COMPREHENSIVE INDEX

of Publications of

The American Association of Petroleum Geologists

1976 - 1980

By

June McFarland
and
Ronald L. Hart

Published by

The American Association of Petroleum Geologists
Tulsa, Oklahoma, U.S.A.

Library of Congress Catalog Card No. 37-9376

ISBN: 0-89181-502-3

Foreword

Essentially all formal publications of the American Association of Petroleum Geologists (except maps) are included in these comprehensive indices.

Indexing of the *Bulletin* was done by June McFarland, Managing Editor, and other special publications of the Association were indexed by Ronald Hart. The two wish to thank E.A. Beaumont, Michael H. Blechner, Eula Matheny, Elizabeth Mitchell, Carol Thompson, Douglas A. White, E.M. Tidwell, Ann Shelby, and Kathryn Stockton for assistance in indexing, preparation and proofing of the computer printouts, and coordination of the project.

June McFarland
Ronald Hart

December, 1984

Introduction: Explanation of Index Format

This comprehensive index is a compilation of the indexes from the Association's *Bulletin* and Special Publications programs for the years 1976 - 1980. In the *Bulletin*, all major paper titles, Geologic Note titles, Discussion titles, and abstract titles are indexed, as well as keywords and author names. For Special Publications, major paper titles, keywords, and author names are included.

Coding in the left-hand columns are as follows:

First Column—Letters signify the publication series: B, Bulletin; M, Memoir; S, Studies in Geology.

Second Column—Numbers signify the volume in the series: four-digit numbers denote *Bulletin* issues by volume year and issue number (ex. 6001 means Volume 60, Issue Number 1—or 1976, January). Two-digit numbers denote the series number of a special publication (ex. M 24 means Memoir Number 24).

Third Column—Numbers signify the page on which the citation is found within the volume or issue.

Fourth Column (optional)—crosses (†) or asterisks (*) signify *Bulletin* codes: Crosses (†) denote the title of a major article within the *Bulletin*; asterisks (*) denote the citation and location of a paper represented only by an abstract. Such abstracts are published in advance of an oral presentation, and likely the paper was not later published in its entire form.

The index does not show multiple authors in any single listing. Each co-author is cited separately.

Only single-line entries are listed. Lines are terminated after the last full word that fits the space.

COMPREHENSIVE INDEX
1976 - 1980

B 6002 246 A AND B POOLS, ALBERTA, RAINBOW LAKE
B 6203 380 A, B, AND C HORIZONS, COOK INLET, ALASKA
B 6102 154 A, B, C, D PRODUCING BLOCKS, CALIFORNIA, VENTURA FIELD
M 26 451† A CASE HISTORY OF GEOSEISMIC MODELING OF BASAL
B 6004 498† A CRETACEOUS TIME SCALE
S 6 269† A CRETACEOUS TIME SCALE
S 6 289† A JURASSIC TIME SCALE
B 6004 489† A JURASSIC TIME SCALE
B 6303 455* A NEW LOOK AT GEOLOGY OF TOGO BY SLAR, ABST.
S 6 213† A NEW PALEOGENE NUMERICAL TIME SCALE
B 6207 1142 A OIL FIELD, IRIAN JAYA
B 6401 107† A PROCESS FOR PRIMARY MIGRATION OF PETROLEUM
B 6202 223 A PROSPECT OF PANGAEA
S 6 179† A RADIOMETRIC TIME SCALE FOR THE NEOGENE OF THE PARATETHYS
B 6206 1063† A REPORT ON THE MEMBERSHIP
M 30 39† A REVIEW OF THE VIKING GAS FIELD
M 24 218 A SANDSTONE ZONE, LOUISIANA
B 6211 2232† A SIGSBEE KNOLL IN EARLY TERTIARY BAY OF BISCAY AND ASSOCIATED
B 6405 704* A TEST OF THE MELANOIDIN HYPOTHESIS, ABST.
B 6304 604 A ZONE, NIOBRARA MEMBER, MANCOS SHALE, NEW MEXICO
B 6104 595 A-1-1 FIELD, LIBYA
M 24 16 A-POOL MARKER, ALASKA, MIDDLE GROUND SHOAL FIELD
B 6102 146 A01 SAND, CALIFORNIA
B 6202 230 AAIUM BASIN, EVAPORITES, MOROCCO
B 6101 85 AAIUN, TARFAYA, BASIN, AFRICAN PLATFORM
B 6305 810 AALTO, K. R., DEEP-WATER SANDSTONE FACIES AND ANCIENT
B 6004 644 AALTO, K. R., SEDIMENTOLOGY OF MELANGE, FRANCISCAN OF
B 6409 1319 AAPG AND API CLASSIFICATION OF WELLS
B 6308 1202 AAPG AND API CLASSIFICATION OF WELLS
B 6108 1155 AAPG AND API CLASSIFICATION OF WELLS
B 6208 1309 AAPG AND API CLASSIFICATION OF WELLS
B 6008 1187 AAPG AND API CLASSIFICATION OF WELLS
B 6205 865 AAPG BOY SCOUT COMMITTEE, GEOLOGY MERIT BADGE BOOKLET
B 6208 1310 AAPG, CSD, INDIVIDUAL WELL TICKET, API-
B 6108 1156 AAPG, CSD, INDIVIDUAL WELL TICKET, API-
B 6008 1188 AAPG, CSD, INDIVIDUAL WELL TICKET, API--
B 6004 694* AAPG DIGITIZED OIL AND GAS FIELD MA OF NORTH AMERICA--STATUS
B 6002 159 AAPG FOUNDATION ISSUE, FEBRUARY 1976 BULLETIN
B 6002 159 AAPG FOUNDATION TRUSTEES
B 6206 1063 AAPG, MEMBERSHIP, AGE
B 6206 1063 AAPG, MEMBERSHIP, BY CLASS
B 6206 1065 AAPG, MEMBERSHIP, DISTRIBUTION BY EMPLOYER, JANUARY 1, 1978
B 6206 1066 AAPG, MEMBERSHIP, NATIONALITIES
B 6206 1063 AAPG, MEMBERSHIP, TOTAL
B 6206 1065 AAPG, MEMBERSHIP, WOMEN
B 6004 716* AAPG OIL AND GAS FIELD DATA BANK PROJECT, UNITED STATES AND
M 24 300† AAPG PUBLICATIONS, BIBLIOGRAPHY OF, NORTH AMERICAN, OIL AND,
B 6311 1996† AAPG RESEARCH CONFERENCE GUIDELINES
B 6303 529 AARON, JOHN A., SLUMPS ON UPPER CONTINENTAL SLOPE,
B 6104 585 AB TEYMUR FIELD, PERSIAN GULF
B 6204 702 ABACO KNOLL, SEISMIC REFLECTION PROFILE
B 6007 1049 ABACO KNOLL, SEISMIC-REFLECTION PROFILE
B 6201 2 ABAKALIKI HIGH, NIGER DELTA
B 6012 2117 ABBOTT, B. M., ORIGIN AND EVOLUTION OF BIOHERMS IN WENLOCK
B 6310 1689 ABBOTT, B. M., PETROLEUM DEVELOPMENT IN CENTRAL AND SOUTHERN
B 6004 671 ABBOTT, P. L., PETROLOGY OF TRIASSIC MARINE SEDIMENTARY ROCKS,
B 6004 707 ABBOTT, PATRICK L., SEDIMENTOLOGIC INDICATION OF EOCENE
B 6101 58 ABD EL AZIZ-MARDIN SWELL, LEVANT, JAUF-, GAARA-
B 6203 488 ABELL, PAUL I.., OXYGEN, AND CARBON ISOTOPIC VARIATION IN
B 6303 409 ABELSON, P. H., SYNTHETIC FUEL FORECASTS. BY TITLE ONLY
B 6004 644 ABELSON, PHILIP H., ENERGY SOURCES, ABST.
B 6101 80 ABENAKI BASINS, OUTER, SCOTIAN SHELF, ORPHEUS AND
B 6004 669* ABENAKI FORMATION, NOVA SCOTIA SHELF, CANADA--DEPOSITIONAL
B 6301 90 ABENAKI FORMATION, SCOTIAN BASIN
B 6301 89 ABENAKI SUBBASIN, SCOTIAN BASIN
B 6107 971 ABIDJAN BASIN OFF. IVORY COAST
B 6008 1403* **ABNORMAL ELECTRIC RESISTIVITY AND FLUID PRESSURE IN BAKKEN**
B 6405 749* ABNORMAL FLUID PRESSURE PRODUCED BY HYDROCARBON
B 6107 1090† ABNORMAL FORMATION PRESSURE. DISCUSSION
B 6007 1124† ABNORMAL FORMATION PRESSURE. DISCUSSION AND REPLY
B 6405 706* ABNORMAL FORMATION PRESSURES, ABST., GLOBAL OCCURRENCE OF
B 6002 228 ABNORMAL PRESSURES
B 6403 415 ABNORMALLY HIGH FLUID PRESSURE, MODELS
B 6403 414† ABNORMALLY HIGH FLUID PRESSURE--SURVEY
B 6412 2179† ABNORMALLY HIGH PORE PRESSURES IN SHALES, MECHANICAL VERSUS,
B 6412 2179 ABNORMALLY HIGH PORE PRESSURES IN SHALES, THERMAL,
B 6205 781 ABO FORMATION, NEW MEXICO
M 26 436 ABO FORMATION, NEW MEXICO, DELAWARE BASIN
M 28 43 ABO SANDSTONE, NEW MEXICO, PERMIAN
M 28 120 ABO SANDSTONE, NEW MEXICO, PERMIAN
M 28 47 ABO SANDSTONE, NEW MEXICO, PERMIAN
M 28 97 ABO SANDSTONE, NEW MEXICO, PERMIAN
M 28 119 ABO SANDSTONE, NEW MEXICO, PERMIAN
M 28 28 ABO SANDSTONE, NEW MEXICO, PERMIAN
M 28 105 ABO SANDSTONE, NEW MEXICO, PERMIAN
S 8 158 ABRA DELTA, PHILIPPINE ISLANDS, SUBMARINE CANYONS
S 8 162 ABRA DELTA, SEAFLOOR TOPOGRAPHY, INDEX MAP
S 8 161 ABRA SEAVALLEY, DESCRIBED, TURBIDITY CURRENTS
S 8 36 ABRA SEAVALLEY, TIME-VELOCITY CURVES, CURRENT FLOW
B 6109 1450 ABRIGO FORMATION, PEDREGOSA BASIN
B 6002 197 ABROLHOS BANK, BRAZIL MARGIN
B 6307 1093 ABROLHOS SUBBASIN, AUSTRALIA
M 26 427 ABRUPT VERTICAL CONTACTS, RESERVOIR-SEAL CONTACT
B 6203 488 ABRY, CLAUDE G., INTERACTIVE GRAPHICS IN SUBSURFACE
B 6405 805* ABSAROKA SHEET, IDAHO AND WYOMING, ABST., THRUST GEOMETRY OF
B 6305 832* ABSAROKA THRUST PLATE, UINTA COUNTY, WYOMING, ABST., PAINTER
B 6307 1040 ABSAROKA THRUST, WYOMING
B 6405 639 ABSAROKA THRUST, WYOMING
B 6408 1286* ABSENCE OF CONNATE WATER IN MICHIGAN REEFS ATTRIBUTED TO
S 6 205 ABSOLUTE DATING, A REVIEW, PALEOGENE BOUNDARIES
S 6 338 ABSOLUTE DATING, AUSTRALIA, DEVONIAN
S 6 338 ABSOLUTE DATING, NORTH AMERICA, DEVONIAN
B 6109 1411 ABSORBING TENDENCIES OF OILS AND AN ASPHALT
B 6109 1411 ABSORBING-TENDENCY DATA, SIGNIFICANCE OF
B 6010 1870 ABU DHABI, 1975, ABU DHABI PETROLEUM CO., LTD., BP PETROLEUM
B 6110 1799 ABU DHABI, 1976, ABU DHABI PETROL. CO., LTD., ABU DHABI OIL CO.,
B 6310 1887 ABU DHABI, 1978, ABU DHABI PETROLEUM CO., ABU DHABI OIL CO., ABU
B 6411 1838 ABU DHABI, 1979, ABU DHABI CO. ET AL, ON DEVELOPMENTS IN
B 6411 1838 ABU DHABI CO. ET AL, ON DEVELOPMENTS IN ABU DHABI, 1979
S 4 384 ABU DHABI COASTAL-CONTINENTAL SABKHA BELT, SECTION
M 28 129 ABU DHABI, HOLOCENE, SABKHA SANDS
B 6310 1887 ABU DHABI MARINE OPERATING CO., ON DEVELOPMENT IN ABU DHABI,
B 6310 1887 ABU DHABI NATIONAL OIL CO., AMERADA HESS, AND ABU DHABI
B 6310 1887 ABU DHABI OIL CO., ABU DHABI NATIONAL OIL CO., AMERADA HESS,
B 6210 2115 ABU DHABI OIL CO., LTD., AMERADA HESS CORP., B. P. TRADING LTD., ET
B 6110 1799 ABU DHABI OIL CO., LTD., AMERADA HESS CORP., ET AL, ON
B 6010 1870 ABU DHABI OIL CO., LTD., SUNNINGDALE OILS, U.K., LTD., AND TOTAL
B 6110 1799 ABU DHABI PETROL. CO., LTD., ABU DHABI OIL CO., LTD., AMERADA HESS
B 6310 1887 ABU DHABI PETROLEUM CO., ABU DHABI OIL CO., ABU DHABI NATIONAL
B 6210 2115 ABU DHABI PETROLEUM CO., LTD., ABU DHABI OIL CO., LTD., AMERADA
B 6010 1870 ABU DHABI PETROLEUM CO., LTD., BP PETROLEUM DEVELOPMENN LTD.,
M 27 190 ABU DHABI, PLEISTOCENE AEOLIANITE
M 27 194 ABU DHABI, PLEISTOCENE AEOLIANITE
B 6210 2130 ABU DHABI, PRODUCTION, 1977
B 6310 1898 ABU DHABI, PRODUCTION, 1978
B 6411 1852 ABU DHABI, PRODUCTION, 1979
M 27 161 ABU DHABI, QUATERNARY CRUSTS
M 27 158 ABU DHABI, QUATERNARY SEDIMENTS
M 27 157 ABU DHABI, QUATERNARY SEDIMENTS
M 27 119 ABU DHABI, QUATERNARY SEDIMENTS
B 6303 505* ABU DHABI SABKHA, ABST., SULFUR AND, OXYGEN, ISOTOPE,
B 6303 496* ABU DHABI, UNITED ARAB EMIRATES, AND ITS RELATION TO DOLOMITE
S 6 146 ABU GROUP, JAPAN, MESOZOIC
B 6104 585 ABU HADNIYA FIELD, PERSIAN GULF
B 6104 585 ABU JIDU, FIELD, PERSIAN GULF, ASAB
B 6104 585 ABU SAFAH FIELD, PERSIAN GULF
B 6004 668* ABYSSAL ORGANISM TRACES ON AN IN MODERN PELAGIC CARBONATE
B 6204 592† ABYSSAL PLAIN, PETROLEUM, POTENTIAL OF, U.S., ATLANTIC, SLOPE,
S 7 15 ABYSSAL PLAIN PROVINCE, DESCRIBED
M 26 273 ABYSSAL PLAIN SYSTEM
S 7 16 ABYSSAL PLAINS MAP, GULF OF MEXICO
B 6108 1246 ACADIA PARISH, LOUISIANA, RAYNE FIELD
B 6101 80 ACADIAN OROGENE, NORTH AMERICA
B 6007 1055 ACANTHOPHORA ZONE, PANAMA, GALETA REEF
B 6205 739† ACCELERATION IN GLOBAL EXPLORATION—REQUIREMENT FOR
S 10 20 ACCOMMODATION, MIGRATION, ENHANCED SOLUBILITY
B 6303 451* ACCRETIONAL SHORELINE PROCESSES AT TIDAL INLETS ALONG SOUTH
M 29 249 ACCRETIONARY BASIN, EASTERN ALEUTIAN TRENCH, PHYSIOGRAPHY
M 29 250 ACCRETIONARY BASIN, EVOLUTION, DIAGRAM
M 29 252 ACCRETIONARY BASIN, EVOLUTION, STRUCTURAL HIGH
B 6008 1398* ACCUMULATION, ABST., REGIONAL, PRESSURE ENVIRONMENTS AND
B 6203 521* ACCUMULATION, ABST., SUBMARINE, FANS—IMPORTANT REGIONS FOR
B 6009 1550† ACCUMULATION AND COMPOSITION IN MEHSANA AREA OF

Index

B 6108 1382* ACCUMULATION AT NORTHEAST LUCIEN FIELD, NOBLE COUNTY,
B 6111 1960 ACCUMULATION, EAST, TEXAS, OIL
B 6104 558† ACCUMULATION IN MAGELLAN BASIN, SOUTH AMERICA, REGIONALIZED,
B 6203 515* ACCUMULATION IN SACRAMENTO VALLEY, CALIFORNIA, ABST.,
B 6408 1219 ACCUMULATION IN SUBTLE TRAPS, PETROLEUM, MIGRATION AND
B 6001 12† ACCUMULATION, LOST SOLDIER AREA, WYOMING, INFLUENCE OF,
B 6004 692* ACCUMULATION OF HYDROCARBONS, ABST., CLAY-, MEMBRANE,
B 6001 108† ACCUMULATION OF HYDROCARBONS IN HELEZ OIL FIELD, ISRAEL,
B 6209 1586 ACCUMULATION OF ORGANIC MATTER, PRODUCTION AND
B 6203 567* ACCUMULATION OF PETROLEUM, ABST., GRAVITY-INDUCED
B 6405 732* ACCUMULATION, PRODUCTION, AND RESIDUAL OIL SHOW. BY TITLE
B 6202 212 ACCUMULATION RATES, SEDIMENT
B 6004 650* ACCUMULATION RELATED TO SAN ANDREAS FAULT SYSTEM AN PLATE
B 6206 1053 ACCUMULATION, URANIUM
B 6108 1376* ACCUMULATIONS AT DRY PINEY, TIP TOP, AND HOGSBACK FIELDS,
B 6203 563* ACCUMULATIONS IN MARINE MUDSTONES OF LATE JURASSIC
B 6408 1278* ACCUMULATIONS IN OVERTHRUST BELT OF ALBERTA, ABST.,
B 6403 316† ACCUMULATIONS, NORTH-CENTRAL MONTANA, COASTAL AND, DELTAIC
B 6104 483† **ACCUMULATIONS OF DIAGENETIC TRAPS--EXPLORATION TARGETS,**
B 6004 714* ACCUMULATIONS--THEIR ORIGIN AND SIGNIFICANCE FOR FUTURE
B 6101 30 ACE MINE, SASKATCHEWAN
B 6303 409 ACEVEDO, J. S., GIANT FIELDS BY SOUTHEASTERN MEXICO. BY TITLE
B 6304 697 ACEVEDO, JOSE SANTIAGO, PETROLEUM GEOLOGY OF CAMPECHE
B 6207 1207 ACHALABHUTI, CHARAN, NATURAL GAS DEPOSITS IN GULF OF
M 25 251 ACHALABHUTI, CHARAN, PETROLEUM GEOLOGY OF THAILAND, GULF O
B 6005 794 ACHARYYA, S. K., CONCEALED MESOZOIC-CENOZOIC ALPINE HIMALAYAN
B 6203 489 ACHAUER, C. W., CARBON AND OXYGEN ISOTOPES AS INDICATORS OF
B 6109 1550 ACHAUER, C. W., CONTRASTS IN CEMENTATION, DISSOLUTION, AND
B 6105 758 ACHAUER, C. W., CONTRASTS IN CEMENTATION, DISSOLUTION, AND
B 6303 325 ACHZIV CANYON, ISRAEL
B 6212 2441† ACID ANIONS IN OIL-FIELD WATERS—IMPLICATIONS FOR ORIGIN OF
M 25 464 ACID VOLCANISM, CHARACTERISTICS OF, FAR EAST, SOUTHERN
M 25 463† ACID VOLCANISM OF SOUTHERN FAR EAST--SUMMARY, SOME
M 25 463 ACID VOLCANISM, USSR, FAR EAST
B 6003 452† ACIDS IN HOLOCENE SEDIMENT, THERMALLY INDUCED DIAGENESIS OF,
S 10 21 ACIDS IN WATER, MIGRATION, SOLUBILITIES OF
B 6405 679 ACKMAN, EDWARD J., INDIAN CREEK FIELD, FALL RIVER COUNTY,
B 6204 633 ACONA RESEARCH CRUISE, R V
B 6110 1587 ACOSTA ESTEVEY, R., ON DEVELOPMENTS IN MEXICO, 1976
B 6412 2149 ACOUSTIC ANOMALIES, BRIGHT SPOTS, O WIPE-OUT ZONES, BERING SEA
M 26 427 ACOUSTIC IMPEDANCE CONTRAST, RESERVOIR-SEAL CONTACT
M 26 6 ACOUSTIC IMPEDANCE, SEISMIC, REFLECTIONS
M 26 455 ACOUSTIC STRATIGRAPHY, SEISMIC, MODELING
M 26 354 ACOUSTIC SYSTEMS, CLASSIFICATION CHART, RESOLUTION SYSTEMS
B 6205 744 ACREAGE HOLDINGS IN NON-COMMUNIST WORLD IN 1970 AND 1976
B 6303 420* ACRITARCH GENUS KWAGUNT FORMATION, LATE PRECAMBRIAN,
M 27 23 ACRITARCHS
B 6405 770* ACRITARCHS IN STUDY OF MARINE KEROGEN, ABST., FLUORESCENCE OF •
S 4 44 ACROPORA ASSEMBLAGE, BELIZE
B 6303 514* ACROPORA CERVICORNIS, ABST., SUBAERIAL DIAGENESIS OF AXIAL,
S 4 144 ACROPORA PALMATA, HYDROMECHANICAL STUDY OF
S 4 159 ACROPORA PALMATA OFF BONAIRE
B 6007 1057 ACROPORA PALMATA, PANAMA, GALETA POINT, REEF
S 4 90 ACTINACIS ASSEMBLAGE, JAMAICA, MULTICOLUMNASTREA-,
S 4 90 ACTINACIS ASSEMBLAGE, JAMAICA, PARACYCLOSERIS-, TROCHOSERIS-
S 4 90 ACTINACIS-DICTUOPHYLLIA-DICHOCOENIA ASSEMBLAGE, JAMAICA,
B 6307 1130 ACTIVATION ENERGIES, ESTIMATES
B 6307 1129 ACTIVATION ENERGIES FOR VARIOUS GEOCHEMICAL AND RELATED
M 29 443† ACTIVE AND PASSIVE CONTINENTAL MARGINS. EXPLOSION. SEISMOLOGY
B 6303 447* **ACTIVE FAULTS IN HOUSTON, TEXAS, AREA AS OBSERVED ON LANDSAT**
B 6301 9 ACTIVE FOREARC SYSTEMS
M 29 306 ACTIVE MARGINS, AEGEAN ARC, MEDITERRANEAN
M 29 305 ACTIVE MARGINS, CALABRIAN ARC, MEDITERRANEAN
M 29 310 ACTIVE MARGINS, CYPRUS ARC, MEDITERRANEAN
M 29 305 ACTIVE MARGINS, MEDITERRANEAN, EASTERN
B 6303 512* ACTIVE SUBMARINE LANDSLIDE PROCESSES, MISSISSIPPI DELTA, ABST.
B 6004 689* ACTIVE SURFACE FAULTING, TEXAS COASTAL ZONE, ABST.
B 6201 88 ADAK ISLAND, ALASKA
B 6207 1227* ADAK ISLAND, ALASKA—A CASE HISTORY, ABST., GEOTHERMAL
B 6404 530 ADAMS BJERG FORMATION, GREENLAND
M 26 277 ADAMS COUNTIES, COLORADO, BOULDER, WELD, AND
B 6405 776* ADAMS COUNTY, COLORADO, SECS. 17-20, T3S, R62W, ABST., ZENITH
B 6409 1552 ADAMS, JOE B., NO. 1 PLAY IN U.S.A.—SOUTH LOUISIANA TUSCALOOSA
B 6203 488 ADAMS, JOHN K., HEAVY-MINERAL RESPONSE TO ENVIRONMENTAL
B 6402 264 ADAMS, R. D., EVOLUTION OF SAND-DOMINANT SUBAERIAL PHASE,
B 6207 1207 ADAMS, R. H., DEVELOPMENT AND RESERVOIR ANALYSIS OF
B 6105 759 ADAMS, S. S., INTERPRETATION OF POSTDEPOSITIONAL PROCESSES
S 4 142 ADAPTIVE STRATEGIES FOR REDUCING HYDRAULIC STRESS
B 6101 114 ADAVILLE FORMATION, WYOMING
B 6105 761 ADDICOTT, WAREN O., POROSITY IN LATE MIOCENE REEFS, SOUTHERN
B 6012 2174 ADDICOTT, WARREN O., NEOGENE MOLLUSCAN STAGES OF OREGON AND
B 6402 188 ADDICOTT, WARREN O., POROSITY EVOLUTION OF UPPER MIOCENE
B 6203 497 ADDINGTON, J. D., DIGITAL IMAGE PROCESSING OF POTENTIAL FIELD
B 6304 668 ADDY, SUNIT K., GAS SEEPS AND SUBSURFAC STRUCTURE OFF PANAMA
B 6303 409 ADDY, SUNIT K.,, SEISMIC STRATIGRAPHY AND STRUCTURE OFF
B 6203 489 ADDY, SUNIT KUMAR, MIXING OF BOTTOM SEDIMENTS IN NORTHWEST
B 6303 538 ADELSECK, C G., MIDDLE CRETACEOUS OXYGEN-DEFICIENT
B 6410 1614 ADENA FIELD, COLORADO
S 4 67 ADEY, WALTER H., HOLOCENE BIOHERMS OF LESSER
M 28 6 ADIRONDACK GNEISS, NEW YORK, PRECAMBRIAN
B 6012 2185 ADKISON, W. L., GEOLOGIC FRAMEWORK OF LOWER COOK INLET,
B 6311 2119 ADLER, FRANK J., CONTROLS ON PENNSYLVANIAN HYDROCARBON
B 6101 55 ADLOUN 1, LOG, LEBANON
B 6101 55 ADLOUN STRUCTURE, LEBANON
B 6101 3 ADMIRALTY BASIN
B 6004 671* **ADMIRE, 650 FT, SANDSTONE, PERMIAN, EL DORADO MICELLAR-POLYMER**
B 6303 541* ADMIRE 650-FOOT SANDSTONE, EL DORADO FIELD, KANSAS, ABST.,
B 6107 1030 ADMIRE GROUP, NEBRASKA
B 6002 319* ADMIRE SANDSTONE, EL DORADO FIELD, BUTLER COUNTY, KANSAS,
M 29 297 ADRIATIC BASINS, SEDIMENTARY SEQUENCE, MEDITERRANEAN
S 5 171 ADRIATIC COAST, SANTA MARGHERITA DI SAVOIA, COMMERCIAL
B 6411 1970 ADRIATIC OFFSHORE AND DINARIC ONSHORE ZONE, YUGOSLAVIA
B 6405 695* ADVANCES IN RADON EXPLORATION TECHNIQUES FOR URANIUM, ABST.
B 6312 2194 ADVENTDALEN GROUP, SVALBARD
M 29 306 AEGEAN ARC, MEDITERRANEAN, ACTIVE MARGINS
M 29 312 AEGEAN BASINS, EVOLUTION, MEDITERRANEAN
M 29 310 AEGEAN BASINS, SEDIMENTARY SEQUENCE, MEDITERRANEAN
B 6106 900 AEGEAN PLATE
B 6411 1964 AEGEAN SEA, GREECE
B 6405 686* AEOLIAN SEDIMENT BY CYANOPHYTES, ABST., TRAPPING AND
B 6105 816* AERIAL PHOTOGRAPH INTERPRETATION OF HIGH-VELOCITY
B 6211 2354* AEROMAGNETIC DATA FOR GEOLOGIC MAPPING ON NORTH SLOPE,
B 6302 245† AEROMAGNETIC DETECTION OF DIAGENETIC MAGNETITE OVER OIL
B 6309 1538† AEROMAGNETIC DETECTION OF DIAGENETIC MAGNETITE OVER OIL
M 29 179 AEROMAGNETIC MAP, BLAKE ESCARPMENT
M 29 90 AEROMAGNETIC SURVEY, ATLANTIC MARGIN, 1964-1966
M 29 86 AEROMAGNETIC SURVEY, ATLANTIC MARGIN, 1975
B 6302 246 AEROMAGNETIC SURVEY, OKLAHOMA, CEMENT OIL FIELD
B 6210 1849 AFARS AND, ISSAS, BOTSWANA, BURUNDI, MALAWI, RHODESIA, RWANDA,
B 6110 1833 AFGHANISTAN, DEVELOPMENTS, 1976
B 6210 1980 AFGHANISTAN, DEVELOPMENTS, 1977
B 6310 1818 AFGHANISTAN, DEVELOPMENTS, 1978
B 6411 1883 AFGHANISTAN, DEVELOPMENTS, 1979
B 6110 1842 AFGHANISTAN, PRODUCTION, 1975-1976
B 6210 1990 AFGHANISTAN, PRODUCTION, 1975-1977
B 6310 1829 AFGHANISTAN, PRODUCTION, 1977-1978
B 6411 1896 AFGHANISTAN, PRODUCTION, 1977-1979
B 6201 18 AFIESERE-OLOMORO MACROSTRUCTURE, NIGER DELTA
B 6201 2 AFIKPO LOW, NIGER DELTA
B 6004 667* AFRICA, ABST., DEVELOPMENT AND, PALEOGEOGRAPHIC EXTENT
B 6303 425* AFRICA, ABST., EARLY, MESOZOIC TECTONIC FRAMEWORK AND
B 6303 544* AFRICA, ABST., MID-, CENOZOIC, FORTUNA FORMATION, EAST-CENTRAL,
B 6303 529* AFRICA, ABST., TERTIARY, SEA-LEVEL MOVEMENTS AROUND SOUTHERN
B 6107 963 AFRICA, AGULHAS FRACTURE ZONE
B 6303 525* AFRICA AND CARRIBEAN, ABST., SIGNIFICANCE OF MIDDLE, TERTIAR
B 6406 858 AFRICA, AOUENAT UPLIFT
B 6107 963 AFRICA, ASCENSION FRACTURE ZONE
B 6107 963 AFRICA, AZORES FRACTURE ZONE
B 6107 972 AFRICA, BENIN SLOPE
B 6406 865 AFRICA, BENUE TROUGH
B 6105 826* AFRICA BETWEEN CANARY AND CAPE VERDE ISLANDS, ABST.,
B 6406 858 AFRICA, CALANSCIO UPLIFT
B 6107 972 AFRICA, CAMEROON VOLCANICS
B 6107 965 AFRICA, CAPE VERDE FRACTURE
B 6406 860 AFRICA, CENOMANIAN-TURONIAN TRANSGRESSIVE DEPOSITS
B 6110 1752 AFRICA, CENTRAL AND SOUTHERN, PRODUCTION, 1975-1976
B 6210 1852 AFRICA, CENTRAL AND SOUTHERN, PRODUCTION, 1976-1977
B 6310 1696 AFRICA, CENTRAL AND SOUTHERN, PRODUCTION, 1977-1978
B 6411 1794 AFRICA, CENTRAL AND SOUTHERN, PRODUCTION, 1978-1979
B 6406 865 AFRICA, CHAD BASIN
B 6107 969 AFRICA, CHAIN FRACTURE ZONE
B 6107 969 AFRICA, CHARCOT FRACTURE
M 26 162 AFRICA, COMPARISON, CYCLE CHARTS, TEXAS, GULF COAST, WEST
B 6406 857† AFRICA CRATON, LATEST, JURASSIC-EARLY, CRETACEOUS, REGRESSIVE
B 6205 780 AFRICA, CRETACEOUS, NORTH
B 6406 861 AFRICA, CYRENAICA PLATFORM
B 6403 429 AFRICA, ECCA HIGH-WAX OIL FIELD

Index

B 6205 777 AFRICA, EOCENE, NORTH
B 6107 967 AFRICA, EQUATORIAL SEGMENT
S 9 17 AFRICA, EVIDENCES OF DRIFT, PALEOMAGNETISM
M 30 577 AFRICA, GABON, GRONDIN FIELD
B 6406 858 AFRICA, GARGAF ARCH
B 6003 389† AFRICA, GEOLOGY AND, OIL, PROSPECTS OF, SOMALIA, EAST
B 6107 965 AFRICA, GUINEA FRACTURE
B 6406 858 AFRICA, HOGGAR MASSIF
S 9 8 AFRICA, ICE FLOW DIRECTION, ORDOVICIAN, SAHARA
B 6010 1813† AFRICA IN 1975, PETROLEUM, DEVELOPMENTS IN CENTRAL AND
B 6010 1767† AFRICA IN 1975, PETROLEUM, DEVELOPMENTS IN, NORTH
B 6110 1746† AFRICA IN 1976, PETROLEUM, DEVELOPMENTS IN CENTRAL AND
B 6110 1698† AFRICA IN 1976, PETROLEUM, DEVELOPMENTS IN, NORTH
B 6210 1844† AFRICA IN 1977, PETROLEUM, DEVELOPMENTS IN CENTRAL AND
B 6210 1796† AFRICA IN 1977, PETROLEUM, DEVELOPMENTS IN, NORTH
B 6310 1689† AFRICA IN 1978, PETROLEUM, DEVELOPMENT IN CENTRAL AND
B 6411 1785† AFRICA IN 1979, PETROLEUM, DEVELOPMENTS IN CENTRAL AND
B 6411 1739† AFRICA IN 1979, PETROLEUM, DEVELOPMENTS IN, NORTH
B 6406 858 AFRICA, MALTA ESCARPMENT
B 6403 429 AFRICA, NIGER DELTA HIGH-WAX OIL FIELD
M 30 565 AFRICA, NIGERIA, NEMBE CREEK FIELD
M 26 190 AFRICA, NORTH, PINNACLE REEF, SEISMIC EXAMPLE
B 6406 859 AFRICA, NUBIA FORMATION
B 6007 1009 AFRICA, NUBIAN, NORTH
B 6406 861 AFRICA, NUBIAN SHIELD
B 6205 778 AFRICA, OFFSHORE, JURASSIC, WEST
B 6205 778 AFRICA, OFFSHORE, SEISMIC DATA, WEST
B 6205 759 AFRICA, OFFSHORE, SEISMIC FACIES
B 6205 780 AFRICA, OLIGOCENE, NORTH
B 6107 979 AFRICA, ORANGE RIVER REGION, PROFILE
B 6205 788 AFRICA, PALEOCENE, NORTH
B 6205 777 AFRICA, PALEOCENE, NORTH
B 6406 865 AFRICA, PELAGIAN BLOCK
B 6406 865 AFRICA, PELUSIUM LINE
B 6310 1636† AFRICA, PETROLEUM, DEVELOPMENTS IN, NORTH
S 6 364 AFRICA, PRECAMBRIAN, RADIOMETRIC DATING
B 6202 226 AFRICA, RIFT VALLEYS
B 6107 969 AFRICA, ROMANCHE-TAKORADI FRACTURES
S 8 40 AFRICA, SEAVALLEYS, CONGO CANYON
B 6107 962 AFRICA SEGMENT, NORTHWEST
B 6205 788 AFRICA, SEISMIC DATA, NORTH
B 6205 777 AFRICA, SEISMIC DATA, NORTH
B 6107 961 AFRICA, SEISMIC PROFILES OFF
M 26 145† AFRICA, SEISMIC, STRATIGRAPHY AND, GLOBAL CHANGES OF SEA LEVEL,
B 6107 965 AFRICA, SENEGAL BASIN
M 26 158 AFRICA, SEQUENCES, JURASSIC, BAJOCIAN
M 26 158 AFRICA, SEQUENCES, JURASSIC, BATHONIAN
M 26 158 AFRICA, SEQUENCES, JURASSIC, CALLOVIAN
B 6107 963 AFRICA, ST. PAUL FRACTURE ZONE
B 6107 963 AFRICA, ST. PAUL-LAS PALMAS FRACTURES
B 6107 961† AFRICA, STRUCTURAL HISTORY OF, ATLANTIC, MARGIN OF
S 8 161 AFRICA, SUBMARINE CANYONS, CONGO CANYON
B 6406 858 AFRICA, TIBESTI MASSIF
M 26 161 AFRICA TO GULF COAST, CYCLE, COMPARISON, WEST
B 6410 1690 AFRICA, TRANSVAAL STROMATOLITIC LIMESTONE
M 29 79 AFRICA, TRIASSIC, RIFTING HISTORY
B 6002 210 AFRICA, WALVIS RIDGE TO BENUE TROUGH, WEST
M 26 158 AFRICA, WEST, CHRONOSTRATIGRAPHIC CORRELATION TO CYCLES
M 26 81 AFRICA, WEST, SEA LEVEL FLUCTUATIONS, CHRONOCORRELATIONS
M 26 191 AFRICA, WEST, SHELF-MARGIN CARBONATE, SEISMIC EXAMPLE
M 25 205 AFRICA, WEST, VELOCITY-STRUCTURE PROBLEM
M 26 155 AFRICA, WESTERN, SEQUENCES, IDENTIFICATION
M 26 159 AFRICA, WESTERN, SEQUENCES, OXFORDIAN-KIMMERIDGIAN
M 26 160 AFRICA, WESTERN, SEQUENCES, TITHONIAN-BERRIASIAN
B 6005 809† AFRICA--SENEGAL TO PORTUGAL, CONTINENTAL, MARGIN OFF WESTERN
B 6303 409* AFRICA, BY TITLE ONLY, OUARGLA REGION, ALGERIA, NORTH
B 6002 211 AFRICAN AND BRAZILIAN SALT EDGES, ALBIAN, MODEL, MATCHING,
B 6002 211 AFRICAN DIAPIR ZONE, GABON TO, ANGOLA, WEST
B 6106 900 AFRICAN PLATE
B 6403 360 AFRICAN PLATE, BOUNDARY
B 6403 360 AFRICAN PLATE, BOUNDARY
B 6101 85 AFRICAN PLATFORM, AAIUN, TARFAYA, BASIN
B 6408 1195 AFRICAN SHELF, BENGUELA CURRENT, ANOXIC LAYER CAUSED BY
S 6 367† AFRICAN STRATIGRAPHIC CLASSIFICATION AND A PROPOSAL FOR THE
B 6409 1563 AGAGU, O. K., DEPOSITIONAL SYSTEMS IN NACATOCH SAND, UPPER
B 6405 668 AGAGU, O. K., PETROLEUM ACCUMULATION TRENDS IN EAST TEXAS
B 6405 668 AGAGU, O. K., PETROLEUM GEOLOGY OF SENONIAN IN SEDIMENTS
B 6004 716 AGATSTON, ROBERT S., AAPG OIL AND GAS FIELD DATA BANK PROJECT,
B 6202 297 AGBADA FORMATION, NIGER DELTA
B 6408 1252 AGBADA FORMATION, NIGERIA
B 6009 1621* AGE AND TECTONICS OF CINCINNATI ARCH IN SOUTH-CENTRAL
B 6104 577 AGE CALCULATIONS OF OILS, EXAMPLES OF
S 5 100 AGE OF REEFS, MICHIGAN BASIN, NIAGARAN SERIES,
B 6205 776 AGE, CARBONATE ROCKS PLOTTED AGAINST GEOLOGIC
B 6005 883† AGE DETERMINATION OF SILURIAN GLAUCONITE. DISCUSSION AND
S 6 363 AGE DETERMINATIONS, A REVIEW, PRECAMBRIAN
M 26 95 AGE ESTIMATION BEFORE DRILLING, GLOBAL CYCLES
B 6101 112† AGE OF FRONTIER FORMATION IN NORTH-CENTRAL UTAH
M 29 392 AGE OF SEDIMENT, GULF OF MEXICO, MISSISSIPPI FAN
B 6211 2352* AGE OF SEDIMENTATION-DIAGENESIS FOR BEDFORD CANYON
M 26 18 AGE-VELOCITY-DEPTH ASPECTS, SEISMIC, INTERPRETATION
B 6203 535 AGENBROAD, LARRY D., SEDIMENTOLOGY OF LATE QUATERNARY
S 10 33† AGENTS FOR PRIMARY HYDROCARBON MIGRATION: A REVIEW
B 6104 573† AGES OF HYDROCARBONS IN OILS--PHYSICAL CHEMISTRY APPLIED
B 6004 722* AGGLUTINATED FORAMINIFERS IN LATE QUATERNARY CONTINENTAL
M 27 34 AGGLUTINATING FORAMS
B 6104 585 AGHA JARI FIELD, PERSIAN GULF
B 6010 1866 AGI S.P.A., AMOCO, AND ATLANTIC RICHFIELD CO., ON DEVELOPMENTS
B 6411 1841 AGIP AND DOME PETROLEUM LTD., ON DEVELOPMENTS IN SOUTH
B 6210 2112 AGIP MINERARIA S.P.A., DEMINEX IRAN OIL CO., TOTAL IRANIAN
B 6310 1885 AGIP MINERARIA S.P.A., ON DEVELOPMENTS IN IRAN, 1978, OIL SERVICE
B 6310 1890 AGIP S.P.A., AND PETROCONSULTANT S.A., ON DEVELOPMENTS IN SOTH
B 6210 2118 AGIP S.P.A., AND PETROCONSULTANTS S.A., ON DEVELOPMENTS IN
B 6407 995 AGLELGAL SANDSTONE MEMBER, MOROCCO
M 25 436 AGNO BATHOLITH, LUZON
B 6405 713 AGTERBERG, FRITZ, APPLICATION OF STATISTICAL MODELS IN
B 6112 2104 AGUA CALIENTE SECTION, PERU
B 6109 1476 AGUA GRANDE FIELD, BRAZIL
B 6307 1033 AGUA GRANDE FIELD, STRUCTURE MAP, BRAZIL
M 30 363 AGUA NUEVA FORMATION, CHIAPAS-TABASCO PROVINCE
B 6202 244 AGUA NUEVA FORMATION, MEXICO
B 6409 1560* AGUA NUEVA FORMATIONS, CRETACEOUS, EAST-CENTRAL MEXICO,
B 6004 644 AGUAYO, J. E. C., SEDIMENTARY ENVIRONMENTS AND DIAGENESIS OF
M 25 535 AGUILAR LEAD-ZINC DEPOSIT, ARGENTINA
B 6109 1476 AGULHA FIELD, BRAZIL
B 6107 963 AGULHAS FRACTURE ZONE, AFRICA
B 6008 1407 AHLBRANDT, THOMAS S., COASTAL SEDIMENTATION MODEL, TENSLEEP
S 4 18 AHOGADA REEF, PUERTO RICO
B 6409 1552 AHR, WAYNE M., ENVIRONMENTS OF DEPOSITION AND DIAGENESIS OF
B 6203 496 AHR, WAYNE M., TURBIDITE ORIGIN FOR ANHYDRITIC DOLOMITES IN
B 6104 585 AHWAZ FIELD, PERSIAN GULF
B 6405 614 AHWAZ SANDSTONE DELTA, MIDDLE EAST
B 6102 297 AIKEN, C. L. V., RESIDUAL BOUGUER GRAVITY ANOMALY ANALYSIS OF
S 6 146 AIOI GROUP, JAPAN, MESOZOIC
M 25 120 AIR POLLUTION FROM HYDROGEN SULFIDE
B 6405 626 AIRCRAFT RADAR IMAGE, CALIFORNIA, SAN ANDREAS FAULT
B 6012 2185* AIRGUN ARRAY, ABST., DESIGN OF HIGH-FREQUENCY-RESOLUTION
M 29 127 AIRGUN-SONOBUOY RESULTS, ARGENTINA, MARGIN
M 29 116 AIRGUN-SONOBUOY RESULTS, ARGENTINA, MARGIN
B 6303 409 AIT-HAMOUDA, M., OUARGLA REGION, ALGERIA, NORTH AFRICA. BY
B 6401 136 AITKEN, JAMES D., AMERICAN COMMISSION ON STRATIGRAPHIC
B 6401 137 AITKEN, JAMES D., NOTE 51--RECORDS OF STRATIGRAPHIC
B 6303 409 AITKEN, R. R., DRILLING AND PRODUCTION PROBLEMS IN
B 6202 296 AJALI FORMATION, NIGERIA
M 27 47 AJAX LIMESTONE, AUSTRALIA, CAMBRIAN
B 6310 1888 AJMAN, 1978, FORMAN EXPLORATION CO., ON DEVELOPMENTS IN
B 6411 1840 AJMAN, 1979, FORMAN EXPLORATION CO., ON DEVELOPMENTS IN
B 6010 1871 AJMAN AND UMM AL QAIWAIN, 1975, UNITED REFINING CO.,
B 6110 1800 AJMAN AND UMM AL QAIWAIN, 1976, HOUSTON OIL AND MINERALS
B 6210 2117 AJMAN AND UMM AL QAIWAIN, 1977, PETROCONSULTANTS S.A.,
B 6202 295 AKATA FORMATION, NIGER DELTA
S 7 117 AKERS, W. H., GULF OF MEXICO CONTINENTAL-SLOPE SEDIMENTS AND
B 6207 1207 AKIL, ISMET, EXPLORATION AND DEVELOPMENT OF GEOTHERMAL
B 6303 325 AKKO EXTENSION, ISRAEL
M 24 54 AKLAVIK ARCH, ALASKA, NORTHWEST TERRITORIES
B 6209 1610 AKRON DOLOMITE, NEW YORK
B 6310 1890 AL-HINAI, ON DEVELOPMENTS IN OMAN, 1978, ASIATIC PETROLEUM
B 6411 1967 AL-KASIM, FAROUK, AND EGIL BERGSAGER, ON DEVELOPMENTS IN
B 6310 1927 AL-KASIM, ON DEVELOPMENTS IN NORWAY NORT OF 62 DEGREES, 1978,
B 6010 1718 AL-KASIM, ON DEVELOPMENTS IN NORWAY NORT OF 62 DEGREES N, 1975,
B 6110 1647 AL-KASIM, ON DEVELOPMENTS IN NORWAY NORTH OF 62 DEGREES N,
B 6210 1918 AL-KASIM, ON DEVELOPMENTS IN NORWAY NORTH OF 62 DEGREES N,
B 6203 502 AL-SHAIEB, ZUHAIR, CALCRETES. BY TITLE ONLY.
B 6203 463 AL-SHAIEB, ZUHAIR,, SECONDARY FERROAN DOLOMITE RHOMBS IN OIL
B 6103 360 AL-SHAIEB, ZUHAIR, URANIUM POTENTIAL OF PERMIAN AND
B 6405 775* ALABAMA, ABST., APPALACHIAN, THRUST BELT BETWEEN TRENTON,
B 6409 1555* ALABAMA, ABST., CHESTER AND, POTTSVILLE DEPOSITIONAL SYSTEMS,
B 6009 1606* ALABAMA, ABST., DEPOSITIONAL ENVIRONMENT AND PETROLEUM,

B 6004 687* ALABAMA, ABST., DISTRIBUTION AND RELATION OF PHYSICAL AND
B 6409 1558* ALABAMA, ABST., EFFECTS OF, HURRICANE, FREDERIC ON MORPHOLOGY
B 6309 1607* ALABAMA, ABST., EOCENE-, OLIGOCENE BOUNDARY IN SOUTHWEST
B 6409 1561* ALABAMA, ABST., JURASSIC GEOLOGY AND HYDROCARBON POTENTIAL
B 6409 1565* ALABAMA, ABST., LATE, PALEOCENE PLANKTONIC, FORAMINIFERAL
B 6409 1559* ALABAMA, ABST., OSTRACODE BIOSTRATIGRAPHY OF LOWER,
B 6409 1563* ALABAMA, ABST., REGIONAL STRATIGRAPHY OF UPPER, JURASSIC,
B 6309 1596* ALABAMA, ABST., UPPER, MISSISSIPPIAN, DELTAS IN, BLACK WARRIOR
B 6004 720* ALABAMA AND MISSISSIPPI, ABST., DEPOSITION, FACIES, PATTERNS, AND,
M 24 278 ALABAMA AND MISSISSIPPI, GILBERTOWN-PICKENS-POLLARD FAULT
M 24 264 ALABAMA, ANNONA CHALK
B 6204 646 ALABAMA, BANGOR SECTION
S 10 248 ALABAMA, BIG ESCAMBIA CREEK FIELD, ETHANE CONCENTRATIONS
B 6204 646 ALABAMA, BLACK WARRIOR BASIN
B 6110 1898 ALABAMA, CAMBRIAN TEST WELLS, MISSISSIPPI AND
S 6 348 ALABAMA, CARTERS LIMESTONE, CORRELATION
M 24 262 ALABAMA, CITRONELLE FIELD, TYPE SECTION
M 24 261 ALABAMA, CITRONELLE FORMATION
M 24 259† ALABAMA, CITRONELLE OIL FIELD, MOBILE COUNTY
M 24 278 ALABAMA, CONECUH RIDGE
B 6008 1282 ALABAMA, DEVELOPMENTS, 1975
B 6108 1264 ALABAMA, DEVELOPMENTS, 1976
B 6208 1394 ALABAMA, DEVELOPMENTS, 1977
B 6308 1401 ALABAMA, DEVELOPMENTS, 1978
B 6411 2029 ALABAMA, DEVELOPMENTS, 1979
M 24 268 ALABAMA, DONOVAN SANDSTONES
M 24 266 ALABAMA, FERRY LAKE ANHYDRITE
S 10 244 ALABAMA, FLOMATON FIELD, ETHANE CONCENTRATIONS
S 10 246 ALABAMA, FLOMATON FIELD, SAND PERCENTAGE CONCENTRATIONS
B 6204 646 ALABAMA, FLOYD SHALE
B 6204 646 ALABAMA, FORT PAYNE FORMATION
B 6101 111 ALABAMA GEOLOGICAL SOCIETY, ESTABLISHED APRIL 25, 1963
B 6101 111† ALABAMA GEOLOGICAL SOCIETY, HISTORY OF
B 6110 1898 ALABAMA, GEORGIA, AND TENNESSEE, CAMBRIAN TYPE LOCALITIES
M 24 259 ALABAMA, GLEN ROSE FORMATION
B 6204 646 ALABAMA, HARTSELLE SANDSTONE
M 24 261 ALABAMA, HATTIESBURG CLAY
M 27 112 ALABAMA, JURASSIC, SMACKOVER FORMATION
M 24 284 ALABAMA, LITTLE ESCAMBIA CREEK FIELD
B 6204 646 ALABAMA, MISSISSIPPIAN SECTION
S 7 99 ALABAMA, MOBILE BAY, SEISMIC FACIES EXAMPLE
B 6204 644 ALABAMA, MONTEAGLE LIMESTONE
B 6204 646 ALABAMA, PARKWOOD FORMATION
B 6204 646 ALABAMA, PENNINGTON SECTION
B 6204 646 ALABAMA, POTTSVILLE FORMATION
B 6204 646 ALABAMA, PRIDE MOUNTAIN FORMATION
B 6008 1284 ALABAMA, PRODUCTION, 1974-1975
B 6108 1266 ALABAMA, PRODUCTION, 1975-1976
B 6208 1396 ALABAMA, PRODUCTION, 1976-1977
B 6308 1404 ALABAMA, PRODUCTION, 1977-1978
B 6107 988 ALABAMA, SELMA CHALK UNIT
M 24 265 ALABAMA, TUSCALOOSA SANDSTONE
B 6204 646 ALABAMA, TUSCUMBIA SECTION
B 6202 229 ALAGOAS BASIN, SEDIMENTARY SEQUENCE, BRAZIL, SERGIPE-
B 6002 197 ALAGOAS COASTAL BASIN, BRAZIL, SERGIPE-
B 6202 230 ALAGOAS, EVAPORITES, BRAZIL, SERGIPE-
B 6302 221 ALAGOAS FRACTURE ZONE, BRAZIL
B 6111 1962 ALAMINOS 1973 CRUISE AL-11, RV
B 6206 1068 ALAMINOS CANYON, GULF OF MEXICO
B 6209 1590 ALAMINOS CANYON, GULF OF MEXICO
B 6206 1067 ALAMINOS CRUISE 11-73, RV
B 6109 1456 ALAMO HUECO BASIN, ARIZONA, NEW MEXICO, AND, MEXICO
B 6104 602 ALAMO HUECO MOUNTAINS, NEW MEXICO
M 30 307 ALAPAH FORMATION, PRUDHOE BAY, LISBURNE GROUP
B 6203 380 ALASKA, A, B, AND C HORIZONS, COOK INLET
B 6405 737* ALASKA, ABST., BARROW GAS FIELDS, NORTH SLOPE
B 6403 447* ALASKA, ABST., BENTHIC, FORAMINIFERS FROM EASTERN, GULF OF
B 6105 815* ALASKA, ABST., CLAY MINERALOGY OF EASTERN, GULF OF
B 6105 817* ALASKA, ABST., COMPLEX, COMPLEX RIDGES AND, TROUGH
B 6103 467* ALASKA, ABST., COMPRESSIVE INTERACTION OF, SIBERIAN AND, NORTH
B 6405 708* ALASKA, ABST., CONTROLS OF, ZEOLITE, CEMENTATION IN UPPER,
B 6405 752* ALASKA, ABST., CRETACEOUS STRATIGRAPHY, CHIGNIK AREA, ALASKA
B 6012 2176* ALASKA, ABST., DISPERSAL, DISTRIBUTION, AND THICKNESS OF,
B 6209 1754* ALASKA, ABST., DISTRIBUTION OF, HOLOCENE, FORAMINIFERA IN, GULF
B 6211 2350* ALASKA, ABST., DISTRIBUTION OF RECENT, FORAMINIFERA IN, GULF OF
B 6012 2175* ALASKA, ABST., ENVIRONMENTAL AND TIME SIGNIFICANCE OF LOCAL
B 6405 754* ALASKA, ABST., EVOLUTION OF, BROOKS RANGE, THRUST BELT AND,
B 6405 766* ALASKA, ABST., GEOLOGY OF CONTINENTAL SLOPE ADJACENT TO, OCS,
B 6004 654* ALASKA, ABST., GEOLOGY OF, SOUTH AUGUSTINE ISLAND, LOWER, COOK
B 6303 411* ALASKA, ABST., GLACIAL ORIGIN OF MEGACHANNELS OF UPPER,
B 6303 411* ALASKA, ABST., GLACIAL-MARINE, TRACE- AND BODY-FOSSIL
B 6403 440* ALASKA, ABST., HOLOCENE, FORAMINIFERAL DISTRIBUTION PATTERNS
B 6403 444* ALASKA, ABST., HOLOCENE, MOLLUSK DISTRIBUTION PATTERNS IN
B 6203 545* ALASKA, ABST., HOLOCENE SEDIMENT VOLUME AND MODERN SEDIMENT
B 6303 417* ALASKA, ABST., HYDROCARBON POTENTIAL OF CENTRAL, GULF OF
B 6405 671* ALASKA, ABST., HYDROCARBON POTENTIAL OF, SITKALIDAK
B 6405 745* ALASKA, ABST., HYDROCARBON PROSPECTS OF, BERING SEA SHELF
B 6203 534* ALASKA, ABST., HYDROCARBONS OF LOW MOLECULAR WEIGHT IN
B 6405 671* ALASKA, ABST., LATE, NEOGENE DEPOSITIONAL AND, CLIMATI CYCLES IN,
B 6004 649* ALASKA, ABST., LISBURNE GROUP--POTENTIAL MAJO OBJECTIVE OF
B 6211 2356* ALASKA, ABST., LITHOFACIES IN LOWER, JURASSIC, VOLCANICLASTIC
B 6004 716* ALASKA, ABST., LITTORAL PROCESSES AND GEOMORPHIC VARIABILITY
B 6405 784* ALASKA, ABST., MESOZOIC ACCRETIONARY TECTONICS OF
B 6205 880* ALASKA, ABST., NEW INFORMATION ON, LISBURNE GROUP,
B 6103 468* ALASKA, ABST., PALEOMAGNETISM AND PALEOGEOGRAPHIC
B 6207 1225* ALASKA, ABST., PETROLEUM GEOLOGY OF, COOK INLET BASIN
B 6203 553* ALASKA, ABST., PRECAMBRIAN-, CAMBRIAN SEDIMENTARY ROCKS FROM
B 6403 446* ALASKA, ABST., QUATERNARY SEDIMENTARY FACIE ON CONTINENTAL
B 6405 792* ALASKA, ABST., RATIONALIZATION OF, KOYUKUK CRUNCH, NORTHERN
B 6405 756* ALASKA, ABST., SEDIMENTARY CHARACTERISTICS AND PROCESSES OF
B 6405 758* ALASKA, ABST., SEDIMENTOLOGY OF FLUVIAL UPPER, DEVONIAN,
B 6405 752* ALASKA, ABST., STRATIGRAPHIC RELATIONS OF, NANUSHUK GROUP,
B 6405 722* ALASKA, ABST., STRATIGRAPHY AND STRUCTURE OF EARLY, TERTIARY,
B 6105 838* ALASKA, ABST., STRUCTURAL INTERPRETATION OF EASTERN, ALEUTIAN
B 6405 706* ALASKA, ABST., STRUCTURE AND HYDROCARBON POTENTIAL OF,
B 6203 513* ALASKA, ABST., STRUCTURE AND TECTONIC SETTING OF, CONTINENTAL
B 6012 2190* ALASKA, ABST., SUBSURFACE GEOLOGY AND SOME IMPLICATIONS
B 6303 443* ALASKA, ABST., TRACE-FOSSIL EVIDENCE FOR, DEEP-WATER
B 6004 672* ALASKA, ABST., TRANSGRESSIVE-, REGRESSIVE, SHELF DEPOSITION,
B 6405 684* **ALASKA, ABST., UNCONFORMITIES—KEY TO MAJOR OIL ACCUMULATIONS,**
B 6405 698* ALASKA, ABST., URANIUM IN, TERTIARY SEDIMENTS IN
B 6012 2188* ALASKA, ABST., USE OF, IR COLOR PHOTOGRAPHY AND CANYON PHOTOS
B 6105 818* ALASKA, ABST., VARIATIONS IN, COASTAL ENVIRONMENTS AND, BEACH
B 6201 88 ALASKA, ADAK ISLAND
B 6407 1019 ALASKA, ALBATROSS BANK
B 6408 1142 ALASKA, ALBATROSS BANK
B 6408 1142 ALASKA, ALBATROSS BASIN
B 6407 1018 ALASKA, ALBATROSS BASIN
B 6201 88 ALASKA, ALEUTIAN BASIN
B 6203 374 ALASKA, ALEUTIAN RANGE
B 6201 88 ALASKA, ALEUTIAN TERRACE
B 6201 88 ALASKA, ALEUTIAN TRENCH
B 6408 1142 ALASKA, ALEUTIAN TRENCH
B 6407 1028 ALASKA, ALEUTIAN TRENCH
M 25 149 ALASKA, ALEUTIAN VOLCANIC ARC, CALDERAS
B 6104 607 ALASKA, ALINCHAK BAY AREA
B 6407 1026 ALASKA, AMATULI TROUGH
B 6002 177 ALASKA, AMERICAN PLATE
B 6201 88 ALASKA, AMLIA BASIN
B 6405 806* ALASKA, AN UPDATE, ABST., NPR-
B 6103 467* ALASKA AND ADJACENT DEEP-SEA AREA, LAT PALEOZOIC THROUGH
M 29 273† ALASKA AND BOUNDARIES OF TECTONIC PLATES, CONTINENTAL,
M 25 512 ALASKA AND BRITISH COLUMBIA, TRANSCURRENT FAULTS, LARGE,
M 24 54 ALASKA, ANDERSON PLAIN
B 6201 88 ALASKA, ANDREANOF ISLANDS
B 6008 1189 ALASKA, ARCTIC NORTH SLOPE, EXPLORATION, 1975
B 6108 1157 ALASKA, ARCTIC NORTH SLOPE, EXPLORATION, 1976
B 6208 1312 ALASKA, ARCTIC NORTH SLOPE, EXPLORATION, 1977
B 6308 1211 ALASKA, ARCTIC NORTH SLOPE, EXPLORATION, 1978
B 6409 1345 ALASKA, ARCTIC NORTH SLOPE, EXPLORATION, 1979
B 6109 1494 ALASKA, ARCTIC WILDLIFE RANGE
B 6203 377 ALASKA, ARKOSE RIDGE FORMATION
B 6201 88 ALASKA, ATKA ISLAND
B 6203 374 ALASKA, AUGUSTINE ISLAND
B 6203 384 ALASKA, AUGUSTINE-SELDOVIA ARCH
B 6209 1697 ALASKA, AVAN HILLS
M 24 52 ALASKA, BARN MOUNTAINS
B 6203 374 ALASKA, BARREN ISLANDS
B 6109 1494 ALASKA, BARROW ARCH
M 24 26 ALASKA, BARROW ARCH
B 6408 1132 ALASKA, BEAVER CREEK OIL FIELD
B 6109 1499 ALASKA, BEECHY POINT SECTION
B 6203 377 ALASKA, BELUGA FORMATION
B 6408 1133 ALASKA, BELUGA FORMATION
B 6408 1132 ALASKA, BELUGA RIVER GAS FIELD
B 6203 394 ALASKA, BENIOFF ZONE

B 6204 635 ALASKA, BERING GLACIER
B 6204 635 ALASKA, BERING TROUGH
M 24 11 ALASKA, BEULGA FORMATION
B 6408 1131† ALASKA, BIOGENIC AND, THERMOGENIC ORIGINS OF, NATURAL, GAS IN,
B 6408 1132 ALASKA, BIRCH HILL GAS FIELD
B 6103 465* ALASKA BLUESCHIST BELTS. BY TITLE ONLY, REGIONA TECTONIC
B 6408 1260 ALASKA, BORDER RANGES FAULT
B 6203 394 ALASKA, BORDER RANGES FAULT SYSTEM
B 6408 1135 ALASKA, BOWSER CREEK FIELD
B 6002 173 ALASKA, BRISTOL BAY
M 25 303 ALASKA, BRISTOL BAY BASIN, SECTION
M 24 52 ALASKA, BRITISH MOUNTAINS
M 24 26 ALASKA, BROOKS RANGE
B 6109 1494 ALASKA, BROOKS RANGE
B 6103 437 ALASKA, BROOKS RANGE SECTION
B 6203 376 ALASKA, BRUIN BAY FAULT
B 6408 1260 ALASKA, BRUIN BAY FAULT
B 6204 635 ALASKA, CAPE SUCKLING
B 6204 635 ALASKA, CAPE YAKATAGA
M 26 356 ALASKA, CAPE YAKATAGA, SIDE-SCAN SONAR RECORD
B 6006 962† ALASKA, CARBONIFEROUS SECTION AND, UNCONFORMITY AT, MOUNT,
B 6408 1260 ALASKA, CASTLE MOUNTAIN FAULT
M 25 293 ALASKA, CASTLE MOUNTAIN FAULT
B 6203 397 ALASKA, CASTLE MOUNTAIN FAULT
M 29 274 ALASKA, CENOZOIC, PLATE BOUNDARIES
B 6408 1142 ALASKA, CENTRAL SHELF UPLIFT
B 6407 1023 ALASKA, CENTRAL SHELF UPLIFT
B 6103 436 ALASKA, CHESTERAN UNCONFORMITY
B 6203 377 ALASKA, CHICKALOON FORMATION
B 6203 374 ALASKA, CHINITNA BAY
B 6203 377 ALASKA, CHINITNA FORMATION
M 24 5 ALASKA, CHINITNA SILTSTONE
B 6408 1142 ALASKA, CHIRIKOF ISLAND
B 6407 1016 ALASKA, CHIRIKOF ISLAND
B 6203 377 ALASKA, CHISIK CONGLOMERATE MEMBER
B 6203 386 ALASKA, CHUGACH BAY FAULT
B 6403 446* ALASKA COASTLINE, CAPE SUCKLING TO CAPE SPENCER, ABST.,
B 6109 1499 ALASKA, COLVILLE DELTA SECTION
B 6109 1494 ALASKA, COLVILLE TROUGH
M 24 26 ALASKA, COLVILLE TROUGH
B 6203 394 ALASKA, CONTACT FAULT
B 6207 1231* ALASKA CONTINENTAL MARGIN, ABST., PLATE TECTONICS IN
B 6209 1593 ALASKA CONTINENTAL MARGIN, ALEUTIAN TRENCH
B 6209 1593 ALASKA CONTINENTAL MARGIN, KODIAK SEAMOUNT
B 6209 1593 ALASKA CONTINENTAL MARGIN, SURVEYOR CHANNEL
B 6204 633† ALASKA CONTINENTAL SHELF, SURFACE SEDIMENTARY UNITS OF
M 29 429 ALASKA, CONVERGENT MARGIN, PACIFIC
B 6303 305 ALASKA, COOK INLET
M 24 4 ALASKA, COOK INLET AREA, FIELDS AND FACILITIES
M 25 304 ALASKA, COOK INLET BASIN
B 6203 377 ALASKA, COOK INLET BASIN, STRATIGRAPHIC COLUMN
B 6408 1134 ALASKA, COOK INLET BAY, RESERVES
B 6408 1260 ALASKA, COOK INLET COST 1 WELL
B 6409 1346 ALASKA, COOK INLET, EXPLORATION
B 6108 1158 ALASKA, COOK INLET, EXPLORATION
B 6208 1312 ALASKA, COOK INLET, EXPLORATION
B 6308 1212 ALASKA, COOK INLET, EXPLORATION
B 6008 1190 ALASKA, COOK INLET, EXPLORATION, 1975
B 6203 376 ALASKA, COOK INLET, GEOLOGIC MAP
B 6203 399 ALASKA, COOK INLET, HYDROCARBON POTENTIAL
B 6104 578 ALASKA, COOK INLET, OIL AGE
B 6408 1132 ALASKA, COOK INLET REGION, OIL AND GAS FIELD
B 6203 380 ALASKA, COOK INLET, SEISMIC DATA
B 6203 395 ALASKA, COOK INLET, TECTONIC-CYCLES
B 6203 395 ALASKA, COOK INLET, TIME-STRATIGRAPHIC CHART
B 6203 379 ALASKA, COOK INLET, WELL DATA
B 6303 306 ALASKA, COOK-SHELIKOF BASIN
B 6203 378 ALASKA, CRETACEOUS
B 6408 1142 ALASKA, CRETACEOUS
B 6407 1016 ALASKA, CRETACEOUS
M 24 4 ALASKA, CRETACEOUS
M 25 295 ALASKA, CRETACEOUS
S 6 264 ALASKA, CRETACEOUS, ANDESITIC VOLCANICS DATA POINT
M 28 117 ALASKA, CRETACEOUS, CORWIN FORMATION
M 28 98 ALASKA, CRETACEOUS, CORWIN FORMATION
M 28 29 ALASKA, CRETACEOUS, CORWIN FORMATION
M 28 31 ALASKA, CRETACEOUS, FORTRESS MOUNTAIN FORMATION
M 28 99 ALASKA, CRETACEOUS, FORTRESS MOUNTAIN FORMATION
M 28 7 ALASKA, CRETACEOUS, KOGOSUKRUK TONGUE
M 28 20 ALASKA, CRETACEOUS, KOGOSUKRUK TONGUE
M 28 132 ALASKA, CRETACEOUS, KOGOSUKRUK TONGUE
M 28 22 ALASKA, CRETACEOUS, KOGOSUKRUK TONGUE
M 28 130 ALASKA, CRETACEOUS, KOGOSUKRUK TONGUE
M 28 19 ALASKA, CRETACEOUS, KOGOSUKRUK TONGUE
M 28 118 ALASKA, CRETACEOUS, KOGOSUKRUK TONGUE
B 6109 1499 ALASKA, CRETACEOUS, LATE, UNCONFORMITY
M 28 20 ALASKA, CRETACEOUS, PRINCE CREEK FORMATION
M 28 22 ALASKA, CRETACEOUS, PRINCE CREEK FORMATION
M 28 7 ALASKA, CRETACEOUS, PRINCE CREEK FORMATION
M 28 118 ALASKA, CRETACEOUS, PRINCE CREEK FORMATION
M 28 130 ALASKA, CRETACEOUS, PRINCE CREEK FORMATION
M 28 132 ALASKA, CRETACEOUS, PRINCE CREEK FORMATION
M 28 19 ALASKA, CRETACEOUS, PRINCE CREEK FORMATION
M 28 160 ALASKA, CRETACEOUS, SHALE
M 28 34 ALASKA, CRETACEOUS, SHUMAGIN FORMATION
M 28 32 ALASKA, CRETACEOUS, TOROK FORMATION
M 28 33 ALASKA, CRETACEOUS, TOROK FORMATION
M 28 30 ALASKA, CRETACEOUS, TUKTU FORMATION
M 28 99 ALASKA, CRETACEOUS, TUKTU FORMATION
M 29 279 ALASKA, CROSS SOUND, CHICHAGOF-BARANOF FAULT
M 29 278 ALASKA, CROSS SOUND, CHICHAGOF-BARANOF FAULT
M 29 282 ALASKA, CROSS SOUND, PLATE BOUNDARY
M 29 276 ALASKA, CROSS SOUND, SEISMIC SECTIONS
B 6408 1142 ALASKA, DANGEROUS CAPE HIGH
B 6407 1021 ALASKA, DANGEROUS CAPE HIGH
B 6201 88 ALASKA, DEEP SEA DRILLING PROJECT, SITE 186
M 25 293 ALASKA, DONNA RIVER FAULT ZONE
B 6203 395 ALASKA, DSDP SITE 178
B 6203 395 ALASKA, DSDP SITE 180
B 6203 395 ALASKA, DSDP SITE 181
B 6203 395 ALASKA, DSDP SITE 182
B 6004 657* ALASKA, DURING LATE CENOZOIC, ABST., CLIMATIC CONTROL OF
B 6203 397 ALASKA, EAGLE RIVER FAULT
M 29 275 ALASKA, EARTHQUAKES AND FAULTS, INDEX
B 6109 1499 ALASKA, EAST MIKKELSEN BAY SECTION
B 6109 1499 ALASKA, EAST UGNU SECTION
B 6109 1496 ALASKA, ECHOOKA FORMATION
M 24 29 ALASKA, ECHOOKA FORMATION
B 6103 438 ALASKA, ECHOOKA SANDSTONE
B 6103 437 ALASKA, ENDICOTT GROUP
B 6109 1496 ALASKA, ENDICOTT GROUP
B 6408 1142 ALASKA, EOCENE
B 6407 1016 ALASKA, EOCENE
M 25 69 ALASKA, ERTS STUDY
B 6108 1158 ALASKA, EXPLORATION, GULF OF
B 6308 1213 ALASKA, EXPLORATION, GULF OF
B 6208 1312 ALASKA, EXPLORATION, GULF OF
B 6012 2186* ALASKA EXPLORATION POTENTIAL, ABST.
M 25 293 ALASKA, FAIRWEATHER FAULT
M 29 276 ALASKA, FAIRWEATHER ISLAND, FAULT ZONE
B 6408 1132 ALASKA, FALLS CREEK GAS FIELD
M 25 293 ALASKA, FAREWELL FAULT
B 6203 386 ALASKA, FITZ CREEK ANTICLINE
B 6203 399 ALASKA, FITZ CREEK SILTSTONE
B 6301 24 ALASKA, FOREARC EVOLUTION
M 29 261† ALASKA, FROM MULTICHANNEL SEISMIC RECORDS, STRUCTURE OF THE
B 6012 2180* ALASKA, FROM SEISMIC DATA, ABST., GEOLOGY OF OFFSHORE LOWER,
B 6203 399 ALASKA, GAIKEMA SANDSTONE
B 6002 164 ALASKA, GEMUK GROUP
B 6203 373† ALASKA, GEOLOGIC FRAMEWORK OF LOWER, COOK INLET
B 6006 952† ALASKA, GEOLOGIC SIGNIFICANCE OF, DOONERAK STRUCTURAL, HIGH,
B 6012 2149† ALASKA GEOLOGICAL SOCIETY HISTORY
M 25 149 ALASKA, GEOTHERMAL ENERGY
B 6303 302 ALASKA, GHOST ROCKS FORMATION
B 6407 1016 ALASKA, GHOST ROCKS FORMATION
B 6408 1142 ALASKA, GHOST ROCKS FORMATION
B 6201 88 ALASKA, GLOMAR CHALLENGER SEISMIC PROFILE
B 6104 595 ALASKA, GRANITE POINT FIELD
B 6408 1132 ALASKA, GRANITE POINT OIL FIELD
M 24 26 ALASKA, GUBIK FORMATION
B 6203 374 ALASKA, GULF OF ALASKA
B 6201 88 ALASKA, HAWLEY RIDGE
B 6203 377 ALASKA, HEMLOCK CONGLOMERATE
B 6408 1133 ALASKA, HEMLOCK CONGLOMERATE
B 6401 26 ALASKA, HEMLOCK FORMATION

M 24 11 ALASKA, HEMLOCK SANDSTONE MEMBER
B 6204 635 ALASKA, HINCHINBROOK ISLAND
B 6204 635 ALASKA, HINCHINBROOK SEA VALLEY
B 6204 637 ALASKA, HOLOCENE
B 6403 376 ALASKA, HOLOCENE FAN DELTA
B 6212 2412† ALASKA, HOLOCENE, SLUMP ON, CONTINENTAL, SHELF OFF, MALASPINA,
B 6204 635 ALASKA, ICY BAY
M 29 285 ALASKA, ICY BAY, REFRACTION DATA
B 6212 2415 ALASKA, ICY BAY-MALASPINA GLACIER SLUMP, SEISMIC DATA
M 25 293 ALASKA, IDITAROD FAULT
B 6203 374 ALASKA, ILIAMNA VOLCANO
B 6008 1189† ALASKA IN 1975, DEVELOPMENTS IN
B 6108 1157† ALASKA IN 1976, DEVELOPMENTS IN
B 6208 1311† ALASKA IN 1977, DEVELOPMENTS IN
B 6308 1211† ALASKA IN 1978, DEVELOPMENTS IN
B 6409 1345† ALASKA IN 1979, DEVELOPMENTS IN
B 6203 374 ALASKA, INISKIN PENINSULA
B 6109 1499 ALASKA, ITKILLIK SECTION
B 6109 1500 ALASKA, ITKILYARIAK FORMATION
B 6103 438 ALASKA, ITKILYARIAK FORMATION
B 6408 1132 ALASKA, IVAN RIVER GAS FIELD
B 6109 1496 ALASKA, IVISHAK FORMATION
M 24 31 ALASKA, IVISHAK SANDSTONE
B 6103 438 ALASKA, IVISHAK SHALE
M 24 4 ALASKA, JURASSIC
M 25 295 ALASKA, JURASSIC
B 6408 1261 ALASKA, JURASSIC
B 6203 378 ALASKA, JURASSIC
B 6407 1016 ALASKA, JURASSIC
B 6104 609 ALASKA, JURASSIC
B 6104 607 ALASKA, JURASSIC, FOSSILS
M 28 160 ALASKA, JURASSIC, SHALE
B 6203 374 ALASKA, KACHEMAK BAY
B 6203 377 ALASKA, KAGUYAK FORMATION
B 6408 1132 ALASKA, KALOA GAS FIELD
B 6109 1499 ALASKA, KALUBIK CREEK SECTION
B 6203 374 ALASKA, KAMISHAK BAY
B 6203 377 ALASKA, KAMISHAK FORMATION
B 6203 374 ALASKA, KAMISHAK HILLS
B 6203 377 ALASKA, KAMISHAK HILLS SECTION
M 25 185 ALASKA, KATALLA OIL FIELD
M 24 32 ALASKA, KAVIK SHALE
B 6204 635 ALASKA, KAYAK ISLAND
M 24 26 ALASKA, KAYAK SHALE
B 6109 1500 ALASKA, KAYAK SHALE
B 6109 1496 ALASKA, KAYAK SHALE
B 6103 437 ALASKA, KAYAK SHALE
B 6204 635 ALASKA, KAYAK TROUGH
B 6103 438 ALASKA, KEKIKTUK CONGLOMERATE
B 6109 1498 ALASKA, KEKIKTUK CONGLOMERATE
B 6408 1132 ALASKA, KENAI FIELD
B 6401 26 ALASKA, KENAI FORMATION
B 6401 26 ALASKA, KENAI FORMATION
B 6203 376 ALASKA, KENAI GAS FIELD
B 6203 377 ALASKA, KENAI GROUP
B 6408 1133 ALASKA, KENAI GROUP
M 24 14 ALASKA, KENAI GROUP
B 6203 383 ALASKA, KENAI LOWLAND
B 6203 374 ALASKA, KENAI LOWLAND
B 6401 27 ALASKA, KENAI MOUNTAIN FAULT ZONE
B 6203 374 ALASKA, KENAI PENINSULA
M 29 271 ALASKA, KENAI PENINSULA, ORCA FORMATION
B 6303 306 ALASKA, KENAI-KODIAK RIDGE
B 6104 608 ALASKA, KIALAGVIK FORMATION
M 24 26 ALASKA, KINGAK FORMATION
B 6203 397 ALASKA, KNIK FAULT
M 25 304 ALASKA, KODIAK BASIN, SECTION
M 29 263 ALASKA, KODIAK, CONTINENTAL MARGIN, BATHYMETRY
M 29 270 ALASKA, KODIAK, CONTINENTAL MARGIN, MODEL
B 6303 302 ALASKA, KODIAK FORMATION
B 6407 1016 ALASKA, KODIAK FORMATION
B 6408 1142 ALASKA, KODIAK FORMATION
B 6408 1142 ALASKA, KODIAK ISLAND, GEOLOGY
B 6407 1016 ALASKA, KODIAK ISLANDS, GEOLOGY
M 29 264 ALASKA, KODIAK, REFLECTION RECORDS (24-CHANNEL)
M 29 268 ALASKA, KODIAK, REFRACTION VELOCITY DATA
B 6408 1151 ALASKA, KODIAK SHELF, GEOTHERMAL GRADIENTS
B 6408 1146 ALASKA, KODIAK SHELF, SOURCE AND RESERVOIR DATA
M 29 268 ALASKA, KODIAK, SLOPE DATA, INTERPRETATION
M 29 267 ALASKA, KODIAK, VELOCITY DATA
M 24 26 ALASKA, KUPARUK RIER FORMATION
B 6109 1499 ALASKA, KUPARUK SECTION
B 6203 397 ALASKA, LAKE CLARK FAULT
B 6408 1260 ALASKA, LAKE CLARK FAULT
B 6204 642 ALASKA, LANDSAT IMAGE, OCTOBER 12, 1972, GULF OF
B 6408 1132 ALASKA, LEWIS RIVER GAS FIELD
M 24 26 ALASKA, LISBURNE GROUP
B 6109 1500 ALASKA, LISBURNE GROUP
B 6103 437 ALASKA, LISBURNE GROUP
B 6109 1493† ALASKA, LISBURNE GROUP, MISSISSIPPIAN AND, PENNSYLVANIAN,
B 6109 1510 ALASKA, LISBURNE GROUP, TRAPS
M 26 370 ALASKA, LOWER COOK INLET, DEPTH SOUNDER RECORD
B 6204 635 ALASKA, MALASPINA GLACIER
B 6103 466* ALASKA MARGIN, ABST., MESOZOIC MAGMATISM AN ACCRETION,
M 29 430 ALASKA, MARGIN, ORGANIC CARBON
M 29 429 ALASKA, MARGIN, ORGANIC CARBON IN SEDIMENTS
B 6103 466* ALASKA MARGIN, BY TITLE ONLY, HISTORIC, EARTHQUAKES AND
B 6209 1592 ALASKA MARGINS, DSDP, SITES 172-182, LEG 18, OREGON AND
B 6203 377 ALASKA, MATANUSKA FORMATION
M 24 4 ALASKA, MATANUSKA FORMATION
M 24 6 ALASKA, MCARTHUR RIVER FIELD
B 6408 1132 ALASKA, MCARTHUR RIVER OIL FIELD
B 6203 395 ALASKA, MCHUGH COMPLEX
B 6103 436 ALASKA, MERAMECIAN, LOWER, UNCONFORMITY
B 6303 302 ALASKA, MESHIK FORMATION
M 24 16 ALASKA, MIDDLE GROUND SHOAL FIELD, A-POOL MARKER
M 24 17 ALASKA, MIDDLE GROUND SHOAL FIELD, B-POOL MARKER
M 24 18 ALASKA, MIDDLE GROUND SHOAL FIELD, E-POOL MARKER
M 24 1† **ALASKA, MIDDLE GROUND SHOAL OIL FIELD**
B 6408 1132 ALASKA, MIDDLE GROUND SHOAL OIL FIELD
M 24 21 ALASKA, MIDDLE GROUND SHOAL OIL FIELD, PRODUCTIVE ZONES A, B, C,
B 6204 635 ALASKA, MIDDLETON ISLAND
B 6109 1499 ALASKA, MIKKELSEN BAY SECTION
B 6109 1499 ALASKA, MILNE POINT SECTION
B 6407 1016 ALASKA, MIOCENE
B 6408 1133 ALASKA, MIOCENE
B 6408 1142 ALASKA, MIOCENE
B 6209 1697 ALASKA, MISHEGUK MOUNTAIN
M 28 40 ALASKA, MISSISSIPPIAN, NUKA FORMATION
M 28 122 ALASKA, MISSISSIPPIAN, NUKA FORMATION
M 28 129 ALASKA, MISSISSIPPIAN, NUKA FORMATION
B 6204 635 ALASKA, MONTAGUE ISLAND
B 6408 1132 ALASKA, MOQUAWKIE GAS FIELD
M 25 301 ALASKA, MOUNT SPURR VOLCANO
M 25 293 ALASKA, MULCHATNA FAULT
B 6408 1261 ALASKA, NAKNEK FORMATION
M 24 4 ALASKA, NAKNEK FORMATION
B 6203 377 ALASKA, NAKNEK FORMATION
B 6406 947 ALASKA, NAKNEK FORMATION SANDSTONE
B 6408 1142 ALASKA, NARROW CAPE FORMATION
B 6407 1016 ALASKA, NARROW CAPE FORMATION
B 6203 395 ALASKA, NARROW CAPE FORMATION
B 6303 302 ALASKA, NARROW CAPE FORMATION
B 6109 1494 ALASKA, NAVAL PETROLEUM RESERVE NO. 4
B 6203 377 ALASKA, NELCHINA LIMESTONE
B 6408 1132 ALASKA, NICOLAI CREEK GAS FIELD
B 6408 1259† ALASKA, NONMARINE UPPER, CRETACEOUS ROCKS, COOK INLET
B 6203 394 ALASKA, NORTH AMERICAN PLATE
M 29 274 ALASKA, NORTH AMERICAN PLATE, BOUNDARY
B 6408 1132 ALASKA, NORTH COOK INLET GAS FIELD
B 6408 1134 ALASKA, NORTH FORK GAS FIELD
B 6408 1132 ALASKA, NORTH MIDDLE GROUND SHOAL GAS FIELD
M 30 298 ALASKA, NORTH SLOPE DISCOVERY WELLS, DATA TABLE
M 30 290 ALASKA, NORTH SLOPE, MAJOR STRUCTURAL ELEMENTS
M 25 303 ALASKA, NORTH SLOPE, SECTION
B 6204 635 ALASKA, NORTHERN, BATHYMETRY, GULF OF
B 6204 634 ALASKA, NORTHERN, SEISMIC PROFILES, GULF OF
M 24 54 ALASKA, NORTHWEST TERRITORIES, AKLAVIK ARCH
B 6407 1078 ALASKA, NORTON BASIN, GAS SEEP
B 6405 715* ALASKA, NPRA, ABST., PETROLEUM EXPLORATION O, NATIONAL
B 6207 1218* ALASKA, NPRA, ABST., PETROLEUM EXPLORATION OF, NATIONAL
B 6408 1133 ALASKA, OLIGOCENE
B 6407 1016 ALASKA, OLIGOCENE
B 6408 1142 ALASKA, OLIGOCENE
B 6203 395 ALASKA, ORCA GROUP
B 6407 1078† ALASKA, ORIGIN OF, GASOLINE-RANGE, HYDROCARBONS AND THEIR,
B 6203 394 ALASKA, PACIFIC PLATE
M 29 274 ALASKA, PACIFIC PLATE, BOUNDARY

1976 - 1980
List of Publications, Showing Abbreviations, used in Index References

BULLETINS

AAPG 6001 through AAPG 6412 (Volumes 60 through 64, 1976-1980)

SPECIAL PUBLICATIONS

Memoirs

M 24 North American Oil and Gas Fields (1976). Edited by Jules Braunstein.

M 25 Circum-Pacific Energy and Mineral Resources (1976). Edited by Michel T. Halbouty, John C. Maher, and Harold M. Lian.

M 26 Seismic Stratigraphy—Applications to Hydrocarbon Exploration (1977). Edited by Charles E. Payton.

M 27 A Color Illustrated Guide to Carbonate Rock Constituents, Textures, Cements, and Porosities (1978). By Peter A. Scholle.

M 28 A Color Illustrated Guide to Constituents, Textures, Cements, and Porosities of Sandstones and Associated Rocks (1979). By Peter A. Scholle.

M 29 Geological and Geophysical Investigations of Continental Margins (1979). Edited by Joel S. Watkins, Lucien Montadert, and Patricia Wood Dickerson.

M 30 Giant Oil and Gas Fields of the Decade 1968-1978 (1980). Edited by Michel T. Halbouty.

Studies in Geology

S 3 Migration of Uranium and Thorium—Exploration Significance (1977). By John W. Gabelman.

S 4 Reefs and Related Carbonates—Ecology and Sedimentology (1977). Edited by Stanley H. Frost, Malcolm P. Weiss, and John B. Saunders.

S 5 Reefs and Evaporites—Concepts and Depositional Models (1977). Edited by James H. Fisher.

S 6 Contributions to the Geologic Time Scale (1978). Edited by George V. Cohee, Martin F. Glaessner, and Hollis D. Hedberg.

S 7 Framework, Facies, and Oil-Trapping Characteristics of the Upper Continental Margin (1978). Edited by Arnold H. Bouma, George T. Moore, and James M. Coleman.

S 8 Currents in Submarine Canyons and Other Seavalleys (1979). By Francis P. Shepard, Neil F. Marshall, Patrick A. McLoughlin, and Gary G. Sullivan.

S 9 Paleoclimate, Paleomagnetism, and Continental Drift (1979). By J.K.A. Habicht.

S 10 Problems of Petroleum Migration (1980). Edited by W.H. Roberts, III, and Robert J. Cordell.

S 11 The Miocene Stratigraphy of California, Revisited (1980). By R.M. Kleinpell. Also Pliocene Biostratigraphy of California by C.R. Haller, with special sections by: G. Hornaday, A.D. Warren, A. Tipton; illustrations by Mary E. Taylor, extra plates by Margaret Hanna.

B 6408 1142 ALASKA, PALEOCENE
B 6408 1133 ALASKA, PALEOCENE
B 6103 467* ALASKA PALEOMAGNETISM, PLATE ROTATION AND TECTONICS, PART 1,
B 6103 467* ALASKA PALEOMAGNETISM, PLATE ROTATION, AND TECTONICS, PART 2,
M 25 295 ALASKA, PALEOZOIC
B 6204 635 ALASKA, PAMPLONA RIDGE
B 6211 2357* ALASKA PENINSULA, ABST., STRATIGRAPHY AND SEDIMENTOLOGY OF,
B 6405 752* ALASKA PENINSULA, ALASKA, ABST., CRETACEOUS STRATIGRAPHY,
B 6207 1228* ALASKA PENINSULA, AND SHUMAGIN KODIAK SHELF, ABST.,
B 6309 1522 ALASKA PENINSULA, BEAR LAKE FORMATION, MIOCENE
B 6309 1522 ALASKA PENINSULA, BELKOFSKI FORMATION, OLIGOCENE
B 6002 164 ALASKA PENINSULA, GEOLOGIC MAP
B 6309 1523 ALASKA PENINSULA, HOODOO FORMATION
B 6105 811 * ALASKA PENINSULA LATE CRETACEOUS FORE-ARC DEPOSITION
B 6309 1523 ALASKA PENINSULA, NAKNEK FORMATION
B 6309 1525 ALASKA PENINSULA, QUATERNARY ROCKS
B 6309 1523 ALASKA PENINSULA, SHUMAGIN FORMATION
B 6104 607† ALASKA PENINSULA, SOME LOWER AND MIDDLE, JURASSI BEDS IN,
B 6309 1523 ALASKA PENINSULA, STEPOVAK FORMATION
B 6309 1524 ALASKA PENINSULA, TACHILNI FORMATION, MIOCENE AND PLIOCENE
B 6309 1522† ALASKA PENINSULA, TERTIARY STRATIGRAPHY AND PETROLEUM
B 6309 1523 ALASKA PENINSULA, TOLSTOI FORMTION
M 28 129 ALASKA, PERMIAN, NUKA FORMATION
M 28 40 ALASKA, PERMIAN, NUKA FORMATION
M 28 122 ALASKA, PERMIAN, NUKA FORMATION
B 6109 1499 ALASKA, PERMIAN UNCONFORMITY
M 24 23† ALASKA, PERMO-, TRIASSIC, RESERVOIRS OF, PRUDHO BAY FIELD, NORTH
B 6408 1140† ALASKA, PETROLEUM GEOLOGY OF, KODIAK SHELF
M 25 298 ALASKA, PETROLEUM PROSPECTS
M 29 277 ALASKA, PLATE BOUNDARIES, JUNCTURE
B 6103 463* ALASKA PLATE TECTONICS, ABST., OVERVIEW OF
B 6103 469* ALASKA PLATE-TECTONIC MODEL BASED ON CONTINENTAL-MARGIN
B 6408 1142 ALASKA, PLEISTOCENE
B 6201 90 ALASKA, PLEISTOCENE
B 6204 637 ALASKA, PLEISTOCENE
B 6201 90 ALASKA, PLIOCENE
B 6408 1142 ALASKA, PLIOCENE
B 6407 1016 ALASKA, PLIOCENE
B 6408 1133 ALASKA, PLIOCENE
B 6303 303 ALASKA, PLUTONIC ROCKS
B 6204 635 ALASKA, POINT RIOU
B 6109 1499 ALASKA, POINT STARKERSEN SECTION
M 25 293 ALASKA, PORCUPINE FAULT
B 6407 1024 ALASKA, PORTLOCK ANTICLINE
B 6408 1142 ALASKA, PORTLOCK ANTICLINE
B 6407 1024 ALASKA, PORTLOCK ANTICLINE
M 24 26 ALASKA, PRINCE CREEK-SCHRADER BLUFF FORMATION
B 6204 635 ALASKA, PRINCE WILLIAM SOUND
B 6008 1191 ALASKA, PRODUCTION, 1974-1975
B 6108 1160 ALASKA, PRODUCTION, 1975-1976
B 6208 1316 ALASKA, PRODUCTION, 1976-1977
B 6308 1216 ALASKA, PRODUCTION, 1977-1978
B 6409 1350 ALASKA, PRODUCTION, 1978-1979
M 30 299 ALASKA, PRUDHOE BAY COMPLEX, HYDROCARBON ACCUMULATIONS
B 6104 594 ALASKA, PRUDHOE BAY FIELD
B 6109 1493 ALASKA, PRUDHOE BAY FIELD
M 30 290 ALASKA, PRUDHOE BAY FIELD, EXPLORATION HISTORY
M 30 297 ALASKA, PRUDHOE BAY FIELD, GEOLOGIC SETTING
M 30 22 ALASKA, PRUDHOE BAY FIELD, LANDSAT IMAGE
M 24 25 ALASKA, PRUDHOE BAY FIELD, LOCATION MAP
B 6103 439 ALASKA, PRUDHOE BAY FIELD, SECTION
B 6401 20 ALASKA, PRUDHOE BAY GIANT OIL AND GAS FIELD, LANDSAT DATA
B 6401 20 ALASKA, PRUDHOE BAY GIANT OIL AND GAS FIELD, LANDSAT DATA
B 6109 1499 ALASKA, PRUDHOE BAY SECTION
M 26 339 ALASKA, PRUDHOE BAY, SEISMIC SECTION
B 6104 607 ALASKA PUALE BAY AREA
B 6109 1499 ALASKA, PUT RIVER 9-11-13 SECTION
B 6203 376 ALASKA, QUATERNARY
B 6204 641 ALASKA, QUATERNARY
M 29 276 ALASKA, QUEEN CHARLOTTE ISLAND, FAULT ZONE
M 29 280 ALASKA, QUEEN CHARLOTTE-FAIRWEATHER, CRUSTAL MODELS
B 6103 465* ALASKA, RANGE, ABST., TECTONICS, MIDDLE PALEOZOIC VOLCANISM,
B 6203 399 ALASKA, RED GLACIER FORMATION
B 6203 374 ALASKA, REDOUBT BAY
B 6408 1132 ALASKA, REDOUBT SHOAL OIL FIELD
B 6203 374 ALASKA, REDOUBT VOLCANO
B 6103 435† ALASKA RELATED TO ANTLER TECTONIC PULSES, MISSISSIPPIAN,
M 24 52 ALASKA, RICHARDSON MOUNTAINS
B 6408 1261 ALASKA, SADDLE MOUNTAIN AREA
B 6401 20 ALASKA, SADLEROCHIT FORMATION
M 24 26 ALASKA, SADLEROCHIT GROUP
B 6109 1496 ALASKA, SADLEROCHIT GROUP
B 6103 438 ALASKA, SADLEROCHIT SANDSTONE
B 6109 1499 ALASKA, SAG DELTA SECTION
M 24 40 ALASKA, SAG RIVER FORMATION
M 24 26 ALASKA, SAGAVANIRTOK FORMATION
B 6303 305 ALASKA, SANAK BASIN
M 24 26 ALASKA, SEABEE FORMATION
B 6204 635 ALASKA, SEAL ROCKS
B 6203 376 ALASKA, SELDOVIA FAULT
B 6304 700* ALASKA, SHELF BASED ON DREDGE SAMPLES FROM ADJACENT
B 6104 608 ALASKA, SHELIKOF FORMATION
B 6203 374 ALASKA, SHELIKOF STRAIT
M 24 38 ALASKA, SHUBLIK FORMATION
M 24 26 ALASKA, SHUBLIK FORMATION
B 6303 305 ALASKA, SHUMAGIN BASIN
B 6407 1016 ALASKA, SHUYAK FORMATION
B 6209 1697 ALASKA, SINIKTAMEYAK MOUNTAIN
B 6408 1142 ALASKA, SITKALIDAK FORMATION
B 6407 1016 ALASKA, SITKALIDAK FORMATION
B 6303 302 ALASKA, SITKALIDAK FORMATION
B 6408 1142 ALASKA, SITKINAK FORMATION
B 6407 1016 ALASKA, SITKINAK FORMATION
B 6303 302 ALASKA, SITKINAK FORMATION
B 6203 395 ALASKA, SITKINAK FORMATION
B 6408 1142 ALASKA, SITKINAK ISLAND
B 6104 594 ALASKA, STARICHKOFF FIELD
B 6408 1133 ALASKA, STERLING FORMATION
B 6203 377 ALASKA, STERLING FORMATION
M 24 11 ALASKA, STERLING FORMATION
B 6408 1132 ALASKA, STERLING GAS FIELD
B 6408 1142 ALASKA, STEVENSON BASIN
B 6407 1023 ALASKA, STEVENSON BASIN
B 6303 301† ALASKA, STRUCTURE AND TECTONIC SETTING OF, CONTINENTAL SHELF
B 6407 1014† ALASKA, STRUCTURE OF UPPER CENOZOIC STRATA BENEATH, KODIAK
M 24 6 ALASKA, SWANSON RIVER FIELD
B 6401 26 ALASKA, SWANSON RIVER GIANT OIL FIELD, LANDSAT DATA
B 6401 26 ALASKA, SWANSON RIVER GIANT OIL FIELD, LANDSAT DATA
M 30 28 ALASKA, SWANSON RIVER, LANDSAT IMAGE
B 6408 1132 ALASKA, SWANSON RIVER OIL FIELD
M 24 61 ALASKA, TAGLU SHALE
M 24 4 ALASKA, TALKEETNA FORMATION
B 6203 376 ALASKA, TALKEETNA FORMATION
B 6204 635 ALASKA, TARR BANK
M 25 291† ALASKA, TECTONIC FRAMEWORK OF PETROLIFEROUS ROCKS IN
B 6209 1696† ALASKA, TECTONICS OF, BROOKS RANGE, OPHIOLITES
B 6204 637 ALASKA, TERTIARY
B 6407 1016 ALASKA, TERTIARY
M 24 4 ALASKA, TERTIARY
M 25 297 ALASKA, TERTIARY
B 6203 380 ALASKA, TERTIARY
B 6408 1133 ALASKA, TERTIARY
B 6408 1142 ALASKA, TERTIARY
B 6407 1016 ALASKA, TERTIARY
B 6303 303 ALASKA, TERTIARY
M 28 172 ALASKA, TERTIARY, KOGUSUKRUK FORMATION
B 6211 2360* ALASKA TERTIARY PROVINCE, ABST., FORAMINIFERAL
B 6405 748* ALASKA TERTIARY PROVINCE, ABST., PRELIMINARY, MOLLUSCAN
B 6004 655* ALASKA TERTIARY PROVINCE, ABST., PRELIMINARY STUDY OF,
B 6012 2176* ALASKA TERTIARY PROVINCE, ABST., STRUCTURAL ELEMENTS OF
M 25 152 ALASKA, THERMAL SPRINGS
B 6303 302 ALASKA, TOLSTOI FORMATION
B 6203 386 ALASKA, TONNIE SYNCLINE
B 6002 220 ALASKA TOPPED OIL
M 25 6 ALASKA, TRADING BAY FIELD
B 6408 1132 ALASKA, TRADING BAY OIL FIELD
B 6407 1016 ALASKA, TRIASSIC
M 25 295 ALASKA, TRIASSIC
B 6203 378 ALASKA, TRIASSIC
B 6104 608 ALASKA, TRIASSIC
B 6303 307 ALASKA, TUGIDAK ANTICLINE
B 6408 1142 ALASKA, TUGIDAK BASIN
B 6303 301 ALASKA, TUGIDAK BASIN, FORE-ARC BASIN
B 6303 305 ALASKA, TUGIDAK BASIN, SEISMIC HORIZONS
B 6303 302 ALASKA, TUGIDAK FORMATION
B 6407 1016 ALASKA, TUGIDAK FORMATION
B 6408 1142 ALASKA, TUGIDAK FORMATION
B 6408 1142 ALASKA, TUGIDAK ISLAND
B 6408 1142 ALASKA, TUGIDAK UPLIFT
M 24 55 ALASKA, TUK FLEXURE ZONE

B 6203 374 ALASKA, TUXEDNI BAY
B 6203 377 ALASKA, TUXEDNI GROUP
M 24 4 ALASKA, TUXEDNI GROUP
B 6104 594 ALASKA, TYONEK FIELD
M 24 10 ALASKA, TYONEK FORMATION
B 6203 377 ALASKA, TYONEK FORMATION
B 6408 1133 ALASKA, TYONEK FORMATION
B 6103 468* ALASKA USING GEOLOGY AND GEOCHEMISTRY. BY TITLE ONLY, LARGE
B 6303 302 ALASKA, UYAK FORMATION
B 6407 1016 ALASKA, UYAK FORMATION
B 6203 395 ALASKA, VALDEZ GROUP
M 25 149 ALASKA, VOLCANIC ROCKS, RECENT
M 25 509 ALASKA, VOLCANIC-SEDIMENTARY TROUGH
B 6103 437 ALASKA, WACHSMUTH LIMESTONE
B 6204 635 ALASKA, WESSELS REEF
B 6408 1261 ALASKA, WEST FORELAND FORMATION
M 24 10 ALASKA, WEST FORELAND FORMATION
B 6203 377 ALASKA, WEST FORELAND FORMATION
B 6408 1133 ALASKA, WEST FORELAND FORMATION
B 6408 1132 ALASKA, WEST FORELAND GAS FIELD
B 6408 1132 ALASKA, WEST FORK GAS FIELD
B 6002 164 ALASKA, WESTERN, GEOLOGIC MAP
B 6203 377 ALASKA, WISHBONE FORMATION
B 6012 2185* ALASKA, WITH EMPHASIS TO ONSHORE GEOLOGY, ABST., GEOLOGIC
M 25 149 ALASKA, WRANGELL MOUNTAINS VOLCANIC PILE
B 6204 635 ALASKA, YAKUTAT BAY
M 26 355 ALASKA, YAKUTAT BAY, DEPTH SOUNDER RECORD
M 29 285 ALASKA, YAKUTAT BAY, REFRACTION DATA
B 6212 2417 ALASKA, YAKUTAT SEA VALLEY
B 6204 635 ALASKA, YAKUTAT SEA VALLEY
B 6002 172 ALASKA, YUKON-KOYUKUK GEOSYNCLINE
M 25 104† ALASKA--SUMMARY, COAL RESOURCES OF
M 25 482† ALASKA--SUMMARY, METAL PROVINCES OF
M 25 149† ALASKA--SUMMARY, POTENTIAL FOR, GEOTHERMAL-, ENERGY
B 6408 1260 ALASKA-ALEUTIAN RANGE BATHOLITH
B 6203 376 ALASKA-ALEUTIAN RANGE BATHOLITH
B 6103 468* ALASKA. BY TITLE ONLY, GEOLOGY OF, AMBLER MINERAL BELT, BROOKS
B 6103 466* ALASKA. BY TITLE ONLY, GEOTHERMAL POTENTIAL OF HIGH-LEVEL
B 6403 438* ALASKA. BY TITLE ONLY, GLACIOMARINE DEPOSITIONAL
B 6211 2351* ALASKA. BY TITLE ONLY, MOBILE BED FORMS AND, VOLCANISM, LOWER,
B 6004 727* ALASKAN ARCTIC PLATE. BY TITLE ONLY, PRUDHOE BAY SUPE, GIANT
B 6103 464* ALASKAN PLATE-TECTONIC HISTORY. BY TITLE ONLY, GRAVITATIONAL
B 6012 2174† ALASKAN SEDIMENTARY BASINS, ABST., GRAVITATIONAL EVIDENCE FOR
B 6405 743* ALASKAS EASTERN INTERIOR, ABST., CONVEYER-BELT TECTONICS AND
B 6207 1227* ALASKA—A CASE HISTORY, ABST., GEOTHERMAL EXPLORATION ON,
B 6209 1591 ALASKA—CONVERGENT MARGINS, OREGON AND
B 6405 744* ALASKA—IMPLICATIONS FOR FUTURE EXPLORATION, ABST., TWO OIL
M 25 466 ALAZEY EUGEOSYNCLINAL SYSTEM, USSR, OLOY-
B 6010 1724 ALBANIA, DEVELOPMENTS, 1975
B 6110 1654 ALBANIA, PRODUCTION, 1975-1976
B 6210 1924 ALBANIA, PRODUCTION, 1976-1977
B 6310 1932 ALBANIA, PRODUCTION, 1977-1978
B 6407 1019 ALBATROSS BANK, ALASKA
B 6408 1142 ALBATROSS BANK, ALASKA
B 6407 1018 ALBATROSS BASIN, ALASKA
B 6405 693 ALBAUGH, DENNIS S., CRYSTALLINE OVERTHRUSTING OF PALEOZOIC
B 6309 1453 ALBEMARLE EMBAYMENT, MID-ATLANTIC
M 29 143 ALBEMARLE SOUND, PROCESSES, ATLANTIC MARGIN
B 6212 2508 ALBERS, JOHN P., TECTONIC FRAMEWORK AND METALLOGENESIS IN
B 6004 644 ALBERS, W. F., USE OF COLOR DISPLAYS IN SEISMIC INTERPRETATION,
B 6004 729 ALBERSTADT, L. P., REGIONAL ENVIRONMENTA STRATIGRAPHY AND OIL
M 25 63 ALBERT, N. R., SURVEYING EARTH
M 6004 726* ALBERTA, ABST., APPLICATION OF, MONTE CARLO METHOD FOR,
B 6303 492* ALBERTA, ABST., BURIAL, DOLOMITIZATION OF UPPER, DEVONIAN,
B 6405 669* ALBERTA, ABST., DEEP BASIN
B 6406 960* ALBERTA, ABST., GEOLOGY OF, IN SITU PILOT PROJECT, WABASCA OIL
B 6405 751* ALBERTA, ABST., GEOMETRY AND MECHANISMS OF TRANSVERSE,
B 6405 673* ALBERTA, ABST., GEOPHYSICAL CASE HISTORY OF, TWO HILLS COLONY
B 6408 1278* ALBERTA, ABST., HYDROCARBON, ACCUMULATIONS IN, OVERTHRUST
B 6105 816* ALBERTA, ABST., IMPORTANCE OF DIAGENESIS IN PREDICTING, POROSITY
B 6004 730* ALBERTA, ABST., POROSITY, REDUCTION IN, REEF, MARGINS BY EARLY,
B 6405 674* ALBERTA, ABST., RETROGRADATIONAL SHELF SEDIMENTATION, VIKING
B 6303 499* ALBERTA, ABST., SEDIMENTOLOGY, PETROLOGY, AND GEOTECHNICAL
B 6405 714* ALBERTA, ABST., SIDERITE TEXTURES IN, CARDIUM FORMATION,
B 6303 469* ALBERTA, ABST., STRACHAN-, RICINUS GAS FIELD
B 6008 1413* ALBERTA, ABST., TRAPPING OF OIL AT MIDDLE, MANNVILLE
B 6008 1391* ALBERTA, ABST., VIKING FORMATION
B 6405 773* ALBERTA AND BOWDOIN DOME AREA, NORTH-CENTRAL MONTANA,
B 6302 152 ALBERTA AND BRITISH COLUMBIA, DEEP BASIN, GAS
B 6302 169 ALBERTA AND BRITISH COLUMBIA, DEEP BASIN, LOG ANALYSIS
B 6107 1014 ALBERTA AND SASKATCHEWAN, COLONY-MCLAREN-WASECA MEMBERS
B 6107 1014 ALBERTA AND SASKATCHEWAN, CUMMINGS MEMBER
B 6107 1014 ALBERTA AND SASKATCHEWAN, DINA MEMBER
B 6107 1014 ALBERTA AND SASKATCHEWAN, GENERAL PETROLEUM MEMBER
B 6107 1014 ALBERTA AND SASKATCHEWAN, LLOYDMINSTER MEMBER
B 6407 981 ALBERTA AND SASKATCHEWAN, MEDICINE HAT FIELD
B 6107 1014 ALBERTA AND SASKATCHEWAN, REX MEMBER
B 6107 1014 ALBERTA AND SASKATCHEWAN, SPARKY MEMBER
B 6107 1010† ALBERTA AND SASKATCHEWAN, TRAPPING OF OIL AT INTRA-,
B 6205 784 ALBERTA, BASHAW COMPLEX
B 6407 972 ALBERTA, BATTLE FORMATION
B 6407 972 ALBERTA, BEARPAW SHALE
B 6112 2135 ALBERTA, BEAVERHILL LAKE FIELD
B 6205 789 ALBERTA, BEAVERHILL LAKE GROUP
M 24 82 ALBERTA, BEAVERHILL LAKE GROUP
B 6111 1954 ALBERTA, BELLSHILL LAKE HEAVY OIL
M 26 429 ALBERTA, BELLY RIVER FORMATION, CANADA
B 6407 972 ALBERTA, BELLY RIVER GROUP
B 6408 1166 ALBERTA, BELLY RIVER SANDSTONE
B 6104 594 ALBERTA, BENTLEY FIELD
B 6112 2135 ALBERTA, BON ACCORD FIELD
B 6407 981 ALBERTA, BOW ISLAND FORMATION
B 6205 785 ALBERTA, CAMBRIAN
B 6303 534* ALBERTA, CANADA, ABST., CLAY-, CARBONATE DIAGENESIS OF, DELTAIC
B 6303 481* ALBERTA, CANADA, ABST., DEPOSITIONAL ENVIRONMENTS AND
B 6405 735* ALBERTA, CANADA, ABST., DIAGENETIC FEATURES OF, GRAND RAPIDS
B 6405 776* ALBERTA, CANADA, ABST., FACIES CONTROL ON, BITUMEN SATURATION,
M 24 79† ALBERTA, CANADA, KAYBOB OIL FIELD
M 30 315† ALBERTA, CANADA, STRACHAN AND, RICINUS WEST, GAS FIELDS
M 26 435 ALBERTA, CARDIUM FORMATION, CANADA
B 6408 1166 ALBERTA, CARDIUM SANDSTONE
B 6107 1086 ALBERTA, CARDIUM SANDSTONE, FOOTHILLS AND FORELAND
B 6109 1421 ALBERTA, CENTRAL, REGIONAL CROSS SECTION
B 6107 1021 ALBERTA, CHAUVIN SOUTH FIELD
B 6207 1117 ALBERTA, CLARESHOLM AREA STUDY
B 6104 597 ALBERTA, CLIVE FIELD
B 6205 782 ALBERTA, COLD LAKE SALT
B 6407 981 ALBERTA, COLORADO GROUP
B 6408 1166 ALBERTA, COLORADO SHALES
B 6205 783 ALBERTA, COOKING LAKE BANK
B 6205 785 ALBERTA, CRETACEOUS
B 6407 981 ALBERTA, CRETACEOUS
B 6302 162 ALBERTA, DEEP BASIN, ELECTRIC LOG SECTION
M 30 315 ALBERTA, DEEP BASIN, EXPLORATION
B 6108 1315 ALBERTA, DEVELOPMENTS, 1976
B 6308 1408 ALBERTA, DEVELOPMENTS, 1978
B 6409 1511 ALBERTA, DEVELOPMENTS, 1979
M 28 161 ALBERTA, DEVONIAN, CAIRN FORMATION, CANADA
M 28 154 ALBERTA, DEVONIAN, CAIRN FORMATION, CANADA
M 28 10 ALBERTA, DEVONIAN, CAIRN FORMATION, CANADA
M 28 153 ALBERTA, DEVONIAN, CAIRN FORMATION, CANADA
M 30 315 ALBERTA, DEVONIAN, RICINUS WEST
M 30 315 ALBERTA, DEVONIAN, STRACHAN FIELD
B 6112 2086 ALBERTA, DOIG FORMATION, KEROGEN
B 6205 783 ALBERTA, DUVERNAY FORMATION
B 6407 972 ALBERTA, EASTEND FORMATION
B 6407 972 ALBERTA, EDMONTON GROUP
S 10 157 ALBERTA, ETZIKOM COULEE, ANALOG FLOW PATTERN
B 6104 594 ALBERTA, FERRIER FIELD
B 6112 2137 ALBERTA, FERRIER OIL FIELD
B 6407 981 ALBERTA, FIRST WHITE SPECKS
B 6407 981 ALBERTA, FISH SCALES SANDSTONE
B 6407 972 ALBERTA, FOREMOST FORMATION
B 6407 972 ALBERTA, FRENCHMAN FORMATION
B 6112 2135 ALBERTA, FT. SASKATCHEWAN FIELD
M 24 82 ALBERTA, FT. VERMILLION FORMATION
B 6104 594 ALBERTA, GILBY FIELD
B 6112 2137 ALBERTA, GILBY OIL FIELD
M 24 73 ALBERTA, GILWOOD SANDSTONE
M 26 196 ALBERTA, GOLDEN SPIKE, CARBONATE BUILDUP, SEISMIC EXAMPLE,
B 6205 783 ALBERTA, GOLDEN SPIKE REEF, SEISMIC DATA
B 6104 597 ALBERTA, HOMEGLEN-RIMBEY FIELD
S 10 65 ALBERTA, HYDRAULIC HEAD DISTRIBUTION
B 6104 596 ALBERTA, INNISFAIL FIELD
B 6205 784 ALBERTA, INNISFAIL FIELD, SEISMIC DATA
M 26 197 ALBERTA, INNISFAIL, REEF BUILDUP, SEISMIC EXAMPLE, CANADA
B 6205 783 ALBERTA, IRETON FORMATION

B	6205	735	ALBERTA, IRETON-DUVERNAY SHALE
B	6109	1421	ALBERTA, JOARCAM FIELD
B	6112	2135	ALBERTA, JOARCAM FIELD
B	6104	594	ALBERTA, JOFFRE FIELD
B	6207	1116	ALBERTA, JOINTING, REGIONAL
B	6104	597	ALBERTA, JUDY CREEK FIELD
B	6112	2135	ALBERTA, JUDY CREEK FIELD
B	6205	739	ALBERTA, JUDY CREEK FIELD, PRODUCTION
B	6205	739	ALBERTA, JUDY CREEK REEF AND BANK COMPLEX, SEISMIC DATA
M	26	232	ALBERTA, JUDY CREEK, SHALLOW-WATER CARBONATE BUILDUP,
M	24	30	ALBERTA, KAYBOB REEF COMPLEX
M	24	33	ALBERTA, KAYBOB REEF COMPLEX
B	6002	246	ALBERTA, KEG RIVER FORMATION
B	6205	732	ALBERTA, KEG RIVER FORMATION
B	6112	2137	ALBERTA, LEAFLAND FIELD
B	6104	594	ALBERTA, LEAFLAND FIELD
B	6104	596	ALBERTA, LEDUC FIELD
B	6109	1421	ALBERTA, LEDUC FIELD
M	30	315	ALBERTA, LEDUC FORMATION, STRACHAN-RICINUS PRODUCTION
B	6205	733	ALBERTA, LEDUC-RIMBEY BARRIER REEF TREND
B	6107	1040	ALBERTA, LLOYDMINSTER AREA, HISTORY
S	10	52	ALBERTA, LOWER CRETACEOUS, ORIGIN OF OIL SANDS
M	24	32	ALBERTA, MAJEAU LAKE FORMATION
B	6104	596	ALBERTA, MALMO FIELD
B	6104	597	ALBERTA, MALMO FIELD
B	6107	1040	ALBERTA, MANNVILLE GROUP
B	6408	1156	ALBERTA, MANNVILLE SANDSTONE
B	6401	29	ALBERTA, MCCONNELL THRUST FAULT
B	6007	1112	ALBERTA, MCMURRAY SANDS, ORIGIN OF OIL IN
B	6007	1110	ALBERTA, MCMURRAY SANDS, ORIGIN OF OIL IN
B	6407	981	ALBERTA, MEDICINE HAT SANDSTONE
B	6408	1154	ALBERTA, MESOZOIC, SANDSTONE WEDGES
B	6107	1080	ALBERTA, MIETTE REEF COMPLEX
B	6407	981	ALBERTA, MILK RIVER EQUIVALENT
B	6302	155	ALBERTA, MILK RIVER GAS FIELD
M	24	72	ALBERTA, MITSUE OIL FIELD
M	24	73	ALBERTA, MUSKEG FORMTION
B	6205	782	ALBERTA, MUSKEG SALT
B	6407	972	ALBERTA, OLDMAN FORMATION
B	6407	981	ALBERTA, PAKOWKI FORMATION
B	6207	1119	ALBERTA, PASKAPOO FORMATION
B	6408	1156	ALBERTA, PASKAPOO-EDMONTON FACIES
M	24	73	ALBERTA, PEACE RIVER UPLANDS
B	6109	1421	ALBERTA, PEMBINA FIELD
B	6104	594	ALBERTA, PEMBINA FIELD
B	6104	486	ALBERTA, PEMBINA FIELD, TRAP
B	6108	1322	ALBERTA, PRODUCTION, 1976
B	6002	246	ALBERTA, RAINBOW LAKE, A AND B POOLS
M	26	195	ALBERTA, RAINBOW-ZAMA AREA, PINNACLE REEFS, SEISMIC EXAMPLE,
B	6205	782	ALBERTA, RAINBOW-ZAMA AREA, SEISMIC DATA
B	6207	1118	ALBERTA, RED DEER AREA STUDY
S	10	141	ALBERTA, RED EARTH REGION, GROUNDWATER FLOW
S	10	125	ALBERTA, RED EARTH REGION, HYDRAULIC DISTRIBUTION
S	10	141	ALBERTA, RED EARTH REGION, HYDROGEOLOGIC CROSS SECTION
S	10	154	ALBERTA, RED EARTH REGION, PETROLEUM MIGRATION
S	10	140	ALBERTA, RED EARTH REGION, STRATIGRAPHY, CANADA
B	6104	597	ALBERTA, REDWATER FIELD
B	6002	246	ALBERTA, REEFS, DEVONIAN
M	30	30	ALBERTA, RICINUS FIELD, LANDSAT IMAGE
B	6401	28	ALBERTA, RICINUS GIANT GAS FIELD, LANDSAT DATA
M	30	327	ALBERTA, RICINUS WEST FIELD, RESERVES AND PRODUCTION
B	6407	981	ALBERTA, SECOND WHITE SPECKS
B	6205	782	ALBERTA, SLAVE POINT FORMATION
M	24	73	ALBERTA, SLAVE POINT FORMATION
M	30	30	ALBERTA, STRACHAN FIELD, LANDSAT IMAGE
M	30	324	ALBERTA, STRACHAN FIELD, LEDUC REEF
M	30	322	ALBERTA, STRACHAN FIELD, RESERVES AND PRODUCTION
M	30	322	ALBERTA, STRACHAN FIELD, SEISMIC SECTION
B	6401	28	ALBERTA, STRACHAN GIANT GAS FIELD, LANDSAT DATA
B	6401	28	ALBERTA, STRACHAN GIANT GAS FIELD, LANDSAT DATA
B	6407	981	ALBERTA, SUFFIELD BLOCK
B	6407	983	ALBERTA, SUFFIELD BLOCK GAS RESERVES
B	6205	782	ALBERTA, SULPHUR POINT FORMATION
M	24	81	ALBERTA, SWAN HILLS FORMATION
M	24	80	ALBERTA, SWAN HILLS PLATFORM
M	24	85	ALBERTA, SWAN HILLS, REEF
B	6112	2086	ALBERTA, TRIASSIC, KEROGEN
B	6109	1420	ALBERTA, VIKING FORMATION, HYDROCARBONS
B	6408	1166	ALBERTA, VIKING SANDSTONE
B	6112	2134	ALBERTA, VIKING SANDSTONES, SUBNORMAL PRESSURES
M	24	82	ALBERTA, WATERWAYS FORMATION
M	24	73	ALBERTA, WATT MOUNTAIN FORMATION
B	6104	597	ALBERTA, WESTEROSE FIELD
B	6112	2135	ALBERTA, WESTLOCK FIELD
B	6407	972	ALBERTA, WHITEMUD FORMATION
M	24	83	ALBERTA, WOODBEND GROUP
M	30	318	ALBERTA, WOODBEND GROUP, FACIES DISTRIBUTION MAP
M	26	198	ALBERTA, YEKAN LAKE, CARBONATE BUILDUP, SEISMIC EXAMPLE,
B	6205	785	ALBERTA, YEKAU LAKE REEF, SEISMIC DATA
M	24	72†	ALBERTA--A RICH STRATIGRAPHIC TRAP, MITSUE OIL FIELD
B	6303	499*	ALBERTA. BY TITLE ONLY, SEDIMENTOLOGY, PETROLOGY, AND
B	6203	565*	ALBERTA. BY TITLE ONLY., HEAVY, OIL AND, TAR, SANDS OF
B	6203	572*	ALBERTA. BY TITLE ONLY., HEAVY, OILS, SASKATCHEWAN AND
B	6204	669	ALBIAN, ALGERIA
B	6303	503*	ALBIAN BLACK SHALES FROM SOUTH ATLANTIC, DSDP LEGS 36,
S	6	279	ALBIAN, FORAMINIFERAL ZONES, CRETACEOUS
S	4	87	ALBIAN, JAMAICA, RIO NUEVO FORMATION
S	4	87	ALBIAN, JAMAICA, TIBER FORMATION
B	6101	44	ALBIAN, LLBANON
B	6002	211	ALBIAN MODEL, MATCHING WEST AFRICAN AND BRAZILIAN SALT EDGES
B	6412	2055	ALBIAN, ORGANIC MATTER, ATLANTIC BASINS, APTIAN-
S	6	132	ALBIAN, POTASSIUM-ARGON, FRANCE, PARIS BASIN
B	6405	614	ALBIAN SEQUENCE, MIDDLE EAST, BAJOCIAN TO
B	6109	1522†	ALBIAN SUBSIDENCE RATES IN EASTERN UNITED STATES, RAPID, APTIAN-
B	6401	75	ALBIAN THROUGH SANTONIAN, ENGLAND, ISLE OF WIGHT
B	6401	75	ALBIAN THROUGH SANTONIAN, ENGLAND, ISLE OF WIGHT
S	6	263	ALBIAN-CENOMANIAN BOUNDARY, NORTH AMERICA, CRETACEOUS
B	6303	429*	ALBIAN-CENOMANIAN, CAMPOS BASIN, OFFSHORE RIO DE JANEIRO,
B	6303	492*	ALBIAN-CENOMANIAN, NORTHERN ALASKA, ABST., DINOFLAGELLATE
B	6109	1480	ALBIAN-SANTONIAN PETROLEUM RESERVOIR SYSTEM, BRAZIL
B	6405	802*	ALBIAN-TURONIAN, DEPOSITIONAL ENVIRONMENTS ALONG A PART OF
B	6002	258	ALBIAN-TURONIAN STRATA, ISRAEL
M	28	22	ALBITE, FELDSPAR, AUTHIGENIC
M	28	17	ALBITE TWINNING, FELDSPAR
M	28	16	ALBITE TWINNING, FELDSPAR
B	6411	1968	ALBORAN SEA, SPAIN
B	6402	219	ALBORZ-5 FIELD, IRAN
B	6402	219	ALBORZ-5 FIELD, IRAN
M	30	173	ALBRIGHT, W. A., NINIAN FIELD, U.K. SECTOR, NORTH SEA
M	25	140	ALCARAZ, ARTURO, GEOTHERMAL EXPLORATION AND DEVELOPMENT IN
S	10	29	ALCOHOL INTERACTION, MIGRATION PROCESSES
S	10	21	ALCOHOLS IN WATER, MIGRATION, SOLUBILITIES OF
B	6008	1389	ALDERMAN, WILLIS H., USE OF INTERACTIVE COMPUTER GRAPHICS FOR
B	6211	2353*	ALDRICH MOUNTAINS INLIER, CENTRAL OREGON, ABST., MESOZOIC
B	6305	804	ALEDO FIELD, OKLAHOMA
B	6010	1770	ALEM, A., AND PETROCONSULTANTS S.A., ON DEVELOPMENTS
B	6110	1701	ALEM, A., AND PETROCONSULTANTS S.A., ON DEVELOPMENTS IN
B	6310	1639	ALEM, AMID, AND PETROCONSULTANTS S.A., ON DEVELOPMENTS IN
B	6210	1799	ALEM, AMID, AND PETROCONSULTANTS S.A., ON DEVELOPMENTS IN
B	6405	668	ALEMAN, ANTENOR, EARLY CRETACEOUS ARC SEDIMENTATION AND
B	6404	530	ALEQATSIAQ FJORD FORMATION, GREENLAND
B	6404	528	ALEQATSIAQ FJORD, GREENLAND
B	6102	193	ALEUTIAN BASIN
B	6201	88	ALEUTIAN BASIN, ALASKA
B	6403	439*	ALEUTIAN BASIN, BERING SEA, ABST., HYDROCARBON POTENTIAL OF
B	6403	439*	ALEUTIAN BASIN, BERING SEA, ABST., HYDROCARBON POTENTIAL OF
B	6311	2070†	ALEUTIAN BASIN, BERING SEA, HYDROCARBON POTENTIAL OF
B	6311	2070†	ALEUTIAN BASIN, BERING SEA, HYDROCARBON POTENTIAL OF
B	6207	1228*	ALEUTIAN ISLANDS, ALASKA PENINSULA, AND SHUMAGIN KODIAK
B	6203	374	ALEUTIAN RANGE, ALASKA
B	6203	376	ALEUTIAN RANGE BATHOLITH, ALASKA-
B	6408	1260	ALEUTIAN RANGE BATHOLITH, ALASKA-
B	6102	193	ALEUTIAN RIDGE
B	6201	87†	ALEUTIAN RIDGE, NEOGENE, VOLCANICLASTIC SEDIMENTS FROM, ATKA
B	6201	88	ALEUTIAN TERRACE, ALASKA
B	6102	193	ALEUTIAN TRENCH
B	6201	88	ALEUTIAN TRENCH, ALASKA
B	6408	1142	ALEUTIAN TRENCH, ALASKA
B	6407	1028	ALEUTIAN TRENCH, ALASKA
B	6209	1593	ALEUTIAN TRENCH, ALASKA CONTINENTAL MARGIN
B	6105	838*	ALEUTIAN TRENCH AND ADJACENT CONTINENTAL MARGIN OFF KODIAK
M	25	485	ALEUTIAN TRENCH, SUBDUCTION
M	25	149	ALEUTIAN VOLCANIC ARC, CALDERAS, ALASKA
M	25	474	ALEUTIAN- CHUKOTSK MEGASTRUCTURE
M	25	280†	ALEUTIAN-BERING SEA REGION--SUMMARY, CENOZOIC AN, MESOZOIC
B	6004	645	ALEXANDER, C. S., SOME PROBLEMATIC LATE CENOZOIC CARIBBEAN
B	**6105**	**759**	**ALEXANDER, PETER, DEVELOPMENT OF GEOPHYSICAL TECHNIQUES FOR**
M	25	509	ALEXANDER TERRANE, CORDILLERAN OROGEN, INSULAR BELT OR
B	6106	875	ALEXANDRIA CANYON, EGYPT

B 6009 1623* ALEXANDRIAN AND EARLY NIAGARAN ROCKS, SILURIAN, IN INDIANA
B 6311 2118* ALFALFA COUNTY, OKLAHOMA, ABST., GEOCHEMISTRY OF SMALL,
B 6203 537* ALGA FROM EOCENE BRIDGER AND GREEN RIVER FORMATIONS,
M 27 11 ALGAE, CHAROPHYTES, GREEN
B 6107 1069 ALGAE, ENDOLITHIC, BORING
M 27 7 ALGAE, HALIMEDA, GREEN
M 27 3 ALGAE, HOLOCENE
S 4 239† ALGAE IN DEVELOPMENT OF DEPOSITIONAL TOPOGRAPHY--REALITY O
B 6107 1069† ALGAE, MICRITIZATION AND, CARBONATE-, GRAIN, BINDING BY,
S 4 243 ALGAE, PENNSYLVANIAN-, PERMIAN LIMESTONES, PHYLLOID
S 4 44 ALGAE-PORITES ASSEMBLAGE, BELIZE
S 4 43 ALGAL ASSEMBLAGES, BELIZE, GLOVERS REEF ATOLL, CORAL-
M 27 3 ALGAL BODIES, BLUE-GREEN
M 27 183 ALGAL BORINGS
M 27 193 ALGAL BORINGS
B 6107 1070 ALGAL BORINGS, INFILLING OF
M 27 127 ALGAL BORINGS, OOIDS
B 6112 2115† ALGAL BUILDUP, SACRAMENTO MOUNTAINS, NEW MEXICO, EVOLUTION
B 6203 527* ALGAL CARBONATE REEFS FROM SNAKE RIVER PLAIN, IDAHO, ABST.,
B 6203 541* ALGAL COLUMNS IN LATE TERTIARY FURNACE CREEK FORMATION,
M 27 47 ALGAL CRUST, BLUE-GREEN
M 27 190 ALGAL CRUSTS
M 27 4 ALGAL ENCRUSTATION, BLUE-GREEN
M 27 13 ALGAL ENCRUSTING, RED
B 6107 1077 ALGAL FILAMENTS AND NATURE OF GIRVANELLA, CALCIFIED
M 27 9 ALGAL FRAGMENT, DASYCLADACEAN, GREEN
M 27 13 ALGAL FRAGMENT, DOLOMITE REPLACEMENT, RED
M 27 12 ALGAL GRAIN, GONIOLITHON, RED
M 27 10 ALGAL GRAINS, PHYLLOID
M 27 131 ALGAL GRAINS, RED
M 27 3 ALGAL MAT, LAMINATED, BLUE-GREEN
S 5 159 ALGAL MOUNDS, CANADA, SVERDRUP BASIN, OTTO FIORD FORMATION,
B 6111 1941 ALGAL MOUNDS, EVOLUTION OF
M 27 6 ALGAL NODULE, TUBULES EXAMPLE
M 27 189 ALGAL ONCOLITE, BLUE-GREEN
B 6004 696* ALGAL, PISOLITES AND PISOLITIC CALICHE IN TEXAS, ABST., CARBON
M 27 6 ALGAL STROMATOLITE, BLUE-GREEN
M 27 4 ALGAL STROMATOLITE, BLUE-GREEN
B 6203 535* ALGAL STROMATOLITES IN EUREKA QUARTZITE, SOUTHEASTERN
B 6004 612† ALGAL STRUCTURES, BURROWERS, AND GRAZERS IN DOGGER, JURASSIC,
B 6203 514* ALGAL TUFA FROM SHORES OF ANCESTRAL DEAD SEA, ABST.
B 6405 767* ALGAL-METAZOAN BIOHERMS OF LOWER ORDOVICIAN AGE--ST. GEORGE
B 6307 1108† ALGAL-RUDIST REEFS, ARIZONA, DEPOSITIONAL, MODEL OF EARLY,
B 6010 1768 ALGERIA, 1975, PETROCONSULTANTS S.A., ON DEVELOPMENTS IN
B 6110 1699 ALGERIA, 1976, PETROCONSULTANTS S.A., ON DEVELOPMENTS IN
B 6210 1797 ALGERIA, 1977, PETROCONSULTANTS S.A., ON DEVELOPMENTS IN
B 6310 1637 ALGERIA, 1978, PETROCONSULTANTS S.A., ON DEVELOPMENTS IN
B 6411 1740 ALGERIA, 1979, PETROCONSULTANTS S.A., ON DEVELOPMENTS IN
B 6204 669 ALGERIA, ALBIAN
B 6204 669 ALGERIA, APTIAN
B 6204 669 ALGERIA, BARREMIAN
B 6204 681 ALGERIA, CAMBRIAN-ORDOVICIAN
B 6204 672 ALGERIA, CARBONIFEROUS
B 6204 669 ALGERIA, CENOMANIAN-TURONIAN
B 6204 672 ALGERIA, DEVONIAN, D UNIT
B 6204 672 ALGERIA, DEVONIAN, F UNIT
B 6412 2111 ALGERIA, DEVONIAN SHALES
B 6204 672 ALGERIA, DEVONIAN-SILURIAN RESERVOIRS, F6 UNITS
B 6204 668 ALGERIA, EDJELEH OIL FIELD
B 6406 860 ALGERIA, GHADAMES BASIN, LIBYA, TUNISIA, AND
B 6204 668 ALGERIA, HASSI MESSAOUD OIL FIELD
B 6204 668 ALGERIA, HASSI RMEL OIL FIELD
B 6204 668 ALGERIA, ILLIZI BASIN
B 6204 669 ALGERIA, ILLIZI BASIN, STRATIGRAPHIC SECTION
B 6204 669 ALGERIA, JURASSIC
B 6204 669 ALGERIA, LIAS
B 6204 670 ALGERIA, MESSAOUD AXIS
B 6204 668 ALGERIA, MZAB BASIN
B 6204 669 ALGERIA, NAMURIAN
B 6204 669 ALGERIA, NEOCOMIAN
B 6303 409* ALGERIA, NORTH AFRICA, BY TITLE ONLY, OUARGLA REGION
B 6204 668 ALGERIA, NORTHEAST SAHARA BASIN
M 30 540 ALGERIA, OUARGLA REGION, BEN-KAHLA FIELD
M 30 540 ALGERIA, OUARGLA REGION, BOU-KHEZANA FIELD
M 30 540 ALGERIA, OUARGLA REGION, DRAA ET TEMRA FIELD
M 30 540 ALGERIA, OUARGLA REGION, GARET ECH CHOUF FIELD
M 30 540 ALGERIA, OUARGLA REGION, GUELLALA FIELD
M 30 540 ALGERIA, OUARGLA REGION, GUELLALA NORD-EST FIELD
M 30 540 ALGERIA, OUARGLA REGION, HANIET-EL-BEIDA FIELD
M 30 540 ALGERIA, OUARGLA REGION, HAOUD-BERKAOUI FIELD
M 30 540 ALGERIA, OUARGLA REGION, KEF EL ARGOUB FIELD
M 30 540 ALGERIA, OUARGLA REGION, N GOUSSA FIELD
M 30 540 ALGERIA, OUARGLA REGION, OUARGLA FIELD
M 30 540 ALGERIA, OUARGLA REGION, TAKHOUKT FIELD
M 30 539† ALGERIA, PETROLEUM POTENTIAL--, OUARGLA REGION, TRIASSIC BASIN
B 6110 1711 ALGERIA, PRODUCTION, 1970-1976
B 6210 1808 ALGERIA, PRODUCTION, 1976-1977
B 6310 1642 ALGERIA, PRODUCTION, 1978
B 6411 1758 ALGERIA, PRODUCTION, 1979
B 6204 669 ALGERIA, SENONIAN
B 6204 668 ALGERIA, SOUTH ATLAS FLEXURE
B 6204 668 ALGERIA, TEN FOUYE OIL FIELD
B 6204 669 ALGERIA, TIGUENTOURINE
B 6204 682 ALGERIA, TIN FOUYE-TABANKORT DEVONIAN OIL FIELD
B 6204 667 ALGERIA, TRIASSIC
B 6204 669 ALGERIA, VISEAN
B 6204 669 ALGERIA, VRACONIAN
B 6204 669 ALGERIA, WESTPHALIAN
M 29 295 ALGERIAN BASIN, CROSS SECTION, MEDITERRANEAN
B 6204 669 ALGERIAN SAHARA, EASTERN, HERCYNIAN UNCONFORMITY
B 6204 667† ALGERIAN SAHARA—INFLUENCE ON HYDROCARBON OCCURRENCE,
S 3 89 ALGONKIAN ALKALIC GRANITE, BRAZIL
B 6004 696 ALI, A. JAWAD, CONTOURING GEOLOGIC DATA FROM ARBITRARY
B 6105 812 **ALI, A. JAWAD, CONTOURING GEOLOGIC DATA FROM ARBITRARY POINTS**
B 6203 370 ALIJOS ROCKS, BAJA CALIFORNIA
B 6104 607† ALINCHAK BAY AREA, ALASKA PENINSULA, SOME LOWER AND MIDDLE,
B 6405 784* ALINE-LAMBERT FIELDS, OKLAHOMA—SOURCE, RESERVOIR, AND TRAP,
B 6212 2441† ALIPHATIC ACID ANIONS IN OIL-FIELD WATERS—IMPLICATIONS FOR
B 6003 452† ALIPHATIC FATTY ACIDS IN HOLOCENE SEDIMENT, THERMALLY
S 6 263 ALISITOS FORMATION, MEXICO, CRETACEOUS
B 6203 535* ALISO CANYON FIELD, EASTERN VENTURA BASIN, CALIFORNIA, ABST.,
S 3 135 ALKALIC VOLCANIC CENTERS, HYDROTHERMAL-VEIN FORMATION
B 6207 1154 ALKANES IN OILS AND EXTRACTS, N-
B 6203 489 ALLAN, J. R., CARBON AND OXYGEN ISOTOPES AS INDICATORS OF
B 6012 2150 ALLAUD, L., VERTICAL NET SANDSTONE DETERMINATION FOR ISOPACH
B 6101 80 ALLEGHANIAN OROGENE, NORTH AMERICA
B 6212 2396 ALLEGHENY FORMATION, WEST VIRGINIA TO, PENNSYLVANIA
B 6304 647† ALLEGHENY FRONTAL ZONE, CONODONT-BASED REINTERPRETATION OF,
B 6307 1059† ALLEN BAY CARBONATE ROCKS AND CONTIGUOUS SHELF AND BASIN
B 6203 489 ALLEN, ELIZABETH A., ANALOGIES BETWEEN HOLOCENE
B 6405 668 ALLEN, ELIZABETH A., EVALUATION OF TEMPERATE ZONE
B 6105 805 * **ALLEN, ELIZABETH A., STRUCTURE OF COASTAL WASHOVER FANS AND**
B 6105 759 **ALLEN, ELIZABETH A., THIN SECTION ANALYSIS OF COASTAL-MARSH**
B 6405 669 ALLEN, FRASER H., DEEP BASIN, ALBERTA, ABST.
B 6101 113 ALLEN HOLLOW SHALE MEMBER, UTAH
B 6203 507 ALLEN, RUSSELL W., GEOLOGIC APPRAISAL OF UNDISCOVERED OIL AND
B 6308 1310 ALLEN, W. W., DEVELOPMENTS IN LOUISIANA GULF COAST OFFSHORE IN
B 6311 2060 ALLENBY FORMATION, BRITISH COLUMBIA
M 27 211 ALLENTOWN DOLOMITE, PENNSYLVANIA, CAMBRIAN
B 6201 126 ALLIGATOR REEF FAULT, FLORIDA
B 6203 553 ALLISON, CAROL W., PRECAMBRIAN-CAMBRIAN SEDIMENTARY ROCKS
B 6405 669 ALLMENDINGER, RICHARD W., TECTONIC SIGNIFICANCE OF
M 25 509 ALLOCHTHON, FORMATION OF, CORDILLERAN OROGEN, PALEOZOIC
B 6203 538* ALLOCHTHON, NORTH-CENTRAL NEVADA, ABST., ELDER SANDSTONE--,
B 6011 2044† **ALLOCHTHONOUS AND AUTOCHTHONOUS NANNOFOSSILS OF MIDDLE**
B 6011 2038† ALLOCHTHONOUS CARBONATE DEBRIS IN MESOZOIC FLYSCH DEPOSITS
B 6303 429* ALLOCHTHONOUS NATURE OF NUMIDIAN FLYSCH COMPLEX, WESTERN
B 6305 828* ALLUVIAL FANS AND THEIR DEPOSIT IN DEVONIAN HORNELEN BASIN,
B 6203 531* ALLUVIAL-DELTAIC-COASTAL SEDIMENTARY-FACIES PATTERNS IN
B 6103 367 ALLUVIAL-FAN DEPOSIT OKLAHOMA WICHITA UPLIFT, ALLUVIAL-FAN
B 6303 437* ALLUVIAL-FAN DEPOSITION AND TECTONIC SIGNIFICANCE OF TWO LATE
B 6303 493* ALLUVIAL-PLAIN DEPOSITS, SIERRA NEVADA FOOTHILLS, CALIFORNIA,
B 6305 829* ALLUVIAL-PLAIN SEDIMENTS OF NUBIA, SOUTHWESTERN EGYPT, ABST.
B 6211 2359* ALLUVIUM BY NATURAL GAS IN PYRAMID LAKE AREA, NEVADA, ABST.,
B 6007 1025 ALMA, RESEARCH LAUNCH
B 6109 1471 ALMADA BASIN, BRAZIL
B 6002 197 ALMADA COASTAL BASIN, BRAZIL
B 6303 324 ALMAGOR, GIDEON AND, SUBMARINE SLUMPING IN CONTINENTAL
B 6402 188† ALMERIA PROVINCE, SOUTHERN SPAIN, POROSITY EVOLUTION OF
B 6004 645 ALMGREN, A. A., TIMING OF TERTIARY SUBMARINE CANYONS AND
M 25 434 ALMOGELA, DOMINADOR H., GEOLOGIC ENVIRONMENT AND ECONOMI
B 6209 1756 ALMON, WILLIAM R., CLAY TECHNOLOGY AND WELL STIMULATION,
B 6108 1393 ALMON, WILLIAM R., DIAGENESIS OF OFFSHORE-BAR SANDSTONES OF
B 6108 1375 ALMON, WILLIAM R., DIAGENESIS OF TERTIARY VOLCANICLASTICS,
B 6109 1535 ALMON, WILLIAM R., EFFECTS OF SANDSTONE COMPOSITION AND
B 6408 1276 ALMON, WILLIAM R., IMPACT OF DIAGENESIS ON EXPLORATION
B 6309 1595 ALMON, WILLIAM R., IMPACT OF DIAGENESIS ON LOG INTERPRETATION,
B 6303 409 ALMON, WILLIAM R., MECHANISMS OF FORMATION DAMAGE IN
B 6203 490 ALMON, WILLIAM R., REGIONAL CONTROLS ON DIAGENESIS OF
B 6108 1371 ALMON, WILLIAM R., REGIONAL DIAGENETIC TRENDS IN LOWER
B 6203 508 ALMON, WILLIAM R., RESERVOIR POTENTIAL OF VOLCANOGENIC

B	**6105**	**784**	**ALMON, WILLIAM R., BEREA, MISSISSIPPIAN, SANDSTONE CORE FROM**
B	6409	1555	ALMON, WILLIAM R., RESERVOIR QUALITY, PLIOCENE-PLEISTOCENE
B	**6105**	**759**	**ALMON, W. R., DIAGENESIS IN FRONTIER FORMATION, SPEARHEAD RANCH**
B	6402	225	ALMOND GRABEN, TEXAS
B	6212	2508	ALONSO ESPINOSA, HECTOR, DEVELOPMENT OF CERRO PRIETO
B	6203	490	ALONZO E., H., RECENT DEVELOPMENTS AT CERRO PRIETO GEOTHERMAL
S	3	49	ALPHA RECOIL PROCESS
B	6303	544*	ALPINE ACTIVITY ON CRATONIC MARGIN OF NORTH AFRICA, ABST.,
B	6005	794†	ALPINE HIMALAYAN GEOSYNCLINE AND ITS PETROLEUM POSSIBILITIES,
B	6004	654*	ALPINE MOBILE BELT AND FORELANDS, ABST., PALEOSEDIMENTOLOGIC,
B	6301	69	ALPINE-CARPATHIAN OROGENE, EUROPE
B	6403	359†	ALPINOTYPE OROGENIES AND CHANGES IN OCEAN-FLOOR SPREADING
B	6403	359†	ALPINOTYPE OROGENIES AND CHANGES IN OCEAN-FLOOR SPREADING
B	6403	366	ALPINOTYPE OROGENIES, WORLDWIDE
B	6301	82	ALPS, ANNOT FORMATION, MARITIME
B	6110	1639	ALPS ZONE, AUSTRIA, FLYSCH-LIMESTONE
B	6411	1964	ALSACE AND JURA, FRANCE
B	6110	1641	ALSACE, FRANCE
B	6406	960*	ALSANDS PROJECT—A CHALLENGE IN PETROLEUM AND MINING
B	6203	533*	ALTAI MOUNTAINS, ABST., LACUSTRINE STROMATOLITES FROM,
B	6111	2002	ALTAMIRA MEMBER, CALIFORNIA
B	6202	259	ALTAMIRA SHALE, CALIFORNIA
B	6111	1956	ALTAMIRA-SUATA HEAVY OIL, VENEZUELA
B	6405	755*	ALTAMONT FIELD, UINTA BASIN, UTAH, ABST., ORIGIN OF SUBSURFACE,
B	6107	1046	ALTAMONT FORMATION, KANSAS
M	24	121†	ALTAMONT-BLUEBELL--A MAJOR, NATURALLY FRACTURED
B	6405	716*	ALTAR DESERT, SONORA, MEXICO, ABST., PETROLEUM POSSIBILITIES IN
M	26	31	ALTERATION, LABORATORY, MEASUREMENT, HEAT
B	6211	2359*	ALTERATION OF ALLUVIUM BY NATURAL GAS IN PYRAMID LAKE AREA,
B	6204	687†	ALTERATION OF KEROGEN FROM DIFFERENT SEDIMENTS,
B	6004	651*	ALTERED VOLCANIC ASH PARTINGS AS STRATIGRAPHIC MARKER BEDS
B	**6105**	**800**	**ALTERNATE GEOLOGIC INTERPRETATION—POTENTIAL USE OF COMPUTER**
B	6004	645	ALTHAUS, V. E., CLASTIC AND STRATIGRAPHIC TRAP ANALYSIS BY
S	6	107	ALTO PARAGUAY GROUP, SEPOTUBA FORMATION, BRAZIL
B	6004	693*	ALTOS CUCHUMATANES MOUNTAINS, WESTERN GUATEMALA, ABST.,
B	6203	490	ALTSCHULER, Z. S., PROFILE OF COAL-FORMING SWAMP—THE
B	6104	595	ALTURITAS FIELD, VENEZUELA
M	25	407	ALUMINA-ALUMINUM INDUSTRY, EXPORT MARKETS, AUSTRALIA,
M	25	460	ALUMINOQUARTZITES, AND FLUORITE, EASTERN ASIATIC
M	25	544	ALUMINUM, AND IRON, TERNARY DIAGRAM, NAZCA PLATE, MANGANESE
M	25	407	ALUMINUM INDUSTRY, DEVELOPMENT AND INTEGRATION, AUSTRALIA
M	25	407	ALUMINUM PRODUCTION COMMISSION, AUSTRALIAN
B	6212	2455	ALVARADO RIDGE
S	11	228	ALVEOLOPHRAGMIUM SCITULUM
S	11	268	ALVEOLOPHRAGMIUM SCITULUM
B	6204	693	ALVIN CRUISE, DSRV
B	6108	1225	ALVIN FIELD, BRAZORIA COUNTY, TEXAS
B	6410	1703	ALVIN, SAMPLING, DSRV
B	6008	1394*	ALVORD VALLEY, OREGON GEOTHERMAL INVESTIGATION, ABST.
B	6103	465*	ALYESKA ENGINEERING. BY TITLE ONLY
M	25	135	ALZWAR, M., POSSIBLE ENERGY SOUUCES IN INDONESIAN VOLCANIC
B	6412	2142	AMAK BASIN, BERING SEA SHELF
B	6002	170	AMAK BASIN, BERINGIAN SHELF
S	6	150	AMAMI-OSHIMA GRANITES, JAPAN, CRETACEOUS
B	6407	1091	AMANINIAN HORIZON, ASIA, NORTHEAST
B	6109	1471	AMAPA BASIN, BRAZIL
B	6405	687*	AMAPA CARBONATE ROCKS, PALEOGENE, OF FOZ DO AMAZONAS BASIN,
B	6306	942	AMAPA FORMATION, AMAZON BASIN
B	6306	940	AMAPA PLATFORM, AMAZON MARGIN AREA
B	6310	1745	AMATO, F. L., PETROLEUM DEVELOPMENTS IN SOUTH AMERICA,
B	6110	1578	AMATO, F. L., PETROLEUM DEVELOPMENTS IN SOUTH AMERICA,
B	6210	2044	AMATO, F. L., PETROLEUM DEVELOPMENTS IN SOUTH AMERICA,
B	**6409**	**1331**	**AMATO, ROGER V., DEVELOPMENTS IN ATLANTIC COASTAL PLAIN IN 1979**
B	6409	1336	AMATO, ROGER V., DEVELOPMENTS IN ATLANTIC OC IN 1978—
B	**6105**	**760**	**AMATO, ROGER V., GEOLOGIC AND OPERATIONAL SUMMARY OF COST B-2**
B	6407	1026	AMATULI TROUGH, ALASKA
B	6306	942	AMAZON BASIN, AMAPA FORMATION
B	6109	1471	AMAZON BASIN, BRAZIL
B	6203	503*	AMAZON BASIN, BRAZIL, ABST., PETROLEUM GEOLOGY OF, PALEOZOIC,
B	6306	942	AMAZON BASIN, JACAREZINHO FORMATION
B	6306	942	AMAZON BASIN, LIMOEIRO FORMATION
B	6306	942	AMAZON BASIN, MARAJO FORMATION
B	6306	942	AMAZON BASIN, PIRARUCU FORMATION
S	6	101	AMAZON BASIN, TROMBETAS FORMATION, BRAZIL
B	6306	942	AMAZON BASIN, TUCURARE FORMATION
M	25	344	AMAZON BASINS, CRETACEOUS
M	25	341	AMAZON BASINS, DEVONIAN
M	25	341	AMAZON BASINS, MISSISSIPPIAN
M	25	339†	AMAZON BASINS OF COLOMBIA, ECUADOR, AND, HYDROCARBON
M	25	347	AMAZON BASINS, OIL FIELDS, WILDCATS, AND SEEPS
M	25	341	AMAZON BASINS. ORDOVICIAN
M	25	341	AMAZON BASINS, PENNSYLVANIA
M	25	344	AMAZON BASINS, PERMIAN
M	25	339	AMAZON BASINS, POTENTIAL, COLOMBIA
M	25	339	AMAZON BASINS, POTENTIAL, ECUADOR
M	25	339	AMAZON BASINS, POTENTIAL, PERU
M	25	341	AMAZON BASINS, SILURIAN
M	25	344	AMAZON BASINS, TERTIARY
M	25	344	AMAZON BASINS, TRIASSIC-JURASSIC
B	6306	936	AMAZON CANYON, BATHYMETRY
B	6306	936	AMAZON CONE, BATHYMETRY
B	6302	219	AMAZON CONE, BRAZIL
B	6202	277	AMAZON CONE OFF, BRAZIL, NORTHERN
B	6306	936	AMAZON CONTINENTAL MARGIN, BATHYMETRY, MAP
B	6306	939	AMAZON CONTINENTAL MARGIN, ST. PAULS FRACTURE ZONE
B	6306	940	AMAZON MARGIN AREA, AMAPA PLATFORM
B	6306	940	AMAZON MARGIN AREA, GURUPA ARCH
B	6306	940	AMAZON MARGIN AREA, MARAJO BASIN
B	6306	940	AMAZON MARGIN AREA, PARA-MARANHAO PLATFORM
B	6306	940	AMAZON MARGIN AREA, SAO LUIS BASIN
B	6306	939	AMAZON MARGIN, MAGNETIC ANOMALIES
M	25	474	AMAZON MEGASTRUCTURE, UPPER
B	6306	940	AMAZON MEXIANA GRABEN
B	6109	1471	AMAZON MOUTH, BRAZIL
B	6306	934	AMAZON RIVER, FAN
B	6206	917	AMAZON SUBMARINE FAN, SOUTH AMERICA
B	6306	934†	AMAZON UPPER CONTINENTAL MARGIN, MORPHOLOGY AND
B	6103	468*	AMBLER MINERAL BELT, BROOKS RANGE, ALASKA. BY TITLE ONLY,
B	6304	689*	AMBROSIA LAKE AREA, ABST., THERMOLUMINESCENCE OF, URANIUM
B	6304	690*	AMBROSIA LAKE AREA, NEW MEXICO, ABST., EFFECTS OF, URANIUM
B	6304	687*	AMBROSIA LAKE DISTRICT, NEW MEXICO, ABST., GEOLOGY AND ORE
B	6209	1674	AMBROSIA LAKE DISTRICT, URANIUM, NEW MEXICO
B	6304	688*	AMBROSIA LAKE TREND AT SANDSTONE MINE, ABST., ANOMALOUS
B	6309	1453	AMCOR, BALTIMORE CANYON TROUGH, USGS, ATLANTIC MARGIN
B	**6105**	**760**	**AMDURER, MICHAEL, GEOCHEMISTRY, HYDROLOGY, AND MINERALOGY OF**
B	6202	296	AMEKI FORMATION, NIGERIA
B	6207	1185†	AMENDMENT OF CODE OF STRATIGRAPHIC NOMENCLATURE TO
B	6310	1887	AMERADA HESS, AND ABU DHABI MARINE OPERATING CO., ON
B	6010	1870	AMERADA HESS CORP., ABU DHABI OIL CO., LTD., SUNNINGDALE OILS,
B	6210	2115	AMERADA HESS CORP., B. P. TRADING LTD., ET AL, ON DEVELOPMENTS IN
B	6110	1799	AMERADA HESS CORP., ET AL, ON DEVELOPMENTS IN ABU DHABI, 1976,
B	6207	1210*	AMERICA TRENCH, ABST., TECTONIC FRAMEWORK OF, MIDDLE
B	6310	1986	AMERICAN COMMISSION OF STRATIGRAPHIC NOMENCLATURE, NOTE 50
B	6002	273	AMERICAN COMMISSION ON STRATIGRAPHIC NOMENCLATURE, NOTE 44
B	6102	248	AMERICAN COMMISSION ON STRATIGRAPHIC NOMENCLATURE, NOTE 45
B	6107	1083	AMERICAN COMMISSION ON STRATIGRAPHIC NOMENCLATURE, NOTE 47
B	6207	1185	AMERICAN COMMISSION ON STRATIGRAPHIC NOMENCLATURE, NOTE 48
B	6207	1187	AMERICAN COMMISSION ON STRATIGRAPHIC NOMENCLATURE, NOTE 49
B	6401	136†	AMERICAN COMMISSION ON STRATIGRAPHIC NOMENCLATURE NOTE
B	6010	1868	AMERICAN INDEPENDENT OIL CO., ARABIAN OIL CO. LTD, JAPAN, AND
B	6110	1798	AMERICAN INDEPENDENT OIL CO., ARABIAN OIL CO. LTD., AND GETTY
B	6002	177	AMERICAN PLATE, ALASKA
B	6403	360	AMERICAN PLATE, BOUNDARY
B	6210	2112	AMERICAN ULTRAMAR LTD., AND ATLANTIC RICHFIELD CO., ON
B	6110	1796	AMERICAN ULTRAMAR LTD., PETROCONSULTANTS S.A., PHILLIPS
B	6002	321*	AMERICUS LIMESTONE, PERMIAN, WOLFCAMPIAN, IN NORTHEAST
S	7	87	AMERY, G. B., RECOGNITION OF CONTINENTAL-SLOPE SEISMIC FACIES,
S	7	141	AMERY, GEORGE B., STRUCTURE OF CONTINENTAL SLOPE, NORTHERN
B	6407	993	AMESKOUDE GRABEN, MOROCCO
S	5	77	AMHERSTBURG FORMATION, DEVONIAN, DETROIT RIVER GROUP,
B	6404	569	AMIAZ PLAIN, ISRAEL
B	6205	827	AMIT, ORA, ORGANOCHEMICAL EVALUATION OF GEVARAM SHALES,
B	6409	1553*	AMITE RIVER, LOUISIANA, ABST., PRELIMINARY OBSERVATIONS OF
B	6201	88	AMLIA BASIN, ALASKA
B	6303	341	AMMERMAN, M. L. AND, DELINEATION OF ROME TROUGH IN EASTERN
B	6009	1621	AMMERMAN, M. L., GEOPHYSICAL AND TECTONIC STUDY OF MOORMAN
S	11	228	AMMOBACULITES AMERICANUS
S	11	227	AMMODISCIDAE
S	11	227	AMMODISCUS
M	27	89	AMMONITES
S	6	309	AMMONOIDEA, PERMIAN, KEY FOSSIL GROUPS
S	6	329	AMMONOIDS, CARBONIFEROUS, BIOSTRATIGRAPHY
B	6010	1866	AMOCO, AND ATLANTIC RICHFIELD CO., ON DEVELOPMENTS IN IRAN,
B	6110	1796	AMOCO INTERNAT. OIL CO., AMERICAN ULTRAMAR LTD.,
B	6411	1841	AMOCO INTERNAT. OIL CO., ET AL, ON DEVELOPMENTS IN OMAN,
B	6411	1840	AMOCO INTERNATIONAL OIL CO. AND FORMAN EXPLORATION CO., ON
B	6310	1889	AMOCO INTERNATIONAL OIL CO., FORMAN EXPLORATION CO., RESERV
B	6210	2112	AMOCO INTERNATIONAL OIL CO., SOCIETE NATIONALE ELF AQUITAINE,
B	6404	569	AMORA FORMATION, ISRAEL

B 6002 220 AMOSEAS LAKE TOPPED OIL, AUSTRALIA
B 6209 1765* AMPHIBOLITES ALONG MOTAGUA FAULT ZONE IN EL PROGRESO
S 4 335 AMPHISTEGINA GRAINSTONES, ST. CROIX
S 4 32 AMPHISTEGINA-SAND FACIES
M 26 482 **AMPLITUDE ANOMALY, BRIGHT SPOT, SEISMIC, MODELING**
M 26 301† AMPLITUDE, FREQUENCY, AND OTHER ATTRIBUTES TO STRATIGRAPHIC
M 26 392 AMPLITUDES, WAVEFORMS, MODELING, USING
B 6303 409 AMSBURY, DAVID L., EDWARDS FORMATION, NORTH-CENTRAL
B 6204 622 AMSDEN FORMATION, WYOMING
B 6303 410 AMSDEN, THOMAS W., LATE ORDOVICIAN--EARLY SILURIAN STRATA IN
B 6105 760 AMSDEN, THOMAS W., MIDDLE PALEOZOIC HISTORY OF ANADARKO AND
B 6307 1135 AMSDEN, THOMAS W., WELLING FORMATION, NEW NAME FOR UPPE
B 6206 1006 AMUND RINGNES ISLAND, ARCTIC, CANADA
B 6101 3 AMUND-CORNWALL ARCH
B 6101 3 AMUNDSEN BASIN
M 25 59 AMUR FOLDED REGION, USSR
B 6407 992 AMZRI FLEXURE, MOROCCO
S 5 169* AN EVAPORITIC LITHOFACIES CONTINUUM: LATEST MIOCENE
B 6004 694* ANACACHO LIMESTONE ASSOCIATED WITH LATE CRETACEOUS
B 6311 2119* ANADARKO BASIN, ABST., DELTAIC DEPOSITS IN UPPER, MORROW
B 6002 325* ANADARKO BASIN, ABST., EVOLUTION OF
B 6311 2116* ANADARKO BASIN, ABST., MAJOR STRUCTURAL AND STRATIGRAPHIC
B 6002 321* ANADARKO BASIN, ABST., SEDIMENTARY, CYCLES AND DEPOSITIONAL
B 6004 711* ANADARKO BASIN, ABST., SEDIMENTARY, CYCLES IN, VIRGILIAN, UPPER,
B 6303 480* ANADARKO BASIN, ABST., SIMPLE, PYROLYSIS TECHNIQU USING WELL
B 6102 297* ANADARKO BASIN, ABST., STATISTICAL SUMMARY OF, WELLS DRILLED
B 6311 2119* ANADARKO BASIN, ABST., UPPER, MORROW, FAN-, DELTA DEPOSITS OF
B 6205 856 ANADARKO BASIN, BASEMENT CONTOUR MAP, TEXAS, PANHANDLE AND,
M 26 447 ANADARKO BASIN, BELLE CITY LIMESTONE, OKLAHOMA
M 26 440 ANADARKO BASIN, HASKELL LIMESTONE, OKLAHOMA
M 26 445 ANADARKO BASIN, MEDRANO SANDSTONE, OKLAHOMA
B 6305 804 ANADARKO BASIN, MILLS RANCH COMPLEX, TEXAS
M 26 452 ANADARKO BASIN, MORROW SANDSTONE, OVERPRESSURING,
B 6002 233 ANADARKO BASIN, OKLAHOMA
B 6103 361 ANADARKO BASIN, OKLAHOMA
B 6203 566* ANADARKO BASIN, OKLAHOMA, ABST., ECONOMIC POTENTIAL OF,
M 26 439† ANADARKO BASIN, OKLAHOMA, SEISMIC, STRATIGRAPHIC, MODEL OF,
B 6109 1437† ANADARKO BASIN, OKLAHOMA, SEISMIC STRATIGRAPHIC, MODEL OF
M 26 439 ANADARKO BASIN, PENNSYLVANIAN, HOXBAR GROUP, OKLAHOMA,
B 6109 1437 ANADARKO BASIN, SEISMIC DATA, OKLAHOMA
M 26 452 ANADARKO BASIN, SPRINGER SANDSTONE OVERPRESSURING,
B 6205 853† ANADARKO BASIN, STATISTICAL SUMMARY OF, WELLS DRILLED BELOW
M 26 440 ANADARKO BASIN, TONKAWA LIMESTONE, OKLAHOMA
B 6303 540* ANADARKO BASIN TREND, ABST., EXPLORATION RESEARCH ALONG,
M 26 445 ANADARKO BASIN, WADE SANDSTONE, OKLAHOMA
M 26 451 ANADARKO BASIN, WATONGA-CHICKASHA TREND, OKLAHOMA
M 26 451 ANADARKO BASIN, WICHITAN UNCONFORMITY, OKLAHOMA
B 6002 327* ANADARKO BASIN. BY TITLE ONLY, MORROW-, SPRINGER-, CHESTER
B 6002 322* ANADARKO BASIN. BY TITLE ONLY, REPORT--
M 26 441 ANADARKO, UPPER HOXBAR GROUP, MISSOURIAN, OKLAHOMA
B 6309 1467† ANADYR AND KHATYRKA BASINS, USSR, REVIEW OF PETROLEUM
B 6309 1473 ANADYR BASIN, APTIAN-TURONIAN SEQUENCE, USSR
B 6002 178 ANADYR BASIN, BERING SEA
B 6309 1474 ANADYR BASIN, MIOCENE-HOLOCENE SEQUENCE, USSR
B 6309 1473 ANADYR BASIN, SENONIAN-OLIGOCENE SEQUENCE, USSR
B 6309 1467 ANADYR BASIN, SIBERIA
M 25 466 ANADYR-KORYAK GEOSYNCLINAL ZONE, USSR
B 6408 1185 ANAEROBIC BACTERIA
B 6202 255 ANAHEIM NOSE, CALIFORNIA
B 6009 1616* ANAHUAC, NORTHEASTERN MEXICO, OPENS UP JURASSIC POSSIBILITIES
B 6408 1169 ANAHUAC SHALE, TEXAS
S 4 236 ANALCIDOMETRA ARMATA POURTALES, BARBADOS, PERICLIMENES OR
S 4 235 ANALCIDOMETRA ARMATA POURTALES, GRAND CAYMAN
S 4 235 ANALCIDOMETRA ARMATA POURTALES, MARTINIQUE
M 28 142 ANALCIME (ZEOLITE), BIREFRINGENCE, ETC.
S 10 157 ANALOG FLOW PATTERN, ALBERTA, ETZIKOM COULEE
M 29 179 ANALOG RECORD OF PROFILE, BLAKE ESCARPMENT
B 6203 489* ANALOGIES BETWEEN HOLOCENE COASTAL-MARSH DEPOSITS AND
B 6203 501* ANALOGS, ABST., CONSTRUCTION OF PROCESS-RESPONSE, MODELS FOR,
B 6408 1223† ANALOGS, CONSTRUCTION OF, PROCESS-RESPONSE MODELS FOR,
B 6408 1240 ANALYSES OF ANCIENT DEPOSITIONAL SYSTEMS
M 26 374 ANALYSES, RESOLUTION SYSTEMS, ENGINEERING
M 26 369 ANALYSES, RESOLUTION SYSTEMS, EXPLORATION
M 26 136 ANALYSIS, INTERPRETATION, FACIES
M 26 135 ANALYSIS, INTERPRETATION, SEQUENCE
M 26 136 ANALYSIS. MAPPING. FACIES
B 6004 659* ANALYSIS OF SUBMARINE DEBRIS-FLOW DEPOSIT, NORTHERN TUNISIA,
B 6004 653* ANALYSIS OF BORERS AND THEIR BORINGS IN BERMUDA REEF
B 6205 879* ANALYSIS OF EFFECT ON GRADED BEDS OF CHANGING GRAVITATIONAL
B 6203 524* ANALYSIS OF GEOCHEMICAL AND SEDIMENTOLOGIC FACTORS
B 6403 400† ANALYSIS OF GRAVITY ANOMALY OVER CORAL-REEF OIL FIELD-- WILFRED
M 26 352† ANALYSIS OF HIGH RESOLUTION SEISMIC DATA
B 6303 459* ANALYSIS OF HYDROCARBON DISTRIBUTION AND FINDING RATES
B 6203 518* ANALYSIS OF OIL-FIELD DISTRIBUTION FOR SEDIMENTARY BASINS OF
M 26 136 ANALYSIS, RECOGNITION, FACIES
M 26 165 ANALYSIS, SEISMIC, FACIES
M 26 53† ANALYSIS, SEISMIC, STRATIGRAPHY AND, GLOBAL CHANGES OF, SEA
M 26 138 ANALYSIS, TOOLS, SEA LEVEL
B 6303 543* ANALYTIC AND INTERPRETIVE PROCEDURES FOR GEOLOGIC
B 6008 1395* ANALYTIC SOLUTIONS APPLIED TO FAULTING IN LARAMIDE ROCKY
B 6405 672* ANAMBRA BASIN, SOUTHERN NIGERIA, ABST., PETROLEUM PROSPECTS
B 6405 668* ANAMBRA SYNCLINE, SOUTHEASTERN NIGERIA, ABST., PETROLEUM
S 3 87 ANATEXIS AND MAGMATION
B 6106 900 ANATOLIAN TRANSFORM
B 6303 462* ANATOMY AND GROWTH HISTORY OF HOLOCENE OOID SHOAL, ABST.
S 4 361 ANATOMY OF SCLERACTINIAN CORALS
M 29 319† ANATOMY OF THE MEXICAN RIDGES, SOUTHWESTERN GULF OF MEXICO
M 25 341 ANAZON BASINS, DEVONIAN
B 6309 1578 ANCELL, KENNETH L., CHARACTERIZATION OF PRODUCTION
B 6004 712* ANCIENT ANALOGUES OF QUATERNARY TIDAL TERRACES OR SUBAERIAL
B 6105 763 * ANCIENT AND MODERN CONTINENTAL MARGINS, PERSPECTIVE FOR FUTURE
B 6004 686* ANCIENT AND MODERN SUBMARINE CANYON-FAN SYSTEMS, ABST.,
B 6203 506* ANCIENT CONTINENTAL SLOPE DEPOSITS AND THEIR USEFULNESS IN
B 6009 1606* ANCIENT DELTA SYSTEMS IN GULF COAST AREA ABST.
B 6408 1240 ANCIENT DEPOSITIONAL SYSTEMS, ANALYSES OF
B 6408 1223† ANCIENT EPICONTINENTAL SEAWAY DEPOSITIONAL SYSTEMS USING
B 6203 501* ANCIENT EPICONTINENTAL SEAWAY DEPOSITIONAL SYSTEMS USING
B 6004 657* ANCIENT ESTUARINE-FILL FACIES, WILLAPA BAY, WASHINGTON, ABST.,
B 6303 448* ANCIENT FLOOD-PLAIN DEPOSITION, ABST., RECENT, ARTHROPO
B 6408 1232 ANCIENT OFFSHORE SANDSTONE BODIES IN EPICONTINENTAL SEAWAYS
S 4 96 ANCIENT REEF ECOSYSTEMS, TROPHIC ANALYSIS OF
B 6309 1545 ANCIENT SEDIMENTARY ENVIRONMENTS, B R. C. SELLEY, KUMAR,
B 6405 709* ANCIENT STRATIGRAPHIC TRAP FOR PETROLEUM IN WEBER
B 6203 542* ANCIENT SUBMARINE CANYON AND CHANNEL MORPHOLOGY, INDUS
B 6301 67† ANCIENT SUBMARINE CANYONS OF TETHYAN CONTINENTAL MARGINS,
B 6206 959 ANCIENT SUBMARINE FAN, ENGLAND, SHALE, GRIT: A COMPLETE
B 6206 950 ANCIENT SUBMARINE FANS, EXAMPLES OF
B 6407 1094† ANCIENT SUBMARINE FANS. DISCUSSION AND REPLIES, MODERN AND
B 6305 810† ANCIENT SUBMARINE FANS. MODELS FOR EXPLORATION FOR
B 6206 932† ANCIENT SUBMARINE FANS: MODELS FOR EXPLORATION FOR
S 9 9 ANCIENT WIND DIRECTION, PALEOCLIMATE INDICATORS
M 25 329 ANCON GROUP, ECUADOR
B 6212 2509* AND BACK-ARC BASINS OF TOHOKU ARC, JAPAN, ABST., CONTRAST OF
B 6004 694* AND FACIES DISTRIBUTIO OF ANACACHO LIMESTONE ASSOCIATED WITH
M 28 48 ANDALUSITE, CRYSTAL
M 28 32 ANDALUSITE SCHIST, NEW HAMPSHIRE, PALEOZOIC
M 28 49 ANDALUSITE SCHIST, NEW HAMPSHIRE, PALEOZOIC
M 28 48 ANDALUSITE SCHIST, NEW HAMPSHIRE, PALEOZOIC
M 25 252 ANDAMAN BASIN, THAILAND
M 29 189† ANDAMAN SEA AND BURMA, TECTONICS OF THE
M 29 190 ANDAMAN SEA, BATHYMETRY, BURMA
M 29 195 ANDAMAN SEA, MAGNETIC PROFILE, BURMA
M 29 192 ANDAMAN SEA, PHYSIOGRAPHIC PROVINCES, BURMA
M 29 194 ANDAMAN SEA, SPREADING EXTENSION, BURMA
M 29 193 ANDAMAN SEA, STRUCTURE SECTIONS, BURMA
M 29 195 ANDAMAN SEA, SYNTHETIC MAGNETICS, BURMA
M 29 191 ANDAMAN SEA, TECTONIC MAP, BURMA
M 29 192 ANDAMAN SEA, TECTONIC PROVINCES, BURMA
M 29 197 ANDAMAN SEA, TECTONICS EVALUATED, BURMA
M 25 251† ANDAMAN SEA--SUMMARY, PETROLEUM GEOLOGY OF, THAILAND, GULF
B 6104 560 ANDEAN BATHOLITH, CHILE
B 6212 2470 ANDEAN CORDILLERAS
S 3 79 ANDEAN MARGIN, SOUTH AMERICAN
M 25 559 ANDEAN MESOZOIC-CENOZOIC GEOSYNCLINAL CYCLE, CHILE
M 25 530 ANDEAN METALLOGENIC PROVINCE, DEPOSITS OF
M 25 532 ANDEAN METALLOGENIC PROVINCE, MAP SHOWING SUBPROVINCES
M 25 529 ANDEAN METALLOGENIC PROVINCE, METALLOGENESIS
M 25 529 ANDEAN OROGENY
M 29 345 ANDEAN OROGENY, COLOMBIA, EVOLUTION
B 6203 559* ANDEAN PLATEAU, NORTHERN ECUADOR, ABST., VOLCANOGENIC
M 25 533 ANDEAN REGION, BASE-METAL DEPOSITS, POLYMETALLIC
M 25 535 ANDEAN REGION, BERYLLIUM
M 25 531 ANDEAN REGION, COPPER
M 25 533 ANDEAN REGION, GOLD, OCCURRENCE AND PRODUCTION
M 25 531 ANDEAN REGION, IRON
M 25 535 ANDEAN REGION, LITHIUM
M 25 527 ANDEAN REGION, MINERAL MINING, FUTURE OF
M 25 527 ANDEAN REGION, OROGENIC BELT, ACTIVE, LONGEST
M 25 528 ANDEAN REGION, PHYSIOGRAPHIC PROVINCES
M 25 533 ANDEAN REGION, PRE-ANDEAN METALLIFEROUS DEPOSITS
M 25 534 ANDEAN REGION, PRECAMBRIAN AND PALEOZOIC METALLIFEROUS
M 25 533 ANDEAN REGION, TIN AND ASSOCIATED DEPOSITS

Index

M 25 535 ANDEAN REGION, TUNGSTEN
M 25 461 ANDEAN-CORDILLERAN METALLIC BELT
B 6303 421 ANDERSON, C. T. J., INFLUENCE OF RATE ON DEMAND AND USE OF
B 6012 2174 ANDERSON, DAVID N., UPDATE OF GEOTHERMAL RESOURCES AN
B 6009 1606 ANDERSON, DON R., ANCIENT DELTA SYSTEMS IN GULF COAST AREA
B 6405 791 ANDERSON, E. G., COMPARISON OF GULF COAST LATE JURASSIC
B 6209 1752 ANDERSON, E. G., RADIOMETRIC INVESTIGATION RELATIVE TO A
B 6409 1552 ANDERSON, E. G., REGIONAL PATTERNS OF WOODBINE-TUSCALOOSA,
B 6309 1545 ANDERSON, FEAZEL, CHARLES T., REVIEW OF, MODERN PETROLEUM, A
B 6309 1579 ANDERSON, G. D., THEORETICAL AND EXPERIMENTAL ANALYSES OF
B 6405 671 ANDERSON, GEORGE A., HYDROCARBON POTENTIAL OF SITKALIDAK
B 6302 258 ANDERSON, HABICHT, J. K. A., REVIEW OF THE STRUCTURE OF WESTERN,
B 6303 410 ANDERSON, JOHN B., ANTARCTIC CONTINENTAL MARGIN, ABST.
B 6004 645 ANDERSON, JOHN B., CYCLIC SEDIMENTATION IN GLACIAL MARINE
B 6303 467 ANDERSON, JOHN B., FLOW OF BOTTOM WATER OU OF WEDDELL SEA,
B 6303 410 ANDERSON, JOHN B.,, GLACIAL-MARINE SEDIMENTS OF WEST
B 6203 490 ANDERSON, JOHN B., SEDIMENTATION ON ANTARCTIC CONTINENTAL
B 6303 483 ANDERSON, JOHN B., SEDIMENTOLOGIC FACIES I MODERN
B 6405 669 ANDERSON, KNUT A., PERMIAN TRACE FOSSILS OF WESTERN WYOMING
B 6304 625 ANDERSON MINE FORMATION, MEMBERS, ARIZONA
B 6304 635 ANDERSON MINE, URANIUM-BEARING INTERVAL, ARIZONA
B 6203 561* ANDERSON MINE, YAVAPAI COUNTY, ARIZONA, ABST., MAJOR, URANIUM
B 6205 879 ANDERSON, PAUL B., ENVIRONMENTAL RECONSTRUCTION OF MIDDLE
M 24 54 ANDERSON PLAIN, ALASKA
B 6004 718 ANDERSON, R. C., SYNTHETIC SEISMIC SECTIONS OF SELECTED
B 6405 683 ANDERSON, RAYMOND R., ASSESSMENT OF POTENTIA FOR FLUID
B 6008 1409 ANDERSON, ROBERT C., SYNTHETIC SEISMIC SECTIONS OF SELECTED
M 29 199 ANDERSON, ROGER N., EXTENSIONAL TECTONICS IN THE OKINAWA
B 6304 625 ANDERSON URANIUM MINE, PRODUCTION, ARIZONA
B 6304 622 ANDERSON URANIUM MINE--ARTILLERY PEAK AREA, ARIZONA
B 6409 1549 ANDES, A GEOLOGICAL REVIEW, BY WERNER ZEIL, VAN HOUTEN,
M 25 458 ANDESITE LINE OR RING, PACIFIDES
S 6 264 ANDESITIC VOLCANICS DATA POINT, ALASKA, CRETACEOUS
M 25 443 ANDONG ANTICLINE, SOUTH KOREA
B 6201 88 ANDREANOF ISLANDS, ALASKA
B 6105 761 **ANDREWS, G. W., MESOZOIC AND CENOZOIC BIOSTRATIGRAPHY OF OAK**
M 25 382 ANDREWS, J., MANGANESE-NODULE STRUCTURES, RELATION TO NODULE
M 25 62 ANDREYEV, S. I., TECTONICS OF CONTINENTAL AND OCEANIC
B 6303 422* ANDROS, BAHAMAS, ABST., HURRICANES AND RAINFALL--KEY FOR,
B 6204 694 ANDROS ISLAND, BAHAMAS
B 6405 788* ANDROS ISLAND, BAHAMAS, ABST., CEMENTATION OF LIME-MUD AND
B 6303 457* ANDROS ISLAND, BAHAMAS, ABST., MIXING ZONE, DOLOMITE IN
S 4 187 ANDROS ISLAND, REEF, BAHAMAS
B 6204 694 ANDROS ISLAND WELL, BAHAMA PLATFORM
M 24 280 ANDROS PLATFORM FACIES, GREAT BAHAMA BANKS
S 4 345 ANEGADA PASSAGE, ST. CROIX
M 30 237 ANGARA SUITE, USSR, LENA-TUNGUSKA AREA
M 30 232 ANGARA-LENA STRUCTURAL TERRACE, USSR
M 25 553 ANGAS, COPPER, ZINC, LEAD, SILVER, AND GOLD DEPOSITS, ECUADOR
M 25 553 ANGAS DEPOSIT, POLYMETALLIC MINERALIZATION, ECUADOR
M 25 237 ANGEL DISCOVERY WELL, AUSTRALIA
B 6201 43 ANGEL FIELD, AUSTRALIA
B 6405 670 ANGEVINE, CHARLES, THERMAL EVOLUTION O SEDIMENTARY BASINS
B 6405 681* ANGOLA, ABST., HYDROCARBON OCCURRENCES IN NONMARINE,
B 6106 930 ANGOLA BASIN
B 6412 2055 ANGOLA BASIN, DSDP SITE 364, LEG 40, GEOCHEMICAL LOG
B 6107 961 ANGOLA BASIN, EVOLUTION OF, GABON-
B 6111 1994 ANGOLA BASIN, POSSIBLE DIAPIRS
B 6110 1747 ANGOLA, DEVELOPMENTS, 1976
B 6210 1845 ANGOLA, DEVELOPMENTS, 1977
B 6310 1690 ANGOLA, DEVELOPMENTS, 1978
B 6411 1786 ANGOLA, DEVELOPMENTS, 1979
B 6202 227 ANGOLA, MOSSAMEDES, EVAPORITES, CUANZA
B 6110 1754 ANGOLA, PRODUCTION, 1975-1976
B 6210 1853 ANGOLA, PRODUCTION, 1976-1977
B 6310 1696 ANGOLA, PRODUCTION, 1978
B 6411 1800 ANGOLA, PRODUCTION, 1979
B 6107 970 ANGOLA SEGMENT, CAMEROON-, GABON-
B 6002 211 ANGOLA, WEST AFRICAN DIAPIR ZONE, GABON TO
M 25 63 ANGSUWATHANA, PRAYONG, SURVEYING EARTH
M 29 36 ANGULAR UNCONFORMITY, BLAKE PLATEAU, STRUCTURE
M 29 34 ANGULAR UNCONFORMITY, BLAKE PLATEAU, TRIASSIC
M 28 102 ANGULARITY, CLASSIFICATION, ROUNDNESS
S 11 280 ANGULOGERINA ANGULOSA
S 11 253 ANGULOGERINA ANGULOSA
S 11 252 ANGULOGERINA HUGHESI
B 6004 688* ANHYDRITE AND PALEOTEMPERATURE DETERMINATION, ABST., SABKHA
S 5 149 ANHYDRITE DEPOSITION, CANADA, SVERDRUP BASIN, OTTO FIORD
S 5 145* ANHYDRITE FACIES RELATIONSHIPS, OTTO FIORD FORMATION
M 28 87 ANHYDRITE, CRYSTALS, IN A SERPULID WORM TUBE
B 6111 1929† ANHYDRITE FACIES RELATIONS IN OTTO FIORD FORMATION,
M 28 129 ANHYDRITE, IN SANDSTONE, CEMENTS
M 28 87 ANHYDRITE, NODULE, IN FORAMINIFERAL SEDIMENTS
M 27 134 ANHYDRITE, REPLACEMENT
S 5 67 ANHYDRITE, SALINA GROUP, RUFF FORMATION, RABBIT EARS
B 6203 496* ANHYDRITIC DOLOMITES IN BELL CANYON FORMATION, DELAWARE
M 25 62 ANIKEYEVA, L. I., TECTONICS OF CONTINENTAL AND OCEANIC
B 6004 708 ANIMA, R. J., BED FORMS AND INTERNAL STRUCTURES IN DEEP ESTUARY
B 6106 904 ANIMAL BASIN, BAJA CALIFORNIA
B 6104 602 ANIMAS MOUNTAINS, NEW MEXICO
B 6212 2441† ANIONS IN OIL-FIELD WATERS—IMPLICATIONS FOR ORIGIN OF NATURAL
M 26 28 ANISOTROPY, LABORATORY, MEASUREMENT
B 6203 498* ANKERITE CEMENTATION IN WILCOX OF SOUTHWEST TEXAS, ABST.,
B 6402 219 ANKLESVAR FIELD, INDIA
B 6403 429 ANKLESWAR HIGH-WAX OIL FIELD, INDIA
B 6303 415* ANNELID BURROWS AS CURRENT-DIRECTION INDICATORS, ABST.,
M 24 264 ANNONA CHALK, ALABAMA
B 6209 1768* ANNONA CHALK AND GOBER CHALK, UPPER CAMPANIAN, AT TYPE
B 6301 82 ANNOT FORMATION, MARITIME, ALPS
B 6205 743 ANNUAL GEOPHYSICAL CREW-MONTHS IN NON-COMMUNIST WORLD,
M 27 60 ANNULARIS, ARAGONITIC
B 6403 299 ANOMALIES, CONTINENTAL, MARGINS OF WORLD CAUSED BY, PLATE
B 6109 1560* ANOMALIES FROM LANDSAT, ABST., DETECTION OF
M 25 472 ANOMALIES, METHOD OF, DIRECTION
S 11 298 ANOMALINA CF. A. SALINASENSIS
S 11 158 ANOMALINA SALINASENSIS
S 11 264 ANOMALINA SALINASENSIS
S 11 264 ANOMALINIDAE
S 11 160 ANOMALINIDS, CALIFORNIA, MIOCENE
S 11 158 ANOMALINIDS, CALIFORNIA, MIOCENE
B 6304 688* ANOMALOUS OREBODY WITHIN AMBROSIA LAKE TREND AT SANDSTONE
B 6405 787* ANOMALOUS THERMOLUMINESCENCE AROUND URANIUM-DEPOSITS,
M 26 482 **ANOMALY, BRIGHT SPOT, SEISMIC, MODELING, AMPLITUDE**
B 6102 297* ANOMALY ANALYSIS OF CHIHUAHUA AND TRANS-PECOS, ABST.,
B 6205 793 ANOMALY, BLAKE SPUR, MAGNETIC
B 6007 1115 ANOMALY, GAMMA CONTOUR OF, EAST COAST, MAGNETIC
B 6007 1101 ANOMALY MAP, BARBADOS RIDGE COMPLEX AND ADJACENT, ATLANTIC
B 6403 400† ANOMALY OVER CORAL-REE OIL FIELD--WILFRED POOL, SULLIVAN
B 6407 1122 ANONYMOUS, REVIEW OF I. SEISMIC, SAFETY AND LAND-USE PLANNING.
B 6405 671* ANOXIC AND CARBONATE CYCLES IN CRETACEOUS ORGANI
B 6412 2060 ANOXIC AND OXIC, OCEANIC, BASINS
B 6408 1186 ANOXIC CONDITIONS IN WATER MASSES
B 6408 1206 ANOXIC DEPOSITIONAL MODELS, CLASSIFICATION
S 7 303† ANOXIC DEPRESSION ON THE CONTINENTAL SLOPE, NORTHWEST GULF
B 6408 1179† ANOXIC ENVIRONMENTS AND OIL SOURCE BED GENESIS
B 6405 697* ANOXIC ENVIRONMENTS AND OIL SOURCE BEDS, ABST.
B 6408 1187 ANOXIC ENVIRONMENTS, CLASSIFICATION OF
B 6306 870† ANOXIC EVENTS AS CAUSAL FACTORS IN DEVELOPMENT OF
B 6408 1187 ANOXIC LAKES, EXAMPLES, LARGE
B 6408 1194 ANOXIC LAYERS CAUSED BY UPWELLING, EXAMPLES
B 6408 1196 ANOXIC LAYERS CAUSED BY UPWELLING, PERU CURRENT
B 6408 1200 ANOXIC OPEN OCEAN, EXAMPLES
B 6408 1189 ANOXIC SILLED BASINS, EXAMPLES
S 6 113 ANSE DU VERYACH SERIES, ISOCHRON DIAGRAM, FRANCE
B 6004 687* ANTAGONIST--PLATE TECTONICS AS A MAJOR EXPLORATION TOOL. BY
M 29 313 ANTALYA BASIN, EVOLUTION, MEDITERRANEAN
M 29 312 ANTALYA BASIN, SEDIMENTARY SEQUENCE, MEDITERRANEAN
B 6303 410* ANTARCTIC CONTINENTAL MARGIN, ABST.
B 6203 490* ANTARCTIC CONTINENTAL SLOPE, ABST., SEDIMENTATION ON
S 4 79 ANTARCTIC GLACIATION
B 6303 543* ANTARCTIC LOGISTICS FOR EARTH SCIENCES, ABST.
M 25 536 ANTARCTIC PENINSULA AND PACIFIC MARGIN, GEOLOGY AND
B 6403 360 ANTARCTIC PLATE, BOUNDARY
B 6303 554* ANTARCTIC--LATE TERTIARY PALEOCLIMATIC HISTORY, ABST., DEEP-SEA,
B 6303 532* ANTARCTICA, ABST., COAL IN
B 6011 2070* ANTARCTICA, ABST., DRILLING IN
B 6303 524* ANTARCTICA, ABST., LEGAL AND POLITICAL STATUS OF
B 6207 1213* ANTARCTICA, ABST., LEGAL ASPECTS OF RESOURCE DEVELOPMENT IN
B 6303 418* ANTARCTICA, ABST., SPECULATIONS ON PETROLEUM, RESOURCE
B 6303 441* ANTARCTICA, ABST., URANIUM RESOURCE
M 25 74† ANTARCTICA, AN UNPROSPECTED, UNEXPLOITED
B 6303 436* ANTARCTICA AND GONDWANALAND, ABST.

B 6303 410 ANTARCTICA AND PUGET SOUND, ABST., GLACIAL-MARINE SEDIMENTS
S 9 18 ANTARCTICA, CENOZOIC, DRIFT MOVEMENTS
S 9 17 ANTARCTICA, EVIDENCES OF DRIFT, PALEOMAGNETISM
M 25 78 ANTARCTICA, GEOTHERMAL ENERGY
M 29 171 ANTARCTICA, MARGIN, COMPARED TO AUSTRALIA
M 25 535 ANTARCTICA, MINERAL OCCURRENCES IN
M 25 77 ANTARCTICA, MINERL RESOURCES
M 25 74 ANTARCTICA, PHYSIOGRAPHIC FEATURES
B 6303 467* ANTARCTICA--700,000 YEARS TO PRESENT, ABST., FLOW OF BOTTOM
B 6303 518* ANTARCTICA--DEVELOPMENT FAR IN FUTURE, ABST., METALLIC
B 6203 351 ANTELOPE HILLS OIL FIELD, CALIFORNIA
B 6211 2245 ANTELOPE SHALE FORMATION, CALIFORNIA
B 6101 80 ANTI-ATLAS
B 6405 724* ANTICIPATING COAL MINING PROBLEMS IN HARTSHORNE FORMATION,
B 6303 495* ANTICLINES AS EXPLORATION OBJECTIVES, ABST., CONTOURITE
B 6108 1389* ANTICLINORIUM, BRAZOS UPLIFT, AND CHAMA BASIN IN
S 4 78 ANTIGUA, CARBONATE SHELF
S 3 98 ANTILLEAN ARC
S 4 84 ANTILLEAN ARC, MESOZOIC, GREATER
S 4 69 ANTILLEAN ISLAND ARC, LESSER
B 6303 528* ANTILLES, ABST., DOLOMITIZATION OF, PLIOCENE-, PLEISTOCENE,
S 4 3† ANTILLES, CORAL, REEFS AND THEIR ZONATION IN, NETHERLANDS
B 6410 1702 ANTILLES CURRENT
S 4 70 ANTILLES, HOLOCENE BIOHERMS, LESSER
B 6007 1079 ANTILLES ISLAND ARC, TRACKLINE COVERAGE EAST OF, LESSER
S 4 3 ANTILLES, LEEWARD GROUP, CORALS, NETHERLANDS
S 4 155† ANTILLES, LOW-WATER EMERGENCE OF, CARIBBEAN, REEFS AND EFFECT
S 4 69 ANTILLES, PLEISTOCENE, LESSER
S 4 69 ANTILLES, PLIOCENE, LESSER
S 4 69 ANTILLES, TERTIARY, LESSER
B 6007 1094 ANTILLES TRENCH, LESSER
S 4 3 ANTILLES, WINDWARD GROUP, CORALS, NETHERLANDS
S 4 67† ANTILLES--GEOLOGIC CONTROL OF DEVELOPMENT, HOLOCENE,
S 3 107 ANTILLES-YUCATAN AXIS
M 25 475 ANTIMONY DEPOSITS, CLASSIFICATION OF
M 25 475† ANTIMONY DEPOSITS OF EASTERN USSR--SUMMARY, EVOLUTIONARY
M 25 518 ANTIMONY, MEXICO
M 25 475 ANTIMONY MINERALIZATION, CORRELATION WITH THICKNESS OF
M 25 558 ANTIMONY MINES, BOLIVIA
M 25 475 ANTIMONY, USSR, KELYAN DEPOSITS
M 25 475 ANTIMONY, USSR, OGNEY ORE DEPOSIT
B 6004 612† ANTIPATHETIC RELATIONS AMONG ALGAL STRUCTURES, BURROWERS,
B 6206 972 ANTIPODE, LEG 1, CALIFORNIA, SCRIPPS CRUISE
B 6212 2509* ANTITRUST LAWS, ABST., WHAT FOREIGN COMPANIES SHOULD KNOW
B 6410 1615 ANTLER FIELD, COLORADO
B 6305 831* ANTLER FORMATION, NORTH TEXAS, ABST., NONMARINE DEPOSITIONAL
B 6405 737* ANTLER OROGENIC BELT OF CENTRAL NEVADA, ABST., TECTONICS AND
B 6405 700* ANTLER OROGENY IN CENTRAL IDAHO—CLARIFICATIONS AND
B 6103 435† ANTLER TECTONIC PULSES, MISSISSIPPIAN, UNCONFORMITIES IN
M 25 170 ANTOFAGASTA HYDROTHERMAL AREA, CHILE
B 6111 1931 ANTOINETTE FORMATION, SVERDRUP BASIN
M 30 379 ANTONIO J. BERMUDEX COMPLEX, CHIAPAS-TABASCO PROVINCE
B 6309 1590* ANTRI SHALE BY IN-SITU PROCESS, ABST., RECOVERY OF, ENERGY FROM,
B 6203 426 ANTRIM BASIN, EUROPE
B 6401 129 ANTRIM SHALE, INDIANA
B 6209 1704 ANTRIM SHALE, UNITED STATES, EASTERN
M 25 462 ANYUY-YUKON FAULT ZONE
M 25 362 AOKI, SHIGERU, LAND SUBSIDENCE IN JAPAN RESULTING FROM FLUID
B 6406 858 AOUENAT UPLIFT, AFRICA
B 6405 696* APACHE COUNTY, ARIZONA ABST., DINEH-BI-KEYAH FIELD
B 6104 602 APACHE HILLS, NEW MEXICO
B 6109 1534* APALACHICOLA EMBAYMENT, ABST., SMACKOVER OIL PROSPECTS IN
M 28 54 APATITE, CRYSTALS, IN BIOTITE
M 25 454 APATITE, FUTURE POTENTIAL, JAPAN, URANIFEROUS, GLAUCONITE AND
B 6004 707* APENNINES, ABST., IRPINIDS--, MODEL OF TECTONICALLY CONTROLLED,
B 6004 666* APENNINES, ITALY, ABST., SPECTRAL AND, MARKOVIAN ANALYSIS OF
B 6112 2076 APENNINES RIDGE, MEDITERRANEAN REGION
B 6112 2070 APENNINES, SOUTH, SECTION
B 6101 30 APHEBIAN CHART, CANADIAN SHIELD
B 6405 785* APHEBIAN WISHART QUARTZITE, KNOB LAKE AREA, QUEBEC AND
B 6308 1202 API CLASSIFICATION OF WELLS, AAPG AND
B 6108 1155 API CLASSIFICATION OF WELLS, AAPG AND
B 6409 1319 API CLASSIFICATION OF WELLS, AAPG AND
B 6008 1187 API CLASSIFICATION OF WELLS, AAPG AND
B 6208 1309 API CLASSIFICATION OF WELLS, AAPG AND
B 6406 923 API GRAVITY OF OILS
B 6008 1188 API--AAPG, CSD, INDIVIDUAL WELL TICKET
B 6308 1203 API--AAPG, CSD, INDIVIDUAL WELL TICKET
B 6208 1310 API-AAPG, CSD, INDIVIDUAL WELL TICKET
B 6108 1156 API-AAPG, CSD, INDIVIDUAL WELL TICKET
B 6409 1320 API—AAPG, CSD, INDIVIDUAL WELL TICKET
B 6305 846* APPALACHIA, ABST., SILURIAN-, DEVONIAN EMBAYMENT INTO
B 6405 761* APPALACHIAN AND EASTERN INTERIOR REGIONS AND ROCKY
B 6405 743* APPALACHIAN AND ILLINOIS BASINS, ABST., RESIN RODS AND WOODY
B 6303 551* APPALACHIAN BASIN, ABST., CORRELATION OF SUBSURFACE MIDDLE
B 6009 1619* APPALACHIAN BASIN, ABST., ENERGY RESOURCE IMPLICATIONS OF, 38TH
B 6305 845* APPALACHIAN BASIN, ABST., FOSSIL FLUIDS IN POST-SILURIAN ROCKS OF
B 6405 692* APPALACHIAN BASIN, ABST., GEOCHEMICAL EFFECTS OF EARLY
B 6405 689* APPALACHIAN BASIN, ABST., GEOLOGIC CONTROLS ON
B 6309 1589* APPALACHIAN BASIN, ABST., HYDROTHERMAL AND UPLIF HISTORIES OF
B 6305 846* APPALACHIAN BASIN, ABST., LATE, ORDOVICIAN STRATIGRAPHY AND
B 6309 1590* APPALACHIAN BASIN, ABST., UPPER, NIAGARAN AND LOWER, CAYUGAN
B 6209 1635 APPALACHIAN BASIN, BRASSFIELD CARBONATE ROCKS
B 6209 1615 APPALACHIAN BASIN, COBBLESKILL FORMATION
B 6209 1638 APPALACHIAN BASIN, DEVONIAN REEFS
B 6405 692* APPALACHIAN BASIN DEVONIAN SHALES—REGIONAL ORGANIC
B 6209 1615 APPALACHIAN BASIN, HELDERBERG GROUP
B 6209 1608 APPALACHIAN BASIN, LOCKPORT FORMATION
B 6209 1635 APPALACHIAN BASIN, MEDINA SANDSTONES AND SHALES
B 6305 847* APPALACHIAN BASIN OF WEST VIRGINIA, ABST., DEVONIAN, SHALE, GAS
B 6209 1615 APPALACHIAN BASIN, ONESQUETHAW STAGE
B 6209 1616 APPALACHIAN BASIN, ONONDAGA LIMESTONE
B 6412 2156† APPALACHIAN BASIN, ORGANIC CONTENT OF, DEVONIAN, SHALE IN
B 6209 1615 APPALACHIAN BASIN, ORISKANY SANDSTONE
B 6209 1607† APPALACHIAN BASIN, PALEOGEOGRAPHY OF SOME, SILURIAN AND,
B 6209 1615 APPALACHIAN BASIN, RONDOUT FORMATION
B 6209 1637 APPALACHIAN BASIN, SALINA EVAPORITE UNITS
B 6209 1611 APPALACHIAN BASIN, SALINA GROUP
B 6211 2308† APPALACHIAN BASIN, SILURIAN EVOLUTION OF, CENTRAL
B 6209 1609 APPALACHIAN BASIN, SILURIAN-DEVONIAN CARBONATE SEQUENCE
B 6209 1711 APPALACHIAN BASIN, THREE LICK BED CORRELATION IN
B 6209 1635 APPALACHIAN BASIN, TUSCARORA SANDSTONE
B 6305 844* APPALACHIAN BASIN WITH MINI-SOSIE, ABST., SHALLOW SEISMIC
B 6303 488* APPALACHIAN BASIN-- PRODELTAIC CLASTIC RAMP, ABST., UPPER,
B 6004 665* APPALACHIAN BASIN--DETERMINATION BY FAUNAL CRITERIA, ABST.,
B 6008 1399* APPALACHIAN BASIN--SOME INITIAL RESULTS, ABST., RECONNAISSANCE
B 6412 2064 APPALACHIAN BASINS, SOUTHERN
B 6203 517* APPALACHIAN BASIN—SOURCE AND RESERVOIR DEPOSITS, ABST.,
B 6004 680* APPALACHIAN CARBONIFEROUS DELTAS, ABST., SANDSTONE, MODELS
B 6405 735* APPALACHIAN COAL BED, ABST., USCHEM, GEOCHEMICAL DATA FILE OF,
B 6203 512* APPALACHIAN COAL FIELD, ABST., COAL-RESOURCE DISTRIBUTION
B 6309 1504† APPALACHIAN DEVONIAN SHALES FROM FORMATION-DENSITY LOGS,
B 6305 847* APPALACHIAN GAS RESERVES--HARD TO FIND OR EASY TO MISS,
B 6303 504* APPALACHIAN NAPPE STRUCTURES AROUND QUEBEC CITY, CANADA,
B 6309 1579* APPALACHIAN OROGENE, ABST., EASTERN PROJECTION O, VALLEY AND
B 6405 770* APPALACHIAN OROGENE, VIRGINIA, ABST., CARBONATE DEPOCENTERS
M 29 86 APPALACHIAN OROGENIC SYSTEM, ATLANTIC MARGIN
B 6412 2166† APPALACHIAN OVERTHRUST BELT, CROSS-STRIKE STRUCTURAL,
B 6105 653 † APPALACHIAN PLATEAU FOLDS, FLOW OF WEAK ROCKS IN
B 6406 938 APPALACHIAN PLATEAU, MAJOR COAL PRODUCING AREA
B 6212 2379† APPALACHIAN REGION, DEPOSITIONAL, MODELS IN, COAL EXPLORATION
B 6405 775* APPALACHIAN THRUST BELT BETWEEN TRENTON, GEORGIA, AND
B 6203 543* APPALACHIANS, ABST., COARSE SEDIMENTS DEPOSITED AT BASE OF,
B 6305 845* APPALACHIANS, ABST., INTEGRATION OF AIRBORNE,
B 6405 798* APPALACHIANS, ABST., PALEOECOLOGIC DETERMINATION OF
B 6303 525* APPALACHIANS, ABST., QUANTITATIVE, BASIN ANALYSIS AND
B 6412 2078 APPALACHIANS, BEAVER VALLEY FAULT
B 6410 1575† APPALACHIANS, CARBONATE, RAMP-TO-BASIN TRANSITIONS AND,
B 6412 2078 APPALACHIANS, CLINCHPORT FAULT
B 6412 2078 APPALACHIANS, COPPER CREEK FAULT
B 6412 2167 APPALACHIANS, CROSS-STRIKE STRUCTURAL DISCONTINUITIES
B 6412 2078 APPALACHIANS, CUMBERLAND MOUNTAIN
B 6412 2078 APPALACHIANS, DORTON BRANCH FAULT ZONE
B 6412 2078 APPALACHIANS, DUMPLIN VALLEY FAULT
B 6412 2078 APPALACHIANS, GUESS CREEK AND GREAT SMOKY FAULTS
B 6412 2078 APPALACHIANS, JACKSBORO FAULT
B 6412 2078 APPALACHIANS, KNOXVILLE FAULT
B 6405 693* APPALACHIANS MAPPED BY COCORP REFLECTION DATA, ABST.,
B 6412 2078 APPALACHIANS, MIDDLESBORO SYNCLINE
B 6203 516* APPALACHIANS OF WEST VIRGINIA, ABST., ORIGIN OF, CYCLICITY IN,
B 6412 2078 APPALACHIANS, PINE MOUNTAIN FAULT
B 6412 2078 APPALACHIANS, POWELL VALLEY ANTICLINE
B 6412 2078 APPALACHIANS, ROCKY FACE FAULT
B 6412 2078 APPALACHIANS, SALTVILLE FAULT
B 6410 1582 APPALACHIANS, SECTIONS, VIRGINIA
B 6203 534* APPALACHIANS, SOUTHERN NOVA SCOTIA, ABST., ORDOVICIAN-,
B 6412 2064 APPALACHIANS, SOUTHERN, VALLEY AND RIDGE AREA, SAR IMAGERY
B 6410 1576 APPALACHIANS, VALLEY AND RIDGE PROVINCE, GEOLOGIC SKETCH
B 6412 2076 APPALACHIANS, VALLEY AND RIDGE PROVINCE, STRUCTURE

B 6412 2078 APPALACHIANS, WARREN VALLEY FAULT
B 6412 2078 APPALACHIANS, WHITE MOUNTAIN FAULT ZONE
B 6303 465* APPENNINES, ITALY, ABST., RESEDIMENTED, CONGLOMERATES IN
B 6309 1580* APPLACHIAN STATES, ABST., NEW ECONOMICS OF, NATURAL, GAS
B 6209 1753 APPLEGATE, A. V., FUTURE OIL POTENTIAL OF LOWER CRETACEOUS
B 6109 1534 APPLEGATE, A. V., SMACKOVER OIL PROSPECTS IN APALACHICOLA
B 6008 1397 APPLEGATE, JAMES K., GEOELECTRICAL INVESTIGATIONS OF BOISE,
B 6008 1389 APPLEGATE, JAMES K., GEOLOGIC AND SEISMIC STUDIES OF BOISE
S 6 93† APPLICABILITY OF THE RUBIDIUM-STRONTIUM METHOD TO SHALES AND
B 6108 1389* APPLICABLE REGULATIONS OF IN-SITU MINING OF URANIUM. BY TITLE
B 6002 273† APPLICATION FOR ADDITION TO CODE CONCERNING
B 6102 248† APPLICATION FOR AMENDMENT CONCERNING TERMINOLOGY FOR
B 6102 252† APPLICATION FOR AMENDMENT OF ARTICLES 8 AND 10 OF CODE,
M 26 301† APPLICATION OF AMPLITUDE, FREQUENCY, AND OTHER ATTRIBUTES TO
B 6405 728* APPLICATION OF CINGULUM INDEX TO DINOGYMNIUM IN HILLIARD
B 6405 738* APPLICATION OF CONODONT AND PALYNOMORPH COLOR ALTERATION
B 6105 801 * APPLICATION OF COMPUTER-MAPPING PROGRAMS, SYMAP AND SYMVU, TO
B 6408 1278* APPLICATION OF DEPOSITIONAL MODELS IN COAL EXPLORATION AND
B 6108 1392* APPLICATION OF DRILLING STATISTICS IN BASIN EVALUATION, ABST.
B 6405 716* APPLICATION OF ECOLOGIC STUDIES OF LIVING, ALGAL
B 6209 1755* APPLICATION OF FRACTURE IDENTIFICATION LOGS IN CRETACEOUS OF
B 6203 551* APPLICATION OF GARNET SYSTEM FOR EVALUATION OF MINERAL AND
B 6405 708* APPLICATION OF INVERSION PROCESSING TO EXPLORATION FOR
B 6207 1222* APPLICATION OF LANDSAT IMAGERY AND RELATED REMOTE SENSING
B 6005 745† APPLICATION OF LANDSAT IMAGERY TO PETROLEUM AND MINERAL
M 6004 726* APPLICATION OF MONTE CARLO METHOD FOR ESTIMATING RESERVES IN
B 6305 828* APPLICATION OF NONMARINE MOLLUSCA TO PALEOENVIRONMENTAL
B 6303 513* APPLICATION OF OIL AND GAS DISCOVERY FUNCTIONS TO RESOURCE
B 6303 502* APPLICATION OF PALYNOLOGY TO AGE DETERMINATION,
B 6303 500* APPLICATION OF POTENTIAL FIELD DAT TO STRUCTURAL
B 6405 695* APPLICATION OF REGIONAL GEOCHEMICAL DATA TO URANIUM
B 6004 677* APPLICATION OF REMOTE SENSING LANDSAT DATA TO PETROLEUM
B 6304 693* APPLICATION OF SOLUTION-MINERAL EQUILIBRIUM CHEMISTRY TO
B 6405 713* APPLICATION OF STATISTICAL MODELS IN CONTINENTAL MARGIN
B 6303 466* APPLICATIONS OF CRAB-BURROW ORIENTATION TO ENVIRONMENTAL
B 6207 1232* APPLICATIONS OF LANDSAT IMAGERY IN CIRCUM-PACIFIC REGION,
B 6002 318* APPLICATIONS OF OIL FIELD WATER GEOCHEMISTRY TO EXPLORATION,
B 6012 2179* APPLICATIONS OF PLATE TECTONICS TO PETROLEUM GEOLOG ALONG
B 6405 677* APPLICATIONS OF PROBABL RANGE CONCEPT TO BIOSTRATIGRAPHY,
B 6303 539* APPLICATIONS OF SCANNING ELECTRON MICROSCOPY TO
B 6309 1599* APPLICATIONS OF SOLAR ENERGY. WHAT IS PRACTICAL IN THE NEAR
B 6012 2181* APPLICATIONS OF SURFACE-CONSISTENT CORRECTIONS, ABST.
B 6108 1371* APPLICATIONS OF VERTICAL SEISMIC PROFILE TO STRATIGRAPHIC
M 26 484 APPLICATIONS, SEISMIC, MODELING
B 6303 509* APPLIED BIOSTRATIGRAPHY IN GULF COAST TERTIARY, ABST.
B 6303 442* APPLIED BIOSTRATIGRAPHY, SOUTH LOUISIANA AND GULF OF MEXICO,
B 6305 822* APPLIED EXPLORATION GEOLOGY AND URANIUM RESOURCE OF GREAT
B 6405 794* APPLIED PETROPHYSICS—CASE HISTORY, ABST.
B 6004 651* APPRAISAL OF POTENTIAL AREAS--CLUSTER AND CORRESPONDENCE
B 6212 2512* APPRAISAL OF PRESENT PETROLEUM-EXPLORATION ACTIVITIES IN
B 6007 1073† APPROXIMATE GEOTHERMAL GRADIENTS IN NIGER DELTA
B 6105 839* APRON IN TRANS-PECOS VOLCANIC FIELD OF TEXAS, ABST., OLIGOCENE,
B 6303 291 APRONS, PROSPECTS, OUTER, MARGIN
B 6204 669 APTIAN, ALGERIA
B 6303 503* APTIAN AND ALBIAN BLACK SHALES FROM SOUTH ATLANTIC, DSDP LEGS
B 6206 1014 APTIAN, ARCTIC, CANADA, MESOZOIC, UPPER, NORIAN TO
S 6 279 APTIAN, FORAMINIFERAL ZONES, CRETACEOUS
S 4 87 APTIAN, JAMAICA, DEVILS RACE COURSE FORMATION, BARREMIAN-
B 6405 610 APTIAN, MIDDLE EAST
B 6109 1477 APTIAN PETROLEUM RESERVOIR SYSTEM, BRAZIL
B 6205 793 APTIAN, PROFILE, BLAKE PLATEAU AREA
B 6401 74 APTIAN THROUGH CENOMANIAN CARBONATES, MEXICO, RANCHO
B 6412 2055 APTIAN-ALBIAN, ORGANIC MATTER, ATLANTIC BASINS
B 6109 1522† APTIAN-ALBIAN SUBSIDENCE RATES IN EASTERN UNITED STATES, RAPID
B 6206 1015 APTIAN-CAMPANIAN TO MAESTRICHTIAN, ARCTIC, CANADA, MESOZOIC
B 6201 63 APTIAN-CENOMANIAN, AUSTRALIA
B 6309 1473 APTIAN-TURONIAN SEQUENCE, USSR, ANADYR BASIN
S 10 40 AQUATHERMAL PRESSURE, MIGRATION, AGENTS
B 6403 417 AQUATHERMAL PRESSURING
B 6309 1527† AQUEOUS SOLUBILITY OF METHANE AT ELEVATED PRESSURES AND
B 6002 213† AQUEOUS SOLUBILITY OF PETROLEUM AS APPLIED
B 6112 2146† AQUEOUS SOLUBILITY OF PETROLEUM AS APPLIED
M 25 366† AQUIFER--SUMMARY, SIMULATION MODEL OF, HONOLULU BASAL
B 6405 711* AQUIFER-OAKVILLE SANDSTONE OF SOUTH TEXAS COASTAL PLAIN,
B 6109 1539* AQUIFERS, HOUSTON-GALVESTON, TEXAS, ABST., GEOMETRY OF, GULF
B 6405 674* AQUIFERS IN MADISON GROUP AND RED RIVER FORMATION, POWDER
B 6004 718* AQUIFERS IN SOUTHEAST POWDER RIVER BASIN, WYOMING, ABST.,
B 6008 1409* AQUIFERS IN SOUTHEAST POWDER RIVER BASIN WYOMING, ABST.,
S 6 209 AQUITAINE BASIN, CORRELATION, PALEOGENE
B 6411 1964 AQUITAINE BASIN, FRANCE
B 6110 1641 AQUITAINE BASIN, FRANCE
B 6310 1919 AQUITAINE BASIN, FRANCE
B 6210 1910 AQUITAINE BASIN, FRANCE
M 29 463 AQUITAINE BASIN, VOLUME OF SEDIMENT PRESERVED, FRANCE
S 10 136 AQUITAINE, CRETACEOUS WATER LEVELS, FRANCE
S 10 136 AQUITAINE, EOCENE WATER LEVELS, FRANCE
S 10 135 AQUITAINE, MIOCENE WATER LEVELS, FRANCE
S 10 134 AQUITAINE, TOPOGRAPHIC SURFACE, FRANCE
S 10 133 AQUITAINE BASIN, GROUNDWATER FLOW, FRANCE
B 6405 614 AQUITANIAN, MIDDLE EAST
S 10 125 AQUITARD, EXTENSIVE, BASIN GEOMETRY
B 6105 827* AR36 RATIOS IN SOIL GAS OVERLYING TWO URANIUM ORE DEPOSITS,
B 6405 616 ARAB C, MIDDLE EAST
B 6405 616 ARAB D, MIDDLE EAST
B 6401 9 ARAB FORMATION, SAUDI ARABIA
B 6411 1838 ARABIAN AMERICAN OIL CO. AND PETROCONSULTANTS S.A., ON
B 6010 1868 ARABIAN AMERICAN OIL CO., AND TENNECO OIL, ON
B 6110 1798 ARABIAN AMERICAN OIL CO., DEVELOPMENTS IN SAUDI ARABIA, 1976,
B 6210 2114 ARABIAN AMERICAN OIL CO., ON DEVELOPMENTS IN SAUDI ARABIA, 1977
B 6010 1868 ARABIAN OIL CO. LTD, JAPAN, AND GETTY OIL CO., ON DEVELOPMENTS
B 6110 1798 ARABIAN OIL CO. LTD., AND GETTY OIL CO., DEVELOPMENTS IN
B 6310 1886 ARABIAN OIL CO., LTD., AND GETTY OIL CO., ON DEVELOPMENTS IN
B 6210 2114 ARABIAN OIL CO., LTD., GETTY OIL CO., AND PETROCONSULTANTS S.A.,
B 6405 719* ARABIAN PENINSULA, ABST., CRETACEOUS SEA LEVEL AND
S 4 382 ARABIAN PENINSULA, COASTAL AND CONTINENTAL SABKHAS
B 6106 900 ARABIAN PLATE
B 6403 360 ARABIAN PLATE, BOUNDARY
B 6406 864 ARABIAN PLATFORM, FACIES
B 6002 215 ARABIAN WHOLE CRUDE OIL, GHAWAR
B 6002 258 ARABO-NUBIAN MASSIF
B 6109 1476 ARACAS PROFUNDO FIELD, BRAZIL
B 6109 1476 ARACAS RASO FIELD, BRAZIL
M 25 241 ARAFURA PLATFORM OR SAHUL SHELF, INDONESIA
M 29 213 ARAFURA SEA SECTION, AUSTRALIA, SHELF
B 6102 208 ARAGON FORMATION, MEXICO
B 6203 534* ARAGONITE, ABST., EGGSHELL DIAGENESIS AND PRECOMPACTIONAL
M 27 158 ARAGONITE CEMENTS
M 27 157 ARAGONITE CEMENTS
M 27 189 ARAGONITE CRUST, CEMENTS
M 28 121 ARAGONITE, IN SANDSTONE, CEMENTS
B 6303 493* ARAGONITE ON PERMIAN SEAFLOOR, ABST., IN-SITU FORMATION OF,
M 27 60 ARAGONITIC ANNULARIS
B 6009 1607* ARANSAS COUNTY, TEXAS ABST., DENSITY-, FLOW, ORIGIN FOR, FRIO,
B 6109 1471 ARARIPE BASIN, BRAZIL
B 6105 774 ARAUJO, M. B., ET AL, CANDEIAS FIELD, RECONCAVO BASIN, BRAZIL—
B 6303 533 ARAUJO, R., GEOCHEMICAL EVALUATION OF ENSENADA DE LA VELA
B 6203 443 ARAVALLI-DELHI OROGENIC BELT, INDIA
B 6405 670 ARBENZ, J. KASPAR, FRESH LOOK AT SOME OUACHITA PROBLEMS, ABST.
B 6103 463 ARBENZ, J. KASPAR, OVERVIEW OF ALASKA PLATE TECTONICS, ABST.
B 6103 361 ARBUCKLE ANTICLINE, OKLAHOMA
B 6305 806 ARBUCKLE DEVELOPMENT, MILLS RANCH COMPLEX, TEXAS
B 6203 516* ARBUCKLE GROUP, ABST., STRATIGRAPHIC, TRAP POSSIBILITIES IN
B 6004 719* ARBUCKLE GROUP, ARBUCKLE MOUNTAINS SOUTH-CENTRAL
B 6108 1375* ARBUCKLE GROUP, CAMBRIAN-ORDOVICIAN, OKLAHOMA, ABST.,
B 6406 960* ARBUCKLE GROUP, CENTRAL KANSAS, ABST., PETROLOGY OF
B 6002 325* ARBUCKLE GROUP. BY TITLE ONLY, STRATIGRAPHIC, TRAP POSSIBILITIES
B 6307 1136 ARBUCKLE MOUNTAINS CORRELATION, OKLAHOMA
B 6209 1767* ARBUCKLE MOUNTAINS, OKLAHOMA, ABST., PERITIDAL, CARBONATE
B 6004 719* ARBUCKLE MOUNTAINS SOUTH-CENTRAL OKLAHOMA, ABST., CHEMICAL
B 6104 582 ARBUCKLE OILS, OIL AGE, OKLAHOMA, SOUTHERN
B 6305 804 ARBUCKLE PRODUCTION, TEXAS
B 6103 370 ARBUCKLE UPLIFT, ALLUVIAL FAN DEPOSITS, OKLAHOMA
B 6301 3 ARC, DEFINITION, MAGMATIC
B 6301 3 ARC, DEFINITION, VOLCANIC
M 25 472 ARC, FORMATION IN RELATION TO ENERGY-GENERATING CENTER,
S 3 82 ARC, ISLAND
B 6301 17 ARC MASSIF
M 25 498 ARC, MINERAL DEPOSITS, CANADIAN, CORDILLERA, ISLAND
B 6312 2175 ARC PROVENANCES, MAGMATIC
M 25 498 ARC, ROCKS OF, CANADIAN, CORDILLERA, ISLAND
B 6406 868† ARC-CONTINENT COLLISION I BANDA SEA REGION
B 6108 1378* ARC-DERIVED SANDSTONES-IMPLICATIONS FOR PETROLEUM
M 25 463 ARC-FAULT-BLOCK MOVEMENT, VOLCANISM, RELATED TO

B 6303 552* ARC-TRENCH GAP, SOUTHWESTERN GUATEMALA ABST.,
B 6207 1225* ARC-TRENCH SYSTEM, ABST., SEISMICITY AND STRUCTURE OF, NEW,
B 6301 3 ARC-TRENCH SYSTEMS
B 6301 4 ARC-TRENCH SYSTEMS, DEFINED BY BACKARC FEATURES
B 6301 8 ARC-TRENCH SYSTEMS, PLATE RUPTURE
B 6301 8 ARC-TRENCH SYSTEMS, POLARITY REVERSAL
B 6009 1623* ARCH-RELATED ORIGIN OF LITTLE NORTH MOUNTAIN STRUCTURAL
B 6004 663* ARCHAELOGIC DATING OF HOLOCENE SHORELINE CHANGES ON
B 6203 509* ARCHAEOLOGICAL EVIDENCE FOR POST-PLEISTOCENE LOW STAND OF
B 6004 650* ARCHAEOLOGY, ABST., NUMERICAL ASPECTS OF, CHEMICAL VARIATION
B 6201 41 ARCHEAN AND PROTEROZOIC, AUSTRALIA
M 25 411 ARCHEAN BANDED IRON FORMATIONS OR BIFS, WESTERN AUSTRALIA
B 6101 30 ARCHEAN CHART, CANADIAN SHIELD
M 25 427 ARCHEAN IRON ORE, WESTERN AUSTRALIA
M 25 428 ARCHEAN, ORE DEPOSITS ASSOCIATED WITH, AUSTRALIA
M 25 401 ARCHEAN, WESTERN AUSTRALIA, NICKEL SULFIDE DEPOSITS IN
S 4 251 ARCHEOLITHOPHYLLUM SEGMENT, TEXAS
B 6109 1536* ARCHEOLOGICAL DATING, GEORGIA COAST, ABST., HISTORY OF,
B 6405 729 ARCHER, ALLEN W., TRACE-MAKERS AS HISTORIANS IN LARGE-SCALE
B 6307 1035 ARCHES, DOMES, AND SAGS, BASEMENT, WARPS--
B 6311 2114 ARCHINAL, BRUCE E., ATOKA FORMATION, PENNSYLVANIAN,
B 6203 491 ARCHINAL, BRUCE E., ATOKA FORMATION, PENNSYLVANIAN,
B 6004 662* ARCHIPELAGO, ABST., EARLY, MARINE AND LATER POSTBURIAL,
M 25 269 ARCHIPELAGO ALIGNMENT, KOREA, NAMHAE
B 6108 1389* ARCHULETA ANTICLINORIUM, BRAZOS UPLIFT, AND CHAMA BASIN IN
B 6104 562 ARCILLA CON PINTAS STRATIGRAPHIC UNIT, CHILE
B 6004 710* ARCS AND IT APPLICATION TO PRETROLEUM GEOLOGY, ABST.,
M 25 448 ARCS, PACIFIC-WARD SHIFTING, JAPAN, ISLAND
B 6004 662* ARCTIC ARCHIPELAGO, ABST., EARLY, MARINE AND LATER POSTBURIAL,
S 5 145* ARCTIC ARCHIPELAGO, CARBONATE- ANHYDRITE FACIES
B 6111 1929† ARCTIC ARCHIPELAGO, CARBONATE-, ANHYDRITE FACIES RELATIONS IN,
S 5 146 ARCTIC ARCHIPELAGO, ELLESMERE ISLAND, OTTO FIORD FORMATION,
B 6307 1060 ARCTIC ARCHIPELAGO, GEOLOGIC PROVINCES, MAP
B 6111 1930 ARCTIC ARCHIPELAGO, OTTO FJORD FORMATION
B 6101 3 ARCTIC BASIN
B 6405 714* ARCTIC BASIN, ABST., IMPLICATIONS OF
B 6105 818* ARCTIC CANADA, ABST., OLD AND NEW HORIZONS IN FOR WESTERN AND
B 6004 697* ARCTIC CANADA, ABST., TERTIARY, DELTAIC, CYCLES ON UNSTABLE,
B 6206 1006 ARCTIC CANADA, AMUND RINGNES ISLAND
B 6206 1005 ARCTIC CANADA, AXEL HEIBERG ISLAND
B 6206 1005 ARCTIC CANADA, BAFFIN ISLAND
B 6206 1005 ARCTIC CANADA, BANKS ISLAND
B 6206 1008 ARCTIC CANADA, BARROW SEGMENT
B 6206 1005 ARCTIC CANADA, BOOTHIA PENINSULA
B 6206 1005 ARCTIC CANADA, BOOTHIA UPLIFT
B 6206 1005 ARCTIC CANADA, BYLOT ISLAND
B 6206 1005 ARCTIC CANADA, DEVON ISLAND
B 6206 1006 ARCTIC CANADA, ELLEF RINGNES ISLAND
B 6206 1005 ARCTIC CANADA, ELLESMERE ISLAND
B 6206 1005 ARCTIC CANADA, ELLESMERIAN OROGENY
B 6206 1004† ARCTIC CANADA, EVOLUTION OF, SVERDRUP BASIN
B 6206 1005 ARCTIC CANADA, FRANKLINIAN GEOSYNCLINE
B 6206 1005 ARCTIC CANADA, GEOLOGIC PROVINCES
B 6206 1005 ARCTIC CANADA, HAZEN TROUGH
B 6206 1005 ARCTIC CANADA, HUDSONIAN OROGENY
B 6206 1006 ARCTIC CANADA, KING CHRISTIAN ISLAND
B 6206 1005 ARCTIC CANADA, MELVILLE ISLAND
B 6206 1015 ARCTIC CANADA, MESOZOIC, APTIAN-CAMPANIAN TO MAESTRICHTIAN
B 6206 1014 ARCTIC CANADA MESOZOIC, GRIESBACHIAN TO UPPER NORIAN
B 6206 1015 ARCTIC CANADA, MESOZOIC TO CENOZOIC, CAMPANIAN TO
B 6206 1014 ARCTIC CANADA, MESOZOIC, UPPER NORIAN TO APTIAN
B 6206 1007 ARCTIC CANADA, PALEOZOIC, VISEAN TO DZHULFIAN
B 6206 1005 ARCTIC CANADA, PEARYA GEANTICLINE
B 6206 1006 ARCTIC CANADA, PRINCE PATRICK ISLAND
B 6206 1005 ARCTIC CANADA, QUEEN ELIZABETH ISLANDS
B 6206 1006 ARCTIC CANADA, SABINE PENINSULA
B 6206 1006 ARCTIC CANADA, SVERDRUP BASIN, OIL AND GAS
B 6206 1005 ARCTIC CANADA, VICTORIA ISLAND
B 6206 1013 ARCTIC, CORNWALL ARCH, CANADA
B 6207 1239* ARCTIC ENERGY—TRANS-ALASKA PIPELINE, ABST., HARNESSING
B 6403 445* ARCTIC FORAMINIFERA--AN OVERVIEW, ABST., RECENT
B 6206 1013 ARCTIC, GRANTLAND UPLIFT, CANADA
B 6412 2111 ARCTIC ISLAND WELL, HYDROCARBONS, CANADIAN
B 6101 7 ARCTIC ISLANDS, CAMBRIAN, CANADIAN
B 6101 20 ARCTIC ISLANDS, CENTRAL ELLESMERE FOLDBELT, CANADIAN
B 6101 19 ARCTIC ISLANDS, CORNWALLIS FOLDBELT, CANADIAN
B 6108 1317 ARCTIC ISLANDS, DEVELOPMENTS, 1976
B 6308 1410 ARCTIC ISLANDS, DEVELOPMENTS, 1978
B 6409 1513 ARCTIC ISLANDS, DEVELOPMENTS, 1979
B 6101 24 ARCTIC ISLANDS, DRAKE POINT, WEST AND EAST, FIELD, CANADIAN
B 6101 20 ARCTIC ISLANDS, EUREKAN FOLDBELT, CANADIAN
B 6101 6 ARCTIC ISLANDS, FRANKLINIAN BASIN, CANADIAN
B 6101 19 ARCTIC ISLANDS, FRANKLINIAN FOLDBELT, CANADIAN
B 6101 1† ARCTIC ISLANDS, GEOLOGY AND, HYDROCARBON DISCOVERIES OF,
B 6101 24 ARCTIC ISLANDS, HECLA FIELD, CANADIAN
B 6101 24 ARCTIC ISLANDS, JACKSON BAY FIELD, CANADIAN
B 6101 24 ARCTIC ISLANDS, KING CHRISTIAN FIELD, CANADIAN
B 6101 24 ARCTIC ISLANDS, KRISTOFFER FIELD, CANADIAN
B 6101 12 ARCTIC ISLANDS, MISSISSIPPIAN, CANADIAN
B 6101 9 ARCTIC ISLANDS, ORDOVICIAN TO DEVONIAN, CANADIAN
B 6101 18 ARCTIC ISLANDS, OROGENIES, CANADIAN
B 6101 10 ARCTIC ISLANDS, PALEOZOIC AND MESOZOIC SEDIMENTS, DEPOCENTERS,
B 6101 19 ARCTIC ISLANDS, PARRY ISLANDS FOLDBELT, CANADIAN
B 6101 13 ARCTIC ISLANDS, PERMO-PENNSYLVANIAN, CANADIAN
B 6101 6 ARCTIC ISLANDS, PRECAMBRIAN, CANADIAN
B 6101 25 ARCTIC ISLANDS, RESERVES, CANADIAN
B 6101 11 ARCTIC ISLANDS, SVERDRUP BASIN, CANADIAN
B 6101 20 ARCTIC ISLANDS, SVERDRUP RIDGE, CANADIAN
B 6101 24 ARCTIC ISLANDS, THOR FIELD, CANADIAN
B 6101 14 ARCTIC ISLANDS, TRIASSIC, CANADIAN
B 6101 15 ARCTIC ISLANDS, TRIASSIC-JURASSIC, CANADIAN
B 6101 24 ARCTIC ISLANDS, WALLIS FIELD, CANADIAN
B 6101 11 ARCTIC ISLANDS, WANDEL SEA BASIN, CANADIAN
B 6206 1013 ARCTIC, LAKE HAZEN FAULT, CANADA
B 6111 1951 ARCTIC, MELVILLE ISLAND HEAVY OIL DEPOSIT, CANADIAN
B 6008 1189 ARCTIC NORTH SLOPE, EXPLORATION, 1975, ALASKA
B 6108 1157 ARCTIC NORTH SLOPE, EXPLORATION, 1976, ALASKA
B 6208 1312 ARCTIC NORTH SLOPE, EXPLORATION, 1977, ALASKA
B 6308 1211 ARCTIC NORTH SLOPE, EXPLORATION, 1978, ALASKA
B 6409 1345 ARCTIC NORTH SLOPE, EXPLORATION, 1979, ALASKA
B 6211 2357* ARCTIC OCEAN, ABST., RECENT, BENTHONIC, FORAMINIFERAL BIOFACIES
B 6211 2355* ARCTIC OUTER CONTINENTAL SHELF. BY TITLE ONLY,
B 6004 727* ARCTIC PLATE. BY TITLE ONLY, PRUDHOE BAY SUPE, GIANT FIELD AND,
B 6206 1013 ARCTIC, PRINCESS MARGARET ARCH, CANADA
B 6405 754* ARCTIC SLOPE, ALASKA, ABST., EVOLUTION OF, BROOKS RANGE, THRUST
B 6004 649* ARCTIC SLOPE, ALASKA, ABST., LISBURNE GROUP--POTENTIAL
B 6109 1493† ARCTIC SLOPE, ALASKA, LISBURNE GROUP, MISSISSIPPIAN AND,
B 6206 1013 ARCTIC, STOLZ FAULT, CANADA
B 6109 1494 ARCTIC WILDLIFE RANGE, ALASKA
B 6207 1172 ARCTURUS GROUP, NEVADA AND, UTAH
B 6401 133 ARCYANA FAMOUS--PHOTOGRAPHIC ATLAS OF THE MID-ATLANTIC
B 6209 1755 ARDEN, DANIEL DOUGLAS, JR., UPPER MESOZOIC AND PALEOGENE
B 6203 507 ARDEN, DANIEL DOUGLAS, MESOZOIC AND CENOZOIC GEOLOGY OF
B 6203 491 ARDILA, LUIS E., POTENTIAL OF MORROW SANDSTONES IN
B 6103 361 ARDMORE BASIN, OKLAHOMA
B 6307 1025 ARDMORE BASIN SEISMIC PROFILE, OKLAHOMA
B 6407 972 ARDMORE BENTONITE BED, GREAT PLAINS
B 6403 325 ARDMORE BENTONITE BED, MONTANA
B 6202 185 ARDMORE BENTONITE BED, WYOMING
B 6301 106† ARDMORE GEOLOGICAL SOCIETY, HISTORY OF
B 6303 540* ARDMORE-ANADARKO BASIN TREND, ABST., EXPLORATION RESEARCH
B 6303 413* ARE PENNSYLVANIAN FRESHWATER LIMESTONES ACTUALLY
B 6004 681* AREAL DISTRIBUTION OF CLAY MINERALS AND THEIR RELATION TO
B 6312 2186 AREAL YIELD, ASSESSING RESOURCES
M 27 33 ARENACEOUS FORAMS
B 6403 438 ARENDS, ROBERT G., DIATOMS AND SILICOFLAGELLATES FROM
B 6104 563 ARENISCA ARCILLOSA STRATIGRAPHIC UNIT, CHILE
B 6407 992 ARGANA FAULT, MOROCCO
B 6407 998 ARGANA GRABEN, MOROCCO
B 6407 993 ARGANA GRABEN, MOROCCO
B 6101 83 ARGANA VALLEY
B 6407 988† ARGANA VALLEY, SOUTHERN MOROCCO, AND THEIR REGIONAL
B 6304 682 ARGAND, TRANSLATED AND EDITED BY ALBERT V. CAROZZI, RIES,
B 6010 1642 ARGENTINA, 1975, LESTA, P. H., AND MARCELO R., YRIGOYEN, ON
B 6110 1579 ARGENTINA, 1976, YRIGOYEN, M. R., AND, PETROCONSULTANTS S.A., ON
B 6210 2045 ARGENTINA, 1977, YRIGOYEN, M. R., AND J. A., SUAREZ LYNCH, ON
B 6310 1746 ARGENTINA, 1978, YRIGOYEN, M. R., AND, ROBBIANO, J. A., ON
M 25 535 ARGENTINA, AGUILAR LEAD-ZINC DEPOSIT
M 29 130 ARGENTINA, ARGENTINE BASIN, BATHYMETRY
M 29 115 ARGENTINA, COLORADO BASIN, GEOLOGY
M 29 119 ARGENTINA, COLORADO BASIN, STRUCTURE SECTION
M 29 118 ARGENTINA, CONTINENT-OCEAN CRUST BOUNDARY
M 29 131 ARGENTINA, FALKLAND PLATEAU
M 29 132 ARGENTINA, FALKLAND PLATEAU, PALEOGENE SEDIMENTS
M 29 135 ARGENTINA, FALKLAND PLATEAU, SEISMIC REFLECTIONS
M 29 133 ARGENTINA, FALKLAND PLATEAU, STRUCTURE SECTION

M 29 132 ARGENTINA, FALKLAND PLATEAU, STRUCTURE SECTIONS
M 29 131 ARGENTINA, FALKLAND TROUGH
M 29 135 ARGENTINA, FALKLAND TROUGH, SEISMIC REFLECTIONS
M 29 126 ARGENTINA, MAGELLAN BASIN, GEOLOGIC MAP
B 6104 558 ARGENTINA, MAGELLAN BASIN, PETROLEUM, CHILE AND
M 29 126 ARGENTINA, MALVINAS BASIN, GEOLOGIC MAP
M 29 131 ARGENTINA, MALVINAS BASIN, STRUCTURE SECTION
M 29 116 ARGENTINA, MARGIN, AIRGUN-SONOBUOY RESULTS
M 29 127 ARGENTINA, MARGIN, AIRGUN-SONOBUOY RESULTS
M 29 114 ARGENTINA, MARGIN, BATHYMETRIC MAP
M 29 123 ARGENTINA MARGIN, GRAPH OF PROFILE VELOCITIES
M 29 118 ARGENTINA, MARGIN, SEISMIC REFRACTIONS
M 29 119 ARGENTINA, MARGIN, STRUCTURE SECTION
B 6403 429 ARGENTINA, MENDOZA HIGH-WAX OIL FIELD
M 29 131 ARGENTINA, NORTH SCOTIA RIDGE
M 29 135 ARGENTINA, NORTH SCOTIA RIDGE, SEISMIC REFLECTIONS
M 29 136 ARGENTINA, OFFSHORE, ISOPACH MAP
B 6110 1594 ARGENTINA, PRODUCTION, 1975-1976
B 6110 1603 ARGENTINA, PRODUCTION, 1976
B 6210 2069 ARGENTINA, PRODUCTION, 1976-1977
B 6310 1770 ARGENTINA, PRODUCTION, 1977-1978
B 6104 563 ARGENTINA, QUATERNARY GLACIOFLUVIAL DEPOSITS, CHILE AND
M 29 113† ARGENTINA, STRUCTURE OF, COLORADO, BASIN AND, CONTINENT-,
M 29 125† ARGENTINA, STRUCTURE OF, FALKLAND PLATEAU AND OFFSHORE,
B 6104 562 ARGENTINA, TERTIARY ZONA GLAUCONITICA STRATIGRAPHIC UNIT,
M 29 126 ARGENTINA, TIERRA DEL FUEGO, OFFSHORE, GEOLOGY
B 6106 930 ARGENTINE BASIN
M 29 130 ARGENTINE BASIN, BATHYMETRY, ARGENTINA
B 6111 1994 ARGENTINE BASIN, POSSIBLE DIAPIRS
B 6409 1557* ARGILLACEOUS SEDIMENTS, UNITED STATES GULF COAST, ABST., WELL
B 6201 41 ARGO ABYSSAL PLAIN, AUSTRALIA
B 6111 1994 ARGO ABYSSAL PLAIN, POSSIBLE DIAPIRS
B 6301 90 ARGO FORMATION, SCOTIAN BASIN
B 6402 244 ARGOLIDE SEQUENCES, GREECE
S 6 207 ARGON DATES, DIAGRAM, PHANEROZOIC, GLAUCONITES
S 6 123 ARGON-ARGON, APPARENT, GAS FRACTION, DIAGRAM
S 6 349 ARGON-ARGON DATING, A REVIEW, ORDOVICIAN
S 6 122 ARGON-ARGON DATING METHOD, A REVIEW
B 6206 968 ARGUELLO CANYON, CALIFORNIA
B 6206 968 ARGUELLO FAN, CALIFORNIA
M 25 458 ARGURI-KERULEN VOLCANIC-PLUTONIC CHAIN, PACIFIDES
B 6307 1031 ARGYLL FIELD, SEISMIC PROFILE, NORTH SEA
S 9 3 ARID ZONES, PRESENT DISTRIBUTION
B 6203 464 ARIDA FORMATION, LIBYA
M 28 51 ARIKAREE FORMATION, NEBRASKA, MIOCENE
M 28 50 ARIKAREE FORMATION, WYOMING, MIOCENE
B 6306 986 ARIZONA, 1978, NEW MEXICO GEOLOGIC SOCIETY WITH ARIZONA
B 6303 420* ARIZONA, ABST., MELAMOCYRILLIUM--NEW, ACRITARCH GENUS FRO,
B 6205 879* ARIZONA, ABST., CARBONIFEROUS, MISSISSIPPIAN, MICROFACIES AND
B 6303 517* ARIZONA, ABST., CONTRASTING, PENNSYLVANIAN, CARBONATE
B 6405 676* ARIZONA, ABST., DEVONIAN PALEOGEOGRAPHY AND
B 6405 696* ARIZONA ABST., DINEH-BI-KEYAH FIELD, APACHE COUNTY
B 6108 1386* ARIZONA, ABST., EARLY, PERMIAN, SLIP AND STRATIGRAPHIC
B 6405 688* ARIZONA, ABST., GENETIC STRATIGRAPHY AND PROVENANCE OF, BACA
B 6203 526* ARIZONA, ABST., GENETIC STRATIGRAPHY AND PROVENANCE OF, BACA
B 6109 1559* ARIZONA, ABST., INTEGRATION OF, LANDSAT AND, GEOPHYSICAL DATA
B 6203 561* ARIZONA, ABST., MAJOR, URANIUM DISCOVERY IN, FRONTIER AREA—,
B 6405 671* ARIZONA, ABST., MISSISSIPPIAN SYSTEM OF, NEW MEXICO AND
B 6008 1401* ARIZONA, ABST., OIL AND GAS POSSIBILITIES, REDWALL LIMESTONE,
B 6405 802* ARIZONA, ABST., URANIUM MINERALIZATION IN, HOPI BUTTES
B 6109 1448 ARIZONA, AND MEXICO, PEDREGOSA BASIN, PETROLEUM POTENTIAL,
B 6207 1159† ARIZONA AND NEVADA DURING LATE MIOCENE TIME, GULF OF
B 6307 1139† ARIZONA AND NEVADA DURING LATE MIOCENE TIME. DISCUSSION AND
B 6207 1160 ARIZONA AND NEVADA, GRAND WASH CLIFFS
B 6207 1159 ARIZONA AND NEVADA, HUALAPAI LIMESTONE MEMBER, MUDDY CREEK
B 6109 1449 ARIZONA AND NEW MEXICO, BURRO-FLORIDA UPLIFT
B 6109 1449 ARIZONA AND NEW MEXICO, ZUNI UPLIFT
B 6304 689* **ARIZONA AND NEW MEXICO**, ABST., DEPOSITIONAL ENVIRONMENTS AS
B 6303 547* ARIZONA AND NORTHEASTERN SONORA, MEXICO, ABST., LOWER,
B 6405 800* ARIZONA AND SONORA, ABST., DOLOMITE AND DEDOLOMITE IN, MURAL
B 6304 625 ARIZONA, ANDERSON MINE FORMATION, MEMBERS
B 6304 635 ARIZONA, ANDERSON MINE, URANIUM-BEARING INTERVAL
B 6304 625 ARIZONA, ANDERSON URANIUM MINE, PRODUCTION
B 6304 622 ARIZONA, ANDERSON URANIUM MINE--ARTILLERY PEAK AREA
B 6304 625 ARIZONA, ARRASTRA VOLCANICS
B 6307 1108 ARIZONA, COCHISE COUNTY, REEFS
S 3 33 ARIZONA, COCONINO SANDSTONE
B 6304 625 ARIZONA, DATE CREEK BASIN
B 6304 626 ARIZONA, DATE CREEK BASIN, MAP
B 6307 1108† ARIZONA, DEPOSITIONAL, MODEL OF EARLY, CRETACEOUS, CORAL-,
B 6008 1239 ARIZONA, DEVELOPMENTS, 1975
B 6108 1204 ARIZONA, DEVELOPMENTS, 1976
B 6208 1360 ARIZONA, DEVELOPMENTS, 1977
B 6308 1206 ARIZONA, DEVELOPMENTS, 1978
B 6409 1340 ARIZONA, DEVELOPMENTS, 1979
B 6405 769* ARIZONA DURING DEPOSITION OF PERMIAN TOROWEAP FORMATION,
B 6304 631 ARIZONA, FLAT TOP FORMATION
B 6109 1449 ARIZONA, FLORENCE UPLIFT
B 6306 986 **ARIZONA GEOLOGICAL SOCIETY FIELD TRIP GUIDEBOOK, EDITED BY J. F.**
B 6109 1449 ARIZONA, GRAHAM UPLIFT
S 3 33 ARIZONA, GRAND CANYON, SOUTH RIM
B 6307 1122 ARIZONA, GRASSY HILL
S 3 33 ARIZONA, HERMIT SHALE
B 6405 717* ARIZONA HINGE LINE—THRUST BELT—POTENTIAL NEW HYDROCARBON
B 6211 2357* ARIZONA HINGE-LINE THRUST BELT—POTENTIAL NEW HYDROCARBON
S 3 119 ARIZONA, HOPI BUTTES
B 6307 1117 ARIZONA, LEE SIDING, REEF, SECTION
B 6304 621† ARIZONA, MAJOR, URANIUM DISCOVERY IN, VOLCANICLASTIC
B 6109 1449 ARIZONA, MAZATZAL UPLIFT
B 6304 624 ARIZONA, MIOCENE
M 28 82 ARIZONA, MISSISSIPPIAN,, REDWALL LIMESTONE
S 3 131 ARIZONA, MONUMENT 2 MINE
B 6307 1122 ARIZONA, MORITA HOGBACK
B 6307 1109 ARIZONA, MURAL LIMESTONE, FIVE DEPOSITIONAL ENVIRONMENTS
B 6307 1123 ARIZONA, MURAL REEFS, MODEL
B 6109 1456 ARIZONA, NEW MEXICO, AND MEXICO, ALAMO HUECO BASIN
B 6307 1114 ARIZONA, PAUL SPUR, REEFS, SECTION
M 28 89 ARIZONA, PLIOCENE, BIG SANDY FORMATION
M 28 155 ARIZONA, PLIOCENE, BIG SANDY FORMATION
M 28 139 ARIZONA, PLIOCENE, BIG SANDY FORMATION
M 28 142 ARIZONA, PLIOCENE, BIG SANDY FORMATION
B 6004 646* ARIZONA, SOUTHWESTERN NEW MEXICO, NORTHEASTERN SONORA, AND
B 6004 723* ARIZONA, SOUTHWESTERN NEW MEXICO, NORTHEASTERN SONORA, AND
S 3 32 ARIZONA, TOROWEAP SANDSTONE, PARASHOUT WASH, SHIVWITS
M 28 97 ARIZONA, TRIASSIC, CHINLE FORMATION
M 28 26 ARIZONA, TRIASSIC, CHINLE FORMATION
M 28 125 ARIZONA, TRIASSIC, MOENKOPI FORMATION
B 6304 640 ARIZONA, URANIUM, GEOCHEMISTRY
B 6209 1680 ARIZONA, VOLCANIC ROCKS, ISOTOPIC AGES
M 25 73† ARIZONA--SUMMARY, SOLAR, DESALINATION EXPERIENCE OF,
B 6108 1386* ARIZONA-EOLIAN OR MARINE IN ORIGIN, ABST., NAVAJO SANDSTONE,
B 6407 1012 ARK-LA-TEX AREA, BETHANY FIELD
B 6407 1005 ARK-LA-TEX AREA, BOSSIER SHALES
B 6407 1005 ARK-LA-TEX AREA, BRYANS MILL FIELD
B 6407 1005 ARK-LA-TEX AREA, BUCKNER ANHYDRITE
B 6407 1005 ARK-LA-TEX AREA, CARTHAGE FIELD
B 6407 1011 ARK-LA-TEX AREA, CASPIANA FIELD
B 6407 1005 ARK-LA-TEX AREA, CHEROKEE TRACE FIELD
B 6407 1004 ARK-LA-TEX AREA, COTTON VALLEY AND SMACKOVER RESERVOIRS
B 6407 1005 ARK-LA-TEX AREA, DAYS CREEK FIELD
B 6407 1005 ARK-LA-TEX AREA, ELM GROVE FIELD
B 6407 1009 ARK-LA-TEX AREA, FALLON FIELD
B 6407 1005 ARK-LA-TEX AREA, FREESTONE FIELD
B 6407 1005 ARK-LA-TEX AREA, HICO SHALES
B 6407 1005 ARK-LA-TEX AREA, KNOWLES LIMESTONE
B 6407 1009 ARK-LA-TEX AREA, LITTLE BUDDY FIELD
B 6407 1005 ARK-LA-TEX AREA, LOUANN SALT
B 6407 1005 ARK-LA-TEX AREA, NORPHLET SILTSTONE
B 6407 1009 ARK-LA-TEX AREA, NORTH SIMSBORO FIELD
B 6407 1005 ARK-LA-TEX AREA, OAKS FIELD
B 6407 1005 ARK-LA-TEX AREA, OVERTON FIELD
B 6407 1005 ARK-LA-TEX AREA, REYNOLDS OOLITE
B 6407 1005 ARK-LA-TEX AREA, ROSEWOOD FIELD
B 6407 1005 ARK-LA-TEX AREA, SCHULER RED BEDS
B 6407 1005 ARK-LA-TEX AREA, TEAGUE FIELD
B 6407 1005 ARK-LA-TEX AREA, TERRYVILLE SANDS
B 6407 1005 ARK-LA-TEX AREA, WASKOM FIELD
B 6209 1757* ARK-LA-TEX COTTON VALLEY FORMATION, ABST., FORMATION
B 6009 1610* ARK-LA-TEX REGION, ABST., CORRELATIONS AND STRATIGRAPHY FROM
B 6004 715* ARKANSAS, ABST., COMBINATION, ENTRAPMENT , SMACKOVER
B 6009 1614* ARKANSAS, ABST., COMBINATION, ENTRAPMENT IN, SMACKOVER
B 6303 495* ARKANSAS, ABST., DEPOSITIONAL ENVIRONMENTS AND THEIR RELATION
B 6409 1563* ARKANSAS, ABST., DEPOSITIONAL SYSTEMS IN, NACATOCH SAND, UPPER,
B 6311 2120* ARKANSAS, ABST., EARLY, PENNSYLVANIAN, BRAIDED STREA
B 6309 1606* ARKANSAS, ABST., OUTER DEEP-SEA FAN DEPOSITIONAL LOBE SEQUENCE
B 6409 1566* ARKANSAS, ABST., PALEOENVIRONMENTAL ANALYSIS OF, JOACHIM

B 6405 753* ARKANSAS, ABST., POST-COMPACTION, SUBSURFACE, SECONDARY,
B 6209 1768* ARKANSAS, ABST., STRATIGRAPHIC RELATIONS OF, ANNONA CHALK
B 6405 721* ARKANSAS, ABST., STRATIGRAPHY AND DEPOSITIONAL HISTORY,
B 6002 322* ARKANSAS, ABST., TRENDS AND ORIGINS OF, ORDOVICIAN SANDSTONES,
B 6203 565* ARKANSAS AND ADJACENT AREAS, ABST., PETROLEUM POTENTIAL AND
B 6203 566* ARKANSAS AND NORTHEASTERN OKLAHOMA, ABST., REEVALUATION OF,
B 6108 1384* ARKANSAS AND OKLAHOMA, ABST., PETROLOGY AND DIAGENESIS OF,
B 6012 2103† ARKANSAS AND OKLAHOMA, NONGLACIAL VARVES IN LOWER MEMBER
B 6303 422* ARKANSAS, ANNONA FORMATION, CHALK MEMBER, SARATOGA
B 6407 1007 ARKANSAS, BASEMENT STRUCTURES, TEXAS, LOUISIANA, AND
M 24 179 ARKANSAS, BUCKNER FORMATION
B 6407 1008 ARKANSAS, CHALYBEAT SPRINGS FIELD
B 6407 1011 ARKANSAS, COTTON VALLEY SANDSTONES, TEXAS, LOUISIANA, AND
M 28 74 ARKANSAS, CRETACEOUS, BAUXITE
B 6208 1487 ARKANSAS, DEVELOPMENTS, 1977
B 6407 1008 ARKANSAS, GRABEN-FAULT STRUCTURES, TEXAS, LOUISIANA, AND
M 24 178 ARKANSAS, JURASSIC
B 6407 1004† ARKANSAS, JURASSIC, COTTON VALLEY AND, SMACKOVER, RESERVOIR
M 24 177 ARKANSAS, LAKE ERLING FIELD
M 24 184 ARKANSAS, LOUANN SALT
B 6108 1234† ARKANSAS, LOUISIANA, AND EAST TEXAS IN 1976, DEVELOPMENTS IN
B 6407 1007 ARKANSAS, MEXIA-TALCO FAULT SYSTEM, TEXAS, LOUISIANA, AND
B 6407 1007 ARKANSAS, MT. VERNON FIELD
M 24 184 ARKANSAS, NORPHLET FORMATION
B 6008 1370† ARKANSAS, NORTH LOUISIANA, AND EAST TEXAS IN 1975, DEVELOPMENT
B 6208 1487† ARKANSAS, NORTH LOUISIANA, AND EAST TEXAS IN 1977,
B 6308 1376† ARKANSAS, NORTH LOUISIANA, AND EAST TEXAS IN 1978, DEVELOPMENT
B 6409 1490† ARKANSAS, NORTH LOUISIANA, AND EAST TEXAS IN 1979,
S 7 30 ARKANSAS, NORTHERN GULF OF MEXICO RIM, EXPOSED
B 6012 2103† ARKANSAS NOVACULITE, DEVONIAN, ARKANSAS AND OKLAHOMA,
B 6405 754* ARKANSAS NOVACULITE FORMATIONS, OKLAHOMA, ABST., OIL
B 6209 1704 ARKANSAS NOVACULITE, UNITED STATES, EASTERN
M 28 113 ARKANSAS, ORDOVICIAN, CALICO ROCK SANDSTONE
M 28 113 ARKANSAS, ORDOVICIAN, EVERTON FORMATION
M 28 112 ARKANSAS, ORDOVICIAN, ST. PETER SANDSTONE
B 6407 1007 ARKANSAS, SALT ANTICLINES, TEXAS, LOUISIANA, AND
B 6002 231 ARKANSAS, SALTS IN OIL-FIELD WATERS VS. DEPTH
M 24 184 ARKANSAS, SMACKOVER, SUBMEMBER, BASAL UPPER
M 28 74 ARKANSAS, TERTIARY, BAUXITE
M 24 177† ARKANSAS, UPPER, SMACKOVER, RESERVOIRS, WALKER CREEK FIELD
M 24 177 ARKANSAS, WALKER CREEK FIELD
B 6407 1008 ARKANSAS, WALKER CREEK FIELD
B 6407 1004 ARKANSAS, WALKER CREEK STRATIGRAPHIC FIELD
B 6407 1004 ARKANSAS, WALKER FIELD
B 6303 544* ARKANSAS--SMACKOVER CASE HISTORY, ABST., WALKER CREEK FIELD
B 6002 321* ARKANSAS, BY TITLE ONLY, EASTERN, OKLAHOMA AND WESTERN
B 6203 491* **ARKOMA BASIN, OKLAHOMA, ABST., ATOKA FORMATION,**
B 6203 543* ARKOMA BASIN, OKLAHOMA, ABST., CONTEMPORANEOUS, FAULTS–
B 6311 2114* ARKOMA BASIN, OKLLHOMA, ABST., ATOKA FORMATION,
M 28 53 ARKOSE, MASSACHUSETTS, TRIASSIC
B 6203 377 ARKOSE RIDGE FORMATION, ALASKA
B 6303 465* ARKOSES, SANTA YNEZ MOUNTAINS, SANTA BARBARA AND VENTURA
B 6108 1392* ARKOSIC AND VITRIC SANDSTONE, CENOZOIC HIGH PLAINS SEQUENCE,
B 6403 438 ARMENTROUT, JOHN, GLACIOMARINE DEPOSITIONAL ENVIRONMENTS
B 6203 491 ARMENTROUT, JOHN M., COMMUNITY SUCCESSION OF PIONEER
B 6303 411 ARMENTROUT, JOHN M., GLACIAL ORIGIN OF MEGACHANNELS OF UPPER
B 6303 411 ARMENTROUT, JOHN M., GLACIAL-MARINE TRACE- AND BODY-FOSSIL
B 6403 440 ARMENTROUT, JOHN M., HOLOCENE FORAMINIFERAL DISTRIBUTION
B 6405 671 ARMENTROUT, JOHN M., HYDROCARBON POTENTIAL OF SITKALIDAK
B 6405 671 ARMENTROUT, JOHN M., LATE NEOGENE DEPOSITIONAL AND CLIMATI
B 6211 2348 ARMENTROUT, JOHN M., NEW MIOCENE FORMATION AT COOS BAY,
B 6405 670 ARMENTROUT, JOHN M., OPHIOMORPHA FROM UPPER BATHYAL EOCENE
B 6310 1823 ARMITAGE, J. H., AND ROBERT DE RUITER, ON DEVELOPMENTS, IN
B 6210 1986 ARMITAGE, J., ON DEVELOPMENTS IN PENINSULAR MALAYSIA, 1977
B 6006 962 ARMSTRONG, A. K., CARBONIFEROUS SECTION AND UNCONFORMITY AT
B 6105 761 ARMSTRONG, A. K., POROSITY IN LATE MIOCENE REEFS, SOUTHERN
B 6105 761 ARMSTRONG, A. K., RELATIONS BETWEEN TRIASSIC CARBONATE SABKHA
B 6205 879 ARMSTRONG, AUGUSTUS K., CARBONIFEROUS, MISSISSIPPIAN,
B 6405 671 ARMSTRONG, AUGUSTUS K., MISSISSIPPIAN SYSTEM OF NEW MEXICO
B 6402 188 ARMSTRONG, AUGUSTUS K., POROSITY EVOLUTION OF UPPER MIOCENE
B 6008 1389 ARMSTRONG, FRANK C., URANIUM IN PLUTONIC ROCKS, ABST.
S 6 73 ARMSTRONG, RICHARD LEE, PRE-CENOZOIC PHANEROZOIC TIME
B 6004 646 ARNAL, R. E., CHANGES IN MIOCENE PALEOBATHYMETR OF CALIFORNIA
B 6012 2174 ARNAL, ROBERT E., LATE MIOCENE PALEOBATHYMETR OF CALIFORNIA
S 10 27 AROMATIC DISTRIBUTION, FRANCE, PARENTIS BASIN
S 10 93 AROMATIC HYDROCARBONS, WEIGHT, CRUDE OIL
S 4 285 AROTHRON MELEAGRIS, PANAMA
B 6304 625 ARRASTRA VOLCANICS, ARIZONA
B 6307 1131 ARRHENIUS PLOTS OVER WIDE TEMPERATURE RANGE, EXTRAPOLATION
B 6410 1645 ARROWIE BASIN, AUSTRALIA
B 6012 2191* ARROYO EL BULITO, SANTA BARBARA COUNTY, CALIFORNIA, ABST.,
B 6211 2262 ARROYO HONDO MEMBER, CALIFORNIA
B 6211 2262 ARROYO HONDO SECTION, CALIFORNIA
B 6109 1515 ARROYO PENASCO FORMATION, NEW MEXICO
B 6109 1515 ARROYO PENASCO GROUP, NEW MEXICO
B 6211 2218 ARROYO SECO TROUGH, CALIFORNIA
M 25 515 ARSENIC, MEXICO, SILVER, LEAD, ZINC, AND
B 6303 474* ARTESIA GROUP. MIDDLE PERMIAN. SANDSTONES OF PERMIAN BASIN,
B 6303 448* **ARTHROPOD LEBENSSPUREN AS INDICATORS OF ANCIENT FLOOD-PLAIN**
S 4 271 ARTHROPODS, BORERS
B 6004 646 ARTHUR, M. A., PRIMARY BEDDING RHYTHMS IN PELAGIC CALCAREOUS
B 6401 67 ARTHUR, MICHAEL A., CARBON ISOTOPE FLUCTUATION IN CRETACEOUS
B 6306 870 ARTHUR, MICHAEL A., CRETACEOUS OCEANIC ANOXIC EVENTS AS
B 6405 680 ARTHUR, MICHAEL A., LOWER CRETACEOUS LACUSTRINE SOURCE BEDS
B 6105 762 ARTHUR, MICHAEL A., MIDDLE CRETACEOUS OCEANIC ANOXIC EVENTS AS
B 6405 671 ARTHUR, MICHAEL A., OXIC-ANOXIC AND CARBONATE CYCLES IN
S 4 88 ARTHURS SEAT FORMATION, SANTONIAN, JAMAICA
B 6107 1083† ARTICLES 27 AND 34 OF STRATIGRAPHIC CODE TO INTRODUCE
B 6102 252† ARTICLES 8 AND 10 OF CODE, CONCERNING SMALLEST FORMAL
M 25 354† ARTIFICIAL RECHARGE OF GR
S 10 117 ARTIFICIAL WATER-ROCK INTERACTION, COAL GASIFICATION
S 10 114 ARTIFICIAL WATER-ROCK INTERACTION, OIL SAND
B 6406 884 ARU, AND SERAM TROUGHS, TIMOR
M 29 216 ARU BASIN, REFLECTION PROFILE, AUSTRALIA
S 4 5 ARUBA ISLAND, NETHERLANDS ANTILLES
M 25 182 ASA ALASKA, WESTERN, CONVERGENT MARGIN, GULF OF
B 6104 585 ASAB, ABU JIDU, FIELD, PERSIAN GULF
B 6212 2509 ASAKAWA, TADASHI, CONTRAST OF HYDROCARBON POTENTIAL
M 25 497 ASBESTOS, CHROMITE, AND NICKEL, CANADA
M 25 518 ASBESTOS, MEXICO
B 6206 968 ASCENSION CANYON, CALIFORNIA
B 6107 963 ASCENSION FRACTURE ZONE, AFRICA
B 6206 918 ASCENSION VALLEY, CALIFORNIA, MONTEREY FAN
B 6203 531 ASCHEN-BRENNER, STANLEY E., HOLOCENE EPOCH
B 6106 929† ASEISMIC RIDGES–EVIDENCE FROM SEDIMENTS ON RIO GRAND RISE
B 6105 771 ASGAARD, U., FRESHWATER ICHNOCOENOSES FROM TRIASSIC RED BEDS OF
M 28 140 ASH FLOW TUFF, NEW MEXICO, TERTIARY
B 6004 651* ASH PARTINGS AS STRATIGRAPHIC MARKER BEDS IN COALS OF ROCKY
B 6201 81 ASHERN CORRELATION, SASKATCHEWAN
B 6104 594 ASHLAND FIELD, KENTUCKY
B 6303 344 ASHLOCK FORMATION, KENTUCKY
B 6311 2024 ASHMORE BLOCK, AUSTRALIA
B 6008 1204 ASHMORE, H. T., OIL AND GAS DEVELOPMENT IN NORTHERN ROCKIES IN
B 6205 828 ASHQELON AREA, CORES, ISRAEL
M 25 8† ASIA, BOUNDARIES OF, SEABED, JURISDICTION OFF, PACIFIC COAST OF
B 6307 1143 ASIA, BY CORAZON MORALES SIDDAYAO, DURKEE, E. F., REVIEW OF,
B 6304 682 ASIA, BY EMILE ARGAND, TRANSLATED AND EDITED BY ALBERT V.
B 6407 1087† ASIA, CHUKOTKA, KAMCHATKA, SAKHALIN, STRATIGRAPHY OF MARINE,
B 6407 1091 ASIA, NORTHEAST, AMANINIAN HORIZON
B 6407 1092 ASIA, NORTHEAST, ENEMTENIAN HORIZON
B 6407 1092 ASIA, NORTHEAST, ERMANOVIAN HORIZON
B 6407 1091 ASIA, NORTHEAST, ILJINIAN HORIZON
B 6407 1092 ASIA, NORTHEAST, KAKERTIAN-ETOLONIAN HORIZON
B 6407 1088 ASIA, NORTHEAST, KARAGINSKY ISLAND
B 6407 1088 ASIA, NORTHEAST, KORYAK UPLAND
B 6407 1091 ASIA, NORTHEAST, KOVACHINIAN HORIZON
B 6407 1091 ASIA, NORTHEAST, NAPANIAN HORIZON OF KAMCHATKA
B 6407 1092 ASIA, NORTHEAST, TUSATUVAYAMIAN HORIZON
B 6407 1092 ASIA, NORTHEAST, UST-LIMIMTEVAYAMIAN HORIZON
B 6407 1091 ASIA, NORTHEAST, UTKHOLOKIAN-VIVENTEKIAN HORIZON
B 6407 1090 ASIA, NORTHEASTERN, CHRONOSTRATIGRAPHIC CLASSIFICATION
M 29 373† ASIA, THE, SULU SEA: A, MARGINAL, BASIN IN, SOUTHEAST
M 25 250† ASIA--SUMMARY, STRUCTURAL EVOLUTION OF, TERTIARY, BASINS OF
B 6209 1669 ASIAN CONTINENTAL MARGIN, CONVERGENCE OF, LUZON ARC WITH
M 25 474 ASIAN MEGASTRUCTURE, EAST
B 6207 1241* ASIAN REGION, ABST., ENERGY, RESOURCES, AND, ENVIRONMENT
M 25 59 ASIAN VOLCANOGENIC BELT, USSR, EAST
B 6110 1802 ASIATIC PETROL. CORP., AND PETROCONSULTANTS S.A., ON
B 6110 1801 ASIATIC PETROL. CORP. AND WINTERSHALL A. G., ON DEVELOPMENTS IN
B 6110 1797 ASIATIC PETROL. CORP. ET AL, ON DEVELOPMENTS IN KUWAIT, 1976
B 6110 1802 ASIATIC PETROL. CORP., ON DEVELOPMENTS IN YEMEN, 1976
B 6110 1801 ASIATIC PETROL. CORP., SOFIRAN, ELF AQUITAINE, ET AL, ON
B 6310 1891 ASIATIC PETROLEUM CO., MOBIL EXPLORATION MEDITERRANEAN INC.,
B 6210 2113 ASIATIC PETROLEUM CO., ON DEVELOPMENTS IN KUWAIT, 1977,
B 6010 1872 ASIATIC PETROLEUM CO., QATAR OIL CO., LTD., AND WINTERSHAL A.G.,
B 6411 1842 ASIATIC PETROLEUM CO., SHELL, AND MOBIL EXPLORATIO
B 6010 1867 ASIATIC PETROLEUM CORP., AN KUWAIT SPANISH PETROLEUM CO., ON
B 6210 2119 ASIATIC PETROLEUM CORP. AND J.P. JAVOGUES, ON DEVELOPMENTS IN
B 6010 1874 ASIATIC PETROLEUM CORP., AND PETROCONSULTANTS S.A., ON
B 6010 1872 ASIATIC PETROLEUM CORP. AND SANTA FE MINERALS, INC., ON
B 6310 1890 ASIATIC PETROLEUM CORP., B.P. PETROLEUM DEVELOPMENT LTD., CLUFF

B 6210 2117 ASIATIC PETROLEUM CORP., ON DEVELOPMENTS IN QATAR, 1977,
B 6210 2118 ASIATIC PETROLEUM CORP., ON DEVELOPMENTS IN YEMEN, 1977
B 6210 2118 ASIATIC PETROLEUM CORP., SOCIETE NATIONALE ELF AQUITAINE, ET AL,
B 6010 1872 ASIATIC PETROLEUM CORP., WINTERSHALL A.G., SUN OIL CO., AN ELF
M 25 460 ASIATIC STANNIFEROUS ZONE, AUSTRALIAN-
B 6412 2104 ASMARI FORMATION, IRAN
B 6206 991 ASMARI LIMESTONE, IRAN
B 6002 196 ASMUS, H., SOUTH ATLANTIC DIAPIRIC STRUCTURES
B 6309 1453 ASP, BALTIMORE CANYON TROUGH, ATLANTIC SLOPE PROJECT
M 26 15† ASPECTS OF ROCK PHYSICS FROM LABORATORY AND LOG DATA THAT
S 6 367† ASPECTS OF THE REVISED SOUTH AFRICAN STRATIGRAPHIC
B 6207 1114 ASPECTS OF TOOL ROTATION ASSOCIATED WITH BREAKOUTS
B 6101 113 ASPEN SHALE, UTAH
B 6101 114 ASPEN SHALE, WYOMING
M 24 115 ASPEN SHALE, WYOMING
B 6109 1411 ASPHALT, ABSORBING TENDENCIES OF OILS AND AN
B 6205 838 ASPHALT LOCATIONS, DEAD SEA BASIN
B 6111 1951 ASPHALT RIDGE HEAVY OIL DEPOSIT, UNITED STATES
B 6205 841 ASPHALT, USES, DEAD SEA
B 6101 100† ASPHALTIC COMPOUNDS IN SEDIMENTARY ROCKS, DISTRIBUTION OF,
B 6101 100 ASPHALTIC COMPOUNDS, TEMPERATURE RANGES, HYDROCARBONS AND
B 6009 1606* ASPHALTIC HARTSELLE SANDSTONE IN NORTH ALABAMA, ABST.,
B 6203 351 ASPHALTO OIL FIELD, CALIFORNIA
B 6205 837† ASPHALTS—HISTORICAL ASPECTS, DEAD SEA
B 6103 361 ASPHALTUM STRUCTURE, OKLAHOMA
B 6008 1389 ASPINALL, WAYNE N., DANGER OF FEDERAL WITHDRAWALS TO
B 6102 295 **ASQUITH, GEORGE B., SEDIMENTOLOGY OF CRETACEOUS MESA RICA**
B 6403 427† ASSAM BASIN, INDIA, DEPOSITIONAL ENVIRONMENT OF, SOURCE BEDS
S 6 301 ASSELIAN STAGE, PERMIAN, BIOCHRONOLOGY
B 6408 1158† ASSESSING OIL AND GAS PLAYS IN FACIES-CYCLE WEDGES
B 6303 552* ASSESSING OIL AND GAS PLAYS IN FACIES-CYCLE WEDGES, ABST.
B 6108 1391* ASSESSMENT OF MAGNETIC MEASUREMENTS IN URANIUM
B 6108 1388* ASSESSMENT OF OIL AND GAS RESOURCES IN IDAHO-WYOMING THRUST
B 6405 683* **ASSESSMENT OF POTENTIAL FOR FLUID HYDROCARBONS IN IOWA,**
B 6312 2183 ASSESSMENT PROBABILITY CURVES
B 6209 1592 ASTORIA FAN SEDIMENTS, OREGON CONTINENTAL MARGIN
S 4 90 ASTROCOENIA-ACTINACIS ASSEMBLAGE, JAMAICA,
S 11 227 ASTRORHIZIDAE
B 6202 296 ASU RIVER GROUP, NIGERIA
B 6307 1021 ASYMMETRIC MASTER TRANSFORM SYSTEM, DEAD SEA
B 6109 1477 ATALAIA FAULT, BRAZIL
B 6301 36 ATASCOSA COUNTY, TEXAS, JOURDANTON FIELD
B 6301 35 ATASCOSA TROUGH, TEXAS
B 6402 270 ATCHAFALAYA BAY, BATHYMETRIC MAPS, LOUISIANA
B 6402 264† ATCHAFALAYA DELTA, EVOLUTION OF, SAND-DOMINANT, SUBAERIAL
B 6402 264† ATCHAFALAYA DELTA, LOUISIANA, EVOLUTION OF, SAND-DOMINANT,
B 6409 1568* ATCHAFALAYA DELTA—LOUISIANAS NEW PROGRADING COAST, ABST.
B 6409 1562* ATCHAFALAYA DIVERSION, ABST., SHOULD WE PERMIT, MISSISSIPPI-
B 6402 268 ATCHAFALAYA RIVER, DISCHARGE AND LOAD, LOUISIANA
M 27 17 ATCO FORMATION, TEXAS, CRETACEOUS
M 28 153 ATCO FORMATION, TEXAS, CRETACEOUS
M 27 18 ATCO FORMATION, TEXAS, CRETACEOUS
M 28 84 ATCO FORMATION, TEXAS, CRETACEOUS
B 6203 566* ATHABASCA BASIN, ABST., COMPUTER APPLICATIONS IN GEOLOGIC
B 6004 673* ATHABASCA BITUMEN, ABST., SIMULATED, GEOTHERMAL, MATURATION
B 6111 1951 ATHABASCA HEAVY OIL DEPOSIT, WESTERN, CANADA
S 10 222 ATHABASCA OIL DEPOSIT, TRAP, CANADA
B 6007 1112† ATHABASCA OIL, ORIGIN OF
B 6007 1110† ATHABASCA OIL, ORIGIN OF
B 6405 742* ATHABASCA OIL SAND AREA, ABST., GEOLOGY OF, SYNCRUDE CANADA
B 6203 547* ATHABASCA OIL SANDS, ABST., DEEP SINUOUS, CHANNELS AND GIANT.
B 6303 499* ATHABASCA OIL SANDS, ALBERTA, ABST., SEDIMENTOLOGY, PETROLOGY,
B 6303 499* ATHABASCA OIL SANDS, ALBERTA. BY TITLE ONLY, SEDIMENTOLOGY,
M 6004 726* ATHABASCA TAR DEPOSITS, ALBERTA, ABST., APPLICATION OF, MONTE
S 10 64 ATHABASCA-PEACE RIVER, CRETACEOUS, GEOLOGIC SETTING, CANADA
B 6109 1560* ATHABASKA REGION BY IMAGE PROCESSING OF LANDSAT DATA, ABST.,
B 6203 522* ATHAPUSCOW AULACOGEN, MIDDLE PRECAMBRIAN, GREAT SLAVE LAKE,
M 27 176 ATHENS LIMESTONE, VIRGINIA, ORDOVICIAN
B 6201 87† ATKA BASIN, ALEUTIAN RIDGE, NEOGENE, VOLCANICLASTIC SEDIMENTS
B 6201 88 ATKA ISLAND, ALASKA
M 25 328 ATLANTA FORMATION, ECUADOR
B 6303 543* **ATLANTIC, ABST., GEOLOGIC SIGNIFICANCE OF SEDIMENTARY**
B 6405 680* ATLANTIC, ABST., LOWER, CRETACEOUS LACUSTRINE SOURCE BEDS
B 6303 475* ATLANTIC, ABST., MESOZOIC-CENOZOIC SEDIMENTARY FORMATIONS
B 6203 489* ATLANTIC, ABST., MIXING OF, BOTTOM SEDIMENTS IN NORTHWEST
B 6405 790* ATLANTIC, ABST., ORGANIC FACIES OF MID-, CRETACEOUS, BLACK
B 6004 649* ATLANTIC, ABST., PRELIMINARY RESULTS FROM, LEG, 44, DEEP SEA,
B 6303 510* ATLANTIC, ABST., TEMPORAL AND SPATIAL DISTRIBUTION OF
B 6007 1078† ATLANTIC AND BARBADOS RIDGE COMPLEX, TECTONICS OF
B 6101 83 ATLANTIC ARCH, CENTRAL
B 6007 1085 ATLANTIC BASIN, BATHYMETRY
B 6204 596 ATLANTIC BASIN, CRETACEOUS, NORTH
B 6204 596 ATLANTIC BASIN, JURASSIC, NORTH
B 6007 1085 ATLANTIC BASIN, MAGNETICS
B 6204 596 ATLANTIC BASIN, QUATERNARY, NORTH
B 6204 596 ATLANTIC BASIN, SECTION, NORTH
B 6007 1088 ATLANTIC BASIN, SEISMIC-REFLECTION DATA
B 6204 596 ATLANTIC BASIN, TERTIARY, NORTH
B 6303 542* ATLANTIC BASINS, ABST., PALEOENVIRONMENT AND PETROLEUM
B 6412 2055 ATLANTIC BASINS, APTIAN-ALBIAN, ORGANIC MATTER
B 6412 2055 ATLANTIC BASINS, CENOMANIAN TO CONIACIAN, ORGANIC MATTER
B 6412 2056 ATLANTIC BASINS, CRETACEOUS, UPPER, ORGANIC MATTER
B 6412 2051 ATLANTIC BASINS, ORGANIC MATTER, TYPES
B 6412 2051† ATLANTIC BASINS, PALEOENVIRONMENT AND PETROLEUM POTENTIAL
B 6412 2059 ATLANTIC BASINS, TURONIAN, ORGANIC MATTER
M 29 181 ATLANTIC, BLAKE-BAHAMA BASIN, SEISMIC REFLECTORS
M 29 79 ATLANTIC COAST, BALTIMORE CANYON, CROSS SECTION
M 29 79 ATLANTIC COAST, BALTIMORE CANYON, SHELF EDGE
M 29 77 ATLANTIC COAST, CAPE COD, CROSS SECTION
M 29 80 ATLANTIC COAST, CAPE MAY, CROSS SECTION
M 29 78 ATLANTIC COAST, CAROLINA PLATFORM, CROSS SECTION
M 29 66 ATLANTIC COAST, COMPRESSIONAL VELOCITIES
M 29 75 ATLANTIC COAST, CONTINENTAL SLOPE, UNCONFORMITY
M 29 81 ATLANTIC COAST, COST B-2 WELL, SUBSIDENCE HISTORY
M 29 72 ATLANTIC COAST, CRETACEOUS, DEPTH SECTIONS
M 29 72 ATLANTIC COAST, CRETACEOUS, SHELF EDGE
M 29 71 ATLANTIC COAST, CRETACEOUS, TIME AND DEPTH SECTIONS
M 29 77 ATLANTIC COAST, CRUST, TRANSITION ZONE
M 29 70 ATLANTIC COAST, DELAWARE, DEPTH SECTIONS
M 29 76 ATLANTIC COAST, INFERRED BASIN STRUCTURE
M 29 66 ATLANTIC COAST, INTERVAL VELOCITY ANALYSIS
M 29 74 ATLANTIC COAST, JURASSIC, TECTONIC DEFORMATION
M 29 72 ATLANTIC COAST, NEW ENGLAND, DEPTH SECTIONS
M 29 78 ATLANTIC COAST, NEW JERSEY, BALTIMORE CANYON, DIAPIR
M 29 67 ATLANTIC COAST, NEW JERSEY, REFRACTING HORIZONS
M 29 81 ATLANTIC COAST, NEW JERSEY, SUBSIDENCE HISTORY
M 29 76 ATLANTIC COAST, OCEAN-CONTINENT TRANSITION ZONE
B 6405 783* ATLANTIC COAST OF EASTERN LONG ISLAND, NEW YORK, ABST.,
M 29 66 ATLANTIC COAST, SHELF INDEX MAP
B 6312 2146 ATLANTIC COAST, TRANSGRESSIVE BARRIERS, DELAWARE
M 29 69 ATLANTIC COAST, VELOCITY SECTIONS, EXAMPLES
B 6008 1323† ATLANTIC COASTAL PLAIN BETWEEN NEW JERSEY AND NORTH
B 6409 1332 ATLANTIC COASTAL PLAIN, EXPLORATORY WELLS, 1979
B 6308 1204† **ATLANTIC COASTAL PLAIN IN 1978, DEVELOPMENTS IN**
B 6409 1331† ATLANTI COASTAL PLAIN IN 1979, DEVELOPMENTS IN
B 6409 1336 ATLANTIC COASTAL PLAIN IN 1978—ADDENDUM, DEVELOPMENTS IN
B 6004 734* ATLANTIC COASTAL PLAIN INTERPRETED FROM RELICT SHORELINES,
B 6001 87† ATLANTIC COASTAL PLAIN OF NEW JERSEY, UPPER, CRETACEOUS
B 6104 519 ATLANTIC COASTAL RIDGE, FLORIDA
B 6203 419 ATLANTIC, CONTINENTAL DRIFT, NORTH
B 6105 821* ATLANTIC CONTINENTAL MARGIN, ABST., PRELIMINARY,
M 29 11† ATLANTIC CONTINENTAL MARGIN, BASED ON MULTICHANNEL SEISMIC
B 6205 792† ATLANTIC CONTINENTAL MARGIN, SEISMIC STRATIGRAPHY AND
M 29 139† ATLANTIC CONTINENTAL MARGIN, VARIATIONS IN, BOTTOM, PROCESSES
B 6405 670* ATLANTIC CONTINENTAL MARGINS OF UNITED STATES, ABST., THERMAL
B 6410 1659 ATLANTIC, CONTINENTAL OFFSHORE STRATIGRAPHIC TEST COST,
B 6405 675* ATLANTIC CONTINENTAL SHELF, ABST., FRACTIONAL PRESERVATION OF
B 6303 481* ATLANTIC CONTINENTAL SHELF AND SLOPE--MULTIVARIATE STUDY,
B 6303 480* ATLANTIC CONTINENTAL SLOPE, ABST., SLUMPING IN INTERCANYON
B 6203 495* ATLANTIC CONTINENTAL SLOPE EAST OF CAPE MAY, NEW JERSEY, ABST.,
B 6203 420 ATLANTIC, CRETACEOUS, NORTH
M 28 123 ATLANTIC, CRETACEOUS, SANDSTONE
B 6309 1608* ATLANTIC, DEEP SEA DRILLING PROJECT LEG 43 SITE 386, ABST.,
B 6410 1659 ATLANTIC, DEEP SEA DRILLING PROJECT, SITES, WESTERN
B 6002 196† ATLANTIC DIAPIRIC STRUCTURES, SOUTH
B 6303 503* ATLANTIC, DSDP LEGS 36, 40--SEDIMENTOLOGIC IMPLICATIONS, ABST.,
B 6004 662* ATLANTIC DURING QUATERNARY, ABST., CLIMATICALLY CONTROLLED,
B 6203 424 ATLANTIC, EOCENE, NORTH
B 6202 226 ATLANTIC, EVAPORITES AROUND
B 6011 2019† ATLANTIC INNER CONTINENTAL SHELF OFF NORTHERN FLORIDA,
B 6205 793 ATLANTIC, JOIDES HOLES, NORTH
B 6409 1662 ATLANTIC JURASSIC INNER MAGNETIC QUIET ZONE, PROFILE, WESTERN
B 6203 420 ATLANTIC, JURASSIC, NORTH
B 6409 1662 ATLANTIC JURASSIC OUTER MAGNETIC QUIET ZONE, PROFILE, WESTERN

B 6312 2222† ATLANTIC KIMMERIDGIAN, COMPOSITION OF, ORGANIC MATTER FROM,
B 6312 2224 ATLANTIC KIMMERIDGIAN SHALES, NORTH
B 6410 1676 ATLANTIC, KNAUSS KNOLL, WESTERN
B 6410 1677 ATLANTIC MAGNETIC QUIET ZONE, JURASSIC SEDIMENTATION RATES,
B 6410 1658† ATLANTIC MAGNETIC QUIET ZONE, JURASSIC, SEISMIC STRATIGRAPH
B 6410 1661 ATLANTIC MAGNETIC QUIET ZONE, SEISMIC DATA
M 29 90 ATLANTIC MARGIN, 1964-1966 AEROMAGNETIC SURVEY
M 29 86 ATLANTIC MARGIN, 1975 AEROMAGNETIC SURVEY
B 6004 686* ATLANTIC MARGIN, ABST., SEDIMENTARY PROCESSES IN, SUBMARINE,
M 29 143 ATLANTIC MARGIN, ALBEMARLE SOUND, PROCESSES
M 29 86 ATLANTIC MARGIN, APPALACHIAN OROGENIC SYSTEM
B 6112 2077 ATLANTIC MARGIN, AUSTRALIA
M 29 100 ATLANTIC MARGIN, BALTIMORE CANYON TROUGH
M 29 85† ATLANTIC MARGIN, BASIN, STRUCTURE OF THE, U.S.
M 29 105 ATLANTIC MARGIN, BLAKE PLATEAU BASIN
M 29 139 ATLANTIC MARGIN, BOTTOM PROCESSES
M 29 102 ATLANTIC MARGIN, CAROLINA PLATFORM
M 29 102 ATLANTIC MARGIN, CAROLINA TROUGH
M 29 94 ATLANTIC MARGIN, COMPARISON OF MAGNETICS
B 6105 822* ATLANTIC MARGIN CORING PROJECT, ABST., SHIPBOARD GEOTECHNICAL
B 6309 1453 ATLANTIC MARGIN CORING PROJECT, AMCOR, BALTIMORE CANYON
B 6105 814* ATLANTIC MARGIN CORING PROJECT—PRELIMINARY RESULTS, ABST.,
B 6105 794 * ATLANTIC MARGIN CORING PROJECT, 1976—SUMMARY OF OPERATIONS AND
M 29 94 ATLANTIC MARGIN, EAST COAST MAGNETIC ANOMALY
M 29 148 ATLANTIC MARGIN, EOCENE, REFLECTORS
M 29 86 ATLANTIC MARGIN, ESSAOUIRA BASIN
B 6004 691* ATLANTIC MARGIN FROM NIGERIA TO CAPE OF GOOD HOPE, ABST.,
M 29 97 ATLANTIC MARGIN, GEORGES BANK BASIN
B 6405 781* ATLANTIC MARGIN IN VICINITY OF CHEVRON COST B-3 WELL, ABST.,
M 29 86 ATLANTIC MARGIN, JURASSIC QUIET ZONE
M 29 98 ATLANTIC MARGIN, LONG ISLAND PLATFORM
M 29 88 ATLANTIC MARGIN, MAGNETIC ANOMALY CONTOURS
M 29 90 ATLANTIC MARGIN, MAGNETIC DEPTH ESTIMATES
M 29 95 ATLANTIC MARGIN, MAGNETICS, SEISMICS
B 6107 961† ATLANTIC MARGIN OF AFRICA, STRUCTURAL HISTORY OF
B 6105 829* ATLANTIC MARGIN OF NORTH AMERICA, ABST., PETROLEUM POTENTIAL
B 6202 223† ATLANTIC MARGIN, ORIGIN AND SIGNIFICANCE OF, EVAPORITES IN,
M 29 143 ATLANTIC MARGIN, REFLECTING HORIZONS
M 29 28 ATLANTIC MARGIN, SEDIMENTARY BASINS
M 29 140 ATLANTIC MARGIN, SEDIMENTATION REGIONAL VARIATIONS
M 29 455 ATLANTIC MARGINS, RIDGE SPREADING, DIAGRAM
M 29 453† ATLANTIC MARGINS, THE EFFECT OF, EUSTATIC, SEA LEVEL CHANGES
B 6203 424 ATLANTIC, MIOCENE, NORTH
B 6409 1336† ATLANTIC OC IN 1978—ADDENDUM, DEVELOPMENTS IN
B 6303 423* ATLANTIC OCEAN, ABST., MID-, CRETACEOUS CALCAREOUS,
B 6004 706* ATLANTIC OCEAN, ABST., NORTH SEA, RIFTING AND OPENING OF, NORTH
B 6203 419† ATLANTIC OCEAN AND THEIR RELATION TO NORTH SEA AREA, RIFT,
B 6408 1202 ATLANTIC OCEAN, ANOXIC OPEN OCEAN
B 6412 2052 ATLANTIC OCEAN, COST B-2 AND GE-1 WELLS
B 6412 2051 ATLANTIC OCEAN, DEEP SEA DRILLING PROJECT, BLACK SHALES
B 6004 695* ATLANTIC OCEAN, EARLY CRETACEOUS THROUGH PLIOCENE, ABST.,
B 6411 2041 ATLANTIC OCEAN, EDITED BY RED H. PILGER, JR., LADD, JOHN W.,
M 25 43 ATLANTIC OCEAN MARGINS, WESTERN, PACIFIC AND
B 6103 334 ATLANTIC OCEAN, PHYSIOGRAPHIC PROVINCES
B 6111 1991 ATLANTIC OCEAN, PIERCEMENT STRUCTURES, SEISMIC PROFILES
B 6403 364 ATLANTIC OCEAN PLATE MOVEMENTS, NORTH
B 6111 1994 ATLANTIC OCEAN, POSSIBLE DIAPIRS
B 6106 929† ATLANTIC OCEAN, SUBSIDENCE OF, ASEISMIC, RIDGES—EVIDENCE FROM
B 6303 453* ATLANTIC OCEANOGRAPHIC AND METEOROLOGICAL LABORATORIES,
B 6205 892* ATLANTIC OCEAN—PALEO-OCEANOGRAPHIC IMPLICATIONS, ABST.,
B 6202 273† ATLANTIC OFF NORTHERN BRAZIL—STRUCTURAL CONTROLS AND
B 6308 1205 ATLANTIC OFFSHORE CONTINENTAL SHELF, DRILLING, 1978
B 6310 1919 ATLANTIC OFFSHORE, FRANCE
B 6203 424 ATLANTIC, OLIGOCENE, NORTH
B 6203 420 ATLANTIC, OPENING OF, NORTH
B 6105 826* ATLANTIC OUTER CONTINENTAL SHELF, ABST., PETROLEUM FRONTIER
B 6405 733* ATLANTIC OUTER CONTINENTAL SHELF, ABST., PROCESSES
B 6105 836* ATLANTIC OUTER CONTINENTAL SHELF, ABST., SEDIMENTARY MODEL
B 6007 1114 ATLANTIC OUTER CONTINENTAL SHELF, CDP MULTICHANNE
B 6007 1112† ATLANTIC OUTER CONTINENTAL SHELF, EVIDENCE OF, POST-,
B 6106 861 ATLANTIC OUTER CONTINENTAL SHELF, SURFICIAL, SAND, SHEET ON,
B 6007 1112† ATLANTIC OUTER CONTINETAL SHELF, EVIDENCE OF, POST-PLEISTOCENE,
B 6007 1113 ATLANTIC OUTER CONTINENTAL SHELF, SHALLOW FAULTS
B 6203 424 ATLANTIC, PALEOCENE, NORTH
S 6 222 ATLANTIC, PALEOGENE, PALEOMAGNETIC REVERSALS
B 6202 274 ATLANTIC, PHYSIOGRAPHIC PROVINCE MAP, WESTERN, EQUATORIAL
B 6409 1662 ATLANTIC, PLANTAGENET FORMATION, WESTERN
B 6404 503 ATLANTIC, PLATE-TECTONIC EVENTS, NORTH
B 6108 1335 ATLANTIC PROVINCES, DEVELOPMENTS, 1976
B 6208 1451 ATLANTIC PROVINCES, DEVELOPMENTS, 1977
B 6308 1344 ATLANTIC PROVINCES, DEVELOPMENTS, 1978
B 6409 1518 ATLANTIC PROVINCES, DEVELOPMENTS, 1979
S 4 127 ATLANTIC REGION, SPECIES OF SPONGES, WESTERN
B 6310 1885 ATLANTIC RICHFIELD CO., AND AGIP MINERARIA S.P.A., ON
B 6010 1866 ATLANTIC RICHFIELD CO., ON DEVELOPMENTS IN IRAN, 1975, OIL
B 6210 2112 ATLANTIC RICHFIELD CO., ON DEVELOPMENTS IN IRAN, 1977, OIL
B 6110 1796 ATLANTIC RICHFIELD, ON DEVELOPMENTS IN IRAN, 1976, OIL SERVICE
B 6203 434 ATLANTIC, RIFT BASINS, HYDROCARBONS, NORTH
B 6101 83 ATLANTIC RIFT VALLEY, CENTRAL
B 6007 1101 ATLANTIC SEA FLOOR, GRAVITY BOUGUER ANOMALY MAP, BARBADOS
B 6007 1100 ATLANTIC SEA FLOOR, GRAVITY FREE-AIR ANOMALY MAP, BARBADOS
B 6203 423 ATLANTIC, SEAFLOOR, AGE, NORTH
B 6405 706* ATLANTIC SHELF, ABST., MODIFICATION OF LINEAR SAND RIDGES BY
B 6405 773* ATLANTIC SHELF, UNITED STATES—FOURIER GRAIN SHAPE ANALYSIS,
B 6204 604 ATLANTIC SLOPE, CARBONATE FACIES, U.S.
B 6204 603 ATLANTIC SLOPE. JURASSIC. U.S.
B 6203 560* ATLANTIC SLOPE OF EASTERN NORTH AMERICA, ABST., STRUCTURE OF
B 6204 600 ATLANTIC SLOPE, PALEOSHELF-EDGE HIGH, POTENTIAL TRAP, U.S.
B 6204 600 ATLANTIC SLOPE, POTENTIAL TRAPS, U.S.
B 6309 1453 ATLANTIC SLOPE PROJECT, ASP, BALTIMORE CANYON TROUGH
B 6204 592† ATLANTIC SLOPE, RISE, AND ABYSSAL PLAIN, PETROLEUM, POTENTIAL
B 6204 596 ATLANTIC SLOPE, SEDIMENT DISTRIBUTION, U.S.
B 6203 564* ATLANTIC STATES, ABST., EROSION VERSUS DEPOSITION ON UPPER,
B 6203 425 ATLANTIC, TRIASSIC, NORTH
M 25 44 ATLANTIC WESTERN BORDERS, PACIFIC AND
M 29 457 ATLANTIC-TYPE MARGIN, MODEL
B 6111 1959 ATLANTIC-TYPE MARGINS, RESERVES, PASSIVE
M 29 455 ATLANTIC-TYPE MARGINS, SEDIMENTARY HISTORY
B 6306 881 ATLANTIC-TYPE PASSIVE MARGIN, POTENTIAL, RESERVOIR IN, RUDIST,
B 6406 871 ATLANTIS II, 1976 GEOPHYSICAL INVESTIGATIONS, RV
B 6109 1483 ATLANTIS II CRUISE 91, JULY-AUGUST 1975, RV
B 6404 501 ATLANTIS II CRUISE 91, RV
B 6106 861 ATLANTIS II CRUISE, JUNE 2-10, 1975, RV
B 6303 521* ATLAS, ABST., STRUCTURE OF, TUNISIAN
B 6101 80 ATLAS, ANTI-
B 6204 668 ATLAS FLEXURE, ALGERIA, SOUTH
B 6101 88 ATLAS FRACTURE ZONE, CORNWALL-, SOUTH
B 6101 80 ATLAS FRACTURE ZONE, SOUTH
M 25 436 ATLAS MINE, CEBU
B 6004 725* ATLAS, MOROCCO, ABST., ARBONATE, PETROLOGY OF BASIN-TO-, REEF,
B 6004 667* ATLAS MOUNTAINS OF MOROCCO, NORTHWEST AFRICA, ABST.,
B 6403 423† ATLAS, PROGRESSIVE, NEOGENE DEFORMATION OF SOUTHERN, HEDIL,
B 6101 81 ATLAS SALT BASIN, SAHARAN
B 6101 85 ATLAS TREND, GUERCIF BASIN
B 6101 84 ATLAS TREND, HIGH, MOROCCO
B 6101 85 ATLAS TREND, KERROUCHEN BASIN
B 6101 84 ATLAS TREND, MIDDLE, MOROCCO
B 6101 85 ATLAS TREND, MOUSA OU SALAH BASIN
B 6101 85 ATLAS TREND, NFIS, TISI NTEST, BASIN
B 6101 85 ATLAS TREND, TAMDAFELT BASIN
M 25 498 ATLIN TERRANE, CANADA
B 6303 507* ATMOSPHERIC CIRCULATION AND UPWELLING IN PALEOZOIC, WITH
B 6301 54 ATOKA CLASTICS, TEXAS
B 6311 2114* ATOKA FORMATION, PENNSYLVANIAN, DEPOSITION AND
B 6203 491* ATOKA FORMATION, PENNSYLVANIAN, DEPOSITIONAL ENVIRONMENT
B 6308 1425* ATOKA GAS IN SOUTHERN COTTLE AND NORTHER KING COUNTIES,
B 6102 294* ATOKA GAS IN SOUTHERN COTTLE AND NORTHERN KING COUNTIES,
M 26 453 ATOKA LIMESTONE, OKLAHOMA, PENNSYLVANIAN
B 6002 327* ATOKA, LOWER PENNSYLVANIAN, CLASTIC ROCKS IN PARTS OF JACK,
B 6301 50† ATOKA, LOWER PENNSYLVANIAN, CLASTIC ROCKS IN PARTS OF JACK,
B 6305 808 ATOKA-MORROW PRODUCTION, TEXAS
B 6203 566* ATOKAN SERIES BOUNDARY IN NORTHWESTERN ARKANSAS AND
S 4 37† ATOLL, BELIZE, PATCH-REEF ECOLOGY AND SEDIMENTOLOGY OF,
B 6405 713* ATOLL REEF, MIDDLE PLEISTOCENE TO HOLOCENE, ABST., DEPOSITION
B 6410 1640 ATOLLS, BAHAMA BANKS AS
S 4 285 ATOLLS, GRAZING AND BROWSING FISH, PACIFIC
S 4 76 ATOLLS, INDO-, PACIFIC AND, CARIBBEAN, MAJOR, REEFS AND
B 6202 302 ATOMIC RATIOS OF BITUMENS AND COALS
M 25 319 ATRATO AREA, POTENTIAL, COLOMBIA
M 25 318† ATRATO VALLEY AND PACIFIC COAST, POTENTIAL FOR EXPLORATION
M 28 72 ATTAPULGITE, SEM, CLAY
M 26 3 ATTENUATING NOISE, SEISMIC, REFLECTIONS
M 26 24 ATTENUATION, FLUID MOTION, LABORATORY, MEASUREMENT
M 26 23 ATTENUATION, LABORATORY, MEASUREMENTS
M 26 25 ATTENUATION, P-WAVE, FLUID MOTION
B 6209 1685 ATTIKA OIL FIELD, KALIMANTAN
B 6203 500 ATTWOOD, PAUL, VADOSE CEMENT PATTERNS IN STRAND-PLAIN
B 6403 438 ATWATER, BRIAN E., HOLOCENE INTERTIDAL DEPOSITS OF

S 11 146 ATWILLINA PSEUDOCOCOAENSIS
S 11 77 ATWILLINA PSEUDOCOCOAENSIS
S 11 78 ATWILLINA PSEUDOCOCOAENSIS
B 6212 2509 AUBOUIN, JEAN, FROM CARIBBEAN TO SOUTH AND NORTH AMERICAN
B 6302 249 AUDLEY-CHARLES, M. G., GEOSYNCLINES AND PLATE TECTONICS IN
B 6203 464 AUGILA FORMATION, LIBYA
B 6203 374 AUGUSTINE ISLAND, ALASKA
B 6203 384 AUGUSTINE-SELDOVIA ARCH, ALASKA
B 6203 523* AULACOGEN, MIDDLE PRECAMBRIAN, GREAT SLAVE LAKE, NORTHWEST
B 6203 522* AULACOGEN, MIDDLE PRECAMBRIAN, GREAT SLAVE LAKE, NORTHWEST
B 6002 327* AULACOGEN, OKLAHOMA, ABST., GEOTECTONIC EVOLUTION OF,
B 6004 730* AULACOGENS AN MEGASHEARS--NATURAL HABITAT FOR OIL AND
B 6011 2067* **AULACOGENS AND MEGASHEARS--NATURAL HABITAT FOR OIL AND**
B 6004 653* AULACOGENS--INITIAL DEPOSITS OF CORDILLERAN GEOSYNCLINE, ABST.,
B 6405 705* AULACOGEN—IMPLICATION FOR PETROLEUM EXPLORATION, ABST.,
B 6405 672 AULT, CURTIS H., NEW EXPLORATION AND EVALUATION OF COAL
S 4 313 AURELIO VICALVI, MARCO, CALCIUM CARBONATE SEDIMENTATION ON
B 6004 684 AUSICH, WILLIAM I., SEARCH FOR SILURANIAN REEF MODEL--GREA
B 6109 1547* AUSTIN AND BUDA FORMATIONS FROM CORE AND FRACTURE ANALYSIS,
B 6405 679* AUSTIN AND TAYLOR GROUPS, UPPER CRETACEOUS, EAST-CENTRAL AND
B 6211 2348 AUSTIN, CARL F., COSO AREA GEOTHERMAL DEVELOPMENTS, ABST.
B 6109 1548* AUSTIN CHALK AND BUDA LIMESTONE IN SOUTH TEXAS, ABST.,
B 6102 296* AUSTIN CHALK, CRETACEOUS, PLAY, ABST., SOUTH, TEXAS
M 26 114 AUSTIN CHALK, TEXAS, POLK COUNTY
B 6107 988 AUSTIN CHALK UNIT, TEXAS
B 6109 1546* AUSTIN CHALK—BUDA TREND OF SOUTH TEXAS, ABST.
B 6409 1356 AUSTIN COUNTY, TEXAS
B 6402 231 AUSTIN FORMATION, TEXAS
B 6109 1556* AUSTIN GROUP AND THEIR CONTROL OF PETROLEUM POTENTIAL, ABST.,
B 6011 2058† AUSTIN GROUP, CRETACEOUS, NORTHEAST TEXAS, GLOBOTRUNCANA
M 27 220 AUSTIN GROUP, TEXAS, CRETACEOUS
M 27 223 AUSTIN GROUP, TEXAS, CRETACEOUS
M 27 193 AUSTIN GROUP, TEXAS, CRETACEOUS
B 6408 1282 AUSTIN, J. M., DEPOSITIONAL TRENDS OF LOWER SILURIAN CLINTON
B 6404 501 AUSTIN, JAMES A., JR., GEOLOGY OF NEW ENGLAND PASSIVE MARGIN
B 6304 686 AUSTIN, S. R., DISSOLUTION AND AUTHIGENESIS IN HOST SANDSTONES,
B 6408 1220 AUSTIN SECTION, TEXAS
B 6203 492 AUSTIN, STEVEN A., UNIFORMITARIANISM—DOCTRINE WHICH NEEDS
B 6107 988 AUSTIN-SAN FELIPE CHALK UNIT, MEXICO
B 6207 1222* AUSTRALASIA, ABST., APPLICATION OF, LANDSAT, IMAGERY AND
B 6307 1093 AUSTRALIA, ABROLHOS SUBBASIN
B 6207 1235* AUSTRALIA, ABST., BAUXITE, MINING AND, STREAM, SALINITY IN,
B 6303 500* AUSTRALIA, ABST., COORONG MODEL FOR PENECONTEMPORANEOUS,
B 6405 746* AUSTRALIA, ABST., DEFINITION AND DEVELOPMENT OF, MACKEREL
B 6203 504* AUSTRALIA, ABST., EXPLORATION CASE HISTORY AND GEOLOGIC
B 6303 539* AUSTRALIA, ABST., GENERATION AND OCCURRENCE OF HYDROCARBONS
B 6212 2511* AUSTRALIA, ABST., GEOCHEMICAL EVALUATION OF, BONAPARTE GULF—,
B 6303 513* AUSTRALIA, ABST., GLACIOMARINE CARBONATE SEDIMENTATION--,
B 6207 1224* AUSTRALIA, ABST., HIGH-RESOLUTION, SEISMIC EXPLORATION FOR,
B 6207 1217* AUSTRALIA, ABST., HYDROGEOLOGY OF, URANIUM DEPOSITS IN,
B 6303 496* AUSTRALIA, ABST., MASSIVE MARINE-SANDSTONE FACIES, MACKEREL
B 6207 1219* AUSTRALIA, ABST., PRODUCTION GEOLOGY OF, NORTH RANKIN GAS
B 6207 1226* AUSTRALIA, ABST., SEISMIC AND GEOLOGIC, FACIES ANALYSES IN FIELD
B 6008 1392* AUSTRALIA, ABST., URANIUM IN, TERTIARY CHANNEL SOUTHEAST OF,
M 25 407 AUSTRALIA, ALUMINUM INDUSTRY, DEVELOPMENT AND INTEGRATION
B 6002 220 AUSTRALIA, AMOSEAS LAKE TOPPED OIL
M 25 229† AUSTRALIA AND SURROUNDING OCEANIC REGION--SUMMARY,
M 25 237 AUSTRALIA, ANGEL DISCOVERY WELL
B 6201 43 AUSTRALIA, ANGEL FIELD
B 6201 63 AUSTRALIA, APTIAN-CENOMANIAN
B 6201 41 AUSTRALIA, ARCHEAN AND PROTEROZOIC
M 25 428 AUSTRALIA, ARCHEAN, ORE DEPOSITS ASSOCIATED WITH
B 6201 41 AUSTRALIA, ARGO ABYSSAL PLAIN
B 6410 1645 AUSTRALIA, ARROWIE BASIN
M 29 216 AUSTRALIA, ARU BASIN, REFLECTION PROFILE
B 6207 1217* AUSTRALIA AS NATURAL GAS EXPORTER, ABST.
B 6311 2024 AUSTRALIA, ASHMORE BLOCK
B 6112 2077 AUSTRALIA, ATLANTIC MARGIN
B 6410 1644 AUSTRALIA, BARRIER RANGES
B 6201 50 AUSTRALIA, BARROW FORMATION
B 6203 457 AUSTRALIA, BARROW GROUP
B 6203 457 AUSTRALIA, BARROW ISLAND OIL FIELD, SECTION
B 6201 41 AUSTRALIA, BARROW SUBBASIN
B 6203 457 AUSTRALIA, BARROW SUBBASIN, WELLS
M 25 427 AUSTRALIA, BASE-METAL DEPOSITS, DISCOVERY AND POTENTIAL
M 25 428 AUSTRALIA, BASE-METAL DEPOSITS, GEOLOGY
M 25 230 AUSTRALIA, BASS BASIN
M 25 407 AUSTRALIA, BAUXITE DEPOSITS, LOCATIONS AND RESERVES
M 25 408 AUSTRALIA, BAUXITE DEPOSITS, TOPOGRAPHY, MINERAL COMPOSITION
M 25 406 AUSTRALIA, BAUXITE, EXPLORATION AND EVALUATION
M 25 405 AUSTRALIA, BAUXITE, HISTORY OF DISCOVERY
M 25 408 AUSTRALIA, BAUXITE MINING, REGENERATION OF MINED AREAS
M 25 408 AUSTRALIA, BAUXITE MINING, SOCIAL BENEFITS
M 25 405 AUSTRALIA, BAUXITE, RESERVES
M 25 405 AUSTRALIA, BAUXITE, RESERVES AND GRADES
M 25 406 AUSTRALIA, BAUXITE, TYPE OF
M 25 407 AUSTRALIA, BAUXITE-ALUMINA-ALUMINUM INDUSTRY, EXPORT
B 6307 1101 AUSTRALIA, BEAGLE RIDGE GEOTHERMAL ANOMALY
M 25 233 AUSTRALIA, BEAGLE SUBBASIN
B 6201 41 AUSTRALIA, BEDOUT SUBBASIN
B 6410 1644 AUSTRALIA, BEEFSTEAK CHANNEL
B 6410 1644 AUSTRALIA, BENAGERIE RIDGE
B 6410 1644 AUSTRALIA, BEVERLEY DEPOSIT
B 6410 1644 AUSTRALIA, BILLEROO CHANNEL
M 25 420 AUSTRALIA, BLACK SANDS
B 6311 2023 AUSTRALIA, BONAPARTE GULF BASIN
B 6210 2152 AUSTRALIA, BONAPARTE GULF BASIN
B 6411 1864 AUSTRALIA, BONAPARTE GULF BASIN
M 25 234 **AUSTRALIA, BONAPARTE GULF BASIN, OFFSHORE**
M 25 426 AUSTRALIA, BOUGAINVILLE COPPER DEPOSITS
M 25 426 AUSTRALIA, BROKEN HILL, MINING CENTER
M 25 234 AUSTRALIA, BROWSE BASIN
B 6411 1864 AUSTRALIA, BROWSE BASIN
B 6210 2152 AUSTRALIA, BROWSE BASIN
B 6311 2023 AUSTRALIA, BROWSE BASIN
B 6201 41 AUSTRALIA, BROWSE BASIN
B 6406 823 AUSTRALIA, BUGLE GAP AREA, GEOLOGIC MAP
B 6406 821 AUSTRALIA, BUGLE GAP FORMATION
B 6307 1093 AUSTRALIA, BUNBURY TROUGH
B 6307 1097 AUSTRALIA, CADDA FORMATION
B 6307 1093 AUSTRALIA, CADDA SHELF
B 6201 52 AUSTRALIA, CALLOVIAN STRATA
M 27 47 AUSTRALIA, CAMBRIAN, AJAX LIMESTONE
B 6410 1682 AUSTRALIA, CAMBRIAN POUND SANDSTONE-QUARTZITE
B 6410 1645 AUSTRALIA, CAMBRIAN RED BEDS
B 6201 41 AUSTRALIA, CANNING BASIN
B 6411 1865 AUSTRALIA, CANNING BASIN
B 6406 836 AUSTRALIA, CANNING BASIN, GEOTHERMAL HISTORY
M 25 233 AUSTRALIA, CANNING BASIN, OFFSHORE
B 6311 2089 AUSTRALIA, CAPE RANGE FRACTURE
B 6207 1154 AUSTRALIA, CARBONIFEROUS
B 6210 2152 AUSTRALIA, CARNARVON BASIN
B 6411 1865 AUSTRALIA, CARNARVON BASIN
B 6201 41 AUSTRALIA, CARNARVON BASIN
B 6307 1096 AUSTRALIA, CARYNGINIA FORMATION
B 6307 1097 AUSTRALIA, CATTAMARRA COAL MEASURES MEMBER
B 6201 67 AUSTRALIA, CENOZOIC
S 9 18 AUSTRALIA, CENOZOIC, DRIFT MOVEMENTS
M 25 81 AUSTRALIA, COAL BASINS AND MINES
B 6307 1094 AUSTRALIA, COCKLESHELL GULLY FORMATION
M 29 152 AUSTRALIA, CONTINENTAL MARGIN, BREMER BASIN
M 29 152 AUSTRALIA, CONTINENTAL MARGIN, DUNTROON BASIN
M 29 152 AUSTRALIA, CONTINENTAL MARGIN, EUCLA BASIN
M 29 152 AUSTRALIA, CONTINENTAL MARGIN, GEOPHYSICAL CHART
M 29 152 AUSTRALIA, CONTINENTAL MARGIN, MURRAY BASIN
M 29 152 AUSTRALIA, CONTINENTAL MARGIN, OTWAY BASIN
B 6307 1094 AUSTRALIA, COOLYENA GROUP
B 6207 1154 AUSTRALIA, COOPER BASIN, STRATIGRAPHY
B 6207 1153 AUSTRALIA, CRETACEOUS
M 25 235 AUSTRALIA, CRETACEOUS
B 6203 460 AUSTRALIA, CRETACEOUS
B 6201 60 AUSTRALIA, CRETACEOUS
B 6410 1644 AUSTRALIA, CURNAMONA CHANNEL
B 6201 41 AUSTRALIA, CUVIER ABYSSAL PLAIN
B 6311 2089 AUSTRALIA, CUVIER ABYSSAL PLAIN
B 6201 50 AUSTRALIA, DAMPIER FORMATION
B 6201 41 AUSTRALIA, DAMPIER SUBBASIN
M 25 232 AUSTRALIA, DAMPIER SUBBASIN
B 6307 1092 AUSTRALIA, DANDARAGAN TROUGH
B 6307 1093 AUSTRALIA, DARLING FAULT
M 25 405 AUSTRALIA, DARLING RANGE, BAUXITE RESERVES
B 6311 2024 AUSTRALIA, DARWIN BLOCK
B 6201 62 AUSTRALIA, DEEP SEA DRILLING PROJECT, SITE 260
B 6201 41 AUSTRALIA, DEEP SEA DRILLING PROJECT, SITE 263
S 6 338 AUSTRALIA, DEVONIAN, ABSOLUTE DATING
B 6406 814† AUSTRALIA, DEVONIAN, GREAT, BARRIER, REEF OF, CANNING BASIN,
B 6203 457 AUSTRALIA, DINGO CLAYSTONE
B 6203 457 AUSTRALIA, DINGO GROUP

B 6410 1643† AUSTRALIA, DISTRIBUTION AND GENESIS OF SEDIMENTARY, URANIU
B 6104 595 AUSTRALIA, DOLPHIN FIELD
B 6307 1093 AUSTRALIA, DONGARA FIELD
B 6307 1095 AUSTRALIA, DONGARA SADDLE
B 6307 1094 AUSTRALIA, DONGARA SANDSTONE
B 6203 457 AUSTRALIA, DUPUY SANDSTONE
M 25 419 AUSTRALIA, EASTERN, HEAVY-MINERAL SAND
B 6410 1682 AUSTRALIA, EDIACARA HILLS, PRECAMBRIAN, SOUTH
B 6307 1093 AUSTRALIA, EDWARDS ISLAND BLOCK
B 6406 823 AUSTRALIA, EMANUEL RANGE
B 6307 1094 AUSTRALIA, ENEABBA MEMBER
B 6207 1153 AUSTRALIA, EOCENE
B 6410 1644 AUSTRALIA, ERUDINA CHANNEL
B 6201 41 AUSTRALIA, EXMOUTH PLATEAU ARCH
B 6411 1865 AUSTRALIA, EXMOUTH PLATEAU AREA
B 6201 70 AUSTRALIA, EXMOUTH PLATEAU, CAPROCKS
B 6201 70 AUSTRALIA, EXMOUTH PLATEAU, RESERVOIRS
B 6201 48 AUSTRALIA, EXMOUTH PLATEAU, SEISMIC DATA
B 6201 69 AUSTRALIA, EXMOUTH PLATEAU, SOURCE ROCKS
B 6201 61 AUSTRALIA, EXMOUTH PLATEAU, STRATIGRAPHIC COLUMN
B 6201 50 AUSTRALIA, EXMOUTH PLATEAU, WELLS
B 6201 41 AUSTRALIA, EXMOUTH SUBBASIN
B 6406 821 AUSTRALIA, FAIRFIELD GROUP
B 6406 826 AUSTRALIA, FAMENNIAN NULLARA REEF CYCLE
B 6201 41 AUSTRALIA, FITZROY TROUGH
B 6203 457 AUSTRALIA, FLINDERS FAULT SYSTEM
B 6410 1644 AUSTRALIA, FLINDERS RANGES
B 6104 594 AUSTRALIA, FLOUNDER FIELD
B 6207 1153 AUSTRALIA, FLOUNDER FORMATION
B 6406 827 AUSTRALIA, FRASNIAN REEF COMPLEXES
B 6410 1644 AUSTRALIA, FROME EMBAYMENT, URANIUM
B 6201 41 AUSTRALIA, GASCOYNE ABYSSAL PLAIN
M 25 229 AUSTRALIA, GASMAN OROGENIC ZONE
B 6203 457 AUSTRALIA, GEARLE SILTSTONE
B 6307 1092† AUSTRALIA, GEOCHEMICAL ANALYSIS OF HYDROCARBON OCCURRENCES
B 6311 2021† AUSTRALIA, GEOCHEMICAL EVALUATION OF PETROLEUM, SOURCE ROCK
B 6201 40† AUSTRALIA, GEOLOGY AND PETROLEUM POTENTIAL OF, EXMOUTH
B 6307 1093 AUSTRALIA, GINGIN FIELD
M 25 230 AUSTRALIA, GIPPSLAND BASIN
B 6207 1153 AUSTRALIA, GIPPSLAND BASIN, STRATIGRAPHY
B 6207 1152 AUSTRALIA, GIPPSLAND BASIN, WELLS
B 6207 1153 AUSTRALIA, GIPPSLAND LIMESTONE
B 6406 826 AUSTRALIA, GIVETIAN-FRASNIAN PILLARA REEF CYCLE
B 6406 823 AUSTRALIA, GLENISTER KNOLLS
B 6406 821 AUSTRALIA, GOGO FORMATION
M 25 237 AUSTRALIA, GOODWYN DISCOVERY WELLS
M 25 237 AUSTRALIA, GOODWYN FAULT BLOCK
B 6201 43 AUSTRALIA, GOODWYN FIELD
B 6410 1646 AUSTRALIA, GOULDS DAM DEPOSIT
M 25 405 AUSTRALIA, GOVE, BAUXITE RESERVES
S 6 114 AUSTRALIA, GREAT ARTESIAN BASIN, ISOCHRON DIAGRAM
M 25 27 AUSTRALIA, GREAT BARRIER REEF
M 25 402 AUSTRALIA, GREENVALE AND ROCKHAMTON, NICKEL-ENRICHED
M 25 423 AUSTRALIA, GROOTE EYLANDT BEDS
M 25 423 AUSTRALIA, GROOTE EYLANDT, LOCATION AND GEOLOGY, MAP
M 25 353† AUSTRALIA GROUNDWATER RESOURCES--SUMMARY, DEVELOPMENT OF
B 6207 1153 AUSTRALIA, GUNARD FORMATION
B 6311 2024 AUSTRALIA, HALLS CREEK MOBILE ZONE
B 6307 1093 AUSTRALIA, HARVEY RIDGE
M 25 420 AUSTRALIA, HEAVY-MINERAL SAND, EXPLORATION
M 25 418 AUSTRALIA, HEAVY-MINERAL SAND, MINING, HISTORY
M 25 418 AUSTRALIA, HEAVY-MINERAL SAND, PRODUCTION FIGURES
M 25 420 AUSTRALIA, HEAVY-MINERAL SAND, PRODUCTION METHODS
M 25 420 AUSTRALIA, HEAVY-MINERAL SAND, REHABILITATION OF MINED LANDS
M 25 421 AUSTRALIA, HEAVY-MINERALS INDUSTRY, FUTURE FOR
M 25 421 AUSTRALIA, HEAVY-MINERALS, MARKETING
B 6307 1096 AUSTRALIA, HOLMWOOD SHALE
B 6207 1151† AUSTRALIA, HYDROCARBON, GENERATION IN, GIPPSLAND BASIN,
B 6010 1957† AUSTRALIA IN 1975, PETROLEUM, DEVELOPMENTS IN
B 6110 1888† AUSTRALIA IN 1976, PETROLEUM, DEVELOPMENTS IN
B 6210 2150† AUSTRALIA IN 1977, PETROLEUM, DEVELOPMENTS IN
B 6310 1804† AUSTRALIA IN 1978, PETROLEUM, DEVELOPMENTS IN
B 6411 1862† AUSTRALIA IN 1979, PETROLEUM, DEVELOPMENTS IN
M 25 415 AUSTRALIA, IRON-ORE RESERVES
B 6307 1096 AUSTRALIA, IRWIN RIVER COAL MEASURES
B 6203 455† AUSTRALIA, ISOTOPIC EVIDENCE FOR, MICROBIOLOGIC ROLE IN GENESIS
B 6201 59 AUSTRALIA, JURASSIC
B 6307 1097 AUSTRALIA, JURASSIC
B 6207 1154 AUSTRALIA, JURASSIC
M 25 235 AUSTRALIA, JURASSIC
B 6203 457 AUSTRALIA, JURASSIC OIL ZONE
B 6201 41 AUSTRALIA, KANGAROO SYNCLINE
B 6311 2089 AUSTRALIA, KANGAROO SYNCLINE
B 6311 2024 AUSTRALIA, KIMBERLEY BLOCK
B 6201 41 AUSTRALIA, KIMBERLEY BLOCK
B 6104 595 AUSTRALIA, KINGFISH FIELD
B 6307 1096 AUSTRALIA, KOCKATEA SHALE
B 6406 823 AUSTRALIA, LAIDLAW RANGE
B 6207 1153 AUSTRALIA, LAKE ENTRANCE FORMATION
B 6410 1644 AUSTRALIA, LAKE NAMBA CHANNEL
M 25 421 AUSTRALIA, LAND USE, COMPETITION FOR
M 25 403 AUSTRALIA, LATERITE NICKEL DEPOSITS
M 25 402 AUSTRALIA, LATERITE NICKEL DEPOSITS
B 6207 1153 AUSTRALIA, LATROBE GROUP
B 6406 823 AUSTRALIA, LAWFORD RANGE
B 6307 1093 AUSTRALIA, LEEUWIN BLOCK
B 6307 1094 AUSTRALIA, LESUEUR SANDSTONE
B 6311 2095 AUSTRALIA, LEWIS TROUGH
B 6406 824 AUSTRALIA, LLOYD HILL PINNACLE REEF-PLATFORM ATOLL
B 6311 2024 AUSTRALIA, LONDONDERRY HIGH
B 6203 457 AUSTRALIA, LONG ISLAND FAULT
B 6311 2024 AUSTRALIA, MALITA GRABEN
M 29 154 AUSTRALIA, MARGIN, BATHYMETRY
M 29 160 AUSTRALIA, MARGIN, CEDUNA PLATEAU, SECTION
M 29 171 AUSTRALIA, MARGIN, COMPARED TO ANTARCTICA
M 29 158 AUSTRALIA, MARGIN, DEPTH-TO-BASEMENT CONTOURS
M 29 153 AUSTRALIA, MARGIN, DIAMANTINA FRACTURE ZONE
M 29 170 AUSTRALIA, MARGIN, EVOLUTION
M 29 160 AUSTRALIA, MARGIN, EYRE PLATEAU, SECTION
M 29 157 AUSTRALIA, MARGIN, ISOSTATIC GRAVITY PROFILES
M 29 155 AUSTRALIA, MARGIN, MAGNETIC LINEATIONS
M 29 159 AUSTRALIA, MARGIN, OTWAY BASIN SECTION
M 29 153 AUSTRALIA, MARGIN, PHYSIOGRAPHIC PROVINCES
M 29 164 AUSTRALIA, MARGIN, SONOBUOY VELOCITIES
B 6207 1153 AUSTRALIA, MARLIN CHANNEL
B 6104 594 AUSTRALIA, MARLIN FIELD
B 6406 831 AUSTRALIA, MCWHAE RIDGE
M 29 171 AUSTRALIA, MESOZOIC, EVOLUTION
M 25 428 AUSTRALIA, MESOZOIC, ORE DEPOSITS ASSOCIATED WITH
M 25 429 AUSTRALIA, METAL DEPOSITS, RESOURCES
B 6207 1153 AUSTRALIA, MIOCENE
B 6307 1093 AUSTRALIA, MONDARRA FIELD
B 6203 457 AUSTRALIA, MONTEBELLO ISLAND
B 6201 44 AUSTRALIA, MONTEBELLO TROUGH
B 6002 240 AUSTRALIA, MOONIE FIELD
B 6410 1682 AUSTRALIA, MOUNT JOHN OSMOND RANGE, LOWER CAMBRIAN FLAGS,
B 6410 1644 AUSTRALIA, MT. PAINTER RIDGE
B 6203 457 AUSTRALIA, MUDERONG GREENSAND MEMBER
B 6203 457 AUSTRALIA, MUDERONG SHALE
M 25 355 AUSTRALIA, MURRAY BASIN
B 6307 1094 AUSTRALIA, NANGETTY FORMATION
B 6406 821 AUSTRALIA, NAPIER FORMATION
B 6201 52 AUSTRALIA, NEOCOMIAN
M 25 400 AUSTRALIA, NICKEL MINERALIZATION, OXIDE ZONE
M 25 400 AUSTRALIA, NICKEL MINERALIZATION, PYRITE-VIOLARITE ZONE
M 25 403 AUSTRALIA, NICKEL, RESERVES AND ECONOMICS
M 25 236 AUSTRALIA, NORTH RANKIN DISCOVERY WELL
B 6201 43 AUSTRALIA, NORTH RANKIN FIELD
B 6307 1093 AUSTRALIA, NORTHAMPTON BLOCK
M 25 427 AUSTRALIA, NORTHERN TERRITORY, EAST ALLIGATOR, URANIUM
M 25 427 AUSTRALIA, NORTHERN TERRITORY, H.Y.C. DEPOSIT, LEAD-ZINC
M 25 427 AUSTRALIA, NORTHERN TERRITORY, RUM JUNGLE, URANIUM, COPPER,
M 25 427 AUSTRALIA, NORTHERN TERRITORY, SHERWIN IRONSTONE, IRON ORE
B 6201 44 AUSTRALIA, NORTHWEST SHELF
B 6311 2089 AUSTRALIA, NORTHWEST SHELF
B 6406 821 AUSTRALIA, NULLARA LIMESTONE
B 6406 821 AUSTRALIA, NULLARA LIMESTONE
M 25 422† AUSTRALIA, OCCURRENCE AND DEVELOPMENT OF SEDIMENTARY,
B 6410 1644 AUSTRALIA, OLARY RANGES
B 6207 1153 AUSTRALIA, OLIGOCENE
M 25 230 AUSTRALIA, OTWAY BASIN
B 6112 2077 AUSTRALIA, PACIFIC MARGIN
B 6406 823 AUSTRALIA, PADDYS VALLEY
B 6207 1153 AUSTRALIA, PALEOCENE
M 25 428 AUSTRALIA, PALEOZOIC, ORE DEPOSITS ASSOCIATED WITH
B 6203 457 AUSTRALIA, PEEDAMULLA SHELF
M 25 400 AUSTRALIA, PENTLANDITE AND PYRRHOTITE
B 6201 50 AUSTRALIA, PERMIAN
B 6207 1154 AUSTRALIA, PERMIAN
M 25 235 AUSTRALIA, PERMIAN

B 6307 1096 AUSTRALIA, PERMIAN
B 6201 41 AUSTRALIA, PERTH ABYSSAL PLAIN
B 6201 41 AUSTRALIA, PERTH BASIN
B 6210 2151 AUSTRALIA, PERTH BASIN
B 6307 1097 AUSTRALIA, PERTH BASIN, GEOTHERMAL GRADIENTS
B 6307 1093 AUSTRALIA, PERTH BASIN, TECTONIC ELEMENTS
B 6311 2024 AUSTRALIA, PETREL SUB-BASIN
B 6406 821 AUSTRALIA, PIKER HILLS FORMATION
B 6201 41 AUSTRALIA, PILBARA BLOCK
B 6406 821 AUSTRALIA, PILLARA LIMESTONE
B 6410 1644 AUSTRALIA, POONTANA STRUCTURE
B 6410 1644 AUSTRALIA, POONTANA SUBBASIN
B 6201 57 AUSTRALIA, PRECAMBRIAN TO CARBONIFEROUS
B 6110 1890 AUSTRALIA, PRODUCTION, 1976
B 6210 2154 AUSTRALIA, PRODUCTION, 1977
B 6310 1807 AUSTRALIA, PRODUCTION, 1978
B 6411 1875 AUSTRALIA, PRODUCTION, 1979
M 25 428 AUSTRALIA, PROTEROZOIC, ORE DEPOSITS ASSOCIATED WITH
M 25 400 AUSTRALIA, PRYITE AND CHALCOPYRITE
M 25 236 AUSTRALIA, QUATERNARY
M 28 74 AUSTRALIA, QUATERNARY SOIL
M 28 66 AUSTRALIA, QUATERNARY SOIL
M 28 131 AUSTRALIA, QUATERNARY, SOIL
M 29 155 AUSTRALIA, QUIET ZONE, ANOMALIES
M 29 170 AUSTRALIA, QUIET ZONE AS HYBRID CRUST
M 29 170 AUSTRALIA, QUIET ZONE AS OCEANIC CRUST
M 29 173 AUSTRALIA, QUIET ZONE, COMPARED TO RIFTS
M 29 166 AUSTRALIA, QUIET ZONE, CRUSTAL ROCKS
M 29 169 AUSTRALIA, QUIET ZONE, EVOLUTION INTERPRETATION
M 29 153 AUSTRALIA, QUIET ZONE, GEOLOGY
M 29 156 AUSTRALIA, QUIET ZONE, GRAVITY
M 29 153 AUSTRALIA, QUIET ZONE, PHYSIOGRAPHY
M 29 163 AUSTRALIA, QUIET ZONE, REFRACTION MEASUREMENTS
M 29 165 AUSTRALIA, QUIET ZONE, SEDIMENTARY ROCKS
M 29 169 AUSTRALIA, QUIET ZONE, SEISMIC INTERPRETATION
M 29 157 AUSTRALIA, QUIET ZONE, SEISMIC REFLECTIONS
M 29 167 AUSTRALIA, QUIET ZONE, SEISMIC SECTIONS
B 6201 41 AUSTRALIA, RANKIN PLATFORM
B 6311 2089 AUSTRALIA, RANKIN PLATFORM
B 6406 820 AUSTRALIA, REEF COMPLEXES, CANNING BASIN
B 6406 823 AUSTRALIA, ROSS HILL
B 6406 821 AUSTRALIA, SADLER FORMATION
B 6311 2024 AUSTRALIA, SAHUL PLATFORM
B 6311 2024 AUSTRALIA, SAHUL SYNCLINE
B 6201 41 AUSTRALIA, SCOTT PLATEAU
M 25 236 AUSTRALIA, SCOTT REEF DISCOVERY WELL
B 6112 2079 AUSTRALIA, SECTION, FLORES SEA ACROSS, BANDA VOLCANIC ARC,
B 6311 2088 † AUSTRALIA, SEDIMENTARY-, WEDGE PROGRADATION FROM,
B 6201 50 AUSTRALIA, SENONIAN-MAESTRICHTIAN
M 29 213 AUSTRALIA, SHELF, ARAFURA SEA SECTION
M 29 212 AUSTRALIA, SHELF, BANDA SEA
M 29 213 AUSTRALIA, SHELF, MELVILLE ISLAND SECTION
M 29 209 AUSTRALIA, SHELF, PHYSIOGRAPHY
M 29 214 AUSTRALIA, SHELF, REFLECTION STUDIES
M 29 210 AUSTRALIA, SHELF, REFRACTION STUDIES
M 29 211 AUSTRALIA, SHELF, VELOCITY SOLUTIONS
B 6104 595 AUSTRALIA, SNAPPER FIELD
B 6307 1104 AUSTRALIA, SOURCE ROCKS, NORTH PERTH BASIN
M 25 80 † AUSTRALIA, STATUS OF, COAL EXPLORATION AND, MINE DEVELOPMENT
M 25 426 AUSTRALIA, STRATIFORM BASE-METAL DEPOSITS, PRODUCTION AND RES
B 6207 1153 AUSTRALIA, STRZELECKI GROUP
B 6311 2024 AUSTRALIA, STURT BLOCK
M 29 215 AUSTRALIA, TANIMBAR TROUGH, REFLECTION PROFILE
B 6406 823 AUSTRALIA, TEICHERT HILLS
M 25 236 AUSTRALIA, TERTIARY
M 25 428 AUSTRALIA, TERTIARY, ORE DEPOSITS ASSOCIATED WITH
M 29 151 † AUSTRALIA, THE, CRUSTAL, STRUCTURE AND, EVOLUTION OF THE AREA
B 6311 2024 AUSTRALIA, TIMOR TROUGH
M 29 215 AUSTRALIA, TIMOR TROUGH, REFLECTION PROFILE
M 29 214 AUSTRALIA, TIMOR-TANIMBAR-ARU TROUGH, CRUSTAL SECTION
B 6203 457 AUSTRALIA, TOOLONGA CALCILUTITE
B 6201 58 AUSTRALIA, TRIASSIC
M 25 235 AUSTRALIA, TRIASSIC
B 6307 1096 AUSTRALIA, TRIASSIC
B 6201 52 AUSTRALIA, TRYAL ROCKS AREA
B 6307 1094 AUSTRALIA, TUMBLAGOODA SANDSTONE
B 6207 1153 AUSTRALIA, TUNA-FLOUNDER CHANNEL
B 6207 1153 AUSTRALIA, TURRUM FORMATION
B 6307 1093 AUSTRALIA, TURTLE DOVE RIDGE
B 6002 220 AUSTRALIA, UNION MOONIE TOPPED OIL
B 6207 1229 * AUSTRALIA URANIUM EXPORT POTENTIAL, ABST.
B 6410 1650 AUSTRALIA, URANIUM MINERALIZATION
B 6307 1093 AUSTRALIA, URELLA FAULT
B 6406 821 AUSTRALIA, VIRGIN HILLS FORMATION
B 6207 1157 AUSTRALIA, VITRINITE REFLECTANCE PROFILES
B 6307 1093 AUSTRALIA, VLAMING SUBBASIN
B 6311 2024 AUSTRALIA, VULCAN SUB-BASIN
B 6406 824 AUSTRALIA, WADE KNOLL PINNACLE REEF
B 6307 1094 AUSTRALIA, WAGINA FORMATION
B 6201 41 AUSTRALIA, WALLABY PLATEAU
B 6307 1093 AUSTRALIA, WALYERING FIELD
B 6307 1094 AUSTRALIA, WARNBRO GROUP
M 25 405 AUSTRALIA, WEIPA, BAUXITE RESERVES
B 6201 43 AUSTRALIA, WEST TRYAL ROCKS FIELD
M 25 29 AUSTRALIA, WESTERNPORT BAY
B 6201 44 AUSTRALIA, WHARTON BASIN
B 6203 457 AUSTRALIA, WINDALIA OIL ZONE
B 6203 457 AUSTRALIA, WINDALIA RADIOLARITE
B 6203 457 AUSTRALIA, WINDALIA SAND MEMBER
B 6406 821 AUSTRALIA, WINDJANA LIMESTONE
B 6201 50 AUSTRALIA, WINNING GROUP
B 6203 457 AUSTRALIA, WINNING GROUP
B 6201 44 AUSTRALIA, WOMBAT PLATEAU
B 6307 1094 AUSTRALIA, WOODADA FORMATION
B 6307 1093 AUSTRALIA, YARDARINO FIELD
B 6307 1097 AUSTRALIA, YARRAGADEE FORMATION
B 6410 1644 AUSTRALIA, YARRAMBA CHANNEL
B 6201 41 AUSTRALIA, YILGARN BLOCK
M 25 410 † AUSTRALIA--GEOLOGY AND DEVELOPMENT, IRON-ORE DEPOSITS OF,
M 25 418 † AUSTRALIA--SUMMARY, MINING HEAVY-MINERAL SAND IN
M 25 399 † AUSTRALIA--SUMMARY, NICKEL RESOURCES OF
M 25 355 † AUSTRALIA--SUMMARY, SALINE, GROUNDWATER INFLOWS TO, MURRAY
M 25 426 † AUSTRALIA--SUMMARY, STRATIFORM AND STRATA-BOUND METAL
M 25 27 † AUSTRALIA. TWO STUDIES--SUMMARY, MARINE ENVIRONMENT IN
B 6111 1994 AUSTRALIAN ABYSSAL PLAIN, POSSIBLE DIAPIRS, SOUTH
M 25 407 AUSTRALIAN ALUMINUM PRODUCTION COMMISSION
M 25 406 AUSTRALIAN BAUXITE DEPOSITS, CONFINED TO TERTIARY LATERITES
M 25 405 † AUSTRALIAN BAUXITE DEPOSITS--DISCOVERY AND DEVELOPMENT
M 25 406 AUSTRALIAN BAUXITE RESERVES AND GRADES, TABLE
B 6112 2070 AUSTRALIAN CONTINENTAL SHELF, SECTION
M 29 209 † AUSTRALIAN CONTINENTAL SHELF, SEISMIC, REFRACTION AND,
B 6112 2066 AUSTRALIAN CRATON, INDONESIA, EASTERN
B 6112 2074 AUSTRALIAN CRATON, SECTION, NEW ISLAND, SAWU GROUP, TO
B 6207 1226 * AUSTRALIAN GEOPHYSICS—CURRENT PERFORMANCE AND FUTURE
B 6207 1239 * AUSTRALIAN MARGINAL PLATEAUS, ABST., PETROLEUM PROSPECTIVITY
B 6207 1208 * AUSTRALIAN MINERAL POLICY, ABST.
M 25 231 † AUSTRALIAN NORTHWEST CONTINENTAL SHELF--RESULTS OF TEN
B 6403 360 AUSTRALIAN PLATE, BOUNDARY, INDO-
B 6207 1221 * AUSTRALIAN SCENE, ABST., LAND-USE PLANNING AND, MINERAL,
B 6406 883 AUSTRALIAN SHELF
B 6008 1407 * AUSTRALIAN URANIUM DEPOSITS, ABST., OVERVIEW OF
M 25 460 AUSTRALIAN-ASIATIC STANNIFEROUS ZONE
B 6207 1224 * AUSTRALIAS ENERGY ECONOMY, ABST., NORTHWEST SHELF, GAS—ITS
B 6207 1151 † AUSTRALIA—COMPARISON WITH COOPER BASIN, AUSTRALIA,
B 6207 1128 AUSTRALO-NEW GUINEAN CONTINENTAL PLATE
B 6010 1707 AUSTRIA, 1975, KROLL, A. J., AND K. F. J., KOLLMANN, ON DEVELOPMENTS
B 6110 1638 AUSTRIA, 1976, KROLL, A. J., AND K. F. J., KOLLMANN, ON DEVELOPMENTS
B 6210 1908 AUSTRIA, 1977, BUDWILL, A.E.H., AND A. J., KROLL, ON DEVELOPMENTS IN
B 6310 1917 AUSTRIA, 1978, KROLL, A.J., AND A. E. H., BUDWILL, ON DEVELOPMENTS
B 6411 1962 AUSTRIA, 1979, BUDWILL, A. E. H., AND A. J., KROLL, ON DEVELOPMENTS
B 6004 687 * AUSTRIA, ABST., TRAILS FROM, THE VIENNA WOODS--,
B 6411 1963 AUSTRIA, BURGENLAND
B 6110 1639 AUSTRIA, BURGENLAND
M 27 205 AUSTRIA, CRETACEOUS, GMUNDEN FLYSCH
M 27 34 AUSTRIA, CRETACEOUS LIMESTONE
M 27 124 AUSTRIA, EOCENE, OBERAUDORF SCHICHTEN
M 27 177 AUSTRIA, EOCENE, OBERAUDORF SCHICHTEN
B 6411 1963 AUSTRIA, FLYSCH-CALCAREOUS ALPS
B 6310 1918 AUSTRIA, FLYSCH-CALCAREOUS ALPS
B 6210 1909 AUSTRIA, FLYSCH-CALCAREOUS ALPS
B 6110 1639 AUSTRIA, FLYSCH-LIMESTONE ALPS ZONE
B 6110 1638 AUSTRIA, INNER ALPINE VIENNA BASIN
B 6210 1908 AUSTRIA, INNER ALPINE VIENNA BASIN
B 6310 1917 AUSTRIA, INNER ALPINE VIENNA BASIN
B 6411 1962 AUSTRIA, INNER ALPINE VIENNA BASIN
B 6411 1963 AUSTRIA, MOLASSE ZONE
B 6310 1918 AUSTRIA, MOLASSE ZONE
B 6210 1909 AUSTRIA, MOLASSE ZONE
B 6110 1639 AUSTRIA, MOLASSE ZONE
B 6110 1656 AUSTRIA, PRODUCTION, 1975-1976

B 6210 1926 AUSTRIA, PRODUCTION, 1976-1977
B 6310 1933 AUSTRIA, PRODUCTION, 1977-1978
B 6411 1971 AUSTRIA, PRODUCTION, 1978-1979
B 6411 1963 AUSTRIA, STYRIAN BASIN
B 6310 1918 AUSTRIA, STYRIAN BASIN
B 6210 1909 AUSTRIA, STYRIAN BASIN
B 6110 1639 AUSTRIA, STYRIAN BASIN
M 27 164 AUSTRIA, TRIASSIC, DACHSTEIN LIMESTONE
M 27 3 AUSTRIA, TRIASSIC DACHSTEIN LIMESTONE
B 6303 460* **AUTHIGENIC ILLITES IN SANDSTONE RESERVOIRS,**
M 28 134 AUTHIGENIC, BIREFRINGENCE, CHLORITE
M 28 66 AUTHIGENIC, CLAY, MONTMORILLONITE-, SMECTITE
B 6303 474* AUTHIGENIC MINERALS IN ARTESIA GROUP, MIDDLE PERMIAN,
B 6108 1384* AUTHIGENIC MINERALS IN ST. PETER SANDSTONE, ABST., PARAGENESIS
B 6303 504* AUTHIGENIC MINERALS--INDICATORS OF PORE-FLUID GEOCHEMISTRY,
M 28 23 AUTHIGENIC POTASSIUM, FELDSPAR
B 6405 720* AUTHIGENIC QUARTZ IN DEVONIAN BLACK SHALE, ABST.
B 6303 456* AUTHIGENIC TOURMALINE CRYSTALS IN PORE SPACE OF UPPER
M 25 101 AUTHIGENIC ZEOLITES IN TERTIARY COALFIELDS, JAPAN
M 25 252 AUTHOCHTHONOUS AND ALLOCHTHONOUS BLOCKS
B 6409 1553 AUTIN, WHITNEY J., PRELIMINARY OBSERVATIONS OF MODERN POINT
B 6011 2044† AUTOCHTHONOUS NANNOFOSSILS OF MIDDLE MIOCENE OF SOUTH
B 6004 697* AUTOKINETI BASINAL LITHOFACIES, ABST., DEVELOPMENT OF VERTICAL
S 5 119 AUTOMATIC STATICS CORRECTION, SEISMIC DATA,
S 5 3 A-0 CARBONATE, SALINA GROUP,
S 5 3 A-1 CARBONATE, SALINA GROUP,
S 5 30 A-1 EVAPORITE, MICHIGAN BASIN, SALINA GROUP,
S 5 3 A-1 EVAPORITE, SALINA GROUP,
S 5 4 A-2 CARBONATE, SALINA GROUP,
S 5 4 A-2 EVAPORITE, SALINA GROUP,
B 6004 683* AUTOPSY OF A DEAD CORAL REEF, ABST.
M 25 453 AUTUNITE, URANIUM ORE MINERAL, JAPAN
B 6004 656* AVAILABILITY AND APPLICATION OF NATIONAL COLE-RESOURCE DATA
B 6207 1222* AVAILABILITY AND QUALITY OF WATER RELATED TO ENERGY IN
B 6205 863† AVAILABILITY OF GEOLOGIC MATERIALS FROM STATE SURVEYS
B 6303 505* AVAILABILITY OF REMOTELY SENSED DATA, ABST.
B 6310 1749 AVALAS, R., ON DEVELOPMENTS IN CHILE, 1978
B 6209 1697 AVAN HILLS, ALASKA
B 6203 492 AVARY, KATHARINE LEE, DISTINCTIVE TURBIDITE SILTSTONE IN
B 6405 672 AVBOVBO, A. A., PETROLEUM PROSPECTS OF ANAMBRA BASIN,
B 6202 295 AVBOVBO, AKPOYOVBIKE ABOGO, TERTIARY LITHOSTRATIGRAPHY OF
B 6004 689* AVELLA, PENNSYLVANIA, ABST., COMPARISON OF DIAGENETIC AND,
B 6405 672 AVIS, LOREN E., PROMISING MORROW SANDSTONE DISCOVERIES IN
M 28 119 AVIS SANDSTONE, TEXAS, PENNSYLVANIAN
B 6104 493 AVON PARK FORMATION, FLORIDA
M 27 219 AVON PARK LIMESTONE, FLORIDA, EOCENE
M 27 131 AVON PARK LIMESTONE, FLORIDA, EOCENE
M 27 194 AVON PARK LIMESTONE, FLORIDA, EOCENE
B 6202 296 AWGU SHALE, NIGERIA
B 6203 488 AWRAMIK, STANLEY M.., OXYGEN, AND CARBON ISOTOPIC VARIATION IN
B 6206 1005 AXEL HEIBERG ISLAND, ARCTIC, CANADA
B 6206 1018 AXEL HEIBERG ISLAND, KANGUK DIAPIR
B 6101 3 AXEL HEIBERG ISLANDS, NORTHERN, ELLESMERE AND
B 6302 195 AXIAL ANTICLINE, COLORADO
M 25 437 AYA-AYA COPPER PROJECT, NEGROS
B 6405 672 AYOOLA, E. O., PETROLEUM PROSPECTS OF ANAMBRA BASIN, SOUTHERN
B 6111 2010 AYRTON, W. G., CANADIAN SOCIETY OF PETROLEUM GEOLOGISTS
B 6107 963 AZORES FRACTURE ZONE, AFRICA
M 25 328 AZUCAR GROUP, ECUADOR
B 6203 380 B, AND C HORIZONS, COOK INLET, ALASKA, A
B 6002 246 B POOLS, ALBERTA, RAINBOW LAKE, A AND
M 24 217 B SANDSTONE ZONE, LOUISIANA
B 6002 319* B, STAFFORD COUNTY, KANSAS, ABST., NUMERICAL ANALYSIS OF, TRAP
M 24 216 B-14 SANDSTONES, LOUISIANA, LOUISIANA
B 6309 1453 B-2, BALTIMORE CANYON TROUGH, CONTINENTAL OFFSHORE
B 6209 1602 B-2 COST WELL, KEROGEN-MATURATION PROFILE, BALTIMORE CANYON
B 6104 595 B-2-1 FIELD, LIBYA
B 6402 218 B-3 FIELD, INDIA
B 6402 218 B-3 FIELD, INDIA
B 6402 218 B-38 FIELD, INDIA
M 24 17 B-POOL MARKER, ALASKA, MIDDLE GROUND SHOAL FIELD
M 25 453 B-URANOPHANE, URANIUM ORE MINERAL, JAPAN
B 6210 2115 B. P. TRADING LTD., ET AL, ON DEVELOPMENTS IN ABU DHABI, 1977, ABU
B 6010 1720 B.P. PETROL. DEVEL. LTD., BRITISH GAS CORP., ET AL, ON DEVELOPMENTS
B 6310 1890 B.P. PETROLEUM DEVELOPMENT LTD., CLUFF OIL LTD.,
B 6009 1618* B.W.I., ABST., MIDDLE, EOCENE, PLANKTONIC,
B 6308 1206 BAARS, D. L., DEVELOPMENTS IN FOUR CORNERS--INTERMOUNTAIN ARE
B 6409 1340 BAARS, D. L., DEVELOPMENTS IN FOUR CORNERS—INTERMOUNTAIN
B 6104 585 BAB, MURBAN, FIELD, PERSIAN GULF
M 25 147 BABA, KENZO, GEOTHERMAL RESOURCES AND ENERGY IN JAPAN
B 6207 1111 BABCOCK, E. A., MEASUREMENT OF SUBSURFACE FRACTURES FROM
B 6203 573 BABCOCK, J. A., SYNDEPOSITIONAL SUBMARINE CEMENTATION OF
B 6203 492 BABCOCK, J. A., SYNDEPOSITIONAL SUBMARINE CEMENTATION OF
B 6209 1661 BABUYAN RIDGE, STRUCTURE, LU TAO-
M 25 158 BACA, CALDERA FIELD, GEOTHERMAL, NEW MEXICO, VALLES
B 6405 688* BACA FORMATION OF NEW MEXICO AND EAGAR FORMATION AND
B 6203 526* BACA FORMATION OF NEW MEXICO AND EAGAR FORMATION OF
M 25 26 BACA, M. F., LATIN AMERICAS OUTLOOK TOWARD ENERGY
B 6301 90 BACCARO MEMBER, SCOTIAN BASIN
B 6104 579 BACHAQUERO FIELD, VENEZUELA
B 6104 579 BACHAQUERO, SHELL, FIELD, VENEZUELA
B 6403 438 BACHMAN, S. B., ET AL, VARIABLES AFFECTING TRENCH AND
B 6211 2349 BACHMAN, STEVEN B., CRETACEOUS AND EARLY TERTIARY SUBDUCTION
B 6405 673 BACHMANN, G. H., EXPLORATION IN CLASSIC THRUST BELT AND ITS
B 6103 416 BACK, WILLIAM, WILDCAT VALLEY SANDSTONE IN SOUTHWEST
B 6009 1621 BACK, WILLIAM, WILDCAT VALLEY SANDSTONE IN SOUTHWEST
B 6212 2510* BACK-ARC BASINS OF BOHOKU ARC, JAPAN, ABST., STRUCTURAL
B 6212 2470 BACK-ARC GEOLOGY
B 6212 2381 BACK-BARRIER, DEPOSITIONAL ENVIRONMENT
B 6203 512* BACK-BARRIER ENVIRONMENT, ABST., DEPOSITIONAL, MODEL OF
B 6203 556* BACK-BARRIER SEDIMENTATION ON TIDE-DOMINATED COASTLINE OF
M 27 59 BACK-REEF BEACH SAND, BELIZE, HOLOCENE
B 6303 519* BACK-REEF CARBONAT ROCKS--EXAMPLE FROM CAPITAN COMPLEX,
M 27 59 BACK-REEF LIMESTONES, BELIZE, PLEISTOCENE
B 6301 4 BACKARC FEATURES, ARC-TRENCH SYSTEMS, DEFINED BY
B 6204 707 BACKBONE LIMESTONE, ILLINOIS BASIN
B 6209 1667 BACKBONE RANGE AND YU SHAN, TAIWAN
M 25 430† BACKGROUND TO PRESENT MINERAL SEARCH IN BRITISH SOLOMON
S 4 19 BACKREEF LAGOON, PUERTO RICO
B 6405 673 BACKUS, MILO M., GEOPHYSICAL EXPLORATION IS JUST BEGINNING,
B 6410 1723 BACKUS, MILO M., REVIEW OF A FIRST COURSE IN GEOPHYSICAL
B 6303 411 BACLAWSKI, A. PAUL, PLEISTOCENE SUBMARINE CANYONS OF
B 6303 412 BACOCCOLI, G., NAMORAD FIELD, MAJOR OIL DISCOVERY IN CAMPOS
B 6303 412 BACOCCOLI, GIUSEPPE,, OIL EXPLORATION IN CAMPOS BASIN, BRAZIL.
M 30 329 BACOCCOLI, GIUSEPPE, THE NAMORADO OIL FIELD--A MAJOR OIL
B 6408 1185 BACTERIA, ANAEROBIC
B 6304 701* BACTERIAL DEGRADATION OF CRUDE OILS--IDENTIFICATION BY
B 6203 513* BACTERIAL MEDIATION OF REDOX POTENTIAL IN URANIUM ORE
B 6207 1208 BACUTA, GEORGE C., GEOLOGY AND ECONOMIC POSSIBILITIES OF SOME
M 30 441 BADAK FIELD, DISCOVERY HISTORY, INDONESIA
B 6303 470* BADAK FIELD, EAST KALIMANTAN, INDONESIA, ABST., DISCOVERY AND
M 30 441† BADAK FIELD, EAST KALIMANTAN, INDONESIA, DISCOVERY AND
M 30 442 BADAK FIELD, REGIONAL GEOLOGY, KUTEI BASIN
M 30 442 BADAK FIELD, STRATIGRAPHY, KUTEI BASIN
M 30 453 BADAK FIELD, STRUCTURAL CROSS SECTION, INDONESIA
M 30 443 BADAK FIELD, ZONE A, RESERVOIR
M 30 456 BADAK FIELD, ZONE B, STRATIGRAPHY
M 30 457 BADAK FIELD, ZONE C, RESERVOIRS
M 30 457 BADAK FIELD, ZONE D, RESERVOIRS
M 30 457 BADAK FIELD, ZONE E, RESERVOIRS
M 30 457 BADAK FIELD, ZONE F, RESERVOIRS
B 6209 1685 BADAK GAS FIELD, KALIMANTAN
B 6109 1476 BADEJO FIELD, BRAZIL
S 6 184 BADENIAN STAGE, PARATETHYS, NEOGENE
B 6310 1819 BADRUDDOJA, M., AND STEWART, R. D., ON DEVELOPMENTS IN
B 6303 412 BAER, CHRIS B., COMPUTER-COMPATIBLE HANDLING OF GEOLOGIC DATA
B 6206 1005 BAFFIN ISLAND, ARCTIC, CANADA
B 6208 1382 BAFFLE POINT FIELD, GALVESTON COUNTY, TEXAS
B 6009 1620 BAGANZ, B. P., CARBONIFEROUS SEDIMENTARY RESPONSES TO
B 6212 2379 BAGANZ, B. P., DEPOSITIONAL MODELS IN COAL EXPLORATION AND
S 6 179 BAGDASARJAN, GEVORG P., A RADIOMETRIC TIME SCALE FOR THE
S 11 162 BAGGINA, CALIFORNIA, MIOCENE
S 11 162 BAGGINA CALIFORNICA
S 11 284 BAGGINA CALIFORNICA
S 11 255 BAGGINA CALIFORNICA
S 11 160 BAGGINA ROBUSTA
S 11 162 BAGGINA ROBUSTA
S 11 162 BAGGINA SUBINEQUALIS
M 25 435 BAGUIO MINERAL DISTRICT, COPPER, LUZON
B 6406 851 BAHAMA ABYSSAL PLAIN, SAND LAYERS, BLAKE-
B 6412 2058 BAHAMA AREA, DSDP SITE 391, LEG 44, GEOCHEMICAL LOG, BLAKE-
B 6203 520* BAHAMA BANK, ABST., OOID, SAND-SHOAL FACIES, JOULTERS CAYS,
B 6105 841* BAHAMA BANK, ABST., POROSITY CONTROLS IN SUBSEA, CEMENTED
B 6203 494* BAHAMA BANK, ABST., PRE-HOLOCENE, CARBONATE ROCKS OF
B 6410 1703 BAHAMA BANK DRIFT, GREAT
B 6410 1703 BAHAMA BANK DRIFT, LITTLE

B	6410	1634†	BAHAMA BANK, FACIES SUCCESSION OF, PLIOCENE-, PLEISTOCENE,
B	6410	1640	BAHAMA BANKS AS ATOLLS
B	6405	674	BAHAMA CHANNEL, NORTH OF CUBA, ABST., EXPLORABLE STRUCTURES
B	6204	693†	BAHAMA CHANNELS IN RELATION TO HYDROCARBON POTENTIAL,
B	6204	694	BAHAMA CHANNELS, ORIGIN OF
B	6410	1701	BAHAMA OUTER RIDGE, ABYSSAL SEDIMENT DRIFT, BLAKE-
B	6004	700*	BAHAMA PLATFORM, ABST., HYDROCARBON, POTENTIAL OF
B	6204	694	BAHAMA PLATFORM, ANDROS ISLAND WELL
B	6204	694	BAHAMA PLATFORM, CAY SAL WELL
B	6204	696	BAHAMA PLATFORM, CENOZOIC
B	6204	696	BAHAMA PLATFORM, CRETACEOUS
B	6204	696	BAHAMA PLATFORM, DSDP 100
B	6204	696	BAHAMA PLATFORM, DSDP 99
B	6204	694	BAHAMA PLATFORM, DSDP SITE 98
B	6204	694	BAHAMA PLATFORM, ELEUTHERA WELL
B	6204	694	BAHAMA PLATFORM, GREAT ISAAC WELL
B	6204	695	BAHAMA PLATFORM, HYDROCARBON POTENTIAL
B	6204	696	BAHAMA PLATFORM, JURASSIC
B	6204	694	BAHAMA PLATFORM, LONG ISLAND WELL
B	6007	1038	BAHAMA PLATFORM, MAP, NORTHWESTERN
B	6204	694	BAHAMA PLATFORM, NEW PROVIDENCE WELL
B	6204	696	BAHAMA PLATFORM, NORTHWESTERN, INTERPRETIVE SECTION
B	6204	694	BAHAMA PLATFORM, SAN SALVADOR WELL
B	6204	694	BAHAMA PLATFORM, SEISMIC REFLECTION PROFILES
B	6204	696	BAHAMA PLATFORM, TRIASSIC
B	6204	694	BAHAMA PLATFORM, WILLIAMS WELL
B	6105	829*	BAHAMA REGION OF ATLANTIC MARGIN OF NORTH AMERICA, ABST.,
B	6010	1642	BAHAMAS, 1975, MINISTRY OF DEVELOPMENT, ON DEVELOPMENTS IN
B	6110	1580	BAHAMAS, 1976, PETROCONSULTANTS S.A., ON DEVELOPMENTS IN
B	6210	2046	BAHAMAS, 1977, PETROCONSULTANTS S.A., ON DEVELOPMENTS IN
B	6310	1747	BAHAMAS, 1978, PETROCONSULTANTS S.A., ON DEVELOPMENTS IN
B	6405	788*	BAHAMAS, ABST., CEMENTATION OF LIME-MUD AND PELLET MU
B	6209	1758*	BAHAMAS, ABST., HOLOCENE MARINE-CEMENTED, SANDS, JOULTERS,
B	6303	422*	BAHAMAS, ABST., HURRICANES AND RAINFALL--KEY FOR,
B	6203	560*	BAHAMAS, ABST., INFLUENCE OF, PLATFORM-DERIVED SEDIMENT ON,
B	6303	457*	BAHAMAS, ABST., MIXING ZONE, DOLOMITE IN TIDAL-FLAT SEDIMENTS
B	6303	500*	BAHAMAS, ABST., SEISMIC FACIES AND DEPOSITIONAL PROCESSES OF
B	6204	694	BAHAMAS, ANDROS ISLAND
S	4	187	BAHAMAS, ANDROS ISLAND, REEF
B	6204	694	BAHAMAS, CAICOS BANK
B	6204	694	BAHAMAS, CAT ISLAND
B	6410	1702	BAHAMAS, CIRCULATION PATTERN OF NORTHERN
B	6204	694	BAHAMAS, COLUMBUS BASIN
B	6410	1641	BAHAMAS, DEFINITION OF, LUCAYAN LIMESTONE, NEW QUATERNARY
B	6204	694	BAHAMAS, ELEUTHERA ISLAND
B	6204	694	BAHAMAS, EXUMA SOUND
B	6204	694	BAHAMAS, GREAT ABACO ISLAND
B	6204	694	BAHAMAS, GREAT INAGUA ISLAND
M	27	189	BAHAMAS, HOLOCENE BEACHROCK
M	27	157	BAHAMAS, HOLOCENE BEACHROCK
M	27	128	BAHAMAS, HOLOCENE SEDIMENTS
M	27	127	BAHAMAS, HOLOCENE SEDIMENTS
M	27	179	BAHAMAS, HOLOCENE SEDIMENTS
M	27	117	BAHAMAS, HOLOCENE SEDIMENTS
M	27	127	BAHAMAS, HOLOCENE SEDIMENTS
M	27	183	BAHAMAS, HOLOCENE SEDIMENTS
M	27	123	BAHAMAS, HOLOCENE SEDIMENTS
M	27	81	BAHAMAS, HOLOCENE SEDIMENTS
B	6004	682*	BAHAMAS ISLAND, ABST., REEF MORPHOLOGY AND SEDIMENT,
B	6404	467	BAHAMAS, JOULTERS CAY, MARINE PHREATIC ARAGONITE CEMENT
S	4	186	BAHAMAS, JOULTERS CAYS, PATCH REEF
B	6204	694	BAHAMAS, LITTLE BAHAMA BANK
B	6204	694	BAHAMAS, LONG ISLAND
S	4	175	BAHAMAS, MONTASTREA ANNULARIS
B	6204	694	BAHAMAS, NORTHEAST PROVIDENCE CHANNEL
B	6204	694	BAHAMAS, OLD BAHAMA CHANNEL
M	27	127	BAHAMAS, PLEISTOCENE DUNES
M	27	115	BAHAMAS, PLEISTOCENE DUNES
M	27	161	BAHAMAS, QUATERNARY SEDIMENTS
B	6103	376†	**BAHAMAS, SHALLOW, CARBONATE-, BANK-, MARGIN GROWTH AND**
B	6007	1037†	BAHAMAS, STRATIGRAPHY AND STRUCTURE OF, NORTHEAST,
B	6204	694	BAHAMAS, TONGUE OF THE OCEAN
B	6103	443†	BAHAMAS. DISCUSSION AND REPLY, JURASSIC LITHOLOGY IN, GREAT
B	6408	1274	BAHAMIAN OOID SHOAL, SEDIMENTA VII, BY PAUL M. HARRIS,
B	6004	653*	BAHAMIAN OOIDS, ABST., ISOTOPIC EVIDENCE BEARING ON, ORIGIN OF
B	6405	755*	**BAHAMIAN SLOPES—POSSIBLE MODEL FOR ORIGIN OF SOME NODULAR**
B	6203	507*	BAHAMIAN TROUGH, ABST., SEDIMENTARY RECORD OF LONGITUDINAL
B	6004	719*	BAHAMIAN TROUGH. BY TITLE ONLY, CARBONATE, MUD TO, CHALK,
M	29	113†	BAHIA BLANCA, ARGENTINA, STRUCTURE OF, COLORADO, BASIN AND,
B	6104	562	BAHIA INUTIL GROUP, CHILE
B	6004	646	BAHLBURG, WILLIAM, DEPOSITIONAL ENVIRONMENTS OF TAMAROA
B	6010	1872	BAHRAIN, 1975, CHEVRON OVERSEAS PETROLEUM, INC., AND, SUPERIOR
B	6110	1801	BAHRAIN, 1976, CHEVRON OVERSEAS PETROL. INC., AND, SUPERIOR OIL
B	6210	2117	BAHRAIN, 1977, CHEVRON OVERSEAS PETROLEUM, INC., SUPERIOR OIL
B	6310	1889	BAHRAIN, 1978, CHEVRON OVERSEAS PETROLEUM, INC., SUPERIOR OIL
B	6411	1838	BAHRAIN, 1979, CHEVRON OVERSEAS PETROLEUM INC. AND, SUPERIOR
B	6310	1892	BAHRAIN, DEVELOPMENTS, 1977-1978
B	6210	2120	BAHRAIN, PRODUCTION, 1976-1977
B	6411	1843	BAHRAIN, PRODUCTION, 1978-1979
S	6	302	BAIGENDZINIAN STAGE, PERMIAN, BIOCHRONOLOGY
S	11	202	BAILEY ASH BED, RADIOACTIVE DATING
B	6407	1035	BAILEY ASH, CALIFORNIA
B	6204	706	BAILEY LIMESTONE, ILLINOIS BASIN
B	6303	413	BAIN, ROGER J., ARE PENNSYLVANIAN FRESHWATER LIMESTONES
B	6109	1534	BAIRD, JAMES A., UNCONFORMITIES—NEGLECTED EXPLORATION TOOL,
B	6106	913	BAJA BORDERLAND
B	6207	1219*	BAJA CALIFORNIA, ABST., PETROLEUM GEOLOGY OF, REFORMA AREA,
B	6004	707*	BAJA CALIFORNIA, ABST., SEDIMENTOLOGIC INDICATION OF, EOCENE,
B	6203	370	BAJA CALIFORNIA, ALIJOS ROCKS
B	6106	904	BAJA CALIFORNIA, ANIMAL BASIN
B	6106	907	BAJA CALIFORNIA, BAJA GAP
B	6106	904	BAJA CALIFORNIA, BLANCA BASIN
B	6106	904	BAJA CALIFORNIA, CEDROS DEEP
B	6203	370	BAJA CALIFORNIA, CEDROS DEEP
B	6106	904	BAJA CALIFORNIA, COLNETT BASIN
B	6106	903†	BAJA CALIFORNIA CONTINENTAL BORDERLAND, MEXICO, MARINE
B	6106	912	BAJA CALIFORNIA, FRANCISCAN BASEMENT
M	25	161	BAJA CALIFORNIA GEOTHERMAL AREA, MEXICO, CERRO PRIETO,
B	6106	904	BAJA CALIFORNIA, GUADALUPE ARRUGADO
M	25	526	BAJA CALIFORNIA, HAWAII-TYPE BASALTS AND HAWAIITES
B	6403	376	BAJA CALIFORNIA, HOLOCENE FAN DELTA
B	6203	365	BAJA CALIFORNIA, MAGDALENA BORDERLAND
M	25	515	BAJA CALIFORNIA METALLOGENETIC PROVINCE, MEXICO
B	6210	2055	BAJA CALIFORNIA, MEXICO
B	6212	2508*	BAJA CALIFORNIA, MEXICO, ABST., DEVELOPMENT OF, CERRO PRIETO,
B	6203	490*	BAJA CALIFORNIA, MEXICO, ABST., RECENT DEVELOPMENTS AT, CERRO
B	6405	758*	BAJA CALIFORNIA, MEXICO, ABST., STRUCTURAL AND SEDIMENTOLOGIC
B	6405	759*	BAJA CALIFORNIA, MEXICO, ABST., STUDY OF DIAGENETIC PROCESSES IN
B	6106	904	BAJA CALIFORNIA, OUTER BASIN
B	6106	907	BAJA CALIFORNIA, POPCORN RIDGE
B	6106	907	BAJA CALIFORNIA, POPCORN RIDGE ESCARPMENT
B	6106	904	BAJA CALIFORNIA, POPCORN RIDGE FAULT
B	6106	911	BAJA CALIFORNIA, POWAY CONGLOMERATE
B	6203	370	BAJA CALIFORNIA, ROSA BANK
B	6106	904	BAJA CALIFORNIA, SAN ISIDRO BASIN
B	6407	1073	BAJA CALIFORNIA, SAN LUCAS FAN
B	6206	922	BAJA CALIFORNIA, SAN LUCAS FAN
B	6106	904	BAJA CALIFORNIA, SAN QUENTIN BASIN
B	6203	362	BAJA CALIFORNIA, SEBASTIAN VIZCAINO BASIN
B	6106	904	BAJA CALIFORNIA, SOLEDAD BASIN
B	6106	904	BAJA CALIFORNIA, SOLEDAD GAP
B	6106	904	BAJA CALIFORNIA, SOLEDAD RIDGE
B	6004	671*	BAJA CALIFORNIA SUR, MEXICO, ABST., PETROLOGY OF, TRIASSIC,
M	25	526	BAJA CALIFORNIA, VOLCANIC ERUPTIONS, RECENT
M	25	526	BAJA CALIFORNIA, VOLCANIC SERIES, CYCLES OF, MEXICO
M	25	526†	BAJA CALIFORNIA--SUMMARY, MAJOR CHEMICAL CHARACTERISTICS OF,
B	6207	1237*	BAJA CALIFORNIA—PROGRESS AND PROSPECTS, ABST.,
B	6106	907	BAJA GAP, BAJA CALIFORNIA
M	26	158	BAJOCIAN, AFRICA, SEQUENCES, JURASSIC
B	6405	614	BAJOCIAN TO ALBIAN SEQUENCE, MIDDLE EAST
M	30	127	BAJOCIAN-BATHONIAN, STRATIGRAPHY, BEATRICE FIELD
S	4	3	BAK, ROLF P. M., CORAL REEFS AND THEIR ZONATION IN NETHERLANDS
B	6111	1986	BAKER CANYON CONGLOMERATE, CALIFORNIA
B	6405	673	BAKER, F. E., GEOPHYSICAL CASE HISTORY OF TWO HILLS COLONY GAS
B	6101	29	BAKER LAKE, PITCHBLENDE DEPOSITS, CANADA
B	6207	1237	BAKER, R. N., SATELLITE-ASSISTED MINERAL EXPLORATION IN BAJA
B	6105	762	**BAKER, RALPH N., DISCRIMINATION OF SURFICIAL MATERIALS IN PART**
B	6311	2114	BAKER, RANDALL K., DEPOSITIONAL ENVIRONMENT OF PENNSYLVANIA
B	6408	1280	BAKER, ROBERT, GEOLOGY OF WHEATON CONSOLIDATED OIL FIELD,
B	6109	1540	BAKER, VICTOR R., LATE QUATERNARY GEOMORPHIC EVOLUTION OF
B	6211	2245	BAKERSFIELD ARCH, CALIFORNIA
B	6307	1036	BAKERSFIELD ARCH, CONTOUR MAP, CALIFORNIA
B	6008	1403*	BAKKEN FORMATION, WILLISTON BASIN, AND ITS RELATION TO
S	10	49	BAKKEN SHALE, MIGRATION, NORTH DAKOTA, WILLISTON BASIN
B	6303	515	BAKKER, M. L., GEOSEISMI MODELING--AN INTERACTIVE COMPUTER
S	10	232	BAKU AREA, GOUSANY FIELD, USSR
S	10	232	BAKU AREA, KARACHUKHUR FIELD, USSR
S	10	203	BAKU AREA, MUD VOLCANO, DISTRIBUTION, USSR

S	10	232	BAKU AREA, PESCHANY FIELD, USSR
S	10	232	BAKU AREA, SURAKHANY FIELD, USSR
S	10	230	BAKU AREA, ZYKH FIELD, HYDROTHERMAL EFFECTS, USSR
B	6108	1371	BALCH, A. H., APPLICATIONS OF VERTICAL SEISMIC PROFILE TO
B	6004	718	BALCH, A. H., SYNTHETIC SEISMIC SECTIONS OF SELECTED
B	6405	674	BALCH, A. H., VERTICAL AN SURFACE SEISMIC PROFILES MAP AQUIFERS
B	6008	1409	BALCH, ALFRED H., SYNTHETIC SEISMIC SECTIONS OF SELECTED
B	6105	762	**BALDWIN, B., FREQUENCY OF TURBIDITE EVENTS IN NORTHERN TACONIC**
B	6105	763	**BALDWIN B., TECTONIC IMPLICATIONS OF CROWN POINT SECTION, NEW**
B	6012	2174	BALDWIN, THOMAS A., SALINAS VALLEY TODAY AND TOMORROW, ABST.
B	6411	1968	BALEARIC OFFSHORE, SPAIN
M	25	138	BALI, THERMAL AREA, PROSPECTIVE
B	6206	1004	BALKWILL, HUGH R., EVOLUTION OF SVERDRUP BASIN, ARCTIC CANADA
B	6303	413	BALKWILL, HUGH R., SVERDRUP BASIN RESPONSE TO CRATONIC EVENTS,
M	28	69	BALL CLAY, MISSOURI, PENNSYLVANIAN
B	6002	295	BALL CLAY NO. 661, GEOTECHNICAL PROPERTIES, ENGLAND
B	6004	647	BALL, DOUGLAS, ENERGY FROM TAR SANDS, ABST.
B	6004	700	BALL, M. M., HYDROCARBON POTENTIAL OF NORTHWESTERN BAHAMA
B	6204	693	BALL, MAHLON M., CHARACTERISTICS OF DEEP BAHAMA CHANNELS IN
B	6405	674	BALL, MAHLON M., EXPLORABLE STRUCTURES IN OLD BAHAMA
B	6203	493	BALL, MAHLON M., PETROLEUM POTENTIAL OF CONTINENTAL SLOPE
S	4	239	BALL, STANTON M., IMPORTANCE OF PHYLLOID ALGAE IN
B	6109	1483	BALLARD, R. D., CONTINENTAL SLOPE AND UPPER RISE OFF WESTERN
B	6404	501	BALLARD, R. D., GEOLOGY OF NEW ENGLAND PASSIVE MARGIN
B	6105	763	**BALLY, A. W., ANCIENT AND MODERN CONTINENTAL MARGINS,**
B	6004	647	BALLY, A. W., PROTAGONIST--PLATE TECTONICS AS A MAJOR
B	6302	257	BALLY, A. W., REVIEW ON SEDIMENTATION IN SUBMARINE CANYONS
B	6408	1271	BALLY, LEHNER, P., REVIEW OF, CONTINENTAL MARGINS—GEOLOGICAL
B	6203	443	BALOL SANTHAL FIELD, INDIA
B	6203	453	BALOL-SANTHAL FIELD, SECTION, INDIA
B	6004	712*	BALSAS OF WESTERN MEXICO, ABST., SLOPE, VALLEYS AND
B	6203	443	BALSASAN FIELD, INDIA
B	6405	724	BALSEY, J. K., CRETACEOUS-TERTIARY VERSUS CARBONIFEROUS
B	6303	410	BALSHAW, KATHERINE, GLACIAL-MARINE SEDIMENTS OF WEST
B	6303	413	BALSHAW, KATHERINE M., CENOZOIC CLAY-MINERAL STRATIGRAPHY I
B	6405	739	BALSLEY, J. K., ET AL, WAVE-DOMINATED DELTAS—IMPORTANT
B	6408	1192	BALTIC SEA, ANOXIC SILLED BASIN
B	6101	80	BALTIMORE CANYON
B	6203	526*	BALTIMORE CANYON, AND ITS RELATION TO SCOTIAN BASIN, ABST.,
B	6209	1602	BALTIMORE CANYON, B-2 COST WELL, KEROGEN-MATURATION PROFILE
M	29	79	BALTIMORE CANYON, CROSS SECTION, ATLANTIC COAST
M	29	104	BALTIMORE CANYON, DEPTH-TO-BASEMENT CONTOURS
M	29	78	BALTIMORE CANYON, DIAPIR, ATLANTIC COAST, NEW JERSEY
B	6004	648*	BALTIMORE CANYON GEOSYNCLINE, ABST., RADIOCARBON, DATING AND
M	29	438	BALTIMORE CANYON, KEROGEN MATURATION IN SEDIMENTS
M	29	102	BALTIMORE CANYON, MAGNETIC ANOMALY CONTOURS
M	29	103	BALTIMORE CANYON, MAGNETIC DEPTH ESTIMATES
M	26	333	BALTIMORE CANYON, SEISMIC LINE
M	29	79	BALTIMORE CANYON, SHELF EDGE, ATLANTIC COAST
B	6309	1465	BALTIMORE CANYON, STRATIGRAPHIC COLUMN, LONG ISLAND
B	6309	1465	BALTIMORE CANYON, STRATIGRAPHIC COLUMN, MARYLAND
B	6309	1465	BALTIMORE CANYON, STRATIGRAPHIC COLUMN, NEW JERSEY
B	6309	1465	BALTIMORE CANYON, STRATIGRAPHIC COLUMN, NORTH CAROLINA
B	6309	1465	BALTIMORE CANYON, STRATIGRAPHIC COLUMN, VIRGINIA
B	6204	593	BALTIMORE CANYON TROUGH
B	6106	862	BALTIMORE CANYON TROUGH AREA, BATHYMETRY
B	6106	918	BALTIMORE CANYON TROUGH AREA, DELAWARE
B	6106	865	BALTIMORE CANYON TROUGH AREA, PROFILES
B	6106	861†	BALTIMORE CANYON TROUGH AREA, THICKNESS AND AGE OF,
B	6106	882	BALTIMORE CANYON TROUGH AREA, TIGER SCARP
B	6106	863	BALTIMORE CANYON TROUGH AREA, VIBRACORE SAMPLES
M	29	100	BALTIMORE CANYON TROUGH, ATLANTIC MARGIN
B	6309	1453	BALTIMORE CANYON TROUGH, ATLANTIC SLOPE PROJECT, ASP
B	6405	715*	BALTIMORE CANYON TROUGH BASED ON MULTIFOLD SEISMIC
B	6309	1455	BALTIMORE CANYON TROUGH, BASEMENT
B	6309	1453	BALTIMORE CANYON TROUGH, CONTINENTAL OFFSHORE STRATIGRAPHI
B	6309	1455	BALTIMORE CANYON TROUGH, CRETACEOUS
B	6309	1453	BALTIMORE CANYON TROUGH, DEEP SEA DRILLING PROJECT, DSDP,
B	6309	1456	BALTIMORE CANYON TROUGH, EOCENE
B	6308	1204	BALTIMORE CANYON TROUGH, EXPLORATORY WELLS, 1978
B	6309	1463	BALTIMORE CANYON TROUGH, HOLOCENE
B	6309	1455	BALTIMORE CANYON TROUGH, JURASSIC
B	6409	1333	BALTIMORE CANYON TROUGH, MID-ATLANTIC OUTER CONTINENTAL
B	6309	1459	BALTIMORE CANYON TROUGH, MIOCENE
B	6309	1458	BALTIMORE CANYON TROUGH, OLIGOCENE
B	6309	1456	BALTIMORE CANYON TROUGH, PALEOCENE
B	6007	1113	BALTIMORE CANYON TROUGH, PETROLEUM EXPLORATION LEASING
B	6309	1461	BALTIMORE CANYON TROUGH, PLEISTOCENE
B	6309	1461	BALTIMORE CANYON TROUGH, PLIOCENE
B	6309	1452†	BALTIMORE CANYON TROUGH, STRATIGRAPHY AND DEPOSITIONAL
B	6309	1455	BALTIMORE CANYON TROUGH, TRIASSIC

B	6309	1453	BALTIMORE CANYON TROUGH, USGS ATLANTIC MARGIN CORING
B	6309	1490†	BALUCHISTAN, PAKISTAN, BY COMPUTER PROCESSIN OF LANDSAT DATA,
B	6207	1208	BAMBRICK, SUSAN C., AUSTRALIAN MINERAL POLICY, ABST.
S	6	109	BAMBUI GROUP, BRAZIL, LATE, PRECAMBRIAN
S	6	109	BAMBUI GROUP, RUBIDIUM-STRONTIUM, BRAZIL
B	6411	1970	BANAT, MURA, DRAVA, AND SAVA SUBBASINS, YUGOSLAVIA
S	7	10	BANCO DE CAMPECHE SLOPE, DESCRIBED
B	6406	890	BANDA ARC, OUTER
B	6112	2063†	BANDA ARCS, EASTERN INDONESIA, GEOSYNCLINES AND, PLATE
B	6302	249†	BANDA ARCS, EASTERN INDONESIA. DISCUSSION AND REPLY,
B	6112	2069	BANDA ARCS, POSTOROGENIC SEQUENCES
B	6112	2064	BANDA ARCS, PREOROGENIC SEQUENCES
B	6112	2075	BANDA ARCS, TECTONIC MODELS, INDONESIA, EASTERN
B	6406	904	BANDA BASIN
M	29	212	BANDA SEA, AUSTRALIA, SHELF
M	25	474	BANDA SEA MEGASTRUCTURE
B	6406	868†	BANDA SEA REGION, ARC-, CONTINENT, COLLISION
B	6406	870	BANDA SEA REGION, BATHYMETRY
B	6406	869	BANDA SEA REGION, PRINCIPAL FEATURES
B	6406	871	BANDA SEA REGION, SEISMIC PROFILING
B	6112	2079	BANDA VOLCANIC ARC, TIMOR, AND NORTHWESTERN SHELF OF
M	25	411	BANDED IRON FORMATIONS OR BIFS, WESTERN AUSTRALIA, ARCHEAN
M	25	411	BANDED IRON FORMATIONS, WESTERN AUSTRALIA, PROTEROZOIC
B	6301	35	BANDERA COUNTY, CRETACEOUS, TEXAS
B	6109	1555*	BANDERA COUNTY, TEXAS, ABST., DIAGENESIS AND, GEOCHEMISTRY OF,
B	6204	688	BANDERAS BAY, KEROGEN, CALIFORNIA
S	7	220	BANDING, WEST FLORIDA SLOPE
M	25	358	BANDUNG VOLCANIC AREA, INDONESIA
B	6304	647†	BANE DOME--STRUCTURAL REEVALUATION OF ALLEGHENY FRONTAL
B	6303	414	BANE, S. C., GEOLOGY AND DEVELOPMENT OF TEAK OIL FIELD,
M	30	387	BANE, S. C., GEOLOGY AND DEVELOPMENT OF THE TEAK OIL FIELD,
B	6402	209	BANERJEE, A., SOURCE AREAS AND MIGRATION TRENDS OF OIL AND GAS
B	6309	1547	BANERJEE, BARR, F. T., REVIEW OF, BIBLIOGRAPHY OF THE GEOLOGY OF,
B	6004	673	BANERJEE, R. C., SIMULATED GEOTHERMAL MATURATION OF
B	6310	1819	BANGDALESH, 1978, BADRUDDOJA, M., AND, STEWART, R. D., ON
B	6206	986	BANGESTAN GROUP, IRAN
B	6110	1833	BANGLADESH, 1976, STEWART, R. D., ON DEVELOPMENTS IN
B	6210	1980	BANGLADESH, 1977, STEWART, R. D., ON DEVELOPMENTS IN
B	6411	1884	BANGLADESH, DEVELOPMENTS, 1979
B	6110	1848	BANGLADESH, PRODUCTION, 1975-1976
B	6210	1990	BANGLADESH, PRODUCTION, 1975-1977
B	6310	1829	BANGLADESH, PRODUCTION, 1977-1978
B	6411	1896	BANGLADESH, PRODUCTION, 1977-1979
B	6004	720*	BANGOR LIMESTONE, MISSISSIPPIAN, NORTHERN
B	6204	646	BANGOR SECTION, ALABAMA
S	4	242	BANK, BUILDUP, OR MOUND, CONCEPT OF
B	6004	688*	BANK COMPLEX EDWARDS LIMESTONE, NORTH-CENTRAL TEXAS, ABST.,
B	6109	1553*	BANK COMPLEX, EDWARDS LIMESTONE, NORTH-CENTRAL TEXAS, ABST.,
B	6205	772	BANK, DEFINITION
B	6004	732*	BANK MARGINS--RESULTS USING MANNED SUBMERSIBLE, ABST.,
B	6203	503*	BANK, SLOPE, AND BASINAL FACIES, ORDOVICIAN TRENTON-UTICA
B	6303	427*	BANK-BARRIER REEF MORPHOGENESIS, ST. CROIX, UNITED STATES
B	6103	376†	BANK-MARGIN GROWTH AND STRUCTURE, LITTLE BAHAMA BANK,
B	6101	3	BANKS BASIN
B	6203	557*	BANKS IN LOWER PROTEROZOIC BELCHER GROUP, HUDSON BAY,
B	6206	1005	BANKS ISLAND, ARCTIC, CANADA
B	6004	697*	BANKS ISLAND, ARCTIC CANADA, ABST., TERTIARY, DELTAIC, CYCLES ON
B	6209	1753	BANKS, JOSEPH E., SOUTHERN FLORIDA SUBSURFACE FEATURES
B	6004	713*	BANKS OF TEXAS OUTER CONTINENTAL SHELF, ABST., CORAL, REEFS
M	26	185	BANKS, SEISMIC, REFLECTIONS, RECOGNITION
B	6303	414	BANNER, F. T., DISTRIBUTION AND FACTORS CONTROLLING
B	6205	882*	BANNOCK AND CARIBOU COUNTIES, IDAHO, ABST., GENERAL GEOLOGY
B	6301	90	BANQUEREAU FORMATION, SCOTIAN BASIN
B	6402	277	BAPTISTE COLLETTE CREVASSE, LOUISIANA
B	6105	827*	BAR MORPHOLOGY IN BRAIDED-TO-MEANDERING TRANSITION ZONE—
B	6203	552*	BAR MORPHOLOGY IN EASTERN NICARAGUA, ABST., EFFECTS OF
B	6006	907†	BAR PROGRADATION, SAN ANDRES FORMATION, MIDLAND BASIN,
B	6108	1393*	BAR SANDSTONES OF FRONTIER FORMATION, SPEARHEAD RANCH FIELD,
B	6104	519†	BAR SEAWARD OF OOID SHOAL COMPLEX NEAR MIAMI, FLORIDA,
B	6004	675*	BAR SEDIMENTATION AND EQUILIBRIUM, KOUCHIBOUGUAC BAY, NEW
B	6204	644†	BAR SEDIMENTATION IN SOUTHERN CUMBERLAND PLATEAU,
B	6211	2330	BARABOO QUARTZITE, WISCONSIN
B	6211	2330	BARABOO RANGE, WISCONSIN
M	30	431	BARAM DELTA, BARONIA FIELD, SARAWAK
B	6010	1643	BARBADOS, 1975, HICKAM, E. E., ON DEVELOPMENTS IN
B	6110	1580	BARBADOS, 1976, PETROCONSULTANTS S.A., ON DEVELOPMENTS IN
B	6210	2046	BARBADOS, 1977, UMIKER, R., ON DEVELOPMENTS IN
B	6310	1747	BARBADOS, 1978, UMIKER, R., ON DEVELOPMENTS IN
B	6007	1094	BARBADOS CENTRAL RIDGE

S 4 235 BARBADOS, CTENANTEDON KINZIEI MEYER
B 6007 1094 BARBADOS FRONTAL HILLS ZONE
S 4 175 BARBADOS, MONTASTREA ANNULARIS
B 6107 1077 BARBADOS, NORTH POINT, MICROGRAPHS
S 4 236 BARBADOS, PERICLIMENES OR ANALCIDOMETRA ARMATA POURTALES
B 6404 472 BARBADOS, PLEISTOCENE, VADOSE WHISKER CALCITE
B 6110 1593 BARBADOS, PRODUCTION, 1973-1976
B 6210 2062 BARBADOS, PRODUCTION, 1973-1977
B 6310 1764 BARBADOS, PRODUCTION, 1974-1978
B 6007 1101 BARBADOS RIDGE COMPLEX AND ADJACENT ATLANTIC SEA FLOOR,
B 6007 1100 BARBADOS RIDGE COMPLEX AND ADJACENT ATLANTIC SEA FLOOR,
B 6007 1089 BARBADOS RIDGE COMPLEX, BATHYMETRY
B 6007 1099 BARBADOS RIDGE COMPLEX, GRAVITY
B 6007 1096 BARBADOS RIDGE COMPLEX, SEISMIC-REFLECTION DATA
B 6007 1078† BARBADOS RIDGE COMPLEX, TECTONICS OF SOUTHWESTERN, NORTH,
B 6107 1075 BARBADOS, RIVER BAY, MICROGRAPHS
B 6007 1094 BARBADOS TRANSVERSE VALLEY, STA. LUCIA-
S 4 236 BARBADOS, TROPIOMETRA CARINATA LAMARCK
B 6007 1094 BARBADOS TROUGH PROVINCE
B 6302 249 BARBER, A. J., GEOSYNCLINES AND PLATE TECTONICS IN BANDA ARCS,
S 4 307 BARBUDA, CODRINGTON LAGOON
S 4 303 BARBUDA, CODRINGTON LIMESTONE
S 4 302 BARBUDA, HIGHLAND LIMESTONE
S 4 304 BARBUDA, HOLOCENE SEDIMENTS, FACIES
S 4 300 BARBUDA, PLEISTOCENE
S 4 301 BARBUDA, REEF, WEST INDIES
S 4 299† BARBUDA, WEST INDIES, FACIES ANALYSIS OF, HOLOCENE, CARBONATE
B 6102 207 BARCENAS P., RODOLFO, ORIGIN AND DISTRIBUTION OF TERTIARY
B 6106 875 BARDAWIL LAGOON, EGYPT
B 6407 1036 BARDSDALE FIELD, CALIFORNIA
B 6312 2193† BARENTS SHELF, SANDSTONE IN LOWER, CRETACEOUS, HELVETIAFJELLET
B 6004 647 BARIA, LAWRENCE R., MIDDLE GUADALUPIAN COASTAL CARBONATE
M 25 496 BARITE AND FLUORITE, CANADIAN, CORDILLERA
M 28 128 BARITE, IN LIMESTONE, BIREFRINGENCE
M 28 128 BARITE, IN SANDSTONE, CEMENTS
B 6112 2146 BARKER, COLIN, AQUEOUS SOLUBILITY OF PETROLEUM AS APPLIED TO
B 6004 647 BARKER, COLIN, DEEP OR SHALLOW PETROLEUM GENERATION. ATTEMPT
B 6311 2115 BARKER, COLIN, DISTRIBUTION OF BITUMENS AND KEROGEN IN SHALE
B 6105 763 BARKER, COLIN, FLOOR FOR GAS
B 6311 2118 BARKER, COLIN, GEOCHEMISTRY OF SMALL LACUSTRINE DELTA, GREAT
B 6303 414 BARKER, COLIN, NATURAL GAS STABILITY IN DEEP SUBSURFACE, ABST.
B 6203 493 BARKER, COLIN, PLATE TECTONICS, ORGANIC MATTER, AND BASIN
S 10 19 BARKER, COLIN, PRIMARY MIGRATION: THE IMPORTANCE OF
B 6303 414 BARKER, COLIN, ROLE OF TEMPERATURE AND BURIAL DEPTH I
B 6203 493 BARKER, COLIN, WATER-ROCK-ORGANIC MATTER INTERACTIONS AT
M 24 52 BARN MOUNTAINS, ALASKA
B 6304 686 BARNARD, RALSTON W., DIRECT MEASUREMENT OF URANIUM BY
B 6102 154 BARNARD THRUST FAULT, CALIFORNIA
B 6102 146 BARNARD THRUST FAULT, CALIFORNIA
B 6102 189 BARNARD THRUST FAULT, CALIFORNIA
B 6405 738 BARNES, CHRISTOPHER R., APPLICATION OF CONODONT AND
B 6012 2174 BARNES, DAVID F., GRAVITATIONAL EVIDENCE FOR MORE ALASKAN
B 6103 464 BARNES. DAVID. GRAVITATIONAL AND OTHER GEOPHYSICAL EVIDENCE
B 6309 1589 BARNES, H. L., HYDROTHERMAL AND UPLIFT HISTORIES OF NORTHERN
M 25 149 BARNES, IVAN, POTENTIAL FOR GEOTHERMAL-ENERGY DEVELOPMENT
B 6207 1208 BARNES, IVAN, PROVENANCE AND GEOCHEMISTRY OF CARBON
B 6203 494 BARNES, JACK R., DETERMINATION OF RESERVOIR FLUID CONTENT BY
B 6105 764 BARNES, PETER, RATES OF ICE GOUGING AND SEDIMENT REWORKING,
B 6003 389 BARNES, SYDNEY U., GEOLOGY AND OIL PROSPECTS OF SOMALIA, EAST
B 6305 833 BARNUM, B. E., WRENCH FAULTING AND HYDROCARBON OCCURRENCE
M 30 431 BARONIA FIELD, SARAWAK, BARAM DELTA
B 6309 1547 BARR, F. T., REVIEW OF BIBLIOGRAPHY OF THE GEOLOGY OF LIBYA BY J.
B 6007 1086 BARRACUDA ABYSSAL PLAIN, BATHYMETRIC MAP
B 6007 1078 BARRACUDA FAULT ZONE
B 6007 1080 BARRACUDA RIDGE, BATHYMETRIC PROFILE
B 6109 1471 BARREIRINHAS BASIN, BRAZIL
M 26 214 BARREIRINHAS BASIN, BRAZIL
B 6405 674 BARRELL, STEVEN S., GRADED BEDS AS PALEOGRAVIMETRIC CAMERAS,
B 6205 879 BARRELL, STEVEN SOLOMON, ANALYSIS OF EFFECT ON GRADED BEDS OF
B 6204 669 BARREMIAN, ALGERIA
B 6405 610 BARREMIAN, MIDDLE EAST
S 4 87 BARREMIAN-APTIAN, JAMAICA, DEVILS RACE COURSE FORMATION
B 6203 374 BARREN ISLANDS, ALASKA
B 6108 1386 BARRETT, JAMES K., USE OF HYDROGEOLOGY, GEOCHEMISTRY, AND
B 6104 519† BARRIER BAR SEAWARD OF OOID SHOAL COMPLEX NEAR MIAMI,
B 6211 2363* BARRIER BEACH-SALINE LAGOON COMPLEX, UPPER SESPE CREEK,
B 6303 439* BARRIER COMPLEXES FROM HOLOCENE MODELS SOUTHEASTERN
B 6212 2381 BARRIER, DEPOSITIONAL ENVIRONMENT
B 6303 554* BARRIER ENVIRONMENTS, ABST., STRATIGRAPHIC MODELS FOR MODERN
B 6203 500* BARRIER ISLAND, ABST., STRATIGRAPHY AND GEOLOGIC HISTORY OF
B 6303 518* BARRIER ISLAND, CAPE ROMAIN, SOUTH CAROLINA, ABST., HOLOCENE
B 6004 711* BARRIER ISLAND COMPLEX IN LOWER SUNDANCE FORMATION OF
B 6303 499* BARRIER ISLAND--KIAWAH ISLAND, SOUTH CAROLINA, ABST., LATE
B 6303 486* BARRIER ISLANDS, ABST., INTERACTION BETWEEN OVERWASH AND
B 6309 1610* BARRIER ISLANDS AND INFLUENCE OF DROWNED VALLEYS ON TIDAL
B 6303 438* BARRIER ISLANDS ON WEST-CENTRAL PENINSULA OF FLORIDA, ABST.,
B 6312 2145† BARRIER, LATERAL AND, VERTICAL, FACIES RELATIONS OF,
B 6410 1644 BARRIER RANGES, AUSTRALIA
B 6406 814† BARRIER REEF OF CANNING BASIN, WESTERN AUSTRALIA, DEVONIAN,
B 6205 783 BARRIER REEF TREND, ALBERTA, LEDUC-RIMBEY
B 6203 556* BARRIER SEDIMENTATION ON TIDE-DOMINATED COASTLINE OF SOUTH
B 6303 415* BARRIER SEQUENCES OF EASTERN KENTUCKY, ABST.,
B 6209 1762* BARRIER TRENDS AND ORIGINS OF LAKE PONTCHARTRAIN, ABST., NEW
M 26 480 BARRIER-BAR SANDSTONE, SEISMIC, MODELING
B 6211 2275† BARRIER-ISLAND COMPLEX—LOWER SUNDANCE FORMATION OF BLACK
B 6004 673* BARRIER-ISLAND FEATURES, ABST., BERMUDAS NORTHWEST, REEFS
B 6303 489* BARRIER-REEF COMPLLX--EVIDENCE OF EXTENSIVE SUBMARINE
B 6303 492* BARRIERS, ABST., DETERMINATION OF STORM OVERWASH PERIODICITY
B 6203 550* BARRIERS OF BEAUFORT SEA COAST, ABST., TRANSGRESSIVE
B 6312 2160 BARRIERS, ORIGIN, TRANSGRESSIVE, WASHOVER
B 6211 2349 BARRINGER, A. R., REMOTE SENSING FOR ENERGY RESOURCES, ABST.
B 6004 648 BARRON, E. J., SOURCE AND DYNAMICS OF SUSPENDED SEDIMENTS IN
B 6012 2175 BARRON, JOHN A., MIDDLE MIOCENE-EARLY PLIOCENE MARINE DIATO
M 24 26 BARROW ARCH, ALASKA
B 6109 1494 BARROW ARCH, ALASKA
B 6201 50 BARROW FORMATION, AUSTRALIA
B 6405 737* BARROW GAS FIELDS, NORTH SLOPE, ALASKA, ABST.
B 6203 457 BARROW GROUP, AUSTRALIA
B 6203 457 BARROW ISLAND OIL FIELD, SECTION, AUSTRALIA
B 6203 455† BARROW ISLAND, WESTERN AUSTRALIA, ISOTOPIC EVIDENCE FOR,
B 6206 1008 BARROW SEGMENT, ARCTIC, CANADA
B 6201 41 BARROW SUBBASIN, AUSTRALIA
B 6203 457 BARROW SUBBASIN, WELLS, AUSTRALIA
B 6405 692 BARROWS, M. H., RELATION OF DEPOSITIONAL FACIES AND HISTORY TO
B 6309 1575 BARROWS, MARY H., PETROLOGY AND MATURATION OF DISPERSED
B 6409 1565* BARS ALONG UNITED STATES GULF COAST, ABST., NEARSHORE
B 6004 676* BARS, NUECES RIVER, TEXAS, ABST., BED FORMS AND STRATIFICATION
M 28 141 BARSTOW FORMATION, CALIFORNIA, MIOCENE
B 6207 1215* BARSTOW POWER TOWER, CALIFORNIA, ABST., SOLAR, ENERGY—,
B 6007 1009 BARTLESVILLE, BLUEJACKET, SANDSTONE, OKLAHOMA
B 6311 2118* BARTLESVILLE SANDSTONE, SALLYARDS FIELD, GREENWOOD COUNTY,
B 6103 374 BARTLESVILLE-BLUEJACKET SANDSTONE, OKLAHOMA
B 6212 2481 BARTLETT AND BENT CRUISES, 1972, USNS
M 25 462 BARTLETT FAULT, HUTANOPAN OR, SUMATRA-, CLIPPERTON-,
B 6105 821 BARTOK, P., CRETACEOUS COGOLLO GROUP, MARACAIBO PLATFORM,
B 6303 482 BARTON, CHRISTOPHER C., FRACTOGRAPHIC DISTINCTION OF
B 6303 482 BARTON, CHRISTOPHER C., NATURE AND FIELD APPLICATION OF
S 6 220 BARTONIAN STAGE DEFINED, EUROPE, EOCENE
B 6105 797 BARWIS, JOHN H., DISTRIBUTION OF SEDIMENTARY STRUCTURES IN EBB—
B 6303 415 BARWIS, JOHN H., MODERN AND ANCIENT EXAMPLES OF ANNELID
B 6303 415 BARWIS, JOHN H., PALEOTIDAL-RANGE INDICATORS IN CARBONIFEROUS
B 6105 764 BARWIS, JOHN H., RELATIVE STABILITIES OF TIDAL-INLET COMPONENT
B 6105 764 BARWIS, JOHN J., STORM-GENERATED SHELL RIDGE ON OPEN LAGOONAL
B 6306 969 BASAL CLASTIC UNIT, EGYPT
B 6107 1031 BASAL PENNSYLVANIAN SAND, PRODUCTION, NEBRASKA
B 6303 344 BASAL SAND, KENTUCKY
B 6405 784* BASALT INTRUSIONS ON KEROGENS IN UNCONSOLIDATED OCEANIC
M 28 69 BASALT WEATHERING, WASHINGTON, TERTIARY
M 25 351† BASALTIC ISLANDS--SUMMARY, DEVELOPING, DIKE-COMPLEX AND
M 25 526 BASALTS AND HAWAIITES, BAJA CALIFORNIA, HAWAII-TYPE
B 6105 828 BASAN, P. B., RHIZOCORALLIUM IN CRETACEOUS DAKOTA SANDSTONE,
B 6305 817 BASAN, PERKINS, BOB F., REVIEW OF, TRACE FOSSIL CONCEPTS, EDITED
M 25 437 BASAY COPPER PROJECT, NEGROS
B 6309 1602 BASCIANO, J. M., HYDROLOGY AND WATER QUALITY OF EOCENE WILCOX
B 6004 671 BASCLE, B., DELTAIC SUBENVIRONMENTS AND PALEOCURRENT ANALYSIS
B 6105 835* BASE MAPS IN EXPLORATION PROCESS, ABST., USE OF,
M 25 485 BASE METALS, COLORADO, PRECIOUS AND
M 25 485 BASE METALS, WASHINGTON, PRECIOUS AND
M 26 120 BASE-DISCORDANT, SEISMIC, REFLECTIONS, PATTERNS
M 25 449 BASE-METAL CONCENTRATION
S 5 181* BASE-METAL CONCENTRATION IN A DENSITY-STRATIFIED EVAPORITE
M 25 533 BASE-METAL DEPOSITS, BOLIVIA
M 25 427 BASE-METAL DEPOSITS, DISCOVERY AND POTENTIAL, AUSTRALIA
M 25 457 BASE-METAL DEPOSITS, DISTRICTS, MAP, JAPAN
M 25 457 BASE-METAL DEPOSITS, EXPLORATION PROJECTS, JAPAN

Index

M 25 456 BASE-METAL DEPOSITS, FUTURE EXPLORATION, JAPAN
M 25 456 BASE-METAL DEPOSITS, GEOLOGIC SURVEY PROJECT, JAPAN
M 25 428 BASE-METAL DEPOSITS, GEOLOGY, AUSTRALIA
M 25 533 BASE-METAL DEPOSITS, POLYMETALLIC, ANDEAN REGION
M 25 426 BASE-METAL DEPOSITS, PRODUCTION AND RES, AUSTRALIA,
M 25 533 BASE-METAL DEPOSITS, PRODUCTION, PERU
M 25 554 BASE-METAL SULFIDES, SURVEYS, ECUADOR, GOLD AND
M 26 57 BASELAP, ONLAP AND DOWNLAP, DEPOSITIONAL SEQUENCES
B 6209 1769* BASEMENT, ABST., REEVALUATING, FLORIDA
B 6405 786* BASEMENT, AND THE DRILL, ABST., BASINS
B 6106 912 BASEMENT, BAJA CALIFORNIA, FRANCISCAN
B 6205 794 BASEMENT, BLAKE PLATEAU AREA
M 25 322 BASEMENT, COLOMBIA
B 6112 2144 BASEMENT, COMPUTER-PROGRAM PACKAGES, SYMAP AND, SYMVU,
B 6205 856 BASEMENT CONTOUR MAP, TEXAS, PANHANDLE AND, OKLAHOMA,
B 6205 855 BASEMENT CONTOUR MAP, WEST, TEXAS AND, NEW MEXICO
B 6206 985 BASEMENT, IRAN, ZAGROS MOUNTAINS
M 29 231 BASEMENT MELANGE, SUMATRA, SUNDA ARC
B 6303 498* BASEMENT MOBILITY IN GULF OF MEXICO, ABST., SPECULATIONS ON
B 6201 3 BASEMENT, NIGER DELTA
B 6205 793 BASEMENT, PROFILE, BLAKE PLATEAU AREA, VOLCANIC
B 6002 199 BASEMENT RIDGE, BRAZIL COASTAL MARGIN
B 6002 177 BASEMENT ROCK, BERINGIAN SHELF AND MARGIN, MESOZOIC
M 25 451 BASEMENT ROCKS, JAPAN
B 6402 189 BASEMENT ROCKS, SPAIN, NEOGENE SEDIMENTARY AND OLDER
B 6206 1033 BASEMENT STRUCTURE MAP, BRAZIL, SOUTHERN
B 6410 1658† BASEMENT STRUCTURE OF WESTERN ATLANTIC MAGNETIC QUIET ZONE,
B 6407 1007 BASEMENT STRUCTURES, TEXAS, LOUISIANA, AND, ARKANSAS
B 6404 511 BASEMENT TECTONIC MAP, NEW ENGLAND, PASSIVE MARGIN
B 6404 508 BASEMENT TECTONICS, NEW ENGLAND
B 6405 710* BASEMENT TECTONICS OF CHINA—CONTINENTAL-SCALE CATACLASTIC
B 6303 550* BASEMENT TECTONICS ON DEPOSITIONAL SYSTEMS AND SEISMIC
B 6102 269† BASEMENT TERRANE IN SIERRA DE OMOA. DISCUSSION AND REPLY,
B 6004 566† BASEMENT TERRANEIN SIERRA DE OMOA, PRE-CRETACEOUS ROCKS OF
B 6307 1026 BASEMENT THRUSTS, COMPRESSIVE, FAULT BLOCKS AND
B 6307 1035 BASEMENT WARPS--ARCHES, DOMES, AND SAGS
B 6212 2489† BASEMENT-BLOCK MOTIONS IN CAMBAY BASIN, INDIA, AND ITS
B 6012 2081† BASEMENT-CONTROLLED DEFORMATION IN SAN ANDREAS FAULT ZONE,
B 6307 1018 BASEMENT-INVOLVED STRUCTURES WRENCH-FAULT ASSEMBLAGES
B 6112 2143† BASEMENT—AN ISOMETRIC VIEW, FLORIDA
B 6205 784 BASHAW COMPLEX, ALBERTA
S 6 324 BASHKIRIAN-MOSCOVIAN BOUNDARY, CARBONIFEROUS
B 6008 1399* BASI AND RANGE PROVINCE, NEVADA AND WESTERN UTAH, ABST., NEW
B 6105 796 * BASIN ANALYSIS OF LOWER MIDDLE MIOCENE STRAT, SANTA ROSA AND
S 10 70 BASIN COMPACTION, DEVELOPMENT MODEL
S 10 125 BASIN GEOMETRY, AQUITARD, EXTENSIVE
S 10 124 BASIN GEOMETRY, DISTRIBUTION OF HYDRAULIC FORCES
S 10 126 BASIN GEOMETRY, HOMOGENEOUS ROCK FRAMEWORK
S 10 127 BASIN GEOMETRY, PARTIAL AQUITARD
S 10 61 BASIN TEMPERATURES, MIGRATION
B 6004 711* BASIN, ABST., SEDIMENTARY, CYCLES IN, VIRGILIAN, UPPER,
B 6203 501* BASIN, ABST., STRUCTURE AND EARLY GEOLOGIC HISTORY OF DEEP
B 6303 525* BASIN ANALYSIS AND EVOLUTION OF DEEP-MARINE SHALE BASIN,
B 6303 465* BASIN ANALYSIS, AND EXPLORATION, ABST., PLATE TECTONICS
M 26 215 BASIN ANALYSIS, DEPOSITIONAL SYSTEMS
M 25 492 BASIN ANALYSIS IN, MINERAL, EXPLORATION, USE OF
B 6405 699* BASIN ANALYSIS—HUMBOLDT BASIN OF CALIFORNIA AS EXAMPLE, ABST.,
M 29 113† BASIN AND CONTINENT-OCEAN CRUST BOUNDARY OFF BAHIA BLANCA,
S 3 79 BASIN AND RANGE PROVINCE
B 6405 799* BASIN AND RANGE PROVINCE ABST., STRATEGY OF EXPLORATION FOR
B 6405 703* BASIN AND RANGE PROVINCE NORTHERN NEVADA, ABST., TERTIARY
B 6405 707* BASIN AND RANGE PROVINCE, WESTERN UNITED STATES, ABST.,
B 6304 621† BASIN AND RANGE PROVINCE, YAVAPAI COUNTY, ARIZONA, MAJOR,
S 4 329† BASIN AND SHELF MODEL, ST. CROIX, U.S VIRGIN ISLANDS, MIOCENE,
S 7 303† BASIN, ANOXIC DEPRESSION ON THE CONTINENTAL SLOPE, NORTHWEST
B 6007 1073† BASIN, APPROXIMATE, GEOTHERMAL, GRADIENTS IN, NIGER, DELTA
S 7 313† BASIN AREAS, NORTHWEST GULF OF MEXICO, HOLOCENE, SEDIMENT,
B 6203 503* BASIN, BRAZIL, ABST., PETROLEUM GEOLOGY OF, PALEOZOIC, CLASTIC
B 6004 705* BASIN, CANTABRIAN MOUNTAINS, NORTH SPAIN, ABST., SEDIMENTATION
B 6306 887 BASIN DAKOTA GAS FIELD, NEW MEXICO
B 6404 550 BASIN DEVELOPMENT MODEL
M 25 282 BASIN DISTRIBUTION, NORTH AMERICA, WESTERN
B 6108 1392* BASIN EVALUATION, ABST., APPLICATION OF, DRILLING, STATISTICS IN
B 6105 833* BASIN EVALUATION, ABST., APPLICATION OF, DRILLING STATISTICS IN
B 6203 493* BASIN EVALUATION FOR PETROLEUM POTENTIAL, ABST., PLATE
B 6305 824* BASIN EVOLUTION OF LATE PALEOZOIC TAOS TROUGH, NORTHERN NEW
B 6004 659* BASIN FILL--NAMURIA OF CENTRAL PENNINE BASIN, ENGLAND, ABST.,
B 6406 854 BASIN FLOORS, RESERVOIR POTENTIAL, FLAT
B 6004 709* BASIN FORMATION. BY TITLE ONLY, INDONESIA-, SOUTH, SUMATRA AND,
B 6203 494* BASIN, FRIO FORMATION, BRAZORIA COUNTY, TEXAS, ABST., DELTAIC
S 7 289† BASIN IN NORTHWEST GULF OF MEXICO, INTRASLOPE
M 29 373† BASIN IN SOUTHEAST ASIA, THE, SULU SEA: A, MARGINAL
B 6004 734* BASIN IN SOUTHERN SOUTH AMERICA, ABST., SUBMARINE-, FAN
B 6203 442† BASIN, INDIA, EARLY, EOCENE SUBAERIAL EROSIONAL VALLEYS IN,
B 6402 209† BASIN, INDIA, SOURCE AREAS AND, MIGRATION TRENDS OF, OIL AND,
M 25 497 BASIN, MINERAL DEPOSITS, CANADIAN, CORDILLERA, OCEANIC
B 6206 946 BASIN PLAIN, SUBMARINE FAN
M 26 194 BASIN PLATFORM, SEISMIC EXAMPLE, NEW MEXICO, LEA COUNTY,
B 6303 297 BASIN REGIONS, PROSPECTS, SMALL, OCEAN
M 25 497 BASIN, ROCKS OF, CANADIAN, CORDILLERA, OCEANIC
B 6002 230 BASIN SALINITIES
B 6402 145† BASIN, SEALING AND, NONSEALING, FAULTS IN, LOUISIANA, GULF
B 6004 675* BASIN SEDIMENTS, ABST., HIGH-RESOLUTION STUDIES OF,
B 6211 2308† BASIN, SILURIAN EVOLUTION OF, CENTRAL, APPALACHIAN
B 6104 558† BASIN, SOUTH AMERICA, REGIONALIZED, VARIABLES FOR EVALUATION
B 6307 1059† BASIN STRATA, CORNWALLIS AND GRIFFITH ISLANDS, NORTHWEST
M 29 85† BASIN STRUCTURE OF THE U.S. ATLANTIC MARGIN
B 6304 702* BASIN SUBSIDENCE, ABST., MECHANISMS OF
M 29 363† BASIN, TECTONIC DEVELOPMENT OF, TRENCH-ARC COMPLEXES ON THE
B 6002 231 BASIN TEMPERATURES, DEEP-
B 6004 554† BASIN, THICKNESS OF, REMOVED SEDIMENTARY ROCKS, PALEOPORE,
B 6004 665* BASIN--DETERMINATION BY FAUNAL CRITERIA, ABST., RELATIVE,
B 6205 890* BASIN-AND-RANGE, ABST., EOCENE STRUCTURING IN EAST-CENTRAL
S 5 29 BASIN-CENTER EVAPORITES, DEPOSITIONAL ENVIRONMENT
S 5 23* BASIN-CENTER EVAPORITES, LOWER SALINA GROUP (UPPER SILURIAN),
S 5 1* BASIN, DEPOSITIONAL ENVIRONMENTS OF PINNACLE REEFS, NIAGARA
B 6403 440* BASIN-EDGE TECTONICS AND STRATIGRAPHIC DEVELOPMENT,
M 26 175 BASIN-FLOOR FACIES, SEISMIC, INTERPRETATION
B 6211 2258 BASIN-MARGIN WEDGING
M 26 246 BASIN-MARGINS, DEPOSITIONAL SYSTEMS, PASSIVE
B 6311 2115* BASIN-SHELF BORDER IN NORTHEAST OKLAHOMA DURING
B 6009 1622* BASIN-SHELF RELATIONS DURING SILURIAN TIME IN WEST VIRGINIA,
B 6205 762 BASIN-SLOPE AND BASIN FLOOR SEISMIC FACIES
M 26 175 BASIN-SLOPE FACIES, SEISMIC, INTERPRETATION
B 6004 725* BASIN-TO-REEF SHOALING SEQUENCE IN DOGGER, MIDDLE JURASSIC, OF
B 6305 812† BASIN. DISCUSSION AND REPLY, THICKNESS OF, REMOVED SEDIMENTARY
B 6203 503* BASINAL FACIES, ORDOVICIAN TRENTON-UTICA SEQUENCE OF NEW
B 6004 697* BASINAL LITHOFACIES, ABST., DEVELOPMENT OF VERTICAL ESCARPMENT
B 6004 517† BASINAL SANDSTONE FACIES, DELAWARE BASIN, WEST TEXAS AND
S 10 69† BASINS, MIGRATION OF, HYDROCARBONS IN, COMPACTING
B 6303 459* BASINS, ABST., ANALYSIS OF HYDROCARBON DISTRIBUTION AND
B 6004 733* BASINS, ABST., CARBONATE, MOUNDS FROM, SHELVES AND
B 6109 1559* BASINS, ABST., EVALUATION OF, LANDSAT DATA IN EXPLORING,
B 6004 693* BASINS, ABST., FACTORS CONTROLLING, SUSPENDED SEDIMENT IN
B 6405 779* BASINS, ABST., GENERATION, MIGRATION, AND ENTRAPMENT OF
B 6012 2174† BASINS, ABST., GRAVITATIONAL EVIDENCE FOR MORE, ALASKAN
B 6405 732* BASINS, ABST., INTEGRATED GEOLOGIC AND GEOPHYSICAL STUDIES AS
B 6203 498* BASINS, ABST., MIGRATION OF, HYDROCARBONS IN, COMPACTING
B 6303 543* BASINS ABST., MODELS FOR EVOLUTION OF INTERIOR
B 6304 697* BASINS, ABST., PROCESS FOR, PRIMARY, MIGRATION OF PETROLEUM IN
B 6004 710* BASINS, ABST., STRUCTURE OF, NEOGENE, MIOCENE, LITTORAL HARD-
B 6405 704* BASINS, ABST., TRACE, FOSSILS AND STAGNATION OF, DEEP-SEA
B 6405 670* BASINS ALONG ATLANTIC CONTINENTAL MARGINS OF UNITED STATES,
B 6004 720* BASINS, ALTERNATIVE MODE OF FORMATION, ABST., DEVELOPMENT OF,
B 6004 732* BASINS AND BANK MARGINS--RESULTS USING MANNED SUBMERSIBLE,
M 25 217† BASINS AND PETROLEUM PROSPECTS, ONSHORE AND OFFSHORE NEW
B 6105 829* BASINS AND THEIR POSSIBLE ROLE IN FORMATION OF PETROLEUM AND
B 6412 2060 BASINS, ANOXIC AND OXIC, OCEANIC
B 6202 223† BASINS AROUND ATLANTIC MARGIN, ORIGIN AND SIGNIFICANCE OF,
B 6405 786* BASINS, BASEMENT, AND THE DRILL, ABST.
B 6002 165 BASINS, BERING SHELF
B 6002 177 BASINS, BERINGIAN SHELF AND, MARGIN, CENOZOIC
B 6002 197 BASINS, BRAZIL, EASTERN, MESOZOIC-, CENOZOIC, MARGIN
B 6406 954 BASINS, BY C. E. B. CONYBEARE, MACKENZIE, D. B., REVIEW OF,
B 6004 717* BASINS CARBON COUNTY, WYOMING, ABST., EOCENE PROVENANCE OF,
M 25 247 BASINS, CLASSIFICATION, INDONESIA
B 6002 197 BASINS, COASTAL, BRAZIL
B 6406 841† BASINS, COMPARISON OF, SAND-LAYER GEOMETRY ON, FLAT FLOORS OF
S 3 136 BASINS, CONCENTRATION IN, EVAPORITE
B 6408 1215 BASINS, EVALUATING, MATURE
B 6203 419† BASINS FLANKING NORTH ATLANTIC OCEAN AND THEIR RELATION TO
B 6112 2135 BASINS, FLUID POTENTIALS COMMON IN SEDIMENTARY
B 6203 434 BASINS, HYDROCARBONS, NORTH, ATLANTIC, RIFT
S 3 136 BASINS, HYPOGENE, LATEROGENESIS IN, GEOPRESSURIZED
M 29 293† BASINS IN THE MEDITERRANEAN, VARIETY OF, MARGINS AND DEEP

B 6203 536* BASINS IN WESTERN GULF OF MAINE, ABST., CARBONATE CATCHMENT
M 25 281† BASINS IN WESTERN NORTH AMERICA, STRUCTURAL EVOLUTION OF
B 6304 573† BASINS, MIGRATION OF, FLUIDS IN, SEDIMENTARY
B 6404 549† BASINS, MIGRATION OF, HYDROCARBONS IN, COMPACTING
M 25 500 BASINS, MINERAL DEPOSITS, CANADA, SUCCESSOR
M 25 493 BASINS, MODELS OF, CANADA, CORDILLERAN
M 25 212 BASINS, NEW ZEALAND
M 25 176† BASINS OF CIRCUM-PACIFIC BELT--SUMMARY, OIL- AND, GAS-BEARING
B 6405 727* BASINS OF EAST INDIA—TECTONO-STRATIGRAPHIC FACIES, ABST.
B 6405 807* BASINS OF EASTERN CHINA, ABST., TECTONIC FEATURES OF OIL AND GAS
M 25 239† BASINS OF INDONESIA, RECENT KNOWLEDGE OF, HYDROCARBON
B 6108 1392* BASINS OF PACIFIC MARGIN, ABST., WIDESPREAD OCCURRENCE OF,
M 25 250† BASINS OF SOUTHEAST ASIA--SUMMARY, STRUCTURAL EVOLUTION OF,
B 6408 1211 BASINS OF THE WORLD, EXPLORATORY STATUS
B 6408 1211 BASINS OF THE WORLD, PRODUCTIVE
B 6203 518* BASINS OF UNITED STATES, ABST., ANALYSIS OF, OIL-FIELD
B 6205 746 BASINS OF WORLD
B 6203 368 BASINS, ONSHORE AND OFFSHORE, CALIFORNIA, NEOGENE
B 6209 1593 BASINS, ORGANIC CONTENT, SMALL, OCEAN
M 25 286 BASINS, PRODUCING, NORTH AMERICA, WESTERN
B 6111 1957 BASINS, RESERVES, FORELAND
B 6111 1957 BASINS, RESERVES, RIFT
M 25 500 BASINS, ROCKS OF, CANADIAN, CORDILLERA, SUCCESSOR
M 26 213† BASINS, SEISMIC-, STRATIGRAPHIC, INTERPRETATION OF, DEPOSITIONAL,
M 25 441 BASINS, SOUTH KOREA, KYONGSANG, SEDIMENTARY
M 25 441 BASINS, SOUTH KOREA, TERTIARY
M 25 466 BASINS, SUPERIMPOSED OR SUCCESSOR, USSR, NORTHEAST
M 29 245† BASINS, THE, EVOLUTION OF, STRUCTURAL HIGHS BORDERING MAJOR,
B 6303 552* BASINS WEST OF UNITED KINGDOM AND EIRE, ABST., HYDROCARBON
B 6205 748 BASINS, WORLD, FRONTIER
B 6205 747 BASINS, WORLD, PRODUCTIVE
B 6004 709* BASINS--MECHANISM OF BASIN FORMATION. BY TITLE ONLY,
M 25 349† BASINS--SUMMARY, HYDROCARBON POTENTIAL OF, BOLIVIAN
B 6108 1372* BASINS-APPLICATION OF OLD IDEAS ON EXPANDED SCALE IN SOUTHERN
B 6303 531* BASINS. BY TITLE ONLY, PETROLEUM ENGINEERING--GEOLOGY,
B 6409 1538† BASINS. DISCUSSION AND REPLY, MIGRATION OF, FLUIDS IN,
B 6203 555* BASINS—POSSIBLE SIGNIFICANT NEW ENERGY SOURCE, ABST., CRUDE OIL
B 6004 727* BASINWARD FACIES CHANGES IN WAPANUCKA LIMESTONE, LOWER
B 6405 806* BASIN—NEW APPROACH, ABST., THREE-DIMENSIONAL SIMULATION OF
B 6203 573* BASIN—QUANTITATIVE APPROACH, ABST., SIMULATION OF GEOLOGIC,
B 6212 2509 BASOEKI, M., RECENT DEVELOPMENT OF KAMOJANG GEOTHERMAL
M 25 230† BASS, GIPPSLAND, AND OTWAY BASINS--SUMMARY, STRUCTURAL STYLE
B 6211 2309 BASS ISLANDS FORMATION, WEST VIRGINIA
B 6402 219 BASSEIN FIELD, INDIA
B 6209 1769* BASSFIELD, JEFFERSON DAVIS COUNTY, MISSISSIPPI, ABST.,
B 6004 735 BASU, A., PALEOCLIMATE INTERPRETATION THROUGH USE OF
B 6203 494 BASU, ABHIJIT, TRACE ELEMENTS IN HOLOCENE FLUVIAL SANDS
B 6402 209 BASU, D. N., SOURCE AREAS AND MIGRATION TRENDS OF OIL AND GAS
M 30 588 BATANGA FORMATION, EWONGUE SHALE, GABON BASIN
M 30 588 BATANGA FORMATION, POINTE CLAIRETTE SHALES, GABON BASIN
M 30 577 BATANGA SANDSTONE, CRETACEOUS, MAASTRICHTIAN
M 30 579 BATANGA SANDSTONE, PETROLEUM, GABON BASIN
B 6207 1131 BATANTA ISLAND, INDONESIA
B 6207 1132 BATANTA ISLAND, KAIS LIMESTONE
B 6207 1132 BATANTA ISLAND, PLIO-PLEISTOCENE CONGLOMERATE FILL
B 6307 1144 BATE AND ERIC ROBINSON, LUNDIN, ROBERT F., REVIEW OF,
B 6207 1194 BATEMAN FORMATION, OREGON AND, WASHINGTON
B 6205 890* BATHOLITH, CALIFORNIA, ABST., FOLD SYSTEM IN SOUTHERN, SIERRA
M 25 449 BATHOLITHS AND MINERAL ZONING, JAPAN, MESOZOIC, LATE, GRANITIC
M 26 158 BATHONIAN, AFRICA, SEQUENCES, JURASSIC
B 6405 608 BATHONIAN, MIDDLE EAST
B 6209 1754* BATHYAL ENVIRONMENTS, ABST., PALEOSTRUCTURAL, MAPPING IN
B 6105 835* BATHYAL SANDSTONES OF CRETACEOUS GREAT VALLEY SEQUENCE,
B 6403 302 BATHYMETRIC AND SEISMIC PROFILES, CONTINENTAL, RISES AND,
B 6203 511* BATHYMETRIC CONTROL OF DEEP-SEA ICHNOFACIES, ABST.
B 6303 476* BATHYMETRIC DISTRIBUTION OF TRACE FOSSILS IN UPPER
B 6007 1050 BATHYMETRIC, FREE-AIR GRAVITY AND MAGNETIC PROFILES,
B 6204 658 BATHYMETRIC MAP, BRAZIL, SAO PAULO PLATEAU
S 7 197 BATHYMETRIC MAP, GULF OF MEXICO
B 6007 1022 BATHYMETRIC MAP, WESTERN, LAKE SUPERIOR
M 26 374 BATHYMETRIC MAPPING, RESOLUTION SYSTEMS
S 7 197 BATHYMETRY AND CORE HOLES, WEST FLORIDA SLOPE
S 7 157 BATHYMETRY AND PHYSIOGRAPHY, EASTERN GULF OF MEXICO
B 6007 1085 BATHYMETRY, ATLANTIC BASIN
B 6106 862 BATHYMETRY, BALTIMORE CANYON TROUGH AREA
B 6007 1089 BATHYMETRY, BARBADOS RIDGE COMPLEX
B 6306 936 BATHYMETRY, MAP, AMAZON, CONTINENTAL, MARGIN
S 7 6 BATHYMETRY MAP, GULF OF MEXICO
S 7 290 BATHYMETRY MAP, GULF OF MEXICO, GYRE BASIN
S 7 304 BATHYMETRY MAP, GULF OF MEXICO, ORCA BASIN
S 7 163 BATHYMETRY MAP, MISSISSIPPI TROUGH
S 7 322 BATHYMETRY MAP, ORCA BASIN
B 6007 1079 BATHYMETRY, MID-ATLANTIC RIDGE
B 6007 1039 BATHYMETRY, NORTHEAST, PROVIDENCE CHANNEL
B 6201 123 BATHYMETRY OF SOUTHERN STRAITS OF FLORIDA
M 29 444 BATHYMETRY PROFILE, PERU TRENCH
B 6202 249 BATHYMETRY, SOUTHERN, CALIFORNIA
S 7 244 BATHYMETRY, TEXAS, SHELF-SLOPE
S 11 268 BATHYSIPHON ARENACEA
S 11 227 BATHYSIPHON ARENACEA
B 6407 972 BATTLE FORMATION, ALBERTA
B 6212 2511* BAUER BASIN, ABST., EVIDENCE FOR METALLOGENESIS ON, OCEANIC,
M 25 542 BAUER DEEP
B 6405 745 BAUER, M. A., DEVELOPMENT OF CONCEPTUAL MODEL TO
B 6209 1760 BAUER, MARY, LIVING SHELLED MICROZOOPLANKTON, RADIOLARIANS,
B 6309 1597 BAUER, MARY, RELICT AND EXPATRIATE RADIOLARIAN FAUNA IN GULF
B 6105 783 BAUMGARDNER, ROBERT, JR., LANDSAT STUDY OF NEARSHORE
M 28 74 BAUXITE, ARKANSAS, CRETACEOUS
M 28 74 BAUXITE, ARKANSAS, TERTIARY
M 25 430 BAUXITE, BRITISH, SOLOMON ISLANDS
M 25 406 BAUXITE DEPOSITS, CONFINED TO TERTIARY LATERITES, AUSTRALIAN
M 25 407 BAUXITE DEPOSITS, LOCATIONS AND RESERVES, AUSTRALIA
M 25 408 BAUXITE DEPOSITS, TOPOGRAPHY, MINERAL COMPOSITION, AUSTRALIA
M 25 405† BAUXITE DEPOSITS--DISCOVERY AND DEVELOPMENT, AUSTRALIAN
M 25 406 BAUXITE, EXPLORATION AND EVALUATION, AUSTRALIA
M 25 405 BAUXITE, GEOLOGIC ENVIRONMENT
M 25 405 BAUXITE, HISTORY OF DISCOVERY, AUSTRALIA
M 28 74 BAUXITE, KAOLINITE, IN
B 6207 1235* BAUXITE MINING AND STREAM SALINITY IN DARLING RANGE, WESTERN
M 25 408 BAUXITE MINING, REGENERATION OF MINED AREAS, AUSTRALIA
M 25 408 BAUXITE MINING, SOCIAL BENEFITS, AUSTRALIA
S 9 5 BAUXITE, PALEOCLIMATE INDICATORS
M 25 405 BAUXITE, RESERVES AND GRADES, AUSTRALIA
M 25 406 BAUXITE RESERVES AND GRADES, TABLE, AUSTRALIAN
M 25 405 BAUXITE, RESERVES, AUSTRALIA
M 25 405 BAUXITE RESERVES, AUSTRALIA, DARLING RANGE
M 25 405 BAUXITE RESERVES, AUSTRALIA, GOVE
M 25 405 BAUXITE RESERVES, AUSTRALIA, WEIPA
M 25 406 BAUXITE, TYPE OF, AUSTRALIA
M 25 407 BAUXITE-ALUMINA-ALUMINUM INDUSTRY, EXPORT MARKETS,
B 6405 673* BAVARIAN ALPS, GERMANY, ABST., EXPLORATION IN CLASSIC, THRUST
B 6303 522 BAWIEC, W., PREDICTING FUTUR OIL USING THREE-DIMENSIONAL
M 24 92 BAXTER SHALE, WYOMING
B 6109 1425 BAXTERVILLE OIL, MISSISSIPPI
S 5 83 BAY COUNTY LOCATION, MICHIGAN,
B 6309 1599* BAY COUNTY, FLORIDA, ABST., SUBSURFACE, NEOGENE STRATIGRAPHY
M 24 231 BAY MARCHAND SALT-RIDGE TREND, LOUISIANA
B 6206 918 BAY OF BENGAL, BENGAL FAN
B 6111 1994 BAY OF BENGAL, POSSIBLE DIAPIRS
B 6211 2232† BAY OF BISCAY AND ASSOCIATED TURBIDITY CURRENTS, A, SIGSBEE,
B 6211 2233 BAY OF BISCAY AREA, CRETACEOUS
B 6211 2234 BAY OF BISCAY AREA, DANIAN
B 6211 2239 BAY OF BISCAY AREA, EOCENE
B 6211 2241 BAY OF BISCAY AREA, EVAPORITES
B 6211 2233 BAY OF BISCAY AREA, KEUPER
B 6211 2234 BAY OF BISCAY AREA, MAESTRICHTIAN
B 6211 2239 BAY OF BISCAY AREA, SECOND SANDSTONE UNIT
B 6211 2233 BAY OF BISCAY AREA, TERTIARY
B 6211 2233 BAY OF BISCAY AREA, TURONIAN
B 6211 2233 BAY OF BISCAY AREA, ZARAUZ DIAPIR
B 6412 2057 BAY OF BISCAY, DSDP SITE 402A, LEG 48, GEOCHEMICAL LOG
B 6411 1968 BAY OF BISCAY, SPAIN
S 7 57 BAY OF CAMPECHE SLOPE, DISCUSSED, GULF OF MEXICO
B 6004 698* BAY OF FUNDY, CANADA, ABST., ORIGIN AND DISTRIBUTION OF,
B 6109 1550 BAY, T. A., JR., SOME LOWER CRETACEOUS STRATIGRAPHIC MODELS
B 6309 1579 BAYER, KENNETH C., EASTERN PROJECTION O VALLEY AND RIDGE
B 6105 811 BAYER, KENNETH C., PETROLEUM POTENTIAL OF POSSIBLE LOWER
B 6008 1390 BAYLIFF, WILLIAM H., PERFORMANCE REVIEW OF TIGER RIDGE AND
B 6405 770 BAYLISS, G. S., DEPOSITIONAL ENVIRONMENTS AND KEROGEN TYPES. BY
B 6405 770 BAYLISS, G. S., FLUORESCENCE OF ACRITARCHS IN STUDY OF MARINE
B 6206 1092 BAYLISS, G. S., ORGANIC GEOCHEMISTRY OF GAS-BEARING SHALES, U.S.
B 6011 2067 BAYLISS, GEOFFREY S., PRINCIPLES OF HYDROCARBON GENERATION
B 6312 2147 BAYMOUTH BARRIER, DELAWARE
B 6101 65 BAYOU CARLIN AREA, GEOTHERMAL PATTERN, LOUISIANA
B 6101 70 BAYOU CARLIN FIELD, LOUISIANA
B 6101 65† BAYOU CARLIN--LAKE SAND AREA, SOUTH LOUISIANA--IMPLICATIONS,
B 6309 1600* BAYOU MIDDLE FORK FIELD, CLAIBORNE PARISH, LOUISIANA--CASE

B 6407 1007 BAYOU MIDDLEFORK FIELD, LOUISIANA
B 6402 165 BAYOU VILLARS FIELD, LOUISIANA
B 6410 1605 BAYS FORMATION, VIRGINIA
M 27 176 BAYS LIMESTONE, VIRGINIA, ORDOVICIAN
M 27 185 BAYS LIMESTONE, VIRGINIA, ORDOVICIAN
M 27 4 BAYS LIMESTONE, VIRGINIA, ORDOVICIAN
B 6302 174 BCF MAPS
B 6203 552* BEACH AND NEARSHORE-BAR MORPHOLOGY IN EASTERN NICARAGUA,
B 6410 1634 BEACH, DAVID K., FACIES SUCCESSION OF PLIOCENE-PLEISTOCENE
B 6303 460 BEACH, DAVID K., POROSITY PRESERVATION AND EARLY FRESHWATER
B 6203 494 BEACH, DAVID K., PRE-HOLOCENE CARBONATE ROCKS OF
B 6105 818* BEACH MORPHOLOGY OF NORTHERN ALASKA, ABST., VARIATIONS IN,
B 6009 1608* BEACH SEDIMENTOLOGY OF MUSTANG AND PADRE
M 28 58 BEACH SEDIMENTS, CONNECTICUT, HOLOCENE
M 28 59 BEACH SEDIMENTS, CONNECTICUT, HOLOCENE
B 6004 704* BEACH TARS WITHIN SOUTHERN CALIFORNIA BORDERLAND, ABST.,
B 6303 468* BEACH-TO-OFFSHORE SEQUENCE, VENTURA--PORT HUENEME AREA,
B 6303 450* BEACHES, ABST., VARIATIONS OF MORPHOLOGY AND SEDIMEN
B 6004 704* BEACHES IN MAGDALEN ISLANDS, GULF OF ST. LAWRENCE, CANADA,
B 6307 1101 BEAGLE RIDGE GEOTHERMAL ANOMALY, AUSTRALIA
M 25 233 BEAGLE SUBBASIN, AUSTRALIA
B 6105 765 **BEALES, FRANK W., COARSE WHITE SPARRY DOLOMITE AND CALCITE**
B 6303 416 BEALES, FRANK W., CRITERIA FOR RECOGNITION OF DIVERSE
B 6303 416 BEALL, A. O., JR., MODERN EXPLORATION STRATEGY IN FRONTIER
B 6202 190 BEAR CREEK AREA, CROSS SECTION, WYOMING
B 6309 1522 BEAR LAKE FORMATION, MIOCENE, ALASKA PENINSULA
S 5 113 BEAR LAKE TOWNSHIP, SEISMIC EXAMPLE, MICHIGAN, MANISTEE
S 11 200 BEAR RIVER AREA, BIOCHRONOLOGY, HUMBOLDT BASIN AND
S 11 193 BEAR RIVER AREA, BIOSTRATIGRAPHY
S 11 222 BEAR RIVER BEDS, FAUNULES
S 11 217 BEAR RIVER BEDS FORAMINIFERA
S 11 191 BEAR RIVER BEDS, HUMBOLDT BASIN
S 11 320 BEAR RIVER COASTAL SYNCLINE, SECTION DESCRIPTION
B 6101 114 BEAR RIVER FORMATON, WYOMING
S 11 187 BEAR RIVER SECTION, CALIFORNIA
S 11 194 BEAR RIVER STRATIGRAPHIC SECTION
S 11 309 BEAR RIVER VALLEY, CALIFORNIA, SAMPLE LOCALITIES
B 6109 1484 BEAR SEAMOUNT, EAST COAST
B 6306 989 BEAR, TED L., REVIEW OF COASTAL SOUTHERN CALIFORNIA, BY ROBERT
B 6101 111 BEARCE, DENNY N., HISTORY OF ALABAMA GEOLOGICAL SOCIETY
B 6008 1398* BEARPAW MOUNTAINS AREA, NORTH-CENTRAL MONTANA, ABST.,
B 6403 317 BEARPAW MOUNTAINS. MONTANA
B 6403 317 BEARPAW MOUNTAINS, MONTANA
B 6305 827* BEARPAW MOUNTAINS, NORTH-CENTRAL MONTANA, ABST.,
B 6407 972 BEARPAW MOUNTAINS, SECTION, MONTANA
B 6407 972 BEARPAW SHALE, ALBERTA
B 6407 971 BEARPAW SHALE, GREAT PLAINS
B 6403 318 BEARPAW SHALE, MONTANA
M 29 337 BEATA RIDGE, STRUCTURE, COLOMBIA
B 6303 487* BEATRIC FIELD, MORAY FIRTH, NORTH SEA, ABST.
M 30 127 BEATRICE FIELD, BAJOCIAN-BATHONIAN, STRATIGRAPHY
M 30 128 BEATRICE FIELD, CALLOVIAN, STRATIGRAPHY
M 30 125 BEATRICE FIELD, CORE SECTION WITH LOGS
M 30 119 BEATRICE FIELD, EXPLORATION HISTORY, NORTH SEA
M 30 124 BEATRICE FIELD, GREAT GLEN FAULT, STRUCTURE
M 30 117† BEATRICE FIELD, INNER MORAY FIRTH, U.K. NORTH SEA, THE
M 30 127 BEATRICE FIELD, RESERVOIR STRATIGRAPHY
M 30 127 BEATRICE FIELD, RHAETIAN-LIAS, STRATIGRAPHY
M 30 126 BEATRICE FIELD, SEISMIC SECTION REFLECTORS, NORTH SEA
M 30 121 BEATRICE FIELD, STRUCTURE MAP, NORTH SEA
M 30 127 BEATRICE FIELD, TOARCIAN-BAJOCIAN, STRATIGRAPHY
B 6103 465* BEAUFORT AND NORTHERN CHUKSHI SEAS, ABST., GEOLOGIC
M 24 51† BEAUFORT BASIN, NORTHWEST TERRITORIES, TAGLU GAS FIELD
B 6103 332 BEAUFORT HIGH, SOUTH CAROLINA
B 6101 3 BEAUFORT HOMOCLINE
B 6203 550* BEAUFORT SEA COAST, ABST., TRANSGRESSIVE, BARRIERS OF
B 6405 675 BEAUMONT, E. A., DEPOSITIONAL FACIES, GEOMETRY, AND GENESIS OF
B 6405 674 BEAUMONT, E.A., RETROGRADATIONAL SHELF SEDIMENTATION, VIKING
B 6105 765 **BEAUMONT, E. C., GEOMETRY OF COAL BEDS IN SAN JUAN BASIN, NEW**
B 6302 194 BEAUMONT, EDWARD A., DEPOSITIONAL ENVIRONMENTS OF FORT
B 6305 836* BEAVER COUNTY, UTAH, ABST., GEOLOGY AND STRUCTURAL CONTROL
B 6408 1132 BEAVER CREEK OIL FIELD, ALASKA
B 6002 281 BEAVER DAM SHALE MEMBER, PENNSYLVANIA
B 6412 2078 BEAVER VALLEY FAULT, APPALACHIANS
B 6112 2135 BEAVERHILL LAKE FIELD, ALBERTA
B 6205 789 BEAVERHILL LAKE GROUP, ALBERTA
M 24 82 BEAVERHILL LAKE GROUP, ALBERTA
B 6101 29 BEAVERLODGE AREA, PITCHBLENDE, SASKATCHEWAN
B 6101 31 BEAVERLODGE AREA, URANIUM, SASKATCHEWAN
B 6009 1606 BEAVERS, WENDELL M., DEPOSITIONAL ENVIRONMENT AND PETROLEUM
B 6109 1541 BEBOUT, D. G., CONSOLIDATION HISTORY OF GULF COAST LOWER
B 6203 494 BEBOUT, D. G., DELTAIC DEPOSITIONAL STYLE IN SALT-WITHDRAWAL
B 6209 1753 BEBOUT, D. G., DISTRIBUTION AND GEOMETRY OF AN OOLITE-SHOAL
B 6009 1606 BEBOUT, D. G., GEOPRESSURED GEOTHERMAL PROSPECT IN FRIO
B 6303 416 BEBOUT, D. G., LOWER CRETACEOUS SHELF-MARGIN CARBONATE FACIES,
B 6303 417 BEBOUT, D. G., POTENTIAL FOR PRODUCTION OF GEOPRESSURED
B 6303 416 BEBOUT, D. G., SECONDARY CARBONATE POROSITY, ZELTEN FIELD LIBYA,
B 6109 1551 BEBOUT, D. G., SLIGO AND HOSSTON DEPOSITIONAL PATTERNS,
B 6309 1595 BEBOUT, D. G., TESTIN GEOPRESSUREDD GEOTHERMAL RESOURCE, FRIO
B 6105 765 **BEBOUT, D. G., TEST SITE FOR GEOPRESSURED GEOTHERMAL WELL—FRIO**
B 6004 648 BEBOUT, D. G., TRANSGRESSIVE SANDSTONE--CARBONATE
B 6303 417 BEBOUT, D. G., WORLDS FIRST GEOPRESSURED GEOTHERMAL TEST WELL,
B 6408 1276 BEBOUT, DON G., ENERGY RESOURCES OF WATER-BEARING
B 6009 1607 BECHER, JACK W., WALKER CREEK FIELD--SMACKOVER DIAGENETIC
B 6303 417 BECKER, ADRIAN A., HYDROCARBON POTENTIAL OF CENTRAL GULF OF
B 6105 815 BECKER, JACK W., CARBONATE RESERVOIRS, ABST., DIAGENETIC
B 6009 1618 BECKER, LEROY E., GENERAL SEDIMENTOLOGIC HISTOR OF LATE
S 11 77 BECKINA FULMERI
S 11 146 BECKINA FULMERI
S 11 78 BECKINA HORNADAYI
S 11 144 BECKINA HORNADAYI
B 6212 2399 BECKLEY COAL AND SURROUNDING LITHOLOGIES, WEST VIRGINIA
B 6303 533* BECKLEY SEAM, ABST., ESTUARINE-COASTAL PLAIN, COAL DEPOSITION IN
B 6303 380 BECKMANN, J. P., REVIEW OF INDIAN OCEAN GEOLOGY AND
M 27 206 BECRAFT LIMESTONE, NEW YORK, DEVONIAN
M 28 166 BECRAFT LIMESTONE, NEW YORK, DEVONIAN
M 27 77 BECRAFT LIMESTONE, NEW YORK, DEVONIAN
B 6105 765 **BECK, L. S., URANIUM DEPOSITS OF NORTHERN SASKATCHEWAN**
B 6004 681* BED CONTINUITY AND PERMEABILITY VARIATIONS OF RECENT DELTAIC
B 6004 700* BED FORMS AN SEDIMENTARY STRUCTURES DEVELOPED UNDER
B 6004 708* BED FORMS AND INTERNAL STRUCTURES IN DEEP ESTUARY CHANNEL
B 6303 509* BED FORMS AND PROCESSES ON ESTUARINE TIDAL-CURRENT RIDGE,
B 6004 676* BED FORMS AND STRATIFICATION TYPES OF MODERN GRAVEL POINT
B 6408 1229 BED FORMS, HYDRAULIC PROCESSES AND RESULTING
B 6004 646* BEDDING RHYTHMS IN PELAGIC CALCAREOUS SEDIMENTS, ABST.,
B 6211 2352* BEDFORD CANYON FORMATION AND SANTA MONICA SLATE IN
B 6111 1986 BEDFORD CANYON FORMATION, CALIFORNIA
B 6209 1705 BEDFORD SHALE, KENTUCKY
B 6209 1710 BEDFORD SHALE, OHIO
B 6201 41 BEDOUT SUBBASIN, AUSTRALIA
B 6004 651* BEDS IN COALS OF ROCKY MOUNTAIN REGION, ABST., ALTERED,
B 6206 943 BEDS, MATRIX-SUPPORTED
B 6301 40 BEE COUNTY, TEXAS, PAWNEE FIELD
B 6309 1575 BEEBE, ROBERT R., LINEAMENTS AND WATER WELLS AS EXPLORATION
B 6109 1499 BEECHY POINT SECTION, ALASKA
B 6410 1644 BEEFSTEAK CHANNEL, AUSTRALIA
B 6410 1577 BEEKMANTOWN FORMATION, VIRGINIA
M 27 174 BEEKMANTOWN LIMESTONE, MARYLAND, ORDOVICIAN
B 6203 495 BEGGS, HUGH R., INTRODUCTION OF COMPUTER APPLICATIONS TO
M 29 85 BEHRENDT, JOHN C., BASIN STRUCTURE OF THE U.S. ATLANTIC MARGIN
B 6105 803 **BEHRENDT, JOHN C., MESOZOIC BASIN DISTRIBUTION ALONG U.S.**
B 6105 742 **BEHRENDT, JOHN C., REGIONAL GEOLOGIC FRAMEWORK OFF NORTHEASTERN**
B 6006 926 BEHRENDT, JOHN C., REGIONAL GEOLOGIC FRAMEWORK OFF
B 6006 926 **BEHRENDT, JOHN C., REGIONAL GEOLOGIC FRAMEWORK OFF NORTHEASTERN**
B 6303 418 BEHRENDT, JOHN C., SPECULATIONS ON PETROLEUM RESOURCE
B 6105 766 **BEHRENDT, JOHN C., STRUCTURE OF BALTIMORE CANYON TROUGH, U.S.**
B 6303 485 BEHRENSMEYER, ANNA K., TRACKS AND SUBSTRATE REWORKING BY
B 6211 2349 BEHRMAN, PHILIP G., PALEOGEOGRAPHIC SIGNIFICANCE OF MIDDLE TO
B 6002 258 BEIN, AMOS, RUDISTID FRINGING REEFS OF CRETACEOUS SHALLOW
B 6002 268 BEIT OREN REEF, ISRAEL
B 6412 2190 BEITRAGE ZUR BIOSTRATIGRAPHIE DER TETHYS-TRIAS, EDITED BY H.
B 6203 506 BEITZEL, J. E., UNDERSTANDING SEISMICALLY DERIVED IMPEDANCE
B 6004 657* BEJUCO FIELD, EASTERN MEXICO, ABST., SUBSURFACE, JURASSIC-,
B 6101 61 BEKAA GRABEN, LEBANON
M 30 464 BEKAPAI FIELD, FRESHWATER SANDSTONE, INDONESIA
M 30 459† BEKAPAI FIELD, GEOLOGY OF THE
M 30 460 BEKAPAI FIELD, INDONESIA, KUTEI BASIN
B 6209 1688 BEKAPAI FIELD, KALIMANTAN
M 30 460 BEKAPAI FIELD, KALIMANTAN, KUTEI BASIN
M 30 401 BEKAPAI FIELD, MIOCENE, MAHAKAM DELTA
M 30 468 BEKAPAI FIELD, RESERVOIR CHARACTERISTICS, INDONESIA
M 30 463 BEKAPAI FIELD, STRUCTURAL GEOLOGY, INDONESIA
B 6111 1931 BELCHER CH FORMATION, SVERDRUP BASIN
B 6203 557* BELCHER GROUP, HUDSON BAY, CANADA, ABST., EXAMPLE OF,
B 6405 685* BELEMNITE ROSTRA, ABST., MICROBIAL AND INVERTEBRATE
M 27 90 BELEMNITE ROSTRUMS

B 6203 351 BELGIAN ANTICLINE OIL FIELD, CALIFORNIA
B 6310 1918 BELGIUM, DEVELOPMENTS, 1978
B 6411 1963 BELGIUM, EXCLUDING NORTH SEA, DEVELOPMENTS, 1979
B 6010 1643 BELIZE, 1975, ZAPPATERRA, E., ON DEVELOPMENTS IN
B 6210 2046 BELIZE, 1977, PETROCONSULTANTS S.A., ON DEVELOPMENTS IN
B 6310 1747 BELIZE, 1978, PETROCONSULTANTS S.A., ON DEVELOPMENTS IN
B 6205 892* BELIZE, ABST., DEPOSITION AND DIAGENESIS OF, HOLOCENE, REEFS
S 4 44 BELIZE, ACROPORA ASSEMBLAGE
S 4 44 BELIZE, ALGAE-PORITES ASSEMBLAGE
B 6404 466 BELIZE AREA, MARINE PHREATIC CEMENT
B 6303 489* BELIZE BARRIER-REEF COMPLLX--EVIDENCE OF EXTENSIVE SUBMARINE
B 6004 677* BELIZE DETERMINED BY DIVER-OPERATED DRILLING DEVICE, ABST.,
S 4 43 BELIZE, GLOVERS REEF ATOLL, CORAL-ALGAL ASSEMBLAGES
S 4 42 BELIZE, GLOVERS REEF ATOLL, FORAMINIFERAL ASSEMBLAGES
M 27 59 BELIZE, HOLOCENE BACK-REEF BEACH SAND
M 27 218 BELIZE, HOLOCENE BEACH SEDIMENTS
M 27 96 BELIZE, HOLOCENE CORAL RUBBLE
M 27 35 BELIZE, HOLOCENE, CORAL-RUBBLE
M 27 18 BELIZE, HOLOCENE, PLANKTON
M 27 95 BELIZE, HOLOCENE SEDIMENTS
M 27 21 BELIZE, HOLOCENE SEDIMENTS
M 27 158 BELIZE, HOLOCENE SEDIMENTS
M 27 32 BELIZE, HOLOCENE SEDIMENTS
M 27 183 BELIZE, HOLOCENE SEDIMENTS
M 27 60 BELIZE, HOLOCENE SEDIMENTS
M 27 83 BELIZE, HOLOCENE SEDIMENTS
M 27 51 BELIZE, HOLOCENE SEDIMENTS
M 27 60 BELIZE, HOLOCENE SEDIMENTS
M 27 25 BELIZE, HOLOCENE SEDIMENTS
M 27 82 BELIZE, HOLOCENE SEDIMENTS
M 27 81 BELIZE, HOLOCENE SEDIMENTS
M 27 61 BELIZE, HOLOCENE SEDIMENTS
M 27 101 BELIZE, HOLOCENE SEDIMENTS
M 27 37 BELIZE, HOLOCENE SEDIMENTS
M 27 159 BELIZE, HOLOCENE SEDIMENTS
M 27 38 BELIZE, HOLOCENE SEDIMENTS, PYRGO SP.
S 4 176 BELIZE, MONTASTREA ANNULARIS AND M. CAVERNOSA
S 4 44 BELIZE, MONTASTREA ASSEMBLAGE
B 6303 528* BELIZE PATC REEFS, ABST., THREE-DIMENSIONAL ASPECTS OF
B 6303 528* BELIZE PATCH REEFS, ABST., THREE-DIMENSIONAL ASPECTS OF
S 4 37† BELIZE, PATCH-REEF ECOLOGY AND SEDIMENTOLOGY OF, GLOVERS,
M 27 59 BELIZE, PLEISTOCENE, BACK-REEF LIMESTONES
M 27 160 BELIZE, QUATERNARY SEDIMENTS
B 6004 648 BELKNAP, D. F., RADIOCARBON DATING AND STRATIGRAPHIC ANALYSIS
B 6405 675 BELKNAP, DANIEL F., FRACTIONAL PRESERVATION OF TRANSGRESSIVE
B 6403 438 BELKNAP, DANIEL F., HOLOCENE INTERTIDAL DEPOSITS OF
B 6403 438 BELKNAP, DANIEL F., HOLOCENE INTERTIDAL DEPOSITS OF
B 6309 1522 BELKOFSKI FORMATION, OLIGOCENE, ALASKA PENINSULA
B 6203 496* BELL CANYON FORMATION, DELAWARE BASIN, NEW MEXICO, ABST.,
B 6203 495* BELL CANYON FORMATION OF DELAWARE BASIN, ABST., COMPARISON
B 6303 553* BELL CANYON FORMATION, PERMIAN, DELAWARE BASIN TEXAS-NEW
M 28 122 BELL CANYON FORMATION, TEXAS, PERMIAN
M 28 22 BELL CANYON FORMATION, TEXAS, PERMIAN
M 28 11 BELL CANYON FORMATION, TEXAS, PERMIAN
M 28 176 BELL CANYON FORMATION, TEXAS, PERMIAN
M 27 69 BELL CANYON FORMATION, TEXAS, PERMIAN
M 28 113 BELL CANYON FORMATION, TEXAS, PERMIAN
B 6104 486 BELL CANYON SANDSTONES TRAPS, TEXAS
M 26 432 BELL CREEK FIELD, MUDDY FORMATION, MONTANA
B 6105 766 BELL, L. H., COMPUTER GRAPHIC DISPLAYS OF DIPMETER INFORMATION
B 6109 1440 BELLE CITY LIMESTONE, OKLAHOMA
M 26 447 BELLE CITY LIMESTONE, OKLAHOMA, ANADARKO BASIN
B 6108 1380* BELLE CREEK-A UNIQUE COMMUNITY CONCEPT, ABST.
B 6404 490 BELLE FOURCHE MEMBER, FRONTIER FORMATION, WYOMING
B 6407 971 BELLE FOURCHE SHALE, GREAT PLAINS
B 6407 976 BELLE FOURCHE SHALE, NORTH DAKOTA
B 6406 943 BELLE FOURCHE SHALE, SOUTH DAKOTA
B 6404 490 BELLE FOURCHE SHALE, WYOMING
B 6107 1091 BELLE ISLE SALT MINE, LOUISIANA
B 6201 76 BELLE PLAINE MEMBER, SASKATCHEWAN
S 5 11 BELLE RIVER MILLS FORMATION, VADOSE DEPOSITS, MICHIGAN,
B 6211 2243 BELLEVUE CYCLE, INFORMAL NAME, CALIFORNIA
B 6211 2264 BELLEVUE FAN, CALIFORNIA
B 6211 2247 BELLEVUE FAN, CALIFORNIA
B 6211 2256 BELLEVUE FIELD, CALIFORNIA
B 6211 2255 BELLEVUE GROWTH FAULT SYSTEM, CALIFORNIA
B 6211 2256 BELLEVUE SAND, CALIFORNIA
B 6211 2247 BELLEVUE TREND, CALIFORNIA, ROSEDALE-
B 6111 1994 BELLINGSHAUSEN BASIN, POSSIBLE DIAPIRS
B 6111 1954 BELLSHILL LAKE HEAVY OIL, ALBERTA
M 26 429 BELLY RIVER FORMATION, CANADA, ALBERTA
B 6303 534* BELLY RIVER FORMATION, UPPER CRETACEOUS, CENTRA ALBERTA,
B 6407 972 BELLY RIVER GROUP, ALBERTA
B 6408 1166 BELLY RIVER SANDSTONE, ALBERTA
B 6304 655 BELPRE BENTONITE, NEW NAME, OHIO
B 6004 653* BELT AN UINTA AULACOGENS--INITIAL DEPOSITS OF CORDILLERAN
B 6303 468* BELT SUPERGROUP, MONTANA, ABST., MICROFOSSILS, MACROFOSSILS,
B 6405 687 BELTRAMI, CARLOS V., ET AL, MICROFACIES AND
B 6408 1133 BELUGA FORMATION, ALASKA
B 6203 377 BELUGA FORMATION, ALASKA
B 6408 1132 BELUGA RIVER GAS FIELD, ALASKA
M 30 236 BELYA SUITE, USSR, LENA-TUNGUSKA AREA
B 6111 1951 BEMOLANGA HEAVY OIL DEPOSIT, MALAGASY
M 30 540 BEN-KAHLA FIELD, ALGERIA, OUARGLA REGION
B 6211 2246 BENA CHANNEL, NEW TERM, CALIFORNIA
B 6211 2245 BENA FAN, CALIFORNIA
B 6410 1644 BENAGERIE RIDGE, AUSTRALIA
B 6410 1577 BENBOLT FORMATION, VIRGINIA
S 4 87 BENBOW INLIER, JAMAICA
S 4 87 BENBOW LIMESTONE, JAMAICA
B 6405 675 BENCH, BERNARD M., DRILLING FOR METHANE GAS IN FISHERS PEA
B 6303 536* BENCH SEQUENCE--SHOOFLY OOLITE OF SOUTHWESTERN IDAHO, ABST.,
B 6301 54 BEND ARCH, ORIGIN, TEXAS
B 6301 54 BEND CONGLOMERATES, TEXAS
B 6207 1209 BENDER, FRIEDRICH, PRESENT AND FUTURE SUPPLY OF ENERGY
B 6303 418 BENDER, GRETCHEN L., MIDDLE PALEOZOIC SEDIMENTATION AND
B 6208 1399 BENDLER, E. PERRY, OIL AND GAS DEVELOPMENTS IN MARYLAND, OHIO,
B 6409 1403 BENDLER, E. PERRY, OIL AND GAS DEVELOPMENTS IN MARYLAND, OHIO,
B 6108 1269 BENDLER, E. PERRY, OIL AND GAS DEVELOPMENTS IN MARYLAND, OHIO,
B 6308 1244 BENDLER, E. PERRY, OIL AND GAS DEVELOPMENTS IN MARYLAND, OHIO,
B 6008 1288 BENDLER, E. PERRY, OIL AND GAS DEVELOPMENTS IN MARYLAND, OHIO,
M 25 411 BENEFICIABLE IRON FORMATION OR TACONITE
B 6004 699* BENGAL FAN AND SWATCH OF NO GROUND CANYON, NORTHEAST
B 6206 918 BENGAL FAN, BAY OF BENGAL
B 6203 495 BENGTSON, C. A., STATISTICAL CURVATURE ANALYSIS TECHNIQUES FOR
B 6211 2350 BENGTSON, C. A., STATISTICAL CURVATURE ANALYSIS TECHNIQUES FOR
B 6408 1195 BENGUELA CURRENT, ANOXIC LAYER CAUSED BY UPWELLING,
B 6209 1646 BENHAM RISE
B 6008 1390 BENICHOU, PIERRE, SEISMIC EXPERIMENTS, CLINKERS AREA, WYOMING,
B 6110 1747 BENIN, DAHOMEY, DEVELOPMENTS, 1976
B 6210 1846 BENIN, DAHOMEY, DEVELOPMENTS, 1977
B 6310 1691 BENIN, DEVELOPMENTS, 1978
B 6411 1786 BENIN, DEVELOPMENTS, 1979
B 6201 2 BENIN FLANK, NIGER DELTA
B 6202 297 BENIN FORMATION, NIGER DELTA
B 6202 296 BENIN FORMATION, NIGERIA
B 6408 1252 BENIN FORMATION, NIGERIA
B 6107 972 BENIN SLOPE, AFRICA
S 3 79 BENIOFF SUBDUCTION ZONES, WELL-DEFINED
B 6203 394 BENIOFF ZONE, ALASKA
M 25 529 BENIOFF ZONE, CALC-ALKALINE MAGMAS FORMED ON
M 25 481 BENIOFF ZONE, MIGRATION OF, PACIFIC
M 25 51 BENIOFF ZONE, PACIFIC REGION
B 6105 788 BENMORE, WILLIAM C., HOLOCENE FRESHWATER DIAGENESIS OF MODERN
B 6303 453 BENNETT, R. H., ENVIRONMENTAL STUDIES AT NOAAS ATLANTIC
B 6303 536 BENNETT, R. L., SOUTH HALLETTSVILLE, GAS FIELD IN LOWER WILCOX,
B 6106 918 BENNETT, RICHARD H., CONTINENTAL SLOPE SEDIMENT INSTABILITY
B 6203 529 BENNETT, RICHARD H., GEOTECHNICAL PROPERTIES OF CONTINENTAL
B 6203 495 BENNETT, RICHARD H., SLOPE MAP DEPICTING MAJOR SUBMARINE SLIDE
B 6311 2115 BENNISON, ALLAN P., MOBILE BASIN-SHELF BORDER IN NORTHEAST
B 6211 2350 BENOIT, WALTER R., DISCOVERY AND GEOLOGY OF DESERT PEAK,
B 6404 572 BENOT LOT SHALE MEMBER, ISRAEL
B 6409 1563 BENSON, D. JOE, REGIONAL STRATIGRAPHY OF UPPER JURASSIC
B 6108 1392 BENSON, L. V., INCIPIENT DIAGENESIS OF ARKOSIC AND VITRIC
B 6305 841* BENSON OIL FIELD, SOUTHEASTERN SASKATCHEWAN, ABST., GEOLOGY
B 6209 1714 BENSON, RICHARD N., RECORRELATION OF EOCENE-MIOCENE
B 6004 649 BENSON, W. E., PRELIMINARY RESULTS FROM LEG 44 DEEP SEA DRILLING
B 6212 2481 BENT CRUISES, 1972, USNS, BARTLETT AND
B 6404 573 BENTHE SALT DOME, SECTIONS, GERMANY
M 26 350 BENTHEIM SANDSTONE
M 26 350 BENTHEIM SANDSTONE, SEISMIC SECTION, NORTH SEA
B 6405 700* BENTHIC FORAMINIFERA AND THEIR BIOSTRATIGRAPHIC POTENTIAL,
S 4 321 BENTHIC FORAMINIFERA, ASSEMBLAGES, BRAZIL, CONTINENTAL, SHELF
S 4 313† BENTHIC FORAMINIFERA, CALCIUM, CARBONATE SEDIMENTATION ON,
B 6203 510* BENTHIC FORAMINIFERA OF CALIFORNIA BORDERLAND, ABST.,
B 6403 444* BENTHIC FORAMINIFERAL BIOFACIES OF EASTERN PACIFIC MARGIN
B 6403 447* BENTHIC FORAMINIFERS FROM EASTERN GULF OF ALASKA, ABST.
B 6403 447* BENTHIC FORAMINIFERS FROM EASTERN GULF OF ALASKA, ABST.
B 6202 259 BENTHIC STAGES, CALIFORNIA
B 6209 1759* BENTHONIC FORAMINIFERA AS INDICATORS OF OCEANOGRAPHIC

B	6211	2357*	BENTHONIC FORAMINIFERAL BIOFACIES IN ARCTIC OCEAN, ABST.,
B	6105	842*	BENTHONIC RATIO IN FORAMINIFERA AS PALEOBATHYMETRIC TOOL—
B	6104	594	BENTLEY FIELD, ALBERTA
M	26	279	BENTON GROUP, CRETACEOUS
M	24	242	BENTONIA FIELD, MISSISSIPPI
M	28	68	BENTONITE, CLAY, SMECTITE, IN A
B	6304	655	BENTONITE, NEW NAME, OHIO, BELPRE
B	6304	656	BENTONITE, OHIO, TIOGA
S	6	344	BENTONITES DISCUSSED, SILURIAN, ISOTOPIC DATING, USE OF
B	6304	655†	BENTONITES IN EASTERN OHIO, DEVONIAN
S	6	347	BENTONITES, ORDOVICIAN, CARADOC SERIES
S	6	348	BENTONITES, ORDOVICIAN, KENTUCKY, TYRONE LIMESTONE
S	6	348	BENTONITES, ORDOVICIAN, SWEDEN, CHASMOPS LIMESTONE
B	6406	865	BENUE TROUGH, AFRICA
B	6201	2	BENUE TROUGH, SOUTHWEST EXTENSION, NIGER DELTA
B	6002	210	BENUE TROUGH, WEST, AFRICA, WALVIS RIDGE TO
B	6401	54	BERAI LIMESTONE, KALIMANTAN
B	6401	54	BERAI LIMESTONE, KALIMANTAN
B	6401	54	BERAI MARL, KALIMANTAN
B	6401	54	BERAI MARL, KALIMANTAN
B	**6105**	**784***	**BEREA, MISSISSIPPIAN, SANDSTONE CORE FROM CSO ISLAND CREEK COAL**
M	26	26	BEREA SANDSTONE, LABORATORY, MEASUREMENTS
B	6209	1709	BEREA SANDSTONE, UNITED STATES, EASTERN
B	6405	675	BERESKIN, S. R., DEVONIAN REEFS EXPOSED ALONG CENTRAL
M	25	513	BERG COPPER DEPOSIT
M	25	482	BERG, HENRY C., METAL PROVINCES OF ALASKA--SUMMARY
B	6311	2115	BERG, J. ROBERT, EXPLORATION FOR PENNSYLVANIAN SANDSTONES,
B	6405	676	BERG, R. R., HYDRODYNAMIC FLOW IN LOWER CRETACEOUS MUDDY
B	6002	324	BERG, R. R., TURBIDITE RESERVOIRS IN CANYON SANDSTONE, VAL
B	6309	1596	BERG, ROBERT R., CHARACTERISTICS OF LOWER VICKSBURG
B	6309	1595	BERG, ROBERT R., CHARACTERISTICS OF LOWER WILCOX RESERVOIRS,
B	6109	1534	BERG, ROBERT R., CHARACTERISTICS OF WILCOX GAS RESERVOIRS,
B	6203	495	BERG, ROBERT R., COMPARISON OF TURBIDITE SANDSTONE
B	6009	1425	BERG, ROBERT R., DEFORMATION OF MESOZOIC SHALES AT HAMILTON
B	6009	1607	BERG, ROBERT R., DENSITY-FLOW ORIGIN FOR FRIO RESERVOIR
B	6409	1553	**BERG, ROBERT R., MORPHOLOGY OF TURBIDITE-CHANNEL RESERVOIRS,**
B	6306	886	BERG, ROBERT R., OIL AND GAS IN DELTA-MARGIN FACIES OF DAKOTA
B	6303	418	BERG, ROBERT R., RESERVOIR MORPHOLOGIES FOR GULF COAST
B	6308	1425	BERG, ROBERT R., TURBIDITE CHANNEL RESERVOIRS IN CANYO
B	6203	496	BERG, ROBERT R., TURBIDITE ORIGIN FOR ANHYDRITIC DOLOMITES IN
B	6405	792	BERG, THOMAS M., INTERPRETATION OF FOSSI FLUVIAL BIVALVE
B	6008	1390	BERGE, CHARLES W., GEOTHERMAL EXPLORATION OF ROOSEVELT KGRA,
B	**6105**	**766**	**BERGE, C. W., EXPLORATION AND EVALUATION OF ROOSEVELT KGRA,**
B	6209	1754	BERGEN, F. W., DISTRIBUTION OF HOLOCENE FORAMINIFERA IN GULF OF
B	6211	2350	BERGEN, F. W., DISTRIBUTION OF RECENT FORAMINIFERA IN GULF OF
B	6309	1545	BERGER AND KENNETH E. ANDERSON, FEAZEL, CHARLES T., REVIEW OF,
B	6309	1576	BERGER, IRVING A., SUGGESTED ORIGIN FOR GAS IN EASTERN
M	25	475	BERGER, V. I., EVOLUTIONARY SERIES OF ANTIMONY DEPOSITS OF
M	25	62	BERGER, V. I., TECTONICS OF CONTINENTAL AND OCEANIC STRUCTURES
B	6004	668	BERGER, W. H., ABYSSAL ORGANISM TRACES ON AN IN MODERN PELAGIC
B	6303	419	BERGER, W. H., DEEP-SEA DRILLING AND MAJOR THEMES OF OCEAN
S	6	213	BERGGREN, W. A., A NEW PALEOGENE NUMERICAL TIME SCALE
S	6	39	BERGGREN, W. A., BIOCHRONOLOGY
S	6	217	BERGGRENS ZONES, A REVIEW, PALEOGENE, FORAMINIFERA
B	6405	771	BERGHORN, CLAUDE E., LOWER AND MIDDLE PALEOZOIC POTENTIAL OF
B	6404	530	BERGLUM RIVER FORMATION, GREENLAND
B	6108	1376	BERGQUIST, L. E., HELIUM DETECTION IN URANIUM EXPLORATION, ABST.
B	6303	419	BERGSAGER, EGIL,, EXPLORATION RESULTS AND POTENTIAL OF
B	6405	676	BERGSTROM, STIG M., EARLY PALEOZOIC CONODONT
B	6405	706*	BERIN SEA AND UNITED STATES ATLANTIC SHELF, ABST., MODIFICATION
B	6212	2483	BERING ABYSSAL PLAIN, VAMPS
B	6207	1211*	BERING AND OKHOTSK SEA ICE COVERS BY MICROWAVE SATELLITE
B	6004	702*	BERING EPICONTINENTAL SHELF, ABST., SEDIMENTARY, FACIES AND
B	6204	635	BERING GLACIER, ALASKA
B	6303	490*	BERING SE CONTINENTAL MARGIN, ABST., SHALLOW-, WATER UPPER,
B	6403	439*	BERING SEA, ABST., HYDROCARBON POTENTIAL OF, ALEUTIAN BASIN
B	**6408**	**1279***	**BERING SEA, ABST., RESOURCE POTENTIAL AND, PLATE-MARGIN**
B	6303	468*	BERING SEA, ABST., SONOBUOY REFRACTION MEASUREMENT FROM,
B	6303	436*	BERING SEA, ABST., THICK SEDIMENT ACCUMULATIONS BENEATH,
B	6303	449*	BERING SEA, ABST., VIDEOTAPES OF ACTIVE PRIMARY PHYSICAL AND
B	6405	756*	BERING SEA, ALASKA, ABST., SEDIMENTARY CHARACTERISTICS AND
B	6002	178	BERING SEA, ANADYR BASIN
B	6212	2481	BERING SEA BASIN, SEISMIC PROFILES
B	6212	2481†	BERING SEA BASIN, VAMPS—POSSIBLE HYDROCARBON-BEARING
B	6311	2071	BERING SEA, BOWERS BASIN
B	6405	694*	BERING SEA CONTINENTAL MARGIN, ABST., MULTIFOLD, SEISMIC DATA
B	6406	945†	BERING SEA CONTINENTAL SHELF, PETROGRAPHY AND GEOLOGIC
B	6002	165	BERING SEA, CONTINENTAL SLOPE, DREDGE SAMPLES
B	6311	2070†	BERING SEA, HYDROCARBON POTENTIAL OF, ALEUTIAN BASIN
B	6102	193	BERING SEA, KOMANDORSKIY BASIN, DEEP SEA DRILLING PROJECT SITE
B	6102	192	BERING SEA, KOMANDORSKIY BASIN, TURBIDITES
B	6311	2071	BERING SEA, KOMANDORSKY BASIN
B	6004	694*	BERING SEA MARGIN AND RELATION TO PLATE TECTONICS. BY TITLE
B	6102	192†	BERING SEA, NEOGENE, TURBIDITE SEDIMENTATION IN, KOMANDORSKIY
B	6412	2109	BERING SEA, ORGANIC CARBON AND BITUMEN
B	6111	1994	BERING SEA, POSSIBLE DIAPIRS
M	25	280†	BERING SEA REGION--SUMMARY, CENOZOIC AN, MESOZOIC PETROLEUM
B	6412	2149	BERING SEA SHELF, ACOUSTIC ANOMALIES, BRIGHT SPOTS, O WIPE-OUT
B	6412	2142	BERING SEA SHELF, AMAK BASIN
B	6412	2142	BERING SEA SHELF, BLACK HILLS RIDGE
B	6412	2142	BERING SEA SHELF, BRISTOL BAY BASIN
B	6412	2139†	BERING SEA SHELF, MESOZOIC AND, CENOZOIC STRUCTURAL TRENDS
B	6412	2145	BERING SEA SHELF, NAKNEK FORMATION
B	6412	2142	BERING SEA SHELF, PRIBILOF BASIN
B	6412	2142	BERING SEA SHELF, PRIBILOF CANYON
B	6412	2142	BERING SEA SHELF, PRIBILOF ISLAND
B	6412	2142	BERING SEA SHELF, PRIBILOF RIDGE
B	6002	179	BERING SEA SHELF, RESERVOIR BEDS
B	6412	2141	BERING SEA SHELF, SEISMIC REFLECTION DATA, PROFILES
B	6002	161†	BERING SEA SHELF SOUTH OF ST LAWRENCE ISLAND, STRUCTURE AND
B	6405	745*	BERING SEA SHELF SOUTH OF ST. LAWRENCE ISLAND, ALASKA, ABST.,
B	6412	2142	BERING SEA SHELF, ST. GEORGE BASIN
B	6002	179	BERING SEA SHELF, TRAPS
B	6412	2142	BERING SEA SHELF, UNIMAK ISLAND
B	6311	2071	BERING SEA, SHIRSHOV RIDGE
B	6311	2077	BERING SEA, SOUNDER RIDGE
B	6405	693*	BERING SEA, SOUTH OF ST. LAWRENCE ISLAND, ABST., PROSPECTIVE AND
B	6311	2071	BERING SHELF
B	6103	465*	BERING SHELF, ABST., GEOLOGY OF
B	6011	2067*	BERING SHELF, ABST., GEOLOGY OF
B	6002	165	BERING SHELF, BASINS
B	6002	165	BERING SHELF, GARDEN BASIN
B	6002	165	BERING SHELF, HALL BASIN
B	6309	1468	BERING SHELF, NAVARIN BASIN
B	6002	165	BERING SHELF, PRIBILOF BASIN
B	6002	165	BERING SHELF, ST. MATTHEW BASIN
B	6103	467*	BERING STRAIT REGION, ALASKA, ABST., COMPRESSIVE INTERACTION OF,
B	6204	635	BERING TROUGH, ALASKA
B	6002	161	BERINGIAN CONTINENTAL MARGIN
B	6412	2152	BERINGIAN MARGIN, CROSS SECTIONS
B	6412	2152	BERINGIAN MARGIN, KULA PLATE
B	6002	175	BERINGIAN MARGIN, KULA PLATE
B	6412	2152	BERINGIAN MARGIN, NORTH AMERICAN PLATE
B	6002	170	BERINGIAN SHELF, AMAK BASIN
B	6002	177	BERINGIAN SHELF AND MARGIN, CENOZOIC BASINS
B	6002	177	BERINGIAN SHELF AND MARGIN, MESOZOIC AND CENOZOIC TECTONIC
B	6002	177	BERINGIAN SHELF AND MARGIN, MESOZOIC BASEMENT ROCK
B	6002	164	BERINGIAN SHELF, BLACK HILLS
B	6002	164	BERINGIAN SHELF, DALNOI BASIN
B	6002	164	BERINGIAN SHELF, DALNOI RIDGE
B	6002	174	BERINGIAN SHELF, DENALI FAULT
B	6002	164	BERINGIAN SHELF, GARDEN RIDGE
B	6002	174	BERINGIAN SHELF, GRAVITY DATA
B	6002	164	BERINGIAN SHELF, INNER BASIN
B	6002	164	BERINGIAN SHELF, KALTAG FAULT
B	6002	172	BERINGIAN SHELF, MAGNETIC DATA
B	6002	175	BERINGIAN SHELF, MESOZOIC FOLDBELT
B	6002	175	BERINGIAN SHELF, MESOZOIC MAGMATIC ARC
B	6002	172	BERINGIAN SHELF, NAVARIN BASIN PROVINCE
B	6002	164	BERINGIAN SHELF, NAVARIN RIDGE
B	6002	164	BERINGIAN SHELF, NELSON ISLAND
B	6002	172	BERINGIAN SHELF, NUNIVAK ARCH PROVINCE
B	6002	164	BERINGIAN SHELF, NUNIVAK ISLAND
B	6002	164	BERINGIAN SHELF, OTTER BASIN
B	6002	178	BERINGIAN SHELF, PETROLEUM GEOLOGY
B	6002	162	BERINGIAN SHELF, PRIBILOF CANYON
B	6002	162	BERINGIAN SHELF, PRIBILOF ISLANDS
B	6002	164	BERINGIAN SHELF, PRIBILOF RIDGE
B	6002	163	BERINGIAN SHELF, PROFILES
B	6002	163	BERINGIAN SHELF, SEISMIC-REFLECTION DATA
B	6002	170	BERINGIAN SHELF, ST. GEORGE BASIN PROVINCE
B	6002	162	BERINGIAN SHELF, ST. MATTHEW ISLAND
B	6002	175	BERINGIAN SHELF, STRIKE-SLIP FAULTS
B	6002	162	BERINGIAN SHELF, UMNAK PLATEAU
B	6002	164	BERINGIAN SHELF, WALRUS BASIN
B	6002	164	BERINGIAN SHELF, ZHEMCHUG BASIN
B	6002	162	BERINGIAN SHELF, ZHEMCHUG CANYON
M	24	205	BERLITZ, R. E., EAST CAMERON BLOCK 270, OFFSHORE LOUISIANA--A

B 6209 1760 BERLITZ, RICHARD, PRE-PLATFORM EXPLORATION OF HIGH ISLAND
B 6405 723* BERMUDA, ABST., INTERTIDAL ZONATION OF RECENT MICROBIAL,
B 6004 657* BERMUDA, ABST., PALEOECOLOGIC INVESTIGATIONS O LATE, HOLOCENE,
B 6004 653* BERMUDA REEF ENVIRONMENTS, ABST., ANALYSIS OF, BORERS AND
B 6412 2058 BERMUDA RISE, DSDP SITE 386, LEG 43, GEOCHEMICAL LOG
M 27 139 BERMUDA RISE, EOCENE DEEP-WATER OOZE
M 27 17 BERMUDA RISE, EOCENE, DEEP-WATER OOZE
S 4 187 BERMUDA, TOBACCO BAY, REEF
B 6303 455* BERMUDAS LAGOONS AND REEFS, ABST., PROBING
B 6004 673* BERMUDAS NORTHWEST REEFS COLONIZE LAST INTERGLACIAL
B 6405 756 BERNARD, BERNIE B., CARBON ISOTOPIC SIGNATURE AS CRITERION FOR
B 6303 419 BERNARD, BERNIE B., ISOTOPIC COMPOSITIONS OF BIOGENIC CARBONATE
B 6303 420 BERNARD, BERNIE B., MODEL FOR INTERSTITIAL SULFATE REDUCTION
B 6405 784 BERNARD, BERNIE B., OSWEGO LIMESTONE, ALINE-LAMBERT FIELDS,
B 6105 780 BERQUIST, L. E., HELIUM DETECTION IN URANIUM EXPLORATION
M 25 319 BERREBERRE ANTICLINE, COLOMBIA
B 6104 585 BERRI FIELDS, PERSIAN GULF
B 6101 83 BERRICHID-RHARB BASINS, MOROCCO, DOUKKALA-
B 6303 513* BERRIEDALE LIMESTONE, PERMIAN, HOBART, TASMANIA, AUSTRALIA,
B 6205 880 BERRONG, BERT H., SEISMIC EXPLORATION LEADING TO DISCOVERY OF
B 6211 2363 BERRONG, BERT H., SEISMIC EXPLORATION LEADING TO DISCOVERY OF
B 6211 2219 BERRY FORMATION, CALIFORNIA
B 6407 1121 BERRY, HOWER, S. GRACE, REVIEW OF THE, PETROLEUM DICTIONARY, BY
B 6012 2179 BERRY, K. D., UPPER CRETACEOUS LITHOFACIES MODEL, SACRAMENTO
B 6004 649 BERRYHILL, H. L., JR., LATE PLEISTOCENE-HOLOCENE EVOLUTION OF
B 6303 420 BERRYHILL, HENRY L., JR.., ENVIRONMENTAL GEOLOGY OF
B 6312 2128 BERTRAND, RUDOLF, COMPILATION AND CORRELATION OF MAJOR
B 6203 496 BERTRAND-SARFATI, JANINE, NONMARINE STROMATOLITES FROM
B 6311 2002 BERYL EMBAYMENT, NORTH SEA
B 6408 1247 BERYL EMBAYMENT, UNITED KINGDOM
M 25 535 BERYLLIUM, ANDEAN REGION
B 6306 934 BESNARD CRUISE, 1973, RV
B 6404 530 BESSEL FJORD FORMATION, GREENLAND
B 6404 528 BESSEL FJORD, GREENLAND
B 6310 1753 BETHANCOURT, H. R., ON DEVELOPMENTS IN GUATEMALA, 1978
B 6407 1012 BETHANY FIELD, ARK-LA-TEX AREA
M 24 11 BEULGA FORMATION, ALASKA
B 6405 676 BEUS, STANLEY S., DEVONIAN PALEOGEOGRAPHY AND
B 6410 1644 BEVERLEY DEPOSIT, AUSTRALIA
M 25 427 BEVERLEY, URANIUM, SOUTH AUSTRALIA
M 24 159 BEXAR SHALE MEMBER, TEXAS
B 6101 43 BEYDOUN, Z. R., PETROLEUM PROSPECTS OF LEBANON--REEVALUATION
B 6305 822 BEYER, L. A., BOREHOLE GRAVITY STUDY OF DENSITY AND POROSITY OF
S 7 1 BEYOND THE SHELF BREAK PAPERS, 1976
M 25 376 BEZRUKOV, P. L., FERROMANGANESE NODULES OF THE PACIFIC
M 25 46 BEZRUKOV, P. L., GEOLOGIC MAP OF PACIFIC MOBILE BELT AND PACIFIC
M 25 58 BEZRUKOV, P. L., GEOLOGIC-STRUCTURAL SYNTHESIS OF PACIFIC AREA
B 6104 585 BIBI HAKIMEH FIELD, PERSIAN GULF
M 28 91 BIBLIOGRAPHY, ASSOCIATED SEDIMENTS
M 28 61 BIBLIOGRAPHY, CARBONATES
M 28 75 BIBLIOGRAPHY, CLAY
M 28 167 BIBLIOGRAPHY, COMPACTION
M 28 145 BIBLIOGRAPHY, DIAGENESIS
M 28 146 BIBLIOGRAPHY, DIAGENESIS AND POROSITY
M 28 147 BIBLIOGRAPHY, DIAGENESIS AND POROSITY, CARBONATES
M 28 147 BIBLIOGRAPHY, DIAGENESIS AND POROSITY, CLAY
M 28 147 BIBLIOGRAPHY, DIAGENESIS AND POROSITY, FELDSPAR
M 28 149 BIBLIOGRAPHY, DIAGENESIS AND POROSITY, IRON
M 28 149 BIBLIOGRAPHY, DIAGENESIS AND POROSITY, MANGANESE
M 28 149 BIBLIOGRAPHY, DIAGENESIS AND POROSITY, PHOSPHATES
M 28 146 BIBLIOGRAPHY, DIAGENESIS AND POROSITY, QUARTZ
M 28 146 BIBLIOGRAPHY, DIAGENESIS AND POROSITY, SILICA
M 28 149 BIBLIOGRAPHY, DIAGENESIS AND POROSITY, ZEOLITES
M 28 91 BIBLIOGRAPHY, EVAPORITES
M 28 24 BIBLIOGRAPHY, FELDSPAR
M 28 167 BIBLIOGRAPHY, FRACTURING
M 28 61 BIBLIOGRAPHY, GLAUCONITE
M 28 109 BIBLIOGRAPHY, GRAIN ORIENTATION
M 28 108 BIBLIOGRAPHY, GRAIN SIZE SORTING
M 28 108 BIBLIOGRAPHY, GRAIN-SHAPE ABRASION
M 28 61 BIBLIOGRAPHY, HEAVY MINERALS
M 28 61 BIBLIOGRAPHY, IRON MINERALS
M 28 75 BIBLIOGRAPHY, MICA
M 24 300† BIBLIOGRAPHY OF NORTH AMERICAN OIL AND GAS FIELDS FROM AAPG
B 6309 1547 BIBLIOGRAPHY OF THE GEOLOGY OF LIBYA BY J. AZZOUZ ETTALHI, D.
M 28 61 BIBLIOGRAPHY, OTHER DETRITAL GRAINS
M 28 91 BIBLIOGRAPHY, PHOSPHATES
M 28 145 BIBLIOGRAPHY, POROSITY
M 28 12 BIBLIOGRAPHY, QUARTZ
M 28 12 BIBLIOGRAPHY, QUARTZ SURFACE TEXTURE
M 28 36 BIBLIOGRAPHY, ROCK FRAGMENTS
M 28 100 BIBLIOGRAPHY, SANDSTONE CLASSIFICATION
M 28 75 BIBLIOGRAPHY, SHALE
M 28 91 BIBLIOGRAPHY, SILICA
M 28 191 BIBLIOGRAPHY, TECHNIQUES
M 28 108 BIBLIOGRAPHY, TEXTURE
M 28 92 BIBLIOGRAPHY, TUFFS
B 6309 1602 BICKER, ALVIN R., JR., PETROGRAPHY OF SOME SUBSURFACE IGNEOUS
B 6407 990 BICOUDINE FORMATION, MOROCCO
B 6209 1754 BIEL, RALPH, PALEOSTRUCTURAL MAPPING IN BATHYAL
M 25 411 BIFS, WESTERN AUSTRALIA, ARCHEAN, BANDED, IRON FORMATIONS OR
B 6012 2177* BIG BLUE FORMATION NEAR CANTUA CREEK, CALIFORNIA, ABST.,
B 6004 692* BIG BROWN LIGNITE DEPOSIT, FREESTONE COUNTY, TEXAS, ABST.,
B 6104 594 BIG CAESAR FIELD, TEXAS
B 6007 1011 BIG DRAW FAULT, NEW MEXICO
S 10 248 BIG ESCAMBIA CREEK FIELD, ETHANE CONCENTRATIONS, ALABAMA
B 6009 1615* BIG ESCAMBIA CREEK-JAY-BLACKJACK CREEK FIELD AREA, ABST.,
B 6104 602 BIG HATCHET MOUNTAINS, NEW MEXICO
B 6409 1503 BIG HORN BASIN, WYOMING
B 6008 1205 BIG HORN BASIN, WYOMING
B 6208 1499 BIG HORN BASIN, WYOMING
B 6108 1347 BIG HORN BASIN, WYOMING
B 6004 723* BIG HORN BASIN, WYOMING, ABST., PALEOENVIRONMENTS OF, FRONTIER
B 6108 1383* BIG HORN BASIN, WYOMING, ABST., DEPOSITION AND DIAGENESIS OF,
S 5 135 BIG MOUNTAIN SHALE, WEST VIRGINIA, KEYSER LIMESTONE,
B 6111 1980 BIG PINE FAULT, CALIFORNIA
B 6203 345 BIG PINE FAULT, CALIFORNIA
B 6107 1079 BIG PINE KEY, MIAMI OOLITE, MICROGRAPHS, FLORIDA KEYS
M 24 119 BIG PINEY ANTICLINE, WYOMING
M 24 96 BIG PINEY FIELD, WYOMING
M 24 91† BIG PINEY--LA BARGE PRODUCING COMPLEX, SUBLETTE AND LINCOLN
M 28 142 BIG SANDY FORMATION, ARIZONA, PLIOCENE
M 28 155 BIG SANDY FORMATION, ARIZONA, PLIOCENE
M 28 89 BIG SANDY FORMATION, ARIZONA, PLIOCENE
M 28 139 BIG SANDY FORMATION, ARIZONA, PLIOCENE
B 6405 631 BIG SIX SAND, WEST VIRGINIA
M 24 145† BIG WELLS FIELD, DIMMIT AND ZAVALA COUNTIES, TEXAS
B 6301 46 BIG WELLS FIELD, TEXAS
M 24 149 BIG WELLS, LO EAST, FIELD, TEXAS
M 24 149 BIG WELLS, SAN MIGUEL FIELD, TEXAS
B 6102 257 BIG WOOD RIVER VALLEY, NOMENCLATURE, IDAHO
M 24 235 BIGENERINA FLORIDANA ZONE, LOUISIANA
M 24 235 BIGENERINA-DISCORBIS ZONE, LOUISIANA
B 6004 651 BIGGS, M. E., COMPUTER-GENERATED SCALE FUNCTIONS FOR
B 6004 651 BIGGS, M. E., COMPUTER-GENERATED SCALE FUNCTIONS FOR
B 6005 879 BIGGS, MAURICE E., TRANSFORMATION OF RESISTIVITY TO
B 6303 439 BIGGS, ROBERT B., INTERPRETATION OF PLEISTOCENE COASTAL-BARRIER
B 6305 825* BIGHORN AN UINTA BASINS, WYOMING AND UTAH, ABST., TERTIARY
B 6307 1028 BIGHORN BASIN, WYOMING
B 6308 1390 BIGHORN BASIN, WYOMING
B 6202 185 BIGHORN BASIN, WYOMING
B 6305 822* BIGHORN BASIN, WYOMING, ABST., BOREHOLE GRAVITY STUDY OF
B 6009 1425† BIGHORN BASIN, WYOMING, DEFORMATION OF, MESOZOIC, SHALES AT,
M 28 7 BIGHORN DOLOMITE, WYOMING, ORDOVICIAN
B 6202 185 BIGHORN MOUNTAINS, WYOMING
B 6110 1746 BIGNELL, ROGER, PETROLEUM DEVELOPMENTS IN CENTRAL AND
B 6407 996 BIGOUDINE FORMATION, MOROCCO
M 29 293 BIJU-DUVAL, B., VARIETY OF MARGINS AND DEEP BASINS IN THE
B 6410 1644 BILLEROO CHANNEL, AUSTRALIA
B 6108 1372 BILLINGSLEY, LEE T., STRATIGRAPHY AND CLAY DIAGENESIS OF
B 6105 767 BILLO, SALEH M., FUTURE PETROLEUM RESOURCES IN SAUDI ARABIA
B 6303 420 BILLO, SALEH M., LITHOFACIES ANALYSIS OF CASTILE FORMATION,
B 6104 585 BINAK FIELD, PERSIAN GULF
B 6203 520* BINARY SIMILARITY COEFFICIENTS AND CONFIDENCE INTERVALS, ABST.
B 6107 1069† BINDING BY ENDOLITHIC ALGAE, MICRITIZATION AND, CARBONATE-,
B 6002 319* BINDLEY FIELD, HODGEMAN COUNTY, KANSAS, ABST., GEOLOGY OF,
B 6103 309† BINDLEY FIELD, HODGEMAN COUNTY, KANSAS, MISSISSIPPIAN
B 6105 767 BINSTOCK, JUTTA H, TURBIDITE SEDIMENTATION IN CAMBRIAN OF NORTH
S 11 207 BIOCHRONOLOGIC SUBDIVISIONS, HUMBOLDT BASIN, NEOGENE
S 6 39† BIOCHRONOLOGY
S 6 301 BIOCHRONOLOGY, ASSELIAN STAGE, PERMIAN
S 6 302 BIOCHRONOLOGY, BAIGENDZINIAN STAGE, PERMIAN
S 6 40 BIOCHRONOLOGY. BIOSTRATIGRAPHY. USED FOR
S 6 305 BIOCHRONOLOGY, DJULFIAN STAGE, PERMIAN
S 6 306 BIOCHRONOLOGY, DORASHAMIAN STAGE, PERMIAN
S 6 40 BIOCHRONOLOGY, FIRST APPEARANCE DATUM
S 6 51 BIOCHRONOLOGY, GLOBIGERINA NEPENTHES, EXAMPLE
S 6 51 BIOCHRONOLOGY, HIPPARION, EXAMPLE
S 11 200 BIOCHRONOLOGY, HUMBOLDT BASIN AND, BEAR RIVER AREA
S 6 304 BIOCHRONOLOGY, KAZANIAN STAGE, PERMIAN
S 6 304 BIOCHRONOLOGY, KUNGURIAN STAGE, PERMIAN
S 6 40 BIOCHRONOLOGY, LAST APPEARANCE DATUM

S 6 299 BIOCHRONOLOGY, MARINE FAUNAS, PERMIAN
S 6 304 BIOCHRONOLOGY, PUNJABIAN STAGE, PERMIAN
S 6 3 BIOCHRONOLOGY, REVIEW OF PRINCIPLE
S 6 301 BIOCHRONOLOGY, SAKMARIAN STAGE, PERMIAN
S 6 310 BIOCHRONOLOGY, STANDARD NOMENCLATURE, PERMIAN
B 6203 563* BIOCLASTIC ACCUMULATIONS IN MARINE MUDSTONES OF LATE
B 6303 424* BIOEROSION IN MASS-WASTING OF PLEISTOCENE OUTCROPS ON
M 27 184 BIOEROSION, SPONGE
B 6002 321* BIOFACIES ANALYSIS OF UPPERMOST HAMLIN SHALE AND AMERICUS
B 6105 839* BIOFACIES RELATIONS IN LOWER DEVONIAN STROMATOPOROID REEF
B 6004 650* BIOGENI CHARACTERISTICS OF SEDIMENTS FROM UPPER HUENEME
B 6408 1131† BIOGENIC AND THERMOGENIC ORIGINS OF NATURAL GAS IN COOK
B 6408 1131† BIOGENIC AND THERMOGENIC ORIGINS OF NATURAL GAS IN COOK
S 7 250 BIOGENIC CALCIUM CARBONATE, TEXAS, SHELF-SLOPE
B 6004 688* BIOGENIC CARBONATE SEDIMENTS IN DSDP CORES FROM PACIFIC
B 6004 679* BIOGENIC SEDIMENTARY STRUCTURES OF SOUTH TEXA OUTER
B 6105 814* BIOGENIC STRUCTURES AND THEIR PALEOENVIRONMENTAL
B 6405 680* BIOGENIC STRUCTURES AS INDICATORS O DEPOSITIONAL RATE, ABST.
B 6303 553* BIOGENOUS SEDIMENTS RECOVERED BY DEEP-SEA DRILLING, ABST.
S 4 181† BIOHERM THROUGH GEOLOGIC TIME, BRYOZOAN CONTRIBUTIONS TO,
B 6004 672* BIOHERMAL DEVELOPMENT IN WOLFCAMPIAN CARBONATES, LEONARD
M 27 68 BIOHERMAL MOUND, KANSAS, MISSISSIPPIAN
B 6203 539* BIOHERMS AND ASSOCIATED FACIES, NOLICHUCKY FORMATION, UPPER
B 6405 629† BIOHERMS AND INTERBIOHERM DEPOSITS OF WESTERN WEST VIRGINIA,
B 6405 629† BIOHERMS AND INTERBIOHERM DEPOSITS OF WESTERN WEST VIRGINIA,
S 4 75 BIOHERMS, CARBONATE FRAMEWORK, CARIBBEAN
B 6012 2117† BIOHERMS IN WENLOCK LIMESTONES, SILURIAN, OF SHROPSHIRE,
S 4 70 BIOHERMS, LESSER, ANTILLES, HOLOCENE
S 4 67† BIOHERMS OF LESSER ANTILLES--GEOLOGIC CONTROL OF DEVELOPMENT,
B 6405 767* BIOHERMS OF LOWER ORDOVICIAN AGE—ST. GEORGE GROUP, WESTERN
B 6405 807* BIOHERMS OF UPPER PALEOZOIC IN YUNNAN-GUIZHOU-GUANGXI
B 6303 421* BIOHERMS, OR NEW MODELS FOR REEFS, ABST., DEVELOPMENTAL PHASES
B 6405 631 BIOHERMS, WEST VIRGINIA, LOCKPORT
M 27 4 BIOLITHITE
M 27 176 BIOLITHITE, RED ALGAL
B 6207 1238* BIOMASS AS ENERGY SOURCE, ABST.
M 27 177 BIOMICRITE
M 27 178 BIOMICRITE, PACKED
M 27 175 BIOMICRITES
M 27 164 BIOSPARITE, CRINOID
M 27 178 BIOSPARITE, POORLY-WASHED
M 27 164 BIOSPARITE, POROUS
M 27 167 BIOSPARITES
M 27 173 BIOSPARITES
B 6405 704* BIOSPHERE—CHEMICAL EVIDENCE FROM DEEP-WEA CORES, ABST.,
B 6009 1562† BIOSTRATIGRAPH AND CLIMATIC STAGES OF PLEISTOCENE BRUNHES
B 6108 1383* BIOSTRATIGRAPHIC CORRELATION OF MESAVERDE GROUP IN
S 6 337 BIOSTRATIGRAPHIC DATA, A REVIEW, DEVONIAN
S 6 341 BIOSTRATIGRAPHIC DATING, A REVIEW, SILURIAN
B 6405 781* BIOSTRATIGRAPHIC PROBLEMS GENERATED BY DEE SEA CORING—
S 6 15 BIOSTRATIGRAPHIC SCALES, REVIEW
B 6409 1558* BIOSTRATIGRAPHIC SEQUENCE OF FOSSIL MOLLUSCA LARVAE, ABST.
B 6409 1558* BIOSTRATIGRAPHIC SIGNIFICANCE OF FOSIL MOLLUSCAN LARVAE, ABST.
S 6 273 BIOSTRATIGRAPHIC ZONES, CRETACEOUS
S 6 216 BIOSTRATIGRAPHY, A REVIEW, PALEOGENE
B 6405 677* BIOSTRATIGRAPHY, ABST., APPLICATIONS OF, PROBABL RANGE CONCEPT
S 6 329 BIOSTRATIGRAPHY, AMMONOIDS, CARBONIFEROUS
S 11 193 BIOSTRATIGRAPHY, BEAR RIVER AREA
S 6 329 BIOSTRATIGRAPHY, BRACHIOPODS, CARBONIFEROUS
S 6 329 BIOSTRATIGRAPHY, CONODONTS, CARBONIFEROUS
S 6 343 BIOSTRATIGRAPHY, CORRELATION WITH ISOTOPIC, SILURIAN
S 11 193 BIOSTRATIGRAPHY, HUMBOLDT BASIN
S 11 201 BIOSTRATIGRAPHY, HUMBOLDT BASIN, NEOGENE
M 30 496 BIOSTRATIGRAPHY, INDIA, BOMBAY HIGH FIELD
S 6 145 BIOSTRATIGRAPHY, JAPAN, CRETACEOUS
S 6 260 BIOSTRATIGRAPHY, NORTH AMERICA, CRETACEOUS
B 6009 1614* **BIOSTRATIGRAPHY OF PLEISTOCENE DEPOCENTER--TEXAS-LOUISIANA**
B 6009 1608* BIOSTRATIGRAPHY OF BOSSIER SHALE AND JURASSIC-CRETACEOUS
S 11 183† BIOSTRATIGRAPHY OF CALIFORNIA, PLIOCENE
B 6303 536* BIOSTRATIGRAPHY OF EARLY AND EARLIEST LATE CRETACEOUS
B 6105 821* BIOSTRATIGRAPHY OF U.S. ATLANTIC CONTINENTAL MARGIN, ABST.,
S 6 330 BIOSTRATIGRAPHY, PLANT MEGAFOSSILS, CARBONIFEROUS
S 6 217 BIOSTRATIGRAPHY, REVIEW, PALEOGENE, PLANKTONIC
S 6 329 BIOSTRATIGRAPHY, SMALLER FORAMINIFERA, CARBONIFEROUS
B 6405 750* BIOSTRATIGRAPHY THROUGH GRAPHIC CORRELATIONS, ABST., NEW
M 25 428 BIOSTRATIGRAPHY TO, METAL CONCENTRATIONS, RELATION OF
S 6 40 BIOSTRATIGRAPHY, USED FOR BIOCHRONOLOGY
B 6404 527† BIOSTROMES OF NORTH GREENLAND, PALEOGEOGRAPHIC AND
M 26 36 BIOT-GEERTSMA EQUATIONS, LABORATORY, MEASUREMENTS
B 6203 559* BIOTA AND BIOLOGIC PROCESSES OF CONTINENTAL SLOPE, ABST.
M 28 45 BIOTITE, DETRITAL
B 6405 685* BIOTURBATION AS FACTOR IN HYDROCARBON GENERATION—EXAMPLE
M 24 99 BIRCH CREEK FIELD, WYOMING
M 24 99 BIRCH CREEK SANDSTONE, WYOMING
B 6303 454* BIRCH CREEK VALLEY, EAST-CENTRAL IDAHO--DIAGENETIC MODEL,
B 6408 1132 BIRCH HILL GAS FIELD, ALASKA
B 6305 844 BIRCH, MICHAEL J., PETROLEUM PRODUCTION FROM BASAL GREENBRIER
B 6012 2175 BIRD, K. J., ENVIRONMENTAL AND TIME SIGNIFICANCE OF LOCAL GUIDE
B 6004 649 BIRD, K. J., LISBURNE GROUP--POTENTIAL MAJO OBJECTIVE OF EASTERN
B 6405 752 BIRD, K. J., STRATIGRAPHIC RELATIONS OF NANUSHUK GROUP, MIDDLE
B 6109 1493 BIRD, KENNETH J., LISBURNE GROUP, MISSISSIPPIAN AND
B 6205 880 BIRD, KENNETH J., NEW INFORMATION ON LISBURNE GROUP,
B 6203 510 BIRD, KENNETH J., SAND TRANSPORT ACROSS EOCENE SHELF AND SLOPE
B 6201 74 BIRDBEAR LIMESTONE-DOLOMITE, SASKATCHEWAN
B 6105 767 BIRDSALL, BARTON, HURRICANE IMPACT ON CONTINENTAL-SHELF
B 6004 650 BIRDSALL, B. C., PHYSICAL AND BIOGENI CHARACTERISTICS OF
M 28 128 BIREFRINGENCE, BARITE, IN LIMESTONE
M 28 137 BIREFRINGENCE, CEMENTS, KAOLINITE
M 28 134 BIREFRINGENCE, CHLORITE, AUTHIGENIC
M 28 142 BIREFRINGENCE, ETC., ANALCIME (ZEOLITE)
M 28 140 BIREFRINGENCE, ETC., CLINOPTILOLITE
M 28 140 BIREFRINGENCE, ETC., ERIONITE
M 28 141 BIREFRINGENCE, ETC., LAUMONTITE
M 28 141 BIREFRINGENCE, ETC., SEARLESITE
M 28 128 BIREFRINGENCE, GLAUCONITE, GRAINS
B 6203 496 BIRKHOLZ, DONALD O., URANIUM DEPOSITS IN VOLCANICLASTIC ROCKS
B 6309 1490 BIRNIE, RICHARD W., RECONNAISSANCE GEOLOGIC MAPPING IN CHAGAI
B 6010 1813 BIRO, PHILIPPE, PETROLEUM DEVELOPMENTS IN CENTRAL AND
B 6311 2059 BIRTISH COLUMBIA, NANAIMO COALFIELD
M 29 373 BISCARRAT, P. A., THE SULU SEA: A MARGINAL BASIN IN SOUTHEAST
B 6211 2232† BISCAY AND ASSOCIATED TURBIDITY CURRENTS, A, SIGSBEE, KNOLL IN
M 28 139 BISEFRINGENCE, ETC., PHILLIPSITE
M 27 34 BISERIAL FORAMS
B 6207 1209 BISHOP, C. P., MINE DEVELOPMENT IN PACIFIC ISLANDS—IMPORTANCE
B 6405 789 BISHOP, CLIVE D., FAN MODELS FOR HYDROCARBON EXPLORATION
B 6306 918 BISHOP, RICHARD S., CALCULATED COMPACTION STATES OF THICK
B 6105 768 BISHOP, RICHARD S., COMPACTION MODEL OF THICK, ABNORMALLY
B 6209 1561 BISHOP, RICHARD S., MECHANISM FOR EMPLACEMENT OF PIERCEMENT
B 6105 768 BISHOP, RICHARD S., NONBUOYANT DIAPIR EMPLACEMENT
B 6008 1390 BISHOP, ROBERT A., SALT SOLUTION-COLLAPSE, WILLISTON
B 6405 730 BISHOP, ROBERT A., WHITNEY CANYON FIELD—POTENTIAL GAS GIANT IN
B 6004 650 BISHOP, RONALD L., NUMERICAL ASPECTS OF CHEMICAL VARIATION
B 6401 37 BISHOP, WILLIAM F., STRUCTURE, STRATIGRAPHY, AND HYDROCARBONS
B 6401 37 BISHOP, WILLIAM F., STRUCTURE, STRATIGRAPHY, AND HYDROCARBONS
B 6309 1582* BISHOP-BRADSHAW CREEK FAULT, ABST.
M 25 474 BISMARCK SEA MEGASTRUCTURE
M 25 558 BISMUTH MINES, BOLIVIA
B 6112 2139 BISTI FIELD, NEW MEXICO
B 6410 1688 BITTER SPRINGS FORMATION, CELLULAR FILAMENTS AND SHEATHS
B 6211 2218 BITTERWATER BASIN, CALIFORNIA
S 10 59 BITUMEN COMPOSITION, IMMATURE ORGANIC MATTER
S 10 56 BITUMEN REDISTRIBUTION IN SOURCE ROCKS, MIGRATION
B 6004 673* BITUMEN, ABST., SIMULATED, GEOTHERMAL, MATURATION OF,
B 6406 921 BITUMEN AND ORGANIC CARBON RATIOS
B 6412 2109 BITUMEN, BERING SEA, ORGANIC CARBON AND
B 6112 2098 BITUMEN GENERATED, CHARACTER OF
B 6312 2135 BITUMEN, ORGANIC, GEOCHEMISTRY PARAMETERS OF
B 6312 2227† BITUMEN SATURATION ON SOURCE-ROCK PYROLYSIS, EFFECT OF, OIL
B 6405 776* BITUMEN SATURATION, PEACE RIVER OIL SANDS DEPOSIT, ALBERTA,
B 6202 302 BITUMENS AND COALS, ATOMIC RATIOS OF
B 6202 301† BITUMENS AND COALS, CHARACTERIZATION OF
B 6311 2115* BITUMENS AND KEROGEN IN SHALE CLAST, ABST., DISTRIBUTION OF
S 10 61 BITUMENS AT HIGH TEMPERATURES, MIGRATION
B 6303 507* BITUMINOU COALS, ABST., PORPHYRINS IN, SUBBITUMINOUS AND
B 6405 761* BITUMINOUS COAL FROM APPALACHIAN AND EASTERN INTERIOR
B 6408 1289* BITUMINOUS COAL RESOURCE, ABST., ILLINOIS COAL—A MAJOR
S 10 180 BITUMINOUS SHALES, METHANE, OCCURRENCES
B 6004 660* BITUMINOUS SUBSTANCES, ABST., PALYNOMORPHS IN PETROLEUM AND
B 6405 702* BIVALVE TRACE FOSSILS IN BRITISH SILESIAN, ABST.
S 4 271 BIVALVED MOLLUSKS, BORERS
B 6304 686 BIVENS, HUGH M., DIRECT MEASUREMENT OF URANIUM BY PROMPT
B 6405 754* BIYAD-WASI SANDSTONES, LOWER MIDDLE CRETACEOUS, OF CENTRAL
B 6303 542* BLAC SHALES IN ATLANTIC BASINS, ABST., PALEOENVIRONMENT AND
B 6108 1372 BLACK, BRUCE A., MEGAEXPLORATION FOR NEW PETROLEUM
B 6305 822 BLACK, BRUCE A.., OIL AND GAS POTENTIAL OF SANTA FE EMBAYMENT,
B 6002 164 BLACK HILLS, BERINGIAN SHELF
B 6211 2280 BLACK HILLS, HULETT SANDSTONE MEMBER
B 6211 2283 BLACK HILLS, LAK MEMBER
B 6405 783* BLACK HILLS, MONTANA AND SOUTH DAKOTA, ABST., GEOMETRY OF
B 6406 943 BLACK HILLS, NORTHERN, UPPER CRETACEOUS, SOUTH DAKOTA

B 6004 711* BLACK HILLS REGION, ABST., LATE, JURASSIC, BARRIER ISLAND COMPLEX
B 6412 2142 BLACK HILLS RIDGE, BERING SEA SHELF
B 6211 2275† BLACK HILLS, SEDIMENTOLOGY OF LATE, JURASSIC, BARRIER-ISLAND
B 6211 2275 BLACK HILLS, SOUTH DAKOTA
S 3 93 BLACK HILLS, SOUTH DAKOTA
B 6211 2280 BLACK HILLS, SPEARFISH FORMATION
B 6211 2279 BLACK HILLS, STOCKADE BEAVER SHALE MEMBER
B 6211 2275 BLACK HILLS, SUNDANCE FORMATION, CALLOVIAN
B 6211 2275 BLACK HILLS, WYOMING
B 6202 184 BLACK HILLS, WYOMING
B 6410 1615 BLACK HOLLOW FIELD, COLORADO
M 28 70 BLACK LEAF FORMATION, MONTANA, CRETACEOUS
B 6008 1401* BLACK MESA BASIN, ARIZONA, ABST., OIL AND GAS POSSIBILITIES,
M 25 448 BLACK ORE DEPOSITS, JAPAN, KUROKO OR
M 27 173 BLACK RIVER LIMESTONE, PENNSYLVANIA, ORDOVICIAN
M 27 72 BLACK RIVER LIMESTONE, PENNSYLVANIA, ORDOVICIAN
M 27 222 BLACK RIVER LIMESTONE, PENNSYLVANIA, ORDOVICIAN
M 27 200 BLACK RIVER LIMESTONE, PENNSYLVANIA, ORDOVICIAN
M 27 206 BLACK RIVER LIMESTONE, PENNSYLVANIA, ORDOVICIAN
M 27 75 BLACK RIVER LIMESTONE, PENNSYLVANIA, ORDOVICIAN
M 25 420 BLACK SANDS, AUSTRALIA
B 6408 1189 BLACK SEA, ANOXIC SILLED BASIN
M 29 305 BLACK SEA BASIN, CROSS SECTION, USSR
M 29 313 BLACK SEA BASIN, EVOLUTION, USSR
B 6306 875 BLACK SEA BASIN, MODEL
M 29 313 BLACK SEA BASIN, SEDIMENTARY SEQUENCE, USSR
B 6408 1191 BLACK SEA, POSITIVE WATER BALANCE BASIN
M 29 431 BLACK SEA, SMALL OCEAN BASIN, ORGANIC SEDIMENTS
B 6405 720* BLACK SHALE, ABST., AUTHIGENIC QUARTZ IN, DEVONIAN
B 6107 1063 BLACK SHALE AND PHOSPHORITE, MID-CONTINENT, PALEOZOIC
B 6107 1058 BLACK SHALE DEPOSITION, MODELS OF, PENNSYLVANIAN
B 6209 1594 BLACK SHALE, DSDP SITE WITH RECOVERY OF, CRETACEOUS
B 6107 1045† BLACK SHALE FACIES IN PENNSYLVANIAN CYCLOTHEMS OF
B 6309 1578* BLACK SHALE FORMATIONS, ABST., NOVEL FORMATION-EVALUATION
B 6405 729* BLACK SHALE IN EAST-CENTRAL KENTUCKY, ABST., TRACE, FOSSILS AND
B 6108 1375* BLACK SHALE MEMBERS OF PERMIAN PHOSPHORIA FORMATION,
B 6201 98† BLACK SHALE MEMBERS OF PHOSPHORIA FORMATION, WESTERN
B 6304 656 BLACK SHALE, OHIO, DEVONIAN
B 6405 692* BLACK SHALES, APPALACHIAN BASIN, ABST., GEOCHEMICAL EFFECTS OF
B 6412 2051 BLACK SHALES, ATLANTIC OCEAN, DEEP SEA DRILLING PROJECT
B 6309 1592* BLACK SHALES, ERIE COUNTY PENNSYLVANIA, ABST., PRESSURE, CORING
B 6303 503* BLACK SHALES FROM SOUTH ATLANTIC, DSDP LEGS 36,
B 6303 542* BLACK SHALES IN ATLANTIC BASINS, ABST., PALEOENVIRONMENT AND
B 6412 2051† BLACK SHALES IN ATLANTIC BASINS, PALEOENVIRONMENT AND
B 6405 790* BLACK SHALES IN DEEP NORTH ATLANTIC, ABST., ORGANIC FACIES OF
B 6303 462* BLACK SHALES IN SUBSURFACE PENNSYLVANIA--STRATIGRAPHY AND
B 6004 720* **BLACK WARRIOR BASIN, ALABAMA AND MISSISSIPPI, ABST., DEPOSITION,**
B 6204 646 BLACK WARRIOR BASIN, ALABAMA
B 6110 1897† BLACK WARRIOR BASIN, CAMBRIAN SYSTEM IN
B 6409 1555* BLACK WARRIOR BASIN, MISSISSIPPI AND ALABAMA, ABST., CHESTER
B 6009 1616* BLACK WARRIOR BASIN, MONROE COUNTY, MISSISSIPPI, ABST.,
B 6309 1596* BLACK WARRIOR BASIN OF MISSISSIPPI AND ALABAMA, ABST., UPPER,
B 6110 1897 BLACK WARRIOR BASIN, PRECAMBRIAN
B 6410 1590 BLACKFORD FORMATION, VIRGINIA
B 6307 1137 BLACKGUM FORMATIONS, OKLAHOMA, TENKILLER-
B 6205 879* BLACKHAWK FORMATION, CRETACEOUS, AND ITS RELATION TO COAL
B 6011 1985 BLACKIE, GARY W., MAGNETIC-REVERSAL STRATIGRAPHY OF
B 6009 1615* BLACKJACK CREEK FIELD AREA, ABST., PALEOENVIRONMENTAL
S 10 249 BLACKJACK CREEK FIELD, METHANE DISTRIBUTION
S 10 244 BLACKJACK CREEK FIELD, ETHANE CONCENTRATIONS, FLORIDA
S 10 252 BLACKJACK CREEK FIELD, METHANE CONCENTRATIONS, FLORIDA
S 10 246 BLACKJACK CREEK FIELD, SAND PERCENTAGE CONCENTRATIONS,
S 10 249 BLACKJACK CREEK FIELD, SUBSURFACE GEOLOGY, FLORIDA
B 6407 972 BLACKLEAF FORMATION, GREAT PLAINS
B 6008 1410 BLACKWELL, DAVID D., HEAT-FLOW STUDIES IN STEAMBOAT
B 6103 346 BLACKWELL FIELD, TEXAS
B 6203 497 BLACKWELL, R. J., DIGITAL IMAGE PROCESSING OF POTENTIAL FIELD
B 6401 105 BLACO FIELD, TEXAS
B 6401 105 BLACO FIELD, TEXAS
B 6303 342 BLAINE FAULT, KENTUCKY
B 6002 320* BLAINE FORMATION AND ASSOCIATED STRATA IN SOUTHWEST
B 6212 2398 BLAINE-WALBRIDGE FAULT, KENTUCKY
B 6309 1479 BLAIR DOLOMITE, TEXTURE AND COMPOSITION DATA
B 6207 1159 BLAIR, WILL N., GULF OF CALIFORNIA IN LAKE MEAD AREA OF ARIZONA
B 6307 1139 BLAIR, WILL N., GULF OF CALIFORNIA IN LAKE MEAD AREA OF ARIZONA
B 6201 74 BLAIRMORE SAND-SHALE-SILT, SASKATCHEWAN
B 6308 1292 BLAISDELL, R. C., DEVELOPMENTS IN WEST COAST AREA IN 1978
B 6409 1393 BLAISDELL, R. C., DEVELOPMENTS IN WEST COAST AREA IN 1979
S 10 190 BLAKE OUTER RIDGE, METHANE-HYDRATE, SEISMIC SKETCH
B 6408 1287 BLAKE, B. MITCHELL, JR., COAL IN WEST VIRGINIA—GEOLOGY AND
B 6403 340 BLAKE BAHAMA BASIN
B 6403 340 BLAKE BAHAMA BASIN
B 6204 597 BLAKE BASIN, JOIDES, DSDP, SITE 391, LEG 44
B 6205 793 BLAKE BASIN, SEISMIC REFLECTION PROFILE, MULTIFOLD
B 6211 2351 BLAKE, CLARK, FRANCISCAN ASSEMBLAGE OF NORTHERN CALIFORNIA, A
B 6403 340 BLAKE ESCARPMENT
B 6403 340 BLAKE ESCARPMENT
M 29 179 BLAKE ESCARPMENT, AEROMAGNETIC MAP
M 29 179 BLAKE ESCARPMENT, ANALOG RECORD OF PROFILE
M 29 177† BLAKE ESCARPMENT BASED ON SEISMIC REFLECTION PROFILES.
M 29 178 BLAKE ESCARPMENT, BOUGUER GRAVITY PROFILE
M 29 178 BLAKE ESCARPMENT, INDEX MAP
M 29 180 BLAKE ESCARPMENT, MIOCENE, CHALK OUTCROPS
M 29 182 BLAKE ESCARPMENT, MIOCENE, SEISMIC REFLECTORS
M 29 179 BLAKE ESCARPMENT, MULTICHANNEL SEISMICS
B 6205 793 BLAKE ESCARPMENT, SEISMIC REFLECTION PROFILE, MULTIFOLD
M 29 182 BLAKE ESCARPMENT, VELOCITY ANALYSES
B 6105 768 **BLAKE, GREGG H., DISTRIBUTION AND TEMPORAL MIGRATION OF BANK-**
B 6012 2175 BLAKE, GREGG H., PALEOCEANOGRAPHIC VARIATIONS AND THEIR
B 6203 344 **BLAKE, M. C., NEOGENE BASIN FORMATION AND HYDROCARBON**
B 6004 650 BLAKE, M. C., NEOGENE BASI FORMATION AND HYDROCARBON
B 6205 793 BLAKE NOSE
B 6312 2206 BLAKE OUTER RIDGE, PROFILE
B 6205 793 BLAKE OUTER RIDGE, SEISMIC REFLECTION PROFILE, MULTIFOLD
M 29 30 BLAKE PLATEAU, AGE OF REFLECTORS
B 6205 792† BLAKE PLATEAU AND ADJACENT WESTERN ATLANTIC CONTINENTAL
B 6205 793 BLAKE PLATEAU AREA, APTIAN, PROFILE
B 6205 794 BLAKE PLATEAU AREA, BASEMENT
B 6205 793 BLAKE PLATEAU AREA, CENOMANIAN, PROFILE
B 6205 793 BLAKE PLATEAU AREA, CENOZOIC, PROFILE
B 6205 793 BLAKE PLATEAU AREA, CRETACEOUS, PROFILE
B 6205 804 BLAKE PLATEAU AREA, DEEP OCEAN, SEISMIC INTERVALS
B 6205 793 BLAKE PLATEAU AREA, HOLOCENE, PROFILE
B 6205 793 BLAKE PLATEAU AREA, MIOCENE, PROFILE
B 6205 793 BLAKE PLATEAU AREA, NEOCOMIAN, PROFILE
B 6205 793 BLAKE PLATEAU AREA, OLIGOCENE, PROFILE
B 6205 793 BLAKE PLATEAU AREA, OXFORDIAN, PROFILE
B 6205 793 BLAKE PLATEAU AREA, PLIOCENE, PROFILE
B 6205 793 BLAKE PLATEAU AREA, VOLCANIC BASEMENT, PROFILE
M 29 105 BLAKE PLATEAU BASIN, ATLANTIC MARGIN
B 6205 803 BLAKE PLATEAU, CENOZOIC
M 29 38 BLAKE PLATEAU, CONTRASTS IN DEVELOPMENT
B 6205 798 BLAKE PLATEAU, CRETACEOUS
M 29 35 BLAKE PLATEAU, DEPTH TO CRETACEOUS, MAP
M 29 37 BLAKE PLATEAU, DEPTH TO JURASSIC REFLECTOR, MAP
M 29 37 BLAKE PLATEAU, DEVELOPMENT
M 29 30 BLAKE PLATEAU, DRILL-SITE SELECTION
B 6205 807 BLAKE PLATEAU, GEOLOGIC HISTORY
M 29 39 BLAKE PLATEAU, GEOLOGIC HISTORY, MODEL
M 29 27 BLAKE PLATEAU GEOLOGIC SETTING
M 29 11 BLAKE PLATEAU, GEORGIA EMBAYMENT, RELATION TO
B 6205 798 BLAKE PLATEAU, JURASSIC
M 29 31 BLAKE PLATEAU, JURASSIC VOLCANICS
M 29 30 BLAKE PLATEAU, MULTICHANNEL SEISMIC PROFILE
M 29 40 BLAKE PLATEAU, NORTHERN, GEOLOGIC HISTORY
M 29 34 BLAKE PLATEAU, REGIONAL STRUCTURE
M 29 40 BLAKE PLATEAU, RELATION TO CRUST
B 6205 798 BLAKE PLATEAU, SEISMIC INTERVALS
B 6205 793 BLAKE PLATEAU, SEISMIC REFLECTION PROFILE, MULTIFOLD
B 6205 799 BLAKE PLATEAU, SEISMIC STRATIGRAPHY
B 6403 347 BLAKE PLATEAU, SEISMIC STUDY, FLORIDA-, HATTERAS SHELF AND
B 6403 347 BLAKE PLATEAU, SEISMIC STUDY, FLORIDA-, HATTERAS SHELF AND
M 29 36 BLAKE PLATEAU, STRUCTURE, ANGULAR UNCONFORMITY
M 29 35 BLAKE PLATEAU, STRUCTURE CONTOURS, REFLECTORS
B 6403 339† BLAKE PLATEAU, STRUCTURE, STRATIGRAPHY, AND GEOLOGIC HISTORY
M 29 34 BLAKE PLATEAU, STRUCTURE, TOP OF CRETACEOUS
B 6205 805 BLAKE PLATEAU, TERTIARY
M 29 30 BLAKE PLATEAU, TOP OF CRETACEOUS
M 29 31 BLAKE PLATEAU, TOP OF NEOCOMIAN
M 29 31 BLAKE PLATEAU, TOP OF TITHONIAN
M 29 34 BLAKE PLATEAU, TRIASSIC, ANGULAR UNCONFORMITY
B 6105 829* BLAKE PLATEAU-BAHAMA REGION OF ATLANTIC MARGIN OF NORTH
M 29 27† BLAKE PLATEAU: PRELIMINARY ANALYSIS, STRUCTURE AND,
B 6403 339† BLAKE PLLTEAU, STRUCTURE, STRATIGRAPHY, AND GEOLOGIC HISTORY
B 6403 340 BLAKE SPUR
B 6205 793 BLAKE SPUR MAGNETIC ANOMALY
B 6410 1659 BLAKE SPUR MAGNETIC ANOMALY
B 6406 851 BLAKE-BAHAMA ABYSSAL PLAIN, SAND LAYERS
B 6412 2058 BLAKE-BAHAMA AREA, DSDP SITE 391, LEG 44, GEOCHEMICAL LOG
B 6303 445* BLAKE-BAHAMA BASIN, ABST., MIOCENE, CARBONATE GRAVITY, FLOW IN

M 29 181 BLAKE-BAHAMA BASIN, SEISMIC REFLECTORS, ATLANTIC
B 6409 1662 BLAKE-BAHAMA FORMATION, WESTERN ATLANTIC
B 6410 1701 BLAKE-BAHAMA OUTER RIDGE, ABYSSAL SEDIMENT DRIFT
B 6004 651 BLAKELY, R. F., COMPUTER-GENERATED SCALE FUNCTIONS FOR
B 6004 651 BLAKELY, R. F., COMPUTER-GENERATED SCALE FUNCTIONS FOR
B 6105 769 BLAKELY, ROBERT F., INFORMATION RETRIVAL SYSTEM FOR WELL DATA
B 6005 879 BLAKELY, ROBERT F., TRANSFORMATION OF RESISTIVITY TO
B 6205 880 BLAKEY, RONALD C., LOWER TRIASSIC HINGE LINE AND SHELF TAR
B 6405 677 BLAKEY, RONALD C., PENNSYLVANIAN AND EARLY PERMIAN
B 6106 904 BLANCA BASIN, BAJA CALIFORNIA
B 6203 494 BLANCHARD, D. P., TRACE ELEMENTS IN HOLOCENE FLUVIAL SANDS
B 6109 1551* BLANCO AND HAYS COUNTIES, TEXAS, ABST., MIDDLE, GLEN ROSE,
B 6008 1392* BLANCO FIELD, SAN JUAN BASIN, RELATED TO SOURCE-ROCK TYPE,
B 6112 2140 BLANCO GAS FIELD, GREEN MARKER HORIZON, NEW MEXICO
B 6009 1621 BLAND, A. E., GEOPHYSICAL AND TECTONIC STUDY OF MOORMAN
B 6405 710 BLAND, F. X., DEEP TUSCALOOSA GAS TREND OF SOUTH LOUISIANA,
B 6405 677 BLANK, RICHARD G., APPLICATIONS OF PROBABL RANGE CONCEPT TO
B 6103 465 BLANTON, SANKEY L., GEOLOGY OF BERING SHELF, ABST.
B 6011 2067 BLANTON, SANKEY L., JR., GEOLOGY OF BERING SHELF, ABST.
B 6410 1726 BLATT, HARVEY, REVIEW OF DIAGENESIS IN SEDIMENTS AND
B 6108 1372 BLATT, HARVEY, DIAGENETIC PROCESSES IN SANDSTONES, ABST.
B 6409 1554 BLAUSER, W. H., CALPIONELLIDS AND NANNOCONIDS OF TARAISES
B 6105 769 BLECHSCHMIDT, GRETCHEN, OLIGOCENE CALCAREOUS NANNOFOSSIL
B 6303 447* BLIGH WATER, FIJI, ABST., DETERMINATION OF DEPTH TO
B 6211 2203 BLIND HORSES COVE GROWTH FAULTS, IRELAND, WESTERN, KILLARD-
B 6104 603 BLISS FORMATION, NEW MEXICO
B 6109 1450 BLISS SANDSTONE, PEDREGOSA BASIN
B 6405 677 BLOCH, S., DISTRIBUTION AND ALTERATION OF OGALLALA
B 6409 1355 BLOCK 154, HIGH ISLAND AREA, TEXAS
B 6002 327* BLOCK 16 FIELD, WARD COUNTY, TEXAS, ABST.
B 6402 155 BLOCK 24 FIELD, LOUISIANA, SOUTH PASS
B 6402 155 BLOCK 24 FIELD, LOUISIANA, SOUTH PASS
B 6409 1569* BLOCK 25 FIELD, CHANDELEUR SOUND, ST. BERNARD PARISH,
B 6402 165 BLOCK 27 FIELD, LOUISIANA, SOUTH PASS
B 6402 165 BLOCK 27 FIELD, LOUISIANA, SOUTH PASS
B 6208 1382 BLOCK 27-L FIELD, HIGH ISLAND AREA, TEXAS
M 24 205† BLOCK 270, OFFSHORE LOUISIANA--A PLEISTOCENE, EAST CAMERON
B 6303 467* BLOCK 330 FIELD, OFFSHOR LOUISIANA, ABST., EUGENE ISLAND
B 6105 834* BLOCK 54 SALT DOME, ABST., PALEONTOLOGIC TECHNIQUES FOR
B 6209 1768* BLOCK 54 SALT DOME AND SALT DOME GROWTH MODELS, ABST.,
B 6109 1429 BLOCK 73 FIELD, LOUISIANA, WEST DELTA
B 6209 1760* BLOCKS A-560 AND A-561, ABST., PRE-PLATFORM EXPLORATION OF, HIGH
B 6303 291 BLOCKS, PROSPECTS, MARGINAL-, PLATEAU
B 6405 678 BLODGETT, ROBERT H., TRIASSIC PALEOCALICHE IN RED BEDS O
B 6303 420 BLOESER, BONNIE, MELAMOCYRILLIUM--NEW ACRITARCH GENUS FRO
B 6405 619 BLOM, RONALD, SEASAT RADAR IMAGE OF SAN ANDREAS FAULT,
B 6303 421 BLOME, CHARLES D., UPPER TRIASSIC RADIOLARIA FROM EASTERN
B 6103 344 BLOOMER, RICHARD R., DEPOSITIONAL ENVIRONMENTS OF A RESERVOIR
B 6211 2309 BLOOMSBURG FORMATION, WEST VIRGINIA
B 6209 1610 BLOOMSBURG RED BEDS, MARYLAND
B 6209 1610 BLOOMSBURG RED BEDS, PENNSYLVANIA
B 6103 467 BLOOMSTEIN, EDWARD, COMPRESSIVE INTERACTION OF SIBERIAN AND
B 6002 318 BLOUSTINE, DONALD A., PRODUCING CHARACTERISTICS OF LOWER
M 27 86 BLOYD FORMATION, OKLAHOMA, PENNSYLVANIAN
M 27 35 BLOYD FORMATION, OKLAHOMA, PENNSYLVANIAN
M 27 118 BLOYD FORMATION, OKLAHOMA, PENNSYLVANIAN
M 27 116 BLOYD FORMATION, OKLAHOMA, PENNSYLVANIAN
M 27 167 BLOYD FORMATION, OKLAHOMA, PENNSYLVANIAN
M 27 96 BLOYD FORMATION, OKLAHOMA, PENNSYLVANIAN
B 6406 943 BLUE HILL MEMBER, MINNESOTA AND, SOUTH DAKOTA
S 4 85 BLUE MOUNTAIN INLIER, JAMAICA
B 6111 1930 BLUE MOUNTAINS, ELLESMERE ISLAND
B 6403 380 BLUE MOUNTAINS, JAMAICA
B 6403 380 BLUE MOUNTAINS, JAMAICA
B 6403 340 BLUE RIDGE PROVINCE
B 6403 340 BLUE RIDGE PROVINCE
B 6410 1576 BLUE RIDGE THRUST, VIRGINIA
M 24 119 BLUE RIM ARCH, WYOMING, MONUMENT-
S 3 130 BLUE WATER FAULT, NEW MEXICO
M 27 3 BLUE-GREEN ALGAL BODIES
M 27 47 BLUE-GREEN ALGAL CRUST
M 27 4 BLUE-GREEN ALGAL ENCRUSTATION
M 27 3 BLUE-GREEN ALGAL MAT, LAMINATED
M 27 189 BLUE-GREEN ALGAL ONCOLITE
M 27 4 BLUE-GREEN ALGAL STROMATOLITE
M 27 6 BLUE-GREEN ALGAL STROMATOLITE
M 24 121† BLUEBELL--A MAJOR, NATURALLY FRACTURED STRATIGRAPHIC TRAP,
B 6304 652 BLUEFIELD FORMATION, VIRGINIA
B 6405 678 BLUEFORD, JOYCE R., MIOCENE SPUMELLARIAN RADIOLARIA FROM
B 6103 374 BLUEJACKET SANDSTONE, OKLAHOMA, BARTLESVILLE-
B 6007 1009 BLUEJACKET, SANDSTONE, OKLAHOMA, BARTLESVILLE
B 6007 1011 BLUEWATER FAULT, NEW MEXICO
B 6111 2002 BLUFF COVE AREA, PALOS VERDES HILLS, CALIFORNIA
B 6007 1013 BLUFF SANDSTONE, NEW MEXICO
B 6007 1011 BLUFF SANDSTONE, NEW MEXICO
S 3 133 BLUFF SANDSTONE, NEW MEXICO
B 6209 1675 BLUFF SANDSTONE, NEW MEXICO
M 25 463 BLUMSTEIN, E. I., SOME CHARACTERISTICS OF ACID VOLCANISM OF
B 6410 1689 BLYTH DOLOMITE, SPHEROIDAL MICROFOSSILS
B 6105 769 BOARDMAN, MARK, LIME MUD DEPOSITION IN ENCLOSED LAGOON, BIGHT
B 6305 822 BOBERG, W. W., APPLIED EXPLORATION GEOLOGY AND URANIUM
M 25 526 BOBIER, CLAUDE, MAJOR CHEMICAL CHARACTERISTICS OF TERTIARY
B 6405 674 BOCK, WAYNE D., ET AL, EXPLORABLE STRUCTURES IN OLD BAHAMA
B 6203 352 BODEGA BASIN, CALIFORNIA
B 6211 2215 BODEGA HEAD, CALIFORNIA
M 28 165 BOEHM LAMELLAE, IN QUARTZ, COMPACTION
M 28 9 BOEHM LAMELLAE, QUARTZ
B 6209 1754 BOELLSTORFF, JOHN, NEED FOR REDEFINITION OF NORTH AMERICAN
B 6303 538 BOERSMA, A., MIDDLE CRETACEOUS OXYGEN-DEFICIENT
B 6008 1391 BOETHLING, FREDERICK C., VIKING FORMATION, ALBERTA, ABST.
B 6104 613 BOFECILLOS MOUNTAINS, TEXAS
B 6203 497 BOGUCHWAL, LAWRENCE A., DYNAMIC SCALE MODELING OF BED
B 6301 69 BOHEMIAN MASSIF, EUROPE
B 6212 2510* BOHOKU ARC, JAPAN, ABST., STRUCTURAL CONTRAST BETWEEN,
B 6004 651 BOHOR, B. F., ALTERED VOLCANIC ASH PARTINGS AS STRATIGRAPHIC
S 4 283 BOILER BAY, ECHINOMETRA LUCUNTER, ST. CROIX
S 4 283 BOILER BAY, PORITES ASTREOIDES, ST. CROIX
B 6209 1610 BOIS BLANC LIMESTONE, OHIO
B 6004 651 BOIS, C., MULTIDIMENSIONAL APPROACHES TO GEOLOGIC CLASSIFICATIO
B 6105 770 BOIS, C., MULTIDIMENSIONAL APPROACHES TO GEOLOGIC
B 6008 1389* BOISE FRONT, IDAHO, FOR GEOTHERMAL RESOURCE EVALUATION, ABST.,
M 26 23 BOISE SANDSTONE, LABORATORY, MEASUREMENTS
S 4 135 BOKER, ALEXA M., STRUCTURAL MODIFICATION OF CORALS IN
B 6105 770 BOKUNIEWICZ, H. J., CONTAINMENT OF PARTICULATE WASTE IN MARINE
B 6004 651 BOLE, GEORGE R., WELL-DATA DISPLAY SYSTEM--INTERACTIVE
M 25 515 BOLEO FORMATION, MEXICO
B 6203 497 BOLES, J. R., CLAY DIAGENESIS IN WILCOX SANDSTONES OF SOUTHWEST
B 6203 498 BOLES, J. R., ORIGIN OF LATE-STAGE IRON-RICH CARBONATE CEMENTS IN
B 6405 678 BOLES, JAMES R., CALCIUM BUDGET IN FRIO SANDSTONES, SOUTHWEST
B 6108 1393 BOLES, JAMES R., DIAGENESIS OF VOLCANOGENIC SANDSTONES, ABST.
B 6310 1981 BOLES, JAMES R., REVIEW OF A COLOR ILLUSTRATED GUIDE TO
M 29 340 BOLIVAR LINEAMENT, COLOMBIA, PALEOTRENCHES
B 6010 1643 BOLIVIA, 1975, OBLITAS GARCIA, J., ON DEVELOPMENTS IN
B 6110 1580 BOLIVIA, 1976, OBLITAS GARCIA, J., ON DEVELOPMENTS IN
B 6210 2046 BOLIVIA, 1977, OBLITAS, J., AND F., JIMENEZ, ON DEVELOPMENTS IN
B 6310 1747 BOLIVIA, 1978, PETROCONSULTANTS S.A., ON DEVELOPMENTS IN
B 6004 684* BOLIVIA, ABST., DEVONIAN STRATIGRAPHY OF
B 6212 2512* BOLIVIA, ABST., EXPLORATION OF, CARBONIFEROUS SYSTEM IN,
M 25 558 BOLIVIA, ANTIMONY MINES
M 25 533 BOLIVIA, BASE-METAL DEPOSITS
M 25 558 BOLIVIA, BISMUTH MINES
M 25 558 BOLIVIA, LEAD MINES
M 25 558 BOLIVIA, LLALLAGUA MINE, TIN PRODUCTION
M 25 558 BOLIVIA, MINERALIZATION, ZONAL ARRANGEMENT
B 6110 1604 BOLIVIA, PRODUCTION, 1975-1976
B 6210 2071 BOLIVIA, PRODUCTION, 1976-1977
B 6310 1772 BOLIVIA, PRODUCTION, 1977-1978
M 25 533 BOLIVIA, TIN
M 25 558 BOLIVIA, TIN, METALLOGENIC CYCLES
M 25 558 BOLIVIA, TIN MINES
M 25 558 BOLIVIA, TIN, RESERVES
B 6303 471* BOLIVIA--CASE STUDY OF MIGRATION-FRACTIONATION, ABST.,
M 25 349† BOLIVIAN SEDIMENTARY BASINS--SUMMARY, HYDROCARBON
M 25 460 BOLIVIAN STANNIFEROUS SYSTEM
M 25 558† BOLIVIAN TIN BELT--SUMMARY, DEPOSITS OF
S 11 132 BOLIVINA ADVENA
S 11 247 BOLIVINA ADVENA
S 11 282 BOLIVINA ADVENA
S 11 39 BOLIVINA BARBARANA ZONE
S 11 132 BOLIVINA BRAMLETTEI
S 11 132 BOLIVINA BREVIOR
S 11 72 BOLIVINA BREVIOR BREVIOR
S 11 72 BOLIVINA BREVIOR DUNLAPI
S 11 136 BOLIVINA CALIFORNICA
S 11 136 BOLIVINA CHURCHI ALMGRENI
S 11 72 BOLIVINA CHURCHI ALMGRENI
S 11 136 BOLIVINA CHURCHI CHURCHI
S 11 72 BOLIVINA CHURCHI CHURCHI
S 11 282 BOLIVINA COCHEI
S 11 247 BOLIVINA COCHEI
S 11 130 BOLIVINA CUNEIFORMIS

S 11 282 BOLIVINA DECUSSATA
S 11 73 BOLIVINA FLORIDANA FLORIDANA
S 11 134 BOLIVINA FLORIDANA FLORIDANA
S 11 73 BOLIVINA FLORIDANA STRIATELLA
S 11 134 BOLIVINA FORAMINATA
S 11 130 BOLIVINA GIRARDENSIS
S 11 136 BOLIVINA GOUDKOFFI
S 11 73 BOLIVINA GOUDKOFFI
S 11 40 BOLIVINA GOUDKOFFI ZONE
S 11 136 BOLIVINA GRANTI
S 11 73 BOLIVINA HOOTSI
S 11 134 BOLIVINA HOOTSI
S 11 132 BOLIVINA HUGHESI
S 11 136 BOLIVINA HUGHESI
S 11 73 BOLIVINA HUGHESI
S 11 130 BOLIVINA IMBRICATA
S 11 282 BOLIVINA INTERJUNCTA
S 11 248 BOLIVINA INTERJUNCTA
S 11 136 BOLIVINA MALAGAENIS
S 11 130 BOLIVINA MARGINATA GRACILLIMA
S 11 132 BOLIVINA MODELOENSIS
S 11 39 BOLIVINA MODELOENSIS ZONE
S 11 130 BOLIVINA MULLERI
S 11 74 BOLIVINA MULLERI
S 11 136 BOLIVINA OBLIQUA
S 11 136 BOLIVINA PARVA
S 11 282 BOLIVINA PSEUDOBEYRICHI
S 11 248 BOLIVINA PSEUDOBEYRICHI
S 11 132 BOLIVINA PSEUDOSPISSA
S 11 134 BOLIVINA RANKINI
B 6405 713* BOLIVINA SEMINUDA, ABST., CALCIFICATION MODEL AND SECONDARY
S 11 134 BOLIVINA SEMINUDA SEMINUDA
S 11 282 BOLIVINA SINUATA
S 11 248 BOLIVINA SINUATA
S 11 134 BOLIVINA SINUATA ALISOENSIS
S 11 74 BOLIVINA SINUATA ALISOENSIS
S 11 282 BOLIVINA SUBADVENA
S 11 248 BOLIVINA SUBADVENA ACUMINATA
S 11 249 BOLIVINA SUBADVENA SPISSA
S 11 132 BOLIVINA SUBADVENA SPISSA
S 11 282 BOLIVINA SULPHURENSIS
S 11 249 BOLIVINA SULPHURENSIS
S 11 74 BOLIVINA WISSLERI
S 11 134 BOLIVINA WISSLERI
S 11 39 BOLIVINA WISSLERI ZONE
S 11 134 BOLIVINIDS, CALIFORNIA, MIOCENE
S 11 132 BOLIVINIDS, CALIFORNIA, MIOCENE
S 11 130 BOLIVINIDS, CALIFORNIA, MIOCENE
S 11 136 BOLIVINIDS, CALIFORNIA, MIOCENE
S 11 327 BOLIVINIDS, KEY TO IDENTIFICATION, CALIFORNIA, NEOGENE
S 11 244 BOLIVINITA QUADRILATERA
S 11 278 BOLIVINITA QUADRILATERA
S 11 132 BOLIVNIA WOODRINGI
B 6303 380 BOLLI, T. A. DAVIES, J. B. SAUNDERS, AND J. G. SCHLATER, BECKMANN, J.
B 6109 1450 BOLSA QUARTZITE, PEDREGOSA BASIN
M 24 242 BOLTON FIELD, MISSISSIPPI
M 30 496 BOMBAY HIGH FIELD, BIOSTRATIGRAPHY, INDIA
B 6402 218 BOMBAY HIGH FIELD, INDIA
B 6402 218 BOMBAY HIGH FIELD, INDIA
M 30 487† BOMBAY HIGH FIELD, INDIA, PETROLEUM, GEOLOGY OF
M 30 488 BOMBAY HIGH FIELD, LOCATION, INDIA
M 30 503 BOMBAY HIGH FIELD, PETROLEUM GEOLOGY, MIOCENE
M 30 488 BOMBAY HIGH FIELD, STRATIGRAPHY, INDIA
M 30 489 BOMBAY OFFSHORE BASIN, INDIA, CAMBAY BASIN, CORRELATION TO
B 6402 209† BOMBAY OFFSHORE BASIN, INDIA, SOURCE AREAS AND, MIGRATION
B 6402 209† BOMBAY OFFSHORE BASIN, INDIA, SOURCE AREAS AND, MIGRATION
B 6402 210 BOMBAY OFFSHORE BASIN, OIL FIELDS, INDIA
B 6402 210 BOMBAY OFFSHORE BASIN, OIL FIELDS, INDIA
M 30 487 BOMBAY OFFSHORE BASIN, REGIONAL GEOLOGY, INDIA
M 30 497 BOMBAY OFFSHORE BASIN, STRUCTURE, INDIA
M 30 495 BOMBAY OFFSHORE BASIN, TECTONIC FRAMEWORK, INDIA
B 6402 210 BOMBAY-RATNAGIRI SHELF, INDIA
B 6402 210 BOMBAY-RATNAGIRI SHELF, INDIA
B 6112 2135 BON ACCORD FIELD, ALBERTA
S 4 159 BONAIRE, ACROPORA PALMATA OFF
S 4 158 BONAIRE, DIPLORIA CLIVOSA ZONE OFF
S 4 160 BONAIRE, DIPLORIA STRIGOSA OFF
S 4 5 BONAIRE ISLAND, CORALS
S 4 233 BONAIRE, NEMASTER RUBIGINOSA POURTALES
S 4 155† BONAIRE NETHERLANDS ANTILLES, LOW-WATER EMERGENCE OF,
B 6311 2023 BONAPARTE GULF BASIN, AUSTRALIA
B 6411 1864 BONAPARTE GULF BASIN, AUSTRALIA
B 6210 2152 BONAPARTE GULF BASIN, AUSTRALIA
M 25 234 BONAPARTE GULF BASIN, GFFSHORE, AUSTRALIA
B 6311 2021† BONAPARTE GULF--TIMOR SEA REGION, NORTHWESTERN AUSTRALIA,
B 6212 2511* BONAPARTE GULF—TIMOR SEA REGION, NORTHWESTERN AUSTRALIA,
B 6203 533 BONE, M. R., FIELD DEVELOPMENT WITH 3-D SEISMIC METHODS—CASE
B 6405 729 BONE, M. R., UNDERSTANDING FIELD DEVELOPMEN HISTORY USING 3-D
B 6211 2351 BONE, MARION R., REVIEW OF 30 SEISMIC CASE HISTORIES. BY TITLE
B 6012 2175 BONE, MARION R., SEISMIC DATA ANALYSIS WITH TIME SEISCROP MAP
B 6205 781 BONE SPRING FORMATION, NEW MEXICO
M 27 78 BONE SPRING LIMESTONE, TEXAS, PERMIAN
M 27 32 BONE SPRING LIMESTONE, TEXAS, PERMIAN
M 28 82 BONE VALLEY FORMATION, FLORIDA, PLIOCENE
M 28 83 BONE VALLEY FORMATION, FLORIDA, PLIOCENE
M 28 126 BONE VALLEY FORMATION, FLORIDA, PLIOCENE
B 6303 421 BONEM, RENA M., DEVELOPMENTAL PHASES IN LAGOONAL PATCH
B 6203 498 BONEM, RENA M., STROMATOPOROIDS—EPIC STRUGGLE FOR SURVIVAL,
B 6008 1391 BONES, MARION R., SEISMIC DATA ANALYSIS WITH TIME-SEISCROP MAP
M 25 443 BONGHWAJAE THRUST, SOUTH KOREA
B 6010 1712 BONGIORNI, D., AND L. SOGARO, ON DEVELOPMENTS IN ITALY, 1975
B 6210 1913 BONGIORNI, D., AND M. PUPILLI, ON DEVELOPMENTS IN ITALY, 1977
B 6110 1643 BONGIORNI, D., AND R. TOCCO, ON DEVELOPMENTS IN ITALY, 1976
B 6310 1922 BONGIORNI, D., M. MARINELLI, AND M. PUPILLI, ON DEVELOPMENTS IN
S 10 69 BONHAM, L. C., MIGRATION OF HYDROCARBONS IN COMPACTING BASINS
B 6404 549 BONHAM, L. C., MIGRATION OF HYDROCARBONS IN COMPACTING BASINS
B 6203 498 BONHAM, L. C., MIGRATION OF HYDROCARBONS IN COMPACTING
S 7 155 BONHAM, L. C., MISSISSIPPI FAN, GULF OF MEXICO—PHYSIOGRAPHY,
B 6212 2478 BONHAM, L. C., SOLUBILITY OF METHANE IN WATER AT ELEVATED
S 6 363 BONHOMME, MICHEL G., NUMERICAL CORRELATION OF MIDDLE AND
B 6405 687 BONILLA, GUIDO, SACHA FIELD OF ECUADORIAN ORIENTE, ABST.
B 6004 662 BONIS, S. B., CONTINENTAL VOLCANIC SEDIMENTS AND THEIR
B 6004 677* BOO BEE PATCH REEF, BELIZE DETERMINED BY DIVER-OPERATED
M 25 415 BOOLGEEDA IRON FORMATION, WESTERN AUSTRALIA
S 5 181 BOON, J. A BASE- METAL CONCENTRATION IN A DENSITY- STRATIFIED
B 6309 1579* BOONE AND NORTHERN LOGAN COUNTIES, WEST VIRGINIA--POSSIBLE
B 6309 1585* BOONE COUNTY, WEST VIRGINIA, ABST., HOW GOOD ARE REGIONAL,
B 6203 499 BOONE, JOHN L., LACUSTRINE DEPOSITIONAL SYSTEMS OF DOCKUM
B 6009 1606 BOONE, PETER A., DEPOSITIONAL ENVIRONMENT AND PETROLEUM
B 6405 672 BOOTH, D. R., PROMISING MORROW SANDSTONE DISCOVERIES IN
B 6405 678 BOOTH, E. A., SIGNATURE PROCESSING OF NORTH SEA AIR-GUN DATA,
B 6405 724 BOOTH, JAMES S., IMPACT OF HOLOCENE TRANSGRESSION ON
B 6203 499 BOOTH, JAMES S., RECENT DEPOSITIONAL HISTORY OF UPPER
B 6101 3 BOOTHIA ARCH
B 6206 1005 BOOTHIA PENINSULA, ARCTIC, CANADA
B 6206 1005 BOOTHIA UPLIFT, ARCTIC, CANADA
B 6209 1756 BOOTHROYD, JON C., STRATIGRAPHIC INTERPRETATION OF A POSSIBLE
B 6004 687* BORDE FORMTION, MISSISSIPPIAN, KENTUCKY AND INDIANA, ABST.,
B 6209 1706 BORDEN FORMATION, KENTUCKY
B 6303 344 BORDEN FORMATION, KENTUCKY
B 6408 1260 BORDER RANGES FAULT, ALASKA
B 6203 394 BORDER RANGES FAULT SYSTEM, ALASKA
B 6004 675* BORDERLAND BASIN SEDIMENTS, ABST., HIGH-RESOLUTION STUDIES OF,
B 6212 2492† BORDERLAND. DISCUSSION AND REPLY, PLIOCENE-, PLEISTOCENE,
B 6405 707* BORE HOLES, ABST., DIPMETER VALIDITY IN, DEVIATED
B 6303 439* BOREHOLE DATA, ABST., KRIGING OF TOP OF, RESERVOIR FROM SEISMIC
B 6012 2176* BOREHOLE GRAVIMETER, ABST., COMMERCIAL APPLICATION AND
B 6305 822* BOREHOLE GRAVITY STUDY OF DENSITY AND POROSITY OF SELECTED
B 6304 686* BOREHOLE LOGGING, ABST., DIRECT MEASUREMENT OF, URANIUM BY,
B 6009 1622 BORELLA, PETER E., LATE MIDDLE AND EARLY LATE ORDOVICIAN
B 6004 732* BORERS, ABST., EVENT SEQUENCES IN GEOLOGIC, REEF
B 6004 653* BORERS AND THEIR BORINGS IN BERMUDA REEF ENVIRONMENTS, ABST.,
S 4 271 BORERS, ARTHROPODS
S 4 271 BORERS, BIVALVED, MOLLUSKS
S 4 263 BORERS, JAMAICA, DISCOVERY BAY, REEF
S 4 268 BORERS, MOROCCO, JEBEL ASSAMEUR NAIT FERGANE, CORALS
S 4 265 BORERS, MOROCCO, JEBEL BOU KENDILL, CORAL
S 4 270 BORERS, POLYCHAETE, WORMS
S 4 270 BORERS, SIPUNCULID, WORMS
S 4 270 BORERS, SPONGES
S 4 262 BORERS, TEXAS, CRETACEOUS LIMESTONE
S 4 261† BORERS--THEIR ROLE IN REEF ECOLOGY AND PRESERVATION,
B 6303 421 BORG, I. Y., INFLUENCE OF RATE ON DEMAND AND USE OF NATURAL
B 6105 770 BORG, I. Y., USE OF RESERVE-RESOURCE ESTIMATES BY NATIONAL ERDA
B 6410 1617 BORIE FIELD, WYOMING
B 6107 1069 BORING, ALGAE, ENDOLITHIC
M 27 183 BORINGS, ALGAL

M 27 193 BORINGS, ALGAL
B 6004 653* BORINGS IN BERMUDA REEF ENVIRONMENTS, ABST., ANALYSIS OF,
B 6107 1070 BORINGS, INFILLING OF, ALGAL
B 6410 1635 BORINGS, LOCATIONS, DEPTHS, RECOVERY, GREAT BAHAMA BANK,
M 27 184 BORINGS, SPONGE
M 27 183 BORINGS, SUBTIDAL
M 27 184 BORINGS VS. BURROWS, EXAMPLE
B 6303 458* BORINGS--NEW APPROACH, ABST., STUDY OF, FOSSI MICROBIAL
M 30 429 BORNEO GEOSYNCLINE, MALAYSIA, NORTHWEST
B 6209 1684† BORNEO, ORGANIC SEDIMENTATION AND, GENESIS OF, PETROLEUM IN,
M 30 430 BORNEO OROGEN, MALAYSIA, NORTHWEST
M 30 426 BORNEO, STRUCTURAL SKETCH
B 6309 1596 BORNHAUSER, MAX A., SUBSURFACE STRATIGRAPHY OF
M 25 460 BORON, ALUMINOQUARTZITES, AND FLUORITE, EASTERN ASIATIC
B 6104 594 BORREGAS FIELD, TEXAS
B 6405 622 BORREGO FORMATION, CALIFORNIA
B 6205 881 BORTZ, LOUIS C., EAGLE SPRINGS OIL FIELD, NYE COUNTY, NEVADA,
B 6405 679 BORTZ, LOUIS C., EXPLORATION IN GREAT SALT LAKE, ABST.
B 6405 679 BORTZ, LOUIS C., INDIAN CREEK FIELD, FALL RIVER COUNTY, SOUTH
S 5 161 BORUP FIORD, ELLESMERE ISLAND,
B 6111 1931 BORUP FJORD FORMATION, SVERDRUP BASIN
M 26 340 BOSCAN FIELD, VENEZUELA
M 26 342 BOSCAN SANDSTONE, VENEZUELA
B 6104 579 BOSEAN FIELD, VENEZUELA
B 6303 551* BOSO PENINSULA, JAPAN, ABST., EARLY, PLEISTOCENE, SUBMARINE
M 24 1 **BOSS, R. F., MIDDLE GROUND SHOAL OIL FIELD, ALASKA**
B 6209 1610 BOSSARDVILLE LIMESTONE, PENNSYLVANIA
B 6009 1609* BOSSIER AND WEBSTER PARISHES, LOUISIANA, ABST., SOUTH SAREPTA
B 6009 1608* BOSSIER SHALE AND JURASSIC-CRETACEOUS BOUNDARY, ABST.,
B 6007 1119 BOSSIER, SHALE, FORMATION, TEXAS
B 6007 1119 BOSSIER, SHALE, FORMATION, TEXAS
B 6407 1005 BOSSIER SHALES, ARK-LA-TEX AREA
B 6405 738 BOSTICK, N. H., EVALUATION OF ORGANIC MATTER AND SUBSURFACE
B 6303 421 BOSTICK, N. H., VITRINITE REFLECTANCE, PRESENT ROCK TEMPERATURE,
B 6410 1577 BOTETOURT FORMATION, VIRGINIA
B 6105 804 **BOTHNER, M. H., FRESH GROUNDWATER BENEATH CONTINENTAL SHELF:**
B 6303 493* BOTRYOIDAL ARAGONITE ON PERMIAN SEAFLOOR, ABST., IN-SITU
B 6210 1849 BOTSWANA, BURUNDI, MALAWI, RHODESIA, RWANDA, SWAZILAND,
B 6411 1786 BOTSWANA, DEVELOPMENTS, 1979
M 29 3 BOTT, M. H. P., SUBSIDENCE MECHANISMS AT PASSIVE CONTINENTAL
B 6405 679 BOTTJER, DAVID J., PALEOENVIRONMENTAL ANALYSIS OF
B 6303 422 BOTTJER, DAVIS J., ICHNOLOGY AND DEPOSITIONAL ENVIRONMENTS OF
B 6004 654* BOTTOM CURRENTS AND FAUNAL ACTIVITY IN LOWER HUDSON
S 7 272 BOTTOM DEFORMATION, COMPLEX SHELF-EDGE GRABEN FAULTS
S 7 272 BOTTOM DEFORMATION, CONCAVE CONVEX FAULT
S 7 273 BOTTOM, DEFORMATION, DEEP-SEATED FAULTS
S 7 278 BOTTOM DEFORMATION, DIFFERENTIAL CONSOLIDATION
S 7 275 BOTTOM DEFORMATION, GENERAL INSTABILITY
S 7 275 BOTTOM DEFORMATION, GRAVITY
S 7 272 BOTTOM DEFORMATION, GROWTH FAULT
S 7 273 BOTTOM, DEFORMATION, HYDRAULIC FORCES
S 7 273 BOTTOM, DEFORMATION, LINEAR FAULT
S 7 273 BOTTOM, DEFORMATION, MONOCLINAL FOLD
S 7 276 BOTTOM DEFORMATION, MUDFLOWS
S 7 269 BOTTOM DEFORMATION PERIPHERAL SLUMPS
S 7 271 BOTTOM DEFORMATION, SHALLOW GRABEN FAULTS
S 7 273 BOTTOM DEFORMATION, SHELF-EDGE SLUMP AND DIAPIRIC INTRUSION
S 7 278 BOTTOM DEFORMATION, SQUEEZING AND DIAPIRISM
S 7 272 BOTTOM DEFORMATION, SUBSURFACE CONVEX FAULT
S 7 271 BOTTOM DEFORMATION, SURFACE MUDFLOW
S 7 273 BOTTOM, DEFORMATION, TECTONIC ACTIVITY
S 7 278 BOTTOM DEFORMATION, TENSIONAL STRESS
S 7 278 BOTTOM DEFORMATION, TURBIDITY FLOWS
M 29 139† BOTTOM PROCESSES ALONG THE U.S. ATLANTIC CONTINENTAL MARGIN,
B 6203 489* BOTTOM SEDIMENTS IN NORTHWEST ATLANTIC, ABST., MIXING OF
B 6203 544* BOTTOM SEDIMENTS OF SAN NICOLAS BASIN, CALIFORNIA, ABST.
S 7 91 BOTTOM-FLOWING CURRENT TRANSPORT, SEDIMENT PROCESSES
S 7 284 BOTTOM-PRESSURE ANOMALY, STRESS, STABILITY
M 30 540 BOU-KHEZANA FIELD, ALGERIA, OUARGLA REGION
M 25 355 BOUCAUT, W. R. P., SALINE GROUNDWATER INFLOWS TO MURRAY RIVER
B 6406 953 BOUCOT, EDS., MEYERHOFF, A. A., REVIEW OF, HISTORICAL
B 6101 85 BOUFEKRANE BASIN, MOROCCAN MESETA
M 25 426 BOUGAINVILLE COPPER DEPOSITS, AUSTRALIA
B 6007 1101 BOUGUER ANOMALY MAP, BARBADOS RIDGE COMPLEX AND ADJACENT,
B 6102 297* BOUGUER GRAVITY ANOMALY ANALYSIS OF CHIHUAHUA AND
B 6212 2435 BOUGUER GRAVITY MAP
B 6007 1076 BOUGUER GRAVITY MAP, NIGER DELTA
M 29 178 BOUGUER GRAVITY PROFILE, BLAKE ESCARPMENT
B 6201 112 BOULDER BATHOLITH, MONTANA
B 6304 599 BOULDER FIELD, NEW MEXICO
M 26 278 BOULDER OIL FIELD, COLORADO
M 26 277 BOULDER, WELD, AND ADAMS COUNTIES, COLORADO
M 24 296 BOULDER ZONE, FLORIDA
B 6201 121 BOULDER ZONES, DEFINITION, FLORIDA
M 26 281 BOULDER-WELD FAULT ZONE, STRATIGRAPHY
M 26 292 BOULDER-WELD FAULT ZONE, STRUCTURE
B 6004 664* BOULDERS AS PALEOCLIMATIC INDICATOR, SOUTH SHORE OF LOWER ST.
B 6207 1216 BOUMA, A. H., SEAFLOOR INSTABILITY ON PACIFIC CONTINENTAL
B 6004 702 BOUMA, A. H., THIN-BEDDED TURBIDITES IN MODERN SUBMARINE
S 7 1 **BOUMA, ARNOLD, EDITED BY, FRAMEWORK, FACIES, AND OIL-TRAPPING**
B 6203 499 BOUMA, ARNOLD H., CONTINENTAL SLOPES, ABST.
S 7 289 BOUMA, ARNOLD H., INTRASLOPE BASIN IN NORTHWEST GULF OF
S 7 1 **BOUMA, ARNOLD H., INTRODUCTION, FRAMEWORK, FACIES, AND OIL-**
S 7 3 BOUMA, ARNOLD H., PHYSIOGRAPHY OF GULF OF MEXICO
B 6407 1118 BOUMA, ARNOLD H., REVIEW OF GUADALUPIAN DELAWARE MOUNTAI
B 6211 2351 BOUMA, ARNOLD, MOBILE BED FORMS AND VOLCANISM, LOWER COOK
B 6206 935 BOUMA MODEL FOR CLASSIC TURBIDITES
B 6207 1220* BOUNDARIES, ABST., GEOMORPHOLOGY OF, OCEAN FLOOR AS NATURAL
B 6207 1232* BOUNDARIES AND STRUCTURE OF BROADLANDS GEOTHERMAL FIELD,
M 26 56 BOUNDARIES, DEPOSITIONAL SEQUENCES, MAGNITUDE AND
M 25 8† BOUNDARIES OF SEABED JURISDICTION OFF PACIFIC COAST OF ASIA
M 29 273† BOUNDARIES OF TECTONIC PLATES, CONTINENTAL, MARGINS OF THE
S 6 171† BOUNDARY BETWEEN PLIOCENE AND PLEISTOCENE, STATUS OF THE
S 11 208 BOUNDARY PROBLEMS
B 6107 1083† BOUNDARY STRATOTYPE CONCEPT, NOTE 47—APPLICATION FOR
B 6410 1701 BOUNDARY UNDERCURRENT, WESTERN
B 6101 39 BOUNDARY-ROLL DEPOSITS, URANIUM
B 6303 422 BOURGEOIS, JOANNE, RETROGRADATIONAL SHELF SEQUENCE--UPPER
B 6305 820 BOURGEOIS, SCHOLLE, PETER A., REVIEW OF, ENCYCLOPEDIA OF,
S 6 323 BOUROZ, A., REPORT ON ISOTOPIC DATING OF ROCKS IN THE
B 6303 422 BOURROUILH-LE, FRANCOISE G., HURRICANES AND RAINFALL--KEY FOR
B 6207 1169 BOUSE FORMATION, GULF OF CALIFORNIA
B 6303 427* BOUSE FORMATION NEAR VIDAL, SOUTHEASTERN CALIFORNIA, ABST.,
B 6203 494 BOUSSA, MEZIANE, DETERMINATION OF RESERVOIR FLUID CONTENT BY
B 6409 1445 BOUTTE, C. BRIAN, DEVELOPMENTS IN LOUISIANA GULF COAST
B 6407 981 BOW ISLAND FORMATION, ALBERTA
B 6303 423 BOWDLER, J. L., MID-CRETACEOUS CALCAREOUS NANNOPLANKTON
B 6205 892 BOWDLER, JAY L., PALEOBIOGEOGRAPHY AND PRESERVATIOON OF
B 6008 1395* BOWDOIN AREA, MONTANA--LOW-PERMEABILITY, SANDY-SHALE,
B 6405 773* BOWDOIN DOME AREA, NORTH-CENTRAL MONTANA, ABST.,
B 6305 838* BOWDOIN DOME AREA, NORTH-CENTRAL MONTANA--EXAMPLE OF
B 6405 760* BOWDOIN DOME AREA, NORTH-CENTRAL MONTANA—EXAMPLE OF
B 6407 972 BOWDOIN DOME, SECTION, MONTANA
B 6407 981 BOWDOIN FIELD, MONTANA
B 6407 984 BOWDOIN FIELD, SECTION, MONTANA
B 6410 1605 BOWEN FORMATION, VIRGINIA
M 25 118 BOWEN, RICHARD G., ENVIRONMENTAL EFFECTS OF GEOTHERMAL
B 6004 652 BOWEN, RICHARD L., EPISODIC UPLIFT AND SEDIMENTATION IN
M 24 51 BOWERMAN, J. N., TAGLU GAS FIELD, BEAUFORT BASIN, NORTHWEST
B 6311 2071 BOWERS BASIN, BERING SEA
B 6211 2351 BOWERSOX, J. R., LATE MIOCENE TURBIDITE DEPOSITION, CROCKER
B 6005 809 BOWIN, C. O., CONTINENTAL MARGIN OFF WESTERN AFRICA--SENEGAL
B 6406 868 **BOWIN, CARL, ARC-CONTINENT COLLISION IN BANDA SEA REGION**
B 6209 1645 BOWIN, CARL, PLATE CONVERGENCE AND ACCRETION IN
B 6407 1120 BOWIN, CARL, REVIEW OF MARINE GRAVITY, BY PETER DEHLINGER
B 6009 1623* BOWLING GREEN FAULT--CASE OF RESURGENT TECTONICS, ABST.
B 6004 652 BOWMAN, J. F., II, TIMING AND PALEOCLIMATE INDICATORS IN
M 25 108 BOWRON RIVER COALFIELD, CANADA
M 25 498 BOWSER BASIN, CANADA
M 25 504 BOWSER BASIN, CANADA
M 25 500 BOWSER BASIN, CROSS SECTION, CANADIAN, CORDILLERA
B 6408 1135 BOWSER CREEK FIELD, ALASKA
B 6405 715 BOWSHER, ARTHUR L., PETROLEUM EXPLORATION O NATIONAL
M 26 398 BOX CAR REEF, MODELING, WAVE THEORY
B 6205 865 BOY SCOUT COMMITTEE, GEOLOGY MERIT BADGE BOOKLET, AAPG
B 6004 653 BOYCE, ROBERT L., LATE PRECAMBRIAN SEDIMENTATION IN BELT AN
B 6303 516 BOYD, DONALD W., MARIN PERMIAN ROCKS OF TUNISIA, ABST.
B 6203 505 BOYER, DAVID L., IRON-SULFIDE MINERALOGY AND MORPHOLOGY IN
B 6303 423 BOYER, LARRY F., CHANGING PERSPECTIVES IN PALEOECOLOGIC USE O
B 6309 1589* BOYLE DOLOMITE, MIDDLE DEVONIAN, IN EAST-CENTRAL KENTUCKY,
B 6303 344 BOYLE FORMATION, KENTUCKY
B 6203 500 BOZANICH, RICHARD G., FLUID DENSITY CURRENT DEPOSITION OF
B 6010 1870 BP PETROLEUM DEVELOPMENN LTD., AMERADA HESS CORP., ABU DHABI
B 6203 426 BRABANT PLATFORM, EUROPE
M 27 76 BRACHIOPOD GRAINS
M 27 76 BRACHIOPOD, PSEUDOPUNCTATE
M 27 76 BRACHIOPOD, PUNCTATE
M 27 78 BRACHIOPOD SPINES
S 6 306 BRACHIOPODA, PERMIAN, KEY FOSSIL GROUPS
M 27 75 BRACHIOPODS

M 27 72 BRACHIOPODS
S 6 329 BRACHIOPODS, CARBONIFEROUS, BIOSTRATIGRAPHY
M 27 77 BRACHIOPODS, PSEUDOPUNCTATE
B 6004 727* BRACKISH-WATE CRETACEOUS OSTRACODA FROM WELLS IN CENTRAL
B 6112 2076 BRADANIC TROUGH, MEDITERRANEAN REGION
B 6307 1139 BRADBURY, J. PLATT, GULF OF CALIFORNIA IN LAKE MEAD AREA OF
B 6007 1124 BRADLEY, J. S., ABNORMAL FORMATION PRESSURE. DISCUSSION AND
B 6012 2176 BRADLEY, J. W., COMMERCIAL APPLICATION AND INTERPRETATION OF
B 6203 556 BRADLEY, JAMES W., BRIGHT SPOTS, MILLIGALS, AND GAMMAS, ABST.
B 6405 789 BRADLEY, RETA, UNITED STATES PROVINCE OVERVIEWS, ABST.
B 6002 318 BRADLEY, WILBUR C., REPORT--EASTERN KANSAS. BY TITLE ONLY
B 6306 955 BRADORE FORMATION, NEWFOUNDLAND
B 6105 793 BRADY, M. J., CARBONATE-SAND WAVES, ISLA MUJERES, YUCATAN.
B 6303 362 BRADY, M. J., STRANDLINE SEDIMENTATION O CARBONATE
B 6303 445 BRADY, MICHAEL J., PROGRESSIVE CEMENTATION FROM MARINE TO
B 6203 500 BRADY, MICHAEL J., VADOSE CEMENT PATTERNS IN STRAND-PLAIN
B 6305 823 BRADY, RAY A., MINERAL EVALUATION OF OIL-SHALE LAND
B 6303 423 BRADY, T. J., INTISAR D OIL FIELD, LIBYA, ABST.
M 30 543 BRADY, T. J., INTISAR D OIL FIELD,LIBYA
B 6405 718* BRAE FIELD AREA, NORTH SEA, ABST.
B 6303 423 BRAGA, JOSE A. E., SEISMIC SEQUENCE ANALYSIS, NEW APPROACH FOR
B 6206 952 BRAIDED CHANNELS ON SUPRAFAN LOBE, ANCIENT, MID-, FAN:
B 6004 655* BRAIDED RIVER, ABST., EFFECTS OF, ICE ON SEDIMENTS ON, SANDY
B 6311 2120* BRAIDED STREA SEDIMENTATION, NORTHWEST ARKANSAS, ABST.,
B 6303 552* BRAIDED-STREAM PROCESSES AND FACIES IN MODERN ARC-TRENCH
B 6105 827* BRAIDED-TO-MEANDERING TRANSITION ZONE—RED RIVER, TEXAS AND
B 6304 652 BRAILLER AND MILLBORO SHALES, VIRGINIA
B 6004 696 BRAITHWAITE, C. J. R., CONTOURING GEOLOGIC DATA FROM ARBITRARY
B 6105 812 BRAITHWAITE, C. J. R., CONTOURING GEOLOGIC DATA FROM ARBITRARY
B 6203 492* BRALLIER FORMATION OF WEST VIRGINIA AND VIRGINIA, ABST.,
B 6303 438 BRAME, JEFFRE W., ORIGIN AND DEVELOPMENT OF BARRIER ISLANDS ON
B 6203 500 BRAME, JEFFREY W., STRATIGRAPHY AND GEOLOGIC HISTORY OF
B 6405 679 BRAMLETT, LARRY B., URANIUM EXPLORATION SYSTEMS CASE STUDY—
B 6008 1391 BRANCH, JERRY L., PALEOGEOMORPHOLOGY OF FRED AND GEORGE
B 6105 771 BRAND, LEONARD, COCONINO SANDSTONE, PERMIAN, FOSSIL VERTEBRATE
B 6304 700 BRANDT, B. M., WAVE-FORM FACTOR ANALYSIS--QUANTITATIVE
B 6405 680 BRANDT, DANITA S., BIOGENIC STRUCTURES AS INDICATORS O
B 6203 494 BRANNON, J. C., TRACE ELEMENTS IN HOLOCENE FLUVIAL SANDS
B 6408 1281 BRANT, RUSSELL A., COAL GEOLOGY AND RESOURCE OF EASTERN
B 6407 1007 BRANTLEY JACKSON FIELD, TEXAS
B 6002 196 BRASILEIRO, NATIONAL SCIENCE FOUNDATION, INTERNATIONAL
B 6209 1635 BRASSFIELD CARBONATE ROCKS, APPALACHIAN BASIN
B 6303 344 BRASSFIELD FORMATION, KENTUCKY
B 6308 1310 BRAUNSTEIN, JULES, DEVELOPMENTS IN LOUISIANA GULF COAST
B 6009 1607 BRAUNSTEIN, JULES, DOOR POINT--BURIED VOLCANO IN SOUTHEAST
M 24 1 BRAUNSTEIN, JULES, EDITED BY, NORTH AMERICAN OIL AND GAS FIELDS
B 6401 107 BRAY, E. E., A PROCESS FOR PRIMARY MIGRATION OF PETROLEUM
B 6304 697 BRAY, E. E., PROCESS FOR PRIMARY MIGRATION OF PETROLEUM IN
B 6209 1610 BRAYMAN SHALE, NEW YORK
S 3 117 BRAZIL
B 6010 1644 BRAZIL, 1975, PETROLEO BRASILEIRO S.A., ON DEVELOPMENTS IN
B 6210 2046 BRAZIL, 1977, MARINHO CAMPOS, C.W., ON DEVELOPMENTS IN
B 6310 1748 BRAZIL, 1978, FERREIRA SOBRAL, L., ON DEVELOPMENTS IN
B 6303 429* BRAZIL, ABST., DEPOSITIONAL-DIAGENETIC HISTORY OF, MACAE,
B 6405 687* BRAZIL, ABST., MICROFACIES AND DEPOSITIONAL-DIAGENETIC MODEL
B 6303 412* BRAZIL, ABST., NAMORAD FIELD, MAJOR OIL DISCOVERY IN, CAMPOS
B 6203 503* BRAZIL, ABST., PETROLEUM GEOLOGY OF, PALEOZOIC, CLASTIC ROCKS
B 6004 655* BRAZIL, ABST., SYNSEDIMENTARY TECTONIC CONTROL OF, PALEOZOIC,
B 6109 1476 BRAZIL, AGUA GRANDE FIELD
B 6307 1033 BRAZIL, AGUA GRANDE FIELD, STRUCTURE MAP
B 6109 1476 BRAZIL, AGULHA FIELD
B 6302 221 BRAZIL, ALAGOAS FRACTURE ZONE
B 6109 1480 BRAZIL, ALBIAN-SANTONIAN PETROLEUM RESERVOIR SYSTEM
S 3 89 BRAZIL, ALGONKIAN ALKALIC GRANITE
B 6109 1471 BRAZIL, ALMADA BASIN
B 6002 197 BRAZIL, ALMADA COASTAL BASIN
S 6 107 BRAZIL, ALTO PARAGUAY GROUP, SEPOTUBA FORMATION
B 6109 1471 BRAZIL, AMAPA BASIN
B 6109 1471 BRAZIL, AMAZON BASIN
S 6 101 BRAZIL, AMAZON BASIN, TROMBETAS FORMATION
B 6302 219 BRAZIL, AMAZON CONE
B 6109 1471 BRAZIL, AMAZON MOUTH
B 6109 1477 BRAZIL, APTIAN PETROLEUM RESERVOIR SYSTEM
B 6109 1476 BRAZIL, ARACAS PROFUNDO FIELD
B 6109 1476 BRAZIL, ARACAS RASO FIELD
B 6109 1471 BRAZIL, ARARIPE BASIN
B 6109 1477 BRAZIL, ATALAIA FAULT
B 6109 1476 BRAZIL, BADEJO FIELD
S 6 109 BRAZIL, BAMBUI GROUP, RUBIDIUM-STRONTIUM
B 6109 1471 BRAZIL, BARREIRINHAS BASIN
M 26 214 BRAZIL, BARREIRINHAS BASIN
B 6106 930 BRAZIL BASIN
B 6002 197 BRAZIL, BASINS, COASTAL
S 6 108 BRAZIL, BRAZILIAN CYCLE, ESTANCIA FORMATION
B 6109 1476 BRAZIL, BREJO GRANDE FIELD
S 3 88 BRAZIL, BREJU DA CRUZ GRANODIORITE BATHOLITH
B 6109 1476 BRAZIL, BURACICA FIELD
B 6109 1479 BRAZIL, CABO FRIO ARCH
B 6302 221 BRAZIL, CABUGI LINEAMENT
B 6109 1476 BRAZIL, CAIOBA FIELD
B 6109 1471 BRAZIL, CAMAMU BASIN
B 6109 1476 BRAZIL, CAMORIM FIELD
B 6109 1471 BRAZIL, CAMPOS BASIN
M 30 330 BRAZIL, CAMPOS BASIN, CABO FRIO ARCH
M 30 330 BRAZIL, CAMPOS BASIN, LOCATION
M 30 333 BRAZIL, CAMPOS BASIN, SHELF-EDGE SEISMIC SECTION
M 30 331 BRAZIL, CAMPOS BASIN, STRATIGRAPHIC FRAMEWORK
M 30 332 BRAZIL, CAMPOS BASIN, STRUCTURAL FRAMEWORK
M 30 330 BRAZIL, CAMPOS BASIN, VITORIA ARCH
B 6002 197 BRAZIL, CAMPOS COASTAL BASIN
M 26 345 BRAZIL, CANDEIAS FIELD
B 6109 1476 BRAZIL, CANDEIAS FIELD
B 6109 1476 BRAZIL, CARMOPOLIS FIELD
B 6109 1479 BRAZIL, CARMOPOLIS MEMBER
B 6109 1478 BRAZIL, CASSARONGONGO FIELD
B 6302 219 BRAZIL, CEARA ABYSSAL PLAIN
B 6109 1471 BRAZIL, CEARA BASIN
B 6302 236 BRAZIL, CEARA GUYOT
B 6302 231 BRAZIL, CEARA PLATEAU
B 6302 219 BRAZIL, CEARA RISE
B 6002 199 BRAZIL COASTAL MARGIN, BASEMENT RIDGE
B 6002 199 BRAZIL COASTAL MARGIN, SALT BOUNDARY
B 6002 206 BRAZIL COASTAL MARGIN, SERRA GERAL LAVA
B 6002 202 BRAZIL, COASTAL WELLS
B 6206 1029 BRAZIL, CONTINENTAL MARGIN
S 4 321 BRAZIL, CONTINENTAL SHELF, BENTHIC FORAMINIFERA, ASSEMBLAGES
S 4 317 BRAZIL, CONTINENTAL SHELF, CARBONATE SEDIMENTATION
S 4 315 BRAZIL, CONTINENTAL SHELF, OCEANOGRAPHY
B 6109 1480 BRAZIL, CRETACEOUS, LATE-TERTIARY PETROLEUM RESERVOIR SYSTEM
M 30 329 BRAZIL, CRETACEOUS, MACAE FORMATION
B 6110 1580 BRAZIL, DEVELOPMENTS, 1976
B 6002 196 BRAZIL, DIAPIRS, COASTAL MARGIN
B 6109 1476 BRAZIL, DOM JOAO FIELD
B 6109 1476 BRAZIL, DOURADO FIELD
B 6002 197 BRAZIL, EASTERN, MESOZOIC-CENOZOIC MARGIN BASINS
B 6109 1480 BRAZIL, EMBORE FORMATION
B 6109 1478 BRAZIL, ENGENHO FURADO FIELD
B 6204 665 BRAZIL, EOCENE
B 6109 1471 BRAZIL, ESPIRITO SANTO BASIN
B 6002 197 BRAZIL, ESPIRITO SANTO COASTAL BASIN
M 26 214 BRAZIL, ESPIRITO-SANTO BASIN
S 6 108 BRAZIL, ESTANCIA FORMATION, RUBIDIUM-STRONTIUM
S 6 105 BRAZIL, ESTRADA NOVA FORMATION, RUBIDIUM-STRONTIUM
B 6109 1476 BRAZIL, FAZENDA CEDRO FIELD
B 6109 1478 BRAZIL, FAZENDA TOMADO FIELD
B 6302 219 BRAZIL, FERNANDO DE NORONHA RIDGE
M 26 214 BRAZIL, FOZ DO AMAZONAS BASIN
B 6109 1476 BRAZIL, FURADO FIELD
B 6109 1476 BRAZIL, GAROUPA FIELD
B 6109 1476 BRAZIL, GUARICEMA FIELD
B 6109 1481 BRAZIL, GUARICEMA SANDSTONE
S 6 104 BRAZIL, IRATI FORMATION, RUBIDIUM-STRONTIUM
B 6109 1471 BRAZIL, JATOBA BASIN
B 6109 1471 BRAZIL, JEQUITINHONHA BASIN
B 6002 197 BRAZIL, JEQUITINHONHA COASTAL BASIN
B 6109 1475 BRAZIL, JURASSIC, LATE, PETROLEUM RESERVOIR SYSTEM
S 9 11 BRAZIL, JURASSIC, WIND DIRECTION
S 6 109 BRAZIL, LATE PRECAMBRIAN, BAMBUI GROUP
B 6109 1480 BRAZIL, MACAE FORMATION
M 30 331 BRAZIL, MACAE FORMATION, NAMORADO SANDSTONE
B 6109 1477 BRAZIL, MALOMBE FIELD
B 6109 1471 BRAZIL, MARANHAO BASIN
B 6002 197 BRAZIL MARGIN, ABROLHOS BANK
B 6002 197 BRAZIL MARGIN, CABO FRIO ARCH
B 6002 198 BRAZIL MARGIN, LAMONT-DOHERTY GEOPHYSICAL CRUISES, MAY 1974
B 6104 546† BRAZIL MARGIN, MARINE, SEISMIC INVESTIGATIONS, SOUTHERN
B 6002 197 BRAZIL MARGIN, RECIFE PLATEAU
B 6002 197 BRAZIL MARGIN, RIO GRANDE RISE
B 6002 197 BRAZIL MARGIN, ROYAL CHARLOTTE BANK
B 6002 197 BRAZIL MARGIN, SAO PAULO PLATEAU

B 6002 197 BRAZIL MARGIN, SEAMOUNTS
B 6002 197 BRAZIL MARGIN, STOCKS
B 6002 197 BRAZIL MARGIN, VEMA CHANNEL
B 6002 197 BRAZIL MARGIN, VITORIA ARCH
B 6002 197 BRAZIL MARGIN, VITORIA-TRINDADE CHAIN
B 6109 1476 BRAZIL, MERO FIELD
B 6204 665 BRAZIL, MIOCENE
B 6109 1476 BRAZIL, MIRANGA FIELD
B 6307 1034 BRAZIL, MIRANGA FIELD, SECTION
B 6109 1476 BRAZIL, MIRANGA NORTE FIELD
B 6206 1029† BRAZIL, MORPHOLOGY AND STRUCTURE OF UPPER, CONTINENTAL,
M 26 214 BRAZIL, MUCURI-CUMURUXATIBA-JEQUITINHONHA BASIN
B 6109 1476 BRAZIL, NAMORADO FIELD
M 30 332 BRAZIL, NAMORADO FIELD, DISCOVERY HISTORY
M 30 338 BRAZIL, NAMORADO FIELD, GEOLOGIC CROSS SECTION
B 6302 219 BRAZIL, NORTH BRAZILIAN RIDGE
B 6202 277 BRAZIL, NORTHERN, AMAZON CONE OFF
B 6202 290 BRAZIL, NORTHERN, CEARA ABYSSAL PLAIN OFF
B 6202 285 BRAZIL, NORTHERN, CEARA RISE OFF
B 6306 935 BRAZIL, NORTHERN, CONTINENTAL MARGIN
B 6202 290 BRAZIL, NORTHERN, DEMERARA ABYSSAL PLAIN OFF
B 6202 286 BRAZIL, NORTHERN, FERNANDO DE NORONHA BASIN OFF
B 6202 286 BRAZIL, NORTHERN, PARA-MARANHAO BASIN OFF
B 6202 282 BRAZIL, NORTHERN, SEISMIC PROFILES OFF
M 26 240 BRAZIL, OFFSHORE BASINS, MODELS
B 6109 1471 BRAZIL, PARAIBA BASIN
B 6302 221 BRAZIL, PARAIBA FRACTURE ZONE
B 6109 1471 BRAZIL, PARANA BASIN
B 6104 547 BRAZIL, PARANA BASIN
B 6206 1033 BRAZIL, PARANA BASIN
S 6 105 BRAZIL, PARANA BASIN, ESTRADA NOVA FORMATION
S 6 104 BRAZIL, PARANA BASIN, IRATI FORMATION
S 6 106 BRAZIL, PARANA BASIN, PONTA GROSSA FORMATION
S 6 106 BRAZIL, PARANA BASIN, RIO BONITO FORMATION
S 6 102 BRAZIL, PARANA BASIN, RIO DO RASTO FORMATION
B 6109 1476 BRAZIL, PARGO FIELD
B 6206 1043 BRAZIL, PATOS ARCH, PROFILES
B 6302 221 BRAZIL, PATOS LINEAMENT
B 6109 1471 BRAZIL, PELOTAS BASIN
B 6206 1042 BRAZIL, PELOTAS BASIN, PROFILES
B 6002 197 BRAZIL, PELOTAS COASTAL BASIN
B 6104 549 BRAZIL, PELOTAS SHELF AND BASIN
B 6109 1471 BRAZIL, PERNAMBUCO BASIN
B 6302 223 BRAZIL, PERNAMBUCO PLATEAU
B 6109 1471 BRAZIL, PIAUI BASIN
B 6204 665 BRAZIL, PLEISTOCENE
B 6204 665 BRAZIL, PLIOCENE
B 6104 547 BRAZIL, PONTA GROSSA ARCH
S 6 106 BRAZIL, PONTA GROSSA FORMATION, RUBIDIUM-STRONTIUM
M 26 240 BRAZIL, POST-RIFT BASIN, MODEL
M 26 214 BRAZIL, POTIGUAR BASIN
B 6109 1471 BRAZIL, POTIGUAR BASIN
B 6302 221 BRAZIL, POTIGUAR BASIN
B 6110 1611 BRAZIL, PRODUCTION, 1975-1976
B 6210 2077 BRAZIL, PRODUCTION, 1977
B 6310 1773 BRAZIL, PRODUCTION, 1977-1978
B 6109 1474 BRAZIL, RECIFE-JOAO PESSOA BASIN
B 6302 221 BRAZIL, RECIFE-JOAO PESSOA BASIN
B 6002 197 BRAZIL, RECIFE-JOAO PESSOA COASTAL BASIN
B 6109 1471 BRAZIL, RECONCAVO BASIN
B 6307 1033 BRAZIL, RECONCAVO BASIN, SANTO AMARO GROUP
B 6307 1034 BRAZIL, RECONCAVO BASIN, SECTION
B 6002 197 BRAZIL, RECONCAVO COASTAL BASIN
B 6109 1481 BRAZIL, RESERVES, RECOVERABLE
B 6109 1476 BRAZIL, RIACHUELO FIELD
M 26 240 BRAZIL, RIFT-BASIN, MODEL
S 6 106 BRAZIL, RIO BONITO FORMATION, RUBIDIUM-STRONTIUM
S 6 102 BRAZIL, RIO DO RASTO FORMATION, RUBIDIUM-STRONTIUM
B 6302 228 BRAZIL, RIO GRANDE DO NORTE PLATEAU
B 6104 547 BRAZIL, RIO GRANDE FAULT
B 6206 1044 BRAZIL, RIO GRANDE-SANTOS FAULT
B 6206 1034 BRAZIL, RIO GRANDE-SANTOS FAULT SYSTEM
B 6109 1479 BRAZIL, RIO PARAIBA PLATFORM
B 6302 219 BRAZIL, ROMANCHE FRACTURE ZONE
B 6109 1476 BRAZIL, S. M. CAMPOS FIELD
B 6109 1478 BRAZIL, SANTIAGO SANDSTONE
B 6109 1471 BRAZIL, SANTOS BASIN
M 26 214 BRAZIL, SANTOS BASIN
B 6206 1035 BRAZIL, SANTOS BASIN, PROFILES
B 6002 197 BRAZIL, SANTOS COASTAL BASIN
B 6206 1033 BRAZIL, SAO GABRIEL ARCH
B 6104 547 BRAZIL, SAO GABRIEL ARCH
B 6109 1476 BRAZIL, SAO MATEUS FIELD
B 6204 658 BRAZIL, SAO PAULO PLATEAU, BATHYMETRIC MAP
B 6204 657 BRAZIL, SAO PAULO PLATEAU, GEOTHERMAL MEASUREMENTS
B 6204 661 BRAZIL, SAO PAULO PLATEAU, SEISMIC REFLECTION PROFILES
B 6109 1471 BRAZIL, SEDIMENTARY BASIN
B 6104 547 BRAZIL, SEISMIC SECTIONS, SOUTHERN MARGIN
S 6 107 BRAZIL, SEPOTUBA FORMATION, RUBIDIUM-STRONTIUM
B 6109 1471 BRAZIL, SERGIPE-ALAGOAS BASIN
B 6202 229 BRAZIL, SERGIPE-ALAGOAS BASIN, SEDIMENTARY SEQUENCE
B 6002 197 BRAZIL, SERGIPE-ALAGOAS COASTAL BASIN
B 6202 230 BRAZIL, SERGIPE-ALAGOAS, EVAPORITES
M 26 214 BRAZIL, SERIGIPE-ALAGOAS BASIN
B 6206 1033 BRAZIL, SERRA GERAL BASALTS
B 6104 551 BRAZIL, SERRA GERAL LAVAS
B 6109 1477 BRAZIL, SERRARIA FORMATION
B 6109 1476 BRAZIL, SIRIRIZINHO FIELD
B 6206 1033 BRAZIL, SOUTHERN, BASEMENT STRUCTURE MAP
B 6302 219 BRAZIL, ST. PAUL FRACTURE ZONE
B 6302 218† BRAZIL, STRUCTURE AND ORIGIN OF THREE, CONTINENTAL-MARGIN
B 6109 1476 BRAZIL, TABULEIRO DOS MARTINS FIELD
B 6109 1476 BRAZIL, TAINHA FIELD
B 6109 1476 BRAZIL, TAQUIPE FIELD
B 6307 1034 BRAZIL, TAQUIPE FIELD, SECTION
M 30 329† BRAZIL, THE, NAMORADO OIL FIELD--A MAJOR, OIL DISCOVERY IN THE,
B 6104 552 BRAZIL, TORRES ARCH
B 6206 1033 BRAZIL, TORRES ARCH
B 6206 1033 BRAZIL, TORRES SYNCLINE
B 6104 547 BRAZIL, TORRES SYNCLINE
S 6 101 BRAZIL, TROMBETAS FORMATION, RUBIDIUM-STRONTIUM
B 6109 1471 BRAZIL, TUCANO BASIN
B 6109 1476 BRAZIL, UBARANA FIELD
B 6109 1480 BRAZIL, UBATUBA MEMBER
S 4 313† BRAZIL, WITH SPECIAL REFERENCE TO BENTHIC FORAMINIFERA,
B 6411 2036 BRAZIL, WITH UNESCO, HEDBERG, HOLLIS D., REVIEW OF, TECTONIC MAP
B 6303 412* BRAZIL. MODEL FOR EXPLORATION IN ATLANTIC-TYPE BASINS, ABST.,
B 6004 690* BRAZILIAN CONTINENTAL MARGIN, ABST., EVOLUTION OF, MARGINAL,
B 6004 656* BRAZILIAN CONTINENTAL MARGIN, ABST., PETROLEUM GEOLOGY OF
B 6109 1470† BRAZILIAN CONTINENTAL MARGIN, PETROLEUM GEOLOGY OF EASTERN
B 6206 1029 BRAZILIAN CONTINENTAL MARGIN, REMAC, COOPERATIVE
B 6204 657† BRAZILIAN CONTINENTAL MARGIN, SALT, DIAPIRS ON, SAO PAULO
B 6204 657† BRAZILIAN CONTINENTAL MARGIN, SALT DIAPIRS ON, SAO PAULO
B 6105 832* BRAZILIAN CRATON—COMPARISON WITH NORTH AMERICA, ABST.,
S 6 108 BRAZILIAN CYCLE, ESTANCIA FORMATION, BRAZIL
B 6202 274 BRAZILIAN RIDGE, NORTH
M 26 213† BRAZILIAN RIFT AND PULL-APART BASINS, SEISMIC-, STRATIGRAPHIC,
B 6002 211 BRAZILIAN SALT EDGES, ALBIAN, MODEL, MATCHING, WEST, AFRICAN
B 6204 657† BRAZILIANCONTINENTAL MARGIN, SALT DIAPIRS ON, SAO PAULO
B 6405 705* BRAZIL—SOURCE ROCK CHARACTERIZATION AND EVALUATION, ABST.,
B 6202 273† BRAZIL—STRUCTURAL CONTROLS AND EVOLUTION, SEDIMENT
B 6203 494* BRAZORIA COUNTY, TEXAS, ABST., DELTAIC DEPOSITIONAL STYLE IN,
B 6303 417* BRAZORIA COUNTY, TEXAS, ABST., WORLDS FIRST GEOPRESSURED,
B 6108 1225 BRAZORIA COUNTY, TEXAS, ALVIN FIELD
B 6208 1382 BRAZORIA COUNTY, TEXAS, DANBURY DOME
B 6208 1382 BRAZORIA COUNTY, TEXAS, EAST OLD OCEAN FIELD
B 6308 1225 BRAZORIA COUNTY, TEXAS, FOUR CORNERS FIELD
B 6108 1226 BRAZORIA COUNTY, TEXAS, SOUTH PEACH POINT
B 6405 709* BRAZORIA, TEXAS, ABST., SHALE MINERALOGY OF, GENERAL CRUDE OIL
B 6208 1382 BRAZOS AREA, TEXAS, MIDDLE BANK REEF
B 6108 1226 BRAZOS COUNTY, TEXAS, KURTEN FIELD
B 6108 1389* BRAZOS UPLIFT, AND CHAMA BASIN IN SOUTHWESTERN COLORADO,
B 6207 1114 BREAKOUTS, ASPECTS OF, TOOL, ROTATION ASSOCIATED WITH
B 6207 1120 BREAKOUTS, CAUSES, ELONGATE
B 6211 2354* BREAKTHROUGHS, BRIGHT SPOTS, AND BEWARE. BY TITLE ONLY
B 6109 1490 BREAKUP OF PANGAEA
B 6312 2222 BREAM FORMATION, NORTH SEA
B 6303 344 BREATHITT FORMATION, KENTUCKY
B 6004 682* BRECCIA, NEWFOUNDLAND, ABST., TIMING OF, CARBONATE AND, CLAY
M 25 438 BRECCIA PIPES AND PORPHYRY COPPER, RELATION OF
B 6309 1586* BRECCIATION--PROPOSED MECHANISM OF FRACTURING, PETROLEUM
B 6211 2268 BRECKENRIDGE FAULT, CALIFORNIA
M 25 352 BREDEHOEFT, J. D., CONTROL OF EARTHQUAKES AT RANGELY,
B 6405 680 BREDEHOEFT, JOHN D., PERMEABILITY OF CLAY AND SHALE, ABST.
S 7 239 BREEN SANDSTONE, ISOPACH, CALIFORNIA, LOS ANGELES BASIN
B 6303 424 BREINER, SHELDON, EXPLORATION GEOPHYSICS IN PEOPLES REPUBLIC
B 6109 1476 BREJO GRANDE FIELD, BRAZIL
S 3 88 BREJU DA CRUZ GRANODIORITE BATHOLITH, BRAZIL
B 6405 798 BRELAND, F. C., PALEOECOLOGIC DETERMINATION OF BATHYMETRIC
M 29 152 BREMER BASIN, AUSTRALIA, CONTINENTAL MARGIN

B 6004 **652** BRENNEKE, J. C., ISOTOPIC EVIDENCE BEARING ON ORIGIN OF BAHAMIAN
S 4 353 BRENNER, I. B., PROGRESSIVE DIAGENETI ELIMINATION OF STRONTIUM
B 6408 1223 BRENNER, ROBERT L., CONSTRUCTION OF PROCESS-RESPONSE MODELS
B 6203 501 BRENNER, ROBERT L., CONSTRUCTION OF PROCESS-RESPONSE MODELS
B 6203 563 BRENNER, ROBERT L., MOLLUSCAN BIOCLASTIC ACCUMULATIONS IN
B 6305 823 BRENNER, ROBERT L., STRATIGRAPHIC AND SEDIMENTOLOGIC RELATION
B 6202 181 BRENNER, ROBERT L., SUSSEX SANDSTONE OF WYOMING—EXAMPLE OF
M 26 430 BRENT FIELD, MODEL, NORTH SEA
M 30 102 BRENT FORMATION, JURASSIC, STATFJORD FIELD
M 30 111 BRENT FORMATION, STRUCTURE CONTOUR, STATFJORD FIELD
M 30 109 BRENT RESERVOIR, STATFJORD FIELD, STRUCTURE
M 26 430 BRENT SANDSTONE, NORTH SEA
B 6310 1920 BRESSE-SAVOIE, FRANCE, JURA-
B 6408 1281 BRETT, CARLTON E., MIDDLE ORDOVICIA SHELF CARBONATE
B 6004 584 BREUNINGER, RAY H., PALAEOAPLYSINA, HYDROZOAN, CARBONAT
B 6102 294 BREZINA, J. L., CAPITAN REEF--EVOLUTION OF A CONCEPT, ABST.
B 6004 661 BREZINA, J. L., CAPITAN REEF--EVOLUTION OF CONCEPT, ABST.
B 6405 681 BRICE, SUZAN E., HYDROCARBON OCCURRENCES IN NONMARINE,
B 6405 680 BRICE, SUZAN E., LOWER CRETACEOUS LACUSTRINE SOURCE BEDS FROM
B 6107 988 BRIDGE CREEK CHALK UNIT, OKLAHOMA
B 6303 464* BRIDGE CREEK MEMBER, GREENHORN LIMESTONE, UPPER CRETACEOUS,
B 6203 537* BRIDGER AND GREEN RIVER FORMATIONS, FREMONT-SWEETWATER
B 6405 639 BRIDGER HILL THRUST, WYOMING
B 6108 1372 BRIDGES, L. W. DAN, RED WING CREEK FIELD, NORTH
B 6303 421 BRIGGS, C., INFLUENCE OF RATE ON DEMAND AND USE OF NATURAL
B 6203 501 BRIGGS, D. Z., COMPUTER SYSTEM FOR LITHOSTRATIGRAPHIC ANALYSIS,
B 6004 653 BRIGGS, D. Z., SET-THEORETIC INFORMATION RETRIEVAL FROM
B 6007 1107 BRIGGS FORMATION, TEXAS
B 6203 501 BRIGGS, L. I., COMPUTER SYSTEM FOR LITHOSTRATIGRAPHIC ANALYSIS,
B 6004 653 BRIGGS, L. I., SET-THEORETIC INFORMATION RETRIEVAL FROM
S 5 53 BRIGGS, LOUIS I. DEPOSITIONAL ENVIRONMENT OF RUFF FORMATION
S 5 1 BRIGGS, LOUIS I. DEPOSITIONAL ENVIRONMENTS OF PINNACLE REEFS,
B 6405 681 BRIGGS, S. R., TIDAL-CURRENT SAND WAVES IN VINEYARD SOUND,
B 6004 716 BRIGHT JERLENE, AAPG OIL AND GAS FIELD DATA BANK PROJECT,
B 6311 2060 BRITISH COLUMBIA, TRANQUILLE FORMATION
M 25 512 BRITISH COLUMBIA, TRANSCURRENT FAULTS, LARGE, ORIGI OF, ALASKA
M 25 308† BRITISH COLUMBIA--SUMMARY, SEDIMENTATION AND TECTONICS OF,
B 6010 1720 BRITISH GAS CORP., ET AL, ON DEVELOPMENTS IN UNITED KINGDOM,
B 6110 1649 BRITISH GAS CORP., ET AL ON DEVELOPMENTS IN UNITED KINGDOM,
S 4 175 BRITISH HONDURAS, MONTASTREA ANNULARIS
S 6 350 BRITISH ISLES, ORDOVICIAN, ISOTOPIC DATING REVIEW
M 24 52 BRITISH MOUNTAINS, ALASKA
B 6307 1144 BRITISH OSTRACODA, EDITED BY RAYMOND H. BATE AND ERIC
B 6110 1649 BRITISH PETROL. DEV. LTD., BRITISH GAS CORP., ET AL ON
B 6405 702* BRITISH SILESIAN, ABST., BIVALVE, TRACE, FOSSILS IN
M 25 430† BRITISH SOLOMON ISLANDS AND FIJI--SUMMARY, BACKGROUND TO
M 25 430 BRITISH SOLOMON ISLANDS, BAUXITE
M 25 430 BRITISH SOLOMON ISLANDS, COPPER
M 25 430 BRITISH SOLOMON ISLANDS, MINERAL EXPLORATION
M 25 430 BRITISH SOLOMON ISLANDS, NICKEL
B 6203 426 BRITTANY PLATFORM, EUROPE
B 6205 759 BROAD, LOW-RELIEF MOUND FACIES
B 6408 1282 BROADAWAY, DIANNE, BURIED STRUCTURES IN MARINE EXPLORATION,
M 26 22 BROADBAND VELOCITY MEASUREMENT, GERMANY, SOLENHOFEN
B 6309 1576 BROADHEAD, R. F., GAS-BEARIN OHIO SHALE ALONG LAKE
M 25 132 BROADLANDS GEOTHERMAL FIELD, NEW ZEALAND
B 6207 1232* BROADLANDS GEOTHERMAL FIELD, NEW ZEALAND, BY GEOPHYSICAL
B 6108 1269 BROCK, SAMUEL M., JR., OIL AND GAS DEVELOPMENTS IN MARYLAND,
B 6304 699 BROCKETT, L. D., PRUDHOE BAY, A TEN-YEAR PERSPECTIVE, ABST.
M 30 289 BROCKETT, L. D., PRUDHOE BAY--A 10-YEAR PERSPECTIVE
B 6102 257 BROCKIE LAKE CONGLOMERATE, IDAHO
B 6104 601† BROCKMAN HILLS, SOUTHWESTERN NEW MEXICO, STRUCTURE O
M 25 411 BROCKMAN IRON FORMATION, WESTERN AUSTRALIA
B 6004 653 BROCOUM, S. J., RELATION OF SEISMICITY AN TECTONIC PROVINCES,
B 6109 1535 BRODGON, LARRY, LITHOFACIES ANALYSIS AND URANIUM
B 6105 771 BROIN, T. LEO, COMPUTER APPROACH TO VERTICAL-VARIABILITY
B 6303 424 BROKAW, RICHARD S., JR., ROLE OF BIOEROSION IN MASS-WASTING OF
M 25 426 BROKEN HILL, MINING CENTER, AUSTRALIA
B 6307 1136 BROMIDE FORMATION, OKLAHOMA
B 6203 538* BROMIDE FORMATION, SOUTH-CENTRAL OKLAHOMA, ABST., SANDSTONE
B 6004 653 BROMLEY, R. G., ANALYSIS OF BORERS AND THEIR BORINGS IN
B 6105 771 BROMLEY, R. G., FRESHWATER ICHNOCOENOSES FROM TRIASSIC RED BEDS
B 6303 446 **BRONDOS, M. D., PETROLEUM POTENTIAL OF SCOTIAN SHELF--CASE**
B 6211 2362 BROOK, CHARLES A., TERTIARY GEOLOGY AND OIL SHALE RESOURCES
B 6006 962† BROOK RANGE, NORTHERN ALASKA, CARBONIFEROUS SECTION AND,
B 6408 1281 BROOKFIELD, MICHAEL E., MIDDLE ORDOVICIA SHELF CARBONATE
B 6304 687 BROOKINS, D. G., GEOCHEMICAL STUDIES OF GRANTS MINERAL BELT,
B 6405 681 BROOKINS, D. G., GEOCHRONOLOGIC EVIDENCE FOR PALEOZOIC
B 6305 823 BROOKINS, D. G., LESS COMMON TRACE ELEMENTS IN SANDSTONE-TYPE
B 6304 686 BROOKINS, D. G., MECHANISMS FOR URANIUM DEPOSITION IN GRANTS

B 6304 692 BROOKINS, D. G., MINERALOGY AND GEOCHEMISTRY OF MARIANO LAKE
B 6304 687 BROOKINS, D. G., PERIODS OF MINERALIZATION IN GRANTS MINERAL
B 6005 883 BROOKINS, D. G., RADIOMETRIC AGE DETERMINATION OF SILURIAN
B 6405 681 BROOKINS, D. G., RB-SR DATING OF DIAGENESIS—EPIGENESIS IN
B 6105 771 BROOKINS, DOUGLAS G., FAVORABLE CRITERIA FOR RADIOACTIVE-WASTE-
B 6304 693 BROOKINS, DOUGLAS G., MOUNT TAYLOR URANIUM DEPOSIT, SAN
B 6305 839 BROOKINS, DOUGLAS G., MOUNT TAYLOR URANIUM
B 6008 1391 BROOKINS, DOUGLAS G., OKLO PHENOMENON--GEOLOGIC AND NUCLEAR
B 6108 1373 BROOKINS, DOUGLAS G., RARE EARTH AND OTHER TRACE ELEMENTS AS
B 6105 772 BROOKINS, DOUGLAS G., RESULTS OF PILOT STUDIES OF PROJECT NURE
B 6410 1718 BROOKINS, DOUGLAS G., RUBIDIUM-STRONTIUM MINIMUM AGES OF
B 6209 1673 BROOKINS, DOUGLAS G., RUBIDIUM-STRONTIUM MINIMUM AGES OF
B 6304 693 BROOKINS, DOUGLAS G., SEM INVESTIGATION OF PARAGENESIS OF
B 6102 296* BROOKS FIELD, SOUTHEAST MIDLAND BASIN, IRION COUNTY, TEXAS,
B 6405 682 BROOKS, GREGG R., UNUSUAL PONDING OF SEDIMENTS ON DEEP-WATER
B 6004 691 BROOKS, H. K., SUBMERGED RELICT FORESTS OFF NORTHWEST FLORIDA
B 6206 1091 BROOKS, R. A., NEW MODEL FOR SOME COLORADO PLATEAU URANIUM
M 24 26 BROOKS RANGE, ALASKA
B 6109 1494 BROOKS RANGE, ALASKA
B 6405 758* BROOKS RANGE, ALASKA, ABST., SEDIMENTOLOGY OF FLUVIAL UPPER,
B 6012 2188* BROOKS RANGE, ALASKA, ABST., USE OF, IR COLOR PHOTOGRAPHY AND
B 6006 952† BROOKS RANGE, ALASKA, GEOLOGIC SIGNIFICANCE OF, DOONERAK
B 6103 468* BROOKS RANGE, ALASKA, BY TITLE ONLY, GEOLOGY OF, AMBLER
B 6012 2177* BROOKS RANGE FOLDBELT, ABST., IPEWICK FORMATION, SIGNIFICANT,
B 6209 1696† BROOKS RANGE OPHIOLITES, ALASKA, TECTONICS OF
B 6103 437 BROOKS RANGE SECTION, ALASKA
B 6405 754* BROOKS RANGE THRUST BELT AND ARCTIC SLOPE, ALASKA, ABST.,
B 6103 469* BROOKS RANGE, BY TITLE ONLY, GEOCHRONOLOGY OF SOUTHWESTERN
B 6108 1373 BROOKS, ROBERT A., GEOLOGICAL, GEOCHEMICAL, AND PETROPHYSICAL
B 6203 501 BROOKS, ROBERT A., NEW MODEL FOR SOME COLORADO PLATEAU
B 6004 653 BROOKS, ROBERT A., PRELIMINARY INVESTIGATIONS OF ELEMENTAL
B 6207 1210 BROOKS, ROBERT, URANIUM RESOURCES IN AMERICAN WEST—
B 6006 962 BROSGE, W. P., CARBONIFEROUS SECTION AND UNCONFORMITY AT
B 6006 952 BROSGE, WILLIAM P., GEOLOGIC SIGNIFICANCE OF DOONERAK
B 6402 237 BROTHERTON ARCH, TEXAS
B 6402 237 BROTHERTON ARCH, TEXAS
B 6008 1392 BROTT, CHARLES A., HEAT-FLOW STUDY OF SNAKE RIVER PLAIN, IDAHO,
B 6409 1555 BROUSSARD, M. C., CHESTER AND POTTSVILLE DEPOSITIONAL SYSTEMS,
B 6309 1596 BROUSSARD, MATTHEW A., UPPER MISSISSIPPIAN DELTAS IN BLACK
M 24 289 BROWARD SYNCLINE, FLORIDA
B 6303 517 BROWER, J. C., STATISTICAL COMPARISON OF MAPPED DATA, ABST.
M 25 503 BROWN, A. SUTHERLAND, METALLIC MINERAL RESOURCES OF
B 6303 424 BROWN, ALISTAIR, R., THREE-DIMENSIONAL SEISMIC INTERPRETATION
B 6305 838 BROWN, CHARLES A., BOWDOIN DOME AREA, NORTH-CENTRAL
B 6405 760 BROWN, CHARLES A., BOWDOIN DOME AREA, NORTH-CENTRAL
B 6207 1210 BROWN, HARRISON, CRITICAL ENERGY PROBLEMS OF PACIFIC BASIN,
B 6105 772 BROWN, JAMES L., JR., PALEOENVIRONMENT OF MOFFAT MOUND, EDWARDS
B 6109 1539 BROWN, L. F., JR., LAND RESOURCES OF TEXAS, ABST.
M 26 213 BROWN, L. F., JR., SEISMIC-STRATIGRAPHIC INTERPRETATION OF
B 6304 698 BROWN, LEWIS R., MICROBIOLOGICAL PROSPECTING FOR
B 6408 1282 BROWN, P. H., DEPOSITIONAL TRENDS OF LOWER SILURIAN CLINTON
B 6303 478 BROWN, PATRICK M., GEOCHEMISTRY OF OIL-FIELD WATERS FROM
B 6108 1374 BROWN, PAUL LARRY, EFFECTIVE USE OF MICROEARTHQUAKE
B 6303 425 BROWN, R. H., EARLY MESOZOIC TECTONIC FRAMEWORK AND
B 6303 425 BROWN, ROGER E., DRANEY LIMESTONE--EARLY CRETACEOUS
B 6407 988 BROWN, ROY H., TRIASSIC ROCKS OF ARGANA VALLEY, SOUTHERN
B 6209 1755 BROWN, RUSSELL O., APPLICATION OF FRACTURE IDENTIFICATION LOGS
B 6303 425 BROWN, STEPHEN W., REGIONAL HYDROCARBON SOURCE-ROCK
B 6004 730 BROWNING, J. M., AULACOGENS AN MEGASHEARS--NATURAL HABITAT
B 6011 2067 BROWNING, JOHN M., AULACOGENS AND MEGASHEARS--NATURAL
B 6307 1143 BROWNLOW, HUNT, JOHN M., REVIEW OF, GEOCHEMISTRY, BY A. H.
M 27 50 BROWNSPORT GROUP, TENNESSEE, SILURIAN
M 25 234 BROWSE BASIN, AUSTRALIA
B 6311 2023 BROWSE BASIN, AUSTRALIA
B 6210 2152 BROWSE BASIN, AUSTRALIA
B 6411 1864 BROWSE BASIN, AUSTRALIA
B 6201 41 BROWSE BASIN, AUSTRALIA
B 6311 1996 BRUCE, CLEMONT H., AAPG RESEARCH CONFERENCE GUIDELINES
B 6311 2002 BRUCE FAN, NORTH SEA
B 6403 439 BRUER, WESLEY G., MIST GAS FIELD, COLUMBIA COUNTY, OREGON, ABST.
B 6203 376 BRUIN BAY FAULT, ALASKA
B 6408 1260 BRUIN BAY FAULT, ALASKA
B 6305 841* BRULE FORMATION, ABST., PALEOSOLS AS ENVIRONMENTAL INDICATORS
B 6012 2142 BRUMBAUGH, D. S., EXPOSE STEP IN LARAMIDE THRUST FAULT,
M 24 250 BRUMFIELD SANDSTONE, MISSISSIPPI
B 6110 1834 BRUNEI, DEVELOPMENTS, 1976
B 6210 1981 BRUNEI, DEVELOPMENTS, 1977
B 6310 1819 BRUNEI, DEVELOPMENTS, 1978
B 6411 1884 BRUNEI, DEVELOPMENTS, 1979
B 6110 1842 BRUNEI, PRODUCTION, 1975-1976
B 6210 1990 BRUNEI, PRODUCTION, 1975-1977
B 6310 1829 BRUNEI, PRODUCTION, 1977-1978
B 6411 1896 BRUNEI, PRODUCTION, 1977-1979
B 6009 1562† BRUNHES EPOCH, NANNOFOSSIL, BIOSTRATIGRAPH AND, CLIMATIC
B 6304 700 BRUNS, T. R., PETROLEUM POTENTIAL OF EASTERN GULF OF ALASKA
B 6207 1231 BRUNS, TERRY, PLATE TECTONICS IN EVOLUTION OF SOUTHERN ALASKA
B 6103 469 BRUNS, TERRY R., LATE TERTIARY GULF OF ALASKA PLATE-TECTONIC
B 6004 655 BRUNS, TERRY R., PRELIMINARY STUDY OF SEA-FLOOR INSTABILITY IN
B 6012 2176 BRUNS, TERRY R., STRUCTURAL ELEMENTS OF OFFSHORE GULF OF
B 6105 773 BRUNS, TERRY R., STRUCTURE OF CONTINENTAL SHELF, EASTERN GULF
M 28 152 BRUNSWICK FORMATION, NEW JERSEY, TRIASSIC
M 28 44 BRUNSWICK FORMATION, NEW JERSEY, TRIASSIC
B 6008 1392 BRUNT, DAVID, URANIUM IN TERTIARY CHANNEL SOUTHEAST OF LAKE
M 30 531 BRUNTINO MARLS, CRETACEOUS, MALOSSA FIELD
B 6104 563 BRUSH LAKE FORMATION, CHILE
B 6104 561 BRUSH LAKE STRATIGRAPHIC UNIT, CHILE
B 6007 1013 BRUSHY BASIN MEMBER, NEW MEXICO
B 6209 1675 BRUSHY BASIN MEMBER, NEW MEXICO
B 6205 781 BRUSHY CANYON FORMATION, NEW MEXICO
M 28 144 BRUSHY CANYON FORMATION, TEXAS, PERMIAN
M 28 131 BRUSHY CANYON FORMATION, TEXAS, PERMIAN
B 6002 196 BRYAN, G., SOUTH ATLANTIC DIAPIRIC STRUCTURES
B 6105 698 BRYAN, G. M., GAS-HYDRATE HORIZONS DETECTED IN SEISMIC-PROFILER
B 6303 342 BRYAN STATION FAULT, KENTUCKY, WEST HICKMAN CREEK-
B 6407 1005 BRYANS MILL FIELD, ARK-LA-TEX AREA
B 6405 682 BRYANT, A. C., ENTRADA OIL FIELDS OF SOUTHERN SAN JUAN BASIN,
B 6205 881 BRYANT, ALAN C., ENTRADA OIL FIELDS OF SOUTHERN SAN JUAN BASIN,
B 6203 511 BRYANT, P. F., BATHYMETRIC CONTROL OF DEEP-SEA ICHNOFACIES,
B 6008 1355 BRYANT, R. G., DEVELOPMENTS IN EASTERN CANADA IN 1975
B 6108 1335 BRYANT, R. G., DEVELOPMENTS IN EASTERN CANADA IN 1976
B 6208 1451 BRYANT, R. G., DEVELOPMENTS IN EASTERN CANADA IN 1977
B 6308 1344 BRYANT, R. G., DEVELOPMENTS IN EASTERN CANADA IN 1978
B 6409 1518 BRYANT, R. G., DEVELOPMENTS IN EASTERN CANADA IN 1979
B 6405 679 BRYANT, W. ANTHONY, PALEOENVIRONMENTAL ANALYSIS OF
B 6004 681 BRYANT, W. R., AREAL DISTRIBUTION OF CLA MINERALS AND THEIR
B 6303 542 BRYANT, W. R., SHALLOW-SUBMARINE SEISMIC STRATIGRAPHY,
S 4 17 BRYANT, W. R., SHELF REEFS, SOUTHWESTERN PUERTO RICO
B 6303 426 BRYANT, W. R., SUBAQUEOUS LANDSLIDES AND DIAPIRS, GULF OF
S 7 243 BRYANT, WILLIAM R., LATE PLEISTOCENE GEOLOGIC HISTORY OF TEXAS
B 6303 449 BRYANT, WILLIAM R., SEDIMENTATION IN MISSISSIPPI TROUGH, GULF OF
B 6109 1545 BRYANT, WILLIAM R., STRUCTURE AND STRATIGRAPHY OF LATE
S 4 181† BRYOZOAN CONTRIBUTIONS TO REEFS AND BIOHERM THROUGH
M 28 39 BRYOZOAN, FRAGMENTS
M 27 71 BRYOZOAN FROND
S 4 183 BRYOZOAN GROUPS, CLASSIFICATION, REEFAL
M 27 67 BRYOZOANS
M 27 70 BRYOZOANS, CRYPTOSOME
M 27 69 BRYOZOANS, FENESTRATE
M 27 68 BRYOZOANS, FENESTRATE
M 27 71 BRYOZOANS, MODERN ENCRUSTERS
B 6104 585 BU HASA FIELD, PERSIAN GULF
M 26 49 BUBB, J. N., SEISMIC STRATIGRAPHY AND GLOBAL CHANGES OF SEA
M 26 185 BUBB, J. N., SEISMIC STRATIGRAPHY AND GLOBAL CHANGES OF SEA
B 6205 772 BUBB, J. N., SEISMIC STRATIGRAPHY AND GLOBAL CHANGES OF SEA
S 11 253 BUCCELLA OREGONENSIS
S 11 253 BUCCELLA TENERRIMA
S 11 284 BUCELLA OREGONENSIS
S 11 284 BUCELLA TENERRIMA
M 25 319 BUCHADO ANTICLINE, COLOMBIA
B 6103 465 BUCHAN, STUART H., PERMIAN-TRIASSIC OF PRUDHOE BAY FIELD, ABST.
B 6205 881 BUCHHEIM, H. PAUL, CARBONATE CYCLES IN GREEN RIVER FORMATION,
B 6010 1719 BUCHI, U. P., AND G. WIENER, ON DEVELOPMENTS IN SWITZERLAND, 1975
B 6110 1649 BUCHI, U. P., AND G. WIENER, ON DEVELOPMENTS IN SWITZERLAND, 1976
B 6210 1919 BUCHI, U. P., AND G. WIENER, ON DEVELOPMENTS IN SWITZERLAND, 1977
B 6310 1928 BUCHI, U. P., AND G. WIENER, ON DEVELOPMENTS IN SWITZERLAND, 1978
B 6411 1968 BUCHI, U. P., AND GABRIEL WIENER, ON DEVELOPMENTS IN
B 6007 1119 BUCHNER FORMATION, TEXAS
B 6209 1754 BUCK, GLENN H., PALEOSTRUCTURAL MAPPING IN BATHYAL
B 6410 1620 BUCKAROO FIELD, COLORADO
B 6206 999 BUCKLE FOLDING, IRAN, SALT, DIAPIRS AND
B 6407 1005 BUCKNER ANHYDRITE, ARK-LA-TEX AREA
B 6008 1408 BUCKNER, DUANE H., ROCKY MOUNTAIN CRETACEOUS SOURCE-ROCK

B 6008 1392 BUCKNER, DUANE H., VARIATIONS IN NATURAL GAS COMPOSITION I
M 24 179 BUCKNER FORMATION, ARKANSAS
B 6007 1119 BUCKNER FORMATION, TEXAS
M 26 150 BUCKNER FORMATION, TEXAS, EAST
M 26 146 BUCKNER FORMATION, TEXAS, EAST TEXAS BASIN
M 24 251 BUCKNER SECTION, MISSISSIPPI, HAYNESVILLE-
B 6405 782 BUCKOVIC, W. A., ETAL, URANIUM DEPOSITS OF PART OF CENTRAL
B 6203 561 BUCKOVIC, W. A., MAJOR URANIUM DISCOVERY IN FRONTIER AREA—
B 6304 621 BUCKOVIC, W. A., MAJOR URANIUM DISCOVERY IN VOLCANICLASTIC
B 6305 824 BUCUREL, HILDRED, STRATIGRAPHY AND COAL DEPOSITS OF UPPER
B 6402 230 BUDA FORMATION, TEXAS
B 6402 230 BUDA FORMATION, TEXAS
B 6109 1547* BUDA FORMATIONS FROM CORE AND FRACTURE ANALYSIS, ABST.,
B 6309 1604* BUDA, GEORGETOWN, MCKNIGHT, AND WEST NUECES CARBONATE
B 6109 1548* BUDA LIMESTONE IN SOUTH TEXAS, ABST., GEOLOGY OF, FRACTURED,
B 6301 35 BUDA LIMESTONE, TEXAS
M 27 21 BUDA LIMESTONE, TEXAS, CRETACEOUS
M 26 114 BUDA LIMESTONE, TEXAS, POLK COUNTY
B 6109 1546* BUDA TREND OF SOUTH TEXAS, ABST., AUSTIN, CHALK—
B 6405 773 BUDAI, ALEXIS, LOW-COST MICROPROCESSOR SYSTEM FOR
B 6405 682 BUDAI, JOYCE M., DIAGENESIS OF LOWER CORALLINE LIMESTONE,
S 6 264 BUDDEN CANYON FORMATION, CALIFORNIA, CRETACEOUS
S 5 53 BUDROS, RON DEPOSITIONAL ENVIRONMENT OF RUFF FORMATION
B 6411 1962 BUDWILL, A. E. H., AND A. J. KROLL, ON DEVELOPMENTS IN AUSTRIA, 1979
B 6210 1908 BUDWILL, A.E.H., AND A. J. KROLL, ON DEVELOPMENTS IN AUSTRIA, 1977
B 6310 1917 BUDWILL, ON DEVELOPMENTS I AUSTRIA, 1978, KROLL, A.J., AND A. E. H.
B 6211 2267 BUENA VISTA HILLS FIELD, CALIFORNIA
B 6203 351 BUENA VISTA OIL FIELD, CALIFORNIA
M 25 318 BUENO S., RAFAEL, POTENTIAL FOR EXPLORATION AND DEVELOPMENT
B 6305 804 BUFFALO WALLOW FIELD, TEXAS
B 6205 862 BUFFALO WALLOW-ULTRADEEP FIELD, PRODUCTION, TEXAS
M 25 280 BUFFINGTON, E. C., CENOZOIC AN MESOZOIC PETROLEUM PROSPECTS,
B 6002 161 BUFFINGTON, EDWIN C., STRUCTURE AND EVOLUTION OF BERING SEA
B 6004 654 BUFFLER, R. T., GEOLOGY OF SOUTH AUGUSTINE ISLAND, LOWER COOK
B 6303 426 BUFFLER, R. T., GRAVITY--SLIDE ORIGIN FOR MEXICAN RIDGES
B 6004 690 BUFFLER, R. T., SEISMIC STRATIGRAPHY AND GEOLOGIC HISTORY OF
M 29 319 BUFFLER, RICHARD T., ANATOMY OF THE MEXICAN RIDGES,
B 6207 1210 BUFFLER, RICHARD T., ET AL, TECTONIC FRAMEWORK OF MIDDLE
B 6408 1277 BUFFLER, RICHARD T., GEOLOGIC HISTORY OF DEEP GULF OF MEXICO
B 6303 544 BUFFLER, RICHARD T., GEOLOGIC HISTORY OF DEEP GULF ON MEXICO
B 6405 682 BUFFLER, RICHARD T., GEOLOGIC HISTORY OF DEEP SOUTHEASTERN
M 29 11 BUFFLER, RICHARD T., GEOLOGY OF THE OFFSHORE SOUTHEAST
B 6303 426 BUFFLER, RICHARD T., MIOCENE SUBMARINE FAN IN DEEP WESTERN
S 7 43 BUFFLER, RICHARD T., OCCURRENCE AND EVOLUTION OF SALT IN DEEP
B 6405 796 BUFFLER, RICHARD T., POST-MIDDLE CRETACEOUS SEISMIC
B 6303 526 BUFFLER, RICHARD T., POST-MIDDLE CRETACEOUS SEISMIC
B 6312 2204 BUFFLER, RICHARD T., SEISMIC EVIDENCE FOR WIDESPREAD POSSIBLE
B 6405 726 BUFFLER, RICHARD T., SEISMIC, STRATIGRAPHIC, AND STRUCTURAL
B 6303 487 BUFFLER, RICHARD T., SEISMIC STRATIGRAPHY AND GEOLOGIC HISTORY
B 6205 792 BUFFLER, RICHARD T., SEISMIC STRATIGRAPHY AND GEOLOGIC HISTORY
B 6105 773 BUFFLER, RICHARD T., SEISMIC STRATIGRAPHY, SEDIMENTARY HISTORY,
M 29 27 BUFFLER, RICHARD T., STRUCTURE AND DEVELOPMENT OF THE
B 6105 781 BUFFLER, RICHARD T., STRUCTURE AND DEVELOPMENT OF SOUTHEAST
B 6203 501 BUFFLER, RICHARD T., STRUCTURE AND EARLY GEOLOGIC HISTORY OF
B 6406 823 BUGLE GAP AREA, GEOLOGIC MAP, AUSTRALIA
B 6406 821 BUGLE GAP FORMATION, AUSTRALIA
B 6205 772 BUILDUP, DEFINITION, CARBONATE
S 4 242 BUILDUP, OR MOUND, CONCEPT OF, BANK
B 6112 2115† BUILDUP, SACRAMENTO MOUNTAINS, NEW MEXICO, EVOLUTION OF,
B 6205 885* BUILDUP-EVAPORITE RELATIONS IN MIDDLE DEVONIAN OF
B 6404 527† BUILDUPS AND BIOSTROMES OF NORTH GREENLAND,
B 6004 667* BUILDUPS AND ORGANIC REEFS DURING JURASSIC TIME IN ATLAS
B 6404 529 BUILDUPS, CLASSIFICATION
B 6004 584† BUILDUPS FROM UPPER PALEOZOIC OF IDAHO, PALAEOAPLYSINA,
B 6203 557* BUILDUPS, MIDDLE ORDOVICIAN, VIRGINIA, ABST., GEOMETRY, FACIES,
M 26 185 BUILDUPS, SEISMIC, REFLECTIONS, RECOGNITION
M 26 185† BUILDUPS, SEISMIC, STRATIGRAPHY AND, GLOBAL CHANGES OF, SEA
B 6205 772† BUILDUPS, SEISMIC STRATIGRAPHY AND, GLOBAL CHANGES OF, SEA
B 6210 2051 BUIS, O. J., ON DEVELOPMENTS IN GUATEMALA, 1977
B 6203 526 BUJAK, J. P., COST B-2 WELL, BALTIMORE CANYON, AND ITS RELATION TO
B 6203 503 BUKHARI, MOHAMMED TAHIR, RECOGNITION OF CRATONAL BANK,
B 6410 1690 BULAWAYAN STROMATOLITIC LIMESTONE, RHODESIA
M 30 237 BULAY SUITE, USSR, LENA-TUNGUSKA AREA
B 6110 1653 BULGARIA, DEVELOPMENTS, 1976
B 6110 1654 BULGARIA, PRODUCTION, 1976-1976
B 6210 1924 BULGARIA, PRODUCTION, 1976-1977
B 6310 1932 BULGARIA, PRODUCTION, 1977-1978
S 11 138 BULIMINA ALLIGATA
S 11 138 BULIMINA BREVIOR
S 11 280 BULIMINA DEFORMATA
S 11 245 BULIMINA DEFORMATA
S 11 245 BULIMINA DENUDATA
S 11 280 BULIMINA FOSSA
S 11 244 BULIMINA FOSSA
S 11 138 BULIMINA MONTEREYANA DELMONTEENSIS
S 11 138 BULIMINA MONTEREYANA MONTEREYANA
S 11 138 BULIMINA PSEUDOAFFINIS
S 11 138 BULIMINA PSEUDOTORTA
S 11 245 BULIMINA SUBACUMINATA
S 11 280 BULIMINA SUBACUMINATA
S 11 278 BULIMINELLA ELEGANTISSIMA
S 11 244 BULIMINELLA ELEGANTISSIMA
S 11 138 BULIMINELLA SEMIHISPIDA
S 11 244 BULIMINELLA SUBFUSIFORMIS
S 11 278 BULIMINELLA SUBFUSIFORMIS
S 11 72 BULIMINIDAE
S 11 243 BULIMINIDAE
S 11 138 BULIMINIDAE, CALIFORNIA, MIOCENE
S 11 146 BULIMINIDS, CALIFORNIA, MIOCENE AND, OLIGOCENE
S 11 144 BULIMINIDS, CALIFORNIA, OLIGOCENE AND, MIOCENE
S 11 62 BULITIAN STAGE
M 26 39 BULK DENSITY, LABORATORY, MEASUREMENTS, SATURATION EFFECTS
M 25 442 BULKUKSA DISTURBANCE, KOREA
M 25 440 BULKUKSA GRANITE, SOUTH KOREA
M 25 444 BULKUKSA GRANITES, SOUTH KOREA
S 4 88 BULLHEAD FORMATION, MAIN RIDGE VOLCANICS, CAMPANIAN,
B 6008 1390* BULLHOOK GAS UNITS, ABST., PERFORMANCE REVIEW OF, TIGER RIDGE
B 6403 332 BULLWACKER FIELD, SECTION, MONTANA
B 6403 332 BULLWACKER FIELD, SECTION, MONTANA
S 11 140 BULMINIDAE, CALIFORNIA, MIOCENE, UVIGERINE
B 6307 1093 BUNBURY TROUGH, AUSTRALIA
B 6007 1009 BUNDSANDSTEIN, GERMANY
B 6103 465 BUNDTZEN, THOMAS K., TECTONICS, MIDDLE PALEOZOIC VOLCANISM,
B 6405 802 BUNKER, BILL J., MIDDLE CRETACEOUS, ALBIAN-TURONIAN,
B 6405 683 BUNKER, BILLY J., ASSESSMENT OF POTENTIA FOR FLUID
B 6405 780 BUNKER, C., HEAT-FLOW AND HEAT-PRODUCTION STUDIES IN NORTH
M 28 121 BUNTER SANDSTONE, NORTHERN IRELAND, TRIASSIC
B 6209 1579 BUOYANCY THEORY, THEORY OF, DIAPIRISM VERSUS
B 6109 1476 BURACICA FIELD, BRAZIL
B 6303 429 BURCHFIEL, B. C., ALLOCHTHONOUS NATURE OF NUMIDIAN FLYSCH
B 6405 683 BURCHFIEL, B. C., FORELAND FOLD AND THRUST BELT OF CARPATHIANS
B 6405 684 BURCHFIEL, B. C., FORELAND FOLD AND THRUST BELTS—REVIEW, ABST.
B 6211 2352 BURCHFIEL, B. C., MESOZOIC CONSTRUCTION OF CORDILLERAN
B 6306 426 BURCKLE, LLOYD H., PAN-PACIFIC NEOGENE DIATOM CORRELATIONS,
M 25 354 BURDEN-JONES, C., ARTIFICIAL RECHARGE OF GR
B 6309 1591* BUREAU OF MINES COAL RESOURCES CLASSIFICATION SYSTEM ABST.,
M 25 463 BUREYA STABLE MASSIF, USSR
M 30 16 BURGAN FIELD, LANDSAT IMAGE, KUWAIT
B 6401 14 BURGAN GIANT OIL FIELD, LANDSAT DATA, KUWAIT
B 6401 14 BURGAN GIANT OIL FIELD, LANDSAT DATA, KUWAIT
B 6405 616 BURGAN SANDSTONE, MIDDLE EAST
B 6411 1963 BURGENLAND, AUSTRIA
B 6110 1639 BURGENLAND, AUSTRIA
B 6405 684 BURGESS, P. A., COAL RESOURCE DATA PROCESSING, OLD AND NEW
B 6009 1608 BURGESS, WILLIAM J., GEOLOGIC EVOLUTION OF MID-CONTINENT AN
B 6309 1597 BURGESS, WILLIAM J., HOW TO CREATE AND SUBMIT A WINNING
B 6110 1901† BURIAL AND DIAGENESIS. DISCUSSION AND REPLY, CLAY, MINERALS
B 6306 923 BURIAL BY PERMEABLE OVERBURDEN, SHALE MASSES BEFORE
B 6303 414* BURIAL DEPTH I DEVELOPMENT OF SUBNORMAL AND ABNORMAL
B 6303 492* BURIAL DOLOMITIZATION OF UPPER DEVONIAN MIETTE BUILDUP,
B 6303 421* BURIAL HISTORY IN LOS ANGELES AND VENTURA BASINS, CALIFORNIA,
B 6401 78 BURIAL RATES, ORGANIC, CARBON
B 6401 78 BURIAL RATES, ORGANIC, CARBON
B 6306 925 BURIAL, SHALE MASSES DURING
B 6001 108† BURIED CANYON--INFLUENCE ON ACCUMULATION OF HYDROCARBONS
B 6203 510* BURIED CREVASSE FILLINGS OF PORT HURON AGE, STOUFFVILLE,
B 6008 1409* BURIED FOSSIL CRATERS, ABST., WHY PETROLEUM EXPLORATIONIST
B 6408 1282* BURIED STRUCTURES IN MARINE EXPLORATION, ABST.
B 6408 1287* BURIED VALLEY NEAR PENINSULA, OHIO, ABST., SEISMIC REFRACTION
B 6009 1607* BURIED VOLCANO IN SOUTHEAST LOUISIANA, ABST., DOOR POINT--
B 6310 1982 BURK AND CHARLES L DRAKE, HEDBERG, HOLLIS D., REVIEW OF THE,
B 6211 2290 BURK, C. A., MARGINS OF GULF OF MEXICO
M 29 403 BURK, C. A., THE MARGINS OF THE GULF OF MEXICO
B 6303 426 BURKART, BURKE, PRE-STRIKE-SLIP POSITIONS OF POLOCHIC FAULT
B 6105 842 BURKE, JACK, PETROS—PETROPHYSICAL LOG AND CORE-ANALYSIS
B 6303 427 BURKE, RANDOLPH B., BANK-BARRIER REEF MORPHOGENESIS, ST. CROIX,
S 4 67 BURKE, RANDOLPH B., HOLOCENE BIOHERMS OF LESSER
B 6303 438 BURKE, RANDOLPH B., ORIGIN AND DEVELOPMENT OF BARRIER ISLANDS
B 6209 1704 BURKET BLACK SHALE, UNITED STATES, EASTERN
B 6002 281 BURKET MEMBER, PENNSYLVANIA

Index

B	6309	1579	BURKHARD, N. R., THEORETICAL AND EXPERIMENTAL ANALYSES OF
M	25	58	BURLIN, YU. K., GEOLOGIC-STRUCTURAL SYNTHESIS OF PACIFIC AREA
M	25	176	BURLIN, YU. K., OIL- AND GAS-BEARING BASINS OF CIRCUM-PACIFIC
M	29	190	BURMA, ANDAMAN SEA, BATHYMETRY
M	29	195	BURMA, ANDAMAN SEA, MAGNETIC PROFILE
M	29	192	BURMA, ANDAMAN SEA, PHYSIOGRAPHIC PROVINCES
M	29	194	BURMA, ANDAMAN SEA, SPREADING EXTENSION
M	29	193	BURMA, ANDAMAN SEA, STRUCTURE SECTIONS
M	29	195	BURMA, ANDAMAN SEA, SYNTHETIC MAGNETICS
M	29	191	BURMA, ANDAMAN SEA, TECTONIC MAP
M	29	192	BURMA, ANDAMAN SEA, TECTONIC PROVINCES
M	29	197	BURMA, ANDAMAN SEA, TECTONICS EVALUATED
B	6110	1834	BURMA, DEVELOPMENTS, 1976
B	6210	1982	BURMA, DEVELOPMENTS, 1977
B	6310	1820	BURMA, DEVELOPMENTS, 1978
B	6411	1885	BURMA, DEVELOPMENTS, 1979
B	6110	1842	BURMA, PRODUCTION, 1975-1976
B	6210	1990	BURMA, PRODUCTION, 1975-1977
B	6310	1829	BURMA, PRODUCTION, 1977-1978
B	6411	1896	BURMA, PRODUCTION 1977-1979
M	29	189†	BURMA, TECTONICS OF THE, ANDAMAN SEA AND
B	6402	219	BURMA, YENANGYAUNG FIELD
B	6402	219	BURMA, YENANGYAUNG FIELD
B	6409	1554*	BURMAH AGATE, WITH CEMENTS ON AVENUES OF OIL TO SEDIMENTS
B	6409	1554*	BURMAH AGATE, WITH COMMENTS ON AVENUE OF OIL TO SEDIMENTS
B	6203	563*	BURNING OF FOSSIL FUELS AND OCEAN CHEMISTRY IN 22D CENTURY,
B	6303	447	BURNS, LARY K., PETROLOGY AND DIAGENESIS OF ARC-DERIVED LITHIC
B	6108	1374	BURNS, LARY K., PETROLOGY AND DIAGENETIC EFFECTS OF
S	5	4	BURNT BLUFF FORMATION, MICHIGAN BASIN, NIAGARA SERIES,
S	5	4	B-SALT, SALINA GROUP,
B	6004	654	BUROLLET, P. F., PALEOSEDIMENTOLOGIC CORRELATION OF OROGENIC
B	6206	1093*	BURRO CANYON FORMATION, AND DAKOTA SANDSTONE OF SAN JUAN
B	6109	1449	BURRO-FLORIDA UPLIFT, ARIZONA AND, NEW MEXICO
B	**6105**	**773**	**BURROUGHS, R. H., OBJECTIVES OF REGULATIONS VERSUS**
B	6304	700	BURROUGHS, W. A., QUANTITATIVE COMPARISONS OF GEOLOGIC MAPS
B	6405	690*	BURROW PATTERNS OF GHOST CRAB OCYPODE CERATOPHTHALMA,
B	6004	612†	BURROWERS, AND GRAZERS IN DOGGER, JURASSIC, CARBONATE ROCKS,
B	6105	816	BURROWES, G., POROSITY IN REEF COMPLEXES—UPPER DEVONIAN
B	6303	433*	BURROWING BEHAVIOR AND REDISTRIBUTION OF OIL, ABST., SPILLE OIL
S	4	132	BURROWING SPONGES AND REEFS
M	27	185	BURROWS, FILLED
M	27	184	BURROWS VS. BORINGS, EXAMPLE
S	7	220	BURROWS, WEST FLORIDA SLOPE
B	6210	1849	BURUNDI, MALAWI, RHODESIA, RWANDA, SWAZILAND, UPPER VOLTA,
B	6211	2361	BUSBY, C., EARLY MESOZOIC PALEOTECTONIC-PALEOGEOGRAPHIC
B	6306	987	BUSCH, DANIEL A., REVIEW OF PRINCIPLES OF SEDIMENTOLOGY, BY
B	6202	235	BUSCH, DANIEL A.,, STRATIGRAPHY AND STRUCTURE OF CHICONTEPEC
B	6405	684	BUSHNELL, HUGH P., UNCONFORMITIES—KEY TO MAJOR OI
B	6002	319*	BUTLER COUNTY, KANSAS, ABST., DELTAIC DEPOSITIONAL
B	6004	671*	BUTLER COUNTY, KANSAS, ABST., DELTAIC SUBENVIRONMENTS AND,
B	6108	1374	BUTLER, DAVID, EFFECTIVE USE OF MICROEARTHQUAKE RECORDINGS
B	6405	684	BUTLER, DAVID R., FUTURE EXPLORATION FOR GEOTHERMAL
B	6004	654	BUTLER, J. C., EXAMINATION OF RELATIONS BETWEEN MODAL
B	6207	1175	BUTTE MOUNTAINS, NEVADA
B	6209	1610	BUTTERMILK FALLS LIMESTONE, PENNSYLVANIA
B	6011	2005†	BUTTES, GYPSUM PLAIN, CULBERSON COUNTY, TEXAS, ORIGIN OF,
B	6205	819	BUTTONWILLOW DEPOCENTER, CALIFORNIA
M	30	14	BUZURGAN FIELD, LANDSAT IMAGE, IRAQ
B	6401	12	BUZURGAN GIANT OIL FIELD, LANDSAT DATA, IRAQ
B	6401	12	BUZURGAN GIANT OIL FIELD, LANDSAT DATA, IRAQ
B	6203	502	BUZZALINI, ARNOLD D., CALCRETES. BY TITLE ONLY.
B	6207	1211	BYBEE, ROBERT W., PETROLEUM DEVELOPMENT IN COASTAL ZONE, ABST.
B	**6105**	**761**	**BYBELL, L. M., MESOZOIC AND CENOZOIC BIOSTRATIGRAPHY OF OAK**
B	**6105**	**773**	**BYBELL, LAUREL, M., PRELIMINARY CALCAREOUS NANNOFOSSIL ZONATION**
B	6009	1531	BYERLEE, JAMES D., EFFECT OF HIGH-PRESSURE DEFORMATION ON
B	**6303**	**354**	**BYERS, CHARLES W. AND, PALEOENVIRONMENTS OF MOWRY SHALE,**
B	6405	685	BYERS, CHARLES W., BIOTURBATION AS FACTOR IN HYDROCARBON
B	6405	685	BYERS, CHARLES W., DEPOSITIONAL ENVIRONMENTS IN UPPER
B	6203	502	BYERS, CHARLES W., ENIGMAS IN WISCONSIN CAMBRIAN AND NEW
M	25	463	BYKOVSKAYA, E. V., SOME CHARACTERISTICS OF ACID VOLCANISM OF
B	6206	1005	BYLOT ISLAND, ARCTIC, CANADA
B	6404	563	BYPASS COMPUTER MODEL
S	10	85	BYPASS COMPUTER MODEL, MIGRATION
S	10	60	BYPASSING WATER, MIGRATION
B	6104	527	BYRNE, J. R., LAVACA BAY--TRANSGRESSIVE DELTAIC SEDIMENTATION IN
M	24	119	C CANYON ANTICLINE, WYOMING
B	6203	380	C HORIZONS, COOK INLET, ALASKA, A, B, AND
B	6202	237	C, MEXICO, CHICONTEPEC HORIZON
S	3	62	C ROLL, URANIUM OREBODY
M	24	214	C SANDSTONE, LOUISIANA
M	28	80	CABALLOS NORACULITE, TEXAS, PALEOZOIC
B	6405	777*	CABIN CREEK FIELD, MONTANA, ABST., MECHANISMS CONTROLLING
B	6303	427*	CABINDA, ABST., OIL EXPLORATION IN PRE-, SALT AND POST-SALT
B	6405	681*	CABINDA, ANGOLA, ABST., HYDROCARBON OCCURRENCES IN
B	6203	502	CABLE, HAROLD, MAPRO—GEOGRAPHIC MAPPING PROGRAM, ABST.
B	6303	441	CABLE, HAROLD W., KOALA--MINICOMPUTER LO ANALYSIS SYSTEM FOR
B	6402	189	CABO DE GATA VOLCANIC COMPLEX, SPAIN
B	6402	189	CABO DE GATA VOLCANIC COMPLEX, SPAIN
B	6109	1479	CABO FRIO ARCH, BRAZIL
M	30	330	CABO FRIO ARCH, BRAZIL, CAMPOS BASIN
B	6002	197	CABO FRIO ARCH, BRAZIL MARGIN
B	6202	255	CABRILLO FAULT, CALIFORNIA
B	6302	221	CABUGI LINEAMENT, BRAZIL
B	6211	2312	CACAPON SANDSTONE MEMBER, WEST VIRGINIA
B	6004	654	CACCHIONE, D. A., SEDIMENT PROCESSES CONTROLLED BY BOTTOM
B	6405	756	CACCHIONE, D. A., SEDIMENTARY CHARACTERISTICS AND PROCESSES OF
B	6405	731	CACCHIONE, D. A., SEDIMENTARY PROCESSES ON CONTINENTAL SHELF
B	6105	817	CACCHIONE, DAVID A., COMPLEX RIDGES AND TROUGH TOPOGRAPHY
M	25	497	CACHE CREEK GROUP, CANADA
M	25	509	CACHE CREEK-SLIDE MOUNTAIN- NICOLA ROCKS, CORDILLERAN
M	30	379	CACTUS FIELD, CHIAPAS-TABASCO PROVINCE
B	6307	1097	CADDA FORMATION, AUSTRALIA
B	6307	1093	CADDA SHELF, AUSTRALIA
B	6103	361	CADDO ANTICLINE, OKLAHOMA
B	6311	2114*	CADDO COUNTY, OKLAHOMA, ABST., DEPOSITIONAL ENVIRONMENT OF,
B	6301	55	CADDO LIMESTONE, TEXAS
B	6408	1171	CADDO LIMESTONE, TEXAS
B	6302	164	CADOTTE SAND, CANADA
B	6303	427	CAFLISCH, L., OIL EXPLORATION IN PRE-SALT AND POST-SALT
B	6204	694	CAICOS BANK, BAHAMAS
B	6406	853	CAICOS BASIN, SAND LAYERS, HISPANIOLA-
B	6304	610	CAIN FORMATION, MICHIGAN BASIN
B	6109	1476	CAIOBA FIELD, BRAZIL
M	28	153	CAIRN FORMATION, CANADA, ALBERTA, DEVONIAN
M	28	10	CAIRN FORMATION, CANADA, ALBERTA, DEVONIAN
M	28	154	CAIRN FORMATION, CANADA, ALBERTA, DEVONIAN
M	28	161	CAIRN FORMATION, CANADA, ALBERTA, DEVONIAN
B	6201	2	CALABAR FLANK, NIGER DELTA
M	29	308	CALABRIAN ARC, EVOLUTION, MEDITERRANEAN
M	29	305	CALABRIAN ARC, MEDITERRANEAN, ACTIVE MARGINS
S	6	172	CALABRIAN, ITALY, BASE OF
S	6	173	CALABRIAN-VILLAFRANCHIAN CORRELATION, ITALY
B	6406	858	CALANSCIO UPLIFT, AFRICA
M	25	529	CALC-ALKALINE MAGMAS FORMED ON, BENIOFF ZONE
B	6309	1576	CALCAGNO, FRANK, JR., FRACTURE PATTERN IN OUTCROP OF
M	27	12	CALCARENITE, MUSCAT, OMAN, TERTIARY
B	6310	1918	CALCAREOUS ALPS, AUSTRIA, FLYSCH-
B	6411	1963	CALCAREOUS ALPS, AUSTRIA, FLYSCH-
B	6210	1909	CALCAREOUS ALPS, AUSTRIA, FLYSCH-
B	6303	478*	CALCAREOUS MICROFOSSIL BIOSTRATIGRAPHY, PALEO-OCEANOGRAPHY,
B	6205	892*	CALCAREOUS NANNOFOSSIL ASSEMBLAGES IN ATLANTIC OCEAN—
B	6004	700*	CALCAREOUS NANNOFOSSILS IN DEEP-SEA CARBONATE
B	6303	443*	CALCAREOUS NANNOPLANKTON BIOSTRATIGRAPHY IN EASTERN
S	11	60†	CALCAREOUS NANNOPLANKTON BIOSTRATIGRAPHY OF CENOZOIC
B	6303	503*	CALCAREOUS NANNOPLANKTON I UPPER APTIAN AND ALBIAN BLACK
B	6004	646*	CALCAREOUS SEDIMENTS, ABST., PRIMARY, BEDDING, RHYTHMS IN,
M	27	49	CALCAREOUS SPONGES
S	10	210	CALCASIEU PARISH, MANCHESTER FIELD, CLAY CONVERSION, LOUISIANA
S	10	210	CALCASIEU PARISH, MANCHESTER FIELD, KEROGENCONVERSION,
S	10	212	CALCASIEU PARISH, MANCHESTER FIELD,SALINITY WITH DEPTH,
B	6303	480*	CALCIFICATION AND SIGNIFICANCE OF SOI FILAMENTOUS
B	6405	713*	CALCIFICATION MODEL AND SECONDARY CALCIFICATION EFFECTS ON
M	27	21	CALCISPHERES
B	6409	1560*	CALCISPHERES FROM CUESTA DEL CURA AND LOWER AGUA NUEVA
M	27	48	CALCISPONGES
B	6405	734*	CALCIT FROM GULF COAST OF LOUISIANA, ABST., EARLY, CEMENTATION
M	28	121	CALCITE (PISOLITIC), IN CALCRETE, CEMENTS
M	28	120	CALCITE (POIKILOTOPIC), IN SANDSTONE, CEMENTS
M	28	120	CALCITE, AFTER QUARTZ OVERGROWTHS, CEMENTS
M	28	90	CALCITE, CEMENT IN SANDSTONE
M	27	159	CALCITE CEMENTS
M	27	164	CALCITE, FIBROUS
M	28	28	CALCITE, FRAGMENTS, POLYMICT
B	6303	450*	CALCITE, HYDROMAGNESITE, AND NESQUEHONITE FROM
M	28	119	CALCITE, IN SANDSTONE, CEMENTS
M	28	138	CALCITE, IN SANDSTONE, CEMENTS
M	28	68	CALCITE, MICRITIC, AS A MATRIX
M	28	155	CALCITE, REPLACEMENT OF DETRITAL PLAGIOCLASE
S	4	379	CALCITE THROUGH METEORIC FRESH WATERS, PRECIPITATION OF LOW-,
B	6404	472	CALCITE, WHISKER

M 28 160 CALCITE-FILLED, IN DARK SHALE, FRACTURES
M 27 217 CALCITE-FILLED INTERPARTICLE POROSITIES
S 3 106 CALCITE-URANIUM ORE, PYRITE-, MARCASITE-, JORDESITE-
M 28 165 CALCITIC OOID, COMPACTION OF
B 6405 678* CALCIUM BUDGET IN FRIO SANDSTONES, SOUTHWEST TEXAS, ABST.
S 4 219† CALCIUM CARBONATE FIXATION BY LARGE REEF-DWELLING
B 6305 825* CALCIUM CARBONATE IN MODERN NORTH TEMPERATE LAKES, ABST.,
S 4 313† CALCIUM CARBONATE SEDIMENTATION ON CONTINENTAL SHELF OFF
B 6203 530* CALCIUM SULFATE DIAGENESIS, MISSISSIPPIAN OF EASTERN WILLISTON
M 27 124 CALCLITHITE, EXTRACLASTS
M 28 98 CALCLITHITE, FRAGMENTS, LIMESTONE, IN
M 27 177 CALCLITHITES, POLYMICT
B 6303 480* CALCRETES, EASTERN SPAIN, ABST., CALCIFICATION AND SIGNIFICANCE
B 6207 1217* CALCRETES OF WESTERN AUSTRALIA, ABST., HYDROGEOLOGY OF,
B 6203 502* CALCRETES. BY TITLE ONLY.
B 6306 918† CALCULATED COMPACTION STATES OF THICK ABNORMALLY PRESSURED
S 10 40 CALCULATED SHALE PRESSURE, MIGRATION
B 6104 573† CALCULATION OF AGES OF HYDROCARBONS IN OILS--PHYSICAL
B 6004 698* CALCULATION OF COAL RESERVES BY COMPUTER METHODS, ABST.
B 6207 1187 CALDWELL, W. G. E., NOTE 49—RECORDS OF STRATIGRAPHIC
B 6203 414 CALEDONIA FORMATION, VIRGIN ISLANDS, ST. CROIX
B 6303 481* CALHOUN COUNTY, FLORIDA, ABST., DIAGENETIC PHENOMENA OF,
B 6409 1561* CALHOUN COUNTY, FLORIDA, ABST., EARLY, CEMENTATION AND
S 5 55 CALHOUN COUNTY, LOCATION, MICHIGAN,
B 6311 2103† CALHOUN COUNTY, MISSISSIPPI, VARDAMAN, TRIASSIC-, JURASSIC, SILL
B 6209 1761* CALHOUN COUNTY, TEXAS, ABST., MIOCENE GEOLOGY OF, MATAGORDA
M 27 119 CALICHE
M 27 189 CALICHE CRUST, TEXAS, QUATERNARY
M 28 121 CALICHE CRUST, TEXAS, QUATERNARY
B 6004 696* CALICHE IN TEXAS, ABST., CARBON AND, OXYGEN, ISOTOPE ANALYSES
M 27 189 CALICHE, LAMINATED
B 6011 2048† CALICHE, VADOSE, PISOLITE AND
M 28 113 CALICO ROCK SANDSTONE, ARKANSAS, ORDOVICIAN
B 6203 350 CALIENTE BASIN, CALIFORNIA
B 6211 2351* CALIFORNIA, A REINTERPRETATION. BY TITLE ONLY, FRANCISCAN
B 6102 146 CALIFORNIA, A01 SAND
B 6203 544* CALIFORNIA, ABST., BOTTOM SEDIMENTS OF, SAN NICOLAS BASIN
B 6012 2186* CALIFORNIA, ABST., CLAST POPULATIONS IN, SESPE AND, POWAY
B 6211 2358* CALIFORNIA, ABST., COAL FOR
B 6211 2349* CALIFORNIA, ABST., CRETACEOUS AND EARLY, TERTIARY, SUBDUCTION
B 6303 427* CALIFORNIA, ABST., DEFORMATIONAL STRUCTURES IN, PLIOCENE, BOUSE
B 6203 515* CALIFORNIA, ABST., DEPOSITIONAL SYSTEMS AND THEIR RELATION TO,
B 6405 790* CALIFORNIA, ABST., DIAGENESIS IN, MONTEREY FORMATION, PISMO
B 6303 465* CALIFORNIA, ABST., DIAGENESIS OF, TERTIARY, ARKOSES, SANTA YNEZ
B 6303 493* CALIFORNIA, ABST., DIAGENETIC ALTERATION IN, EOCENE,
B 6403 449* CALIFORNIA, ABST., DIAGENETIC REACTIONS IN, MONTEREY
B 6403 449* CALIFORNIA, ABST., DIAGENETIC REACTIONS IN, MONTEREY
B 6309 1586* CALIFORNIA, ABST., DILATION, BRECCIATION--PROPOSED MECHANISM
B 6211 2360* CALIFORNIA, ABST., EARLY, MIOCENE SUBMARINE DUNE FIELDS—
B 6405 784* CALIFORNIA, ABST., EFFECTS OF, BASALT INTRUSIONS ON, KEROGENS IN
B 6012 2185* CALIFORNIA, ABST., ELEMENTS OF TECTONISM DEDUCED FROM,
B 6205 890* CALIFORNIA, ABST., FOLD SYSTEM IN SOUTHERN, SIERRA NEVADA,
B 6012 2190* CALIFORNIA, ABST., FORAMINIFERAL ZONATION OF, REFUGIAN STAGE,
B 6108 1385* CALIFORNIA, ABST., GEOLOGY AND, RESERVOIR ANALYSIS OF, EAST
B 6203 503* CALIFORNIA, ABST., GEOTHERMAL RESOURCE INVESTIGATIONS AND
B 6012 2188* CALIFORNIA, ABST., HOLOCENE SHELF SEDIMENTATION PATTERNS
B 6405 740* CALIFORNIA, ABST., HYDROCARBON POTENTIAL OF, MATILIJA
B 6207 1231* CALIFORNIA, ABST., HYDROGEOLOGY OF, LONG VALLEY, GEOTHERMAL
B 6405 770* CALIFORNIA, ABST., HYPOTHESIS INVOLVING DILATION AND NATURA
B 6303 421* CALIFORNIA, ABST., INFLUENCE OF RATE ON DEMAND AND USE OF,
B 6211 2363* CALIFORNIA, ABST., LATE, CRETACEOUS DEPOSITIONAL ENVIRONMENTS,
B 6211 2351* CALIFORNIA, ABST., LATE, CRETACEOUS SEDIMENTATION IN, SANTA
B 6012 2182* CALIFORNIA, ABST., LATE, CRETACEOUS TURBIDITE SEDIMENTATIO AND
B 6211 2351* CALIFORNIA, ABST., LATE, MIOCENE, TURBIDITE DEPOSITION, CROCKER
B 6012 2182* CALIFORNIA, ABST., LATE, NEOGENE PLAEOBATHYMETRY AND
B 6012 2180* CALIFORNIA, ABST., LATE, NEOGENE-, QUATERNARY TECTONICS AN
B 6303 473* CALIFORNIA, ABST., LATERAL DIAGENESIS IN, MONTEREY SHALE, SANTA
B 6211 2352* CALIFORNIA, ABST., MESOZOIC CONSTRUCTION OF, CORDILLERAN
B 6012 2175† CALIFORNIA, ABST., MIDDLE, MIOCENE-EARLY. PLIOCENE MARINE. DIATO
B 6403 440* CALIFORNIA, ABST., MODERN AND ANCIENT COASTAL SEDIMENTARY
B 6403 444 * CALIFORNIA, ABST., MONTEREY ROCKS ALONG, SANTA BARBARA COAST
B 6012 2191* CALIFORNIA, ABST., NANNOPLANKTON BIOSTRATIGRAPHY
B 6004 650* CALIFORNIA, ABST., NEOGENE BASI FORMATION AND, HYDROCARBON,
B 6012 2192* CALIFORNIA, ABST., NEOGENE TECTONICS OF CENTRAL, VENTURA BASIN
B 6207 1230* CALIFORNIA, ABST., OIL AND GAS EXPLORATION, OFFSHORE CENTRAL
B 6207 1238* CALIFORNIA, ABST., OIL AND GAS EXPLORATION, OFFSHORE SOUTHERN
B 6207 1237* CALIFORNIA, ABST., OIL AND GAS FROM, SUBMARINE, CHANNELS IN,
B 6405 698* CALIFORNIA, ABST., PALEOGENE DEPOSITIONAL SYSTEMS, WESTERN,
B 6211 2349* CALIFORNIA, ABST., PALEOGEOGRAPHIC SIGNIFICANCE OF MIDDLE TO
B 6012 2187* CALIFORNIA, ABST., PALEONTOLOGIC STUDIES OF MIDDLE, TERTIARY,
B 6211 2359* CALIFORNIA, ABST., PETROLEUM ACCUMULATIONS IN, SANTA
B 6211 2353* CALIFORNIA, ABST., PLATE-TECTONIC EVOLUTION OF, SACRAMENTO
B 6012 2189* CALIFORNIA, ABST., PLIOCENE BIOSTRATIGRAPHY AND DEPOSITIONAL
B 6403 445* CALIFORNIA, ABST., PORE-FILLING, CEMENTS--PRODUCTS OF SHALE
B 6403 445* CALIFORNIA, ABST., PORE-FILLING, CEMENTS—PRODUCTS OF SHALE
B 6405 803* CALIFORNIA, ABST., POTENTIAL PETROLEUM RESERVOIRS ON, DEEP-SEA,
B 6012 2192* CALIFORNIA, ABST., RESERVOIR AND SOURCE-BED HISTORY IN, GREAT
B 6105 844* CALIFORNIA, ABST., RESERVOIR AND SOURCE-BED HISTORY IN, GREAT
B 6012 2188* CALIFORNIA, ABST., SATELLITE LOOK AT REGIONAL GEOLOGY OF
B 6405 731* CALIFORNIA, ABST., SEDIMENTARY PROCESSES ON CONTINENTAL SHELF
B 6012 2177* CALIFORNIA, ABST., SEDIMENTARY SERPENTINITE OF THE MIOCENE, BIG
B 6303 468* CALIFORNIA, ABST., SEDIMENTARY STRUCTURES OF, HIGH-ENERGY,
B 6004 707* CALIFORNIA, ABST., SEDIMENTOLOGIC INDICATION OF, EOCENE,
B 6105 835* CALIFORNIA, ABST., SEDIMENTOLOGY OF, FLUVIAL TO, BATHYAL,
B 6004 644* CALIFORNIA, ABST., SEDIMENTOLOGY OF, MELANGE, FRANCISCAN OF,
B 6207 1215* CALIFORNIA, ABST., SOLAR, ENERGY—, CENTRAL RECEIVER PROGRAM,
B 6203 541* CALIFORNIA, ABST., STEEPLY INCLINED, ALGAL COLUMNS IN LATE,
B 6203 535* CALIFORNIA, ABST., STRUCTURE OF, ALISO CANYON FIELD, EASTERN,
B 6305 844* CALIFORNIA, ABST., STRUCTURE OF, SAN CAYETANO AND, OAK RIDGE
B 6303 522* CALIFORNIA, ABST., SUBSURFACE GEOLOGY OF, HONOR RANCHO AREA,
B 6212 2508* CALIFORNIA, ABST., TECTONIC FRAMEWORK AND, METALLOGENESIS IN
B 6211 2360* CALIFORNIA, ABST., TERTIARY, PRINCETON SUBMARINE VALLEY SYSTEM
B 6012 2181* CALIFORNIA, ABST., TERTIARY STRATIGRAPHIC AND DEPOSITIONAL
B 6004 645* CALIFORNIA, ABST., TIMING OF, TERTIARY, SUBMARINE, CANYONS AND,
B 6004 685* CALIFORNIA, ABST., TIMING--GEOLOGIC AN POLITICAL--IN, URBAN, OIL
B 6012 2179* CALIFORNIA, ABST., UPPER, CRETACEOUS LITHOFACIES MODEL.
B 6211 2363* CALIFORNIA, ABST., UPPER, MIOCENE, BARRIER BEACH-, SALINE, LAGOON
B 6203 545* CALIFORNIA, ABST., URANIUM IN, TUFFACEOUS, FLUVIAL SANDSTONE,
B 6303 421* CALIFORNIA, ABST., VITRINITE REFLECTANCE, PRESENT ROCK,
B 6011 2038† CALIFORNIA, ALLOCHTHONOUS, CARBONATE DEBRIS IN, MESOZOIC,
B 6111 2002 CALIFORNIA, ALTAMIRA MEMBER
B 6202 259 CALIFORNIA, ALTAMIRA SHALE
B 6211 2355* CALIFORNIA, AN ENIGMA, ABST., SALINIAN BLOCK OF
B 6202 255 CALIFORNIA, ANAHEIM NOSE
B 6004 668* CALIFORNIA AND ITS PALINSPASTIC IMPLICATIONS, ABST., MAGNITUDE
B 6407 1034† CALIFORNIA, AND MODEL OF LONGITUDINAL TRANSPORT OF
B 6012 2174† CALIFORNIA AND NORTHERN MEXICO, ABST., UPDATE OF, GEOTHERMAL
B 6004 707* CALIFORNIA AND NORTHWESTERN BAJA CALIFORNIA, ABST.,
B 6205 883* CALIFORNIA AND POSSIBLE ANALOGS IN CRETACEOUS OF UTAH, ABST.,
B 6207 1223* CALIFORNIA AND SOUTHWESTERN OREGON, ABST., LATE, MESOZOIC
B 6211 2354* CALIFORNIA AND WESTERN MEXICO, ABST., MESOZOIC GEOLOGY OF
B 6203 351 CALIFORNIA, ANTELOPE HILLS OIL FIELD
B 6211 2245 CALIFORNIA, ANTELOPE SHALE FORMATION
B 6206 968 CALIFORNIA, ARGUELLO CANYON
B 6206 968 CALIFORNIA, ARGUELLO FAN
B 6211 2262 CALIFORNIA, ARROYO HONDO MEMBER
B 6211 2262 CALIFORNIA, ARROYO HONDO SECTION
B 6211 2218 CALIFORNIA, ARROYO SECO TROUGH
B 6405 699* CALIFORNIA AS EXAMPLE, ABST., PALEOECOLOGY IN, BASIN ANALYSIS—,
B 6206 968 CALIFORNIA, ASCENSION CANYON
B 6203 351 CALIFORNIA, ASPHALTO OIL FIELD
B 6407 1035 CALIFORNIA, BAILEY ASH
B 6111 1986 CALIFORNIA, BAKER CANYON CONGLOMERATE
B 6211 2245 CALIFORNIA, BAKERSFIELD ARCH
B 6307 1036 CALIFORNIA, BAKERSFIELD ARCH, CONTOUR MAP
B 6204 688 CALIFORNIA, BANDERAS BAY, KEROGEN
B 6407 1036 CALIFORNIA, BARDSDALE FIELD
B 6102 146 CALIFORNIA, BARNARD THRUST FAULT
B 6102 189 CALIFORNIA, BARNARD THRUST FAULT
B 6102 154 CALIFORNIA, BARNARD THRUST FAULT
B 6012 2180* CALIFORNIA BASINS, ABST., TOMORROWS OIL FROM SOUTHERN
B 6004 732* CALIFORNIA BASINS AND BANK MARGINS--RESULTS USING MANNED
B 6202 249 CALIFORNIA, BATHYMETRY, SOUTHERN
S 11 187 CALIFORNIA, BEAR RIVER SECTION
B 6111 1986 CALIFORNIA, BEDFORD CANYON FORMATION
B 6203 351 CALIFORNIA, BELGIAN ANTICLINE OIL FIELD
B 6211 2243 CALIFORNIA, BELLEVUE CYCLE, INFORMAL NAME
B 6211 2264 CALIFORNIA, BELLEVUE FAN
B 6211 2247 CALIFORNIA, BELLEVUE FAN
B 6211 2256 CALIFORNIA, BELLEVUE FIELD
B 6211 2255 CALIFORNIA, BELLEVUE GROWTH FAULT SYSTEM
B 6211 2256 CALIFORNIA, BELLEVUE SAND
B 6211 2246 CALIFORNIA, BENA CHANNEL, NEW TERM
B 6211 2245 CALIFORNIA, BENA FAN
B 6202 259 CALIFORNIA, BENTHIC STAGES
B 6211 2219 CALIFORNIA, BERRY FORMATION
B 6111 1980 CALIFORNIA, BIG PINE FAULT
B 6203 345 CALIFORNIA, BIG PINE FAULT
B 6211 2218 CALIFORNIA, BITTERWATER BASIN
B 6111 2002 CALIFORNIA, BLUFF COVE AREA, PALOS VERDES HILLS

B 6203 352 CALIFORNIA, BODEGA BASIN
B 6211 2215 CALIFORNIA, BODEGA HEAD
B 6403 441* CALIFORNIA BORDERLAND, ABST., MASS MOVEMENT EFFECTS ON A
B 6403 442* CALIFORNIA BORDERLAND, ABST., SEDIMENTARY PROCESSES ACTIVE ON
B 6403 442* CALIFORNIA BORDERLAND, ABST., SEDIMENTOLOGIC HISTORY AND
B 6203 510* CALIFORNIA BORDERLAND, ABST., SLOPE-BASIN, BENTHIC,
B 6405 792* CALIFORNIA BORDERLAND, ABST., SLOPE-CENTERED PROCESSES IN,
B 6004 704* CALIFORNIA BORDERLAND, ABST., SOURCE IDENTIFICATION OF, BEACH,
B 6403 440* CALIFORNIA BORDERLAND, ABST., SPECULATIONS ON HYDROCARBON
B 6403 440* CALIFORNIA BORDERLAND, ABST., SPECULATIONS ON HYDROCARBON
B 6004 663* CALIFORNIA BORDERLAND FROM SATELLITE IMAGERY, ABST.,
B 6403 442* CALIFORNIA BORDERLAND—QUATERNARY TECTONICS, SEISMIC
B 6405 622 CALIFORNIA, BORREGO FORMATION
B 6207 1196 CALIFORNIA BOUNDARY, PLATE BOUNDARIES IN WESTERN, NORTH
B 6211 2268 CALIFORNIA, BRECKENRIDGE FAULT
B 6211 2267 CALIFORNIA, BUENA VISTA HILLS FIELD
B 6203 351 CALIFORNIA, BUENA VISTA OIL FIELD
B 6205 819 CALIFORNIA, BUTTONWILLOW DEPOCENTER
B 6306 989 CALIFORNIA, BY ROBERT SHARP, BEAR, TED L., REVIEW OF, COASTAL
B 6202 255 CALIFORNIA, CABRILLO FAULT
S 11 60† CALIFORNIA, CALCAREOUS, NANNOPLANKTON BIOSTRATIGRAPHY OF,
B 6203 350 CALIFORNIA, CALIENTE BASIN
B 6205 848 CALIFORNIA, CANTUA CREEK FIELD
B 6211 2262 CALIFORNIA, CANTUA CREEK SECTION
B 6211 2262 CALIFORNIA, CANTUA SAND
B 6206 955 CALIFORNIA, CAPISTRANO FORMATION
B 6206 968 CALIFORNIA, CARMEL CANYON
B 6211 2224 CALIFORNIA, CARRIZO PLAIN
B 6203 345 CALIFORNIA, CASCADE RANGE
B 6203 355 CALIFORNIA, CASMALIA OIL FIELD
B 6203 355 CALIFORNIA, CAT CANYON GROUP
B 6202 248 CALIFORNIA, CATALINA BASIN
B 6203 363 CALIFORNIA, CATALINA BASIN
B 6111 1981 CALIFORNIA, CATALINA BASIN
B 6111 1984 CALIFORNIA, CATALINA SCHIST
B 6203 362 CALIFORNIA, CATALINA SCHIST
S 11 62 CALIFORNIA, CENOZOIC MARINE STAGES
B 6202 259 CALIFORNIA, CENOZOIC, UPPER, STRATIGRAPHIC COLUMN
M 25 314 CALIFORNIA, CENTRAL, STRATIGRAPHIC TRAPS
B 6211 2262 CALIFORNIA, CERROS HILL SECTION
B 6211 2262 CALIFORNIA, CERROS SHALE MEMBER
B 6211 2220 CALIFORNIA, CHALONE CREEK FAULT
B 6211 2245 CALIFORNIA, CHANAC FORMATION
B 6206 967† CALIFORNIA, CHANNEL, SANDS AND, PETROLEUM, POTENTIAL OF,
S 11 91 CALIFORNIA, CHECK LIST OF, FORAMINIFERA, MONTEREY FORMATION,
S 11 94 CALIFORNIA, CHECK LIST OF, FORAMINIFERA, MONTEREY SHALE,
S 11 82 CALIFORNIA, CHECK LIST OF, FORAMINIFERA, TEMBLOR-MONTEREY
S 11 85 CALIFORNIA, CHECK LIST OF, FORAMINIFERA, TEMBLOR-MONTEREY
S 11 88 CALIFORNIA, CHECK LIST OF, FORAMINIFERA, TEMBLOR-MONTEREY
S 11 99 CALIFORNIA, CHECK LIST OF, MIOCENE, FORAMINIFERA, SHALE POINT
B 6203 351 CALIFORNIA, CHICO-MARTINEZ OIL FIELD
B 6111 1982 CALIFORNIA, CHINO FAULT
B 6211 2220 CALIFORNIA, CHUPINES FAULT
B 6211 2220 CALIFORNIA, CHURCH CREEK FAULT
B 6211 2261 CALIFORNIA, CIMA SANDSTONE
B 6103 371 CALIFORNIA, CLOUD CHIEF FORMATION
B 6304 649 CALIFORNIA CO. F. P. STRADER NO. 1, SECTION, VIRGINIA
B 6203 368 CALIFORNIA, COACHELLA-IMPERIAL VALLEYS
B 6205 848 CALIFORNIA, COALINGA EAST EXTENSION FIELD
B 6004 650* CALIFORNIA COAST, ABST., PHYSICAL AND, BIOGENI CHARACTERISTICS
B 6301 21 CALIFORNIA, COAST RANGE THRUST
B 6211 2244 CALIFORNIA, COAST RANGES
B 6203 345 CALIFORNIA, COAST RANGES
M 25 485 CALIFORNIA, COAST RANGES, MERCURY
B 6403 445* CALIFORNIA COASTAL LAGOON, ABST., LATE, PLEISTOCENE-, HOLOCENE
B 6211 2247 CALIFORNIA, COLES LEVEE TREND
S 5 178 CALIFORNIA, COMMERCIAL SALINA
B 6104 595 CALIFORNIA, CONCEPTION FIELD
B 6403 438* CALIFORNIA CONTINENTAL BORDERLAND, ABST., DIATOMS AND,
B 6403 438* CALIFORNIA CONTINENTAL BORDERLAND, ABST., DIATOMS AND,
B 6203 548* CALIFORNIA CONTINENTAL BORDERLAND, ABST., HYDROCARBON,
B 6203 511* CALIFORNIA CONTINENTAL BORDERLAND, ABST., NEW EVIDENCE OF,
B 6403 440* CALIFORNIA CONTINENTAL BORDERLAND, ABST., PALYNOMORPHS
B 6403 440* CALIFORNIA CONTINENTAL BORDERLAND, ABST., PALYNOMORPHS
B 6004 670* CALIFORNIA CONTINENTAL BORDERLAND, ABST., SHALLOW STRUCTURE
B 6004 675* CALIFORNIA CONTINENTAL BORDERLAND BASIN SEDIMENTS, ABST.,
B 6004 646* CALIFORNIA CONTINENTAL BORDERLAND NORTH OF 32 DEGREES, ABST.,
B 6012 2174† CALIFORNIA CONTINENTAL BORDERLAND NORTH OF 32 DEGREES, ABST.,
B 6202 247† CALIFORNIA CONTINENTAL BORDERLAND, PLIOCENE-, PLEISTOCENE,
B 6012 2181* CALIFORNIA CONTINENTAL BORDERLAND SEDIMENTATION DERIVED
B 6212 2492† CALIFORNIA CONTINENTAL BORDERLAND, DISCUSSION AND REPLY,
B 6405 706* CALIFORNIA CONTINENTAL MARGIN, ABST., DIAPIR-LIKE RIDGES AND
B 6405 736* CALIFORNIA CONTINENTAL MARGIN, ABST., THERMOGENIC
B 6202 249 CALIFORNIA, CORES, JET AND DART
B 6102 152 CALIFORNIA, CORONADO BANK
B 6111 1981 CALIFORNIA, CORTES BANK
B 6203 367 CALIFORNIA, CORTES BANK
B 6111 1983 CALIFORNIA, CORTES BANK, STRATIGRAPHIC COLUMN
B 6211 2243 CALIFORNIA, COULTER CYCLE, INFORMAL NAME
B 6211 2247 CALIFORNIA, COULTER FAN
B 6211 2264 CALIFORNIA, COULTER FAN
B 6211 2256 CALIFORNIA, COULTER SAND
B 6211 2261 CALIFORNIA, CRETACEOUS
S 6 264 CALIFORNIA, CRETACEOUS, BUDDEN CANYON FORMATION
B 6205 819 CALIFORNIA, CRETACEOUS, ISOPACH MAP
S 6 264 CALIFORNIA, CRETACEOUS, SHASTA BALLY BATHOLITH
B 6111 1989 CALIFORNIA, CRETACEOUS SHORELINE
B 6206 934 CALIFORNIA, CRETACEOUS, TURBIDITE FACIES
B 6211 2217 CALIFORNIA, CRETACEOUS-PALEOCENE
B 6111 1987 CALIFORNIA, CRETACEOUS-TERTIARY UNCONFORMITY
B 6111 1981 CALIFORNIA, CUCAMONGA FAULT
B 6203 361 CALIFORNIA, CUCAMONGA-SIERRA MADRE FAULT
B 6203 353 CALIFORNIA, CUYAMA OIL DISTRICT
B 6203 351 CALIFORNIA, CYMRIC OIL FIELD
B 6102 146 CALIFORNIA, DA SAND
B 6206 968 CALIFORNIA, DELGADA CHANNEL
B 6206 968 CALIFORNIA, DELGADA FAN
B 6211 2246 CALIFORNIA, DELMONTIAN STAGE
B 6202 259 CALIFORNIA, DELMONTIAN STAGE
S 11 43 CALIFORNIA, DELMONTIAN STAGE
B 6205 819 CALIFORNIA, DELTA DEPOCENTER
B 6008 1196 CALIFORNIA, DEVELOPMENTS, 1975
B 6108 1166 CALIFORNIA, DEVELOPMENTS, 1976
B 6208 1322 CALIFORNIA, DEVELOPMENTS, 1977
B 6308 1292 CALIFORNIA, DEVELOPMENTS, 1978
B 6409 1394 CALIFORNIA, DEVELOPMENTS, 1979
B 6211 2244 CALIFORNIA, DIABLO RANGE
B 6203 369 CALIFORNIA, DILIGENCIA BASIN
B 6211 2262 CALIFORNIA, DOMENGINE CREEK SECTION
B 6211 2261 CALIFORNIA, DOMENGINE SAND
B 6205 849 CALIFORNIA, DOMENGINE SHALE
B 6203 362 CALIFORNIA, DOMINGUEZ OIL FIELD
B 6211 2261 CALIFORNIA, DOS PALOS SHALE
B 6211 2261 CALIFORNIA, DOS PALOS SHALE
B 6103 371 CALIFORNIA, DOXEY FORMATION
B 6405 621 CALIFORNIA, DURMID HILLS, SEASAT IMAGE
B 6211 2215 CALIFORNIA, EAGLE REST PEAK
B 6111 1981 CALIFORNIA, EAST CORTES BASIN
B 6212 2493 CALIFORNIA, EAST WILMINGTON FIELD, REPETTO STRUCTURE, UPPER
B 6211 2246 CALIFORNIA, EDISON HIGH, NEW TERM
B 6203 346 CALIFORNIA, EEL RIVER BASIN
B 6205 818 CALIFORNIA, ELK HILLS FIELD
B 6203 351 CALIFORNIA, ELK HILLS OIL FIELD
B 6205 816 CALIFORNIA, ELK HILLS POROSITY VALUES
B 6111 1982 CALIFORNIA, ELSINORE FAULT ZONE
B 6111 1982 CALIFORNIA, ELSINORE TROUGH
B 6211 2217 CALIFORNIA, EOCENE
B 6205 820 CALIFORNIA, EOCENE
B 6102 237† CALIFORNIA, EOCENE, PALEOCURRENTS AND SEDIMENTATION, SA
B 6102 237 CALIFORNIA, EOCENE, SAN NICOLAS ISLAND
S 11 4 CALIFORNIA, EPOCH BOUNDARIES, EARLY PROBLEMS
B 6211 2267 CALIFORNIA, ETCHEGOIN FORMATION
B 6206 968 CALIFORNIA, FARALLON CANYON
B 6203 356 CALIFORNIA, FARALLON RIDGE
M 25 526 CALIFORNIA FAULT SYSTEM, MOVEMENTS, SAN ANDREAS-GULF OF
B 6111 2001† CALIFORNIA, FAULTED UPPER, PLEISTOCENE, MARINE, TERRACE, PALOS
B 6104 594 CALIFORNIA, FILLMORE FIELD
B 6205 848 CALIFORNIA, FIVE POINTS FIELD
S 11 188 CALIFORNIA, FORAMINIFERAL LOCALITIES
B 6211 2246 CALIFORNIA, FORAMINIFERAL STAGE
B 6211 2218 CALIFORNIA, FORAMINIFERAL STAGES
B 6301 20 CALIFORNIA, FOREARC EVOLUTION
B 6301 21 CALIFORNIA, FOREARC REGIONS
B 6212 2444 CALIFORNIA, FORMATION WATERS
B 6102 238 CALIFORNIA, FORTY MILE BANK
B 6301 21 CALIFORNIA, FRANCISCAN COMPLEX
B 6211 2217 CALIFORNIA, FRANCISCAN COMPLEX
B 6203 349 CALIFORNIA, FRANCISCAN COMPLEX
B 6111 1980 CALIFORNIA, FRANCISCAN FORMATION
B 6211 2245 CALIFORNIA, FRUITVALE FAN

B 6211 2217 CALIFORNIA, GABILAN MESA
B 6211 2217 CALIFORNIA, GABILAN RANGE
B 6211 2224 CALIFORNIA, GARLOCK FAULT
B 6203 345 CALIFORNIA, GARLOCK FAULT
B 6211 2270 CALIFORNIA, GARLOCK FAULT
B 6205 852 CALIFORNIA, GATCHELL SAND, KAOLINIZATION
B 6205 848 CALIFORNIA, GATCHELL SAND, TRAPS
B 6211 2263 CALIFORNIA, GATCHELL TURBIDITE
M 25 41 CALIFORNIA, GEOLOGIC MAP
B 6006 988† CALIFORNIA GEOLOGICAL SOCIETY, SHORT, HISTORY OF, NORTHERN
B 6008 1199 CALIFORNIA, GEOTHERMAL PROSPECTING, 1975
B 6108 1169 CALIFORNIA, GEOTHERMAL PROSPECTING, 1976
B 6208 1326 CALIFORNIA, GEOTHERMAL PROSPECTING, 1977
B 6308 1296 CALIFORNIA, GEOTHERMAL PROSPECTING, 1978
B 6409 1398 CALIFORNIA, GEOTHERMAL PROSPECTING, 1979
B 6211 2243 CALIFORNIA, GOSFORD CYCLE, INFORMAL NAME
B 6211 2247 CALIFORNIA, GOSFORD FAN
B 6211 2256 CALIFORNIA, GOSFORD SAND
B 6203 345 CALIFORNIA, GREAT BASIN
B 6203 345 CALIFORNIA, GREAT VALLEY
B 6211 2244 CALIFORNIA, GREAT VALLEY
B 6211 2217 CALIFORNIA, GREAT VALLEY SEQUENCE
B 6301 20 CALIFORNIA, GREAT VALLEY SEQUENCE
B 6211 2264 CALIFORNIA, GREELEY FIELD
B 6211 2245 CALIFORNIA, GREELEY HIGH
B 6102 156 CALIFORNIA, GRUBB FAULT
B 6203 355 CALIFORNIA, GUADALUPE OIL FIELD
B 6205 848 CALIFORNIA, GUIJARRAL HILLS FIELD
B 6205 850 CALIFORNIA, GUIJARRAL HILLS SAND
B 6203 368 CALIFORNIA, GULF OF SANTA CATALINA
B 6407 1037 CALIFORNIA, HALL CANYON
B 6202 259 CALIFORNIA, HALLIAN STAGE
B 6211 2219 CALIFORNIA, HAMES MEMBER
B 6211 2261 CALIFORNIA, HELM FIELD
B 6202 250 CALIFORNIA, HOLOCENE FAULT SCARP
M 28 10 CALIFORNIA, HOLOCENE, GRUS
B 6102 152 CALIFORNIA, HOLOCENE SAND
B 6111 1986 CALIFORNIA, HOLZ SHALE
B 6205 849 CALIFORNIA, HONDO SILT
B 6203 357 CALIFORNIA, HOSGRI FAULT
B 6203 349 CALIFORNIA, HUASNA-PISMO BASIN
S 11 183 CALIFORNIA, HUMBOLDT BASIN
B 6102 137† CALIFORNIA, I--FACIES GEOMETRY AND GENESIS OF LOWER PLIOCENE
B 6102 169† CALIFORNIA, II--LITHOLOGY, COMPACTION AND PERMEABILITY OF
B 6307 1139† CALIFORNIA IN LAKE MEAD AREA OF ARIZONA AND NEVADA DURING
S 11 8 CALIFORNIA, INFRA-SERIES SUBDIVISIONS
S 11 105 CALIFORNIA, INTERPRETATION, FIGURE, FORAM SEQUENCE
B 6211 2220 CALIFORNIA, JOLON FAULT
B 6211 2219 CALIFORNIA, JUNIPERO FORMATION
M 27 25 CALIFORNIA, JURASSIC, FRANCISCAN FORMATION
M 27 24 CALIFORNIA, JURASSIC, POINT SAL OPHIOLITE
M 27 23 CALIFORNIA, JURASSIC, POINT SAL OPHIOLITE
M 28 79 CALIFORNIA, JURASSIC, POINT SAL OPHIOLITE
B 6111 1982 CALIFORNIA, JURUPA HILLS
B 6211 2245 CALIFORNIA, KERN ARCH
B 6211 2268 CALIFORNIA, KERN CANYON FAULT
B 6211 2268 CALIFORNIA, KERN RIVER, ANCESTRAL
B 6211 2270 CALIFORNIA, KERNSUMNER FIELD
M 30 26 CALIFORNIA, KETTLEMAN HILLS FIELD, LANDSAT IMAGE
B 6401 24 CALIFORNIA, KETTLEMAN HILLS GIANT OIL FIELD, LANDSAT DATA
B 6401 24 CALIFORNIA, KETTLEMAN HILLS GIANT OIL FIELD, LANDSAT DATA
B 6212 2446 CALIFORNIA, KETTLEMAN NORTH DOME
B 6205 848 CALIFORNIA, KETTLEMAN NORTH DOME FIELD
B 6211 2263 CALIFORNIA, KETTLEMAN NORTH DOME FIELD
B 6203 345 CALIFORNIA, KLAMATH MOUNTAINS
B 6301 21 CALIFORNIA, KLAMATH MOUNTAINS
B 6205 849 CALIFORNIA, KREYENHAGEN SHALE
B 6203 349 CALIFORNIA, LA HONDA BASIN
B 6102 152 CALIFORNIA, LA JOLLA CANYON
B 6206 921 CALIFORNIA, LA JOLLA FAN VALLEY
B 6211 2220 CALIFORNIA, LA PANZA FAULT
B 6211 2218 CALIFORNIA, LA PANZA RANGE
B 6111 1986 CALIFORNIA, LA VIDA SILTSTONE
B 6405 622 CALIFORNIA, LAKE CAHUILLA DEPOSITS
B 6202 255 CALIFORNIA, LASUEN ANTICLINORIUM
B 6202 248 CALIFORNIA, LASUEN KNOLL
B 6203 349 CALIFORNIA, LIVERMORE BASIN
B 6211 2218 CALIFORNIA, LOCKWOOD HIGH
B 6203 369 CALIFORNIA, LOCKWOOD VALLEY BASIN
B 6211 2262 CALIFORNIA, LODO GULCH SECTION
B 6211 2262 CALIFORNIA, LODO MEMBER
B 6203 355 CALIFORNIA, LOMPOC OIL FIELD
B 6203 355 CALIFORNIA, LOS ALAMOS OIL FIELD
B 6111 1988 CALIFORNIA, LOS ANGELES ARCH
B 6202 248 CALIFORNIA, LOS ANGELES BASIN
B 6203 356 CALIFORNIA, LOS ANGELES BASIN
S 7 239 CALIFORNIA, LOS ANGELES BASIN, BREEN SANDSTONE, ISOPACH
S 7 239 CALIFORNIA, LOS ANGELES BASIN, CHAPMAN SANDSTONE, ISOPACH
S 7 233 CALIFORNIA, LOS ANGELES BASIN, DISTAL TURBIDITE FACIES
S 7 234 CALIFORNIA, LOS ANGELES BASIN, E-LOG SECTION
S 7 240 CALIFORNIA, LOS ANGELES BASIN, MOHNIAN PALEOGEOGRAPHY
S 7 238 CALIFORNIA, LOS ANGELES BASIN, PROXIMAL SANDSTONE FACIES
S 7 236 CALIFORNIA, LOS ANGELES BASIN, PROXIMAL TURBIDITE FACIES
S 7 240 CALIFORNIA, LOS ANGELES BASIN, RECONSTRUCTION
S 10 50 CALIFORNIA, LOS ANGELES BASIN,MIGRATION
S 10 51 CALIFORNIA, LOS ANGELES BASIN, PLAYA DEL REYFIELD
B 6002 220 CALIFORNIA, LOS ANGELES DOWNTOWN FIELD, FLOWER COMMUNITY
B 6111 1981 CALIFORNIA, LOS ANGELES LOWLANDS AND BORDERLAND
B 6211 2220 CALIFORNIA, LOS LOBOS THRUST
B 6211 2220 CALIFORNIA, LUCIA FAULT
B 6211 2219 CALIFORNIA, LUCIA FORMATION
B 6206 968 CALIFORNIA, LUCIA-PARTINGTON CANYON
B 6211 2246 CALIFORNIA, LUISIAN STAGE
B 6202 259 CALIFORNIA, LUISIAN STAGE
B 6211 2247 CALIFORNIA, MACOMA SHALE
B 6202 249 CALIFORNIA, MAGNETIC PROFILES, SOUTHERN
B 6011 1985† CALIFORNIA, MAGNETIC-REVERSAL STRATIGRAPHY OF, PLIOCENE-,
B 6202 259 CALIFORNIA, MALAGA MUDSTONE MEMBER
B 6203 359 CALIFORNIA, MALIBU COAST FAULT
B 6211 2246 CALIFORNIA, MARICOPA SUBBASIN, INFORMAL NAME
B 6206 934 CALIFORNIA, MATILIJA FORMATION, TURBIDITES
B 6203 351 CALIFORNIA, MCDONALD ANTICLINE OIL FIELD
B 6211 2267 CALIFORNIA, MCDONALD SHALE
B 6203 351 CALIFORNIA, MCKITTRICK OIL FIELD
B 6206 951 CALIFORNIA, MEGANOS CHANNEL
B 6203 356 CALIFORNIA, MENDOCINO FAULT
S 11 4 CALIFORNIA, MICROPALEONTOLOGY, PIONEERS
B 6203 351 CALIFORNIA, MIDWAY-SUNSET OIL FIELD
B 6211 2220 CALIFORNIA, MILLER CREEK FAULT
S 11 52 CALIFORNIA, MIO-PLIOCENE, DIATOMS
B 6205 820 CALIFORNIA, MIOCENE
B 6211 2223 CALIFORNIA, MIOCENE
B 6211 2269 CALIFORNIA, MIOCENE
B 6206 951 CALIFORNIA, MIOCENE
B 6111 1985 CALIFORNIA, MIOCENE
B 6102 153 CALIFORNIA, MIOCENE
S 11 146 CALIFORNIA MIOCENE AND OLIGOCENE BULIMINIDS
S 11 160 CALIFORNIA MIOCENE ANOMALINIDS
S 11 158 CALIFORNIA MIOCENE ANOMALINIDS
S 11 162 CALIFORNIA MIOCENE BAGGINA
M 28 141 CALIFORNIA, MIOCENE, BARSTOW FORMATION
S 11 134 CALIFORNIA MIOCENE BOLIVINIDS
S 11 136 CALIFORNIA MIOCENE BOLIVINIDS
S 11 130 CALIFORNIA MIOCENE BOLIVINIDS
S 11 132 CALIFORNIA MIOCENE BOLIVINIDS
S 11 138 CALIFORNIA MIOCENE BULIMINIDAE
S 11 154 CALIFORNIA MIOCENE CASSIDULINA
M 28 66 CALIFORNIA, MIOCENE, CIERBO SANDSTONE
M 28 133 CALIFORNIA, MIOCENE, CIERBO SANDSTONE
S 11 107 CALIFORNIA, MIOCENE CORRELATION
S 11 156 CALIFORNIA MIOCENE EPISTOMINELLIDS AND DISCORBINELLA
M 28 142 CALIFORNIA, MIOCENE, MONTEREY FORMATION
M 27 26 CALIFORNIA, MIOCENE, MONTEREY SHALE
M 28 81 CALIFORNIA, MIOCENE, MONTEREY SHALE
M 28 66 CALIFORNIA, MIOCENE, NEROLOY SANDSTONE
M 28 133 CALIFORNIA, MIOCENE, NEROLOY SANDSTONE
S 11 122 CALIFORNIA, MIOCENE, NONIONIDAE
S 11 124 CALIFORNIA MIOCENE, NONIONIDAE, MOHNIAN AND, DELMONTIAN
S 11 8 CALIFORNIA, MIOCENE, SALINAS VALLEY SEQUENCE
S 11 148 CALIFORNIA MIOCENE SIPHOGENERINA
S 11 150 CALIFORNIA MIOCENE SIPHOGENERINA
S 11 152 CALIFORNIA MIOCENE SIPHOGENERINA
S 11 142 CALIFORNIA MIOCENE UVIGERINA
S 11 140 CALIFORNIA MIOCENE UVIGERINE BULMINIDAE
S 11 170 CALIFORNIA MIOCENE VALVULINERIA
S 11 168 CALIFORNIA MIOCENE VALVULINERIA
S 11 166 CALIFORNIA MIOCENE VALVULINERIA
B 6203 345 CALIFORNIA, MODOC PLATEAU
B 6211 2221 CALIFORNIA, MOHNIAN PALEOGEOGRAPHY
B 6211 2246 CALIFORNIA, MOHNIAN STAGE

B 6202 259 CALIFORNIA, MOHNIAN STAGE
B 6203 345 CALIFORNIA, MOJAVE DESERT
S 11 186 CALIFORNIA, MOLLUSCAN LOCALITIES
B 6104 594 CALIFORNIA, MONTALVO FIELD
B 6203 362 CALIFORNIA, MONTEBELLO OIL FIELD
B 6211 2220 CALIFORNIA, MONTEREY BAY FAULT ZONE
B 6206 968 CALIFORNIA, MONTEREY CANYON
B 6206 968 CALIFORNIA, MONTEREY FAN
B 6206 918 CALIFORNIA, MONTEREY FAN, ASCENSION VALLEY
B 6206 974 CALIFORNIA, MONTEREY FAN, RESERVOIR BEDS AND TRAPS
B 6206 970 CALIFORNIA, MONTEREY FAN, SEISMIC REFLECTION DATA
B 6211 2219 CALIFORNIA, MONTEREY FORMATION
B 6111 2002 CALIFORNIA, MONTEREY FORMATION
B 6202 259 CALIFORNIA, MONTEREY SHALE
B 6206 916 CALIFORNIA, MONTEREY SUBMARINE FAN
B 6203 353 CALIFORNIA, MORALES AND TAYLOR CANYON TREND
B 6211 2261 CALIFORNIA, MORENO FORMATION
B 6211 2270 CALIFORNIA, MOUNTAIN VIEW FAULT
B 6211 2245 CALIFORNIA, MOUNTAIN VIEW PLATFORM
B 6111 1980 CALIFORNIA, MT. DIABLO RANGE
B 6211 2259 CALIFORNIA, N-CHERT
B 6102 240 CALIFORNIA, NARIZIAN ROCKS
B 6206 916 CALIFORNIA, NAVY SUBMARINE FAN
B 6203 344† CALIFORNIA, NEOGENE BASINFORMATION IN RELATION TO,
S 11 327 CALIFORNIA, NEOGENE BOLIVINIDS, KEY TO IDENTIFICATION
B 6203 368 CALIFORNIA, NEOGENE DEPOSITIONAL BASINS, ONSHORE AND OFFSHORE
S 11 205 CALIFORNIA NEOGENE, FORAM LINEAGES
S 11 213 CALIFORNIA NEOGENE, FORAM NAMES, PRIOR USAGE
B 6111 1982 CALIFORNIA, NEWPORT-INGLEWOOD FAULT ZONE
B 6202 255 CALIFORNIA, NEWPORT-INGLEWOOD ZONE
S 11 126 CALIFORNIA, NONION MULTICAMERATUM, MOHNIAN
B 6202 265 CALIFORNIA, NORTH AMERICAN AND PACIFIC PLATES, DISPLACEMENT
B 6211 2215 CALIFORNIA, NORTH AMERICAN PLATE
B 6203 351 CALIFORNIA, NORTH ANTELOPE HILLS OIL FIELD
B 6203 351 CALIFORNIA, NORTH BELRIDGE OIL FIELD
B 6211 2245 CALIFORNIA, NORTH TEJON HIGH
B 6203 366 CALIFORNIA, NORTHEAST BANK
M 25 65 CALIFORNIA, NORTHERN, ERTS PROSPECTING
B 6211 2270 CALIFORNIA, NOZU SAND
B 6102 171 CALIFORNIA, OAK RIDGE FAULT
B 6102 139 CALIFORNIA, OAK RIDGE FAULT
B 6407 1036 CALIFORNIA, OAK RIDGE FAULT
B 6203 356 CALIFORNIA, OCONOSTATA FAULT
B 6102 170 CALIFORNIA, OJAI VALLEY
B 6102 138 CALIFORNIA, OJAI VALLEY
B 6211 2270 CALIFORNIA, OLCESE SAND
B 6211 2223 CALIFORNIA, OLIGOCENE
B 6205 820 CALIFORNIA, OLIGOCENE
S 11 144 CALIFORNIA OLIGOCENE AND MIOCENE BULIMINIDS
S 7 229 CALIFORNIA, ONSHORE AND OFFSHORE BASINS, MAP
B 6111 1982 CALIFORNIA, ONTARIO BASIN
B 6203 355 CALIFORNIA, ORCUTT OIL FIELD
B 6211 2364* CALIFORNIA, OREGON, AND WASHINGTON, ABST., SYNTHESIS OF, OCS
B 6203 352 CALIFORNIA, OUTER SANTA CRUZ BASIN
B 6104 594 CALIFORNIA, OXNARD FIELD
B 6407 1036 CALIFORNIA, OXNARD FIELD
B 6203 360 CALIFORNIA, OXNARD OIL FIELD
B 6203 360 CALIFORNIA, OXNARD SHELF
B 6211 2215 CALIFORNIA, PACIFIC PLATE
B 6102 167 CALIFORNIA, PADRE CANYON FIELD
B 6102 165 CALIFORNIA, PADRE JUAN FAULT
B 6211 2261 CALIFORNIA, PALEOCENE
B 6205 820 CALIFORNIA, PALEOCENE
B 6206 951 CALIFORNIA, PALEOCENE
B 6211 2220 CALIFORNIA, PALO COLORADO FAULT
B 6102 177 CALIFORNIA, PALOMA FIELD, STEVENS SAND
B 6202 252 CALIFORNIA, PALOS VERDES FAULT
B 6202 248 CALIFORNIA, PALOS VERDES HILLS
B 6202 255 CALIFORNIA, PALOS VERDES HILLS ANTICLINORIUM
B 6202 259 CALIFORNIA, PALOS VERDES HILLS AREA, COLUMN
B 6203 361 CALIFORNIA, PALOS VERDES HILLS FAULT
B 6202 259 CALIFORNIA, PALOS VERDES SAND
B 6211 2218 CALIFORNIA, PASO ROBLES HIGH
B 6111 1980 CALIFORNIA, PATTON ESCARPMENT
B 6203 363 CALIFORNIA, PATTON ESCARPMENT
B 6111 1981 CALIFORNIA, PATTON RIDGE
B 6102 238 CALIFORNIA, PATTON RIDGE
B 6203 368 CALIFORNIA, PATTON RIDGE A BASIN
B 6203 368 CALIFORNIA, PATTON RIDGE B BASIN
B 6111 1980 CALIFORNIA, PENINSULAR RANGES
B 6203 345 CALIFORNIA, PENINSULAR RANGES
B 6205 848† CALIFORNIA, PERMEABILITY, TRAPS IN, GATCHELL, EOCENE, SAND OF
B 6111 1981 CALIFORNIA, PERRIS BLOCK
B 6211 2224 CALIFORNIA, PESCADERO POINT
B 6202 259 CALIFORNIA, PICO FORMATION
B 6102 177 CALIFORNIA, PICO SAND
B 6203 359 CALIFORNIA, PINE MOUNTAIN FAULT
B 6211 2224 CALIFORNIA, PINNACLES VOLCANIC VENT AREA
B 6206 968 CALIFORNIA, PIONEER CANYON
S 11 54 CALIFORNIA, PLANKTONIC FORAMINIFERA
B 6202 255 CALIFORNIA, PLAYA DEL REY ANTICLINAL STRUCTURE
B 6205 850 CALIFORNIA, PLEASANT VALLEY OIL FIELD, SECTION
B 6211 2267 CALIFORNIA, PLEISTOCENE
B 6205 824 CALIFORNIA, PLEISTOCENE
B 6202 250 CALIFORNIA, PLEISTOCENE
B 6102 139 CALIFORNIA, PLEISTOCENE
B 6102 171 CALIFORNIA, PLEISTOCENE
M 28 42 CALIFORNIA, PLEISTOCENE SANDSTONE
M 28 141 CALIFORNIA, PLEISTOCENE, SEDIMENTS
M 28 140 CALIFORNIA, PLEISTOCENE, SEDIMENTS
B 6102 171 CALIFORNIA, PLIOCENE
B 6102 139 CALIFORNIA, PLIOCENE
B 6211 2247 CALIFORNIA, PLIOCENE
B 6205 824 CALIFORNIA, PLIOCENE
S 11 1 CALIFORNIA, PLIOCENE BIOSTRATIGRAPHY OF
S 11 183† CALIFORNIA, PLIOCENE, BIOSTRATIGRAPHY OF
M 28 142 CALIFORNIA, PLIOCENE, MONTEREY FORMATION
M 28 81 CALIFORNIA, PLIOCENE, MONTEREY SHALE
B 6407 1037 CALIFORNIA, PLIOCENE-PLEISTOCENE CONTACT
B 6211 2217 CALIFORNIA, PLIOCENE-RECENT
B 6211 2215 CALIFORNIA, POINT ARENA
B 6203 352 CALIFORNIA, POINT ARENA BASIN
B 6008 1198 CALIFORNIA, PRODUCTION, 1974-1975
B 6108 1168 CALIFORNIA, PRODUCTION, 1975-1976
B 6208 1325 CALIFORNIA, PRODUCTION, 1976-1977
B 6308 1295 CALIFORNIA, PRODUCTION, 1977-1978
B 6409 1401 CALIFORNIA, PRODUCTION, 1978-1979
B 6203 356 CALIFORNIA, PT. REYES FAULT
B 6111 1986 CALIFORNIA, PUENTE FORMATION
B 6111 1981 CALIFORNIA, PUENTE HILLS
B 6202 251 CALIFORNIA, QUATERNARY
B 6211 2220 CALIFORNIA, RED HILLS FAULT
B 6102 170 CALIFORNIA, RED MOUNTAIN THRUST
B 6102 138 CALIFORNIA, RED MOUNTAIN THRUST
B 6202 248 CALIFORNIA, REDONDO CANYON
B 6202 250 CALIFORNIA, REDONDO CANYON FAULT SCARP
B 6202 248 CALIFORNIA, REDONDO PLATFORM
B 6202 253 CALIFORNIA, REDONDO PLATFORM ANTICLINORIUM
B 6202 248 CALIFORNIA, REDONDO SUBMARINE FAN
B 6211 2256 CALIFORNIA, REEF RIDGE SAND
B 6211 2247 CALIFORNIA, REEF RIDGE SHALE
B 6211 2219 CALIFORNIA, RELIZ CANYON FORMATION
B 6211 2219 CALIFORNIA, RELIZ CANYON SECTION
B 6211 2246 CALIFORNIA, RELIZIAN STAGE
B 6202 259 CALIFORNIA, RELIZIAN STAGE
B 6211 2221 CALIFORNIA, RELIZIAN-LUISIAN BOUNDARY PALEOGEOGRAPHY
B 6202 254 CALIFORNIA, REPETTIAN STRATA
B 6206 958 CALIFORNIA, REPETTO FIELD
B 6202 259 CALIFORNIA, REPETTO FORMATION
B 6102 138 CALIFORNIA, REPETTO FORMATION
B 6102 177 CALIFORNIA, REPETTO SAND
B 6205 813† CALIFORNIA, RESERVOIR AND, SOURCE-BED HISTORY OF, GREAT VALLEY
S 11 1 CALIFORNIA REVISITED BY R. J. KLEINPELL
B 6203 369 CALIFORNIA, RIDGE BASIN
B 6102 166 CALIFORNIA, RINCON ANTICLINE
B 6104 594 CALIFORNIA, RINCON FIELD
B 6407 1039 CALIFORNIA, RINCON FIELD
B 6203 353 CALIFORNIA, RINCONADA FAULT
B 6211 2218 CALIFORNIA, RINCONADA-RELIZ FAULT SYSTEM
B 6102 177 CALIFORNIA, RIO BRAVO SAND
B 6211 2214† CALIFORNIA, ROLE OF, SALINIAN BLOCK IN EVOLUTION OF, SAN
B 6206 951 CALIFORNIA, ROSEDALE CHANNEL
B 6211 2243 CALIFORNIA, ROSEDALE CYCLE, INFORMAL NAME
B 6211 2247 CALIFORNIA, ROSEDALE FAN
B 6211 2245 CALIFORNIA, ROSEDALE FAN
B 6211 2256 CALIFORNIA, ROSEDALE SAND
B 6211 2247 CALIFORNIA, ROSEDALE-BELLEVUE TREND
B 6203 353 CALIFORNIA, RUSSELL RANCH OIL FIELD
B 6206 951 CALIFORNIA, SACRAMENTO VALLEY
B 6211 2244 CALIFORNIA, SACRAMENTO VALLEY
B 6301 23 CALIFORNIA, SACRAMENTO VALLEY
B 6301 27 CALIFORNIA, SACRAMENTO VALLEY PETROLEUM OCCURRENCE

B 6205 818 CALIFORNIA, SACRAMENTO VALLEY, SOURCE ROCKS
B 6203 349 CALIFORNIA, SALINAS BASIN
B 6211 2217 CALIFORNIA, SALINAS VALLEY
B 6203 349 CALIFORNIA, SALINIAN BLOCK
B 6211 2262 CALIFORNIA, SALT CREEK SECTION
B 6203 345 CALIFORNIA, SALTON TROUGH
B 6111 1980 CALIFORNIA, SALTON TROUGH
B 6111 1980 CALIFORNIA, SAN ANDREAS FAULT
B 6205 819 CALIFORNIA, SAN ANDREAS FAULT
B 6211 2244 CALIFORNIA, SAN ANDREAS FAULT
M 25 41 CALIFORNIA, SAN ANDREAS FAULT
B 6405 626 CALIFORNIA, SAN ANDREAS FAULT, AIRCRAFT RADAR IMAGE
B 6405 624 CALIFORNIA, SAN ANDREAS FAULT, LANDSAT IMAGE
B 6405 626 CALIFORNIA, SAN ANDREAS FAULT, SKYLAB PHOTOGRAPH
B 6202 265 CALIFORNIA, SAN ANDREAS FAULT, SLIP
B 6401 25 CALIFORNIA, SAN ANDREAS FAULT ZONE
B 6401 25 CALIFORNIA, SAN ANDREAS FAULT ZONE
B 6211 2220 CALIFORNIA, SAN ANTONIO FAULT
B 6211 2219 CALIFORNIA, SAN ARDO OIL FIELD
B 6203 349 CALIFORNIA, SAN BENITO BASIN
B 6102 170 CALIFORNIA, SAN CAYENTO FAULT
B 6102 138 CALIFORNIA, SAN CAYENTO FAULT
B 6111 1981 CALIFORNIA, SAN CLEMENTE BASIN
B 6203 363 CALIFORNIA, SAN CLEMENTE ESCARPMENT
B 6111 1981 CALIFORNIA, SAN CLEMENTE ISLAND
B 6102 238 CALIFORNIA, SAN CLEMENTE ISLAND
B 6203 367 CALIFORNIA, SAN CLEMENTE ISLAND PLATFORM
B 6111 1984 CALIFORNIA, SAN CLEMENTE RIDGE
B 6203 364 CALIFORNIA, SAN DIEGO TROUGH
B 6111 1981 CALIFORNIA, SAN DIEGO TROUGH
B 6102 238 CALIFORNIA, SAN DIEGO TROUGH
B 6102 152 CALIFORNIA, SAN DIEGO TROUGH
B 6211 2270 CALIFORNIA, SAN EMIDIO MOUNTAINS
B 6212 2447 CALIFORNIA, SAN EMIDIO NOSE
B 6111 1982 CALIFORNIA, SAN GABRIEL BASIN
B 6202 248 CALIFORNIA, SAN GABRIEL CANYON
B 6203 359 CALIFORNIA, SAN GABRIEL FAULT
B 6111 1982 CALIFORNIA, SAN GABRIEL MOUNTAINS
B 6203 357 CALIFORNIA, SAN GREGORIO FAULT
B 6211 2218 CALIFORNIA, SAN GREGORIO-HOSGRI FAULT
B 6111 1981 CALIFORNIA, SAN JACINTO FAULT
B 6203 350 CALIFORNIA, SAN JOAQUIN BASIN
M 26 74 CALIFORNIA, SAN JOAQUIN BASIN, FACIES SEQUENCES, STRATIGRAPHY
M 26 72 CALIFORNIA, SAN JOAQUIN BASIN, FRUITVALE FORMATION
B 6211 2267 CALIFORNIA, SAN JOAQUIN FORMATION
B 6111 1982 CALIFORNIA, SAN JOAQUIN HILLS
B 6211 2224 CALIFORNIA, SAN JOAQUIN VALLEY
B 6211 2244 CALIFORNIA, SAN JOAQUIN VALLEY
B 6212 2442 CALIFORNIA, SAN JOAQUIN VALLEY
B 6307 1036 CALIFORNIA, SAN JOAQUIN VALLEY BASIN, EAST, CONTOUR MAP
M 26 173 CALIFORNIA, SAN JOAQUIN VALLEY, MIOCENE
B 6205 820 CALIFORNIA, SAN JOAQUIN VALLEY, SOURCE ROCKS
B 6111 1982 CALIFORNIA, SAN JOSE FAULT
B 6111 1982 CALIFORNIA, SAN JOSE HILLS
B 6102 238 CALIFORNIA, SAN MIGUEL ISLAND, EOCENE
B 6102 167 CALIFORNIA, SAN MIGUELITO FIELD
B 6104 594 CALIFORNIA, SAN MIGUELITTO FIELD
B 6111 1981 CALIFORNIA, SAN NICOLAS BASIN
B 6203 363 CALIFORNIA, SAN NICOLAS BASIN
B 6102 238 CALIFORNIA, SAN NICOLAS BASIN
B 6111 1981 CALIFORNIA, SAN NICOLAS ISLAND
B 6102 237 CALIFORNIA, SAN NICOLAS ISLAND, EOCENE
B 6102 238 CALIFORNIA, SAN PEDRO BASIN
B 6203 363 CALIFORNIA, SAN PEDRO BASIN
B 6111 1984 CALIFORNIA, SAN PEDRO BASIN
B 6202 248 CALIFORNIA, SAN PEDRO BASIN
B 6202 255 CALIFORNIA, SAN PEDRO BAY ANTICLINORIUM
B 6202 248 CALIFORNIA, SAN PEDRO ESCARPMENT
B 6202 255 CALIFORNIA, SAN PEDRO ESCARPMENT FAULT ZONE
B 6202 259 CALIFORNIA, SAN PEDRO FORMATION
B 6202 248 CALIFORNIA, SAN PEDRO SEA VALLEY
B 6202 249 CALIFORNIA, SAN PEDRO SHELF, PROFILES
B 6202 249 CALIFORNIA, SAN PEDRO SHELF, SEISMIC DATA
B 6203 355 CALIFORNIA, SAN RAFAEL MOUNTAINS
B 6211 2219 CALIFORNIA, SANDHOLDT MEMBER
B 6202 248 CALIFORNIA, SANTA ANA MOUNTAINS
B 6111 1982 CALIFORNIA, SANTA ANA MOUNTAINS
B 6111 1983 CALIFORNIA, SANTA ANA MOUNTAINS, STRATIGRAPHIC COLUMN
B 6203 364 CALIFORNIA, SANTA BARBARA BASIN
B 6111 1981 CALIFORNIA, SANTA BARBARA BASIN
B 6206 973 CALIFORNIA, SANTA BARBARA BASIN
B 6102 238 CALIFORNIA, SANTA BARBARA BASIN
B 6204 688 CALIFORNIA, SANTA BARBARA BASIN, KEROGEN
B 6102 238 CALIFORNIA, SANTA BARBARA ISLAND
B 6111 1981 CALIFORNIA, SANTA BARBARA ISLAND
B 6102 238 CALIFORNIA, SANTA CATALINA BASIN
B 6206 973 CALIFORNIA, SANTA CATALINA BASIN
B 6102 238 CALIFORNIA, SANTA CATALINA ISLAND
B 6111 1981 CALIFORNIA, SANTA CATALINA ISLAND
B 6202 248 CALIFORNIA, SANTA CATALINA ISLAND
B 6203 359 CALIFORNIA, SANTA CLARA FAULT
B 6203 359 CALIFORNIA, SANTA CLARA TROUGH
B 6203 346 CALIFORNIA, SANTA CRUZ BASIN
B 6111 1981 CALIFORNIA, SANTA CRUZ BASIN
B 6102 238 CALIFORNIA, SANTA CRUZ BASIN
B 6203 356 CALIFORNIA, SANTA CRUZ HIGH
B 6102 238 CALIFORNIA, SANTA CRUZ ISLAND, EOCENE
B 6211 2224 CALIFORNIA, SANTA CRUZ MOUNTAINS
B 6203 357 CALIFORNIA, SANTA LUCIA BANK FAULT
B 6211 2217 CALIFORNIA, SANTA LUCIA RANGE
B 6211 2245 CALIFORNIA, SANTA MARGARITA FORMATION
B 6203 350 CALIFORNIA, SANTA MARIA BASIN
B 6203 355 CALIFORNIA, SANTA MARIA OIL DISTRICT AND BASIN
B 6203 355 CALIFORNIA, SANTA MARIA VALLEY OIL FIELD
B 6203 364 CALIFORNIA, SANTA MONICA BASIN
B 6111 1981 CALIFORNIA, SANTA MONICA BASIN
B 6202 248 CALIFORNIA, SANTA MONICA BASIN
B 6102 238 CALIFORNIA, SANTA MONICA BASIN
B 6406 853 CALIFORNIA, SANTA MONICA BASIN, SAND LAYERS
B 6202 248 CALIFORNIA, SANTA MONICA CANYON
B 6202 255 CALIFORNIA, SANTA MONICA FAULT ZONE
B 6202 248 CALIFORNIA, SANTA MONICA MOUNTAINS
B 6102 266 CALIFORNIA, SANTA MONICA MOUNTAINS, NORTHWESTERN, GEOLOGIC
B 6202 248 CALIFORNIA, SANTA MONICA SHELF
B 6202 249 CALIFORNIA, SANTA MONICA SHELF, PROFILES
B 6202 249 CALIFORNIA, SANTA MONICA SHELF, SEISMIC DATA
B 6203 361 CALIFORNIA, SANTA MONICA-RAYMOND HILL FAULT
B 6102 238 CALIFORNIA, SANTA ROSA ISLAND, EOCENE
B 6001 66 CALIFORNIA, SANTA ROSA VALLEY
B 6203 363 CALIFORNIA, SANTA ROSA-CORTES RIDGE
B 6111 1981 CALIFORNIA, SANTA ROSA-CORTES RIDGE
B 6203 359 CALIFORNIA, SANTA SUSANA FAULT
B 6203 359 CALIFORNIA, SANTA YNEZ FAULT
B 6203 360 CALIFORNIA, SANTA YNEZ UPLIFT
B 6111 1986 CALIFORNIA, SANTIAGO PEAK INTRUSIVE
B 6106 906 CALIFORNIA, SANTO TOMAS FAULT, OFFSHORE
B 6102 139 CALIFORNIA, SATICOY FIELD
B 6102 171 CALIFORNIA, SATICOY FIELD
B 6407 1036 CALIFORNIA, SATICOY FIELD
B 6102 177 CALIFORNIA, SATICOY FIELD, PICO SAND
B 6203 360 CALIFORNIA, SATICOY OIL FIELD
B 6211 2221 CALIFORNIA, SAUCESIAN PALEOGEOGRAPHY
B 6211 2221 CALIFORNIA, SAUCESIAN-RELIZIAN BOUNDARY PALEOGEOGRAPHY
B 6202 255 CALIFORNIA, SCHIST RIDGE
B 6206 972 CALIFORNIA, SCRIPPS CRUISE 7 TOW, LEG 10
B 6206 972 CALIFORNIA, SCRIPPS CRUISE ANTIPODE, LEG 1
B 6206 972 CALIFORNIA, SCRIPPS CRUISE MONTCANYON II
B 6405 619† CALIFORNIA, SEASAT, RADAR IMAGE OF, SAN ANDREAS FAULT
B 6211 2243† CALIFORNIA, SEDIMENTATION AND, TRAPPING MECHANISM IN UPPER,
B 6202 249 CALIFORNIA, SEISMIC REFLECTION, SOUTHERN
M 25 209 CALIFORNIA, SEISMIC STUDY
S 11 7 CALIFORNIA, SERIES SUBDIVISIONS, FORMATIONAL APPROACH
B 6111 1986 CALIFORNIA, SESPE FORMATION
B 6407 1037 CALIFORNIA, SEXTON CANYON
B 6102 138 CALIFORNIA, SEXTON CONGLOMERATE OUTCROPS
B 6102 170 CALIFORNIA, SEXTON CONGLOMERATE OUTCROPS
B 6407 1042 CALIFORNIA, SEXTON LAKE CONGLOMERATE
B 6102 156 CALIFORNIA, SHELL TAYLOR LEASE, ISOPACH MAP
B 6211 2218 CALIFORNIA, SIERRA DE SALINAS SCHIST-TYPE BASEMENT
B 6211 2244 CALIFORNIA, SIERRA NEVADA
B 6203 345 CALIFORNIA, SIERRA NEVADA
B 6111 1982 CALIFORNIA, SIERRA PEAK
B 6211 2217 CALIFORNIA, SIERRAN-SALINIAN BASEMENT
B 6102 170 CALIFORNIA, SISAR FAULT
B 6102 138 CALIFORNIA, SISAR FAULT
B 6203 369 CALIFORNIA, SOLEDAD BASIN
B 6111 1986 CALIFORNIA, SOQUEL SANDSTONE
B 6203 351 CALIFORNIA, SOUTH BELRIDGE OIL FIELD
B 6203 353 CALIFORNIA, SOUTH CUYAMA OIL FIELD
B 6301 21 CALIFORNIA, SOUTH FORK MOUNTAIN THRUST

B 6211 2268 CALIFORNIA, SOUTH FORK VALLEY
B 6407 1036 CALIFORNIA, SOUTH MOUNTAIN FIELD
B 6203 360 CALIFORNIA, SOUTH MOUNTAIN OIL FIELD
B 6203 346 CALIFORNIA, SOUTHERN CALIFORNIA BASIN
S 11 61 CALIFORNIA STAGES VERSUS, NANNOPLANKTON ZONES
B 6211 2271 CALIFORNIA, STENDERUP SAND
B 6102 177 CALIFORNIA, STEVENS SAND
B 6211 2246 CALIFORNIA, STEVENS SANDSTONE
B 6211 2264 CALIFORNIA, STEVENS TURBIDITE FAN
B 6211 2257 CALIFORNIA, STEVENS TURBIDITES
B 6211 2244 CALIFORNIA, STOCKTON ARCH
B 6102 176 CALIFORNIA, STRAND FIELD, MIOCENE SANDS
B 6102 177 CALIFORNIA, STRAND FIELD, RIO BRAVO SAND
B 6211 2260 CALIFORNIA, STRAND OIL FIELD, SECTIONS
S 8 78 CALIFORNIA, SUBMARINE CANYONS, CARMEL CANYON
S 8 67 CALIFORNIA, SUBMARINE CANYONS, HUENEME CANYON
S 8 58 CALIFORNIA, SUBMARINE CANYONS, LA JOLLA CANYON
S 8 87 CALIFORNIA, SUBMARINE CANYONS, MONTEREY CANYON
S 8 61 CALIFORNIA, SUBMARINE CANYONS, NEWPORT CANYON
S 8 64 CALIFORNIA, SUBMARINE CANYONS, REDONDO CANYON
S 8 91 CALIFORNIA, SUBMARINE CANYONS, SAN CLEMENTE RIFT
S 8 73 CALIFORNIA, SUBMARINE CANYONS, SANTA CRUZ CANYON
S 8 65 CALIFORNIA, SUBMARINE CANYONS, SANTA MONICA CANYON
S 8 58 CALIFORNIA, SUBMARINE CANYONS, SCRIPPS CANYON
B 6403 443* CALIFORNIA SUBMARINE FANS, ABST., QUATERNARY STYLES OF
B 6407 1037 CALIFORNIA, SULPHUR CANYON
B 6203 357 CALIFORNIA, SUR BASIN
B 6206 968 CALIFORNIA, SUR CANYON
B 6211 2222 CALIFORNIA, SUR-NACIMIENTO FAULT
B 6203 345 CALIFORNIA, SUR-NACIMIENTO FAULT
B 6111 1986 CALIFORNIA, SYCAMORE CANYON SANDSTONE AND CONGLOMERATE
B 6203 351 CALIFORNIA, TAFT OIL DISTRICT
B 6203 366 CALIFORNIA, TANNER BANK
B 6102 238 CALIFORNIA, TANNER BANK
B 6102 238 CALIFORNIA, TANNER BASIN
B 6111 1981 CALIFORNIA, TANNER BASIN
B 6203 368 CALIFORNIA, TANNER BASIN
B 6104 504 CALIFORNIA, TANNER BASIN, KEROGEN
B 6204 688 CALIFORNIA, TANNER BASIN, KEROGEN
B 6102 154 CALIFORNIA, TAYLOR THRUST FAULT
B 6102 189 CALIFORNIA, TAYLOR THRUST FAULT
B 6003 356† CALIFORNIA, TECTONIC SIGNIFICANCE AND, HYDROCARBON, TRAPPING
B 6012 2081† CALIFORNIA, TECTONIC TRANSPRESSION AND, BASEMENT-CONTROLLED
B 6211 2245 CALIFORNIA, TEHACHAPI MOUNTAINS
B 6205 819 CALIFORNIA, TEJON DEPOCENTER
B 6211 2246 CALIFORNIA, TEJON PLATFORM
B 6211 2247 CALIFORNIA, TEN SECTION-STRAND TREND
B 6211 2223 CALIFORNIA, TERTIARY
B 6111 1989 CALIFORNIA, TERTIARY METAVOLCANIC CLASTS
S 11 15 CALIFORNIA, TERTIARY, STAGE BOUNDARIES
S 11 164 CALIFORNIA TERTIARY VALVULINERIDS
B 6203 369 CALIFORNIA, TEXAS CANYON BASIN
M 25 158 CALIFORNIA, THE GEYSERS FIELD, GEOTHERMAL
B 6211 2219 CALIFORNIA, THE ROCKS MEMBER
B 6203 363 CALIFORNIA, THIRTY MILE BANK
B 6102 152 CALIFORNIA, THIRTY MILE BANK
B 6102 238 CALIFORNIA, THIRTY MILE BANK
B 6111 1986 CALIFORNIA, TOPANGA FORMATION
B 6203 360 CALIFORNIA, TOPATOPA UPLIFT
B 6203 362 CALIFORNIA, TORRANCE OIL FIELD
B 6202 255 CALIFORNIA, TORRANCE-WILMINGTON ANTICLINAL STRUCTURE
B 6203 346 CALIFORNIA, TORTUGAS BASIN
B 6203 345 CALIFORNIA, TRANSVERSE RANGES
B 6111 1980 CALIFORNIA, TRANSVERSE RANGES
B 6211 2220 CALIFORNIA, TULARCITOS FAULT
B 6211 2267 CALIFORNIA, TULARE FORMATION
B 6211 2262 CALIFORNIA, TUMEY GULCH SECTION
B 6303 548* CALIFORNIA TURBIDITE AND CANYON-RELATED POOLS, ABST.,
B 6102 240 CALIFORNIA, ULATISIAN ROCKS
B 6203 360 CALIFORNIA, UPPER PIRU-SAN RAFAEL UPLIFT
B 6202 259 CALIFORNIA, VALMONTE DIATOMITE MEMBER
B 6211 2245 CALIFORNIA, VALPREDO FAN
B 6211 2220 CALIFORNIA, VAQUEROS FAULT
B 6211 2219 CALIFORNIA, VAQUEROS FORMATION
B 6102 165 CALIFORNIA, VENTURA ANTICLINE
B 6102 182 CALIFORNIA, VENTURA ANTICLINE
B 6104 594 CALIFORNIA, VENTURA AVENUE FIELD
B 6407 1036 CALIFORNIA, VENTURA AVENUE FIELD
B 6203 360 CALIFORNIA, VENTURA AVENUE OIL FIELD
B 6203 356 CALIFORNIA, VENTURA BASIN
S 7 232 CALIFORNIA, VENTURA BASIN, CROSS SECTION
S 7 235 CALIFORNIA, VENTURA BASIN, DISTAL TURBIDITE FACIES
B 6407 1036 CALIFORNIA, VENTURA BASIN, OIL FIELDS
S 7 236 CALIFORNIA, VENTURA BASIN, OXNARD SLOPE, SECTION
B 6206 958 CALIFORNIA, VENTURA FIELD
B 6102 154 CALIFORNIA, VENTURA FIELD, A, B, C, D PRODUCING BLOCKS
B 6102 176 CALIFORNIA, VENTURA FIELD, PLIOCENE SANDS
B 6102 177 CALIFORNIA, VENTURA FIELD, REPETTO AND PICO SANDS
B 6102 172 CALIFORNIA, VENTURA FIELD, TYPE ELECTRIC LOG
B 6102 140 CALIFORNIA, VENTURA FIELD, TYPE ELECTRIC LOG
B 6102 140 CALIFORNIA, VENTURA FIELD, ZONES 1 TO 8
B 6102 172 CALIFORNIA, VENTURA FIELD, ZONES 1 TO 8
B 6102 149 CALIFORNIA, VENTURA FIELD, ZONES 5, 6, 7, 8
B 6202 259 CALIFORNIA, VENTURIAN STAGE
B 6211 2217 CALIFORNIA, VERGELLES FAULT
B 6211 2268 CALIFORNIA, WALKER VALLEY
B 6102 153 CALIFORNIA, WELDON CANYON
B 6111 1984 CALIFORNIA, WEST CORTES BASIN
B 6104 594 CALIFORNIA, WEST MONTALVO FIELD
B 6407 1036 CALIFORNIA, WEST MONTALVO FIELD
B 6407 1036 CALIFORNIA, WEST MOUNTAIN FIELD
B 6102 138 CALIFORNIA, WESTERN, VENTURA BASIN, GEOLOGIC MAP
B 6102 170 CALIFORNIA, WESTERN, VENTURA BASIN, GEOLOGIC MAP
B 6212 2447 CALIFORNIA, WHEELER RIDGE
B 6202 259 CALIFORNIA, WHEELERIAN STAGE
B 6211 2245 CALIFORNIA, WHITE WOLF FAULT
B 6203 361 CALIFORNIA, WHITTIER FAULT
B 6202 255 CALIFORNIA, WHITTIER FAULT
B 6111 1982 CALIFORNIA, WHITTIER FAULT ZONE
B 6111 1988 CALIFORNIA, WHITTIER-ELSINORE FAULT ZONE
S 11 185 CALIFORNIA, WILDCAT SERIES
B 6111 1986 CALIFORNIA, YORBA SILTSTONE
B 6203 355 CALIFORNIA, ZACA OIL FIELD
S 11 54† CALIFORNIA, ZEMORRIAN THROUGH DELMONTIAN, PLANKTONIC AND
B 6403 446* CALIFORNIA--CONFIGURATION AND IMPLICATIONS FOR PETROLEUM
B 6012 2192* CALIFORNIA--PROBLEMS AND SOLUTIONS, ABST., NEOGENE, DIATOM
M 25 311† CALIFORNIA--SUMMARY, HYDROCARBON POTENTIAL OF OFFSHORE
M 25 313† CALIFORNIA--SUMMARY, PETROLEUM POTENTIAL OF, CONTINENTAL,
B 6403 440* CALIFORNIA. BY TITLE ONLY, BASIN-EDGE TECTONICS AND
B 6211 2356* CALIFORNIA. VIEWPOINT FROM ISOTOPE GEOLOGY. BY TITLE ONLY,
B 6211 2352* CALIFORNIA; RB-SR EVALUATION, ABST., AGE OF
B 6004 727* CALIFORNIAS DEVELOPED AND UNDEVELOPED BORDERLAND, ABST.
B 6403 446* CALIFORNIA—CONFIGURATION AND IMPLICATIONS FOR PETROLEUM
B 6405 727* CALIFORNIA—INFLUENCE OF BASIN EVOLUTION AND TECTONICS, ABST.,
B 6405 790* CALIFORNIA—ISOTOPIC AND CHEMICAL DATA, ABST., DIAGENESIS IN,
B 6212 2513* CALIFORNIA—TECTONICS AND OIL ACCUMULATION, ABST., SOUTHERN
M 25 321 CALIMA ANTICLINE, COLOMBIA
B 6304 598 CALLENDER, J. F., FRACTURES IN CRETACEOUS ROCKS FROM SELECTED
B 6306 986 CALLENDER, JAN C. WILT, AND R. E. CLEMONS, KING, PHILIP B., REVIEW
B 6206 1053 CALLIHAN SANDSTONE MEMBER, TEXAS
M 26 158 CALLOVIAN, AFRICA, SEQUENCES, JURASSIC
B 6211 2275 CALLOVIAN, BLACK HILLS, SUNDANCE FORMATION
M 30 143 CALLOVIAN, NORTH SEA, MORAY FIRTH BASIN
B 6201 52 CALLOVIAN STRATA, AUSTRALIA
M 30 128 CALLOVIAN, STRATIGRAPHY, BEATRICE FIELD
B 6303 344 CALLOWAY CREEK LIMESTONE, KENTUCKY
M 27 167 CALOOSAHATCHEE FORMATION, FLORIDA, PLIOCENE
M 27 199 CALOOSAHATCHEE FORMATION, FLORIDA, PLIOCENE
M 27 175 CALOOSAHATCHEE MARL, FLORIDA, PLIOCENE
M 27 84 CALOOSAHATCHEE MARL, FLORIDA, PLIOCENE
M 27 178 CALOOSAHATCHEE MARL, FLORIDA, PLIOCENE
M 27 83 CALOOSAHATCHEE MARL, FLORIDA, PLIOCENE
M 27 212 CALOOSAHATCHEE MARL, FLORIDA, PLIOCENE
M 27 221 CALOOSAHATCHEE MARL, FLORIDA, PLIOCENE
M 27 211 CALOOSAHATCHEE MARL, FLORIDA, PLIOCENE
M 27 19 CALPIONELLID LIMESTONE, ITALY, JURASSIC
M 27 20 CALPIONELLID LIMESTONE, ITALY, JURASSIC
M 27 177 CALPIONELLID LIMESTONE, ITALY, JURASSIC
B 6409 1554* CALPIONELLIDS AND NANNOCONIDS OF TARAISES FORMATION, EARLY
B 6303 384 CALVERT, WARREN L., REVIEW OF GEOLOGY OF MANITOULIN AREA,
B 6203 502 CALVERT, WARREN L., STRATIGRAPHIC CONCEPTS AND TERMINOLOGY,
B 6405 797* CALVIN FIELD—DEEP BASINAL JURASSIC PLAY IN NORTH LOUISIANA
B 6303 427 CALZIA, J. P., DEFORMATIONAL STRUCTURES IN PLIOCENE BOUSE
B 6109 1471 CAMAMU BASIN, BRAZIL
M 30 489 CAMBAY BASIN, CORRELATION TO BOMBAY OFFSHORE BASIN, INDIA
B 6212 2489† CAMBAY BASIN, INDIA, AND ITS IMPORTANCE IN PETROLEUM

B 6203 442† CAMBAY BASIN, INDIA, EARLY, EOCENE SUBAERIAL EROSIONAL VALLEYS
B 6009 1550† CAMBAY BASIN, INDIA, PATTERN OF, ACCUMULATION AND
M 30 495 CAMBAY BASIN, TECTONIC FRAMEWORK, INDIA
B 6402 219 CAMBAY FIELD, INDIA
B 6402 219 CAMBAY FIELD, INDIA
B 6402 210 CAMBAY GULF SHELF, INDIA
B 6402 210 CAMBAY GULF SHELF, INDIA
B 6403 429 CAMBAY HIGH-WAX OIL FIELD, INDIA
B 6403 429 CAMBAY HIGH-WAX OIL FIELD, INDIA
B 6203 444 CAMBAY SHALE, INDIA
B 6110 1835 CAMBODIA, DEVELOPMENTS, 1976
M 27 47 CAMBRIAN, AJAX LIMESTONE, AUSTRALIA
B 6205 785 CAMBRIAN, ALBERTA
M 27 211 CAMBRIAN, ALLENTOWN DOLOMITE, PENNSYLVANIA
B 6003 414† CAMBRIAN AND LOWE ORDOVICIAN ROCKS OF WESTERN
B 6203 502* CAMBRIAN AND NEW DEPOSITIONAL MODEL FOR TYPE ST. CROIXAN,
B 6101 7 CAMBRIAN, CANADIAN, ARCTIC ISLANDS
S 9 22 CAMBRIAN, CARBONATE REEF BELT DISTRIBUTION
B 6004 697* CAMBRIAN CARBONATE SHELF EDGE AND ITS CONTRIBUTION TO
M 28 160 CAMBRIAN, CONOCOCHEAGUE FORMATION, VIRGINIA
M 28 154 CAMBRIAN, CONOCOCHEAGUE LIMESTONE, VIRGINIA
M 27 141 CAMBRIAN, CONOCOCHEAGUE LIMESTONE, VIRGINIA
M 27 139 CAMBRIAN, CONOCOCHEAGUE LIMESTONE, VIRGINIA
M 27 140 CAMBRIAN, COPPER RIDGE DOLOMITE, VIRGINIA
M 27 139 CAMBRIAN, COPPER RIDGE DOLOMITE, VIRGINIA
M 27 141 CAMBRIAN, COPPER RIDGE DOLOMITE, VIRGINIA
B 6009 1570† CAMBRIAN, EAST-CENTRAL NEW YORK, DEFINITION OF, TYPE, LITTLE
B 6410 1682 CAMBRIAN FLAGS, WESTERN, AUSTRALIA, MOUNT JOHN OSMOND
S 9 20 CAMBRIAN, FOLDOUT, PALEOCLIMATES
B 6101 7 CAMBRIAN, FRANKLINIAN BASIN
M 27 118 CAMBRIAN, GALLATIN FORMATION, WYOMING
M 28 113 CAMBRIAN, GATESBURG FORMATION, PENNSYLVANIA
M 28 165 CAMBRIAN, GATESBURG FORMATION, PENNSYLVANIA
M 28 79 CAMBRIAN, GATESBURG FORMATION, PENNSYLVANIA
M 28 5 CAMBRIAN, GATESBURG FORMATION, PENNSYLVANIA
M 27 207 CAMBRIAN, GATESBURG FORMATION, PENNSYLVANIA
M 27 199 CAMBRIAN, GATESBURG FORMATION, PENNSYLVANIA
M 28 52 CAMBRIAN, GATESBURG FORMATION, PENNSYLVANIA
M 28 116 CAMBRIAN, GATESBURG FORMATION, PENNSYLVANIA
M 28 2 CAMBRIAN, GATESBURG FORMATION, PENNSYLVANIA
B 6101 8 CAMBRIAN, GREENLAND, NORTHERN
M 28 8 CAMBRIAN, HARRISON SUMMIT QUARTZITE, IDAHO
M 28 21 CAMBRIAN, HARRISON SUMMIT QUARTZITE, IDAHO
M 28 164 CAMBRIAN, HICKORY SANDSTONE, TEXAS
M 28 126 CAMBRIAN, HICKORY SANDSTONE, TEXAS
M 28 106 CAMBRIAN, HICKORY SANDSTONE, TEXAS
B 6303 508* CAMBRIAN, IN SOUTHERN LABRADOR, ABST., ICHNOLOG OF, LABRADOR
B 6405 685* CAMBRIAN JORDAN SANDSTONE IN WISCONSIN, ABST., DEPOSITIONAL
M 27 117 CAMBRIAN, KITTATINNY LIMESTONE, NEW JERSEY
M 27 199 CAMBRIAN, KITTATINNY LIMESTONE, NEW JERSEY
M 27 154 CAMBRIAN, MEAGHER FORMATION, MONTANA
M 27 6 CAMBRIAN, MEAGLER LIMESTONE, MONTANA
M 27 141 CAMBRIAN MINES DOLOMITE, PENNSYLVANIA
M 28 116 CAMBRIAN, MINES DOLOMITE, PENNSYLVANIA
B 6306 952 CAMBRIAN, NEWFOUNDLAND
M 28 143 CAMBRIAN, OLD FORT ISLAND SANDSTONE, CANADA, N.W.T.
M 28 179 CAMBRIAN, OLD FORT ISLAND SANDSTONE, CANADA, N.W.T.
S 6 358 CAMBRIAN, ORDOVICIAN BOUNDARY, TREMADOC AGE
M 28 52 CAMBRIAN, POTSDAM SANDSTONE, NEW YORK
B 6410 1682 CAMBRIAN POUND SANDSTONE-QUARTZITE, AUSTRALIA
S 6 358 CAMBRIAN, PRECAMBRIAN BOUNDARY, TOMMOTIAN AGE
S 6 358 CAMBRIAN, RADIOMETRIC DATES, RANGES
S 6 356 CAMBRIAN, RADIOMETRIC DATES, TABLE
S 6 355 CAMBRIAN, RADIOMETRIC DATING, A REVIEW
S 6 359 CAMBRIAN, RADIOMETRIC DATING, CONSTANTS USED
B 6410 1645 CAMBRIAN RED BEDS, AUSTRALIA
B 6203 531* CAMBRIAN REEFS, LABRADOR, ABST., CAVITY-DWELLING ORGANISMS,
M 28 126 CAMBRIAN, RILEY FORMATION, TEXAS
M 28 164 CAMBRIAN, RILEY FORMATION, TEXAS
M 28 106 CAMBRIAN, RILEY FORMATION, TEXAS
B 6203 553* CAMBRIAN SEDIMENTARY ROCKS FROM EAST-CENTRAL ALASKA, ABST.,
B 6203 532* CAMBRIAN SEKWI FORMATION, MACKENZIE MOUNTAINS, NORTHWEST
B 6004 712* CAMBRIAN SHALLOW SUBTIDAL ENVIRONMENT--TIME AND PLACE FOR
B 6405 770* CAMBRIAN SHELF AND MIDDLE ORDOVICIAN FORELAND BASIN,
M 28 65 CAMBRIAN, SILLERY GROUP, CANADA, QUEBEC
M 28 162 CAMBRIAN, SILLERY GROUP, CANADA, QUEBEC
B 6004 659* CAMBRIAN, SOUTHERN GREAT BASIN, ABST., STRATIGRAPHY AND
B 6203 539* CAMBRIAN, SOUTHWEST VIRGINIA, ABST., DIGITATE, STROMATOLITIC,
M 28 161 CAMBRIAN, ST. ROCH FORMATION, CANADA, QUEBEC
B 6110 1899 CAMBRIAN STRATIGRAPHIC SUMMARY, MISSISSIPPI
B 6110 1897† CAMBRIAN SYSTEM IN BLACK WARRIOR BASIN
S 6 355† CAMBRIAN SYSTEM, THE
B 6110 1898 CAMBRIAN TEST WELLS, MISSISSIPPI AND, ALABAMA
B 6402 222 CAMBRIAN, TEXAS
B 6402 222 CAMBRIAN, TEXAS
B 6402 223 CAMBRIAN THROUGH PERMIAN, TEXAS
B 6402 223 CAMBRIAN THROUGH PERMIAN, TEXAS
B 6405 701* CAMBRIAN TIDAL SAND BODY, ABST., SHALLOW-MARINE, TRACE
B 6004 678* CAMBRIAN TO LOWER ORDOVICIAN, QUEBEC, ABST., SUBMARINE-, FAN
B 6004 703* CAMBRIAN TO PRESENT, ABST., CAUSALITY AND, TIMING IN, NORTH
B 6405 750* CAMBRIAN TRANSGRESSION IN WYOMING, ABST., SEDIMENTOLOGY AND
B 6110 1898 CAMBRIAN TYPE LOCALITIES, ALABAMA, GEORGIA, AND, TENNESSEE
M 28 162 CAMBRIAN, UNICOI FORMATION, VIRGINIA
B 6203 550* CAMBRIAN, UPPER MISSISSIPPI VALLEY, ABST., LITHOSTRATIGRAPHY,
B 6304 652 CAMBRIAN, VIRGINIA
M 28 46 CAMBRIAN, WATERBURY GNEISS, CONNECTICUT
M 28 55 CAMBRIAN, WATERBURY GNEISS, CONNECTICUT
M 27 212 CAMBRIAN, WILBERNS FORMATION, TEXAS
B 6204 681 CAMBRIAN-ORDOVICIAN, ALGERIA
B 6004 717* CAMBRIAN-ORDOVICIAN OF EASTERN NEW YORK--TIMING AND
B 6108 1375* CAMBRIAN-ORDOVICIAN, OKLAHOMA, ABST., PETROLEUM,
B 6004 682* CAMBRIAN-ORDOVIDIAN COW HEA BRECCIA, NEWFOUNDLAND, ABST.,
B 6107 1030 CAMBRIDGE ARCH, NEBRASKA
S 10 209 CAMERON COUNTY, TEMPERATURES SHOWINGGEOPRESSURE, TEXAS
M 25 308 CAMERON, B. E. B., SEDIMENTATION AND TECTONICS OF PACIFIC
B 6105 801 CAMERON, BARRY, APPLICATION OF COMPUTER-MAPPING PROGRAMS, SYMAP
B 6004 692 CAMERON, BARRY, MARINE COLD-WATER CARBONATE DEPOSITS AS
B 6405 685 CAMERON, BARRY, MICROBIAL AND INVERTEBRATE ENDOLITHIC
B 6004 728 CAMERON, BARRY, MIOCENE TRANSGRESSIONS AND REGRESSIONS OF
B 6203 503 CAMERON, BARRY, RECOGNITION OF CRATONAL BANK, SLOPE, AND
B 6105 723 CAMERON, BARRY, RECORRELATION OF EOCENE-MIOCENE BOUNDARY OF
B 6209 1714 CAMERON, BARRY, RECORRELATION OF EOCENE-MIOCENE BOUNDARY
B 6409 1558* CAMERON PARISH, LOUISIANA, ABST., PRESSURE, TEMPERATURE,
B 6411 1786 CAMEROON, DEVELOPMENTS, 1979
B 6401 60 CAMEROON, DOUALA BASIN, CRETACEOUS, KEROGEN
B 6411 1794 CAMEROON, PRODUCTION, 1978-1979
B 6107 972 CAMEROON VOLCANICS, AFRICA
B 6107 970 CAMEROON-GABON-ANGOLA SEGMENT
B 6110 1747 CAMEROUN, DEVELOPMENTS, 1976
B 6210 1846 CAMEROUN, DEVELOPMENTS, 1977
B 6310 1691 CAMEROUN, DEVELOPMENTS, 1978
B 6401 60 CAMEROUN, DOUALA BASIN, CRETACEOUS, KEROGEN
B 6210 1855 CAMEROUN, PRODUCTION, 1976-1977
B 6310 1698 CAMEROUN, PRODUCTION, 1978
B 6109 1537* CAMINADA-MOREAU CHENIER-BEACH SYSTEM, ABST.,
B 6309 1609* CAMINADA-MOREAU COAST LOUISIANA, ABST., VARIATIONS IN
B 6109 1476 CAMORIM FIELD, BRAZIL
B 6209 1768* CAMPANIAN, AT TYPE LOCALITIES IN NORTHEAST TEXAS AND
B 6202 182 CAMPANIAN INTERIOR SEAWAY, WYOMING
S 4 88 CAMPANIAN, JAMAICA, BULLHEAD FORMATION, MAIN RIDGE
B 6405 613 CAMPANIAN, MIDDLE EAST
B 6408 1233 CAMPANIAN SEAWAY, NORTH AMERICA
B 6206 1015 CAMPANIAN TO MAESTRICHTIAN, ARCTIC, CANADA, MESOZOIC, APTIAN-
B 6202 184 CAMPANIAN, WYOMING
S 6 261 CAMPANIAN-MAASTRICHTIAN BOUNDARY, NORTH AMERICA,
B 6303 422* CAMPANIAN-MAESTRICHTIAN, CAPE SEBASTIAN SANDSTONE
B 6007 1009 CAMPBELL, C. V., RESERVOIR GEOMETRY OF A FLUVIAL SHEET
B 6203 503 CAMPBELL, D. A., GEOTHERMAL RESOURCE INVESTIGATIONS AND
B 6405 685 CAMPBELL, JACK H., CARBONATE MINERAL REACTIONS DURING
B 6309 1538 CAMPBELL, JOCK A., AEROMAGNETIC DETECTION OF DIAGENETIC
B 6205 882 CAMPBELL, JOCK A., CARBON DIOXIDE RESOURCES OF UTAH—A FIRST
B 6405 686 CAMPBELL, JOCK A., MAJOR PETROLEUM-IMPREGNATED ROCK DEPOSITS
B 6008 1392 CAMPBELL, JOCK A., STRUCTURAL GEOLOGY AND PETROLEUM
B 6405 686 CAMPBELL, JOHN A., LOWER PERMIAN DEPOSITIONAL SYSTEMS
B 6004 653 CAMPBELL, JOHN A., PRELIMINARY INVESTIGATIONS OF ELEMENTAL
B 6008 1411 CAMPBELL, JOHN N., MULTI-DIMENSIONAL SEISMIC DISPLAYS, ABST.
B 6203 532 CAMPBELL, KERRY, GEOTECHNICAL ENGINEERING PROBLEMS OF UPPER
B 6303 423 CAMPBELL, N. D. ., INTISAR D OIL FIELD, LIBYA, ABST.
M 30 543 CAMPBELL, N. D. J., INTISAR D OIL FIELD,LIBYA
M 25 220 CAMPBELL PLATEAU BASIN O GREAT SOUTH BASIN, NEW ZEALAND,
B 6004 650 CAMPBELL, R. H., NEOGENE BASIN FORMATION AND HYDROCARBON
B 6203 344 CAMPBELL, R. H., NEOGENE BASIN FORMATION IN RELATION TO
B 6102 263 CAMPBELL, R. H., SANTA MONICA AND SANTA ANA
B 6405 686 CAMPBELL, S. E., TRAPPING AND ACCRETION OF AEOLIAN SEDIMENT BY
B 6303 458 CAMPBELL, S., STUDY OF FOSSIL MICROBIAL BORINGS--NEW APPROACH,
B 6207 1211 CAMPBELL, W. J., STRUCTURE AND MOTION OF BERING AND OKHOTSK
M 29 418 CAMPECHE BANK, REFLECTION PROFILE, GULF OF MEXICO
B 6210 2055 CAMPECHE BAY, MEXICO, SOUTHEASTERN ZONE
S 7 10 CAMPECHE ESCARPMENT, DESCRIBED
B 6305 783 CAMPECHE ESCARPMENT, GULF OF MEXICO

B 6405 726* CAMPECHE ESCARPMENT, GULF OF MEXICO, ABST., SEISMIC,
B 6303 487* CAMPECHE ESCARPMENT, GULF OF MEXICO, ABST., SEISMIC
M 29 395 CAMPECHE ESCARPMENT, MISSISSIPPI FAN, RELATIONS TO
M 26 272 CAMPECHE FAN SYSTEM
S 7 62 CAMPECHE KNOLLS, CHALLENGER SALT FLOW, DIAGRAM
B 6305 794 CAMPECHE KNOLLS, GULF OF MEXICO
M 30 384 CAMPECHE MARINE PLATFORM, HYDROCARBON PRODUCTION, MEXICO
B 6304 697* CAMPECHE OFFSHOR AREA, SOUTHEASTERN MEXICO, ABST., PETROLEUM
S 7 63 CAMPECHE PLATFORM, GULF OF MEXICO
B 6211 2294 CAMPECHE SCARP, GULF OF MEXICO
S 7 49 CAMPECHE SCARP, GULF OF MEXICO, SALT
M 29 407 CAMPECHE SCARP, SEISMIC SECTION, GULF OF MEXICO
S 7 10 CAMPECHE TERRACE, DESCRIBED
B 6211 2291 CAMPECHE UNIT, GULF OF MEXICO
B 6008 1393 CAMPEN, E. B., TED,, WELL-LOG ANALYSIS IN CRETACEOUS GAS SAND IN
B 6402 189 CAMPO DE NIJAR SECTION, SPAIN
B 6402 189 CAMPO DE NIJAR SECTION, SPAIN
B 6109 1471 CAMPOS BASIN, BRAZIL
B 6303 412* CAMPOS BASIN, BRAZIL, ABST., NAMORAD FIELD, MAJOR OIL DISCOVERY
M 30 329† CAMPOS BASIN, BRAZIL, THE, NAMORADO OIL FIELD--A MAJOR, OIL
B 6303 412* CAMPOS BASIN, BRAZIL. MODEL FOR EXPLORATION IN ATLANTIC-TYPE
M 30 330 CAMPOS BASIN, CABO FRIO ARCH, BRAZIL
M 30 330 CAMPOS BASIN, LOCATION, BRAZIL
B 6303 429* CAMPOS BASIN, OFFSHORE RIO DE JANEIRO, BRAZIL, ABST.,
M 30 333 CAMPOS BASIN, SHELF-EDGE SEISMIC SECTION, BRAZIL
M 30 331 CAMPOS BASIN, STRATIGRAPHIC FRAMEWORK, BRAZIL
M 30 332 CAMPOS BASIN, STRUCTURAL FRAMEWORK, BRAZIL
M 30 330 CAMPOS BASIN, VITORIA ARCH, BRAZIL
B 6002 197 CAMPOS COASTAL BASIN, BRAZIL
B 6303 412 CAMPOS, ODIMAR A. J., NAMORAD FIELD, MAJOR OIL DISCOVERY IN
M 30 329 CAMPOS, ODIMAR A. J., THE NAMORADO OIL FIELD--A MAJOR OIL
B 6303 441* CAN GEOCHEMISTRY FIND OIL. BY TITLE ONLY
B 6004 554† CANAD BASIN, THICKNESS OF, REMOVED SEDIMENTARY ROCKS,
M 27 5 CANADA (N.W.T.), DEVONIAN, JEAN MARIE FORMATION
B 6004 716* CANADA, ABST., AAPG, OIL AND, GAS, FIELD, DATA BANK PROJECT,
B 6004 716* CANADA, ABST., AAPG, OIL AND, GAS, FIELD, DATA BANK PROJECT,
B 6203 542* CANADA, ABST., COARSE-GRAINED DEPOSITS BY EPHEMERAL, STREAMS
B 6004 704* CANADA, ABST., COMPARISON OF, COASTAL DYNAMICS ON WEST- AND
B 6203 532* CANADA, ABST., CONTINENTAL SLOPE-RISE DEPOSITS FROM LOWER,
B 6303 481* CANADA, ABST., DEPOSITIONAL ENVIRONMENTS AND DIAGENESIS O,
B 6405 735* CANADA, ABST., DIAGENETIC FEATURES OF, GRAND RAPIDS FORMATION,
B 6405 793* CANADA, ABST., EARLY, PALEOZOIC, CONODONT BIOSTRATIGRAPHY AND
B 6405 759* CANADA, ABST., EASTERN, CORDILLERAN FOLDBELT AND FORELAND OF
B 6203 557* CANADA, ABST., EXAMPLE OF, CARBONATE, BANKS IN LOWER,
B 6405 776* CANADA, ABST., FACIES CONTROL ON, BITUMEN SATURATION, PEACE
B 6008 1398* CANADA, ABST., GEOLOGY AND HYDROCARBON POTENTIAL OF,
B 6004 701* CANADA, ABST., LOWER, CRETACEOUS, NEOCOMIAN, SANDSTONE
B 6004 675* CANADA, ABST., MARINE-, BAR SEDIMENTATION AND EQUILIBRIUM,
B 6105 818* CANADA, ABST., OLD AND NEW HORIZONS IN FOR WESTERN AND,
B 6004 698* CANADA, ABST., ORIGIN AND DISTRIBUTION OF, SUSPENDED SEDIMENT
B 6105 818* CANADA, ABST., PALEOGEOGRAPHY AND PALEOENVIRONMENTS OF,
B 6203 539* CANADA, ABST., REGRESSIVE SEDIMENTATION, MODEL FOR, JAMES BAY
B 6405 747* CANADA, ABST., TERRESTRIAL SEDIMENTATION ASSOCIATED WITH
B 6004 697* CANADA, ABST., TERTIARY, DELTAIC, CYCLES ON UNSTABLE, CRATO,
B 6303 504* CANADA, ABST., THERMA MATURATION OF LOWER PALEOZOIC, SHALES
M 26 429 CANADA, ALBERTA, BELLY RIVER FORMATION
M 26 435 CANADA, ALBERTA, CARDIUM FORMATION
M 28 161 CANADA, ALBERTA, DEVONIAN, CAIRN FORMATION
M 28 10 CANADA, ALBERTA, DEVONIAN, CAIRN FORMATION
M 28 153 CANADA, ALBERTA, DEVONIAN, CAIRN FORMATION
M 28 154 CANADA, ALBERTA, DEVONIAN, CAIRN FORMATION
M 26 196 CANADA, ALBERTA, GOLDEN SPIKE, CARBONATE BUILDUP, SEISMIC
M 26 197 CANADA, ALBERTA, INNISFAIL, REEF BUILDUP, SEISMIC EXAMPLE
M 26 202 CANADA, ALBERTA, JUDY CREEK, SHALLOW-WATER CARBONATE
M 26 195 CANADA, ALBERTA, RAINBOW-ZAMA AREA, PINNACLE REEFS, SEISMIC
S 10 140 CANADA, ALBERTA, RED EARTH REGION,STRATIGRAPHY
M 26 198 CANADA, ALBERTA, YEKAN LAKE, CARBONATE BUILDUP, SEISMIC
B 6206 1006 CANADA, AMUND RINGNES ISLAND, ARCTIC
B 6206 1013 CANADA ARCTIC, CORNWALL ARCH
B 6206 1013 CANADA ARCTIC, GRANTLAND UPLIFT
B 6206 1013 CANADA ARCTIC, LAKE HAZEN FAULT
B 6206 1013 CANADA ARCTIC, PRINCESS MARGARET ARCH
B 6206 1013 CANADA ARCTIC, STOLZ FAULT
S 5 146 CANADA, ARCTIC ARCHIPELAGO, ELLESMERE ISLAND, OTTO FIORD
M 25 497 CANADA, ASBESTOS, CHROMITE, AND NICKEL
B 6111 1951 CANADA, ATHABASCA HEAVY OIL DEPOSIT, WESTERN
S 10 222 CANADA, ATHABASCA OIL DEPOSIT, TRAP
S 10 64 CANADA, ATHABASCA-PEACE RIVER, CRETACEOUS,GEOLOGIC SETTING
M 25 498 CANADA, ATLIN TERRANE
B 6206 1005 CANADA, AXEL HEIBERG ISLAND, ARCTIC
B 6206 1005 CANADA, BAFFIN ISLAND, ARCTIC
B 6101 29 CANADA, BAKER LAKE, PITCHBLENDE DEPOSITS
B 6206 1005 CANADA, BANKS ISLAND, ARCTIC
B 6206 1008 CANADA, BARROW SEGMENT, ARCTIC
B 6103 434† CANADA BASIN, RATIO OF PETROLEUM TO, WATER DURING, PRIMARY,
B 6305 812† CANADA BASIN. DISCUSSION AND REPLY, THICKNESS OF, REMOVED
B 6206 1005 CANADA, BOOTHIA PENINSULA, ARCTIC
B 6206 1005 CANADA, BOOTHIA UPLIFT, ARCTIC
M 25 108 CANADA, BOWRON RIVER COALFIELD
M 25 504 CANADA, BOWSER BASIN
M 25 498 CANADA, BOWSER BASIN
B 6206 1005 CANADA, BYLOT ISLAND, ARCTIC
M 25 497 CANADA, CACHE CREEK GROUP
B 6302 164 CANADA, CADOTTE SAND
M 25 108 CANADA, CARMACKS COAL AREA
M 25 105 CANADA, COAL RESERVES
M 25 106 CANADA, COALFIELDS
B 6111 1951 CANADA, COLD LAKE HEAVY OIL DEPOSIT, WESTERN
M 25 493 CANADA, CORDILLERAN BASINS, MODELS OF
M 25 493 CANADA, CORDILLERAN OROGEN
M 25 500 CANADA, CORDILLERAN OROGENY
M 28 177 CANADA, CRETACEOUS, LOGAN CANYON FORMATION
M 28 124 CANADA, CRETACEOUS, LOGAN CANYON FORMATION
M 28 172 CANADA, CRETACEOUS, LOGAN CANYON SANDSTONE
S 6 263 CANADA, CRETACEOUS, PEACE RIVER FORMATION
M 28 172 CANADA, CRETACEOUS, SANDSTONE
B 6403 318 CANADA, CRETACEOUS, UPPER, CORRELATION CHART, MONTANA AND
B 6403 318 CANADA, CRETACEOUS, UPPER, CORRELATION CHART, MONTANA AND
B 6302 152† CANADA, DEEP BASIN, GAS, TRAP, WESTERN
B 6307 1059† CANADA, DEPOSITIONAL, FACIES OF LOWER, PALEOZOIC, ALLEN BAY,
B 6206 1005 CANADA, DEVON ISLAND, ARCTIC
S 10 114 CANADA, DEVONIAN, DISTRIBUTION OFLEAD-ZINC
M 27 135 CANADA, DEVONIAN, KEG RIVER EQUIVALENT
B 6409 1519 CANADA, EASTERN, DEVELOPMENTS, 1979
B 6108 1335 CANADA, EASTERN, OFFSHORE, DEVELOPMENTS, 1976
B 6208 1451 CANADA, EASTERN, OFFSHORE, DEVELOPMENTS, 1977
B 6308 1344 CANADA, EASTERN, OFFSHORE, DEVELOPMENTS 1978
B 6108 1340 CANADA, EASTERN, PRODUCTION, 1973-1976
B 6208 1456 CANADA, EASTERN, PRODUCTION, 1974-1977
B 6308 1349 CANADA, EASTERN, PRODUCTION, 1975-1978
B 6409 1522 CANADA, EASTERN, PRODUCTION, 1976-1979
B 6205 763 CANADA, EASTERN, SEISMIC FACIES, OFFSHORE
S 3 90 CANADA, ELDORADO VEIN DISTRICT
B 6206 1006 CANADA, ELLEF RINGNES ISLAND, ARCTIC
B 6404 535 CANADA, ELLESMERE ISLAND
B 6206 1005 CANADA, ELLESMERE ISLAND, ARCTIC
B 6206 1005 CANADA, ELLESMERIAN OROGENY, ARCTIC
B 6202 178 CANADA, ENERGY RESOURCES
B 6206 1004† CANADA, EVOLUTION OF, SVERDRUP BASIN, ARCTIC
B 6008 1156 CANADA, EXPLORATORY DRILLING, 1975
B 6108 1124 CANADA, EXPLORATORY DRILLING, 1976
B 6208 1278 CANADA, EXPLORATORY DRILLING, 1977
B 6308 1171 CANADA, EXPLORATORY DRILLING, 1978
B 6409 1298 CANADA, EXPLORATORY DRILLING, 1979
M 25 496 CANADA, FARO MINE
B 6302 166 CANADA, FISH SCALES STRUCTURE
B 6302 180 CANADA, FOOTHILLS, GAS TESTS
B 6206 1005 CANADA, FRANKLINIAN GEOSYNCLINE, ARCTIC
B 6206 1005 CANADA, GEOLOGIC PROVINCES, ARCTIC
M 26 60 CANADA, GETHING FORMATION
M 25 497 CANADA, GOLD
B 6101 29 CANADA, GREAT BEAR LAKE, PITCHBLENDE DEPOSITS
M 25 108 CANADA, GROUNDHOG COALFIELD
M 25 506 CANADA, HART RIVER DEPOSIT
M 25 107 CANADA, HAT CREEK COALFIELDS
B 6206 1005 CANADA, HAZEN TROUGH, ARCTIC
B 6111 1953 CANADA, HEAVY OIL, WESTERN
B 6206 1005 CANADA, HUDSONIAN OROGENY, ARCTIC
B 6008 1355† CANADA IN 1975, DEVELOPMENTS IN, EASTERN
B 6008 1343† CANADA IN 1975, DEVELOPMENTS IN, WESTERN
B 6108 1335† CANADA IN 1976, DEVELOPMENTS IN, EASTERN
B 6108 1313† CANADA IN 1976, DEVELOPMENTS IN, WESTERN
B 6208 1451† CANADA IN 1977, DEVELOPMENTS IN, EASTERN
B 6308 1344† CANADA IN 1978, DEVELOPMENTS IN, EASTERN
B 6308 1407† CANADA IN 1978, DEVELOPMENTS IN, WESTERN
B 6409 1518† CANADA IN 1979, DEVELOPMENTS IN, EASTERN
B 6409 1510† CANADA IN 1979, DEVELOPMENTS IN, WESTERN
M 25 500 CANADA, INTERMONTANE BELT
M 24 79† CANADA, KAYBOB OIL FIELD, ALBERTA

B 6206 1006 CANADA, KING CHRISTIAN ISLAND, ARCTIC
M 25 507 CANADA, KOOTENAY ARC
M 25 108 CANADA, KYNOCKS COAL AREAS
M 25 108 CANADA, LIARD RIVER, SHEEP CREEK, AND DAWSON COALFIELDS
M 25 497 CANADA, LIARD TROUGH
S 10 226 CANADA, LLOYDMINSTER HEAVY OIL DEPOSIT,PROFILE
B 6111 1954 CANADA, LLOYDMINSTER HEAVY OIL, WESTERN
B 6206 1005 CANADA, MELVILLE ISLAND, ARCTIC
B 6206 1015 CANADA, MESOZOIC, APTIAN-CAMPANIAN TO MAESTRICHTIAN, ARCTIC
B 6206 1014 CANADA, MESOZOIC, GRIESBACHIAN TO UPPER NORIAN, ARCTIC
B 6206 1015 CANADA, MESOZOIC TO CENOZOIC, CAMPANIAN TO MAESTRICHTIAN,
B 6206 1014 CANADA, MESOZOIC, UPPER NORIAN TO APTIAN, ARCTIC
B 6302 166 CANADA, MILK RIVER SAND
M 27 133 CANADA, MISSISSIPPIAN, DEBOLT FORMATION
M 28 84 CANADA, MISSISSIPPIAN, DEBOLT FORMATION
M 28 143 CANADA, N.W.T., CAMBRIAN, OLD FORT ISLAND SANDSTONE
M 28 179 CANADA, N.W.T., CAMBRIAN, OLD FORT ISLAND SANDSTONE
M 28 180 CANADA, N.W.T., CRETACEOUS, SANDSTONE
M 28 178 CANADA, N.W.T., CRETACEOUS, SANDSTONE
M 28 163 CANADA, N.W.T., CRETACEOUS, SANDSTONE
M 25 500 CANADA, OMINECA CRYSTALLINE BELT
B 6206 1007 CANADA, PALEOZOIC, VISEAN TO DZHULFIAN, ARCTIC
B 6111 1951 CANADA, PEACE RIVER HEAVY OIL DEPOSIT, WESTERN
B 6206 1005 CANADA, PEARYA GEANTICLINE, ARCTIC
M 25 496 CANADA, PINE POINT DEPOSIT, LEAD-ZINC
M 25 506 CANADA, POTATO HILLS TUNGSTEN DEPOSIT
B 6206 1006 CANADA, PRINCE PATRICK ISLAND, ARCTIC
M 25 107 CANADA, PRINCETON-MERRITT COALFIELDS
M 28 162 CANADA, QUEBEC, CAMBRIAN, SILLERY GROUP
M 28 65 CANADA, QUEBEC, CAMBRIAN, SILLERY GROUP
M 28 161 CANADA, QUEBEC, CAMBRIAN, ST. ROCH FORMATION
M 25 109 CANADA, QUEEN CHARLOTTE ISLANDS, COAL
B 6206 1005 CANADA, QUEEN ELIZABETH ISLANDS, ARCTIC
B 6101 29 CANADA, RABBIT LAKE, PITCHBLENDE DEPOSITS
M 25 496 CANADA, RAPITAN GROUP, IRON DEPOSITS
B 6206 1006 CANADA, SABINE PENINSULA, ARCTIC
M 25 507 CANADA, SALMO TYPE OF LEAD-ZINC DEPOSIT
M 25 506 CANADA, SAM GOOSLY DEPOSIT
B 6107 988 CANADA, SCOTIAN SHELF, WYANDOT CHALK UNIT
S 8 40 CANADA, SEAVALLEYS, FRASER DELTA
M 25 493 CANADA, SELWYN BASIN
M 25 497 CANADA, SELWYN BASIN, ZINC-LEAD
M 25 507 CANADA, SHUSWAP TYPE OF LEAD-ZINC DEPOSIT
M 25 500 CANADA, SIFTON BASIN
M 25 500 CANADA, SILVER-RICH VEINS
M 25 500 CANADA, SKEENA ARCH
M 25 504 CANADA, SKEENA ARCH
M 25 500 CANADA, SLOKO BASIN
M 25 501 CANADA, SLOKO, SUSTUT, AND SIFTON BASINS, CROSS SECTION
M 25 508 CANADA, SNAKE RIVER IRON DEPOSIT
M 26 60 CANADA, SPIRIT RIVER FORMATION
M 25 504 CANADA, STIKINE ARCH
M 25 500 CANADA, STIKINE ARCH
M 30 315† CANADA, STRACHAN AND, RICINUS WEST, GAS FIELDS, ALBERTA
S 8 97 CANADA, SUBMARINE CANYONS, FRASER SEAVALLEY
M 25 500 CANADA, SUCCESSOR BASINS, MINERAL DEPOSITS
M 25 507 CANADA, SULLIVAN DEPOSIT
M 25 496 CANADA, SULLIVAN MINE
M 25 500 CANADA, SUSTUT BASIN
M 25 507 CANADA, SUSTUT COPPER DEPOSIT
B 6206 1021 CANADA, SVERDRUP BASIN, EUREKAN OROGENY
B 6206 1016 CANADA, SVERDRUP BASIN, HALOKINETIC STRUCTURES
B 6206 1025 CANADA, SVERDRUP BASIN, HYDROCARBONS
B 6206 1019 CANADA, SVERDRUP BASIN, MAFIC INTRUSIVE AND VOLCANIC ROCKS
B 6206 1006 CANADA, SVERDRUP BASIN, OIL AND GAS, ARCTIC
B 6206 1007 CANADA, SVERDRUP BASIN, STRATIGRAPHY
S 5 147 CANADA, SVERDRUP BASIN, OTTO FIORD FORMATION, STRATIGRAPHY
S 5 159 CANADA, SVERDRUP BASIN, OTTO FIORD FORMATION, ALGAL MOUNDS
S 5 149 CANADA, SVERDRUP BASIN, OTTO FIORD FORMATION, ANHYDRITE
M 25 497 CANADA, SYLVESTER TROUGH
M 25 500 CANADA, TANTALUS BASIN
M 25 108 CANADA, TELKWA COALFIELD
M 25 500 CANADA, TYAUGHTON TROUGH
M 25 109 CANADA, VANCOUVER ISLAND, COAL
B 6206 1005 CANADA, VICTORIA ISLAND, ARCTIC
B 6111 1951 CANADA, WABASCA HEAVY OIL DEPOSIT, WESTERN
B 6111 1955 CANADA, WAINWRIGHT HYDROCARBONS, WESTERN
M 25 485 CANADA, WESTERN, COPPER AND MOLYBDENUM
B 6407 980 CANADA, WESTERN, CRETACEOUS RESERVOIRS
M 25 108 CANADA, WHITEHORSE COAL AREAS
M 25 499 CANADA, WHITEHORSE TROUGH
B 6004 669* CANADA--DEPOSITIONAL AND DIAGENETIC MODEL FOR MESOZOIC
B 6303 492* CANADA. BY TITLE ONLY, ENGINEERING, GEOLOGY IN, DEEP BASIN
S 5 145* CANADIAN ARCTIC ARCHIPELAGO, CARBONATE- ANHYDRITE FACIES
B 6111 1929† CANADIAN ARCTIC ARCHIPELAGO, CARBONATE-, ANHYDRITE FACIES
B 6111 1932 CANADIAN ARCTIC, HOODOO DIAPIR
B 6412 2111 CANADIAN ARCTIC ISLAND WELL, HYDROCARBONS
B 6101 7 CANADIAN ARCTIC ISLANDS, CAMBRIAN
B 6101 20 CANADIAN ARCTIC ISLANDS, CENTRAL ELLESMERE FOLDBELT
B 6101 19 CANADIAN ARCTIC ISLANDS, CORNWALLIS FOLDBELT
B 6101 24 CANADIAN ARCTIC ISLANDS, DRAKE POINT, WEST AND EAST, FIELD
B 6101 20 CANADIAN ARCTIC ISLANDS, EUREKAN FOLDBELT
B 6101 6 CANADIAN ARCTIC ISLANDS, FRANKLINIAN BASIN
B 6101 19 CANADIAN ARCTIC ISLANDS, FRANKLINIAN FOLDBELT
B 6101 1† CANADIAN ARCTIC ISLANDS, GEOLOGY AND, HYDROCARBON
B 6101 24 CANADIAN ARCTIC ISLANDS, HECLA FIELD
B 6101 24 CANADIAN ARCTIC ISLANDS, JACKSON BAY FIELD
B 6101 24 CANADIAN ARCTIC ISLANDS, KING CHRISTIAN FIELD
B 6101 24 CANADIAN ARCTIC ISLANDS, KRISTOFFER FIELD
B 6101 12 CANADIAN ARCTIC ISLANDS, MISSISSIPPIAN
B 6101 9 CANADIAN ARCTIC ISLANDS, ORDOVICIAN TO DEVONIAN
B 6101 18 CANADIAN ARCTIC ISLANDS, OROGENIES
B 6101 10 CANADIAN ARCTIC ISLANDS, PALEOZOIC AND MESOZOIC SEDIMENTS,
B 6101 19 CANADIAN ARCTIC ISLANDS, PARRY ISLANDS FOLDBELT
B 6101 13 CANADIAN ARCTIC ISLANDS, PERMO-PENNSYLVANIAN
B 6101 6 CANADIAN ARCTIC ISLANDS, PRECAMBRIAN
B 6101 25 CANADIAN ARCTIC ISLANDS, RESERVES
B 6101 11 CANADIAN ARCTIC ISLANDS, SVERDRUP BASIN
B 6101 20 CANADIAN ARCTIC ISLANDS, SVERDRUP RIDGE
B 6101 24 CANADIAN ARCTIC ISLANDS, THOR FIELD
B 6101 14 CANADIAN ARCTIC ISLANDS, TRIASSIC
B 6101 15 CANADIAN ARCTIC ISLANDS, TRIASSIC-JURASSIC
B 6101 24 CANADIAN ARCTIC ISLANDS, WALLIS FIELD
B 6101 11 CANADIAN ARCTIC ISLANDS, WANDEL SEA BASIN
B 6111 1951 CANADIAN ARCTIC, MELVILLE ISLAND HEAVY OIL DEPOSIT
B 6111 1930 CANADIAN ARCTIC, SVERDRUP BASIN
B 6111 1930 CANADIAN ARCTIC, VAN HAUEN PASS
B 6207 1214* CANADIAN CORDILLERA, ABST., METALLOGENIC ZONES AND EPOCHS IN
M 25 496 CANADIAN CORDILLERA, BARITE AND FLUORITE
M 25 500 CANADIAN CORDILLERA, BOWSER BASIN, CROSS SECTION
M 25 497 CANADIAN CORDILLERA, CLASTIC WEDGE, MINERAL DEPOSITS
M 25 503 CANADIAN CORDILLERA, COAST CRYSTALLINE BELT
M 25 503 CANADIAN CORDILLERA, COLUMBIAN OROGEN
M 25 492 CANADIAN CORDILLERA, COMPLEXITY OF GEOLOGY
M 25 496 CANADIAN CORDILLERA, COPPER
M 25 505 CANADIAN CORDILLERA, COPPER, PRODUCTION AND RESERVES
M 25 499 CANADIAN CORDILLERA, CROSS SECTION
M 25 497 CANADIAN CORDILLERA, CROSS SECTION
M 25 498 CANADIAN CORDILLERA, CROSS SECTION
M 25 495 CANADIAN CORDILLERA, CROSS SECTION, RESTORED
M 25 492† CANADIAN CORDILLERA DEPOSITIONAL BASINS, ENVIRONMENTS OF
M 25 503 CANADIAN CORDILLERA, EASTERN MARGINAL BELT
M 25 503 CANADIAN CORDILLERA, GEOLOGIC FRAMEWORK
M 25 494 CANADIAN CORDILLERA, GEOLOGIC PROVINCES, MAP
M 25 503 CANADIAN CORDILLERA, INSULAR BELT
M 25 503 CANADIAN CORDILLERA, INTERMONTANE BELT
M 25 505 CANADIAN CORDILLERA, IRON, PRODUCTION
M 25 498 CANADIAN CORDILLERA, ISLAND ARC, MINERAL DEPOSITS
M 25 498 CANADIAN CORDILLERA, ISLAND ARC, ROCKS OF
M 25 499 CANADIAN CORDILLERA, JURASSIC SUCCESSOR BASIN, CROSS SECTION
M 25 505 CANADIAN CORDILLERA, LEAD, PRODUCTION AND RESERVES
M 25 505 CANADIAN CORDILLERA, MAGMATIC DEPOSITS OF METALS
M 25 506 CANADIAN CORDILLERA, MASSIVE SULFIDE DEPOSITS OF METALS
M 25 505 CANADIAN CORDILLERA, METAL PRODUCTION AND RESERVES, TABLE
M 25 503 CANADIAN CORDILLERA, METAL PRODUCTION, ANNUAL
M 25 503† CANADIAN CORDILLERA, METALLIC, MINERAL RESOURCES OF
M 25 504 CANADIAN CORDILLERA, METALS, BACKGROUND ABUNDANCES
M 25 504 CANADIAN CORDILLERA, METALS, CLASSES OF
M 25 504 CANADIAN CORDILLERA, METALS, ZONAL PATTERN OF DISTRIBUTION
M 25 496 CANADIAN CORDILLERA, MID-PALEOZOIC CLASTIC WEDGE
M 25 506 CANADIAN CORDILLERA, MINERAL DEPOSITS, DISTRIBUTION BY CLASSES
M 25 493 CANADIAN CORDILLERA, MIOGEOSYNCLINE, GEOLOGIC HISTORY
M 25 496 CANADIAN CORDILLERA, MIOGEOSYNCLINE, MINERAL DEPOSITS
M 25 505 CANADIAN CORDILLERA, MOLYBDENUM, PRODUCTION AND RESERVES
M 25 497 CANADIAN CORDILLERA, OCEANIC BASIN, MINERAL DEPOSITS
M 25 497 CANADIAN CORDILLERA, OCEANIC BASIN, ROCKS OF
M 25 503 CANADIAN CORDILLERA, OMINECA CRYSTALLINE BELT
M 25 503 CANADIAN CORDILLERA, PACIFIC OROGEN
M 25 506 CANADIAN CORDILLERA, PORPHYRY DEPOSITS OF METALS
M 25 496 CANADIAN CORDILLERA, QUARTZ MONZONITE PLUTONS

M 25 503 CANADIAN CORDILLERA, SEDIMENTATION AND MINERALIZATION,
M 25 505 CANADIAN CORDILLERA, SILVER, PRODUCTION AND RESERVES
M 25 496 CANADIAN CORDILLERA, SILVER-LEAD-ZINC
M 25 505 CANADIAN CORDILLERA, SKARN DEPOSITS OF METALS
M 25 507 CANADIAN CORDILLERA, STRATIFORM DEPOSITS OF METALS
M 25 500 CANADIAN CORDILLERA, SUCCESSOR BASINS, ROCKS OF
M 25 503 CANADIAN CORDILLERA, TECTONIC BELTS
M 25 504 CANADIAN CORDILLERA, TECTONIC BELTS, MAP
M 25 496 CANADIAN CORDILLERA, TUNGSTEN
M 25 507 CANADIAN CORDILLERA, VEIN DEPOSITS OF METALS
M 25 505 CANADIAN CORDILLERA, ZINC, PRODUCTION AND RESERVES
M 25 105† CANADIAN CORDILLERA--SUMMARY, COAL RESOURCES OF
B 6002 322* CANADIAN COUNTIES, OKLAHOMA, ABST., DIBBLE--, MUSTANG,
B 6109 1437 CANADIAN COUNTY, ANADARKO BASIN, OKLAHOMA
M 26 439 CANADIAN COUNTY, ANADARKO BASIN, PENNSYLVANIAN, HOXBAR
B 6405 767* CANADIAN FOOTHILLS PROVINCE ANALOG, ABST., OIL AND GAS
B 6207 1236* CANADIAN GEOTHERMAL RESEARCH PROGRAM, ABST.
B 6405 721* CANADIAN, IN NORTHERN ARKANSAS, ABST., STRATIGRAPHY AND
B 6203 560* CANADIAN OIL SANDS—GEOLOGY AND PRODUCTION. BY TITLE ONLY.,
B 6405 768* CANADIAN ROCKIES SEGMENT OF CORDILLERAN FORELAND THRUST
B 6304 701* CANADIAN ROCKY MOUNTAINS, ABST., STRUCTURAL ANALYSIS OF
B 6207 1228* CANADIAN SEGMENT OF CORDILLERA, ABST., EVOLUTION OF
B 6101 3 CANADIAN SHIELD
S 3 118 CANADIAN SHIELD
S 3 118 CANADIAN SHIELD
B 6101 30 CANADIAN SHIELD, APHEBIAN CHART
B 6101 30 CANADIAN SHIELD, ARCHEAN CHART
B 6101 30 CANADIAN SHIELD, HELIKIAN CHART
B 6101 34 CANADIAN SHIELD, HELIKIAN URANIUM PROVINCES, NORTHWESTERN
B 6111 2010† CANADIAN SOCIETY OF PETROLEUM GEOLOGISTS
B 6210 2117 CANADIAN SUPERIOR OIL LTD., ON DEVELOPMENTS IN AJMAN AND UMM
B 6004 710* CANADIAN WESTERN INTERIOR, ABST., PROVENANCE AND, DISPERSAL OF
M 25 519 CANANEA GRANITE, MEXICO
M 25 519 CANANEA MINE, COPPER, MEXICO, SONORA
B 6105 826* CANARY AND CAPE VERDE ISLANDS, ABST., GEOCHEMISTRY OF RECENT,
B 6111 1993 CANARY BASIN, VEMA-30, SEISMIC PROFILE
B 6111 2005 CANARY BASIN-CAPE VERDE RISE
S 11 160 CANCRIS BROGNIARTI
S 4 195 CANCUN, MOLLUSCA, MEXICO, ISLA
B 6109 1476 CANDEIAS FIELD, BRAZIL
M 26 345 CANDEIAS FIELD, BRAZIL
B 6105 774 * CANDEIAS FIELD, RECONCAVO BASIN, BRAZIL—NEW SEDIMENTOLOGIC
M 25 522 CANDU TYPE, NUCLEAR REACTORS, HEAVY-WATER OR
M 25 556 CANEPA, C., MINERAL RESOURCES POTENTIAL O PERU--SUMMARY
B 6010 1650 CANFIELD, R. W., ON DEVELOPMENTS IN ECUADOR, 1975
B 6110 1584 CANFIELD, R. W., ON DEVELOPMENTS IN ECUADOR, 1976
B 6405 687 CANFIELD, ROBERT W., SACHA FIELD OF ECUADORIAN ORIENTE, ABST.
M 25 411 CANGA, SCREE ORE, CONGLOMERATE ORE, DETRITAL ORE, OR
B 6411 1865 CANNING BASIN, AUSTRALIA
B 6201 41 CANNING BASIN, AUSTRALIA
B 6406 836 CANNING BASIN, GEOTHERMAL HISTORY, AUSTRALIA
M 25 233 CANNING BASIN, OFFSHORE, AUSTRALIA
B 6406 814† CANNING BASIN, WESTERN AUSTRALIA, DEVONIAN, GREAT, BARRIER,
B 6004 655 CANT, D. J., EFFECTS OF ICE ON SEDIMENTS ON SANDY BRAIDED RIVER,
B 6004 705* CANTABRIAN MOUNTAINS, NORTH SPAIN, ABST., SEDIMENTATION
S 6 325 CANTABRIAN STAGE, CARBONIFEROUS, RADIOMETRIC DATING
B 6405 675* CANTABRIC COAST, NORTHERN SPAIN, ABST., DEVONIAN, REEFS
M 25 220 CANTERBURY-EAST OTAGO-CAMPBELL PLATEAU BASIN O GREAT SOUTH
B 6009 1620 CANTRELL, C. L., CARBONIFEROUS SEDIMENTARY RESPONSES TO
B 6012 2177* CANTUA CREEK, CALIFORNIA, ABST., SEDIMENTARY SERPENTINITE OF
B 6205 848 CANTUA CREEK FIELD, CALIFORNIA
B 6211 2262 CANTUA CREEK SECTION, CALIFORNIA
B 6211 2262 CANTUA SAND, CALIFORNIA
B 6308 1425* CANYO SANDSTONE, ROUNDTOP AREA, FISHER COUNTY, TEXAS, ABST.,
B 6004 654* CANYON, ABST., SEDIMENT PROCESSES CONTROLLED BY, BOTTOM,
B 6004 725* CANYON AND FAN-VALLEY SEDIMENTATION, ABST., LATERAL, INFILL AS
B 6004 721* CANYON AND THEIR RELATION TO DEEP-SEA FANS, ABST., CURRENTS IN,
B 6004 650* CANYON, CALIFORNIA COAST, ABST., PHYSICAL AND, BIOGENI
B 6203 495* CANYON GROUP OF VAL VERDE BASIN, WILCOX GROUP OF SOUTH
B 6303 314 CANYON RANGE THRUST, UTAH
B 6408 1171 CANYON REEF, TEXAS
B 6002 324* CANYON SANDSTONE, VAL VERDE BASIN, TEXAS, ABST., TURBIDITE,
B 6205 787 CANYON SEISMIC DATA, TEXAS, CISCO-
B 6211 2277 CANYON SPRINGS SANDSTONE MEMBER, WYOMING
B 6004 695* CANYON TRIBUTARY, ABST., LARGE SEDIMENT, SLUMP REOPENS
B 6001 108† CANYON--INFLUENCE ON ACCUMULATION OF HYDROCARBONS IN HELEZ
B 6004 686* CANYON-FAN SYSTEMS, ABST., SIGNIFICANCE OF, CLAST SIZE, SHAPE, AN
B 6203 566* CANYON-FAN UNIT IN ANADARKO BASIN, OKLAHOMA, ABST., ECONOMIC
B 6303 548* CANYON-RELATED POOLS, ABST., ENTRAPMENT FACTORS IN,
B 6004 667* CANYONS, ABST., CONTRASTS IN CONCENTRATION OF, SUSPENDED
B 6004 702* CANYONS AND FANS, ABST., THIN-BEDDED, TURBIDITES IN, MODERN,
B 6004 645* CANYONS AND MARINE CYCLES OF DEPOSITION IN SOUTHERN
B 6004 686* CANYONS OF NORTHWESTERN ATLANTIC MARGIN, ABST., SEDIMENTARY
B 6407 1064 CAP CHAT CHANNEL, QUEBEC
B 6407 1054 CAP DES ROSIERS GROUP, QUEBEC
B 6206 938 CAP ENRAGE FORMATION, MASSIVE SANDSTONE FACIES, QUEBEC
B 6004 678* CAP ENRAGE FORMATION MIDDLE CAMBRIAN TO LOWER ORDOVICIAN,
B 6109 1544* CAP ROCK, ABST., SUBSALT ORIGIN OF, SALT-, DOME
B 6012 2162† CAP ROCK, ORIGIN OF, GULF COAST, SALT-, DOME
S 10 59 CAP ROCKS, MIGRATION IN AQUEOUSSOLUTION
B 6407 1062 CAP STE. ANNE, QUEBEC
B 6409 1556* CAP-ROCK FORMATION AND DIAGENESIS, GYP HILL SALT DOME, SOUTH
B 6112 2105 CAPAHUARI FIELD, PERU
B 6106 930 CAPE BASIN
B 6412 2055 CAPE BASIN, DSDP SITE 361, LEG 40, GEOCHEMICAL LOG
B 6111 1994 CAPE BASIN, POSSIBLE DIAPIRS
B 6203 491* CAPE BLANCO, OREGON, ABST., COMMUNITY SUCCESSION OF, PIONEER
B 6404 530 CAPE CALHOUN FORMATION, GREENLAND
M 29 77 CAPE COD, CROSS SECTION, ATLANTIC COAST
M 29 65† CAPE COD, MULTICHANNEL, SEISMIC, DEPTH, SECTIONS AND, INTERVAL,
B 6303 444* CAPE COD TO FLORIDA, ABST., DISTRIBUTION, MORPHOLOGY,
M 25 222 CAPE EGMONT FIELD, NEW ZEALAND
B 6403 340 CAPE FEAR ARCH
B 6403 340 CAPE FEAR ARCH
B 6109 1523 CAPE FEAR ARCH, ACCUMULATION RATES, EAST COAST
B 6103 336 CAPE FEAR ARCH, GREAT CAROLINA RIDGE
B 6107 975 CAPE GOOD HOPE SEGMENT, WALVIS RIDGE—
B 6405 746* CAPE HATTERAS, ABST., GEOLOGIC SETTING AND OIL AND GAS
M 29 65† CAPE HATTERAS AND CAPE COD, MULTICHANNEL, SEISMIC, DEPTH,
B 6402 184 CAPE HATTERAS, SUBSIDENCE AND HEAT FLOW
B 6402 184 CAPE HATTERAS, SUBSIDENCE AND HEAT FLOW
B 6402 184 CAPE HATTERAS, TIME, TEMPERATURE, AND DEPTH PLOT
B 6402 184 CAPE HATTERAS, TIME, TEMPERATURE, AND DEPTH PLOT
B 6203 529* CAPE HATTERASTO HYDROGRAPHER CANYON, ABST., GEOTECHNICAL
B 6312 2150 CAPE HENLOPEN SPIT, DELAWARE
B 6102 193 CAPE KAMCHATKA, MYS KAMCHATSKIY
B 6402 256 CAPE MATAPAN SUBMARINE PENINSULA, SEISMIC REFLECTION, GREECE
B 6402 256 CAPE MATAPAN SUBMARINE PENINSULA, SEISMIC REFLECTION, GREECE
M 29 80 CAPE MAY, CROSS SECTION, ATLANTIC COAST
B 6203 495* CAPE MAY, NEW JERSEY, ABST., SLOPE, MAP DEPICTING MAJOR
B 6004 691* CAPE OF GOOD HOPE, ABST., PETROLEUM HABITAT OF, ATLANTIC,
B 6102 193 CAPE OLYUTORSKY, MYS OLYUTORSKIY
B 6307 1061 CAPE PHILLIPS FORMATION, NORTHWEST TERRITORIES
B 6311 2089 CAPE RANGE FRACTURE, AUSTRALIA
B 6403 340 CAPE ROMAIN, SOUTH CAROLINA
B 6403 340 CAPE ROMAIN, SOUTH CAROLINA
B 6303 518* CAPE ROMAIN, SOUTH CAROLINA, ABST., HOLOCENE STRATIGRAPHY OF
S 8 101 CAPE SAN LUCAS CANYON, MEXICO, SUBMARINE CANYONS
B 6404 530 CAPE SCHUCHERT FORMATION, GREENLAND
B 6303 422* CAPE SEBASTIAN SANDSTONE SOUTHWESTERN OREGON, ABST.,
B 6403 446* CAPE SPENCER, ABST., TWENTIETH-CENTURY HISTORY OF, GULF OF,
B 6204 635 CAPE SUCKLING, ALASKA
B 6403 446* CAPE SUCKLING TO CAPE SPENCER, ABST., TWENTIETH-CENTURY
S 6 369 CAPE SYSTEM, SOUTH AFRICA
B 6412 2056 CAPE VERDE BASIN, DSDP SITE 367, LEG 41, GEOCHEMICAL LOG
B 6107 965 CAPE VERDE FRACTURE, AFRICA
B 6105 826* CAPE VERDE ISLANDS, ABST., GEOCHEMISTRY OF RECENT,
B 6111 2004† CAPE VERDE RISE AND ITS RELATION TO PIERCEMENT STRUCTURES,
B 6111 2005 CAPE VERDE RISE, CANARY BASIN-
B 6111 2008 CAPE VERDE RISE, MIOCENE NANNO-FORAM CHALK INCLUSION
B 6111 2006 CAPE VERDE RISE, SEISMIC PROFILE
B 6111 1993 CAPE VERDE RISE, VEMA-30, SEISMIC PROFILE
B 6111 2005 CAPE VERDE-MADEIRA ABYSSAL PLAIN
B 6204 635 CAPE YAKATAGA, ALASKA
M 26 356 CAPE YAKATAGA, SIDE-SCAN SONAR RECORD, ALASKA
B 6002 245† CAPILLARY PRESSURE, PORE, GEOMETRY OF, CARBONATE ROCKS AS
B 6305 738 CAPILLARY PRESSURE TESTS, MERCURY
B 6305 742 CAPILLARY PROPERTIES OF DRILL CUTTINGS
B 6206 955 CAPISTRANO FORMATION, CALIFORNIA
B 6303 519* CAPITAN COMPLEX, ABST., DIAGENESIS OF, BACK-, REEF, CARBONAT
B 6205 781 CAPITAN FORMATION, NEW MEXICO
M 27 5 CAPITAN FORMATION, NEW MEXICO, PERMIAN
M 27 4 CAPITAN FORMATION, NEW MEXICO, PERMIAN
M 27 6 CAPITAN LIMESTONE, NEW MEXICO, PERMIAN
M 27 67 CAPITAN LIMESTONE, NEW MEXICO, PERMIAN
M 27 174 CAPITAN LIMESTONE, NEW MEXICO, PERMIAN
M 27 77 CAPITAN LIMESTONE, NEW MEXICO, PERMIAN
M 27 47 CAPITAN LIMESTONE, NEW MEXICO, PERMIAN
B 6203 573* CAPITAN LIMESTONE, PERMIAN, GUADALUPE MOUNTAINS, WEST TEXAS

B 6203 492* CAPITAN LIMESTONE, PERMIAN, GUADALUPE MOUNTAINS, WEST TEXAS
M 27 9 CAPITAN LS.-TANSILL FM. TRANSITION, NEW MEXICO, PERMIAN
B 6105 843* CAPITAN MASSIVE LIMESTONE, PERMIAN, OF GUADALUPE MOUNTAINS,
B 6004 670* CAPITAN REEF COMPLEX, PERMIAN, GUADALUPE MOUNTAIN, NEW
B 6109 1449 CAPITAN REEF, NEW MEXICO AND, TEXAS
B 6203 541* CAPITAN REEF, PERMIAN, GUADALUPE MOUNTAINS, NEW MEXICO, ABST.,
B 6303 509* CAPITAN REEF, PERMIAN, WEST TEXAS, ABST., TRACE-ELEMENT,
B 6102 294* CAPITAN REEF--EVOLUTION OF A CONCEPT, ABST.
B 6004 661* CAPITAN REEF--EVOLUTION OF CONCEPT, ABST.
B 6203 504* CAPLES TERRANE, SOUTH ISLAND, NEW ZEALAND, ABST.,
M 25 519 CAPOTE BASIN, MEXICO
M 25 519 CAPOTE BRECCIA, MEXICO
M 25 519 CAPOTE QUARTZITE, MEXICO
B 6203 553* CAPRINID REEFS AND ASSOCIATED SEDIMENTS, ABST., EFFECTS OF,
M 27 85 CAPRINID RUDISTS
B 6304 479 CAPROCK, GEOCHEMICAL CONCEPT
B 6201 70 CAPROCKS, AUSTRALIA, EXMOUTH PLATEAU
M 25 426 CAPTAINS FLAT, LEAD-ZINC-COPPER DEPOSIT, NEW SOUTH WALES
M 25 462 CARACAS-MARACAIBO FRACTURE ZONE
S 6 347 CARADOC SERIES, BENTONITES, ORDOVICIAN
B 6401 74 CARBO ISOTOPE SYSTEM, RELIABILITY OF, ISOTOPE DATA--EFFECTS OF
B 6304 701* CARBO ISOTOPES, ABST., ORIGIN OF, GASES, ADSORBED IN
B 6105 774 CARBO, SALVATORE, SUNY, DISTRIBUTTION AND SIGNIFICANCE OF TRACE
B 6004 696* CARBON AND OXYGEN ISOTOPE ANALYSES OF PLEISTOCENE FLUVIAL,
B 6203 489* CARBON AND OXYGEN ISOTOPES AS INDICATORS OF METEORIC AND
B 6105 810 * CARBON AND OXYGEN ISOTOPIC COMPOSITIONS AS INDICATORS OF
B 6109 1555* CARBON AND OXYGEN ISOTOPIC EVOLUTION OF WHOLE ROCK AND
B 6101 100† CARBON AS HYDROCARBONS AND ASPHALTIC COMPOUNDS IN
B 6004 717* CARBON BASINS CARBON COUNTY, WYOMING, ABST., EOCENE
B 6401 78 CARBON BURIAL RATES, ORGANIC
B 6401 78 CARBON BURIAL RATES, ORGANIC
B 6209 1763* CARBON CONTENT AND PETROLEUM SOURCE ROCK POTENTIAL OF
B 6008 1408* CARBON COUNTY, MONTANA, ABST., MESAVERDE GAS POTENTIAL IN,
B 6205 879* CARBON COUNTY, UTAH, ABST., ENVIRONMENTAL RECONSTRUCTION OF
B 6004 717* CARBON COUNTY, WYOMING, ABST., EOCENE PROVENANCE OF, HANNA
B 6303 479* CARBON DIOXIDE, ABST., STABLE, CARBON, ISOTOPES IN, OIL-FIELD,
B 6407 1078† CARBON DIOXIDE IN NORTON BASIN, ALASKA, ORIGIN OF,
B 6401 112 CARBON DIOXIDE IN PORE WATER OF SOURCE ROCKS, CONCENTRATION
B 6401 112 CARBON DIOXIDE IN PORE WATER OF SOURCE ROCKS, CONCENTRATION
B 6205 882* CARBON DIOXIDE RESOURCES OF UTAH—A FIRST APPRAISAL, ABST.
B 6401 108 CARBON DIOXIDE, SANTA CRUZ BASIN RECENT SEDIMENTS, GENERATION
B 6401 108 CARBON DIOXIDE, SANTA CRUZ BASIN RECENT SEDIMENTS, GENERATION
B 6207 1208* CARBON DIOXIDE-RICH GROUND WATERS, CIRCUM-PACIFIC REGION,
B 6112 2098 CARBON DURING CATAGENESIS, LOSS OF
B 6405 704* CARBON FLUX FROM OCEAN TO BIOSPHERE—CHEMICAL EVIDENCE FROM
B 6209 1589 CARBON IN MARINE ENVIRONMENTS, ORGANIC
S 10 258 CARBON ISOTOPE DATA, METHANE,NEAR-SURFACE
B 6401 67† CARBON ISOTOPE FLUCTUATION IN CRETACEOUS PELAGIC
B 6401 67† CARBON ISOTOPE FLUCTUATIONS IN CRETACEOUS PELAGIC
B 6304 698* CARBON ISOTOPE MEASUREMENTS OF HYDROCARBONS ADSORBED IN
B 6401 74 CARBON ISOTOPE SYSTEM
B 6401 76 CARBON ISOTOPE VARIATIONS, PRIMARY CONTROLS ON
B 6401 76 CARBON ISOTOPE VARIATIONS, PRIMARY CONTROLS ON
B 6303 519* CARBON ISOTOPES A PALEOCLIMATOLOGICAL TOOL, ABST., STABLE
B 6304 701* CARBON ISOTOPES, ABST., BACTERIAL, DEGRADATION OF, CRUDE,
B 6303 479* CARBON ISOTOPES IN OIL-FIELD WATERS AND ORIGIN OF CARBON
B 6303 532* CARBON ISOTOPES--NEW TOOL IN HYDROCARBON CORRELATION AND
B 6303 532* CARBON ISOTOPES. BY TITLE ONLY, ORIGIN OF, GASES ADSORBED IN
B 6401 71 CARBON ISOTOPIC DATA, MEXICO, PEREGRINA CANYON
B 6401 71 CARBON ISOTOPIC DATA, MEXICO, PEREGRINA CANYON
B 6405 756* CARBON ISOTOPIC SIGNATURE AS CRITERION FOR INTERPRETING
B 6203 488* CARBON ISOTOPIC VARIATION IN PLEISTOCENE LACUSTRINE
B 6406 921 CARBON PREFERENCE INDEX, CPI
B 6406 921 CARBON RATIOS, BITUMEN AND, ORGANIC
S 6 165 CARBON-14, QUATERNARY DATING
B 6306 872 CARBON-RICH SEDIMENTS O CRETACEOUS, GLOBAL DEPOSITION OF.
B 6002 185 CARBONACEOUS DETRITUS IN DITCH CUTTINGS, NORTH SEA
S 10 180 CARBONACEOUS SHALES, METHANE,OCCURRENCES
B 6004 584† CARBONAT BUILDUPS FROM UPPER PALEOZOIC OF IDAHO,
B 6004 669* CARBONAT PLATFORM, ABST., ABENAKI FORMATION, NOVA SCOTIA
B 6303 519* CARBONAT ROCKS--EXAMPLE FROM CAPITAN COMPLEX, ABST.,
B 6004 674* CARBONATE AND CLASTIC DEPOSITS, ABST., SEDIMENT, FACIES OF, SALT
B 6004 682* CARBONATE AND CLAY DIAGENESIS IN CAMBRIAN-ORDOVIDIAN COW
B 6303 462* CARBONATE AND EVAPORITE SEDIMENTATION PATTERNS IN LOWER
B 6303 489* CARBONATE AND EVAPORITE SOURCE ROCKS, ABST., ENVIRONMENTAL
B 6409 1562* CARBONATE AND EVAPORITE SOURCE ROCKS, ABST., ENVIRONMENTAL
B 6203 569* CARBONATE AND SILICICLASTIC DEPOSITIONAL PATTERNS AND
B 6209 1767* CARBONATE BANK DEPOSITION, COZUMEL ISLAND, QUINTANA ROO,
B 6303 546* CARBONATE BANK OF YUCATAN AND ITS PETROLEUM POTENTIAL,
B 6203 557* CARBONATE BANKS IN LOWER PROTEROZOIC BELCHER GROUP, HUDSON
S 4 329† CARBONATE BASIN AND SHELF MODEL, ST. CROIX, U.S VIRGIN ISLANDS,
S 4 261† CARBONATE BORERS--THEIR ROLE IN REEF ECOLOGY AND
B 6205 772 CARBONATE BUILDUP, DEFINITION
M 26 196 CARBONATE BUILDUP, SEISMIC EXAMPLE, CANADA, ALBERTA, GOLDEN
M 26 202 CARBONATE BUILDUP, SEISMIC EXAMPLE, CANADA, ALBERTA, JUDY
M 26 198 CARBONATE BUILDUP, SEISMIC EXAMPLE, CANADA, ALBERTA, YEKAN
B 6205 885* CARBONATE BUILDUP-EVAPORITE RELATIONS IN MIDDLE DEVONIAN OF
B 6404 527† CARBONATE BUILDUPS AND BIOSTROMES OF NORTH GREENLAND,
M 26 186 CARBONATE BUILDUPS, MODEL
M 26 185† CARBONATE BUILDUPS, SEISMIC, STRATIGRAPHY AND, GLOBAL
B 6205 772† CARBONATE BUILDUPS, SEISMIC STRATIGRAPHY AND, GLOBAL CHANGES
B 6203 536* CARBONATE CATCHMENT AND EXCHANGE BASINS IN WESTERN GULF OF
B 6303 419* CARBONATE CEMENTS, ABST., ISOTOPIC COMPOSITIONS OF BIOGENIC
B 6303 454* CARBONATE CEMENTS IN QUATERNARY ALLUVIAL-FAN DEPOSITS, BIRCH
B 6203 498* CARBONATE CEMENTS IN SANDSTONES—ACTIVE ANKERITE
B 6303 428* CARBONATE CHEMISTRY OF BRINES FROM ORCA BASIN AND DEAD SEA,
B 6004 647* CARBONATE COMPLEX--CO3-SO4 DIAGENESIS AND TIMING OF
B 6004 693* CARBONATE COMPLEX--STRATIGRAPHY AND ENVIRONMENTAL
B 6109 1554* CARBONATE COMPLEXES--SUBSURFACE PEARSALLFORMATION, LOWER
S 9 2 CARBONATE CONTENT, TABLE, MARINE MUD
B 6405 671* CARBONATE CYCLES IN CRETACEOUS ORGANI CARBON-RICH MARINE
B 6205 881* CARBONATE CYCLES IN GREEN RIVER FORMATION, WYOMING, ABST.
B 6011 2038† CARBONATE DEBRIS IN MESOZOIC FLYSCH DEPOSITS IN SANTA ANA
S 4 172 CARBONATE, DENSITY VARIATIONS IN, SKELETAL
B 6405 770* CARBONATE DEPOCENTERS AND FACIES DISTRIBUTION ON PASSIVE
B 6303 425* CARBONATE DEPOSITION IN WESTERN WYOMING AND SOUTHEASTERN
B 6304 609 CARBONATE DEPOSITION, MICHIGAN BASIN
B 6405 798* CARBONATE DEPOSITIONAL ENVIRONMENT, ABST., SEISMIC
B 6004 692* CARBONATE DEPOSITS AS POSSIBLE PALEOCLIMATIC INDICATORS, ABST.,
B 6303 472* CARBONATE DEPOSITS IN HANNET SHALE, LOWER CRETACEOUS, TEXAS,
B 6105 840* CARBONATE DEPOSITS IN NORTH-CENTRAL NEVADA, ABST., GOODWIN
B 6004 709* CARBONATE DEPOSITS ON MESOZOIC CONTINENTAL MARGIN, OTHRIS,
M 25 439 CARBONATE DEPOSITS, TAIWAN
B 6004 668* CARBONATE DEPOSITS, WESTERN EQUATORIAL PACIFIC, ABST., ABYSSAL,
B 6405 765* CARBONATE DIAGENESIS, ABST., DIFFERENTIAL DIFFUSION DURING
B 6004 694* CARBONATE DIAGENESIS AND FACIES DISTRIBUTIO OF ANACACHO
B 6408 1289* CARBONATE DIAGENESIS IN A QUARTZARENITE—DOLOMITIC KEEFER
B 6303 534* CARBONATE DIAGENESIS OF DELTAIC SANDSTONES--BASAL BELLY RIVER
B 6405 764* CARBONATE DIAGENESIS—A NEW MODEL, ABST., PHANEROZOIC
B 6404 461† CARBONATE DIAGENETIC TEXTURES FROM NEASURFACE DIAGENETIC
B 6004 731* CARBONATE DUNE ROCKS, ABST., INDICATORS OF, CLIMATE IN
B 6203 514* CARBONATE ENVIRONMENTS OF MIDDLE ORDOVICIAN INTERVAL,
B 6002 328* CARBONATE ENVIRONMENTS USING WELL CUTTINGS, ABST., STUDY OF
B 6109 1554* CARBONATE FACIES DISTRIBUTION ANDDIAGENESIS ASSOCIATED WITH
B 6303 416* CARBONATE FACIES, NORTHWESTERN GULF OF MEXICO, ABST., LOWER,
B 6405 712* CARBONATE FACIES OF PENAS ALTAS FORMATION, VENEZUELA—CASE
B 6204 604 CARBONATE FACIES, U.S., ATLANTIC SLOPE
B 6303 501* CARBONATE FACIES VARIATION ON GUADALUPIAN SHEL CREST, UPPER
B 6404 530 CARBONATE FORMATION, GREENLAND
M 29 242 CARBONATE FOSSILS, NORTH PHILIPPINE SEA, EOCENE
B 6303 533* CARBONATE GEOLOGY OF PENA BLANCA URANIUM DISTRICT,
B 6303 362† CARBONATE GRAINSTONES, UPPER PLEISTOCENE, YUCATAN PENINSULA,
B 6303 445* CARBONATE GRAVITY FLOW IN BLAKE-BAHAMA BASIN, ABST., MIOCENE
M 28 28 CARBONATE, IN CALICHE, FRAGMENTS
B 6203 506* CARBONATE LIQUEFIED-FLOW DEPOSIT, NORTHERN TUNISIA—LINK IN
B 6004 678* CARBONATE MASS BALANCE--CYCLING AND DEPOSITION ON SHELVES
B 6209 1767* CARBONATE MATERIAL AND EVIDENCE FOR VANISHED EVAPORITES IN
B 6004 723* CARBONATE MATERIALS, ABST., RADIOCARBON, DATING OF, ESTUARINE
B 6405 685* CARBONATE MINERAL REACTIONS DURING COMBUSTION RETORTING OF
B 6004 733* CARBONATE MOUNDS FROM SHELVES AND BASINS, ABST.
B 6405 764* CARBONATE MOUNDS IN PARADOX BASIN, UTAH, ABST., SEISMIC
B 6004 733* CARBONATE MOUNDS, LITHOHERMS, IN NORTHERN STRAITS OF
B 6405 743* CARBONATE MOUNDS OF FORT PAYNE FORMATION, LOWER
B 6303 555* CARBONATE MUD MOUNDS FROM LOWER ORDOVICIAN WAH WAH
B 6004 719* CARBONATE MUD TO CHALK, PLEISTOCENE, IN BAHAMIAN TROUGH. BY
S 4 272 CARBONATE NONREEF SURFACES, HARDGROUNDS AND OTHER
B 6004 700* CARBONATE OOZE--EMPIRICAL OBSERVATIONS AND EXPERIMENTAL
M 28 119 CARBONATE OVERGROWTHS, CEMENTS
B 6004 725' CARBONATE PETROLOGY OF BASIN-TO-REEF SHOALING SEQUENCE IN
B 6203 562* CARBONATE PETROLOGY WITH EXAMPLE FROM SILURIAN TONOLOWAY
B 6109 1552* CARBONATE PLATFORM, ABST., EXPLORATION HISTORY OF SOUTH,
B 6301 32† CARBONATE PLATFORM, EXPLORATION HISTORY OF SOUTH, TEXAS
B 6303 508* CARBONATE PLATFORM, GULF COAST, MEXICO, ABST., FACIES,
B 6004 731* CARBONATE PLATFORM, MEXICO, ABST., DIAGENETIC SEQUENCE OF
M 26 243 CARBONATE PLATFORM, MODEL, DEPOSITIONAL SYSTEMS, POST-RIFT
B 6002 258† CARBONATE PLATFORM OF ISRAEL, RUDISTID, FRINGING, REEFS OF,
B 6105 803 * CARBONATE POROSITY MODIFICATION BY HALITE IMPREGNATION AND
B 6004 694* CARBONATE POROSITY TO DEPOSITIONAL AND DIAGENETIC FACIES,
B 6303 416* CARBONATE POROSITY, ZELTEN FIELD LIBYA, ABST., SECONDARY
B 6410 1575† CARBONATE RAMP-TO-BASIN TRANSITIONS AND FORELAND BASIN
M 26 203 CARBONATE RECOGNITION, CRITERIA, CHART, SEISMIC, REFLECTIONS
M 26 203 CARBONATE RECOGNITION, EXAMPLES, SEISMIC, REFLECTIONS

M	26	187	CARBONATE RECOGNITION, GEOPHYSICAL CRITERIA, SEISMIC,
M	26	187	CARBONATE RECOGNITION, PROCEDURE, SEISMIC, REFLECTIONS
S	9	22	CARBONATE REEF BELT DISTRIBUTION, CAMBRIAN
S	9	22	CARBONATE REEF BELT DISTRIBUTION, CARBONIFEROUS
S	9	22	CARBONATE REEF BELT DISTRIBUTION, CRETACEOUS
S	9	22	CARBONATE REEF BELT DISTRIBUTION, DEVONIAN
S	9	22	CARBONATE REEF BELT DISTRIBUTION, HOLOCENE
S	9	22	CARBONATE REEF BELT DISTRIBUTION, JURASSIC
S	9	22	CARBONATE REEF BELT DISTRIBUTION, ORDOVICIAN
S	9	22	CARBONATE REEF BELT DISTRIBUTION, PERMIAN
S	9	22	CARBONATE REEF BELT DISTRIBUTION, SILURIAN
S	9	22	CARBONATE REEF BELT DISTRIBUTION, TERTIARY
S	9	22	CARBONATE REEF BELT DISTRIBUTION, TRIASSIC
S	9	22	CARBONATE REEF DISTRIBUTION, PALEOCLIMATE
B	6203	527*	CARBONATE REEFS FROM SNAKE RIVER PLAIN, IDAHO, ABST., TERTIARY,
B	6303	512*	CARBONATE REPLACEMENT OF SULFATE--NEW MECHANISM FOR
B	6405	777*	CARBONATE RESERVOIR, CABIN CREEK FIELD, MONTANA, ABST.,
B	6104	484	CARBONATE RESERVOIRS
B	6105	815*	CARBONATE RESERVOIRS, ABST., DIAGENETIC CONSTRAINTS ON USE OF
B	6303	474*	CARBONATE RESERVOIRS BY DICKITE, ABST., OCCLUSION OF POROSITY
B	6207	1230*	CARBONATE RESERVOIRS IN INDONESIA, ABST., TERTIARY
B	6002	325*	CARBONATE RESERVOIRS OF PERMIAN BASIN, ABST., FACTORS
B	6205	886*	CARBONATE RESERVOIRS OF SOUTHERN SASKATCHEWAN, ABST.,
B	6303	551*	CARBONATE RESERVOIRS, UNITED STATES GULF COAST, ABST., PORE
B	6303	429*	CARBONATE RESERVOIRS, ALBIAN-CENOMANIAN, CAMPOS BASIN,
M	27	1	**CARBONATE ROCK CONSTITUENTS, TEXTURES, CEMENTS,**
B	6004	665*	CARBONATE ROCKS, ABST., EXPERIMENTAL DEVELOPMENT OF,
B	6203	525*	CARBONATE ROCKS, ABST., OCCURRENCES OF, TERTIARY, POROSITY IN
B	6109	1541*	CARBONATE ROCKS, ABST., PLATE TECTONICS AND SEDIMENTARY,
B	6105	840*	CARBONATE ROCKS, ABST., POROSITY PRESERVATION IN,
B	6307	1059†	CARBONATE ROCKS AND CONTIGUOUS SHELF AND BASIN STRATA,
B	6004	662*	CARBONATE ROCKS, ARCTIC ARCHIPELAGO, ABST., EARLY, MARINE AND
B	6002	245†	CARBONATE ROCKS AS REVEALED BY PORE CASTS AND CAPILLARY
B	6309	1478†	CARBONATE ROCKS, EFFECTS OF, TEXTURE AND COMPOSITION ON,
B	6303	500*	CARBONATE ROCKS IN NORTHERN BAHAMAS, ABST., SEISMIC FACIES
B	6205	884*	CARBONATE ROCKS IN THIN SECTION, ABST., ESTIMATING, PORE AND,
B	6004	676*	CARBONATE ROCKS, MIDDLE ORDOVICIAN, VIRGINIA, ABST., FENESTRAL,
B	6109	1557*	CARBONATE ROCKS OF CENTRAL SOUTH TEXAS, ABST., LOWER,
B	6203	541*	CARBONATE ROCKS OF EASTERN NETHERLANDS FROM SEISMIC DATA,
B	6004	720*	CARBONATE ROCKS OF EVAPORITIC BASINS, ALTERNATIVE MODE OF
B	6309	1604*	CARBONATE ROCKS OF MAVERICK BASIN, ABST., FACIES, DIAGENESIS,
B	6303	469*	CARBONATE ROCKS OF MONTOYA GROUP, MIDDLE AN UPPER
B	6203	494*	CARBONATE ROCKS OF NORTHWESTERN GREAT BAHAMA BANK, ABST.,
B	6004	723*	CARBONATE ROCKS, PEDREGOSA BASIN, SOUTHEASTERN ARIZONA,
B	6205	776	CARBONATE ROCKS PLOTTED AGAINST GEOLOGIC AGE
B	6209	1763*	CARBONATE ROCKS, SOUTH FLORIDA BASIN, ABST., PRELIMINARY
B	6004	612†	CARBONATE ROCKS, SOUTHEAST OF PARIS, FRANCE, ANTIPATHETIC
B	6404	480	CARBONATE ROCKS, STAGES, DIAGENESIS IN
B	6405	715*	CARBONATE ROCKS, VIRGINIA, ABST., CEMENT TYPES AND
B	6004	712*	CARBONATE ROCKS, VIRGINIA--ANCIENT ANALOGUES OF QUATERNARY
B	6303	488*	CARBONATE ROCKS--CAUSES AND CURES, ABST., ERROR IN, X-RAY
B	6105	838*	CARBONATE SANDS, ABST., EFFECT OF SHAPE SORTING ON PREDICTION
B	6303	460*	CARBONATE SANDS, ABST., POROSITY PRESERVATION AND EARLY
B	**6105**	**793***	**CARBONATE-SAND WAVES, ISLA MUJERES, YUCATAN, MEXICO**
M	29	20	CARBONATE SECTION, GEORGIA EMBAYMENT
B	6410	1715	CARBONATE SEDIMENT DRIFTS, CRITERIA FOR RECOGNITION OF
B	6410	1701†	CARBONATE SEDIMENT DRIFTS IN NORTHER STRAITS OF FLORIDA
B	6303	442*	CARBONATE SEDIMENT IN HANSON CREEK FORMATION,
S	4	313†	CARBONATE SEDIMENTATION ON CONTINENTAL SHELF OFF SOUTHERN
B	6303	517*	CARBONATE SEDIMENTATION, SOUTHEASTERN ARIZONA, ABST.,
B	6203	555*	CARBONATE SEDIMENTOLOGY AND A RETURN TO THE ROCKS, ABST.
B	6105	827*	CARBONATE SEDIMENTOLOGY, DIAGENESIS, AND PETROLEUM
B	6105	826*	CARBONATE SEDIMENTS, ABST., MESOGENETIC, POROSITY ALTERATION
B	6303	445*	CARBONATE SEDIMENTS, ABST., PORE SPACE IN, HOLOCENE
S	4	299†	CARBONATE SEDIMENTS AND TERTIARY-PLEISTOCENE LIMESTONES ON
B	6303	528*	CARBONATE SEDIMENTS, BONAIRE, NETHERLANDS ANTILLES, ABST.,
B	6004	657*	CARBONATE SEDIMENTS, HARRINGTON SOUND, BERMUDA, ABST.,
B	6004	688*	CARBONATE SEDIMENTS IN DSDP CORES FROM PACIFIC OCEAN, ABST.,
M	26	191	CARBONATE, SEISMIC EXAMPLE, AFRICA, WEST, SHELF-MARGIN
B	6209	1609	CARBONATE SEQUENCE, APPALACHIAN BASIN, SILURIAN-, DEVONIAN
B	6004	686*	CARBONATE SEQUENCES, ABST., SIMILARITIES IN EARLY DIAGENETIC
B	6405	700*	CARBONATE SHELF, ABST., SANDS AND SAND TRANSPORT ON,
B	6004	697*	CARBONATE SHELF EDGE AND ITS CONTRIBUTION TO AUTOKINETI
B	6303	547*	CARBONATE SHELF IN SOUTHEASTERN ARIZONA AND NORTHEASTERN
B	6405	782*	CARBONATE SHELF IN UTAH AND NEVADA, ABST., PALEOGEOGRAPHY
B	6305	828*	CARBONATE SHELF MARGIN ALONG OVERTHRUST BELT FROM
M	26	435	CARBONATE SHELF MARGIN TRAPS, SEISMIC, MODELS
B	6004	724*	CARBONATE SHELF, SILURIAN MCKENZIE FORMATION, WEST VIRGINIA,
B	6104	492†	CARBONATE SHORELINE SEQUENCES, CENTRAL FLORIDA, DIAGENESIS OF
B	6203	549*	CARBONATE SLOPES, ABST.
B	6405	744*	CARBONATE SOURCE ROCKS, ABST., ENVIRONMENTAL AND DIAGENETIC
B	6203	558*	CARBONATE SUCCESSION IN NORTHWESTERN BRITISH COLUMBIA, ABST.,
B	6109	1552*	CARBONATE SUCCESSIONS OF PLEISTOCENE AND CRETACEOUS AGE,
S	7	219	CARBONATE TEXTURES, WEST FLORIDA SLOPE
S	7	217	CARBONATE TEXTURES, WEST FLORIDA SLOPE
B	6004	674*	CARBONATE TIDAL FLATS, ABST., SEDIMENTARY RECORD OF,
M	26	334	CARBONATE TRAPS
B	6408	1167	CARBONATE WEDGES, EXAMPLES
B	6408	1168	CARBONATE WEDGES, TEXAS, WEST, PALEOZOIC
B	6408	1168	CARBONATE WEDGES, UTAH AND, COLORADO, PALEOZOIC
B	6408	1168	CARBONATE WEDGES, UTAH AND, COLORADO, PALEOZOIC
B	6004	685*	CARBONATE WELL-CUTTING SAMPLES, ABST., INTERPRETATION OF
B	6111	1929†	CARBONATE-ANHYDRITE FACIES RELATIONS IN OTTO FIORD
S	5	145*	CARBONATE-ANHYDRITE FACIES RELATIONSHIPS, OTTO FIORD
B	6303	547*	CARBONATE-BANK GROWTH, ABST., ROLE OF PHYSICAL SEDIMENTATION
B	6103	376†	CARBONATE-BANK-MARGIN GROWTH AND STRUCTURE, LITTLE BAHAMA
B	6203	530*	CARBONATE-EVAPORITE RELATIONS IN MIDDLEDEVONIAN OF
B	6203	516*	CARBONATE-FACIES DISTRIBUTION, ABST., PREDICTIVE, MODELS FOR
B	6107	1069†	CARBONATE-GRAIN BINDING BY ENDOLITHIC ALGAE, MICRITIZATION
B	6105	819*	CARBONATE-MUDBANK BUILDUPS, PENNSYLVANIAN AND PERMIAN,
B	6303	431*	CARBONATE-ROCK FABRIC, ABST., DOLOMITE, NONSTOICHIOMETRY, ITS
B	6011	2068*	CARBONATE-ROCK SEQUENCES--SEDIMENTARY OR METAMORPHIC
S	4	281†	CARBONATE-SEDIMENT PRODUCTION BY PARROT FISH AN SEA URCHINS
B	6303	461*	CARBONATE-SHELF MARGINS AND DELTAIC SANDSTONNS PALO DURO
B	6303	461*	CARBONATE-SHELF MARGINS, PALO DURO BASINS, TEXAS PANHANDLE,
B	6105	823*	CARBONATE-SHELF MARGINS, WESTERN UNITED STATES, ABST.,
B	6206	1092*	CARBONATE-SLOPE DEPOSITS, NORTHEASTERN MEXICO, ABST., MIDDLE,
B	6405	787*	CARBONATE-SULFATE MINERAL REPLACEMENTS IN DIAGENESIS OF
B	6206	1058	CARBONATES
M	28	61	CARBONATES, BIBLIOGRAPHY
M	28	147	CARBONATES, BIBLIOGRAPHY, DIAGENESIS AND POROSITY
M	26	188	CARBONATES, DIRECT RECOGNITION, SEISMIC, REFLECTIONS
S	9	2	CARBONATES, DISTRIBUTION, MARINE
S	4	1	**CARBONATES--ECOLOGY AND SEDIMENTOLOGY**
M	26	188	CARBONATES, INDIRECT RECOGNITION, SEISMIC, REFLECTIONS
B	6004	672*	**CARBONATES, LEONARD MOUNTAIN AREA, GLASS MOUNTAIN, WEST**
B	6410	1634†	CARBONATES, NORTHWESTERN GREAT BAHAMA BANK, FACIES
M	26	203	CARBONATES, SEISMIC, REFLECTIONS, RECOGNITION, INTERVAL
S	3	135	CARBONATITES, DIAPIRIC EMPLACEMENT OF, HYDROTHERMAL
B	6204	672	CARBONIFEROUS, ALGERIA
B	6205	880*	CARBONIFEROUS AND PERMIAN, IN NAVAL PETROLEUM RESERVE
B	6207	1154	CARBONIFEROUS, AUSTRALIA
B	6201	57	CARBONIFEROUS, AUSTRALIA, PRECAMBRIAN TO
B	6212	2385	CARBONIFEROUS BACK-BARRIER DEPOSITS, KENTUCKY AND, WEST
B	6303	415*	CARBONIFEROUS BARRIER SEQUENCES OF EASTERN KENTUCKY, ABST.,
S	6	324	CARBONIFEROUS, BASHKIRIAN-MOSCOVIAN BOUNDARY
B	6004	705*	CARBONIFEROUS, BASIN, CANTABRIAN MOUNTAINS, NORTH SPAIN, ABST.,
S	6	329	CARBONIFEROUS, BIOSTRATIGRAPHY, AMMONOIDS
S	6	329	CARBONIFEROUS, BIOSTRATIGRAPHY, BRACHIOPODS
S	6	329	CARBONIFEROUS, BIOSTRATIGRAPHY, CONODONTS
S	6	330	CARBONIFEROUS, BIOSTRATIGRAPHY, PLANT MEGAFOSSILS
S	6	329	CARBONIFEROUS, BIOSTRATIGRAPHY, SMALLER FORAMINIFERA
S	6	311	CARBONIFEROUS BOUNDARY, PERMIAN, RADIOMETRIC
S	6	324	CARBONIFEROUS BOUNDARY, TABLE, RADIOMETRIC DATING
S	9	22	CARBONIFEROUS, CARBONATE REEF BELT DISTRIBUTION
S	6	331	CARBONIFEROUS, COMPARISON, NORTH AMERICA
S	6	333	CARBONIFEROUS, COMPARISON, USSR
B	6405	737*	CARBONIFEROUS CONODONTS, ABST.
M	27	63	CARBONIFEROUS, CORAL LIMESTONE, ENGLAND
B	6204	644	CARBONIFEROUS, CUMBERLAND PLATEAU, SOUTHERN
B	6004	680*	CARBONIFEROUS DELTAS, ABST., SANDSTONE, MODELS FROM,
S	6	330	CARBONIFEROUS DEPOSITION, STRUCTURAL CONTROL
B	6405	724*	CARBONIFEROUS DEPOSITIONAL SETTINGS—FACTORS AFFECTING COAL
S	6	324	CARBONIFEROUS, DEVONIAN-CARBONIFEROUS BOUNDARY
S	9	23	CARBONIFEROUS, FOLDOUT, PALEOCLIMATES
S	6	326	CARBONIFEROUS, GEOCHRONOLOGIC SCALE, SUMMARY
S	6	323	CARBONIFEROUS, ISOTOPIC DATING, A REVIEW
S	6	324	CARBONIFEROUS, ISOTOPIC DATING, DEVONIAN BOUNDARY
S	6	330	CARBONIFEROUS, LACK OF FORAMINIFERAL ZONE 19
B	6211	2192	CARBONIFEROUS LIMESTONE, IRELAND, WESTERN
B	6203	544*	CARBONIFEROUS LIMESTONES OF SOUTHERN KENTUCKY AND
B	6405	603	CARBONIFEROUS, MIDDLE EAST
S	6	327	**CARBONIFEROUS, MISSISSIPPIAN-PENNSYLVANIAN, A REVIEW**
B	6205	879*	CARBONIFEROUS, MISSISSIPPIAN, MICROFACIES AND STRATIGRAPHY,
S	6	324	CARBONIFEROUS, MISSISSIPPIAN-PENNSYLVANIAN BOUNDARY
B	6306	952	CARBONIFEROUS, NEWFOUNDLAND
B	6203	542*	CARBONIFEROUS OF MARITIME PROVINCES, CANADA, ABST.,
B	6004	695*	CARBONIFEROUS, OF NORTHERN ENGLAND--SLOPE INFLUENCED BY
B	6405	747*	CARBONIFEROUS OF NOVA SCOTIA, CANADA, ABST., TERRESTRIAL
B	6203	515*	CARBONIFEROUS OF SOUTHERN IRELAND, ABST., REGRESSIVE-, DELTA,
B	6211	2191†	CARBONIFEROUS OF WESTERN IRELAND, GROWTH, FAULTS IN
S	6	334	CARBONIFEROUS, RADIOMETRIC DATING, A REVIEW
S	6	325	CARBONIFEROUS, RADIOMETRIC DATING, CANTABRIAN STAGE
B	6405	749*	CARBONIFEROUS ROCKS OF CENTRAL UNITED STATES, ABST.

B 6006 962† CARBONIFEROUS SECTION AND UNCONFORMITY AT MOUNT DOONERAK,
B 6009 1620* CARBONIFEROUS SEDIMENTARY RESPONSES TO CONTEMPORANEOUS
B 6405 730* CARBONIFEROUS SEQUENCE EXPOSED NEAR JOGGINS, NOVA SCOTIA,
S 9 8 CARBONIFEROUS, SOUTH AFRICA, ICE FLOW DIRECTION
M 28 30 CARBONIFEROUS, SULZBACHER, GERMANY
B 6212 2512* CARBONIFEROUS SYSTEM IN ORIENTAL BASINS OF BOLIVIA, ABST.,
S 6 323† CARBONIFEROUS SYSTEM, REPORT ON, ISOTOPIC DATING OF ROCKS IN
B 6405 605 CARBONIFEROUS TO LIASSIC, MIDDLE EAST
S 6 328 CARBONIFEROUS, TYPE-MISSISSIPPIAN REGION
S 6 329 CARBONIFEROUS, TYPE-MORROWAN SERIES
S 6 328 CARBONIFEROUS, TYPE-PENNSYLVANIAN REGION
S 6 325 CARBONIFEROUS, WESTPHALIAN-STEPHANIAN BOUNDARY
S 6 326 CARBONIFEROUS-PERMIAN BOUNDARY
B 6112 2085 CARBONIZATION, TEMPERATURE VERSUS
B 6201 125 CARD SOUND DOLOMITE, FLORIDA
B 6103 465 CARDEN, JOHN, REGIONA TECTONIC IMPLICATIONS OF ALASKA
B 6203 430 CARDIGAN BAY BASIN, BRITAIN, WEST
B 6203 430 CARDIGAN BAY, RIFT SYSTEM
M 26 435 CARDIUM FORMATION, CANADA, ALBERTA
B 6405 714* CARDIUM FORMATION, FERRIER FIELD, ALBERTA, ABST., SIDERITE
B 6408 1166 CARDIUM SANDSTONE, ALBERTA
B 6107 1086 CARDIUM SANDSTONE, FOOTHILLS AND FORELAND, ALBERTA
B 6108 1375 CARDWELL, A. LYNN, PETROLEUM SOURCE-ROCK POTENTIAL OF
B 6303 532 CAREY, B. D., CARBON ISOTOPES--NEW TOOL IN HYDROCARBON
B 6303 532 CAREY, B. D., ORIGIN OF GASES ADSORBED IN NEAR-SURFACE SEDIMENT
B 6304 701 CAREY, B. D., ORIGIN OF GASES, ADSORBED IN NEAR-SURFACE
B 6303 431 CAREY, BYRL D., JR., REGIONAL HYDROCARBON SOURC ROCK AND
B 6211 2351 CAREY, SEAN M., LATE CRETACEOUS SEDIMENTATION IN SANTA MONICA
B 6203 551 CARGILL, SIMON M., APPLICATION OF GARNET SYSTEM FOR
B 6004 703 CARGILL, SIMON M., INTERACTIVE COMPUTER PROGRAM TO ANALYZE
B 6408 1192 CARIACO TRENCH, SILLED DEPRESSION IN OXIC OPEN OCEAN
B 6412 2125 CARIACO TRENCH, SITE 147, GEOCHEMISTRY
M 29 368 CARIB BEDS, VENEZUELA BASIN, VENEZUELA TRENCH
S 4 102 CARIBBEAN AREA, CENOZOIC FOSSIL, REEFS OF
B 6010 1640† CARIBBEAN AREA IN 1975, PETROLEUM, DEVELOPMENTS IN, SOUTH
B 6110 1578† CARIBBEAN AREA IN 1976, PETROLEUM, DEVELOPMENTS IN, SOUTH
B 6210 2044† CARIBBEAN AREA IN 1977, PETROLEUM, DEVELOPMENTS IN, SOUTH
B 6310 1745† CARIBBEAN AREA IN 1978, PETROLEUM, DEVELOPMENTS IN, SOUTH
S 4 75 CARIBBEAN BIOHERMS, CARBONATE FRAMEWORK
S 4 105 CARIBBEAN, CORAL BANKS
S 4 5 CARIBBEAN CORAL REEFS, LITERATURE
B 6303 538* CARIBBEAN CORES--PALEO-OCEANOGRAPHIC IMPLICATIONS, ABST., LATE,
S 4 73 CARIBBEAN, EASTERN AND NORTHEASTERN, WIND-SPEED INDEX
B 6007 1054† CARIBBEAN FRINGING REEF, GALETA POINT, PANAMA, EVOLUTION OF,
B 6007 1054† CARIBBEAN FRINGING REEF, GALETA POINT, PANAMA, EVOLUTION OF,
S 4 76 CARIBBEAN, MAJOR REEFS AND ATOLLS, INDO-, PACIFIC AND
S 4 106 CARIBBEAN, MIOCENE
S 4 106 CARIBBEAN, NEOGENE REEFS
S 4 105 CARIBBEAN, OLIGOCENE
S 4 103 CARIBBEAN, PALEOCENE-EARLY EOCENE
S 4 102 CARIBBEAN, PALEOGENE REEFS
B 6004 645* CARIBBEAN PALEOSOLS AND THEIR GEOTECTONIC AND PALEOCLIMATIC
B 6403 360 CARIBBEAN PLATE, BOUNDARY
B 6403 360 CARIBBEAN PLATE, BOUNDARY
S 4 107 CARIBBEAN, PLEISTOCENE-HOLOCENE REEFS
S 4 106 CARIBBEAN, PLIOCENE
S 4 171† CARIBBEAN REEF CORALS MONTASTREA ANNULARIS AND MONTASTREA
S 4 158 CARIBBEAN REEF CORALS OBSERVED DURING EMERGENCE, LIVING
S 4 155† CARIBBEAN REEFS AND EFFECT OF EXPOSURE ON CORAL
S 4 281† CARIBBEAN REEFS, CARBONATE-SEDIMENT PRODUCTION BY, PARROT,
B 6212 2509* CARIBBEAN TO SOUTH AND NORTH AMERICAN CORDILLERAS, ABST.,
S 4 93† CARIBBEAN--PROSPECTS FOR PALEOECOLOGIC SYNTHESIS, CENOZOIC,
B 6205 882* CARIBOU COUNTIES, IDAHO, ABST., GENERAL GEOLOGY OF NORTHERN,
B 6303 363 CARILLO PUERTO FORMATION, MEXICO
B 6410 1622 CARLILE SHALE, DENVER BASIN
B 6407 971 CARLILE SHALE, GREAT PLAINS
B 6407 976 CARLILE SHALE, NORTH DAKOTA
B 6406 942† CARLILE SHALE, UPPER CRETACEOUS, AT LAKE TRAVERSE, WESTERN
B 6404 490 CARLILE SHALE, WYOMING, POOL CREEK MEMBER
B 6404 490 CARLILE SHALE, WYOMING, TURNER SANDY MEMBER
B 6101 65† CARLIN--LAKE SAND AREA, SOUTH LOUISIANA--IMPLICATIONS,
B 6209 1610 CARLISLE CENTER SHALE, NEW YORK
S 11 196 CARLOTTA FORMATION
S 11 224 CARLOTTA FORMATION, ASSEMBLAGES, SCOTIA BLUFFS AND
B 6203 544 CARLSON, KURT H., NEW DEVELOPMENTS IN METHODS OF RESOURCE
B 6304 700 CARLSON, P. R., PETROLEUM POTENTIAL OF EASTERN GULF OF ALASKA
B 6207 1216 CARLSON, P. R., SEAFLOOR INSTABILITY ON PACIFIC CONTINENTAL
B 6004 702 CARLSON, P. R., THIN-BEDDED TURBIDITES IN MODERN SUBMARINE
B 6403 447 CARLSON, PAUL R., BENTHIC FORAMINIFERS FROM EASTERN GULF OF
B 6012 2176 CARLSON, PAUL R., DISPERSAL, DISTRIBUTION, AND THICKNESS OF
B 6203 545 CARLSON, PAUL R., HOLOCENE SEDIMENT VOLUME AND MODERN
B 6212 2412 CARLSON, PAUL R., HOLOCENE SLUMP ON CONTINENTAL SHELF OFF
B 6105 806 **CARLSON, PAUL R., HYDROCARBON GASES IN SEDIMENTS OF EASTERN**
B 6004 655 CARLSON, PAUL R., PRELIMINARY STUDY OF SEA-FLOOR INSTABILITY IN
B 6403 446 **CARLSON, PAUL R., QUATERNARY SEDIMENTARY FACIES ON**
B 6105 774 **CARLSON, PAUL R., SUBMARINE-SLUMP FEATURES SEAWARD OF ICY BAY**
B 6204 633 CARLSON, PAUL R., SURFACE SEDIMENTARY UNITS OF NORTHERN GULF
B 6303 428 CARLSON, TONALEE, COMPARISON OF CARBONATE CHEMISTRY OF
M 25 108 CARMACKS COAL AREA, CANADA
B 6303 428 CARMALT, S. W., GEOLOGIC DISTRIBUTION OF OIL ABST.
B 6206 968 CARMEL CANYON, CALIFORNIA
S 8 78 CARMEL CANYON, CALIFORNIA, SUBMARINE CANYONS
S 8 30 CARMEL CANYON, DEPTH AGREEMENT, CURRENT FLOW
S 8 83 CARMEL CANYON, TIME-VELOCITY CURVES, CURRENT FLOW
S 8 81 CARMEL CANYON, TOPOGRAPHY, INDEX MAP
B 6405 712* CARMEL FORMATION, SOUTHERN UTAH, ABST., FACIES AND
B 6002 266 CARMEL REEF, ISRAEL, SOUTHERN
B 6002 268 CARMEL REGION, ISRAEL
B 6303 325 CARMEL STRUCTURE, ISRAEL
B 6112 2105 CARMEN FIELD, PERU
B 6109 1476 CARMOPOLIS FIELD, BRAZIL
B 6109 1479 CARMOPOLIS MEMBER, BRAZIL
B 6201 41 CARNARVON BASIN, AUSTRALIA
B 6411 1865 CARNARVON BASIN, AUSTRALIA
B 6210 2152 CARNARVON BASIN, AUSTRALIA
S 3 126 CARNATION MINE, COLORADO
B 6212 2459 CARNEGIE RIDGE
B 6004 655 CARNEIRO, R. G., SYNSEDIMENTARY TECTONIC CONTROL OF PALEOZOIC
B 6405 606 CARNIAN, MIDDLE EAST, LADINIAN TO
B 6101 88 CARNIAN PREDRIFT FRAMEWORK
M 25 453 CARNOTITE, URANIUM ORE MINERAL, JAPAN
B 6403 340 CAROLINA PLATFORM
B 6403 340 CAROLINA PLATFORM
M 29 102 CAROLINA PLATFORM, ATLANTIC MARGIN
M 29 78 CAROLINA PLATFORM, CROSS SECTION, ATLANTIC COAST
M 29 108 CAROLINA PLATFORM, DEPTH-TO-BASEMENT CONTOURS
M 29 106 CAROLINA PLATFORM, MAGNETIC ANOMALY CONTOURS
M 29 107 CAROLINA PLATFORM, MAGNETIC DEPTH ESTIMATES
M 29 102 CAROLINA TROUGH, ATLANTIC MARGIN
B 6312 2194 CAROLINEFJELLET FORMATION, SVALBARD
B 6004 665 CAROTHERS, MARSHAL C., RELATIVE MARINE TRANSGRESSIVE AND
B 6212 2441 CAROTHERS, WILLIAM W., ALIPHATIC ACID ANIONS IN OIL-FIELD
B 6303 478 CAROTHERS, WILLIAM W., GEOCHEMISTRY OF OIL-FIELD WATERS FROM
B 6303 428 CAROTHERS, WILLIAM W., ORGANIC ACID ANIONS IN OIL-FIELD WATERS
B 6303 479 CAROTHERS, WILLIAM W., STABLE CARBON ISOTOPES IN OIL-FIELD
B 6105 774 **CAROZZI, A. V., CANDEIAS FIELD, RECONCAVO BASIN, BRAZIL—NEW**
B 6203 503 CAROZZI, A. V., PETROLEUM GEOLOGY OF PALEOZOIC CLASTIC ROCKS OF
B 6004 655 CAROZZI, A. V., SYNSEDIMENTARY TECTONIC CONTROL OF PALEOZOIC
B 6303 429 CAROZZI, ALBERT V., DEPOSITIONAL-DIAGENETIC HISTORY OF MACAE
B 6004 665 CAROZZI, ALBERT V., EXPERIMENTAL DEVELOPMENT OF
B 6105 781 **CAROZZI, ALBERT V., FABRIC-SELECTIVE PERMOPOROSITY**
B 6405 687 CAROZZI, ALBERT V., MICROFACIES AND DEPOSITIONAL-DIAGENETIC
B 6304 682 CAROZZI, RIES, EDWARD R., REVIEW OF, TECTONICS OF, ASIA, BY EMILE,
B 6301 69 CARPATHIAN OROGENE, EUROPE, ALPINE-
B 6307 1052 CARPATHIAN THRUST-FOLD BELT, PLOESTI AREA, SALT FEATURES,
B 6405 683* CARPATHIANS AND ITS RELATION TO PANNONIAN AND RELATED
S 3 99 CARPATHIANS, COMPRESSIONAL FOLDING
B 6301 82 CARPATHIANS, KROSNO FORMATION
B 6303 429 **CARPENTER, A. B., ROLE OF BRINES AND FERROUS IRON IN FORMATION**
B 6105 775 **CARPENTER, ALDEN B., PETROLEUM AND MISSISSIPPI VALLEY TYPE ORE**
B 6105 769 **CARPENTER, GERALD L., INFORMATION RETRIVAL SYSTEM FOR WELL DATA**
B 6409 1462 CARPENTER, G. L., DEVELOPMENT IN EAST-CENTRAL STATES IN 1979
B 6008 1222 CARPENTER, G. L., DEVELOPMENTS IN EAST-CENTRAL STATES IN 1975
B 6108 1183 CARPENTER, G. L., DEVELOPMENTS IN EAST-CENTRAL STATES IN 1976
B 6208 1340 CARPENTER, G. L., DEVELOPMENTS IN EAST-CENTRAL STATES IN 1977
B 6308 1324 CARPENTER, G. L., DEVELOPMENTS IN EAST-CENTRAL STATES IN 1978
M 25 402 CARR BOYD AREA, NICKEL SULFIDE, WESTERN AUSTRALIA, SCOTIA-
M 25 401 CARR BOYD NICKEL DEPOSIT, KALGOORLIE
B 6405 672 CARR, DONALD D., NEW EXPLORATION AND EVALUATION OF COAL
B 6303 429 CARR, MICHAEL D., ALLOCHTHONOUS NATURE OF NUMIDIAN FLYSCH
M 27 201 CARRARA MARBLE, ITALY, JURASSIC
B 6303 430 CARRASCO, V. V. BALDOMERO,, LOWER CRETACEOUS VOLCANIC ROCKS
B 6303 525* CARRIBEAN, ABST., SIGNIFICANCE OF MIDDLE, TERTIAR LARGE,
B 6309 1606* CARRILLO PUERTO FORMATION OF NORTHEASTERN QUINTANA ROO,
B 6304 689* CARRIZO MOUNTAINS, ARIZONA AND NEW MMXICO, ABST.,
B 6211 2224 CARRIZO PLAIN, CALIFORNIA
B 6402 168 CARRIZO SANDSTONE, TEXAS
B 6402 168 CARRIZO SANDSTONE, TEXAS
B 6105 823 CARROLL, S., GRAY-LEVEL MAPPING OF GEOLOGIC DATA, ABST.

Index

B 6303 430 CARROLL, W. KIPP, DOLOMITIZATION IN UPPER RE RIVER FORMATION,
B 6211 2209 CARROWMORE POINT GROWTH FAULTS, IRELAND, WESTERN
M 25 433 CARRUTHERS, D. S., PORPHYRY COPPER DEPOSITS OF SOUTHWEST
B 6203 428 CARSON BASIN, GRAND BASIN
B 6303 430 CARSTENS, HALFDAN,, GEOTHERMAL INVESTIGATION OF NORTHERN
B 6402 234 CARTA VALLEY FAULT ZONE, TEXAS
B 6402 234 CARTA VALLEY FAULT ZONE, TEXAS
B 6402 222 CARTA VALLEY QUADRANGLE, TEXAS
B 6402 222 CARTA VALLEY QUADRANGLE, TEXAS
B 6110 1583 CARTAGENA OFFSHORE, COLOMBIA
S 10 195 CARTAGENA, GAS BUBBLE IN MUD VOLCANO,COLOMBIA
S 10 196 CARTAGENA, TOTUMO MUD VOLCANO,COLOMBIA
B 6302 249 CARTER, D. J., GEOSYNCLINES AND PLATE TECTONICS IN BANDA ARCS,
B 6004 656 CARTER, M. DEVEREUX, AVAILABILITY AND APPLICATION OF NATIONAL
B 6109 1559 CARTER, W. D., EVALUATION OF LANDSAT DATA IN EXPLORING
M 25 63 CARTER, W. D., SURVEYING EARTH
B 6103 361 CARTER-KNOX STRUCTURE, OKLAHOMA
S 6 348 CARTERS LIMESTONE, CORRELATION, ALABAMA
S 6 348 CARTERS LIMESTONE, CORRELATION, TENNESSEE
B 6407 1005 CARTHAGE FIELD, ARK-LA-TEX AREA
B 6407 1011 CARTHAGE FIELD, TEXAS
B 6104 595 CARTHAGE POINT FIELD, MISSISSIPPI
B 6104 594 CARTHAGE POINT FIELD, MISSISSIPPI
B 6009 1520 CARTMILL, JOHN C., OBSCURE NATURE OF PETROLEUM MIGRATION AND
B 6212 2379 CARUCCIO, F. T., DEPOSITIONAL MODELS IN COAL EXPLORATION AND
B 6110 1584 CARVAJAL, J., ON DEVELOPMENTS IN DOMINICAN REPUBLIC, 1976
B 6110 1590 CARVAJAL, J., ON DEVELOPMENTS IN EL SALVADOR, 1976
B 6110 1587 CARVAJAL, J., ON DEVELOPMENTS IN HAITI, 1976
B 6310 1891 CARVAJAL, J., ON DEVELOPMENTS IN JORDAN, 1978
B 6110 1802 CARVAJAL, J., ON DEVELOPMENTS IN LEBANON, 1976
B 6210 2119 CARVAJAL, J., ON DEVELOPMENTS IN LEBANON, 1977
B 6110 1590 CARVAJAL, J., ON DEVELOPMENTS IN PANAMA, 1976
B 6110 1802 CARVAJAL, J., PETROCONSULTANTS S.A., AND NATIONAL RES.
B 6307 1096 CARYNGINIA FORMATION, AUSTRALIA
B 6405 687 CASBY, SUSAN, GEOLOGIC AND GEOPHYSICAL INVESTIGATION OF PART
B 6203 345 CASCADE RANGE, CALIFORNIA
B 6105 784 * CASE HISTORY OF DISCOVERY OF NEW URANIUM DEPOSIT, AMBROSIA LAKE
M 26 451† CASE HISTORY OF GEOSEISMIC MODELING OF BASAL MORROW-SPRINGER
B 6408 1281 CASEROTTI, PHILLIP M., GEOLOGY AND GEOPHYSICS OF
B 6305 824 CASEY, J. MICHAEL, BASIN EVOLUTION OF LATE PALEOZOIC TAOS
B 6405 687 CASEY, J. MICHAEL, PALEOGEOGRAPHIC EVOLUTION OF LATE PALEOZOIC
B 6305 834 CASEY, J. MICHAEL, PENNSYLVANIAN COARSE-GRAINED MEANDERING
B 6405 688 CASEY, J. MICHAEL, SPECTRUM OF DELTAIC SYSTEMS IN
B 6209 1759 CASEY, RICHARD, DISTRIBUTION OF LIVING BENTHONIC FORAMINIFERA
B 6303 514 CASEY, RICHARD E., NEOGENE RADIOLARIAN BIOSTRATIGRAPHY AND
B 6303 531 CASEY, RICHARD E., RADIOLARIA--PRESENT AND PAST INDICATORS OF
B 6405 688 CASEY, RICHARD E., RADIOLARIAN CHARACTERISTICS OF MAJOR
B 6303 483 CASEY, RICHARD E., RADIOLARIAN PRESERVATIO IN PRESENT AND PAST
B 6309 1597 CASEY, RICHARD, ECOLOGIC NICHES OF RADIOLARIANS, PLANKTONIC
B 6209 1760 CASEY, RICHARD, LIVING SHELLED MICROZOOPLANKTON,
B 6409 1554 CASEY, RICHARD, PRELIMINARY REPORT ON MICROPLANKTON AND
B 6309 1598 CASEY, RICHARD, RADIOLARIAN DISTRIBUTION, DIVERSITY, AND
B 6309 1597 CASEY, RICHARD, RELICT AND EXPATRIATE RADIOLARIAN FAUNA IN
B 6012 2177 CASEY, TOM ANN L., SEDIMENTARY SERPENTINITE OF THE MIOCENE BIG
B 6303 421 CASHMAN, S. M., VITRINITE REFLECTANCE, PRESENT ROCK
B 6207 1236* CASING PROGRAMS IN INDONESIA, ABST., GEOTHERMAL, WELL
B 6203 355 CASMALIA OIL FIELD, CALIFORNIA
B 6202 184 CASPER ARCH, WYOMING
S 10 200 CASPIAN AREA, GARNIARIK TEPE MUD VOLCANO,IRAN
S 10 200 CASPIAN AREA, NAFTLI DJEH MUD VOLCANO,IRAN
B 6407 1011 CASPIANA FIELD, ARK-LA-TEX AREA
B 6109 1478 CASSARONGONGO FIELD, BRAZIL
B 6208 1322 CASSELL, J. K., DEVELOPMENTS IN WEST COAST AREA IN 1977
B 6303 430 CASSELL, J. K., MAKING PUBLIC LANDS PRIVATE, ABST.
B 6211 2364 CASSELL, J. K., SYNTHESIS OF OCS WELL INFORMATION, OFFSHORE
S 11 154 CASSIDULINA, CALIFORNIA, MIOCENE
S 11 290 CASSIDULINA CALIFORNICA
S 11 259 CASSIDULINA CALIFORNICA
S 11 290 CASSIDULINA CARINATA
S 11 259 CASSIDULINA CARINATA
S 11 290 CASSIDULINA CRASSA
S 11 259 CASSIDULINA CRASSA
S 11 259 CASSIDULINA DELICATA
S 11 290 CASSIDULINA DELICATA
S 11 154 CASSIDULINA DELICATA
S 11 259 CASSIDULINA LIMBATA
S 11 292 CASSIDULINA LIMBATA
S 11 260 CASSIDULINA MINUTA
S 11 290 CASSIDULINA MINUTA
S 11 154 CASSIDULINA MODELOENSIS
S 11 154 CASSIDULINA MONICANA
S 11 154 CASSIDULINA PANZANA
S 11 292 CASSIDULINA TORTUOSA
S 11 260 CASSIDULINA TORTUOSA
S 11 260 CASSIDULINA TRANSLUCENS
S 11 292 CASSIDULINA TRANSLUCENS
S 11 257 CASSIDULINIDAE
S 11 261 CASSIDULINOIDES BRADYI
S 11 292 CASSIDULINOIDES BRADYI
S 11 261 CASSIDULINOIDES CORNUTA
S 11 292 CASSIDULINOIDES CORNUTA
M 25 460 CASSITERITE, EASTERN ASIATIC VOLCANOGENIC BELT, MERCURY,
M 25 470 CASSITERITE, USSR, NORTHEAST
M 25 467 CASSITERITE, USSR, NORTHEAST, GOLD AND
B 6303 420* CASTILE FORMATION, DELAWARE BASIN, TEXAS AND NEW MEXICO,
S 5 32 CASTILE FORMATION, NEW MEXICO,
M 27 133 CASTILE FORMATION, NEW MEXICO, PERMIAN
M 28 84 CASTILE FORMATION, NEW MEXICO, PERMIAN
S 5 32 CASTILE FORMATION, TEXAS,
B 6405 775* CASTILLO GAS FIELD, SPAIN, ABST.
B 6203 504 CASTLE, JAMES W., SEDIMENTOLOGY OF CAPLES TERRANE, SOUTH
M 25 293 CASTLE MOUNTAIN FAULT, ALASKA
B 6408 1260 CASTLE MOUNTAIN FAULT, ALASKA
B 6203 397 CASTLE MOUNTAIN FAULT, ALASKA
B 6405 687 CASTRO, JOEL C., MICROFACIES AND DEPOSITIONAL-DIAGENETIC MODEL
B 6002 245† CASTS AND CAPILLARY PRESSURE, PORE, GEOMETRY OF, CARBONATE
B 6203 355 CAT CANYON GROUP, CALIFORNIA
B 6407 973 CAT CREEK ANTICLINE, MONTANA
B 6204 694 CAT ISLAND, BAHAMAS
B 6405 710* CATACLASTIC FLOW, ABST., BASEMENT TECTONICS OF, CHINA—
B 6002 288 CATAGENESIS
B 6101 101 CATAGENESIS, DEFINITION
B 6112 2098 CATAGENESIS, LOSS OF, CARBON DURING
B 6112 2082 CATAGENESIS OF KEROGEN TYPES, NATURAL
B 6303 493* CATAHOULA FORMATION, CENTRAL TEXAS COASTAL PLAIN, ABST.,
B 6309 1607* CATAHOULA FORMATION IN PARTS OF LAVACA, GONZALES, AND
B 6108 1379* CATAHOULA FORMATION, TEXAS COASTAL PLAIN, ABST., RELATIONS
B 6405 738* CATAHOULA FORMATION—IMPLICATIONS ON SOURCE OF SILICA
S 3 98 CATAHOULA GROUP, MEXICO
B 6109 1546* CATAHOULA SANDSTONES IN EAST TEXAS, ABST., PETROLOGY OF
M 28 116 CATAHOULA TUFF, TEXAS, OLIGOCENE
M 28 105 CATAHOULA TUFF, TEXAS, OLIGOCENE
M 28 107 CATAHOULA TUFF, TEXAS, OLIGOCENE
B 6206 1050 CATAHOULA TUFFACEOUS ROCKS, URANIUM SOURCE, TEXAS
S 5 169 CATALANO, RAIMONDO AN EVAPORITIC LITHOFACIES CONTINUUM:
B 6111 1981 CATALINA BASIN, CALIFORNIA
B 6202 248 CATALINA BASIN, CALIFORNIA
B 6203 363 CATALINA BASIN, CALIFORNIA
B 6203 362 CATALINA SCHIST, CALIFORNIA
B 6111 1984 CATALINA SCHIST, CALIFORNIA
B 6311 2054 CATATUMBO RIVER DELTA, CHARACTERISTICS, TABLE, VENEZUELA
B 6311 2042† CATATUMBO RIVER, LAKE MARACAIBO, VENEZUELA, STRATIGRAPHY OF,
B 6411 2026 CATE, P. D., DEVELOPMENTS IN SOUTHEASTERN STATES
B 6008 1277 CATE, P. D., DEVELOPMENTS IN SOUTHEASTERN STATES IN 1975
B 6108 1259 CATE, P. D., DEVELOPMENTS IN SOUTHEASTERN STATES IN 1976
B 6208 1390 CATE, P. D., DEVELOPMENTS IN SOUTHEASTERN STATES IN 1977
B 6308 1399 CATE, P. D., DEVELOPMENTS IN SOUTHEASTERN STATES IN 1978
B 6405 688 CATHER, STEVEN M., GENETIC STRATIGRAPHY AND PROVENANCE OF
B 6203 553* CATHODO-LUMINESCENCE OF DOLOMITE BY IRON AND MANGANESE,
M 28 114 CATHODOLUMINESCENCE, CEMENTS, QUARTZ OVERGROWTHS
M 28 186 CATHODOLUMINESCENCE, EXPLANATION, TECHNIQUES
B 6405 792* CATSKILL FORMATION, BASED ON ANALOGY WITH MARGARITIFERA
B 6405 765* CATSKILL FORMATION IN EASTERN PENNSYLVANIA, ABST., DIAGENETIC
B 6307 1097 CATTAMARRA COAL MEASURES MEMBER, AUSTRALIA
S 5 175 CATTOLICA BASIN, ITALY, SICILY,
B 6009 1616 CAUGHEY, CHARLES A., FORM AND COMPOSITION OF MISSISSIPPI FAN,
M 26 249 CAUGHEY, CHARLES A., SEISMIC FACIES AND SEDIMENTOLOGY OF
B 6004 703* CAUSALITY AND TIMING IN NORTH AMERICAN
M 26 92 CAUSES, SEA LEVEL FLUCTUATIONS, GLOBAL CYCLES
B 6405 689 CAVAROC, VICTOR V., JR., VARIATIONS I CRETACEOUS COAL-BEARING
B 6203 531* CAVITY-DWELLING ORGANISMS, COELOBITES, IN LOWER CAMBRIAN
M 25 432 CAVITY-FILLING TYPE, THAILAND, FLUORITE OF
B 6408 1252 CAWTHORNE CHANNEL ONSHORE, NIGER DELTA
B 6204 694 CAY SAL WELL, BAHAMA PLATFORM
B 6303 431 CAYCE, P. W., REGIONAL HYDROCARBON SOURC ROCK AND THERMAL
S 4 121 CAYO DE LOS MUERTOS, VENEZUELA
S 4 123 CAYO PERAZA, VENEZUELA
S 4 122 CAYO SAL, VENEZUELA
S 5 130 CAYUGAN GROUP, SALINA FORMATION, TONOLOWAY LIMESTONE,
S 5 127 CAYUGAN GROUP, SALINA FORMATION, TONOLOWAY LIMESTONE, WEST

S 5 4 CAYUGAN SERIES, SALINA GROUP, MICHIGAN BASIN,
S 5 54 CAYUGAN SERIES, SALINA GROUP, STRATIGRAPHY, MICHIGAN,
B 6309 1590* CAYUGAN STRATIGRAPHY AND DEPOSITIONAL ENVIRONMENTS OF
S 5 95 **CAYUGAN SERIES, CORRELATION, SILURIAN,**
B 6401 47 CD LIMESTONE, KALIMANTAN
B 6401 47 CD LIMESTONE, KALIMANTAN
B 6007 1114 CDP MULTICHANNE REFLECTION PROFILE, ATLANTIC, OUTER,
B 6306 936 CEARA ABYSSAL PLAIN, BATHYMETRY
B 6302 219 CEARA ABYSSAL PLAIN, BRAZIL
B 6202 290 CEARA ABYSSAL PLAIN OFF, BRAZIL, NORTHERN
B 6111 1994 CEARA ABYSSAL PLAIN, POSSIBLE DIAPIRS
B 6109 1471 CEARA BASIN, BRAZIL
B 6302 236 CEARA GUYOT, BRAZIL
B 6302 231 CEARA PLATEAU, BRAZIL
B 6106 930 CEARA RISE
B 6306 936 CEARA RISE, BATHYMETRY
B 6302 219 CEARA RISE, BRAZIL
B 6202 285 CEARA RISE OFF, BRAZIL, NORTHERN
M 25 436 CEBU, ATLAS MINE
M 25 436 CEBU, TOLEDO COPPER OREBODIES
B 6105 775 **CECIL, BLAINE C., GEOLOGIC FACTORS CONTROLLING COALIFICATION**
B 6405 788 CECIL, C. B., FORMS AND ASSOCIATIONS OF PYRITE IN UPPER FREEPORT
B 6405 689 CECIL, C. B., GEOLOGIC CONTROLS ON MINERAL-MATTER CONTENT OF
B 6405 689 CECIL, C. B., GEOLOGIC CONTROLS ON SULFUR CONTENT IN COAL, ABST.
B 6405 702 CECIL, C. B., X-RAY MINERALOGY OF UPPER FREEPORT COAL, ABST.
B 6204 633 CECIL GREEN RESEARCH CRUISE, MV
B 6405 793 CECILE, M. P., EARLY PALEOZOIC CONODONT BIOSTRATIGRAPHY AND
B 6011 2067* CEDA CREEK, ABST., GEOLOGIC HISTORY--KEY TO ACCUMULATION AT
B 6407 973 CEDAR CREEK ANTICLINE, MONTANA
B 6008 1394* CEDAR CREEK ANTICLINE, PAST AND PRESENT, ABST.
B 6212 2408 CEDAR GROVE COAL, WEST VIRGINIA
B 6201 125 CEDAR KEYS FORMATION, FLORIDA
B 6205 886* CEDAR MOUNTAIN AND DAKOTA FORMATIONS, EARLY CRETACEOUS,
B 6104 602 CEDAR MOUNTAINS, NEW MEXICO
M 28 114 CEDAR VALLEY LIMESTONE, ILLINOIS, DEVONIAN
B 6106 904 CEDROS DEEP, BAJA CALIFORNIA
B 6203 370 CEDROS DEEP, BAJA CALIFORNIA
M 29 160 CEDUNA PLATEAU, SECTION, AUSTRALIA, MARGIN
M 28 84 CELESTITE, IN A CARBONATE MATRIX
M 27 135 CELESTITE, REPLACEMENT
B 6203 567* CELLS IN FORMATION OF PALEOSURFACE-RELATED VEIN-TYPE URANIUM
B 6410 1688 CELLULAR FILAMENTS AND SHEATHS, BITTER SPRINGS FORMATION
B 6109 1470 CELSO PONTE, F., PETROLEUM GEOLOGY OF EASTERN BRAZILIAN
B 6004 656 CELSO-PONTE, F., PETROLEUM GEOLOGY OF EASTERN BRAZILIAN
B 6203 426 CELTIC SEA BASIN, EUROPE, NORTH
B 6203 426 CELTIC SEA BASIN, EUROPE, SOUTH
B 6107 988 CELTIC SEA, CHALK UNIT
B 6411 1963 CELTIC SEA, EIRE
B 6305 844 CEMEN, IBRAHIM, STRUCTURE OF SAN CAYETANO AND OAK RIDGE
B 6104 511† CEMENT IN OOID RESERVOIRS IN DOGGER, MIDDLE JURASSIC, OF FRANCE,
B 6405 735* CEMENT, ABST., EARLY POROSITY AND PERMEABILITY REDUCTION IN,
B 6103 361 CEMENT ANTICLINE, OKLAHOMA
B 6203 573* CEMENT EVOLUTION, ABST., SYNDEPOSITIONAL, SUBMARINE,
B 6304 676† **CEMENT IN OOID RESERVOIRS IN DOGGER, MIDDLE JURASSIC, OR FRANCE,**
B 6302 246 CEMENT OIL FIELD, AEROMAGNETIC SURVEY, OKLAHOMA
B 6302 247 CEMENT OIL FIELD, MAGNETITE, OKLAHOMA
B 6203 500* CEMENT PATTERNS IN STRAND-PLAIN LIMESTONES OF EASTERN
B 6405 715* CEMENT TYPES AND CEMENTATION PATTERNS O MIDDLE ORDOVICIAN
B 6103 360 CEMENT URANIUM DEPOSIT, OKLAHOMA
B 6205 884* CEMENT VOLUMES OF CARBONATE ROCKS IN THIN SECTION, ABST.,
B 6203 497* CEMENTATION, ABST., CLAY DIAGENESIS IN, WILCOX SANDSTONES OF
B 6203 554* CEMENTATION, ABST., INTERSTITIAL, WATER CHEMISTRY OF, JAMAICAN,
B 6102 174 CEMENTATION AND COMPACTION ON PERMEABILITY, EFFECT OF
B 6004 662* CEMENTATION AND DIAGENESIS OF PALEOZOIC CARBONATE ROCKS,
B 6409 1561* CEMENTATION AND MINERAL STABILITY OF CHIPOLA FORMATION,
B 6405 734* CEMENTATION BY HIGH-MAGNESIUM CALCIT FROM GULF COAST OF
B 6405 754* CEMENTATION, DIAGENESIS AND PARAGENETIC SEQUENCE IN
B 6109 1550* CEMENTATION, DISSOLUTION, AND POROSITY DEVELOPMENT BETWEEN
M 27 161 CEMENTATION, FRESHWATER
B 6303 445* CEMENTATION FROM MARINE TO DEEP SUBSURFACE ENVIRONMENTS,
B 6004 730* CEMENTATION, GOLDEN SPIKE OIL FIELD, ALBERTA, ABST., POROSITY,
B 6405 708* CEMENTATION IN UPPER JURASSIC SANDSTONES, LOWER COOK INLET,
B 6203 498* CEMENTATION IN WILCOX OF SOUTHWEST TEXAS, ABST., ORIGIN OF
B 6304 592 CEMENTATION MAPS, GEOCHEMICAL TOOL
B 6203 573* CEMENTATION OF CAPITAN LIMESTONE, PERMIAN, GUADALUPE
B 6203 492* CEMENTATION OF CAPITAN LIMESTONE, PERMIAN, GUADALUPE
M 27 142 CEMENTATION OF CARBONATE GRAINS, CHERT
B 6405 719* CEMENTATION OF HOLOCENE AND JURASSIC GRAINSTONES, ABST.,
M 27 160 CEMENTATION OF LARGE VOIDS
B 6405 788* CEMENTATION OF LIME-MUD AND PELLET MU BENEATH TIDAL FLATS OF
B 6405 766* CEMENTATION ON PHYSICAL PROPERTIE OF SANDSTONES, ABST., EFFECT
B 6107 990 CEMENTATION TERMED HARDGROUNDS, SYNSEDIMENTARY SEAFLOOR
M 27 166 CEMENTATIONS, MULTIPLE-GENERATION
B 6004 733* CEMENTED CARBONATE MOUNDS, LITHOHERMS, IN NORTHERN STRAITS
B 6105 841* CEMENTED ROCKS FROM DEEP-FLANK ENVIRONMENT OF LITTLE
B 6405 738* CEMENTED SANDS IN CATAHOULA FORMATION—IMPLICATIONS ON
B 6303 419* CEMENTS, ABST., ISOTOPIC COMPOSITIONS OF BIOGENIC, CARBONATE
S 4 372 CEMENTS AND DECOMPOSITION OF ORGANIC TISSUE IN SUBMERGENT
B 6203 541* CEMENTS AND FABRIC-RETENTIVE DIAGENESIS, CAPITAN REEF,
M 27 1 **CEMENTS, AND POROSITIES, A COLOR ILLUSTRATED**
M 28 1 **CEMENTS, AND POROSITIES OF SANDSTONES AND ASSOCIATED**
B 6404 462 CEMENTS AND POROSITY, TYPES AND TEXTURES OF
M 28 129 CEMENTS, ANHYDRITE, IN SANDSTONE
M 28 121 CEMENTS, ARAGONITE, IN SANDSTONE
M 28 128 CEMENTS, BARITE, IN SANDSTONE
M 28 121 CEMENTS, CALCITE (PISOLITIC), IN CALCRETE
M 28 120 CEMENTS, CALCITE (POIKILOTOPIC), IN SANDSTONE
M 28 120 CEMENTS, CALCITE, AFTER QUARTZ OVERGROWTHS
M 28 138 CEMENTS, CALCITE, IN SANDSTONE
M 28 119 CEMENTS, CALCITE, IN SANDSTONE
M 28 119 CEMENTS, CARBONATE OVERGROWTHS
M 28 118 CEMENTS, CHALCEDONY, IN OOLITIC CARBONATES
M 28 135 CEMENTS, CHLORITE (AUTHIGENIC), IN SANDSTONE
M 28 134 CEMENTS, CHLORITE, IN SANDSTONE
M 28 131 CEMENTS, CLAY COATINGS
M 28 133 CEMENTS, CLAY, OCCLUSION OF POROSITY
M 28 130 CEMENTS, CLAYS, IN SANDSTONES
M 28 144 CEMENTS, COMPLEX, CHLORITE AND KAOLINITE
M 28 143 CEMENTS, COMPLEX, QUARTZ WITH BITUMEN
M 28 144 CEMENTS, COMPLEX, QUARTZ WITH CALCITE
M 28 121 CEMENTS, DOLOMITE, IN HEMATITIC SANDSTONE
M 28 122 CEMENTS, DOLOMITE, IN SANDSTONE
M 28 118 CEMENTS, FELDSPAR OVERGROWTHS
B 6109 1555* CEMENTS FROM STUART CITY TREND, LOWER CRETACEOUS,
M 28 111 CEMENTS, GENERAL
M 28 129 CEMENTS, GLAUCONITE, IN SANDSTONE
M 28 126 CEMENTS, GOETHITE, IN SANDSTONE
M 28 129 CEMENTS, GYPSUM, IN SANDSTONE
M 28 124 CEMENTS, HEMATITE, IN SANDSTONE
M 28 126 CEMENTS, HEMATITE, IN SANDSTONE
M 28 143 CEMENTS, HYDROCARBON, IN SANDSTONE
M 28 136 CEMENTS, ILLITE-SERICITE, IN SANDSTONE
B 6303 533* CEMENTS IN HIGH PLAINS SEQUENCE OF WYOMING AND NEBRASKA,
B 6303 454* CEMENTS IN QUATERNARY ALLUVIAL-FAN DEPOSITS, BIRCH CREEK
B 6203 498* CEMENTS IN SANDSTONES—ACTIVE ANKERITE CEMENTATION IN WILCOX
M 28 137 CEMENTS, KAOLINITE, BIREFRINGENCE
M 28 136 CEMENTS, KAOLINITE, IN SANDSTONE
M 28 128 CEMENTS, KAOLINITE, IN SANDSTONE
M 28 141 CEMENTS, LAUMONTITE, IN ARKOSIC SANDSTONE
M 28 126 CEMENTS, LIMONITE, IN SANDSTONE
M 28 117 CEMENTS, MEGAQUARTZ, WITH CHERT
M 27 160 CEMENTS, MENISCUS
M 28 144 CEMENTS, MONTMORILLONITE COATING DETRITAL
M 28 116 CEMENTS, OPAL, IN SANDSTONE
B 6007 1066 CEMENTS, PANAMA, GALETA REEF
B 6405 756* CEMENTS, PATCH REEF FACIES, ENEWETAK ATOLL, ABST., CARBON,
M 28 127 CEMENTS, PHOSPHATE, IN SANDSTONE
M 28 126 CEMENTS, PHOSPHATE, IN SANDSTONE
M 28 117 CEMENTS, QUARTZ GROWTH
M 28 116 CEMENTS, QUARTZ, IN OOLITIC LIMESTONE
M 28 115 CEMENTS, QUARTZ, MENISCUS
M 28 112 CEMENTS, QUARTZ OVERGROWTH
M 28 173 CEMENTS, QUARTZ OVERGROWTH, IN SANDSTONE
M 28 113 CEMENTS, QUARTZ OVERGROWTH, SEM
M 28 143 CEMENTS, QUARTZ, OVERGROWTHS
M 28 114 CEMENTS, QUARTZ OVERGROWTHS, CATHODOLUMINESCENCE
M 28 114 CEMENTS, QUARTZ OVERGROWTHS, DUST RIMS
M 28 115 CEMENTS, QUARTZ OVERGROWTHS, RE-WORKED
M 28 123 CEMENTS, SIDERITE, IN SANDSTONE
M 28 124 CEMENTS, SIDERITE, IN SANDSTONE
M 28 134 CEMENTS, SMECTITE (AUTHIGENIC), IN SANDSTONE
M 28 138 CEMENTS, ZEOLITE, IN SANDSTONE
B 6403 445* CEMENTS--PRODUCTS OF SHALE DEWATERING IN UPPER MIOCENE
B 6403 445* CEMENTS—PRODUCTS OF SHALE DEWATERING IN UPPER MIOCENE
B 6201 63 CENOMANIAN, AUSTRALIA, APTIAN-
B 6303 429* CENOMANIAN, CAMPOS BASIN, OFFSHORE RIO DE JANEIRO, BRAZIL,
B 6401 74 CENOMANIAN CARBONATES, MEXICO, RANCHO JACALITOS, APTIAN
B 6401 74 CENOMANIAN CARBONATES, MEXICO, RANCHO JACALITOS, APTIAN
B 6205 829 CENOMANIAN, ISRAEL
S 6 278 CENOMANIAN, MACROFOSSIL ZONES, CRETACEOUS

Index

B	6405	612	CENOMANIAN, MIDDLE EAST
S	6	160	CENOMANIAN, MIFUNE GROUP, JAPAN
B	6303	492*	CENOMANIAN, NORTHERN ALASKA, ABST., DINOFLAGELLATE
S	6	132	CENOMANIAN, POTASSIUM-ARGON, FRANCE, PARIS BASIN
B	6205	793	CENOMANIAN, PROFILE, BLAKE PLATEAU AREA
S	6	131	CENOMANIAN, RUBIDIUM-STRONTIUM, ISOCHRON DIAGRAM
B	6412	2055	CENOMANIAN TO CONIACIAN, ORGANIC MATTER, ATLANTIC BASINS
B	6204	669	CENOMANIAN-TURONIAN, ALGERIA
S	6	263	CENOMANIAN-TURONIAN BOUNDARY, NORTH AMERICA, CRETACEOUS
B	6101	44	CENOMANIAN-TURONIAN, LEBANON
B	6406	860	CENOMANIAN-TURONIAN TRANSGRESSIVE DEPOSITS, AFRICA
B	6004	657*	CENOZOIC, ABST., CLIMATIC CONTROL OF DEPOSITION AT, PRUDHOE
B	6303	538*	CENOZOIC, ABST., OCEANS AND, CLIMATE DURING
B	6005	794†	CENOZOIC ALPINE HIMALAYAN GEOSYNCLINE AND ITS PETROLEUM
M	25	280†	CENOZOIC AN MESOZOIC PETROLEUM PROSPECTS, ALEUTIAN-BERING
B	6203	568*	CENOZOIC AND RECENT NONMARINE STROMATOLITES, ABST., ORGANIC,
B	6201	67	CENOZOIC, AUSTRALIA
B	6303	456*	CENOZOIC BACK-ARC LIMESTONE-TUFF SEQUENCE, CENTRAL LUZON,
B	6204	696	CENOZOIC, BAHAMA PLATFORM
B	6002	177	CENOZOIC BASINS, BERINGIAN SHELF AND, MARGIN
M	25	281†	CENOZOIC BASINS IN WESTERN NORTH AMERICA, STRUCTURAL
B	6205	803	CENOZOIC, BLAKE PLATEAU
S	6	251	CENOZOIC BOUNDARY, MESOZOIC, ISOTOPIC DATING
B	6303	478*	CENOZOIC CALCAREOUS MICROFOSSIL BIOSTRATIGRAPHY,
B	6206	1015	CENOZOIC, CAMPANIAN TO MAESTRICHTIAN, ARCTIC, CANADA,
B	6004	645*	CENOZOIC CARIBBEAN PALEOSOLS AND THEIR GEOTECTONIC AND
M	26	87	CENOZOIC, CHRONOCORRELATION, GLOBAL CYCLES
B	6303	413*	CENOZOIC CLAY-MINERAL STRATIGRAPHY I SOUTH PHILIPPINE SEA,
B	6405	703*	CENOZOIC CRUSTAL DEFORMATION IN WESTERN UNITED STATES, ABST.,
S	9	18	CENOZOIC, DRIFT MOVEMENTS, ANTARCTICA
S	9	18	CENOZOIC, DRIFT MOVEMENTS, AUSTRALIA
M	29	223†	CENOZOIC EVOLUTION OF THE SUNDA ARC IN THE CENTRAL SUMATRA
B	6405	791*	CENOZOIC FORELAND SEDIMENTATION IN SOUTHWEST MONTANA,
B	6303	544*	CENOZOIC FORTUNA FORMATION, EAST-CENTRAL TUNISIA, RECORD OF
S	4	102	CENOZOIC FOSSIL REEFS OF CARIBBEAN AREA
S	6	43	CENOZOIC, GEOCHRONOLOGIC SYSTEMS, RESOLUTION
S	6	225	CENOZOIC, GEOMAGNETIC SCALE, DIAGRAM
M	25	559	CENOZOIC GEOSYNCLINAL CYCLE, CHILE, ANDEAN MESOZOIC-
M	29	472	CENOZOIC, GLOBAL CYCLES OF CHANGE
B	6108	1392*	CENOZOIC HIGH PLAINS SEQUENCE, WYOMING AND NEBRASKA, ABST.,
M	29	235	CENOZOIC, HISTORY, SUMATRA
M	29	23	CENOZOIC INTERVAL, GEORGIA EMBAYMENT
S	6	251	CENOZOIC, ISOTOPIC DATING, MESOZOIC BOUNDARY
B	6002	197	CENOZOIC MARGIN BASINS, BRAZIL, EASTERN, MESOZOIC-
S	11	62	CENOZOIC MARINE STAGES, CALIFORNIA
S	11	113	CENOZOIC MARINE STAGES, CORRELATIONS
S	11	60†	CENOZOIC MARINE STAGES IN CALIFORNIA, CALCAREOUS,
S	7	28	CENOZOIC, NORTHERN GULF OF MEXICO, STRATA
S	9	26	CENOZOIC, OXYGEN ISOTOPE OCEAN TEMPERATURES
B	6303	521*	CENOZOIC PALEO-OCEANOGRAPHY AND CORRELATED EVENTS IN
M	29	274	CENOZOIC, PLATE BOUNDARIES, ALASKA
S	10	212	CENOZOIC, POROSITY RELATED TO DEPTH,LOUISIANA
B	6205	793	CENOZOIC, PROFILE, BLAKE PLATEAU AREA
S	4	93†	CENOZOIC REEF SYSTEMS OF CARIBBEAN--PROSPECTS FOR
B	6207	1228*	CENOZOIC ROCKS OF ALEUTIAN ISLANDS, ALASKA PENINSULA, AND
B	6012	2185*	CENOZOIC SAL LAKE BEDS IN NORTHEASTERN GREAT BASIN, ABST., NEW
B	6103	465*	CENOZOIC SEDIMENTATION AND TECTONICS OF HOPE BASIN, SOUTHERN
B	6407	1014†	CENOZOIC STRATA BENEATH KODIAK SHELF, ALASKA, STRUCTURE OF
B	6407	1014†	CENOZOIC STRATA BENEATH KODIAK SHELF, ALASKA, STRUCTURE OF
B	6412	2139†	CENOZOIC STRUCTURAL TRENDS UNDER SOUTHERN BERING SEA SHELF,
M	29	224	CENOZOIC, SUBDUCTION HISTORY, SUMATRA
M	29	225	CENOZOIC, SUNDA CONVERGENCE ZONE, SUMATRA
B	6002	177	CENOZOIC TECTONIC SETTING, BERINGIAN SHELF AND, MARGIN,
S	9	24	CENOZOIC, TEMPERATURE CHANGES
B	6402	224	CENOZOIC, TEXAS
B	6402	224	CENOZOIC, TEXAS
B	6202	259	CENOZOIC, UPPER, STRATIGRAPHIC COLUMN, CALIFORNIA
B	6004	661*	CENOZOIC--MODEL FOR APPLICATION OF STRATIGRAPHIC CONCEPTS TO
B	6209	1755*	CENOZOIC—MODEL FOR APPLICATION OF STRATIGRAPHIC CONCEPTS TO
S	11	310	CENTERVILLE, CALIFORNIA, SAMPLE LOCALITIES
S	11	320	CENTERVILLE COASTAL SECTION DESCRIPTION
S	11	194	CENTERVILLE COASTAL STRATIGRAPHIC SECTION
B	6010	1655	CENTONZIO, A., ON DEVELOPMENTS IN PUERTO RICO, 1975
B	6210	1846	CENTRAL AFRICAN EMPIRE, DEVELOPMENTS, 1977
B	6310	1691	CENTRAL AFRICAN EMPIRE, DEVELOPMENTS, 1978
B	6110	1747	CENTRAL AFRICAN REPUBLIC, DEVELOPMENTS, 1976
B	6411	1787	CENTRAL AFRICAN REPUBLIC, DEVELOPMENTS, 1979
B	6207	1229*	CENTRAL AMERICA, ABST., GEOLOGY AND PETROLEUM POTENTIAL OF
B	6010	1640†	CENTRAL AMERICA, AND CARIBBEAN AREA IN 1975, PETROLEUM,
M	29	253	CENTRAL AMERICA, GUATEMALA, TECTONIC MAP
B	6110	1578†	CENTRAL AMERICA, MEXICO, AND CARIBBEAN AREA IN 1976,
B	6210	2044†	CENTRAL AMERICA, MEXICO, AND CARIBBEAN AREA IN 1977,
B	6310	1745†	CENTRAL AMERICA, MEXICO, AND CARIBBEAN AREA IN 1978,
M	25	460	CENTRAL AMERICA, SILVER-COPPER-POLYMETALLIC PROFILE, PACIFIC,
M	25	244	CENTRAL AND SOUTH SUMATRA BASINS, INDONESIA
B	6211	2308†	CENTRAL APPALACHIAN BASIN, SILURIAN EVOLUTION OF
B	6101	83	CENTRAL ATLANTIC ARCH
B	6101	83	CENTRAL ATLANTIC RIFT VALLEY
B	6205	781	CENTRAL BASIN PLATFORM, SEISMIC DATA, NEW MEXICO
B	6205	787	CENTRAL BASIN PLATFORM, TEXAS
B	6209	1666	CENTRAL BASIN RIDGE, PHILIPPINE BASIN
B	6006	952†	CENTRAL BROOKS RANGE, ALASKA, GEOLOGIC SIGNIFICANCE OF,
B	6211	2194	CENTRAL CLARE GROUP, IRELAND, WESTERN
B	6101	20	CENTRAL ELLESMERE FOLDBELT, CANADIAN, ARCTIC ISLANDS
B	6203	436	CENTRAL GRABEN, DANISH SECTOR, RIFT SYSTEM, NORTH SEA
B	6203	426	CENTRAL GRABEN, EUROPE
M	30	196	CENTRAL GRABEN, LOCATION, NORTH SEA
B	6203	436	CENTRAL GRABEN, NETHERLANDS SECTOR, RIFT SYSTEM, NORTH SEA
B	6107	1002	CENTRAL GRABEN, NORTH SEA
B	6307	1031	CENTRAL GRABEN, ROTLIEGENDES SANDSTONE, SEISMIC PROFILE,
S	4	88	CENTRAL INLIER, JAMAICA
M	25	241	CENTRAL IRIAN MIOGEOSYNCLINE, INDONESIA, CENTRAL, IRIAN
M	25	241	CENTRAL IRIAN UNSTABLE SHELF AREA, CENTRAL IRIAN
B	6103	311	CENTRAL KANSAS UPLIFT, KANSAS
B	6104	581	CENTRAL LAKE HIGH, VENEZUELA
B	6203	404	CENTRAL LIMESTONE PLAIN, VIRGIN ISLANDS, ST. CROIX
M	30	436	CENTRAL LUCONIA FIELD, MALAYSIA, SARAWAK
B	6405	775*	CENTRAL MEDITERRANEAN THRUST BELTS, ABST.
B	6407	973	CENTRAL MONTANA UPLIFT, MONTANA
B	6303	288	CENTRAL OCEAN REGION, PROSPECTS
B	6004	659*	CENTRAL PENNINE BASIN, ENGLAND, ABST., DELTAIC EVOLUTION
B	6106	901	CENTRAL PLATE
S	10	136	CENTRAL PLATEAU, GROUNDWATER FLOW,IRAN
S	10	139	CENTRAL PLATEAU, GROUNDWATER FLOW PATTERN,IRAN
B	6209	1667	CENTRAL RANGE PROVINCE, TAIWAN
B	6207	1215*	CENTRAL RECEIVER PROGRAM, BARSTOW POWER TOWER, CALIFORNIA,
B	6408	1142	CENTRAL SHELF UPLIFT, ALASKA
B	6407	1023	CENTRAL SHELF UPLIFT, ALASKA
S	7	141	CENTRAL SLOPE, NORTHERN GULF OF MEXICO, CONTINENTAL SLOPE
B	6004	653*	CENTRAL STABLE REGION, ABST., RELATION OF, SEISMICITY AN
B	6104	527†	CENTRAL TEXAS ESTUARY, LAVACA BAY--, TRANSGRESSIVE, DELTAIC
B	6302	183	CENTRAL TROUGH, NORTH SEA
B	6401	133	CENTRE NATIONAL POUR LEXPLOITATION DES OCEANS, MOORE, GEORGE
B	6104	579	CENTRO FIELD, VENEZUELA
B	6105	775	CENTURY, J. R., WESTERN CANADIAN OIL AND GAS DATA BASE AS
B	6303	538	CEPEK, P., MIDDLE CRETACEOUS OXYGEN-DEFICIENT
M	27	89	CEPHALOPODS
M	27	90	CEPHALOPODS, NAUTILOID
B	6112	2069	CERAM ISLAND, SECTION
M	25	543	CERIUM DEPLETION, SIGNIFICANCE OF
B	6303	431	CERNOCK, P. J., SOURCE ROCKS IN GULF COAST AREA--THEIR
B	6303	425	CERNOCK, PAUL J., REGIONAL HYDROCARBON SOURCE-ROCK
B	6405	758*	CERR PRIETO GEOTHERMAL FIELD, BAJA CALIFORNIA, MEXICO, ABST.,
B	6111	1956	CERRO NEGRO-MORICHAL-JOBO HEAVY OIL, VENEZUELA
M	25	159	CERRO PRIETO AREA, GEOTHERMAL, MEXICO
M	25	161	CERRO PRIETO GEOTHERMAL AREA, BAJA CALIFORNIA GEOTHERMAL
B	6203	490*	CERRO PRIETO GEOTHERMAL FIELD, BAJA CALIFORNIA, MEXICO, ABST.,
B	6405	759*	CERRO PRIETO GEOTHERMAL FIELD, BAJA CALIFORNIA, MEXICO, ABST.,
B	6212	2508*	CERRO PRIETO GEOTHERMAL FIELD, BAJA CALIFORNIA, MEXICO, ABST.,
B	6211	2262	CERROS HILL SECTION, CALIFORNIA
B	6211	2262	CERROS SHALE MEMBER, CALIFORNIA
B	6109	1417	CHA CHA FIELD, NEW MEXICO
M	28	142	CHABAZITE, CRYSTALLIZED, IN TUFF
S	3	99	CHACAHOULA DOME, LOUISIANA
B	6406	865	CHAD BASIN, AFRICA
B	6110	1747	CHAD, DEVELOPMENTS, 1976
B	6210	1846	CHAD, DEVELOPMENTS, 1977
B	6310	1691	CHAD, DEVELOPMENTS, 1978
B	6411	1787	CHAD, DEVELOPMENTS, 1979
B	6406	863	CHAD, MURZUK BASIN, LIBYA AND
B	6203	463†	CHADRA SANDS, GIALO FIELD, LIBYA, SECONDARY, FERROAN,
B	6008	1393	CHADWICK, ROBERT A, GEOLOGIC CONTROLS OVER HOT WATER
B	6309	1490†	CHAGAI HILLS, BALUCHISTAN, PAKISTAN, BY COMPUTER PROCESSIN OF
B	6309	1494	CHAGAI HILLS, HYDROTHERMAL ALTERATION, PAKISTAN
B	6309	1494	CHAGAI HILLS, VOLCANIC FLOAT, PAKISTAN
B	6309	1494	CHAGAI INTRUSIVES, PAKISTAN
B	6303	504	CHAGNON, A., THERMA MATURATION OF LOWER PALEOZOIC SHALES IN
B	6312	2128	CHAGNON, ANDRE, COMPILATION AND CORRELATION OF MAJOR
B	6004	704*	CHAGRIN FORMATION, UPPER DEVONIAN, NORTHEAST OHIO, ABST.,
M	25	321	CHAGUII ANTICLINE, COLOMBIA

Index

B 6104 512 CHAILLY OIL FIELD, FRANCE
B 6104 515 CHAILLY WELLS, FRANCE
B 6306 934 CHAIN 115, CRUISE, 1974
B 6410 1660 CHAIN 57 AND 73, WOODS HOLE OCEANOGRAPHIC INSTITUTION CRUISES
B 6106 872 CHAIN CRUISE 119, FEBRUARY-APRIL 1975, RV
B 6412 2126 CHAIN CRUISE 161, RV
B 6302 218 CHAIN CRUISE, 1974, RV
B 6107 969 CHAIN FRACTURE ZONE, AFRICA
B 6305 838* CHAINMA SHALE OF EAST-CENTRAL NEVADA, ABST., EVIDENCE FOR TWO
B 6405 690 CHAKRABARTI, A., BURROW PATTERNS OF GHOST CRAB OCYPODE
M 28 81 CHALCEDONY, IN A CAVITY, QUARTZ
M 28 27 CHALCEDONY, IN CARBONATE CEMENT, FRAGMENTS
M 28 118 CHALCEDONY, IN OOLITIC CARBONATES, CEMENTS
M 28 157 CHALCEDONY, REPLACEMENT OF A FOSSIL
M 28 81 CHALCEDONY, ZEBRAIC, IN A CAVITY
M 25 400 CHALCOPYRITE, AUSTRALIA, PRYITE AND
M 27 17 CHALK
B 6109 1548* CHALK AND BUDA LIMESTONE IN SOUTH TEXAS, ABST., GEOLOGY OF,
B 6107 990 CHALK BURIAL DIAGENESIS
B 6101 113 CHALK CREEK MEMBER, UTAH
B 6107 988 CHALK, DENMARK, MAESTRICHTIAN
B 6107 982† CHALK DIAGENESIS AND ITS RELATION TO PETROLEUM EXPLORATION—
B 6303 422* CHALK MEMBER, SARATOGA FORMATION, ABST., ICHNOLOGY AND
B 6107 988 CHALK, NETHERLANDS, MAESTRICHTIAN
M 28 153 CHALK, NORTH SEA, CRETACEOUS
M 29 180 CHALK OUTCROPS, BLAKE ESCARPMENT, MIOCENE
B 6004 719* CHALK, PLEISTOCENE, IN BAHAMIAN TROUGH. BY TITLE ONLY,
B 6107 983 CHALK, PRODUCTION FROM, COLORADO
B 6107 1004 CHALK, PRODUCTION FROM, GULF COAST
B 6107 1001 CHALK, PRODUCTION FROM, NORTH SEA
B 6303 448* CHALK RESERVOIRS, ABST., GENERATIO AND OCCLUSION OF POROSITY
B 6303 522* CHALK RESERVOIRS, ABST., POROSITY RELATIONS IN
B 6107 986 CHALK SECTION, LOWER, ENGLAND
B 6107 988 CHALK UNIT, ALABAMA, SELMA
B 6107 988 CHALK UNIT, CANADA, SCOTIAN SHELF, WYANDOT
B 6107 988 CHALK UNIT, CELTIC SEA
B 6107 988 CHALK UNIT, FRANCE
B 6107 988 CHALK UNIT, GERMANY
B 6107 988 CHALK UNIT, IRELAND, WHITE LIMESTONE
B 6107 988 CHALK UNIT, MEXICO, AUSTIN-, SAN FELIPE
B 6107 988 CHALK UNIT, MIDDLE AND UPPER, ENGLAND
B 6107 988 CHALK UNIT, OKLAHOMA, BRIDGE CREEK
B 6107 988 CHALK UNIT, OKLAHOMA, HARTLAND
B 6107 988 CHALK UNIT, TEXAS, AUSTIN
B 6209 1767* CHALK, UPPER CRETACEOUS, MAESTRICHTIAN, AND OF PINE BARREN
B 6107 982† CHALKS, A MODERN MIRACLE, CHALK DIAGENESIS AND ITS RELATION
B 6004 719* CHALKS, ABST., DIAGENETIC, PATTERNS OF
B 6107 1001 CHALKS, HYDROCARBON, PRODUCTION FROM
B 6205 892* CHALKS OF WESTERN INTERIOR, ABST., DIAGENESIS AND,
B 6107 984 CHALKS, PHYSICAL AND CHEMICAL PROPERTIES OF
B 6303 422* CHALKS, SOUTHWESTERN ARKANSAS, ANNONA FORMATION, CHALK
B 6107 988 CHALKS, TABLE, NORTH AMERICAN AND, EUROPEAN
B 6109 1546* CHALK—BUDA TREND OF SOUTH TEXAS, ABST., AUSTIN
B 6211 2294 CHALLENGER KNOLL, GULF OF MEXICO
S 7 49 CHALLENGER KNOLL, GULF OF MEXICO, SALT
S 7 46 CHALLENGER KNOLL, REFLECTION PROFILE, SALT, GULF OF MEXICO
M 29 407 CHALLENGER KNOLL, SEISMIC SECTION, GULF OF MEXICO
S 7 62 CHALLENGER SALT FLOW, DIAGRAM, CAMPECHE KNOLLS
S 7 48 CHALLENGER UNIT, DESCRIBED, DEEP, GULF OF MEXICO
B 6211 2291 CHALLENGER UNIT, GULF OF MEXICO
S 7 53 CHALLENGER UNIT, GULF OF MEXICO, SIGSBEE KNOLLS
B 6211 2220 CHALONE CREEK FAULT, CALIFORNIA
B 6407 1008 CHALYBEAT SPRINGS FIELD, ARKANSAS
B 6009 1614* CHALYBEAT SPRINGS FIELD, JURASSIC OF SOUT ARKANSAS, ABST.,
B 6004 715* CHALYBEAT SPRINGS FIELD, JURASSIC OF SOUTH ARKANSAS, ABST.,
B 6108 1389* CHAMA BASIN IN SOUTHWESTERN COLORADO, ABST., HYDROCARBON
B 6108 1373 CHAMBERLAIN, ALAN K., SAGEBRUSH AND GEOCHEMICAL EXPLORATION
B 6405 690 CHAMBERLAIN, C. KENT, SYNONYMY, ETHOLOGY, AND ENVIRONMENTAL
S 4 135 CHAMBERLAIN, JOHN A., JR., STRUCTURAL MODIFICATION OF CORALS IN
B 6008 1257 CHAMBERS COUNTY, TEXAS, DEVILLIER FIELD
B 6308 1225 CHAMBERS COUNTY, TEXAS, ELWOOD FIELD
B 6208 1383 CHAMBERS COUNTY, TEXAS, SHIP CHANNEL FIELD
B 6412 2100 CHAMPEH MEMBER, IRAN
B 6207 1211 CHAMPENY, JON D., PETROLEUM POTENTIAL IN FAR EAST, ABST.
B 6008 1215 **CHAMPION, BEVERLY L., DEVELOPMENTS IN MICHIGAN IN 1975**
B 6303 447 CHAN, K. C., DETERMINATION OF DEPTH TO HYDROCARBON
B 6211 2245 CHANAC FORMATION, CALIFORNIA
B 6008 1393 CHANCELLOR, ROBERT E., PETROLEUM GEOLOGY OF FORT UNION AND
B 6409 1569* CHANDELEUR SOUND, ST. BERNARD PARISH, LOUISIANA, ABST., BLOCK
B 6211 2351 CHANDLER, SUSAN M., LATE MIOCENE TURBIDITE DEPOSITION, CROCKER
B 6002 281 CHANEYSVILLE SILTSTONE MEMBER, PENNSYLVANIA
B 6405 690 CHANG, YI-MAW, DECOMPACTION TECHNIQUE HELPS CORRELATION,
S 4 209† CHANGE IN OSTRACOD STANDING CROP, LAGUNA MANDINGA,
B 6004 646* CHANGES IN MIOCENE PALEOBATHYMETR OF CALIFORNIA
M 29 456 CHANGES IN SEA LEVEL, COMPUTATION OF SHORELINE
M 29 469† CHANGES OF SEA LEVEL FROM SEISMIC STRATIGRAPHY, GLOBAL, CYCLES
M 26 51† CHANGES OF SEA LEVEL, PART 1: OVERVIEW, SEISMIC, STRATIGRAPHY
M 26 49† CHANGES OF SEA LEVEL, SEISMIC, STRATIGRAPHY AND, GLOBAL
B 6303 423* CHANGING PERSPECTIVES IN PALEOECOLOGIC USE O TRACE FOSSILS,
B 6008 1407* CHANGING SITES AND STYLES OF MESOZOIC AND CENOZOIC TECTONISM,
B 6007 1037† CHANNEL, BAHAMAS, STRATIGRAPHY AND STRUCTURE OF, NORTHEAST,
B 6007 1037† CHANNEL, BAHAMAS, STRATIGRAPHY AND STRUCTURE OF, NORTHEAST,
B 6203 426 CHANNEL BASIN, EUROPE
B 6105 776 * **CHANNEL-FILL AND SPILLOVER FACIES OF NORTHWEST ATLANTIC MID-**
B 6202 237 CHANNEL FILL, MEXICO, CHICONTEPEC
B 6012 2185* CHANNEL ISLANDS, CALIFORNIA, ABST., ELEMENTS OF TECTONISM
B 6203 542* CHANNEL MORPHOLOGY, INDUS CONE, OFFSHORE PAKISTAN, ABST.,
B 6209 1756* CHANNEL OF ANCIENT NUECES RIVER, SOUTH TEXAS, ABST.,
B 6206 967† CHANNEL SANDS AND PETROLEUM POTENTIAL OF MONTEREY DEEP-SEA
B 6311 2120* CHANNEL SANDSTONE OIL RESERVOIRS OF PENNSYLVANIAN AGE,
B 6103 366 CHANNEL SANDSTONES, OKLAHOMA, PERMIAN
M 26 479 CHANNEL SANDSTONES, SEISMIC, MODELING
B 6007 1015 CHANNEL SYSTEM, SEQUENCE, SEPARATE
B 6004 708* CHANNEL SYSTEM, WILLAPA BAY, WASHINGTON, ABST., BED FORMS AND,
B 6206 950 CHANNEL, TERM, FEEDER
B 6004 707* CHANNELIZED SEDIMENTATION IN SOUTHERN APENNINES, ABST.,
B 6206 912† CHANNELS, AND DEPOSITIONAL LOBES ON MODERN SUBMARINE FANS:
B 6203 547* CHANNELS AND GIANT CROSS-BEDS IN ATHABASCA OIL SANDS, ABST.,
B 6206 978 CHANNELS AND SUPRAFANS AS STRATIGRAPHIC TRAPS
B 6207 1237* CHANNELS IN GREAT VALLEY, CALIFORNIA, ABST., OIL AND GAS FROM,
B 6204 693† CHANNELS IN RELATION TO HYDROCARBON POTENTIAL,
B 6206 952 CHANNELS ON SUPRAFAN LOBE, ANCIENT, MID-, FAN:, BRAIDED
B 6204 694 CHANNELS, ORIGIN OF, BAHAMA
B 6007 1015 CHANNELS, SEQUENCE, INDIVIDUAL
B 6206 947 CHANNELS, SUBMARINE FAN, FEEDER
B 6303 414 CHANPONG, R. R., GEOLOGY AND DEVELOPMENT OF TEAK OIL FIELD,
M 30 387 CHANPONG, R. R., GEOLOGY AND DEVELOPMENT OF THE TEAK OIL
M 25 252 CHAO PHRAYA DEPRESSION, THAILAND
B 6408 1211 CHAO-SHUI BASIN, CHINA
S 7 260 CHAOTIC FACIES FROM UPLIFT, TEXAS, SHELF-SLOPE
S 7 112 CHAOTIC PATTERNS, SEDIMENT PROCESSES, SEISMIC IDENTIFICATION
S 7 258 CHAOTIC SEISMIC FACIES, TEXAS, SHELF-SLOPE
M 26 128 CHAOTIC, SEISMIC, REFLECTIONS, CONFIGURATION PATTERNS
B 6205 770 CHAOTIC-FILL FACIES
M 26 183 CHAOTIC-FILL FACIES, SEISMIC, INTERPRETATION
B 6004 657 CHAPA, ABELARDO CANTU, SUBSURFACE JURASSIC-CRETACEOUS
B 6004 677 CHAPMAN, R. P., SOME APPROACHES TO CLUSTER-ANALYSIS
B 6412 2179 CHAPMAN, RICHARD E., MECHANICAL VERSUS THERMAL CAUSE OF
B 6409 1538 CHAPMAN, RICHARD E., MIGRATION OF FLUIDS IN SEDIMENTARY
S 7 239 CHAPMAN SANDSTONE, ISOPACH, CALIFORNIA, LOS ANGELES BASIN
B 6303 435 CHAPMAN, W. L., SURFACE SEISMIC METHODS APPLIED TO COAL-MINING
B 6102 208 CHAPOPOTE FORMATION, MEXICO
B 6303 548* CHAPPEL LIMESTONE, FORT WORTH BASIN, TEXAS, ABST., DEPOSITION
B 6208 1364 CHAPPELLE, RAY L., DEVELOPMENTS IN WEST TEXAS AND EASTERN NEW
B 6105 653 **CHAPPLE, W. M., FLOW OF WEAK ROCKS IN APPALACHIAN PLATEAU FOLDS**
M 24 103 CHAPPO MEMBER, WYOMING
B 6112 2098 CHARACTER OF BITUMEN GENERATED
B 6204 693† CHARACTERISTICS OF DEEP BAHAMA CHANNELS IN RELATION TO
B 6002 289 CHARACTERISTICS OF DEGASSED SEDIMENTS
B 6303 537* CHARACTERISTICS OF LANDSAT SYSTEM FO GEOLOGIC APPLICATIONS,
B 6309 1596* CHARACTERISTICS OF LOWER VICKSBURG RESERVOIRS, MCALLEN
B 6309 1595* CHARACTERISTICS OF LOWER WILCOX RESERVOIRS, VALENTINE FIELD,
B 6303 506* CHARACTERISTICS OF REMOTELY SENSED DATA ACQUIRED FROM
B 6303 498* CHARACTERISTICS OF SHALLOW SUBSURFACE FAULTS ON TEXAS INNER
B 6008 1396* CHARACTERISTICS OF UPPER CRETACEOUS LANCE CHANNELS, NIOBRARA
B 6109 1534* CHARACTERISTICS OF WILCOX GAS RESERVOIRS, NORTHEAST
B 6202 301† CHARACTERIZATION OF BITUMENS AND COALS
B 6004 692* CHARACTERIZATION OF INSOLUBLE ORGANIC MATTER ASSOCIATED
B 6102 297* CHARACTERIZATION OF OIL TYPES IN PERMIAN BASIN BY TITLE ONLY
B 6309 1578* CHARACTERIZATION OF PRODUCTION MECHANISM IN DEVONIAN SHALE
B 6405 794* CHARACTERIZATION OF REGULARLY INTERSTRATIFIED
B 6303 501* CHARACTERIZATION OF ROCK MINERAL AND PORE SPACE PROPERTIES
B 6107 969 CHARCOT FRACTURE, AFRICA
B 6008 1398* CHARLES FORMATION, WILLISTON BASIN, NORTH-CENTRAL NORTH
B 6201 102 CHARLESTON OVERTHRUST FAULT, UTAH
B 6303 555 CHARLESWORTH, H. A. K., STRUCTURE-CONTOUR MAPS COMPUTER
S 5 55 CHARLEVOIX COUNTY, LOCATION, MICHIGAN,
M 30 92 CHARLIE SAND, DEPOSITIONAL HISTORY, FORTIES FIELD
M 30 89 CHARLIE SAND, ISOPACH, FORTIES FIELD
M 30 90 CHARLIE SAND UNIT, FORTIES FIELD, FORTIES FORMATION
B 6203 518* CHARLOTTE HARBOR, FLORIDA, ABST., PHOSPHATE IN BOTTOM
M 27 11 CHAROPHYTES, GREEN, ALGAE
M 26 88 CHART CONSTRUCTION, SEA LEVEL FLUCTUATIONS, GLOBAL CYCLES
M 26 166 CHART, SEISMIC, INTERPRETATION, DEPOSITIONAL FRAMEWORK,
M 26 203 CHART, SEISMIC, REFLECTIONS, CARBONATE, RECOGNITION, CRITERIA
M 26 186 CHART, SEISMIC, REFLECTIONS, INTERPRETATION, PROCEDURE
M 26 162 CHARTS, TEXAS, GULF COAST, WEST, AFRICA, COMPARISON, CYCLE
B 6107 1030 CHASE GROUP, NEBRASKA
B 6206 967 CHASE, T. E., CHANNEL SANDS AND PETROLEUM POTENTIAL OF
M 25 313 CHASE, T. E., PETROLEUM POTENTIAL OF CONTINENTAL RISE OFF
B 6405 803 CHASE, T. E., POTENTIAL PETROLEUM RESERVOIRS ON DEEP-SEA FANS
B 6104 585 CHASHMEH KHUSH FIELD, PERSIAN GULF
S 6 348 CHASMOPS LIMESTONE, BENTONITES, ORDOVICIAN, SWEDEN
B 6410 1577 CHATHAM HILL FORMATION, VIRGINIA
S 10 198 CHATHAM MUD ISLAND, WEST INDIES,TRINIDAD
M 25 395† **CHATHAM RISE NEAR NEW ZEALAND--SUMMARY, SUBMARINE,**
M 25 397 CHATHAM RISE, PHOSPHATE NODULE RESERVES
M 25 395 CHATHAM RISE, PHOSPHORITE OCCURRENCE
B 6209 1704 CHATTANOOGA SHALE, UNITED STATES, EASTERN
B 6405 682* CHATTIAN, MALTESE ISLANDS, ABST., DIAGENESIS OF LOWER, CORALLINE
S 6 221 CHATTIAN STAGE DEFINED, EUROPE, OLIGOCENE

M 25 553 CHAUCHA COPPER-MOLYBDENUM DEPOSIT, ECUADOR
M 25 553 CHAUCHA DEPOSIT, COPPER-MOLYBDENUM, ECUADOR
B 6405 691 CHAUDHURI, ASRU K., EARLY DIAGENETIC GLAUCONITIZATION OF
B 6405 691 CHAUDHURI, ASRU K., TRACE FOSSILS FROM PROTEROZOIC SEDIMENTS
B 6107 1021 CHAUVIN SOUTH FIELD, ALBERTA
B 6203 519 CHAVANNE, R. E., CHEMICAL ANALYSES OF GEOTHERMAL WATERS FROM
B 6303 344 CHAZY GROUP, KENTUCKY
M 27 173 CHAZY LIMESTONE, PENNSYLVANIA, ORDOVICIAN
M 27 200 CHAZY LIMESTONE, PENNSYLVANIA, ORDOVICIAN
S 11 91 CHECK LIST OF FORAMINIFERA, MONTEREY FORMATION, NAPLES,
S 11 94 CHECK LIST OF FORAMINIFERA, MONTEREY SHALE, NAPLES, CALIFORNIA
S 11 88 CHECK LIST OF FORAMINIFERA, TEMBLOR-MONTEREY SEQUENCE,
S 11 82 CHECK LIST OF FORAMINIFERA, TEMBLOR-MONTEREY SEQUENCE,
S 11 85 CHECK LIST OF FORAMINIFERA, TEMBLOR-MONTEREY SEQUENCE,
S 11 99 CHECK LIST OF MIOCENE FORAMINIFERA, SHALE POINT, CALIFORNIA
B 6105 775 CHEEMA, MUJEEB R., GAS-PRODUCING SUBMARINE-FAN DEPOSITS OF
B 6102 294 CHEESEMAN, RAYMOND J., OIL AND GAS EXPLORATION AND
B 6412 2100 CHEHEL MEMBER, IRAN
M 25 269 CHEJU VOLCANIC BELT, KOREA
M 25 480 CHEMEKOV, YU. F., STRUCTURAL CLASSIFICATION OF NORTH PACIFIC
B 6203 519* CHEMICAL ANALYSES OF GEOTHERMAL WATERS FROM SUTH LOUISIANA
B 6203 529* CHEMICAL ANALYSIS OF GAS DISSOLVED IN GEOTHERMAL WATERS IN
B 6409 1566* CHEMICAL AND ISOTOPIC EVIDENCE OF ORIGINS OF NATURAL GASES IN
B 6004 719* **CHEMICAL AND ISOTOPIC INVESTIGATION OF STRATIGRAPHIC AND**
S 10 89† CHEMICAL AND PHYSICAL CONSTRAINTS, OIL AND,GAS, MIGRATION:
B 6309 1604* CHEMICAL DIFFERENTIATION OF TEMPERATURE AND TROPICAL
B 6004 650* CHEMICAL VARIATION AND TEXTURE--CONTRIBUTION FROM MAYA
B 6104 573† **CHEMISTRY APPLIED TO PETROLEUM GEOCHEMISTRY I, CALCULATION OF,**
B 6109 1407† CHEMISTRY APPLIED TO PETROLEUM GEOCHEMISTRY, II, DISTRIBUTION
S 7 325 CHEMISTRY, BRINE CLASSIFICATION, GYRE BASIN
B 6207 1211* CHEMISTRY OF PHILIPPINE SEA SEDIMENTS, PARTICULARLY MANGANESE
B 6412 2169 CHEMUNG GROUP, WEST VIRGINIA
B 6207 1211 CHEN, JU.-CHIN, MINERALOGY AND CHEMISTRY OF PHILIPPINE SEA
B 6304 698 CHEN SI ZHONG,, GEOLOGY OF GU-DAO OIL FIELD AN ADJACENT AREAS,
S 11 62 CHENEYAN STAGE
B 6207 1221 CHENG, WENG-TSE, EXPLORATION AND DEVELOPMENT OF GEOTHERMAL
M 25 146 CHENG, WENG-TSE, ORIGIN AND POTENTIAL OF SOM GEOTHERMAL
B 6109 1537* CHENIER, LOUISIANA, ABST., INTERIOR STRUCTURE OF, OAK GROVE
B 6409 1560* CHENIER SYSTEM—SOME PRELIMINARY REINTERPRETATIONS AND
B 6109 1537* CHENIER-BEACH SYSTEM, ABST., SEDIMENT-DISPERSAL TRENDS OF,
B 6304 687 CHENOWETH, WILLIAM L., EXPLORATION IN GRANTS URANIUM REGION
B 6304 649 CHEPULTEPEC DOLOMITE, VIRGINIA
B 6002 322* CHEROKE RESERVOIR SANDS, KINCAID OIL FIELD, SOUTHEAST KANSAS,
B 6103 373 CHEROKEE DELTAIC OIL-PRODUCTIVE SANDSTONES, OKLAHOMA
B 6311 2116* CHEROKEE, DESMOINESIAN, SANDSTONES,
B 6107 1030 CHEROKEE GROUP, NEBRASKA
B 6407 1005 CHEROKEE TRACE FIELD, ARK-LA-TEX AREA
B 6007 1120 CHEROKEE TRACE FIELD, TEXAS
B 6112 2086 CHERRY CANYON FORMATION, KEROGEN, TEXAS
B 6205 781 CHERRY CANYON FORMATION, NEW MEXICO
B 6207 1172 CHERRY CREEK RANGE, SECTION, NEVADA
B 6107 1046 CHERRYVALE FORMATION, KANSAS, LOWER
M 28 105 CHERT ARENITE, OPAL-CEMENTED
M 27 142 CHERT CEMENTATION OF CARBONATE GRAINS
B 6109 1554* CHERT, CENTRAL TEXAS,ABST., HYDROGEN AND, OXYGEN, ISOTOPIC
M 28 78 CHERT, DETRITAL, FRAGMENTS
M 28 117 CHERT, DETRITAL, FRAGMENTS
M 28 98 CHERT, DETRITAL, IN QUARTZ ARENITE
M 28 80 CHERT, FILLING, IN BRECCIA
M 28 27 CHERT, FRAGMENTS
M 28 78 CHERT MATRIX, SPONGE SPICULES, IN
M 28 78 CHERT, NODULE, REPLACEMENT
B 6203 529* CHERT PROBLEM, ABST., DEEP-SEA SEDIMENT AND THE
M 27 139 CHERT, REPLACEMENT
M 28 156 CHERT, REPLACEMENT OF DOLOMITE
M 27 139 CHERTS
B 6405 754* CHERTS OF WOODFORD AND ARKANSAS NOVACULITE FORMATIONS,
B 6203 426 CHESHIRE BASIN, EUROPE
B 6409 1555* CHESTER AND POTTSVILLE DEPOSITIONAL SYSTEMS, OUTCROP AND
B 6002 327* CHESTER RELATIONS IN ANADARKO BASIN. BY TITLE ONLY, MORROW-,
M 26 453 CHESTER SHALE, OKLAHOMA, MISSISSIPPIAN
B 6103 436 CHESTERAN UNCONFORMITY, ALASKA
B 6009 1619* CHESTERIAN SHORELINE SEQUENCES IN EASTERN KENTUCKY, ABST.
B 6408 1282* CHESTERIAN, STRATIGRAPHIC ENTRAPMENT OF PETROLEUM, ABST.,
B 6405 781* CHEVRON COST B-3 WELL, ABST., SEISMIC STRATIGRAPHY OF, ATLANTIC
B 6110 1801 CHEVRON OVERSEAS PETROL. INC., AND SUPERIOR OIL CO., ON
B 6010 1872 CHEVRON OVERSEAS PETROLEUM, INC., AND SUPERIOR OIL CO., ON
B 6411 1838 CHEVRON OVERSEAS PETROLEUM INC. AND SUPERIOR OIL CO., ON
B 6210 2117 CHEVRON OVERSEAS PETROLEUM, INC., SUPERIOR OIL CO., AND
B 6310 1889 CHEVRON OVERSEAS PETROLEUM, INC., SUPERIOR OIL CO., ET AL, ON
B 6311 2119* CHEYENNE AREA, WESTERN OKLAHOMA, ABST., STIMULATION DESIGN
M 30 362 CHIAPAS-TABASCO MESOZOIC PROVINCE, MEXICO
M 30 363 CHIAPAS-TABASCO PROVINCE, AGUA NUEVA FORMATION
M 30 379 CHIAPAS-TABASCO PROVINCE, ANTONIO J. BERMUDEX COMPLEX
M 30 379 CHIAPAS-TABASCO PROVINCE, CACTUS FIELD
M 30 363 CHIAPAS-TABASCO PROVINCE, MENDEZ FORMATION
M 30 365 CHIAPAS-TABASCO PROVINCE, PARAJE SOLO FORMATION
M 30 363 CHIAPAS-TABASCO PROVINCE, SAN FELIPE FORMATION
M 30 365 CHIAPAS-TABASCO PROVINCE, SITIO GRANDE FIELD
B 6408 1245 CHIARELLI, A., PRESSURE ORIGIN AND DISTRIBUTION IN JURASSIC OF
B 6204 667 CHIARELLI, ANDRE, HYDRODYNAMIC FRAMEWORK OF EASTERN
M 29 278 CHICHAGOF-BARANOF FAULT, ALASKA, CROSS SOUND
M 29 279 CHICHAGOF-BARANOF FAULT, ALASKA, CROSS SOUND
M 25 449 CHICHIBU GEOSYNCLINAL UNITS, MANGANESE DEPOSITS, JAPAN
S 4 113 CHICHIRIVICHE AREA, COASTAL REEFS, VENEZUELA
S 4 113 CHICHIRIVICKE AREA, COASTAL REEFS, VENEZUELA
B 6410 1688 CHICHKAN FORMATION, CELLULAR FILAMENTS AND SHEATHS
B 6203 377 CHICKALOON FORMATION, ALASKA
B 6103 361 CHICKASHA ANTICLINE, OKLAHOMA
M 26 451† CHICKASHA TREND: GEARY, OKLAHOMA—T13N, R10W, A, CASE HISTORY
B 6312 2216 CHICKEN-WIRE FRACTURES
S 4 243 CHICO RIDGE AREA, TEXAS
B 6203 351 CHICO-MARTINEZ OIL FIELD, CALIFORNIA
B 6202 237 CHICONTEPEC CHANNEL FILL, MEXICO
B 6202 240 CHICONTEPEC FORMATION, MEXICO
B 6202 237 CHICONTEPEC HORIZON C, MEXICO
B 6202 236 CHICONTEPEC OIL POOLS AND WELLS, MEXICO
B 6202 237 CHICONTEPEC TURBIDITES, CROSS SECTIONS, MEXICO
B 6202 235† CHICONTEPEC TURBIDITES, SOUTHEASTERN TAMPICO-MISANTLA BASIN,
B 6405 752* CHIGNIK AREA, ALASKA PENINSULA, ALASKA, ABST., CRETACEOUS
B 6004 646* CHIHUAHUA, ABST., DEPOSITIONAL ENVIRONMENTS OF, TAMAROA
B 6004 723* CHIHUAHUA, ABST., DIAGENETIC HISTORY OF, MISSISSIPPIAN,
B 6102 297* CHIHUAHUA AND TRANS-PECOS, ABST., RESIDUAL, BOUGUER, GRAVITY,
B 6303 533* CHIHUAHUA, MEXICO, ABST., CARBONATE GEOLOGY OF, PENA BLANCA,
B 6109 1449 CHIHUAHUA TECTONIC BELT, TEXAS
M 30 360 CHILAPILLA FIELD, MEXICO, MACUSPANA BASIN
B 6311 2070 CHILDS, JONATHAN R., HYDROCARBON POTENTIAL OF ALEUTIAN BASIN,
B 6010 1646 CHILE, 1975, MARIO MARINO P., ON DEVELOPMENTS IN
B 6110 1580 CHILE, 1976, MARINO P., M., ON DEVELOPMENTS IN
B 6210 2048 CHILE, 1977, MARINO, M., ON DEVELOPMENTS IN
B 6310 1749 CHILE, 1978, AVALAS, R., ON DEVELOPMENTS IN
B 6104 558 CHILE AND ARGENTINA, MAGELLAN BASIN, PETROLEUM
B 6104 563 CHILE AND ARGENTINA, QUATERNARY GLACIOFLUVIAL DEPOSITS
B 6104 562 CHILE AND ARGENTINA, TERTIARY ZONA GLAUCONITICA
B 6104 560 CHILE, ANDEAN BATHOLITH
M 25 559 CHILE, ANDEAN MESOZOIC-CENOZOIC GEOSYNCLINAL CYCLE
M 25 170 CHILE, ANTOFAGASTA HYDROTHERMAL AREA
B 6104 562 CHILE, ARCILLA CON PINTAS STRATIGRAPHIC UNIT
B 6104 563 CHILE, ARENISCA ARCILLOSA STRATIGRAPHIC UNIT
B 6104 562 CHILE, BAHIA INUTIL GROUP
B 6104 563 CHILE, BRUSH LAKE FORMATION
B 6104 561 CHILE, BRUSH LAKE STRATIGRAPHIC UNIT
M 25 531 CHILE, COPPER DEPOSITS AND PRODUCTION
M 25 559 CHILE, COPPER, SILVER, AND GOLD
B 6104 562 CHILE, CRETA DURA STRATIGRAPHIC UNIT
B 6104 562 CHILE, CRETACEOUS
B 6104 561 CHILE, CULLEN FIELD, COLUMN
M 25 531 CHILE, EL LACO, MAGNETITE INTERPRETED AS A VOLCANIC FLOW
M 25 114 CHILE, EL TATIO FIELD, GEOTHERMAL
M 25 170 CHILE, EL TATIO GEOTHERMAL FIELD
B 6104 561 CHILE, EOCENE
B 6104 562 CHILE, ESTRATOS CON FAVRELLA STEINMANNI STRATIGRAPHIC UNIT
B 6104 561 CHILE, ESTRATOS CON FAVRELLA STRATIGRAPHIC UNIT
B 6104 560 CHILE, FELIPE BAY
B 6104 563 CHILE, FILARET FORMATION
B 6104 561 CHILE, FILARET STRATIGRAPHIC UNIT
B 6104 560 CHILE, FITZ ROY CHANNEL
B 6104 560 CHILE, GAJARDO CHANNEL
M 25 559 CHILE, IRON AND MANGANESE
B 6104 562 CHILE, JURASSIC
B 6104 562 CHILE, LUTITAS ARENOSAS STRATIGRAPHIC UNIT
B 6104 562 CHILE, LUTITAS GRIS VERDOSAS STRATIGRAPHIC UNIT
M 25 533 CHILE, MANGANESE
M 25 559 CHILE, METALLOGENESIS AND MINERALIZATION
M 25 559 CHILE, METALLOGENIC-TECTONIC EVENTS
M 25 559 CHILE, MINERALS
B 6104 561 CHILE, MIOCENE
M 25 559 CHILE, NICKEL AND CHROMIUM
M 25 169 CHILE, NORTHERN, GEOTHERMAL AREAS

B 6104 561 CHILE, OLIGOCENE
B 6104 561 CHILE, PALEOCENE
M 25 559 CHILE, PALEOZOIC GEOSYNCLINAL CYCLE
B 6104 563 CHILE, PALOMARES FORMATION
B 6104 561 CHILE, PALOMARES STRATIGRAPHIC UNIT
B 6104 561 CHILE, PLIOCENE
B 6110 1611 CHILE, PRODUCTION, 1975-1976
B 6210 2077 CHILE, PRODUCTION, 1976-1977
B 6310 1774 CHILE, PRODUCTION, 1978
M 25 173 CHILE, PUCHULDIZA GEOTHERMAL FIELD
B 6104 562 CHILE, SERIE TOBIFERA STRATIGRAPHIC UNIT
B 6104 562 CHILE, SPRINGHILL FORMATION
B 6212 2470 CHILE SUBDUCTION ZONE, PERU-
M 25 559 CHILE, SULFUR
M 25 174 CHILE, SURIRE THERMAL AREA
M 25 170 CHILE, TARAPACA HYDROTHERMAL AREA
M 25 171 CHILE TRENCH, PERU-
M 25 539 CHILE TRENCH, PERU-
M 25 529 CHILE TRENCH, PERU-
B 6212 2462 CHILE TRENCH, PERU-
M 25 541 CHILE TRENCH, RATE OF, NAZCA RIDGE, CONVERGENCE WITH, PERU-
B 6207 1239* CHILE TRENCH—RELATION TO MINERAL DEPOSITS, ABST., TECTONIC
M 25 169 CHILE, VOLCANISM
M 25 114 CHILE, WAIRAKEI FIELD, GEOTHERMAL
M 25 169† CHILE--SUMMARY, GEOTHERMAL EXPLORATION IN NORTHERN
M 25 559† CHILEAN CIRCUM-PACIFIC CONTINENTAL BORDER--SUMMARY,
B 6410 1726 CHILINGAR, BLATT, HARVER, REVIEW OF, DIAGENESIS IN SEDIMENTS
S 11 262 CHILOSTOMELLA CZIZEKI
S 11 294 CHILOSTOMELLA CZIZEKI
S 11 262 CHILOSTOMELLIDAE
B 6303 431 CHIMAHUSKY, JOHN S., DOLOMITE NONSTOICHIOMETRY, ITS RELATION
M 24 177 CHIMENE, CALVIN A., UPPER SMACKOVER RESERVOIRS, WALKER CREEK
B 6405 804 CHIMNEY, PETER C., PROPOSED MODEL FOR DEVELOPMENT OF RED
B 6210 1988 CHINA, 1977, YANG, Y. F., ON DEVELOPMENTS IN, TAIWAN, REPUBLIC OF
B 6303 424* CHINA, ABST., EXPLORATION GEOPHYSICS IN, PEOPLES REPUBLIC OF
B 6405 807* CHINA, ABST., LITHOFACIES AND, BIOHERMS OF UPPER, PALEOZOIC IN,
B 6303 524* CHINA ABST., OIL AND, GAS IN, PEOPLES REPUBLIC OF
B 6207 1226* CHINA, ABST., PETROLEUM GEOLOGY—, PEOPLES REPUBLIC OF
B 6405 699* CHINA, ABST., SEG DELEGATION VISIT TO, PEOPLES REPUBLIC OF
B 6405 807* CHINA, ABST., TECTONIC FEATURES OF OIL AND GAS, BASINS OF
B 6209 1654 CHINA BASIN, MAGNETIC ANOMALIES, SOUTH
B 6209 1646 CHINA BASIN RIDGE
B 6408 1211 CHINA, CHAO-SHUI BASIN
B 6210 1982 CHINA, DEVELOPMENTS, 1977, PEOPLES REPUBLIC OF
B 6310 1825 CHINA, DEVELOPMENTS, 1978, PEOPLES REPUBLIC OF
B 6411 1891 CHINA, DEVELOPMENTS, 1979, PEOPLES REPUBLIC OF
B 6408 1211 CHINA, DZUNGARIA BASIN
S 10 156 CHINA, DZUNGARIAN ARTESIAN BASIN, PETROLEUMMIGRATION
B 6408 1213 CHINA, GEOLOGIC BASINS, PEOPLES REPUBLIC OF
M 30 480 CHINA, GUADAO FIELD, SEISMIC PROFILES
M 30 482 CHINA, GUDAO FIELD, CHROMATOGRAPHS OF OIL SAMPLES
M 30 474 CHINA, GUDAO FIELD, STRUCTURE
B 6408 1213 CHINA, KIANGSU BASIN
B 6408 1211 CHINA, KWANGSI-KWEICHOW BASIN
B 6408 1211 CHINA, MIN-HO BASIN
M 30 471 CHINA, MIOCENE, GUDAO FIELD
B 6408 1211 CHINA, NORTH, BASINS
B 6408 1211 CHINA, ORDOS BASIN
B 6408 1211 CHINA, PO HAI BASIN
B 6408 1211 CHINA, PRE-NAN-SHAN BASIN
B 6110 1842 CHINA, PRODUCTION, 1975-1976
B 6210 1990 CHINA, PRODUCTION, 1975-1977, PEOPLES REPUBLIC OF
B 6411 1896 CHINA, PRODUCTION, 1977-1979, PEOPLES REPUBLIC OF
M 25 13 CHINA SEA, SEABED JURISDICTION, EAST
M 25 15 CHINA SEA SEABED JURISDICTION, SOUTH
B 6401 9 CHINA, SONGLIAO BASIN
B 6401 9 CHINA, SONGLIAO BASIN
B 6408 1211 CHINA, SUNG-LIAO BASIN
B 6408 1211 CHINA, SZECHWAN BASIN
B 6401 9 CHINA, TACHING, DAQING, GIANT OIL FIELD, LANDSAT DATA
B 6401 9 CHINA, TACHING, DAQING, GIANT OIL FIELD, LANDSAT DATA
M 30 10 CHINA, TACHING FIELD, LANDSAT IMAGE
B 6403 429 CHINA, TACHING HIGH-WAX OIL FIELD
B 6403 429 CHINA, TACHING HIGH-WAX OIL FIELD
B 6408 1211 CHINA, TARIM BASIN
B 6408 1211 CHINA, TECTONISM IN PRINCIPAL ONSHORE REGIONS
B 6408 1211 CHINA, TSAIDAM BASIN
B 6408 1211 CHINA, TUNGTING-HU BASIN
B 6408 1211 CHINA, TURFAN BASIN
M 30 476 CHINA, ZHANHUA BASIN, GUANTAO FORMATION
M 30 472 CHINA, ZHANHUA BASIN, SHAHEJIE FORMATION
M 30 473 CHINA, ZHANHUA BASIN, STRATIGRAPHIC COLUMN
B 6305 841* CHINA. BY TITLE ONLY, OIL AND, GAS IN
B 6405 710* CHINA—CONTINENTAL-SCALE CATACLASTIC FLOW, ABST., BASEMENT
B 6410 1701 CHINBURG, SUSAN J., CARBONATE SEDIMENT DRIFTS IN NORTHER
M 25 474 CHINESE MEGASTRUCTURE, SOUTH
M 25 467 CHINESE VOLCANIC BELT, EAST
B 6104 613 CHINITI MOUNTAINS, TEXAS
B 6203 374 CHINITNA BAY, ALASKA
B 6203 377 CHINITNA FORMATION, ALASKA
M 24 5 CHINITNA SILTSTONE, ALASKA
M 28 97 CHINLE FORMATION, ARIZONA, TRIASSIC
M 28 26 CHINLE FORMATION, ARIZONA, TRIASSIC
B 6007 1013 CHINLE FORMATION, NEW MEXICO
B 6007 1009 CHINLE FORMATION, ROCKY MOUNTAINS
B 6101 33 CHINLE FORMATION, URANIUM ORES
M 28 20 CHINLE FORMATION, UTAH, TRIASSIC
B 6111 1982 CHINO FAULT, CALIFORNIA
M 25 319 CHINTADO ANTICLINE, COLOMBIA
S 4 205 CHIONE CANCELLATA, YUCATAN
B 6303 481* CHIPOLA FORMATION, CALHOUN COUNTY, FLORIDA, ABST., DIAGENETIC
B 6409 1561* CHIPOLA FORMATION, CALHOUN COUNTY, FLORIDA, ABST., EARLY,
B 6408 1142 CHIRIKOF ISLAND, ALASKA
B 6407 1016 CHIRIKOF ISLAND, ALASKA
B 6203 377 CHISIK CONGLOMERATE MEMBER, ALASKA
B 6103 361 CHITWOOD STRUCTURE, OKLAHOMA
B 6209 1647 CHIU LIEN, TAIJUNT CRUISE, 1975, RV
M 28 135 CHLORITE (AUTHIGENIC), IN SANDSTONE, CEMENTS
M 28 46 CHLORITE, AFTER BIOTITE
M 28 134 CHLORITE, AUTHIGENIC, BIREFRINGENCE
M 28 135 CHLORITE, AUTHIGENIC, IN ROSETTES
M 28 71 CHLORITE, AUTHIGENIC, IN SANDSTONE
M 28 71 CHLORITE, AUTHIGENIC, IN SHALE
M 28 135 CHLORITE, AUTHIGENIC, VERMICULAR
M 28 47 CHLORITE, DETRITAL
M 28 67 CHLORITE, IN ARKOSE, CLAY
M 28 134 CHLORITE, IN SANDSTONE, CEMENTS
B 6405 794* CHLORITE-SMECTITE MIXED-LAYERED CLAY USING COMBINED
B 6212 2444 CHOCOLATE BAYOU, TEXAS
B 6303 486* CHOCOLAY GROUP--EARLY PROTEROZOIC CRATONI SEQUENCE, ABST.
B 6109 1538* CHOCTAWHATCHEE FORMATION, FLORIDA PANHANDLE, ABST., LATE,
M 25 268 CHOI, S. O., MARIN PETROLEUM EXPLORATION OF HUKSAN PLATFORM,
B 6405 775 CHONG, SHUANG-LING, SEDIMENTARY ASPECTS O ORGANIC MATERIAL
B 6112 2104 CHONTA FORMATION, PERU
B 6112 2104 CHONTA LIMESTONE, PERU
B 6303 431 CHOQUETTE, PHILIP W., MISSISSIPPIAN NON-SUPRATIDAL DOLOMITE STE.
B 6105 793 CHOQUETTE, P. W., CARBONATE-SAND WAVES, ISLA MUJERES, YUCATAN,
M 25 252 CHOU, J. T., PETROLIFEROUS TAIWAN BASINS IN FRAMEWOR OF
B 6309 1577 CHOU, MEI-IN M., PYROLYSIS OF EASTERN GAS SHALES--EFFECTS OF
B 6105 776 CHOUGH, S. K., CHANNEL-FILL AND SPILLOVER FACIES OF NORTHWEST
B 6203 442 CHOWDHARY, L. R., EARLY EOCENE SUBAERIAL EROSIONAL VALLEYS IN
B 6203 504 CHOWNS, T. M., PRE-CRETACEOUS GEOLOGY BENEATH GEORGIA
B 6108 1382 CHRISTENSEN. DAVID K., EVIDENCE FOR SOURCE AREAS AND
B 6012 2179 CHRISTENSEN, E. W., UPPER CRETACEOUS LITHOFACIES MODEL,
M 25 154 CHRISTIANSEN, ROBERT L., DEEP RESEARCH DRILLHOLE IN KILAUEAU
S 8 142 CHRISTIANSTED CANYON, TIME-VELOCITY CURVES, CURRENT FLOW
S 8 140 CHRISTIANSTED CANYON, TOPOGRAPHY, INDEX MAP
S 8 139 CHRISTIANSTED CANYON, WEST INDIES, ST. CROIX
S 8 139 CHRISTIANSTED CANYON, WEST INDIES, SUBMARINE CANYONS
B 6105 776 CHRISTIE, D. L., OTTO FIORD FORMATION—EVIDENCE OF SUBAQUEOUS
M 24 72 CHRISTIE, HAL H., MITSUE OIL FIELD, ALBERTA--A RICH STRATIGRAPHIC
B 6405 781 CHRISTIE, PHILLIP A. F., CONTINENTAL STRETCHING—EXPLANATION OF
B 6309 1598 CHRISTINA, CARLO C., YOU AINT SEEN NOTHING TILL YOUVE SEEN THE
B 6105 761 CHRISTOPHER, R. A., MESOZOIC AND CENOZOIC BIOSTRATIGRAPHY OF
B 6109 1432 CHROMATOGRAPH COLUMNS, GAS-
B 6201 108 CHROMATOGRAPHIC ANALYSES, MEADE PEAK MEMBER, GAS
B 6201 108 CHROMATOGRAPHIC ANALYSES, RETORT MEMBER, GAS
M 25 497 CHROMITE, AND NICKEL, CANADA, ASBESTOS
B 6207 1208* CHROMITE DEPOSITS IN PHILIPPINES, ABST., GEOLOGY AND ECONOMIC
M 25 460 CHROMITE, PHILIPPINES AND, INDONESIA
M 25 436 CHROMITE, REFRACTORY, LUZON
B 6303 443* CHRONICLE OF MIOCENE, PHASE III, MIDDLE MIOCENE EVENTS, ABST.
M 26 87 CHRONOCORRELATION, GLOBAL CYCLES, CENOZOIC
M 26 81 CHRONOCORRELATIONS, AFRICA, WEST, SEA LEVEL FLUCTUATIONS
S 6 38 CHRONOHORIZONS, NEED FOR RECOGNITION
S 6 369 CHRONOLITHOSTRATIGRAPHIC COLUMN, SOUTH AFRICA
S 6 13 CHRONOLOGIC SCALE, REVIEW
B 6002 276 CHRONOLOGIC UNITS, POLARITY-
S 6 21 CHRONOMETRIC SCALE, DIVISIONS OF UNITS
S 6 20 CHRONOMETRIC SCALE, NOMENCLATURE

S 6 13 CHRONOMETRIC SCALE, REVIEW
S 6 21 CHRONOSTRATIC SCALE, DIVISIONS OF UNITS
S 6 21 CHRONOSTRATIC SCALE, NOMENCLATURE
S 6 13 CHRONOSTRATIC SCALE, REVIEW
M 26 115 CHRONOSTRATIGRAPHIC CORRELATION, ACCURACY, SEISMIC,
M 26 158 CHRONOSTRATIGRAPHIC CORRELATION TO CYCLES, AFRICA, WEST
M 26 153 CHRONOSTRATIGRAPHIC CORRELATION TO CYCLES, TEXAS, GULF COAST
M 26 116 CHRONOSTRATIGRAPHIC CORRELATIONS, PROBLEMS, SEISMIC,
S 6 35 CHRONOSTRATIGRAPHIC HIERARCHY OF TERMS
M 26 99† CHRONOSTRATIGRAPHIC SIGNIFICANCE OF SEISMIC REFLECTIONS,
M 26 55 CHRONOSTRATIGRAPHIC SIGNIFICANCE, SEISMIC ANALYSIS,
S 6 367† CHRONOSTRATIGRAPHIC SUBDIVISION OF THE PRECAMBRIAN, ASPECTS
B 6002 276 CHRONOSTRATIGRAPHIC UNITS, POLARITY-
M 26 139 CHRONOSTRATIGRAPHIC WITH SEA LEVEL, CHART, CORRELATION
S 7 198 CHRONOSTRATIGRAPHY, CHART, WEST FLORIDA SLOPE
S 6 299† CHRONOSTRATIGRAPHY FOR THE WORLD PERMIAN
M 26 100 CHRONOSTRATIGRAPHY, SEISMIC, REFLECTIONS, UNCONFORMITIES
S 6 376 CHRONOSTRATIGRAPHY, SOUTH AFRICA, PRECAMBRIAN
M 26 109 CHRONOSTRATIGRAPHY, STRATIGRAPHIC TRAP EXPLORATION, SEISMIC,
B 6303 420* CHUAR GROUP, GRAND CANYON SUPERGROUP ARIZONA, AABST.,
B 6309 1598 CHUBER, STEWART, EXPLORATION METHODS OF DISCOVERY AND
B 6203 386 CHUGACH BAY FAULT, ALASKA
M 25 440 CHUGARYONG RIFT VALLEY, KOREA
M 25 443 CHUGARYONG RIFT VALLEY, SOUTH KOREA
B 6103 465* CHUKCHI SEA, ABST., CENOZOIC SEDIMENTATION AND TECTONICS OF,
B 6002 163 CHUKCHI VOLCANIC AND PLUTONIC ZONE, SIBERIA, OKHOTSK-
B 6407 1087† CHUKOTKA, KAMCHATKA, SAKHALIN, STRATIGRAPHY OF MARINE,
M 25 466 CHUKOTSK FOLDED AREA, USSR, VERKHOYANSK-
M 25 59 CHUKOTSK FOLDED REGION, USSR, VERKHOYANSK-
M 25 474 CHUKOTSK MEGASTRUCTURE, ALEUTIAN-
M 25 466 CHUKOTSK VOLCANIC BELT, USSR, OKHOTSK-
B 6103 465* CHUKSHI SEAS, ABST., GEOLOGIC FRAMEWORK AN RELATED
M 25 440 CHUNSONG GROUP, SOUTH KOREA
B 6211 2220 CHUPINES FAULT, CALIFORNIA
B 6211 2220 CHURCH CREEK FAULT, CALIFORNIA
B 6209 1674 CHURCHROCK MINES, NEW MEXICO
B 6206 1053 CHUSA TUFF MEMBER, TEXAS
B 6209 1674 CHUSKA DISTRICT, URANIUM, NEW MEXICO
S 11 160 CIBICIDES AMERICANUS AMERICANUS
S 11 160 CIBICIDES AMERICANUS CRASSISEPTUS
S 11 266 CIBICIDES FLORIDANUS
S 11 300 CIBICIDES FLORIDANUS
S 11 160 CIBICIDES FLORIDANUS
S 11 164 CIBICIDES HODGEI HODGEI
S 11 158 CIBICIDES ILLINGI
S 11 302 CIBICIDES LOBATULUS
S 11 266 CIBICIDES LOBATULUS
S 11 302 CIBICIDES MCKANNAI
S 11 266 CIBICIDES MCKANNAI
S 11 266 CIBICIDES MCKANNAI SPIRALIS
S 11 300 CIBICIDINA WASHINGTONENSIS
S 11 266 CIBICIDINA WASHINGTONENSIS
M 28 133 CIERBO SANDSTONE, CALIFORNIA, MIOCENE
M 28 66 CIERBO SANDSTONE, CALIFORNIA, MIOCENE
M 29 313 CILICIA-ADANA BASIN, EVOLUTION, MEDITERRANEAN
M 29 312 CILICIA-ADANA BASIN, SEDIMENTARY SEQUENCE, MEDITERRANEAN
B 6211 2261 CIMA SANDSTONE, CALIFORNIA
B 6311 2118* CIMARRON COUNTY, OKLAHOMA, ABST., ENVIRONMENTS AND
M 30 61 CIMMERIAN OROGENY, EFFECTS ON VIKING BASIN
S 5 175 CIMMINA BASIN, ITALY, SICILY,
B 6009 1621* CINCINNATI ARCH, ABST., GEOPHYSICAL AND TECTONIC STUDY OF,
B 6009 1621* CINCINNATI ARCH IN SOUTH-CENTRAL KENTUCKY, ABST., AGE AND
B 6009 1622* CINCINNATI ARCH PROVINCE, CENTRAL KENTUCKY T CENTRAL
B 6405 744* CINCINNATIAN KOPE FORMATION AND SOME PALEOECOLOGIC
B 6211 2291 CINCO DE MAYO UNIT, GULF OF MEXICO
M 30 343 CINCO PRESIDENTES FIELD, MEXICO, SALINE BASIN
M 25 477 CINNABAR-FLUORITE-ANTIMONITE JASPEROID
M 25 475 CINNABAR-FLUORITE-ANTIMONITE, USSR, EASTERN, STRATIFORM
B 6111 1951 CIRCLE CLIFFS HEAVY OIL DEPOSIT, UNITED STATES
B 6004 663* CIRCULATION AN SUSPENDED-SEDIMENT TRANSPORT OVER PART OF
B 6410 1702 CIRCULATION PATTERN OF NORTHERN BAHAMAS
B 6207 1234* CIRCUM-PACIFIC, AREA, ABST., ROLE OF, NUCLEAR, ENERGY IN FUTURE
M 25 176† CIRCUM-PACIFIC BELT--SUMMARY, OIL- AND, GAS-BEARING, BASINS OF
M 25 559† CIRCUM-PACIFIC CONTINENTAL BORDER--SUMMARY, METALLOGENIC
M 25 1† **CIRCUM-PACIFIC ENERGY AND MINERAL RESOURCES CONFERENCE, PAPERS,**
M 25 50 CIRCUM-PACIFIC GEOTHERMAL PHENOMENA
B 6212 2510* CIRCUM-PACIFIC MAP PROJECT, ABST.
B 6207 1232* CIRCUM-PACIFIC REGION, ABST., APPLICATIONS OF, LANDSAT, IMAGERY
B 6303 510* CIRCUM-PACIFIC REGION, ABST., DEDUCTION OF PAST, GEOTHERMAL
B 6207 1234* CIRCUM-PACIFIC REGION, ABST., NICKELIFEROUS, LATERITE DEPOSITS IN
B 6207 1222* CIRCUM-PACIFIC REGION, ABST., ORIGIN, DISTRIBUTION, AND
B 6207 1234* CIRCUM-PACIFIC REGION, ABST., PHOSPHOGENIC PROVINCES OF
B 6207 1212* CIRCUM-PACIFIC REGION, ABST., POSSIBLE RELATIONS BETWEEN, NORTH,
B 6207 1208* CIRCUM-PACIFIC REGION, ABST., PROVENANCE AND, GEOCHEMISTRY OF,
B 6207 1238* CIRCUM-PACIFIC REGION AS RELATED TO MINERAL AND ENERGY
M 25 32 CIRCUM-PACIFIC REGION, BATHYMETRY
M 25 48† CIRCUM-PACIFIC REGION, PLATE TECTONICS AND, MINERAL RESOURCES
M 25 37 CIRCUM-PACIFIC REGION, REGMATIC SHEAR PATTERN
M 25 33 CIRCUM-PACIFIC REGION, TECTONIC MAP
M 25 18† CIRCUM-PACIFIC REGION--SUMMARY, MINERAL TRADE IN
B 6408 1171 CISCO FACIES, TEXAS
B 6205 787 CISCO-CANYON SEISMIC DATA, TEXAS
M 24 262 CITRONELLE FIELD, TYPE SECTION, ALABAMA
M 24 261 CITRONELLE FORMATION, ALABAMA
M 24 259† CITRONELLE OIL FIELD, MOBILE COUNTY, ALABAMA
B 6008 1256 CITY DIKE FIELD, GALVESTON COUNTY, TEXAS
B 6004 681* CLA MINERALS AND THEIR RELATION TO PHYSICAL PROPERTIES, GULF
B 6012 2177 CLAERBOUT, JON F., SLANT-WAVE STACK, ABST.
B 6407 971 CLAGGETT SHALE, GREAT PLAINS
B 6403 325 CLAGGETT SHALE, MONTANA
B 6403 325 CLAGGETT SHALE, MONTANA
B 6309 1599* CLAIBORNE PARISH, LOUISIANA, ABST., CLASTICITY INDEX--KEY TO
B 6309 1600* CLAIBORNE PARISH, LOUISIANA--CASE HISTORY FROM DISCOVERY TO
B 6408 1220 CLAIBORNE SECTION, TEXAS
B 6303 432 CLANTON, UEL S., FAULTS OFFSETTING LAND SURFACE IN
B 6207 1117 CLARESHOLM AREA STUDY, ALBERTA
M 25 482 CLARK, ALLEN L., METAL PROVINCES OF ALASKA--SUMMARY
B 6105 776 CLARK, D. A., SYSTEM FOR REGIONAL LITHOFACIES MAPPING BY
B 6304 676 CLARK, D. N., PATTERNS O POROSITY AND CEMENT IN OOID RESERVOIRS
B 6405 691 CLARK, DAVID L., RISE AND FALL OF TRIASSIC CONODONTS, ABST.
B 6004 657 CLARK, G. R., II, PALEOECOLOGIC INVESTIGATIONS O LATE HOLOCENE
B 6309 1599 CLARK, GENE, APPLICATIONS OF SOLAR ENERGY. WHAT IS PRACTICAL IN
B 6303 432 CLARK, GEORGE R., II, PYRITIZATION IN SHELLS OF LIVING BIVALVES,
B 6004 566 CLARK, GEORGE S., PRE-CRETACEOUS ROCKS OF NORTHWESTERN
B 6309 1599 CLARK, MURLENE WIGGS,, SUBSURFACE NEOGENE STRATIGRAPHY OF
B 6207 1212 CLARK, ROBERT A., POSSIBLE RELATIONS BETWEEN NORTH PACIFIC
B 6408 1282 CLARK, ROBEY H., BURIED STRUCTURES IN MARINE EXPLORATION,
B 6403 439 CLARKE, DONALD D., COMPUTER SAND VOLUME DETERMINATIONS IN
B 6303 432 CLARKE, JAMES W., WEST SIBERIAN BASIN, ABST.
B 6004 657 CLARKE, R. H., CLIMATIC CONTROL OF DEPOSITION AT PRUDHOE BAY,
B 6305 799 CLARKE, R. H., RESERVOIR PROPERTIES OF CONGLOMERATES AND
B 6004 672 CLARKE, R. H., TRANSGRESSIVE-REGRESSIVE SHELF DEPOSITION,
B 6207 1216 CLARKE, S. H., JR., SEAFLOOR INSTABILITY ON PACIFIC CONTINENTAL
B 6405 706 CLARKE, SAMUEL H., JR., DIAPIR-LIKE RIDGES AND POSSIBLE
B 6105 783 CLARKE, SAMUEL H., SUBMARINE SLUMPS AND SEDIMENT FLOWS OFF SAN
B 6405 736 CLARKE, SAMUEL H., THERMOGENIC HYDROCARBON GASES IN
B 6206 933 CLASSIC TURBIDITES
B 6206 935 CLASSIC TURBIDITES, BOUMA, MODEL FOR
B 6004 651* CLASSIFICATIO OF PETROLEUM ZONES FOR APPRAISAL OF POTENTIAL
B 6403 297† CLASSIFICATION AND PETROLEUM PROSPECTS, CONTINENTAL, MARGINS--
S 6 367† CLASSIFICATION AND A PROPOSAL FOR THE CHRONOSTRATIGRAPHIC
B 6403 297† CLASSIFICATION AND PETROLEUM PROSPECTS, CONTINENTAL,
M 26 32 CLASSIFICATION CHART, LABORATORY, MEASUREMENTS, TEMPERATURE
M 26 354 CLASSIFICATION CHART, RESOLUTION SYSTEMS, ACOUSTIC, SYSTEMS
M 28 170 CLASSIFICATION, COMPARISONS, POROSITY
M 28 64 CLASSIFICATION, FINE-GRAINED ROCKS
M 28 93 CLASSIFICATION, GENERAL, SANDSTONE
B 6008 1158 CLASSIFICATION GUIDELINES, WELL-
B 6409 1301 CLASSIFICATION GUIDELINES, WELL-
B 6208 1281 CLASSIFICATION GUIDELINES, WELL-
B 6308 1174 CLASSIFICATION GUIDELINES, WELL-
B 6108 1126 CLASSIFICATION GUIDELINES, WELL-
B 6312 2214† CLASSIFICATION, NATURAL, FRACTURE SYSTEMS, DESCRIPTION AND
B 6004 663* CLASSIFICATION O SANDSTONE PORE STRUCTURE, ABST., NUMERICAL
B 6408 1187 CLASSIFICATION OF ANOXIC ENVIRONMENTS
B 6004 702* CLASSIFICATION OF CLASTIC SEDIMENTS, ABST.
B 6409 1529 CLASSIFICATION OF NATURAL GAS RESOURCES
M 25 480† CLASSIFICATION OF NORTH PACIFIC SHELVES--SUMMARY, STRUCTURAL
B 6409 1319 CLASSIFICATION OF WELLS, AAPG AND, API
B 6108 1155 CLASSIFICATION OF WELLS, AAPG AND, API
B 6308 1202 CLASSIFICATION OF WELLS, AAPG AND, API
B 6208 1309 CLASSIFICATION OF WELLS, AAPG AND, API
B 6008 1187 CLASSIFICATION OF WELLS, AAPG AND, API
B 6008 1187 CLASSIFICATION OF WELLS, AS APPLIED BY CSD, LAHEE
B 6208 1309 CLASSIFICATION OF WELLS, AS APPLIED BY CSD, LAHEE
B 6308 1202 CLASSIFICATION OF WELLS, AS APPLIED BY CSD, LAHEE
B 6108 1155 CLASSIFICATION OF WELLS, AS APPLIED BY CSD, LAHEE
B 6409 1319 CLASSIFICATION OF WELLS, AS APPLIED BY CSD, LAHEE
B 6309 1591* CLASSIFICATION SYSTEM ABST., REVISION OF, U.S GEOLOGICAL SURVEY
M 28 94 CLASSIFICATIONS, DIAGRAMS, SANDSTONE

B 6311 2115* CLAST, ABST., DISTRIBUTION OF, BITUMENS AND, KEROGEN IN, SHALE
B 6012 2186* CLAST POPULATIONS IN SESPE AND POWAY CONGLOMERATES,
B 6004 686* CLAST SIZE, SHAPE, AN COMPOSITION IN ANCIENT AND MODERN
B 6206 941 CLAST-SUPPORTED CONGLOMERATES
B 6004 709* **CLASTIC CARBONATE DEPOSITS ON MESOZOIC CONTINENTAL MARGIN,**
B 6409 1557* CLASTIC AND ARGILLACEOUS SEDIMENTS, UNITED STATES GULF COAST,
B 6004 645* CLASTIC AND STRATIGRAPHIC TRAP ANALYSIS BY COMPUTER, ABST.
B 6105 814* CLASTIC CORRELATIVES, DEVONIAN, EAST-CENTRAL NEW YORK, ABST.,
B 6004 655* CLASTIC DEPOSITION IN MARANHAO BASIN, NORTHEAST BRAZIL, ABST.,
M 26 165† CLASTIC DEPOSITIONAL FACIES, SEISMIC, STRATIGRAPHY AND, GLOBAL
B 6205 752† CLASTIC DEPOSITIONAL FACIES, SEISMIC STRATIGRAPHY AND, GLOBAL
B 6004 674* CLASTIC DEPOSITS, ABST., SEDIMENT, FACIES OF, SALT RIVER, ESTUARY,
B 6405 726* CLASTIC DIAGENESIS IN BOTH OPEN AND CLOSED SYSTEMS, ABST.,
B 6206 933 CLASTIC FAMILY, RESEDIMENTED COARSE-
M 28 64 CLASTIC NOMENCLATURE, TABLE, FINE-GRAINED
B 6104 484 CLASTIC RESERVOIRS
B 6002 327* CLASTIC ROCKS IN PARTS OF JACK, PALO PINTO, PARKER, AND WISE
B 6203 503* CLASTIC ROCKS OF MIDDLE AMAZON BASIN, BRAZIL, ABST., PETROLEUM
B 6002 186 CLASTIC SEDIMENTARY ROCKS, FOUR CHARACTERISTIC, GAMMA-, LOG
B 6004 702* CLASTIC SEDIMENTS, ABST., CLASSIFICATION OF
B 6002 324* CLASTIC SEDIMENTS, ABST., LIQUID, HYDROCARBON, MIGRATION PATHS
B 6203 506* CLASTIC SEDIMENTS, ABST., MIGRATION PATHWAYS IN, COMPACTING
B 6004 543† CLASTIC SEDIMENTS DURING COMPACTION, WATER, EXPULSION FROM
B 6205 753 CLASTIC SEISMIC FACIES UNITS, SEISMIC CHARACTERISTICS AND,
B 6303 464* CLASTIC STRANDLINE SYSTEMS, ABST., SPECTRUM OF DEPOSITIONAL
M 25 496 CLASTIC WEDGE, CANADIAN, CORDILLERA, MID-, PALEOZOIC
M 25 497 CLASTIC WEDGE, MINERAL DEPOSITS, CANADIAN, CORDILLERA
B 6309 1599* CLASTICITY INDEX--KEY TO CORRELATING DEPOSITIONAL AND
B 6303 501* CLASTICS IN SOUTHERN WEST VIRGINIA, ABST., SUBSURFACE
B 6208 1511 CLAUGHTON, JAMES L., DEVELOPMENTS IN SOUTH TEXAS IN 1977
B 6308 1338 CLAUGHTON, JAMES L., DEVELOPMENTS IN SOUTH TEXAS IN 1978
B 6405 680* CLAY AND SHALE, ABST., PERMEABILITY OF
S 10 210 CLAY CONVERSION, LOUISIANA, CALCASIEUPARISH, MANCHESTER FIELD
S 10 41 CLAY MINERAL CONVERSION, MIGRATION, AGENTS
M 28 72 CLAY, ATTAPULGITE, SEM
M 28 97 CLAY, AUTHIGENIC, IN QUARTZARENITE
M 28 122 CLAY, AUTHIGENIC, IN SANDSTONE
M 28 75 CLAY, BIBLIOGRAPHY
M 28 147 CLAY, BIBLIOGRAPHY, DIAGENESIS AND POROSITY
M 28 67 CLAY, CHLORITE, IN ARKOSE
M 28 131 CLAY COATINGS, CEMENTS
B 6002 293 CLAY, COMPACTIONAL RESPONSE OF, KAOLINITIC
M 28 65 CLAY, DETRITAL, IN PORE SPACE
M 28 136 CLAY, DETRITAL OCCURING WITH ALTERATION
B 6004 682* CLAY DIAGENESIS IN CAMBRIAN-ORDOVIDIAN COW HEA BRECCIA,
B 6203 497* CLAY DIAGENESIS IN WILCOX SANDSTONES OF SOUTHWEST TEXAS—
B 6108 1372* CLAY DIAGENESIS OF TRINIDAD SANDSTONE AND VERMEJO FORMATION,
S 7 117 CLAY FRACTION, GULF OF MEXICO, CONTINENTAL SLOPE
B 6002 293† CLAY, HIGH-, PRESSURE, CONSOLIDATION OF, KAOLINITIC
B 6303 541* CLAY IN RECENT SANDS, ABST., DISTRIBUTION OF
M 28 65 CLAY, INTERSTITIAL
M 28 67 CLAY, KAOLINITE, IN ARKOSE
B 6405 741* CLAY MINERAL SYSTEMS, ABST., PROPERTIES OF WATER IN
B 6105 815* CLAY MINERALOGY OF EASTERN GULF OF ALASKA, ABST.
B 6206 1056 CLAY MINERALS
B 6110 1901† CLAY, MINERALS AND PETROLEUM-FORMING REACTIONS DURING
B 6409 1557* CLAY MINERALS AS INDICATORS FOR DEPOSITIONAL ENVIRONMENT IN
B 6405 753* CLAY MINERALS FROM NORTHEAST GULF OF ALASKA, ABST.,
S 7 300 CLAY MINERALS, GYRE BASIN
B 6311 2036 CLAY MINERALS, X-RAY ANALYSIS OF
M 28 66 CLAY, MONTMORILLONITE-SMECTITE, AUTHIGENIC
B 6002 295 CLAY NO. 661, GEOTECHNICAL PROPERTIES, ENGLAND, BALL
M 28 69 CLAY, NONTRONITE, SEM VIEW
M 28 133 CLAY, OCCLUSION OF POROSITY, CEMENTS
M 28 72 CLAY, PALYGORSKITE, SEM
B 6002 297 CLAY, SCANNING-ELECTRON MICROGRAPH
B 6009 1615* CLAY SEDIMENTATION I PEARL RIVER DELTA, LOUISIANA-MISSISSIPPI,
M 28 67 CLAY, SERICITE, IN ARKOSE
M 28 65 CLAY, SHALE CLASTS
M 28 68 CLAY, SMECTITE, BIREFRINGENCE
M 28 68 CLAY, SMECTITE, IN A BENTONITE
B 6209 1756* CLAY TECHNOLOGY AND WELL STIMULATION, ABST.
B 6403 417 CLAY TRANSFORMATION
B 6403 417 CLAY TRANSFORMATION
B 6002 224 CLAY TRANSFORMATION, MONTMORILLONITE-TO-, ILLITE
B 6405 794* CLAY USING COMBINED SCANNING ELECTRON MICROSCOPY, X-RAY
B 6303 534* CLAY-CARBONATE DIAGENESIS OF DELTAIC SANDSTONES--BASAL BELLY
B 6004 692* CLAY-MEMBRANE FILTRATION AS MECHANISM FOR PRIMARY
B 6303 413* CLAY-MINERAL STRATIGRAPHY I SOUTH PHILIPPINE SEA, DSDP LEGS 59
B 6004 689* CLAY-MINERAL VARIABILITY IN SUSPENDED SEDIMENTS OF SAN
B 6404 488 CLAYPOOL, G. E., ORGANIC COMPOSITION OF SOME UPPER CRETACEOUS
B 6207 1225 CLAYPOOL, G. E., PETROLEUM GEOLOGY OF COOK INLET BASIN, ALASKA,
B 6405 692 CLAYPOOL, GEORGE E., APPALACHIAN BASIN DEVONIAN SHALES—
B 6408 1131 CLAYPOOL, GEORGE E., BIOGENIC AND THERMOGENIC ORIGINS OF
B 6305 838 CLAYPOOL, GEORGE E., EVIDENCE FOR TWO MAJOR CYCLES OF
B 6405 692 CLAYPOOL, GEORGE E., GEOCHEMICAL EFFECTS OF EARLY DIAGENESIS
B 6105 785 **CLAYPOOL, GEORGE E., NEWLY RECOGNIZED PETROLEUM SOURCE-ROCK**
B 6201 98 CLAYPOOL, GEORGE E., ORGANIC GEOCHEMISTRY, INCIPIENT
B 6108 1375 CLAYPOOL, GEORGE E., ORGANIC GEOCHEMISTRY, INCIPIENT
B 6103 466 CLAYPOOL, GEORGE E., ORIGIN OF COOK INLET OIL, ABST.
B 6407 1078 CLAYPOOL, GEORGE E., ORIGIN OF GASOLINE-RANGE HYDROCARBONS
B 6206 1075 CLAYPOOL, GEORGE E., PETROLEUM SOURCE-BED EVALUATION OF
B 6004 608 **CLAYPOOL, GEORGE E., THERMAL-ANALYSIS TECHNIQUE FOR**
B 6405 744 CLAYPOOL, GEORGE E., TWO OIL TYPES ON NORTH SLOPE OF ALASKA—
B 6405 766 CLAYPOOL, GEORGE, GEOLOGY OF CONTINENTAL SLOPE ADJACENT TO
B 6304 700 CLAYPOOL, GEORGE, PETROLEUM POTENTIAL OF EASTERN GULF OF
M 28 63 CLAYS AND SHALES, GENERAL
M 25 553 CLAYS AND TRAVERTINE MARBLE, SURVEYS, ECUADOR
B 6002 233 CLAYS, CORRELATION BETWEEN, PETROLEUM PRODUCTIVITY AND,
B 6406 922 CLAYS, EXPANDABLE
B 6303 344 CLAYS FERRY FORMATION, KENTUCKY
B 6006 973† CLAYS, FLOW OF, WATER AN, SOLUTES THROUGH, COMPACTED
B 6303 469* CLAYS IN EOCENE INTERLAMINATE SHALES AND SANDSTONES, ABST.,
M 28 130 CLAYS, IN SANDSTONES, CEMENTS
B 6203 548* CLAYS ON URANIUM DISTRIBUTION IN SOUTH TEXAS, ABST., EFFECT OF
M 28 69 CLAYSTONE, KAOLINITE, IN
M 28 70 CLAYSTONE, SEM, KAOLINITE, IN
B 6209 1767* CLAYTON FORMATION, LOWER PALEOCENE, DANIAN, ABST., OSTRACODA
B 6303 433 CLAYTON, J. L., PETROLEUM GENERATION AN MIGRATION IN DENVER
B 6410 1613 CLAYTON, J. L., PETROLEUM GENERATION AND MIGRATION IN DENVER
B 6309 1610 CLAYTON, JERRY L., ORGANIC GEOCHEMISTRY OF DEEP WELL IN HINDS
B 6008 1412 CLAYTON, JERRY L., SOURCE BEDS OF PETROLEUM IN DENVER BASIN,
B 6203 513* CLEAR CREEK AREA, NORTHEASTERN WYOMING, ABST., STRATIGRAPHY
B 6204 706 CLEAR CREEK CHERT, ILLINOIS BASIN
B 6305 832* CLEAR CREEK--RYCKMA CREEK NUGGET SANDSTONE TREND IN
B 6112 2091 CLEARFORK OIL, KEROGEN, TEXAS
B 6002 281 CLEARVILLE SILTSTONE, PENNSYLVANIA
B 6311 2060 CLEARWATER FORMATION, BRITISH COLUMBIA
B 6008 1394 CLEARY, JOHN G., ALVORD VALLEY, OREGON GEOTHERMAL
B 6406 841 CLEARY, WILLIA J., COMPARISON OF SAND-LAYER GEOMETRY ON FLAT
B 6105 777 **CLEARY, W. J., SAND PETROLOGY AND BIOCLASTIC COMPONENTS OF**
B 6409 1555 CLEAVES, A. W., CHESTER AND POTTSVILLE DEPOSITIONAL SYSTEMS,
B 6409 1554 CLEAVES, A. W., DEPOSITIONAL SYSTEMS AND LIGNITE PROSPECTING
B 6002 324 CLEAVES, ARTHUR, UPPER STRAWN DEPOSITIONAL SYSTEMS, LATE
B 6109 1551 CLEAVES, ARTHUR W., MIDDLE GLEN ROSE, CRETACEOUS, FACIES
B 6309 1601 CLEAVES, ARTHUR W., MIDDLE STRAWN, DESMOINESIAN, CRATONIC
B 6309 1596 CLEAVES, ARTHUR W., UPPER MISSISSIPPIAN DELTAS IN BLACK WARRIOR
B 6008 1394 CLEMENT, JAMES H., CEDAR CREEK ANTICLINE, PAST AND PRESENT,
B 6011 2067 CLEMENT, JAMES H., GEOLOGIC HISTORY--KEY TO ACCUMULATION AT
M 26 451 CLEMENT, WILLIAM A., A CASE HISTORY OF GEOSEISMIC MODELING OF
B 6312 2227 CLEMENTZ, DAVID M., EFFECT OF OIL AND BITUMEN SATURATION ON
M 25 4 CLEVELAND, HARLAN, THE MACROPROBLEM OF MODERNIZATION
B 6209 1706 CLEVELAND MEMBER, KENTUCKY
B 6209 1710 CLEVELAND MEMBER, OHIO
B 6303 433 CLIFFORD, HAROLD J., GEOLOGY OF A STRATIGRAPHIC GIANT--MESSLA
M 30 507 CLIFFORD, HAROLD J., GEOLOGY OF A STRATIGRAPHIC GIANT--MESSLA
B 6004 708 CLIFTON, H. E., BED FORMS AND INTERNAL STRUCTURES IN DEEP
B 6403 440 CLIFTON, H. E., MODERN AND ANCIENT COASTAL SEDIMENTARY FACIES,
B 6004 657 CLIFTON, H. E., MODERN AND ANCIENT ESTUARINE-FILL FACIES.
B 6303 433 **CLIFTON, H. EDWARD, SPILLED OIL AND INFAUNAL**
B 6404 692 CLIFTON, H. EDWARD, STRATIFICATION TYPES IN INTERTIDAL
B 6405 763* CLIMATE ASYMMETRY AND BIOGEOGRAPHIC DISTRIBUTIONS, ABST.
B 6403 447* CLIMATE CHANGE AND TECTONISM, PINE VALLEY NEVADA, ABST.,
B 6303 538* CLIMATE DURING CENOZOIC, ABST., OCEANS AND
B 6004 731* CLIMATE IN CARBONATE DUNE ROCKS, ABST., INDICATORS OF
M 25 120 CLIMATE MODIFICATION FROM RELEASE OF VAPORS
B 6403 434 CLIMATE, PAMPE, WILLIAM R., REVIEW OF, ENERGY AND
B 6405 671* CLIMATI CYCLES IN YAKATAGA FORMATION, GULF OF ALASKA, ABST.,
B 6004 707* CLIMATIC CHANGE, SOUTHWESTERN CALIFORNIA, ABST.,
B 6004 707* CLIMATIC CHANGE, SOUTHWESTERN CALIFORNIA AND NORTHWESTERN
S 9 17 CLIMATIC CHANGES, CAUSES
S 9 19 CLIMATIC CHANGES, DISTRIBUTION OF CONTINENTS
S 9 19 CLIMATIC CHANGES, ICE AGES
S 9 17 CLIMATIC CHANGES, MOUNTAIN BUILDING
S 9 19 CLIMATIC CHANGES, SEA-LEVEL CHANGES
B 6004 657* CLIMATIC CONTROL OF DEPOSITION AT PRUDHOE BAY, ALASKA,
B 6004 695* CLIMATIC CYCLES, ABST., PALEOGENE, RADIOLARIAN,
B 6009 1562† CLIMATIC STAGES OF PLEISTOCENE BRUNHES EPOCH, NANNOFOSSIL,
B 6004 662* CLIMATICALLY CONTROLLED, CYCLIC SEDIMENTATION IN WESTERN
B 6203 504 CLIMIE, J. A., EXPLORATION CASE HISTORY AND GEOLOGIC

B 6412 2078 CLINCHPORT FAULT, APPALACHIANS
B 6103 417 CLINCHPORT FAULT, VIRGINIA
B 6008 1390* CLINKERS AREA, WYOMING, ABST., SEISMIC EXPERIMENTS
M 26 127 CLINOFORMS, SEISMIC, REFLECTIONS, PATTERNS
M 26 125 CLINOFORMS, SEISMIC, REFLECTIONS, PATTERNS
M 28 140 CLINOPTILOLITE, BIREFRINGENCE, ETC.
M 28 140 CLINOPTILOLITE, CRYSTALS, PRISMATIC TO PLATY
M 28 141 CLINOPTILOLITE, CRYSTALS, SEM
M 28 56 CLINOZOISITE, BIREFRINGENCE
M 28 56 CLINOZOISITE, CRYSTALS, IN SCHIST
M 27 147 CLINTON FORMATION, PENNSYLVANIA, SILURIAN
M 27 144 CLINTON FORMATION, PENNSYLVANIA, SILURIAN
M 28 45 CLINTON FORMATION, PENNSYLVANIA, SILURIAN
S 5 54 CLINTON FORMATION, SILURIAN, NIAGARA GROUP,
M 28 106 CLINTON FORMATION, VIRGINIA, SILURIAN
M 28 5 CLINTON FORMATION, VIRGINIA, SILURIAN
S 5 55 CLINTON INLET, LOCATION, MICHIGAN,
B 6408 1282* CLINTON SANDSTONE, NORTHEASTERN OHIO, ABST., DEPOSITIONAL
M 25 462 CLIPPERTON-TEHUANTEPEC-BARTLETT FAULT, HUTANOPAN OR,
B 6104 597 CLIVE FIELD, ALBERTA
M 29 428 CLOSED BASINS, GULF OF MEXICO, GENERAL BATHYMETRY
B 6209 1763* CLOSED GULF OF MEXICO, PRE-ATLANTIC OCEANPLATE
B 6103 371 CLOUD CHIEF FORMATION, CALIFORNIA
B 6109 1556 CLOUD, KELTON, DIAGENETIC PATTERNS OF AUSTIN GROUP AND THEIR
B 6103 465 CLUFF, LLOYD S., ALYESKA ENGINEERING. BY TITLE ONLY
B 6310 1890 CLUFF OIL LTD., PETROCONSULTANTS S.A., QUINTANA OVERSEAS, INC.
B 6405 692 CLUFF, R. M., RELATION OF DEPOSITIONAL FACIES AND HISTORY TO
B 6408 1282 CLUFF, ROBERT M., PALEOCHANNEL ACROSS LOUDEN ANTICLINE,
B 6309 1575 CLUFF, ROBERT M., PETROLOGY AND MATURATION OF DISPERSED
B 6004 651* CLUSTER AND CORRESPONDENCE ANALYSIS. BY TITLE ONLY.,
B 6004 677* CLUSTER-ANALYSIS INTERPRETATION OF SUBSURFACE DATA. BY TITLE
B 6004 647* CO3-SO4 DIAGENESIS AND TIMING OF HYDROCARBON MIGRATION,
B 6203 368 COACHELLA-IMPERIAL VALLEYS, CALIFORNIA
B 6409 1564* COAHUILA, MEXICO, ABST., DIAGENESIS OF, DELTAIC SANDSTONE--,
B 6109 1548* COAHUILA, MEXICO, ABST., GEOLOGY OF, SIERRA BALUARTES, DIAPIR
B 6303 495* COAHUILA PENINSULA, POTRERO DE LA MULA QAND SIERRA DEL FUSTE,
S 10 180 COAL, METHANE, OCCURRENCES
S 10 117 COAL GASIFICATION, ARTIFICIAL WATER-ROCK INTERACTION
B 6405 689* COAL, ABST., GEOLOGIC CONTROLS ON, SULFUR CONTENT IN
B 6408 1286* **COAL, ABST., GEOLOGY AND, MINING OF, TENNESSEE**
B 6405 702* COAL, ABST., X-RAY MINERALOGY OF, UPPER, FREEPORT
B 6408 1290* COAL AND COAL DEVELOPMENTS IN WESTERN KENTUCKY, ABST.
M 25 391† COAL, AND STEEL PRODUCTION IN NEW ZEALAND--SUMMARY, IRON ORE
B 6212 2399 COAL AND SURROUNDING LITHOLOGIES, WEST VIRGINIA, BECKLEY
M 25 81 COAL BASINS AND MINES, AUSTRALIA
B 6305 839* COAL BED, HIGH-SULFUR, RADIOACTIVE COAL OF PALUDAL-LACUSTRINE
B 6405 788* COAL BED, HOMER CITY, PENNSYLVANIA, ABST., FORMS AND
B 6405 774* COAL BEDS, ABST., RESOURCE EVALUATION OF GAS-BEARING
B 6008 1395* COAL BEDS IN WASATCH FORMATION, WESTERN POWDER RIVER BASIN,
B 6405 743* COAL BEDS OF APPALACHIAN AND ILLINOIS BASINS, ABST., RESIN RODS
B 6405 756* COAL CHARACTERIZATION FOR OPTIMUM UTILIZATION, ABST.
B 6203 548* COAL CLEANING, DESULFURIZATION, BY HIGH-INTENSITY MAGNETIC
B 6105 831* COAL CONVERSION, ABST., WEST VIRGINIAS, NORTHERN, COALFIELD--
B 6405 732* COAL DEPOSITION ACROSS POWDER RIVER BASIN IN NORTHEASTERN
B 6303 533* COAL DEPOSITION IN SOUTHERN WEST VIRGINIA, PENNSYLVANIAN
B 6302 194† COAL, DEPOSITIONAL ENVIRONMENTS OF, FORT UNION SEDIMENTS,
B 6305 846* COAL DEPOSITS IN SOUTHEASTERN SALINE COUNTY, ILLINOIS, ABST.,
B 6102 295* **COAL DEPOSITS IN THURBER AREA, TEXAS. BY TITLE ONLY**
B 6305 824* COAL DEPOSITS OF UPPER CRETACEOUS MESAVERDE GROUP IN
B 6408 1283* COAL DEPOSITS, SOUTHERN ILLINOIS, ABST., FACIES, PALEOECOLOGY,
M 25 467 COAL DEPOSITS, USSR, NORTHEAST
B 6408 1283* COAL DEVELOPMEN IN MICHIGAN BASIN, ABST., PROSPECTS FOR
B 6305 827* **COAL DEVELOPMENT IN POWDER RIVER BASIN OF WYOMING,**
M 25 85 COAL DISTRIBUTION, INDONESIA
B 6004 658* COAL ENVIRONMENTS, ABST.
B 6303 438* COAL EXPLORATION, ABST., HIGH-RESOLUTION, SEISMIC METHODS IN
B 6405 803* COAL EXPLORATION, ABST., NEW AND DEVELOPING TECHNIQUES IN
B 6207 1239* COAL EXPLORATION AND DEVELOPMENT, ABST., QUEENSLAND
B 6108 1380* COAL EXPLORATION AND DEVELOPMENT, ABST., UTILIZATION OF,
M 25 80† COAL EXPLORATION AND MINE DEVELOPMENT IN AUSTRALIA, STATUS
B 6408 1278* COAL EXPLORATION AND MINE PLANNING, ABST., APPLICATION OF
B 6212 2379† COAL EXPLORATION AND MINE PLANNING IN APPALACHIAN REGION,
B 6008 1403* COAL EXPLORATION IN SOUTHEASTERN MONTANA AND
B 6405 689* COAL FIELD, NEW MEXICO, ABST., VARIATIONS I, CRETACEOUS
B 6211 2358* COAL FOR CALIFORNIA, ABST.
B 6405 761* COAL FROM APPALACHIAN AND EASTERN INTERIOR REGIONS AND
B 6408 1281* COAL GEOLOGY AND RESOURCE OF EASTERN KENTUCKY, ABST.
B 6303 532* COAL IN ANTARCTICA, ABST.
B 6207 1224* COAL IN AUSTRALIA, ABST., HIGH-RESOLUTION, SEISMIC EXPLORATION
B 6405 689* COAL IN CENTRAL APPALACHIAN BASIN, ABST., GEOLOGIC CONTROLS ON
M 25 98† COAL IN JAPAN--SUMMARY, INFLUENCE OF, TEMPERATURE ON,
B 6408 1284* COAL IN PENNSYLVANIA--GEOLOGY, CURRENT PRODUCTION, AND
M 25 94† COAL IN REPUBLIC OF KOREA, PALEOZOIC AND, MESOZOIC
B 6408 1287* COAL IN WEST VIRGINIA--GEOLOGY AND CURRENT MINING TRENDS,
M 25 553 COAL INVESTIGATIONS, ECUADOR
B 6112 2094 COAL, KEROGEN, PENNSYLVANIAN-AGE
S 9 14 COAL, LATITUDE DISTRIBUTION, PALEOCLIMATE
B 6203 523* COAL MINES, ABST., USE OF DEPOSITIONAL, MODELS IN PREDICTING,
B 6405 724* COAL MINING PROBLEMS IN HARTSHORNE FORMATION, EAST-CENTRAL
B 6212 2511* COAL MINING--SIMULATION MODELS, ABST., ECONOMICS OF
B 6305 839* COAL OF PALUDAL-LACUSTRINE ORIGIN IN WASATCH FORMATION OF
S 9 6 COAL, PALEOCLIMATE INDICATORS
B 6405 724* COAL PARAMETERS, ABST., CRETACEOUS-, TERTIARY VERSUS,
B 6311 2058† COAL PETROLOGY OF TULAMEEN COALFIELD, SOUTH-CENTRAL BRITISH
B 6406 938 COAL PRODUCING AREA, APPALACHIAN PLATEAU, MAJOR
M 25 391 COAL PRODUCTION, NEW ZEALAND
B 6405 727* COAL PROPERTIES, ABST., GEOLOGIC FRAMEWORK FOR DEVELOPMENT,
B 6303 523* COAL PROPERTIES ON COMBUSTION CHARACTERISTICS, ABST., IMPACT
B 6105 819* COAL PROPERTY AND MAXIMIZING TOTAL RECOVERABLE RESERVE,
B 6212 2400 COAL QUALITY, SULFUR PROBLEMS
B 6408 1285* COAL RANK IN PART OF WESTERN KENTUCKY COALFIELD, ABST.
B 6008 1404* COAL RESEARCH ABST., NEW FRONTIERS IN, COLORADO
B 6309 1585* COAL RESERVE FIGURES. CASE STUDY OF WINIFREDE COAL, BOONE
B 6004 698* COAL RESERVES BY COMPUTER METHODS, ABST., CALCULATION OF
M 25 105 COAL RESERVES, CANADA
M 25 86 COAL RESERVES, INDONESIA
M 25 99 COAL RESERVES, JAPAN
M 25 96 COAL RESERVES, KOREA, SOUTHERN
M 25 92 COAL RESERVES, TAIWAN
B 6408 1289* COAL RESOURCE, ABST., ILLINOIS COAL--A MAJOR, BITUMINOUS
B 6405 762* COAL RESOURCE AND RESERVE EVALUATIONS IN UNITED STATES, ABST.,
B 6405 684* COAL RESOURCE DATA PROCESSING, OLD AND NEW TECHNIQUES, ABST.
B 6405 745* COAL RESOURCE ESTIMATES, ABST., DEVELOPMENT OF CONCEPTUAL
B 6405 705* COAL RESOURCE ESTIMATION, ABST., GEOSTATISTICAL APPROAC IN
B 6408 1284* COAL RESOURCE STUDIES IN VIRGINIA ABST.
M 25 84† COAL RESOURCES AND POTENTIAL IN INDONESIA--SUMMARY
B 6309 1591* COAL RESOURCES CLASSIFICATION SYSTEM ABST., REVISION OF, U.S
B 6405 735* COAL RESOURCES DATA SYSTEM, NCRDS, APPLIED TO STUDY OF
B 6405 672* COAL RESOURCES IN COMPLEXLY FAULTED AREA CONTAINING OLD
B 6408 1289* COAL RESOURCES IN OHIO, ABST.
M 25 104† COAL RESOURCES OF ALASKA--SUMMARY
M 25 105† COAL RESOURCES OF CANADIAN CORDILLERA--SUMMARY
B 6105 806 * **COAL RESOURCES OF COLORADO AND WYOMING**
B 6203 552* COAL RESOURCES OF ILLINOIS, ABST.
B 6408 1284* COAL RESOURCES OF INDIANA AND POTENTIAL GEOLOGIC PROBLEMS IN
B 6303 503* COAL RESOURCES OF PART OF SOUTHERN ILLINOIS BASIN, ABST.
M 25 89† COAL RESOURCES OF TAIWAN--SUMMARY
B 6310 1983 COAL RESOURCES OF THE AMERICAS--SELECTED PAPERS, EDITED BY
B 6205 879* COAL RESOURCES, PINE CANYON QUADRANGLE, CARBON COUNTY,
B 6405 725* COAL SEAM, ABST., PETROGRAPHIC VARIATION IN WESTERN, KENTUCKY,
B 6406 939 COAL SEAM, WEST VIRGINIA, LOWER, STOCKTON
B 6108 1390* COAL SEAMS, ABST., HIGH-RESOLUTION, SEISMIC EXPLORATION FOR
B 6405 800* COAL SEAMS, UPPER PENNSYLVANIAN OF WEST VIRGINIA, ABST.,
B 6305 826* **COAL VARIATIONS IN FLUVIAL DEPOSITION OF PALEOCENE TONGUE**
B 6212 2408 COAL, WEST VIRGINIA, CEDAR GROVE
B 6405 730* COAL-BEARING CARBONIFEROUS SEQUENCE EXPOSED NEAR JOGGINS,
B 6105 830* COAL-BEARING NARRAGANSETT BASIN, MASSACHUSETTS AND RHODE
B 6405 796* COAL-BEARING STRATA, NORTHERN KAIPAROWITS PLATEAU, UTAH,
B 6004 703* COAL-ENERGY RESOURCES, PACER, INTERACTIVE, COMPUTER PROGRAM
B 6203 490* COAL-FORMING SWAMP--THE EVERGLADES, ABST., PROFILE OF
B 6203 564* COAL-GAS WELLS, ABST., FOAM, STIMULATION FOR, ENHANCED
B 6303 435* COAL-MINING PROBLEMS, ABST., SURFACE SEISMIC METHODS APPLIED
B 6209 1586 COAL-RANK SCALE WITH VARIOUS MATURATION INDICES AND ZONES
B 6009 1613* **COAL-RESOURCE ASSESSMENT AND UTILIZATION, ABST., EVALUATION OF,**
B 6203 512* COAL-RESOURCE DISTRIBUTION TRENDS IN APPALACHIAN COAL FIELD,
B 6008 1412* COAL-SEA CORRELATIONS IN UPPER CRETACEOUS SWAMP COMPLEX,
M 25 99 COALFIELD, JAPAN
B 6311 2058† COALFIELD, SOUTH-CENTRAL BRITISH COLUMBIA, DEPOSITIONAL
M 25 106 COALFIELDS, CANADA
M 25 95 COALFIELDS, KOREA, SOUTHERN
M 25 89 COALFIELDS, TAIWAN
B 6105 831* COALFIELD--DIRECTION OF RESEARCH TO SUPPORT COAL CONVERSION,
M 25 98† COALIFICATION OF TERTIARY COAL IN JAPAN--SUMMARY, INFLUENCE
B 6309 1577* COALIFICATION PATTERNS IN EASTERN KENTUCKY, VIRGINIA, WEST
B 6205 848 COALINGA EAST EXTENSION FIELD, CALIFORNIA
B 6108 1381* COALMONT FORMATION NEAR COALMONT, JACKSON, COUNTY,
S 10 58 COALS AS SOURCE ROCKS, MIGRATION IN AQUEOUS SOLUTION
B 6309 1581* COALS, ABST., GEOLOGIC AGE VERSUS MAJOR MINOR, AND
B 6303 507* COALS, ABST., PORPHYRINS IN, SUBBITUMINOUS AND, BITUMINOU
B 6203 489* COALS AND SHALES, ABST., ANALOGIES BETWEEN, HOLOCENE, COASTAL-,

B 6202 302 COALS, ATOMIC RATIOS OF, BITUMENS AND
B 6202 301† COALS, CHARACTERIZATION OF, BITUMENS AND
B 6309 1578* COALS FOR COAL UTILIZATION SCHEMES, ABST., PETROGRAPHIC
B 6212 2401 COALS, FORMS, PYRITE IN
B 6408 1288* COALS IN NORTHERN WEST VIRGINIA, ABST., POTTSVILLE ALLUVIAL
B 6309 1579* COALS IN SOUTHERN BOONE AND NORTHERN LOGAN COUNTIES, WEST
B 6305 840* COALS IN UPPE CRETACEOUS FERON SANDSTONE MEMBER, CENTRAL
B 6212 2387 COALS, KENTUCKY, ELKHORN
B 6303 519* COALS OF FERRON SANDSTONE MEMBER OF MANCOS
B 6004 651* COALS OF ROCKY MOUNTAIN REGION, ABST., ALTERED, VOLCANIC, ASH
B 6406 938† COALS, SOUTHWEST WEST VIRGINIA, EXPLORATION MODEL FOR,
B 6101 113 COALVILLE MEMBER, UTAH
B 6101 114 COALVILLE, TO CUMBERLAND GAP, WYOMING, CORRELATION, UTAH
B 6203 540* COAL—IMPORTANT ENERGY RESOURCE, ABST., MONTANAS
B 6405 795* COAL—MAJOR UNEXPLOITED SEAM IN ILLINOIS, ABST., SEELYVILLE
M 27 131 COAMO SPRINGS LIMESTONE, PUERTO RICO, EOCENE
M 27 13 COAMO SPRINGS LIMESTONE, PUERTO RICO, EOCENE
M 27 33 COAMO SPRINGS LIMESTONE, PUERTO RICO, EOCENE
M 27 140 COAMO SPRINGS LIMESTONE, PUERTO RICO, EOCENE
M 27 12 COAMO SPRINGS LIMESTONE, PUERTO RICO, EOCENE
B 6203 543* COARSE SEDIMENTS DEPOSITED AT BASE OF CONTINENTAL SLOPE BY
B 6203 542* COARSE-GRAINED DEPOSITS BY EPHEMERAL STREAMS IN
B 6405 785* COARSENING-UPWARD SHELF SEQUENCES IN APHEBIAN WISHART
B 6105 765 * COARSE WHITE SPARRY DOLOMITE AND CALCITE CEMENTS—INDICATORS OF
M 25 503 COAST CRYSTALLINE BELT, CANADIAN, CORDILLERA
B 6403 325 COAST, HYDRODYNAMIC ZONES, MODERN INTERDELTAIC
B 6403 325 COAST, HYDRODYNAMIC ZONES, MODERN INTERDELTAIC
B 6004 704* COAST PLAIN, ABST., CRETACEOUS AND EARLY, TERTIARY,
M 25 511 COAST RANGE BATHOLITH
M 25 509 COAST RANGE BATHOLITH
B 6301 21 COAST RANGE THRUST, CALIFORNIA
B 6211 2244 COAST RANGES, CALIFORNIA
B 6203 345 COAST RANGES, CALIFORNIA
M 25 485 COAST RANGES, MERCURY, CALIFORNIA
B 6405 790* COAST RANGES OF CALIFORNIA, ABST., DIAGENESIS IN, MONTEREY
B 6403 316† COASTAL AND DELTAIC SEDIMENTATION OF UPPER CRETACEOUS EAGLE
B 6411 1842 COASTAL CORP., MARATHON OIL CORP.,AND PETROCONSULTANTS S.A.
B 6004 724* COASTAL DEPOSITS, ABST., MODELS USING, SUBSURFACE DATA FO
B 6105 789 * COASTAL-DUNE GEOMORPHOLOGY AND CLASSIFICATION
B 6004 704* COASTAL DYNAMICS ON WEST- AND EAST-FACING BEACHES IN
B 6105 818* COASTAL ENVIRONMENTS AND BEACH MORPHOLOGY OF NORTHERN
M 26 66 COASTAL ONLAP, SEA LEVEL FLUCTUATIONS
M 26 63† COASTAL ONLAP, SEISMIC, STRATIGRAPHY AND, GLOBAL CHANGES OF,
B 6203 507* COASTAL PLAIN, ABST., MESOZOIC AND CENOZOIC GEOLOGY OF,
B 6001 87† COASTAL PLAIN OF NEW JERSEY, UPPER, CRETACEOUS SUBSURFACE
B 6209 1714† COASTAL PLAIN: DISCUSSION AND REPLY, RECORRELATION OF, EOCENE-,
B 6209 1667 COASTAL RANGE PROVINCE, TAIWAN
B 6405 668* COASTAL RANGES, CENTRAL PERU, ABST., EARLY, CRETACEOUS ARC
S 4 113 COASTAL REEFS, VENEZUELA, CHICHIRIVICHE AREA
S 4 113 COASTAL REEFS, VENEZUELA, CHICHIRIVICKE AREA
B 6203 531* COASTAL SEDIMENTARY-FACIES PATTERNS IN GRABEN EMBAYMENTS IN
B 6008 1407* COASTAL SEDIMENTATION MODEL, TENSLEEP SANDSTONE,
B 6306 989 COASTAL SOUTHERN CALIFORNIA, BY ROBERT SHARP, BEAR, TED L.,
B 6405 787* COASTAL SWAMP ORIGIN OF SAN MIGUEL LIGNITE DEPOSIT, JACKSON
B 6002 202 COASTAL WELLS, BRAZIL
B 6002 202 COASTAL WELLS, SOUTH ATLANTIC
B 6009 1617* COASTAL ZON MANAGEMENT IN LOUISIANA, ABST., THE, OIL, INDUSTRY
B 6009 1609* COASTAL ZONE, ABST., INTERPRETATION OF UNENHANCED, LANDSAT,
B 6207 1211* COASTAL ZONE, ABST., PETROLEUM DEVELOPMENT IN
B 6303 501* COASTAL-BOUNDARY LAYER, ABST., DISPERSAL OF FINE-GRAINED
B 6203 489* COASTAL-MARSH DEPOSITS AND PRE-QUATERNARY COALS AND SHALES,
B 6009 1611* COASTAL-PLAIN SEDIMENTS, ABST., MULTIVARIATE, MINERAL ANALYSIS
B 6203 556* COASTLINE OF SOUTH CAROLINA, ABST., BACK-, BARRIER
S 4 83 COATES, ANTHONY G., JAMAICAN CORAL-RUDIST FRAMEWORKS AND
B 6409 1555 COATES, E. J., SUBSURFACE WILCOX LIGNITE IN WEST-CENTRAL
M 28 132 COATINGS, ILLITE, AUTHIGENIC
M 28 127 COATINGS, PHOSPHATE, OOLITIC
M 28 133 COATINGS, SMECTITE, ON DETRITAL SAND
M 25 460 COBALT, PHILIPPINES, INDONESIA, AND, NEW CALEDONIA, NICKEL AND
M 25 426 COBAR, COPPER, ZINC, AND GOLD, NEW SOUTH WALES
M 25 104 COBB, E. H., COAL RESOURCES OF ALASKA--SUMMARY
B 6408 1283 COBB, LAVERNE B., DEVONIAN SHALE EXPLORATION RATIONALE TESTED
B 6209 1615 COBBLESKILL FORMATION, APPALACHIAN BASIN
B 6004 697* COBEQUID BAY, NOVA SCOTIA, ABST., FACIES, MODE FOR, MACROTIDAL
B 6209 1610 COBLESKILL LIMESTONE, NEW YORK
M 27 55 COBLESKILL LIMESTONE, NEW YORK, SILURIAN
M 27 19 COCCOLITH, RHABDOLITH
M 27 168 COCCOLITHS
M 27 17 COCCOLITHS, GOLDEN-BROWN ALGAE
B 6012 2191* COCCOLITHUS PELAGICUS--CLARIFICATION. ABST.
M 27 18 COCCOSPHERE
M 30 199 COCCOSPHERE, SEM PHOTO, EKOFISK FORMATION
B 6307 1108 COCHISE COUNTY, REEFS, ARIZONA
B 6306 986 COCHISE, SOUTHEASTERN ARIZONA, 1978, NEW MEXICO GEOLOGIC
B 6307 1094 COCKLESHELL GULLY FORMATION, AUSTRALIA
B 6107 975 COCOBEACH SERIES, GABON
S 3 33 COCONINO SANDSTONE, ARIZONA
B 6105 771 * COCONINO SANDSTONE, PERMIAN, FOSSIL VERTEBRATE FOOTPRINTS—
B 6004 653 COCONOWER, A. B., SET-THEORETIC INFORMATION RETRIEVAL FROM
B 6405 693* COCORP REFLECTION DATA, ABST., CRYSTALLINE, OVERTHRUSTING OF
B 6405 742* COCORP WIND RIVER, WYOMING, SEISMIC REFLECTION DATA, ABST.,
B 6403 360 COCOS PLATE, BOUNDARY
B 6403 360 COCOS PLATE, BOUNDARY
B 6002 273† CODE CONCERNING MAGNETOSTRATIGRAPHIC UNITS, NOTE 44--,
B 6102 252† CODE, CONCERNING SMALLEST FORMAL ROCK-STRATIGRAPHIC UNIT,
B 6207 1185† CODE OF STRATIGRAPHIC NOMENCLATURE TO PROVIDE GUIDELINES
B 6107 1083† CODE TO INTRODUCE POINT-BOUNDARY STRATOTYPE CONCEPT, NOTE
B 6406 943 CODELL SANDSTONE MEMBER, MINNESOTA AND, SOUTH DAKOTA
S 4 307 CODRINGTON LAGOON, BARBUDA
S 4 303 CODRINGTON LIMESTONE, BARBUDA
B 6202 185 CODY SHALE, WYOMING
B 6404 490 CODY SHALE, WYOMING, SAGE BREAKS MEMBER
B 6203 531* COELOBITES, IN LOWER CAMBRIAN REEFS, LABRADOR, ABST.,
B 6209 1610 COEYMANS FORMATION, PENNSYLVANIA
B 6209 1610 COEYMANS LIMESTONE, MARYLAND
M 25 420 COFFEE ROCK
B 6302 259 COFFEEN, HUH, JOHN M., REVIEW ON, SEISMIC EXPLORATION
M 28 157 COFFINITE, REPLACEMENT OF QUARTZ
B 6206 1059 COFFINITE, URANINITE OR
M 25 453 COFFINITE, URANIUM ORE MINERAL, JAPAN
M 24 51 COFFMAN, R C., TAGLU GAS FIELD, BEAUFORT BASIN, NORTHWEST
B 6105 821* COGOLLO GROUP, MARACAIBO PLATFORM, VENEZUELA, ABST.,
S 6 1 COHEE, GEORGE V., EDITED BY, THE GEOLOGIC TIME SCALE
B 6105 796 COHEN, A. D., DEPOSITIONAL MODELS IN COAL EXPLORATION AND MINE
B 6105 759 COHEN, A. D., THIN SECTION ANALYSIS OF COASTAL-MARSH SEDIMENTS
B 6303 434 COHEN, A. S., ENVIRONMENTAL SIGNIFICANCE OF FOSSIL ROOT CAST
B 6403 423 COHEN, CURTIS R., PROGRESSIVE NEOGENE DEFORMATION OF
B 6203 505 COHEN, STAN, SPEAKEASY—COMPUTER LANGUAGE FOR GEOLOGISTS,
B 6001 108 COHEN, ZFANIA, EARLY CRETACEOUS BURIED CANYON--INFLUENCE ON
B 6012 2184* COILING RATIOS IN PLANKTONIC FORAMINIFERA--ADAPTIVE STRATEGY
B 6303 457* COKE COUNTY, TEXAS, ABST., POROSITY EXAMPLES FROM, IAB FIELD
B 6105 787 * COKING-COAL RESERVES OF OKLAHOMA
B 6403 439 COLAZAS, XENOPHON C., COMPUTER SAND VOLUME DETERMINATIONS
B 6211 2351 COLBURN, IVAN P., LATE CRETACEOUS SEDIMENTATION IN SANTA
B 6211 2357* COLD BAY AREA OF ALASKA PENINSULA, ABST., STRATIGRAPHY AND
B 6309 1522† COLD BAY--FALSE PASS AREA, ALASKA PENINSULA, TERTIARY
B 6111 1951 COLD LAKE HEAVY OIL DEPOSIT, WESTERN, CANADA
B 6205 782 COLD LAKE SALT, ALBERTA
B 6004 692* COLD-WATER CARBONATE DEPOSITS AS POSSIBLE PALEOCLIMATIC
B 6309 1577 COLE, GARY A., REGIONAL COALIFICATION PATTERNS IN EASTERN
B 6102 237 COLE, MARK R., EOCENE PALEOCURRENTS AND SEDIMENTATION, SA
B 6104 616 COLE, MARK R., SANTA MONICA AND SANTA ANA
B 6303 410 COLE, MARY LOU, GLACIAL-MARINE SEDIMENTS OF WEST ANTARCTICA
B 6303 434 COLE, MARY LOU, NEARSHORE GLACIAL-MARINE SEDIMENTATION
B 6305 824 COLE, R. D., DEFINITION OF DEPOSITIONAL FACIES OF PARACHUTE
B 6203 550 COLE, R. D., DEPOSITIONAL LITHOFACIES OF PARACHUTE CREEK
B 6203 505 COLE, R. D., IRON-SULFIDE MINERALOGY AND MORPHOLOGY IN OIL
B 6305 836 COLE, R. D., PETROLOGY OF TERTIARY SANDSTONES OF SOUTHERN
B 6004 656* COLE-RESOURCE DATA SYSTEM, ABST., AVAILABILITY AND APPLICATION
B 6004 658 COLEMAN, D. D., IDENTIFICATION OF LEAKAGE GAS FROM
B 6303 434 COLEMAN, D. D., ISOTOPE GEOCHEMISTRY OF NATURA GAS IN ILLINOIS
B 6203 505 COLEMAN, J. M., CONTEMPORARY GRAVITY TECTONICS—AN EVERYDAY
B 6303 426 COLEMAN, J. M., SUBAQUEOUS LANDSLIDES AND DIAPIRS, GULF OF
B 6303 512 COLEMAN, JAMES M., ACTIVE SUBMARINE LANDSLIDE PROCESSES,
B 6004 658 COLEMAN, JAMES M., COAL ENVIRONMENTS, ABST.
S 7 1 COLEMAN, JAMES M., EDITED BY, FRAMEWORK, FACIES, AND OIL-
S 7 1 COLEMAN, JAMES M., INTRODUCTION, FRAMEWORK, FACIES, AND OIL-
B 6011 2068 COLEMAN, JAMES M., MODERN RIVER DELTAS--VARIABILITY OF
B 6303 550 COLEMAN, JAMES M., SEDIMENT SUSPENSION, DEWATERING, AND
B 6011 2068 COLEMAN, JAMES M., SUBMARINE-SLOPE FAILURES AND DOWNSLOPE
B 6009 1613 COLEMAN, S. L., EVALUATION OF QUALITY AND QUANTITY CONTROLS
B 6211 2247 COLES LEVEE TREND, CALIFORNIA
B 6303 555* COLIMA, MEXICO, ABST., IMMANITAS BED, PASO DEL RIO
B 6104 603 COLINA LIMESTONE, NEW MEXICO
B 6203 523* COLLAPSE IN ATHAPUSCOW AULACOGEN, MIDDLE PRECAMBRIAN,
B 6205 864 COLLECTIONS AVAILABLE FROM FORTY-EIGHT STATES, MINERAL, ROCK,
B 6009 1613* COLLIER COUNTIES, FLORIDA, ABST., SUNOCO-, FELDA AND, WEST
M 24 287† COLLIER COUNTIES, FLORIDA, SUNOCO-FELDA FIELD, HENDRY AND
B 6303 493* COLLIER COUNTY, FLORIDA, ABST., SHELL BORINGS IN UPPER, MIOCENE-,
B 6410 1577 COLLIERSTOWN BEDS, VIRGINIA
B 6002 318 COLLINS, A. GENE, APPLICATIONS OF OIL FIELD WATER GEOCHEMISTRY

B 6308 1376 COLLINS, DOUGLAS J., DEVELOPMENT IN ARKANSAS, NORTH LOUISIANA,
B 6108 1234 COLLINS, DOUGLAS J., DEVELOPMENTS IN ARKANSAS, LOUISIANA, AND
B 6409 1490 COLLINS, DOUGLAS J., DEVELOPMENTS IN ARKANSAS, NORTH
B 6208 1487 COLLINS, DOUGLAS J., DEVELOPMENTS IN ARKANSAS, NORTH
B 6309 1603 COLLINS, E. W., QUEEN CITY FORMATION IN EAST TEXAS
B 6304 655 COLLINS, HORACE R., DEVONIAN BENTONITES IN EASTERN OHIO
B 6407 1004 COLLINS, STEPHEN E., JURASSIC COTTON VALLEY AND SMACKOVER
M 25 73 COLLINS, WAYNE L., SOLAR DESALINATION EXPERIENCE OF UNIVERSITY
B 6004 659 COLLINSON, J. D., DELTAIC EVOLUTION DURING BASIN FILL--NAMURIA
B 6303 311 COLLINSON, JAMES W., DEPOSITIONAL HISTORY OF PALEOCENE--LOWER
B 6207 1171 COLLINSON, JAMES W., STRATIGRAPHIC RELATIONS OF PARK CITY
B 6406 868† COLLISION I BANDA SEA REGION, ARC-, CONTINENT
M 27 153 COLLOPHANE FRAGMENTS
B 6206 984 COLMAN-SADD, S. P., FOLD DEVELOPMENT IN ZAGROS SIMPLY FOLDED
B 6408 1286* COLMAR-PLYMOUTH CONUNDRUM, ABST.
B 6106 904 COLNETT BASIN, BAJA CALIFORNIA
B 6412 2189 COLOMBIA 1959-1978, BY COLOMBIAN SOCIETY OF PETROLEUM
B 6010 1658 COLOMBIA, 1975, ROWLINSON, N. R., ON DEVELOPMENTS IN
B 6110 1582 COLOMBIA, 1976, ROWLINSON, N. R., ON DEVELOPMENTS IN
B 6210 2049 COLOMBIA, 1977, VALDERRAMA, B., AND J., VELANDIA, ON
B 6310 1749 COLOMBIA, 1978, VALDERRAMA, R., AND, VELANDIA, J., ON
B 6212 2510* COLOMBIA, ABST., GEOLOGY AND, HYDROCARBON PROSPECTS OF, SINU
M 25 339 COLOMBIA, AMAZON BASINS, POTENTIAL
M 25 319 COLOMBIA, ATRATO AREA, POTENTIAL
M 25 322 COLOMBIA, BASEMENT
M 29 337 COLOMBIA, BEATA RIDGE, STRUCTURE
M 25 319 COLOMBIA, BERREBERRE ANTICLINE
M 25 319 COLOMBIA, BUCHADO ANTICLINE
M 25 321 COLOMBIA, CALIMA ANTICLINE
B 6110 1583 COLOMBIA, CARTAGENA OFFSHORE
S 10 195 COLOMBIA, CARTAGENA, GAS BUBBLE IN MUD VOLCANO
S 10 196 COLOMBIA, CARTAGENA, TOTUMO MUD VOLCANO
M 25 321 COLOMBIA, CHAGUII ANTICLINE
M 25 319 COLOMBIA, CHINTADO ANTICLINE
M 29 330 COLOMBIA, COLOMBIAN BASIN, BATHYMETRY
M 25 321 COLOMBIA, COPOMA ANTICLINE
M 25 323 COLOMBIA, CRETACEOUS
M 29 341 COLOMBIA, CRETACEOUS, EVOLUTION
M 25 339† COLOMBIA, ECUADOR, AND, HYDROCARBON POTENTIAL OF, AMAZON
M 25 324 COLOMBIA, EOCENE
M 29 341 COLOMBIA, EOCENE, EVOLUTION
M 29 345 COLOMBIA, EVOLUTION, ANDEAN OROGENY
M 29 345 COLOMBIA, EVOLUTION, OLIGOCENE-MIOCENE DIASTROPHISM
M 29 341 COLOMBIA, EVOLUTION, PRE-ANDEAN OROGENY
M 25 326 COLOMBIA, EXPLORATION, HISTORY
M 25 533 COLOMBIA, GOLD
B 6004 652* COLOMBIA GROUP OF NEW JERSEY COSTAL PLAIN ABST., TIMING AND,
B 6110 1583 COLOMBIA, GUAJIRA BASIN
M 29 341 COLOMBIA LINEAMENT, COLOMBIA, PALEOTRENCHES
B 6110 1583 COLOMBIA, LLANOS BASIN
B 6110 1583 COLOMBIA, LOWER MAGDALENA VALLEY
M 29 329† COLOMBIA, MAJOR, STRUCTURAL ELEMENTS AND, EVOLUTION OF
M 25 551 COLOMBIA, METALLOGENIC PROVINCES
M 25 551 COLOMBIA, METALLOGENIC PROVINCES
B 6110 1582 COLOMBIA, MIDDLE MAGDALENA VALLEY
M 25 325 COLOMBIA, MIOCENE
M 29 345 COLOMBIA, MIOCENE, EVOLUTION
S 10 192 COLOMBIA, MUD DIAPIRS, ECHOGRAM
M 25 319 COLOMBIA, NAPIPI ANTICLINE
S 4 234 COLOMBIA, NEMASTER DISCOIDEA CARPENTER
S 4 233 COLOMBIA, NEMASTER GRANDIS CLARK
M 29 342 COLOMBIA, NORTHWESTERN, CRETACEOUS-PALEOCENE
M 29 343 COLOMBIA, NORTHWESTERN, EOCENE PALEOGEOGRAPHY
M 29 331 COLOMBIA, NORTHWESTERN, GENERAL GEOLOGY
M 29 346 COLOMBIA, NORTHWESTERN, MIOCENE PALEOGEOGRAPHY
M 29 344 COLOMBIA, NORTHWESTERN, OLIGOCENE PALEOGEOGRAPHY
M 29 332 COLOMBIA, NORTHWESTERN, PHYSIOGRAPHY
M 29 348 COLOMBIA, NORTHWESTERN, PLEISTOCENE PALEOGEOGRAPHY
M 29 347 COLOMBIA, NORTHWESTERN, PLIOCENE PALEOGEOGRAPHY
M 29 334 COLOMBIA, NORTHWESTERN, REGIONAL STRATIGRAPHY
M 29 335 COLOMBIA, NORTHWESTERN, SAN JACINTO FRAGMENTED BELT
M 29 339 COLOMBIA, NORTHWESTERN, SINU BELT
M 29 349 COLOMBIA, NORTHWESTERN, TECTONIC EVOLUTION
M 29 333 COLOMBIA, NORTHWESTERN, TECTONIC MAP
M 25 324 COLOMBIA, OLIGOCENE
M 29 341 COLOMBIA, OLIGOCENE, EVOLUTION
M 25 319 COLOMBIA, OPOGADO ANTICLINE
M 29 341 COLOMBIA, PALEOCENE, EVOLUTION
M 29 340 COLOMBIA, PALEOTRENCHES, BOLIVAR LINEAMENT
M 29 341 COLOMBIA, PALEOTRENCHES, COLOMBIA LINEAMENT
M 29 340 COLOMBIA, PALEOTRENCHES, ROMERAL LINEAMENT
M 29 340 COLOMBIA, PALEOTRENCHES, SINU LINEAMENT
M 25 533 COLOMBIA, PLATINUM
M 29 345 COLOMBIA, PLEISTOCENE, EVOLUTION
M 25 325 COLOMBIA, PLIOCENE
M 29 345 COLOMBIA, PLIOCENE, EVOLUTION
M 25 551 COLOMBIA, PORPHYRY COPPER MINERALIZATION DISCOVERED
B 6110 1613 COLOMBIA, PRODUCTION, 1975-1976
B 6210 2078 COLOMBIA, PRODUCTION, 1976-1977
B 6310 1775 COLOMBIA, PRODUCTION, 1977-1978
B 6110 1582 COLOMBIA, PUTUMAYO BASIN
M 25 325 COLOMBIA, QUATERNARY
M 25 321 COLOMBIA, REMOLINO GRANDE ANTICLINE
M 29 337 COLOMBIA, SAN JACINTO BELT, STRUCTURE
M 25 319 COLOMBIA, SAN JUAN AREA, POTENTIAL
M 25 319 COLOMBIA, SATAGUI ANTICLINE
M 29 448 COLOMBIA, SEISMIC EXPERIMENT, CONTINENTAL MARGINS
M 25 321 COLOMBIA, SIGUIRISUA ANTICLINE
M 25 319 COLOMBIA, TAGACHI ANTICLINE
M 29 330 COLOMBIA, TECTONICS, MAJOR ELEMENTS
M 25 323 COLOMBIA, TERTIARY
S 4 233 COLOMBIA, TROPIOMETRA CARINATA LAMARCK
M 25 321 COLOMBIA, TUMACO AREA, POTENTIAL
M 25 319 COLOMBIA, URODO ANTICLINE
S 10 195 COLOMBIA, VOLCANCITOS, MUD VOLCANO
M 25 551† COLOMBIA--SUMMARY, POSSIBILITIES FOR DEVELOPMENT OF, MINERAL
M 29 330 COLOMBIAN BASIN, BATHYMETRY, COLOMBIA
B 6412 2189 COLOMBIAN SOCIETY OF PETROLEUM GEOLOGISTS AND GEOPHYSICISTS,
B 6107 1014 COLONY-MCLAREN-WASECA MEMBERS, ALBERTA AND, SASKATCHEWAN
M 26 303 COLOR DISPLAY, SEISMIC APPLICATION, USE OF
M 26 304 COLOR DISPLAYS, EXAMPLES, SEISMIC, INTERPRETATION
B 6004 644* COLOR DISPLAYS IN SEISMIC INTERPRETATION, ABST., USE OF
B 6105 829* COLOR DISPLAYS OF COMPLEX SEISMIC TRACE, ABST., GEOLOGIC
M 27 1 **COLOR ILLUSTRATED GUIDE TO CARBONATE ROCK**
M 28 1 **COLOR ILLUSTRATED GUIDE TO CONSTITUENTS.**
B 6310 1981 COLOR ILLUSTRATED GUIDE TO CONSTITUENTS, TEXTURES, CEMENTS
B 6104 505 COLOR OF HEATED KEROGEN, ELEMENTARY COMPOSITION AND
B 6104 504† COLOR OF KEROGEN AS INDEX OF ORGANIC MATURITY
B 6405 686 COLORADO, ABST., LOWER, PERMIAN DEPOSITIONAL SYSTEMS,
B 6108 1383* COLORADO, ABST., BIOSTRATIGRAPHIC, CORRELATION OF, MESAVERDE
B 6203 550* COLORADO, ABST., DEPOSITIONAL LITHOFACIES OF, PARACHUTE CREEK
B 6303 510* COLORADO, ABST., DIAGENESIS IN, HYGIENE AND, TERRY SANDSTONES,
B 6405 675* COLORADO, ABST., DRILLING FOR, METHANE, GAS IN, FISHERS PEA AREA,
B 6108 1381* COLORADO, ABST., ENVIRONMENTS OF DEPOSITION OF, COALMONT
B 6405 709* COLORADO, ABST., EOLIAN-, FLUVIATILE, CONTINENTAL, ORIGIN OF,
B 6205 882* COLORADO, ABST., EXPLORATION FOR NONELECTRIC, GEOTHERMAL
B 6108 1373* COLORADO, ABST., GEOLOGICAL, GEOCHEMICAL, AND PETROPHYSICAL
B 6108 1387* COLORADO, ABST., GEOLOGY AND, SHALE-OIL RESOURCES OF, WHITE
B 6108 1389* COLORADO, ABST., GEOTHERMAL EXPLORATION IN CENTRAL
B 6305 834* COLORADO, ABST., HOT-SPRING-GENERATED, KARST FEATURES NEAR
B 6108 1389* COLORADO, ABST., HYDROCARBON POTENTIAL OF, ARCHULETA,
B 6203 505* COLORADO, ABST., IRON-, SULFIDE MINERALOGY AND MORPHOLOGY IN,
B 6205 885* COLORADO, ABST., MAESTRICHTIAN CONGLOMERATES IN
B 6108 1388* COLORADO, ABST., MARINE-SHELF MODEL FOR UPPER, CRETACEOUS,
B 6405 801* COLORADO, ABST., METHANE MEASUREMENT FROM, SALINE ZONE, OIL
B 6405 714* COLORADO, ABST., OIL AND GAS POTENTIAL OF, SAN LUIS BASIN,
B 6108 1385* COLORADO, ABST., OIL POTENTIAL OF MESOZOIC SEDIMENTS BENEATH,
B 6105 828* COLORADO, ABST., ORIGIN OF, RHIZOCORALLIUM IN, CRETACEOUS,
B 6008 1413* COLORADO, ABST., OVERPRESSURED, SHALE MASSES AND, GROWTH,
B 6305 826* COLORADO, ABST., PALEOKARST CONTROLS O LOCALIZATION OF,
B 6405 733* COLORADO, ABST., PATTERNS OF SHALLOW-MARIN DEPOSITION, UPPER,
B 6008 1393* COLORADO, ABST., PETROLEUM GEOLOGY OF, FORT UNION AND,
B 6203 491* COLORADO, ABST., POTENTIAL OF, MORROW SANDSTONES IN
B 6108 1382* COLORADO, ABST., PREVENTION OF DAMAGE TO, WATERFOWL ON,
B 6405 672* COLORADO, ABST., PROMISING, MORROW SANDSTONE DISCOVERIES IN
B 6405 796* COLORADO, ABST., REEVALUATION OF DEPOSITIONAL ENVIRONMENTS
B 6108 1372* COLORADO, ABST., STRATIGRAPHY AND, CLAY DIAGENESIS OF,
B 6205 886* COLORADO, ABST., STRATIGRAPHY OF, CEDAR MOUNTAIN AND, DAKOTA
B 6205 887* COLORADO, ABST., STRATIGRAPHY OF, MESAVERDE GROUP AND
B 6405 678* COLORADO, ABST., TRIASSIC, PALEOCALICHE IN, RED BEDS O, DOLORES
B 6405 755* COLORADO, ABST., URANIUM OCCURRENCE IN, LEADVILLE DOLOMITE A,
B 6305 839* COLORADO, ABST., VERMILLION CREEK, COAL BED, HIGH-, SULFUR,
B 6405 801* COLORADO, ABST., WATTENBERG AND, SPINDLE FIELDS—
B 6410 1614 COLORADO, ADENA FIELD
B 6302 158 COLORADO AND NEW MEXICO, HOGBACK MONOCLINE
B 6302 159 COLORADO AND NEW MEXICO, MESAVERDE SANDSTONE
S 3 94 COLORADO AND NEW MEXICO, MORRISON FORMATION
B 6302 194† COLORADO, AND THEIR RELATION TO COAL, DEPOSITIONAL
B 6206 1093* COLORADO, AND UTAH, ABST., PALEOCURRENTS OF, JACKPILE
B 6405 728* COLORADO AND UTAH, ABST., STRATIGRAPHIC RELATIONS OF, PERMIAN
B 6302 198 COLORADO AND WYOMING, SAND WASH BASIN

B 6410 1615 COLORADO, ANTLER FIELD
B 6302 195 COLORADO, AXIAL ANTICLINE
M 29 113† COLORADO BASIN AND CONTINENT-OCEAN CRUST BOUNDARY OFF
M 29 118 COLORADO BASIN, FORMATIONS, AGES GIVEN
M 29 115 COLORADO BASIN, GEOLOGY, ARGENTINA
M 29 119 COLORADO BASIN, STRUCTURE SECTION, ARGENTINA
B 6410 1615 COLORADO, BLACK HOLLOW FIELD
M 26 278 COLORADO, BOULDER OIL FIELD
M 26 277 COLORADO, BOULDER, WELD, AND, ADAMS COUNTIES
B 6410 1620 COLORADO, BUCKAROO FIELD
S 3 126 COLORADO, CARNATION MINE
B 6107 983 COLORADO, CHALK, PRODUCTION FROM
B 6405 761* COLORADO, CLASSIC AREA OF STRATIGRAPHIC EXPLORATION, ABST.,
B 6008 1404* COLORADO COAL RESEARCH ABST., NEW FRONTIERS IN
B 6108 1226 COLORADO COUNTY, TEXAS, SOUTH DUBINA FIELD
M 27 107 COLORADO, CRETACEOUS, DAKOTA GROUP
M 27 108 COLORADO, CRETACEOUS, DAKOTA SANDSTONE
M 27 177 COLORADO, CRETACEOUS, GREENHORN LIMESTONE
M 27 36 COLORADO, CRETACEOUS, GREENHORN LIMESTONE
M 28 41 COLORADO, CRETACEOUS, MESAVERDE GROUP
B 6408 1172 COLORADO, CUTLER ARKOSE, UTAH AND
M 24 139 COLORADO, D SANDSTONE
M 24 139 COLORADO, DAKOTA FORMATION
M 24 139 COLORADO, DAKOTA FORMATION
S 7 246 COLORADO DELTA BULGE, TEXAS, SHELF-SLOPE
B 6402 277 COLORADO DELTA, TEXAS
B 6402 277 COLORADO DELTA, TEXAS
B 6410 1614 COLORADO, DENVER BASIN
M 24 137 COLORADO, DENVER BASIN, MAP
M 24 139 COLORADO, DENVER-JULESBURG BASIN
B 6008 1234† COLORADO, DEVELOPMENTS IN EASTERN AND NORTHWESTERN
M 28 60 COLORADO, EOCENE, GREEN RIVER FORMATION
M 28 59 COLORADO, EOCENE, GREEN RIVER FORMATION
M 28 57 COLORADO, EOCENE, GREEN RIVER FORMATION
M 28 67 COLORADO, EOCENE, HUERFANO FORMATION
M 28 138 COLORADO, EOCENE, HUERFANO FORMATION
B 6107 997 COLORADO, FORT HAYS LIMESTONE MEMBER, NIOBRARA LIMESTONE
B 6302 210 COLORADO, FORT UNION COALS
B 6302 212 COLORADO, FORT UNION FORMATION
B 6410 1614 COLORADO, FT. COLLINS FIELD
B 6410 1621 COLORADO, GRANEROS SHALE
B 6407 981 COLORADO GROUP, ALBERTA
S 3 129 COLORADO, GYPSUM VALLEY
B 6408 1172 COLORADO, HERMOSA CARBONATE ROCK, UTAH AND
B 6408 1238 COLORADO, HYGIENE SANDSTONE BAR COMPLEXES
B 6108 1199† COLORADO IN 1976, DEVELOPMENTS IN EASTERN AND NORTHWESTERN
B 6208 1355† COLORADO IN 1977, DEVELOPMENTS IN EASTERN AND NORTHWESTERN
B 6308 1371† COLORADO IN 1978, DEVELOPMENTS IN EASTERN AND NORTHWESTERN
B 6409 1484† COLORADO IN 1979, DEVELOPMENTS IN EASTERN AND NORTHWESTERN
M 24 136 COLORADO, J SANDSTONE
M 28 65 COLORADO, JURASSIC, MORRISON FORMATION
M 28 26 COLORADO, JURASSIC, MORRISON FORMATION
M 28 133 COLORADO, JURASSIC, MORRISON FORMATION
M 28 166 COLORADO, JURASSIC, MORRISON FORMATION
M 28 157 COLORADO, JURASSIC, MORRISON FORMATION
B 6302 196 COLORADO, LANCE FORMATION
B 6410 1620 COLORADO, LINDON FIELD
B 6410 1615 COLORADO, LOVELAND FIELD
B 6201 81 COLORADO, LOWER, CORRELATION, SASKATCHEWAN
S 3 129 COLORADO, MARY L. MINE
B 6112 2141 COLORADO, MESAVERDE SANDSTONES, NEW MEXICO AND
B 6408 1172 COLORADO, MOLAS CARBONATE ROCK, UTAH AND
B 6410 1621 COLORADO, MOWRY SHALE
B 6107 983 COLORADO, NIOBRARA LIMESTONE
B 6107 997 COLORADO, NIOBRARA LIMESTONE
B 6108 1376* COLORADO OIL-SHALE DEPOSITS, ABST., DISTRIBUTION AND, ORIGIN OF,
M 27 105 COLORADO, OLIGOCENE, CREEDE FORMATION
M 28 54 COLORADO, OLIGOCENE, FISH CANYON TUFF
M 28 141 COLORADO, PALEOCENE, POISON CANYON FORMATION
B 6408 1168 COLORADO, PALEOZOIC, CARBONATE WEDGES, UTAH AND
B 6408 1168 COLORADO, PALEOZOIC, CARBONATE WEDGES, UTAH AND
B 6408 1172 COLORADO, PARADOX SALT, UTAH AND
B 6302 195 COLORADO, PARK RANGE
M 27 10 COLORADO, PENNSYLVANIAN, MINTURN FORMATION
M 27 11 COLORADO, PENNSYLVANIAN, MINTURN FORMATION
B 6410 1614 COLORADO, PIERCE FIELD
B 6405 686* COLORADO PLATEAU, ABST., MAJOR PETROLEUM-IMPREGNATED ROCK
B 6105 843* COLORADO PLATEAU, ABST., SANDSTONE DEPOSITIONAL SYSTEMS IN,
B 6405 677* COLORADO PLATEAU AND VICINITY, ABST., PENNSYLVANIAN AND
B 6101 35 COLORADO PLATEAU, URANIUM DEPOSITS
B 6206 1091* COLORADO PLATEAU URANIUM DEPOSITS, ABST., NEW, MODEL FOR
B 6203 501* COLORADO PLATEAU URANIUM DEPOSITS. BY TITLE ONLY., NEW, MODEL
B 6112 2141 COLORADO, POINT LOOKOUT SANDSTONE, NEW MEXICO AND
B 6403 429 COLORADO, POWDER WASH HIGH-WAX OIL FIELD
B 6403 429 COLORADO, POWDER WASH HIGH-WAX OIL FIELD
M 25 485 COLORADO, PRECIOUS AND BASE METALS
S 10 227 COLORADO, RANGELY FIELD, HYDROTHERMAL EFFECTS
B 6307 1026 COLORADO, RANGELY OIL FIELD STRUCTURE, SEISMIC PROFILE
B 6408 1172 COLORADO, REDWALL FACIES, UTAH AND
B 6109 1540* COLORADO RIVER, INNER TEXAS COASTAL PLAIN, ABST., LATE,
S 3 65 COLORADO, SALT WASH SANDSTONE MEMBER, MORRISON FORMATION
B 6112 2137 COLORADO, SAN JUAN BASIN, SUBNORMAL PRESSURES, NEW MEXICO
S 3 30 COLORADO, SAN JUAN MOUNTAINS
B 6405 776* COLORADO, SECS. 17-20, T3S, R62W, ABST., ZENITH FIELD—SIGNIFICANT,
B 6408 1166 COLORADO SHALES, ALBERTA
B 6107 983 COLORADO, SMOKY HILL CHALK MEMBER, NIOBRARA LIMESTONE
S 3 124 COLORADO, SOUTH PARK, EOCENE SANDSTONE
B 6108 1204 COLORADO, SOUTHWEST, DEVELOPMENTS, 1976
B 6008 1239 COLORADO, SOUTHWESTERN, DEVELOPMENTS, 1975
B 6208 1360 COLORADO, SOUTHWESTERN, DEVELOPMENTS, 1977
B 6308 1206 COLORADO, SOUTHWESTERN, DEVELOPMENTS, 1978
B 6409 1340 COLORADO, SOUTHWESTERN, DEVELOPMENTS, 1979
B 6410 1615 COLORADO, SPINDLE FIELD
M 26 278 COLORADO SPINDLE OIL FIELD
M 26 277† COLORADO, STRATIGRAPHIC AND, SEISMIC EVIDENCE FOR LATE,
M 24 141 COLORADO, SUSSEX-SHANNON HYGIENE DISCOVERY WELLS
S 3 129 COLORADO, TALLAHASSEE CREEK
M 28 41 COLORADO, TERTIARY, CREEDE FORMATION
B 6302 195 COLORADO, UINTA MOUNTAINS
S 3 124 COLORADO, URAVAN MINERAL BELT, SALT WASH SANDSTONE
B 6302 197 COLORADO, WASATCH FORMATION
B 6302 155 COLORADO, WATTENBERG FIELD
M 24 136 COLORADO, WATTENBERG FIELD
B 6410 1631 COLORADO, WATTENBERG FIELD
M 24 136† COLORADO, WATTENBERG FIELD, DENVER BASIN
M 26 278 COLORADO, WATTENBERG OIL FIELD
B 6408 1172 COLORADO, WHITE RIM TAR, UTAH AND
B 6302 195 COLORADO, WHITE RIVER PLATEAU
M 26 281 COLORADO WHITE ROCKS AREA, CROSS SECTION
B 6410 1614 COLORADO, ZORICHAK FIELD
B 6305 824* COLORADO--EVIDENCE FROM SULFUR ISOTOPES AND WHOLE-ROCK
B 6305 836* COLORADO--IMPLICATION FOR PROVENANCE AND DEPOSITIONAL
M 25 352† COLORADO--SUMMARY, CONTROL OF, EARTHQUAKES AT, RANGELY
B 6305 830* COLORADO-WYOMING KIMBERLITE PROVINCE, ABST., UPDATE ON
B 6203 570* COLORADO. BY TITLE ONLY., URANIUM DEPOSITS OF, PITCH MINE
M 25 124 COLP, JOHN L., MAGMA TAP, THE ULTIMATE GEOTHERMAL ENERGY
B 6207 1212 COLP, JOHN L., UTILIZATION OF MAGMA ENERGY, ABST.
B 6303 313 COLTON FORMATION, UTAH
M 24 177† COLUMBIA COUNTIES, ARKANSAS, UPPER, SMACKOVER, RESERVOIRS,
B 6403 439* COLUMBIA COUNTY, OREGON, ABST., MIST GAS FIELD
M 27 9 COLUMBIA, CRETACEOUS, LIMESTONE
B 6105 833* COLUMBIA RIVER MOUTH, ABST., STABILITY OF, DREDGED MATERIAL
M 25 509 COLUMBIA VOLCANIC PLATEAU
M 25 503 COLUMBIAN OROGEN, CANADIAN, CORDILLERA
B 6209 1614 COLUMBIANA COUNTY, LOG, OHIO
B 6204 694 COLUMBUS BASIN, BAHAMAS
B 6406 852 COLUMBUS BASIN, SAND LAYERS
B 6209 1610 COLUMBUS LIMESTONE, OHIO
B 6109 1499 COLVILLE DELTA SECTION, ALASKA
B 6109 1494 COLVILLE TROUGH, ALASKA
M 24 26 COLVILLE TROUGH, ALASKA
S 4 235 COMACTINIA ECHINOPTERA MULLER, GRAND BAHAMA ISLAND
S 4 235 COMACTINIA ECHINOPTERA MULLER, JAMAICA
B 6301 34 COMANCHE SERIES, NOMENCLATURE, TEXAS
B 6103 361 COMANCHE STRUCTURE, OKLAHOMA
B 6209 1684 COMBAZ, A., ORGANIC SEDIMENTATION AND GENESIS OF PETROLEUM IN
B 6303 542 COMBAZ, A., PALEOENVIRONMENT AND PETROLEUM POTENTIAL OF
B 6412 2051 COMBAZ, A., PALEOENVIRONMENT AND PETROLEUM POTENTIAL OF
B 6004 715* COMBINATION ENTRAPMENT I SMACKOVER FORMATION AT CHALYBEAT
B 6009 1614* COMBINATION ENTRAPMENT IN SMACKOVER FORMATION AT
B 6002 319* COMBINATION TRAP, BINDLEY FIELD, HODGEMAN COUNTY, KANSAS,
B 6004 552 COMBINED-MIGRATION MODEL
B 6303 523* COMBUSTION CHARACTERISTICS, ABST., IMPACT OF, COAL PROPERTIES
B 6405 685* COMBUSTION RETORTING OF OIL SHALE, ABST., CARBONATE MINERAL
S 10 180 COMBUSTION, METHANE, OCCURRENCES
B 6102 193 COMMANDER ISLANDS, KOMANDORSKIYE OSTROVA
B 6004 689 COMMEAU, J. A., CLAY-MINERAL VARIABILITY IN SUSPENDED SEDIMENTS
B 6012 2176* COMMERCIAL APPLICATION AND INTERPRETATION OF BOREHOLE
S 5 177 COMMERCIAL SALINAS, ITALY,
S 5 178 COMMERCIAL SALINA, CALIFORNIA.

S 5 171 COMMERCIAL SALINA, ITALY, ADRIATIC COAST, SANTA MARGHERITA
B 6411 2036 COMMISSION FOR GEOLOGICAL MAP OF THE WORLD, NATIONAL
B 6002 273 COMMISSION ON STRATIGRAPHIC NOMENCLATURE, NOTE 44, AMERICAN
B 6102 248 COMMISSION ON STRATIGRAPHIC NOMENCLATURE, NOTE 45, AMERICAN
B 6308 1173 COMMITTEE ON STATISTICS OF DRILLING, STRUCTURE
B 6409 1300 COMMITTEE ON STATISTICS OF DRILLING, STRUCTURE
B 6208 1280 COMMITTEE ON STATISTICS OF DRILLING, STRUCTURE
B 6108 1126 COMMITTEE ON STATISTICS OF DRILLING, STRUCTURE
B 6008 1158 COMMITTEE ON STATISTICS OF DRILLING, STRUCTURE
B 6405 774* COMMON FACTORS AMONG ATYPICAL FIELDS, ABST.
B 6203 491* COMMUNITY SUCCESSION OF PIONEER TERRACE MOLLUSCAN FAUNA,
B 6006 973† COMPACTED CLAYS, FLOW OF, WATER AN, SOLUTES THROUGH
S 10 69† COMPACTING BASINS, MIGRATION OF, HYDROCARBONS IN
B 6203 498* COMPACTING BASINS, ABST., MIGRATION OF, HYDROCARBONS IN
B 6404 549† COMPACTING BASINS, MIGRATION OF, HYDROCARBONS IN
B 6203 506* COMPACTING CLASTIC SEDIMENTS, ABST., MIGRATION PATHWAYS IN
B 6406 927† COMPACTING SEDIMENTS—RELATION TO HYDROCARBON-MATURATION
S 10 39 COMPACTION, MIGRATION, AGENTS
B 6209 1761* COMPACTION, ABST., POROSITY LOSS IN, SANDSTONE BY DUCTILE GRAIN
M 27 207 COMPACTION AFTER CEMENTATION
B 6304 682 COMPACTION AND FLUID MIGRATION, PRACTICAL PETROLEUM
B 6102 169† COMPACTION AND PERMEABILITY OF SANDS, STUDIES OF, VENTURA
M 28 167 COMPACTION, BIBLIOGRAPHY
M 28 165 COMPACTION, BOEHM LAMELLAE, IN QUARTZ
B 6002 294 COMPACTION CELL, VERTICAL SECTION THROUGH
B 6403 415 COMPACTION, DEFINITION
B 6403 415 COMPACTION, DEFINITION
B 6306 918 COMPACTION, DEFINITION, MECHANICAL
S 10 75 COMPACTION, DEPTH VERSUS POROSITY
B 6105 768 * COMPACTION MODEL OF THICK, ABNORMALLY PRESSURED SHALES
S 10 83 COMPACTION OF HIGH PRESSURE SHALES, MIGRATION
S 10 4 COMPACTION STAGES, GULF COAST, SHALE POROSITY
S 10 4 COMPACTION STAGES, MIOCENE, GULF COAST, SHALE POROSITY
S 10 3 COMPACTION STAGES, SHALE POROSITY
B 6102 188 COMPACTION, EFFECT OF, OVERBURDEN AND, FOLDING ON
B 6303 530* COMPACTION EFFECTS IN FUSULINID LIMESTONE ABST.
B 6404 550 COMPACTION, FLUID MOVEMENT DURING
B 6304 573 COMPACTION, GEOCHEMICAL CONCEPT, SHALE
B 6105 841* COMPACTION IN LIME GRAINSTONES, ABST., EVIDENCE FOR SOLUTION
B 6306 922 COMPACTION, MECHANISM OF, SHALE
B 6403 415 COMPACTION, NONEQUILIBRIUM
B 6403 415 COMPACTION, NONEQUILIBRIUM
M 28 165 COMPACTION OF CALCITIC OOID
M 28 163 COMPACTION, OF GLAUCONITE PELLETS
M 28 163 COMPACTION, OF GRAINS IN SANDY LIMESTONE
B 6405 783* COMPACTION OF LIME SEDIMENT, ABST., EXPERIMENTAL
M 28 164 COMPACTION, OF PLAGIOCLASE
M 28 162 COMPACTION, OF SHALE CLASTS, IN SAND
M 28 163 COMPACTION, OF SIDERITE NODULES AND PELLETS
B 6007 1126 COMPACTION ON FLUID PRESSURES, EFFECT OF
B 6102 174 COMPACTION ON PERMEABILITY, EFFECT OF, CEMENTATION AND
B 6002 297 COMPACTION, PORE, FLUID EXPELLED DURING
B 6306 919 COMPACTION PROFILES, DEFINITION
B 6404 551 COMPACTION, RELATIVE POSITIONS OF, ISOTHERMS DURING
B 6007 1124 COMPACTION, SHALE
B 6306 918† COMPACTION STATES OF THICK ABNORMALLY PRESSURED SHALES,
B 6004 543† COMPACTION, WATER, EXPULSION FROM, CLASTIC SEDIMENTS DURING
B 6002 293 COMPACTIONAL RESPONSE OF KAOLINITIC CLAY
M 26 162 COMPARISON, CYCLE CHARTS, TEXAS, GULF COAST, WEST, AFRICA
B 6405 743* COMPARISON OF SHALLOW-MARINE SHELF CARBONATE MOUNDS OF
B 6303 428* COMPARISON OF CARBONATE CHEMISTRY OF BRINES FROM ORCA BASIN
B 6004 704* COMPARISON OF COASTAL DYNAMICS ON WEST- AND EAST-FACING
B 6004 689* COMPARISON OF DIAGENETIC AND WEATHERING MICROTEXTURES ON
B 6105 793 * COMPARISON OF ENDOLITH FLORAS FROM HOLOCENE-PLEISTOCENE, BAHAMA
B 6004 683* COMPARISON OF EOLIAN AND SUBAQUEOUS SAND-FLOW CROSS-STRATA,
B 6405 791* COMPARISON OF GULF COAST LATE JURASSIC OSTRACODA WITH THOSE
B 6303 550* COMPARISON OF PORE-FILLING MATERIAL IN SOME PENNSYLVANIAN
B 6209 1577 COMPARISON OF SALT TO SHALE DIAPIRISM
B 6406 841† COMPARISON OF SAND-LAYER GEOMETRY ON FLAT FLOORS OF 10
B 6105 771 * COMPUTER APPROACH TO VERTICAL-VARIABILITY MAPPING
B 6203 495* COMPARISON OF TURBIDITE SANDSTONE RESERVOIRS—CANYON GROUP
B 6211 2240 COMPARISON OF ZARAUZ DIAPIR WITH SALT DOMES IN PRESENT
M 26 161 COMPARISON, WEST AFRICA TO GULF COAST, CYCLE
S 6 377 COMPARISONS, DIAGRAM, PRECAMBRIAN, TIME-ROCK
S 6 377 COMPARISONS, DIAGRAM, PRECAMBRIAN, TIME-ROCK
B 6206 987 COMPETENT GROUP, IRAN, ZAGROS MOUNTAINS
B 6009 1613* COMPETITION AND QUEST FOR OFFSHORE LEASES, ABST.
B 6312 2128† COMPILATION AND CORRELATION OF MAJOR THERMAL MATURATION
B 6007 1015 COMPLETE FLUVIAL SEQUENCE
B 6003 379† COMPLEX FAULT STRUCTURES IN VERACRUZ PROVINCE OF MEXICO
S 7 109 COMPLEX PATTERNS, OBLIQUE-PROGRADATIONAL SEISMIC FACIES
S 7 108 COMPLEX PATTERNS, SEDIMENT PROCESSES, SEISMIC IDENTIFICATION
S 7 108 COMPLEX PATTERNS, SIGMOID-PROGRADATIONAL SEISMIC FACIES
B 6105 817* COMPLEX RIDGES AND TROUGH TOPOGRAPHY ON SHALLOW,
S 7 272 COMPLEX SHELF-EDGE GRABEN FAULTS, BOTTOM, DEFORMATION
B 6312 2222† COMPOSITION OF ORGANIC MATTER FROM NORTH ATLANTIC
B 6004 735* COMPOSITIONAL TRENDS IN SANDS AN SANDSTONES, ABST.,
B 6309 1581* COMPOSITIONAL VARIATION IN FABRIC-ELEMENT-BASED LITHOTYPE
M 29 66 COMPRESSIONAL VELOCITIES, ATLANTIC COAST
B 6307 1026 COMPRESSIVE FAULT BLOCKS AND BASEMENT THRUSTS
B 6103 467* COMPRESSIVE INTERACTION OF SIBERIAN AND NORTH AMERICAN
B 6303 549* COMPUTATION OF INITIAL WELL PRODUCTIVITIES IN EOLIAN
S 10 73 COMPUTER MODELING, DEPTH-POROSITY HISTORY CURVES
S 10 73 COMPUTER MODELING, GEOLOGICAL RESTORATION
S 10 76 COMPUTER MODELING, GEOTHERMAL RESTORATION
S 10 78 COMPUTER MODELING, TIMING OF FLUID SOURCES
S 10 72 COMPUTER MODELING OF GULF COAST, MIGRATION
B 6004 645* COMPUTER, ABST., CLASTIC AND, STRATIGRAPHIC, TRAP ANALYSIS BY
B 6405 696* COMPUTER, ABST., GRAPHIC REPRESENTATION OF SUBSURFACE DATA BY
B 6203 528* COMPUTER, ABST., REGIONAL GEOLOGIC, MAPPING OF, NORTH SEA BY
B 6004 714* COMPUTER ANALYSIS OF LINEAMENTS TO INFER EXISTENCE OF
B 6203 508* COMPUTER ANALYSIS OF REGIONAL WELL-DENSITY DATA FOR
B 6203 566* COMPUTER APPLICATIONS IN GEOLOGIC INTERPRETATION OF
B 6203 528* COMPUTER APPLICATIONS IN REGIONAL STUDY OF POWDER RIVER
B 6303 542* COMPUTER APPLICATIONS OF PETROLEUM DATA SYSTEM, ABST.
B 6203 495* COMPUTER APPLICATIONS TO IDENTIFY AND FACILITATE EFFICIENT
B 6004 706* COMPUTER APPROACH TO FACIES STUDIES, ABST.
B 6303 515* COMPUTER APPROACH TO STRATIGRAPHIC AND
B 6405 774* COMPUTER ASSISTED PALEOECOLOGIC ANALYSES AND APPLICATION TO
S 7 291 COMPUTER BATHYMETRY, GYRE BASIN
B 6303 555* COMPUTER CONSTRUCTED FROM ORIENTATIONAL, STRATIGRAPHIC,
B 6303 494* COMPUTER DRAFTING--ITS APPLICATIO IN PETROLEUM EXPLORATION,
S 6 73† COMPUTER FILE OF CRITICAL DATES AND CONSEQUENCES OF NEW AND
B 6105 834* COMPUTER FOR TIME-TO-DEATH CONVERSION AND STRUCTURE
B 6105 824* COMPUTER GEOLOGY IN SMALL COMPANY, ABST., PRACTICALITY OF
B 6105 766 * COMPUTER GRAPHIC DISPLAYS OF DIPMETER INFORMATION AS AIDS TO
B 6008 1389* COMPUTER GRAPHICS FOR EXPLORATION, ABST., USE OF INTERACTIVE
B 6203 505* COMPUTER LANGUAGE FOR GEOLOGISTS, ABST., SPEAKEASY—
B 6004 698* COMPUTER METHODOLOGIES USED BY U.S. GEOLOGICAL SURVEY FOR
B 6004 698* COMPUTER METHODS, ABST., CALCULATION OF, COAL, RESERVES BY
B 6404 563 COMPUTER MODEL, BYPASS
B 6303 345 COMPUTER MODELING OF GRAVITY DATA, KENTUCKY, ROME TROUGH
B 6105 808 * COMPUTER PLOTTING OF LAND LEASES ON DIGITIZED BASE MAP
B 6303 518* COMPUTER PREDICTION OF CHANGES IN STRATIGRAPHIC INTERVALS,
B 6302 172 COMPUTER PRINT OUT OF LOG ANALYSIS FROM DIGITIZED LOGS
B 6309 1490† COMPUTER PROCESSING OF LANDSAT DATA, RECONNAISSANCE
B 6303 537* COMPUTER PROCESSING OF LANDSAT DATA
B 6105 777 * COMPUTER PROCESSING OF LANDSAT DATA FOR GEOLOGIC INVESTIGATION
B 6004 703* COMPUTER PROGRAM TO ANALYZE COAL-ENERGY RESOURCES, PACER,
B 6403 439* COMPUTER SAND VOLUME DETERMINATIONS IN LONG BEACH UNIT,
B 6203 501* COMPUTER SYSTEM FOR LITHOSTRATIGRAPHIC ANALYSIS, ABST.
B 6303 509* COMPUTER WELL-DATA SYSTEM, ABST., SMALL
B 6309 1585* COMPUTER-AIDED GRAPHICS FOR ANALYSIS OF COTTAGEVILLE FIELD,
B 6105 836* COMPUTER-ASSISTED COMPOSITE MAPPING SYSTEM TO ENERGY
B 6209 1764* COMPUTER-ASSISTED PALEOECOLOGIC ANALYSES AND APPLICATION TO
B 6303 454* COMPUTER-ASSISTED STRUCTURAL ANALYSIS, ABST.
B 6303 412* COMPUTER-COMPATIBLE HANDLING OF GEOLOGIC DATA ABST.
B 6105 835* COMPUTER-GENERATED BASE MAPS IN EXPLORATION PROCESS, ABST.,
B 6004 651† COMPUTER-GENERATED SCALE FUNCTIONS FOR CONVERSION OF
B 6212 2428 COMPUTER-MADE SLOPE MAPS
B 6112 2144 COMPUTER-PROGRAM PACKAGES, SYMAP AND SYMVU, USED FOR
B 6004 728* COMPUTERIZED MODELING IN RECONSTRUCTION
B 6301 54 COMYN LIMESTONE, TEXAS
B 6110 1899 CONASAUGA FORMATION, MISSISSIPPI
B 6303 344 CONASAUGA SHALE, KENTUCKY
S 7 272 CONCAVE CONVEX FAULT, BOTTOM, DEFORMATION
B 6005 794† CONCEALED MESOZOIC-CENOZOIC ALPINE HIMALAYAN GEOSYNCLINE
S 5 181* CONCENTRATION IN A DENSITY-STRATIFIED EVAPORITE PAN, BASE-
S 3 136 CONCENTRATION IN EVAPORITE BASINS
B 6401 112 CONCENTRATION OF CARBON DIOXIDE IN PORE WATER OF SOURCE
B 6401 112 CONCENTRATION OF CARBON DIOXIDE IN PORE WATER OF SOURCE
S 5 184 CONCENTRATIONS IN BRINE, VENEZUELA, LAGO PUEBLO,
M 26 215 CONCEPT, DEPOSITIONAL SYSTEMS
S 4 242 CONCEPT OF BANK, BUILDUP, OR MOUND
S 6 12 CONCEPT OF SEQUENCE, TIME
B 6104 595 CONCEPTION FIELD, CALIFORNIA
B 6011 2070* CONCEPTS AND TECHNIQUES UTILIZED IN SUCCESSFUL EXPLORATION
B 6209 1571 CONCEPTUAL MODEL OF FORCEFULLY INTRUSIVE DIAPIRISM

B 6104 603 CONCHA LIMESTONE, NEW MEXICO
B 6306 966 CONCHE TROUGH, NEWFOUNDLAND
B 6301 52 CONCHO ARCH, TEXAS
B 6309 1601* CONCHO PLATFORM OF NORTH-CENTRAL TEXAS, ABST., MIDDLE,
M 25 211† CONDENSATE FIELDS OF NEW ZEALAND--SUMMARY, KAPUNI AND, MAUI,
B 6303 480* CONDENSATE MATURITY, AND ABNORMAL FLUID PRESSURES
B 6305 730 CONDENSATE, RESERVOIR DENSITY OF, GAS
B 6004 686* CONDENSED DEEP-WATER PELAGIC AND SHALLOW-WATER EPERIC
B 6004 716* CONE, ABST., RECENT SEDIMENTATION ON, NILE, DELTA AND
B 6106 872† CONE AREA, STRUCTURE AND SEDIMENTARY HISTORY OF
B 6203 542* CONE, OFFSHORE PAKISTAN, ABST., ANCIENT, SUBMARINE CANYON AND,
B 6203 506 CONE, R. M., UNDERSTANDING SEISMICALLY DERIVED IMPEDANCE LOGS,
M 24 278 CONECUH RIDGE, ALABAMA
B 6203 516* CONEMAUGH FORMATION IN APPALACHIANS OF WEST VIRGINIA, ABST.,
B 6109 1419 CONERLY OIL, MISSISSIPPI
B 6303 293 CONES, PROSPECTS, CONTINENT-RELATED, DEEP-WATER, FANS AND
M 26 128 CONFIGURATION PATTERNS, CHAOTIC, SEISMIC, REFLECTIONS
M 26 123 CONFIGURATION PATTERNS, DIVERGENT, SEISMIC, REFLECTIONS
M 26 123 CONFIGURATION PATTERNS, PARALLEL, SEISMIC, REFLECTIONS
M 26 123 CONFIGURATION PATTERNS, PROGRADING, SEISMIC, REFLECTIONS
M 26 128 CONFIGURATION PATTERNS, REFLECTION-FREE AREAS, SEISMIC,
M 26 123 CONFIGURATION PATTERNS, SUB-PARALLEL, SEISMIC, REFLECTIONS
M 26 123 CONFIGURATION PATTERNS, TYPES, SEISMIC, REFLECTIONS
M 26 122 CONFIGURATION, SEISMIC, REFLECTIONS
M 26 132 CONFIGURATIONS, MIGRATING-WAVE, SEISMIC, REFLECTIONS
B 6102 229 CONFINING PRESSURE IN FRACTURE PERMEABILITY
M 25 556 CONFIRMED RESOURCES, MINERALS
M 26 56 CONFORMITIES, DEPOSITIONAL SEQUENCES
B 6101 105† CONFUSION FLATS--AN EXPLORATION GAME
B 6207 1172 CONFUSION RANGE, SECTION, UTAH
B 6305 802 CONGLOMERATE AND PEBBLY SANDSTONES, PERMEABILITY VERSUS,
M 25 411 CONGLOMERATE ORE, DETRITAL ORE, OR CANGA, SCREE ORE
B 6305 799† CONGLOMERATES AND CONGLOMERATIC SANDSTONES, RESERVOIR
B 6004 730* CONGLOMERATES AND SANDSTONES, JURASSIC OF SOUTHWESTERN
B 6206 941 CONGLOMERATES, CLAST-SUPPORTED
B 6303 465* CONGLOMERATES IN SUBMARINE CHANNEL, MIOCENE FLYSCH OF
B 6305 846* CONGLOMERATES OF VIRGINIA, ABST., PETROLOGY OF, TRIASSIC-,
B 6203 520* CONGLOMERATES, PEBBLY AND MASSIVE SANDSTONES, ABST., FACIES
B 6102 207† CONGLOMERATES, VERACRUZ BASIN, MEXICO, ORIGIN AND
B 6203 506* CONGLOMERATIC CARBONATE LIQUEFIED-FLOW DEPOSIT, NORTHERN
B 6303 462* CONGLOMERATIC UNITS, ABST., SEA GULLIES AND DEVELOPMENT OF
S 8 40 CONGO CANYON, AFRICA, SEAVALLEYS
S 8 161 CONGO CANYON, AFRICA, SUBMARINE CANYONS
S 8 166 CONGO CANYON, TIME-VELOCITY CURVES, CURRENT FLOW
S 8 165 CONGO CANYON, TOPOGRAPHY, TIME-VELOCITY CURVES
B 6202 227 CONGO, EVAPORITES
B 6206 920 CONGO FAN
B 6110 1756 CONGO, PRODUCTION, 1975-1976
B 6210 1856 CONGO, PRODUCTION, 1976-1977
B 6310 1700 CONGO, PRODUCTION, 1978
B 6411 1805 CONGO, PRODUCTION, 1979
B 6110 1747 CONGO, REPUBLIC OF, DEVELOPMENTS, 1976
B 6210 1846 CONGO, REPUBLIC OF, DEVELOPMENTS, 1977
B 6310 1691 CONGO, REPUBLIC OF, DEVELOPMENTS, 1978
B 6411 1787 CONGO, REPUBLIC OF, DEVELOPMENTS, 1979
B 6202 229 CONGO, SEDIMENTARY SEQUENCE
S 3 90 CONGO, SHINKOLOBWE VEIN DISTRICT
B 6202 184 CONIACIAN, WYOMING
S 6 261 CONIACIAN-SANTONIAN BOUNDARY, NORTH AMERICA, CRETACEOUS
B 6109 1551 CONKLIN, JACK, PALEOENVIRONMENTAL ANALYSIS OF LOWER
B 6004 709 CONLEY, C. D., SPATIAL VARIATION IN POROSITY--NEW CONCEPT, ABST.
B 6406 960 CONLEY, CURTIS D., PETROLOGY OF ARBUCKLE GROUP, CENTRAL
B 6403 450 CONLEY, J. N., OIL AND GAS EXPLORATION WELLS IN PEDREGOSA BASIN,
B 6209 1755 CONNALLY, TOM, SAND BODY GEOMETRY AND RESERVOIR POTENTIAL
B 6005 884 CONNAN, J., TIME-TEMPERATURE RELATION IN OIL GENESIS. DISCUSSION
B 6408 1286* CONNATE WATER IN MICHIGAN REEFS ATTRIBUTED TO ANHYDRITE,
M 28 55 CONNECTICUT, CAMBRIAN, WATERBURY GNEISS
M 28 46 CONNECTICUT, CAMBRIAN, WATERBURY GNEISS
M 28 51 CONNECTICUT, DEVONIAN, META-DIABASE
M 28 52 CONNECTICUT, DEVONIAN, META-DIABASE
M 28 59 CONNECTICUT, HOLOCENE, BEACH SEDIMENTS
M 28 58 CONNECTICUT, HOLOCENE, BEACH SEDIMENTS
M 28 44 CONNECTICUT, HOLOCENE SEDIMENT
M 28 45 CONNECTICUT, ORDOVICIAN, GARNET SCHIST
M 28 46 CONNECTICUT, ORDOVICIAN, GARNET SCHIST
M 28 6 CONNECTICUT, ORDOVICIAN, KYANITE SCHIST
M 28 16 CONNECTICUT, ORDOVICIAN, NEWTOWN GNEISS
M 28 17 CONNECTICUT, ORDOVICIAN, NEWTOWN GNEISS
M 28 47 CONNECTICUT, ORDOVICIAN SCHIST
M 28 49 CONNECTICUT, ORDOVICIAN, SCHIST
M 28 50 CONNECTICUT, ORDOVICIAN, STRAITS SCHIST
M 28 17 CONNECTICUT, PRECAMBRIAN, WATERBURY GNEISS
M 28 135 CONNECTICUT, TRIASSIC, NEW HAVEN ARKOSE
M 28 136 CONNECTICUT, TRIASSIC, NEW HAVEN ARKOSE
M 28 67 CONNECTICUT, TRIASSIC, NEW HAVEN ARKOSE
M 28 164 CONNECTICUT, TRIASSIC, NEW HAVEN ARKOSE
M 28 5 CONNECTICUT, TRIASSIC, NEW HAVEN ARKOSE
M 28 55 CONNECTICUT, TRIASSIC, NEW HAVEN ARKOSE
M 28 6 CONNECTICUT, TRIASSIC, NEW HAVEN ARKOSE
M 28 124 CONNECTICUT, TRIASSIC, NEW HAVEN ARKOSE
M 28 32 CONNECTICUT, TRIASSIC, NEW HAVEN ARKOSE
B 6101 80 CONNECTICUT-MASSACHUSETTS, HARTFORD-DEERFIELD BASIN
B 6103 466 CONNELLY, WILLIAM, MESOZOIC MAGMATISM AN ACCRETION,
B 6103 465 **CONNELLY, WILLIAM, REGIONAL TECTONIC IMPLICATIONS OF ALASKA**
M 28 160 CONOCOCHEAGUE FORMATION, VIRGINIA, CAMBRIAN
M 27 139 CONOCOCHEAGUE LIMESTONE, VIRGINIA, CAMBRIAN
M 27 141 CONOCOCHEAGUE LIMESTONE, VIRGINIA, CAMBRIAN
M 28 154 CONOCOCHEAGUE LIMESTONE, VIRGINIA, CAMBRIAN
B 6405 738* CONODONT AND PALYNOMORPH COLOR ALTERATION STUDIES TO
B 6405 772* CONODONT ANIMAL—HYPOTHESES AND SPECULATIONS, ABST.
B 6405 799* CONODONT BIOSTRATIGRAPHY AND PALEOECOLOGY, ABST., PERMIAN
B 6405 793* CONODONT BIOSTRATIGRAPHY AND PALEOGEOGRAPHY OF
B 6405 676* CONODONT BIOSTRATIGRAPHY, BIOGEOGRAPHY, AND PALEOECOLOGY,
B 6203 518* CONODONT BIOSTRATIGRAPHY OF WAPANUCKA FORMATION,
B 6107 1045 CONODONT DISTRIBUTION AND THE CYCLIC ROCK SEQUENCE IN UPPER
B 6405 791* CONODONT TAXONOMY, ABST., STATE OF THE ART IN
B 6405 719* CONODONT-BASED ASSESSMENT OF THERMAL MATURITY IN PALEOZOIC
B 6304 647† CONODONT-BASED REINTERPRETATION OF BANE DOME--STRUCTURAL
M 27 26 CONODONTS
M 27 27 CONODONTS
B 6405 737* CONODONTS, ABST., CARBONIFEROUS
B 6405 691* CONODONTS, ABST., RISE AND FALL OF, TRIASSIC
S 6 329 CONODONTS, CARBONIFEROUS, BIOSTRATIGRAPHY
M 28 40 CONODONTS, FRAGMENTS, PHOSPHATE
B 6405 780* CONODONTS IN PETROLEUM EXPLORATION, WESTERN UNITED STATES,
S 6 310 CONODONTS, PERMIAN, KEY FOSSIL GROUPS
B 6304 650 CONODONTS, VIRGINIA, ORDOVICIAN
B 6405 718* CONODONTS—MODELS OF PRAGMATIC PALEONTOLOGY, ABST.
B 6004 689 CONOMOS, T. J., CLAY-MINERAL VARIABILITY IN SUSPENDED SEDIMENTS
B 6303 520* CONQUISTA PROJECT, URANIUM MINING AN MILLING IN SOUTH TEXAS,
B 6303 484 CONRAD, BARBARA U., DETAILED TEMPERATURE LOGGING AS USEFUL
B 6007 1050 CONRAD, CRUISE 10--LEG 12, RV
B 6204 660 CONRAD CRUISE 15, RV
B 6104 546 CONRAD CRUISES 15 AND 16, RV
B 6002 196 CONRAD CRUISES THROUGH MAY 1974, RV, ROBERT D.
B 6109 1559 CONRAD, J. F., LANDSAT—USEFUL PETROLEUM EXPLORATION TOOL IN
B 6410 1658 CONRAD SEISMIC REFLECTION PROFILES, RV
B 6405 693 CONROY, B. W., URANIUM SOLUTION MINING—INTEGRATION OF
M 30 530 CONSESIO FORMATION, JURASSIC, MALOSSA FIELD
B 6109 1541* CONSOLIDATION HISTORY OF GULF COAST LOWER TERTIARY
B 6002 293† CONSOLIDATION OF KAOLINITIC CLAY, HIGH-, PRESSURE
B 6409 1558* CONSTANCE BAYOU, DEEP LAKE, AND SOUTHEAST LITTLE PECAN LAKE
M 25 427 CONSTANCE RANGE, IRON ORE, QUEENSLAND
B 6009 1618 CONSTANS, RICHARD E., MIDDLE EOCENE PLANKTONIC
B 6208 1364 CONSTANT, FRANK L., DEVELOPMENTS IN WEST TEXAS AND EASTERN
S 6 70 CONSTANTS, RECOMMENDED VALUES, DECAY
S 6 359 CONSTANTS USED, CAMBRIAN, RADIOMETRIC DATING
B 6209 1569 CONSTRAINED DIAPIRISM, UNCONSTRAINED AND
B 6309 1586* CONSTRAINTS AND PROBLEMS IN PRODUCING GAS FROM EASTERN
S 10 89† CONSTRAINTS, OIL AND, GAS, MIGRATION:, CHEMICAL AND, PHYSICAL
S 10 90 CONSTRAINTS, SOLUBILITIES IN WATER, MIGRATION
S 10 47† CONSTRAINTS ON MIGRATION MECHANISMS, SOME, MASS BALANCE AND,
M 29 252 CONSTRUCTED BASIN, EVOLUTION, STRUCTURAL HIGH
M 29 251 CONSTRUCTED BASIN, IQUIQUE BASIN, PERU-CHILE TRENCH
B 6408 1223† CONSTRUCTION OF PROCESS-RESPONSE MODELS FOR ANCIENT
B 6203 501* CONSTRUCTION OF PROCESS-RESPONSE MODELS FOR ANCIENT
M 26 77 CONSTRUCTION PROCEDURE, SEA LEVEL FLUCTUATIONS, CORRELATION
M 26 88 CONSTRUCTION, SEA LEVEL FLUCTUATIONS, GLOBAL CYCLES, CHART
B 6303 471* CONSULTING GEOLOGIST IN URBAN-ENVIRONMENTAL CONTEXT, ABST.,
B 6203 394 CONTACT FAULT, ALASKA
B 6105 770 * CONTAINMENT OF PARTICULATE WASTE IN MARINE ENVIRONMENT
B 6203 543* CONTEMPORANEOUS FAULTS—MECHANISM FOR CONTROL OF
B 6405 800* CONTEMPORANEOUS RECONSTRUCTION AND CEMENTATION OF LIVING
B 6203 505* CONTEMPORARY GRAVITY TECTONICS—AN EVERYDAY CATASTROPHE,
B 6406 961 CONTI, LOUIS J., RED WING CREEK FIELD—COSMIC IMPACT STRUCTURE,
B 6303 505* CONTINENNAL MARGIN, ABST., OLIGOCENE TRANSGRESSIV SEDIMENTS
B 6004 705* CONTINENNAL SHELF OFF NEW JERSEY INTERPRETED FROM R- AND
M 25 231† CONTINENNAL SHELF--RESULTS OF TEN YEARS OF EXPLORATION,
B 6406 868† CONTINENT COLLISION I BANDA SEA REGION, ARC-
B 6409 1437† CONTINENT IN 1979, OIL AND, GAS, DEVELOPMENTS IN NORTH, MID-
M 29 118 CONTINENT-OCEAN CRUST BOUNDARY, ARGENTINA
M 29 113† CONTINENT-OCEAN CRUST BOUNDARY OFF BAHIA BLANCA, ARGENTINA,

B 6303 293 CONTINENT-RELATED DEEP-WATER FANS AND CONES, PROSPECTS
B 6312 2165 CONTINENTAL BLOCK PROVENANCES, TABLE
B 6203 548* CONTINENTAL BORDERLAND, ABST., HYDROCARBON, SEEPAGE IN,
B 6203 511* CONTINENTAL BORDERLAND, ABST., NEW EVIDENCE OF, CURRENT,
B 6004 670* CONTINENTAL BORDERLAND, ABST., SHALLOW STRUCTURE OF
B 6004 675* CONTINENTAL BORDERLAND BASIN SEDIMENTS, ABST.,
B 6106 903† CONTINENTAL BORDERLAND, MEXICO, MARINE GEOLOGY OF, BAJA
B 6004 646* CONTINENTAL BORDERLAND NORTH OF 32 DEGREES, ABST., CHANGES
B 6202 247† CONTINENTAL BORDERLAND, PLIOCENE-, PLEISTOCENE, DIASTROPHISM
B 6212 2492† CONTINENTAL BORDERLAND. DISCUSSION AND REPLY, PLIOCENE-,
B 6202 223 CONTINENTAL DRIFT
B 6203 419 CONTINENTAL DRIFT, NORTH, ATLANTIC
S 9 1 **CONTINENTAL DRIFT, PALEOCLIMATE, PALEOMAGNETISM, AND**
B 6101 79 CONTINENTAL DRIFT, TRANS-ATLANTIC SIMILARITIES
S 4 381† CONTINENTAL GROUNDWATER SOURCES IN A PERSIAN GULF COASTAL
B 6004 708* CONTINENTAL MARGIN, 1926-1976, ABST., SEDIMENTOLOGY AND,
B 6303 410* CONTINENTAL MARGIN, ABST., ANTARCTIC
B 6405 706* CONTINENTAL MARGIN, ABST., DIAPIR-LIKE RIDGES AND POSSIBLE
B 6004 690* CONTINENTAL MARGIN, ABST., EVOLUTION OF, MARGINAL,
B 6004 690* CONTINENTAL MARGIN, ABST., EVOLUTION OF, MARGINAL,
B 6004 656* CONTINENTAL MARGIN, ABST., PETROLEUM GEOLOGY OF EASTERN,
B 6207 1231* CONTINENTAL MARGIN, ABST., PLATE TECTONICS IN EVOLUTION OF
B 6105 821* CONTINENTAL MARGIN, ABST., PRELIMINARY, BIOSTRATIGRAPHY OF,
B 6004 708* CONTINENTAL MARGIN, ABST., SEDIMENTOLOGY AND, EASTERN
B 6303 490* CONTINENTAL MARGIN, ABST., SHALLOW-, WATER UPPER, JURASSIC
B 6405 736* CONTINENTAL MARGIN, ABST., THERMOGENIC HYDROCARBON, GASES IN
B 6106 881 CONTINENTAL MARGIN AND ADJACENT DEEP SEA, MAGNETIC
M 29 11† CONTINENTAL MARGIN, BASED ON MULTICHANNEL SEISMIC
M 29 263 CONTINENTAL MARGIN, BATHYMETRY, ALASKA, KODIAK
B 6306 936 CONTINENTAL MARGIN, BATHYMETRY, MAP, AMAZON
B 6002 161 CONTINENTAL MARGIN, BERINGIAN
B 6405 713* CONTINENTAL MARGIN BIOSTRATIGRAPHY, ABST., APPLICATION OF
B 6306 935 CONTINENTAL MARGIN, BRAZIL, NORTHERN
M 29 152 CONTINENTAL MARGIN, BREMER BASIN, AUSTRALIA
B 6209 1669 CONTINENTAL MARGIN, CONVERGENCE OF, LUZON ARC WITH, ASIAN
M 29 152 CONTINENTAL MARGIN, DUNTROON BASIN, AUSTRALIA
M 29 152 CONTINENTAL MARGIN, EUCLA BASIN, AUSTRALIA
B 6109 1489 CONTINENTAL MARGIN, EUGEOCLINAL WEDGE, EAST COAST
S 7 1 **CONTINENTAL MARGIN, FRAMEWORK, FACIES, AND OIL-TRAPPING**
B 6404 516 CONTINENTAL MARGIN, GEOLOGY, NEW ENGLAND
M 29 152 CONTINENTAL MARGIN, GEOPHYSICAL CHART, AUSTRALIA
S 3 96 CONTINENTAL MARGIN, INCIPIENT, OROGENIC
B 6209 1588 CONTINENTAL MARGIN, INDIA-, PAKISTAN
B 6109 1486 CONTINENTAL MARGIN, MIOGEOCLINAL WEDGE, EAST COAST
M 29 270 CONTINENTAL MARGIN, MODEL, ALASKA, KODIAK
B 6306 934† CONTINENTAL MARGIN, MORPHOLOGY AND STRUCTURE OF, AMAZON
M 29 152 CONTINENTAL MARGIN, MURRAY BASIN, AUSTRALIA
B 6209 1603 CONTINENTAL MARGIN, NORTH AMERICA, EASTERN
B 6405 746* CONTINENTAL MARGIN NORTH OF CAPE HATTERAS, ABST., GEOLOGIC
M 25 308† CONTINENTAL MARGIN OF BRITISH COLUMBIA--SUMMARY,
B 6303 324† CONTINENTAL MARGIN OF ISRAEL AND NORTHERN SINAI, SUBMARINE,
B 6303 436* CONTINENTAL MARGIN OF OUTER BERING SEA, ABST., THICK SEDIMENT
B 6303 420* CONTINENTAL MARGIN OFF CENTRAL AND SOUTH TEXAS IN ATLAS
B 6105 838* CONTINENTAL MARGIN OFF KODIAK ISLAND, ALASKA, ABST.,
B 6206 1029† CONTINENTAL MARGIN OFF SOUTHERN BRAZIL, MORPHOLOGY AND
B 6005 809† CONTINENTAL MARGIN OFF WESTERN AFRICA--SENEGAL TO PORTUGAL
M 29 427 CONTINENTAL MARGIN, ORGANIC CONTENT, SOURCE BEDS
B 6004 709* CONTINENTAL MARGIN, OTHRIS, GREECE, ABST., SEDIMENTATION AND
M 29 152 CONTINENTAL MARGIN, OTWAY BASIN, AUSTRALIA
B 6109 1470† CONTINENTAL MARGIN, PETROLEUM GEOLOGY OF EASTERN, BRAZILIAN
B 6303 289 CONTINENTAL MARGIN REGION, PROSPECTS
B 6204 657† CONTINENTAL MARGIN, SALT DIAPIRS ON, SAO PAULO PLATEAU,
B 6204 657† CONTINENTAL MARGIN, SALT DIAPIRS ON, SAO PAULO
B 6209 1595 CONTINENTAL MARGIN SEDIMENTS, THERMAL MATURITY OF
B 6205 792† CONTINENTAL MARGIN, SEISMIC STRATIGRAPHY AND GEOLOGIC
B 6402 242† CONTINENTAL MARGIN, SEISMIC STRUCTURE OF SOUTHWESTERN,
B 6402 242† CONTINENTAL MARGIN, SEISMIC STRUCTURE OF SOUTHWESTERN,
B 6402 173† CONTINENTAL MARGIN SUBSIDENCE AND HEAT FLOW--IMPORTANT
M 29 432 CONTINENTAL MARGIN, THERMAL MATURITY, SOURCE BEDS
M 29 139† CONTINENTAL MARGIN, VARIATIONS IN, BOTTOM, PROCESSES ALONG
S 7 21† CONTINENTAL MARGIN: STRATIGRAPHIC AND STRUCTURAL
B 6207 1216* CONTINENTAL MARGINS, ABST., SEAFLOOR INSTABILITY ON, PACIFIC
M 25 178 CONTINENTAL MARGINS, CLASSIFICATION
M 29 448 CONTINENTAL MARGINS, COLOMBIA, SEISMIC EXPERIMENT
B 6301 67† CONTINENTAL MARGINS, CZECHOSLOVAKIA, ANCIENT, SUBMARINE
M 29 443† CONTINENTAL MARGINS, EXPLOSION, SEISMOLOGY STUDIES OF, ACTIVE
B 6202 208 CONTINENTAL MARGINS. GEOHISTORY OF
M 29 1 **CONTINENTAL MARGINS, GEOLOGICAL AND**
B 6403 300 CONTINENTAL MARGINS IDENTIFIABLE FROM CONTINUOUS SEISMIC
M 29 446 CONTINENTAL MARGINS, MEXICO, SEISMIC EXPERIMENT
M 29 273† CONTINENTAL MARGINS OF THE EASTERN GULF OF ALASKA AND
B 6405 670* CONTINENTAL MARGINS OF UNITED STATES, ABST., THERMAL
B 6403 299 CONTINENTAL MARGINS OF WORLD CAUSED BY PLATE MOVEMENTS
M 29 443 CONTINENTAL MARGINS, PERU, SEISMIC EXPERIMENT
B 6009 1463† CONTINENTAL MARGINS, PLATE TECTONICS IN, OIL AND, GAS,
M 25 177† CONTINENTAL MARGINS, PLATE TECTONICS IN PETROLEUM
B 6403 307 CONTINENTAL MARGINS, SEDIMENT DAMS BENEATH
B 6403 307 CONTINENTAL MARGINS, SEDIMENT DAMS BENEATH
M 29 3† CONTINENTAL MARGINS, SUBSIDENCE MECHANISMS AT, PASSIVE
M 29 448 CONTINENTAL MARGINS, U.S. EAST COAST, SEISMIC EXPERIMENT
B 6403 303 CONTINENTAL MARGINS, WORLD DISTRIBUTION OF STAGES OF
B 6403 303 CONTINENTAL MARGINS, WORLD DISTRIBUTION OF STAGES OF
B 6403 297† CONTINENTAL MARGINS--CLASSIFICATION AND PETROLEUM PROSPECTS
B 6408 1271 CONTINENTAL MARGINS—GEOLOGICAL AND GEOPHYSICAL RESEARCH
B 6105 842* CONTINENTAL MARGIN—U.S. GULF COAST, ABST.
S 9 15 CONTINENTAL MOBILITY, PALEOCLIMATE
B 6309 1453 **CONTINENTAL OFFSHORE STRATIGRAPHIC TEST, COST, WELL B-2,**
B 6410 1659 CONTINENTAL OFFSHORE STRATIGRAPHIC TEST COST, WESTERN,
B 6404 512 CONTINENTAL OFFSHORE STRATIGRAPHIC TEST WELLS, COST—
B 6010 1870 CONTINENTAL OIL CO., AND TEXAS PACIFIC OIL CO., ON DEVELOPMENT
B 6207 1128 CONTINENTAL PLATE, AUSTRALO-, NEW GUINEAN
S 7 14 CONTINENTAL RISE MAP, GULF OF MEXICO
B 6106 877 CONTINENTAL RISE, NILE CONE, LEVANT PLATFORM, AND
B 6303 438* CONTINENTAL RISE OF EASTERN SOUTH AMERICA, ABST., MASS WASTING
M 25 313† CONTINENTAL RISE OFF CENTRAL CALIFORNIA--SUMMARY, PETROLEUM
B 6303 467* CONTINENTAL RISE OFF NORTH AMERICA, ABST., SEDIMENTATION
S 7 15 CONTINENTAL RISE PROVINCE, DESCRIBED
M 26 271 CONTINENTAL RISE-FAN SYSTEMS
B 6403 302 CONTINENTAL RISES AND DEEP-OCEAN TRENCHES INFERRED FROM
B 6403 302 CONTINENTAL RISES AND DEEP-OCEAN TRENCHES INFERRED FROM
B 6004 679* CONTINENTAL SHELF, ABST., BIOGENIC SEDIMENTARY STRUCTURES OF
B 6009 1614* CONTINENTAL SHELF, ABST., BIOSTRATIGRAPHY O, PLEISTOCENE
B 6004 713* CONTINENTAL SHELF, ABST., CORAL, REEFS AND, BANKS OF, TEXAS,
B 6209 1759* CONTINENTAL SHELF, ABST., DISTRIBUTION OF LIVING, BENTHONIC,
B 6303 419* CONTINENTAL SHELF, ABST., EXPLORATION RESULTS AND POTENTIAL
B 6405 675* CONTINENTAL SHELF, ABST., FRACTIONAL PRESERVATION OF
B 6004 649* CONTINENTAL SHELF, ABST., LATE, PLEISTOCENE-, HOLOCENE
B 6105 826* CONTINENTAL SHELF, ABST., PETROLEUM FRONTIER AREAS ON,
B 6405 733* CONTINENTAL SHELF, ABST., PROCESSES CONTROLLING
B 6004 722* CONTINENTAL SHELF, ABST., SEDIMENT-, DISPERSAL, PATTERNS OF
B 6105 836* CONTINENTAL SHELF, ABST., SEDIMENTARY MODEL PREDICTION OF
B 6109 1547* CONTINENTAL SHELF, ABST., STRUCTURE AND STRATIGRAPHY OF
S 7 267† CONTINENTAL SHELF AND SLOPE OFF LOUISIANA AND TEXAS:
B 6303 481* CONTINENTAL SHELF AND SLOPE--MULTIVARIATE STUDY, ABT ST.,
B 6209 1758* CONTINENTAL SHELF AND SOURCE OF TURBID BOTTOM WATER LAYER,
M 29 65† CONTINENTAL SHELF AND UPPER CONTINENTAL SLOPE BETWEEN CAPE
B 6405 687* CONTINENTAL SHELF AND UPPER CONTINENTAL SLOPE, NORTHWEST
S 4 321 CONTINENTAL SHELF, BENTHIC FORAMINIFERA, ASSEMBLAGES, BRAZIL
S 4 317 CONTINENTAL SHELF, CARBONATE SEDIMENTATION, BRAZIL
B 6007 1114 CONTINENTAL SHELF, CDP MULTICHANNE REFLECTION PROFILE,
B 6308 1205 CONTINENTAL SHELF, DRILLING, 1978, ATLANTIC, OFFSHORE
B 6007 1112† CONTINENTAL SHELF, EVIDENCE OF, POST-, PLEISTOCENE, FAULTS ON,
B 6109 1542* CONTINENTAL SHELF, GULF OF MEXICO, ABST., INTERNAL, WAVES AND,
B 6303 527* CONTINENTAL SHELF, GULF OF MEXICO-TIME-SEQUENCE STUDY, ABST.,
B 6004 715* CONTINENTAL SHELF NORTH OF 62 DEGREES NORTH, ABST., GEOLOGY
B 6303 466* CONTINENTAL SHELF, NORTH SEA, ABST., FORTIES FIELD, UNITED
M 30 81† CONTINENTAL SHELF, NORTH SEA, GEOLOGY OF THE, FORTIES FIELD,
B 6012 2176* CONTINENTAL SHELF, NORTHERN GULF OF ALASKA, ABST., DISPERSAL,
B 6105 830* CONTINENTAL SHELF, NORTHWEST GULF OF MEXICO, ABST.,
S 4 315 CONTINENTAL SHELF, OCEANOGRAPHY, BRAZIL
B 6209 1760* CONTINENTAL SHELF OF SOUTH TEXAS, ABST., LIVING, SHELLED,
B 6212 2412† CONTINENTAL SHELF OFF MALASPINA GLACIER, GULF OF ALASKA,
B 6011 2019† CONTINENTAL SHELF OFF NORTHERN FLORIDA, NEOGENE SEDIMENTS
B 6004 687* CONTINENTAL SHELF OFF NORTHWESTERN FLORIDA AND ALABAMA,
S 4 313† CONTINENTAL SHELF OFF SOUTHERN BRAZIL, WITH SPECIAL REFERENCE
B 6406 945† CONTINENTAL SHELF, PETROGRAPHY AND GEOLOGIC SIGNIFICANCE OF
S 7 4 CONTINENTAL SHELF PROVINCE, DESCRIPTION
B 6105 814* CONTINENTAL SHELF REGION—INITIAL SURVEY, ABST., HYDROCARBONS
B 6006 932 CONTINENTAL SHELF, SECTION, NEW JERSEY
M 29 209† CONTINENTAL SHELF, SEISMIC, REFRACTION AND, REFLECTION STUDIES
B 6007 1113 CONTINENTAL SHELF, SHALLOW FAULTS, ATLANTIC, OUUER
B 6203 513* CONTINENTAL SHELF SOUTHWEST OF KODIAK ISLAND, ALASKA, ABST.,
B 6303 301† CONTINENTAL SHELF SOUTHWEST OF KODIAK ISLAND, ALASKA,
B 6204 633† CONTINENTAL SHELF, SURFACE SEDIMENTARY UNITS OF NORTHERN,
B 6106 861 CONTINENTAL SHELF, SURFICIAL, SAND, SHEET ON, MIDDLE, ATLANTIC,
B 6203 426 CONTINENTAL SHELF, TECTONICS, EUROPE, NORTHWEST
S 7 279 CONTINENTAL SHELF-SLOPE, STABILITY EVALUATION
B 6211 2355* CONTINENTAL SHELF. BY TITLE ONLY, GEOENVIRONMENTAL STUDIES

B 6203 495* CONTINENTAL SHELF. BY TITLE ONLY., INTRODUCTION OF, COMPUTER
B 6203 559* CONTINENTAL SLOPE, ABST., BIOTA AND BIOLOGIC PROCESSES OF
B 6304 700* CONTINENTAL SLOPE, ABST., PETROLEUM POTENTIAL OF EASTERN, GULF
B 6203 490* CONTINENTAL SLOPE, ABST., SEDIMENTATION ON, ANTARCTIC
B 6203 510* CONTINENTAL SLOPE, ABST., SEDIMENTATION ON EASTERN, UNITED
B 6303 480* CONTINENTAL SLOPE, ABST., SLUMPING IN INTERCANYON AREAS,
S 7 313† CONTINENTAL SLOPE AND INTRASLOPE BASIN AREAS, NORTHWEST GULF
B 6109 1483† CONTINENTAL SLOPE AND UPPER RISE OFF WESTERN NOVA SCOTIA AND
M 29 65† CONTINENTAL SLOPE BETWEEN CAPE HATTERAS AND CAPE COD,
B 6203 543* CONTINENTAL SLOPE BY SEDIMENT-GRAVITY FLOWS, EXAMPLES FROM
S 7 141 CONTINENTAL SLOPE, CENTRAL SLOPE, NORTHERN GULF OF MEXICO
S 7 117 CONTINENTAL SLOPE, CLAY FRACTION, GULF OF MEXICO
B 6203 493* CONTINENTAL SLOPE DEPOSITS, ABST., PETROLEUM POTENTIAL OF
B 6203 506* CONTINENTAL SLOPE DEPOSITS AND THEIR USEFULNESS IN
B 6203 529* CONTINENTAL SLOPE DEPOSITS, CAPE HATTERASTO HYDROGRAPHER
B 6002 165 CONTINENTAL SLOPE, DREDGE SAMPLES, BERING SEA
B 6203 495* CONTINENTAL SLOPE EAST OF CAPE MAY, NEW JERSEY, ABST., SLOPE,
S 7 291 CONTINENTAL SLOPE, GYRE BASIN, TEXAS
S 7 243† CONTINENTAL SLOPE, LATE, PLEISTOCENE, GEOLOGIC HISTORY OF,
S 7 141 CONTINENTAL SLOPE, LOWER SLOPE, NORTHERN GULF OF MEXICO
S 7 141 CONTINENTAL SLOPE, MISSISSIPPI SLOPE, NORTHERN GULF OF MEXICO
B 6203 542* CONTINENTAL SLOPE, NORTHEAST GULF OF MEXICO, ABST., LATE,
B 6303 529* CONTINENTAL SLOPE, NORTHEASTERN UNITED STATES--OBSERVATIONS
B 6004 683* CONTINENTAL SLOPE NORTHERN GULF OF MEXICO, ABST., SALT
B 6203 499* CONTINENTAL SLOPE, NORTHERN GULF OF MEXICO AS INTERPRETED
B 6305 782† CONTINENTAL SLOPE NORTHERN GULF OF MEXICO, SALT MOVEMENT
S 7 69† CONTINENTAL SLOPE, NORTHERN GULF OF MEXICO, SALT MOVEMENT
S 7 141† CONTINENTAL SLOPE, NORTHERN GULF OF MEXICO, STRUCTURE OF
S 7 303† CONTINENTAL SLOPE, NORTHWEST GULF OF MEXICO, ORCA, BASIN,
S 7 141 CONTINENTAL SLOPE, NORTHWEST SLOPE, NORTHERN GULF OF MEXICO
B 6203 564* CONTINENTAL SLOPE OFF MID-ATLANTIC STATES, ABST., EROSION
B 6203 559* CONTINENTAL SLOPE OFF NEW ENGLAND, ABST., SEDIMENTARY
S 7 9 CONTINENTAL SLOPE PROVINCE, DESCRIPTION
S 7 72 CONTINENTAL SLOPE, SALT DOME GROWTH
S 7 76 CONTINENTAL SLOPE, SALT DOME GROWTH, DIAGRAM
B 6305 785 CONTINENTAL SLOPE, SALT, DOME GROWTH ON
S 7 150 CONTINENTAL SLOPE, SALT THICKNESS, NORTHERN GULF OF MEXICO
S 7 117 CONTINENTAL SLOPE, SAND FRACTION, GULF OF MEXICO
S 7 70 CONTINENTAL SLOPE, SEDIMENT DEPOSITION, SALT
B 6106 918† CONTINENTAL SLOPE SEDIMENT INSTABILITY NORTHEAST OF
S 7 150 CONTINENTAL SLOPE, SEDIMENT LOADING, NORTHERN GULF OF
B 6203 557* CONTINENTAL SLOPE SEDIMENTS, EASTERN UNITED STATES, ABST.,
B 6004 722* CONTINENTAL SLOPE SEDIMENTS FROM NORTHWEST GULF OF MEXICO,
B 6303 549* CONTINENTAL SLOPE SEDIMENTS RESULTING FROM LARGE-SCALE
S 7 151 CONTINENTAL SLOPE, STRUCTURAL GENESIS, NORTHERN GULF OF
S 7 142 CONTINENTAL SLOPE, STRUCTURAL PROVINCES, NORTHERN GULF OF
M 26 270 CONTINENTAL SLOPE SYSTEMS
M 29 75 CONTINENTAL SLOPE, UNCONFORMITY, ATLANTIC COAST
B 6203 532* CONTINENTAL SLOPE-RISE DEPOSITS FROM LOWER CAMBRIAN SEKWI
B 6203 499* CONTINENTAL SLOPES, ABST.
B 6209 1584† CONTINENTAL SLOPES AND RISES, PETROLEUM, SOURCE BEDS ON
M 29 423† CONTINENTAL SLOPES AND RISES, PETROLEUM, SOURCE BEDS ON
B 6312 2204† CONTINENTAL SLOPES AND RISES, SEISMIC EVIDENCE FOR WIDESPREAD
B 6312 2211 CONTINENTAL SLOPES AND RISES, THERMAL GRADIENTS
B 6406 957 CONTINENTAL SLOPES, EDITED BY LARRY J. DOYLE AND ORRIN H.
B 6203 533* CONTINENTAL SLOPE—ACTIVE-MARGIN ENVIRONMENT, ABST.,
B 6004 696* CONTINENTAL SODA-RICH WATERS WITH CONCENTRATED SEA WATER,
B 6405 781* CONTINENTAL STRETCHING—EXPLANATION OF POST-MID-CRETACEOUS
B 6411 2038 CONTINENTAL TECTONICS, STUDIE IN GEOPHYSICS RESEARCH BOARD,
B 6403 305 CONTINENTAL TERRACES CLASSIFIED USING CONTINUOUS SEISMIC
B 6004 662* CONTINENTAL VOLCANIC SEDIMENTS AND THEIR SEDIMENTATION,
B 6303 473* CONTINENTAL-MARGIN AND DEEP-SEA SEQUENCES--NEOGENE EXAMPLES
S 3 82 CONTINENTAL-MARGIN HISTORY, OROGENIC
S 3 115 CONTINENTAL-MARGIN HISTORY, TAPHROGENIC
M 25 460 CONTINENTAL-MARGIN METALLOGENIC ZONES, PACIFIDES
S 3 93 CONTINENTAL-MARGIN MOBILE BELT, DEFINITION
B 6302 218† CONTINENTAL-MARGIN PLATEAUS, NORTHEASTERN BRAZIL, STRUCTURE
B 6105 826* CONTINENTAL-MARGIN SEDIMENTS OFF NORTHWEST AFRICA BETWEEN
B 6209 1588 CONTINENTAL-MARGIN SEDIMENTS, ORGANIC CONTENT OF
B 6105 842* CONTINENTAL-MARGIN TYPES RELATED TO PLATE TECTONICS AND
B 6408 1224 CONTINENTAL-SHELF ENVIRONMENTS AND PROCESSES, REVIEW
S 7 117† CONTINENTAL-SLOPE SEDIMENTS AND SEDIMENTATION, GULF OF
S 7 87† CONTINENTAL-SLOPE SEISMIC FACIES, OFFSHORE TEXAS-LOUISIANA,
S 3 36 CONTINENTS, MANTLE MAGMAS
B 6007 1112† CONTINETAL SHELF, EVIDENCE OF, POST-PLEISTOCENE, FAULTS ON, NEW
S 10 48 CONTINUOUS PHASE, MIGRATION
B 6403 305 CONTINUOUS SEISMIC REFLECTION PROFILES, CONTINENTAL, TERRACES
M 26 100 CONTINUOUS, SEISMIC, REFLECTIONS
S 5 169* CONTINUUM: LATEST MIOCENE (MESSINIAN) DEPOSITS OF SALEMI BASIN
B 6307 1036 CONTOUR MAP, CALIFORNIA, SAN JOAQUIN VALLEY BASIN, EAST
M 26 371 CONTOUR MAPPING, RESOLUTION SYSTEMS. STRUCTURE FORM
B 6105 812* CONTOURING GEOLOGIC DATA FROM ARBITRARY POINTS
B 6004 696* CONTOURING GEOLOGIC DATA FROM ARBITRARY POINTS, ABST.
B 6303 495* CONTOURITE ANTICLINES AS EXPLORATION OBJECTIVES, ABST.
S 4 195 CONTOY, MOLLUSCA, MEXICO, ISLA
B 6312 2215 CONTRACTIONAL FRACTURES
B 6312 2216 CONTRACTIONAL FRACTURES, THERMAL
B 6212 2509* CONTRAST OF HYDROCARBON POTENTIAL BETWEEN FORE-ARC AND
B 6303 517* CONTRASTING PENNSYLVANIAN CARBONATE SEDIMENTATION,
B 6109 1550* CONTRASTS IN CEMENTATION, DISSOLUTION, AND POROSITY
B 6105 758* CONTRASTS IN CEMENTATION, DISSOLUTION, AND POROSITY DEVELOPMENT
B 6004 667* CONTRASTS IN CONCENTRATION OF SUSPENDED MATTER IN
B 6004 655 CONTREIRAS, C. A., SYNSEDIMENTARY TECTONIC CONTROL OF
B 6203 503 CONTREIRAS, C. J. A., PETROLEUM GEOLOGY OF PALEOZOIC CLASTIC
B 6109 1561* CONTRIBUTION OF LANDSAT IMAGERY TO GEOLOGIC PROSPECTING
B 6211 2358* CONTRIBUTION OF SHEAR WAVES TO STRATIGRAPHIC INTERPRETATION,
B 6203 553* CONTROL OF CATHODO-LUMINESCENCE OF DOLOMITE BY IRON AND
M 25 352† CONTROL OF EARTHQUAKES AT RANGELY, COLORADO--SUMMARY
B 6004 701* CONTROL OF SEDIMENT DISTRIBUTION BY CURRENT DYNAMICS IN WEST
B 6303 487* CONTROLS OF SEDIMENTATIO OF LOWER MISSISSIPPIAN
B 6405 708* CONTROLS OF ZEOLITE CEMENTATION IN UPPER JURASSIC SANDSTONES,
B 6303 546* CONTROLS ON LATE CRETACEOUS-PALEOCENE SEDIMENTATION IN
B 6311 2119* CONTROLS ON PENNSYLVANIAN HYDROCARBON ACCUMULATIONS IN
S 3 80 CONVECTION, CRUSTAL
S 3 25 CONVECTION, MANTLE
S 3 79 CONVECTION, MANTLE
B 6209 1645† CONVERGENCE AND ACCRETION IN TAIWAN-LUZON REGION, PLATE
B 6209 1669 CONVERGENCE OF LUZON ARC WITH ASIAN CONTINENTAL MARGIN
B 6212 2456 CONVERGENCE OF NAZCA AND SOUTH AMERICAN PLATES
M 29 369 CONVERGENCE, VENEZUELA BASIN, TRENCH-ARC COMPLEXES
M 25 177† CONVERGENT CONTINENTAL MARGINS, PLATE TECTONICS IN
M 29 261† CONVERGENT MARGIN OFF KODIAK ISLAND, ALASKA, FROM
M 29 429 CONVERGENT MARGIN, PACIFIC, ALASKA
M 29 429 CONVERGENT MARGIN, PACIFIC, OREGON
B 6209 1591 CONVERGENT MARGINS, OREGON AND, ALASKA—
M 29 435 CONVERGENT MARGINS, THERMAL MATURITY OF SEDIMENTS
M 25 51 CONVERGENT PLATE BOUNDARIES, METAL DEPOSITS AT OR NEAR
M 25 179 CONVERGENT PLATES AND LITHOSPHERIC SUBDUCTION
B 6008 1406* CONVERSE COUNTY WYOMING, ABST., SPEARHEAD RANCH FIELD
S 10 169† CONVERSION OF ORGANIC COMPOUNDS IN WATER, ORIGIN OF,
B 6004 651* CONVERSION OF RESISTIVITY TO PSEUDOVELOCITY LOGS, ABST.,
S 7 272 CONVEX FAULT, BOTTOM, DEFORMATION, CONCAVE
B 6405 743* CONVEYER-BELT TECTONICS AND GEOLOGIC EVOLUTION OF ALASKAS
M 26 8 CONVOLUTION, SEISMIC, DATA
B 6406 954 CONYBEARE, MACKENZIE, D. B., REVIEW OF, LITHOSTRATIGRAPHIC
B 6408 1282 COOGAN, A. H., DEPOSITIONAL TRENDS OF LOWER SILURIAN CLINTON
B 6105 838 COOGAN, A. H., POROSITY IN CARBONATE SANDS, ABST., EFFECT OF
B 6109 1551 COOGAN, ALAN H., EARLY AND MIDDLE CRETACEOUS HIPPURITACEA,
B 6309 1583 COOGAN, ALAN H., OIL- AND GAS-PRODUCING CONFIGURATIONS IN
B 6303 434 COOGAN, ALAN H., SOURCES OF LIABILITY FOR GEOLOGIC HAZARDS,
B 6206 951 COOK CHANNEL, TEXAS
B 6103 352 COOK DELTAS, TEXAS
B 6303 435 COOK, EARL, EXPLOITATION OF VERY LOW-GRADE URANIUM DEPOSITS,
B 6405 693 COOK, FREDERICK A., CRYSTALLINE OVERTHRUSTING OF PALEOZOIC
B 6203 506 COOK, HARRY E., ANCIENT CONTINENTAL SLOPE DEPOSITS AND THEIR
B 6303 435 COOK, HARRY E., GENERATION OF DEBRIS FLOWS AND
B 6303 305 COOK INLET, ALASKA
B 6203 380 COOK INLET, ALASKA, A, B, AND, C HORIZONS
B 6405 708* COOK INLET, ALASKA, ABST., CONTROLS OF, ZEOLITE, CEMENTATION IN
B 6004 654* COOK INLET, ALASKA, ABST., GEOLOGY OF, SOUTH AUGUSTINE ISLAND,
B 6012 2180* COOK INLET ALASKA, FROM SEISMIC DATA, ABST., GEOLOGY OF
B 6203 373† COOK INLET, ALASKA, GEOLOGIC FRAMEWORK OF LOWER
B 6408 1259† COOK INLET, ALASKA, NONMARINE UPPER, CRETACEOUS ROCKS
B 6012 2185* COOK INLET, ALASKA, WITH EMPHASIS TO ONSHORE GEOLOGY, ABST.,
B 6211 2351* COOK INLET, ALASKA. BY TITLE ONLY, MOBILE BED FORMS AND,
M 24 4 COOK INLET AREA, FIELDS AND FACILITIES, ALASKA
M 25 304 COOK INLET BASIN, ALASKA
B 6408 1131† COOK INLET BASIN, ALASKA, BIOGENIC AND, THERMOGENIC ORIGINS OF,
B 6203 377 COOK INLET BASIN, STRATIGRAPHIC COLUMN, ALASKA
B 6408 1134 COOK INLET BAY, RESERVES, ALASKA
B 6408 1260 COOK INLET COST 1 WELL, ALASKA
M 26 370 COOK INLET, DEPTH SOUNDER RECORD, ALASKA, LOWER
B 6409 1346 COOK INLET, EXPLORATION, ALASKA
B 6108 1158 COOK INLET, EXPLORATION, ALASKA
B 6208 1312 COOK INLET, EXPLORATION, ALASKA
B 6308 1212 COOK INLET, EXPLORATION, ALASKA
B 6203 376 COOK INLET, GEOLOGIC MAP, ALASKA
B 6203 399 COOK INLET, HYDROCARBON POTENTIAL, ALASKA
B 6103 466* COOK INLET OIL, ABST., ORIGIN OF
B 6104 578 COOK INLET, OIL AGE, ALASKA

Index

B 6408 1132 COOK INLET REGION, OIL AND GAS FIELD, ALASKA
B 6203 380 COOK INLET, SEISMIC DATA, ALASKA
B 6211 2356* COOK INLET, SOUTHWESTERN KENAI PENINSULA, ALASKA, ABST.,
B 6203 395 COOK INLET, TECTONIC-CYCLES, ALASKA
B 6203 395 COOK INLET, TIME-STRATIGRAPHIC CHART, ALASKA
B 6203 379 COOK INLET, WELL DATA, ALASKA
B 6103 345 COOK PALEODRAINAGE SYSTEMS, TEXAS
B 6412 2118 COOK, PETER J., ORGANIC GEOCHEMISTRY OF PLIOCENE-PLEISTOCENE
B 6103 348 COOK POINT-BAR STRATIGRAPHIC TRAP, TEXAS
B 6103 346 COOK SANDSTONE, TEXAS
B 6109 1552 COOK, T. D., EXPLORATION HISTORY OF SOUTH TEXAS LOWER
B 6301 32 COOK, THEODORE D., EXPLORATION HISTORY OF SOUTH TEXAS LOWER
B 6303 306 COOK-SHELIKOF BASIN, ALASKA
B 6205 783 COOKING LAKE BANK, ALBERTA
B 6209 1767* COOL CREEK FORMATION, ARBUCKLE MOUNTAINS, OKLAHOMA, ABST.,
B 6307 1094 COOLYENA GROUP, AUSTRALIA
B 6303 435 COON, J.B., SURFACE SEISMIC METHODS APPLIED TO COAL-MINING
B 6405 745 COOPER, A. K., HYDROCARBON PROSPECTS OF BERING SEA SHELF SOUTH
B 6303 490 COOPER, A. K., SHALLOW-WATER UPPER JURASSIC ROCKS DREDGED
B 6103 467 COOPER, ALAN A., INDICATOR OF PLATE TECTONICS IN SOUTHERN
B 6311 2070 COOPER, ALAN K., HYDROCARBON POTENTIAL OF ALEUTIAN BASIN,
B 6403 439 COOPER, ALAN K., HYDROCARBON POTENTIAL OF ALEUTIAN BASIN,
B 6412 2139 COOPER, ALAN K., MESOZOIC AND CENOZOIC STRUCTURAL TRENDS
B 6405 694 COOPER, ALAN K., MULTIFOLD SEISMIC DATA ACROSS OUTER BERING
B 6405 693 COOPER, ALAN K., PROSPECTIVE AND FUTURE HYDROCARBON
B 6002 161 COOPER, ALAN K., STRUCTURE AND EVOLUTION OF BERING SEA SHELF
B 6303 436 COOPER, ALAN K., THICK SEDIMENT ACCUMULATIONS BENEATH
B 6212 2481 COOPER, ALAN K., VAMPS—POSSIBLE HYDROCARBON-BEARING
B 6207 1151† COOPER BASIN, AUSTRALIA, HYDROCARBON, GENERATION IN,
B 6207 1154 COOPER BASIN, STRATIGRAPHY, AUSTRALIA
B 6004 659 COOPER, J. D., STRATIGRAPHY AND DEPOSITIONAL ENVIRONMENTS OF
B 6211 2363 COOPER, JOHN D., LATE CRETACEOUS DEPOSITIONAL ENVIRONMENTS,
B 6405 786 COOPER, MARY BETH, CRETACEOUS NIOBRARA GAS PLAY, ABST.
B 6004 684 COOPER, W. A., STRATIGRAPHY AND SEDIMENTATION OF FRESHWATER,
B 6011 2044 COOPER, W. W., DIFFERENTIATION OF SHALE TYPES BY ALLOCHTHONOU
B 6009 1608 COOPER, W. W., NANNOFOSSIL BIOSTRATIGRAPHY OF BOSSIER SHALE
B 6311 2042 COOPER, WILLIAM A., STRATIGRAPHY OF INTERMONTANE, LACUSTRINE
B 6303 500* COORONG MODEL FOR PENECONTEMPORANEOUS DOLOMITE
B 6211 2348* COOS BAY, OREGON, ABST., NEW, MIOCENE FORMATION AT
B 6303 436 COPELIN, EDWARD C., CRUDE OIL-LIKE HYDROCARBONS IN RECENT
B 6305 820 COPELIN, EDWARD C., REVIEW OF OIL SAND AND OIL SHALE CHEMISTRY,
M 25 321 COPOMA ANTICLINE, COLOMBIA
B 6004 659 COPP, E. A., RESPONSE OF RESERVES TO PRICE. BY TITLE ONLY
S 10 227 COPPER IN SAME TRAP, USSR, FERGHANA, OIL AND
B 6004 694 COPPER, A. E., STRUCTURE AN EVOLUTION OF BERING SEA MARGIN AND
M 25 461 COPPER AND GOLD METALLOGENIC SYSTEMS, TRANS-, PACIFIC
M 25 427 COPPER, AND LEAD, AUSTRALIA, NORTHERN TERRITORY, RUM JUNGLE,
M 25 485 COPPER AND MOLYBDENUM, CANADA, WESTERN
M 25 554 COPPER AND MOLYBDENUM DEPOSITS, ECUADOR, LOS LINDEROS
M 25 554 COPPER AND MOLYBDENUM DEPOSITS, ECUADOR, RIO PLAYAS
M 25 545 COPPER AND NICKEL, CONCENTRATIONS OF, NAZCA PLATE
M 25 554 COPPER AND RELATED MINERALS, ECUADOR, FIERRO URCO DEPOSIT
M 25 531 COPPER, ANDEAN REGION
M 25 436 COPPER, ASSOCIATED MINERALS, PHILIPPINES
B 6102 256† COPPER BASIN GROUP, PIONEER MOUNTAINS, SOUTH-CENTRAL IDAHO,
M 25 430 COPPER, BRITISH SOLOMON ISLANDS
M 25 496 COPPER, CANADIAN, CORDILLERA
B 6410 1580 COPPER CREEK BLOCK, VIRGINIA
B 6412 2078 COPPER CREEK FAULT, APPALACHIANS
B 6103 417 COPPER CREEK FAULT, VIRGINIA
B 6311 2060 COPPER CREEK INTRUSIONS, BRITISH COLUMBIA
B 6410 1576 COPPER CREEK THRUST, VIRGINIA
M 25 513 COPPER DEPOSIT, BERG
M 25 507 COPPER DEPOSIT, CANADA, SUSTUT
M 25 513 COPPER DEPOSIT, DIAPIRIC-TYPE, PORPHYRY
M 25 513 COPPER DEPOSIT, GRANISLE
M 25 426 COPPER DEPOSIT, NEW SOUTH WALES, CAPTAINS FLAT, LEAD-, ZINC-
M 25 426 COPPER DEPOSIT, TASMANIA, MOUNT LYELL
M 25 426 COPPER DEPOSIT, TASMANIA, ROSEBERY, LEAD-, ZINC-
M 25 434 COPPER DEPOSIT, USAGE DEFINED, PORPHYRY
M 25 531 COPPER DEPOSITS AND PRODUCTION, CHILE
M 25 433 COPPER DEPOSITS, ASSOCIATED ROCKS, SOUTHWEST, PACIFIC,
M 25 426 COPPER DEPOSITS, AUSTRALIA, BOUGAINVILLE
M 25 512 COPPER DEPOSITS, DIORITE-MODEL, PORPHYRY
M 25 433 COPPER DEPOSITS, EXPLOITATION, SOCIAL AND POLITICAL I, PORPHYRY
M 25 486 COPPER DEPOSITS, GENETIC DISTINCTIONS, STRATA-BOUND
M 25 435 COPPER DEPOSITS, GEOLOGIC ENVIRONMENT, PHILIPPINES, PORPHYRY
M 25 509 COPPER DEPOSITS, GEOLOGY OF, NORTH AMERICA, CORDILLERAN
M 25 512 COPPER DEPOSITS, GRANITIC-PLUTON-TYPE, PORPHYRY
M 25 519† COPPER DEPOSITS IN SONORA, MEXICO--SUMMARY, PRINCIPAL
M 25 434† COPPER DEPOSITS IN THE PHILIPPINES--SUMMARY, GEOLOGIC
M 25 519 COPPER DEPOSITS, MAIN, MEXICO, SONORA, PORPHYRY
M 25 510 COPPER DEPOSITS, MAP, CORDILLERAN OROGEN, NORTHERN, PORPHYRY
M 25 520 COPPER DEPOSITS, MAP, MEXICO, SONORA, PORPHYRY
M 25 433† COPPER DEPOSITS OF SOUTHWEST PACIFIC, DISCOVERY AND
M 25 509 COPPER DEPOSITS, PETROLOGIC MODELS, PORPHYRY
M 25 512 COPPER DEPOSITS, QUARTZ MONZONITE-QUARTZ DIORITE MODEL OF,
M 25 511 COPPER DEPOSITS, TABLE, CORDILLERAN OROGEN, NORTHERN,
M 25 439 COPPER DEPOSITS, TAIWAN, GOLD-, SILVER-
M 25 509 COPPER DEPOSITS, TYPES OF, QUARTZ MONZONIE
M 25 486 COPPER DISTRICT, NEVADA, ELY
M 25 438 COPPER, ECONOMIC POSSIBILITIES, PHILIPPINES, PORPHYRY
M 25 552 COPPER, ECUADOR, MACUCHI MINE
M 25 552 COPPER, ECUADOR, TOACHI DEPOSITS, LA PLATA MINE, GOLD, SILVER,
M 25 430 COPPER, FIJI
M 25 469 COPPER, LEAD, AND ZINC, USSR, NORTHEAST
S 4 87 COPPER LIMESTONE, JAMAICA
M 25 435 COPPER, LUZON, BAGUIO MINERAL DISTRICT
M 25 515 COPPER, MEXICO
M 25 437 COPPER, MINDANAO, DAVAO DISTRICT
M 25 520 COPPER MINERALIZATION, MEXICO, SONORA
B 6203 573* COPPER MOUNTAIN, WYOMING, URANIUM DEPOSIT REDISCOVERED. BY
M 25 487 COPPER OF OPHIOLITIC ORIGIN
M 25 457 COPPER ORE, JAPAN
M 25 521 COPPER, ORE MINERALS, MEXICO
M 25 434 COPPER, ORE RESERVE, PHILIPPINES, PORPHYRY
B 6303 507* COPPER PORPHYRINS IN DEEP-SEA SEDIMENTS, ABST.
M 25 505 COPPER, PRODUCTION AND RESERVES, CANADIAN, CORDILLERA
M 25 531 COPPER PRODUCTION, PERU
B 6207 1221* COPPER PROJECT, PAPUA NEW GUINEA, ABST., OK TEDI
M 25 433 COPPER PROJECTS, CAPITAL INTENSIVE, PORPHYRY
M 25 509† COPPER PROVINCE OF NORTHERN CORDILLERAN OROGEN--SUMMARY,
M 25 486 COPPER PROVINCES, WESTERN CORDILLERA
M 25 427 COPPER, QUEENSLAND, LADY ANNIE
M 25 427 COPPER, QUEENSLAND, MOUNT ISA, LEAD-, SILVER-, ZINC-
M 25 427 COPPER, QUEENSLAND, MOUNT MORGAN, GOLD-
M 25 438 COPPER, RELATION OF, BRECCIA PIPES AND, PORPHYRY
M 27 140 COPPER RIDGE DOLOMITE, VIRGINIA, CAMBRIAN
M 27 141 COPPER RIDGE DOLOMITE, VIRGINIA, CAMBRIAN
M 27 139 COPPER RIDGE DOLOMITE, VIRGINIA, CAMBRIAN
M 25 559 COPPER, SILVER, AND GOLD, CHILE
M 25 444 COPPER, SOUTH KOREA
M 25 434 COPPER SULFIDE DEPOSIT, HYPOGENE
M 25 426 COPPER, ZINC, AND GOLD, NEW SOUTH WALES, COBAR
M 25 554 COPPER, ZINC, LEAD, AND MOLYBDENU DEPOSITS, ECUADOR, FIERRO
M 25 554 COPPER, ZINC, LEAD, AND MOLYBDENUM DEPOSITS, ECUADOR, FIERRO
M 25 553 COPPER, ZINC, LEAD, SILVER, AND GOLD DEPOSITS, ECUADOR, ANGAS
M 25 460 COPPER-BEARING RING, GREAT, PACIFIC
M 25 426 COPPER-LEAD-ZINC DEPOSIT, NEW SOUTH WALES, WOODLAWN
M 25 553 COPPER-MOLYBDENUM DEPOSIT, ECUADOR, CHAUCHA
M 25 553 COPPER-MOLYBDENUM, ECUADOR, CHAUCHA DEPOSIT
M 25 554 COPPER-MOLYBDENUM, ECUADOR, RIO PLAYAS
M 25 460 COPPER-POLYMETALLIC PROFILE, PACIFIC, MEXICO-, CENTRAL AMERICA,
M 25 390† COPPER-RICH NODULES OF EQUATORIAL NORTH PACIFIC--SUMMARY,
S 4 135 CORAL ADAPTATION, HYDRAULIC MODEL FOR
S 4 105 CORAL BANKS, CARIBBEAN
B 6004 732* CORAL BORERS, ABST., EVENT SEQUENCES IN GEOLOGIC, REEF
S 4 265 CORAL, BORERS, MOROCCO, JEBEL BOU KENDILL
S 4 155† CORAL DIVERSITY--OBSERVATIONS OFF BONAIRE NETHERLANDS
S 4 289 CORAL, GALAXEA CLAVUS DANA, HERMATYPIC, SCLERACTINIAN
M 27 63 CORAL LIMESTONE, ENGLAND, CARBONIFEROUS
M 27 164 CORAL LIMESTONE, ENGLAND, MISSISSIPPIAN
B 6004 683* CORAL REEF, ABST., AUTOPSY OF A DEAD
B 6306 877 CORAL REEFAL BUILDUPS, DISTRIBUTION, CRETACEOUS, RUDIST-
B 6004 713* CORAL REEFS AND BANKS OF TEXAS OUTER CONTINENTAL SHELF, ABST.
S 4 3† CORAL REEFS AND THEIR ZONATION IN NETHERLANDS ANTILLES
S 4 231† CORAL REEFS, CRINOIDS OF, WEST, INDIAN
S 4 5 CORAL REEFS, LITERATURE, CARIBBEAN
S 4 353† CORAL REEFS OF RED SEA--SEQUENCE AND TIME SCALE, PROGRESSIVE,
S 9 3 CORAL REEFS, PRESENT DISTRIBUTION
S 4 3 CORAL REEFS, SABA BANK
S 4 357† CORAL REEFS, SEQUENCE OF PROGRESSIVE, DIAGENESIS IN
B 6303 482* CORAL REEFS, WESTERN GULF OF MEXICO--ROLE OF ECOLOGY AND
B 6309 1605* CORAL REEFS, WESTERN GULF OF MEXICO--ROLE OF ECOLOGY AND
S 4 178 CORAL, SAIPAN, MIOCENE, REEF
M 27 36 CORAL SEA, HOLOCENE OOZE
M 27 90 CORAL SEA, HOLOCENE SEDIMENTS
B 6401 129 CORAL ZONES, INDIANA, STROMATOPOROID-
B 6401 129 CORAL ZONES, INDIANA, STROMATOPOROID-
S 4 43 CORAL-ALGAL ASSEMBLAGES, BELIZE, GLOVERS REEF ATOLL

B 6307 1108† CORAL-ALGAL-RUDIST REEFS, ARIZONA, DEPOSITIONAL, MODEL OF
B 6403 400† CORAL-REEF OIL FIELD—WILFRED POOL, SULLIVAN COUNTY, INDIANA,
M 27 35 CORAL-RUBBLE, BELIZE, HOLOCENE
S 4 83† CORAL-RUDIST FRAMEWORKS AND THEI GEOLOGIC SETTING, JAMAICAN
M 27 41 CORALLIAN SERIES, ENGLAND, JURASSIC
B 6405 682* CORALLINE LIMESTONE, CHATTIAN, MALTESE ISLANDS, ABST.,
B 6303 514* CORALLITE OF ACROPORA CERVICORNIS, ABST., SUBAERIAL DIAGENESIS
B 6303 540* CORALS, ABST., EFFECTS OF AN OFFSHORE, DRILLING, MUD ON SELECTED
B 6303 521* CORALS, ABST., TROPICAL, CENOZOIC PALEO-OCEANOGRAPHY AND
S 4 361 CORALS, ANATOMY OF, SCLERACTINIAN
S 4 5 CORALS, BONAIRE ISLAND
S 4 268 CORALS, BORERS, MOROCCO, JEBEL ASSAMEUR NAIT FERGANE
M 27 62 CORALS, COLONIAL, RUGOSE
S 4 5 CORALS, CURACAO ISLAND
S 4 371 CORALS, DENSITY, POROSITY, AND PERMEABILITY
S 4 95 CORALS, DIVERSITY AND ABUNDANCE OF, SCLERACTINIAN
S 4 177 CORALS, GROWTH RATES FOR, FOSSIL
S 4 289† CORALS IN DETERMINING GLACIAL-INTERGLACIAL CHANGES IN
S 4 135† CORALS IN RELATION TO WAVES AND CURRENTS, STRUCTURAL
S 4 171† CORALS MONTASTREA ANNULARIS AND MONTASTREA
S 4 158 CORALS OBSERVED DURING EMERGENCE, LIVING, CARIBBEAN, REEF
S 4 5 CORALS, PISCADERA BAY
S 4 223 CORALS, REEF-BUILDING
M 27 59 CORALS, SCLERACTINIAN
M 27 61 CORALS, SCLERACTINIAN
M 27 61 CORALS, SILICIFIED, TABULATE
M 27 62 CORALS, SOLITARY, RUGOSE
M 27 63 CORALS, SOLITARY, RUGOSE
B 6205 882 CORBETT, MARSHALL K., GENERAL GEOLOGY OF NORTHERN PORTNEUF
B 6104 601 **CORBITT, L. L., STRUCTURE OF BROCKMAN HILLS, SOUTHWESTERN NEW**
B 6405 694 CORBO, SALVATORE, PALEOENVIRONMENT CONTROL OF BIOGENIC
S 6 93 CORDANI, UMBERTO G., APPLICABILITY OF THE RUBIDIUM-STRONTIUM
S 10 1 **CORDELL, R. J., INTRODUCTION, PROBLEMS OF PETROLEUM MIGRATION**
S 10 1 **CORDELL, ROBERT J., EDITED BY, PROBLEMS OF PETROLEUM MIGRATION**
B 6405 694 CORDELL, ROBERT J., INTRODUCTION TO SEPM RESEARCH SYMPOSIUM—
B 6002 324 CORDELL, ROBERT J., LIQUID HYDROCARBON MIGRATION PATHS IN
B 6203 506 CORDELL, ROBERT J., MIGRATION PATHWAYS IN COMPACTING CLASTIC
B 6207 1228* CORDILLERA, ABST., EVOLUTION OF, CANADIAN SEGMENT OF
B 6004 715* CORDILLERA, ABST., FUSULINACEAN PROVINCES OF, NORTH AMERICAN
B 6207 1214* CORDILLERA, ABST., METALLOGENIC ZONES AND EPOCHS IN, CANADIAN
M 25 496 CORDILLERA, BARITE AND FLUORITE, CANADIAN
M 25 500 CORDILLERA, BOWSER BASIN, CROSS SECTION, CANADIAN
M 25 497 CORDILLERA, CLASTIC WEDGE, MINERAL DEPOSITS, CANADIAN
M 25 503 CORDILLERA, COAST CRYSTALLINE BELT, CANADIAN
M 25 503 CORDILLERA, COLUMBIAN OROGEN, CANADIAN
M 25 492 CORDILLERA, COMPLEXITY OF GEOLOGY, CANADIAN
M 25 496 CORDILLERA, COPPER, CANADIAN
M 25 505 CORDILLERA, COPPER, PRODUCTION AND RESERVES, CANADIAN
M 25 497 CORDILLERA, CROSS SECTION, CANADIAN
M 25 499 CORDILLERA, CROSS SECTION, CANADIAN
M 25 498 CORDILLERA, CROSS SECTION, CANADIAN
M 25 495 CORDILLERA, CROSS SECTION, RESTORED, CANADIAN
M 25 492† CORDILLERA DEPOSITIONAL BASINS, ENVIRONMENTS OF, CANADIAN
M 25 503 CORDILLERA, EASTERN MARGINAL BELT, CANADIAN
M 25 503 CORDILLERA, GEOLOGIC FRAMEWORK, CANADIAN
M 25 494 CORDILLERA, GEOLOGIC PROVINCES, MAP, CANADIAN
M 25 503 CORDILLERA, INTERMONTANE BELT, CANADIAN
M 25 505 CORDILLERA, IRON, PRODUCTION, CANADIAN
M 25 498 CORDILLERA, ISLAND ARC, MINERAL DEPOSITS, CANADIAN
M 25 498 CORDILLERA, ISLAND ARC, ROCKS OF, CANADIAN
M 25 499 CORDILLERA, JURASSIC SUCCESSOR BASIN, CROSS SECTION, CANADIAN
M 25 505 CORDILLERA, MAGMATIC DEPOSITS OF METALS, CANADIAN
M 25 506 CORDILLERA, MASSIVE SULFIDE DEPOSITS OF METALS, CANADIAN
M 25 505 CORDILLERA, METAL PRODUCTION AND RESERVES, TABLE, CANADIAN
M 25 503 CORDILLERA, METAL PRODUCTION, ANNUAL, CANADIAN
M 25 503† CORDILLERA, METALLIC, MINERAL RESOURCES OF, CANADIAN
M 25 483 CORDILLERA, METALLOGENIC PROVINCES, NORTH AMERICA, WESTERN
M 25 504 CORDILLERA, METALS, BACKGROUND ABUNDANCES, CANADIAN
M 25 504 CORDILLERA, METALS, CLASSES OF, CANADIAN
M 25 496 CORDILLERA, MID-PALEOZOIC CLASTIC WEDGE, CANADIAN
M 25 506 CORDILLERA, MINERAL DEPOSITS, DISTRIBUTION BY CLASSES,
M 25 493 CORDILLERA, MIOGEOSYNCLINE, GEOLOGIC HISTORY, CANADIAN
M 25 496 CORDILLERA, MIOGEOSYNCLINE, MINERAL DEPOSITS, CANADIAN
M 25 505 CORDILLERA, MOLYBDENUM, PRODUCTION AND RESERVES, CANADIAN
M 25 497 CORDILLERA, OCEANIC BASIN, MINERAL DEPOSITS, CANADIAN
M 25 497 CORDILLERA, OCEANIC BASIN, ROCKS OF, CANADIAN
M 25 503 CORDILLERA, OMINECA CRYSTALLINE BELT, CANADIAN
M 25 503 CORDILLERA, PACIFIC OROGEN, CANADIAN
M 25 506 CORDILLERA, PORPHYRY DEPOSITS OF METALS, CANADIAN
M 25 496 CORDILLERA, QUARTZ MONZONITE PLUTONS, CANADIAN
M 25 503 CORDILLERA, SEDIMENTATION AND MINERALIZATION, CYCLES OF,
M 25 505 CORDILLERA, SILVER, PRODUCTION AND RESERVES, CANADIAN
M 25 496 CORDILLERA, SILVER-LEAD-ZINC, CANADIAN
M 25 505 CORDILLERA, SKARN DEPOSITS OF METALS, CANADIAN
M 25 507 CORDILLERA, STRATIFORM DEPOSITS OF METALS, CANADIAN
M 25 500 CORDILLERA, SUCCESSOR BASINS, ROCKS OF, CANADIAN
M 25 503 CORDILLERA, TECTONIC BELTS, CANADIAN
M 25 504 CORDILLERA, TECTONIC BELTS, MAP, CANADIAN
M 25 496 CORDILLERA, TUNGSTEN, CANADIAN
M 25 507 CORDILLERA, VEIN DEPOSITS OF METALS, CANADIAN
M 25 505 CORDILLERA, ZINC, PRODUCTION AND RESERVES, CANADIAN
M 25 105† CORDILLERA--SUMMARY, COAL RESOURCES OF, CANADIAN
M 25 493 CORDILLERAN BASINS, MODELS OF, CANADA
B 6211 2352* CORDILLERAN COLLAGE, CENTRAL BRITISH COLUMBIA TO CENTRAL
B 6405 759* CORDILLERAN FOLDBELT AND FORELAND OF NORTHERN CANADA,
B 6104 602 CORDILLERAN FOLDBELT, NEW MEXICO, LARAMIDE THRUSTS
B 6405 768* CORDILLERAN FORELAND THRUST BELT, ABST., HYDROCARBON-,
B 6004 653* CORDILLERAN GEOSYNCLINE, ABST., LATE, PRECAMBRIAN
M 25 461 CORDILLERAN METALLIC BELT, ANDEAN-
M 25 493 CORDILLERAN OROGEN, CANADA
M 25 510 CORDILLERAN OROGEN, GRAVINA-NUTZOTIN BELT
M 25 509 CORDILLERAN OROGEN, INSULAR BELT OR ALEXANDER TERRANE
M 25 511 CORDILLERAN OROGEN, NORTHERN, PALEOZOIC CRUST FILLS RIFT ZONE
M 25 510 CORDILLERAN OROGEN, NORTHERN, PORPHYRY COPPER DEPOSITS, MAP
M 25 511 CORDILLERAN OROGEN, NORTHERN, PORPHYRY COPPER DEPOSITS,
M 25 509 CORDILLERAN OROGEN, PALEOZOIC ALLOCHTHON, FORMATION OF
M 25 509 CORDILLERAN OROGEN, PORPHYRY COPPER DEPOSITS, GEOLOGY OF,
M 25 509 CORDILLERAN OROGEN, TAKLA-HAZELTON-CACHE CREEK-SLIDE
M 25 509† CORDILLERAN OROGEN--SUMMARY, PORPHYRY, COPPER PROVINCE OF
M 25 500 CORDILLERAN OROGENY, CANADA
M 25 488 CORDILLERAN , SOURCES
B 6204 609 CORDINER, F. S., TENSLEEP RESERVOIR, OREGON BASIN FIELD, WYOMING
B 6105 816 CORDINER, F. S., TENSLEEP RESERVOIR STUDY, OREGON BASIN FIELD,
M 25 525 **CORDOBA, D. A., MINERAL RESOURCES OF STATE OF SINALOA.**
M 25 526 CORDOBA, DIEG A., MAJOR CHEMICAL CHARACTERISTICS OF TERTIARY
B 6209 1764* CORE 77B OF LEG 9, ABST., COSMOPOLITAN BIOZONATION FOR LATE
S 7 315 CORE ANALYSES, DIAGRAMS, NORMAL SLOPE SEQUENCE
S 7 316 CORE ANALYSIS, EXAMPLE, NORMAL SLOPE SEQUENCE
S 7 306 CORE ANALYSIS, ORCA BASIN
B 6109 1547* CORE AND FRACTURE ANALYSIS, ABST., EVALUATION OF, AUSTIN AND,
B 6203 547* CORE BANKS, NORTH CAROLINA—EFFECTS ON DEPOSITIONAL PATTERNS
B 6410 1635 CORE BORINGS, LOCATIONS, DEPTHS, RECOVERY, GREAT BAHAMA BANK,
B 6308 1376 CORE, EUGENE H., DEVELOPMENT IN ARKANSAS, NORTH LOUISIANA,
B 6108 1234 CORE, EUGENE H., DEVELOPMENTS IN ARKANSAS, LOUISIANA, AND EAST
B 6409 1490 CORE, EUGENE H., DEVELOPMENTS IN ARKANSAS, NORTH LOUISIANA,
B 6208 1487 CORE, EUGENE H., DEVELOPMENTS IN ARKANSAS, NORTH LOUISIANA,
B 6102 214 CORE, MEXICO, NOVILLERO 9
B 6303 463* CORE STUDY, ABST., KEY LARGO--SUBSURFACE
B 6209 1766* CORE STUDY, ABST., SUBMARINE-FAN DEPOSITION OF, WOODBINE
B 6009 1615* CORE STUDY, FRIERSON FIELD, LOUISIANA, ABST., INTERPRETATION OF,
B 6105 842* CORE-ANALYSIS SYSTEM, ABST., PETROS—PETROPHYSICAL, LOG AND
B 6405 704* CORES, ABST., CARBON FLUX FROM, OCEAN TO, BIOSPHERE—CHEMICAL
B 6101 102 CORES, DEEP SEA DRILLING PROJECT
B 6004 688* CORES FROM PACIFIC OCEAN, ABST., RESEDIMENTATION OF, BIOGENIC,
B 6202 249 CORES, JET AND DART, CALIFORNIA
B 6007 1025 CORES, LAKE SUPERIOR, SEDIMENT
B 6401 128 CORES, LOGS, INDIANA GEOLOGICAL SURVEY
B 6401 128 CORES, LOGS, INDIANA GEOLOGICAL SURVEY
B 6007 1061 CORES, PANAMA, GALETA
B 6303 538* CORES--PALEO-OCEANOGRAPHIC IMPLICATIONS, ABST., LATE,
B 6405 711* CORING IN EQUATORIAL PACIFIC—PRELIMINARY RESULTS FROM DSDP
B 6309 1592* CORING OF GAS-BEARING DEVONIAN BLACK SHALES, ERIE COUNTY
B 6105 822* CORING PROJECT, ABST., SHIPBOARD GEOTECHNICAL RESULTS FROM,
B 6303 482* CORING-INDUCED FRACTURES FROM NATURA CORED FRACTURES,
B 6405 781* CORING—BIOSYSTEMATIC ANALYSIS, EVOLUTIONARY SPECIES, AND
B 6304 689 CORKEN, R. JAMES., DEPOSITIONAL ENVIRONMENTS AS ORE CONTROLS
B 6105 778 **CORLISS, J. B., MARINE METALLIFEROUS SEDIMENTS FROM NAZCA PLATE**
M 25 539 CORLISS, JOHN B., METALLOGENESIS IN SOUTHEAST PACIFIC OCEAN,
B 6307 1139 CORNELL, WILLIAM C., GULF OF CALIFORNIA IN LAKE MEAD AREA OF
B 6203 426 CORNUBIAN PLATFORM, EUROPE
B 6206 1013 CORNWALL ARCH, CANADA, ARCTIC
B 6101 88 CORNWALL-SOUTH ATLAS FRACTURE ZONE
B 6307 1059† CORNWALLIS AND GRIFFITH ISLANDS, NORTHWEST TERRITORIES,
B 6101 19 CORNWALLIS FOLDBELT, CANADIAN, ARCTIC ISLANDS
B 6307 1064 CORNWALLIS GROUP, NORTHWEST TERRITORIES
B 6105 765 **CORON, CYNTHIA R., COARSE WHITE SPARRY DOLOMITE AND CALCITE**
B 6102 152 CORONADO BANK, CALIFORNIA
B 6206 920 CORONADO FAN
S 4 341 CORONOCYCLUS, ST. CROIX
B 6305 844 CORRE, PIERRE, SHALLOW SEISMIC REFLECTION IN APPALACHIAN BASIN

S 5 119 CORRECTION, SEISMIC DATA, AUTOMATIC STATICS
S 5 121 CORRECTION, SEISMIC DATA, HAND- STATICS
M 26 115 CORRELATION, ACCURACY, SEISMIC, REFLECTIONS,
B 6211 2277 CORRELATION AND NOMENCLATURE, WESTERN INTERIOR, JURASSIC
B 6303 474* CORRELATION AND TIMING OF PLATFORM-MARGIN MEGABRECCIA
B 6002 233 CORRELATION BETWEEN PETROLEUM PRODUCTIVITY AND SWELLING
B 6401 119 CORRELATION CHART, GREAT LAKES AREA, SILURIAN
B 6401 119 CORRELATION CHART, GREAT LAKES AREA, SILURIAN
B 6007 1120 CORRELATION CHART, TEXAS, JURASSIC, UPPER
M 26 139 CORRELATION, CHRONOSTRATIGRAPHIC WITH SEA LEVEL, CHART
M 26 77 CORRELATION CURVE CONSTRUCTION PROCEDURE, SEA LEVEL
S 6 174 CORRELATION, DIAGRAM, NEOGENE-, QUATERNARY
S 6 17 CORRELATION METHOD, REVIEW
B 6311 2116* CORRELATION OF CHEROKEE, DESMOINESIAN, SANDSTONES,
B 6209 1586 CORRELATION OF COAL-RANK SCALE WITH VARIOUS MATURATION
B 6303 473* CORRELATION OF CONTINENTAL-MARGIN AND DEEP-SEA
B 6207 1224* CORRELATION OF MAJOR UNCONFORMITIES IN WESTERN INDONESIA,
B 6108 1383* CORRELATION OF MESAVERDE GROUP IN SOUTHWESTERN WYOMING
S 6 363† CORRELATION OF MIDDLE AND UPPER PRECAMBRIAN SEDIMENTS,
B 6004 654* CORRELATION OF OROGENIC AND EPEIROGENIC EVENTS IN WESTERN
B 6205 867† CORRELATION OF PINNACLES AND NEENACH VOLCANIC FORMATIONS
B 6012 2128† CORRELATION OF PINNACLES AND NEENACH VOLCANIC FORMATIONS
B 6303 551* CORRELATION OF SUBSURFACE MIDDLE AND UPPER DEVONIAN ROCKS
B 6406 921 CORRELATION OF TIME-TEMPERATURE INDEX GEOCHEMICAL DATA
M 26 151 CORRELATION OF UNITS TO A CYCLE CHART, TEXAS, EAST TEXAS BASIN
M 26 90 CORRELATION, REGIONAL, GLOBAL CYCLES
M 26 106 CORRELATION, SEISMIC, REFLECTIONS, PALEONTOLOGIC
M 26 158 CORRELATION TO CYCLES, AFRICA, WEST, CHRONOSTRATIGRAPHIC
M 26 153 CORRELATION TO CYCLES, TEXAS, GULF COAST,
S 5 95 **CORRELATION, SILURIAN, CAYUGAN SERIES,**
S 5 95 CORRELATION, SILURIAN, LUDLOVIAN SERIES,
S 5 95 CORRELATION, SILURIAN, NIAGARAN SERIES,
S 5 95 CORRELATION, SILURIAN, PRIDOLIAN SERIES,
S 5 95 CORRELATION, SILURIAN, WENLOCKIAN SERIES,
B 6101 114 CORRELATION, UTAH, COALVILLE, TO, CUMBERLAND GAP, WYOMING
M 26 84 CORRELATION WITH FIRST-ORDER, GLOBAL CYCLES, SECOND-ORDER
M 26 84 CORRELATION WITH SECOND-ORDER, GLOBAL CYCLES, FIRST-ORDER
S 6 36 CORRELATION, WORLDWIDE, NEED FOR
B 6009 1610* CORRELATIONS AND STRATIGRAPHY FROM BASE OF PINE ISLAND TO
M 26 408 CORRELATIONS, MODELING, STRATIGRAPHIC
M 26 116 CORRELATIONS, PROBLEMS, SEISMIC, REFLECTIONS,
B 6009 1618* CORRELATIONS--NORTHERN ITALY-JAMAICA, B.W.I., ABST., MIDDLE,
B 6004 651* CORRESPONDENCE ANALYSIS. BY TITLE ONLY., MULTIDIMENSIONAL
B 6112 2104 CORRIENTES FIELD, PERU
B 6112 2105 CORRIENTES FIELD, PERU
M 28 177 CORROSION, OF HEAVY MINERALS
M 28 177 CORROSION, OF HORNBLENDE, IN SANDSTONE
B 6109 1484 CORSAIR CANYON, EAST COAST
B 6203 367 CORTES BANK, CALIFORNIA
B 6111 1981 CORTES BANK, CALIFORNIA
B 6111 1983 CORTES BANK, STRATIGRAPHIC COLUMN, CALIFORNIA
B 6203 363 CORTES RIDGE, CALIFORNIA, SANTA ROSA-
B 6004 670* CORTES RIDGE, SOUTHERN CALIFORNIA CONTINENTAL BORDERLAND,
B 6207 1213 CORVALAN, JOSE, REGIONAL GEOLOGIC, STRUCTURAL, AND MINERAL
M 28 98 CORWIN FORMATION, ALASKA, CRETACEOUS
M 28 117 CORWIN FORMATION, ALASKA, CRETACEOUS
M 28 29 CORWIN FORMATION, ALASKA, CRETACEOUS
B 6008 1412* CORWIN SPRINGS KNOWN GEOTHERMAL RESOURCE AREA, PARK
B 6008 1411* CORWIN SPRINGS--GARDINER AREA, MONTANA, ABST., PROPOSED,
B 6105 834* CORWIN SPRINGS—GARDINER AREA, MONTANA—STRUCTURAL AND
B 6209 1764* COSMOPOLITAN BIOZONATION FOR LATE CENOZOIC RADIOLARIANS AND
B 6211 2348* COSO AREA GEOTHERMAL DEVELOPMENTS, ABST.
B 6004 659 COSSEY, S. P. J., ANALYSIS O SUBMARINE DEBRIS-FLOW DEPOSIT,
B 6203 506 COSSEY, STEPHEN P. J., CONGLOMERATIC CARBONATE LIQUEFIED-FLOW
B 6408 1260 COST 1 WELL, ALASKA, COOK INLET
B 6412 2052 COST B-2 AND GE-1 WELLS, ATLANTIC OCEAN
B 6203 526* COST B-2 WELL, BALTIMORE CANYON, AND ITS RELATION TO SCOTIAN
M 29 70 COST B-2, WELL DATA, SONIC VELOCITIES
M 29 81 COST B-2 WELL, SUBSIDENCE HISTORY, ATLANTIC COAST
B 6405 781* COST B-3 WELL, ABST., SEISMIC STRATIGRAPHY OF, ATLANTIC MARGIN
B 6412 2059 COST GE-1 WELL, GEOCHEMICAL LOG, GEORGIA EMBAYMENT
B 6303 510* COST GE-1 WELL, SOUTHEAST GEORGIA EMBAYMENT, ABST., IMPORTANT
B 6309 1453 COST, WELL B-2, BALTIMORE CANYON TROUGH, CONTINENTAL
B 6410 1659 COST, WESTERN, ATLANTIC, CONTINENTAL OFFSHORE STRATIGRAPHIC
B 6010 1649 COSTA RICA, 1975, PRADAL, L. J., ON DEVELOPMENTS IN
B 6110 1583 COSTA RICA, 1976, PETROCONSULTANTS S.A., ON DEVELOPMENTS IN
B 6210 2049 COSTA RICA, 1977, PETROCONSULTANTS S.A., ON DEVELOPMENTS IN
B 6310 1750 COSTA RICA, 1978, PETROCONSULTANTS S.A., ON DEVELOPMENTS IN
B 6312 2207 COSTA RICA, NICOYA PENINSULA, PROFILE
B 6303 484 COSTAIN, JOHN K., DETAILED TEMPERATURE LOGGING AS USEFUL TOOL
B 6205 862 COSTS, PER FOOT DRILLED, UNITED STATES, WELL
B 6205 862 COSTS, ULTRADEEP, WELLS, AVERAGE
B 6404 512 COST—CONTINENTAL OFFSHORE STRATIGRAPHIC TEST WELLS
B 6101 66 COTE BLANCHE ISLAND, LOUISIANA
B 6101 67 COTE BLANCHE ISLAND SALT DOME, LOUISIANA
B 6309 1580* **COTTAGEVILLE FIELD, DEVONIAN SHALE, ABST., UNCONVENTIONAL**
B 6309 1585* COTTAGEVILLE FIELD, JACKSON COUNTY, WEST VIRGINIA, ABST.,
B 6412 2168 COTTAGEVILLE FIELD, WEST VIRGINIA
B 6309 1588* COTTAGEVILLE GAS FIELD CORRELATION ANALYSES FOR RESERVOIR
B 6308 1425* **COTTLE AND NORTHERN KING COUNTIES, TEXAS, ABST., ATOKA GAS IN**
B 6102 294* COTTLE AND NORTHERN KING COUNTIES, TEXAS, ABST., ATOKA, GAS IN
B 6003 401 COTTON FORMATION, SOMALIA
B 6407 1004† COTTON VALLEY AND SMACKOVER RESERVOIR TRENDS, EAST TEXAS,
B 6009 1615* COTTON VALLEY DEPOSITIONAL ENVIRONMENT FROM CORE STUDY,
B 6209 1757* COTTON VALLEY FORMATION, ABST., FORMATION EVALUATION WITH,
M 26 147 COTTON VALLEY FORMATION, TEXAS, EAST TEXAS BASIN
B 6407 1008 COTTON VALLEY LIMESTONE, TEXAS
B 6007 1119 COTTON VALLEY LIMESTONE, TEXAS
M 24 242 COTTON VALLEY PRODUCING SAND, MISSISSIPPI
B 6407 1011 COTTON VALLEY SANDSTONES, TEXAS, LOUISIANA, AND, ARKANSAS
B 6108 1381* COTTONWOOD CREEK, WYOMING, ABST., REGIONAL IMPLICATIONS OF
B 6402 165 COUBA ISLAND FIELD, LOUISIANA
B 6211 2243 COULTER CYCLE, INFORMAL NAME, CALIFORNIA
B 6211 2247 COULTER FAN, CALIFORNIA
B 6211 2264 COULTER FAN, CALIFORNIA
B 6211 2256 COULTER SAND, CALIFORNIA
B 6107 1030 COUNCIL GROVE GROUP, NEBRASKA
B 6211 2206 COUNTY CLARE, GULF COAST, AND NIGER DELTA, GROWTH-FAULT
B 6211 2194 COUNTY CLARE, NAMURIAN, IRELAND, WESTERN
B 6008 1395 COUPLES, GARY, ANALYTIC SOLUTIONS APPLIED TO FAULTING IN
B 6203 507 COURY, ANNY B., GEOLOGIC APPRAISAL OF UNDISCOVERED OIL AND
B 6105 777 **COURY, ANNY B., PROSPECTIVE HYDROCARBON PROVINCES OF THE WORLD**
B 6405 736 COUSINS, JOHN, EXPLORING FOR NIAGARAN PINNACLE REEF IN
B 6303 316 COVE MOUNTAIN MEMBER, UTAH, FLAGSTAFF LIMESTONE
B 6004 694* COW CREEK LIMESTONE, LOWER CRETACEOUS SOUTH TEXAS, ABST.,
B 6004 682* **COW HEAD BRECCIA, NEWFOUNDLAND, ABST., TIMING OF, CARBONATE**
B 6303 474* COW HEAD AND RELATED GROUPS, WESTERN NEWFOUNDLAND, ABST.,
S 3 132 COW SPRINGS SANDSTONE, NEW MEXICO
B 6004 657 COWIE, A. M., PALEOECOLOGIC INVESTIGATIONS O LATE HOLOCENE
S 6 355 COWIE, J. W., THE CAMBRIAN SYSTEM
B 6103 363 COX CITY DEPOSIT, OKLAHOMA
B 6408 1282 COX J. C., DEPOSITIONAL TRENDS OF LOWER SILURIAN CLINTON
B 6112 2134 COX, WAYNE C., OIL AND GAS IN RESERVOIRS WITH SUBNORMAL
B 6104 602 COYOTE HILLS, NEW MEXICO
B 6308 1244 COZART, CHERYL L., OIL AND GAS DEVELOPMENTS IN MARYLAND, OHIO,
B 6208 1399 COZART, CHERYL L., OIL AND GAS DEVELOPMENTS IN MARYLAND, OHIO,
B 6209 1767* COZUMEL ISLAND, QUINTANA ROO, MEXICO, ABST., LATE, PLEISTOCENE,
B 6406 921 CPI, CARBON PREFERENCE INDEX
B 6405 690* CRAB OCYPODE CERATOPHTHALMA, PALLAS, AS POSSIBLE INDICATORS
B 6209 1705 CRAB ORCHARD FORMATION, KENTUCKY
B 6303 344 CRAB ORCHARD FORMATION, KENTUCKY
B 6405 751* CRAB ORCHARD GROUP, PENNSYLVANIAN, NORTH-CENTRAL TENNESSEE,
B 6303 466* CRAB-BURROW ORIENTATION TO ENVIRONMENTAL ANALYSIS, ABST.,
B 6203 571* CRACKING, AND MIGRATION, ABST., SHALE, DEHYDRATION, SALINITY
B 6303 436 CRADDOCK, CAMPBELL,, ANTARCTICA AND GONDWANALAND, ABST.
B 6109 1547 CRAFT, J. MILTON, EVALUATION OF AUSTIN AND BUDA FORMATIONS
M 28 82 CRAIE GRISE, NETHERLANDS, CRETACEOUS
M 27 143 CRAIE GRISE, NETHERLANDS, CRETACEOUS
B 6309 1607 CRAIG, LIS E., FACIES DISTRIBUTION OF HOSSTON SAND, CENTRAL
B 6404 582 CRAIG, R. MICHAEL, GEOCHEMICAL AND SEDIMENTOLOGIC PROBLEMS
B 6203 507 CRAMER, HOWARD ROSS, MESOZOIC AND CENOZOIC GEOLOGY OF
B 6209 1755 CRAMER, HOWARD ROSS, UPPER MESOZOIC AND PALEOGENE GEOLOGY
B 6203 573 CRAMER, RONALD T., COPPER MOUNTAIN, WYOMING, URANIUM
B 6105 777 **CRANE, DAVID C., COMPUTER PROCESSING OF LANDSAT DATA FOR**
B 6012 2177 CRANE, R. C., IPEWICK FORMATION, SIGNIFICANT JURASSIC-NEOCOMIAN
B 6008 1409* CRATERS, ABST., WHY PETROLEUM EXPLORATIONIST SHOULD BE ON
B 6004 697* CRATO MARGIN, BANKS ISLAND, ARCTIC CANADA, ABST., TERTIARY,
S 3 118 CRATON, DEFINITION
S 3 118 CRATON, DEFINITION
M 29 461† CRATON, GLOBAL, SEA LEVEL CHANGE: A VIEW FROM THE
S 3 118 CRATON HISTORY
S 3 118 CRATON HISTORY
B 6203 503* CRATONAL BANK, SLOPE, AND BASINAL FACIES, ORDOVICIAN
B 6303 437 CRAWFORD, G. A., ALLUVIAL-FAN DEPOSITION AND TECTONIC
B 6405 768 CRAWFORD, G. A., EARLY GUADALUPIAN, PERMIAN, BANK MARGIN
B 6303 437 CRAWFORD, G. A., SEDIMENTOLOGY OF GOAT SEEP DOLOMITE
B 6004 725 CRAWFORD, W. A., GLAUCONITE-MAGHEMITE-HEMATITE
B 6008 1397* CRAZY MOUNTAIN BASIN, ABST., GEOTHERMAL POTENTIAL OF
B 6008 1400* CRAZY MOUNTAINS AREA--GEOLOGY, EXPLORATION HISTORY, AND
B 6008 1408* CRAZY MOUNTAINS BASIN, ABST., ROCKY MOUNTAIN, CRETACEOUS
B 6105 833 CREAGER, J. S., DREDGED MATERIAL DEPOSITED AT OCEAN DUMPING

B 6303 494* CREATION AND APPLICATION OF VARIABL DENSITY GRIDS TO OIL
M 27 105 CREEDE FORMATION, COLORADO, OLIGOCENE
M 28 41 CREEDE FORMATION, COLORADO, TERTIARY
B 6303 549* CREEP, ABST., DEFORMATION OF, CONTINENTAL, SLOPE SEDIMENTS
S 7 276 CREEP BEHAVIOR IN CLAY SOILS, STABILITY
S 7 283 CREEP, DIAGRAM, STABILITY, STRUCTURAL LOADING AND
M 26 259 CREEP MODIFICATION
B 6303 513* CREEP OF ACTIVE FAULTS IN NORTHWESTERN HOUSTON, HARRIS
B 6104 579 CREOLE CENTRAL LAKE FIELD, VENEZUELA
B 6104 579 CREOLE LA ROSA FIELD, VENEZUELA
B 6104 579 CREOLE LAGUNILLAS FIELD, VENEZUELA
B 6104 579 CREOLE LAGUNILLAS FIELD, VENEZUELA
B 6104 579 CREOLE PUNTA BENITEZ FIELD, VENEZUELA
B 6104 579 CREOLE PUNTA BENITEZ FIELD, VENEZUELA
B 6104 579 CREOLE TIA JUANA FIELD, VENEZUELA
B 6402 158 CRESCENT FARMS FIELD, LOUISIANA
B 6402 158 CRESCENT FARMS FIELD, LOUISIANA
B 6207 1194 CRESCENT FORMATION, OREGON AND, WASHINGTON
M 28 160 CRESCENT FORMATION, WASHINGTON, EOCENE
B 6207 1196 CRESCENT STRATIGRAPHY, OREGON AND, WASHINGTON, SILETZ RIVER-
B 6104 562 CRETA DURA STRATIGRAPHIC UNIT, CHILE
B 6008 1398* CRETACEOUS EAGLE FORMATION, BEARPAW MOUNTAINS AREA,
B 6102 295* CRETACEOUS MESA RICA SANDSTONE, TUCUMCARI BASIN, NEW MEXICO,
B 6004 699* CRETACEOUS MUDDY FORMATION OF GRIEVE FIELD, WYOMING, ABST.,
B 6304 700* CRETACEOUS OCEAN TEMPERATURES, ABST., TERTIARY AND
M 27 153 CRETACEOUS , SAN CARLOS FORMATION, TEXAS
B 6303 460* CRETACEOUS, ABST., PORE TYPES IN, SUNNILAND LIMESTONE, LOWER
B 6109 1552* CRETACEOUS AGE, ABST., SIGNIFICANCE OF, FRESHWATER, LIMESTONES
B 6111 1979† CRETACEOUS AGE, LOS ANGELES, EROSION SURFACE OF MIDDLE
B 6203 378 CRETACEOUS, ALASKA
M 24 4 CRETACEOUS, ALASKA
M 25 295 CRETACEOUS, ALASKA
B 6407 1016 CRETACEOUS, ALASKA
B 6408 1142 CRETACEOUS, ALASKA
B 6407 981 CRETACEOUS, ALBERTA
B 6205 785 CRETACEOUS, ALBERTA
S 6 279 CRETACEOUS, ALBIAN, FORAMINIFERAL ZONES
S 6 263 CRETACEOUS, ALBIAN-CENOMANIAN BOUNDARY, NORTH AMERICA
B 6405 802* CRETACEOUS, ALBIAN-TURONIAN, DEPOSITIONAL ENVIRONMENTS
S 6 263 CRETACEOUS, ALISITOS FORMATION, MEXICO
S 6 150 CRETACEOUS, AMAMI-OSHIMA GRANITES, JAPAN
M 25 344 CRETACEOUS, AMAZON BASINS
B 6405 752* CRETACEOUS, AND ASSOCIATED STRATA, NORTH SLOPE OF ALASKA,
B 6004 704* CRETACEOUS AND EARLY TERTIARY PALEOBATHYMETRIC HISTORY OF
B 6211 2349* CRETACEOUS AND EARLY TERTIARY SUBDUCTION COMPLEX,
B 6205 879* CRETACEOUS, AND ITS RELATION TO COAL RESOURCES, PINE CANYON
B 6209 1763* CRETACEOUS AND LOWER TERTIARY CARBONATE ROCKS, SOUTH
B 6405 738* CRETACEOUS AND LOWER TERTIARY SANDSTONES IN PACIFIC CREEK
B 6201 121† CRETACEOUS AND PALEOCENE, IN SOUTH FLORIDA, REBECCA, SHOAL,
B 6004 710* CRETACEOUS AND PALEOCENE SANDSTONES OF CANADIAN WESTERN
B 6004 704* CRETACEOUS AND TERTIARY OF NEW JERSEY, ABST., TIMING OF,
B 6203 536* CRETACEOUS AND TERTIARY STRATA FROM NUGSSUAQ EMBAYMENT,
S 6 264 CRETACEOUS, ANDESITIC VOLCANICS DATA POINT, ALASKA
B 6305 831* CRETACEOUS ANTLER FORMATION, NORTH TEXAS, ABST., NONMARINE
S 6 279 CRETACEOUS, APTIAN, FORAMINIFERAL ZONES
B 6405 668* CRETACEOUS ARC SEDIMENTATION AND VOLCANISM IN COASTAL
B 6303 443* CRETACEOUS ARC-TRENCH GAP, SOUTH-CENTRAL ALASKA, ABST.,
B 6405 800* CRETACEOUS, ARIZONA AND SONORA, ABST., DOLOMITE AND
B 6406 942† CRETACEOUS, AT LAKE TRAVERSE, WESTERN MINNESOTA, EXPOSURES
M 27 17 CRETACEOUS, ATCO FORMATION, TEXAS
M 28 153 CRETACEOUS, ATCO FORMATION, TEXAS
M 27 18 CRETACEOUS, ATCO FORMATION, TEXAS
M 28 84 CRETACEOUS, ATCO FORMATION, TEXAS
M 27 220 CRETACEOUS, AUSTIN GROUP, TEXAS
M 27 193 CRETACEOUS, AUSTIN GROUP, TEXAS
M 27 223 CRETACEOUS, AUSTIN GROUP, TEXAS
B 6207 1153 CRETACEOUS, AUSTRALIA
B 6201 60 CRETACEOUS, AUSTRALIA
B 6203 460 CRETACEOUS, AUSTRALIA
M 25 235 CRETACEOUS, AUSTRALIA
B 6204 696 CRETACEOUS, BAHAMA PLATFORM
B 6309 1455 CRETACEOUS, BALTIMORE CANYON TROUGH
B 6109 1553* CRETACEOUS BANK COMPLEX, EDWARDS LIMESTONE, NORTH-CENTRAL
B 6004 688* CRETACEOUS BANK COMPLEX EDWARDS LIMESTONE, NORTH-CENTRAL
M 28 74 CRETACEOUS, BAUXITE, ARKANSAS
B 6211 2233 CRETACEOUS, BAY OF BISCAY AREA
B 6405 685* CRETACEOUS BELEMNITE ROSTRA, ABST., MICROBIAL AND
M 26 279 CRETACEOUS, BENTON GROUP
S 6 273 CRETACEOUS, BIOSTRATIGRAPHIC ZONES
S 6 145 CRETACEOUS, BIOSTRATIGRAPHY, JAPAN
S 6 260 CRETACEOUS, BIOSTRATIGRAPHY, NORTH AMERICA
M 28 70 CRETACEOUS, BLACK LEAF FORMATION, MONTANA
B 6209 1594 CRETACEOUS BLACK SHALE, DSDP SITE WITH RECOVERY OF
B 6412 2051† CRETACEOUS BLACK SHALES IN ATLANTIC BASINS, PALEOENVIRONMENT
B 6405 790* CRETACEOUS BLACK SHALES IN DEEP NORTH ATLANTIC, ABST., ORGANIC
B 6205 798 CRETACEOUS, BLAKE PLATEAU
M 29 34 CRETACEOUS, BLAKE PLATEAU, STRUCTURE, TOP OF
M 29 30 CRETACEOUS, BLAKE PLATEAU, TOP OF
S 6 272 CRETACEOUS, BOUNDARIES, A REVIEW
B 6009 1608* CRETACEOUS BOUNDARY, ABST., NANNOFOSSIL, BIOSTRATIGRAPHY OF,
B 6004 657* CRETACEOUS BOUNDARY, NEOCOMIAN STRATIGRAPHY, AND
B 6404 514 CRETACEOUS BOUNDARY, NEW ENGLAND, JURASSIC-
M 27 21 CRETACEOUS, BUDA LIMESTONE, TEXAS
S 6 264 CRETACEOUS, BUDDEN CANYON FORMATION, CALIFORNIA
B 6001 108† CRETACEOUS BURIED CANYON--INFLUENCE ON ACCUMULATION OF
B 6205 892* CRETACEOUS CALCAREOUS NANNOFOSSIL ASSEMBLAGES IN ATLANTIC
B 6303 423* CRETACEOUS CALCAREOUS NANNOPLANKTON PALEOBIOGEOGRAPHY
B 6211 2261 CRETACEOUS, CALIFORNIA
S 6 261 CRETACEOUS, CAMPANIAN-MAASTRICHTIAN BOUNDARY, NORTH
B 6303 422* CRETACEOUS, CAMPANIAN-MAESTRICHTIAN, CAPE SEBASTIAN
B 6109 1552* CRETACEOUS CARBONATE PLATFORM, ABST., EXPLORATION HISTORY OF
B 6301 32† CRETACEOUS CARBONATE PLATFORM, EXPLORATION HISTORY OF
B 6004 731* CRETACEOUS CARBONATE PLATFORM, MEXICO, ABST., DIAGENETIC
S 9 22 CRETACEOUS, CARBONATE REEF BELT DISTRIBUTION
B 6109 1557* CRETACEOUS CARBONATE ROCKS OF CENTRAL SOUTH TEXAS, ABST.,
B 6303 547* CRETACEOUS CARBONATE SHELF IN SOUTHEASTERN ARIZONA AND
B 6206 1092* CRETACEOUS CARBONATE-SLOPE DEPOSITS, NORTHEASTERN MEXICO,
S 6 278 CRETACEOUS, CENOMANIAN, MACROFOSSIL ZONES
S 6 263 CRETACEOUS, CENOMANIAN-TURONIAN BOUNDARY, NORTH AMERICA
B 6303 534* CRETACEOUS, CENTRA ALBERTA, CANADA, ABST., CLAY-, CARBONATE
B 6405 674* CRETACEOUS, CENTRAL ALBERTA, ABST., RETROGRADATIONAL SHELF
B 6209 1766* CRETACEOUS, CENTRAL TYLER COUNTY, TEXAS—SUBSURFACE CORE
M 27 208 CRETACEOUS CHALK, ENGLAND
M 27 132 CRETACEOUS CHALK, FRANCE
M 27 168 CRETACEOUS CHALK, NORTH SEA
M 28 153 CRETACEOUS, CHALK, NORTH SEA
M 27 149 CRETACEOUS CHALK, NORTH SEA
M 27 21 CRETACEOUS CHALK, NORTH SEA
B 6107 996 CRETACEOUS CHALK, UPPER, SAMPLES, NORTH SEA
B 6205 892* CRETACEOUS CHALKS OF WESTERN INTERIOR, ABST., DIAGENESIS AND,
B 6303 422* CRETACEOUS CHALKS, SOUTHWESTERN ARKANSAS, ANNONA
B 6407 971 CRETACEOUS, CHART, GREAT PLAINS
B 6104 562 CRETACEOUS, CHILE
B 6405 689* CRETACEOUS COAL-BEARING STRATA, GALLUP COAL FIELD, NEW
B 6405 796* CRETACEOUS COAL-BEARING STRATA, NORTHERN KAIPAROWITS
B 6303 519* CRETACEOUS COALS OF WESTERN INTERIOR, ABST., DELTAIC, COALS OF,
B 6105 821* CRETACEOUS COGOLLO GROUP, MARACAIBO PLATFORM, VENEZUELA,
M 25 323 CRETACEOUS, COLOMBIA
S 6 261 CRETACEOUS, CONIACIAN-SANTONIAN BOUNDARY, NORTH AMERICA
B 6307 1108† CRETACEOUS CORAL-ALGAL-RUDIST REEFS, ARIZONA, DEPOSITIONAL,
M 28 117 CRETACEOUS, CORWIN FORMATION, ALASKA
M 28 29 CRETACEOUS, CORWIN FORMATION, ALASKA
M 28 98 CRETACEOUS, CORWIN FORMATION, ALASKA
M 28 82 CRETACEOUS, CRAIE GRISE, NETHERLANDS
M 27 143 CRETACEOUS, CRAIE GRISE, NETHERLANDS
M 27 86 CRETACEOUS, CUPIDIO FORMATION, MEXICO
M 27 123 CRETACEOUS, CUPIDO FORMATION, MEXICO
M 27 87 CRETACEOUS, CUPIDO FORMATION, MEXICO
M 27 88 CRETACEOUS, CUPIDO FORMATION, MEXICO
M 27 31 CRETACEOUS, CUPIDO FORMATION, MEXICO
B 6109 1551* CRETACEOUS CUPIDO FORMATION, NORTHEAST MEXICO, ABST.,
M 26 279 CRETACEOUS, DAKOTA GROUP
M 27 107 CRETACEOUS, DAKOTA GROUP, COLORADO
M 27 108 CRETACEOUS, DAKOTA SANDSTONE, COLORADO
B 6105 828* CRETACEOUS DAKOTA SANDSTONE, COLORADO, ABST., ORIGIN OF,
M 27 33 CRETACEOUS, DEL RIO CLAY, TEXAS
B 6009 1614* CRETACEOUS DELTAS OF SOUTH MISSISSIPPI AND PRODUCING TRENDS,
M 26 279 CRETACEOUS, DENVER-ARAPAHOE
B 6211 2363* CRETACEOUS DEPOSITIONAL ENVIRONMENTS, NORTHERN SANTA ANA
B 6109 1557* CRETACEOUS DEPOSITIONAL SYSTEMS, WEST TEXAS, ABST., LOWER
M 29 72 CRETACEOUS, DEPTH SECTIONS, ATLANTIC COAST
S 9 12 CRETACEOUS, DIAGRAM, POLAR WANDERING
B 6107 1010† CRETACEOUS, DISCONFORMITY IN LLOYDMINSTER AREA, ALBERTA AND
S 7 82 CRETACEOUS, DISCUSSED, GULF OF MEXICO, END OF
B 6303 451* CRETACEOUS DISTRIBUTARY-CHANNEL, DELTA-FRONT, AND BEACH-BAR
B 6403 316† **CRETACEOUS EAGLE FORMATION, BEARPAW MOUNTAINS AREA,**
B 6403 316† CRETACEOUS EAGLE SANDSTONE—RELATION TO SHALLOW GAS
M 25 446 CRETACEOUS, EARLY, METALLOGENI PROVINCES, SOUTH KOREA,
B 6409 1563* CRETACEOUS, EAST TEXAS BASIN AN SOUTHWEST ARKANSAS, ABST.,
B 6405 679* CRETACEOUS, EAST-CENTRAL AND NORTHEASTERN TEXAS, ABST.,
B 6409 1560* CRETACEOUS, EAST-CENTRAL MEXICO, ABST., FORAMINIFERS AND,
M 25 328 CRETACEOUS, ECUADOR

B 6109 1554* CRETACEOUS, EDWARDS GROUP, CHERT, CENTRAL TEXAS,ABST.,
B 6405 749* CRETACEOUS EDWARDS LIMESTONE, SAN ANTONIO AREA, TEXAS, ABST.,
M 27 205 CRETACEOUS, EDWARDS LIMESTONE, TEXAS
B 6306 976 CRETACEOUS, EGYPT
B 6306 970 CRETACEOUS, EGYPT
M 27 56 CRETACEOUS, EL ABRA FORMATION, MEXICO
M 27 161 CRETACEOUS, EL ABRA FORMATION, MEXICO
M 27 193 CRETACEOUS, EL ABRA LIMESTONE, MEXICO
M 27 167 CRETACEOUS, EL ABRA LIMESTONE, MEXICO
M 27 85 CRETACEOUS, EL ABRA LIMESTONE, MEXICO
M 29 341 CRETACEOUS, EVOLUTION, COLOMBIA
B 6109 1551* CRETACEOUS, FACIES MOSAIC, BLANCO AND HAYS COUNTIES, TEXAS,
B 6305 840* CRETACEOUS FERON SANDSTONE MEMBER, CENTRAL UTAH, ABST.,
M 27 134 CRETACEOUS, FERRY LAKE ANHYDRITE, TEXAS
M 28 87 CRETACEOUS, FERRY LAKE ANHYDRITE, TEXAS
S 9 23 CRETACEOUS, FOLDOUT, PALEOCLIMATES
M 27 166 CRETACEOUS, FOLKESTONE BEDS, ENGLAND
B 6008 1413* CRETACEOUS FORELAND BASIN, COLORADO, ABST., OVERPRESSURED,
M 28 31 CRETACEOUS, FORTRESS MOUNTAIN FORMATION, ALASKA
M 28 99 CRETACEOUS, FORTRESS MOUNTAIN FORMATION, ALASKA
M 26 279 CRETACEOUS, FOX HILLS SANDSTONE
M 27 89 CRETACEOUS, FOX HILLS SANDSTONE, SOUTH DAKOTA
M 28 174 CRETACEOUS, FRONTIER FORMATION, WYOMING
M 28 144 CRETACEOUS, FRONTIER FORMATION, WYOMING
M 28 21 CRETACEOUS, FRONTIER FORMATION, WYOMING
M 28 134 CRETACEOUS, FRONTIER FORMATION, WYOMING
M 28 130 CRETACEOUS, FRONTIER FORMATION, WYOMING
B 6008 1402* CRETACEOUS FRONTIER SANDSTONE, SOUTHEASTERN WYOMING AND
B 6405 794* CRETACEOUS GAS RESERVOIRS, ROCKY MOUNTAINS, ABST., EVOLUTION
B 6008 1393* CRETACEOUS GAS SAND IN NORTHERN MONTANA, ABST., WELL-LOG
B 6205 890* CRETACEOUS GAS-BEARING SANDSTONES, ABST., PERMEABILITY
B 6303 456* CRETACEOUS GAS-PRODUCTIVE SANDSTONE, NORTH-CENTRL MONTANA,
M 28 127 CRETACEOUS, GAULT FORMATION, ENGLAND
M 28 128 CRETACEOUS, GAULT FORMATION, ENGLAND
S 10 64 CRETACEOUS, GEOLOGIC SETTING, CANADA, ATHABASCA-PEACE RIVER
S 6 273 CRETACEOUS, GEOMAGNETIC REVERSAL SCALE
S 6 280 CRETACEOUS, GEOMAGNETIC REVERSAL SCALES
M 29 30 CRETACEOUS, GEORGIA EMBAYMENT, TOP OF
S 6 136 CRETACEOUS, GLAUCONITES, EXAMPLES, EUROPE
M 27 135 CRETACEOUS, GLEN ROSE LIMESTONE, TEXAS
M 27 87 CRETACEOUS, GLEN ROSE LIMESTONE, TEXAS
M 27 41 CRETACEOUS, GLEN ROSE LIMESTONE, TEXAS
M 27 144 CRETACEOUS, GLEN ROSE LIMESTONE, TEXAS
M 27 147 CRETACEOUS, GLEN ROSE LIMESTONE, TEXAS
M 27 124 CRETACEOUS, GLEN ROSE LIMESTONE, TEXAS
M 27 134 CRETACEOUS, GLEN ROSE LIMESTONE, TEXAS
B 6306 872 CRETACEOUS, GLOBAL DEPOSITION OF, ORGANIC, CARBON-RICH
M 27 205 CRETACEOUS, GMUNDEN FLYSCH, AUSTRIA
B 6105 835* CRETACEOUS GREAT VALLEY SEQUENCE, CALIFORNIA, ABST.,
M 27 177 CRETACEOUS, GREENHORN LIMESTONE, COLORADO
M 27 36 CRETACEOUS, GREENHORN LIMESTONE, COLORADO
M 26 277† CRETACEOUS GROWTH FAULTING, DENVER BASIN, COLORADO,
S 7 82 CRETACEOUS, GULF OF MEXICO, DISCUSSED
S 6 158 CRETACEOUS, HARACHIYAMA FORMATION, JAPAN
B 6312 2193† CRETACEOUS HELVETIAFJELLET FORMATION, SVALBARD, BEARING ON
B 6109 1551* CRETACEOUS HIPPURITACEA, RUDISTS, OF GULF COAST, ABST., EARLY
B 6108 1388* CRETACEOUS HYGIENE MEMBER OF PIERRE SHALE, WEST DENVER BASIN,
M 26 279 CRETACEOUS, HYGIENE SANDSTONE
S 6 150 CRETACEOUS, IGNEOUS GEOCHRONOLOGY, JAPAN
M 28 33 CRETACEOUS, ILDEFONSO FORMATION, PUERTO RICO
M 28 99 CRETACEOUS, ILDEFONSO FORMATION, PUERTO RICO
B 6105 832* CRETACEOUS, IN SOUTH DAKOTA AND WYOMING AS DETERMINED BY
M 29 16 CRETACEOUS INTERVAL, GEORGIA EMBAYMENT
B 6207 1132 CRETACEOUS, IRIAN JAYA
B 6205 819 CRETACEOUS, ISOPACH MAP, CALIFORNIA
B 6206 951 CRETACEOUS, ISRAEL
B 6205 839 CRETACEOUS, ISRAEL
B 6205 827† CRETACEOUS, ISRAEL, AS POSSIBLE OIL SOURCE ROCK,
S 4 84 CRETACEOUS, JAMAICA
S 4 84 CRETACEOUS, JAMAICA
S 6 155 CRETACEOUS, KITAKAMI MOUNTAINS, FORMATION, JAPAN
M 28 118 CRETACEOUS, KOGOSUKRUK TONGUE, ALASKA
M 28 20 CRETACEOUS, KOGOSUKRUK TONGUE, ALASKA
M 28 130 CRETACEOUS, KOGOSUKRUK TONGUE, ALASKA
M 28 7 CRETACEOUS, KOGOSUKRUK TONGUE, ALASKA
M 28 132 CRETACEOUS, KOGOSUKRUK TONGUE, ALASKA
M 28 19 CRETACEOUS, KOGOSUKRUK TONGUE, ALASKA
M 28 22 CRETACEOUS, KOGOSUKRUK TONGUE, ALASKA
M 27 84 CRETACEOUS, KRONER REEF LIMESTONE, GERMANY
S 6 145 CRETACEOUS, KWANMON GROUP, JAPAN
B 6303 425* CRETACEOUS LACUSTRINE CARBONATE DEPOSITION IN WESTERN
B 6405 680* CRETACEOUS LACUSTRINE SOURCE BEDS FROM EARLY RIFTING PHASES
B 6008 1396* CRETACEOUS LANCE CHANNELS, NIOBRARA COUNTY, WYOMING, ABST.,
M 26 279 CRETACEOUS, LARAMIE
B 6202 182 CRETACEOUS, LATE, PALEOGEOGRAPHY, WYOMING
M 25 443 CRETACEOUS, LATE, TO TERTIARY DEFORMATION, SOUTH KOREA
M 25 446 CRETACEOUS, LATE, TO TERTIARY, EARLY, METALLOGENIC PROVINCES,
B 6109 1499 CRETACEOUS, LATE, UNCONFORMITY, ALASKA
B 6109 1480 CRETACEOUS, LATE-TERTIARY PETROLEUM RESERVOIR SYSTEM, BRAZIL
M 27 34 CRETACEOUS LIMESTONE, AUSTRIA
S 4 262 CRETACEOUS, LIMESTONE, BORERS, TEXAS
M 27 9 CRETACEOUS, LIMESTONE, COLUMBIA
B 6012 2179* CRETACEOUS LITHOFACIES MODEL, SACRAMENTO VALLEY, CALIFORNIA,
M 28 124 CRETACEOUS, LOGAN CANYON FORMATION, CANADA
M 28 177 CRETACEOUS, LOGAN CANYON FORMATION, CANADA
M 28 172 CRETACEOUS, LOGAN CANYON SANDSTONE, CANADA
M 27 34 CRETACEOUS, LOWER CHALK, ENGLAND
M 27 20 CRETACEOUS, LOWER CHALK, ENGLAND
M 27 49 CRETACEOUS, LOWER GREENSAND, ENGLAND
M 30 577 CRETACEOUS, MAASTRICHTIAN, BATANGA SANDSTONE
M 30 329 CRETACEOUS, MACAE FORMATION, BRAZIL
B 6209 1767* CRETACEOUS, MAESTRICHTIAN, AND OF PINE BARREN MEMBER OF
M 29 129 CRETACEOUS, MAGELLAN BASIN, ONSHORE
M 30 531 CRETACEOUS, MALOSSA FIELD, BRUNTINO MARLS
M 30 531 CRETACEOUS, MALOSSA FIELD, SASS DE LA LUNA FORMATION
B 6203 566* CRETACEOUS, MARATHON REGION, WEST TEXAS, ABST., DEPOSITIONAL
B 6004 724* CRETACEOUS MARINE AND COASTAL DEPOSITS, ABST., MODELS USING,
M 28 175 CRETACEOUS, MESAVERDE FORMATION, WYOMING
M 28 176 CRETACEOUS, MESAVERDE FORMATION, WYOMING
M 28 178 CRETACEOUS, MESAVERDE FORMATION, WYOMING
M 28 179 CRETACEOUS, MESAVERDE FORMATION, WYOMING
M 28 41 CRETACEOUS, MESAVERDE GROUP, COLORADO
B 6305 824* CRETACEOUS MESAVERDE GROUP IN SOUTHERN WASATCH PLATEAU,
M 30 507 CRETACEOUS, MESSLA FIELD, SARIR SANDSTONE
B 6102 208 CRETACEOUS, MEXICO
B 6405 675* CRETACEOUS MID-SHELF SAND COMPLEX—SUSSEX SANDSTONE AT
M 28 70 CRETACEOUS, MIDDENDORF FORMATION, SOUTH CAROLINA
S 6 160 CRETACEOUS, MIDDLE YEZO GROUP, JAPAN
B 6406 943 CRETACEOUS, MINNESOTA
M 24 250 CRETACEOUS, MISSISSIPPI
B 6407 972 CRETACEOUS, MONTANA, GLACIER NATIONAL PARK
B 6407 972 CRETACEOUS, MONTANA, WILLISTON BASIN
M 27 148 CRETACEOUS, MONTE ANTOLA FORMATION, ITALY
M 28 73 CRETACEOUS, MONTE ANTOLA FORMATION, ITALY
M 28 22 CRETACEOUS, MONTE ANTOLA FORMATION, ITALY
M 27 185 CRETACEOUS, MONTE ANTOLA FORMATION, ITALY
M 28 155 CRETACEOUS, MONTE ANTOLA FORMATION, ITALY
M 28 45 CRETACEOUS, MONTE ANTOLA FORMATION, ITALY
M 28 162 CRETACEOUS, MONTE ANTOLA FORMATION, ITALY
M 28 38 CRETACEOUS, MONTE ANTOLA FORMATION, ITALY
M 27 200 CRETACEOUS, MONTE ANTOLA FORMATION, ITALY
M 27 143 CRETACEOUS, MONTE ANTOLA FORMATION, ITALY
M 28 152 CRETACEOUS, MONTE ANTOLA FORMATION, ITALY
B 6405 772* CRETACEOUS MOSBY SANDSTONE, CENTRAL MONTANA—EXAMPLE OF
M 27 22 CRETACEOUS, MOUNT LAUREL SANDSTONE, NEW JERSEY
M 28 107 CRETACEOUS, MOWRY SHALE, UTAH
M 28 72 CRETACEOUS, MOWRY SHALE, UTAH
M 28 68 CRETACEOUS, MOWRY SHALE, WYOMING
B 6405 676* CRETACEOUS MUDDY FORMATION, NORTHEAST POWDER RIVER BASIN,
B 6108 1384* **CRETACEOUS MUDDY FORMATION OF GRIEVE FIELD, WYOMING,**
B 6108 1371* CRETACEOUS MUDDY SANDSTONE, POWDER RIVER BASIN, WYOMING,
M 28 136 CRETACEOUS, MUDDY SANDSTONE, WYOMING
M 28 137 CRETACEOUS, MUDDY SANDSTONE, WYOMING
M 28 38 CRETACEOUS, NAVESINK FORMATION, NEW JERSEY
M 27 154 CRETACEOUS, NAVESINK FORMATION, NEW JERSEY
B 6009 1614* CRETACEOUS NEAR KERRVILLE, TEXAS, ABST., NEW, PELECYPOD SPECIES
B 6004 701* CRETACEOUS NEOCOMIAN, SANDSTONE SEQUENCE, MACKENZIE DELTA
B 6404 516 CRETACEOUS, NEW ENGLAND CONTINENTAL MARGIN
B 6007 1013 CRETACEOUS, NEW MEXICO
M 26 102 CRETACEOUS, NEW MEXICO, SAN JUAN BASIN
M 27 11 CRETACEOUS, NEWARK CANYON FORMATION, NEVADA
B 6201 2 CRETACEOUS, NIGER DELTA
M 26 279 CRETACEOUS, NIOBRARA
B 6405 786* CRETACEOUS NIOBRARA GAS PLAY, ABST.
B 6205 780 CRETACEOUS, NORTH, AFRICA
B 6105 824* CRETACEOUS, NORTH AMERICA, ABST., MODELS OF SHALLOW-MARINE,
M 25 283 CRETACEOUS, NORTH AMERICA, WESTERN
B 6203 420 CRETACEOUS, NORTH, ATLANTIC
B 6204 596 CRETACEOUS, NORTH, ATLANTIC BASIN
B 6011 2058† CRETACEOUS, NORTHEAST TEXAS, GLOBOTRUNCANA FORNICATA ZONE
B 6409 1564* CRETACEOUS, NORTHERN RIO ESCONDIDO BASIN, COAHUILA, MEXICO,
B 6004 723* CRETACEOUS, NORTHWESTERN BIG HORN BASIN, WYOMING, ABST.,

B 6205 886* CRETACEOUS, NORTHWESTERN COLORADO, ABST., STRATIGRAPHY OF,
S 6 273 CRETACEOUS, NUMERIC TIME SCALES
S 6 282 CRETACEOUS, OCEAN BASEMENT AGES
B 6306 870† CRETACEOUS OCEANIC ANOXIC EVENTS AS CAUSAL FACTORS IN
B 6405 754* CRETACEOUS, OF CENTRAL SAUDI ARABIA, ABST., CEMENTATION,
B 6209 1755* CRETACEOUS OF NORTH LOUISIANA AND MISSISSIPPI, ABST.,
B 6209 1761* CRETACEOUS, OF NORTH-CENTRAL TEXAS, ABST., FORAMINIFERAL
B 6203 514* CRETACEOUS OF NORTHEASTERN SPAIN, ABST., ONCOLITES FROM,
B 6405 733* CRETACEOUS OF NORTHERN COLORADO, ABST., PATTERNS OF
B 6109 1541* CRETACEOUS, OF TEXAS, ABST., DEPOSITIONAL ENVIRONMENT OF,
B 6205 883* CRETACEOUS OF UTAH, ABST., HOLOCENE, ICHNOFACIES FROM
B 6405 739* CRETACEOUS OF WESTERN INTERIOR, ABST., WAVE-DOMINATED,
B 6405 793* CRETACEOUS, OFFSHORE BAR FACIES DISTRIBUTION, SALT CREEK AREA,
B 6202 181† CRETACEOUS OFFSHORE SEDIMENTATION, SUSSEX SANDSTONE OF,
B 6405 671* CRETACEOUS ORGANIC CARBON-RICH MARINE STRATA, ABST., OXIC-,
B 6303 536* CRETACEOUS OSTRACODA FROM PENINSULAR FLORIDA, ABST.,
B 6004 727* CRETACEOUS OSTRACODA FROM WELLS IN CENTRAL AND SOUTHERN
B 6303 538* CRETACEOUS OXYGEN-DEFICIENT PALEOENVIRONMENTS IN MID-PACIFIC
S 6 127† CRETACEOUS, PALEOGENE SEDIMENTS, EUROPE, RESULTS OF DATING
M 28 120 CRETACEOUS, PANTHER TONGUE, UTAH
M 28 9 CRETACEOUS, PATULA ARKOSE, MEXICO
M 28 165 CRETACEOUS, PATULA ARKOSE, MEXICO
M 28 134 CRETACEOUS, PATULA ARKOSE, MEXICO
M 28 18 CRETACEOUS, PATULA ARKOSE, MEXICO
S 6 263 CRETACEOUS, PEACE RIVER FORMATION, CANADA
B 6109 1461 CRETACEOUS, PEDREGOSA BASIN
B 6401 67† CRETACEOUS PELAGIC LIMESTONES--POTENTIAL STRATIGRAPHIC AND
B 6401 67† CRETACEOUS PELAGIC LIMMSTONES--POTENTIAL STRATIGRAPHIC AND
B 6203 517* CRETACEOUS PETERSON LIMESTONE, WESTERN WYOMING, ABST.,
B 6101 56 CRETACEOUS PETROLEUM PROSPECTS, LEBANON, JURASSIC-
S 6 269 CRETACEOUS, PHANEROZOIC TIME SCALE, A REVIEW
M 26 279 CRETACEOUS, PIERRE SHALE, ZONES
B 6102 296* CRETACEOUS, PLAY, ABST., SOUTH, TEXAS, AUSTIN CHALK
S 10 157 CRETACEOUS, POTENTIOMETRIC SURFACE, PERSIAN GULF
S 10 136 CRETACEOUS WATER LEVELS, FRANCE, AQUITAINE
M 28 178 CRETACEOUS, PRICE RIVER FORMATION, UTAH
M 28 130 CRETACEOUS, PRINCE CREEK FORMATION, ALASKA
M 28 20 CRETACEOUS, PRINCE CREEK FORMATION, ALASKA
M 28 132 CRETACEOUS, PRINCE CREEK FORMATION, ALASKA
M 28 118 CRETACEOUS, PRINCE CREEK FORMATION, ALASKA
M 28 7 CRETACEOUS, PRINCE CREEK FORMATION, ALASKA
M 28 22 CRETACEOUS, PRINCE CREEK FORMATION, ALASKA
M 28 19 CRETACEOUS, PRINCE CREEK FORMATION, ALASKA
B 6205 793 CRETACEOUS, PROFILE, BLAKE PLATEAU AREA
S 6 138 CRETACEOUS, PROPOSED RADIOMETRIC SCALE, EUROPE
S 6 266 CRETACEOUS, PROPOSED TIME SCALE, NORTH AMERICA
M 30 304 CRETACEOUS, PUT RIVER SANDSTONE, PRUDHOE BAY
S 6 275 CRETACEOUS, RADIOMETRIC SCALE, GLAUCONITE
S 6 274 CRETACEOUS, RADIOMETRIC SCALE, NON-GLAUCONITE
M 27 106 CRETACEOUS, RED BANK SANDSTONE, NEW JERSEY
M 27 22 CRETACEOUS, RED BANK SANDSTONE, NEW JERSEY
B 6405 782* CRETACEOUS REEF COMMUNITIES IN GULF COAST, ABST., EARLY
B 6109 1550* CRETACEOUS REEFS OF TEXAS, ABST., CONTRASTS IN, CEMENTATION,
B 6406 857† CRETACEOUS REGRESSIVE FACIES, NORTHEAST AFRICA CRATON, LATEST,
B 6303 550* CRETACEOUS RESERVOIR AND NONRESERVOIR SANDSTONES, ABST.,
B 6407 980 CRETACEOUS, RESERVOIRS, CANADA, WESTERN
M 28 28 CRETACEOUS, RIESELBERGER SANDSTONE, GERMANY
M 27 84 CRETACEOUS, RIO YAUCO FORMATION, PUERTO RICO
B 6408 1259† CRETACEOUS ROCKS, COOK INLET, ALASKA, NONMARINE UPPER
B 6304 598† CRETACEOUS ROCKS FROM SELECTED AREAS OF SAN JUAN BASIN, NEW
B 6203 546* CRETACEOUS ROCKS OF CENTRAL SAUDI ARABIA, ABST., STRATIGRAPHY,
B 6109 1557* CRETACEOUS ROCKS OF MEXICO AND TEXAS, ABST., MIDDLE
B 6305 832* CRETACEOUS ROCKS OF SOUTHMAN CANYON GAS FIELD, UINTA BASIN,
B 6306 876 CRETACEOUS RUDIST BUILDUPS AS STRATIGRAPHIC TRAPS
B 6306 877 CRETACEOUS RUDIST-CORAL REEFAL BUILDUPS. DISTRIBUTION
M 27 82 CRETACEOUS SAN CARLOS FORMATION, TEXAS
B 6309 1611* CRETACEOUS SAN MIGUEL FORMATION, MAVERICK BASIN, SOUTH
B 6408 1236 CRETACEOUS SAND BARS, UTAH
B 6408 1233 CRETACEOUS SAND FIELD, WYOMING, POWDER RIVER BASIN
B 6004 661* CRETACEOUS SAND FLAT-SHALLOW SUBTIDAL ENVIRONMENT, ABST.,
B 6405 694* CRETACEOUS SAND UNITS, DELAWARE AND NEW JERSEY, ABST., TRACE,
M 28 123 CRETACEOUS, SANDSTONE, ATLANTIC
M 28 172 CRETACEOUS, SANDSTONE, CANADA
M 28 180 CRETACEOUS, SANDSTONE, CANADA, N.W.T.
M 28 178 CRETACEOUS, SANDSTONE, CANADA, N.W.T.
M 28 163 CRETACEOUS, SANDSTONE, CANADA, N.W.T.
B 6405 748* CRETACEOUS SANDSTONES, ABST., EASTERN, GREEN RIVER BASIN—A
B 6409 1554* CRETACEOUS, SANTA ROSA CANYON, SIERRA DE SANTA ROSA, NUEVO
S 6 261 CRETACEOUS, SANTONIAN-CAMPANIAN BOUNDARY, NORTH AMERICA
B 6201 74 CRETACEOUS, SASKATCHEWAN
M 28 156 CRETACEOUS, SCHISTI GALESTRINI, ITALY
B 6405 719* CRETACEOUS SEA LEVEL AND STRATIGRAPHY, EASTERN ARABIAN
B 6009 1618* CRETACEOUS SECTION IN SOUTH FLORIDA, ABST., SIX PROPOSED
B 6304 599 CRETACEOUS, SECTION, NEW MEXICO
B 6109 1555* CRETACEOUS SEDIMENTARYFACIES AND SEA-LEVEL CHANGES, GULF
B 6303 550* CRETACEOUS SEDIMENTATION AND PETROLEUM OCCURRENCES IN
B 6211 2351* CRETACEOUS SEDIMENTATION IN SANTA MONICA MOUNTAINS,
M 27 90 CRETACEOUS SEDIMENTS, NEW JERSEY
M 26 145† CRETACEOUS SEISMIC SEQUENCES IN GULF OF MEXICO AND OFFSHORE
B 6303 526* CRETACEOUS SEISMIC STRATIGRAPHY AND GEOLOGIC HISTORY OF DEEP
B 6405 796* CRETACEOUS SEISMIC STRATIGRAPHY OF ABYSSAL SOUTHWESTERN
S 6 277 CRETACEOUS, SENONIAN, FORAMINIFERAL SCHEME
M 28 160 CRETACEOUS, SHALE, ALASKA
B 6404 488† CRETACEOUS SHALE, POWDER RIVER BASIN, WYOMING, ORGANIC
B 6002 258† CRETACEOUS SHALLOW CARBONATE PLATFORM OF ISRAEL, RUDISTID,
B 6405 712* CRETACEOUS SHALLOW-MARINE SHELF, ABST., CARBONATE FACIES OF,
S 6 264 CRETACEOUS, SHASTA BALLY BATHOLITH, CALIFORNIA
M 29 72 CRETACEOUS, SHELF EDGE, ATLANTIC COAST
M 26 169 CRETACEOUS, SHELF FACIES, SAN JUAN BASIN
B 6109 1557* CRETACEOUS SHELF MARGIN IN NORTHERN MEXICO, ABST., LOWER
B 6009 1617* CRETACEOUS SHELF MARGIN IN NORTHERN MEXICO AND SOUTH TEXAS,
B 6405 723* CRETACEOUS SHELF STORM DEPOSITS, NORTH TEXAS, ABST., LOWER
B 6303 416* CRETACEOUS SHELF-MARGIN CARBONATE FACIES, NORTHWESTERN
B 6111 1989 CRETACEOUS SHORELINE, CALIFORNIA
M 28 34 CRETACEOUS, SHUMAGIN FORMATION, ALASKA
B 6004 687* CRETACEOUS SIEVERING FORMATION, VIENNA, AUSTRIA, ABST., TRAILS
B 6209 1753* CRETACEOUS SLIGO FORMATION, SOUTH TEXAS, ABST., DISTRIBUTION
B 6008 1408* CRETACEOUS SOURCE-ROCK TYPES, MATURITY, AND RELATION TO
B 6406 943 CRETACEOUS, SOUTH DAKOTA, BLACK HILLS, NORTHERN, UPPER
B 6109 1554* CRETACEOUS, SOUTH TEXAS, ABST., POROSITY DEVELOPMENT AND
B 6004 694* CRETACEOUS SOUTH TEXAS, ABST., RELATION OF EARLY, CARBONATE,
B 6109 1555* CRETACEOUS, SOUTH-CENTRAL TEXAS, ABST., CARBON AND, OXYGEN,
B 6203 548* CRETACEOUS, SOUTH-CENTRAL TEXAS, ABST., LITHOFACIES AND
B 6409 1552* CRETACEOUS, SPEARY GAS FIELD, KARNES COUNTY, TEXAS, ABST.,
B 6303 510* CRETACEOUS, SPINDLE FIELD, COLORADO, ABST., DIAGENESIS IN,
M 28 120 CRETACEOUS, STAR POINT SANDSTONE, UTAH
M 28 156 CRETACEOUS, STAR POINT SANDSTONE, UTAH
M 28 180 CRETACEOUS, STAR POINT SANDSTONE, UTAH
M 30 105 CRETACEOUS, STATFJORD FIELD, CROMER KNOLL GROUP
M 30 105 CRETACEOUS, STATFJORD FIELD, SHETLAND GROUP
B 6405 729* CRETACEOUS STRATA, ABST., TRACE-MAKERS AS HISTORIANS IN
B 6108 1375* CRETACEOUS STRATA IN WESTERN NEBRASKA AND ADJACENT STATES,
B 6109 1550* CRETACEOUS STRATIGRAPHIC MODELS FROM TEXAS AND MEXICO,
B 6405 752* CRETACEOUS STRATIGRAPHY, CHIGNIK AREA, ALASKA PENINSULA,
B 6405 781* CRETACEOUS SUBSIDENCE OF CENTRA NORTH SEA BASIN, ABST.,
B 6001 87† CRETACEOUS SUBSURFACE STRATIGRAPHY OF ATLANTIC COASTAL
B 6405 728* CRETACEOUS SUBTIDAL SAND COMPLEX, WOBURN SANDS, SOUTHERN
B 6303 528* CRETACEOUS, SUGAR CREEK FIELD, TYLER COUNTY, TEXAS, ABST.,
B 6209 1753* CRETACEOUS SUNNILAND FORMATION IN SOUTH FLORIDA, ABST.,
B 6101 16 CRETACEOUS, SVERDRUP BASIN
B 6101 15 CRETACEOUS, SVERDRUP BASIN, JURASSIC-
B 6008 1412* CRETACEOUS SWAMP COMPLEX, JOHN HENRY MEMBER OF STRAIGHT
M 27 195 CRETACEOUS, TAMABRA LIMESTONE, MEXICO
M 27 212 CRETACEOUS, TAMABRA LIMESTONE, MEXICO
M 27 190 CRETACEOUS, TAMABRA LIMESTONE, MEXICO
M 27 14 CRETACEOUS, TAMABRA LIMESTONE, MEXICO
M 27 85 CRETACEOUS, TAMABRA LIMESTONE, MEXICO
M 27 13 CRETACEOUS, TAMABRA LIMESTONE, MEXICO
M 27 9 CRETACEOUS, TAMABRA LIMESTONE, MEXICO
M 28 178 CRETACEOUS, TEAPOT SANDSTONE, WYOMING
M 28 176 CRETACEOUS, TEAPOT SANDSTONE, WYOMING
M 28 179 CRETACEOUS, TEAPOT SANDSTONE, WYOMING
M 28 175 CRETACEOUS, TEAPOT SANDSTONE, WYOMING
M 26 279 CRETACEOUS, TERRY SANDSTONE
S 6 260 CRETACEOUS, TERTIARY BOUNDARY, A REVIEW
B 6402 225 CRETACEOUS, TEXAS
B 6402 225 CRETACEOUS, TEXAS
B 6303 472* CRETACEOUS, TEXAS, ABST., DOLOMITIZATION OF OFFSHORE,
B 6301 35 CRETACEOUS, TEXAS, BANDERA COUNTY
B 6301 35 CRETACEOUS, TEXAS, DEWITT COUNTY
S 7 203 CRETACEOUS THICKNESS, WEST FLORIDA SLOPE
B 6004 695* CRETACEOUS THROUGH PLIOCENE, ABST., SEDIMENT, LITHOFACIES IN,
B 6405 778* CRETACEOUS TIDES, WESTERN INTERIOR BASIN, NORTH AMERICA, ABST.,
B 6405 731* CRETACEOUS TIDES, WESTERN INTERIOR BASIN, NORTH AMERICA, ABST.,
M 29 71 CRETACEOUS, TIME AND DEPTH SECTIONS, ATLANTIC COAST
S 6 269† CRETACEOUS TIME SCALE, A
B 6004 498† CRETACEOUS TIME SCALE, A
S 6 259† CRETACEOUS TIME SCALE FROM NORTH AMERICA
S 6 259 CRETACEOUS, TIME SCALE STUDIES, REVIEW
S 6 270 CRETACEOUS, TIME SCALES, COMPARISON
S 6 262 CRETACEOUS, TIME-SCALES, COMPARISON, NORTH AMERICA
M 28 38 CRETACEOUS, TINTON SAND, NEW JERSEY

Index

B 6108 1378* CRETACEOUS TO EOCENE, AND ASSOCIATED CLOSED-BASIN DEPOSITS,
M 28 32 CRETACEOUS, TOROK FORMATION, ALASKA
M 28 33 CRETACEOUS, TOROK FORMATION, ALASKA
B 6303 495* CRETACEOUS TRANSGRESSION OF COAHUILA PENINSULA, POTRERO DE
M 28 27 CRETACEOUS, TRAVIS PEAK CONGLOMERATE, TEXAS
M 28 78 CRETACEOUS, TRAVIS PEAK FORMATION, TEXAS
M 28 9 CRETACEOUS, TRAVIS PEAK FORMATION, TEXAS
M 28 117 CRETACEOUS, TRAVIS PEAK FORMATION, TEXAS
M 28 115 CRETACEOUS, TRAVIS PEAK FORMATION, TEXAS
S 6 260 CRETACEOUS, TRICERATOPS, NORTH AMERICA
M 28 30 CRETACEOUS, TUKTU FORMATION, ALASKA
M 28 99 CRETACEOUS, TUKTU FORMATION, ALASKA
B 6403 424 CRETACEOUS, TUNISIA
B 6403 424 CRETACEOUS, TUNISIA
B 6206 934 CRETACEOUS, TURBIDITE FACIES, CALIFORNIA
B 6012 2182* CRETACEOUS TURBIDITE SEDIMENTATIO AND SUBMARINE-FAN FACIES
S 6 278 CRETACEOUS, TURONIAN, MACROFOSSIL ZONES
S 6 261 CRETACEOUS, TURONIAN-CONIACIAN BOUNDARY, NORTH AMERICA
M 28 70 CRETACEOUS, TUSCALOOSA FORMATION, LOUISIANA
M 28 144 CRETACEOUS, TUSCALOOSA FORMATION, LOUISIANA
M 28 138 CRETACEOUS, TUSCALOOSA FORMATION, LOUISIANA
M 28 135 CRETACEOUS, TUSCALOOSA FORMATION, LOUISIANA
M 28 176 CRETACEOUS, TUSCALOOSA FORMATION, LOUISIANA
M 30 178 CRETACEOUS UNCONFORMITY, NINIAN FIELD, MAP
M 27 95 CRETACEOUS, UPPER CHALK, ENGLAND
M 27 82 CRETACEOUS, UPPER CHALK, ENGLAND
M 28 78 CRETACEOUS, UPPER CHALK, ENGLAND
M 27 35 CRETACEOUS, UPPER CHALK, ENGLAND
B 6201 125 CRETACEOUS, UPPER, COLUMNAR SECTION, FLORIDA
B 6403 318 CRETACEOUS, UPPER, CORRELATION CHART, MONTANA AND, CANADA
B 6403 318 CRETACEOUS, UPPER, CORRELATION CHART, MONTANA AND, CANADA
B 6412 2056 CRETACEOUS, UPPER, ORGANIC MATTER, ATLANTIC BASINS
B 6404 471 CRETACEOUS, VADOSE PENDANT CEMENTS, TEXAS
B 6104 581 CRETACEOUS, VENEZUELA
B 6203 416 CRETACEOUS, VIRGIN ISLANDS, ST. CROIX
M 25 269 CRETACEOUS VOLCANIC ACTIVITY, KOREA
B 6303 430* CRETACEOUS VOLCANIC ROCKS CLOSE TO GULF COAST, SIERRA MADRE
S 6 153 CRETACEOUS VOLCANICS, TABLE, JAPAN
B 6004 694* CRETACEOUS VOLCANO, DIMMI COUNTY, TEXAS, ABST., CARBONATE
S 7 200 CRETACEOUS, WEST FLORIDA SLOPE, NORTHERN AREA, LOWER
S 7 202 CRETACEOUS, WEST FLORIDA SLOPE, NORTHERN AREA, UPPER
B 6303 354† CRETACEOUS, WESTERN AND CENTRAL WYOMING,
S 6 260 CRETACEOUS, WESTERN INTERIOR, NORTH AMERICA
B 6303 464* CRETACEOUS, WESTERN INTERIOR UNITED STATES, ABST., REGIONAL
M 27 205 CRETACEOUS, WHITE LIMESTONE, IRELAND
M 27 184 CRETACEOUS, WHITE LIMESTONE, IRELAND
M 27 51 CRETACEOUS, WHITE LIMESTONE, IRELAND
M 27 160 CRETACEOUS, WHITE LIMESTONE, IRELAND
M 28 78 CRETACEOUS, WHITE LIMESTONE, NORTHERN IRELAND
B 6004 675* CRETACEOUS, WIND RIVER BASIN, WYOMING, ABST., INFLUENC OF
M 28 124 CRETACEOUS, WOODBINE FORMATION, TEXAS
M 24 109 CRETACEOUS, WYOMING
B 6309 1601* **CRETACEOUS--LOWER EOCENE STRATA, HAINESVILLE, KEECHI, AND**
B 6109 1549* CRETACEOUS-EOCENE FORMATIONS IN FLORIDA, ABST., TYPE AND
B 6203 536* CRETACEOUS-HOLOCENE SEDIMENTATIONIN EGYPTIAN RED SEA AND
B 6211 2217 CRETACEOUS-PALEOCENE, CALIFORNIA
B 6303 437* CRETACEOUS-PALEOCENE CONGLOMERATES IN NORTH-CENTRAL UTAH
B 6004 732* CRETACEOUS-PALEOCENE, PARRAS BASIN, MEXICO, ABST.,
B 6303 512* CRETACEOUS-PALEOCENE RESERVOIR SANDSTONES IN WEST-CENTRAL
B 6303 546* CRETACEOUS-PALEOCENE SEDIMENTATION IN WYOMING, ABST.,
B 6401 72 CRETACEOUS-TERTIARY BOUNDARY, CARBON ISOTOPE RATIOS, NORTH
B 6401 72 CRETACEOUS-TERTIARY BOUNDARY, ISOTOPE RATIOS, NORTH
B 6111 1987 CRETACEOUS-TERTIARY UNCONFORMITY, CALIFORNIA
B 6405 724* CRETACEOUS-TERTIARY VERSUS CARBONIFEROUS DEPOSITIONAL
B 6101 115 CRETACEOUS-UPPER CRETACEOUS BOUNDARY, UTAH, LOWER
B 6203 510* CREVASSE FILLINGS OF PORT HURON AGE, STOUFFVILLE, ONTARIO,
B 6203 560* CREVASSE-SPLAY AND REWORKED DELTAIC-SANDSTONE RESERVOIRS,
B 6203 507 CREVELLO, PAUL D., SEDIMENTARY RECORD OF LONGITUDINAL
B 6205 743 CREW-MONTHS IN NON-COMMUNIST WORLD, 1970-1975, ANNUAL,
S 6 355 CRIBB, S. J., THE CAMBRIAN SYSTEM
S 11 270 CRIBROGOESELLA PACIFICA
S 11 230 CRIBROGOESELLA PACIFICA
B 6305 816 CRIMES AND J. C. HARPER, WARME, JOHN E., REVIEW OF, TRACE FOSSILS
B 6104 582 CRINER HILLS, CROSS SECTION, OKLAHOMA, SOUTHERN
B 6103 361 CRINER HILLS, OKLAHOMA
M 27 164 CRINOID BIOSPARITE
M 27 93 CRINOID COLUMNALS
M 28 39 CRINOID, FRAGMENTS
M 27 76 CRINOID PLATES
M 27 93 CRINOIDS
S 4 231† CRINOIDS OF WEST INDIAN CORAL REEFS
S 4 231 CRINOIDS, WEST, INDIAN
B 6211 2352 CRISCIONE, JOSEPH J., AGE OF SEDIMENTATION-DIAGENESIS FOR
M 24 235 CRISTELLARIA K ZONE, LOUISIANA
M 28 82 CRISTOBALITE, SPHERULES
B 6405 738* CRISTOBALITE-CEMENTED SANDS IN CATAHOULA FORMATION—
M 26 203 CRITERIA, CHART, SEISMIC, REFLECTIONS, CARBONATE, RECOGNITION
B 6410 1715 CRITERIA FOR RECOGNITION OF CARBONATE SEDIMENT DRIFTS
B 6303 416* CRITERIA FOR RECOGNITION OF DIVERSE DOLOSTONE TYPES FROM
B 6407 1119 CRITERIA IN CORRELATION—RELEVANT PRINCIPLES OF SCIENCE, BY
B 6207 1210* CRITICAL ENERGY PROBLEMS OF PACIFIC BASIN, ABST.
M 26 418 CRITICAL RESOLUTION THICKNESS, SEISMIC RESOLUTION
S 6 235† CRITICAL REVIEW OF ISOTOPIC DATES IN RELATION TO PALEOGENE
B 6109 1537* CRITICAL WATER ESTIMATES FOR GULF COAST SANDSTONES, ABST.
M 29 233 CRITICAL WELL RESULTS, SUMATRA, OFFSHORE
B 6211 2351* CROCKER CANYON AREA, SOUTHWEST SAN JOAQUIN VALLEY,
M 30 105 CROMER KNOLL GROUP, CRETACEOUS, STATFJORD FIELD
B 6207 1213 CROMMELIN, MICHAEL, LEGAL ASPECTS OF RESOURCE DEVELOPMENT IN
B 6207 1213 CRONAN, D. S., GEOCHEMICAL EXPLORATION FOR SUBMARINE
B 6203 508 CRONIN, THOMAS C., GEOLOGY AND DEVELOPMENT OF LEEDY FIELD,
B 6105 790 **CRONIN, WILLIAM B., DREDGING AND SPOIL DISPOSAL IN CHESAPEAK**
B 6002 278† CROS SECTION OF HAMILTON GROUP, DEVONIAN, AND ADJACENT
M 26 200 CROSBY COUNTY, PERMIAN BASIN, EASTERN-SHELF BUILDUPS, SEISMIC
B 6205 787 CROSBY COUNTY, SEISMIC DATA, TEXAS
B 6105 766 **CROSBY, G. W., EXPLORATION AND EVALUATION OF ROOSEVELT KGRA,**
B 6008 1390 CROSBY, GARY W., GEOTHERMAL EXPLORATION OF ROOSEVELT KGRA,
B 6310 1983 CROSS, AND A. A. MEYERHOFF, MEYERHOFF, HOWARD A., REVIEW OF,
B 6004 660 CROSS, AUREAL T., PALYNOMORPHS IN PETROLEUM AND OTHER
B 6408 1283 CROSS, AUREAL T., PROSPECTS FOR COAL DEVELOPMEN IN MICHIGAN
M 29 79 CROSS SECTION, ATLANTIC COAST, BALTIMORE CANYON
M 29 77 CROSS SECTION, ATLANTIC COAST, CAPE COD
M 29 80 CROSS SECTION, ATLANTIC COAST, CAPE MAY
M 29 78 CROSS SECTION, ATLANTIC COAST, CAROLINA PLATFORM
M 29 295 CROSS SECTION, MEDITERRANEAN, ALGERIAN BASIN
M 29 295 CROSS SECTION, MEDITERRANEAN, NORTH BALEARIC BASIN
M 29 295 CROSS SECTION, MEDITERRANEAN, SOUTH BALEARIC BASIN
M 29 305 CROSS SECTION, USSR, BLACK SEA BASIN
M 29 279 CROSS SOUND, CHICHAGOF-BARANOF FAULT, ALASKA
M 29 278 CROSS SOUND, CHICHAGOF-BARANOF FAULT, ALASKA
M 29 282 CROSS SOUND, PLATE BOUNDARY, ALASKA
M 29 276 CROSS SOUND, SEISMIC SECTIONS, ALASKA
B 6405 760* CROSS-BEDDED SANDSTONE RESERVOIRS—GEOLOGIC MODELING OF
B 6203 547* CROSS-BEDS IN ATHABASCA OIL SANDS, ABST., DEEP SINUOUS, CHANNELS
S 10 121† CROSS-FORMATIONAL GRAVITY-FLOW OF GROUNDWATER
B 6412 2167 CROSS-STRIKE STRUCTURAL DISCONTINUITIES, APPALACHIANS
B 6309 1591* CROSS-STRIKE STRUCTURAL DISCONTINUITIES--EXPLORATION
B 6412 2166† CROSS-STRIKE STRUCTURAL DISCONTINUITIES—POSSIBLE EXPLORATION
S 8 42 CROSS-VALLEY CURRENTS, SEAVALLEYS
M 25 483 CROSSROADS CONCEPT, TECTONIC
B 6112 2063 CROSTELLA, A., GEOSYNCLINES AND PLATE TECTONICS IN BANDA ARCS,
B 6302 249 CROSTELLA, ANGELO, GEOSYNCLINES AND PLATE TECTONICS IN BANDA
B 6403 440 CROUCH, JAMES K., SPECULATIONS ON HYDROCARBON POTENTIAL OF
B 6008 1395 CROUCH, MARSHALL C., III, EXPLOITATION OF GAS RESERVES AT
B 6002 318 CROUCH, MARSHALL C., III, OIL AND GAS POTENTIAL OF WESTER
B 6012 2177 CROWE, BRUCE M., MIOCENE VOLCANIC AND VOLCANICLASTIC ROCKS
B 6004 660 CROWE, C. T., FOSSIL SALT STRUCTURES IN VACHERIE SALT DOM AREA,
B 6009 1612 CROWE, CLIFFORD T., SALT MOVEMENT, HALOKINETICS, IN NORTH
B 6108 1204 CROWLEY, CHARLES J., DEVELOPMENTS IN FOUR
B 6208 1360 CROWLEY, CHARLES J., DEVELOPMENTS IN FOUR CORNERS—
B 6004 660 CROWLEY, DONALD J., INTERTIDAL AND SUBTIDAL STROMATOLITES IN
B 6311 2059 CROWSNEST COALFIELD, BRITISH COLUMBIA
B 6203 555* CRUDE OIL AND NATURAL GAS DISSOLVED IN DEEP, HOT GEOTHERMAL
B 6202 179 CRUDE OIL CONSUMPTION VERSUS TIME, WORLD, POPULATION AND
B 6203 455† CRUDE OIL FROM BARROW ISLAND, WESTERN AUSTRALIA, ISOTOPIC
S 10 94 CRUDE OIL, DISSOLVED HYDROCARBONS
S 10 93 CRUDE OIL, AROMATIC HYDROCARBONS, WEIGHT
S 10 94 CRUDE OIL, COMPOSITION
S 10 94 CRUDE OIL, GASOLINE-RANGE HYDROCARBONS
B 6002 215 CRUDE OIL, GHAWAR, ARABIAN, WHOLE
B 6305 763 CRUDE OIL, HYDROCARBONS DISSOLVED FROM
S 10 93 CRUDE OIL, NAPHTHA FRACTION, ANALYSIS
S 10 91 CRUDE OIL, SOLUBILITY
B 6303 436* CRUDE OIL-LIKE HYDROCARBONS IN RECENT SEDIMENTS OF SANTA
B 6303 547* CRUDE OILS, ABST., EFFECTS OF SOURCE MATERIAL AND, THERMAL,
B 6303 467* CRUDE OILS IN MOLASSE BASIN, SOUTHERN GERMANY, ABST., ORIGIN
B 6304 701* CRUDE OILS--IDENTIFICATION BY CARBON ISOTOPES, ABST., BACTERIAL,
B 6002 198 CRUISES, MAY 1974, BRAZIL MARGIN, LAMONT-DOHERTY GEOPHYSICAL
B 6002 196 CRUISES THROUGH MAY 1974, RV, ROBERT D., CONRAD
B 6002 196 CRUISES THROUGH MAY 1974, RV, VEMA
B 6202 303 CRUSE-FOREST OIL SANDSTONES, TRINIDAD

B 6303 452* CRUST, ABST., OCEANIC
M 25 541 CRUST, AGE OF, NAZCA PLATE, OCEANIC
M 29 113† CRUST BOUNDARY OFF BAHIA BLANCA, ARGENTINA, STRUCTURE OF,
M 25 511 CRUST FILLS RIFT ZONE, CORDILLERAN OROGEN, NORTHERN,
S 3 22 CRUST, MECHANISMS OF, URANIUM AND, THORIUM TRANSFER TO THE
S 3 36 CRUST, MIGRATION OF, MANTLE-DERIVED, URANIUM AND, THORIUM
M 29 77 CRUST, TRANSITION ZONE, ATLANTIC COAST
S 3 80 CRUSTAL CONVECTION
M 29 6 CRUSTAL FLOW, DIAGRAM, SUBSIDENCE
S 7 22 CRUSTAL FRAMEWORK, NORTHERN, GULF OF MEXICO
M 29 280 CRUSTAL MODELS, ALASKA, QUEEN CHARLOTTE-FAIRWEATHER
S 3 41 CRUSTAL ROCKS, MECHANISMS OF, RADIOELEMENT, MOBILIZATION IN
M 29 214 CRUSTAL SECTION, AUSTRALIA, TIMOR-TANIMBAR-ARU TROUGH
M 29 454 CRUSTAL SHORTENING AND THERMAL WELTS, MECHANISMS
M 29 151† CRUSTAL STRUCTURE AND EVOLUTION OF THE AREA UNDERLYING THE
M 29 7 CRUSTAL THINNING, NECKING
M 27 190 CRUSTS, ALGAL
B 6102 207 CRUZ HELU, PABLO, ORIGIN AND DISTRIBUTION OF TERTIARY
M 25 551 CRUZ, J. A., POSSIBILITIES FOR DEVELOPMENT OF MINERAL RESOURCES
M 25 424 CRYPTOMELANE, GROOTE EYLANDT
M 27 70 CRYPTOSOME BRYOZOANS
B 6103 348 CRYSTAL FALLS LIMESTONE, TEXAS
B 6010 1871 CRYSTAL OIL CO., RESERVE OIL AND GAS CO., AND PETROCONSULTANTS
B 6405 693* CRYSTALLINE OVERTHRUSTING OF PALEOZOIC SHELF IN SOUTHERN
S 3 135 CRYSTALLIZATION DISSEMINATION OF THORIUM-URANIUM MINERALS
B 6008 1188 CSD, INDIVIDUAL WELL TICKET, API--, AAPG
B 6308 1203 CSD, INDIVIDUAL WELL TICKET, API--AAPG
B 6108 1156 CSD, INDIVIDUAL WELL TICKET, API-, AAPG
B 6208 1310 CSD, INDIVIDUAL WELL TICKET, API-, AAPG
B 6409 1320 CSD, INDIVIDUAL WELL TICKET, API—AAPG
S 4 235 CTENANTEDON KINZIEI MEYER, BARBADOS
B 6202 227 CUANZA, ANGOLA, MOSSAMEDES, EVAPORITES
B 6202 229 CUANZA, SEDIMENTARY SEQUENCE
B 6010 1649 CUBA, 1975, PETROCONSULTANTS S.A., ON DEVELOPMENTS IN
B 6110 1583 CUBA, 1976, MERMINOD, F., ON DEVELOPMENTS IN
B 6210 2049 CUBA, 1977, PETROCONSULTANTS S.A., ON DEVELOPMENTS IN
B 6310 1750 CUBA, 1978, PETROCONSULTANTS S.A., ON DEVELOPMENTS IN
B 6405 674 CUBA, ABST., EXPLORABLE STRUCTURES IN, OLD, BAHAMA CHANNEL,
B 6110 1593 CUBA, PRODUCTION, 1972-1976
B 6210 2062 CUBA, PRODUCTION, 1973-1977
B 6310 1764 CUBA, PRODUCTION, 1974-1978
B 6410 1723 CUBITT AND STEPHEN HENLEY, FONT, ROBERT G., REVIEW OF,
B 6111 1981 CUCAMONGA FAULT, CALIFORNIA
B 6203 361 CUCAMONGA-SIERRA MADRE FAULT, CALIFORNIA
B 6004 693* CUCHUMATANES MOUNTAINS, WESTERN GUATEMALA, ABST.,
B 6409 1560* CUESTA DEL CURA AND LOWER AGUA NUEVA FORMATIONS,
B 6402 189 CUEVAS VIEJAS FORMATION, SPAIN
B 6402 189 CUEVAS VIEJOS FORMATION, SPAIN
S 4 181 CUFFEY, ROGER J., BRYOZOAN CONTRIBUTIONS TO REEFS AND BIOHERM
M 25 519 CUITACA, BATHOLITHIC GRANODIORITE, MEXICO
B 6011 2005† CULBERSON COUNTY, TEXAS, ORIGIN OF, LIMESTONE, BUTTES, GYPSUM
B 6008 1395 CULBERTSON, WILLIAM C., THICK COAL BEDS IN WASATCH FORMATION,
B 6104 561 CULLEN FIELD, COLUMN, CHILE
B 6101 80 CULPEPER BASIN, VIRGINIA
B 6203 537* CULPEPER BASIN, VIRGINIA, ABST., TECTONIC CONTROL OF
B 6305 846* CULPEPER GROUP CONGLOMERATES OF VIRGINIA, ABST., PETROLOGY
B 6107 986 CULVER CLIFFS, CHALK, ENGLAND
B 6311 2059 CUMBERLAND COALFIELD, BRITISH COLUMBIA
B 6101 114 CUMBERLAND GAP, WYOMING, CORRELATION, UTAH, COALVILLE, TO
B 6412 2078 CUMBERLAND MOUNTAIN, APPALACHIANS
B 6103 417 CUMBERLAND OVERTHRUST, VIRGINIA
B 6204 644† CUMBERLAND PLATEAU, MONTEAGLE LIMESTONE, UPPER,
B 6204 644 CUMBERLAND PLATEAU, SOUTHERN, CARBONIFEROUS
B 6204 645 CUMBERLAND PLATEAU, SOUTHERN, MONTEAGLE LIMESTONE, SIX
B 6103 417 CUMBERLAND PLATEAU, VIRGINIA
B 6107 1014 CUMMINGS MEMBER, ALBERTA AND, SASKATCHEWAN
B 6409 1529 CUMULATIVE PRODUCTION, DEFINITION
B 6402 264 CUNNINGHAM, R. H. W., EVOLUTION OF SAND-DOMINANT SUBAERIAL
B 6304 690 CUNNINGHAM, S. B., DEPOSITIONAL ENVIRONMENT OF POISON CANYON
M 27 31 CUPIDO FORMATION, MEXICO, CRETACEOUS
M 27 87 CUPIDO FORMATION, MEXICO, CRETACEOUS
M 27 123 CUPIDO FORMATION, MEXICO, CRETACEOUS
M 27 88 CUPIDO FORMATION, MEXICO, CRETACEOUS
B 6109 1551* CUPIDO FORMATION, NORTHEAST MEXICO, ABST.,
B 6405 745 CUPPS, C. Q., RECENT ACTIVITIES IN UNITED STATES TAR SAND, ABST.
M 25 439 CUPRIFEROUS IRON SULFIDE DEPOSITS, BEDDED, TAIWAN
S 4 5 CURACAO ISLAND, CORALS
M 29 364 CURACAO RIDGE, INTERPRETATION, VENEZUELA BASIN
M 29 367 CURACAO RIDGE, REFLECTING HORIZONS, VENEZUELA BASIN
M 29 365 CURACAO RIDGE, SEISMIC SECTION, VENEZUELA BASIN
B 6311 2117 CURIALE, JOSEPH A., INVESTIGATION OF DESMOINESIAN ROCKS IN
B 6410 1644 CURNAMONA CHANNEL, AUSTRALIA
B 6410 1643† CURNAMONA, LAKE FROME REGION, SOUTH AUSTRALIA, DISTRIBUTION
B 6105 777 **CURRAN, H. A., SAND PETROLOGY AND BIOCLASTIC COMPONENTS OF**
B 6405 694 CURRAN, H. ALLEN, TRACE FOSSIL ASSEMBLAGES OF UPPER CRETACEOUS
B 6004 661 CURRAN, H. ALLEN., TRACE-FOSSIL ASSEMBLAGE OF EARLY CRETACEOUS
B 6105 778 **CURRAY, JOSEPH R., MODES OF EMPLACEMENT OF PROSPECTIVE**
B 6004 699 CURRAY, J. R., DYNAMIC PROCESSES OF UPPER BENGAL FAN AND
M 29 189 CURRAY, J. R., TECTONICS OF THE ANDAMAN SEA AND BURMA
B 6203 500* CURRENT DEPOSITION OF PERMIAN DELAWARE MOUNTAIN GROUP.
B 6004 701* CURRENT DYNAMICS IN WEST MISSISSIPPI DELTA AREA. ABST., CONTROL
B 6303 459* CURRENT ESTIMATES AND METHODS OF POTENTIAL GAS COMMITTEE.
S 8 36 CURRENT FLOW, ABRA SEAVALLEY, TIME-VELOCITY CURVES
S 8 15 CURRENT FLOW, ALTERNATING DIRECTIONS
S 8 30 CURRENT FLOW, CARMEL CANYON, DEPTH AGREEMENT
S 8 83 CURRENT FLOW, CARMEL CANYON, TIME-VELOCITY CURVES
S 8 142 CURRENT FLOW, CHRISTIANSTED CANYON, TIME-VELOCITY CURVES
S 8 166 CURRENT FLOW, CONGO CANYON, TIME-VELOCITY CURVES
S 8 22 CURRENT FLOW, CONSISTENCY OF SPEED
S 8 45 CURRENT FLOW, CURRENT-METER READINGS, TABLE
S 8 17 CURRENT FLOW, CYCLES, COMPOSITE DIAGRAM
S 8 15 CURRENT FLOW, DEPTH VS. TIDE
S 8 57 CURRENT FLOW, DESCRIPTION OF CURRENTS
S 8 23 CURRENT FLOW, EFFECTS OF TIDE IN BAY ENVIRONMENT
S 8 102 CURRENT FLOW, FRASER SEAVALLEY, TIME-VELOCITY CURVES
S 8 20 CURRENT FLOW, FRASER SEAVALLEY, TIME-VELOCITY CURVES
S 8 152 CURRENT FLOW, HANAKAPIAI CANYON, TIME-VELOCITY CURVES
S 8 124 CURRENT FLOW, HUDSON CANYON, TIME-VELOCITY CURVES
S 8 27 CURRENT FLOW, HYDROGRAPHER CANYON, TIME-VELOCITY CURVES
S 8 116 CURRENT FLOW, HYDROGRAPHER CANYON, TIME-VELOCITY CURVES
S 8 155 CURRENT FLOW, KAULAKAHI CHANNEL, TIME-VELOCITY CURVES
S 8 57 CURRENT FLOW, LA JOLLA CANYON, AVERAGE FLOW TIME
S 8 33 CURRENT FLOW, LA JOLLA CANYON, DOWNCANYON VELOCITIES
S 8 60 CURRENT FLOW, LA JOLLA CANYON, TIDAL RELATIONS
S 8 62 CURRENT FLOW, LA JOLLA CANYON, TIME-VELOCITY CURVES
S 8 10 CURRENT FLOW, LA JOLLA CANYON, TIME-VELOCITY CURVES
S 8 40 CURRENT FLOW, LAKE WALENSEE, SWITZERLAND, DOWNSLOPE FLOW
S 8 28 CURRENT FLOW, MAX DOWNCANYON FLOW, COMPARISON
S 8 26 CURRENT FLOW, MAX SPEED AT 3 M ABOVE BOTTOM
S 8 93 CURRENT FLOW, MONTEREY CANYON, TIME-VELOCITY CURVES
S 8 21 CURRENT FLOW, NET FLOW IN SEAVALLEYS
S 8 160 CURRENT FLOW, NOHILI SEAVALLEY, TIME-VELOCITY CURVES
S 8 18 CURRENT FLOW, PINGET SEAVALLEY, TIME-VELOCITY CURVES
S 8 66 CURRENT FLOW, REDONDO CANYON, TIME-VELOCITY CURVES
S 8 25 CURRENT FLOW, RELATION TO HEIGHT ABOVE BOTTOM
S 8 110 CURRENT FLOW, RIO BALSAS CANYON, TIME-VELOCITY CURVES
S 8 31 CURRENT FLOW, RIO BALSAS CANYON, TIME-VELOCITY CURVES
S 8 34 CURRENT FLOW, RIO BALSAS CANYON, TIME-VELOCITY RECORD
S 8 35 CURRENT FLOW, RIO BALSAS CANYON, VELOCITY AVERAGES
S 8 16 CURRENT FLOW, RIO BALSAS DELTA, TIME-VELOCITY CURVES
S 8 38 CURRENT FLOW, RIO DE LA PLATA CANYON, TIME-VELOCITY CURVES
S 8 19 CURRENT FLOW, SALT RIVER CANYON, TIME-VELOCITY CURVES
S 8 136 CURRENT FLOW, SALT RIVER CANYON, TIME-VELOCITY CURVES
S 8 96 CURRENT FLOW, SAN CLEMENTE RIFT, TIME-VELOCITY CURVES
S 8 106 CURRENT FLOW, SAN LUCAS CANYON, TIME-VELOCITY CURVES
S 8 24 CURRENT FLOW, SANTA CRUZ CANYON, TIDAL EFFECTS
S 8 79 CURRENT FLOW, SANTA CRUZ CANYON, TIME-VELOCITY CURVES
S 8 70 CURRENT FLOW, SANTA MONICA CANYON, TIME-VELOCITY CURVES
S 8 22 CURRENT FLOW, SEDIMENT TRANSPORT, SEAVALLEYS
S 8 32 CURRENT FLOW, TURBIDITY CURRENTS, LA JOLLA CANYON
S 8 29 CURRENT FLOW, TURBIDITY CURRENTS, REVIEW
S 8 18 CURRENT FLOW, UNIDIRECTIONAL FLOW
S 8 21 CURRENT FLOW, VAR CANYON, TURBIDITY CURRENT
S 8 128 CURRENT FLOW, WILMINGTON CANYON, TIME-VELOCITY CURVES
S 8 8 CURRENT METER, DEPLOYMENT METHOD, FIGURE
S 8 6 CURRENT METER, SCHICK-SAVONIUS, MODEL
S 8 9 CURRENT METER, TIME-RELEASE MECHANISM
B 6007 1025 CURRENT METERS, LAKE SUPERIOR
B 6203 511* CURRENT WINNOWING ACTIVITY ON HUENEME SILL, CALIFORNIA
B 6105 817* CURRENT-DOMINATED SHELF, NORTHWEST ALASKA, ABST., COMPLEX,
S 8 45 CURRENT-METER READINGS, TABLE, CURRENT FLOW
B 6105 816* CURRENTS, ABST., AERIAL, PHOTOGRAPH INTERPRETATION OF
B 6004 695* CURRENTS, ABST., KINDERSCOUTIAN, DELTA, CARBONIFEROUS, OF
B 6004 654* CURRENTS AND FAUNAL ACTIVITY IN LOWER HUDSON SUBMARINE
B 6004 721* CURRENTS IN SUBMARINE CANYON AND THEIR RELATION TO DEEP-SEA
B 6405 649 CURRENTS IN SUBMARINE CANYONS AND OTHER SEAVALLEYS, BY
B 6203 561* CURRENTS IN SUBMARINE CANYONS AND OTHER TYPES OF SLOPE
S 4 135† CURRENTS, STRUCTURAL MODIFICATION OF, CORALS IN RELATION TO,
B 6407 1007 CURRIE FIELD, TEXAS
B 6107 1086 CURRIE, J. B., FRACTURE POROSITY, SIGNIFICANT GEOLOGIC PROCESSES
B 6405 755 CURRIE, JOHN B., ORIGIN OF SUBSURFACE FRACTURE SYSTEMS—
B 6105 778 **CURRIE, PHILLIP W., DINOSAUR TRACKS FROM CRETACEOUS SEDIMENTS**

B 6207 1175 CURRIE RAVINE, NEVADA
B 6105 671 CURRY, WILLIAM H., JR., TEAPOT DOME--PAST, PRESENT, AND FUTURE
B 6008 1395 CURRY, WILLIAM H., III, TEAPOT DELTA OF WESTERN POWDER RIVE
B 6305 825 CURRY, WILLIAM H., III, TERTIARY CORRELATIONS IN BIGHORN AN
B 6203 403 CURTH, PATRICK J., STRATIGRAPHY AND DEPOSITIONAL SETTING,
B 6305 845 CURTIS, C. E., INTEGRATION OF AIRBORNE, MAGNETI AND GAMMA-RAY
B 6004 661 CURTIS, D. M., GULF COAST CENOZOIC--MODEL FOR APPLICATION OF
B 6209 1755 CURTIS, D. M., GULF COAST CENOZOIC--MODEL FOR APPLICATION OF
B 6303 509 CURTIS, DORIS M., APPLIED BIOSTRATIGRAPHY IN GULF COAST
B 6303 443 CURTIS, DORIS M., CHRONICLE OF MIOCENE, PHASE III, MIDDLE MIOCENE
B 6303 437 CURTIS, DORIS M., SOURCE OF OILS IN GULF COAST TERTIARY. WHY LOO
M 28 114 CURTIS FORMATION, UTAH, JURASSIC
M 28 27 CURTIS FORMATION, UTAH, JURASSIC
M 28 39 CURTIS FORMATION, UTAH, JURASSIC
M 28 115 CURTIS FORMATION, UTAH, JURASSIC
M 28 7 CURTIS FORMATION, UTAH, JURASSIC
B 6205 882 CURTIS, GRAHAM R., HISTORY OF EARLY EXPLORATION IN GREAT
B 6112 2104 CUSHABATAY SECTION, PERU
B 6004 612 CUSSEY, R., ANTIPATHETIC RELATIONS AMONG ALGAL STRUCTURES,
B 6304 676 CUSSEY, R., PATTERNS OF POROSITY AND CEMENT IN OOID RESERVOIRS
B 6104 511 CUSSEY, R., PATTERNS OF POROSITY AND CEMENT IN OOID RESERVOIRS IN
B 6405 708 CUSTARD, H. C., SMECTITE-ILLITE TRANSFORMATION--ROLE IN
B 6407 971 CUT BANK SANDSTONE MEMBER, GREAT PLAINS
B 6408 1172 CUTLER ARKOSE, UTAH AND, COLORADO
S 3 58 CUTLER FORMTION, UTAH
B 6002 328* CUTTINGS, ABST., STUDY OF, CARBONATE ENVIRONMENTS USING, WELL
B 6305 742 CUTTINGS, CAPILLARY PROPERTIES OF, DRILL
B 6311 2089 CUVIER ABYSSAL PLAIN, AUSTRALIA
B 6201 41 CUVIER ABYSSAL PLAIN, AUSTRALIA
B 6203 353 CUYAMA OIL DISTRICT, CALIFORNIA
B 6405 686* CYANOPHYTES, ABST., TRAPPING AND ACCRETION OF, AEOLIAN
S 11 228 CYCLAMMINA
S 11 268 CYCLAMMINA
S 11 268 CYCLAMMINA CANCELLATA
S 11 228 CYCLAMMINA CANCELLATA
S 5 93* CYCLC EVAPORITES OF MICHIGAN BASIN, SYNCHRONIZATION OF
M 26 151 CYCLE CHART, TEXAS, EAST TEXAS BASIN, CORRELATION OF UNITS TO A
M 26 162 CYCLE CHARTS, TEXAS, GULF COAST, WEST, AFRICA, COMPARISON
M 26 161 CYCLE, COMPARISON, WEST AFRICA TO GULF COAST
B 6004 703* CYCLE FROM CAMBRIAN TO PRESENT, ABST., CAUSALITY AND, TIMING
S 5 73* CYCLES AND LITHOFACIES IN LUCAS FORMATION, DETROIT RIVER
B 6004 695* CYCLES, ABST., PALEOGENE, RADIOLARIAN,
B 6004 699* CYCLES, ABST., REGIONAL, SEISMIC INTERPRETATION USING SEQUENCES
M 26 158 CYCLES, AFRICA, WEST, CHRONOSTRATIGRAPHIC, CORRELATION TO
B 6002 321* CYCLES AND DEPOSITIONAL ENVIRONMENTS IN VIRGILIAN,
B 6109 1541* CYCLES IN CARBONATE ROCKS, ABST., PLATE TECTONICS AND
B 6205 881* CYCLES IN GREEN RIVER FORMATION, WYOMING, ABST., CARBONATE
B 6004 711* CYCLES IN VIRGILIAN, UPPER PENNSYLVANIAN, ROCKS IN ANADARKO
B 6004 645* CYCLES OF DEPOSITION IN SOUTHERN SACRAMENTO VALLEY,
M 29 469† CYCLES OF RELATIVE CHANGES OF SEA LEVEL FROM SEISMIC
S 3 83 CYCLES OF URANIUM AND THORIUM, SECTION
S 3 77 CYCLES OF URANIUM AND THORIUM, SECTION
B 6004 697* CYCLES ON UNSTABLE CRATO MARGIN, BANKS ISLAND, ARCTIC
M 26 64 CYCLES, SEA LEVEL FLUCTUATIONS
M 26 153 CYCLES, TEXAS, GULF COAST, CHRONOSTRATIGRAPHIC, CORRELATION
B 6004 696* CYCLIC INTERACTION OF CONTINENTAL SODA-RICH WATERS WITH
B 6004 696* CYCLIC INTERACTION OF CONTINENTAL SODA-RICH WATERS WITH
B 6107 1045 CYCLIC ROCK SEQUENCE IN UPPER PENNSYLVANIAN ROCKS OF EASTERN
B 6311 2115* CYCLIC SEDIMENTATION, ABST., MOBILE, BASIN-SHELF BORDER IN
B 6004 645* CYCLIC SEDIMENTATION IN GLACIAL MARINE ENVIRONMENTS, ABST.
B 6004 662* CYCLIC SEDIMENTATION IN WESTERN EQUATORIAL ATLANTIC DURING
B 6303 512* CYCLIC SEDIMENTATION OF CRETACEOUS-PALEOCENE RESERVOIR
B 6105 824* CYCLIC SEDIMENTATION, WESTERN INTERIOR CRETACEOUS, NORTH
B 6203 570* CYCLIC SEDIMENTS IN NORTHERN KANSAS AND SOUTHERN NEBRASKA,
S 4 339 CYCLICARGOLITHUS FLORIDANUS, ST. CROIX
B 6203 516* CYCLICITY IN CONEMAUGH FORMATION IN APPALACHIANS OF WEST
B 6004 678* CYCLING AND DEPOSITION ON SHELVES AND IN DEEP SEA, ABST.,
B 6405 721* CYCLOTHEMS, ABST., PALEOGEOGRAPHY OF EUSTATIC MODEL FOR
B 6211 2194 CYCLOTHEMS, IRELAND, WESTERN, DELTAIC
B 6107 1045† CYCLOTHEMS OF MID-CONTINENT NORTH AMERICA, ORIGIN OF,
B 6107 1060 CYCLOTHEMS, POSSIBLE RELATIONS OF, KANSAS AND, ILLINOIS
B 6203 351 CYMRIC OIL FIELD, CALIFORNIA
B 6408 1282* CYPRESS, CHESTERIAN, STRATIGRAPHIC ENTRAPMENT OF PETROLEUM,
M 29 311 CYPRUS ARC, EVOLUTION, MEDITERRANEAN
M 29 310 CYPRUS ARC, MEDITERRANEAN, ACTIVE MARGINS
B 6106 875 CYPRUS BASIN
M 28 85 CYPRUS, MIOCENE, UNIT
B 6406 861 CYRENAICA PLATFORM, AFRICA
M 28 39 CYRENENSCHICHTEN, GERMANY, OLIGOCENE
M 29 304 CYRENIAN PLATEAU, MAGNETIC ANOMALIES, MEDITERRANEAN
B 6102 294 CYS, J. M., CAPITAN REEF--EVOLUTION OF A CONCEPT, ABST.
B 6004 661 CYS, J. M., CAPITAN REEF--EVOLUTION OF CONCEPT, ABST.
B 6303 493 CYS, JOHN M., IN-SITU FORMATION OF BOTRYOIDAL ARAGONITE ON
B 6103 431 CYS, JOHN M., NEAL RANCH-LENOX HILLS CONTACT IN GLASS
B 6203 541 CYS, JOHN M., SUBMARINE CEMENTS AND FABRIC-RETENTIVE
B 6007 1107 CYS, JOHN M., WILKIE RANCH FORMATION, NEW PERMIAN UNIT IN
B 6405 695 CZARNECKI, ROBERT F., ADVANCES IN RADON EXPLORATION
B 6301 67† CZECHOSLOVAKIA, ANCIENT, SUBMARINE CANYONS OF TETHYAN,
B 6010 1724 CZECHOSLOVAKIA, DEVELOPMENTS
B 6110 1653 CZECHOSLOVAKIA, DEVELOPMENTS, 1976
S 3 90 CZECHOSLOVAKIA, JOACHIMSTHAL VEIN DISTRICT
B 6301 70 CZECHOSLOVAKIA, NESVACILKA CANYON
B 6301 68 CZECHOSLOVAKIA, PALEOGENE CANYONS
B 6301 70 CZECHOSLOVAKIA, POUZDRANY UNIT
B 6110 1654 CZECHOSLOVAKIA, PRODUCTION, 1975-1976
B 6210 1924 CZECHOSLOVAKIA, PRODUCTION, 1976-1977
B 6310 1932 CZECHOSLOVAKIA, PRODUCTION, 1977-1978
B 6301 69 CZECHOSLOVAKIA, VARISCAN PLATFORM
B 6301 70 CZECHOSLOVAKIA, VRANOVICE CANYON
B 6301 75 CZECHOSLOVAKIA, ZDANICE NAPPE
B 6203 344 D G., NEOGENE BASINFORMATIN IN RELATION TO PLATE-TECTONIC
B 6104 585 D RESERVOIR, FIELD, PERSIAN GULF, GHAWAR-AIN DAR
B 6104 585 D RESERVOIR, FIELD, PERSIAN GULF, GHAWAR-HARADH
B 6104 585 D RESERVOIR, FIELD, PERSIAN GULF, GHAWAR-UTHMANIYAH
B 6410 1615 D SAND, DENVER BASIN
M 24 139 D SANDSTONE, COLORADO
B 6405 776* D SANDSTONE DISCOVERY, ADAMS COUNTY, COLORADO, SECS. 17-20, T3S,
B 6303 889 D SANDSTONE, NEW MEXICO, DAKOTA
B 6204 672 D UNIT, ALGERIA, DEVONIAN
B 6402 218 D-1A FIELD, INDIA
B 6102 146 DA SAND, CALIFORNIA
M 27 3 DACHSTEIN LIMESTONE, AUSTRIA, TRIASSIC
M 27 164 DACHSTEIN LIMESTONE, AUSTRIA, TRIASSIC
M 25 440 DAEBO GRANITE, SOUTH KOREA
M 25 444 DAEBO GRANITES, SOUTH KOREA
M 25 442 DAEBO OROGENY, KOREA
M 25 443 DAEBO OROGENY, SOUTH KOREA, JURASSIC DEFORMATION
M 25 443 DAEDONG SEDIMENTARY ROCKS, SOUTH KOREA
B 6402 210 DAHANU DEPRESSION, INDIA
B 6402 218 DAHANU FIELD, INDIA
B 6101 85 DAHAR NSOUR, BASIN, MOROCCAN MESITA, PRE-RIF
B 6203 508 DAHLBERG, E. C., COMPUTER ANALYSIS OF REGIONAL WELL-DENSITY
B 6110 1747 DAHOMEY, DEVELOPMENTS, 1976, BENIN
B 6210 1846 DAHOMEY, DEVELOPMENTS, 1977, BENIN
B 6009 1543 DAILLY, GERARD C., PENDULUM EFFECT AND NIGER DELTA PROLIFI
B 6305 825 DAILY, MICHEAL, LINEAMENT MAPPING WITH ORBITING IMAGING
M 29 239† DAITO RIDGE AND ASSOCIATED BASINS, NORTH PHILIPPINE SEA,
M 29 240 DAITO RIDGE AREA, PHYSIOGRAPHY, NORTH PHILIPPINE SEA
M 29 240 DAITO RIDGE, GEOLOGY, NORTH PHILIPPINE SEA
M 29 240 DAITO RIDGE, REFLECTION STUDIES, NORTH PHILIPPINE SEA
M 29 241 DAITO RIDGE, REFLECTION SURVEY, FOLDOUT
B 6306 969 DAKHLA BASIN, EGYPT
B 6306 970 DAKHLA SHALE, EGYPT
B 6008 1405* DAKOTA AND SUBJACENT SANDSTONES, SAN JUAN BASIN, NEW MEXICO,
B 6303 889 DAKOTA D SANDSTONE, NEW MEXICO
M 24 139 DAKOTA FORMATION, COLORADO
B 6306 887 DAKOTA FORMATION SECTION, NEW MEXICO
B 6205 886* DAKOTA FORMATIONS, EARLY CRETACEOUS, NORTHWESTERN
M 27 107 DAKOTA GROUP, COLORADO, CRETACEOUS
M 26 279 DAKOTA GROUP, CRETACEOUS
B 6410 1624 DAKOTA GROUP, DENVER BASIN
B 6405 776* DAKOTA, MUDDY, D SANDSTONE DISCOVERY, ADAMS COUNTY,
B 6105 828* DAKOTA SANDSTONE, COLORADO, ABST., ORIGIN OF, RHIZOCORALLIUM
M 27 108 DAKOTA SANDSTONE, COLORADO, CRETACEOUS
B 6306 886† DAKOTA SANDSTONE, LONE PINE FIELD, NEW MEXICO, OIL AND, GAS IN,
B 6209 1675 DAKOTA SANDSTONE, NEW MEXICO
B 6007 1013 DAKOTA SANDSTONE, NEW MEXICO
B 6206 1093* DAKOTA SANDSTONE OF SAN JUAN BASIN REGION, NEW MEXICO,
M 24 95 DAKOTA SANDSTONE, WYOMING
M 25 366 DALE, R. H., SIMULATION MODEL OF HONOLULU BASAL
M 25 411 DALES GORGE MEMBER, WESTERN AUSTRALIA
B 6308 1427* DALHART BASINS, TEXAS, ABST., SAN ANDRES FACIES PATTERNS, PALO
B 6312 2194 DALKJEGLA MEMBER, SVALBARD
B 6204 686 DALLAS GEOLOGICAL SOCIETY, GEOLOGICAL INFORMATION LIBRARY OF
B 6105 728 † DALLAS GEOLOGICAL SOCIETY, HISTORY OF
B 6002 164 DALNOI BASIN, BERINGIAN SHELF
B 6002 164 DALNOI RIDGE, BERINGIAN SHELF
B 6002 273 DALRYMPLE G. BRENT, NOTE 44--APPLICATION FOR ADDITION TO CODE
B 6004 697 DALRYMPLE, R. W., FACIES MODE FOR MACROTIDAL ENVIRONMENT,
B 6405 696 DALRYMPLE, ROBERT W., INTERNAL STRUCTURES OF SHALLOW-MARINE

B 6303 438 DALY, THOMAS E., HIGH-RESOLUTION SEISMIC METHODS IN COAL
B 6308 1225 DAMASCUS FIELD, POLK COUNTY, TEXAS
B 6403 440 DAMASSA, SARAH PIERCE, PALYNOMORPHS FROM HOLOCENE SEDIMENTS
B 6105 778 DAME, R., MICROCONTINENTAL MARGINS
M 25 556 DAMIANI, O., MINERAL RESOURCES POTENTIAL OF PERU--SUMMARY
B 6106 880 DAMIETTA FAN, EGYPT
B 6104 585 DAMMAN FIELD, PERSIAN GULF
B 6209 1757* DAMON MOUND, TEXAS, ABST., OLIGOCENE, REEF COMMUNITY
B 6201 50 DAMPIER FORMATION, AUSTRALIA
B 6201 41 DAMPIER SUBBASIN, AUSTRALIA
M 25 232 DAMPIER SUBBASIN, AUSTRALIA
B 6204 657 DAMUTH, J. E., SALT DIAPIRS ON SAO PAULO PLATEAU, SOUTHEASTERN
B 6004 662 DAMUTH, JOHN E., CLIMATICALLY CONTROLLED, CYCLIC
B 6303 438 DAMUTH, JOHN E., MASS WASTING ON CONTINENTAL RISE OF EASTERN
B 6405 696 DANA, G. F., PETE, SHALLOW OIL SHALE DEPOSITS OF SOUTHERN UINTA
B 6403 400 DANA, STEPHEN W., ANALYSIS OF GRAVITY ANOMALY OVER CORAL-REE
B 6208 1382 DANBURY DOME, BRAZORIA COUNTY, TEXAS
B 6307 1092 DANDARAGAN TROUGH, AUSTRALIA
B 6405 695 DANDREA, RALPH D., JR., APPLICATION OF REGIONAL GEOCHEMICAL
B 6008 1389* DANGER OF FEDERAL WITHDRAWALS TO NATURAL RESOURCES
B 6407 1021 DANGEROUS CAPE HIGH, ALASKA
B 6408 1142 DANGEROUS CAPE HIGH, ALASKA
B 6209 1767* DANIAN, ABST., OSTRACODA OF, PRAIRIE BLUFF, CHALK, UPPER,
B 6211 2234 DANIAN, BAY OF BISCAY AREA
M 30 67 DANIAN, FRIGG FIELD, PALEOGEOGRAPHY
S 6 217 DANIAN STAGE DEFINED, EUROPE, PALEOCENE
B 6405 696 DANIE, T. C., DINEH-BI-KEYAH FIELD, APACHE COUNTY, ARIZONA ABST.
B 6102 294 DANIELS, JOHN, EVOLUTION OF PRESENT-DAY SEISMIC EXPLORATION IN
B 6302 182† DANISH BASIN, NORTH SEA, PETROLOGY AND DIAGENESIS OF, JURASSIC
B 6302 183 DANISH BASIN, TECTONIC MAP, NORWEGIAN
B 6203 426 DANISH TROUGH, EUROPE
B 6401 9 DAQING, GIANT OIL FIELD, LANDSAT DATA, CHINA, TACHING
B 6307 1040 DARBY-PROSPECT THRUST, WYOMING
B 6312 2119 DARIUS FIELD, YAMAMA LIMESTONE, PERSIAN GULF
B 6307 1093 DARLING FAULT, AUSTRALIA
M 25 405 DARLING RANGE, BAUXITE RESERVES, AUSTRALIA
B 6207 1235* DARLING RANGE, WESTERN AUSTRALIA, ABST., BAUXITE, MINING AND,
B 6301 38 DARST CREEK FIELD, TEXAS
M 27 89 DARTRY LIMESTONE, IRELAND, MISSISSIPPIAN
B 6311 2024 DARWIN BLOCK, AUSTRALIA
B 6105 778 DASCH, E. J., MARINE METALLIFEROUS SEDIMENTS FROM NAZCA PLATE
M 25 539 DASCH, E. JULIUS, METALLOGENESIS IN SOUTHEAST PACIFIC OCEAN,
B 6303 484 DASHEVSKY, SAMUEL S., DETAILED TEMPERATURE LOGGING AS USEFUL
B 6206 991 DASHT-E ARJAN GRABEN, IRAN
M 27 9 DASYCLADACEAN, GREEN, ALGAL FRAGMENT
M 26 352† DATA, ANALYSIS OF, HIGH, RESOLUTION, SEISMIC
B 6004 716* DATA BANK PROJECT, UNITED STATES AND CANADA, ABST., AAPG, OIL
S 6 75 DATA BASE, COMPUTER FILE FORMAT
B 6004 653* DATA BASES, ABST., SET-THEORETIC, INFORMATION, RETRIEVAL FROM,
M 26 8 DATA, CONVOLUTION, SEISMIC
S 6 79 DATA FILE, DECAY CONSTANTS
S 6 76 DATA FILE, GEOLOGIC TIME SCALE, BASED ON
M 26 3 DATA, MODELING, SEISMIC
B 6303 385 DATA ON THE GEOCHEMISTRY OF OIL FIELD WATERS, BY KARL
B 6405 684* DATA PROCESSING, OLD AND NEW TECHNIQUES, ABST., COAL RESOURCE
M 26 10 DATA, SEISMIC LOG PROCESS, SEISMIC
M 26 9 DATA, SEISMIC TRACES, SEISMIC
M 26 11 DATA, SPATIAL RESOLUTION, SEISMIC
M 26 9 DATA, SYNTHETIC SEISMOGRAMS, , SEISMIC
M 26 3 DATA, SYNTHETIC SEISMOGRAMS, SEISMIC
B 6004 656* DATA SYSTEM, ABST., AVAILABILITY AND APPLICATION OF NATIONAL,
M 26 15† DATA THAT ARE IMPORTANT TO SEISMIC INTERPRETATION, ASPECTS OF,
S 5 111* DATA-PROCESSING TECHNIQUES IN EXPLORATION FOR REEFS,
B 6212 2429 DATACOLOR SLOPE MAPS
B 6304 625 DATE CREEK BASIN, ARIZONA
B 6304 626 DATE CREEK BASIN, MAP, ARIZONA
S 6 1† DATING AND CORRELATION, A REVIEW
B 6004 648* DATING AND STRATIGRAPHIC ANALYSIS OF HOLOCENE TRANSGRESSION,
B 6109 1536* DATING, GEORGIA COAST, ABST., HISTORY OF, SHORELINE CHANGES BY,
B 6004 723* DATING OF ESTUARINE CARBONATE MATERIALS, ABST., RADIOCARBON
B 6004 663* DATING OF HOLOCENE SHORELINE CHANGES ON GEORGIA COAST, ABST.,
B 6405 696 DAUB, GERALD J., GRAPHIC REPRESENTATION OF SUBSURFACE DATA BY
B 6209 1756 DAUB, GERALD J., STRATIGRAPHIC INTERPRETATION OF A POSSIBLE
M 25 143 DAUIN GEOTHERMAL PROSPECT, PHILIPPINES, PALIMPINON-
B 6409 1558* DAUPHIN ISLAND, ALABAMA, ABST., EFFECTS OF, HURRICANE, FREDERIC
M 25 437 DAVAO DISTRICT, COPPER, MINDANAO
B 6201 74 DAVIDSON EVAPORITES, SASKATCHEWAN
B 6004 675 DAVIDSON-ARNOTT, G. D., MARINE-BAR SEDIMENTATION AND
B 6004 662 DAVIES, D. K., CONTINENTAL VOLCANIC SEDIMENTS AND THEIR
B 6303 460 DAVIES, DAVID K., AUTHIGENI ILLITES IN SANDSTONE RESERVOIRS,
B 6209 1756 DAVIES, DAVID K., CLAY TECHNOLOGY AND WELL STIMULATION, ABST.
B 6108 1375 DAVIES, DAVID K., DIAGENESIS OF TERTIARY VOLCANICLASTICS,
B 6309 1603 DAVIES, DAVID K., DISTRIBUTION OF VOLCANIC ACTIVITY IN TIME AND
B 6109 1535 DAVIES, DAVID K., EFFECTS OF SANDSTONE COMPOSITION AND
B 6311 2118 DAVIES, DAVID K., ENVIRONMENTS AND DIAGENESIS, MORROW
B 6108 1371 DAVIES, DAVID K., REGIONAL DIAGENETIC TRENDS IN LOWER
B 6203 508 DAVIES, DAVID K., RESERVOIR POTENTIAL OF VOLCANOGENIC
B 6409 1555 DAVIES, DAVID K., RESERVOIR QUALITY, PLIOCENE-PLEISTOCENE
B 6305 843 DAVIES, DAVID K., SEDIMENTOLOGY OF VOLCANICLASTIC DEPOSITS
S 5 145 DAVIES, GRAHAM R. CARBONATE- ANHYDRITE FACIES RELATIONSHIPS,
B 6111 1929 DAVIES, GRAHAM R., CARBONATE-ANHYDRITE FACIES RELATIONS IN
B 6004 662 DAVIES, GRAHAM R., EARLY MARINE AND LATER POSTBURIAL
B 6105 776 DAVIES, GRAHAM R., OTTO FIORD FORMATION—EVIDENCE OF SUBAQUEOUS
B 6303 380 DAVIES, J. B. SAUNDERS, AND J. G. SCHLATER, BECKMANN, J. P., REVIEW
B 6303 499 DAVIES, JOHN, LATE HOLOCENE DEPOSITIONAL HISTORY OF REGRESSIVE
B 6303 438 DAVIES, THOMAS A., PALEOENVIRONMENTAL IMPLICATIONS OF OCEANIC
B 6004 663 DAVIS, C. C., OBSERVATIONS OF CIRCULATION AN SUSPENDED-SEDIMENT
B 6206 1092 DAVIS, D., INSTRUMENTATION FOR EVALUATION OF IN-SITU RECOVERY
B 6002 318 DAVIS, D. V., JR., GEOLOGY OF PANOMA GAS AREA, SOUTHWESTER
B 6207 1213 DAVIS, DAN A., WATER SUPPLIES IN LIMESTONE TERRANES OF
B 6211 2352 DAVIS, G. A., MESOZOIC CONSTRUCTION OF CORDILLERAN COLLAGE.
B 6004 663 DAVIS, J. C., NUMERICAL CLASSIFICATION OF SANDSTONE PORE
B 6004 714 DAVIS, J. H., COMPUTER ANALYSIS OF LINEAMENTS TO INFER EXISTENCE
B 6002 318 DAVIS, JOHN C., PROBABILITY METHODS OF OIL EXPLORATION IN
B 6104 558 DAVIS, JOHN C., REGIONALIZED VARIABLES FOR EVALUATION OF
B 6004 675 DAVIS, K. RENEE, INFLUENC OF SOURCE AND PROCESSES IN
B 6002 159 DAVIS, MORGAN J., IN THE PUBLIC INTEREST
B 6104 613 DAVIS MOUNTAINS, TEXAS
B 6009 1608 DAVIS, RICHARD A., JR., BEACH SEDIMENTOLOGY OF MUSTANG AND
B 6303 438 DAVIS, RICHARD A., JR., ORIGIN AND DEVELOPMENT OF BARRIER
B 6105 779 DAVIS, RICHARD A., JR., PERMIAN INFILLED SUBGLACIAL GULLY FROM
B 6105 779 DAVIS, RICHARD A., JR., SEDIMENTATION IN LAKE REEVE, VICTORIA.
M 26 277 DAVIS, T. L., STRATIGRAPHIC AND SEISMIC EVIDENCE FOR LATE
B 6211 2352 DAVIS, TERRY E., AGE OF SEDIMENTATION-DIAGENESIS FOR BEDFORD
B 6403 440 DAVIS, THOM, BASIN-EDGE TECTONICS AND STRATIGRAPHIC
B 6008 1413 DAVIS, TOM L., OVERPRESSURED SHALE MASSES AND GROWTH
B 6201 74 DAWSON BAY LIMESTONE, SASKATCHEWAN
B 6301 90 DAWSON CANYON FORMATION, SCOTIAN BASIN
M 25 108 DAWSON COALFIELDS, CANADA, LIARD RIVER, SHEEP CREEK, AND
B 6207 1214 DAWSON, KENNETH M., METALLOGENIC ZONES AND EPOCHS IN
B 6203 508 DAWSON, WILLIAM C., IMPROVEMENT OF SANDSTONE POROSITY DURING
M 25 427 DAWSONVALE, IRON ORE, QUEENSLAND
B 6105 779 DAY, PETER C., FACTORS INFLUENCING SEDIMENT DISTRIBUTIONS AND
B 6407 1005 DAYS CREEK FIELD, ARK-LA-TEX AREA
B 6402 219 DCS AREA, INDIA
B 6305 835 DE AZCONA, C. LOPEZ, RECENT SALT SEDIMENTATION IN PLAYA LAKES
B 6203 503 DE CASTRO, J. C., PETROLEUM GEOLOGY OF PALEOZOIC CLASTIC ROCKS
B 6303 412 DE CASTRO, PAULO JACKSON MORGADO, OIL EXPLORATION IN CAMPOS
B 6002 293 DE FREITAS, M. H., HIGH-PRESSURE CONSOLIDATION OF KAOLINITIC
B 6105 770 DE JEKHOWSKY, B., MULTIDIMENSIONAL APPROACHES TO GEOLOGIC
B 6303 439 DE MARSILY, G., KRIGING OF TOP OF RESERVOIR FROM SEISMIC AND
B 6209 1684 DE MATHAREL, M., ORGANIC SEDIMENTATION AND GENESIS OF
B 6004 663 DE PRATTER, C. B., ARCHAELOGIC DATING OF HOLOCENE SHORELINE
B 6310 1823 DE RUITER, ON DEVELOPMENTS, IN MALAYSIA, ARMITAGE, J. H., AND
B 6107 961 DE RUITER, P. A. C., ATLANTIC MARGIN OF AFRICA, STRUCTURAL
B 6004 691 DE RUITER, P. A. C., PETROLEUM HABITAT OF ATLANTIC MARGIN FROM
B 6105 780 DE RUITER, P. A. C., PETROLEUM POTENTIAL OFF GABON
B 6004 664 DE VOTO, R. H., ENVIRONMENTS OF FORMATION OF URANIUM DEPOSITS.
B 6105 780 DE VOTO, R. H., HELIUM DETECTION IN URANIUM EXPLORATION
B 6108 1376 DE VOTO, RICHARD H., HELIUM DETECTION IN URANIUM EXPLORATION,
B 6203 514* DEAD SEA, ABST., ALGAL, TUFA FROM SHORES OF ANCESTRAL
B 6303 428* DEAD SEA, ABST., COMPARISON OF, CARBONATE CHEMISTRY OF, BRINES
B 6205 841 DEAD SEA ASPHALT, USES
B 6205 837† DEAD SEA ASPHALTS—HISTORICAL ASPECTS
B 6307 1021 DEAD SEA, ASYMMETRIC MASTER TRANSFORM SYSTEM
B 6205 838 DEAD SEA BASIN, ASPHALT LOCATIONS
B 6205 838 DEAD SEA BASIN, SOLID HYDROCARBONS
B 6404 569 DEAD SEA RIFT VALLEY, ISRAEL
B 6404 568 DEAD SEA TROUGH, ISRAEL
B 6306 987 DEAN AND B. CHARLOTTE SCHREIBER, EVANS, ROBERT, REVIEW OF,
B 6408 1283 DEAN, CLAUDE S., DEVONIAN SHALE EXPLORATION RATIONALE TESTED
B 6309 1577 DEAN, CLAUDE S., EXPLORATION STRATEGY FOR UNCONVENTIONAL
B 6205 781 DEAN FORMATION, NEW MEXICO
B 6303 482 DEAN, STUART L., FRACTOGRAPHIC DISTINCTION OF CORING-INDUCED
B 6303 482 DEAN, STUART L., NATURE AND FIELD APPLICATION OF PLUMOSE
B 6303 538 DEAN, W. E., MIDDLE CRETACEOUS OXYGEN-DEFICIENT
B 6305 825 DEAN, WALTER E., DEPOSITION AND DIAGENESIS OF ORGANIC MATTER
B 6203 509 DEAN, WALTER E., FRESHWATER ONCOLITES, ONONDAGA LAKE, NEW
B 6405 671 DEAN, WALTER E., OXIC-ANOXIC AND CARBONATE CYCLES IN

B 6303 439 DEAN, WALTER E., STATISTICAL ZONATION OF OIL SHALE ON BASIS OF
B 6203 541* DEATH VALLEY, CALIFORNIA, ABST., STEEPLY INCLINED, ALGAL
M 28 84 DEBOLT FORMATION, CANADA, MISSISSIPPIAN
M 27 133 DEBOLT FORMATION, CANADA, MISSISSIPPIAN
B 6308 1244 DEBROSSE, THEODORE A., OIL AND GAS DEVELOPMENTS IN MARYLAND,
B 6208 1399 DEBROSSE, THEODORE A., OIL AND GAS DEVELOPMENTS IN MARYLAND,
B 6108 1269 DEBROSSE, THEODORE A., OIL AND GAS DEVELOPMENTS IN MARYLAND,
B 6008 1288 DEBROSSE, THEODORE A., OIL AND GAS DEVELOPMENTS IN MARYLAND,
B 6409 1403 DEBROSSE, THEODORE H., OIL AND GAS DEVELOPMENTS IN MARYLAND,
M 26 364 DEBUBBLING, DEREVERBERATION, DIGITAL PROCESSING
B 6204 707 DECATUR LIMESTONE, ILLINOIS BASIN
B 6203 524* DECAY CHAIN—PERSPECTIVE ON MULTIELEMENT URANIUM
S 6 73 DECAY CONSTANT CALIBRATION, PHANEROZOIC
S 6 74 DECAY CONSTANT, DATA BASE
S 6 67† DECAY CONSTANTS IN GEOCHRONOLOGY AND COSMOCHRONOLOGY,
S 6 124 DECAY CONSTANTS, POTASSIUM-ARGON
S 6 70 DECAY CONSTANTS, RECOMMENDED VALUES
S 6 67 DECAY CONSTANTS, REPORT ON CURRENT USAGE
S 6 77 DECAY CONSTANTS, USAGE, COMPARISONS
S 6 73† DECAY-CONSTANT REVISIONS, PRE-CENOZOIC, PHANEROZOIC, TIME
B 6203 443 DECCAN TRAP, INDIA
B 6402 210 DECCAN TRAP LAVA FLOWS, INDIA
B 6209 1610 DECEW DOLOMITE, NEW YORK
B 6110 1901 DECKER, EDWARD R., CLAY, MINERALS AND PETROLEUM-FORMING
B 6209 1610 DECKER FORMATION, PENNSYLVANIA
B 6307 1038 DECOLLEMENT THRUST-FOLD ASSEMBLAGES
B 6405 690* DECOMPACTION TECHNIQUE HELPS CORRELATION, ABST.
S 5 118 DECONVOLUTION, SEISMIC DATA,
M 26 477 DEDMAN, E. V., PRACTICAL STRATIGRAPHIC MODELING AND
B 6203 537 DEDMAN, E. V., SEISMIC CALIBRATION TECHNIQUES FOR IMPROVED
B 6207 1214 DEDRICK, CLAIRE T., EFFECTIVE ENVIRONMENTAL REGULATION, ABST.
B 6303 510* DEDUCTION OF PAST GEOTHERMAL GRADIENTS IN NEOGENE SILICEOUS
B 6405 781* DEE SEA CORING—BIOSYSTEMATIC ANALYSIS, EVOLUTIONARY SPECIES,
B 6405 669* DEEP BASIN, ALBERTA, ABST.
B 6303 492* DEEP BASIN CANADA, BY TITLE ONLY, ENGINEERING, GEOLOGY IN
B 6302 162 DEEP BASIN, ELECTRIC LOG SECTION, ALBERTA
B 6302 152 DEEP BASIN, GAS, ALBERTA AND, BRITISH COLUMBIA
B 6302 152† DEEP BASIN GAS TRAP, WESTERN CANADA
B 6302 169 DEEP BASIN, LOG ANALYSIS, ALBERTA AND, BRITISH COLUMBIA
B 6405 703* DEEP DRILLING AND CURRENT MODELS OF CENOZOIC CRUSTAL
B 6303 341† DEEP DRILLING DATA, DELINEATION OF, ROME TROUGH IN EASTERN,
B 6004 708* DEEP ESTUARY CHANNEL SYSTEM, WILLAPA BAY, WASHINGTON, ABST.,
S 4 53† DEEP FOREREEF AND UPPER ISLAND SLOPE, NORTH JAMAICA
S 7 48 DEEP GULF OF MEXICO, CHALLENGER UNIT, DESCRIBED
S 7 47 DEEP GULF OF MEXICO, DESCRIBED
B 6409 1558* DEEP LAKE, AND SOUTHEAST LITTLE PECAN LAKE FIELDS, CAMERON
B 6205 804 DEEP OCEAN, SEISMIC INTERVALS, BLAKE PLATEAU AREA
B 6111 1991† DEEP OCEANS, PIERCEMENT STRUCTURES IN
B 6303 297 DEEP OFFSHORE, DRILLING TECHNOLOGY
B 6303 286† DEEP OFFSHORE, PETROLEUM, PROSPECT OF THE
B 6004 647* DEEP OR SHALLOW PETROLEUM GENERATION. ATTEMPT TO RESOLVE
B 6002 227 DEEP POROSITY MEASUREMENTS
M 25 154† DEEP REASEA
B 6101 80 DEEP RIVER BASIN, NORTH CAROLINA AND, VIRGINIA, DURHAM-
B 6104 594 DEEP RIVER FIELD, MICHIGAN
B 6004 678* DEEP SEA, ABST., CARBONATE MASS BALANCE--, CYCLING AND
B 6303 380 DEEP SEA DRILLING LEGS 22-29, EDITED J R. HEIRTZLER, H. M. BOLLI, T. A.
B 6402 245 DEEP SEA DRILLING PROJECT SITES 127 AND 128, PELOPONNESUS
B 6412 2051 DEEP SEA DRILLING PROJECT, BLACK SHALES, ATLANTIC OCEAN
B 6209 1764* DEEP SEA DRILLING PROJECT CORE 77B OF LEG 9, ABST., COSMOPOLITAN
B 6101 102 DEEP SEA DRILLING PROJECT CORES
B 6309 1453 DEEP SEA DRILLING PROJECT, DSDP, CORE HOLE 108, BALTIMORE
B 6303 287 DEEP SEA DRILLING PROJECT, DSDP, INTERNATIONAL PHASE OF OCEAN
B 6303 290 DEEP SEA DRILLING PROJECT HOLES, BASE MAP, 1975
B 6004 649* DEEP SEA DRILLING PROJECT IN WESTERN NORTH ATLANTIC, ABST.,
B 6303 287 DEEP SEA DRILLING PROJECT, IPOD INITIAL REPORTS OF THE
B 6211 2294 DEEP SEA DRILLING PROJECT, LEG 1, HOLE 1
B 6106 931 DEEP SEA DRILLING PROJECT LEG 3 RESULTS
B 6106 931 DEEP SEA DRILLING PROJECT LEG 39 RESULTS
B 6309 1608* DEEP SEA DRILLING PROJECT LEG 43 SITE 386, ABST., MERIDIONALLY
B 6303 464* DEEP SEA DRILLING PROJECT ON PALEO-OCEANOGRAPHY, ABST., IMPACT
B 6201 88 DEEP SEA DRILLING PROJECT, SITE 186, ALASKA
B 6102 193 DEEP SEA DRILLING PROJECT SITE 191, BERING SEA, KOMANDORSKIY
B 6201 62 DEEP SEA DRILLING PROJECT, SITE 260, AUSTRALIA
B 6112 2068 DEEP SEA DRILLING PROJECT SITE 262
B 6201 41 DEEP SEA DRILLING PROJECT, SITE 263, AUSTRALIA
B 6211 2306 DEEP SEA DRILLING PROJECT, SITE 97
B 6007 1041 DEEP SEA DRILLING PROJECT, SITE 98
B 6402 245 DEEP SEA DRILLING PROJECTS SITES 127 AND 128, PELOPONNESUS
B 6410 1659 DEEP SEA DRILLING PROJECT, SITES, WESTERN, ATLANTIC
B 6111 1992 DEEP SEA DRILLING SITE 141, SEISMIC PROFILE
B 6106 930 DEEP SEA DRILLING SITES ON RIO GRANDE RISE, WALVIS RIDGE
B 6203 547* DEEP SINUOUS CHANNELS AND GIANT CROSS-BEDS IN ATHABASCA OIL
B 6004 705* DEEP SMACKOVER OF MISSISSIPPI, ABST., GEOPRESSURES AND,
B 6405 715* DEEP STRATIGRAPHY AND EVOLUTION OF BALTIMORE CANYON
B 6303 445* DEEP SUBSURFACE ENVIRONMENTS, UPPER SMACKOVER FORMATION,
B 6405 771* DEEP TERRESTRIAL HEAT-FLOW STUDIES IN SOUTHWESTERN UNITED
B 6405 710* DEEP TUSCALOOSA GAS TREND OF SOUTH LOUISIANA, ABST.
B 6004 690* DEEP WESTERN GULF OF MEXICO, ABST., SEISMIC STRATIGRAPHY AND
B 6002 231 DEEP-BASIN TEMPERATURES
B 6403 302 DEEP-OCEAN TRENCHES INFERRED FROM BATHYMETRIC AND SEISMIC
B 6403 302 DEEP-OCEAN TRENCHES INFERRED FROM BATHYMETRIC AND SEISMIC
B 6405 704* DEEP-SEA BASINS, ABST., TRACE, FOSSILS AND STAGNATION OF
B 6405 700* DEEP-SEA BENTHIC FORAMINIFERA AND THEIR BIOSTRATIGRAPHIC
B 6004 700* DEEP-SEA CARBONATE OOZE--EMPIRICAL OBSERVATIONS AND
B 6203 520* DEEP-SEA CONGLOMERATES, PEBBLY AND MASSIVE SANDSTONES, ABST.,
B 6303 553* DEEP-SEA DEPOSITION IN DENSITY-STRATIFIED CRATONIC BASIN, BELL
B 6303 553* DEEP-SEA DRILLING, ABST., BIOGENOUS SEDIMENTS RECOVERED BY
B 6303 537* DEEP-SEA DRILLING AND GLOBAL TECTONICS, ABST.
B 6303 419* DEEP-SEA DRILLING AND MAJOR THEMES OF OCEAN EVOLUTION, ABST.
B 6303 554* DEEP-SEA DRILLING IN ANTARCTIC--LATE TERTIARY PALEOCLIMATIC
B 6303 484* DEEP-SEA DRILLING, NEW DIMENSION IN OUR APPROACH TO OCEANIC
B 6405 740* DEEP-SEA FAN AND SHALLOW-MARINE COMPLEX, CALIFORNIA, ABST.,
B 6206 967† DEEP-SEA FAN, CALIFORNIA, CHANNEL, SANDS AND, PETROLEUM,
B 6303 535* DEEP-SEA FAN MODELS, FINE-GRAINED SEDIMENT DISTRIBUTION, AND
B 6405 735* DEEP-SEA FAN SANDSTONE AND SHALE BY AUTHIGENIC SMECTITE AND
B 6004 721* DEEP-SEA FANS, ABST., CURRENTS IN, SUBMARINE, CANYON AND THEIR
B 6405 758* DEEP-SEA FANS FROM VARIOUS TECTONIC SETTINGS, ABST., GEOMETRY
B 6206 913 DEEP-SEA FANS, GROWTH PATTERNS OF
B 6405 803* DEEP-SEA FANS OFF CENTRAL CALIFORNIA, ABST., POTENTIAL
B 6203 511* DEEP-SEA ICHNOFACIES, ABST., BATHYMETRIC CONTROL OF
B 6207 1216* DEEP-SEA MINING , ABST., WHAT LAW SHALL GOVERN
B 6306 908 DEEP-SEA SANDS, DIPMETER MOTIFS OF
B 6306 906 DEEP-SEA SANDS, GAMMA, LOG MOTIFS OF
B 6203 529* DEEP-SEA SEDIMENT AND THE CHERT PROBLEM, ABST.
B 6111 1994 DEEP-SEA SEDIMENT DIAPIRS FROM LAMONT-DOHERTY SEISMIC PROFILE
B 6303 507* DEEP-SEA SEDIMENTS, ABST., COPPER, PORPHYRINS IN
B 6303 473* DEEP-SEA SEQUENCES--NEOGENE EXAMPLES FROM PACIFIC, ABST.,
S 7 273 DEEP-SEATED FAULTS, BOTTOM, DEFORMATION
B 6403 440* DEEP-WATER BASINS IN OUTER SOUTHERN CALIFORNIA BORDERLAND,
B 6105 827* DEEP-WATER CARBONATE SEDIMENTOLOGY, DIAGENESIS, AND
S 10 235 DEEP-WATER DISCHARGE, SUBSURFACE OBSERVATION
S 10 235 DEEP-WATER DISCHARGE, SURFACE OBSERVATIONS
B 6306 911 DEEP-WATER FACIES, LOGS, NORTH SEA
B 6303 293 DEEP-WATER FANS AND CONES, PROSPECTS, CONTINENT-RELATED
M 27 17 DEEP-WATER OOZE, BERMUDA RISE, EOCENE
B 6004 686* DEEP-WATER PELAGIC AND SHALLOW-WATER EPERIC CARBONATE
B 6408 1280* DEEP-WATER RESERVOIRS—SUBMARINE FANS AND FANTASIES, ABST.
B 6305 810† DEEP-WATER SANDSTONE FACIES AND ANCIENT SUBMARINE FANS.
B 6206 932† DEEP-WATER SANDSTONE FACIES AND ANCIENT SUBMARINE FANS:
B 6108 1384* DEEP-WATER SANDSTONES, OUACHITA MOUNTAINS, ARKANSAS AND
B 6303 443* DEEP-WATER SEDIMENTATIO IN CRETACEOUS ARC-TRENCH GAP,
B 6105 828* DEEP-WATER SEDIMENTATION AND TECTONICS IN MIDDLE ORDOVICIAN
B 6009 1616* DEEP-WELL DATA, ABST., GEOTHERMAL EXPLORATION FROM
B 6108 1126 DEEPER POOL TEST
B 6208 1281 DEEPER POOL TEST
B 6308 1174 DEEPER POOL TEST
B 6008 1159 DEEPER POOL TEST
B 6409 1301 DEEPER-POOL TEST
M 27 111 DEEPKILL SHALE, NEW YORK, ORDOVICIAN
B 6107 1046 DEER CREEK FORMATION, KANSAS
B 6412 2169 DEER PARK ANTICLINE, WEST VIRGINIA
B 6101 80 DEERFIELD BASIN, CONNECTICUT-, MASSACHUSETTS, HARTFORD-
B 6104 595 DEERFIELD FIELD, MICHIGAN
B 6105 779 DEERY, J. R., PHYSICAL CHARACTERISTICS OF WASHOVER FANS OF
B 6109 1535 DEERY, JOHN R., ORIGIN AND CHARACTER OF WASHOVER FANS ON
B 6203 506 DEES, J. L., UNDERSTANDING SEISMICALLY DERIVED IMPEDANCE LOGS.
B 6105 811 DEETER, THOMAS M., ALASKA PENINSULA LATE CRETACEOUS FORE-ARC
B 6207 1232* DEFINING BOUNDARIES AND STRUCTURE OF BROADLANDS
B 6405 746* DEFINITION AND DEVELOPMENT OF MACKEREL FIELD, GIPPSLAND
B 6305 824* DEFINITION OF DEPOSITIONAL FACIES OF PARACHUTE CREEK MEMBER
B 6007 1119† DEFINITION OF GILMER LIMESTONE, UPPER JURASSIC FORMATION.
B 6410 1641 DEFINITION OF LUCAYAN LIMESTONE, NEW QUATERNARY FORMATION,
B 6209 1584 DEFINITION OF SOURCE BEDS
B 6009 1570† DEFINITION OF TYPE LITTLE FALLS DOLOSTONE, LATE CAMBRIAN,
S 7 272 DEFORMATION, COMPLEX SHELF-EDGE GRABEN FAULTS, BOTTOM
S 7 272 DEFORMATION, CONCAVE CONVEX FAULT, BOTTOM
S 7 273 DEFORMATION, DEEP-SEATED FAULTS, BOTTOM
S 7 278 DEFORMATION, DIFFERENTIAL CONSOLIDATION, BOTTOM
M 28 159 DEFORMATION FABRICS, GENERAL

S 3 41 DEFORMATION, FLUID, MOBILIZATION BY
S 7 275 DEFORMATION, GENERAL INSTABILITY, BOTTOM
S 7 275 DEFORMATION, GRAVITY, BOTTOM
S 7 272 DEFORMATION, GROWTH FAULT, BOTTOM
S 7 273 DEFORMATION, HYDRAULIC FORCES, BOTTOM
S 7 273 DEFORMATION, LINEAR FAULT, BOTTOM
S 7 273 DEFORMATION, MONOCLINAL FOLD, BOTTOM
S 7 276 DEFORMATION, MUDFLOWS, BOTTOM
B 6303 549* DEFORMATION OF CONTINENTAL SLOPE SEDIMENTS RESULTING FROM
B 6009 1425† DEFORMATION OF MESOZOIC SHALES AT HAMILTON DOME, BIGHORN
B 6009 1531† DEFORMATION ON PERMEABILITY OF OTTAWA SAND, EFFECT OF, HIGH-,
S 7 269 DEFORMATION PERIPHERAL SLUMPS, BOTTOM
S 7 271 DEFORMATION, SHALLOW GRABEN FAULTS, BOTTOM
S 7 273 DEFORMATION, SHELF-EDGE SLUMP AND DIAPIRIC INTRUSION, BOTTOM
S 7 278 DEFORMATION, SQUEEZING AND DIAPIRISM, BOTTOM
S 7 272 DEFORMATION, SUBSURFACE CONVEX FAULT, BOTTOM
S 7 271 DEFORMATION, SURFACE MUDFLOW, BOTTOM
S 7 273 DEFORMATION, TECTONIC ACTIVITY, BOTTOM
S 7 278 DEFORMATION, TENSIONAL STRESS, BOTTOM
S 7 278 DEFORMATION, TURBIDITY FLOWS, BOTTOM
S 7 269 DEFORMATIONAL FEATURES, SHELF, STABILITY
S 7 269 DEFORMATIONAL FEATURES, SLOPE, STABILITY
B 6303 427* DEFORMATIONAL STRUCTURES IN PLIOCENE BOUSE FORMATION NEAR
M 27 208 DEFORMATIONS, NEW LIMESTONE
B 6309 1478† DEFORMED CARBONATE ROCKS, EFFECTS OF, TEXTURE AND
B 6002 289 DEGASSED SEDIMENTS, CHARACTERISTICS OF
B 6304 585 DEGRADATION, GEOCHEMICAL CONCEPT
B 6304 701* DEGRADATION OF CRUDE OILS--IDENTIFICATION BY CARBON ISOTOPES,
B 6111 1953 DEGRADATION, OIL, GENERATION, MIGRATION, ENTRAPMENT, AND
B 6108 1375 DEGRAW, HAROLD M., GEOMETRY AND PALEOENVIRONMENTS OF
B 6407 1120 DEHLINGER, BOWIN, CARL, REVIEW OF, MARINE GRAVITY, BY PETER
B 6203 571* DEHYDRATION, SALINITY, CRACKING, AND MIGRATION, ABST., SHALE
B 6405 749 DEIKE, RUTH G., STABLE ISOTOPE EVIDENCE FOR MODERN FRESHWATER
B 6103 416 DEIKE, RUTH G., WILDCAT VALLEY SANDSTONE IN SOUTHWEST
B 6009 1621 DEIKE, RUTH G., WILDCAT VALLEY SANDSTONE IN SOUTHWEST
B 6305 835 DEL CURA, A. GARCIA, RECENT SALT SEDIMENTATION IN PLAYA LAKES
M 27 33 DEL RIO CLAY, TEXAS, CRETACEOUS
B 6402 230 DEL RIO FORMATION, TEXAS
B 6303 492* DELAWAR COASTAL BARRIERS, ABST., DETERMINATION OF STORM
B 6303 439* DELAWARE, ABST., INTERPRETATION OF, PLEISTOCENE COASTAL-,
B 6405 694* DELAWARE AND NEW JERSEY, ABST., TRACE, FOSSIL ASSEMBLAGES OF
B 6312 2146 DELAWARE, ATLANTIC COAST, TRANSGRESSIVE BARRIERS
B 6106 918 DELAWARE, BALTIMORE CANYON TROUGH AREA
M 26 436 DELAWARE BASIN, ABO FORMATION, NEW MEXICO
B 6203 495* DELAWARE BASIN, ABST., COMPARISON OF, TURBIDITE, SANDSTONE,
B 6405 803* DELAWARE BASIN, ABST., RESERVOIR ROCK, SOURCE ROCK, AND
B 6203 496* DELAWARE BASIN, NEW MEXICO, ABST., TURBIDITE ORIGIN FOR,
M 26 437 DELAWARE BASIN, SAN ANDRES FORMATION, NEW MEXICO
B 6109 1449 DELAWARE BASIN, TEXAS AND, NEW MEXICO
B 6303 420* DELAWARE BASIN, TEXAS AND NEW MEXICO, ABST., LITHOFACIES
B 6303 553* DELAWARE BASIN TEXAS-NEW MEXICO, ABST., DEEP-SEA DEPOSITION IN
B 6303 530* DELAWARE BASIN, WEST TEXAS ABST., ELLENBURGER GROUP
B 6004 517† DELAWARE BASIN, WEST TEXAS AND SOUTHEAST NEW MEXICO, BASINAL,
B 6312 2147 DELAWARE, BAYMOUTH BARRIER
B 6312 2150 DELAWARE, CAPE HENLOPEN SPIT
B 6004 728* DELAWARE COASTAL PLAIN, ABST., MIOCENE, TRANSGRESSIONS AND,
B 6209 1714† DELAWARE COASTAL PLAIN: DISCUSSION AND REPLY, RECORRELATION
B 6105 723 † DELAWARE COASTAL PLAIN, RECORRELATION OF EOCENE-MIOCENE BOUNDAR
M 29 70 DELAWARE, DEPTH SECTIONS, ATLANTIC COAST
B 6209 1610 DELAWARE LIMESTONE, OHIO
B 6407 1118 DELAWARE MOUNTAIN GROUP OF WEST TEXAS AND SOUTHEAST NEW
B 6203 500* DELAWARE MOUNTAIN GROUP, WEST TEXAS—SUBSURFACE EVIDENCE,
B 6106 918 DELAWARE, WILMINGTON CANYON AREA, SUBMARINE SLIDE
B 6303 439 DELFINER, P., KRIGING OF TOP OF RESERVOIR FROM SEISMIC AND
B 6206 968 DELGADA CHANNEL, CALIFORNIA
B 6206 968 DELGADA FAN, CALIFORNIA
B 6405 697 DELGADO, DAVID J., SUBMARINE DIAGENESIS, ARAGONITE DISSOLUTION,
B 6305 835 DELGADO, S. ORDONEZ, RECENT SALT SEDIMENTATION IN PLAYA LAKES
B 6303 439 DELHOMME, J. P., KRIGING OF TOP OF RESERVOIR FROM SEISMIC AND
B 6405 725* DELINEATION OF JACKSONIAN-VICKSBURGIAN BOUNDARY IN
B 6303 341† DELINEATION OF ROME TROUGH IN EASTERN KENTUCKY BY GRAVITY
B 6304 692 DELLA VALLE, R., MINERALOGY AND GEOCHEMISTRY OF MARIANO LAKE
B 6304 687 DELLA VALLE, R. S., GEOCHEMICAL STUDIES OF GRANTS MINERAL BELT,
B 6305 823 DELLA VALLE, R. S., LESS COMMON TRACE ELEMENTS IN
B 6405 681 DELLA VALLE, R. S., RB-SR DATING OF DIAGENESIS—EPIGENESIS IN
B 6304 693 DELLA VALLE, RICHARD, SEM INVESTIGATION OF PARAGENESIS OF
B 6303 455 DELLWIG, L. F., A NEW LOOK AT GEOLOGY OF TOGO BY SLAR, ABST.
B 6002 325 DELLWIG, LOUIS F., SLAR IMAGERY IS NOT JUST A PICTURE, ABST.
S 11 54† DELMONTIAN, PLANKTONIC AND LARGE, FORAMINIFERA IN MIDDLE,
S 11 40 DELMONTIAN STAGE
S 11 67 DELMONTIAN STAGE
S 11 43 DELMONTIAN STAGE, CALIFORNIA
B 6211 2246 DELMONTIAN STAGE, CALIFORNIA
B 6202 259 DELMONTIAN STAGE, CALIFORNIA
S 11 40 DELMONTIAN STAGE, CONTROVERSY
S 11 50 DELMONTIAN STAGE, ZONATION
S 11 124 DELMONTIAN STAGES, CALIFORNIA MIOCENE, NONIONIDAE, MOHNIAN
B 6303 542 DELPEIL, J. R., PALEOENVIRONMENT AND PETROLEUM POTENTIAL OF
B 6312 2185 DELPHI, ASSESSING RESOURCES
B 6401 115† DELPHI, INDIANA, STRATIGRAPHY, STRUCTURE, AND, ZONATION OF
B 6401 127 DELPHI REEF, INDIANA
B 6303 512* DELTA, ABST., ACTIVE SUBMARINE, LANDSLIDE PROCESSES, MISSISSIPPI
B 6203 488* DELTA, ABST., HEAVY-MINERAL RESPONSE TO ENVIRONMENTAL
B 6004 684* DELTA, ABST., STRATIGRAPHY AND SEDIMENTATION OF, FRESHWATER,
B 6004 716* DELTA AND CONE, ABST., RECENT SEDIMENTATION ON, NILE
B 6004 701* DELTA AND RICHARDSON MOUNTAINS, NORTHWEST TERRITORIES,
B 6004 701* DELTA AREA, ABST., CONTROL OF SEDIMENT DISTRIBUTION BY,
B 6202 207 DELTA AREA, GEOHISTORY, LOUISIANA, MISSISSIPPI
B 6209 1684† DELTA, BORNEO, ORGANIC SEDIMENTATION AND, GENESIS OF,
B 6004 695* DELTA, CARBONIFEROUS, OF NORTHERN ENGLAND--SLOPE INFLUENCED
B 6311 2042† DELTA, CATATUMBO RIVER, LAKE MARACAIBO, VENEZUELA,
B 6311 2054 DELTA, CHARACTERISTICS, TABLE, MISSISSIPPI
B 6311 2054 DELTA, CHARACTERISTICS, TABLE, VENEZUELA, CATATUMBO RIVER
B 6409 1556* DELTA COMPLEX—UNSTABLE, SHELF-EDGE DELTA IN DEEP WILCOX
B 6205 819 DELTA DEPOCENTER, CALIFORNIA
B 6311 2119* DELTA DEPOSITS OF ANADARKO BASIN, ABST., UPPER, MORROW, FAN-
B 6402 264† DELTA, EVOLUTION OF, SAND-DOMINANT, SUBAERIAL PHASE,
B 6303 542* DELTA FRONT, ABST., SHALLOW-SUBMARINE SEISMIC STRATIGRAPHY,
B 6007 1074 DELTA, GEOTHERMAL GRADIENT MAP, NIGER
B 6311 2118* DELTA, GREAT SALT PLAINS, ALFALFA COUNTY, OKLAHOMA, ABST.,
B 6201 1† DELTA, HYDROCARBON HABITAT OF, TERTIARY, NIGER
B 6403 376 DELTA, JAMAICA, HOLOCENE, FAN
B 6007 1024 DELTA, LAKE SUPERIOR, RESERVE, FAN AND
B 6002 280 DELTA LOBES, MASON AND DIXON LINE, LOCATION OF
B 6402 264† DELTA, LOUISIANA, EVOLUTION OF, SAND-DOMINANT, SUBAERIAL
B 6009 1615* DELTA, LOUISIANA-MISSISSIPPI, ABST., CLAY SEDIMENTATION I, PEARL
B 6409 1559* DELTA MARGIN EMBAYMENT—UPPER WOODBINE FORMATION OF
B 6203 515* DELTA MODEL WITH EXAMPLE FROM UPPER DEVONIAN—LOWER
M 26 224 DELTA MODELS, DEPOSITIONAL SYSTEMS
B 6408 1251† DELTA, PETROLEUM, SOURCE-BED, EVALUATION OF, TERTIARY, NIGER
B 6406 938† DELTA PLAIN COALS, SOUTHWEST WEST VIRGINIA, EXPLORATION
B 6212 2381 DELTA PLAIN, FLUVIAL AND UPPER
B 6212 2381 DELTA PLAIN, LOWER
B 6212 2381 DELTA PLAIN, TRANSITIONAL LOWER
B 6009 1543† DELTA PROLIFI BELT, PENDULUM EFFECT AND, NIGER
B 6311 2116* DELTA SANDSTONES, PAL DURO BASIN, TEXAS, ABST., PENNSYLVANIAN,
B 6007 1073† DELTA SEDIMENTARY BASIN, APPROXIMATE, GEOTHERMAL, GRADIENTS
B 6403 374† DELTA SEDIMENTOLOGY AND TECTONIC SETTING—YALLAHS FAN
B 6303 551* DELTA SEDIMENTS, ABST., EFFECT OF BIOGENIC, METHANE ON SEDIMENT
B 6203 572* DELTA, SOUTHEASTERN JAMAICA, ABST., DEPOSITIONAL ENVIRONMENTS
B 6405 707* DELTA STRATIFICATION AND ITS RELATION TO TIDAL INLET PROCESSES,
B 6405 703* DELTA SYSTEM, SOUTH TEXAS, ABST., GROWTH, FAULTING AND
B 6202 303 DELTA SYSTEM, TRINIDAD, PROTO-ORINOCO
B 6202 303 DELTA SYSTEM, VENEZUELA, ORINOCO
B 6309 1601* DELTA SYSTEMS, CONCHO PLATFORM OF NORTH-CENTRAL TEXAS, ABST.,
B 6009 1606* DELTA SYSTEMS IN GULF COAST AREA ABST., ANCIENT
B 6311 2116* DELTA SYSTEMS, WOLFCAMPIAN, SOUTHEASTERN PALO DURO BASIN,
B 6202 295† DELTA, TERTIARY LITHOSTRATIGRAPHY OF, NIGER
B 6402 277 DELTA, TEXAS, COLORADO
B 6402 277 DELTA, TEXAS, GUADALUPE
B 6402 277 DELTA, TEXAS, TRINITY
B 6304 660† DELTA--APRIL 9, 1976--FROM LANDSAT 2, MISSISSIPPI RIVER
B 6409 1567* DELTA-FRONT DEPOSITS, ABST., SEDIMENTARY STRUCTURES
B 6306 886† DELTA-MARGIN FACIES OF DAKOTA SANDSTONE, LONE PINE FIELD, NEW
B 6303 519* DELTAIC COALS OF FERRON SANDSTONE MEMBER OF MANCOS
B 6203 571* DELTAIC COMPONENTS OF VALMY AND VININI FORMATIONS,
B 6004 697* DELTAIC CYCLES ON UNSTABLE CRATO MARGIN, BANKS ISLAND, ARCTIC
B 6211 2194 DELTAIC CYCLOTHEMS, IRELAND, WESTERN
B 6002 319* DELTAIC DEPOSITIONAL ENVIRONMENTS OF PERMIAN ADMIRE
B 6203 494* DELTAIC DEPOSITIONAL STYLE IN SALT-WITHDRAWAL BASIN, FRIO
B 6311 2119* DELTAIC DEPOSITS IN UPPER MORROW FORMATION OF ANADARKO
B 6004 659* DELTAIC EVOLUTION DURING BASIN FILL--NAMURIA OF CENTRAL
B 6303 514* DELTAIC FACIES RECOGNITIO ON SEISMIC DATA, ABST.
B 6209 1589 DELTAIC MARGIN, LOUISIANA, GULF COAST—, DIVERGENT
B 6009 1502† DELTAIC OIL HABITAT, TRINIDAD, MIOCENE
B 6303 541* DELTAIC SANDSTONE, ADMIRE 650-FOOT SANDSTONE, EL DORADO FIELD,
B 6105 825* DELTAIC SANDSTONE RESERVOIRS—EXPLORATION MODELS AND
B 6109 1536* DELTAIC SANDSTONE, STRAWN SERIES, PENNSYLVANIAN,
B 6303 516* DELTAIC SANDSTONES, LAKE MARACAIBO, VENEZUELA, ABST., FACIES

Index

B 6303 534* DELTAIC SANDSTONES--BASAL BELLY RIVER FORMATION, UPPER
B 6405 780* DELTAIC SANDSTONES—EXCELLENT SUBSURFACE RESERVOIR, ABST.,
B 6409 1564* DELTAIC SANDSTONE—OLMOS SAN MIGUEL, AND UPSON FORMATIONS,
B 6104 527† DELTAIC SEDIMENTATION IN CENTRAL TEXAS ESTUARY, LAVACA BAY--,
B 6403 316† DELTAIC SEDIMENTATION OF UPPER CRETACEOUS EAGLE
B 6403 316† DELTAIC SEDIMENTATION OF UPPER CRETACEOUS EAGLE SANDSTONE—
B 6004 681* DELTAIC SEDIMENTS, ABST., BED CONTINUITY AND, PERMEABILITY
B 6203 538* DELTAIC SEDIMENTS OF ROBERTS MOUNTAIN ALLOCHTHON,
B 6003 341† DELTAIC SEDIMENTS, SVALBARD, GROWTH, FAULTS IN UPPER, TRIASSIC
B 6004 668* DELTAIC SEDIMENTS, SVALBARD, NORWAY, ABST., GROWTH, FAULTS
B 6307 1054 DELTAIC SEQUENCE, SEISMIC EXPRESSION OF, SHALE, DIAPIRS WITHIN
B 6004 671* DELTAIC SUBENVIRONMENTS AND PALEOCURRENT ANALYSIS OF
B 6405 717* DELTAIC SYSTEMS AND ASSOCIATED GROWTH FAULTING OF VICKSBURG
B 6405 688* DELTAIC SYSTEMS IN PENNSYLVANIAN TECTONICALLY ACTIVE
M 26 431 DELTAIC TRAPS, SEISMIC, MODELS
B 6203 531* DELTAIC-COASTAL SEDIMENTARY-FACIES PATTERNS IN GRABEN
B 6203 560* DELTAIC-SANDSTONE RESERVOIRS, ABST., EXPLORATION MODELS AND
B 6004 680* DELTAS, ABST., SANDSTONE, MODELS FROM, APPALACHIAN,
B 6309 1596* DELTAS IN BLACK WARRIOR BASIN OF MISSISSIPPI AND ALABAMA, ABST.,
B 6009 1614* DELTAS OF SOUTH MISSISSIPPI AND PRODUCING TRENDS, ABST., PRE-,
M 26 224 DELTAS, SEISMIC, DEPOSITIONAL SYSTEMS
B 6103 352 DELTAS, TEXAS, COOK
B 6011 2068* DELTAS--VARIABILITY OF PROCESSES AND SAND BODIES, ABST., MODERN
B 6405 739* DELTAS—IMPORTANT ECONOMIC DEPOSITIONAL SETTING IN UPPER
B 6105 778 DELTEIL, J. R., MICROCONTINENTAL MARGINS
B 6412 2051 DELTEIL, J. R., PALEOENVIRONMENT AND PETROLEUM POTENTIAL OF
B 6105 780 DELTEIL, J. R., PETROLEUM POTENTIAL OFF GABON
B 6408 1179 DEMAISON, G. J., ANOXIC ENVIRONMENTS AND OIL SOURCE BED GENESIS
B 6405 697 DEMAISON, G. J., ANOXIC ENVIRONMENTS AND OIL SOURCE BEDS, ABST.
B 6405 729 DEMAISON, G. J., ORGANIC FACIES—STRATIGRAPHIC CONCEPT AND
B 6111 1950 DEMAISON, G. J., TAR SANDS AND SUPERGIANT OIL FIELDS
B 6012 2178 DEMAISON, G. J., TAR SANDS GIVE CLUES TO SEARCH FOR SUPERGIANT
B 6303 542 DEMAISON, G., PALEOENVIRONMENT AND PETROLEUM POTENTIAL OF
B 6412 2051 DEMAISON, G., PALEOENVIRONMENT AND PETROLEUM POTENTIAL OF
M 25 526 DEMANT, ALAIN, MAJOR CHEMICAL CHARACTERISTICS OF TERTIARY
B 6303 439 DEMAREST, JAMES M., INTERPRETATION OF PLEISTOCENE
M 30 459 DEMATHAREL, M., GEOLOGY OF THE BEKAPAI FIELD
B 6405 697 DEMATHIEU, GEORGES R., USE OF TRACE FOSSILS FOR INTERPRETATION
B 6405 776 DEMBICKI, HARRY, JR., ET AL, OIL-TO-SOURCE CORRELATION—PINEVIEW
B 6007 1086 DEMERARA ABYSSAL PLAIN, BATHYMETRIC MAP
B 6202 290 DEMERARA ABYSSAL PLAIN OFF, BRAZIL, NORTHERN
B 6111 1994 DEMERARA ABYSSAL PLAIN, POSSIBLE DIAPIRS
B 6010 1866 DEMINEX IRA OIL CO., HORMOZ PETROLEUM CO., SOFIRAN, IRAN NIPPON
B 6110 1796 DEMINEX IRAN OIL CO., TOTAL IRAN PETROL. CO., MOBIL IRAN
B 6210 2112 DEMINEX IRAN OIL CO., TOTAL IRANIAN PETROLEUM CO., AMERICAN
B 6002 174 DENALI FAULT, BERINGIAN SHELF
B 6405 698 DENHAM, L. R., HOW GEOLOGIC OBJECTIVES SHOULD DETERMINE
B 6002 325 DENISON, RODGER E., EVOLUTION OF ANADARKO BASIN, ABST.
B 6110 1640 DENMARK, DEVELOPMENTS, 1976
B 6210 1909 DENMARK, DEVELOPMENTS, 1977
B 6310 1918 DENMARK, DEVELOPMENTS, 1978
B 6411 1963 DENMARK, EXCLUDING NORTH SEA, DEVELOPMENTS, 1979
B 6104 597 DENMARK FIELD, OHIO
B 6107 988 DENMARK, MAESTRICHTIAN CHALK
M 27 19 DENMARK, MAESTRICHTIAN CHALK
B 6411 1871 DENMARK, NORTH SEA, PRODUCTION, 1978-1979
M 27 60 DENMARK, PALEOCENE, FAKSE LIMESTONE
M 27 201 DENMARK, PALEOCENE, FAXEKALK
B 6210 1924 DENMARK-NORTH SEA, PRODUCTION, 1976-1977
B 6310 1932 DENMARK-NORTH SEA, PRODUCTION, 1977-1978
B 6107 1046 DENNIS FORMATION, KANSAS
S 4 243 DENNIS LIMESTONE, KANSAS
B 6408 1285 DENNISON, JOHN M., DEVONIA OIL SHALE OF EASTERN UNITED STATES,
B 6203 492 DENNISON, JOHN M., DISTINCTIVE TURBIDITE SILTSTONE IN DEVONIAN
B 6009 1619 DENNISON, JOHN M., ENERGY RESOURCE IMPLICATIONS OF 38TH
B 6002 278 DENNISON, JOHN M., STRATIGRAPHIC CROSS SECTION OF HAMILTON
B 6304 692* DENNISON-BUNN CLAIM AND POSSIBILITY OF URANIUM DEPOSITS IN
B 6007 1128 DENSITY AND POROSITY FOR OFFSHORE WELL, LOUISIANA, SHALE
B 6004 695* DENSITY CURRENTS, ABST., KINDERSCOUTIAN, DELTA, CARBONIFEROUS,
B 6303 494* DENSITY GRIDS TO OIL EXPLORATION DATA, ABST., CREATION AND
M 26 39 DENSITY, LABORATORY, MEASUREMENTS, SATURATION EFFECTS, BULK
B 6309 1504† DENSITY LOGS, DETERMINATION OF, ORGANIC CONTENT OF,
S 4 371 DENSITY, POROSITY, AND PERMEABILITY, CORALS
S 4 172 DENSITY VARIATIONS IN SKELETAL CARBONATE
B 6007 1125 DENSITY VERSUS DEPTH RELATION FOR SHALES
B 6009 1607* DENSITY-FLOW ORIGIN FOR FRIO RESERVOIR SANDSTONES, NINE MILE
M 26 18 DENSITY-LITHOLOGY ASPECTS, SEISMIC, INTERPRETATION, VELOCITY-
S 5 181* DENSITY-STRATIFIED EVAPORITE PAN, BASE- METAL CONCENTRATION IN
B 6212 2509 DENT, RICHARD J., WHAT FOREIGN COMPANIES SHOULD KNOW ABOUT
S 11 272 DENTALINA QUADRULATA
S 11 234 DENTALINA QUADRULATA
S 11 274 DENTALINA SOLUTA
S 11 234 DENTALINA SOLUTA
B 6303 433* DENVER BASIN, ABST., PETROLEUM GENERATION AN, MIGRATION IN
B 6008 1412* DENVER BASIN, ABST., SOURCE BEDS OF PETROLEUM IN
M 26 280 DENVER BASIN, BOULDER-WELD FAULT ZONE
B 6410 1622 DENVER BASIN, CARLILE SHALE
B 6410 1614 DENVER BASIN, COLORADO
B 6108 1388* DENVER BASIN, COLORADO, ABST., MARINE-SHELF MODEL FOR UPPER,
B 6405 801* DENVER BASIN COLORADO, ABST., WATTENBERG AND, SPINDLE FIELDS—
M 26 277† DENVER BASIN, COLORADO, STRATIGRAPHIC AND, SEISMIC EVIDENCE
M 24 136† DENVER BASIN, COLORADO, WATTENBERG FIELD
B 6410 1614 DENVER BASIN, CRUDE-OIL GEOCHEMISTRY
B 6410 1615 DENVER BASIN, D SAND
B 6410 1624 DENVER BASIN, DAKOTA GROUP
B 6410 1615 DENVER BASIN, FLORENCE FIELD
B 6410 1615 DENVER BASIN, FORT HAYS LIMESTONE
B 6410 1622 DENVER BASIN, GREENHORN LIMESTONE
M 26 293 DENVER BASIN, GROWTH FAULTING
B 6410 1623 DENVER BASIN, HUNTSMAN OF DRILLERS
B 6410 1617 DENVER BASIN, HYGIENE SANDSTONE
M 26 289 DENVER BASIN HYGIENE STRUCTURE CONTOUR MAP
B 6410 1615 DENVER BASIN, J SAND OIL
B 6410 1615 DENVER BASIN, LYONS FORMATION
B 6410 1615 DENVER BASIN, LYTLE FORMATION
M 24 137 DENVER BASIN, MAP, COLORADO
B 6410 1614 DENVER BASIN, NEBRASKA
B 6410 1618 DENVER BASIN, NIOBRARA FORMATION
M 26 288 DENVER BASIN, NIOBRARA STRUCTURE CONTOUR MAP
M 26 283 DENVER BASIN, NIOBRARA-PIERRE SEISMIC LINE
B 6410 1613† DENVER BASIN, PETROLEUM, GENERATION AND, MIGRATION IN
B 6410 1615 DENVER BASIN, PIERRE SHALE
M 26 290 DENVER BASIN, PIERRE SHALE STRUCTURE CONTOUR
M 26 296 DENVER BASIN, PIERRE SHALE, SUBSURFACE PRESSURE GRADIENT
M 26 286 DENVER BASIN, PIERRE SHALE VELOCITY-CONDUCTIVITY RELATION
M 26 287 DENVER BASIN, PRECAMBRIAN STRUCTURE CONTOUR MAP
B 6410 1615 DENVER BASIN, SINGLETREE FIELD
B 6410 1623 DENVER BASIN, SKULL CREEK EQUIVALENT
B 6410 1619 DENVER BASIN, SOURCE-ROCK EVALUATION
M 26 278 DENVER BASIN TECTONIC AND GEOLOGIC MAP
B 6410 1615 DENVER BASIN, TERRY SANDSTONE
M 26 279 DENVER BASIN, WEST FLANK
M 26 281 DENVER BASIN, WEST FLANK, STRATIGRAPHY
B 6409 1503 DENVER BASIN, WYOMING
B 6410 1614 DENVER BASIN, WYOMING
B 6308 1393 DENVER BASIN, WYOMING
B 6208 1500 DENVER BASIN, WYOMING
B 6008 1206 DENVER BASIN, WYOMING
B 6108 1347 DENVER BASIN, WYOMING, HANNA-, LARAMIE-
M 26 279 DENVER-ARAPAHOE, CRETACEOUS
B 6305 842* DENVER-JULESBURG BASIN, ABST., NEW LOOK AT, OIL AND, GAS,
M 24 139 DENVER-JULESBURG BASIN, COLORADO
B 6405 709* DEPARTMENT OF ENERGY 1 PLEASANT BAYOU
B 6405 732* DEPARTMENT OF ENERGY DRILLING PROGRAMS AND THE GEOSCIENCES,
B 6411 1838 DEPARTMENT OF PETROLEUM AFFAIRS, ON DEVELOPMENTS IN QATAR,
B 6409 1556 DEPAUL, GILBERT J., ENVIRONMENT OF DEPOSITION OF UPPER WILCOX
B 6206 1093 DEPETRO, LYNNE M., PALEOCURRENTS OF JACKPILE SANDSTONE, BURRO
B 6303 468* DEPLETION AND GEOCHEMICAL SIGNAL DECAY. B TITLE ONLY, POOL
B 6303 468* DEPLETION ON NEAR-SURFACE HYDROCARBON ANOMALIES, ABST.,
B 6109 1553* DEPOSITION AND DIAGENESIS OF FORT TERRETT FORMATION, EDWARDS
B 6109 1553* DEPOSITION AND DIAGENESIS OF FORTFACIES AND SEA-LEVEL CHANGES,
B 6205 892* DEPOSITION AND DIAGENESIS OF HOLOCENE REEFS, BELIZE, ABST.
B 6303 548* DEPOSITION AND DIAGENESIS OF MISSISSIPPIAN PINNACLE REEFS OF
B 6305 825* DEPOSITION AND DIAGENESIS OF ORGANIC MATTER AND CALCIUM
B 6108 1383* DEPOSITION AND DIAGENESIS OF TENSLEEP SANDSTONE, EASTERN BIG
B 6304 689* DEPOSITION AND EARLY HYDROLOGIC EVOLUTION OF WESTWATER
B 6004 734* DEPOSITION, DEFORMATION, AND HEAVY MINERAL DISTRIBUTION ON
S 6 216 DEPOSITION, EUROPE, PALEOGENE
B 6004 720* DEPOSITION, FACIES PATTERNS, AND HYDROCARBON POTENTIAL OF
M 26 273 DEPOSITION, GULF OF MEXICO, PLEISTOCENE
B 6405 713* DEPOSITION OF ENEWETA ATOLL REEF, MIDDLE PLEISTOCENE TO
B 6303 479* DEPOSITION OF MARBLE FALLS FORMATION, CENTRAL TEXAS, ABST.
B 6405 717* DEPOSITIONAL AND TECTONIC EVOLUTION OF A BASEMENT-BOUNDED,
B 6012 2178* DEPOSITIONAL AND TECTONIC HISTORY OF SAN FRANCISCO BAY
B 6008 1412* DEPOSITIONAL ENVIRONMENNSTS OF OIL SHALE, ABST.
B 6009 1606* DEPOSITIONAL ENVIRONMENT AND PETROLEUM POTENTIAL OF
B 6311 2118* DEPOSITIONAL ENVIRONMENT OF BARTLESVILLE SANDSTONE,
B 6405 753* DEPOSITIONAL ENVIRONMENT OF CLAY MINERALS FROM NORTHEAST
B 6109 1541* DEPOSITIONAL ENVIRONMENT OF GRAYSON FORMATION, UPPER
B 6405 744* DEPOSITIONAL ENVIRONMENT OF LOWER CINCINNATIAN KOPE
B 6004 699* DEPOSITIONAL ENVIRONMENT OF LOWER CRETACEOUS MUDDY

B 6108 1384* DEPOSITIONAL ENVIRONMENT OF LOWER CRETACEOUS MUDDY
B 6311 2114* DEPOSITIONAL ENVIRONMENT OF PENNSYLVANIA UPPER MARCHAND
B 6304 690* DEPOSITIONAL ENVIRONMENT OF POISON CANYON SANDSTONE IN GULF
S 5 53* DEPOSITIONAL ENVIRONMENT OF RUFF FORMATION (UPPER SILURIAN)
B 6403 427† DEPOSITIONAL ENVIRONMENT OF SOURCE BEDS OF HIGH-WAX OILS IN
B 6309 1601* DEPOSITIONAL ENVIRONMENT OF WOODBINE SANDSTONES, POLK,
B 6405 757* DEPOSITIONAL ENVIRONMENTAL ANALYSIS OF KAIBAB AND TOROWEAP
B 6405 751* DEPOSITIONAL ENVIRONMENTS AND BIOGENIC STRUCTURES,
B 6203 518* DEPOSITIONAL ENVIRONMENTS AND CONODONT BIOSTRATIGRAPHY OF
B 6303 481* **DEPOSITIONAL ENVIRONMENTS AND DIAGENESIS OF GRAND RAPIDS**
B 6205 887* DEPOSITIONAL ENVIRONMENTS AND DIAGENESIS OF MADISON
B 6405 770* DEPOSITIONAL ENVIRONMENTS AND KEROGEN TYPES. BY TITLE ONLY
B 6303 495* DEPOSITIONAL ENVIRONMENTS AND THEIR RELATION TO POROSITY IN
B 6304 689* DEPOSITIONAL ENVIRONMENTS AS ORE CONTROLS IN SALT WASH
B 6405 685* DEPOSITIONAL ENVIRONMENTS IN UPPER CAMBRIAN JORDAN
B 6103 344† DEPOSITIONAL ENVIRONMENTS OF A RESERVOIR SANDSTONE IN
S 5 23* DEPOSITIONAL ENVIRONMENTS OF BASIN-CENTER EVAPORITES, LOWER
B 6002 322* DEPOSITIONAL ENVIRONMENTS OF CHEROKE RESERVOIR SANDS,
B 6008 1408* DEPOSITIONAL ENVIRONMENTS OF EAGLE SANDSTONE, NORTH-CENTRAL
B 6302 194† DEPOSITIONAL ENVIRONMENTS OF FORT UNION SEDIMENTS, TERTIARY,
B 6209 1768* DEPOSITIONAL ENVIRONMENTS OF GULF OF MEXICO SOUTH TIMBALIER
B 6405 800* DEPOSITIONAL ENVIRONMENTS OF LACUSTRINE LIMESTONES AND THEIR
B 6203 566* DEPOSITIONAL ENVIRONMENTS OF MAXON FORMATION, LOWER
S 5 1* DEPOSITIONAL ENVIRONMENTS OF PINNACLE REEFS, NIAGARA AND
B 6409 1560* DEPOSITIONAL ENVIRONMENTS OF SOME NABORTON—DOLET HILLS
B 6004 646* DEPOSITIONAL ENVIRONMENTS OF TAMAROA SEQUENCE, UPPER
B 6408 1287* DEPOSITIONAL ENVIRONMENTS OF ULLIN LIMESTONE AND FORT PAYNE
B 6203 572* DEPOSITIONAL ENVIRONMENTS OF YALLAHS FAN DELTA,
B 6403 448* DEPOSITIONAL ENVIRONMENTS WITHIN HIGH-ENERGY TIDALLY
S 5 125* DEPOSITIONAL ENVIRONMENTS, TONOLOWAY LIMESTONE, AND
S 5 29 DEPOSITIONAL ENVIRONMENT, BASIN-CENTER EVAPORITES,
S 5 126 DEPOSITIONAL ENVIRONMENT, WEST VIRGINIA, TONOLOWAY
B 6203 538* DEPOSITIONAL FACIES AND OCCURRENCE OF POROSITY IN LOWER
B 6305 835* DEPOSITIONAL FACIES AND URANIUM OCCURRENCE TRIASSIC, DOCKUM
B 6405 675* DEPOSITIONAL FACIES, GEOMETRY, AND GENESIS OF UPPER
B 6307 1059† DEPOSITIONAL FACIES OF LOWER PALEOZOIC ALLEN BAY CARBONATE
M 26 165† DEPOSITIONAL FACIES, SEISMIC, STRATIGRAPHY AND, GLOBAL CHANGES
B 6109 1542* DEPOSITIONAL FRAMEWORK OF LOWER DOCKUM GROUP, TRIASSIC,
B 6407 1052† DEPOSITIONAL FRAMEWORK OF SANDY MID-FAN COMPLEXES OF
M 26 166 DEPOSITIONAL FRAMEWORK, REFLECTION CONFIGURATION, CHART,
B 6311 2116* DEPOSITIONAL HISTORY OF HIGH-CONSTRUCTIVE DELTA SYSTEMS,
B 6303 311† DEPOSITIONAL HISTORY OF PALEOCENE--LOWER EOCENE FLAGSTAFF
B 6203 558* DEPOSITIONAL HISTORY OF UPPER PALEOZOIC CARBONATE SUCCESSION
S 5 66 DEPOSITIONAL HISTORY, SILURIAN, SALINA GROUP, RUFF FORMATION,
S 5 93* DEPOSITION: SILURIAN REEF-BEARING ROCKS ON WABASH PLATFORM
S 5 149 DEPOSITION, CANADA, SVERDRUP BASIN, OTTO FIORD FORMATION,
B 6203 550* DEPOSITIONAL LITHOFACIES OF PARACHUTE CREEK MEMBER OF GREEN
B 6307 1108† DEPOSITIONAL MODEL OF EARLY CRETACEOUS CORAL-ALGAL-RUDIST
B 6203 512* DEPOSITIONAL MODEL OF MESOTIDAL BACK-BARRIER ENVIRONMENT,
B 6308 1425* DEPOSITIONAL MODELS AND RESOURCE POTENTIAL OF PENNSYLVANIAN
B 6405 787* DEPOSITIONAL MODELS FOR FINE-GRAINED SEDIMENT IN WESTERN
B 6105 796 * DEPOSITIONAL MODELS IN COAL EXPLORATION AND MINE PLANNING
B 6212 2379† DEPOSITIONAL MODELS IN COAL EXPLORATION AND MINE PLANNING IN
M 26 73 DEPOSITIONAL PATTERNS, HIGHSTAND OF SEA LEVEL
M 26 73 DEPOSITIONAL PATTERNS, LOWSTAND OF SEA LEVEL
M 26 439† DEPOSITIONAL PLATFORM MARGIN, EASTERN ANADARKO BASIN,
M 26 53† DEPOSITIONAL SEQUENCE AS A BASIC UNIT FOR STRATIGRAPHIC
M 26 55 DEPOSITIONAL SEQUENCE, CHRONOSTRATIGRAPHIC SIGNIFICANCE,
B 6303 514* DEPOSITIONAL SEQUENCE IN SUBARCTIC SANDY BEACH FACE, CENTRAL
M 26 53 DEPOSITIONAL SEQUENCE, SEISMIC ANALYSIS
M 26 57 DEPOSITIONAL SEQUENCES, BASELAP, ONLAP AND DOWNLAP
M 26 56 DEPOSITIONAL SEQUENCES, CONFORMITIES
M 26 57 DEPOSITIONAL SEQUENCES, HIATUSES
M 26 56 DEPOSITIONAL SEQUENCES, MAGNITUDE AND BOUNDARIES
M 26 117† DEPOSITIONAL SEQUENCES, SEISMIC, STRATIGRAPHY AND, GLOBAL
M 26 58 DEPOSITIONAL SEQUENCES, TOPLAP
M 26 56 DEPOSITIONAL SEQUENCES, UNCONFORMITIES
B 6311 2058† DEPOSITIONAL SETTING AND COAL PETROLOGY OF TULAMEEN
B 6305 840* DEPOSITIONAL SETTING OF COALS IN UPPE CRETACEOUS FERON
B 6405 718* DEPOSITIONAL SETTING OF MIDDLE DOLOMITE UNIT IN METALINE
M 26 165 DEPOSITIONAL SETTINGS, SEISMIC, INTERPRETATION
M 29 43† DEPOSITIONAL STRUCTURES OF THE SOUTHWEST ICELAND INSULAR
B 6409 1554* DEPOSITIONAL SYSTEMS AND LIGNITE PROSPECTING MODELS—WILCOX
B 6308 1426* DEPOSITIONAL SYSTEMS AND PETROLEU POTENTIAL OF LOWER
B 6203 515* DEPOSITIONAL SYSTEMS AND THEIR RELATION TO GAS ACCUMULATION
M 26 215 DEPOSITIONAL SYSTEMS, BASIN ANALYSIS
M 26 215 DEPOSITIONAL SYSTEMS, CONCEPT
M 26 224 DEPOSITIONAL SYSTEMS, DELTA MODELS
M 26 224 DEPOSITIONAL SYSTEMS, DELTAS, SEISMIC
M 26 236 DEPOSITIONAL SYSTEMS, EXAMPLES
M 26 227 DEPOSITIONAL SYSTEMS. FACIES ASSOCIATIONS. STRANDPLAIN DEPOSITS
M 26 227 DEPOSITIONAL SYSTEMS, FACIES ASSOCIATIONS, TIDAL-FLAT DEPOSITS
M 26 217 DEPOSITIONAL SYSTEMS, FORELAND BASIN, LATE PALEOZOIC
M 26 217 DEPOSITIONAL SYSTEMS, FORT WORTH BASIN, LATE PALEOZOIC
M 26 215 DEPOSITIONAL SYSTEMS, GULF BASIN
B 6002 325* DEPOSITIONAL SYSTEMS IN MATADOR UPLIFT AREA. BY TITLE ONLY
B 6409 1563* DEPOSITIONAL SYSTEMS IN NACATOCH SAND, UPPER CRETACEOUS, EAST
B 6109 1544* DEPOSITIONAL SYSTEMS IN SPARTA FORMATION, EOCENE GULF COAST
M 26 217 DEPOSITIONAL SYSTEMS, MICHIGAN BASIN, LATE PALEOZOIC
M 26 235 DEPOSITIONAL SYSTEMS, OFFLAP
M 26 235 DEPOSITIONAL SYSTEMS, ONLAP
M 26 246 DEPOSITIONAL SYSTEMS, PASSIVE BASIN-MARGINS
M 26 244 DEPOSITIONAL SYSTEMS, PASSIVE OFFLAP, MODEL
M 26 243 DEPOSITIONAL SYSTEMS, POST-RIFT CARBONATE PLATFORM, MODEL
M 26 220 DEPOSITIONAL SYSTEMS, SEISMIC STRATIGRAPHIC ANALYSIS
M 26 231 DEPOSITIONAL SYSTEMS, SHELF MODELS
M 26 231 DEPOSITIONAL SYSTEMS, SHELFS, SEISMIC
M 26 232 DEPOSITIONAL SYSTEMS, SLOPE MODELS
M 26 232 DEPOSITIONAL SYSTEMS, SLOPES, SEISMIC
M 26 59 DEPOSITIONAL SYSTEMS, TRUNCATION
M 26 219 DEPOSITIONAL SYSTEMS, WEST TEXAS BASIN
M 26 213† DEPOSITIONAL SYSTEMS: EXAMPLES FROM BRAZILIAN RIFT AND
B 6408 1282* DEPOSITIONAL TRENDS OF LOWER SILURIAN CLINTON SANDSTONE,
B 6303 429* DEPOSITIONAL-DIAGENETIC HISTORY OF MACAE CARBONATE
M 25 558† DEPOSITS OF BOLIVIAN TIN BELT--SUMMARY
B 6405 649 DEPOWSKI, RUHLE, EDWARD, REVIEW OF,
B 6203 509 DEPRATTER, CHESTER B., ARCHAEOLOGICAL EVIDENCE FOR
B 6109 1536 DEPRATTER, CHESTER B., HISTORY OF SHORELINE CHANGES BY
S 10 241† DEPTH, NEAR-, SURFACE, EVIDENCE OF, HYDROCARBON, MOVEMENT
S 10 75 DEPTH VERSUS POROSITY, COMPACTION
B 6203 524* DEPTH, ABST., NEAR-SURFACE EVIDENCE OF, HYDROCARBON MOVEMENT
M 26 18 DEPTH ASPECTS, SEISMIC, INTERPRETATION, AGE-, VELOCITY-
B 6105 790 * DEPTH CONVERSION OF MULTICHANNEL SEISMIC-REFLECTION PROFILES
B 6202 201 DEPTH DETERMINATION, PALEO-, WATER-
B 6205 813 DEPTH IN SANDSTONE RESERVOIRS, POROSITY VERSUS
B 6405 742* DEPTH MIGRATION AND INTERPRETATION OF COCORP WIND RIVER,
B 6303 527* DEPTH MIGRATION OF SEISMIC DATA, ABST.
B 6007 1125 DEPTH RELATION FOR SHALES, DENSITY VERSUS
B 6004 544 DEPTH RELATION, SHALE, POROSITY-
M 29 65† DEPTH SECTIONS AND INTERVAL VELOCITIES OVER OUTER
M 29 72 DEPTH SECTIONS, ATLANTIC COAST, CRETACEOUS
M 29 70 DEPTH SECTIONS, ATLANTIC COAST, DELAWARE
M 29 72 DEPTH SECTIONS, ATLANTIC COAST, NEW ENGLAND
M 26 370 DEPTH SOUNDER RECORD, ALASKA, LOWER, COOK INLET
M 26 355 DEPTH SOUNDER RECORD, ALASKA, YAKUTAT BAY
B 6404 554 DEPTH VERSUS POROSITY, GULF COAST SHALES
S 10 73 DEPTH-POROSITY HISTORY CURVES, COMPUTER MODELING
M 29 158 DEPTH-TO-BASEMENT CONTOURS, AUSTRALIA, MARGIN
M 29 104 DEPTH-TO-BASEMENT CONTOURS, BALTIMORE CANYON
M 29 108 DEPTH-TO-BASEMENT CONTOURS, CAROLINA PLATFORM
M 29 95 DEPTH-TO-BASEMENT CONTOURS, GEORGES BANK BASIN
M 29 99 DEPTH-TO-BASEMENT CONTOURS, LONG ISLAND PLATFORM
M 29 91 DEPTH-TO-SOURCE ESTIMATES, MAGNETIC DATA
B 6206 962 DERBYSHIRE MASSIF, ENGLAND
M 26 364 DEREVERBERATION, DIGITAL PROCESSING, DEBUBBLING
B 6107 1032 DES MOINES, MARKER IN, NEBRASKA
M 25 73† DESALINATION EXPERIENCE OF UNIVERSITY OF ARIZONA--SUMMARY,
B 6405 767 DESBOROUGH, GEORGE A., OIL AND METALS IN ORDOVICIAN AND
B 6303 528* DESCRIPTION AND SEDIMENTOLOGY O SUBMARINE-FAN GAS RESERVOIR
M 26 141 DESCRIPTION, SEISMIC, REFLECTIONS, SECTION
M 26 302 DESCRIPTION, SEISMIC TRACE, COMPLEX
S 5 58 DESCRIPTIONS, SILURIAN, SALINA GROUP, RUFF FORMATION,
B 6405 779* DESERET STARVED BASIN, MISSISSIPPIAN, WESTERN UNITED STATES,
B 6211 2350* DESERT PEAK, NEVADA, GEOTHERMAL FIELD, ABST., DISCOVERY AND
B 6306 972 DESERT ROSE UNIT, EGYPT
B 6408 1283 DESHOWITZ, MARC P., FACIES, PALEOECOLOGY, AND DEPOSITIONAL
B 6312 2215 DESICCATION FRACTURES, MUD CRACKS
B 6308 1428* DESIGN AND FUNCTION OF OIL AND GAS TRAPS, ABST.
B 6203 558* DESIGN AND FUNCTION OF OIL AND GAS TRAPS AS CLUES TO
S 10 217† DESIGN AND FUNCTION OF OIL AND GAS TRAPS
B 6012 2185* DESIGN OF HIGH-FREQUENCY-RESOLUTION AIRGUN ARRAY, ABST.
B 6309 1601* DESMOINESIAN, CRATONIC DELTA SYSTEMS, CONCHO PLATFORM OF
B 6311 2115* DESMOINESIAN CYCLIC SEDIMENTATION, ABST., MOBILE, BASIN-SHELF
B 6311 2116* DESMOINESIAN DEPOSITIONAL SYSTEMS IN KNOX-BAYLOR TROUGH,
B 6002 324* DESMOINESIAN, OF NORTH-CENTRAL TEXAS, ABST., UPPER, STRAWN
B 6105 819* DESMOINESIAN, PENNSYLVANIA, AGE IN ILLINOIS, ABST.,
B 6002 322* DESMOINESIAN, PENNSYLVANIAN, SANDSTONE RESERVOIR AT SLEEPY
B 6107 1029† DESMOINESIAN, PENNSYLVANIAN, SANDSTONE RESERVOIR, SLEEPY
B 6311 2117* DESMOINESIAN ROCKS IN NORTHEASTERN OKLAHOMA FOR HEAVY-OIL

B 6311 2120* DESMOINESIAN-MISSOURIAN, PENNSYLVANIAN SERIES BOUNDARY IN
S 7 213 DESOTO CANYON AREA, WEST FLORIDA SLOPE
S 7 216 DESOTO CANYON, DEVELOPMENT STAGES
S 7 215 DESOTO CANYON, SEISMIC PROFILE
S 7 199 DESOTO CANYON, SEISMIC PROFILE
M 29 391 DESOTO DEEP SEA CHANNEL, SEISMIC PROFILE, MISSISSIPPI FAN
M 29 389 DESOTO DEEP-SEA CHANNEL, GULF OF MEXICO, MISSISSIPPI FAN
S 7 186 DESOTO DEEP-SEA CHANNEL, HIGH-RESOLUTION PROFILES
S 7 13 DESOTO SLOPE, DESCRIBED
B 6411 1785 DESTEFANO, E. R., PETROLEUM DEVELOPMENTS IN CENTRAL AND
B 6304 671 DESTIN DOME, FLORIDA
B 6203 548* DESULFURIZATION, BY HIGH-INTENSITY MAGNETIC SEPARATION, ABST.,
B 6307 1043 DETACHED NORMAL-FAULT ASSEMBLAGES
M 26 3† DETAIL DERIVABLE FROM THEM, LIMITATIONS ON, RESOLUTION OF,
B 6303 440* DETAILED STRATIGRAPHY AND SEDIMENTOLOGY OF WHITSETT
B 6303 484* DETAILED TEMPERATURE LOGGING AS USEFUL TOOL FOR LITHOLOGIC
B 6109 1560* DETECTION OF PETROLEUM-RELATED SOIL ANOMALIES FROM LANDSAT,
M 26 301 DETECTION, WAVES AS SIGNALS, SEISMIC
M 26 301† DETERMINATION, APPLICATION OF, AMPLITUDE, FREQUENCY, AND
B 6303 447* DETERMINATION OF DEPTH TO HYDROCARBON MATURATION
B 6007 1073 DETERMINATION OF GEOTHERMAL GRADIENTS
B 6309 1504† DETERMINATION OF ORGANIC CONTENT OF APPALACHIAN DEVONIAN
B 6203 494* DETERMINATION OF RESERVOIR FLUID CONTENT BY SEISMIC
B 6303 492* DETERMINATION OF STORM OVERWASH PERIODICITY FROM
M 28 1 DETRITAL GRAINS, QUARTZ, GENERAL
M 25 411 DETRITAL ORE, OR CANGA, SCREE ORE, CONGLOMERATE ORE
S 5 77 DETROIT RIVER GROUP, AMHERSTBURG FORMATION, DEVONIAN,
S 5 73* DETROIT RIVER GROUP, DEVONIAN, MIDLAND, MICHIGAN, EVAPORITE
S 5 79 DETROIT RIVER GROUP, LUCAS FORMATION, STRATIGRAPHY,
B 6104 607 DETTERMAN, R. L., SOME LOWER AND MIDDLE JURASSI BEDS IN PUALE
S 3 135 DEUTERIC CRYSTALLIZATION OF THORIUM-URANIUM MINERALS
B 6411 1840 DEUTSCHE SCHACHTBAU AND TIEFBOHRGESELLSCHAFT MBH, ON
M 25 351† DEVELOPING DIKE-COMPLEX AND BASAL-LENS WATER SUPPLIES ON
B 6004 667* DEVELOPMENT AND PALEOGEOGRAPHIC EXTENT O BUILDUPS AND
B 6207 1207* DEVELOPMENT AND RESERVOIR ANALYSIS OF MOMOTOMBO
B 6303 458* DEVELOPMENT AND USE OF FORMATION-TOPS FIL SYSTEM, ABST.
M 30 167 DEVELOPMENT DRILLING, MORAY FIRTH BASIN, PIPER FIELD
B 6008 1370† DEVELOPMENT IN ARKANSAS, NORTH LOUISIANA, AND EAST TEXAS IN
B 6308 1376† DEVELOPMENT IN ARKANSAS, NORTH LOUISIANA, AND EAST TEXAS IN
B 6310 1689† DEVELOPMENT IN CENTRAL AND SOUTHERN AFRICA IN 1978,
B 6409 1462† DEVELOPMENT IN EAST-CENTRAL STATES IN 1979
B 6008 1204† DEVELOPMENT IN NORTHERN ROCKIES IN 1975, OIL AND, GAS
M 25 353† DEVELOPMENT OF AUSTRALIA GROUNDWATER RESOURCES--SUMMARY
B 6405 773* DEVELOPMENT OF BIOGENIC GAS FROM SHALLOW, LOW-PERMEABILITY
B 6212 2508* DEVELOPMENT OF CERRO PRIETO GEOTHERMAL FIELD, BAJA
B 6405 745* DEVELOPMENT OF CONCEPTUAL MODEL TO CHARACTERIZE
B 6105 800 * DEVELOPMENT OF CONTINENTAL-SLOPE BASINS ALONG CONVERGENCE ZONE
B 6004 688* DEVELOPMENT OF EARLY CRETACEOUS BANK COMPLEX EDWARDS
B 6105 759 * DEVELOPMENT C GEOPHYSICAL TECHNIQUES FOR URANIUM ASSESSMENT
B 6203 526* DEVELOPMENT OF MILLS RANCH COMPLEX, TEXAS AND OKLAHOMA,
B 6303 497* DEVELOPMENT OF PETROLEUM RESOURCE APPRAISAL METHODS IN U.S.
B 6004 720* DEVELOPMENT OF POROSITY IN CARBONATE ROCKS OF EVAPORITIC
B 6405 737* DEVELOPMENT OF SECONDARY POROSITY IN RESERVOIR SANDSTONES B
M 29 27† DEVELOPMENT OF THE SOUTHEAST GEORGIA EMBAYMENT AND
B 6004 697* DEVELOPMENT OF VERTICAL ESCARPMENT AT MIDDLE CAMBRIAN
B 6010 1637† DEVELOPMENT PAPERS, 1975, FOREIGN
B 6110 1575† DEVELOPMENT PAPERS, 1976, FOREIGN
B 6210 1793† DEVELOPMENT PAPERS, 1977, FOREIGN
B 6310 1632† DEVELOPMENT PAPERS, 1978, FOREIGN
B 6411 1736† DEVELOPMENT PAPERS, 1979, FOREIGN
B 6008 1159 DEVELOPMENT WELL
B 6308 1174 DEVELOPMENT WELL
B 6208 1281 DEVELOPMENT WELL
B 6108 1127 DEVELOPMENT WELL
B 6409 1301 DEVELOPMENT WELL
S 5 6 DEVELOPMENT, PINNACLE REEF
M 25 118† DEVELOPMENT--SUMMARY, ENVIRONMENTAL EFFECTS OF,
B 6303 421* DEVELOPMENTAL PHASES IN LAGOONAL PATCH REEFS--IMPLICATIONS
B 6110 1658 DEVELOPMENTS, 1975-1976, GERMAN FEDERAL REPUBLIC
B 6108 1186 DEVELOPMENTS, 1976, INDIANA
B 6308 1206 DEVELOPMENTS, 1978, COLORADO, SOUTHWESTERN
B 6308 1206 DEVELOPMENTS, 1978, NEW MEXICO, WESTERN
B 6008 1189† DEVELOPMENTS IN ALASKA IN 1975
B 6108 1157† DEVELOPMENTS IN ALASKA IN 1976
B 6208 1311† DEVELOPMENTS IN ALASKA IN 1977
B 6308 1211† DEVELOPMENTS IN ALASKA IN 1978
B 6409 1345† DEVELOPMENTS IN ALASKA IN 1979
B 6108 1234† DEVELOPMENTS IN ARKANSAS, LOUISIANA, AND EAST TEXAS IN 1976
B 6409 1490† DEVELOPMENTS IN ARKANSAS, NORTH LOUISIANA, AND EAST TEXAS IN
B 6208 1487† DEVELOPMENTS IN ARKANSAS, NORTH LOUISIANA, AND EAST TEXAS IN
B 6409 1331† DEVELOPMENTS IN ATLANTIC COASTAL PLAIN IN 1979
B 6308 1204† DEVELOPMENTS IN ATLANTIC COASTAL PLAIN IN 1978
B 6409 1336 DEVELOPMENTS IN ATLANTIC COASTAL PLAIN IN 1978—ADDENDUM
B 6409 1336† DEVELOPMENTS IN ATLANTIC OC IN 1978—ADDENDUM
B 6010 1957† DEVELOPMENTS IN AUSTRALIA IN 1975, PETROLEUM
B 6110 1888† DEVELOPMENTS IN AUSTRALIA IN 1976, PETROLEUM
B 6210 2150† DEVELOPMENTS IN AUSTRALIA IN 1977, PETROLEUM
B 6310 1804† DEVELOPMENTS IN AUSTRALIA IN 1978, PETROLEUM
B 6411 1862† DEVELOPMENTS IN AUSTRALIA IN 1979, PETROLEUM
B 6010 1813† DEVELOPMENTS IN CENTRAL AND SOUTHERN AFRICA IN 1975,
B 6110 1746† DEVELOPMENTS IN CENTRAL AND SOUTHERN AFRICA IN 1976,
B 6210 1844† DEVELOPMENTS IN CENTRAL AND SOUTHERN AFRICA IN 1977,
B 6411 1785† DEVELOPMENTS IN CENTRAL AND SOUTHERN AFRICA IN 1979,
B 6008 1222† DEVELOPMENTS IN EAST-CENTRAL STATES IN 1975
B 6108 1183† DEVELOPMENTS IN EAST-CENTRAL STATES IN 1976
B 6208 1340† DEVELOPMENTS IN EAST-CENTRAL STATES IN 1977
B 6308 1324† DEVELOPMENTS IN EAST-CENTRAL STATES IN 1978
B 6008 1234† DEVELOPMENTS IN EASTERN AND NORTHWESTERN COLORADO
B 6108 1199† DEVELOPMENTS IN EASTERN AND NORTHWESTERN COLORADO IN 1976
B 6208 1355† DEVELOPMENTS IN EASTERN AND NORTHWESTERN COLORADO IN 1977
B 6308 1371† DEVELOPMENTS IN EASTERN AND NORTHWESTERN COLORADO IN 1978
B 6409 1484† DEVELOPMENTS IN EASTERN AND NORTHWESTERN COLORADO IN 1979
B 6008 1355† DEVELOPMENTS IN EASTERN CANADA IN 1975
B 6108 1335† DEVELOPMENTS IN EASTERN CANADA IN 1976
B 6208 1451† DEVELOPMENTS IN EASTERN CANADA IN 1977
B 6308 1344† DEVELOPMENTS IN EASTERN CANADA IN 1978
B 6409 1518† DEVELOPMENTS IN EASTERN CANADA IN 1979
B 6411 1880† DEVELOPMENTS IN FAR EAST, 1979, PETROLEUM
B 6010 1908† DEVELOPMENTS IN FAR EAST IN 1975, PETROLEUM
B 6110 1832† DEVELOPMENTS IN FAR EAST IN 1976, PETROLEUM
B 6210 1978† DEVELOPMENTS IN FAR EAST IN 1977, PETROLEUM
B 6310 1816† DEVELOPMENTS IN FAR EAST IN 1978, PETROLEUM
B 6008 1239† DEVELOPMENTS IN FOUR CORNERS--INTERMOUNTAIN AREA IN 1975
B 6308 1206† DEVELOPMENTS IN FOUR CORNERS--INTERMOUNTAIN AREA IN 1978
B 6108 1204† DEVELOPMENTS IN FOUR CORNERS-INTERMOUNTAIN AREA IN 1976
B 6208 1360† DEVELOPMENTS IN FOUR CORNERS—INTERMOUNTAIN AREA IN 1977
B 6409 1340† DEVELOPMENTS IN FOUR CORNERS—INTERMOUNTAIN AREA IN 1979
B 6108 1326† DEVELOPMENTS IN GREENLAND IN 1976
B 6010 1905† DEVELOPMENTS IN ISRAEL IN 1975, PETROLEUM
B 6110 1830† DEVELOPMENTS IN ISRAEL IN 1976, PETROLEUM
B 6210 1898† DEVELOPMENTS IN ISRAEL IN 1977, PETROLEUM
B 6310 1743† DEVELOPMENTS IN ISRAEL IN 1978, PETROLEUM
B 6008 1265† DEVELOPMENTS IN LOUISIAN GULF COAST IN 1975
B 6108 1244† DEVELOPMENTS IN LOUISIANA GULF COAST IN 1976
B 6208 1519† DEVELOPMENTS IN LOUISIANA GULF COAST OFFSHORE IN 1977
B 6308 1310† DEVELOPMENTS IN LOUISIANA GULF COAST OFFSHORE IN 1978
B 6409 1445† DEVELOPMENTS IN LOUISIANA GULF COAST OFFSHORE IN 1979
B 6208 1462† DEVELOPMENTS IN LOUISIANA GULF COAST ONSHORE IN 1977
B 6308 1357† DEVELOPMENTS IN LOUISIANA GULF COAST ONSHORE IN 1978
B 6411 2012† DEVELOPMENTS IN LOUISIANA GULF COAST ONSHORE IN 1979
B 6308 1244† DEVELOPMENTS IN MARYLAND, OHIO, PENNSYLVANIA, VIRGINIA, AND
B 6208 1399† DEVELOPMENTS IN MARYLAND, OHIO, PENNSYLVANIA, VIRGINIA, AND
B 6409 1403† DEVELOPMENTS IN MARYLAND, OHIO, PENNSYLVANIA, VIRGINIA, AND
B 6108 1269† DEVELOPMENTS IN MARYLAND, OHIO, PENNSYLVANIA, VIRGINIA, AND
B 6008 1288† DEVELOPMENTS IN MARYLAND, OHIO, PENNSYLVANIA, VIRGINIA, AND
B 6008 1215† DEVELOPMENTS IN MICHIGA IN 1975
B 6108 1173† DEVELOPMENTS IN MICHIGAN IN 1976
B 6208 1330† DEVELOPMENTS IN MICHIGAN IN 1977
B 6308 1278† DEVELOPMENTS IN MICHIGAN IN 1978
B 6409 1377† DEVELOPMENTS IN MICHIGAN IN 1979
B 6010 1865 † DEVELOPMENTS IN MIDDLE EAST COUNTRIES IN 1975, PETROLEUM
B 6110 1795† DEVELOPMENTS IN MIDDLE EAST COUNTRIES IN 1976, PETROLEUM
B 6210 2111† DEVELOPMENTS IN MIDDLE EAST COUNTRIES IN 1977, PETROLEUM
B 6310 1884† DEVELOPMENTS IN MIDDLE EAST COUNTRIES IN 1978, PETROLEUM
B 6411 1836† DEVELOPMENTS IN MIDDLE EAST COUNTRIES IN 1979, PETROLEUM
B 6008 1324† DEVELOPMENTS IN NEW YORK IN 1975
B 6108 1328† DEVELOPMENTS IN NEW YORK IN 1976
B 6208 1445† DEVELOPMENTS IN NEW YORK IN 1977
B 6308 1318† DEVELOPMENTS IN NEW YORK IN 1978
B 6409 1455† DEVELOPMENTS IN NEW YORK IN 1979
B 6310 1680† DEVELOPMENTS IN NEW ZEALAND AND SOUTHWEST PACIFIC ISLAND
B 6210 1900† DEVELOPMENTS IN NEW ZEALAND AND SOUTHWEST PACIFIC ISLAND
B 6110 1880† DEVELOPMENTS IN NEW ZEALAND AND SOUTHWEST PACIFIC ISLAND
B 6010 1947† DEVELOPMENTS IN NEW ZEALAND AND SOUTHWEST PACIFIC ISLAND
B 6411 1776† DEVELOPMENTS IN NEW ZEALAND AND SOUTHWEST PACIFIC ISLAND
B 6010 1767† DEVELOPMENTS IN NORTH AFRICA IN 1975, PETROLEUM
B 6110 1698† DEVELOPMENTS IN NORTH AFRICA IN 1976, PETROLEUM
B 6210 1796† DEVELOPMENTS IN NORTH AFRICA IN 1977, PETROLEUM
B 6411 1739† DEVELOPMENTS IN NORTH AFRICA IN 1979, PETROLEUM
B 6310 1636† DEVELOPMENTS IN NORTH AFRICA, PETROLEUM

B 6108 1305† DEVELOPMENTS IN NORTH MID-CONTINENT IN 1976, OIL AND, GAS
B 6208 1477† DEVELOPMENTS IN NORTH MID-CONTINENT IN 1977, OIL AND, GAS
B 6308 1301† DEVELOPMENTS IN NORTH MID-CONTINENT IN 1978, OIL AND, GAS
B 6409 1437† DEVELOPMENTS IN NORTH MID-CONTINENT IN 1979, OIL AND, GAS
B 6008 1340† DEVELOPMENTS IN NORTH-CENTRAL TEXAS IN 1975
B 6108 1220† DEVELOPMENTS IN NORTH-CENTRAL TEXAS IN 1976
B 6208 1377† DEVELOPMENTS IN NORTH-CENTRAL TEXAS IN 1977
B 6308 1233† DEVELOPMENTS IN NORTH-CENTRAL TEXAS IN 1978
B 6409 1366† DEVELOPMENTS IN NORTH-CENTRAL TEXAS IN 1979
B 6108 1346† DEVELOPMENTS IN NORTHERN ROCKIES IN 1976, OIL AND, GAS
B 6308 1386† DEVELOPMENTS IN NORTHERN ROCKIES IN 1978
B 6409 1501† DEVELOPMENTS IN NORTHERN ROCKIES IN 1979
B 6108 1216† DEVELOPMENTS IN OKLAHOMA AND PANHANDLE OF TEXAS
B 6008 1251† DEVELOPMENTS IN OKLAHOMA AND PANHANDLE OF TEXAS IN 1975
B 6208 1372† DEVELOPMENTS IN OKLAHOMA AND PANHANDLE OF TEXAS IN 1977
B 6308 1286† DEVELOPMENTS IN OKLAHOMA AND PANHANDLE OF TEXAS IN 1978
B 6409 1387† DEVELOPMENTS IN OKLAHOMA AND PANHANDLE OF TEXAS IN 1979
B 6010 1640† DEVELOPMENTS IN SOUTH AMERICA, CENTRAL AMERICA, AND
B 6310 1745† DEVELOPMENTS IN SOUTH AMERICA, CENTRAL AMERICA, MEXICO, AND
B 6210 2044† DEVELOPMENTS IN SOUTH AMERICA, CENTRAL AMERICA, MEXICO, AND
B 6110 1578† DEVELOPMENTS IN SOUTH AMERICA, CENTRAL AMERICA, MEXICO, AND
B 6008 1365† **DEVELOPMENTS IN SOUTH TEXAS IN 1975**
B 6108 1356† DEVELOPMENTS IN SOUTH TEXAS IN 1976
B 6208 1511† DEVELOPMENTS IN SOUTH TEXAS IN 1977
B 6308 1338† DEVELOPMENTS IN SOUTH TEXAS IN 1978
B 6409 1476† DEVELOPMENTS IN SOUTH TEXAS IN 1979
B 6002 321* DEVELOPMENTS IN SOUTH-CENTRAL OKLAHOMA, ABST.
B 6411 2026† DEVELOPMENTS IN SOUTHEASTERN STATES
B 6008 1277† DEVELOPMENTS IN SOUTHEASTERN STATES IN 1975
B 6108 1259† DEVELOPMENTS IN SOUTHEASTERN STATES IN 1976
B 6208 1390† DEVELOPMENTS IN SOUTHEASTERN STATES IN 1977
B 6308 1399† DEVELOPMENTS IN SOUTHEASTERN STATES IN 1978
B 6008 1254† DEVELOPMENTS IN UPPER GULF COAST OF TEXAS IN 1975
B 6108 1223† DEVELOPMENTS IN UPPER GULF COAST OF TEXAS IN 1976
B 6208 1380† DEVELOPMENTS IN UPPER GULF COAST OF TEXAS IN 1977
B 6308 1223† DEVELOPMENTS IN UPPER GULF COAST OF TEXAS IN 1978
B 6409 1354† DEVELOPMENTS IN UPPER GULF COAST OF TEXAS IN 1979
B 6008 1196† DEVELOPMENTS IN WEST COAST AREA IN 1975
B 6108 1166† DEVELOPMENTS IN WEST COAST AREA IN 1976
B 6208 1322† DEVELOPMENTS IN WEST COAST AREA IN 1977
B 6308 1292† DEVELOPMENTS IN WEST COAST AREA IN 1978
B 6409 1393† DEVELOPMENTS IN WEST COAST AREA IN 1979
B 6008 1329† DEVELOPMENTS IN WEST GREENLAND IN 1975
B 6008 1243† DEVELOPMENTS IN WEST TEXAS AND EASTERN NEW MEXICO IN 1975
B 6108 1208† DEVELOPMENTS IN WEST TEXAS AND EASTERN NEW MEXICO IN 1976
B 6208 1364† DEVELOPMENTS IN WEST TEXAS AND EASTERN NEW MEXICO IN 1977
B 6308 1236† **DEVELOPMENTS IN WEST TEXAS AND EASTERN NEW MEXICO IN 1978**
B 6409 1369† DEVELOPMENTS IN WEST TEXAS AND EASTERN NEW MEXICO IN 1979
B 6008 1343† DEVELOPMENTS IN WESTERN CANADA IN 1975
B 6108 1313† DEVELOPMENTS IN WESTERN CANADA IN 1976
B 6308 1407† DEVELOPMENTS IN WESTERN CANADA IN 1978
B 6409 1510† DEVELOPMENTS IN WESTERN CANADA IN 1979
B 6208 1441† DEVELOPMENTS OFFSHORE WEST GREENLAND IN 1977
B 6308 1210† DEVELOPMENTS OFFSHORE WEST GREENLAND IN 1978
B 6008 1323† DEVELOPMENTS ON ATLANTIC COASTAL PLAIN BETWEEN NEW JERSEY
B 6009 1619 DEVER, G. R., JR., ENERGY RESOURCE IMPLICATIONS OF 38TH PARALLEL
S 6 316 DEVIANT RESULTS, RADIOMETRIC DATING, PERMIAN
B 6405 707* DEVIATED BORE HOLES, ABST., DIPMETER VALIDITY IN
B 6205 845† DEVIATED WELLS, GRAPHIC DETERMINATION OF, STRATIGRAPHIC AND,
B 6008 1257 DEVILLIER FIELD, CHAMBERS COUNTY, TEXAS
B 6402 222 DEVILS LAKE QUADRANGLE, TEXAS
S 4 87 DEVILS RACE COURSE FORMATION, BARREMIAN-APTIAN, JAMAICA
B 6402 230 DEVILS RIVER FORMATION, TEXAS
B 6402 221† DEVILS RIVER UPLIFT--SOUTHERN VAL VERDE BASIN, SOUTHWEST
B 6312 2218 DEVILS TOWER, COLUMNAR JOINTING, WYOMING
B 6206 1005 DEVON ISLAND, ARCTIC, CANADA
B 6305 846* **DEVONIAN EMBAYMENT INTO APPALACHIA, ABST., SILURIAN-**
B 6408 1285† **DEVONIAN OIL SHALE OF EASTERN UNITED STATES,**
S 6 337† DEVONIAN
S 6 338 DEVONIAN, ABSOLUTE DATING, AUSTRALIA
S 6 338 DEVONIAN, ABSOLUTE DATING, NORTH AMERICA
B 6002 246 DEVONIAN, ALBERTA, REEFS
M 25 341 DEVONIAN, ANAZON BASINS
B 6002 278† DEVONIAN, AND ADJACENT STRATA ALONG SOUTH BORDER OF
B 6004 646* DEVONIAN AND MISSISSIPPIAN, PEDREGOSA BASIN, SOUTHEASTERN
B 6012 2103† DEVONIAN, ARKANSAS AND OKLAHOMA, NONGLACIAL, VARVES IN
M 27 206 DEVONIAN, BECRAFT LIMESTONE, NEW YORK
M 28 166 DEVONIAN, BECRAFT LIMESTONE, NEW YORK
M 27 77 DEVONIAN, BECRAFT LIMESTONE, NEW YORK
B 6304 655† DEVONIAN BENTONITES IN EASTERN OHIO
S 6 337 DEVONIAN, BIOSTRATIGRAPHIC DATA, A REVIEW
B 6405 720* DEVONIAN BLACK SHALE, ABST., AUTHIGENIC QUARTZ IN
B 6405 729* DEVONIAN BLACK SHALE IN EAST-CENTRAL KENTUCKY, ABST., TRACE,
B 6304 656 DEVONIAN BLACK SHALE, OHIO
B 6405 692* DEVONIAN BLACK SHALES, APPALACHIAN BASIN, ABST., GEOCHEMICAL
B 6309 1592* DEVONIAN BLACK SHALES, ERIE COUNTY PENNSYLVANIA, ABST.,
S 6 324 DEVONIAN BOUNDARY, CARBONIFEROUS, ISOTOPIC DATING
S 6 324 DEVONIAN BOUNDARY, TABLE, RADIOMETRIC DATING
B 6203 492* DEVONIAN BRALLIER FORMATION OF WEST VIRGINIA AND VIRGINIA,
M 28 154 DEVONIAN, CAIRN FORMATION, CANADA, ALBERTA
M 28 10 DEVONIAN, CAIRN FORMATION, CANADA, ALBERTA
M 28 153 DEVONIAN, CAIRN FORMATION, CANADA, ALBERTA
M 28 161 DEVONIAN, CAIRN FORMATION, CANADA, ALBERTA
B 6101 9 DEVONIAN, CANADIAN, ARCTIC ISLANDS, ORDOVICIAN TO
S 9 22 DEVONIAN, CARBONATE REEF BELT DISTRIBUTION
B 6209 1609 DEVONIAN CARBONATE SEQUENCE, APPALACHIAN BASIN, SILURIAN-
B 6209 1609 DEVONIAN CARBONATE SEQUENCE, KENTUCKY, SILURIAN-
B 6209 1609 DEVONIAN CARBONATE SEQUENCE, MARYLAND, SILURIAN-
B 6209 1610 DEVONIAN CARBONATE SEQUENCE, NOMENCLATURE, MARYLAND,
B 6209 1610 DEVONIAN CARBONATE SEQUENCE, NOMENCLATURE, NEW YORK,
B 6209 1610 DEVONIAN CARBONATE SEQUENCE, NOMENCLATURE, OHIO, SILURIAN-
B 6209 1610 DEVONIAN CARBONATE SEQUENCE, NOMENCLATURE, PENNSYLVANIA,
B 6209 1609 DEVONIAN CARBONATE SEQUENCE, VIRGINIA, SILURIAN-
M 28 114 DEVONIAN, CEDAR VALLEY LIMESTONE, ILLINOIS
B 6303 501* DEVONIAN CLASTICS IN SOUTHERN WEST VIRGINIA, ABST., SUBSURFACE
B 6405 780* DEVONIAN CONODONTS IN PETROLEUM EXPLORATION, WESTERN
B 6204 672 DEVONIAN, D UNIT, ALGERIA
B 6303 478* DEVONIAN DOLOMITIZATION IN WILLISTON BASIN, ABST., ORIGIN OF
S 5 77 DEVONIAN, DETROIT RIVER GROUP, AMHERSTBURG FORMATION
S 5 79 DEVONIAN, DETROIT RIVER GROUP, LUCAS FORMATION, STRATIGRAPHY
S 10 114 DEVONIAN, DISTRIBUTION OF LEAD-ZINC, CANADA
B 6105 814* DEVONIAN, EAST-CENTRAL NEW YORK, ABST., BIOGENIC STRUCTURES
B 6009 1618* DEVONIAN EVENTS IN SOUTHWESTERN INDIANA, ABST., GENERAL
B 6105 816* DEVONIAN EXAMPLES FROM ALBERTA, ABST., IMPORTANCE OF
B 6204 672 DEVONIAN, F UNIT, ALGERIA
S 9 23 DEVONIAN, FOLDOUT, PALEOCLIMATES
B 6101 9 DEVONIAN, FRANKLINIAN BASIN
B 6101 8 DEVONIAN, FRANKLINIAN BASIN, ORDOVICIAN TO
M 27 42 DEVONIAN, GENUNDEWA LIMESTONE, NEW YORK
M 28 8 DEVONIAN, GNEISS, MASSACHUSETTS
B 6406 814† DEVONIAN GREAT BARRIER REEF OF CANNING BASIN, WESTERN
M 28 114 DEVONIAN, HOING SANDSTONE, ILLINOIS
B 6305 828* DEVONIAN HORNELEN BASIN, NORWAY, ABST., ALLUVIAL, FANS AND
B 6309 1589* DEVONIAN, IN EAST-CENTRAL KENTUCKY, ABST., PETROLOGY AND
B 6309 1591* DEVONIAN, IN SOUTH-CENTRAL NEW YORK STATE, ABST., STRUCTURAL
B 6401 128 DEVONIAN, INDIANA
B 6403 401 DEVONIAN, INDIANA
B 6401 128 DEVONIAN, INDIANA
B 6403 401 DEVONIAN, INDIANA
M 27 5 DEVONIAN, JEAN MARIE FORMATION, CANADA (N.W.T.)
B 6405 758* DEVONIAN KANAYUT CONGLOMERATE, BROOKS RANGE, ALASKA, ABST.,
M 27 135 DEVONIAN, KEG RIVER EQUIVALENT, CANADA
B 6405 767* DEVONIAN KEROGENOUS MARINE STRATA OF CENTRAL NEVADA, ABST.,
M 27 55 DEVONIAN LIMESTONE, IOWA
M 27 62 DEVONIAN LIMESTONES, IOWA
B 6407 992 DEVONIAN LIMESTONES, MOROCCO
M 28 51 DEVONIAN, META-DIABASE, CONNECTICUT
M 28 52 DEVONIAN, META-DIABASE, CONNECTICUT
B 6303 492* DEVONIAN MIETTE BUILDUP, ALBERTA, ABST., BURIAL,
S 5 73* DEVONIAN, MIDLAND, MICHIGAN, EVAPORITE CYCLES AND LITHOFACIES
B 6004 704* DEVONIAN, NORTHEAST OHIO, ABST., TRACE, FOSSILS OF, CHAGRIN
B 6203 498* DEVONIAN OF MIGHICAN, PENNSYLVANIAN OF OKLAHOMA, AND
B 6203 530* DEVONIAN OF SASKATCHEWAN, ABST., CARBONATE-, EVAPORITE
B 6205 885* DEVONIAN OF SASKATCHEWAN—REINTERPRETATION, ABST.,
B 6309 1590* DEVONIAN OF VIRGINIA AN WEST VIRGINIA, ABST., PETROLOGIC
B 6309 1576* DEVONIAN, OHIO, SHALE IN NORTHEASTERN OHIO, ABST., FRACTURE
M 28 19 DEVONIAN, OLD RED SANDSTONE, NORTHERN, IRELAND
B 6303 462* DEVONIAN ORGANIC-RIC BLACK SHALES IN SUBSURFACE
M 28 65 DEVONIAN, ORISKANY SANDSTONE, VIRGINIA
M 28 97 DEVONIAN, ORISKANY SANDSTONE, VIRGINIA
M 28 165 DEVONIAN, ORISKANY SANDSTONE, VIRGINIA
B 6405 676* DEVONIAN PALEOGEOGRAPHY AND PALEOENVIRONMENTS IN
B 6405 694* DEVONIAN PRODELTAIC TURBIDITE DEPOSIT, ABST.,
S 6 337 DEVONIAN, RADIOMETRIC DATA, A REVIEW
S 6 338 DEVONIAN, RADIOMETRIC DATING, NEED FOR MORE
S 6 339 DEVONIAN, RADIOMETRIC SCALE AGAINST ORGANIC
B 6209 1607† DEVONIAN REEF TRENDS, CENTRAL APPALACHIAN BASIN,
B 6209 1638 DEVONIAN REEFS, APPALACHIAN BASIN
B 6405 675* DEVONIAN REEFS EXPOSED ALONG CENTRAL CANTABRIC COAST,
S 6 338 DEVONIAN, RELATIVE PALEONTOLOGIC UNITS, INDEX

M 30 315 DEVONIAN, RICINUS WEST, ALBERTA
B 6303 551* DEVONIAN ROCKS IN APPALACHIAN BASIN, ABST., CORRELATION OF
B 6303 542* DEVONIAN SANDSTONES OF UPPER MISSISSIPPI VALLEY, ABST.,
B 6209 1612 DEVONIAN SECTION, NEW YORK, SILURIAN-
B 6209 1622 DEVONIAN SECTION, NEW YORK TO, OHIO, SILURIAN-
B 6209 1618 DEVONIAN SECTION, NEW YORK TO, PENNSYLVANIA, SILURIAN-
B 6209 1624 DEVONIAN SECTION, OHIO TO, PENNSYLVANIA, SILURIAN-
B 6209 1626 DEVONIAN SECTION, OHIO TO, WEST VIRGINIA, SILURIAN-
B 6309 1580* DEVONIAN SHALE, ABST., UNCONVENTIONAL EVALUATION OF,
B 6309 1578* DEVONIAN SHALE AND ITS SENSITIVITY TO CHANGE IN VARIOUS
B 6309 1584* DEVONIAN SHALE CHARACTERIZATION--EVALUATION OF SOME
B 6408 1283* DEVONIAN SHALE EXPLORATION RATIONALE TESTED IN NORTHEAST
B 6412 2163 DEVONIAN SHALE FACIES, KENTUCKY
B 6412 2163 DEVONIAN SHALE FACIES, NEW YORK
B 6412 2163 DEVONIAN SHALE FACIES, OHIO
B 6412 2163 DEVONIAN SHALE FACIES, PENNSYLVANIA
B 6412 2163 DEVONIAN SHALE FACIES, WEST VIRGINIA
B 6412 2163 DEVONIAN SHALE FACIES, WEST VIRGINIA
B 6305 847* DEVONIAN SHALE GAS PRODUCTION, PHOTOLINEAMENTS, AND
B 6309 1581* DEVONIAN SHALE GAS-RESERVOIR PERFORMANCE, ABST., MODELING OF
B 6412 2156† DEVONIAN SHALE IN WESTERN APPALACHIAN BASIN, ORGANIC
B 6309 1581* DEVONIAN SHALES, ABST., COMPOSITIONAL VARIATION IN
B 6309 1577* DEVONIAN SHALES, ABST., EXPLORATION STRATEGY FOR
B 6309 1576* DEVONIAN SHALES, ABST., SUGGESTED ORIGIN FOR, GAS IN EASTERN
B 6412 2111 DEVONIAN SHALES, ALGERIA
B 6309 1504† DEVONIAN SHALES FROM FORMATION-DENSITY LOGS, DETERMINATION
B 6309 1580* DEVONIAN SHALES IN SOUTHERN WEST VIRGINIA, ABST., ELEMENTAL
B 6408 1285* DEVONIAN SHALES OF KENTUCKY AND WEST VIRGINIA—STATISTICAL
B 6405 692* DEVONIAN SHALES—REGIONAL ORGANIC GEOCHEMISTRY AND
B 6203 517* DEVONIAN SLOPE SEDIMENTS IN CENTRAL APPALACHIAN BASIN—
M 30 315 DEVONIAN, STRACHAN FIELD, ALBERTA
M 30 317 DEVONIAN, STRACHAN-RICINUS WEST FIELDS, REEF HORIZONS
B 6309 1577* DEVONIAN STRATIGRAPHY IN NORTHWESTERN WEST VIRGINIA, ABST.,
B 6004 684* DEVONIAN STRATIGRAPHY OF BOLIVIA, ABST.
B 6105 839* DEVONIAN STROMATOPOROID REEF COMPLEX, ABST., BIOFACIES
B 6309 1510† DEVONIAN SUBSURFACE UNITS IN PERMIAN BASIN, WEST TEXAS,
M 27 42 DEVONIAN, TENTACULITEN KNOTTENKALK, GERMANY
B 6408 1288* DEVONIAN TIDE-DOMINATED DELTAIC-INTRADELTAIC SEDIMENTATION
B 6102 257 DEVONIAN TO PERMIAN, NOMENCLATURE, IDAHO
B 6303 488* DEVONIAN TURBIDITES OF CENTRAL AND SOUTHERN APPALACHIAN
B 6209 1614 DEVONIAN TYPE LOG, OHIO, SILURIAN-
B 6304 652 DEVONIAN, VIRGINIA
B 6309 1584* DEVONIAN WELL CUTTINGS IN EASTERN KENTUCKY GAS FIELD, ABST.,
S 6 324 DEVONIAN-CARBONIFEROUS BOUNDARY, CARBONIFEROUS
B 6405 692* DEVONIAN-MISSISSIPPIAN, OF ILLINOIS, ABST., RELATION OF
B 6204 672 DEVONIAN-SILURIAN RESERVOIRS, F6 UNITS, ALGERIA
B 6203 515* DEVONIAN—LOWER CARBONIFEROUS OF SOUTHERN IRELAND, ABST.,
B 6008 1396 DEVOTO, RICHARD H., HELIUM DETECTION EXPLORATION GUIDE TO
B 6309 1599 DEVOTO, RICHARD H., NUCLEAR POWER AND GEOLOGY OF URANIUM,
S 3 100 DEWATERING AND GROWTH FAULTING
B 6206 1053 DEWEESVILLE SANDSTONE MEMBER, TEXAS
B 6203 508* DEWEY COUNTY, OKLAHOMA. BY TITLE ONLY., GEOLOGY AND
B 6301 35 DEWITT COUNTY, CRETACEOUS, TEXAS
B 6203 561 DEWITT, D. B., MAJOR URANIUM DISCOVERY IN FRONTIER AREA—
B 6304 621 DEWITT, D. B., MAJOR URANIUM DISCOVERY IN VOLCANICLASTIC
B 6009 1621 DEYLING, T. H., RELATION BETWEE SALINA, UPPER SILURIAN, OIL
B 6010 1872 DHOFAR, 1975, ASIATIC PETROLEUM CORP., WINTERSHALL A.G., SUN OIL
B 6110 1801 DHOFAR, 1976, ASIATIC PETROL. CORP., SOFIRAN, ELF AQUITAINE, ET AL,
B 6210 2118 DHOFAR, 1977, ASIATIC PETROLEUM CORP., SOCIETE NATIONALE ELF
B 6411 1841 DHOFAR, 1979, AMOCO INTERNAT. OIL CO., ET AL, ON DEVELOPMENTS IN,
B 6405 602 DHOFAR ARCH, MIDDLE EAST, HUQF-
B 6405 600 DHRUMA FORMATION, MIDDLE EAST
B 6211 2244 DIABLO RANGE, CALIFORNIA
B 6109 1449 DIABLO UPLIFT, TEXAS
S 4 284 DIADEMA ANTILLARUM EROSION RATES
S 4 283 DIADEMA ANTILLARUM, VIRGIN ISLANDS
B 6109 1555* DIAGENESIS AND GEOCHEMISTRY OF GLEN ROSE PATCH-REEF
B 6205 892* DIAGENESIS AND HYDROCARBON POTENTIAL OF UPPER CRETACEOUS
M 28 146 DIAGENESIS AND POROSITY, BIBLIOGRAPHY
M 28 147 DIAGENESIS AND POROSITY, CARBONATES, BIBLIOGRAPHY
M 28 147 DIAGENESIS AND POROSITY, CLAY, BIBLIOGRAPHY
B 6109 1536* DIAGENESIS AND POROSITY DISTRIBUTION IN DELTAIC SANDSTONE,
M 28 147 DIAGENESIS AND POROSITY, FELDSPAR, BIBLIOGRAPHY
M 28 149 DIAGENESIS AND POROSITY, IRON, BIBLIOGRAPHY
M 28 149 DIAGENESIS AND POROSITY, MANGANESE, BIBLIOGRAPHY
B 6105 802 * DIAGENESIS AND POROSITY OF RODESSA LIMESTONE, RUNNING DUKE
M 28 149 DIAGENESIS AND POROSITY, PHOSPHATES, BIBLIOGRAPHY
M 28 146 DIAGENESIS AND POROSITY, QUARTZ, BIBLIOGRAPHY
M 28 146 DIAGENESIS AND POROSITY, SILICA, BIBLIOGRAPHY
M 28 149 DIAGENESIS AND POROSITY, ZEOLITES, BIBLIOGRAPHY
S 10 41 DIAGENESIS, MIGRATION, EXPANDABLE CLAY VERSUS DEPTH
B 6108 1387* DIAGENESIS, AND PRODUCTIVE CAPABILITY OF SANDSTONE RESERVOIRS,
M 28 145 DIAGENESIS, BIBLIOGRAPHY
B 6104 483 DIAGENESIS BY HYDROCARBONS, INHIBITION OF
B 6002 288 DIAGENESIS, EARLY
B 6404 480 DIAGENESIS IN CARBONATE ROCKS, STAGES
S 4 357† DIAGENESIS IN CORAL REEFS, SEQUENCE OF PROGRESSIVE
B 6105 759 † DIAGENESIS IN FRONTIER FORMATION, SPEARHEAD RANCH FIELD,
B 6303 510* DIAGENESIS IN HYGIENE AND TERRY SANDSTONES, UPPER CRETACEOUS,
B 6405 790* DIAGENESIS IN MONTEREY FORMATION, PISMO SYNCLINE, COAST
B 6410 1726 DIAGENESIS IN SEDIMENTS AND SEDIMENTARY ROCKS, EDITED BY G.
B 6405 712* DIAGENESIS IN SHALLOW CONVENTIONAL AND LOW-PERMEABILITY
B 6405 790* DIAGENESIS IN VOLCANOGENIC ROCKS OF GREAT VALLEY SEQUENCE,
B 6002 288 DIAGENESIS, LATE
M 27 200 DIAGENESIS, MICRITE
B 6408 1274 DIAGENESIS OF A BAHAMIAN OOID SHOAL, SEDIMENTA VII, BY PAUL M.
B 6109 1552* DIAGENESIS OF A GIANT—POZA RICA TREND, MEXICO, ABST.
B 6303 519* DIAGENESIS OF BACK-REEF CARBONAT ROCKS--EXAMPLE FROM CAPITAN
B 6409 1564* DIAGENESIS OF DELTAIC SANDSTONE—OLMOS SAN MIGUEL, AND UPSON
B 6405 682* DIAGENESIS OF LOWER CORALLINE LIMESTONE, CHATTIAN, MALTESE
B 6104 492† DIAGENESIS OF MIDDLE AND UPPER EOCENE CARBONATE SHORELINE
B 6409 1557* DIAGENESIS OF MIDDLE MIOCENE SANDS IN GEOPRESSURE ZONE OF
B 6108 1393* DIAGENESIS OF OFFSHORE-BAR SANDSTONES OF FRONTIER FORMATION,
B 6303 465* DIAGENESIS OF TERTIARY ARKOSES, SANTA YNEZ MOUNTAINS, SANTA
B 6108 1375* DIAGENESIS OF TERTIARY VOLCANICLASTICS, GUATEMALA, ABST.
B 6108 1393* DIAGENESIS OF VOLCANOGENIC SANDSTONES, ABST.
B 6408 1276* DIAGENESIS ON EXPLORATION STRATEGY AND RESERVOIR
M 27 195 DIAGENESIS, VADOSE
S 4 353† DIAGENETI ELIMINATION OF STRONTIUM IN QUATERNARY TO LATE
B 6303 493* DIAGENETIC ALTERATION IN EOCENE ALLUVIAL-PLAIN DEPOSITS,
B 6004 653* DIAGENETIC ASPECTS OF TABULAR URANIUM DEPOSIT LA SAL MINE,
B 6108 1378* DIAGENETIC CONTROL OF RESERVOIR QUALITY IN ARC-DERIVED
B 6404 461† DIAGENETIC ENVIRONMENTS, CARBONATE DIAGENETIC TEXTURES
B 6405 735* DIAGENETIC FEATURES OF GRAND RAPIDS FORMATION,
B 6002 287† DIAGENETIC GENERATION OF HYDROCARBON GASES AND THEIR
M 30 204 DIAGENETIC HISTORY, EKOFISK FIELD, EKOFISK CHALK
B 6004 723* DIAGENETIC HISTORY OF MISSISSIPPIAN CARBONATE ROCKS,
B 6302 245† DIAGENETIC MAGNETITE OVER OIL FIELDS, AEROMAGNETIC DETECTION
B 6309 1538† DIAGENETIC MAGNETITE OVER OIL FIELDS. DISCUSSION, AEROMAGNETIC
B 6405 765* DIAGENETIC MOBILIZATION OF URANIUM AND IRON FROM RED BEDS OF
B 6109 1556* DIAGENETIC PATTERNS OF AUSTIN GROUP AND THEIR CONTROL OF
B 6004 719* DIAGENETIC PATTERNS OF CHALKS, ABST.
B 6303 481* DIAGENETIC PHENOMENA OF CHIPOLA FORMATION, CALHOUN COUNTY,
B 6108 1372* DIAGENETIC PROCESSES IN SANDSTONES, ABST.
B 6403 449* DIAGENETIC REACTIONS IN MONTEREY FORMATION, PISMO SYNCLINE,
B 6403 449* DIAGENETIC REACTIONS IN MONTEREY FORMATION, PISMO SYNCLINE,
B 6004 731* DIAGENETIC SEQUENCE OF MIDDLE CRETACEOUS CARBONATE
B 6009 1607* DIAGENETIC TRAP, ABST., WALKER CREEK FIELD--, SMACKOVER
B 6104 486 DIAGENETIC TRAP, DEFINITION
B 6104 485 DIAGENETIC TRAPS AS STRATIGRAPHIC TRAPS, MISINTERPRETATION OF
B 6104 483† DIAGENETIC TRAPS--EXPLORATION TARGETS, FROZEN-IN,
B 6009 1611* DIAGNOSTIC CENTRAL GULF COAST VICKSBURGIAN OSTRACODS IN H. V.
B 6409 1550 DIAGRAPHIES DIFFEREES—BASES DE L'INTERPRETATION, TOME 1—
M 29 153 DIAMANTINA FRACTURE ZONE, AUSTRALIA, MARGIN
B 6305 830* DIAMONDS IN COLORADO-WYOMING KIMBERLITE PROVINCE, ABST.,
B 6109 1548* DIAPIR AND SIERRA DE PAJAROS AZULES, COAHUILA, MEXICO, ABST.,
M 29 78 DIAPIR, ATLANTIC COAST, NEW JERSEY, BALTIMORE CANYON
B 6211 2233 DIAPIR, BAY OF BISCAY AREA, ZARAUZ
M 26 267 DIAPIR GROWTH, EXAMPLE, GULF OF MEXICO
B 6206 997 DIAPIR, IRAN, HORMUZ, SALT
B 6404 568† DIAPIR, ISRAEL, STRAIN MEASUREMENTS IN EASTERN MARGINAL SHEAR
B 6211 2232 DIAPIR, PYRENEES
M 26 264 DIAPIR UPLIFT, LOUISIANA, WESTERN
B 6211 2240 DIAPIR WITH SALT DOMES IN PRESENT DEEP-SEA ENVIRONMENTS,
B 6002 211 DIAPIR ZONE, GABON TO, ANGOLA, WEST, AFRICAN
B 6405 706* DIAPIR-LIKE RIDGES AND POSSIBLE HYDROCARBON OCCURRENCE,
S 3 135 DIAPIRIC EMPLACEMENT OF HYDROTHERMAL MODIFICATION OF
S 7 81 DIAPIRIC SALT FEATURES, MAP, GULF OF MEXICO
B 6305 790 DIAPIRIC SALT, GULF OF MEXICO
B 6002 196† DIAPIRIC STRUCTURES, SOUTH, ATLANTIC
M 25 513 DIAPIRIC-TYPE PORPHYRY COPPER DEPOSIT
S 3 99 DIAPIRISM
B 6209 1577 DIAPIRISM, COMPARISON OF, SALT TO, SHALE
B 6209 1571 DIAPIRISM, CONCEPTUAL, MODEL OF FORCEFULLY, INTRUSIVE
S 7 278 DIAPIRISM, DIAGRAM, STABILITY, SQUEEZING AND
B 6209 1564 DIAPIRISM, EMPLACEMENT HISTORIES OF, PIERCEMENT
B 6003 458† DIAPIRISM IN SOUTHERN IRAN. DISCUSSION, SALT
S 3 136 DIAPIRISM, MOBILIZATION OR LOCALIZATION BY, SALT OR, SHALE
B 6209 1569 DIAPIRISM, UNCONSTRAINED AND, CONSTRAINED
B 6209 1579 DIAPIRISM VERSUS BUOYANCY THEORY, THEORY OF
B 6206 999 DIAPIRS AND BUCKLE FOLDING, IRAN, SALT

B 6002 196 DIAPIRS, COASTAL MARGIN, BRAZIL
B 6209 1567 DIAPIRS, EXTRUSION AND INTRUSION, INCIPIENT
B 6111 1994 DIAPIRS FROM LAMONT-DOHERTY SEISMIC PROFILE LIBRARY, DEEP-SEA
B 6209 1566 DIAPIRS, GROWTH OF, INCIPIENT
S 7 27 DIAPIRS, JURASSIC, NORTHERN, GULF OF MEXICO, SALT AND
B 6209 1561† DIAPIRS, MECHANISM FOR, EMPLACEMENT OF, PIERCEMENT
B 6204 657† DIAPIRS ON SAO PAULO PLATEAU, SOUTHEASTERN BRAZILIAN
B 6209 1570 DIAPIRS, RADIAL AND, TRAP-DOOR, FAULTS ASSOCIATED WITH, SALT
S 10 191 DIAPIRS, SEISMIC PROFILE, OVERPRESSURED SHALES
B 6311 2097† DIAPIRS, SEISMIC RECOGNITION OF, SALT
B 6209 1576 DIAPIRS, TRAP EXISTENCE IN
B 6307 1054 DIAPIRS WITHIN DELTAIC SEQUENCE, SEISMIC EXPRESSION OF, SHALE
B 6202 247† DIASTROPHISM OF SANTA MONICA AND SAN PEDRO SHELVES,
B 6212 2492† DIASTROPHISM OF SANTA MONICA AND SAN PEDRO SHELVES,
B 6306 426* DIATOM CORRELATIONS, ABST., PAN-, PACIFIC, NEOGENE
B 6012 2192* DIATOM CORRELATIONS IN CALIFORNIA--PROBLEMS AND SOLUTIONS,
B 6207 1222* DIATOMACEOUS SEDIMENTS IN CIRCUM-PACIFIC REGION, ABST., ORIGIN,
M 28 81 DIATOMITE, WITH OPALLINE SILICA
B 6403 438* DIATOMS AND SILICOFLAGELLATES FROM HOLOCENE SEDIMENTS,
S 11 52 DIATOMS, CALIFORNIA, MIO-PLIOCENE
B 6303 554* DIATOMS IN PETROLEUM EXPLORATION, ABST., UTILITY OF
M 27 25 DIATOMS, SILICEOUS ALGAE
M 27 26 DIATOMS, SILICEOUS ALGAE
S 3 135 DIATREME FEEDING OF URANIFEROUS MANTLE VOLATILES
S 3 30 DIATREME FEEDING OF URANIUM
B 6203 464 DIBA FORMATION, LIBYA
B 6405 602 DIBBA LINE, MIDDLE EAST
B 6002 322* DIBBLE--MUSTANG, CURRENT HUNTON PLAY IN MCCLAIN, GRADY, AND
B 6203 344 DIBBLEE, T. W., JR., NEOGENE BASINFORMATION IN RELATION TO
B 6012 2178 DIBBLEE, THOMAS W., JR., DEPOSITIONAL AND TECTONIC HISTORY OF
S 4 90 DICHOCOENIA ASSEMBLAGE, JAMAICA, MULTICOLUMNASTREA-,
S 4 90 DICHOCOENIA TRECHMANNI ASSEMBLAGE, JAMAICA
B 6205 882 DICK JAY D., EXPLORATION FOR NONELECTRIC GEOTHERMAL
B 6309 1577 DICKERSON, D. R., PYROLYSIS OF EASTERN GAS SHALES--EFFECTS OF
M 29 1 DICKERSON, PAT, FOREWORD, GEOLOGICAL AND GEOPHYSICAL
M 29 1 DICKERSON, PATRICIA WOOD, EDITED BY, GEOLOGICAL AND GEOPHYSICAL
B 6311 2042 DICKEY, PARK A., STRATIGRAPHY OF INTERMONTANE, LACUSTRINE
B 6007 1124 DICKEY, PARKE A., ABNORMAL FORMATION PRESSURE, DISCUSSION AND
B 6105 780 DICKEY, PARKE A., OIL AND GAS IN RESERVOIRS WITH SUBNORMAL
B 6112 2134 DICKEY, PARKE A., OIL AND GAS IN RESERVOIRS WITH SUBNORMAL
B 6411 2040 DICKEY, SHANMUGAM, G., REVIEW OF, PETROLEUM DEVELOPMENT
B 6008 1401* DICKINSON AREA, NORTH DAKOTA, ABST., STRATIGRAPHY AND
B 6303 485* DICKINSON AREA, NORTH DAKOTA--A 24-MILLION BARREL SOIL-ZONE
B 6303 440 DICKINSON, KENDALL A., DETAILED STRATIGRAPHY AND
B 6405 698 DICKINSON, KENDALL A., URANIUM IN TERTIARY SEDIMENTS IN ALASKA,
B 6301 2 DICKINSON, W. R., STRUCTURE AND STRATIGRAPHY OF FOREARC
B 6012 2179 DICKINSON, WILLIAM R., APPLICATIONS OF PLATE TECTONICS TO
B 6211 2353 DICKINSON, WILLIAM R., MESOZOIC PALEOGEOGRAPHY AND
B 6405 698 DICKINSON, WILLIAM R., PALEOGENE DEPOSITIONAL SYSTEMS, WESTERN
B 6312 2164 DICKINSON, WILLIAM R., PLATE TECTONIC AND SANDSTONE
B 6211 2353 DICKINSON, WILLIAM R., PLATE-TECTONIC EVOLUTION OF SACRAMENTO
B 6403 449 DICKINSON, WILLIAM R., SANDSTONE COMPOSITIONS RELATED TO PLATE
B 6012 2177 DICKINSON, WILLIAM R., SEDIMENTARY SERPENTINITE OF THE MIOCENE
B 6105 781 DICKINSON, WILLIAM R., STRATIGRAPHY AND STRUCTURE OF
B 6303 474* DICKITE, ABST., OCCLUSION OF POROSITY IN, CARBONATE, RESERVOIRS
M 28 138 DICKITE, WITH KAOLINITE, SEM
B 6007 1021 DICKSON, FRANKLIN H., SUBLACUSTRINE FAN MORPHOLOGY IN LAKE
B 6308 1421 DICKSON, MORAN, W. R., REVIEW OF ATLAS OF, ECONOMIC, MINERAL
S 4 90 DICTUOPHYLLIA-DICHOCOENIA ASSEMBLAGE, JAMAICA,
B 6205 890 DIDERIKSEN, DARRELL JULIAN, VOLCANIC SANDSTONE IN GREEN RIVER
B 6409 1369 DIEMER, R. A., DEVELOPMENTS IN WEST TEXAS AND EASTERN NEW
B 6410 1722 DIENES AND THEODORE SHABAD, MEYERHOFF, A. A., REVIEW OF THE,
B 6004 719 DIESTER-HAAS, L., EOLIAN-SAND TURBIDITES, ABST.
B 6409 1548 DIETRICH AND BRIAN J. SKINNER, SUCZEK, CHIRSTOPHER A., REVIEW OF,
S 7 278 DIFFERENTIAL CONSOLIDATION, BOTTOM, DEFORMATION
B 6405 765* DIFFERENTIAL DIFFUSION DURING CARBONATE DIAGENESIS, ABST.
B 6305 757 DIFFERENTIAL ENTRAPMENT
B 6304 608† DIFFERENTIAL ENTRAPMENT OF OIL AND GAS IN NIAGARAN
B 6109 1546* DIFFERENTIAL SUBSIDENCE OF MARSHALND PEAT AS GEOLOGIC
M 29 453 DIFFERENTIATION OF MANTLE, MECHANISMS
B 6105 785 * DIFFERENTIATION OF SANDSTONE DEPOSITIONAL ENVIRONMENTS IN
B 6011 2044† DIFFERENTIATION OF SHALE TYPES BY ALLOCHTHONOU AND
B 6002 215 DIFFUSION HYPOTHESIS
B 6002 215 DIFFUSION HYPOTHESIS, KEROGEN
S 10 17 DIFFUSION OF WATER AND DISSOLVED SOLIDS, PORE WATER
B 6004 732* DIFUNTA GROUP, UPPER CRETACEOUS-PALEOCENE, PARRAS BASIN,
B 6402 219 DIGBOI FIELD, INDIA
B 6403 428 DIGBOI FIELD, INDIA
S 10 180 DIGESTIVE TRACTS OF ANIMALS, METHANE, OCCURRENCES
B 6203 497* DIGITAL IMAGE PROCESSING OF POTENTIAL FIELD DATA, ABST.
M 26 364 DIGITAL PROCESSING, DEBUBBLING, DEREVERBERATION
B 6203 546* DIGITAL PROCESSING OF LANDSAT DATA, ABST., SMALL ROCK
M 26 364 DIGITAL PROCESSING, RESOLUTION SYSTEMS
B 6203 539* DIGITATE STROMATOLITIC BIOHERMS AND ASSOCIATED FACIES,
B 6302 172 DIGITIZED LOGS, COMPUTER PRINT OUT OF, LOG ANALYSIS FROM
B 6004 694* DIGITIZED OIL AND GAS FIELD MA OF NORTH AMERICA--STATUS
B 6409 1393 DIGNES, T. W., DEVELOPMENTS IN WEST COAST AREA IN 1979
M 25 351† DIKE-COMPLEX AND BASAL-LENS WATER SUPPLIES ON BASALTIC
B 6012 2187 DIKOFF, C., PROCESSING OF NONSTRAIGHT SEISMIC LINES USING
B 6406 932 DILATANCY, KINETICS OF HYDROCARBON, MATURATION AND
B 6309 1586* DILATION BRECCIATION--PROPOSED MECHANISM OF FRACTURING,
B 6203 369 DILIGENCIA BASIN, CALIFORNIA
B 6203 509 DILL, R. F., GEOLOGY SEEN WHILE LIVING IN A SUBMARINE CANYON
B 6004 725 DILL, ROBERT F., LATERAL INFILL AS MAJOR FACTOR IN
B 6105 826 DILLON, WILLIAM P., ATLANTIC OUTER CONTINENTAL SHELF, ABST..
B 6105 782 DILLON, WILLIAM P., GEOLOGIC DEVELOPMENT OF CENOZOIC
M 29 11 DILLON, WILLIAM P., GEOLOGY OF THE OFFSHORE SOUTHEAST GEORGIA
M 29 27 DILLON, WILLIAM P., STRUCTURE AND DEVELOPMENT OF THE
B 6105 781 DILLON, WILLIAM P., STRUCTURE AND DEVELOPMENT OF SOUTHEAST
B 6203 560 DILLON, WILLIAM P., STRUCTURE OF ATLANTIC SLOPE OF EASTERN
B 6403 339 DILLON, WILLIAM P., STRUCTURE, STRATIGRAPHY, AND GEOLOGIC
B 6206 1053 DILWORTH URANIUM HOST ROCK, TEXAS
B 6004 694* DIMMI COUNTY, TEXAS, ABST., CARBONATE DIAGENESIS, AND FACIES
M 24 145† DIMMIT AND ZAVALA COUNTIES, TEXAS, BIG WELLS FIELD
B 6109 1538* DIMMIT COUNTY, TEXAS—SOME UNIQUE ASPECTS OF
B 6107 1014 DINA MEMBER, ALBERTA AND, SASKATCHEWAN
B 6411 1970 DINARIC ONSHORE ZONE, YUGOSLAVIA, ADRIATIC OFFSHORE AND
B 6405 696* DINEH-BI-KEYAH FIELD, APACHE COUNTY, ARIZONA ABST.
B 6303 529 DINGLE, RICHARD V., TERTIARY SEA-LEVEL MOVEMENTS AROUND
B 6203 457 DINGO CLAYSTONE, AUSTRALIA
B 6203 457 DINGO GROUP, AUSTRALIA
B 6303 492* DINOFLAGELLATE ASSEMBLAGES FROM SURFACE AN SUBSURFACE
M 27 22 DINOFLAGELLATES
B 6405 728* DINOGYMNIUM IN HILLIARD SHALE, SOUTHWESTERN WYOMING, ABST..
B 6105 778 * DINOSAUR TRACKS FROM CRETACEOUS SEDIMENTS OF PEACE RIVER CANYON
B 6405 714 DINTER, DAVID A., IMPLICATIONS OF GEOLOGI STRUCTURE AND
B 6201 100 DINWOODY FORMATION, IDAHO AND, WYOMING
B 6004 664 DIONNE, J. C., ICE-DRIFTED BOULDERS AS PALEOCLIMATIC INDICATOR,
M 25 512 DIORITE-MODEL PORPHYRY COPPER DEPOSITS
M 25 435 DIORITES, INTRUSIONS OF, PHILIPPINES
M 25 437 DIORITIC MASSES, RELATION OF, PHILIPPINE RIFT SYSTEM AND
B 6002 188 DIP LOGS, NORTH SEA
S 4 158 DIPLORIA CLIVOSA ZONE OFF BONAIRE
S 4 160 DIPLORIA STRIGOSA OFF BONAIRE
S 4 31 DIPLORIA STRIGOSA ON LIVING REEF
S 4 283 DIPLORIA STRIGOSA, VIRGIN ISLANDS
S 4 33 DIPLORIA-MONTASTREA-PORITES ASSEMBLAGE, WEST FLOWER GARDEN
S 4 262 DIPLOTHYRA SMYTHI, TEXAS
B 6306 905† DIPMETER AND LOG MOTIFS IN NORTH SEA SUBMARINE-FAN SANDS
B 6211 2350* DIPMETER DATA, ABST., STATISTICAL CURVATURE ANALYSIS
B 6203 495* DIPMETER DATA, ABST., STATISTICAL CURVATURE ANALYSIS
B 6207 1112 DIPMETER LOG, SCHLUMBERGER, 4-ARM HIGH-RESOLUTION
B 6207 1111† DIPMETER LOGS, MEASUREMENT OF, SUBSURFACE, FRACTURES FROM
B 6306 908 DIPMETER MOTIFS OF DEEP-SEA SANDS
B 6303 504* DIPMETER RESULTS, ABST., EOLIAN SEDIMENTOLOGY INTERPRETED
B 6405 707* DIPMETER VALIDITY IN DEVIATED BORE HOLES, ABST.
B 6211 2362* DIPPING EVENTS, ABST., EQUALIZING, STACKING VELOCITIES OF
B 6310 1927 DIRECCION GENERAL DE LA ENERGIA, ON DEVELOPMENTS IN SPAIN, 1978
B 6304 686* DIRECT MEASUREMENT OF URANIUM BY PROMPT FISSION NEUTRON
M 26 188 DIRECT RECOGNITION, SEISMIC, REFLECTIONS, CARBONATES
M 25 472 DIRECTION ANOMALIES, METHOD OF
B 6304 689* DISCONFORMITIES IN GRANTS MINERAL BELT AND THEIR RELATION TO
B 6107 1010† DISCONFORMITY IN LLOYDMINSTER AREA, ALBERTA AND
B 6412 2167 DISCONTINUITIES, APPALACHIANS, CROSS-STRIKE STRUCTURAL
B 6403 366 DISCONTINUITIES OF OCEAN-FLOOR SPREADING AND UNCONFORMITIES
B 6412 2166† DISCONTINUITIES—POSSIBLE EXPLORATION TOOL FOR NATURAL GAS IN
M 26 100 DISCONTINUOUS, SEISMIC, REFLECTIONS
S 11 156 DISCORBINELLA, CALIFORNIA, MIOCENE, EPISTOMINELLIDS AND
S 11 156 DISCORBINELLA VALMONTEENSIS
S 11 253 DISCORBIS CAMPANULATA
S 11 284 DISCORBIS CAMPANULATA
B 6007 1078 DISCOVERER, CRUISES, 1971-1972, NOAA SHIPS, RESEARCHER AND
B 6204 633 DISCOVERER RESEARCH CRUISE, NOAA SHIP
M 30 1† DISCOVERIES 1968-1978: AN OVERVIEW, GIANT FIELD
B 6208 1282 DISCOVERIES, RESERVE ESTIMATES OF, NEW-FIELD
B 6108 1128 DISCOVERIES, RESERVE ESTIMATES OF, NEW-FIELD
B 6308 1175 DISCOVERIES, RESERVE ESTIMATES OF, NEW-FIELD
B 6008 1160 DISCOVERIES, RESERVE ESTIMATES OF, NEW-FIELD
B 6409 1303 DISCOVERIES, RESERVE ESTIMATES OF, NEW-FIELD
B 6303 470* DISCOVERY AND DEVELOPMENT OF BADAK FIELD, EAST KALIMANTAN,

B 6303 502* DISCOVERY AND DEVELOPMENT OF NEMBE CREEK OIL FIELD, NIGERIA,
M 30 441† DISCOVERY AND DEVELOPMENT OF THE BADAK FIELD, EAST
B 6211 2350* DISCOVERY AND GEOLOGY OF DESERT PEAK, NEVADA, GEOTHERMAL
S 4 53 DISCOVERY BAY CANYON, JAMAICA
B 6107 1073 DISCOVERY BAY, MICROGRAPHS, JAMAICA
S 4 263 DISCOVERY BAY, REEF, BORERS, JAMAICA
S 4 129 DISCOVERY BAY, REEF, JAMAICA
M 30 3 DISCOVERY HISTORY, SINCE 1900
B 6205 744 DISCOVERY HISTORY, WORLD
B 6312 2189 DISCOVERY RATES, ASSESSING RESOURCES, EXTRAPOLATION OF
M 30 2 DISCOVERY STATISTICS, GIANT FIELDS, HISTORY
B 6105 762 * DISCRIMINATION OF SURFICIAL MATERIALS IN PART OF MISSISSIPPI
B 6407 1094† DISCUSSION AND REPLIES, MODERN AND, ANCIENT, SUBMARINE, FANS.
B 6410 1688 DISMAL LAKES GROUP, CELLULAR FILAMENTS AND SHEATHS
M 27 176 DISMICRITES
B 6012 2176* DISPERSAL, DISTRIBUTION, AND THICKNESS OF HOLOCENE SEDIMENTS
B 6303 501* DISPERSAL OF FINE-GRAINED SEDIMENT IN STRONGLY STRATIFIED
B 6004 710* DISPERSAL OF UPPER CRETACEOUS AND PALEOCENE SANDSTONES OF
B 6004 722* DISPERSAL PATTERNS OF SOUTH TEXAS OUTER CONTINENTAL SHELF,
B 6004 668* DISPLACEMENT OF SAN ANDREAS FAULT IN SOUTHERN CALIFORNIA AND
B 6004 651* DISPLAY SYSTEM--INTERACTIVE EXPLORATION TOOL, ABST., WELL-DATA
B 6004 670* DISPLAYS AS INTERPRETATION AID, ABST., THREE-DIMENSIONAL,
B 6004 644* DISPLAYS IN SEISMIC INTERPRETATION, ABST., USE OF, COLOR
B 6004 724* DISPOSAL, ABST., HYDROLOGIC STABILIT OF NORTH, LOUISIANA, SALT,
M 25 365† DISPOSAL IN HONOLULU--SUMMARY, IMPACT OF, ECONOMIC
B 6105 828* DISPOSAL IN LITTORAL ZONE, NEW RIVER INLET, NORTH CAROLINA,
B 6303 475* DISPOSAL OF INDUSTRIAL WASTES IN OKLAHOMA, ABST., POTENTIAL
B 6108 1382* DISPOSAL PONDS IN NORTHEAST COLORADO, ABST., PREVENTION OF
M 25 371† DISPOSAL PRACTICE ON GROUNDWATER RESOURCES IN HAWAIIAN
B 6105 818* DISPOSAL SITE, GEORGIA, ABST., DEPOSITIONAL CHARACTERISTICS OF
B 6304 686* DISSOLUTION AND AUTHIGENESIS IN HOST SANDSTONES, ABST.
B 6109 1550* DISSOLUTION, AND POROSITY DEVELOPMENT BETWEEN TWO LOWER
B 6002 323* DISSOLUTION IN OIL AND GAS TEST HOLE IN CENTRAL KANSAS, ABST.,
M 28 173 DISSOLUTION, PARTIAL, OF DETRITAL GRAINS
S 3 47 DISSOLUTION, SELECTIVE
B 6004 700* DISSOLUTION SEQUENCES FOR CALCAREOUS NANNOFOSSILS IN DEEP-SEA
B 6212 2478 DISSOLVED GAS CONTENT OF GEOTHERMAL GEOPRESSURED WATERS
B 6305 766 DISSOLVED HYDROCARBONS AS POSSIBLE PRIMARY MIGRATION
B 6304 699* DISSOLVED HYDROCARBONS IN COASTAL WATERS OF NORTH AMERICA,
S 10 94 DISSOLVED HYDROCARBONS, CRUDE OIL
S 10 95 DISSOLVED HYDROCARBONS, NATURAL GAS
S 7 233 DISTAL TURBIDITE FACIES, CALIFORNIA, LOS ANGELES BASIN
S 7 235 DISTAL TURBIDITE FACIES, CALIFORNIA, VENTURA BASIN
B 6203 492* DISTINCTIVE TURBIDITE SILTSTONE IN DEVONIAN BRALLIER
B 6408 1276 DISTINGUISHED LECTURE TOUR ABSTRACTS, 1979-80
B 6004 691* DISTORTIONS INHERENT IN SOME MAP PROJECTIONS, ABST.,
B 6105 819* DISTRIBUTARY-CHANNEL SYSTEMS OF DESMOINESIAN, PENNSYLVANIA,
B 6309 1608* DISTRIBUTION AN SIGNIFICANCE OF COARSE BIOGENIC AND CLASTIC
B 6405 677* DISTRIBUTION AND ALTERATION OF OGALLALA VOLCANIC-ASH
B 6303 414* DISTRIBUTION AND FACTORS CONTROLLING FORAMINIFERAL
B 6410 1643† DISTRIBUTION AND GENESIS OF SEDIMENTARY URANIU NEAR
B 6209 1753* DISTRIBUTION AND GEOMETRY OF AN OOLITE-SHOAL COMPLEX—
B 6108 1376* DISTRIBUTION AND ORIGIN OF SULFUR IN COLORADO OIL-SHALE
B 6004 687* DISTRIBUTION AND RELATION OF PHYSICAL AND BIOGENIC
B 6105 774 * DISTRIBUTION AND SIGNIFICANCE OF TRACE FOSSILS IN UPPER
B 6105 768 * DISTRIBUTION AND TEMPORAL MIGRATION OF BANK-TOP FORAMINIFERA
S 5 125* DISTRIBUTION OF EVAPORITES IN THE SALINE FORMATION, WEST
B 6405 757* DISTRIBUTION, DIAGENESIS, AND DEPOSITIONAL HISTORY OF POROUS
B 6303 444* DISTRIBUTION, MORPHOLOGY, MECHANISMS, AND AGES OF SEDIMENT
B 6311 2115* DISTRIBUTION OF BITUMENS AND KEROGEN IN SHALE CLAST, ABST.
B 6101 100† DISTRIBUTION OF CARBON AS HYDROCARBONS AND ASPHALTIC
B 6303 454* DISTRIBUTION OF CARBONATE CEMENTS IN QUATERNARY
B 6303 541* DISTRIBUTION OF CLAY IN RECENT SANDS, ABST.
S 9 19 DISTRIBUTION OF CONTINENTS, CLIMATIC CHANGES
B 6209 1754* DISTRIBUTION OF HOLOCENE FORAMINIFERA IN GULF OF ALASKA, ABST.
B 6209 1759* DISTRIBUTION OF LIVING BENTHONIC FORAMINIFERA AS INDICATORS
B 6211 2350* DISTRIBUTION OF RECENT FORAMINIFERA IN GULF OF ALASKA, ABST.
B 6303 491* DISTRIBUTION OF SALT STRUCTURES IN GULF O MEXICO, ABST.
B 6105 797 * DISTRIBUTION OF SEDIMENTARY STRUCTURES IN EBB—TIDAL DELTA
B 6105 802 * DISTRIBUTION OF SULFUR IN CONTINENTAL AND NEARSHORE COALS FROM
B 6105 804 * DISTRIBUTION OF SUSPENDED PARTICULATE MATTER NEAR SEWAGE
B 6008 1403* DISTRIBUTION OF URANIUM RESOURCES IN NONCOMMUNIST WORLD,
B 6309 1603* DISTRIBUTION OF VOLCANIC ACTIVITY IN TIME AND SPACE IN GULF
B 6405 757* DISTURBED BELT, MONTANA, ABST., DISTRIBUTION, DIAGENESIS, AND
B 6305 836* DISTURBED BELT OF MONTANA, ABST., STRUCTURAL PATTERN OF
M 24 93 DISTURBED BELT, WYOMING
B 6004 677* DIVER-OPERATED DRILLING DEVICE, ABST., RECENT AND RELICT
M 29 437 DIVERGENT CONTINENTAL MARGINS, THERMAL MATURITY
M 29 428 DIVERGENT DELTAIC MARGIN, GULF OF MEXICO, LOUISIANA
B 6209 1589 DIVERGENT DELTAIC MARGIN, LOUISIANA, GULF COAST—
M 25 53 DIVERGENT PLATE BOUNDARIES, METAL DEPOSITS AT
M 25 178 DIVERGENT PLATES AND SEAFLOOR SPREADING
M 26 123 DIVERGENT, SEISMIC, REFLECTIONS, CONFIGURATION PATTERNS
S 4 95 DIVERSITY AND ABUNDANCE OF SCLERACTINIAN CORALS
B 6209 1703† DIVISION OF BLACK OHIO SHALE IN EASTERN KENTUCKY
B 6206 1063 DIX, F. A., JR., A REPORT ON THE MEMBERSHIP
B 6309 1592 DIXON, J. A., PRESSURE CORING OF GAS-BEARING DEVONIAN BLACK
B 6309 1591 DIXON, J. A., PROJECTION OF FAVORABLE GAS-PRODUCING AREAS FROM
M 25 436 DIZON COPPER PROSPECT, LUZON
B 6403 424 DJEBEL EL HARA, TUNISIA
B 6403 424 DJEBEL HASSANE, TUNISIA
B 6403 424 DJEBEL TIOUR EL LIL-ANNTRA, TUNISIA
S 6 305 DJULFIAN STAGE, PERMIAN, BIOCHRONOLOGY
B 6104 595 DKM-2 WELL, SENEGAL
B 6104 597 DKM-2 WELL, SENEGAL
M 26 329 DOBRIN, MILTON B., SEISMIC EXPLORATION FOR STRATIGRAPHIC TRAPS
B 6012 2179 DOBRIN, MILTON B., SEISMIC STRATIGRAPHY--NEW KIND OF
M 28 98 DOCKUM FORMATION, NEW MEXICO, TRIASSIC
B 6203 499* DOCKUM GROUP IN TULE CANYON, TEXAS PANHANDLE, ABST.,
B 6305 835* DOCKUM GROUP, TEXAS PANHANDLE, ABST., DEPOSITIONAL FACIES
M 28 173 DOCKUM GROUP, TEXAS, TRIASSIC
M 28 28 DOCKUM GROUP, TEXAS, TRIASSIC
M 28 137 DOCKUM GROUP, TEXAS, TRIASSIC
B 6109 1542* DOCKUM GROUP, TRIASSIC, TEXAS PANHANDLE, ABST., DEPOSITIONAL
B 6012 2189 DODD, J. R., PLIOCENE BIOSTRATIGRAPHY AND DEPOSITIONAL
B 6405 699 DODD, J. ROBERT, PALEOECOLOGY IN BASIN ANALYSIS—HUMBOLDT
B 6008 1396 DODGE, HARRY W., JR., CHARACTERISTICS OF UPPER CRETACEOUS
B 6405 699 DODGE, M. M., EVALUATION OF VOLUME OF ENTRAINED METHANE IN
B 6309 1606 DODGE, M. M., IMPORTANCE OF SECONDARY LEACHED POROSITY IN
B 6303 440 DODGE, M. M., MINERALOGIC COMPOSITION AND DIAGENESIS OF
B 6303 488 DODGE, M. M., RESERVOIR QUALITY IN TERTIARY SANDSTONES ALONG
B 6405 741 DODGE, M. M., SANDSTONE DIAGENESIS IN GEOPRESSURED TERTIARY
B 6302 183 DOGGER HIGH, NORTH SEA
B 6203 436 DOGGER HIGH, NORTH SEA
B 6004 612† DOGGER, JURASSIC, CARBONATE ROCKS, SOUTHEAST OF PARIS, FRANCE,
B 6004 725* DOGGER, MIDDLE JURASSIC, OF CENTRAL HIGH ATLAS, MOROCCO, ABST.,
B 6104 511† DOGGER, MIDDLE JURASSIC, OF FRANCE, PATTERNS OF, POROSITY AN,
B 6304 676† DOGGER, MIDDLE JURASSIC, OF FRANCE. DISCUSSIN AND REPLY,
B 6304 678 DOGGER, MIDDLE JURASSIC, RESERVOIR, FRANCE
M 26 337 DOGGER SANDSTONE, HOHNE FIELD, WEST, GERMANY
B 6405 673 DOHR, G., EXPLORATION IN CLASSIC THRUST BELT AND ITS
B 6112 2086 DOIG FORMATION, KEROGEN, ALBERTA
B 6202 274 DOLDRUMS FRACTURE ZONE
B 6212 2427 DOLE, W. E., SLOPE MAPPING
B 6004 665 DOLE, W. E., TECHNIQUES AND USES OF SLOPE MAPPING, ABST.
B 6108 1225 DOLLAR POINT FIELD, GALVESTON COUNTY, TEXAS
B 6205 883 DOLLY, EDWARD D., GEOLOGIC TECHNIQUES USED IN TRAP SPRING
B 6405 707 DOLLY, EDWARD D., PETROLEUM POTENTIAL OF BASIN AND RANGE
B 6403 442 DOLLY, EDWARD D., PETROLEUM POTENTIAL OF GREAT BASIN, ABST.
M 30 537 DOLOMIA PRINCIPALE FORMATION, ITALY, PO BASIN
B 6401 32 DOLOMIA PRINCIPATE, ITALY
B 6303 490* DOLOMITE, ABST., POSSIBLE BIOGENIC ORIGIN FOR SOM SEDIMENTARY
B 6405 800* DOLOMITE AND DEDOLOMITE IN MURAL LIMESTONE, LOWER
M 28 122 DOLOMITE, AUTHIGENIC, IN SANDSTONE
B 6203 553* DOLOMITE BY IRON AND MANGANESE, ABST., CONTROL OF,
M 28 28 DOLOMITE, DETRITAL, FRAGMENTS
B 6303 500* DOLOMITE FORMATION IN MIDDLE PROTEROZOIC MCARTHUR GROUP,
B 6002 253 DOLOMITE GROWTH, INHIBITION OF
B 6002 248 DOLOMITE GROWTH, MECHANISM OF
B 6303 488* DOLOMITE IN CARBONATE ROCKS--CAUSES AND CURES, ABST., ERROR IN,
M 28 156 DOLOMITE, IN CHERT CLAST, REPLACEMENT
M 28 121 DOLOMITE, IN HEMATITIC SANDSTONE, CEMENTS
M 28 122 DOLOMITE, IN SANDSTONE, CEMENTS
B 6303 457* DOLOMITE IN TIDAL-FLAT SEDIMENTS OF CENTRAL-WEST ANDRŌS
B 6303 453* DOLOMITE IS EVAPORITE MINERAL--EVIDENCE FROM ROCK RECORD
B 6105 833* DOLOMITE LOCALITY IN FLORIDA BAY, ABST., HOLOCENE
B 6303 431* DOLOMITE NONSTOICHIOMETRY, ITS RELATION TO CARBONATE-ROCK
M 27 132 DOLOMITE, REPLACEMENT
M 27 176 DOLOMITE REPLACEMENT
M 27 13 DOLOMITE REPLACEMENT, RED, ALGAL FRAGMENT
M 27 199 DOLOMITE REPLACEMENTS
M 28 29 DOLOMITE RHOMBS, FRAGMENTS
B 6203 463† DOLOMITE RHOMBS IN OIL RESERVOIRS, CHADRA SANDS, GIALO FIELD,
S 7 218 DOLOMITE, WEST FLORIDA SLOPE
B 6004 696* DOLOMITE--MODEL BASED ON CYCLIC INTERACTION OF CONTINENTAL
B 6303 485* DOLOMITE--STATE OF THE ART, ABST., ISOTOPIC AND, TRACE-ELEMENT,
B 6004 719* DOLOMITES IN ARBUCKLE GROUP, ARBUCKLE MOUNTAINS
B 6203 496* DOLOMITES IN BELL CANYON FORMATION, DELAWARE BASIN, NEW
B 6303 503* DOLOMITES IN GREAT BASIN, ABST., FIELD RELATIONS OF SOME
B 6002 249 DOLOMITES, PORE SYSTEMS IN

B 6002 245 DOLOMITES, PORE, THROATS IN
M 25 475 DOLOMITES, USSR, LENA STAGE
B 6303 499* DOLOMITES--PROBLEMS OF ORIGIN, ABST., LATE-STAGE SUBSURFACE
B 6303 422* DOLOMITIZATION IN TIDAL FLATS OF WESTERN ANDROS, BAHAMAS,
B 6303 430* DOLOMITIZATION IN UPPER RE RIVER FORMATION, UPPER ORDOVICIAN,
B 6303 478* DOLOMITIZATION IN WILLISTON BASIN, ABST., ORIGIN OF REGIONAL
B 6303 446* DOLOMITIZATION MODELS FROM EARLY PRECAMBRIAN SEQUENCE IN
B 6303 524* DOLOMITIZATION OF NORTHERN MICHIGAN NIAGARAN REEFS BY BRINE
B 6303 472* DOLOMITIZATION OF OFFSHORE CARBONATE DEPOSITS IN HANNET
B 6303 528* DOLOMITIZATION OF PLIOCENE-PLEISTOCENE CARBONATE SEDIMENTS,
B 6303 442* DOLOMITIZATION OF SUBTIDALLY DEPOSITED CARBONATE SEDIMENT IN
B 6303 492* DOLOMITIZATION OF UPPER DEVONIAN MIETTE BUILDUP, ALBERTA,
B 6008 1400* DOLOMITIZATION PATTERNS IN PALEOZOIC O WILLISTON BASIN, ABST.
M 27 131 DOLOMITIZATION, REPLACEMENT
B 6303 455* DOLOMITIZATION--RECENT EXPERIMENTAL APPROACHES, ABST.
B 6405 678* DOLORES FORMATION, SOUTHWESTERN COLORADO, ABST., TRIASSIC,
B 6212 2458 DOLORES-GUAYAQUIL MEGASHEAR, ECUADOR
B 6303 429* DOLOSTONE, ABST., ROL OF, BRINES AND, FERROUS IRON IN FORMATION
B 6009 1570† DOLOSTONE, LATE CAMBRIAN, EAST-CENTRAL NEW YORK, DEFINITION
B 6303 416* DOLOSTONE TYPES FROM HOST ROCKS FOR MISSISSIPPI VALLEY--TYPE
B 6104 595 DOLPHIN FIELD, AUSTRALIA
B 6203 509 DOLTON, GORDON L., GEOLOGIC APPRAISAL OF UNDISCOVERED OIL AND
B 6004 665 DOLTON, GORDON L., NE APPROACH TO GEOLOGIC ESTIMATES OF OIL
B 6004 660* DOM AREA, LOUISIANA, ABST., FOSSIL, SALT STRUCTURES IN, VACHERIE,
B 6109 1476 DOM JOAO FIELD, BRAZIL
B 6303 410 DOMACK, EUGENE, GLACIAL-MARINE SEDIMENTS OF WEST ANTARCTICA
B 6105 834* DOME, ABST., PALEONTOLOGIC TECHNIQUES FOR DETERMINING
B 6405 668* DOME AREA, ABST., PETROLEUM ACCUMULATION TRENDS IN, EAST,
B 6009 1425† DOME, BIGHORN BASIN, WYOMING, DEFORMATION OF, MESOZOIC,
B 6109 1544* DOME CAP ROCK, ABST., SUBSALT ORIGIN OF, SALT-
B 6012 2162† DOME CAP ROCK, ORIGIN OF, GULF COAST, SALT-
B 6305 785 DOME GROWTH ON CONTINENTAL SLOPE, SALT
B 6101 67 DOME, LOUISIANA, COTE BLANCHE ISLAND, SALT
B 6209 1759* DOME OF LOUISIANA—GEOTEMPERATURES AND HALOKINETICS, ABST.,
B 6411 1841 DOME PETROLEUM LTD., ON DEVELOPMENTS IN SOUTH YEMEN, PEOPLES
B 6409 1556* DOME, SOUTH TEXAS, ABST., CAP-ROCK FORMATION AND DIAGENESIS,
M 25 472 DOME STRUCTURES
B 6109 1540* DOME, WOOD COUNTY, TEXAS, ABST., GROWTH HISTORY OF,
M 25 472 DOME-RING STRUCTURES
B 6211 2262 DOMENGINE CREEK SECTION, CALIFORNIA
B 6211 2261 DOMENGINE SAND, CALIFORNIA
B 6205 849 DOMENGINE SHALE, CALIFORNIA
B 6406 927 DOMENICO, P. A., MICROFRACTURE DEVELOPMENT IN COMPACTING
B 6211 2353 DOMENICO, S. N., MULTIPLE REFLECTIONS AND HEAD WAVES IN GULF OF
B 6405 699 DOMENICO, S. N., SEG DELEGATION VISIT TO PEOPLES REPUBLIC OF
B 6307 1035 DOMES, AND SAGS, BASEMENT, WARPS--, ARCHES
B 6309 1605* DOMES IN EAST TEXAS BASIN FOR GEOLOGIC ISOLATION OF NUCLEAR
B 6211 2240 DOMES IN PRESENT DEEP-SEA ENVIRONMENTS, COMPARISON OF,
B 6209 1574 DOMES, ONSHORE, GULF COAST, SALT
B 6004 724* DOMES--POTENTIAL SITES FOR RADIOACTIVE WASTE DISPOSAL, ABST.,
B 6303 497* DOMESTIC OIL RESERVES FORECASTING METHOD AND ASSESSMENT OF
B 6203 563* DOMING, ABST., REGIONAL VARIATION OF DEPOSITIONAL
B 6203 490 DOMINGUEZ A., B., RECENT DEVELOPMENTS AT CERRO PRIETO
B 6203 362 DOMINGUEZ OIL FIELD, CALIFORNIA
B 6007 1094 DOMINICA TRANSVERSE VALLEY
B 6010 1649 DOMINICAN REPUBLIC, 1975, PETROCONSULTANTS S.A., ON
B 6110 1584 DOMINICAN REPUBLIC, 1976, CARVAJAL, J., ON DEVELOPMENTS IN
B 6210 2050 DOMINICAN REPUBLIC, 1977, PETROCONSULTANTS S.A., ON
B 6310 1750 DOMINICAN REPUBLIC, 1978, PETROCONSULTANTS S.A., ON
S 6 369 DOMINION REEF SYSTEM, SOUTH AFRICA
S 10 233 DON MEDVEDICA AREA (VOLGA), HYDROTHERMAL EFFECTS, USSR
B 6004 689 DONAHUE, J. D., COMPARISON OF DIAGENETIC AND WEATHERING
B 6405 730 DONAHUE, J., SEDIMENTOLOGIC DESCRIPTION OF PART OF
B 6004 665 DONAHUE, JACK., RELATIVE MARINE TRANSGRESSIVE AND REGRESSIVE
B 6105 775 DONALDSON, ALAN C., GAS-PRODUCING SUBMARINE-FAN DEPOSITS OF
B 6008 1397 DONALDSON, PAUL R., GEOELECTRICAL INVESTIGATIONS OF BOISE,
B 6008 1389 DONALDSON, PAUL R., GEOLOGIC AND SEISMIC STUDIES OF BOISE
B 6004 665 DONATH, FRED A., EXPERIMENTAL DEVELOPMENT OF PERMOPOROSITY
B 6105 781 DONATH, FRED A., FABRIC-SELECTIVE PERMOPOROSITY EXPERIMENTALLY
B 6305 825 DONDANVILLE, RICHARD F., GEOLOGIC CHARACTERISTICS OF VALLE
B 6411 1963 DONEGAL AND FASTNET BASINS, EIRE, STYNE TROUGH
B 6211 2201 DONEGAL POINT GROWTH FAULTS, IRELAND, WESTERN
B 6209 1757 DONELLAN, MONICA S., MINERALOGY OF PLEISTOCENE TERRACE
B 6307 1093 DONGARA FIELD, AUSTRALIA
B 6307 1095 DONGARA SADDLE, AUSTRALIA
B 6307 1094 DONGARA SANDSTONE, AUSTRALIA
M 25 445 DONGJOM QUARTZITE, SOUTH KOREA
M 25 293 DONNA RIVER FAULT ZONE, ALASKA
B 6004 666 DONNELL, JOHN R., ENVIRONMENTS OF DEPOSITION OF OIL SHALE, ABST.
B 6405 700 DONNELL, JOHN R., POTENTIAL CONTRIBUTION OF OIL SHALE TO
B 6008 1397 DONNERSTAG, PHILIP, URANIUM EXPLORATION IN PRECAMBRIAN
B 6303 440 DONOGHUE, JOSEPH F., SEDIMENTATION RATES IN COASTAL-PLAIN
M 24 268 DONOVAN SANDSTONES, ALABAMA
B 6302 245 DONOVAN, TERRENCE J., AEROMAGNETIC DETECTION OF DIAGENETIC
B 6006 962† DOONERAK, BROOK RANGE, NORTHERN ALASKA, CARBONIFEROUS
B 6006 952† DOONERAK STRUCTURAL HIGH, CENTRAL BROOKS RANGE, ALASKA,
B 6009 1607* DOOR POINT--BURIED VOLCANO IN SOUTHEAST LOUISIANA, ABST.
B 6303 325 DOR DISTURBANCE, ISRAEL
S 6 306 DORASHAMIAN STAGE, PERMIAN, BIOCHRONOLOGY
B 6012 2179 DORMAN, J. H., NEW OIL AND GAS FROM MATURE PROVINCES, ABST.
B 6412 2078 DORTON BRANCH FAULT ZONE, APPALACHIANS
B 6211 2261 DOS PALOS SHALE, CALIFORNIA
B 6109 1470 DOS REIS FONSECA, J., PETROLEUM GEOLOGY OF EASTERN BRAZILIAN
B 6303 437 DOTT, R. H., JR., ALLUVIAL-FAN DEPOSITION AND TECTONIC
B 6303 441 DOTT, R. H., JR., NUGGET-NAVAJO SANDSTONE ENVIRONMENTAL
B 6308 1424 DOTT, R. H., JR., NUGGET-NAVAJO SANDSTONE ENVIRONMENTAL
B 6004 666 DOTT, R. H., JR., PALEOLATITUDE AND PALEOCLIMATE, ABST.
B 6203 510 DOTT, R. H., JR., SAND TRANSPORT ACROSS EOCENE SHELF AND SLOPE IN
B 6004 734 DOTT, R. H., JR., SUBMARINE-FAN SEDIMENTATION IN MESOZOIC
B 6405 685 DOTT, ROBERT H., JR., DEPOSITIONAL ENVIRONMENTS IN UPPER
B 6401 60 DOUALA BASIN, CRETACEOUS, KEROGEN, CAMEROON
B 6401 60 DOUALA BASIN, CRETACEOUS, KEROGEN, CAMEROUN
B 6405 788 DOUGHERTY, TOM, WHATS NEW IN U.S. INDUSTRY ACTIVITY, ABST.
B 6203 550* DOUGLAS CREEK ARCH, COLORADO, ABST., DEPOSITIONAL LITHOFACIES
B 6107 1030 DOUGLAS GROUP, NEBRASKA
B 6405 700 DOUGLAS, R. G., DEEP-SEA BENTHIC FORAMINIFERA AND THEIR
B 6012 2175 DOUGLAS, R. G., PALEOCEANOGRAPHIC VARIATIONS AND THEIR
B 6203 510 DOUGLAS, ROBERT G., SLOPE-BASIN BENTHIC FORAMINIFERA OF
B 6304 700 DOUGLAS, ROBERT G., TERTIARY AND CRETACEOU OCEAN
B 6212 2510 DOUGLASS, R. H., PROGRESS AND PROSPECTS IN OCEAN THERMAL
B 6405 730 DOUGLASS, RAYMOND C., PALEOGEOGRAPHY AND STRUCTURAL
B 6101 83 DOUKKALA-BERRICHID-RHARB BASINS, MOROCCO
B 6109 1476 DOURADO FIELD, BRAZIL
B 6207 1215 DOUST. H., GEOLOGY AND EXPLORATION HISTORY OF OFFSHORE
B 6401 68 DOVER AREA, SAMPLED SECTION, ENGLAND
B 6405 700 DOVER, JAMES H., STATUS OF ANTLER OROGENY IN CENTRAL IDAHO—
B 6004 666 DOVETON, J. H., SPECTRAL AND MARKOVIAN ANALYSIS OF BED
B 6303 441 DOVETON, JOHN H., KOALA--MINICOMPUTER LO ANALYSIS SYSTEM FOR
B 6002 319 DOVETON, JOHN H., NUMERICAL ANALYSIS OF TRAP LOCATIONS IN
B 6303 441 DOW, WALLACE, CAN GEOCHEMISTRY FIND OIL. BY TITLE ONLY
B 6209 1584 DOW, WALLACE G., PETROLEUM SOURCE BEDS ON CONTINENTAL SLOPES
M 29 423 DOW, WALLACE G., PETROLEUM SOURCE BEDS ON CONTINENTAL SLOPES
B 6105 781 DOW, WALLACE G., PETROLEUM SOURCE BEDS ON CONTINENTAL SLOPES
B 6303 441 DOW, WALLACE G., VITRINITE REFLECTANCE--WHAT, HOW, AND WHY,
B 6303 495* DOWNCURRENT CHANGE IN MIOCENE CANYON-CHANNEL SYSTEMS IN
B 6302 168 DOWNDIP WATER FLOW, OR WATER BLOCK
M 26 57 DOWNLAP, DEPOSITIONAL SEQUENCES, BASELAP, ONLAP AND
B 6309 1577 DOWSE, MARY E., MIDDLE AND UPPER DEVONIAN STRATIGRAPHY IN
B 6205 893 DOWSE, MARY E., SUBSURFACE THRUSTS ALONG WASATCH HINGE LINE
B 6103 371 DOXEY FORMATION, CALIFORNIA
B 6406 957 DOYLE AND ORRIN H. PILKEY, SUMMERHAYES, COLIN P., REVIEW OF,
B 6106 903 DOYLE, LARRY J., MARINE GEOLOGY OF BAJA CALIFORNIA
B 6405 700 DOYLE, LARRY J., SANDS AND SAND TRANSPORT ON PALIMPSEST
B 6203 510 DOYLE, LARRY J., SEDIMENTATION ON EASTERN UNITED STATES
M 30 540 DRAA ET TEMRA FIELD, ALGERIA, OUARGLA REGION
B 6303 494* DRAFTING--ITS APPLICATIO IN PETROLEUM EXPLORATION, ABST.,
B 6312 2222 DRAGNESET FORMATION, NORTH SEA
B 6303 463* DRAINAGE ANOMALIES IN GULF COAST TERTIARY SANDSTONES, ABST.
B 6004 667 DRAKE, D. E., CONTRASTS IN CONCENTRATION OF SUSPENDED MATTER
B 6405 731 DRAKE, D. E., ET AL, SEDIMENTARY PROCESSES ON CONTINENTAL SHELF
B 6405 756 DRAKE, D. E., SEDIMENTARY CHARACTERISTICS AND PROCESSES OF
B 6303 344 DRAKE FORMATION, KENTUCKY
B 6310 1982 DRAKE, HEDBERG, HOLLIS D., REVIEW OF THE, IMPACT OF THE,
B 6101 24 DRAKE POINT, WEST AND EAST, FIELD, CANADIAN, ARCTIC ISLANDS
B 6303 425* DRANEY LIMESTONE--EARLY CRETACEOUS LACUSTRINE CARBONATE
M 26 258 DRAPE UNCONFORMITIES
B 6411 1970 DRAVA, AND SAVA SUBBASINS, YUGOSLAVIA, BANAT, MURA
B 6004 667 DRAVIS, JEFF, WIDESPREAD GENERATION OF HARDGROUNDS IN
B 6105 828* DREDGE DISPOSAL IN LITTORAL ZONE, NEW RIVER INLET, NORTH
B 6105 833* DREDGED MATERIAL DEPOSITED AT OCEAN DUMPING SITES SEAWARD
B 6105 831* DREDGED MATERIAL—MAJOR ENVIRONMENTAL CONSIDERATIONS IN
B 6105 790 * DREDGING AND SPOIL DISPOSAL IN CHESAPEAKE BAY, 1975-2025
B 6303 441 DRESCHHOFF, GISELA, URANIUM RESOURCE EVALUATION IN
B 6303 442 DRESHER, WILLIAM H., FEDERAL BARRIERS TO ENERGY RESOURCE
B 6004 727 DRESSER, H., SUBTIDAL TRACE FOSSILS, MUDDY AND FRONTIER
B 6303 522 DREW, L. J., PREDICTING FUTUR OIL USING THREE-DIMENSIONAL
B 6105 782 DREW, L. J., STATISTICAL HISTORY OF PETROLEUM EXPLORATION IN
B 6203 558 DREW, ROY H., GEOLOGIC APPRAISAL OF UNDISCOVERED OIL AND GAS
M 24 121 DREXLER, JAMES M., ALTAMONT-BLUEBELL--A MAJOR, NATURALLY
B 6405 784 DREXLER, TIMOTHY, OSWEGO LIMESTONE, ALINE-LAMBERT FIELDS,
B 6405 701 DRIESE, STEVEN G., SHALLOW-MARINE TRACE ASSEMBLAGES IN A

B 6410 1701 DRIFT, BLAKE-, BAHAMA, OUTER RIDGE, ABYSSAL SEDIMENT
B 6202 223 DRIFT, CONTINENTAL
B 6410 1703 DRIFT, GREAT, BAHAMA BANK
B 6410 1703 DRIFT, LITTLE, BAHAMA BANK
S 9 18 DRIFT MOVEMENTS, ANTARCTICA, CENOZOIC
S 9 18 DRIFT MOVEMENTS, AUSTRALIA, CENOZOIC
B 6203 419 DRIFT, NORTH, ATLANTIC, CONTINENTAL
S 9 1 DRIFT, PALEOCLIMATE, PALEOMAGNETISM, AND CONTINENTAL
B 6101 79 DRIFT, TRANS-ATLANTIC SIMILARITIES, CONTINENTAL
B 6410 1715 DRIFTS, CRITERIA FOR RECOGNITION OF, CARBONATE SEDIMENT
B 6410 1701† DRIFTS IN NORTHER STRAITS OF FLORIDA, CARBONATE SEDIMENT
B 6405 786* DRILL, ABST., BASINS, BASEMENT, AND THE
B 6405 785* DRILL AS TOOL FOR STUDYING MINERAL DEPOSITS AND
B 6305 742 DRILL CUTTINGS, CAPILLARY PROPERTIES OF
B 6008 1156 DRILLING, 1975, CANADA, EXPLORATORY
B 6008 1157 DRILLING, 1975, MEXICO, EXPLORATORY
B 6108 1124 DRILLING, 1976, CANADA, EXPLORATORY
B 6108 1124 DRILLING, 1976, MEXICO, EXPLORATORY
B 6208 1278 DRILLING, 1977, CANADA, EXPLORATORY
B 6208 1279 DRILLING, 1977, MEXICO, EXPLORATORY
B 6308 1171 DRILLING, 1978, CANADA, EXPLORATORY
B 6308 1172 DRILLING, 1978, MEXICO, EXPLORATORY
B 6409 1298 DRILLING, 1979, CANADA, EXPLORATORY
B 6409 1299 DRILLING, 1979, MEXICO, EXPLORATORY
B 6303 553* DRILLING, ABST., BIOGENOUS SEDIMENTS RECOVERED BY, DEEP-SEA
B 6008 1153† DRILLING ACTIVITY IN 1975, NORTH AMERICAN
B 6108 1121† DRILLING ACTIVITY IN 1976, NORTH AMERICAN
B 6208 1275† DRILLING ACTIVITY IN 1977, NORTH AMERICAN
B 6308 1168† DRILLING ACTIVITY IN 1978, NORTH AMERICAN
B 6409 1295† DRILLING ACTIVITY IN 1979, NORTH AMERICAN
B 6205 742 DRILLING ACTIVITY, WORLD, EXPLORATORY
B 6303 537* DRILLING AND GLOBAL TECTONICS, ABST., DEEP-SEA
B 6303 409* DRILLING AND PRODUCTION PROBLEMS IN UNDISCOVERED RESERVOIRS,
B 6303 341† DRILLING DATA, DELINEATION OF, ROME TROUGH IN EASTERN,
B 6004 677* DRILLING DEVICE, ABST., RECENT AND RELICT TOPOGRAPHY OF, BOO
B 6405 675* DRILLING FOR METHANE GAS IN FISHERS PEA AREA, LAS ANIMAS
M 26 374 DRILLING HAZARDS, RESOLUTION SYSTEMS, POTENTIAL
B 6303 554* DRILLING IN ANTARCTIC--LATE TERTIARY PALEOCLIMATIC HISTORY,
B 6011 2070* DRILLING IN ANTARCTICA, ABST.
B 6405 769* DRILLING IN SAN ANDREAS FAULT, ABST.
B 6101 50 DRILLING, LEBANON
B 6303 540* DRILLING MUD ON SELECTED CORALS, ABST., EFFECTS OF AN OFFSHORE
B 6303 546* DRILLING ON CONVERGENT OCEAN MARGINS, ABST., IPOD
B 6004 649* DRILLING PROJECT IN WESTERN NORTH ATLANTIC, ABST., PRELIMINARY
B 6007 1041 DRILLING PROJECT, SITE 98, DEEP SEA
B 6409 1300 DRILLING REPORTING, DEFINITIONS
B 6308 1173 DRILLING REPORTING, DEFINITIONS
B 6208 1280 DRILLING REPORTING, DEFINITIONS
B 6108 1126 DRILLING REPORTING, DEFINITIONS
B 6008 1158 DRILLING REPORTING, DEFINITIONS
B 6105 833* DRILLING STATISTICS IN BASIN EVALUATION, ABST., APPLICATION OF
B 6108 1392* DRILLING STATISTICS IN BASIN EVALUATION, ABST., APPLICATION OF
B 6008 1154 DRILLING SUMMARY--1975, U.S.
B 6308 1169 DRILLING SUMMARY--1978, U.S.
B 6108 1122 DRILLING SUMMARY—1976, U.S.
B 6208 1277 DRILLING SUMMARY—1977, U.S.
B 6409 1297 DRILLING SUMMARY—1979, U.S.
B 6303 297 DRILLING TECHNOLOGY, DEEP, OFFSHORE
S 10 23 DROPLET DEFORMATION, MIGRATION, MECHANISMS
S 10 97 DROPLET EXPULSION, MIGRATION MECHANISMS
B 6305 770 DROPLET EXPULSION, OIL
S 5 93 DROSTE, JOHN B. SYNCHRONIZATION OF DEPOSITION: SILURIAN
S 5 4 D-SALT, SALINA GROUP,
B 6009 1618 DROSTE, JOHN B., GENERAL SEDIMENTOLOGIC HISTOR OF LATE
B 6004 684 DROSTE, JOHN B., SEARCH FOR SILURANIAN REEF MODEL--GREA LAKES
B 6408 1283 DROSTE, JOHN B., STRUCTURAL INTERPRETATION OF BURIED SILURIAN
B 6304 699 DROZD, R. J., PREDICTIONS OF OIL OR GA POTENTIAL BY NEAR-SURFACE
B 6405 753 DRUCKMAN, YEHEZKEEL, POST-COMPACTION, SUBSURFACE SECONDARY
B 6107 1046 DRUM, QUIVIRA, FORMATION, KANSAS
B 6012 2179 DRUMMOND, K. F., UPPE CRETACEOUS LITHOFACIES MODEL,
B 6102 257 DRUMMOND MINE LIMESTONE, IDAHO
B 6112 2116 DRY CANYON, NEW MEXICO
B 6008 1408* DRY CREEK AREA, CARBON COUNTY, MONTANA, ABST., MESAVERDE GAS
B 6402 222 DRY DEVIL QUADRANGLE, TEXAS
B 6101 113 DRY HOLLOW MEMBER, UTAH
B 6104 595 DRY MESA FIELD, FOUR CORNERS
M 24 114 DRY PINEY FIELD, WYOMING
B 6108 1376* DRY PINEY, TIP TOP, AND HOGSBACK FIELDS, SUBLETTE COUNTY,
B 6204 696 DSDP 100, BAHAMA PLATFORM
B 6205 797 DSDP 102, JOIDES
B 6205 797 DSDP 104, JOIDES
B 6204 696 DSDP 99, BAHAMA PLATFORM
B 6309 1453 DSDP, CORE HOLE 108, BALTIMORE CANYON TROUGH, DEEP SEA
M 28 89 DSDP CORE, MEDITERRANEAN, MIOCENE
M 28 88 DSDP CORE, MEDITERRANEAN, MIOCENE
M 28 87 DSDP CORE, MEDITERRANEAN, MIOCENE
B 6004 688* DSDP CORES FROM PACIFIC OCEAN, ABST., RESEDIMENTATION OF,
M 25 218 DSDP DRILLING SITES, NEW ZEALAND, GLOMAR CHALLENGER
B 6205 793 DSDP HOLE 105, JOIDES
B 6204 601 DSDP HOLE 108, JOIDES
B 6401 72 DSDP HOLES 390A, 398D, CRETACEOUS-TERTIARY BOUNDARY, CARBON
B 6401 72 DSDP HOLES 390A, 398D, CRETACEOUS-TERTIARY BOUNDARY, ISOTOPE
B 6303 287 DSDP, INTERNATIONAL PHASE OF OCEAN DRILLING,, JOIDES, DEEP SEA
B 6303 503* DSDP LEGS 36, 40--SEDIMENTOLOGIC IMPLICATIONS, ABST., CALCAREOUS,
B 6303 413* DSDP LEGS 59 AND 60, ABST., CENOZOIC, CLAY-MINERAL STRATIGRAPHY
B 6204 596 DSDP SITE 100, JOIDES
B 6204 596 DSDP SITE 101, JOIDES
B 6204 596 DSDP SITE 105, JOIDES
B 6412 2058 DSDP SITE 105, LEG 11, GEOCHEMICAL LOG, HATTERAS ABYSSAL PLAIN
B 6204 596 DSDP SITE 106, JOIDES
B 6202 274 DSDP SITE 142
B 6203 395 DSDP SITE 178, ALASKA
B 6203 395 DSDP SITE 180, ALASKA
B 6203 395 DSDP SITE 181, ALASKA
B 6203 395 DSDP SITE 182, ALASKA
B 6212 2486 DSDP SITE 190
B 6412 2118† DSDP SITE 262, TIMOR TROUGH, ORGANIC, GEOCHEMISTRY OF
B 6202 274 DSDP SITE 354
B 6412 2055 DSDP SITE 361, LEG 40, GEOCHEMICAL LOG, CAPE BASIN
B 6412 2055 DSDP SITE 364, LEG 40, GEOCHEMICAL LOG, ANGOLA BASIN
B 6412 2056 DSDP SITE 367, LEG 41, GEOCHEMICAL LOG, CAPE VERDE BASIN
B 6412 2056 DSDP SITE 370, LEG 41, GEOCHEMICAL LOG, MOROCCO
B 6412 2058 DSDP SITE 386, LEG 43, GEOCHEMICAL LOG, BERMUDA RISE
B 6204 597 DSDP SITE 391, LEG 44, BLAKE BASIN, JOIDES
B 6412 2058 DSDP SITE 391, LEG 44, GEOCHEMICAL LOG, BLAKE-, BAHAMA AREA
B 6209 1605 DSDP SITE 391 OFF, NEW JERSEY
B 6209 1601 DSDP SITE 391, PROFILE, NORTH ATLANTIC
B 6412 2057 DSDP SITE 398D, LEG 47, GEOCHEMICAL LOG, PORTUGAL
B 6412 2057 DSDP SITE 402A, LEG 48, GEOCHEMICAL LOG, BAY OF BISCAY
B 6405 711* DSDP SITE 503, LEG 68—INDICATE CONTINUOUS SECTION OF
B 6204 694 DSDP SITE 98, BAHAMA PLATFORM
B 6007 1042 DSDP SITE 98, RV, GLOMAR CHALLENGER
B 6204 596 DSDP SITE 99, CAT GAP, JOIDES
B 6209 1594 DSDP SITE WITH RECOVERY OF CRETACEOUS BLACK SHALE
B 6209 1592 DSDP, SITES 172-182, LEG 18, OREGON AND, ALASKA MARGINS
B 6401 72 DSDP SITES 367 AND 391C, CARBON ISOTOPE RATIOS, NORTH ATLANTIC
B 6401 72 DSDP SITES 367 AND 391C, CARBON ISOTOPE RATIOS, NORTH ATLANTIC
B 6209 1594 DSDP SITES, MID-ATLANTIC RIDGE
B 6303 483* DSDP-IPOD, ABST., ORGANIC, GEOCHEMISTRY OF SEDIMENTS RECOVERE
B 6303 293 DSDP-IPOD HOLES, RESERVOIR SANDS, ORGANIC CARBON CONTENT, AND
B 6303 516* DSDPBIOSTRATIGRAPHY, ABST., RECENT AND POTENTIAL ADVANCES IN
B 6203 531* DSDP—DEEP DRILLING IN SHIKOKU BASIN AND NORTHEASTERN
B 6204 693 DSRV ALVIN CRUISE
B 6410 1703 DSRV ALVIN, SAMPLING
B 6004 667 DU DRESNAY, R., DEVELOPMENT AND PALEOGEOGRAPHIC EXTENT O
B 6303 522* DUAL ORIGIN OF NATURAL GASES IN SUBALPINE TERTIARY BASINS,
B 6309 1579 DUBA, A. G., THEORETICAL AND EXPERIMENTAL ANALYSES OF
B 6010 1870 DUBAI, 1975, CONTINENTAL OIL CO., AND, TEXAS PACIFIC OIL CO., ON
B 6110 1799 DUBAI, 1976, DUBAI PETROL. CO. AND, TEXAS PACIFIC OIL CO., ON
B 6310 1888 DUBAI, 1978, DUBAI PETROLEUM CO., AND, SEDCO ENERGY CO., ON
B 6411 1840 DUBAI, 1979, DUBAI PETROLEUM CO. AND, SEDCO, ENERGY CORP., ON
B 6110 1799 DUBAI PETROL. CO. AND TEXAS PACIFIC OIL CO., ON DEVELOPMENTS IN
B 6310 1888 DUBAI PETROLEUM CO., AND SEDCO ENERGY CO., ON DEVELOPMENTS IN
B 6411 1840 DUBAI PETROLEUM CO. AND SEDCO ENERGY CORP., ON DEVELOPMENTS
B 6210 2116 DUBAI PETROLEUM CO. AND TEXAS PACIFIC OIL CO., INC., ON
B 6210 2120 DUBAI, PRODUCTION, 1976-1977
B 6310 1898 DUBAI, PRODUCTION, 1978
B 6411 1861 DUBAI PRODUCTION, 1979
B 6207 1215 DUBEY, MICHAEL, ENERGY FROM WIND—OLD IDEA WITH NEW
B 6207 1225 DUBOIS, JACQUES, SEISMICITY AND STRUCTURE OF NEW HEBRIDES
B 6405 701 DUBOIS, MARTIN K., PALEOTOPOGRAPHYS INFLUENCE ON POROSITY
B 6109 1513 DUCHENE, HARVEY R., OSHA CANYON FORMATION, PENNSYLVANIAN,
M 25 427 DUCHESS AREA, PHOSPHATES, QUEENSLAND
B 6203 510 DUCKWORTH, PETER B., BURIED CREVASSE FILLINGS OF PORT HURON
B 6309 1602 DUEX, T. W., HYDROLOGY AND WATER QUALITY OF EOCENE WILCOX
B 6205 883 DUEY, HERBERT D., TRAP SPRING OIL FIELD, NYE COUNTY, NEVADA,
B 6211 2354 DUEY, HERBERT D., TRAP SPRING OIL FIELD, NYE COUNTY, NEVADA,
B 6405 701 DUEY, HERBERT D., TRAP SPRING OIL FIELD, NYE COUNTY, NEVADA
B 6408 1245 DUFFAUD, F., PRESSURE ORIGIN AND DISTRIBUTION IN JURASSIC OF
B 6309 1578 DUFFY, JULIAN, PAST AND CURRENT DEVELOPMENTS FOR DRILLING
M 25 427 DUGALD RIVER, ZINC-LEAD, QUEENSLAND
B 6309 1597 DUIS, THEO, ECOLOGIC NICHES OF RADIOLARIANS, PLANKTONIC

B 6309 1598 DUIS, THEO, RADIOLARIAN DISTRIBUTION, DIVERSITY AND DENSITY IN
B 6007 1039 DUKE UNIVERSITY RV EASTWOOD CRUISE, JANUARY, 1974
B 6303 460 DUKE, W. L., HUMMOCKY CROSS-STRATIFICATION--INDICATOR OF
B 6104 585 DUKHAN FIELDS, PERSIAN GULF
B 6405 689 DULONG, F. T., ET AL, GEOLOGIC CONTROLS ON MINERAL-MATTER
B 6405 689 DULONG, F. T., ET AL, GEOLOGIC CONTROLS ON SULFUR CONTENT IN
B 6405 788 DULONG, F. T., FORMS AND ASSOCIATIONS OF PYRITE IN UPPER
B 6405 702 DULONG, F. T., X-RAY MINERALOGY OF UPPER FREEPORT COAL, ABST.
B 6105 831* DUMPING OF DREDGED MATERIAL—MAJOR ENVIRONMENTAL
B 6105 841* DUMPING ON NEW YORK BIGHT INNER SHELF, ABST., GEOLOGIC EFFECTS
B 6105 833* DUMPING SITES SEAWARD OF COLUMBIA RIVER MOUTH, ABST.,
B 6412 2078 DUMPLIN VALLEY FAULT, APPALACHIANS
B 6109 1557 DUNAWAY, W. E., LOWER CRETACEOUS CARBONATE ROCKS OF CENTRAL
B 6103 332 DUNBARTON BASIN, SOUTH CAROLINA
B 6105 791 * DUNE DYNAMICS IN COASTAL ZONE—PAST, PRESENT, AND FUTURE
B 6004 731* DUNE ROCKS, ABST., INDICATORS OF, CLIMATE IN, CARBONATE
M 28 85 DUNES, UTAH, HOLOCENE
B 6303 442 DUNHAM, JOHN B., SHALLOW SUBSURFACE DOLOMITIZATION OF
M 25 460 DUNITE-HARZBURGITE FACIES, PACIFIDES, PERIDOTITE BELT
B 6212 2396 DUNKARD BASIN, SECTION
B 6303 442 DUNLAP, JOHN B., JR., APPLIED BIOSTRATIGRAPHY, SOUTH LOUISIANA
B 6303 542 DUNLAP, W., SHALLOW-SUBMARINE SEISMIC STRATIGRAPHY, MISSISSIPPI
M 30 102 DUNLIN FORMATION, JURASSIC, STATFJORD FIELD
B 6105 803 DUNN, C. E., CARBONATE POROSITY MODIFICATION BY HALITE
B 6108 1376 DUNNEWALD, JOHN B., NUGGET OIL ACCUMULATIONS AT DRY PINEY,
B 6303 435 DUNSTER, D. E., SURFACE SEISMIC METHODS APPLIED TO COAL-MINING
M 29 152 DUNTROON BASIN, AUSTRALIA, CONTINENTAL MARGIN
B 6201 74 DUPEROW LIMESTONE-DOLOMITE, SASKATCHEWAN
B 6405 704 DUPLESSY, JEAN CLAUDE, CARBON FLUX FROM OCEAN TO BIOSPHERE—
B 6403 440 DUPRE, W. R., MODERN AND ANCIENT COASTAL SEDIMENTARY FACIES,
B 6305 826 DUPREE, J. ANN, PALEOKARST CONTROLS O LOCALIZATION OF URANIUM
B 6203 457 DUPUY SANDSTONE, AUSTRALIA
M 29 329 DUQUE-CARO, HERMAN, MAJOR STRUCTURAL ELEMENTS AND
B 6212 2510 DUQUE-CARO, HERMANN, GEOLOGY AND HYDROCARBON PROSPECTS OF
B 6207 1215 DUQUETTE, FRANK F., SOLAR ENERGY—CENTRAL RECEIVER PROGRAM,
B 6405 704 DURAZZI, JOSEPH T. ET AL, CARBON FLUX FROM OCEAN TO BIOSPHERE—
B 6101 80 DURHAM-DEEP RIVER BASIN, NORTH CAROLINA AND, VIRGINIA
B 6010 1957 DURKEE, E. F., PETROLEUM DEVELOPMENTS IN AUSTRALIA IN 1975
B 6110 1888 DURKEE, E. F., PETROLEUM DEVELOPMENTS IN AUSTRALIA IN 1976
B 6210 2150 DURKEE, E. F., PETROLEUM DEVELOPMENTS IN AUSTRALIA IN 1977
B 6310 1804 DURKEE, E. F., PETROLEUM DEVELOPMENTS IN AUSTRALIA IN 1978
B 6411 1862 DURKEE, E. F., PETROLEUM DEVELOPMENTS IN AUSTRALIA IN 1979
B 6307 1143 DURKEE, E. F., REVIEW OF OFF-SHORE PETROLEUM RESOURCES OF
B 6408 1273 DURKEE, E. F., REVIEW OF TURKIYE DORDUNCU PETROL KONGRESI,
B 6405 621 DURMID HILLS, SEASAT IMAGE, CALIFORNIA
B 6303 499 DUSSEAULT, M. B., SEDIMENTOLOGY, PETROLOGY, AND GEOTECTONIC
M 28 114 DUST RIMS, CEMENTS, QUARTZ OVERGROWTHS
B 6203 490* DUTCH SECTOR OF NORTH SEA, ABST., REGIONAL CONTROLS ON
B 6412 2189 DUTCHER, MITCHLER, JOHN, REVIEW OF DEPOSITIONAL AND
B 6405 758 DUTRO, J. T., JR., ET AL, SEDIMENTOLOGY OF FLUVIAL UPPER DEVONIAN
B 6006 952 DUTRO, J. T., JR., GEOLOGIC SIGNIFICANCE OF DOONERAK STRUCTURAL
B 6203 535 DUTROW, BARBARA L., SEDIMENTOLOGY OF LATE QUATERNARY
B 6403 427 DUTTA, T. K., DEPOSITIONAL ENVIRONMENT OF SOURC BEDS OF
B 6403 427 DUTTA, T. K., DEPOSITIONAL ENVIRONMENT OF SOURCE BEDS OF
B 6303 511 DUTTON, S. P., PALO DURO BASIN ANALYSIS, ABST.
B 6409 1556 DUTTON, SHIRLEY P., CAP-ROCK FORMATION AND DIAGENESIS, GYP
B 6405 717 DUTTON, SHIRLEY P., DEPOSITIONAL AND TECTONIC EVOLUTION OF A
B 6308 1425 DUTTON, SHIRLEY P., DEPOSITIONAL MODELS AND RESOURCE
B 6303 461 DUTTON, SHIRLEY P., EXPLORATION POTENTIA OF
B 6303 442 DUTTON, SHIRLEY P., FACIES PATTERNS AND DEPOSITIONAL MODELS,
B 6311 2116 DUTTON, SHIRLEY P., PENNSYLVANIAN FAN-DELTA SANDSTONES, PAL
B 6401 88 DUTTON, SHIRLEY P., PENNSYLVANIAN-EARLY PERMIAN DEPOSITIONAL
B 6405 702 DUTTON, SHIRLEY P., SOURCE-ROCK QUALITY AND THERMAL
B 6109 1536 DUTTON, SHIRLEY PETERSON, DIAGENESIS AND POROSITY
B 6205 783 DUVERNAY FORMATION, ALBERTA
B 6205 785 DUVERNAY SHALE, ALBERTA, IRETON-
B 6306 977 DUWI FORMATION, EGYPT
B 6306 970 DUWI PHOSPHATE, EGYPT
B 6207 1207 DYKSTRA, H., DEVELOPMENT AND RESERVOIR ANALYSIS OF
B 6105 778 DYMOND, J., MARINE METALLIFEROUS SEDIMENTS FROM NAZCA PLATE
B 6309 1490 DYKSTRA, JON D., RECONNAISSANCE GEOLOGIC MAPPING IN CHAGAI
M 25 539 DYMOND, JACK R., METALLOGENESIS IN SOUTHEAST PACIFIC OCEAN,
M 26 30 DYNAMIC MODULI, LABORATORY, MEASUREMENT
B 6004 699* DYNAMIC PROCESSES OF UPPER BENGAL FAN AND SWATCH OF NO
B 6203 497* DYNAMIC SCALE MODELING OF BED CONFIGURATIONS, ABST.
S 10 11 DYNAMIC STRUCTURING, WATER, MODEL
B 6004 701* DYNAMICS IN WEST MISSISSIPPI DELTA AREA, ABST., CONTROL OF
B 6109 1550* DYNAMICS OF NATURAL SYSTEMS—GENETIC BASIS FOR ASSESSING
B 6405 801* DYNAMICS OF UNVEGETATED TIDAL-FLAT MUDS, ABST.
B 6108 1376 DYNI, JOHN R., DISTRIBUTION AND ORIGIN OF SULFUR IN COLORADO
B 6303 439 DYNI, JOHN R., STATISTICAL ZONATION OF OIL SHALE ON BASIS OF
B 6302 182 DYPVIK, HENNING AND, PETROLOGY AND DIAGENESIS OF JURASSIC
B 6312 2222 DYPVIK, HENNING, COMPOSITION OF ORGANIC MATTER FROM NORTH
B 6408 1211 DZUNGARIA BASIN, CHINA
S 10 156 DZUNGARIAN ARTESIAN BASIN, PETROLEUM MIGRATION, CHINA
B 6004 644* E ABRA LIMESTONE AT ITS TYPE LOCALITY, EASTERN MEXICO, ABST.,
B 6004 671* E DORADO MICELLAR-POLYMER PROJECT AREA, BUTLER COUNTY,
S 7 234 E-LOG SECTION, CALIFORNIA, LOS ANGELES BASIN
M 24 18 E-POOL MARKER, ALASKA, MIDDLE GROUND SHOAL FIELD
B 6405 688* EAGAR FORMATION AND MOGOLLON RIM GRAVELS OF ARIZONA, ABST.,
B 6203 526* EAGAR FORMATION OF ARIZONA, ABST., GENETIC STRATIGRAPHY AND
B 6405 702 EAGAR, R. M. C., BIVALVE TRACE FOSSILS IN BRITISH SILESIAN, ABST.
B 6402 230 EAGLE FORD GROUP, TEXAS
B 6303 528* EAGLE FORD INTERVAL, UPPER CRETACEOUS, SUGAR CREEK FIELD,
M 26 114 EAGLE FORD SHALE, TEXAS, POLK COUNTY
B 6008 1398* EAGLE FORMATION, BEARPAW MOUNTAINS AREA, NORTH-CENTRAL
S 3 100 EAGLE HILL SILICIFIED MUD VOLCANO, TEXAS
M 26 150 EAGLE MILLS FORMATION, TEXAS, EAST
M 26 147 EAGLE MILLS FORMATION, TEXAS, EAST TEXAS BASIN
B 6211 2215 EAGLE REST PEAK, CALIFORNIA
B 6203 397 EAGLE RIVER FAULT, ALASKA
B 6407 971 EAGLE SANDSTONE, GREAT PLAINS
B 6403 337 EAGLE SANDSTONE, MAJOR RESERVOIR SANDSTONES, MONTANA
B 6405 712* EAGLE SANDSTONE, MONTANA, ABST., DIAGENESIS IN SHALLOW
B 6008 1408* EAGLE SANDSTONE, NORTH-CENTRAL MONTANA--AID FOR
B 6305 827* EAGLE SANDSTONE OF BEARPAW MOUNTAINS, NORTH-CENTRAL
B 6403 316† EAGLE SANDSTONE—RELATION TO SHALLOW GAS ACCUMULATIONS,
B 6205 881* EAGLE SPRINGS OIL FIELD, NYE COUNTY, NEVADA, ABST.
B 6408 1220 EAGLEFORD SECTION, TEXAS
B 6109 1551* EARLY AND MIDDLE CRETACEOUS HIPPURITACEA, RUDISTS, OF GULF
B 6004 712* EARLY CAMBRIAN SHALLOW SUBTIDAL ENVIRONMENT--TIME AND
B 6409 1561* EARLY CEMENTATION AND MINERAL STABILITY OF CHIPOLA
B 6405 734* EARLY CEMENTATION BY HIGH-MAGNESIUM CALCIT FROM GULF COAST
B 6405 668* EARLY CRETACEOUS ARC SEDIMENTATION AND VOLCANISM IN COASTAL
B 6001 108† EARLY CRETACEOUS BURIED CANYON--INFLUENCE ON ACCUMULATION
B 6405 782* EARLY CRETACEOUS REEF COMMUNITIES IN GULF COAST, ABST.
B 6009 1617* EARLY CRETACEOUS SHELF MARGIN IN NORTHERN MEXICO AND SOUTH
B 6002 288 EARLY DIAGENESIS
B 6002 287† EARLY DIAGENETIC GENERATION OF HYDROCARBON GASES AND THEIR
B 6405 691* EARLY DIAGENETIC GLAUCONITIZATION OF FELDSPAR GRAINS IN A
B 6203 442† EARLY EOCENE SUBAERIAL EROSIONAL VALLEYS IN CAMBAY BASIN,
B 6405 768* EARLY GUADALUPIAN, PERMIAN, BANK MARGIN EROSION SURFACES,
B 6004 662* EARLY MARINE AND LATER POSTBURIAL CEMENTATION AND
B 6211 2361* EARLY MESOZOIC PALEOTECTONIC-PALEOGEOGRAPHIC
B 6303 425* EARLY MESOZOIC TECTONIC FRAMEWORK AND SEDIMENTATION,
B 6211 2360* EARLY MIOCENE SUBMARINE DUNE FIELDS—CROSS-BEDDING IN
B 6405 793* EARLY PALEOZOIC CONODONT BIOSTRATIGRAPHY AND
B 6405 676* EARLY PALEOZOIC CONODONT BIOSTRATIGRAPHY, BIOGEOGRAPHY,
B 6311 2120* EARLY PENNSYLVANIAN BRAIDED STREA SEDIMENTATION, NORTHWEST
B 6009 1624* EARLY PENNSYLVANIAN DEPOSITIONAL SYSTEMS IN NORTHEASTERN
B 6108 1386* EARLY PERMIAN SLIP AND STRATIGRAPHIC CONSEQUENCES OF
B 6303 551* EARLY PLEISTOCENE SUBMARINE CANYON, BOSO PENINSULA, JAPAN,
B 6405 735* EARLY POROSITY AND PERMEABILITY REDUCTION IN DEEP-SEA FAN
B 6203 570* EARLY SUBAERIAL EXPOSURE AND FRESHWATER DIAGENESIS OF UPPER
B 6309 1611* EARLY TERTIARY LITHOSTRATIGRAPHIC INTERPRETATION OF
B 6207 1193† EARLY TERTIARY RIFTING IN WESTERN OREGON-WASHINGTON
B 6209 1765* EARLY TERTIARY TERRESTRIAL PALEOGEOGRAPHY OF NORTH
B 6405 710* EARLY TRANSFORMATION REACTION OF STEROIDS IN MARINE
B 6104 603 EARP FORMATION, NEW MEXICO
M 25 473 EARTH, MOON, MARS, AND VENUS, RING, STRUCTURES, DISTINCT ON
B 6109 1560* EARTH RESOURCES OBSERVATORY—A CONCEPT, ABST.
M 25 63 EARTH RESOURCES TECHNOLOGY SATELLITE ERTS, DEVELOPMENT
B 6304 578 EARTH TEMPERATURES AND OCCURRENCE OF OIL AND GAS
B 6212 2413 EARTHQUAKE EPICENTERS, GULF OF ALASKA
M 29 444 EARTHQUAKE HYPOCENTERS, PERU TRENCH
B 6012 2182* EARTHQUAKE PREDICTION, ABST., RECENT RESULTS IN
B 6103 466* EARTHQUAKES AND ENGINEERING--REVIEW OF CURRENT PRACTICES,
M 29 275 EARTHQUAKES AND FAULTS, INDEX, ALASKA
B 6103 466* EARTHQUAKES AND FUTURE EARTHQUAKE HAZARDS ALONG GULF OF
M 25 352† EARTHQUAKES AT RANGELY, COLORADO--SUMMARY, CONTROL OF
M 25 427 EAST ALLIGATOR, URANIUM, AUSTRALIA, NORTHERN TERRITORY
B 6304 603 EAST AND WEST PUERTO CHIQUITO FIELDS, FRACTURE RESERVOIRS,
M 25 474 EAST ASIAN MEGASTRUCTURE
M 25 473 EAST ASIAN MEGASTRUCTURE
M 25 59 EAST ASIAN VOLCANOGENIC BELT, USSR
M 25 467 EAST ASIATIC SYSTEM, CONTINENT-OCEAN ZONE
S 10 262 EAST BERNARD FIELD, LEVERIDGE ESTATE WELL, LOG, TEXAS
B 6307 1047 EAST CAMERON AREA, SECTION, GULF COAST
M 24 205† EAST CAMERON BLOCK 270, OFFSHORE LOUISIANA--A PLEISTOCENE
B 6302 245 EAST CEMENT DOME, OKLAHOMA

M 25 263 EAST CHINA SEA, KOREA OFFSHORE, SOUTH SEA
M 25 13 EAST CHINA SEA, SEABED JURISDICTION
M 25 467 EAST CHINESE VOLCANIC BELT
B 6404 506 EAST COAST ANOMALY
B 6109 1484 EAST COAST, BEAR SEAMOUNT
B 6109 1523 EAST COAST, CAPE FEAR ARCH, ACCUMULATION RATES
B 6109 1489 EAST COAST CONTINENTAL MARGIN, EUGEOCLINAL WEDGE
B 6109 1486 EAST COAST CONTINENTAL MARGIN, MIOGEOCLINAL WEDGE
B 6109 1484 EAST COAST, CORSAIR CANYON
M 25 222 EAST COAST FOLD BELT, NEW ZEALAND
B 6109 1484 EAST COAST, GEORGES BANK BASIN
B 6109 1484 EAST COAST, GEORGES BASIN
B 6109 1483 EAST COAST GEOSYNCLINE
B 6109 1484 EAST COAST, GULF OF MAINE HIGH
B 6109 1484 EAST COAST, LAHAVE PLATFORM
B 6109 1484 EAST COAST, LONG ISLAND PLATFORM
B 6410 1659 EAST COAST MAGNETIC ANOMALY
M 29 94 EAST COAST MAGNETIC ANOMALY, ATLANTIC MARGIN
B 6007 1115 EAST COAST MAGNETIC ANOMALY, GAMMA CONTOUR OF
B 6109 1484 EAST COAST, NEW ENGLAND SEAMOUNT CHAIN
B 6106 919 EAST COAST, NORFOLK CANYON
B 6109 1484 EAST COAST, NORTHEAST CHANNEL
B 6109 1523 EAST COAST, PENINSULAR ARCH, ACCUMULATION RATES
B 6109 1484 EAST COAST, SCOTIAN BASIN
B 6109 1523 EAST COAST, SOUTHEAST GEORGIA EMBAYMENT, ACCUMULATION
B 6109 1522 EAST COAST, SUBSIDENCE RATES
B 6402 174 EAST COAST, SUBSIDENCE, UNITED STATES
B 6106 919 EAST COAST, WASHINGTON CANYON
B 6109 1484 EAST COAST, YARMOUTH ARCH
B 6303 349 EAST CONTINENT GRAVITY HIGH, KENTUCKY
B 6111 1981 EAST CORTES BASIN, CALIFORNIA
B 6203 414 EAST END RANGE, VIRGIN ISLANDS, ST. CROIX
B 6311 2002 EAST FRIGG FAN, NORTH SEA
B 6010 1724 EAST GERMANY, DDR, DEVELOPMENTS, 1975
B 6210 1923 EAST GERMANY, DEVELOPMENTS, 1977
B 6210 1923 EAST GERMANY, DEVELOPMENTS, 1977, POLAND-
B 6110 1654 EAST GERMANY, PRODUCTION, 1975-1976
B 6210 1924 EAST GERMANY, PRODUCTION, 1976-1977
B 6310 1932 EAST GERMANY, PRODUCTION, 1977-1978
M 25 246 EAST INDONESIA, IRIAN JAYA, BASINS
M 25 246 EAST JAVA BASIN, INDONESIA
B 6401 43 EAST JAVA SEA REGION, TERMINOLOGY
M 25 246 EAST KALIMANTAN BASINS, INDONNSIA
B 6303 545* EAST KALIMANTAN, INDONESIA, ABST., HANDIL FIELD
M 30 441† EAST KALIMANTAN, INDONESIA, DISCOVERY AND DEVELOPMENT OF
M 30 399† EAST KALIMANTAN, INDONESIA, GEOLOGY OF THE, HANDIL FIELD
B 6008 1408* EAST KEITH GAS AREA, LIBERTY COUNTY, MONTANA, ABST.
B 6101 70 EAST LAKE SAND FIELD, LOUISIANA
B 6210 1985 EAST MALAYSIA, DEVELOPMENTS, 1977
B 6303 526* EAST MALAYSIA OVER PAST DECADE, ABST., EXPLORATION IN
M 30 423† EAST MALAYSIA OVER THE PAST DECADE, EXPLORATION IN
B 6203 503* EAST MESA, CALIFORNIA, ABST., GEOTHERMAL RESOURCE
B 6108 1385* EAST MESA GEOTHERMAL FIELD, IMPERIAL COUNTY, CALIFORNIA, ABST.,
S 7 9 EAST MEXICO SHELF, DESCRIBED
S 7 11 EAST MEXICO SLOPE, DESCRIBED
B 6109 1499 EAST MIKKELSEN BAY SECTION, ALASKA
B 6208 1382 EAST OLD OCEAN FIELD, BRAZORIA COUNTY, TEXAS
M 25 220 EAST OTAGO-CAMPBELL PLATEAU BASIN O GREAT SOUTH BASIN, NEW
M 25 539 EAST PACIFIC RISE
S 3 79 EAST PACIFIC RISE
B 6212 2511* EAST PACIFIC RISE AND BAUER BASIN, ABST., EVIDENCE FOR
M 25 487 EAST PACIFIC RISE AND MID-ATLANTIC RIDGE, METAL
M 25 382 EAST PACIFIC RISE, NODULES
M 25 485 EAST PACIFIC RISE, SEAFLOOR SPREADING
M 25 460 EAST PACIFIC RISE, ULTRABASITES, METAL CONCENTRATIONS IN
S 4 129 EAST PALISADES, REEF, JAMAICA
B 6203 404 EAST RANGE, VIRGIN ISLANDS, ST. CROIX
M 25 263 EAST SEA, SEA OF JAPAN, KOREAN OFFSHORE
M 30 99 EAST SHETLAND BASIN, CROSS SECTION, STATFJORD FIELD
M 30 547 EAST SIRTE BASIN, STRATIGRAPHY, LIBYA
B 6409 1356 EAST SOUR LAKE FIELD, HARDIN COUNTY, TEXAS
B 6409 1563* **EAST TEXAS BASIN AND SOUTHWEST ARKANSAS, ABST., DEPOSITIONAL**
M 26 146 EAST TEXAS BASIN, BUCKNER FORMATION, TEXAS
M 26 151 EAST TEXAS BASIN, CORRELATION OF UNITS TO A CYCLE CHART, TEXAS
M 26 147 EAST TEXAS BASIN, COTTON VALLEY FORMATION, TEXAS
B 6408 1220 EAST TEXAS BASIN, CROSS SECTION
M 26 147 EAST TEXAS BASIN, EAGLE MILLS FORMATION, TEXAS
M 26 147 EAST TEXAS BASIN, GILMER FORMATION, TEXAS
M 26 147 EAST TEXAS BASIN, HOSSTON FORMATION, TEXAS
M 26 147 EAST TEXAS BASIN, LOUANN FORMATION, TEXAS
M 26 147 EAST TEXAS BASIN, SMACKOVER FORMATION, TEXAS
M 26 147 EAST TEXAS BASIN, WERNER FORMATION, TEXAS
B 6408 1220 EAST TEXAS FIELD, SECTION, TEXAS
S 10 223 EAST TEXAS FIELD, TRAP, TEXAS
B 6111 1960 EAST TEXAS, OIL ACCUMULATION
B 6407 1006 EAST TEXAS SALT BASIN, TEXAS
B 6405 668* EAST TEXAS SALT DOME AREA, ABST., PETROLEUM ACCUMULATION
B 6108 1373† EAST TINTIC DISTRICT, UTAH, ABST., SAGEBRUSH AND, GEOCHEMICAL
B 6303 434* EAST TURKANA, KENYA, ENVIRONMENTAL SIGNIFICANCE OF, FOSSIL,
B 6109 1499 EAST UGNU SECTION, ALASKA
B 6212 2493 EAST WILMINGTON FIELD, REPETTO STRUCTURE, UPPER, CALIFORNIA
B 6008 1222† EAST-CENTRAL STATES IN 1975, DEVELOPMENTS IN
B 6108 1183† EAST-CENTRAL STATES IN 1976, DEVELOPMENTS IN
B 6208 1340† EAST-CENTRAL STATES IN 1977, DEVELOPMENTS IN
B 6308 1324† EAST-CENTRAL STATES IN 1978, DEVELOPMENTS IN
B 6409 1462† EAST-CENTRAL STATES IN 1979, DEVELOPMENT IN
B 6407 972 EASTEND FORMATION, ALBERTA
M 29 249 EASTERN ALEUTIAN TRENCH, PHYSIOGRAPHY, ACCRETIONARY BASIN
M 25 458 EASTERN ASIATIC BELT OR RHYOLITE LINE, PACIFIDES
M 25 460 EASTERN ASIATIC VOLCANOGENIC BELT, BORON, ALUMINOQUARTZITES,
M 25 460 EASTERN ASIATIC VOLCANOGENIC BELT, GOLD AND SILVER
M 25 460 EASTERN ASIATIC VOLCANOGENIC BELT, MERCURY, POLYMETALLIC
B 6008 1355† EASTERN CANADA IN 1975, DEVELOPMENTS IN
B 6108 1335† EASTERN CANADA IN 1976, DEVELOPMENTS IN
B 6208 1451† EASTERN CANADA IN 1977, DEVELOPMENTS IN
B 6308 1344† EASTERN CANADA IN 1978, DEVELOPMENTS IN
B 6409 1518† EASTERN CANADA IN 1979, DEVELOPMENTS IN
B 6004 708* EASTERN CONTINENTAL MARGIN, 1926-1976, ABST., SEDIMENTOLOGY
B 6004 708* EASTERN CONTINENTAL MARGIN, ABST., SEDIMENTOLOGY AND
B 6405 759* EASTERN CORDILLERAN FOLDBELT AND FORELAND OF NORTHERN
B 6306 976 EASTERN DESERT, CENTRAL, GEOLOGIC MAP, EGYPT
B 6306 975 EASTERN DESERT, CENTRAL, NUBIA FORMATION, EGYPT
B 6306 975† EASTERN DESERT, EGYPT--MAJOR SUBDIVISIONS AND DEPOSITIONAL
B 6405 748* EASTERN GREEN RIVER BASIN—A DEVELOPING GIAN GAS SUPPLY FROM
S 7 157 EASTERN GULF OF MEXICO, BATHYMETRY AND PHYSIOGRAPHY
S 7 21† EASTERN GULF OF MEXICO CONTINENTAL MARGIN: STRATIGRAPHIC
S 7 26 EASTERN GULF OF MEXICO, DISTRIBUTION MAP, JURASSIC
S 7 23 EASTERN GULF OF MEXICO, GEOLOGIC MAP
S 7 36 EASTERN GULF OF MEXICO MARGIN, DESCRIBED
S 7 29 EASTERN GULF OF MEXICO, TECTONIC MAP
B 6206 1092* EASTERN INTERIOR BASIN, ABST., ORGANIC, GEOCHEMISTRY OF,
B 6405 761* EASTERN INTERIOR REGIONS AND ROCKY MOUNTAIN COAL PROVINCE—
B 6309 1584* EASTERN KENTUCKY GAS FIELD AND COMPARISON WITH GAS
M 30 396 EASTERN OFFSHORE BASIN, STRATIGRAPHIC COLUMN, TRINIDAD
M 30 393 EASTERN OFFSHORE BASIN, STRUCTURAL BOUNDARIES, TRINIDAD
B 6002 321* EASTERN OKLAHOMA AND WESTERN ARKANSAS, BY TITLE ONLY
B 6309 1579* **EASTERN PROJECTION OF VALLEY AND RIDGE BENEATH METAMORPHIC**
B 6205 787 EASTERN SHELF AREA, SEISMIC DATA, TEXAS
B 6103 346 EASTERN SHELF, TEXAS
M 30 225† EASTERN SIBERIA, USSR, GEOLOGY AND, PETROLEUM FIELDS IN,
B 6209 1667 EASTERN SUBPROVINCE, TAIWAN
B 6405 686 EASTERN UTAH AND SOUTHWESTERN COLORADO, ABST., LOWER,
M 26 200 EASTERN-SHELF BUILDUPS, SEISMIC EXAMPLE, TEXAS, CROSBY COUNTY,
B 6204 698 EASTWARD CRUISE E-16A-74, RV
B 6204 701 EASTWARD CRUISE E-1B-74, RV
B 6204 702 EASTWARD CRUISE E-1C-77, RV
B 6410 1703 EASTWARD, SEISMIC REFLECTION PROFILES, RV
S 4 54 EASTWARD SOUNDINGS, RV
B 6007 1039 EASTWOOD CRUISE, JANUARY, 1974, DUKE UNIVERSITY, RV
B 6405 703 EATON, GORDON P., DEEP DRILLING AND CURRENT MODELS OF
M 24 259 EAVES, EVERETT, CITRONELLE OIL FIELD, MOBILE COUNTY, ALABAMA
B 6311 2116 EBANKS, W. J., JR., CORRELATION OF CHEROKEE, DESMOINESIAN,
B 6002 319 EBANKS, W. J., JR.,, GEOLOGY OF MISSISSIPPIAN COMBINATION TRAP,
B 6103 309 EBANKS, W. J., JR., MISSISSIPPIAN COMBINATION TRAP, BINDLEY FIELD,
B 6203 511 EBANKS, W. J., JR., SUBSURFACE HEAVY-OIL-BEARING SANDSTONES IN
B 6203 570 EBANKS, WILLIAM J., EARLY SUBAERIAL EXPOSURE AND FRESHWATER
B 6405 707* EBB-TIDAL DELTA STRATIFICATION AND ITS RELATION TO TIDAL INLET
B 6305 835* EBRO BASIN, SPAIN, ABST., RECENT, SALT SEDIMENTATION IN, PLAYA
B 6309 1599 EBY, DAVID E., CLASTICITY INDEX--KEY TO CORRELATING
B 6209 1767 EBY, DAVID E., PERITIDAL CARBONATE MATERIAL AND EVIDENCE FOR
B 6308 1426 EBY, DAVID E., SEDIMENTOLOGY AND SOURCE-BED GEOCHEMISTRY OF
B 6403 429 ECCA HIGH-WAX OIL FIELD, AFRICA
B 6303 475 **ECE, OMER I., MICROSCOPIC DISTRIBUTION OF URANIUM IN OAKVILLE**
M 27 94 ECHINODERM FRAGMENTS
M 28 39 ECHINODERM, GLAUCONITE AFTER CALCITE, FRAGMENTS
B 6401 130 ECHINODERMAL DEBRIS ZONES, INDIANA
B 6401 130 ECHINODERMAL DEBRIS ZONES, INDIANA
M 27 96 ECHINODERMS
M 27 95 ECHINOIDS
S 4 271 ECHINOIDS, BORERS

S 4 286 ECHINOMETRA LUCUNTER EROSION RATES
S 4 283 ECHINOMETRA LUCUNTER, ST. CROIX, BOILER BAY
S 4 285 ECHINOMETRA MATHEI, PERSIAN GULF
B 6303 443 ECHOLS, DOROTHY J., CHRONICLE OF MIOCENE, PHASE III, MIDDLE
B 6403 440 ECHOLS, R. J., HOLOCENE FORAMINIFERAL DISTRIBUTION PATTERNS ON
M 24 29 ECHOOKA FORMATION, ALASKA
B 6109 1496 ECHOOKA FORMATION, ALASKA
B 6103 438 ECHOOKA SANDSTONE, ALASKA
B 6309 1597* ECOLOGIC NICHES OF RADIOLARIANS, PLANKTONIC FORAMINIFERS, AND
S 10 109† ECONOMIC ASPECTS OF WATER-ROCK INTERACTION, SOME
M 25 365† ECONOMIC DEVELOPMENT ON DEMANDS FOR GROUNDWATER AND
B 6308 1421 ECONOMIC MINERAL DEPOSITS, BY COLIN J. DICKSON, MORAN, W. R.,
B 6203 566* ECONOMIC POTENTIAL OF CANYON-FAN UNIT IN ANADARKO BASIN,
M 25 388 ECONOMIC POTENTIAL OF NODULES
B 6008 1405* ECONOMIC VALUES USIN DISTRIBUTED RANGES OF VALUES OF RESERVES
B 6004 711* ECONOMICS, ABST., IMPACT OF, OIL AND, GA, EXPLORATION
M 25 24† ECONOMICS OF ENERGY AND MINERAL RESOURCES--SUMMARY
B 6212 2511* ECONOMICS OF LARGE-SCALE SURFACE COAL MINING—SIMULATION
B 6209 1764* ECOPHENOTYPES IN GULF COAST ESTUARIES—ECOLOGIC AND
B 6010 1650 ECUADOR, 1975, CANFIELD, R. W., ON DEVELOPMENTS IN
B 6110 1584 ECUADOR, 1976, CANFIELD, R. W., ON DEVELOPMENTS IN
B 6210 2050 ECUADOR, 1977, FASSETT, B. D., ON DEVELOPMENTS IN
B 6310 1750 ECUADOR, 1978, FASSETT, B. D., ON DEVELOPMENTS IN
B 6203 559* ECUADOR, ABST., VOLCANOGENIC SEDIMENTATION ON, ANDEAN
M 25 339 ECUADOR, AMAZON BASINS, POTENTIAL
M 25 329 ECUADOR, ANCON GROUP
M 25 339† ECUADOR, AND, HYDROCARBON POTENTIAL OF, AMAZON BASINS OF,
M 25 553 ECUADOR, ANGAS, COPPER, ZINC, LEAD, SILVER, AND GOLD DEPOSITS
M 25 553 ECUADOR, ANGAS DEPOSIT, POLYMETALLIC MINERALIZATION
M 25 328 ECUADOR, ATLANTA FORMATION
M 25 328 ECUADOR, AZUCAR GROUP
M 25 553 ECUADOR, CHAUCHA COPPER-MOLYBDENUM DEPOSIT
M 25 553 ECUADOR, CHAUCHA DEPOSIT, COPPER-MOLYBDENUM
M 25 553 ECUADOR, CLAYS AND TRAVERTINE MARBLE, SURVEYS
M 25 553 ECUADOR, COAL INVESTIGATIONS
M 25 328 ECUADOR, CRETACEOUS
B 6212 2458 ECUADOR, DOLORES-GUAYAQUIL MEGASHEAR
M 25 329 ECUADOR, EOCENE
M 25 554 ECUADOR, FIERRO URCO COPPER, ZINC, LEAD, AND MOLYBDENU
M 25 554 ECUADOR, FIERRO URCO, COPPER, ZINC, LEAD, AND MOLYBDENUM
M 25 554 ECUADOR FIERRO URCO DEPOSIT, COPPER AND RELATED MINERALS
M 25 554 ECUADOR, GOLD AND BASE-METAL SULFIDES, SURVEYS
M 25 554 ECUADOR, LOS LINDEROS COPPER AND MOLYBDENUM DEPOSITS
M 25 554 ECUADOR, LOS LINDEROS DEPOSIT, COPPER-MOLYBDENUM
M 25 552 ECUADOR, MACUCHI MINE, COPPER
B 6212 2468 ECUADOR, MANABI BASIN
M 25 552 ECUADOR, MINERALS, EXPLORATION
M 25 555 ECUADOR, MINING PROSPECTS
M 25 328 ECUADOR, MIOCENE
B 6112 2086 ECUADOR, NAPO FORMATION, KEROGEN
M 25 328 ECUADOR, PALEOCENE
B 6212 2468 ECUADOR, PORTOVIEJO FAULT
B 6110 1617 ECUADOR, PRODUCTION, 1975-1976
B 6210 2084 ECUADOR, PRODUCTION, 1976-1977
B 6310 1780 ECUADOR, PRODUCTION, 1977-1978
M 25 554 ECUADOR, RIO PLAYAS, COPPER AND MOLYBDENUM DEPOSITS
M 25 554 ECUADOR, RIO PLAYAS, COPPER-MOLYBDENUM
M 25 553 ECUADOR, SAN BARTOLOME DEPOSIT, SILVER-ZINC-LEAD
M 25 553 ECUADOR, SAN BARTOLOME, SILVER, ZINC, AND LEAD DEPOSITS
M 25 329 ECUADOR, SAN EDUARDO FORMATION
M 25 328 ECUADOR, SAN JOSE FORMATION
M 25 329 ECUADOR, SAN MATEO FORMATION
B 6212 2461 ECUADOR, SANTA ELENA OIL FIELD
M 25 330 ECUADOR, SANTIAGO AND ZAPALLO FORMATIONS
M 25 554 ECUADOR, SARAGURD DEPOSIT, TIN AND TUNGSTEN
M 25 554 ECUADOR, SARAGURO TIN AND TUNGSTEN DEPOSITS
M 25 329 ECUADOR, SECE FORMATION
B 6212 2458 ECUADOR, SEISMICITY
M 25 329 ECUADOR, SOCORRO FORMATION
B 6212 2473 ECUADOR SUBDUCTION SYSTEM, PETROLEUM IN
M 25 330 ECUADOR, SUBIBAJA FORMATION
M 25 552† ECUADOR, THEIR DEVELOPMENT AND PROSPECTS FOR NEW
M 25 552 ECUADOR, TOACHI DEPOSITS, LA PLATA MINE, GOLD, SILVER, AND
B 6212 2455 ECUADOR TRENCH
M 25 554 ECUADOR, URITOHUASER DEPOSIT, LEAD-ZINC
M 25 554 ECUADOR, URITOHUASER, LEAD AND ZINC DEPOSITS
M 25 328† ECUADOR--SUMMARY, EXPLORATION AND DEVELOPMENT OF
B 6405 687* ECUADORIAN ORIENTE, ABST., SACHA FIELD OF
B 6212 2454† ECUADORIAN SUBDUCTION SYSTEM
B 6402 280† ECUADORIAN SUBDUCTION SYSTEM. DISCUSSION AND REPLY
B 6206 961 EDALE SHALES, ENGLAND
B 6002 326* EDDY COUNTY, NEW MEXICO, ABST., ROCK TANK, MORROW SANDSTONE,
B 6204 706 EDGEWOOD LIMESTONE, ILLINOIS BASIN
B 6410 1682 EDIACARA HILLS, PRECAMBRIAN, SOUTH, AUSTRALIA
B 6410 1586 EDINBURG BEDS, VIRGINIA
B 6211 2246 EDISON HIGH, NEW TERM, CALIFORNIA
B 6303 476 EDISON, T. A., INTEGRATION OF MICROSCOPIC ORGANIC ANALYSIS AND
B 6204 668 EDJELEH OIL FIELD, ALGERIA
M 25 231 EDMOND, G. M., AUSTRALIAN NORTHWEST CONTINENNAL
B 6403 441 EDMONDSON, WILLIAM F., MEMBER SANDS OF WINTERS FORMATION,
B 6408 1166 EDMONTON FACIES, ALBERTA, PASKAPOO-
B 6407 972 EDMONTON GROUP, ALBERTA
B 6105 782 EDSALL, DOUGLAS W., GEOLOGIC DEVELOPMENT OF CENOZOIC
B 6203 544 EDWARDS, B. D., BOTTOM SEDIMENTS OF SAN NICOLAS BASIN,
B 6405 703 EDWARDS, BRIAN D., LEBENSSPUREN OF DYSAEROBIC, BATHYAL BASIN.
B 6403 441 EDWARDS, BRIAN D., MASS MOVEMENT EFFECTS ON A BATHYAL
B 6203 511 EDWARDS, BRIAN D., NEW EVIDENCE OF CURRENT WINNOWING
B 6403 442 EDWARDS, BRIAN D., SEDIMENTARY PROCESSES ACTIVE ON SLOPES OF
S 11 88 EDWARDS CANYON, CALIFORNIA, CHECK LIST OF, FORAMINIFERA,
B 6402 221 EDWARDS COUNTY, MAPPING, TEXAS
B 6303 409* EDWARDS FORMATION, NORTH-CENTRAL TEXAS--STRATIGRAPHY.
B 6109 1554* EDWARDS GROUP, CHERT, CENTRAL TEXAS,ABST., HYDROGEN AND,
B 6109 1553* EDWARDS GROUP, IN VICINITY OF JUNCTION, TEXAS, ABST., DEPOSITION
B 6102 294 EDWARDS, H. S., ATOKA GAS IN SOUTHERN COTTLE AND NORTHERN
B 6308 1425 EDWARDS, H. S. GUS, ATOKA GAS IN SOUTHERN COTTLE AND NORTHER
B 6307 1093 EDWARDS ISLAND BLOCK, AUSTRALIA
B 6108 1269 EDWARDS, J., JR., OIL AND GAS DEVELOPMENTS IN MARYLAND, OHIO,
B 6008 1288 EDWARDS, JONATHAN, JR., OIL AND GAS DEVELOPMENTS IN MARYLAND,
B 6004 688* EDWARDS LIMESTONE, NORTH-CENTRAL TEXAS, ABST., DEVELOPMENT
B 6109 1553* EDWARDS LIMESTONE, NORTH-CENTRAL TEXAS, ABST., FACIES,
B 6405 749* EDWARDS LIMESTONE, SAN ANTONIO AREA, TEXAS, ABST., STABLE,
B 6301 37 EDWARDS LIMESTONE, TEXAS
M 27 205 EDWARDS LIMESTONE, TEXAS, CRETACEOUS
B 6004 668 EDWARDS, M. B., GROWTH FAULTS AND ASSOCIATED STRUCTURES IN
B 6405 703 EDWARDS, MARC B., GROWTH FAULTING AND SEDIMENTATION IN
B 6003 341 EDWARDS, MARC B., GROWTH FAULTS IN UPPER TRIASSIC DELTAIC
B 6409 1556 EDWARDS, MARC B., LIVE OAK DELTA COMPLEX—UNSTABLE,
B 6312 2193 EDWARDS, MARC B., SANDSTONE IN LOWER CRETACEOUS
B 6402 238 EDWARDS PLATEAU, ERTS IMAGE, TEXAS
B 6402 238 EDWARDS PLATEAU, ERTS IMAGE, TEXAS
B 6203 571* EDWARDS PLATEAU REGION, TEXAS, ABST., STRUCTURAL GEOLOGY OF
M 26 310 EDWARDS REEF, TEXAS, EAST
B 6109 1538* EDWARDS-MCKNIGHT DIAGENESIS, ABST., JFS FIELD, DIMMIT COUNTY,
B 6203 346 EEL RIVER BASIN, CALIFORNIA
S 11 318 EEL RIVER, CALIFORNIA, SAMPLE LOCALITIES, SCOTIA-
S 11 196 EEL RIVER FORMATION
S 11 222 EEL RIVER FORMATION, ASSEMBLAGES
S 11 218 EEL RIVER FORMATION, FOSSIL AGGREGATE
S 11 323 EEL RIVER SECTION DESCRIPTION, SCOTIA-
B 6303 551* EFFECT OF BIOGENIC METHANE ON SEDIMENT INSTABILITY IN MODERN
B 6102 174 EFFECT OF CEMENTATION AND COMPACTION ON PERMEABILITY
B 6405 766* EFFECT OF CEMENTATION ON PHYSICAL PROPERTIE OF SANDSTONES,
B 6203 548* EFFECT OF CLAYS ON URANIUM DISTRIBUTION IN SOUTH TEXAS, ABST.
B 6007 1126 EFFECT OF COMPACTION ON FLUID PRESSURES
B 6303 468* EFFECT OF DEPLETION ON NEAR-SURFACE HYDROCARBON ANOMALIES,
B 6203 555* EFFECT OF FAULT-RELATED GRANULATION ON POROSITY AND
B 6009 1531† EFFECT OF HIGH-PRESSURE DEFORMATION ON PERMEABILITY OF
B 6401 65 EFFECT OF MINERAL MATRIX ON FREE HYDROCARBONS
B 6401 62 EFFECT OF MINERAL MATRIX ON HYDROCARBON YIELD
B 6312 2227† EFFECT OF OIL AND BITUMEN SATURATION ON SOURCE-ROCK
B 6102 188 EFFECT OF OVERBURDEN AND FOLDING ON COMPACTION
M 25 371† EFFECT OF SUBSURFAC WASTE-DISPOSAL PRACTICE ON GROUNDWATER
B 6207 1214* EFFECTIVE ENVIRONMENTAL REGULATION, ABST.
B 6405 766* EFFECTIVE STRESS AND ABNORMALLY HIGH FLUID PRESSURE, ABST.
B 6403 414 EFFECTIVE STRESS, DEFINITION
B 6108 1374* EFFECTIVE USE OF MICROEARTHQUAKE RECORDINGS AS EXPLORATION
B 6303 540* EFFECTS OF AN OFFSHORE DRILLING MUD ON SELECTED CORALS, ABST.
B 6405 784* EFFECTS OF BASALT INTRUSIONS ON KEROGENS IN UNCONSOLIDATED
B 6105 803 * EFFECTS OF BURIAL STAGES ON POROSITY IN RESERVOIR ROCKS IN
B 6105 795 * EFFECTS OF DREDGING AND HANDLING TECHNIQUES ON SEDIMENT TEXTURE
B 6409 1558* EFFECTS OF HURRICANE FREDERIC ON MORPHOLOGY OF DAUPHIN
B 6004 655* EFFECTS OF ICE ON SEDIMENTS ON SANDY BRAIDED RIVER, ABST.
B 6109 1535* EFFECTS OF SANDSTONE COMPOSITION AND DIAGENESIS ON RESERVOIR
B 6303 547* EFFECTS OF SOURCE MATERIAL AND THERMAL MATURATION ON
B 6203 553* EFFECTS OF SUBMARINE DIAGENESIS ON PRIMARY POROSITY IN
B 6309 1478† EFFECTS OF TEXTURE AND COMPOSITION ON MECHANICAL BEHAVIOR
B 6304 690* EFFECTS OF URANIUM MINING ON GROUNDWATER IN AMBROSIA LAKE
B 6203 552* EFFECTS OF VARIATIONS IN WAVE ENERGY ON BEACH AND
B 6405 703 EFFIMOFF, I., TERTIARY STRUCTURAL DEVELOPMENT OF SELECTED
B 6410 1577 EFFNA BEDS, VIRGINIA
B 6205 886* EGAN RANGE, NEVADA, ABST., TERTIARY STRUCTURAL DEVELOPMENT
B 6408 1259 EGBERT, ROBERT M., NONMARINE UPPER CRETACEOUS ROCKS, COOK
S 6 180 EGERIAN STAGE, PARATETHYS, NEOGENE

Index

B 6302 183 EGERSUND SUBBASIN, NORTH SEA
S 6 180 EGGENBURGIAN STAGE, PARATETHYS, NEOGENE
S 11 230 EGGERELLA BRADYI
S 11 270 EGGERELLA BRADYI
B 6410 1602 EGGLESTON FORMATION, VIRGINIA
B 6203 509 EGGLESTON, JANE R., FRESHWATER ONCOLITES, ONONDAGA LAKE, NEW
B 6203 534* EGGSHELL DIAGENESIS AND PRECOMPACTIONAL DISSOLUTION OF
M 25 62 EGIAZAROV, B. KH., TECTONICS OF CONTINENTAL AND OCEANIC
B 6411 1967 EGIL BERGSAGER, ON DEVELOPMENTS IN NORWAY NORTH OF 62
B 6101 3 EGLINTON GRABEN
M 29 43 EGLOFF, JULIUS, EROSIONAL AND DEPOSITIONAL STRUCTURES OF THE
B 6010 1769 EGYPT, 1975, PETROCONSULTANTS S.A., ON DEVELOPMENTS IN
B 6110 1700 EGYPT, 1976, PETROCONSULTANTS S.A., ON DEVELOPMENTS IN
B 6210 1797 EGYPT, 1977, PETROCONSULTANTS S.A., ON DEVELOPMENTS IN
B 6310 1637 EGYPT, 1978, PETROCONSULTANTS S.A., ON DEVELOPMENTS IN
B 6411 1740 EGYPT, 1979, PETROCONSULTANTS S.A., ON DEVELOPMENTS IN
B 6305 829* EGYPT, ABST., ALLUVIAL-PLAIN SEDIMENTS OF, NUBIA, SOUTHWESTERN
B 6303 416* EGYPT, ABST., MODERN EXPLORATION STRATEGY IN FRONTIER AREAS,
B 6304 594 EGYPT, ABU MADI FIELD, TEMPERATURE MAP
B 6106 875 EGYPT, ALEXANDRIA CANYON
B 6306 974 EGYPT AND SUDAN, NUBIA GROUP NOMENCLATURE AND CORRELATION
B 6106 875 EGYPT, BARDAWIL LAGOON
B 6306 969 EGYPT, BASAL CLASTIC UNIT
B 6306 976 EGYPT, CRETACEOUS
B 6306 970 EGYPT, CRETACEOUS
B 6306 969 EGYPT, DAKHLA BASIN
B 6306 970 EGYPT, DAKHLA SHALE
B 6106 880 EGYPT, DAMIETTA FAN
B 6306 972 EGYPT, DESERT ROSE UNIT
B 6306 977 EGYPT, DUWI FORMATION
B 6306 970 EGYPT, DUWI PHOSPHATE
B 6306 976 EGYPT, EASTERN DESERT, CENTRAL, GEOLOGIC MAP
B 6306 975 EGYPT, EASTERN DESERT, CENTRAL, NUBIA FORMATION
B 6001 34† EGYPT, EVOLUTION OF, EOCENE-, MIOCENE SEDIMENTATION, PATTERN
B 6306 971 EGYPT, GILF KEBIR UNIT
B 6307 1032 EGYPT, GULF OF SUEZ GRABEN, FAULTS
B 6306 970 EGYPT, JURASSIC
B 6306 971 EGYPT, LINGULA SHALE UNIT
B 6306 967† EGYPT, MAJOR SUBDIVISIONS AND DEPOSITIONAL ENVIRONMENTS OF,
B 6106 873 EGYPT, MESSINIAN EVAPORITES
B 6106 877 EGYPT, NILE VALLEY AND DELTA
B 6406 861 EGYPT, NORTHERN BASIN
B 6306 975 EGYPT, NUBIA FORMATION, FIVE LITHOLOGIC UNITS
B 6306 977 EGYPT, NUBIA FORMATION, STRATIGRAPHIC COLUMN
B 6106 877 EGYPT, PHYSIOGRAPHY
B 6306 972 EGYPT, PLANT BED UNIT
B 6306 976 EGYPT, PRECAMBRIAN BASEMENT COMPLEX
B 6110 1719 EGYPT, PRODUCTION, 1976
B 6210 1815 EGYPT, PRODUCTION, 1977
B 6310 1653 EGYPT, PRODUCTION, 1978
B 6411 1764 EGYPT, PRODUCTION, 1979
B 6106 875 EGYPT, QATTARA DEPRESSION
B 6306 976 EGYPT, QUATERNARY
B 6106 880 EGYPT, ROSETTA FAN
B 6406 862 EGYPT, SOUTHERN BASIN
B 6306 968 EGYPT, SOUTHWESTERN, OUTCROP MAP
B 6306 973 EGYPT, TAREF SANDSTONE UNIT
B 6306 976 EGYPT, TERTIARY
B 6306 973 EGYPT, VARIEGATED SHALE UNIT
B 6306 975† EGYPT--MAJOR SUBDIVISIONS AND DEPOSITIONAL SETTING, NUBIA
B 6106 881 EGYPTIAN CONTINENTAL MARGIN AND ADJACENT DEEP SEA, MAGNETIC
B 6203 536* EGYPTIAN RED SEA AND GULF OF SUEZ AREA—TALE OF FOUR TROUGHS,
B 6203 563* EGYPT—RESPONSE TO PRERIFT CRUSTAL DOMING, ABST., REGIONAL
B 6004 668 EHLIG, P. L., MAGNITUDE AND TIMING OF DISPLACEMENT ON SA
B 6211 2352 EHLIG, PERRY, AGE OF SEDIMENTATION-DIAGENESIS FOR BEDFORD
S 11 261 EHRENBERGINA COMPRESSA
S 11 294 EHRENBERGINA COMPRESSA
B 6004 659 **EHRLICH, R., ANALYSIS OF SUBMARINE DEBRIS-FLOW DEPOSIT,**
B 6303 477 EHRLICH, R., REGIONAL TECTONICS AND PETROLEUM GEOLOGY OF
B 6303 477 EHRLICH, R., TUNISIA--PLATE TECTONIC AND HYDROCARBON
B 6203 506 EHRLICH, ROBERT, CONGLOMERATIC CARBONATE LIQUEFIED-FLOW
B 6203 519 EHRLICH, ROBERT, HIGH LEVELS OF VARIATION IN SIZE-FREQUENCY
B 6405 773 EHRLICH, ROBERT, LOCATION OF LITTORAL ENERGY FENCE AND
B 6002 268 EIN SHARQYA REEF, ISRAEL
B 6303 552* EIRE, ABST., HYDROCARBON PROSPECTS IN, BASINS WEST OF, UNITED
B 6411 1963 EIRE, CELTIC SEA
B 6010 1709 EIRE, DEVELOPMENTS, 1975
B 6110 1640 EIRE, DEVELOPMENTS, 1976
B 6210 1909 EIRE, DEVELOPMENTS, 1977
B 6310 1918 EIRE, DEVELOPMENTS, 1978
B 6411 1963 EIRE, DEVELOPMENTS, 1979
B 6411 1963 EIRE, GOBAN SPUR
B 6411 1963 EIRE, KISH BANK BASIN
B 6411 1963 EIRE, PORCUPINE TROUGH
B 6310 1932 EIRE, PRODUCTION, 1977-1978
B 6411 1971 EIRE, PRODUCTION, 1978-1979
B 6411 1963 EIRE, STYNE TROUGH, DONEGAL AND FASTNET BASINS
B 6405 693 EISEL, J. D., URANIUM SOLUTION MINING—INTEGRATION OF
M 27 148 EISENOOLITH, GERMANY, JURASSIC
M 28 44 EISENOOLITH, GERMANY, JURASSIC
B 6103 465 EITTREIM, STEPHEN L., CENOZOIC SEDIMENTATION AND TECTONICS OF
B 6103 465 EITTREIM, STEPHEN L., GEOLOGIC FRAMEWORK AN RELATED
M 25 518 EJE NEO-VOLCANICO METALLOGENETIC PROVINCE, MEXICO
B 6002 246 EJECTION CURVES, MERCURY-, INJECTION AND
B 6207 1241 EJIMA, YASUHIKO, PRESENT STATUS OF GEOTHERMAL DEVELOPMENT
B 6203 511 EKDALE, A. A., BATHYMETRIC CONTROL OF DEEP-SEA ICHNOFACIES,
B 6205 883 EKDALE, A. A., HOLOCENE ICHNOFACIES FROM NORTHERN GULF OF
S 4 195 EKDALE, A. A., QUANTITATIVE PALEOECOLOGICAL ASPECTS OF MODERN
B 6405 704 EKDALE, A. A., TRACE FOSSILS AND STAGNATION OF DEEP-SEA BASINS,
B 6303 443 EKDALE, A. A., TRACE-FOSSIL EVIDENCE FOR DEEP-WATER
B 6004 668 **EKDALE, A. A., ABYSSAL ORGANISM TRACES ON AND IN MODERN PELAGIC**
M 30 204 EKOFISK CHALK, DIAGENETIC HISTORY, EKOFISK FIELD
M 30 203 EKOFISK CHALK, SEM PHOTO, EKOFISK FORMATION
M 30 217 EKOFISK FIELD, BRIGHT SPOT EXAMPLE, NORTH SEA
M 30 208 EKOFISK FIELD, CROSS SECTION, NORTH SEA
M 30 221 EKOFISK FIELD, DEVELOPMENT PROGRAM, NORTH SEA
M 30 204 EKOFISK FIELD, EKOFISK CHALK, DIAGENETIC HISTORY
M 30 200 EKOFISK FIELD, EKOFISK FORMATION, NORTH SEA
M 30 213 EKOFISK FIELD, GEOCHEMICAL DATA, NORTH SEA
M 30 197 EKOFISK FIELD, GEOLOGIC SETTING, NORTH SEA
M 30 213 EKOFISK FIELD, GEOPRESSURES, NORTH SEA
M 30 198 EKOFISK FIELD, LITHOSTRATIGRAPHY, NORTH SEA
M 30 196 EKOFISK FIELD, LOCATION, NORWAY
B 6107 1002 EKOFISK FIELD, NORTH SEA
M 30 211 EKOFISK FIELD, SOURCE ROCK, NORTH SEA
M 30 200 EKOFISK FIELD, STRATIGRAPHY, NORTH SEA
M 30 211 EKOFISK FIELD, STRUCTURE, NORTH SEA
M 30 218 EKOFISK FIELD, THERMAL ALTERATION EXAMPLE
M 30 200 EKOFISK FIELD, TOR FORMATION, NORTH SEA
M 30 199 EKOFISK FORMATION, COCCOSPHERE, SEM PHOTO
M 30 203 EKOFISK FORMATION, EKOFISK CHALK, SEM PHOTO
M 30 200 EKOFISK FORMATION, NORTH SEA, EKOFISK FIELD
B 6401 68 EKOFISK WELL, SAMPLED SECTION, NORWEGIAN, NORTH SEA
M 30 195† EKOFISK--FIRST OF THE GIANT OIL FIELDS IN WESTERN EUROPE
B 6304 701* EKOFISK--FIRST OF WESTERN EUROPEAN GIANT OIL FIELDS, ABST.
B 6405 668 EKWEOZOR, C. M., PETROLEUM GEOLOGY OF SENONIAN IN SEDIMENTS
B 6408 1251 EKWEOZOR, CHUKWUEMEKA M., PETROLEUM SOURCE-BED EVALUATION
M 27 56 EL ABRA FORMATION, MEXICO, CRETACEOUS
M 27 161 EL ABRA FORMATION, MEXICO, CRETACEOUS
M 27 85 EL ABRA LIMESTONE, MEXICO, CRETACEOUS
M 27 193 EL ABRA LIMESTONE, MEXICO, CRETACEOUS
M 27 167 EL ABRA LIMESTONE, MEXICO, CRETACEOUS
M 30 20 EL BORMA FIELD, LANDSAT IMAGE, TUNISIA
B 6401 18 EL BORMA GIANT OIL FIELD, LANDSAT DATA, TUNISIA
B 6401 18 EL BORMA GIANT OIL FIELD, LANDSAT DATA, TUNISIA
B 6002 319* EL DORADO FIELD, BUTLER COUNTY, KANSAS, ABST., DELTAIC
B 6303 541* EL DORADO FIELD, KANSAS, ABST., SEDIMENTOLOGY AND SYNERGY OF,
M 30 548 EL GIZA MEMBER, SIRTE BASIN, GIALO FORMATION
B 6403 424 EL HARA, TUNISIA, DJEBEL
M 25 531 EL LACO, MAGNETITE INTERPRETED AS A VOLCANIC FLOW, CHILE
B 6104 603 EL PASO FORMATION, NEW MEXICO
B 6102 253† EL PASO GEOLOGICAL SOCIETY, HISTORY OF
B 6102 255 EL PASO GEOLOGICAL SOCIETY, OFFICERS, 1967-1976
M 27 48 EL PASO GROUP, TEXAS, ORDOVICIAN
B 6109 1451 EL PASO-ELLENBURGER UNIT, PEDREGOSA BASIN
M 30 342 EL PLAN FIELD, MEXICO, SALINE BASIN
B 6209 1765* EL PROGRESO QUADRANGLE, GUATEMALA, ABST., STRUCTURAL FABRIC
B 6101 54 EL QAA 1, LOG, LEBANON
B 6101 54 EL QAA STRUCTURE, LEBANON
B 6010 1655 EL SALVADOR, 1975, PETROCONSULTANTS S.A., ON DEVELOPMENTS IN
B 6110 1590 EL SALVADOR, 1976, CARVAJAL, J., ON DEVELOPMENTS IN
B 6210 2059 EL SALVADOR, 1977, PETROCONSULTANTS S.A., ON DEVELOPMENTS IN
B 6310 1761 EL SALVADOR, 1978, PETROCONSULTANTS S.A., ON DEVELOPMENTS IN
B 6004 669 EL SHAZLY, E. M., GEOLOGY OF RED SEA MARGIN, ABST
M 25 114 EL TATIO FIELD, GEOTHERMAL, CHILE
M 25 170 EL TATIO GEOTHERMAL FIELD, CHILE
M 25 519 EL TORRE SYENITE, MEXICO
B 6207 1216 EL-HINNAWI, ESSAM E., ENVIRONMENTAL IMPACT OF EXPLOITATION OF
B 6405 619 ELACHI, CHARLES, SEASAT RADAR IMAGE OF SAN ANDREAS FAULT,
B 6410 1578 ELBROOK CARBONATE ROCKS, VIRGINIA
B 6203 538* ELDER SANDSTONE—DELTAIC SEDIMENTS OF ROBERTS MOUNTAIN
S 3 90 ELDORADO VEIN DISTRICT, CANADA
B 6102 172 ELECTRIC LOG, CALIFORNIA, VENTURA FIELD, TYPE

B 6102 140 ELECTRIC LOG, CALIFORNIA, VENTURA FIELD, TYPE
M 26 106 ELECTRIC LOG SECTIONS, NEW MEXICO, SAN JUAN BASIN
B 6304 595 ELECTRIC LOGS IN MIGRATION STUDIES, POTENTIAL USE OF
B 6412 2191 ELECTRICAL LOG INTERPRETATION, BY DOUGLAS W. HILCHIE, EVANS,
M 25 119 ELECTRICITY, GEOTHERMAL ENERGY FOR PRODUCTION OF
M 25 523 ELECTRICITY, PROJECTED INCREASES IN DEMAND FOR, TABLE
B 6203 567* ELECTROCHEMICAL CELLS IN FORMATION OF PALEOSURFACE-RELATED
M 26 357 ELECTROMECHANICAL SUBBOTTOM PROFILING SYSTEMS, RESOLUTION
M 28 190 ELECTRON MICROPROBE ANALYSIS, TECHNIQUES
B 6007 1125 ELECTRON-SCANNING PHOTOMICROGRAPH, OKLAHOMA, PALEOZOIC
B 6408 1285* ELEMENTAL ABUNDANCES IN DEVONIAN SHALES OF KENTUCKY AND
B 6309 1580* ELEMENTAL ANALYSES OF DEVONIAN SHALES IN SOUTHERN WEST
B 6004 653* ELEMENTAL VARIATION AND DIAGENETIC ASPECTS OF TABULAR
B 6104 505 ELEMENTARY COMPOSITION AND COLOR OF HEATED KEROGEN
B 6012 2185* ELEMENTS OF TECTONISM DEDUCED FROM MIOCENE STRATA,
B 6004 667* ELEUTHERA BANK, BAHAMAS, ABST., WIDESPREAD GENERATION OF,
B 6204 694 ELEUTHERA ISLAND, BAHAMAS
B 6204 694 ELEUTHERA WELL, BAHAMA PLATFORM
S 5 118 ELEVATION CORRECTIONS, SEISMIC DATA,
B 6010 1872 ELF R.E., ON DEVELOPMENTS IN OMAN, INCLUDING DHOFAR, 1975,
S 11 138 ELIPSOGLANDULINA FRAGILIS
B 6305 794 ELISHA KANE SPARKER PROFILES, USNS
B 6004 669 ELIUK, L. S., ABENAKI FORMATION, NOVA SCOTIA SHELF,
B 6103 373 ELK CITY FORMATION, OKLAHOMA
B 6205 818 ELK HILLS FIELD, CALIFORNIA
B 6403 445* ELK HILLS FIELD, KERN COUNTY, CALIFORNIA, ABST., PORE-FILLING,
B 6203 351 ELK HILLS OIL FIELD, CALIFORNIA
B 6205 816 ELK HILLS POROSITY VALUES, CALIFORNIA
M 29 465 ELK POINT BASIN, AVERAGE THICKNESS OF SEDIMENTS
S 11 312 ELK RIVER VALLEY, CALIFORNIA, SAMPLE LOCALTIES
B 6212 2387 ELKHORN COALS, KENTUCKY
B 6012 2149 ELKINS, REGINALD W., ALASKA GEOLOGICAL SOCIETY HISTORY
B 6412 2169 ELKINS VALLEY ANTICLINE, WEST VIRGINIA
B 6211 2362* ELKO BASIN, NEVADA, ABST., TERTIARY GEOLOGY AND, OIL, SHALE
B 6206 1006 ELLEF RINGNES ISLAND, ARCTIC, CANADA
B 6206 1017 ELLEF RINGNES ISLAND, ISACHSEN DOME
B 6111 1930 ELLEF RINGNES ISLAND, OTTO FIORD FORMATION
B 6402 233 ELLENBURGER DOLOMITE, TEXAS
B 6303 488* ELLENBURGER FORMATION, PUCKETT FIELD, PECOS COUNTY, TEXAS,
B 6303 530* ELLENBURGER GROUP, DELAWARE BASIN, WEST TEXAS ABST.
M 27 132 ELLENBURGER LIMESTONE, TEXAS, ORDOVICIAN
B 6205 862 ELLENBURGER, ULTRADEEP FIELD, PRODUCTION, TEXAS, GOMEZ
B 6109 1451 ELLENBURGER UNIT, PEDREGOSA BASIN, EL PASO-
B 6101 3 ELLESMERE AND AXEL HEIBERG ISLANDS, NORTHERN
B 6101 20 ELLESMERE FOLDBELT, CANADIAN, ARCTIC ISLANDS, CENTRAL
B 6206 1005 ELLESMERE ISLAND, ARCTIC, CANADA
B 6111 1930 ELLESMERE ISLAND, BLUE MOUNTAINS
S 5 161 ELLESMERE ISLAND, BORUP FIORD
B 6404 535 ELLESMERE ISLAND, CANADA
B 6111 1930 ELLESMERE ISLAND, GIRTY CREEK
B 6111 1930 ELLESMERE ISLAND, GREELY FJORD
B 6111 1944 ELLESMERE ISLAND, HARE FIORD
S 5 161 ELLESMERE ISLAND, HARE FIORD
B 6111 1930 ELLESMERE ISLAND, HARE FJORD
B 6111 1930 ELLESMERE ISLAND, HARE FJORD DIAPIR
B 6111 1930 ELLESMERE ISLAND, NANSEN SOUND
B 6111 1930 ELLESMERE ISLAND, NORTHWESTERN, OTTO FIORD FORMATION
S 5 146 ELLESMERE ISLAND, OTTO FIORD FORMATION, CANADA, ARCTIC
B 6111 1930 ELLESMERE ISLAND, OTTO FJORD
B 6111 1930 ELLESMERE ISLAND, STEPANOW CREEK
B 6111 1930 ELLESMERE ISLAND, SVARTFJELD PENINSULA
B 6111 1931 ELLESMERE ISLAND, SVERDRUP BASIN, SECTION
B 6206 1005 ELLESMERIAN OROGENY, ARCTIC, CANADA
B 6405 704 ELLIOTT, DAVID, HOW DO THRUST BELTS FORM, ABST.
B 6101 28 ELLIOTT LAKE, URANIUM DEPOSITS, ONTARIO
B 6206 1092 ELLIOTT, T. L., MIDDLE CRETACEOUS CARBONATE-SLOPE DEPOSITS,
B 6405 677 ELLIS, C. HOWARD, APPLICATIONS OF PROBABL RANGE CONCEPT TO
B 6303 443 ELLIS, C. HOWARD, NEOGENE CALCAREOUS NANNOPLANKTON
B 6002 318* ELLIS COUNTY, OKLAHOMA, ABST., PRODUCING CHARACTERISTICS OF
B 6410 1643 ELLIS, G. K., DISTRIBUTION AND GENESIS OF SEDIMENTARY URANIU
B 6211 2277 ELLIS GROUP, MONTANA
B 6109 1483 ELLIS, J. P., CONTINENTAL SLOPE AND UPPER RISE OFF WESTERN NOVA
B 6009 1609 ELLISON, LUTHER, SOUTH SAREPTA FIELD, GRAY SANDSTONE,
B 6008 1215 **ELLS, GARLAND D., DEVELOPMENTS IN MICHIGAN IN 1975**
B 6108 1173 ELLS, GARLAND D., DEVELOPMENTS IN MICHIGAN IN 1976
B 6208 1330 ELLS, GARLAND D., DEVELOPMENTS IN MICHIGAN IN 1977
B 6308 1278 ELLS, GARLAND D., DEVELOPMENTS IN MICHIGAN IN 1978
B 6409 1377 ELLS, GARLAND D., DEVELOPMENTS IN MICHIGAN IN 1979
B 6101 4 ELLSMERIAN OROGENY
B 6101 5 ELLSMERIAN OROGENY, SVERDRUP BASIN
B 6009 1619 ELLSWORTH, GEORGE, CHESTERIAN SHORELINE SEQUENCES IN EASTERN
B 6407 1005 ELM GROVE FIELD, ARK-LA-TEX AREA
B 6303 444 ELMORE, R. DOUGLAS, MIOCENE-PLIOCENE MOLASSE, ZAGROS
B 6105 777 ELMORE, R. D., SAND PETROLOGY AND BIOCLASTIC COMPONENTS OF
B 6207 1120 ELONGATE BREAKOUTS, CAUSES
S 11 278 ELPHIDIELLA HANNAI
S 11 241 ELPHIDIELLA HANNAI
S 11 278 ELPHIDIELLA OREGONENSIS
S 11 242 ELPHIDIELLA OREGONENSIS
S 11 276 ELPHIDIUM ACUTUM
S 11 239 ELPHIDIUM ACUTUM
S 4 221 ELPHIDIUM CRISPUM LINNE AND MARGINOPORA VERTEBRALIS QUOY
S 11 239 ELPHIDIUM FORAMINOSUM
S 11 276 ELPHIDIUM FORAMINOSUM
S 11 276 ELPHIDIUM GRANTI
S 11 240 ELPHIDIUM GRANTI
S 11 240 ELPHIDIUM HUGHESI
S 11 276 ELPHIDIUM HUMBOLDTENSIS
S 11 240 ELPHIDIUM HUMBOLDTENSIS
S 11 241 ELPHIDIUM PEOYANUM
S 11 278 ELPHIDIUM POEYANUM
B 6111 1982 ELSINORE FAULT ZONE, CALIFORNIA
B 6111 1988 ELSINORE FAULT ZONE, CALIFORNIA, WHITTIER-
B 6111 1982 ELSINORE TROUGH, CALIFORNIA
B 6004 669 ELVERS, DOUGLAS, J., NEW ENVIRONMENTAL MAPS FOR GULF OF
B 6410 1590 ELWAY FORMATION, VIRGINIA
B 6308 1225 ELWOOD FIELD, CHAMBERS COUNTY, TEXAS
M 25 486 ELY COPPER DISTRICT, NEVADA
M 25 8 ELY, NORTHCUTT, BOUNDARIES OF SEABED JURISDICTION OFF PACIFIC
B 6207 1216 ELY, NORTHCUTT, WHAT LAW SHALL GOVERN DEEP-SEA MINING , ABST.
B 6406 823 EMANUEL RANGE, AUSTRALIA
M 29 27† EMBAYMENT AND NORTHERN BLAKE PLATEAU: PRELIMINARY
B 6111 1991 EMBLEY, R. W., PIERCEMENT STRUCTURES IN DEEP OCEANS
B 6303 444 EMBLEY, ROBERT W., DISTRIBUTION, MORPHOLOGY, MECHANISMS, AND
B 6111 2004 EMBLEY, ROBERT W., EXOTIC MIDDLE MIOCENE SEDIMENT FROM CAPE
B 6303 438 EMBLEY, ROBERT W., MASS WASTING ON CONTINENTAL RISE OF
B 6109 1480 EMBORE FORMATION, BRAZIL
B 6211 2358* EMERGENCE OF NATIONAL OIL COMPANIES, ABST.
B 6005 809 EMERY, K. O., CONTINENTAL MARGIN OFF WESTERN AFRICA--SENEGAL
B 6403 297 EMERY, K. O., CONTINENTAL MARGINS--CLASSIFICATION AND
B 6105 782 EMERY, K. O., STRATIGRAPHY AND STRUCTURE OF PULL-APART MARGINS
B 6405 679 EMILIA, DAVID A., URANIUM EXPLORATION SYSTEMS CASE STUDY—
B 6009 1562 EMILIANI, CESARE, NANNOFOSSIL BIOSTRATIGRAPH AND CLIMATIC
B 6211 2209 EMLAGH POINT GROWTH FAULT, IRELAND, WESTERN
B 6004 699 EMMEL, F. J., DYNAMIC PROCESSES OF UPPER BENGAL FAN AND SWATCH
M 29 189 EMMEL, F. J., TECTONICS OF THE ANDAMAN SEA AND BURMA
B 6103 361 EMPIRE STRUCTURE, OKLAHOMA
B 6401 105 EMPIRE-ABO FIELD, NEW MEXICO
B 6004 700* EMPIRICAL OBSERVATIONS AND EXPERIMENTAL RESULTS, ABST.,
B 6209 1564 EMPLACEMENT HISTORIES OF PIERCEMENT DIAPIRISM
B 6209 1561† EMPLACEMENT OF PIERCEMENT DIAPIRS, MECHANISM FOR
B 6401 9 EN NALA ANTICLINAL AXIS, SAUDI ARABIA
B 6102 208 ENCANTO DEPOSITO FORMATION, MEXICO
M 27 35 ENCRUSTING FORAMS
B 6407 1117 ENCYCLOPEDIA O PALEONTOLOGY, EDITED BY RHODES W. FAIRBRIDGE
B 6305 820 ENCYCLOPEDIA OF SEDIMENTOLOGY, EDITED BY RHODES W.
B 6103 437 ENDICOTT GROUP, ALASKA
B 6109 1496 ENDICOTT GROUP, ALASKA
B 6107 1069† ENDOLITHIC ALGAE, MICRITIZATION AND, CARBONATE-, GRAIN,
B 6107 1069 ENDOLITHIC, BORING, ALGAE
B 6405 723* ENDOLITHS, BERMUDA, ABST., INTERTIDAL ZONATION OF RECENT
B 6307 1094 ENEABBA MEMBER, AUSTRALIA
B 6407 1092 ENEMTENIAN HORIZON, ASIA, NORTHEAST
B 6108 1374* ENERGY, ABST., EFFECTIVE USE OF MICROEARTHQUAKE RECORDINGS AS
B 6212 2510* ENERGY, ABST., PROGRESS AND PROSPECTS IN, OCEAN, THERMAL
B 6108 1379* ENERGY, ABST., QUATERNARY GEOLOGIC PROCESSES ASSOCIATED WITH,
B 6207 1212* ENERGY, ABST., UTILIZATION OF, MAGMA
M 25 1† ENERGY AN MINERAL RESOURCES TO WORLD, IMPORTANCE OF,
B 6403 434 ENERGY AND CLIMATE, PAMPE, WILLIAM R., REVIEW OF
B 6207 1227* ENERGY AND MINERAL DEVELOPMENT, ABST., ENVIRONMENTAL
M 25 24† ENERGY AND MINERAL RESOURCES--SUMMARY, ECONOMICS OF
B 6004 731* ENERGY COASTAL ENVIRONMENNTS, SOUTHWES FLORIDA, ABST.,
B 6411 1840 ENERGY CORP., ON DEVELOPMENTS IN DUBAI, 1979, DUBAI PETROLEUM
B 6202 176† ENERGY CRISIS—ENERGY SOLUTION: TWO CASE HISTORIES
B 6405 771* ENERGY DEPOSITS, ABST., RECENT ADVANCES IN, HELIUM ANALYSIS AS
B 6105 836* ENERGY DEVELOPMENT AND ENVIRONMENTAL-IMPACT ASSESSMENT,
M 25 149† ENERGY DEVELOPMENT IN ALASKA--SUMMARY, POTENTIAL FOR,
B 6207 1225* ENERGY DEVELOPMENT THROUGH PLANNING, ABST., OVERCOMING
M 25 118† ENERGY DEVELOPMENT--SUMMARY, ENVIRONMENTAL EFFECTS OF,
B 6405 761* ENERGY DEVELOPMENTS, ABST., OVERVIEW OF, GEOTHERMAL
B 6009 1613* ENERGY DILEMMA, CHOICES AND CONSEQUENCES, ABST.,
M 28 189 ENERGY DISPERSIVE ANALYSIS, TECHNIQUES

B 6303 556* ENERGY DISSIPATION OVER SALT MARSH AND IT EFFECT ON ESTUARINE
B 6207 1224* ENERGY ECONOMY, ABST., NORTHWEST SHELF, GAS—ITS ROLE IN,
B 6207 1233* ENERGY EXPLORATION AND DEVELOPMENT IN PHILIPPINES, ABST.,
B 6004 688* ENERGY EXPLORATION. BY TITLE ONLY, TIMING FACTOR IN
B 6405 773* ENERGY FENCE AND RESOLUTION OF RELICT FEATURES ON ATLANTIC
M 25 473 ENERGY FLOW IN LITHOSPHERE
B 6309 1590* ENERGY FROM MICHIGAN ANTRI SHALE BY IN-SITU PROCESS, ABST.,
B 6004 647* ENERGY FROM TAR SANDS, ABST.
B 6207 1215* ENERGY FROM WIND—OLD IDEA WITH NEW POTENTIAL, ABST.
B 6207 1233* ENERGY IN CONTIGUOUS UNITED STATES, ABST., FUTURE POTENTIAL
B 6207 1234* ENERGY IN FUTURE OF CIRCUM-PACIFIC, AREA, ABST., ROLE OF,
B 6207 1235* ENERGY IN HAWAII—PRESENT AND FUTURE, ABST., GEOTHERMAL
M 25 147† ENERGY IN JAPAN ESTIMATED FROM GEOPHYSICAL DATA--SUMMARY,
M 25 130† ENERGY IN NEW ZEALAND--SUMMARY, GEOTHERMAL
M 25 163† ENERGY IN NICARAGUA--SUMMARY, EXPLORATION FOR, GEOTHERMAL
B 6207 1222* ENERGY IN WESTERN UNITED STATES, ABST., AVAILABILITY AND
B 6105 820* ENERGY MINERALS—INTRODUCTION, ABST., RESOURCE-, RESERVE
B 6405 700* ENERGY NEEDS OF UNITED STATES AND OTHER PARTS OF THE WORLD,
B 6203 552* ENERGY ON BEACH AND NEARSHORE-BAR MORPHOLOGY IN EASTERN
B 6105 835* ENERGY OUTLOOK FOR 1980S, ABST., U.S. AND WORLD
B 6305 842* ENERGY OUTLOOK FOR THE 1980S, ABST., UNITED STATES AND, WORLD
B 6303 473* ENERGY PERSPECTIVE, ABST., UNITED STATES PETROLEUM IN, WORLD
B 6207 1238* ENERGY POLICY, ABST., PHILIPPINE, NATIONAL
B 6203 549* ENERGY PROBLEM, ABST., HOW TO HELP SOLVE THE
B 6207 1210* ENERGY PROBLEMS OF PACIFIC BASIN, ABST., CRITICAL
B 6209 1760* ENERGY PRODUCTION IN VICKSBURG FORMATION ALONG TEXAS GULF
M 25 124† ENERGY PROGRAM--SUMMARY, MAGMA TAP, THE ULTIMATE,
B 6207 1217* ENERGY RESOURCE, ABST., FUELS FROM, MAGMA—POTENTIAL
B 6105 813 * ENERGY RESOURCE DATA—SOURCES, ACQUISITION, USES
B 6303 442* ENERGY RESOURCE DEVELOPMENT, ABST., FEDERAL BARRIERS TO
B 6009 1619* ENERGY RESOURCE IMPLICATIONS OF 38TH PARALLEL LINEAMENT
B 6203 551* ENERGY RESOURCES, ABST., APPLICATION OF, GARNET SYSTEM FOR,
B 6207 1216* ENERGY RESOURCES, ABST., ENVIRONMENTAL IMPACT OF
B 6207 1230* ENERGY RESOURCES, ABST., FUNDAMENTAL STRUCTURAL PATTERN IN
B 6211 2349* ENERGY RESOURCES, ABST., REMOTE SENSING FOR
B 6207 1238* ENERGY RESOURCES, ABST., SLOPE-STABILITY PROBLEMS OF,
B 6207 1241* ENERGY, RESOURCES, AND ENVIRONMENT MANAGEMENT—BASE FOR
B 6202 178 ENERGY RESOURCES, CANADA
B 6310 1982 ENERGY RESOURCES, EDITED BY CREIGHTON A. BURK AND CHARLES L
B 6202 178 ENERGY RESOURCES, MIDDLE EAST
B 6207 1209* ENERGY RESOURCES OF THE WORLD, ABST., PRESENT AND FUTURE
B 6408 1276* ENERGY RESOURCES OF WATER-BEARING GEOPRESSURED RESERVOIRS,
B 6202 178 ENERGY RESOURCES, RUSSIA, CHINA, ET AL
B 6202 178 ENERGY RESOURCES, UNITED STATES
B 6202 178 ENERGY RESOURCES, WORLD
M 25 26† ENERGY RESOURCES--SUMMARY, LATIN AMERICAS OUTLOOK TOWARD
B 6408 1283* ENERGY SHALE MEMBER, PENNSYLVANIAN, AND THEIR RELATION T
B 6207 1238* ENERGY SOURCE, ABST., BIOMASS AS
B 6203 555* ENERGY SOURCE, ABST., CRUDE OIL AND NATURAL, GAS DISSOLVED IN
B 6004 644* ENERGY SOURCES, ABST.
M 25 135† ENERGY SOURCES IN INDONESIAN VOLCANIC AREAS--SUMMARY,
B 6004 711* ENERGY SUPPLIES AND ECONOMICS, ABST., IMPACT OF, OIL AND, GA,
B 6207 1218* ENERGY SYMPOSIUM, INTRODUCTION, ABST., RENEWABLE
B 6105 837* ENERGY TECHNOLOGY TO CREATE SURGE IN DEMAND FOR LITHIUM,
B 6008 1410* ENERGY--ROCKY ROAD ABST., LEASE OF FEDERAL, GEOTHERMAL
M 25 472 ENERGY-GENERATING CENTER, ISLAND, ARC, FORMATION IN RELATION
B 6105 829* ENERGY-TRANSPORT PROCESSES IN SEDIMENTARY BASINS AND THEIR
B 6309 1599* ENERGY. WHAT IS PRACTICAL IN THE NEAR TERM, ABST., APPLICATIONS
B 6207 1215* ENERGY—CENTRAL RECEIVER PROGRAM, BARSTOW POWER TOWER,
B 6207 1239* ENERGY—TRANS-ALASKA PIPELINE, ABST., HARNESSING, ARCTIC
B 6208 1441 ENEVOLDSEN, SOREN, DEVELOPMENTS OFFSHORE WEST GREENLAND IN
B 6405 713* ENEWETAK ATOLL REEF, MIDDLE PLEISTOCENE TO HOLOCENE,
B 6405 756* ENEWETAK ATOLL, ABST., CARBON, ISOTOPIC SIGNATURE AS CRITERION
B 6303 458* ENEWETAK ATOLL, ABST., PLEISTOCENE, METEORIC-VADOSE DIAGENESIS
B 6405 800* ENEWETAK ATOLL, MARSHALL ISLANDS, ABST., CONTEMPORANEOUS
S 5 54 ENGADINE FORMATION, SILURIAN, NIAGARA GROUP,
B 6303 445 ENGELMANN, WILLIAM H., ENVIRONMENTAL REGULATIONS FOR IN-SITU
B 6109 1478 ENGENHO FURADO FIELD, BRAZIL
M 26 374 ENGINEERING ANALYSES, RESOLUTION SYSTEMS
B 6002 323* ENGINEERING AND GEOLOGIC CHALLENGES IN HEAVY-OIL RECOVERY,
B 6303 492* ENGINEERING GEOLOGY IN DEEP BASIN CANADA. BY TITLE ONLY
B 6004 659* ENGLAND, ABST., DELTAIC EVOLUTION DURING, BASIN, FILL--, NAMURIA
B 6405 740* ENGLAND, ABST., HYDROCARBON EXPLORATION IN, WESTERN
B 6405 728* ENGLAND, ABST., SEDIMENTOLOGY OF LOWER, CRETACEOUS SUBTIDAL
B 6002 295 ENGLAND, BALL CLAY NO. 661, GEOTECHNICAL PROPERTIES
M 27 63 ENGLAND, CARBONIFEROUS, CORAL LIMESTONE
B 6107 986 ENGLAND, CHALK SECTION, LOWER
B 6107 988 ENGLAND, CHALK UNIT, MIDDLE AND UPPER
M 27 208 ENGLAND, CRETACEOUS CHALK
M 27 166 ENGLAND, CRETACEOUS, FOLKESTONE BEDS
M 28 127 ENGLAND, CRETACEOUS, GAULT FORMATION
M 28 128 ENGLAND, CRETACEOUS, GAULT FORMATION
M 27 20 ENGLAND, CRETACEOUS, LOWER CHALK
M 27 34 ENGLAND, CRETACEOUS, LOWER CHALK
M 27 49 ENGLAND, CRETACEOUS, LOWER GREENSAND
M 27 95 ENGLAND, CRETACEOUS, UPPER CHALK
M 27 35 ENGLAND, CRETACEOUS, UPPER CHALK
M 27 82 ENGLAND, CRETACEOUS, UPPER CHALK
M 28 78 ENGLAND, CRETACEOUS, UPPER CHALK
B 6107 986 ENGLAND, CULVER CLIFFS, CHALK
B 6206 962 ENGLAND, DERBYSHIRE MASSIF
B 6401 68 ENGLAND, DOVER AREA, SAMPLED SECTION
B 6206 961 ENGLAND, EDALE SHALES
B 6002 297 ENGLAND, FALMOUTH CLAY
B 6206 960 ENGLAND, GRINDSLOW SHALES
B 6403 376 ENGLAND, HOLOCENE FAN DELTA
B 6401 75 ENGLAND, ISLE OF WIGHT, ALBIAN THROUGH SANTONIAN
M 27 41 ENGLAND, JURASSIC, CORALLIAN SERIES
M 27 190 ENGLAND, JURASSIC, FOREST MARBLE BEDS
M 27 184 ENGLAND, JURASSIC, FOREST MARBLE BEDS
B 6206 960 ENGLAND, KINDERSCOUT GRIT
B 6206 960 ENGLAND, MAM TOR SANDSTONES
M 27 164 ENGLAND, MISSISSIPPIAN, CORAL LIMESTONE
B 6401 68 ENGLAND, NORFOLK REGION, SAMPLED SECTION
B 6012 2117† ENGLAND, ORIGIN AND EVOLUTION OF, BIOHERMS IN, WENLOCK
B 6206 960 ENGLAND, PENNINE BASIN
B 6206 962 ENGLAND, PENNINE DELTA
B 6002 295 ENGLAND, RIVER FAL KAOLINITIC DEPOSITS
B 6206 959 ENGLAND, SHALE GRIT: A COMPLETE ANCIENT SUBMARINE FAN
B 6004 695* ENGLAND--SLOPE INFLUENCED BY DENSITY CURRENTS, ABST.,
B 6303 445 ENGLEMAN, MARY, PROGRESSIVE CEMENTATION FROM MARINE TO DEEP
B 6411 1964 ENGLISH CHANNEL-WESTERN APPROACHES, FRANCE
B 6009 1613 ENGLUND, K. J., EVALUATION OF QUALITY AND QUANTITY CONTROLS
B 6203 512 ENGLUND, KENNETH J., COAL-RESOURCE DISTRIBUTION TRENDS IN
B 6309 1603* ENHANCED OIL RECOVERY, ABST.
B 6203 564* ENHANCED RECOVERY FROM COAL-GAS WELLS, ABST., FOAM,
S 10 20 ENHANCED SOLUBILITY, ACCOMMODATION, MIGRATION
S 10 21 ENHANCED SOLUBILITY, MICELLES, MIGRATION
S 10 20 ENHANCED SOLUBILITY, PRECURSORS, MIGRATION
B 6203 502* ENIGMAS IN WISCONSIN CAMBRIAN AND NEW DEPOSITIONAL MODEL
B 6004 649 ENOS, P., PRELIMINARY RESULTS FROM LEG 44 DEEP SEA DRILLING
B 6109 1552 ENOS, PAUL, DIAGENESIS OF A GIANT—POZA RICA TREND, MEXICO, ABST.
B 6004 669 ENOS, PAUL, EVOLUTION OF FLORIDA BAY INTERPRETED FROM ISLAND
B 6303 445 ENOS, PAUL, MIOCENE CARBONATE GRAVITY FLOW IN BLAKE-BAHAMA
B 6303 445 ENOS, PAUL, PORE SPACE IN HOLOCENE CARBONATE SEDIMENTS, ABST.
S 4 20 ENRIQUE REEF ENVIRONMENT, PUERTO RICO
B 6303 533* ENSENADA DE LA VELA BASIN, OFFSHORE FALCON STATE, VENEZUELA,
B 6205 881* ENTRADA OIL FIELDS OF SOUTHERN SAN JUAN BASIN, NEW MEXICO,
B 6405 682* ENTRADA OIL FIELDS OF SOUTHERN SAN JUAN BASIN, NEW MEXICO,
B 6007 1013 ENTRADA SANDSTONE, NEW MEXICO
M 28 106 ENTRADA SANDSTONE, UTAH, JURASSIC
M 28 107 ENTRADA SANDSTONE, UTAH, JURASSIC
B 6111 1953 ENTRAPMENT, AND DEGRADATION, OIL, GENERATION, MIGRATION
B 6108 1381* ENTRAPMENT AT COTTONWOOD CREEK, WYOMING, ABST., REGIONAL
M 26 347 ENTRAPMENT BY, FACIES CHANGES
B 6305 757 ENTRAPMENT, DIFFERENTIAL
B 6303 548* ENTRAPMENT FACTORS IN CALIFORNIA TURBIDITE AND
B 6004 715* ENTRAPMENT IN SMACKOVER FORMATION AT CHALYBEAT SPRINGS
B 6009 1614* ENTRAPMENT IN SMACKOVER FORMATION AT CHALYBEAT SPRINGS
B 6305 723† ENTRAPMENT, MECHANICS OF, SECONDARY, HYDROCARBON,
B 6305 752 ENTRAPMENT MODEL, MIGRATION AND
B 6009 1520† ENTRAPMENT, OBSCURE NATURE OF, PETROLEUM, MIGRATION AND
B 6008 1398* ENTRAPMENT OF GAS IN UPPER CRETACEOU EAGLE FORMATION,
B 6108 1391* ENTRAPMENT OF HYDROCARBONS, ABST., MECHANICS OF, SECONDARY,
B 6109 1540* ENTRAPMENT OF HYDROCARBONS IN SAN MIGUEL SANDSTONES OF
B 6304 608† ENTRAPMENT OF OIL AND GAS IN NIAGARAN PINNACLE-REEF BELT OF
B 6203 551* ENVIRONMENT CREATED BY VERTICALLY MOVING FLUIDS, ABST.
B 6207 1241* ENVIRONMENT MANAGEMENT—BASE FOR REGIONAL COOPERATION FOR
B 6409 1556* ENVIRONMENT OF DEPOSITION OF UPPER WILCOX SANDSTONES, KATY
S 5 53* ENVIRONMENT OF RUFF FORMATION (UPPER SILURIAN) IN SOUTHEAST
B 6009 1613* ENVIRONMENT-ENERGY DILEMMA, CHOICES AND CONSEQUENCES, ABST.
B 6002 184† ENVIRONMENTAL ANALYSIS OF NORT SEA SEDIMENTS, SUBSURFACE
B 6303 489* ENVIRONMENTAL AND DIAGENETIC CONTROLS O CARBONATE AND
B 6409 1562* ENVIRONMENTAL AND DIAGENETIC CONTROLS OF CARBONATE AND
B 6405 744* ENVIRONMENTAL AND DIAGENETIC CONTROLS OF CARBONATE SOURCE
B 6012 2175* ENVIRONMENTAL AND TIME SIGNIFICANCE OF LOCAL GUIDE FOSSILS,
B 6207 1227* ENVIRONMENTAL ASPECTS OF INTERNATIONAL ENERGY AND MINERAL
B 6207 1240* ENVIRONMENTAL ASSESSMENT INTO EXPLORATION-DEVELOPMENT
B 6303 538* ENVIRONMENTAL ASSESSMENT OF IN-SITU LEACHING OF URANIUM,
B 6108 1379* ENVIRONMENTAL CASE STUDY-HENDERSON MINE MILL, ABST.
B 6207 1226* ENVIRONMENTAL CONSIDERATIONS IN DEVELOPMENT PLANNING, ABST.,

B 6405 733* ENVIRONMENTAL CONTROL OF TRACE FOSSIL MORPHOLOGY, ABST.
S 7 230 ENVIRONMENTAL CYCLE CONCEPT, SEDIMENTATION
M 25 118† ENVIRONMENTAL EFFECTS OF GEOTHERMAL ENERGY
B 6205 753 ENVIRONMENTAL FACIES INTERPRETATION OF CLASTIC SEISMIC FACIES
B 6303 420* ENVIRONMENTAL GEOLOGY OF CONTINENTAL MARGIN OFF CENTRAL
B 6207 1216* ENVIRONMENTAL IMPACT OF EXPLOITATION OF MINERAL AND ENERGY
B 6109 1550* ENVIRONMENTAL IMPACTS, ABST., STATICS AND, DYNAMICS OF
B 6004 669* ENVIRONMENTAL MAPS FOR GULF OF MEXICO OCS PROGRAMS, ABST.,
B 6205 879* ENVIRONMENTAL RECONSTRUCTION OF MIDDLE BLACKHAWK
B 6207 1214* ENVIRONMENTAL REGULATION, ABST., EFFECTIVE
B 6303 445* ENVIRONMENTAL REGULATIONS FOR IN-SITU URANIUM MINING, FROM
B 6303 434* ENVIRONMENTAL SIGNIFICANCE OF FOSSIL ROOT CAST FROM KOOBI
B 6303 463* ENVIRONMENTAL STABILITY--A DIFFERENT APPROACH, ABST.,
B 6303 453* ENVIRONMENTAL STUDIES AT NOAAS ATLANTIC OCEANOGRAPHIC AND
B 6109 1545* ENVIRONMENTAL-INFORMATION AND IMPACT-ASSESSMENT SYSTEM,
B 6311 2118* ENVIRONMENTS AND DIAGENESIS, MORROW SANDSTONES, CIMARRON
B 6305 827* ENVIRONMENTS FOR SEDIMENTARY URANIUM IN TRIASSIC-JURASSIC
S 5 23* ENVIRONMENTS OF BASIN-CENTER EVAPORITES, LOWER SALINA GROUP
S 5 1* ENVIRONMENTS OF PINNACLE REEFS, NIAGARA AND SALINA GROUPS,
B 6006 981† ENVIRONMENTS, HYDRATE FORMATION IN, SUBSURFACE
M 25 492† ENVIRONMENTS OF CANADIAN CORDILLERA DEPOSITIONAL BASINS
B 6409 1552* ENVIRONMENTS OF DEPOSITION AND DIAGENESIS OF STUART CITY
B 6108 1381* ENVIRONMENTS OF DEPOSITION OF COALMONT FORMATION NEAR
B 6004 666* ENVIRONMENTS OF DEPOSITION OF OIL SHALE, ABST.
B 6102 295* ENVIRONMENTS OF EVAPORITI SEDIMENTATION--SHALLOWS, SABKHAS,
B 6004 664* ENVIRONMENTS OF FORMATION OF URANIUM DEPOSITS, ABST.
S 5 125* ENVIRONMENTS, TONOLOWAY LIMESTONE, AND DISTRIBUTION OF
S 5 29 ENVIRONMENT, BASIN-CENTER EVAPORITES, DEPOSITIONAL
S 10 136 EOCENE WATER LEVELS, FRANCE, AQUITAINE
B 6408 1142 EOCENE, ALASKA
B 6407 1016 EOCENE, ALASKA
B 6303 493* EOCENE ALLUVIAL-PLAIN DEPOSITS, SIERRA NEVADA FOOTHILLS,
B 6108 1378* EOCENE, AND ASSOCIATED CLOSED-BASIN DEPOSITS, EOCENE AND
B 6108 1378* EOCENE AND OLIGOCENE, IN EAST-CENTRAL NEVADA-IMPLICATIONS FOR
B 6207 1153 EOCENE, AUSTRALIA
M 27 219 EOCENE, AVON PARK LIMESTONE, FLORIDA
M 27 194 EOCENE, AVON PARK LIMESTONE, FLORIDA
M 27 131 EOCENE, AVON PARK LIMESTONE, FLORIDA
B 6309 1456 EOCENE, BALTIMORE CANYON TROUGH
S 6 220 EOCENE, BARTONIAN STAGE DEFINED, EUROPE
B 6211 2239 EOCENE, BAY OF BISCAY AREA
B 6204 665 EOCENE, BRAZIL
B 6203 537* EOCENE BRIDGER AND GREEN RIVER FORMATIONS,
B 6205 820 EOCENE, CALIFORNIA
B 6211 2217 EOCENE, CALIFORNIA
M 29 242 EOCENE, CARBONATE FOSSILS, NORTH PHILIPPINE SEA
B 6104 492† EOCENE CARBONATE SHORELINE SEQUENCES, CENTRAL FLORIDA,
S 4 103 EOCENE, CARIBBEAN, PALEOCENE-EARLY
B 6104 561 EOCENE, CHILE
B 6004 707* EOCENE CLIMATIC CHANGE, SOUTHWESTERN CALIFORNIA, ABST.,
B 6004 707* EOCENE CLIMATIC CHANGE, SOUTHWESTERN CALIFORNIA AND
M 27 140 EOCENE, COAMO SPRINGS LIMESTONE, PUERTO RICO
M 27 131 EOCENE, COAMO SPRINGS LIMESTONE, PUERTO RICO
M 27 33 EOCENE, COAMO SPRINGS LIMESTONE, PUERTO RICO
M 27 12 EOCENE, COAMO SPRINGS LIMESTONE, PUERTO RICO
M 27 13 EOCENE, COAMO SPRINGS LIMESTONE, PUERTO RICO
M 25 324 EOCENE, COLOMBIA
M 28 160 EOCENE, CRESCENT FORMATION, WASHINGTON
M 27 139 EOCENE DEEP-WATER OOZE, BERMUDA RISE
M 27 17 EOCENE, DEEP-WATER OOZE, BERMUDA RISE
B 6303 516* EOCENE DELTAIC SANDSTONES, LAKE MARACAIBO, VENEZUELA, ABST.,
B 6004 692* EOCENE DEPOSITIONAL ENVIRONMENTS AND RELATED SULFUR
S 6 238 EOCENE, EARLY-MIDDLE BOUNDARY, PALEOGENE
M 25 329 EOCENE, ECUADOR
M 29 341 EOCENE, EVOLUTION, COLOMBIA
B 6303 311† EOCENE FLAGSTAFF LIMESTONE AND COEVAL ROCKS, CENTRA UTAH,
B 6109 1549* EOCENE FORMATIONS IN FLORIDA, ABST., TYPE AND COTYPE, WELLS FOR
M 30 67 EOCENE, FRIGG FIELD, SEDIMENTATION HISTORY
B 6203 565* EOCENE GREEN RIVER FORMATION, ABST., STROMATOLITES IN
M 28 59 EOCENE, GREEN RIVER FORMATION, COLORADO
M 28 60 EOCENE, GREEN RIVER FORMATION, COLORADO
M 28 57 EOCENE, GREEN RIVER FORMATION, COLORADO
M 27 6 EOCENE, GREEN RIVER FORMATION, WYOMING
M 27 100 EOCENE, GREEN RIVER FORMATION, WYOMING
M 28 118 EOCENE, GREEN RIVER FORMATION, WYOMING
M 27 142 EOCENE, GREEN RIVER FORMATION, WYOMING
B 6109 1544* EOCENE GULF COAST BASIN OF TEXAS, ABST., DEPOSITIONAL SYSTEMS
M 28 138 EOCENE, HUERFANO FORMATION, COLORADO
M 28 67 EOCENE, HUERFANO FORMATION, COLORADO
B 6402 219 EOCENE, INDIA
M 27 189 EOCENE, INDIAN MEADOWS FORMATION, WYOMING
B 6303 469* EOCENE INTERLAMINATE SHALES AND SANDSTONES, ABST.,
B 6205 829 EOCENE, ISRAEL
B 6109 1547* EOCENE, KARNES COUNTY, TEXAS, ABST., SECONDARY, POROSITY IN,
B 6303 535* EOCENE LAKES GOSIUTE AND UINTA, ABST., EVOLUTION OF
M 27 13 EOCENE LIMESTONE, GERMANY
S 6 220 EOCENE, LUTETIAN STAGE DEFINED, EUROPE
S 6 252 EOCENE, LUTETIAN STAGE, ISOTOPIC DATES, PALEOGENE
B 6102 208 EOCENE, MEXICO
B 6405 613 EOCENE, MIDDLE EAST, PALEOCENE TO
S 6 241 EOCENE, MIDDLE-LATE BOUNDARY, PALEOGENE
B 6201 3 EOCENE, NIGER DELTA
B 6205 777 EOCENE, NORTH, AFRICA
B 6203 424 EOCENE, NORTH, ATLANTIC
B 6409 1554* EOCENE, NORTHERN MISSISSIPPI, ABST., DEPOSITIONAL SYSTEMS AND,
M 27 33 EOCENE, NUMMULITE LIMESTONE, YUGOSLAVIA
M 27 124 EOCENE, OBERAUDORF SCHICHTEN, AUSTRIA
M 27 177 EOCENE, OBERAUDORF SCHICHTEN, AUSTRIA
M 27 217 EOCENE, OCALA FORMATION, FLORIDA
M 27 223 EOCENE, OCALA FORMATION, FLORIDA
M 27 164 EOCENE, OCALA FORMATION, FLORIDA
M 27 95 EOCENE, OCALA FORMATION, FLORIDA
M 27 219 EOCENE, OCALA FORMATION, FLORIDA
M 27 142 EOCENE, OCALA LIMESTONE, FLORIDA
M 27 31 EOCENE, OCALA LIMESTONE, FLORIDA
M 27 220 EOCENE, OCALA LIMESTONE, FLORIDA
M 28 119 EOCENE, OCALA LIMESTONE, FLORIDA
B 6012 2190* EOCENE OF CALIFORNIA, ABST., FORAMINIFERAL ZONATION OF,
B 6203 563* EOCENE, OF EASTERN EGYPT—RESPONSE TO PRERIFT CRUSTAL DOMING,
B 6405 802* EOCENE OF NORTH-CENTRAL PAKISTAN, ABST., NEARSHORE MARINE
B 6207 1164 EOCENE, OPALINE CLAYSTONES, SOUTH CAROLINA
B 6102 237† EOCENE PALEOCURRENTS AND SEDIMENTATION, SA NICOLAS ISLAND,
B 6009 1618* EOCENE PLANKTONIC CORRELATIONS--NORTHERN ITALY-JAMAICA,
M 28 68 EOCENE, PLUM BENTONITE, TEXAS
S 6 221 EOCENE, PRIABONIAN STAGE DEFINED, EUROPE
B 6004 717* EOCENE PROVENANCE OF HANNA AND CARBON BASINS CARBON
S 6 227 EOCENE, RADIOMETRIC AGES, CORRELATIONS
M 29 148 EOCENE, REFLECTORS, ATLANTIC MARGIN
B 6207 1194 EOCENE, RIFT HYPOTHESIS, OREGON AND, WASHINGTON
M 30 59† EOCENE ROCKS OF THE VIKING GRABEN, NORTH SEA, FRIGG
B 6102 237 EOCENE, SAN NICOLAS ISLAND, CALIFORNIA
B 6205 848† EOCENE, SAND OF CALIFORNIA, PERMEABILITY, TRAPS IN, GATCHELL
B 6405 740* EOCENE SAND-RICH, DEEP-SEA FAN AND SHALLOW-MARINE COMPLEX,
S 3 124 EOCENE SANDSTONE, COLORADO, SOUTH PARK
B 6004 725* EOCENE, SANDSTONE, KARNES COUNTY TEXAS, ABST., FACTORS
B 6203 521* EOCENE SANDSTONES OF NORTH SEA VIKING GRABEN, ABST., FRIGG
B 6203 510* EOCENE SHELF AND SLOPE IN SOUTHWESTERN OREGON, ABST., SAND,
S 6 228 EOCENE, STAGE CORRELATIONS, RADIOMETRIC
B 6309 1601* EOCENE STRATA, HAINESVILLE, KEECHI, AND OAKWOOD SALT DOMES,
B 6211 2358* EOCENE STRATIGRAPHY OF SOUTHWEST OREGON, ABST., HOLOTHURIAN
B 6205 890* EOCENE STRUCTURING IN EAST-CENTRAL BASIN-AND-RANGE, ABST.
B 6203 442† EOCENE SUBAERIAL EROSIONAL VALLEYS IN CAMBAY BASIN, INDIA,
B 6405 671* EOCENE SUBMARINE FAN COMPLEX, KODIAK ISLAND ARCHIPELAGO,
B 6405 670* EOCENE SUBSEA FAN FACIES, NORTHWESTERN WASHINGTON, ABST.,
M 29 242 EOCENE, SUBSIDENCE, NORTH PHILIPPINE SEA
B 6206 951 EOCENE, TEXAS
B 6403 424 EOCENE, TUNISIA
S 6 229 EOCENE, U.S.A.-EUROPE CORRELATIONS, DIAGRAM
B 6108 1374* EOCENE UMPQUA FORMATION, SOUTHWEST OREGON, ABST., PETROLOGY
B 6004 657* EOCENE UNCONFORMITY IN BEJUCO FIELD, EASTERN MEXICO, ABST.,
B 6104 581 EOCENE, VENEZUELA
B 6309 1602* EOCENE WILCOX GROUP IN RELATION TO LIGNITE DEVELOPMENT IN
B 6303 530* EOCENE, WYOMING, ABST., SEDIMENTARY SUBENVIRONMENTS OF,
S 6 220 EOCENE, YPRESIAN STAGE DEFINED, EUROPE
B 6303 494* EOCENE ZONATIONS OF WEST COAST, ABST., PALEOECOLOGIC
S 7 202 EOCENE-LOWER MIOCENE, WEST FLORIDA SLOPE, NORTHERN AREA
B 6209 1714† EOCENE-MIOCENE BOUNDARY OF DELAWARE COASTAL PLAIN:
B 6001 34† EOCENE-MIOCENE SEDIMENTATION PATTERN IN PARTS OF NORTHERN
S 6 254 EOCENE-OLIGOCENE BOUNDARY, DIAGRAM, PALEOGENE
B 6309 1607* EOCENE-OLIGOCENE BOUNDARY IN SOUTHWEST ALABAMA, ABST.
S 6 253 EOCENE-OLIGOCENE BOUNDARY, PALEOGENE
S 6 242 EOCENE-OLIGOCENE BOUNDARY, PALEOGENE
B 6004 683* EOLIAN AND SUBAQUEOUS SAND-FLOW CROSS-STRATA, ABST.,
B 6108 1386* EOLIAN OR MARINE IN ORIGIN, ABST., NAVAJO SANDSTONE, JURASSIC,
M 30 50 EOLIAN SANDS, VIKING FIELD, ROTLIEGENDES SANDSTONE
B 6303 504* EOLIAN SEDIMENTOLOGY INTERPRETED FROM DIPMETER RESULTS,
B 6405 709* EOLIAN-FLUVIATILE, CONTINENTAL, ORIGIN OF ANCIENT
B 6004 719* EOLIAN-SAND TURBIDITES, ABST.

B 6004 654* EPEIROGENIC EVENTS IN WESTERN ALPINE MOBILE BELT AND
B 6004 686* EPERIC CARBONATE SEQUENCES, ABST., SIMILARITIES IN EARLY
B 6408 1223† EPICONTINENTAL SEAWAY DEPOSITIONAL SYSTEMS USING PARTIAL
B 6408 1232 EPICONTINENTAL SEAWAYS, ANCIENT, OFFSHORE, SANDSTONE BODIES IN
B 6004 702* EPICONTINENTAL SHELF, ABST., SEDIMENTARY, FACIES AND
B 6303 449* EPICONTINENTAL SHELF, NORTHERN BERING SEA, ABST., VIDEOTAPES OF
M 28 57 EPIDOTE, CRYSTALS
M 28 53 EPIDOTE, CRYSTALS, IN QUARTZ
M 28 8 EPIDOTE MICROLITE INCLUSION, QUARTZ
B 6403 362 EPISODIC READJUSTMENTS OF SEAFLOOR SPREADING AND PLATE
B 6004 652* EPISODIC UPLIFT AND SEDIMENTATION IN POST-PENNSYLVANIAN
S 11 156 EPISOTOMINELLA CAPITANENSIS
S 11 254 EPISTOMINELLA EXIGUA
S 11 286 EPISTOMINELLA EXIGUA
S 11 156 EPISTOMINELLA GYROIDINAFORMIS
S 11 128 EPISTOMINELLA PACIFICA
S 11 254 EPISTOMINELLA PACIFICA
S 11 286 EPISTOMINELLA PACIFICA
S 11 156 EPISTOMINELLA PURISIMA
S 11 286 EPISTOMINELLA SUBPERUVIANA
S 11 254 EPISTOMINELLA SUBPERUVIANA
S 11 156 EPISTOMINELLIDS AND DISCORBINELLA, CALIFORNIA, MIOCENE
B 6104 603 EPITAPH DOLOMITE, NEW MEXICO
S 11 4 EPOCH BOUNDARIES, EARLY PROBLEMS, CALIFORNIA
S 11 256 EPONIDES HEALDI
S 11 288 EPONIDES HEALDI
S 11 256 EPONIDES REPANDUS
S 11 288 EPONIDES REPANDUS
B 6107 1016 EPPING FIELD, SASKATCHEWAN
B 6211 2362* EQUALIZING STACKING VELOCITIES OF DIPPING EVENTS, ABST.
B 6007 1073 EQUATION, MODIFIED, LACHENBRUCH-BREWER
M 26 36 EQUATIONS, LABORATORY, MEASUREMENTS, BIOT-, GEERTSMA
B 6004 662* EQUATORIAL ATLANTIC DURING QUATERNARY, ABST., CLIMATICALLY
B 6202 274 EQUATORIAL ATLANTIC, PHYSIOGRAPHIC PROVINCE MAP, WESTERN
B 6110 1748 EQUATORIAL GUINEA, RIO MUNI-FERNANDO POO, DEVELOPMENTS, 1976
B 6310 1691 EQUATORIAL GUINEA, RIO MUNI-FERNANDO POO, DEVELOPMENTS, 1978
B 6411 1787 EQUATORIAL GUINEA, RIO MUNI-FERNANDO POO, DEVELOPMENTS, 1979
B 6210 1846 EQUATORIAL GUINEA, RIO-MUNI-FERNANDO POO, DEVELOPMENTS, 1977
B 6004 668* EQUATORIAL PACIFIC, ABST., ABYSSAL, ORGANISM, TRACES ON AN IN
B 6109 1408 EQUILIBRIUM CONCEPT, DEVELOPMENT OF THE
B 6109 1430 EQUILIBRIUM EQUATIONS, DERIVATION OF
B 6309 1605* EQUILIBRIUM IN MODERN CORAL REEFS, WESTERN GULF OF
B 6303 482* EQUILIBRIUM IN MODERN CORAL REEFS, WESTERN GULF OF
M 28 86 ERACLEA MINOA MASS FLOW, ITALY, MIOCENE
M 29 304 ERATHOSTHENES PLATEAU, MAGNETIC ANOMALIES, MEDITERRANEAN
B 6106 880 ERATOSTHENES ABYSSAL PLAIN
B 6106 877 ERATOSTHENES SEAMOUNT, CONTINENTAL RISE, NILE CONE, LEVANT
B 6303 446 ERDMAN, J. G., PETROLEUM POTENTIA OF SCOTIAN SHELF--CASE STUDY,
B 6004 697 EREMENKO, N. A., EURASIAN PLATE BOUNDARIES IN USSR, ABST.
M 25 527 ERICKSEN, GEORGE E., METALLOGENIC PROVINCES OF SOUTHEASTERN
B 6109 1559 ERICKSON, DON, LANDSAT—USEFUL PETROLEUM EXPLORATION TOOL IN
B 6303 447 ERICKSON, J., DETERMINATION OF DEPTH TO HYDROCARBON
B 6207 1216 ERICKSON, R. C., SUBSURFACE GEOTECHNICAL HAZARDS IN PETROLEUM
B 6405 775 ERICSON, ERIC K., CASTILLO GAS FIELD, SPAIN, ABST.
B 6309 1592* ERIE COUNTY PENNSYLVANIA, ABST., PRESSURE, CORING OF,
B 6303 446 ERIKSSON, KENNETH A., DOLOMITIZATION MODELS FROM EARLY
M 28 140 ERIONITE, BIREFRINGENCE, ETC.
M 28 140 ERIONITE, CRYSTALS, IN RHYOLITE TUFF
M 28 140 ERIONITE, CRYSTALS, SEM
M 25 62 ERMAKOV, B. V., TECTONICS OF CONTINENTAL AND OCEANIC
B 6407 1092 ERMANOVIAN HORIZON, ASIA, NORTHEAST
M 26 263 EROSION, MODIFICATIONS, STRUCTURAL
B 6111 1979† EROSION SURFACE OF MIDDLE CRETACEOUS AGE, LOS ANGELES
B 6203 564* EROSION VERSUS DEPOSITION ON UPPER CONTINENTAL SLOPE OFF
S 3 48 EROSION-TRANSPORT-DEPOSITION CYCLE
M 29 43† EROSIONAL AND DEPOSITIONAL STRUCTURES OF THE SOUTHWEST
S 6 113 ERQUY FORMATION, ISOCHRON DIAGRAM, FRANCE
B 6303 446 ERRICO, G., MALOSSA FIELD, DEEP DISCOVERY IN PO VALLEY, ITALY,
M 30 525 ERRICO, G., MALOSSA FIELD--A DEEP DISCOVERY IN THE PO VALLEY,
B 6303 488* ERROR IN X-RAY DIFFRACTION ESTIMATES O DOLOMITE IN CARBONATE
B 6307 1128† ERRORS IN EXTRAPOLATION OF EXPERIMENTA KINETIC PARAMETERS
B 6303 447 ERSKINE, M. C., JR., DETERMINATION OF DEPTH TO HYDROCARBON
B 6405 704 ERTEL, JOHN R., A TEST OF THE MELANOIDIN HYPOTHESIS, ABST.
B 6002 321* ERTS, ABST., NEBRASKA OIL FIELDS AND
M 25 63 ERTS, DEVELOPMENT, EARTH RESOURCES TECHNOLOGY SATELLITE
M 25 65 ERTS IMAGES, UNITED STATES
M 25 63 ERTS, NOW LANDSAT
M 25 65 ERTS PROSPECTING, CALIFORNIA, NORTHERN
M 25 69 ERTS STUDY, ALASKA
B 6002 320* ERTS-1, ABST., KANSAS GEOLOGY AS SEEN BY
B 6203 515* ERTS-LANDSAT IMAGERY IN URANIUM EXPLORATION, ABST., TEST USE
B 6410 1644 ERUDINA CHANNEL, AUSTRALIA
S 6 221 ERUOPE, OLIGOCENE, RUPELIAN STAGE DEFINED
B 6201 100 ERVAY CARBONATE ROCK MEMBER, WYOMING
B 6309 1599 ERWIN, CHARLES R., CLASTICITY INDEX--KEY TO CORRELATING
M 24 287 ERWIN, W. L., SUNOCO-FELDA FIELD, HENDRY AND COLLIER COUNTIES,
B 6104 603 ESCABROSA GROUP, NEW MEXICO
M 29 177† ESCARPMENT BASED ON SEISMIC REFLECTION PROFILES, STRUCTURE
B 6408 1273 ESELLER, DURKEE, E. F., REVIEW OF, TURKIYE DORDUNCU PETROL
M 30 529 ESINO FORMATION, TRIASSIC, MALOSSA FIELD
B 6209 1610 ESOPUS SHALE, NEW YORK
B 6209 1610 ESOPUS SHALE, PENNSYLVANIA
B 6301 90 ESPERANTO BEDS, SCOTIAN BASIN
B 6109 1471 ESPIRITO SANTO BASIN, BRAZIL
B 6002 197 ESPIRITO SANTO COASTAL BASIN, BRAZIL
M 26 214 ESPIRITO-SANTO BASIN, BRAZIL
B 6303 456 ESPIRITU, E., PETROLOGY, SEDIMENTOLOGY, AND PETROLEUM
B 6109 1515 ESPIRITU SANTO FORMATION, NEW MEXICO
B 6401 59 ESPITALIE, J., ROLE OF MINERAL MATRIX IN KEROGEN
M 29 86 ESSAOUIRA BASIN, ATLANTIC MARGIN
B 6101 86 ESSAOUIRA BASIN, MOROCCO
B 6202 228 ESSAOUIRA BASIN, SEDIMENTARY SEQUENCE, MOROCCO
B 6303 288 ESSENTIAL REQUISITES FOR FAVORABLE PETROLEUM PROSPECTS
B 6202 304 ESSEQUIBO DRAINAGE BASIN, GUYANA
M 25 32 ESSER, R. W., TECTONIC FRAMEWORK OF PACIFIC REGION
S 6 108 ESTANCIA FORMATION, BRAZIL, BRAZILIAN CYCLE
S 6 108 ESTANCIA FORMATION, RUBIDIUM-STRONTIUM, BRAZIL
B 6011 2048 ESTEBAN C. MATEO, VADOSE PISOLITE AND CALICHE
B 6004 670 ESTEBAN, MATEO, NONVADOSE ORIGIN OF PISOLITIC FACIES, CAPITAN
B 6004 655 ESTEVES, F R., SYNSEDIMENTARY TECTONIC CONTROL OF PALEOZOIC
B 6012 2177 ESTEVEZ, RAUL, SLANT-WAVE STACK, ABST.
B 6004 608† ESTIMATE OF ORGANIC RICHNESS AND EFFECT OF LITHOLOGIC
B 6004 608† ESTIMATE OF SOURCE-ROCK EVALUATION--QUANTITATIVE ESTIMATE OF
B 6303 470* ESTIMATES, ABST., TWENTY YEARS OF, UNITED STATES PETROLEUM
B 6303 459* ESTIMATES AND METHODS OF POTENTIAL GAS COMMITTEE, ABST.,
B 6409 1531 ESTIMATES, EVALUATION OF, RESOURCE
B 6004 665* ESTIMATES OF OIL AND GAS RESOURCES BY U.S. GEOLOGICAL SURVEY,
B 6004 698* ESTIMATES OF UNDISCOVERED OIL AND GAS RESOURCES IN UNITED
B 6004 691* ESTIMATES RESULTING FROM AREAL DISTORTIONS INHERENT IN SOME
B 6312 2183† ESTIMATING OIL AND GAS RESOURCES, METHODS OF
B 6205 884* ESTIMATING PORE AND CEMENT VOLUMES OF CARBONATE ROCKS IN
M 6004 726* ESTIMATING RESERVES IN GREGOIRE LAKE AREA, ATHABASCA TAR
M 26 89 ESTIMATION, SEA LEVEL FLUCTUATIONS, EUSTATIC
B 6105 798 * ESTIMATION OF ORE RESERVES FOR IN-SITU SLUTION MINING
S 11 79 ESTORFFINA MAYI
S 11 146 ESTORFFINA MAYI
S 11 144 ESTORFFINA MAYI
S 6 105 ESTRADA NOVA FORMATION, BRAZIL, PARANA BASIN
S 6 105 ESTRADA NOVA FORMATION, RUBIDIUM-STRONTIUM, BRAZIL
B 6104 562 ESTRATOS CON FAVRELLA STEINMANNI STRATIGRAPHIC UNIT, CHILE
B 6104 561 ESTRATOS CON FAVRELLA STRATIGRAPHIC UNIT, CHILE
B 6004 723* ESTUARINE CARBONATE MATERIALS, ABST., RADIOCARBON, DATING OF
B 6303 448* ESTUARINE SEDIMENTS, ABST., STICKY, MUDS--, VISCOSITY OF
B 6303 509* ESTUARINE TIDAL-CURRENT RIDGE, WILLAPA BAY, WASHINGTON, ABST.,
B 6303 533* ESTUARINE-COASTAL PLAIN COAL DEPOSITION IN SOUTHERN WEST
B 6004 657* ESTUARINE-FILL FACIES, WILLAPA BAY, WASHINGTON, ABST., MODERN
B 6004 664* ESTUARY, ABST., ICE-DRIFTED, BOULDERS AS, PALEOCLIMATIC
B 6004 720* ESTUARY, ABST., SEDIMENTARY ENVIRONMENTS OF, TIDAL, FLAT
B 6004 703* ESTUARY, ABST., SEDIMENTATION, RATES IN, HUDSON RIVER
B 6004 708* ESTUARY CHANNEL SYSTEM, WILLAPA BAY, WASHINGTON, ABST., BED
B 6104 527† ESTUARY, LAVACA BAY--, TRANSGRESSIVE, DELTAIC SEDIMENTATION IN,
B 6004 674* ESTUARY, ST. CROIX, U.S.V.I.--MIXED CARBONATE AND CLASTIC
B 6412 2169 ETAM ANTICLINE, WEST VIRGINIA
B 6211 2267 ETCHEGOIN FORMATION, CALIFORNIA
S 10 243 ETHANE, MIGRATION, SURFACE EVIDENCE
S 10 248 ETHANE CONCENTRATIONS, ALABAMA, BIG ESCAMBIA CREEK FIELD
S 10 244 ETHANE CONCENTRATIONS, ALABAMA, FLOMATON FIELD
S 10 244 ETHANE CONCENTRATIONS, FLORIDA, BLACKJACK CREEK FIELD
S 10 244 ETHANE CONCENTRATIONS, FLORIDA, JAY FIELD
S 10 245 ETHANE INTERPRETATION, MIGRATION, SURFACE EVIDENCE
S 10 254 ETHANE MEASUREMENTS, LOUISIANA, OFFSHORE
B 6303 431* ETHIOPIA, ABST., REGIONAL HYDROCARBON, SOURC ROCK AND,
B 6110 1748 ETHIOPIA, DEVELOPMENTS, 1976
B 6210 1846 ETHIOPIA, DEVELOPMENTS, 1977
B 6411 1787 ETHIOPIA, DEVELOPMENTS, 1979
B 6004 681 ETHRIDGE, F. G., TRANSPORTATIONAL AND DEPOSITIONAL HISTORY OF
B 6203 572 ETHRIDGE, FRANK G., DEPOSITIONAL ENVIRONMENTS OF YALLAHS FAN
B 6403 374 ETHRIDGE, FRANK G., FAN-DELTA SEDIMENTOLOGY AND TECTONIC
B 6305 831 ETHRIDGE, FRANK G., FLOOD-PLAIN SEQUENCES OF FINE-GRAINED
B 6004 650 ETHRIDGE, FRANK G., NUMERICAL ASPECTS OF CHEMICAL VARIATION

B 6303 447 ETHRIDGE, FRANK G., PETROLOGY AND DIAGENESIS OF ARC-DERIVED
B 6108 1374 ETHRIDGE, FRANK G., PETROLOGY AND DIAGENETIC EFFECTS OF
B 6405 796 ETHRIDGE, FRANK G., REEVALUATION OF DEPOSITIONAL
B 6203 525 ETHRIDGE, FRANK G., TRANSITION BETWEEN A SANDY AND MUDDY
B 6407 1092 ETOLONIAN HORIZON, ASIA, NORTHEAST, KAKERTIAN-
B 6309 1547 ETTALHI, D. KROKOVIC, AND S. BANERJEE, BARR, F. T., REVIEW OF,
B 6009 1619 ETTENSOHN, FRANK R., STRATIGRAPHIC, PALEOENVIRONMENTAL, AND
S 10 157 ETZIKOM COULEE, ANALOG FLOW PATTERN, ALBERTA
M 29 152 EUCLA BASIN, AUSTRALIA, CONTINENTAL MARGIN
M 30 262 EUGENE ISLAND 330 FIELD, COMPOSITE TYPE LOG
M 30 275 EUGENE ISLAND 330 FIELD, GA SANDSTONE, NET OIL ISOPACH
M 30 264 EUGENE ISLAND 330 FIELD, GA SANDSTONE, RESERVOIR
M 30 271 EUGENE ISLAND 330 FIELD, GEOCHEMISTRY
M 30 255 EUGENE ISLAND 330 FIELD, GEOLOGIC SETTING
M 30 266 EUGENE ISLAND 330 FIELD, JD SANDSTONE, RESERVOIR
M 30 266 EUGENE ISLAND 330 FIELD, LF SANDSTONE, RESERVOIR
M 30 254 EUGENE ISLAND 330 FIELD, LOCATION, GULF COAST
M 30 267 EUGENE ISLAND 330 FIELD, OI SANDSTONE, RESERVOIR
M 30 262 EUGENE ISLAND 330 FIELD, PETROGRAPHY
M 30 253 EUGENE ISLAND 330 FIELD, PLEISTOCENE, LOUISIANA
M 30 256 EUGENE ISLAND 330 FIELD, PRODUCTION HISTORY
M 30 263 EUGENE ISLAND 330 FIELD, SANDSTONE COMPOSITION
M 30 257 EUGENE ISLAND 330 FIELD, SEISMIC SECTION
M 30 268 EUGENE ISLAND 330 FIELD, SEM PETROGRAPHY
M 30 257 EUGENE ISLAND 330 FIELD, STRATIGRAPHY
B 6303 467* EUGENE ISLAND BLOCK 330 FIELD,
M 30 253† EUGENE ISLAND BLOCK 330 FIELD,
B 6104 587 EUGENE ISLAND BLOCK FIELDS, LOUISIANA
B 6109 1489 EUGEOCLINAL WEDGE, EAST COAST, CONTINENTAL, MARGIN
S 9 17 EURASIA, EVIDENCES OF DRIFT, PALEOMAGNETISM
S 9 16 EURASIA, MAGNETIC POLE POSITIONS, PALEOZOIC
B 6106 900 EURASIAN PLATE
M 25 434 EURASIAN PLATE
B 6004 697* EURASIAN PLATE BOUNDARIES IN USSR, ABST.
B 6403 360 EURASIAN PLATE, BOUNDARY
B 6203 535* EUREKA QUARTZITE, SOUTHEASTERN NEVADA, ABST., ALGAL,
B 6101 20 EUREKAN FOLDBELT, CANADIAN, ARCTIC ISLANDS
B 6206 1021 EUREKAN OROGENY, CANADA, SVERDRUP BASIN
B 6101 5 EUREKAN OROGENY, SVERDRUP BASIN
B 6405 743* EUROPE, ABST., COMPARISON OF SHALLOW-MARINE SHELF, CARBONATE,
B 6405 791* EUROPE, ABST., COMPARISON OF, GULF COAST LATE, JURASSIC,
B 6301 69 EUROPE, ALPINE-CARPATHIAN OROGENE
B 6203 426 EUROPE, ANTRIM BASIN
B 6301 69 EUROPE, BOHEMIAN MASSIF
B 6203 426 EUROPE, BRABANT PLATFORM
B 6203 426 EUROPE, BRITTANY PLATFORM
B 6302 258 EUROPE, BY J. G. C. ANDERSON, HABICHT, J. K. A., REVIEW OF THE
B 6203 426 EUROPE, CENTRAL GRABEN
B 6203 426 EUROPE, CHANNEL BASIN
B 6203 426 EUROPE, CHESHIRE BASIN
B 6203 426 EUROPE, CORNUBIAN PLATFORM
S 6 136 EUROPE, CRETACEOUS, GLAUCONITES, EXAMPLES
S 6 138 EUROPE, CRETACEOUS, PROPOSED RADIOMETRIC SCALE
B 6203 426 EUROPE, DANISH TROUGH
S 6 220 EUROPE, EOCENE, BARTONIAN STAGE DEFINED
S 6 220 EUROPE, EOCENE, LUTETIAN STAGE DEFINED
S 6 221 EUROPE, EOCENE, PRIABONIAN STAGE DEFINED
S 6 220 EUROPE, EOCENE, YPRESIAN STAGE DEFINED
B 6203 426 EUROPE, FYN-RINGKSBING HIGH
B 6010 1704† EUROPE IN 1975, PETROLEUM, EXPLORATION AND, PRODUCTION IN
B 6110 1636† EUROPE IN 1976, PETROLEUM, EXPLORATION AND, PRODUCTION IN
B 6210 1906† EUROPE IN 1977, PETROLEUM, EXPLORATION AND, PRODUCTION IN
B 6310 1915† EUROPE IN 1978, PETROLEUM, EXPLORATION AND, PRODUCTION IN
B 6411 1961† EUROPE IN 1979, PETROLEUM DEVELOPMENTS IN
B 6203 426 EUROPE, IRISH PLATFORM
S 6 134 EUROPE, LUTETIAN BOUNDARIES, GLAUCONITES
B 6203 426 EUROPE, MANX FURNESS BASIN
B 6203 426 EUROPE, MID NORTH SEA PLATFORM
B 6203 426 EUROPE, MINCH BASIN
S 6 221 EUROPE, MIOCENE-OLIGOCENE BOUNDARY DEFINED
B 6203 426 EUROPE, MORAY FIRTH GRABEN
B 6301 69 EUROPE, NEOGENE FOREDEEPS
B 6203 426 EUROPE, NORTH CELTIC SEA BASIN
B 6203 426 EUROPE, NORTHWEST, CONTINENTAL SHELF, TECTONICS
B 6203 426 EUROPE, NORWEGIAN PLATFORM
S 6 221 EUROPE, OLIGOCENE, CHATTIAN STAGE DEFINED
S 6 221 EUROPE, OLIGOCENE-MIOCENE BOUNDARY DEFINED
S 6 217 EUROPE, PALEOCENE, DANIAN STAGE DEFINED
S 6 220 EUROPE, PALEOCENE, THANETIAN STAGE DEFINED
S 6 216 EUROPE, PALEOGENE, DEPOSITION
S 6 139 EUROPE, PALEOGENE, PROPOSED RADIOMETRIC SCALE
S 6 215 EUROPE, PALEOGENE, STAGE STRATOTYPES
S 6 179 EUROPE, PARATETHYS, DESCRIPTION OF AREA
S 9 10 EUROPE, PERMIAN, WIND DIRECTION
B 6301 69 EUROPE, PIENINY KLIPPEN BELT
B 6203 426 EUROPE, PORCUPINE SEABIGHT
B 6110 1654 EUROPE, PRODUCTION, 1975-1976
B 6210 1924 EUROPE, PRODUCTION, 1976-1977
B 6310 1932 EUROPE, PRODUCTION, 1977-1978
B 6411 1971 EUROPE, PRODUCTION, 1978-1979
S 6 127† EUROPE, RESULTS OF DATING, CRETACEOUS, PALEOGENE SEDIMENTS
B 6203 426 EUROPE, SHETLAND PLATFORM
B 6203 426 EUROPE, SOUTH CELTIC SEA BASIN
B 6203 426 EUROPE, VIKING GRABEN
B 6203 426 EUROPE, WELSH PLATFORM
B 6203 426 EUROPE, WEST NETHERLANDS BASIN
B 6203 426 EUROPE, WEST SHETLAND BASIN
B 6203 426 EUROPE, WEST SOLE TROUGH
B 6203 426 EUROPE, WESTERN APPROACHES
B 6111 1994 EUROPEAN BASIN, POSSIBLE DIAPIRS
B 6004 710* EUROPEAN BASINS, ABST., STRUCTURE OF, NEOGENE, MIOCENE,
B 6107 988 EUROPEAN CHALKS, TABLE, NORTH AMERICAN AND
B 6203 495* EUROPEAN CONTINENTAL SHELF. BY TITLE ONLY., INTRODUCTION OF,
B 6304 701* EUROPEAN GIANT OIL FIELDS, ABST., EKOFISK--FIRST OF WESTERN
S 11 7 EUROPEAN, TERTIARY STAGES AND ZONES
B 6301 90 EURYDICE FORMATION, SCOTIAN BASIN
B 6004 699* EUSTATIC CYCLES, ABST., REGIONAL, SEISMIC INTERPRETATION USING
M 26 89 EUSTATIC ESTIMATION, SEA LEVEL FLUCTUATIONS
B 6109 1538* EUSTATIC LOWERING OF SEA LEVEL—EVIDENCE FROM
M 29 453† EUSTATIC SEA LEVEL CHANGES ON STRATIGRAPHIC SEQUENCES AT
M 24 250 EUTAW SANDSTONES, MISSISSIPPI
B 6103 309 EUWER, R. M., MISSISSIPPIAN COMBINATION TRAP, BINDLEY FIELD,
B 6002 319 EUWER, ROBERT M., GEOLOGY OF MISSISSIPPIAN COMBINATION TRAP,
B 6408 1215 EVALUATING MATURE BASINS
B 6109 1547* EVALUATION OF AUSTIN AND BUDA FORMATIONS FROM CORE AND
B 6105 799 * EVALUATION OF FLUVIAL-FACIES MODELS
B 6203 529* EVALUATION OF IN SITU RECOVERY PROJECTS. BY TITLE ONLY.,
B 6206 1092* EVALUATION OF IN-SITU RECOVERY PROJECTS, ABST.,
B 6206 1075† EVALUATION OF JURASSIC TWIN CREEK LIMESTONE, IDAHO-WYOMING
B 6109 1559* EVALUATION OF LANDSAT DATA IN EXPLORING EVAPORITE BASINS,
B 6105 806 * EVALUATION OF MECHANISMS OF CALCITE GROWTH INHIBITION AND
B 6203 551* EVALUATION OF MINERAL AND ENERGY RESOURCES, ABST.,
B 6405 738* EVALUATION OF ORGANIC MATTER AND SUBSURFACE TEMPERATURE
B 6303 287 EVALUATION OF PROSPECTS, OFFSHORE, GUIDELINES TO
B 6009 1613* EVALUATION OF QUALITY AND QUANTITY CONTROLS ON
B 6409 1531 EVALUATION OF RESOURCE ESTIMATES
B 6405 668* EVALUATION OF TEMPERATE ZONE COASTAL-MARSH SEDIMENTS AS
B 6408 1251† EVALUATION OF TERTIARY NIGER DELTA, PETROLEUM, SOURCE-BED
B 6405 798* EVALUATION OF URANIUM POTENTIAL OF FRONTIER AREAS BY
B 6405 699* EVALUATION OF VOLUME OF ENTRAINED METHANE IN DEEP, TERTIARY
B 6004 608† EVALUATION--QUANTITATIVE ESTIMATE OF ORGANIC RICHNESS AND
B 6004 608† EVALUATION--QUANTITATIVE ESTIMATE OF SOURCE-ROCK
B 6201 1 EVAMY, B. D., HYDROCARBON HABITAT OF TERTIARY NIGER DELTA
B 6402 280 EVANS, CHRISTOPHER D. R., ECUADORIAN SUBDUCTION SYSTEM.
M 25 405 EVANS, H. J., AUSTRALIAN BAUXITE DEPOSITS--DISCOVERY AND
B 6412 2191 EVANS, HILTON B., REVIEW OF OLD ELECTRICAL LOG INTERPRETATION,
B 6311 2116 EVANS, JAMES L., MAJOR STRUCTURAL AND STRATIGRAPHIC FEATURES
B 6102 295 EVANS, ROBERT, ENVIRONMENTS OF EVAPORITI
B 6202 223 EVANS, ROBERT, ORIGIN AND SIGNIFICANCE OF EVAPORITES IN BASINS
B 6105 783 EVANS, ROBERT, ORIGIN AND SIGNIFICANCE OF EVAPORITES IN BASINS
B 6011 2005 EVANS, ROBERT, ORIGIN OF LIMESTONE BUTTES, GYPSUM PLAIN,
B 6306 987 EVANS, ROBERT, REVIEW OF MARINE EVAPORITES, EDITED BY WALTER E.
B 6405 733 EVANS, ROBERT, SOURCE-ROCK POTENTIAL OF EVAPORITIC
B 6405 640 EVANSTON FORMATION, WYOMING
S 5 73* EVAPORITE CYCLES AND LITHOFACIES IN LUCAS FORMATION, DETROIT
S 5 181* EVAPORITE PAN, BASE- METAL CONCENTRATION IN A DENSITY-
B 6408 1288* EVAPORITE, AND CARBONATE FACIES ASSOCIATIONS IN INTERIOR
B 6107 975 EVAPORITE BASIN, GABON
S 9 4 EVAPORITE BASIN MODEL, CIRCULATION, ARID CONDITIONS
S 9 4 EVAPORITE BASIN MODEL, CIRCULATION, HUMID CONDITIONS
S 9 4 EVAPORITE BASIN MODEL, GENERAL CIRCULATION
B 6303 512* EVAPORITE BASINS, ABST., CARBONATE REPLACEMENT OF,
B 6109 1559* EVAPORITE BASINS, ABST., EVALUATION OF, LANDSAT DATA IN
S 3 136 EVAPORITE BASINS, CONCENTRATION IN
B 6004 718* EVAPORITE DEPOSITION, PERMIAN SHELF SEVEN RIVERS FORMATION,
B 6207 1223* EVAPORITE DEPOSITS OF NORTHEAST THAILAND, ABST.
B 6205 885* EVAPORITE RELATIONS IN MIDDLE DEVONIAN OF SASKATCHEWAN—
B 6203 530* EVAPORITE RELATIONS IN MIDDLEDEVONIAN OF SASKATCHEWAN, ABST.,
B 6303 462* EVAPORITE SEDIMENTATION PATTERNS IN LOWER AND MIDDLE
B 6203 523* EVAPORITE SOLUTION COLLAPSE IN ATHAPUSCOW AULACOGEN, MIDDLE

B 6303 489* EVAPORITE SOURCE ROCKS, ABST., ENVIRONMENTAL AND DIAGENETIC
B 6409 1562* EVAPORITE SOURCE ROCKS, ABST., ENVIRONMENTAL AND DIAGENETIC
B 6209 1637 EVAPORITE UNITS, APPALACHIAN BASIN, SALINA
B 6004 674* EVAPORITE-ENCASED REEF RESERVOIRS, ABST., SOURCE-, RESERVOIR-,
B 6303 511* EVAPORITES AND RED BEDS OF PALO DURO BASIN, TEXAS--FACIES
B 6202 226 EVAPORITES AROUND ATLANTIC
S 5 125* EVAPORITES IN THE SALINE FORMATION, WEST VIRGINIA, RELATIONS
S 5 93* EVAPORITES OF MICHIGAN BASIN, SYNCHRONIZATION OF DEPOSITION:
B 6211 2241 EVAPORITES, BAY OF BISCAY AREA
B 6202 230 EVAPORITES, BRAZIL, SERGIPE-, ALAGOAS
S 5 1 EVAPORITES--CONCEPTS AND DEPOSITIONAL MODELS, REEFS AND
B 6202 227 EVAPORITES, CONGO
B 6202 227 EVAPORITES, CUANZA, ANGOLA, MOSSAMEDES
B 6306 987 EVAPORITES, EDITED BY WALTER E. DEAN AND B. CHARLOTTE
B 6106 873 EVAPORITES, EGYPT, MESSINIAN
B 6202 226 EVAPORITES, GABON
B 6202 230 EVAPORITES, GRAND BANKS
B 6202 231 EVAPORITES, GULF OF MEXICO
B 6202 223† EVAPORITES IN BASINS AROUND ATLANTIC MARGIN, ORIGIN AND
B 6209 1767* EVAPORITES IN LOWER ORDOVICIAN COOL CREEK FORMATION,
M 29 306 EVAPORITES, MAP, MEDITERRANEAN, MESSINIAN
B 6202 230 EVAPORITES, MOROCCO
B 6202 230 EVAPORITES, MOROCCO, AAIUM BASIN
B 6202 230 EVAPORITES, SCOTIAN SHELF
B 6202 230 EVAPORITES, SENEGAL
B 6203 544* EVAPORITES, SILICIFIED, FROM CARBONIFEROUS LIMESTONES OF
B 6011 2069* EVAPORITES, SULFIDES, AND PETROLEUM, ABST.
S 5 189* EVAPORITES: A SUMMARY, REEFS AND
S 5 29 EVAPORITES, DEPOSITIONAL ENVIRONMENT, BASIN-CENTER
S 5 23* EVAPORITES, LOWER SALINA GROUP (UPPER SILURIAN), MICHIGAN
B 6102 295* EVAPORITIC SEDIMENTATION--SHALLOWS, SABKHAS, SHELVES, AND
B 6004 720* EVAPORITIC BASINS, ALTERNATIVE MODE OF FORMATION, ABST.,
B 6405 733* EVAPORITIC ENVIRONMENT, ABST., SOURCE-ROCK POTENTIAL OF
S 5 169* EVAPORITIC LITHOFACIES CONTINUUM: LATEST MIOCENE (MESSINIAN)
B 6004 732* EVENT SEQUENCES IN GEOLOGIC REEF PRESERVATION--ROL OF CORAL
B 6303 447 EVERETT, JOHN R., ACTIVE FAULTS IN HOUSTON, TEXAS, AREA A
B 6203 490* EVERGLADES, ABST., PROFILE OF, COAL-FORMING SWAMP—THE
M 28 113 EVERTON FORMATION, ARKANSAS, ORDOVICIAN
B 6107 1025 EVESHAM AREA, SASKATCHEWAN
B 6405 724 EVEY, R. A., CRETACEOUS-TERTIARY VERSUS CARBONIFEROUS
B 6012 2188* EVIDENCE FOR LOW FREE-GAS SATURATION IN WATER-BEARING
B 6212 2511* EVIDENCE FOR METALLOGENESIS ON OCEANIC PLATES—EAST PACIFIC
B 6309 1610* EVIDENCE FOR POST-JURASSIC TECTONISM IN EASTERN NORTH
B 6108 1382* EVIDENCE FOR SOURCE AREAS AND DEPOSITIONAL ENVIRONMENTS OF
B 6305 838* EVIDENCE FOR TWO MAJOR CYCLES OF PETROLEUM GENERATION IN
B 6308 1428* EVIDENCE FOR WRENCH FAULTING, SOUTHERN VAL VERDE BASIN,
S 10 241† EVIDENCE OF HYDROCARBON MOVEMENT FROM DEPTH, NEAR-,
B 6007 1112† EVIDENCE OF POST-PLEISTOCENE FAULTS ON NEW JERSEY ATLANTIC
B 6412 2108† EVIDENCES OF PRIMARY OIL MIGRATION
B 6305 840* EVOLUTION AND RESPONSE OF FLUVIAL SYSTEM--SEDIMENTOLOGIC
B 6405 794* EVOLUTION AND STRESS HISTORY OF LOW-PERMEABILITY UPPER
B 6203 564* EVOLUTION AND UNIFORMITARIANISM, ABST.
M 29 345 EVOLUTION, ANDEAN OROGENY, COLOMBIA
M 29 341 EVOLUTION, COLOMBIA, CRETACEOUS
M 29 341 EVOLUTION, COLOMBIA, EOCENE
M 29 345 EVOLUTION, COLOMBIA, MIOCENE
M 29 341 EVOLUTION, COLOMBIA, OLIGOCENE
M 29 341 EVOLUTION, COLOMBIA, PALEOCENE
M 29 345 EVOLUTION, COLOMBIA, PLEISTOCENE
M 29 345 EVOLUTION, COLOMBIA, PLIOCENE
M 29 250 EVOLUTION, DIAGRAM, ACCRETIONARY BASIN
M 29 259 EVOLUTION DIAGRAM, FOREARC BASIN, RESIDUAL
M 29 246 EVOLUTION, FOREARC BASIN, MODEL
M 29 312 EVOLUTION, MEDITERRANEAN, AEGEAN BASINS
M 29 313 EVOLUTION, MEDITERRANEAN, ANTALYA BASIN
M 29 308 EVOLUTION, MEDITERRANEAN, CALABRIAN ARC
M 29 313 EVOLUTION, MEDITERRANEAN, CILICIA-ADANA BASIN
M 29 311 EVOLUTION, MEDITERRANEAN, CYPRUS ARC
M 29 304 EVOLUTION, MEDITERRANEAN, EASTERN BASINS
M 29 313 EVOLUTION, MEDITERRANEAN, NORTH LEVANTINE BASIN
M 29 307 EVOLUTION, MEDITERRANEAN, SOUTH BALEARIC-ALGERIAN BASIN
M 29 297 EVOLUTION, MEDITERRANEAN, TYRRHENIAN BASIN
B 6111 1941 EVOLUTION OF ALGAL MOUNDS
B 6002 325* EVOLUTION OF ANADARKO BASIN, ABST.
B 6009 1622* EVOLUTION OF BASIN-SHELF RELATIONS DURING SILURIAN TIME IN
B 6405 754* EVOLUTION OF BROOKS RANGE THRUST BELT AND ARCTIC SLOPE,
B 6207 1228* EVOLUTION OF CANADIAN SEGMENT OF CORDILLERA, ABST.
B 6004 724* EVOLUTION OF CARBONATE SHELF, SILURIAN MCKENZIE FORMATION,
B 6105 790 * EVOLUTION OF CONTINENTAL-MARGIN BASINS
B 6001 34† EVOLUTION OF EOCENE-MIOCENE SEDIMENTATION PATTERN IN PARTS
B 6004 669* EVOLUTION OF FLORIDA BAY INTERPRETED FROM ISLAND
B 6303 535* EVOLUTION OF HYDROGRAPHIC BASINS AND LIMNOLOGY OF EOCENE
B 6004 690* EVOLUTION OF MARGINAL PLATEAU--DSDP DATA FROM SITE 356 ON SAO
B 6007 1054† EVOLUTION OF MODERN CARIBBEAN FRINGING REEF, GALETA POINT,
M 29 329† EVOLUTION OF NORTHWESTERN COLOMBIA, MAJOR, STRUCTURAL
B 6102 294* EVOLUTION OF PRESENT-DAY SEISMIC EXPLORATION IN CENTRAL
S 7 43† EVOLUTION OF SALT IN DEEP GULF OF MEXICO, OCCURRENCE AND
B 6402 264† EVOLUTION OF SAND-DOMINANT SUBAERIAL PHASE, ATCHAFALAYA
M 29 245† EVOLUTION OF STRUCTURAL HIGHS BORDERING MAJOR FOREARC
M 29 229 EVOLUTION OF STRUCTURE, SUMATRA, SUNDA ARC
B 6206 1004† EVOLUTION OF SVERDRUP BASIN, ARCTIC CANADA
M 29 151† EVOLUTION OF THE AREA UNDERLYING THE MAGNETIC QUIET ZONE ON
M 29 223† EVOLUTION OF THE SUNDA ARC IN THE CENTRAL SUMATRA REGION,
B 6112 2115† EVOLUTION OF YUCCA MOUND COMPLEX, LATE PENNSYLVANIAN
M 29 345 EVOLUTION, OLIGOCENE-MIOCENE DIASTROPHISM, COLOMBIA
M 29 341 EVOLUTION, PRE-ANDEAN OROGENY, COLOMBIA
S 5 7 EVOLUTION, STAGES, NIAGARA GROUP, PINNACLE REEFS,
S 5 7 EVOLUTION, STAGES, SALINA GROUP, PINNACLE REEFS,
M 29 252 EVOLUTION, STRUCTURAL HIGH, ACCRETIONARY BASIN
M 29 252 EVOLUTION, STRUCTURAL HIGH, CONSTRUCTED BASIN
M 29 253 EVOLUTION, STRUCTURAL HIGH, RESIDUAL BASIN
M 29 313 EVOLUTION, USSR, BLACK SEA BASIN
M 25 475† EVOLUTIONARY SERIES OF ANTIMONY DEPOSITS OF EASTERN
B 6408 1281* EWING AREA, JEFFERSON AND FRANKLIN COUNTIES, ILLINOIS, ABST.,
B 6105 698 GAS-HYDRATE HORIZONS DETECTED IN SEISMIC-PROFILER DATA
M 29 177 EWING, J. I., STRUCTURE AND STRATIGRAPHY OF THE BLAKE
M 29 113 EWING, J. I., STRUCTURE OF COLORADO BASIN AND CONTINENT-OCEAN
M 29 125 EWING, J. I., STRUCTURE OF FALKLAND PLATEAU AND OFFSHORE
M 30 588 EWONGUE SHALE, GABON BASIN, BATANGA FORMATION
B 6004 654* EXAMINATION OF RELATIONS BETWEEN MODAL COMPONENTS USING
M 26 178 EXAMPLE, GULF OF MEXICO, SHEET-DRAPE FACIES
B 6203 557* EXAMPLE OF CARBONATE BANKS IN LOWER PROTEROZOIC BELCHER
M 26 236 EXAMPLES, DEPOSITIONAL SYSTEMS
B 6405 749* EXAMPLES OF ABNORMAL FLUID PRESSURE PRODUCED BY
B 6104 577 EXAMPLES OF AGE CALCULATIONS OF OILS
B 6206 950 EXAMPLES OF ANCIENT SUBMARINE FANS
M 26 304 EXAMPLES, SEISMIC, INTERPRETATION, COLOR DISPLAYS
M 26 203 EXAMPLES, SEISMIC, REFLECTIONS, CARBONATE, RECOGNITION
B 6107 1046 EXCELLO FORMATION, KANSAS
S 3 83 EXHALOGENIC, DEFINITION
B 6201 41 EXMOUTH PLATEAU ARCH, AUSTRALIA
B 6411 1865 EXMOUTH PLATEAU AREA, AUSTRALIA
B 6201 40† EXMOUTH PLATEAU AREA OFF WESTERN AUSTRALIA, GEOLOGY AND
B 6201 70 EXMOUTH PLATEAU, CAPROCKS, AUSTRALIA
B 6201 70 EXMOUTH PLATEAU, RESERVOIRS, AUSTRALIA
B 6201 48 EXMOUTH PLATEAU, SEISMIC DATA, AUSTRALIA
B 6201 69 EXMOUTH PLATEAU, SOURCE ROCKS, AUSTRALIA
B 6201 61 EXMOUTH PLATEAU, STRATIGRAPHIC COLUMN, AUSTRALIA
B 6201 50 EXMOUTH PLATEAU, WELLS, AUSTRALIA
B 6311 2088† EXMOUTH PLATEAU, WESTERN AUSTRALIA, SEDIMENTARY-, WEDGE
B 6201 41 EXMOUTH SUBBASIN, AUSTRALIA
B 6201 40 EXON, N. F., GEOLOGY AND PETROLEUM POTENTIAL OF EXMOUTH
B 6111 2004† EXOTIC MIDDLE MIOCENE SEDIMENT FROM CAPE VERDE RISE AND ITS
S 10 41 EXPANDABLE CLAY VERSUS DEPTH, DIAGENESIS, MIGRATION
B 6406 922 EXPANDABLE CLAYS
B 6203 564* EXPANDING FIELD OF SEISLOG APPLICATIONS, ABST.
B 6209 1580 EXPANSION OF PRESSURIZED HOLES
B 6011 1993† EXPECTATIONS FROM URANIUM EXPLORATION
B 6405 783* EXPERIMENTAL COMPACTION OF LIME SEDIMENT, ABST.
B 6004 665* EXPERIMENTAL DEVELOPMENT OF PERMOPOROSITY IN CARBONATE
B 6401 60 EXPERIMENTAL EVOLUTION OF SOURCE ROCK AND ISOLATED KEROGEN
B 6401 60 EXPERIMENTAL EVOLUTION OF SOURCE ROCK AND ISOLATED KEROGEN
B 6102 227† EXPERIMENTAL STUDY OF FRACTURE PERMEABILITY IN POROUS ROCK
B 6108 1379* EXPERIMENTAL STUDY OF PYRITE OXIDATION AT PH 5-9.5,
B 6008 1395* EXPLOITATION OF GAS RESERVES AT BOWDOIN AREA,
M 25 386† EXPLOITATION OF MANGANESE NODULES IN SOUTH PACIFIC--SUMMARY
B 6303 435* EXPLOITATION OF VERY LOW-GRADE URANIUM DEPOSITS, ABST.
B 6405 674 EXPLORABLE STRUCTURES IN OLD BAHAMA CHANNEL, NORTH OF CUBA,
S 5 111* EXPLORATION FOR REEFS, NORTHERN MICHIGAN, SEISMIC
S 5 4 E-UNIT, SALINA GROUP,
B 6004 714* EXPLORATION, ABST., SHALLOW, GAS, ACCUMULATIONS--THEIR ORIGIN
B 6105 820* EXPLORATION, ABST., TECHNOLOGICAL FOREFRONT OF
M 26 369 EXPLORATION ANALYSES, RESOLUTION SYSTEMS
B 6102 294† EXPLORATION AND DEVELOPMENT IN OIL-POTASH AREA OF
B 6207 1207* EXPLORATION AND DEVELOPMENT OF GEOTHERMAL FIELDS IN
B 6207 1221* EXPLORATION AND DEVELOPMENT OF GEOTHERMAL RESOURCES IN
M 25 328† EXPLORATION AND DEVELOPMENT OF HYDROCARBON RESOURCES IN
B 6105 766 * EXPLORATION AND EVALUATION OF ROOSEVELT KGRA, UTAH

B 6010 1704† EXPLORATION AND PRODUCTION IN EUROPE IN 1975, PETROLEUM
B 6110 1636† EXPLORATION AND PRODUCTION IN EUROPE IN 1976, PETROLEUM
B 6210 1906† EXPLORATION AND PRODUCTION IN EUROPE IN 1977, PETROLEUM
B 6310 1915† EXPLORATION AND PRODUCTION IN EUROPE IN 1978, PETROLEUM
B 6004 711* EXPLORATION ASSUMPTIONS ON PROJECTIONS OF LONG-RANGE
B 6203 504* EXPLORATION CASE HISTORY AND GEOLOGIC ENVIRONMENT OF
B 6101 105 EXPLORATION CLASSROOM GAME FOR STUDENTS, PETROLEUM-
B 6203 569* EXPLORATION CYCLE MODELING, ABST., WHY ARE LARGE DEPOSITS
M 25 163† EXPLORATION FOR GEOTHERMAL ENERGY IN NICARAGUA--SUMMARY
B 6205 882* EXPLORATION FOR NONELECTRIC GEOTHERMAL RESOURCES IN
B 6311 2115* EXPLORATION FOR PENNSYLVANIAN SANDSTONES, SOUTH-CENTRAL
M 26 329† EXPLORATION FOR STRATIGRAPHIC TRAPS, SEISMIC
B 6304 688* EXPLORATION FOR URANIUM DEPOSITS IN GRANT MINERAL BELT, NEW
B 6101 105† EXPLORATION GAME, CONFUSION FLATS--AN
B 6303 424* EXPLORATION GEOPHYSICS IN PEOPLES REPUBLIC OF CHINA, ABST.
B 6001 3† EXPLORATION GEOPHYSICS--RECENT BREAKTHROUGHS AND
M 25 326 EXPLORATION, HISTORY, COLOMBIA
B 6109 1552* EXPLORATION HISTORY OF SOUTH TEXAS LOWER CRETACEOUS
B 6301 32† EXPLORATION HISTORY OF SOUTH TEXAS LOWER CRETACEOUS
B 6405 673* EXPLORATION IN CLASSIC THRUST BELT AND ITS FORELAND--BAVARIAN
B 6303 526* EXPLORATION IN EAST MALAYSIA OVER PAST DECADE, ABST.
M 30 423† EXPLORATION IN EAST MALAYSIA OVER THE PAST DECADE
B 6304 687* EXPLORATION IN GRANTS URANIUM REGION SINCE 1963, ABST.
B 6405 679* EXPLORATION IN GREAT SALT LAKE, ABST.
B 6211 2356* EXPLORATION IN RUGGED TERRAIN IN OVERTHRUST BELT, ABST.
B 6309 1598* EXPLORATION METHODS OF DISCOVERY AND DEVELOPMEN OF LOWER
B 6408 1216 EXPLORATION, METHODS USED
B 6406 938† EXPLORATION MODEL FOR PENNSYLVANIAN UPPER DELTA PLAIN
B 6203 560* EXPLORATION MODELS AND CRITERIA FOR SUBSURFACE RECOGNITION
B 6212 2512* EXPLORATION OF CARBONIFEROUS SYSTEM IN ORIENTAL BASINS OF
B 6009 1463† EXPLORATION OF CONTINENTAL MARGINS, PLATE TECTONICS IN, OIL
B 6303 461* **EXPLORATION POTENTIAL OF PENNSYLVANIAN-PERMIAN**
B 6206 963 EXPLORATION PREDICTIONS FROM FAN MODEL
B 6409 1557* EXPLORATION RAMIFICATIONS OF SUBSURFACE FLUID MIGRATIONS IN
M 26 330 EXPLORATION, REFLECTION PATTERNS IN
B 6303 540* EXPLORATION RESEARCH ALONG ARDMORE-ANADARKO BASIN TREND,
B 6303 419* EXPLORATION RESULTS AND POTENTIAL OF NORWEGIAN CONTINENTAL
B 6303 520* EXPLORATION RESULTS OFFSHORE SOUTH NEW ZEALAND, ABST.
B 6211 2364* EXPLORATION RISK ANALYSIS—DEVELOPMENT DECISION FOR FRONTIER
B 6004 713* EXPLORATION, ROCKY MOUNTAIN REGION, ABST., INTEGRATION OF
M 26 109 EXPLORATION, SEISMIC, REFLECTIONS, CHRONOSTRATIGRAPHY,
B 6309 1577* EXPLORATION STRATEGY FOR UNCONVENTIONAL NATURAL GAS
B 6104 483† EXPLORATION TARGETS, FROZEN-IN, HYDROCARBON, ACCUMULATIONS
B 6105 808 * EXPLORATION TECHNIQUES OF CONTINENTAL SLOPES AND RISES
M 25 492 EXPLORATION, USE OF BASIN ANALYSIS IN, MINERAL
B 6205 739† EXPLORATION—REQUIREMENT FOR SURVIVAL, ACCELERATION IN,
B 6008 1156 EXPLORATORY DRILLING, 1975, CANADA
B 6008 1157 EXPLORATORY DRILLING, 1975, MEXICO
B 6108 1124 EXPLORATORY DRILLING, 1976, CANADA
B 6108 1124 EXPLORATORY DRILLING, 1976, MEXICO
B 6208 1278 EXPLORATORY DRILLING, 1977, CANADA
B 6208 1279 EXPLORATORY DRILLING, 1977, MEXICO
B 6308 1171 EXPLORATORY DRILLING, 1978, CANADA
B 6308 1172 EXPLORATORY DRILLING, 1978, MEXICO
B 6409 1298 EXPLORATORY DRILLING, 1979, CANADA
B 6409 1299 EXPLORATORY DRILLING, 1979, MEXICO
B 6205 742 EXPLORATORY DRILLING ACTIVITY, WORLD
B 6308 1204 EXPLORATORY WELLS, 1978, BALTIMORE CANYON TROUGH
B 6205 743 EXPLORATORY WELLS DRILLED ANNUALLY IN NON-COMMUNIST
M 25 485 EXPLORER RISE, SEAFLOOR SPREADING
B 6405 736* EXPLORING FOR NIAGARAN PINNACLE REEF IN SOUTHERN MICHIGAN
M 29 443† EXPLOSION SEISMOLOGY STUDIES OF ACTIVE AND PASSIVE
B 6406 942† EXPOSURES OF GREENHORN FORMATION AND CARLILE SHALE, UPPER
B 6302 241 EXPULSION EFFICIENCY IN SOURCE BED EVALUATION
B 6004 543† EXPULSION FROM CLASTIC SEDIMENTS DURING COMPACTION, WATER
B 6002 213 EXPULSION HYPOTHESIS, GLOBULE
S 10 12 EXPULSION MECHANISMS, HYDROCARBON GENERATION
S 10 13 EXPULSION MECHANISMS, OSMOTIC IMBALANCES
S 10 12 EXPULSION MECHANISMS, THERMAL EXPANSION
B 6305 770 EXPULSION, OIL, DROPLET
B 6308 1174 EXTENSION TEST, OUTPOST OR
B 6409 1301 EXTENSION TEST, OUTPOST OR
B 6008 1159 EXTENSION TEST, OUTPOST OR
B 6208 1281 EXTENSION TEST, OUTPOST OR
B 6108 1126 EXTENSION TEST, OUTPOST OR
B 6405 779* EXTENSIONAL BASINS, ABST., GENERATION, MIGRATION, AND
B 6307 1031 EXTENSIONAL FAULT BLOCKS
M 29 199† EXTENSIONAL TECTONICS IN THE OKINAWA TROUGH
M 26 131 EXTERNAL FORMS, FILL TYPES, SEISMIC, FACIES
M 26 131 EXTERNAL FORMS, MOUND TYPES, SEISMIC, FACIES
M 26 130 EXTERNAL FORMS, TYPES, SEISMIC, REFLECTIONS, FACIES UNITS
M 27 124 EXTRACLASTS, CALCLITHITE
M 25 362† EXTRACTION--SUMMARY, LAND, SUBSIDENCE IN, JAPAN RESULTING
B 6307 1131 EXTRAPOLATION OF ARRHENIUS PLOTS OVER WIDE TEMPERATURE
B 6312 2189 EXTRAPOLATION OF DISCOVERY RATES, ASSESSING RESOURCES
B 6209 1567 EXTRUSION AND INTRUSION, INCIPIENT, DIAPIRS
B 6204 694 EXUMA SOUND, BAHAMAS
B 6107 1024 EYEHILL FIELD, SASKATCHEWAN
M 29 160 EYRE PLATEAU, SECTION, AUSTRALIA, MARGIN
B 6202 296 EZE AKU SHALE, NIGERIA
B 6204 672 F UNIT, ALGERIA, DEVONIAN
B 6304 649 F. P. STRADER NO. 1, SECTION, VIRGINIA, CALIFORNIA CO.
B 6204 672 F6 UNITS, ALGERIA, DEVONIAN-, SILURIAN RESERVOIRS
B 6203 512 FAAS, RICHARD W., MASS PHYSICAL AND ENGINEERING PROPERTIES OF
B 6105 817 FAAS, RICHARD W., PROBLEM OF MISPLACED SEDIMENT, ABST.
B 6303 448 FAAS, RICHARD W., STICKY MUDS--VISCOSITY OF ESTUARINE SEDIMENTS,
B 6303 532 FABER, E., CARBON ISOTOPES--NEW TOOL IN HYDROCARBON
B 6303 532 FABER, E., ORIGIN OF GASES ADSORBED IN NEAR-SURFACE SEDIMENT
B 6304 701 FABER, E., ORIGIN OF GASES, ADSORBED IN NEAR-SURFACE SEDIMENTS
B 6304 698 FABER, ECKHARD, CARBON ISOTOPE MEASUREMENTS OF
M 30 315 FABER, J. M., STRACHAN AND RICINUS WEST GAS FIELDS, ALBERTA,
B 6303 469 FABER, J. M., STRACHAN-RICINUS GAS FIELD, ALBERTA, ABST.
B 6203 520* FABRIC ANALYSES OF DEEP-SEA CONGLOMERATES, PEBBLY AND MASSIVE
B 6309 1585* FABRIC-ELEMENT-BASED CLASSIFICATION FOR LOW-POROSITY-SHALE
B 6004 676* FABRICS IN NONCYCLIC LOW SALINITY TIDAL-FLAT CARBONATE ROCKS,
M 28 151 FABRICS, REPLACEMENT OR DISPLACEMENT, GENERAL
B 6105 781 * FABRIC-SELECTIVE PERMOPOROSITY EXPERIMENTALLY DEELOPED IN
B 6207 1226* FACIES ANALYSES IN FIELD DEVELOPMENT, MACKEREL FIELD,
B 6205 752 FACIES ANALYSIS, DEFINITION, SEISMIC
M 26 136 FACIES ANALYSIS, INTERPRETATION
M 26 136 FACIES ANALYSIS, MAPPING
S 4 299† FACIES ANALYSIS OF HOLOCENE CARBONATE SEDIMENTS AND
M 26 136 FACIES ANALYSIS, RECOGNITION
S 5 145* FACIES RELATIONSHIPS, OTTO FIORD FORMATION
M 26 165 FACIES, ANALYSIS, SEISMIC
B 6408 1274 **FACIES ANATOMY AND DIAGENESIS OF A BAHAMIAN OOID SHOAL,**
B 6405 712* FACIES AND DEPOSITIONAL TECTONICS OF MIDDLE JURASSIC CARMEL
B 6303 491* FACIES AND RESERVOIR CHARACTERISTICS OF SHELF SANDSTONE,
M 26 249† FACIES AND SEDIMENTOLOGY OF TERRIGENOUS PLEISTOCENE DEPOSITS
B 6004 702* FACIES AND STRUCTURES OF BERING EPICONTINENTAL SHELF, ABST.,
B 6203 520* FACIES ASSOCIATIONS AND FABRIC ANALYSES OF DEEP-SEA
M 26 227 FACIES ASSOCIATIONS, STRANDPLAIN DEPOSITS, DEPOSITIONAL SYSTEMS
M 26 227 FACIES ASSOCIATIONS, TIDAL-FLAT DEPOSITS, DEPOSITIONAL SYSTEMS
B 6205 762 FACIES, BASIN-SLOPE AND BASIN FLOOR, SEISMIC
B 6306 953 FACIES BELTS, NEWFOUNDLAND, WESTERN
B 6205 759 FACIES, BROAD, LOW-RELIEF MOUND
B 6004 670* FACIES, CAPITAN REEF COMPLEX, PERMIAN, GUADALUPE MOUNTAIN,
M 26 347 FACIES CHANGES, ENTRAPMENT BY
B 6004 727* FACIES CHANGES IN WAPANUCKA LIMESTONE, LOWER PENNSYLVANIAN,
B 6205 770 FACIES, CHAOTIC-FILL
B 6405 776* FACIES CONTROL ON BITUMEN SATURATION, PEACE RIVER OIL SANDS
M 26 121 FACIES DEFINITION, SEISMIC, REFLECTIONS
B 6004 517† FACIES, DELAWARE BASIN, WEST TEXAS AND SOUTHEAST NEW MEXICO,
B 6203 560* FACIES, DIAGENESIS, AND DEFORMATION IN SLOPE AND BASINAL
B 6109 1553* FACIES, DIAGENESIS, AND POROSITY DEVELOPMENT IN LOWER
B 6309 1604* FACIES, DIAGENESIS, AND POROSITY RELATIONS OF BUDA,
B 6309 1607* FACIES DISTRIBUTION OF HOSSTON SAND, CENTRAL
B 6303 508* FACIES, ENVIRONMENTS, AND DEVELOPMENT OF TUXPAN-TECOLUTLA
M 26 131 FACIES, EXTERNAL FORMS, FILL TYPES, SEISMIC
M 26 131 FACIES, EXTERNAL FORMS, MOUND TYPES, SEISMIC
B 6102 137† FACIES GEOMETRY AND GENESIS OF LOWER PLIOCENE TURBIDITES,
B 6105 786 * FACIES GEOMETRY FROM CONVENTIONAL SEISMIC DATA-SEISLOG
B 6205 756 FACIES, HIGH-CONTINUITY AND, HIGH-AMPLITUDE
B 6205 770 FACIES INTERPRETATION, FUTURE OF, SEISMIC
B 6205 753 FACIES INTERPRETATION OF CLASTIC SEISMIC FACIES UNITS, SEISMIC
B 6205 757 FACIES, LOW-AMPLITUDE
B 6205 757 FACIES, LOW-CONTINUITY AND, VARIABLE-AMPLITUDE
M 26 142 FACIES, MAPPING, SEISMIC
B 6004 697* FACIES MODE FOR MACROTIDAL ENVIRONMENT, COBEQUID BAY, NOVA
B 6404 585 FACIES MODELS, EDITED B ROGER G. WALKER, VAN HOUTEN, FRANKLYN
B 6109 1551* FACIES MOSAIC, BLANCO AND HAYS COUNTIES, TEXAS, ABST., MIDDLE,
B 6205 767 FACIES, MOUNDED ONLAPPING-FILL
B 6205 764 FACIES, MOUNDED, SEISMIC
B 6205 760 FACIES, OBLIQUE-PROGRADATIONAL
B 6307 1059† FACIES OF LOWER PALEOZOIC ALLEN BAY CARBONATE ROCKS AND
B 6004 674* FACIES OF SALT RIVER ESTUARY, ST. CROIX, U.S.V.I.--MIXED CARBONATE
B 6205 764 FACIES, ONLAPPING-FILL, SEISMIC
B 6408 1283* FACIES, PALEOECOLOGY, AND DEPOSITIONAL ENVIRONMENTS OF
M 26 121 FACIES, PARAMETERS, SEISMIC
B 6303 442* FACIES PATTERNS AND DEPOSITIONAL MODELS, PENNSYLVANIAN
B 6004 720* FACIES PATTERNS, AND HYDROCARBON POTENTIAL OF BANGOR
B 6303 516* FACIES RECOGNITION FROM WELL LOGS TO PREDICT PERMEABILITY OF
B 6312 2145† FACIES RELATIONS OF TRANSGRESSIVE BARRIER, LATERAL AND,

M 26 165† FACIES, SEISMIC, STRATIGRAPHY AND, GLOBAL CHANGES OF, SEA LEVEL,
B 6205 752† FACIES, SEISMIC STRATIGRAPHY AND, GLOBAL CHANGES OF, SEA LEVEL,
M 26 74 FACIES SEQUENCES, STRATIGRAPHY, CALIFORNIA, SAN JOAQUIN BASIN
B 6205 763 FACIES, SHEET-DRAPE
B 6205 756 FACIES, SHELF, SEISMIC
B 6205 759 FACIES, SHELF-MARGIN AND PROGRADED-, SLOPE, SEISMIC
B 6205 762 FACIES, SIGMOID-PROGRADATIONAL
B 6205 764 FACIES, SLOPE-FRONT FILL
B 6004 706* FACIES STUDIES, ABST., COMPUTER APPROACH TO
B 6004 694* FACIES, SUBSURFACE COW CREEK LIMESTONE, LOWER CRETACEOUS
B 6410 1634† FACIES SUCCESSION OF PLIOCENE-PLEISTOCENE CARBONATES,
M 26 130 FACIES UNITS, EXTERNAL FORMS, TYPES, SEISMIC, REFLECTIONS
B 6205 753 FACIES UNITS, SEISMIC CHARACTERISTICS AND, ENVIRONMENTAL,
M 26 122 FACIES UNITS, SEISMIC, REFLECTIONS
B 6004 657* FACIES, WILLAPA BAY, WASHINGTON, ABST., MODERN AND, ANCIENT,
B 6408 1158† FACIES-CYCLE WEDGES, ASSESSING, OIL AND, GAS, PLAYS IN
B 6203 552* FACTORS CONTROLLING GEOCHEMISTRY AND EARLY DIAGENESIS OF
B 6105 799 * FACTORS CONTROLLING POROSITY RELATIONS IN LIMESTONE RESERVOIRS
B 6002 325* FACTORS CONTROLLING POROSITY RELATIONS IN PENNSYLVANIAN AND
B 6004 693* FACTORS CONTROLLING SUSPENDED SEDIMENT IN STREAMS DRAINING
B 6004 725* **FACTORS INFLUENCING POROSITY AND PERMEABILITY OF LOWER**
B 6105 779 * FACTORS INFLUENCING SEDIMENT DISTRIBUTIONS AND PATTERNS ON
B 6203 429 FAEROE TROUGH, BRITAIN, WEST, SHETLAND-
B 6203 423 FAEROE-ICELAND RIDGE
B 6303 448 FAGERSTROM, J. A., RECENT ARTHROPO LEBENSSPUREN AS INDICATORS
B 6305 846 FAGIN, STUART W., PETROLOGY OF TRIASSIC-JURASSIC CULPEPER
B 6203 550 FAHNESTOCK, ROBERT K., TRANSGRESSIVE BARRIERS OF BEAUFORT SEA
M 29 27 FAIL, JEAN-PIERRE, STRUCTURE AND DEVELOPMENT OF THE SOUTHEAST
B 6302 218 FAINSTEIN, ROBERTO AND, STRUCTURE AND ORIGIN OF THREE
B 6405 704 FAIRBANKS, RICHARD G., CARBON FLUX FROM OCEAN TO BIOSPHERE—
B 6407 1117 FAIRBRIDGE AND DAVID JABLONSKI, LINTZ, JOSEPH, JR., REVIEW OF THE,
B 6305 820 FAIRBRIDGE AND JEANNE BOURGEOIS, SCHOLLE, PETER A., REVIEW OF,
B 6406 821 FAIRFIELD GROUP, AUSTRALIA
B 6406 943 FAIRPORT MEMBER, MINNESOTA AND, SOUTH DAKOTA
M 24 157† FAIRWAY FIELD, EAST TEXAS, GEOLOGY OF
M 24 157 FAIRWAY FIELD, TEXAS
M 25 293 FAIRWEATHER FAULT, ALASKA
M 25 485 FAIRWEATHER FAULT ZONE, TRANSCURRENT MOVEMENTS
M 29 276 FAIRWEATHER ISLAND, FAULT ZONE, ALASKA
M 27 60 FAKSE LIMESTONE, DENMARK, PALEOCENE
B 6002 295 FAL KAOLINITIC DEPOSITS, ENGLAND, RIVER
B 6303 533* FALCON STATE, VENEZUELA, ABST., GEOCHEMICAL EVALUATION OF,
B 6004 655 FALKENHEIN, F. U. H., SYNSEDIMENTARY TECTONIC CONTROL OF
B 6303 429 FALKENHEIN, FRANK U. H., DEPOSITIONAL-DIAGENETIC HISTORY OF
B 6405 742* FALKLAND PLATEAU, ABST., SEISMIC STRATIGRAPH AND STRUCTURE OF
M 29 125† FALKLAND PLATEAU AND OFFSHORE TIERRA DEL FUEGO, ARGENTINA,
M 29 131 FALKLAND PLATEAU, ARGENTINA
M 29 132 FALKLAND PLATEAU, PALEOGENE SEDIMENTS, ARGENTINA
M 29 135 FALKLAND PLATEAU, SEISMIC REFLECTIONS, ARGENTINA
M 29 133 FALKLAND PLATEAU, STRUCTURE SECTION, ARGENTINA
M 29 132 FALKLAND PLATEAU, STRUCTURE SECTIONS, ARGENTINA
B 6402 182 FALKLAND PLATEAU, SUBSIDENCE AND HEAT FLOW
B 6402 182 FALKLAND PLATEAU, TIME, TEMPERATURE, AND DEPTH PLOT
M 29 131 FALKLAND TROUGH, ARGENTINA
M 29 135 FALKLAND TROUGH, SEISMIC REFLECTIONS, ARGENTINA
B 6304 687 FALKOWSKI, S. K., GEOLOGY AND ORE DEPOSITS OF JOHNNY M MINE,
B 6405 679* FALL RIVER COUNTY, SOUTH DAKOTA, ABST., INDIAN CREEK FIELD
B 6407 971 FALL RIVER SANDSTONE, GREAT PLAINS
B 6403 443* FALLON BASIN, WESTERN NEVADA, ABST., RESULTS OF EXPLORATORY
B 6407 1009 FALLON FIELD, ARK-LA-TEX AREA
B 6408 1132 FALLS CREEK GAS FIELD, ALASKA
M 26 71 FALLS, SEA LEVEL FLUCTUATIONS, RELATIVE
B 6405 749 FALLS, W FRED, ORGANOPHOSPHORITES IN CARBONIFEROUS ROCKS OF
B 6002 297 FALMOUTH CLAY, ENGLAND
B 6309 1522† FALSE PASS AREA, ALASKA PENINSULA, TERTIARY STRATIGRAPHY AND
B 6405 720* FALSE RIVER FIELD, ABST.
B 6108 1246 FALSE RIVER FIELD, POINTE COUPEE PARISH, LOUISIANA
B 6406 826 FAMENNIAN NULLARA REEF CYCLE, AUSTRALIA
B 6401 133 FAMOUS--PHOTOGRAPHIC ATLAS OF THE MID-ATLANTIC RIDGE, RIFT
B 6303 495* FAN, ABST., DOWNCURRENT CHANGE IN, MIOCENE CANYON-CHANNEL
B 6009 1616* FAN, ABST., FORM AND COMPOSITION OF, MISSISSIPPI
B 6405 734* FAN, ABST., LITHOLOGY AND STRUCTURES OF, QUATERNARY
B 6306 934 FAN, AMAZON RIVER
B 6007 1024 FAN AND DELTA, LAKE SUPERIOR, RESERVE
B 6405 740* FAN AND SHALLOW-MARINE COMPLEX, CALIFORNIA, ABST.,
B 6004 699* FAN AND SWATCH OF NO GROUND CANYON, NORTHEAST INDIAN
B 6004 687* FAN APRON IN BORDE FORMTION, MISSISSIPPIAN, KENTUCKY AND
B 6206 967† FAN, CALIFORNIA, CHANNEL, SANDS AND, PETROLEUM, POTENTIAL OF,
B 6004 678* FAN COMPLEX, CAP ENRAGE FORMATION MIDDLE CAMBRIAN TO LOWER
B 6405 671* FAN COMPLEX, KODIAK ISLAND ARCHIPELAGO, ALASKA, ABST.,
B 6004 730* FAN CONGLOMERATES AND SANDSTONES, JURASSIC OF SOUTHWESTERN
B 6403 376 FAN DELTA, JAMAICA, HOLOCENE
B 6203 572* FAN DELTA, SOUTHEASTERN JAMAICA, ABST., DEPOSITIONAL
B 6303 437* FAN DEPOSITION AND TECTONIC SIGNIFICANCE OF TWO LATE
B 6206 959 FAN, ENGLAND, SHALE, GRIT: A COMPLETE, ANCIENT, SUBMARINE
B 6405 670* FAN FACIES, NORTHWESTERN WASHINGTON, ABST., OPHIOMORPHA
B 6303 528* FAN GAS RESERVOIR IN WOODBINE--EAGLE FORD INTERVAL, UPPER
B 6203 546* FAN, GULF OF MEXICO, ABST., MISSISSIPPI
M 29 383† FAN, GULF OF MEXICO, INVESTIGATION OF, MISSISSIPPI
B 6303 426* FAN IN DEEP WESTERN GULF OF MEXICO AS INTERPRETED FROM
B 6206 927 FAN, LOWER, BASIN PLAIN, GROWTH PATTERNS
B 6206 946 FAN, LOWER, SUBMARINE FAN
B 6206 921 FAN, MIDDLE, GROWTH PATTERNS
B 6206 963 FAN MODEL, EXPLORATION PREDICTIONS FROM
B 6407 1096 FAN MODELS AND SUPRAFAN CONCEPT, SUBMARINE
B 6303 535* FAN MODELS, FINE-GRAINED SEDIMENT DISTRIBUTION, AND
B 6405 789* FAN MODELS FOR HYDROCARBON EXPLORATION WITH EXAMPLES FROM
B 6007 1021† FAN MORPHOLOGY IN LAKE SUPERIOR, SUBLACUSTRINE
B 6007 1023 FAN, NEW TERM, LAKE SUPERIOR, RESERVE
B 6306 905† FAN SANDS, DIPMETER AND, LOG MOTIFS IN, NORTH SEA, SUBMARINE-
B 6405 735* FAN SANDSTONE AND SHALE BY AUTHIGENIC SMECTITE AND
B 6004 734* FAN SEDIMENTATION IN MESOZOIC ARC-REAR MARGINAL BASIN IN
B 6304 689* FAN SYSTEM, ABST., DEPOSITION AND EARLY HYDROLOGIC EVOLUTION
B 6004 686* FAN SYSTEMS, ABST., SIGNIFICANCE OF, CLAST SIZE, SHAPE, AN
B 6203 521* FAN TRAP IN LOWER EOCENE SANDSTONES OF NORTH SEA VIKING
B 6203 566* FAN UNIT IN ANADARKO BASIN, OKLAHOMA, ABST., ECONOMIC
B 6206 950 FAN, UPPER, ANCIENT
B 6206 913 FAN, UPPER, GROWTH PATTERNS
B 6206 947 FAN, UPPER, SUBMARINE FAN
B 6206 912† FAN VALLEYS, CHANNELS, AND DEPOSITIONAL LOBES ON MODERN
B 6007 1033 FAN VALLEYS, LAKE SUPERIOR
B 6311 2119* FAN-DELTA DEPOSITS OF ANADARKO BASIN, ABST., UPPER, MORROW
B 6311 2116* FAN-DELTA SANDSTONES, PAL DURO BASIN, TEXAS, ABST.,
B 6403 374† FAN-DELTA SEDIMENTOLOGY AND TECTONIC SETTING—YALLAHS FAN
B 6004 725* FAN-VALLEY SEDIMENTATION, ABST., LATERAL, INFILL AS MAJOR
B 6206 952 FAN: BRAIDED CHANNELS ON SUPRAFAN LOBE, ANCIENT, MID-
B 6405 705 **FANG, J. H., GEOSTATISTICAL APPROACH IN COAL RESOURCE ESTIMATION,**
B 6004 721* FANS, ABST., CURRENTS IN, SUBMARINE, CANYON AND THEIR RELATION
B 6405 722* FANS, ABST., POSSIBLE GEOMETRIES OF, SANDSTONE BODIES AS
B 6403 443* FANS, ABST., QUATERNARY STYLES OF, CALIFORNIA, SUBMARINE
B 6004 702* FANS, ABST., THIN-BEDDED, TURBIDITES IN, MODERN, SUBMARINE,
B 6303 293 FANS AND CONES, PROSPECTS, CONTINENT-RELATED, DEEP-WATER
B 6408 1280* FANS AND FANTASIES, ABST., DEEP-WATER, RESERVOIRS—SUBMARINE
B 6305 828* FANS AND THEIR DEPOSIT IN DEVONIAN HORNELEN BASIN, NORWAY,
B 6206 950 FANS, EXAMPLES OF, ANCIENT, SUBMARINE
B 6405 758* FANS FROM VARIOUS TECTONIC SETTINGS, ABST., GEOMETRY AND
B 6206 913 FANS, GROWTH PATTERNS OF, DEEP-SEA
B 6306 907 FANS, NOMENCLATURE, SUBMARINE
B 6211 2243† FANS OF SOUTHEASTERN SAN JOAQUIN VALLEY, CALIFORNIA,
B 6405 803* FANS OFF CENTRAL CALIFORNIA, ABST., POTENTIAL PETROLEUM
B 6109 1535* FANS ON GEORGIA COAST, ABST., ORIGIN AND CHARACTER OF,
B 6407 1052 FANS, QUEBEC, SUBMARINE
B 6407 1094† FANS. DISCUSSION AND REPLIES, MODERN AND, ANCIENT, SUBMARINE
B 6305 810† FANS. MODELS FOR EXPLORATION FOR STRATIGRAPHIC TRAPS.
B 6206 912† FANS: CHARACTERS FOR RECOGNITION OF SANDY TURBIDITE
B 6206 932† FANS: MODELS FOR EXPLORATION FOR STRATIGRAPHIC TRAPS, DEEP-,
B 6008 1397 FANSHAWE, JOHN R., GEOTHERMAL POTENTIAL OF CRAZY MOUNTAIN
B 6203 521* FANS—IMPORTANT REGIONS FOR OIL AND GAS GENESIS AND
B 6206 1053 FANT TUFF MEMBER, TEXAS
B 6411 1880† FAR EAST, 1979, PETROLEUM, DEVELOPMENTS IN
B 6207 1211* FAR EAST, ABST., PETROLEUM POTENTIAL IN
B 6212 2513* FAR EAST, ABST., PLATE-TECTONIC EVOLUTION OF THE
M 25 463 FAR EAST, ACID VOLCANISM, USSR
B 6303 459* FAR EAST AND MIDDLE EAST ABST., GAS VERSUS OIL IN
B 6010 1908† FAR EAST IN 1975, PETROLEUM, DEVELOPMENTS IN
B 6110 1832† FAR EAST IN 1976, PETROLEUM, DEVELOPMENTS IN
B 6210 1978† FAR EAST IN 1977, PETROLEUM, DEVELOPMENTS IN
B 6310 1816† FAR EAST IN 1978, PETROLEUM, DEVELOPMENTS IN
B 6110 1842 FAR EAST, PRODUCTION, 1975-1976
B 6210 1990 FAR EAST, PRODUCTION, 1975-1977
B 6310 1829 FAR EAST, PRODUCTION, 1976-1978
B 6411 1896 FAR EAST, PRODUCTION, 1977-1979
M 25 464 FAR EAST, SOUTHERN, ACID VOLCANISM, CHARACTERISTICS OF
M 25 465 FAR EAST, SOUTHERN, MINERALIZATION
B 6203 346 FARALLON AND NORTH AMERICAN PLATES, INTERACTION OF, PACIFIC
B 6206 968 FARALLON CANYON, CALIFORNIA
B 6207 1195 FARALLON PLATE

B 6207 1193 FARALLON PLATE BOUNDARIES, SEAFLOOR SPREADING ALONG, KULA-
B 6203 356 FARALLON RIDGE, CALIFORNIA
M 25 293 FAREWELL FAULT, ALASKA
B 6209 1706 FARMERS MEMBER, HENLEY BED, KENTUCKY
B 6002 220 FARMERS WHOLE OIL, WYOMING
M 25 496 FARO MINE, CANADA
B 6303 448 FARR, JOHN B., HIGH-RESOLUTION SEISMIC WORK AS TOOL TO LOCATE
B 6004 670 FARRELL, R. C., THREE-DIMENSIONAL SEISMIC DISPLAYS AS
B 6008 1411 FARRELL, RAYMOND C., MULTI-DIMENSIONAL SEISMIC DISPLAYS, ABST.
B 6412 2095† FARS GROUP, MIOCENE, SOUTH-SOUTHWEST IRAN, STRATIGRAPHY AND
B 6301 47 FASHING CARBONATE PRODUCTION, TEXAS
B 6210 2050 FASSETT, B. D., ON DEVELOPMENTS IN ECUADOR, 1977
B 6310 1750 FASSETT, B. D., ON DEVELOPMENTS IN ECUADOR, 1978
B 6405 648 FASSETT, NICK D. THOMAIDIS, AND MARVIN L. MATHENY, WENGERD,
B 6411 1963 FASTNET BASINS, EIRE, STYNE TROUGH, DONEGAL AND
B 6003 452† FATTY ACIDS IN HOLOCENE SEDIMENT, THERMALLY INDUCED
B 6303 545* FAULT ACTIVITY IN TEXAS GULF COAST, ABST., QUATERNARY
B 6307 1043 FAULT ASSEMBLAGES, DETACHED, NORMAL-
B 6307 1026 FAULT BLOCKS AND BASEMENT THRUSTS, COMPRESSIVE
B 6307 1031 FAULT BLOCKS, EXTENSIONAL
B 6402 146 FAULT ENTRAPMENT OF HYDROCARBONS UNDER HYDROSTATIC
M 25 182 FAULT HYPOTHESIS FOR GENERATION OF OIL AND GAS, THRUST-
B 6108 1385* FAULT, NORTH PARK, COLORADO, ABST., OIL POTENTIAL OF MESOZOIC
B 6003 379† FAULT STRUCTURES IN VERACRUZ PROVINCE OF MEXICO, COMPLEX
B 6105 843* FAULT STRUCTURES, STYLES, MECHANISMS, AND RELATIONS TO
B 6211 2214† FAULT SYSTEM, CALIFORNIA, ROLE OF, SALINIAN BLOCK IN EVOLUTION
M 29 276 FAULT ZONE, ALASKA, FAIRWEATHER ISLAND
M 29 276 FAULT ZONE, ALASKA, QUEEN CHARLOTTE ISLAND
B 6402 167 FAULT ZONE MATERIALS, GULF COAST
B 6402 167 FAULT ZONES, GULF COAST
M 25 463 FAULT-BLOCK MOVEMENT, VOLCANISM, RELATED TO, ARC-
B 6203 555* FAULT-RELATED GRANULATION ON POROSITY AND PERMEABILITY OF
B 6105 834* FAULTED AREAS, ABST., USING, COMPUTER FOR, TIME-TO-DEATH
B 6111 2001† FAULTED UPPER PLEISTOCENE MARINE TERRACE, PALOS VERDES HILLS,
B 6405 804* FAULTING AND GEOTHERMAL RESERVOIR QUALITY, ABST., LOWER,
B 6305 833* FAULTING AND HYDROCARBON OCCURRENCE IN NORTHWESTERN
B 6405 703* FAULTING AND SEDIMENTATION IN UPPER WILCOX DELTA SYSTEM,
M 26 277† FAULTING, DENVER BASIN, COLORADO, STRATIGRAPHIC AND, SEISMIC
M 26 293 FAULTING, DENVER BASIN, GROWTH
S 3 100 FAULTING, DEWATERING AND, GROWTH
B 6004 707* FAULTING IN EASTERN WEST VIRGINIA, ABST., POSSIBILITY OF
B 6008 1395* FAULTING IN LARAMIDE ROCKY MOUNTAIN FORELANDS. ABST..
M 26 391 FAULTING, MODELING, OVERTHRUST
B 6303 545* FAULTING NEA HOUSTON, TEXAS, ABST., LAND DEVELOPMENT AND
B 6008 1413* FAULTING, NEW TECTONIC STYLE IN CRETACEOUS FORELAND BASIN,
B 6405 717* FAULTING OF VICKSBURG FORMATION, OLIGOCENE, SOUTH TEXAS,
M 29 203 FAULTING, OKINAWA TROUGH, TECTONICS
B 6405 751* FAULTING, ROCKY MOUNTAIN FRONT RANGES, CANMORE, ALBERTA,
B 6003 356† FAULTING, SAN JOAQUIN VALLEY, CALIFORNIA, TECTONIC
B 6308 1428* FAULTING, SOUTHERN VAL VERDE BASIN, SOUTHWEST TEXAS, ABST.,
B 6203 491* FAULTING, SOUTHWESTERN ARKOMA BASIN, OKLAHOMA, ABST., ATOKA
B 6004 689* FAULTING, TEXAS COASTAL ZONE, ABST., ACTIVE, SURFACE
B 6004 668* FAULTS AND ASSOCIATED STRUCTURES IN UPPER TRIASSIC DELTAIC
B 6209 1570 FAULTS ASSOCIATED WITH SALT DIAPIRS, RADIAL AND, TRAP-DOOR
B 6007 1113 **FAULTS, ATLANTIC, OUTER, CONTINENTAL, SHELF, SHALLOW**
B 6002 175 FAULTS, BERINGIAN SHELF, STRIKE-SLIP
B 6002 229 FAULTS, FLUID, MOVEMENT UP
B 6003 414† **FAULTS IN CAMBRIAN AND LOWER ORDOVICIAN ROCKS OF WESTERN**
B 6211 2191† FAULTS IN CARBONIFEROUS OF WESTERN IRELAND, GROWTH
B 6108 1390* FAULTS IN COAL SEAMS, ABST., HIGH-RESOLUTION, SEISMIC
B 6303 529* FAULTS IN GULF COAST SALT BASIN, ABST., SEALING AND NONSEALING
B 6303 447* FAULTS IN HOUSTON, TEXAS, AREA A OBSERVED ON LANDSAT IMAGERY,
B 6402 145† FAULTS IN LOUISIANA GULF COAST SALT BASIN, SEALING AND,
B 6303 513* FAULTS IN NORTHWESTERN HOUSTON, HARRIS COUNTY, TEXAS, ABST.,
B 6003 341† FAULTS IN UPPER TRIASSIC DELTAIC SEDIMENTS, SVALBARD, GROWTH
B 6211 2207 FAULTS, IRELAND, WESTERN, NAMURIAN, GROWTH
M 25 512 FAULTS, LARGE, ORIGI OF, ALASKA AND, BRITISH COLUMBIA,
B 6108 1386* FAULTS, NORTHWEST ARIZONA, ABST., EARLY, PERMIAN, SLIP AND
B 6303 432* FAULTS OFFSETTING LAND SURFACE IN SOUTHEASTERN HOUSTON
B 6007 1112† FAULTS ON NEW JERSEY ATLANTIC OUTER CONTINENTAL SHELF,
B 6303 498* FAULTS ON TEXAS INNER SHELF, ABST., CHARACTERISTICS OF SHALLOW
B 6203 543* FAULTS—MECHANISM FOR CONTROL OF SEDIMENTATION IN
B 6004 727* FAUNA OF SOME NIAGARAN REEFS OF WESTERN OHIO, ABST.,
S 9 7 FAUNA, PALEOCLIMATE INDICATORS, FLORA AND
B 6004 654* FAUNAL ACTIVITY IN LOWER HUDSON SUBMARINE CANYON, ABST.,
M 30 64 FAUNAL ASSEMBLAGES, FRIGG FIELD, TERTIARY
B 6004 665* FAUNAL CRITERIA, ABST., RELATIVE, MARINE, TRANSGRESSIVE AND,
B 6105 824* FAUNAS IN BIOGEOGRAPHY AND BIOSTRATIGRAPHY, ABST., DISPERSAL,
B 6303 533 FAURE, GUNTER, ISOTOPIC COMPOSITION AN SOURCES OF STRONTIUM
B 6303 423 **FAVERA, JORGE C. DELLA, SEISMIC SEQUENCE ANALYSIS, NEW APPROACH**
S 4 362 FAVIA SP., MICROGRAPH
B 6105 771 * **FAVORABLE CRITERIA FOR RADIOACTIVE-WASTE-DISPOSAL SITES IN**
B 6108 1389 FAWCETT, DONALD W., GEOTHERMAL EXPLORATION IN CENTRAL
M 27 201 FAXEKALK, DENMARK, PALEOCENE
B 6403 339 FAY CRUISES 017, 018, AND 025, RV
B 6410 1660 FAY CRUISES 19, 20, 21, AND 25, USGS, RV
B 6101 30 FAY MINE, SASKATCHEWAN
B 6309 1607* FAYETTE COUNTIES, TEXAS ABST., FIELD RELATIONS AND PETROLOGY
B 6408 1282* FAYETTE COUNTY, ILLINOIS—ITS RELATION TO CYPRESS, CHESTERIAN,
B 6109 1476 FAZENDA CEDRO FIELD, BRAZIL
B 6109 1478 FAZENDA TOMADO FIELD, BRAZIL
B 6303 448 **FEAZEL, CHARLES T., GENERATION AND OCCLUSION OF POROSITY IN**
B 6309 1545 FEAZEL, CHARLES T., REVIEW OF MODERN PETROLEUM, A BASIC PRIME
M 27 111 FECAL PELLETS
B 6303 442* FEDERAL BARRIERS TO ENERGY RESOURCE DEVELOPMENT, ABST.
B 6008 1256 FEDERAL BLOCK A-302, TEXAS, HIGH ISLAND AREA EAST ADDITION
B 6108 1225 FEDERAL BLOCK A-492, TEXAS, HIGH ISLAND SOUTH
B 6008 1256 FEDERAL BLOCK A-532, TEXAS, HIGH ISLAND AREA SOUTH
B 6108 1225 FEDERAL BLOCK A-560, TEXAS, HIGH ISLAND SOUTH
B 6008 1256 FEDERAL BLOCK A-561, TEXAS, HIGH ISLAND SOUTH
B 6010 1711 FEDERAL REPUBLIC, 1975, HARK, H. U., ON DEVELOPMENTS IN, GERMAN
B 6310 1920 FEDERAL REPUBLIC OF GERMANY, 1978, SCHRODER, L., ON
B 6310 1936 FEDERAL REPUBLIC OF GERMANY, PRODUCTION, 1977-1978
B 6008 1389* FEDERAL WITHDRAWALS TO NATURAL RESOURCES DEVELOPMENT.
B 6206 950 FEEDER CHANNEL, TERM
B 6206 947 FEEDER CHANNELS, SUBMARINE FAN
S 3 135 FEEDING, AT TECTONIC INTERSECTIONS, TAPHROGENIC,
B 6102 295 FEEMSTER, LEE, COAL DEPOSITS IN THURBER AREA, TEXAS. BY TITL
B 6405 705 FEINSTEIN, SHIMON, SUBSIDENCE AND THERMAL HISTORY OF
B 6406 960 FEIR, J. DOUGLAS, THE ALSANDS PROJECT—A CHALLENGE IN
B 6012 2180 FEISTER, GEORGE H., TOMORROWS OIL FROM SOUTHERN CALIFORNIA
B 6009 1613* FELDA AND WEST SUNOCO-FELDA FIELDS HENDRY AND COLLIER
M 28 16 FELDSPAR, ALBITE TWINNING
M 28 17 FELDSPAR, ALBITE TWINNING
M 28 19 FELDSPAR, ALTERATION
M 28 22 FELDSPAR, AUTHIGENIC ALBITE
M 28 23 FELDSPAR, AUTHIGENIC POTASSIUM
M 28 155 FELDSPAR, AUTHIGENIC, REPLACEMENT OF ANALCIME
M 28 155 FELDSPAR, AUTHIGENIC, REPLACEMENT OF LIMESTONE
M 28 24 FELDSPAR, BIBLIOGRAPHY
M 28 147 FELDSPAR, BIBLIOGRAPHY, DIAGENESIS AND POROSITY
M 28 20 FELDSPAR, DETRITAL, LEACHING OF
M 28 35 FELDSPAR, FRAGMENTS, VOLCANIC
M 28 174 FELDSPAR GROWTH AND LEACHING, POROSITY, COMPLEX
M 28 20 FELDSPAR, LEACHING
M 28 17 FELDSPAR, MICROCLINE, SPINDLE TWINNING
M 28 17 FELDSPAR, MICROCLINE TWINNING
M 28 18 FELDSPAR, ORTHOCLASE, UNTWINNED
M 28 118 FELDSPAR OVERGROWTHS, CEMENTS
M 28 22 FELDSPAR, PARTIAL DISSOLUTION
M 28 16 FELDSPAR, PLAGIOCLASE TWINNING
M 27 143 FELDSPAR, REPLACEMENT
M 28 17 FELDSPAR, SANIDINE
M 28 18 FELDSPAR, SERICITIZATION
M 28 21 FELDSPAR, UNTWINNED AND UNSTAINED
M 28 18 FELDSPAR, VACUOLIZATION
B 6206 1058 FELDSPARS
M 28 175 FELDSPARS, FROM VOLCANIC ROCK, LEACHING
M 28 15 FELDSPARS, GENERAL
B 6104 560 FELIPE BAY, CHILE
B 6004 676* FENESTRAL FABRICS IN NONCYCLIC LOW SALINITY TIDAL-FLAT
M 27 69 FENESTRATE BRYOZOANS
M 27 68 FENESTRATE BRYOZOANS
S 7 87 FENNESSY, W. J., RECOGNITION OF CONTINENTAL-SLOPE SEISMIC FACIES,
M 25 477 FERBERITE, USSR, EASTERN, LODE DEPOSITS OF ANTIMONITE AN
B 6303 449 FEREBEE, THOMAS W., JR., SEDIMENTATION IN MISSISSIPPI TROUGH,
B 6308 1338 FERGESON, WILLIAM G., DEVELOPMENTS IN SOUTH TEXAS IN 1978
B 6409 1476 FERGESON, WILLIAM G., DEVELOPMENTS IN SOUTH TEXAS IN 1979
S 10 227 FERGHANA, OIL AND COPPER IN SAME TRAP, USSR
S 10 227 FERGHANA FIELD, TRAP, USSR
B 6009 1620 FERM, J. C., CARBONIFEROUS SEDIMENTARY RESPONSES TO
B 6105 796 **FERM, J. C., DEPOSITIONAL MODELS IN COAL EXPLORATION AND MINE**
B 6212 2379 FERM, J. C., DEPOSITIONAL MODELS IN COAL EXPLORATION AND MINE
B 6303 533 FERM, JOHN C., ESTUARINE-COASTAL PLAIN COAL DEPOSITION IN
B 6406 938 FERM, JOHN C., EXPLORATION MODEL FOR PENNSYLVANIAN UPPER
B 6004 680 FERM, JOHN C., SANDSTONE MODELS FROM APPALACHIAN
B 6203 523 FERM, JOHN C., USE OF DEPOSITIONAL MODELS IN PREDICTING ROOF
B 6310 1751 FERNANDEZ, J., ON DEVELOPMENTS IN GUATEMALA, 1978
B 6208 1355 FERNANDEZ, SHARYL M., DEVELOPMENTS IN EASTERN AND
B 6202 286 FERNANDO DE NORONHA BASIN OFF, BRAZIL, NORTHERN
B 6302 219 FERNANDO DE NORONHA RIDGE, BRAZIL
B 6110 1748 FERNANDO POO, DEVELOPMENTS, 1976, EQUATORIAL GUINEA, RIO MUNI-

Index

B 6210 1846 FERNANDO POO, DEVELOPMENTS, 1977, EQUATORIAL GUINEA, RIO-MUNI-
B 6310 1691 FERNANDO POO, DEVELOPMENTS, 1978, EQUATORIAL GUINEA, RIO MUNI-
B 6411 1787 FERNANDO POO, DEVELOPMENTS, 1979, EQUATORIAL GUINEA, RIO MUNI-
B 6309 1600 FERNS, C. KIPP, BAYOU MIDDLE FORK FIELD, CLAIBORNE PARISH,
B 6307 1135† FERNVALE, WELLING FORMATION, NEW NAME FOR UPPE, ORDOVICIAN
B 6305 840* FERON SANDSTONE MEMBER, CENTRAL UTAH, ABST., DEPOSITIONAL
B 6405 705 FERREIRA, JUSTO CAMEJO. SERGIPE-ALAGOAS BASIN. BRAZIL—SOURCE
B 6310 1748 FERREIRA SOBRAL, L., ON DEVELOPMENTS IN BRAZIL, 1978
B 6309 1609 FERRELL, R. E., JR., GEOPRESSURE IN HOUMA AND HOLLYWOOD FIELDS,
B 6409 1557 FERRELL, RAY E., JR., DIAGENESIS OF MIDDLE MIOCENE SANDS IN
B 6209 1757 FERRELL, RAY E., JR., MINERALOGY OF PLEISTOCENE TERRACE DEPOSITS,
B 6207 1229 FERREY, CARLOS, GEOLOGY AND DEVELOPMENT OF MOMOTOMBO
B 6104 594 FERRIER FIELD, ALBERTA
B 6405 714* FERRIER FIELD, ALBERTA, ABST., SIDERITE TEXTURES IN, CARDIUM
B 6112 2137 FERRIER OIL FIELD, ALBERTA
B 6203 512 FERRIS, CLINTON S., URANIUM IN GRANITES. BY TITLE ONLY.
B 6203 463† FERROAN DOLOMITE RHOMBS IN OIL RESERVOIRS, CHADRA SANDS,
M 25 545 FERROMANGANESE NODULES, IRON-NICKEL RATIOS, NAZCA PLATE
M 25 376† FERROMANGANESE NODULES OF THE PACIFIC
M 25 535 FERROMANGANESE NODULES, PACIFIC, SOUTHEASTERN
B 6303 316 FERRON MOUNTAIN MEMBER, UTAH, FLAGSTAFF LIMESTONE
B 6303 519* FERRON SANDSTONE MEMBER OF MANCOS SHALE--PREDICTIVE MODEL
B 6408 1238 FERRON SANDSTONES, UTAH
B 6303 429* FERROUS IRON IN FORMATION OF DOLOSTONE, ABST., ROL OF, BRINES
M 24 266 FERRY LAKE ANHYDRITE, ALABAMA
M 24 159 FERRY LAKE ANHYDRITE, TEXAS
M 27 134 FERRY LAKE ANHYDRITE, TEXAS, CRETACEOUS
M 28 87 FERRY LAKE ANHYDRITE, TEXAS, CRETACEOUS
B 6009 1614* FERRY LAKE EARLY CRETACEOUS DELTAS OF SOUTH MISSISSIPPI AND
B 6405 705 FERTL, W. H., GEOTHERMAL RESOURCES EVALUATED BY WELL LOGS,
B 6405 774 FERTL, W. H., RESOURCE EVALUATIO OF GAS-BEARING COAL BEDS, ABST.
B 6309 1578 FERTL, W. J., NOVEL FORMATION-EVALUATION CONCEPT FOR SHALE
B 6303 449 FERTL, WALTER H., GAMMA-RAY SPECTRAL LOGGING DATA ASSIST I
B 6405 706 FERTL, WALTER H., GLOBAL OCCURRENCE OF ABNORMAL FORMATION
B 6309 1600 FERTL, WALTER H., INTERPRETIVE WELL-LOGGING CONCEPTS SOLVE
B 6409 1557 FERTL, WALTER H., WELL LOG ANALYSIS CONCEPTS IN CLASTIC AND
B 6312 2194 FESTNINGEN MEMBER, SVALBARD
B 6309 1600 FETT, T. H.., LOG EVALUATION OF TIGHT ROCKS OF SOUTH TEXAS, ABST
M 25 463 FIAMME, PUMICE LENTILS OR
M 28 48 FIBROLITE, VARIETY OF SILLIMANITE
B 6111 2015 FIELD CONFERENCE GUIDEBOOKS, NEW MEXICO GEOLOGICAL SOCIETY
M 25 539 FIELD, CYRUS W., METALLOGENESIS IN SOUTHEAST PACIFIC OCEAN,
B 6004 716* FIELD DATA BANK PROJECT, UNITED STATES AND CANADA, ABST., AAPG,
B 6205 749 FIELD, DEFINITION, GIANT, GAS
B 6205 749 FIELD, DEFINITION, GIANT, OIL
B 6203 533* FIELD DEVELOPMENT WITH 3-D SEISMIC METHODS—CASE HISTORY, ABST.
B 6008 1411* FIELD EVIDENCE FOR STRUCTURAL STYLE OF LARAMIDE TRAPS IN
B 6401 10 FIELD, LANDSAT DATA, IRAN, KANGAN GIANT GAS
B 6405 733 FIELD, M. E., PATTERNS OF SHALLOW-MARIN DEPOSITION, UPPER
B 6207 1216 FIELD, M. E., SEAFLOOR INSTABILITY ON PACIFIC CONTINENTAL
B 6403 442 FIELD, M. E., SEDIMENTARY PROCESSES ACTIVE ON SLOPES OF
B 6105 817 FIELD, MICHAEL E., COMPLEX RIDGES AND TROUGH TOPOGRAPHY ON
B 6405 706 FIELD, MICHAEL E., DIAPIR-LIKE RIDGES AND POSSIBLE HYDROCARBON
B 6405 706 FIELD, MICHAEL E., MODIFICATION OF LINEAR SAND RIDGES BY
B 6011 2019 FIELD, MICHAEL E., NEOGENE SEDIMENTS OF ATLANTIC INNER
B 6004 670 FIELD, MICHAEL E., SHALLOW STRUCTURE OF NORTHERN SANT
B 6105 783 FIELD, MICHAEL E., SUBMARINE SLUMPS AND SEDIMENT FLOWS OFF SAN
B 6405 736 FIELD, MICHAEL E., THERMOGENIC HYDROCARBON GASES IN
B 6303 449 FIELD, MICHAEL E., VIDEOTAPES OF ACTIVE PRIMARY PHYSICAL AND
B 6306 887 FIELD, NEW MEXICO, BASIN DAKOTA GAS
B 6312 2188 FIELD NUMBER AND SIZE, ASSESSING RESOURCES
B 6309 1607* FIELD RELATIONS AND PETROLOGY OF CATAHOULA FORMATION IN
B 6303 503* FIELD RELATIONS OF SOME PRE-TERTIAR DOLOMITES IN GREAT BASIN,
B 6002 318* FIELD WATER GEOCHEMISTRY TO EXPLORATION, ABST., APPLICATIONS
B 6405 774* FIELDS, ABST., COMMON FACTORS AMONG ATYPICAL
B 6009 1611* FIELDS, ABST., NEW, OIL FROM, OLD
B 6405 778* FIELDS CONTAINING SANDSTONE RESERVOIRS IN ROCKY MOUNTAIN
M 24 300† FIELDS FROM AAPG PUBLICATIONS, BIBLIOGRAPHY OF, NORTH
B 6401 8† FIELDS, GEOLOGIC SIGNIFICANCE OF, LANDSAT DATA FOR 15, GIANT, OIL
B 6109 1543* FIELDS IN GULF COAST, ABST., TOPOGRAPHIC EXPRESSION OF, OIL AND,
B 6408 1210† FIELDS IN HIGHLY EXPLORED, MATURE, AREAS, METHODS USED, AND
S 11 314 FIELDS LANDING, CALIFORNIA, SAMPLE LOCALITIES
M 24 1 FIELDS, NORTH AMERICAN OIL AND GAS
M 30 1 FIELDS OF THE DECADE 1968-1978, GIANT OIL
M 30 1† FIELDS OF THE DECADE 1968-1978, GIANT, OIL AND, GAS
B 6201 30 FIELDS, SHELL-BP, NIGER DELTA
B 6111 1950† FIELDS, TAR, SANDS AND, SUPERGIANT, OIL
M 25 554 FIERRO URCO, COPPER, ZINC, LEAD, AND MOLYBDENUM DEPOSITS,
M 25 554 FIERRO URCO DEPOSIT, COPPER AND RELATED MINERALS, ECUADOR
B 6009 1609 FIETZ, R. P., RECENT DEVELOPMENTS IN SUNNILAND EXPLORATION O
B 6007 1089 FIFTEEN TWENTY FRACTURE ZONE
B 6303 447* FIJI, ABST., DETERMINATION OF DEPTH TO HYDROCARBON,
M 25 430 FIJI, COPPER
B 6110 1881 FIJI, DEVELOPMENTS, 1976
B 6210 1901 FIJI, DEVELOPMENTS, 1977
B 6310 1681 FIJI, DEVELOPMENTS, 1978
B 6411 1778 FIJI, DEVELOPMENTS, 1979
M 25 430 FIJI, MINERAL EXPLORATION
S 4 291 FIJI, VITI LEVU, GALAXEA
M 25 430† FIJI--SUMMARY, BACKGROUND TO PRESENT MINERAL SEARCH IN,
B 6405 715* FILAMENT-PRODUCING HYDROCARBONS IN PALYNOLOGY
B 6104 563 FILARET FORMATION, CHILE
B 6104 561 FILARET STRATIGRAPHIC UNIT, CHILE
S 6 93 FILHO, ANTONIO THOMAZ, APPLICABILITY OF THE
B 6004 657* FILL FACIES, WILLAPA BAY, WASHINGTON, ABST., MODERN AND,
B 6104 612† FILL IN MARFA BASIN, TEXAS, THE
M 26 131 FILL TYPES, SEISMIC, FACIES, EXTERNAL FORMS
B 6004 659* FILL--NAMURIA OF CENTRAL PENNINE BASIN, ENGLAND, ABST., DELTAIC
S 3 70 FILL-SPILL BRAIDED-STREAM SYSTEMS, DEFINITION
B 6303 291 FILLINGS, PROSPECTS, MARGINAL-, TRENCH
B 6104 594 FILLMORE FIELD, CALIFORNIA
B 6010 1873 FILON EXPLORATION CORP., AND PETROCONSULTANTS S.A., ON
S 3 24 FILTER PRESSING
S 5 118 FILTERING, SEISMIC DATA,
B 6004 692* FILTRATION AS MECHANISM FOR PRIMARY ACCUMULATION OF
B 6108 1383* FINDING RATE METHODOLOGIES TO APPRAISAL OF UNDISCOVERED OIL
B 6405 725 FINE, BRIAN, HIGH-RESOLUTION LANDSAT FOR GEOPHYSICAL STUDIES,
B 6109 1407† FINE-GRAINED SEDIMENTARY ROCKS—PHYSICAL CHEMISTRY APPLIED
B 6004 671 FINK, J. W., PETROLOGY OF TRIASSIC MARINE SEDIMENTARY ROCKS,
B 6405 743 FINKELMAN, ROBERT B., RESIN RODS AND WOODY ROD-LIKE
B 6303 450 FINKELSTEIN, KENNETH, VARIATIONS OF MORPHOLOGY AND SEDIMEN
B 6007 1107† FINLAY MOUNTAINS, WEST TEXAS, WILKIE RANCH FORMATION, NEW,
B 6303 511 FINLEY, R. J., PALO DURO BASIN ANALYSIS, ABST.
B 6009 1609 FINLEY, ROBERT J., INTERPRETATION OF UNENHANCED LANDSAT
B 6105 893 FINLEY, ROBERT J, LANDSAT STUDY OF NEARSHORE CIRCULATION OFF
B 6303 430 FINSTAD, K. G., GEOTHERMAL INVESTIGATION OF NORTHERN NORTH
B 6004 671 FINSTAD, K. G., NORWAYS OIL AND GAS POTENTIAL BASED ON PRESENT
S 6 40 FIRST APPEARANCE DATUM, BIOCHRONOLOGY
S 6 43 FIRST APPEARANCE DATUM, EXAMPLES
S 6 41 FIRST APPEARANCE DATUM, REVIEW OF PRINCIPLE
B 6201 83 FIRST RED BED, SASKATCHEWAN
B 6201 74 FIRST RED CLAYSTONE, SASKATCHEWAN
B 6407 981 FIRST WHITE SPECKS, ALBERTA
M 26 84 FIRST-ORDER, CORRELATION WITH SECOND-ORDER, GLOBAL CYCLES
M 29 470 FIRST-ORDER CYCLES, PHANEROZOIC, GLOBAL CYCLES OF CHANGE
M 26 84 FIRST-ORDER, GLOBAL CYCLES, SECOND-ORDER, CORRELATION WITH
B 6303 450 FISCHBECK, REINHARD,, SYNTHESIS OF PROTODOLOMITE,
B 6203 512 FISCHER, IAN A., DEPOSITIONAL MODEL OF MESOTIDAL BACK-BARRIER
B 6105 783 FISCHER, IAN A. TIDAL-FLAT SEDIMENTATION ON TIDE-DOMINATED
B 6203 550 FISCHER, IAN A., TRANSGRESSIVE BARRIERS OF BEAUFORT SEA COAST,
B 6012 2180 FISCHER, P. J., LATE NEOGENE-QUATERNARY TECTONICS AN
B 6403 442 FISCHER, PETER J., INTERMARGIN OF SOUTHERN CALIFORNIA
M 25 63 FISCHER, W. A., SURVEYING EARTH
S 4 281† FISH AND SEA URCHINS ON CARIBBEAN REEFS, CARBONATE-SEDIMENT
S 4 271 FISH, BORERS
M 28 54 FISH CANYON TUFF, COLORADO, OLIGOCENE
B 6008 1332 FISH, JACK T., OIL AND GAS DEVELOPMENTS IN NORTH MID-CONTINENT
B 6409 1437 FISH, JACK T., OIL AND GAS DEVELOPMENTS IN NORTH MID-CONTINENT
B 6208 1477 FISH, JACK T., OIL AND GAS DEVELOPMENTS IN NORTH MID-CONTINENT
B 6108 1305 FISH, JACK T., OIL AND GAS DEVELOPMENTS IN NORTH MID-CONTINENT
B 6308 1301 FISH, JACK T., OIL AND GAS DEVELOPMENTS IN NORTH MID-CONTINENT
B 6407 981 FISH SCALES SANDSTONE, ALBERTA
B 6302 166 FISH SCALES STRUCTURE, CANADA
B 6105 819 FISHEL, KEN, COAL PROPERTY AND MAXIMIZING TOTAL RECOVERABLE
B 6203 513 FISHER, CARLA, BACTERIAL MEDIATION OF REDOX POTENTIAL IN
B 6002 326* FISHER COUNTIES, TEXAS, ABST., NEILL SOUTH, ODOM LIME, FIELD,
B 6308 1425* FISHER COUNTY, TEXAS, ABST., TURBIDITE CHANNEL, RESERVOIRS IN,
B 6101 105 FISHER, JAMES H., CONFUSION FLATS--AN EXPLORATION GAME
S 5 1 FISHER, JAMES H., EDITED BY. REEFS AND EVAPORITES--CONCEPTS
S 5 1 FISHER, JAMES H., INTRODUCTION, REEFS AND EVAPORITES
B 6303 450 FISHER, JAMES H., STRUCTURAL EVOLUTION OF MICHIGAN BASIN AND
B 6303 468 FISHER, M. A., SONOBUOY REFRACTION MEASUREMENT FROM NORTON
B 6203 373 FISHER, MICHAEL A., GEOLOGIC FRAMEWORK OF LOWER COOK INLET,
B 6012 2180 FISHER, MICHAEL A., GEOLOGY OF OFFSHORE LOWER COOK INLET
B 6408 1140 FISHER, MICHAEL A., PETROLEUM GEOLOGY OF KODIAK SHELF, ALASKA
B 6405 706 FISHER, MICHAEL A., STRUCTURE AND HYDROCARBON POTENTIAL OF
B 6303 301 FISHER, MICHAEL A., STRUCTURE AND TECTONIC SETTING OF
B 6203 513 FISHER, MICHAEL A., STRUCTURE AND TECTONIC SETTING OF
B 6407 1014 FISHER, MICHAEL A., STRUCTURE OF UPPER CENOZOIC STRATA BENEATH

M 26 213 FISHER, W. L., SEISMIC-STRATIGRAPHIC INTERPRETATION OF
B 6405 675* FISHERS PEA AREA, LAS ANIMAS COUNTY, COLORADO, ABST., DRILLING
B 6202 185 FISHTOOTH SANDSTONE, WYOMING
M 25 431 FISSURE-VEIN TYPE, THAILAND, FLUORITE OF
S 11 238 FISSURINA LUCIDA
S 11 276 FISSURINA LUCIDA
S 11 238 FISSURINA SEMIMARGINATA
S 11 276 FISSURINA SEMIMARGINATA
B 6105 784 FITCH, DAVID C., CASE HISTORY OF DISCOVERY OF NEW URANIUM
B 6304 688 FITCH, DAVID C., EXPLORATION FOR URANIUM DEPOSITS IN GRANT
B 6307 1135 FITE LIMESTONE, OKLAHOMA
B 6203 386 FITZ CREEK ANTICLINE, ALASKA
B 6203 399 FITZ CREEK SILTSTONE, ALASKA
B 6104 560 FITZ ROY CHANNEL, CHILE
B 6405 707 FITZGERALD, DON D., DIPMETER VALIDITY IN DEVIATED BORE HOLES,
B 6303 451 FITZGERALD, DUNCAN M., ACCRETIONAL SHORELINE PROCESSES AT
B 6405 707 FITZGERALD, DUNCAN M., EBB-TIDAL DELTA STRATIFICATION AND ITS
B 6004 682 FITZGERALD, DUNCAN M., REEF MORPHOLOGY AND SEDIMENT
B 6303 451 FITZGERALD, T. A., GIANT FIELD DISCOVERIES, 1967-77--AN OVERVIEW,
M 30 1 FITZGERALD, T. A., GIANT FIELD DISCOVERIES 1968-1978: AN OVERVIEW
B 6201 41 FITZROY TROUGH, AUSTRALIA
B 6101 65 FIVE ISLANDS TREND, LOUISIANA
B 6410 1590 FIVE OAKS FORMATION, VIRGINIA
B 6205 848 FIVE POINTS FIELD, CALIFORNIA
S 10 130 FIVE-AQUIFER SYSTEM, EXAMPLE, GROUNDWATER FLOW
S 3 75 FIXATION OF URANIUM
S 3 136 FIXATION OF URANIUM BY NATURAL FLUID OR GASEOUS
B 6302 183 FJERRITSELV FAULT, NORTH SEA
B 6111 1931 FJORD FORMATION, SVERDRUP BASIN, CANYON
B 6203 547 FLACH, PETER D., DEEP SINUOUS CHANNELS AND GIANT CROSS-BEDS IN
B 6311 2001 FLADEN SPUR, NORTH SEA
B 6303 311 FLAGSTAFF LACUSTRINE-ALLUVIAL, UTAH, LAKE
B 6303 311† FLAGSTAFF LIMESTONE AND COEVAL ROCKS, CENTRA UTAH,
B 6303 316 FLAGSTAFF LIMESTONE, COVE MOUNTAIN MEMBER, UTAH
B 6303 316 FLAGSTAFF LIMESTONE FERRON MOUNTAIN MEMBER, UTAH
B 6303 318 FLAGSTAFF LIMESTONE, MUSINIA PEAK MEMBER, UTAH
B 6406 854 FLAT BASIN FLOORS, RESERVOIR POTENTIAL
B 6004 676* FLAT CARBONATE ROCKS, MIDDLE ORDOVICIAN, VIRGINIA, ABST.,
B 6004 720* FLAT CREATED BY SHORELINE EROSION IN TIDAL ESTUARY, ABST.,
B 6406 841† FLAT FLOORS OF 10 MODERN DEPOSITIONAL BASINS, COMPARISON OF,
B 6402 222 FLAT ROCK QUADRANGLE, TEXAS
B 6004 661* FLAT-SHALLOW SUBTIDAL ENVIRONMENT, ABST., TRACE-, FOSSIL
B 6004 674* FLATS, ABST., SEDIMENTARY RECORD OF, PALEOCLIMATE IN,
B 6304 688 FLEISCHER, ROBERT L., RADON EMANATION OVER OREBODY, HAS
B 6302 183 FLEKKEFJORD HIGH, NORTH SEA
B 6408 1169 FLEMING TONGUE, TEXAS
B 6004 671 FLESCH, G. A., DELTAIC SUBENVIRONMENTS AND PALEOCURRENT
B 6105 784 FLESCH, GARY A., BEREA, MISSISSIPPIAN, SANDSTONE CORE FROM CSO
B 6002 319 FLESCH, GARY, DELTAIC DEPOSITIONAL ENVIRONMENTS OF PERMIAN
B 6411 1880 FLETCHER, G. L., PETROLEUM DEVELOPMENTS IN FAR EAST, 1979
B 6210 1978 FLETCHER, G. L., PETROLEUM DEVELOPMENTS IN FAR EAST IN 1977
B 6310 1816 FLETCHER, G. L., PETROLEUM DEVELOPMENTS IN FAR EAST IN 1978
B 6310 1827 FLETCHER, ON DEVELOPMENTS IN TAIWAN, 1978, YANG, S., AND G. L.
B 6405 742 FLEWITT, W. E., GEOLOGY OF SYNCRUDE CANADA LIMITED MINE SITE,
B 6203 457 FLINDERS FAULT SYSTEM, AUSTRALIA
B 6410 1644 FLINDERS RANGES, AUSTRALIA
B 6103 348 FLIPPEN LIMESTONE, TEXAS
S 10 244 FLOMATON FIELD, ETHANE CONCENTRATIONS, ALABAMA
S 10 246 FLOMATON FIELD, SAND PERCENTAGE CONCENTRATIONS, ALABAMA
B 6401 30 FLOMATON STRUCTURE, FLORIDA
S 10 243 FLOMATON-JAY-BLACKJACK CREEK, MIGRATION, SURFACE EVIDENCE
B 6205 892* FLOOD DEPOSITION AND DEPOSITS, ABST., SHEET-
B 6411 1894 FLOOD, LLOYD, ON DEVELOPMENTS IN VIETNAM, 1979
B 6407 972 FLOOD MEMBER, MONTANA
B 6305 831* FLOOD-PLAIN SEQUENCES OF FINE-GRAINED MEANDER-BELT SYSTEM,
B 6004 700* FLOODING, ABST., LARGE-SCALE, BED FORMS AN SEDIMENTARY
B 6105 763* FLOOR FOR GAS
S 9 7 FLORA AND FAUNA, PALEOCLIMATE INDICATORS
M 24 242 FLORA FIELD, MISSISSIPPI
S 4 269 FLORAL MICROBORERS
B 6309 1583 FLORENCE, ALEXANDER L., INTERACTION OF PROPPANTS WITH CRACK
B 6103 332 FLORENCE BASIN, SOUTH CAROLINA
B 6410 1615 FLORENCE FIELD, DENVER BASIN
B 6109 1449 FLORENCE UPLIFT, ARIZONA
B 6004 672 FLORES, R. M., BIOHERMAL DEVELOPMENT IN WOLFCAMPIAN
B 6305 826 FLORES, ROMEO M., COAL VARIATIONS IN FLUVIAL DEPOSITION
B 6105 785 FLORES, ROMEO M., DIFFERENTIATION OF SANDSTONE DEPOSITIONAL
B 6303 451 FLORES, ROMEO M., PICTURED CLIFFS SANDSTONE--UPPER CRETACEOUS
B 6203 513 FLORES, ROMEO M., STRATIGRAPHY AND DEPOSITIONAL ENVIRONMENTS
B 6405 689 FLORES, ROMEO M., VARIATIONS I CRETACEOUS COAL-BEARING STRATA,
B 6112 2079 FLORES SEA ACROSS BANDA VOLCANIC ARC, TIMOR, AND
B 6303 536* FLORIDA, ABST., BIOSTRATIGRAPHY OF EARLY AND EARLIEST LATE,
B 6303 481* FLORIDA, ABST., DIAGENETIC PHENOMENA OF, CHIPOLA FORMATION,
B 6303 444* FLORIDA, ABST., DISTRIBUTION, MORPHOLOGY, MECHANISMS, AND AGES
B 6409 1561* FLORIDA, ABST., EARLY, CEMENTATION AND MINERAL STABILITY OF,
B 6209 1753* FLORIDA, ABST., FUTURE, OIL POTENTIAL OF LOWER, CRETACEOUS,
B 6209 1768* FLORIDA, ABST., HIGH-LEVEL, PLIO-, PLEISTOCENE, SHORELINE IN SOUTH
B 6004 727* FLORIDA, ABST., MARINE AND, BRACKISH-, WATE, CRETACEOUS,
B 6303 438* FLORIDA, ABST., ORIGIN AND DEVELOPMENT OF, BARRIER ISLANDS ON
B 6303 548* FLORIDA, ABST., PALEOECOLOGIC IMPLICATION OF, FORAMINIFERAL
B 6004 733* FLORIDA, ABST., PETROLOGY OF, SUBSE, CEMENTED, CARBONATE,
B 6203 518* FLORIDA, ABST., PHOSPHATE IN BOTTOM SEDIMENTS IN, MYAKKA AND,
B 6009 1609* FLORIDA, ABST., RECENT DEVELOPMENTS IN, SUNNILAND EXPLORATION
B 6309 1605* FLORIDA, ABST., REVISION OF, TAMPA FORMATION, WEST-CENTRAL
B 6303 508* FLORIDA, ABST., SEDIMENT DISPERSAL AT, FORT GEORGE INLET
B 6303 493* FLORIDA, ABST., SHELL BORINGS IN UPPER, MIOCENE-, PLIOCENE
B 6009 1618* FLORIDA, ABST., SIX PROPOSED FORMATIONS IN UNDESCRIBED PART OF
B 6004 648* FLORIDA, ABST., SOURCE AND DYNAMICS OF, SUSPENDED SEDIMENTS IN,
B 6309 1609* FLORIDA, ABST., STRATIGRAPHY AND PALEOECOLOGY OF, TAMIAMI
B 6309 1599* FLORIDA, ABST., SUBSURFACE, NEOGENE STRATIGRAPHY OF, BAY
B 6009 1613* FLORIDA, ABST., SUNOCO-, FELDA AND, WEST SUNOCO-FELDA FIELDS,
B 6004 731* FLORIDA, ABST., TIMING, CHARACTER, AND PRESERVABILIT OF
B 6109 1549* FLORIDA, ABST., TYPE AND COTYPE, WELLS FOR SOME, CRETACEOUS-,
B 6201 126 FLORIDA, ALLIGATOR REEF FAULT
B 6004 687* FLORIDA AND ALABAMA, ABST., DISTRIBUTION AND RELATION OF
B 6104 519 FLORIDA, ATLANTIC COASTAL RIDGE
B 6104 493 FLORIDA, AVON PARK FORMATION
B 6211 2306 FLORIDA BANKS, GULF OF MEXICO, WEST
B 6209 1769* FLORIDA BASEMENT, ABST., REEVALUATING
B 6112 2144 FLORIDA BASEMENT, COMPUTER-PROGRAM PACKAGES, SYMAP AND,
B 6112 2143† FLORIDA BASEMENT—AN ISOMETRIC VIEW
B 6209 1763* FLORIDA BASIN, ABST., PRELIMINARY ASSESSMENT OF, ORGANIC,
B 6201 123 FLORIDA, BATHYMETRY OF SOUTHERN, STRAITS OF
B 6105 833* FLORIDA BAY, ABST., HOLOCENE, DOLOMITE LOCALITY IN
B 6004 669* FLORIDA BAY INTERPRETED FROM ISLAND STRATIGRAPHY, ABST.,
B 6102 176 FLORIDA BEACHES, RECENT SANDS
S 10 244 FLORIDA, BLACKJACK CREEK FIELD, ETHANE CONCENTRATIONS
S 10 252 FLORIDA, BLACKJACK CREEK FIELD, METHANE CONCENTRATIONS
S 10 246 FLORIDA, BLACKJACK CREEK FIELD, SAND PERCENTAGE
S 10 249 FLORIDA, BLACKJACK CREEK FIELD, SUBSURFACE GEOLOGY
M 24 296 FLORIDA, BOULDER ZONE
B 6201 121 FLORIDA, BOULDER ZONES, DEFINITION
M 24 289 FLORIDA, BROWARD SYNCLINE
B 6410 1701† FLORIDA, CARBONATE SEDIMENT, DRIFTS IN NORTHERN, STRAITS OF
B 6201 125 FLORIDA, CARD SOUND DOLOMITE
B 6201 125 FLORIDA, CEDAR KEYS FORMATION
B 6004 691* FLORIDA COAST--PALEOENVIRONMENTAL SIGNIFICANCE, ABST.,
B 6201 125 FLORIDA, CRETACEOUS, UPPER, COLUMNAR SECTION
B 6410 1702 FLORIDA CURRENT
B 6304 671 FLORIDA, DESTIN DOME
B 6008 1282 FLORIDA, DEVELOPMENTS, 1975
B 6108 1264 FLORIDA, DEVELOPMENTS, 1976
B 6208 1394 FLORIDA, DEVELOPMENTS, 1977
B 6308 1401 FLORIDA, DEVELOPMENTS, 1978
B 6411 2029 FLORIDA, DEVELOPMENTS, 1979
B 6104 492† FLORIDA, DIAGENESIS OF MIDDLE AND UPPER, EOCENE, CARBONATE,
M 27 194 FLORIDA, EOCENE, AVON PARK LIMESTONE
M 27 131 FLORIDA, EOCENE, AVON PARK LIMESTONE
M 27 219 FLORIDA, EOCENE, AVON PARK LIMESTONE
B 6104 492 FLORIDA, EOCENE CARBONATE SHORELINE SEQUENCES
M 27 219 FLORIDA, EOCENE, OCALA FORMATION
M 27 223 FLORIDA, EOCENE, OCALA FORMATION
M 27 217 FLORIDA, EOCENE, OCALA FORMATION
M 27 164 FLORIDA, EOCENE, OCALA FORMATION
M 27 95 FLORIDA, EOCENE, OCALA FORMATION
M 27 31 FLORIDA, EOCENE, OCALA LIMESTONE
M 27 142 FLORIDA, EOCENE, OCALA LIMESTONE
M 27 220 FLORIDA, EOCENE, OCALA LIMESTONE
M 28 119 FLORIDA, EOCENE, OCALA LIMESTONE
B 6305 783 FLORIDA ESCARPMENT, GULF OF MEXICO
M 29 417 FLORIDA ESCARPMENT, REFLECTION SECTION, GULF OF MEXICO
B 6401 30 FLORIDA, FLOMATON STRUCTURE
B 6304 668† FLORIDA, GAS, SEEPS AND SUBSURFAC STRUCTURE OFF, PANAMA CITY
B 6403 339 FLORIDA, GEORGIA, AND SOUTH CAROLINA, FLORIDA, HATTERAS SHELF
B 6203 500* FLORIDA GULF COAST BARRIER ISLAND, ABST., STRATIGRAPHY AND
B 6109 1549* FLORIDA GULF COAST INTERPRETED FROM RELICT SHORELINES, ABST.,
B 6403 339 FLORIDA, HATTERAS SHELF, FLORIDA, GEORGIA, AND SOUTH CAROLINA

B 6103 335 FLORIDA HATTERAS SLOPE
M 27 12 FLORIDA, HOLOCENE BOTTOM SEDIMENT
M 27 7 FLORIDA, HOLOCENE BOTTOM SEDIMENTS
M 27 96 FLORIDA, HOLOCENE SEDIMENTS
M 27 42 FLORIDA, HOLOCENE SEDIMENTS
M 27 61 FLORIDA, HOLOCENE SEDIMENTS
M 27 115 FLORIDA, HOLOCENE SEDIMENTS
M 27 222 FLORIDA, HOLOCENE SOIL CRUSTS
B 6104 494 FLORIDA, INGLIS FORMATION
S 10 244 FLORIDA, JAY FIELD, ETHANE CONCENTRATIONS
M 30 32 FLORIDA, JAY FIELD, LANDSAT IMAGE
S 10 252 FLORIDA, JAY FIELD, METHANE CONCENTRATIONS
S 10 246 FLORIDA, JAY FIELD, SAND PERCENTAGE CONCENTRATIONS
S 10 249 FLORIDA, JAY FIELD, SUBSURFACE GEOLOGY
B 6401 30 FLORIDA, JAY GIANT OIL AND GAS FIELD, LANDSAT DATA
S 4 187 FLORIDA, KEY LARGO LIMESTONE, REEF
B 6201 121 FLORIDA, KEYS AREA WELLS
B 6107 1079 FLORIDA KEYS, BIG PINE KEY, MIAMI OOLITE, MICROGRAPHS
B 6201 122 FLORIDA, MARQUESAS KEYS, WELLS
B 6104 522 FLORIDA, MIAMI LIMESTONE
B 6404 475 FLORIDA, MIAMI OOLITE, FRESHWATER PHREATIC ZONE
M 28 72 FLORIDA, MIOCENE, HAWTHORN FORMATION
B 6201 123 FLORIDA, MITCHELL ESCARPMENT
B 6201 126 FLORIDA, MITCHELL ESCARPMENT FAULT
S 4 175 FLORIDA, MONTASTREA ANNULARIS
B 6405 681* FLORIDA MOUNTAINS, NEW MEXICO, ABST., GEOCHRONOLOGIC
B 6201 125 FLORIDA, NAPLES BAY GROUP
B 6011 2019† FLORIDA, NEOGENE SEDIMENTS OF, ATLANTIC, INNER, CONTINENTAL,
M 24 284 FLORIDA, NORPHLET SANDSTONES
B 6104 493 FLORIDA, OCALA ARCH
M 24 289 FLORIDA, OCALA UPLIFT
B 6201 125 FLORIDA, OLDSMAR FORMATION
M 27 218 FLORIDA, OLIGOCENE, SUWANNEE FORMATION
M 27 7 FLORIDA, OLIGOCENE, SUWANNEE LIMESTONE
M 27 223 FLORIDA, OLIGOCENE, SUWANNEE LIMESTONE
M 27 165 FLORIDA, OLIGOCENE, SUWANNEE LIMESTONE
M 27 173 FLORIDA, OLIGOCENE, SUWANNEE LIMESTONE
M 27 128 FLORIDA, OLIGOCENE, SUWANNEE LIMESTONE
B 6112 2144 FLORIDA, OSCEOLA LOW
B 6201 125 FLORIDA, PALEOCENE, COLUMNAR SECTION
B 6109 1538* FLORIDA PANHANDLE, ABST., LATE, MIOCENE, GLACIO-, EUSTATIC
B 6112 2144 FLORIDA, PENINSULAR ARCH
M 24 289 FLORIDA, PENINSULAR ARCH
B 6201 125 FLORIDA, PINE KEY FORMATION
S 7 17 FLORIDA PLAIN, DESCRIBED
B 6201 125 FLORIDA, PLANTATION TONGUE
B 6403 340 FLORIDA PLATFORM, PENINSULAR ARCH
B 6104 519† FLORIDA, PLEISTOCENE, BARRIER, BAR SEAWARD OF, OOID, SHOAL
M 27 59 FLORIDA, PLEISTOCENE, KEY LARGO LIMESTONE
M 27 71 FLORIDA, PLEISTOCENE, KEY LARGO LIMESTONE
M 27 176 FLORIDA, PLEISTOCENE, KEY LARGO LIMESTONE
M 27 7 FLORIDA, PLEISTOCENE, KEY LARGO LIMESTONE
M 27 224 FLORIDA, PLEISTOCENE, KEY LARGO LIMESTONE
M 27 218 FLORIDA, PLEISTOCENE, KEY LARGO LIMESTONE
M 27 162 FLORIDA, PLEISTOCENE, KEY LARGO LIMESTONE
M 27 219 FLORIDA, PLEISTOCENE, MIAMI OOLITES
M 27 36 FLORIDA, PLEISTOCENE OOZE
M 28 126 FLORIDA, PLIOCENE, BONE VALLEY FORMATION
M 28 83 FLORIDA, PLIOCENE, BONE VALLEY FORMATION
M 28 82 FLORIDA, PLIOCENE, BONE VALLEY FORMATION
M 27 199 FLORIDA, PLIOCENE, CALOOSAHATCHEE FORMATION
M 27 167 FLORIDA, PLIOCENE, CALOOSAHATCHEE FORMATION
M 27 221 FLORIDA, PLIOCENE, CALOOSAHATCHEE MARL
M 27 84 FLORIDA, PLIOCENE, CALOOSAHATCHEE MARL
M 27 178 FLORIDA, PLIOCENE, CALOOSAHATCHEE MARL
M 27 212 FLORIDA, PLIOCENE, CALOOSAHATCHEE MARL
M 27 83 FLORIDA, PLIOCENE, CALOOSAHATCHEE MARL
M 27 211 FLORIDA, PLIOCENE, CALOOSAHATCHEE MARL
M 27 175 FLORIDA, PLIOCENE, CALOOSAHATCHEE MARL
M 27 166 FLORIDA, PLIOCENE, TAMIAMI FORMATION
M 27 178 FLORIDA, PLIOCENE, TAMIAMI FORMATION
B 6201 123 FLORIDA, POURTALES ESCARPMENT
B 6201 126 FLORIDA, POURTALES ESCARPMENT FAULT
B 6008 1284 FLORIDA, PRODUCTION, 1974-1975
B 6108 1266 FLORIDA, PRODUCTION, 1975-1976
B 6208 1396 FLORIDA, PRODUCTION, 1976-1977
B 6308 1404 FLORIDA, PRODUCTION, 1977-1978
B 6201 124 FLORIDA, REBECCA SHOAL DOLOMITE
B 6201 124 FLORIDA, REBECCA SHOAL DOLOMITE TONGUES
B 6201 121† FLORIDA, REBECCA, SHOAL, REEF COMPLEX, UPPER, CRETACEOUS AND,
M 24 292 FLORIDA, ROBERTS LIMESTONE PRODUCING ZONE
B 6211 2303 FLORIDA SCARP, GULF OF MEXICO, WEST
B 6205 793 FLORIDA SHELF, SEISMIC REFLECTION PROFILE, MULTIFOLD
S 7 103 FLORIDA, SHELF-UPPER SLOPE, SEISMIC FACIES EXAMPLE
B 6401 30 FLORIDA, SMACKOVER FORMATION
M 24 289 FLORIDA, SOUTH FLORIDA EMBAYMENT
M 24 289 FLORIDA, SOUTH FLORIDA SHELF
B 6211 2293 FLORIDA STRAIT, MULTICHANNEL PROFILE LINES
B 6204 694 FLORIDA STRAITS
M 29 418 FLORIDA STRAITS, REFLECTION PROFILE, GULF OF MEXICO
B 6209 1753* FLORIDA SUBSURFACE FEATURES RELATED TO OIL EXPLORATION, ABST.,
M 24 287† FLORIDA, SUNOCO-FELDA FIELD, HENDRY AND, COLLIER COUNTIES
B 6112 2144 FLORIDA, SUWANNEE STRAITS
B 6201 125 FLORIDA, TAVERNIER TONGUE
B 6109 1523 FLORIDA TO LONG ISLAND, SEDIMENT ACCUMULATION RATES
M 24 298 FLORIDA, TRINITY GROUP
B 6109 1449 FLORIDA UPLIFT, ARIZONA AND, NEW MEXICO, BURRO-
B 6404 483 FLORIDA, WINDLEY KEY QUARRY, PHREATIC ENVIRONMENT
M 24 276† FLORIDA--A JURASSIC STRATIGRAPHIC TRAP, JAY FIELD
B 6403 347 FLORIDA-HATTERAS SHELF AND INNER BLAKE PLATEAU. SEISMIC STUDY
B 6403 339† FLORIDA-HATTERAS SHELF AND INNER BLAKE PLLTEAU, STRUCTURE,
B 6403 339 FLORIDA-HATTERAS SHELF, FLORIDA, GEORGIA, AND SOUTH CAROLINA
B 6403 340 FLORIDA-HATTERAS SLOPE
B 6309 1583* FLORIDA-HATTERAS SLOPE, ABST., PETROLEUM, GEOCHEMISTRY AND
B 6303 409* FLORIDA. ABST., SEISMIC STRATIGRAPHY AND STRUCTURE OFF, PANAMA
B 6006 992† FLORIDAS OCALA UPLIFT IS NOT AN UPLIFT
B 6104 594 FLOUNDER FIELD, AUSTRALIA
B 6207 1153 FLOUNDER FORMATION, AUSTRALIA
B 6207 1194 FLOURNOY FORMATION, OREGON AND, WASHINGTON
B 6409 1557 FLOURNOY, LEIGH ANNE, DIAGENESIS OF MIDDLE MIOCENE SANDS IN
B 6004 700* FLOW DURING HURRICANE FLOODING, ABST., LARGE-SCALE, BED FORMS
B 6303 445* FLOW IN BLAKE-BAHAMA BASIN, ABST., MIOCENE, CARBONATE GRAVITY
B 6303 467* **FLOW OF BOTTOM WATER OUT OF WEDDELL SEA, ANTARCTICA--700,000**
B 6006 973† **FLOW OF WATER AND SOLUTES THROUGH COMPACTED CLAYS**
B 6105 653† **FLOW OF WEAK ROCKS IN APPALACHIAN PLATEAU FOLDS**
B 6009 1607* FLOW ORIGIN FOR FRIO RESERVOIR SANDSTONES, NINE MILE POINT
B 6105 827* FLOW PATTERNS AND BAR MORPHOLOGY IN BRAIDED-TO-MEANDERING
B 6203 542* FLOW PHENOMENA—IN SITU OBSERVATIONS, ABST., SHELF SEDIMENT
B 6404 563 FLOWER COMMUNITY CRUDE
B 6002 220 FLOWER COMMUNITY WHOLE OIL, CALIFORNIA, LOS ANGELES
B 6211 2354 FLOWERS, B. S., BREAKTHROUGHS, BRIGHT SPOTS, AND BEWARE. BY
B 6001 3 FLOWERS, B. S., OVERVIEW OF EXPLORATION GEOPHYSICS--RECENT
M 25 203 **FLOWERS, B. S., OVERVIEW OF EXPLORATION GEOPHYSICS--RECENT**
B 6008 1397 FLOWERS, BILLY S., OVERVIEW OF EXPLORATION GEOPHYSICS--RECENT
B 6303 435* FLOWS AND TURBIDITY-CURRENT FLOWS FROM SUBMARINE SLIDES,
B 6203 543* FLOWS, EXAMPLES FROM NORTHERN APPALACHIANS, ABST., COARSE
B 6203 567* FLOW—POSSIBLE MECHANISM FOR TRANSPORT AND ACCUMULATION OF
B 6204 646 FLOYD SHALE, ALABAMA
M 26 115 FLUID CONTACTS, HYDRATE ZONES, SEISMIC, REFLECTIONS
B 6203 494* FLUID CONTENT BY SEISMIC MODELING, ABST., DETERMINATION OF,
B 6203 500* FLUID DENSITY CURRENT DEPOSITION OF PERMIAN DELAWARE
B 6002 297 FLUID EXPELLED DURING COMPACTION, PORE
M 25 362† FLUID EXTRACTION--SUMMARY, LAND, SUBSIDENCE IN, JAPAN
M 28 187 FLUID INCLUSION STUDIES, EXPLANATION, TECHNIQUES
S 3 41 FLUID MOBILIZATION BY DEFORMATION
M 26 25 FLUID MOTION, ATTENUATION, P-WAVE
M 26 24 FLUID MOTION, LABORATORY, MEASUREMENT, ATTENUATION
B 6404 550 FLUID MOVEMENT DURING COMPACTION
S 10 71 FLUID MOVEMENT DURING COMPACTION, MODEL
B 6002 229 FLUID MOVEMENT UP FAULTS
B 6112 2135 FLUID POTENTIALS COMMON IN SEDIMENTARY BASINS
B 6405 766* FLUID PRESSURE, ABST., EFFECTIVE STRESS AND ABNORMALLY HIGH
M 26 37 FLUID PRESSURE, LABORATORY, MEASUREMENTS
B 6403 415 FLUID PRESSURE, MODELS, ABNORMALLY HIGH
B 6403 415 FLUID PRESSURE, MODELS, NORMAL
B 6405 749* FLUID PRESSURE PRODUCED BY HYDROCARBON GENERATION, ABST.,
B 6403 414† FLUID PRESSURE--SURVEY O SOME BASIC PRINCIPLES, ABNORMALLY
B 6004 707* FLUID PRESSURES DURING FOLDING AND FAULTING IN EASTERN WEST
B 6007 1126 FLUID PRESSURES, EFFECT OF, COMPACTION ON
B 6406 929 FLUID PRESSURES IN ACCUMULATING SEDIMENTS, MODEL FOR EXCESS
B 6403 414† FLUID PRESSURE—SURVEY OF SOME BASIC PRINCIPLES, ABNORMALLY
S 10 48 FLUID PROPERTIES, MIGRATION
M 25 120 FLUID REINJECTION, INDUCTION OF, SEISMIC ACTIVITY FROM
B 6404 558 FLUID VOLUME CALCULATIONS
S 10 79 FLUID VOLUME CALCULATIONS, MIGRATION
B 6211 2362* FLUID-FLOW RATE VARIATIONS IN NORTHERN EAST MESA GEOTHERMAL
B 6203 551* FLUIDS, ABST., ENVIRONMENT CREATED BY VERTICALLY MOVING
B 6304 573† FLUIDS IN SEDIMENTARY BASINS, MIGRATION OF
B 6409 1538† FLUIDS IN SEDIMENTARY BASINS. DISCUSSION AND REPLY, MIGRATION
B 6405 770* FLUORESCENCE OF ACRITARCHS IN STUDY OF MARINE KEROGEN, ABST.
S 3 23 FLUORINE, FRANCE, MASSIF CENTRAL
M 25 496 FLUORITE, CANADIAN, CORDILLERA, BARITE AND

M 25 431 FLUORITE DEPOSITS, OCCURRENCE, THAILAND
M 25 460 FLUORITE, EASTERN ASIATIC VOLCANOGENIC BELT, BORON,
M 25 432 FLUORITE, GEOPHYSICAL EXPLORATION, THAILAND
M 25 431 FLUORITE, MINERALS ASSOCIATED WITH
M 25 432 FLUORITE, MINING AND PROCESSING, THAILAND
M 25 432 FLUORITE OF CAVITY-FILLING TYPE, THAILAND
M 25 431 FLUORITE OF FISSURE-VEIN TYPE, THAILAND
M 25 431 FLUORITE OF PEGMATITE TYPE, THAILAND
M 25 432 FLUORITE OF REPLACEMENT TYPE, THAILAND
M 25 432 FLUORITE, RESERVES, THAILAND
M 25 431† FLUORITE RESOURCES OF THAILAND--SUMMARY
M 25 444 FLUORITE, SOUTH KOREA
M 25 431 FLUORITE, USE FOR GEMSTONES, THAILAND
M 25 477 FLUORITE-ANTIMONITE JASPEROID, CINNABAR-
M 25 475 FLUORITE-ANTIMONITE, USSR, EASTERN, STRATIFORM DEPOSITS O,
B 6004 696* FLUVIAL, ALGAL, PISOLITES AND PISOLITIC CALICHE IN TEXAS, ABST.,
B 6212 2381 FLUVIAL AND UPPER DELTA PLAIN
B 6305 834* FLUVIAL DEPOSITS, TAOS TROUGH, NORTH-CENTRAL NEW MEXICO
B 6004 729* FLUVIAL MODELS--WET SLOPES AND RAIN SHADOWS, ABST.
B 6203 494* FLUVIAL SANDS DERIVED FROM GRANITIC PLUTONS, ABST., TRACE
B 6007 1017 FLUVIAL SANDSTONE, ORIGIN OF, SHEETS OF
B 6203 545* FLUVIAL SANDSTONE, SONORA PASS, CALIFORNIA, ABST., URANIUM IN,
B 6307 1144 FLUVIAL SEDIMENTOLOGY, EDITED BY ANDREW D. MIALL, LE BLANC,
B 6007 1015 FLUVIAL SEQUENCE, COMPLETE
B 6007 1015 FLUVIAL SEQUENCES, INCOMPLETE
B 6007 1010 FLUVIAL SHEET SANDSTONE, NEW MEXICO, WESTWATER CANYON
B 6007 1009† FLUVIAL SHEET SANDSTONE, RESERVOIR, GEOMETRY OF A
B 6305 840* FLUVIAL SYSTEM--SEDIMENTOLOGIC IMPLICATIONS, ABST., EVOLUTION
B 6105 835* FLUVIAL TO BATHYAL SANDSTONES OF CRETACEOUS GREAT VALLEY
S 7 89 FLUVIAL TRANSPORT, SEDIMENT PROCESSES
M 26 269 FLUVIAL-DELTAIC SYSTEMS, SHELF-PLATFORM
B 6405 709* FLUVIATILE, CONTINENTAL, ORIGIN OF ANCIENT STRATIGRAPHIC TRAP
B 6011 2038† FLYSCH DEPOSITS IN SANTA ANA MOUNTAINS, CALIFORNIA,
B 6411 1963 FLYSCH-CALCAREOUS ALPS, AUSTRIA
B 6310 1918 FLYSCH-CALCAREOUS ALPS, AUSTRIA
B 6210 1909 FLYSCH-CALCAREOUS ALPS, AUSTRIA
B 6110 1639 FLYSCH-LIMESTONE ALPS ZONE, AUSTRIA
B 6203 564* FOAM STIMULATION FOR ENHANCED RECOVERY FROM COAL-GAS
B 6405 673 FOCHT, G. W., GEOPHYSICAL CASE HISTORY OF TWO HILLS COLONY GAS
B 6206 984† FOLD DEVELOPMENT IN ZAGROS SIMPLY FOLDED BELT, SOUTHWEST
B 6205 890* FOLD SYSTEM IN SOUTHERN SIERRA NEVADA BATHOLITH, CALIFORNIA,
B 6206 984† FOLDED BELT, SOUTHWEST IRAN, FOLD DEVELOPMENT IN, ZAGROS
B 6004 707* FOLDING AND FAULTING IN EASTERN WEST VIRGINIA, ABST.,
B 6206 999 FOLDING, IRAN, SALT, DIAPIRS AND, BUCKLE
B 6102 188 FOLDING ON COMPACTION, EFFECT OF, OVERBURDEN AND
B 6003 356† FOLDING SYNCHRONOUS WITH SAN ANDREAS FAULTING, SAN JOAQUIN
B 6004 672 FOLGER, D. W., MESOZOIC STRATIGRAPHIC TEST, NANTUCKET ISLAND,
B 6105 785 FOLK, ROBERT L., POISSON BED-THICKNESS DISTRIBUTION AND
M 27 166 FOLKESTONE BEDS, ENGLAND, CRETACEOUS
B 6105 774 FONSECA, J. R., CANDEIAS FIELD, RECONCAVO BASIN, BRAZIL—NEW
B 6410 1723 FONT, ROBERT G., REVIEW OF STATISTICAL ANALYSIS IN GEOLOGY, BY
B 6409 1553 FONTANA, CHRIS M., PRELIMINARY OBSERVATIONS OF MODERN POINT
B 6302 180 FOOTHILLS, GAS TESTS, CANADA
B 6405 767* FOOTHILLS PROVINCE ANALOG, ABST., OIL AND GAS POTENTIAL OF,
B 6004 677* FOOTPRINTS, ABST., PALEOENVIRONMENT OF EARLY, PERMIAN,
B 6004 592 FORAM FACIES, IDAHO, TUBULAR-
S 11 205 FORAM LINEAGES, CALIFORNIA, NEOGENE
S 11 213 FORAM NAMES, PRIOR USAGE, CALIFORNIA, NEOGENE
S 11 105 FORAM SEQUENCE, CALIFORNIA, INTERPRETATION, FIGURE
B 6004 678* FORAMINIFERA, ABST., PALEOCLIMATIC SIGNIFICANCE OF,
B 6405 700* FORAMINIFERA AND THEIR BIOSTRATIGRAPHIC POTENTIAL, ABST.,
B 6209 1759* FORAMINIFERA AS INDICATORS OF OCEANOGRAPHIC PROCESSES ON
B 6105 842* FORAMINIFERA AS PALEOBATHYMETRIC TOOL—QUANTITATIVE
S 4 321 FORAMINIFERA, ASSEMBLAGES, BRAZIL, CONTINENTAL, SHELF, BENTHIC
S 11 217 FORAMINIFERA, BEAR RIVER BEDS
S 6 217 FORAMINIFERA, BERGGRENS ZONES, A REVIEW, PALEOGENE
S 4 219† FORAMINIFERA, CALCIUM CARBONATE FIXATION BY LARGE,
S 4 313† FORAMINIFERA, CALCIUM, CARBONATE SEDIMENTATION ON,
S 11 54 FORAMINIFERA, CALIFORNIA, PLANKTONIC
S 6 329 FORAMINIFERA, CARBONIFEROUS, BIOSTRATIGRAPHY, SMALLER
B 6303 525* FORAMINIFERA COMMON TO WEST AFRICA AND CARRIBEAN, ABST.,
B 6209 1754* FORAMINIFERA IN GULF OF ALASKA, ABST., DISTRIBUTION OF,
B 6211 2350* FORAMINIFERA IN GULF OF ALASKA, ABST., DISTRIBUTION OF RECENT
S 11 54† FORAMINIFERA IN MIDDLE TERTIARY OF CALIFORNIA, ZEMORRIAN
S 11 57 FORAMINIFERA, INTERREGIONAL CORRELATIONS
S 11 58 FORAMINIFERA, LARGE
S 11 55 FORAMINIFERA, MIOCENE PLANKTONIC
S 11 91 FORAMINIFERA, MONTEREY FORMATION, NAPLES, CALIFORNIA, CHECK
S 11 94 FORAMINIFERA, MONTEREY SHALE, NAPLES, CALIFORNIA, CHECK LIST
S 11 77 FORAMINIFERA, NEW GENERA, WEST COAST, SIPHOGENERINID
S 11 222 FORAMINIFERA, NUMBER OF, HUMBOLDT BASIN, CALIFORNIA
B 6203 510* FORAMINIFERA OF CALIFORNIA BORDERLAND, ABST., SLOPE-BASIN,
B 6403 448* FORAMINIFERA OF SANGAMON ESTUARY, CENTRAL SAN FRANCISCO
S 11 217 FORAMINIFERA, PERCENT LIVING, HUMBOLDT BASIN
S 11 201 FORAMINIFERA, RELATIVE ABUNDANCE, PLANKTONIC
S 11 99 FORAMINIFERA, SHALE POINT, CALIFORNIA, CHECK LIST OF, MIOCENE
S 11 88 FORAMINIFERA, TEMBLOR-MONTEREY SEQUENCE, EDWARDS CANYON,
S 11 82 FORAMINIFERA, TEMBLOR-MONTEREY SEQUENCE, INDIAN CREEK,
S 11 85 FORAMINIFERA, TEMBLOR-MONTEREY SEQUENCE, QUAILWATER CREEK,
B 6405 716* FORAMINIFERA TO PALEOECOLOGIC INTERPRETATION, ABST.,
B 6012 2184* FORAMINIFERA--ADAPTIVE STRATEGY AND PALEOENVIRONMENTAL
B 6403 445* FORAMINIFERA--AN OVERVIEW, ABST., RECENT, ARCTIC
S 4 42 FORAMINIFERAL ASSEMBLAGES, BELIZE, GLOVERS REEF ATOLL
B 6211 2357* FORAMINIFERAL BIOFACIES IN ARCTIC OCEAN, ABST., RECENT,
B 6403 444* FORAMINIFERAL BIOFACIES OF EASTERN PACIFIC MARGIN BETWEEN 32S
B 6211 2360* FORAMINIFERAL BIOSTRATIGRAPHY IN GULF OF ALASKA TERTIARY
B 6203 572* FORAMINIFERAL BIOSTRATIGRAPHY OF NORTHEASTERN TUNISIA, ABST.,
B 6409 1565* FORAMINIFERAL BISTRATIGRAPHY OF TUSCAHOMA MARLS, SOUTHWEST
B 6403 440* FORAMINIFERAL DISTRIBUTION PATTERNS ON SHELF AND SLOPE,
B 6303 548* FORAMINIFERAL DISTRIBUTIONAL PATTERNS OFF BIG PINE KEY,
B 6209 1764* FORAMINIFERAL ECOPHENOTYPES IN GULF COAST ESTUARIES—
S 11 188 FORAMINIFERAL LOCALITIES, CALIFORNIA
B 6209 1761* FORAMINIFERAL PALEOECOLOGY OF GRAYSON FORMATION, UPPER
B 6303 538* FORAMINIFERAL RECORD IN EASTERN CARIBBEAN
S 6 277 FORAMINIFERAL SCHEME, CRETACEOUS, SENONIAN
B 6211 2246 FORAMINIFERAL STAGE, CALIFORNIA
B 6211 2218 FORAMINIFERAL STAGES, CALIFORNIA
B 6012 2190* FORAMINIFERAL ZONATION OF REFUGIAN STAGE, UPPERMOST EOCENE
S 6 332 FORAMINIFERAL ZONE 19, LOCATIONS, MISSISSIPPIAN
S 11 206 FORAMINIFERAL ZONES
S 6 279 FORAMINIFERAL ZONES, CRETACEOUS, ALBIAN
S 6 279 FORAMINIFERAL ZONES, CRETACEOUS, APTIAN
M 25 278 FORAMINIFERAL ZONULES, JAPAN
B 6403 445* FORAMINIFERA—AN OVERVIEW ABST., RECENT, ARCTIC
B 6409 1560* FORAMINIFERS AND CALCISPHERES FROM CUESTA DEL CURA AND
B 6209 1760* FORAMINIFERS, AND PTEROPODS, AS INDICATORS OF OCEANOGRAPHIC
B 6309 1597* FORAMINIFERS, AND PTEROPODS INFERRED FROM STUDIES ON LIVING
B 6403 447* FORAMINIFERS FROM EASTERN GULF OF ALASKA, ABST., BENTHIC
B 6309 1608* FORAMINIFERS FROM WESTERN ATLANTIC, DEEP SEA DRILLING PROJECT
B 6004 722* FORAMINIFERS IN LATE QUATERNARY CONTINENTAL SLOPE SEDIMENTS
B 6309 1609* FORAMINIFERS OF WEST FLOWER GARDEN BANK, NORTHERNMOST
M 27 34 FORAMS, AGGLUTINATING
M 27 33 FORAMS, ARENACEOUS
M 27 34 FORAMS, BISERIAL
M 27 35 FORAMS, ENCRUSTING
M 27 32 FORAMS, FUSULINID
M 27 36 FORAMS, GLOBIGERINID
M 27 36 FORAMS, GLOBOROTALIC
M 27 31 FORAMS, MILIOLID
M 27 34 FORAMS, ORBITOLINID
M 27 32 FORAMS, PENEROPLID
M 27 35 FORAMS, PLANKTONIC
M 27 34 FORAMS, UNISERIAL
B 6103 465 FORBES, ROBERT B., REGIONAL TECTONIC IMPLICATIONS OF ALASKA
B 6103 469 FORBES, ROBERT F., GEOCHRONOLOGY OF SOUTHWESTERN BROOKS
B 6303 451 FORCE, LUCY M, RIVER NEAR CHARLESTON SOUTH CAROLINA, ABST.
B 6211 2354 FORD, DON, ONSHORE WASHINGTON AND OREGON, BY TITLE ONLY
B 6412 2064 FORD, J. P., SEASAT ORBITAL RADAR IMAGERY FOR GEOLOGIC
B 6309 1578 FORD, W. KENT, CHARACTERIZATION OF PRODUCTION MECHANISM IN
B 6212 2510* FORE-ARC AND BACK-ARC BASINS OF BOHOKU ARC, JAPAN, ABST.,
B 6212 2509* FORE-ARC AND BACK-ARC BASINS OF TOHOKU ARC, JAPAN, ABST.,
B 6303 301 FORE-ARC BASIN, ALASKA, TUGIDAK BASIN
B 6301 16 FOREARC BASIN
M 29 246 FOREARC BASIN, MODEL, EVOLUTION
B 6405 727* FOREARC BASIN OF CALIFORNIA—INFLUENCE OF BASIN EVOLUTION AND
M 29 259 FOREARC BASIN, RESIDUAL, EVOLUTION DIAGRAM
M 29 245† FOREARC BASINS, THE, EVOLUTION OF, STRUCTURAL HIGHS BORDERING
B 6301 24 FOREARC EVOLUTION, ALASKA
B 6301 20 FOREARC EVOLUTION, CALIFORNIA
B 6301 5 FOREARC MODEL, TERMINOLOGIES
M 29 248 FOREARC MORPHOLOGIES, MODELS
M 29 250 FOREARC, MORPHOLOGY, RIDGED
M 29 250 FOREARC, MORPHOLOGY, SHELVED
M 29 250 FOREARC, MORPHOLOGY, TERRACED
B 6301 24 FOREARC PETROLEUM POTENTIAL
B 6301 21 FOREARC REGIONS, CALIFORNIA
B 6301 2† FOREARC REGIONS, STRUCTURE AND STRATIGRAPHY O
B 6301 9 FOREARC SYSTEMS, ACTIVE
B 6301 3 FOREARC TECTONICS
B 6301 6 FOREARC TYPES
B 6303 497* FORECASTING METHOD AND ASSESSMENT OF REGIONAL POTENTIALS,

B 6303 539* FORECASTING--FACT OR FOLLY, ABST.
B 6010 1637† FOREIGN DEVELOPMENT PAPERS, 1975
B 6110 1575† FOREIGN DEVELOPMENT PAPERS, 1976
B 6210 1793† FOREIGN DEVELOPMENT PAPERS, 1977
B 6310 1632† FOREIGN DEVELOPMENT PAPERS, 1978
B 6411 1736† FOREIGN DEVELOPMENT PAPERS, 1979
B 6405 716* FOREIGN FRONTIERS—AN OVERVIEW, ABST.
B 6405 770* FORELAND BASIN, APPALACHIAN OROGENE, VIRGINIA, ABST.,
B 6008 1413* FORELAND BASIN, COLORADO, ABST., OVERPRESSURED, SHALE MASSES
B 6410 1575† FORELAND BASIN EVOLUTION, MIDDLE ORDOVICIAN, VIRGINIA
M 26 217 FORELAND BASIN, LATE PALEOZOIC, DEPOSITIONAL SYSTEMS
B 6111 1957 FORELAND BASINS, RESERVES
S 3 93 FORELAND, DEFINITION
B 6405 683* FORELAND FOLD AND THRUST BELT OF CARPATHIANS AND ITS
B 6405 684* FORELAND FOLD AND THRUST BELTS—REVIEW, ABST.
S 3 93 FORELAND MARGIN, OROGENIC
B 6407 972 FOREMOST FORMATION, ALBERTA
S 4 53† FOREREEF AND UPPER ISLAND SLOPE, NORTH JAMAICA, DEEP
S 4 21 FOREREEF, PUERTO RICO
M 27 190 FOREST MARBLE BEDS, ENGLAND, JURASSIC
M 27 184 FOREST MARBLE BEDS, ENGLAND, JURASSIC
B 6202 303 FOREST OIL SANDSTONES, TRINIDAD, CRUSE-
B 6004 691* FORESTS OFF NORTHWEST FLORIDA COAST--PALEOENVIRONMENTAL
B 6302 245 FORGEY, RANDAL L. AND, AEROMAGNETIC DETECTION OF DIAGENETIC
B 6009 1610 FORGOTSON, J. M. JR., CORRELATIONS AND STRATIGRAPHY FROM BASE
B 6007 1119 FORGOTSON, J. M., JR., DEFINITION OF GILMER LIMESTONE, UPPER
B 6009 1610 FORGOTSON, JAMES M., CORRELATIONS AND STRATIGRAPHY FROM BASE
B 6007 1119 FORGOTSON, JAMES M., DEFINITION OF GILMER LIMESTONE, UPPER
B 6309 1534 FORGOTSON, JAMES M., SR., SUPER OVERPRESSURED GAS
B 6009 1616* FORM AND COMPOSITION OF MISSISSIPPI FAN, ABST.
B 6102 252 FORMAL ROCK-STRATIGRAPHIC TERM, VOLCANIC ROCKS
B 6102 252† FORMAL ROCK-STRATIGRAPHIC UNIT, NOTE 46--, APPLICATION FOR
B 6207 1185† FORMAL TERMINOLOGY FOR OCEANIC ROCKS, NOTE 48—APPLICATION
B 6310 1888 FORMAN EXPLORATION CO., ON DEVELOPMENTS IN AJMAN, 1978
B 6411 1840 FORMAN EXPLORATION CO., ON DEVELOPMENTS IN AJMAN, 1979
B 6411 1840 FORMAN EXPLORATION CO., ON DEVELOPMENTS IN SHARJAH, 1979,
B 6209 1757* FORMATION EVALUATION WITH LOGS IN ARK-LA-TEX COTTON VALLEY
B 6206 993 FORMATION, IRAN, GACHSARAN
B 6107 1090† FORMATION PRESSURE. DISCUSSION, ABNORMAL
B 6007 1124† FORMATION PRESSURE. DISCUSSION AND REPLY, ABNORMAL
B 6212 2444 FORMATION WATERS, CALIFORNIA
B 6212 2444 FORMATION WATERS, TEXAS
B 6309 1506 FORMATION-DENSITY LOG STUDY, KENTUCKY
B 6309 1506 FORMATION-DENSITY LOG STUDY, NEW YORK
B 6309 1506 FORMATION-DENSITY LOG STUDY, OHIO
B 6309 1506 FORMATION-DENSITY LOG STUDY, VIRGINIA
B 6309 1506 FORMATION-DENSITY LOG STUDY, WEST VIRGINIA
B 6309 1504† FORMATION-DENSITY LOGS, DETERMINATION OF, ORGANIC CONTENT
B 6303 458* FORMATION-TOPS FIL SYSTEM, ABST., DEVELOPMENT AND USE OF
S 11 7 FORMATIONAL APPROACH, CALIFORNIA, SERIES SUBDIVISIONS
B 6405 788* FORMS AND ASSOCIATIONS OF PYRITE IN UPPER FREEPORT COAL BED,
B 6009 1615 FORSTHOFF, GARY M., CLAY SEDIMENTATION I PEARL RIVER DELTA.
B 6409 1356 FORT BEND COUNTY, TEXAS, PASTOR FIELD
B 6303 508* FORT GEORGE INLET, FLORIDA, ABST., SEDIMENT DISPERSAL AT
B 6410 1615 FORT HAYS LIMESTONE, DENVER BASIN
B 6107 997 FORT HAYS LIMESTONE MEMBER, NIOBRARA LIMESTONE, COLORADO
B 6107 997 FORT HAYS LIMESTONE MEMBER, NIOBRARA LIMESTONE, KANSAS
B 6402 222 FORT LANCASTER FORMATION, TEXAS
S 3 61 FORT PAYNE CHERT, TENNESSEE
B 6204 646 FORT PAYNE FORMATION, ALABAMA
B 6405 743* FORT PAYNE FORMATION, LOWER MISSISSIPPIAN, OF TENNESSEE WITH
B 6408 1287* FORT PAYNE FORMATION, MISSISSIPPIAN, ILLINOIS BASIN, ABST.,
B 6303 487* FORT PAYNE FORMATION, NORTHERN TENNESSEE, ABST., CONTROLS OF
B 6107 1046 FORT SCOTT FORMATION, KANSAS
B 6103 313 FORT SCOTT LIMESTONE, KANSAS
B 6109 1553* FORT TERRETT FORMATION, EDWARDS GROUP, IN VICINITY OF
B 6402 222 FORT TERRETT FORMATION, TEXAS
B 6008 1393* FORT UNION AND MESAVERDE FORMATIONS--ITS RELATION TO
B 6302 210 FORT UNION COALS, COLORADO
B 6302 212 FORT UNION FORMATION, COLORADO
B 6203 513* FORT UNION FORMATION, EOCENE, POWDER RIVER-CLEAR CREEK AREA,
B 6407 971 FORT UNION FORMATION, GREAT PLAINS
B 6305 826* FORT UNION FORMATION, POWDER RIVER AREA, WYOMIN AND
B 6105 821* FORT UNION FORMATION, WEST-CENTRAL NORTH DAKOTA, ABST.,
B 6311 2052 FORT UNION FORMATION, WYOMING
B 6305 831* FORT UNION FORMATIONS, CENTRAL POWDER RIVER BASIN, WYOMING
M 24 106 FORT UNION MARKER, WYOMING
B 6302 194† FORT UNION SEDIMENTS, TERTIARY, NORTHWEST COLORADO, AND
B 6308 1427* FORT WORTH BASIN, ABST., STRATIGRAPHY AND EXPLORATION
M 26 217 FORT WORTH BASIN, LATE PALEOZOIC, DEPOSITIONAL SYSTEMS
B 6301 50 FORT WORTH BASIN, TEXAS
B 6303 548* FORT WORTH BASIN, TEXAS, ABST., DEPOSITION AND DIAGENESIS OF,
B 6306 955 FORTEAU FORMATION, NEWFOUNDLAND
M 25 411 FORTESCUE GROUP, WESTERN AUSTRALIA
B 6303 474 FORTEY, RICHARD A., CORRELATION AND TIMING OF
M 30 92 FORTIES FIELD, CHARLIE SAND, DEPOSITIONAL HISTORY
M 30 89 FORTIES FIELD, CHARLIE SAND, ISOPACH
M 30 82 FORTIES FIELD, DEVELOPMENT PLAN, NORTH SEA
M 30 86 FORTIES FIELD, FACIES DISTRIBUTION
M 30 85 FORTIES FIELD, FIVE COMMON LOG PATTERNS
M 30 90 FORTIES FIELD, FORTIES FORMATION, CHARLIE SAND UNIT
M 30 84 FORTIES FIELD, FORTIES FORMATION, DEPOSITIONAL ENVIRONMENT
M 30 82 FORTIES FIELD, LITHOFACIES ANALYSIS, CORE DATA, NORTH SEA
M 30 82 FORTIES FIELD, LITHOSTRATIGRAPHY, NORTH SEA
M 30 82 FORTIES FIELD, LOCATION MAP, NORTH SEA
M 30 85 FORTIES FIELD, LOG DATA, SEQUENTIAL FACIES
M 30 82 FORTIES FIELD, STRUCTURAL SETTING, NORTH SEA
M 30 86 FORTIES FIELD, SUBMARINE FAN DEPOSITIONAL MODEL
M 30 81† FORTIES FIELD, U.K. CONTINENTAL SHELF, NORTH SEA, GEOLOGY OF
B 6303 466* **FORTIES FIELD, UNITED KINGDOM CONTINENTAL SHELF, NORTH SEA,**
M 30 90 FORTIES FORMATION, CHARLIE SAND UNIT, FORTIES FIELD
M 30 84 FORTIES FORMATION, DEPOSITIONAL ENVIRONMENT, FORTIES FIELD
M 28 31 FORTRESS MOUNTAIN FORMATION, ALASKA, CRETACEOUS
M 28 99 FORTRESS MOUNTAIN FORMATION, ALASKA, CRETACEOUS
B 6303 544* FORTUNA FORMATION, EAST-CENTRAL TUNISIA, RECORD OF LATE
B 6102 238 FORTY MILE BANK, CALIFORNIA
B 6309 1601 FOSS, DEANE C., DEPOSITIONAL ENVIRONMENT OF WOODBINE
M 25 448 FOSSA MAGNA, JAPAN
B 6303 458* **FOSSIL MICROBIAL BORINGS--NEW APPROACH, ABST., STUDY OF**
B 6405 740* FOSSIL ACCRETIONARY PRISM, GREECE, ABST., SYNSEDIMENTARY
B 6004 661* FOSSIL ASSEMBLAGE OF EARLY CRETACEOUS SAND FLAT-SHALLOW
B 6405 694* FOSSIL ASSEMBLAGES OF UPPER CRETACEOUS SAND UNITS, DELAWARE
B 6205 864 FOSSIL COLLECTIONS AVAILABLE FROM FORTY-EIGHT STATES, MINERAL,
S 4 177 FOSSIL CORALS, GROWTH RATES FOR
B 6410 1682 FOSSIL EVIDENCE, PRECAMBRIAN
B 6305 845* FOSSIL FLUIDS IN POST-SILURIAN ROCKS OF APPALACHIAN BASIN, ABST.
B 6203 563* FOSSIL FUELS AND OCEAN CHEMISTRY IN 22D CENTURY, ABST., BURNING
B 6105 815* FOSSIL GOLD, URANIUM, AND PYRITE CONCENTRATIONS IN
S 5 99 FOSSIL HILL FORMATION, ONTARIO,
B 6405 733* FOSSIL MORPHOLOGY, ABST., ENVIRONMENTAL CONTROL OF, TRACE
B 6109 1540* FOSSIL RECORD, ABST., GASTROPOD, RADULAE—THEIR POTENTIAL IN
B 6303 434* FOSSIL ROOT CAST FROM KOOBI FORA FORMATION,
B 6004 660* FOSSIL SALT STRUCTURES IN VACHERIE SALT DOM AREA, LOUISIANA,
B 6405 690* FOSSILS, ABST., SYNONYMY, ETHOLOGY, AND ENVIRONMENTAL
B 6405 704* FOSSILS AND STAGNATION OF DEEP-SEA BASINS, ABST., TRACE
B 6405 729* FOSSILS AND STRATIGRAPHY OF DEVONIAN BLACK SHALE IN
B 6405 697* FOSSILS FOR INTERPRETATION OF TRIASSIC DEPOSITIONAL
B 6405 691* FOSSILS FROM PROTEROZOIC SEDIMENTS OF GODAVARI VALLEY, SOUTH
B 6405 702* FOSSILS IN BRITISH SILESIAN, ABST., BIVALVE, TRACE
B 6303 476* FOSSILS IN UPPER PENNSYLVANIAN AND LOWER PERMIAN, WESTERN
B 6012 2175* FOSSILS, LISBURNE GROUP, NORTHEAST ALASKA, ABST.,
B 6004 727* FOSSILS, MUDDY AND FRONTIER FORMATIONS, WYOMING AND
B 6004 704* FOSSILS OF CHAGRIN FORMATION, UPPER DEVONIAN, NORTHEAST OHIO,
B 6004 687* FOSSILS OF UPPER CRETACEOUS SIEVERING FORMATION, VIENNA,
B 6405 669* FOSSILS OF WESTERN WYOMING AND ADJACENT AREAS, ABST., PERMIAN,
B 6304 688 FOSTER, JOHN F., ANOMALOUS OREBODY WITHIN AMBROSIA LAKE
B 6011 2070 FOSTER, NORMAN H., CONCEPTS AND TECHNIQUES UTILIZED IN
B 6205 884 FOSTER, NORMAN H., GEOMORPHIC EXPLORATION USED IN DISCOVERY
B 6405 707 FOSTER, NORMAN H., PETROLEUM POTENTIAL OF BASIN AND RANGE
B 6403 442 FOSTER, NORMAN H., PETROLEUM POTENTIAL OF GREAT BASIN, ABST.
B 6207 1217 FOSTER, ROBERT J., AUSTRALIA AS NATURAL GAS EXPORTER, ABST.
M 25 27 FOSTER, ROBERT J., MARINE ENVIRONMENT IN AUSTRALIA. TWO
B 6401 107 FOSTER, W. R., A PROCESS FOR PRIMARY MIGRATION OF PETROLEUM
B 6304 697 FOSTER, W. R., PROCESS FOR PRIMARY MIGRATION OF PETROLEUM IN
B 6405 708 FOSTER, W. R., SMECTITE-ILLITE TRANSFORMATION—ROLE IN
B 6205 891 FOUCH, T. D., MINERALOGIC CHARACTERISTICS OF SOME LOWER
B 6305 838 FOUCH, THOMAS D., EVIDENCE FOR TWO MAJOR CYCLES OF PETROLEUM
B 6305 832 FOUCH, THOMAS D., INFLUENCE OF DIAGENETIC REACTIONS ON
B 6108 1377 FOUCH, THOMAS D., INTERDISCIPLINARY ANALYSIS OF SOME POTENTIAL
B 6105 785 **FOUCH, THOMAS D., NEWLY RECOGNIZED PETROLEUM SOURCE-ROCK UNITS**
B 6108 1378 FOUCH, THOMAS D., SHEEP PASS, CRETACEOUS TO EOCENE, AND
B 6405 798 FOULKES, D. E., SEISMIC STRATIGRAPHY IN LOW-ENERGY CARBONATE
B 6303 544 FOULKES, D. E., WALKER CREEK FIELD, ARKANSAS--SMACKOVER CASE
B 6405 708 FOULKES, DONALD E., APPLICATION OF INVERSION PROCESSING TO
B 6105 786 **FOULKES, DONALD E., FACIES GEOMETRY FROM CONVENTIONAL SEISMIC**
B 6002 159 FOUNDATION ISSUE, FEBRUARY 1976 BULLETIN, AAPG
B 6002 159 FOUNDATION TRUSTEES, AAPG
B 6203 404 FOUNTAIN VALLEY, VIRGIN ISLANDS, ST. CROIX
B 6002 186 FOUR CHARACTERISTIC GAMMA-LOG PATTERNS IN CLASTIC
B 6405 648 FOUR CORNERS AREA, EDITED BY JAMES E. FASSETT, NICK D.
B 6104 595 FOUR CORNERS, DRY MESA FIELD

B 6308 1225 FOUR CORNERS FIELD, BRAZORIA COUNTY, TEXAS
B 6104 596 FOUR CORNERS, RATTLESNAKE FIELD
B 6104 597 FOUR CORNERS, TOCITO DOME FIELD
B 6104 594 FOUR CORNERS, TOCITO DOME FIELD
B 6104 594 FOUR CORNERS, WALKER CREEK FIELD
B 6008 1239† **FOUR CORNERS--INTERMOUNTAIN AREA IN 1975, DEVELOPMENTS IN**
B 6308 1206† **FOUR CORNERS-INTERMOUNTAIN AREA IN 1978, DEVELOPMENTS IN**
B 6108 1204 FOUR CORNERS-INTERMOUNTAIN AREA, GEOTHERMAL ACTIVITY, 1976
B 6108 1204† FOUR CORNERS-INTERMOUNTAIN AREA IN 1976, DEVELOPMENTS IN
B 6208 1361 FOUR CORNERS—INTERMOUNTAIN AREA, GEOTHERMAL ACTIVITY, 1977
B 6208 1360† FOUR CORNERS—INTERMOUNTAIN AREA IN 1977, DEVELOPMENTS IN
B 6409 1340† FOUR CORNERS—INTERMOUNTAIN AREA IN 1979, DEVELOPMENTS IN
B 6405 773* FOURIER GRAIN SHAPE ANALYSIS, ABST., LOCATION OF LITTORAL,
B 6009 1616* FOURMILE CREEK AND SPLUNGE FIELDS, BLACK WARRIOR BASIN,
B 6405 675 FOWLER, M. L., DEPOSITIONAL FACIES, GEOMETRY, AND GENESIS OF
B 6008 1398 FOWLER, WILLIAM A., JR., REGIONAL PRESSURE ENVIRONMENTS AND
M 26 279 FOX HILLS SANDSTONE, CRETACEOUS
B 6407 971 FOX HILLS SANDSTONE, GREAT PLAINS
B 6403 318 FOX HILLS SANDSTONE, MONTANA
M 27 89 FOX HILLS SANDSTONE, SOUTH DAKOTA, CRETACEOUS
B 6008 1407 FOX, JAMES E, COASTAL SEDIMENTATION MODEL, TENSLEEP SANDSTONE,
B 6303 429 FOX, L. A., ROL OF BRINES AND FERROUS IRON IN FORMATION OF
B 6105 797 FOX, PAUL J., LONG-DISTANCE TURBITITE CORRELATIONS IN HORSESHOE
B 6303 452 FOX, PAUL J., OCEANIC CRUST, ABST.
B 6103 361 FOX-GRAHAM STRUCTURE, OKLAHOMA
M 26 214 FOZ DO AMAZONAS BASIN, BRAZIL
B 6405 687* FOZ DO AMAZONAS BASIN, OFFSHORE NE BRAZIL, ABST., MICROFACIES
S 3 45 FRACTIONAL HYDROLYSIS
S 3 45 FRACTIONAL MELTING
S 3 45 FRACTIONAL METAMORPHISM
B 6405 675* FRACTIONAL PRESERVATION OF TRANSGRESSIVE COASTAL LITHOSOMES
S 3 45 FRACTIONAL VOLATILIZATION
B 6303 482* FRACTOGRAPHIC DISTINCTION OF CORING-INDUCED FRACTURES FROM
B 6109 1547* FRACTURE ANALYSIS, ABST., EVALUATION OF, AUSTIN AND, BUDA
B 6209 1755* FRACTURE IDENTIFICATION LOGS IN CRETACEOUS OF NORTH
B 6309 1576* FRACTURE PATTERN IN OUTCROP OF GAS-PRODUCING DEVONIAN, OHIO,
B 6102 227† FRACTURE PERMEABILITY IN POROUS ROCK, EXPERIMENTAL STUDY OF
B 6107 1086† FRACTURE POROSITY, SIGNIFICANT GEOLOGIC PROCESSES IN
B 6303 480* FRACTURE RESERVOIRS, EXAMPLE FRO ANADARKO BASIN, ABST.,
B 6312 2214† FRACTURE SYSTEMS, DESCRIPTION AND CLASSIFICATION, NATURAL
B 6405 755* FRACTURE SYSTEMS—EXAMPLE FROM ALTAMONT FIELD, UINTA BASIN,
B 6109 1548* FRACTURED AUSTIN CHALK AND BUDA LIMESTONE IN SOUTH TEXAS,
B 6405 754* FRACTURED CHERTS OF WOODFORD AND ARKANSAS NOVACULITE
S 10 54 FRACTURED SHALE ACCUMULATIONS, MIGRATION
M 24 121† FRACTURED STRATIGRAPHIC TRAP, UINTA BASIN, UTAH, ALTAMONT-,
M 27 208 FRACTURES
M 28 160 FRACTURES, CALCITE-FILLED, IN DARK SHALE
B 6312 2216 FRACTURES, CHICKEN-WIRE
B 6312 2215 FRACTURES, CONTRACTIONAL
B 6207 1111† FRACTURES FROM DIPMETER LOGS, MEASUREMENT OF, SUBSURFACE
B 6303 482* FRACTURES FROM NATURA CORED FRACTURES, ABST., FRACTOGRAPHIC
B 6304 598† FRACTURES IN CRETACEOUS ROCKS FROM SELECTED AREAS OF SAN
M 28 160 FRACTURES, IN DOLOMITIC CHERT
M 28 160 FRACTURES, IN LIMESTONE
B 6309 1590* FRACTURES IN LOW-POROSITY SHALE GAS RESERVOIRS--RESULTS OF
B 6312 2217 FRACTURES, MINERAL PHASE-CHANGE
B 6312 2215 FRACTURES, MUD CRACKS, DESICCATION
M 28 164 FRACTURES OF QUARTZ GRAINS
M 28 180 FRACTURES, OIL STAINING, EXAMPLE
M 28 161 FRACTURES, QUARTZ AND CALCITE FILLED
M 28 161 FRACTURES, QUARTZ FILLED, IN LIMESTONE
M 28 161 FRACTURES, QUARTZ FILLED, IN SANDSTONE
B 6312 2216 FRACTURES, SYNERESIS
B 6312 2216 FRACTURES, THERMAL, CONTRACTIONAL
B 6309 1583* FRACTURING, ABST., INTERACTION OF, PROPPANTS WITH CRACK
B 6309 1579* FRACTURING AND SONIC LOGGING IN GAS SHALES, ABST., THEORETICAL
M 28 167 FRACTURING, BIBLIOGRAPHY
B 6309 1586* FRACTURING, PETROLEUM EXPULSION, AND DOLOMITIZATION IN
B 6309 1579* **FRACTURING AND SONIC LOGGING IN GAS SHALES, ABST., THEORETICAL**
M 28 39 FRAGMENTS, BRYOZOAN
M 28 28 FRAGMENTS, CARBONATE, IN CALICHE
M 28 27 FRAGMENTS, CHALCEDONY, IN CARBONATE CEMENT
M 28 27 FRAGMENTS, CHERT
M 28 117 FRAGMENTS, CHERT, DETRITAL
M 28 78 FRAGMENTS, CHERT, DETRITAL
M 28 39 FRAGMENTS, CRINOID
M 28 28 FRAGMENTS, DOLOMITE, DETRITAL
M 28 29 FRAGMENTS, DOLOMITE RHOMBS
M 28 39 FRAGMENTS, ECHINODERM, GLAUCONITE AFTER CALCITE
M 28 38 FRAGMENTS, GLAUCONITE, DETRITAL
M 28 42 FRAGMENTS, GYPSUM CRYSTALS
M 28 43 FRAGMENTS, IRON OXIDES
M 28 98 FRAGMENTS, LIMESTONE, IN CALCLITHITE
M 28 31 FRAGMENTS, METAMORPHIC
M 28 30 FRAGMENTS, METAMORPHIC
M 28 40 FRAGMENTS, MOLLUSCAN
M 28 34 FRAGMENTS, OPALLINE SILICA
M 28 42 FRAGMENTS, ORGANIC MATTER
M 28 41 FRAGMENTS, ORGANIC MATTER
M 28 39 FRAGMENTS, PELECYPOD
M 28 40 FRAGMENTS, PHOSPHATE, CONODONTS
M 28 34 FRAGMENTS, PLAGIOCLASE
M 28 28 FRAGMENTS, POLYMICT CALCITE
M 28 32 FRAGMENTS, SCHIST
M 28 31 FRAGMENTS, SCHIST, IN MUSCOVITE
M 28 29 FRAGMENTS, SILTSTONE AND SANDSTONE
M 28 30 FRAGMENTS, SLATE
M 28 39 FRAGMENTS, TRILOBITE
M 28 34 FRAGMENTS, VOLCANIC
M 28 35 FRAGMENTS, VOLCANIC FELDSPAR
B 6309 1511 FRAME MEMBER, NEW NAME, TEXAS
B 6002 281 FRAME SHALE MEMBER, RESTRICTED, PENNSYLVANIA
S 7 1 **FRAMEWORK, FACIES, AND OIL-TRAPPING CHARACTERISTICS OF**
S 7 21† FRAMEWORK, NORTHERN AND, EASTERN, GULF OF MEXICO,
B 6006 926† FRAMEWORK OFF NORTHEASTERN UNITED STATES, REGIONAL
B 6203 496* FRANCE (MICROSTRUCTURES, ALGAL REMAINS)—COMPARISON WITH
B 6010 1710 FRANCE, 1975, SERVICE CONS. GISEMENTS, ON DEVELOPMENTS IN
B 6110 1640 FRANCE, 1976, SERVICE CONS. GISEMENTS, ON DEVELOPMENTS IN
B 6210 1910 FRANCE, 1977, SERVICE CONS. GISEMENTS, ON DEVELOPMENTS IN
B 6310 1919 FRANCE, 1978, SERVICE CONS. GISEMENTS, ON DEVELOPMENTS IN
B 6405 697* FRANCE, ABST., USE OF, TRACE, FOSSILS FOR INTERPRETATION OF,
B 6110 1641 FRANCE, ALSACE
B 6411 1964 FRANCE, ALSACE AND JURA
S 6 113 FRANCE, ANSE DU VERYACH SERIES, ISOCHRON DIAGRAM
B 6004 612† FRANCE, ANTIPATHETIC RELATIONS AMON, ALGAL STRUCTURES,
B 6411 1964 FRANCE, AQUITAINE BASIN
B 6110 1641 FRANCE, AQUITAINE BASIN
B 6310 1919 FRANCE, AQUITAINE BASIN
B 6210 1910 FRANCE, AQUITAINE BASIN
M 29 463 FRANCE, AQUITAINE BASIN, VOLUME OF SEDIMENT PRESERVED
S 10 136 FRANCE, AQUITAINE, CRETACEOUS WATER LEVELS
S 10 136 FRANCE, AQUITAINE, EOCENE WATER LEVELS
S 10 135 FRANCE, AQUITAINE, MIOCENE WATER LEVELS
S 10 134 FRANCE, AQUITAINE, TOPOGRAPHIC SURFACE
S 10 133 FRANCE, AQUITAINE BASIN, GROUNDWATER FLOW
B 6310 1919 FRANCE, ATLANTIC OFFSHORE
B 6211 2232 FRANCE, BAY OF BISCAY AREA
B 6104 512 FRANCE, CHAILLY OIL FIELD
B 6104 515 FRANCE, CHAILLY WELLS
B 6107 988 FRANCE, CHALK UNIT
M 27 132 FRANCE, CRETACEOUS CHALK
B 6411 1964 FRANCE, DEVELOPMENTS, 1979
B 6304 678 FRANCE, DOGGER, MIDDLE JURASSIC, RESERVOIR
B 6411 1964 FRANCE, ENGLISH CHANNEL-WESTERN APPROACHES
S 6 113 FRANCE, ERQUY FORMATION, ISOCHRON DIAGRAM
B 6310 1920 FRANCE, JURA-BRESSE-SAVOIE
B 6110 1641 FRANCE, JURA-BRESSE-SAVOIE
S 6 113 FRANCE, KERITY FORMATION, ISOCHRON DIAGRAM
B 6104 516 FRANCE, MAROLLES WELLS
S 3 23 FRANCE, MASSIF CENTRAL, FLUORINE
B 6310 1920 FRANCE, MEDITERRANEAN, DEVELOPMENTS
S 10 154 FRANCE, PARENTIS ARTESIAN BASIN, PETROLEUM MIGRATION
S 10 27 FRANCE, PARENTIS BASIN, AROMATIC DISTRIBUTION
S 10 27 FRANCE, PARENTIS BASIN, NSO DISTRIBUTION
S 10 27 FRANCE, PARENTIS BASIN, SATURATE DISTRIBUTION
B 6411 1964 FRANCE, PARIS BASIN
B 6210 1911 FRANCE, PARIS BASIN
S 6 132 FRANCE, PARIS BASIN, ALBIAN, POTASSIUM-ARGON
S 6 132 FRANCE, PARIS BASIN, CENOMANIAN, POTASSIUM-ARGON
B 6104 511 FRANCE, PARIS BASIN, CHAILLY OIL FIELD
B 6202 202 FRANCE, PARIS BASIN, QUALITATIVE GEOHISTORY DIAGRAM
B 6401 61 FRANCE, PARIS BASIN, TOARCIAN, KEROGEN
B 6110 1641 FRANCE, PARIS BASIN-WESTERN APPROACHES
B 6104 511† FRANCE, PATTERNS OF, POROSITY AN, CEMEN IN, OOID, RESERVOIRS IN,
B 6110 1657 FRANCE, PRODUCTION, 1975-1976
B 6210 1928 FRANCE, PRODUCTION, 1976-1977
B 6310 1935 FRANCE, PRODUCTION, 1977-1978
B 6411 1971 FRANCE, PRODUCTION, 1978-1979
B 6411 1964 FRANCE, PROVENCE-MEDITERRANEAN
B 6104 513 FRANCE, SAINT BRICE WELL

Index

B 6104 511 FRANCE, ST. BRICE-VILLENEUVE WELLS
S 6 114 FRANCE, STEIGE SERIES, ISOCHRON DIAGRAM
S 8 21 FRANCE, TURBIDITY CURRENTS, VAR CANYON
S 6 114 FRANCE, VILLE SERIES, ISOCHRON DIAGRAM
B 6304 676† FRANCE. DISCUSSIN AND REPLY, PATTERNS O, POROSITY AND, CEMENT
B 6211 2351* FRANCISCAN ASSEMBLAGE OF NORTHERN CALIFORNIA, A
B 6106 912 FRANCISCAN BASEMENT, BAJA CALIFORNIA
B 6203 349 FRANCISCAN COMPLEX, CALIFORNIA
B 6211 2217 FRANCISCAN COMPLEX, CALIFORNIA
B 6301 21 FRANCISCAN COMPLEX, CALIFORNIA
B 6111 1980 FRANCISCAN FORMATION, CALIFORNIA
M 27 25 FRANCISCAN FORMATION, CALIFORNIA, JURASSIC
B 6004 644* FRANCISCAN OF TRINIDAD, CALIFORNIA, ABST., SEDIMENTOLOGY OF,
B 6104 546 FRANCISCONI, O., MARINE SEISMIC INVESTIGATIONS, SOUTHERN BRAZIL
B 6209 1757 FRANK, R. W., FORMATION EVALUATION WITH LOGS IN ARK-LA-TEX
B 6303 452 FRANK, W. M., GEOLOGY OF SOUTHEAST GEORGIA EMBAYMENT, ABST.
B 6303 429 FRANKE, MILTON R., DEPOSITIONAL-DIAGENETIC HISTORY OF MACAE
B 6408 1281* FRANKLIN COUNTIES, ILLINOIS, ABST., GEOLOGY AND GEOPHYSICS OF
B 6101 7 FRANKLINIAN BASIN, CAMBRIAN
B 6101 6 FRANKLINIAN BASIN, CANADIAN, ARCTIC ISLANDS
B 6101 9 FRANKLINIAN BASIN, DEVONIAN
B 6101 7 FRANKLINIAN BASIN, ORDOVICIAN
B 6101 8 FRANKLINIAN BASIN, ORDOVICIAN TO DEVONIAN
B 6101 6 FRANKLINIAN BASIN, PROTEROZOIC
B 6101 4 FRANKLINIAN BASIN, SILURIAN
B 6101 4 FRANKLINIAN BASIN, STRATIGRAPHIC COLUMN
B 6307 1061 FRANKLINIAN BASIN, STRATIGRAPHIC COLUMN
B 6101 3 FRANKLINIAN FOLDBELT
B 6101 19 FRANKLINIAN FOLDBELT, CANADIAN, ARCTIC ISLANDS
B 6206 1005 FRANKLINIAN GEOSYNCLINE, ARCTIC, CANADA
B 6203 497 FRANKS, S. G., CLAY DIAGENESIS IN WILCOX SANDSTONES OF
B 6105 786 FRANKS, S. G., INVESTIGATION OF FACTORS AFFECTING HYDROCARBON
B 6405 708 FRANKS, STEPHEN G., CONTROLS OF ZEOLITE CEMENTATION IN UPPER
B 6201 100 FRANSON MEMBER, IDAHO AND, WYOMING
S 8 40 FRASER DELTA, CANADA, SEAVALLEYS
B 6004 672 FRASER, G. S., TRANSGRESSIVE-REGRESSIVE SHELF DEPOSITION, SHUBLIK
S 8 97 FRASER SEAVALLEY, CANADA, SUBMARINE CANYONS
S 8 102 FRASER SEAVALLEY, TIME-VELOCITY CURVES, CURRENT FLOW
S 8 20 FRASER SEAVALLEY, TIME-VELOCITY CURVES, CURRENT FLOW
S 8 100 FRASER SEAVALLEY, TOPOGRAPHY, INDEX MAP
B 6406 826 FRASNIAN PILLARA REEF CYCLE, AUSTRALIA, GIVETIAN-
B 6406 827 FRASNIAN REEF COMPLEXES, AUSTRALIA
B 6405 722* FRAUNHOFER LINE DISCRIMINATOR, ABST., MARINE PETROLEUM
S 7 87 FRAZIER, D. E., RECOGNITION OF CONTINENTAL-SLOPE SEISMIC FACIES,
M 25 268 FRAZIER, S. B., MARIN PETROLEUM EXPLORATION OF HUKSAN
B 6211 2204 FREAGH POINT AND GREEN ISLAND, GROWTH FAULTS BETWEEN,
B 6008 1391* FRED AND GEORGE CREEK AREA, TOOLE COUNTY, MONTANA, ABST.,
B 6409 1558* FREDERIC ON MORPHOLOGY OF DAUPHIN ISLAND, ALABAMA, ABST.,
B 6404 528 FREDERICK E. HYDE FJORD, GREENLAND
B 6402 222 FREDERICKSBURG GROUP, TEXAS
M 24 242 FREDERICKSBURG PRODUCING SAND, MISSISSIPPI
B 6205 786 FREDERICKSBURG RUDISTID REEFS AND BANKS, TEXAS, WASHITA-
B 6408 1220 FREDERICKSBURG SECTION, TEXAS
M 29 355 FREDERICKSTED CANYON, VIRGIN ISLANDS, ST. CROIX
M 29 360 FREDERICKSTED PLATEAU, BATHYMETRY, VIRGIN ISLANDS
M 29 362 FREDERICKSTED PLATEAU, GEOMORPHOLOGIC DEVELOPMENT, VIRGIN
M 29 358 FREDERICKSTED PLATEAU, SEISMIC SECTIONS, VIRGIN ISLANDS
B 6007 1050 FREE-AIR GRAVITY AND MAGNETIC PROFILES, NORTHWEST AND,
B 6002 288 FREE-GAS PHASE OF HYDROCARBONS, STUDY OF, WHOLE
B 6409 1557 FREED, ROBERT L., CLAY MINERALS AS INDICATORS FOR DEPOSITIONAL
B 6405 709 FREED, ROBERT L., SHALE MINERALOGY OF GENERAL CRUDE OIL AND
B 6303 453 FREELAND, GEORGE L., ENVIRONMENTAL STUDIES AT NOAAS ATLANTIC
B 6105 786 FREELAND, G. J., NOAAS WASTE-DISPOSAL STUDIES IN NEW YORK BIGHT
B 6205 884 FREEMAN, KEVIN J., TERTIARY VOLCANIC STRATIGRAPHY OF TRAP
B 6209 1762 FREEMAN, ROBERT E., LOBO TREND OF SOUTH LAREDO AREA, WEBB AND
B 6405 695 FREEMAN, ROBERT W., ADVANCES IN RADON EXPLORATION
B 6203 514 FREEMAN, TOM, ONCOLITES FROM LACUSTRINE SEDIMENTS IN
B 6105 775 FREEMAN, TOM, PETROLEUM AND MISSISSIPPI VALLEY TYPE ORE
B 6405 702* FREEPORT COAL, ABST., X-RAY MINERALOGY OF, UPPER
B 6405 788* FREEPORT COAL BED, HOMER CITY, PENNSYLVANIA, ABST., FORMS AND
B 6305 804 FREEPORT OIL CO. 1 SIDNEY FABIAN, MILLS RANCH FIELD, TEXAS
B 6004 692* FREESTONE COUNTY, TEXAS, ABST., EOCENE DEPOSITIONAL
B 6407 1005 FREESTONE FIELD, ARK-LA-TEX AREA
B 6408 1165 FREITES SHALE, VENEZUELA
B 6203 537* FREMONT-SWEETWATER COUNTIES, WYOMING, ABST.,
B 6205 884 FRENCH, DON E., TERTIARY VOLCANIC STRATIGRAPHY OF TRAP SPRING
B 6110 1586 FRENCH GUIANA, 1976, PETROCONSULTANTS S.A., ON DEVELOPMENTS IN
B 6210 2052 FRENCH GUIANA, 1977, JACOBSEN, P., JR., ON DEVELOPMENTS OF
B 6310 1753 FRENCH GUIANA, 1978, JACOBSEN, P., JR., ON DEVELOPMENTS IN
B 6010 1652 FRENCH GUIANA, DEVELOPMENTS, 1975
B 6409 1563* FRENCH ISLAND AREAS, JEFFERSON COUNTY, TEXAS, ABST., GEOLOGY
B 6405 697* FRENCH MASSIF, CENTRAL FRANCE, ABST., USE OF, TRACE, FOSSILS FOR
B 6407 972 FRENCHMAN FORMATION, ALBERTA
B 6305 831* FRENCHMAN FORMATION, MAESTRICHTIAN, OF SOUTHERN
M 26 301† FREQUENCY, AND OTHER ATTRIBUTES TO STRATIGRAPHIC AND
B 6105 762 * FREQUENCY OF TURBIDITE EVENTS IN NORTHERN TACONIC KLIPPE
M 26 303 FREQUENCY, SEISMIC APPLICATION
B 6405 670* FRESH LOOK AT SOME OUACHITA PROBLEMS, ABST.
S 4 379 FRESH WATERS, PRECIPITATION OF LOW-, MAGNESIAN, CALCITE
B 6105 804 * FRESH GROUNDWATER BENEATH CONTINENTAL SHELF: FINDINGS OF
M 27 161 FRESHWATER CEMENTATION
B 6405 719* FRESHWATER CEMENTATION OF HOLOCENE AND JURASSIC
S 11 314 FRESHWATER CREEK, CALIFORNIA, SAMPLE LOCALITIES
B 6203 570* FRESHWATER DIAGENESIS OF UPPER PENNSYLVANIAN CYCLIC
B 6105 771 * FRESHWATER ICHNOCOENOSES FROM TRIASSIC RED BEDS OF EAST
B 6004 684* FRESHWATER, INTERMONTANE DELTA, ABST., STRATIGRAPHY AND
B 6303 413* FRESHWATER LIMESTONES ACTUALLY INTERTIDAL AND SUPRATIDAL
B 6109 1552* FRESHWATER LIMESTONES IN MARINE CARBONATE SUCCESSIONS OF
B 6203 509* FRESHWATER ONCOLITES, ONONDAGA LAKE, NEW YORK, ABST.
M 30 464 FRESHWATER SANDSTONE, INDONESIA, BEKAPAI FIELD
M 26 396 FRESNEL ZONE ILLUSTRATION, MODELING
M 26 11 FRESNEL ZONE, SEISMIC, RESOLUTION
B 6404 528 FREUCHIN LAND, GREENLAND
B 6404 568 FREUND, RAPHAEL, STRAIN MEASUREMENTS IN EASTERN MARGINAL
B 6004 681 FREY, R. W., OPHIOMORPHA--GOOD NEWS AND BAD NEWS, ABST.
B 6203 514 FREY, ROBERT W., ICHNOLOGY AND UNIFORMITARIANISM, ABST.
B 6203 496 FREYTET, PIERRE, NONMARINE STROMATOLITES FROM CENTRAL AND
B 6203 558 FREZON, SHERWOOD E., GEOLOGIC APPRAISAL OF UNDISCOVERED OIL
B 6105 842 FRICK, ALAN P., STRATIGRAPHIC MODELING, ABST., INTERPRETIVE
M 26 26 FRICTION LOSS, LABORATORY, MEASUREMENTS
B 6211 2354 FRIEDBERG, JEFFERY L., USE OF VERTICAL-GRADIENT AEROMAGNETIC
B 6306 987 FRIEDMAN AND JOHN E. SANDERS, BUSCH, DANIEL A., REVIEW OF,
B 6004 672 FRIEDMAN, G. M., 50 YEARS OF SEDIMENTOLOGY--RETROSPECT AN
B 6203 514 FRIEDMAN, G. M., ALGAL TUFA FROM SHORES OF ANCESTRAL DEAD SEA,
B 6303 453 FRIEDMAN, G. M., SUBAQUEOUS GRAVITY-DISPLACEMENT PRODUCTS,
B 6004 717 FRIEDMAN, G. M., THREE DIAGENETIC FEATURES FROM WHITEHALL
B 6004 612 FRIEDMAN, GERALD M., ANTIPATHETIC RELATIONS AMONG ALGAL
B 6405 713 FRIEDMAN, GERALD M., DEPOSITION OF ENEWETAK ATOLL REEF, MIDDLE
B 6303 453 FRIEDMAN, GERALD M., DOLOMITE IS EVAPORITE MINERAL--EVIDENCE
B 6105 786 FRIEDMAN, GERALD M., KINDS OF PORES IN CENOZOIC REEFS OF
B 6405 777 FRIEDMAN, GERALD M., MECHANISMS CONTROLLING POROSITY IN RED
B 6304 676 FRIEDMAN, GERALD M., PATTERNS OF POROSITY AND CEMENT IN OOID
B 6104 511 FRIEDMAN, GERALD M., PATTERNS OF POROSITY AND CEMENT IN OOID
B 6303 458 FRIEDMAN, GERALD M., PLEISTOCENE METEORIC-VADOSE DIAGENESIS OF
S 4 353 FRIEDMAN, GERALD M., PROGRESSIVE DIAGENETIC ELIMINATION OF
S 5 23 FRIEDMAN, GERALD M. SEDIMENTOLOGY AND DEPOSITIONAL
S 4 357 FRIEDMAN, GERALD M., SEQUENCE OF PROGRESSIVE DIAGENESIS IN
B 6309 1478 FRIEDMAN, MELVIN, EFFECTS OF TEXTURE AND COMPOSITION ON
B 6105 787 FRIEDMAN, S. A., COKING-COAL RESERVES OF OKLAHOMA
B 6009 1615* FRIERSON FIELD, LOUISIANA, ABST., INTERPRETATION OF, COTTON
B 6105 769 FRIESEN, PAUL, INFORMATION RETRIVAL SYSTEM FOR WELL DATA IN
B 6311 2003 FRIGG 25-1-1 DISCOVERY WELL, NORTH SEA
B 6311 2002 FRIGG FAN, NORTH SEA
M 30 72 FRIGG FIELD, CROSS SECTIONS
M 30 65 FRIGG FIELD, FRIGG SANDSTONE, TOP OF STRUCTURE
M 30 59 FRIGG FIELD, HISTORY OF DISCOVERY, NORTH SEA
M 30 71 FRIGG FIELD, HYDROCARBON COMPOSITION AND ORIGIN
M 30 60 FRIGG FIELD, LOCATION MAP, NORTH SEA
M 30 67 FRIGG FIELD, PALEOGEOGRAPHY, DANIAN
M 30 67 FRIGG FIELD, PALEOGEOGRAPHY, MONTIAN, THANETIAN
M 30 71 FRIGG FIELD, PALEOGEOGRAPHY, SPARNACIAN
M 30 71 FRIGG FIELD, PALEOGEOGRAPHY, YPRESIAN
M 30 78 FRIGG FIELD, RESERVOIR CHARACTERISTICS
M 30 67 FRIGG FIELD, SEDIMENTATION HISTORY, EOCENE
M 30 67 FRIGG FIELD, SEDIMENTATION HISTORY, PALEOCENE
M 30 67 FRIGG FIELD, SEDIMENTATION HISTORY, TERTIARY
M 30 68 FRIGG FIELD, SEISMIC LINE FLAT SPOT
M 30 63 FRIGG FIELD, STRATIGRAPHIC COLUMN, NORTH SEA
M 30 64 FRIGG FIELD, STRATIGRAPHY
M 30 65 FRIGG FIELD STRUCTURE, DEEP-SEA FAN ORIGIN
M 30 65 FRIGG FIELD, STRUCTURE, NORTH SEA
M 30 64 FRIGG FIELD, TERTIARY, FAUNAL ASSEMBLAGES
B 6408 1247 FRIGG FIELD, UNITED KINGDOM
B 6311 1999† FRIGG FIELD--LARGE SUBMARINE-FAN TRAP IN LOWER EOCENE ROCKS
M 30 59† FRIGG FIELD--LARGE SUBMARINE-FAN TRAP IN LOWER EOCENE ROCKS
B 6203 521* FRIGG FIELD--LARGE SUBMARINE-FAN TRAP IN LOWER EOCENE
M 30 65 FRIGG SANDSTONE, TOP OF STRUCTURE, FRIGG FIELD
B 6311 2002 FRIGG SUBBASIN, NORTH SEA
B 6408 1247 FRIGG SUBBASIN, UNITED KINGDOM

B 6007 1054† FRINGING REEF, GALETA POINT, PANAMA, EVOLUTION OF, MODERN,
B 6002 258† FRINGING REEFS OF CRETACEOUS SHALLOW CARBONATE PLATFORM OF
B 6206 1053 FRIO CLAY, TEXAS
B 6203 494* FRIO FORMATION, BRAZORIA COUNTY, TEXAS, ABST., DELTAIC
B 6009 1606* FRIO FORMATION OF TEXAS GULF COAST--IDEAL VERSUS ACTUAL
B 6009 1612* FRIO FORMATION, OLIGOCENE, SOUTH TEXAS, ABST., LEACHED,
B 6309 1595* FRIO FORMATION, TEXAS GULF COAST, ABST., TESTIN GEOPRESSUREDD,
M 28 173 FRIO FORMATION, TEXAS, OLIGOCENE
M 28 67 FRIO FORMATION, TEXAS, OLIGOCENE
M 28 174 FRIO FORMATION, TEXAS, OLIGOCENE
M 28 154 FRIO FORMATION, TEXAS, OLIGOCENE
M 28 175 FRIO FORMATION, TEXAS, OLIGOCENE
M 28 135 FRIO FORMATION, TEXAS, OLIGOCENE
M 28 20 FRIO FORMATION, TEXAS, OLIGOCENE
M 28 71 FRIO FORMATION, TEXAS, OLIGOCENE
B 6009 1607* FRIO RESERVOIR SANDSTONES, NINE MILE POINT FIELD, ARANSAS
B 6408 1169 FRIO SAND, TEXAS
B 6405 678* FRIO SANDSTONES, SOUTHWEST TEXAS, ABST., CALCIUM BUDGET IN
B 6203 514 FRISCHMANN, PETER S., CARBONATE ENVIRONMENTS OF MIDDLE
B 6405 760 FRISINGER, RUSTY, CROSS-BEDDED SANDSTONE RESERVOIRS—GEOLOGIC
B 6203 515 FRITSCHE, A. EUGENE, INTERPRETATION OF TECTONIC AND
B 6211 2360 FRITSCHE, EUGENE A., EARLY MIOCENE SUBMARINE DUNE FIELDS—
B 6207 1127 FROIDEVAUX, CLAUDE M., TERTIARY TECTONIC HISTORY OF SALAWATI
B 6212 2509* FROM CARIBBEAN TO SOUTH AND NORTH AMERICAN CORDILLERAS,
B 6410 1644 FROME EMBAYMENT, URANIUM, AUSTRALIA
B 6410 1643† FROME REGION, SOUTH AUSTRALIA, DISTRIBUTION AND GENESIS OF
B 6008 1392* FROME, SOUTH AUSTRALIA, ABST., URANIUM IN, TERTIARY CHANNEL
S 11 120 FRONDICULARIA FOLIACEA
B 6410 1614 FRONT RANGE AREA, GENERALIZED STRATIGRAPHY COLUMN
B 6405 751* FRONT RANGES, CANMORE, ALBERTA, ABST., GEOMETRY AND
B 6112 2076 FRONT RANGES, ROCKY MOUNTAINS
B 6304 701* FRONT RANGES, SOUTHERN CANADIAN ROCKY MOUNTAINS, ABST.,
B 6203 561* FRONTIER AREA—ANDERSON MINE, YAVAPAI COUNTY, ARIZONA, ABST.,
B 6205 748 FRONTIER BASINS, WORLD
B 6307 1039 FRONTIER FORMATION, IDAHO-, WYOMING, THRUST-FOLD BELT
B 6101 112† FRONTIER FORMATION IN NORTH-CENTRAL UTAH, AGE OF
B 6108 1393* FRONTIER FORMATION, SPEARHEAD RANCH FIELD, WYOMING, ABST.,
B 6004 723* FRONTIER FORMATION, UPPER CRETACEOUS, NORTHWESTERN BIG
B 6101 113 FRONTIER FORMATION, UTAH
B 6101 114 FRONTIER FORMATION, WYOMING
M 24 111 FRONTIER FORMATION, WYOMING
B 6404 490 FRONTIER FORMATION, WYOMING, BELLE FOURCHE MEMBER
M 28 130 FRONTIER FORMATION, WYOMING, CRETACEOUS
M 28 134 FRONTIER FORMATION, WYOMING, CRETACEOUS
M 28 174 FRONTIER FORMATION, WYOMING, CRETACEOUS
M 28 144 FRONTIER FORMATION, WYOMING, CRETACEOUS
M 28 21 FRONTIER FORMATION, WYOMING, CRETACEOUS
B 6404 490 FRONTIER FORMATION, WYOMING, WALL CREEK MEMBER
B 6004 727* FRONTIER FORMATIONS, WYOMING AND MONTANA, ABST., SUBTIDAL,
B 6008 1402* FRONTIER SANDSTONE, SOUTHEASTERN WYOMING AND WESTERN
B 6305 822* FRONTIER, TENSLEEP, AND MADISON RESERVOIRS IN BIGHORN BASIN,
B 6212 2510 FROSCH, ROBERT A., WHICH WAY IS UP IN SPACE, ABST.
B 6405 719 FROST, S. H., CRETACEOUS SEA LEVEL AND STRATIGRAPHY, EASTERN
S 4 329 FROST, S. H., MIOCENE KINGSHILL SEAWAY--A DYNAMIC CARBONATE
B 6303 525 FROST, S. H., SIGNIFICANCE OF MIDDLE TERTIAR LARGE FORAMINIFERA
S 4 93 FROST, STANLEY H., CENOZOIC REEF SYSTEMS OF
S 4 1 FROST, STANLEY H., EDITED BY, REEFS AND RELATED CARBONATES--
S 4 1 FROST, STANLEY H., INTRODUCTION, REEFS AND RELATED CARBONATES--
B 6209 1757 FROST, STANLEY H., OLIGOCENE REEF COMMUNITY SUCCESSION, DAMON
S 4 1 FROST, STANLEY H., PREFACE, REEFS AND RELATED CARBONATES--
B 6203 403 FROST, STANLEY H., STRATIGRAPHY AND DEPOSITIONAL SETTING,
B 6303 521 FROST, STANLEY H., TROPICAL CENOZOIC PALEO-OCEANOGRAPHY AND
B 6405 709 FROST, STANLEY H., UNIFORMITARIANISM AND TERTIARY REEF
B 6104 483† FROZEN-IN HYDROCARBON ACCUMULATIONS O DIAGENETIC
B 6112 2086 FRUITLAND FORMATION, KEROGEN, NEW MEXICO
B 6211 2245 FRUITVALE FAN, CALIFORNIA
M 26 72 FRUITVALE FORMATION, CALIFORNIA, SAN JOAQUIN BASIN
B 6004 665 FRUTH, LESTER S., JR., EXPERIMENTAL DEVELOPMENT OF
B 6105 781 FRUTH, LESTER S., JR., FABRIC-SELECTIVE PERMOPOROSITY
M 25 559 FRUTOS, J., JOSE, METALLOGENIC BELTS AND TECTONIC EVOLUTION OF
B 6405 675 FRY, S. A., DEVONIAN REEFS EXPOSED ALONG CENTRAL CANTABRIC
B 6405 709 FRYBERGER, STEVEN G., EOLIAN-FLUVIATILE, CONTINENTAL, ORIGIN OF
B 6105 829 FRYE, D., COLOR DISPLAYS OF COMPLEX SEISMIC TRACE, ABST.,
B 6012 2181 FRYE, DON W., APPLICATIONS OF SURFACE-CONSISTENT CORRECTIONS,
B 6409 1355 FT. BEND COUNTY, TEXAS, TAVENER FIELD
B 6410 1614 FT. COLLINS FIELD, COLORADO
B 6112 2135 FT. SASKATCHEWAN FIELD, ALBERTA
M 24 82 FT. VERMILLION FORMATION, ALBERTA
B 6105 787 FUCHS, ROBERT L., RESERVE INFORMATION—INVALUABLE TOOL OF
B 6305 843* FUEGO, GUATEMALA, ABST., SEDIMENTOLOGY OF, VOLCANICLASTIC
M 25 524 FUEL COST, TABLE, MEXICO
B 6303 409* FUEL FORECASTS, BY TITLE ONLY, SYNTHETIC
B 6105 826* FUEL-RESERVE INFORMATION, ABST., ELECTRIC UTILITIES NEED FOR
B 6207 1217* FUELS FROM MAGMA—POTENTIAL ENERGY RESOURCE, ABST.
B 6010 1871 FUJAIRAH, 1975, RESERVE OIL AND GAS CO., ON DEVELOPMENTS IN
B 6110 1800 FUJAIRAH, 1976, RESERVE OIL AND GAS CO., ON DEVELOPMENTS IN
B 6210 2117 FUJAIRAH, 1977, RESERVE OIL AND GAS CO., ON DEVELOPMENTS IN
B 6310 1888 FUJAIRAH, 1978, RESERVE OIL, INC., ON DEVELOPMENTS IN
B 6411 1840 FUJAIRAH, 1979, NORDIC PETROLEUMS, INC., ON DEVELOPMENTS IN
B 6303 538 FUJII, N., MIDDLE CRETACEOUS OXYGEN-DEFICIENT
M 25 441 FUKIEN-REINAN MASSIF, SOUTH KOREA
M 25 261 FUKIEN-REINAN MASSIF, SOUTH, KOREA
M 25 269 FUKUE BASIN, KOREA
B 6304 689 FULLER, G. E., THERMOLUMINESCENCE OF URANIUM HOST ROCKS IN
B 6105 815 FULLER, PAUL T., CLAY MINERALOGY OF EASTERN GULF OF ALASKA,
B 6002 280 FULTON LOBE, PENNSYLVANIA
M 25 454 FUNATSU GRANITE, JAPAN
S 6 150 FUNATSU GRANITE, JAPAN
S 3 10 FUNDAMENTAL SOURCES OF URANIUM AND THORIUM
B 6207 1230* FUNDAMENTAL STRUCTURAL PATTERN IN DISPOSITION OF MINERAL
B 6101 80 FUNDY BASIN, NOVA SCOTIA
B 6303 454 FUNK, JAMES M., DISTRIBUTION OF CARBONATE CEMENTS IN
B 6405 710 FUNKHOUSER, L. W., DEEP TUSCALOOSA GAS TREND OF SOUTH
B 6109 1476 FURADO FIELD, BRAZIL
B 6105 760 FURBUSH, MALCOLM A., GEOLOGIC AND OPERATIONAL SUMMARY OF COST
B 6203 541* FURNACE CREEK FORMATION, DEATH VALLEY, CALIFORNIA, ABST.,
S 11 247 FURSENKOINA
B 6109 1452 FUSSELMAN DOLOMITE, PEDREGOSA BASIN
B 6303 454 FUSSO, JOSEPH W., JR., COMPUTER-ASSISTED STRUCTURAL ANALYSIS,
S 6 309 FUSULINACEA, PERMIAN, KEY FOSSIL GROUPS
B 6004 715* FUSULINACEAN PROVINCES OF NORTH AMERICAN CORDILLERA, ABST.
M 27 32 FUSULINID FORAMS
B 6303 530* FUSULINID LIMESTONE ABST., COMPACTION EFFECTS IN
B 6405 684* FUTURE EXPLORATION FOR GEOTHERMAL RESOURCES, ABST.
B 6303 526* FUTURE EXPLORATION OPPORTUNITIES IN GULF OF MEXICO, ABST.
B 6002 319* FUTURE GAS SUPPLIES, ABST., GAS, RESERVES, GAS RESOURCE BASE, AND
B 6207 1220* FUTURE GEOLOGIC REMOTE SENSING FROM SPACE—POTENTIAL AND
B 6203 543* FUTURE NATURAL GAS SUPPLY, ABST.
B 6205 770 FUTURE OF SEISMIC FACIES INTERPRETATION
B 6303 503* FUTURE OF URANIUM, ABST.
B 6209 1753* FUTURE OIL POTENTIAL OF LOWER CRETACEOUS SUNNILAND
B 6409 1564* FUTURE PETROLEUM PROVINCES OF GULF OF MEXICO REGION, ABST.
B 6105 767 * FUTURE PETROLEUM RESOURCES IN SAUDI ARABIA
B 6207 1233* FUTURE POTENTIAL FOR DEVELOPMENT OF GEOTHERMAL ENERGY IN
B 6405 762* FUTURE TRENDS IN COAL RESOURCE AND RESERVE EVALUATIONS IN
B 6305 826 FUZESY, LASZLO M., GEOLOGY OF PALEOZOIC STRATA IN WEST-CENTRAL
B 6008 1398 FUZESY, LAZSLO M., GEOLOGY AND HYDROCARBON POTENTIAL OF
B 6203 436 FYN-RINGKOBING HIGH, NORTH SEA
B 6203 426 FYN-RINGKSBING HIGH, EUROPE
M 24 209 G, J, K, O, Q SANDSTONE, LOUISIANA
M 24 209 G SANDSTONE, LOUISIANA
M 24 212 G THROUGH K SANDSTONES, LOUISIANA
B 6204 630 G ZONE, WYOMING, TENSLEEP
B 6004 711* GA EXPLORATION ASSUMPTIONS ON PROJECTIONS OF LONG-RANGE
B 6305 847* GA RESERVES--HARD TO FIND OR EASY TO MISS, ABST., PROFITABLE
M 30 275 GA SANDSTONE, NET OIL ISOPACH, EUGENE ISLAND 330 FIELD
M 30 264 GA SANDSTONE, RESERVOIR, EUGENE ISLAND 330 FIELD
B 6101 58 GAARA-ABD EL AZIZ-MARDIN SWELL, LEVANT, JAUF-
B 6011 1993 GABELMAN, JOHN W., EXPECTATIONS FROM URANIUM EXPLORATION
S 3 1 GABELMAN, JOHN W., MIGRATION OF URANIUM AND THORIUM--
S 3 1 GABELMAN, JOHN W., PREFACE, MIGRATION OF URANIUM AND THORIUM--
B 6203 515 GABELMAN, JOHN W., TEST USE OF ERTS-LANDSAT IMAGERY IN
B 6303 534* GABES BASIN, TUNISIA, ABST., GEOLOGY OF
B 6211 2217 GABILAN MESA, CALIFORNIA
B 6211 2217 GABILAN RANGE, CALIFORNIA
B 6303 546* GABON, ABST., GRONDIN OIL FIELD
M 30 588 GABON BASIN, BATANGA FORMATION, EWONGUE SHALE
M 30 588 GABON BASIN, BATANGA FORMATION, POINTE CLAIRETTE SHALES
M 30 579 GABON BASIN, BATANGA SANDSTONE, PETROLEUM
M 30 580 GABON BASIN, GRONDIN FIELD, CROSS SECTION
M 30 577 GABON BASIN, GRONDIN FIELD, GABON
M 30 579 GABON BASIN, POST-SALT STRATIGRAPHY
B 6107 975 GABON, COCOBEACH SERIES
B 6110 1748 GABON, DEVELOPMENTS, 1976
B 6210 1847 GABON, DEVELOPMENTS, 1977
B 6310 1691 GABON, DEVELOPMENTS, 1978
B 6411 1787 GABON, DEVELOPMENTS, 1979
B 6107 975 GABON EVAPORITE BASIN
B 6202 226 GABON, EVAPORITES
M 30 577 GABON, GABON BASIN, GRONDIN FIELD
B 6107 975 GABON, GAMBA FORMATION

Index

M 30 577 GABON, GRONDIN FIELD, AFRICA
M 30 589 GABON, GRONDIN FIELD, DEVELOPMENT HISTORY
M 30 578 GABON, GRONDIN FIELD, LOCATION
B 6107 975 GABON, MADIELA FORMATION
B 6311 2098 GABON, OFFSHORE, SEISMOGRAM
B 6110 1757 GABON, PRODUCTION, 1975-1976
B 6210 1859 GABON, PRODUCTION, 1976-1977
B 6310 1701 GABON, PRODUCTION, 1978
B 6411 1815 GABON, PRODUCTION, 1979
B 6202 229 GABON, SEDIMENTARY SEQUENCE
B 6002 211 GABON TO ANGOLA, WEST AFRICAN DIAPIR ZONE
B 6107 961 GABON-ANGOLA BASIN, EVOLUTION OF
B 6107 970 GABON-ANGOLA SEGMENT, CAMEROON-
M 25 492 GABRIELSE, H., ENVIRONMENTS OF CANADIAN CORDILLERA
B 6303 454 GABRYSCH, R. K., LAND-SURFACE SUBSIDENCE IN HOUSTON-GALVESTON
S 6 205 GABUNIA, L., ON DATING OF THE PALEOGENE
B 6206 985 GACHSARAN FIELD, IRAN
B 6206 993 GACHSARAN FORMATION, IRAN
B 6412 2098 GACHSARAN FORMATION, IRAN
B 6102 295 GAGE, JOHN E., SEDIMENTOLOGY OF CRETACEOUS MESA RICA
M 30 39 GAGE, MICHAEL, A REVIEW OF THE VIKING GAS FIELD
B 6405 705 GAGLIANONE, PAULO CESAR, SERGIPE-ALAGOAS BASIN, BRAZIL—SOURCE
B 6405 710 GAGOSIAN, ROBERT B., EARLY TRANSFORMATION REACTION OF
B 6209 1646 GAGUA EXTINCT SPREADING AXIS
B 6209 1646 GAGUA RIDGE
B 6203 399 GAIKEMA SANDSTONE, ALASKA
B 6303 455 GAINES, ALAN, M., DOLOMITIZATION--RECENT EXPERIMENTAL
S 4 249 GAINES BEND MOUND, TEXAS
B 6104 560 GAJARDO CHANNEL, CHILE
M 25 539 GALAPAGOS RISE
S 4 289 GALAXEA CLAVUS DANA, HERMATYPIC, SCLERACTINIAN, CORAL
S 4 292 GALAXEA, DISTRIBUTION
S 4 291 GALAXEA, FIJI, VITI LEVU
S 4 291 GALAXEA, GUAM
S 4 291 GALAXEA, HERON ISLAND
S 4 291 GALAXEA, HOOK ISLAND
S 4 291 GALAXEA, LADY MUSGRAVE REEF
S 4 291 GALAXEA, MARIANA ISLANDS, SAIPAN
S 4 291 GALAXEA, NEW BRITAIN, RABAUL
S 4 291 GALAXEA, NEW CALEDONIA, NOUMEA
S 4 291 GALAXEA, PALEOCLIMATIC INFORMATION OBTAINED FROM FOSSIL
S 4 291 GALAXEA, PAPUA, PORT MORESBY
S 4 291 GALAXEA, SAMOA, PAGO PAGO
B 6403 428 GALEKI FIELD, INDIA
B 6405 697* GALENA GROUP, UPPER MISSISSIPPI VALLEY, ABST., SUBMARINE
B 6211 2329 GALESVILLE QUARTZ ARENITE, WISCONSIN
B 6211 2331 GALESVILLE SANDSTONE, WISCONSIN
B 6007 1061 GALETA CORES, PANAMA
B 6007 1055 GALETA POINT, DRILL HOLES, PANAMA
B 6007 1054† GALETA POINT, PANAMA, EVOLUTION OF, MODERN, CARIBBEAN,
B 6007 1057 GALETA POINT, REEF ACROPORA PALMATA, PANAMA
B 6007 1056 GALETA POINT, REEF ASSEMBLAGES, PANAMA
B 6007 1055 GALETA REEF, ACANTHOPHORA ZONE, PANAMA
B 6007 1061 GALETA REEF, ACCUMULATION RATES, PANAMA
B 6007 1066 GALETA REEF, CEMENTS, PANAMA
B 6007 1055 GALETA REEF, LAURENCIA ZONE, PANAMA
B 6007 1069 GALETA REEF, OTHER COMPONENTS, PANAMA
B 6007 1062 GALETA REEF, RADIOCARBON-DATED SAMPLES, PANAMA
B 6007 1059 GALETA REEF, SECTIONS, PANAMA
B 6007 1055 GALETA REEF, THALASSIA ZONE, PANAMA
B 6007 1055 GALETA REEF, ZOANTHUS ZONE, PANAMA
B 6405 710 GALLAGHER, J. J., BASEMENT TECTONICS OF CHINA—
B 6405 727 GALLAGHER, J. J., BASINS OF EAST INDIA—TECTONO-STRATIGRAPHIC
M 27 118 GALLATIN FORMATION, WYOMING, CAMBRIAN
B 6012 2187* GALLAWAY FORMATIONS AT POINT ARENA, CALIFORNIA, ABST.,
B 6112 2139 GALLEGOS FIELD, NEW MEXICO
S 11 75 GALLIHERINA DELREYENSIS
S 11 140 GALLIHERINA DELREYENSIS
S 11 75 GALLIHERINA UVIGERINAFORMIS
S 11 75 GALLIHERINA UVIGERINAFORMIS DOANEI
S 11 140 GALLIHERINA UVIGERINAFORMIS DOANEI
S 11 140 GALLIHERINA UVIGERINAFORMIS UVIGERINAFORMIS
S 11 76 GALLIHERINA UVIGERINAFORMIS UVIGERINAFORMIS
S 11 76 GALLIHERINA UVIGERINAFORMIS WARRENI
S 11 140 GALLIHERINA UVIGERINAFORMIS WARRENI
B 6008 1393 GALLOWAY, MICHAEL J., GEOLOGIC CONTROLS OVER HOT WATER
B 6109 1541 GALLOWAY, W. E., CONSOLIDATION HISTORY OF GULF COAST LOWER
B 6304 689 GALLOWAY, W. E., DEPOSITION AND EARLY HYDROLOGIC EVOLUTION OF
B 6309 1606 GALLOWAY, W. E., IMPORTANCE OF SECONDARY LEACHED POROSITY IN
B 6303 511 GALLOWAY, W. E., PALO DURO BASIN ANALYSIS, ABST.
B 6303 488 GALLOWAY, W. E., RESERVOIR QUALITY IN TERTIARY SANDSTONES
B 6109 1536 GALLOWAY, W. E., URANIUM HOST DEPOSITIONAL SYSTEMS OF TEXAS
B 6108 1378 GALLOWAY, WILLIAM E., DIAGENETIC CONTROL OF RESERVOIR QUALITY
B 6405 798 GALLOWAY, WILLIAM E., EVALUATION OF URANIUM POTENTIAL OF
B 6405 711 GALLOWAY, WILLIAM E., HYDROSTRATIGRAPHIC FRAMEWORK AND
B 6303 455 GALLOWAY, WILLIAM E., OAKVILLE FORMATION OF TEXAS COASTAL
B 6405 722 GALLOWAY, WILLIAM E., OXIDATION-REDUCTION IN OAKVILLE
B 6108 1379 GALLOWAY, WILLIAM E., RELATIONS AMONG DEPOSITIONAL-SYSTEM,
M 26 439 GALLOWAY, WILLIAM E., SEISMIC STRATIGRAPHIC MODEL OF
B 6109 1437 GALLOWAY, WILLIAM E., SEISMIC STRATIGRAPHIC MODEL OF
B 6405 689* GALLUP COAL FIELD, NEW MEXICO, ABST., VARIATIONS I, CRETACEOUS
B 6209 1674 GALLUP DISTRICT, URANIUM, NEW MEXICO
B 6112 2139 GALLUP SANDSTONE, NEW MEXICO
B 6109 1417 GALLUP SANDSTONES AND SHALES, NEW MEXICO
B 6208 1382 GALVESTON AREA, TEXAS, SHIPWRECK FIELD
B 6208 1382 GALVESTON COUNTY, TEXAS, BAFFLE POINT FIELD
B 6008 1256 GALVESTON COUNTY, TEXAS, CITY DIKE FIELD
B 6108 1225 GALVESTON COUNTY, TEXAS, DOLLAR POINT FIELD
B 6108 1225 GALVESTON COUNTY, TEXAS, HALF MOON SHOAL FIELD
B 6208 1383 GALVESTON COUNTY, TEXAS, HANNAH ISLAND FIELD
B 6108 1225 GALVESTON COUNTY, TEXAS, NORTH TEXAS CITY DIKE FIELD
B 6308 1225 GALVESTON COUNTY, TEXAS, PELICAN ISLAND FIELD
B 6303 454* GALVESTON REGION, TEXAS, ABST., LAND-SURFACE, SUBSIDENCE IN,
B 6109 1539* GALVESTON, TEXAS, ABST., GEOMETRY OF, GULF COAST, AQUIFERS,
M 25 522 GALVEZ, L., URANIUM AND ITS PROJECTED USE IN NUCLEAR
B 6109 1470 GAMARRA MORALES, R., PETROLEUM GEOLOGY OF EASTERN BRAZILIAN
B 6004 656 GAMARRA MORALES, R., PETROLEUM GEOLOGY OF EASTERN BRAZILIAN
M 30 329 GAMARRA MORALES, ROBERTO, THE NAMORADO OIL FIELD--A MAJOR
B 6107 975 GAMBA FORMATION, GABON
B 6009 1550 GAMBHIR, S. C., PATTERN OF ACCUMULATION AND COMPOSITION OF OI
B 6212 2489 GAMBHIR, S. C., REVERSAL OF BASEMENT-BLOCK MOTIONS IN CAMBAY
B 6110 1748 GAMBIA, DEVELOPMENTS, 1976
B 6210 1847 GAMBIA, DEVELOPMENTS, 1977
B 6310 1692 GAMBIA, DEVELOPMENTS, 1978
B 6411 1788 GAMBIA, DEVELOPMENTS, 1979
B 6004 690 GAMBOA, L., EVOLUTION OF MARGINAL PLATEAU--DSDP DATA FROM
B 6306 906 GAMMA LOG MOTIFS OF DEEP-SEA SANDS
B 6002 186 GAMMA-LOG PATTERNS IN CLASTIC SEDIMENTARY ROCKS, FOUR
B 6209 1752* GAMMA-RAY LOG, AND ITS GEOLOGIC IMPLICATIONS, ABST.,
B 6002 185 GAMMA-RAY LOGS, NORTH SEA
B 6303 449* GAMMA-RAY SPECTRAL LOGGING DATA ASSIST I GEOLOGIC STUDIES,
B 6203 556* GAMMAS, ABST., BRIGHT SPOTS, MILLIGALS, AND
B 6105 788 GAMMISCH, R. A., WAVE-CLIMATE STUDIES IN BALTIMORE CANYON
B 6202 185 GAMMON FERRUGINOUS SHALE, WYOMING
B 6407 976 GAMMON MEMBER, NORTH DAKOTA
B 6407 971 GAMMON SHALE, GREAT PLAINS
B 6111 1979 GANDER, CRAIG, LOS ANGELES EROSION SURFACE OF MIDDLE
B 6002 281 GANDER RUN SHALE MEMBER, PENNSYLVANIA
M 25 480 GANESHIN, G. S., STRUCTURAL CLASSIFICATION OF NORTH PACIFIC
B 6206 1056 GANGUE MINERALS
B 6405 640 GANNETT GROUP, WYOMING
M 24 115 GANNETT GROUP, WYOMING
B 6104 529 GARCETAS RIVER VALLEY, GULF OF MEXICO
B 6203 515 GARCIA, ROBERTO, DEPOSITIONAL SYSTEMS AND THEIR RELATION TO
B 6002 165 GARDEN BASIN, BERING SHELF
B 6002 164 GARDEN RIDGE, BERINGIAN SHELF
B 6008 1411* GARDINER AREA, MONTANA, ABST., PROPOSED, GEOTHERMAL
B 6105 834* GARDINER AREA, MONTANA—STRUCTURAL AND LITHOLOGIC
B 6203 515 GARDINER, PIERS R., REGRESSIVE-DELTA MODEL WITH EXAMPLE FROM
B 6405 711 GARDNER, G. H. F., NEW DIMENSIONS IN SEISMIC EXPLORATION, ABST.
B 6303 455 GARDNER, J. V., A NEW LOOK AT GEOLOGY OF TOGO BY SLAR, ABST.
B 6405 711 GARDNER, J. V., ET AL, HYDRAULIC PISTON CORING IN EQUATORIAL
B 6406 945 GARDNER, J. V., PETROGRAPHY AND GEOLOGIC SIGNIFICANCE OF UPPER
M 30 540 GARET ECH CHOUF FIELD, ALGERIA, OUARGLA REGION
B 6303 324 GARFUNKEL, ZVI, SUBMARINE SLUMPING IN CONTINENTAL MARGIN OF
B 6406 858 GARGAF ARCH, AFRICA
B 6211 2224 GARLOCK FAULT, CALIFORNIA
B 6203 345 GARLOCK FAULT, CALIFORNIA
B 6211 2270 GARLOCK FAULT, CALIFORNIA
B 6109 1539 GARNER, L. E., LAND RESOURCES OF TEXAS, ABST.
M 28 55 GARNET, DETRITAL, (ALMANDITE)
M 28 58 GARNET, ETCHING
M 28 57 GARNET, EXAMPLE
M 28 55 GARNET, GROSSULARITE, CRYSTAL
M 28 45 GARNET SCHIST, CONNECTICUT, ORDOVICIAN
M 28 46 GARNET SCHIST, CONNECTICUT, ORDOVICIAN
B 6203 551* GARNET SYSTEM FOR EVALUATION OF MINERAL AND ENERGY
M 25 479 GARNETS, USSR
S 10 200 GARNIARIK TEPE MUD VOLCANO, IRAN, CASPIAN AREA
B 6109 1476 GAROUPA FIELD, BRAZIL
B 6303 344 GARRARD SILTSTONE, KENTUCKY
B 6004 673 GARRETT, PETER, BERMUDAS NORTHWEST REEFS COLONIZE LAST

B 6303 457 GARRETT, PETER, MIXING ZONE DOLOMITE IN TIDAL-FLAT SEDIMENTS
B 6303 455 GARRETT, PETER,, PROBING BERMUDAS LAGOONS AND REEFS, ABST.
B 6303 512 GARRISON, LOUIS E., ACTIVE SUBMARINE LANDSLIDE PROCESSES,
B 6303 456 GARRISON, R. E., PETROLOGY, SEDIMENTOLOGY, AND PETROLEUM
B 6203 563 GARRISON, R. E., REGIONAL VARIATION OF DEPOSITIONAL
B 6207 1222 GARRISON, ROBERT E., ORIGIN, DISTRIBUTION, AND DIAGENETIC
S 7 243 GARTNER, STEFAN, LATE PLEISTOCENE GEOLOGIC HISTORY OF TEXAS
B 6009 1562 **GARTNER, STEFAN, NANNOFOSSIL BIOSTRATIGRAPHY AND CLIMATIC**
B 6309 1587 GARY, C., IMPLICATIONS OF GASEOUS HYDROCARBON GEOCHEMISTRY
B 6309 1583 GARY, C., PETROLEUM GEOCHEMISTRY AND GEOLOGY OF SOUTHEAST
S 10 89† GAS MIGRATION: CHEMICAL AND PHYSICAL CONSTRAINTS, OIL AND
S 10 217† GAS TRAPS, DESIGN AND FUNCTION OF, OIL AND
B 6009 1612* GAS, ABST., INDEPENDENT AND MAJOR ROLES IN EXPLORATION FOR, OIL
B 6303 428* GAS, ABST., ORGANIC ACID ANIONS IN OIL-FIELD, WATERS AND, ORIGIN
B 6303 487* GAS ACCUMULATION, GULF OF MEXICO, ABST., SEISMIC STRATIGRAPHIC,
B 6203 515* GAS ACCUMULATION IN SACRAMENTO VALLEY, CALIFORNIA, ABST.,
B 6403 316† GAS ACCUMULATIONS, NORTH-CENTRAL MONTANA, COASTAL AND,
B 6303 515* **GAS ACCUMULATIONS, WESTERN GULF OF MEXICO, ABST., RELATION OF**
B 6004 714* GAS ACCUMULATIONS--THEIR ORIGIN AND SIGNIFICANCE FOR FUTURE
B 6212 2441† GAS, ALIPHATIC, ACID, ANIONS IN, OIL-FIELD, WATERS—IMPLICATIONS
M 26 482 GAS CAP, SEISMIC, MODELING, WATER SANDSTONE
B 6201 108 GAS CHROMATOGRAPHIC ANALYSES, MEADE PEAK MEMBER
B 6201 108 GAS CHROMATOGRAPHIC ANALYSES, RETORT MEMBER
B 6409 1529† GAS COMMITTEE, POTENTIAL RESOURCES OF, NATURAL, GAS IN, UNITED
B 6008 1392* GAS COMPOSITION I BLANCO FIELD, SAN JUAN BASIN, RELATED TO
M 30 41 GAS COMPOSITION, NORTH SEA, VIKING FIELD
B 6002 290 GAS COMPOSITION OF RECENT SEDIMENTS
B 6305 730 GAS CONDENSATE, RESERVOIR DENSITY OF
B 6207 1207* GAS DEPOSITS IN GULF OF THAILAND, ABST., NATURAL
B 6008 1204† GAS DEVELOPMENT IN NORTHERN ROCKIES IN 1975, OIL AND
B 6308 1301† GAS DEVELOPMENTS I NORTH MID-CONTINENT IN 1978, OIL AND
B 6308 1244† GAS DEVELOPMENTS IN MARYLAND, OHIO, PENNSYLVANIA, VIRGINIA,
B 6008 1288† GAS DEVELOPMENTS IN MARYLAND, OHIO, PENNSYLVANIA, VIRGINIA,
B 6208 1399† GAS DEVELOPMENTS IN MARYLAND, OHIO, PENNSYLVANIA, VIRGINIA,
B 6409 1403† GAS DEVELOPMENTS IN MARYLAND, OHIO, PENNSYLVANIA, VIRGINIA,
B 6108 1269† GAS DEVELOPMENTS IN MARYLAND, OHIO, PENNSYLVANIA, VIRGINIA,
B 6008 1332† GAS DEVELOPMENTS IN NORTH MID-CONTINENT IN 1975, OIL AND
B 6108 1305† GAS DEVELOPMENTS IN NORTH MID-CONTINENT IN 1976, OIL AND
B 6208 1477† GAS DEVELOPMENTS IN NORTH MID-CONTINENT IN 1977, OIL AND
B 6308 1301† GAS DEVELOPMENTS IN NORTH MID-CONTINENT IN 1978, OIL AND
B 6409 1437† GAS DEVELOPMENTS IN NORTH MID-CONTINENT IN 1979, OIL AND
B 6108 1346† GAS DEVELOPMENTS IN NORTHERN ROCKIES IN 1976, OIL AND
B 6208 1498† GAS DEVELOPMENTS IN NORTHERN ROCKIES IN 1977, OIL AND
B 6303 520* GAS DISCOVERIES IN KANGAN AREA, IRAN, ABST., RECENT HUGE
B 6303 513* GAS DISCOVERY FUNCTIONS TO RESOURCE ASSESSMENT, ABST.,
B 6203 555* GAS DISSOLVED IN DEEP, HOT GEOTHERMAL WATERS OF PETROLEUM
B 6203 529* GAS DISSOLVED IN GEOTHERMAL WATERS IN SOUTH LOUISIANA WELL,
B 6102 294† GAS EXPLORATION AND DEVELOPMENT IN OIL-POTASH AREA OF
B 6009 1463† GAS EXPLORATION OF CONTINENTAL MARGINS, PLATE TECTONICS IN,
B 6207 1217* GAS EXPORTER, ABST., AUSTRALIA AS, NATURAL
M 30 39† GAS FIELD, A REVIEW OF THE, VIKING
B 6004 716* GAS FIELD DATA BANK PROJECT, UNITED STATES AND CANADA, ABST.,
B 6205 749 GAS FIELD, DEFINITION, GIANT
B 6303 488* GAS FIELD, ELLENBURGER FORMATION, PUCKETT FIELD, PECOS COUNTY,
B 6303 536* GAS FIELD IN LOWER WILCOX, LAVACA COUNTY, TEXAS, ABST., SOUTH
M 30 12 GAS FIELD, LANDSAT IMAGE, IRAN, KANGAN
B 6303 502* GAS FIELDS, ABST., RELATIONS OF GAS OCCURRENCE TO GEOLOGIC
M 30 315† GAS FIELDS, ALBERTA, CANADA, STRACHAN AND, RICINUS WEST
B 6401 8† GAS FIELDS, GEOLOGIC SIGNIFICANCE OF, LANDSAT DATA FOR 15,
M 30 7† GAS FIELDS, GEOLOGIC SIGNIFICANCE OF, LANDSAT DATA FOR 15, GIANT
B 6109 1543* GAS FIELDS IN GULF COAST, ABST., TOPOGRAPHIC EXPRESSION OF, OIL
B 6408 1210† GAS FIELDS IN HIGHLY EXPLORED, MATURE, AREAS, METHODS USED,
B 6302 153 GAS FIELDS, NORTH AMERICA
M 24 1 GAS FIELDS, NORTH AMERICN OIL AND
M 30 1 GAS FIELDS OF THE DECADE 1968-1978, OIL AND
M 30 1† GAS FIELDS OF THE DECADE 1968-1978, GIANT, OIL AND
B 6002 292 GAS FIELDS OF THE USSR
B 6309 1586* GAS FROM EASTERN SHALES, ABST., CONSTRAINTS AND PROBLEMS IN
B 6012 2179* GAS FROM MATURE PROVINCES, ABST., NEW, OIL AND
B 6004 658* GAS FROM UNDERGROUND NATURAL GAS STORAGE RESERVOIRS BY
B 6112 2082† GAS GENERATION BY LABORATORY PYROLYSIS OF KEROGEN, OIL AND
B 6401 108 GAS GENERATION FROM RECENT SEDIMENTS, GULF OF MEXICO SHELF
B 6305 832* GAS GIANT IN WYOMING THRUST BELT, ABST., WHITNEY CANYO
B 6405 730* GAS GIANT IN WYOMING THRUST BELT, ABST., WHITNEY CANYON
B 6312 2204† GAS HYDRATE HORIZONS ON CONTINENTAL SLOPES AND RISES, SEISMIC
B 6412 2166† GAS IN APPALACHIAN OVERTHRUST BELT, CROSS-STRIKE STRUCTURAL,
B 6402 209† GAS IN BOMBAY OFFSHORE BASIN, INDIA, SOURCE AREAS AND,
B 6303 421* GAS IN CALIFORNIA, ABST., INFLUENCE OF RATE ON DEMAND AND USE
B 6305 841* GAS IN CHINA, BY TITLE ONLY, OIL AND
B 6408 1131† GAS IN COOK INLET BASIN, ALASKA, BIOGENIC AND, THERMOGENIC
B 6306 886† GAS IN DELTA-MARGIN FACIES OF DAKOTA SANDSTONE, LONE PINE
B 6309 1576* GAS IN EASTERN DEVONIAN SHALES, ABST., SUGGESTED ORIGIN FOR
B 6405 675* GAS IN FISHERS PEA AREA, LAS ANIMAS COUNTY, COLORADO, ABST.,
B 6303 434* GAS IN ILLINOIS BASIN, ABST., ISOTOPE, GEOCHEMISTRY OF, NATURA
B 6205 887* GAS IN NORTHWEST COLORADO, ABST., STRATIGRAPHY OF, MESAVERDE
B 6303 524* GAS IN PEOPLES REPUBLIC OF CHINA ABST., OIL AND
B 6401 107 GAS IN PORES OF SOURCE ROCKS, GENERATION OF
B 6211 2359* GAS IN PYRAMID LAKE AREA, NEVADA, ABST., ALTERATION OF,
B 6112 2134† GAS IN RESERVOIRS WITH SUBNORMAL PRESSURES, OIL AND
B 6102 294* GAS IN SOUTHERN COTTLE AND NORTHERN KING COUNTIES, TEXAS,
B 6303 478* GAS IN UNITED STATES--CASE HISTORY OF POTENTIAL GAS COMMITTEE,
B 6409 1529† GAS IN UNITED STATES—CASE HISTORY OF POTENTIAL GAS COMMITTEE,
B 6309 1578* GAS IN WESTERN KENTUCKY, ABST., PAST AND CURRENT
B 6210 2118 GAS LTD., AGIP S.P.A., AND PETROCONSULTANTS S.A., ON DEVELOPMENTS
S 7 259 GAS MIGRATION, SEISMIC, TEXAS, SHELF-SLOPE
B 6305 761† GAS MIGRATION--CHEMICAL AN PHYSICAL CONSTRAINTS, OIL AND
B 6312 2136 GAS, ORGANIC, GEOCHEMISTRY PARAMETERS OF
B 6002 288 GAS PHASE OF HYDROCARBONS, STUDY OF, WHOLE, FREE-
B 6303 552* GAS PLAYS IN FACIES-CYCLE WEDGES, ABST., ASSESSING, OIL AND
B 6408 1158† GAS PLAYS IN FACIES-CYCLE WEDGES, ASSESSING, OIL AND
B 6004 671* GAS POTENTIAL BASED ON PRESENT EXPLORATION, ABST., NORWAYS,
B 6002 318* **GAS POTENTIAL OF WESTERN KANSAS, ABST., OIL AND**
B 6305 829* GAS POTENTIAL OF WYOMING, ABST., OIL AND
B 6004 729* GAS POTENTIALS OF MIDDLE AND UPPER ORDOVICIAN OF TENNESSEE,
B 6305 767 GAS, PRIMARY MIGRATION MECHANISM, NATURAL
B 6405 760* GAS PRODUCTION FROM LOWER-PERMEABILITY RESERVOIRS, ABST.,
B 6102 297* GAS PRODUCTION FROM TIGHT SANDSTONES OF
B 6309 1580* GAS PRODUCTION IN APPLACHIAN STATES, ABST., NEW ECONOMICS OF,
B 6305 847* GAS PRODUCTION, PHOTOLINEAMENTS, AND GEOLOGY IN CENTRAL
B 6401 133 GAS PROSPECTING, BY N. I. MARKOVSKII, MCGOOKEY, D. P., REVIEW OF,
B 6008 1395* GAS RESERVES AT BOWDOIN AREA, MONTANA--LOW-PERMEABILITY,
B 6002 319* GAS RESERVES, GAS RESOURCE BASE, AND PROSPECT FOR FUTURE GAS
B 6303 528* GAS RESERVOIR IN WOODBINE--EAGLE FORD INTERVAL, UPPER
B 6309 1585* GAS RESERVOIRS, ABST., FABRIC-ELEMENT-BASED CLASSIFICATION FOR
B 6109 1534* GAS RESERVOIRS, NORTHEAST THOMPSONVILLE FIELD, JIM HOGG AND
B 6405 794* GAS RESERVOIRS, ROCKY MOUNTAINS, ABST., EVOLUTION AND STRESS
B 6309 1590* GAS RESERVOIRS--RESULTS OF INVESTIGATION IN UPPER DEVONIAN OF
B 6303 540* GAS RESOURCE APPRAISAL--STATE OF THE ART, ABST., OIL AND
B 6309 1577* GAS RESOURCE--DEVONIAN SHALES, ABST., EXPLORATION STRATEGY
B 6004 665* GAS RESOURCES BY U.S. GEOLOGICAL SURVEY, ABST., NE APPROACH TO
B 6409 1529 GAS RESOURCES, CLASSIFICATION OF, NATURAL
B 6203 561* GAS RESOURCES IN GULF OF MEXICO, ABST., GEOLOGIC APPRAISAL OF
B 6205 891* GAS RESOURCES IN SHALLOW, LOW-PERMEABILITY RESERVOIRS OF
B 6004 698* GAS RESOURCES IN UNITED STATES, ABST., PROBABILISTIC AND,
B 6312 2183† GAS RESOURCES, METHODS OF, ESTIMATING, OIL AND
B 6305 842* GAS RESOURCES OF DENVER-JULESBURG BASIN, ABST., NEW LOOK AT,
B 6407 969† GAS RESOURCES, SHALLOW, LOW-PERMEABILITY, RESERVOIRS OF
M 24 242 GAS ROCK PRODUCING SAND, MISSISSIPPI
B 6008 1393* GAS SAND IN NORTHERN MONTANA, ABST., WELL-LOG ANALYSIS IN,
B 6206 1067† GAS SEEP AREA, SOUTH TEXAS OFFSHORE, SERENDIPITY
B 6304 668† GAS SEEPS AND SUBSURFAC STRUCTURE OFF PANAMA CITY, FLORIDA
B 6309 1579* GAS SHALES, ABST., THEORETICAL AND EXPERIMENTAL ANALYSES OF,
B 6309 1577* GAS SHALES--EFFECTS OF TEMPERATURE AND ATMOSPHERE, ABST.,
B 6303 414* GAS STABILITY IN DEEP SUBSURFACE, ABST., NATURAL
B 6004 658* GAS STORAGE RESERVOIRS BY ISOTOPIC ANALYSIS OF METHANE, ABST.,
B 6309 1534† GAS, SUPER OVERPRESSURED
B 6203 543* GAS SUPPLY, ABST., FUTURE, NATURAL
B 6302 152† GAS TRAP, WESTERN CANADA, DEEP BASIN
B 6308 1428* GAS TRAPS, ABST., DESIGN AND FUNCTION OF, OIL AND
B 6203 554* GAS TRAPS, ABST., PHYSICAL AND CHEMICAL SIGNALS AROUND, OIL AND
B 6203 558* GAS TRAPS AS CLUES TO MIGRATION, ABST., DESIGN AND FUNCTION OF,
B 6405 710* GAS TREND OF SOUTH LOUISIANA, ABST., DEEP, TUSCALOOSA
B 6303 459* GAS VERSUS OIL IN FAR EAST AND MIDDLE EAST ABST.
B 6203 564* GAS WELLS, ABST., FOAM, STIMULATION FOR, ENHANCED RECOVERY
B 6111 1973 GAS-ASSOCIATED FEATURES, SIGSBEE SCARP
B 6309 1576* GAS-BEARIN OHIO SHALE ALONG LAKE ERIE--STRATIGRAPHY,
M 25 176† GAS-BEARING BASINS OF CIRCUM-PACIFIC BELT--SUMMARY, OIL- AND
B 6309 1592* GAS-BEARING DEVONIAN BLACK SHALES, ERIE COUNTY PENNSYLVANIA,
B 6303 502* GAS-BEARING, NONMARINE ROCKS IN CENTRAL ROCKY MOUNTAINS,
B 6205 890* GAS-BEARING SANDSTONES, ABST., PERMEABILITY CONTROL IN,
B 6206 1092* GAS-BEARING SHALES, U.S. EASTERN INTERIOR BASIN, ABST., ORGANIC,
B 6109 1432 GAS-CHROMATOGRAPH COLUMNS
M 25 211† GAS-CONDENSATE FIELDS OF NEW ZEALAND--SUMMARY, KAPUNI AND.
B 6105 698 † GAS-HYDRATE HORIZONS DETECTED IN SEISMIC-PROFILER DATA FROM
B 6309 1591* GAS-PRODUCING AREAS FROM PALEOENVIRONMENTAL DATA, ABST.,
B 6309 1583* GAS-PRODUCING CONFIGURATIONS IN TRENTON LIMESTONE,
B 6105 775 * GAS-PRODUCING SUBMARINE-FAN DEPOSITS OF CATSKILL DELTA COMPLEX
B 6309 1581* GAS-RESERVOIR PERFORMANCE, ABST., MODELING OF, DEVONIAN, SHALE
B 6305 827* GAS-RESERVOIR QUALITY IN EAGLE SANDSTONE OF BEARPAW

B 6401 110 GAS-SATURATED WATER, OIL-CARRYING CAPACITY OF
B 6201 41 GASCOYNE ABYSSAL PLAIN, AUSTRALIA
B 6002 214 GASEOUS SOLUTION HYPOTHESIS
B 6303 532* GASES ADSORBED IN NEAR-SURFACE SEDIMENT IDENTIFIED BY CARBON
B 6304 701* GASES, ADSORBED IN NEAR-SURFACE SEDIMENTS IDENTIFIED BY CARBO
B 6002 287† GASES AND THEIR VARIATIONS DEPENDENT ON INITIAL ORGANIC
B 6409 1566* GASES IN OFFSHORE GULF OF MEXICO, ABST., CHEMICAL AND ISOTOPIC
B 6303 522* GASES IN SUBALPINE TERTIARY BASINS, ABST., DUAL ORIGIN OF,
B 6407 1083 GASES IN SUBMARINE SEEPS, ORIGIN OF, HYDROCARBON
B 6405 736* GASES IN UNCONSOLIDATED SEAFLOOR DEPOSITS, NORTHERN
B 6207 1217 GASKIN, A. J., HYDROGEOLOGY OF URANIUM DEPOSITS IN CALCRETES
M 25 229 GASMAN OROGENIC ZONE, AUSTRALIA
B 6407 1078† GASOLINE-RANGE HYDROCARBONS AND THEIR MIGRATION BY
S 10 94 GASOLINE-RANGE HYDROCARBONS, CRUDE OIL
S 10 198 GAS-CUT MUDFLOW, WEST INDIES, TRINIDAD, LABREA FORMATION
M 26 33 GASSMANNS THEORY, LABORATORY, MEASUREMENTS
B 6211 2354 GASTIL, GORDON, MESOZOIC GEOLOGY OF PENINSULAR CALIFORNIA
B 6105 805 * **GASTROPOD RADULAE—THEIR POTENTIAL IN FOSSIL RECORD**
B 6109 1540* GASTROPOD RADULAE—THEIR POTENTIAL IN THE FOSSIL RECORD, ABST.
B 6401 130 GASTROPOD ZONES, INDIANA
M 27 87 GASTROPODS, FRESHWATER
M 27 88 GASTROPODS, HIGH SPIRED
M 27 167 GASTROPODS, VERMATID
B 6207 1224* GAS—ITS ROLE IN WESTERN AUSTRALIAS ENERGY ECONOMY, ABST.,
B 6205 848† GATCHELL, EOCENE, SAND OF CALIFORNIA, PERMEABILITY, TRAPS IN
B 6211 2263 GATCHELL TURBIDITE, CALIFORNIA
B 6409 1557 GATENBY, GLEN M., EXPLORATION RAMIFICATIONS OF SUBSURFACE
M 28 165 GATESBURG FORMATION, PENNSYLVANIA, CAMBRIAN
M 28 2 GATESBURG FORMATION, PENNSYLVANIA, CAMBRIAN
M 28 79 GATESBURG FORMATION, PENNSYLVANIA, CAMBRIAN
M 27 207 GATESBURG FORMATION, PENNSYLVANIA, CAMBRIAN
M 27 199 GATESBURG FORMATION, PENNSYLVANIA, CAMBRIAN
M 28 5 GATESBURG FORMATION, PENNSYLVANIA, CAMBRIAN
M 28 52 GATESBURG FORMATION, PENNSYLVANIA, CAMBRIAN
M 28 116 GATESBURG FORMATION, PENNSYLVANIA, CAMBRIAN
M 28 113 GATESBURG FORMATION, PENNSYLVANIA, CAMBRIAN
B 6203 516 GATEWOOD, LLOYD E., STRATIGRAPHIC TRAP POSSIBILITIES IN
B 6002 325 GATEWOOD, LLOYD, STRATIGRAPHIC TRAP POSSIBILITIES IN ARBUCKLE
B 6007 1058 GATUN FORMATION, PANAMA
B 6303 439 GAUDEMET, L. M., KRIGING OF TOP OF RESERVOIR FROM SEISMIC AND
B 6303 456 GAUDETTE, HENRI E., PROVENANC STUDIES IN TUNISIA BY U-PB AGES OF
S 11 268 GAUDRYINA ATLANTICA
S 11 229 GAUDRYINA ATLANTICA
S 11 268 GAUDRYINA PLIOCENICA
S 11 229 GAUDRYINA PLIOCENICA
M 28 128 GAULT FORMATION, ENGLAND, CRETACEOUS
M 28 127 GAULT FORMATION, ENGLAND, CRETACEOUS
B 6401 133 GAUTHIER-VILLARS AND CENTRE NATIONAL POUR LEXPLOITATION DES
B 6303 456 GAUTIER, DONALD L., AUTHIGENIC TOURMALINE CRYSTALS IN PORE
B 6405 773 GAUTIER, DONALD L., DEVELOPMENT OF BIOGENIC GAS FROM SHALLOW,
B 6405 712 GAUTIER, DONALD L., DIAGENESIS IN SHALLOW CONVENTIONAL AND
B 6405 711 GAUTIER, DONALD L., PHYSICAL CHARACTERISTICS OF SHALLOW
B 6305 827 GAUTIER, DONALD L., POST-DEPOSITIONAL CONTROL OF GAS-RESERVOIR
B 6012 2191* GAVIOTA FORMATIONS, ARROYO EL BULITO, SANTA BARBARA COUNTY,
B 6402 244 GAVROVO ZONE, GREECE
B 6303 325 GAZA DISTURBANCE, ISRAEL AND SINAI
B 6209 1757 GAZZIER, CONRAD A., PLEISTOCENE SHORELINES AND
B 6412 2052 GE-1 WELLS, ATLANTIC OCEAN, COST B-2 AND
B 6203 457 GEARLE SILTSTONE, AUSTRALIA
M 26 460 GEARY AREA, GEOSEISMIC MODELING, OKLAHOMA
M 26 451† GEARY, OKLAHOMA—T13N, R10W, A, CASE HISTORY OF, GEOSEISMIC,
B 6105 788 **GEBELEIN, CONRAD D., HOLOCENE FRESHWATER DIAGENESIS OF MODERN**
B 6303 457 GEBELEIN, CONRAD D., MIXING ZONE DOLOMITE IN TIDAL-FLAT
B 6105 787 **GEBELEIN, CONRAD D., MIXING-ZONE DOLOMITIZATION OF HOLOCENE**
B 6203 516 GEBELEIN, CONRAD D., PREDICTIVE MODELS FOR CARBONATE-FACIES
M 26 36 GEERTSMA EQUATIONS, LABORATORY, MEASUREMENTS, BIOT-
B 6405 712 GEESAMAN, RICHARD C., FACIES AND DEPOSITIONAL TECTONICS OF
B 6105 788 **GEFFEN, TED M., METHODS FOR RECOVERY OF MORE OIL FROM KNOWN**
B 6312 2183 GEHMAN, HARRY M., METHODS OF ESTIMATING OIL AND GAS
B 6304 589 GELA OIL FIELD, SICILY
B 6002 164 GEMUK GROUP, ALASKA
B 6201 73 GENDZWILL, D. J., WINNIPEGOSIS MOUNDS AND PRAIRIE EVAPORITE
B 6405 709* GENERAL CRUDE OIL AND DEPARTMENT OF ENERGY 1 PLEASANT
B 6205 882* GENERAL GEOLOGY OF NORTHERN PORTNEUF RANGE, BANNOCK AND
B 6107 1014 GENERAL PETROLEUM MEMBER, ALBERTA AND, SASKATCHEWAN
B 6009 1618* **GENERAL SEDIMENTOLOGIC HISTORY OF LATE SILURIAN AND EARLY**
B 6303 448* **GENERATION AND OCCLUSION OF POROSITY IN CHALK RESERVOIRS,**
S 10 179† GENERATION AND PETROLEUM MIGRATION, METHANE
B 6408 1279* GENERATION, ABST., OIL EXPULSION—A CONSEQUENCE OF OIL
B 6011 2067* GENERATION AND APPLICATION TO OIL AND GAS EXPLORATION, ABST.,
B 6209 1586 GENERATION AND DESTRUCTION, CORRELATION OF, COAL-RANK SCALE
B 6410 1613† GENERATION AND MIGRATION IN DENVER BASIN, PETROLEUM
B 6304 577 GENERATION AND MIGRATION OF HYDROCARBONS, GEOCHEMICAL
B 6401 59† GENERATION AND MIGRATION, ROLE OF, MINERAL, MATRIX IN,
B 6401 59† GENERATION AND MIGRATION, ROLE OF MINERAL MATRIX IN KEROGEN
B 6303 539* GENERATION AND OCCURRENCE OF HYDROCARBONS RELATED TO
B 6112 2082† GENERATION BY LABORATORY PYROLYSIS OF KEROGEN, OIL AND, GAS
B 6204 687† GENERATION BY THERMAL ALTERATION OF KEROGEN FROM DIFFERENT
B 6201 98† GENERATION IN BLACK SHALE MEMBERS OF PHOSPHORIA FORMATION,
B 6207 1151† GENERATION IN GIPPSLAND BASIN, AUSTRALIA—COMPARISON WITH
B 6305 838* GENERATION IN MISSISSIPPIAN CHAINMA SHALE OF EAST-CENTRAL
B 6405 779* GENERATION, MIGRATION, AND ENTRAPMENT OF PETROLEUM IN
B 6111 1953 GENERATION, MIGRATION, ENTRAPMENT, AND DEGRADATION, OIL
B 6405 753* GENERATION OF ABNORMAL PRESSURES THROUG ORGANIC MATTER
B 6303 435* GENERATION OF DEBRIS FLOWS AND TURBIDITY-CURRENT FLOWS FROM
B 6401 107 GENERATION OF GAS IN PORES OF SOURCE ROCKS
B 6002 287† GENERATION OF HYDROCARBON GASES AND THEIR VARIATIONS
B 6205 818 GENERATION, SOURCE ROCKS AND, HYDROCARBON
B 6112 2095 GENERATION VERSUS H-C RATIO, HYDROCARBON
B 6004 647* GENERATION. ATTEMPT TO RESOLVE PARADOX, ABST., DEEP OR,
B 6209 1704 GENESEO SHALE, UNITED STATES, EASTERN
B 6002 322* GENESIS AND DISTRIBUTION OF DESMOINESIAN, PENNSYLVANIAN,
B 6303 508* GENESIS, OCCURRENCE, AND CAUSES OF SEDIMENT DISTRIBUTION IN
B 6209 1684† GENESIS OF PETROLEUM IN MAHAKAM DELTA, BORNEO, ORGANIC
B 6405 792* GENESIS OF SAND RIDGES ON STORM-DOMINATED SHELVES—STATUS
B 6005 884† GENESIS. DISCUSSION AND REPLY, TIME-, TEMPERATURE RELATION IN,
B 6405 688* GENETIC STRATIGRAPHY AND PROVENANCE OF BACA FORMATION OF
B 6203 526* GENETIC STRATIGRAPHY AND PROVENANCE OF BACA FORMATION OF
B 6303 551 GENSAMER, A. R., PORE SYSTEMS IN JURASSI CARBONATE RESERVOIRS,
M 27 42 GENUNDEWA LIMESTONE, NEW YORK, DEVONIAN
B 6106 931 GEOBRESIL CRUISES, JEAN CHARCOT
B 6309 1584* GEOCHEMICA STUDIES OF EASTERN KENTUCKY GAS FIELD AND
B 6307 1092† GEOCHEMICAL ANALYSIS OF HYDROCARBON OCCURRENCES IN
S 10 110 GEOCHEMICAL CYCLE, COMPARED TO THE HYDROLOGIC CYCLE
B 6108 1373* GEOCHEMICAL, AND PETROPHYSICAL FEATURES OF A ROLL WITHIN A
B 6307 1129 GEOCHEMICAL AND RELATED REACTIONS, TIME, TEMPERATURE, AND,
B 6203 524* GEOCHEMICAL AND SEDIMENTOLOGIC FACTORS CONTROLLING
B 6404 582† GEOCHEMICAL AND SEDIMENTOLOGIC PROBLEMS OF URANIUM
B 6206 1049† GEOCHEMICAL AND SEDIMENTOLOGIC PROBLEMS OF URANIUM
S 3 61 GEOCHEMICAL CELL
B 6410 1650 GEOCHEMICAL CELL CONCEPT, WYOMING
M 25 148† GEOCHEMICAL CHARACTERISTICS OF GEOTHERMAL RESOURCES IN
B 6301 87† GEOCHEMICAL CHARACTERISTICS OF SEDIMENTARY ROCKS IN SCOTIAN
B 6304 479 GEOCHEMICAL CONCEPT, CAPROCK
B 6304 585 GEOCHEMICAL CONCEPT, DEGRADATION
B 6304 577 GEOCHEMICAL CONCEPT, GENERATION AND, MIGRATION OF,
B 6304 578 GEOCHEMICAL CONCEPT, PRIMARY, MIGRATION
B 6304 581 GEOCHEMICAL CONCEPT, SECONDARY, MIGRATION
B 6304 573 GEOCHEMICAL CONCEPT, SHALE, COMPACTION
B 6203 545* GEOCHEMICAL CONSIDERATIONS, ABST., OIL, MIGRATION LIMITATIONS
B 6004 722* GEOCHEMICAL CONTROLS ON VERTICAL DISTRIBUTION OF IRON-RICH
B 6211 2357* GEOCHEMICAL DATA, ABST., MODERN METHODS OF ACQUIRING AND
B 6406 921 GEOCHEMICAL DATA, CORRELATION OF, TIME-, TEMPERATURE INDEX
B 6405 735* GEOCHEMICAL DATA FILE OF NATIONAL COAL RESOURCES DATA
B 6405 695* GEOCHEMICAL DATA TO URANIUM EXPLORATION, ABST., APPLICATION
B 6206 1077 GEOCHEMICAL DATA, UTAH, TWIN CREEK LIMESTONE, SAMPLES AND
B 6206 1077 GEOCHEMICAL DATA, WYOMING, TWIN CREEK LIMESTONE, SAMPLES
S 3 121 GEOCHEMICAL DISTINCTION OF URANIUM MINERALIZATION PROCESSES
B 6405 692* GEOCHEMICAL EFFECTS OF EARLY DIAGENESIS OF ORGANIC MATTER,
B 6212 2511* GEOCHEMICAL EVALUATION OF BONAPARTE GULF—TIMOR SEA REGION,
B 6303 533* GEOCHEMICAL EVALUATION OF ENSENADA DE LA VELA BASIN,
B 6311 2021† GEOCHEMICAL EVALUATION OF PETROLEUM SOURCE ROCK IN
B 6207 1213* GEOCHEMICAL EXPLORATION FOR SUBMARINE METALLIFEROUS
B 6108 1373† GEOCHEMICAL EXPLORATION IN EAST TINTIC DISTRICT, UTAH, ABST.,
B 6303 497* GEOCHEMICAL EXPLORATION IN POWDER RIVER BASIN, NORTHEASTERN
B 6309 1587* GEOCHEMICAL INFORMATION, ABST., INTERPRETATIO AND STATISTICAL
B 6412 2055 GEOCHEMICAL LOG, ANGOLA BASIN, DSDP SITE 364, LEG 40
B 6412 2057 GEOCHEMICAL LOG, BAY OF BISCAY, DSDP SITE 402A, LEG 48
B 6412 2058 GEOCHEMICAL LOG, BERMUDA RISE, DSDP SITE 386, LEG 43
B 6412 2058 GEOCHEMICAL LOG, BLAKE-, BAHAMA AREA, DSDP SITE 391, LEG 44
B 6412 2055 GEOCHEMICAL LOG, CAPE BASIN, DSDP SITE 361, LEG 40
B 6412 2056 GEOCHEMICAL LOG, CAPE VERDE BASIN, DSDP SITE 367, LEG 41
B 6412 2059 GEOCHEMICAL LOG, GEORGIA EMBAYMENT, COST GE-1 WELL
B 6412 2058 GEOCHEMICAL LOG, HATTERAS ABYSSAL PLAIN, DSDP SITE 105, LEG 11
B 6412 2056 GEOCHEMICAL LOG, MOROCCO, DSDP SITE 370, LEG 41
B 6412 2057 GEOCHEMICAL LOG, PORTUGAL, DSDP SITE 398D, LEG 47
B 6312 2187 GEOCHEMICAL MATERIAL BALANCE, ASSESSING RESOURCES
B 6303 476* GEOCHEMICAL MEASUREMENTS IN EVALUATION OF SOUUCE ROCKS,
B 6206 1059 GEOCHEMICAL MECHANISMS FOR URANIUM ACCUMULATION,
B 6311 2117* GEOCHEMICAL PROSPECTING FOR STRATIGRAPHIC TRAPS, ABST.
B 6303 468* GEOCHEMICAL SIGNAL DECAY. B TITLE ONLY, POOL, DEPLETION AND

B 6304 687* GEOCHEMICAL STUDIES OF GRANTS MINERAL BELT, NEW MEXICO, ABST.
B 6307 1128† GEOCHEMICAL SYSTEMS, ERRORS IN EXTRAPOLATION OF EXPERIMENTA,
B 6304 592 GEOCHEMICAL TOOL, CEMENTATION, MAPS
B 6304 590 GEOCHEMICAL TOOL, HYDRAULIC POTENTIAL, MAPS
B 6304 590 GEOCHEMICAL TOOL, ISOTOPE ANALYSES
B 6304 592 GEOCHEMICAL TOOL, MATURITY, MAPS
B 6304 590 GEOCHEMICAL TOOL, SALINITY, MAPS
B 6304 591 GEOCHEMICAL TOOL, TEMPERATURE, MAPS
B 6304 694* GEOCHEMICAL ZONATION ACROSS ROLL-TYPE URANIUM
B 6405 754* GEOCHEMISTRY, ABST., GETTING THE MOST OUT OF, RADON
B 6303 504* GEOCHEMISTRY, ABST., PARAGENETIC AND STABILITY RELATIONS
B 6304 699* GEOCHEMISTRY, ABST., PREDICTIONS OF OIL OR GAS POTENTIAL BY
B 6303 509* GEOCHEMISTRY AND DIAGENESIS IN CAPITAN REEF, PERMIAN, WEST
B 6203 552* GEOCHEMISTRY AND EARLY DIAGENESIS OF RECENT SEDIMENTS IN
B 6406 954 GEOCHEMISTRY AND GEOLOGY, BY JOHN M. HUNT, KVENVOLDEN,
B 6309 1583* GEOCHEMISTRY AND GEOLOGY OF SOUTHEAST GEORGIA EMBAYMENT
B 6108 1386* GEOCHEMISTRY, AND GEOTHERMOMETER MODELS IN RECONNAISSANCE
B 6405 692* GEOCHEMISTRY AND HYDROCARBON GENESIS, ABST., APPALACHIAN
B 6304 691* GEOCHEMISTRY AND URANIUM IN GRANTS MINERAL BELT, ABST.,
B 6304 640 GEOCHEMISTRY, ARIZONA, URANIUM
B 6004 717* GEOCHEMISTRY AS AID TO EXPLORATION IN EASTERN JAVA SEA, ABST.,
B 6303 472* GEOCHEMISTRY AS EXPLORATION TOOL, ABST., OIL
B 6410 1725 GEOCHEMISTRY, BY A. A. LEVINSON, WITH CHAPTERS BY R. B.
B 6307 1143 GEOCHEMISTRY, BY A. H. BROWNLOW, HUNT, JOHN M., REVIEW OF
B 6412 2125 GEOCHEMISTRY, CARIACO TRENCH, SITE 147
B 6410 1614 GEOCHEMISTRY, DENVER BASIN, CRUDE-OIL
B 6303 441* GEOCHEMISTRY FIND OIL. BY TITLE ONLY, CAN
B 6105 760 * GEOCHEMISTRY, HYDROLOGY, AND MINERALOGY OF NONCARBONATE COASTAL
B 6104 573† GEOCHEMISTRY I, CALCULATION OF, AGES OF, HYDROCARBONS IN,
B 6109 1407† GEOCHEMISTRY, II, DISTRIBUTION OF, HYDROCARBONS BETWEEN OILS
B 6411 2037 GEOCHEMISTRY IN MINERAL EXPLORATION BY ARTHUR W. ROSE,
B 6201 98† GEOCHEMISTRY, INCIPIENT METAMORPHISM, AND OIL GENERATION IN
B 6108 1375* GEOCHEMISTRY, INCIPIENT METAMORPHISM, AND OIL GENERATION IN
B 6303 505* GEOCHEMISTRY OF ABU DHABI SABKHA, ABST., SULFUR AND, OXYGEN,
B 6405 763* GEOCHEMISTRY OF ARTIFICIALLY HEATED HUMIC AND SAPROPELIC
B 6207 1208* GEOCHEMISTRY OF CARBON DIOXIDE-RICH GROUND WATERS,
B 6203 568* GEOCHEMISTRY OF CENOZOIC AND RECENT NONMARINE
S 7 313† GEOCHEMISTRY OF CONTINENTAL SLOPE AND INTRASLOPE BASIN
B 6303 485* GEOCHEMISTRY OF DOLOMITE--STATE OF THE ART, ABST., ISOTOPIC
B 6206 1092* GEOCHEMISTRY OF GAS-BEARING SHALES, U.S. EASTERN INTERIOR
B 6109 1555* GEOCHEMISTRY OF GLEN ROSE PATCH-REEF COMPLEX—BANDERA
B 6105 814* GEOCHEMISTRY OF LIGHT HYDROCARBONS IN SHALLOW HOLES,
B 6303 434* GEOCHEMISTRY OF NATURA GAS IN ILLINOIS BASIN, ABST., ISOTOPE
B 6303 385 GEOCHEMISTRY OF OIL FIELD WATERS, BY KARL KREJCI-GRAF, OSTROFF,
B 6303 478* GEOCHEMISTRY OF OIL-FIELD WATERS FROM NORTHERN GULF OF
B 6112 2103† GEOCHEMISTRY OF OILS FROM MARANON BASIN, PERU, HYDROCARBON
B 6303 471* GEOCHEMISTRY OF OILS FROM SANTA CRUZ BASIN, BOLIVIA--CASE
B 6411 2039 GEOCHEMISTRY OF ORGANIC MOLECULES, EDITED BY KEITH A.
B 6412 2118† GEOCHEMISTRY OF PLIOCENE-PLEISTOCENE CALCAREOUS SEDIMENTS,
B 6410 1690 GEOCHEMISTRY OF PRECAMBRIAN ROCKS, ORGANIC
B 6105 826* GEOCHEMISTRY OF RECENT CONTINENTAL-MARGIN SEDIMENTS OFF
B 6405 750* GEOCHEMISTRY OF REGIONALLY EXTENSIVE CALCITE CEMENT ZONES IN
B 6303 483* GEOCHEMISTRY OF SEDIMENTS RECOVERE BY DSDP-IPOD, ABST.,
B 6311 2118* GEOCHEMISTRY OF SMALL LACUSTRINE DELTA, GREAT SALT PLAINS,
B 6303 472* GEOCHEMISTRY OF SOUTH TEXAS URANIUM DEPOSITS, ABST.
B 6308 1426* GEOCHEMISTRY OF SPRABERRY SANDSTONE, MIDLAND BASIN, WEST
S 3 19 GEOCHEMISTRY OF URANIUM AND THORIUM
B 6312 2135 GEOCHEMISTRY PARAMETERS OF BITUMEN, ORGANIC
B 6312 2136 GEOCHEMISTRY PARAMETERS OF GAS, ORGANIC
B 6312 2135 GEOCHEMISTRY PARAMETERS OF KEROGEN, ORGANIC
B 6412 2125 GEOCHEMISTRY, RED SEA, SITE 229
B 6002 318* GEOCHEMISTRY TO EXPLORATION, ABST., APPLICATIONS OF, OIL, FIELD,
B 6103 468* GEOCHEMISTRY. BY TITLE ONLY, LARGE NEW PORPHYRY MOLYBDENUM
B 6405 681* GEOCHRONOLOGIC EVIDENCE FOR PALEOZOIC PLUTONIC ACTIVITY IN
S 6 35 GEOCHRONOLOGIC HIERARCHY OF TERMS
B 6004 492 GEOCHRONOLOGIC SCALE, JURASSIC TIME SCALE
S 6 33† GEOCHRONOLOGIC SCALE, STRATOTYPES AND AN, INTERNATIONAL
S 6 326 GEOCHRONOLOGIC SCALE, SUMMARY, CARBONIFEROUS
S 6 9† GEOCHRONOLOGIC SCALES
S 6 9 GEOCHRONOLOGIC SCALES, REVIEW
S 6 43 GEOCHRONOLOGIC SYSTEMS, RESOLUTION, CENOZOIC
S 6 237 GEOCHRONOLOGY, COMPARISON DIAGRAM, PALEOGENE
S 6 25 GEOCHRONOLOGY, GLOSSARY OF USAGE
B 6103 469* GEOCHRONOLOGY OF SOUTHWESTERN BROOKS RANGE. BY TITLE ONLY
S 6 347† GEOCHRONOLOGY, ORDOVICIAN
B 6209 1676 GEOCHRONOLOGY, RB-, SR
S 6 352 GEOCHRONOLOGY, TIME SCALE CHART, ORDOVICIAN
S 6 67† GEOCHRONOLOGY--CONVENTION ON THE USE OF DECAY CONSTANTS IN
B 6403 359† GEODYNAMIC PEAKS IN ALPINOTYPE OROGENIES AND CHANGES IN
B 6008 1397* GEOELECTRICAL INVESTIGATIONS OF BOISE, IDAHO, GEOTHERMAL
B 6211 2355* GEOENVIRONMENTAL STUDIES OF PACIFIC ARCTIC OUTER
B 6202 201† GEOHISTORY ANALYSIS—APPLICATION OF MICROPALEONTOLOGY IN
B 6202 202 GEOHISTORY DIAGRAM, FRANCE, PARIS BASIN, QUALITATIVE
B 6202 204 GEOHISTORY DIAGRAMS, TUNISIA, CENTRAL, QUALITATIVE
B 6202 207 GEOHISTORY, LOUISIANA, MISSISSIPPI, DELTA AREA
B 6202 208 GEOHISTORY OF CONTINENTAL MARGINS
M 25 451† GEOLOGIC ASPECTS OF URANIUM RESOURCES OF JAPAN--SUMMARY
B 6309 1581* GEOLOGIC AGE VERSUS MAJOR MINOR, AND TRACE-ELEMENT
B 6312 2185 GEOLOGIC ANALOGY, ASSESSING RESOURCES
B 6405 687* GEOLOGIC AND GEOPHYSICAL INVESTIGATION OF PART OF OUTER
B 6105 760 * GEOLOGIC AND OPERATIONAL SUMMARY OF COST B-2 WELL—APPRAISAL OF
B 6008 1389* GEOLOGIC AND SEISMIC STUDIES OF BOISE FRONT, IDAHO, FOR
B 6405 797* GEOLOGIC AND STRUCTURAL HISTORY OF ZAGROS FOLDBELT, IRAN,
B 6203 561* GEOLOGIC APPRAISAL OF UNDISCOVERED OIL AND GAS RESOURCES IN
B 6203 558* GEOLOGIC APPRAISAL OF UNDISCOVERED OIL AND GAS RESOURCES IN
B 6203 507* GEOLOGIC APPRAISAL OF UNDISCOVERED OIL AND GAS RESOURCES IN
B 6203 509* GEOLOGIC APPRAISAL OF UNDISCOVERED OIL AND GAS RESOURCES IN
B 6305 825* GEOLOGIC CHARACTERISTICS OF VALLE CALDERA GEOTHERMAL SYSTEM
M 26 417† GEOLOGIC CONSIDERATIONS FOR STRATIGRAPHIC MODELING AND
M 25 439† GEOLOGIC CONTROLS OF MINERAL DEPOSITS IN TAIWAN--SUMMARY
B 6405 689* GEOLOGIC CONTROLS ON MINERAL-MATTER CONTENT OF COAL IN
B 6405 689* GEOLOGIC CONTROLS ON SULFUR CONTENT IN COAL, ABST.
B 6008 1393* GEOLOGIC CONTROLS OVER HOT WATER MIGRATION AT SELECTED HOT
B 6105 782 * GEOLOGIC DEVELOPMENT OF CENOZOIC CONTINENTAL MARGIN OF
B 6303 428* GEOLOGIC DISTRIBUTION OF OIL ABST.
M 25 434† GEOLOGIC ENVIRONMENT AND ECONOMIC POSSIBILITIES OF PORPHYRY
B 6004 689* GEOLOGIC ENVIRONMENTS OF GEOTHERMAL RESOURCES ABST.
B 6012 2186* GEOLOGIC ESTIMATES OF UNDISCOVERED OIL AND GAS RESOURCES IN
B 6009 1608* GEOLOGIC EVOLUTION OF MID-CONTINENT AN GULF COAST
B 6004 732* GEOLOGIC EXPLORATION OF SOUTHERN CALIFORNIA BASINS AND BANK
B 6105 775 * GEOLOGIC FACTORS CONTROLLING COALIFICATION AND HYDROCARBON
B 6103 465* GEOLOGIC FRAMEWORK AN RELATED HYDROCARBON POTENTIAL OF
B 6405 727* GEOLOGIC FRAMEWORK FOR DEVELOPMENT, PRODUCTION, AND
B 6203 373† GEOLOGIC FRAMEWORK OF LOWER COOK INLET, ALASKA
B 6012 2185* GEOLOGIC FRAMEWORK OF LOWER COOK INLET, ALASKA, WITH
M 25 466† GEOLOGIC FRAMEWORK OF METALLIC MINERAL DEPOSITS OF
B 6303 434* GEOLOGIC HAZARDS, ABST., SOURCES OF LIABILITY FOR
M 25 47† GEOLOGIC HISTORY AND PALEOZOOGEOGRAPHY OF PACIFIC
B 6004 727† GEOLOGIC HISTORY AND TECTONICS OF ZAGROS MOUNTAINS IN
M 29 40 GEOLOGIC HISTORY, BLAKE PLATEAU, NORTHERN
M 29 24 GEOLOGIC HISTORY, GEORGIA EMBAYMENT
M 29 39 GEOLOGIC HISTORY, MODEL, BLAKE PLATEAU
B 6408 1277* GEOLOGIC HISTORY OF DEEP GULF OF MEXICO BASIN, ABST.
B 6303 544* GEOLOGIC HISTORY OF DEEP GULF ON MEXICO BASED ON REGIONAL
B 6405 682* GEOLOGIC HISTORY OF DEEP SOUTHEASTERN GULF OF MEXICO BASIN—
S 7 243† GEOLOGIC HISTORY OF TEXAS OUTER CONTINENTAL SHELF AND UPPER
B 6011 2067* GEOLOGIC HISTORY--KEY TO ACCUMULATION AT CEDA CREEK, ABST.
M 29 13 GEOLOGIC INTERPRETATION, GEORGIA EMBAYMENT
B 6203 562* GEOLOGIC INTERPRETATION OF SEISMIC-ATTRIBUTE MEASUREMENTS,
B 6211 2362* GEOLOGIC INTERPRETATIONS FOR RESERVOIR AND FLUID-FLOW RATE
M 29 254 GEOLOGIC MAP, GUATEMALA, MAGNETIC BASEMENT
M 25 46† GEOLOGIC MAP OF PACIFIC MOBILE BELT AND PACIFIC OCEAN, GENERAL
B 6205 863† GEOLOGIC MATERIALS FROM STATE SURVEYS, AVAILABILITY OF
M 26 405 GEOLOGIC MODEL, NORTH SEA
B 6207 1236* GEOLOGIC REMOTE SENSING, ABST., INTRODUCTION TO
B 6405 746* GEOLOGIC SETTING AND OIL AND GAS POTENTIAL OF EASTERN UNITED
S 10 64 GEOLOGIC SETTING, CANADA, ATHABASCA-PEACE RIVER, CRETACEOUS
M 29 27 GEOLOGIC SETTING, BLAKE PLATEAU
M 29 12 GEOLOGIC SETTING, GEORGIA EMBAYMENT
B 6006 952† GEOLOGIC SIGNIFICANCE OF DOONERAK STRUCTURAL HIGH, CENTRAL
B 6401 8† GEOLOGIC SIGNIFICANCE OF LANDSAT DATA FOR 15 GIANT OIL AND GAS
M 30 7† GEOLOGIC SIGNIFICANCE OF LANDSAT DATA FOR 15 GIANT OIL AND GAS
B 6304 698* GEOLOGIC SIGNIFICANCE OF LANDSAT DATA ON SOME KNOWN GIANT
B 6303 543* GEOLOGIC SIGNIFICANCE OF SEDIMENTAR REFLECTORS IN DEEP
M 26 122 GEOLOGIC SIGNIFICANCE, REFLECTION, PARAMETERS
B 6306 986 GEOLOGIC SOCIETY WITH ARIZONA GEOLOGICA SOCIETY FIELD TRIP
B 6004 714* GEOLOGIC SYNERGISM. BY TITLE ONLY, GEOPHYSICAL-
B 6205 883* GEOLOGIC TECHNIQUES USED IN TRAP SPRING FIELD DISCOVERY,
S 6 76 GEOLOGIC TIME SCALE, BASED ON DATA FILE
B 6405 780* GEOLOGIC-SEISMIC EXPLORATION MODEL FOR REWORKED DELTAIC
M 25 58† GEOLOGIC-STRUCTURAL SYNTHESIS OF PACIFIC AREA
M 25 148† GEOLOGICAL AND GEOCHEMICAL CHARACTERISTICS OF GEOTHERMAL
M 29 1 GEOLOGICAL AND GEOPHYSICAL INVESTIGATIONS
B 6403 448* GEOLOGICAL ASSESSMENT OF PETROLEUM POTENTIAL OF TERTIARY
S 10 47† GEOLOGICAL CONSTRAINTS ON MIGRATION MECHANISMS, SOME, MASS
B 6412 2189 GEOLOGICAL FIELD TRIPS COLOMBIA 1959-1978, BY COLOMBIAN SOCIETY
S 10 73 GEOLOGICAL RESTORATION, COMPUTER MODELING
B 6108 1373* GEOLOGICAL, GEOCHEMICAL, AND PETROPHYSICAL FEATURES OF A
B 6204 686 GEOLOGICAL INFORMATION LIBRARY OF DALLAS, DALLAS GEOLOGICAL
B 6411 2036 GEOLOGICAL MAP OF THE WORLD, NATIONAL DEPARTMENT OF
B 6102 253† GEOLOGICAL SOCIETY, HISTORY OF, EL PASO

B 6007 1117 GEOLOGICAL SOCIETY, HISTORY, SAN ANTONIO
B 6007 1119 GEOLOGICAL SOCIETY, HONORARY MEMBERS, 1935-1975, SOUTH, TEXAS
B 6102 255 GEOLOGICAL SOCIETY, OFFICERS, 1967-1976, EL PASO
B 6006 988† GEOLOGICAL SOCIETY, SHORT, HISTORY OF, NORTHERN, CALIFORNIA
B 6007 1117† GEOLOGICAL SOCIETY--A FRAGMENTARY INFORMAL HISTORY, SOUTH,
S 6 11 GEOLOGICAL TIME
S 6 1 GEOLOGICAL TIME SCALE SYMPOSIUM 106.6, AUSTRALIA, AUGUST 1976
B 6303 531* GEOLOGISTS AND POLITICS--A MATTER OF SURVIVAL, ABST.
B 6308 1427* GEOLOGISTS--ENDANGERED SPECIES, ABST., INDEPENDENT
B 6303 491* GEOLOGISTS--ENDANGERED SPECIES, ABST., INDEPENDENT
B 6305 843* GEOLOGY AND CURRENT DEVELOPMENT ACTIVITY AT LITTLE KNIFE
B 6203 508* GEOLOGY AND DEVELOPMENT OF LEEDY FIELD, DEWEY COUNTY,
B 6304 691* GEOLOGY AND DEVELOPMENT OF MARQUEZ, NE MEXICO, URANIUM
B 6305 804† GEOLOGY AND DEVELOPMENT OF MILLS RANCH COMPLEX--WORLDS
B 6207 1229* GEOLOGY AND DEVELOPMENT OF MOMOTOMBO GEOTHERMAL FIELD,
B 6303 414* GEOLOGY AND DEVELOPMENT OF TEAK OIL FIELD, TRINIDAD, WEST
M 30 387† GEOLOGY AND DEVELOPMENT OF THE TEAK OIL FIELD, TRINIDAD, WEST
B 6207 1208* GEOLOGY AND ECONOMIC POSSIBILITIES OF SOME CHROMITE DEPOSITS
B 6207 1215* GEOLOGY AND EXPLORATION HISTORY OF OFFSHORE CENTRAL
B 6105 792 * GEOLOGY AND EXPLORATION OF URAVAN MINERAL BELT
B 6408 1281* GEOLOGY AND GEOPHYSICS OF MIDDLEMISSISSIPPIAN, VALMEYERAN,
B 6205 889* GEOLOGY AND GEOPOLITICS OF MEXICAN PETROLEUM DISCOVERIES,
B 6305 841* GEOLOGY AND HYDROCARBON ACCUMULATION IN MISSISSIPPIAN
B 6101 1† GEOLOGY AND HYDROCARBON DISCOVERIES OF CANADIAN ARCTIC
B 6008 1398* GEOLOGY AND HYDROCARBON POTENTIAL OF MISSISSIPPIAN OUNGRE
B 6212 2510* GEOLOGY AND HYDROCARBON PROSPECTS OF SINU BELT,
B 6403 434 GEOLOGY AND HYDROGEOLOGY OF NORTHEASTERN YUCATAN, EDITED
B 6408 1286* **GEOLOGY AND MINING OF TENNESSEE COAL,**
B 6003 389† GEOLOGY AND OIL PROSPECTS OF SOMALIA, EAST AFRICA
B 6304 687* GEOLOGY AND ORE DEPOSITS OF JOHNNY M MINE, AMBROSIA LAKE
M 30 225† GEOLOGY AND PETROLEUM FIELDS IN PROTEROZOIC AND LOWER
B 6207 1229* GEOLOGY AND PETROLEUM POTENTIAL OF CENTRAL AMERICA, ABST.
B 6201 40† GEOLOGY AND PETROLEUM POTENTIAL OF EXMOUTH PLATEAU AREA
B 6004 682* GEOLOGY AND POTENTIAL USES OF GEOPRESSURE RESOURCES OF GULF
B 6305 845* GEOLOGY AND PRODUCTION POTENTIAL OF MONTEAGLE LIMESTONE,
B 6108 1385* GEOLOGY AND RESERVOIR ANALYSIS OF EAST MESA GEOTHERMAL
B 6108 1387* GEOLOGY AND SHALE-OIL RESOURCES OF WHITE RIVER-GRAY HILLS
B 6305 836* GEOLOGY AND STRUCTURAL CONTROL OF GEOTHERMAL SYSTEM AT
B 6405 720* GEOLOGY IN 1980S, ABST., PETROLEUM
B 6303 492* GEOLOGY IN DEEP BASIN CANADA. BY TITLE ONLY, ENGINEERING
B 6303 380 GEOLOGY IN ENVIRONMENTAL PLANNING. BY ARTHUR D. HOWARD
B 6105 790 * GEOLOGY IN GEOTHERMAL ENERGY EXPLORATION
B 6205 865 GEOLOGY MERIT BADGE BOOKLET, AAPG, BOY SCOUT COMMITTEE
M 29 240 GEOLOGY, NORTH PHILIPPINE SEA, DAITO RIDGE
M 29 243 GEOLOGY, NORTH PHILIPPINE SEA, KITA-DAITO BASIN
M 29 242 GEOLOGY, NORTH PHILIPPINE SEA, MINAMI-DAITO BASIN
M 29 240 GEOLOGY, NORTH PHILIPPINE SEA, OKI-DAITO RIDGE
M 30 507† GEOLOGY OF A STRATIGRAPHIC GIANT--MESSLA OIL FIELD, LIBYA
B 6303 433* GEOLOGY OF A STRATIGRAPHIC GIANT--MESSLA OIL FIELD, LIBYA, ABST.
B 6103 468* GEOLOGY OF AMBLER MINERAL BELT, BROOKS RANGE, ALASKA. BY
B 6103 465* GEOLOGY OF BERING SHELF, ABST.
B 6011 2067* GEOLOGY OF BERING SHELF, ABST.
M 30 487† GEOLOGY OF BOMBAY HIGH FIELD, INDIA, PETROLEUM
B 6405 797* GEOLOGY OF CALVIN FIELD—DEEP BASINAL JURASSIC PLAY IN NORTH
B 6305 846* GEOLOGY OF COAL DEPOSITS IN SOUTHEASTERN SALINE COUNTY,
B 6405 766* GEOLOGY OF CONTINENTAL SLOPE ADJACENT TO OCS LEASE SALE 55,
B 6406 957 GEOLOGY OF CONTINENTAL SLOPES, EDITED BY LARRY J. DOYLE AND
B 6304 693* GEOLOGY OF EASTERN SMITH LAKE ORE TREND, ABST.
M 24 157† GEOLOGY OF FAIRWAY FIELD, EAST TEXAS
B 6109 1548* GEOLOGY OF FRACTURED AUSTIN CHALK AND BUDA LIMESTONE IN
B 6303 534* GEOLOGY OF GABES BASIN, TUNISIA, ABST.
M 30 577† GEOLOGY OF GRONDIN FIELD
B 6304 698* GEOLOGY OF GU-DAO OIL FIELD AN ADJACENT AREAS, ABST.
M 30 471† GEOLOGY OF GUDAO OIL FIELD AND SURROUNDING AREAS
B 6409 1563* GEOLOGY OF GUM ISLAND NORTH AND FRENCH ISLAND AREAS,
B 6105 794 * GEOLOGY OF HEAVY-OIL DEPOSIT AT COLD LAKE, ALBERTA
B 6406 960* GEOLOGY OF IN SITU PILOT PROJECT, WABASCA OIL SANDS DEPOSIT,
B 6105 809 * GEOLOGY OF JURASSIC-TRIASSIC CULPEPER BASIN, NORTH OF
B 6011 2069* GEOLOGY OF LABRADOR SEA AND ITS PETROLEUM POTENTIAL, ABST.
B 6303 384 GEOLOGY OF MANITOULIN AREA, INCLUDING THE ROAD LOG TO THE
B 6002 319* GEOLOGY OF MISSISSIPPIAN COMBINATION TRAP, BINDLEY FIELD,
B 6404 501† GEOLOGY OF NEW ENGLAND PASSIVE MARGIN
B 6004 715* GEOLOGY OF NORWEGIAN CONTINENTAL SHELF NORTH OF 62 DEGREES
B 6012 2180* GEOLOGY OF OFFSHORE LOWER COOK INLET ALASKA, FROM SEISMIC
B 6305 826* GEOLOGY OF PALEOZOIC STRATA IN WEST-CENTRAL SASKATCHEWAN,
B 6002 318* GEOLOGY OF PANOMA GAS AREA, SOUTHWESTER KANSAS, ABST.
B 6002 320* GEOLOGY OF PERMIAN BLAINE FORMATION AND ASSOCIATED STRATA
B 6008 1402* GEOLOGY OF PINEVIEW FIELD AREA, SUMMI COUNTY, UTAH, ABST.
B 6004 669* GEOLOGY OF RED SEA MARGIN, ABST
B 6004 728* GEOLOGY OF SALTON SEA GEOTHERMAL FIELD, ABST.
B 6109 1548* GEOLOGY OF SIERRA BALUARTES DIAPIR AND SIERRA DE PAJAROS
B 6004 654* GEOLOGY OF SOUTH AUGUSTINE ISLAND, LOWER COOK INLET, ALASKA,
B 6303 452* GEOLOGY OF SOUTHEAST GEORGIA EMBAYMENT, ABST.
B 6405 742* GEOLOGY OF SYNCRUDE CANADA LIMITED MINE SITE, ATHABASCA OIL
M 30 459† GEOLOGY OF THE BEKAPAI FIELD
M 30 81† GEOLOGY OF THE FORTIES FIELD, U.K. CONTINENTAL SHELF, NORTH SEA
M 30 399† GEOLOGY OF THE HANDIL FIELD, EAST KALIMANTAN, INDONESIA
M 29 11† GEOLOGY OF THE OFFSHORE SOUTHEAST GEORGIA EMBAYMENT, U.S.
B 6408 1280* GEOLOGY OF WHEATON CONSOLIDATED OIL FIELD, GIBSON COUNTY,
B 6203 509* GEOLOGY SEEN WHILE LIVING IN A SUBMARINE CANYON NEAR ST.
M 29 353† GEOLOGY WEST OF ST. CROIX, U.S. VIRGIN ISLANDS, GEOMORPHOLOGY
S 6 295 GEOMAGNETIC REVERSAL SCALE, COMPARISON, JURASSIC
S 6 273 GEOMAGNETIC REVERSAL SCALE, CRETACEOUS
S 6 293 GEOMAGNETIC REVERSAL SCALE, PROPOSED, JURASSIC
S 6 280 GEOMAGNETIC REVERSAL SCALES, CRETACEOUS
S 6 225 GEOMAGNETIC SCALE, DIAGRAM, CENOZOIC
S 4 313 GEOMAR, GEOLOGIA MARINHA IN 1971-73, PROJETO
B 6405 758* GEOMETRY AND DISPERSAL PATTERNS OF DEEP-SEA FANS FROM
B 6405 751* GEOMETRY AND MECHANISMS OF TRANSVERSE FAULTING, ROCKY
B 6108 1375* GEOMETRY AND PALEOENVIRONMENTS OF UPPERMOST CRETACEOUS
B 6203 557* GEOMETRY, FACIES, AND DEVELOPMENT OF LARGE NONREEFAL SHELF
B 6007 1009† GEOMETRY OF A FLUVIAL SHEET SANDSTONE, RESERVOIR
B 6002 245† GEOMETRY OF CARBONATE ROCKS AS REVEALED BY PORE CASTS AND
B 6105 765 * GEOMETRY OF COAL BEDS IN SAN JUAN BASIN, NEW MEXICO
B 6109 1539* GEOMETRY OF GULF COAST AQUIFERS, HOUSTON-GALVESTON, TEXAS,
B 6405 783* GEOMETRY OF SHELF SANDSTONE BODIES IN SHANNON-EQUIVALENT
B 6205 884* GEOMORPHIC EXPLORATION USED IN DISCOVERY OF TRAP SPRING OIL
M 29 362 GEOMORPHOLOGIC DEVELOPMENT, VIRGIN ISLANDS, FREDERICKSTED
M 29 353† GEOMORPHOLOGY AND SUBSURFACE GEOLOGY WEST OF ST. CROIX, U.S.
B 6207 1220* GEOMORPHOLOGY OF OCEAN FLOOR AS NATURAL GUIDE TO
S 7 305 GEOMORPHOLOGY, ORCA BASIN
M 27 211 GEOPETAL FILLS
M 27 212 GEOPETAL FILLS
M 27 211 GEOPETAL OOIDS
B 6205 742 GEOPHYSICAL ACTIVITY, WORLD
B 6211 2357* GEOPHYSICAL AND GEOCHEMICAL DATA, ABST., MODERN METHODS OF
B 6203 518* GEOPHYSICAL AND GEOLOGICAL SIGNIFICANCE OF VELOCITY
B 6009 1621* GEOPHYSICAL AND TECTONIC STUDY OF MOORMAN SYNCLINE AND
B 6405 673* GEOPHYSICAL CASE HISTORY OF TWO HILLS COLONY GAS FIELD OF
B 6205 743 GEOPHYSICAL CREW-MONTHS IN NON-COMMUNIST WORLD, 1970-1975,
M 26 187 GEOPHYSICAL CRITERIA, SEISMIC, REFLECTIONS, CARBONATE,
B 6109 1559* GEOPHYSICAL DATA IN NORTHEASTERN ARIZONA, ABST., INTEGRATION
B 6304 695* GEOPHYSICAL EXPERIMENTS AT THE MARIANO LAK URANIUM
B 6410 1723 GEOPHYSICAL EXPLORATION AND INTERPRETATION, BY ROBERT E.
B 6405 786* GEOPHYSICAL EXPLORATION FOR PRECAMBRIAN-RELATED URANIUM
B 6405 673* GEOPHYSICAL EXPLORATION IS JUST BEGINNING, ABST.
B 6211 2362* GEOPHYSICAL EXPLORATION METHOD—A REVIEW, ABST.,
M 25 432 GEOPHYSICAL EXPLORATION, THAILAND, FLUORITE
B 6303 465* GEOPHYSICAL INVESTIGATIONS FOR PILOT PLANT FOR NUCLEAR-WASTE
M 29 1 GEOPHYSICAL INVESTIGATIONS OF CONTINENTAL MARGIN
B 6207 1232* GEOPHYSICAL METHODS, ABST., DEFINING, BOUNDARIES AND
M 26 389† GEOPHYSICAL PRINCIPLES AND TECHNIQUES, STRATIGRAPHIC,
M 29 43† GEOPHYSICAL PROFILES, EROSIONAL AND, DEPOSITIONAL, STRUCTURES
B 6209 1659 GEOPHYSICAL PROFILES, MANILA TRENCH
B 6009 1624* GEOPHYSICAL PROVINCES IN INDIANA, ABST., PRECAMBRIAN
B 6004 714* GEOPHYSICAL-GEOLOGIC SYNERGISM. BY TITLE ONLY
B 6012 2192* GEOPHYSICISTS, ABST., SATELLITE EXPLORATION BY
M 26 13 GEOPHYSICS, GLOSSARY
B 6108 1380* GEOPHYSICS IN COAL EXPLORATION AND DEVELOPMENT, ABST.,
B 6101 50 GEOPHYSICS, LEBANON
B 6411 2038 GEOPHYSICS RESEARCH BOARD, RODGERS, DONALD A., REVIEW OF,
B 6008 1397* GEOPHYSICS--RECENT BREAKTHROUGHS AND CHALLENGING NEW
M 25 203† GEOPHYSICS--RECENT BREAKTHROUGHS AND CHALLENGING NEW
B 6001 3† GEOPHYSICS--RECENT BREAKTHROUGHS AND CHALLENGING NEW
B 6207 1226* GEOPHYSICS—CURRENT PERFORMANCE AND FUTURE PROSPECTS, ABST.,
B 6205 889* GEOPOLITICS OF MEXICAN PETROLEUM DISCOVERIES, ABST., GEOLOGY
B 6405 708* GEOPRESSURE, ABST., SMECTITE-, ILLITE TRANSFORMATION—ROLE IN
B 6309 1609* GEOPRESSURE IN HOUMA AND HOLLYWOOD FIELDS, LOUISIANA, ABST.
B 6101 65† GEOPRESSURE PATTERNS OF BAYOU CARLIN--LAKE SAND AREA, SOUTH
B 6004 682* GEOPRESSURE RESOURCES OF GULF COAST, ABST., GEOLOGY AND
B 6409 1557* GEOPRESSURE ZONE OF LIRETTE FIELD, TERREBONNE PARISH,
S 10 208 GEOPRESSURE, ORIGIN AND PRESERVATION
S 10 207† GEOPRESSURE IN THE HYDROCARBON AND WATER SYSTEM, ROLE OF
S 10 213 GEOPRESSURE WITH DEPTH OF REVERSAL OF SHALE DENSITY, GULF OF
S 10 209 GEOPRESSURE ZONE, GEOTHERMAL REGIME
S 10 214 GEOPRESSURE ZONE, HYDROCARBON REGIME
B 6209 1760* GEOPRESSURED GEOTHERMAL ENERGY PRODUCTION IN VICKSBURG
B 6009 1606* GEOPRESSURED GEOTHERMAL PROSPECT IN FRIO FORMATION OF TEXAS
B 6212 2478 GEOPRESSURED WATERS, DISSOLVED GAS CONTENT OF, GEOTHERMAL
B 6405 709* GEOPRESSURED-GEOTHERMAL TEST WELL, BRAZORIA, TEXAS, ABST.,

B 6004 705* GEOPRESSURES AND SECONDARY POROSITY IN DEEP SMACKOVER OF
M 30 213 GEOPRESSURES, NORTH SEA, EKOFISK FIELD
S 3 102 GEOPRESSURIZATION
S 3 136 GEOPRESSURIZED BASINS, HYPOGENE, LATEROGENESIS IN
B 6310 1984 GEOREF THESAURUS AND GUIDE T INDEXING, BY CAROL HECKMAN AND
B 6004 673 GEORGE, A. E., SIMULATED GEOTHERMAL MATURATION OF ATHABASCA
B 6303 457 GEORGE, C. J., POROSITY EXAMPLES FROM IAB FIELD, COKE COUNTY,
B 6008 1391* GEORGE CREEK AREA, TOOLE COUNTY, MONTANA, ABST.,
B 6303 547 GEORGE, DANIEL T., EFFECTS OF SOURCE MATERIAL AND THERMAL
B 6008 1204 GEORGE, G. R., OIL AND GAS DEVELOPMENT IN NORTHERN ROCKIES IN
B 6303 429 GEORGE, M. A., ROL OF BRINES AND FERROUS IRON IN FORMATION OF
B 6101 80 GEORGES BANK
M 29 97 GEORGES BANK BASIN, ATLANTIC MARGIN
B 6404 505 GEORGES BANK BASIN, CORRELATION
M 29 95 GEORGES BANK BASIN, DEPTH-TO-BASEMENT CONTOURS
B 6109 1484 GEORGES BANK BASIN, EAST COAST
M 29 93 GEORGES BANK BASIN, MAGNETIC ANOMALY CONTOURS
M 29 94 GEORGES BANK BASIN, MAGNETIC DEPTH CONTOURS
B 6109 1483† GEORGES BANK, CONTINENTAL, SLOPE AND UPPER, RISE OFF WESTERN,
B 6109 1484 GEORGES BASIN, EAST COAST
B 6309 1604* GEORGETOWN, MCKNIGHT, AND WEST NUECES CARBONATE ROCKS OF
B 6105 818* GEORGIA, ABST., DEPOSITIONAL CHARACTERISTICS OF SEDIMENTS AT,
B 6309 1611* GEORGIA, ABST., EARLY, TERTIARY LITHOSTRATIGRAPHIC
B 6209 1755* GEORGIA, ABST., UPPER, MESOZOIC AND, PALEOGENE GEOLOGY AND,
B 6403 339 GEORGIA, AND SOUTH CAROLINA, FLORIDA, HATTERAS SHELF, FLORIDA
B 6110 1898 GEORGIA, AND TENNESSEE, CAMBRIAN TYPE LOCALITIES, ALABAMA
B 6405 775* GEORGIA, AND TUSCALOOSA, ALABAMA, ABST., APPALACHIAN, THRUST
B 6004 663* GEORGIA COAST, ABST., ARCHAELOGIC, DATING OF, HOLOCENE,
B 6109 1536* GEORGIA COAST, ABST., HISTORY OF, SHORELINE CHANGES BY,
B 6109 1535* GEORGIA COAST, ABST., ORIGIN AND CHARACTER OF, WASHOVER, FANS
B 6303 424* GEORGIA COAST, ABST., ROLE OF, BIOEROSION IN MASS-WASTING OF,
B 6203 507* GEORGIA COASTAL PLAIN, ABST., MESOZOIC AND CENOZOIC GEOLOGY
B 6203 504* GEORGIA COASTAL PLAIN, ABST., PRE-CRETACEOUS GEOLOGY BENEATH
B 6008 1282 GEORGIA, DEVELOPMENTS, 1975
B 6108 1264 GEORGIA, DEVELOPMENTS, 1976
B 6303 452* GEORGIA EMBAYMENT, ABST., GEOLOGY OF, SOUTHEAST
B 6303 510* GEORGIA EMBAYMENT, ABST., IMPORTANT STRATIGRAPHIC BREAKS IN,
B 6303 425* GEORGIA EMBAYMENT, ABST., REGIONAL HYDROCARBON, SOURCE-ROCK
M 29 30 GEORGIA EMBAYMENT, AGE OF REFLECTORS
M 29 27† GEORGIA EMBAYMENT AND NORTHERN BLAKE PLATEAU: PRELIMINARY
M 29 20 GEORGIA EMBAYMENT, CARBONATE SECTION
M 29 23 GEORGIA EMBAYMENT, CENOZOIC INTERVAL
B 6412 2059 GEORGIA EMBAYMENT, COST GE-1 WELL, GEOCHEMICAL LOG
M 29 16 GEORGIA EMBAYMENT, CRETACEOUS INTERVAL
M 29 23 GEORGIA EMBAYMENT, GENERAL CROSS SECTION
M 29 24 GEORGIA EMBAYMENT, GEOLOGIC HISTORY
M 29 13 GEORGIA EMBAYMENT, GEOLOGIC INTERPRETATION
M 29 12 GEORGIA EMBAYMENT, GEOLOGIC SETTING
M 29 13 GEORGIA EMBAYMENT, INTERPRETIVE CROSS SECTION
M 29 17 GEORGIA EMBAYMENT, MAJOR SEISMIC INTERVALS
M 29 18 GEORGIA EMBAYMENT, MULTICHANNEL SEISMIC PROFILES
M 29 21 GEORGIA EMBAYMENT, MULTICHANNEL SEISMIC PROFILES
M 29 19 GEORGIA EMBAYMENT, NEAR-TRACE SHIPBOARD MONITOR
M 29 11 GEORGIA EMBAYMENT, RELATION TO BLAKE PLATEAU
M 29 16 GEORGIA EMBAYMENT, SEISMIC STRATIGRAPHY
M 29 13 GEORGIA EMBAYMENT, SEISMIC VELOCITIES
B 6205 793 GEORGIA EMBAYMENT, SOUTHEAST
M 29 30 GEORGIA EMBAYMENT, TOP OF CRETACEOUS
M 29 11† GEORGIA EMBAYMENT, U.S. ATLANTIC CONTINENTAL MARGIN, BASED
M 29 19 GEORGIA EMBAYMENT, UPPER CRETACEOUS INTERVAL
B 6204 644 GEORGIA, MONTEAGLE LIMESTONE
S 5 55 GEORGIAN INLET, LOCATION, MICHIGAN,
B 6109 1559* GEOSAT—REMOTE-SENSING SYSTEMS OF FUTURE, ABST.
B 6310 1982 GEOSCIENCES ON CRITICAL ENERGY RESOURCES, EDITED BY CREIGHTON
S 6 11 GEOSCIENTIST TIME
B 6004 720* GEOSCIENTISTS IN PETROLEUM INDUSTRY, ABST., WOMEN
B 6303 515* GEOSEISMI MODELING--AN INTERACTIVE COMPUTER APPROACH TO
M 26 468 GEOSEISMIC MODELING, MAPPING
M 26 451† GEOSEISMIC MODELING OF BASAL MORROW-SPRINGER SANDSTONES,
M 26 460 GEOSEISMIC MODELING, OKLAHOMA, GEARY AREA
B 6405 705* GEOSTATISTICAL APPROAC IN COAL RESOURCE ESTIMATION, ABST.
M 25 466 GEOSYNCLINAL DEVELOPMENT, USSR, MESOZOIC
B 6004 648* GEOSYNCLINE, ABST., RADIOCARBON, DATING AND STRATIGRAPHIC
B 6005 794† GEOSYNCLINE AND ITS PETROLEUM POSSIBILITIES, CONCEALED,
B 6302 249† GEOSYNCLINES AND PLATE TECTONICS IN BANDA ARCS, EASTERN
B 6112 2063† GEOSYNCLINES AND PLATE TECTONICS IN BANDA ARCS, EASTERN
B 6303 289 GEOSYNCLINES, PROSPECTS, MARGINAL
S 7 267† GEOTECHNICAL ASPECTS, STABILITY OF, CONTINENTAL, SHELF AND,
B 6203 532* GEOTECHNICAL ENGINEERING PROBLEMS OF UPPER SLOPE SITES IN
B 6207 1216* GEOTECHNICAL HAZARDS IN PETROLEUM DEVELOPMENT, ABST.,
B 6203 529* GEOTECHNICAL PROPERTIES OF CONTINENTAL SLOPE DEPOSITS, CAPE
B 6203 557* GEOTECHNICAL PROPERTIES OF UPPER CONTINENTAL SLOPE SEDIMENTS,
B 6002 327* GEOTECTONIC EVOLUTION OF WICHITA AULACOGEN, OKLAHOMA, ABST.
B 6209 1759* GEOTEMPERATURES AND HALOKINETICS, ABST., REDDELL, SALT, DOME
B 6108 1204 GEOTHERMAL ACTIVITY, 1976, FOUR CORNERS-, INTERMOUNTAIN AREA
B 6308 1227 GEOTHERMAL ACTIVITY, 1978, TEXAS
B 6101 65† GEOTHERMAL AND GEOPRESSURE PATTERNS OF BAYOU CARLIN--LAKE
B 6307 1101 GEOTHERMAL ANOMALY, AUSTRALIA, BEAGLE RIDGE
M 25 161 GEOTHERMAL AREA, BAJA CALIFORNIA GEOTHERMAL AREA, MEXICO,
B 6207 1231* GEOTHERMAL AREA, CALIFORNIA, ABST., HYDROGEOLOGY OF, LONG
M 25 161 GEOTHERMAL AREA, MEXICO, CERRO PRIETO, GEOTHERMAL AREA, BAJA
M 25 160 GEOTHERMAL AREA, MEXICO, MICHOACAN
M 25 161 GEOTHERMAL AREA, MEXICO, PATHE
M 25 169 GEOTHERMAL AREAS, CHILE, NORTHERN
M 25 146† GEOTHERMAL AREAS IN TAIWAN--SUMMARY, ORIGIN AND POTENTIAL
S 10 230 GEOTHERMAL PROFILE, USSR, SIBERIA, KOLPASHEVO
S 10 209 GEOTHERMAL REGIME, GEOPRESSURE ZONE
S 10 76 GEOTHERMAL RESTORATION, COMPUTER MODELING
M 25 158 GEOTHERMAL, CALIFORNIA, THE, GEYSERS FIELD
M 25 114 GEOTHERMAL, CHILE, EL TATIO FIELD
M 25 114 GEOTHERMAL, CHILE, WAIRAKEI FIELD
B 6008 1411* GEOTHERMAL CIRCULATION PATTERN, CORWIN SPRINGS--GARDINER
B 6207 1241* GEOTHERMAL DEVELOPMENT AND UTILIZATION IN JAPAN, ABST.,
B 6211 2348* GEOTHERMAL DEVELOPMENTS, ABST., COSO AREA
B 6108 1374* GEOTHERMAL ENERGY, ABST., EFFECTIVE USE OF MICROEARTHQUAKE
B 6108 1379* GEOTHERMAL ENERGY, ABST., QUATERNARY GEOLOGIC PROCESSES
M 25 78 GEOTHERMAL ENERGY, ANTARCTICA
M 25 118† GEOTHERMAL ENERGY DEVELOPMENT--SUMMARY, ENVIRONMENTAL
B 6405 761* GEOTHERMAL ENERGY DEVELOPMENTS, ABST., OVERVIEW OF
B 6207 1233* GEOTHERMAL ENERGY EXPLORATION AND DEVELOPMENT IN
M 25 119 GEOTHERMAL ENERGY FOR PRODUCTION OF ELECTRICITY
B 6207 1233* GEOTHERMAL ENERGY IN CONTIGUOUS UNITED STATES, ABST., FUTURE
B 6207 1235* GEOTHERMAL ENERGY IN HAWAII—PRESENT AND FUTURE, ABST.
M 25 130† GEOTHERMAL ENERGY IN NEW ZEALAND--SUMMARY
M 25 163† GEOTHERMAL ENERGY IN NICARAGUA--SUMMARY, EXPLORATION FOR
B 6209 1760* GEOTHERMAL ENERGY PRODUCTION IN VICKSBURG FORMATION
M 25 124† GEOTHERMAL ENERGY PROGRAM--SUMMARY, MAGMA TAP, THE
B 6303 417* GEOTHERMAL ENERGY, TEXAS GULF COAST, ABST., POTENTIAL FOR
B 6008 1410* GEOTHERMAL ENERGY--ROCKY ROAD ABST., LEASE OF FEDERAL
M 25 140† GEOTHERMAL EXPLORATION AND DEVELOPMENT IN THE
B 6207 1223* GEOTHERMAL EXPLORATION AND EXPLOITATION DEVELOPMENTS,
B 6009 1616* GEOTHERMAL EXPLORATION FROM DEEP-WELL DATA, ABST.
B 6108 1389* GEOTHERMAL EXPLORATION IN CENTRAL COLORADO, ABST.
M 25 169† GEOTHERMAL EXPLORATION IN NORTHERN CHILE--SUMMARY
M 25 135 GEOTHERMAL EXPLORATION, INDONESIA
B 6008 1390* GEOTHERMAL EXPLORATION OF ROOSEVELT KGRA, UTAH, ABST.
B 6207 1227* GEOTHERMAL EXPLORATION ON ADAK ISLAND, ALASKA—A CASE
B 6409 1356 GEOTHERMAL EXPLORATION, TEXAS
B 6211 2350* GEOTHERMAL FIELD, ABST., DISCOVERY AND GEOLOGY OF, DESERT
B 6004 728* GEOTHERMAL FIELD, ABST., GEOLOGY OF, SALTON SEA
B 6212 2508* GEOTHERMAL FIELD, BAJA CALIFORNIA, MEXICO, ABST., DEVELOPMENT
B 6203 490* GEOTHERMAL FIELD, BAJA CALIFORNIA, MEXICO, ABST., RECENT
B 6405 758* GEOTHERMAL FIELD, BAJA CALIFORNIA, MEXICO, ABST., STRUCTURAL
B 6405 759* GEOTHERMAL FIELD, BAJA CALIFORNIA, MEXICO, ABST., STUDY OF
B 6108 1385* GEOTHERMAL FIELD, IMPERIAL COUNTY, CALIFORNIA, ABST., GEOLOGY
B 6207 1232* GEOTHERMAL FIELD, NEW ZEALAND, BY GEOPHYSICAL METHODS, ABST.,
B 6207 1229* GEOTHERMAL FIELD, NICARAGUA, ABST., GEOLOGY AND DEVELOPMENT
B 6207 1207* GEOTHERMAL FIELDS IN INDONESIA, ABST., EXPLORATION AND
M 25 113 GEOTHERMAL FIELDS, NEW ZEALAND, NORTH ISLAND
M 25 141 GEOTHERMAL FIELDS, PHILIPPINES
M 25 121 GEOTHERMAL FIELDS, SUBSIDENCE IN
B 6212 2478 GEOTHERMAL GEOPRESSURED WATERS, DISSOLVED GAS CONTENT OF
B 6007 1074 GEOTHERMAL GRADIENT MAP, NIGER, DELTA
M 25 465 GEOTHERMAL GRADIENT, SIKHOTE ALIN
B 6004 714* GEOTHERMAL GRADIENTS, ABST., SCOTIAN, SHELF
B 6408 1151 GEOTHERMAL GRADIENTS, ALASKA, KODIAK SHELF
B 6307 1097 GEOTHERMAL GRADIENTS, AUSTRALIA, PERTH BASIN
B 6007 1073 GEOTHERMAL GRADIENTS, DETERMINATION OF
B 6002 233 GEOTHERMAL GRADIENTS FOR DIFFERENT BASINS
B 6303 510* GEOTHERMAL GRADIENTS IN NEOGENE SILICEOUS ROCKS IN
B 6007 1073† GEOTHERMAL GRADIENTS IN NIGER DELTA SEDIMENTARY BASIN,
B 6406 836 GEOTHERMAL HISTORY, AUSTRALIA, CANNING BASIN
B 6303 539* GEOTHERMAL HISTORY OF ONSHORE PERTH BASIN, WESTER AUSTRALIA,
B 6008 1394* GEOTHERMAL INVESTIGATION, ABST., ALVORD VALLEY, OREGON
B 6303 430* GEOTHERMAL INVESTIGATION OF NORTHERN NORTH SEA, ABST.
B 6004 673* GEOTHERMAL MATURATION OF ATHABASCA BITUMEN, ABST.,
B 6311 2033 GEOTHERMAL MATURATION OF POTENTIAL SOURCE ROCK
B 6204 657 GEOTHERMAL MEASUREMENTS, BRAZIL, SAO PAULO PLATEAU
M 25 159 GEOTHERMAL, MEXICO, CERRO PRIETO AREA
M 25 158 GEOTHERMAL, NEW MEXICO, VALLES, BACA, CALDERA FIELD
B 6101 65 GEOTHERMAL PATTERN, LOUISIANA, BAYOU CARLIN AREA

B 6101 65 GEOTHERMAL PATTERN, LOUISIANA, LAKE SAND AREA
M 25 50 GEOTHERMAL PHENOMENA, CIRCUM-PACIFIC
M 25 137 GEOTHERMAL POTENTIAL, INDONESIA, VOLCANIC
B 6008 1397* GEOTHERMAL POTENTIAL OF CRAZY MOUNTAIN BASIN, ABST.
B 6103 466* GEOTHERMAL POTENTIAL OF HIGH-LEVEL MAGM CHAMBERS IN ALASKA.
M 25 160† GEOTHERMAL POTENTIAL OF MEXICO
B 6105 798 * **GEOTHERMAL POTENTIAL OF THREE AREAS IN SOUTHEASTERN IDAHO AND**
M 25 156† GEOTHERMAL POTENTIAL OF WESTERN UNITED STATES--SUMMARY
M 25 157 GEOTHERMAL POTENTIAL, UNITED STATES, SNAKE RIVER DOWNWARP
B 6207 1207* GEOTHERMAL PROJECT, NICARAGUA, ABST., DEVELOPMENT AND,
B 6212 2509* GEOTHERMAL PROJECT, WEST JAVA, INDONESIA, ABST., RECENT
B 6009 1606* GEOTHERMAL PROSPECT IN FRIO FORMATION OF TEXAS GULF
M 25 164 GEOTHERMAL PROSPECT, NICARAGUA, MOMOTOMBO
M 25 163 GEOTHERMAL PROSPECT, NICARAGUA, SAN JACINTO, TISATE
M 25 144 GEOTHERMAL PROSPECT, PHILIPPINES, MAMBUCAL
M 25 141 GEOTHERMAL PROSPECT, PHILIPPINES, MT. MAQUILING--, MT. BANAHAO,
M 25 143 GEOTHERMAL PROSPECT, PHILIPPINES, PALIMPINON-, DAUIN
M 25 143 GEOTHERMAL PROSPECT, PHILIPPINES, TONGONAN
B 6008 1199 GEOTHERMAL PROSPECTING, 1975, CALIFORNIA
B 6108 1169 GEOTHERMAL PROSPECTING, 1976, CALIFORNIA
B 6208 1326 GEOTHERMAL PROSPECTING, 1977, CALIFORNIA
B 6409 1398 GEOTHERMAL PROSPECTING, 1979, CALIFORNIA
M 25 110† GEOTHERMAL PROSPECTS AROUND THE PACIFIC
M 25 144 GEOTHERMAL PROSPECTS, MINDANAO
M 25 143 GEOTHERMAL PROSPECTS, PHILIPPINES, NEGROS ISLAND
M 25 156 GEOTHERMAL PROVINCES, UNITED STATES, WESTERN
B 6207 1236* GEOTHERMAL RESEARCH PROGRAM, ABST., CANADIAN
B 6405 804* GEOTHERMAL RESERVOIR QUALITY, ABST., LOWER, WILCOX SHELF EDGE
B 6008 1412* GEOTHERMAL RESOURCE AREA, PARK COUNTY, MONTANA, ABST.,
B 6008 1389* GEOTHERMAL RESOURCE EVALUATION, ABST., GEOLOGIC AND, SEISMIC
B 6309 1595* GEOTHERMAL RESOURCE, FRIO FORMATION, TEXAS GULF COAST, ABST.,
B 6203 503* GEOTHERMAL RESOURCE INVESTIGATIONS AND DEVELOPMENT PLANS
B 6405 684* GEOTHERMAL RESOURCES, ABST., FUTURE EXPLORATION FOR
B 6004 689* GEOTHERMAL RESOURCES ABST., GEOLOGIC ENVIRONMENTS OF
B 6409 1569* GEOTHERMAL RESOURCES, ABST., REGIONAL TECTONIC FEATURES OF
B 6012 2174† GEOTHERMAL RESOURCES AN DEVELOPMENTS IN CALIFORNIA AND
M 25 147† GEOTHERMAL RESOURCES AND ENERGY IN JAPAN ESTIMATED FROM
B 6405 705* GEOTHERMAL RESOURCES EVALUATED BY WELL LOGS, ABST.
B 6205 882* GEOTHERMAL RESOURCES IN COLORADO, ABST., EXPLORATION FOR
M 25 148† GEOTHERMAL RESOURCES IN JAPAN--SUMMARY, GEOLOGICAL AND,
B 6207 1221* GEOTHERMAL RESOURCES IN TAIWAN, ABST., EXPLORATION AND
B 6404 556 GEOTHERMAL RESTORATION
B 6008 1397* GEOTHERMAL SYSTEM, ABST., GEOELECTRICAL INVESTIGATIONS OF
B 6305 836* GEOTHERMAL SYSTEM AT ROOSEVELT HOT SPRINN KGRA, BEAVER
B 6305 825* GEOTHERMAL SYSTEM IN NEW MEXICO, ABST., GEOLOGIC
B 6105 834* GEOTHERMAL SYSTEMS OF CORWIN SPRINGS—GARDINER AREA,
B 6303 417* GEOTHERMAL TEST WELL, BRAZORIA COUNTY, TEXAS, ABST., WORLDS
B 6405 709* GEOTHERMAL TEST WELL, BRAZORIA, TEXAS, ABST., SHALE MINERALOGY
B 6309 1607* GEOTHERMAL WATERS, ABST., FACIES DISTRIBUTION OF, HOSSTON SAND,
B 6203 519* GEOTHERMAL WATERS FROM SUTH LOUISIANA WELL, ABST., CHEMICAL
B 6203 529* GEOTHERMAL WATERS IN SOUTH LOUISIANA WELL, ABST., CHEMICAL
B 6203 555* GEOTHERMAL WATERS OF PETROLEUM BASINS—POSSIBLE SIGNIFICANT
B 6207 1236* GEOTHERMAL WELL COMPLETION AND CASING PROGRAMS IN
B 6211 2362* GEOTHERMAL WELLS, ABST., GEOLOGIC INTERPRETATIONS FOR,
B 6008 1411* GEOTHERMAL WELLS--SITING, DRILLING, AND TESTING, ABST., RAFT
M 25 149† GEOTHERMAL-ENERGY DEVELOPMENT IN ALASKA--SUMMARY,
B 6108 1386* GEOTHERMOMETER MODELS IN RECONNAISSANCE EXPLORATION FOR A
M 25 439 GEOTHERMS, TAIWAN
B 6409 1558 GERDES, BOB, ET AL, EFFECTS OF HURRICANE FREDERIC ON
S 4 329 GERHARD, L. C., MIOCENE KINGSHILL SEAWAY--A DYNAMIC CARBONATE
B 6303 430 GERHARD, LEE C., DOLOMITIZATION IN UPPER RE RIVER FORMATION,
B 6004 674 GERHARD, LEE C., SEDIMENT FACIES OF SALT RIVER ESTUARY, ST. CROIX,
B 6203 403 GERHARD, LEE C., STRATIGRAPHY AND DEPOSITIONAL SETTING,
B 6207 1217 GERLACH, TERRENCE M., FUELS FROM MAGMA—POTENTIAL ENERGY
B 6010 1711 GERMAN FEDERAL REPUBLIC, 1975, HARK, H. U., ON DEVELOPMENTS IN
B 6110 1642 GERMAN FEDERAL REPUBLIC, 1976, SCHRODER, L., ON DEVELOPMENTS IN
B 6210 1911 GERMAN FEDERAL REPUBLIC, 1977, SCHRODER, L., ON DEVELOPMENTS IN
B 6110 1658 GERMAN FEDERAL REPUBLIC, DEVELOPMENTS, 1975-1976
B 6210 1929 GERMAN FEDERAL REPUBLIC, PRODUCTION, 1976-1977
B 6310 1920 GERMANY, 1978, SCHRODER, L., ON DEVELOPMENTS IN, FEDERAL
B 6411 1969 GERMANY, 1979, SCHRODER, L., AND DEVELOPMENTS IN, WEST
B 6405 673* GERMANY, ABST., EXPLORATION IN CLASSIC, THRUST BELT AND ITS
B 6303 467* GERMANY, ABST., ORIGIN AND OCCURRENCE OF, CRUDE, OILS IN,
B 6405 759* GERMANY, ABST., SAND DISPERSAL AT, NORDERNEYER SEEGAT, WEST
B 6404 573 GERMANY, BENTHE SALT DOME, SECTIONS
B 6007 1009 GERMANY, BUNDSANDSTEIN
M 28 30 GERMANY, CARBONIFEROUS, SULZBACHER
B 6107 988 GERMANY, CHALK UNIT
M 27 84 GERMANY, CRETACEOUS, KRONER REEF LIMESTONE
M 28 28 GERMANY, CRETACEOUS, RIESELBERGER SANDSTONE
B 6010 1724 GERMANY, DDR, DEVELOPMENTS, 1975, EAST
B 6210 1923 GERMANY, DEVELOPMENTS, 1977, EAST
B 6210 1923 GERMANY, DEVELOPMENTS, 1977, POLAND-, EAST
M 27 42 GERMANY, DEVONIAN, TENTACULITEN KNOTTENKALK
M 26 337 GERMANY, DOGGER SANDSTONE, HOHNE FIELD, WEST
M 27 13 GERMANY, EOCENE LIMESTONE
B 6401 34 GERMANY, GRONINGEN GIANT GAS FIELD LANDSAT DATA,
M 26 338 GERMANY, HOHNE FIELD, SEISMIC SECTION
M 27 148 GERMANY, JURASSIC, EISENOOLITH
M 28 44 GERMANY, JURASSIC, EISENOOLITH
M 27 208 GERMANY, JURASSIC, HIERLATZKALK
M 27 128 GERMANY, JURASSIC, SOLNHOFEN LIMESTONE
M 27 222 GERMANY, JURASSIC, SOLNHOFEN LIMESTONE
M 27 175 GERMANY, JURASSIC, SOLNHOFEN LIMESTONE
M 27 87 GERMANY, MIOCENE, HYDROBIEN LIMESTONE
M 28 39 GERMANY, OLIGOCENE, CYRENENSCHICHTEN
M 27 81 GERMANY, OLIGOCENE, MOLASSE
M 28 71 GERMANY, ORDOVICIAN, GRIFFEL SCHIEFER
M 28 73 GERMANY, ORDOVICIAN, PHYCODEN SCHICHTEN
M 28 166 GERMANY, ORDOVICIAN, PHYCODEN SCHICHTEN
B 6401 68 GERMANY, OTHFRESEN SAMPLED SECTION
M 28 66 GERMANY, PENNSYLVANIAN, TEUSCHNITZER CONGLOMERATE
M 28 105 GERMANY, PENNSYLVANIAN, TEUSCHNITZER CONGLOMERATE
B 6210 1924 GERMANY, PRODUCTION, 1976-1977, EAST
B 6310 1936 GERMANY, PRODUCTION, 1977-1978, FEDERAL REPUBLIC OF
B 6401 34 GERMANY, ROTLIEGENDES FORMATION, NETHERLANDS AND, WEST
M 26 22 GERMANY, SOLENHOFEN LIMESTONE, S-WAVE MEASUREMENT
M 27 86 GERMANY, TRIASSIC, HALLSTATTER LIMESTONE
B 6401 68 GERMANY, WOLTWIESCHE SAMPLED SECTION
B 6207 1174 GERSTER LIMESTONE, NEVADA AND, UTAH
M 28 43 GESSO SOLFIFERA, ITALY, MIOCENE
M 28 85 GESSO SOLFIFERA, ITALY, MIOCENE
B 6205 781 GETAWAY FORMATION, NEW MEXICO
M 27 142 GETAWAY LIMESTONE, TEXAS, PERMIAN
M 27 48 GETAWAY LIMESTONE, TEXAS, PERMIAN
M 26 60 GETHING FORMATION, CANADA
B 6405 754* GETTING THE MOST OUT OF RADON GEOCHEMISTRY, ABST.
B 6210 2114 GETTY OIL CO., AND PETROCONSULTANTS S.A., ON DEVELOPMENTS IN
B 6110 1798 GETTY OIL CO., DEVELOPMENTS IN KUWAIT-SAUDI ARABIA DIVIDED
B 6310 1886 GETTY OIL CO., ON DEVELOPMENTS IN KUWAIT-SAUDI ARABIA DIVIDED
B 6010 1868 GETTY OIL CO., ON DEVELOPMENTS IN KUWAIT-SAUDI ARABIA DIVIDED
B 6411 1838 GETTY OIL CO., ON DEVELOPMENTS IN KUWAIT-SAUDI ARABIA DIVIDED
B 6101 80 GETTYSBURG BASIN, PENNSYLVANIA
B 6002 271 GEVAR-AM FORMATION, ISRAEL
B 6205 828 GEVARAM CANYON, ISRAEL
B 6206 951 GEVARAM CHANNEL, ISRAEL
B 6205 829 GEVARAM FORMATION, ISRAEL
B 6205 827† GEVARAM SHALES, LOWER CRETACEOUS, ISRAEL, AS POSSIBLE OIL
B 6209 1759 GEVIRTZ, JOEL, DISTRIBUTION OF LIVING BENTHONIC FORAMINIFERA AS
B 6109 1545 GEVIRTZ, JOEL L., NATURAL ENVIRONMENTAL-INFORMATION AND
B 6309 1597 GEVIRTZ, JOEL, RELICT AND EXPATRIATE RADIOLARIAN FAUNA IN GULF
M 25 158 GEYSERS FIELD, GEOTHERMAL, CALIFORNIA, THE
B 6406 860 GHADAMES BASIN, LIBYA, TUNISIA, AND, ALGERIA
B 6401 18 GHADAMES BASIN, TUNISIA
B 6110 1748 GHANA, DEVELOPMENTS, 1976
B 6210 1847 GHANA, DEVELOPMENTS, 1977
B 6310 1692 GHANA, DEVELOPMENTS, 1978
B 6411 1788 GHANA, DEVELOPMENTS, 1979
B 6310 1703 GHANA, PRODUCTION, 1978
B 6411 1794 GHANA, PRODUCTION, 1978-1979
B 6002 215 GHAWAR ARABIAN WHOLE CRUDE OIL
M 30 9 GHAWAR FIELD, LANDSAT IMAGE, SAUDI ARABIA
B 6401 9 GHAWAR GIANT OIL FIELD, LANDSAT DATA, SAUDI ARABIA
B 6104 585 GHAWAR-AIN DAR, D RESERVOIR, FIELD, PERSIAN GULF
B 6104 585 GHAWAR-HARADH, D RESERVOIR, FIELD, PERSIAN GULF
B 6104 585 GHAWAR-UTHMANIYAH, D RESERVOIR, FIELD, PERSIAN GULF
B 6303 512 GHORI, K. A. R., CYCLIC SEDIMENTATION OF CRETACEOUS-PALEOCENE
B 6203 516 GHOSH, SANTOSH K., ORIGIN OF CYCLICITY IN CONEMAUGH
B 6405 712 GHOSH, SANTOSH KUMAR, CARBONATE FACIES OF PENAS ALTAS
B 6405 690* GHOST CRAB OCYPODE CERATOPHTHALMA, PALLAS, AS POSSIBLE
B 6407 1016 GHOST ROCKS FORMATION, ALASKA
B 6408 1142 GHOST ROCKS FORMATION, ALASKA
B 6303 302 GHOST ROCKS FORMATION, ALASKA
S 3 76 GHOST ROLL FRONTS, WYOMING
B 6203 463† GIALO FIELD, LIBYA, SECONDARY, FERROAN, DOLOMITE, RHOMBS IN,
M 30 548 GIALO FORMATION, EL GIZA MEMBER, SIRTE BASIN
B 6405 748* GIAN GAS SUPPLY FROM DEEP OVERPRESSURED UPPER CRETACEOUS
B 6303 479* GIANT, ABST., STATFJORD FIELD, NORTH SEA
B 6004 727* GIANT FIELD AND ALASKAN ARCTIC PLATE, BY TITLE ONLY, PRUDHOE
B 6303 451* GIANT FIELD DISCOVERIES, 1967-77--AN OVERVIEW, ABST.
M 30 1† GIANT FIELD DISCOVERIES 1968-1978: AN OVERVIEW
B 6303 409* GIANT FIELDS BY SOUTHEASTERN MEXICO, BY TITLE ONLY
B 6304 698* GIANT FIELDS, GEOLOGIC SIGNIFICANCE OF, LANDSAT DATA ON SOME

M 30 339† GIANT FIELDS OF THE SOUTHERN ZONE--MEXICO
B 6205 749 GIANT FIELDS, WORLD
B 6205 749 GIANT GAS FIELD, DEFINITION
B 6303 488* GIANT GAS FIELD, ELLENBURGER FORMATION, PUCKETT FIELD, PECOS
B 6305 832 **GIANT IN WYOMING THRUST BELT, ABST., WHITNEY CANYON**
B 6405 730* GIANT IN WYOMING THRUST BELT, ABST., WHITNEY CANYON FIELD—
B 6405 638† GIANT IN WYOMING THRUST BELT, PAINTER RESERVOIR FIELD—
B 6303 532* GIANT MUDFLOW DEPOSITS IN SUBMARINE TRENCHES, HELLENIC BASINS
B 6401 20 GIANT OIL AND GAS FIELD, LANDSAT DATA, ALASKA, PRUDHOE BAY
B 6401 8† GIANT OIL AND GAS FIELDS, GEOLOGIC SIGNIFICANCE OF, LANDSAT
M 30 7† GIANT OIL AND GAS FIELDS, GEOLOGIC SIGNIFICANCE OF, LANDSAT
M 30 1 **GIANT OIL AND GAS FIELDS OF THE DECADE 1968-1978**
M 30 1† GIANT OIL AND GAS FIELDS OF THE DECADE 1968-1978
B 6205 888* GIANT OIL FIELD, ABST., HARTZOG DRAW, NEW
B 6205 749 GIANT OIL FIELD, DEFINITION
B 6304 701* GIANT OIL FIELDS, ABST., EKOFISK--FIRST OF WESTERN, EUROPEAN
B 6306 870† GIANT OIL FIELDS, CRETACEOUS, OCEANIC, ANOXIC EVENTS AS CAUSAL
B 6405 723* GIANT PETROLEUM PROVINCE, ABST., WILLISTON BASIN—REAWAKENING
B 6405 806* GIANT—ONLY SMAL PART OF ROCKY MOUNTAIN OVERTHRUST BELT,
B 6109 1552* GIANT—POZA RICA TREND, MEXICO, ABST., DIAGENESIS OF A
B 6204 686 GIBBS, JAMES A., GILD—A SOURCE OF GEOLOGIC DATA
M 25 512 GIBRALTAR DEPOSIT, BRITISH COLUMBIA
B 6101 81 GIBRALTAR FRACTURE ZONE
B 6101 88 GIBRALTAR FRACTURE ZONE, GLOSSCAP
B 6303 458 GIBSON, A. BRYCE, DEVELOPMENT AND USE OF FORMATION-TOPS FIL
B 6408 1280* GIBSON COUNTY, INDIANA, ABST., GEOLOGY OF, WHEATON
B 6408 1286* GIBSON COUNTY, INDIANA, ABST., ST. LOUIS AND, SALEM
B 6402 158 GIBSON FIELD, LOUISIANA
B 6012 2186 GIBSON, KEITH N., CLAST POPULATIONS IN SESPE AND POWAY
B 6308 1236 **GIBSON, WAYNE R., DEVELOPMENTS IN WEST TEXAS AND EASTERN NEW**
B 6409 1369 GIBSON, WAYNE R., DEVELOPMENTS IN WEST TEXAS AND EASTERN NEW
B 6103 465 GILBERT, WYATT G., TECTONICS, MIDDLE PALEOZOIC VOLCANISM, AND
M 24 278 GILBERTOWN-PICKENS-POLLARD FAULT SYSTEM, ALABAMA AND,
B 6104 594 GILBY FIELD, ALBERTA
B 6112 2137 GILBY OIL FIELD, ALBERTA
B 6405 777 GILCHRIST, KAREN A., PALEOCENE, MIDWAY, CONTINENTAL SHELF
B 6204 686† GILD—A SOURCE OF GEOLOGIC DATA
B 6405 668 GILES, A. B., PETROLEUM ACCUMULATION TRENDS IN EAST TEXAS SALT
B 6309 1601 GILES, ALICE B., UPPER CRETACEOUS--LOWE EOCENE STRATA,
B 6203 533 GILES, BEN F., FIELD DEVELOPMENT WITH 3-D SEISMIC METHODS—CASE
B 6012 2175 GILES, BEN F., SEISMIC DATA ANALYSIS WITH TIME SEISCROP MAP
B 6008 1391 GILES, BEN F., SEISMIC DATA ANALYSIS WITH TIME-SEISCROP MAP
B 6304 652 GILES COUNTY, CROSS SECTIONAL MODEL, VIRGINIA
B 6306 971 GILF KEBIR UNIT, EGYPT
S 5 1 GILL, DAN DEPOSITIONAL ENVIRONMENTS OF PINNACLE REEFS,
B 6304 608 GILL, DAN, DIFFERENTIAL ENTRAPMENT OF OIL AND GAS IN NIAGARAN
B 6004 674 GILL, DAN, SOURCE-RESERVOIR-SEAL RELATIONS IN PRODUCTIVE
B 6008 1370 GILLEY, JERRY D., DEVELOPMENT IN ARKANSAS, NORTH LOUISIANA,
B 6108 1379 GILLILAND, JIM, ENVIRONMENTAL CASE STUDY-HENDERSON MINE MILL,
B 6204 700 GILLIS CRUISE, RV
B 6407 1004 GILMER FIELD, TEXAS
M 26 150 GILMER FORMATION, TEXAS, EAST
M 26 147 GILMER FORMATION, TEXAS, EAST TEXAS BASIN
B 6007 1119 GILMER, INDIAN ROCK, FIELD, TEXAS
B 6309 1535 GILMER LIMESTONE GAS-CONDENSATE FIELD, TEXAS
B 6007 1120 GILMER LIMESTONE, TYPE SECTION, TEXAS
B 6007 1119† GILMER LIMESTONE, UPPER JURASSIC FORMATION. NORTHEASTERN
B 6007 1119 GILMER SOUTH FIELD, TEXAS
B 6308 1301 GILMORE, JACK, OIL AND GAS DEVELOPMENTS I NORTH MID-CONTINENT
B 6208 1477 GILMORE, JACK, OIL AND GAS DEVELOPMENTS IN NORTH
B 6108 1305 GILMORE, JACK, OIL AND GAS DEVELOPMENTS IN NORTH
M 24 205 GILREATH, J. A., EAST CAMERON BLOCK 270, OFFSHORE LOUISIANA--A
B 6209 1760 GILREATH, J. A., PRE-PLATFORM EXPLORATION OF HIGH ISLAND BLOCKS
B 6205 840 GILSONITE, SOLID HYDROCARBON, UTAH
M 24 73 GILWOOD SANDSTONE, ALBERTA
B 6203 528 GIMLIN, R. W., COMPUTER APPLICATIONS IN REGIONAL STUDY OF
B 6407 1005 GINGER FIELD, TEXAS
B 6307 1093 GINGIN FIELD, AUSTRALIA
B 6203 560 GINSBURG, R. N., INFLUENCE OF PLATFORM-DERIVED SEDIMENT ON
B 6004 674 GINSBURG, R. N., SEDIMENTARY RECORD OF PALEOCLIMATE IN
B 6410 1634 GINSBURG, ROBERT N., FACIES SUCCESSION OF PLIOCENE-PLEISTOCENE
B 6203 494 GINSBURG, ROBERT N., PRE-HOLOCENE CARBONATE ROCKS OF
M 25 230† GIPPSLAND, AND OTWAY BASINS--SUMMARY, STRUCTURAL STYLE AND
B 6405 746* GIPPSLAND BASIN, AUSTRALIA, ABST., DEFINITION AND DEVELOPMENT
B 6303 496* GIPPSLAND BASIN, AUSTRALIA, ABST., MASSIVE MARINE-SANDSTONE
B 6207 1226* GIPPSLAND BASIN, AUSTRALIA, ABST., SEISMIC AND GEOLOGIC, FACIES
B 6207 1151† GIPPSLAND BASIN, AUSTRALIA—COMPARISON WITH COOPER BASIN,
B 6207 1153 GIPPSLAND BASIN, STRATIGRAPHY, AUSTRALIA
B 6207 1152 GIPPSLAND BASIN, WELLS, AUSTRALIA
B 6207 1153 GIPPSLAND LIMESTONE, AUSTRALIA
B 6204 592 GIRARD, O. W., JR., PETROLEUM POTENTIAL OF U.S. ATLANTIC SLOPE,
B 6303 432 GIRARD, OSWALD W., JR., WEST SIBERIAN BASIN, ABST.
B 6111 1930 GIRTY CREEK, ELLESMERE ISLAND
B 6107 1077 GIRVANELLA, CALCIFIED, ALGAL FILAMENTS AND NATURE OF
B 6004 675 GIVENS, M. M., STRATIGRAPHY OF NOVA SCOTIA SHELF, ABST.
B 6406 826 GIVETIAN-FRASNIAN PILLARA REEF CYCLE, AUSTRALIA
B 6004 645* GLACIAL MARINE ENVIRONMENTS, ABST., CYCLIC SEDIMENTATION IN
B 6303 411* GLACIAL ORIGIN OF MEGACHANNELS OF UPPER YAKATAGA
S 9 19 GLACIAL PERIODS, PALEOCLIMATE
B 6203 512* GLACIAL SEDIMENTS OF VORING PLATEAU, NORWEGIAN SEA, ABST.,
B 6004 716* GLACIAL SHORELINE, MALASPINA FORELAND, GULF OF ALASKA, ABST.,
S 4 289† GLACIAL-INTERGLACIAL CHANGES IN TEMPERATURE AND ISOTOPIC
B 6303 483* GLACIAL-MARINE ENVIRONMENT, ABST., SEDIMENTOLOGIC FACIES I
B 6303 434* GLACIAL-MARINE SEDIMENTATION PRELIMINARY MODELING BASED ON
B 6303 410* GLACIAL-MARINE SEDIMENTS OF WEST ANTARCTICA AND PUGET
B 6303 411* GLACIAL-MARINE TRACE- AND BODY-FOSSIL ASSOCIATIONS, UPPER
B 6004 693* GLACIATED BASINS, ABST., FACTORS CONTROLLING, SUSPENDED
S 4 79 GLACIATION, ANTARCTIC
S 6 317 GLACIATION DATING, SOLAR CHRONOLOGY
B 6212 2412† GLACIER, GULF OF ALASKA, HOLOCENE, SLUMP ON, CONTINENTAL,
B 6407 972 GLACIER NATIONAL PARK, CRETACEOUS, MONTANA
B 6109 1538* GLACIO-EUSTATIC LOWERING OF SEA LEVEL—EVIDENCE FROM
B 6203 534* GLACIO-EUSTATIC SEA-LEVEL CHANGES IN EASTERNMOST
B 6303 513* GLACIOMARINE CARBONATE SEDIMENTATION--BERRIEDALE LIMESTONE,
B 6403 438* GLACIOMARINE DEPOSITIONAL ENVIRONMENTS AND BIOFACIES,
B 6407 1087 GLADENKOV, YURI B., STRATIGRAPHY OF MARINE PALEOGENE AND
S 5 78 GLADWIN COUNTY LOCATION, MICHIGAN,
B 6203 517 GLAESER, J. DOUGLAS, DEVONIAN SLOPE SEDIMENTS IN CENTRAL
B 6305 827 GLAESER, J. DOUGLAS, ENVIRONMENTS FOR SEDIMENTARY URANIUM IN
S 6 1 **GLAESSNER, MARTIN F., EDITED BY, THE GEOLOGIC TIME SCALE**
S 11 237 GLANDULINA COMATULA
S 11 274 GLANDULINA COMATULA
S 11 274 GLANDULINA LAEVIGATA
S 11 237 GLANDULINA LAEVIGATA
M 25 386 GLASBY, G. P., EXPLOITATION OF MANGANESE NODULES IN SOUTH
B 6207 1218 GLASKI, FRED A., RENEWABLE ENERGY SYMPOSIUM, INTRODUCTION,
B 6305 827 GLASS, GARY B., COAL DEVELOPMENT IN POWDER RIVER BASIN O
B 6004 672* GLASS MOUNTAIN, WEST TEXAS, ABST., BIOHERMAL DEVELOPMENT IN,
B 6103 431† GLASS MOUNTAINS, WEST TEXAS, NEAL RANCH-, LENOX HILLS CONTACT
B 6203 517 GLASS, STEVEN W., LACUSTRINE PSEUDOVARVES FROM CRETACEOUS
B 6105 833 GLASSLEY, W., DREDGED MATERIAL DEPOSITED AT OCEAN DUMPING
M 28 39 GLAUCONITE AFTER CALCITE, FRAGMENTS, ECHINODERM
M 25 454 GLAUCONITE AND APATITE, FUTURE POTENTIAL, JAPAN, URANIFEROUS
M 27 154 GLAUCONITE, ANGULAR GRAINS
M 28 39 GLAUCONITE, AUTHIGENIC
M 28 61 GLAUCONITE, BIBLIOGRAPHY
S 6 275 GLAUCONITE, CRETACEOUS, RADIOMETRIC SCALE
M 28 38 GLAUCONITE, CRYSTALS, BROWN
S 6 130 GLAUCONITE DATING, RELIABILITY
M 28 38 GLAUCONITE, DETRITAL, FRAGMENTS
S 6 128 GLAUCONITE, FRACTIONATION
M 28 128 GLAUCONITE, GRAINS, BIREFRINGENCE
B 6002 185 GLAUCONITE IN DITCH CUTTINGS, NORTH SEA
M 28 129 GLAUCONITE, IN SANDSTONE, CEMENTS
M 27 154 GLAUCONITE, MARL
M 28 163 GLAUCONITE PELLETS, COMPACTION, OF
S 6 128 GLAUCONITE, SAMPLE SELECTION, REVIEW
B 6004 725* GLAUCONITE-MAGHEMITE-HEMATITE TRANSFORMATION AND ITS
B 6005 883† GLAUCONITE. DISCUSSION AND REPLY, RADIOMETRIC, AGE
S 6 207 GLAUCONITES, ARGON DATES, DIAGRAM, PHANEROZOIC
S 6 134 GLAUCONITES, EUROPE, LUTETIAN BOUNDARIES
S 6 136 GLAUCONITES, EXAMPLES, EUROPE, CRETACEOUS
S 6 206 GLAUCONITES, INTERPRETATION
S 6 206 GLAUCONITES, LOSS OF AUTHIGENIC MINERALS
S 6 131 GLAUCONITES, REPRODUCIBILITY
S 6 133 GLAUCONITES, VERTICAL REPRODUCIBILITY, DIAGRAM
B 6405 691* GLAUCONITIZATION OF FELDSPAR GRAINS IN A MIDDLE PROTEROZOIC
B 6109 1551* GLEN ROSE, CRETACEOUS, FACIES MOSAIC, BLANCO AND HAYS
M 24 259 GLEN ROSE FORMATION, ALABAMA
B 6402 232 GLEN ROSE FORMATION, TEXAS
B 6301 33 GLEN ROSE LIMESTONE, TEXAS
M 27 41 GLEN ROSE LIMESTONE, TEXAS, CRETACEOUS
M 27 144 GLEN ROSE LIMESTONE, TEXAS, CRETACEOUS
M 27 124 GLEN ROSE LIMESTONE, TEXAS, CRETACEOUS
M 27 87 GLEN ROSE LIMESTONE, TEXAS, CRETACEOUS
M 27 135 GLEN ROSE LIMESTONE, TEXAS, CRETACEOUS
M 27 147 GLEN ROSE LIMESTONE, TEXAS, CRETACEOUS
M 27 134 GLEN ROSE LIMESTONE, TEXAS, CRETACEOUS
B 6109 1555* GLEN ROSE PATCH-REEF COMPLEX—BANDERA COUNTY, TEXAS, ABST.,
B 6408 1220 GLEN ROSE SECTION, TEXAS

M 27 77 GLENCAR LIMESTONE, IRELAND, MISSISSIPPIAN
M 27 78 GLENCAR LIMESTONE, IRELAND, MISSISSIPPIAN
B 6406 823 GLENISTER KNOLLS, AUSTRALIA
B 6105 843* GLIDE-PLANE FAULT STRUCTURES, STYLES, MECHANISMS, AND
B 6203 494 GLINKMAN, JACK V., DETERMINATION OF RESERVOIR FLUID CONTENT
B 6312 2194 GLITREFJELLET MEMBER, SVALBARD
M 29 463 GLOBAL CHANGE, AVERAGE SUBSIDENCE RATES
M 29 461 GLOBAL CHANGE, MARGINS VERSUS INTERIORS
M 29 462 GLOBAL CHANGE, MESOZOIC-CENOZOIC BASINS, ANALYSIS
M 29 464 GLOBAL CHANGE, MESOZOIC-CENOZOIC BASINS, INTERPRETATION
M 29 464 GLOBAL CHANGE, MID-PALEOZOIC BASINS, ANALYSIS
M 26 185† GLOBAL CHANGES OF SEA LEVEL, PART 10: SEISMIC RECOGNITION OF
B 6205 772† GLOBAL CHANGES OF SEA LEVEL, PART 10: SEISMIC RECOGNITION OF
M 26 205† GLOBAL CHANGES OF SEA LEVEL, PART 11: GLOSSARY OF TERMS USED IN
M 26 51† GLOBAL CHANGES OF SEA LEVEL, PART 1: OVERVIEW, SEISMIC,
M 26 53† GLOBAL CHANGES OF SEA LEVEL, PART 2: THE DEPOSITIONAL SEQUENCE
M 26 63† GLOBAL CHANGES OF SEA LEVEL, PART 3: RELATIVE CHANGES OF SEA
M 26 83† GLOBAL CHANGES OF SEA LEVEL, PART 4: GLOBAL CYCLS OF RELATIVE
M 26 99† GLOBAL CHANGES OF SEA LEVEL, PART 5: CHRONOSTRATIGRAPHIC
M 26 117† GLOBAL CHANGES OF SEA LEVEL, PART 6: STRATIGRAPHIC
M 26 135† GLOBAL CHANGES OF SEA LEVEL, PART 7: SEISMIC STRATIGRAPHIC
M 26 145† GLOBAL CHANGES OF SEA LEVEL, PART 8: IDENTIFICATION OF UPPER
M 26 165† GLOBAL CHANGES OF SEA LEVEL, PART 9: SEISMIC INTERPRETATION OF
B 6205 752† GLOBAL CHANGES OF SEA LEVEL, PART 9: SEISMIC INTERPRETATION OF
M 26 49† GLOBAL CHANGES OF SEA LEVEL, SEISMIC, STRATIGRAPHY AND
M 26 95 GLOBAL CYCLES, AGE ESTIMATION BEFORE DRILLING
M 26 92 GLOBAL CYCLES, CAUSES, SEA LEVEL FLUCTUATIONS
M 26 87 GLOBAL CYCLES, CENOZOIC, CHRONOCORRELATION
M 26 88 GLOBAL CYCLES, CHART CONSTRUCTION, SEA LEVEL FLUCTUATIONS
M 26 90 GLOBAL CYCLES, CORRELATION, REGIONAL
M 26 84 GLOBAL CYCLES, FIRST-ORDER, CORRELATION WITH SECOND-ORDER
M 26 93 GLOBAL CYCLES, INTERREGIONAL UNCONFORMITIES
M 26 85 GLOBAL CYCLES, JURASSIC
M 29 472 GLOBAL CYCLES OF CHANGE, CENOZOIC
M 29 470 GLOBAL CYCLES OF CHANGE, FIRST-ORDER CYCLES, PHANEROZOIC
M 29 471 GLOBAL CYCLES OF CHANGE, JURASSIC-TERTIARY
M 29 470 GLOBAL CYCLES OF CHANGE, SECOND-ORDER CYCLES, PHANEROZOIC
M 29 469† **GLOBAL CYCLES OF RELATIVE CHANGES OF SEA LEVEL, SEISMIC,**
M 26 83† GLOBAL CYCLES OF RELATIVE CHANGES OF SEA LEVEL, SEISMIC,
M 26 83 GLOBAL CYCLES, SEA LEVEL FLUCTUATIONS
M 26 84 GLOBAL CYCLES, SECOND-ORDER, CORRELATION WITH FIRST-ORDER
M 26 85 GLOBAL CYCLES, TERTIARY
B 6306 872 GLOBAL DEPOSITION OF ORGANIC CARBON-RICH SEDIMENTS O
B 6205 739† GLOBAL EXPLORATION—REQUIREMENT FOR SURVIVAL, ACCELERATION
M 26 92 GLOBAL HIGHSTANDS, SEA LEVEL FLUCTUATIONS
M 26 92 GLOBAL LOWSTANDS, SEA LEVEL FLUCTUATIONS
B 6405 706* GLOBAL OCCURRENCE OF ABNORMAL FORMATION PRESSURES, ABST.
M 29 461† GLOBAL SEA LEVEL CHANGE: A VIEW FROM THE CRATON
B 6303 537* GLOBAL TECTONICS, ABST., DEEP-SEA, DRILLING AND
B 6203 562* GLOBAL TECTONICS, ABST., UNIFORMITARIANISM AND
S 11 296 GLOBIGERINA
S 11 263 GLOBIGERINA BULLOIDES
S 11 296 GLOBIGERINA BULLOIDES
S 6 51 GLOBIGERINA NEPENTHES, EXAMPLE, BIOCHRONOLOGY
M 28 86 GLOBIGERINA OOZE, IN GYPSUM CEMENT
S 11 263 GLOBIGERINA PACHYDERMA
S 11 296 GLOBIGERINA PACHYDERMA
M 27 36 GLOBIGERINID FORAMS
S 11 263 GLOBIGERINIDAE
B 6004 690* GLOBIGERINOIDES DATUM, ABST., UNRELIABILITY OF
B 6009 1564† GLOBIGERINOIDES DATUM, UNRELIABILITY
S 11 280 GLOBOBULIMINA AFFINIS
S 11 246 GLOBOBULIMINA AFFINIS
S 11 246 GLOBOBULIMINA AURICULATA
S 11 280 GLOBOBULIMINA AURICULATA
S 11 246 GLOBOBULIMINA PACIFICA
S 11 280 GLOBOBULIMINA PACIFICA
S 11 263 GLOBOROTALIA CRASSAFORMIS
S 11 296 GLOBOROTALIA CRASSAFORMIS
S 11 264 GLOBOROTALIA CRASSULA
S 11 296 GLOBOROTALIA INFLATA
S 11 264 GLOBOROTALIA INFLATA
S 4 341 GLOBOROTALIA MAYERI, ST. CROIX
S 11 298 GLOBOROTALIA SCITULA
S 11 264 GLOBOROTALIA SCITULA
M 27 36 GLOBOROTALIC FORAMS
S 11 263 GLOBOROTALIIDAE
B 6011 2058† GLOBOTRUNCANA FORNICATA ZONE OF UPPER AUSTIN GROUP,
B 6002 213 GLOBULE EXPULSION HYPOTHESIS
M 25 218 GLOMAR CHALLENGER DSDP DRILLING SITES, NEW ZEALAND
B 6106 931 GLOMAR CHALLENGER DSDP LEG 3
B 6106 931 GLOMAR CHALLENGER DSDP LEG 39
B 6007 1042 GLOMAR CHALLENGER, DSDP SITE 98, RV
B 6201 88 GLOMAR CHALLENGER SEISMIC PROFILE, ALASKA
B 6102 195 GLOMAR CHALLENGER SEISMIC PROFILE, KOMANDORSKIY BASIN
B 6305 828 GLOPPEN, TOR GUNNAR, ALLUVIAL FANS AND THEIR DEPOSIT IN
M 26 13 GLOSSARY, GEOPHYSICS
M 26 205† GLOSSARY OF TERMS USED IN SEISMIC STRATIGRAPHY, SEISMIC,
S 6 25 GLOSSARY OF USAGE, GEOCHRONOLOGY
B 6412 2092 GLOSSARY, REMOTE SENSING
M 26 379 GLOSSARY, RESOLUTION SYSTEMS
M 26 205 GLOSSARY, SEISMIC, STRATIGRAPHY, TERMINOLOGY
B 6101 88 GLOSSCAP GIBRALTAR FRACTURE ZONE
B 6408 1284 GLOVER, ALBERT D., COAL IN PENNSYLVANIA—GEOLOGY, CURRENT
S 4 37† GLOVERS REEF ATOLL, BELIZE, PATCH-REEF ECOLOGY AND
S 4 43 GLOVERS REEF ATOLL, CORAL-ALGAL ASSEMBLAGES, BELIZE
S 4 42 GLOVERS REEF ATOLL, FORAMINIFERAL ASSEMBLAGES, BELIZE
B 6103 331 GLOWACZ, M. E., PETROLEUM GEOLOGY AND OIL AND GAS POTENTIAL
B 6007 1054 GLYNN, P. W., EVOLUTION OF MODERN CARIBBEAN FRINGING REEF,
M 27 205 GMUNDEN FLYSCH, AUSTRIA, CRETACEOUS
M 28 8 GNEISS, MASSACHUSETTS, DEVONIAN
B 6105 801 **GOAT, C. G., TEXAS LIGNITE**
B 6007 1011 GOAT MOUNTAIN FAULT, NEW MEXICO
B 6303 437* GOAT SEEP DOLOMITE GUADALUPIAN, PERMIAN, GUADALUPE
B 6205 781 GOAT SEEP FORMATION, NEW MEXICO
B 6411 1963 GOBAN SPUR, EIRE
B 6402 225 GOBBLE GRABEN, TEXAS
B 6209 1768* GOBER CHALK, UPPER CAMPANIAN, AT TYPE LOCALITIES IN NORTHEAST
B 6405 691* GODAVARI VALLEY, SOUTH INDIA, ABST., TRACE, FOSSILS FROM
S 4 111 GODDARD, D. A., MANS IMPACT ON COASTAL REEFS--A EXAMPLE FROM
B 6110 1587 GODEL, M., ON DEVELOPMENTS IN LESSER ANTILLES, 1976
S 5 34 GODERICH SALT MINE, ONTARIO, HURON COUNTY,
B 6002 319 GOEBEL, EDWIN D., GAS RESERVES, GAS RESOURCE BASE, AND PROSPECT
B 6008 1393 **GOERING, JOHN D., GEOLOGIC CONTROLS OVER HOT WATER MIGRATION**
M 28 126 GOETHITE, IN SANDSTONE, CEMENTS
B 6406 821 GOGO FORMATION, AUSTRALIA
M 25 554 GOLD AND BASE-METAL SULFIDES, SURVEYS, ECUADOR
M 25 467 GOLD AND CASSITERITE, USSR, NORTHEAST
M 25 460 GOLD AND SILVER, EASTERN ASIATIC VOLCANOGENIC BELT
B 6303 555 GOLD, C. M., STRUCTURE-CONTOUR MAPS COMPUTER CONSTRUCTED
M 25 497 GOLD, CANADA
M 25 533 GOLD, COLOMBIA
M 25 553 GOLD DEPOSITS, ECUADOR, ANGAS, COPPER, ZINC, LEAD, SILVER, AND
M 25 468 GOLD DEPOSITS, USSR, NORTHEAST
B 6207 1172 GOLD HILL, SECTION, UTAH
M 25 461 GOLD METALLOGENIC SYSTEMS, TRANS-, PACIFIC, COPPER AND
M 25 518 GOLD, MEXICO, PACHUCA Y REAL DEL MONTE MINE, SILVER AND
M 25 518 GOLD, MEXICO, TAYOLTITA-, SAN DIMAS MINE
M 25 426 GOLD, NEW SOUTH WALES, COBAR, COPPER, ZINC, AND
M 25 533 GOLD, OCCURRENCE AND PRODUCTION, ANDEAN REGION
M 25 552 GOLD, SILVER, AND COPPER, ECUADOR, TOACHI DEPOSITS, LA PLATA
B 6105 815* GOLD, URANIUM, AND PYRITE CONCENTRATIONS IN PROTEROZOIC
M 25 476 GOLD-ANTIMONITE MINERALIZATION, TIME OF, USSR, EASTERN
M 25 476 GOLD-ANTIMONITE, USSR, EASTERN, LODE DEPOSITS OF
M 25 461 GOLD-BEARING RING, GREAT, PACIFIC
M 25 427 GOLD-COPPER, QUEENSLAND, MOUNT MORGAN
M 25 444 GOLD-SILVER, SOUTH KOREA
M 25 439 GOLD-SILVER-COPPER DEPOSITS, TAIWAN
M 29 321 GOLDEN LANE CARBONATE BANK, MEXICO, SLOPE
B 6202 236 GOLDEN LANE FIELD, MEXICO
B 6301 48 GOLDEN LANE, SECTION, MEXICO
B 6004 545 GOLDEN MEADOW FAULT SYSTEM, LOUISIANA
M 26 196 GOLDEN SPIKE, CARBONATE BUILDUP, SEISMIC EXAMPLE, CANADA,
B 6004 730* GOLDEN SPIKE OIL FIELD, ALBERTA, ABST., POROSITY, REDUCTION IN,
B 6205 783 GOLDEN SPIKE REEF, SEISMIC DATA, ALBERTA
M 27 17 GOLDEN-BROWN ALGAE, COCCOLITHS
B 6405 772 GOLDHABER, M. B., HISTORY OF SULFIDIZATION OF SOUTH TEXAS
B 6108 1379 GOLDHABER, MARTIN B., EXPERIMENTAL STUDY OF PYRITE OXIDATION
B 6405 692 GOLDHABER, MARTIN B., GEOCHEMICAL EFFECTS OF EARLY DIAGENESIS
B 6303 514 GOLDHABER, MARTIN B., MARCOSITE IN ROLL-TYPE URANIUM DEPOSITS,
B 6108 1388 GOLDHABER, MARTIN B., RECOGNITION OF OXIDIZED SULFIDE
B 6010 1653 GOLDSMITH, L. H., ON DEVELOPMENTS IN NICARAGUA, 1975
B 6105 789 **GOLDSMITH, V., COASTAL-DUNE GEOMORPHOLOGY AND CLASSIFICATION**
B 6105 791 **GOLDSMITH, V., DUNE DYNAMICS IN COASTAL ZONE—PAST, PRESENT, AND**
B 6105 788 **GOLDSMITH, V., WAVE-CLIMATE STUDIES IN BALTIMORE CANYON TROUGH**
S 7 11 GOLFO DE CAMPECHE SLOPE, DESCRIBED
M 29 121 GOLFO SAN JORGE, SEISMIC PROFILE
B 6109 1535* GOLIAD FORMATIONS OF SOUTH TEXAS COASTAL PLAIN, ABST.,
B 6206 1053 GOLIAD SAND, TEXAS
B 6405 685 GOLUBIC, S., ET AL, MICROBIAL AND INVERTEBRATE ENDOLITHIC
B 6405 713 GOLUBIC, S., MORPHOMETRY OF LATE ORDOVICIAN MICROBIAL

B 6303 458 GOLUBIC, S., STUDY OF FOSSI MICROBIAL BORINGS--NEW APPROACH,
B 6205 862 GOMEZ, ELLENBURGER, ULTRADEEP FIELD, PRODUCTION, TEXAS
B 6112 2086 GOMEZ FIELD, KEROGEN, TEXAS
B 6303 436* GONDWANALAND, ABST., ANTARCTICA AND
B 6112 2079 GONDWANALAND MARGINS
B 6408 1285 GONG, ZHICONG, COAL RANK IN PART OF WESTERN KENTUCKY
M 27 12 GONIOLITHON, RED, ALGAL GRAIN
B 6309 1607* GONZALES, AND FAYETTE COUNTIES, TEXAS ABST., FIELD RELATIONS
M 25 333 GONZALES, GROVER, HYDROCARBON POTENTIAL OF COASTAL BASINS OF
B 6301 39 GONZALES TROUGH, TEXAS
B 6402 148 GOOD HOPE FIELD, LOUISIANA
B 6107 975 GOOD HOPE SEGMENT, WALVIS RIDGE--, CAPE
M 24 149 GOOD LUCK FIELD, TEXAS
B 6303 533 GOODELL, PHILIP C., CARBONATE GEOLOGY OF PENA BLANCA URANIUM
B 6303 458 GOODELL, PHILIP C., PRELIMINARY ANALYSES OF BRINES FROM
B 6303 509 GOODELL, PHILIP C., TRACE-ELEMENT GEOCHEMISTRY AND DIAGENESIS
B 6211 2361 GOODIN, S., EARLY MESOZOIC PALEOTECTONIC-PALEOGEOGRAPHIC
B 6303 409 GOODWIN, DANIEL A., SEISMIC STRATIGRAPHY AND STRUCTURE OFF
B 6105 840* GOODWIN LIMESTONE--LOWER ORDOVICIAN SHELF-SLOPE CARBONATE
B 6303 415 GOODWIN, PETER W., MODERN AND ANCIENT EXAMPLES OF ANNELID
B 6004 653 GOODWIN, R. H., RELATION OF SEISMICITY AN TECTONIC PROVINCES,
M 25 237 GOODWYN DISCOVERY WELLS, AUSTRALIA
M 25 237 GOODWYN FAULT BLOCK, AUSTRALIA
B 6201 43 GOODWYN FIELD, AUSTRALIA
B 6306 955 GOOSE COVE FORMATION, NEWFOUNDLAND
B 6201 100 GOOSE EGG FORMATION, WYOMING
B 6306 955 GOOSE TICKLE FORMATION, NEWFOUNDLAND
B 6004 675 GOPINATH, TUMKUR R., INFLUENC OF SOURCE AND PROCESSES IN
M 25 485 GORDA RISE, SEAFLOOR SPREADING
B 6409 1437 GORDON, DONIVAN L., OIL AND GAS DEVELOPMENTS IN NORTH
B 6008 1332 GORDON DONIVAN L., OIL AND GAS DEVELOPMENTS IN NORTH
S 6 327 GORDON, MACKENZIE, JR., THE MISSISSIPPIAN-PENNSYLVANIAN
B 6105 770 GORDON, R. B., CONTAINMENT OF PARTICULATE WASTE IN MARINE
B 6408 1278 GORDY, PETER L., HYDROCARBON ACCUMULATIONS IN OVERTHRUST
B 6304 598 GORHAM, FRANK D., JR., FRACTURES IN CRETACEOUS ROCKS FROM
B 6203 569* GORHAM OIL FIELD, RUSSELL COUNTY, KANSAS, ABST.
B 6111 1956 GORRIN-MACHETE HEAVY OIL, VENEZUELA
B 6005 834 GORRINGE BANK
B 6203 544 GORSLINE, D. S., BOTTOM SEDIMENTS OF SAN NICOLAS BASIN,
B 6004 675 GORSLINE, D. S., HIGH-RESOLUTION STUDIES OF PALEOCLIMATIC AND
B 6012 2181 GORSLINE, D. S., PATTERNS OF CALIFORNIA CONTINENTAL BORDERLAND
B 6403 442 GORSLINE, D. S., SEDIMENTOLOGIC HISTORY AND CHARACTERISTICS OF
B 6105 789 GORSLINE, D. S., SHELF-SEDIMENT DYNAMICS OF SOLID-WASTE
B 6106 903 GORSLINE, DONN S., MARINE GEOLOGY OF BAJA CALIFORNIA
B 6203 511 GORSLINE, DONN S., NEW EVIDENCE OF CURRENT WINNOWING ACTIVITY
B 6108 1376 GORTON, KENNETH A., NUGGET OIL ACCUMULATIONS AT DRY PINEY,
B 6211 2243 GOSFORD CYCLE, INFORMAL NAME, CALIFORNIA
B 6211 2247 GOSFORD FAN, CALIFORNIA
B 6211 2256 GOSFORD SAND, CALIFORNIA
B 6402 242 GOT, HENRI, SEISMIC STRUCTURE OF SOUTHWESTERN PELOPONNESUS
B 6405 713 GOTER, EDWIN R., DEPOSITION OF ENEWETA ATOLL REEF, MIDDLE
B 6303 458 GOTER, EDWIN R., PLEISTOCENE METEORIC-VADOSE DIAGENESIS OF
B 6004 718* GOTLAND, SWEDEN, ABST., SEDIMENTOLOGY OF SOME, SILURIA ROCKS
B 6203 455 GOULD, K. W.,, ISOTOPIC EVIDENCE FOR MICROBIOLOGIC ROLE IN
B 6410 1646 GOULDS DAM DEPOSIT, AUSTRALIA
S 10 232 GOUSANY FIELD, USSR, BAKU AREA
M 25 405 GOVE, BAUXITE RESERVES, AUSTRALIA
M 25 318 GOVEA R., CARLOS, POTENTIAL FOR EXPLORATION AND DEVELOPMENT
B 6405 713 GOVEAN, FRANCES M., CALCIFICATION MODEL AND SECONDARY
B 6202 235 GOVELA S., AMADO, STRATIGRAPHY AND STRUCTURE OF CHICONTEPEC
B 6105 773 GOVONI, DAVID L., PRELIMINARY CALCAREOUS NANNOFOSSIL ZONATION
M 25 354† GR, ARTIFICIAL RECHARGE OF
B 6203 531* GRABEN EMBAYMENTS IN SOUTHERN GREECE, ABST., HOLOCENE EPOCH,
B 6407 1008 GRABEN-FAULT STRUCTURES, TEXAS, LOUISIANA, AND, ARKANSAS
M 29 7 GRABENS, NORMAL FAULT MECHANISMS
B 6203 517 GRABOWSKI, GEORGE, TIDELESS MIDDLE ORDOVICIAN SEDIMENTS, ABST.
B 6203 519 GRACE, JAMES T., HIGH LEVELS OF VARIATION IN SIZE-FREQUENCY
B 6105 766 GRACE, L. M., COMPUTER GRAPHIC DISPLAYS OF DIPMETER INFORMATION
B 6203 518 GRACE, SCOTT R., PHOSPHATE IN BOTTOM SEDIMENTS IN MYAKKA AND
B 6405 674* GRADED BEDS AS PALEOGRAVIMETRIC CAMERAS, ABST.
B 6205 879* GRADED BEDS OF CHANGING GRAVITATIONAL ACCELERATION WITH
S 7 220 GRADED TURBIDITES( ), WEST FLORIDA SLOPE
B 6007 1074 GRADIENT MAP, NIGER, DELTA, GEOTHERMAL
B 6004 714* GRADIENTS, ABST., SCOTIAN, SHELF, GEOTHERMAL
B 6007 1073 GRADIENTS, DETERMINATION OF, GEOTHERMAL
B 6007 1073† GRADIENTS IN NIGER DELTA SEDIMENTARY BASIN, APPROXIMATE,
B 6105 824* GRADING IN GRAIN FLOWS, ABST., DEVELOPMENT OF INVERSE
B 6203 536* GRADING IN OVERWASH DEPOSITS, ABST., SEDIMENT
B 6405 713 GRADSTEIN, FELIX, APPLICATION OF STATISTICAL MODELS IN
B 6002 322* GRADY, AND CANADIAN COUNTIES, OKLAHOMA, ABST., DIBBLE--,
B 6309 1578 GRADY, WILLIAM C., PETROGRAPHIC EVALUATION OF WEST VIRGINI
M 27 48 GRAFORD FORMATION, TEXAS, PENNSYLVANIAN
M 27 70 GRAHAM FORMATION, TEXAS, PENNSYLVANIAN
M 27 62 GRAHAM FORMATION, TEXAS, PENNSYLVANIAN
M 28 119 GRAHAM FORMATION, TEXAS, PENNSYLVANIAN
B 6012 2179 GRAHAM, STEPHAN A., APPLICATIONS OF PLATE TECTONICS TO
B 6211 2214 GRAHAM, STEPHAN A., ROLE OF SALINIAN BLOCK IN EVOLUTION OF SAN
B 6012 2181 GRAHAM, STEPHEN ALAN, TERTIARY STRATIGRAPHIC AND
B 6109 1449 GRAHAM UPLIFT, ARIZONA
B 6107 1069† GRAIN BINDING BY ENDOLITHIC ALGAE, MICRITIZATION AND,
M 27 205 GRAIN COMPACTION
M 27 205 GRAIN DEFORMATION
B 6105 824* GRAIN FLOWS, ABST., DEVELOPMENT OF INVERSE, GRADING IN
M 28 109 GRAIN ORIENTATION, BIBLIOGRAPHY
M 28 108 GRAIN SIZE SORTING, BIBLIOGRAPHY
M 28 108 GRAIN-SHAPE ABRASION, BIBLIOGRAPHY
S 3 124 GRAIN-SIZE ANALYSIS, URANIUM HOST SANDSTONES
B 6002 184 GRAIN-SIZE PROFILES
S 7 298 GRAIN-SORTING CURVE, GYRE BASIN
B 6105 841* GRAINSTONES, ABST., EVIDENCE FOR SOLUTION, COMPACTION IN, LIME
B 6405 757* GRAINSTONES AT TOP OF MADISON GROUP, DISTURBED BELT,
B 6303 362† GRAINSTONES, UPPER PLEISTOCENE, YUCATAN PENINSULA, MEXICO,
S 5 182 GRAN ROQUE, LAGO PUEBLO, LOCATION, VENEZUELA,
B 6109 1542 GRANATA, G. E., DEPOSITIONAL FRAMEWORK OF LOWER DOCKUM
B 6109 1539 GRANATA, GEORGE, GEOMETRY OF GULF COAST AQUIFERS,
B 6109 1537 GRANBERRY, RAYMOND J., CRITICAL WATER ESTIMATES FOR GULF
B 6402 267 GRAND AND SIX MILE LAKES, LACUSTRINE DELTA FILL, LOUISIANA
S 4 235 GRAND BAHAMA ISLAND, COMACTINIA ECHINOPTERA MULLER
B 6004 682* GRAND BAHAMAS ISLAND, ABST., REEF MORPHOLOGY AND SEDIMENT,
B 6101 81 GRAND BANKS
B 6202 230 GRAND BANKS, EVAPORITES
B 6105 837* GRAND BANKS, NEWFOUNDLAND, ABST., GEOLOGY AND PETROLEUM
B 6203 428 GRAND BANKS, SECTIONS
B 6202 228 GRAND BANKS, SEDIMENTARY SEQUENCE
B 6203 428 GRAND BASIN, CARSON BASIN
B 6203 428 GRAND BASIN, HORSESHOE BASIN
B 6203 428 GRAND BASIN, JEANNE DARC BASIN
B 6203 428 GRAND BASIN, WHALE BASIN
S 3 33 GRAND CANYON, SOUTH RIM, ARIZONA
B 6303 420* GRAND CANYON SUPERGROUP ARIZONA, AABST.,
S 4 235 GRAND CAYMAN, ANALCIDOMETRA ARMATA POURTALES
M 24 232 GRAND ISLE ASH, LOUISIANA
M 24 235 GRAND ISLE ASH ZONE, LOUISIANA
M 24 229† GRAND ISLE BLOCK 16 FIELD, OFFSHORE LOUISIANA
B 6104 587 GRAND ISLE BLOCK FIELDS, LOUISIANA
B 6303 481* GRAND RAPIDS FORMATION, NORTH-CENTRAL ALBERTA, CANADA,
B 6405 735* GRAND RAPIDS FORMATION, NORTH-CENTRAL ALBERTA, CANADA,
B 6007 1112 GRAND RAPIDS OIL, MASS SPECTROMETRIC ANALYSIS
S 10 199 GRAND RAVINE FIELD, MUDFLOWS, WEST INDIES, TRINIDAD
B 6204 706 GRAND TOWER LIMESTONE, ILLINOIS BASIN
S 5 121 GRAND TRAVERSE COUNTY, EXAMPLE, MICHIGAN,
S 5 115 GRAND TRAVERSE COUNTY, LOCATION, MICHIGAN,
B 6207 1160 GRAND WASH CLIFFS, ARIZONA AND, NEVADA
B 6108 1386* GRAND WASH FAULTS, NORTHWEST ARIZONA, ABST., EARLY, PERMIAN,
B 6410 1621 GRANEROS SHALE, COLORADO
B 6406 943 GRANEROS SHALE, MINNESOTA AND, SOUTH DAKOTA
B 6410 1621 GRANEROS SHALE, WYOMING
M 25 513 GRANISLE COPPER DEPOSIT
M 25 440 GRANITE GNEISS SYSTEM, SOUTH KOREA
M 25 454 GRANITE, JAPAN, FUNATSU
B 6003 455† GRANITE, LAFAYETTE COUNTY, MISSISSIPPI, SUBSURFACE, PRECAMBRIAN
B 6201 100 GRANITE MOUNTAINS, WYOMING
B 6104 595 GRANITE POINT FIELD, ALASKA
B 6408 1132 GRANITE POINT OIL FIELD, ALASKA
B 6305 808 GRANITE WASH PRODUCTION, TEXAS
M 25 475 GRANITE-METAMORPHIC LAYER, ANTIMONY, MINERALIZATION,
S 6 149 GRANITES, DIAGRAM, JAPAN, POTASSIUM-ARGON AGES
S 3 135 GRANITES, HYPOTHERMAL REPLACEMENT OF LATE-PHASE ALKALIC
S 6 147 GRANITES, JAPAN, POTASSIUM-ARGON DATING
M 25 443 GRANITES, SOUTH KOREA
M 25 440 GRANITES, SOUTH, KOREA, MESOZOIC
B 6203 512* GRANITES. BY TITLE ONLY., URANIUM IN
M 25 449 GRANITIC BATHOLITHS AND MINERAL ZONING, JAPAN, MESOZOIC, LATE
M 25 452 GRANITIC BODIES, DISTRIBUTION, JAPAN, URANIUM CONCENTRATION
B 6203 494* GRANITIC PLUTONS, ABST., TRACE ELEMENTS IN, HOLOCENE, FLUVIAL,
B 6405 789* GRANITIC ROCKS AND IMPLICATIONS FOR URANIUM EXPLORATION,
M 25 512 GRANITIC-PLUTON-TYPE PORPHYRY COPPER DEPOSITS
B 6405 739* GRANT COUNTY, INDIANA, ABST., POROSITY EVOLUTION OF, NIAGARAN,
B 6008 1405 GRANT, KEITH, PALEOCURRENT ANALYSIS OF DAKOTA AND SUBJACENT
B 6304 688* GRANT MINERAL BELT, NEW MEXICO, ABST., EXPLORATION FOR,
B 6206 1013 GRANTLAND UPLIFT, CANADA, ARCTIC
S 3 4 GRANTS DISTRICT, MOUNT TAYLOR VOLCANIC FIELD, NEW MEXICO

B 6304 686* GRANTS MINERAL BELT, ABST., MECHANISMS FOR, URANIUM
B 6304 691* GRANTS MINERAL BELT, ABST., ORGANIC, GEOCHEMISTRY AND,
B 6304 689* GRANTS MINERAL BELT AND THEIR RELATION TO URANIUM
B 6304 687* GRANTS MINERAL BELT, NEW MEXICO, ABST., GEOCHEMICAL STUDIES OF
B 6304 687* GRANTS MINERAL BELT, NEW MEXICO, ABST., PERIODS OF
B 6108 1373† GRANTS MINERAL BELT, NEW MEXICO, ABST., RARE EARTH AND OTHER,
B 6304 693* GRANTS MINERAL BELT, NEW MEXICO, ABST., SEM INVESTIGATION OF
B 6209 1673† GRANTS MINERAL BELT, NEW MEXICO, RUBIDIUM-, STRONTIUM
B 6410 1718† GRANTS MINERAL BELT, NEW MEXICO. DISCUSSION AND REPLY,
S 3 124 GRANTS, POISON CANYON SANDSTONE, NEW MEXICO
S 3 76 GRANTS, TREND AND STACK ORES, NEW MEXICO
B 6101 38 GRANTS URANIUM REGION, NEW MEXICO
B 6105 825* GRANTS URANIUM REGION, NEW MEXICO, ABST., EXPLORATION AND
B 6304 687* GRANTS URANIUM REGION SINCE 1963, ABST., EXPLORATION IN
S 3 124 GRANTS, WESTWATER SANDSTONE, NEW MEXICO
B 6103 465 GRANTZ, ARTHUR, CENOZOIC SEDIMENTATION AND TECTONICS OF HOPE
B 6103 465 GRANTZ, ARTHUR, GEOLOGIC FRAMEWORK AN RELATED
B 6405 714 GRANTZ, ARTHUR, IMPLICATIONS OF GEOLOGI STRUCTURE AND
M 25 291 GRANTZ, ARTHUR, TECTONIC FRAMEWORK OF PETROLIFEROUS ROCKS
B 6203 555* GRANULATION ON POROSITY AND PERMEABILITY OF
B 6004 705* GRANULOMETRIC RELATIONS WITH RIDGE-AND-SWALE TOPOGRAPHY
M 27 123 GRAPESTONE INTRACLASTS
B 6205 845† GRAPHIC DETERMINATION OF STRATIGRAPHIC AND VERTICAL
B 6405 696* GRAPHIC REPRESENTATION OF SUBSURFACE DATA BY COMPUTER, ABST.
B 6203 488* GRAPHICS IN SUBSURFACE STRATIGRAPHIC ANALYSIS, ABST.,
M 25 444 GRAPHITE, SOUTH KOREA
B 6101 113 GRASS CREEK MEMBER, UTAH
B 6307 1122 GRASSY HILL, ARIZONA
B 6204 706 GRASSY KNOB CHERT, ILLINOIS BASIN
B 6105 789 GRAUCH, RICHARD I., POSSIBLE PRESENCE OF ECONOMIC URANIUM
S 4 135 GRAUS, RICHARD R., STRUCTURAL MODIFICATION OF CORALS IN
B 6012 2176* GRAVIMETER, ABST., COMMERCIAL APPLICATION AND INTERPRETATION
M 25 510 GRAVINA-NUTZOTIN BELT, CORDILLERAN OROGEN
B 6205 879* GRAVITATIONAL ACCELERATION WITH TIME—A PROTOTYPE
B 6103 464* GRAVITATIONAL AND OTHER GEOPHYSICAL EVIDENCE OF ALASKAN
B 6012 2174† GRAVITATIONAL EVIDENCE FOR MORE ALASKAN SEDIMENTARY BASINS,
B 6303 341† GRAVITY AND DEEP DRILLING DATA, DELINEATION OF, ROME TROUGH
B 6007 1050 GRAVITY AND MAGNETIC PROFILES, NORTHWEST AND, NORTHEAST,
B 6303 476* GRAVITY AND MAGNETICS DATA, ABST., INTERACTIVE MODELING AND
B 6102 297* GRAVITY ANOMALY ANALYSIS OF CHIHUAHUA AND TRANS-PECOS,
B 6403 400† GRAVITY ANOMALY OVER CORAL-REEF OIL FIELD—WILFRED POOL,
B 6007 1099 GRAVITY, BARBADOS RIDGE COMPLEX
S 7 275 GRAVITY, BOTTOM, DEFORMATION
B 6007 1101 GRAVITY BOUGUER ANOMALY MAP, BARBADOS RIDGE COMPLEX AND
B 6002 174 GRAVITY DATA, BERINGIAN SHELF
B 6303 345 GRAVITY DATA, KENTUCKY, ROME TROUGH, COMPUTER MODELING OF
B 6007 1100 GRAVITY FREE-AIR ANOMALY MAP, BARBADOS RIDGE COMPLEX AND
B 6212 2435 GRAVITY MAP, BOUGUER
B 6303 343 GRAVITY MAP, KENTUCKY, ROME TROUGH
B 6007 1076 GRAVITY MAP, NIGER DELTA, BOUGUER
B 6406 923 GRAVITY OF OILS, API
S 7 230 GRAVITY SEDIMENTS, SEDIMENTATION
B 6203 505* GRAVITY TECTONICS—AN EVERYDAY CATASTROPHE, ABST.,
B 6303 426* GRAVITY--SLIDE ORIGIN FOR MEXICAN RIDGES FOLDBELT,
B 6303 453* GRAVITY-DISPLACEMENT PRODUCTS, ABST., SUBAQUEOUS
S 10 121† GRAVITY-FLOW OF GROUNDWATER, CROSS-FORMATIONAL
B 6203 567* GRAVITY-INDUCED CROSS-FORMATIONAL WATER FLOW—POSSIBLE
B 6406 953 GRAY AND ARTHUR J. BOUCOT, EDS., MEYERHOFF, A. A., REVIEW OF,
B 6108 1387* GRAY HILLS AREA, PICEANCE CREEK BASIN, NORTHWESTERN
B 6409 1565* GRAY SAND, JURASSIC, NORTH LOUISIANA, ABST., PETROGRAPHIC,
B 6009 1609* GRAY SANDSTONE, SMACKOVER, BOSSIER AND WEBSTER PARISHES,
M 28 112 GRAY SANDSTONE, TEXAS, PENNSYLVANIAN
M 28 128 GRAY SANDSTONE, TEXAS, PENNSYLVANIAN
M 28 157 GRAY SANDSTONE, TEXAS, PENNSYLVANIAN
B 6004 685 GRAY, T. L., INTERPRETATION OF DEPOSITIONAL ENVIRONMENTS,
B 6105 823* GRAY-LEVEL MAPPING OF GEOLOGIC DATA, ABST.
B 6205 781 GRAYBURG FORMATION, NEW MEXICO
B 6209 1761* GRAYSON FORMATION, UPPER CRETACEOUS, OF NORTH-CENTRAL
B 6109 1541* GRAYSON FORMATION, UPPER CRETACEOUS, OF TEXAS, ABST.,
B 6004 727 GRAYSON, ROBERT C., JR., BASINWARD FACIES CHANGES IN WAPANUCKA
B 6203 518 GRAYSON, ROBERT C., JR., DEPOSITIONAL ENVIRONMENTS AND
B 6203 566 GRAYSON, ROBERT C., JR., REEVALUATION OF MORROWAN-ATOKAN
B 6108 1374* GRAYWACKE SANDSTONES-EOCENE UMPQUA FORMATION, SOUTHWEST
B 6004 612† GRAZERS IN DOGGER, JURASSIC, CARBONATE ROCKS, SOUTHEAST OF
B 6004 684* GREAT LAKES AREA, ABST., SEARCH FOR, SILURANIAN, REEF, MODEL--
B 6204 694 GREAT ABACO ISLAND, BAHAMAS
S 6 114 GREAT ARTESIAN BASIN, ISOCHRON DIAGRAM, AUSTRALIA
M 25 474 GREAT ARTESIAN BASIN, MEGASTRUCTURE
B 6204 694 GREAT BAHAMA BANK
B 6203 520* GREAT BAHAMA BANK, ABST., OOID, SAND-SHOAL FACIES, JOULTERS
B 6203 494* GREAT BAHAMA BANK, ABST., PRE-HOLOCENE, CARBONATE ROCKS OF
B 6410 1703 GREAT BAHAMA BANK DRIFT
B 6410 1634† GREAT BAHAMA BANK, FACIES SUCCESSION OF, PLIOCENE-, PLEISTOCENE,
B 6410 1638 GREAT BAHAMA BANK, LUCAYAN LIMESTONE
B 6410 1635 GREAT BAHAMA BANK, NORTHWESTERN, CORE BORINGS, LOCATIONS,
M 24 280 GREAT BAHAMA BANKS, ANDROS PLATFORM FACIES
M 24 280 GREAT BAHAMA BANKS, JOULTERS BAY AREA, FACIES
M 25 27 GREAT BARRIER REEF, AUSTRALIA
B 6406 814† GREAT BARRIER REEF OF CANNING BASIN, WESTERN AUSTRALIA,
B 6405 719* GREAT BASIN, ABST., CONODONT-BASED ASSESSMENT OF THERMAL
B 6303 503* GREAT BASIN, ABST., FIELD RELATIONS OF SOME PRE-TERTIAR,
B 6205 882* GREAT BASIN, ABST., HISTORY OF EARLY EXPLORATION IN
B 6303 418* GREAT BASIN, ABST., MIDDLE PALEOZOIC SEDIMENTATION AND
B 6012 2185* GREAT BASIN, ABST., NEW EVIDENCE FOR AGE OF, CENOZOIC, SAL LAKE
B 6403 442* GREAT BASIN, ABST., PETROLEUM POTENTIAL OF
B 6205 888* GREAT BASIN, ABST., PROBABLE, PERMIAN, SOURCE ROCKS,
B 6205 885* GREAT BASIN, ABST., REVIEW OF, HYDROCARBON, POTENTIAL OF,
B 6004 659* GREAT BASIN, ABST., STRATIGRAPHY AND DEPOSITIONAL
B 6203 345 GREAT BASIN, CALIFORNIA
M 25 474 GREAT BASIN, MEGASTRUCTURE
M 25 485 GREAT BASIN, METAL PROVINCE, UNITED STATES
B 6205 888* GREAT BASIN, WESTERN UTAH AND NEVADA, ABST., PETROLEUM
B 6101 29 GREAT BEAR LAKE, PITCHBLENDE DEPOSITS, CANADA
B 6203 537* GREAT BLUE FORMATION AT WELLSVILLE MOUNTAIN, UTAH, ABST.,
B 6408 1194 GREAT BRITAIN, LIASSIC FACIES
B 6203 560* GREAT CANADIAN OIL SANDS—GEOLOGY AND PRODUCTION, BY TITLE
B 6303 546* GREAT CARBONATE BANK OF YUCATAN AND ITS PETROLEUM
B 6103 336 GREAT CAROLINA RIDGE, CAPE FEAR ARCH
B 6305 822* GREAT DIVIDE BASIN, WYOMING, ABST., APPLIED EXPLORATION
B 6405 782* GREAT DIVIDE BASIN, WYOMING, ABST., URANIUM DEPOSITS OF PART OF
M 30 124 GREAT GLEN FAULT, STRUCTURE, BEATRICE FIELD
B 6204 694 GREAT INAGUA ISLAND, BAHAMAS
B 6103 443† GREAT ISAA 1 WELL, BAHAMAS. DISCUSSION AND REPLY, JURASSIC
B 6204 694 GREAT ISAAC WELL, BAHAMA PLATFORM
B 6401 116 GREAT LAKES AREA, REEF ARCHIPELAGO
B 6401 119 GREAT LAKES AREA, SILURIAN CORRELATION CHART
B 6209 1633 GREAT LAKES REGION, NIAGARAN PALEOGEOGRAPHY
B 6209 1633 GREAT LAKES REGION, REEFS
S 10 134 GREAT LOWLAND ARTESIAN BASIN, GROUNDWATER FLOW, HUNGARY
S 10 155 GREAT LOWLAND ARTESIAN BASIN, PETROLEUM MIGRATION, HUNGARY
S 10 138 GREAT LOWLAND ARTESIAN BASIN, WATER LEVELS, HUNGARY
B 6306 952 GREAT NORTHERN PENINSULA, GEOLOGIC MAP, NEWFOUNDLAND
B 6109 1560* GREAT NORTHERN PENINSULA, NEWFOUNDLAND, ABST., SATELLITE
B 6306 951† GREAT NORTHERN PENINSULA, NEWFOUNDLAND, SATELLITE IMAGE
M 25 460 GREAT PACIFIC COPPER-BEARING RING
M 25 461 GREAT PACIFIC GOLD-BEARING RING
B 6004 678* GREAT PATAGONIAN OIL SPILL, ABST.
B 6205 891* GREAT PLAINS, ABST., NATURAL, GAS RESOURCES IN SHALLOW, LOW-,
B 6405 711* GREAT PLAINS, ABST., PHYSICAL CHARACTERISTICS OF SHALLOW,
B 6407 972 GREAT PLAINS, ARDMORE BENTONITE BED
B 6407 971 GREAT PLAINS, BEARPAW SHALE
B 6407 971 GREAT PLAINS, BELLE FOURCHE SHALE
B 6407 972 GREAT PLAINS, BLACKLEAF FORMATION
B 6407 971 GREAT PLAINS, CARLILE SHALE
B 6407 971 GREAT PLAINS, CLAGGETT SHALE
B 6407 971 GREAT PLAINS, CRETACEOUS, CHART
B 6407 971 GREAT PLAINS, CUT BANK SANDSTONE MEMBER
B 6407 971 GREAT PLAINS, EAGLE SANDSTONE
B 6407 971 GREAT PLAINS, FALL RIVER SANDSTONE
B 6407 971 GREAT PLAINS, FORT UNION FORMATION
B 6407 971 GREAT PLAINS, FOX HILLS SANDSTONE
B 6407 971 GREAT PLAINS, GAMMON SHALE
B 6407 971 GREAT PLAINS, GREENHORN FORMATION
B 6407 971 GREAT PLAINS, HELL CREEK FORMATION
B 6407 972 GREAT PLAINS, HORSETHIEF SANDSTONE
B 6407 971 GREAT PLAINS, JUDITH RIVER FORMATION
B 6407 971 GREAT PLAINS, KOOTENAI FORMATION
B 6008 1410* GREAT PLAINS LIGNITES, ABST., LIQUEFACTION OF, NORTHERN
B 6407 972 GREAT PLAINS, MARIAS RIVER SHALE
B 6407 971 GREAT PLAINS, MORRISON FORMATION
B 6407 971 GREAT PLAINS, MOWRY SHALE
B 6407 971 GREAT PLAINS, MUDDY SANDSTONE
B 6407 971 GREAT PLAINS, NIOBRARA FORMATION
B 6407 971 GREAT PLAINS, PIERRE SHALE
B 6407 971 GREAT PLAINS, SKULL CREEK SHALE
B 6407 971 GREAT PLAINS, SUNBURST SANDSTONE MEMBER
B 6407 971 GREAT PLAINS, TELEGRAPH CREEK FORMATION
B 6407 972 GREAT PLAINS, VIRGELLE SANDSTONE
B 6407 969† GREAT PLAINS—ASSESSMENT OF THEIR NATURAL GAS RESOURCES,
B 6405 679* GREAT SALT LAKE, ABST., EXPLORATION IN

B 6205 888* GREAT SALT LAKE BASIN, ABST., HYDROCARBON EXPLORATION IN
B 6311 2118* GREAT SALT PLAINS, ALFALFA COUNTY, OKLAHOMA, ABST.,
M 25 483 GREAT SILVER CHANNEL
B 6203 523* GREAT SLAVE LAKE, NORTHWEST TERRITORIES, ABST., SPELEOTHEMS
B 6203 522* GREAT SLAVE LAKE, NORTHWEST TERRITORIES, ABST., TECTONIC
B 6412 2078 GREAT SMOKY FAULTS, APPALACHIANS, GUESS CREEK AND
M 25 220 GREAT SOUTH BASIN, NEW ZEALAND, CANTERBURY-, EAST OTAGO-,
B 6404 502 GREAT SOUTH CHANNEL, SEISMIC REFLECTION PROFILES
B 6211 2244 GREAT VALLEY, CALIFORNIA
B 6203 345 GREAT VALLEY, CALIFORNIA
B 6207 1237* GREAT VALLEY, CALIFORNIA, ABST., OIL AND GAS FROM, SUBMARINE,
B 6105 844* GREAT VALLEY, CALIFORNIA, ABST., RESERVOIR AND SOURCE-BED
B 6012 2192* GREAT VALLEY, CALIFORNIA, ABST., RESERVOIR AND SOURCE-BED
B 6205 813† GREAT VALLEY, CALIFORNIA, RESERVOIR AND, SOURCE-BED HISTORY OF
B 6405 727* GREAT VALLEY FOREARC BASIN OF CALIFORNIA—INFLUENCE OF BASIN
B 6211 2217 GREAT VALLEY SEQUENCE, CALIFORNIA
B 6301 20 GREAT VALLEY SEQUENCE, CALIFORNIA
B 6105 835* GREAT VALLEY SEQUENCE, CALIFORNIA, ABST., SEDIMENTOLOGY OF,
B 6012 2182* GREAT VALLEY SEQUENCE, NORTHERN AND CENTRAL CALIFORNIA,
B 6405 790* GREAT VALLEY SEQUENCE, NORTHERN CALIFORNIA—ISOTOPIC AND
B 6111 1956 GREATER ANACO OIL FIELD, VENEZUELA
S 4 84 GREATER ANTILLEAN ARC, MESOZOIC
B 6401 14 GREATER BURGAN FIELD, IRAN
B 6208 1499 GREATER GREEN RIVER BASIN, GREATER, WYOMING
B 6111 1956 GREATER MERCEDES OIL FIELD, VENEZUELA
B 6111 1956 GREATER OFICINA OIL FIELD, VENEZUELA
B 6201 18 GREATER UGHELLI MEGASTRUCTURE, NIGER DELTA
B 6010 1712 GREECE, 1975, LALECHOS, N., ON DEVELOPMENTS IN
B 6110 1643 GREECE, 1976, LALECHOS, N., ON DEVELOPMENTS IN
B 6210 1912 GREECE, 1977, LALECHOS, N., ON DEVELOPMENTS IN
B 6310 1921 GREECE, 1978, LALECHOS, N., ON DEVELOPMENTS IN
B 6411 1964 GREECE, 1979, LALECHOS, N., ON DEVELOPMENTS IN
B 6203 531* GREECE, ABST., HOLOCENE EPOCH, ALLUVIAL-, DELTAIC-, COASTAL
B 6004 709* GREECE, ABST., SEDIMENTATION AND RESEDIMENTATION OF, CLASTI,
B 6405 740* GREECE, ABST., SYNSEDIMENTARY DEFORMATION IN, FOSSIL
B 6411 1964 GREECE, AEGEAN SEA
B 6402 244 GREECE, ARGOLIDE SEQUENCES
B 6402 256 GREECE, CAPE MATAPAN SUBMARINE PENINSULA, SEISMIC REFLECTION
B 6402 244 GREECE, GAVROVO ZONE
B 6402 244 GREECE, IONIAN BASIN
B 6411 1964 GREECE, IONIAN SEA
B 6402 244 GREECE, IONIAN ZONE
B 6402 246 GREECE, KEPHALONIA-ZAKINTHOS PLATEAU, SEISMIC REFLECTIONS
B 6402 256 GREECE, KITHIRA-ANTIKITHIRA CONTINENTAL MARGIN, SEISMI
B 6402 256 GREECE, LACONIA SUBMARINE BASIN, SEISMIC REFLECTION
B 6402 245 GREECE, MATAPAN SEAMOUNTS
B 6402 249 GREECE, MESSENIA BASIN, SEISMIC REFLECTION
B 6402 244 GREECE, NEOGENE
B 6402 249 GREECE, NORTH MATAPAN TROUGH, SEISMIC REFLECTION
B 6402 244 GREECE, OLYMPIA BASIN
B 6402 244 GREECE, PELOPONNESUS AND ADJACENT ISLANDS, GEOLOGY
B 6402 244 GREECE, PINDUS ZONE
B 6402 245 GREECE, PYLOS BASIN
B 6402 256 GREECE, SOUTH MATAPAN TROUGH, SEISMIC REFLECTION
B 6402 249 GREECE, STROPHADES RISE, SEISMIC REFLECTION
B 6211 2264 GREELEY FIELD, CALIFORNIA
B 6211 2245 GREELEY HIGH, CALIFORNIA
B 6111 1930 GREELY FJORD, ELLESMERE ISLAND
S 10 53 GREEN RIVER FORMATION, MIGRATION, UTAH, UINTA BASIN
B 6207 1222 GREEN, A. A., APPLICATION OF LANDSAT IMAGERY AND RELATED
M 27 11 GREEN ALGAE, CHAROPHYTES
M 27 8 GREEN ALGAE, DASYCLAD
M 27 7 GREEN ALGAE, HALIMEDA
M 27 20 GREEN ALGAE, WALLED CALCISPHERES
M 27 9 GREEN ALGAL FRAGMENT, DASYCLADACEAN
B 6105 790 GREEN, A. R., EVOLUTION OF CONTINENTAL-MARGIN BASINS
B 6211 2204 GREEN ISLAND, GROWTH FAULTS BETWEEN, IRELAND, WESTERN,
S 4 85 GREEN ISLAND INLIER, JAMAICA
B 6111 1963 GREEN KNOLL, GULF OF MEXICO
B 6102 257 GREEN LAKE LIMESTONE MEMBER, IDAHO
B 6112 2140 GREEN MARKER HORIZON, NEW MEXICO, BLANCO GAS FIELD
B 6304 689 GREEN, MORRIS W., DISCONFORMITIES IN GRANTS MINERAL BELT AND
B 6205 890* GREEN RIVER AND UINTA FORMATIONS, UINTA BASIN, UTAH, ABST.,
B 6308 1391 GREEN RIVER BASIN, GREATER, WYOMING
B 6108 1347 GREEN RIVER BASIN, GREATER, WYOMING
B 6409 1503 GREEN RIVER BASIN, GREATER, WYOMING
B 6008 1205 GREEN RIVER BASIN, WYOMING, GREATER
B 6405 748* **GREEN RIVER BASIN—A DEVELOPING GIANT GAS SUPPLY FROM DEEP**
B 6203 565* GREEN RIVER FORMATION, ABST., STROMATOLITES IN, EOCENE
M 28 60 GREEN RIVER FORMATION, COLORADO, EOCENE
M 28 59 GREEN RIVER FORMATION, COLORADO, EOCENE
M 28 57 GREEN RIVER FORMATION, COLORADO, EOCENE
B 6305 824* GREEN RIVER FORMATION, COLORADO--EVIDENCE FROM SULFUR
B 6203 550* GREEN RIVER FORMATION, DOUGLAS CREEK ARCH, COLORADO, ABST.,
B 6303 530* GREEN RIVER FORMATION, EOCENE, WYOMING, ABST., SEDIMENTARY
B 6203 505* GREEN RIVER FORMATION, PICEANCE CREEK BASIN, COLORADO, ABST.,
B 6205 885* GREEN RIVER FORMATION, PRICE RIVER CANYON AREA,
M 24 123 GREEN RIVER FORMATION, UTAH
B 6303 314 GREEN RIVER FORMATION, UTAH
M 24 92 GREEN RIVER FORMATION, WYOMING
B 6205 881* GREEN RIVER FORMATION, WYOMING, ABST., CARBONATE, CYCLES IN
M 27 142 GREEN RIVER FORMATION, WYOMING, EOCENE
M 27 6 GREEN RIVER FORMATION, WYOMING, EOCENE
M 27 100 GREEN RIVER FORMATION, WYOMING, EOCENE
M 28 118 GREEN RIVER FORMATION, WYOMING, EOCENE
B 6203 537* GREEN RIVER FORMATIONS, FREMONT-SWEETWATER COUNTIES,
B 6405 640 GREEN RIVER FORMATIONS, WYOMING, WASATCH-
B 6405 775* GREEN RIVER SHALE, ABST., SEDIMENTARY ASPECTS O, ORGANIC
M 25 448 GREEN TUFF AREA, JAPAN, MIOCENE SUBMARINE, VOLCANIC OR
B 6405 696 GREENALL, ROGER K., JR., GRAPHIC REPRESENTATION OF SUBSURFACE
B 6305 844* GREENBRIER FORMATION IN HYDEN WEST POOL, EASTERN KENTUCKY,
B 6304 652 GREENBRIER LIMESTONE, VIRGINIA
B 6103 417 GREENDALE SYNCLINE, VIRGINIA
B 6207 1216 GREENE, H. G., SEAFLOOR INSTABILITY ON PACIFIC CONTINENTAL
B 6406 942† GREENHORN FORMATION AND CARLILE SHALE, UPPER CRETACEOUS, AT
B 6407 971 GREENHORN FORMATION, GREAT PLAINS
B 6407 976 GREENHORN FORMATION, NORTH DAKOTA
B 6404 490 GREENHORN FORMATION, WYOMING
M 27 177 GREENHORN LIMESTONE, COLORADO, CRETACEOUS
M 27 36 GREENHORN LIMESTONE, COLORADO, CRETACEOUS
B 6410 1622 GREENHORN LIMESTONE, DENVER BASIN
B 6306 887 GREENHORN LIMESTONE MARKER, NEW MEXICO
B 6303 464* GREENHORN LIMESTONE, UPPER CRETACEOUS, WESTERN INTERIOR
B 6203 536* GREENLAND, ABST., HYDROCARBON GENERATION POTENTIAL OF,
B 6404 530 GREENLAND, ADAMS BJERG FORMATION
B 6404 528 GREENLAND, ALEQATSIAQ FJORD
B 6404 530 GREENLAND, ALEQATSIAQ FJORD FORMATION
B 6101 3 GREENLAND BASIN, NORTHEAST
B 6404 530 GREENLAND, BERGLUM RIVER FORMATION
B 6404 528 GREENLAND, BESSEL FJORD
B 6404 530 GREENLAND, BESSEL FJORD FORMATION
B 6404 530 GREENLAND, CAPE CALHOUN FORMATION
B 6404 530 GREENLAND, CAPE SCHUCHERT FORMATION
B 6404 530 GREENLAND, CARBONATE FORMATION
B 6411 1964 GREENLAND, DEVELOPMENTS, 1979
B 6404 528 GREENLAND, FREDERICK E. HYDE FJORD
B 6404 528 GREENLAND, FREUCHIN LAND
B 6404 530 GREENLAND, HALL LAND, STRATIGRAPHIC CHART
B 6404 528 GREENLAND, HAUGE BJERGE
B 6404 530 GREENLAND, HAUGE BJERGE FORMATION
B 6008 1329† GREENLAND IN 1975, DEVELOPMENTS IN WEST
B 6108 1326† GREENLAND IN 1976, DEVELOPMENTS IN
B 6208 1441† GREENLAND IN 1977, DEVELOPMENTS, OFFSHORE WEST
B 6308 1210† GREENLAND IN 1978, DEVELOPMENTS, OFFSHORE WEST
B 6404 528 GREENLAND, INGLEFIELD LAND
B 6404 528 GREENLAND, KANE BASIN
B 6404 528 GREENLAND, KAP CONSTITUTION
B 6404 536 GREENLAND, KAP GODFRED HANSEN
B 6404 530 GREENLAND, KAP GODFRED HANSEN FORMATION
B 6404 528 GREENLAND, KAP JEFFERSON
B 6404 528 GREENLAND, KAP LUCIE MARIE
B 6404 530 GREENLAND, KAP LUCIE MARIE FORMATION
B 6404 530 GREENLAND, KAP MAYNARD FORMATION
B 6404 528 GREENLAND, KAP MORRIS JESUP
B 6404 530 GREENLAND, KAP MORTON FORMATION
B 6404 528 GREENLAND, KAP TYSON
B 6404 529 GREENLAND, LLANDOVERY CONTACT AND BUILDUPS
B 6404 528 GREENLAND, NORTH, GEOLOGIC MAP
B 6101 8 GREENLAND, NORTHERN, CAMBRIAN
B 6101 1 GREENLAND, NORTHERN, GEOLOGY
B 6404 528 GREENLAND, NYEBOE LAND
B 6404 530 GREENLAND, OFFLEY ISLAND FORMATION
B 6404 530 GREENLAND, ORDOVICIAN
B 6404 527† GREENLAND, PALEOGEOGRAPHIC AND STRATIGRAPHIC DIFFERENTIATIO
B 6404 530 GREENLAND, PEARY LAND, STRATIGRAPHIC CHART
B 6404 528 GREENLAND, PENTAMERUS BJERGE
B 6404 530 GREENLAND, PENTAMERUS BJERGE FORMATION
B 6404 530 GREENLAND, PETERMANN HALVO FORMATION

B	6101	3	GREENLAND SHIELD
B	6404	528	GREENLAND, VICTORIA FJORD
B	6404	528	GREENLAND, WASHINGTON LAND
B	6404	530	GREENLAND, WASHINGTON LAND, STRATIGRAPHIC CHART
B	6404	530	GREENLAND, WULFF LAND, STRATIGRAPHIC CHART
M	25	411	GREENSTONE BELTS, WESTERN AUSTRALIA
M	25	402	GREENVALE AND ROCKHAMTON, NICKEL-ENRICHED LATERITES,
B	6004	675	GREENWOOD, BRIAN, MARINE-BAR SEDIMENTATION AND EQUILIBRIUM,
B	6311	2118*	GREENWOOD COUNTY, KANSAS, ABST., DEPOSITIONAL ENVIRONMENT
B	6102	294	GREENWOOD, E., CAPITAN REEF--EVOLUTION OF A CONCEPT, ABST.
B	6004	661	GREENWOOD, E., CAPITAN REEF--EVOLUTION OF CONCEPT, ABST.
B	6109	1448	GREENWOOD, EUGENE, POTENTIAL AND STRATIGRAPHY OF PEDREGOSA
B	6004	675	GREENWOOD, ROBIN, MARINE-BAR SEDIMENTATION AND EQUILIBRIUM,
B	6304	598	GREER, ALBERT R., FRACTURES IN CRETACEOUS ROCKS FROM SELECTED
M	6004	726*	GREGOIRE LAKE AREA, ATHABASCA TAR DEPOSITS, ALBERTA, ABST.,
M	26	15	GREGORY, A. R., ASPECTS OF ROCK PHYSICS FROM LABORATORY AND
B	6405	699	GREGORY, A. R., EVALUATION OF VOLUME OF ENTRAINED METHANE IN
B	6303	417	GREGORY, A. R., POTENTIAL FOR PRODUCTION OF GEOPRESSURED
B	6309	1595	**GREGORY, A. R., TESTING GEOPRESSURED GEOTHERMAL RESOURCE**
B	**6105**	**765**	**GREGORY, A. R., TEST SITE FOR GEOPRESSURED GEOTHERMAL WELL—FRIO**
B	6303	417	GREGORY, A. R., WORLDS FIRST GEOPRESSURED GEOTHERMAL TEST
B	6307	1035	GREGORY RIFTS, MAP, KENYA, KAVIRONDO AND
B	6309	1601	GREIMEL, THOMAS C., MIDDLE STRAWN, DESMOINESIAN, CRATONIC
B	6109	1537	GREMILLION, RICKY P., INTERIOR STRUCTURE OF OAK GROVE RIDGE
B	6007	1094	GRENADA TROUGH
B	6203	518	GRENDER, G. C., ANALYSIS OF OIL-FIELD DISTRIBUTION FOR
B	6004	711	GRENDER, G. C., IMPACT OF OIL AND GA EXPLORATION ASSUMPTIONS
B	6303	459	GRENDER, GORDON C., ANALYSIS OF HYDROCARBON DISTRIBUTION
B	6303	513	GRENDER, GORDON C., APPLICATION OF OIL AND GAS DISCOVERY
B	6303	459	GRENON, MICHEL, WORLD OIL RESOURCE ASSESSMENTS AND RELATION
M	28	86	GRESSO SOLFIFERA, ITALY, MIOCENE
B	6203	518	GRETENER, P. E., GEOPHYSICAL AND GEOLOGICAL SIGNIFICANCE OF
B	6303	429	**GRETHEN, B. L., ROLE OF BRINES AND FERROUS IRON IN FORMATION OF**
B	6205	867	GRIBI, EDWARD A., JR., CORRELATION OF PINNACLES AND NEENACH
B	6303	494*	GRIDS TO OIL EXPLORATION DATA, ABST., CREATION AND APPLICATION
B	6405	714	GRIES, R. R., OIL AND GAS POTENTIAL OF SAN LUIS BASIN,
B	6408	1259	GRIESBACH, FREDERICK B., NONMARINE UPPER CRETACEOUS ROCKS,
B	6206	1014	GRIESBACHIAN TO UPPER NORIAN, ARCTIC, CANADA, MESOZOIC
B	6004	699*	GRIEVE FIELD, WYOMING, ABST., DEPOSITIONAL ENVIRONMENT OF
B	6108	1384*	GRIEVE FIELD, WYOMING, ABST., DEPOSITIONAL ENVIRONMENT OF
M	28	71	GRIFFEL SCHIEFER, GERMANY, ORDOVICIAN
B	6209	1758	GRIFFIN, GEORGE M., WATER CLARITY NEAR OIL PRODUCTION
B	6307	1059†	GRIFFITH ISLANDS, NORTHWEST TERRITORIES, CANADA, DEPOSITIONAL,
B	6405	714	GRIFFITH, L. A., SIDERITE TEXTURES IN CARDIUM FORMATION, FERRIER
B	6212	2455	GRIJALVA SCARP
M	25	425	GRINDALL METAMORPHICS, GROOTE EYLANDT
B	6206	960	GRINDSLOW SHALES, ENGLAND
B	6206	959	GRIT: A COMPLETE ANCIENT SUBMARINE FAN, ENGLAND, SHALE
B	6409	1555	GROAT, C. G., SUBSURFACE WILCOX LIGNITE IN WEST-CENTRAL
B	**6105**	**801**	**GROAT, C. G., TEXAS LIGNITE**
B	6407	1005	GROGAN FIELD, LOUISIANA
M	30	577	GRONDIN FIELD, AFRICA, GABON
M	30	580	GRONDIN FIELD, CROSS SECTION, GABON BASIN
M	30	589	GRONDIN FIELD, DEVELOPMENT HISTORY, GABON
M	30	577	GRONDIN FIELD, GABON, GABON BASIN
M	30	577†	GRONDIN FIELD, GEOLOGY OF
M	30	578	GRONDIN FIELD, LOCATION, GABON
B	6303	546*	GRONDIN OIL FIELD, GABON, ABST.
M	30	36	GRONINGEN FIELD, LANDSAT IMAGE, NETHERLANDS
M	30	36	GRONINGEN FIELD, LANDSAT IMAGE, WEST GERMANY
B	6401	34	GRONINGEN GIANT GAS FIELD LANDSAT DATA, NETHERLANDS AND,
M	25	423	GROOTE EYLANDT BEDS, AUSTRALIA
M	25	424	GROOTE EYLANDT, CRYPTOMELANE
M	25	425	GROOTE EYLANDT, GRINDALL METAMORPHICS
M	25	423	GROOTE EYLANDT, LOCATION AND GEOLOGY, MAP, AUSTRALIA
M	25	427	GROOTE EYLANDT, MANGANESE
M	25	425	GROOTE EYLANDT, MANGANESE DEPOSITS, GENESIS OF
M	25	424	GROOTE EYLANDT, MANGANESE ORE, CHARACTERISTICS OF
M	25	424	GROOTE EYLANDT, MANGANESE ORE, CONSTITUENT MINERALS
M	25	423	GROOTE EYLANDT, MANGANESE ORE, EVALUATION OF RESOURCES
M	25	423	GROOTE EYLANDT, MANGANESE ORE, PRODUCTION
M	25	424	GROOTE EYLANDT, MANGANESE ORE, PRODUCTION AND TREATMENT
M	25	425	GROOTE EYLANDT, MCARTHUR GROUP
M	25	423	GROOTE EYLANDT, MULLAMAN BEDS
M	25	422†	GROOTE EYLANDT, NORTHERN AUSTRALIA, OCCURRENCE AND
M	25	424	GROOTE EYLANDT, PYROLUSITE
M	25	423	GROOTE EYLANDT, STRATIGRAPHY
B	6303	446	GROPPI, G., MALOSSA FIELD, DEEP DISCOVERY IN PO VALLEY, ITALY,
M	30	525	GROPPI, G., MALOSSA FIELD--A DEEP DISCOVERY IN THE PO VALLEY,
M	30	395	GROS MORNE FORMATION, PLIOCENE, TEAK FIELD
B	**6105**	**790**	**GROSE, L. T., GEOLOGY OF GEOTHERMAL ENERGY EXPLORATION**
B	6108	1379	GROSE, L. T., QUATERNARY GEOLOGIC PROCESSES ASSOCIATED WITH
B	6105	835	GROSHONG, R. H., JR., FLUVIAL TO BATHYAL SANDSTONES OF
B	6409	1548	GROSHONG, RICHARD H., JR., REVIEW OF TECTONOMECHANICS—AN
B	**6105**	**790**	**GROSS, M GRANT, DREDGING AND SPOIL DISPOSAL IN CHESAPEAK BAY**
B	6206	952	GROSSES ROCHES CONGLOMERATE, QUEBEC
B	6305	818	GROSSLING, MEYERHOFF, A. A., REVIEW OF, WINDOW ON OIL, A SURVEY
M	28	55	GROSSULARITE, CRYSTAL, GARNET
B	6405	715	GROTH, PETER K. H., FILAMENT-PRODUCING HYDROCARBONS IN
B	6203	519	GROTHAUS, BRIAN T., HIGH LEVELS OF VARIATION IN SIZE-FREQUENCY
B	6207	1208*	GROUND WATERS, CIRCUM-PACIFIC REGION, ABST., PROVENANCE AND,
B	6311	2059	GROUNDHOG COALFIELD, BRITISH COLUMBIA
M	25	108	GROUNDHOG COALFIELD, CANADA
M	25	365†	GROUNDWATER AND WASTEWATER DISPOSAL IN
B	6108	1379*	GROUNDWATER FLOW HISTORY AND ORIGIN, MIGRATION AND
M	25	360†	GROUNDWATER IN VOLCANIC AREAS IN JAPAN--SUMMARY
B	6212	2512*	GROUNDWATER IN VOLCANIC ISLANDS, ABST.
M	25	355†	GROUNDWATER INFLOWS TO MURRAY RIVER IN SOUTH
S	10	121†	GROUNDWATER, CROSS-FORMATIONAL, GRAVITY-FLOW OF
S	10	141	GROUNDWATER FLOW, ALBERTA, RED EARTH REGION
S	10	133	GROUNDWATER FLOW, FRANCE, AQUITAINE BASIN
S	10	134	GROUNDWATER FLOW, HUNGARY, GREAT LOWLAND ARTESIAN BASIN
S	10	136	GROUNDWATER FLOW, IRAN, CENTRAL PLATEAU
S	10	133	GROUNDWATER FLOW, BASIN PROPERTIES, EXAMPLE
S	10	147	GROUNDWATER FLOW, DELAYED RESPONSE OF WATER LEVELS
S	10	131	GROUNDWATER FLOW, FIELD PHENOMENA, MIGRATION
S	10	130	GROUNDWATER FLOW, FIVE-AQUIFER SYSTEM, EXAMPLE
S	10	123	GROUNDWATER FLOW, IN A STRATIFIED FRAMEWORK
S	10	147	GROUNDWATER FLOW, LEAKAGE ACROSS AQUITARD
S	10	130	GROUNDWATER FLOW, NONSTEADY-STATE, MIGRATION
S	10	154	GROUNDWATER FLOW, PETROLEUM TRANSPORT AND ACCUMULATION
S	10	122	GROUNDWATER FLOW, PRINCIPLES, MEASUREMENT EQUATIONS
S	10	132	GROUNDWATER FLOW, REGIONAL, TRANSIENT DEVELOPMENT
S	10	151	GROUNDWATER FLOW, REGIONAL FIELD PHENOMENA
S	10	148	GROUNDWATER FLOW, REGIONAL HYDRAULIC CONTINUITY
S	10	151	GROUNDWATER FLOW, REGIONAL PATTERNS
S	10	123	GROUNDWATER FLOW, STEADY-STATE, MIGRATION
S	10	139	GROUNDWATER FLOW PATTERN, IRAN, CENTRAL PLATEAU
S	10	122	GROUNDWATER MOVEMENT, REGIONAL, MIGRATION
M	25	363	GROUNDWATER, JAPAN
B	6303	442*	GROUNDWATER MIXING, ABST., SHALLOW SUBSURFACE,
M	25	357†	**GROUNDWATER POTENTIAL OF AREAS UNDERLAIN BY VOLCANICLASTIC**
M	25	356†	GROUNDWATER POTENTIAL OF LIMESTONE TERRANES, EXAMPLES FROM
M	25	371†	GROUNDWATER RESOURCES IN HAWAIIAN ISLANDS, EFFECT OF
M	25	353†	GROUNDWATER RESOURCES--SUMMARY, DEVELOPMENT OF, AUSTRALIA
S	4	381†	GROUNDWATER SOURCES IN A PERSIAN GULF COASTAL SABKHA,
S	3	54	GROUNDWATER SYSTEM, TRANSPORTATION IN, METEORIC
B	6103	346	GROUP 4000 FIELD, TEXAS
B	6405	715	GROVER, GEORGE, JR., CEMENT TYPES AND CEMENTATION PATTERNS O
B	6004	676	GROVER, GEORGE, JR., FENESTRAL FABRICS IN NONCYCLIC LOW
B	6004	712	GROVER, GEORGE, JR., SCALLOPED SOLUTION SURFACES, MIDDLE
M	25	430	GROVER, J. C., BACKGROUND TO PRESENT MINERAL SEARCH IN BRITISH
B	6004	672	GROVES, D. C., BIOHERMAL DEVELOPMENT IN WOLFCAMPIAN
B	6303	459	GROW, GEORGE C., JR., CURRENT ESTIMATES AND METHODS OF
B	6204	592	GROW, J. A., PETROLEUM POTENTIAL OF U.S. ATLANTIC SLOPE, RISE, AND
B	6303	527	GROW, J. A., RECENT ADVANCES IN PASSIVE-MARGIN RESEARCH, ABST.
B	6405	715	GROW, JOHN A., DEEP STRATIGRAPHY AND EVOLUTION OF BALTIMORE
B	**6105**	**790**	**GROW, JOHN A., DEPTH CONVERSION OF MULTICHANNEL SEISMIC-**
B	6410	1658	GROW, JOHN A., JURASSIC SEISMIC STRATIGRAPH AND BASEMENT
M	29	65	GROW, JOHN A., MULTICHANNEL SEISMIC DEPTH SECTIONS AND
B	6006	926	GROW, JOHN A., REGIONAL GEOLOGIC FRAMEWORK OFF
B	**6105**	**742**	**GROW, JOHN A., REGIONAL GEOLOGIC FRAMEWORK OFF NORTHEASTERN**
B	6203	560	GROW, JOHN A., STRUCTURE OF ATLANTIC SLOPE OF EASTERN NORTH
S	7	272	GROWTH FAULT, BOTTOM, DEFORMATION
B	6405	804*	GROWTH FAULTING AND GEOTHERMAL RESERVOIR QUALITY, ABST.,
B	6405	703*	GROWTH FAULTING AND SEDIMENTATION IN UPPER WILCOX DELTA
M	26	293	GROWTH FAULTING, DENVER BASIN
M	26	277†	GROWTH FAULTING, DENVER BASIN, COLORADO, STRATIGRAPHIC AND,
S	3	100	GROWTH FAULTING, DEWATERING AND
B	6008	1413*	GROWTH FAULTING, NEW TECTONIC STYLE IN CRETACEOUS FORELAND
B	6405	717*	GROWTH FAULTING OF VICKSBURG FORMATION, OLIGOCENE, SOUTH
B	6203	491*	GROWTH FAULTING, SOUTHWESTERN ARKOMA BASIN, OKLAHOMA,
B	6004	668*	GROWTH FAULTS AND ASSOCIATED STRUCTURES IN UPPER TRIASSIC
S	7	247	GROWTH FAULTS, DIAGRAM, TEXAS, SHELF-SLOPE
B	6003	414†	GROWTH FAULTS IN CAMBRIAN AND LOWER ORDOVICIAN ROCKS OF
B	6211	2191†	GROWTH FAULTS IN CARBONIFEROUS OF WESTERN IRELAND
B	6003	341†	GROWTH FAULTS IN UPPER TRIASSIC DELTAIC SEDIMENTS, SVALBARD
B	6211	2207	GROWTH FAULTS, IRELAND, WESTERN, NAMURIAN
B	6109	1540*	GROWTH HISTORY OF HAINESVILLE SALT DOME, WOOD COUNTY, TEXAS,
B	6209	1566	GROWTH OF INCIPIENT DIAPIRS
B	6206	913	GROWTH PATTERNS OF DEEP-SEA FANS
S	4	177	GROWTH RATES FOR FOSSIL CORALS

B 6211 2208 GROWTH-FAULT DEVELOPMENT, IDEALIZED
B 6211 2206 GROWTH-FAULT FEATURES FROM COUNTY CLARE, GULF COAST, AND
S 3 103 GROWTH-FAULT SYSTEMS, TEXAS
B 6102 156 GRUBB FAULT, CALIFORNIA
B 6102 256 GRUBER, DAVID P., LITTLE COPPER FORMATION--NEW NAM FOR LOWEST
B 6303 459 GRUNAU, HANS R., GAS VERSUS OIL IN FAR EAST AND MIDDLE EAST
B 6303 433 GRUND, ROGER, GEOLOGY OF A STRATIGRAPHIC GIANT--MESSLA OIL
M 30 507 GRUND, ROGER, GEOLOGY OF A STRATIGRAPHIC GIANT--MESSLA OIL
M 28 10 GRUS, CALIFORNIA, HOLOCENE
M 28 11 GRUS, TEXAS, QUATERNARY
M 25 104 GRYC, GEORGE, COAL RESOURCES OF ALASKA--SUMMARY
B 6405 715 **GRYC, GEORGE, PETROLEUM EXPLORATION OF NATIONAL PETROLEUM**
B 6207 1218 GRYC, GEORGE, PETROLEUM EXPLORATION OF NATIONAL PETROLEUM
B 6106 904 GUADALUPE ARRUGADO, BAJA CALIFORNIA
B 6402 277 GUADALUPE DELTA, TEXAS
B 6408 1171 GUADALUPE FACIES, TEXAS
B 6004 670* GUADALUPE MOUNTAIN, NEW MEXIC AND WEST TEXAS, ABST.,
B 6303 501* GUADALUPE MOUNTAINS, NEW MEXICO, ABST., CARBONATE FACIES
B 6303 471* GUADALUPE MOUNTAINS, NEW MEXICO, ABST., SEAWARD PRIMARY DIP
B 6203 541* GUADALUPE MOUNTAINS, NEW MEXICO, ABST., SUBMARINE, CEMENTS
B 6105 843* GUADALUPE MOUNTAINS, NEW MEXICO AND WEST TEXAS, ABST.,
B 6405 768* GUADALUPE MOUNTAINS, TEXAS, ABST., EARLY, GUADALUPIAN,
B 6303 437* GUADALUPE MOUNTAINS, WEST TEXAS AND NEW MEXICO, ABST.,
B 6203 573* GUADALUPE MOUNTAINS, WEST TEXAS AND NEW MEXICO—CEMENT
B 6203 492* GUADALUPE MOUNTAINS, WEST TEXAS AND NEW MEXICO—NATURE
B 6203 355 GUADALUPE OIL FIELD, CALIFORNIA
B 6004 647* GUADALUPIAN COASTAL CARBONATE COMPLEX--CO3-SO4 DIAGENESIS
B 6004 693* GUADALUPIAN COASTAL CARBONATE COMPLEX--STRATIGRAPHY AND
B 6407 1118 GUADALUPIAN DELAWARE MOUNTAI GROUP OF WEST TEXAS AND
B 6405 768* GUADALUPIAN, PERMIAN, BANK MARGIN EROSION SURFACES,
B 6303 437* GUADALUPIAN, PERMIAN, GUADALUPE MOUNTAINS, WEST TEXAS AND
B 6303 501* GUADALUPIAN SHEL CREST, UPPER YATES AND LOWER TANSILL
M 30 480 GUADAO FIELD, SEISMIC PROFILES, CHINA
S 4 234 GUADELOUPE, NEMASTER DISCOIDEA CARPENTER
S 4 233 GUADELOUPE, NEMASTER RUBIGINOSA POURTALES
B 6110 1583 GUAJIRA BASIN, COLOMBIA
S 4 291 GUAM, GALAXEA
B 6405 807* GUANGXI PROVINCES, CHINA, ABST., LITHOFACIES AND, BIOHERMS OF
M 30 476 GUANTAO FORMATION, CHINA, ZHANHUA BASIN
B 6109 1476 GUARICEMA FIELD, BRAZIL
B 6109 1481 GUARICEMA SANDSTONE, BRAZIL
B 6111 1956 GUARICO OIL FIELD, VENEZUELA
B 6010 1651 GUATEMALA, 1975, HALL, W. K., ON DEVELOPMENTS IN
B 6110 1585 GUATEMALA, 1976, VAN HART, D., ON DEVELOPMENTS IN
B 6210 2051 GUATEMALA, 1977, BUIS, O. J., ON DEVELOPMENTS IN
B 6310 1753 GUATEMALA, 1978, BETHANCOURT, H. R., ON DEVELOPMENTS IN
B 6310 1751 GUATEMALA, 1978, FERNANDEZ, J., ON DEVELOPMENTS IN
B 6310 1753 GUATEMALA, 1978, PETROCONSULTANTS S.A., ON DEVELOPMENTS IN
B 6303 552* GUATEMALA ABST., BRAIDED-STREAM PROCESSES AND FACIES IN
B 6108 1375* GUATEMALA, ABST., DIAGENESIS OF, TERTIARY, VOLCANICLASTICS
B 6303 426* GUATEMALA, ABST., PRE-STRIKE-SLIP POSITIONS OF, POLOCHIC FAULT
B 6305 843* GUATEMALA, ABST., SEDIMENTOLOGY OF, VOLCANICLASTIC DEPOSITS
B 6209 1765* GUATEMALA, ABST., STRUCTURAL FABRIC OF, SERPENTINITES AND,
B 6004 693* GUATEMALA, ABST., STRUCTURE AND, TIMING OF, OROGENIC EVENTS,
M 29 254 GUATEMALA, MAGNETIC BASEMENT, GEOLOGIC MAP
B 6310 1764 GUATEMALA, PRODUCTION, 1974-1978
M 29 258 GUATEMALA, SHELF, SEISMIC SECTION
M 29 253 GUATEMALA, TECTONIC MAP, CENTRAL AMERICA
B 6102 208 GUAYABAL FORMATION, MEXICO
B 6405 784* GUAYMAS BASIN, GULF OF CALIFORNIA, ABST., EFFECTS OF, BASALT
B 6401 68 GUBBIO AREA, SAMPLED SECTION, ITALY
M 24 26 GUBIK FORMATION, ALASKA
M 30 471 GUDAO FIELD, CHINA, MIOCENE
M 30 482 GUDAO FIELD, CHROMATOGRAPHS OF OIL SAMPLES, CHINA
M 30 474 GUDAO FIELD, STRUCTURE, CHINA
M 30 471† GUDAO OIL FIELD AND SURROUNDING AREAS, GEOLOGY OF
M 30 540 GUELLALA FIELD, ALGERIA, OUARGLA REGION
M 30 540 GUELLALA NORD-EST FIELD, ALGERIA, OUARGLA REGION
B 6209 1610 GUELPH DOLOMITES, OHIO, LOCKPORT AND
S 5 4 GUELPH FORMATION, MICHIGAN BASIN, NIAGARAN SERIES, NIAGARA
S 5 54 GUELPH FORMATION, SILURIAN, NIAGARA GROUP,
M 25 269 GUEONGSANG, BASIN, KOREA, KYONGSAN
B 6101 85 GUERCIF BASIN, ATLAS TREND
B 6303 439 GUERRINI Y., KRIGING OF TOP OF RESERVOIR FROM SEISMIC AND
B 6412 2078 GUESS CREEK AND GREAT SMOKY FAULTS, APPALACHIANS
B 6309 1601 **GUEVARA, EDGAR H., UPPER CRETACEOUS--LOWER EOCENE STRATA,**
B 6109 1539 GUEVERA, EDGAR H., GEOMETRY OF GULF COAST AQUIFERS,
B 6111 1994 GUIANA BASIN, POSSIBLE DIAPIRS
M 25 513 GUICHON BATHOLITH, BRITISH COLUMBIA
B 6405 650 GUIDEBOOK OF SANTA FE COUNTRY EDITED BY RAYMOND V.
B 6111 2015 GUIDEBOOKS, NEW MEXICO GEOLOGICAL SOCIETY, FIELD CONFERENCE
B 6303 287 GUIDELINES TO EVALUATION OF PROSPECTS, OFFSHORE
B 6205 848 GUIJARRAL HILLS FIELD, CALIFORNIA
B 6205 850 GUIJARRAL HILLS SAND, CALIFORNIA
B 6309 1602 GUILLOTTE, J., SMACKOVER RESERVOIR--INTERPRETATION CASE STUDY
B 6110 1748 GUINEA BISSAU, DEVELOPMENTS, 1976
B 6210 1847 GUINEA BISSAU, DEVELOPMENTS, 1977
B 6310 1692 GUINEA BISSAU, DEVELOPMENTS, 1978
B 6411 1788 GUINEA BISSAU, DEVELOPMENTS, 1979
B 6210 1847 GUINEA CONAKRY, DEVELOPMENTS, 1977
S 4 88 GUINEA CORN FORMATION, MAESTRICHTIAN, JAMAICA
B 6110 1748 GUINEA, DEVELOPMENTS, 1976
B 6310 1692 GUINEA, DEVELOPMENTS, 1978
B 6411 1788 GUINEA, DEVELOPMENTS, 1979
B 6107 965 GUINEA PLATEAU AND ADJACENT CONTINENTAL SHELF
B 6211 2356 GUION, D. J., EXPLORATION IN RUGGED TERRAIN IN OVERTHRUST BELT,
B 6405 807* GUIZHOU-GUANGXI PROVINCES, CHINA, ABST., LITHOFACIES AND,
B 6007 1109 GULF 1 BURNER-STATE B WELL, TEXAS
M 26 215 GULF BASIN, DEPOSITIONAL SYSTEMS
B 6105 842* GULF COAST, ABST., CONTINENTAL, MARGIN—, U.S.
B 6109 1553* GULF COAST, ABST., DEPOSITION AND DIAGENESIS OF, FORTFACIES AND,
B 6109 1551* GULF COAST, ABST., EARLY AND MIDDLE, CRETACEOUS, HIPPURITACEA,
B 6405 782* GULF COAST, ABST., EARLY, CRETACEOUS, REEF COMMUNITIES IN
B 6209 1765* GULF COAST, ABST., EARLY, TERTIARY TERRESTRIAL PALEOGEOGRAPHY
B 6405 699* GULF COAST, ABST., EVALUATION OF VOLUME OF ENTRAINED, METHANE
B 6004 682* GULF COAST, ABST., GEOLOGY AND POTENTIAL USES OF, GEOPRESSURE
B 6109 1543* GULF COAST, ABST., HISTORICAL, SHORELINE CHANGES AND THEIR
B 6309 1606* GULF COAST, ABST., IMPORTANCE OF SECONDARY, LEACHED POROSITY
B 6109 1555* GULF COAST, ABST., LOWER, CRETACEOUS SEDIMENTARYFACIES AND,
B 6303 440* GULF COAST, ABST., MINERALOGIC COMPOSITION AND DIAGENESIS OF,
B 6409 1565* GULF COAST, ABST., NEARSHORE, BARS ALONG, UNITED STATES
B 6309 1610* GULF COAST, ABST., ORIGINS OF PERMANEN INLETS SEPARATING,
B 6209 1757* GULF COAST, ABST., PLEISTOCENE, SHORELINES AND PALEOGEOGRAPHIC
B 6303 551* GULF COAST, ABST., PORE SYSTEMS IN, JURASSI, CARBONATE,
B 6303 417* GULF COAST, ABST., POTENTIAL FOR PRODUCTION OF GEOPRESSURED,
B 6303 545* GULF COAST, ABST., QUATERNARY, FAULT ACTIVITY IN, TEXAS
B 6303 488* GULF COAST, ABST., RESERVOIR QUALITY IN, TERTIARY SANDSTONES
B 6209 1760* GULF COAST, ABST., SANDSTONE DISTRIBUTION AND POTENTIAL FOR,
B 6309 1602* GULF COAST, ABST., SEDIMENTATION ON TRAILING, PLATE
B 6309 1595* **GULF COAST, ABST., TESTIN GEOPRESSURED, GEOTHERMAL RESOURCE,**
B 6109 1543* GULF COAST, ABST., TOPOGRAPHIC EXPRESSION OF, OIL AND, GAS,
B 6409 1557* GULF COAST, ABST., WELL LOG ANALYSIS CONCEPTS IN, CLASTIC AND,
B 6211 2206 GULF COAST, AND NIGER DELTA, GROWTH-FAULT FEATURES FROM,
B 6109 1539* GULF COAST AQUIFERS, HOUSTON-GALVESTON, TEXAS, ABST.,
B 6009 1606* GULF COAST AREA ABST., ANCIENT, DELTA SYSTEMS IN
B 6404 552 GULF COAST AREA, COMPUTER MODELING
B 6404 549 GULF COAST AREA, PETROLEUM MIGRATION STUDY
B 6303 431* GULF COAST AREA--THEIR IDENTIFICATION AND EXPLORATION
B 6009 1608* GULF COAST AREAS--PLATE TECTONICS VIEW, ABST., GEOLOGIC
B 6203 500* GULF COAST BARRIER ISLAND, ABST., STRATIGRAPHY AND GEOLOGIC
B 6405 741* GULF COAST BASIN, ABST., SANDSTONE DIAGENESIS IN GEOPRESSURED,
B 6409 1569* GULF COAST BASIN AND MISSISSIPPI EMBAYMENT—IMPLICATIONS FOR
S 7 31 GULF COAST BASIN, CROSS SECTIONS, QUATERNARY
S 7 31 GULF COAST BASIN, CROSS SECTIONS, TERTIARY
B 6109 1544* GULF COAST BASIN OF TEXAS, ABST., DEPOSITIONAL SYSTEMS IN,
S 7 30 GULF COAST BASIN, PALEOSHELF EDGES, MAP
B 6307 1045 GULF COAST BASIN, SEISMIC PROFILE, LOUISIANA AND, TEXAS
B 6307 1043 GULF COAST BASIN, SEISMIC PROFILE, TEXAS
M 29 463 GULF COAST BASIN, VOLUME OF SEDIMENT PRESERVED
B 6004 661* GULF COAST CENOZOIC--MODEL FOR APPLICATION OF STRATIGRAPHIC
B 6209 1755* GULF COAST CENOZOIC—MODEL FOR APPLICATION OF STRATIGRAPHIC
B 6107 1004 GULF COAST, CHALK, PRODUCTION FROM
M 26 153 GULF COAST, CHRONOSTRATIGRAPHIC CORRELATION TO CYCLES, TEXAS
B 6303 547* **GULF COAST CRUDE OILS, ABST., EFFECTS OF SOURCE MATERIAL AND,**
M 26 161 GULF COAST, CYCLE, COMPARISON, WEST, AFRICA TO
B 6307 1047 GULF COAST, EAST CAMERON AREA, SECTION
B 6209 1767* GULF COAST ECONOMIC MICROPALEONTOLOGY, ABST., MILESTONES IN
B 6103 337 GULF COAST EQUIVALENTS, STRATIGRAPHIC LIMITS, SOUTH CAROLINA
B 6209 1764* GULF COAST ESTUARIES—ECOLOGIC AND PALEOECOLOGIC
M 30 254 GULF COAST, EUGENE ISLAND 330 FIELD, LOCATION
B 6109 1538* GULF COAST EVALUATED BY MEANS OF OSTRACODA, ABST.,
B 6402 167 GULF COAST FAULT ZONE MATERIALS
B 6402 167 GULF COAST FAULT ZONES
S 3 96 GULF COAST GEOSYNCLINE
B 6008 1265† **GULF COAST IN 1975, DEVELOPMENTS IN, LOUISIANA**
B 6108 1244† GULF COAST IN 1976, DEVELOPMENTS IN, LOUISIANA
B 6109 1549* GULF COAST INTERPRETED FROM RELICT SHORELINES, ABST., PLIOCENE-,
B 6002 234 GULF COAST, LACK OF ALTERNATIVE, MIGRATION MECHANISM IN
B 6405 791* GULF COAST LATE JURASSIC OSTRACODA WITH THOSE OF WESTERN
B 6309 1604* GULF COAST LIGNITE--STATUS REPORT, ABST.
B 6209 1759* GULF COAST LIMESTONE SOILS, ABST., MINERALOGIC AND PHYSICAL
B 6305 795 GULF COAST, LOUANN DEPOSITION

B 6109 1541* GULF COAST LOWER TERTIARY SANDSTONE RESERVOIRS—POROSITY
B 6309 1527 GULF COAST, METHANE IN GEOPRESSURED WATERS
B 6303 508* GULF COAST, MEXICO, ABST., FACIES, ENVIRONMENTS, AND
B 6405 734* GULF COAST OF LOUISIANA, ABST., EARLY, CEMENTATION BY
B 6008 1254† GULF COAST OF TEXAS IN 1975, DEVELOPMENTS IN UPPER
B 6108 1223† GULF COAST OF TEXAS IN 1976, DEVELOPMENTS IN UPPER
B 6208 1380† GULF COAST OF TEXAS IN 1977, DEVELOPMENTS IN UPPER
B 6308 1223† GULF COAST OF TEXAS IN 1978, DEVELOPMENTS IN UPPER
B 6409 1354† GULF COAST OF TEXAS IN 1979, DEVELOPMENTS IN UPPER
B 6208 1519† GULF COAST OFFSHORE IN 1977, DEVELOPMENTS IN, LOUISIANA
B 6308 1310† GULF COAST OFFSHORE IN 1978, DEVELOPMENTS IN, LOUISIANA
B 6409 1445† GULF COAST OFFSHORE IN 1979, DEVELOPMENTS IN, LOUISIANA
B 6208 1521 GULF COAST, OFFSHORE, PRODUCTION, 1974-1977, LOUISIANA
B 6009 1610* GULF COAST OIL, ABST., ORGANIC DIAGENESIS--ITS RELATION TO
B 6208 1462† GULF COAST ONSHORE IN 1977, DEVELOPMENTS IN, LOUISIANA
B 6308 1357† GULF COAST ONSHORE IN 1978, DEVELOPMENTS IN, LOUISIANA
B 6411 2012† GULF COAST ONSHORE IN 1979, DEVELOPMENTS IN, LOUISIANA
B 6411 2025 GULF COAST ONSHORE, PRODUCTION, 1979, LOUISIANA
M 29 436 GULF COAST, PETROLEUM, PRODUCTION INTERVALS
B 6305 792 GULF COAST, PLEISTOCENE THICKNESS MAP
B 6209 1598 GULF COAST, PROFILES
B 6109 1535* GULF COAST REGION, ABST., EFFECTS OF, SANDSTONE COMPOSITION
B 6303 534* GULF COAST REGION, ABST., SMACKOVER FORMATION OF
B 6303 529* GULF COAST SALT BASIN, ABST., SEALING AND NONSEALING, FAULTS IN
B 6402 145† GULF COAST SALT BASIN, SEALING AND, NONSEALING, FAULTS IN,
B 6209 1574 GULF COAST SALT DOMES, ONSHORE
B 6009 1434† GULF COAST SALT STOCK HELP TO DELINEATE SPINES OF MOVEMENT,
B 6107 1090† GULF COAST SALT STOCKS HELP TO DELINEATE SPINES OF MOVEMENT.
B 6012 2162† GULF COAST SALT-DOME CAP ROCK, ORIGIN OF
B 6109 1537* GULF COAST SANDSTONES, ABST., CRITICAL, WATER ESTIMATES FOR
M 25 208 GULF COAST, SEISMIC STUDY
M 26 153 GULF COAST SEQUENCES, TEXAS
B 6404 554 GULF COAST SHALES, DEPTH VERSUS POROSITY
B 6409 1568 GULF COAST STRATIGRAPHIC TRAPS, ABST., SEISMIC-STRATIGRAPHIC,
B 6303 509* GULF COAST TERTIARY, ABST., APPLIED BIOSTRATIGRAPHY IN
S 10 10 GULF COAST, SHALE COMPACTION SEQUENCE
S 10 4 GULF COAST, SHALE POROSITY, COMPACTION STAGES, MIOCENE
S 10 4 GULF COAST, SHALE POROSITY, COMPACTION STAGES
S 10 60 GULF COAST SOLUBILITY REQUIREMENTS, MIGRATION
S 10 61 GULF COAST TERTIARY, ORIGIN OF PETROLEUM
S 10 1† GULF COAST TERTIARY SECTION, THE NATURE OF, SHALES AND THE
B 6408 1167 GULF COAST, TERTIARY, SANDSTONE WEDGES, TEXAS
B 6303 463* GULF COAST TERTIARY SANDSTONES, ABST., DRAINAGE ANOMALIES IN
B 6203 521* GULF COAST TERTIARY SECTION, ABST., NATURE OF, SHALES AND
B 6303 437* **GULF COAST TERTIARY. WHY LOOK A GIFT HORSE IN THE MOUTH,**
B 6303 418* GULF COAST TURBIDITE SANDSTONES, TEXAS, ABST., RESERVOIR
B 6405 725* GULF COAST USING EVOLUTIONARY SERIES IN OSTRACODS, ABST.,
B 6009 1611* GULF COAST VICKSBURGIAN OSTRACODS IN H. V. HOWE COLLECTION,
B 6002 227 GULF COAST WELLS, LOUISIANA
M 26 162 GULF COAST, WEST AFRICA, COMPARISON, CYCLE CHARTS, TEXAS
B 6009 1606* GULF COAST--IDEAL VERSUS ACTUAL MODELS, ABST., GEOPRESSURED,
B 6303 501* GULF COAST. BY TITLE ONLY, CHARACTERIZATION OF ROCK MINERAL
B 6303 470* GULF COASTA PLAIN--TREND, EXPLORATION, AND PRODUCTION, ABST.,
B 6203 524* GULF COASTAL PLAIN, ABST., ANALYSIS OF, GEOCHEMICAL AND
B 6206 1049† GULF COASTAL PLAIN, GEOCHEMICAL AND SEDIMENTOLOGIC PROBLEMS
B 6404 582† GULF COASTAL PLAIN. DISCUSSION, GEOCHEMICAL AND,
B 6109 1536* GULF COASTAL PLAIN—INFLUENCE OF GENETIC FACIES ON
B 6309 1603* GULF COASTAL PROVINCE, ABST., DISTRIBUTION OF, VOLCANIC
B 6209 1589 GULF COAST—DIVERGENT DELTAIC MARGIN, LOUISIANA
B 6309 1535 GULF ENERGY AND MINERALS CO., U.S.A., D. A. COLLIE GAS UNIT, TEXAS,
B 6310 1888 GULF EXPLORATION AND PRODUCTION CO. AND PETROCONSULTANTS
B 6304 690* GULF MARIANO LAKE MINE, ABST., DEPOSITIONAL ENVIRONMENT OF.
B 6403 447* GULF OF ALASKA, ABST., BENTHIC, FORAMINIFERS FROM EASTERN
B 6105 815* GULF OF ALASKA, ABST., CLAY MINERALOGY OF EASTERN
B 6405 753* GULF OF ALASKA, ABST., DEPOSITIONAL ENVIRONMENT OF, CLAY
B 6012 2176* GULF OF ALASKA, ABST., DISPERSAL, DISTRIBUTION, AND THICKNESS OF,
B 6209 1754* GULF OF ALASKA, ABST., DISTRIBUTION OF, HOLOCENE, FORAMINIFERA
B 6211 2350* GULF OF ALASKA, ABST., DISTRIBUTION OF RECENT, FORAMINIFERA IN
B 6405 766* GULF OF ALASKA, ABST., GEOLOGY OF CONTINENTAL SLOPE ADJACENT
B 6303 411* GULF OF ALASKA, ABST., GLACIAL ORIGIN OF MEGACHANNELS OF
B 6303 411* GULF OF ALASKA, ABST., GLACIAL-MARINE, TRACE- AND BODY-FOSSIL
B 6403 440* GULF OF ALASKA, ABST., HOLOCENE, FORAMINIFERAL DISTRIBUTION
B 6403 444* GULF OF ALASKA, ABST., HOLOCENE, MOLLUSK DISTRIBUTION PATTERNS
B 6203 545* GULF OF ALASKA, ABST., HOLOCENE SEDIMENT VOLUME AND MODERN
B 6303 417* GULF OF ALASKA, ABST., HYDROCARBON POTENTIAL OF CENTRAL
B 6405 671* GULF OF ALASKA, ABST., LATE, NEOGENE DEPOSITIONAL AND, CLIMATI
B 6004 716* GULF OF ALASKA, ABST., LITTORAL PROCESSES AND GEOMORPHIC
B 6403 446* GULF OF ALASKA, ABST., QUATERNARY SEDIMENTARY FACIE ON
B 6203 374 GULF OF ALASKA, ALASKA
M 29 273† GULF OF ALASKA AND BOUNDARIES OF TECTONIC PLATES.
B 6403 446* GULF OF ALASKA COASTLINE, CAPE SUCKLING TO CAPE SPENCER, ABST.,
B 6204 633† GULF OF ALASKA CONTINENTAL SHELF, SURFACE SEDIMENTARY UNITS
B 6212 2415 GULF OF ALASKA, DATA-COLLECTION CRUISES
B 6212 2413 GULF OF ALASKA, EARTHQUAKE EPICENTERS
B 6208 1312 GULF OF ALASKA, EXPLORATION
B 6108 1158 GULF OF ALASKA, EXPLORATION
B 6308 1213 GULF OF ALASKA, EXPLORATION
B 6212 2412† GULF OF ALASKA, HOLOCENE, SLUMP ON, CONTINENTAL, SHELF OFF,
B 6204 642 GULF OF ALASKA, LANDSAT IMAGE, OCTOBER 12, 1972
B 6212 2413 GULF OF ALASKA, LEASE TRACTS
B 6103 466* GULF OF ALASKA MARGIN, BY TITLE ONLY, HISTORIC, EARTHQUAKES
B 6204 635 GULF OF ALASKA, NORTHERN, BATHYMETRY
B 6204 634 GULF OF ALASKA, NORTHERN, SEISMIC PROFILES
M 29 284 GULF OF ALASKA, OBLIQUE CONVERGENT MARGIN, SEISMIC TRACES
M 29 282 GULF OF ALASKA, OBLIQUE CONVERGENT PLATE BOUNDARY
B 6103 469* GULF OF ALASKA PLATE-TECTONIC MODEL BASED ON
M 29 284 GULF OF ALASKA, REFRACTION STATION DATA
B 6304 700* **GULF OF ALASKA, SHELF BASED ON DREDGE SAMPLES FROM ADJACENT**
B 6212 2413 GULF OF ALASKA, SLUMP AREA
M 29 287 GULF OF ALASKA, STRUCTURES (MAJOR), MAP
B 6211 2360* GULF OF ALASKA TERTIARY PROVINCE, ABST., FORAMINIFERAL
B 6405 748* GULF OF ALASKA TERTIARY PROVINCE, ABST., PRELIMINARY,
B 6004 655* GULF OF ALASKA TERTIARY PROVINCE, ABST., PRELIMINARY STUDY OF,
B 6012 2176* GULF OF ALASKA TERTIARY PROVINCE, ABST., STRUCTURAL ELEMENTS
B 6212 2423 GULF OF ALASKA, TSUNAMIS
B 6403 438* GULF OF ALASKA. BY TITLE ONLY, GLACIOMARINE DEPOSITIONAL
B 6403 376 GULF OF AQABA, HOLOCENE FAN DELTA
M 25 182 GULF OF ASA ALASKA, WESTERN, CONVERGENT MARGIN
B 6411 1968 GULF OF CADIZ, SPAIN
B 6405 784* GULF OF CALIFORNIA, ABST., EFFECTS OF, BASALT INTRUSIONS ON,
B 6205 883* GULF OF CALIFORNIA AND POSSIBLE ANALOGS IN CRETACEOUS OF
B 6408 1201 GULF OF CALIFORNIA, ANOXIC OPEN OCEAN
B 6207 1169 GULF OF CALIFORNIA, BOUSE FORMATION
B 6307 1139† GULF OF CALIFORNIA IN LAKE MEAD AREA OF ARIZONA AND NEVADA
B 6207 1159† GULF OF CALIFORNIA IN LAKE MEAD AREA OF ARIZONA AND NEVADA
B 6207 1169 GULF OF CALIFORNIA, MIOCENE
B 6307 1021 GULF OF CALIFORNIA PARALLEL TRANSFORM SET
B 6207 1169 GULF OF CALIFORNIA, PLIOCENE
B 6101 83 GULF OF MAINE
B 6203 536* GULF OF MAINE, ABST., CARBONATE CATCHMENT AND EXCHANGE,
B 6109 1484 GULF OF MAINE HIGH, EAST COAST
B 6404 511 GULF OF MAINE PLATFORM
S 7 80 GULF OF MEXICO, (RECONSTRUCTION) DISCUSSED, MESOZOIC
B 6303 442* GULF OF MEXICO, ABST., APPLIED BIOSTRATIGRAPHY, SOUTH,
B 6004 681* **GULF OF MEXICO, ABST., AREAL DISTRIBUTION OF, CLAY, MINERALS AND**
B 6409 1566* GULF OF MEXICO, ABST., CHEMICAL AND ISOTOPIC EVIDENCE OF
B 6303 491* GULF OF MEXICO, ABST., DISTRIBUTION OF, SALT STRUCTURES IN
B 6303 526* GULF OF MEXICO, ABST., FUTURE EXPLORATION OPPORTUNITIES IN
B 6004 722* GULF OF MEXICO, ABST., GEOCHEMICAL CONTROLS ON VERTICAL
B 6405 687* GULF OF MEXICO, ABST., GEOLOGIC AND GEOPHYSICAL INVESTIGATION
B 6203 561* GULF OF MEXICO, ABST., GEOLOGIC APPRAISAL OF UNDISCOVERED, OIL
B 6203 532* GULF OF MEXICO, ABST., GEOTECHNICAL ENGINEERING PROBLEMS OF
B 6303 426* GULF OF MEXICO, ABST., GRAVITY--SLIDE ORIGIN FOR, MEXICAN RIDGES
B 6109 1542* GULF OF MEXICO, ABST., INTERNAL, WAVES AND, NEPHELOID LAYER ON,
B 6203 542* GULF OF MEXICO, ABST., LATE, QUATERNARY SEDIMENTATION ON
B 6303 520* GULF OF MEXICO, ABST., LATE, TRIASSIC-, JURASSIC PALEOGEOGRAPHY
B 6309 1609* GULF OF MEXICO, ABST., LIVING, FORAMINIFERS OF, WEST FLOWER
B 6303 416* GULF OF MEXICO, ABST., LOWER, CRETACEOUS SHELF-MARGIN,
B 6203 546* GULF OF MEXICO, ABST., MISSISSIPPI, FAN
B 6409 1562* GULF OF MEXICO, ABST., MODERN, LAGOONAL, OSTRACODES AND SPECIE
B 6303 515* GULF OF MEXICO, ABST., RELATION OF SEDIMENTARY HISTORY AND
B 6409 1555* GULF OF MEXICO, ABST., RESERVOIR QUALITY, PLIOCENE-, PLEISTOCENE
B 6004 683* GULF OF MEXICO, ABST., SALT MOVEMENT ON, CONTINENTAL, SLOPE
B 6303 449* GULF OF MEXICO, ABST., SEDIMENTATION IN, MISSISSIPPI TROUGH
B 6303 487* GULF OF MEXICO, ABST., SEISMIC STRATIGRAPHIC, AND ECONOMIC
B 6405 726* GULF OF MEXICO, ABST., SEISMIC, STRATIGRAPHIC, AND STRUCTURAL
B 6303 487* GULF OF MEXICO, ABST., SEISMIC STRATIGRAPHY AND GEOLOGIC
B 6004 690* GULF OF MEXICO, ABST., SEISMIC STRATIGRAPHY AND GEOLOGIC
B 6303 498* GULF OF MEXICO, ABST., SPECULATIONS ON, BASEMENT MOBILITY IN
B 6105 830* GULF OF MEXICO, ABST., SUSPENDED-SEDIMENT DISTRIBUTION ON
S 7 16 GULF OF MEXICO, ABYSSAL PLAINS MAP
B 6209 1590 GULF OF MEXICO, ALAMINOS CANYON
B 6206 1068 GULF OF MEXICO, ALAMINOS CANYON
M 29 319† GULF OF MEXICO, ANATOMY OF THE, MEXICAN RIDGES, SOUTHWESTERN
B 6309 1597* GULF OF MEXICO AND ADJACENT WATERS, ABST., ECOLOGIC NICHES OF,
B 6309 1598* GULF OF MEXICO AND ADJACENT WATERS, ABST., RADIOLARIAN
M 26 145† GULF OF MEXICO AND OFFSHORE WEST AFRICA, SEISMIC,
B 6203 544* GULF OF MEXICO AND PERMIAN BASIN, WEST TEXAS AND
B 6411 2041 GULF OF MEXICO AND THE EARLY OPENING OF THE CENTRAL NORTH
B 6203 499* GULF OF MEXICO AS INTERPRETED FROM CONSOLIDATION HISTORY OF
B 6303 426* GULF OF MEXICO AS INTERPRETED FROM SEISMIC REFLECTIO PROFILES,
B 6303 478* GULF OF MEXICO BASIN, ABST., GEOCHEMISTRY OF, OIL-FIELD, WATERS
B 6408 1277* GULF OF MEXICO BASIN, ABST., GEOLOGIC HISTORY OF DEEP

B 6405 796* GULF OF MEXICO BASIN, ABST., POST-MIDDLE, CRETACEOUS, SEISMIC
B 6203 501* GULF OF MEXICO BASIN, ABST., STRUCTURE AND EARLY GEOLOGIC
B 6405 682* GULF OF MEXICO BASIN—SEISMIC STRATIGRAPHIC INTERPRETATION
S 7 197 GULF OF MEXICO, BATHYMETRIC MAP
S 7 6 GULF OF MEXICO, BATHYMETRY MAP
S 7 57 GULF OF MEXICO, BAY OF CAMPECHE SLOPE, DISCUSSED
M 29 418 GULF OF MEXICO, CAMPECHE BANK, REFLECTION PROFILE
B 6305 783 GULF OF MEXICO, CAMPECHE ESCARPMENT
B 6305 794 GULF OF MEXICO, CAMPECHE KNOLLS
S 7 63 GULF OF MEXICO, CAMPECHE PLATFORM
B 6211 2294 GULF OF MEXICO, CAMPECHE SCARP
M 29 407 GULF OF MEXICO, CAMPECHE SCARP, SEISMIC SECTION
B 6211 2291 GULF OF MEXICO, CAMPECHE UNIT
M 29 321 GULF OF MEXICO, CENTRAL GULF, SEISMIC STRATIGRAPHY
S 7 57 GULF OF MEXICO, CENTRAL SIGSBEE ESCARPMENT, DISCUSSED
B 6211 2294 GULF OF MEXICO, CHALLENGER KNOLL
S 7 46 GULF OF MEXICO, CHALLENGER KNOLL, REFLECTION PROFILE, SALT
M 29 407 GULF OF MEXICO, CHALLENGER KNOLL, SEISMIC SECTION
B 6211 2291 GULF OF MEXICO, CHALLENGER UNIT
S 7 48 GULF OF MEXICO, CHALLENGER UNIT, DESCRIBED, DEEP
S 7 113 GULF OF MEXICO, CHAOTIC SEISMIC FACIES, THICKNESS, EXAMPLE
B 6211 2291 GULF OF MEXICO, CINCO DE MAYO UNIT
B 6405 724* GULF OF MEXICO CONTINENTAL MARGIN, ABST., IMPACT OF, HOLOCENE
S 7 21† GULF OF MEXICO CONTINENTAL MARGIN: STRATIGRAPHIC AND
S 7 14 GULF OF MEXICO, CONTINENTAL RISE MAP
S 7 117 GULF OF MEXICO, CONTINENTAL SLOPE, CLAY FRACTION
S 7 117 GULF OF MEXICO, CONTINENTAL SLOPE, SAND FRACTION
S 7 117† GULF OF MEXICO CONTINENTAL-SLOPE SEDIMENTS AND
S 7 129 GULF OF MEXICO, CORE HOLE LOG
S 7 119 GULF OF MEXICO, CORE HOLE LOG
S 7 133 GULF OF MEXICO, CORE HOLE LOG
S 7 131 GULF OF MEXICO, CORE HOLE LOG
B 6305 793 GULF OF MEXICO, CROSS SECTION
S 7 22 GULF OF MEXICO CRUSTAL FRAMEWORK, NORTHERN
S 7 47 GULF OF MEXICO, DESCRIBED, DEEP
S 7 22 GULF OF MEXICO, DESCRIBED, NORTHERN
S 7 33 GULF OF MEXICO, DESCRIBED, NORTHERN
M 26 267 GULF OF MEXICO, DIAPIR GROWTH, EXAMPLE
B 6305 790 GULF OF MEXICO, DIAPIRIC SALT
S 7 81 GULF OF MEXICO, DIAPIRIC SALT FEATURES, MAP
S 7 82 GULF OF MEXICO, DISCUSSED, CRETACEOUS
S 7 26 GULF OF MEXICO, DISTRIBUTION MAP, JURASSIC, EASTERN
M 30 261 GULF OF MEXICO, DISTRIBUTION OF SALT TECTONICS
S 7 82 GULF OF MEXICO, END OF CRETACEOUS, DISCUSSED
S 7 84 GULF OF MEXICO, END OF MIOCENE, TERTIARY
B 6202 231 GULF OF MEXICO, EVAPORITES
S 7 82 GULF OF MEXICO, EVOLUTION DISCUSSED
S 7 51 GULF OF MEXICO, EVOLUTION DISCUSSED, DIAGRAM
B 6305 783 GULF OF MEXICO, FLORIDA ESCARPMENT
M 29 417 GULF OF MEXICO, FLORIDA ESCARPMENT, REFLECTION SECTION
M 29 418 GULF OF MEXICO, FLORIDA STRAITS, REFLECTION PROFILE
B 6104 529 GULF OF MEXICO, GARCETAS RIVER VALLEY
M 29 428 GULF OF MEXICO, GENERAL BATHYMETRY, CLOSED BASINS
S 7 23 GULF OF MEXICO, GEOLOGIC MAP, EASTERN
S 7 23 GULF OF MEXICO, GEOLOGIC MAP, NORTHERN
S 7 44 GULF OF MEXICO, GEOLOGIC SETTING, DESCRIBED
S 10 213 GULF OF MEXICO, GEOPRESSURE WITH DEPTH OF REVERSAL OF SHALE
B 6111 1963 GULF OF MEXICO, GREEN KNOLL
S 7 290 GULF OF MEXICO, GYRE BASIN, BATHYMETRY MAP
S 7 313† GULF OF MEXICO, HOLOCENE, SEDIMENT, GEOCHEMISTRY OF,
S 7 289† GULF OF MEXICO, INTRASLOPE, BASIN IN, NORTHWEST
M 29 383† GULF OF MEXICO, INVESTIGATION OF, MISSISSIPPI, FAN
S 7 16 GULF OF MEXICO, KNOLLS MAP
S 7 82 GULF OF MEXICO, LOUANN DEPOSITION, JURASSIC
M 29 428 GULF OF MEXICO, LOUISIANA, DIVERGENT DELTAIC MARGIN
M 29 427 GULF OF MEXICO, LOUISIANA, ORGANIC-CARBON BY DEPTH
M 29 433 GULF OF MEXICO, LOUISIANA, THERMAL MATURITY
S 7 14 GULF OF MEXICO, LOWER MISSISSIPPI FAN MAP
S 7 24 GULF OF MEXICO MARGIN, CROSS SECTION, NORTHERN
S 7 36 GULF OF MEXICO MARGIN, DESCRIBED, EASTERN
M 29 405 GULF OF MEXICO, MARGINS, CENTRAL GULF
M 29 418 GULF OF MEXICO, MARGINS, EASTERN GULF
M 29 410 GULF OF MEXICO, MARGINS, NORTHERN GULF
B 6211 2290† GULF OF MEXICO, MARGINS OF
M 29 403 GULF OF MEXICO, MARGINS, SEISMIC STRATIGRAPHY
M 29 418 GULF OF MEXICO, MARGINS, SOUTHEASTERN GULF
M 29 410 GULF OF MEXICO, MARGINS, SOUTHERN GULF
M 29 407 GULF OF MEXICO, MARGINS, WESTERN GULF
B 6305 793 GULF OF MEXICO, MESOZOIC RECONSTRUCTION
S 10 191 GULF OF MEXICO, METHANE SEEPS, ECHOGRAM
S 10 189 GULF OF MEXICO, METHANE-HYDRATE HORIZON, SEISMIC LINE
M 26 260 GULF OF MEXICO, MEXICAN RIDGE, SOUTH TEXAS
S 7 54 GULF OF MEXICO, MEXICAN RIDGES, DISCUSSED
M 29 320 GULF OF MEXICO, MEXICAN RIDGES, INDEX MAP
S 7 55 GULF OF MEXICO, MEXICAN RIDGES, REFLECTION PROFILE
M 29 407 GULF OF MEXICO, MEXICAN RIDGES, SEISMIC SECTION
M 29 323 GULF OF MEXICO, MEXICAN RIDGES, SEISMIC SECTION (24-FOLD)
M 29 324 GULF OF MEXICO, MEXICAN RIDGES, SEISMIC SECTION (FOLDOUT)
M 29 326 GULF OF MEXICO, MEXICAN RIDGES, STRUCTURE EXPLAINED
B 6211 2291 GULF OF MEXICO, MEXICAN RIDGES UNIT
S 7 124 GULF OF MEXICO, MICROFAUNA, PALEOBATHYMETRY
M 29 409 GULF OF MEXICO, MISSISSIPPI CONE, REFLECTION SECTION
S 7 56 GULF OF MEXICO, MISSISSIPPI CONE REGION, DISCUSSED
S 7 58 GULF OF MEXICO, MISSISSIPPI CONE SALT
M 29 392 GULF OF MEXICO, MISSISSIPPI FAN, AGE OF SEDIMENT
M 29 389 GULF OF MEXICO, MISSISSIPPI FAN, DESOTO DEEP-SEA CHANNEL
M 29 387 GULF OF MEXICO, MISSISSIPPI FAN, GEOMORPHIC UNITS
M 29 400 GULF OF MEXICO, MISSISSIPPI FAN, HISTORY
M 29 385 GULF OF MEXICO, MISSISSIPPI FAN, INDEX
M 29 400 GULF OF MEXICO, MISSISSIPPI FAN, PETROLEUM POTENTIAL
M 29 385 GULF OF MEXICO, MISSISSIPPI FAN, PHYSIOGRAPHY
M 29 395 GULF OF MEXICO, MISSISSIPPI FAN, SEDIMENT DISTRIBUTION
M 29 393 GULF OF MEXICO, MISSISSIPPI FAN, SEISMIC STRATIGRAPHY
M 29 393 GULF OF MEXICO, MISSISSIPPI FAN, STRATIGRAPHY
B 6404 467 GULF OF MEXICO, MISSISSIPPI, MARINE PHREATIC ARAGONITE CEMENT
S 7 134 GULF OF MEXICO, MISSISSIPPI SUBMARINE CHANNEL, ISOPACH
M 26 261 GULF OF MEXICO, MISSISSIPPI TROUGH, SEISMIC LINE
S 10 194 GULF OF MEXICO, MUD DIAPIR WITH METHANE-HYDRATE, SEISMIC
B 6211 2292 GULF OF MEXICO, MULTICHANNEL PROFILE LINES
S 7 80 GULF OF MEXICO, NORTH-TO-SOUTH CROSS SECTION
B 6111 1962† GULF OF MEXICO, NORTHEASTERN EXTENSION OF, SIGSBEE SCARP
M 26 145 GULF OF MEXICO, NORTHERN, SEQUENCES
S 7 43† GULF OF MEXICO, OCCURRENCE AND, EVOLUTION OF, SALT IN DEEP
B 6004 669* GULF OF MEXICO OCS PROGRAMS, ABST., NEW, ENVIRONMENTAL, MAPS
B 6409 1554* GULF OF MEXICO OIL SPILLS, IXTOC I AND BURMAH AGATE, WITH
M 26 216 GULF OF MEXICO, OLIGOCENE, HACKBERRY SLOPE SYSTEM
B 6209 1590 GULF OF MEXICO, ORCA BASIN
S 7 303† GULF OF MEXICO, ORCA, BASIN, ANOXIC DEPRESSION ON THE,
S 7 304 GULF OF MEXICO, ORCA BASIN, BATHYMETRY MAP
S 7 303 GULF OF MEXICO, ORCA BASIN, GEOLOGIC SETTING
S 7 121 GULF OF MEXICO, PALEOBATHYMETRIC LOG
B 6202 206 GULF OF MEXICO, PALEOBATHYMETRY CURVES AND NUMERICAL AGE
S 7 249 GULF OF MEXICO, PALEOCLIMATIC ZONES, PLEISTOCENE
B 6206 1068 GULF OF MEXICO, PERDIDO SCARP
S 7 31 GULF OF MEXICO, PERIPHERAL FAULT ZONES, DESCRIBED, NORTHERN
S 7 3† GULF OF MEXICO, PHYSIOGRAPHY OF
M 26 273 GULF OF MEXICO, PLEISTOCENE DEPOSITION
M 26 251 GULF OF MEXICO, PLEISTOCENE ISOPACH
S 7 79 GULF OF MEXICO, PLEISTOCENE THICKNESS MAP
B 6209 1763* GULF OF MEXICO, PRE-ATLANTIC OCEANPLATE RECONSTRUCTION—
S 10 214 GULF OF MEXICO, PROFILE SHOWING SALT DIAPIRS
M 26 252 GULF OF MEXICO, PROVINCES
S 7 7 GULF OF MEXICO, PROVINCES MAP
S 7 82 GULF OF MEXICO, RED SEA MODEL
B 6305 795 GULF OF MEXICO, RED SEA MODEL FOR
B 6409 1564* GULF OF MEXICO REGION, ABST., FUTURE PETROLEUM PROVINCES OF
B 6009 1617* GULF OF MEXICO REGION ABST., OIL, PROSPECTS IN
S 7 70 GULF OF MEXICO, REGIONAL MAP (HUMPHRIS)
S 7 30 GULF OF MEXICO RIM, EXPOSED, ARKANSAS, NORTHERN
S 7 30 GULF OF MEXICO RIM, EXPOSED, OKLAHOMA, NORTHERN
S 7 30 GULF OF MEXICO RIM, EXPOSED, PALEOZOIC, NORTHERN
B 6206 1068 GULF OF MEXICO, RIO PERDIDO CANYON
S 7 27 GULF OF MEXICO, SALT AND DIAPIRS, JURASSIC, NORTHERN
S 7 49 GULF OF MEXICO, SALT, CAMPECHE SCARP
S 7 49 GULF OF MEXICO, SALT, CHALLENGER KNOLL
S 7 50 GULF OF MEXICO, SALT DEPOSITION, DISCUSSED
S 7 47 GULF OF MEXICO, SALT IDENTIFICATION, CRITERIA
S 7 69† GULF OF MEXICO, SALT MOVEMENT ON, CONTINENTAL, SLOPE,
B 6305 782† GULF OF MEXICO, SALT MOVEMENT ON, CONTINENTAL, SLOPE
S 7 50 GULF OF MEXICO, SALT STRUCTURES, DISCUSSED
B 6205 765 GULF OF MEXICO, SEISMIC FACIES
M 26 249† GULF OF MEXICO, SEISMIC, FACIES AND, SEDIMENTOLOGY OF,
S 7 193† GULF OF MEXICO, SEISMIC, STRATIGRAPHIC INVESTIGATION OF, WEST
S 7 48 GULF OF MEXICO, SEISMIC UNITS, TABLE, WESTERN
S 7 84 GULF OF MEXICO, SEQUENTIAL DEVELOPMENT
B 6305 797 GULF OF MEXICO, SEQUENTIAL DEVELOPMENT
S 7 35 GULF OF MEXICO, SHALE STRUCTURES, DESCRIBED, NORTHERN
M 26 178 GULF OF MEXICO, SHEET-DRAPE FACIES, EXAMPLE
B 6401 108 GULF OF MEXICO SHELF, GAS GENERATION FROM RECENT SEDIMENTS
S 7 8 GULF OF MEXICO, SHELF SUBPROVINCES MAP

B 6209 1590 GULF OF MEXICO, SIGSBEE ESCARPMENT
S 7 75 GULF OF MEXICO, SIGSBEE ESCARPMENT, ORIGIN
S 7 78 GULF OF MEXICO, SIGSBEE ESCARPMENT, REFLECTION PROFILE
S 7 60 GULF OF MEXICO, SIGSBEE ESCARPMENT, REFLECTION PROFILE
S 7 73 GULF OF MEXICO, SIGSBEE ESCARPMENT, SPARKER PROFILE
S 7 71 GULF OF MEXICO, SIGSBEE ESCARPMENT, SPARKER PROFILE
S 7 74 GULF OF MEXICO, SIGSBEE ESCARPMENT, SPARKER PROFILE
B 6305 783 GULF OF MEXICO, SIGSBEE KNOLLS
M 26 258 GULF OF MEXICO, SIGSBEE KNOLLS
S 7 53 GULF OF MEXICO, SIGSBEE KNOLLS, CHALLENGER UNIT
S 7 52 GULF OF MEXICO, SIGSBEE KNOLLS, REFLECTION PROFILE
B 6211 2294 GULF OF MEXICO, SIGSBEE KNOLLS REGION
M 29 406 GULF OF MEXICO, SIGSBEE KNOLLS, SEISMIC SECTION
M 29 408 GULF OF MEXICO, SIGSBEE SALT DOME PROVINCE, SECTION
B 6206 1068 GULF OF MEXICO, SIGSBEE SCARP
B 6211 2291 GULF OF MEXICO, SIGSBEE UNIT
S 7 10 GULF OF MEXICO, SLOPE SUBPROVINCES MAP
B 6209 1768* GULF OF MEXICO SOUTH TIMBALIER BLOCK 54 SALT DOME AND SALT
M 29 324 GULF OF MEXICO, SOUTHWESTERN, SEISMIC CHARACTERISTICS
M 29 405 GULF OF MEXICO, SOUTHWESTERN, SEISMIC REFLECTORS
M 29 324 GULF OF MEXICO, SOUTHWESTERN, SEISMIC UNITS
B 6211 2291 GULF OF MEXICO, SOUTHWESTERN, SEISMIC UNITS
B 6305 783 GULF OF MEXICO, SPARKER AND SEISMIC PROFILES
S 7 28 GULF OF MEXICO, STRATA, CENOZOIC, NORTHERN
S 7 25 GULF OF MEXICO, STRATA, JURASSIC, NORTHERN
S 7 22 GULF OF MEXICO STRATIGRAPHIC FRAMEWORK, NORTHERN
S 7 28 GULF OF MEXICO STRUCTURAL FRAMEWORK, NORTHERN
S 7 141† GULF OF MEXICO, STRUCTURE OF, CONTINENTAL SLOPE, NORTHERN
S 7 5 GULF OF MEXICO, SUBMARINE TOPOGRAPHY, OUTLINE CHART
S 7 29 GULF OF MEXICO, TECTONIC MAP, EASTERN
S 7 29 GULF OF MEXICO, TECTONIC MAP, NORTHERN
S 7 56 GULF OF MEXICO, TEXAS-LOUISIANA SLOPE, DISCUSSED
M 29 403† GULF OF MEXICO, THE, MARGINS OF THE
S 7 49 GULF OF MEXICO, VIEJO UNIT
B 6211 2306 GULF OF MEXICO, WEST FLORIDA BANKS
M 29 418 GULF OF MEXICO, WEST FLORIDA BANKS, REFLECTION PROFILE
S 7 63 GULF OF MEXICO, WEST FLORIDA PLATFORM
B 6211 2303 GULF OF MEXICO, WEST FLORIDA SCARP
S 4 27† GULF OF MEXICO, WEST FLOWER GARDEN BANK
S 4 33 GULF OF MEXICO, WEST FLOWER GARDEN BANK, REEFS
S 7 88 GULF OF MEXICO, WESTERN CLASTIC REGION
S 7 57 GULF OF MEXICO, WESTERN SIGSBEE ESCARPMENT, DISCUSSED
B 6104 529 GULF OF MEXICO, WISCONSIN LOWSTAND OF
B 6309 1597* GULF OF MEXICO--IMPLICATIONS, ABST., RELICT AND EXPATRIATE,
B 6309 1605* GULF OF MEXICO--ROLE OF ECOLOGY AND OSTRACOD MICROFAUNA,
B 6303 482* GULF OF MEXICO--ROLE OF ECOLOGY AND OSTRACOD MICROFAUNA.
B 6303 411* GULF OF MEXICO--STUDY IN SEISMIC STRATIGRAPHY, ABST.,
B 6303 527* GULF OF MEXICO-TIME-SEQUENCE STUDY, ABST., SURFACE, TURBIDITY
B 6303 426* GULF OF MEXICO. BY TITLE ONLY, SUBAQUEOUS LANDSLIDES AND
S 7 155† GULF OF MEXICO—PHYSIOGRAPHY, STRATIGRAPHY, AND
B 6303 526* **GULF OF MEXICO BASIN, ABST., POST-MIDDLE, CRETACEOUS SEISMIC**
B 6205 779 GULF OF PAPUA, MIOCENE
B 6205 779 GULF OF PAPUA, SEISMIC DATA
M 26 192 GULF OF PAPUA, SHELF-MARGIN REEF, SEISMIC EXAMPLE
B 6203 368 GULF OF SANTA CATALINA, CALIFORNIA
B 6004 704* GULF OF ST. LAWRENCE, CANADA, ABST., COMPARISON OF, COASTAL
B 6211 2353* GULF OF SUEZ, ABST., MULTIPLE REFLECTIONS AND HEAD WAVES IN
B 6203 536* GULF OF SUEZ AREA—TALE OF FOUR TROUGHS, ABST., CRETACEOUS-,
B 6307 1032 GULF OF SUEZ GRABEN, FAULTS, EGYPT
M 25 251† **GULF OF THAILAND, AND ANDAMAN SEA--SUMMARY, PETROLEUM**
B 6207 1207* GULF OF THAILAND, ABST., NATURAL, GAS DEPOSITS IN
M 25 252 GULF OF THAILAND, CHUMPHON BASIN
M 25 16 GULF OF THAILAND, SEABED JURISDICTION
M 25 252 GULF OF THAILAND, STRATIGRAPHIC SECTION
B 6411 1968 GULF OF VALENCIA, SPAIN
B 6411 1840 GULF OIL EXPLORATION AND PRODUCTION CO. AND DEUTSCHE
B 6303 544* GULF ON MEXICO BASED ON REGIONAL MULTIFOLD SEISMIC LINES,
B 6408 1272 GULF REGION, UNITED STATES AND MEXICO, BY MICHEL T. HALBOUTY,
B 6007 1113 GULF SEAL, CRUISE, 1973, MV
M 25 189 GULFREX SEISMIC SURVEY, RF
B 6004 685* GULL RIVER LIMESTONE--TRANSGRESSIVE SEQUENCE OF SUPRATIDAL TO
B 6409 1563* GUM ISLAND NORTH AND FRENCH ISLAND AREAS, JEFFERSON COUNTY,
B 6207 1153 GUNARD FORMATION, AUSTRALIA
B 6405 795 GUNDERSEN, L. C., SEDIMENTOLOGY OF WESTWATER CANYON MEMBER,
B 6004 678 GUNDLACH, ERICH R., GREAT PATAGONIAN OIL SPILL, ABST.
B 6410 1688 GUNFLINT IRON FORMATION, CELLULAR FILAMENTS AND SHEATHS
B 6002 325 GUNN, ROBERT D., DEPOSITIONAL SYSTEMS IN MATADOR UPLIFT AREA.
B 6311 2116 GUNN, ROBERT D., DESMOINESIAN DEPOSITIONAL SYSTEMS IN
M 27 20 GUNSIGHT LIMESTONE, TEXAS, PENNSYLVANIAN
B 6308 1399 GUNTER, C. P., DEVELOPMENTS IN SOUTHEASTERN STATES IN 1978
B 6206 993 GURPI MARL, IRAN
B 6306 940 GURUPA ARCH, AMAZON MARGIN AREA
B 6211 2356* GUS-BUS, ABST., THE
B 6309 1597 GUST, LINDA, ECOLOGIC NICHES OF RADIOLARIANS, PLANKTONIC
B 6305 845 GUSTAVSON, JOHN B., GEOLOGY AND PRODUCTION POTENTIAL OF
B 6303 511 GUSTAVSON, T. C., PALO DURO BASIN ANALYSIS, ABST.
B 6004 676 GUSTAVSON, THOMAS C., BED FORMS AND STRATIFICATION TYPES OF
B 6303 552 GUSTAVSON, THOMAS C., MAP DISPLAY FORMATS FOR ENVIRONMENTAL
B 6105 789 **GUTMAN, A., COASTAL-DUNE GEOMORPHOLOGY AND CLASSIFICATION**
B 6105 791 **GUTMAN, A., DUNE DYNAMICS IN COASTAL ZONE—PAST, PRESENT, AND**
B 6109 1559 GUTMAN, S. I., INTEGRATION OF LANDSAT AND GEOPHYSICAL DATA IN
B 6311 2120 GUTRU, ROBERT J., CHANNEL SANDSTONE OIL RESERVOIRS OF
B 6305 828 GUTSCHICK, RAYMOND C., MISSISSIPPIAN CARBONATE SHELF MARGIN
B 6405 716 GUTSCHICK, RAYMOND C., MISSISSIPPIAN SHELF MARGIN AND
B 6405 779 GUTSCHICK, RAYMOND C., SEDIMENTATION, BIOSTRATIGRAPHY, AND
B 6303 460 **GUVEN, NECIP, AUTHIGENIC ILLITES IN SANDSTONE RESERVOIRS,**
B 6303 474 GUVEN, NECIP, OCCLUSION OF POROSITY IN CARBONATE RESERVOIRS BY
B 6110 1586 GUYANA, 1976, PETROCONSULTANTS S.A., ON DEVELOPMENTS IN
B 6210 2052 GUYANA, 1977, PETROCONSULTANTS S.A., ON DEVELOPMENTS IN
B 6310 1754 GUYANA, 1978, PETROCONSULTANTS S.A., ON DEVELOPMENTS IN
B 6010 1652 GUYANA, DEVELOPMENTS, 1975
B 6202 304 GUYANA, ESSEQUIBO DRAINAGE BASIN
B 6008 1397* GUYANA, SOUTH AMERICA, ABST., URANIUM EXPLORATION IN,
B 6405 716 GUZMAN, A. E., PETROLEUM POSSIBILITIES IN ALTAR DESERT, SONORA,
B 6207 1219 GUZMAN, ALFREDO E., PETROLEUM GEOLOGY OF REFORMA AREA,
B 6207 1219 GUZMAN, EDUARDO J., PETROLEUM GEOLOGY OF REFORMA AREA,
B 6403 434 GUZMAN, EDUARDO J., REVIEW OF GEOLOGY AND HYDROGEOLOGY OF
S 4 357 GVIRTZMAN, GEDALIA, SEQUENCE OF PROGRESSIVE DIAGENESIS IN
B 6303 523* GWNA MELANGE, UPPER PRECAMBRIAN OLISTOSTROMAL SEQUENCE,
M 25 269 GYEONGGI, MASSIF, KOREA, KYONGGI
B 6409 1556* GYP HILL SALT DOME, SOUTH TEXAS, ABST., CAP-ROCK FORMATION AND
S 5 181 GYPSUM, VENEZUELA, LAGO PUEBLO, PRECIPITATION OF
S 5 4 G-UNIT, SALINA GROUP,
M 27 133 GYPSUM, BEDDED
M 28 84 GYPSUM, BEDDED AND VARVED
M 28 84 GYPSUM, BIREFRINGENCE
M 28 86 GYPSUM CEMENT, GLOBIGERINA OOZE, IN
M 28 42 GYPSUM CRYSTALS, FRAGMENTS
M 28 87 GYPSUM, CRYSTALS, GROWTH WITHIN SEDIMENTS
M 28 85 GYPSUM, CRYSTALS, IN A CARBONATE MATRIX
M 28 85 GYPSUM, DETRITAL (WINDBLOWN), IN DUNES
M 28 42 GYPSUM DUNES, UTAH, QUATERNARY
M 28 129 GYPSUM, IN SANDSTONE, CEMENTS
M 28 85 GYPSUM, LATH-LIKE CRYSTALS
M 28 86 GYPSUM, OOLITIC GROWTHS
B 6011 2005† GYPSUM PLAIN, CULBERSON COUNTY, TEXAS, ORIGIN OF, LIMESTONE,
M 27 133 GYPSUM, REPLACEMENT
B 6211 2277 GYPSUM SPRING FORMATION, WYOMING
M 28 43 GYPSUM, TWINNED CRYSTALS
S 3 129 GYPSUM VALLEY, COLORADO
M 28 85 GYPSUM, WIND-DEPOSITED
S 7 290 GYRE BASIN, BATHYMETRY MAP, GULF OF MEXICO
S 7 325 GYRE BASIN, CHEMISTRY, BRINE CLASSIFICATION
S 7 300 GYRE BASIN, CLAY MINERALS
S 7 291 GYRE BASIN, COMPUTER BATHYMETRY
S 7 291 GYRE BASIN, GEOLOGIC SETTING
S 7 298 GYRE BASIN, GRAIN-SORTING CURVE
S 7 300 GYRE BASIN, HEAVY MINERALS
S 7 295 GYRE BASIN, HIGH-RESOLUTION PROFILES
S 7 296 GYRE BASIN, HIGH-RESOLUTION PROFILES
S 7 321 GYRE BASIN, MANGANESE IN PORE WATER
S 7 291 GYRE BASIN, MORPHOLOGY
S 7 319 GYRE BASIN, NORTHWEST GULF OF MEXICO, INTRASLOPE BASINS
S 7 318 GYRE BASIN, SCANNING-ELECTRON MICROGRAPHS
S 7 317 GYRE BASIN, SCANNING-ELECTRON MICROGRAPHS
S 7 299 GYRE BASIN, SEDIMENT ACCUMULATION, HOLOCENE
S 7 297 GYRE BASIN, SEDIMENT ANOMALY MAP
S 7 293 GYRE BASIN, SEDIMENT CORE
S 7 294 GYRE BASIN, SEDIMENT DISTRIBUTION
S 7 294 GYRE BASIN, SUBBOTTOM CHARACTERISTICS, PROFILING
S 7 320 GYRE BASIN, TEMPERATURE-SALINITY DIAGRAM
S 7 291 GYRE BASIN, TEXAS, CONTINENTAL SLOPE
S 7 319 GYRE BASIN, WATER COLUMN ANALYSIS
S 11 288 GYROIDINA SOLDANII
S 11 256 GYROIDINA SOLDANII
B 6112 2095 H-C RATIO, HYDROCARBON, GENERATION VERSUS
B 6009 1611* H. V. HOWE COLLECTION, ABST., DIAGNOSTIC CENTRAL, GULF COAST,
M 25 427 H.Y.C. DEPOSIT, LEAD-ZINC, AUSTRALIA, NORTHERN TERRITORY
S 9 1 **HABICHT, J. K. A., PALEOCLIMATE, PALEOMAGNETISM, AND**
B 6402 218 H1-5 FIELD, INDIA
B 6302 258 HABICHT, J. K. A., REVIEW OF THE STRUCTURE OF WESTERN EUROPE, BY
B 6203 568* HABITAT OF OIL—A RECONSIDERATION OF OLD PRINCIPLES, ABST.
B 6206 951 HACKBERRY CHANNEL, LOUISIANA

B 6409 1553* HACKBERRY, OLIGOCENE, SOUTHEAST TEXAS, ABST., MORPHOLOGY O,
M 26 216 HACKBERRY SLOPE SYSTEM, GULF OF MEXICO, OLIGOCENE
B 6105 791 HACQUEBARD, PETER A., RANK OF COAL AS INDEX OF ORGANIC
B 6108 1380 HADENHORST, FRED, BELLE CREEK-A UNIQUE COMMUNITY CONCEPT,
M 25 135 HADIKUSUMO, DJAYADI, POSSIBLE ENERGY SOUUCES IN INDONESIAN
B 6203 559 HAEDRICH, RICHARD L., BIOTA AND BIOLOGIC PROCESSES OF
B 6104 585 HAFT KEL FIELD, PERSIAN GULF
B 6310 1927 HAGEMANN, F., AND F. AL-KASIM, ON DEVELOPMENTS IN NORWAY NORT
B 6010 1718 HAGEMANN, F., AND F. AL-KASIM, ON DEVELOPMENTS IN NORWAY NORT
B 6110 1647 HAGEMANN, F., AND F. AL-KASIM, ON DEVELOPMENTS IN NORWAY
B 6210 1918 HAGEMANN, F., AND FAROUK AL-KASIM, ON DEVELOPMENTS IN
B 6105 791 HAGEMANN, RICHARD F., HIGH-RESOLUTION REFLECTION-SEISMIC
B 6303 467 HAHN-WEINHEIMER, P., ORIGIN AND OCCURRENCE OF CRUDE OILS IN
B 6307 1050 HAINESVILLE DOME, SEISMIC SECTION, TEXAS
B 6309 1601* HAINESVILLE, KEECHI, AND OAKWOOD SALT DOMES, EAST TEXAS, ABST.,
B 6109 1540* HAINESVILLE SALT DOME, WOOD COUNTY, TEXAS, ABST., GROWTH
S 4 249 HAIRPIN CURVE EXPOSURE, TEXAS
B 6010 1652 HAITI, 1975, PETROCONSULTANTS S.A., ON DEVELOPMENTS IN
B 6110 1587 HAITI, 1976, CARVAJAL, J., ON DEVELOPMENTS IN
B 6210 2052 HAITI, 1977, PETROCONSULTANTS S.A., ON DEVELOPMENTS IN
B 6310 1754 HAITI, 1978, PETROCONSULTANTS S.A., ON DEVELOPMENTS IN
B 6405 747* HAITI, ABST., RADIOLARIAN BIOSTRATIGRAPHY AND PALEOECOLOGY OF
B 6209 1768 HAJISHAFIE, M., HIGH-LEVEL PLIO-PLEISTOCENE SHORELINE IN SOUTH
B 6008 1398 HALBERTSMA, HENK L., STRATIGRAPHY OF MISSISSIPPIAN CHARLES
B 6205 739 HALBOUTY, MICHEL T., ACCELERATION IN GLOBAL EXPLORATION—
B 6005 745 HALBOUTY, MICHEL T., APPLICATION OF LANDSAT IMAGERY TO
B 6004 677 HALBOUTY, MICHEL T., APPLICATION OF REMOTE SENSING LANDSAT
B 6212 2510 HALBOUTY, MICHEL T., CIRCUM-PACIFIC MAP PROJECT, ABST.
M 25 1 HALBOUTY, MICHEL T., EDITED BY CIRCUM-PACIFIC ENERGY AND
B 6405 716 HALBOUTY, MICHEL T., FOREIGN FRONTIERS—AN OVERVIEW, ABST.
M 25 1 HALBOUTY, MICHEL T., FOREWORD, CIRCUM-PACIFIC ENERGY AND
B 6401 8 HALBOUTY, MICHEL T., GEOLOGIC SIGNIFICANCE OF LANDSAT DATA
M 30 7 HALBOUTY, MICHEL T., GEOLOGIC SIGNIFICANCE OF LANDSAT DATA
B 6401 8 HALBOUTY, MICHEL T., GEOLOGIC SIGNIFICANCE OF LANDSAT DATA
B 6304 698 HALBOUTY, MICHEL T., GEOLOGIC SIGNIFICANCE OF LANDSAT DATA ON
M 30 1 HALBOUTY, MICHEL T., GIANT OIL AND GAS FIELDS OF THE DECADE
M 25 1 HALBOUTY, MICHEL T., INTRODUCTION, CIRCUM-PACIFIC ENERGY AND
M 30 1 HALBOUTY, MICHEL T., INTRODUCTION, GIANT OIL AND GAS FIELDS OF
B 6002 324 HALBOUTY, MICHEL T., KEYNOTE ADDRESS AAPG SOUTHWEST SEC
B 6408 1210 HALBOUTY, MICHEL T., METHODS USED, AND EXPERIENCE GAINED, IN
B 6105 792 HALBOUTY, MICHEL T., TOMORROWS ENERGY—WHERE WILL IT COME FROM
B 6408 1272 HALBOUTY, SHELTON, JOHN W., REVIEW OF, SALT DOMES OF, GULF
B 6302 184 HALDAGER FORMATION, NORTH SEA
M 27 120 HALE FORMATION, OKLAHOMA, PENNSYLVANIAN
M 27 166 HALE FORMATION, OKLAHOMA, PENNSYLVANIAN
B 6108 1225 HALF MOON SHOAL FIELD, GALVESTON COUNTY, TEXAS
B 6311 2001 HALIBUT HORST, NORTH SEA
M 27 7 HALIMEDA, GREEN, ALGAE
S 4 18 HALIMEDA-THALASSIA MEADOWS, PUERTO RICO
M 28 88 HALITE, CEMENTING ANHYDRITE
M 28 89 HALITE, HOPPER BEDS
M 28 88 HALITE, HOPPER CRYSTALS
M 27 135 HALITE, REPLACEMENT
B 6002 165 HALL BASIN, BERING SHELF
B 6405 790 HALL, C. A., DIAGENESIS IN MONTEREY FORMATION, PISMO SYNCLINE,
B 6403 449 HALL, C. A., DIAGENETIC REACTIONS IN MONTEREY FORMATION, PISMO
B 6407 1037 HALL CANYON, CALIFORNIA
B 6012 2180 HALL, EDWARD A., TOMORROWS OIL FROM SOUTHERN CALIFORNIA
B 6105 792 HALL, JOHN B., GEOLOGY AND EXPLORATION OF URAVAN MINERAL BELT
B 6404 530 HALL LAND, STRATIGRAPHIC CHART, GREENLAND
B 6405 785 HALL, W. E., UPPER PALEOZOIC PALEOGEOGRAPHY OF IDAHO, ABST.
B 6010 1651 HALL, W. K., ON DEVELOPMENTS IN GUATEMALA, 1975
S 11 183 HALLER, C. R., PLIOCENE BIOSTRATIGRAPHY OF CALIFORNIA
B 6205 892 HALLEY, R. B., DEPOSITION AND DIAGENESIS OF HOLOCENE REEFS,
B 6004 683 HALLEY, ROBERT B., AUTOPSY OF A DEAD CORAL REEF, ABST.
B 6105 793 HALLEY, ROBERT B., COMPARISON OF ENDOLITH FLORAS FROM HOLOCENE-
B 6105 833 HALLEY, ROBERT B., HOLOCENE DOLOMITE LOCALITY IN FLORIDA BAY,
B 6105 792 HALLEY, ROBERT B., HOLOCENE FRESHWATER CEMENTS, JOULTERS CAYS,
B 6303 463 HALLEY, ROBERT B., KEY LARGO--SUBSURFACE CORE STUDY, ABST.
B 6104 519 HALLEY, ROBERT B., PLEISTOCENE BARRIER BAR SEAWARD OF OOID
B 6303 460 HALLEY, ROBERT B., PORE TYPES IN SUNNILAND LIMESTONE, LOWER
B 6303 460 HALLEY, ROBERT B., POROSITY PRESERVATION AND EARLY FRESHWATER
B 6004 677 HALLEY, ROBERT B., RECENT AND RELICT TOPOGRAPHY OF BOO BEE
B 6109 1552 HALLEY, ROBERT B., SIGNIFICANCE OF FRESHWATER LIMESTONES IN
B 6303 528 HALLEY, ROBERT B., THREE-DIMENSIONAL ASPECTS OF BELIZE PATCH
B 6205 884 HALLEY, ROBERT E., ESTIMATING PORE AND CEMENT VOLUMES OF
S 11 68 HALLIAN STAGE
S 11 217 HALLIAN STAGE
B 6202 259 HALLIAN STAGE, CALIFORNIA
B 6405 716 HALLOCK, PAMELA, APPLICATION OF ECOLOGIC STUDIES OF LIVING,
B 6311 2024 HALLS CREEK MOBILE ZONE, AUSTRALIA
M 27 86 HALLSTATTER LIMESTONE, GERMANY, TRIASSIC
B 6206 1016 HALOKINETIC STRUCTURES, CANADA, SVERDRUP BASIN
B 6209 1759* HALOKINETICS, ABST., REDDELL, SALT, DOME OF, LOUISIANA—,
B 6009 1612* HALOKINETICS, IN NORTH LOUISIANA, ABST., SALT MOVEMENT
B 6211 2355 HALPIN, DAVID, MAGNETOTELLURIC EXPLORATION—AN UPDATE, ABST.
B 6207 1219 HALUNEN, A. JOHN, MINERAL RESOURCES IN SOUTHWEST PACIFIC, ABST.
B 6203 519 HAM, R. A., CHEMICAL ANALYSES OF GEOTHERMAL WATERS FROM SUTH
B 6111 1956 HAMACA-SANTA CLARA HEAVY OIL, VENEZUELA
S 6 143 HAMAMOTO, REIKO, ISOTOPIC AGES AND STRATIGRAPHIC CONTROL OF
B 6303 460 HAMBLIN, A. P., HUMMOCKY CROSS-STRATIFICATION--INDICATOR OF
M 25 413 HAMERSLEY BASIN, IRON FORMATIONS, MINERALS PRESENT IN
M 25 413 HAMERSLEY BASIN, IRON FORMATIONS, STRATIGRAPHIC COLUMNS
M 25 411 HAMERSLEY BASIN, IRON, WESTERN AUSTRALIA
M 25 411 HAMERSLEY GROUP, WESTERN AUSTRALIA
M 25 427 HAMERSLEY IRON PROVINCE, BANDED IRO FORMATIONS, WESTERN
B 6211 2219 HAMES MEMBER, CALIFORNIA
B 6009 1425† HAMILTON DOME, BIGHORN BASIN, WYOMING, DEFORMATION OF,
B 6002 278† HAMILTON GROUP, DEVONIAN, AND ADJACENT STRATA ALONG SOUTH
B 6002 279 HAMILTON GROUP, PENNSYLVANIA
B 6309 1546 HAMILTON, THOMAS D., REVIEW OF ICE AGES, SOLVING THE MYSTERY
B 6211 2355 HAMILTON, WARREN, MESOZOIC TECTONICS OF WESTERN UNITED
B 6002 321* HAMLIN SHALE AND AMERICUS LIMESTONE, PERMIAN, WOLFCAMPIAN,
B 6002 159 HAMM, W. DOW, IN THE PUBLIC INTEREST
B 6008 1410 HAMMOND, PAUL E. HEAT-FLOW STUDIES IN STEAMBOAT
M 30 539 HAMOUDA, AIT, PETROLEUM POTENTIAL--OUARGLA REGION TRIASSIC
B 6412 2169 HAMPSHIRE FORMATION, WEST VIRGINIA
B 6207 1216 HAMPTON, M. A., SEAFLOOR INSTABILITY ON PACIFIC CONTINENTAL
B 6211 2355 HAMPTON, MONTE A., GEOENVIRONMENTAL STUDIES OF PACIFIC ARCTIC
B 6203 414 HAMS BLUFF, VIRGIN ISLANDS, ST. CROIX
B 6405 717 HAN, JOHN HWAN, DELTAIC SYSTEMS AND ASSOCIATED GROWTH
S 8 152 HANAKAPIAI CANYON, HAWAII, SUBMARINE CANYONS
S 8 152 HANAKAPIAI CANYON, TIME-VELOCITY CURVES, CURRENT FLOW
B 6209 1763* HANCOCK COUNTY, MISSISSIPPI, ABST., LATE, PLEISTOCENE
B 6006 981 HAND, J. H., HYDRATE FORMATION IN SUBSURFACE ENVIRONMENTS
S 5 121 HAND-STATICS CORRECTION, SEISMIC DATA,
B 6303 511 HANDFORD, C. R., PALO DURO BASIN ANALYSIS, ABST.
B 6405 717 HANDFORD, C. ROBERTSON, DEPOSITIONAL AND TECTONIC EVOLUTION
B 6311 2116 HANDFORD, C. ROBERTSON, DEPOSITIONAL HISTORY OF
B 6308 1426 HANDFORD, C. ROBERTSON, DEPOSITIONAL SYSTEMS AND PETROLEU
B 6303 461 HANDFORD, C. ROBERTSON, EXPLORATION POTENTIAL OF
B 6303 461 HANDFORD, C. ROBERTSON, LOWER PERMIAN FACIES TRACTS AND
B 6204 644 HANDFORD, C. ROBERTSON, MONTEAGLE LIMESTONE, UPPER
B 6105 793 HANDFORD, C. ROBERTSON, MONTEAGLE LIMESTONE, UPPER
B 6303 545* HANDIL FIELD, EAST KALIMANTAN, INDONESIA, ABST.
M 30 399† HANDIL FIELD, EAST KALIMANTAN, INDONESIA, GEOLOGY OF THE
M 30 419 HANDIL FIELD, GEOCHEMISTRY, INDONESIA
B 6209 1687 HANDIL FIELD, KALIMANTAN
M 30 420 HANDIL FIELD, MIGRATION HISTORY
M 30 401 HANDIL FIELD, MIOCENE, MAHAKAM DELTA
M 30 414 HANDIL FIELD, RESERVOIR 19-7, INDONESIA
M 30 413 HANDIL FIELD, RESERVOIR CHARACTERISTICS, INDONESIA
M 30 403 HANDIL FIELD, STRATIGRAPHIC COLUMN, MIOCENE
M 30 402 HANDIL FIELD, STRUCTURE, INDONESIA
B 6102 227 HANDIN, JOHN, EXPERIMENTAL STUDY OF FRACTURE PERMEABILITY IN
B 6303 462 HANER, BARBARA E., SEA GULLIES AND DEVELOPMENT OF LINEAR
B 6009 1620 HANEY, DONALD C., STRUCTURAL CONTROL ON PALEOZOIC
B 6303 471 HANEY, F. R., GEOCHEMISTRY OF OILS FROM SANTA CRUZ BASIN,
B 6112 2103 HANEY, F. R., HYDROCARBON GEOCHEMISTRY OF OILS FROM MARANON
B 6303 472 HANEY, F. R., OIL GEOCHEMISTRY AS EXPLORATION TOOL, ABST.
B 6405 600 HANIFA FORMATION, MIDDLE EAST
B 6211 2232 HANISCH, JORG, A SIGSBEE KNOLL IN EARLY TERTIARY BAY OF BISCAY
B 6203 519 HANKINS, B. E., CHEMICAL ANALYSES OF GEOTHERMAL WATERS FROM
B 6203 529 HANKINS, B. E., CHEMICAL ANALYSIS OF GAS DISSOLVED IN
B 6305 828 HANLEY, JOHN H., APPLICATION OF NONMARINE MOLLUSCA TO
B 6108 1377 HANLEY, JOHN H., INTERDISCIPLINARY ANALYSIS OF SOME POTENTIAL
B 6105 785 HANLEY, JOHN H., NEWLY RECOGNIZED PETROLEUM SOURCE-ROCK UNITS
B 6004 717* HANNA AND CARBON BASINS CARBON COUNTY, WYOMING, ABST.,
B 6008 1206 HANNA BASINS, WYOMING, LARAMIE-
B 6308 1391 HANNA-LARAMIE BASIN, WYOMING
B 6208 1500 HANNA-LARAMIE BASIN, WYOMING
B 6409 1503 HANNA-LARAMIE BASIN, WYOMING
B 6108 1347 HANNA-LARAMIE-DENVER BASIN, WYOMING
S 11 1 HANNA, MARGARET, MIOCENE STRATIGRAPHY OF CALIFORNIA
B 6305 838* HANNA-SHIRLEY BASINS--FUTURE PETROLEUM POTENTIAL, ABST.,
B 6208 1383 HANNAH ISLAND FIELD, GALVESTON COUNTY, TEXAS
B 6412 2169 HANNASVILLE SYNCLINE, WEST VIRGINIA
B 6303 472* HANNET SHALE, LOWER CRETACEOUS, TEXAS, ABST., DOLOMITIZATION
B 6104 594 HANOVER FIELD, MICHIGAN

S 4 85 HANOVER INLIER, JAMAICA
B 6010 1905 HANRECK, R. G. S., PETROLEUM DEVELOPMENTS IN ISRAEL IN 1975
B 6110 1830 HANRECK, R. G. S., PETROLEUM DEVELOPMENTS IN ISRAEL IN 1976
B 6210 1898 HANRECK, R. G. S., PETROLEUM DEVELOPMENTS IN ISRAEL IN 1977
B 6310 1743 HANRECK, R. G. S., PETROLEUM DEVELOPMENTS IN ISRAEL IN 1978
B 6405 717 HANSEN, ALAN R., UTAH-ARIZONA HINGE LINE—THRUST BELT—
B 6211 2357 HANSEN, ALAN R., UTAH-ARIZONA HINGE-LINE THRUST BELT—
B 6303 507 HANSEN, KIRK S., ATMOSPHERIC CIRCULATION AND UPWELLING IN
B 6409 1558 HANSEN, THOR, BIOSTRATIGRAPHIC SEQUENCE OF FOSSIL MOLLUSCA
B 6409 1558 HANSEN, THOR, BIOSTRATIGRAPHIC SIGNIFICANCE OF FOSIL
B 6303 442* HANSON CREEK FORMATION, ORDOVICIAN-SILURIAN, CENTRAL
B 6309 1579 HANSON, M. E., THEORETICAL AND EXPERIMENTAL ANALYSES OF
B 6103 360 HANSON, RICHARD E., URANIUM POTENTIAL O PERMIAN AND
M 30 540 HAOUD-BERKAOUI FIELD, ALGERIA, OUARGLA REGION
S 6 158 HARACHIYAMA FORMATION, JAPAN, CRETACEOUS
B 6405 718 HARBOUR, JERRY L., DEPOSITIONAL SETTING OF MIDDLE DOLOMITE
B 6405 718 HARBRIDGE, CONSTANCE B., WAMSUTTER ARCH TIGHT GAS PLAY,
B 6103 366 HARDEMAN-HOLLIS BASIN, OKLAHOMA
S 6 213 HARDENBOL, J., A NEW PALEOGENE NUMERICAL TIME SCALE
B 6410 1707 HARDGROUND, DEFINITION
S 4 272 HARDGROUNDS AND OTHER CARBONATE NONREEF SURFACES
B 6004 667* HARDGROUNDS IN HIGH-ENERGY OOLITIC ENVIRONMENT ON
B 6107 990 HARDGROUNDS, SYNSEDIMENTARY SEAFLOOR, CEMENTATION TERMED
B 6409 1356 HARDIN COUNTY, TEXAS, EAST SOUR LAKE FIELD
B 6303 462 HARDING, ANDREW G., SHALLOW-WATER CARBONATE AND EVAPORITE
B 6303 544 HARDING, RICHARD, HYDROCARBON OCCURRENCE A FUNCTION OF
B 6307 1016 HARDING, T. P., STRUCTURAL STYLES, THEIR PLATE-TECTONIC
B 6003 356 HARDING, T. P., TECTONIC SIGNIFICANCE AND HYDROCARBON
B 6301 87 HARDY, I. A., GEOCHEMICAL CHARACTERISTICS OF SEDIMENTARY ROCKS
B 6105 765 HARDY, JENNA L., COARSE WHITE SPARRY DOLOMITE AND CALCITE
B 6303 416 HARDY, JENNA L., CRITERIA FOR RECOGNITION OF DIVERSE DOLOSTONE
S 5 161 HARE FIORD, ELLESMERE ISLAND,
B 6111 1944 HARE FIORD, ELLESMERE ISLAND
S 5 162 HARE FIORD, NANSEN FORMATION, SVERDRUP BASIN,
B 6111 1930 HARE FJORD DIAPIR, ELLESMERE ISLAND
B 6111 1930 HARE FJORD, ELLESMERE ISLAND
B 6111 1931 HARE FJORD FORMATION, SVERDRUP BASIN
B 6201 1 HAREMBOURE, J., HYDROCARBON HABITAT OF TERTIARY NIGER DELTA
B 6203 519 HARGRAVES, R. B., UNIFORMITY AND THE PRECAMBRIAN, ABST.
B 6010 1711 HARK, H. U., ON DEVELOPMENTS IN GERMAN FEDERAL REPUBLIC, 1975
B 6303 410 HARLAN, JANIS, GLACIAL-MARINE SEDIMENTS OF WEST ANTARCTICA
S 6 9 HARLAND, W. B., GEOCHRONOLOGIC SCALES
B 6305 845 HARMON, DAVID E., JR., PETROLEUM GEOLOGY OF WHIRLPOOL MEMBER,
B 6305 829 HARMS, J. C., ALLUVIAL-PLAIN SEDIMENTS OF NUBIA, SOUTHWESTERN
B 6405 718 HARMS, J. C., BRAE FIELD AREA, NORTH SEA, ABST.
B 6105 793 HARMS, J. C., CARBONATE-SAND WAVES, ISLA MUJERES, YUCATAN,
B 6306 967 HARMS, J. C., MAJOR SUBDIVISIONS AND DEPOSITIONAL ENVIRONMENTS
B 6407 1117 HARMS, J. C., REVIEW OF SEDIMENTARY ENVIRONMENTS AND FACIES,
B 6207 1239* HARNESSING ARCTIC ENERGY—TRANS-ALASKA PIPELINE, ABST.
B 6408 1284 HARPER, DENVER, COAL RESOURCES OF INDIANA AND POTENTIAL
B 6004 677 HARPER, J. D., SOME APPROACHES TO CLUSTER-ANALYSIS
B 6303 462 HARPER, JOHN A., DEVONIAN ORGANIC-RIC BLACK SHALES IN
B 6109 1537 HARPER, JOHN, SEDIMENT-DISPERSAL TRENDS OF CAMINADA-MOREAU
B 6305 816 HARPER, WARME, JOHN E., REVIEW OF, TRACE FOSSILS 2, EDITED BY T. P,
B 6002 285 HARRELL SHALE, PENNSYLVANIA
B 6309 1602 HARRELSON, DANNY W., PETROGRAPHY OF SOME SUBSURFACE IGNEOUS
B 6004 657* HARRINGTON SOUND, BERMUDA, ABST., PALEOECOLOGIC
B 6405 719 HARRIS, ANITA G., CONODONT-BASED ASSESSMENT OF THERMAL
B 6304 647 HARRIS, ANITA G., CONODONT-BASED REINTERPRETATION OF BANE
B 6405 718 HARRIS, ANITA G., CONODONTS—MODELS OF PRAGMATIC
B 6303 513* HARRIS COUNTY, TEXAS, ABST., RATES OF, CREEP OF ACTIVE, FAULTS IN
B 6207 1219 HARRIS, E. J., PRODUCTION GEOLOGY OF NORTH RANKIN GAS FIELD,
B 6211 2359 HARRIS, J. R., SUMMARY OF RECENT EXPLORATION AND EVALUATION
B 6304 647 HARRIS, LEONARD D., CONODONT-BASED REINTERPRETATION OF BANE
B 6309 1579 HARRIS, LEONARD D., EASTERN PROJECTION OF VALLEY AND RIDGE
B 6405 768 HARRIS, M. T., ET AL, EARLY GUADALUPIAN, PERMIAN, BANK MARGIN
B 6408 1274 HARRIS, MILLIMAN, JOHN D., REVIEW OF, FACIES ANATOMY AND,
B 6405 719 HARRIS, P. M., CRETACEOUS SEA LEVEL AND STRATIGRAPHY, EASTERN
B 6303 462 HARRIS, PAUL M., ANATOMY AND GROWTH HISTORY OF HOLOCENE
B 6105 793 HARRIS, PAUL M., COMPARISON OF ENDOLITH FLORAS FROM HOLOCENE-
B 6405 719 HARRIS, PAUL M., FRESHWATER CEMENTATION OF HOLOCENE AND
B 6105 792 HARRIS, PAUL M., HOLOCENE FRESHWATER CEMENTS, JOULTERS CAYS,
B 6209 1758 HARRIS, PAUL M., HOLOCENE MARINE-CEMENTED SANDS, JOULTERS
B 6203 520 HARRIS, PAUL M., OOID SAND-SHOAL FACIES, JOULTERS CAYS, GREAT
B 6105 794 HARRISON, DONALD B., GEOLOGY OF HEAVY-OIL DEPOSIT AT COLD LAKE,
B 6409 1558 HARRISON, FRANK W., III, PRESSURE, TEMPERATURE, SALINITY,
B 6009 1610 HARRISON, FRANK W., IVE GOT A NEW PRODUCER IN SOUTH
B 6405 720 HARRISON, FRANK W., JR., FALSE RIVER FIELD, ABST.
B 6409 1558 HARRISON, FRANK W., JR., LOUISIANA TUSCALOOSA VERSUS SOUTHEAST
B 6409 1552 HARRISON, FRANK W., JR., NO. 1 PLAY IN U.S.A.—SOUTH LOUISIANA
B 6303 463 HARRISON, RAND S., KEY LARGO--SUBSURFACE CORE STUDY, ABST.
M 28 21 HARRISON SUMMIT QUARTZITE, IDAHO, CAMBRIAN
M 28 8 HARRISON SUMMIT QUARTZITE, IDAHO, CAMBRIAN
B 6405 744 HARRISON, WILLIAM B., III, DEPOSITIONAL ENVIRONMENT OF LOWER
B 6303 463 HARRISON, WILLIAM B., III, PALEOECOLOGIC INTERPRETATION OF
B 6311 2117 HARRISON, WILLIAM E., GEOCHEMICAL PROSPECTING FOR
B 6311 2117 HARRISON, WILLIAM E., INVESTIGATION OF DESMOINESIAN ROCKS IN
B 6009 1610 HARRISON, WILLIAM E., ORGANIC DIAGENESIS--ITS RELATION TO GULF
B 6405 705 HARRISON, WILLIAM E., SUBSIDENCE AND THERMAL HISTORY OF
B 6003 452 HARRISON, WILLIAM E., THERMALLY INDUCED DIAGENESIS OF
B 6409 1555 HART, GEORGE F., SUBSURFACE WILCOX LIGNITE IN WEST-CENTRAL
B 6209 1758 HART, JEANNE, REMOTELY SENSED SEDIMENTATION PATTERNS IN LAKE
M 25 506 HART RIVER DEPOSIT, CANADA
B 6101 80 HARTFORD-DEERFIELD BASIN, CONNECTICUT-, MASSACHUSETTS
B 6107 988 HARTLAND CHALK UNIT, OKLAHOMA
B 6303 463 HARTMAN, JAMES A., DRAINAGE ANOMALIES IN GULF COAST TERTIARY
B 6009 1611 HARTMAN, JAMES A., NEW OIL FROM OLD FIELDS, ABST.
B 6305 845 HARTMAN, R. R., INTEGRATION OF AIRBORNE, MAGNETI AND
S 4 127 HARTMAN, WILLARD D., SPONGES AS REEF BUILDERS AND SHAPERS
B 6406 868 HARTONO, H. M. S., ARC-CONTINENT COLLISION I BANDA SEA REGION
B 6204 646 HARTSELLE SANDSTONE, ALABAMA
B 6009 1606* HARTSELLE SANDSTONE IN NORTH ALABAMA, ABST., DEPOSITIONAL
B 6405 724* HARTSHORNE FORMATION, EAST-CENTRAL OKLAHOMA, USING
B 6305 830* HARTVILLE UPLIFT--NEW LOOK AT AN OLD AREA, ABST.
B 6409 1354 HARTZELL, STEPHEN P., DEVELOPMENTS IN UPPER GULF COAST OF
B 6308 1223 HARTZELL, STEPHEN P., DEVELOPMENTS IN UPPER GULF COAST OF
B 6303 491* HARTZOG DRAW FIELD, POWDER RIVER BASIN, WYOMING, ABST., FACIES
B 6203 540* HARTZOG DRAW, NEW GIANT OIL FIELD, ABST.
B 6205 888* HARTZOG DRAW, NEW GIANT OIL FIELD, ABST.
B 6309 1575 HARVEY, RICHARD D., PETROLOGY AND MATURATION OF DISPERSED
B 6307 1093 HARVEY RIDGE, AUSTRALIA
B 6112 2082 HARWOOD, ROBERT J., OIL AND GAS GENERATION BY LABORATORY
M 25 460 HARZBURGITE FACIES, PACIFIDES, PERIDOTITE BELT, DUNITE-
B 6108 1380 HASBROUCK, WILFRED P., UTILIZATION OF GEOPHYSICS IN COAL
B 6210 2119 HASHEMITE KINGDOM OF JORDAN, AND PETROCONSULTANTS S.A., ON
B 6010 1873 HASHEMITE KINGDOM OF JORDAN FILON EXPLORATION CORP., AND
B 6107 1046 HASKELL FORMATION, KANSAS
B 6109 1440 HASKELL LIMESTONE, OKLAHOMA
M 26 440 HASKELL LIMESTONE, OKLAHOMA, ANADARKO BASIN
B 6309 1479 HASMARK DOLOMITE, TEXTURE AND COMPOSITION DATA
B 6403 424 HASSANE, TUNISIA, DJEBEL
B 6407 996 HASSEINE MUDSTONE MEMBER, MOROCCO
B 6407 1000 HASSI EL BEGUER FAULT, MOROCCO
B 6204 668 HASSI MESSAOUD OIL FIELD, ALGERIA
B 6204 668 HASSI RMEL OIL FIELD, ALGERIA
B 6002 278 HASSON, KENNETH O., STRATIGRAPHIC CROSS SECTION OF HAMILTON
B 6010 1865 HASSON, RICHARD C., PETROLEUM
B 6210 2111 HASSON, RICHARD C., PETROLEUM DEVELOPMENTS IN MIDDLE EAST
B 6310 1884 HASSON, RICHARD C., PETROLEUM DEVELOPMENTS IN MIDDLE EAST
B 6110 1795 HASSON, RICHARD C., PETROLEUM DEVELOPMENTS IN MIDDLE EAST
B 6403 443 HASTINGS, DOUGLAS D., RESULTS OF EXPLORATORY DRILLING,
B 6311 2059 HAT CREEK COALFIELD, BRITISH COLUMBIA
M 25 107 HAT CREEK COALFIELDS, CANADA
B 6411 1836 HATCH, GREGORY C., PETROLEUM DEVELOPMENTS IN MIDDLE EAST
B 6004 651 HATCH, J. R., ALTERED VOLCANIC ASH PARTINGS AS STRATIGRAPHIC
B 6004 667 HATCHER, P., CONTRASTS IN CONCENTRATION OF SUSPENDED MATTER
B 6104 602 HATCHET GAP, NEW MEXICO
B 6104 603 HATCHITA FORMATION, NEW MEXICO
M 25 354 HATFIELD, F. A., ARTIFICIAL RECHARGE OF GR
M 25 353 HATFIELD, F. A., DEVELOPMENT OF AUSTRALIA GROUNDWATER
B 6010 1640 HATFIELD, L. E., PETROLEUM DEVELOPMENTS IN SOUTH AMERICA,
B 6004 672 HATHAWAY, J. C., MESOZOIC STRATIGRAPHIC TEST, NANTUCKET ISLAND,
B 6105 794 HATHAWAY, JOHN C., ATLANTIC MARGIN CORING PROJECT, 1976—SUMMARY
B 6405 720 HATHON, CHRIS, AUTHIGENIC QUARTZ IN DEVONIAN BLACK SHALE,
M 26 49 HATLELID, W. G., SEISMIC STRATIGRAPHY AND GLOBAL CHANGES OF SEA
M 26 185 HATLELID, W. G., SEISMIC STRATIGRAPHY AND GLOBAL CHANGES OF SEA
B 6205 772 HATLELID, W. G., SEISMIC STRATIGRAPHY AND GLOBAL CHANGES OF SEA
M 24 51 HATLELID, W. G., TAGLU GAS FIELD, BEAUFORT BASIN, NORTHWEST
B 6405 746* HATTERAS, ABST., GEOLOGIC SETTING AND OIL AND GAS POTENTIAL OF
B 6205 793 HATTERAS ABYSSAL PLAIN
B 6412 2058 HATTERAS ABYSSAL PLAIN, DSDP SITE 105, LEG 11, GEOCHEMICAL LOG
B 6402 184 HATTERAS ABYSSAL PLAIN, SUBSIDENCE AND HEAT FLOW
B 6402 184 HATTERAS ABYSSAL PLAIN, TIME, TEMPERATURE, AND DEPTH PLOT
B 6406 851 HATTERAS ABYSSAL PLAINS, SAND LAYERS, SOHM AND
B 6403 347 HATTERAS SHELF AND INNER BLAKE PLATEAU, SEISMIC STUDY,
B 6403 339† HATTERAS SHELF AND INNER BLAKE PLATEAU, STRUCTURE,
B 6403 339 HATTERAS SHELF, FLORIDA, GEORGIA, AND SOUTH CAROLINA, FLORIDA
B 6309 1583* HATTERAS SLOPE, ABST., PETROLEUM, GEOCHEMISTRY AND GEOLOGY
B 6403 340 HATTERAS SLOPE, FLORIDA-
B 6103 335 HATTERAS SLOPE, FLORIDA

B 6402 184 HATTERAS, SUBSIDENCE AND HEAT FLOW, CAPE
B 6402 184 HATTERAS, TIME, TEMPERATURE, AND DEPTH PLOT, CAPE
B 6203 529* HATTERASTO HYDROGRAPHER CANYON, ABST., GEOTECHNICAL
M 24 261 HATTIESBURG CLAY, ALABAMA
B 6303 464 HATTIN, DONALD E., REGIONAL STRATIGRAPHY OF LIMESTONE MARKER
B 6405 729 HATTIN, DONALD E., TRACE-MAKERS AS HISTORIANS IN LARGE-SCALE
B 6105 807 HATTON, L., WAVE-EQUATION MIGRATION—TWO APPROACHES
B 6012 2184 HATTON, L., WAVE-EQUATION MIGRATION--TWO APPROACHES, ABST.
M 27 17 HATTON-ROCKALL BASIN, MIOCENE
B 6004 677 HAUBOLD, H., PALEOENVIRONMENT OF EARLY PERMIAN, ROTLIEGENDE
B 6404 530 HAUGE BJERGE FORMATION, GREENLAND
B 6404 528 HAUGE BJERGE, GREENLAND
B 6311 2117 HAUN, JOHN D., HOW MANY WILDCATS MUST WE DRILL, ABST.
B 6305 829 HAUN, JOHN D., OIL AND GAS POTENTIAL OF WYOMING, ABST.
B 6405 720 HAUN, JOHN D., PETROLEUM GEOLOGY IN 1980S, ABST.
M 30 537 HAUPT DOLOMITE FORMATION, TRIASSIC, MALOSSA FIELD
B 6305 830 HAUSEL, W. D., POWDER RIVER BASIN URANIUM DEPOSITS, HISTORY AND
B 6305 830 HAUSEL, W. D., UPDATE ON EXPLORATION FOR DIAMONDS IN
S 6 159 HAUTERIVIAN, KANAIGAURA FORMATION, JAPAN
S 6 159 HAUTERIVIAN, YAMADORI FORMATION, JAPAN
B 6004 657* HAUTERIVIAN-EOCENE UNCONFORMITY IN BEJUCO FIELD, EASTERN
M 25 351 HAWAII, BASALTIC ISLANDS
M 25 371 HAWAII, GROUNDWATER
M 25 365 HAWAII, GROUNDWATER AND WASTEWATER DISPOSAL
M 28 139 HAWAII, HOLOCENE, HONOLULU VOLCANICS
M 25 366 HAWAII, HONOLULU BASIN
M 25 366 HAWAII, HONOLULU, GROUNDWATER
M 25 154 HAWAII, KILAUEA VOLCANO
M 28 139 HAWAII, PLEISTOCENE, HONOLULU VOLCANICS
S 8 152 HAWAII, SUBMARINE CANYONS, HANAKAPIAI CANYON
S 8 152 HAWAII, SUBMARINE CANYONS, KALALAU CANYON
S 8 147 HAWAII, SUBMARINE CANYONS, KAUAI CANYONS
S 8 152 HAWAII, SUBMARINE CANYONS, NOHILI SEAVALLEY
M 25 351 HAWAII, WATER SOURCES, RAINFALL
M 25 526 HAWAII-TYPE BASALTS AND HAWAIITES, BAJA CALIFORNIA
M 25 371† HAWAIIAN ISLANDS, EFFECT OF SUBSURFAC, WASTE-, DISPOSAL
M 25 526 HAWAIITES, BAJA CALIFORNIA, HAWAII-TYPE, BASALTS AND
B 6207 1235* HAWAII—PRESENT AND FUTURE, ABST., GEOTHERMAL, ENERGY IN
B 6306 955 HAWKE BAY FORMATION, NEWFOUNDLAND
B 6411 2037 HAWKES, AND JOHN S. WEBB, BRIGHT, M. J., REVIEW OF, GEOCHEMISTRY
M 24 51 HAWKINGS, T. J., TAGLU GAS FIELD, BEAUFORT BASIN, NORTHWEST
B 6201 88 HAWLEY RIDGE, ALASKA
M 28 72 HAWTHORN FORMATION, FLORIDA, MIOCENE
B 6308 1407 HAY, PETER W., DEVELOPMENTS IN WESTERN CANADA IN 1978
B 6409 1510 HAY, PETER W., DEVELOPMENTS IN WESTERN CANADA IN 1979
B 6105 794 HAY, PETER W., REGIONAL LOWER MANNVILLE LITHOFACIES COMPUTER
B 6004 678 HAY, W. W., CARBONATE MASS BALANCE--CYCLING AND DEPOSITION ON
B 6203 520 HAY, WILLIAM W., BINARY SIMILARITY COEFFICIENTS AND CONFIDENCE
B 6303 464 HAY, WILLIAM W., IMPACT OF DEEP SEA DRILLING PROJECT ON
B 6105 794 HAY, WILLIAM W., POST-PALEOZOIC CHANGES IN VOLUMES AND ORGANIC-
M 25 147 HAYAKAWA, MASAMI, GEOTHERMAL RESOURCES AND ENERGY IN JAPAN
B 6108 1380 HAYES, JOHN B., SANDSTONE DIAGENESIS-THE HOLE TRUTH, ABST.
B 6004 716 HAYES, M. O., LITTORAL PROCESSES AND GEOMORPHIC VARIABILITY ON
B 6105 797 HAYES, MILES O., DISTRIBUTION OF SEDIMENTARY STRUCTURES IN EBB—
B 6004 678 HAYES, MILES O., GREAT PATAGONIAN OIL SPILL, ABST.
B 6303 464 HAYES, MILES O., SPECTRUM OF DEPOSITIONAL MODELS FOR CLASTIC
B 6105 783 HAYES, MILES O., TIDAL-FLAT SEDIMENTATION ON TIDE-DOMINATED
B 6303 425 HAYKUS, JOSEPH A., REGIONAL HYDROCARBON SOURCE-ROCK
M 28 23 HAYNER RANCH FORMATION, NEW MEXICO, MIOCENE
M 28 177 HAYNER RANCH FORMATION, NEW MEXICO, MIOCENE
M 28 125 HAYNER RANCH FORMATION, NEW MEXICO, MIOCENE
M 28 174 HAYNER RANCH FORMATION, NEW MEXICO, MIOCENE
B 6407 1005 HAYNESVILLE FIELD, LOUISIANA
B 6007 1119 HAYNESVILLE LIMESTONE, TEXAS
M 24 251 HAYNESVILLE-BUCKNER SECTION, MISSISSIPPI
B 6109 1551* HAYS COUNTIES, TEXAS, ABST., MIDDLE, GLEN ROSE, CRETACEOUS,
B 6304 689 HAYSLIP, D. L., THERMOLUMINESCENCE OF URANIUM HOST ROCKS IN
B 6303 380 HAYWARD, O. T., REVIEW OF GEOLOGY IN ENVIRONMENTAL PLANNING.
B 6303 434* HAZARDS, ABST., SOURCES OF LIABILITY FOR, GEOLOGIC
M 26 374 HAZARDS, RESOLUTION SYSTEMS, POTENTIAL, DRILLING
B 6409 1559 HAZEL, JOSEPH E., OSTRACODE BIOSTRATIGRAPHY OF LOWER
M 25 509 HAZELTON-CACHE CREEK-SLIDE MOUNTAIN- NICOLA ROCKS,
B 6206 1005 HAZEN TROUGH, ARCTIC, CANADA
B 6305 846 HAZLETT, JEAN M., PETROLOGY OF TRIASSIC-JURASSIC CULPEPER GROUP
M 25 1 HAZZARD, JOHN CHARLES, 1903-1973, DEDICATION TO
B 6203 561* HE4, RN222, AND PB210 OVER STRATA-BOUND URANIUM DEPOSIT, RED
B 6105 827* HE4-AR36 RATIOS IN SOIL GAS OVERLYING TWO URANIUM ORE
B 6004 718 HEAD, W. J., SYNTHETIC SEISMIC SECTIONS OF SELECTED STRATIGRAPHIC
B 6105 807 HEALD, M. T., IMPORTANCE OF TEXTURAL AND DIAGENETIC PROPERTIES
B 6105 775 HEALD, MILTON T., GAS-PRODUCING SUBMARINE-FAN DEPOSITS OF
B 6103 361 HEALDTON ANTICLINE, OKLAHOMA
B 6104 583 HEALDTON POOL, OKLAHOMA
M 25 352 HEALY, J. H., CONTROL OF EARTHQUAKES AT RANGELY,
M 25 130 HEALY, JAMES, GEOTHERMAL ENERGY IN NEW ZEALAND--SUMMARY
M 25 110 HEALY, JAMES, GEOTHERMAL PROSPECTS AROUND THE PACIFIC
B 6012 2182 HEALY, JOHN H., RECENT RESULTS IN EARTHQUAKE PREDICTION, ABST.
M 26 31 HEAT ALTERATION, LABORATORY, MEASUREMENT
M 25 448 HEAT FLOW, HIGH, JAPAN
M 25 443 HEAT FLOW, HIGH, SOUTH KOREA, TONGRAE FAULT
M 25 443 HEAT FLOW, HIGH, SOUTH KOREA, YANGSAN FAULT
B 6405 735* HEAT FLOW REGIME IN OCEANS AS RELATED TO HYDROCARBON
B 6402 173† HEAT FLOW--IMPORTANT PARAMETERS IN FORMATION OF PETROLEUM
B 6405 780* HEAT-FLOW AND HEAT-PRODUCTION STUDIES IN NORTH DAKOTA, ABST.
S 3 12 HEAT-FLOW MODEL FOR RADIOELEMENT DISTRIBUTION
B 6405 771* HEAT-FLOW STUDIES IN SOUTHWESTERN UNITED STATES, ABST., DEEP
B 6008 1410* HEAT-FLOW STUDIES IN STEAMBOAT MOUNTAIN-LEMEI ROCK AREA,
B 6008 1392* HEAT-FLOW STUDY OF SNAKE RIVER PLAIN, IDAHO, ABST.
S 3 14 HEAT-PRODUCING CAPACITIES OF RADIOELEMENTS
M 25 539 HEATH, G. ROSS, METALLOGENESIS IN SOUTHEAST PACIFIC OCEAN,
B 6002 320 HEATH, LARMAN J., STIMULATING RECOVERY FROM HEAVY-OIL
M 30 104 HEATHER FORMATION, JURASSIC, STATFJORD FIELD
M 25 467 HEAVY METALS, USSR, NORTHEAST
B 6004 734* HEAVY MINERAL DISTRIBUTION ON SOUTHERN ATLANTIC COASTAL
B 6206 1058 HEAVY MINERALS
M 28 61 HEAVY MINERALS, BIBLIOGRAPHY
S 7 300 HEAVY MINERALS, GYRE BASIN
B 6004 675* HEAVY MINERALS, MUDDY SANDSTONE, LOWER CRETACEOUS, WIND
B 6203 565* HEAVY OIL AND TAR SANDS OF ALBERTA. BY TITLE ONLY.
B 6405 740* HEAVY OIL AREA, SASKATCHEWAN, ABST., NEARSHORE LITHOFACIES OF,
B 6002 323* HEAVY OIL IN WESTERN MISSOURI--UPDATE 1975, ABST.
B 6111 1955 HEAVY OIL, USSR, VOLGA-, URAL
B 6111 1955 HEAVY OIL, UTAH
B 6111 1953 HEAVY OIL, VENEZUELA, EASTERN
B 6111 1953 HEAVY OIL, WESTERN, CANADA
B 6203 572* HEAVY OILS, SASKATCHEWAN AND ALBERTA. BY TITLE ONLY.
B 6012 2188* HEAVY-MINERAL DATA, MONTEREY BAY, CALIFORNIA, ABST., HOLOCENE
B 6203 488* HEAVY-MINERAL RESPONSE TO ENVIRONMENTAL CONDITIONS IN A
M 25 419 HEAVY-MINERAL SAND, AUSTRALIA, EASTERN
M 25 420 HEAVY-MINERAL SAND, EXPLORATION, AUSTRALIA
M 25 418 HEAVY-MINERAL SAND, MINING, HISTORY, AUSTRALIA
M 25 418 HEAVY-MINERAL SAND, PRODUCTION FIGURES, AUSTRALIA
M 25 420 HEAVY-MINERAL SAND, PRODUCTION METHODS, AUSTRALIA
M 25 420 HEAVY-MINERAL SAND, REHABILITATION OF MINED LANDS, AUSTRALIA
M 25 419 HEAVY-MINERAL SAND, WESTERN AUSTRALIA
M 25 421 HEAVY-MINERALS INDUSTRY, FUTURE FOR, AUSTRALIA
M 25 421 HEAVY-MINERALS, MARKETING, AUSTRALIA
B 6111 1951 HEAVY-OIL DEPOSITS OF WORLD, LOCATION
B 6311 2117* HEAVY-OIL POTENTIAL, ABST., INVESTIGATION OF, DESMOINESIAN
B 6002 323* HEAVY-OIL RECOVERY, ABST., ENGINEERING AND GEOLOGIC
B 6002 320* HEAVY-OIL RESOURCES, MID-CONTINENT AREA, ABST., STIMULATING
B 6203 511* HEAVY-OIL-BEARING SANDSTONES IN SOUTHEASTERN KANSAS, ABST.,
B 6203 430 HEBRIDEAN PLATFORM, BRITAIN, WEST
B 6207 1225* HEBRIDES ARC-TRENCH SYSTEM, ABST., SEISMICITY AND STRUCTURE OF,
B 6004 678 HECHT, A. D., PALEOCLIMATIC SIGNIFICANCE OF MORPHOLOGIC
B 6305 845 HECK, E. T., FOSSIL FLUIDS IN POST-SILURIAN ROCKS OF APPALACHIAN
B 6207 1220 HECKARD, JOHN M., PHYSICAL CONSTRAINTS ON MINERAL
B 6405 721 HECKEL, PHILIP H., PALEOGEOGRAPHY OF EUSTATIC MODEL FOR
B 6107 1045 HECKEL, PHILIP H., PHOSPHATIC BLACK SHALE FACIES IN
B 6310 1984 HECKMAN AND OTHERS, RICE, PEGGY J., REVIEW OF, GEOREF
B 6109 1559 HECKMANN, G. A., INTEGRATION OF LANDSAT AND GEOPHYSICAL DATA
B 6101 24 HECLA FIELD, CANADIAN, ARCTIC ISLANDS
B 6303 286 HEDBERG, H. D., PETROLEUM PROSPECT OF THE DEEP OFFSHORE
B 6003 458 HEDBERG, H. D., SALT DIAPIRISM IN SOUTHERN IRAN. DISCUSSION
S 6 1 HEDBERG, HOLLIS D., EDITED BY, THE GEOLOGIC TIME SCALE
B 6207 1220 HEDBERG, HOLLIS D., GEOMORPHOLOGY OF OCEAN FLOOR AS NATURAL
S 6 1 HEDBERG, HOLLIS D., INTRODUCTION, THE GEOLOGIC TIME SCALE
S 10 179 HEDBERG, HOLLIS D., METHANE GENERATION AND PETROLEUM
B 6203 520 HEDBERG, HOLLIS D., METHANE GENERATION AND PETROLEUM
B 6105 795 HEDBERG, HOLLIS D., PETROLEUM PROSPECTS OF CONTINENTAL SLOPES
B 6411 2036 HEDBERG, HOLLIS D., REVIEW OF TECTONIC MAP OF SOUTH AMERICA BY
B 6310 1982 HEDBERG, HOLLIS D., REVIEW OF THE IMPACT OF THE GEOSCIENCES ON
S 6 33 HEDBERG, HOLLIS D., STRATOTYPES AND AN INTERNATIONAL
B 6303 286 HEDBERG, R. M., PETROLEUM PROSPECT OF THE DEEP OFFSHORE
B 6405 721 HEDDEN, WILLIAM J., STRATIGRAPHY AND DEPOSITIONAL HISTORY O
B 6405 704 HEDGES, JOHN, A TEST OF THE MELANOIDIN HYPOTHESIS, ABST.
B 6403 423† HEDIL, TUNISIAN ATLAS, PROGRESSIVE, NEOGENE DEFORMATION OF
B 6404 507 HEEZEN CANYON, NEW ENGLAND MARGIN
B 6305 830 HEGNA, EARL T., HARTVILLE UPLIFT--NEW LOOK AT AN OLD AREA, ABST.
M 29 223 HEHANUSSA, P. E., STRUCTURE AND CENOZOIC EVOLUTION OF THE

B 6101 3 HEIBERG ISLANDS, NORTHERN, ELLESMERE AND, AXEL
B 6203 520 HEIN, F. J., FACIES ASSOCIATIONS AND FABRIC ANALYSES OF DEEP-SEA
B 6004 678 HEIN, F. J., SUBMARINE-FAN COMPLEX, CAP ENRAGE FORMATION
B 6303 380 HEIRTZLER, H. M. BOLLI, T. A. DAVIES, J. B. SAUNDERS, AND J. G.
B 6405 721 HEIRTZLER, J. R., SCIENTIFIC OCEAN DRILLING NEAR UNITED STATES.
B 6409 1331 HEISE, BRUCE A., DEVELOPMENTS IN ATLANTIC COASTAL PLAIN IN 1979
B 6409 1336 HEISE, BRUCE A., DEVELOPMENTS IN ATLANTIC OC IN 1978—ADDENDUM
B 6203 510 HEITMAN, HAL, SLOPE-BASIN BENTHIC FORAMINIFERA OF CALIFORNIA
B 6309 1589* HELDERBER GROUP OF WEST VIRGINIA AND VIRGINIA, ABST., PATTERNS
B 6209 1615 HELDERBERG GROUP, APPALACHIAN BASIN
B 6209 1610 HELDERBERG GROUP, MARYLAND
B 6209 1610 HELDERBERG GROUP, NEW YORK
B 6209 1610 HELDERBERG GROUP, PENNSYLVANIA
S 5 130 HELDERBERG GROUP, KEYSER LIMESTONE, WEST VIRGINIA, UPPER
B 6205 829 HELEZ FORMATION, ISRAEL
B 6205 827 HELEZ OIL FIELD, ISRAEL
B 6001 108† HELEZ OIL FIELD, ISRAEL, EARLY, CRETACEOUS, BURIED,
B 6002 266 HELEZ REEF, ISRAEL
B 6203 435 HELGELAND BASIN, NORTH SEA
B 6101 30 HELIKIAN CHART, CANADIAN SHIELD
B 6101 33 HELIKIAN URANIUM DEPOSITS, SASKATCHEWAN
B 6101 34 HELIKIAN URANIUM PROVINCES, NORTHWESTERN, CANADIAN SHIELD
B 6405 771* HELIUM ANALYSIS AS EXPLORATION TOOL FOR ENERGY DEPOSITS,
B 6008 1396* HELIUM DETECTION EXPLORATION GUIDE TO URANIUM SOURCES, ABST.
B 6105 780 * HELIUM DETECTION IN URANIUM EXPLORATION
B 6108 1376* HELIUM DETECTION IN URANIUM EXPLORATION, ABST.
B 6405 766* HELIUM SURVEYING FOR DEEPLY BURIED URANIUM DEPOSITS, ABST.
B 6407 971 HELL CREEK FORMATION, GREAT PLAINS
B 6403 318 HELL CREEK FORMATION, MONTANA
B 6104 603 HELL-TO-FINISH FORMATION, NEW MEXICO
B 6106 880 HELLENIC ARC, MEDITERRANEAN RIDGE AND
B 6303 532* HELLENIC BASINS AND SLOPES IN EASTERN MEDITERRANEAN, ABST.,
B 6405 787* HELLENIC TRENCH, EASTERN MEDITERRANEAN, ABST., DEPOSITIONAL
B 6203 561 HELLINGER, T. S., MAJOR URANIUM DISCOVERY IN FRONTIER AREA—
B 6304 621 HELLINGER T. S., MAJOR URANIUM DISCOVERY IN VOLCANICLASTIC
B 6211 2261 HELM FIELD, CALIFORNIA
B 6303 470 HELMIG, H. M., DISCOVERY AND DEVELOPMENT OF BADAK FIELD, EAST
M 30 441 HELMIG, H. M., DISCOVERY AND DEVELOPMENT OF THE BADAK FIELD,
B 6303 465 HELMOLD, KENNETH P., DIAGENESIS OF TERTIARY ARKOSES, SANTA
B 6306 905 HELMSDALE FAULT, SCOTLAND
M 29 443 HELSLEY, C., EXPLOSION SEISMOLOGY STUDIES OF ACTIVE AND PASSIVE
B 6312 2194 HELVETIAFJELLET FORMATION, SVALBARD
B 6312 2193† HELVETIAFJELLET FORMATION, SVALBARD, BEARING ON RESERVOIR
B 6405 690 HELWIG, JAMES, DECOMPACTION TECHNIQUE HELPS CORRELATION,
B 6303 465 HELWIG, JAMES, PLATE TECTONICS, BASIN ANALYSIS, AND
B 6405 722 HELWIG, JAMES, STRATIGRAPHY AND STRUCTURE OF EARLY TERTIARY
M 28 125 HEMATITE, AUTHIGENIC, ON SAND GRAINS
M 27 144 HEMATITE CEMENT
M 27 147 HEMATITE CEMENT
M 27 144 HEMATITE COATINGS
M 25 413 HEMATITE ENRICHMENT ORE, GENESIS OF, IRON
M 25 411 HEMATITE ENRICHMENT ORE, WESTERN AUSTRALIA, IRON
M 27 147 HEMATITE FILLINGS
M 27 144 HEMATITE FILLINGS
M 28 124 HEMATITE, IN SANDSTONE, CEMENTS
M 28 126 HEMATITE, IN SANDSTONE, CEMENTS
M 28 126 HEMATITE, MYLONITIZATION
M 27 147 HEMATITE OOLITIC COATINGS
B 6004 725* HEMATITE TRANSFORMATION AND ITS POSSIBLE SIGNIFICANCE FOR
M 25 411 HEMATITE-GOETHITE ORE
M 28 44 HEMATITIC OOIDS
M 28 45 HEMATITIC OOIDS
B 6010 1900 HEMER, D. O., SELECTED REFERENCES ON MIDDLE EAST, 1970 TO 1975
B 6411 1836 HEMER, DARWIN O., PETROLEUM DEVELOPMENTS IN MIDDLE EAST
B 6310 1884 HEMER, DARWIN O., PETROLEUM DEVELOPMENTS IN MIDDLE EAST
B 6210 2111 HEMER, DARWIN O., PETROLEUM DEVELOPMENTS IN MIDDLE EAST
B 6110 1795 HEMER, DARWIN O., PETROLEUM DEVELOPMENTS IN MIDDLE EAST
S 11 120 HEMICRISTELLARIA BEALI
M 25 480 HEMISHELVES, DEFINED
B 6203 377 HEMLOCK CONGLOMERATE, ALASKA
B 6408 1133 HEMLOCK CONGLOMERATE, ALASKA
B 6401 26 HEMLOCK FORMATION, ALASKA
M 24 11 HEMLOCK SANDSTONE MEMBER, ALASKA
B 6109 1559 HENDERSON, F. B., III, GEOSAT—REMOTE-SENSING SYSTEMS OF FUTURE,
B 6207 1220 HENDERSON, FREDERICK B., III, FUTURE GEOLOGIC REMOTE SENSING
B 6208 1441 HENDERSON, GILROY, DEVELOPMENTS OFFSHORE WEST GREENLAND IN
B 6308 1210 HENDERSON, GILROY, DEVELOPMENTS OFFSHORE WEST GREENLAND IN
B 6406 961 HENDERSON, GORDON W., RED WING CREEK FIELD—COSMIC IMPACT
B 6408 1284 HENDERSON, JAMES A., JR., COAL RESOURCE STUDIES IN VIRGINIA ABST.
B 6008 1398 HENDERSON, L. BROOKE,, ENTRAPMENT OF GAS IN UPPER CRETACEOU
B 6108 1379* HENDERSON MINE MILL, ABST., ENVIRONMENTAL CASE STUDY-
B 6402 280 HENDERSON, WILLIAM G., ECUADORIAN SUBDUCTION SYSTEM.
B 6108 1381 HENDRICKS, MICHAEL L., ENVIRONMENTS OF DEPOSITION OF
B 6105 777 HENDRICKS, THOMAS A., PROSPECTIVE HYDROCARBON PROVINCES OF THE
B 6009 1613* HENDRY AND COLLIER COUNTIES, FLORIDA, ABST., SUNOCO-, FELDA
M 24 287† HENDRY AND COLLIER COUNTIES, FLORIDA, SUNOCO-FELDA FIELD
B 6309 1609* HENDRY COUNTIES, FLORIDA, ABST., STRATIGRAPHY AND
B 6305 831 HENDRY, H. E., RELATIONS BETWEEN SEDIMENTARY FACIES AND
B 6303 465 HENDRY, HUGH E., RESEDIMENTED CONGLOMERATES IN SUBMARINE
B 6101 113 HENEFER FORMATION, UTAH
B 6309 1602 HENK, FLOYD H., JR., SEDIMENTATION ON TRAILING PLATE
B 6309 1579 HENKLE, WILLIAM R., JR., NEW RIVER FORMATION COALS IN SOUTHERN
B 6209 1706 HENLEY BED, KENTUCKY, FARMERS MEMBER
B 6410 1723 HENLEY, FONT, ROBERT G., REVIEW OF, STATISTICAL ANALYSIS IN
B 6403 424 HENNCHIR EN NEMRA, TUNISIA
B 6105 789 HENNIGAR, H., COASTAL-DUNE GEOMORPHOLOGY AND CLASSIFICATION
B 6105 791 HENNIGAR, H., DUNE DYNAMICS IN COASTAL ZONE—PAST, PRESENT, AND
B 6309 1580 HENNINGTON, W. M., UNCONVENTIONAL EVALUATION OF COTTAGEVILL
M 25 18 HENRIE, THOMAS A., MINERAL TRADE IN CIRCUM-PACIFIC
B 6405 763* HENRY BASIN, UTAH, ABST., MUDSTONES AS EXPLORATION GUIDES TO
B 6309 1602 HENRY, C. D., HYDROLOGY AND WATER QUALITY OF EOCENE WILCOX
B 6405 798 HENRY, CHRISTOPHER D., EVALUATION OF URANIUM POTENTIAL OF
B 6405 711 HENRY, CHRISTOPHER D., HYDROSTRATIGRAPHIC FRAMEWORK AND
B 6405 722 HENRY, CHRISTOPHER D., OXIDATION-REDUCTION IN OAKVILLE
B 6008 1340 HENRY, GARY E., DEVELOPMENTS IN NORTH-CENTRAL TEXAS IN 1975
B 6108 1220 HENRY, GARY E., DEVELOPMENTS IN NORTH-CENTRAL TEXAS IN 1976
B 6208 1377 HENRY, GARY E., DEVELOPMENTS IN NORTH-CENTRAL TEXAS IN 1977
B 6309 1603 HENRY, JAMES D., ENHANCED OIL RECOVERY, ABST.
M 29 189 HENRY, M., TECTONICS OF THE ANDAMAN SEA AND BURMA
B 6405 722 HENRY, MITCHELL E., MARINE PETROLEUM PROSPECTING WITH
B 6208 1519 HENTON, J. M., DEVELOPMENTS IN LOUISIANA GULF COAST OFFSHORE IN
B 6212 2492 HENYEY, T. L., PLIOCENE-PLEISTOCENE DIASTROPHISM OF SANTA
B 6202 247 HENYEY, THOMAS L., PLIOCENE-PLEISTOCENE DIASTROPHISM OF SANTA
B 6303 504 HEPP, V. R., EOLIAN SEDIMENTOLOGY INTERPRETED FROM DIPMETER
B 6204 669 HERCYNIAN UNCONFORMITY, ALGERIAN, SAHARA, EASTERN
B 6110 1838 HERGENRODER, J. D., ON DEVELOPMENTS IN PAKISTAN, 1976
B 6311 1999 HERITIER, F. E., FRIGG FIELD--LARGE SUBMARINE-FAN TRAP IN LOWER
M 30 59 HERITIER, F. E., FRIGG FIELD--LARGE SUBMARINE-FAN TRAP IN LOWER
B 6203 521 HERITIER, F. E., FRIGG FIELD—LARGE SUBMARINE-FAN TRAP IN LOWER
M 29 199 HERMAN, BRUCE M., EXTENSIONAL TECTONICS IN THE OKINAWA
B 6203 521 HERMAN, YVONNE, SUBMARINE FANS—IMPORTANT REGIONS FOR OIL
B 6208 1399 HERMANN, JEFFREY B., OIL AND GAS DEVELOPMENTS IN MARYLAND,
B 6008 1288 HERMANN, JEFFREY B., OIL AND GAS DEVELOPMENTS IN MARYLAND,
B 6108 1269 HERMANN, JEFFREY B., OIL AND GAS DEVELOPMENTS IN MARYLAND,
B 6308 1244 HERMANN, JEFFREY B., OIL AND GAS DEVELOPMENTS IN MARYLAND,
B 6409 1403 HERMANN, JEFFREY B., OIL AND GAS DEVELOPMENTS IN MARYLAND,
S 4 289 HERMATYPIC SCLERACTINIAN CORAL, GALAXEA CLAVUS DANA
B 6101 54 HERMEL STRUCTURE, LEBANON
S 3 33 HERMIT SHALE, ARIZONA
B 6408 1172 HERMOSA CARBONATE ROCK, UTAH AND, COLORADO
S 4 243 HERMOSA GROUP, PARODOX BASIN
B 6303 465 HERN, JOHN L., GEOPHYSICAL INVESTIGATIONS FOR PILOT PLANT FOR
B 6106 880 HERODOTUS ABYSSAL PLAIN
B 6106 875 HERODOTUS BASIN
M 29 299 HERODOTUS BASIN, SEDIMENTARY SEQUENCE, MEDITERRANEAN
B 6203 547 HERON, DUNCAN, RATES OF LANDWARD MIGRATION OF CORE BANKS,
S 4 291 HERON ISLAND, GALAXEA
B 6303 504 HEROUX, Y., THERMA MATURATION OF LOWER PALEOZOIC SHALES IN
B 6312 2128 HEROUX, YVON, COMPILATION AND CORRELATION OF MAJOR THERMAL
B 6108 1157 HERRERA, ROGER C., DEVELOPMENTS IN ALASKA IN 1976
B 6208 1311 HERRERA, ROGER C., DEVELOPMENTS IN ALASKA IN 1977
B 6211 2356 HERRING, A. T., EXPLORATION IN RUGGED TERRAIN IN OVERTHRUST
B 6009 1611 HERRMAN, LEO A., JAMES LIMESTONE IN WINN AND NATCHITOCHES
B 6108 1381 HERROD, WILSON H., REGIONAL IMPLICATIONS OF OIL ENTRAPMENT AT
B 6107 1046 HERTHA FORMATION, KANSAS
B 6405 722 HESS, GORDON R., POSSIBLE GEOMETRIES OF SANDSTONE BODIES AS
B 6403 443 HESS, GORDON R., QUATERNARY STYLES OF CALIFORNIA SUBMARINE
B 6405 762 HESS, PAUL D., MONDAK MISSISSIPPIAN OIL FIELD, WILLISTON BASIN,
B 6303 538* HESS RISE, CENTRAL NORTH PACIFIC OCEAN, ABST., MIDDLE,
B 6105 776 HESSE, R., CHANNEL-FILL AND SPILLOVER FACIES OF NORTHWEST
B 6303 504 HESSE, R., THERMA MATURATION OF LOWER PALEOZOIC SHALES IN
B 6009 1612 HESSENBRUCH, JOHN M., SALT MOVEMENT, HALOKINETICS, IN NORTH
B 6004 679 HESSENBRUCH, JOHN M., SALT TECTONICS OF WINN PARISH, LOUISIANA,
B 6408 1281 HESTER, NORMAN C., COAL GEOLOGY AND RESOURCE OF EASTERN
B 6105 795 HEWARD, ALAN P., STEPHANIAN A AND B ALLUVIAL-FAN AND LACUSTRINE
B 6104 583 HEWITT FIELD, OKLAHOMA
B 6208 1399 HEYMAN, LOUIS, OIL AND GAS DEVELOPMENTS IN MARYLAND, OHIO,
M 26 57 HIATUSES, DEPOSITIONAL SEQUENCES
B 6010 1643 HICKAM, E. E., ON DEVELOPMENTS IN BARBADOS, 1975

B 6403 444 HICKMAN, CAROLE S., HOLOCENE MOLLUSK DISTRIBUTION PATTERNS IN
B 6402 225 HICKMAN GRABEN, TEXAS
B 6402 225 HICKMAN GRABEN, TEXAS
M 28 164 HICKORY SANDSTONE, TEXAS, CAMBRIAN
M 28 126 HICKORY SANDSTONE, TEXAS, CAMBRIAN
M 28 106 HICKORY SANDSTONE, TEXAS, CAMBRIAN
B 6110 1839 HICKS, G. W., ON DEVELOPMENTS IN PHILIPPINES, 1976
B 6407 1005 HICO SHALES, ARK-LA-TEX AREA
B 6309 1596* HIDALGO COUNTY, TEXAS, ABST., CHARACTERISTICS OF LOWER,
B 6409 1566* HIDALGO COUNTY, TEXAS, ABST., MINERALOGY, DIAGENESIS, AND
B 6104 603 HIDALGO FORMATION, NEW MEXICO
M 27 208 HIERLATZKALK, GERMANY, JURASSIC
B 6207 1221 HIERN, M. N., LAND-USE PLANNING AND MINERAL INDUSTRY—
M 25 279 HIGASHI-KASHIWATZAKI GAS FIELD, JAPAN
B 6004 725* HIGH ATLAS, MOROCCO, ABST., CARBONATE, PETROLOGY O, BASIN-TO-,
B 6303 344 HIGH BRIDGE GROUP, KENTUCKY
B 6006 952† HIGH, CENTRAL BROOKS RANGE, ALASKA, GEOLOGIC SIGNIFICANCE OF,
B 6008 1256 HIGH ISLAND AREA EAST ADDITION SOUTH EXTENSION, FEDERAL BLOCK
B 6008 1256 HIGH ISLAND AREA SOUTH, FEDERAL BLOCK A-532, TEXAS
B 6409 1355 HIGH ISLAND AREA, TEXAS, BLOCK 154
B 6208 1382 HIGH ISLAND AREA, TEXAS, BLOCK 27-L FIELD
B 6209 1760* HIGH ISLAND BLOCKS A-560 AND A-561, ABST., PRE-PLATFORM
B 6008 1256 HIGH ISLAND EAST ADDITION, SOUTH EXTENSION FEDERAL BLOCK 341,
B 6108 1225 HIGH ISLAND SOUTH, FEDERAL BLOCK A-492, TEXAS
B 6108 1225 HIGH ISLAND SOUTH, FEDERAL BLOCK A-560, TEXAS
B 6008 1256 HIGH ISLAND SOUTH, FEDERAL BLOCK A-561, TEXAS
B 6305 837 HIGH, LEE R., JR., STRATIGRAPHY OF LACUSTRINE DEPOSITS, ABST.
B 6203 519* HIGH LEVELS OF VARIATION IN SIZE-FREQUENCY DISTRIBUTIONS AT
B 6303 533* HIGH PLAINS SEQUENCE OF WYOMING AND NEBRASKA, ABST., ISOTOPIC
B 6108 1392* HIGH PLAINS SEQUENCE, WYOMING AND NEBRASKA, ABST., INCIPIENT
M 26 352† HIGH RESOLUTION SEISMIC DATA, ANALYSIS OF
B 6403 429 HIGH--WAX OIL FIELDS, TABLE
B 6205 756 HIGH-AMPLITUDE FACIES, HIGH-CONTINUITY AND
M 26 169 HIGH-AMPLITUDE FACIES, SEISMIC, INTERPRETATION
B 6205 756 HIGH-CONTINUITY AND HIGH-AMPLITUDE FACIES
M 26 169 HIGH-CONTINUITY FACIES, SEISMIC, INTERPRETATION
B 6303 468* HIGH-ENERGY BEACH-TO-OFFSHORE SEQUENCE, VENTURA--PORT
S 7 91 HIGH-ENERGY TURBIDITY-CURRENT TRANSPORT, SEDIMENT PROCESSES
B 6209 1768* HIGH-LEVEL PLIO-PLEISTOCENE SHORELINE IN SOUTH FLORIDA, ABST.
B 6002 293† HIGH-PRESSURE CONSOLIDATION OF KAOLINITIC CLAY
B 6009 1531† HIGH-PRESSURE DEFORMATION ON PERMEABILITY OF OTTAWA SAND,
B 6304 699* HIGH-RESOLUTION LANDSAT FOR GEOLOGIC STUDIES, ABST.
B 6405 725* HIGH-RESOLUTION LANDSAT FOR GEOPHYSICAL STUDIES, ABST.
S 7 296 HIGH-RESOLUTION PROFILES, GYRE BASIN
S 7 295 HIGH-RESOLUTION PROFILES, GYRE BASIN
B 6105 791 * HIGH-RESOLUTION REFLECTION-SEISMIC EXPLORATION FOR COAL MINING
B 6207 1224* HIGH-RESOLUTION SEISMIC EXPLORATION FOR COAL IN AUSTRALIA,
B 6108 1390* HIGH-RESOLUTION SEISMIC EXPLORATION FOR GEOLOGIC FAULTS IN
B 6303 438* HIGH-RESOLUTION SEISMIC METHODS IN COAL EXPLORATION, ABST.
B 6303 448* HIGH-RESOLUTION SEISMIC WORK AS TOOL TO LOCATE STRATIGRAPHIC
B 6004 675* HIGH-RESOLUTION STUDIES OF PALEOCLIMATIC AND
B 6403 429 HIGH-WAX OIL FIELDS, TABLE 1
S 4 302 HIGHLAND LIMESTONE, BARBUDA
M 26 73 HIGHSTAND OF SEA LEVEL, DEPOSITIONAL PATTERNS
M 26 92 HIGHSTANDS, SEA LEVEL FLUCTUATIONS, GLOBAL
B 6412 2191 HILCHIE, EVANS, HILTON B., REVIEW OF OLD, ELECTRICAL, LOG
B 6111 1951 HILL CREEK HEAVY OIL DEPOSIT, UNITED STATES
B 6004 651 HILL, D. J., ALTERED VOLCANIC ASH PARTINGS AS STRATIGRAPHIC
B 6211 2355 HILL, D. R., UNION ISLAND GAS FIELD, ABST.
B 6211 2355 HILL, F. L., WORKSHOP ON URBAN OIL AND GAS DEVELOPMENT, ABST.
B 6105 798 HILL, G. W., NEARSHORE CURRENT PATTERN OFF SOUTH TEXAS—
B 6303 466 HILL, GARY W., APPLICATIONS OF CRAB-BURROW ORIENTATION TO
B 6004 679 HILL, GARY W., BIOGENIC SEDIMENTARY STRUCTURES OF SOUTH TEXA
B 6303 466 HILL, J. RANDEL, RESPONSIBILITIES OF TEXAS RAILROAD COMMISSION
B 6207 1214 HILL, NORMAN E., EFFECTIVE ENVIRONMENTAL REGULATION, ABST.
B 6303 466 HILL, P. J., FORTIES FIELD, UNITED KINGDO CONTINENTAL SHELF,
M 30 81 HILL, P. J., GEOLOGY OF THE FORTIES FIELD, U.K. CONTINENTAL SHELF,
B 6306 951 HILL, PATRICK ARTHUR, SATELLITE IMAGE ANALYSI IN MINERAL
B 6109 1560 HILL, PATRICK ARTHUR, SATELLITE IMAGE ANALYSIS IN MINERAL
B 6211 2356 HILLHOUSE, J. W., MESOZOIC CONTINENTAL ACCRETION OF NORTH
B 6101 114 HILLIARD FORMATION, WYOMING
B 6307 1039 HILLIARD SHALE, IDAHO-, WYOMING, THRUST-FOLD BELT
B 6405 728* HILLIARD SHALE, SOUTHWESTERN WYOMING, ABST., APPLICATION OF
M 24 105 HILLIARD SHALE, WYOMING
B 6303 471* HILLS BEACH-FLETCHER NECK TOMBOLO SYSTEM, BIDDEFORD, MAINE,
B 6303 458 HILLS, JOHN M., PRELIMINARY ANALYSES OF BRINES FROM PERMIAN
B 6309 1510 HILLS, JOHN M., PROPOSED TYPE SECTION FOR UPPER SILURIAN AND
B 6012 2187 HILTERMAN, F. J., INTERPRETIV USES OF SEISMIC MODELING WITH
B 6203 525 HILTERMAN, FRED, USE OF SEISMIC MODELING FOR REEF EXPLORATION
M 25 427 HILTON DEPOSIT, LEAD-ZINC-SILVER, QUEENSLAND
B 6409 1567 HILTON, NEAL, OCCURRENCE OF HIGH GRAVITY OIL IN AN OLIGOCENE
B 6005 794† HIMALAYAN GEOSYNCLINE AND ITS PETROLEUM POSSIBILITIES,
B 6403 428 HIMALAYAN THRUST SHEET, INDIA
B 6008 1234 HINAMAN, SUSAN L., DEVELOPMENTS IN EASTERN AND NORTHWESTERN
B 6203 521 HINCH, HENRY H., NATURE OF SHALES AND DYNAMICS OF
S 10 1 HINCH, HENRY H., THE NATURE OF SHALES AND THE DYNAMICS OF
B 6204 635 HINCHINBROOK ISLAND, ALASKA
B 6204 635 HINCHINBROOK SEA VALLEY, ALASKA
M 27 100 HINDSVILLE LIMESTONE, OKLAHOMA, MISSISSIPPIAN
M 27 211 HINDSVILLE LIMESTONE, OKLAHOMA, MISSISSIPPIAN
M 27 76 HINDSVILLE LIMESTONE, OKLAHOMA, MISSISSIPPIAN
B 6410 1701 HINE, ALBERT C., CARBONATE SEDIMENT DRIFTS IN NORTHER STRAITS
B 6303 455 HINE, ALBERT C., PROBING BERMUDAS LAGOONS AND REEFS, ABST.
B 6004 679 HINE, ALBERT C., RESPONSE OF ACTIVE OOLITE SAND SHOAL TO ITS
B 6103 376 HINE, ALBERT C., SHALLOW CARBONATE-BANK-MARGIN GROWTH AND
B 6205 880* HINGE LINE AND SHELF TAR SAND DEPOSITS, CENTRAL AND SOUTHERN
B 6205 893* HINGE LINE IN CENTRAL UTAH, ABST., SUBSURFACE, THRUSTS ALONG,
B 6205 889* HINGE LINE—OVERTHRUST BELT, UTAH, WYOMING, AND IDAHO, ABST.,
B 6405 717* HINGE LINE—THRUST BELT—POTENTIAL NEW HYDROCARBON
B 6211 2357* HINGE-LINE THRUST BELT—POTENTIAL NEW HYDROCARBON PROVINCE,
B 6008 1404* HINGELINE AREA, CENTRAL UTAH, ABST., LOWER MESOZOIC AND UPPER
B 6405 771 HINKLE, M. E., RECENT ADVANCES IN HELIUM ANALYSIS AS
M 25 437 HINOBAAN COPPER PROJECT, NEGROS
B 6203 503 HINRICHS, T. C., GEOTHERMAL RESOURCE INVESTIGATIONS AND
S 3 93 HINTERLAND, DEFINITION
M 26 65 HINTERLAND SEQUENCES, SEA LEVEL FLUCTUATIONS
B 6304 652 HINTON FORMATION, VIRGINIA
B 6303 527 HINZ, K., RECENT ADVANCES I PASSIVE-MARGIN RESEARCH, ABST.
B 6009 1623 HINZE, W. J., BOWLING GREEN FAULT--CASE OF RESURGENT TECTONICS,
S 6 51 HIPPARION, EXAMPLE, BIOCHRONOLOGY
B 6109 1551* HIPPURITACEA, RUDISTS, OF GULF COAST, ABST., EARLY AND MIDDLE,
B 6303 467 HIRNER, A., ORIGIN AND OCCURRENCE OF CRUDE OILS IN MOLASSE
B 6008 1323 HIRSCH, ALFRED M., DEVELOPMENTS ON ATLANTIC COASTAL PLAIN
B 6407 1052 HISCOTT, RICHARD N., DEPOSITIONAL FRAMEWORK OF SANDY MID-FAN
B 6406 853 HISPANIOLA-CAICOS BASIN, SAND LAYERS
B 6103 466* HISTORIC EARTHQUAKES AND FUTURE EARTHQUAKE HAZARDS ALONG
B 6406 953 HISTORICAL BIOGEOGRAPHY, PLATE TECTONICS, AND THE CHANGING
B 6109 1543* HISTORICAL SHORELINE CHANGES AND THEIR CAUSES, TEXAS GULF
S 5 66 HISTORY, SILURIAN, SALINA GROUP, RUFF FORMATION, DEPOSITIONAL
B 6012 2149† HISTORY, ALASKA GEOLOGICAL SOCIETY
B 6101 111† HISTORY OF ALABAMA GEOLOGICAL SOCIETY
B 6301 106† HISTORY OF ARDMORE GEOLOGICAL SOCIETY
B 6105 728 † HISTORY OF DALLAS GEOLOGICAL SOCIETY
B 6205 882* HISTORY OF EARLY EXPLORATION IN GREAT BASIN, ABST.
B 6102 253† HISTORY OF EL PASO GEOLOGICAL SOCIETY
B 6105 796 * HISTORY OF EXPLOITATION AT SALTON SEA GEOTHERMAL FIELD,
B 6111 2012† HISTORY OF NEW MEXICO GEOLOGICAL SOCIETY, 1947-1976
B 6006 988† HISTORY OF NORTHERN CALIFORNIA GEOLOGICAL SOCIETY, SHORT
B 6109 1536* HISTORY OF SHORELINE CHANGES BY ARCHEOLOGICAL DATING,
S 11 4† HISTORY OF STRATIGRAPHIC PALEONTOLOGY OF WEST COAST
B 6405 772* HISTORY OF SULFIDIZATION OF SOUTH TEXAS ROLL-TYPE URANIUM
B 6007 1117 HISTORY, SAN ANTONIO, GEOLOGICAL, SOCIETY
B 6007 1117† HISTORY, SOUTH, TEXAS, GEOLOGICAL, SOCIETY--A FRAGMENTARY
B 6405 701* HITCHCOCK COUNTY, NEBRASKA, ABST., PALEOTOPOGRAPHYS
S 10 109 HITCHON, BRIAN, SOME ECONOMIC ASPECTS OF WATER-ROCK
B 6203 522 HITCHON, BRIAN, SOME ECONOMIC ASPECTS OF WATER-ROCK
B 6410 1725 HITCHON, MORAN, W. R., REVIEW OF INTRODUCTION TO EXPLORATION,
B 6405 708 HITE, DAVID M., CONTROLS OF ZEOLITE CEMENTATION IN UPPER
B 6405 600 HITH FORMATION, MIDDLE EAST
B 6405 776 HO, THOMAS T. Y., OIL-TO-SOURCE CORRELATION—PINEVIEW FIELD,
M 24 103 HOBACK FORMATION, WYOMING
B 6309 1603 HOBDAY, D. K., QUEEN CITY FORMATION IN EAST TEXAS
B 6405 723 HOBDAY, DAVID K., LOWER CRETACEOUS SHELF STORM DEPOSITS,
B 6305 831 HOBDAY, DAVID K., NONMARINE DEPOSITIONAL ENVIRONMENTS AND
B 6409 1559 HOBDAY, DAVID K., PALEOENVIRONMENTS AND TRACE FOSSILS OF
B 6107 1090 HOBSON, G. D., ABNORMAL FORMATION PRESSURE. DISCUSSION
B 6405 675 HOBSON, J. P., JR., DEPOSITIONAL FACIES, GEOMETRY, AND GENESIS OF
B 6307 1059 HOBSON, JOHN P., JR., DEPOSITIONAL FACIES OF LOWER PALEOZOIC
B 6105 795 HOBSON, RICHARD D., EFFECTS OF DREDGING AND HANDLING TECHNIQUES
B 6002 319* HODGEMAN COUNTY, KANSAS, ABST., GEOLOGY OF, MISSISSIPPIAN,
B 6103 309† HODGEMAN COUNTY, KANSAS, MISSISSIPPIAN COMBINATION, TRAP,
M 25 73 HODGES, CARL N., SOLAR DESALINATION EXPERIENCE OF UNIVERSITY
M 25 230 HODGSON, E. A., STRUCTURAL STYLE AND HYDROCARBON OF BASS,
B 6203 522 HODGSON, G. W., ORIGIN OF PETROLEUM—IN-TRANSIT CONVERSION OF
S 10 169 HODGSON, G. W., ORIGIN OF PETROLEUM: IN-TRANSIT CONVERSION OF
B 6109 1560 HODGSON, R. A., EARTH RESOURCES OBSERVATORY—A CONCEPT, ABST.
B 6309 1510 HOENIG, MARGARET ALLEN, PROPOSED TYPE SECTION FOR UPPER
B 6404 572 HOF SHALE AND SALT MEMBER, ISRAEL
B 6405 723 HOFFMAN, DAVID R., WILLISTON BASIN—REAWAKENING OF GIANT

B 6405 723 HOFFMAN, E., INTERTIDAL ZONATION OF RECENT MICROBIAL
B 6303 458 HOFFMAN, E. J., STUDY OF FOSSI MICROBIAL BORINGS--NEW APPROACH,
B 6405 685 HOFFMAN, E., MICROBIAL AND INVERTEBRATE ENDOLITHIC
B 6303 457 HOFFMAN, EMILY J., MIXING ZONE DOLOMITE IN TIDAL-FLAT
B 6203 523 HOFFMAN, PAUL F., SPELEOTHEMS AND EVAPORITE SOLUTION COLLAPSE
B 6203 522 HOFFMAN, PAUL F., TECTONIC HISTORY OF ATHAPUSCOW AULACOGEN,
B 6302 158 HOGBACK MONOCLINE, COLORADO AND, NEW MEXICO
B 6304 599 HOGBACK MONOCLINE, NEW MEXICO
B 6409 1560 HOGENSON, D. D., ET AL, DEPOSITIONAL ENVIRONMENTS OF SOME
B 6406 858 HOGGAR MASSIF, AFRICA
S 11 257 HOGLUNDINA ELEGANS
S 11 290 HOGLUNDINA ELEGANS
B 6108 1376* HOGSBACK FIELDS, SUBLETTE COUNTY, WYOMING, ABST., NUGGET OIL,
B 6307 1039 HOGSBACK THRUST, SEISMIC PROFILE, WYOMING
B 6405 639 HOGSBACK THRUST, WYOMING
B 6309 1581 HOHN, MICHAEL E., COMPOSITIONAL VARIATION IN
B 6309 1580 HOHN, MICHAEL E., ELEMENTAL ANALYSES OF DEVONIAN SHALES IN
B 6408 1285 HOHN, MICHAEL ED., ELEMENTAL ABUNDANCES IN DEVONIAN SHALES
M 26 338 HOHNE FIELD, SEISMIC SECTION, GERMANY
M 26 337 HOHNE FIELD, WEST, GERMANY, DOGGER SANDSTONE
M 28 114 HOING SANDSTONE, ILLINOIS, DEVONIAN
B 6303 410 HOKANSON, CLAUDIA, GLACIAL-MARINE SEDIMENTS OF WEST
B 6303 467 HOKANSON, CLAUDIA L., FLOW OF BOTTOM WATER OU OF WEDDELL
B 6203 531* HOKKAIDO ISLAND, JAPAN, ABST., SLOPE SEDIMENTATION PROCESSES IN,
M 29 353 HOLCOMBE, TROY L., GEOMORPHOLOGY AND SUBSURFACE GEOLOGY
B 6405 679 HOLDEN, JOHN F., ET AL, URANIUM EXPLORATION SYSTEMS CASE
S 4 243 HOLDER FORMATION, NEW MEXICO
B 6006 981 HOLDER, G. D., HYDRATE FORMATION IN SUBSURFACE ENVIRONMENTS
B 6309 1453 HOLE 108, BALTIMORE CANYON TROUGH, DEEP SEA DRILLING PROJECT,
B 6002 323* HOLE IN CENTRAL KANSAS, ABST., SALT, DISSOLUTION IN OIL AND GAS,
B 6405 707* HOLES, ABST., DIPMETER VALIDITY IN, DEVIATED, BORE
B 6209 1580 HOLES, EXPANSION OF, PRESSURIZED
M 24 205 HOLLAND, D. S., EAST CAMERON BLOCK 270, OFFSHORE LOUISIANA--A
B 6303 467 HOLLAND, D. S., EUGENE ISLAND BLOCK 330 FIELD, OFFSHOR LOUISIANA,
M 30 253 HOLLAND, D. S., EUGENE ISLAND BLOCK 330 FIELD, OFFSHORE
B 6103 366 HOLLIS BASIN, OKLAHOMA, HARDEMAN-
B 6303 467 HOLLISTER, CHARLES D., SEDIMENTATION PROCESSES AND
M 25 509 HOLLISTER, V. F., PORPHYRY COPPER PROVINCE OF NORTHERN
B 6203 523 HOLLRAH, TERRY L., SUBSURFACE LITHOSTRATIGRAPHY OF HUNTON
B 6303 516* HOLLY SPRINGS NATIONAL FORESTS, MISSISSIPPI, ABST., LIGNITES OF,
B 6309 1609* HOLLYWOOD FIELDS, LOUISIANA, ABST., GEOPRESSURE IN, HOUMA AND
B 6004 649 HOLMES, C. W., LATE PLEISTOCENE-HOLOCENE EVOLUTION OF SOUTH
B 6405 724 HOLMES, CHARLES W., IMPACT OF HOLOCENE TRANSGRESSION ON
B 6105 795 HOLMES, KENNETH H., SOUTH TEXAS MINERAL BELT—REVIEW
B 6303 468 HOLMES, MARK L., SONOBUOY REFRACTION MEASUREMENT FROM
B 6008 1393 HOLMES, MICHAEL, PETROLEUM GEOLOGY OF FORT UNION AND
M 25 32 HOLMGREN, D. A., TECTONIC FRAMEWORK OF PACIFIC REGION
B 6307 1096 HOLMWOOD SHALE, AUSTRALIA
B 6405 713* HOLOCENE, ABST., DEPOSITION OF, ENEWETA, ATOLL, REEF, MIDDLE,
B 6204 637 HOLOCENE, ALASKA
M 27 3 HOLOCENE ALGAE
B 6405 719* HOLOCENE AND JURASSIC GRAINSTONES, ABST., FRESHWATER,
M 27 59 HOLOCENE BACK-REEF BEACH SAND, BELIZE
B 6309 1463 HOLOCENE, BALTIMORE CANYON TROUGH
B 6209 1762* HOLOCENE BARRIER TRENDS AND ORIGINS OF LAKE PONTCHARTRAIN,
M 27 218 HOLOCENE BEACH SEDIMENTS, BELIZE
M 28 59 HOLOCENE, BEACH SEDIMENTS, CONNECTICUT
M 28 58 HOLOCENE, BEACH SEDIMENTS, CONNECTICUT
M 27 189 HOLOCENE BEACHROCK, BAHAMAS
M 27 157 HOLOCENE BEACHROCK, BAHAMAS
S 4 70 HOLOCENE BIOHERMS, LESSER, ANTILLES
S 4 67† HOLOCENE BIOHERMS OF LESSER ANTILLES--GEOLOGIC CONTROL OF
M 27 12 HOLOCENE BOTTOM SEDIMENT, FLORIDA
M 27 7 HOLOCENE BOTTOM SEDIMENTS, FLORIDA
B 6309 1607* HOLOCENE BOUNDARY, ABST., RADIOLARIAN RATIOS AN, PLEISTOCENE-
S 9 22 HOLOCENE, CARBONATE REEF BELT DISTRIBUTION
B 6303 445* HOLOCENE CARBONATE SEDIMENTS, ABST., PORE SPACE IN
S 4 299† HOLOCENE CARBONATE SEDIMENTS AND TERTIARY-PLEISTOCENE
B 6004 657* HOLOCENE CARBONATE SEDIMENTS, HARRINGTON SOUND, BERMUDA,
B 6203 489* HOLOCENE COASTAL-MARSH DEPOSITS AND PRE-QUATERNARY COALS
M 27 96 HOLOCENE CORAL RUBBLE, BELIZE
M 27 35 HOLOCENE, CORAL-RUBBLE, BELIZE
B 6105 833* HOLOCENE DOLOMITE LOCALITY IN FLORIDA BAY, ABST.
M 27 160 HOLOCENE DUNE SANDS, MEXICO
M 28 85 HOLOCENE, DUNES, UTAH
B 6203 531* HOLOCENE EPOCH ALLUVIAL-DELTAIC-COASTAL SEDIMENTARY-FACIES
B 6004 649* HOLOCENE EVOLUTION OF SOUTH TEXAS OUTER CONTINENTAL SHELF,
B 6403 376 HOLOCENE FAN DELTA, ALASKA
B 6403 376 HOLOCENE FAN DELTA, BAJA CALIFORNIA
B 6403 376 HOLOCENE FAN DELTA, ENGLAND
B 6403 376 HOLOCENE FAN DELTA, GULF OF AQABA
B 6403 376 HOLOCENE FAN DELTA, JAMAICA
B 6403 376 HOLOCENE FAN DELTA, JAPAN
B 6403 376 HOLOCENE FAN DELTA, NEW ZEALAND
B 6403 376 HOLOCENE FAN DELTA, PUERTO RICO
B 6403 376 HOLOCENE FAN DELTA, RED SEA
B 6403 376 HOLOCENE FAN DELTA, TEXAS
B 6202 250 HOLOCENE FAULT SCARP, CALIFORNIA
B 6203 494* HOLOCENE FLUVIAL SANDS DERIVED FROM GRANITIC PLUTONS, ABST.,
B 6209 1754* HOLOCENE FORAMINIFERA IN GULF OF ALASKA, ABST., DISTRIBUTION
B 6403 440* HOLOCENE FORAMINIFERAL DISTRIBUTION PATTERNS ON SHELF AND
B 6105 792* HOLOCENE FRESHWATER CEMENTS, JOULTERS CAYS, BAHAMAS
B 6105 788* HOLOCENE FRESHWATER DIAGENESIS OF MODERN OOLITES, JOULTERS CAY,
M 28 10 HOLOCENE, GRUS, CALIFORNIA
S 7 299 HOLOCENE, GYRE BASIN, SEDIMENT ACCUMULATION
M 28 139 HOLOCENE, HONOLULU VOLCANICS, HAWAII
B 6205 883* HOLOCENE ICHNOFACIES FROM NORTHERN GULF OF CALIFORNIA AND
B 6403 438* HOLOCENE INTERTIDAL DEPOSITS OF SACRAMENTO--SAN JOAQUIN
B 6112 2086 HOLOCENE, KEROGEN, LOUISIANA
B 6209 1758* HOLOCENE MARINE-CEMENTED SANDS, JOULTERS OOID SHOAL,
B 6303 439* HOLOCENE MODELS SOUTHEASTERN DELAWARE, ABST..
B 6403 444* HOLOCENE MOLLUSK DISTRIBUTION PATTERNS IN NORTHERN GULF OF
B 6203 507* HOLOCENE OF BAHAMIAN TROUGH, ABST., SEDIMENTARY RECORD OF
B 6203 525* HOLOCENE OF SOUTHEAST TEXAS, ABST., TRANSITION BETWEEN A
B 6303 462* HOLOCENE OOID SHOAL, ABST., ANATOMY AND GROWTH HISTORY OF
M 27 217 HOLOCENE OOLITES, UTAH
M 27 36 HOLOCENE OOZE, CORAL SEA
M 27 18 HOLOCENE, PLANKTON, BELIZE
B 6205 793 HOLOCENE, PROFILE, BLAKE PLATEAU AREA
M 27 49 HOLOCENE REEF SEDIMENT, JAMAICA
B 6205 892* HOLOCENE REEFS, BELIZE, ABST., DEPOSITION AND DIAGENESIS OF
S 4 107 HOLOCENE REEFS, CARIBBEAN, PLEISTOCENE-
S 4 94 HOLOCENE REEFS, PLEISTOCENE-
M 28 129 HOLOCENE, SABKHA SANDS, ABU DHABI
B 6102 152 HOLOCENE SAND, CALIFORNIA
B 6004 679* HOLOCENE SEA-LEVEL RISE, ABST., RESPONSE OF ACTIVE, OOLITE, SAND,
M 28 44 HOLOCENE SEDIMENT, CONNECTICUT
S 7 313† HOLOCENE SEDIMENT GEOCHEMISTRY OF CONTINENTAL SLOPE AND
M 27 26 HOLOCENE SEDIMENT, MINNESOTA
M 28 58 HOLOCENE, SEDIMENT, MINNESOTA
B 6003 452† HOLOCENE SEDIMENT, THERMALLY INDUCED DIAGENESIS OF,
M 29 400 HOLOCENE, SEDIMENT THICKNESS, MISSISSIPPI FAN
B 6203 545* HOLOCENE SEDIMENT VOLUME AND MODERN SEDIMENT YIELD,
B 6403 445* HOLOCENE SEDIMENTARY INFILLING AND FAUNAL CHANGE IN A
M 29 398 HOLOCENE, SEDIMENTATION, MISSISSIPPI FAN
B 6203 536* HOLOCENE SEDIMENTATIONIN EGYPTIAN RED SEA AND GULF OF SUEZ
M 27 183 HOLOCENE SEDIMENTS, BAHAMAS
M 27 117 HOLOCENE SEDIMENTS, BAHAMAS
M 27 123 HOLOCENE SEDIMENTS, BAHAMAS
M 27 81 HOLOCENE SEDIMENTS, BAHAMAS
M 27 127 HOLOCENE SEDIMENTS, BAHAMAS
M 27 179 HOLOCENE SEDIMENTS, BAHAMAS
M 27 128 HOLOCENE SEDIMENTS, BAHAMAS
M 27 82 HOLOCENE SEDIMENTS, BELIZE
M 27 51 HOLOCENE SEDIMENTS, BELIZE
M 27 37 HOLOCENE SEDIMENTS, BELIZE
M 27 21 HOLOCENE SEDIMENTS, BELIZE
M 27 60 HOLOCENE SEDIMENTS, BELIZE
M 27 25 HOLOCENE SEDIMENTS, BELIZE
M 27 158 HOLOCENE SEDIMENTS, BELIZE
M 27 183 HOLOCENE SEDIMENTS, BELIZE
M 27 95 HOLOCENE SEDIMENTS, BELIZE
M 27 159 HOLOCENE SEDIMENTS, BELIZE
M 27 61 HOLOCENE SEDIMENTS, BELIZE
M 27 101 HOLOCENE SEDIMENTS, BELIZE
M 27 32 HOLOCENE SEDIMENTS, BELIZE
M 27 81 HOLOCENE SEDIMENTS, BELIZE
M 27 60 HOLOCENE SEDIMENTS, BELIZE
M 27 83 HOLOCENE SEDIMENTS, BELIZE
M 27 90 HOLOCENE SEDIMENTS, CORAL SEA
S 4 304 HOLOCENE SEDIMENTS, FACIES, BARBUDA
M 27 115 HOLOCENE SEDIMENTS, FLORIDA
M 27 61 HOLOCENE SEDIMENTS, FLORIDA
M 27 42 HOLOCENE SEDIMENTS, FLORIDA
M 27 96 HOLOCENE SEDIMENTS, FLORIDA
M 27 111 HOLOCENE SEDIMENTS, MEXICO
M 27 25 HOLOCENE SEDIMENTS, MEXICO
M 28 57 HOLOCENE, SEDIMENTS, MINNESOTA
M 27 119 HOLOCENE SEDIMENTS, NEW MEXICO

B 6309 1598* HOLOCENE SEDIMENTS OF GULF OF MEXICO AND ADJACENT WATERS,
B 6012 2176* HOLOCENE SEDIMENTS ON CONTINENTAL SHELF, NORTHERN GULF OF
M 27 38 HOLOCENE SEDIMENTS. PYRGO SP., BELIZE
B 6403 438* HOLOCENE SEDIMENTS, SOUTHERN CALIFORNIA CONTINENTAL
M 27 116 HOLOCENE SEDIMENTS, TEXAS
M 27 115 HOLOCENE SEDIMENTS, UTAH
M 27 159 HOLOCENE SEDIMENTS, VIRGIN ISLANDS
B 6309 1474 HOLOCENE SEQUENCE, USSR, ANADYR BASIN, MIOCENE-
B 6012 2188* HOLOCENE SHELF SEDIMENTATION PATTERNS DETERMINED BY
B 6004 663* HOLOCENE SHORELINE CHANGES ON GEORGIA COAST, ABST.,
B 6212 2412† HOLOCENE SLUMP ON CONTINENTAL SHELF OFF MALASPINA GLACIER,
M 27 222 HOLOCENE SOIL CRUSTS, FLORIDA
B 6303 518* HOLOCENE STRATIGRAPHY OF RAPIDLY TRANSGRESSIVE BARRIER
M 28 121 HOLOCENE, SUBMARINE SANDS, PERSIAN GULF
B 6004 648* HOLOCENE TRANSGRESSION, NORTHWEST FLANK OF BALTIMORE
B 6405 724* HOLOCENE TRANSGRESSION ON DEPOSITIONAL ENVIRONMENT OF
B 6004 733* HOLOCOCCOLITHS, ABST., MESOZOIC, ORIGIN OF
B 6211 2358* HOLOTHURIAN MICROFOSSILS AND EOCENE STRATIGRAPHY OF
M 27 96 HOLOTHURIAN, MODERN
S 4 186 HOLSTON LIMESTONE, REEF, TENNESSEE
B 6111 1986 HOLZ SHALE, CALIFORNIA
B 6004 652 HOLZER, ROBERT, TIMING AND PALEOCLIMATE INDICATORS IN
B 6104 597 HOMEGLEN-RIMBEY FIELD, ALBERTA
B 6304 652 HONAKER DOLOMITE, VIRGINIA
B 6205 849 HONDO SILT, CALIFORNIA
B 6110 1587 HONDURAS, 1976, PETROCONSULTANTS S.A., ON DEVELOPMENTS IN
B 6210 2053 HONDURAS, 1977, JACOBSEN, P., JR., ON DEVELOPMENTS IN
B 6310 1754 HONDURAS, 1978, JACOBSEN, P., JR., ON DEVELOPMENTS IN
B 6010 1653 HONDURAS, DEVELOPMENTS, 1975
B 6102 269† HONDURAS--BASEMENT TERRANE IN SIERRA DE OMOA. DISCUSSION AND
B 6004 566† HONDURAS--BASEMENT TERRANEIN SIERRA DE OMOA, PRE-CRETACEOUS
M 28 176 HONEYCOMBING OF GRAINS, POROSITY
M 25 366† HONOLULU BASAL AQUIFER--SUMMARY, SIMULATION MODEL OF
M 25 366 HONOLULU BASIN, HAWAII
M 25 366 HONOLULU, GROUNDWATER, HAWAII
M 28 139 HONOLULU VOLCANICS, HAWAII, HOLOCENE
M 28 139 HONOLULU VOLCANICS, HAWAII, PLEISTOCENE
M 25 365† HONOLULU--SUMMARY, IMPACT OF, ECONOMIC DEVELOPMENT ON
B 6303 522* HONOR RANCHO AREA, VENTURA COUNTY, CALIFORNIA, ABST.,
M 25 467 HONSHU-KOREAN VOLCANIC BELT
B 6111 1932 HOODOO DIAPIR, CANADIAN ARCTIC
B 6309 1523 HOODOO FORMATION, ALASKA PENINSULA
S 4 291 HOOK ISLAND, GALAXEA
S 11 197 HOOKTON FORMATION
S 11 224 HOOKTON FORMATION, ASSEMBLAGES
S 11 220 HOOKTON FORMATION, FOSSIL ASSEMBLAGE
B 6207 1227 HOOVER, DONALD B., GEOTHERMAL EXPLORATION ON ADAK ISLAND,
B 6103 465* HOPE BASIN, SOUTHERN CHUKCHI SEA, ABST., CENOZOIC
S 3 119 HOPI BUTTES, ARIZONA
B 6405 802* HOPI BUTTES, ARIZONA, ABST., URANIUM MINERALIZATION IN
S 11 138 HOPKINSINA MAGNIFICA
B 6303 456 HORAN, L. J., PETROLOGY, SEDIMENTOLOGY, AND PETROLEUM
B 6102 208 HORCONES FORMATION, MEXICO
B 6311 2001 HORDA BASIN, NORTH SEA
B 6203 436 HORDA PLATFORM, NORTH SEA
S 6 35 HORIZONS, NEED FOR DEFINITION
B 6312 2204† HORIZONS ON CONTINENTAL SLOPES AND RISES, SEISMIC EVIDENCE FOR
M 30 361 HORMIGUERO FIELD, MEXICO, MACUSPANA BASIN
S 11 227 HORMOSINA GLOBULIFERA
S 11 268 HORMOSINA GLOBULIFERA
B 6010 1866 HORMOZ PETROLEUM CO., SOFIRAN, IRAN NIPPON PETROLEUM CO., AGI
B 6206 997 HORMUZ SALT DIAPIR, IRAN
M 25 390 **HORN, B. M., NICKEL- AND COPPER-RICH NODULES OF EQUATORIAL NORTH**
B 6203 535 HORN, BERYL, ALGAL STROMATOLITES IN EUREKA QUARTZITE,
M 25 390 HORN, D. R., NICKEL- AND COPPER-RICH NODULES OF EQUATORIAL
B 6302 183 HORN GRABEN, NORTH SEA
S 11 54 HORNADAY, GORDON, PLANKTONIC AND LARGE FORAMINIFERA IN
M 28 58 HORNBLENDE, BIREFRINGENCE
M 28 50 HORNBLENDE, CRYSTALS
M 28 51 HORNBLENDE, DETRITAL GRAINS
M 28 58 HORNBLENDE, FROM METAMORPHICS
B 6102 269 HORNE, G. G., PRE-CRETACEOUS ROCKS OF NORTHWESTERN
B 6004 566 HORNE, GREGORY S., PRE-CRETACEOUS ROCKS OF NORTHWESTERN
B 6009 1620 HORNE, J. C., CARBONIFEROUS SEDIMENTARY RESPONSES TO
B 6105 796 HORNE, J. C., DEPOSITIONAL MODELS IN COAL EXPLORATION AND MINE
B 6405 724 HORNE, J. C., CRETACEOUS-TERTIARY VERSUS CARBONIFEROUS
B 6212 2379 HORNE, J. C., DEPOSITIONAL MODELS IN COAL EXPLORATION AND MINE
B 6405 739 HORNE, J. C., WAVE-DOMINATED DELTAS--IMPORTANT ECONOMIC
B 6408 1278 HORNE, JOHN C., APPLICATION OF DEPOSITIONAL MODELS IN COAL
B 6303 415 HORNE, JOHN C., PALEOTIDAL-RANGE INDICATORS IN CARBONIFEROUS
B 6004 680 HORNE, JOHN C., SANDSTONE MODELS FROM APPALACHIAN
B 6203 523 HORNE, JOHN C., USE OF DEPOSITIONAL MODELS IN PREDICTING ROOF
B 6305 828* HORNELEN BASIN, NORWAY, ABST., ALLUVIAL, FANS AND THEIR DEPOSIT
B 6303 468 HORODYSKI, ROBERT J., MICROFOSSILS, MACROFOSSILS, AND
B 6104 603 HORQUILLA LIMESTONE, NEW MEXICO
B 6410 1617 HORSE CREEK FIELD, WYOMING
M 28 34 HORSE SPRINGS FORMATION, NEVADA, TERTIARY
M 28 90 HORSE SPRINGS FORMATION, NEVADA, TERTIARY
M 28 4 HORSE SPRINGS FORMATION, NEVADA, TERTIARY
M 26 336 HORSESHOE ATOLL, SEISMIC SECTION, WEST, TEXAS
B 6205 787 HORSESHOE ATOLL, TEXAS
B 6203 428 HORSESHOE BASIN, GRAND BASIN
B 6304 599 HORSESHOE FIELD, SECTION, NEW MEXICO
B 6407 972 HORSETHIEF SANDSTONE, GREAT PLAINS
M 26 404 HORST-FAULTING, MODELING, NORTH SEA
B 6405 724 HORSTINK, J., OIL EXPLORATION IN SOUTHEAST TURKEY THRUST BELT,
B 6303 468 HORVITZ, L., POOL DEPLETION AND GEOCHEMICAL SIGNAL DECAY. B
B 6303 468 HORVITZ, LEO, EFFECT OF DEPLETION ON NEAR-SURFACE
S 10 241 HORVITZ, LEO, NEAR- SURFACE EVIDENCE OF HYDROCARBON
B 6203 524 HORVITZ, LEO, NEAR-SURFACE EVIDENCE OF HYDROCARBON
B 6203 357 HOSGRI FAULT, CALIFORNIA
B 6211 2218 HOSGRI FAULT, CALIFORNIA, SAN GREGORIO-
M 25 63 HOSHINO, KAZUO,, SURVEYING EARTH
B 6306 887 HOSPAH-LONE PINE AREA, NEW MEXICO
B 6109 1551* HOSSTON DEPOSITIONAL PATTERNS, SUBSURFACE OF SOUTH TEXAS,
B 6203 548* HOSSTON FORMATION, LOWER CRETACEOUS, SOUTH-CENTRAL TEXAS,
M 24 159 HOSSTON FORMATION, TEXAS
M 26 147 HOSSTON FORMATION, TEXAS, EAST TEXAS BASIN
M 24 242 HOSSTON PRODUCING SANDS, MISSISSIPPI
B 6309 1607* HOSSTON SAND, CENTRAL TEXAS--IMPLICATIONS T LOW-TEMPERATURE
B 6209 1769* HOSSTON SANDSTONE RESERVOIRS AT BASSFIELD, JEFFERSON DAVIS
B 6101 65 HOT BELT, LOUISIANA
B 6203 535* HOT SPRINGS, SOUTH DAKOTA, ABST., SEDIMENTOLOGY OF LATE,
B 6008 1393* HOT SPRINGS. SOUTHWESTERN MONTANA, ABST., GEOLOGIC CONTROLS
B 6305 836* **HOT SPRINGS, KGRA, BEAVER COUNTY, UTAH, ABST., GEOLOGY AND**
B 6305 834* HOT-SPRING-GENERATED KARST FEATURES NEAR SALIDA. COLORADO.
B 6004 680 HOTTMAN, W. E., AREAL DISTRIBUTION OF CLAY MINERALS AND THEIR
B 6004 681 HOTTMAN, W. E., AREAL DISTRIBUTION OF CLAY MINERALS AND THEIR
B 6308 1426 HOUDE, RICHARD F., SEDIMENTOLOGY AND SOURCE-BED
B 6309 1609* HOUMA AND HOLLYWOOD FIELDS, LOUISIANA, ABST., GEOPRESSURE IN
B 6405 675* HOUSE CREEK FIELD, POWDER RIVER BASIN, WYOMING, ABST.,
B 6202 184 HOUSE CREEK FIELD, WYOMING
B 6004 682 HOUSE, P. A., GEOLOGY AND POTENTIAL USES OF GEOPRESSURE
B 6004 681 HOUSEKNECHT, D. W., TRANSPORTATIONAL AND DEPOSITIONAL
B 6405 724 HOUSEKNECHT, DAVID W., ANTICIPATING COAL MINING PROBLEMS IN
B 6004 681 HOUSER, J. F., BED CONTINUITY AND PERMEABILITY VARIATIONS OF
B 6206 1051 HOUSTON EMBAYMENT, TEXAS
B 6303 513* HOUSTON, HARRIS COUNTY, TEXAS, ABST., RATES OF, CREEP OF ACTIVE,
S 7 43 HOUSTON, MARK H., OCCURRENCE AND EVOLUTION OF SALT IN DEEP
B 6312 2204 HOUSTON, MARK H., SEISMIC EVIDENCE FOR WIDESPREAD POSSIBLE GAS
B 6303 432* HOUSTON METROPOLITAN AREA, TEXAS, ABST., FAULTS OFFSETTING
B 6310 1888 HOUSTON OIL AND MINERALS CORP AND ZAPEX, U.K., LTD., ON
B 6110 1800 HOUSTON OIL AND MINERALS CORP. AND PETROCONSULTANTS S.A., ON
B 6411 1841 HOUSTON OIL AND MINERALS OF DUBAI, INC., AND ZAPEX, U.K., ON
B 6409 1567* HOUSTON TREND, MISSISSIPPI, ABST.
B 6303 454* HOUSTON-GALVESTON REGION, TEXAS, ABST., LAND-SURFACE,
B 6109 1539* HOUSTON-GALVESTON, TEXAS, ABST., GEOMETRY OF, GULF COAST,
M 29 125 HOUTZ, R. E., STRUCTURE OF FALKLAND PLATEAU AND OFFSHORE
M 29 151 HOUTZ, ROBERT, THE CRUSTAL STRUCTURE AND EVOLUTION OF THE
B 6405 704* HOW DO THRUST BELTS FORM, ABST.
B 6405 698* HOW GEOLOGIC OBJECTIVES SHOULD DETERMINE SEISMIC FIELD
B 6309 1585* HOW GOOD ARE REGIONAL COAL RESERVE FIGURES. CASE STUDY OF
B 6311 2117* HOW MANY WILDCATS MUST WE DRILL, ABST.
B 6309 1597* HOW TO CREATE AND SUBMIT A WINNING PROSPECT, ABST.
B 6203 549* HOW TO HELP SOLVE THE ENERGY PROBLEM, ABST.
B 6303 380 HOWARD AND IRWIN REMSON, HAYWARD, O. T., REVIEW OF, GEOLOGY
B 6008 1399 HOWARD, EUGENE L., NEW PALEOSTRUCTURAL INTERPRETATION OF
B 6205 885 HOWARD, EUGENE L., REVIEW OF HYDROCARBON POTENTIAL OF
B 6004 682 HOWARD. H. L. GEOLOGY AND POTENTIAL USES OF GEOPRESSURE
B 6004 681 **HOWARD, J. C., PHYSICAL-BIOGENIC BREAK IN NEARSHORE SEDIMENTARY**
B 6004 663 HOWARD, J. D., ARCHAELOGIC DATING OF HOLOCENE SHORELINE

B 6004 734 HOWARD, J. D., DEPOSITION, DEFORMATION, AND HEAVY MINERAL
B 6004 687 HOWARD, J. D., DISTRIBUTION AND RELATION OF PHYSICAL AND
B 6004 681 HOWARD, J. D., OPHIOMORPHA--GOOD NEWS AND BAD NEWS, ABST.
B 6105 779 HOWARD, J. D., PHYSICAL CHARACTERISTICS OF WASHOVER FANS OF
B 6203 503 HOWARD, J. H., GEOTHERMAL RESOURCE INVESTIGATIONS AND
B 6105 796 HOWARD, J. H., HISTORY OF EXPLOITATION AT SALTON SEA GEOTHERMAL
B 6203 509 HOWARD, JAMES D., ARCHAEOLOGICAL EVIDENCE FOR
B 6109 1536 HOWARD, JAMES D., HISTORY OF SHORELINE CHANGES BY
B 6109 1535 HOWARD, JAMES D., ORIGIN AND CHARACTER OF WASHOVER FANS ON
B 6109 1549 HOWARD, JAMES D., PLIOCENE-PLEISTOCENE PALEOGEOGRAPHY OF
B 6303 424 HOWARD, JAMES D., ROLE OF BIOEROSION IN MASS-WASTING OF
B 6004 720 HOWARD, JAMES D., SEDIMENTARY ENVIRONMENTS OF TIDAL FLAT
B 6303 468 HOWARD, JAMES D., SEDIMENTARY STRUCTURES OF HIGH-ENERGY
B 6303 469 HOWARD, JAMES J., MIXED-LAYER CLAYS IN EOCENE INTERLAMINATE
B 6405 725 HOWARD, JAMES J., SEDIMENTATION RATES AND ILLITE-SMECTITE
B 6303 469 HOWE, HERBERT J., CARBONATE ROCKS OF MONTOYA GROUP, MIDDLE
B 6405 725 HOWE, HERBERT J., DELINEATION OF JACKSONIAN-VICKSBURGIAN
B 6009 1611 HOWE, HERBERT J., DIAGNOSTIC CENTRAL GULF COAST VICKSBURGIAN
B 6109 1538 HOWE, HERBERT J., JACKSONIAN-VICKSBURGIAN BOUNDARY IN
B 6105 796 HOWELL, D. G., BASIN ANALYSIS OF LOWER MIDDLE MIOCENE STRAT,
B 6211 2355 HOWELL, D. G., SALINIAN BLOCK OF CALIFORNIA, AN ENIGMA, ABST.
B 6403 450 HOWELL, D. G., STRATIGRAPHIC AND STRUCTURAL RELATIONS TO
B 6012 2185 HOWELL, DAVID G., ELEMENTS OF TECTONISM DEDUCED FROM MIOCENE
B 6012 2177 HOWELL, DAVID G., MIOCENE VOLCANIC AND VOLCANICLASTIC ROCKS
B 6012 2182 HOWELL, DAVID G., REVIEW OF ESTIMTES OF RADIOMETRIC AGES FOR
B 6406 938 HOWELL, DAVID J., EXPLORATION MODEL FOR PENNSYLVANIAN UPPER
B 6405 780 HOWELL, F., HEAT-FLOW AND HEAT-PRODUCTION STUDIES IN NORTH
B 6203 344 HOWELL,, NEOGENE BASINFORMATION IN RELATION TO
B 6207 1221 HOWELL, W. J. S., OK TEDI COPPER PROJECT, PAPUA NEW GUINEA, ABST.
B 6408 1285 HOWER, JAMES C., COAL RANK IN PART OF WESTERN KENTUCKY
B 6405 725 HOWER, JAMES C., PETROGRAPHIC VARIATION IN WESTERN KENTUCKY
B 6407 1121 HOWER, S. GRACE, REVIEW OF THE PETROLEUM DICTIONARY, BY DAVI F.
B 6303 460 HOWER, WAYNE F., AUTHIGENI ILLITES IN SANDSTONE RESERVOIRS,
B 6302 246 HOXBAR FORMATION, OKLAHOMA
B 6109 1439 HOXBAR GROUP FACIES CHANGE, OKLAHOMA
M 26 441 HOXBAR GROUP, MISSOURIAN, OKLAHOMA, ANADARKO, UPPER
M 26 439 HOXBAR GROUP, OKLAHOMA, CANADIAN COUNTY, ANADARKO BASIN,
B 6105 797 HOYT, WILLIAM H., LONG-DISTANCE TURBIDITE CORRELATIONS IN
M 30 315 HRISKEVICH, M. E., STRACHAN AND RICINUS WEST GAS FIELDS, ALBERTA,
B 6303 469 HRISKEVICH, M. E., STRACHAN-RICINUS GAS FIELD, ALBERTA, ABST.
B 6303 496 HSU, KENNETH J., MOVEMENT OF SUBSURFACE WATERS UNDER SABKHA,
B 6407 1034 HSU, KENNETH J., RESEDIMENTED FACIES IN VENTURA BASIN,
B 6102 137 HSU, KENNETH J., STUDIES OF VENTURA FIELD, CALIFORNIA, I--FACIES
B 6102 169 HSU, KENNETH J., STUDIES OF VENTURA FIELD, CALIFORNIA,
B 6209 1667 HSUEH SHAN RANGE, TAIWAN
B 6207 1159 HUALAPAI LIMESTONE MEMBER, MUDDY CREEK FORMATION, ARIZONA
M 25 533 HUANCAVELICA DISTRICT, MERCURY, PERU
B 6207 1221 HUANG, KE-KONG, EXPLORATION AND DEVELOPMENT OF GEOTHERMAL
B 6105 797 HUANG, T. C., DEPOSITION AND RESEDIMENTATION OF ABYSSAL TEPHRA
B 6203 524 HUANG, WEN H., ANALYSIS OF GEOCHEMICAL AND SEDIMENTOLOGIC
B 6206 1049 HUANG, WEN H., GEOCHEMICAL AND SEDIMENTOLOGIC PROBLEMS OF
B 6303 470 HUANG, WEN H., URANIUM DEPOSITS OF TEXAS GULF COASTA
B 6203 349 HUASNA-PISMO BASIN, CALIFORNIA
B 6304 699 HUBBARD, CLYDE, HIGH-RESOLUTION LANDSAT FOR GEOLOGIC STUDIES,
B 6405 725 HUBBARD, CLYDE, HIGH-RESOLUTION LANDSAT FOR GEOPHYSICAL
B 6004 682 HUBBARD, DENNIS K., REEF MORPHOLOGY AND SEDIMENT TRANSPORT,
B 6105 797 HUBBARD, DENNIS L., DISTRIBUTION OF SEDIMENTARY STRUCTURES IN
B 6303 470 HUBBERT, M. KING, TWENTY YEARS OF UNITED STATES PETROLEUM
B 6004 682 HUBERT, JOHN F., TIMING OF CARBONATE AND CLAY DIAGENESIS IN
B 6405 726 HUC, ALAIN Y., PETROLEUM SOURCE-ROCK EVALUATION BY THERMAL
B 6109 1538 HUDDLESTUN, PAUL F., LATE MIOCENE GLACIO-EUSTATIC LOWERING OF
S 5 181 HUDEC, P.P. BASE- METAL CONCENTRATION IN A DENSITY- STRATIFIED
B 6004 667* HUDSON, AND WILMINGTON SUBMARINE CANYONS, ABST., CONTRASTS
B 6303 535* HUDSON BAY BASIN, ABST.
B 6203 557* HUDSON BAY, CANADA, ABST., EXAMPLE OF, CARBONATE, BANKS IN
S 8 124 HUDSON CANYON, TIME-VELOCITY CURVES, CURRENT FLOW
S 8 122 HUDSON CANYON, TOPOGRAPHY, INDEX MAP
S 8 121 HUDSON CANYON, U.S. EAST COAST, SUBMARINE CANYONS
B 6207 1222 HUDSON, HUGH H., AVAILABILITY AND QUALITY OF WATER RELATED
B 6205 892 HUDSON, J. H., DEPOSITION AND DIAGENESIS OF HOLOCENE REEFS,
B 6004 683 HUDSON, J. HAROLD, AUTOPSY OF A DEAD CORAL REEF, ABST.
B 6104 519 HUDSON, J. HAROLD, PLEISTOCENE BARRIER BAR SEAWARD OF OOID
B 6004 677 HUDSON, J. HAROLD, RECENT AND RELICT TOPOGRAPHY OF BOO BEE
B 6303 528 HUDSON, J. HAROLD, THREE-DIMENSIONAL ASPECTS OF BELIZE PATCH
B 6108 1199 HUDSON, JON P., DEVELOPMENTS IN EASTERN AND NORTHWESTERN
B 6008 1234 HUDSON, JON P., DEVELOPMENTS IN EASTERN AND NORTHWESTERN
B 6004 703* HUDSON RIVER ESTUARY, ABST., SEDIMENTATION, RATES IN
B 6004 654* HUDSON SUBMARINE CANYON, ABST., SEDIMENT PROCESSES
B 6206 1005 HUDSONIAN OROGENY, ARCTIC, CANADA
B 6007 1107 HUECO LIMESTONE, TEXAS
B 6109 1449 HUECO UPLIFT, TEXAS
S 8 67 HUENEME CANYON, CALIFORNIA, SUBMARINE CANYONS
S 8 69 HUENEME CANYON, LOCATION, INDEX MAP
S 8 75 HUENEME CANYON, PROGRESSIVE-VECTOR DIAGRAMS
B 6203 511* HUENEME SILL, CALIFORNIA CONTINENTAL BORDERLAND, ABST., NEW
B 6004 650* HUENEME SUBMARINE CANYON, CALIFORNIA COAST, ABST., PHYSICAL
B 6209 1759 HUENI, CAMILLE, DISTRIBUTION OF LIVING BENTHONIC FORAMINIFERA
M 28 67 HUERFANO FORMATION, COLORADO, EOCENE
M 28 138 HUERFANO FORMATION, COLORADO, EOCENE
B 6405 726 HUERTA, RAUL, SEISMIC, STRATIGRAPHIC, AND STRUCTURAL ANALYS
B 6203 542 HUFF, DAVID W., SHELF SEDIMENT AND LOCAL FLOW PHENOMENA—IN
B 6409 1559 HUFF, WILLIAM J., OSTRACODE BIOSTRATIGRAPHY OF LOWER
B 6303 470 HUFFINGTON, ROY M., DISCOVERY AND DEVELOPMENT OF BADAK
M 30 441 HUFFINGTON, ROY M., DISCOVERY AND DEVELOPMENT OF THE BADAK
B 6304 689 HUFFMAN, A. CURTIS, JR., DEPOSITIONAL ENVIRONMENTS AS ORE
B 6008 1204 HUGHES, J. H., OIL AND GAS DEVELOPMENT IN NORTHERN ROCKIES IN
B 6303 471 HUGHES, LYNN N., LEGAL IMPLICATIONS OF CONSULTING
B 6309 1478 HUGMAN, R. H. H., EFFECTS OF TEXTURE AND COMPOSITION ON
S 6 367 HUGO, P. J., ASPECTS OF THE REVISED SOUTH AFRICAN STRATIGRAPHIC
S 5 1 HUH, JOHN M. DEPOSITIONAL ENVIRONMENTS OF PINNACLE REEFS,
B 6302 259 HUH, JOHN M., REVIEW ON SEISMIC EXPLORATION FUNDAMENTALS,
M 25 268† HUKSAN PLATFORM, KOREA, MARIN PETROLEUM EXPLORATION OF
B 6211 2280 HULETT SANDSTONE MEMBER, BLACK HILLS
B 6211 2277 HULETT SANDSTONE MEMBER, WYOMING
B 6108 1382 HULKE, STEVEN D., ORDOVICIAN OIL ACCUMULATION AT NORTHEAST
B 6303 471 HULMES, LEITA J., ORIGIN OF HILLS BEACH-FLETCHER NECK TOMBOLO
B 6311 2118 HULSE, WILLIAM J., DEPOSITIONAL ENVIRONMENT OF BARTLESVILLE
B 6305 843* HUMAT MODEL--SEDIMENTOLOGIC AND GEOCHEMICAL MODEL FOR
M 30 104 HUMBER GROUP, JURASSIC, STATFJORD FIELD
S 11 200 HUMBOLDT BASIN AND BEAR RIVER AREA, BIOCHRONOLOGY
S 11 191 HUMBOLDT BASIN, BEAR RIVER BEDS
S 11 193 HUMBOLDT BASIN, BIOSTRATIGRAPHY
S 11 183 HUMBOLDT BASIN, CALIFORNIA
S 11 222 HUMBOLDT BASIN, CALIFORNIA, FORAMINIFERA, NUMBER OF
S 11 223 HUMBOLDT BASIN, CALIFORNIA, MOLLUSCA, NUMBER OF
S 11 217 HUMBOLDT BASIN, FORAMINIFERA, PERCENT LIVING
S 11 190 HUMBOLDT BASIN, GEOLOGY OF
S 11 207 HUMBOLDT BASIN NEOGENE, BIOCHRONOLOGIC SUBDIVISIONS
S 11 201 HUMBOLDT BASIN NEOGENE, BIOSTRATIGRAPHY
B 6012 2182* HUMBOLDT BASIN, NORTHERN CALIFORNIA, ABST., LATE, NEOGENE
B 6405 699* HUMBOLDT BASIN OF CALIFORNIA AS EXAMPLE, ABST., PALEOECOLOGY
S 11 201 HUMBOLDT BASIN PLIOCENE AND PLEISTOCENE , BIOSTRATIGRAPHY
S 11 191 HUMBOLDT BASIN PULLEN FORMATION
S 11 305 HUMBOLDT COUNTY, CALIFORNIA, SAMPLE LOCALITIES
B 6303 460* HUMMOCKY CROSS-STRATIFICATION--INDICATOR OF
B 6405 710 HUMPHRIS, C. C., JR., DEEP TUSCALOOSA GAS TREND OF SOUTH
S 7 69 HUMPHRIS, C. C., JR., SALT MOVEMENT ON CONTINENTAL SLOPE,
B 6004 683 HUMPHRIS, C. C., JR., SALT MOVEMENT ON CONTINENTAL SLOPE
B 6305 782 HUMPHRIS, C. C., JR., SALT MOVEMENT ON CONTINENTAL SLOPE
B 6110 1652 HUNGARY, DEVELOPMENTS, 1976
B 6210 1923 HUNGARY, DEVELOPMENTS, 1977
S 10 134 HUNGARY, GREAT LOWLAND ARTESIAN BASIN, GROUNDWATER FLOW
S 10 155 HUNGARY, GREAT LOWLAND ARTESIAN BASIN, PETROLEUM MIGRATION
S 10 138 HUNGARY, GREAT LOWLAND ARTESIAN BASIN, WATER LEVELS
B 6110 1654 HUNGARY, PRODUCTION, 1975-1976
B 6210 1924 HUNGARY, PRODUCTION, 1976-1977
B 6310 1932 HUNGARY, PRODUCTION, 1977-1978
B 6002 233 HUNGARY, WELLS, BOTTOM-HOLE TEMPERATURES
B 6105 798 HUNKIN, GEOFFREY G., ESTIMATION OF ORE RESERVES FOR IN-SITU
B 6309 1592 HUNT, A E., PRESSURE CORING OF GAS-BEARING DEVONIAN BLACK
B 6203 524 HUNT, C. WARREN, UP THE DECAY CHAIN—PERSPECTIVE ON
B 6309 1602 HUNT, E. R., SMACKOVER RESERVOIR--INTERPRETATION CASE STUDY OF
B 6202 301 HUNT, JOHN M., CHARACTERIZATION OF BITUMENS AND COALS
B 6101 100 HUNT, JOHN M., DISTRIBUTION OF CARBON AS HYDROCARBONS AND
B 6007 1112 HUNT, JOHN M., ORIGIN OF ATHABASCA OIL
B 6405 726 HUNT, JOHN M., PETROLEUM SOURCE-ROCK EVALUATION BY THERMAL
B 6103 434 HUNT, JOHN M., RATIO OF PETROLEUM TO WATER DURING PRIMARY
B 6307 1143 HUNT, JOHN M. REVIEW OF GEOCHEMISTRY, BY A. H. BROWNLOW
B 6406 954 HUNT, KVENVOLDEN, KEITH A., REVIEW OF, PETROLEUM,
B 6309 1603 HUNTER, BRUCE E., DISTRIBUTION OF VOLCANIC ACTIVITY IN TIME AND
B 6004 683 HUNTER, R. E., COMPARISON OF EOLIAN AND SUBAQUEOUS SAND-FLOW
B 6403 440 HUNTER, R. E., MODERN AND ANCIENT COASTAL SEDIMENTARY FACIES.
B 6105 798 HUNTER, R. E., NEARSHORE CURRENT PATTERN OFF SOUTH TEXAS—
B 6103 417 HUNTER VALLEY FAULT, VIRGINIA
B 6209 1610 HUNTERSVILLE CHERT, MARYLAND
B 6002 280 HUNTERSVILLE CHERT, PENNSYLVANIA
B 6304 652 HUNTERSVILLE CHERT, VIRGINIA
B 6207 1222 HUNTINGTON, J. F., APPLICATION OF LANDSAT IMAGERY AND RELATED
B 6305 805 HUNTON GEOLOGY, MILLS RANCH COMPLEX, TEXAS

B 6203 523* HUNTON GROUP IN PARTS OF PAYNE, LINCOLN, AND LOGAN COUNTIES,
B 6109 1452 HUNTON GROUP, PEDREGOSA BASIN
B 6002 322* HUNTON PLAY IN MCCLAIN, GRADY, AND CANADIAN COUNTIES,
B 6410 1623 HUNTSMAN OF DRILLERS, DENVER BASIN
B 6008 1410 HUNTTING, MARSHALL T., HEAT-FLOW STUDIES IN STEAMBOAT
B 6406 956 HUNZIKER, MCDOWELL, FRED W., REVIEW OF, LECTURES IN, ISOTOPE
B 6405 602 HUQF-DHOFAR ARCH, MIDDLE EAST
B 6303 471 HURLEY, NEIL F., SEAWARD PRIMARY DIP OF FALL-IN BEDS, LOWER
B 6303 456 HURLEY, PATRICK M., PROVENANC STUDIES IN TUNISIA BY U-PB AGES
S 5 34 HURON COUNTY, GODERICH SALT MINE, ONTARIO,
B 6203 552* HURON, ABST., FACTORS CONTROLLING, GEOCHEMISTRY AND EARLY
B 6209 1706 HURON MEMBER, KENTUCKY
B 6209 1710 HURON MEMBER, OHIO
B 6108 1386* HURRICANE AND GRAND WASH FAULTS, NORTHWEST ARIZONA, ABST.,
B 6004 700* HURRICANE FLOODING, ABST., LARGE-SCALE, BED FORMS AN
B 6409 1558* HURRICANE FREDERIC ON MORPHOLOGY OF DAUPHIN ISLAND,
B 6105 767 * HURRICANE IMPACT ON CONTINENTAL-SHELF STRATIGRAPHY, GULF OF
B 6303 422* HURRICANES AND RAINFALL--KEY FOR DOLOMITIZATION IN TIDAL
B 6404 527 HURST, JOHN M., PALEOGEOGRAPHIC AND STRATIGRAPHIC
M 25 462 HUTANOPAN OR SUMATRA-CLIPPERTON-TEHUANTEPEC-BARTLETT
B 6405 726 HUTCHEON, IAN, WATER-ROCK INTERACTION DURING CLASTIC
B 6402 225 HUTTO GRABEN, TEXAS
B 6105 798 HUTTRER, GERALD W., GEOTHERMAL POTENTIAL OF THREE AREAS IN
M 25 441 HWANGGANGRI FORMATION, TILLITIC ORIGIN, SOUTH KOREA
B 6305 844* HYDEN WEST POOL, EASTERN KENTUCKY, ABST., PETROLEUM
B 6006 981† HYDRATE FORMATION IN SUBSURFACE ENVIRONMENTS
B 6312 2204† HYDRATE HORIZONS ON CONTINENTAL SLOPES AND RISES, SEISMIC
M 26 115 HYDRATE ZONES, SEISMIC, REFLECTIONS, FLUID CONTACTS
S 10 125 HYDRAULIC DISTRIBUTION, ALBERTA, RED EARTH REGION
S 10 131 HYDRAULIC DISTRIBUTION, ONE-DIMENSIONAL DECAY
S 10 124 HYDRAULIC FORCES, BASIN GEOMETRY, DISTRIBUTION OF
S 7 273 HYDRAULIC FORCES, BOTTOM, DEFORMATION
B 6309 1583* HYDRAULIC FRACTURING, ABST., INTERACTION OF, PROPPANTS WITH
B 6309 1579* HYDRAULIC FRACTURING AND SONIC LOGGING IN GAS SHALES, ABST.,
S 4 135 HYDRAULIC MODEL FOR CORAL ADAPTATION
B 6405 711* HYDRAULIC PISTON CORING IN EQUATORIAL PACIFIC—PRELIMINARY
B 6304 590 HYDRAULIC POTENTIAL MAPS, GEOCHEMICAL TOOL
B 6408 1229 HYDRAULIC PROCESSES AND RESULTING BED FORMS
S 4 142 HYDRAULIC STRESS, ADAPTIVE STRATEGIES FOR REDUCING
S 10 162 HYDRAULIC THEORY OF OIL AND GAS ACCUMULATION, MUNN
S 10 121† HYDRAULIC THEORY OF PETROLEUM MIGRATION, THE GENERALIZED
M 27 87 HYDROBIEN LIMESTONE, GERMANY, MIOCENE
B 6004 650* HYDROCARBON ACCUMULATION RELATED TO SAN ANDREAS FAULT
B 6408 1278* HYDROCARBON ACCUMULATIONS IN OVERTHRUST BELT OF ALBERTA,
B 6104 483† HYDROCARBON ACCUMULATIONS OF DIAGENETIC TRAPS--EXPLORATION
B 6202 197 HYDROCARBON ACCUMULATIONS, WYOMING, SUSSEX SANDSTONE
S 10 207† HYDROCARBON AND WATER SYSTEM, ROLE OF, GEOPRESSURE IN THE
M 26 301† HYDROCARBON DETERMINATION, APPLICATION OF, AMPLITUDE,
B 6101 1† HYDROCARBON DISCOVERIES OF CANADIAN ARCTIC ISLANDS, GEOLOGY
S 10 56 HYDROCARBON DISTRIBUTION IN SOURCE ROCKS, MIGRATION
B 6205 888* HYDROCARBON EXPLORATION IN GREAT SALT LAKE BASIN, ABST.
B 6108 1377* HYDROCARBON EXPLORATION IN NONMARINE ROCKS OF WESTERN
B 6405 740* HYDROCARBON EXPLORATION IN WESTERN APPROACHES, OFFSHORE
B 6405 789* HYDROCARBON EXPLORATION WITH EXAMPLES FROM THE NORTH SEA,
M 25 189† HYDROCARBON EXPLORATIONIST, SOME SHALLOW TECTONIC
S 10 1† HYDROCARBON EXPULSION IN THE GULF COAST TERTIARY SECTION,
B 6002 287† HYDROCARBON GASES AND THEIR VARIATIONS DEPENDENT ON INITIAL
B 6105 806 * HYDROCARBON GASES IN SEDIMENTS OF EASTERN GULF OF ALASKA
B 6407 1083 HYDROCARBON GASES IN SUBMARINE SEEPS, ORIGIN OF
B 6011 2067* HYDROCARBON GENERATION AND APPLICATION TO OIL AND GAS
B 6204 687† HYDROCARBON GENERATION BY THERMAL ALTERATION OF KEROGEN
B 6207 1151† HYDROCARBON GENERATION IN GIPPSLAND BASIN, AUSTRALIA—
B 6203 536* HYDROCARBON GENERATION POTENTIAL OF CRETACEOUS AND
S 10 12 HYDROCARBON GENERATION, EXPULSION MECHANISMS
B 6205 818 HYDROCARBON GENERATION, SOURCE ROCKS AND
B 6112 2095 HYDROCARBON GENERATION VERSUS H-C RATIO
B 6112 2103† HYDROCARBON GEOCHEMISTRY OF OILS FROM MARANON BASIN, PERU
B 6201 1† HYDROCARBON HABITAT OF TERTIARY NIGER DELTA
S 10 33† HYDROCARBON MIGRATION: A REVIEW, AGENTS FOR, PRIMARY
S 10 241† HYDROCARBON MOVEMENT FROM DEPTH, NEAR-, SURFACE, EVIDENCE
S 10 214 HYDROCARBON REGIME, GEOPRESSURE ZONE
S 10 59 HYDROCARBON SOLUBILITIES, MIGRATION IN AQUEOUS SOLUTION
M 28 143 HYDROCARBON, IN SANDSTONE, CEMENTS
B 6112 2103 HYDROCARBON, ISOPRENOID-TYPE
B 6004 647* HYDROCARBON MIGRATION, ABST., MIDDLE, GUADALUPIAN COASTAL,
B 6305 723† HYDROCARBON MIGRATION AND ENTRAPMENT, MECHANICS OF,
B 6002 324* HYDROCARBON MIGRATION PATHS IN THICK, RAPIDLY DEPOSITED
B 6004 699* HYDROCARBON MIGRATION--RESULT OF GEOLOGIC-EVENT TIMING,
B 6203 524* HYDROCARBON MOVEMENT FROM DEPTH, ABST., NEAR-SURFACE
B 6109 1493† HYDROCARBON OBJECTIVE OF ARCTIC SLOPE, ALASKA, LISBURNE
B 6303 544* HYDROCARBON OCCURRENCE A FUNCTION OF THERMAL ALTERATION
B 6204 667† HYDROCARBON OCCURRENCE, HYDRODYNAMIC FRAMEWORK OF
B 6405 681* HYDROCARBON OCCURRENCES IN NONMARINE, PRE-SALT SEQUENCE OF
B 6204 695 HYDROCARBON POTENTIAL, BAHAMA PLATFORM
B 6204 693† HYDROCARBON POTENTIAL, CHARACTERISTICS OF DEEP, BAHAMA,
B 6311 2070† HYDROCARBON POTENTIAL OF ALEUTIAN BASIN, BERING SEA
B 6403 439* HYDROCARBON POTENTIAL OF ALEUTIAN BASIN, BERING SEA, ABST.
M 25 339† HYDROCARBON POTENTIAL OF AMAZON BASINS OF COLOMBIA,
B 6108 1389* HYDROCARBON POTENTIAL OF ARCHULETA ANTICLINORIUM, BRAZOS
B 6004 720* HYDROCARBON POTENTIAL OF BANGOR LIMESTONE, MISSISSIPPIAN,
M 25 349† HYDROCARBON POTENTIAL OF BOLIVIAN SEDIMENTARY
B 6303 417* HYDROCARBON POTENTIAL OF CENTRAL GULF OF ALASKA, ABST.
M 25 333† HYDROCARBON POTENTIAL OF COASTAL BASINS OF PERU
B 6405 740* HYDROCARBON POTENTIAL OF MATILIJA SANDSTONE, AN EOCENE
B 6004 700* HYDROCARBON POTENTIAL OF NORTHWESTERN BAHAMA PLATFORM,
M 25 311† HYDROCARBON POTENTIAL OF OFFSHORE CALIFORNIA--SUMMARY
B 6205 885* HYDROCARBON POTENTIAL OF PALEOZOIC ROCKS IN EASTERN GREAT
B 6405 671* HYDROCARBON POTENTIAL OF SITKALIDAK FORMATION, EOCENE
B 6205 892* HYDROCARBON POTENTIAL OF UPPER CRETACEOUS CHALKS OF
B 6105 816* HYDROCARBON POTENTIAL PREDICTED FOR OUACHITA MOUNTAINS,
B 6412 2134 HYDROCARBON POTENTIAL, TIMOR TROUGH
M 25 239† HYDROCARBON POTENTIALS IN SEDIMENTARY BASINS OF INDONESIA,
B 6107 1001 HYDROCARBON PRODUCTION FROM CHALKS
B 6303 552* HYDROCARBON PROSPECTS IN BASINS WEST OF UNITED KINGDOM AND
B 6405 745* HYDROCARBON PROSPECTS OF BERING SEA SHELF SOUTH OF ST.
B 6212 2510* HYDROCARBON PROSPECTS OF SINU BELT, NORTHWESTERN COLOMBIA,
B 6012 2150† HYDROCARBON RESERVOIRS, VERTICAL, NET, SANDSTONE
B 6105 823* HYDROCARBON RESOURCES AND RESERVES ESTIMATES, ABST., DATA
B 6203 548* HYDROCARBON SEEPAGE IN SANTA MONICA BAY—CALIFORNIA
B 6301 89 HYDROCARBON SHOWS, SCOTIAN BASIN
B 6305 840* HYDROCARBON SHOWS--PROXIMITY INDICATORS FOR STRATIGRAPHIC
B 6003 356† HYDROCARBON TRAPPING CONSEQUENCES OF SEQUENTIAL FOLDING
M 26 334 HYDROCARBON TRAPS, DELINEATION
B 6307 1016† HYDROCARBON TRAPS IN PETROLEUM PROVINCES, STRUCTURAL
B 6007 1076 HYDROCARBON-LIQUID WINDOW, NIGER DELTA
B 6406 927† HYDROCARBON-MATURATION KINETICS, MICROFRACTURE
B 6405 768* HYDROCARBON-TRAPPING STRUCTURES IN SOUTHERN CANADIAN
M 28 143 HYDROCARBONS, "DEAD," BRIDGING PORES
B 6004 692* HYDROCARBONS, ABST., CLAY-, MEMBRANE, FILTRATION AS MECHANISM
B 6108 1391* HYDROCARBONS, ABST., MECHANICS OF, SECONDARY, MIGRATION AND,
B 6304 698* HYDROCARBONS, ABST., MICROBIOLOGICAL, PROSPECTING FOR
B 6101 100† HYDROCARBONS AND ASPHALTIC COMPOUNDS IN SEDIMENTARY
B 6101 100 HYDROCARBONS AND ASPHALTIC COMPOUNDS, TEMPERATURE RANGES
B 6407 1078† HYDROCARBONS AND THEIR MIGRATION BY SOLUTION IN CARBON
B 6305 766 HYDROCARBONS AS POSSIBLE PRIMARY MIGRATION MECHANISM,
B 6109 1407† HYDROCARBONS BETWEEN OILS AND ASSOCIATED FINE-GRAINED
B 6304 701* HYDROCARBONS BY MEASUREMENTS DERIVED FROM SEISMIC DATA,
B 6402 173† HYDROCARBONS, CONTINENTAL, MARGIN, SUBSIDENCE AND, HEAT
B 6305 763 HYDROCARBONS DISSOLVED FROM CRUDE OIL
S 10 100 HYDROCARBONS, DISTRIBUTION IN NONRESERVOIR ROCKS
B 6304 577 HYDROCARBONS, GEOCHEMICAL CONCEPT, GENERATION AND,
B 6304 699* HYDROCARBONS IN COASTAL WATERS OF NORTH AMERICA, ABST.,
B 6203 498* HYDROCARBONS IN COMPACTING BASINS, ABST., MIGRATION OF
B 6404 549† HYDROCARBONS IN COMPACTING BASINS, MIGRATION OF
B 6001 108† HYDROCARBONS IN HELEZ OIL FIELD, ISRAEL, EARLY, CRETACEOUS,
B 6104 573† HYDROCARBONS IN OILS--PHYSICAL CHEMISTRY APPLIED T PETROLEUM
B 6308 1426* HYDROCARBONS IN SAN MIGUEL SANDS OF SOUTHWEST TEXAS, ABST.,
B 6109 1540* HYDROCARBONS IN SAN MIGUEL SANDSTONES OF SOUTHWEST TEXAS,
B 6105 814* HYDROCARBONS IN SHALLOW HOLES, ATLANTIC MARGIN CORING
B 6105 814* HYDROCARBONS IN SURFACE SEDIMENTS OF MID-ATLANTIC
B 6008 1404* HYDROCARBONS IN UNITED STATES PETROLEUM PROVINCES, ABST.,
B 6305 762 HYDROCARBONS IN WATER, SOLUBILITIES OF
B 6104 483 HYDROCARBONS, INHIBITION OF, DIAGENESIS BY
B 6101 52 HYDROCARBONS, LEBANON
B 6203 434 HYDROCARBONS, NORTH, ATLANTIC, RIFT, BASINS
B 6203 534* HYDROCARBONS OF LOW MOLECULAR WEIGHT IN SEDIMENTS FROM
B 6205 740 HYDROCARBONS, POLITICAL DISTRIBUTION OF
M 28 142 HYDROCARBONS, PORE-FILLING, IN SANDSTONES
S 10 112 HYDROCARBONS, RELATIONS WITH NONHYDROCARBONS IN BASINS
S 10 69† HYDROCARBONS IN COMPACTING BASINS, MIGRATION OF
S 10 104 HYDROCARBONS IN SOLUTION, MIGRATION, SECONDARY
B 6002 288 HYDROCARBONS, STUDY OF, WHOLE, FREE-, GAS PHASE OF
M 28 142 HYDROCARBONS, TAR SANDS, WITH
B 6002 236 HYDROCARBONS, THERMAL DESTRUCTION OF
B 6112 2148 HYDROCARBONS, THERMAL DESTRUCTION OF
S 10 111 HYDROCHEMISTRY, RELATIONS WITH HYDRODYNAMICS
S 10 152 HYDRODYNAMIC TYPES, EXAMPLES, SEDIMENTARY BASINS
B 6203 573* HYDRODYNAMIC, AND THERMODYNAMIC DEVELOPMENT OF A
B 6405 676* HYDRODYNAMIC FLOW IN LOWER CRETACEOUS MUDDY FORMATION,
B 6204 667† HYDRODYNAMIC FRAMEWORK OF EASTERN ALGERIAN SAHARA—

B 6403 325 HYDRODYNAMIC ZONES, MODERN INTERDELTAIC, COAST
B 6104 486 HYDRODYNAMICS CONNRIBUTING TO TILTED OIL-WATER CONTACTS
S 10 111 HYDRODYNAMICS, HYDROCHEMISTRY, RELATIONS WITH
B 6109 1554* HYDROGEN AND OXYGEN ISOTOPIC COMPOSITION OF CRETACEOUS,
S 10 24 HYDROGEN BONDING, STRUCTURED WATER, PORE WATER
M 25 120 HYDROGEN SULFIDE, AIR, POLLUTION FROM
B 6406 922 HYDROGEN-CARBON RATIOS, KEROGEN
B 6305 841* HYDROGEOLOGIC SIGNIFICANCE OF OGALLALA FLUVIAL
B 6108 1386* HYDROGEOLOGY, GEOCHEMISTRY, AND GEOTHERMOMETER MODELS IN
B 6207 1231* HYDROGEOLOGY OF LONG VALLEY GEOTHERMAL AREA, CALIFORNIA,
M 25 350† HYDROGEOLOGY OF THE PACIFIC--SUMMARY
B 6207 1217* HYDROGEOLOGY OF URANIUM DEPOSITS IN CALCRETES OF WESTERN
B 6108 1386* HYDROGEOTHERMAL RESOURCE, ABST., USE OF, HYDROGEOLOGY,
B 6203 529* HYDROGRAPHER CANYON, ABST., GEOTECHNICAL PROPERTIES OF,
S 8 120 HYDROGRAPHER CANYON, FLOW RELATED TO STORMS
S 8 27 HYDROGRAPHER CANYON, TIME-VELOCITY CURVES, CURRENT FLOW
S 8 116 HYDROGRAPHER CANYON, TIME-VELOCITY CURVES, CURRENT FLOW
S 8 114 HYDROGRAPHER CANYON, TOPOGRAPHY, INDEX MAP
S 8 111 HYDROGRAPHER CANYON, U.S. EAST COAST, SUBMARINE CANYONS
B 6004 667* HYDROGRAPHER, HUDSON, AND WILMINGTON SUBMARINE CANYONS,
S 10 110 HYDROLOGIC CYCLE, COMPARED TO THE GEOCHEMICAL CYCLE
B 6004 724* **HYDROLOGIC STABILITY OF NORTH LOUISIANA SALT DOMES--POTENTIAL**
B 6309 1602* HYDROLOGY AND WATER QUALITY OF EOCENE WILCOX GROUP IN
B 6303 450* HYDROMAGNESITE, AND NESQUEHONITE FROM CALCIUM-MAGNESIU
S 4 144 HYDROMECHANICAL STUDY OF ACROPORA PALMATA
B 6402 146 HYDROSTATIC CONDITIONS, FAULT ENTRAPMENT OF HYDROCARBONS
B 6405 711* HYDROSTRATIGRAPHIC FRAMEWORK AND FLOW DYNAMICS OF
B 6309 1494 HYDROTHERMAL ALTERATION, PAKISTAN, CHAGAI HILLS
B 6309 1589* HYDROTHERMAL AND UPLIF HISTORIES OF NORTHERN APPALACHIAN
S 10 227 HYDROTHERMAL EFFECTS, COLORADO, RANGELY FIELD
S 10 228 HYDROTHERMAL EFFECTS, MIGRATION, BRINE FLOW
S 10 230 HYDROTHERMAL EFFECTS, USSR, BAKU AREA, ZYKH FIELD
S 10 233 HYDROTHERMAL EFFECTS, USSR, DON MEDVEDICA AREA (VOLGA)
S 10 229 HYDROTHERMAL EFFECTS, USSR, SIBERIA, OB RIVER AREA
S 10 229 HYDROTHERMAL EFFECTS, USSR, TRANS CAUCASUS AND BAKU
S 3 135 HYDROTHERMAL FEEDING, AT TECTONIC INTERSECTIONS,
S 3 135 HYDROTHERMAL MODIFICATION OF CARBONATITES, DIAPIRIC
B 6405 785* HYDROTHERMAL SYSTEMS, ABST., DRILL AS TOOL FOR STUDYING
B 6405 799* HYDROTHERMAL SYSTEMS IN BASIN AND RANGE PROVINCE ABST.,
S 3 135 HYDROTHERMAL-VEIN FORMATION AROUND TAPHROGENIC ALKALIC
B 6004 584† HYDROZOAN, CARBONAT BUILDUPS FROM UPPER PALEOZOIC OF IDAHO,
B 6303 510* HYGIENE AND TERRY SANDSTONES, UPPER CRETACEOUS, SPINDLE FIELD,
M 24 141 HYGIENE DISCOVERY WELLS, COLORADO, SUSSEX-, SHANNON
B 6108 1388* HYGIENE MEMBER OF PIERRE SHALE, WEST DENVER BASIN, COLORADO,
B 6408 1238 HYGIENE SANDSTONE BAR COMPLEXES, COLORADO
M 26 279 HYGIENE SANDSTONE, CRETACEOUS
B 6410 1617 HYGIENE SANDSTONE, DENVER BASIN
M 26 289 HYGIENE STRUCTURE CONTOUR MAP, DENVER BASIN
B 6004 684 HYNE, N. J., STRATIGRAPHY AND SEDIMENTATION OF FRESHWATER,
B 6311 2042 HYNE, NORMAN J., STRATIGRAPHY OF INTERMONTANE, LACUSTRINE
B 6408 1272 HYNE, STEWART, G. F., REVIEW OF, PENNSYLVANIAN SANDSTONES OF
M 25 434 HYPOGENE COPPER SULFIDE DEPOSIT
S 3 136 HYPOGENE LATEROGENESIS IN GEOPRESSURIZED BASINS
S 3 135 HYPOTHERMAL REPLACEMENT OF LATE-PHASE ALKALIC GRANITES
B 6002 213 HYPOTHESES, PRIMARY, PETROLEUM, MIGRATION
B 6002 215 HYPOTHESIS, DIFFUSION
B 6412 2179 HYPOTHESIS FOR ABNORMALLY HIGH PORE PRESSURES IN SHALES,
M 25 182 HYPOTHESIS FOR GENERATION OF OIL AND GAS, THRUST-, FAULT
B 6002 214 HYPOTHESIS, GASEOUS, SOLUTION
B 6002 213 HYPOTHESIS, GLOBULE, EXPULSION
B 6405 770* HYPOTHESIS INVOLVING DILATION AND NATURA HYDRAULIC
B 6002 215 HYPOTHESIS, KEROGEN, DIFFUSION
B 6002 214 HYPOTHESIS, MICELLAR, SOLUTION
B 6002 215 HYPOTHESIS, MOLECULAR, SOLUTION
M 25 178 HYPOTHESIS OF PLATE TECTONICS
B 6104 594 I-1-6 FIELD, LIBYA
B 6303 457* IAB FIELD, COKE COUNTY, TEXAS, ABST., POROSITY EXAMPLES FROM
B 6405 724 IANNACCHIONE, ANTHONY T., ANTICIPATING COAL MINING PROBLEMS
B 6101 81 IBERIAN MESETA
B 6109 1561* IBERIAN PENINSULA, ABST., CONTRIBUTION OF, LANDSAT, IMAGERY TO
B 6303 555* IBEX AREA, WESTERN MILLARD COUNTY, UTAH, ABST., CARBONATE,
S 9 19 ICE AGES, CLIMATIC CHANGES
B 6309 1546 ICE AGES, SOLVING THE MYSTERY BY JOHN IMBRIE AND KATHERINE
B 6207 1211* ICE COVERS BY MICROWAVE SATELLITE IMAGERY, ABST., STRUCTURE
S 9 8 ICE FLOW DIRECTION, CARBONIFEROUS, SOUTH AFRICA
S 9 8 ICE FLOW DIRECTION, ORDOVICIAN, SAHARA, AFRICA
S 9 7 ICE FLOW DIRECTIONS, PALEOCLIMATE INDICATORS
B 6004 655* ICE ON SEDIMENTS ON SANDY BRAIDED RIVER, ABST., EFFECTS OF
B 6409 1560 ICE, R. GEOFF, FORAMINIFERS AND CALCISPHERES FROM CUESTA DEL
B 6405 779* ICE SHELF, ABST., ORGANIC MATTER IN SEDIMENTS UNDERLYING, ROSS

B 6004 664* ICE-DRIFTED BOULDERS AS PALEOCLIMATIC INDICATOR, SOUTH SHORE
M 29 44 ICELAND, BATHYMETRIC CONTOURS SURROUNDING
M 29 46 ICELAND, INSULAR MARGIN, STRUCTURAL DIAGRAM
M 29 49 ICELAND, INSULAR MARGIN, THIRTEEN PROFILES
M 29 43† ICELAND INSULAR MARGIN: THIRTEEN GEOPHYSICAL PROFILES,
M 29 47 ICELAND, INSULAR RISE, EROSION
M 29 47 ICELAND, INSULAR SLOPE
M 29 48 ICELAND, MARGIN PROFILES, DESCRIPTIONS
M 29 45 ICELAND, OUTER INSULAR SHELF
M 29 45 ICELAND, PALEOSHELF
M 29 45 ICELAND, PHYSIOGRAPHIC PROVINCES SURROUNDING
M 29 45 ICELAND, PLIOCENE, OUTER INSULAR SHELF
M 29 45 ICELAND, REYKJANES RIDGE
B 6203 423 ICELAND RIDGE, FAEROE-
M 29 46 ICELAND, SNAEFELLSNES ARCH
B 6107 1070 ICELAND SPAR SUBSTRATES, MICRITIZATION OF
B 6004 710* ICHNOCOENOSES--THEIR SIGNIFICANCE FOR ENVIRONMENTAL ANALYSIS
B 6203 511* ICHNOFACIES, ABST., BATHYMETRIC CONTROL OF, DEEP-SEA
B 6205 883* ICHNOFACIES FROM NORTHERN GULF OF CALIFORNIA AND POSSIBLE
B 6303 508* ICHNOLOG OF LABRADOR GROUP, LOWER CAMBRIAN, IN SOUTHERN
B 6303 422* ICHNOLOGY AND DEPOSITIONAL ENVIRONMENTS OF UPPER
B 6203 514* ICHNOLOGY AND UNIFORMITARIANISM, ABST.
B 6104 581 ICOTEA HIGH, VENEZUELA
B 6204 635 ICY BAY, ALASKA
M 29 285 ICY BAY, REFRACTION DATA, ALASKA
B 6212 2415 ICY BAY-MALASPINA GLACIER SLUMP, SEISMIC DATA, ALASKA
B 6205 794 IDA GREEN CRUISE, RV
B 6206 1067 IDA GREEN CRUISES 1G-7 AND 1G-8, RV
B 6008 1413* IDAHO A DEDUCED FROM REGIONAL FRAMEWORK EXPLORATION, ABST.,
B 6303 425* IDAHO, ABST., DRANEY LIMESTONE--EARLY, CRETACEOUS LACUSTRINE,
B 6205 882* IDAHO, ABST., GENERAL GEOLOGY OF NORTHERN, PORTNEUF RANGE,
B 6008 1392* IDAHO, ABST., HEAT-FLOW STUDY OF, SNAKE RIVER PLAIN
B 6303 536* IDAHO, ABST., NORMALLY AND REVERSELY GRADED BEDS WITHIN
B 6305 836* IDAHO, ABST., PETROLEUM EXPLORATION I YOUNGER OVER OLDER,
B 6203 527* IDAHO, ABST., TERTIARY, ALGAL, CARBONATE, REEFS FROM, SNAKE
B 6205 889* IDAHO, ABST., TWIN CREEK FORMATION AND EQUIVALENT UNITS,
B 6405 785* IDAHO, ABST., UPPER, PALEOZOIC PALEOGEOGRAPHY OF
B 6108 1389* IDAHO, AND NORTHERN UTAH-STRUCTURAL GEOMETRY AND RELATED
B 6405 805* IDAHO AND WYOMING, ABST., THRUST GEOMETRY OF NORTHERN,
B 6201 100 IDAHO AND WYOMING, DINWOODY FORMATION
B 6201 100 IDAHO AND WYOMING, FRANSON MEMBER
B 6201 100 IDAHO AND WYOMING, REX CHERT MEMBER
M 25 509 IDAHO BATHOLITH
B 6201 112 IDAHO BATHOLITH, MONTANA AND, IDAHO
B 6102 257 IDAHO, BIG WOOD RIVER VALLEY, NOMENCLATURE
B 6102 257 IDAHO, BROCKIE LAKE CONGLOMERATE
M 28 8 IDAHO, CAMBRIAN, HARRISON SUMMIT QUARTZITE
M 28 21 IDAHO, CAMBRIAN, HARRISON SUMMIT QUARTZITE
B 6008 1207 IDAHO, DEVELOPMENTS, 1975
B 6308 1386 IDAHO, DEVELOPMENTS, 1978
B 6409 1501 IDAHO, DEVELOPMENTS, 1979
B 6102 257 IDAHO, DEVONIAN TO PERMIAN, NOMENCLATURE
B 6102 257 IDAHO, DRUMMOND MINE LIMESTONE
B 6008 1389* IDAHO, FOR GEOTHERMAL RESOURCE EVALUATION, ABST., GEOLOGIC
B 6008 1397* IDAHO, GEOTHERMAL SYSTEM, ABST., GEOELECTRICAL INVESTIGATIONS
B 6008 1411* IDAHO, GEOTHERMAL WELLS--SITING, DRILLING, AND TESTING, ABST.,
B 6102 257 IDAHO, GREEN LAKE LIMESTONE MEMBER
B 6201 112 IDAHO, IDAHO BATHOLITH, MONTANA AND
B 6102 257 IDAHO, IRON BOG CREEK FORMATION
B 6102 257 IDAHO, JEFFERSON FORMATION
B 6102 259 IDAHO, LITTLE COPPER FORMATION
B 6102 256† IDAHO, LITTLE COPPER FORMATION--NEW
B 6102 257 IDAHO, LOST RIVER RANGE, NOMENCLATURE
B 6102 257 IDAHO, MCGOWAN CREEK FORMATION
B 6201 99 IDAHO, MEADE PEAK PHOSPHATIC SHALE MEMBER
B 6102 257 IDAHO, MILLIGEN FORMATION
B 6102 257 IDAHO, MISSISSIPPIAN, NOMENCLATURE
B 6102 257 IDAHO, MULDOON CANYON FORMATION
B 6405 799* IDAHO, NEVADA, MONTANA, UTAH, AND WYOMING, ABST.,
B 6405 767* IDAHO OVERTHRUST BELT—RELATION TO CANADIAN FOOTHILLS
B 6004 584† **IDAHO, PALAEOAPLYSINA, HYDROZOAN, CARBONATE BUILDUPS FROM**
B 6201 100 IDAHO, PARK CITY FORMATION
B 6102 257 IDAHO, PENNSYLVANIAN, NOMENCLATURE
M 28 127 IDAHO, PERMIAN, PHOSPHORIA FORMATION
M 27 120 IDAHO, PERMIAN, PHOSPHORIA FORMATION
M 28 83 IDAHO, PERMIAN, PHOSPHORIA FORMATION
B 6102 257 IDAHO, PERMIAN TO DEVONIAN, NOMENCLATURE
B 6102 257 IDAHO, PIONEER MOUNTAINS, NOMENCLATURE
B 6201 100 IDAHO, PREUSS RANGE
B 6201 99 IDAHO, RETORT PHOSPHATIC SHALE MEMBER

B 6102 257 IDAHO, SCORPION MOUNTAIN FORMATION
B 6201 100 IDAHO, SUBLETT RANGE
B 6102 257 IDAHO, THREE FORKS FORMATION
B 6004 592 IDAHO, TUBULAR-FORAM FACIES
B 6201 100 IDAHO, WELLS FORMATION
B 6102 257 IDAHO, WHITE KNOB LIMESTONE GROUP
B 6102 257 IDAHO, WOOD RIVER FORMATION
B 6303 454* IDAHO--DIAGENETIC MODEL, ABST., DISTRIBUTION OF, CARBONATE,
B 6405 736* IDAHO-WYOMIN THRUST BELT, AS COMPARED WITH OTHER
B 6303 500* IDAHO-WYOMING THRUST BELT, ABST., APPLICATION OF POTENTIAL
B 6108 1388* IDAHO-WYOMING THRUST BELT, ABST., ASSESSMENT OF OIL AND GAS
B 6105 821* IDAHO-WYOMING THRUST BELT, ABST., ASSESSMENT OF OIL AND GAS
B 6109 1559* IDAHO-WYOMING THRUST BELT, ABST., LANDSAT—USEFUL PETROLEUM
B 6405 669* IDAHO-WYOMING THRUST BELT AND HINTERLAND, ABST., TECTONIC
B 6206 1075 IDAHO-WYOMING THRUST BELT, NUGGET SANDSTONE
B 6206 1075† IDAHO-WYOMING THRUST BELT, PETROLEUM, SOURCE-BED,
B 6305 833* IDAHO-WYOMING THRUST BELT--PRELIMINARY REPORT, ABST.,
B 6307 1039 IDAHO-WYOMING THRUST-FOLD BELT, FRONTIER FORMATION
B 6307 1039 IDAHO-WYOMING THRUST-FOLD BELT, HILLIARD SHALE
B 6307 1038 IDAHO-WYOMING, THRUST-FOLD BELT, MAP
B 6307 1039 IDAHO-WYOMING THRUST-FOLD BELT, NUGGET SANDSTONE
B 6307 1040 IDAHO-WYOMING THRUST-FOLD BELT SEQUENTIAL EVOLUTION
B 6405 700* IDAHO—CLARIFICATIONS AND CONSTRAINTS FROM PIONEER
B 6211 2208 IDEALIZED GROWTH-FAULT DEVELOPMENT
B 6111 1934 IDEALIZED SABKHA CYCLE
M 26 155 IDENTIFICATION, AFRICA, WESTERN, SEQUENCES
S 7 47 IDENTIFICATION, CRITERIA, GULF OF MEXICO, SALT
B 6004 658* IDENTIFICATION OF LEAKAGE GAS FROM UNDERGROUND NATURAL
M 26 145† IDENTIFICATION OF UPPER TRIASSIC, JURASSIC, AND LOWER
M 25 293 IDITAROD FAULT, ALASKA
B 6405 674 IDRIS, FAISAL M., EXPLORABLE STRUCTURES IN OLD BAHAMA CHANNEL,
B 6407 992 IFERD FAULT, MOROCCO
B 6101 16 IGNEOUS ACTIVITY, SVERDRUP BASIN
B 6102 248† IGNEOUS AND HIGH-GRADE METAMORPHIC ROCKS, NOTE 45--,
S 6 150 IGNEOUS GEOCHRONOLOGY, JAPAN, CRETACEOUS
S 6 144 IGNEOUS LOCALITIES, MAP, JAPAN, MESOZOIC
S 6 143† IGNEOUS ROCKS IN JAPAN, ISOTOPIC AGES AND STRATIGRAPHIC
B 6309 1602* IGNEOUS ROCKS OF MISSISSIPPI, ABST., PETROGRAPHY OF SOME
M 25 486 IGNEOUS-TECTONIC PROVINCES, MOLYBDENUM DEPOSITS
M 25 465 IGNIMBRITE VOLCANISM, JAPAN, NORTHEAST
M 25 463 IGNIMBRITE VOLCANISM, LESSER, KHINGAN RANGE
M 25 526 IGNIMBRITES, MEXICO
B 6007 1076 IHENDIAGU AREA, NIGER DELTA
M 25 98 IIJIMA, AZUMA, INFLUENCE OF TEMPERATURE ON COALIFICATION OF
B 6407 991 IKAKERN FORMATION, MOROCCO
B 6212 2510 IKEBE, YUTAKA, STRUCTURAL CONTRAST BETWEEN FORE-ARC AND
B 6209 1670 ILAN PLAIN, TAIWAN
M 28 33 ILDEFONSO FORMATION, PUERTO RICO, CRETACEOUS
M 28 99 ILDEFONSO FORMATION, PUERTO RICO, CRETACEOUS
B 6101 60 ILE DU PALMIER STRUCTURE, LEBANON
B 6203 374 ILIAMNA VOLCANO, ALASKA
B 6407 1091 ILJINIAN HORIZON, ASIA, NORTHEAST
B 6303 471 ILLICH, H. A., GEOCHEMISTRY OF OILS FROM SANTA CRUZ BASIN,
B 6112 2103 ILLICH, H. A., HYDROCARBON GEOCHEMISTRY OF OILS FROM MARANON
B 6303 472 ILLICH, H. A., OIL GEOCHEMISTRY AS EXPLORATION TOOL, ABST.
B 6203 552* ILLINOIS, ABST., COAL RESOURCES OF
B 6408 1283* ILLINOIS, ABST., FACIES, PALEOECOLOGY, AND DEPOSITIONAL
B 6408 1281* ILLINOIS, ABST., GEOLOGY AND GEOPHYSICS OF MIDDLE, MISSISSIPPIAN,
B 6305 846* ILLINOIS, ABST., GEOLOGY OF, COAL DEPOSITS IN SOUTHEASTERN,
B 6105 819* ILLINOIS, ABST., IDENTIFICATION OF MAJOR, DISTRIBUTARY-CHANNEL
B 6405 748* ILLINOIS, ABST., POROSITY EVOLUTION OF, NIAGARAN,
B 6405 692* ILLINOIS, ABST., RELATION OF DEPOSITIONAL FACIES AND HISTORY TO
B 6405 795* ILLINOIS, ABST., SEELYVILLE, COAL—MAJOR UNEXPLOITED, SEAM IN
B 6303 503* ILLINOIS BASIN, ABST., COAL RESOURCES OF PART OF SOUTHERN
B 6408 1287* ILLINOIS BASIN, ABST., DEPOSITIONAL ENVIRONMENTS OF, ULLIN
B 6303 434* ILLINOIS BASIN, ABST., ISOTOPE, GEOCHEMISTRY OF,
B 6303 431* ILLINOIS BASIN, ABST., MISSISSIPPIAN NON-SUPRATIDAL DOLOMITE, STE.
B 6303 506* ILLINOIS BASIN, ABST., ORIGIN AND CHARACTERTICS OF,
B 6309 1575* **ILLINOIS BASIN, ABST., PETROLOGY AND, MATURATION OF DISPERSED,**
B 6204 707 ILLINOIS BASIN, BACKBONE LIMESTONE
B 6204 706 ILLINOIS BASIN, BAILEY LIMESTONE
B 6204 706 ILLINOIS BASIN, CLEAR CREEK CHERT
B 6204 707 ILLINOIS BASIN, DECATUR LIMESTONE
B 6204 706 ILLINOIS BASIN, EDGEWOOD LIMESTONE
B 6412 2189 ILLINOIS BASIN, EDITED BY JAMES E. PALMER AND RUSSELL R. DUTCHER,
B 6204 706 ILLINOIS BASIN, ELECTRIC-LOG CROSS SECTION
B 6204 706 ILLINOIS BASIN, GRAND TOWER LIMESTONE
B 6204 706 ILLINOIS BASIN, GRASSY KNOB CHERT
B 6408 1287* ILLINOIS BASIN, ITS FUTURE PETROLEUM PROSPECTS, AND NUMBERS,
B 6204 706 ILLINOIS BASIN, MAQUOKETA SHALE
B 6204 706 ILLINOIS BASIN, MOCCASIN SPRINGS FORMATION
B 6204 707 ILLINOIS BASIN, OZORA LIMESTONE
B 6204 705† ILLINOIS BASIN, POSITION OF, SUB-KASKASKIA UNCONFORMITY IN
B 6204 706 ILLINOIS BASIN, POST-EMSIAN UNCONFORMITY
B 6204 706 ILLINOIS BASIN, SEXTON CREEK LIMESTONE
B 6204 706 ILLINOIS BASIN, ST. CLAIR LIMESTONE
B 6204 707 ILLINOIS BASIN, SUB-KASKASKIA-POST-TIPPECANOE UNCONFORMITY
B 6209 1711 ILLINOIS BASIN, THREE LICK BED CORRELATION IN
B 6405 743* ILLINOIS BASINS, ABST., RESIN RODS AND WOODY ROD-LIKE
B 6408 1289* ILLINOIS COAL—A MAJOR BITUMINOUS COAL RESOURCE, ABST.
B 6107 1060 ILLINOIS CYCLOTHEMS, POSSIBLE RELATIONS OF, KANSAS AND
B 6008 1223 ILLINOIS, DEVELOPMENTS, 1975
B 6108 1185 ILLINOIS, DEVELOPMENTS, 1976
B 6208 1341 ILLINOIS, DEVELOPMENTS, 1977
B 6308 1325 ILLINOIS, DEVELOPMENTS, 1978
B 6409 1463 ILLINOIS, DEVELOPMENTS, 1979
M 28 114 ILLINOIS, DEVONIAN, CEDAR VALLEY LIMESTONE
M 28 114 ILLINOIS, DEVONIAN, HOING SANDSTONE
S 5 99 ILLINOIS, KANKAKEE DOLOMITE
B 6009 1623* ILLINOIS, KENTUCKY AND OHIO, ABST., LITHOFACIES RELATIONS IN,
S 10 149 ILLINOIS, MAQUOKETA AQUITARD, PIEZOMETRIC SURFACE
M 27 105 ILLINOIS, PENNSYLVANIAN, MAZON CREEK COAL BALLS
B 6008 1223 ILLINOIS, PRODUCTION, 1975
B 6108 1185 ILLINOIS, PRODUCTION, 1976
B 6208 1341 ILLINOIS, PRODUCTION, 1977
B 6308 1325 ILLINOIS, PRODUCTION, 1978
B 6409 1463 ILLINOIS, PRODUCTION, 1979
S 5 2 ILLINOIS, SILURIAN, NIAGARA GROUP
B 6408 1282* ILLINOIS—ITS RELATION TO CYPRESS, CHESTERIAN, STRATIGRAPHIC
M 28 132 ILLITE, AUTHIGENIC, COATINGS
M 28 71 ILLITE, AUTHIGENIC, IN SANDSTONE
B 6002 224 ILLITE CLAY TRANSFORMATION, MONTMORILLONITE-TO-
B 6203 497* ILLITE REACTION FOR SANDSTONE CEMENTATION, ABST., CLAY
B 6405 708* ILLITE TRANSFORMATION—ROLE IN GENERATING AND MAINTAINING
M 28 136 ILLITE-SERICITE, IN SANDSTONE, CEMENTS
B 6405 725* ILLITE-SMECTITE DIAGENESIS, ABST., SEDIMENTATION RATES AND
B 6303 460* ILLITES IN SANDSTONE RESERVOIRS, ABST., AUTHIGENI
B 6204 668 ILLIZI BASIN, ALGERIA
B 6204 669 ILLIZI BASIN, STRATIGRAPHIC SECTION, ALGERIA
M 25 418 ILMENITE, COMPOSITION AND USES
M 25 443 ILWOLSAN UPTHRUST, SOUTH KOREA
B 6109 1560* IMAGE PROCESSING OF LANDSAT DATA, ABST., URANIUM SEARCH IN,
B 6004 663* IMAGERY, ABST., OBSERVATIONS OF, CIRCULATION AN,
B 6207 1211* IMAGERY, ABST., STRUCTURE AND MOTION OF, BERING AND, OKHOTSK
B 6207 1222* IMAGERY AND RELATED REMOTE SENSING TECHNIQUES TO MINERAL
B 6412 2064† IMAGERY FOR GEOLOGIC MAPPING—TENNESSEE-KENTUCKY-VIRGINIA,
B 6009 1609* IMAGERY FOR WETLAND AND LAND-USE DELINEATION IN TEXAS
B 6207 1232* IMAGERY IN CIRCUM-PACIFIC REGION, ABST., APPLICATIONS OF,
B 6002 325* IMAGERY IS NOT JUST A PICTURE, ABST., SLAR
B 6004 709* IMAGERY STUDIES TO TIME AND ORIGIN OF MICHIGAN BASIN LINEAR
B 6109 1561* IMAGERY TO GEOLOGIC PROSPECTING OVER IBERIAN PENINSULA, ABST.,
B 6005 745† IMAGERY TO PETROLEUM AND MINERAL EXPLORATION, APPLICATION
B 6309 1546 IMBRIE AND KATHERINE PALMER IMBRIE, HAMILTON, THOMAS D.,
B 6309 1546 IMBRIE, HAMILTON, THOMAS D., REVIEW OF, ICE AGES, SOLVING THE
B 6407 993 IMI TANOUTE HORST
B 6104 607 **IMLAY, R. W., SOME LOWER AND MIDDLE JURASSIC BEDS IN PUALE**
B 6303 555* IMMANITAS BED, PASO DEL RIO, COLIMA, MEXICO, ABST.
S 10 59 IMMATURE ORGANIC MATTER, BITUMEN COMPOSITION
M 28 105 IMMATURE SEDIMENTARY, TEXTURE
B 6202 296 IMO FORMATION, NIGERIA
B 6301 47 IMOGENE CARBONATE PRODUCTION, TEXAS
B 6303 523* IMPACT OF COAL PROPERTIES ON COMBUSTION CHARACTERISTICS,
B 6303 464* IMPACT OF DEEP SEA DRILLING PROJECT ON PALEO-OCEANOGRAPHY,
B 6408 1276* IMPACT OF DIAGENESIS ON EXPLORATION STRATEGY AND RESERVOIR
B 6309 1595* IMPACT OF DIAGENESIS ON LOG INTERPRETATION, ABST.
M 25 365† IMPACT OF ECONOMIC DEVELOPMENT ON DEMANDS FOR
B 6405 724* IMPACT OF HOLOCENE TRANSGRESSION ON DEPOSITIONAL
B 6004 711* IMPACT OF OIL AND GAS EXPLORATION ASSUMPTIONS ON PROJECTIONS
B 6207 1228* IMPACT OF SHORELINE PROCESSES ON RESOURCE DEVELOPMENT, ABST.
B 6310 1982 IMPACT OF THE GEOSCIENCES ON CRITICAL ENERGY RESOURCES, EDITED
B 6109 1545* IMPACT-ASSESSMENT SYSTEM, ABST., NATURAL,
B 6203 506* IMPEDANCE LOGS, ABST., UNDERSTANDING, SEISMICALLY DERIVED
B 6108 1385* IMPERIAL COUNTY, CALIFORNIA, ABST., GEOLOGY AND, RESERVOIR
B 6203 368 IMPERIAL VALLEYS, CALIFORNIA, COACHELLA-
B 6309 1587* IMPLICATIONS OF GASEOUS HYDROCARBON GEOCHEMISTRY OF
B 6405 714* IMPLICATIONS OF GEOLOGI STRUCTURE AND REGIONAL
B 6004 709* IMPLICATIONS OF IMAGERY STUDIES TO TIME AND ORIGIN OF
M 25 1† IMPORTANCE OF CIRCUM-PACIFIC ENERGY AN MINERAL RESOURCES TO
S 4 239† IMPORTANCE OF PHYLLOID ALGAE IN DEVELOPMENT OF DEPOSITIONAL
B 6309 1606* IMPORTANCE OF SECONDARY LEACHED POROSITY IN LOWER TERTIARY

B 6305 834* **IMPORTANCE OF SECONDARY POROSITY IN SANDSTONES TO**
B 6405 747* IMPORTANCE OF SECONDARY POROSITY IN SANDSTONES TO
B 6303 510* IMPORTANT STRATIGRAPHIC BREAKS IN COST GE-1 WELL, SOUTHEAST
B 6105 807* IMPORTANCE OF TEXTURAL AND DIAGENETIC PROPERTIES OF BEREA
B 6303 498* IN SEARCH OF RESOLUTION, ABST.
B 6406 960* IN SITU PILOT PROJECT, WABASCA OIL SANDS DEPOSIT, ALBERTA, ABST.,
B 6203 529* IN SITU RECOVERY PROJECTS. BY TITLE ONLY., INSTRUMENTATION FOR,
B 6002 159† IN THE PUBLIC INTEREST
B 6303 493* IN-SITU FORMATION OF BOTRYOIDAL ARAGONITE ON PERMIAN
B 6303 538* IN-SITU LEACHING OF URANIUM, ABST., ENVIRONMENTAL ASSESSMENT
B 6105 840* IN-SITU LEACHING OF URANIUM—DEVELOPING TECHNOLOGY, ABST.
B 6108 1389* IN-SITU MINING OF URANIUM. BY TITLE ONLY., APPLICABLE
B 6309 1590* IN-SITU PROCESS, ABST., RECOVERY OF, ENERGY FROM, MICHIGAN,
B 6206 1092* IN-SITU RECOVERY PROJECTS, ABST., INSTRUMENTATION FOR,
B 6309 1587* IN-SITU TESTING OF WELL-SHOOTING CONCEPTS, ABST.
B 6303 519* IN-SITU URANIUM SOLUTION MINING--ENVIRONMENTAL AND
B 6108 1392* INCIPIENT DIAGENESIS OF ARKOSIC AND VITRIC SANDSTONE, CENOZOIC
B 6209 1567 INCIPIENT DIAPIRS, EXTRUSION AND INTRUSION
B 6209 1566 INCIPIENT DIAPIRS, GROWTH OF
B 6201 98† INCIPIENT METAMORPHISM, AND OIL GENERATION IN BLACK SHALE
B 6108 1375* INCIPIENT METAMORPHISM, AND OIL GENERATION IN BLACK SHALE
S 3 96 INCIPIENT OROGENIC CONTINENTAL MARGIN
M 26 258 INCLINED-REFLECTOR PATTERNS
B 6105 809* INCOMPLETE CARBONATE CEMENTATION—PRINCIPAL FACTOR IN
B 6007 1015 INCOMPLETE FLUVIAL SEQUENCES
B 6004 691* INCONSISTENCIES IN RESOURCE-APPRAISAL ESTIMATES RESULTING
B 6303 472 INDEN, RICHARD F., DOLOMITIZATION OF OFFSHORE CARBONATE
B 6108 1385* INDEPENDENCE MOUNTAIN THRUST FAULT, NORTH PARK, COLORADO,
B 6009 1612* INDEPENDENT AND MAJOR ROLES IN EXPLORATION FOR OIL AND GAS,
B 6303 491* INDEPENDENT GEOLOGISTS--ENDANGERED SPECIES, ABST.
B 6308 1427* INDEPENDENT GEOLOGISTS--ENDANGERED SPECIES, ABST.
B 6012 2187* INDEPENDENT IN TOMORROWS OIL, ABST., ROLE OF SMALL
B 6310 1820 INDIA, 1978, RAO, K. L. N., ON DEVELOPMENTS IN
B 6405 691* INDIA, ABST., TRACE, FOSSILS FROM PROTEROZOIC SEDIMENTS OF,
B 6212 2489† INDIA, AND ITS IMPORTANCE IN PETROLEUM EXPLORATION.
B 6306 875 INDIA AND PAKISTAN, BASIN MODEL OFF
B 6402 219 INDIA, ANKLESVAR FIELD
B 6403 429 INDIA, ANKLESWAR HIGH-WAX OIL FIELD
B 6403 429 INDIA, ANKLESWAR HIGH-WAX OIL FIELD
B 6203 443 INDIA, ARAVALLI-DELHI OROGENIC BELT
B 6402 218 INDIA, B-3 FIELD
B 6402 218 INDIA, B-38 FIELD
B 6203 443 INDIA, BALOL SANTHAL FIELD
B 6203 453 INDIA, BALOL-SANTHAL FIELD, SECTION
B 6203 443 INDIA, BALSASAN FIELD
B 6402 219 INDIA, BASSEIN FIELD
B 6402 218 INDIA, BOMBAY HIGH FIELD
M 30 496 INDIA, BOMBAY HIGH FIELD, BIOSTRATIGRAPHY
M 30 488 INDIA, BOMBAY HIGH FIELD, LOCATION
M 30 488 INDIA, BOMBAY HIGH FIELD, STRATIGRAPHY
B 6402 210 INDIA, BOMBAY OFFSHORE BASIN, OIL FIELDS
M 30 487 INDIA, BOMBAY OFFSHORE BASIN, REGIONAL GEOLOGY
M 30 497 INDIA, BOMBAY OFFSHORE BASIN, STRUCTURE
M 30 495 INDIA, BOMBAY OFFSHORE BASIN, TECTONIC FRAMEWORK
B 6402 210 INDIA, BOMBAY-RATNAGIRI SHELF
M 30 489 INDIA, CAMBAY BASIN, CORRELATION TO BOMBAY OFFSHORE BASIN
M 30 495 INDIA, CAMBAY BASIN, TECTONIC FRAMEWORK
B 6402 219 INDIA, CAMBAY FIELD
B 6402 210 INDIA, CAMBAY GULF SHELF
B 6403 429 INDIA, CAMBAY HIGH-WAX OIL FIELD
B 6203 444 INDIA, CAMBAY SHALE
B 6402 218 INDIA, D-1 FIELD
B 6402 218 INDIA, D-1A FIELD
B 6402 210 INDIA, DAHANU DEPRESSION
B 6402 218 INDIA, DAHANU FIELD
B 6402 219 INDIA, DCS AREA
B 6203 443 INDIA, DECCAN TRAP
B 6402 210 INDIA, DECCAN TRAP LAVA FLOWS
B 6403 427† INDIA, DEPOSITIONAL ENVIRONMENT OF, SOURCE BEDS OF HIGH-, WAX,
B 6110 1835 INDIA, DEVELOPMENTS, 1976
B 6210 1983 INDIA, DEVELOPMENTS, 1977
B 6411 1886 INDIA, DEVELOPMENTS, 1979
B 6403 428 INDIA, DIGBOI FIELD
B 6402 219 INDIA, DIGBOI FIELD
B 6403 428 INDIA, DIGBOI FIELD
B 6402 219 INDIA, DIGBOI FIELD
B 6203 442† INDIA, EARLY, EOCENE SUBAERIAL EROSIONAL VALLEYS IN, CAMBAY,
B 6402 219 INDIA, EOCENE
B 6403 428 INDIA, GALEKI FIELD
B 6402 218 INDIA, H1-5 FIELD
B 6403 428 INDIA, HIMALAYAN THRUST SHEET
B 6203 444 INDIA, KADI FORMATION
B 6203 444 INDIA, KALOL FORMATION
B 6203 448 INDIA, KALOL FORMATION, MSP MARKER
B 6203 449 INDIA, KALOL FORMATION, USP MARKER
B 6403 428 INDIA, LAKWA FIELD
B 6203 443 INDIA, LANWA FIELD
M 29 427 INDIA, MARGIN, OXYGEN IN SEAWATER
B 6203 444 INDIA, MEHSANA HORST, SECTION
B 6203 453 INDIA, MIOCENE
B 6203 444 INDIA, MIOCENE
B 6402 219 INDIA, MIOCENE
B 6403 428 INDIA, MORAN FIELD
B 6402 219 INDIA, NAHAR-KATIYA AND MORAN FIELD
B 6403 428 INDIA, NAHARKATIYA FIELD
B 6402 213 INDIA, NEOGENE
B 6203 443 INDIA, NORTH KADI FIELD
B 6203 453 INDIA, NORTH KADI OIL FIELD, SECTIONS
B 6402 219 INDIA, OLIGOCENE
B 6402 213 INDIA, PALEOGENE
B 6009 1550† INDIA, PATTERN OF, ACCUMULATION AND COMPOSITION OF, OI IN,
M 30 487† INDIA, PETROLEUM, GEOLOGY OF, BOMBAY HIGH FIELD
B 6203 443 INDIA, PRECAMBRIAN
B 6110 1842 INDIA, PRODUCTION, 1975-1976
B 6210 1990 INDIA, PRODUCTION, 1975-1977
B 6310 1829 INDIA, PRODUCTION, 1977-1978
B 6411 1896 INDIA, PRODUCTION, 1977-1979
B 6203 443 INDIA, QUATERNARY
B 6403 428 INDIA, RUDRASAGAR FIELD
B 6402 209† INDIA, SOURCE AREAS AND, MIGRATION TRENDS OF, OIL AND, GAS IN,
B 6402 218 INDIA, SOUTH BASSEIN FIELD
B 6402 218 INDIA, SOUTH TAPTI FIELD
B 6203 451 INDIA, SURAJ EROSIONAL VALLEYS
B 6402 218 INDIA, TARAPUR FIELD
B 6203 444 INDIA, TARAPUR SHALE
B 6203 443 INDIA, TERTIARY
B 6402 219 INDIA, TERTIARY
B 6209 1588 INDIA-PAKISTAN CONTINENTAL MARGIN
S 4 231† INDIAN CORAL REEFS, CRINOIDS OF, WEST
S 11 82 INDIAN CREEK, CALIFORNIA, CHECK LIST OF, FORAMINIFERA,
B 6405 679* INDIAN CREEK FIELD, FALL RIVER COUNTY, SOUTH DAKOTA, ABST.
S 4 231 INDIAN CRINOIDS, WEST
M 27 189 INDIAN MEADOWS FORMATION, WYOMING, EOCENE
B 6004 727* INDIAN NATIONS TURNPIKE, OUACHITA MOUNTAINS, OKLAHOMA,
B 6004 699* INDIAN OCEAN, ABST., DYNAMIC PROCESSES OF UPPER, BENGAL, FAN
B 6408 1201 INDIAN OCEAN, ANOXIC OPEN OCEAN
M 29 226 INDIAN OCEAN BASIN, RELATIONS, SUMATRA
B 6303 380 INDIAN OCEAN GEOLOGY AND BIOSTRATIGRAPHY. STUDIES FOLLOWING
B 6207 1164 INDIAN OCEAN, KERGUELEN PLATEAU, CRISTOBALITE CHERT
B 6111 1991 INDIAN OCEAN, PIERCEMENT STRUCTURES, SEISMIC PROFILES
B 6111 1994 INDIAN OCEAN, POSSIBLE DIAPIRS
M 25 186 INDIAN OCEANS, CONVERGENT MARGINS, PACIFIC AND
B 6007 1119 INDIAN ROCK, FIELD, TEXAS, GILMER
B 6003 428 INDIAN UNIVERSITY PALEONTOLOGY SEMINAR, SILURIAN REEF
B 6009 1618* INDIANA, ABST., GENERAL SEDIMENTOLOGIC HISTOR OF LATE, SILURIAN
B 6408 1280* INDIANA, ABST., GEOLOGY OF, WHEATON CONSOLIDATED OIL FIELD,
B 6405 739* INDIANA, ABST., POROSITY EVOLUTION OF, NIAGARAN, PIPE CREE JR.,
B 6405 739* INDIANA, ABST., POROSITY EVOLUTION OF, NIAGARAN, PIPE CREE JR.,
B 6009 1624* INDIANA, ABST., PRECAMBRIAN, GEOPHYSICAL PROVINCES IN
B 6004 687* INDIANA, ABST., PRODELTA, TURBIDITE-, FAN, APRON IN, BORDE
B 6408 1286* INDIANA, ABST., ST. LOUIS AND, SALEM STRATIGRAPHY AND OIL
B 6408 1283* INDIANA, ABST., STRUCTURAL INTERPRETATION OF BURIED, SILURIAN,
B 6405 790* INDIANA, ABST., SUBMARINE CARBONATE CEMENTATION AND, PISOLITH
B 6309 1582* INDIANA, ABST., SUBMARINE, SLUMPING IN, REEF-FLANK DEPOSITS,
B 6403 400† INDIANA, ANALYSIS OF, GRAVITY, ANOMALY OVER, CORAL-, REEF OIL
B 6009 1623* INDIANA AND PARTS OF ILLINOIS, KENTUCKY AND OHIO, ABST.,
B 6408 1284* INDIANA AND POTENTIAL GEOLOGIC PROBLEMS IN THEIR
B 6401 129 INDIANA, ANTRIM SHALE
B 6401 127 INDIANA, DELPHI REEF
B 6008 1224 INDIANA, DEVELOPMENTS, 1975
B 6108 1186 INDIANA, DEVELOPMENTS, 1976
B 6208 1342 INDIANA, DEVELOPMENTS, 1977
B 6308 1325 INDIANA, DEVELOPMENTS, 1978
B 6409 1464 INDIANA, DEVELOPMENTS, 1979
B 6401 128 INDIANA, DEVONIAN
B 6403 401 INDIANA, DEVONIAN
B 6401 128 INDIANA, DEVONIAN
B 6403 401 INDIANA, DEVONIAN
B 6401 130 INDIANA, ECHINODERMAL DEBRIS ZONES
B 6401 130 INDIANA, GASTROPOD ZONES
B 6401 128 INDIANA GEOLOGICAL SURVEY, CORES, LOGS

B 6401 129 INDIANA, LIMBERLOST DOLOMITE
B 6405 796* INDIANA LIMESTONE AND YULE MARBLE, ABST., NATURE OF
B 6309 1479 INDIANA LIMESTONE, TEXTURE AND COMPOSITION DATA
S 5 96 INDIANA, LIMBERLOST DOLOMITE
B 6401 128 INDIANA, LOUISVILLE LIMESTONE
B 6401 129 INDIANA, MAQUOKETA GROUP
B 6403 404 INDIANA, MISSISSIPPIAN
M 27 90 INDIANA, MISSISSIPPIAN, SALEM LIMESTONE
M 27 163 INDIANA, MISSISSIPPIAN, STE. GENEVIEVE LIMESTONE
B 6401 129 INDIANA, ORDOVICIAN
B 6401 130 INDIANA, PENTAMERID ZONES
B 6008 1224 INDIANA, PRODUCTION, 1975
B 6108 1186 INDIANA, PRODUCTION, 1976
B 6208 1342 INDIANA, PRODUCTION, 1977
B 6308 1326 INDIANA, PRODUCTION, 1978
B 6409 1464 INDIANA, PRODUCTION, 1979
B 6401 129 INDIANA, SALAMONIE DOLOMITE
S 5 96 INDIANA, SALAMONIE DOLOMITE
B 6401 129 INDIANA, SALINA FORMATION, KENNETH LIMESTONE MEMBER
B 6401 128 INDIANA, SALINA FORMATION, KOKOMA LIMESTONE MEMBER
B 6401 129 INDIANA, SEXTON CREEK LIMESTONE
B 6401 128 INDIANA, SILURIAN
B 6401 121 INDIANA, SILURIAN REEF COMPLEX, FOSSILS
S 5 2 INDIANA, SILURIAN, NIAGARA GROUP
B 6401 115† INDIANA, STRATIGRAPHY, STRUCTURE, AND, ZONATION OF LARGE,
B 6401 129 INDIANA, STROMATOPOROID-CORAL ZONES
B 6403 400 INDIANA, SULLIVAN COUNTY, WILFRED POOL
B 6401 128 INDIANA, TRAVERSE FORMATION
B 6401 118 INDIANA, TRENTON LIMESTONE
B 6004 684 INDIANA UNIVERSITY PALEONTOLOGY SEMINAR,, SEARCH FOR
B 6003 428 INDIANA UNIVERSITY PALEONTOLOGY SEMINAR: SILURIAN REEF COMPLEX
B 6401 115 INDIANA UNIVERSITY PALEONTOLOGY SEMINAR, STRATIGRAPHY,
B 6401 128 INDIANA, WABASH FORMATION
S 5 99 INDIANA, WABASH FORMATION, MISSISSINEWA SHALE
B 6401 128 INDIANA, WALDRON FORMATION
B 6403 400 INDIANA, WILFRED POOL, SULLIVAN COUNTY
B 6012 2181* INDIANS RANCH, MONTEREY COUNTY, CALIFORNIA, ABST., TERTIARY
B 6405 727* INDIA—TECTONO-STRATIGRAPHIC FACIES, ABST., BASINS OF EAST
B 6103 467* INDICATOR OF PLATE TECTONICS IN SOUTHERN ALASKA AND ADJACENT
B 6004 731* INDICATORS OF CLIMATE IN CARBONATE DUNE ROCKS, ABST.
B 6410 1681† INDIGENOUS PRECAMBRIAN PETROLEUM REVISITED
B 6203 547* INDIGENOUS PRECAMBRIAN PETROLEUM REVISITED, ABST.
B 6109 1543* INDIGENOUS PRECAMBRIAN PETROLEUM REVISITED, ABST.
M 26 188 INDIRECT RECOGNITION, SEISMIC, REFLECTIONS, CARBONATES
B 6007 1015 INDIVIDUAL CHANNELS, SEQUENCE
B 6403 360 INDO-AUSTRALIAN PLATE, BOUNDARY
S 4 76 INDO-PACIFIC AND CARIBBEAN, MAJOR REEFS AND ATOLLS
B 6404 467 INDO-PACIFIC, MARINE PHREATIC ARAGONITE CEMENT
M 30 443 INDOENSIA, KUTAI BASIN, TECTONIC FRAMEWORK
B 6110 1836 INDONESIA, 1976, SWEENEY, PAT, ON DEVELOPMENTS IN
B 6210 1984 INDONESIA, 1977, SWEENEY, P. J., ON DEVELOPMENTS IN
B 6310 1821 INDONESIA, 1978, JAUNCEY, W., AND, ZELIFF, C. W., ON DEVELOPMENTS IN
B 6411 1887 INDONESIA, 1979, JAMES, R., AND W., WILLIAMS, ON DEVELOPMENTS IN
B 6011 2070* INDONESIA ABST., CONCEPTS AND TECHNIQUES UTILIZED IN
B 6207 1224* INDONESIA, ABST., CORRELATION OF MAJOR, UNCONFORMITIES IN,
B 6303 470* INDONESIA, ABST., DISCOVERY AND DEVELOPMENT OF, BADAK FIELD,
B 6207 1207* INDONESIA, ABST., EXPLORATION AND DEVELOPMENT OF, GEOTHERMAL
B 6207 1236* INDONESIA, ABST., GEOTHERMAL, WELL COMPLETION AND, CASING
B 6303 545* INDONESIA, ABST., HANDIL FIELD, EAST, KALIMANTAN
B 6212 2509* INDONESIA, ABST., RECENT DEVELOPMENT OF, KAMOJANG,
B 6207 1230* INDONESIA, ABST., TERTIARY, CARBONATE, RESERVOIRS IN
M 25 460 INDONESIA, AND NEW CALEDONIA, NICKEL AND COBALT, PHILIPPINES
M 25 241 INDONESIA, ARAFURA PLATFORM OR SAHUL SHELF
M 30 441 INDONESIA, BADAK FIELD, DISCOVERY HISTORY
M 30 453 INDONESIA, BADAK FIELD, STRUCTURAL CROSS SECTION
M 25 358 INDONESIA, BANDUNG VOLCANIC AREA
M 25 247 INDONESIA, BASINS, CLASSIFICATION
B 6207 1131 INDONESIA, BATANTA ISLAND
M 30 464 INDONESIA, BEKAPAI FIELD, FRESHWATER SANDSTONE
M 30 468 INDONESIA, BEKAPAI FIELD, RESERVOIR CHARACTERISTICS
M 30 463 INDONESIA, BEKAPAI FIELD, STRUCTURAL GEOLOGY
M 25 244 INDONESIA, CENTRAL AND SOUTH SUMATRA BASINS
M 25 241 INDONESIA, CENTRAL IRIAN UNSTABLE SHELF AREA, CENTRAL IRIAN
M 25 460 INDONESIA, CHROMITE, PHILIPPINES AND
M 25 85 INDONESIA, COAL DISTRIBUTION
M 25 86 INDONESIA, COAL RESERVES
M 25 187 INDONESIA, CONVERGENT MARGIN
M 30 441† INDONESIA, DISCOVERY AND DEVELOPMENT OF THE, BADAK FIELD,
M 25 246 INDONESIA, EAST JAVA BASIN
B 6112 2066 INDONESIA, EASTERN, AUSTRALIAN CRATON
B 6112 2075 INDONESIA, EASTERN, BANDA ARCS, TECTONIC MODELS
B 6112 2066 INDONESIA, EASTERN, METAMORPHIC BELT
B 6112 2066 INDONESIA, EASTERN, MIOCENE, MIDDLE, OROGENIC BELT
B 6112 2066 INDONESIA, EASTERN, PLIOCENE, LATE, OROGENIC BELT
B 6112 2065 INDONESIA, EASTERN, TECTONIC MAP
B 6112 2066 INDONESIA, EASTERN, VOLCANIC ARC
M 30 399† INDONESIA, GEOLOGY OF THE, HANDIL FIELD, EAST KALIMANTAN
B 6112 2063† INDONESIA, GEOSYNCLINES AND, PLATE TECTONICS IN, BANDA ARCS,
M 25 135 INDONESIA, GEOTHERMAL EXPLORATION
M 25 357 INDONESIA, GROUNDWATER POTENTIAL, VOLCANICLASTIC ROCKS
M 30 419 INDONESIA, HANDIL FIELD, GEOCHEMISTRY
M 30 414 INDONESIA, HANDIL FIELD, RESERVOIR 19-7
M 30 413 INDONESIA, HANDIL FIELD, RESERVOIR CHARACTERISTICS
M 30 402 INDONESIA, HANDIL FIELD, STRUCTURE
M 25 246 INDONESIA, IRIAN JAYA, BASINS, EAST
M 25 246 INDONESIA, KUTAI BASIN
M 30 460 INDONESIA, KUTEI BASIN, BEKAPAI FIELD
M 30 400 INDONESIA, KUTEI BASIN, LOCATION MAP
B 6307 1021 INDONESIA, LONGITUDINAL WRENCH FAULTS, WESTERN
M 30 408 INDONESIA, MAHAKAM DELTA, DEPOSITIONAL ENVIRONMENTS
M 25 242 INDONESIA, MOLUCCAS GEOLOGIC PROVINCE
M 25 359 INDONESIA, NGANJUK-KERTOSONO VOLCANIC AREA
M 25 244 INDONESIA, NORTH SUMATRA BASIN
M 25 243 INDONESIA, NORTH SUMATRA BASIN
M 25 244 INDONESIA, NORTHWEST JAVA BASIN AREA
B 6009 1448† INDONESIA, OIL-BEARING, REEFS IN, SALAWATI BASIN OF, IRIAN JAYA
B 6110 1856 INDONESIA, PRODUCTION, 1975-1976
B 6210 1990 INDONESIA, PRODUCTION, 1975-1977
B 6310 1829 INDONESIA, PRODUCTION, 1977-1978
B 6411 1896 INDONESIA, PRODUCTION, 1977-1979
M 25 239† INDONESIA, RECENT KNOWLEDGE OF, HYDROCARBON POTENTIALS IN
B 6401 38 INDONESIA, SEBUKU PLATFORM
M 25 242 INDONESIA, SORONG FAULT ZONE
B 6401 37† INDONESIA, STRUCTURE, STRATIGRAPHY, AND HYDROCARBONS,
M 25 239 INDONESIA, TECTONIC FEATURES
B 6207 1127† INDONESIA, TERTIARY TECTONIC HISTORY OF, SALAWATI AREA, IRIAN
M 25 137 INDONESIA, VOLCANIC GEOTHERMAL POTENTIAL
M 25 357 INDONESIA, VOLCANOES
M 25 136 INDONESIA, VOLCANOES, ACTIVE
M 25 357 INDONESIA, YOGYAKARTA VOLCANIC AREA
M 25 84† INDONESIA--SUMMARY, COAL RESOURCES AND POTENTIAL IN
M 25 356† INDONESIA--SUMMARY, GROUNDWATER POTENTIAL OF, LIMESTONE
B 6004 709* INDONESIA-SOUTH SUMATRA AND JAV SEA BASINS--MECHANISM OF
B 6302 249† INDONESIA, DISCUSSION AND REPLY, GEOSYNCLINES AND, PLATE
B 6406 903 INDONESIAN ISLAND INNER VOLCANIC ARC
M 25 135† INDONESIAN VOLCANIC AREAS--SUMMARY, POSSIBLE, ENERGY SOUUCES
M 25 246 INDONNSIA, EAST KALIMANTAN BASINS
M 25 357† INDONNSIA--SUMMARY, GROUNDWATER POTENTIAL O AREAS
B 6303 495* INDU SUBMARINE FAN, ABST., DOWNCURRENT CHANGE IN, MIOCENE
M 25 120 INDUCTION OF SEISMIC ACTIVITY FROM FLUID REINJECTION
B 6203 542* INDUS CONE, OFFSHORE PAKISTAN, ABST., ANCIENT, SUBMARINE
B 6405 734* INDUS FAN, ABST., LITHOLOGY AND STRUCTURES OF, QUATERNARY
B 6004 720* INDUSTRY, ABST., WOMEN, GEOSCIENTISTS IN PETROLEUM
B 6009 1617* INDUSTRY AND COASTAL ZON MANAGEMENT IN LOUISIANA, ABST., THE,
B 6207 1221* INDUSTRY—PRACTICES AND PROBLEMS, AUSTRALIAN SCENE, ABST.,
B 6209 1585 INERTINITE, DEFINITION
M 29 76 INFERRED BASIN STRUCTURE, ATLANTIC COAST
B 6004 528† INFERRING STRATIGRAPHY FROM SEISMIC DATA
B 6004 725* INFILL AS MAJOR FACTOR IN SUBMARINE-CANYON AND FAN-VALLEY
B 6004 675* INFLUENC OF SOURCE AND PROCESSES IN DISTRIBUTION OF HEAVY
B 6303 550* INFLUENCE OF BASEMENT TECTONICS ON DEPOSITIONAL SYSTEMS AND
B 6205 886* INFLUENCE OF DIAGENETIC AND POSTLITHIFICATION PROCESSES ON
B 6305 832* INFLUENCE OF DIAGENETIC REACTIONS ON NONMARINE UPPER
B 6203 560* INFLUENCE OF PLATFORM-DERIVED SEDIMENT ON FACIES, DIAGENESIS,
B 6303 421* INFLUENCE OF RATE ON DEMAND AND USE OF NATURAL GAS IN
B 6001 12† INFLUENCE OF RECURRENT LARAMIDE STRUCTURAL GROWTH ON
M 25 98† INFLUENCE OF TEMPERATURE ON COALIFICATION OF TERTIARY COAL
B 6303 550* INFLUENCE OF TRANSCONTINENTAL ARCH O CRETACEOUS
B 6004 653* INFORMATION RETRIEVAL FROM MEGA-BYTE DATA BASES, ABST.,
B 6105 769 * INFORMATION RETRIVAL SYSTEM FOR WELL DATA IN INDIANA
B 6405 602 INFRA-CAMBRIAN SALT, MIDDLE EAST
S 11 8 INFRA-SERIES SUBDIVISIONS, CALIFORNIA
B 6312 2223 INFRARED ANALYSES, KEROGEN
B 6108 1385* INFRARED SURVEYS OF URANIUM EXPLORATION IN SOUTH TEXAS,
B 6405 650 INGERSOLL, LEE A. WOODWARD, AND H. L. JAMES, SCHILLING, JOHN,
B 6012 2179 INGERSOLL, RAYMOND V., APPLICATIONS OF PLATE TECTONICS TO
B 6012 2182 INGERSOLL, RAYMOND V., LATE CRETACEOUS TURBIDITE SEDIMENTATIO
B 6211 2353 INGERSOLL, RAYMOND V., PLATE-TECTONIC EVOLUTION OF
B 6405 727 INGERSOLL, RAYMOND V., TEMPORAL CHANGES IN DEPOSITIONAL
B 6303 472 INGERSON, EARL, GEOCHEMISTRY OF SOUTH TEXAS URANIUM DEPOSITS,
B 6403 444 INGLE, JAMES C., BENTHIC FORAMINIFERAL BIOFACIES OF EASTERN

B 6303 473 INGLE, JAMES C., JR., CORRELATION OF CONTINENTAL-MARGIN AND
B 6012 2182 INGLE, JAMES C., JR., LATE NEOGENE PLAEOBATHYMETRY AND
B 6207 1222 INGLE, JAMES C., JR., ORIGIN, DISTRIBUTION, AND DIAGENETIC
B 6404 528 INGLEFIELD LAND, GREENLAND
B 6202 255 INGLEWOOD ZONE, CALIFORNIA, NEWPORT-
B 6104 494 INGLIS FORMATION, FLORIDA
S 10 73 INHIBITED COMPACTION, MIGRATION
B 6104 483 INHIBITION OF DIAGENESIS BY HYDROCARBONS
B 6002 253 INHIBITION OF DOLOMITE GROWTH
B 6203 374 INISKIN PENINSULA, ALASKA
B 6002 246 INJECTION AND EJECTION CURVES, MERCURY-
B 6405 727 INKOLLU, S. N. MURTY, BASINS OF EAST INDIA—
B 6211 2353* INLIER, CENTRAL OREGON, ABST., MESOZOIC PALEOGEOGRAPHY AND
B 6411 1962 INNER ALPINE VIENNA BASIN, AUSTRIA
B 6110 1638 INNER ALPINE VIENNA BASIN, AUSTRIA
B 6310 1917 INNER ALPINE VIENNA BASIN, AUSTRIA
B 6210 1908 INNER ALPINE VIENNA BASIN, AUSTRIA
B 6002 164 INNER BASIN, BERINGIAN SHELF
B 6011 2019† INNER CONTINENTAL SHELF OFF NORTHERN FLORIDA, NEOGENE
B 6203 430 INNER HEBRIDES BASIN, BRITAIN, WEST
M 29 376 INNER SULU SEA, OCEANIC BASIN, SOUTHEAST ASIA
M 25 454 INNER ZONE, JAPAN
S 4 17 INNER-SHELF PROVINCE, PUERTO RICO
B 6104 596 INNISFAIL FIELD, ALBERTA
B 6205 784 INNISFAIL FIELD, SEISMIC DATA, ALBERTA
M 26 197 INNISFAIL, REEF BUILDUP, SEISMIC EXAMPLE, CANADA, ALBERTA
B 6312 2194 INNKJEGLA MEMBER, SVALBARD
M 25 436 INO-CAPAYANG COPPER PROPERTY, MARINDUQUE ISLAND
M 27 82 INOCERAMUS
B 6110 1837 INOMATA, T., ON DEVELOPMENTS IN JAPAN, 1976
B 6210 1985 INOMATA, T., ON DEVELOPMENTS IN JAPAN, 1977
B 6004 729* INORGANIC AND ORGANIC METAMORPHISM IN SILICICLASTIC ROCKS,
B 6309 1584* INORGANIC GEOCHEMICA STUDIES OF EASTERN KENTUCKY GAS FIELD
M 26 453 INPUT PARAMETERS, SEISMIC, MODELING
B 6004 692* INSOLUBLE ORGANIC MATTER ASSOCIATED WITH URANIUM ORES, ABST.,
M 26 302 INSTANTANEOUS PHASE, SEISMIC APPLICATION
B 6203 529* INSTRUMENTATION FOR EVALUATION OF IN SITU RECOVERY PROJECTS.
B 6206 1092* INSTRUMENTATION FOR EVALUATION OF IN-SITU RECOVERY PROJECTS,
B 6401 33 INSUBRIC FAULT ZONE, ITALY
M 25 503 INSULAR BELT, CANADIAN CORDILLERA
M 25 509 INSULAR BELT OR ALEXANDER TERRANE, CORDILLERAN OROGEN
M 29 46 INSULAR MARGIN, STRUCTURAL DIAGRAM, ICELAND
M 29 49 INSULAR MARGIN, THIRTEEN PROFILES, ICELAND
M 29 43† INSULAR MARGIN: THIRTEEN GEOPHYSICAL PROFILES, EROSIONAL AND,
M 29 47 INSULAR RISE, EROSION, ICELAND
M 29 47 INSULAR SLOPE, ICELAND
B 6211 2356* INTEGRATED EXPLORATION, ABST.
B 6405 732* INTEGRATED GEOLOGIC AND GEOPHYSICAL STUDIES AS EXPLORATION
B 6207 1240* INTEGRATING ENVIRONMENTAL ASSESSMENT INTO
B 6207 1226* INTEGRATING ENVIRONMENTAL CONSIDERATIONS IN DEVELOPMENT
B 6305 845* INTEGRATION OF AIRBORNE, MAGNETI AND GAMMA-RAY
B 6109 1559* INTEGRATION OF LANDSAT AND GEOPHYSICAL DATA IN
B 6303 476* INTEGRATION OF MICROSCOPIC ORGANIC ANALYSIS AND GEOCHEMICAL
B 6004 713* INTEGRATION OF SEDIMENTOLOGY, PALEONTOLOGY, AND STRUCTURAL
B 6203 527* INTERACTING DYNAMICS OF PRESSURE, TEMPERATURE, AND WATER
B 6303 486* INTERACTION BETWEEN OVERWASH AND EOLIAN PROCESSES ON
B 6203 346 INTERACTION OF PACIFIC FARALLON AND NORTH AMERICAN PLATES
B 6309 1583* INTERACTION OF PROPPANTS WITH CRACK FORMATION AND
S 10 109† INTERACTION, SOME, ECONOMIC ASPECTS OF, WATER-ROCK
S 10 169† IN-TRANSIT CONVERSION OF ORGANIC COMPOUNDS IN WATER, ORIGIN
S 10 172 IN-TRANSIT GENERATION, ESTIMATION, MIGRATION
S 10 173 IN-TRANSIT GENERATION, MODEL, MIGRATION
B 6004 703* INTERACTIVE COMPUTER PROGRAM TO ANALYZE COAL-ENERGY
B 6203 488* INTERACTIVE GRAPHICS IN SUBSURFACE STRATIGRAPHIC ANALYSIS,
B 6303 476* INTERACTIVE MODELING AND INTERPRETATION OF TWO-DIMENSIONAL
B 6405 629† INTERBIOHERM DEPOSITS OF WESTERN WEST VIRGINIA, NIAGARAN,
B 6002 249 INTERBOUNDARY-SHEET PORES
B 6108 1377* INTERDISCIPLINARY ANALYSIS OF SOME POTENTIAL PETROLEUM
M 26 6 INTERFERENCE, SEISMIC, REFLECTIONS
S 4 289† INTERGLACIAL CHANGES IN TEMPERATURE AND ISOTOPIC
B 6004 673* INTERGLACIAL STORM-MODIFIED BARRIER-ISLAND FEATURES, ABST.,
M 28 166 INTERGRANULAR SOLUTION, OF FOSSILS
B 6109 1537* INTERIOR STRUCTURE OF OAK GROVE RIDGE CHENIER, LOUISIANA,
B 6201 81 INTERLAKE CORRELATION, SASKATCHEWAN
B 6403 442* INTERMARGIN OF SOUTHERN CALIFORNIA BORDERLAND—
B 6403 442* INTERMARGIN OF SOUTHERN CALIFORNII BORDERLAND--QUATERNARY
B 6203 436 INTERMEDIATE BLOCK, NORTH SEA
M 25 500 INTERMONTANE BELT, CANADA
M 25 503 INTERMONTANE BELT, CANADIAN, CORDILLERA
B 6004 684* INTERMONTANE DELTA, ABST., STRATIGRAPHY AND SEDIMENTATION
B 6311 2042† INTERMONTANE, LACUSTRINE DELTA, CATATUMBO RIVER, LAKE
B 6108 1204 INTERMOUNTAIN AREA, GEOTHERMAL ACTIVITY, 1976, FOUR CORNERS-
B 6208 1361 INTERMOUNTAIN AREA, GEOTHERMAL ACTIVITY, 1977, FOUR CORNERS—
B 6008 1239† INTERMOUNTAIN AREA IN 1975, DEVELOPMENTS IN, FOUR CORNERS--
B 6108 1204† INTERMOUNTAIN AREA IN 1976, DEVELOPMENTS IN, FOUR CORNERS-
B 6208 1360† INTERMOUNTAIN AREA IN 1977, DEVELOPMENTS IN, FOUR CORNERS—
B 6308 1206† INTERMOUNTAIN AREA IN 1978, DEVELOPMENTS IN, FOUR CORNERS--
B 6409 1340† INTERMOUNTAIN AREA IN 1979, DEVELOPMENTS IN, FOUR CORNERS—
B 6105 799 * INTERNAL STRUCTURE OF LOWER CAMBRIAN ARCHAEOCYATHID REEFS OF
B 6004 708* INTERNAL STRUCTURES IN DEEP ESTUARY CHANNEL SYSTEM, WILLAPA
B 6405 696* INTERNAL STRUCTURES OF SHALLOW-MARINE TIDAL SAND WAVES,
B 6109 1542* INTERNAL WAVES AND NEPHELOID LAYER ON CONTINENTAL SHELF,
B 6105 802 * INTERNAL WAVES—IMPLICATIONS FOR DISPERSAL OF SUSPENDED
B 6002 196 INTERNATIONAL DECADE OF OCEAN EXPLORATION AND REMAC
M 25 539 INTERNATIONAL DECADE OF OCEAN EXPLORATION PROGRAM
S 6 1 INTERNATIONAL GEOLOGICAL CONGRESS, 25TH, SYDNEY, AUSTRALIA,
S 6 33† INTERNATIONAL GEOCHRONOLOGIC SCALE, STRATOTYPES AND AN
B 6303 287 INTERNATIONAL PHASE OF OCEAN DRILLING,, JOIDES, DEEP SEA
B 6205 794 INTERNATIONAL PHASE OF OCEAN DRILLING SITE-LOCATION SURVEY
B 6309 1587* INTERPRETATION AND STATISTICAL ANALYSIS OF KENTUCKY AND WEST
M 26 18 INTERPRETATION, AGE-VELOCITY-DEPTH ASPECTS, SEISMIC
M 26 15† INTERPRETATION, ASPECTS OF, ROCK, PHYSICS FROM, LABORATORY
M 26 175 INTERPRETATION, BASIN-FLOOR FACIES, SEISMIC
M 26 175 INTERPRETATION, BASIN-SLOPE FACIES, SEISMIC
M 26 183 INTERPRETATION, CHAOTIC-FILL FACIES, SEISMIC
M 26 304 INTERPRETATION, COLOR DISPLAYS, EXAMPLES, SEISMIC
M 26 166 INTERPRETATION, DEPOSITIONAL FRAMEWORK, REFLECTION
M 26 165 INTERPRETATION, DEPOSITIONAL SETTINGS, SEISMIC
M 26 136 INTERPRETATION, FACIES, ANALYSIS
M 26 417† INTERPRETATION, GEOLOGIC CONSIDERATIONS FOR, STRATIGRAPHIC,
M 29 13 INTERPRETATION, GEORGIA EMBAYMENT, GEOLOGIC
M 26 169 INTERPRETATION, HIGH-AMPLITUDE FACIES, SEISMIC
M 26 169 INTERPRETATION, HIGH-CONTINUITY FACIES, SEISMIC
M 26 170 INTERPRETATION, LOW-AMPLITUDE FACIES, SEISMIC
M 26 170 INTERPRETATION, LOW-CONTINUITY FACIES, SEISMIC
M 26 177 INTERPRETATION, MOUNDED FACIES, SEISMIC
M 26 180 INTERPRETATION, MOUNDED ONLAPPING-FILL FACIES, SEISMIC
M 26 173 INTERPRETATION, OBLIQUE-PROGRADATIONAL FACIES, SEISMIC
B 6011 2068* INTERPRETATION OF CARBONATE-ROCK SEQUENCES--SEDIMENTARY OR
M 26 165† INTERPRETATION OF CLASTIC DEPOSITIONAL FACIES, SEISMIC,
B 6009 1615* INTERPRETATION OF COTTON VALLEY DEPOSITIONAL ENVIRONMENT
B 6004 685* INTERPRETATION OF DEPOSITIONAL ENVIRONMENTS, POROSITY
M 26 213† INTERPRETATION OF DEPOSITIONAL SYSTEMS: EXAMPLES FROM
B 6405 792* INTERPRETATION OF FOSSIL FLUVIAL BIVALVE BURROWS IN CATSKILL
B 6303 439* INTERPRETATION OF PLEISTOCENE COASTAL-BARRIER COMPLEXES
B 6105 759 * INTERPRETATION OF POSTDEPOSITIONAL PROCESSES RELATED TO
M 26 117† INTERPRETATION OF SEISMIC REFLECTION PATTERNS IN DEPOSITIONAL
B 6203 515* INTERPRETATION OF TECTONIC AND SEDIMENTATION RATES FROM
B 6405 789* INTERPRETATION OF THORIUM TO URANIUM RATIO IN GRANITIC
B 6009 1609* INTERPRETATION OF UNENHANCED LANDSAT IMAGERY FOR WETLAND
M 26 177 INTERPRETATION, ONLAPPING-FILL FACIES, SEISMIC
M 26 18 INTERPRETATION, PERMEABILITY ASPECTS, SEISMIC
M 26 477† INTERPRETATION, PRACTICAL, STRATIGRAPHIC, MODELING AND
M 26 186 INTERPRETATION PROCEDURE, CHART, SEISMIC, REFLECTIONS
M 26 500 INTERPRETATION PROCEDURE, SEISMIC, MODELING
M 26 135† INTERPRETATION PROCEDURE, SEISMIC, STRATIGRAPHY AND, GLOBAL
M 26 172 INTERPRETATION, PROGRADED-SLOPE FACIES, SEISMIC
M 26 249 INTERPRETATION, SEISMIC DATA
M 26 118 INTERPRETATION, SEISMIC FACIES, PARAMETERS
M 26 138 INTERPRETATION, SEISMIC, REFLECTIONS, PATTERNS, STRATIGRAPHIC
M 26 135 INTERPRETATION, SEQUENCE, ANALYSIS
M 26 176 INTERPRETATION, SHEET-DRAPE FACIES, SEISMIC
M 26 169 INTERPRETATION, SHELF FACIES, SEISMIC
M 26 172 INTERPRETATION, SHELF-MARGIN FACIES, SEISMIC
M 26 175 INTERPRETATION, SIGMOID-PROGRADATIONAL FACIES, SEISMIC
M 26 177 INTERPRETATION, SLOPE-FRONT FILL FACIES, SEISMIC
M 26 170 INTERPRETATION, VARIABLE-AMPLITUDE FACIES, SEISMIC
M 26 18 INTERPRETATION, VELOCITY-DENSITY-LITHOLOGY ASPECTS, SEISMIC
M 26 389† INTERPRETATION—GEOPHYSICAL PRINCIPLES AND TECHNIQUES.
M 29 13 INTERPRETIVE CROSS SECTION, GEORGIA EMBAYMENT
B 6309 1600* INTERPRETIVE WELL-LOGGING CONCEPTS SOLVE SOUTH TEXAS
M 26 92 INTERREGIONAL UNCONFORMITIES, SEA LEVEL FLUCTUATIONS
B 6203 554* INTERSTITIAL WATER CHEMISTRY OF JAMAICAN REEF SEDIMENTS—
B 6004 660* INTERTIDAL AND SUBTIDAL STROMATOLITES IN LOCKPORT
B 6405 723* INTERTIDAL ZONATION OF RECENT MICROBIAL ENDOLITHS, BERMUDA,
M 29 65† INTERVAL VELOCITIES OVER OUTER CONTINENTAL SHELF AND UPPER
M 29 66 INTERVAL VELOCITY ANALYSIS, ATLANTIC COAST
M 26 203 INTERVAL VELOCITY, CARBONATES, SEISMIC, REFLECTIONS,
M 30 546 INTISAR D FIELD, CROSS SECTION, LIBYA
M 30 557 INTISAR D FIELD, DEPOSITIONAL HISTORY, LIBYA

M 30 552 INTISAR D FIELD, INTISAR D REEF, SEISMIC SECTION
M 30 544 INTISAR D FIELD, LOCATION, LIBYA
M 30 543 INTISAR D FIELD, PALEOCENE, SIRTE BASIN
M 30 564 INTISAR D FIELD, PRODUCTION HISTORY
M 30 548 INTISAR D FIELD, RESERVOIR CHARACTERISTICS, LIBYA
M 30 564 INTISAR D FIELD, TRAPPING MECHANISM
B 6303 423* INTISAR D OIL FIELD, LIBYA, ABST.
M 30 543† INTISAR D OIL FIELD,LIBYA
M 30 552 INTISAR D REEF, SEISMIC SECTION, INTISAR D FIELD
M 27 124 INTRACLASTS
B 6203 519* INTRALAMINAR LEVELS SUGGEST REVISION OF CONVENTIONAL
M 27 174 INTRAMICRITES
S 7 313† INTRASLOPE BASIN AREAS, NORTHWEST GULF OF MEXICO, HOLOCENE,
S 7 289† INTRASLOPE BASIN IN NORTHWEST GULF OF MEXICO
S 7 319 INTRASLOPE BASINS, GYRE BASIN, NORTHWEST GULF OF MEXICO
S 7 314 INTRASLOPE BASINS, MAP, TEXAS-LOUISIANA SLOPE
S 7 318 INTRASLOPE BASINS, NORTHWEST GULF OF MEXICO
S 7 320 INTRASLOPE BASINS, ORCA BASIN, NORTHWEST GULF OF MEXICO
M 27 178 INTRASPARITE, POORLY-SORTED
M 27 173 INTRASPARITES
B 6203 495* INTRODUCTION OF COMPUTER APPLICATIONS TO IDENTIFY AND
B 6105 803 * **INTRODUCTION OF SYMPOSIUM ON PREDICTIVE MODELS FOR EARTH**
B 6207 1236* INTRODUCTION TO GEOLOGIC REMOTE SENSING, ABST.
B 6405 694* INTRODUCTION TO SEPM RESEARCH SYMPOSIUM—SOURCE BEDS—
B 6209 1567 INTRUSION, INCIPIENT, DIAPIRS, EXTRUSION AND
B 6209 1571 INTRUSIVE DIAPIRISM, CONCEPTUAL, MODEL OF FORCEFULLY
M 28 106 INVERSION, BIMODAL GRAIN SIZE DISTRIBUTION, TEXTURE
B 6405 708* INVERSION PROCESSING TO EXPLORATION FOR POINT-BAR SANDSTONES,
M 28 106 INVERSION, WELL-ROUNDED BUT POORLY SORTED, TEXTURE
B 6311 2117* INVESTIGATION OF DESMOINESIAN ROCKS IN NORTHEASTERN
B 6105 786 * **INVESTIGATION OF FACTORS AFFECTING HYDROCARBON POTENTIAL OF**
B 6405 735* INVESTIGATION OF MEASUREMENTS OF HEAT FLOW REGIME IN OCEANS
M 29 383† INVESTIGATION OF MISSISSIPPI FAN, GULF OF MEXICO
M 28 54 INWOOD LIMESTONE, NEW YORK, PRECAMBRIAN
B 6107 1046 IOLA FORMATION, KANSAS
S 10 41 ION FILTRATION, MIGRATION, AGENTS
B 6303 473 ION, D. C., UNITED STATES PETROLEUM IN WORLD ENERGY PERSPECTIVE,
B 6402 244 IONIAN BASIN, GREECE
M 29 299 IONIAN BASIN, SEDIMENTARY SEQUENCE, MEDITERRANEAN
B 6106 900 IONIAN PLATE
B 6411 1964 IONIAN SEA, GREECE
B 6402 260 IONIAN SEA, MEDITERRANEAN RIDGE
B 6402 261 IONIAN SHELF, MATAPAN TROUGH
B 6402 244 IONIAN ZONE, GREECE
B 6405 683* IOWA, ABST., ASSESSMENT OF POTENTIA FOR FLUID HYDROCARBONS IN
B 6208 1477 IOWA, DEVELOPMENTS, 1977
M 27 55 IOWA, DEVONIAN LIMESTONE
M 27 62 IOWA, DEVONIAN LIMESTONES
B 6402 154 IOWA FIELD, LOUISIANA
B 6311 2104 IOWA LIMESTONE, MISSISSIPPI
B 6012 2177* IPEWICK FORMATION, SIGNIFICANT JURASSIC-NEOCOMIAN MAP UNIT IN
B 6303 546* IPOD DRILLING ON CONVERGENT OCEAN MARGINS, ABST.
B 6303 293 IPOD HOLES, RESERVOIR SANDS, ORGANIC CARBON CONTENT, AND OIL
B 6303 287 IPOD INITIAL REPORTS OF THE DEEP SEA DRILLING PROJECT
M 29 251 IQUIQUE BASIN, PERU-CHILE TRENCH, CONSTRUCTED BASIN
B 6012 2188* IR COLOR PHOTOGRAPHY AND CANYON PHOTOS IN PHOTOGEOLOGIC
B 6010 1866 IRAN, 1975, OIL SERVICE CO. OF IRAN, PHILLIPS PETROLEUM CO.,
B 6110 1796 IRAN, 1976, OIL SERVICE CO. OF IRAN, AMOCO INTERNAT. OIL CO.,
B 6210 2112 IRAN, 1977, OIL SERVICE CO. OF IRAN, AMOCO INTERNATIONAL OIL CO.,
B 6310 1885 IRAN, 1978, OIL SERVICE CO. OF IRAN, ATLANTIC RICHFIELD CO., AND,
B 6411 1837 IRAN, 1979, PETROCONSULTANTS S.A., ON DEVELOPMENTS IN
B 6405 797* IRAN, ABST., GEOLOGIC AND STRUCTURAL HISTORY OF, ZAGROS
B 6303 444* IRAN, ABST., MIOCENE-, PLIOCENE MOLASSE, ZAGROS MOUNTAINS
B 6303 444* IRAN, ABST., MIOCENE-, PLIOCENE SYNDEPOSITIONAL, TAR DEPOSI IN
B 6303 520* IRAN, ABST., RECENT HUGE, GAS DISCOVERIES IN, KANGAN AREA
B 6402 219 IRAN, ALBORZ-5 FIELD
B 6412 2104 IRAN, ASMARI FORMATION
B 6206 991 IRAN, ASMARI LIMESTONE
B 6206 986 IRAN, BANGESTAN GROUP
S 10 200 IRAN, CASPIAN AREA, GARNIARIK TEPE MUD VOLCANO
S 10 200 IRAN, CASPIAN AREA, NAFTLI DJEH MUD VOLCANO
S 10 136 IRAN, CENTRAL PLATEAU, GROUNDWATER FLOW
S 10 139 IRAN, CENTRAL PLATEAU, GROUNDWATER FLOW PATTERN
B 6412 2100 IRAN, CHAMPEH MEMBER
B 6412 2100 IRAN, CHEHEL MEMBER
B 6206 991 IRAN, DASHT-E ARJAN GRABEN
B 6206 984† IRAN, FOLD DEVELOPMENT IN, ZAGROS SIMPLY, FOLDED BELT,
B 6206 985 IRAN, GACHSARAN FIELD
B 6206 993 IRAN, GACHSARAN FORMATION
B 6412 2098 IRAN, GACHSARAN FORMATION
B 6401 14 IRAN, GREATER BURGAN FIELD
B 6206 993 IRAN, GURPI MARL
B 6206 997 IRAN, HORMUZ SALT DIAPIR
M 30 12 IRAN, KANGAN GAS FIELD, LANDSAT IMAGE
B 6401 10 IRAN, KANGAN GIANT GAS FIELD, LANDSAT DATA
B 6401 10 IRAN, KHUFF FORMATION
B 6206 985 IRAN, KUH-E ASMARI ANTICLINE
B 6206 985 IRAN, KUH-E BAMU ANTICLINE
B 6206 985 IRAN, KUH-E DASHTAK ANTICLINE
B 6206 985 IRAN, KUH-E MISH ANTICLINE
B 6206 985 IRAN, KUH-E MUNGASHT ANTICLINE
B 6206 985 IRAN, KUH-E PAHN ANTICLINE
B 6206 985 IRAN, KUH-E SHAH NISHIN ANTICLINE
B 6206 985 IRAN, KUH-E SURMEH ANTICLINE
B 6206 985 IRAN, LALI FIELD
B 6206 985 IRAN, MASJED-E SULEYMAN FIELD
S 10 200 IRAN, MEKRAN COAST, NAPAG MUD VOLCANO
B 6412 2100 IRAN, MOL MEMBER
B 6412 2100 IRAN, NAMAKI FORMATION, THREE MEMBER
B 6010 1866 IRAN NIPPON PETROLEUM CO., AGI S.P.A., AMOCO, AND ATLANTIC
B 6307 1021 IRAN, OBLIQUE WRENCH FAULTS
B 6206 993 IRAN, PABDEH MARL
B 6110 1804 IRAN, PRODUCTION, 1976
B 6210 2120 IRAN, PRODUCTION, 1977
B 6310 1892 IRAN, PRODUCTION, 1978
B 6411 1843 IRAN, PRODUCTION, 1978-1979
B 6412 2100 IRAN, QESHM FORMATION
B 6412 2099 IRAN, RAZAK FORMATION
B 6206 999 IRAN, SALT DIAPIRS AND BUCKLE FOLDING
B 6412 2095† IRAN, STRATIGRAPHY AND ENVIRONMENTAL SEDIMENTOLOG OF
B 6401 10 IRAN, ZAGROS FOOTHILLS THRUST BELT
B 6412 2096 IRAN, ZAGROS GEOSYNCLINE, MAP
B 6412 2098 IRAN, ZAGROS GEOSYNCLINE, STRATIGRAPHIC CHART
B 6206 985 IRAN, ZAGROS MOUNTAINS, BASEMENT
B 6206 987 IRAN, ZAGROS MOUNTAINS, COMPETENT GROUP
B 6206 987 IRAN, ZAGROS MOUNTAINS, INCOMPETENT GROUP
B 6206 987 IRAN, ZAGROS MOUNTAINS, MOBILE GROUP
B 6307 1022 IRAN, ZAGROS OROGENIC BELT
B 6206 986 IRAN, ZAGROS SIMPLY FOLDED BELT, STRATIGRAPHIC COLUMN
B 6004 727† IRAN. BY TITLE ONLY, GEOLOGIC HISTORY AND TECTONICS OF, ZAGROS
B 6003 458† IRAN. DISCUSSION, SALT, DIAPIRISM IN SOUTHERN
B 6008 1379† IRANIAN PETROLEUM INSTITUTE
B 6010 1865 IRAQ, 1975, PETROCONSULTANTS S.A., ON DEVELOPMENTS IN
B 6110 1795 IRAQ, 1976, JAVOGUES, J.-P., ON DEVELOPMENTS IN
B 6210 2112 IRAQ, 1977, JAVOGUES, J.-P., ON DEVELOPMENTS IN
B 6310 1885 IRAQ, 1978, JAVOGUES, J. P., ON DEVELOPMENTS IN
B 6411 1836 IRAQ, 1979, PETROCONSULTANTS S.A., ON DEVELOPMENTS IN
M 30 14 IRAQ, BUZURGAN FIELD, LANDSAT IMAGE
B 6401 12 IRAQ, BUZURGAN GIANT OIL FIELD, LANDSAT DATA
B 6310 1892 IRAQ, DEVELOPMENTS, 1977-1978
B 6401 12 IRAQ, MISHRIF LIMESTONE FORMATION
B 6210 2120 IRAQ, PRODUCTION, 1976-1977
B 6411 1843 IRAQ, PRODUCTION, 1978-1979
B 6401 12 IRAQ, ZAGROS FOOTHILLS THRUST BELT

B 6401 12 IRAQ, ZUBAIR SANDSTONE FORMATION
S 6 104 IRATI FORMATION, BRAZIL, PARANA BASIN
S 6 104 IRATI FORMATION, RUBIDIUM-STRONTIUM, BRAZIL
M 27 184 IRELAND, CRETACEOUS, WHITE LIMESTONE
M 27 160 IRELAND, CRETACEOUS, WHITE LIMESTONE
M 27 51 IRELAND, CRETACEOUS, WHITE LIMESTONE
M 27 205 IRELAND, CRETACEOUS, WHITE LIMESTONE
M 28 19 IRELAND, DEVONIAN, OLD RED SANDSTONE, NORTHERN
B 6211 2191† IRELAND, GROWTH, FAULTS IN, CARBONIFEROUS OF WESTERN
M 27 89 IRELAND, MISSISSIPPIAN, DARTRY LIMESTONE
M 27 77 IRELAND, MISSISSIPPIAN, GLENCAR LIMESTONE
M 27 78 IRELAND, MISSISSIPPIAN, GLENCAR LIMESTONE
M 28 128 IRELAND, MISSISSIPPIAN, LIMESTONE
M 27 101 IRELAND, MISSISSIPPIAN LIMESTONE
M 27 179 IRELAND, MISSISSIPPIAN, WAUL SORTIAN LIMESTONE
M 27 69 IRELAND, MISSISSIPPIAN, WAULSORTIAN LIMESTONE
B 6211 2192 IRELAND, WESTERN, CARBONIFEROUS LIMESTONE
B 6211 2209 IRELAND, WESTERN, CARROWMORE POINT GROWTH FAULTS
B 6211 2194 IRELAND, WESTERN, CENTRAL CLARE GROUP
B 6211 2194 IRELAND, WESTERN, COUNTY CLARE, NAMURIAN
B 6211 2194 IRELAND, WESTERN, DELTAIC CYCLOTHEMS
B 6211 2201 IRELAND, WESTERN, DONEGAL POINT GROWTH FAULTS
B 6211 2209 IRELAND, WESTERN, EMLAGH POINT GROWTH FAULT
B 6211 2204 IRELAND, WESTERN, FREAGH POINT AND GREEN ISLAND, GROWTH
B 6211 2195 IRELAND, WESTERN, KILKEE CYCLOTHEM
B 6211 2203 IRELAND, WESTERN, KILLARD-BLIND HORSES COVE GROWTH FAULTS
B 6211 2192 IRELAND, WESTERN, NAMURIAN BASIN
B 6211 2207 IRELAND, WESTERN, NAMURIAN GROWTH FAULTS
B 6211 2194 IRELAND, WESTERN, NAMURIAN STRATIGRAPHY
B 6211 2209 IRELAND, WESTERN, NORTH GOLEEN GROWTH FAULT
B 6211 2192 IRELAND, WESTERN, OLD RED SANDSTONE
B 6211 2209 IRELAND, WESTERN, PULEEN BAY GROWTH FAULTS
B 6211 2194 IRELAND, WESTERN, SHANNON GROUP
B 6211 2198 IRELAND, WESTERN, SLUMP FEATURES
B 6211 2199 IRELAND, WESTERN, SPANISH POINT GROWTH FAULT
B 6107 988 IRELAND, WHITE LIMESTONE, CHALK UNIT
B 6307 1061 IRENE BAY FORMATION, NORTHWEST TERRITORIES
B 6205 783 IRETON FORMATION, ALBERTA
B 6205 785 IRETON-DUVERNAY SHALE, ALBERTA
B 6207 1142 IRIAN JAYA, A OIL FIELD
M 25 246 IRIAN JAYA, BASINS, EAST, INDONESIA
B 6207 1132 IRIAN JAYA, CRETACEOUS
B 6009 1448† IRIAN JAYA, INDONESIA, OIL-BEARING, REEFS IN, SALAWATI BASIN OF
B 6207 1127† IRIAN JAYA, INDONESIA, TERTIARY TECTONIC HISTORY OF, SALAWATI
B 6207 1132 IRIAN JAYA, KAIS FORMATION
B 6207 1131 IRIAN JAYA, KASIM ISLAND
B 6207 1142 IRIAN JAYA, KASIM OIL FIELD
B 6207 1132 IRIAN JAYA, KLAMONO HIGH
B 6207 1132 IRIAN JAYA, KLASAMAN FORMATION
B 6207 1129 IRIAN JAYA, LENGGURU FOLDBELT
B 6207 1136 IRIAN JAYA, MIOCENE
B 6207 1129 IRIAN JAYA, MISOOL-ONIN ANTICLINORIUM FLEXURE
B 6207 1129 IRIAN JAYA, OETA ANOMALY
B 6207 1144 IRIAN JAYA, RIFT HYPOTHESIS, TESTING
B 6207 1131 IRIAN JAYA, SAILEEN BAY
M 25 248 IRIAN JAYA, SALAWATI BASIN
B 6207 1138 IRIAN JAYA, SALAWATI ISLAND
B 6112 2069 IRIAN JAYA, SECTION
B 6207 1132 IRIAN JAYA, SELE HIGH
B 6207 1129 IRIAN JAYA, SELE STRAIT—MISOOL FAULT
B 6207 1127 IRIAN JAYA, SORONG FAULT ZONE
B 6207 1127 IRIAN JAYA, TARERA-AIDUNA FAULT ZONE
B 6207 1136 IRIAN JAYA, TERTIARY
B 6207 1128 IRIAN JAYA, TRIASSIC
B 6207 1132 IRIAN JAYA, WALIO OIL FIELD
M 25 241 IRIAN MIOGEOSYNCLINE, INDONESIA, CENTRAL, IRIAN UNSTABLE SHELF
M 25 241 IRIAN UNSTABLE SHELF AREA, CENTRAL IRIAN MIOGEOSYNCLINE,
B 6102 296* IRION COUNTY, TEXAS, ABST., BROOKS FIELD, SOUTHEAST, MIDLAND
B 6203 426 IRISH PLATFORM, EUROPE
B 6203 430 IRISH SEA, RIFT SYSTEM
M 30 241 IRKUTSK AMPHITHEATER, PETROLEUM POTENTIAL, USSR
B 6410 1691 IRKUTSK, BASIN, OIL AND GAS FIELDS, USSR
M 25 427 IRO FORMATIONS, WESTERN AUSTRALIA, HAMERSLEY, IRON PROVINCE,
B 6407 995 IROHALENE MUDSTONE MEMBER, MOROCCO
B 6203 553* IRON AND MANGANESE, ABST., CONTROL OF, CATHODO-LUMINESCENCE
M 25 543 IRON, AND MANGANESE, TERNARY DIAGRAM, NAZCA PLATE, SILICON
M 25 531 IRON, ANDEAN REGION
M 28 149 IRON, BIBLIOGRAPHY, DIAGENESIS AND POROSITY
B 6102 257 IRON BOG CREEK FORMATION, IDAHO
M 25 508 IRON DEPOSIT. CANADA. SNAKE RIVER
M 25 496 IRON DEPOSITS, CANADA, RAPITAN GROUP
M 25 411 IRON FORMATION OR TACONITE, BENEFICIABLE
B 6303 529* IRON FORMATIONS, ABST., SEDIMENTOLOGY OF SOME, PRECAMBRIAN
M 25 413 IRON FORMATIONS, MINERALS PRESENT IN, HAMERSLEY BASIN
M 25 411 IRON FORMATIONS OR BIFS, WESTERN AUSTRALIA, ARCHEAN, BANDED
M 25 413 IRON FORMATIONS, STRATIGRAPHIC COLUMNS, HAMERSLEY BASIN
M 25 411 IRON FORMATIONS, WESTERN AUSTRALIA, PROTEROZOIC, BANDED
B 6405 765* IRON FROM RED BEDS OF CATSKILL FORMATION IN EASTERN
M 25 413 IRON, HEMATITE ENRICHMENT ORE, GENESIS OF
M 25 411 IRON, HEMATITE ENRICHMENT ORE, WESTERN AUSTRALIA
M 28 61 IRON MINERALS, BIBLIOGRAPHY
M 25 427 IRON ORE, AUSTRALIA, NORTHERN TERRITORY, SHERWIN IRONSTONE
M 25 391† IRON ORE, COAL, AND STEEL PRODUCTION IN NEW ZEALAND--SUMMARY
M 25 415 IRON ORE, ECONOMICS, WESTERN AUSTRALIA
M 25 411 IRON ORE, GEOLOGY AND DISTRIBUTION, WESTERN AUSTRALIA
M 25 391 IRON ORE PRODUCTION, NEW ZEALAND
M 25 415 IRON ORE, PRODUCTION, WESTERN AUSTRALIA
M 25 427 IRON ORE, QUEENSLAND, CONSTANCE RANGE
M 25 427 IRON ORE, QUEENSLAND, DAWSONVALE
M 25 427 IRON ORE, SOUTH AUSTRALIA, MIDDLEBACK RANGES
M 25 411 IRON ORE, TYPES OF, WESTERN AUSTRALIA
M 25 427 IRON ORE, WESTERN AUSTRALIA, ARCHEAN
M 25 427 IRON ORE, WESTERN AUSTRALIA, YAMPI SOUND
B 6004 725* IRON ORES, ABST., GLAUCONITE-, MAGHEMITE-, HEMATITE
M 28 43 IRON OXIDES, FRAGMENTS
M 25 414 IRON, PISOLITIC LIMONITE ORE, WESTERN AUSTRALIA
M 25 505 IRON, PRODUCTION, CANADIAN, CORDILLERA
M 25 427 IRON PROVINCE, BANDED IRO FORMATIONS, WESTERN AUSTRALIA,
M 25 414 IRON, SEDIMENTARY ORE, WESTERN AUSTRALIA
M 25 444 IRON, SOUTH KOREA
M 25 439 IRON SULFIDE DEPOSITS, BEDDED, TAIWAN, CUPRIFEROUS
M 25 544 IRON, TERNARY DIAGRAM, NAZCA PLATE, MANGANESE, ALUMINUM,
M 25 411 IRON, WESTERN AUSTRALIA, HAMERSLEY BASIN
M 25 411 IRON, WESTERN AUSTRALIA, PILBARA AND, YILGARN BLOCKS
M 25 460 IRON-MANGANESE CONCRETIONS, ORIGIN OF, PACIFIC
M 25 415 IRON-ORE DEPOSITS, EXPLORATION, WESTERN AUSTRALIA
M 25 412 IRON-ORE DEPOSITS, MAP, WESTERN AUSTRALIA
M 25 410† IRON-ORE DEPOSITS OF WESTERN AUSTRALIA--GEOLOGY AND
M 25 416 IRON-ORE PRODUCTION BY COMPANIES, WESTERN AUSTRALIA
M 25 416 IRON-ORE PRODUCTION, FUTURE OUTLOOK, WESTERN AUSTRALIA
M 25 415 IRON-ORE RESERVES, AUSTRALIA
B 6004 722* IRON-RICH AGGLUTINATED FORAMINIFERS IN LATE QUATERNARY
B 6203 498* IRON-RICH CARBONATE CEMENTS IN SANDSTONES—ACTIVE ANKERITE
B 6203 505* IRON-SULFIDE MINERALOGY AND MORPHOLOGY IN OIL SHALE AND
B 6301 90 IROQUOIS FORMATION, SCOTIAN BASIN
B 6004 707* IRPINIDS--MODEL OF TECTONICALLY CONTROLLED CHANNELIZED
B 6303 342 IRVINE--PAINT CREEK FAULT ZONE, KENTUCKY
B 6004 691 IRWIN, D. D., STATISTICAL MAPS AS AID TO GEOLOGIC INTERPRETATION,
B 6307 1096 IRWIN RIVER COAL MEASURES, AUSTRALIA
S 4 145 ISAAC BAY, REEF, ST. CROIX
B 6303 473 ISAACS, CAROLINE M., LATERAL DIAGENESIS IN MONTEREY SHALE,
B 6403 444 ISAACS, CAROLINE M., MONTEREY ROCKS ALONG SANTA BARBARA
B 6308 1204 ISAACS, KALMAN N., DEVELOPMENTS IN ATLANTIC COASTAL PLAIN IN
B 6004 684 ISAACSON, PETER E., DEVONIAN STRATIGRAPHY OF BOLIVIA, ABST.
B 6206 1017 ISACHSEN DOME, ELLEF RINGNES ISLAND
B 6207 1225 ISACKS, BRYAN, SEISMICITY AND STRUCTURE OF NEW HEBRIDES
B 6212 2509 ISHIWADA, YASUFUMI, CONTRAST OF HYDROCARBON POTENTIAL
B 6212 2510 ISHIWADA, YASUFUMI, STRUCTURAL CONTRAST BETWEEN FORE-ARC
B 6104 504 ISHIWATARI, R., COLOR OF KEROGEN AS INDEX OF ORGANIC MATURITY
B 6204 687 ISHIWATARI, R., HYDROCARBON GENERATION BY THERMAL
S 4 195 ISLA CONTOY, MOLLUSCA, MEXICO
S 4 195 ISLA MUJERES, MOLLUSCA, MEXICO
S 3 82 ISLAND ARC
M 25 472 ISLAND ARC, FORMATION IN RELATION TO ENERGY-GENERATING
M 25 498 ISLAND ARC, MINERAL DEPOSITS, CANADIAN, CORDILLERA
M 25 498 ISLAND ARC, ROCKS OF, CANADIAN, CORDILLERA
B 6004 710* ISLAND ARCS AND IT APPLICATION TO PRETROLEUM GEOLOGY, ABST.,
B 6302 259 ISLAND ARCS, DEEP SEA TRENCHES, AND BACK-ARC BASINS, SERIES 1,
M 25 448 ISLAND, ARCS, PACIFIC-WARD SHIFTING, JAPAN
B 6004 669* ISLAND STRATIGRAPHY, ABST., EVOLUTION OF, FLORIDA BAY
S 4 83 ISLAND-ARC SEDIMENTATION, MODEL
M 25 351† ISLANDS--SUMMARY, DEVELOPING, DIKE-COMPLEX AND BASAL-LENS,
B 6107 1080 ISLE LAMOTTE, ORDOVICIAN, VERMONT
B 6401 75 ISLE OF WIGHT, ALBIAN THROUGH SANTONIAN, ENGLAND
S 6 99 ISOCHRONIC MODEL I, EXAMPLE
S 6 98 ISOCHRONIC MODEL I, FINE FRACTION, HAND SPECIMEN
S 6 99 ISOCHRONIC MODEL II, EXAMPLE
S 6 98 ISOCHRONIC MODEL II, WHOLE-ROCK, HAND SPECIMEN
S 6 99 ISOCHRONIC MODEL III, EXAMPLE
S 6 98 ISOCHRONIC MODEL III, FINE FRACTION, LITHOSTRATIGRAPHIC UNIT

S 6 99 ISOCHRONIC MODEL IV, EXAMPLE
S 6 98 ISOCHRONIC MODEL IV, WHOLE-ROCK, LITHOSTRATIGRAPHIC UNIT
S 6 97 ISOCHRONS, SAMPLE, USE OF
B 6112 2143† ISOMETRIC VIEW, FLORIDA, BASEMENT—AN
M 26 251 ISOPACH, GULF OF MEXICO, PLEISTOCENE
M 29 136 ISOPACH MAP, ARGENTINA, OFFSHORE
B 6205 845 ISOPACH MAPPING OF HYDROCARBON RESERVOIRS
B 6012 2150† ISOPACH MAPPING OF HYDROCARBON RESERVOIRS, VERTICAL, NET,
M 26 371 ISOPACH MAPPING, SHALLOW, RESOLUTION SYSTEMS
B 6112 2103 ISOPRENOID-TYPE HYDROCARBON
M 29 157 ISOSTATIC GRAVITY PROFILES, AUSTRALIA, MARGIN
B 6404 551 ISOTHERMS DURING COMPACTION, RELATIVE POSITIONS OF
S 10 71 ISOTHERMS, POSITIONS DURING COMPACTION
B 6304 590 ISOTOPE ANALYSES, GEOCHEMICAL TOOL
B 6004 696* ISOTOPE ANALYSES OF PLEISTOCENE FLUVIAL, ALGAL, PISOLITES AND
B 6401 74 ISOTOPE DATA--EFFECTS OF DIAGENESIS, CARBO ISOTOPE SYSTEM,
B 6401 74 ISOTOPE DATA--EFFECTS OF DIAGENESIS, RELIABILITY OF
S 6 3 ISOTOPE DATING, REVIEW OF PRINCIPLE
B 6405 749* ISOTOPE EVIDENCE FOR MODERN FRESHWATER DIAGENESIS OF
B 6401 67† ISOTOPE FLUCTUATION IN CRETACEOUS PELAGIC
B 6401 67† ISOTOPE FLUCTUATIONS IN CRETACEOUS PELAGIC
B 6303 505* ISOTOPE GEOCHEMISTRY OF ABU DHABI SABKHA, ABST., SULFUR AND,
B 6303 434* ISOTOPE GEOCHEMISTRY OF NATURA GAS IN ILLINOIS BASIN, ABST.
B 6406 956 ISOTOPE GEOLOGY EDITED BY E. JAGER AND J. C. HUNZIKER,
B 6211 2356* ISOTOPE GEOLOGY. BY TITLE ONLY, MESOZOIC PALEOGEOGRAPHY OF,
B 6304 698* ISOTOPE MEASUREMENTS OF HYDROCARBONS ADSORBED IN
B 6401 74 ISOTOPE SYSTEM, CARBON
B 6401 76 ISOTOPE SYSTEM, OXYGEN
B 6401 74 ISOTOPE SYSTEM, RELIABILITY OF, ISOTOPE DATA--EFFECTS OF
B 6401 76 ISOTOPE VARIATIONS, PRIMARY CONTROLS ON, CARBON
B 6303 519* ISOTOPES A PALEOCLIMATOLOGICAL TOOL, ABST., STABLE, CARBON
B 6304 701* ISOTOPES, ABST., BACTERIAL, DEGRADATION OF, CRUDE,
B 6304 701* ISOTOPES, ABST., ORIGIN OF, GASES, ADSORBED IN NEAR-SURFACE
B 6305 824* ISOTOPES AND WHOLE-ROCK MINERALOGIC DATA, ABST., DEFINITION OF
B 6203 489* ISOTOPES AS INDICATORS OF METEORIC AND MIXING-ZONE DIAGENESIS
B 6303 479* ISOTOPES IN OIL-FIELD WATERS AND ORIGIN OF CARBON DIOXIDE,
B 6303 532* ISOTOPES--NEW TOOL IN HYDROCARBON CORRELATION AND
B 6303 532* ISOTOPES. BY TITLE ONLY, ORIGIN OF, GASES ADSORBED IN
S 6 143† ISOTOPIC AGES AND STRATIGRAPHIC CONTROL OF MESOZOIC IGNEOUS
B 6004 658* ISOTOPIC ANALYSIS OF METHANE, ABST., IDENTIFICATION OF, LEAKAGE,
B 6303 485* ISOTOPIC AND TRACE-ELEMENT GEOCHEMISTRY OF DOLOMITE--STATE
B 6303 533* ISOTOPIC COMPOSITION AN SOURCES OF STRONTIUM IN SANDSTONE
B 6109 1554* ISOTOPIC COMPOSITION OF CRETACEOUS, EDWARDS GROUP, CHERT,
S 4 289† ISOTOPIC COMPOSITION OF SEAWATER, USE OF, CORALS IN
B 6303 419* ISOTOPIC COMPOSITIONS OF BIOGENIC CARBONATE CEMENTS, ABST.
S 6 247† ISOTOPIC DATES FOR A PALEOGENE TIME SCALE
S 6 235† ISOTOPIC DATES IN RELATION TO PALEOGENE STRATOTYPES, CRITICAL
S 6 252 ISOTOPIC DATES, PALEOGENE, EOCENE, LUTETIAN STAGE
S 6 323 ISOTOPIC DATING, A REVIEW, CARBONIFEROUS
S 6 166 ISOTOPIC DATING, A REVIEW, QUATERNARY
S 6 342 ISOTOPIC DATING, A REVIEW, SILURIAN
S 6 251 ISOTOPIC DATING, CENOZOIC BOUNDARY, MESOZOIC
S 6 235 ISOTOPIC DATING, CHOICE OF A SPECIMEN, PALEOGENE
S 6 236 ISOTOPIC DATING, CONSIDERATIONS, PALEOGENE
S 6 343 ISOTOPIC DATING, CORRELATION WITH BIOSTRATIGRAPHY, SILURIAN
S 6 236 ISOTOPIC DATING, DECAY OF CONSTANTS, PALEOGENE
S 6 324 ISOTOPIC DATING, DEVONIAN BOUNDARY, CARBONIFEROUS
S 6 251 ISOTOPIC DATING, MESOZOIC BOUNDARY, CENOZOIC
S 6 119† ISOTOPIC DATING METHOD AND ITS APPLICATION TO PHYSICAL
S 6 323† ISOTOPIC DATING OF ROCKS IN THE CARBONIFEROUS SYSTEM, REPORT
S 6 350 ISOTOPIC DATING REVIEW, BRITISH ISLES, ORDOVICIAN
S 6 344 ISOTOPIC DATING, USE OF BENTONITES DISCUSSED, SILURIAN
B 6004 653* ISOTOPIC EVIDENCE BEARING ON ORIGIN OF BAHAMIAN OOIDS, ABST.
B 6203 455† ISOTOPIC EVIDENCE FOR MICROBIOLOGIC ROLE IN GENESIS OF CRUDE
B 6109 1555* ISOTOPIC EVOLUTION OF WHOLE ROCK AND CEMENTS FROM STUART
B 6004 719* ISOTOPIC INVESTIGATION OF STRATIGRAPHIC AND TECTONIC DOLOMITES
S 6 165 ISOTOPIC METHODS, A REVIEW, QUATERNARY
S 6 165† ISOTOPIC METHODS IN QUATERNARY GEOLOGY
M 25 488 ISOTOPIC MODELING
B 6405 756* ISOTOPIC SIGNATURE AS CRITERION FOR INTERPRETING ORIGIN OF
B 6203 488* ISOTOPIC VARIATION IN PLEISTOCENE LACUSTRINE STROMATOLITES
B 6309 1604 ISPHORDING, W. C., CHEMICAL DIFFERENTIATION OF TEMPERATURE
B 6209 1759 ISPHORDING, WAYNE C., MINERALOGIC AND PHYSICAL PROPERTIES OF
B 6009 1611 ISPHORDING, WAYNE C., MULTIVARIATE MINERAL ANALYSIS OF
B 6303 325 ISRAEL, ACHZIV CANYON
B 6303 325 ISRAEL, AKKO EXTENSION
B 6002 258 ISRAEL, ALBIAN-TURONIAN STRATA
B 6404 569 ISRAEL, AMIAZ PLAIN
B 6404 569 ISRAEL, AMORA FORMATION
B 6303 324† ISRAEL AND NORTHERN SINAI, SUBMARINE, SLUMPING IN,
B 6303 325 ISRAEL AND SINAI, GAZA DISTURBANCE
B 6303 326 ISRAEL AND SINAI, SEISMIC PROFILES
B 6205 827† ISRAEL, AS POSSIBLE OIL SOURCE ROCK, ORGANOCHEMICAL
B 6205 828 ISRAEL, ASHQELON AREA, CORES
B 6002 268 ISRAEL, BEIT OREN REEF
B 6404 572 ISRAEL, BENOT LOT SHALE MEMBER
B 6002 268 ISRAEL, CARMEL REGION
B 6303 325 ISRAEL, CARMEL STRUCTURE
B 6205 829 ISRAEL, CENOMANIAN
B 6205 839 ISRAEL, CRETACEOUS
B 6206 951 ISRAEL, CRETACEOUS
B 6404 569 ISRAEL, DEAD SEA RIFT VALLEY
B 6404 568 ISRAEL, DEAD SEA TROUGH
B 6303 325 ISRAEL, DOR DISTURBANCE
B 6001 108† ISRAEL, EARLY, CRETACEOUS, BURIED, CANYON--INFLUENCE ON,
B 6002 268 ISRAEL, EIN SHARQYA REEF
B 6205 829 ISRAEL, EOCENE
B 6002 271 ISRAEL, GEVAR-AM FORMATION
B 6205 828 ISRAEL, GEVARAM CANYON
B 6206 951 ISRAEL, GEVARAM CHANNEL
B 6205 829 ISRAEL, GEVARAM FORMATION
B 6205 829 ISRAEL, HELEZ FORMATION
B 6205 827 ISRAEL, HELEZ OIL FIELD
B 6002 266 ISRAEL, HELEZ REEF
B 6404 572 ISRAEL, HOF SHALE AND SALT MEMBER
B 6010 1905† ISRAEL IN 1975, PETROLEUM, DEVELOPMENTS IN
B 6110 1830† ISRAEL IN 1976, PETROLEUM, DEVELOPMENTS IN
B 6210 1898† ISRAEL IN 1977, PETROLEUM, DEVELOPMENTS IN
B 6310 1743† ISRAEL IN 1978, PETROLEUM, DEVELOPMENTS IN
B 6002 259 ISRAEL, JUDEA LIMESTONE
B 6205 829 ISRAEL, JURASSIC
B 6404 572 ISRAEL, KARBOLET CAPROCK
B 6404 572 ISRAEL, KARBOLET SALT AND SHALE MEMBER
B 6303 324 ISRAEL, LEVANTINE PLATFORM
B 6404 569 ISRAEL, LISAN FORMATION
B 6404 572 ISRAEL, LOT CAPROCK
B 6404 572 ISRAEL, LOT SALT MEMBER
B 6404 572 ISRAEL, MEARAT SEDOM CAPROCK
B 6404 572 ISRAEL, MEARAT SEDOM SALT MEMBER
B 6311 2091 ISRAEL, MEDITERRANEAN MARGIN, SEISMIC SECTION
B 6303 337 ISRAEL, MESSINIAN EVAPORITIC SERIES
B 6002 261 ISRAEL, NAHAL HAMEAROT REEF COMPLEX
B 6205 829 ISRAEL, NEOGENE
B 6002 271 ISRAEL, NEOGENE FAULT SYSTEM
B 6303 325 ISRAEL, PALMAHIM DISTURBANCE
B 6205 839 ISRAEL, PLEISTOCENE
B 6404 569 ISRAEL, PLEISTOCENE
B 6404 569 ISRAEL, PLIOCENE
B 6110 1831 ISRAEL, PRODUCTION, 1976
B 6210 1899 ISRAEL, PRODUCTION, 1976-1977
B 6310 1743 ISRAEL, PRODUCTION, 1978
B 6303 325 ISRAEL, QISHON CANYON
B 6002 258 ISRAEL, REEFS
B 6002 258† ISRAEL, RUDISTID, FRINGING, REEFS OF, CRETACEOUS, SHALLOW,
B 6404 571 ISRAEL, SEDOM COVE AND QUARRY
B 6404 569 ISRAEL, SEDOM FORMATION
B 6205 829 ISRAEL, SENONIAN
B 6002 266 ISRAEL, SOUTHERN CARMEL REEF
B 6404 568† ISRAEL, STRAIN MEASUREMENTS IN EASTERN MARGINAL SHEAR ZONE
B 6002 259 ISRAEL, TALME YAFE FORMATION
B 6002 263 ISRAEL, ZIKHRON REEF
B 6210 1849 ISSAS, BOTSWANA, BURUNDI, MALAWI, RHODESIA, RWANDA,
S 6 1 ISSC, INVOLVEMENT IN TIME SCALES
B 6102 207 ISTHMUS SALT BASIN, MEXICO
B 6010 1712 ITALY, 1975, BONGIORNI, D., AND L., SOGARO, ON DEVELOPMENTS IN
B 6110 1643 ITALY, 1976, BONGIORNI, D., AND R., TOCCO, ON DEVELOPMENTS IN
B 6210 1913 ITALY, 1977, BONGIORNI, D., AND M., PUPILLI, ON DEVELOPMENTS IN
B 6310 1922 ITALY, 1978, BONGIORNI, D., M., MARINELLI, AND M., PUPILLI, ON
B 6411 1964 ITALY, 1979, MINISTERO DELL-INDUSTRIA, DEL COMMERCIO E
B 6303 446* ITALY, ABST., MALOSSA FIELD, DEEP DISCOVERY IN, PO VALLEY
B 6303 465* ITALY, ABST., RESEDIMENTED, CONGLOMERATES IN SUBMARINE
B 6004 701* ITALY, ABST., SEDIMENTOLOGY OF, OUTER-FAN DEPOSITIONAL, LOBES
B 6004 666* ITALY, ABST., SPECTRAL AND, MARKOVIAN ANALYSIS OF BED,
S 5 171 ITALY, ADRIATIC COAST, SANTA MARGHERITA DI SAVOIA, COMMERCIAL
S 6 172 ITALY, BASE OF CALABRIAN
S 6 173 ITALY, CALABRIAN-VILLAFRANCHIAN CORRELATION
S 5 177 ITALY, COMMERCIAL SALINAS
B 6304 591 ITALY, CORTEMAGGIORE FIELD
M 27 143 ITALY, CRETACEOUS, MONTE ANTOLA FORMATION
M 28 73 ITALY, CRETACEOUS, MONTE ANTOLA FORMATION

M 28 152 ITALY, CRETACEOUS, MONTE ANTOLA FORMATION
M 28 38 ITALY, CRETACEOUS, MONTE ANTOLA FORMATION
M 27 185 ITALY, CRETACEOUS, MONTE ANTOLA FORMATION
M 27 200 ITALY, CRETACEOUS, MONTE ANTOLA FORMATION
M 28 22 ITALY, CRETACEOUS, MONTE ANTOLA FORMATION
M 28 45 ITALY, CRETACEOUS, MONTE ANTOLA FORMATION
M 28 162 ITALY, CRETACEOUS, MONTE ANTOLA FORMATION
M 28 155 ITALY, CRETACEOUS, MONTE ANTOLA FORMATION
M 27 148 ITALY, CRETACEOUS, MONTE ANTOLA FORMATION
M 28 156 ITALY, CRETACEOUS, SCHISTI GALESTRINI
B 6401 32 ITALY, DOLOMIA PRINCIPATE
B 6401 68 ITALY, GUBBIO AREA, SAMPLED SECTION
B 6401 33 ITALY, INSUBRIC FAULT ZONE
M 27 177 ITALY, JURASSIC, CALPIONELLID LIMESTONE
M 27 19 ITALY, JURASSIC, CALPIONELLID LIMESTONE
M 27 20 ITALY, JURASSIC, CALPIONELLID LIMESTONE
M 27 201 ITALY, JURASSIC, CARRARA MARBLE
M 27 163 ITALY, JURASSIC, PISOLITIC LIMESTONE
M 28 81 ITALY, JURASSIC, RADIOLARITI
B 6401 32 ITALY, MAIOLICA LIMESTONE
M 30 525 ITALY, MALOSSA FIELD, GEOLOGIC SETTING
M 30 34 ITALY, MALOSSA FIELD, LANDSAT IMAGE
M 30 526 ITALY, MALOSSA FIELD, LOCATION
M 30 537 ITALY, MALOSSA FIELD, RESERVOIR CHARACTERISTICS
M 30 525† ITALY, MALOSSA FIELD--A DEEP DISCOVERY IN THE, PO VALLEY
B 6401 32 ITALY, MALOSSA GIANT GAS-CONDENSATE FIELD, LANDSAT
B 6401 32 ITALY, MALOSSA GIANT GAS-CONDENSATE FIELD, LANDSAT DATA
M 28 86 ITALY, MIOCENE, ERACLEA MINOA MASS FLOW
M 28 43 ITALY, MIOCENE, GESSO SOLFIFERA
M 28 85 ITALY, MIOCENE, GESSO SOLFIFERA
S 6 173 ITALY, NEOGENE-QUATERNARY BOUNDARY
M 28 161 ITALY, OLIGOCENE, TONGRIANO CONGLOMERATE
M 28 31 ITALY, OLIGOCENE, TONGRIANO FORMATION
M 28 87 ITALY, PLEISTOCENE, MONTALLEGRO FORMATION
M 27 88 ITALY, PLIOCENE LIMESTONE
M 27 89 ITALY, PLIOCENE LIMESTONES
S 5 175 ITALY, PLIOCENE, TRUBI FORMATION
M 30 529 ITALY, PO BASIN, CROSS SECTION
M 30 537 ITALY, PO BASIN, DOLOMIA PRINCIPALE FORMATION
M 30 527 ITALY, PO BASIN, PRODUCTION HISTORY
M 30 531 ITALY, PO BASIN, STRATIGRAPHIC COLUMN
M 30 527 ITALY, PO BASIN, STRUCTURAL OUTLINE
B 6310 1922 ITALY, PO VALLEY
B 6411 1964 ITALY, PO VALLEY
B 6210 1913 ITALY, PO VALLEY ONSHORE
B 6110 1643 ITALY, PO VALLEY ONSHORE
B 6110 1654 ITALY, PRODUCTION, 1975-1976
B 6210 1924 ITALY, PRODUCTION, 1976-1977
B 6310 1932 ITALY, PRODUCTION, 1977-1978
B 6411 1971 ITALY, PRODUCTION, 1978-1979
M 26 343 ITALY, SERGNANO FIELD, PO VALLEY
S 5 175 ITALY, SICILY, CATTOLICA BASIN
S 5 175 ITALY, SICILY, CIMMINA BASIN
S 5 174 ITALY, SICILY, MESSINIAN STRATIGRAPHY
S 5 175 ITALY, SICILY, SALEMI BASIN
B 6401 32 ITALY, ZANDOBBIO DOLOMITE
B 6009 1618* ITALY-JAMAICA, B.W.I., ABST., MIDDLE, EOCENE, PLANKTONIC,
B 6109 1499 ITKILLIK SECTION, ALASKA
B 6109 1500 ITKILYARIAK FORMATION, ALASKA
B 6103 438 ITKILYARIAK FORMATION, ALASKA
M 25 62 ITSIKSEN, M. I., TECTONICS OF CONTINENTAL AND OCEANIC
M 25 458 ITSIKSON, M. I., METALLOGENIC SYSTEMS OF PACIFIDES--SUMMARY
S 6 1 IUGS, INVOLVEMENT IN TIME SCALES
B 6408 1132 IVAN RIVER GAS FIELD, ALASKA
B 6009 1610* IVE GOT A NEW PRODUCER IN SOUT LOUISIANA--NOW WHAT, ABST.
B 6405 727 IVEY, JOHN B., GEOLOGIC FRAMEWORK FOR DEVELOPMENT,
B 6109 1496 IVISHAK FORMATION, ALASKA
M 24 31 IVISHAK SANDSTONE, ALASKA
M 30 301 IVISHAK SANDSTONE, PRUDHOE BAY, SADLEROCHIT GROUP
B 6103 438 IVISHAK SHALE, ALASKA
B 6107 971 IVORY COAST, ABIDJAN BASIN OFF
B 6110 1748 IVORY COAST, DEVELOPMENTS, 1976
B 6210 1847 IVORY COAST, DEVELOPMENTS, 1977
B 6310 1692 IVORY COAST, DEVELOPMENTS, 1978
B 6411 1788 IVORY COAST, DEVELOPMENTS, 1979
M 25 160 IXTLAN DE LOS HERVORES GEOTHERMAL AREA, MEXICO
B 6409 1554* IXTOC I AND BURMAH AGATE, WITH COMMENTS ON AVENUE OF OIL TO
M 24 209 J, K, O, Q SANDSTONE, LOUISIANA, G
B 6410 1615 J SAND OIL, DENVER BASIN
M 24 136 J SANDSTONE, COLORADO
B 6104 595 J-1-1 FIELD, LIBYA
B 6104 596 J-2-6 FIELD, LIBYA
B 6407 1117 JABLONSKI, LINTZ, JOSEPH, JR., REVIEW OF THE, ENCYCLOPEDIA O,
B 6012 2189* JACALITOS CANYON AREA, CALIFORNIA, ABST., PLIOCENE
B 6306 942 JACAREZINHO FORMATION, AMAZON BASIN
B 6004 685 JACK, A. D., INTERPRETATION OF DEPOSITIONAL ENVIRONMENTS,
B 6301 50† JACK, PALO PINTO, PARKER, AND WISE COUNTIES, NORTH-CENTRAL
B 6002 327* JACK, PALO PINTO, PARKER, AND WISE COUNTIES, NORTH-CENTRAL
B 6109 1553 JACKA, ALONZO D., DEPOSITION AND DIAGENESIS OF FORT TERRETT
B 6303 548 JACKA, ALONZO D., DEPOSITION AND DIAGENESIS OF MISSISSIPPIAN
B 6309 1604 JACKA, ALONZO D., FACIES, DIAGENESIS, AND POROSITY RELATIONS OF
B 6105 799 **JACKA, ALONZO D., FACTORS CONTROLLING POROSITY RELATIONS IN**
B 6109 1538 JACKA, ALONZO D., JFS FIELD, DIMMIT COUNTY, TEXAS—SOME UNIQUE
B 6303 474 JACKA, ALONZO D., OCCLUSION OF POROSITY IN CARBONATE
B 6203 525 JACKA, ALONZO D., OCCURRENCES OF TERTIARY POROSITY IN
B 6303 474 JACKA, ALONZO D., PRECIPITATION OF AUTHIGENIC MINERALS IN
B 6102 295 JACKA, ALONZO D., PREDICTING POROSITY RELATIONS IN
B 6309 1606* JACKFORK GROUP OF SOUTHERN ARKANSAS, ABST., OUTER DEEP-SEA
B 6209 1674 JACKPILE MINE, NEW MEXICO
B 6206 1093* JACKPILE SANDSTONE, BURRO CANYON FORMATION, AND DAKOTA
B 6209 1675 JACKPILE SANDSTONE, NEW MEXICO
S 3 70 JACKPILE SANDSTONE, NEW MEXICO
B 6412 2078 JACKSBORO FAULT, APPALACHIANS
B 6101 24 JACKSON BAY FIELD, CANADIAN, ARCTIC ISLANDS
B 6108 1381* JACKSON, COUNTY, COLORADO, ABST., ENVIRONMENTS OF DEPOSITION
B 6309 1585* JACKSON COUNTY, WEST VIRGINIA, ABST., COMPUTER-AIDED GRAPHICS
M 24 242 JACKSON FIELD, MISSISSIPPI
B 6405 787* JACKSON GROUP, SOUTH TEXAS, ABST., COASTAL SWAMP ORIGIN OF,
B 6009 1612 JACKSON, J. R., JR., INDEPENDENT AND MAJOR ROLES IN EXPLORATION
B 6105 799 **JACKSON, ROSCOE G., II, EVALUATION OF FLUVIAL-FACIES MODELS**
B 6112 2103 JACKSON, T. J., HYDROCARBON GEOCHEMISTRY OF OILS FROM
B 6303 472 JACKSON, T. J., OIL GEOCHEMISTRY AS EXPLORATION TOOL, ABST.
B 6305 831 JACKSON, TIMOTHY J., FLOOD-PLAIN SEQUENCES OF FINE-GRAINED
B 6203 525 JACKSON, TIMOTHY J., TRANSITION BETWEEN A SANDY AND MUDDY
B 6203 525 JACKSON, WALDO, USE OF SEISMIC MODELING FOR REEF EXPLORATION
B 6109 1538* JACKSONIAN-VICKSBURGIAN BOUNDARY IN EAST-CENTRAL GULF COAST
B 6405 725* JACKSONIAN-VICKSBURGIAN BOUNDARY IN EAST-CENTRAL GULF COAST
B 6008 1399 JACOB, ARTHUR F., RECONNAISSANCE FOR URANIUM IN APPALACHIAN
B 6111 2004 JACOBI, ROBERT D., EXOTIC MIDDLE MIOCENE SEDIMENT FROM CAPE
B 6103 443 JACOBS, CYRIL, JURASSIC LITHOLOGY IN GREAT ISAA 1 WELL, BAHAMAS.
B 6310 1753 JACOBSEN, P., JR., ON DEVELOPMENTS IN FRENCH GUIANA, 1978
B 6210 2053 JACOBSEN, P., JR., ON DEVELOPMENTS IN HONDURAS, 1977
B 6310 1754 JACOBSEN, P., JR., ON DEVELOPMENTS IN HONDURAS, 1978
B 6010 1654 JACOBSEN, P., JR., ON DEVELOPMENTS IN PARAGUAY, 1975
B 6310 1758 JACOBSEN, P., JR., ON DEVELOPMENTS IN PARAGUAY, 1978
B 6210 2059 JACOBSEN, P., JR., ON DEVELOPMENTS IN SURINAM, 1977
B 6310 1761 JACOBSEN, P., JR., ON DEVELOPMENTS IN SURINAM, 1978
B 6210 2052 JACOBSEN, P., JR., ON DEVELOPMENTS OF FRENCH GUIANA, 1977
B 6405 728 JACOBSEN, S. R., APPLICATION OF CINGULUM INDEX TO DINOGYMNIUM
B 6004 685 JACOBSON, J. B., TIMING--GEOLOGIC AN POLITICAL--IN URBAN OIL
M 29 209 JACOBSON, R. S., SEISMIC REFRACTION AND REFLECTION STUDIES IN THE
B 6303 506 **JACOBSON, RUSSELL J., ORIGIN AND CHARACTERISTICS OF**
B 6405 728 JADO, ABDUL RAOF, STRATIGRAPHIC RELATIONS OF PERMIAN
B 6302 183 JAEREN HIGH, NORTH SEA
B 6406 956 JAGER AND J. C. HUNZIKER, MCDOWELL, FRED W., REVIEW OF, LECTURES
S 6 67 JAGER, E., SUBCOMMISSION ON GEOCHRONOLOGY--CONVENTION ON THE
M 25 483 JAHNS, RICHARD H., METALLOGENIC PROVINCES OF NORTHEAST PACIFIC
B 6110 1587 JAMAICA, 1976, PETROCONSULTANTS S.A., ON DEVELOPMENTS IN
B 6210 2053 JAMAICA, 1977, WRIGHT, R. M., ON DEVELOPMENTS IN
B 6310 1754 JAMAICA, 1978, PETROCONSULTANTS S.A., ON DEVELOPMENTS IN
B 6203 572* JAMAICA, ABST., DEPOSITIONAL ENVIRONMENTS OF, YALLAHS, FAN,
B 6303 447* JAMAICA, ABST., PETROLOGY AND DIAGENESIS OF ARC-DERIVED LITHIC,
S 4 88 JAMAICA, ARTHURS SEAT FORMATION, SANTONIAN
B 6009 1618* JAMAICA, B.W.I., ABST., MIDDLE, EOCENE, PLANKTONIC,
S 4 87 JAMAICA, BENBOW INLIER
S 4 87 JAMAICA, BENBOW LIMESTONE
S 4 85 JAMAICA, BLUE MOUNTAIN INLIER
B 6403 380 JAMAICA, BLUE MOUNTAINS
S 4 88 JAMAICA, BULLHEAD FORMATION, MAIN RIDGE VOLCANICS,
S 4 88 JAMAICA, CENTRAL INLIER
S 4 235 JAMAICA, COMACTINIA ECHINOPTERA MULLER
S 4 87 JAMAICA, COPPER LIMESTONE
S 4 84 JAMAICA, CRETACEOUS
S 4 53† JAMAICA, DEEP, FOREREEF AND UPPER ISLAND SLOPE, NORTH
B 6010 1653 JAMAICA, DEVELOPMENTS, 1975
S 4 87 JAMAICA, DEVILS RACE COURSE FORMATION, BARREMIAN-APTIAN
S 4 90 JAMAICA, DICHOCOENIA TRECHMANNI ASSEMBLAGE
S 4 53 JAMAICA, DISCOVERY BAY CANYON

B 6107 1073 JAMAICA, DISCOVERY BAY, MICROGRAPHS
S 4 129 JAMAICA, DISCOVERY BAY, REEF
S 4 263 JAMAICA, DISCOVERY BAY, REEF, BORERS
S 4 129 JAMAICA, EAST PALISADES, REEF
B 6403 374† JAMAICA, FAN-, DELTA SEDIMENTOLOG AND TECTONIC SETTING--.
S 4 85 JAMAICA, GREEN ISLAND INLIER
S 4 88 JAMAICA, GUINEA CORN FORMATION, MAESTRICHTIAN
S 4 85 JAMAICA, HANOVER INLIER
B 6403 376 JAMAICA, HOLOCENE FAN DELTA
M 27 49 JAMAICA, HOLOCENE REEF SEDIMENT
S 4 85 JAMAICA, JERUSALEM MT. INLIER
S 4 87 JAMAICA, JUBILEE LIMESTONE
S 4 54 JAMAICA, KAISER CHANNEL
S 4 85 JAMAICA, MALDON INLIER
S 4 85 JAMAICA, MARCHMONT INLIER
S 4 175 JAMAICA, MONTASTREA ANNULARIS
S 4 90 JAMAICA, MULTICOLUMNASTREA-ASTROCOENIA-ACTINACIS
S 4 234 JAMAICA, NEMASTER DISCOIDEA CARPENTER
S 4 90 JAMAICA, PARACYCLOSERIS-TROCHOSERIS-ACTINACIS ASSEMBLAGE
S 4 88 JAMAICA, PETERS HILL FORMATION, UPPER SANTONIAN
S 4 87 JAMAICA, PHILLIPSBURG LIMESTONE
B 6403 380 JAMAICA, PORT ROYAL MOUNTAINS
S 4 87 JAMAICA, RIO NUEVO FORMATION, ALBIAN
S 4 128 JAMAICA, RUNAWAY BAY, REEF
S 4 87 JAMAICA, SEAFIELD LIMESTONE
S 4 88 JAMAICA, SLIPPERY ROCK FORMATION, MAESTRICHTIAN
S 4 85 JAMAICA, ST. ANNS INLIER
S 4 88 JAMAICA, SUMMERFIELD FORMATION, MAESTRICHTIAN
S 4 85 JAMAICA, SUNDERLAND INLIER
S 4 87 JAMAICA, TIBER FORMATION, ALBIAN
B 6403 380 JAMAICA, WAGWATER TROUGH
S 4 83† JAMAICAN CORAL-RUDIST FRAMEWORKS AND THEI GEOLOGIC SETTING
B 6203 554* JAMAICAN REEF SEDIMENTS—REDOX MODEL FOR SUBMARINE
B 6203 498* JAMAICAN REEFS, ABST., STROMATOPOROIDS—EPIC STRUGGLE FOR
B 6311 2120 JAMES, ALFRED, III, CHANNEL SANDSTONE OIL RESERVOIRS OF
B 6203 539* JAMES BAY LOWLANDS, ONTARIO, CANADA, ABST., REGRESSIVE
B 6203 531 JAMES,, CAVITY-DWELLING ORGANISMS, COELOBITES, IN LOWER
B 6009 1611* JAMES LIMESTONE IN WINN AND NATCHITOCHES PARISHES, LOUISIANA,
M 24 158 JAMES LIMESTONE MEMBER, TEXAS
M 24 159 JAMES LIMESTONE MEMBER, TEXAS
B 6303 474 JAMES, NOEL P., CORRELATION AND TIMING OF PLATFORM-MARGIN
B 6303 508 JAMES, NOEL P., ICHNOLOG OF LABRADOR GROUP, LOWER CAMBRIAN. IN
B 6105 799 JAMES, N. P., INTERNAL STRUCTURE OF LOWER CAMBRIAN
B 6411 1887 JAMES, R., AND W. WILLIAMS, ON DEVELOPMENTS IN INDONESIA, 1979
M 25 130 JAMES, RUSSELL, GEOTHERMAL ENERGY IN NEW ZEALAND--SUMMARY
M 25 354 JAMES, S. E., ARTIFICIAL RECHARGE OF GR
B 6405 650 JAMES, SCHILLING, JOHN, REVIEW OF, GUIDEBOOK OF, SANTA FE
B 6405 791 JAMES, W. CALVIN, LATE MESOZOIC TO EARLY CENOZOIC FORELAND
B 6303 475 JAMES, WILLIAM R., U.S. GEOLOGICAL SURVEY METHODS FOR OCS TRACT
B 6103 355 JAMESON STRAWN SANDSTONE FIELD, TEXAS
B 6304 699 JAMISON, H. C., PRUDHOE BAY, A TEN-YEAR PERSPECTIVE, ABST.
M 30 289 JAMISON, H. C., PRUDHOE BAY--A 10-YEAR PERSPECTIVE
B 6412 2189 JAMISON, W. R., REVIEW OF GEOLOGICAL FIELD TRIPS COLOMBIA
B 6403 445 JANES, STEPHEN A., LATE PLEISTOCENE-HOLOCENE SEDIMENTARY
M 25 440 JANGRAK GROUP, SOUTH KOREA
B 6408 1285 JANKA, JOHN C., DEVONIA OIL SHALE OF EASTERN UNITED STATES, ABST.
B 6105 799 JANSA, L., STRATIGRAPHIC SYNTHESIS AND HYDROCARBON POTENTIAL
B 6203 526 JANSA, L. F., COST B-2 WELL, BALTIMORE CANYON, AND ITS RELATION TO
B 6303 475 JANSA, LUBOMIR F., MESOZOIC-CENOZOIC SEDIMENTARY FORMATIONS O
B 6009 1621 JANSSENS, A., RELATION BETWEE SALINA, UPPER SILURIAN, OIL
B 6312 2222 JANUSFJELLET FORMATION, NORTH SEA
B 6312 2194 JANUSFJELLET FORMATION, SVALBARD
B 6207 1223 JAPAKASETR, THAWAT, EVAPORITE DEPOSITS OF NORTHEAST THAILAND,
B 6110 1837 JAPAN, 1976, INOMATA, T., ON DEVELOPMENTS IN
B 6210 1985 JAPAN, 1977, INOMATA, T., ON DEVELOPMENTS IN
B 6310 1822 JAPAN, 1978, KUJIRAOKA, A., ON DEVELOPMENTS IN
B 6411 1888 JAPAN, 1979, KUJIRAOKA, AKIRA, ON DEVELOPMENTS IN
B 6212 2509* JAPAN, ABST., CONTRAST OF HYDROCARBON POTENTIAL BETWEEN,
B 6303 551* JAPAN, ABST., EARLY, PLEISTOCENE, SUBMARINE CANYON, BOSO
B 6207 1231* JAPAN, ABST., LANDSLIDE PROBLEMS IN
B 6207 1241* JAPAN, ABST., PRESENT STATUS OF, GEOTHERMAL DEVELOPMENT AND
B 6203 531* JAPAN, ABST., SLOPE SEDIMENTATION PROCESSES IN, NEOGENE
B 6212 2510* JAPAN, ABST., STRUCTURAL CONTRAST BETWEEN, FORE-ARC AND,
M 25 101 JAPAN, AUTHIGENIC ZEOLITES IN TERTIARY COALFIELDS
M 25 453 JAPAN, AUTUNITE, URANIUM ORE MINERAL
M 25 453 JAPAN, B-URANOPHANE, URANIUM ORE MINERAL
M 25 457 JAPAN, BASE-METAL DEPOSITS, DISTRICTS, MAP
M 25 457 JAPAN, BASE-METAL DEPOSITS, EXPLORATION PROJECTS
M 25 456 JAPAN, BASE-METAL DEPOSITS, FUTURE EXPLORATION
M 25 456 JAPAN, BASE-METAL DEPOSITS, GEOLOGIC SURVEY PROJECT
M 25 456 JAPAN, BASE-METAL ORES, DEMAND AND PRODUCTION, DISCREPANCY
M 25 451 JAPAN, BASEMENT ROCKS
M 25 453 JAPAN, CARNOTITE, URANIUM ORE MINERAL
S 6 160 JAPAN, CENOMANIAN, MIFUNE GROUP
M 25 449 JAPAN, CHICHIBU GEOSYNCLINAL UNITS, MANGANESE DEPOSITS
M 25 99 JAPAN, COAL RESERVES
M 25 99 JAPAN, COALFIELD
M 25 453 JAPAN, COFFINITE, URANIUM ORE MINERAL
M 25 457 JAPAN, COPPER ORE
S 6 150 JAPAN, CRETACEOUS, AMAMI-OSHIMA GRANITES
S 6 145 JAPAN, CRETACEOUS, BIOSTRATIGRAPHY
S 6 158 JAPAN, CRETACEOUS, HARACHIYAMA FORMATION
S 6 150 JAPAN, CRETACEOUS, IGNEOUS GEOCHRONOLOGY
S 6 155 JAPAN, CRETACEOUS, KITAKAMI MOUNTAINS, FORMATION
S 6 145 JAPAN, CRETACEOUS, KWANMON GROUP
S 6 160 JAPAN, CRETACEOUS, MIDDLE YEZO GROUP
S 6 153 JAPAN, CRETACEOUS VOLCANICS, TABLE
M 25 147† JAPAN ESTIMATED FROM GEOPHYSICAL DATA--SUMMARY, GEOTHERMAL
M 25 448 JAPAN, FAULTS, TRANSVERSE
M 25 278 JAPAN, FORAMINIFERAL ZONULES
M 25 448 JAPAN, FOSSA MAGNA
M 25 454 JAPAN, FUNATSU GRANITE
S 6 150 JAPAN, FUNATSU GRANITE
M 25 147 JAPAN, GEOTHERMAL ENERGY
M 25 363 JAPAN, GROUNDWATER
S 6 159 JAPAN, HAUTERIVIAN, KANAIGAURA FORMATION
S 6 159 JAPAN, HAUTERIVIAN, YAMADORI FORMATION
M 25 448 JAPAN, HEAT FLOW, HIGH
M 25 279 JAPAN, HIGASHI-KASHIWATZAKI GAS FIELD
B 6403 376 JAPAN, HOLOCENE FAN DELTA
M 25 454 JAPAN, INNER ZONE
M 25 448 JAPAN, ISLAND, ARCS, PACIFIC-WARD SHIFTING
S 6 143† JAPAN, ISOTOPIC AGES AND STRATIGRAPHIC CONTROL OF, MESOZOIC,
S 6 154 JAPAN, KOSHIMEGURI FORMATIONS
M 25 456 JAPAN, KUROKO DEPOSITS
M 25 448 JAPAN, KUROKO OR BLACK ORE DEPOSITS
M 25 362 JAPAN, KWANTO REGION, LAND SUBSIDENCE
M 25 457 JAPAN, LEAD-ZINC ORES
S 6 154 JAPAN, MAGISAWA FORMATIONS
S 6 146 JAPAN, MESOZOIC, ABU GROUP
S 6 146 JAPAN, MESOZOIC, AIOI GROUP
S 6 144 JAPAN, MESOZOIC, IGNEOUS LOCALITIES, MAP
M 25 449 JAPAN, MESOZOIC, LATE, GRANITIC BATHOLITHS AND MINERAL ZONING
S 6 146 JAPAN, MESOZOIC, PALEOMAGNETIC EVIDENCE
M 25 456 JAPAN, METAL MINING AGENCY OF
M 25 449 JAPAN, METAMORPHISM
M 25 448 JAPAN, MINERAL CONCENTRATION AND SUBDUCTION ZONE, RELATION
M 25 456 JAPAN, MINING INDUSTRY, GOVERNMENT POLICY TOWARD
M 25 448 JAPAN, MIOCENE SUBMARINE VOLCANIC OR GREEN TUFF AREA
B 6412 2113 JAPAN, MITI-YOSHIDA WELL, HYDROCARBONS
S 6 152 JAPAN, MIYAKO GRANITE, RUBIDIUM-STRONTIUM DATA
S 6 156 JAPAN, MIYAKO GRANITE, RUBIDIUM-STRONTIUM ISOCHRON DIAGRAM
M 25 360 JAPAN, MT. FUJI STRATOVOLCANO
M 22 279 JAPAN, MYOHOJI FIELDS
M 25 448 JAPAN, NEOGENE MINERAL DISTRIBUTION
S 6 159 JAPAN, NIITSUKI FORMATION
M 25 451 JAPAN, NINGYO-TOGE URANIUM DEPOSITS
M 25 453 JAPAN, NINGYOITE, URANIUM ORE MINERAL
M 25 465 JAPAN, NORTHEAST, IGNIMBRITE VOLCANISM
B 6407 1090 JAPAN, NORTHERN, REGIONAL HORIZONS
M 25 278 JAPAN, OIL AND GAS FIELDS, MAJOR
S 6 154 JAPAN, OMOTO FORMATION
S 6 154 JAPAN, OUTCROP MAP
M 25 454 JAPAN, OUTER ZONE
M 25 449 JAPAN, PERMIAN GEOSYNCLINAL SEDIMENTATION AND MINERA
M 25 448 JAPAN, PLATE MOVEMENT, DIRECTION OF
S 6 149 JAPAN, POTASSIUM-ARGON AGES, GRANITES, DIAGRAM
S 6 147 JAPAN, POTASSIUM-ARGON DATING, GRANITES
B 6110 1857 JAPAN, PRODUCTION, 1975-1976
B 6210 1990 JAPAN, PRODUCTION, 1975-1977
B 6310 1829 JAPAN, PRODUCTION, 1977-1978
B 6411 1896 JAPAN, PRODUCTION, 1977-1979
M 25 456 JAPAN, PROVINCIAL BUREAU OF THE MINISTRY OF INTERNATIONAL
M 25 362† JAPAN RESULTING FROM FLUID EXTRACTION--SUMMARY, LAND,
M 25 449 JAPAN, SAMBAGAWA SYSTEM, STRATA-BOUND SULFIDE DEPOSITS
S 6 160 JAPAN, SENNAN ACID PYROCLASTICS
M 25 463 JAPAN, SOUTHWEST, VOLCANIC BELTS, CHARACTERISTICS OF
M 25 68 JAPAN, STRUCTURAL ANALYSIS
M 25 448 JAPAN, SULFUR AND PYRITE ORE DEPOSITS, RECENT

M 25 448 JAPAN, SULFUR DEPOSITS, RELATION TO VOLCANIC FRONT
S 6 152 JAPAN, TARO GRANITE, RUBIDIUM-STRONTIUM DATA
S 6 157 JAPAN, TARO GRANITE, RUBIDIUM-STRONTIUM ISOCHRON PLOT
M 25 448 JAPAN, THOLEIITIC SUITE
M 25 452 JAPAN, TOKI LIGNITE-BEARING FORMATION, URANIUM HOST ROCKS
M 25 454 JAPAN, TONO AREA, URANIUM CONTENT OF GROUNDWATERS
M 25 452 JAPAN, TONO URANIUM DEPOSITS
M 25 449 JAPAN, TUNGSTEN AND MOLYBDENUM DEPOSITS, ZONAL DISTRIBUTION
M 25 454 JAPAN, URANIFEROUS GLAUCONITE AND APATITE, FUTURE POTENTIAL
M 25 453 JAPAN, URANINITE, URANIUM ORE MINERAL
M 25 453 JAPAN, URANIUM ADSORBED ON ZEOLITES
M 25 451 JAPAN, URANIUM AND GRANITIC MASSES, RELATION OF
M 25 454 JAPAN, URANIUM AND GRANITIC ROCKS
M 25 452 JAPAN, URANIUM CONCENTRATION AND GRANITIC BODIES,
M 25 453 JAPAN, URANIUM CONCENTRATION, GEOCHEMISTRY OF
M 25 451 JAPAN, URANIUM, DISTRIBUTION AND OCCURRENCE
M 25 453 JAPAN, URANIUM ORE RESERVES
M 25 451 JAPAN, URANIUM, RESERVES
M 25 276 JAPAN, VOLCANIC-ROCK RESERVOIRS
M 25 276† **JAPAN WITH VOLCANIC-ROCK RESERVOIRS--SUMMARY, PETROLEUM**
M 25 279 JAPAN, YOSHII GAS FIELD
M 25 451† JAPAN--SUMMARY, GEOLOGI ASPECTS OF, URANIUM RESOURCES OF
M 25 148† JAPAN--SUMMARY, GEOLOGICAL AND, GEOCHEMICAL CHARACTERISTICS
M 25 360† JAPAN--SUMMARY, GROUNDWATER IN, VOLCANIC AREAS IN
M 25 98† JAPAN--SUMMARY, INFLUENCE OF, TEMPERATURE ON, COALIFICATION
M 25 456† JAPAN--SUMMARY, NATIONAL EXPLORATION PROGRAM FOR BASE-META
M 25 448† JAPAN--SUMMARY, RELATION BETWEEN, TECTONICS AND,
M 25 463 JAPANESE ISLAND ARC, KURIL-
B 6105 800 JARDINE, D., RESERVOIR DESCRIPTION AND POOL PERFORMANCE OF JUDY
M 25 328 JARRIN, ALFONSO, EXPLORATION AND DEVELOPMENT OF
M 25 475 JASPERIZATION
B 6109 1471 JATOBA BASIN, BRAZIL
B 6101 58 JAUF-GAARA-ABD EL AZIZ-MARDIN SWELL, LEVANT
B 6405 602 JAUF-GAARA-MOSIL ARCH, MIDDLE EAST
B 6310 1821 JAUNCEY, W., AND ZELIFF, C. W., ON DEVELOPMENTS IN INDONESIA, 1978
B 6004 709* **JAVA SEA BASINS--MECHANISM OF BASIN FORMATION. BY TITLE ONLY,**
M 25 244 JAVA BASIN AREA, INDONESIA, NORTHWEST
M 25 246 JAVA BASIN, INDONESIA, EAST
B 6004 717* JAVA SEA, ABST., ORGANIC, GEOCHEMISTRY AS AID TO EXPLORATION IN
B 6401 56 JAVA SEA, POLENG FIELD
B 6401 56 JAVA SEA, POLENG FIELD, KALIMANTAN
B 6401 43 JAVA SEA REGION, TERMINOLOGY, EAST
M 25 138 JAVA, THERMAL AREAS, PROSPECTIVE
B 6406 889 JAVA TRENCH
B 6310 1891 JAVOGUES, J. P., ET AL, ON DEVELOPMENTS IN SYRIA, 1978
B 6310 1885 JAVOGUES, J. P., ON DEVELOPMENTS IN IRAQ, 1978
B 6110 1795 JAVOGUES, J.-P., ON DEVELOPMENTS IN IRAQ, 1976
B 6210 2112 JAVOGUES, J.-P., ON DEVELOPMENTS IN IRAQ, 1977
B 6210 2119 JAVOGUES, J.-P., ON DEVELOPMENTS IN SYRIA, 1977
B 6210 2119 JAVOGUES, ON DEVELOPMENTS IN TURKEY, 1977. ASIATIC PETROLEUM
B 6110 1802 **JAVOGUES, J.-P., ON DEVELOPMENTS IN SYRIA, 1976**
B 6309 1580 JAWOREK, W. GIBSON, NEW ECONOMICS OF NATURAL GAS PRODUCTION
S 10 249 JAY FIELD, METHANE DISTRIBUTION
S 10 244 JAY FIELD, ETHANE CONCENTRATIONS, FLORIDA
M 24 276† JAY FIELD, FLORIDA--A JURASSIC STRATIGRAPHIC TRAP
M 30 32 JAY FIELD, LANDSAT IMAGE, FLORIDA
S 10 252 JAY FIELD, METHANE CONCENTRATIONS, FLORIDA
S 10 246 JAY FIELD, SAND PERCENTAGE CONCENTRATIONS, FLORIDA
S 10 249 JAY FIELD, SUBSURFACE GEOLOGY, FLORIDA
B 6401 30 JAY GIANT OIL AND GAS FIELD, LANDSAT DATA, FLORIDA
B 6009 1615* JAY-BLACKJACK CREEK FIELD AREA, ABST., PALEOENVIRONMENTAL
M 30 266 JD SANDSTONE, RESERVOIR, EUGENE ISLAND 330 FIELD
S 4 331 JEALOUSY FORMATION, ST. CROIX
B 6203 406 JEALOUSY FORMATION, VIRGIN ISLANDS
B 6106 931 JEAN CHARCOT GEOBRESIL CRUISES
M 27 5 JEAN MARIE FORMATION, CANADA (N.W.T.), DEVONIAN
B 6203 428 JEANNE DARC BASIN, GRAND BASIN
B 6203 526 JEARY, GENE L., LEONARDIAN STRATA IN NORTHERN MIDLAND BASIN
B 6407 992 JEBEL AMRHOUZ, MOROCCO
S 4 268 JEBEL ASSAMEUR NAIT FERGANE, CORALS, BORERS, MOROCCO
S 4 265 JEBEL BOU KENDILL, CORAL, BORERS, MOROCCO
B 6408 1281* JEFFERSON ANDFRANKLIN COUNTIES, ILLINOIS, ABST., GEOLOGY AND
B 6409 1563* JEFFERSON COUNTY, TEXAS, ABST., GEOLOGY OF, GUM ISLAND NORTH
B 6208 1383 JEFFERSON COUNTY, TEXAS, NORTH GUM ISLAND
B 6209 1769* JEFFERSON DAVIS COUNTY, MISSISSIPPI, ABST., PETROGRAPHY AND
B 6102 257 JEFFERSON FORMATION, IDAHO
S 7 313 JEFFREY, L. M., HOLOCENE SEDIMENT GEOCHEMISTRY OF CONTINENTAL
B 6211 2282 JELM FORMATION, WYOMING
B 6305 804 JEMISON, R. M., JR., GEOLOGY AND DEVELOPMENT OF MILLS RANCH
B 6203 526 JEMISON, ROBERT M., JR., DEVELOPMENT OF MILLS RANCH COMPLEX.
B 6304 690 JENKINS, J. T., JR., DEPOSITIONAL ENVIRONMENT OF POISON CANYON
B 6012 2183 JENKINSON, W. D., SIDE-SCAN SONAR--APPLICATIONS IN OIL
B 6308 1399 JENNINGS, S. P., DEVELOPMENTS IN SOUTHEASTERN STATES IN 1978
B 6304 686 JENSEN, DAL H., DIRECT MEASUREMENT OF URANIUM BY PROMPT
B 6109 1471 JEQUITINHONHA BASIN, BRAZIL
B 6002 197 JEQUITINHONHA COASTAL BASIN, BRAZIL
S 4 85 JERUSALEM MT. INLIER, JAMAICA
B 6406 868 JEZEK, PETER, ARC-CONTINENT COLLISION I BANDA SEA REGION
B 6109 1538* JFS FIELD, DIMMIT COUNTY, TEXAS—SOME UNIQUE ASPECTS OF
B 6008 1410* JIM COULEE-LITTLE WALL FIELD, MUSSELSHELL COUNTY, MONTANA,
B 6109 1534* JIM HOGG AND WEBB COUNTIES, TEXAS, ABST., CHARACTERISTICS OF,
B 6409 1567* JIM HOGG COUNTY, TEXAS, ABST., OCCURRENCE OF HIGH GRAVITY OIL
B 6210 2046 JIMENEZ, ON DEVELOPMENTS IN BOLIVIA, 1977, OBLITAS, J., AND F.
B 6409 1566* JOACHIM DOLOMITE, MIDDLE ORDOVICIAN, NORTH ARKANSAS, ABST.,
S 3 90 JOACHIMSTHAL VEIN DISTRICT, CZECHOSLOVAKIA
B 6302 221 JOAO PESSOA BASIN, BRAZIL, RECIFE-
B 6002 197 JOAO PESSOA COASTAL BASIN, BRAZIL, RECIFE-
B 6112 2135 JOARCAM FIELD, ALBERTA
B 6109 1421 JOARCAM FIELD, ALBERTA
B 6104 594 JOFFRE FIELD, ALBERTA
M 25 411 JOFFRE MEMBER, WESTERN AUSTRALIA
B 6303 417 JOHANNSON, THORA M., HYDROCARBON POTENTIAL OF CENTRAL GULF
B 6103 466 JOHANSEN, NILS I., EARTHQUAKES AND ENGINEERING--REVIEW OF
B 6208 1441 JOHANSEN, POUL, DEVELOPMENTS OFFSHORE WEST GREENLAND IN 1977
B 6105 805 JOHN, CHACKO, STRUCTURE OF COASTAL WASHOVER FANS
B 6303 492 JOHN, CHACKO J., DETERMINATION OF STORM OVERWASH PERIODICITY
B 6312 2145 JOHN, CHACKO J., LATERAL AND VERTICAL FACIES RELATIONS OF
B 6008 1412* JOHN HENRY MEMBER OF STRAIGHT CLIFFS FORMATION, SOUTHERN
B 6304 687* JOHNNY M MINE, AMBROSIA LAKE DISTRICT, NEW MEXICO, ABST.,
B 6110 1901 JOHNS, W. D., CLAY, MINERALS AND PETROLEUM-FORMING REACTIONS
B 6004 714 JOHNSON, A. C., COMPUTER ANALYSIS OF LINEAMENTS TO INFER
B 6409 1445 JOHNSON, ARTHUR H., DEVELOPMENTS IN LOUISIANA GULF COAST
B 6203 526 JOHNSON, BRUCE D., GENETIC STRATIGRAPHY AND PROVENANCE OF
B 6405 688 JOHNSON, BRUCE D., GENETIC STRATIGRAPHY AND PROVENANCE OF
B 6303 342 JOHNSON CREEK FAULT, KENTUCKY
B 6403 800 JOHNSON, C. R., ALTERNATE GEOLOGIC INTERPRETATION—POTENTIAL USE
B 6004 645 JOHNSON, D. L., SOME PROBLEMATIC LATE CENOZOIC CARIBBEAN
M 29 43 JOHNSON, G. LEONARD, EROSIONAL AND DEPOSITIONAL STRUCTURES OF
B 6405 728 JOHNSON, HOWARD D., SEDIMENTOLOGY OF LOWER CRETACEOUS
B 6105 833 JOHNSON, J., DREDGED MATERIAL DEPOSITED AT OCEAN DUMPING SITES
B 6405 729 JOHNSON, J. P., UNDERSTANDING FIELD DEVELOPMEN HISTORY USING
B 6405 677 JOHNSON, KENNETH S., DISTRIBUTION AND ALTERATION OF OGALLALA
B 6002 320 JOHNSON, KENNETH S., GEOLOGY OF PERMIAN BLAINE FORMATION AND
B 6203 526 JOHNSON, KENNETH S., PERMIAN SALT DEPOSITS IN TEXAS PANHANDLE
B 6303 475 JOHNSON, KENNETH S., POTENTIAL FOR SUBSURFACE DISPOSAL OF
B 6405 729 JOHNSON, MARCUS W., TRACE-MAKERS AS HISTORIANS IN LARGE-SCALE
B 6303 417 JOHNSON, MICHAEL L., HYDROCARBON POTENTIAL OF CENTRAL GULF
B 6108 1382 JOHNSON, MICHAEL S., NESSON ANTICLINE-PAST, PRESENT, AND FUTURE.
B 6004 682 JOHNSON, P. M., GEOLOGY AND POTENTIAL USES OF GEOPRESSURE
B 6308 1386 JOHNSON, R. P., DEVELOPMENTS IN NORTHERN ROCKIES IN 1978
B 6208 1498 JOHNSON, R. P., OIL AND GAS DEVELOPMENTS IN NORTHERN ROCKIES IN
B 6108 1346 JOHNSON, ROBERT P., OIL AND GAS DEVELOPMENTS IN NORTHERN
B 6002 326 JOHNSON, RON, LITHOFACIES AND DEPOSITIONAL ENVIRONMENTS OF
B 6205 885 JOHNSON, RONALD C., MAESTRICHTIAN CONGLOMERATES IN
B 6405 796 JOHNSON, SAMUEL R., MULTIPLE BARRIER-ISLAND AND DELTAIC
B 6105 800 JOHNSON, STEPHEN H., DEVELOPMENT OF CONTINENTAL-SLOPE BASINS
B 6406 868 JOHNSTON, CHRIS, ARC-CONTINENT COLLISION I BANDA SEA REGION
B 6008 1153 JOHNSTON, ROBERT R., NORTH AMERICAN DRILLING ACTIVITY IN 1975
B 6108 1121 JOHNSTON, ROBERT R., NORTH AMERICAN DRILLING ACTIVITY IN 1976
B 6208 1275 JOHNSTON, ROBERT R., NORTH AMERICAN DRILLING ACTIVITY IN 1977
B 6308 1168 JOHNSTON, ROBERT R., NORTH AMERICAN DRILLING ACTIVITY IN 1978
B 6409 1295 JOHNSTON, ROBERT R., NORTH AMERICAN DRILLING ACTIVITY IN 1979
B 6105 801 JOHNSTON, JOHN F., NEED FOR BETTER COAL-RESOURCES ASSESSMENTS
B 6107 976 JOIDES CORE HOLES 365 AND 364
B 6303 287 JOIDES DEEP SEA DRILLING PROJECT, DSDP, INTERNATIONAL PHASE OF
B 6410 1713 JOIDES, DRILL DATA, 1965
B 6205 797 JOIDES, DSDP 102
B 6205 797 JOIDES, DSDP 104
B 6205 793 JOIDES, DSDP HOLE 105
B 6204 601 JOIDES, DSDP HOLE 108
B 6204 596 JOIDES, DSDP SITE 100
B 6204 596 JOIDES, DSDP SITE 101
B 6204 596 JOIDES, DSDP SITE 105
B 6204 596 JOIDES, DSDP SITE 106
B 6204 597 JOIDES, DSDP SITE 391, LEG 44, BLAKE BASIN
B 6204 596 JOIDES, DSDP SITE 99, CAT GAP
B 6107 979 JOIDES HOLE 361
B 6205 793 JOIDES HOLES, NORTH, ATLANTIC
B 6207 1116 JOINTING, REGIONAL, ALBERTA
B 6211 2220 JOLON FAULT, CALIFORNIA
M 25 443 JOMCHON THRUST, SOUTH KOREA

B 6303 475 JONAS, EDWARD, MICROSCOPIC DISTRIBUTION OF URANIU IN OAKVILLE
B 6303 548 JONES, A. V., DEPOSITION AND DIAGENESIS OF MISSISSIPPIAN PINNACLE
B 6308 1211 JONES, B. C., DEVELOPMENTS IN ALASKA IN 1978
B 6409 1345 JONES, B. C., DEVELOPMENTS IN ALASKA IN 1979
B 6207 1223 JONES, D. L., LATE MESOZOIC TECTONIC HISTORY OF NORTHERN
B 6211 2356 JONES, D. L., MESOZOIC CONTINENTAL ACCRETION OF NORTH AMERICA.
B 6406 945 JONES, D. L., PETROGRAPHY AND GEOLOGIC SIGNIFICANCE OF UPPER
B 6303 490 JONES, D. L., SHALLOW-WATER UPPER JURASSIC ROCKS DREDGED FROM
S 6 259 JONES, DAVID L., CRETACEOUS TIME SCALE FROM NORTH AMERICA
B 6405 784 JONES, DAVID L., MESOZOIC ACCRETIONARY TECTONICS OF ALASKA,
B 6305 832 JONES, EARL V., JR., PAINTER RESERVOIR--CLEAR CREEK--RYCKMA CREEK
B 6203 527 JONES, FRANCES G., TERTIARY ALGAL CARBONATE REEFS FROM SNAKE
M 24 23 JONES, H. P., PERMO-TRIASSIC RESERVOIRS OF PRUDHO BAY FIELD,
B 6105 801 JONES, J., RICHARD, APPLICATION OF COMPUTER-MAPPING PROGRAMS
B 6008 1400 JONES, M. KAY, CRAZY MOUNTAINS AREA--GEOLOGY, EXPLORATION
M 25 354 JONES, N. O., ARTIFICIAL RECHARGE OF GR
M 25 353 JONES, N. O., DEVELOPMENT OF AUSTRALIA GROUNDWATER
B 6303 495 JONES, NORRIS W., CRETACEOUS TRANSGRESSION OF COAHUILA
B 6203 527 JONES, PAUL H., INTERACTING DYNAMICS OF PRESSURE, TEMPERATURE,
S 10 207 JONES, PAUL H., ROLE OF GEOPRESSURE IN THE HYDROCARBON AND
B 6303 476 JONES, R. W., INTEGRATION OF MICROSCOPIC ORGANIC ANALYSIS AND
B 6405 729 JONES, R. W., ORGANIC FACIES—STRATIGRAPHIC CONCEPT AND
S 10 47 JONES, R. W., SOME MASS BALANCE AND GEOLOGICAL CONSTRAINTS ON
B 6203 528 JONES, R. W., SOME MASS BALANCE AND GEOLOGIC CONSTRAINTS ON
B 6304 699 JONES, VICTOR T., PREDICTIONS OF OIL OR GA POTENTIAL BY
B 6002 326 JONS, RICHARD D., ROCK TANK, MORROW SANDSTONE, FIELD, EDDY
B 6211 2356 JOOSTEN, WESLEY L., THE GUS-BUS, ABST.
B 6010 1873 JORDAN, 1975, NATURAL RESOURCES AUTHORITY, HASHEMITE KINGDOM
B 6110 1802 JORDAN, 1976, CARVAJAL, J., PETROCONSULTANTS S.A., AND, NATIONAL
B 6210 2119 JORDAN, 1977, NATURAL RESOURCES AUTHORITY, THE, HASHEMITE
B 6310 1891 JORDAN, 1978, CARVAJAL, J., ON DEVELOPMENTS IN
B 6411 1841 JORDAN, 1979, PETROCONSULTANTS S.A., ON DEVELOPMENTS IN
B 6004 649 JORDAN, C. F., LISBURNE GROUP--POTENTIAL MAJO OBJECTIVE OF
B 6109 1493 JORDAN, CLIFTON F., LISBURNE GROUP, MISSISSIPPIAN AND
B 6105 801 JORDAN, CLIFTON F., REVIEW OF RECENT CARBONATE LAGOONAL
B 6203 550* JORDAN FORMATION, CAMBRIAN, UPPER MISSISSIPPI VALLEY, ABST.,
B 6212 2427 JORDAN, N. F., SLOPE MAPPING
B 6004 665 JORDAN, N. F., TECHNIQUES AND USES OF SLOPE MAPPING, ABST.
B 6303 476 JORDAN, NEAL F., INTERACTIVE MODELING AND INTERPRETATION OF
B 6209 1714 JORDAN, ROBERT R., RECORRELATION OF EOCENE-MIOCENE BOUNDARY
B 6405 685* JORDAN SANDSTONE IN WISCONSIN, ABST., DEPOSITIONAL
B 6012 2179 JORDAN, TERESA E., APPLICATIONS OF PLATE TECTONICS TO PETROLEUM
B 6303 476 JORDAN, TERESA E., BATHYMETRIC DISTRIBUTION OF TRACE FOSSILS IN
B 6405 730 JORDAN, TERESA E., PALEOGEOGRAPHY AND STRUCTURAL
S 3 106 JORDESITE-CALCITE-URANIUM ORE, PYRITE-, MARCASITE-
M 30 360 JOSE COLOMO FIELD, MEXICO, MACUSPANA BASIN
M 24 280 JOULTERS BAY AREA, FACIES, GREAT BAHAMA BANKS
B 6404 467 JOULTERS CAY, MARINE PHREATIC ARAGONITE CEMENT, BAHAMAS
B 6203 520* JOULTERS CAYS, GREAT BAHAMA BANK, ABST., OOID, SAND-SHOAL
S 4 186 JOULTERS CAYS, PATCH REEF, BAHAMAS
B 6209 1758* JOULTERS OOID SHOAL, BAHAMAS, ABST., HOLOCENE
B 6301 36 JOURDANTON FIELD, ATASCOSA COUNTY, TEXAS
B 6207 1195 JUAN DE FUCA PLATE
M 25 485 JUAN DE FUCA RISE, SEAFLOOR SPREADING
B 6405 600 JUBAILA FORMATION, MIDDLE EAST
B 6401 9 JUBAILA FORMATION, SAUDI ARABIA
S 4 87 JUBILEE LIMESTONE, JAMAICA
B 6305 832 JUDD, JAMES B., WHITNEY CANYO FIELD--POTENTIAL GAS GIANT IN
B 6405 730 JUDD, JAMES B., WHITNEY CANYON FIELD—POTENTIAL GAS GIANT IN
B 6101 113 JUDD SHALE MEMBER, UTAH
B 6002 259 JUDEA LIMESTONE, ISRAEL
B 6203 414 JUDITH FANCY FORMATION, VIRGIN ISLANDS, ST. CROIX
B 6407 971 JUDITH RIVER FORMATION, GREAT PLAINS
B 6403 318 JUDITH RIVER FORMATION, MONTANA
B 6303 527 JUDSON, DAVID, DEPTH MIGRATION OF SEISMIC DATA, ABST.
B 6211 2362 JUDSON, DAVID, EQUALIZING STACKING VELOCITIES OF DIPPING
B 6012 2183 JUDSON, DAVID,, REGIONAL VELOCITY REDUCTION, ABST.
B 6112 2135 JUDY CREEK FIELD, ALBERTA
B 6104 597 JUDY CREEK FIELD, ALBERTA
B 6205 789 JUDY CREEK FIELD, PRODUCTION, ALBERTA
B 6205 789 JUDY CREEK REEF AND BANK COMPLEX, SEISMIC DATA, ALBERTA
M 26 202 JUDY CREEK, SHALLOW-WATER CARBONATE BUILDUP, SEISMIC
B 6305 842* JULESBURG BASIN, ABST., NEW LOOK AT, OIL AND, GAS, RESOURCES OF,
M 24 139 JULESBURG BASIN, COLORADO, DENVER-
B 6105 777 JUNEMANN, PAUL M., COMPUTER PROCESSING OF LANDSAT DATA FOR
B 6211 2309 JUNIATA FORMATION, WEST VIRGINIA
B 6211 2219 JUNIPERO FORMATION, CALIFORNIA
B 6411 1964 JURA, FRANCE, ALSACE AND
B 6411 1968 JURA MOUNTAINS, SWITZERLAND
B 6310 1920 JURA-BRESSE-SAVOIE, FRANCE
B 6110 1641 JURA-BRESSE-SAVOIE, FRANCE
B 6104 607† **JURASSIC BEDS IN PUALE BAY--ALINCHAK BAY AREA, ALASKA PENINSULA,**
B 6303 551* **JURASSIC CARBONATE RESERVOIRS, UNITED STATES GULF COAST,**
S 7 83 JURASSIC (RECONSTRUCTION) DIAGRAM, NORTH ATLANTIC
B 6104 609 JURASSIC, ALASKA
B 6203 378 JURASSIC, ALASKA
B 6407 1016 JURASSIC, ALASKA
M 24 4 JURASSIC, ALASKA
M 25 295 JURASSIC, ALASKA
B 6408 1261 JURASSIC, ALASKA
B 6204 669 JURASSIC, ALGERIA
M 25 344 JURASSIC, AMAZON BASINS, TRIASSIC-
M 26 145† JURASSIC, AND LOWER CRETACEOUS SEISMIC SEQUENCES IN GULF OF
M 24 178 JURASSIC, ARKANSAS
M 25 235 JURASSIC, AUSTRALIA
B 6207 1154 JURASSIC, AUSTRALIA
B 6201 59 JURASSIC, AUSTRALIA
B 6307 1097 JURASSIC, AUSTRALIA
B 6204 696 JURASSIC, BAHAMA PLATFORM
M 26 158 JURASSIC, BAJOCIAN, AFRICA, SEQUENCES
B 6309 1455 JURASSIC, BALTIMORE CANYON TROUGH
B 6004 711* JURASSIC BARRIER ISLAND COMPLEX IN LOWER SUNDANCE FORMATION
B 6211 2275† JURASSIC BARRIER-ISLAND COMPLEX—LOWER SUNDANCE FORMATION
B 6305 827* JURASSIC BASINS, EASTERN NORTH AMERICA, ABST., ENVIRONMENTS
M 26 158 JURASSIC, BATHONIAN, AFRICA, SEQUENCES
B 6205 798 JURASSIC, BLAKE PLATEAU
M 26 158 JURASSIC, CALLOVIAN, AFRICA, SEQUENCES
M 27 20 JURASSIC, CALPIONELLID LIMESTONE, ITALY
M 27 19 JURASSIC, CALPIONELLID LIMESTONE, ITALY
M 27 177 JURASSIC, CALPIONELLID LIMESTONE, ITALY
B 6101 15 JURASSIC, CANADIAN, ARCTIC ISLANDS, TRIASSIC-
S 9 22 JURASSIC, CARBONATE REEF BELT DISTRIBUTION
B 6004 612† JURASSIC, CARBONATE ROCKS, SOUTHEAST OF PARIS, FRANCE,
B 6405 712* JURASSIC CARMEL FORMATION, SOUTHERN UTAH, ABST., FACIES AND
M 27 201 JURASSIC, CARRARA MARBLE, ITALY
B 6304 689* JURASSIC, CARRIZO MOUNTAINS, ARIZONA AND NEW MMXICO, ABST.,
B 6104 562 JURASSIC, CHILE
B 6007 1013 JURASSIC, COLUMN, NEW MEXICO
M 27 41 JURASSIC, CORALLIAN SERIES, ENGLAND
B 6211 2277 JURASSIC CORRELATION AND NOMENCLATURE, WESTERN INTERIOR
B 6407 1004† JURASSIC COTTON VALLEY AND SMACKOVER RESERVOIR TRENDS, EAST
B 6203 537* JURASSIC CULPEPER BASIN, VIRGINIA, ABST., TECTONIC CONTROL OF
B 6305 846* JURASSIC CULPEPER GROUP CONGLOMERATES OF VIRGINIA, ABST.,
M 28 114 JURASSIC, CURTIS FORMATION, UTAH
M 28 27 JURASSIC, CURTIS FORMATION, UTAH
M 28 7 JURASSIC, CURTIS FORMATION, UTAH
M 28 115 JURASSIC, CURTIS FORMATION, UTAH
M 28 39 JURASSIC, CURTIS FORMATION, UTAH
M 25 443 JURASSIC DEFORMATION, DAEBO OROGENY, SOUTH KOREA
S 7 26 JURASSIC, EASTERN GULF OF MEXICO, DISTRIBUTION MAP
B 6306 970 JURASSIC, EGYPT
M 28 44 JURASSIC, EISENOOLITH, GERMANY
M 27 148 JURASSIC, EISENOOLITH, GERMANY
M 28 106 JURASSIC, ENTRADA SANDSTONE, UTAH
M 28 107 JURASSIC, ENTRADA SANDSTONE, UTAH
B 6203 563* JURASSIC EPICONTINENTAL SEAWAY, WESTERN UNITED STATES, ABST.,
S 9 23 JURASSIC, FOLDOUT, PALEOCLIMATES
M 27 190 JURASSIC, FOREST MARBLE BEDS, ENGLAND
M 27 184 JURASSIC, FOREST MARBLE BEDS, ENGLAND
B 6007 1119† JURASSIC FORMATION, NORTHEASTERN TEXAS, DEFINITION OF, GILMER
M 27 25 JURASSIC, FRANCISCAN FORMATION, CALIFORNIA
S 6 291 JURASSIC, GEOLOGICAL SOCIETY OF LONDON
B 6409 1561* JURASSIC GEOLOGY AND HYDROCARBON POTENTIAL OF
S 6 295 JURASSIC, GEOMAGNETIC REVERSAL SCALE, COMPARISON
S 6 293 JURASSIC, GEOMAGNETIC REVERSAL SCALE, PROPOSED
M 26 85 JURASSIC, GLOBAL CYCLES
B 6405 719* JURASSIC GRAINSTONES, ABST., FRESHWATER, CEMENTATION OF,
B 6209 1673† JURASSIC, GRANTS MINERAL BELT, NEW MEXICO, RUBIDIUM-,
B 6410 1718† JURASSIC, GRANTS MINERAL BELT, NEW MEXICO. DISCUSSION AND
S 7 82 JURASSIC, GULF OF MEXICO, LOUANN DEPOSITION
B 6405 763* JURASSIC, HENRY BASIN, UTAH, ABST., MUDSTONES AS EXPLORATION
M 27 208 JURASSIC, HIERLATZKALK, GERMANY
B 6409 1662 JURASSIC INNER MAGNETIC QUIET ZONE, PROFILE, WESTERN, ATLANTIC
B 6205 829 JURASSIC, ISRAEL
B 6109 1475 JURASSIC, LATE, PETROLEUM RESERVOIR SYSTEM, BRAZIL

B 6103 443† JURASSIC LITHOLOGY IN GREAT ISAA 1 WELL, BAHAMAS. DISCUSSION
B 6410 1659 JURASSIC MAGNETIC QUIET ZONE, BASEMENT AND TECTONIC MAP
M 30 530 JURASSIC, MALOSSA FIELD, CONSESIO FORMATION
M 30 531 JURASSIC, MALOSSA FIELD, SELCIFERO FORMATION
M 30 530 JURASSIC, MALOSSA FIELD, ZANDOBBIO FORMATION
B 6102 217 JURASSIC, MEXICO
B 6211 2277 JURASSIC, MONTANA, WILLISTON BASIN
M 26 279 JURASSIC, MORRISON FORMATION
M 28 65 JURASSIC, MORRISON FORMATION, COLORADO
M 28 26 JURASSIC, MORRISON FORMATION, COLORADO
M 28 157 JURASSIC, MORRISON FORMATION, COLORADO
M 28 166 JURASSIC, MORRISON FORMATION, COLORADO
M 28 133 JURASSIC, MORRISON FORMATION, COLORADO
M 28 57 JURASSIC, MORRISON FORMATION, UTAH
M 28 56 JURASSIC, MORRISON FORMATION, UTAH
B 6404 516 JURASSIC, NEW ENGLAND CONTINENTAL MARGIN
B 6203 420 JURASSIC, NORTH, ATLANTIC
B 6204 596 JURASSIC, NORTH, ATLANTIC BASIN
B 6409 1565* JURASSIC, NORTH LOUISIANA, ABST., PETROGRAPHIC, STRATIGRAPHY,
S 7 27 JURASSIC, NORTHERN GULF OF MEXICO, SALT AND DIAPIRS
S 7 25 JURASSIC, NORTHERN GULF OF MEXICO, STRATA
S 6 290 JURASSIC, NUMERIC TIME SCALE, COMPARISON
S 6 295 JURASSIC, OCEANIC BASEMENT AGES
B 6004 725* JURASSIC, OF CENTRAL HIGH ATLAS, MOROCCO, ABST., CARBONATE,
B 6104 511† JURASSIC, OF FRANCE, PATTERNS OF, POROSITY AN, CEMEN IN, OOID,
B 6304 676† JURASSIC, OF FRANCE. DISCUSSIN AND REPLY, PATTERNS O, POROSITY
B 6009 1614* JURASSIC OF SOUT ARKANSAS, ABST., COMBINATION, ENTRAPMENT IN,
B 6004 715* JURASSIC OF SOUTH ARKANSAS, ABST., COMBINATION, ENTRAPMENT I,
B 6305 837* JURASSIC, OF SOUTH-CENTRA UTAH, ABST., SEDIMENTARY AND
B 6004 730* JURASSIC OF SOUTHWESTERN OREGON, NOT ASSOCIATED WITH
B 6303 553* JURASSIC OF TUNISIA, ABST., SEDIMENTOLOGIC STUDIES IN
B 6108 1386* JURASSIC, OF UTAH AND NORTHERN ARIZONA-EOLIAN OR MARINE IN
B 6408 1245† JURASSIC OF VIKING BASIN, UNITED KINGDOM-NORWAY, PRESSURE
B 6203 457 JURASSIC OIL ZONE, AUSTRALIA
B 6405 791* JURASSIC OSTRACODA WITH THOSE OF WESTERN EUROPE, ABST.,
B 6409 1662 JURASSIC OUTER MAGNETIC QUIET ZONE, PROFILE, WESTERN, ATLANTIC
M 30 281 JURASSIC, PAINTER RESERVOIR FIELD, NUGGET SANDSTONE
B 6303 520* JURASSIC PALEOGEOGRAPHY AND ORIGIN OF GULF OF MEXICO, ABST.,
S 6 289 JURASSIC, PALEOMAGNETIC SCALES, A REVIEW
B 6104 489 JURASSIC PALEOSTRUCTURE TRAPS, TILTED-OUT, SALT-GENERATED
B 6303 495* JURASSIC, PAUP SPUR FIELD, MILLER COUNTY, ARKANSAS, ABST.,
B 6109 1460 JURASSIC, PEDREGOSA BASIN
M 27 208 JURASSIC, PENNINE BÜNDNERSCHIEFER, SWITZERLAND
B 6408 1245 JURASSIC, PIEZOMETRIC MAP, UNITED KINGDOM AND, NORWAY
M 27 163 JURASSIC, PISOLITIC LIMESTONE, ITALY
B 6405 797* JURASSIC PLAY IN NORTH LOUISIANA SALT BASIN, ABST., GEOLOGY OF,
M 28 79 JURASSIC, POINT SAL OPHIOLITE, CALIFORNIA
M 27 24 JURASSIC, POINT SAL OPHIOLITE, CALIFORNIA
M 27 23 JURASSIC, POINT SAL OPHIOLITE, CALIFORNIA
B 6009 1616* JURASSIC POSSIBILITIES IN RIO GRANDE EMBAYMENT, ABST., NEW,
S 6 293 JURASSIC, PROPOSED TIME SCALE
S 6 291 JURASSIC, PROPOSED TIME SCALE
M 29 86 JURASSIC QUIET ZONE, ATLANTIC MARGIN
M 27 50 JURASSIC, RADIOLARITE, MUSCAT, OMAN
M 27 24 JURASSIC, RADIOLARITE, MUSCAT, OMAN
M 28 81 JURASSIC, RADIOLARITI, ITALY
S 6 290 JURASSIC, RADIOMETRIC SCALE, COMPARISON
B 6304 678 JURASSIC, RESERVOIR, FRANCE, DOGGER, MIDDLE
B 6303 490* JURASSIC ROCKS DREDGED FROM BERING SE CONTINENTAL MARGIN,
B 6406 945† JURASSIC ROCKS DREDGED NEAR PRIBILOF ISLANDS, SOUTHERN BERING
B 6303 462* JURASSIC ROCKS OF SOUTHERN TUNISIA, ABST., SHALLOW-WATER,
M 27 194 JURASSIC, RONDA UNIT, SPAIN
M 27 88 JURASSIC, RONDA UNIT, SPAIN
M 27 221 JURASSIC, RONDA UNIT, SPAIN
M 27 111 JURASSIC, RONDA UNIT, SPAIN
B 6408 1232 JURASSIC SAND RIDGES, NORTH AMERICA
M 28 179 JURASSIC, SANDSTONE, NORTH SEA
B 6302 182† JURASSIC SANDSTONES FROM NORWEGIAN DANISH BASIN, NORTH SEA,
B 6405 708* JURASSIC SANDSTONES, LOWER COOK INLET, ALASKA, ABST., CONTROLS
B 6405 793* JURASSIC SEA-LEVEL CHANGES FROM SEISMIC STRATIGRAPHY, ABST.
B 6101 58 JURASSIC, SEDIMENTARY ROCKS, LEVANT
B 6211 2349* JURASSIC SEDIMENTARY SEQUENCE, SIERRA NEVADA FOOTHILLS,
B 6105 825* JURASSIC SEDIMENTATION, EASTERN SIERRA MADRE ORIENTAL,
B 6410 1677 JURASSIC SEDIMENTATION RATES, WESTERN, ATLANTIC MAGNETIC
B 6410 1658† JURASSIC SEISMIC STRATIGRAPH AND BASEMENT STRUCTURE OF
M 28 160 JURASSIC, SHALE, ALASKA
B 6311 2103† JURASSIC SILL, CALHOUN COUNTY, MISSISSIPPI, VARDAMAN, TRIASSIC-
B 6409 1563* JURASSIC SMACKOVER CARBONATE ROCKS OF SOUTHWEST ALABAMA,
M 27 112 JURASSIC, SMACKOVER FORMATION, ALABAMA
B 6405 753* JURASSIC SMACKOVER IN SOUTHERN ARKANSAS, ABST..
M 27 175 JURASSIC, SOLNHOFEN LIMESTONE, GERMANY
M 27 128 JURASSIC, SOLNHOFEN LIMESTONE, GERMANY
M 27 222 JURASSIC, SOLNHOFEN LIMESTONE, GERMANY
B 6004 708* JURASSIC, SOUTHWESTERN WYOMING AND NORTHEASTER UTAH, ABST.,
M 30 102 JURASSIC, STATFJORD FIELD, BRENT FORMATION
M 30 102 JURASSIC, STATFJORD FIELD, DUNLIN FORMATION
M 30 104 JURASSIC, STATFJORD FIELD, HEATHER FORMATION
M 30 104 JURASSIC, STATFJORD FIELD, HUMBER GROUP
M 30 105 JURASSIC, STATFJORD FIELD, KIMMERIDGE CLAY FORMATION
M 30 101 JURASSIC, STATFJORD FIELD, STATFJORD FORMATION
M 24 276† JURASSIC STRATIGRAPHIC TRAP, JAY FIELD, FLORIDA--A
M 25 499 JURASSIC SUCCESSOR BASIN, CROSS SECTION, CANADIAN, CORDILLERA
B 6101 15 JURASSIC, SVERDRUP BASIN
B 6101 13 JURASSIC, SVERDRUP BASIN, TRIASSIC-
M 29 74 JURASSIC, TECTONIC DEFORMATION, ATLANTIC COAST
B 6004 667* JURASSIC TIME IN ATLAS MOUNTAINS OF MOROCCO, NORTHWEST
S 6 289† JURASSIC TIME SCALE, A
B 6004 489† JURASSIC TIME SCALE, A
B 6004 492 JURASSIC TIME SCALE, GEOCHRONOLOGIC SCALE
M 25 446 JURASSIC TO CRETACEOUS, EARLY, METALLOGENI PROVINCES, SOUTH
B 6403 366 JURASSIC TO LATE TERTIARY TIME, TIME RELATIONS BETWEEN,
B 6405 606 JURASSIC TO TURONIAN, MIDDLE EAST
B 6206 949 JURASSIC, TURBIDITES, OREGON
B 6206 1075† JURASSIC TWIN CREEK LIMESTONE, IDAHO-WYOMING THRUST BELT,
B 6204 603 JURASSIC, U.S., ATLANTIC SLOPE
B 6007 1120 JURASSIC, UPPER, CORRELATION CHART, TEXAS
B 6211 2356* JURASSIC VOLCANICLASTIC ROCKS ADJACENT TO LOWER COOK INLET,
M 29 31 JURASSIC VOLCANICS, BLAKE PLATEAU
B 6205 778 JURASSIC, WEST, AFRICA, OFFSHORE
S 9 11 JURASSIC, WIND DIRECTION, BRAZIL
M 27 141 JURASSIC, ZULOAGO FORMATION, MEXICO
B 6009 1608* JURASSIC-CRETACEOUS BOUNDARY, ABST., NANNOFOSSIL,
B 6004 657* JURASSIC-CRETACEOUS BOUNDARY, NEOCOMIAN STRATIGRAPHY, AND
B 6404 514 JURASSIC-CRETACEOUS BOUNDARY, NEW ENGLAND
B 6101 56 JURASSIC-CRETACEOUS PETROLEUM PROSPECTS, LEBANON
B 6101 15 JURASSIC-CRETACEOUS, SVERDRUP BASIN
B 6406 857† JURASSIC-EARLY CRETACEOUS REGRESSIVE FACIES, NORTHEAST AFRICA
B 6403 359† JURASSIC-LATE TERTIARY TIME, GEODYNAMIC, PEAKS IN, ALPINOTYPE,
B 6012 2177* JURASSIC-NEOCOMIAN MAP UNIT IN NORTHERN BROOKS RANGE
M 29 471 JURASSIC-TERTIARY, GLOBAL CYCLES OF CHANGE
B 6109 1548* JURASSIC—PERRY AND STONE COUNTIES, MISSISSIPPI, ABST., PETROLOGY
B 6104 585 JURAYBIAT FIELDS, PERSIAN GULF
M 25 8† JURISDICTION OFF PACIFIC COAST OF ASIA, BOUNDARIES OF, SEABED
B 6111 1982 JURUPA HILLS, CALIFORNIA
B 6201 7 K BLOCK, OFFSHORE, NIGER DELTA
M 24 209 K, O, Q SANDSTONE, LOUISIANA, G, J
M 24 212 K SANDSTONES, LOUISIANA, G THROUGH
B 6404 514 K UNCONFORMITY, NEW ENGLAND
B 6203 374 KACHEMAK BAY, ALASKA
B 6203 547 KACZOR, MICHAEL J., INDIGENOUS PRECAMBRIAN PETROLEUM
B 6410 1681 KACZOR, MICHAEL J., INDIGENOUS PRECAMBRIAN PETROLEUM
B 6409 1560 KACZOROWSKI, R. T., LOUISIANA CHENIER SYSTEM—SOME PRELIMINARY
B 6203 444 KADI FORMATION, INDIA
M 29 239 KAGAMI, HIDEO, SUBSIDENCE OF THE DAITO RIDGE AND ASSOCIATED
B 6203 377 KAGUYAK FORMATION, ALASKA
B 6203 528 KAHLE, CHARLES F., SUBAERIAL EXPOSURE OF SILURIAN SHELF-MARGIN
B 6405 757* KAIBAB AND TOROWEAP FORMATIONS IN SOUTHWESTERN UTAH, ABST.,
B 6207 1171 KAIBAB LIMESTONE, NEVADA AND, UTAH
B 6301 42 KAINER FORMATION, TEXAS
B 6405 796* KAIPAROWITS PLATEAU, UTAH, ABST., MULTIPLE BARRIER-ISLAND AND
B 6008 1412* KAIPAROWITS REGION, UTAH, ABST., SEDIMENTATION AND MULTIPLE,
B 6207 1132 KAIS FORMATION, IRIAN JAYA
B 6207 1132 KAIS LIMESTONE, BATANTA ISLAND
S 4 54 KAISER CHANNEL, JAMAICA
B 6309 1604 KAISER, W. R., GULF COAST LIGNITE--STATUS REPORT, ABST.
B 6105 801 KAISER, W. R., TEXAS LIGNITE
S 8 152 KALALAU CANYON, HAWAII, SUBMARINE CANYONS
M 25 401 KALGOORLIE, CARR BOYD NICKEL DEPOSIT
M 25 401 KALGOORLIE, NICKEL DEPOSITS, WESTERN AUSTRALIA
M 25 400 KALGOORLIE, NICKEL MINERALIZATION
B 6008 1379 KALHOR, R., IRANIAN PETROLEUM INSTITUTE
B 6209 1685 KALIMANTAN, ATTIKA OIL FIELD
B 6209 1685 KALIMANTAN, BADAK GAS FIELD
M 25 246 KALIMANTAN BASINS, INDONNSIA, EAST
B 6209 1688 KALIMANTAN, BEKAPAI FIELD
B 6401 54 KALIMANTAN, BERAI LIMESTONE
B 6401 47 KALIMANTAN, CD LIMESTONE
B 6209 1687 KALIMANTAN, HANDIL FIELD
B 6303 470* KALIMANTAN, INDONESIA, ABST., DISCOVERY AND DEVELOPMENT OF,
B 6303 545* KALIMANTAN, INDONESIA, ABST., HANDIL FIELD, EAST

M 30 441† KALIMANTAN, INDONESIA, DISCOVERY AND DEVELOPMENT OF THE,
B 6401 37† KALIMANTAN, INDONESIA, STRUCTURE, STRATIGRAPHY, AND
B 6401 56 KALIMANTAN, JAVA SEA, POLENG FIELD
B 6401 51 KALIMANTAN, KUDJUNG UNIT 1
B 6401 49 KALIMANTAN, KUDJUNG UNIT 2
B 6401 48 KALIMANTAN, KUDJUNG UNIT 3
M 30 460 KALIMANTAN, KUTEI BASIN, BEKAPAI FIELD
B 6209 1685 KALIMANTAN, MAHAKAM DELTA, HYDROCARBON ACCUMULATIONS
B 6209 1685 KALIMANTAN, MAHAKAM DELTA, MORPHOLOGY
B 6401 44 KALIMANTAN, PRE-CD LIMESTONE
B 6401 40 KALIMANTAN, SOUTHERN, OFFSHORE, COLUMNAR SECTION
B 6401 41 KALIMANTAN, SOUTHERN, OFFSHORE, KUDJUNG LIMESTONE
B 6401 45 KALIMANTAN, SOUTHERN, OFFSHORE, SECTIONS
B 6401 39 KALIMANTAN, SOUTHERN, OFFSHORE, STRUCTURAL FEATURES
B 6401 56 KALIMANTAN, TANDJUNG FIELD
M 25 479 KALIMANTAN, ULTRAMAFIC VOLCANIC ROCKS
B 6401 56 KALIMANTAN, WARUKIN FIELD
S 5 121 KALKASKA COUNTY, EXAMPLE, MICHIGAN,
S 5 115 KALKASKA COUNTY, LOCATION, MICHIGAN,
S 5 3 KALKASKA COUNTY, LOCATION, MICHIGAN,
B 6209 1610 KALKBERG LIMESTONE, NEW YORK
B 6405 786 KALLIOKOSKI, J., GEOPHYSICAL EXPLORATION FOR
B 6408 1132 KALOA GAS FIELD, ALASKA
B 6203 444 KALOL FORMATION, INDIA
B 6203 448 KALOL FORMATION, MSP MARKER, INDIA
B 6203 449 KALOL FORMATION, USP MARKER, INDIA
B 6002 164 KALTAG FAULT, BERINGIAN SHELF
B 6109 1499 KALUBIK CREEK SECTION, ALASKA
B 6004 685 KAMAL, R. A., GULL RIVER LIMESTONE--TRANSGRESSIVE SEQUENCE OF
M 25 401 KAMBALDA, NICKEL DEPOSITS, WESTERN AUSTRALIA
M 25 479† KAMCHATKA AND KIMBERLITE PROBLEMS OF PACIFIC MOBILE
M 25 481 KAMCHATKA ARC, MIGRATION OF, KURIL-
M 25 481 KAMCHATKA ARC, MINERALS, KURIL-
S 3 23 KAMCHATKA FUMAROLES
M 25 466 KAMCHATKA GEOSYNCLINAL ZONE, USSR, OLYUTOR-
B 6102 193 KAMCHATKA PENINSULA, KAMCHATSKIY POLUOSTROV
B 6407 1087† KAMCHATKA, SAKHALIN, STRATIGRAPHY OF MARINE, PALEOGENE AND,
B 6102 193 KAMCHATKA TRENCH, KURIL-
M 25 479 KAMCHATKA, ULTRAMAFIC VOLCANIC ROCKS, USSR
B 6002 163 KAMCHATKA-KORYAK EUGEOSYNCLINAL COMPLEX, SIBERIA
B 6309 1468 KAMCHATKA-KORYAK FOLDBELT, USSR
B 6102 193 KAMCHATSKIY POLUOSTROV, KAMCHATKA PENINSULA
B 6201 1 KAMERLING, P., HYDROCARBON HABITAT OF TERTIARY NIGER DELTA
B 6109 1560 KAMINSKY, BARNEY, DETECTION OF PETROLEUM-RELATED SOIL
B 6203 374 KAMISHAK BAY, ALASKA
B 6203 377 KAMISHAK FORMATION, ALASKA
B 6203 374 KAMISHAK HILLS, ALASKA
B 6203 377 KAMISHAK HILLS SECTION, ALASKA
M 25 451 KAMIYAMA, TEIJI, GEOLOGI ASPECTS OF URANIUM RESOURCES OF
B 6311 2060 KAMLOOPS GROUP, BRITISH COLUMBIA
B 6212 2509* KAMOJANG GEOTHERMAL PROJECT, WEST JAVA, INDONESIA, ABST.,
B 6210 1985 KAMPUCHEA, DEVELOPMENTS, 1977
B 6310 1823 KAMPUCHEA, DEVELOPMENTS, 1978
B 6411 1888 KAMPUCHEA, DEVELOPMENTS, 1979
S 6 159 KANAIGAURA FORMATION, JAPAN, HAUTERIVIAN
B 6406 938 KANAWHA FORMATION, WEST VIRGINIA
B 6405 758* KANAYUT CONGLOMERATE, BROOKS RANGE, ALASKA, ABST.,
B 6404 528 KANE BASIN, GREENLAND
B 6303 477 KANES, W. H., REGIONAL TECTONICS AND PETROLEUM GEOLOGY OF
B 6303 477 KANES, W. H., TUNISIA--PLATE TECTONIC AND HYDROCARBON
B 6303 520* KANGAN AREA, IRAN, ABST., RECENT HUGE, GAS DISCOVERIES IN
M 30 12 KANGAN GAS FIELD, LANDSAT IMAGE, IRAN
B 6401 10 KANGAN GIANT GAS FIELD, LANDSAT DATA, IRAN
B 6311 2089 KANGAROO SYNCLINE, AUSTRALIA
B 6201 41 KANGAROO SYNCLINE, AUSTRALIA
B 6206 1018 KANGUK DIAPIR, AXEL HEIBERG ISLAND
S 5 99 KANKAKEE DOLOMITE, ILLINOIS,
B 6002 321* KANSAS, ABST., BIOFACIES ANALYSIS OF UPPERMOST, HAMLIN SHALE
B 6311 2120* KANSAS, ABST., CHANNEL, SANDSTONE OIL RESERVOIRS OF,
B 6002 319* KANSAS, ABST., DELTAIC DEPOSITIONAL ENVIRONMENTS OF, PERMIAN,
B 6004 671* KANSAS, ABST., DELTAIC SUBENVIRONMENTS AND, PALEOCURRENT
B 6311 2118* KANSAS, ABST., DEPOSITIONAL ENVIRONMENT OF, BARTLESVILLE
B 6002 322* KANSAS, ABST., DEPOSITIONAL ENVIRONMENTS OF, CHEROKE,
B 6311 2115* KANSAS, ABST., EXPLORATION FOR, PENNSYLVANIAN SANDSTONES,
B 6002 319* KANSAS, ABST., GEOLOGY OF, MISSISSIPPIAN, COMBINATION, TRAP,
B 6002 318* KANSAS, ABST., GEOLOGY OF, PANOMA GAS AREA, SOUTHWESTER
B 6203 569* KANSAS, ABST., GORHAM OIL FIELD, RUSSELL COUNTY
B 6004 722* KANSAS, ABST., LINEAMENT AND STRUCTURAL ANALYSES OF
B 6002 319* KANSAS, ABST., NUMERICAL ANALYSIS OF, TRAP LOCATIONS IN,
B 6002 318* KANSAS, ABST., OIL AND, GAS, POTENTIAL OF WESTER
B 6406 960* KANSAS, ABST., PETROLOGY OF, ARBUCKLE GROUP, CENTRAL
B 6002 318* KANSAS, ABST., PROBABILITY METHODS OF OIL EXPLORATION IN
B 6405 788* KANSAS, ABST., PROFILE OF UNUSUAL, OOLITE DEPOSIT—DRUM
B 6002 323* KANSAS, ABST., SALT, DISSOLUTION IN OIL AND GAS, TEST, HOLE IN
B 6303 541* KANSAS, ABST., SEDIMENTOLOGY AND SYNERGY OF, DELTAIC,
B 6203 511* KANSAS, ABST., SUBSURFACE, HEAVY-OIL-BEARING, SANDSTONES IN
B 6002 320* KANSAS, ABST., YARNELL TREND, SUMNER COUNTY
B 6107 1046 KANSAS, ALTAMONT FORMATION
B 6107 1060 KANSAS AND ILLINOIS CYCLOTHEMS, POSSIBLE RELATIONS OF
B 6203 570* KANSAS AND SOUTHERN NEBRASKA, ABST., EARLY SUBAERIAL
B 6103 311 KANSAS, CENTRAL KANSAS UPLIFT
S 4 243 KANSAS CITY GROUP, KANSAS
B 6107 1030 KANSAS CITY GROUP, NEBRASKA
B 6105 839* KANSAS CITY GROUPS IN NORTHWEST KANSAS-SOUTHWEST NEBRASKA,
B 6107 1045 KANSAS, CONODONT DISTRIBUTION AND THE, CYCLIC ROCK SEQUENCE
B 6107 1046 KANSAS, DEER CREEK FORMATION
B 6107 1046 KANSAS, DENNIS FORMATION
S 4 243 KANSAS, DENNIS LIMESTONE
B 6108 1305 KANSAS, DEVELOPMENTS, 1976
B 6208 1477 KANSAS, DEVELOPMENTS, 1977
B 6308 1301 KANSAS, DEVELOPMENTS, 1978
B 6409 1437 KANSAS, DEVELOPMENTS, 1979
B 6107 1046 KANSAS, DRUM, QUIVIRA, FORMATION
B 6107 1046 KANSAS, EASTERN, PENNSYLVANIAN, STRATIGRAPHIC COLUMN
B 6107 1046 KANSAS, EXCELLO FORMATION
B 6107 997 KANSAS, FORT HAYS LIMESTONE MEMBER, NIOBRARA LIMESTONE
B 6107 1046 KANSAS, FORT SCOTT FORMATION
B 6103 313 KANSAS, FORT SCOTT LIMESTONE
B 6002 320* KANSAS GEOLOGY AS SEEN BY ERTS-1, ABST.
B 6107 1046 KANSAS, HASKELL FORMATION
B 6107 1046 KANSAS, HERTHA FORMATION
B 6107 1046 KANSAS, IOLA FORMATION
S 4 243 KANSAS, KANSAS CITY GROUP
B 6107 1046 KANSAS, LECOMPTON FORMATION
B 6107 1046 KANSAS, LENAPAH FORMATION
B 6107 1046 KANSAS, LOWER CHERRYVALE FORMATION
B 6212 2431 KANSAS, MAGNETIC INTENSITY
M 27 68 KANSAS, MISSISSIPPIAN BIOHERMAL MOUND
B 6103 309† KANSAS, MISSISSIPPIAN COMBINATION, TRAP, BINDLEY FIELD,
B 6107 987 KANSAS, NIOBRARA LIMESTONE
B 6107 997 KANSAS, NIOBRARA LIMESTONE
B 6107 1046 KANSAS, OREAD FORMATION
B 6107 1046 KANSAS, PAWNEE FORMATION
B 6107 1046 KANSAS, PLATTSBURG FORMATION
S 4 243 KANSAS, PLATTSBURG LIMESTONE
B 6108 1305 KANSAS, PRODUCTION, 1976
B 6208 1477 KANSAS, PRODUCTION, 1977
B 6107 987 KANSAS, SMOKY HILL CHALK MEMBER, NIOBRARA LIMESTONE
B 6107 1046 KANSAS, SOUTH BEND FORMATION
B 6107 1046 KANSAS, STANTON FORMATION
B 6107 1046 KANSAS, SWOPE FORMATION
B 6107 1046 KANSAS, TOPEKA FORMATION
B 6103 312 KANSAS, WARSAW FORMATION
S 4 243 KANSAS, WYANDOTTE FORMATION
B 6107 1046 KANSAS, WYANDOTTE FORMATION
B 6311 2116* KANSAS-OKLAHOMA TRI-STATE AREA, ABST., CORRELATION OF,
B 6105 839* KANSAS-SOUTHWEST NEBRASKA, ABST., LITHOFACIES OF UPPER,
B 6002 318 KANSAS. BY TITLE ONLY, REPORT--EASTERN
B 6002 322* KANSAS. BY TITLE ONLY, REPORT--WESTERN
B 6405 742 KANTER, LISA R., ET AL, DEPTH MIGRATION AND INTERPRETATION OF
M 25 444 KAOLIN, SOUTH KOREA
M 28 70 KAOLINITE, AUTHIGENIC, IN SANDSTONE
M 28 69 KAOLINITE, BIREFRINGENCE
M 28 137 KAOLINITE, BIREFRINGENCE, CEMENTS
M 28 67 KAOLINITE, IN ARKOSE, CLAY
M 28 74 KAOLINITE, IN BAUXITE
M 28 69 KAOLINITE, IN CLAYSTONE
M 28 70 KAOLINITE, IN CLAYSTONE, SEM
M 28 128 KAOLINITE, IN SANDSTONE, CEMENTS
M 28 136 KAOLINITE, IN SANDSTONE, CEMENTS
M 28 157 KAOLINITE, REPLACEMENT OF CLAY
M 28 67 KAOLINITE, VERMICULAR, AUTHIGENIC
M 28 138 KAOLINITE, VERMICULAR, IN CALCITE CEMENT
M 28 137 KAOLINITE, VERMICULAR, PORE FILLING
B 6002 293 KAOLINITIC CLAY, COMPACTIONAL RESPONSE OF
B 6002 293† KAOLINITIC CLAY, HIGH-, PRESSURE, CONSOLIDATION OF
B 6002 295 KAOLINITIC DEPOSITS, ENGLAND, RIVER, FAL
B 6205 852 KAOLINIZATION, CALIFORNIA, GATCHELL SAND
B 6404 528 KAP CONSTITUTION. GREENLAND

B 6404 530 KAP GODFRED HANSEN FORMATION, GREENLAND
B 6404 536 KAP GODFRED HANSEN, GREENLAND
B 6404 528 KAP JEFFERSON, GREENLAND
B 6404 530 KAP LUCIE MARIE FORMATION, GREENLAND
B 6404 528 KAP LUCIE MARIE, GREENLAND
B 6404 530 KAP MAYNARD FORMATION, GREENLAND
B 6404 528 KAP MORRIS JESUP, GREENLAND
B 6404 530 KAP MORTON FORMATION, GREENLAND
B 6404 528 KAP TYSON, GREENLAND
B 6203 528 KAPLAN, A. E., COMPUTER APPLICATIONS IN REGIONAL STUDY OF
B 6203 528 KAPLAN, A. E., REGIONAL GEOLOGIC MAPPING OF NORTH SEA BY
B 6104 504 KAPLAN, I. R., COLOR OF KEROGEN AS INDEX OF ORGANIC MATURITY
B 6405 763 KAPLAN, I. R., GEOCHEMISTRY OF ARTIFICIALLY HEATED HUMIC AND
B 6204 687 KAPLAN, I. R., HYDROCARBON GENERATION BY THERMAL ALTERATION
B 6405 730 KAPLAN, SANFORD S., SEDIMENTOLOGIC DESCRIPTION OF PART OF
B 6105 802 KAPLAN, SANFORD SANDY, DISTRIBUTION OF SULFUR IN CONTINENTAL
M 25 211† KAPUNI AND MAUI GAS-CONDENSATE FIELDS OF NEW
M 25 213 KAPUNI FIELD, NEW ZEALAND
M 25 214 KAPUNI FORMATION, NEW ZEALAND
M 25 218 KAPUNI GAS-CONDENSATE FIELD, NEW ZEALAND
S 10 232 KARACHUKHUR FIELD, USSR, BAKU AREA
B 6407 1088 KARAGINSKY ISLAND, ASIA, NORTHEAST
B 6102 193 KARAGINSKY ISLAND, OSTROV KARAGINSKIY
B 6404 572 KARBOLET CAPROCK, ISRAEL
B 6404 572 KARBOLET SALT AND SHALE MEMBER, ISRAEL
M 29 223 KARIG, D. E., STRUCTURE AND CENOZOIC EVOLUTION OF THE SUNDA
B 6203 519 KARKALITS, O. C., CHEMICAL ANALYSES OF GEOTHERMAL WATERS FROM
B 6203 529 KARKALITS, O. C., CHEMICAL ANALYSIS OF GAS DISSOLVED IN
B 6405 731 KARL, H. A., SEDIMENTARY PROCESSES ON CONTINENTAL SHELF OFF
B 6105 802 KARL, HERMAN A., INTERNAL WAVES—IMPLICATION FOR DISPERSAL OF
B 6303 440* KARNES COUNTY, TEXAS, ABST., DETAILED STRATIGRAPHY AND
B 6409 1552* KARNES COUNTY, TEXAS, ABST., ENVIRONMENTS OF DEPOSITION AND
B 6004 725* KARNES COUNTY TEXAS, ABST., FACTORS INFLUENCING POROSITY AND,
B 6109 1547* KARNES COUNTY, TEXAS, ABST., SECONDARY, POROSITY IN,
B 6301 36 KARNES COUNTY, TEXAS, PERSON FIELD
B 6301 35 KARNES TROUGH, TEXAS
M 25 395 KARNS, A. WESLEY, SUBMARINE PHOSPHORITE DEPOSITS OF CHATHAM
S 6 369 KAROO SYSTEM, SOUTH AFRICA
S 6 184 KARPATHIAN STAGE, PARATETHYS, NEOGENE
B 6409 1561 KARPOVICH, RAYMOND P., JURASSIC GEOLOGY AND HYDROCARBON
S 11 230 KARRERIELLA MILLERI
S 11 270 KARRERIELLA MILLERI
B 6004 712* KARST, ABST., SCALLOPED, SOLUTION SURFACES, MIDDLE, ORDOVICIAN,
B 6305 834* KARST FEATURES NEAR SALIDA, COLORADO, ABST.,
B 6206 1092 KASAMEYER, P., INSTRUMENTATION FOR EVALUATION OF IN-SITU
B 6203 529 KASAMEYER, P. W., INSTRUMENTATION FOR EVALUATION OF IN SITU
B 6412 2095 KASHFI, MANSOUR S., STRATIGRAPHY AND ENVIRONMENTAL
M 25 279 KASHIWATZAKI GAS FIELD, JAPAN, HIGASHI-
B 6207 1131 KASIM ISLAND, IRIAN JAYA
B 6207 1142 KASIM OIL FIELD, IRIAN JAYA
B 6311 2118 KASINO, RAYMOND E., ENVIRONMENTS AND DIAGENESIS, MORROW
B 6303 445 KASPER, DENNIS R., ENVIRONMENTAL REGULATIONS FOR IN-SITU
B 6303 477 KASSLER, P., OIL GENERATION AND MIGRATION IN LUSITANIAN BASIN,
B 6411 1961 KAT, CYNTHIA, PETROLEUM DEVELOPMENTS IN EUROPE IN 1979
B 6408 1252 KATA FORMATION, NIGERIA
M 25 276 KATAHIRA, T., PETROLEUM FIELDS OF JAPAN WIT VOLCANIC-ROCK
M 25 185 KATALLA OIL FIELD, ALASKA
B 6002 164 KATTAG FAULT, BERINGIAN SHELF
B 6409 1556* KATY GAS FIELD, WALLER COUNTY, TEXAS, ABST., ENVIRONMENT OF
B 6006 981 KATZ, D. L., HYDRATE FORMATION IN SUBSURFACE ENVIRONMENTS
B 6408 1286 KATZ, DONALD L., ABSENCE OF CONNATE WATER IN MICHIGAN REEFS
B 6010 1947 KATZ, H. R., DEVELOPMENTS IN NEW ZEALAND AND SOUTHWEST PACIFIC
B 6210 1900 KATZ, H. R., DEVELOPMENTS IN NEW ZEALAND AND SOUTHWEST PACIFIC
B 6110 1880 KATZ, H. R., DEVELOPMENTS IN NEW ZEALAND AND SOUTHWEST PACIFIC
B 6310 1680 KATZ, H. R., DEVELOPMENTS IN NEW ZEALAND AND SOUTHWEST PACIFIC
B 6411 1776 KATZ, H. R., DEVELOPMENTS IN NEW ZEALAND AND SOUTHWEST PACIFIC
M 25 217 KATZ, H. R., SEDIMENTARY BASINS AND PETROLEUM PROSPECTS,
S 8 147 KAUAI CANYONS, HAWAII, SUBMARINE CANYONS
S 8 148 KAUAI CANYONS, TOPOGRAPHY, INDEX MAP
B 6405 731 KAUFFMAN, ERLE G., PALEOBIOLOGIC EVIDENCE FOR CRETACEOUS
B 6405 778 KAUFFMAN, ERLE G., PHYSICAL EVIDENCE FOR CRETACEOUS TIDES,
S 8 155 KAULAKAHI CHANNEL, TIME-VELOCITY CURVES, CURRENT FLOW
M 24 32 KAVIK SHALE, ALASKA
B 6307 1035 KAVIRONDO AND GREGORY RIFTS, MAP, KENYA
M 25 468 KAVRALYAN COMPLEX, USSR
S 6 93 KAWASHITA, KOJI, APPLICABILITY OF THE RUBIDIUM-STRONTIUM
M 25 131 KAWERAU GEOTHERMAL FIELD, NEW ZEALAND
B 6204 635 KAYAK ISLAND, ALASKA
M 24 26 KAYAK SHALE, ALASKA
B 6109 1496 KAYAK SHALE, ALASKA
B 6109 1500 KAYAK SHALE, ALASKA
B 6103 437 KAYAK SHALE, ALASKA
B 6204 635 KAYAK TROUGH, ALASKA
M 24 79† KAYBOB OIL FIELD, ALBERTA, CANADA
M 24 80 KAYBOB REEF COMPLEX, ALBERTA
M 24 83 KAYBOB REEF COMPLEX, ALBERTA
B 6410 1682 KAZAKSTAN, PRECAMBRIAN MICROFOSSILS, USSR, SOUTH
S 6 304 KAZANIAN STAGE, PERMIAN, BIOCHRONOLOGY
B 6303 448 KEANY, JOHN, GENERATIO AND OCCLUSION OF POROSITY IN CHALK
B 6207 1223 KEAR, D., NEW ZEALAND GEOTHERMAL EXPLORATION AND
B 6303 501* KEATHLEY ANOMALY, MAGNETOSTRATIGRAPHY OF TUNISIA, ABST.,
B 6111 1962 KEATHLEY T-AGS 35 CRUISE, USNS
B 6104 603 KEATING FORMATION, NEW MEXICO
B 6309 1601* KEECHI, AND OAKWOOD SALT DOMES, EAST TEXAS, ABST., UPPER,
B 6408 1289* KEEFER SANDSTONE OF WEST VIRGINIA, ABST., CARBONATE DIAGENESIS
B 6304 652 KEEFER SANDSTONE, VIRGINIA
B 6405 629 KEEFER SANDSTONE, WEST VIRGINIA
B 6211 2309 KEEFER SANDSTONE, WEST VIRGINIA
B 6307 1136 KEEL FORMATION, OKLAHOMA
B 6109 1537 KEELAN, DARE K., CRITICAL WATER ESTIMATES FOR GULF COAST
B 6405 732 KEELAN, DARE K., ROCK PROPERTIES—INFLUENCE ON HYDROCARBON
B 6004 735 KEELING, J. A., PETROLEUM GEOLOGY AND PRODUCTIVE POTENTIAL OF
B 6203 529 KEENE, JOHN B., DEEP-SEA SEDIMENT AND THE CHERT PROBLEM, ABST.
B 6303 531* KEF ANTICLINE--BOX FOLDING IN TUNISIAN ATLAS, CLUE TO REGIONAL
M 30 540 KEF EL ARGOUB FIELD, ALGERIA, OUARGLA REGION
B 6403 424 KEF EL KATTOUS, TUNISIA
M 27 135 KEG RIVER EQUIVALENT, CANADA, DEVONIAN
B 6002 246 KEG RIVER FORMATION, ALBERTA
B 6205 782 KEG RIVER FORMATION, ALBERTA
B 6109 1540 KEHLE, R. O., GROWTH HISTORY OF HAINESVILLE SALT DOME, WOOD
B 6108 1382 KEHLE, RALPH O., ORDOVICIAN OIL ACCUMULATION AT NORTHEAST
B 6303 487 KEHLE, RALPH O., SEISMIC STRATIGRAPHY AND GEOLOGIC HISTORY OF
B 6309 1591 KEHN, THOMAS M., REVISION OF U.S GEOLOGICAL SURVEY AND U.S.
B 6405 732 KEIGHIN, C. W., SOME RELATIONS BETWEEN DIAGENESIS AND POROSITY
B 6305 832 KEIGHIN, C. WILLIAM, INFLUENCE OF DIAGENETIC REACTIONS ON
B 6205 885 KEIGHIN, C. WILLIAM, SOME PETROGRAPHIC CHARACTERISTICS OF
B 6105 802 KEITH, BRIAN D., DIAGENESIS AND POROSITY OF RODESSA LIMESTONE,
B 6103 438 KEKIKTUK CONGLOMERATE, ALASKA
B 6109 1498 KEKIKTUK CONGLOMERATE, ALASKA
M 29 443 KELLER, B., EXPLOSION SEISMOLOGY STUDIES OF ACTIVE AND PASSIVE
B 6004 667 KELLER, G. H., CONTRASTS IN CONCENTRATION OF SUSPENDED MATTER
B 6004 686 KELLER, G. H., SEDIMENTARY PROCESSES IN SUBMARINE CANYONS OF
B 6303 341 KELLER, G. R., DELINEATION OF ROME TROUGH IN EASTERN KENTUCKY
B 6009 1621 KELLER G. R., GEOPHYSICAL AND TECTONIC STUDY OF MOORMAN
B 6405 732 KELLER, G. R., INTEGRATED GEOLOGIC AND GEOPHYSICAL STUDIES AS
B 6203 529 KELLER, GEORGE H., GEOTECHNICAL PROPERTIES OF CONTINENTAL
M 25 154 KELLER, GEORGE V., DEEP RESEARCH DRILLHOLE AT KILAUEA
B 6403 444 KELLER, GERTA, BENTHIC FORAMINIFERAL BIOFACIES OF EASTERN
B 6408 1286 KELLER, STANLEY J., ST. LOUIS AND SALEM STRATIGRAPHY AND OIL
B 6008 1288 KELLEY, WILLIAM W., JR., OIL AND GAS DEVELOPMENTS IN MARYLAND,
B 6108 1269 KELLEY, WILLIAM W., JR., OIL AND GAS DEVELOPMENTS IN MARYLAND,
B 6302 257 KELLING, BALLY, A. W., REVIEW ON SEDIMENTATION IN, SUBMARINE
B 6004 686 KELLING, GILBERT, SIGNIFICANCE OF CLAST SIZE, SHAPE, AN
B 6205 886 KELLOGG, HAROLD E., TERTIARY STRUCTURAL DEVELOPMENT OF
B 6211 2356 KELLY, J. S., LITHOFACIES IN LOWER JURASSIC VOLCANICLASTIC ROCKS
B 6304 690 KELLY, T. E., EFFECTS OF URANIUM MINING ON GROUNDWATER IN
B 6108 1382 KELLY, T. E., NURE PROGRAM-A NATIONWIDE APPROACH TO URANIUM
B 6208 1399 KELLY, WILLIAM W., JR., OIL AND GAS DEVELOPMENTS IN MARYLAND,
B 6409 1403 KELLY WILLIAM W., JR., OIL AND GAS DEVELOPMENTS IN MARYLAND,
B 6308 1244 KELLY, WILLIAM W., OIL AND GAS DEVELOPMENTS IN MARYLAND, OHIO,
B 6405 680 KELTS, KERRY R., LOWER CRETACEOUS LACUSTRINE SOURCE BEDS FROM
B 6407 1034 KELTS, KERRY, RESEDIMENTED FACIES IN VENTURA BASIN, CALIFORNIA,
B 6101 113 KELVIN FORMATION, UTAH
M 25 475 KELYAN DEPOSITS, ANTIMONY, USSR
B 6401 105 KEMNITZ FIELD, NEW MEXICO
B 6105 763 KEMP, MARWIN K., FLOOR FOR GAS
B 6303 414 KEMP, MARWIN K., NATURAL GAS STABILITY IN DEEP SUBSURFACE,
B 6408 1132 KENAI FIELD, ALASKA
B 6401 26 KENAI FORMATION, ALASKA
B 6203 376 KENAI GAS FIELD, ALASKA
B 6203 377 KENAI GROUP, ALASKA
B 6408 1133 KENAI GROUP, ALASKA
M 24 14 KENAI GROUP, ALASKA
B 6203 383 KENAI LOWLAND, ALASKA
B 6203 374 KENAI LOWLAND, ALASKA
B 6401 27 KENAI MOUNTAIN FAULT ZONE, ALASKA

B 6203 374 KENAI PENINSULA, ALASKA
B 6211 2356* KENAI PENINSULA, ALASKA, ABST., LITHOFACIES IN LOWER, JURASSIC,
M 29 271 KENAI PENINSULA, ORCA FORMATION, ALASKA
B 6303 306 KENAI-KODIAK RIDGE, ALASKA
B 6205 885 KENDAL, ALAN C., CARBONATE BUILDUP-EVAPORITE RELATIONS IN
B 6105 803 KENDALL, A. C., CARBONATE POROSITY MODIFICATION BY HALITE
B 6303 478 KENDALL, A. C., ORIGIN OF REGIONAL PRE-MIDDLE DEVONIAN
B 6004 686 KENDALL, A. C., SIMILARITIES IN EARLY DIAGENETIC HISTORIES OF
B 6203 530 KENDALL, ALAN C., CALCIUM SULFATE DIAGENESIS, MISSISSIPPIAN OF
B 6203 530 KENDALL, ALAN C., CARBONATE-EVAPORITE RELATIONS IN
B 6008 1400 KENDALL, ALAN C., DOLOMITIZATION PATTERNS IN PALEOZOIC O
B 6405 719 KENDALL, C., ET AL, CRETACEOUS SEA LEVEL AND STRATIGRAPHY,
B 6105 840 KENDALL, C. G., POROSITY PRESERVATION IN SHALLOW-MARINE
B 6203 573 KENDALL, ROBERT G., COPPER MOUNTAIN, WYOMING, URANIUM
B 6401 129 KENNETH LIMESTONE MEMBER, INDIANA, SALINA FORMATION
S 5 94 KENNETH LIMESTONE, SALINA FORMATION,
B 6303 478 KENNETT, JAMES P., LATE CENOZOIC CALCAREOUS MICROFOSSIL
B 6405 732 KENT, BION H., PATTERNS OF THICK COAL DEPOSITION ACROSS POWDER
B 6105 803 KENT, D. M., EFFECTS OF BURIAL STAGES ON POROSITY IN RESERVOIR
B 6205 886 KENT, DONALD M., INFLUENCE OF DIAGENETIC AND POSTLITHIFICATION
B 6004 687 KENT, H. C., DISTRIBUTION AND RELATION OF PHYSICAL AND BIOGENIC
B 6409 1529 KENT, HARRY C., POTENTIAL RESOURCES OF NATURAL GAS IN UNITED
B 6303 478 KENT, HARRY C., POTENTIAL RESOURCES OF NATURAL GAS IN UNITED
S 6 367 KENT, L. E., ASPECTS OF THE REVISED SOUTH AFRICAN STRATIGRAPHIC
B 6008 1401 KENT, NORMAN, OIL AND GAS POSSIBILITIES REDWALL LIMESTONE,
B 6004 687 KENT, P. E., ANTAGONIST--PLATE TECTONICS AS A MAJOR EXPLORATION
B 6003 458 KENT, P. E., SALT DIAPIRISM IN SOUTHERN IRAN. DISCUSSION
B 6009 1621* KENTUCKY, ABST., AGE AND TECTONICS OF, CINCINNATI ARCH IN
B 6009 1620* KENTUCKY, ABST., CARBONIFEROUS SEDIMENTARY RESPONSES TO
B 6009 1619* KENTUCKY, ABST., CHESTERIAN SHORELINE SEQUENCES IN EASTERN
B 6408 1290* KENTUCKY, ABST., COAL AND COAL DEVELOPMENTS IN WESTERN
B 6408 1281* KENTUCKY, ABST., COAL GEOLOGY AND RESOURCE OF EASTERN
B 6303 415* KENTUCKY, ABST., PALEOTIDAL-RANGE INDICATORS IN,
B 6309 1578* KENTUCKY, ABST., PAST AND CURRENT DEVELOPMENTS FOR DRILLING,
B 6305 844* KENTUCKY, ABST., PETROLEUM PRODUCTION FROM BASAL, GREENBRIER
B 6309 1589* KENTUCKY, ABST., PETROLOGY AND DEPOSITIONAL ENVIRONMENTS OF,
B 6009 1619* KENTUCKY, ABST., STRATIGRAPHIC, PALEOENVIRONMENTAL, AND
B 6009 1620* KENTUCKY, ABST., STRUCTURAL CONTROL ON, PALEOZOIC
B 6405 729* KENTUCKY, ABST., TRACE, FOSSILS AND STRATIGRAPHY OF, DEVONIAN,
B 6004 687* KENTUCKY AND INDIANA, ABST., PRODELTA, TURBIDITE-, FAN, APRON
B 6203 544* KENTUCKY AND NORTHERN TENNESSEE, ABST., NON-SABKHA,
B 6009 1623* KENTUCKY AND OHIO, ABST., LITHOFACIES RELATIONS IN,
B 6212 2385 KENTUCKY AND WEST VIRGINIA, CARBONIFEROUS BACK-BARRIER
B 6309 1587* KENTUCKY AND WEST VIRGINIA OPEN-FLOW DATA, INCORPORATING
B 6408 1285* KENTUCKY AND WEST VIRGINIA—STATISTICAL COMPARISON AND
B 6104 594 KENTUCKY, ASHLAND FIELD
B 6303 344 KENTUCKY, ASHLOCK FORMATION
B 6212 2384 KENTUCKY, BARRIER AND BACK-BARRIER ENVIRONMENTS
B 6303 344 KENTUCKY, BASAL SAND
B 6209 1705 KENTUCKY, BEDFORD SHALE
B 6303 342 KENTUCKY, BLAINE FAULT
B 6212 2398 KENTUCKY, BLAINE-WALBRIDGE FAULT
B 6303 344 KENTUCKY, BORDEN FORMATION
B 6209 1706 KENTUCKY, BORDEN FORMATION
B 6303 344 KENTUCKY, BOYLE FORMATION
B 6303 344 KENTUCKY, BRASSFIELD FORMATION
B 6303 344 KENTUCKY, BREATHITT FORMATION
B 6303 341† KENTUCKY BY GRAVITY AND DEEP DRILLING DATA, DELINEATION OF,
B 6303 344 KENTUCKY, CALLOWAY CREEK LIMESTONE
B 6303 344 KENTUCKY, CHAZY GROUP
B 6303 344 KENTUCKY, CLAYS FERRY FORMATION
B 6209 1706 KENTUCKY, CLEVELAND MEMBER
B 6408 1285* KENTUCKY COALFIELD, ABST., COAL RANK IN PART OF WESTERN
B 6303 344 KENTUCKY, CONASAUGA SHALE
B 6303 344 KENTUCKY, CRAB ORCHARD FORMATION
B 6209 1705 KENTUCKY, CRAB ORCHARD FORMATION
B 6008 1224 KENTUCKY, DEVELOPMENTS, 1975
B 6108 1186 KENTUCKY, DEVELOPMENTS, 1976
B 6208 1343 KENTUCKY, DEVELOPMENTS, 1977
B 6308 1327 KENTUCKY, DEVELOPMENTS, 1978
B 6409 1464 KENTUCKY, DEVELOPMENTS, 1979
B 6412 2163 KENTUCKY, DEVONIAN SHALE FACIES
B 6212 2387 KENTUCKY, DISTRIBUTARY-MOUTH BAR SANDSTONE
B 6209 1703† KENTUCKY, DIVISION OF BLACK, OHIO SHALE IN EASTERN
B 6303 344 KENTUCKY, DRAKE FORMATION
B 6303 349 KENTUCKY, EAST CONTINENT GRAVITY HIGH
B 6212 2387 KENTUCKY, ELKHORN COALS
B 6209 1706 KENTUCKY, FARMERS MEMBER, HENLEY BED
B 6212 2389 KENTUCKY, FLUVIAL AND UPPER DELTA PLAIN
B 6309 1506 KENTUCKY, FORMATION-DENSITY LOG STUDY
B 6303 344 KENTUCKY, GARRARD SILTSTONE
B 6309 1584* KENTUCKY GAS FIELD, ABST., LITHOLOGY STUDIES OF UPPER,
B 6309 1584* KENTUCKY GAS FIELD AND COMPARISON WITH GAS PRODUCTION,
B 6303 502* KENTUCKY GAS FIELDS, ABST., RELATIONS OF GAS OCCURRENCE TO
B 6303 344 KENTUCKY, HIGH BRIDGE GROUP
B 6209 1706 KENTUCKY, HURON MEMBER
B 6303 342 KENTUCKY, IRVINE--PAINT CREEK FAULT ZONE
B 6303 342 KENTUCKY, JOHNSON CREEK FAULT
B 6303 342 KENTUCKY, KENTUCKY RIVER FAULT ZONE
B 6303 344 KENTUCKY, KNOX GROUP
B 6303 344 KENTUCKY, LEE FORMATION
B 6303 349 KENTUCKY, LEXINGTON FAULT ZONE
B 6303 344 KENTUCKY, LEXINGTON LIMESTONE
B 6303 342 KENTUCKY, LEXINGTON LINEAMENT
B 6212 2386 KENTUCKY, LOWER DELTA PLAIN DEPOSITS
B 6303 344 KENTUCKY, NEW ALBANY SHALE
B 6303 344 KENTUCKY, NEWMAN LIMESTONE
B 6405 725* KENTUCKY NO. 11 COAL SEAM, ABST., PETROGRAPHIC VARIATION IN
B 6209 1710 KENTUCKY, OLENTANGY SHALE
B 6212 2397 KENTUCKY, PIKEVILLE TO LOUISA, SECTION
B 6008 1228 KENTUCKY, PRODUCTION, 1974-1975
B 6108 1193 KENTUCKY, PRODUCTION, 1975-1976
B 6208 1349 KENTUCKY, PRODUCTION, 1976-1977
B 6308 1331 KENTUCKY, PRODUCTION, 1977-1978
B 6409 1470 KENTUCKY, PRODUCTION, 1977-1979
B 6303 342 KENTUCKY RIVER FAULT ZONE, KENTUCKY
B 6303 344 KENTUCKY, ROME FORMATION
B 6303 345 KENTUCKY, ROME TROUGH, COMPUTER MODELING OF GRAVITY DATA
B 6303 343 KENTUCKY, ROME TROUGH, GRAVITY MAP
B 6303 344 KENTUCKY, ROME TROUGH, STRATIGRAPHY
B 6303 344 KENTUCKY, ROSE RUN SANDSTONE
B 6303 344 KENTUCKY, SHADY TOMSTOWN FORMATION
B 6209 1609 KENTUCKY, SILURIAN-DEVONIAN CARBONATE SEQUENCE
B 6209 1705 KENTUCKY, SUNBURY SHALE
B 6009 1622* KENTUCKY TO CENTRAL TENNESSEE, ABST., LATE, MIDDLE AND EARLY
B 6209 1705 KENTUCKY, THREE LICK BED, NEW MARKER
B 6212 2392 KENTUCKY, TRANSITIONAL LOWER DELTA PLAIN
S 6 348 KENTUCKY, TYRONE LIMESTONE, BENTONITES, ORDOVICIAN
B 6309 1577* KENTUCKY, VIRGINIA, WEST VIRGINIA OHIO, MARYLAND, AND
B 6303 342 KENTUCKY, WEST HICKMAN CREEK-BRYAN STATION FAULT
B 6009 1624* KENTUCKY--THEIR RELATION TO LOCAL STRUCTURE, ABST., EARLY,
B 6412 2064† KENTUCKY-VIRGINIA, SEASAT, ORBITAL, RADAR, IMAGERY FOR
B 6303 414* KENYA, ABST., DISTRIBUTION AND FACTORS CONTROLLING
B 6303 508* KENYA, ABST., GENESIS, OCCURRENCE, AND CAUSES OF SEDIMENT
B 6203 488* KENYA, ABST., OXYGEN, AND, CARBON, ISOTOPIC VARIATION IN,
B 6303 485* KENYA, ABST., TRACKS AND SUBSTRATE REWORKING BY, TERRESTRIAL,
B 6110 1748 KENYA, DEVELOPMENTS, 1976
B 6210 1847 KENYA, DEVELOPMENTS, 1977
B 6310 1693 KENYA, DEVELOPMENTS, 1978
B 6411 1788 KENYA, DEVELOPMENTS, 1979
B 6303 434* KENYA, ENVIRONMENTAL SIGNIFICANCE OF, FOSSIL, ROOT CAST FROM,
B 6307 1035 KENYA, KAVIRONDO AND GREGORY RIFTS, MAP
B 6309 1583 KEOUGH, DOUGLAS D., INTERACTION OF PROPPANTS WITH CRACK
B 6309 1576 KEPFERLE, R. C., GAS-BEARIN OHIO SHALE ALONG LAKE
B 6004 687 KEPFERLE, R. C., PRODELTA TURBIDITE-FAN APRON IN BORDE
B 6209 1703 KEPFERLE, ROY C., DIVISION OF BLACK OHIO SHALE IN EASTERN
B 6402 246 KEPHALONIA-ZAKINTHOS PLATEAU, SEISMIC REFLECTIONS, GREECE
B 6210 1985 KERGUELEN ISLANDS, DEVELOPMENTS, 1977
B 6110 1837 KERGUELEN ISLANDS, DEVELOPMENTS, 1976
B 6310 1823 KERGUELEN ISLANDS, DEVELOPMENTS, 1978
B 6411 1889 KERGUELEN ISLANDS, DEVELOPMENTS, 1979
B 6207 1164 KERGUELEN PLATEAU, CRISTOBALITE CHERT, INDIAN OCEAN
S 6 113 KERITY FORMATION, ISOCHRON DIAGRAM, FRANCE
B 6211 2245 KERN ARCH, CALIFORNIA
B 6211 2268 KERN CANYON FAULT, CALIFORNIA
B 6403 445* KERN COUNTY, CALIFORNIA, ABST., PORE-FILLING, CEMENTS--PRODUCTS
B 6004 687 KERN, J. P., TRAILS FROM THE VIENNA WOODS--PALEOENVIRONMENTS
B 6211 2268 KERN RIVER, ANCESTRAL, CALIFORNIA
B 6211 2270 KERNSUMNER FIELD, CALIFORNIA
S 10 210 KEROGEN CONVERSION, LOUISIANA, CALCASIEU PARISH, MANCHESTER
S 10 48 KEROGEN PROPERTIES, MIGRATION
S 10 98 KEROGEN SAMPLES, SEM PHOTOS IN SHALE SAMPLES
B 6405 770* KEROGEN, ABST., FLUORESCENCE OF, ACRITARCHS IN STUDY OF MARINE
B 6112 2086 KEROGEN, ALBERTA, DOIG FORMATION
B 6112 2086 KEROGEN, ALBERTA, TRIASSIC
B 6104 504† KEROGEN AS INDEX OF ORGANIC MATURITY, COLOR OF
B 6204 688 KEROGEN, CALIFORNIA, BANDERAS BAY
B 6204 688 KEROGEN, CALIFORNIA, SANTA BARBARA BASIN
B 6204 688 KEROGEN, CALIFORNIA, TANNER BASIN
B 6104 504 KEROGEN, CALIFORNIA, TANNER BASIN

B 6401 60 KEROGEN, CAMEROUN, DOUALA BASIN, CRETACEOUS
M 29 424 KEROGEN CONCENTRATION, SOURCE BEDS, PETROLEUM
B 6002 215 KEROGEN DIFFUSION HYPOTHESIS
B 6112 2086 KEROGEN, ECUADOR, NAPO FORMATION
B 6104 505 KEROGEN, ELEMENTARY COMPOSITION AND, COLOR OF HEATED
B 6401 60 KEROGEN, EXPERIMENTAL EVOLUTION OF, SOURCE ROCK AND
B 6401 61 KEROGEN, FRANCE, PARIS BASIN, TOARCIAN
B 6204 687† KEROGEN FROM DIFFERENT SEDIMENTS, HYDROCARBON, GENERATION
B 6406 922 KEROGEN HYDROGEN-CARBON RATIOS
B 6311 2115* KEROGEN IN SHALE CLAST, ABST., DISTRIBUTION OF, BITUMENS AND
B 6209 1584 KEROGEN IN SOURCE BEDS
B 6312 2223 KEROGEN, INFRARED ANALYSES
B 6112 2086 KEROGEN, LOUISIANA, HOLOCENE
B 6112 2086 KEROGEN, LOUISIANA, MIOCENE
B 6112 2086 KEROGEN, LOUISIANA, SOUTH PECAN LAKE FIELD
B 6305 771 KEROGEN, MATRIX, PETROLEUM MIGRATION FROM, SOURCE ROCKS
M 29 438 KEROGEN MATURATION IN SEDIMENTS, BALTIMORE CANYON
M 29 437 KEROGEN MATURATION, NORTH ATLANTIC, CONTINENTAL RISE
B 6209 1585 KEROGEN MATURITY
M 29 424 KEROGEN MATURITY, SOURCE BEDS, PETROLEUM
B 6412 2053 KEROGEN, METHODS FOR CHARACTERIZATION
B 6112 2086 KEROGEN, NEW MEXICO, FRUITLAND FORMATION
B 6112 2086 KEROGEN, NEW MEXICO, NAVAJO COAL MINE
B 6112 2085 KEROGEN, NORMAL OIL-TYPE
B 6112 2082† KEROGEN, OIL AND, GAS, GENERATION BY LABORATORY, PYROLYSIS OF
B 6112 2086 KEROGEN, OKLAHOMA, PENNSYLVANIAN
B 6312 2135 KEROGEN, ORGANIC, GEOCHEMISTRY PARAMETERS OF
B 6112 2094 KEROGEN, PENNSYLVANIAN-AGE, COAL
B 6201 108 KEROGEN, PHOSPHORIA FORMATION
B 6401 59† KEROGEN PYROLYSIS--INFLUENCE ON PETROLEUM GENERATION AND
B 6312 2223 KEROGEN SAMPLE, NORTH SEA, KIMMERIDGIAN
B 6112 2086 KEROGEN, TEXAS, CHERRY CANYON FORMATION
B 6112 2091 KEROGEN, TEXAS, CLEARFORK OIL
B 6112 2086 KEROGEN, TEXAS, GOMEZ FIELD
B 6412 2121 KEROGEN, TIMOR TROUGH
M 29 424 KEROGEN TYPE, SOURCE BEDS, PETROLEUM
B 6412 2053 KEROGEN TYPES, I, II, AND III
B 6112 2083 KEROGEN TYPES, LABORATORY, PYROLYSIS OF FOUR
B 6112 2082 KEROGEN TYPES, NATURAL, CATAGENESIS OF
B 6405 770* KEROGEN TYPES. BY TITLE ONLY, DEPOSITIONAL ENVIRONMENTS AND
B 6401 61 KEROGEN, UTAH, UINTA BASIN, EOCENE
B 6112 2091 KEROGEN, VERY HIGH HYDROGEN
B 6112 2092 KEROGEN, WET-GAS-GENERATING
B 6405 767* KEROGENOUS MARINE STRATA OF CENTRAL NEVADA, ABST., OIL AND
B 6405 784* KEROGENS IN UNCONSOLIDATED OCEANIC SEDIMENTS IN GUAYMAS
B 6302 240 KEROGENS, THERMAL MATURITY OF
B 6405 732 KERR, DONALD M., DEPARTMENT OF ENERGY DRILLING PROGRAMS AND
B 6004 688 KERR, R. S., DEVELOPMENT OF EARLY CRETACEOUS BANK COMPLEX
B 6109 1553 KERR, RALPH S., FACIES, DIAGENESIS, AND POROSITY DEVELOPMENT IN
B 6101 85 KERROUCHEN BASIN, ATLAS TREND
B 6009 1614* KERRVILLE, TEXAS, ABST., NEW, PELECYPOD SPECIES FROM LOWER,
B 6405 787 KERSEY, DAVID G., COASTAL SWAMP ORIGIN OF SAN MIGUEL LIGNITE
B 6405 699 KERSEY, DAVID G., PALEOECOLOGY IN BASIN ANALYSIS—HUMBOLDT
M 25 458 KERULEN VOLCANIC-PLUTONIC CHAIN, PACIFIDES, ARGURI-
B 6203 542 KESSLER, L. G., ANCIENT SUBMARINE CANYON AND CHANNEL
B 6303 495 KESSLER, L. GIFFORD, II, DOWNCURRENT CHANGE IN MIOCENE
M 30 26 KETTLEMAN HILLS FIELD, LANDSAT IMAGE, CALIFORNIA
B 6401 24 KETTLEMAN HILLS GIANT OIL FIELD, LANDSAT DATA, CALIFORNIA
B 6212 2446 KETTLEMAN NORTH DOME, CALIFORNIA
B 6211 2263 KETTLEMAN NORTH DOME FIELD, CALIFORNIA
B 6205 848 KETTLEMAN NORTH DOME FIELD, CALIFORNIA
B 6211 2233 KEUPER, BAY OF BISCAY AREA
B 6303 371 KEWANEE INTERAMERICAN OIL CO. 4CPR TEST WELL, PUERTO RICO
M 27 59 KEY LARGO LIMESTONE, FLORIDA, PLEISTOCENE
M 27 7 KEY LARGO LIMESTONE, FLORIDA, PLEISTOCENE
M 27 218 KEY LARGO LIMESTONE, FLORIDA, PLEISTOCENE
M 27 176 KEY LARGO LIMESTONE, FLORIDA, PLEISTOCENE
M 27 162 KEY LARGO LIMESTONE, FLORIDA, PLEISTOCENE
M 27 71 KEY LARGO LIMESTONE, FLORIDA, PLEISTOCENE
M 27 224 KEY LARGO LIMESTONE, FLORIDA, PLEISTOCENE
S 4 187 KEY LARGO LIMESTONE, REEF, FLORIDA
B 6303 463* KEY LARGO--SUBSURFACE CORE STUDY, ABST.
S 4 175 KEY WEST, MONTASTREA ANNULARIS
M 24 276 KEYES, P. L., JAY FIELD, FLORIDA--A JURASSIC STRATIGRAPHIC TRAP
B 6201 121 KEYS AREA WELLS, FLORIDA
S 5 130 **KEYSER LIMESTONE, WEST VIRGINIA, UPPER SILURIAN, HELDERBURG**
S 5 135 KEYSER LIMESTONE, BIG MOUNTAIN SHALE, WEST VIRGINIA,
S 5 135 KEYSER LIMESTONE, MANDATA SHALE, WEST VIRGINIA,
B 6209 1610 KEYSER LIMESTONE, MARYLAND
M 28 157 KEYSER LIMESTONE, PENNSYLVANIA, SILURIAN
B 6211 2309 KEYSER LIMESTONE, WEST VIRGINIA
B 6307 1029 KEYSTONE FIELD, SECTION, TEXAS
B 6305 836* KGRA, BEAVER COUNTY, UTAH, ABST., GEOLOGY AND STRUCTURAL
B 6008 1390* KGRA, UTAH, ABST., GEOTHERMAL EXPLORATION OF, ROOSEVELT
B 6104 585 KHAFJI RESERVOIR, FIELD, PERSIAN GULF, MARJAN
B 6105 723 KHALIFA, H., RECORRELATION OF EOCENE-MIOCENE BOUNDARY OF
B 6209 1714 KHALIFA, H., RECORRELATION OF EOCENE-MIOCENE BOUNDARY OF
B 6212 2441 KHARAKA, YOUSIF K., ALIPHATIC ACID ANIONS IN OIL-FIELD WATERS—
B 6006 973 KHARAKA, YOUSIF K., FLOW OF WATER AND SOLUTES THROUGH
B 6303 478 KHARAKA, YOUSIF K., GEOCHEMISTRY OF OIL-FIELD WATERS FROM
B 6303 428 KHARAKA, YOUSIF K., ORGANIC ACID ANIONS IN OIL-FIELD WATERS
B 6303 479 KHARAKA, YOUSIF K., STABLE CARBON ISOTOPES IN OIL-FIELD WATERS
B 6104 585 KHARG FIELD, PERSIAN GULF
B 6309 1467 KHATYRKA BASIN, SIBERIA
B 6309 1467† KHATYRKA BASINS, USSR, REVIEW OF PETROLEUM GEOLOGY OF,
M 30 550 KHEIR FORMATION, LIBYA, SIRTE BASIN
S 6 369 KHEIS SYSTEM, SOUTH AFRICA
B 6101 83 KHEMISSET BASIN, MOROCCO
M 25 463 KHINGAN RANGE, IGNIMBRITE VOLCANISM, LESSER
M 25 458 KHINGAN VOLCANIC-PLUTONIC CHAIN, PACIFIDES, GREAT
M 25 252 KHORAT PLATEAU, THAILAND
M 25 47 KHUDOLEY, K. M., GEOLOGIC HISTORY AND PALEOZOOGEOGRAPHY OF
B 6401 10 KHUFF FORMATION, IRAN
B 6104 585 KHURSANIYA, A B RESERVOIRS, FIELD, PERSIAN GULF
B 6104 608 KIALAGVIK FORMATION, ALASKA
B 6408 1213 KIANGSU BASIN, CHINA
B 6303 499* KIAWAH ISLAND, SOUTH CAROLINA, ABST., LATE HOLOCENE
B 6109 1557 KIDSON, E. J., LOWER CRETACEOUS DEPOSITIONAL SYSTEMS, WEST
M 29 209 KIECKHEFER, R. M., SEISMIC REFRACTION AND REFLECTION STUDIES IN
M 29 189 KIECKHEFER, R., TECTONICS OF THE ANDAMAN SEA AND BURMA
B 6303 479 KIER, R. S., DEPOSITION OF MARBLE FALLS FORMATION, CENTRAL TEXAS,
B 6109 1539 KIER, R. S., LAND RESOURCES OF TEXAS, ABST.
B 6104 594 KILAR KURIM FIELD, PERSIAN GULF
B 6211 2195 KILKEE CYCLOTHEM, IRELAND, WESTERN
M 25 156 KILKENNY, JOHN E., GEOTHERMAL POTENTIAL OF WESTERN UNITED
B 6012 2183 KILKENNY, JOHN E., KEYNOTE ADDRESS--YESTERDAY, TODAY, AN
B 6004 688 KILKENNY, JOHN E., TIMING FACTOR IN ENERGY EXPLORATION. BY
B 6211 2203 KILLARD-BLIND HORSES COVE GROWTH FAULTS, IRELAND, WESTERN
B 6104 585 KILUR KARIM FIELD, PERSIAN GULF
M 25 268 KIM, B. K., MARIN PETROLEUM EXPLORATION OF HUKSAN PLATFORM,
M 25 94 KIM, BONG KYUN, PALEOZOIC AND MESOZOIC COAL IN REPUBLIC OF
M 125 261 KIM, CHONG SU, PETROLEUM POTENTIAL OF KOREAN OFFSHORE
M 25 440 KIM, OK JOON, MINERAL RESOURCES OF KOREA
M 25 414 KIMBERLEY BASIN, WESTERN AUSTRALIA
B 6311 2024 KIMBERLEY BLOCK, AUSTRALIA
B 6201 41 KIMBERLEY BLOCK, AUSTRALIA
M 25 414 KIMBERLEY GROUP, WESTERN AUSTRALIA
B 6306 988 KIMBERLEY, MARRALL, GERALD E., REVIEW OF, URANIUM DEPOSITS,
M 25 479† KIMBERLITE PROBLEMS OF PACIFIC MOBILE BELT--SUMMARY,
B 6305 830* KIMBERLITE PROVINCE, ABST., UPDATE ON EXPLORATION FOR,
M 30 105 KIMMERIDGE CLAY FORMATION, JURASSIC, STATFJORD FIELD
B 6312 2222 KIMMERIDGE CLAY, NORTH SEA
M 30 137 KIMMERIDGE SHALE ISOPACH, NORTH SEA, PIPER FIELD
M 30 139 KIMMERIDGE SHALE, NORTH SEA, PIPER FIELD
B 6312 2222† KIMMERIDGIAN, COMPOSITION OF, ORGANIC MATTER FROM, NORTH,
B 6312 2223 KIMMERIDGIAN KEROGEN SAMPLE, NORTH SEA
B 6405 608 KIMMERIDGIAN, MIDDLE EAST, OXFORDIAN TO
B 6312 2224 KIMMERIDGIAN SHALES, NORTH, ATLANTIC
B 6004 716* KINCAID FORMATION, LOWER MIDWAY GROUP, PALEOCENE,
B 6002 322* KINCAID OIL FIELD, SOUTHEAST KANSAS, ABST., DEPOSITIONAL
B 6206 960 KINDERSCOUT GRIT, ENGLAND
B 6004 695* KINDERSCOUTIAN DELTA, CARBONIFEROUS, OF NORTHERN
B 6105 786 * KINDS OF PORES IN CENOZOIC REEFS OF INDONESIAN ARCHIPELAGO
B 6403 365 KINEMATIC CONSIDERATIONS IN PLATE MOTION
B 6403 365 KINEMATIC CONSIDERATIONS IN PLATE MOTION AND OROGENY
M 29 224 KINEMATIC FRAMEWORK, SUMATRA, MARGIN
B 6307 1128† KINETIC PARAMETERS TO ORGANIC GEOCHEMICAL SYSTEMS, ERRORS IN
B 6406 927† KINETICS, MICROFRACTURE DEVELOPMENT IN, COMPACTING
B 6406 932 KINETICS OF HYDROCARBON MATURATION AND DILATANCY
B 6101 24 KING CHRISTIAN FIELD, CANADIAN, ARCTIC ISLANDS
B 6206 1006 KING CHRISTIAN ISLAND, ARCTIC, CANADA
B 6102 294* KING COUNTIES, TEXAS, ABST., ATOKA, GAS IN SOUTHERN, COTTLE AND
B 6308 1425* KING COUNTIES, TEXAS, ABST., ATOKA GAS IN SOUTHERN, COTTLE AND
B 6207 1224 KING, DAVID W., HIGH-RESOLUTION SEISMIC EXPLORATION FOR COAL
M 25 426 KING, HADDON F., STRATIFORM AND STRATA-BOUND METAL
B 6309 1581 KING, HOBART M., GEOLOGIC AGE VERSUS MAJOR MINOR, AND
M 25 426 KING ISLAND SCHEELITE, TUNGSTEN, TASMANIA
B 6209 1760 KING, JOHN S., PRE-PLATFORM EXPLORATION OF HIGH ISLAND BLOCKS
B 6309 1605 KING, KEITH C., REVISION OF TAMPA FORMATION, WEST-CENTRAL
B 6306 986 KING, PHILIP B., REVIEW OF LAND OF COCHISE, SOUTHEASTERN

B 6310 1915 KING, ROBERT E., PETROLEUM EXPLORATION AND PRODUCTION IN
B 6210 1906 KING, ROBERT E., PETROLEUM EXPLORATION AND PRODUCTION IN
B 6010 1704 KING, ROBERT E., PETROLEUM EXPLORATION AND PRODUCTION IN
B 6110 1636 KING, ROBERT E., PETROLEUM EXPLORATION AND PRODUCTION IN
M 24 26 KINGAK FORMATION, ALASKA
B 6104 595 KINGFISH FIELD, AUSTRALIA
B 6307 1046 KINGS BAYOU SAND, LOUISIANA
B 6203 407 KINGSHILL LIMESTONE, FIVE MAJOR FACIES, VIRGIN ISLANDS
B 6203 403† KINGSHILL LIMESTONE, MIOCENE, ST. CROIX, U.S. VIRGIN ISLANDS,
B 6203 417 KINGSHILL LIMESTONE, RESERVOIR POTENTIAL, VIRGIN ISLANDS
B 6203 405 KINGSHILL LIMESTONE, STRATIGRAPHIC COLUMN, VIRGIN ISLANDS
B 6203 407 KINGSHILL LIMESTONE, TYPE SECTION, VIRGIN ISLANDS
S 4 332 KINGSHILL MARL, ST. CROIX
B 6203 403 KINGSHILL MARL, VIRGIN ISLANDS, ST. CROIX
S 4 329† KINGSHILL SEAWAY--A DYNAMIC CARBONATE BASIN AND SHELF MODEL,
B 6304 649 KINGSPORT DOLOMITE, VIRGINIA
B 6402 221 KINNEY COUNTY, MAPPING, TEXAS
B 6004 720 KINSMAN, D. J. J., DEVELOPMENT OF POROSITY IN CARBONATE ROCKS OF
S 4 381 KINSMAN, DAVID J. J., MARINE AND CONTINENTAL GROUNDWATER
B 6004 688 KINSMAN, DAVID J. J., SABKHA FACIES NODULAR ANHYDRITE AND
B 6401 18 KIRCHAOU SANDSTONE FORMATION, TUNISIA
M 25 59 KIRIKOV, D. A., STRUCTURAL-FORMATION ANALYSIS OF
B 6304 689 KIRK, ALLAN R., DEPOSITIONAL ENVIRONMENTS AS ORE CONTROLS IN
B 6303 479 KIRK, R. H., STATFJORD FIELD, NORTH SEA GIANT, ABST.
M 30 95 KIRK, R. H., STATFJORD FIELD--A NORTH SEA GIANT
B 6011 2005 KIRKLAND, DOUGLAS W., ORIGIN OF LIMESTONE BUTTES, GYPSUM
B 6405 733 KIRKLAND, DOUGLAS W., SOURCE-ROCK POTENTIAL OF EVAPORITIC
B 6303 532 KIRKSEY, D. L., ORIGIN OF GASES ADSORBED IN NEAR-SURFACE
B 6304 701 KIRKSEY, D. L., ORIGIN OF GASES, ADSORBED IN NEAR-SURFACE
B 6207 1224 KIRKWOOD, J. BRUCE, NORTHWEST SHELF GAS—ITS ROLE IN WESTERN
B 6205 886 KIRKWOOD, STEVEN G., STRATIGRAPHY OF CEDAR MOUNTAIN AND
M 25 291 KIRSCHNER, C. E., TECTONIC FRAMEWORK OF PETROLIFEROUS ROCKS IN
B 6411 1963 KISH BANK BASIN, EIRE
B 6207 1224 KISMORO, H., CORRELATION OF MAJOR UNCONFORMITIES IN WESTERN
B 6211 2356 KISTLER, R. W., MESOZOIC PALEOGEOGRAPHY OF CALIFORNIA.
M 29 243 KITA-DAITO BASIN, GEOLOGY, NORTH PHILIPPINE SEA
M 29 244 KITA-DAITO BASIN, STRATIGRAPHY, NORTH PHILIPPINE SEA
S 6 155 KITAKAMI MOUNTAINS, FORMATION, JAPAN, CRETACEOUS
B 6201 21 KITCHENS, OIL AND GAS
B 6405 733 KITELEY, L. W., PATTERNS OF SHALLOW-MARIN DEPOSITION, UPPER
B 6205 887 KITELEY, L. W., STRATIGRAPHY OF MESAVERDE GROUP AND
B 6402 256 KITHIRA-ANTIKITHIRA CONTINENTAL MARGIN, SEISMI REFLECTION,
B 6402 256 KITHIRA-ANTIKITHIRA CONTINENTAL MARGIN, SEISMI REFLECTION,
M 27 199 KITTATINNY LIMESTONE, NEW JERSEY, CAMBRIAN
M 27 117 KITTATINNY LIMESTONE, NEW JERSEY, CAMBRIAN
B 6408 1188 KIVU, LARGE ANOXIC LAKE, LAKE
B 6203 345 KLAMATH MOUNTAINS, CALIFORNIA
B 6301 21 KLAMATH MOUNTAINS, CALIFORNIA
B 6207 1132 KLAMONO HIGH, IRIAN JAYA
B 6309 1590 KLANDERMAN, DAVID, PETROLOGIC EVALUATION OF SIGNIFICANCE OF
B 6309 1581 KLANDERMAN, DAVID S., COMPOSITIONAL VARIATION IN
B 6303 480 KLAPPA, COLIN F., CALCIFICATION AND SIGNIFICANCE OF SOI
B 6207 1132 KLASAMAN FORMATION, IRIAN JAYA
B 6008 1365 KLATT, JACK G., DEVELOPMENTS IN SOUTH TEXAS IN 1975
B 6108 1356 KLATT, JACK G., DEVELOPMENTS IN SOUTH TEXAS IN 1976
B 6105 803 KLEIN, GEORGE DEVRIES, INTRODUCTION OF SYMPOSIUM ON PREDICTIVE
B 6203 531 KLEIN, GEORGE DEVRIES, LEG 58, DSDP—DEEP DRILLING IN SHIKOKU
B 6004 688 KLEIN, GEORGE DEVRIES,, RESEDIMENTATION OF BIOGENIC CARBONATE
B 6203 531 KLEIN, GEORGE DEVRIES, SLOPE SEDIMENTATION PROCESSES IN
B 6407 1119 KLEINPELL, MORAN, W. R., REVIEW OF, CRITERIA IN CORRELATION—
S 11 4 KLEINPELL, R. M., HISTORY OF STRATIGRAPHIC PALEONTOLOGY OF
S 11 1 KLEINPELL, R. M., MIOCENE STRATIGRAPHY OF CALIFORNIA
S 11 70 KLEINPELL, R. M., TAXONOMY
B 6008 1401 KLIPPING, ROBERT S., STRUCTURAL INTERPRETATION AND POROSITY
M 29 85 KLITGORD, KIM D., BASIN STRUCTURE OF THE U.S. ATLANTIC MARGIN
B 6410 1658 KLITGORD, KIM D., JURASSIC SEISMIC STRATIGRAPH AND BASEMENT
B 6105 KLITGORD, KIM D., MESOZOIC BASIN DISTRIBUTION ALONG U.S.
B 6405 715 KLITGORD, KIM, DEEP STRATIGRAPHY AND EVOLUTION OF BALTIMORE
B 6306 967 KLITZSCH, E., MAJOR SUBDIVISIONS AND DEPOSITIONAL ENVIRONMENTS
B 6104 602 KLONDIKE HILLS, NEW MEXICO
M 25 252 KLONG MARUI FAULT, THAILAND
S 3 125 KM-24 MINE, NEW MEXICO
B 6201 1 KNAAP, W. A., HYDROCARBON HABITAT OF TERTIARY NIGER DELTA
B 6108 1223 KNAPPE, R., JR., DEVELOPMENTS IN UPPER GULF COAST OF TEXAS IN 1976
B 6008 1254 KNAPPE, ROY, JR., DEVELOPMENTS IN UPPER GULF COAST OF TEXAS IN
B 6410 1676 KNAUSS KNOLL, WESTERN, ATLANTIC
B 6004 689 KNEBEL, H. J., CLAY-MINERAL VARIABILITY IN SUSPENDED SEDIMENTS
B 6007 1112 KNEBEL, H. J., EVIDENCE OF POST-PLEISTOCENE FAULTS ON NEW JERSEY
B 6405 733 KNEBEL, HARLEY J., PROCESSES CONTROLLING CHARACTERISTICS OF
B 6303 480 KNEBEL, HARLEY J., SLUMPING IN INTERCANYON AREAS, MIDDLE
B 6106 861 KNEBEL, HARLEY J., SURFICIAL SAND SHEET, BALTIMORE CANYON
B 6105 795 KNIGHT, JOHN A., STEPHANIAN A AND B ALLUVIAL-FAN AND LACUSTRINE
B 6004 697 KNIGHT, R. J., FACIES MODE FOR MACROTIDAL ENVIRONMENT,
B 6303 532 KNIGHT, R. JOHN, GIANT MUDFLOW DEPOSITS IN SUBMARINE
B 6002 320 KNIGHT, WILLIAM V., YARNELL TREND, SUMNER COUNTY, KANSAS,
B 6203 397 KNIK FAULT, ALASKA
B 6405 785* KNOB LAKE AREA, QUEBEC AND NEWFOUNDLAND, ABST.,
B 6405 713 KNOLL, A. H., MORPHOMETRY OF LATE ORDOVICIAN MICROBIAL
B 6211 2232† KNOLL IN EARLY TERTIARY BAY OF BISCAY AND ASSOCIATED
S 7 16 KNOLLS MAP, GULF OF MEXICO
B 6203 550 KNOTH, JEFFREY S., TRANSGRESSIVE BARRIERS OF BEAUFORT SEA
B 6407 1005 KNOWLES LIMESTONE, ARK-LA-TEX AREA
B 6311 2104 KNOX DOLOMITE, MISSISSIPPI
B 6304 652 KNOX DOLOMITE OR GROUP, VIRGINIA
B 6303 344 KNOX GROUP, KENTUCKY
B 6410 1577 KNOX GROUP, VIRGINIA
B 6408 1286 KNOX, LARRY M., GEOLOGY AND MINING O TENNESSEE COAL, ABST.
B 6405 751 KNOX, LARRY W., DEPOSITIONAL ENVIRONMENTS AND BIOGENIC
B 6405 733 KNOX, LARRY W., ENVIRONMENTAL CONTROL OF TRACE FOSSIL
B 6311 2116* KNOX-BAYLOR TROUGH, ABST., DESMOINESIAN DEPOSITIONAL SYSTEMS
B 6412 2078 KNOXVILLE FAULT, APPALACHIANS
M 25 252 KO KRA RIDGE, THAILAND
B 6303 441* KOALA--MINICOMPUTER LO ANALYSIS SYSTEM FOR GEOLOGISTS, ABST.
M 25 448 KOBAYASHI, ISAMU, RELATION BETWEEN TECTONICS AND
B 6203 531 KOBAYASHI, KAZUO, LEG 58, DSDP—DEEP DRILLING IN SHIKOKU BASIN
B 6109 1560 KOBER, CARL L., URANIUM SEARCH IN LAKE ATHABASKA REGION BY
B 6105 799 KOBLUK, D. R., INTERNAL STRUCTURE OF LOWER CAMBRIAN
B 6203 553 KOBLUK, D. R., SILURIAN MACROBORING ORGANISMS, SOUTHERN
B 6203 531 KOBLUK, DAVID R., CAVITY-DWELLING ORGANISMS, COELOBITES, IN
B 6303 508 KOBLUK, DAVID R., ICHNOLOG OF LABRADOR GROUP, LOWER
B 6107 1069 KOBLUK, DAVID R., MICRITIZATION AND CARBONATE-GRAIN BINDING
B 6108 1384 KOCH, MICHAEL R., PETROLOGY AND DIAGENESIS OF DEEP-WATER
B 6303 480 KOCH, W. JERRY, SIMPLE PYROLYSIS TECHNIQU USING WELL CUTTINGS
B 6303 496 KOCH, W. JERRY, TEST-TUBE PYROLYSIS--SIMPLE TECHNIQUE FOR
B 6307 1096 KOCKATEA SHALE, AUSTRALIA
B 6207 1225 KOCKELMAN, WILLIAM J., OVERCOMING EARTH RESOURCE AND HAZARD
B 6105 804 KOCURKO, M. J., SILICIFICATION OF CALCIUM PHOSPHATE IN
B 6303 481 KOCURKO, M. JOHN, DIAGENETIC PHENOMENA OF CHIPOLA FORMATION,
B 6409 1561 KOCURKO, M. JOHN, EARLY CEMENTATION AND MINERAL STABILITY OF
B 6405 734 KOCURKO, M. JOHN, EARLY CEMENTATION BY HIGH-MAGNESIUM
M 25 304 KODIAK BASIN, SECTION, ALASKA
M 29 263 KODIAK, CONTINENTAL MARGIN, BATHYMETRY, ALASKA
M 29 270 KODIAK, CONTINENTAL MARGIN, MODEL, ALASKA
M 29 262 KODIAK, CONVERGENCE MODEL, GENERAL
B 6407 1016 KODIAK FORMATION, ALASKA
B 6408 1142 KODIAK FORMATION, ALASKA
B 6303 302 KODIAK FORMATION, ALASKA
B 6105 838* KODIAK ISLAND, ALASKA, ABST., STRUCTURAL INTERPRETATION OF
M 29 261† KODIAK ISLAND, ALASKA, FROM MULTICHANNEL SEISMIC RECORDS,
B 6303 301† KODIAK ISLAND, ALASKA, STRUCTURE AND TECTONIC SETTING OF,
B 6405 671* KODIAK ISLAND ARCHIPELAGO, ALASKA, ABST., HYDROCARBON
B 6408 1142 KODIAK ISLAND, GEOLOGY, ALASKA
B 6407 1016 KODIAK ISLANDS, GEOLOGY, ALASKA
M 29 264 KODIAK, REFLECTION RECORDS (24-CHANNEL), ALASKA
M 29 268 KODIAK, REFRACTION VELOCITY DATA, ALASKA
B 6303 306 KODIAK RIDGE, ALASKA, KENAI-
B 6209 1593 KODIAK SEAMOUNT, ALASKA CONTINENTAL MARGIN
B 6207 1228* KODIAK SHELF, ABST., PLATE-TECTONIC INTERPRETATION OF, MESOZOIC
B 6405 706* KODIAK SHELF, ALASKA, ABST., STRUCTURE AND HYDROCARBON
B 6408 1140† KODIAK SHELF, ALASKA, PETROLEUM GEOLOGY OF
B 6407 1014† KODIAK SHELF, ALASKA, STRUCTURE OF UPPER CENOZOIC STRATA
B 6408 1151 KODIAK SHELF, GEOTHERMAL GRADIENTS, ALASKA
B 6408 1146 KODIAK SHELF, SOURCE AND RESERVOIR DATA, ALASKA
M 29 268 KODIAK, SLOPE DATA, INTERPRETATION, ALASKA
M 29 267 KODIAK, VELOCITY DATA, ALASKA
B 6105 829 KOEHLER, F., COLOR DISPLAYS OF COMPLEX SEISMIC TRACE, ABST.,
B 6303 472 KOEHN, HENRY, DOLOMITIZATION OF OFFSHORE CARBONATE DEPOSITS
B 6004 689 KOENIG, J. B., GEOLOGIC ENVIRONMENTS OF GEOTHERMAL RESOURCES
M 28 20 KOGOSUKRUK TONGUE, ALASKA, CRETACEOUS
M 28 19 KOGOSUKRUK TONGUE, ALASKA, CRETACEOUS
M 28 118 KOGOSUKRUK TONGUE, ALASKA, CRETACEOUS
M 28 132 KOGOSUKRUK TONGUE, ALASKA, CRETACEOUS
M 28 22 KOGOSUKRUK TONGUE, ALASKA, CRETACEOUS
M 28 7 KOGOSUKRUK TONGUE, ALASKA, CRETACEOUS
M 28 130 KOGOSUKRUK TONGUE, ALASKA, CRETACEOUS
M 28 172 KOGUSUKRUK FORMATION, ALASKA, TERTIARY
B 6405 774 KOHL, BARRY, COMPUTER ASSISTED PALEOECOLOGIC ANALYSES AND
B 6209 1764 KOHL, BARRY, COMPUTER-ASSISTED PALEOECOLOGIC ANALYSES AND
B 6105 804 KOHOUT, F. A., FRESH GROUNDWATER BENEATH CONTINENTAL SHELF:

B 6401 128 KOKOMA LIMESTONE MEMBER, INDIANA, SALINA FORMATION
S 5 94 KOKOMO LIMESTONE, SALINA FORMATION,
B 6201 18 KOKORI MACROSTRUCTURE, NIGER DELTA
B 6409 1562 KOLB, CHARLES R., SHOULD WE PERMIT MISSISSIPPI-ATCHAFALAYA
B 6112 2070 KOLBANO AREA, SECTION, TIMOR
B 6112 2068 KOLBANO AREA, SECTION, TIMOR ISLAND, NIKINIKI-
B 6405 734 KOLLA, VENKATARATHNAM, LITHOLOGY AND STRUCTURES OF
B 6010 1707 **KOLLMANN, ON DEVELOPMENTS IN AUSTRIA, 1975, KROLL, A. J., AND K. F.**
B 6110 1638 KOLLMANN, ON DEVELOPMENTS IN AUSTRIA, 1976, KROLL, A. J., AND K.
B 6105 804 **KOLPACK, RONALD L., DISTRIBUTION OF SUSPENDED PARTICULATE**
S 10 230 KOLPASHEVO, GEOTHERMAL PROFILE, USSR, SIBERIA
M 25 474 KOLYMA MEGASTRUCTURE, YANA-
M 25 466 KOLYMA MIOGEOSYNCLINAL SYSTEM, USSR, YANA-
M 25 466 KOLYNA MASSIF, MIDDLE, RECLASSIFIED, USSR
B 6102 193 KOMANDORSKIY BASIN, DEEP SEA DRILLING PROJECT SITE 191, BERING
B 6102 195 KOMANDORSKIY BASIN, GLOMAR CHALLENGER SEISMIC PROFILE
B 6102 201 KOMANDORSKIY BASIN, MIOCENE
B 6102 201 KOMANDORSKIY BASIN, PLEISTOCENE
B 6102 201 KOMANDORSKIY BASIN, PLIOCENE
B 6102 196 KOMANDORSKIY BASIN, TURBIDITE SAND AND SILT LAYERS
B 6102 192 KOMANDORSKIY BASIN, TURBIDITES, BERING SEA
B 6102 192† KOMANDORSKIY BASIN, WESTERN BERING SEA, NEOGENE, TURBIDITE
B 6102 193 KOMANDORSKIYE OSTROVA, COMMANDER ISLANDS
B 6311 2071 KOMANDORSKY BASIN, BERING SEA
B 6309 1587 KOMAR, C. A., INTERPRETATIO AND STATISTICAL ANALYSIS OF
M 29 151 KONIG, MICHAEL, THE CRUSTAL STRUCTURE AND EVOLUTION OF THE
B 6303 434* KOOBI FORA FORMATION, PLIOCENE-PLEISTOCENE, EAST TURKANA,
B 6203 504* KOONGARRA URANIUM DEPOSIT, NORTHERN TERRITORY, AUSTRALIA,
B 6407 971 KOOTENAI FORMATION, GREAT PLAINS
M 25 507 KOOTENAY ARC, CANADA
B 6405 744* KOPE FORMATION AND SOME PALEOECOLOGIC IMPLICATION, ABST.,
B 6303 538 KOPORULIN, V. I., MIDDLE CRETACEOUS OXYGEN-DEFICIENT
M 25 62 KOPYLOVA, T. N., TECTONICS OF CONTINENTAL AND OCEANIC
M 25 442 KOREA, BULKUKSA DISTURBANCE
M 25 269 KOREA, CHEJU VOLCANIC BELT
M 25 440 KOREA, CHUGARYONG RIFT VALLEY
M 25 269 KOREA, CRETACEOUS VOLCANIC ACTIVITY
M 25 442 KOREA, DAEBO OROGENY
B 6110 1837 KOREA, DEVELOPMENTS, 1976, SOUTH
B 6210 1986 KOREA, DEVELOPMENTS, 1977, NORTH
B 6210 1987 KOREA, DEVELOPMENTS, 1977, SOUTH
B 6310 1824 KOREA, DEVELOPMENTS, 1978, NORTH
B 6310 1826 KOREA, DEVELOPMENTS, 1978, SOUTH
B 6411 1890 KOREA, DEVELOPMENTS, 1979, NORTH
B 6411 1893 KOREA, DEVELOPMENTS, 1979, SOUTH
M 25 261 KOREA, FUKIEN-REINAN MASSIF, SOUTH
M 25 269 KOREA, FUKUE BASIN
M 25 440 KOREA, GEOLOGIC SETTING, SOUTH
M 25 269 KOREA, KUNSAN BASIN
M 25 269 KOREA, KWANGJU-POHANG ALIGNMENT
M 25 269 KOREA, KYONGGI, GYEONGGI, MASSIF
M 25 269 KOREA, KYONGSAN, GUEONGSANG, BASIN
M 25 268† KOREA, MARIN PETROLEUM EXPLORATION OF, HUKSAN PLATFORM
M 25 440 KOREA, MESOZOIC GRANITES, SOUTH
M 25 440† KOREA, MINERAL RESOURCES OF
M 25 269 KOREA, NAMHAE ARCHIPELAGO ALIGNMENT
M 25 263 KOREA OFFSHORE, SOUTH SEA, EAST CHINA SEA
M 25 269 KOREA, OGCHEON, OROGENIC ZONE
M 25 94† KOREA, PALEOZOIC AND, MESOZOIC, COAL IN, REPUBLIC OF
M 25 269 KOREA, POHANG SUBBASIN
M 25 269 KOREA, SOBAEKSAN, SOBAEGON OR YONGNAM, REINAN, MASSIF
M 25 269 KOREA, SOCOTRA SUBBASIN
M 25 442 KOREA, SONGRIM DISTURBANCE
M 25 96 KOREA, SOUTHERN, COAL RESERVES
M 25 95 KOREA, SOUTHERN, COALFIELDS
M 25 94 KOREA, SOUTHERN, STRATIGRAPHY AND PALEONTOLOGY
M 25 440 KOREA, STRATIGRAPHY, SOUTH
M 25 269 KOREA, TSUSHIMA BASIN AND CONTINENTAL BORDERLAND
M 25 269 KOREA, WEST YELLOW SEA, NORTH KIANGSU, AND SHANGHAI BASINS
M 25 269 KOREA, WESTERN SUBBASIN
M 25 442 KOREA, YONIL DISTURBANCE
M 25 263 KOREAN OFFSHORE, EAST SEA, SEA OF JAPAN
M 125 261† KOREAN OFFSHORE, PETROLEUM POTENTIAL OF
M 25 262 KOREAN OFFSHORE, WEST SEA, YELLOW SEA
M 25 261 KOREAN PENINSULA, GEOLOGY
M 25 467 KOREAN VOLCANIC BELT, HONSHU-
B 6002 163 KORYAK EUGEOSYNCLINAL COMPLEX, SIBERIA, KAMCHATKA-
M 25 466 KORYAK GEOSYNCLINAL ZONE, USSR, ANADYR-
B 6309 1468 KORYAK HIGHLANDS, PIKULNEY SERIES, USSR
B 6102 193 KORYAK MOUNTAINS, KORYAKSKIY KHREBET
B 6407 1088 KORYAK UPLAND, ASIA, NORTHEAST
B 6102 193 KORYAKSKIY KHREBET, KORYAK MOUNTAINS
S 6 154 KOSHIMEGURI FORMATIONS, JAPAN
B 6204 657 KOSTACK,J. I., SALT DIAPIRS ON SAO PAULO PLATEAU, SOUTHEASTERN
B 6405 734 KOSTECKI, JOHN A., LITHOLOGY AND STRUCTURES OF QUATERNARY
B 6008 1405 KOSTURA, JOHN R., PALEOCURRENT ANALYSIS OF DAKOTA AND
B 6006 988 KOTICK, OTTMAR F., SHORT HISTORY OF NORTHERN CALIFORNIA
B 6310 1983 KOTTLOWSKI, AUREAL T. CROSS, AND A. A. MEYERHOFF, MEYERHOFF,
B 6109 1448 KOTTLOWSKI, FRANK E., POTENTIAL AND STRATIGRAPHY OF
M 25 221 KOTUKU OIL SEEPAGE, NEW ZEALAND

B 6004 675* KOUCHIBOUGUAC BAY, NEW BRUNSWICK, CANADA, ABST., MARINE-, BAR
B 6407 1091 KOVACHINIAN HORIZON, ASIA, NORTHEAST
B 6105 792 **KOVSCHAK, ANTHONY A., JR., GEOLOGY AND EXPLORATION OF URAVAN**
B 6104 546 KOWSMANN, R., MARINE SEISMIC INVESTIGATIONS, SOUTHERN BRAZIL
B 6405 792* KOYUKUK CRUNCH, NORTHERN AND CENTRAL ALASKA, ABST.,
B 6002 172 KOYUKUK GEOSYNCLINE, ALASKA, YUKON-
B 6405 735 KOZEY, KATHLEEN M., USCHEM, GEOCHEMICAL DATA FILE OF NATIONAL
B 6004 648 KRAFT, J. C., RADIOCARBON DATING AND STRATIGRAPHIC ANALYSIS OF
B 6303 492 **KRAFT, JOHN C., DETERMINATION OF STORM OVERWASH PERIODICITY**
B 6303 439 **KRAFT, JOHN C., INTERPRETATION OF PLEISTOCENE COASTAL-BARRIER**
B 6405 675 KRAFT, JOHN C., FRACTIONAL PRESERVATION OF TRANSGRESSIVE
B 6203 531 KRAFT, JOHN C., HOLOCENE EPOCH ALLUVIAL-DELTAIC-COASTAL
B 6312 2145 KRAFT, JOHN C., LATERAL AND VERTICAL FACIES RELATIONS OF
B 6105 805 **KRAFT, JOHN C., STRUCTURE OF COASTAL WASHOVER FANS AND THEIR**
B 6203 532 KRAFT, LELAND M., JR., GEOTECHNICAL ENGINEERING PROBLEMS OF
S 7 267 KRAFT, LELAND M., JR., STABILITY OF CONTINENTAL SHELF AND SLOPE
B 6303 481 KRAMERS, JOHN W., DEPOSITIONAL ENVIRONMENTS AND DIAGENESIS O
B 6405 735 KRAMERS, JOHN W., DIAGENETIC FEATURES OF GRAND RAPIDS
M 25 46 KRASNY, L. I., GEOLOGIC MAP OF PACIFIC MOBILE BELT AND PACIFIC
M 25 58 KRASNY, L. I., GEOLOGIC-STRUCTURAL SYNTHESIS OF PACIFIC AREA
B 6212 2511 KRAUS, G. P., GEOCHEMICAL EVALUATION OF BONAPARTE GULF—TIMOR
B 6311 2021 KRAUS, GREGORY P., GEOCHEMICAL EVALUATION OF PETROLEUM
B 6203 532 KRAUSE, F. F., CONTINENTAL SLOPE-RISE DEPOSITS FROM LOWER
B 6303 481 KRECOW, F. C., MEGASCALE VARIABILITY OF SEDIMENT ON UNITED
B 6004 689 KREITLER, C. W., ACTIVE SURFACE FAULTING, TEXAS COASTAL ZONE,
B 6409 1556 KREITLER, CHARLES W., CAP-ROCK FORMATION AND DIAGENESIS, GYP
B 6109 1539 KREITLER, CHARLES W., GEOMETRY OF GULF COAST AQUIFERS,
B 6309 1605 KREITLER, CHARLES W., SUITABILITY STUDIES OF SALT DOMES IN EAST
B 6303 385 KREJCI-GRAF, OSTROFF, A. G., REVIEW OF, DATA ON THE,
B 6403 442 KREUTZER, PAUL A., INTERMARGIN OF SOUTHERN CALIFORNIA
B 6205 849 KREYENHAGEN SHALE, CALIFORNIA
B 6303 439* KRIGING OF TOP OF RESERVOIR FROM SEISMIC AND BOREHOLE DATA,
B 6004 689 KRINSLEY, D. H., COMPARISON OF DIAGENETIC AND WEATHERING
B 6203 532 KRINSLEY, DAVID H., ORIGIN OF NAVAJO SANDSTONE, ABST.
B 6101 24 KRISTOFFER FIELD, CANADIAN, ARCTIC ISLANDS
B 6309 1547 KROKOVIC, AND S. BANERJEE, BARR, F. T., REVIEW OF, BIBLIOGRAPHY OF
B 6010 1707 KROLL, A. J., AND K. F. J. KOLLMANN, ON DEVELOPMENTS I AUSTRIA,
B 6110 1638 KROLL, A. J., AND K. F. J. KOLLMANN, ON DEVELOPMENTS IN AUSTRIA,
B 6310 1917 KROLL, A.J., AND A. E. H. BUDWILL, ON DEVELOPMENTS I AUSTRIA, 1978
B 6210 1908 KROLL, ON DEVELOPMENTS IN AUSTRIA, 1977, BUDWILL, A.E.H., AND A. J.
B 6411 1962 KROLL, ON DEVELOPMENTS IN AUSTRIA, 1979, BUDWILL, A. E. H., AND A.
M 27 84 KRONER REEF LIMESTONE, GERMANY, CRETACEOUS
B 6301 82 KROSNO FORMATION, CARPATHIANS
B 6105 805 **KRUEGER, W. C., JR., PERCENTAGE ESTIMATE OF DISCOVERABLE**
B 6405 735 KRUEGER, WM. C., JR., INVESTIGATION OF MEASUREMENTS OF HEAT
S 4 209 KRUTAK, PAUL R., CHANGE IN OSTRACOD STANDING CROP, LAGUNA
B 6303 482 KRUTAK, PAUL R., EQUILIBRIUM IN MODERN CORAL REEFS, WESTERN
B 6309 1605 KRUTAK, PAUL R., EQUILIBRIUM IN MODERN CORAL REEFS, WESTERN
B 6105 805 **KRUTAK, PAUL R., GASTROPOD RADULAE—THEIR POTENTIAL IN FOSSIL**
B 6109 1540 KRUTAK, PAUL R., GASTROPOD RADULAE—THEIR POTENTIAL IN THE
B 6409 1562 KRUTAK, PAUL R., MODERN LAGOONAL OSTRACODES AND SPECIE
B 6203 533 KRYLOV, I. N., LACUSTRINE STROMATOLITES FROM KYZYLGIR
B 6405 735 KRYSTINIK, LEE F., EARLY POROSITY AND PERMEABILITY REDUCTION IN
B 6403 445 KRYSTINIK, LEE F., PORE-FILLING CEMENTS--PRODUCTS OF SHALE
M 25 451 KUBO, KYOSUKE, GEOLOGI ASPECTS OF URANIUM RESOURCES OF
B 6309 1581 KUCUK, FIKRI, MODELING OF DEVONIAN SHALE GAS-RESERVOIR
B 6401 41 KUDJUNG LIMESTONE, KALIMANTAN, SOUTHERN, OFFSHORE
B 6401 51 KUDJUNG UNIT 1, KALIMANTAN
B 6401 49 KUDJUNG UNIT 2, KALIMANTAN
B 6401 48 KUDJUNG UNIT 3, KALIMANTAN
B 6303 552 KUENZI, W. DAVID, BRAIDED-STREAM PROCESSES AND FACIES IN
B 6109 1513 KUES, BARRY S., OSHA CANYON FORMATION, PENNSYLVANIAN, NEW
B 6406 863 KUFRA BASIN, LIBYA AND, SUDAN
B 6008 1189 KUGLER, HARRY W., DEVELOPMENTS IN ALASKA IN 1975
B 6206 985 KUH-E ASMARI ANTICLINE, IRAN
B 6206 985 KUH-E BAMU ANTICLINE, IRAN
B 6206 985 KUH-E DASHTAK ANTICLINE, IRAN
B 6206 985 KUH-E MISH ANTICLINE, IRAN
B 6206 985 KUH-E MUNGASHT ANTICLINE, IRAN
B 6206 985 KUH-E PAHN ANTICLINE, IRAN
B 6206 985 KUH-E SHAH NISHIN ANTICLINE, IRAN
B 6206 985 KUH-E SURMEH ANTICLINE, IRAN
B 6405 762 KUHNEL, CLARENCE, ROCK-EVAL PYROLYSIS AS SOURC ROCK
B 6310 1822 KUJIRAOKA, A., ON DEVELOPMENTS IN JAPAN, 1978
B 6411 1888 KUJIRAOKA, AKIRA, ON DEVELOPMENTS IN JAPAN, 1979
B 6207 1195 KULA PLATE
B 6412 2152 KULA PLATE, BERINGIAN MARGIN
B 6002 175 KULA PLATE, BERINGIAN MARGIN
B 6207 1193 KULA-FARALLON PLATE BOUNDARIES, SEAFLOOR SPREADING ALONG
B 6303 482 KULANDER, BYRON R., FRACTOGRAPHIC DISTINCTION OF
B 6303 482 KULANDER, BYRON R., NATURE AND FIELD APPLICATION OF PLUMOSE
B 6105 800 **KULM, L. D., DEVELOPMENT OF CONTINENTAL-SLOPE BASINS ALONG**
B 6203 533 KULM, L. D., QUATERNARY SEDIMENTATION AND TECTONICS ON
B 6101 65 KUMAR, MADHURENDU BHUSHAN, GEOTHERMAL AND GEOPRESSURE
B 6209 1759 KUMAR, MADHURENDU BHUSHAN, REDDELL SALT DOME OF
B 6004 690 KUMAR, N., EVOLUTION OF MARGINAL PLATEAU--DSDP DATA FROM SITE
B 6309 1545 KUMAR, NARESH, REVIEW OF ANCIENT SEDIMENTARY ENVIRONMENTS, B
B 6202 273 KUMAR, NARESH, SEDIMENT DISTRIBUTION IN WESTERN ATLANTIC OFF
B 6305 812 KUMAR, NARESH, THICKNESS OF REMOVED SEDIMENTARY ROCKS,
S 6 304 KUNGURIAN STAGE, PERMIAN, BIOCHRONOLOGY
M 25 269 KUNSAN BASIN, KOREA
B 6303 531 KUNZE, FLORENCE R, RADIOLARIA--PRESENT AND PAST INDICATORS OF
B 6303 483 KUNZE, FLORENCE R., RADIOLARIAN PRESERVATION IN PRESENT AND
B 6309 1598 KUNZE, FLORENCE, RADIOLARIAN DISTRIBUTION, DIVERSITY, AND
B 6008 1411 KUNZE, JAY F., RAFT RIVER, IDAHO, GEOTHERMAL WELLS--SITING,
M 24 26 KUPARUK RIER FORMATION, ALASKA
M 30 310 KUPARUK RIVER FORMATION, LOWER SAND, PRUDHOE BAY
M 30 310 KUPARUK RIVER FORMATION, MIDDLE SAND, PRUDHOE BAY
M 30 310 KUPARUK RIVER FORMATION, UPPER SAND, PRUDHOE BAY
B 6109 1499 KUPARUK SECTION, ALASKA
B 6009 1612 KUPFER, DONALD H., SALT MOVEMENT, HALOKINETICS, IN NORTH
B 6009 1434 KUPFER, DONALD H., SHEAR ZONES INSIDE GULF COAST SALT STOCK
B 6107 1090 KUPFER, DONALD H., SHEAR ZONES INSIDE GULF COAST SALT STOCKS
B 6203 533 KURFESS, JAMES A., FIELD DEVELOPMENT WITH 3-D SEISMIC METHODS—
M 25 463 KURIL-JAPANESE ISLAND ARC
M 25 481 KURIL-KAMCHATKA ARC, MIGRATION OF
M 25 481 KURIL-KAMCHATKA ARC, MINERALS
M 25 481† KURIL-KAMCHATKA ARC--SUMMARY, OUTLINE OF
B 6102 193 KURIL-KAMCHATKA TRENCH
B 6207 1231 KURODA, KAZUO, LANDSLIDE PROBLEMS IN JAPAN, ABST.
M 25 456 KUROKO DEPOSITS, JAPAN
M 25 448 KUROKO OR BLACK ORE DEPOSITS, JAPAN
B 6108 1226 KURTEN FIELD, BRAZOS COUNTY, TEXAS
B 6409 1566* KURTEN FIELD—DISCOVERED BY STRATIGRAPHIC PROSPECTING, ABST.
B 6303 483 KURTZ, DENNIS D., SEDIMENTOLOGIC FACIES I MODERN
B 6303 410 KURTZ, DENNIS,, GLACIAL-MARINE SEDIMENTS OF WEST ANTARCTICA
B 6203 490 KURTZ, DENNIS, SEDIMENTATION ON ANTARCTIC CONTINENTAL SLOPE,
M 25 246 KUTAI BASIN, INDONESIA
M 30 443 KUTAI BASIN, TECTONIC FRAMEWORK, INDOENSIA
M 30 442 KUTEI BASIN, BADAK FIELD, REGIONAL GEOLOGY
M 30 442 KUTEI BASIN, BADAK FIELD, STRATIGRAPHY
M 30 460 KUTEI BASIN, BEKAPAI FIELD, INDONESIA
M 30 460 KUTEI BASIN, BEKAPAI FIELD, KALIMANTAN
M 30 400 KUTEI BASIN, LOCATION MAP, INDONESIA

B 6207 1178 KUVELOUSIA LEPTOSA ZONE, NEVADA AND, UTAH
B 6010 1867 KUWAIT, 1975, KUWAIT OIL CO., ASIATIC PETROLEUM CORP., AN, KUWAIT
B 6110 1797 KUWAIT, 1976, ASIATIC PETROL. CORP. ET AL, ON DEVELOPMENTS IN
B 6210 2113 KUWAIT, 1977, MERMINOD, F., PETROCONSULTANTS S.A., AND, ASIATIC
B 6310 1886 KUWAIT, 1978, MERMINOD, F., AND, KUWAIT OIL CO., IN
B 6411 1837 KUWAIT, 1979, KUWAIT OIL CO. AND, PETROCONSULTANTS S.A., ON
M 30 16 KUWAIT, BURGAN FIELD, LANDSAT IMAGE
B 6401 14 KUWAIT, BURGAN GIANT OIL FIELD, LANDSAT DATA
B 6411 1837 KUWAIT OIL CO. AND PETROCONSULTANTS S.A., ON DEVELOPMENTS IN
B 6010 1867 KUWAIT OIL CO., ASIATIC PETROLEUM CORP., AN KUWAIT SPANISH
B 6310 1886 KUWAIT OIL CO., IN KUWAIT, 1978, MERMINOD, F., AND
B 6210 2120 KUWAIT, PRODUCTION, 1976-1977
B 6310 1895 KUWAIT, PRODUCTION, 1978
B 6411 1847 KUWAIT, PRODUCTION, 1979
B 6010 1867 KUWAIT SPANISH PETROLEUM CO., ON DEVELOPMENTS IN KUWAIT, 1975,
B 6010 1868 KUWAIT-SAUDI ARABIA DIVIDED NEUTRAL ZONE, 1975, SAUDI ARABIA
B 6110 1798 KUWAIT-SAUDI ARABIA DIVIDED NEUTRAL ZONE, 1976, AMERICAN
B 6210 2114 KUWAIT-SAUDI ARABIA DIVIDED NEUTRAL ZONE, 1977, ARABIAN OIL CO.,
B 6310 1886 KUWAIT-SAUDI ARABIA DIVIDED NEUTRAL ZONE, 1978, ARABIAN OIL CO.,
B 6411 1838 KUWAIT-SAUDI ARABIA DIVIDED NEUTRAL ZONE, 1979,
B 6210 2126 KUWAIT-SAUDI ARABIA DIVIDED NEUTRAL ZONE, PRODUCTION, 1977
B 6310 1895 KUWAIT-SAUDI ARABIA DIVIDED NEUTRAL ZONE, PRODUCTION, 1978
B 6411 1852 KUWAIT-SAUDI ARABIA DIVIDED NEUTRAL ZONE, PRODUCTION, 1979
B 6311 2070 KVENVOLDEN, KEITH A., HYDROCARBON POTENTIAL OF ALEUTIAN
B 6105 806 **KVENVOLDEN, KEITH A., HYDROCARBON GASES IN SEDIMENTS OF EASTERN**
B 6203 534 KVENVOLDEN, KEITH A., HYDROCARBONS OF LOW MOLECULAR WEIGHT
B 6303 483 KVENVOLDEN, KEITH A., ORGANIC GEOCHEMISTRY OF SEDIMENTS
B 6407 1078 KVENVOLDEN, KEITH A., ORIGIN OF GASOLINE-RANGE HYDROCARBONS
B 6406 954 KVENVOLDEN, KEITH A., REVIEW OF PETROLEUM GEOCHEMISTRY AND
B 6303 433 KVENVOLDEN, KEITH A., SPILLE OIL AND INFAUNAL
B 6405 736 KVENVOLDEN, KEITH A., THERMOGENIC HYDROCARBON GASES IN
B 6405 706 KVENVOLDEN, KEITH, DIAPIR-LIKE RIDGES AND POSSIBLE
B 6411 2039 KVENVOLDEN, WAPLES, DOUGLAS, REVIEW OF, GEOCHEMISTRY OF
M 25 252 KWAE NOI FAULT, THAILAND
B 6303 420* KWAGUNT FORMATION, LATE PRECAMBRIAN, CHUAR GROUP, GRAND
M 25 269 KWANGJU-POHANG ALIGNMENT, KOREA
B 6408 1211 KWANGSI-KWEICHOW BASIN, CHINA
S 6 145 KWANMON GROUP, JAPAN, CRETACEOUS
M 25 362 KWANTO REGION, LAND SUBSIDENCE, JAPAN
B 6303 518 KWON, B. D., COMPUTER PREDICTION OF CHANGES IN STRATIGRAPHIC
M 28 47 KYANITE, CRYSTAL
M 28 59 KYANITE, CRYSTALS, IN GRAIN MOUNT
M 28 6 KYANITE SCHIST, CONNECTICUT, ORDOVICIAN
M 25 108 KYNOCKS COAL AREAS, CANADA
M 25 269 KYONGGI, GYEONGGI, MASSIF, KOREA
M 25 441 KYONGGI MASSIF, SOUTH KOREA
M 25 269 KYONGSAN, GUEONGSANG, BASIN, KOREA
M 25 447 KYONGSANG BASIN, MINERALS, SOUTH KOREA
M 25 441 KYONGSANG SEDIMENTARY BASINS, SOUTH KOREA
B 6203 533* KYZYLGIR FORMATION, PLIOCENE, ALTAI MOUNTAINS, ABST.,
M 24 106 LA BARGE ANTICLINE, WYOMING
M 24 92 LA BARGE AREA, GEOLOGIC MAP, WYOMING
M 24 105 LA BARGE FIELD, WYOMING
M 24 103 LA BARGE MEMBER, WYOMING
M 24 91† LA BARGE PRODUCING COMPLEX, SUBLETTE AND LINCOLN COUNTIES,
M 25 521 LA CARIDAD MINE, COPPER, MEXICO, SONORA
M 25 519 LA COLORADA QUARTZ PORPHTYRY, COPPER, MEXICO
B 6203 349 LA HONDA BASIN, CALIFORNIA
S 8 57 LA JOLLA CANYON, AVERAGE FLOW TIME, CURRENT FLOW
B 6102 152 LA JOLLA CANYON, CALIFORNIA
S 8 58 LA JOLLA CANYON, CALIFORNIA, SUBMARINE CANYONS
S 8 54 LA JOLLA CANYON, CROSS-CANYON FLOW, SPEED CURVE
S 8 32 LA JOLLA CANYON, CURRENT FLOW, TURBIDITY CURRENTS
S 8 61 LA JOLLA CANYON, DESCRIBED, TURBIDITY CURRENTS
S 8 33 LA JOLLA CANYON, DOWNCANYON VELOCITIES, CURRENT FLOW
S 8 55 LA JOLLA CANYON, PROGRESSIVE-VECTOR DIAGRAM
S 8 60 LA JOLLA CANYON, TIDAL RELATIONS, CURRENT FLOW
S 8 62 LA JOLLA CANYON, TIME-VELOCITY CURVES, CURRENT FLOW
S 8 10 LA JOLLA CANYON, TIME-VELOCITY CURVES, CURRENT FLOW
S 8 59 LA JOLLA CANYON, TOPOGRAPHY, INDEX MAP
B 6206 921 LA JOLLA FAN VALLEY, CALIFORNIA
B 6102 208 LA LAJA FORMATION, MEXICO
B 6112 2116 LA LUZ ANTICLINE, NEW MEXICO
B 6211 2220 LA PANZA FAULT, CALIFORNIA
B 6211 2218 LA PANZA RANGE, CALIFORNIA
B 6104 579 LA PAZ FIELD, VENEZUELA
B 6408 1165 LA PICA SANDSTONE, VENEZUELA
M 25 552 LA PLATA MINE, GOLD, SILVER, AND COPPER, ECUADOR, TOACHI
B 6004 653* LA SAL MINE, SAN JUAN COUNTY, UTAH, ABST., PRELIMINARY
M 25 161 LA SOLEDAD GEOTHERMAL AREA, MEXICO
B 6111 1986 LA VIDA SILTSTONE, CALIFORNIA
B 6405 736 LABO, J., EXPLORING FOR NIAGARAN PINNACLE REEF IN SOUTHERN
M 26 15† LABORATORY AND LOG DATA THAT ARE IMPORTANT TO SEISMIC
M 26 28 LABORATORY MEASUREMENT, ANISOTROPY
M 26 24 LABORATORY MEASUREMENT ATTENUATION, FLUID MOTION
M 26 30 LABORATORY MEASUREMENT, DYNAMIC MODULI
M 26 31 LABORATORY MEASUREMENT, HEAT ALTERATION
M 26 30 LABORATORY MEASUREMENT, STATIC MODULI
M 26 23 LABORATORY MEASUREMENTS, ATTENUATION
M 26 26 LABORATORY MEASUREMENTS, BEREA SANDSTONE
M 26 36 LABORATORY MEASUREMENTS, BIOT-GEERTSMA EQUATIONS
M 26 23 LABORATORY MEASUREMENTS, BOISE SANDSTONE
M 26 37 LABORATORY MEASUREMENTS, FLUID PRESSURE
M 26 26 LABORATORY MEASUREMENTS, FRICTION LOSS
M 26 33 LABORATORY MEASUREMENTS, GASSMANNS THEORY
M 26 37 LABORATORY MEASUREMENTS, OVERBURDEN
M 26 32 LABORATORY MEASUREMENTS, ROCK PROPERTIES, TEMPERATURE
M 26 39 LABORATORY MEASUREMENTS, SATURATION EFFECTS, BULK DENSITY
M 26 39 LABORATORY MEASUREMENTS, SATURATION EFFECTS, PRESSURE
M 26 39 LABORATORY MEASUREMENTS, SATURATION EFFECTS, ROCK
M 26 39 LABORATORY MEASUREMENTS, SATURATION EFFECTS, SHEAR VELOCITY
M 26 32 LABORATORY MEASUREMENTS, TEMPERATURE EFFECTS CLASSIFICATION
B 6112 2083 LABORATORY PYROLYSIS OF FOUR KEROGEN TYPES
M 26 22 LABORATORY SYSTEMS, BROADBAND VELOCITY MEASUREMENT,
M 26 19 LABORATORY SYSTEMS, VELOCITY MEASUREMENT
B 6203 531* LABRADOR, ABST., CAVITY-DWELLING ORGANISMS, COELOBITES, IN
B 6303 514* LABRADOR, ABST., DEPOSITIONAL SEQUENCE IN SUBARCTIC SANDY
B 6303 508* LABRADOR, ABST., ICHNOLOG OF, LABRADOR GROUP, LOWER,
B 6405 771* LABRADOR, ABST., SEABED CHARACTERISTICS AND SAND DISPERSAL ON
B 6303 508* LABRADOR GROUP, LOWER CAMBRIAN, IN SOUTHERN LABRADOR, ABST.,
B 6306 952 LABRADOR GROUP, NEWFOUNDLAND
B 6011 2069* LABRADOR SEA AND ITS PETROLEUM POTENTIAL, ABST., GEOLOGY OF
B 6111 1994 LABRADOR SEA, POSSIBLE DIAPIRS
S 10 198 LABREA FORMATION, GAS-CUT MUDFLOW, WEST INDIES, TRINIDAD
B 6303 514 LACAZE, J., DELTAIC FACIES RECOGNITIO ON SEISMIC DATA, ABST.
B 6007 1073 LACHENBRUCH-BREWER EQUATION, MODIFIED
B 6002 234 LACK OF ALTERNATIVE MIGRATION MECHANISM IN GULF COAST
B 6402 256 LACONIA SUBMARINE BASIN, SEISMIC REFLECTION, GREECE
B 6311 2042† LACUSTRINE DELTA, CATATUMBO RIVER, LAKE MARACAIBO,
B 6311 2118* LACUSTRINE DELTA, GREAT SALT PLAINS, ALFALFA COUNTY,
B 6203 499* LACUSTRINE DEPOSITIONAL SYSTEMS OF DOCKUM GROUP IN TULE
B 6305 837* LACUSTRINE DEPOSITS, ABST., STRATIGRAPHY OF
B 6403 447* LACUSTRINE, FLUVIAL, AND FAN SEDIMENTATION--QUATERNARY
B 6403 447* LACUSTRINE, FLUVIAL, AND FAN SEDIMENTATION—QUATERNARY
B 6305 843* LACUSTRINE HUMAT MODEL--SEDIMENTOLOGIC AND GEOCHEMICAL
B 6305 839* LACUSTRINE ORIGIN IN WASATCH FORMATION OF VERMILLION CREEK
B 6203 517* LACUSTRINE PSEUDOVARVES FROM CRETACEOUS PETERSON LIMESTONE,
B 6105 836* LACUSTRINE SEDIMENTATION IN NEWARK BASIN, PENNSYLVANIA-NEW
B 6203 514* LACUSTRINE SEDIMENTS IN CRETACEOUS OF NORTHEASTERN SPAIN,
B 6203 533* LACUSTRINE STROMATOLITES FROM KYZYLGIR FORMATION, PLIOCENE,
B 6203 488* LACUSTRINE STROMATOLITES FROM LAKE TURKANA, KENYA, ABST.,
B 6203 551* LACUSTRINE STROMATOLITES, WALKER LAKE, NEVADA, ABST., MODERN
B 6004 690 LADD, J. W., SEISMIC STRATIGRAPHY AND GEOLOGIC HISTORY OF DEEP
S 7 43 LADD, JOHN W., OCCURRENCE AND EVOLUTION OF SALT IN DEEP GULF
B 6411 2041 LADD, JOHN W., REVIEW OF PROCEEDINGS OF THE SYMPOSIUM ON THE
B 6312 2204 LADD, JOHN W., SEISMIC EVIDENCE FOR WIDESPREAD POSSIBLE GAS
M 29 363 LADD, JOHN W., TECTONIC DEVELOPMENT OF TRENCH-ARC COMPLEXES
B 6405 606 LADINIAN TO CARNIAN, MIDDLE EAST
B 6203 534 LADOR, J. M., STATISTICAL APPROACH FOR FUTURE RESERVES
M 25 427 LADY ANNIE, COPPER, QUEENSLAND
B 6107 961 LADY GLORITA CRUISES, 1969-1974
M 25 427 LADY LORETTA, LEAD-ZINC-SILVER, QUEENSLAND
S 4 291 LADY MUSGRAVE REEF, GALAXEA
B 6003 455† LAFAYETTE COUNTY, MISSISSIPPI, SUBSURFACE, PRECAMBRIAN, GRANITE
M 24 177† LAFAYETTE AND COLUMBIA COUNTIES, ARKANSAS, UPPER, SMACKOVER,
B 6211 2356 LAFEHR, T. R., EXPLORATION IN RUGGED TERRAIN IN OVERTHRUST
B 6211 2356 LAFEHR, T. R., INTEGRATED EXPLORATION, ABST.
S 11 274 LAGENA ALCOCKI
S 11 235 LAGENA ALCOCKI
S 11 236 LAGENA AMPHORA
S 11 274 LAGENA CLYPEATA
S 11 236 LAGENA CLYPEATA
S 11 274 LAGENA FOVEOLATA
S 11 236 LAGENA FOVEOLATA
S 11 236 LAGENA MELO
S 11 274 LAGENA MELO
S 11 274 LAGENA PLIOCENICA
S 11 236 LAGENA PLIOCENICA
S 11 237 LAGENA STRIATA
S 11 236 LAGENA STRIATA

S 11 274 LAGENA STRIATA
S 11 237 LAGENA SULCATA
S 11 233 LAGENIDAE
S 11 120 LAGENIDAE, MIOCENE, ROTALIA, PULLENIA
B 6205 887 LAGESON, DAVID R., DEPOSITIONAL ENVIRONMENTS AND DIAGENESIS OF
B 6405 736 LAGESON, DAVID R., STEWART PEAK CULMINATION, IDAHO-WYOMIN
B 6305 833 LAGESON, DAVID R., STRUCTURAL GEOLOGY OF NORTHERN SALT RIVE
S 5 184 LAGO PUEBLO, CONCENTRATIONS IN BRINE, VENEZUELA,
S 5 182 LAGO PUEBLO, LOCATION, VENEZUELA, GRAN ROQUE,
S 5 181 LAGO PUEBLO, PRECIPITATION OF GYPSUM, VENEZUELA,
B 6403 445 LAGOE, MARTIN B., RECENT ARCTIC FORAMINIFERA--AN OVERVIEW.
B 6211 2357 LAGOE, MARTIN B., RECENT BENTHONIC FORAMINIFERAL BIOFACIES IN
B 6403 445* LAGOON, ABST., LATE, PLEISTOCENE-, HOLOCENE SEDIMENTARY
B 6211 2363* LAGOON COMPLEX, UPPER SESPE CREEK, VENTURA COUNTY,
B 6409 1562* LAGOONAL OSTRACODES AND SPECIE DIVERSITY GRADIENTS, GULF OF
B 6303 421* LAGOONAL PATCH REEFS--IMPLICATIONS FOR PALEOZOIC BIOHERMS, OR
B 6303 455* LAGOONS AND REEFS, ABST., PROBING, BERMUDAS
B 6209 1680 LAGUNA DISTRICT, URANIUM, NEW MEXICO
S 4 209† LAGUNA MANDINGA, VERACRUZ, MEXICO, CHANGE IN, OSTRACOD
B 6105 806 LAHANN, RICHARD, SUNY, EVALUATION OF MECHANISMS OF CALCITE
B 6303 483 LAHANN, RICHARD W., SILICA DISSOLUTION FROM MONTMORILLONITE,
B 6404 505 LAHAVE PLATFORM, CORRELATION
B 6109 1484 LAHAVE PLATFORM, EAST COAST
B 6301 89 LAHAVE PLATFORM, SCOTIAN BASIN
B 6108 1155 LAHEE CLASSIFICATION OF WELLS, AS APPLIED BY CSD
B 6208 1309 LAHEE CLASSIFICATION OF WELLS, AS APPLIED BY CSD
B 6409 1319 LAHEE CLASSIFICATION OF WELLS, AS APPLIED BY CSD
B 6308 1202 LAHEE CLASSIFICATION OF WELLS, AS APPLIED BY CSD
B 6008 1187 LAHEE CLASSIFICATION OF WELLS, AS APPLIED BY CSD
B 6103 466 LAHR, JOHN C., HISTORIC EARTHQUAKES AND FUTURE EARTHQUAKE
M 25 169 LAHSEN, ALFREDO, GEOTHERMAL EXPLORATION IN NORTHERN
B 6406 823 LAIDLAW RANGE, AUSTRALIA
S 11 80 LAIMINGINA SMITHI
S 11 146 LAIMINGINA SMITHI
B 6105 806 LAING, W. E., STRAT-SEIS—SIMPLE EXPLANATION WITH FIELD EXAMPLES
B 6303 456 LAJMI, T., PROVENANC STUDIES IN TUNISIA BY U-PB AGES OF DETRITAL
B 6211 2283 LAK MEMBER, BLACK HILLS
B 6211 2277 LAK MEMBER, WYOMING
B 6109 1560* LAKE ATHABASKA REGION BY IMAGE PROCESSING OF LANDSAT DATA,
B 6409 1557* LAKE BORGNE—VALENTINE AREA OF SOUTHEASTERN LOUISIANA, ABST.,
B 6405 622 LAKE CAHUILLA DEPOSITS, CALIFORNIA
B 6408 1260 LAKE CLARK FAULT, ALASKA
B 6203 397 LAKE CLARK FAULT, ALASKA
B 6207 1153 LAKE ENTRANCE FORMATION, AUSTRALIA
B 6309 1576* LAKE ERIE--STRATIGRAPHY, PETROLOGY, AND SEDIMENTOLOGY, ABST.,
M 24 177 LAKE ERLING FIELD, ARKANSAS
B 6303 311 LAKE FLAGSTAFF LACUSTRINE-ALLUVIAL, UTAH
B 6410 1643† LAKE FROME REGION, SOUTH AUSTRALIA, DISTRIBUTION AND GENESIS
B 6008 1392* LAKE FROME, SOUTH AUSTRALIA, ABST., URANIUM IN, TERTIARY
B 6004 545 LAKE HATCH FAULT SYSTEM, LOUISIANA
B 6206 1013 LAKE HAZEN FAULT, CANADA, ARCTIC
B 6203 552* LAKE HURON, ABST., FACTORS CONTROLLING, GEOCHEMISTRY AND
B 6408 1188 LAKE KIVU, LARGE ANOXIC LAKE
B 6109 1424 LAKE MARACAIBO AREA, RESERVOIRS, VENEZUELA
B 6408 1191 LAKE MARACAIBO, POSITIVE WATER BALANCE BASIN, VENEZUELA
B 6303 516* LAKE MARACAIBO, VENEZUELA, ABST., FACIES RECOGNITION FROM,
B 6311 2042† LAKE MARACAIBO, VENEZUELA, STRATIGRAPHY OF, INTERMONTANE,
B 6207 1159† LAKE MEAD AREA OF ARIZONA AND NEVADA DURING LATE MIOCENE
B 6307 1139† LAKE MEAD AREA OF ARIZONA AND NEVADA DURING LATE MIOCENE
B 6410 1644 LAKE NAMBA CHANNEL, AUSTRALIA
B 6209 1762* LAKE PONTCHARTRAIN, ABST., NEW ORLEANS—, SOUTH HANCOCK,
B 6209 1758* LAKE PONTCHARTRAIN, ABST., REMOTELY SENSED SEDIMENTATION
B 6102 234 LAKE POWELL AREA, TESTS, UTAH
B 6101 65 LAKE SAND AREA, GEOTHERMAL PATTERN, LOUISIANA
B 6101 65† LAKE SAND AREA, SOUTH LOUISIANA--IMPLICATIONS, GEOTHERMAL
B 6101 69 LAKE SAND FAULT, LOUISIANA
B 6101 70 LAKE SAND FIELD, LOUISIANA
B 6007 1022 LAKE SUPERIOR, BATHYMETRIC MAP, WESTERN
B 6007 1025 LAKE SUPERIOR, CURRENT METERS
B 6007 1033 LAKE SUPERIOR, FAN VALLEYS
B 6007 1023 LAKE SUPERIOR, REFLECTION PROFILES
B 6206 918 LAKE SUPERIOR, RESERVE FAN
B 6007 1024 LAKE SUPERIOR, RESERVE FAN AND DELTA
B 6007 1023 LAKE SUPERIOR, RESERVE FAN, NEW TERM
B 6007 1025 LAKE SUPERIOR, SEDIMENT CORES
B 6007 1021† LAKE SUPERIOR, SUBLACUSTRINE, FAN MORPHOLOGY IN
B 6007 1022 LAKE SUPERIOR, TAILINGS DEPOSITION
B 6408 1187 LAKE TANGANYIKA, LARGE ANOXIC LAKE
B 6406 942† LAKE TRAVERSE, WESTERN MINNESOTA, EXPOSURES OF, GREENHORN
B 6203 488* LAKE TURKANA, KENYA, ABST., OXYGEN, AND, CARBON, ISOTOPIC
S 8 40 LAKE WALENSEE, SWITZERLAND, DOWNSLOPE FLOW, CURRENT FLOW
B 6104 587 LAKE WASHINGTON FIELD, LOUISIANA
B 6104 595 LAKE WASHINGTON FIELD, LOUISIANA
B 6305 825* LAKES, ABST., DEPOSITION AND DIAGENESIS OF, ORGANIC MATTER AND,
B 6105 832* LAKOTA FORMATION, LOWER CRETACEOUS, IN SOUTH DAKOTA AND
M 24 95 LAKOTA SANDSTONE, WYOMING
B 6403 428 LAKWA FIELD, INDIA
B 6010 1712 LALECHOS, N., ON DEVELOPMENTS IN GREECE, 1975
B 6110 1643 LALECHOS, N., ON DEVELOPMENTS IN GREECE, 1976
B 6210 1912 LALECHOS, N., ON DEVELOPMENTS IN GREECE, 1977
B 6310 1921 LALECHOS, N., ON DEVELOPMENTS IN GREECE, 1978
B 6411 1964 LALECHOS, N., ON DEVELOPMENTS IN GREECE, 1979
B 6206 985 LALI FIELD, IRAN
B 6104 579 LAMA FIELD, VENEZUELA
B 6104 581 LAMA TREND, VENEZUELA
B 6104 579 LAMAR FIELD, VENEZUELA
B 6104 581 LAMAR HIGH, VENEZUELA
B 6205 887 LAMB, C. F., PAINTER RESERVOIR FIELD, UINTA COUNTY, WYOMING,
B 6303 484 LAMB, C. F., PAINTER RESERVOIR FIELD--GIANT IN WYOMING THRUST
M 30 281 LAMB, CHARLES F., PAINTER RESERVOIR FIELD--GIANT IN THE WYOMING
B 6405 638 LAMB, CHARLES F., PAINTER RESERVOIR FIELD—GIANT IN WYOMING
B 6004 690 LAMB, J. L., UNRELIABILITY OF GLOBIGERINOIDES DATUM, ABST.
B 6009 1564 LAMB, JAMES L., UNRELIABILITY OF GLOBIGERINOIDES DATUM
B 6203 529 LAMBERT, DOUGLAS N., GEOTECHNICAL PROPERTIES OF CONTINENTAL
B 6405 784* LAMBERT FIELDS, OKLAHOMA—SOURCE, RESERVOIR, AND TRAP, ABST.,
B 6008 1407 LAMBERT, PAUL W., COASTAL SEDIMENTATION MODEL, TENSLEEP
S 6 347 LAMBERT, RICHARD S., ORDOVICIAN GEOCHRONOLOGY
B 6203 564 LAMBERT, STEPHEN W., FOAM STIMULATION FOR ENHANCED RECOVERY
B 6303 484 LAMBIASE, JOSEPH J., DETAILED TEMPERATURE LOGGING AS USEFUL
B 6105 806 LAMBIASE, JOSEPH J., SEDIMENT DYNAMICS IN MACROTIDAL AVON RIVER
S 5 89 LAMBTON COUNTY, SARNIA, CORE SECTION, ONTARIO,
B 6203 519* LAMINAR AND INTRALAMINAR LEVELS SUGGEST REVISION OF
M 30 253 LAMMLEIN, D. R., EUGENE ISLAND BLOCK 330 FIELD, OFFSHORE
B 6410 1660 LAMONT-DOHERTY GEOLOGICAL OBSERVATORY CRUISES VEMA 18, 19
B 6002 198 LAMONT-DOHERTY GEOPHYSICAL CRUISES, MAY 1974, BRAZIL MARGIN
B 6111 1994 LAMONT-DOHERTY SEISMIC PROFILE LIBRARY, DEEP-SEA SEDIMENT,
B 6004 681* LAMOTTE SANDSTONE OF SOUTHEASTERN MISSOURI, ABST.,
B 6203 430 LANCASHIRE COAST, BRITAIN, WEST
B 6008 1396* LANCE CHANNELS, NIOBRARA COUNTY, WYOMING, ABST.,
B 6111 1991 LANCELOT, Y., PIERCEMENT STRUCTURES IN DEEP OCEANS
B 6303 484 LANCELOT, YVES, DEEP-SEA DRILLING, NEW DIMENSION IN OUR
B 6303 550 LAND, COOPER B., COMPARISON OF PORE-FILLING MATERIAL IN SOME
B 6008 1401 LAND, COOPER B., STRATIGRAPHY AND PETROLEUM ACCUMULATION,
B 6303 485 LAND, COOPFR B., TYLER SANDSTONES, PENNSYLVANIAN, DICKINSON
B 6303 545* LAND DEVELOPMENT AND FAULTING NEAR HOUSTON, TEXAS,
S 4 53 LAND, LYNTON S., DEEP FOREREEF AND UPPER ISLAND SLOPE, NORTH
B 6109 1554 LAND, LYNTON S., HYDROGEN AND OXYGEN ISOTOPIC COMPOSITION OF
B 6303 485 LAND, LYNTON S., ISOTOPIC AND TRACE-ELEMENT GEOCHEMISTRY OF
B 6306 986 LAND OF COCHISE, SOUTHEASTERN ARIZONA, 1978, NEW MEXICO
B 6109 1539* LAND RESOURCES OF TEXAS, ABST.
M 25 362† LAND SUBSIDENCE IN JAPAN RESULTING FROM FLUID
M 25 421 LAND USE, COMPETITION FOR, AUSTRALIA
B 6303 454* LAND-SURFACE SUBSIDENCE IN HOUSTON-GALVESTON REGION, TEXAS,
B 6009 1609* LAND-USE DELINEATION IN TEXAS COASTAL ZONE, ABST.,
B 6207 1221* LAND-USE PLANNING AND MINERAL INDUSTRY—PRACTICES AND
B 6407 1122 LAND-USE PLANNING. 2. FLATLAND DEPOSITS—THEIR GEOLOGY AND
M 28 7 LANDER SANDSTONE, WYOMING, ORDOVICIAN
B 6203 534 LANDING, ED, EGGSHELL DIAGENESIS AND PRECOMPACTIONAL
B 6105 806 LANDIS, E. R., COAL RESOURCES OF COLORADO AND WYOMING
B 6108 1199 LANDON, SUSAN M., DEVELOPMENTS IN EASTERN AND NORTHWESTERN
B 6304 660† LANDSAT 2, MISSISSIPPI RIVER, DELTA--APRIL 9, 1976--FROM
B 6109 1560* LANDSAT, ABST., DETECTION OF PETROLEUM-RELATED, SOIL,
B 6109 1559* LANDSAT AND GEOPHYSICAL DATA IN NORTHEASTERN ARIZONA, ABST.,
B 6203 546* LANDSAT DATA, ABST., SMALL ROCK OUTCROPS IN, VEGETATED,
B 6109 1560* LANDSAT DATA, ABST., URANIUM SEARCH IN, LAKE, ATHABASKA
B 6401 20 LANDSAT DATA, ALASKA, PRUDHOE BAY, GIANT OIL AND GAS FIELD
B 6401 26 LANDSAT DATA, ALASKA, SWANSON RIVER GIANT OIL FIELD
B 6401 28 LANDSAT DATA, ALBERTA, RICINUS GIANT GAS FIELD
B 6401 28 LANDSAT DATA, ALBERTA, STRACHAN GIANT GAS FIELD
B 6401 24 LANDSAT DATA, CALIFORNIA, KETTLEMAN HILLS GIANT OIL FIELD
B 6401 9 LANDSAT DATA, CHINA, TACHING, DAQING, GIANT OIL FIELD
B 6401 30 LANDSAT DATA, FLORIDA, JAY GIANT OIL AND GAS FIELD
B 6303 537* LANDSAT DATA FO GEOLOGIC APPLICATIONS, ABST., COMPUTER
B 6401 8† LANDSAT DATA FOR 15 GIANT OIL AND GAS FIELDS, GEOLOGIC
M 30 7† LANDSAT DATA FOR 15 GIANT OIL AND GAS FIELDS, GEOLOGIC
B 6109 1559* LANDSAT DATA IN EXPLORING EVAPORITE BASINS, ABST., EVALUATION
B 6401 10 LANDSAT DATA, IRAN, KANGAN GIANT GAS, FIELD
B 6401 12 LANDSAT DATA, IRAQ, BUZURGAN GIANT, OIL FIELD
B 6401 32 LANDSAT DATA, ITALY, MALOSSA GIANT GAS-CONDENSATE FIELD
B 6401 14 LANDSAT DATA, KUWAIT, BURGAN GIANT OIL FIELD
B 6401 16 LANDSAT DATA, LIBYA, MESSLA GIANT OIL FIELD

B 6401 34 LANDSAT DATA, NETHERLANDS AND, WEST, GERMANY, GRONINGEN
B 6304 698* LANDSAT DATA ON SOME KNOWN GIANT FIELDS, GEOLOGIC
B 6309 1490† LANDSAT DATA, RECONNAISSANCE GEOLOGIC, MAPPING IN, CHAGAI
B 6401 9 LANDSAT DATA, SAUDI ARABIA, GHAWAR GIANT OIL FIELD
B 6004 677* LANDSAT DATA TO PETROLEUM EXPLORATION, ABST., APPLICATION OF,
B 6401 18 LANDSAT DATA, TUNISIA, EL BORMA GIANT OIL FIELD
B 6401 22 LANDSAT DATA, U.S.S.R., SAMOTLOR, SAMOTLORSKOYE, GIANT OIL
M 25 63 LANDSAT, ERTS, NOW
B 6304 699* LANDSAT FOR GEOLOGIC STUDIES, ABST., HIGH-RESOLUTION
B 6405 725* LANDSAT FOR GEOPHYSICAL STUDIES, ABST., HIGH-RESOLUTION
M 30 22 LANDSAT IMAGE, ALASKA, PRUDHOE BAY FIELD
M 30 28 LANDSAT IMAGE, ALASKA, SWANSON RIVER
M 30 30 LANDSAT IMAGE, ALBERTA, RICINUS FIELD
M 30 30 LANDSAT IMAGE, ALBERTA, STRACHAN FIELD
M 30 26 LANDSAT IMAGE, CALIFORNIA, KETTLEMAN HILLS FIELD
B 6405 624 LANDSAT IMAGE, CALIFORNIA, SAN ANDREAS FAULT
M 30 10 LANDSAT IMAGE, CHINA, TACHING FIELD
M 30 32 LANDSAT IMAGE, FLORIDA, JAY FIELD
M 30 12 LANDSAT IMAGE, IRAN, KANGAN, GAS FIELD
M 30 14 LANDSAT IMAGE, IRAQ, BUZURGAN FIELD
M 30 34 LANDSAT IMAGE, ITALY, MALOSSA FIELD
M 30 16 LANDSAT IMAGE, KUWAIT, BURGAN FIELD
M 30 18 LANDSAT IMAGE, LIBYA, MESSLA FIELD
M 30 36 LANDSAT IMAGE, NETHERLANDS, GRONINGEN FIELD
B 6204 642 LANDSAT IMAGE, OCTOBER 12, 1972, GULF OF, ALASKA
M 30 9 LANDSAT IMAGE, SAUDI ARABIA, GHAWAR FIELD
M 30 24 LANDSAT IMAGE, SIBERIA, SAMOTLOR FIELD
M 30 20 LANDSAT IMAGE, TUNISIA, EL BORMA FIELD
M 30 24 LANDSAT IMAGE, USSR, SAMOTLOR FIELD
M 30 36 LANDSAT IMAGE, WEST GERMANY, GRONINGEN FIELD
B 6303 521* LANDSAT IMAGERY, ABST., SOIL REGIMES OF, TAZIRBU REGION OF
B 6207 1222* LANDSAT IMAGERY AND RELATED REMOTE SENSING TECHNIQUES TO
B 6009 1609* LANDSAT IMAGERY FOR WETLAND AND LAND-USE DELINEATION IN
B 6207 1232* LANDSAT IMAGERY IN CIRCUM-PACIFIC REGION, ABST., APPLICATIONS
B 6203 515* LANDSAT IMAGERY IN URANIUM EXPLORATION, ABST., TEST USE OF,
B 6109 1561* LANDSAT IMAGERY TO GEOLOGIC PROSPECTING OVER IBERIAN
B 6005 745† LANDSAT IMAGERY TO PETROLEUM AND MINERAL EXPLORATION,
B 6008 1414* LANDSAT IMAGERY--EVOLVING EXPLORATION TOOL FOR GEOLOGISTS,
B 6306 951 LANDSAT IMAGES, NEWFOUNDLAND
B 6401 32 LANDSAT, ITALY, MALOSSA GIANT GAS-CONDENSATE FIELD
B 6105 783 * LANDSAT STUDY OF NEARSHORE CIRCULTION OFF ARANSAS PASS, TEXAS
B 6303 537* LANDSAT SYSTEM FOR GEOLOGIC APPLICATIONS, ABST.,
B 6109 1559* LANDSAT—USEFUL PETROLEUM EXPLORATION TOOL IN
B 6207 1231* LANDSLIDE PROBLEMS IN JAPAN, ABST.
B 6303 512* LANDSLIDE PROCESSES, MISSISSIPPI DELTA, ABST., ACTIVE SUBMARINE
B 6405 737 LANE, H. RICHARD, CARBONIFEROUS CONODONTS, ABST.
B 6203 534 LANE, TOM E., ORDOVICIAN-SILURIAN GLACIO-EUSTATIC SEA-LEVEL
B 6310 1981 LANGENHEIM, R. L., JR., REVIEW OF A GEOLOGIST IN THE SERVICE OF
B 6203 535 LANGENHEIM, RALPH L., JR., ALGAL STROMATOLITES IN EUREKA
B 6408 1286 LANGENHEIM, RALPH L., JR., COLMAR-PLYMOUTH CONUNDRUM, ABST.
B 6303 542 LANGENHEIM, RALPH L., JR., PALEOENVIRONMENT OF MIDDLE
B 6309 1582 LANGENHEIM, RALPH L., JR., SUBMARINE SLUMPING IN REEF-FLANK
B 6304 690 LANGFORD, FRED F., STRATIGRAPHIC CONTROL OF URANIUM DEPOSITS
B 6101 28 LANGFORD, FRED F., SURFICIAL ORIGIN OF NORTH AMERICAN
M 30 315 LANGTON, J. R., STRACHAN AND RICINUS WEST GAS FIELDS, ALBERTA,
B 6303 469 LANGTON, J. R., STRACHAN-RICINUS GAS FIELD, ALBERTA, ABST.
B 6006 952 LANPHERE, M. A., GEOLOGIC SIGNIFICANCE OF DOONERAK STRUCTURAL
S 6 259 LANPHERE, MARVIN A., CRETACEOUS TIME SCALE FROM NORTH
S 4 256 LANSING GROUP, MID-CONTINENT
B 6107 1030 LANSING GROUP, NEBRASKA
B 6405 701* LANSING-KANSAS CITY E ZONE, HITCHCOCK COUNTY, NEBRASKA, ABST.,
B 6105 839* LANSING-KANSAS CITY GROUPS IN NORTHWEST KANSAS-SOUTHWEST
B 6203 535 LANT, KEVIN J., STRUCTURE OF ALISO CANYON FIELD, EASTERN
B 6410 1577 LANTZ MILLS BEDS, VIRGINIA
B 6405 737 LANTZ, ROBERT J., BARROW GAS FIELDS, NORTH SLOPE, ALASKA, ABST.
B 6105 810 LANTZY, RONALD J., MANS IMPACT ON GLOBAL SEDIMENTARY CYCLES OF
B 6203 443 LANWA FIELD, INDIA
B 6108 1382 LAPOINT, DENNIS J., EVIDENCE FOR SOURCE AREAS AND DEPOSITIONAL
B 6303 485 LAPORTE, LEO F., TRACKS AND SUBSTRATE REWORKING BY
S 11 315 LARABEE CREEK, CALIFORNIA, SAMPLE LOCALITIES
B 6008 1395* LARAMIDE ROCKY MOUNTAIN FORELANDS, ABST., ANALYTIC
B 6001 12† LARAMIDE STRUCTURAL GROWTH ON SEDIMENTATION ON PETROLEUM
B 6104 602 LARAMIDE THRUSTS, CORDILLERAN FOLDBELT, NEW MEXICO
B 6008 1411* LARAMIDE TRAPS IN ROCKY MOUNTAIN FORELANDS, ABST., FIELD
B 6008 1409* LARAMIDE TRAPS IN ROCKY MOUNTAIN FORELANDS, ABST., SEISMIC
B 6409 1503 LARAMIE BASIN, WYOMING, HANNA-
B 6308 1391 LARAMIE BASIN, WYOMING, HANNA-
B 6208 1500 LARAMIE BASIN, WYOMING, HANNA-
M 26 279 LARAMIE, CRETACEOUS
B 6108 1347 LARAMIE-DENVER BASIN, WYOMING, HANNA-
B 6008 1206 LARAMIE-HANNA BASINS, WYOMING
B 6305 838* LARAMIE-HANNA-SHIRLEY BASINS--FUTURE PETROLEUM POTENTIAL,
B 6405 676 LARBERG, G. M., HYDRODYNAMIC FLOW IN LOWER CRETACEOUS
B 6209 1762* LAREDO AREA, WEBB AND ZAPATA COUNTIES, TEXAS, ABST., LOBO
B 6105 807 LARESE, R. E., IMPORTANCE OF TEXTURAL AND DIAGENETIC PROPERTIES
B 6405 737 LARESE, RICHARD E., DEVELOPMENT OF SECONDARY POROSITY IN
B 6408 1187 LARGE ANOXIC LAKES, EXAMPLES
B 6103 468* LARGE NEW PORPHYRY MOLYBDENUM DISCOVERY IN SOUTHEASTERN
B 6408 1188 LARGE OXIC LAKES, EXAMPLES
B 6004 695* LARGE SEDIMENT SLUMP REOPENS UNKNOWN, SCRIPPS SUBMARINE
B 6004 700* LARGE-SCALE BED FORMS AN SEDIMENTARY STRUCTURES DEVELOPED
B 6105 807 LARNER, K., WAVE-EQUATION MIGRATION—TWO APPROACHES
B 6012 2184 LARNER, K., WAVE-EQUATION MIGRATION--TWO APPROACHES, ABST.
B 6410 1726 LARSEN AND G. V. CHILINGAR, BLATT, HARVER, REVIEW OF, DIAGENESIS
B 6303 354 LARSON, DAVID W., PALEOENVIRONMENTS O MOWRY SHALE, LOWER
B 6008 1402 LARSON, ROBERT T., JR., SOUTHWEST RAGGED POINT FIELD,
B 6303 486 LARSON, ROGER L., LATE MESOZOIC PLATE TECTONICS, ABST.
B 6105 808 LARSON, T. L. L., COMPUTER PLOTTING OF LAND LEASES ON DIGITIZED
B 6303 486 LARSON, WILLIAM C., URANIUM IN-SITU LEACH MINING--A THIRD
B 6303 486 LARUE, DAVID K., CHOCOLAY GROUP--EARLY PROTEROZOIC CRATONI
B 6405 675* LAS ANIMAS COUNTY, COLORADO, ABST., DRILLING FOR, METHANE, GAS
B 6107 963 LAS PALMAS FRACTURES, AFRICA, ST. PAUL-
B 6408 1165 LAS PIEDRAS SANDSTONE, VENEZUELA
B 6301 38 LASALLE COUNTY, TEXAS, STUART CITY FIELD
B 6108 1384 LASSIN, R. J., PARAGENESIS OF AUTHIGENIC MINERALS IN ST. PETER
S 6 40 LAST APPEARANCE DATUM, BIOCHRONOLOGY
S 6 44 LAST APPEARANCE DATUM, EXAMPLES
S 6 41 LAST APPEARANCE DATUM, REVIEW OF PRINCIPLE
B 6202 255 LASUEN ANTICLINORIUM, CALIFORNIA
B 6202 248 LASUEN KNOLL, CALIFORNIA
B 6303 478* LATE CENOZOIC CALCAREOUS MICROFOSSIL BIOSTRATIGRAPHY,
B 6211 2363* LATE CRETACEOUS DEPOSITIONAL ENVIRONMENTS, NORTHERN SANTA
B 6211 2351* LATE CRETACEOUS SEDIMENTATION IN SANTA MONICA MOUNTAINS,
B 6012 2182* LATE CRETACEOUS TURBIDITE SEDIMENTATION AND SUBMARINE-FAN
B 6002 288 LATE DIAGENESIS
B 6303 499* LATE HOLOCENE DEPOSITIONAL HISTORY OF REGRESSIVE BARRIER
B 6105 708 † LATE HOLOCENE SEDIMENTARY PROVINCES, SOUTH TEXAS OUTER
B 6004 711* LATE JURASSIC BARRIER ISLAND COMPLEX IN LOWER SUNDANCE
B 6303 486* LATE MESOZOIC PLATE TECTONICS, ABST.
B 6207 1223* LATE MESOZOIC TECTONIC HISTORY OF NORTHERN CALIFORNIA AND
B 6405 791* LATE MESOZOIC TO EARLY CENOZOIC FORELAND SEDIMENTATION IN
B 6009 1622* LATE MIDDLE AND EARLY LATE ORDOVICIAN HISTORY OF CINCINNATI
B 6109 1538* LATE MIOCENE GLACIO-EUSTATIC LOWERING OF SEA LEVEL—EVIDENCE
B 6012 2174† LATE MIOCENE PALEOBATHYMETR OF CALIFORNIA CONTINENTAL
B 6211 2351* LATE MIOCENE TURBIDITE DEPOSITION, CROCKER CANYON AREA,
B 6405 671* LATE NEOGENE DEPOSITIONAL AND CLIMATI CYCLES IN YAKATAGA
B 6012 2182* LATE NEOGENE PLAEOBATHYMETRY AND PALEOENVIRONMENNS OF
B 6012 2180* LATE NEOGENE-QUATERNARY TECTONICS AN DEPOSITIONAL
B 6305 846* LATE ORDOVICIAN STRATIGRAPHY AND SEDIMENTATION IN NORTHERN
B 6303 410* LATE ORDOVICIAN--EARLY SILURIAN STRATA IN MID-CONTINENT AREA,
B 6409 1565* LATE PALEOCENE PLANKTONIC FORAMINIFERAL BISTRATIGRAPHY OF
B 6209 1767* LATE PLEISTOCENE CARBONATE BANK DEPOSITION, COZUMEL ISLAND,
B 6309 1611* LATE PLEISTOCENE FLUVIAL-DELTAIC DEPOSITION TEXAS COASTAL
S 7 243† LATE PLEISTOCENE GEOLOGIC HISTORY OF TEXAS OUTER CONTINENTAL
B 6209 1763* LATE PLEISTOCENE STRATIGRAPHY OF SOUTHERN HANCOCK COUNTY,
B 6004 649* LATE PLEISTOCENE-HOLOCENE EVOLUTION OF SOUTH TEXAS OUTER
B 6403 445* LATE PLEISTOCENE-HOLOCENE SEDIMENTARY INFILLING AND FAUNAL
B 6004 653* LATE PRECAMBRIAN SEDIMENTATION IN BELT AN UINTA
B 6303 538* LATE QUATERNARY FORAMINIFERAL RECORD IN EASTERN CARIBBEAN
B 6109 1540* LATE QUATERNARY GEOMORPHIC EVOLUTION OF COLORADO RIVER,
B 6203 542* LATE QUATERNARY SEDIMENTATION ON UPPER CONTINENTAL SLOPE,
B 6103 469* LATE TERTIARY GULF OF ALASKA PLATE-TECTONIC MODEL BASED ON
B 6303 520* LATE TRIASSIC-JURASSIC PALEOGEOGRAPHY AND ORIGIN OF GULF OF
B 6303 499* LATE-STAGE SUBSURFACE DOLOMITES--PROBLEMS OF ORIGIN, ABST.
B 6312 2145† LATERAL AND VERTICAL FACIES RELATIONS OF TRANSGRESSIVE
M 26 427 LATERAL CONTACTS, RESERVOIR-SEAL CONTACT
B 6303 473* LATERAL DIAGENESIS IN MONTEREY SHALE, SANTA BARBARA COAST,
B 6004 725* LATERAL INFILL AS MAJOR FACTOR IN SUBMARINE-CANYON AND
B 6207 1234* LATERITE DEPOSITS IN CIRCUM-PACIFIC REGION, ABST., NICKELIFEROUS
M 25 403 LATERITE NICKEL DEPOSITS, AUSTRALIA
M 25 402 LATERITE NICKEL DEPOSITS, AUSTRALIA
M 25 402 LATERITE NICKEL DEPOSITS, SILICATE AND LIMONITE ORE
M 25 402 LATERITES, AUSTRALIA, GREENVALE AND, ROCKHAMTON,
M 25 406 LATERITES, AUSTRALIAN, BAUXITE DEPOSITS, CONFINED TO, TERTIARY
S 3 7 LATEROGENE, LATERAL SECRETION
S 3 136 LATEROGENESIS AT THE LOW-INTENSITY END OF THE
S 3 136 LATEROGENESIS IN GEOPRESSURIZED BASINS, HYPOGENE
B 6406 857† LATEST JURASSIC-EARLY CRETACEOUS REGRESSIVE FACIES, NORTHEAST
B 6207 1225 LATHAM, GARY V., SEISMICITY AND STRUCTURE OF NEW HEBRIDES
M 25 63 LATHRAM, E. H., SURVEYING EARTH

S 11 286 LATICARININA HALOPHORA
S 11 255 LATICARININA HALOPHORA
M 25 26† LATIN AMERICAS OUTLOOK TOWARD ENERGY RESOURCES--SUMMARY
S 9 14 LATITUDE DISTRIBUTION, PALEOCLIMATE, COAL
S 9 13 LATITUDES, PALEOCLIMATE INDICATORS, PALEOMAGNETIC
M 25 105 LATOUR, B. A., COAL RESOURCES OF CANADIAN CORDILLERA--SUMMARY
B 6207 1153 LATROBE GROUP, AUSTRALIA
B 6309 1606 LAUDERDALE, R. W., CARRILLO PUERTO FORMATION OF NORTHEASTERN
B 6303 495 LAUDON, THOMAS S., CRETACEOUS TRANSGRESSION OF COAHUILA
B 6105 807 LAUGHLIN, A. W., HOT, DRY ROCK, HDR, GEOTHERMAL ENERGY—
B 6405 737 LAULE, SUSAN WUNDERLICH, TECTONICS AND SEDIMENTATION ALONG
M 28 141 LAUMONTITE, BIREFRINGENCE, ETC.
M 28 141 LAUMONTITE, IN ARKOSIC SANDSTONE, CEMENTS
B 6108 1392* LAUMONTITE IN LATE MESOZOIC AND TERTIARY BASINS OF PACIFIC
M 28 141 LAUMONTITE, IN NON-TUFF
B 6303 421* LAUMONTITE OCCURRENCE, AND BURIAL HISTORY IN LOS ANGELES
B 6007 1055 LAURENCIA ZONE, PANAMA, GALETA REEF
B 6206 920 LAURENTIAN FAN
B 6303 535* LAURENTIAN FAN--DEEP-SEA FAN MODELS, FINE-GRAINED SEDIMENT
B 6301 90 LAURENTIAN FORMATION, SCOTIAN BASIN
B 6203 535 LAURY, ROBERT L., SEDIMENTOLOGY OF LATE QUATERNARY
B 6104 527† LAVACA BAY--TRANSGRESSIVE DELTAIC SEDIMENTATION IN CENTRAL
B 6309 1595* LAVACA COUNTY, TEXAS, ABST., CHARACTERISTICS OF LOWER, WILCOX,
B 6409 1557* LAVACA COUNTY, TEXAS, ABST., CLAY MINERALS AS INDICATORS FOR
B 6309 1598* LAVACA COUNTY, TEXAS, ABST., EXPLORATION METHODS OF DISCOVERY
B 6303 536* LAVACA COUNTY, TEXAS, ABST., SOUTH HALLETTSVILLE, GAS FIELD IN
B 6309 1607* LAVACA, GONZALES, AND FAYETTE COUNTIES, TEXAS ABST., FIELD
B 6104 528 LAVACA VALLEY, TEXAS
B 6105 786 LAVELLE, J. W., NOAA'S WASTE DISPOSAL STUDIES IN NEW YORK BIGHT
B 6304 691 LAVERY, N. G., MAGNITUDE AND VARIABILITY OF DISEQUILIBRIUM IN
B 6405 738 LAW, B. E., EVALUATION OF ORGANIC MATTER AND SUBSURFACE
B 6305 833 LAW, B. E., WRENCH FAULTING AND HYDROCARBON OCCURRENCE IN
B 6406 823 LAWFORD RANGE, AUSTRALIA
B 6203 536 LAWRENCE, D. R., CRETACEOUS-HOLOCENE SEDIMENTATIONIN
B 6004 691 LAWRENCE, MARC A., SUBMERGED RELICT FORESTS OFF NORTHWEST
B 6006 742 LAWSON, BARBARA J., REGIONAL GEOLOGIC FRAMEWORK OFF
B 6006 926 LAWSON, BARBARA J., REGIONAL GEOLOGIC FRAMEWORK OFF
B 6203 532 LAWSON, DANIEL, ORIGIN OF NAVAJO SANDSTONE, ABST.
M 29 189 LAWVER, L. A., TECTONICS OF THE ANDAMAN SEA AND BURMA
B 6406 868 LAWVER, LAWRENCE, ARC-CONTINENT COLLISION I BANDA SEA REGION
M 24 145 LAYDEN, R. L., BIG WELLS FIELD, DIMMIT AND ZAVALA COUNTIES, TEXAS
B 6303 477 LAZREG, HABIB, REGIONAL TECTONICS AND PETROLEUM GEOLOGY OF
B 6307 1144 LE BLANC, RUFUS J., REVIEW OF FLUVIAL SEDIMENTOLOGY, EDITED BY
B 6402 242 LE QUELLEC, PATRICK, SEISMIC STRUCTURE OF SOUTHWESTERN
M 26 194 LEA COUNTY, CENTRAL BASIN PLATFORM, SEISMIC EXAMPLE, NEW
B 6205 781 LEA COUNTY, SEISMIC DATA, NEW MEXICO
B 6004 691 LEA, J. D., STATISTICAL MAPS AS AID TO GEOLOGIC INTERPRETATION,
B 6303 486* LEACH MINING--A THIRD ALTERNATIVE, ABST., URANIUM IN-SITU
B 6207 1175 LEACH MOUNTAINS, NEVADA
M 27 194 LEACHED FOSSILS
B 6309 1606* LEACHED POROSITY IN LOWER TERTIARY SANDSTONE RESERVOIRS
B 6009 1612* LEACHED POROSITY IN OVERPRESSURED SANDSTONES--FRIO
M 27 195 LEACHING
M 28 20 LEACHING, FELDSPAR
M 28 175 LEACHING, FELDSPARS, FROM VOLCANIC ROCK
S 4 375 LEACHING OF SCLERODERMITES UNDER SUBAERIAL CONDITIONS
B 6303 538* LEACHING OF URANIUM, ABST., ENVIRONMENTAL ASSESSMENT OF,
B 6105 840* LEACHING OF URANIUM—DEVELOPING TECHNOLOGY, ABST., IN-SITU
B 6206 1051 LEACHING, URANIUM
M 25 554 LEAD, AND MOLYBDENUM DEPOSITS, ECUADOR, FIERRO URCO, COPPER,
M 25 554 LEAD AND ZINC DEPOSITS, ECUADOR, URITOHUASER
M 25 469 LEAD, AND ZINC, USSR, NORTHEAST, COPPER
M 25 427 LEAD, AUSTRALIA, NORTHERN TERRITORY, RUM JUNGLE, URANIUM,
M 25 497 LEAD, CANADA, SELWYN BASIN, ZINC-
M 25 553 LEAD DEPOSITS, ECUADOR, SAN BARTOLOME, SILVER, ZINC, AND
M 25 553 LEAD, ECUADOR, SAN BARTOLOME DEPOSIT, SILVER-, ZINC-
M 25 505 LEAD, PRODUCTION AND RESERVES, CANADIAN CORDILLERA
M 25 427 LEAD, QUEENSLAND, DUGALD RIVER, ZINC-
M 25 553 LEAD, SILVER, AND GOLD DEPOSITS, ECUADOR, ANGAS, COPPER, ZINC
M 25 444 LEAD, SOUTH KOREA
M 25 515 LEAD, ZINC, AND ARSENIC, MEXICO, SILVER
M 25 427 LEAD-SILVER-ZINC DEPOSIT, NEW SOUTH WALES, BROKEN HILL
M 25 427 LEAD-SILVER-ZINC-COPPER, QUEENSLAND, MOUNT ISA
S 10 114 LEAD-ZINC, CANADA, DEVONIAN, DISTRIBUTION OF
M 25 427 LEAD-ZINC, AUSTRALIA, NORTHERN TERRITORY, H.Y.C. DEPOSIT
M 25 496 LEAD-ZINC, CANADIAN, CORDILLERA, SILVER-
M 25 535 LEAD-ZINC DEPOSIT, ARGENTINA, AGUILAR
M 25 507 LEAD-ZINC DEPOSIT, CANADA, SALMO TYPE OF
M 25 507 LEAD-ZINC DEPOSIT, CANADA, SHUSWAP TYPE OF
M 25 426 LEAD-ZINC DEPOSIT, NEW SOUTH WALES, WOODLAWN, COPPER-
M 25 554 LEAD-ZINC, ECUADOR, URITOHUASER DEPOSIT
M 25 457 LEAD-ZINC ORES, JAPAN
M 25 426 LEAD-ZINC-COPPER DEPOSIT, NEW SOUTH WALES, CAPTAINS FLAT
M 25 426 LEAD-ZINC-COPPER DEPOSIT, TASMANIA, ROSEBERY
M 25 427 LEAD-ZINC-SILVER, QUEENSLAND, HILTON DEPOSIT
M 25 427 LEAD-ZINC-SILVER, QUEENSLAND, LADY LORETTA
B 6405 755* LEADVILLE DOLOMITE A PITCH MINE, SAGUACHE COUNTY, COLORADO,
B 6112 2137 LEAFLAND FIELD, ALBERTA
B 6104 594 LEAFLAND FIELD, ALBERTA
B 6004 658* LEAKAGE GAS FROM UNDERGROUND NATURAL GAS STORAGE
B 6008 1410* LEASE OF FEDERAL GEOTHERMAL ENERGY--ROCKY ROAD ABST.
B 6405 766* LEASE SALE 55, NEAR YAKUTAT, EASTER GULF OF ALASKA, ABST.,
B 6212 2413 LEASE TRACTS, GULF OF ALASKA
B 6009 1613* LEASES, ABST., COMPETITION AND QUEST FOR, OFFSHORE
B 6303 486 LEATHERMAN, STEPHEN P., INTERACTION BETWEEN OVERWASH AND
B 6203 536 LEATHERMAN, STEPHEN P., SEDIMENT GRADING IN OVERWASH
B 6309 1597 LEAVESLEY, ANN, ECOLOGIC NICHES OF RADIOLARIANS, PLANKTONIC
B 6209 1760 LEAVESLEY, ANN, LIVING SHELLED MICROZOOPLANKTON,
B 6010 1873 LEBANON, 1975, PETROCONSULTANTS S.A., ON DEVELOPMENTS IN
B 6110 1802 LEBANON, 1976, CARVAJAL, J., ON DEVELOPMENTS IN
B 6210 2119 LEBANON, 1977, CARVAJAL, J., ON DEVELOPMENTS IN
B 6101 55 LEBANON, ADLOUN 1, LOG
B 6101 55 LEBANON, ADLOUN STRUCTURE
B 6101 61 LEBANON, BEKAA GRABEN
B 6101 44 LEBANON, CENOMANIAN-TURONIAN
B 6310 1891 LEBANON, DEVELOPMENTS, 1978
B 6101 50 LEBANON, DRILLING
B 6101 54 LEBANON, EL QAA 1, LOG
B 6101 54 LEBANON, EL QAA STRUCTURE
B 6101 49 LEBANON, EXPLORATORY WELLS
B 6101 44 LEBANON, GEOLOGICAL MAP
B 6101 50 LEBANON, GEOPHYSICS
B 6101 54 LEBANON, HERMEL STRUCTURE
B 6101 52 LEBANON, HYDROCARBONS
B 6101 60 LEBANON, ILE DU PALMIER STRUCTURE
B 6101 56 LEBANON, JURASSIC-CRETACEOUS PETROLEUM PROSPECTS
B 6101 48 LEBANON, MAPPING
B 6101 60 LEBANON, QARTABA STRUCTURE
B 6101 44 LEBANON, QUATERNARY
B 6101 47 LEBANON, ROCKS EXPOSED
B 6101 44 LEBANON, SENONIAN
B 6101 61 LEBANON, SIR ED DANIEH DOME
B 6101 53 LEBANON, SOHMOR 1, LOG
B 6101 53 LEBANON, SOHMOR ANTICLINE
B 6101 51 LEBANON, TELL ZNOUB 1B, LOG
B 6101 51 LEBANON, TERBOL, 1A, LOG
B 6101 61 LEBANON, TERBOL NO. 1, SECTION
B 6101 44 LEBANON, TERTIARY
B 6101 44 LEBANON, YAMMOUNEH FAULT
B 6101 53 LEBANON, YOHMOR 1, LOG
B 6101 43† LEBANON--REEVALUATION, PETROLEUM, PROSPECTS OF
B 6405 703* LEBENSSPUREN OF DYSAEROBIC, BATHYAL BASIN, CALIFORNIA
B 6107 1046 LECOMPTON FORMATION, KANSAS
B 6406 956 LECTURES IN ISOTOPE GEOLOGY EDITED BY E. JAGER AND J. C.
B 6405 738 LEDGER, ERNEST B., OPAL-CRISTOBALITE-CEMENTED SANDS IN
B 6109 1421 LEDUC FIELD, ALBERTA
B 6104 596 LEDUC FIELD, ALBERTA
M 30 315 LEDUC FORMATION, STRACHAN-RICINUS PRODUCTION, ALBERTA
B 6007 1112 LEDUC OIL, MASS SPECTROMETRIC ANALYSIS
M 30 324 LEDUC REEF, ALBERTA, STRACHAN FIELD
B 6205 783 LEDUC-RIMBEY BARRIER REEF TREND, ALBERTA
B 6309 1609* LEE AND HENDRY COUNTIES, FLORIDA, ABST., STRATIGRAPHY AND
B 6209 1645 LEE, CHAO-SHING, PLATE CONVERGENCE AND ACCRETION IN
B 6303 344 LEE FORMATION, KENTUCKY
B 6108 1371 LEE, M. W., APPLICATIONS OF VERTICAL SEISMIC PROFILE TO
B 6004 718 LEE, M. W., SYNTHETIC SEISMIC SECTIONS OF SELECTED STRATIGRAPHIC
B 6405 674 **LEE, M. W., VERTICAL AND SURFACE SEISMIC PROFILES MAP AQUIFERS IN**
B 6209 1673 LEE, MOON J., RUBIDIUM-STRONTIUM MINIMUM AGES OF
B 6410 1718 LEE, MOON J., RUBIDIUM-STRONTIUM MINIMUM AGES OF
B 6405 778 LEE, MYUNG W., SEISMIC MODEL OF 15 STRATIGRAPHICALLY
B 6305 840 LEE, MYUNG W., SEISMIC MODEL OF FIFTEEN STRATIGRAPHICALLY
B 6008 1409 LEE, MYUNG W., SYNTHETIC SEISMIC SECTIONS OF SELECTED
B 6105 808 LEE, P. J., COMPUTER PLOTTING OF LAND LEASES ON DIGITIZED BASE
B 6307 1117 LEE SIDING, REEF, SECTION, ARIZONA
B 6203 571 LEE, T. D., DELTAIC COMPONENTS OF VALMY AND VININI FORMATIONS,
B 6203 538 LEE, T. D., ELDER SANDSTONE—DELTAIC SEDIMENTS OF ROBERTS
B 6203 508* LEEDY FIELD, DEWEY COUNTY, OKLAHOMA. BY TITLE ONLY., GEOLOGY
B 6105 760 LEES, JOHN A., GEOLOGIC AND OPERATIONAL SUMMARY OF COST B-2
B 6307 1093 LEEUWIN BLOCK, AUSTRALIA
S 4 3 LEEWARD GROUP, CORALS, NETHERLANDS, ANTILLES
B 6103 395 LEEWARD-BANK MARGINS, LITTLE BAHAMA BANK
B 6211 2294 LEG 1, HOLE 1, DEEP SEA DRILLING PROJECT
B 6209 1592 LEG 18, OREGON AND, ALASKA MARGINS, DSDP, SITES 172-182

B 6106 931 LEG 3 RESULTS, DEEP SEA DRILLING PROJECT
B 6106 931 LEG 39 RESULTS, DEEP SEA DRILLING PROJECT
B 6309 1608* LEG 43 SITE 386, ABST., MERIDIONALLY ORNAMENTED HEDBERGELLID,
B 6204 597 LEG 44, BLAKE BASIN, JOIDES, DSDP, SITE 391
B 6004 649* LEG 44 DEEP SEA DRILLING PROJECT IN WESTERN NORTH ATLANTIC,
B 6203 531* LEG 58, DSDP—DEEP DRILLING IN SHIKOKU BASIN AND NORTHEASTERN
B 6405 711* LEG 68—INDICATE CONTINUOUS SECTION OF UNDISTURBED LATE
B 6209 1764* LEG 9, ABST., COSMOPOLITAN BIOZONATION FOR LATE CENOZOIC,
B 6303 524* LEGAL AND POLITICAL STATUS OF ANTARCTICA, ABST.
B 6207 1213* LEGAL ASPECTS OF RESOURCE DEVELOPMENT IN ANTARCTICA, ABST.
B 6303 471* LEGAL IMPLICATIONS OF CONSULTING RELATIONSHIP--PROFESSIONAL
B 6405 738 LEGALL, FRANKLYN D., APPLICATION OF CONODONT AND
B 6012 2189* LEGISLATION AND TOMORROWS OIL FROM TODAYS PROVINCES, ABST.
B 6303 380 LEGS 22-29, EDITED J R. HEIRTZLER, H. M. BOLLI, T. A. DAVIES, J. B.
B 6303 503* LEGS 36, 40--SEDIMENTOLOGIC IMPLICATIONS, ABST., CALCAREOUS,
B 6303 413* LEGS 59 AND 60, ABST., CENOZOIC, CLAY-MINERAL STRATIGRAPHY I,
B 6004 669 LEHMAN, J., NEW ENVIRONMENTAL MAPS FOR GULF OF MEXICO OCS
M 30 459 LEHMANN, P., GEOLOGY OF THE BEKAPAI FIELD
B 6405 739 LEHMANN, PATRICK J., POROSITY EVOLUTION OF NIAGARAN PIPE CREE
B 6107 961 LEHNER, P., ATLANTIC MARGIN OF AFRICA, STRUCTURAL HISTORY OF
B 6004 691 LEHNER, P., PETROLEUM HABITAT OF ATLANTIC MARGIN FROM NIGERIA
B 6408 1271 LEHNER, P., REVIEW OF CONTINENTAL MARGINS—GEOLOGICAL AND
B 6105 808 LEHNER, PETER, EXPLORATION TECHNIQUES OF CONTINENTAL SLOPES AND
B 6306 967 LEJAL-NICOL, A., MAJOR SUBDIVISIONS AND DEPOSITIONAL
B 6405 739 LEKAS, MITCHELL A., LOFRECO PROCESS—TAILORING SHALE OIL
B 6405 616 LEKHWAIR FORMATION, MIDDLE EAST
B 6303 549* LEMAN GAS FIELD, UNITED KINGDOM ABST., COMPUTATION OF INITIAL
M 30 41 LEMAN SANDSTONE FORMATION, NORTH SEA, VIKING FIELD
B 6008 1410* LEMEI ROCK AREA, SKAMANIA COUNTY, WASHINGTON, ABST.,
M 25 475 LENA STAGE, DOLOMITES, USSR
M 30 237 LENA-TUNGUSKA AREA, ANGARA SUITE, USSR
M 30 236 LENA-TUNGUSKA AREA, BELYA SUITE, USSR
M 30 237 LENA-TUNGUSKA AREA, BULAY SUITE, USSR
M 30 237 LENA-TUNGUSKA AREA, LITVINTSEV SUITE, USSR
M 30 226 LENA-TUNGUSKA AREA, LOCATION MAP, USSR
M 30 235 LENA-TUNGUSKA AREA, MOTY SUITE, USSR
M 30 233 LENA-TUNGUSKA AREA, STRATIGRAPHY, USSR
M 30 235 LENA-TUNGUSKA AREA, USHAKOVA SUITE, USSR
M 30 236 LENA-TUNGUSKA AREA, USOLYE SUITE, USSR
M 30 237 LENA-TUNGUSKA AREA, VERKHOLENSK SUITE, USSR
M 30 225† LENA-TUNGUSKA PETROLEUM PROVINCE, EASTERN SIBERIA, USSR,
M 30 231 LENA-TUNGUSKA PROVINCE, EXPLORATION HISTORY, USSR
M 30 228 LENA-TUNGUSKA PROVINCE, STRUCTURE CONTOURS, USSR
B 6107 1046 LENAPAH FORMATION, KANSAS
B 6207 1129 LENGGURU FOLDBELT, IRIAN JAYA
B 6004 691 LENNARTZ, F. F., INCONSISTENCIES IN RESOURCE-APPRAISAL ESTIMATES
M 24 1 LENNON, R. B., MIDDLE GROUND SHOAL OI FIELD, ALASKA
B 6406 960 LENNOX, T. R., GEOLOGY OF IN SITU PILOT PROJECT, WABASCA OIL
B 6103 431† LENOX HILLS CONTACT IN GLASS MOUNTAINS. WEST TEXAS. NEAL
B 6103 432 LENOX HILLS FORMATION, TEXAS
B 6205 888 LENT, R. L., HYDROCARBON EXPLORATION IN GREAT SALT LAKE BASIN,
S 11 272 LENTICULINA
S 11 233 LENTICULINA CALCAR
S 11 272 LENTICULINA CALCAR
S 11 272 LENTICULINA CUSHMANI
S 11 233 LENTICULINA CUSHMANI
S 11 272 LENTICULINA NIKOBARENSIS
S 11 233 LENTICULINA NIKOBARENSIS
S 11 272 LENTICULINA POLITA
S 11 233 LENTICULINA POLITA
B 6004 692 LENTZ, R. C., EOCENE DEPOSITIONAL ENVIRONMENTS AND RELATED
B 6105 766 LENZER, R. C., EXPLORATION AND EVALUATION OF ROOSEVELT KGRA,
B 6008 1390 LENZER, R. C., GEOTHERMAL EXPLORATION OF ROOSEVELT KGRA, UTAH,
B 6408 1171 LEONARD DOLOMITE, TEXAS
B 6203 536 LEONARD, JAY E., CARBONATE CATCHMENT AND EXCHANGE BASINS IN
B 6004 692 LEONARD, JAY E., MARINE COLD-WATER CARBONATE DEPOSITS AS
B 6303 481 LEONARD, JAY E., MEGASCALE VARIABILITY OF SEDIMENT ON UNITED
B 6303 487 LEONARD, LEROY J.. SEISMIC STRATIGRAPHIC, AND ECONOMIC ANALYSIS
B 6004 672* LEONARD MOUNTAI AREA, GLASS MOUNTAIN, WEST TEXAS, ABST.,
B 6408 1171 LEONARD REEF, TEXAS
B 6007 1107 LEONARD SERIES, UNDIVIDED, TEXAS
B 6203 526* LEONARDIAN STRATA IN NORTHERN MIDLAND BASIN OF WEST TEXAS,
B 6112 2119 LEOPARD KNOB, NEW MEXICO
S 11 59 LEPIDOCYCLINA
M 27 105 LEPIDODENDRALES
M 27 143 LEPISPHERES
B 6406 960 LERAND, M. M., GEOLOGY OF IN SITU PILOT PROJECT, WABASCA OIL
B 6004 710 LERBEKMO, J. F., PROVENANCE AND DISPERSAL OF UPPER CRETACEOUS
B 6309 1587 LERCH, H., IMPLICATION OF GASEOUS HYDROCARBON GEOCHEMISTRY
B 6309 1583 LERCH, H., PETROLEUM GEOCHEMISTRY AND GEOLOGY OF SOUTHEAST
B 6403 331 LEROY FIELD, CROSS SECTION, MONTANA
B 6004 677 LESKIW, K., SOME APPROACHES TO CLUSTER-ANALYSIS INTERPRETATION
B 6110 1749 LESOTHO, DEVELOPMENTS, 1976
B 6210 1848 LESOTHO, DEVELOPMENTS, 1977
B 6310 1693 LESOTHO, DEVELOPMENTS, 1978
B 6411 1789 LESOTHO, DEVELOPMENTS, 1979
B 6305 823* LESS COMMON TRACE ELEMENTS IN SANDSTONE-TYPE URANIUM
S 4 69 LESSER ANTILLEAN ISLAND ARC
B 6010 1653 LESSER ANTILLES, 1975, PETROCONSULTANTS S.A., ON DEVELOPMENTS IN
B 6110 1587 LESSER ANTILLES, 1976, GODEL, M., ON DEVELOPMENTS IN
B 6210 2054 LESSER ANTILLES, 1977, UMIKER, R., ON DEVELOPMENTS IN
B 6310 1755 LESSER ANTILLES, 1978, UMIKER, R., ON DEVELOPMENTS IN
S 4 70 LESSER ANTILLES, HOLOCENE BIOHERMS
B 6007 1079 LESSER ANTILLES ISLAND ARC, TRACKLINE COVERAGE EAST OF
S 4 69 LESSER ANTILLES, PLEISTOCENE
S 4 69 LESSER ANTILLES, PLIOCENE
S 4 69 LESSER ANTILLES, TERTIARY
B 6007 1094 LESSER ANTILLES TRENCH
S 4 67† LESSER ANTILLES--GEOLOGIC CONTROL OF DEVELOPMENT, HOLOCENE,
M 25 463 LESSER KHINGAN RANGE, IGNIMBRITE VOLCANISM
B 6309 1582 LESSING, PETER, BISHOP-BRADSHAW CREEK FAULT, ABST.
B 6010 1642 LESTA, P. H., AND MARCELO R. YRIGOYEN, ON DEVELOPMENTS IN
B 6108 1382 LESTINA, RONALD F., PREVENTION OF DAMAGE TO WATERFOWL ON
B 6307 1094 LESUEUR SANDSTONE, AUSTRALIA
M 29 293 LETOUZEY, J., VARIETY OF MARGINS AND DEEP BASINS IN THE
B 6101 58 LEVANT, JAUF-GAARA-ABD EL AZIZ-MARDIN SWELL
B 6101 58 LEVANT, JURASSIC, SEDIMENTARY ROCKS
B 6106 877 LEVANT PLATFORM, AND ERATOSTHENES SEAMOUNT, CONTINENTAL
B 6311 2094 LEVANT, TOPOGRAPHIC PROFILES
B 6101 58 LEVANT, TRIASSIC, SEDIMENTARY ROCKS
M 29 299 LEVANTINE BASIN, SEDIMENTARY SEQUENCE, MEDITERRANEAN
B 6106 900 LEVANTINE PLATE
B 6303 324 LEVANTINE PLATFORM, ISRAEL

B 6405 728 LEVELL, BRUCE K., SEDIMENTOLOGY OF LOWER CRETACEOUS SUBTIDAL
B 6004 692 LEVENTHAL, JOEL S., CHARACTERIZATION OF INSOLUBLE ORGANIC
B 6405 692 LEVENTHAL, JOEL S., GEOCHEMICAL EFFECTS OF EARLY DIAGENESIS OF
B 6304 691 LEVENTHAL, JOEL S., ORGANIC GEOCHEMISTRY AND URANIUM IN
S 10 261 LEVERIDGE ESTATE WELL, HYDROCARBON LOG, TEXAS, WHARTON
S 10 262 LEVERIDGE ESTATE WELL, LOG, TEXAS, EAST BERNARD FIELD
B 6405 739 LEVEY, R. A., WAVE-DOMINATED DELTAS—IMPORTANT ECONOMIC
B 6405 740 LEVIN, DAVID M., HYDROCARBON EXPLORATION IN WESTERN
B 6410 1725 LEVINSON, WITH CHAPTERS BY R. B. MACCAMMON AND BRIAN HITCHON,
B 6304 689 LEVY, P. W., THERMOLUMINESCENCE OF URANIUM HOST ROCKS IN
B 6203 545 LEVY, WILLIAM P., HOLOCENE SEDIMENT VOLUME AND MODERN
M 29 443 LEWIS, B. T. R., EXPLOSION SEISMOLOGY STUDIES OF ACTIVE AND
B 6004 692 LEWIS, C. R., CLAY-MEMBRANE FILTRATION AS MECHANISM FOR
B 6109 1540 LEWIS, J. O., STRATIGRAPHY AND ENTRAPMENT OF HYDROCARBONS IN
B 6308 1426 LEWIS, JAMES O., STRATIGRAPHY AND ENTRAPMENT OF
B 6009 1621 LEWIS, R. Q., JR., AGE AND TECTONICS OF CINCINNATI ARCH IN
B 6105 808 LEWIS, R. S., SHALLOW SEDIMENTARY FRAMEWORK OF GEORGES BANK
B 6408 1132 LEWIS RIVER GAS FIELD, ALASKA
B 6205 757 LEWIS SHALE, SAN JUAN BASIN
B 6304 701* LEWIS THRUST, FRONT RANGES, SOUTHERN CANADIAN ROCKY
B 6311 2095 LEWIS TROUGH, AUSTRALIA
B 6303 349 LEXINGTON FAULT ZONE, KENTUCKY
B 6303 344 LEXINGTON LIMESTONE, KENTUCKY
B 6303 342 LEXINGTON LINEAMENT, KENTUCKY
B 6104 546 LEYDEN, R., MARINE SEISMIC INVESTIGATIONS, SOUTHERN BRAZIL
B 6204 657 LEYDEN, R., SALT DIAPIRS ON SAO PAULO PLATEAU, SOUTHEASTERN
B 6002 196 LEYDEN, R., SOUTH ATLANTIC DIAPIRIC STRUCTURES
B 6203 536 LEYTHAEUSER, D., HYDROCARBON GENERATION POTENTIAL OF
M 30 266 LF SANDSTONE, RESERVOIR, EUGENE ISLAND 330 FIELD
B 6303 471* LIABILITY OF CONSULTING GEOLOGIST IN URBAN-ENVIRONMENTAL
M 25 1 LIAN, HAROLD M, EDITED BY CIRCUM-PACIFIC ENERGY AND MINERAL
M 25 1 LIAN, HAROLD M., FOREWORD, CIRCUM-PACIFIC ENERGY AND MINERAL
M 25 1 LIAN, HAROLD M., INTRODUCTION, CIRCUM-PACIFIC ENERGY AND
M 25 108 LIARD RIVER, SHEEP CREEK, AND DAWSON COALFIELDS, CANADA
M 25 497 LIARD TROUGH, CANADA
B 6204 669 LIAS, ALGERIA
B 6101 83 LIASSIC CORRELATIONS, TENTATIVE, NORTH AMERICA AND, MOROCCO,
B 6101 82 LIASSIC DEPOSITS, MOROCCO, TRIASSIC-
B 6101 79† LIASSIC DEPOSITS OF MOROCCO AND EASTERN NORTH
B 6408 1194 LIASSIC FACIES, GREAT BRITAIN
B 6405 605 LIASSIC, MIDDLE EAST, CARBONIFEROUS TO
B 6008 1404* LIBERATED, PRODUCED, RECYCLED, OR CONTAMINATION, ABST.
B 6210 1848 LIBERIA, DEVELOPMENTS, 1977
B 6310 1693 LIBERIA, DEVELOPMENTS, 1978
B 6411 1789 LIBERIA, DEVELOPMENTS, 1979
B 6008 1408* LIBERTY COUNTY, MONTANA, ABST., EAST KEITH GAS AREA
B 6008 1257 LIBERTY COUNTY, TEXAS, WEST RAYWOOD FIELD
B 6410 1577 LIBERTY HALL BEDS, VIRGINIA
B 6204 686 LIBRARY OF DALLAS, DALLAS GEOLOGICAL SOCIETY, GEOLOGICAL
B 6104 595 LIBYA, 0-1-6 FIELD
B 6010 1769 LIBYA, 1975, PETROCONSULTANTS S.A., ON DEVELOPMENTS IN
B 6110 1701 LIBYA, 1976, PETROCONSULTANTS S.A., ON DEVELOPMENTS IN
B 6210 1798 LIBYA, 1977, PETROCONSULTANTS S.A., ON DEVELOPMENTS IN
B 6310 1638 LIBYA, 1978, PETROCONSULTANTS S.A., ON DEVELOPMENTS IN
B 6411 1741 LIBYA, 1979, PETROCONSULTANTS S.A., ON DEVELOPMENTS IN
B 6104 595 LIBYA, A-1-1 FIELD
B 6303 433* LIBYA, ABST., GEOLOGY OF A STRATIGRAPHIC GIANT--, MESSLA OIL
B 6303 423* LIBYA, ABST., INTISAR D OIL FIELD
B 6303 416* LIBYA, ABST., SECONDARY, CARBONATE POROSITY, ZELTEN FIELD
B 6406 863 LIBYA AND CHAD, MURZUK BASIN
B 6406 863 LIBYA AND SUDAN, KUFRA BASIN
B 6203 464 LIBYA, ARIDA FORMATION
B 6203 464 LIBYA, AUGILA FORMATION
B 6104 595 LIBYA, B-2-1 FIELD
B 6309 1547 LIBYA BY J. AZZOUZ ETTALHI, D. KROKOVIC, AND S. BANERJEE, BARR, F.
B 6303 521* LIBYA DETERMINED FROM LANDSAT IMAGERY, ABST., SOIL REGIMES OF,
B 6203 464 LIBYA, DIBA FORMATION
M 30 547 LIBYA, EAST SIRTE BASIN, STRATIGRAPHY
M 30 507† LIBYA, GEOLOGY OF A, STRATIGRAPHIC GIANT--MESSLA OIL FIELD
B 6104 594 LIBYA, I-1-6 FIELD
M 30 546 LIBYA, INTISAR D FIELD, CROSS SECTION
M 30 557 LIBYA, INTISAR D FIELD, DEPOSITIONAL HISTORY
M 30 544 LIBYA, INTISAR D FIELD, LOCATION
M 30 548 LIBYA, INTISAR D FIELD, RESERVOIR CHARACTERISTICS
M 30 543† LIBYA, INTISAR D OIL FIELD
B 6104 595 LIBYA, J-1-1 FIELD
B 6104 596 LIBYA, J-2-6 FIELD
B 6104 595 LIBYA, MABRUK FIELD
B 6104 594 LIBYA, MABRUK FIELD
M 30 18 LIBYA, MESSLA FIELD, LANDSAT IMAGE
M 30 516 LIBYA, MESSLA FIELD, STRATIGRAPHIC COLUMN
B 6401 16 LIBYA, MESSLA GIANT OIL FIELD, LANDSAT DATA
B 6203 463 LIBYA, OLIGOCENE
B 6104 597 LIBYA, P-8-6 WELL
B 6110 1725 LIBYA, PRODUCTION, 1976
B 6210 1823 LIBYA, PRODUCTION, 1976-1977
B 6310 1660 LIBYA, PRODUCTION, 1978
B 6411 1770 LIBYA, PRODUCTION, 1979
B 6104 597 LIBYA, R-2-47 WELL
B 6104 597 LIBYA, RUAGA FIELD
B 6401 16 LIBYA, SARIR SANDSTONE
B 6203 463† LIBYA, SECONDARY, FERROAN, DOLOMITE, RHOMBS IN, OIL, RESERVOIRS,
B 6203 464 LIBYA, SIRTE BASIN
B 6406 860 LIBYA, SIRTE BASIN
B 6401 16 LIBYA, SIRTE BASIN
M 30 511 LIBYA, SIRTE BASIN, GEOLOGIC HISTORY
M 30 523 LIBYA, SIRTE BASIN, INTRA-CRETACEOUS UNCONFORMITY
M 30 550 LIBYA, SIRTE BASIN, KHEIR FORMATION
M 30 508 LIBYA, SIRTE BASIN, LOCATION MAP
M 30 546 LIBYA, SIRTE BASIN, MARSA BREGA TROUGH
M 30 507 LIBYA, SIRTE BASIN, MESSLA FIELD
M 30 550 LIBYA, SIRTE BASIN, SABIL FORMATION
M 30 549 LIBYA, SIRTE BASIN, SHETERAT SHALE
B 6406 860 LIBYA, TUNISIA, AND ALGERIA, GHADAMES BASIN
B 6104 597 LIBYA, ZELTEN FIELD
B 6203 551 LICARI, GERALD R., MODERN LACUSTRINE STROMATOLITES, WALKER
B 6203 537 LICARI, GERALD R., PALEOLIMNOPHYCUS—STROMATOLITE-FORMING
B 6203 537 LICARI, JOAN P., PALEOLIMNOPHYCUS—STROMATOLITE-FORMING ALGA
M 30 175 LICENSE BLOCKS, NORTH SEA, NINIAN FIELD
B 6205 892 LIDZ, B., DEPOSITION AND DIAGENESIS OF HOLOCENE REEFS, BELIZE,
B 6004 683 LIDZ BARBARA H., AUTOPSY OF A DEAD CORAL REEF, ABST.
B 6104 519 LIDZ BARBARA H., PLEISTOCENE BARRIER BAR SEAWARD OF OOID SHOAL
B 6004 677 LIDZ BARBARA H., RECENT AND RELICT TOPOGRAPHY OF BOO BEE
B 6303 528 LIDZ, BARBARA, THREE-DIMENSIONAL ASPECTS OF BELIZE PATCH REEFS,
B 6303 487 LIEBER, ROBERT B., CONTROLS OF SEDIMENTATIO OF LOWER
S 11 231 LIEBUSELLA PLIOCENICA
S 11 268 LIEBUSELLA PLIOCENICA
B 6303 540* LIGHT HYDROCARBONS OF PETROLEUM, INTERNAL EVIDENCE OF
B 6105 808 LIGHTY, ROBIN G., SUBMARINE DIAGENESIS IN RELICT HOLOCENE CORAL
B 6409 1560* LIGNITE, ABST., DEPOSITIONAL ENVIRONMENTS OF SOME, NABORTON—
B 6004 692* LIGNITE DEPOSIT, FREESTONE COUNTY, TEXAS, ABST., EOCENE
B 6309 1602* LIGNITE DEVELOPMENT IN EAST TEXAS, ABST., HYDROLOGY AND,
B 6409 1555* LIGNITE IN WEST-CENTRAL LOUISIANA, ABST., SUBSURFACE, WILCOX
B 6409 1554* LIGNITE PROSPECTING MODELS—WILCOX GROUP AND MERIDIAN
B 6105 821* LIGNITE RESOURCES OF FORT UNION FORMATION, WEST-CENTRAL
B 6309 1604* LIGNITE--STATUS REPORT, ABST., GULF COAST
M 25 452 LIGNITE-BEARING FORMATION, URANIUM HOST ROCKS, JAPAN, TOKI
B 6008 1410* LIGNITES, ABST., LIQUEFACTION OF, NORTHERN, GREAT PLAINS
B 6303 516* LIGNITES OF TOMBIGBEE AND HOLLY SPRINGS NATIONAL FORESTS,
B 6309 1587 LIGON, D., IMPLICATION OF GASEOUS HYDROCARBON GEOCHEMISTRY
B 6309 1583 LIGON, D., PETROLEUM GEOCHEMISTRY AND GEOLOGY OF SOUTHEAST
B 6103 387 LILY BANK OOLITE SHOAL, LITTLE BAHAMA BANK
S 5 96 LIMBERLOST DOLOMITE, INDIANA,

B 6401 129 LIMBERLOST DOLOMITE, INDIANA
B 6105 841* LIME GRAINSTONES, ABST., EVIDENCE FOR SOLUTION. COMPACTION IN
B 6105 769 * LIME MUD DEPOSITION IN ENCLOSED LAGOON, BIGHT OF ABACO, BAHAMAS
B 6405 783* LIME SEDIMENT, ABST., EXPERIMENTAL, COMPACTION OF
M 27 159 LIME SILT, CEMENTED
B 6011 2005† LIMESTONE BUTTES, GYPSUM PLAIN, CULBERSON COUNTY, TEXAS,
M 28 98 LIMESTONE, IN CALCLITHITE, FRAGMENTS
M 28 128 LIMESTONE, IRELAND, MISSISSIPPIAN
B 6303 464* LIMESTONE MARKER BED IN BRIDGE CREEK MEMBER, GREENHORN
M 26 447 LIMESTONE, OKLAHOMA, ANADARKO BASIN, BELLE CITY
M 26 440 LIMESTONE, OKLAHOMA, ANADARKO BASIN, HASKELL
M 26 440 LIMESTONE, OKLAHOMA, ANADARKO BASIN, TONKAWA
B 6002 255 LIMESTONE, PORES IN
B 6209 1759* LIMESTONE SOILS, ABST., MINERALOGIC AND PHYSICAL PROPERTIES OF,
M 25 356† LIMESTONE TERRANES, EXAMPLES FROM INDONESIA--SUMMARY,
B 6207 1213* LIMESTONE TERRANES OF MICRONESIA, ABST., WATER SUPPLIES IN
B 6203 489* LIMESTONES, ABST., CARBON AND, OXYGEN, ISOTOPES AS INDICATORS
B 6203 554* LIMESTONES, ABST., MODEL FOR, TRACE-ELEMENT PARTITIONING IN
B 6203 569* LIMESTONES, ABST., THREE TYPES OF, PRESSURE SOLUTION IN
B 6109 1552* LIMESTONES IN MARINE CARBONATE SUCCESSIONS OF PLEISTOCENE AND
B 6203 500* LIMESTONES OF EASTERN YUCATAN PENINSULA, MEXICO, ABST.,
S 4 299† LIMESTONES ON AND AROUND BARBUDA, WEST INDIES, FACIES
M 26 203 LIMITATIONS AND QUALIFICATIONS, SEISMIC, REFLECTIONS,
M 26 3† LIMITATIONS ON RESOLUTION OF SEISMIC REFLECTIONS AND GEOLOGIC
B 6303 535* LIMNOLOGY OF EOCENE LAKES GOSIUTE AND UINTA, ABST., EVOLUTION
B 6306 942 LIMOEIRO FORMATION, AMAZON BASIN
M 28 126 LIMONITE, IN SANDSTONE, CEMENTS
M 25 402 LIMONITE ORE, LATERITE, NICKEL DEPOSITS, SILICATE AND
M 25 414 LIMONITE ORE, WESTERN AUSTRALIA, IRON, PISOLITIC
B 6405 676 LIN, J. T., HYDRODYNAMIC FLOW IN LOWER CRETACEOUS MUDDY
B 6303 527 LIN, JIN, DEPTH MIGRATION OF SEISMIC DATA, ABST.
B 6203 523* LINCOLN, AND LOGAN COUNTIES, OKLAHOMA, ABST., SUBSURFACE
M 24 91† LINCOLN COUNTIES, WYOMING, BIG PINEY--, LA BARGE PRODUCING
B 6410 1577 LINCOLNSHIRE LIMESTONE, VIRGINIA
B 6409 1462 LINDAU, ROBERT D., DEVELOPMENT IN EAST-CENTRAL STATES IN 1979
B 6308 1324 LINDAU, ROBERT D., DEVELOPMENTS IN EAST-CENTRAL STATES IN 1978
B 6004 685 LINDBLOM, R. G., TIMING--GEOLOGIC AN POLITICAL--IN URBAN OIL
B 6105 809 LINDHOLM, RAY C., GEOLOGY OFJRUASSIC-TRIASSIC CULPEPER BASIN,
B 6305 846 LINDHOLM, ROY C., PETROLOGY OF TRIASSIC-JURASSIC CULPEPER
B 6203 537 LINDHOLM, ROY C., TECTONIC CONTROL OF SEDIMENTATION IN
B 6105 809 LINDNER, A. W., RUNDLE OIL SHALE DEPOSIT, QUEENSLAND, AUSTRALIA
B 6410 1620 LINDON FIELD, COLORADO
B 6009 1612 LINDQUIST, SANDRA, LEACHED POROSITY IN OVERPRESSURED
B 6203 537 LINDSAY, ROBERT F., PETROLOGY AND PETROGRAPHY OF GREAT BLUF
B 6105 809 LINDSETH, R. O., MAPPING STRATIGRAPHIC TRAPS WITH SEISLOGS
B 6012 2184 LINDSETH, ROY O., STRATIGRAPHIC EXPLORATION WITH SEISLOG
B 6004 693 LINDSEY, DOUGLAS D., STRUCTURE AND TIMING OF OROGENIC EVENTS,
B 6012 2187 LINDSEY, J. P., INTERPRETIV USES OF SEISMIC MODELING WITH EMPHASIS
M 26 477 LINDSEY, J. P., PRACTICAL STRATIGRAPHIC MODELING AND
B 6203 537 LINDSEY, J. P., SEISMIC CALIBRATION TECHNIQUES FOR IMPROVED
B 6405 764* LINEAMENT ANALYSIS FOR OIL AND GAS EXPLORATIO AND
B 6004 722* LINEAMENT AND STRUCTURAL ANALYSES OF SOUTHEASTERN KANSAS,
B 6305 825* LINEAMENT MAPPING WITH ORBITING IMAGING RADAR, ABST.
B 6309 1575* LINEAMENTS AND WATER WELLS AS EXPLORATION TOOLS IN
B 6004 714* LINEAMENTS TO INFER EXISTENCE OF SUBSURFACE GEOLOGIC
B 6105 840* LINEAMENTS, WEST VIRGINIA, ABST., FIELD STUDIES OF, PARSONS AND,
S 3 62 LINEAR AND PLANAR SHAPES, URANIUM OREBODIES
S 7 273 LINEAR FAULT, BOTTOM, DEFORMATION
B 6202 201 LINEAR NUMERICAL TIME
B 6004 709* LINEAR STRUCTURES, ABST., IMPLICATIONS OF, IMAGERY STUDIES TO,
B 6408 1287 LINEBACK, JERRY A., DEPOSITIONAL ENVIRONMENTS OF ULLIN
B 6302 183 LING SUBBASIN, NORTH SEA
B 6306 971 LINGULA SHALE UNIT, EGYPT
B 6004 693 LINK, D. A., FACTORS CONTROLLING SUSPENDED SEDIMENT IN STREAMS
B 6405 740 LINK, MARTIN H., HYDROCARBON POTENTIAL OF MATILIJA SANDSTONE,
B 6203 551 LINK, MARTIN H., MODERN LACUSTRINE STROMATOLITES, WALKER
B 6304 690 LINK, R. L., EFFECTS OF URANIUM MINING ON GROUNDWATER IN
B 6305 846 LINN, EARL H., LATE ORDOVICIAN STRATIGRAPHY AND SEDIMENTATION
S 4 221 LINNE AND MARGINOPORA VERTEBRALIS QUOY AN GAIMARD, GROWTH
B 6303 487 LINSLEY, P. N., BEATRIC FIELD, MORAY FIRTH, NORTH SEA, ABST.
M 30 117 LINSLEY, PHILIP N., THE BEATRICE FIELD, INNER MORAY FIRTH, U.K.
B 6407 1117 LINTZ, JOSEPH, JR., REVIEW OF THE ENCYCLOPEDIA O PALEONTOLOGY,
B 6203 490 LIPPMANN, M. J., RECENT DEVELOPMENTS AT CERRO PRIETO
B 6012 2184 LIPPS, JERE H., COILING RATIOS IN PLANKTONIC
B 6311 2118 LIPSTREUER, KENNETH A., GEOCHEMISTRY OF SMALL LACUSTRINE
B 6008 1410* LIQUEFACTION OF NORTHERN GREAT PLAINS LIGNITES, ABST.
B 6002 324* LIQUID HYDROCARBON MIGRATION PATHS IN THICK, RAPIDLY
B 6409 1557* LIRETTE FIELD, TERREBONNE PARISH, LOUSIANA, ABST., DIAGENESIS OF
B 6104 594 LIRIK FIELD, SUMATRA
B 6404 569 LISAN FORMATION, ISRAEL
S 3 58 LISBON VALLEY URANIUM DISTRICT, UTAH
S 3 129 LISBON VALLEY, UTAH
M 30 307 LISBURNE GROUP, ALAPAH FORMATION, PRUDHOE BAY
B 6103 437 LISBURNE GROUP, ALASKA
M 24 26 LISBURNE GROUP, ALASKA
B 6109 1500 LISBURNE GROUP, ALASKA
B 6205 880* LISBURNE GROUP, CARBONIFEROUS AND PERMIAN, IN NAVAL
B 6109 1493† LISBURNE GROUP, MISSISSIPPIAN AND PENNSYLVANIAN, POTENTIAL
B 6012 2175* LISBURNE GROUP, NORTHEAST ALASKA, ABST., ENVIRONMENTAL AND
M 30 305 LISBURNE GROUP, STRATIGRAPHY, PRUDHOE BAY
B 6109 1510 LISBURNE GROUP, TRAPS, ALASKA
M 30 307 LISBURNE GROUP, WAHOO FORMATION, PRUDHOE BAY
B 6004 649* LISBURNE GROUP--POTENTIAL MAJO OBJECTIVE OF EASTERN ARCTIC
B 6306 967 LIST, F. K., MAJOR SUBDIVISIONS AND DEPOSITIONAL ENVIRONMENTS OF
B 6302 183 LISTA NOSE, NORTH SEA
B 6105 837* LITHIUM, ABST., DEVELOPMENT OF, ENERGY TECHNOLOGY TO CREATE
M 25 535 LITHIUM, ANDEAN REGION
B 6105 838* LITHIUM IN BRINES AND SEDIMENTARY ROCKS, ABST., POTENTIAL NEW
S 5 169* LITHOFACIES CONTINUUM: LATEST MIOCENE (MESSINIAN) DEPOSITS OF
S 5 58 LITHOFACIES DESCRIPTIONS, SILURIAN, SALINA GROUP, RUFF
S 5 73* LITHOFACIES IN LUCAS FORMATION, DETROIT RIVER GROUP, DEVONIAN,
S 5 24 LITHOFACIES, MICHIGAN BASIN, LOWER SALINA GROUP,
B 6004 697* LITHOFACIES, ABST., DEVELOPMENT OF VERTICAL ESCARPMENT AT
B 6109 1535* LITHOFACIES ANALYSIS AND URANIUM FAVORABILITY OF OAKVILLE
M 30 82 LITHOFACIES ANALYSIS, CORE DATA, NORTH SEA, FORTIES FIELD
B 6303 420* LITHOFACIES ANALYSIS OF CASTILE FORMATION, DELAWARE BASIN,
B 6405 807* LITHOFACIES AND BIOHERMS OF UPPER PALEOZOIC IN
B 6203 548* LITHOFACIES AND DEPOSITIONAL ENVIRONMENTS OF HOSSTON
B 6002 326* LITHOFACIES AND DEPOSITIONAL ENVIRONMENTS OF RUSH CREEK
B 6211 2356* LITHOFACIES IN LOWER JURASSIC VOLCANICLASTIC ROCKS ADJACENT
B 6004 695* LITHOFACIES IN SOUTH ATLANTIC OCEAN, EARLY CRETACEOUS
B 6009 1623* LITHOFACIES RELATIONS IN ALEXANDRIAN AND EARLY NIAGARAN
B 6405 649 LITHOFACIES-PALEOGEOGRAPHICAL ATLAS OF THE PERMIAN OF
S 6 16 LITHOGENETIC SCALES, REVIEW
B 6004 733* LITHOHERMS, IN NORTHERN STRAITS OF FLORIDA, ABST., PETROLOGY
M 26 453 LITHOLOGIC CHARACTER, SEISMIC, MODELING
M 29 389 LITHOLOGIC LOG, CORE EXAMPLE, MISSISSIPPI FAN
B 6405 734* LITHOLOGY AND STRUCTURES OF QUATERNARY SEDIMENTS OF INDUS
M 26 18 LITHOLOGY ASPECTS, SEISMIC, INTERPRETATION, VELOCITY-, DENSITY-
B 6102 169† LITHOLOGY, COMPACTION AND PERMEABILITY OF SANDS, STUDIES OF,
B 6303 537* LITHOLOGY DATA SYSTEMS--ROCKS TO APPLICATIONS, ABST.
M 26 421 LITHOLOGY, EFFECTS OF, SEISMIC
B 6309 1584* LITHOLOGY STUDIES OF UPPER DEVONIAN WELL CUTTINGS IN EASTERN
S 3 12 LITHOPHILE ELEMENTS, URANIUM AND, THORIUM
B 6405 675* LITHOSOMES ON ATLANTIC CONTINENTAL SHELF, ABST., FRACTIONAL
M 25 473 LITHOSPHERE, ENERGY FLOW IN
M 25 48 LITHOSPHERIC PLATE BOUNDARIES, PACIFIC
M 25 546 LITHOSPHERIC PLATE CONVERGENCE, RELATION, TECTONIC BELTS,
M 25 179 LITHOSPHERIC SUBDUCTION, CONVERGENT, PLATES AND
B 6406 954 LITHOSTRATIGRAPHIC ANALYSIS OF SEDIMENTARY BASINS, BY C. E. B.
B 6004 721* LITHOSTRATIGRAPHY OF NORWEGIAN NORTH SEA, ABST.
B 6203 550* LITHOSTRATIGRAPHY, PETROLOGY, AND DEPOSITIONAL ENVIRONMENTS
M 26 100 LITHOSTRATIGRAPHY, SEISMIC, REFLECTIONS
B 6403 340 LITTLE BAHAMA BANK
B 6105 841* LITTLE BAHAMA BANK, ABST., POROSITY CONTROLS IN SUBSEA,
B 6204 694 LITTLE BAHAMA BANK, BAHAMAS
B 6103 376† LITTLE BAHAMA BANK, BAHAMAS, SHALLO, CARBONATE-, BANK-,
B 6410 1703 LITTLE BAHAMA BANK DRIFT
B 6103 395 LITTLE BAHAMA BANK, LEEWARD-BANK MARGINS
B 6103 387 LITTLE BAHAMA BANK, LILY BANK OOLITE SHOAL
B 6103 387 LITTLE BAHAMA BANK, MATANILLA REEF EAST
B 6103 382 LITTLE BAHAMA BANK, SEISMIC PROFILES
B 6103 384 LITTLE BAHAMA BANK, WINDWARD-BANK MARGINS
B 6407 1009 LITTLE BUDDY FIELD, ARK-LA-TEX AREA
B 6102 259 LITTLE COPPER FORMATION, IDAHO
B 6102 256† LITTLE COPPER FORMATION--NEW NAM FOR LOWEST FORMATION OF
B 6104 595 LITTLE CREEK FIELD, MISSISSIPPI
M 24 284 LITTLE ESCAMBIA CREEK FIELD, ALABAMA
B 6009 1570† LITTLE FALLS DOLOSTONE, LATE CAMBRIAN, EAST-CENTRAL NEW YORK,
B 6104 602 LITTLE HATCHET MOUNTAINS, NEW MEXICO
B 6305 843* LITTLE KNIFE FIELD, NORTH DAKOTA, ABST., GEOLOGY AND CURRENT
B 6203 430 LITTLE MINCH BASIN, BRITAIN, WEST
B 6009 1623* LITTLE NORTH MOUNTAIN STRUCTURAL FRONT, VIRGINI AND WEST
B 6407 973 LITTLE ROCKY MOUNTAINS, MONTANA
M 25 225 LITTLE TOTARA SAND, NEW ZEALAND
B 6008 1410* LITTLE WALL FIELD, MUSSELSHELL COUNTY, MONTANA, ABST., JIM
B 6004 710* LITTORAL HARD- AND SOFT-BOTTOM ICHNOCOENOSES--THEIR
B 6004 716* LITTORAL PROCESSES AND GEOMORPHIC VARIABILITY ON
S 11 228 LITUOLIDAE

M 30 237 LITVINTSEV SUITE, USSR, LENA-TUNGUSKA AREA
B 6409 1556* LIVE OAK DELTA COMPLEX—UNSTABLE, SHELF-EDGE DELTA IN DEEP
B 6203 349 LIVERMORE BASIN, CALIFORNIA
S 4 158 LIVING CARIBBEAN REEF CORALS OBSERVED DURING EMERGENCE
B 6309 1609* LIVING FORAMINIFERS OF WEST FLOWER GARDEN BANK,
B 6209 1760* LIVING SHELLED MICROZOOPLANKTON, RADIOLARIANS, FORAMINIFERS,
B 6204 609 LIVINGSTON, A. R., TENSLEEP RESERVOIR, OREGON BASIN FIELD,
B 6105 816 LIVINGSTON, A. R., TENSLEEP RESERVOIR STUDY, OREGON BASIN FIELD,
B 6304 691 LIVINGSTON, B. A., GEOLOGY AND DEVELOPMENT OF MARQUEZ, NE
B 6203 561 LJUNG, J. R., MAJOR URANIUM DISCOVERY IN FRONTIER AREA—
B 6404 529 LLANDOVERY CONTACT AND BUILDUPS, GREENLAND
M 28 3 LLANO SERIES, TEXAS, PRECAMBRIAN
B 6308 1427* LLANO UPLIFT TO SOUTHERN PART OF FORT WORTH BASIN, ABST.,
B 6110 1583 LLANOS BASIN, COLOMBIA
B 6101 44 LLBANON, ALBIAN
B 6406 824 LLOYD HILL PINNACLE REEF-PLATFORM ATOLL, AUSTRALIA
B 6008 1413* LLOYDMINISTER AREA, SASKATCHEWAN AND ALBERTA, ABST.,
B 6107 1010† LLOYDMINSTER AREA, ALBERTA AND SASKATCHEWAN, TRAPPING OF
B 6405 740* LLOYDMINSTER HEAVY OIL AREA, SASKATCHEWAN, ABST., NEARSHORE
S 10 226 LLOYDMINSTER HEAVY OIL DEPOSIT, PROFILE, CANADA
B 6111 1954 LLOYDMINSTER HEAVY OIL, WESTERN, CANADA
B 6107 1014 LLOYDMINSTER MEMBER, ALBERTA AND, SASKATCHEWAN
B 6007 1112 LLOYDMINSTER OIL, MASS SPECTROMETRIC ANALYSIS
B 6105 809 **LLOYD, R. M., POROSITY REDUCTION BY CHEMICAL COMPACTION—STABLE-**
B 6303 441* LO ANALYSIS SYSTEM FOR GEOLOGISTS, ABST., KOALA--MINICOMPUTER
M 24 149 LO EAST, FIELD, TEXAS, BIG WELLS
B 6405 742 LOBB, G. D., GEOLOGY OF SYNCRUDE CANADA LIMITED MINE SITE,
B 6206 952 LOBE, ANCIENT, MID-, FAN:, BRAIDED, CHANNELS ON, SUPRAFAN
B 6002 280 LOBES, MASON AND DIXON LINE, LOCATION OF, DELTA
B 6206 912† LOBES ON MODERN SUBMARINE FANS: CHARACTERS FOR RECOGNITION
B 6004 701* **LOBES ON UPPER MIOCENE-PLIOCENE LAGA FORMATION, EAST-CENTRAL**
B 6206 947 LOBES, SUBMARINE FAN, SUPRAFAN
B 6209 1762* LOBO TREND OF SOUTH LAREDO AREA, WEBB AND ZAPATA COUNTIES,
M 26 371 LOCATING SEEPS AND SHALLOW HYDROCARBONS, RESOLUTION
S 10 61 LOCATION OF EXSOLUTION, MIGRATION
S 5 83 LOCATION, MICHIGAN, BAY COUNTY
S 5 55 LOCATION, MICHIGAN, CALHOUN COUNTY,
S 5 55 LOCATION, MICHIGAN, CHARLEVOIX COUNTY,
S 5 55 LOCATION, MICHIGAN, CLINTON INLET,
S 5 55 LOCATION, MICHIGAN, GEORGIAN INLET,
S 5 78 LOCATION, MICHIGAN, GLADWIN COUNTY
S 5 115 LOCATION, MICHIGAN, GRAND TRAVERSE COUNTY,
S 5 115 LOCATION, MICHIGAN, KALKASKA COUNTY,
S 5 55 LOCATION, MICHIGAN, MACOMB COUNTY,
S 5 115 LOCATION, MICHIGAN, MANISTEE COUNTY,
S 5 78 LOCATION, MICHIGAN, MASON COUNTY
S 5 83 LOCATION, MICHIGAN, MIDLAND COUNTY
S 5 55 LOCATION, MICHIGAN, MIDLAND COUNTY,
S 5 55 LOCATION, MICHIGAN, NEWAGO COUNTY,
S 5 55 LOCATION, MICHIGAN, OCEANA COUNTY,
S 5 80 LOCATION, MICHIGAN, SAGINAW COUNTY
S 5 55 LOCATION, MICHIGAN, SANILAC COUNTY,
S 5 55 LOCATION, MICHIGAN, ST. CLAIR COUNTY,
M 30 330 LOCATION, BRAZIL, CAMPOS BASIN
B 6405 773* LOCATION OF LITTORAL ENERGY FENCE AND RESOLUTION OF RELICT
B 6106 931 LOCATION OF RIO GRANDE RISE SAMPLES
B 6309 1606 LOCK, BRIAN E., OUTER DEEP-SEA FAN DEPOSITIONAL LOBE SEQUENCE
B 6303 500 LOCK, DAVID, COORONG MODEL FOR PENECONTEMPORANEOUS
B 6406 841 LOCKER, STANLEY D., COMPARISON OF SAND-LAYER GEOMETRY ON
B 6209 1610 LOCKPORT AND GUELPH DOLOMITES, OHIO
B 6405 631 LOCKPORT BIOHERMS, WEST VIRGINIA
B 6405 635 LOCKPORT DOLOMITE, GAS POTENTIAL, WEST VIRGINIA
S 5 4 LOCKPORT FORMATION, MICHIGAN BASIN, NIAGARAN SERIES, NIAGARA
S 5 54 LOCKPORT FORMATION, SILURIAN, NIAGARA GROUP,
B 6209 1608 LOCKPORT FORMATION, APPALACHIAN BASIN
B 6004 660* LOCKPORT FORMATION, MIDDLE SILURIAN, OF NEW YORK STATE, ABST.,
S 5 96 LOCKPORT GROUP, OHIO,
B 6209 1610 LOCKPORT GROUP, NEW YORK
S 5 96 LOCKPORT GROUP, OUTCROP RELATIONS OF REEFS, OHIO, SILURIAN,
S 5 96 LOCKPORT GROUP, STRUCTURE, OHIO, SILURIAN,
S 5 96 LOCKPORT GROUP, THICKNESS, OHIO, SILURIAN,
B 6008 1402 LOCKRIDGE, JOHN P., CRETACEOUS FRONTIER SANDSTONE,
B 6205 862 LOCKRIDGE ULTRADEEP FIELD, PRODUCTION, TEXAS
B 6211 2218 LOCKWOOD HIGH, CALIFORNIA
B 6203 369 LOCKWOOD VALLEY BASIN, CALIFORNIA
B 6103 361 LOCO STRUCTURE, OKLAHOMA
B 6004 652 LODDING, WILLIAM, TIMING AND PALEOCLIMATE INDICATORS IN
M 25 477 LODE DEPOSITS OF ANTIMONITE AN ANTIMONITE-FERBERITE, USSR,
M 25 476 LODE DEPOSITS OF GOLD-ANTIMONITE, USSR, EASTERN
M 27 164 LODGEPOLE LIMESTONE, MONTANA, MISSISSIPPIAN
M 27 26 LODGEPOLE LIMESTONE, MONTANA, MISSISSIPPIAN
M 27 27 LODGEPOLE LIMESTONE, MONTANA, MISSISSIPPIAN
M 28 40 LODGEPOLE LIMESTONE, MONTANA, MISSISSIPPIAN
B 6211 2262 LODO GULCH SECTION, CALIFORNIA
B 6211 2262 LODO MEMBER, CALIFORNIA
B 6405 739* LOFRECO PROCESS—TAILORING SHALE OIL EXTRACTION METHOD TO
B 6302 169 LOG ANALYSIS, ALBERTA AND, BRITISH COLUMBIA, DEEP BASIN
B 6302 173 LOG ANALYSIS BY FORMATION, SUMMARY OF
B 6302 172 LOG ANALYSIS FROM DIGITIZED LOGS, COMPUTER PRINT OUT OF
B 6105 842* LOG AND CORE-ANALYSIS SYSTEM, ABST., PETROS—PETROPHYSICAL
B 6209 1752* LOG, AND ITS GEOLOGIC IMPLICATIONS, ABST., RADIOMETRIC
B 6102 140 LOG, CALIFORNIA, VENTURA FIELD, TYPE, ELECTRIC
B 6102 172 LOG, CALIFORNIA, VENTURA FIELD, TYPE, ELECTRIC
M 26 15† LOG DATA THAT ARE IMPORTANT TO SEISMIC INTERPRETATION,
B 6002 189 LOG, ENVIRONMENTAL ANALYSES, NORTH SEA, WELL
B 6309 1600* LOG EVALUATION OF TIGHT ROCKS OF SOUTH TEXAS, ABST
B 6409 1568* LOG EVALUATION OF WELLS IN TUSCALOOSA TREND OF SOUTH
B 6309 1595* LOG INTERPRETATION, ABST., IMPACT OF DIAGENESIS ON
B 6412 2191 LOG INTERPRETATION, BY DOUGLAS W. HILCHIE, EVANS, HILTON B.,
B 6306 905† LOG MOTIFS IN NORTH SEA SUBMARINE-FAN SANDS, DIPMETER AND
B 6306 906 LOG MOTIFS OF DEEP-SEA SANDS, GAMMA
B 6002 186 LOG PATTERNS IN CLASTIC SEDIMENTARY ROCKS, FOUR
B 6002 184 LOG PATTERNS, NORTH SEA, WIRELINE-
B 6007 1073 LOG RECORDS, TEMPERATURES, NIGER DELTA PETROLIFEROUS
B 6207 1112 LOG, SCHLUMBERGER, 4-ARM HIGH-RESOLUTION, DIPMETER
B 6109 1515 LOG SPRINGS FORMATION, NEW MEXICO
B 6011 2068 LOGAN, BRIAN W., INTERPRETATION OF CARBONATE-ROCK
M 28 177 LOGAN CANYON FORMATION, CANADA, CRETACEOUS
M 28 124 LOGAN CANYON FORMATION, CANADA, CRETACEOUS
B 6301 90 LOGAN CANYON FORMATION, SCOTIAN BASIN
M 28 172 LOGAN CANYON SANDSTONE, CANADA, CRETACEOUS
B 6203 523* LOGAN COUNTIES, OKLAHOMA, ABST., SUBSURFACE
B 6309 1579* LOGAN COUNTIES, WEST VIRGINIA--POSSIBLE NEW COAL AND-OR
B 6104 594 LOGAN FIELD, OHIO
B 6304 686* LOGGING, ABST., DIRECT MEASUREMENT OF, URANIUM BY, PROMPT
B 6303 484* LOGGING AS USEFUL TOOL FOR LITHOLOGIC INTERPRETATION, ABST.,
B 6303 449* LOGGING DATA ASSIST I GEOLOGIC STUDIES, ABST., GAMMA-RAY
B 6309 1579* LOGGING IN GAS SHALES, ABST., THEORETICAL AND EXPERIMENTAL
M 30 168 LOGGING PROCEDURES, PIPER FIELD
B 6309 1579* **LOGGING IN GAS SHALES, ABST., THEORETICAL AND EXPERIMENTAL**
B 6004 651† LOGS, ABST., COMPUTER-GENERATED, SCALE FUNCTIONS FOR,
B 6405 760* LOGS, ABST., CROSS-BEDDED SANDSTONE, RESERVOIRS—GEOLOGIC
B 6405 705* LOGS, ABST., GEOTHERMAL RESOURCES EVALUATED BY, WELL
B 6203 506* LOGS, ABST., UNDERSTANDING, SEISMICALLY DERIVED, IMPEDANCE
B 6309 1504† LOGS, DETERMINATION OF, ORGANIC CONTENT OF, APPALACHIAN,
B 6405 806* LOGS FROM EXPLORATIONISTS VIEWPOINT, ABST., SONIC
B 6209 1757* LOGS IN ARK-LA-TEX COTTON VALLEY FORMATION, ABST., FORMATION
B 6209 1755* LOGS IN CRETACEOUS OF NORTH LOUISIANA AND MISSISSIPPI, ABST.,
B 6304 595 LOGS IN MIGRATION STUDIES, POTENTIAL USE OF, ELECTRIC
B 6207 1111† LOGS, MEASUREMENT OF, SUBSURFACE, FRACTURES FROM, DIPMETER
B 6002 188 LOGS, NORTH SEA, DIP
B 6002 185 LOGS, NORTH SEA, GAMMA-RAY
B 6002 188 LOGS, NORTH SEA, WELL
B 6207 1115 LOGS, REPLOTTED
B 6303 516* LOGS TO PREDICT PERMEABILITY OF EOCENE DELTAIC SANDSTONES,
B 6005 879† LOGS, TRANSFORMATION OF, RESISTIVITY TO, PSEUDOVELOCITY
B 6303 443 LOHMAN, WILLIAM H., NEOGENE CALCAREOUS NANNOPLANKTON
B 6311 2097 LOHMANN, HANS H., SEISMIC RECOGNITION OF SALT DIAPIRS
B 6405 750 LOHMANN, KYGER C., GEOCHEMISTRY OF REGIONALLY EXTENSIVE
B 6403 445 LOHMAR, JOHN M., LATE PLEISTOCENE-HOLOCENE SEDIMENTARY
B 6008 1405 LOHRENZ, JOHN, ECONOMIC VALUES USIN DISTRIBUTED RANGES OF
B 6203 355 LOMPOC OIL FIELD, CALIFORNIA
M 29 113 LONARDI, A. G., STRUCTURE OF COLORADO BASIN AND
B 6311 2024 LONDONDERRY HIGH, AUSTRALIA
B 6306 887 LONE PINE AREA, NEW MEXICO, HOSPAH-
B 6306 886† LONE PINE FIELD, NEW MEXICO, OIL AND, GAS IN, DELTA-, MARGIN
B 6306 899 LONE PINE FIELD, RESERVOIR, NEW MEXICO
B 6403 439* LONG BEACH UNIT, WILMINGTON OIL FIELD, ABST., COMPUTER SAND
B 6207 1227 LONG, CARL, GEOTHERMAL EXPLORATION ON ADAK ISLAND, ALASKA—A
B 6405 749 LONG, DAVID T., ORGANOPHOSPHORITES IN CARBONIFEROUS ROCKS OF
B 6105 797* **LONG-DISTANCE TURBIDITE CORRELATIONS IN HORSESHOE ABYSSAL PLAIN**
B 6204 694 LONG ISLAND, BAHAMAS
B 6309 1465 LONG ISLAND, BALTIMORE CANYON, STRATIGRAPHIC COLUMN
B 6203 457 LONG ISLAND FAULT, AUSTRALIA
B 6405 783* LONG ISLAND, NEW YORK, ABST., SEDIMENTARY FACIES RELATIONS AND
M 29 98 LONG ISLAND PLATFORM, ATLANTIC MARGIN
B 6404 505 LONG ISLAND PLATFORM, CORRELATION
M 29 99 LONG ISLAND PLATFORM, DEPTH-TO-BASEMENT CONTOURS
B 6109 1484 LONG ISLAND PLATFORM, EAST COAST
M 29 97 LONG ISLAND PLATFORM, MAGNETIC ANOMALY CONTOURS

M 29 98 LONG ISLAND PLATFORM, MAGNETIC DEPTH ESTIMATES
B 6109 1523 LONG ISLAND, SEDIMENT ACCUMULATION RATES, FLORIDA TO
B 6204 694 LONG ISLAND WELL, BAHAMA PLATFORM
B 6303 487 LONG, JOHN M., SEISMIC STRATIGRAPHY AND GEOLOGIC HISTORY OF
S 4 186 LONG POINT FORMATION, REEF, NEWFOUNDLAND
B 6306 952 LONG RANGE MOUNTAINS, NEWFOUNDLAND
B 6207 1231* LONG VALLEY GEOTHERMAL AREA, CALIFORNIA, ABST.,
B 6004 693 LONGACRE, SUSAN A., MIDDLE GUADALUPIAN COASTAL CARBONATE
B 6408 1229 LONGITUDINAL SAND RIDGES
B 6307 1021 LONGITUDINAL WRENCH FAULTS, WESTERN, INDONESIA
B 6404 461 LONGMAN, MARK W., CARBONATE DIAGENETIC TEXTURES FROM
B 6203 538 LONGMAN, MARK W., SANDSTONE DIAGENESIS IN BROMIDE FORMATION,
B 6101 113 LONGWALL SANDSTONE MEMBER, UTAH
M 25 422 LONIE, W. M., OCCURRENCE AND DEVELOPMENT OF SEDIMENTARY
B 6212 2454 LONSDALE, PETER, ECUADORIAN SUBDUCTION SYSTEM
B 6402 280 LONSDALE, PETER, ECUADORIAN SUBDUCTION SYSTEM. DISCUSSION
B 6109 1540 LOOCKE, J. E., GROWTH HISTORY OF HAINESVILLE SALT DOME, WOOD
B 6207 1194 LOOKINGGLASS FORMATION, OREGON AND, WASHINGTON
B 6109 1540 LOONEY, R. MICHAEL, LATE QUATERNARY GEOMORPHIC EVOLUTION OF
B 6205 819 LOPATIN CALCULATIONS ON SOURCE ROCKS
B 6406 918 LOPATINS METHOD, THEORY OF
B 6406 916† LOPATINS METHOD TO PETROLEUM EXPLORATION, TIME AND,
M 25 519 LOPEZ AVILA, JAVIER, PRINCIPAL CHARACTERISTICS OF SOME PORPHYR
B 6409 1563 LOPEZ, CYNTHIA M., DEPOSITIONAL SYSTEMS IN NACATOCH SAND,
M 25 410 LORD, J. H., IRON-ORE DEPOSITS OF WESTERN AUSTRALIA--GEOLOGY
S 6 2 LORD KELVIN, TIME SCALE
M 24 242 LORING FIELD, MISSISSIPPI
B 6405 740 LORSONG, J. A., NEARSHORE LITHOFACIES OF MANNVILLE GROUP,
B 6405 740 LORSONG, J. A., SYNSEDIMENTARY DEFORMATION IN FOSSIL
B 6203 355 LOS ALAMOS OIL FIELD, CALIFORNIA
B 6303 421* LOS ANGELES AND VENTURA BASINS, CALIFORNIA, ABST., VITRINITE
B 6111 1988 LOS ANGELES ARCH, CALIFORNIA
S 7 239 LOS ANGELES BASIN, BREEN SANDSTONE, ISOPACH, CALIFORNIA
B 6203 356 LOS ANGELES BASIN, CALIFORNIA
B 6202 248 LOS ANGELES BASIN, CALIFORNIA
S 7 239 LOS ANGELES BASIN, CHAPMAN SANDSTONE, ISOPACH, CALIFORNIA
S 7 233 LOS ANGELES BASIN, DISTAL TURBIDITE FACIES, CALIFORNIA
S 7 234 LOS ANGELES BASIN, E-LOG SECTION, CALIFORNIA
S 10 50 LOS ANGELES BASIN, MIGRATION, CALIFORNIA
S 7 240 LOS ANGELES BASIN, MOHNIAN PALEOGEOGRAPHY, CALIFORNIA
S 10 51 LOS ANGELES BASIN, PLAYA DEL REY FIELD, CALIFORNIA
S 7 238 LOS ANGELES BASIN, PROXIMAL SANDSTONE FACIES, CALIFORNIA
S 7 236 LOS ANGELES BASIN, PROXIMAL TURBIDITE FACIES, CALIFORNIA
S 7 240 LOS ANGELES BASIN, RECONSTRUCTION, CALIFORNIA
B 6004 685* LOS ANGELES, CALIFORNIA, ABST., TIMING--GEOLOGIC AN POLITICAL--IN,
B 6002 220 LOS ANGELES DOWNTOWN FIELD, FLOWER COMMUNITY WHOLE OIL,
B 6111 1979† LOS ANGELES EROSION SURFACE OF MIDDLE CRETACEOUS AGE
B 6111 1981 LOS ANGELES LOWLANDS AND BORDERLAND, CALIFORNIA
M 25 160 LOS AZUFRIS GEOTHERMAL AREA, MEXICO
M 25 141 LOS BANOS, GEOTHERMAL PROSPECT, PHILIPPINES, MT. MAQUILING--,
M 25 554 LOS LINDEROS COPPER AND MOLYBDENUM DEPOSITS, ECUADOR
M 25 554 LOS LINDEROS DEPOSIT, COPPER-MOLYBDENUM, ECUADOR
B 6211 2220 LOS LOBOS THRUST, CALIFORNIA
M 25 160 LOS NEGRITOS GEOTHERMAL AREA, MEXICO
M 29 366 LOS ROQUES TRENCH, SEISMIC SECTION, VENEZUELA BASIN
B 6112 2098 LOSS OF CARBON DURING CATAGENESIS
B 6311 1999 LOSSEL, P., FRIGG FIELD--LARGE SUBMARINE-FAN TRAP IN LOWER
M 30 59 LOSSEL, P., FRIGG FIELD--LARGE SUBMARINE-FAN TRAP IN LOWER
B 6203 521 LOSSEL, P., FRIGG FIELD—LARGE SUBMARINE-FAN TRAP IN LOWER
B 6102 257 LOST RIVER RANGE, NOMENCLATURE, IDAHO
B 6001 12† LOST SOLDIER AREA, WYOMING, INFLUENCE OF, RECURRENT, LARAMIDE
B 6404 572 LOT CAPROCK, ISRAEL
B 6404 572 LOT SALT MEMBER, ISRAEL
B 6305 795 LOUANN DEPOSITION, GULF COAST
S 7 82 LOUANN DEPOSITION, JURASSIC, GULF OF MEXICO
M 26 147 LOUANN FORMATION, TEXAS, EAST TEXAS BASIN
S 3 99 LOUANN MOTHER SALT
M 24 184 LOUANN SALT, ARKANSAS
B 6307 1050 LOUANN SALT, TEXAS
M 26 150 LOUANN-WERNER FORMATION, TEXAS, EAST
B 6007 1119 LOUARK GROUP, TEXAS
B 6109 1541 LOUCKS, R. G., CONSOLIDATION HISTORY OF GULF COAST LOWER
B 6203 494 LOUCKS R. G., DELTAIC DEPOSITIONAL STYLE IN SALT-WITHDRAWAL
B 6203 538 LOUCKS, R. G., DEPOSITIONAL FACIES AND OCCURRENCE OF POROSITY
B 6009 1606 LOUCKS, R. G., GEOPRESSURED GEOTHERMAL PROSPECT IN FRIO
B 6309 1606 LOUCKS, R. G., IMPORTANCE OF SECONDARY LEACHED POROSITY IN
B 6303 440 LOUCKS, R. G., MINERALOGIC COMPOSITION AND DIAGENESIS OF
B 6109 1554 LOUCKS, R. G., POROSITY DEVELOPMENT AND DISTRIBUTION
B 6303 488 LOUCKS, R. G., POROSITY IN GIANT GAS FIELD, ELLENBURGER
B 6303 417 LOUCKS, R. G., POTENTIAL FOR PRODUCTION OF GEOPRESSURED
B 6303 488 LOUCKS, R. G., RESERVOIR QUALITY IN TERTIARY SANDSTONES ALONG
B 6405 741 LOUCKS, R. G., SANDSTONE DIAGENESIS IN GEOPRESSURED TERTIARY
B 6209 1760 LOUCKS, R. G., SANDSTONE DISTRIBUTION AND POTENTIAL FOR
B 6309 1595 LOUCKS, R. G., TESTIN GEOPRESSUREDD GEOTHERMAL RESOURCE, FRIO
B 6105 765 **LOUCKS, R. G., TEST SITE FOR GEOPRESSURED GEOTHERMAL WELL—FRIO**
B 6303 417 LOUCKS, R. G., WORLDS FIRST GEOPRESSURED GEOTHERMAL TEST WELL,
B 6105 809 **LOUCKS, ROBERT G., INCOMPLETE CARBONATE CEMENTATION—PRINCIPAL**
B 6004 694 LOUCKS, ROBERT G., RELATION OF EARLY CARBONATE POROSITY TO
B 6408 1282* LOUDEN ANTICLINE, FAYETTE COUNTY, ILLINOIS—ITS RELATION TO
B 6309 1602 LOUDON, EDDIE J., SEDIMENTATION ON TRAILING PLATE
B 6101 3 LOUGHEED ARCH
B 6303 514 LOUIS, P. R., DELTAIC FACIES RECOGNITIO ON SEISMIC DATA, ABST.
M 24 218 LOUISIANA, A SANDSTONE ZONE
B 6409 1569* LOUISIANA, ABST., BLOCK 25 FIELD, CHANDELEUR SOUND, ST. BERNARD
B 6309 1599* LOUISIANA, ABST., CLASTICITY INDEX--KEY TO CORRELATING
B 6405 710* LOUISIANA, ABST., DEEP, TUSCALOOSA, GAS TREND OF SOUTH
B 6109 1546* LOUISIANA, ABST., DIFFERENTIAL, SUBSIDENCE OF, MARSHALND, PEAT
B 6009 1607* LOUISIANA, ABST., DOOR POINT--, BURIED, VOLCANO IN SOUTHEAST
B 6405 734* LOUISIANA, ABST., EARLY, CEMENTATION BY HIGH-MAGNESIUM, CALCIT
B 6303 467* LOUISIANA, ABST., EUGENE ISLAND, BLOCK 330 FIELD, OFFSHOR
B 6409 1557* LOUISIANA, ABST., EXPLORATION RAMIFICATIONS OF SUBSURFACE
B 6004 660* LOUISIANA, ABST., FOSSIL, SALT STRUCTURES IN, VACHERIE, SALT, DOM
B 6309 1609* LOUISIANA, ABST., GEOPRESSURE IN, HOUMA AND, HOLLYWOOD FIELDS
B 6109 1537* LOUISIANA, ABST., INTERIOR STRUCTURE OF, OAK GROVE RIDGE,
B 6009 1615* LOUISIANA, ABST., INTERPRETATION OF, COTTON VALLEY DEPOSITIONAL
B 6009 1611* LOUISIANA, ABST., JAMES LIMESTONE IN, WINN AND, NATCHITOCHES
B 6409 1568* LOUISIANA, ABST., LOG EVALUATION OF WELLS IN, TUSCALOOSA TREND
B 6209 1757* LOUISIANA, ABST., MINERALOGY OF, PLEISTOCENE, TERRACE DEPOSITS
B 6309 1611* LOUISIANA, ABST., ORIGIN OF POROSITY IN DEEP, WOODBINE-,
B 6409 1565* LOUISIANA, ABST., PETROGRAPHIC, STRATIGRAPHY, AND STRUCTURAL
B 6409 1553* LOUISIANA, ABST., PRELIMINARY OBSERVATIONS OF MODERN, POINT
B 6409 1558* LOUISIANA, ABST., PRESSURE, TEMPERATURE, SALINITY, LITHOLOGY,
B 6009 1612* LOUISIANA, ABST., SALT MOVEMENT, HALOKINETICS, IN NORTH
B 6004 679* LOUISIANA, ABST., SALT, TECTONICS OF, WINN PARISH
B 6109 1539* LOUISIANA, ABST., SLICE TECHNIQUE APPLIED TO, URANIUM
B 6009 1609* LOUISIANA, ABST., SOUTH SAREPTA FIELD, GRAY SANDSTONE,
B 6409 1555* LOUISIANA, ABST., SUBSURFACE, WILCOX, LIGNITE IN WEST-CENTRAL
B 6009 1617* LOUISIANA, ABST., THE, OIL, INDUSTRY AND, COASTAL ZON,
B 6309 1609* LOUISIANA, ABST., VARIATIONS IN LITTORAL SAND FEATURES WITH
B 6407 1007 LOUISIANA, AND ARKANSAS, BASEMENT STRUCTURES, TEXAS
B 6407 1011 LOUISIANA, AND ARKANSAS, COTTON VALLEY SANDSTONES, TEXAS
B 6407 1008 LOUISIANA, AND ARKANSAS, GRABEN-FAULT STRUCTURES, TEXAS
B 6407 1007 LOUISIANA, AND ARKANSAS, MEXIA-TALCO FAULT SYSTEM, TEXAS
B 6407 1007 LOUISIANA, AND ARKANSAS, SALT ANTICLINES, TEXAS
B 6008 1370† LOUISIANA, AND EAST TEXAS IN 1975, DEVELOPMENT IN, ARKANSAS,
B 6108 1234† LOUISIANA, AND EAST TEXAS IN 1976, DEVELOPMENTS IN, ARKANSAS
B 6208 1487† LOUISIANA, AND EAST TEXAS IN 1977, DEVELOPMENTS IN, ARKANSAS,
B 6308 1376† LOUISIANA, AND EAST TEXAS IN 1978, DEVELOPMENT IN, ARKANSAS,
B 6409 1490† LOUISIANA, AND EAST TEXAS IN 1979, DEVELOPMENTS IN, ARKANSAS,
B 6303 442* LOUISIANA AND GULF OF MEXICO, ABST., APPLIED BIOSTRATIGRAPHY,
B 6209 1755* LOUISIANA AND MISSISSIPPI, ABST., APPLICATION OF, FRACTURE
B 6407 1004† LOUISIANA, AND SOUTH ARKANSAS, JURASSIC, COTTON VALLEY AND,
B 6307 1045 LOUISIANA AND TEXAS, GULF COAST BASIN, SEISMIC PROFILE
S 7 267† LOUISIANA AND TEXAS: GEOTECHNICAL ASPECTS, STABILITY OF,
B 6409 1560* LOUISIANA, AND THEIR RELATION TO LIGNITE, ABST., DEPOSITIONAL
B 6402 270 LOUISIANA, ATCHAFALAYA BAY, BATHYMETRIC MAPS
B 6402 268 LOUISIANA, ATCHAFALAYA RIVER, DISCHARGE AND LOAD
B 6402 268 LOUISIANA, ATCHAFALAYA RIVER, DISCHARGE AND LOAD
M 24 217 LOUISIANA, B SANDSTONE ZONE
M 24 216 LOUISIANA, B-14 SANDSTONES, LOUISIANA
B 6402 277 LOUISIANA, BAPTISTE COLLETTE CREVASSE
M 24 231 LOUISIANA, BAY MARCHAND SALT-RIDGE TREND
B 6101 65 LOUISIANA, BAYOU CARLIN AREA, GEOTHERMAL PATTERN
B 6101 70 LOUISIANA, BAYOU CARLIN FIELD
B 6407 1007 LOUISIANA, BAYOU MIDDLEFORK FIELD
B 6402 165 LOUISIANA, BAYOU VILLARS FIELD
B 6107 1091 LOUISIANA, BELLE ISLE SALT MINE
M 24 235 LOUISIANA, BIGENERINA FLORIDANA ZONE
M 24 235 LOUISIANA, BIGENERINA-DISCORBIS ZONE
M 24 214 LOUISIANA, C SANDSTONE
S 10 212 LOUISIANA, CALCASIEU PARISH, MANCHESTER FIELD, SALINITY WITH
S 10 210 LOUISIANA, CALCASIEU PARISH, MANCHESTER FIELD, CLAY CONVERSION
S 10 210 LOUISIANA, CALCASIEU PARISH, MANCHESTER FIELD, KEROGEN
S 10 212 LOUISIANA, CENOZOIC, POROSITY RELATED TO DEPTH
S 3 99 LOUISIANA, CHACAHOULA DOME
B 6409 1560* LOUISIANA CHENIER SYSTEM—SOME PRELIMINARY REINTERPRETATIONS
B 6009 1614* LOUISIANA CONTINENTAL SHELF, ABST., BIOSTRATIGRAPHY O,
B 6109 1547* LOUISIANA CONTINENTAL SHELF, ABST., STRUCTURE AND
B 6209 1758* LOUISIANA CONTINENTAL SHELF AND SOURCE OF TURBID BOTTOM
B 6101 66 LOUISIANA, COTE BLANCHE ISLAND

B 6101 67 LOUISIANA, COTE BLANCHE ISLAND SALT DOME
B 6402 165 LOUISIANA, COUBA ISLAND FIELD
B 6402 158 LOUISIANA, CRESCENT FARMS FIELD
M 28 138 LOUISIANA, CRETACEOUS, TUSCALOOSA FORMATION
M 28 135 LOUISIANA, CRETACEOUS, TUSCALOOSA FORMATION
M 28 70 LOUISIANA, CRETACEOUS, TUSCALOOSA FORMATION
M 28 144 LOUISIANA, CRETACEOUS, TUSCALOOSA FORMATION
M 28 176 LOUISIANA, CRETACEOUS, TUSCALOOSA FORMATION
M 24 235 LOUISIANA, CRISTELLARIA K ZONE
B 6011 2044† LOUISIANA, DIFFERENTIATION OF, SHALE TYPES BY, ALLOCHTHONOU
M 29 428 LOUISIANA, DIVERGENT DELTAIC MARGIN, GULF OF MEXICO
B 6101 70 LOUISIANA, EAST LAKE SAND FIELD
M 30 253 LOUISIANA, EUGENE ISLAND 330 FIELD, PLEISTOCENE
M 30 253† LOUISIANA, EUGENE ISLAND BLOCK 330 FIELD, OFFSHORE
B 6104 587 LOUISIANA, EUGENE ISLAND BLOCK FIELDS
B 6108 1246 LOUISIANA, FALSE RIVER FIELD, POINTE COUPEE PARISH
B 6101 65 LOUISIANA, FIVE ISLANDS TREND
M 24 209 LOUISIANA, G, J, K, O, Q SANDSTONE
M 24 209 LOUISIANA, G SANDSTONE
M 24 212 LOUISIANA, G THROUGH K SANDSTONES
B 6402 158 LOUISIANA, GIBSON FIELD
B 6004 545 LOUISIANA, GOLDEN MEADOW FAULT SYSTEM
B 6402 148 LOUISIANA, GOOD HOPE FIELD
B 6402 267 LOUISIANA, GRAND AND SIX MILE LAKES, LACUSTRINE DELTA FILL
M 24 232 LOUISIANA, GRAND ISLE ASH
M 24 235 LOUISIANA, GRAND ISLE ASH ZONE
M 24 229† LOUISIANA, GRAND ISLE BLOCK 16 FIELD, OFFSHORE
B 6104 587 LOUISIANA, GRAND ISLE BLOCK FIELDS
B 6407 1005 LOUISIANA, GROGAN FIELD
B 6008 1265 † LOUISIANA GULF COAST IN 1975, DEVELOPMENTS IN
B 6108 1244† LOUISIANA GULF COAST IN 1976, DEVELOPMENTS IN
B 6208 1519† LOUISIANA GULF COAST OFFSHORE IN 1977, DEVELOPMENTS IN
B 6308 1310† LOUISIANA GULF COAST OFFSHORE IN 1978, DEVELOPMENTS IN
B 6409 1445† LOUISIANA GULF COAST OFFSHORE IN 1979, DEVELOPMENTS IN
B 6208 1521 LOUISIANA, GULF COAST, OFFSHORE, PRODUCTION, 1974-1977
B 6208 1462† LOUISIANA GULF COAST ONSHORE IN 1977, DEVELOPMENTS IN
B 6308 1357† LOUISIANA GULF COAST ONSHORE IN 1978, DEVELOPMENTS IN
B 6411 2012† LOUISIANA GULF COAST ONSHORE IN 1979, DEVELOPMENTS IN
B 6411 2025 LOUISIANA, GULF COAST ONSHORE, PRODUCTION, 1979
B 6108 1258 LOUISIANA GULF COAST, PRODUCTION, 1967-1976
B 6402 145† LOUISIANA GULF COAST SALT BASIN, SEALING AND, NONSEALING,
B 6402 145† LOUISIANA GULF COAST SALT BASIN, SEALING AND, NONSEALING,
B 6002 227 LOUISIANA, GULF COAST WELLS
B 6209 1589 LOUISIANA GULF COAST—DIVERGENT DELTAIC MARGIN
B 6206 951 LOUISIANA, HACKBERRY CHANNEL
B 6407 1005 LOUISIANA, HAYNESVILLE FIELD
B 6112 2086 LOUISIANA, HOLOCENE, KEROGEN
B 6101 65 LOUISIANA, HOT BELT
B 6402 154 LOUISIANA, IOWA FIELD
B 6307 1046 LOUISIANA, KINGS BAYOU SAND
B 6004 545 LOUISIANA, LAKE HATCH FAULT SYSTEM
B 6101 65 LOUISIANA, LAKE SAND AREA, GEOTHERMAL PATTERN
B 6101 69 LOUISIANA, LAKE SAND FAULT
B 6101 70 LOUISIANA, LAKE SAND FIELD
B 6104 595 LOUISIANA, LAKE WASHINGTON FIELD
B 6104 587 LOUISIANA, LAKE WASHINGTON FIELD
M 24 216 LOUISIANA, LOUISIANA, B-14 SANDSTONES
B 6002 239 LOUISIANA, MAIN PASS BLOCK 41 FIELD
B 6104 587 LOUISIANA, MAIN PASS BLOCK FIELDS
B 6002 227 LOUISIANA, MANCHESTER FIELD
S 7 96 LOUISIANA, MARSH ISLAND, SEISMIC FACIES EXAMPLE
S 7 100 LOUISIANA, MIDSLOPE, SEISMIC FACIES DIAGRAM
S 7 101 LOUISIANA, MIDSLOPE SEISMIC FACIES EXAMPLE
S 7 107 LOUISIANA, MIDSLOPE, SEISMIC FACIES EXAMPLE
B 6112 2086 LOUISIANA, MIOCENE, KEROGEN
B 6206 951 LOUISIANA, MISSISSIPPI CHANNEL
B 6202 207 LOUISIANA, MISSISSIPPI DELTA AREA, GEOHISTORY
B 6402 158 LOUISIANA, N SANDSTONE
B 6104 586 LOUISIANA, OFFSHORE, BIOZONES
S 10 211 LOUISIANA, OFFSHORE, FORMATION PRESSURE GRADIENT
S 10 257 LOUISIANA, OFFSHORE, PENTANE CONCENTRATIONS
S 10 254 LOUISIANA, OFFSHORE ETHANE MEASUREMENTS
B 6308 1312 LOUISIANA, OFFSHORE, PRODUCTION, 1978
B 6409 1446 LOUISIANA, OFFSHORE, PRODUCTION, 1979
B 6109 1427 LOUISIANA, OFFSHORE, TERTIARY BEDS
B 6206 951 LOUISIANA, OLIGOCENE
M 29 427 LOUISIANA, ORGANIC-CARBON BY DEPTH, GULF OF MEXICO
S 7 102 LOUISIANA, OUTER SHELF EDGE, SEISMIC FACIES EXAMPLE
B 6402 149 LOUISIANA, P SANDSTONE
B 6206 951 LOUISIANA, PLEISTOCENE
B 6402 265 LOUISIANA, POINT AU FER SHELL REEF
B 6108 1246 LOUISIANA, RAYNE FIELD, ACADIA PARISH
B 6402 156 LOUISIANA, REALTY A ZONE
B 6407 1006 LOUISIANA, SABINE UPLIFT, TEXAS AND
B 6407 1006 LOUISIANA, SALT BASIN
B 6405 797* LOUISIANA SALT BASIN, ABST., GEOLOGY OF, CALVIN FIELD—DEEP
B 6004 724* LOUISIANA SALT DOMES--POTENTIAL SITES FOR RADIOACTIVE WASTE
B 6002 231 LOUISIANA, SALTS IN OIL-FIELD WATERS VS. DEPTH
B 6402 165 LOUISIANA, SATURDAY ISLAND FIELD
B 6007 1128 LOUISIANA, SHALE DENSITY AND POROSITY FOR OFFSHORE WELL
B 6109 1545* LOUISIANA SHELF, ABST., STRUCTURE AND STRATIGRAPHY OF LATE,
B 6104 586 LOUISIANA SHELF, TERTIARY RESERVOIRS, OIL AGE
S 10 6 LOUISIANA, SHALE MINERALOGY EXAMPLE, MIOCENE
B 6104 587 LOUISIANA, SHIP SHOAL BLOCK FIELD
B 6307 1046 LOUISIANA, SIPHONINA DAVISI ZONE
B 6404 466 LOUISIANA, SMACKOVER FORMATION, CEMENTS
B 6104 587 LOUISIANA, SOUTH MARSH ISLAND FIELDS
B 6308 1368 LOUISIANA, SOUTH, ONSHORE, PRODUCTION, 1977-1978
B 6402 155 LOUISIANA, SOUTH PASS BLOCK 24 FIELD
B 6402 165 LOUISIANA, SOUTH PASS BLOCK 27 FIELD
B 6112 2086 LOUISIANA, SOUTH PECAN LAKE FIELD, KEROGEN
B 6104 587 LOUISIANA, SOUTH PLASS BLOCK FIELD
B 6105 834* LOUISIANA SOUTH TIMBALIER BLOCK 54 SALT DOME, ABST.,
B 6104 587 LOUISIANA, SOUTHEAST PASS FIELD
B 6004 545 LOUISIANA, SOUTHEASTERN, GEOLOGIC CROSS SECTION
S 3 99 LOUISIANA, STARKS DOME
S 3 107 LOUISIANA, STARKS DOME
B 6002 220 LOUISIANA STATE WHOLE OIL
B 6402 152 LOUISIANA, T ZONE
B 6004 545 LOUISIANA, TEPETATE-BATON ROUGE FAULT SYSTEM
M 24 231 LOUISIANA, TERREBONNE TROUGH
M 24 235 LOUISIANA, TEXTULARIA L ZONE
M 29 433 LOUISIANA, THERMAL MATURITY, GULF OF MEXICO
M 26 256 LOUISIANA, TIMBALIER CHANNEL, SEISMIC SECTION
B 6409 1552* LOUISIANA TUSCALOOSA TREND, 1975-80, ABST., NO. 1 PLAY IN U.S.A.—
B 6409 1558* LOUISIANA TUSCALOOSA VERSUS SOUTHEAST TEXAS WOODBINE, ABST.
B 6104 587 LOUISIANA, VERMILION BLOCK FIELD
B 6203 519* LOUISIANA WELL, ABST., CHEMICAL ANALYSES OF, GEOTHERMAL,
B 6203 529* LOUISIANA WELL, ABST., CHEMICAL ANALYSIS OF, GAS DISSOLVED IN,
B 6104 587 LOUISIANA, WEST CAMERON BLOCK FIELD
B 6109 1429 LOUISIANA, WEST DELTA BLOCK 73 FIELD
B 6104 587 LOUISIANA, WEST DELTA BLOCK FIELDS
B 6402 159 LOUISIANA, WEST LAKE VERRET FIELD
M 26 264 LOUISIANA, WESTERN, DIAPIR UPLIFT
M 24 205† LOUISIANA--A PLEISTOCENE, EAST CAMERON, BLOCK 270, OFFSHORE
B 6309 1600* LOUISIANA--CASE HISTORY FROM DISCOVERY TO WATERFLOOD, ABST.,
B 6101 65† LOUISIANA--IMPLICATIONS, GEOTHERMAL AND, GEOPRESSURE
B 6009 1610* LOUISIANA--NOW WHAT, ABST., IVE GOT A NEW, PRODUCER IN SOUT
B 6009 1615* LOUISIANA-MISSISSIPPI, ABST., CLAY SEDIMENTATION I, PEARL RIVER,
B 6209 1759* LOUISIANA—GEOTEMPERATURES AND HALOKINETICS, ABST., REDDELL,
B 6401 128 LOUISVILLE LIMESTONE, INDIANA
B 6409 1557* LOUSIANA, ABST., DIAGENESIS OF MIDDLE, MIOCENE SANDS IN,
B 6108 1375 LOVE, ALONZA H., ORGANIC GEOCHEMISTRY, INCIPIENT
B 6201 98 LOVE, ALONZA H., ORGANIC GEOCHEMISTRY, INCIPIENT
B 6409 1550 LOVEJOY, WILSON, JAMES E., REVIEW OF, SIERRA DE JUAREZ,
B 6410 1615 LOVELAND FIELD, COLORADO
B 6405 741 LOW, PHILIP F., PROPERTIES OF WATER IN CLAY MINERAL SYSTEMS,
B 6205 757 LOW-AMPLITUDE FACIES
M 26 170 LOW-AMPLITUDE FACIES, SEISMIC, INTERPRETATION
B 6205 757 LOW-CONTINUITY AND VARIABLE-AMPLITUDE FACIES
B 6405 773* LOW-COST MICROPROCESSOR SYSTEM FOR PALEONTOLOGIC
B 6004 731* LOW-ENERGY COASTAL ENVIRONMENNTS, SOUTHWES FLORIDA, ABST.,
S 7 90 LOW-ENERGY TURBIDITY-CURRENT TRANSPORT, SEDIMENT PROCESSES
M 29 133 LOW-VELOCITY SEDIMENTS, MALVINAS BASIN, TERTIARY
S 4 155† LOW-WATER EMERGENCE OF CARIBBEAN REEFS AND EFFECT OF
B 6012 2103 LOWE, DONALD R., NONGLACIAL VARVES IN LOWER MEMBER OF
B 6003 414† LOWE ORDOVICIAN ROCKS OF WESTERN PENNSYLVANIAN, GROWTH,
B 6307 1016 LOWELL, JAMES D., STRUCTURAL STYLES, THEIR PLATE-TECTONIC
B 6405 771* LOWER AND MIDDLE PALEOZOIC POTENTIAL OF PARADOX BASIN, ABST.
M 30 225† LOWER CAMBRIAN STRATA, LENA-TUNGUSKA PETROLEUM PROVINCE,
M 27 20 LOWER CHALK, ENGLAND, CRETACEOUS
M 27 34 LOWER CHALK, ENGLAND, CRETACEOUS
B 6107 1046 LOWER CHERRYVALE FORMATION, KANSAS
B 6004 654* LOWER COOK INLET, ALASKA, ABST., GEOLOGY OF, SOUTH AUGUSTINE
M 26 370 LOWER COOK INLET, DEPTH SOUNDER RECORD, ALASKA
B 6109 1557* LOWER CRETACEOUS CARBONATE ROCKS OF CENTRAL SOUTH TEXAS,
B 6303 547* LOWER CRETACEOUS CARBONATE SHELF IN SOUTHEASTERN ARIZONA
B 6109 1557* LOWER CRETACEOUS DEPOSITIONAL SYSTEMS, WEST TEXAS, ABST.

B 6405 680* LOWER CRETACEOUS LACUSTRINE SOURCE BEDS FROM EARLY RIFTING
B 6004 701* LOWER CRETACEOUS NEOCOMIAN, SANDSTONE SEQUENCE, MACKENZIE
B 6109 1555* LOWER CRETACEOUS SEDIMENTARYFACIES AND SEA-LEVEL CHANGES,
B 6109 1557* LOWER CRETACEOUS SHELF MARGIN IN NORTHERN MEXICO, ABST.
B 6405 723* LOWER CRETACEOUS SHELF STORM DEPOSITS, NORTH TEXAS, ABST.
B 6303 416* LOWER CRETACEOUS SHELF-MARGIN CARBONATE FACIES,
B 6303 430* LOWER CRETACEOUS VOLCANIC ROCKS CLOSE TO GULF COAST, SIERRA
S 10 62 LOWER CRETACEOUS, ORIGIN OF OIL SANDS, ALBERTA
B 6212 2381 LOWER DELTA PLAIN
M 27 49 LOWER GREENSAND, ENGLAND, CRETACEOUS
B 6104 579 LOWER LAGUNILLAS FIELD, VENEZUELA
B 6110 1583 LOWER MAGDALENA VALLEY, COLOMBIA
B 6008 1404* LOWER MESOZOIC AND UPPER PALEOZOIC PETROLEUM POTENTIAL OF
B 6004 716* LOWER MIDWAY GROUP, PALEOCENE, SOUTH-CENTRAL TEXAS, ABST.,
S 7 14 LOWER MISSISSIPPI FAN MAP, GULF OF MEXICO
B 6004 727* LOWER PENNSYLVANIAN, INDIAN NATIONS TURNPIKE, OUACHITA
B 6405 686 LOWER PERMIAN DEPOSITIONAL SYSTEMS UNCOMPAHGRE BASIN,
B 6303 461* LOWER PERMIAN FACIES TRACTS AND EVOLUTIO OF CARBONATE-SHELF
M 30 310 LOWER SAND, PRUDHOE BAY, KUPARUK RIVER FORMATION
S 7 141 LOWER SLOPE, NORTHERN GULF OF MEXICO, CONTINENTAL SLOPE
B 6406 939 LOWER STOCKTON COAL SEAM, WEST VIRGINIA
B 6004 711* LOWER SUNDANCE FORMATION OF BLACK HILLS REGION, ABST., LATE,
B 6211 2277 LOWER SUNDANCE FORMATION, WYOMING
B 6205 880* LOWER TRIASSIC HINGE LINE AND SHELF TAR SAND DEPOSITS, CENTRAL
B 6405 804* LOWER WILCOX SHELF EDGE IN TEXAS—RELATION TO GROWTH
B 6405 741 LOWRIE, ALLEN, MULTI-PARAMETER REGIONAL MAPPING AND
M 26 73 LOWSTAND OF SEA LEVEL, DEPOSITIONAL PATTERNS
M 26 92 LOWSTANDS, SEA LEVEL FLUCTUATIONS, GLOBAL
B 6209 1645 LU, RICHARD S., PLATE CONVERGENCE AND ACCRETION IN
B 6209 1661 LU TAO-BABUYAN RIDGE, STRUCTURE
S 5 74 LUCAS FORMATION, MICHIGAN
S 5 73* LUCAS FORMATION, DETROIT RIVER GROUP, DEVONIAN, MIDLAND,
S 5 79 LUCAS FORMATION, STRATIGRAPHY, DEVONIAN, DETROIT RIVER
M 24 121 LUCAS, PETER T., ALTAMONT-BLUEBELL--A MAJOR, NATURALLY
B 6004 682* LUCAYA, GRAND BAHAMAS ISLAND, ABST., REEF MORPHOLOGY AND
B 6410 1638 LUCAYAN LIMESTONE, GREAT BAHAMA BANK
B 6410 1641 LUCAYAN LIMESTONE, NEW QUATERNARY FORMATION, BAHAMAS,
B 6303 429 LUCCHESI, CELSO F., DEPOSITIONAL-DIAGENETIC HISTORY OF MACAE
B 6303 524 LUCIA, F. J., DOLOMITIZATION OF NORTHERN MICHIGAN NIAGARAN
B 6211 2220 LUCIA FAULT, CALIFORNIA
B 6211 2219 LUCIA FORMATION, CALIFORNIA
B 6206 968 LUCIA-PARTINGTON CANYON, CALIFORNIA
B 6108 1382* LUCIEN FIELD, NOBLE COUNTY, OKLAHOMA, ABST., ORDOVICIAN OIL,
S 5 95 LUDLOVIAN SERIES, CORRELATION, SILURIAN,
B 6405 802 LUDVIGSON, GREG A., MIDDLE CRETACEOUS, ALBIAN-TURONIAN,
B 6405 772 LUDWIG, K. R., ET AL, HISTORY OF SULFIDIZATION OF SOUTH TEXAS
B 6410 1718 LUDWIG, K. R., RUBIDIUM-STRONTIUM MINIMUM AGES OF
M 29 113 LUDWIG, W. J., STRUCTURE OF COLORADO BASIN AND
M 29 125 LUDWIG, W. J., STRUCTURE OF FALKLAND PLATEAU AND OFFSHORE
B 6405 742 LUDWIG, WILLIAM J., SEISMIC STRATIGRAPH AND STRUCTURE OF
B 6008 1265 LUECK, EVERETT W., DEVELOPMENTS IN LOUISIAN GULF COAST IN 1975
B 6211 2221 LUISIAN BOUNDARY PALEOGEOGRAPHY, CALIFORNIA, RELIZIAN-
S 11 24 LUISIAN STAGE
S 11 66 LUISIAN STAGE
B 6211 2246 LUISIAN STAGE, CALIFORNIA
B 6202 259 LUISIAN STAGE, CALIFORNIA
S 11 26 LUISIAN STAGE, ZONATION
B 6105 793 LUKAS, KAREN J., COMPARISON OF ENDOLITH FLORAS FROM HOLOCENE-
B 6301 34 LULING FAULT ZONES, TEXAS
B 6405 742 LULMAN, P. S., GEOLOGY OF SYNCRUDE CANADA LIMITED MINE SITE,
B 6109 1483 LULU CRUISE 33, SEPTEMBER 1970, RV
M 25 351 LUM, D., DEVELOPING DIKE-COMPLEX AND BASAL-LENS WATER SUPPLIES
B 6303 431 LUMSDEN, DAVID N., DOLOMITE NONSTOICHIOMETRY, ITS RELATION TO
B 6303 488 LUMSDEN, DAVID N., ERROR IN X-RAY DIFFRACTION ESTIMATES O
B 6209 1760 LUND, JEFFREY W., PRE-PLATFORM EXPLORATION OF HIGH ISLAND
B 6303 488 LUNDEGARD, PAUL D., UPPER DEVONIAN TURBIDITES OF CENTRAL AND
B 6203 565 LUNDELL, L. L., STROMATOLITES IN EOCENE GREEN RIVER FORMATION,
B 6008 1412 LUNDELL, LESLIE L., DEPOSITIONAL ENVIRONMENNSTS OF OIL SHALE,
B 6307 1144 LUNDIN, ROBERT F., REVIEW OF STRATIGRAPHIC INDEX OF BRITISH
B 6303 477* LUSITANIAN BASIN, PORTUGAL, ABST., OIL GENERATION AND,
B 6004 653 LUSK, J. D., SET-THEORETIC INFORMATION RETRIEVAL FROM
S 6 134 LUTETIAN BOUNDARIES, GLAUCONITES, EUROPE
S 6 220 LUTETIAN STAGE DEFINED, EUROPE, EOCENE
S 6 252 LUTETIAN STAGE, ISOTOPIC DATES, PALEOGENE, EOCENE
B 6104 562 LUTITAS ARENOSAS STRATIGRAPHIC UNIT, CHILE
B 6104 562 LUTITAS GRIS VERDOSAS STRATIGRAPHIC UNIT, CHILE
B 6109 1554 LUTTRELL, P. L., CARBONATE FACIES DISTRIBUTION ANDDIAGENESIS
B 6004 694 LUTTRELL, PAMELA E., CARBONATE DIAGENESIS AND FACIES
B 6303 432 LUTZ, RICHARD A., PYRITIZATION IN SHELLS OF LIVING BIVALVES, ABST.
M 25 436 LUZON, AGNO BATHOLITH
B 6209 1662 LUZON ARC, EAST MARGIN, STRUCTURE
B 6209 1669 LUZON ARC WITH ASIAN CONTINENTAL MARGIN, CONVERGENCE OF
M 25 435 LUZON, BAGUIO MINERAL DISTRICT, COPPER
M 25 436 LUZON, CHROMITE, REFRACTORY
M 25 436 LUZON, DIZON COPPER PROSPECT
M 25 436 LUZON, PHILEX MINE
B 6303 456* LUZON, PHILIPPINES, ABST., PETROLOGY, SEDIMENTOLOGY, AND
B 6209 1646 LUZON REGION, ACTIVE AND EXTINCT PLATES, TAIWAN-
B 6209 1645† LUZON REGION, PLATE, CONVERGENCE AND ACCRETION IN, TAIWAN-
B 6209 1646 LUZON REGION, VOLCANOES, TAIWAN-
M 25 436 LUZON, STO. NINO MINE
B 6209 1646 LUZON TROUGH, NORTH
B 6209 1646 LUZON TROUGH, WEST
B 6303 546 LYDAY, JOHN REED, CONTROLS ON LATE CRETACEOUS-PALEOCENE
B 6410 1660 LYNCH 73-1 AND 73-2, U.S. NAVAL OCEANOGRAPHIC OFFICE CRUISES
B 6405 742 LYNN, HELOISE BLOXSOM, DEPTH MIGRATION AND INTERPRETATION OF
B 6004 700 LYNTS, G. W., HYDROCARBON POTENTIAL OF NORTHWESTERN BAHAMA
B 6204 693 LYNTS, GEORGE W., CHARACTERISTICS OF DEEP BAHAMA CHANNELS IN
B 6007 1037 LYNTS, GEORGE W., STRATIGRAPHY AND STRUCTURE OF NORTHEAST
B 6410 1615 LYONS FORMATION, DENVER BASIN
B 6405 743 LYONS, PAUL C., RESIN RODS AND WOODY ROD-LIKE STRUCTURES IN
B 6410 1615 LYTLE FORMATION, DENVER BASIN
B 6206 1092 LYTLE, J., INSTRUMENTATION FOR EVALUATION OF IN-SITU RECOVERY
B 6203 529 LYTLE, P. J., INSTRUMENTATION FOR EVALUATION OF IN SITU
B 6008 1288 LYTLE, WILLIAM S., OIL AND GAS DEVELOPMENTS IN MARYLAND, OHIO,
B 6108 1269 LYTLE, WILLIAM S., OIL AND GAS DEVELOPMENTS IN MARYLAND, OHIO,
M 30 577 MAASTRICHTIAN, BATANGA SANDSTONE, CRETACEOUS
B 6104 595 MABRUK FIELD, LIBYA
B 6104 594 MABRUK FIELD, LIBYA
B 6303 429* MACAE CARBONATE RESERVOIRS, ALBIAN-CENOMANIAN, CAMPOS BASIN,
B 6109 1480 MACAE FORMATION, BRAZIL
M 30 329 MACAE FORMATION, BRAZIL, CRETACEOUS
M 30 331 MACAE FORMATION, NAMORADO SANDSTONE, BRAZIL
B 6410 1725 MACCAMMON AND BRIAN HITCHON, MORAN, W. R., REVIEW OF
B 6304 652 MACCRADY FORMATION, VIRGINIA
B 6403 445 MACDONALD, KEITH B., LATE PLEISTOCENE-HOLOCENE SEDIMENTARY
B 6203 559 MACILVAINE, JOSEPH C., SEDIMENTARY PROCESSES ON CONTINENTAL
B 6007 1054 MACINTYRE, I. G., EVOLUTION OF MODERN CARIBBEAN FRINGING REEF,
B 6105 810 MACINTYRE, IAN G., POROSITY DISTRIBUTION THROUGHOUT MODERN
B 6303 489 MACINTYRE, IAN G., PSEUDOSTALACTITES FROM SUBMARINE CAVE NEAR
B 6303 456 MACK, L E., PETROLOGY, SEDIMENTOLOGY, AND PETROLEUM POTENTIAL
B 6406 954 MACKENZIE, D. B., REVIEW OF LITHOSTRATIGRAPHIC ANALYSIS OF
B 6004 701* MACKENZIE DELTA AND RICHARDSON MOUNTAINS, NORTHWEST
M 24 53 MACKENZIE DELTA AREA, NORTHWEST TERRITORIES
B 6105 810 MACKENZIE, FRED, T., MANS IMPACT ON GLOBAL SEDIMENTARY CYCLES
B 6405 764 MACKENZIE, FRED T., PHANEROZOIC CARBONATE DIAGENESIS—A NEW
B 6203 532* MACKENZIE MOUNTAINS, NORTHWEST TERRITORIES, CANADA, ABST.,
B 6405 746* MACKEREL FIELD, GIPPSLAND BASIN, AUSTRALIA, ABST., DEFINITION
B 6303 496* MACKEREL FIELD, GIPPSLAND BASIN, AUSTRALIA, ABST., MASSIVE
B 6207 1226* MACKEREL FIELD, GIPPSLAND BASIN, AUSTRALIA, ABST., SEISMIC AND
B 6105 761 MACKEVETT, E. M., JR., RELATIONS BETWEEN TRIASSIC CARBONATE
B 6010 1657 MACKINNON, J. L., ON DEVELOPMENTS IN URUGUAY, 1975
B 6211 2247 MACOMA SHALE, CALIFORNIA
S 5 55 MACOMB COUNTY, LOCATION, MICHIGAN,
B 6211 2243 MACPHERSON, BRUCE A., SEDIMENTATION AND TRAPPING MECHANISM
B 6002 273 MACQUEEN, ROGER W., NOTE 44--APPLICATION FOR ADDITION TO CODE
B 6107 1083 MACQUEEN, ROGER W., NOTE 47—APPLICATION FOR AMENDMENT OF
B 6405 743 MACQUOWN, WILLIAM C., COMPARISON O SHALLOW-MARINE SHELF
B 6303 487 MACQUOWN, WILLIAM C., CONTROLS OF SEDIMENTATIO OF LOWER
B 6305 844 MACQUOWN, WILLIAM C., PETROLEUM PRODUCTION FROM BASAL
B 6203 553* MACROBORING ORGANISMS, SOUTHERN ONTARIO, ABST., SILURIAN
S 6 278 MACROFOSSIL ZONES, CRETACEOUS, CENOMANIAN
S 6 278 MACROFOSSIL ZONES, CRETACEOUS, TURONIAN
B 6004 697* MACROTIDAL ENVIRONMENT, COBEQUID BAY, NOVA SCOTIA, ABST.,
M 25 552 MACUCHI MINE, COPPER, ECUADOR
S 4 231 MACURDA, DONALD B., JR., CRINOIDS OF WEST INDIAN CORAL REEFS
M 30 360 MACUSPANA BASIN, CHILAPILLA FIELD, MEXICO
M 30 361 MACUSPANA BASIN, HORMIGUERO FIELD, MEXICO
M 30 360 MACUSPANA BASIN, JOSE COLOMO FIELD, MEXICO
B 6111 1994 MADAGASCAR BASIN, POSSIBLE DIAPIRS
B 6110 1749 MADAGASCAR, MALAGASY REPUBLIC, DEVELOPMENTS, 1976
B 6210 1848 MADAGASCAR, MALAGASY REPUBLIC, DEVELOPMENTS, 1977
B 6310 1693 MADAGASCAR, MALAGASY REPUBLIC, DEVELOPMENTS, 1978
B 6411 1789 MADAGASCAR, MALAGASY REPUBLIC, DEVELOPMENTS, 1979
B 6401 59 MADEC, M., ROLE OF MINERAL MATRIX IN KEROGEN
B 6111 2005 MADEIRA ABYSSAL PLAIN, CAPE VERDE-
B 6109 1515 MADERA FORMATION, NEW MEXICO
B 6107 975 MADIELA FORMATION, GABON
B 6309 1479 MADISON FORMATION, TEXTURE AND COMPOSITION DATA
B 6405 674* MADISON GROUP AND RED RIVER FORMATION, POWDER RIVER BASIN,

B 6405 757* MADISON GROUP, DISTURBED BELT, MONTANA, ABST., DISTRIBUTION,
B 6205 887* MADISON LIMESTONE, MISSISSIPPIAN, NORTHERN MEDICINE BOW
B 6305 822* MADISON RESERVOIRS IN BIGHORN BASIN, WYOMING, ABST., BOREHOLE
B 6210 1987 MADRID, A. P., ON DEVELOPMENTS IN PHILIPPINES, 1977
B 6310 1826 MADRID, A. P., ON DEVELOPMENTS IN PHILIPPINES, 1978
B 6401 68 MAESTRICHT AREA, SAMPLED SECTION, NETHERLANDS
B 6209 1767* MAESTRICHTIAN, AND OF PINE BARREN MEMBER OF CLAYTON
B 6206 1015 MAESTRICHTIAN, ARCTIC, CANADA, MESOZOIC, APTIAN-, CAMPANIAN TO
B 6206 1015 MAESTRICHTIAN, ARCTIC, CANADA, MESOZOIC TO, CENOZOIC,
B 6201 50 MAESTRICHTIAN, AUSTRALIA, SENONIAN-
B 6211 2234 MAESTRICHTIAN, BAY OF BISCAY AREA
B 6303 422* MAESTRICHTIAN, CAPE SEBASTIAN SANDSTONE SOUTHWESTERN
B 6107 988 MAESTRICHTIAN CHALK, DENMARK
M 27 19 MAESTRICHTIAN CHALK, DENMARK
B 6107 988 MAESTRICHTIAN CHALK, NETHERLANDS
B 6205 885* MAESTRICHTIAN CONGLOMERATES IN SOUTHWESTERN PICEANCE CREEK
S 4 88 MAESTRICHTIAN, JAMAICA, GUINEA CORN FORMATION
S 4 88 MAESTRICHTIAN, JAMAICA, SLIPPERY ROCK FORMATION
S 4 88 MAESTRICHTIAN, JAMAICA, SUMMERFIELD FORMATION
B 6305 831* MAESTRICHTIAN, OF SOUTHERN SASKATCHEWAN, ABST., RELATIONS
B 6206 1019 MAFIC INTRUSIVE AND VOLCANIC ROCKS, CANADA, SVERDRUP BASIN
M 30 345 MAGALLANES FIELD, MEXICO, SALINE BASIN
B 6203 538 MAGARA, **KINJI**, PRIMARY MIGRATION AGENTS, ABST.
S 10 33 MAGARA, KINJI, AGENTS FOR PRIMARY HYDROCARBON MIGRATION: A
B 6412 2108 MAGARA, KINJI, EVIDENCES OF PRIMARY OIL MIGRATION
B 6004 554 MAGARA, KINJI, THICKNESS OF REMOVED SEDIMENTARY ROCKS,
B 6305 812 MAGARA, KINJI, THICKNESS OF REMOVED SEDIMENTARY ROCKS.
B 6004 543 MAGARA, KINJI, WATER EXPULSION FROM CLASTIC SEDIMENTS DURING
B 6304 682 MAGARA, MOMPER, JAMES A., REVIEW OF, COMPACTION AND FLUID,
B 6105 810 MAGARITZ, M., CARBON AND OXYGEN ISOTOPIC COMPOSITIONS AS
B 6405 743 MAGATHAN, ESTHER R., CONVEYER-BELT TECTONICS AND GEOLOGIC
B 6203 365 MAGDALENA BORDERLAND, BAJA CALIFORNIA
M 29 126 MAGELLAN BASIN, GEOLOGIC MAP, ARGENTINA
M 29 129 MAGELLAN BASIN, ONSHORE, CRETACEOUS
B 6104 558 MAGELLAN BASIN, PETROLEUM, CHILE AND, ARGENTINA
B 6104 558† MAGELLAN BASIN, SOUTH AMERICA, REGIONALIZED, VARIABLES FOR
M 29 126 MAGELLAN BASIN, SPRINGHILL FORMATION, PETROLEUM PRODUCTION
B 6004 725* MAGHEMITE-HEMATITE TRANSFORMATION AND ITS POSSIBLE
S 6 154 MAGISAWA FORMATIONS, JAPAN
B 6207 1212* MAGMA ENERGY, ABST., UTILIZATION OF
S 3 36 MAGMA, OCEAN BASINS, MANTLE-DERIVED
M 25 124† MAGMA TAP, THE ULTIMATE GEOTHERMAL ENERGY
S 3 36 MAGMAS, CONTINENTS, MANTLE
M 25 529 MAGMAS FORMED ON, BENIOFF ZONE, CALC-ALKALINE
B 6301 3 MAGMATIC ARC, DEFINITION
B 6312 2175 MAGMATIC ARC PROVENANCES
M 25 505 MAGMATIC DEPOSITS OF METALS, CANADIAN, CORDILLERA
S 3 44 MAGMATION
S 3 87 MAGMATION, ANATEXIS AND
S 3 23 MAGMATION, CONVERSION OF SOLID-EARTH MATERIAL TO MAGMA
M 25 467 MAGMATISM, USSR, NORTHEAST
B 6207 1217* MAGMA—POTENTIAL ENERGY RESOURCE, ABST., FUELS FROM
S 4 379 MAGNESIAN CALCITE THROUGH METEORIC FRESH WATERS,
B 6305 845* MAGNETI AND GAMMA-RAY SPECTROMETER DATA FOR URANIUM
B 6306 939 MAGNETIC ANOMALIES, AMAZON MARGIN
B 6403 299 MAGNETIC ANOMALIES, CONTINENTAL, MARGINS OF WORLD CAUSED
M 29 304 MAGNETIC ANOMALIES, MEDITERRANEAN, CYRENIAN PLATEAU
M 29 304 MAGNETIC ANOMALIES, MEDITERRANEAN, ERATHOSTHENES PLATEAU
M 29 304 MAGNETIC ANOMALIES, MEDITERRANEAN, SYRTE GULF
B 6106 881 MAGNETIC ANOMALIES OF EGYPTIAN CONTINENTAL MARGIN AND
B 6209 1654 MAGNETIC ANOMALIES, SOUTH, CHINA BASIN
B 6205 793 MAGNETIC ANOMALY, BLAKE SPUR
B 6410 1659 MAGNETIC ANOMALY, BLAKE SPUR
M 29 88 MAGNETIC ANOMALY CONTOURS, ATLANTIC MARGIN
M 29 102 MAGNETIC ANOMALY CONTOURS, BALTIMORE CANYON
M 29 106 MAGNETIC ANOMALY CONTOURS, CAROLINA PLATFORM
M 29 93 MAGNETIC ANOMALY CONTOURS, GEORGES BANK BASIN
M 29 97 MAGNETIC ANOMALY CONTOURS, LONG ISLAND PLATFORM
B 6410 1659 MAGNETIC ANOMALY, EAST COAST
B 6007 1115 MAGNETIC ANOMALY, GAMMA CONTOUR OF, EAST COAST
S 6 58 MAGNETIC ANOMALY NUMBER, DESCRIBED
M 29 254 MAGNETIC BASEMENT, GEOLOGIC MAP, GUATEMALA
M 29 91 MAGNETIC DATA, ANALYSIS METHODS
B 6002 172 MAGNETIC DATA, BERINGIAN SHELF
B 6303 447* MAGNETIC DATA, BLIGH WATER, FIJI, ABST., DETERMINATION OF DEPTH
M 29 91 MAGNETIC DATA, DEPTH-TO-SOURCE ESTIMATES
M 29 94 MAGNETIC DEPTH CONTOURS, GEORGES BANK BASIN
M 29 90 MAGNETIC DEPTH ESTIMATES, ATLANTIC MARGIN
M 29 103 MAGNETIC DEPTH ESTIMATES, BALTIMORE CANYON
M 29 107 MAGNETIC DEPTH ESTIMATES, CAROLINA PLATFORM
M 29 98 MAGNETIC DEPTH ESTIMATES, LONG ISLAND PLATFORM
B 6212 2431 MAGNETIC INTENSITY, KANSAS
M 29 155 MAGNETIC LINEATIONS, AUSTRALIA, MARGIN
S 7 82 MAGNETIC LINEATIONS, NORTH ATLANTIC, SALT DIAPIRS
B 6108 1391* MAGNETIC MEASUREMENTS IN URANIUM EXPLORATION, ABST.,
S 6 57 MAGNETIC POLARITY, STRATIGRAPHIC CORRELATION
S 6 57† MAGNETIC POLARITY TIME SCALE--PROSPECTS AND POSSIBILITIES IN
S 9 16 MAGNETIC POLE POSITIONS, PALEOZOIC, EURASIA
M 29 195 MAGNETIC PROFILE, BURMA, ANDAMAN SEA
B 6007 1050 MAGNETIC PROFILES, NORTHWEST AND, NORTHEAST, PROVIDENCE
B 6202 249 MAGNETIC PROFILES, SOUTHERN, CALIFORNIA
B 6410 1659 MAGNETIC QUIET ZONE, BASEMENT AND TECTONIC MAP, JURASSIC
B 6410 1658† MAGNETIC QUIET ZONE, JURASSIC, SEISMIC STRATIGRAPH AND,
M 29 151† MAGNETIC QUIET ZONE ON THE MARGIN SOUTH OF AUSTRALIA, THE,
B 6409 1662 MAGNETIC QUIET ZONE, PROFILE, WESTERN, ATLANTIC, JURASSIC INNER
M 29 151 MAGNETIC QUIET ZONE, SEE QUIET ZONE
S 6 63 MAGNETIC REVERSAL PATTERNS, PHANEROZOIC
S 6 59 MAGNETIC REVERSAL, USED IN GEOCHRONOLOGY
S 6 62 MAGNETIC REVERSALS, EXAMPLE, PHANEROZOIC
B 6203 548* MAGNETIC SEPARATION, ABST., COAL CLEANING, DESULFURIZATION, BY
S 6 59 MAGNETIC ZONES, CORRELATIONS, NEOGENE
B 6011 1985† MAGNETIC-REVERSAL STRATIGRAPHY OF PLIOCENE-PLEISTOCENE
B 6007 1085 MAGNETICS, ATLANTIC BASIN
M 29 94 MAGNETICS, ATLANTIC MARGIN, COMPARISON OF
B 6303 476* MAGNETICS DATA, ABST., INTERACTIVE MODELING AND
B 6007 1084 MAGNETICS, MID-ATLANTIC RIDGE
M 29 133 MAGNETICS, SCOTIA SEA BASIN, SEAFLOOR SPREADING
M 29 95 MAGNETICS, SEISMICS, ATLANTIC MARGIN
M 28 44 MAGNETITE GRAINS
B 6302 245† MAGNETITE OVER OIL FIELDS, AEROMAGNETIC DETECTION OF,
B 6309 1538† MAGNETITE OVER OIL FIELDS. DISCUSSION, AEROMAGNETIC DETECTION
M 25 426 MAGNETITE-PYRITE, TASMANIA, SAVAGE RIVER
B 6002 275 MAGNETOPOLARITY
B 6002 273† MAGNETOSTRATIGRAPHIC UNITS, NOTE 44--, APPLICATION FOR
B 6303 501* MAGNETOSTRATIGRAPHY OF TUNISIA, ABST., STRATIGRAPHIC
S 6 57† MAGNETOSTRATIGRAPHY, THE, MAGNETIC, POLARITY TIME
B 6211 2355* MAGNETOTELLURIC EXPLORATION—AN UPDATE, ABST.
B 6211 2362* MAGNETOTELLURIC GEOPHYSICAL EXPLORATION METHOD—A REVIEW,
S 6 58 MAGNETOZONE, LOCAL MAGNETIC UNIT
M 26 56 MAGNITUDE AND BOUNDARIES, DEPOSITIONAL SEQUENCES
B 6004 668* MAGNITUDE AND TIMING OF DISPLACEMENT ON SA ANDREAS FAULT IN
B 6304 691* MAGNITUDE AND VARIABILITY OF DISEQUILIBRIUM IN SAN ANTONIO
B 6207 1225 MAGOON, L. B., PETROLEUM GEOLOGY OF COOK INLET BASIN, ALASKA,
B 6408 1131 MAGOON, LESLIE B., BIOGENIC AND THERMOGENIC ORIGINS OF
B 6012 2185 MAGOON, LESLIE B., GEOLOGIC FRAMEWORK OF LOWER COOK INLET,
B 6203 373 MAGOON, LESLIE B., GEOLOGIC FRAMEWORK OF LOWER COOK INLET,
B 6408 1259 MAGOON, LESLIE B., NONMARINE UPPER CRETACEOUS ROCKS, COOK
B 6103 466 MAGOON, LESLIE B., ORIGIN OF COOK INLET OIL, ABST.
B 6405 744 MAGOON, LESLIE B., TWO OIL TYPES ON NORTH SLOPE OF ALASKA—
S 4 20 MAGUEYES LIMESTONE, PUERTO RICO
M 30 401 MAHAKAM DELTA, BEKAPAI FIELD, MIOCENE
B 6209 1684† MAHAKAM DELTA, BORNEO, ORGANIC SEDIMENTATION AND, GENESIS
M 30 408 MAHAKAM DELTA, DEPOSITIONAL ENVIRONMENTS, INDONESIA
M 30 401 MAHAKAM DELTA, HANDIL FIELD, MIOCENE
B 6209 1685 MAHAKAM DELTA, HYDROCARBON ACCUMULATIONS, KALIMANTAN
B 6209 1685 MAHAKAM DELTA, MORPHOLOGY, KALIMANTAN
B 6405 744 MAHAN, THOMAS K., DEPOSITIONAL ENVIRONMENT OF LOWER
B 6002 279 MAHANTANGO FORMATION, PENNSYLVANIA
B 6303 423 MAHER, C. E., INTISAR D OIL FIELD, LIBYA, ABST.
M 30 543 MAHER, C. E., INTISAR D OIL FIELD,LIBYA
M 30 131 MAHER, C. E., PIPER OIL FIELD
B 6303 489 MAHER, C. E., PIPER OIL FIELD, ABST.
M 25 1 MAHER, JOHN C., EDITED BY. CIRCUM-PACIFIC ENERGY AND MINERAL
M 25 1 MAHER, JOHN C., FOREWORD, CIRCUM-PACIFIC ENERGY AND MINERAL
M 25 1 MAHER, JOHN C., INTRODUCTION, CIRCUM-PACIFIC ENERGY AND MINERAL
B 6008 1402 MAHER, PATRICK D., GEOLOGY OF PINEVIEW FIELD AREA, SUMMI
M 25 214 MAHOENUI FORMATION, NEW ZEALAND
B 6306 955 MAIDEN POINT FORMATION, NEWFOUNDLAND
B 6105 835* MAIN COAST, ABST., PHOTOGRAPHIC, MAPPING OF RECENT COASTAL
B 6203 436 MAIN GRABEN, NORTH SEA
B 6002 239 MAIN PASS BLOCK 41 FIELD, LOUISIANA
B 6104 587 MAIN PASS BLOCK FIELDS, LOUISIANA
S 4 88 MAIN RIDGE VOLCANICS, CAMPANIAN, JAMAICA, BULLHEAD
B 6203 536* MAINE, ABST., CARBONATE CATCHMENT AND EXCHANGE, BASINS IN
B 6303 471* MAINE, ABST., ORIGIN OF, HILLS BEACH-FLETCHER NECK TOMBOLO
B 6108 1346 MAIO, CHARLES G., OIL AND GAS DEVELOPMENTS IN NORTHERN
B 6401 32 MAIOLICA LIMESTONE, ITALY
M 24 82 MAJEAU LAKE FORMATION, ALBERTA

M 25 526† MAJOR CHEMICAL CHARACTERISTICS OF TERTIARY AND QUATERNARY
B 6405 686* MAJOR PETROLEUM-IMPREGNATED ROCK DEPOSITS OF WESTERN
B 6311 2116* MAJOR STRUCTURAL AND STRATIGRAPHIC FEATURES OF ANADARKO
M 29 329† MAJOR STRUCTURAL ELEMENTS AND EVOLUTION OF NORTHWESTERN
B 6306 967† MAJOR SUBDIVISIONS AND DEPOSITIONAL ENVIRONMENTS OF NUBIA
B 6203 561* MAJOR URANIUM DISCOVERY IN FRONTIER AREA—ANDERSON MINE,
B 6304 621† MAJOR URANIUM DISCOVERY IN VOLCANICLASTIC SEDIMENTS, BASIN
B 6303 430* MAKING PUBLIC LANDS PRIVATE, ABST.
B 6409 1548 MAKIYAMA, GROSHONG, RICHARD H., JR., REVIEW OF,
B 6203 571 MAKURATH, J. H., DELTAIC COMPONENTS OF VALMY AND VININI
B 6203 538 MAKURATH, J. H., ELDER SANDSTONE—DELTAIC SEDIMENTS OF ROBERTS
B 6105 840 MAKURATH, JOSEPH H., GOODWIN LIMESTONE—LOWER ORDOVICIAN
B 6105 764 MAKURATH, JOSEPH H., STORM-GENERATED SHELL RIDGE ON OPEN
B 6202 259 MALAGA MUDSTONE MEMBER, CALIFORNIA
B 6111 1951 MALAGASY, BEMOLANGA HEAVY OIL DEPOSIT
B 6110 1749 MALAGASY REPUBLIC, DEVELOPMENTS, 1976, MADAGASCAR
B 6210 1848 MALAGASY REPUBLIC, DEVELOPMENTS, 1977, MADAGASCAR
B 6310 1693 MALAGASY REPUBLIC, DEVELOPMENTS, 1978, MADAGASCAR
B 6411 1789 MALAGASY REPUBLIC, DEVELOPMENTS, 1979, MADAGASCAR
B 6004 654 MALAHOFF, A., SEDIMENT PROCESSES CONTROLLED BY BOTTOM
B 6303 444 MALAHOFF, ALEXANDER, DISTRIBUTION, MORPHOLOGY, MECHANISMS,
M 25 479 MALAITA, ULTRAMAFIC VOLCANIC ROCKS
B 6004 716* MALASPINA FORELAND, GULF OF ALASKA, ABST., LITTORAL PROCESSES
B 6204 635 MALASPINA GLACIER, ALASKA
B 6212 2412† MALASPINA GLACIER, GULF OF ALASKA, HOLOCENE, SLUMP ON,
B 6212 2415 MALASPINA GLACIER SLUMP, SEISMIC DATA, ALASKA, ICY BAY-
B 6210 1849 MALAWI, RHODESIA, RWANDA, SWAZILAND, UPPER VOLTA, ZAMBIA,
M 25 252 MALAY BASIN, THAILAND
M 25 431 MALAYA GEOSYNCLINE, THAI-
B 6110 1837 MALAYSIA, 1976, NIJHUIS, H. J., ON DEVELOPMENTS IN
B 6210 1986 MALAYSIA, 1977, ARMITAGE, J., ON DEVELOPMENTS IN, PENINSULAR
B 6310 1823 MALAYSIA, ARMITAGE, J. H., AND ROBERT, DE RUITER, ON
B 6210 1985 MALAYSIA, DEVELOPMENTS, 1977, EAST
B 6411 1889 MALAYSIA, DEVELOPMENTS, 1979
M 30 424 MALAYSIA, DISCOVERIES AND PRODUCTION, LOCATION
B 6210 1990 MALAYSIA, EAST, PRODUCTION, 1975-1977
B 6310 1829 MALAYSIA, EAST, PRODUCTION, 1977-1978
B 6411 1896 MALAYSIA EAST, SABAH, PRODUCTION, 1977-1979
B 6411 1896 MALAYSIA EAST, SARAWAK, PRODUCTION, 1977-1979
M 30 429 MALAYSIA, NORTHWEST BORNEO GEOSYNCLINE
M 30 430 MALAYSIA, NORTHWEST BORNEO OROGEN
B 6303 526* MALAYSIA OVER PAST DECADE, ABST., EXPLORATION IN, EAST
M 30 423† MALAYSIA OVER THE PAST DECADE, EXPLORATION IN, EAST
B 6310 1829 MALAYSIA, PENINSULAR, PRODUCTION, 1977-1978
B 6411 1896 MALAYSIA, PENINSULAR, PRODUCTION, 1977-1979
B 6110 1842 MALAYSIA, PRODUCTION, 1975-1976
M 30 439 MALAYSIA, RESERVOIR CHARACTERISTICS, MIOCENE
M 30 436 MALAYSIA, SARAWAK, CENTRAL LUCONIA FIELD
M 30 433 MALAYSIA, SARAWAK SHELF, RECENT SEDIMENTATION
B 6402 219 MALAYSIA, SERIA FIELD
M 30 425 MALAYSIAN SHELF, EVOLUTION OF TECTONIC FRAMEWORK
B 6405 787 MALDANADO, ANDRES, DEPOSITIONAL MODELS FOR FINE-GRAINED
B 6110 1838 MALDIVE ISLANDS, DEVELOPMENTS, 1976
B 6210 1986 MALDIVE ISLANDS, DEVELOPMENTS, 1977
B 6310 1824 MALDIVE ISLANDS, DEVELOPMENTS, 1978
B 6411 1890 MALDIVE ISLANDS, DEVELOPMENTS, 1979
S 4 85 MALDON INLIER, JAMAICA
B 6105 832 MALDONADO, ANDRES, MUD-FACIES DEFINITION FOR
B 6409 1562 MALEK-ASLANI, M, ENVIRONMENTAL AND DIAGENETIC CONTROLS OF
B 6303 489 MALEK-ASLANI, M., ENVIRONMENTAL AND DIAGENETIC CONTROLS O
B 6405 744 MALEK-ASLANI, M., ENVIRONMENTAL AND DIAGENETIC CONTROLS OF
B 6109 1541 MALEK-ASLANI, M., PLATE TECTONICS AND SEDIMENTARY CYCLES IN
B 6110 1749 MALI, DEVELOPMENTS, 1976
B 6210 1848 MALI, DEVELOPMENTS, 1977
B 6310 1693 MALI, DEVELOPMENTS, 1978
B 6411 1789 MALI, DEVELOPMENTS, 1979
B 6203 359 MALIBU COAST FAULT, CALIFORNIA
B 6311 2024 MALITA GRABEN, AUSTRALIA
B 6105 779 MALLETT, C. W., PERMIAN INFILLED SUBGLACIAL GULLY FROM BACCHUS
B 6004 694 MALLORY, WILLIAM W., USGS-AAPG DIGITIZED OIL AND GAS FIELD MA
B 6104 597 MALMO FIELD, ALBERTA
B 6104 596 MALMO FIELD, ALBERTA
B 6109 1477 MALOMBE FIELD, BRAZIL
M 30 531 MALOSSA FIELD, BRUNTINO MARLS, CRETACEOUS
M 30 530 MALOSSA FIELD, CONSESIO FORMATION, JURASSIC
B 6303 446* MALOSSA FIELD, DEEP DISCOVERY IN PO VALLEY, ITALY, ABST.
M 30 529 MALOSSA FIELD, ESINO FORMATION, TRIASSIC
M 30 525 MALOSSA FIELD, GEOLOGIC SETTING, ITALY
M 30 537 MALOSSA FIELD, HAUPT DOLOMITE FORMATION, TRIASSIC
M 30 34 MALOSSA FIELD, LANDSAT IMAGE, ITALY
M 30 526 MALOSSA FIELD, LOCATION, ITALY
M 30 525 MALOSSA FIELD, PO BASIN, MESOZOIC
M 30 537 MALOSSA FIELD, RESERVOIR CHARACTERISTICS, ITALY
M 30 530 MALOSSA FIELD, S. GIOVANNI BIANCO FORMATION, TRIASSIC
M 30 531 MALOSSA FIELD, SASS DE LA LUNA FORMATION, CRETACEOUS
M 30 531 MALOSSA FIELD, SELCIFERO FORMATION, JURASSIC
M 30 530 MALOSSA FIELD, ZANDOBBIO FORMATION, JURASSIC
M 30 525† MALOSSA FIELD--A DEEP DISCOVERY IN THE PO VALLEY, ITALY
B 6401 32 MALOSSA GIANT GAS-CONDENSATE FIELD, LANDSAT DATA, ITALY
B 6401 32 MALOSSA GIANT GAS-CONDENSATE FIELD, LANDSAT, ITALY
B 6110 1645 MALTA, DEVELOPMENTS, 1976
B 6406 858 MALTA ESCARPMENT, AFRICA
B 6405 682* MALTESE ISLANDS, ABST., DIAGENESIS OF LOWER, CORALLINE
M 29 126 MALVINAS BASIN, GEOLOGIC MAP, ARGENTINA
M 29 131 MALVINAS BASIN, STRUCTURE SECTION, ARGENTINA
M 29 133 MALVINAS BASIN, TERTIARY, LOW-VELOCITY SEDIMENTS
B 6206 960 MAM TOR SANDSTONES, ENGLAND
M 25 144 MAMBUCAL GEOTHERMAL PROSPECT, PHILIPPINES
B 6006 962 MAMET, B. L., CARBONIFEROUS SECTION AND UNCONFORMITY AT
S 6 327 MAMET, B. L., THE MISSISSIPPIAN-PENNSYLVANIAN BOUNDARY
B 6205 879 MAMET, BERNARD L., CARBONIFEROUS, MISSISSIPPIAN, MICROFACIES
B 6405 671 MAMET, BERNARD L., MISSISSIPPIAN SYSTEM OF NEW MEXICO AND
B 6203 535* MAMMOTH-BEARING SINKHOLE DEPOSITS, HOT SPRINGS, SOUTH
B 6202 296 MAMU FORMATION, NIGERIA
B 6212 2468 MANABI BASIN, ECUADOR
B 6009 1617* MANAGEMENT IN LOUISIANA, ABST., THE, OIL, INDUSTRY AND, COASTAL
M 25 222 MANAIA FAULT, NEW ZEALAND
S 10 210 MANCHESTER FIELD, CLAY CONVERSION, LOUISIANA, CALCASIEU PARISH
S 10 210 MANCHESTER FIELD, KEROGEN CONVERSION, LOUISIANA, CALCASIEU
B 6002 227 MANCHESTER FIELD, LOUISIANA
S 10 212 MANCHESTER FIELD, SALINITY WITH DEPTH, LOUISIANA, CALCASIEU
B 6105 811 MANCINI, ERNEST A., ALASKA PENINSULA LATE CRETACEOUS FORE-ARC
B 6109 1541 MANCINI, ERNEST A., DEPOSITIONAL ENVIRONMENT OF GRAYSON
B 6309 1607 MANCINI, ERNEST A., EOCENE-OLIGOCENE BOUNDARY IN SOUTHWEST
B 6209 1761 MANCINI, ERNEST A., FORAMINIFERAL PALEOECOLOGY OF GRAYSON
B 6409 1565 MANCINI, ERNEST A., LATE PALEOCENE PLANKTONIC FORAMINIFERAL
B 6303 490 MANCINI, ERNEST A., MODELS FOR INTERPRETATION OF MICROMORPH
B 6409 1563 MANCINI, ERNEST A., REGIONAL STRATIGRAPHY OF UPPER JURASSIC
B 6304 604 MANCOS SHALE, NEW MEXICO, A ZONE, NIOBRARA MEMBER
B 6205 757 MANCOS SHALE, SAN JUAN BASIN
B 6303 519* MANCOS SHALE--PREDICTIVE MODEL FOR CRETACEOUS COALS OF
B 6302 183 MANDAL HIGH, NORTH SEA
B 6209 1610 MANDATA FORMATION, MARYLAND
S 5 135 MANDATA SHALE, WEST VIRGINIA, KEYSER LIMESTONE,
S 4 209† MANDINGA, VERACRUZ, MEXICO, CHANGE IN, OSTRACOD STANDING
B 6203 553* MANGANESE, ABST., CONTROL OF, CATHODO-LUMINESCENCE OF,
M 25 544 MANGANESE, ALUMINUM, AND IRON, TERNARY DIAGRAM, NAZCA PLATE
M 28 149 MANGANESE, BIBLIOGRAPHY, DIAGENESIS AND POROSITY
M 25 533 MANGANESE, CHILE
M 25 460 MANGANESE CONCRETIONS, ORIGIN OF, PACIFIC, IRON-
M 25 425 MANGANESE DEPOSITS, GENESIS OF, GROOTE EYLANDT
M 25 449 MANGANESE DEPOSITS, JAPAN, CHICHIBU GEOSYNCLINAL UNITS
M 25 427 MANGANESE, GROOTE EYLANDT
S 7 321 MANGANESE IN PORE WATER, GYRE BASIN
B 6207 1211* MANGANESE NODULES, ABST., MINERALOGY AND, CHEMISTRY OF,
M 25 386† MANGANESE NODULES IN SOUTH PACIFIC--SUMMARY, EXPLOITATION OF
M 25 424 MANGANESE ORE, CHARACTERISTICS OF, GROOTE EYLANDT
M 25 424 MANGANESE ORE, CONSTITUENT MINERALS, GROOTE EYLANDT
M 25 423 MANGANESE ORE, EVALUATION OF RESOURCES, GROOTE EYLANDT
M 25 422† MANGANESE ORE, GROOTE EYLANDT, NORTHERN AUSTRALIA,
M 25 424 MANGANESE ORE, PRODUCTION AND TREATMENT, GROOTE EYLANDT
M 25 423 MANGANESE ORE, PRODUCTION, GROOTE EYLANDT
M 25 543 MANGANESE, TERNARY DIAGRAM, NAZCA PLATE, SILICON, IRON, AND
M 25 382† MANGANESE-NODULE STRUCTURES, RELATION TO NODULE
S 4 17 MANGROVE-DOMINATED COAST AND SHALLOWS, PUERTO RICO
B 6408 1287 MANGUN, MARK B., SEISMIC REFRACTION STUDY OF BURIED VALLEY
B 6301 90 MANHASSET BEDS, SCOTIAN BASIN
B 6105 804 MANHEIM, F. T., FRESH GROUNDWATER BENEATH CONTINENTAL SHELF:
B 6209 1658 MANILA SUBDUCTION ZONE
B 6209 1659 MANILA TRENCH, GEOPHYSICAL PROFILES
M 25 434 MANILA TRENCH, PHILIPPINES
S 5 113 MANISTEE COUNTY, BEAR LAKE TOWNSHIP, SEISMIC EXAMPLE,
S 5 115 MANISTEE COUNTY, LOCATION, MICHIGAN,
S 5 4 MANISTIQUE FORMATION, MICHIGAN BASIN, NIAGARAN SERIES,
B 6108 1316 MANITOBA, DEVELOPMENTS, 1976
B 6308 1409 MANITOBA, DEVELOPMENTS, 1978
B 6409 1512 MANITOBA, DEVELOPMENTS, 1979
B 6108 1322 MANITOBA, PRODUCTION, 1976
B 6303 384 MANITOULIN AREA, INCLUDING THE ROAD LOG TO THE MICHIGAN
B 6303 503 MANIVIT, HELENE, CALCAREOUS NANNOPLANKTON I UPPER APTIAN

B 6205 840 MANJAK, SOLID HYDROCARBON, TRINIDAD
B 6108 1383 MANKIEWICZ, DAVID, DEPOSITION AND DIAGENESIS OF TENSLEEP
B 6408 1286 MANN, C. JOHN, COLMAR-PLYMOUTH CONUNDRUM, ABST.
B 6408 1287 MANN, C. JOHN, ILLINOIS BASIN, ITS FUTURE PETROLEUM PROSPECTS,
B 6309 1582 MANN, C. JOHN, METEORITIC IMPACT--RESERVOIR-FORMING PROCESS,
B 6309 1582 MANN, C. JOHN, SUBMARINE SLUMPING IN REEF-FLANK DEPOSITS,
B 6004 732* MANNED SUBMERSIBLE, ABST., GEOLOGIC EXPLORATION OF SOUTHERN,
B 6105 765 MANNS, FRANCIS T., COARSE WHITE SPARRY DOLOMITE AND CALCITE
B 6107 1010 MANNVILLE GROUP, ALBERTA
B 6405 740* MANNVILLE GROUP, LLOYDMINSTER HEAVY OIL AREA, SASKATCHEWAN,
B 6107 1010 MANNVILLE GROUP, SASKATCHEWAN
B 6107 1010† MANNVILLE, LOWER CRETACEOUS, DISCONFORMITY IN LLOYDMINSTER
B 6408 1166 MANNVILLE SANDSTONE, ALBERTA
B 6008 1413* MANNVILLE UNCONFORMITY IN LLOYDMINISTER AREA,
B 6203 490 MANON M., A., RECENT DEVELOPMENTS AT CERRO PRIETO GEOTHERMAL
M 25 160 MANON MERCADO, ALFREDO, GEOTHERMAL POTENTIAL OF MEXICO
S 4 111† MANS IMPACT ON COASTAL REEFS--A EXAMPLE FROM VENEZUELA
B 6105 810* MANS IMPACTS ON GLOBAL SEDIMENTARY CYCLES OF METALS
B 6303 490 MANSFIELD, CHARLES F., POSSIBLE BIOGENIC ORIGIN FOR SOM
M 29 1 MANTADERT, LUCIEN, EDITED BY, GEOLOGICAL AND GEOPHYSICAL
S 3 79 MANTLE CONVECTION
S 3 25 MANTLE CONVECTION
M 25 487 MANTLE INHOMOGENEITIES, ROLE IN METALLIZATION
S 3 36 MANTLE MAGMAS, CONTINENTS
S 3 28 MANTLE PLUMES
S 3 135 MANTLE VOLATILES, DIATREME FEEDING OF, URANIFEROUS
S 3 36 MANTLE-DERIVED MAGMA, OCEAN BASINS
S 3 36 MANTLE-DERIVED URANIUM AND THORIUM WITHIN CRUST, MIGRATION
B 6412 2191 MANUAL FOR TEPHROCHRONOLOGY, BY VIRGINIA STEEN-MCINTYRE,
B 6212 2427 MANUAL SLOPE MAPS
B 6203 426 MANX FURNESS BASIN, EUROPE
B 6203 430 MANX PLATFORM, BRITAIN, WEST
B 6303 290 MAP, 1975 DEEP SEA DRILLING PROJECT HOLES, BASE
B 6212 2435 MAP, BOUGUER, GRAVITY
B 6102 266 MAP, CALIFORNIA, SANTA MONICA MOUNTAINS, NORTHWESTERN,
B 6212 2427 MAP CONCEPT, SLOPE
B 6203 495* MAP DEPICTING MAJOR SUBMARINE SLIDE ON ATLANTIC CONTINENTAL
B 6303 552* MAP DISPLAY FORMATS FOR ENVIRONMENTAL GEOLOGIC AND RELATED
B 6304 594 MAP, EGYPT, ABU MADI FIELD, TEMPERATURE
B 6303 343 MAP, KENTUCKY, ROME TROUGH, GRAVITY
B 6101 44 MAP, LEBANON, GEOLOGICAL
B 6007 1076 MAP, NIGER DELTA, BOUGUER, GRAVITY
B 6007 1074 MAP, NIGER, DELTA, GEOTHERMAL, GRADIENT
M 25 46† MAP OF PACIFIC MOBILE BELT AND PACIFIC OCEAN, GENERAL
B 6212 2510* MAP PROJECT, ABST., CIRCUM-PACIFIC
B 6004 691* MAP PROJECTIONS, ABST., INCONSISTENCIES IN, RESOURCE-APPRAISAL,
B 6012 2175* MAP TECHNIQUES, ABST., SEISMIC DATA ANALYSIS WITH, TIME, SEISCROP
B 6008 1391* MAP TECHNIQUES, ABST., SEISMIC DATA ANALYSIS WITH, TIME-,
B 6007 1022 MAP, WESTERN, LAKE SUPERIOR, BATHYMETRIC
B 6303 517* MAPPED DATA, ABST., STATISTICAL COMPARISON OF
B 6004 665* MAPPING, ABST., TECHNIQUES AND USES OF, SLOPE
B 6405 741* MAPPING AND INTERPRETATIONS, ABST., MULTI-PARAMETER REGIONAL
B 6012 2188* MAPPING, CENTRAL BROOKS RANGE, ALASKA, ABST., USE OF, IR COLOR
M 26 136 MAPPING, FACIES, ANALYSIS
M 26 468 MAPPING, GEOSEISMIC, MODELING
B 6209 1754* MAPPING IN BATHYAL ENVIRONMENTS, ABST., PALEOSTRUCTURAL
B 6309 1490† MAPPING IN CHAGAI HILLS, BALUCHISTAN, PAKISTAN, BY COMPUTER
B 6105 834* MAPPING IN COMPLEXLY FAULTED AREAS, ABST., USING, COMPUTER
B 6101 48 MAPPING, LEBANON
B 6105 823* MAPPING OF GEOLOGIC DATA, ABST., GRAY-LEVEL
B 6409 1568 MAPPING OF GULF COAST STRATIGRAPHIC TRAPS, ABST.,
B 6205 845 MAPPING OF HYDROCARBON RESERVOIRS, ISOPACH
B 6012 2150† MAPPING OF HYDROCARBON RESERVOIRS, VERTICAL, NET, SANDSTONE
B 6203 528* MAPPING OF NORTH SEA BY COMPUTER, ABST., REGIONAL GEOLOGIC
B 6105 835* MAPPING OF RECENT COASTAL MARINE SEDIMENTARY ENVIRONMENTS
B 6211 2354* MAPPING ON NORTH SLOPE, ABST., USE OF, VERTICAL-GRADIENT,
B 6203 502* MAPPING PROGRAM, ABST., MAPRO—GEOGRAPHIC
M 26 374 MAPPING, RESOLUTION SYSTEMS, BATHYMETRIC
M 26 369 MAPPING, RESOLUTION SYSTEMS, SEAFLOOR
M 26 371 MAPPING, RESOLUTION SYSTEMS, STRUCTURE FORM, CONTOUR
M 26 371 MAPPING, RESOLUTION SYSTEMS, SUBCROP
B 6105 812* MAPPING RESOURCES BY MINICOMPUTER
M 26 142 MAPPING, SEISMIC, FACIES
M 26 371 MAPPING, SHALLOW, RESOLUTION SYSTEMS, ISOPACH
B 6212 2427† MAPPING, SLOPE
B 6105 809* MAPPING STRATIGRAPHIC TRAPS WITH SEISLOGS
B 6105 836* MAPPING SYSTEM TO ENERGY DEVELOPMENT AND
M 25 67 MAPPING, THAILAND, GEOLOGIC
B 6305 825* MAPPING WITH ORBITING IMAGING RADAR, ABST., LINEAMENT
B 6412 2064† MAPPING—TENNESSEE-KENTUCKY-VIRGINIA, SEASAT, ORBITAL, RADAR,
B 6203 502* MAPRO—GEOGRAPHIC MAPPING PROGRAM, ABST.
B 6004 691* MAPS AS AID TO GEOLOGIC INTERPRETATION, ABST., STATISTICAL
B 6302 174 MAPS, BCF
B 6303 555* MAPS COMPUTER CONSTRUCTED FROM ORIENTATIONAL,
B 6212 2428 MAPS, COMPUTER-MADE, SLOPE
B 6212 2429 MAPS, DATACOLOR, SLOPE
B 6004 669* MAPS FOR GULF OF MEXICO OCS PROGRAMS, ABST., NEW,
B 6304 592 MAPS, GEOCHEMICAL TOOL, CEMENTATION
B 6304 590 MAPS, GEOCHEMICAL TOOL, HYDRAULIC POTENTIAL
B 6304 592 MAPS, GEOCHEMICAL TOOL, MATURITY
B 6304 590 MAPS, GEOCHEMICAL TOOL, SALINITY
B 6304 591 MAPS, GEOCHEMICAL TOOL, TEMPERATURE
B 6105 835* MAPS IN EXPLORATION PROCESS, ABST., USE OF,
B 6212 2427 MAPS, MANUAL, SLOPE
B 6304 700* MAPS TO PETROLEUM PROSPECTS, ABST., QUANTITATIVE COMPARISONS
S 10 149 MAQUOKETA AQUITARD, PIEZOMETRIC SURFACE, ILLINOIS
B 6401 129 MAQUOKETA GROUP, INDIANA
B 6204 706 MAQUOKETA SHALE, ILLINOIS BASIN
B 6104 579 MARA FIELD, VENEZUELA
B 6109 1424 MARACAIBO AREA, RESERVOIRS, VENEZUELA, LAKE
B 6104 579 MARACAIBO BASIN, OIL AGE, VENEZUELA
M 25 462 MARACAIBO FRACTURE ZONE, CARACAS-
B 6105 821* MARACAIBO PLATFORM, VENEZUELA, ABST., DEPOSITIONAL PATTERNS
B 6408 1191 MARACAIBO, POSITIVE WATER BALANCE BASIN, VENEZUELA, LAKE
B 6303 516* MARACAIBO, VENEZUELA, ABST., FACIES RECOGNITION FROM, WELL,
B 6311 2042† MARACAIBO, VENEZUELA, STRATIGRAPHY OF, INTERMONTANE,
B 6306 940 MARAJO BASIN, AMAZON MARGIN AREA
B 6306 942 MARAJO FORMATION, AMAZON BASIN
B 6109 1471 MARANHAO BASIN, BRAZIL
B 6004 655* MARANHAO BASIN, NORTHEAST BRAZIL, ABST., SYNSEDIMENTARY
B 6202 286 MARANHAO BASIN OFF, BRAZIL, NORTHERN, PARA-
B 6112 2104 MARANON BASIN OILS, CLASSIFICATION, PERU
B 6112 2103† MARANON BASIN, PERU, HYDROCARBON, GEOCHEMISTRY OF OILS FROM
S 3 107 MARATHON BELT, OUACHITA-
S 3 98 MARATHON MOBILE BELTS, OUACHITA AND
B 6411 1842 MARATHON OIL CORP.,AND PETROCONSULTANTS S.A. ON
B 6203 566* MARATHON REGION, WEST TEXAS, ABST., DEPOSITIONAL
B 6109 1449 MARATHON-OUACHITA TECTONIC BELT, TEXAS
B 6303 479* MARBLE FALLS FORMATION, CENTRAL TEXAS, ABST., DEPOSITION OF
B 6308 1427* MARBLE FALLS GROUP, LLANO UPLIFT TO SOUTHERN PART OF FORT
B 6301 55 MARBLE FALLS LIMESTONE, TEXAS
B 6408 1171 MARBLE FALLS LIMESTONE, TEXAS
M 27 50 MARBLE FALLS LIMESTONE, TEXAS, PENNSYLVANIAN
M 27 206 MARBLE FALLS LIMESTONE, TEXAS, PENNSYLVANIAN
M 27 93 MARBLE FALLS LIMESTONE, TEXAS, PENNSYLVANIAN
M 25 553 MARBLE, SURVEYS, ECUADOR, CLAYS AND, TRAVERTINE
S 3 106 MARCASITE-JORDESITE-CALCITE-URANIUM ORE, PYRITE-
B 6002 284 MARCELLUS SHALE, PENNSYLVANIA
B 6311 2114* MARCHAND SANDSTONES, NORTHERN CADDO COUNTY, OKLAHOMA,
B 6405 745 MARCHANT, L. C., RECENT ACTIVITIES IN UNITED STATES TAR SAND,
B 6203 571 MARCHEL, R. J., DELTAIC COMPONENTS OF VALMY AND VININI
B 6203 538 MARCHEL, R. J., ELDER SANDSTONE—DELTAIC SEDIMENTS OF ROBERTS
B 6105 840 MARCHEL, RONALD, GOODWIN LIMESTONE—LOWER ORDOVICIAN
S 4 85 MARCHMONT INLIER, JAMAICA
B 6303 514* MARCOSITE IN ROLL-TYPE URANIUM DEPOSITS, ABST.
B 6101 58 MARDIN SWELL, LEVANT, JAUF-, GAARA-, ABD EL AZIZ-
B 6104 612† MARFA BASIN, TEXAS, THE, FILL IN
B 6405 792* MARGARITIFERA MARGARITIFERA L., ABST., INTERPRETATION OF FOSSI
B 6004 708* MARGIN, 1926-1976, ABST., SEDIMENTOLOGY AND, EASTERN,
B 6004 669* MARGIN, ABST, GEOLOGY OF, RED SEA
B 6303 410* MARGIN, ABST., ANTARCTIC, CONTINENTAL
B 6405 706* MARGIN, ABST., DIAPIR-LIKE RIDGES AND POSSIBLE HYDROCARBON
B 6004 690* MARGIN, ABST., EVOLUTION OF, MARGINAL, PLATEAU--DSDP DATA
B 6004 690* MARGIN, ABST., EVOLUTION OF, MARGINAL, PLATEAU--DSDP DATA
B 6303 505* MARGIN, ABST., OLIGOCENE TRANSGRESSIVE SEDIMENTS OF, NEW JERSEY,
B 6004 656* MARGIN, ABST., PETROLEUM GEOLOGY OF EASTERN, BRAZILIAN,
B 6207 1231* MARGIN, ABST., PLATE TECTONICS IN EVOLUTION OF SOUTHERN,
B 6105 821* MARGIN, ABST., PRELIMINARY, BIOSTRATIGRAPHY OF, U.S., ATLANTIC,
B 6004 686* MARGIN, ABST., SEDIMENTARY PROCESSES IN, SUBMARINE, CANYONS OF
B 6004 708* MARGIN, ABST., SEDIMENTOLOGY AND, EASTERN, CONTINENTAL
B 6303 490* MARGIN, ABST., SHALLOW-, WATER UPPER, JURASSIC ROCKS DREDGED
B 6405 736* MARGIN, ABST., THERMOGENIC HYDROCARBON, GASES IN
B 6108 1392* MARGIN, ABST., WIDESPREAD OCCURRENCE OF, LAUMONTITE IN LATE,
B 6305 828* MARGIN ALONG OVERTHRUST BELT FROM MONTANA TO NEVADA,
B 6106 881 MARGIN AND ADJACENT DEEP SEA, MAGNETIC ANOMALIES OF,
B 6004 694* MARGIN AND RELATION TO PLATE TECTONICS. BY TITLE ONLY,
B 6303 291 MARGIN APRONS, PROSPECTS, OUTER
B 6004 697* MARGIN, BANKS ISLAND, ARCTIC CANADA, ABST., TERTIARY, DELTAIC,
M 29 11† MARGIN, BASED ON MULTICHANNEL SEISMIC REFLECTION PROFILES,
B 6002 197 MARGIN BASINS, BRAZIL, EASTERN, MESOZOIC-, CENOZOIC

B 6306 936 MARGIN, BATHYMETRY, MAP, AMAZON, CONTINENTAL
B 6002 161 MARGIN, BERINGIAN, CONTINENTAL
B 6405 713* MARGIN BIOSTRATIGRAPHY, ABST., APPLICATION OF STATISTICAL
B 6306 935 MARGIN, BRAZIL, NORTHERN, CONTINENTAL
B 6002 177 MARGIN, CENOZOIC BASINS, BERINGIAN SHELF AND
B 6209 1669 MARGIN, CONVERGENCE OF, LUZON ARC WITH, ASIAN, CONTINENTAL
B 6105 822* MARGIN CORING PROJECT, ABST., SHIPBOARD GEOTECHNICAL RESULTS
B 6105 814* MARGIN CORING PROJECT—PRELIMINARY RESULTS, ABST.,
S 7 36 MARGIN, DESCRIBED, EASTERN, GULF OF MEXICO
B 6109 1437† MARGIN, EASTERN ANADARKO BASIN, OKLAHOMA, SEISMIC
B 6109 1489 MARGIN, EUGEOCLINAL WEDGE, EAST COAST, CONTINENTAL
B 6401 88† MARGIN EVOLUTION, PALO DURO BASIN, TEXAS,
B 6306 886† MARGIN FACIES OF DAKOTA SANDSTONE, LONE PINE FIELD, NEW
S 7 1 **MARGIN, FRAMEWORK, FACIES, AND OIL-TRAPPING CHARACTERISTICS**
B 6004 691* MARGIN FROM NIGERIA TO CAPE OF GOOD HOPE, ABST., PETROLEUM
B 6404 516 MARGIN, GEOLOGY, NEW ENGLAND, CONTINENTAL
B 6404 501† MARGIN, GEOLOGY OF, NEW ENGLAND, PASSIVE
B 6103 376† MARGIN GROWTH AND STRUCTURE, LITTLE BAHAMA BANK, BAHAMAS,
S 3 82 MARGIN HISTORY, OROGENIC, CONTINENTAL-
S 3 115 MARGIN HISTORY, TAPHROGENIC, CONTINENTAL-
B 6109 1557* MARGIN IN NORTHERN MEXICO, ABST., LOWER, CRETACEOUS, SHELF
B 6009 1617* MARGIN IN NORTHERN MEXICO AND SOUTH TEXAS, ABST., EARLY,
S 3 96 MARGIN, INCIPIENT, OROGENIC, CONTINENTAL
B 6209 1588 MARGIN, INDIA-, PAKISTAN, CONTINENTAL
B 6209 1589 MARGIN, LOUISIANA, GULF COAST—, DIVERGENT, DELTAIC
B 6104 546† MARGIN, MARINE, SEISMIC INVESTIGATIONS, SOUTHERN, BRAZIL
B 6002 177 MARGIN, MESOZOIC AND CENOZOIC TECTONIC SETTING, BERINGIAN
B 6109 1486 MARGIN, MIOGEOCLINAL WEDGE, EAST COAST, CONTINENTAL
B 6306 934† MARGIN, MORPHOLOGY AND STRUCTURE OF, AMAZON UPPER,
M 29 226 MARGIN, MORPHOLOGY, SUMATRA
B 6209 1603 MARGIN, NORTH AMERICA, EASTERN, CONTINENTAL
B 6405 746* MARGIN NORTH OF CAPE HATTERAS, ABST., GEOLOGIC SETTING AND OIL
B 6107 961† MARGIN OF AFRICA, STRUCTURAL HISTORY OF, ATLANTIC
M 25 308† MARGIN OF BRITISH COLUMBIA--SUMMARY, SEDIMENTATION AND
B 6303 324† MARGIN OF ISRAEL AND NORTHERN SINAI, SUBMARINE, SLUMPING IN,
B 6303 436* MARGIN OF OUTER BERING SEA, ABST., THICK SEDIMENT
B 6303 420* MARGIN OFF CENTRAL AND SOUTH TEXAS IN ATLAS FORMAT, ABST.,
B 6105 838* MARGIN OFF KODIAK ISLAND, ALASKA, ABST., STRUCTURAL
M 29 261† MARGIN OFF KODIAK ISLAND, ALASKA, FROM MULTICHANNEL SEISMIC
B 6206 1029† MARGIN OFF SOUTHERN BRAZIL, MORPHOLOGY AND STRUCTURE OF
B 6005 809† MARGIN OFF WESTERN AFRICA--SENEGAL TO PORTUGAL, CONTINENTAL
B 6202 223† MARGIN, ORIGIN AND SIGNIFICANCE OF, EVAPORITES IN, BASINS
S 3 93 MARGIN, OROGENIC, FORELAND
B 6004 709* MARGIN, OTHRIS, GREECE, ABST., SEDIMENTATION AND
B 6109 1470† MARGIN, PETROLEUM GEOLOGY OF EASTERN, BRAZILIAN, CONTINENTAL
M 29 48 MARGIN PROFILES, DESCRIPTIONS, ICELAND
B 6303 289 MARGIN REGION, PROSPECTS, CONTINENTAL
B 6204 657† MARGIN, SALT, DIAPIRS ON, SAO PAULO PLATEAU, SOUTHEASTERN,
B 6204 657† MARGIN, SALT DIAPIRS ON, SAO PAULO PLATEAU, SOUTHEASTERN,
B 6105 826* MARGIN SEDIMENTS OFF NORTHWEST AFRICA BETWEEN CANARY AND
B 6209 1588 MARGIN SEDIMENTS, ORGANIC CONTENT OF, CONTINENTAL-
B 6209 1595 MARGIN SEDIMENTS, THERMAL MATURITY OF, CONTINENTAL
B 6205 792† MARGIN, SEISMIC STRATIGRAPHY AND GEOLOGIC HISTORY OF, BLAKE
B 6402 242† MARGIN, SEISMIC STRUCTURE OF SOUTHWESTERN, PELOPONNESUS,
M 29 151† MARGIN SOUTH OF AUSTRALIA, THE, CRUSTAL, STRUCTURE AND,
M 29 226 MARGIN, STRUCTURE, SUMATRA
B 6402 173† MARGIN SUBSIDENCE AND HEAT FLOW--IMPORTANT PARAMETERS IN
B 6105 842* MARGIN TYPES RELATED TO PLATE TECTONICS AND EVOLUTION OF
M 29 139† MARGIN, VARIATIONS IN, BOTTOM, PROCESSES ALONG THE, U.S.,
B 6303 444* MARGIN--CAPE COD TO FLORIDA, ABST., DISTRIBUTION, MORPHOLOGY,
M 29 43† MARGIN: THIRTEEN GEOPHYSICAL PROFILES, EROSIONAL AND,
M 29 373† MARGINAL BASIN IN SOUTHEAST ASIA, THE, SULU SEA: A
B 6004 734* MARGINAL BASIN IN SOUTHERN SOUTH AMERICA, ABST., SUBMARINE-,
B 6303 289 MARGINAL GEOSYNCLINES, PROSPECTS
B 6004 690* MARGINAL PLATEAU--DSDP DATA FROM SITE 356 ON SAO PAULO
B 6207 1239* MARGINAL PLATEAUS, ABST., PETROLEUM PROSPECTIVITY OF,
B 6303 291 MARGINAL-PLATEAU BLOCKS, PROSPECTS
B 6303 291 MARGINAL-TRENCH FILLINGS, PROSPECTS
S 4 221 MARGINOPORA VERTEBRALIS QUOY AN GAIMARD, GROWTH RATES,
B 6004 661* MARGINS, ABST., GULF COAST, CENOZOIC--, MODEL FOR APPLICATION OF
B 6209 1755* MARGINS, ABST., GULF COAST, CENOZOIC—MODEL FOR APPLICATION OF
B 6303 546* MARGINS, ABST., IPOD, DRILLING ON CONVERGENT, OCEAN
B 6207 1216* MARGINS, ABST., SEAFLOOR INSTABILITY ON, PACIFIC, CONTINENTAL
M 29 293† MARGINS AND DEEP BASINS IN THE MEDITERRANEAN, VARIETY OF
B 6004 730* MARGINS BY EARLY CEMENTATION, GOLDEN SPIKE OIL FIELD, ALBERTA,
M 29 453 MARGINS, CHANGES IN SEA LEVEL, SEE MECHANISMS
M 25 178 MARGINS, CLASSIFICATION, CONTINENTAL
B 6301 67† MARGINS, CZECHOSLOVAKIA, ANCIENT, SUBMARINE CANYONS OF
M 29 443† MARGINS, EXPLOSION, SEISMOLOGY STUDIES OF, ACTIVE AND, PASSIVE,
B 6202 208 MARGINS, GEOHISTORY OF, CONTINENTAL
M 29 1 **MARGINS, GEOLOGICAL AND GEOPHYSICAL INVESTIGATIONS**
B 6403 300 MARGINS IDENTIFIABLE FROM CONTINUOUS SEISMIC REFLECTION
B 6211 2290† MARGINS OF GULF OF MEXICO
M 29 273† MARGINS OF THE EASTERN GULF OF ALASKA AND BOUNDARIES OF
M 29 403† MARGINS OF THE GULF OF MEXICO, THE
M 29 363† MARGINS OF THE VENEZUELA BASIN, TECTONIC DEVELOPMENT OF,
B 6405 670* MARGINS OF UNITED STATES, ABST., THERMAL EVOLUTION
B 6403 299 MARGINS OF WORLD CAUSED BY PLATE MOVEMENTS ACCORDING TO
B 6209 1591 MARGINS, OREGON AND, ALASKA—, CONVERGENT
S 3 115 MARGINS, OROGENIC-, TAPHROGENIC
B 6009 1463† MARGINS, PLATE TECTONICS IN, OIL AND, GAS, EXPLORATION OF,
M 25 177† MARGINS, PLATE TECTONICS IN PETROLEUM EXPLORATION OF,
S 3 115 MARGINS, PURELY, TAPHROGENIC
B 6111 1959 MARGINS, RESERVES, PASSIVE, ATLANTIC-TYPE
B 6403 307 MARGINS, SEDIMENT DAMS BENEATH, CONTINENTAL
M 29 403 MARGINS, SEISMIC STRATIGRAPHY, GULF OF MEXICO
M 29 292 MARGINS, STRUCTURAL MAP, MEDITERRANEAN
M 29 3† MARGINS, SUBSIDENCE MECHANISMS AT, PASSIVE, CONTINENTAL
M 29 453† MARGINS, THE EFFECT OF, EUSTATIC, SEA LEVEL CHANGES ON,
M 25 43 MARGINS, WESTERN, PACIFIC AND, ATLANTIC OCEAN
B 6105 823* MARGINS, WESTERN UNITED STATES, ABST., MISSISSIPPIAN, CARBONATE-,
B 6403 303 MARGINS, WORLD DISTRIBUTION OF STAGES OF, CONTINENTAL
B 6403 297† MARGINS--CLASSIFICATION AND PETROLEUM PROSPECTS, CONTINENTAL
B 6004 732* MARGINS--RESULTS USING MANNED SUBMERSIBLE, ABST., GEOLOGIC
B 6105 842* MARGINS. BY TITLE ONLY, CONTINENTAL-, MARGIN TYPES RELATED TO,
B 6403 297† MARGINS—CLASSIFICATION AN PETROLEUM PROSPECTS, CONTINENTAL
B 6105 842* MARGIN—U.S. GULF COAST, ABST., CONTINENTAL
M 25 382 MARGOLIS, S., MANGANESE-NODULE STRUCTURES, RELATION TO
S 4 291 MARIANA ISLANDS, SAIPAN, GALAXEA
B 6304 695* MARIANO LAKE URANIUM OREBODY, ABST., GEOPHYSICAL EXPERIMENTS
B 6304 694* MARIANO LAKE TYPE, ABST., MINERALOGICAL AND, GEOCHEMICAL
B 6304 692* MARIANO LAKE URANIUM DEPOSIT, SMITH LAKE DISTRICT, NEW
B 6407 972 MARIAS RIVER SHALE, GREAT PLAINS
B 6211 2246 MARICOPA SUBBASIN, INFORMAL NAME, CALIFORNIA
B 6103 361 MARIETTA BASIN, OKLAHOMA
B 6303 516* MARIN PERMIAN ROCKS OF TUNISIA, ABST.
M 25 268† MARIN PETROLEUM EXPLORATION OF HUKSAN PLATFORM, KOREA
B 6405 748 MARINCOVICH, L. N., JR., PRELIMINARY MOLLUSCAN BIOSTRATIGRAPHY
M 25 436 MARINDUQUE ISLAND, INO-CAPAYANG COPPER PROPERTY
M 25 436 MARINDUQUE ISLAND, MARCOPPER MINE
B 6004 727* MARINE AND BRACKISH-WATE CRETACEOUS OSTRACODA FROM WELLS
B 6004 724* MARINE AND COASTAL DEPOSITS, ABST., MODELS USING, SUBSURFACE
S 4 381† MARINE AND CONTINENTAL GROUNDWATER SOURCES IN A PERSIAN
B 6004 662* MARINE AND LATER POSTBURIAL CEMENTATION AND DIAGENESIS OF
B 6109 1552* MARINE CARBONATE SUCCESSIONS OF PLEISTOCENE AND CRETACEOUS
S 9 2 MARINE CARBONATES, DISTRIBUTION
S 4 372 MARINE CEMENTS AND DECOMPOSITION OF ORGANIC TISSUE IN
B 6004 692* MARINE COLD-WATER CARBONATE DEPOSITS AS POSSIBLE
B 6004 645* MARINE CYCLES OF DEPOSITION IN SOUTHERN SACRAMENTO VALLEY,
B 6303 483* MARINE ENVIRONMENT, ABST., SEDIMENTOLOGIC FACIES I MODERN,
M 25 27† MARINE ENVIRONMENT IN AUSTRALIA. TWO STUDIES--SUMMARY
B 6004 645* MARINE ENVIRONMENTS, ABST., CYCLIC SEDIMENTATION IN, GLACIAL
S 11 221 MARINE ENVIRONMENTS, CLASSIFICATION
B 6306 987 MARINE EVAPORITES, EDITED BY WALTER E. DEAN AND B. CHARLOTTE
S 6 299 MARINE FAUNAS, PERMIAN, BIOCHRONOLOGY
B 6106 903† MARINE GEOLOGY OF BAJA CALIFORNIA CONTINENTAL BORDERLAND,
B 6004 714* MARINE GEOTECHNOLOGY TO SEDIMENTOLOGY, ABST., OSLOFJORD,
B 6407 1120 MARINE GRAVITY, BY PETER DEHLINGER, BOWIN, CARL, REVIEW OF
B 6108 1386* MARINE IN ORIGIN, ABST., NAVAJO SANDSTONE, JURASSIC, OF, UTAH
M 29 426 MARINE, INDEX MAP, ORGANIC PRODUCTIVITY
S 6 302 MARINE, INTERCORRELATIONS, PERMIAN
B 6105 778* **MARINE METALLIFEROUS SEDIMENTS FROM NAZCA PLATE**
S 4 195† MARINE MOLLUSK DISTRIBUTION, NORTHEAST YUCATAN COAST,
S 9 2 MARINE MUD, CARBONATE CONTENT, TABLE
B 6405 722* MARINE PETROLEUM PROSPECTING WITH AIRBORNE FRAUNHOFER LINE
B 6404 463 MARINE PHREATIC ZONE
B 6004 671* MARINE SEDIMENTARY ROCKS, VIZCAINO PENINSULVA, BAJA
B 6104 546† MARINE SEISMIC INVESTIGATIONS, SOUTHERN BRAZIL MARGIN
S 6 300 MARINE, SUBDIVISIONS, PERMIAN
B 6111 2001† MARINE TERRACE, PALOS VERDES HILLS, CALIFORNIA, FAULTED UPPER,
B 6303 445* MARINE TO DEEP SUBSURFACE ENVIRONMENTS, UPPER SMACKOVER
B 6004 665* MARINE TRANSGRESSIVE AND REGRESSIVE RATES IN PENNSYLVANIAN
B 6004 675* MARINE-BAR SEDIMENTATION AND EQUILIBRIUM, KOUCHIBOUGUAC
B 6108 1388* MARINE-SHELF MODEL FOR UPPER CRETACEOUS HYGIENE MEMBER OF
B 6310 1922 MARINELLI, AND M. PUPILLI, ON DEVELOPMENTS IN ITALY, 1978,
B 6210 2046 MARINHO CAMPOS, C.W., ON DEVELOPMENTS IN BRAZIL, 1977
B 6210 2048 MARINO, M., ON DEVELOPMENTS IN CHILE, 1977
B 6110 1580 MARINO P., M., ON DEVELOPMENTS IN CHILE, 1976
B 6010 1646 MARIO MARINO P., ON DEVELOPMENTS IN CHILE, 1975
B 6301 82 MARITIME ALPS. ANNOT FORMATION

B 6203 542* MARITIME PROVINCES, CANADA, ABST., COARSE-GRAINED DEPOSITS BY
M 26 65 MARITIME SEQUENCES, SEA LEVEL FLUCTUATIONS
B 6104 585 MARJAN, KHAFJI RESERVOIR, FIELD, PERSIAN GULF
B 6405 770 MARKELLO, J. R., CARBONATE DEPOCENTERS AND FACIES DISTRIBUTION
B 6203 539 MARKELLO, JAMES R., DIGITATE STROMATOLITIC BIOHERMS AND
B 6004 651* MARKER BEDS IN COALS OF ROCKY MOUNTAIN REGION, ABST.,
B 6203 544 MARKEWICH, WILLIAM, NEW DEVELOPMENTS IN METHODS OF RESOURCE
M 30 243 MARKOVA FIELD, EXPLORATION HISTORY, USSR
B 6004 666* MARKOVIAN ANALYSIS OF BED THICKNESS AND LITHOLOGIC
B 6410 1691 MARKOVO DISCOVERY, USSR
M 30 234 MARKOVO FIELD, CROSS SECTION, USSR
M 30 232 MARKOVO FIELD, LITHOLOGIC COLUMN, USSR
M 30 245 MARKOVO FIELD, PARFENOVA HORIZON, USSR
B 6401 133 MARKOVSKII, MCGOOKEY, D. P., REVIEW OF, PALEOGEOGRAPHIC
M 25 479 MARKOVSKY, B. A., MEIMECHITES OF KAMCHATKA AND KIMBERLITE
B 6207 1153 MARLIN CHANNEL, AUSTRALIA
B 6104 594 MARLIN FIELD, AUSTRALIA
M 25 280 MARLOW, M. S., CENOZOIC AN MESOZOIC PETROLEUM PROSPECTS,
B 6405 745 MARLOW, M. S., HYDROCARBON PROSPECTS OF BERING SEA SHELF
B 6004 694 MARLOW, M. S., STRUCTURE AN EVOLUTION OF BERING SEA MARGIN
B 6303 436 MARLOW, M. S., THICK SEDIMENT ACCUMULATIONS BENEATH
B 6403 439 MARLOW, MICHAEL S., ET AL, HYDROCARBON POTENTIAL OF ALEUTIAN
B 6311 2070 MARLOW, MICHAEL S., HYDROCARBON POTENTIAL OF ALEUTIAN BASIN,
B 6103 467 MARLOW, MICHAEL S., INDICATOR OF PLATE TECTONICS IN SOUTHERN
B 6412 2139 MARLOW, MICHAEL S., MESOZOIC AND CENOZOIC STRUCTURAL TRENDS
B 6405 694 MARLOW, MICHAEL S., MULTIFOLD SEISMIC DATA ACROSS OUTER
B 6405 693 MARLOW, MICHAEL S., PROSPECTIVE AND FUTURE HYDROCARBON
B 6303 490 MARLOW, MICHAEL S., SHALLOW-WATER UPPER JURASSIC ROCKS
B 6002 161 MARLOW, MICHAEL S., STRUCTURE AND EVOLUTION OF BERING SEA
B 6203 505* MARLSTONE, GREEN RIVER FORMATION, PICEANCE CREEK BASIN,
B 6107 1030 MARMATON GROUP, NEBRASKA
B 6104 516 MAROLLES WELLS, FRANCE
B 6002 320 MARONEY, DAVE, REPORT--CENTRALNEBRASKA, ABST.
B 6201 122 MARQUESAS KEYS, WELLS, FLORIDA
B 6303 495 MARQUEZ D., BENJAMIN, CRETACEOUS TRANSGRESSION OF COAHUILA
B 6304 691* MARQUEZ, NE MEXICO, URANIUM DEPOSIT, ABST., GEOLOGY AND
M 25 411 MARRA MAMBA IRON FORMATION, WESTERN AUSTRALIA
B 6306 988 MARRALL, GERALD E., REVIEW OF URANIUM DEPOSITS, THEIR
B 6109 1560 MARRS, RONALD W., DETECTION OF PETROLEUM-RELATED SOIL
M 25 473 MARS, AND VENUS, RING, STRUCTURES, DISTINCT ON, EARTH, MOON
M 30 546 MARSA BREGA TROUGH, LIBYA, SIRTE BASIN
B 6203 570* MARSH, ABST., PHYSICAL AND SEDIMENTOLOGIC PROCESSES IN,
B 6303 556* MARSH AND EFFECT ON ESTUARINE SEDIMENT TRANSPORT AND
B 6203 489* MARSH DEPOSITS AND PRE-QUATERNARY COALS AND SHALES, ABST.,
S 7 96 MARSH ISLAND, SEISMIC FACIES EXAMPLE, LOUISIANA
B 6405 668* MARSH SEDIMENTS AS HYDROCARBON SOURCE BEDS, ABST.,
B 6405 800* MARSHALL ISLANDS, ABST., CONTEMPORANEOUS RECONSTRUCTION
B 6004 721 MARSHALL, N. F., CURRENTS IN SUBMARINE CANYON AND THEIR
S 8 1 MARSHALL, NEIL F., CURRENTS IN SUBMARINE CANYONS AND
B 6004 695 MARSHALL, NEIL F., LARGE SEDIMENT SLUMP REOPENS UNKNOWN,
B 6405 649 MARSHALL, PATRICK A. MCLOUGHLIN, AND GARY G. SULLIVAN, POAG,
M 25 391 MARSHALL, THOMAS, IRON ORE, COAL, AND STEEL PRODUCTION IN NEW
B 6309 1596 MARSHALL, WILLIAM D., CHARACTERISTICS OF LOWER VICKSBURG
B 6109 1546* MARSHLAND PEAT AS GEOLOGIC HAZARD IN GREATER NEW ORLEANS
B 6109 1454 MARTIN FORMATION, PEDREGOSA BASIN
B 6101 30 MARTIN FORMATION, SASKATCHEWAN
B 6209 1761 MARTIN, GENE B., PALEOECOLOGY AND POROSITY TRENDS OF
B 6108 1376 MARTIN, JOSEPH P., HELIUM DETECTION IN URANIUM EXPLORATION,
B 6105 780 MARTIN, J. P., HELIUM DETECTION IN URANIUM EXPLORATION
B 6309 1598 MARTIN, KEN G., YOU AINT SEEN NOTHING TILL YOUVE SEEN THE
B 6004 649 MARTIN, R. G., JR., LATE PLEISTOCENE-HOLOCENE EVOLUTION OF SOUTH
B 6303 491 MARTIN, RAY G., DISTRIBUTION OF SALT STRUCTURES IN GULF O
S 7 21 MARTIN, RAY G., NORTHERN AND EASTERN GULF OF MEXICO
S 7 3 MARTIN, RAY G., PHYSIOGRAPHY OF GULF OF MEXICO
B 6206 1092 MARTIN, S. J., ORGANIC GEOCHEMISTRY OF GAS-BEARING SHALES, U.S.
B 6004 675 MARTIN, WAYNE D., INFLUENC OF SOURCE AND PROCESSES IN
B 6004 724 MARTINEZ, J. D., HYDROLOGIC STABILIT OF NORTH LOUISIANA SALT
B 6009 1613 MARTINEZ, JOSEPH D., ENVIRONMENT-ENERGY DILEMMA, CHOICES AND
B 6004 666 MARTINI, I. P., SPECTRAL AND MARKOVIAN ANALYSIS OF BED
B 6203 539 MARTINI, I. PETER, REGRESSIVE SEDIMENTATION MODEL FOR JAMES BAY
S 4 235 MARTINIQUE, ANALCIDOMETRA ARMATA POURTALES
S 4 78 MARTINIQUE, CARBONATE SHELF
S 4 236 MARTINIQUE, PERICLIMENES ON NEMASTER RUBIGINOSA POURTALES
B 6007 1094 MARTINIQUE TRANSVERSE VALLEY
B 6405 694 MARTINO, RONALD L., TRACE FOSSIL ASSEMBLAGES OF UPPER
S 11 268 MARTINOTTIELLA COMMUNIS
S 11 230 MARTINOTTIELLA COMMUNIS
B 6410 1578 MARTINSBURG FORMATION, VIRGINIA
M 27 71 MARTINSBURG FORMATION, VIRGINIA, ORDOVICIAN
B 6304 652 MARTINSBURG SHALE, VIRGINIA
B 6205 888 MARTINSEN, R. S., HARTZOG DRAW, NEW GIANT OIL FIELD, ABST.
B 6203 540 MARTINSEN, R. S., HARTZOG DRAW, NEW GIANT OIL FIELD, ABST.
B 6405 793 MARTINSEN, R. S., SHANNON SANDSTONE, UPPER CRETACEOUS,
B 6303 491 MARTINSEN, RANDI S., FACIES AND RESERVOIR CHARACTERISTICS OF
B 6012 2185 MARTY, T. A., DESIGN OF HIGH-FREQUENCY-RESOLUTION AIRGUN
B 6104 585 MARUN FIELD, PERSIAN GULF
M 25 427 MARY KATHLEEN, URANIUM-RARE EARTHS, QUEENSLAND
S 3 129 MARY L MINE, COLORADO
B 6303 440* MARYLAND, ABST., SEDIMENTATION RATES IN COASTAL-PLAIN ESTUARY,
B 6309 1577* MARYLAND, AND SOUTHERN PENNSYLVANIA, ABST., REGIONAL,
B 6309 1465 MARYLAND, BALTIMORE CANYON, STRATIGRAPHIC COLUMN
B 6209 1610 MARYLAND, BLOOMSBURG RED BEDS
S 5 130 MARYLAND, CAYUGAN GROUP, SALINA FORMATION, TONOLOWAY
B 6209 1610 MARYLAND, COEYMANS LIMESTONE
B 6008 1288 MARYLAND, DEVELOPMENTS, 1975
B 6108 1269 MARYLAND, DEVELOPMENTS, 1976
B 6208 1400 MARYLAND, DEVELOPMENTS, 1977
B 6308 1245 MARYLAND, DEVELOPMENTS, 1978
B 6409 1404 MARYLAND, DEVELOPMENTS, 1979
B 6209 1610 MARYLAND, HELDERBERG GROUP
B 6209 1610 MARYLAND, HUNTERSVILLE CHERT
B 6209 1610 MARYLAND, KEYSER LIMESTONE
B 6209 1610 MARYLAND, MANDATA FORMATION
B 6209 1610 MARYLAND, MCKENZIE FORMATION
B 6209 1610 MARYLAND, NEEDMORE SHALE
B 6209 1610 MARYLAND, NEW SCOTLAND LIMESTONE
B 6409 1403† MARYLAND, OHIO, PENNSYLVANIA, VIRGINIA, AND WEST VIRGINIA IN
B 6308 1244† MARYLAND, OHIO, PENNSYLVANIA, VIRGINIA, AND WEST VIRGINIA IN
B 6208 1399† MARYLAND, OHIO, PENNSYLVANIA, VIRGINIA, AND WEST VIRGINIA, OIL
B 6108 1269† MARYLAND, OHIO, PENNSYLVANIA, VIRGINIA, AND WEST VIRGINIA, OIL
B 6008 1288† MARYLAND, OHIO, PENNSYLVANIA, VIRGINIA, AND WEST VIRGINIA, OIL
M 27 174 MARYLAND, ORDOVICIAN, BEEKMANTOWN LIMESTONE
B 6008 1298 MARYLAND, PRODUCTION, 1974-1975
B 6108 1281 MARYLAND, PRODUCTION, 1975-1976
B 6208 1410 MARYLAND, PRODUCTION, 1976-1977
B 6308 1247 MARYLAND, PRODUCTION, 1977-1978
B 6409 1412 MARYLAND, PRODUCTION, 1978-1979
B 6209 1610 MARYLAND, RIDGELEY SANDSTONE
B 6209 1610 MARYLAND, ROCHESTER SHALE
B 6209 1610 MARYLAND, SHRIVER CHERT
B 6209 1609 MARYLAND, SILURIAN-DEVONIAN CARBONATE SEQUENCE
B 6209 1610 MARYLAND, SILURIAN-DEVONIAN CARBONATE SEQUENCE,
B 6209 1610 MARYLAND, TONOLOWAY LIMESTONE
B 6209 1610 MARYLAND, WILLS CREEK SHALE
M 29 373 MASCLE, A., THE SULU SEA: A MARGINAL BASIN IN SOUTHEAST ASIA
B 6402 242 MASCLE, JEAN, SEISMIC STRUCTURE OF SOUTHWESTERN PELOPONNESUS
B 6304 649 MASCOT DOLOMITE, VIRGINIA
B 6405 602 MASIRAH LINE, MIDDLE EAST
B 6206 985 MASJED-E SULEYMAN FIELD, IRAN
B 6301 90 MASKONOMET BEDS, SCOTIAN BASIN
B 6305 834 MASLYN, R. MARK, HOT-SPRING-GENERATED KARST FEATURES NEAR
B 6305 826 MASLYN, R. MARK, PALEOKARST CONTROLS OF LOCALIZATION OF
B 6002 280 MASON AND DIXON LINE, LOCATION OF DELTA LOBES
S 5 78 MASON COUNTY LOCATION, MICHIGAN,
S 5 3 MASON COUNTY, LOCATION, MICHIGAN,
B 6010 1865 MASON, JOHN F., PETROLEUM DEVELOPMENTS IN MIDDLE EAST
B 6310 1884 MASON, JOHN F., PETROLEUM DEVELOPMENTS IN MIDDLE EAST
B 6210 2111 MASON, JOHN F., PETROLEUM DEVELOPMENTS IN MIDDLE EAST
B 6411 1836 MASON, JOHN F., PETROLEUM DEVELOPMENTS IN MIDDLE EAST
B 6110 1795 MASON, JOHN F., PETROLEUM DEVELOPMENTS IN MIDDLE EAST
B 6010 1900 MASON, JOHN F., SELECTED REFERENCES ON MIDDLE EAST, 1970 TO 1975
B 6308 1427 MASON, JOHN W., INDEPENDENT GEOLOGISTS--ENDANGERED SPECIES,
B 6303 491 MASON, JOHN W., INDEPENDENT GEOLOGISTS--ENDANGERED SPECIES,
B 6405 797* MASON LAKE FIELD, MUSSELSHELL COUNTY, MONTANA, ABST.
B 6409 1563 MASON, ROBERT C., GEOLOGY OF GUM ISLAND NORTH AND FRENCH
S 10 47† MASS BALANCE AND GEOLOGICAL CONSTRAINTS ON MIGRATION
B 6403 441* MASS MOVEMENT EFFECTS ON A BATHYAL MACROFAUNAL POPULATION,
B 6203 512* MASS PHYSICAL AND ENGINEERING PROPERTIES OF GLACIAL SEDIMENTS
B 6007 1112 MASS SPECTROMETRIC ANALYSIS, GRAND RAPIDS OIL
B 6007 1112 MASS SPECTROMETRIC ANALYSIS, LEDUC OIL
B 6007 1112 MASS SPECTROMETRIC ANALYSIS, LLOYDMINSTER OIL
B 6007 1112 MASS SPECTROMETRIC ANALYSIS, MCMURRAY OIL
S 7 91 MASS TRANSPORT, SEDIMENT PROCESSES
B 6303 438* MASS WASTING ON CONTINENTAL RISE OF EASTERN SOUTH AMERICA,
B 6002 239 MASS-BALANCE CALCULATIONS
S 7 114 MASS-TRANSPORT SCAR, SEISMIC EXAMPLE, TEXAS
B 6004 672* MASSACHUSETTS, ABST., MESOZOIC, STRATIGRAPHIC, TEST, NANTUCKET
B 6405 681* MASSACHUSETTS, ABST., TIDAL-CURRENT, SAND, WAVES IN, VINEYARD
B 6105 830* MASSACHUSETTS AND RHODE ISLAND, ABST., RENEWED EXPLORATION
M 28 8 MASSACHUSETTS, DEVONIAN, GNEISS

B 6101 80 MASSACHUSETTS, HARTFORD-DEERFIELD BASIN, CONNECTICUT-
M 28 48 MASSACHUSETTS, ORDOVICIAN, SCHIST
M 28 56 MASSACHUSETTS, PALEOZOIC, SCHIST
M 28 48 MASSACHUSETTS, SILURIAN, SCHIST
M 28 53 MASSACHUSETTS, TRIASSIC, ARKOSE
S 3 23 MASSIF CENTRAL, FLUORINE, FRANCE
B 6105 843* MASSIVE LIMESTONE, PERMIAN, OF GUADALUPE MOUNTAINS, NEW
B 6303 496* MASSIVE MARINE-SANDSTONE FACIES, MACKEREL FIELD, GIPPSLAND
B 6206 936 MASSIVE SANDSTONES
M 25 506 MASSIVE SULFIDE DEPOSITS OF METALS, CANADIAN, CORDILLERA
B 6303 542 MASSON, P., PALEOENVIRONMENT AND PETROLEUM POTENTIAL OF
B 6412 2051 MASSON, P., PALEOENVIRONMENT AND PETROLEUM POTENTIAL OF
B 6302 152 MASTERS, JOHN A., DEEP BASIN GAS TRAP, WESTERN CANADA
B 6303 492 MASTERS, JOHN A., ENGINEERING GEOLOGY IN DEEP BASIN CANADA. BY
B 6203 540 MASURSKY, HAROLD, TERRESTRIAL EVOLUTION—INSIGHTS FROM THE
B 6002 325* MATADOR UPLIFT AREA. BY TITLE ONLY, DEPOSITIONAL SYSTEMS IN
B 6104 544 MATAGORDA BAY, SEDIMENTATION, TEXAS
B 6308 1225 MATAGORDA COUNTY, TEXAS, S. W. RUGELY FIELD
B 6209 1761* MATAGORDA ISLAND, CALHOUN COUNTY, TEXAS, ABST., MIOCENE
M 25 214 MATAMATEAONGA FORMATION, NEW ZEALAND
B 6103 387 MATANILLA REEF EAST, LITTLE BAHAMA BANK
M 24 4 MATANUSKA FORMATION, ALASKA
B 6203 377 MATANUSKA FORMATION, ALASKA
B 6402 245 MATAPAN SEAMOUNTS, GREECE
B 6402 261 MATAPAN TROUGH, IONIAN SHELF
B 6405 648 MATHENY, WENGERD, SHERMAN A., REVIEW OF, OIL AND GAS FIELDS OF
B 6405 745 MATHEW, DAVID, DEVELOPMENT OF CONCEPTUAL MODEL TO
B 6203 503 MATHIAS, K. E., GEOTHERMAL RESOURCE INVESTIGATIONS AND
B 6006 926 MATICK, ROBERT E., REGIONAL GEOLOGIC FRAMEWORK OFF
B 6206 934 MATILIJA FORMATION, TURBIDITES, CALIFORNIA
B 6405 740* MATILIJA SANDSTONE, AN EOCENE SAND-RICH, DEEP-SEA FAN AND
B 6401 59† MATRIX IN KEROGEN PYROLYSIS--INFLUENCE ON PETROLEUM
B 6401 65 MATRIX ON FREE HYDROCARBONS, EFFECT OF, MINERAL
B 6401 62 MATRIX ON HYDROCARBON YIELD, EFFECT OF, MINERAL
B 6206 943 MATRIX-SUPPORTED BEDS
B 6203 540 MATSON, ROBERT E., MONTANAS COAL—IMPORTANT ENERGY
B 6008 1403 MATSON, ROBERT E., RECENT DEVELOPMENTS IN COAL EXPLORATION IN
S 6 143 MATSUMOTO, TATSURO, ISOTOPIC AGES AND STRATIGRAPHIC CONTROL
B 6405 705 MATTERN, JOE, GEOSTATISTICAL APPROAC IN COAL RESOURCE
B 6303 492 MATTES, BRET, BURIAL DOLOMITIZATION OF UPPER DEVONIAN MIETTE
S 5 73 MATTHEWS, DAVID R. EVAPORITE CYCLES AND LITHOFACIES IN LUCAS
B 6109 1537 MATTHEWS, MICHELE, INTERIOR STRUCTURE OF OAK GROVE RIDGE
B 6408 1285 **MATTHEWS, R. DAVID, DEVONIAN OIL SHALE OF EASTERN UNITED STATES,**
B 6203 489 MATTHEWS, R. K., CARBON AND OXYGEN ISOTOPES AS INDICATORS OF
B 6205 867 MATTHEWS, VINCENT, III, CORRELATION OF PINNACLES AND NEENACH
B 6012 2128 MATTHEWS, VINCENT, III, CORRELATION OF PINNACLES AND NEENACH
B 6207 1226 MATTHEWS, WILLIAM H., INTEGRATING ENVIRONMENTAL
B 6105 826 MATTICK, R. E., ATLANTIC OUTER CONTINENTAL SHELF, ABST.,
B 6204 592 MATTICK, R. E., PETROLEUM POTENTIAL OF U.S. ATLANTIC SLOPE, RISE,
B 6105 790 MATTICK, ROBERT E., DEPTH CONVERSION OF MULTICHANNEL SEISMIC-
B 6405 746 MATTICK, ROBERT E., GEOLOGIC SETTING AND OIL AND GAS POTENTIAL
M 29 65 MATTICK, ROBERT E., MULTICHANNEL SEISMIC DEPTH SECTIONS AND
B 6105 811 MATTICK, ROBERT E., PETROLEUM POTENTIAL OF POSSIBLE LOWER
B 6006 926 MATTICK, ROBERT E., REGIONAL GEOLOGIC FRAMEWORK OFF
B 6105 742 MATTICK, ROBERT E., REGIONAL GEOLOGIC FRAMEWORK OFF
B 6406 932 MATURATION AND DILATANCY, KINETICS OF HYDROCARBON
B 6312 2128† MATURATION INDICATORS, COMPILATION AND CORRELATION OF
B 6312 2139 MATURATION INDICATORS, GLOSSARY, THERMAL
B 6209 1586 MATURATION INDICES AND ZONES OF PETROLEUM GENERATION AND
B 6406 927† MATURATION KINETICS, MICROFRACTURE DEVELOPMENT IN,
B 6004 673* MATURATION OF ATHABASCA BITUMEN, ABST., SIMULATED,
B 6309 1575* MATURATION OF DISPERSED ORGANIC MATTER IN NEW ALBANY SHALE
B 6311 2033 MATURATION OF POTENTIAL SOURCE ROCK, GEOTHERMAL
B 6303 547* MATURATION ON CHEMICAL COMPOSITION OF GULF COAS CRUDE OILS,
B 6303 447* MATURATION TEMPERATURE FROM MAGNETIC DATA, BLIGH WATER,
B 6408 1215 MATURE BASINS, EVALUATING
M 28 105 MATURE SEDIMENTARY, TEXTURE
M 29 441 MATURE SEDIMENTS, NEW JERSEY, MARGIN
M 29 439 MATURE SEDIMENTS, NORTH AMERICA, EASTERN MARGIN
B 6104 504† MATURITY, COLOR OF, KEROGEN AS INDEX OF, ORGANIC
B 6304 592 MATURITY MAPS, GEOCHEMICAL TOOL
M 24 136 MATUSZCZAK, R. A., WATTENBERG FIELD, DENVER BASIN, COLORADO
B 6309 1586* MAUCH CHUNK ALLUVIAL PLAIN AND MUD-FLA SEDIMENTS IN
B 6405 746 MAUGHAN, DAVID M., DEFINITION AND DEVELOPMENT OF MACKEREL
B 6201 98 MAUGHAN, EDWIN K., ORGANIC GEOCHEMISTRY, INCIPIENT
B 6108 1375 MAUGHAN, EDWIN K., ORGANIC GEOCHEMISTRY, INCIPIENT
B 6405 746 MAUGHAN, EDWIN K., PHOSPHORITE, ORGANIC CARBON, AND
B 6205 888 MAUGHAN, EDWIN K., PROBABLE PERMIAN SOURCE ROCKS,
M 25 215 MAUI FIELD, NEW ZEALAND
M 25 211† MAUI GAS-CONDENSATE FIELDS OF NEW ZEALAND--SUMMARY, KAPUNI
B 6203 541 MAUREAU, G. T., PREDICTION OF POROSITY IN PERMIAN, ZECHSTEIN 2.
M 29 131 MAURICE EWING BANK, DESCRIBED, NORTH SCOTIA RIDGES
B 6110 1749 MAURITANIA, DEVELOPMENTS, 1976
B 6210 1848 MAURITANIA, DEVELOPMENTS, 1977
B 6310 1693 MAURITANIA, DEVELOPMENTS, 1978
B 6411 1789 MAURITANIA, DEVELOPMENTS, 1979
B 6110 1749 MAURITIUS, DEVELOPMENTS, 1976
B 6210 1848 MAURITIUS, DEVELOPMENTS, 1977
B 6310 1693 MAURITIUS, DEVELOPMENTS, 1978
B 6411 1789 MAURITIUS, DEVELOPMENTS, 1979
S 4 291 MAURITIUS, GALAXEA
B 6303 492 MAURMEYER, EVELYN M., DETERMINATION OF STORM OVERWASH
B 6105 805 MAURMEYER, EVELYN M., STRUCTURE OF COASTAL WASHOVER FANS AND
B 6004 695 MAURRASSE, F. J-M. R., PALEOGENE RADIOLARIAN
B 6105 811 MAURRASSE, FLORENTIN J. M. R., PALEOTECTONIC IMPLICATIONS OF
B 6405 747 MAURRASSE, FLORENTIN, RADIOLARIAN BIOSTRATIGRAPHY AND
B 6309 1604* MAVERICK BASIN, ABST., FACIES, DIAGENESIS, AND POROSITY RELATIONS
B 6309 1611* MAVERICK BASIN, SOUTH TEXAS, ABST., WAVE-DOMINATED DELTAIC
B 6301 41 MAVERICK BASIN, TEXAS
B 6309 1583 MAXEY, MARILYN H., OIL- AND GAS-PRODUCING CONFIGURATIONS IN
B 6004 724* MAXIANALYSIS, ABST., MINICOMPUTERS FOR
B 6002 287 MAXIMOV, S. P., EARLY DIAGENETIC GENERATION OF HYDROCARBON
B 6203 566* MAXON FORMATION, LOWER CRETACEOUS, MARATHON REGION, WEST
B 6303 492 MAY, FRED E., DINOFLAGELLATE ASSEMBLAGES FROM SURFACE AN
B 6205 885 MAY, FRED, MAESTRICHTIAN CONGLOMERATES IN SOUTHWESTERN
B 6203 495* MAY, NEW JERSEY, ABST., SLOPE, MAP DEPICTING MAJOR SUBMARINE,
B 6103 360 MAY, RICHARD T., URANIUM POTENTIAL OF PERMIAN AND
B 6004 650* MAYA ARCHAEOLOGY, ABST., NUMERICAL ASPECTS OF, CHEMICAL
M 30 395 MAYARO FORMATION, PLIOCENE, TEAK FIELD
B 6405 759 MAYNARD, AMY, SAND DISPERSAL AT NORDERNEYER SEEGAT, WEST
B 6405 769 MAYNARD, J BARRY, SHALE—AN OVERVIEW, ABST.
B 6303 493 MAYNARD, JAMES I., DIAGENETIC ALTERATION IN EOCENE
B 6303 493 MAYOU, TAYLOR V., SHELL BORINGS IN UPPER MIOCENE-PLIOCENE
B 6109 1449 MAZATZAL UPLIFT, ARIZONA
M 25 548 MAZCA AND SOUTH AMERICAN PLATES, PLATE-TECTONIC SUMMARY
M 27 105 MAZON CREEK COAL BALLS, ILLINOIS, PENNSYLVANIAN
B 6409 1561 MAZZULLO, S. J., JURASSIC GEOLOGY AND HYDROCARBON POTENTIAL
B 6203 541 MAZZULLO, S. J., SUBMARINE CEMENTS AND FABRIC-RETENTIVE
B 6303 493 MAZZULLO, S.J., IN-SITU FORMATION OF BOTRYOIDAL ARAGONITE ON
B 6309 1596* MCALLEN RANCH FIELD, HIDALGO COUNTY, TEXAS, ABST.,
B 6409 1566* MCALLEN RANCH FIELD, HIDALGO COUNTY, TEXAS, ABST.,
B 6203 541 MCALLISTER, JAMES F., STEEPLY INCLINED ALGAL COLUMNS IN LATE
B 6102 253 MCANULTY, W. N., HISTORY OF EL PASO GEOLOGICAL SOCIETY
M 25 425 MCARTHUR GROUP, GROOTE EYLANDT
B 6303 500* MCARTHUR GROUP, NORTHERN TERRITORY, AUSTRALIA, ABST.,
B 6203 547 MCARTHUR, RICHARD E., INDIGENOUS PRECAMBRIAN PETROLEUM
B 6410 1681 MCARTHUR, RICHARD E., INDIGENOUS PRECAMBRIAN PETROLEUM
M 24 6 MCARTHUR RIVER FIELD, ALASKA
B 6408 1132 MCARTHUR RIVER OIL FIELD, ALASKA
B 6203 541 MCAULIFFE, C. D., ROLE OF SOLUBILITY IN MIGRATION OF PETROLEUM
S 10 89 MCAULIFFE, CLAYTON D., OIL AND GAS MIGRATION: CHEMICAL AND
B 6305 761 MCAULIFFE, CLAYTON D., OIL AND GAS MIGRATION--CHEMICAL AN
M 25 211 MCBEATH, D. M., KAPUNI AND MAUI GAS-CONDENSATE FIELDS OF NEW
B 6402 225 MCBEE GRABEN, TEXAS
B 6404 585 MCBIRNEY, MORAN, W. R., REVIEW OF, VOLCANOLOGY, BY HOWEL,
B 6209 1761 MCBRIDE, E. F., POROSITY LOSS IN SANDSTONE BY DUCTILE GRAIN
B 6004 725 MCBRIDE, EARLE F., FACTORS INFLUENCIN POROSITY AND
B 6305 834 MCBRIDE, EARLE F., IMPORTANCE OF SECONDARY POROSITY I
B 6405 747 MCBRIDE, EARLE F., IMPORTANCE OF SECONDARY POROSITY IN
B 6105 785 MCBRIDE, EARLE F., POISSON BED-THICKNESS DISTRIBUTION AND
B 6004 732 MCBRIDE, EARLE F., TRANSGRESSIVE-REGRESSIVE DEPOSITS OF DIFUNTA
B 6309 1607 MCBRIDE, MARY W., FACIES DISTRIBUTION OF HOSSTON SAND, CENTRAL
B 6409 1366 MCBROOM, R. L., DEVELOPMENTS IN NORTH-CENTRAL TEXAS IN 1979
B 6305 834 MCBRYDE, JOHN C., PENNSYLVANIAN COARSE-GRAINED MEANDERING
B 6004 655 MCCABE, P. J., EFFECTS OF ICE ON SEDIMENTS ON SANDY BRAIDED
B 6004 695 MCCABE, P. J., KINDERSCOUTIAN DELTA, CARBONIFEROUS, OF
B 6203 542 MCCABE, PETER J., COARSE-GRAINED DEPOSITS BY EPHEMERAL STREAMS
B 6405 747 MCCABE, PETER J., TERRESTRIAL SEDIMENTATION ASSOCIATED WITH
B 6305 830 MCCALLUM, M. E., UPDATE ON EXPLORATION FOR DIAMONDS IN
B 6105 811 MCCARTHY, BRUCE, TRACE FOSSILS IN RELATION TO WATER DEPTH IN
B 6209 1761 MCCARTHY, JAMES A., MIOCENE GEOLOGY OF MATAGORDA ISLAND,
B 6405 747 MCCARTY, KEVIN L., TERRESTRIAL SEDIMENTATION ASSOCIATED WITH
B 6102 296 MCCASLAND, B. C., JR., TEXAS URANIUM, ABST.
B 6002 320 MCCAULEY, JAMES R., KANSAS GEOLOGY AS SEEN BY ERTS-1, ABST.
B 6002 322* MCCLAIN, GRADY, AND CANADIAN COUNTIES, OKLAHOMA, ABST.,
B 6309 1585 MCCLELAND, STEVEN, HOW GOOD ARE REGIONAL COAL RESERVE
B 6012 2185 MCCLELLAN, PATRICK H., NEW EVIDENCE FOR AGE OF CENOZOIC SAL
S 5 111 MCCLINTOCK, P. L. SEISMIC DATA-PROCESSING TECHNIQUES IN
S 5 169* (MESSINIAN) DEPOSITS OF SALEMI BASIN (SICILY) AND A MODERN

B 6408 1287 MCCOLLOCH, GAYLE H., JR., COAL IN WEST VIRGINIA—GEOLOGY AND
B 6401 29 MCCONNELL THRUST FAULT, ALBERTA
B 6305 845* MCCONNELSVILLE FIELD, MORGAN COUNTY, OHIO, ABST., PETROLEUM
B 6004 695 MCCOY, F. W., SEDIMENT LITHOFACIES IN SOUTH ATLANTIC OCEAN,
B 6405 748 MCCOY, S., JR, PRELIMINARY MOLLUSCAN BIOSTRATIGRAPHY OF GULF
B 6207 1226 MCCRACKEN, K. G., AUSTRALIAN GEOPHYSICS—CURRENT PERFORMANCE
B 6309 1607 MCCRACKEN, WILLARD A., FIELD RELATIONS AND PETROLOGY OF
B 6303 493 MCCRACKEN, WILLARD A., PETROLOGY OF MIOCENE CATAHOULA
B 6303 494 MCCULLAGH, MICHAEL J., CREATION AND APPLICATION OF VARIABL
B 6105 812 MCCULLAGH, M. J., MAPPING RESOURCES BY MINICOMPUTER
B 6303 421 MCCULLOH, T. H., VITRINITE REFLECTANCE, PRESENT ROCK
B 6403 446 MCCULLOH, THANE H., SUBSURFACE INTERFACE BETWEEN ZEOLITIZED
B 6108 1392 MCCULLOH, THANE H., WIDESPREAD OCCURRENCE OF LAUMONTITE IN
B 6002 326 MCCULLOUGH, RICHARD, NEILL SOUTH, ODOM LIME, FIELD, NOLAN AND
B 6004 696 MCCUNN, H. J., ORIGIN OF DOLOMITE--MODEL BASED ON CYCLIC
B 6004 696 MCCUNN, H. J. , ORIGIN OF DOLOMITE--MODEL BASED ON CYCLIC
B 6203 351 MCDONALD ANTICLINE OIL FIELD, CALIFORNIA
B 6004 696 MCDONALD, D. R., CONTOURING GEOLOGIC DATA FROM ARBITRARY
B 6105 812 MCDONALD, D. R., CONTOURING GEOLOGIC DATA FROM ARBITRARY POINTS
B 6108 1390 MCDONALD, DAVID A., ROLE OF SECONDARY POROSITY IN SANDSTONE
M 24 97 MCDONALD DRAW FIELD, WYOMING
B 6211 2357 MCDONALD, JOHN A., MODERN METHODS OF ACQUIRING AND
B 6306 975 MCDONALD, K. C., NUBIA FORMATION OF CENTRAL EASTERN DESERT,
M 24 91 MCDONALD, ROBERT E., BIG PINEY--LA BARGE PRODUCING COMPLEX,
B 6211 2267 MCDONALD SHALE, CALIFORNIA
S 6 119 MCDOUGALL, IAN, POTASSIUM-ARGON ISOTOPIC DATING METHOD AND
B 6303 494 MCDOUGALL, KRISTIN,, PALEOECOLOGIC EVALUATION OF LATE EOCENE
B 6406 956 MCDOWELL, FRED W., REVIEW OF LECTURES IN ISOTOPE GEOLOGY
S 6 57 MCELHINNY, M. W., THE MAGNETIC POLARITY TIME SCALE--PROSPECTS
B 6303 494 MCELROY, M. N., COMPUTER DRAFTING--ITS APPLICATIO IN PETROLEUM
B 6109 1555 MCFARLAN, EDWARD, JR., LOWER CRETACEOUS SEDIMENTARYFACIES
B 6004 696 MCGANNON, DONALD E., JR., CARBON AND OXYGEN ISOTOPE ANALYSES
B 6002 159 MCGEE, DEAN A., IN THE PUBLIC INTEREST
B 6303 552 MCGEHEE, R. V., BRAIDED-STREAM PROCESSES AND FACIES IN MODERN
B 6303 511 MCGILLIS, K. A., PALO DURO BASIN ANALYSIS, ABST.
B 6408 1288 MCGILLIS, KATHY A., RED-BED EVAPORITE, AND CARBONATE FACIES
B 6203 528 MCGLEW, R. J., COMPUTER APPLICATIONS IN REGIONAL STUDY OF
B 6303 495 MCGOOKEY, D. P., CONTOURITE ANTICLINES AS EXPLORATION
B 6401 133 MCGOOKEY, D. P., REVIEW OF PALEOGEOGRAPHIC PRINCIPLES OF OI AND
B 6405 748 MCGOVNEY, JAMES E., POROSITY EVOLUTION OF NIAGARAN
B 6102 257 MCGOWAN CREEK FORMATION, IDAHO
B 6305 835 MCGOWEN, J. H., DEPOSITIONAL FACIES AND URANIUM OCCURRENCE
B 6109 1542 MCGOWEN, J. H., DEPOSITIONAL FRAMEWORK OF LOWER DOCKUM
B 6004 700 MCGOWEN, J. H., LARGE-SCALE BED FORMS AN SEDIMENTARY
B 6303 552 MCGOWEN, J. H., MAP DISPLAY FORMATS FOR ENVIRONMENTAL
B 6004 716 MCGOWEN, J. H., SEDIMENTOLOGY AND PALEOECOLOGY OF TYPE
B 6409 1563 MCGOWEN, MARY K., DEPOSITIONAL SYSTEMS IN NACATOCH SAND,
B 6405 792 MCGRAIL, DAVID W., ET AL, GENESIS OF SAND RIDGES ON
B 6109 1542 MCGRAIL, DAVID W., INTERNAL WAVES AND NEPHELOID LAYER ON
B 6203 542 MCGRAIL, DAVID W., SHELF SEDIMENT AND LOCAL FLOW PHENOMENA—
B 6303 495 MCGRAW, M. M., DEPOSITIONAL ENVIRONMENTS AND THEIR RELATION
M 24 250 MCGRAW SANDSTONE, MISSISSIPPI
B 6106 918 MCGREGOR, BONNIE A., CONTINENTAL SLOPE SEDIMENT INSTABILITY
M 29 139 MCGREGOR, BONNIE A., VARIATIONS IN BOTTOM PROCESSES ALONG THE
B 6203 542 MCHARGUE, T. R., ANCIENT SUBMARINE CANYON AND CHANNEL
B 6303 495 MCHARGUE, TIMOTHY R., DOWNCURRENT CHANGE IN MIOCENE
B 6309 1582 MCHONE, JOHN F., JR., METEORITIC IMPACT--RESERVOIR-FORMING
B 6203 395 MCHUGH COMPLEX, ALASKA
B 6309 1583 MCHUGH, STUART L., INTERACTION OF PROPPANTS WITH CRACK
B 6209 1752 MCILHENNY, ROBERT C., RADIOMETRIC INVESTIGATION RELATIVE TO A
B 6004 697 MCILREATH, IAN, DEVELOPMENT OF VERTICAL ESCARPMENT AT MIDDLE
B 6207 1194 MCINTOSH FORMATION, OREGON AND, WASHINGTON
M 25 422 MCINTOSH, J. L., OCCURRENCE AND DEVELOPMENT OF SEDIMENTARY
B 6304 699 MCINTOSH, R. A., PRUDHOE BAY, A TEN-YEAR PERSPECTIVE, ABST.
M 30 289 MCINTOSH, R. A., PRUDHOE BAY--A 10-YEAR PERSPECTIVE
B 6206 1092 MCIVER, R. D., ORGANIC GEOCHEMISTRY OF GAS-BEARING SHALES, U.S.
B 6309 1592 MCIVER, R. D., PRESSURE CORING OF GAS-BEARING DEVONIAN BLACK
B 6309 1591 MCIVER, R. D., PROJECTION OF FAVORABLE GAS-PRODUCING AREAS
B 6109 1407 MCIVER, RICHARD D., HYDROCARBONS BETWEEN OILS AND ASSOCIATED
B 6109 1539 MCKALIPS, DAWN G., GEOMETRY OF GULF COAST AQUIFERS,
B 6305 831 MCKALIPS, DAWN G., NONMARINE DEPOSITIONAL ENVIRONMENTS AND
B 6211 2362 MCKEE, EDWIN H., TERTIARY GEOLOGY AND OIL SHALE RESOURCES OF
B 6303 495 MCKEE, JAMES W., CRETACEOUS TRANSGRESSION OF COAHUILA
B 6004 722 MCKEE, T. R., GEOCHEMICAL CONTROLS ON VERTICAL DISTRIBUTION OF
S 7 313 MCKEE, T. R., HOLOCENE SEDIMENT GEOCHEMISTRY OF CONTINENTAL
S 7 289 MCKEE, THOMAS R., INTRASLOPE BASIN IN NORTHWEST GULF OF
B 6209 1610 MCKENZIE FORMATION, MARYLAND
B 6211 2309 MCKENZIE FORMATION, WEST VIRGINIA
B 6405 634 MCKENZIE FORMATION, WEST VIRGINIA
B 6004 724* MCKENZIE FORMATION, WEST VIRGINIA, ABST., EVOLUTION OF,
B 6309 1585* MCKENZIE FORMATION, WEST VIRGINIA, ABST., PATCH, REEFS AND
B 6303 496 MCKENZIE, JUDITH ANN, MOVEMENT OF SUBSURFACE WATERS UNDER
B 6203 546 MCKEON, JOHN B., SMALL ROCK OUTCROPS IN VEGETATED TERRAIN
B 6305 836 MCKINNEY, D. BROOKS, GEOLOGY AND STRUCTURAL CONTROL OF
B 6412 2118 MCKIRDY, DAVID M., ORGANIC GEOCHEMISTRY OF
B 6203 351 MCKITTRICK OIL FIELD, CALIFORNIA
B 6309 1604* MCKNIGHT, AND WEST NUECES CARBONATE ROCKS OF MAVERICK
B 6109 1538* MCKNIGHT DIAGENESIS, ABST., JFS FIELD, DIMMIT COUNTY, TEXAS—
B 6402 222 MCKNIGHT FORMATION, TEXAS
S 6 1 MCLAREN, D. J., DATING AND CORRELATION, A REVIEW
B 6107 1014 MCLAREN-WASECA MEMBERS, ALBERTA AND, SASKATCHEWAN, COLONY-
B 6105 796 MCLEAN, H., BASIN ANALYSIS OF LOWER MIDDLE MIOCENE STRAT, SANTA
B 6012 2185 MCLEAN, HUGH, ELEMENTS OF TECTONISM DEDUCED FROM MIOCENE
B 6012 2177 MCLEAN, HUGH, MIOCENE VOLCANIC AND VOLCANICLASTIC ROCKS OF
B 6309 1467 MCLEAN, HUGH, REVIEW OF PETROLEUM GEOLOGY OF ANADYR AND
B 6403 450 MCLEAN, HUGH, STRATIGRAPHIC AND STRUCTURAL RELATIONS TO
B 6211 2357 MCLEAN, HUGH, STRATIGRAPHY AND SEDIMENTOLOGY OF TERTIARY
B 6309 1522 MCLEAN, HUGH, TERTIARY STRATIGRAPHY AND PETROLEUM
B 6012 2184 MCLEAY R. D., STRATIGRAPHIC EXPLORATION WITH SEISLOG METHOD,
B 6405 649 MCLOUGHLIN, AND GARY G. SULLIVAN, POAG, C. WYLIE, REVIEW OF,
S 8 1 MCLOUGHLIN, PATRICK A., CURRENTS IN SUBMARINE CANYONS AND
B 6009 1613 MCMICHAEL, C. E., COMPETITION AND QUEST FOR OFFSHORE LEASES,
B 6009 1607 MCMICHAEL, CLAUDE E., DOOR POINT--BURIED VOLCANO IN SOUTHEAST
B 6011 2069 MCMILLAN, N. J., GEOLOGY OF LABRADOR SEA AND ITS PETROLEUM
B 6004 672 MCMILLAN, T. L., BIOHERMAL DEVELOPMENT IN WOLFCAMPIAN
B 6209 1760 MCMILLEN, KEN, LIVING SHELLED MICROZOOPLANKTON,
B 6309 1598 MCMILLEN, KEN, RADIOLARIAN DISTRIBUTION, DIVERSITY AND DENSITY
B 6309 1597 MCMILLEN, KEN, RELICT AND EXPATRIATE RADIOLARIAN FAUNA IN
B 6203 542 MCMILLEN, KENNETH J., LATE QUATERNARY SEDIMENTATION ON UPPER
B 6303 426 MCMILLEN, KENNETH J., MIOCENE SUBMARINE FAN IN DEEP WESTERN
B 6105 812 MCMILLEN, KENNETH J., PRESERVATION AND DISSOLUTION OF
B 6303 531 MCMILLEN, KENNETH J., RADIOLARIA--PRESENT AND PAST INDICATORS
B 6303 483 MCMILLEN, KENNETH J., RADIOLARIAN PRESERVATION IN PRESENT AND
B 6309 1607 MCMILLEN, KENNETH J., RADIOLARIAN RATIOS AN
B 6312 2204 MCMILLEN, KENNETH J., SEISMIC EVIDENCE FOR WIDESPREAD POSSIBLE
B 6007 1112 MCMURRAY OIL, MASS SPECTROMETRIC ANALYSIS
B 6007 1110 MCMURRAY SANDS, ORIGIN OF OIL IN, ALBERTA
B 6007 1112 MCMURRAY SANDS, ORIGIN OF OIL IN, ALBERTA
B 6212 2511 MCMURTRY, GARY M., EVIDENCE FOR METALLOGENESIS ON OCEANIC
B 6303 487 MCNAB, G., BEATRIC FIELD, MORAY FIRTH, NORTH SEA, ABST.
M 30 117 MCNAB, GREG, THE BEATRICE FIELD, INNER MORAY FIRTH, U.K. NORTH
B 6409 1554 MCNULTY, C. L., CALPIONELLIDS AND NANNOCONIDS OF TARAISES
B 6011 2058 MCNULTY, C. L., JR., GLOBOTRUNCANA FORNICATA ZONE OF UPPER
B 6409 1560 MCNULTY, CHARLES L., FORAMINIFERS AND CALCISPHERES FROM
B 6309 1608 MCNULTY, CHARLES L., MERIDIONALLY ORNAMENTED HEDBERGELLID
B 6405 748 MCPEEK, L. A., EASTERN GREEN RIVER BASIN—A DEVELOPING GIAN GAS
B 6203 543 MCQUILLAN, MICHAEL WILLIAM, CONTEMPORANEOUS FAULTS—
B 6406 831 MCWHAE RIDGE, AUSTRALIA
B 6207 1193 MCWILLIAMS, ROBERT G., EARLY TERTIARY RIFTING IN WESTERN
B 6207 1159† MEAD AREA OF ARIZONA AND NEVADA DURING LATE MIOCENE TIME,
B 6307 1139† MEAD AREA OF ARIZONA AND NEVADA DURING LATE MIOCENE TIME.
B 6009 1624 MEAD, JUDSON, PRECAMBRIAN GEOPHYSICAL PROVINCES IN INDIANA,
B 6008 1396 MEAD, RICHARD H., HELIUM DETECTION EXPLORATION GUIDE TO
B 6108 1376 MEAD, RICHARD H., HELIUM DETECTION IN URANIUM EXPLORATION,
B 6201 108 MEADE PEAK MEMBER, GAS CHROMATOGRAPHIC ANALYSES
B 6105 780 MEAD, R. H., HELIUM DETECTION IN URANIUM EXPLORATION
B 6201 99 MEADE PEAK PHOSPHATIC SHALE MEMBER, IDAHO
B 6207 1172 MEADE PEAK PHOSPHATIC SHALE MEMBER, PHOSPHORIA FORMATION,
B 6201 99 MEADE PEAK PHOSPHATIC SHALE MEMBER, UTAH
B 6201 105 MEADE PEAK PHOSPHATIC SHALE MEMBER, WESTERN INTERIOR, UNITED
B 6201 99 MEADE PEAK PHOSPHATIC SHALE MEMBER, WYOMING
B 6307 1040 MEADE THRUST, WYOMING
B 6012 2183 MEADOURS, BRUCE, REGIONAL VELOCITY REDUCTION, ABST.
B 6004 689* MEADOWCROFT ROCK SHELTER, AVELLA, PENNSYLVANIA, ABST.,
M 27 154 MEAGHER FORMATION, MONTANA, CAMBRIAN
M 27 6 MEAGLER LIMESTONE, MONTANA, CAMBRIAN
B 6305 831* MEANDER-BELT SYSTEM, LOWE WASATCH AND UPPER FORT UNION
B 6305 834* MEANDERING FLUVIAL DEPOSITS, TAOS TROUGH, NORTH-CENTRAL NEW
B 6105 827* MEANDERING TRANSITION ZONE—RED RIVER, TEXAS AND OKLAHOMA,
B 6009 1613 MEANS, JOHN A., SUNOCO-FELDA AND WEST SUNOCO-FELDA FIELDS
B 6404 572 MEARAT SEDOM CAPROCK, ISRAEL
B 6404 572 MEARAT SEDOM SALT MEMBER, ISRAEL
S 9 9 MEASURED PALEOTEMPERATURES, PALEOCLIMATE INDICATORS
S 10 122 MEASUREMENT EQUATIONS, GROUNDWATER FLOW, PRINCIPLES
M 26 28 MEASUREMENT, ANISOTROPY, LABORATORY
M 26 24 MEASUREMENT ATTENUATION, FLUID MOTION, LABORATORY
M 26 30 MEASUREMENT, DYNAMIC MODULI, LABORATORY
M 26 22 MEASUREMENT, GERMANY, SOLENHOFEN LIMESTONE, S-WAVE,
M 26 31 MEASUREMENT, HEAT ALTERATION, LABORATORY
M 26 22 MEASUREMENT LABORATORY SYSTEMS. BROADBAND VELOCITY

M 26 19 MEASUREMENT, LABORATORY, SYSTEMS, VELOCITY
B 6207 1111† MEASUREMENT OF SUBSURFACE FRACTURES FROM DIPMETER LOGS
M 26 30 MEASUREMENT, STATIC MODULI, LABORATORY
B 6105 817* MEASUREMENT TECHNIQUE FOR PROCESS-RESPONSE STUDIES IN
M 26 23 MEASUREMENTS, ATTENUATION, LABORATORY
M 26 26 MEASUREMENTS, BEREA SANDSTONE, LABORATORY
M 26 36 MEASUREMENTS, BIOT-GEERTSMA EQUATIONS, LABORATORY
M 26 23 MEASUREMENTS, BOISE SANDSTONE, LABORATORY
M 26 37 MEASUREMENTS, FLUID PRESSURE, LABORATORY
M 26 26 MEASUREMENTS, FRICTION LOSS, LABORATORY
M 26 33 MEASUREMENTS, GASSMANNS THEORY, LABORATORY
B 6312 2223 MEASUREMENTS OF VITRINITE REFLECTANCE
M 26 37 MEASUREMENTS, OVERBURDEN, LABORATORY
M 26 32 MEASUREMENTS, ROCK PROPERTIES, TEMPERATURE EFFECTS,
M 26 39 MEASUREMENTS, SATURATION EFFECTS, BULK DENSITY, LABORATORY
M 26 39 MEASUREMENTS, SATURATION EFFECTS, PRESSURE, LABORATORY
M 26 39 MEASUREMENTS, SATURATION EFFECTS, ROCK PROPERTIES,
M 26 39 MEASUREMENTS, SATURATION EFFECTS, SHEAR VELOCITY, LABORATORY
M 26 32 MEASUREMENTS, TEMPERATURE EFFECTS CLASSIFICATION CHART,
B 6207 1226 MEBBERSON, A. J., SEISMIC AND GEOLOGIC FACIES ANALYSES IN FIELD
B 6303 496 MEBBERSON, ANDREW J., MASSIVE MARINE-SANDSTONE FACIES,
B 6309 1478† MECHANICAL BEHAVIOR OF EXPERIMENTALLY DEFORMED CARBONATE
B 6306 918 MECHANICAL COMPACTION, DEFINITION
B 6412 2179† MECHANICAL VERSUS THERMAL CAUSE OF ABNORMALLY HIGH PORE
B 6305 723† MECHANICS OF SECONDARY HYDROCARBON MIGRATION AND
B 6108 1391* MECHANICS OF SECONDARY MIGRATION AND ENTRAPMENT OF
B 6209 1561† MECHANISM FOR EMPLACEMENT OF PIERCEMENT DIAPIRS
B 6002 248 MECHANISM OF DOLOMITE GROWTH
B 6306 922 MECHANISM OF SHALE COMPACTION
B 6405 777* MECHANISMS CONTROLLING POROSITY IN RED RIVER, UPPER
S 10 47† MECHANISMS, SOME, MASS BALANCE AND, GEOLOGICAL, CONSTRAINTS
M 29 454 MECHANISMS, CRUSTAL SHORTENING AND THERMAL WELTS
M 29 453 MECHANISMS, DIFFERENTIATION OF MANTLE
B 6304 686* MECHANISMS FOR URANIUM DEPOSITION IN GRANTS MINERAL BELT,
B 6304 702* MECHANISMS OF BASIN SUBSIDENCE, ABST.
B 6303 409* MECHANISMS OF FORMATION DAMAGE IN DIAGENETICALLY ALTERED
S 3 41 MECHANISMS OF RADIOELEMENT MOBILIZATION IN CRUSTAL ROCKS
S 3 22 MECHANISMS OF URANIUM AND THORIUM TRANSFER TO THE CRUST
M 29 454 MECHANISMS, SEDIMENT INFLUX AND REMOVAL
M 29 454 MECHANISMS, VOLUME CHANGES IN MID-OCEAN RIDGE
M 26 417 MECKEL, L. D., JR., GEOLOGIC CONSIDERATIONS FOR STRATIGRAPHIC
S 4 22 MEDIA LUNA REEF, PUERTO RICO
B 6205 887* MEDICINE BOW MOUNTAINS, WYOMING, ABST., DEPOSITIONAL
B 6405 639 MEDICINE BUTTE THRUST, WYOMING
B 6407 981 MEDICINE HAT FIELD, ALBERTA AND, SASKATCHEWAN
B 6407 981 MEDICINE HAT SANDSTONE, ALBERTA
B 6207 1172 MEDICINE RANGE, SECTION, NEVADA
B 6209 1635 MEDINA SANDSTONES AND SHALES, APPALACHIAN BASIN
B 6405 787* MEDITERRANEAN, ABST., DEPOSITIONAL MODELS FOR FINE-GRAINED
B 6303 532* MEDITERRANEAN, ABST., GIANT, MUDFLOW DEPOSITS IN SUBMARINE
B 6303 443* MEDITERRANEAN, ABST., NEOGENE, CALCAREOUS, NANNOPLANKTON
M 29 306 MEDITERRANEAN, ACTIVE MARGINS, AEGEAN ARC
M 29 305 MEDITERRANEAN, ACTIVE MARGINS, CALABRIAN ARC
M 29 310 MEDITERRANEAN, ACTIVE MARGINS, CYPRUS ARC
M 29 297 MEDITERRANEAN, ADRIATIC BASINS, SEDIMENTARY SEQUENCE
M 29 312 MEDITERRANEAN, AEGEAN BASINS, EVOLUTION
M 29 310 MEDITERRANEAN, AEGEAN BASINS, SEDIMENTARY SEQUENCE
M 29 295 MEDITERRANEAN, ALGERIAN BASIN, CROSS SECTION
M 29 313 MEDITERRANEAN, ANTALYA BASIN, EVOLUTION
M 29 312 MEDITERRANEAN, ANTALYA BASIN, SEDIMENTARY SEQUENCE
M 29 293 MEDITERRANEAN BASIN, SEDIMENTARY SEQUENCE, WESTERN
M 29 308 MEDITERRANEAN, CALABRIAN ARC, EVOLUTION
M 29 313 MEDITERRANEAN, CILICIA-ADANA BASIN, EVOLUTION
M 29 312 MEDITERRANEAN, CILICIA-ADANA BASIN, SEDIMENTARY SEQUENCE
M 29 311 MEDITERRANEAN, CYPRUS ARC, EVOLUTION
M 29 304 MEDITERRANEAN, CYRENIAN PLATEAU, MAGNETIC ANOMALIES
B 6310 1920 MEDITERRANEAN, DEVELOPMENTS, FRANCE
M 29 305 MEDITERRANEAN, EASTERN, ACTIVE MARGINS
M 29 304 MEDITERRANEAN, EASTERN BASINS, EVOLUTION
M 29 304 MEDITERRANEAN, ERATHOSTHENES PLATEAU, MAGNETIC ANOMALIES
B 6411 1964 MEDITERRANEAN, FRANCE, PROVENCE-
M 29 299 MEDITERRANEAN, HERODOTUS BASIN, SEDIMENTARY SEQUENCE
M 29 299 MEDITERRANEAN, IONIAN BASIN, SEDIMENTARY SEQUENCE
M 29 299 MEDITERRANEAN, LEVANTINE BASIN, SEDIMENTARY SEQUENCE
B 6311 2091 MEDITERRANEAN MARGIN, SEISMIC SECTION, ISRAEL
M 29 292 MEDITERRANEAN, MARGINS, STRUCTURAL MAP
M 29 306 MEDITERRANEAN, MESSINIAN EVAPORITES, MAP
M 28 89 MEDITERRANEAN, MIOCENE, DSDP CORE
M 28 88 MEDITERRANEAN, MIOCENE, DSDP CORE
M 28 87 MEDITERRANEAN, MIOCENE, DSDP CORE
M 29 295 MEDITERRANEAN, NORTH BALEARIC BASIN, CROSS SECTION
M 29 313 MEDITERRANEAN, NORTH LEVANTINE BASIN, EVOLUTION
M 29 312 MEDITERRANEAN, NORTH LEVANTINE BASIN, SEDIMENTARY SEQUENCE
M 29 294 MEDITERRANEAN, PROVENCAL BASIN, CROSS SECTION
B 6112 2076 MEDITERRANEAN REGION, APENNINES RIDGE
B 6112 2076 MEDITERRANEAN REGION, BRADANIC TROUGH
B 6106 880 MEDITERRANEAN RIDGE AND HELLENIC ARC
B 6402 260 MEDITERRANEAN RIDGE, IONIAN SEA
B 6408 1190 MEDITERRANEAN SEA, OXIC NEGATIVE WATER BALANCE BASIN
B 6106 901 MEDITERRANEAN SEA, PLATE TECTONICS, SOUTHEASTERN
B 6106 884 MEDITERRANEAN SEA, SEISMIC PROFILES, SOUTHEASTERN
B 6311 2098 MEDITERRANEAN SEA, SEISMOGRAM
M 29 431 MEDITERRANEAN SEA, SMALL OCEAN BASIN, ORGANIC SEDIMENTS
B 6106 872† MEDITERRANEAN SEA-NILE CONE AREA, STRUCTURE AND SEDIMENTARY
M 29 295 MEDITERRANEAN, SOUTH BALEARIC BASIN, CROSS SECTION
M 29 307 MEDITERRANEAN, SOUTH BALEARIC-ALGERIAN BASIN, EVOLUTION
M 29 304 MEDITERRANEAN, SYRTE GULF, MAGNETIC ANOMALIES
B 6405 775* MEDITERRANEAN THRUST BELTS, ABST., CENTRAL
M 29 297 MEDITERRANEAN, TYRRHENIAN BASIN, EVOLUTION
M 29 297 MEDITERRANEAN, TYRRHENIAN BASIN, SEDIMENTARY SEQUENCE
M 29 293† MEDITERRANEAN, VARIETY OF, MARGINS AND DEEP, BASINS IN THE
B 6405 761 MEDLIN, A. L., ET AL, TRACE-ELEMENT CONTENT OF BITUMINOUS COAL
B 6405 735 MEDLIN, A. L., ET AL, USCHEM, GEOCHEMICAL DATA FILE OF NATIONAL
B 6004 656 MEDLIN, ANTOINETTE L., AVAILABILITY AND APPLICATION OF
B 6009 1613 MEDLIN, J. H., EVALUATION OF QUALITY AND QUANTITY CONTROLS ON
B 6203 490 MEDLIN, JACK H., PROFILE OF COAL-FORMING SWAMP—THE
B 6109 1440 MEDRANO SANDSTONE, OKLAHOMA
M 26 445 MEDRANO SANDSTONE, OKLAHOMA, ANADARKO BASIN
M 29 443 MEEDER, C., EXPLOSION SEISMOLOGY STUDIES OF ACTIVE AND PASSIVE
B 6008 1403 MEEHAN, ROBERT J., DISTRIBUTION OF URANIUM RESOURCES IN
B 6004 658 MEENTS, W. F., IDENTIFICATION OF LEAKAGE GAS FROM UNDERGROUND
B 6303 434 MEENTS, W. F., ISOTOPE GEOCHEMISTRY OF NATURA GAS IN ILLINOIS
B 6004 653* MEGA-BYTE DATA BASES, ABST., SET-THEORETIC, INFORMATION,
B 6303 474* MEGABRECCIA DEPOSITION, COW HEAD AND RELATED GROUPS,
B 6108 1372* MEGAEXPLORATION FOR NEW PETROLEUM BASINS-APPLICATION OF OLD
S 6 330 MEGAFOSSILS, CARBONIFEROUS, BIOSTRATIGRAPHY, PLANT
B 6206 951 MEGANOS CHANNEL, CALIFORNIA
M 28 117 MEGAQUARTZ, WITH CHERT, CEMENTS
B 6303 481* MEGASCALE VARIABILITY OF SEDIMENT ON UNITED STATES ATLANTIC
B 6004 730* MEGASHEARS--NATURAL HABITAT FOR OIL AND MINERAL DEPOSITS,
B 6011 2067* MEGASHEARS--NATURAL HABITAT FOR OIL AND MINERAL DEPOSITS,
M 25 474 MEGASTRUCTURES AND METALLOGENIC PROSPECTING, RELATION OF,
M 25 473 MEGASTRUCTURES, CENTRAL-TYPE, PACIFIC BASIN
M 25 474 MEGASTRUCTURES, MAP, PACIFIC
M 25 472† MEGASTRUCTURES OF THE PACIFIC--SUMMARY, RING
M 25 473 MEGASTRUCTURES, TECTONOMAGMATIC PROCESSES WITHIN
B 6203 536 MEGUID, A. A., CRETACEOUS-HOLOCENE SEDIMENTATIONIN EGYPTIAN
B 6203 563 MEGUID, A. A., REGIONAL VARIATION OF DEPOSITIONAL ENVIRONMENTS
B 6101 80 MEGUMA MESETA
B 6009 1550† MEHSANA AREA OF CAMBAY BASIN, INDIA, PATTERN OF,
B 6203 444 MEHSANA HORST, SECTION, INDIA
M 25 479† MEIMECHITES OF KAMCHATKA AND KIMBERLITE PROBLEMS OF PACIFIC
B 6011 2019 MEISBURGER, EDWARD P., NEOGENE SEDIMENTS OF ATLANTIC INNER
B 6105 812 MEISBURGER, EDWARD P., SHALLOW FRAMEWORK OF INNER CONTINENTAL
B 6405 749 MEISSNER, FRED F., EXAMPLES OF ABNORMAL FLUID PRESSURE
B 6205 888 MEISSNER, FRED F., PETROLEUM GEOLOGY OF GREAT BASIN, WESTERN
B 6303 480 MEISSNER, FRED F., SIMPLE PYROLYSIS TECHNIQU USING WELL
B 6303 496 MEISSNER, FRED F., TEST-TUBE PYROLYSIS--SIMPLE TECHNIQUE FOR
S 10 199 MEKRAN COAST, MUD VOLCANO, PAKISTAN
S 10 200 MEKRAN COAST, NAPAG MUD VOLCANO, IRAN
B 6303 420* MELAMOCYRILLIUM--NEW ACRITARCH GENUS FRO KWAGUNT
B 6004 644* MELANGE, FRANCISCAN OF TRINIDAD, CALIFORNIA, ABST.,
B 6405 704* MELANOIDIN HYPOTHESIS, ABST., A TEST OF THE
B 6105 764 MELCHIOR, JOHN, RATES OF ICE GOUGING AND SEDIMENT REWORKING,
B 6111 1951 MELEKESS HEAVY OIL DEPOSIT, USSR
B 6110 1897 MELLEN, FREDERIC F., CAMBRIAN SYSTEM IN BLACK WARRIOR BASIN
B 6311 2103 MELLEN, FREDERIC F., VARDAMAN TRIASSIC-JURASSIC SILL, CALHOUN
S 11 265 MELONIS BARLEEANUS
S 11 298 MELONIS POMPILIOIDES
S 11 265 MELONIS POMPILIOIDES
S 3 22 MELTING, COMPLETE, RADIOELEMENT TRANSFER BY
S 3 45 MELTING, FRACTIONAL
S 3 23 MELTING, PARTIAL, RADIOELEMENT TRANSFER BY
B 6008 1365 MELTON, EARL, DEVELOPMENTS IN SOUTH TEXAS I 1975
B 6203 523 MELTON, ROBERT A., USE OF DEPOSITIONAL MODELS IN PREDICTING
B 6206 1005 MELVILLE ISLAND, ARCTIC, CANADA
B 6111 1951 MELVILLE ISLAND HEAVY OIL DEPOSIT, CANADIAN, ARCTIC
M 29 213 MELVILLE ISLAND SECTION, AUSTRALIA, SHELF
B 6101 5 MELVILLIAN DISTURBANCE, SVERDRUP BASIN

B 6403 441* MEMBER SANDS OF WINTERS FORMATION, ABST.
B 6206 1063† MEMBERSHIP, A REPORT ON THE
B 6004 692* MEMBRANE FILTRATION AS MECHANISM FOR PRIMARY ACCUMULATION
B 6203 538 MENCH, P. A., DEPOSITIONAL FACIES AND OCCURRENCE OF POROSITY IN
B 6303 488 MENCH, P. A., POROSITY IN GIANT GAS FIELD, ELLENBURGER
B 6405 749 MENCH, PATRICIA A., STABLE ISOTOPE EVIDENCE FOR MODERN
M 30 363 MENDEZ FORMATION, CHIAPAS-TABASCO PROVINCE
B 6202 244 MENDEZ FORMATION, MEXICO
B 6211 2349* MENDOCINO COAST, NORTHERN CALIFORNIA, ABST., CRETACEOUS AND
B 6203 356 MENDOCINO FAULT, CALIFORNIA
M 25 384 MENDOCINO FRACTURE ZONE, NODULES, PACIFIC, NORTH
B 6403 429 MENDOZA HIGH-WAX OIL FIELD, ARGENTINA
B 6303 471 MENDOZA, M., GEOCHEMISTRY OF OILS FROM SANTA CRUZ BASIN,
B 6104 579 MENE GRANDE FIELD, VENEZUELA
B 6104 595 MENEG-CENTA FIELD, VENEZUELA
B 6104 579 MENEG. CUETA FIELD, VENEZUELA
B 6104 579 MENEG. LAMAR FIELD, VENEZUELA
M 25 252 MENG, C. Y., PETROLIFEROUS TAIWAN BASINS IN FRAMEWOR OF
M 27 160 MENISCUS CEMENTS
B 6407 992 MENIZLA INLIER, MOROCCO
B 6309 1598* MENKING FIELDS, LAVACA COUNTY, TEXAS, ABST., EXPLORATION
B 6103 436 MERAMECIAN, LOWER, UNCONFORMITY, ALASKA
B 6008 1404 MERCER, R. F., LIBERATED, PRODUCED, RECYCLED, OR CONTAMINATION,
B 6002 327 MERCURIO, RICHARD N., BLOCK 16 FIELD, WARD COUNTY, TEXAS, ABST.
M 25 485 MERCURY, CALIFORNIA, COAST RANGES
B 6305 738 MERCURY CAPILLARY PRESSURE TESTS
M 25 486 MERCURY DEPOSITS, UNITED STATES, WESTERN
M 25 518 MERCURY, MEXICO
M 25 475 MERCURY MINERALIZATION, TIME OF, USSR, EASTERN, ANTIMONY-
M 25 533 MERCURY, PERU, HUANCAVELICA DISTRICT
M 25 460 MERCURY, POLYMETALLIC ORES, CASSITERITE, EASTERN ASIATIC
M 25 470 MERCURY, USSR, NORTHEAST
B 6002 246 MERCURY-INJECTION AND EJECTION CURVES
B 6405 801 MEREDITH, JAMES A., METHANE MEASUREMENT FROM SALINE ZONE OIL
B 6404 488 MEREWETHER, E. A., ORGANIC COMPOSITION OF SOME UPPER
B 6307 1039 MERIDIAN ANTICLINE, SEISMIC PROFILE, WYOMING
B 6409 1554* MERIDIAN SANDSTONE, EOCENE, NORTHERN MISSISSIPPI, ABST.,
B 6309 1608* MERIDIONALLY ORNAMENTED HEDBERGELLID FORAMINIFERS FROM
B 6205 865 MERIT BADGE BOOKLET, AAPG, BOY SCOUT COMMITTEE, GEOLOGY
B 6310 1886 MERMINOD, F., AND KUWAIT OIL CO., IN KUWAIT, 1978
B 6110 1583 MERMINOD, F., ON DEVELOPMENTS IN CUBA, 1976
B 6210 2113 MERMINOD, F., PETROCONSULTANTS S.A., AND ASIATIC PETROLEUM CO.,
B 6109 1476 MERO FIELD, BRAZIL
B 6304 700 MERRIAM, D. F., QUANTITATIVE COMPARISONS OF GEOLOGIC MAPS TO
B 6303 517 MERRIAM, D. F., STATISTICAL COMPARISON OF MAPPED DATA, ABST.
B 6405 749 MERRILL, GLEN K., ORGANOPHOSPHORITES IN CARBONIFEROUS ROCKS
B 6311 2059 MERRITT COAL FIELD, BRITISH COLUMBIA
M 25 107 MERRITT COALFIELDS, CANADA, PRINCETON-
B 6409 1564 MERRITT, LINDA C., DIAGENESIS OF DELTAIC SANDSTONE—OLMOS SAN
M 25 518 MESA CENTRAL METALLOGENETIC PROVINCE, MEXICO
B 6108 1373* MESA COUNTY, COLORADO, ABST., GEOLOGICAL, GEOCHEMICAL, AND
B 6102 295* MESA RICA SANDSTONE, TUCUMCARI BASIN, NEW MEXICO, ABST.,
B 6302 158 MESAVERDE FIELD, NEW MEXICO
M 24 109 MESAVERDE FORMATION, WYOMING
B 6202 185 MESAVERDE FORMATION, WYOMING
M 28 178 MESAVERDE FORMATION, WYOMING, CRETACEOUS
M 28 176 MESAVERDE FORMATION, WYOMING, CRETACEOUS
M 28 179 MESAVERDE FORMATION, WYOMING, CRETACEOUS
M 28 175 MESAVERDE FORMATION, WYOMING, CRETACEOUS
B 6008 1393* MESAVERDE FORMATIONS--ITS RELATION TO FRACTURE STIMULATION
B 6008 1408* MESAVERDE GAS POTENTIAL IN DRY CREEK AREA, CARBON COUNTY,
B 6205 887* MESAVERDE GROUP AND OCCURRENCE OF NATURAL GAS IN
M 28 41 MESAVERDE GROUP, COLORADO, CRETACEOUS
B 6305 824* MESAVERDE GROUP IN SOUTHERN WASATCH PLATEAU, UTAH, ABST.,
B 6108 1383* MESAVERDE GROUP IN SOUTHWESTERN WYOMING AND
B 6304 600 MESAVERDE GROUP, NEW MEXICO, POINT LOOKOUT SANDSTONE
B 6205 885* MESAVERDE GROUP, NORTH HORN FORMATION AND LOWER PART OF
B 6405 732* MESAVERDE GROUP, UINTA BASIN, UTAH, ABST., SOME RELATIONS
B 6302 159 MESAVERDE SANDSTONE, COLORADO AND, NEW MEXICO
B 6112 2141 MESAVERDE SANDSTONES, NEW MEXICO AND, COLORADO
B 6303 302 MESHIK FORMATION, ALASKA
B 6209 1607 MESOLELLA, KENNETH J., PALEOGEOGRAPHY OF SOME SILURIAN AND
B 6203 570* MESOTIDAL SALT MARSH, ABST., PHYSICAL AND SEDIMENTOLOGIC
S 6 146 MESOZOIC, ABU GROUP, JAPAN
B 6405 784* MESOZOIC ACCRETIONARY TECTONICS OF ALASKA, ABST.
S 6 146 MESOZOIC, AIOI GROUP, JAPAN
M 25 281† MESOZOIC AND CENOZOIC BASINS IN WESTERN NORTH AMERICA,
B 6105 761* **MESOZOIC AND CENOZOIC BIOSTRATIGRAPHY OF OAK GROVE CORE,**
B 6203 507* MESOZOIC AND CENOZOIC GEOLOGY OF GEORGIA COASTAL PLAIN, ABST.
B 6207 1228* MESOZOIC AND CENOZOIC ROCKS OF ALEUTIAN ISLANDS, ALASKA
B 6412 2139† MESOZOIC AND CENOZOIC STRUCTURAL TRENDS UNDER SOUTHERN
B 6002 177 MESOZOIC AND CENOZOIC TECTONIC SETTING, BERINGIAN SHELF AND,
B 6209 1755* MESOZOIC AND PALEOGENE GEOLOGY AND UNCONFORMITIES, COASTAL
B 6108 1392* MESOZOIC AND TERTIARY BASINS OF PACIFIC MARGIN, ABST.,
B 6206 1015 MESOZOIC, APTIAN-CAMPANIAN TO MAESTRICHTIAN, ARCTIC, CANADA
B 6004 734* MESOZOIC ARC-REAR MARGINAL BASIN IN SOUTHERN SOUTH AMERICA,
B 6002 177 MESOZOIC BASEMENT ROCK, BERINGIAN SHELF AND MARGIN
B 6105 803* **MESOZOIC BASIN DISTRIBUTION ALONG U.S. EASTERN CONTINENTAL**
M 30 118 MESOZOIC BASINS, LOCATION MAP, NORTH SEA
S 6 251 MESOZOIC BOUNDARY, CENOZOIC, ISOTOPIC DATING
B 6004 669* **MESOZOIC CARBONATE PLATFORM, ABST., ABENAKI FORMATION, NOVA**
M 25 94† MESOZOIC COAL IN REPUBLIC OF KOREA, PALEOZOIC AND
B 6211 2352* MESOZOIC CONSTRUCTION OF CORDILLERAN COLLAGE, CENTRAL
B 6211 2356* MESOZOIC CONTINENTAL ACCRETION OF NORTH AMERICA. BY TITLE
B 6004 709* MESOZOIC CONTINENTAL MARGIN, OTHRIS, GREECE, ABST.,
B 6408 1232 MESOZOIC EPICONTINENTAL SEAWAYS, NORTH AMERICA, WESTERN
M 29 171 MESOZOIC, EVOLUTION, AUSTRALIA
B 6011 2038† MESOZOIC FLYSCH DEPOSITS IN SANTA ANA MOUNTAINS, CALIFORNIA,
B 6002 175 MESOZOIC FOLDBELT, BERINGIAN SHELF
B 6211 2354* MESOZOIC GEOLOGY OF PENINSULAR CALIFORNIA AND WESTERN
M 25 466 MESOZOIC GEOSYNCLINAL DEVELOPMENT, USSR
M 25 440 MESOZOIC GRANITES, SOUTH, KOREA
S 4 84 MESOZOIC GREATER ANTILLEAN ARC
B 6206 1014 MESOZOIC, GRIESBACHIAN TO UPPER NORIAN, ARCTIC, CANADA
S 7 80 MESOZOIC, GULF OF MEXICO, (RECONSTRUCTION) DISCUSSED
M 30 134 MESOZOIC HISTORY, NORTH SEA, MORAY FIRTH BASIN
S 6 144 MESOZOIC, IGNEOUS LOCALITIES, MAP, JAPAN
S 6 143† MESOZOIC IGNEOUS ROCKS IN JAPAN, ISOTOPIC AGES AND
S 6 251 MESOZOIC, ISOTOPIC DATING, CENOZOIC BOUNDARY
M 25 449 MESOZOIC, LATE, GRANITIC BATHOLITHS AND MINERAL ZONING, JAPAN
B 6002 175 MESOZOIC MAGMATIC ARC, BERINGIAN SHELF
B 6103 466* **MESOZOIC MAGMATISM ARC ACCRETION, SOUTHWEST ALASKA MARGIN,**
M 30 525 MESOZOIC, MALOSSA FIELD, PO BASIN
M 25 461 MESOZOIC MOBILE BELTS OR MESOZOIDES
M 25 428 MESOZOIC, ORE DEPOSITS ASSOCIATED WITH, AUSTRALIA
B 6004 733* MESOZOIC ORIGIN OF HOLOCOCCOLITHS, ABST.
B 6211 2353* MESOZOIC PALEOGEOGRAPHY AND PALEOTECTONICS, SUPLEE-IZEE AND
B 6211 2356* MESOZOIC PALEOGEOGRAPHY OF CALIFORNIA. VIEWPOINT FROM
S 6 146 MESOZOIC, PALEOMAGNETIC EVIDENCE, JAPAN
B 6211 2361* MESOZOIC PALEOTECTONIC-PALEOGEOGRAPHIC RECONSTRUCTION OF
M 25 280† MESOZOIC PETROLEUM PROSPECTS, ALEUTIAN-BERING SEA
B 6303 486* MESOZOIC PLATE TECTONICS, ABST., LATE
M 30 362 MESOZOIC PROVINCE, MEXICO, CHIAPAS-TABASCO
B 6305 793 MESOZOIC RECONSTRUCTION, GULF OF MEXICO
B 6405 681* MESOZOIC ROCKS, SAN JUAN BASIN, NEW MEXICO, ABST., RB-, SR DATING
B 6408 1164 MESOZOIC, SANDSTONE WEDGES, ALBERTA
B 6101 10 MESOZOIC SEDIMENTS, DEPOCENTERS, CANADIAN, ARCTIC ISLANDS,
B 6009 1425† MESOZOIC SHALES AT HAMILTON DOME, BIGHORN BASIN, WYOMING,
B 6004 672* MESOZOIC STRATIGRAPHIC TEST, NANTUCKET ISLAND, MASSACHUSETTS,
B 6101 12 MESOZOIC, SVERDRUP BASIN
M 25 466 MESOZOIC TAPHROGENIC OR RIFTOGENIC ZONES, USSR, NORTHEAST
B 6303 425* MESOZOIC TECTONIC FRAMEWORK AND SEDIMENTATION, NORTHWEST
B 6207 1223* MESOZOIC TECTONIC HISTORY OF NORTHERN CALIFORNIA AND
B 6211 2355* MESOZOIC TECTONICS OF WESTERN UNITED STATES. BY TITLE ONLY
B 6206 1015 MESOZOIC TO CENOZOIC, CAMPANIAN TO MAESTRICHTIAN, ARCTIC,
B 6405 791* MESOZOIC TO EARLY CENOZOIC FORELAND SEDIMENTATION IN
M 25 466 MESOZOIC TROUGHS, USSR, NORTHEAST
B 6206 1014 MESOZOIC, UPPER NORIAN TO APTIAN, ARCTIC, CANADA
B 6005 794† MESOZOIC-CENOZOIC ALPINE HIMALAYAN GEOSYNCLINE AND ITS
M 29 462 MESOZOIC-CENOZOIC BASINS, ANALYSIS, GLOBAL CHANGE
M 29 464 MESOZOIC-CENOZOIC BASINS, INTERPRETATION, GLOBAL CHANGE
B 6002 197 MESOZOIC-CENOZOIC MARGIN BASINS, BRAZIL, EASTERN
B 6303 475* MESOZOIC-CENOZOIC SEDIMENTARY FORMATIONS O NORTH AMERICAN
M 25 461 MESOZOIDES, MESOZOIC MOBILE BELTS OR
M 25 466 MESOZOIDES, USAGE OF TERM
B 6204 670 MESSAOUD AXIS, ALGERIA
B 6402 249 MESSENIA BASIN, SEISMIC REFLECTION, GREECE
B 6106 873 MESSINIAN EVAPORITES, EGYPT
M 29 306 MESSINIAN EVAPORITES, MAP, MEDITERRANEAN
B 6303 337 MESSINIAN EVAPORITIC SERIES, ISRAEL
S 5 174 MESSINIAN STRATIGRAPHY, ITALY, SICILY,
M 30 18 MESSLA FIELD, LANDSAT IMAGE, LIBYA
M 30 507 MESSLA FIELD, LIBYA, SIRTE BASIN
M 30 517 MESSLA FIELD, SARIR GROUP, STRATIGRAPHY
M 30 507 MESSLA FIELD, SARIR SANDSTONE, CRETACEOUS
M 30 521 MESSLA FIELD, SARIR SANDSTONE, NET PAY ISOPACH
M 30 514 MESSLA FIELD, SARIR SANDSTONE, SEISMIC LINE
M 30 523 MESSLA FIELD, SOURCE ROCKS

M 30 516 MESSLA FIELD, STRATIGRAPHIC COLUMN, LIBYA
M 30 515 MESSLA FIELD, STRUCTURE, SIRTE BASIN
B 6401 16 MESSLA GIANT OIL FIELD, LANDSAT DATA, LIBYA
B 6401 16 MESSLA GIANT OIL FIELD, LANDSAT DATA, LIBYA
B 6303 433* MESSLA OIL FIELD, LIBYA, ABST., GEOLOGY OF A STRATIGRAPHIC
M 28 51 META-DIABASE, CONNECTICUT, DEVONIAN
M 28 52 META-DIABASE, CONNECTICUT, DEVONIAN
M 25 484 METAL AND METALLOGENIC PROVINCES, GEOLOGIC FRAMEW,
M 25 449 METAL CONCENTRATION AND PETROCHEMICAL ZONING, CORRELATION
S 5 181* METAL CONCENTRATION IN A DENSITY-STRATIFIED EVAPORITE PAN,
M 25 449 METAL CONCENTRATION, BASE-
M 25 460 METAL CONCENTRATIONS IN, EAST PACIFIC RISE, ULTRABASITES
M 25 488 METAL CONCENTRATIONS, MULTISTAGE ORIGIN
M 25 428 METAL CONCENTRATIONS, RELATION OF BIOSTRATIGRAPHY TO
M 25 387 METAL CONTENTS OF NODULES
M 25 53 METAL DEPOSITS AT DIVERGENT PLATE BOUNDARIES
M 25 51 METAL DEPOSITS AT OR NEAR CONVERGENT PLATE BOUNDARIES
M 25 533 METAL DEPOSITS, BOLIVIA, BASE-
M 25 427 METAL DEPOSITS, DISCOVERY AND POTENTIAL, AUSTRALIA, BASE-
M 25 457 METAL DEPOSITS, DISTRICTS, MAP, JAPAN, BASE-
M 25 457 METAL DEPOSITS, EXPLORATION PROJECTS, JAPAN, BASE-
M 25 456 METAL DEPOSITS, FUTURE EXPLORATION, JAPAN, BASE-
M 25 456 METAL DEPOSITS, GEOLOGIC SURVEY PROJECT, JAPAN, BASE-
M 25 428 METAL DEPOSITS, GEOLOGY, AUSTRALIA, BASE-
M 25 533 METAL DEPOSITS, POLYMETALLIC, ANDEAN REGION, BASE-
M 25 426 METAL DEPOSITS, PRODUCTION AND RES, AUSTRALIA, STRATIFORM,
M 25 533 METAL DEPOSITS, PRODUCTION, PERU, BASE-
M 25 429 METAL DEPOSITS, RESOURCES, AUSTRALIA
M 25 456 METAL MINING AGENCY OF JAPAN
M 25 486 METAL OCCURRENCE, SAN ANDREAS FAULT, APPARENT RELATION TO
M 25 485 METAL OCCURRENCES, RAYLIKE DISTRIBUTION
M 25 456 METAL ORES, DEMAND AND PRODUCTION, DISCREPANCY BE, JAPAN,
M 25 505 METAL PRODUCTION AND RESERVES, TABLE, CANADIAN, CORDILLERA
M 25 503 METAL PRODUCTION, ANNUAL, CANADIAN, CORDILLERA
M 25 482 METAL PROVINCE, USAGE DEFINED
M 25 485 METAL PROVINCES, NORTHEAST, PACIFIC
M 25 482† METAL PROVINCES OF ALASKA--SUMMARY
M 25 483 METAL PROVINCES, USAGE OF TERM
M 25 554 METAL SULFIDES, SURVEYS, ECUADOR, GOLD AND, BASE-
M 25 535 METAL-RICH UNCONSOLIDATED PELAGIC SEDIM, PACIFIC,
B 6405 718* METALINE DISTRICT, WASHINGTON, ABST., DEPOSITIONAL SETTING OF
B 6405 718* METALINE FORMATION, METALINE DISTRICT, WASHINGTON, ABST.,
M 25 525 METALLIC CONCENTRATION, TIMES OF, MEXICO, SINALOA
M 25 536 METALLIC MINERAL OCCURRENCES, ANTARCTIC PENINSULA AND,
B 6303 518* METALLIC MINERAL RESOURCES OF ANTARCTICA--DEVELOPMENT FAR
M 25 503† METALLIC MINERAL RESOURCES OF CANADIAN CORDILLERA
M 25 534 METALLIFEROUS DEPOSIT, ANDEAN REGION, PRECAMBRIAN AND,
M 25 533 METALLIFEROUS DEPOSITS, ANDEAN REGION, PRE-ANDEAN
B 6207 1213* METALLIFEROUS DEPOSITS IN SOUTHWEST PACIFIC, ABST.,
M 25 483 METALLIFEROUS PROVINCES
M 25 546 METALLIFEROUS SEDIMENTS, GENESIS, NAZCA PLATE
M 25 541 METALLIFEROUS SEDIMENTS, NAZCA PLATE REGION
M 25 544 METALLIFEROUS SEDIMENTS, STRONTIUM-ISOTOPE ABUNDANCE DATA
M 25 529 METALLIZATION, AGE OF DEPOSITS ASSOCIAT, PACIFIC, SOUTHEASTERN
M 25 485 METALLIZATION, EPOCHS OF, NORTHEAST, PACIFIC
M 25 485 METALLIZATION, FOCI OF
M 25 487 METALLIZATION, MANTLE INHOMOGENEITIES, ROLE IN
B 6212 2508* METALLOGENESIS IN CALIFORNIA, ABST., TECTONIC FRAMEWORK AND
M 25 448† METALLOGENESIS IN PERIPHERAL-SEA - ISLAND-AR COMPLEX,
M 25 539† METALLOGENESIS IN SOUTHEAST PACIFIC OCEAN, NAZCA PLATE
M 25 471 METALLOGENESIS, MAJOR, TIME OF, USSR, NORTHEAST
M 25 516 METALLOGENETIC CHART OF MEXICO, ILLUSTRATION
M 25 515† METALLOGENETIC CHART OF MEXICO--SUMMARY
M 25 515 METALLOGENETIC CHART OF THE WORLD
M 25 515 METALLOGENETIC PROVINCE, USAGE DEFINED
M 25 483 METALLOGENETIC PROVINCES
M 25 515 METALLOGENETIC PROVINCES OF, MEXICO
M 25 559† METALLOGENIC BELTS AND TECTONIC EVOLUTION OF CHILEAN
M 25 486 METALLOGENIC DEPOSITS, METHODS OF DEFINING
M 25 445 METALLOGENIC EPOCHS, SOUTH KOREA
M 25 488 METALLOGENIC ORES, FUTURE SEARCH FOR
M 25 474 METALLOGENIC PROSPECTING, RELATION OF, RING, MEGASTRUCTURES
M 25 532 METALLOGENIC PROVINCE, MAP SHOWING SUBPROVINCES, ANDEAN
M 25 529 METALLOGENIC PROVINCE, METALLOGENESIS, ANDEAN
M 25 551 METALLOGENIC PROVINCES, COLOMBIA
M 25 484 METALLOGENIC PROVINCES, DETERMINATION OF
M 25 484 METALLOGENIC PROVINCES, GEOLOGIC FRAMEW, NORTHEAST PACIFIC,
M 25 486 METALLOGENIC PROVINCES, NEVADA
M 25 483 METALLOGENIC PROVINCES, NORTH AMERICA, WESTERN, CORDILLERA
M 25 486 METALLOGENIC PROVINCES, NORTHEAST, PACIFIC
M 25 483† METALLOGENIC PROVINCES OF NORTHEAST PACIFIC
M 25 527† METALLOGENIC PROVINCES OF SOUTHEASTERN PACIFIC REGION
M 25 483 METALLOGENIC PROVINCES, OTHER TERMS USED FOR
M 25 446 METALLOGENIC PROVINCES, SOUTH KOREA
M 25 446 METALLOGENIC PROVINCES, SOUTH KOREA, CRETACEOUS, LATE, TO,
M 25 445 METALLOGENIC PROVINCES, SOUTH KOREA, PALEOZOIC AND OTHER
M 25 444 METALLOGENIC PROVINCES, SOUTH KOREA, PRECAMBRIAN
M 25 483 METALLOGENIC PROVINCES, USAGE OF TERM
M 25 462 METALLOGENIC SYSTEM, PACIFIDES, STANNIFEROUS
M 25 460 METALLOGENIC SYSTEMS, LARGE-SCALE, PACIFIDES
M 25 458† METALLOGENIC SYSTEMS OF PACIFIDES--SUMMARY
M 25 461 METALLOGENIC SYSTEMS, TRANS-, PACIFIC, COPPER AND, GOLD
M 25 459 METALLOGENIC ZONATION, MAP, PACIFIC SEGMENT
M 25 461 METALLOGENIC ZONATION, PACIFIC
M 25 462 METALLOGENIC ZONATION, SINGLE-ELEMENT, TIN
B 6207 1214* METALLOGENIC ZONES AND EPOCHS IN CANADIAN CORDILLERA, ABST.
M 25 459 METALLOGENIC ZONES, PACIFIDES
M 25 460 METALLOGENIC ZONES, PACIFIDES, CONTINENTAL-MARGIN
M 25 460 METALLOGENIC ZONES, PACIFIDES, OCEAN-MARGIN
M 25 459 METALLOGENIC ZONES, PACIFIDES, OCEANIC
M 25 483 METALLOGRAPHIC PROVINCES
S 3 91 METALLOTECTONIC GRADIENT, OROGENIC, MINERALIZATION BELTS
M 25 504 METALS, BACKGROUND ABUNDANCES, CANADIAN, CORDILLERA
M 25 505 METALS, CANADIAN, CORDILLERA, MAGMATIC DEPOSITS OF
M 25 506 METALS, CANADIAN, CORDILLERA, MASSIVE, SULFIDE DEPOSITS OF
M 25 506 METALS, CANADIAN, CORDILLERA, PORPHYRY DEPOSITS OF
M 25 505 METALS, CANADIAN, CORDILLERA, SKARN DEPOSITS OF
M 25 507 METALS, CANADIAN, CORDILLERA, STRATIFORM DEPOSITS OF
M 25 507 METALS, CANADIAN, CORDILLERA, VEIN DEPOSITS OF
M 25 504 METALS, CLASSES OF, CANADIAN, CORDILLERA
M 25 485 METALS, COLORADO, PRECIOUS AND, BASE
M 25 487 METALS, OCCURRENCE OF, GENETIC ORDER DISCUSSED
M 25 487 METALS, STRATA-BOUND DEPOSITS
M 25 541 METALS, TRANSITION, NAZCA PLATE REGION
M 25 460 METALS, TYPES OF, PACIFIDES
M 25 467 METALS, USSR, NORTHEAST, HEAVY
M 25 487 METALS, VOLCANOGENIC SULFIDE ASSOCIATIONS
M 25 485 METALS, WASHINGTON, PRECIOUS AND, BASE
M 25 504 METALS, ZONAL PATTERN OF DISTRIBUTION, CANADIAN CORDILLERA
M 25 475 METAMORPHIC LAYER, ANTIMONY, MINERALIZATION, CORRELATION
S 3 136 METAMORPHIC MOBILIZATION, LATEROGENESIS AT THE LOW-INTENSITY
B 6102 248† METAMORPHIC ROCKS, NOTE 45--, APPLICATION FOR AMENDMENT
B 6102 248 METAMORPHIC ROCKS, TERMINOLOGY, NOTE 45
B 6108 1375* METAMORPHISM, AND OIL GENERATION IN BLACK SHALE MEMBERS OF
B 6201 98† METAMORPHISM, AND OIL GENERATION IN BLACK SHALE MEMBERS OF
S 3 45 METAMORPHISM, FRACTIONAL
B 6004 729* METAMORPHISM IN SILICICLASTIC ROCKS, ABST., INORGANIC AND,
M 25 449 METAMORPHISM, JAPAN
B 6402 180 METAMORPHISM OF ORGANIC SEDIMENTS, THERMAL
S 10 116 METAMORPHISM, NATURAL, PRESSURE-TEMPERATURE DIAGRAM
M 25 469 METASOMATIC REWORKING, USSR, NORTHEAST
M 25 469 METASOMATITES
B 6405 767* METAZOAN BIOHERMS OF LOWER ORDOVICIAN AGE—ST. GEORGE
B 6108 1372* METEOR-IMPACT STRUCTURE, ABST., RED WING CREEK FIELD, NORTH
B 6203 489* METEORIC AND MIXING-ZONE DIAGENESIS IN LIMESTONES, ABST.,
S 4 379 METEORIC FRESH WATERS, PRECIPITATION OF LOW-, MAGNESIAN,
S 3 54 METEORIC GROUNDWATER SYSTEM, TRANSPORTATION IN
B 6303 458* METEORIC-VADOSE DIAGENESIS OF ENEWETAK ATOLL, ABST.,
B 6309 1582* METEORITIC IMPACT--RESERVOIR-FORMING PROCESS, ABST.
S 3 10 METEORITIC MODEL FOR RADIOELEMENT DISTRIBUTION
B 6004 658* METHANE, ABST., IDENTIFICATION OF, LEAKAGE, GAS FROM,
B 6309 1527† METHANE AT ELEVATED PRESSURES AND TEMPERATURES, AQUEOUS,
B 6405 675* METHANE GAS IN FISHERS PEA AREA, LAS ANIMAS COUNTY, COLORADO,
B 6203 520* METHANE GENERATION AND PETROLEUM MIGRATION, ABST.
B 6405 699* METHANE IN DEEP, TERTIARY SANDSTONE RESERVOIRS ALONG
B 6309 1527 METHANE IN GEOPRESSURED WATERS, GULF COAST
B 6212 2478† METHANE IN WATER AT ELEVATED TEMPERATURES AND PRESSURES,
B 6305 776 METHANE IN WATER AT ELEVATED TEMPERATURES AND PRESSURES,
B 6405 801* METHANE MEASUREMENT FROM SALINE ZONE OIL SHALE, PICEANCE
B 6303 551* METHANE ON SEDIMENT INSTABILITY IN MODERN DELTA SEDIMENTS,
B 6303 420* METHANE PRODUCTION, ABST., MODEL FOR INTERSTITIAL SULFATE
B 6405 712* METHANE RESERVOIRS OF EAGLE SANDSTONE, MONTANA, ABST.,
B 6405 711* METHANE RESERVOIRS OF NORTHERN GREAT PLAINS, ABST., PHYSICAL
B 6309 1579* METHANE RESOURCE, ABST., NEW RIVER FORMATION, COALS IN
S 10 181 METHANE, BIOCHEMICAL ZONES BENEATH OCEAN FLOOR
S 10 182 METHANE, GENERATION RELATED TO DEPTH OF BURIAL
S 10 181 METHANE, GENESIS, BIOCHEMICALLY GENERATED
S 10 181 METHANE, GENESIS, IN SEDIMENTARY ROCKS
S 10 182 METHANE, GENESIS, THERMOCHEMICALLY GENERATED
S 10 258 METHANE, NEAR-SURFACE, CARBON ISOTOPE DATA

S 10 180 METHANE, OCCURRENCES, BITUMINOUS SHALES
S 10 180 METHANE, OCCURRENCES, CARBONACEOUS SHALES
S 10 180 METHANE, OCCURRENCES, COAL
S 10 180 METHANE, OCCURRENCES, COMBUSTION
S 10 180 METHANE, OCCURRENCES, DIGESTIVE TRACTS OF ANIMALS
S 10 179 METHANE, OCCURRENCES, MARSH GAS
S 10 180 METHANE, OCCURRENCES, PETROLEUM
S 10 180 METHANE, OCCURRENCES, SEDIMENTS BENEATH THE OCEANS
S 10 180 METHANE, OCCURRENCES, SEWAGE SLUDGE
S 10 180 METHANE, OCCURRENCES, SOLID GAS HYDRATES
S 10 180 METHANE, OCCURRENCES, SUBSURFACE WATERS
S 10 180 METHANE, OCCURRENCES, SURFICIAL WATERS
S 10 180 METHANE, OCCURRENCES, VOLCANIC ACTIVITY
S 10 192 METHANE, PRESSURES, MUD VOLCANOES
S 10 192 METHANE, PRESSURES, SHALE DIAPIRS
S 10 179 METHANE, PROPERTIES
S 10 215 METHANE, SOLUBILITY IN FRESH WATER AT PRESSURE
S 10 103 METHANE, SOLUBILITY IN WATER AT TEMPERATURE
S 10 187 METHANE BUBBLES, BARRIERS TO EXPULSION IN PORES
S 10 252 METHANE CONCENTRATIONS, FLORIDA, BLACKJACK CREEK FIELD
S 10 252 METHANE CONCENTRATIONS, FLORIDA, JAY FIELD
S 10 249 METHANE DISTRIBUTION, BLACKJACK CREEK FIELD
S 10 249 METHANE DISTRIBUTION, JAY FIELD
S 10 179† METHANE GENERATION AND PETROLEUM MIGRATION
S 10 191 METHANE SEEPS, ECHOGRAM, GULF OF MEXICO
S 10 188 METHANE-HYDRATE, RELATION TO PRESSURE, TEMPERATURE
S 10 190 METHANE-HYDRATE, SEISMIC SKETCH, BLAKE OUTER RIDGE
S 10 187 METHANE-HYDRATE, SOLID, BARRIERS TO MIGRATION
S 10 189 METHANE-HYDRATE HORIZON, SEISMIC LINE, GULF OF MEXICO
B 6312 2183† METHODS OF ESTIMATING OIL AND GAS RESOURCES
B 6105 788* METHODS FOR RECOVERY OF MORE OIL FROM KNOWN FIELDS
B 6408 1210† METHODS USED, AND EXPERIENCE GAINED, IN EXPLORATION FOR NEW
B 6407 1007 MEXIA-TALCO FAULT SYSTEM, TEXAS, LOUISIANA, AND, ARKANSAS
B 6306 940 MEXIANA GRABEN, AMAZON
B 6205 889* MEXICAN PETROLEUM DISCOVERIES, ABST., GEOLOGY AND, GEOPOLITICS
M 26 260 MEXICAN RIDGE, SOUTH TEXAS, GULF OF MEXICO
S 7 54 MEXICAN RIDGES, DISCUSSED, GULF OF MEXICO
M 29 320 MEXICAN RIDGES FOLD BELT, EVOLUTION HYPOTHESIZED
B 6303 426* MEXICAN RIDGES FOLDBELT, SOUTHWESTERN GULF OF MEXICO, ABST.,
M 29 320 MEXICAN RIDGES, INDEX MAP, GULF OF MEXICO
S 7 55 MEXICAN RIDGES, REFLECTION PROFILE, GULF OF MEXICO
M 29 411 MEXICAN RIDGES, REFLECTION SECTION, MEXICO
M 29 323 MEXICAN RIDGES, SEISMIC SECTION (24-FOLD), GULF OF MEXICO
M 29 324 MEXICAN RIDGES, SEISMIC SECTION (FOLDOUT), GULF OF MEXICO
M 29 407 MEXICAN RIDGES, SEISMIC SECTION, GULF OF MEXICO
M 29 319† MEXICAN RIDGES, SOUTHWESTERN GULF OF MEXICO, ANATOMY OF THE
M 29 326 MEXICAN RIDGES, STRUCTURE EXPLAINED, GULF OF MEXICO
B 6211 2291 MEXICAN RIDGES UNIT, GULF OF MEXICO
B 6110 1587 MEXICO, 1976, ACOSTA ESTEVEY, R., ON DEVELOPMENTS IN
B 6210 2054 MEXICO, 1977, PETROCONSULTANTS S.A., ON DEVELOPMENTS IN
B 6310 1755 MEXICO, 1978, PETROCONSULTANTS S.A., ON DEVELOPMENTS IN
B 6409 1554* MEXICO, ABST., CALPIONELLIDS AND, NANNOCONIDS OF, TARAISES
B 6303 533* MEXICO, ABST., CARBONATE GEOLOGY OF, PENA BLANCA, URANIUM
B 6309 1606* MEXICO, ABST., CARRILLO PUERTO FORMATION OF NORTHEASTERN,
B 6303 495* MEXICO, ABST., CRETACEOUS TRANSGRESSION OF, COAHUILA
B 6212 2508* MEXICO, ABST., DEVELOPMENT OF, CERRO PRIETO, GEOTHERMAL FIELD,
B 6109 1552* MEXICO, ABST., DIAGENESIS OF A, GIANT—, POZA RICA TREND
B 6409 1564* MEXICO, ABST., DIAGENESIS OF, DELTAIC SANDSTONE—, OLMOS, SAN
B 6004 731* MEXICO, ABST., DIAGENETIC SEQUENCE OF MIDDLE, CRETACEOUS,
B 6303 508* MEXICO, ABST., FACIES, ENVIRONMENTS, AND DEVELOPMENT OF,
B 6409 1560* MEXICO, ABST., FORAMINIFERS AND, CALCISPHERES FROM, CUESTA DEL
B 6109 1548* MEXICO, ABST., GEOLOGY OF, SIERRA BALUARTES, DIAPIR AND, SIERRA
B 6303 555* MEXICO, ABST., IMMANITAS BED, PASO DEL RIO, COLIMA
B 6105 825* MEXICO, ABST., LATE, JURASSIC SEDIMENTATION, EASTERN, SIERRA
B 6209 1767* MEXICO, ABST., LATE, PLEISTOCENE, CARBONATE BANK DEPOSITION,
B 6303 547* MEXICO, ABST., LOWER, CRETACEOUS, CARBONATE SHELF IN
B 6109 1557* MEXICO, ABST., LOWER, CRETACEOUS, SHELF, MARGIN IN NORTHERN
B 6211 2354* MEXICO, ABST., MESOZOIC GEOLOGY OF PENINSULAR, CALIFORNIA AND
B 6206 1092* MEXICO, ABST., MIDDLE, CRETACEOUS, CARBONATE-, SLOPE DEPOSITS,
B 6109 1551* MEXICO, ABST., PALEOENVIRONMENTAL ANALYSIS OF LOWER,
B 6304 697* MEXICO, ABST., PETROLEUM GEOLOGY OF, CAMPECHE OFFSHOR AREA,
B 6405 716* MEXICO, ABST., PETROLEUM POSSIBILITIES IN, ALTAR DESERT, SONORA
B 6004 671* MEXICO, ABST., PETROLOGY OF, TRIASSIC, MARINE SEDIMENTARY
B 6203 490* MEXICO, ABST., RECENT DEVELOPMENTS AT, CERRO PRIETO,
B 6004 644* MEXICO, ABST., SEDIMENTARY ENVIRONMENTS AND DIAGENESIS OF, E
B 6004 712* MEXICO, ABST., SLOPE, VALLEYS AND SEDIMENTARY STRUCTURES OFF,
B 6109 1550* MEXICO, ABST., SOME LOWER, CRETACEOUS STRATIGRAPHIC, MODELS
B 6405 758* MEXICO, ABST., STRUCTURAL AND SEDIMENTOLOGIC STUDY OF, CERR
B 6405 759* MEXICO, ABST., STUDY OF DIAGENETIC PROCESSES IN SANDSTONE FROM,
B 6004 657* MEXICO, ABST., SUBSURFACE, JURASSIC-, CRETACEOUS BOUNDARY,
B 6004 732* MEXICO, ABST., TRANSGRESSIVE-, REGRESSIVE DEPOSITS OF, DIFUNTA
B 6012 2174† MEXICO, ABST., UPDATE OF, GEOTHERMAL RESOURCES AN
B 6203 500* MEXICO, ABST., VADOSE, CEMENT PATTERNS IN, STRAND-PLAIN,
B 6202 244 MEXICO, AGUA NUEVA FORMATION
B 6109 1456 MEXICO, ALAMO HUECO BASIN, ARIZONA, NEW MEXICO, AND
M 29 319† MEXICO, ANATOMY OF THE, MEXICAN RIDGES, SOUTHWESTERN, GULF
B 6110 1578† MEXICO, AND CARIBBEAN AREA IN 1976, PETROLEUM, DEVELOPMENTS
B 6210 2044† MEXICO, AND CARIBBEAN AREA IN 1977, PETROLEUM, DEVELOPMENTS
B 6310 1745† MEXICO, AND CARIBBEAN AREA IN 1978, PETROLEUM, DEVELOPMENTS
B 6207 1219* MEXICO, AND EXPLORATORY EFFORT IN BAJA CALIFORNIA, ABST.,
B 6009 1617* MEXICO AND SOUTH TEXAS, ABST., EARLY, CRETACEOUS, SHELF,
B 6109 1557* MEXICO AND TEXAS, ABST., MIDDLE, CRETACEOUS ROCKS OF
M 25 518 MEXICO, ANTIMONY
B 6102 208 MEXICO, ARAGON FORMATION
M 25 518 MEXICO, ASBESTOS
B 6107 988 MEXICO, AUSTIN-SAN FELIPE CHALK UNIT
B 6210 2055 MEXICO, BAJA CALIFORNIA
M 25 515 MEXICO, BAJA CALIFORNIA METALLOGENETIC PROVINCE
M 25 526 MEXICO, BAJA CALIFORNIA, VOLCANIC SERIES, CYCLES OF
M 25 515 MEXICO, BOLEO FORMATION
B 6408 1272 MEXICO, BY MICHEL T. HALBOUTY, SHELTON, JOHN W., REVIEW OF, SALT
M 30 384 MEXICO, CAMPECHE MARINE PLATFORM, HYDROCARBON PRODUCTION
M 25 519 MEXICO, CANANEA GRANITE
M 25 519 MEXICO, CAPOTE BASIN
M 25 519 MEXICO, CAPOTE BRECCIA
M 25 519 MEXICO, CAPOTE QUARTZITE
B 6303 363 MEXICO, CARILLO PUERTO FORMATION
S 3 98 MEXICO, CATAHOULA GROUP
M 25 159 MEXICO, CERRO PRIETO AREA, GEOTHERMAL
M 25 161 MEXICO, CERRO PRIETO GEOTHERMAL AREA, BAJA CALIFORNIA
S 4 209† MEXICO, CHANGE IN, OSTRACOD STANDING CROP, LAGUNA, MANDINGA,
B 6102 208 MEXICO, CHAPOPOTE FORMATION
M 30 362 MEXICO, CHIAPAS-TABASCO MESOZOIC PROVINCE
B 6202 237 MEXICO, CHICONTEPEC CHANNEL FILL
B 6202 240 MEXICO, CHICONTEPEC FORMATION
B 6202 237 MEXICO, CHICONTEPEC HORIZON C
B 6202 236 MEXICO, CHICONTEPEC OIL POOLS AND WELLS
B 6202 237 MEXICO, CHICONTEPEC TURBIDITES, CROSS SECTIONS
B 6003 379† MEXICO, COMPLEX, FAULT STRUCTURES IN, VERACRUZ PROVINCE OF
M 25 515 MEXICO, COPPER
M 25 521 MEXICO, COPPER, ORE MINERALS
B 6102 208 MEXICO, CRETACEOUS
S 6 263 MEXICO, CRETACEOUS, ALISITOS FORMATION
M 27 86 MEXICO, CRETACEOUS, CUPIDIO FORMATION
M 27 88 MEXICO, CRETACEOUS, CUPIDO FORMATION
M 27 31 MEXICO, CRETACEOUS, CUPIDO FORMATION
M 27 87 MEXICO, CRETACEOUS, CUPIDO FORMATION
M 27 123 MEXICO, CRETACEOUS, CUPIDO FORMATION
M 27 56 MEXICO, CRETACEOUS, EL ABRA FORMATION
M 27 161 MEXICO, CRETACEOUS, EL ABRA FORMATION
M 27 85 MEXICO, CRETACEOUS, EL ABRA LIMESTONE
M 27 193 MEXICO, CRETACEOUS, EL ABRA LIMESTONE
M 27 167 MEXICO, CRETACEOUS, EL ABRA LIMESTONE
M 28 18 MEXICO, CRETACEOUS, PATULA ARKOSE
M 28 134 MEXICO, CRETACEOUS, PATULA ARKOSE
M 28 165 MEXICO, CRETACEOUS, PATULA ARKOSE
M 28 9 MEXICO, CRETACEOUS, PATULA ARKOSE
M 27 13 MEXICO, CRETACEOUS, TAMABRA LIMESTONE
M 27 195 MEXICO, CRETACEOUS, TAMABRA LIMESTONE
M 27 85 MEXICO, CRETACEOUS, TAMABRA LIMESTONE
M 27 190 MEXICO, CRETACEOUS, TAMABRA LIMESTONE
M 27 9 MEXICO, CRETACEOUS, TAMABRA LIMESTONE
M 27 14 MEXICO, CRETACEOUS, TAMABRA LIMESTONE
M 27 212 MEXICO, CRETACEOUS, TAMABRA LIMESTONE
M 25 519 MEXICO, CUITACA, BATHOLITHIC GRANODIORITE
M 25 518 MEXICO, EJE NEO-VOLCANICO METALLOGENETIC PROVINCE
M 25 519 MEXICO, EL TORRE SYENITE
B 6102 208 MEXICO, ENCANTO DEPOSITO FORMATION
B 6102 208 MEXICO, EOCENE
B 6008 1157 MEXICO, EXPLORATORY DRILLING, 1975
B 6108 1124 MEXICO, EXPLORATORY DRILLING, 1976
B 6208 1279 MEXICO, EXPLORATORY DRILLING, 1977
B 6308 1172 MEXICO, EXPLORATORY DRILLING, 1978
B 6409 1299 MEXICO, EXPLORATORY DRILLING, 1979
M 25 524 MEXICO, FUEL COST, TABLE
M 25 160† MEXICO, GEOTHERMAL POTENTIAL OF
M 30 339† MEXICO, GIANT FIELDS OF THE SOUTHERN ZONE--
B 6202 236 MEXICO, GOLDEN LANE FIELD
B 6301 48 MEXICO, GOLDEN LANE, SECTION
B 6102 208 MEXICO, GUAYABAL FORMATION

M 25 474 MEXICO, GULF OF, MEGASTRUCTURE
M 27 160 MEXICO, HOLOCENE DUNE SANDS
M 27 111 MEXICO, HOLOCENE SEDIMENTS
M 27 25 MEXICO, HOLOCENE SEDIMENTS
B 6102 208 MEXICO, HORCONES FORMATION
M 25 526 MEXICO, IGNIMBRITES
M 25 516 MEXICO, ILLUSTRATION, METALLOGENETIC CHART OF
S 4 195 MEXICO, ISLA CANCUN, MOLLUSCA
S 4 195 MEXICO, ISLA CONTOY, MOLLUSCA
S 4 195 MEXICO, ISLA MUJERES, MOLLUSCA
B 6102 207 MEXICO, ISTHMUS SALT BASIN
M 25 160 MEXICO, IXTLAN DE LOS HERVORES GEOTHERMAL AREA
B 6102 217 MEXICO, JURASSIC
M 27 141 MEXICO, JURASSIC, ZULOAGO FORMATION
M 25 519 MEXICO, LA COLORADA QUARTZ PORPHYRY, COPPER
B 6102 208 MEXICO, LA LAJA FORMATION
M 25 161 MEXICO, LA SOLEDAD GEOTHERMAL AREA
M 25 160 MEXICO, LOS AZUFRIS GEOTHERMAL AREA
M 25 160 MEXICO, LOS NEGRITOS GEOTHERMAL AREA
M 30 360 MEXICO, MACUSPANA BASIN, CHILAPILLA FIELD
M 30 361 MEXICO, MACUSPANA BASIN, HORMIGUERO FIELD
M 30 360 MEXICO, MACUSPANA BASIN, JOSE COLOMO FIELD
B 6106 903† MEXICO, MARINE GEOLOGY OF, BAJA CALIFORNIA, CONTINENTAL
B 6202 244 MEXICO, MENDEZ FORMATION
M 25 518 MEXICO, MERCURY
M 25 518 MEXICO, MESA CENTRAL METALLOGENETIC PROVINCE
M 25 515 MEXICO, METALLOGENETIC PROVINCES OF
M 29 411 MEXICO, MEXICAN RIDGES, REFLECTION SECTION
M 25 160 MEXICO, MICHOACAN GEOTHERMAL AREA
M 29 446 MEXICO, MIDDLE AMERICAS TRENCH, SEISMICITY
M 25 515 MEXICO, MINERALIZATION, AGE OF
B 6102 208 MEXICO, MIOCENE
M 25 526 MEXICO, MIOCENE SUBDUCTION
B 6109 1449 MEXICO, MOYOTES UPLIFT
B 6210 2055 MEXICO, NORTHEASTERN ZONE
B 6102 214 MEXICO, NOVILLERO 9, CORE
B 6102 208 MEXICO, OLIGOCENE
B 6009 1616* MEXICO, OPENS UP JURASSIC POSSIBILITIES IN RIO GRANDE
B 6102 207† MEXICO, ORIGIN AND DISTRIBUTION OF, TERTIARY, CONGLOMERATES,
M 25 518 MEXICO, PACHUCA Y REAL DEL MONTE MINE, SILVER AND GOLD
B 6102 208 MEXICO, PALEOCENE
B 6310 1756 MEXICO, PAPALOAPAN BASIN
B 6102 222 MEXICO, PAPALOAPAN CANYON
M 25 161 MEXICO, PATHE GEOTHERMAL AREA
B 6109 1448 MEXICO, PEDREGOSA BASIN, PETROLEUM POTENTIAL, NEW MEXICO,
B 6401 71 MEXICO, PEREGRINA CANYON, CARBON ISOTOPIC DATA
M 27 163 MEXICO, PLEISTOCENE DUNES
M 27 162 MEXICO, PLEISTOCENE DUNES
B 6202 236 MEXICO, PRESIDENTE ALEMAN POOL
B 6110 1593 MEXICO, PRODUCTION, 1976
B 6210 2088 MEXICO, PRODUCTION, 1977
S 4 195† MEXICO, QUANTITATIVE PALEOECOLOGICAL ASPECTS OF MODERN,
B 6401 74 MEXICO, RANCHO JACALITOS, APTIAN THROUGH CENOMANIAN
M 25 519 MEXICO, RICKETTS FAULT
M 30 343 MEXICO, SALINE BASIN, CINCO PRESIDENTES FIELD
M 30 342 MEXICO, SALINE BASIN, EL PLAN FIELD
M 30 345 MEXICO, SALINE BASIN, MAGALLANES FIELD
M 30 345 MEXICO, SALINE BASIN, OGARRIO FIELD
M 30 342 MEXICO, SALINE BASIN, TONALA-EL BURRO FIELD
B 6202 244 MEXICO, SAN ANDRES FORMATION
B 6102 207 MEXICO, SANTA ANA UPLIFT
M 29 446 MEXICO, SEISMIC EXPERIMENT, CONTINENTAL MARGINS
B 6102 207 MEXICO, SIERRA DE CHIAPAS
M 25 518 MEXICO, SIERRA MADRE DEL SUR METALLOGENETIC PROVINCE
M 25 515 MEXICO, SIERRA MADRE OCCIDENTAL METALLOGENETIC PROVINCE
B 6102 207 MEXICO, SIERRA MADRE ORIENTAL
M 25 518 MEXICO, SIERRA MADRE ORIENTAL METALLOGENETIC PROVINCE
M 25 515 MEXICO, SILVER, LEAD, ZINC, AND ARSENIC
M 25 525 MEXICO, SINALOA, METALLIC CONCENTRATION, TIMES OF
M 25 525 MEXICO, SINALOA, MINERAL DEPOSITS, GROUPS OF
M 29 321 MEXICO, SLOPE, GOLDEN LANE CARBONATE BANK
M 25 519 MEXICO, SONORA, CANANEA MINE, COPPER
M 25 520 MEXICO, SONORA, COPPER MINERALIZATION
M 25 521 MEXICO, SONORA, LA CARIDAD MINE, COPPER
M 25 520 MEXICO, SONORA, PILARES DE NACOZARI MINE, COPPER
M 25 519 MEXICO, SONORA, PORPHYRY COPPER DEPOSITS, MAIN
M 25 520 MEXICO, SONORA, PORPHYRY COPPER DEPOSITS, MAP
B 6310 1756 MEXICO, SOUTHEASTERN ZONE
B 6210 2055 MEXICO, SOUTHEASTERN ZONE, CAMPECHE BAY
B 6303 362† MEXICO, STRANDLINE SEDIMENTATION CARBONATE, GRAINSTONES,
B 6202 235† MEXICO, STRATIGRAPHY AND STRUCTURE OF, CHICONTEPEC,
S 8 101 MEXICO, SUBMARINE CANYONS, CAPE SAN LUCAS CANYON
S 8 105 MEXICO, SUBMARINE CANYONS, PETACALCO CANYON
S 8 105 MEXICO, SUBMARINE CANYONS, RIO BALSAS DELTA CANYON
S 8 101 MEXICO, SUBMARINE CANYONS, SAN JOSE CANYON
B 6202 244 MEXICO, TAMABRA FORMATION
B 6202 244 MEXICO, TAMAULIPAS FORMATION
B 6102 207 MEXICO, TAMPICO MISANTLA BASIN
M 25 518 MEXICO, TAYOLTITA-SAN DIMAS MINE, GOLD
B 6102 208 MEXICO, TERTIARY
M 25 518 MEXICO, TIN
B 6102 207 MEXICO, TUXTLAS UPLIFT
M 25 523 MEXICO, URANIUM DEMAND AND SERVICES
M 25 524 MEXICO, URANIUM, NEEDS OF, TABLE
M 25 522 MEXICO, URANIUM ORE RESERVES, TABLE
M 25 523 MEXICO, URANIUM SUPPLY AND SERVICES
B 6102 208 MEXICO, VELASCO FORMATION
B 6202 244 MEXICO, VELASCO FORMATION
B 6102 210 MEXICO, VERACRUZ BASIN, SEISMIC PROFILES
M 25 526 MEXICO, VOLCANIC PROVINCES, PETROGENIC INTERPRETATION
M 25 523 MEXICO, YELLOW CAKE, DEMAND FOR
M 25 515† MEXICO--SUMMARY, METALLOGENETIC CHART OF
M 25 525† MEXICO--SUMMARY, MINERAL RESOURCES OF STATE O, SINALOA
M 25 519† MEXICO--SUMMARY, PRINCIPAL CHARACTERISTICS OF SOME PORPHYR,
M 25 522† MEXICO--SUMMARY, URANIUM AND ITS PROJECTED USE IN, NUCLEAR
M 25 460 MEXICO-CENTRAL AMERICA, SILVER-COPPER-POLYMETALLIC PROFILE,
B 6303 409* MEXICO. BY TITLE ONLY, GIANT FIELDS BY SOUTHEASTERN
B 6203 573* MEXICO—CEMENT EVOLUTION, ABST., SYNDEPOSITIONAL, SUBMARINE,
S 4 231 MEYER, DAVID L., CRINOIDS OF WEST INDIAN CORAL REEFS
M 29 443 MEYER, R. P., EXPLOSION SEISMOLOGY STUDIES OF ACTIVE AND PASSIVE
B 6105 813 MEYER, RICHARD F., ENERGY RESOURCE DATA—SOURCES, ACQUISITION,
B 6203 543 MEYER, RICHARD F.., FUTURE NATURAL GAS SUPPLY, ABST.
B 6004 697 MEYERHOFF, A. A., EURASIAN PLATE BOUNDARIES IN USSR, ABST.
B 6409 1564 MEYERHOFF, A. A., FUTURE PETROLEUM PROVINCES OF GULF OF
B 6205 889 MEYERHOFF, A. A., GEOLOGY AND GEOPOLITICS OF MEXICAN
M 30 225 MEYERHOFF, A. A., GEOLOGY AND PETROLEUM FIELDS IN PROTEROZOIC
B 6303 546 MEYERHOFF, A. A., GREAT CARBONATE BANK OF YUCATAN AND ITS
B 6103 443 MEYERHOFF, A. A., JURASSIC LITHOLOGY IN GREAT ISAA 1 WELL,
B 6207 1226 MEYERHOFF, A. A., PETROLEUM GEOLOGY—PEOPLES REPUBLIC OF
B 6406 953 MEYERHOFF, A. A., REVIEW OF HISTORICAL BIOGEOGRAPHY, PLATE
B 6410 1722 MEYERHOFF, A. A., REVIEW OF THE SOVIET ENERGY SYSTEM, BY LESLIE
B 6305 818 MEYERHOFF, A. A., REVIEW OF WINDOW ON OIL, A SURVEY OF WORLD
B 6310 1983 MEYERHOFF, HOWARD A., REVIEW OF COAL RESOURCES OF THE
B 6310 1983 MEYERHOFF, MEYERHOFF, HOWARD A., REVIEW OF, COAL RESOURCES OF
B 6004 717 MEYERS, W. C., ORGANIC GEOCHEMISTRY AS AID TO EXPLORATION IN
B 6405 750 MEYERS, WILLIAM J., GEOCHEMISTRY OF REGIONALLY EXTENSIVE
B 6105 813 MEYERS, WILLIAM J., REGIONAL CEMENTATION PATTERNS IN
B 6004 697 MIALL, A. D., TERTIARY DELTAIC CYCLES ON UNSTABLE CRATO MARGIN,
B 6307 1144 MIALL, LE BLANC, RUFUS J., REVIEW OF, FLUVIAL, SEDIMENTOLOGY,
B 6104 519† MIAMI, FLORIDA, PLEISTOCENE, BARRIER, BAR SEAWARD OF, OOID,
B 6104 522 MIAMI LIMESTONE, FLORIDA
B 6404 475 MIAMI OOLITE, FRESHWATER PHREATIC ZONE, FLORIDA
B 6107 1079 MIAMI OOLITE, MICROGRAPHS, FLORIDA KEYS, BIG PINE KEY
M 27 219 MIAMI OOLITES, FLORIDA, PLEISTOCENE
S 3 29 MIASKITE, DEFINITION
B 6301 90 MIC MAC FORMATION, SCOTIAN BASIN
M 28 75 MICA, BIBLIOGRAPHY
B 6305 778 MICELLAR MIGRATION
S 10 105 MICELLAR MIGRATION, MIGRATION, SECONDARY
B 6002 214 MICELLAR SOLUTION HYPOTHESIS
B 6004 671* MICELLAR-POLYMER PROJECT AREA, BUTLER COUNTY, KANSAS, ABST.,
S 10 96 MICELLE CONCENTRATIONS, SURFACTANT SOLUTIONS
S 10 23 MICELLE CONFIGURATION, MIGRATION, MECHANISMS
S 10 21 MICELLES, MIGRATION, ENHANCED SOLUBILITY
S 10 59 MICELLES, MIGRATION IN AQUEOUS SOLUTION
S 10 95 MICELLES, MIGRATION MECHANISMS, SURFACTANT
B 6305 768 MICELLES, OIL SOLUBILIZED IN, SURFACTANT
B 6305 775 MICELLES, SECONDARY, PETROLEUM MIGRATION IN SOLUTION OR BY
B 6009 1502 MICHELSON, JAMES E., MIOCENE DELTAIC OIL HABITAT, TRINIDAD
B 6309 1590* MICHIGAN ANTRIM SHALE BY IN-SITU PROCESS, ABST., RECOVERY OF,
S 5 1* MICHIGAN BASIN, DEPOSITIONAL ENVIRONMENTS OF PINNACLE REEFS,
S 5 23* MICHIGAN BASIN, SEDIMENTOLOGY AND DEPOSITIONAL ENVIRONMENTS
S 5 93* MICHIGAN BASIN, SYNCHRONIZATION OF DEPOSITION: SILURIAN
B 6405 736* MICHIGAN BASIN, ABST., EXPLORING FOR, NIAGARAN, PINNACLE, REEF
B 6405 768* MICHIGAN BASIN, ABST., PETROLEUM EXPLORATION AND, WRENCHING
B 6408 1283* MICHIGAN BASIN, ABST., PROSPECTS FOR, COAL DEVELOPMEN IN
B 6405 759* MICHIGAN BASIN, ABST., THERMAL CONTRACTION AND PETROLEUM
B 6203 525* MICHIGAN BASIN, ABST., USE OF, SEISMIC, MODELING FOR, REEF
B 6303 450* MICHIGAN BASIN AND ITS PETROLEUM POTENTIAL, ABST., STRUCTURAL
M 29 465 MICHIGAN BASIN, AVERAGE THICKNESS OF SEDIMENTS

B 6304 610 MICHIGAN BASIN, CAIN FORMATION
B 6304 609 MICHIGAN BASIN, CARBONATE DEPOSITION
B 6303 384 MICHIGAN BASIN GEOLOGICAL SOCIETY FIELD TRIP, SEPTEMBER 29, 30
S 5 4 MICHIGAN BASIN, CAYUGAN SERIES, SALINA GROUP
M 26 217 MICHIGAN BASIN, LATE PALEOZOIC, DEPOSITIONAL SYSTEMS
B 6004 709* MICHIGAN BASIN LINEAR STRUCTURES, ABST., IMPLICATIONS OF,
S 5 24 MICHIGAN BASIN, LOWER SALINA GROUP, LITHOFACIES
S 5 4 MICHIGAN BASIN, NIAGARA GROUP, PINNACLE REEFS
S 5 4 MICHIGAN BASIN, NIAGARA SERIES, NIAGARA GROUP, BURNT BLUFF
B 6304 609 MICHIGAN BASIN, NIAGARAN PALEOGEOGRAPHY
S 5 100 MICHIGAN BASIN, NIAGARAN SERIES, AGE OF REEFS
S 5 4 MICHIGAN BASIN, NIAGARAN SERIES, NIAGARA GROUP, GUELPH
S 5 4 MICHIGAN BASIN, NIAGARAN SERIES, NIAGARA GROUP, MANISTIQUE
S 5 4 MICHIGAN BASIN, NIAGARAN SERIES, NIAGARA GROUP, LOCKPORT
S 5 4 MICHIGAN BASIN, PINNACLE REEF BELT, STRATIGRAPHY
S 5 30 MICHIGAN BASIN, SALINA GROUP, A-1 EVAPORITE
S 5 16 MICHIGAN BASIN, SALINA GROUP, RUFF FORMATION
B 6209 1611 MICHIGAN BASIN, SALINA GROUP, UNITS
B 6304 610 MICHIGAN BASIN, SILURIAN FORMATIONS
M 25 204 MICHIGAN BASIN, SILURIAN REEF ANOMALY
M 25 203 MICHIGAN BASIN VELOCITY-STRUCTURE PROBLEM
S 5 74 MICHIGAN LUCAS FORMATION
S 5 53* MICHIGAN, DEPOSITIONAL ENVIRONMENT OF RUFF FORMATION (UPPER
S 5 73* MICHIGAN, EVAPORITE CYCLES AND LITHOFACIES IN LUCAS
S 5 111* MICHIGAN, SEISMIC DATA-PROCESSING TECHNIQUES IN EXPLORATION
S 5 83 MICHIGAN, BAY COUNTY LOCATION
S 5 11 MICHIGAN, BELLE RIVER MILLS FORMATION, VADOSE DEPOSITS
S 5 55 MICHIGAN, CALHOUN COUNTY, LOCATION
S 5 55 MICHIGAN, CHARLEVOIX COUNTY, LOCATION
S 5 55 MICHIGAN, CLINTON INLET, LOCATION
B 6104 594 MICHIGAN, DEEP RIVER FIELD
B 6104 595 MICHIGAN, DEERFIELD FIELD
B 6304 608† MICHIGAN, DIFFERENTIAL, ENTRAPMENT OF OIL AND GAS IN,
S 5 55 MICHIGAN, GEORGIAN INLET, LOCATION
S 5 78 MICHIGAN, GLADWIN COUNTY LOCATION
S 5 121 MICHIGAN, GRAND TRAVERSE COUNTY, EXAMPLE
S 5 115 MICHIGAN, GRAND TRAVERSE COUNTY, LOCATION
B 6104 594 MICHIGAN, HANOVER FIELD
B 6008 1215† MICHIGAN IN 1975, DEVELOPMENTS IN
B 6108 1173† MICHIGAN IN 1976, DEVELOPMENTS IN
B 6208 1330† MICHIGAN IN 1977, DEVELOPMENTS IN
B 6308 1278† MICHIGAN IN 1978, DEVELOPMENTS IN
B 6409 1377† MICHIGAN IN 1979, DEVELOPMENTS IN
B 6303 524* MICHIGAN NIAGARAN REEFS BY BRINE REFLUXION AND MIXING OF
S 5 121 MICHIGAN, KALKASKA COUNTY, EXAMPLE
S 5 115 MICHIGAN, KALKASKA COUNTY, LOCATION
S 5 3 MICHIGAN, KALKASKA COUNTY, LOCATION
S 5 55 MICHIGAN, MACOMB COUNTY, LOCATION
S 5 113 MICHIGAN, MANISTEE COUNTY, BEAR LAKE TOWNSHIP, SEISMIC
S 5 115 MICHIGAN, MANISTEE COUNTY, LOCATION
S 5 78 MICHIGAN, MASON COUNTY LOCATION
S 5 3 MICHIGAN, MASON COUNTY, LOCATION
S 5 83 MICHIGAN, MIDLAND COUNTY LOCATION
S 5 55 MICHIGAN, MIDLAND COUNTY, LOCATION
S 5 84 MICHIGAN, MIDLAND COUNTY, MIDLAND SALT FIELD
S 5 55 MICHIGAN, NEWAGO COUNTY, LOCATION
B 6410 1690 MICHIGAN, NONESUCH SHALE, CRUDE OIL
B 6104 596 MICHIGAN, NORTHVIEW FIELD
S 5 55 MICHIGAN, OCEANA COUNTY, LOCATION
B 6304 619 MICHIGAN, ONOWAY OIL FIELD
B 6203 498* MICHIGAN, PENNSYLVANIAN OF OKLAHOMA, AND MODERN JAMAICAN
S 5 3 MICHIGAN, PRESQUE ISLE COUNTY, LOCATION
B 6008 1217 MICHIGAN, PRODUCTION, 1974-1975
B 6108 1176 MICHIGAN, PRODUCTION, 1975-1976
B 6208 1333 MICHIGAN, PRODUCTION 1976-1977
B 6308 1281 MICHIGAN, PRODUCTION, 1977-1978
B 6409 1381 MICHIGAN, PRODUCTION, 1977-1978
B 6104 597 MICHIGAN, PULASKI FIELD
B 6408 1286* MICHIGAN REEFS ATTRIBUTED TO ANHYDRITE, ABST., ABSENCE OF,
S 5 80 MICHIGAN, SAGINAW COUNTY LOCATION
S 5 55 MICHIGAN, SANILAC COUNTY, LOCATION
S 5 54 MICHIGAN, SILURIAN, CAYUGAN SERIES, SALINA GROUP, STRATIGRAPHY
S 5 2 MICHIGAN, SILURIAN, NIAGARA GROUP
S 5 54 MICHIGAN, SILURIAN, NIAGARAN SERIES, NIAGARA GROUP,
M 28 88 MICHIGAN, SILURIAN, SALINA FORMATION
S 5 3 MICHIGAN, SILURIAN, SALINA GROUP
S 5 55 MICHIGAN, ST. CLAIR COUNTY, LOCATION
M 25 160 MICHOACAN GEOTHERMAL AREA, MEXICO
B 6409 1565 MICIOTTO, SAMUEL A., PETROGRAPHIC, STRATIGRAPHY, AND
M 27 200 MICRITE DIAGENESIS
M 27 194 MICRITE ENVELOPES
M 27 190 MICRITE ENVELOPES
M 27 193 MICRITE ENVELOPES
M 27 177 MICRITE, FOSSILIFEROUS
M 27 175 MICRITES
B 6107 1069† MICRITIZATION AND CARBONATE-GRAIN BINDING BY ENDOLITHIC
M 27 128 MICRITIZED GRAINS
B 6409 1554* MICROBENTHON RESPONSES TO 1979 GULF OF MEXICO OIL SPILLS, IXTOC
B 6409 1554* MICROBENTHON RESPONSES TO 1979 GULF OF MEXICO OIL SPILLS, IXTOC
B 6405 685* MICROBIAL AND INVERTEBRATE ENDOLITHIC ASSEMBLAGES FROM LATE
B 6203 455† MICROBIOLOGIC ROLE IN GENESIS OF CRUDE OIL FROM BARROW
B 6304 698* MICROBIOLOGICAL PROSPECTING FOR HYDROCARBONS, ABST.
S 4 269 MICROBORERS, FLORAL
B 6004 706* MICROBORING ORGANISMS, ABST., PALEOECOLOGIC, POTENTIAL OF
M 28 17 MICROCLINE, SPINDLE TWINNING, FELDSPAR
M 28 17 MICROCLINE TWINNING, FELDSPAR
B 6105 778* MICROCONTINENTAL MARGINS
B 6405 687* MICROFACIES AND DEPOSITIONAL-DIAGENETIC MODEL OF AMAPA
S 7 124 MICROFAUNA, PALEOBATHYMETRY, GULF OF MEXICO
B 6303 468* MICROFOSSILS, MACROFOSSILS, AND STROMATOLITES FROM MIDDLE
B 6406 927† MICROFRACTURE DEVELOPMENT IN COMPACTING SEDIMENTS—
M 28 179 MICROFRACTURE, IN SANDSTONE, POROSITY
B 6303 490* MICROMORPH FAUNAS IN WASHITA GROUP, ABST., MODELS FOR
B 6207 1213* MICRONESIA, ABST., WATER SUPPLIES IN, LIMESTONE TERRANES OF
S 11 10 MICROPALEONTOLOGIC BIOSTRATIGRAPHY, EARLY PROBLEMS
S 11 11 MICROPALEONTOLOGIC BIOSTRATIGRAPHY, MID-TERTIARY STAGES
B 6209 1767* MICROPALEONTOLOGY, ABST., MILESTONES IN, GULF COAST ECONOMIC
B 6202 201† MICROPALEONTOLOGY IN EXPLORATION GEOLOGY, GEOHISTORY
S 11 4 MICROPALEONTOLOGY, PIONEERS, CALIFORNIA
B 6409 1554* MICROPLANKTON AND MICROBENTHON RESPONSES TO 1979 GULF OF
B 6405 773* MICROPROCESSOR SYSTEM FOR PALEONTOLOGIC INFORMATION,
B 6303 475* MICROSCOPIC DISTRIBUTION OF URANIU IN OAKVILLE SANDSTONE,
M 27 161 MICROSTALACTITIC CEMENTS
B 6004 689* MICROTEXTURES ON QUARTZ SAND FROM MEADOWCROFT ROCK
B 6207 1211* MICROWAVE SATELLITE IMAGERY, ABST., STRUCTURE AND MOTION OF,
B 6209 1760* MICROZOOPLANKTON, RADIOLARIANS, FORAMINIFERS, AND
B 6302 183 MID NORTH SEA HIGH, NORTH SEA
B 6203 436 MID NORTH SEA HIGH, NORTH SEA
B 6203 426 MID NORTH SEA PLATFORM, EUROPE
B 6309 1453 MID-ATLANTIC, ALBEMARLE EMBAYMENT
B 6312 2159 MID-ATLANTIC BIGHT, MAP
B 6105 814* MID-ATLANTIC CONTINENTAL SHELF REGION—INITIAL SURVEY, ABST.,
B 6409 1333 MID-ATLANTIC OUTER CONTINENTAL SHELF, EXPLORATORY WELLS,
B 6309 1453 MID-ATLANTIC, RARITAN EMBAYMENT
B 6202 274 MID-ATLANTIC RIDGE
B 6106 930 MID-ATLANTIC RIDGE
B 6007 1079 MID-ATLANTIC RIDGE, BATHYMETRY
B 6209 1594 MID-ATLANTIC RIDGE, DSDP SITES
B 6007 1084 MID-ATLANTIC RIDGE, MAGNETICS
M 25 487 MID-ATLANTIC RIDGE, METAL CONCENTRATIONS, EAST PACIFIC RISE
B 6401 133 MID-ATLANTIC RIDGE, RIFT AND TRANSFORM FAULT AT 3000 METERS
S 3 79 MID-ATLANTIC RIDGE, SPREADING
B 6101 79 MID-ATLANTIC RIFT
B 6309 1453 MID-ATLANTIC, SALISBURY EMBAYMENT
B 6203 564* MID-ATLANTIC STATES, ABST., EROSION VERSUS DEPOSITION ON UPPER,
B 6303 544* MID-CENOZOIC FORTUNA FORMATION, EAST-CENTRAL TUNISIA, RECORD
B 6311 2119* MID-CONTINENT, ABST., CONTROLS ON, PENNSYLVANIAN
B 6009 1608* MID-CONTINENT AN GULF COAST AREAS--PLATE TECTONICS VIEW, ABST.,
B 6303 410* MID-CONTINENT AREA, ABST., LATE, ORDOVICIAN--EARLY, SILURIAN
B 6002 320* MID-CONTINENT AREA, ABST., STIMULATING RECOVERY FROM,
B 6107 1045 MID-CONTINENT, CYCLOTHEMS
B 6408 1272 MID-CONTINENT, EDITED BY NORMAN J. HYNE, STEWART, G. F., REVIEW
B 6008 1332† MID-CONTINENT IN 1975, OIL AND, GAS DEVELOPMENTS IN NORTH
B 6108 1305† MID-CONTINENT IN 1976, OIL AND, GAS, DEVELOPMENTS IN NORTH
B 6208 1477† MID-CONTINENT IN 1977, OIL AND, GAS, DEVELOPMENTS IN NORTH
B 6308 1301† MID-CONTINENT IN 1978, OIL AND, GAS, DEVELOPMENTS I NORTH
B 6308 1301† MID-CONTINENT IN 1978, OIL AND, GAS, DEVELOPMENTS IN NORTH
B 6409 1437† MID-CONTINENT IN 1979, OIL AND, GAS, DEVELOPMENTS IN NORTH
S 4 256 MID-CONTINENT, LANSING GROUP
B 6107 1062 MID-CONTINENT MIDDLE AND LATE PENNSYLVANIAN DEPOSITION,
B 6107 1050 MID-CONTINENT, MISSOURIAN STAGE
B 6107 1045† MID-CONTINENT NORTH AMERICA, ORIGIN OF, PHOSPHATIC, BLACK,
B 6107 1063 MID-CONTINENT PALEOZOIC BLACK SHALE AND PHOSPHORITE
B 6107 1055 MID-CONTINENT SEA, PROBABLE CIRCULATION IN
B 6405 721* MID-CONTINENT UPPER PENNSYLVANIAN CYCLOTHEMS, ABST.,
B 6303 542* MID-CRETACEOUS BLACK SHALES IN ATLANTIC BASINS, ABST.,
B 6303 423* MID-CRETACEOUS CALCAREOUS NANNOPLANKTON
B 6407 1052† MID-FAN COMPLEXES OF TOURELLE FORMATION, ORDOVICIAN, QUEBEC,
B 6206 952 MID-FAN: BRAIDED CHANNELS ON SUPRAFAN LOBE, ANCIENT
B 6303 538* MID-PACIFIC MOUNTAINS AND ON HESS RISE, CENTRAL NORTH PACIFIC

M 29 464 MID-PALEOZOIC BASINS, ANALYSIS, GLOBAL CHANGE
S 11 11 MID-TERTIARY STAGES, MICROPALEONTOLOGIC BIOSTRATIGRAPHY
B 6305 841* MIDALE BEDS, BENSON OIL FIELD, SOUTHEASTERN SASKATCHEWAN,
M 28 70 MIDDENDORF FORMATION, SOUTH CAROLINA, CRETACEOUS
B 6207 1210* MIDDLE AMERICA TRENCH, ABST., TECTONIC FRAMEWORK OF
B 6312 2208 MIDDLE AMERICA TRENCH, PROFILE
M 25 485 MIDDLE AMERICA TRENCH, SUBDUCTION
M 29 446 MIDDLE AMERICAS TRENCH, SEISMICITY, MEXICO
B 6009 1622* MIDDLE AND EARLY LATE ORDOVICIAN HISTORY OF CINCINNATI ARCH
B 6309 1577* MIDDLE AND UPPER DEVONIAN STRATIGRAPHY IN NORTHWESTERN
B 6405 805* MIDDLE AND UPPER ORDOVICIAN PALEOGEOGRAPHY OF REGION
B 6106 861 MIDDLE ATLANTIC OUTER CONTINENTAL SHELF, SURFICIAL, SAND,
B 6208 1382 MIDDLE BANK REEF, BRAZOS AREA, TEXAS
B 6405 802* MIDDLE CRETACEOUS, ALBIAN-TURONIAN, DEPOSITIONAL
B 6206 1092* MIDDLE CRETACEOUS CARBONATE-SLOPE DEPOSITS, NORTHEASTERN
B 6105 762* MIDDLE CRETACEOUS OCEANIC ANOXIC EVENTS AS CAUSAL FACTORS IN
B 6303 538* MIDDLE CRETACEOUS OXYGEN-DEFICIENT PALEOENVIRONMENTS IN
B 6109 1557* MIDDLE CRETACEOUS ROCKS OF MEXICO AND TEXAS, ABST.
B 6010 1900† MIDDLE EAST, 1970 TO 1975, SELECTED REFERENCES ON
B 6303 459* MIDDLE EAST ABST., GAS VERSUS OIL IN, FAR EAST AND
B 6405 614 MIDDLE EAST, AHWAZ SANDSTONE DELTA
B 6405 610 MIDDLE EAST, APTIAN
B 6405 614 MIDDLE EAST, AQUITANIAN
B 6405 616 MIDDLE EAST, ARAB C
B 6405 616 MIDDLE EAST, ARAB D
B 6405 614 MIDDLE EAST, BAJOCIAN TO ALBIAN SEQUENCE
B 6405 610 MIDDLE EAST, BARREMIAN
B 6405 608 MIDDLE EAST, BATHONIAN
B 6405 616 MIDDLE EAST, BURGAN SANDSTONE
B 6405 613 MIDDLE EAST, CAMPANIAN
B 6405 603 MIDDLE EAST, CARBONIFEROUS
B 6405 605 MIDDLE EAST, CARBONIFEROUS TO LIASSIC
B 6405 612 MIDDLE EAST, CENOMANIAN
B 6405 615 MIDDLE EAST, CENTRAL GULF AREA, SECTIONS
B 6405 614 MIDDLE EAST, CENTRAL GULF OIL HABITAT
B 6010 1865† MIDDLE EAST COUNTRIES IN 1975, PETROLEUM
B 6110 1795† MIDDLE EAST COUNTRIES IN 1976, PETROLEUM, DEVELOPMENTS IN
B 6210 2111† MIDDLE EAST COUNTRIES IN 1977, PETROLEUM, DEVELOPMENTS IN
B 6310 1884† MIDDLE EAST COUNTRIES IN 1978, PETROLEUM, DEVELOPMENTS IN
B 6411 1836† MIDDLE EAST COUNTRIES IN 1979, PETROLEUM, DEVELOPMENTS IN
B 6405 600 MIDDLE EAST, DHRUMA FORMATION
B 6405 602 MIDDLE EAST, DIBBA LINE
B 6303 526* MIDDLE EAST DURING MIDDLE TRIASSIC TIME, ABST., REGIONAL ASPECT
B 6202 178 MIDDLE EAST, ENERGY RESOURCES
B 6405 600 MIDDLE EAST, HANIFA FORMATION
B 6405 600 MIDDLE EAST, HITH FORMATION
B 6405 602 MIDDLE EAST, HUQF-DHOFAR ARCH
B 6405 602 MIDDLE EAST, INFRA-CAMBRIAN SALT
B 6405 602 MIDDLE EAST, JAUF-GAARA-MOSIL ARCH
B 6405 600 MIDDLE EAST, JUBAILA FORMATION
B 6405 606 MIDDLE EAST, JURASSIC TO TURONIAN
B 6405 606 MIDDLE EAST, LADINIAN TO CARNIAN
B 6405 616 MIDDLE EAST, LEKHWAIR FORMATION
B 6405 602 MIDDLE EAST, MASIRAH LINE
B 6405 616 MIDDLE EAST, NAHR UMR SHALE
B 6405 608 MIDDLE EAST, OXFORDIAN TO KIMMERIDGIAN
B 6405 613 MIDDLE EAST, PALEOCENE TO EOCENE
B 6405 602 MIDDLE EAST, PALMYRA-SINJAR FOLDBELT
B 6405 605 MIDDLE EAST, PERMIAN
S 10 156 MIDDLE EAST, PERSIAN GULF, PETROLEUM MIGRATION
B 6405 600 MIDDLE EAST, QATAR AREA, COLUMN
B 6405 600 MIDDLE EAST, QATAR FORMATION
B 6405 602 MIDDLE EAST, QATAR-SOUTH FARS ARCH
B 6405 607 MIDDLE EAST, RHAETIAN
B 6405 610 MIDDLE EAST, SENONIAN TO RECENT
B 6405 616 MIDDLE EAST, SHUAIBA FORMATION
B 6405 616 MIDDLE EAST, SHUAIBA LIMESTONE
B 6405 598 MIDDLE EAST, STRATIGRAPHIC COLUMN
B 6405 609 MIDDLE EAST, TITHONIAN
B 6405 606 MIDDLE EAST, TRIASSIC
B 6405 600 MIDDLE EAST, TUWAIQ FORMATION
B 6405 600 MIDDLE EAST, UWAINAT MEMBER
B 6405 609 MIDDLE EAST, VALANGINIAN
B 6405 602 MIDDLE EAST, ZAGROS FOLDBELT
B 6405 597† MIDDLE EAST—STRATIGRAPHIC EVOLUTION AND OIL HABITAT
B 6009 1618* MIDDLE EOCENE PLANKTONIC CORRELATIONS--NORTHERN
B 6109 1551* MIDDLE GLEN ROSE, CRETACEOUS, FACIES MOSAIC, BLANCO AND HAYS
M 24 16 MIDDLE GROUND SHOAL FIELD, A-POOL MARKER, ALASKA
M 24 17 MIDDLE GROUND SHOAL FIELD, B-POOL MARKER, ALASKA
M 24 18 MIDDLE GROUND SHOAL FIELD, E-POOL MARKER, ALASKA
M 24 1† MIDDLE GROUND SHOAL OIL FIELD, ALASKA
B 6408 1132 MIDDLE GROUND SHOAL OIL FIELD, ALASKA
M 24 21 MIDDLE GROUND SHOAL OIL FIELD, PRODUCTIVE ZONES A, B, C, D E, F,
B 6004 647* MIDDLE GUADALUPIAN COASTAL CARBONATE COMPLEX--CO3-SO4
B 6004 693* MIDDLE GUADALUPIAN COASTAL CARBONATE
B 6110 1582 MIDDLE MAGDALENA VALLEY, COLOMBIA
B 6405 778* MIDDLE MEMBER OF MINNELUSA FORMATION, MIDDLE AND UPPER
B 6203 572* MIDDLE MIOCENE TO EARLY PLIOCENE SEDIMENTATION AND
B 6012 2175† MIDDLE MIOCENE-EARLY PLIOCENE MARINE DIATO AND
B 6408 1281* MIDDLE ORDOVICIA SHELF CARBONATE SEDIMENTATION AROUND
B 6105 760* MIDDLE PALEOZOIC HISTORY OF ANADARKO AND ARKOMA BASINS,
B 6303 418* MIDDLE PALEOZOIC SEDIMENTATION AND PALEOGEOGRAPHY OF
M 30 310 MIDDLE SAND, PRUDHOE BAY, KUPARUK RIVER FORMATION
B 6309 1601* MIDDLE STRAWN, DESMOINESIAN, CRATONIC DELTA SYSTEMS, CONCHO
B 6405 799* MIDDLE-LATE PERMIAN PALEOGEOGRAPHY OF IDAHO, NEVADA,
S 4 22 MIDDLE-SHELF PROVINCE, PUERTO RICO
B 6412 2078 MIDDLESBORO SYNCLINE, APPALACHIANS
B 6103 417 MIDDLESBORO SYNCLINE, VIRGINIA
B 6004 697 MIDDLETON, G. V., FACIES MODE FOR MACROTIDAL ENVIRONMENT,
B 6203 543 MIDDLETON, GERARD V., COARSE SEDIMENTS DEPOSITED AT BASE OF
B 6204 635 MIDDLETON ISLAND, ALASKA
B 6405 750 MIDDLETON, LARRY, SEDIMENTOLOGY AND PALEOGEOGRAPHY OF
B 6102 296* MIDLAND BASIN, IRION COUNTY, TEXAS, ABST., BROOKS FIELD,
B 6203 526* MIDLAND BASIN OF WEST TEXAS, ABST., LEONARDIAN STRATA IN
B 6008 1407* MIDLAND BASIN, SUBSIDENCE VERSUS SEDIMENTATION PHENOMENON,
M 26 218 MIDLAND BASIN, SWEETWATER SLOPE SYSTEM
B 6006 907† MIDLAND BASIN, TEXAS, OOLITE-, BAR PROGRADATION, SAN ANDRES
B 6308 1426* MIDLAND BASIN, WEST TEXAS, ABST., SEDIMENTOLOGY AND
S 5 83 MIDLAND COUNTY LOCATION, MICHIGAN,
S 5 55 MIDLAND COUNTY, LOCATION, MICHIGAN,
S 5 84 MIDLAND COUNTY, MIDLAND SALT FIELD, MICHIGAN,
S 5 84 MIDLAND SALT FIELD, MICHIGAN, MIDLAND COUNTY,
S 5 73* MIDLAND, MICHIGAN, EVAPORITE CYCLES AND LITHOFACIES IN LUCAS
S 7 101 MIDSLOPE SEISMIC FACIES EXAMPLE, LOUISIANA
B 6405 777* MIDWAY, CONTINENTAL SHELF DEPOSITS, RIO GRANDE EMBAYMENT,
B 6409 1569* MIDWAY GROUP IN NORTHEAST TEXAS, ABST., PALEOECOLOGY OF
B 6004 716* MIDWAY GROUP, PALEOCENE, SOUTH-CENTRAL TEXAS, ABST.,
B 6408 1220 MIDWAY SECTION, TEXAS
M 24 244 MIDWAY SHALE, MISSISSIPPI
B 6309 1575* MIDWAY-EXTRA GAS FIELD, WEST VIRGINIA, ABST., LINEAMENTS AND
B 6203 351 MIDWAY-SUNSET OIL FIELD, CALIFORNIA
B 6309 1596* MIDWAY-WILCOX, ZAPATA COUNTY, TEXAS, ABST., SUBSURFACE
B 6402 225 MIERS GRABEN, TEXAS
B 6303 492* MIETTE BUILDUP, ALBERTA, ABST., BURIAL, DOLOMITIZATION OF UPPER,
B 6107 1080 MIETTE REEF COMPLEX, ALBERTA
S 6 160 MIFUNE GROUP, JAPAN, CENOMANIAN
M 26 132 MIGRATING-WAVE, SEISMIC, REFLECTIONS, CONFIGURATIONS
S 10 50 MIGRATION, CALIFORNIA, LOS ANGELES BASIN
S 10 179† MIGRATION, METHANE, GENERATION AND, PETROLEUM
S 10 49 MIGRATION, NORTH DAKOTA, WILLISTON BASIN, BAKKEN SHALE
S 10 121† MIGRATION, THE GENERALIZED, HYDRAULIC THEORY OF, PETROLEUM
S 10 53 MIGRATION, UTAH, UINTA BASIN, GREEN RIVER FORMATION
B 6203 558* MIGRATION, ABST., DESIGN AND FUNCTION OF, OIL AND, GAS, TRAPS AS
B 6203 520* MIGRATION, ABST., METHANE GENERATION AND, PETROLEUM
B 6004 647* MIGRATION, ABST., MIDDLE, GUADALUPIAN COASTAL, CARBONATE
B 6203 571* MIGRATION, ABST., SHALE, DEHYDRATION, SALINITY, CRACKING, AND
B 6203 538* MIGRATION AGENTS, ABST., PRIMARY
B 6408 1219 MIGRATION AND ACCUMULATION IN SUBTLE TRAPS, PETROLEUM
B 6108 1379* MIGRATION AND CONCENTRATION OF URANIUM-CATAHOULA
B 6305 723† MIGRATION AND ENTRAPMENT, MECHANICS OF, SECONDARY,
B 6305 752 MIGRATION AND ENTRAPMENT MODEL
B 6405 774* MIGRATION AND ENTRAPMENT OF PETROLEUM—EXAMPLES FROM UTAH
B 6009 1520† MIGRATION AND ENTRAPMENT, OBSCURE NATURE OF, PETROLEUM
B 6108 1391* MIGRATION AND ENTRAPMENT OF HYDROCARBONS, ABST., MECHANICS
S 10 40 MIGRATION, AGENTS, AQUATHERMAL PRESSURE
S 10 41 MIGRATION, AGENTS, CLAY MINERAL CONVERSION
S 10 39 MIGRATION, AGENTS, COMPACTION
S 10 41 MIGRATION, AGENTS, ION FILTRATION
S 10 41 MIGRATION, AGENTS, OSMOSIS
B 6002 213† MIGRATION, AQUEOUS, SOLUBILITY OF, PETROLEUM AS APPLIED T ITS,
B 6002 232 MIGRATION BY MOLECULAR SOLUTION, MODEL OF, PRIMARY,
B 6407 1078† MIGRATION BY SOLUTION IN CARBON DIOXIDE IN NORTON BASIN,
S 10 189 MIGRATION, BARRIERS, OVERPRESSURED SHALE
S 10 61 MIGRATION, BASIN TEMPERATURES
S 10 56 MIGRATION, BITUMEN REDISTRIBUTION IN SOURCE ROCKS
S 10 61 MIGRATION, BITUMENS AT HIGH TEMPERATURES
S 10 228 MIGRATION, BRINE FLOW, HYDROTHERMAL EFFECTS
S 10 85 MIGRATION, BYPASS COMPUTER MODEL
S 10 60 MIGRATION, BYPASSING WATER

S 10 40 MIGRATION, CALCULATED SHALE PRESSURE
S 10 72 MIGRATION, CHANGE OF GEOTHERMAL GRADIENT
S 10 42 MIGRATION, COMBINED AGENTS
S 10 83 MIGRATION, COMPACTION OF HIGH PRESSURE SHALES
S 10 72 MIGRATION, COMPUTER MODELING OF GULF COAST
S 10 90 MIGRATION, CONSTRAINTS, SOLUBILITIES IN WATER
S 10 48 MIGRATION, CONTINUOUS PHASE
S 10 97 MIGRATION, CRUDE OIL IN GAS PHASE
S 10 42 MIGRATION, DIRECTION OF MOVEMENT
S 10 20 MIGRATION, ENHANCED SOLUBILITY, ACCOMMODATION
S 10 21 MIGRATION, ENHANCED SOLUBILITY, MICELLES
S 10 20 MIGRATION, ENHANCED SOLUBILITY, PRECURSORS
B 6111 1953 MIGRATION, ENTRAPMENT, AND DEGRADATION, OIL, GENERATION
B 6412 2108† MIGRATION, EVIDENCES OF, PRIMARY, OIL
S 10 41 MIGRATION, EXPANDABLE CLAY VERSUS DEPTH, DIAGENESIS
S 10 48 MIGRATION, FLUID PROPERTIES
S 10 79 MIGRATION, FLUID VOLUME CALCULATIONS
S 10 35 MIGRATION, FORM OF HYDROCARBONS
S 10 54 MIGRATION, FRACTURED SHALE ACCUMULATIONS
S 10 160 MIGRATION, GENERALIZED THEORY, PETROLEUM
B 6304 578 MIGRATION, GEOCHEMICAL CONCEPT, PRIMARY
B 6304 581 MIGRATION, GEOCHEMICAL CONCEPT, SECONDARY
M 30 420 MIGRATION HISTORY, HANDIL FIELD
S 3 79 MIGRATION HISTORY OF URANIUM AND THORIUM, OCEANIC
S 10 131 MIGRATION, GROUNDWATER FLOW, FIELD PHENOMENA
S 10 130 MIGRATION, GROUNDWATER FLOW, NONSTEADY-STATE
S 10 123 MIGRATION, GROUNDWATER FLOW, STEADY-STATE
S 10 122 MIGRATION, GROUNDWATER MOVEMENT, REGIONAL
S 10 60 MIGRATION, GULF COAST SOLUBILITY REQUIREMENTS
S 10 56 MIGRATION, HYDROCARBON DISTRIBUTION IN SOURCE ROCKS
B 6002 213 MIGRATION, HYPOTHESES, PRIMARY, PETROLEUM
B 6303 433* MIGRATION IN DENVER BASIN, ABST., PETROLEUM GENERATION AN
B 6410 1613† MIGRATION IN DENVER BASIN, PETROLEUM, GENERATION AND
B 6303 477* MIGRATION IN LUSITANIAN BASIN, PORTUGAL, ABST., OIL GENERATION
B 6103 434† MIGRATION IN WESTERN CANADA BASIN, RATIO OF PETROLEUM TO,
B 6203 545* MIGRATION LIMITATIONS SUGGESTED BY GEOLOGIC AND GEOCHEMICAL
B 6305 766 MIGRATION MECHANISM, DISSOLVED, HYDROCARBONS AS POSSIBLE,
B 6002 234 MIGRATION MECHANISM IN GULF COAST, LACK OF ALTERNATIVE
B 6203 528* MIGRATION MECHANISMS, ABST., SOME MASS BALANCE AND GEOLOGIC
S 10 73 MIGRATION, INHIBITED COMPACTION
S 10 172 MIGRATION, IN-TRANSIT GENERATION, ESTIMATION
S 10 173 MIGRATION, IN-TRANSIT GENERATION, MODEL
S 10 48 MIGRATION, KEROGEN PROPERTIES
S 10 61 MIGRATION, LOCATION OF EXSOLUTION
S 10 82 MIGRATION, MECHANISM, MODEL
S 10 23 MIGRATION, MECHANISMS, DROPLET DEFORMATION
S 10 23 MIGRATION, MECHANISMS, MICELLE CONFIGURATION
S 10 21 MIGRATION, MECHANISMS, SEPARATE OIL PHASE
S 10 20 MIGRATION, MECHANISMS, SOLUTION
B 6305 778 MIGRATION, MICELLAR
B 6004 552 MIGRATION MODEL, COMBINED-
S 10 42 MIGRATION, MODEL
B 6201 19 MIGRATION, NIGER DELTA
B 6304 573† MIGRATION OF FLUIDS IN SEDIMENTARY BASINS
B 6409 1538† MIGRATION OF FLUIDS IN SEDIMENTARY BASINS. DISCUSSION AND
B 6304 577 MIGRATION OF HYDROCARBONS, GEOCHEMICAL CONCEPT, GENERATION
B 6404 549† MIGRATION OF HYDROCARBONS IN COMPACTING BASINS
B 6203 498* MIGRATION OF HYDROCARBONS IN COMPACTING BASINS, ABST.
B 6109 1428 MIGRATION OF HYDROCARBONS, PRIMARY
S 3 36 MIGRATION OF MANTLE-DERIVED URANIUM AND THORIUM WITHIN
B 6401 107† MIGRATION OF PETROLEUM, A PROCESS FOR, PRIMARY
B 6203 556* MIGRATION OF PETROLEUM, ABST., NEW EVIDENCE FOR HOT, DEEP,
B 6203 541* MIGRATION OF PETROLEUM FROM SOURCE, ABST., ROLE OF, SOLUBILITY
B 6304 697* MIGRATION OF PETROLEUM IN SEDIMENTARY BASINS, ABST., PROCESS
B 6305 770 MIGRATION OF SOLUBLE NONHYDROCARBON ORGANIC COMPOUNDS
B 6002 324* MIGRATION PATHS IN THICK, RAPIDLY DEPOSITED CLASTIC SEDIMENTS,
B 6203 506* MIGRATION PATHWAYS IN COMPACTING CLASTIC SEDIMENTS, ABST.
S 10 54 MIGRATION, ORGANIC CONTENT OF SOURCE ROCKS
S 10 34 MIGRATION, PEAK ACCUMULATION VERSUS PEAK GENERATION
S 10 60 MIGRATION, PEAK GENERATION
S 10 44 MIGRATION, PERMEABILITY VERSUS COMPACTION
S 10 183 MIGRATION, PETROLEUM, ROLE OF METHANE
S 10 187 MIGRATION, PETROLEUM, ROLE OF METHANE IN IMPEDING
S 10 72 MIGRATION, PETROLEUM IN WATER SOLUTION
S 10 175 MIGRATION, PETROLEUM PORPHYRINS, GENERATION
S 10 25 MIGRATION, PORE-CENTER NETWORK WITH DEPTH
S 10 40 MIGRATION, POROSITY VERSUS DEPTH OF CLAYS
B 6304 682 MIGRATION, PRACTICAL PETROLEUM GEOLOGY, BY KINJI MAGARA,
S 10 1 MIGRATION, PROBLEMS OF PETROLEUM
B 6409 1544 MIGRATION, REDEFINITION OF, PRIMARY AND, SECONDARY

S 10 47 MIGRATION, ROCK PROPERTIES
B 6401 59† MIGRATION, ROLE OF, MINERAL, MATRIX IN, KEROGEN,
B 6401 59† MIGRATION, ROLE OF MINERAL MATRIX IN KEROGEN
B 6304 595 MIGRATION STUDIES, POTENTIAL USE OF, ELECTRIC, LOGS IN
B 6402 209† MIGRATION TRENDS OF OIL AND GAS IN BOMBAY OFFSHORE BASIN,
S 10 104 MIGRATION, SECONDARY, HYDROCARBONS IN SOLUTION
S 10 105 MIGRATION, SECONDARY, MICELLAR MIGRATION
S 10 102 MIGRATION, SECONDARY, NATURAL GAS IN SOLUTION
S 10 101 MIGRATION, SECONDARY, RESIDUAL OIL SATURATION
S 10 43 MIGRATION, SHALE POROSITY-CHLORINITY RELATION
S 10 25 MIGRATION, SIGNIFICANCE OF CORRELATIONS
S 10 21 MIGRATION, SOLUBILITIES OF ACIDS IN WATER
S 10 21 MIGRATION, SOLUBILITIES OF ALCOHOLS IN WATER
S 10 21 MIGRATION, SOLUBILITIES OF PARAFFINS IN WATER
S 10 37 MIGRATION, SOLUBILITY AS TEMPERATURE FUNCTION
S 10 38 MIGRATION, SOLUBILITY OF GAS IN FORMATION WATER
S 10 20 MIGRATION, SOLUBILITY OF HYDROCARBON IN WATER
S 10 170 MIGRATION, SOLUBLE ORGANIC MATTER
S 10 57 MIGRATION, SOURCE ROCK DRAINAGE EFFICIENCY
S 10 36 MIGRATION, SOURCE-ROCK ANALYSIS, COMPARISON
S 10 41 MIGRATION, SPECIFIC VOLUME AT DEPTH
S 10 174 MIGRATION, SULFUR COMPOUNDS, GENERATION
S 10 243 MIGRATION, SURFACE EVIDENCE, ETHANE
S 10 245 MIGRATION, SURFACE EVIDENCE, ETHANE INTERPRETATION
S 10 243 MIGRATION, SURFACE EVIDENCE, FLOMATON-JAY-BLACKJACK CREEK
S 10 73 MIGRATION, TEMPERATURE EFFECTS ON PORE FLUIDS
S 10 42 MIGRATION, TEMPERATURE VERSUS DEPTH
S 10 57 MIGRATION, TIME OF PRIMARY MIGRATION
S 10 86 MIGRATION, TRANSPORT BY BYPASSING GEOPRESSURED SHALE
S 10 260 MIGRATION, VERTICAL, DATA FROM WELL CUTTINGS
S 10 222 MIGRATION, VERTICAL, PRESSURE CHANGE IMPOSED
S 10 222 MIGRATION, VERTICAL, TEMPERATURE CHANGE IMPOSED
S 10 39 MIGRATION, VISCOSITY VERSUS QUANTITY OF WATER IN CLAY
S 10 162 MIGRATION BY GROUNDWATER FLOW, CONDITIONS
S 10 163 MIGRATION BY GROUNDWATER FLOW, KINDS OF TRAPS
S 10 163 MIGRATION BY GROUNDWATER FLOW, STABILITY OF ACCUMULATIONS
S 10 163 MIGRATION BY GROUNDWATER FLOW, TIME AND DIRECTION
S 10 59 MIGRATION IN AQUEOUS SOLUTION, CAP ROCKS
S 10 58 MIGRATION IN AQUEOUS SOLUTION, COALS AS SOURCE ROCKS
S 10 59 MIGRATION IN AQUEOUS SOLUTION, HYDROCARBON SOLUBILITIES
S 10 59 MIGRATION IN AQUEOUS SOLUTION, MICELLES
S 10 59 MIGRATION IN AQUEOUS SOLUTION, OIL COMPOSITION
S 10 59 MIGRATION IN AQUEOUS SOLUTION, SECONDARY MIGRATION
S 10 59 MIGRATION IN AQUEOUS SOLUTION, TOTAL ORGANIC CONTENT
S 10 58 MIGRATION IN AQUEOUS SOLUTION, WATER MOVEMENT
S 10 47† MIGRATION MECHANISMS, SOME, MASS BALANCE AND, GEOLOGICAL,
S 10 97 MIGRATION MECHANISMS, DROPLET EXPULSION
S 10 97 MIGRATION MECHANISMS, NONHYDROCARBON ORGANIC COMPOUNDS
S 10 95 MIGRATION MECHANISMS, SURFACTANT MICELLES
S 10 69† MIGRATION OF HYDROCARBONS IN COMPACTING BASINS
S 10 29 MIGRATION PROCESSES, ALCOHOL INTERACTION
S 10 99 MIGRATION THROUGH ORGANIC MATTER NETWORK
B 6305 761† MIGRATION--CHEMICAL AN PHYSICAL CONSTRAINTS, OIL AND, GAS
B 6004 699* MIGRATION--RESULT OF GEOLOGIC-EVENT TIMING, ABST., ORIENTED,
B 6303 471* MIGRATION-FRACTIONATION, ABST., GEOCHEMISTRY OF OILS FROM,
B 6112 2146† MIGRATION. DISCUSSION AND REPLY, AQUEOUS, SOLUBILITY OF,
S 10 33† MIGRATION: A REVIEW, AGENTS FOR, PRIMARY, HYDROCARBON
S 10 89† MIGRATION: CHEMICAL AND PHYSICAL CONSTRAINTS, OIL AND, GAS
S 10 19† MIGRATION: THE IMPORTANCE OF WATER-MINERAL-ORGANIC MATTER
B 6409 1557* MIGRATIONS IN LAKE BORGNE—VALENTINE AREA OF SOUTHEASTERN
B 6405 706* MIGRATION—BERIN SEA AND UNITED STATES ATLANTIC SHELF, ABST.,
B 6209 1766* MIGRATION—NEW TACK ON AN OLD THEME, ABST., PRIMARY
B 6109 1499 MIKKELSEN BAY SECTION, ALASKA
B 6303 410 MILAM, ROBERT, GLACIAL-MARINE SEDIMENTS OF WEST ANTARCTICA
B 6004 706 MILES, A. E., COMPUTER APPROACH TO FACIES STUDIES, ABST.
B 6211 2358 MILES, G. A., HOLOTHURIAN MICROFOSSILS AND EOCENE STRATIGRAPHY
B 6209 1767* MILESTONES IN GULF COAST ECONOMIC MICROPALEONTOLOGY, ABST.
M 27 31 MILIOLID FORAMS
S 11 231 MILIOLIDAE
B 6407 981 MILK RIVER EQUIVALENT, ALBERTA
B 6302 155 MILK RIVER GAS FIELD, ALBERTA
B 6302 166 MILK RIVER SAND, CANADA
B 6303 555* MILLARD COUNTY, UTAH, ABST., CARBONATE, MUD, MOUNDS FROM
B 6002 286 MILLBORO SHALE, PENNSYLVANIA
B 6209 1704 MILLBORO SHALE, UNITED STATES, EASTERN
B 6304 652 MILLBORO SHALES, VIRGINIA, BRAILLER AND
B 6004 691 MILLER, B. M., INCONSISTENCIES IN RESOURCE-APPRAISAL ESTIMATES
B 6303 476 MILLER, BETTY K., INTERACTIVE MODELING AND INTERPRETATION OF
B 6303 497 MILLER, BETTY M., DEVELOPMENT OF PETROLEUM RESOURCE
B 6012 2186 MILLER, BETTY M., GEOLOGIC ESTIMATES OF UNDISCOVERED OIL AND

B 6008 1404 MILLER, BETTY M., NEW CONCEPTS IN METHODOLOGY FOR PROCESSING
B 6203 544 MILLER, BETTY M., NEW DEVELOPMENTS IN METHODS OF RESOURCE
B 6305 842 MILLER, BETTY M., NEW LOOK AT OIL AND GAS RESOURCES OF
B 6004 698 MILLER, BETTY M., PROBABILISTIC AND COMPUTER METHODOLOGIES
B 6105 813 MILLER, BETTY M., REVIEW AND APPLICATION OF PETROLEUM
B 6105 813 **MILLER, BETTY M., REVIEW AND APPLICATION OF PETROLEUM FINDING-**
B 6108 1383 MILLER, BETTY M., REVIEW AND APPLICATION OF PETROLEUM-FINDING
B 6203 571 MILLER, CALVIN F., STRUCTURAL GEOLOGY OF SOUTHWEST EDWARDS
B 6303 495* MILLER COUNTY, ARKANSAS, ABST., DEPOSITIONAL ENVIRONMENTS AND
B 6211 2220 MILLER CREEK FAULT, CALIFORNIA
B 6308 1386 MILLER, D. D., DEVELOPMENTS IN NORTHERN ROCKIES IN 1978
B 6409 1501 MILLER, D. D., DEVELOPMENTS IN NORTHERN ROCKIES IN 1979
B 6211 2351 MILLER, ED, LATE MIOCENE TURBIDITE DEPOSITION, CROCKER CANYON
B 6303 429 MILLER, ELIZABETH L., ALLOCHTHONOUS NATURE OF NUMIDIAN
B 6108 1383 MILLER, F. X., BIOSTRATIGRAPHIC CORRELATION OF MESAVERDE GROUP
B 6405 750 MILLER, F. X., NEW PRECISION IN BIOSTRATIGRAPHY THROUGH
B 6409 1369 MILLER, HARRY A., JR., DEVELOPMENTS IN WEST TEXAS AND EASTERN
B 6004 727 MILLER, J. A., MARINE AND BRACKISH-WATE CRETACEOUS OSTRACODA
B 6405 674 MILLER, J. J., ET AL, VERTICAL AN SURFACE SEISMIC PROFILES MAP
B 6303 536 MILLER, JAMES A., BIOSTRATIGRAPHY OF EARLY AND EARLIEST LATE
B 6004 698 MILLER, JAMES A., ORIGIN AND DISTRIBUTION OF SUSPENDED SEDIMENT
B 6303 505 MILLER, K. G., OLIGOCENE TRANSGRESSIV SEDIMENTS OF NEW JERSEY
B 6008 1411 MILLER, LOWELL G., RAFT RIVER, IDAHO, GEOTHERMAL WELLS--SITING,
B 6405 733 MILLER, MOLLY F., ENVIRONMENTAL CONTROL OF TRACE FOSSIL
B 6105 814 MILLER, MOLLY FRITZ, BIOGENIC STRUCTURES AND THEIR
B 6405 751 MILLER, MOLLY FRITZ, DEPOSITIONAL ENVIRONMENTS AND BIOGENIC
B 6105 814 MILLER, R. E., GEOCHEMISTRY OF LIGHT HYDROCARBONS IN SHALLOW
B 6105 814 MILLER, R. E., HYDROCARBONS IN SURFACE SEDIMENTS OF
B 6309 1587 MILLER, R. E., IMPLICATION OF GASEOUS HYDROCARBON
B 6309 1583 MILLER, R. E., PETROLEUM GEOCHEMISTRY AND GEOLOGY OF
B 6004 659 MILLER, R. H., STRATIGRAPHY AND DEPOSITIONAL ENVIRONMENTS OF
B 6009 1621 MILLER, RALPH L., WILDCAT VALLEY SANDSTONE IN SOUTHWEST
B 6103 416 MILLER, RALPH L., WILDCAT VALLEY SANDSTONE IN SOUTHWEST
B 6303 418 MILLER, RICHARD H., MIDDLE PALEOZOIC SEDIMENTATION AND
B 6405 751 MILLER, RONALD J., ORIGIN OF BIOGENIC CARBONATE SANDS OF
B 6207 1227 MILLER, THOMAS P., GEOTHERMAL EXPLORATION ON ADAK ISLAND,
B 6103 466 MILLER, THOMAS P., GEOTHERMAL POTENTIAL OF HIGH-LEVEL MAGM
M 25 149 MILLER, THOMAS P., POTENTIAL FOR GEOTHERMAL-ENERGY
B 6004 698 MILLER, W. G., CALCULATION OF COAL RESERVES BY COMPUTER
B 6203 556* MILLIGALS, AND GAMMAS, ABST., BRIGHT SPOTS
M 25 80 MILLIGAN, E. N., STATUS OF COAL EXPLORATION AND MINE
B 6102 257 MILLIGEN FORMATION, IDAHO
B 6203 544 MILLIKEN, KITTY L., NON-SABKHA EVAPORITES, SILICIFIED, FROM
S 4 313 MILLIMAN, JOHN D., CALCIUM CARBONATE SEDIMENTATION ON
B 6306 934 MILLIMAN, JOHN D., MORPHOLOGY AND STRUCTURE OF AMAZON UPPER
B 6206 1029 MILLIMAN, JOHN D., MORPHOLOGY AND STRUCTURE OF UPPER
B 6408 1274 MILLIMAN, JOHN D., REVIEW OF FACIES ANATOMY AND DIAGENESIS O A
B 6302 218 MILLIMAN, JOHN D., STRUCTURE AND ORIGIN OF THREE
B 6111 2001 MILLS, MICHAEL F., FAULTED UPPER PLEISTOCENE MARINE TERRACE,
B 6305 805 MILLS RANCH COMPLEX, TEXAS AND, OKLAHOMA
B 6203 526* MILLS RANCH COMPLEX, TEXAS AND OKLAHOMA, ABST., DEVELOPMENT
B 6305 804† MILLS RANCH COMPLEX--WORLDS DEEPEST FIELD, GEOLOGY AND
B 6305 804 MILLS RANCH FIELD, DISCOVERED 1972, TEXAS
B 6305 804 MILLS RANCH FIELD, TEXAS, FREEPORT OIL CO. 1 SIDNEY FABIAN
B 6405 789 MILLS, STEPHEN J., FAN MODELS FOR HYDROCARBON EXPLORATION
B 6110 1841 MILNE, A. D., ON DEVELOPMENTS IN THAILAND, 1976
B 6210 1988 MILNE, A. D., ON DEVELOPMENTS IN THAILAND, 1977
B 6109 1499 MILNE POINT SECTION, ALASKA
B 6103 361 MILROY STRUCTURE, OKLAHOMA
B 6408 1211 MIN-HO BASIN, CHINA
M 29 242 MINAMI-DAITO BASIN, GEOLOGY, NORTH PHILIPPINE SEA
M 29 243 MINAMI-DAITO BASIN, STRATIGRAPHY, NORTH PHILIPPINE SEA
B 6402 219 MINAS FIELD, SUMATRA
B 6203 426 MINCH BASIN, EUROPE
B 6012 2186 MINCH, JOHN A., CLAST POPULATIONS IN SESPE AND POWAY
M 25 437 MINDANAO, DAVAO DISTRICT, COPPER
M 25 144 MINDANAO, GEOTHERMAL PROSPECTS
M 25 434 MINDANAO TRENCH, PHILIPPINES
M 25 80† MINE DEVELOPMENT IN AUSTRALIA, STATUS OF, COAL EXPLORATION
B 6207 1209* MINE DEVELOPMENT IN PACIFIC ISLANDS—IMPORTANCE OF
B 6408 1278* MINE PLANNING, ABST., APPLICATION OF DEPOSITIONAL MODELS IN,
B 6212 2379† MINE PLANNING IN APPALACHIAN REGION, DEPOSITIONAL, MODELS IN,
B 6004 653* MINE, SAN JUAN COUNTY, UTAH, ABST., PRELIMINARY INVESTIGATIONS
M 25 449 MINERA CONCENTRATION, JAPAN, PERMIAN GEOSYNCLINAL
B 6009 1611* MINERAL ANALYSIS OF MIOCENE-PLIOCENE COASTAL-PLAIN SEDIMENTS,
B 6207 1225* MINERAL AND ENERGY DEVELOPMENT THROUGH PLANNING, ABST.,
B 6207 1216* MINERAL AND ENERGY RESOURCES, ABST., ENVIRONMENTAL IMPACT OF
B 6207 1230* MINERAL AND ENERGY RESOURCES, ABST., FUNDAMENTAL
B 6207 1238* MINERAL AND ENERGY RESOURCES, ABST., SLOPE-STABILITY PROBLEMS
M 25 448 MINERAL CONCENTRATION AND SUBDUCTION ZONE, RELATION
B 6004 730* MINERAL DEPOSITS, ABST., AULACOGENS AN, MEGASHEARS--NATURAL
B 6105 829* MINERAL DEPOSITS, ABST., ENERGY-TRANSPORT PROCESSES IN
B 6207 1239* MINERAL DEPOSITS, ABST., TECTONIC SETTING OF, PERU-, CHILE
M 25 487 MINERAL DEPOSITS AND SEAFLOOR SPREADING, RELATION OF
B 6308 1421 MINERAL DEPOSITS, BY COLIN J. DICKSON, MORAN, W. R., REVIEW OF
M 25 500 MINERAL DEPOSITS, CANADA, SUCCESSOR, BASINS
M 25 497 MINERAL DEPOSITS, CANADIAN, CORDILLERA, CLASTIC WEDGE
M 25 498 MINERAL DEPOSITS, CANADIAN, CORDILLERA, ISLAND, ARC
M 25 496 MINERAL DEPOSITS. CANADIAN. CORDILLERA. MIOGEOSYNCLINE
M 25 497 MINERAL DEPOSITS, CANADIAN, CORDILLERA, OCEANIC, BASIN
M 25 506 MINERAL DEPOSITS, DISTRIBUTION BY CLASSES, CANADIAN,
M 25 525 MINERAL DEPOSITS, GROUPS OF, MEXICO, SINALOA
M 25 466† MINERAL DEPOSITS OF NORTHEAST USSR, GEOLOGIC FRAMEWORK OF
M 25 486 MINERAL DEPOSITS, ORIGIN OF
M 25 530 MINERAL DEPOSITS, PACIFIC, SOUTHEASTERN
M 25 529 MINERAL DEPOSITS, TECTONIC AND GEOLOGIC, PACIFIC, SOUTHEASTERN
M 25 468 MINERAL DEPOSITS, USSR, NORTHEAST
M 25 447 MINERAL DEPOSITS, ZONAL DISTRIBUTION, SOUTH KOREA
B 6207 1227* MINERAL DEVELOPMENT, ABST., ENVIRONMENTAL ASPECTS OF
B 6207 1220* MINERAL DEVELOPMENT, ABST., PHYSICAL CONSTRAINTS ON
M 25 448 MINERAL DISTRIBUTION, JAPAN, NEOGENE
B 6004 734* MINERAL DISTRIBUTION ON SOUTHERN ATLANTIC COASTAL PLAIN
B 6305 823* MINERAL EVALUATION OF OIL-SHALE LAND EXCHANGES, ABST.
B 6005 745† MINERAL EXPLORATION, APPLICATION OF, LANDSAT, IMAGERY TO
M 25 430 MINERAL EXPLORATION, BRITISH SOLOMON ISLANDS
M 25 430 MINERAL EXPLORATION, FIJI
B 6207 1237* MINERAL EXPLORATION IN BAJA CALIFORNIA—PROGRESS AND
B 6109 1560* MINERAL EXPLORATION IN COMPLICATED STRUCTURAL TERRANES,
B 6306 951† MINERAL EXPLORATION IN COMPLICATED STRUCTURAL TERRANES,
M 25 492 MINERAL EXPLORATION, USE OF BASIN ANALYSIS IN
B 6207 1221* MINERAL INDUSTRY—PRACTICES AND PROBLEMS, AUSTRALIAN SCENE,
B 6401 59† MINERAL MATRIX IN KEROGEN PYROLYSIS--INFLUENCE ON PETROLEUM
B 6401 65 MINERAL MATRIX ON FREE HYDROCARBONS, EFFECT OF
B 6401 62 MINERAL MATRIX ON HYDROCARBON YIELD, EFFECT OF
M 25 527 MINERAL MINING, FUTURE OF, ANDEAN REGION
M 25 536 MINERAL OCCURRENCES, ANTARCTIC PENINSULA AND, PACIFIC
M 25 535 MINERAL OCCURRENCES IN, ANTARCTICA
M 25 535 MINERAL OCCURRENCES, SCOTIA ARC
B 6312 2217 MINERAL PHASE-CHANGE FRACTURES
B 6207 1208* MINERAL POLICY, ABST., AUSTRALIAN
B 6207 1222* MINERAL RESOURCES EXPLORATION IN AUSTRALASIA, ABST.,
B 6207 1219* MINERAL RESOURCES IN SOUTHWEST PACIFIC, ABST.
M 25 503† MINERAL RESOURCES OF CANADIAN CORDILLERA, METALLIC
M 25 48† MINERAL RESOURCES OF CIRCUM-PACIFIC REGION, PLATE TECTONICS
M 25 551† MINERAL RESOURCES OF COLOMBIA--SUMMARY, POSSIBILITIES FOR
M 25 552† MINERAL RESOURCES OF ECUADOR, THEIR DEVELOPMENT AND
M 25 440† MINERAL RESOURCES OF KOREA
M 25 525† **MINERAL RESOURCES OF STATE OF SINALOA, MEXICO--SUMMARY**
M 25 556† **MINERAL RESOURCES POTENTIAL OF PERU--SUMMARY**
M 25 24† MINERAL RESOURCES--SUMMARY, ECONOMICS OF, ENERGY AND
B 6205 864 MINERAL, ROCK, OR FOSSIL COLLECTIONS AVAILABLE FROM
M 25 419 MINERAL SAND, AUSTRALIA, EASTERN, HEAVY-
M 25 420 MINERAL SAND, EXPLORATION, AUSTRALIA, HEAVY-
M 25 418 MINERAL SAND, PRODUCTION FIGURES, AUSTRALIA, HEAVY-
M 25 420 MINERAL SAND, PRODUCTION METHODS, AUSTRALIA, HEAVY-
M 25 420 MINERAL SAND, REHABILITATION OF MINED LANDS, AUSTRALIA,
M 25 419 MINERAL SAND, WESTERN AUSTRALIA, HEAVY-
M 25 18† MINERAL TRADE IN CIRCUM-PACIFIC REGION--SUMMARY
M 25 18 MINERAL TRADE, WORLD
B 6207 1213* MINERAL TRENDS IN WESTERN SOUTH AMERICA AS RELATED TO
M 28 143 MINERAL WELLS FORMATION, TEXAS, PENNSYLVANIAN
M 28 112 MINERAL WELLS FORMATION, TEXAS, PENNSYLVANIAN
M 25 449 MINERAL ZONING, JAPAN, MESOZOIC, LATE, GRANITIC, BATHOLITHS AND
M 25 515 MINERALIZATION, AGE OF, MEXICO
S 3 91 MINERALIZATION BELTS AND METALLOTECTONIC GRADIENT, OROGENIC
M 25 475 MINERALIZATION, CORRELATION WITH THICKNESS OF
M 25 503 MINERALIZATION, CYCLES OF, CANADIAN, CORDILLERA,
M 25 465 MINERALIZATION, FAR EAST, SOUTHERN
M 25 520 MINERALIZATION, MEXICO, SONORA, COPPER
B 6109 1536* MINERALIZATION PATTERN, ABST., URANIUM HOST DEPOSITIONAL
S 3 121 MINERALIZATION PROCESSES, GEOCHEMICAL DISTINCTION OF,
M 25 429 MINERALIZATION, STRUCTURAL CONTROLS OF
M 25 475 MINERALIZATION, TIME OF, USSR, EASTERN, ANTIMONY-, MERCURY
M 25 476 MINERALIZATION, TIME OF, USSR, EASTERN, GOLD-ANTIMONITE
M 25 480 MINERALIZATION TYPICAL OF, SHELVES
S 3 135 MINERALIZATION, UNDEREMPHASIZED PROCESSES OF, URANIUM
B 6209 1759* MINERALOGIC AND PHYSICAL PROPERTIES OF GULF COAST LIMESTONE
B 6205 891* MINERALOGIC CHARACTERISTICS OF SOME LOWER TERTIARY
B 6303 440* MINERALOGIC COMPOSITION AND DIAGENESIS OF TERTIARY
B 6304 694* MINERALOGICAL AND GEOCHEMICAL ZONATION ACROSS ROLL-TYPE
B 6207 1211* MINERALOGY AND CHEMISTRY OF PHILIPPINE SEA SEDIMENTS,
B 6304 692* MINERALOGY AND GEOCHEMISTRY OF MARIANO LAKE URANIUM
B 6409 1566* MINERALOGY, DIAGENESIS, AND POROSITY IN VICKSBURG SANDSTONES,
B 6209 1757* MINERALOGY OF PLEISTOCENE TERRACE DEPOSITS, LOUISIANA, ABST.

S 7 217 MINERALOGY, WEST FLORIDA SLOPE
B 6110 1901† MINERALS AND PETROLEUM-FORMING REACTIONS DURING BURIAL AND
B 6004 681* MINERALS AND THEIR RELATION TO PHYSICAL PROPERTIES, GULF OF
M 25 559 MINERALS, CHILE
B 6206 1056 MINERALS, CLAY
M 25 556 MINERALS, CONFIRMED RESOURCES
M 25 552 MINERALS, EXPLORATION, ECUADOR
B 6206 1056 MINERALS, GANGUE
B 6206 1058 MINERALS, HEAVY
B 6108 1384* MINERALS IN ST. PETER SANDSTONE, ABST., PARAGENESIS OF,
M 25 421 MINERALS INDUSTRY, FUTURE FOR, AUSTRALIA, HEAVY-
M 25 481 MINERALS, KURIL-, KAMCHATKA ARC
M 25 421 MINERALS, MARKETING, AUSTRALIA, HEAVY-
B 6206 1055 MINERALS, ORE
M 25 535 MINERALS, PACIFIC, SOUTHEASTERN, SEAFLOOR DEPOSITS BEARING
M 25 444 MINERALS, SOUTH KOREA
M 25 447 MINERALS, SOUTH KOREA, KYONGSANG BASIN
M 25 447 MINERALS, SOUTH KOREA, SILLA SERIES, UPPER
B 6206 1055 MINERALS, URANIUM
M 25 77 MINERL RESOURCES, ANTARCTICA
M 25 483 MINEROGENETIC PROVINCES
B 6203 523* MINES, ABST., USE OF DEPOSITIONAL, MODELS IN PREDICTING, ROOF
M 27 141 MINES DOLOMITE, PENNSYLVANIA, CAMBRIAN
M 28 116 MINES DOLOMITE, PENNSYLVANIA, CAMBRIAN
B 6212 2403 MINES, ROOF CONDITIONS
B 6305 835 MINGARRO, MARTIN F., RECENT SALT SEDIMENTATION IN PLAYA LAKES
B 6305 844* MINI-SOSIE, ABST., SHALLOW SEISMIC REFLECTION IN, APPALACHIAN
B 6004 724* MINICOMPUTERS FOR MAXIANALYSIS, ABST.
B 6309 1610 **MINIHAN, E. D., WAVELAND FIELD, UNIQUE STRUCTURAL AND**
B 6308 1428 **MINIHAN, E. D., WAVELAND FIELD, UNIQUE STRUCTURAL AND**
B 6303 520* MINING AN MILLING IN SOUTH TEXAS, ABST., CONQUISTA PROJECT,
M 25 432 MINING AND PROCESSING, THAILAND, FLUORITE
B 6207 1235* MINING AND STREAM SALINITY IN DARLING RANGE, WESTERN
M 25 527 MINING, FUTURE OF, ANDEAN REGION, MINERAL
B 6406 960* MINING GEOLOGY, ABST., THE, ALSANDS PROJECT—A CHALLENGE IN
M 25 418† MINING HEAVY-MINERAL SAND IN AUSTRALIA--SUMMARY
M 25 456 MINING INDUSTRY, GOVERNMENT POLICY TOWARD, JAPAN
B 6408 1286* **MINING OF TENNESSEE COAL, ABST., GEOLOGY AND**
B 6206 1093* MINING OF OIL SANDS, ABST., NEW GEOLOGIC FRAMEWORK FOR
B 6305 843* MINING OF URANIUM AND ITS EFFECT ON EXPLORATION, ABST.,
B 6304 693* MINING OF URANIUM ORES, ABST., APPLICATION OF SOLUTION-MINERAL
B 6108 1389* MINING OF URANIUM. BY TITLE ONLY., APPLICABLE REGULATIONS OF,
B 6207 1240* MINING PROBLEMS, ABST., OFFSHORE, TIN DEPOSITS OF, THAILAND AND
B 6303 466* MINING PURSUANT TO SURFACE MINING AND RECLAMATION ACT OF
B 6408 1287* MINING TRENDS, ABST., COAL IN, WEST VIRGINIA—GEOLOGY AND
B 6303 486* MINING--A THIRD ALTERNATIVE, ABST., URANIUM IN-SITU, LEACH
B 6303 519* MINING--ENVIRONMENTAL AND LICENSING REGULATIONS, ABST.,
B 6207 1216* MINING , ABST., WHAT LAW SHALL GOVERN, DEEP-SEA
B 6212 2511* MINING—SIMULATION MODELS, ABST., ECONOMICS OF LARGE-SCALE
B 6209 1610 MINISINK LIMESTONE, PENNSYLVANIA
B 6010 1718 MINISTERIO DE INDUSTRIA, ON DEVELOPMENTS IN SPAIN, 1975
B 6110 1648 MINISTERIO DE INDUSTRIA, ON DEVELOPMENTS IN SPAIN, 1976
B 6210 1918 MINISTERIO DE INDUSTRIA, ON DEVELOPMENTS IN SPAIN, 1977
B 6411 1964 MINISTERO DELL-INDUSTRIA, DEL COMMERCIO E DELL'ARTIGIANATO,
B 6010 1642 MINISTRY OF DEVELOPMENT, ON DEVELOPMENTS IN BAHAMAS, 1975
B 6110 1798 MINISTRY OF PETROL. AND MINERAL RESOURCES, TENNECO, AND
B 6010 1656 MINISTRY OF PETROLEUM AND MINES, ON DEVELOPMENTS IN TRINIDAD
M 25 350 MINK, JOHN F., HYDROGEOLOGY OF THE PACIFIC--SUMMARY
B 6008 1389 MINK, LELAND L., GEOLOGIC AND SEISMIC STUDIES OF BOISE FRONT,
B 6207 1227 MINK, PATSY T., ENVIRONMENTAL ASPECTS OF INTERNATIONAL ENERGY
B 6405 778* MINNELUSA FORMATION, MIDDLE AND UPPER PENNSYLVANIAN—
B 6305 839* MINNELUSA--POWDER RIVER BASIN--PAST, PRESENT AND FUTURE, ABST.
B 6406 943 MINNESOTA AND SOUTH DAKOTA, BLUE HILL MEMBER
B 6406 943 MINNESOTA AND SOUTH DAKOTA, CODELL SANDSTONE MEMBER
B 6406 943 MINNESOTA AND SOUTH DAKOTA, FAIRPORT MEMBER
B 6406 943 MINNESOTA AND SOUTH DAKOTA, GRANEROS SHALE
B 6406 943 MINNESOTA AND SOUTH DAKOTA, NIOBRARA FORMATION
B 6406 943 MINNESOTA AND SOUTH DAKOTA, PRECAMBRIAN
B 6406 943 MINNESOTA, CRETACEOUS
B 6108 1307 MINNESOTA, DEVELOPMENTS, 1976
B 6208 1480 MINNESOTA, DEVELOPMENTS, 1977
B 6406 942† MINNESOTA, EXPOSURES OF, GREENHORN FORMATION AND, CARLILE
M 27 26 MINNESOTA, HOLOCENE SEDIMENT
M 28 58 MINNESOTA, HOLOCENE, SEDIMENT
M 28 57 MINNESOTA, HOLOCENE, SEDIMENTS
B 6007 1021 MINNESOTA, RESERVE MINING CO., TACONITE ORE TAILINGS,
B 6007 1023 MINNESOTA, SILVER BAY, TAILINGS, DEPOSITION
B 6105 815 MINTER, W. E. LAWRENCE, FOSSIL GOLD, URANIUM, AND PYRITE
B 6101 3 MINTO ARCH
M 27 10 MINTURN FORMATION, COLORADO, PENNSYLVANIAN
M 27 11 MINTURN FORMATION, COLORADO, PENNSYLVANIAN
B 6109 1543 MINTURN, L. W., TOPOGRAPHIC EXPRESSION OF OIL AND GAS FIELDS IN
S 11 52 MIO-PLIOCENE, DIATOMS, CALIFORNIA

S 5 169* MIOCENE (MESSINIAN) DEPOSITS OF SALEMI BASIN (SICILY) AND A
B 6407 1016 MIOCENE, ALASKA
B 6408 1142 MIOCENE, ALASKA
B 6408 1133 MIOCENE, ALASKA
B 6309 1522 MIOCENE, ALASKA PENINSULA, BEAR LAKE FORMATION
S 11 146 MIOCENE AND OLIGOCENE BULIMINIDS, CALIFORNIA
B 6309 1524 MIOCENE AND PLIOCENE, ALASKA PENINSULA, TACHILNI FORMATION
S 11 160 MIOCENE ANOMALINIDS, CALIFORNIA
S 11 158 MIOCENE ANOMALINIDS, CALIFORNIA
M 28 51 MIOCENE, ARIKAREE FORMATION, NEBRASKA
M 28 50 MIOCENE, ARIKAREE FORMATION, WYOMING
B 6304 624 MIOCENE, ARIZONA
B 6207 1153 MIOCENE, AUSTRALIA
S 11 162 MIOCENE BAGGINA, CALIFORNIA
B 6309 1459 MIOCENE, BALTIMORE CANYON TROUGH
B 6211 2363* MIOCENE BARRIER BEACH-SALINE LAGOON COMPLEX, UPPER SESPE
M 28 141 MIOCENE, BARSTOW FORMATION, CALIFORNIA
S 11 130 MIOCENE BOLIVINIDS, CALIFORNIA
S 11 136 MIOCENE BOLIVINIDS, CALIFORNIA
S 11 134 MIOCENE BOLIVINIDS, CALIFORNIA
S 11 132 MIOCENE BOLIVINIDS, CALIFORNIA
M 30 503 MIOCENE, BOMBAY HIGH FIELD, PETROLEUM GEOLOGY
S 6 255 MIOCENE BOUNDARY, DIAGRAM, PALEOGENE, OLIGOCENE-
B 6209 1714† MIOCENE BOUNDARY OF DELAWARE COASTAL PLAIN: DISCUSSION AND
B 6204 665 MIOCENE, BRAZIL
S 11 138 MIOCENE BULIMINIDAE, CALIFORNIA
S 11 144 MIOCENE BULIMINIDS, CALIFORNIA, OLIGOCENE AND
B 6211 2223 MIOCENE, CALIFORNIA
B 6206 951 MIOCENE, CALIFORNIA
B 6205 820 MIOCENE, CALIFORNIA
B 6211 2269 MIOCENE, CALIFORNIA
B 6102 153 MIOCENE, CALIFORNIA
B 6111 1985 MIOCENE, CALIFORNIA
M 26 173 MIOCENE, CALIFORNIA, SAN JOAQUIN VALLEY
B 6303 495* MIOCENE CANYON-CHANNEL SYSTEMS IN INDU SUBMARINE FAN, ABST.,
B 6303 445* MIOCENE CARBONATE GRAVITY FLOW IN BLAKE-BAHAMA BASIN, ABST.
S 4 106 MIOCENE, CARIBBEAN
S 11 154 MIOCENE CASSIDULINA, CALIFORNIA
B 6303 493* MIOCENE CATAHOULA FORMATION, CENTRAL TEXAS COASTAL PLAIN,
M 29 180 MIOCENE, CHALK OUTCROPS, BLAKE ESCARPMENT
B 6104 561 MIOCENE, CHILE
M 28 66 MIOCENE, CIERBO SANDSTONE, CALIFORNIA
M 28 133 MIOCENE, CIERBO SANDSTONE, CALIFORNIA
M 25 325 MIOCENE, COLOMBIA
S 11 107 MIOCENE CORRELATION, CALIFORNIA
B 6009 1502† MIOCENE DELTAIC OIL HABITAT, TRINIDAD
M 28 89 MIOCENE, DSDP CORE, MEDITERRANEAN
M 28 88 MIOCENE, DSDP CORE, MEDITERRANEAN
M 28 87 MIOCENE, DSDP CORE, MEDITERRANEAN
M 25 328 MIOCENE, ECUADOR
S 11 156 MIOCENE EPISTOMINELLIDS AND DISCORBINELLA, CALIFORNIA
M 28 86 MIOCENE, ERACLEA MINOA MASS FLOW, ITALY
M 29 345 MIOCENE, EVOLUTION, COLOMBIA
B 6303 465* MIOCENE FLYSCH OF NORTHERN APPENNINES, ITALY, ABST.,
S 11 99 MIOCENE FORAMINIFERA, SHALE POINT, CALIFORNIA, CHECK LIST OF
B 6211 2348* MIOCENE FORMATION AT COOS BAY, OREGON, ABST., NEW
B 6209 1761* MIOCENE GEOLOGY OF MATAGORDA ISLAND, CALHOUN COUNTY,
M 28 85 MIOCENE, GESSO SOLFIFERA, ITALY
M 28 43 MIOCENE, GESSO SOLFIFERA, ITALY
B 6109 1538* MIOCENE GLACIO-EUSTATIC LOWERING OF SEA LEVEL—EVIDENCE FROM
M 28 86 MIOCENE, GRESSO SOLFIFERA, ITALY
M 30 471 MIOCENE, GUDAO FIELD, CHINA
S 10 4 MIOCENE, GULF COAST, SHALE POROSITY, COMPACTION STAGES
B 6207 1169 MIOCENE, GULF OF CALIFORNIA
B 6205 779 MIOCENE, GULF OF, PAPUA
M 30 403 MIOCENE, HANDIL FIELD, STRATIGRAPHIC COLUMN
M 28 72 MIOCENE, HAWTHORN FORMATION, FLORIDA
M 28 125 MIOCENE, HAYNER RANCH FORMATION, NEW MEXICO
M 28 23 MIOCENE, HAYNER RANCH FORMATION, NEW MEXICO
M 28 177 MIOCENE, HAYNER RANCH FORMATION, NEW MEXICO
M 28 174 MIOCENE, HAYNER RANCH FORMATION, NEW MEXICO
M 27 87 MIOCENE, HYDROBIEN LIMESTONE, GERMANY
B 6203 453 MIOCENE, INDIA
B 6203 444 MIOCENE, INDIA
B 6402 219 MIOCENE, INDIA
B 6402 219 MIOCENE, INDIA
B 6207 1136 MIOCENE, IRIAN JAYA
B 6112 2086 MIOCENE, KEROGEN, LOUISIANA
S 4 329† MIOCENE KINGSHILL SEAWAY--A DYNAMIC CARBONATE BASIN AND
B 6102 201 MIOCENE, KOMANDORSKIY BASIN
B 6004 710* MIOCENE, LITTORAL HARD- AND SOFT-BOTTOM ICHNOCOENOSES--THEIR
S 10 6 MIOCENE, LOUISIANA, SHALE MINERALOGY EXAMPLE
S 10 135 MIOCENE WATER LEVELS, FRANCE, AQUITAINE

M 30 401 MIOCENE, MAHAKAM DELTA, BEKAPAI FIELD
M 30 401 MIOCENE, MAHAKAM DELTA, HANDIL FIELD
M 30 439 MIOCENE, MALAYSIA, RESERVOIR CHARACTERISTICS
B 6102 208 MIOCENE, MEXICO
B 6112 2066 MIOCENE, MIDDLE, OROGENIC BELT, INDONESIA, EASTERN
M 28 142 MIOCENE, MONTEREY FORMATION, CALIFORNIA
M 27 26 MIOCENE, MONTEREY SHALE, CALIFORNIA
M 28 81 MIOCENE, MONTEREY SHALE, CALIFORNIA
B 6111 2008 MIOCENE NANNO-FORAM CHALK INCLUSION, CAPE VERDE RISE
M 30 565 MIOCENE, NEMBE CREEK FIELD, NIGERIA
M 28 133 MIOCENE, NEROLOY SANDSTONE, CALIFORNIA
M 28 66 MIOCENE, NEROLOY SANDSTONE, CALIFORNIA
B 6201 3 MIOCENE, NIGER DELTA
S 11 122 MIOCENE, NONIONIDAE, CALIFORNIA
S 11 124 MIOCENE, NONIONIDAE, MOHNIAN AND, DELMONTIAN STAGES,
B 6203 424 MIOCENE, NORTH, ATLANTIC
B 6011 2044† MIOCENE OF SOUTH LOUISIANA, DIFFERENTIATION OF, SHALE TYPES BY,
B 6202 303 MIOCENE OIL-BEARING SANDSTONES, TRINIDAD
B 6004 646* **MIOCENE PALEOBATHYMETRY OF CALIFORNIA CONTINENTAL**
B 6012 2174† **MIOCENE PALEOBATHYMETRY OF CALIFORNIA CONTINENTAL**
B 6303 443* MIOCENE, PHASE III, MIDDLE MIOCENE EVENTS, ABST., CHRONICLE OF
S 11 55 MIOCENE PLANKTONIC FORAMINIFERA
B 6205 793 MIOCENE, PROFILE, BLAKE PLATEAU AREA
S 4 178 MIOCENE, REEF CORAL, SAIPAN
B 6402 188† MIOCENE REEFS, ALMERIA PROVINCE, SOUTHERN SPAIN, POROSITY
S 11 120 MIOCENE ROTALIA, PULLENIA, LAGENIDAE
S 11 8 MIOCENE, SALINAS VALLEY SEQUENCE, CALIFORNIA
B 6102 176 MIOCENE SANDS, CALIFORNIA, STRAND FIELD
B 6409 1557* MIOCENE SANDS IN GEOPRESSURE ZONE OF LIRETTE FIELD,
B 6111 2004† MIOCENE SEDIMENT FROM CAPE VERDE RISE AND ITS RELATION TO
B 6001 34† MIOCENE SEDIMENTATION PATTERN IN PARTS OF NORTHERN EGYPT,
M 29 182 MIOCENE, SEISMIC REFLECTORS, BLAKE ESCARPMENT
S 11 148 MIOCENE SIPHOGENERINA, CALIFORNIA
S 11 150 MIOCENE SIPHOGENERINA, CALIFORNIA
S 11 152 MIOCENE SIPHOGENERINA, CALIFORNIA
B 6412 2095† MIOCENE, SOUTH-SOUTHWEST IRAN, STRATIGRAPHY AND
B 6405 678* MIOCENE SPUMELLARIAN RADIOLARIA FROM SOUTH PACIFIC, ABST.
B 6203 403† MIOCENE, ST. CROIX, U.S. VIRGIN ISLANDS, STRATIGRAPHY AND
B 6211 2243† MIOCENE STEVENS AND OLDER TURBIDITE FANS OF SOUTHEASTERN SAN
B 6403 445* MIOCENE STEVENS SANDSTONE, ELK HILLS FIELD, KERN COUNTY,
B 6012 2185* MIOCENE STRATA, NORTHERN CHANNEL ISLANDS, CALIFORNIA, ABST.,
S 11 1 **MIOCENE STRATIGRAPHY OF CALIFORNIA REVISITED**
M 25 526 MIOCENE SUBDUCTION, MEXICO
B 6211 2360* MIOCENE SUBMARINE DUNE FIELDS—CROSS-BEDDING IN VAQUEROS
B 6303 426* MIOCENE SUBMARINE FAN IN DEEP WESTERN GULF OF MEXICO AS
M 25 448 MIOCENE SUBMARINE VOLCANIC OR GREEN TUFF AREA, JAPAN
S 7 84 MIOCENE, TERTIARY, GULF OF MEXICO, END OF
S 7 207 MIOCENE THICKNESS, WEST FLORIDA SLOPE
B 6207 1159† MIOCENE TIME, GULF OF CALIFORNIA IN, LAKE, MEAD AREA OF,
B 6307 1139† MIOCENE TIME. DISCUSSION AND REPLY, GULF OF, CALIFORNIA IN,
B 6203 572* MIOCENE TO EARLY PLIOCENE SEDIMENTATION AND FORAMINIFERAL
B 6004 728* MIOCENE TRANSGRESSIONS AND REGRESSIONS OF DELAWARE COASTAL
B 6403 424 MIOCENE, TUNISIA
B 6211 2351* MIOCENE TURBIDITE DEPOSITION, CROCKER CANYON AREA,
M 28 85 MIOCENE, UNIT, CYPRUS
B 6205 761 MIOCENE, UPPER, SEISMIC FACIES
S 11 142 MIOCENE UVIGERINA, CALIFORNIA
S 11 140 MIOCENE UVIGERINE BULMINIDAE, CALIFORNIA
S 11 166 MIOCENE VALVULINERIA, CALIFORNIA
S 11 170 MIOCENE VALVULINERIA, CALIFORNIA
S 11 168 MIOCENE VALVULINERIA, CALIFORNIA
B 6104 581 MIOCENE, VENEZUELA
B 6012 2177* MIOCENE VOLCANIC AND VOLCANICLASTIC ROCKS OF SANTA CRUZ
S 7 205 MIOCENE, WEST FLORIDA SLOPE, NORTHERN AREA, MIDDLE AND UPPER
B 6012 2175† MIOCENE-EARLY PLIOCENE MARINE DIATO AND SILICOFLAGELLATE
B 6309 1474 MIOCENE-HOLOCENE SEQUENCE, USSR, ANADYR BASIN
S 6 221 MIOCENE-OLIGOCENE BOUNDARY DEFINED, EUROPE
B 6009 1611* MIOCENE-PLIOCENE COASTAL-PLAIN SEDIMENTS, ABST., MULTIVARIATE,
B 6004 701* MIOCENE-PLIOCENE LAGA FORMATION, EAST-CENTRAL ITALY, ABST.,
B 6303 444* MIOCENE-PLIOCENE MOLASSE, ZAGROS MOUNTAINS, IRAN, ABST.
B 6303 444* MIOCENE-PLIOCENE SYNDEPOSITIONAL TAR DEPOSI IN IRAN, ABST.
B 6303 493* MIOCENE-PLIOCENE TAMIAMI FORMATION, COLLIER COUNTY, FLORIDA,
B 6109 1486 MIOGEOCLINAL WEDGE, EAST COAST, CONTINENTAL, MARGIN
M 25 493 MIOGEOSYNCLINE, GEOLOGIC HISTORY, CANADIAN, CORDILLERA
M 25 496 MIOGEOSYNCLINE, MINERAL DEPOSITS, CANADIAN, CORDILLERA
S 11 59 MIOGYPSINA
B 6109 1476 MIRANGA FIELD, BRAZIL
B 6307 1034 MIRANGA FIELD, SECTION, BRAZIL
B 6109 1476 MIRANGA NORTE FIELD, BRAZIL
B 6301 90 MISAINE MEMBER, SCOTIAN BASIN
B 6202 235† MISANTLA BASIN, MEXICO, STRATIGRAPHY AND STRUCTURE OF,
B 6102 207 MISANTLA BASIN, MEXICO, TAMPICO
B 6209 1697 MISHEGUK MOUNTAIN, ALASKA
B 6401 12 MISHRIF LIMESTONE FORMATION, IRAQ
B 6104 485 MISINTERPRETATION OF DIAGENETIC TRAPS AS STRATIGRAPHIC TRAPS
B 6007 1009 MISOA-MIRADOR FORMATION, VENEZUELA
B 6207 1129 MISOOL FAULT, IRIAN JAYA, SELE STRAIT—
B 6112 2069 MISOOL ISLAND, SECTION
B 6207 1129 MISOOL-ONIN ANTICLINORIUM FLEXURE, IRIAN JAYA
B 6301 90 MISSISAUGA FORMATION, SCOTIAN BASIN
S 5 99 MISSISSINEWA SHALE, INDIANA, WABASH FORMATION,
B 6409 1568 MISSISSIPPI, ABST., PACHUTA CREEK, SMACKOVER, FIELD
B 6209 1755* MISSISSIPPI, ABST., APPLICATION OF, FRACTURE IDENTIFICATION, LOGS
B 6009 1615* MISSISSIPPI, ABST., CLAY SEDIMENTATION I, PEARL RIVER, DELTA,
B 6004 720* MISSISSIPPI, ABST., DEPOSITION, FACIES, PATTERNS, AND,
B 6409 1554* MISSISSIPPI, ABST., DEPOSITIONAL SYSTEMS AND, LIGNITE, PROSPECTING
B 6009 1616* MISSISSIPPI, ABST., FOURMILE CREEK AND, SPLUNGE FIELDS, BLACK
B 6004 705* MISSISSIPPI, ABST., GEOPRESSURES AND, SECONDARY, POROSITY IN,
B 6209 1763* MISSISSIPPI, ABST., LATE, PLEISTOCENE STRATIGRAPHY OF SOUTHERN,
B 6303 516* MISSISSIPPI, ABST., LIGNITES OF, TOMBIGBEE AND, HOLLY SPRINGS
B 6209 1769* MISSISSIPPI, ABST., PETROGRAPHY AND DIAGENESIS OF, HOSSTON
B 6309 1602* MISSISSIPPI, ABST., PETROGRAPHY OF SOME SUBSURFACE, IGNEOUS
B 6109 1548* MISSISSIPPI, ABST., PETROLOGY OF, SMACKOVER FORMATION,
B 6409 1555* MISSISSIPPI AND ALABAMA, ABST., CHESTER AND, POTTSVILLE
B 6409 1559* **MISSISSIPPI AND ALABAMA, ABST., OSTRACODE BIOSTRATIGRAPHY OF**
B 6309 1596* MISSISSIPPI AND ALABAMA, ABST., UPPER, MISSISSIPPIAN, DELTAS IN,
B 6110 1898 MISSISSIPPI AND ALABAMA, CAMBRIAN TEST WELLS
B 6009 1614* MISSISSIPPI AND PRODUCING TRENDS, ABST., PRE-, FERRY LAKE EARLY,
B 6109 1425 MISSISSIPPI, BAXTERVILLE OIL
M 24 242 MISSISSIPPI, BENTONIA FIELD
M 24 242 MISSISSIPPI, BOLTON FIELD
M 24 250 MISSISSIPPI, BRUMFIELD SANDSTONE
B 6110 1899 MISSISSIPPI CAMBRIAN STRATIGRAPHIC SUMMARY
B 6104 595 MISSISSIPPI, CARTHAGE POINT FIELD
B 6206 951 MISSISSIPPI CHANNEL, LOUISIANA
B 6110 1899 MISSISSIPPI, CONASAUGA FORMATION
B 6111 1963 MISSISSIPPI CONE
S 7 59 MISSISSIPPI CONE, BURIED SALT RIDGES, DIAGRAM
B 6211 2293 MISSISSIPPI CONE, MULTICHANNEL PROFILE LINES
M 29 409 MISSISSIPPI CONE, REFLECTION SECTION, GULF OF MEXICO
S 7 56 MISSISSIPPI CONE REGION, DISCUSSED, GULF OF MEXICO
S 7 58 MISSISSIPPI CONE SALT, GULF OF MEXICO
B 6109 1419 MISSISSIPPI, CONERLY OIL
M 24 242 MISSISSIPPI, COTTON VALLEY PRODUCING SAND
M 24 250 MISSISSIPPI, CRETACEOUS
B 6303 512* MISSISSIPPI DELTA, ABST., ACTIVE SUBMARINE, LANDSLIDE PROCESSES
B 6004 701* MISSISSIPPI DELTA AREA, ABST., CONTROL OF SEDIMENT DISTRIBUTION
B 6202 207 MISSISSIPPI DELTA AREA, GEOHISTORY, LOUISIANA
B 6311 2054 MISSISSIPPI DELTA, CHARACTERISTICS, TABLE
S 7 96 MISSISSIPPI DELTA, SEISMIC FACIES EXAMPLE
B 6008 1282 MISSISSIPPI, DEVELOPMENTS, 1975
B 6108 1264 MISSISSIPPI, DEVELOPMENTS, 1976
B 6208 1394 MISSISSIPPI, DEVELOPMENTS, 1977
B 6308 1401 MISSISSIPPI, DEVELOPMENTS, 1978
B 6411 2029 MISSISSIPPI, DEVELOPMENTS, 1979
B 6409 1569* MISSISSIPPI EMBAYMENT—IMPLICATIONS FOR POTENTIAL
M 24 250 MISSISSIPPI, EUTAW SANDSTONES
B 6206 920 MISSISSIPPI FAN
B 6009 1616* MISSISSIPPI FAN, ABST., FORM AND COMPOSITION OF
M 29 392 MISSISSIPPI FAN, AGE OF SEDIMENT, GULF OF MEXICO
S 7 166 MISSISSIPPI FAN, AGE OF SEDIMENTS
S 7 13 MISSISSIPPI FAN, DESCRIBED
M 29 391 MISSISSIPPI FAN, DESOTO DEEP SEA CHANNEL, SEISMIC PROFILE
M 29 389 MISSISSIPPI FAN, DESOTO DEEP-SEA CHANNEL, GULF OF MEXICO
S 7 168 MISSISSIPPI FAN, DSDP SITES RELATIVE TO TABLE
S 7 179 MISSISSIPPI FAN, FAN CHANNELS
S 7 159 MISSISSIPPI FAN, GEOLOGIC SETTING
M 29 387 MISSISSIPPI FAN, GEOMORPHIC UNITS, GULF OF MEXICO
B 6203 546* MISSISSIPPI FAN, GULF OF MEXICO, ABST.
M 29 383† MISSISSIPPI FAN, GULF OF MEXICO, INVESTIGATION OF
S 7 155† MISSISSIPPI FAN, GULF OF MEXICO—PHYSIOGRAPHY, STRATIGRAPHY,
S 7 164 MISSISSIPPI FAN, HIGH-RESOLUTION PROFILE
M 29 400 MISSISSIPPI FAN, HISTORY, GULF OF MEXICO
M 29 400 MISSISSIPPI FAN, HOLOCENE, SEDIMENT THICKNESS
M 29 398 MISSISSIPPI FAN, HOLOCENE, SEDIMENTATION
M 29 385 MISSISSIPPI FAN, INDEX, GULF OF MEXICO
M 29 397 MISSISSIPPI FAN, ISOPACH MAP
M 29 389 MISSISSIPPI FAN, LITHOLOGIC LOG, CORE EXAMPLE
S 7 171 MISSISSIPPI FAN, LONGITUDINAL PROFILE
M 29 390 MISSISSIPPI FAN, MIDDLE, HIGH-RESOLUTION PROFILE
S 7 184 MISSISSIPPI FAN, PETROLEUM POTENTIAL
M 29 400 MISSISSIPPI FAN, PETROLEUM POTENTIAL, GULF OF MEXICO
S 7 159 MISSISSIPPI FAN, PHYSIOGRAPHY
M 29 385 MISSISSIPPI FAN, PHYSIOGRAPHY, GULF OF MEXICO
M 29 401 MISSISSIPPI FAN, PLEISTOCENE, PETROLEUM POTENTIAL
M 29 396 MISSISSIPPI FAN, PLEISTOCENE, SEDIMENTATION

S 7 169 MISSISSIPPI FAN, REGIONAL PROFILE, DIAGONAL
S 7 174 MISSISSIPPI FAN, RELATION TO CAMPECHE ESCARPMENT
M 29 395 MISSISSIPPI FAN, RELATIONS TO CAMPECHE ESCARPMENT
S 7 184 MISSISSIPPI FAN, RESERVOIR POTENTIAL
S 7 172 MISSISSIPPI FAN, SEDIMENT DISTRIBUTION
M 29 395 MISSISSIPPI FAN, SEDIMENT DISTRIBUTION, GULF OF MEXICO
S 7 165 MISSISSIPPI FAN, SEDIMENT OVERLAP, PROFILE
S 7 188 MISSISSIPPI FAN, SEDIMENTATION HISTORY
S 7 176 MISSISSIPPI FAN, SEDIMENTATION RATES
S 7 178 MISSISSIPPI FAN, SEDIMENTATION RATES, TABLE
M 29 393 MISSISSIPPI FAN, SEISMIC PROFILE, (FOLDOUT)
S 7 181 MISSISSIPPI FAN, SEISMIC PROFILES, PLEISTOCENE
M 29 393 MISSISSIPPI FAN, SEISMIC STRATIGRAPHY, GULF OF MEXICO
S 7 170 MISSISSIPPI FAN, SEISMIC-STRATIGRAPHIC UNITS
S 7 187 MISSISSIPPI FAN, SOURCE ROCKS
M 29 393 MISSISSIPPI FAN, STRATIGRAPHY, GULF OF MEXICO
S 7 170 MISSISSIPPI FAN, STRATIGRAPHY, SEISMIC INTERPRETATION
S 7 185 MISSISSIPPI FAN, SUPRAFAN GROWTH, STAGES
M 26 271 MISSISSIPPI FAN SYSTEM
M 29 397 MISSISSIPPI FAN, TIME-INTERVAL MAP
S 7 173 MISSISSIPPI FAN, TRANSVERSE PROFILE
S 7 187 MISSISSIPPI FAN, TRAPS
S 7 177 MISSISSIPPI FAN, UNIT ISOPACHS
M 24 242 MISSISSIPPI, FLORA FIELD
M 24 242 MISSISSIPPI, FREDERICKSBURG PRODUCING SAND
M 24 242 MISSISSIPPI, GAS ROCK PRODUCING SAND
M 24 278 MISSISSIPPI, GILBERTOWN-PICKENS-POLLARD FAULT SYSTEM, ALABAMA
B 6209 1757* MISSISSIPPI GULF COAST, ABST., PLEISTOCENE, SHORELINES AND
M 24 251 MISSISSIPPI, HAYNESVILLE-BUCKNER SECTION
M 24 242 MISSISSIPPI, HOSSTON PRODUCING SANDS
B 6311 2104 MISSISSIPPI, IOWA LIMESTONE
M 24 242 MISSISSIPPI, JACKSON FIELD
B 6311 2104 MISSISSIPPI, KNOX DOLOMITE
B 6104 595 MISSISSIPPI, LITTLE CREEK FIELD
M 24 242 MISSISSIPPI, LORING FIELD
B 6404 467 MISSISSIPPI, MARINE PHREATIC ARAGONITE CEMENT, GULF OF MEXICO
M 24 250 MISSISSIPPI, MCGRAW SANDSTONE
M 24 244 MISSISSIPPI, MIDWAY SHALE
M 24 242 MISSISSIPPI, MOORINGSPORT PRODUCING SAND
M 24 242 MISSISSIPPI, PALUXY PRODUCING SAND
M 24 247 MISSISSIPPI, PERRY SANDSTONE
M 24 242 MISSISSIPPI, PICKENS FIELD
M 24 250 MISSISSIPPI, POWELL SANDSTONE
B 6008 1284 MISSISSIPPI, PRODUCTION, 1974-1975
B 6108 1266 MISSISSIPPI, PRODUCTION, 1975-1976
B 6208 1396 MISSISSIPPI, PRODUCTION, 1976-1977
B 6308 1404 MISSISSIPPI, PRODUCTION, 1977-1978
B 6002 240 MISSISSIPPI, REEDY CREEK FIELD
B 6002 220 MISSISSIPPI, REEDY CREEK TOPPED OIL
B 6303 542* MISSISSIPPI RIVER DELTA FRONT, ABST., SHALLOW-SUBMARINE SEISMIC
S 7 99 MISSISSIPPI RIVER DELTA, SEISMIC FACIES EXAMPLE
B 6304 660† MISSISSIPPI RIVER DELTA--APRIL 9, 1976--FROM LANDSAT 2
B 6409 1567* MISSISSIPPI RIVER DELTA-FRONT DEPOSITS, ABST., SEDIMENTARY
B 6304 665 MISSISSIPPI RIVER DISCHARGE JANUARY 2 TO MAY 3, 1976
B 6302 213 MISSISSIPPI RIVER, RECENT HISTORY
B 6102 176 MISSISSIPPI RIVER, RECENT SAND
M 24 242 MISSISSIPPI, RODESSA PRODUCING SAND
B 6110 1899 MISSISSIPPI, ROME FORMATION
M 24 242 MISSISSIPPI, SARTARTIA FIELD
M 24 242 MISSISSIPPI, SELMA CHALK
M 24 242 MISSISSIPPI, SLIGO PRODUCING SANDS
S 7 141 MISSISSIPPI SLOPE, NORTHERN GULF OF MEXICO, CONTINENTAL SLOPE
M 24 252 MISSISSIPPI, SMACKOVER FORMATION
B 6104 594 MISSISSIPPI, SMITHDALE FIELD
M 24 259 MISSISSIPPI, SOSO FIELD
B 6111 1963 MISSISSIPPI SPUR
M 24 250 MISSISSIPPI, STEVENS SANDSTONE
S 7 134 MISSISSIPPI SUBMARINE CHANNEL, ISOPACH, GULF OF MEXICO
B 6003 455† MISSISSIPPI, SUBSURFACE, PRECAMBRIAN, GRANITE,
B 6111 1963 MISSISSIPPI TROUGH
S 7 163 MISSISSIPPI TROUGH, BATHYMETRY MAP
B 6303 449* MISSISSIPPI TROUGH, GULF OF MEXICO, ABST., SEDIMENTATION IN
S 7 164 MISSISSIPPI TROUGH, HIGH-RESOLUTION PROFILE
M 29 389 MISSISSIPPI TROUGH, LONGITUDINAL SEISMIC PROFILE
M 26 261 MISSISSIPPI TROUGH, SEISMIC LINE, GULF OF MEXICO
M 26 270 MISSISSIPPI TROUGH SYSTEM
S 7 182 MISSISSIPPI TROUGH-FAN, LONGITUDINAL PROFILE
B 6109 1421 MISSISSIPPI, TUSCALOOSA FORMATION, HYDROCARBONS
M 24 242 MISSISSIPPI, TUSCALOOSA PRODUCING SAND
B 6203 550* MISSISSIPPI VALLEY, ABST., LITHOSTRATIGRAPHY, PETROLOGY, AND
B 6303 542* MISSISSIPPI VALLEY, ABST., PALEOENVIRONMENT OF MIDDLE,
B 6405 697* MISSISSIPPI VALLEY, ABST., SUBMARINE DIAGENESIS, ARAGONITE
M 24 242 MISSISSIPPI, VALLEY PARK FIELD
B 6303 416* MISSISSIPPI VALLEY--TYPE ORE DEPOSITS, ABST., CRITERIA FOR
B 6311 2103† MISSISSIPPI, VARDAMAN, TRIASSIC-, JURASSIC, SILL, CALHOUN COUNTY
M 24 242 MISSISSIPPI, WASHITA PRODUCING SAND
B 6110 1899 MISSISSIPPI, WEISNER SANDSTONE
M 24 243 MISSISSIPPI, WOODRUFF SANDSTONE
S 7 8 MISSISSIPPI-ALABAMA SHELF, DESCRIBED
B 6409 1562* MISSISSIPPI-ATCHAFALAYA DIVERSION, ABST., SHOULD WE PERMIT
M 25 341 MISSISSIPPIAN, AMAZON BASINS
B 6109 1493† MISSISSIPPIAN AND PENNSYLVANIAN, POTENTIAL MAJOR
B 6002 319* MISSISSIPPIAN B, STAFFORD COUNTY, KANSAS, ABST., NUMERICAL
M 27 68 MISSISSIPPIAN BIOHERMAL MOUND, KANSAS
B 6101 12 MISSISSIPPIAN, CANADIAN, ARCTIC ISLANDS
B 6205 886* MISSISSIPPIAN CARBONATE RESERVOIRS OF SOUTHERN SASKATCHEWAN,
B 6004 723* MISSISSIPPIAN CARBONATE ROCKS, PEDREGOSA BASIN, SOUTHEASTERN
B 6305 828* MISSISSIPPIAN CARBONATE SHELF MARGIN ALONG OVERTHRUST BELT
B 6105 823* MISSISSIPPIAN CARBONATE-SHELF MARGINS, WESTERN UNITED STATES,
B 6305 838* MISSISSIPPIAN CHAINMAN SHALE OF EAST-CENTRAL NEVADA,
B 6008 1398* MISSISSIPPIAN CHARLES FORMATION, WILLISTON BASIN,
M 26 453 MISSISSIPPIAN, CHESTER SHALE, OKLAHOMA
B 6103 309† MISSISSIPPIAN COMBINATION TRAP, BINDLEY FIELD, HODGEMAN
B 6002 319* MISSISSIPPIAN COMBINATION TRAP, BINDLEY FIELD, HODGEMAN
B 6102 256† MISSISSIPPIAN COPPER BASIN GROUP, PIONEER MOUNTAINS,
M 27 164 MISSISSIPPIAN, CORAL LIMESTONE, ENGLAND
M 27 89 MISSISSIPPIAN, DARTRY LIMESTONE, IRELAND
M 27 133 MISSISSIPPIAN, DEBOLT FORMATION, CANADA
M 28 84 MISSISSIPPIAN, DEBOLT FORMATION, CANADA
B 6309 1596* MISSISSIPPIAN DELTAS IN BLACK WARRIOR BASIN OF MISSISSIPPI AND
S 6 331 MISSISSIPPIAN DEPOSITION, STRUCTURAL CONTROLS
S 6 332 MISSISSIPPIAN, FORAMINIFERAL ZONE 19, LOCATIONS
M 27 77 MISSISSIPPIAN, GLENCAR LIMESTONE, IRELAND
M 27 78 MISSISSIPPIAN, GLENCAR LIMESTONE, IRELAND
M 27 211 MISSISSIPPIAN, HINDSVILLE LIMESTONE, OKLAHOMA
M 27 76 MISSISSIPPIAN, HINDSVILLE LIMESTONE, OKLAHOMA
M 27 100 MISSISSIPPIAN, HINDSVILLE LIMESTONE, OKLAHOMA
B 6408 1287* MISSISSIPPIAN, ILLINOIS BASIN, ABST., DEPOSITIONAL ENVIRONMENTS
B 6403 404 MISSISSIPPIAN, INDIANA
B 6004 687* MISSISSIPPIAN, KENTUCKY AND INDIANA, ABST., PRODELTA, TURBIDITE-,
B 6204 654 MISSISSIPPIAN, LATE, UNITED STATES, SOUTHEASTERN
M 28 128 MISSISSIPPIAN, LIMESTONE, IRELAND
M 27 101 MISSISSIPPIAN LIMESTONE, IRELAND
M 27 27 MISSISSIPPIAN, LODGEPOLE LIMESTONE, MONTANA
M 27 26 MISSISSIPPIAN, LODGEPOLE LIMESTONE, MONTANA
M 28 40 MISSISSIPPIAN, LODGEPOLE LIMESTONE, MONTANA
M 27 164 MISSISSIPPIAN, LODGEPOLE LIMESTONE, MONTANA
B 6205 879* MISSISSIPPIAN, MICROFACIES AND STRATIGRAPHY, NEW MEXICO AND
B 6305 841* MISSISSIPPIAN MIDALE BEDS, BENSON OIL FIELD, SOUTHEASTERN
B 6102 257 MISSISSIPPIAN, NOMENCLATURE, IDAHO
B 6303 431* MISSISSIPPIAN NON-SUPRATIDAL DOLOMITE STE. GENEVIEVE
B 6004 720* MISSISSIPPIAN, NORTHERN BLACK WARRIOR BASIN, ALABAMA AND
B 6205 887* MISSISSIPPIAN, NORTHERN MEDICINE BOW MOUNTAINS, WYOMING,
M 28 40 MISSISSIPPIAN, NUKA FORMATION, ALASKA
M 28 129 MISSISSIPPIAN, NUKA FORMATION, ALASKA
M 28 122 MISSISSIPPIAN, NUKA FORMATION, ALASKA
B 6203 530* MISSISSIPPIAN OF EASTERN WILLISTON BASIN, ABST., CALCIUM, SULFATE
B 6405 692* MISSISSIPPIAN, OF ILLINOIS, ABST., RELATION OF DEPOSITIONAL FACIES
B 6305 845* MISSISSIPPIAN, OF MORGAN COUNTY, TENNESSEE, ABST., GEOLOGY AND
B 6405 743* MISSISSIPPIAN, OF TENNESSEE WITH WAULSORTIAN MOUNDS OF
B 6405 762* MISSISSIPPIAN OIL FIELD, WILLISTON BASIN, ABST., MONDAK
B 6008 1398* MISSISSIPPIAN OUNGRE TREND IN SOUTHEASTERN SASKATCHEWAN,
M 26 453 MISSISSIPPIAN, PARVIN LIMESTONE, OKLAHOMA
B 6109 1454 MISSISSIPPIAN, PEDREGOSA BASIN
B 6004 646* MISSISSIPPIAN, PEDREGOSA BASIN, SOUTHEASTERN ARIZONA,
B 6303 548* MISSISSIPPIAN PINNACLE REEFS OF CHAPPEL LIMESTONE, FORT WORTH
M 27 217 MISSISSIPPIAN, PITKIN LIMESTONE, OKLAHOMA
M 27 118 MISSISSIPPIAN, PITKIN LIMESTONE, OKLAHOMA
M 28 82 MISSISSIPPIAN,, REDWALL LIMESTONE, ARIZONA
B 6009 1619* MISSISSIPPIAN ROCKS NEWMAN AND PENNINGTON FORMATIONS,
B 6203 558* MISSISSIPPIAN ROCKS OF PERMIAN BASIN, WEST TEXAS AND
M 27 90 MISSISSIPPIAN, SALEM LIMESTONE, INDIANA
B 6204 646 MISSISSIPPIAN SECTION, ALABAMA
B 6405 716* MISSISSIPPIAN SHELF MARGIN AND CARBONATE PLATFORM FROM
B 6405 750* MISSISSIPPIAN SKELETAL LIMESTONES, NEW MEXICO, ABST.,
M 27 163 MISSISSIPPIAN, STE. GENEVIEVE LIMESTONE, INDIANA
M 27 207 MISSISSIPPIAN, STE. GENEVIEVE LIMESTONE, TENNESSEE
B 6101 11 MISSISSIPPIAN, SVERDRUP BASIN
B 6405 671* MISSISSIPPIAN SYSTEM OF NEW MEXICO AND SOUTHERN ARIZONA, ABST.
B 6109 1515 MISSISSIPPIAN, TERMINOLOGY, NEW MEXICO
B 6103 435† MISSISSIPPIAN UNCONFORMITIES IN NORTHER ALASKA RELATED TO
B 6408 1281* MISSISSIPPIAN, VALMEYERAN, EWING AREA, JEFFERSON ANDFRANKLIN
B 6304 652 MISSISSIPPIAN, VIRGINIA
M 27 179 MISSISSIPPIAN, WAUL SORTIAN LIMESTONE, IRELAND
M 27 69 MISSISSIPPIAN, WAULSORTIAN LIMESTONE, IRELAND
B 6303 487* MISSISSIPPIAN WAULSORTIAN-TYPE MOUNDS OF FORT PAYNE

B 6405 779* MISSISSIPPIAN, WESTERN UNITED STATES, ABST., SEDIMENTATION,
S 6 324 MISSISSIPPIAN-PENNSYLVANIAN BOUNDARY, CARBONIFEROUS
S 6 327† MISSISSIPPIAN-PENNSYLVANIAN BOUNDARY, THE
B 6111 1929† MISSISSIPPIAN-PENNSYLVANIAN, CANADIAN ARCTIC ARCHIPELAGO,
S 5 145* MISSISSIPPIAN-PENNSYLVANIAN, CANADIAN ARCTIC ARCHIPELAGO,
B 6204 644† MISSISSIPPIAN—OOLITIC TIDAL-BAR SEDIMENTATION IN SOUTHERN
B 6004 681* MISSOURI, ABST., TRANSPORTATIONAL AND DEPOSITIONAL HISTORY OF,
B 6108 1307 MISSOURI, DEVELOPMENTS, 1976
B 6208 1480 MISSOURI, DEVELOPMENTS, 1977
B 6308 1305 MISSOURI, DEVELOPMENTS, 1978
B 6409 1440 MISSOURI, DEVELOPMENTS, 1979
M 28 69 MISSOURI, PENNSYLVANIAN, BALL CLAY
B 6002 323* MISSOURI--UPDATE 1975, ABST., HEAVY, OIL IN WESTERN
B 6311 2116* MISSOURI-KANSAS-OKLAHOMA TRI-STATE AREA, ABST., CORRELATION
B 6405 788* MISSOURIAN, MONTGOMERY COUNTY, KANSAS, ABST., PROFILE OF
M 26 441 MISSOURIAN, OKLAHOMA, ANADARKO, UPPER, HOXBAR GROUP
B 6311 2120* MISSOURIAN, PENNSYLVANIAN SERIES BOUNDARY IN NORTHEASTERN
B 6107 1050 MISSOURIAN STAGE, MID-CONTINENT
B 6403 439* MIST GAS FIELD, COLUMBIA COUNTY, OREGON, ABST.
B 6201 126 MITCHELL ESCARPMENT FAULT, FLORIDA
B 6201 123 MITCHELL ESCARPMENT, FLORIDA
B 6004 699 MITCHELL, G. C., DEPOSITIONAL ENVIRONMENT OF LOWER CRETACEOU
B 6108 1384 MITCHELL, GARY C., DEPOSITIONAL ENVIRONMENT OF LOWER
B 6205 889 MITCHELL, GARY C., TWIN CREEK FORMATION AND EQUIVALENT UNITS,
B 6008 1392 MITCHELL, JOHN C., HEAT-FLOW STUDY OF SNAKE RIVER PLAIN, IDAHO,
B 6002 324 MITCHELL, M. H., TURBIDITE RESERVOIRS IN CANYON SANDSTONE, VAL
B 6412 2189 MITCHLER, JOHN, REVIEW OF DEPOSITIONAL AND STRUCTURAL
M 29 469 MITCHUM, R. M., JR., GLOBAL CYCLES OF RELATIVE CHANGES OF SEA
B 6004 699 MITCHUM, R. M., JR., REGIONAL SEISMIC INTERPRETATION USING
S 7 193 MITCHUM, R. M., JR., SEISMIC STRATIGRAPHIC INVESTIGATION OF WEST
M 26 49 MITCHUM, R. M., JR., SEISMIC STRATIGRAPHY AND GLOBAL CHANGES OF
M 26 205 MITCHUM, R. M., JR., SEISMIC STRATIGRAPHY AND GLOBAL CHANGES OF
M 26 51 MITCHUM, R. M., JR., SEISMIC STRATIGRAPHY AND GLOBAL CHANGES OF
M 26 145 MITCHUM, R. M., JR., SEISMIC STRATIGRAPHY AND GLOBAL CHANGES OF
M 26 63 MITCHUM, R. M., JR., SEISMIC STRATIGRAPHY AND GLOBAL CHANGES OF
M 26 117 MITCHUM, R. M., JR., SEISMIC STRATIGRAPHY AND GLOBAL CHANGES OF
M 26 135 MITCHUM, R. M., JR., SEISMIC STRATIGRAPHY AND GLOBAL CHANGES OF
M 26 83 MITCHUM, R. M., JR., SEISMIC STRATIGRAPHY AND GLOBAL CHANGES OF
M 26 53 MITCHUM, R. M., JR., SEISMIC STRATIGRAPHY AND GLOBAL CHANGES OF
B 6412 2113 MITI-YOSHIDA WELL, HYDROCARBONS, JAPAN
M 24 72† MITSUE OIL FIELD, ALBERTA--A RICH STRATIGRAPHIC TRAP
B 6203 531 MITSUI, KIYOHIRO, SLOPE SEDIMENTATION PROCESSES IN NEOGENE
B 6203 544 MITSUSHIO, HIROMI, BOTTOM SEDIMENTS OF SAN NICOLAS BASIN,
B 6202 187 MITTEN BLACK SHALE MEMBER, WYOMING
B 6303 469* MIXED-LAYER CLAYS IN EOCENE INTERLAMINATE SHALES AND
B 6203 489* MIXING OF BOTTOM SEDIMENTS IN NORTHWEST ATLANTIC, ABST.
B 6105 787* MIXING-ZONE DOLOMITIZATION OF HOLOCENE TIDAL-FLAT SEDIMENTS,
B 6303 457* MIXING ZONE DOLOMITE IN TIDAL-FLAT SEDIMENTS OF CENTRAL-WEST
B 6105 822 MIXON, ROBERT B., OAK GROVE CORE, NORTHEASTERN VIRGINIA, ABST.,
B 6004 734* MIXTITE-BEARINN SEQUENCES, ABST., PROBLEMM OF,
S 6 152 MIYAKO GRANITE, RUBIDIUM-STRONTIUM DATA, JAPAN
S 6 156 MIYAKO GRANITE, RUBIDIUM-STRONTIUM ISOCHRON DIAGRAM, JAPAN
M 29 239 MIZUNO, ATSUYUKI, SUBSIDENCE OF THE DAITO RIDGE AND
B 6104 594 MO-2 WELL, MOROCCO
B 6104 596 MO-2 WELL, MOROCCO
B 6104 596 MO-3 WELL, MOROCCO
B 6401 105 MOBEETIE FIELD, TEXAS
B 6411 1842 MOBIL EXPLORATION MEDITERRANEAN INC., ON EXPLORATION IN
B 6310 1891 MOBIL EXPLORATION MEDITERRANEAN INC., TURKIYE PETROLLERI A.O.,
B 6110 1796 MOBIL IRAN ONSHORE, INC., AND ATLANTIC RICHFIELD, ON
B 6311 2115* MOBILE BASIN-SHELF BORDER IN NORTHEAST OKLAHOMA DURING
S 7 99 MOBILE BAY, SEISMIC FACIES EXAMPLE, ALABAMA
B 6211 2351* MOBILE BED FORMS AND VOLCANISM, LOWER COOK INLET, ALASKA. BY
M 25 46† MOBILE BELT AND PACIFIC OCEAN, GENERAL PRINCIPLES AND
M 25 479† MOBILE BELT--SUMMARY, MEIMECHITES OF, KAMCHATKA AND,
M 25 62† MOBILE BELT--SUMMARY, TECTONICS OF CONTINENTAL AND OCEANIC
S 3 91 MOBILE BELTS, OROGENIC ZONES ACROSS
M 24 259† MOBILE COUNTY, ALABAMA, CITRONELLE OIL FIELD
B 6206 987 MOBILE GROUP, IRAN, ZAGROS MOUNTAINS
S 3 41 MOBILIZATION BY DEFORMATION, FLUID
S 3 41 MOBILIZATION IN CRUSTAL ROCKS, MECHANISMS OF, RADIOELEMENT
S 3 136 MOBILIZATION, LATEROGENESIS AT THE LOW-INTENSITY END OF THE,
S 3 136 MOBILIZATION OR LOCALIZATION BY SALT OR SHALE DIAPIRISM
S 3 40 MOBILIZATION PROCESSES, SHALLOW, URANIUM
S 3 46 MOBILIZATION, SELECTIVE, OXIDATION AND
B 6410 1605 MOCCASIN FORMATION, VIRGINIA
B 6204 706 MOCCASIN SPRINGS FORMATION, ILLINOIS BASIN
B 6004 654* MODAL COMPONENTS USING SIMULATION TECHNIQUES, ABST.,
B 6004 697* MODE FOR MACROTIDAL ENVIRONMENT, COBEQUID BAY, NOVA SCOTIA,
B 6004 721* MODEL, ABST., VARIATIONS IN, TRENCH-SLOPE
B 6004 696* MODEL BASED ON CYCLIC INTERACTION OF CONTINENTAL SODA-RICH
M 26 240 MODEL, BRAZIL, POST-RIFT BASIN
M 26 240 MODEL, BRAZIL, RIFT-BASIN
M 26 186 MODEL, CARBONATE BUILDUPS
B 6004 552 MODEL, COMBINED-, MIGRATION
M 26 244 MODEL, DEPOSITIONAL SYSTEMS, PASSIVE OFFLAP
M 26 243 MODEL, DEPOSITIONAL SYSTEMS, POST-RIFT, CARBONATE PLATFORM
B 6206 963 MODEL, EXPLORATION PREDICTIONS FROM, FAN
B 6004 661* MODEL FOR APPLICATION OF STRATIGRAPHIC CONCEPTS TO
B 6206 935 MODEL FOR CLASSIC TURBIDITES, BOUMA
B 6406 929 MODEL FOR EXCESS FLUID PRESSURES IN ACCUMULATING SEDIMENTS
B 6303 420* MODEL FOR INTERSTITIAL SULFATE REDUCTION AND METHANE
B 6203 539* MODEL FOR JAMES BAY LOWLANDS, ONTARIO, CANADA, ABST.,
B 6004 669* MODEL FOR MESOZOIC CARBONAT PLATFORM, ABST., ABENAKI
B 6107 1062 MODEL FOR MID-CONTINENT MIDDLE AND LATE PENNSYLVANIAN
B 6206 1091* MODEL FOR SOME COLORADO PLATEAU URANIUM DEPOSITS, ABST., NEW
B 6203 501* MODEL FOR SOME COLORADO PLATEAU URANIUM DEPOSITS. BY TITLE
B 6203 554* MODEL FOR SUBMARINE CEMENTATION, ABST., INTERSTITIAL, WATER
B 6203 554* MODEL FOR TRACE-ELEMENT PARTITIONING IN LIMESTONES, ABST.
B 6203 502* MODEL FOR TYPE ST. CROIXAN, ABST., ENIGMAS IN, WISCONSIN,
B 6002 211 MODEL, MATCHING WEST AFRICAN AND BRAZILIAN SALT EDGES, ALBIAN
M 26 430 MODEL, NORTH SEA, BRENT FIELD
B 6002 184 MODEL, NORTH SEA, SEDIMENTARY
B 6405 805* MODEL OF AUTOCEMENTATION OF QUARTZ SANDS AS SUGGESTED BY
M 26 439† MODEL OF DEPOSITIONAL PLATFORM MARGIN, EASTERN ANADARKO
B 6109 1437† MODEL OF DEPOSITIONAL PLATFORM MARGIN, EASTERN ANADARKO
B 6307 1108† MODEL OF EARLY CRETACEOUS CORAL-ALGAL-RUDIST REEFS, ARIZONA,
B 6209 1571 MODEL OF FORCEFULLY INTRUSIVE DIAPIRISM, CONCEPTUAL
B 6203 512* MODEL OF MESOTIDAL BACK-BARRIER ENVIRONMENT, ABST.,
B 6002 232 MODEL OF PRIMARY PETROLEUM MIGRATION BY MOLECULAR
B 6004 707* MODEL OF TECTONICALLY CONTROLLED CHANNELIZED SEDIMENTATION
B 6203 515* MODEL WITH EXAMPLE FROM UPPER DEVONIAN—LOWER
B 6004 684* MODEL--GREA LAKES AREA, ABST., SEARCH FOR, SILURANIAN, REEF
B 6203 494* MODELING, ABST., DETERMINATION OF, RESERVOIR, FLUID CONTENT BY,
B 6105 842* MODELING, ABST., INTERPRETIVE METHODS OF, STRATIGRAPHIC
M 26 455 MODELING, ACOUSTIC STRATIGRAPHY, SEISMIC
M 26 482 MODELING, AMPLITUDE ANOMALY, SEISMIC
M 26 417† MODELING AND INTERPRETATION, GEOLOGIC CONSIDERATIONS FOR,
M 26 477† MODELING AND INTERPRETATION, PRACTICAL, STRATIGRAPHIC
M 26 389† MODELING AND INTERPRETATION—GEOPHYSICAL PRINCIPLES AND
M 26 484 MODELING, APPLICATIONS, SEISMIC
M 26 480 MODELING, BARRIER-BAR SANDSTONE, SEISMIC
M 26 479 MODELING, CHANNEL SANDSTONES, SEISMIC
M 26 394 MODELING, CONTRIBUTIONS
B 6203 525* MODELING FOR REEF EXPLORATION IN MICHIGAN BASIN, ABST., USE OF,
M 26 396 MODELING, FRESNEL ZONE ILLUSTRATION
B 6004 728* MODELING IN RECONSTRUCTION OF OBJECTS FROM SERIAL SECTIONS,
M 26 453 MODELING, INPUT PARAMETERS, SEISMIC
M 26 500 MODELING, INTERPRETATION PROCEDURE, SEISMIC
M 26 453 MODELING, LITHOLOGIC CHARACTER, SEISMIC
M 26 468 MODELING, MAPPING, GEOSEISMIC
M 26 404 MODELING, NORTH SEA HORST-FAULTING
M 26 451† MODELING OF BASAL MORROW-SPRINGER SANDSTONES,
B 6203 497* MODELING OF BED CONFIGURATIONS, ABST., DYNAMIC, SCALE
B 6309 1581* MODELING OF DEVONIAN SHALE GAS-RESERVOIR PERFORMANCE, ABST.
M 26 460 MODELING, OKLAHOMA, GEARY AREA, GEOSEISMIC
M 26 440 MODELING, ONE-DIMENSIONAL, SEISMIC
M 26 391 MODELING, OVERTHRUST FAULTING
M 26 403 MODELING, PINNACLE REEF ILLUSTRATION
M 26 455 MODELING PRESSURE AND PORE FLUIDS, SEISMIC
M 26 465 MODELING, PROCEDURE, SEISMIC
M 26 455 MODELING, REFLECTION COEFFICIENTS, SEISMIC
M 26 389 MODELING, RELATING DATA TO STRATIGRAPHY
M 26 453 MODELING, RESERVOIR GEOMETRY, SEISMIC
M 26 3 MODELING, SEISMIC, DATA
M 26 390 MODELING, SEISMOGRAM SYNTHESIS
M 26 480 MODELING, SHORELINE SANDSTONES, SEISMIC
M 26 408 MODELING, STRATIGRAPHIC CORRELATIONS
M 26 493 MODELING, STRATIGRAPHIC TRAP, QUANTIFICATION, SEISMIC
M 26 444 MODELING, TWO-DIMENSIONAL, SEISMIC
M 26 491 MODELING, UPDIP PERMEABILITY BARRIER, SEISMIC
M 26 392 MODELING, USING AMPLITUDES, WAVEFORMS
M 26 478 MODELING, VERTICAL LITHOLOGY, SEISMIC RESPONSE, SEISMIC
M 26 482 MODELING, WATER SANDSTONE, GAS CAP, SEISMIC
M 26 398 MODELING, WAVE THEORY, BOX CAR REEF
B 6009 1606* MODELS, ABST., GEOPRESSURED, GEOTHERMAL, PROSPECT IN, FRIO
B 6203 515* MODELS, ABST., INTERPRETATION OF TECTONIC AND, SEDIMENTATION
M 26 240 MODELS, BRAZIL, OFFSHORE BASINS
M 26 435 MODELS, CARBONATE SHELF MARGIN TRAPS, SEISMIC
M 26 431 MODELS, DELTAIC TRAPS, SEISMIC

M 26 224 MODELS, DEPOSITIONAL SYSTEMS, DELTA
M 26 231 MODELS, DEPOSITIONAL SYSTEMS, SHELF
M 26 232 MODELS, DEPOSITIONAL SYSTEMS, SLOPE
B 6203 501* MODELS FOR ANCIENT EPICONTINENTAL SEAWAY DEPOSITIONAL
B 6203 516* MODELS FOR CARBONATE-FACIES DISTRIBUTION, ABST., PREDICTIVE
B 6303 543* MODELS FOR EVOLUTION OF INTERIOR BASINS ABST.
B 6206 932† MODELS FOR EXPLORATION FOR STRATIGRAPHIC TRAPS, DEEP-, WATER,
B 6305 810† MODELS FOR EXPLORATION FOR STRATIGRAPHIC TRAPS. DISCUSSION
B 6303 490* MODELS FOR INTERPRETATION OF MICROMORPH FAUNAS IN WASHITA
B 6004 680* MODELS FROM APPALACHIAN CARBONIFEROUS DELTAS, ABST.,
B 6109 1550* MODELS FROM TEXAS AND MEXICO, ABST., SOME LOWER, CRETACEOUS
B 6212 2379† MODELS IN COAL EXPLORATION AND MINE PLANNING IN APPALACHIAN
B 6203 523* MODELS IN PREDICTING ROOF PROBLEMS IN COAL MINES, ABST., USE OF
B 6107 1058 MODELS OF PENNSYLVANIAN BLACK SHALE DEPOSITION
M 26 435 MODELS, PLATFORM TRAPS, SEISMIC
M 26 101 MODELS, SEISMIC, REFLECTIONS, SEISMIC
M 26 428 MODELS, SUBSURFACE TRAPS, SEISMIC
M 26 432 MODELS, THIN RESERVOIR SYSTEMS, SEISMIC
M 26 430 MODELS, UNCONFORMITY TRAP, SEISMIC
B 6004 724* **MODELS USING SUBSURFACE DATA FOR INTERPRETATION OF**
B 6004 729* MODELS--WET SLOPES AND RAIN SHADOWS, ABST., FLUVIAL
B 6403 440* MODERN AND ANCIENT COASTAL SEDIMENTARY FACIES, MONTEREY
B 6004 657* MODERN AND ANCIENT ESTUARINE-FILL FACIES, WILLAPA BAY,
B 6303 415* MODERN AND ANCIENT EXAMPLES OF ANNELID BURROWS AS
B 6407 1094† MODERN AND ANCIENT SUBMARINE FANS. DISCUSSION AND REPLIES
B 6007 1054† MODERN CARIBBEAN FRINGING REEF, GALETA POINT, PANAMA,
M 27 71 MODERN ENCRUSTERS, BRYOZOANS
B 6303 416* MODERN EXPLORATION STRATEGY IN FRONTIER AREAS, CASE HISTORY
B 6403 325 MODERN INTERDELTAIC COAST, HYDRODYNAMIC ZONES
B 6203 551* MODERN LACUSTRINE STROMATOLITES, WALKER LAKE, NEVADA, ABST.
B 6409 1562* MODERN LAGOONAL OSTRACODES AND SPECIE DIVERSITY GRADIENTS,
B 6211 2357* MODERN METHODS OF ACQUIRING AND PROCESSING GEOPHYSICAL AND
B 6309 1545 MODERN PETROLEUM, A BASIC PRIME OF THE INDUSTRY, BY BILL D.
B 6011 2068* MODERN RIVER DELTAS--VARIABILITY OF PROCESSES AND SAND BODIES,
B 6004 686* MODERN SUBMARINE CANYON-FAN SYSTEMS, ABST., SIGNIFICANCE OF,
B 6004 702* MODERN SUBMARINE CANYONS AND FANS, ABST., THIN-BEDDED,
B 6206 912† MODERN SUBMARINE FANS: CHARACTERS FOR RECOGNITION OF SANDY
M 25 4† MODERNIZATION, THE MACROPROBLEM OF
B 6105 778* **MODES OF EMPLACEMENT OF PROSPECTIVE HYDROCARBON RESERVOIR ROCKS**
M 26 259 MODIFICATION, CREEP
B 6405 706* MODIFICATION OF LINEAR SAND RIDGES BY BED-FORM MIGRATION—
M 26 263 MODIFICATION, SLIDE
M 26 259 MODIFICATION, SLUMP
M 26 263 MODIFICATIONS, STRUCTURAL, EROSION
B 6007 1073 MODIFIED LACHENBRUCH-BREWER EQUATION
B 6203 345 MODOC PLATEAU, CALIFORNIA
M 26 30 MODULI, LABORATORY, MEASUREMENT, DYNAMIC
M 26 30 MODULI, LABORATORY, MEASUREMENT, STATIC
M 28 125 MOENKOPI FORMATION, ARIZONA, TRIASSIC
B 6405 751 MOFFAT, I. W., GEOMETRY AND MECHANISMS OF TRANSVERSE
B 6303 429* MOGO MOUNTAINS, NORTHERN TUNISIA, ABST., ALLOCHTHONOUS
B 6405 688* MOGOLLON RIM GRAVELS OF ARIZONA, ABST., GENETIC STRATIGRAPHY
B 6304 688 MOGROCAMPERO, ANTONIO, RADON EMANATION OVER OREBODY, HAS
M 25 214 MOHAKATINO FORMATION, NEW ZEALAND
B 6301 90 MOHICAN FORMATION, SCOTIAN BASIN
S 11 124 MOHNIAN AND DELMONTIAN STAGES, CALIFORNIA MIOCENE,
S 11 126 MOHNIAN, CALIFORNIA, NONION MULTICAMERATUM
B 6211 2221 MOHNIAN PALEOGEOGRAPHY, CALIFORNIA
S 11 28 MOHNIAN STAGE
S 11 67 MOHNIAN STAGE
B 6211 2246 MOHNIAN STAGE, CALIFORNIA
B 6202 259 MOHNIAN STAGE, CALIFORNIA
S 11 37 MOHNIAN STAGE, ZONATION
B 6104 603 MOJADO FORMATION, NEW MEXICO
B 6203 345 MOJAVE DESERT, CALIFORNIA
M 25 214 MOKAU FORMATION, NEW ZEALAND
B 6305 835 MOKLESTAD, TOM C., YELLOW CREEK FIELD, UINTA COUNTY, WYOMING,
B 6412 2100 MOL MEMBER, IRAN
B 6408 1172 MOLAS CARBONATE ROCK, UTAH AND, COLORADO
B 6411 1968 MOLASSE BASIN, NORTHERN, SWITZERLAND
B 6303 467* MOLASSE BASIN, SOUTHERN GERMANY, ABST., ORIGIN AND
M 27 81 MOLASSE, GERMANY, OLIGOCENE
B 6411 1963 MOLASSE ZONE, AUSTRIA
B 6310 1918 MOLASSE ZONE, AUSTRIA
B 6110 1639 MOLASSE ZONE, AUSTRIA
B 6210 1909 MOLASSE ZONE, AUSTRIA
M 27 220 MOLDIC POROSITIES
M 27 219 MOLDIC POROSITIES
M 27 221 MOLDIC POROSITIES
B 6002 215 MOLECULAR SOLUTION HYPOTHESIS
B 6002 232 MOLECULAR SOLUTION, MODEL OF, PRIMARY, PETROLEUM, MIGRATION
B 6405 752 MOLENAAR, C. M., CRETACEOUS STRATIGRAPHY, CHIGNIK AREA,
B 6405 752 MOLENAAR, C. M., STRATIGRAPHIC RELATIONS OF NANUSHUK GROUP,
B 6201 1 MOLLOY, F. A., HYDROCARBON HABITAT OF TERTIARY NIGER DELTA
S 11 223 MOLLUSCA, NUMBER OF, HUMBOLDT BASIN, CALIFORNIA
B 6305 828* MOLLUSCA TO PALEOENVIRONMENTAL INTERPRETATION AND
S 4 195 MOLLUSCA, YUCATAN
B 6203 563* MOLLUSCAN BIOCLASTIC ACCUMULATIONS IN MARINE MUDSTONES OF
B 6405 748* MOLLUSCAN BIOSTRATIGRAPHY OF GULF OF ALASKA TERTIARY
B 6203 491* MOLLUSCAN FAUNA, CAPE BLANCO, OREGON, ABST., COMMUNITY
M 28 40 MOLLUSCAN, FRAGMENTS
S 11 186 MOLLUSCAN LOCALITIES, CALIFORNIA
B 6012 2174† MOLLUSCAN STAGES OF OREGON AND WASHINGTON, ABST., NEOGENE
S 4 195† MOLLUSK DISTRIBUTION, NORTHEAST YUCATAN COAST, MEXICO,
B 6403 444* MOLLUSK DISTRIBUTION PATTERNS IN NORTHERN GULF OF ALASKA,
M 27 81 MOLLUSKS
S 4 271 MOLLUSKS, BORERS, BIVALVED
M 27 83 MOLLUSKS, OSTREID
M 27 84 MOLLUSKS, RUDISTID
B 6207 1216 MOLNIA, B. F., SEAFLOOR INSTABILITY ON PACIFIC CONTINENTAL
B 6403 447 MOLNIA, BRUCE F., BENTHIC FORAMINIFERS FROM EASTERN GULF OF
B 6203 536 MOLNIA, BRUCE F., CARBONATE CATCHMENT AND EXCHANGE BASINS IN
B 6105 815 MOLNIA, BRUCE F., CLAY MINERALOGY OF EASTERN GULF OF ALASKA,
B 6405 753 MOLNIA, BRUCE F., DEPOSITIONAL ENVIRONMENT OF CLAY MINERALS
B 6012 2176 MOLNIA, BRUCE F., DISPERSAL, DISTRIBUTION, AND THICKNESS OF
B 6203 545 MOLNIA, BRUCE F., HOLOCENE SEDIMENT VOLUME AND MODERN
B 6207 1228 MOLNIA, BRUCE F., IMPACT OF SHORELINE PROCESSES ON RESOURCE
B 6004 655 MOLNIA, BRUCE F., PRELIMINARY STUDY OF SEA-FLOOR INSTABILITY IN
B 6403 446 MOLNIA, BRUCE F., QUATERNARY SEDIMENTARY FACIE ON
B 6204 633 MOLNIA, BRUCE F., SURFACE SEDIMENTARY UNITS OF NORTHERN GULF
B 6403 446 MOLNIA, BRUCE F., TWENTIETH-CENTURY HISTORY OF GULF OF ALASKA
M 25 242 MOLUCCAS GEOLOGIC PROVINCE, INDONESIA
M 25 554 MOLYBDENU DEPOSITS, ECUADOR, FIERRO URCO, COPPER, ZINC, LEAD,
M 25 485 MOLYBDENUM, CANADA, WESTERN, COPPER AND
M 25 553 MOLYBDENUM DEPOSIT, ECUADOR, CHAUCHA, COPPER-
M 25 554 MOLYBDENUM DEPOSITS, ECUADOR, FIERRO URCO, COPPER, ZINC, LEAD,
M 25 554 MOLYBDENUM DEPOSITS, ECUADOR, LOS LINDEROS, COPPER AND
M 25 554 MOLYBDENUM DEPOSITS, ECUADOR, RIO PLAYAS, COPPER AND
M 25 486 MOLYBDENUM DEPOSITS, IGNEOUS-TECTONIC PROVINCES
M 25 449 MOLYBDENUM DEPOSITS, ZONAL DISTRIBUTION, JAPAN, TUNGSTEN AND
M 25 553 MOLYBDENUM, ECUADOR, CHAUCHA DEPOSIT, COPPER-
M 25 554 MOLYBDENUM, ECUADOR, LOS LINDEROS DEPOSIT, COPPER-
M 25 554 MOLYBDENUM, ECUADOR, RIO PLAYAS, COPPER-
M 25 533 MOLYBDENUM, PERU
M 25 505 MOLYBDENUM, PRODUCTION AND RESERVES, CANADIAN, CORDILLERA
M 25 444 MOLYBDENUM, SOUTH KOREA
B 6207 1229* MOMOTOMBO GEOTHERMAL FIELD, NICARAGUA, ABST., GEOLOGY AND
B 6207 1207* MOMOTOMBO GEOTHERMAL PROJECT, NICARAGUA, ABST.,
M 25 164 MOMOTOMBO GEOTHERMAL PROSPECT, NICARAGUA
B 6303 497 MOMPER, JAMES A., DOMESTIC OIL RESERVES FORECASTING METHOD
B 6405 753 MOMPER, JAMES A., GENERATION OF ABNORMAL PRESSURES THROUG
B 6303 497 MOMPER, JAMES A., GEOCHEMICAL EXPLORATION IN POWDER RIVER
B 6408 1279 MOMPER, JAMES A., OIL EXPULSION—A CONSEQUENCE OF OIL
B 6203 545 MOMPER, JAMES A., OIL MIGRATION LIMITATIONS SUGGESTED BY
B 6304 682 MOMPER, JAMES A., REVIEW OF COMPACTION AND FLUID MIGRATION,
B 6309 1548 MOMPER, JAMES A., REVIEW OF PETROLEUM FORMATION AN
B 6104 573 MONAGHAN, PATRICK H., CALCULATION OF AGES OF HYDROCARBONS
M 25 418 MONAZITE, COMPOSITION AND USES
B 6405 762* MONDAK MISSISSIPPIAN OIL FIELD, WILLISTON BASIN, ABST.
B 6307 1093 MONDARRA FIELD, AUSTRALIA
B 6207 1228 MONGER, J. W. H., EVOLUTION OF CANADIAN SEGMENT OF CORDILLERA,
B 6211 2352 MONGER, J. W. H., MESOZOIC CONSTRUCTION OF CORDILLERAN
B 6002 327 MONLEY, LAWRENCE E., BLOCK 16 FIELD, WARD COUNTY, TEXAS, ABST.
S 7 273 MONOCLINAL FOLD, BOTTOM, DEFORMATION
B 6009 1616* MONROE COUNTY, MISSISSIPPI, ABST., FOURMILE CREEK AND, SPLUNGE
B 6209 1759 MONROE, JANE, DISTRIBUTION OF LIVING BENTHONIC FORAMINIFERA
B 6105 778 **MONTADERT, L., ET AL, MICROCONTINENTAL MARGINS**
B 6303 527 MONTADERT, L., RECENT ADVANCES IN PASSIVE-MARGIN RESEARCH,
M 29 293 MONTADERT, L., VARIETY OF MARGINS AND DEEP BASINS IN THE
M 29 1 **MONTADERT, LUCIEN, EDITED BY, GEOLOGICAL AND GEOPHYSICAL**
M 29 1 **MONTADERT, LUCIEN, FOREWORD, GEOLOGICAL AND GEOPHYSICAL**
B 6204 635 MONTAGUE ISLAND, ALASKA
M 28 87 MONTALLEGRO FORMATION, ITALY, PLEISTOCENE
B 6104 594 MONTALVO FIELD, CALIFORNIA
B 6303 456* MONTANA, ABST., AUTHIGENIC, TOURMALINE CRYSTALS IN PORE SPACE
B 6305 826* MONTANA, ABST., COAL VARIATIONS IN FLUVIAL DEPOSITION O,
B 6008 1412* MONTANA, ABST., CORWIN SPRINGS KNOWN, GEOTHERMAL RESOURCE
B 6008 1400* MONTANA, ABST., CRAZY MOUNTAINS AREA--GEOLOGY, EXPLORATION
B 6405 773* MONTANA, ABST., DEVELOPMENT OF BIOGENIC GAS FROM SHALLOW,
B 6405 712* MONTANA, ABST., DIAGENESIS IN SHALLOW CONVENTIONAL AND
B 6405 757* MONTANA, ABST., DISTRIBUTION, DIAGENESIS, AND DEPOSITIONAL
B 6008 1408* MONTANA, ABST., EAST KEITH GAS AREA, LIBERTY COUNTY
B 6008 1398* MONTANA, ABST., ENTRAPMENT OF GAS IN UPPER, CRETACEOU, EAGLE
B 6303 497* MONTANA, ABST., GEOCHEMICAL EXPLORATION IN, POWDER RIVER

B 6008 1393* MONTANA, ABST., GEOLOGIC CONTROLS OVER HOT WATER MIGRATION
B 6405 676* MONTANA, ABST., HYDRODYNAMIC FLOW IN LOWER, CRETACEOUS,
B 6008 1410* MONTANA, ABST., JIM COULEE-, LITTLE WALL FIELD, MUSSELSHELL
B 6405 791* MONTANA, ABST., LATE, MESOZOIC TO EARLY, CENOZOIC FORELAND
B 6405 777* MONTANA, ABST., MECHANISMS CONTROLLING POROSITY IN, RED RIVER,
B 6008 1408* MONTANA, ABST., MESAVERDE GAS POTENTIAL IN, DRY CREEK AREA,
B 6303 468* MONTANA, ABST., MICROFOSSILS, MACROFOSSILS, AND STROMATOLITES
B 6008 1391* MONTANA, ABST., PALEOGEOMORPHOLOGY OF, FRED AND, GEORGE
B 6305 827* MONTANA, ABST., POST-DEPOSITIONAL CONTROL OF, GAS-, RESERVOIR
B 6008 1411* MONTANA, ABST., PROPOSED, GEOTHERMAL CIRCULATION PATTERN,
B 6008 1401* MONTANA, ABST., STRUCTURAL INTERPRETATION AND POROSITY
B 6305 836* MONTANA, ABST., STRUCTURAL PATTERN OF EASTERN PART O,
B 6004 727* MONTANA, ABST., SUBTIDAL, TRACE, FOSSILS, MUDDY AND, FRONTIER
B 6405 674* MONTANA, ABST., VERTICAL AN SURFACE, SEISMIC PROFILES MAP,
B 6008 1393* MONTANA, ABST., WELL-LOG ANALYSIS IN, CRETACEOUS, GAS SAND IN
B 6403 318 MONTANA AND CANADA, CRETACEOUS, UPPER, CORRELATION CHART
B 6201 112 MONTANA AND IDAHO, IDAHO BATHOLITH
B 6008 1403* MONTANA AND NORTHEASTERN WYOMING, ABST., RECENT
B 6405 783* MONTANA AND SOUTH DAKOTA, ABST., GEOMETRY OF SHELF
B 6405 804* MONTANA AND WESTERN NORTH DAKOTA, ABST., PROPOSED MODEL
B 6305 833* MONTANA AND WYOMING, ABST., WRENCH, FAULTING AND
B 6403 325 MONTANA, ARDMORE BENTONITE BED
B 6403 317 MONTANA, BEARPAW MOUNTAINS
B 6407 972 MONTANA, BEARPAW MOUNTAINS, SECTION
B 6403 318 MONTANA, BEARPAW SHALE
M 26 432 MONTANA, BELL CREEK FIELD, MUDDY FORMATION
B 6201 112 MONTANA, BOULDER BATHOLITH
B 6407 972 MONTANA, BOWDOIN DOME, SECTION
B 6407 981 MONTANA, BOWDOIN FIELD
B 6407 984 MONTANA, BOWDOIN FIELD, SECTION
B 6403 332 MONTANA, BULLWACKER FIELD, SECTION
M 27 154 MONTANA, CAMBRIAN, MEAGHER FORMATION
M 27 6 MONTANA, CAMBRIAN, MEAGLER LIMESTONE
B 6407 973 MONTANA, CAT CREEK ANTICLINE
B 6407 973 MONTANA, CEDAR CREEK ANTICLINE
B 6407 973 MONTANA, CENTRAL MONTANA UPLIFT
B 6403 325 MONTANA, CLAGGETT SHALE
B 6403 316† MONTANA, COASTAL AND, DELTAIC SEDIMENTATION OF UPPER,
M 28 70 MONTANA, CRETACEOUS, BLACK LEAF FORMATION
B 6008 1207 MONTANA, DEVELOPMENTS, 1975
B 6108 1348 MONTANA, DEVELOPMENTS, 1976
B 6208 1500 MONTANA, DEVELOPMENTS, 1977
B 6308 1386 MONTANA, DEVELOPMENTS, 1978
B 6409 1501 MONTANA, DEVELOPMENTS, 1979
B 6403 337 MONTANA, EAGLE SANDSTONE, MAJOR RESERVOIR SANDSTONES
B 6211 2277 MONTANA, ELLIS GROUP
B 6407 972 MONTANA, FLOOD MEMBER
B 6403 318 MONTANA, FOX HILLS SANDSTONE
B 6407 972 MONTANA, GLACIER NATIONAL PARK, CRETACEOUS
B 6403 318 MONTANA, HELL CREEK FORMATION
B 6403 318 MONTANA, JUDITH RIVER FORMATION
B 6403 331 MONTANA, LEROY FIELD, CROSS SECTION
B 6407 973 MONTANA, LITTLE ROCKY MOUNTAINS
M 27 164 MONTANA, MISSISSIPPIAN, LODGEPOLE LIMESTONE
M 27 27 MONTANA, MISSISSIPPIAN, LODGEPOLE LIMESTONE
M 28 40 MONTANA, MISSISSIPPIAN, LODGEPOLE LIMESTONE
M 27 26 MONTANA, MISSISSIPPIAN, LODGEPOLE LIMESTONE
B 6211 2277 MONTANA, MORRISON FORMATION
B 6407 973 MONTANA, NESSON ANTICLINE
B 6403 318 MONTANA, NIOBRARA FORMATION
B 6403 318 MONTANA, PARKMAN SANDSTONE MEMBER
B 6211 2277 MONTANA, PIPER FORMATION
B 6008 1213 MONTANA, PRODUCTION, 1974-1975
B 6108 1354 MONTANA, PRODUCTION, 1974-1976
B 6208 1507 MONTANA, PRODUCTION, 1974-1977
B 6308 1391 MONTANA, PRODUCTION, 1975-1978
B 6409 1506 MONTANA, PRODUCTION, 1975-1979
B 6201 99 MONTANA, RETORT PHOSPHATIC SHALE MEMBER
B 6211 2277 MONTANA, RIERDON FORMATION
B 6403 317 MONTANA, SHERARD FIELD
B 6407 972 MONTANA, ST. MARY RIVER FORMATION
B 6407 973 MONTANA, SWEETGRASS HILLS
B 6211 2277 MONTANA, SWIFT FORMATION
B 6403 319 MONTANA, TELEGRAPH CREEK FORMATION
B 6403 317 MONTANA, TIGER RIDGE FIELD
B 6407 981 MONTANA, TIGER RIDGE FIELD
B 6305 828* MONTANA TO NEVADA, ABST., MISSISSIPPIAN, CARBONATE, SHELF,
B 6405 716* MONTANA TO NEVADA, ABST., MISSISSIPPIAN SHELF MARGIN AND
B 6407 972 MONTANA, TWO MEDICINE FORMATION
B 6405 799* MONTANA, UTAH, AND WYOMING, ABST., MIDDLE-LATE, PERMIAN
B 6407 972 MONTANA, VAUGHN MEMBER
B 6403 319 MONTANA, VIRGELLE SANDSTONE MEMBER, FIVE LITHOFACIES
B 6403 324 MONTANA, VIRGELLE SANDSTONE MEMBER, UPPER MEMBER
B 6403 320 MONTANA, VIRGELLE SANDSTONE, MIDDLE MEMBER
B 6407 972 MONTANA, WILLISTON BASIN, CRETACEOUS
B 6211 2277 MONTANA, WILLISTON BASIN, JURASSIC
B 6407 972 MONTANA, WILLOW CREEK FORMATION
B 6403 317 MONTANA, WINIFRED FIELD
B 6008 1408* MONTANA--AID FOR HYDROCARBON EXPLORATION, ABST.,
B 6305 838* MONTANA--EXAMPLE OF SHALLOW BIOGENIC GAS PRODUCTION FROM
B 6008 1402* MONTANA--FOOD FOR THOUGHT, ABST., SOUTHWEST RAGGED POINT
B 6008 1395* MONTANA--LOW-PERMEABILITY, SANDY-SHALE, BIOGENIC-GAS
B 6203 540* MONTANAS COAL—IMPORTANT ENERGY RESOURCE, ABST.
B 6405 760* MONTANA—EXAMPLE OF SHALLOW BIOGENIC GAS PRODUCTION FROM
B 6405 772* MONTANA—EXAMPLE OF THIN, WIDESPREAD STORM-GENERATED
B 6105 834* MONTANA—STRUCTURAL AND LITHOLOGIC CONTROLS, ABST.,
S 4 176 MONTASTREA ANNULARIS AND M. CAVERNOSA, BELIZE
S 4 176 MONTASTREA ANNULARIS AND M. CAVERNOSA, PANAMA
S 4 171† MONTASTREA ANNULARIS AND MONTASTREA CAVERNOSA--LONG-TERM
S 4 175 MONTASTREA ANNULARIS, BAHAMAS
S 4 175 MONTASTREA ANNULARIS, BARBADOS
S 4 175 MONTASTREA ANNULARIS, BRITISH HONDURAS
S 4 175 MONTASTREA ANNULARIS, FLORIDA
S 4 175 MONTASTREA ANNULARIS, JAMAICA
S 4 175 MONTASTREA ANNULARIS, KEY WEST
S 4 163 MONTASTREA ANNULARIS, REDUCTION
S 4 44 MONTASTREA ASSEMBLAGE, BELIZE
S 4 171† MONTASTREA CAVERNOSA--LONG-TERM GROWTH DATA AS
S 4 33 MONTASTREA-PORITES ASSEMBLAGE, WEST FLOWER GARDEN BANK,
B 6206 972 MONTCANYON II, CALIFORNIA, SCRIPPS CRUISE
M 27 185 MONTE ANTOLA FORMATION, ITALY, CRETACEOUS
M 27 200 MONTE ANTOLA FORMATION, ITALY, CRETACEOUS
M 28 162 MONTE ANTOLA FORMATION, ITALY, CRETACEOUS
M 27 148 MONTE ANTOLA FORMATION, ITALY, CRETACEOUS
M 28 22 MONTE ANTOLA FORMATION, ITALY, CRETACEOUS
M 28 155 MONTE ANTOLA FORMATION, ITALY, CRETACEOUS
M 28 152 MONTE ANTOLA FORMATION, ITALY, CRETACEOUS
M 28 38 MONTE ANTOLA FORMATION, ITALY, CRETACEOUS
M 28 45 MONTE ANTOLA FORMATION, ITALY, CRETACEOUS
M 27 143 MONTE ANTOLA FORMATION, ITALY, CRETACEOUS
M 28 73 MONTE ANTOLA FORMATION, ITALY, CRETACEOUS
B 6004 666* MONTE ANTOLA FORMATION, NORTHERN APENNINES, ITALY, ABST.,
M 6004 726* MONTE CARLO METHOD FOR ESTIMATING RESERVES IN GREGOIRE LAKE
B 6312 2188 MONTE CARLO SUMMATION OF PLAY ASSESSMENTS
B 6204 644 MONTEAGLE LIMESTONE, ALABAMA
B 6204 644 MONTEAGLE LIMESTONE, GEORGIA
B 6204 645 MONTEAGLE LIMESTONE, SIX LITHOFACIES, CUMBERLAND PLATEAU,
B 6204 644 MONTEAGLE LIMESTONE, TENNESSEE
B 6105 793* MONTEAGLE LIMESTONE, UPPER MISSISSIPPIAN,—CASE FOR OOLITIC
B 6305 845* MONTEAGLE LIMESTONE, UPPER MISSISSIPPIAN, OF MORGAN COUNTY,
B 6204 644† MONTEAGLE LIMESTONE, UPPER MISSISSIPPIAN—OOLITIC TIDAL-BAR
B 6203 457 MONTEBELLO ISLAND, AUSTRALIA
B 6203 362 MONTEBELLO OIL FIELD, CALIFORNIA
B 6201 44 MONTEBELLO TROUGH, AUSTRALIA
M 25 189 MONTECCHI, PETER A., SOME SHALLOW TECTONIC CONSEQUENCES O
B 6012 2188* MONTEREY BAY, CALIFORNIA, ABST., HOLOCENE SHELF SEDIMENTATION
B 6403 440* MONTEREY BAY, CALIFORNIA, ABST., MODERN AND ANCIENT COASTAL
B 6211 2220 MONTEREY BAY FAULT ZONE, CALIFORNIA
B 6206 968 MONTEREY CANYON, CALIFORNIA
S 8 87 MONTEREY CANYON, CALIFORNIA, SUBMARINE CANYONS
S 8 90 MONTEREY CANYON, PROGRESSIVE VECTOR DIAGRAMS
S 8 93 MONTEREY CANYON, TIME-VELOCITY CURVES, CURRENT FLOW
S 8 82 MONTEREY CANYON, TOPOGRAPHY, INDEX MAP
B 6012 2181* MONTEREY COUNTY, CALIFORNIA, ABST., TERTIARY STRATIGRAPHIC
B 6206 967† MONTEREY DEEP-SEA FAN, CALIFORNIA, CHANNEL, SANDS AND,
B 6206 918 MONTEREY FAN, ASCENSION VALLEY, CALIFORNIA
B 6206 968 MONTEREY FAN, CALIFORNIA
B 6206 974 MONTEREY FAN, RESERVOIR BEDS AND TRAPS, CALIFORNIA
B 6206 970 MONTEREY FAN, SEISMIC REFLECTION DATA, CALIFORNIA
B 6211 2219 MONTEREY FORMATION, CALIFORNIA
B 6111 2002 MONTEREY FORMATION, CALIFORNIA
B 6309 1586* MONTEREY FORMATION, CALIFORNIA, ABST., DILATION,
M 28 142 MONTEREY FORMATION, CALIFORNIA, MIOCENE
S 11 9 MONTEREY FORMATION, FACIES PROBLEMS
S 11 91 MONTEREY FORMATION, NAPLES, CALIFORNIA, CHECK LIST OF,
B 6403 449* MONTEREY FORMATION, PISMO SYNCLINE, CALIFORNIA, ABST.,
B 6405 790* MONTEREY FORMATION, PISMO SYNCLINE, COAST RANGES OF
B 6403 444* MONTEREY ROCKS ALONG SANTA BARBARA COAST, CALIFORNIA, ABST.
B 6202 259 MONTEREY SHALE, CALIFORNIA
M 27 26 MONTEREY SHALE, CALIFORNIA, MIOCENE

M 28 81 MONTEREY SHALE, CALIFORNIA, MIOCENE
M 28 81 MONTEREY SHALE, CALIFORNIA, PLIOCENE
S 11 94 MONTEREY SHALE, NAPLES, CALIFORNIA, CHECK LIST OF,
B 6303 473* MONTEREY SHALE, SANTA BARBARA COAST, CALIFORNIA, ABST.,
B 6405 770* MONTEREY SHALE, SANTA MARIA AREA, CALIFORNIA, ABST.,
B 6206 916 MONTEREY SUBMARINE FAN, CALIFORNIA
B 6405 788* MONTGOMERY COUNTY, KANSAS, ABST., PROFILE OF UNUSUAL, OOLITE
B 6203 545 MONTGOMERY, K. JOEL, URANIUM IN TUFFACEOUS FLUVIAL
M 30 67 MONTIAN, THANETIAN, FRIGG FIELD, PALEOGEOGRAPHY
M 28 144 MONTMORILLONITE COATING DETRITAL, CEMENTS
B 6303 483* MONTMORILLONITE, EFFECT OF SOLUTION CHEMISTRY, ABST., SILICA
M 28 66 MONTMORILLONITE-SMECTITE, AUTHIGENIC, CLAY
B 6002 224 MONTMORILLONITE-TO-ILLITE CLAY TRANSFORMATION
B 6104 603 MONTOYA DOLOMITE, NEW MEXICO
B 6303 469* MONTOYA GROUP, MIDDLE AN UPPER ORDOVICIAN, OF TRANS-PECOS
B 6109 1452 MONTOYA GROUP, PEDREGOSA BASIN
S 3 131 MONUMENT 2 MINE, ARIZONA
S 3 124 MONUMENT VALLEY, SHINARUMP CONGLOMERATE, UTAH
S 3 129 MONUMENT VALLEY, UTAH
M 24 119 MONUMENT-BLUE RIM ARCH, WYOMING
S 6 369 MOODIES SYSTEM, SOUTH AFRICA
B 6105 795 MOODY, J. D., PETROLEUM PROSPECTS OF CONTINENTAL SLOPES AND
B 6303 286 MOODY, J. D., PETROLEUM PROSPECT OF THE DEEP OFFSHORE
M 25 32 MOODY, J. D., TECTONIC FRAMEWORK OF PACIFIC REGION
B 6202 176 MOODY, JOHN D., ENERGY CRISIS—ENERGY SOLUTION: TWO CASE
B 6203 540* MOON AND PLANETS, ABST., TERRESTRIAL EVOLUTION—INSIGHTS FROM
B 6108 1373 MOON, J. LEE, RARE EARTH AND OTHER TRACE ELEMENTS AS
M 25 473 MOON, MARS, AND VENUS, RING, STRUCTURES, DISTINCT ON, EARTH
B 6002 240 MOONIE FIELD, AUSTRALIA
B 6105 815 MOORE, CLYDE H., JR., CARBONATE RESERVOIRS, ABST., DIAGENETIC
S 4 53 MOORE, CLYDE H., JR., DEEP FOREREEF AND UPPER ISLAND SLOPE,
B 6405 753 MOORE, CLYDE H., POST-COMPACTION, SUBSURFACE SECONDARY
B 6009 1607 MOORE, CLYDE H., WALKER CREEK FIELD--SMACKOVER DIAGENETIC
B 6109 1551 MOORE, CLYDE, PALEOENVIRONMENTAL ANALYSIS OF LOWER
B 6303 498 MOORE, CRAIG E., SPECULATIONS ON BASEMENT MOBILITY IN GULF OF
B 6004 699 MOORE, D. G., DYNAMIC PROCESSES OF UPPER BENGAL FAN AND
M 29 189 MOORE, D. G., TECTONICS OF THE ANDAMAN SEA AND BURMA
B 6303 498 MOORE, DAVID G., SUBMARINE SLIDES, ABST.
B 6004 699 MOORE, G. A., ORIENTED HYDROCARBON MIGRATION--RESULT OF
M 29 223 MOORE, G. F., STRUCTURE AND CENOZOIC EVOLUTION OF THE SUNDA
B 6408 1179 MOORE, G. T., ANOXIC ENVIRONMENTS AND OI SOURCE BED GENESIS
B 6405 697 MOORE, G. T., ANOXIC ENVIRONMENTS AND OIL SOURCE BEDS, ABST.
M 29 383 MOORE, G. T., INVESTIGATION OF MISSISSIPPI FAN, GULF OF MEXICO
B 6203 546 MOORE, G. T., MISSISSIPPI FAN, GULF OF MEXICO, ABST.
S 7 155 MOORE, G. T., MISSISSIPPI FAN, GULF OF MEXICO—PHYSIOGRAPHY,
B 6205 863 MOORE, GEORGE T., AVAILABILITY OF GEOLOGIC MATERIALS FROM
S 7 1 MOORE, GEORGE T., EDITED BY, FRAMEWORK, FACIES, AND OIL-
S 7 1 MOORE, GEORGE T., INTRODUCTION, FRAMEWORK, FACIES, AND OIL-
B 6304 660 MOORE, GEORGE T., MISSISSIPPI RIVER DELTA--APRIL 9, 1976--FROM
B 6401 133 MOORE, GEORGE T., REVIEW OF ARCYANA FAMOUS--PHOTOGRAPHIC
B 6103 466 MOORE, J. CASEY, MESOZOIC MAGMATISM AN ACCRETION, SOUTHWEST
B 6207 1228 MOORE, J. CASEY, PLATE-TECTONIC INTERPRETATION OF MESOZOIC AND
B 6207 1229 MOORE, JAMES L., JR., GEOLOGY AND DEVELOPMENT OF MOMOTOMBO
B 6010 1865 MOORE, QUENTIN M., PETROLEUM DEVELOPMENTS IN MIDDLE EAST
B 6210 2111 MOORE, QUENTIN M., PETROLEUM DEVELOPMENTS IN MIDDLE EAST
B 6110 1795 MOORE, QUENTIN M., PETROLEUM DEVELOPMENTS IN MIDDLE EAST
B 6310 1884 MOORE QUENTIN M., PETROLEUM DEVELOPMENTS IN MIDDLE EAST
B 6004 669 MOORE, R., NEW ENVIRONMENTAL MAPS FOR GULF OF MEXICO OCS
B 6304 691 MOORE, STEPHEN C., MAGNITUDE AND VARIABILITY OF
B 6303 498 MOORE, T. C., JR., IN SEARCH OF RESOLUTION, ABST.
B 6405 758 MOORE, THOMAS E., SEDIMENTOLOGY OF FLUVIAL UPPER DEVONIAN
B 6405 753 MOORE, W. RICHARD, POWDER RIVER BASIN HIGH AND ITS
M 25 34 MOOREA VOLCANIC ISLAND, PACIFIC REGION
M 24 242 MOORINGSPORT PRODUCING SAND, MISSISSIPPI
B 6009 1621* MOORMAN SYNCLINE AND ROME TROUGH AND THEIR RELATION TO
B 6408 1132 MOQUAWKIE GAS FIELD, ALASKA
M 25 338 MOQUEGUA BASIN, PERU
B 6203 353 MORALES AND TAYLOR CANYON TREND, CALIFORNIA
B 6402 189 MORALES FORMATION, SPAIN
B 6303 412 MORALES, R. G., NAMORAD FIELD, MAJOR OIL DISCOVERY IN CAMPOS
B 6303 412 MORALES, ROBERTO GAMARRA, OIL EXPLORATION IN CAMPOS BASIN,
B 6011 2038 MORAN, ANDREW I., ALLOCHTHONOUS CARBONATE DEBRIS IN
B 6403 428 MORAN FIELD, INDIA
B 6402 219 MORAN FIELD, INDIA, NAHAR-KATIYA AND
B 6308 1421 MORAN, W. R. REVIEW OF ATLAS OF ECONOMIC MINERAL DEPOSITS, BY
B 6407 1119 MORAN, W. R., REVIEW OF CRITERIA IN CORRELATION—RELEVANT
B 6410 1725 MORAN, W. R., REVIEW OF INTRODUCTION TO EXPLORATION
B 6404 585 MORAN, W. R., REVIEW OF VOLCANOLOGY, BY HOWEL WILLIAMS AND
M 30 143 MORAY FIRTH BASIN, CALLOVIAN, NORTH SEA
M 30 134 MORAY FIRTH BASIN, MESOZOIC HISTORY, NORTH SEA
M 30 143 MORAY FIRTH BASIN, OXFORDIAN, NORTH SEA
M 30 167 MORAY FIRTH BASIN, PIPER FIELD, DEVELOPMENT DRILLING
M 30 165 MORAY FIRTH BASIN, PIPER FIELD, DEVELOPMENT GEOLOGY
M 30 133 MORAY FIRTH BASIN, PIPER FIELD, NORTH SEA
M 30 162 MORAY FIRTH BASIN, PIPER FIELD, RESERVOIR PERFORMANCE
M 30 159 MORAY FIRTH BASIN, PIPER FIELD, SOURCE ROCK
M 30 132 MORAY FIRTH BASIN, PIPER FIELD, TECTONIC SETTING
M 30 157 MORAY FIRTH BASIN, PIPER SANDSTONE, PETROGRAPHY
M 30 143 MORAY FIRTH BASIN, PIPER SANDSTONE, STRATIGRAPHY
M 30 119 MORAY FIRTH BASIN, REGIONAL STRATIGRAPHY, NORTH SEA
M 30 123 MORAY FIRTH BASIN, REGIONAL STRUCTURE, NORTH SEA
M 30 133 MORAY FIRTH BASIN, RIFTING, NORTH SEA
M 30 139 MORAY FIRTH BASIN, TERTIARY HISTORY, NORTH SEA
B 6203 426 MORAY FIRTH GRABEN, EUROPE
B 6303 487* MORAY FIRTH, NORTH SEA, ABST., BEATRIC FIELD
M 30 120 MORAY FIRTH, REGIONAL GEOLOGY MAP, NORTH SEA
M 30 123 MORAY FIRTH, STRATIGRAPHIC COLUMN, NORTH SEA
M 30 117† MORAY FIRTH, U.K. NORTH SEA, THE, BEATRICE FIELD, INNER
M 28 139 MORDENITE, IN SEDIMENTS, SEM
B 6203 435 MORE-LOFOTEN AREA RIFT SYSTEM, NORTH SEA
B 6109 1537* MOREAU CHENIER-BEACH SYSTEM, ABST., SEDIMENT-DISPERSAL TRENDS
S 4 17 MORELOCK, JACK, SHELF REEFS, SOUTHWESTERN PUERTO RICO
B 6211 2261 MORENO FORMATION, CALIFORNIA
B 6212 2511 MOREY, PHILIP G., ECONOMICS OF LARGE-SCALE SURFACE COAL
B 6305 845* MORGAN COUNTY, OHIO, ABST., PETROLEUM GEOLOGY OF, WHIRLPOOL
B 6305 845* MORGAN COUNTY, TENNESSEE, ABST., GEOLOGY AND PRODUCTION
B 6211 2354 MORGAN, GEORGE, MESOZOIC GEOLOGY OF PENINSULAR CALIFORNIA
B 6204 609 MORGAN, J. T., TENSLEEP RESERVOIR, OREGON BASIN FIELD, WYOMING
B 6105 816 MORGAN, J. T., TENSLEEP RESERVOIR STUDY, OREGON BASIN FIELD,
B 6307 1122 MORITA HOGBACK, ARIZONA
B 6101 80 MOROCCAN MESETA
B 6101 85 MOROCCAN MESETA, BOUFEKRANE BASIN
B 6407 989 MOROCCAN MESETA, MOROCCO
B 6101 85 MOROCCAN MESITA, PRE-RIF, DAHAR NSOUR, BASIN
B 6010 1770 MOROCCO, 1975, ALEM, A., AND, PETROCONSULTANTS S.A., ON
B 6110 1701 MOROCCO, 1976, ALEM, A., AND, PETROCONSULTANTS S.A., ON
B 6210 1799 MOROCCO, 1977, ALEM, AMID, AND, PETROCONSULTANTS S.A., ON
B 6310 1639 MOROCCO, 1978, ALEM, AMID, AND, PETROCONSULTANTS S.A., ON
B 6411 1742 MOROCCO, 1979, PETROCONSULTANTS S.A., ON DEVELOPMENTS IN
B 6202 230 MOROCCO, AAIUM BASIN, EVAPORITES
B 6004 725* MOROCCO, ABST., CARBONATE, PETROLOGY O, BASIN-TO-, REEF,
B 6407 995 MOROCCO, AGLELGAL SANDSTONE MEMBER
B 6407 993 MOROCCO, AMESKOUDE GRABEN
B 6407 992 MOROCCO, AMZRI FLEXURE
B 6101 79† MOROCCO AND EASTERN NORTH AMERICA--COMPARISON, TRIASSIC-,
B 6407 988† MOROCCO, AND THEIR REGIONAL STRUCTURAL IMPLICATIONS, TRIASSIC
B 6407 992 MOROCCO, ARGANA FAULT
B 6407 993 MOROCCO, ARGANA GRABEN
B 6407 998 MOROCCO, ARGANA GRABEN
B 6101 84 MOROCCO, ATLAS TREND, HIGH
B 6101 84 MOROCCO, ATLAS TREND, MIDDLE
B 6407 990 MOROCCO, BICOUDINE FORMATION
B 6407 996 MOROCCO, BIGOUDINE FORMATION
B 6407 992 MOROCCO, DEVONIAN LIMESTONES
B 6101 83 MOROCCO, DOUKKALA-BERRICHID-RHARB BASINS
B 6407 991 MOROCCO, DRISS RIVER CONGLOMERATE MEMBER
B 6412 2056 MOROCCO, DSDP SITE 370, LEG 41, GEOCHEMICAL LOG
B 6101 86 MOROCCO, ESSAOUIRA BASIN
B 6202 228 MOROCCO, ESSAOUIRA BASIN, SEDIMENTARY SEQUENCE
B 6202 230 MOROCCO, EVAPORITES
B 6407 996 MOROCCO, HASSEINE MUDSTONE MEMBER
B 6407 1000 MOROCCO, HASSI EL BEGUER FAULT
B 6407 992 MOROCCO, IFERD FAULT
B 6407 991 MOROCCO, IKAKERN FORMATION
B 6407 995 MOROCCO, IROHALENE MUDSTONE MEMBER
B 6407 992 MOROCCO, JEBEL AMRHOUZ
S 4 268 MOROCCO, JEBEL ASSAMEUR NAIT FERGANE, CORALS, BORERS
S 4 265 MOROCCO, JEBEL BOU KENDILL, CORAL, BORERS
B 6101 83 MOROCCO, KHEMISSET BASIN
B 6407 992 MOROCCO, MENIZLA INLIER
B 6104 596 MOROCCO, MO-2 WELL
B 6104 594 MOROCCO, MO-2 WELL
B 6104 596 MOROCCO, MO-3 WELL
B 6407 989 MOROCCO, MOROCCAN MESETA
B 6004 667* MOROCCO, NORTHWEST AFRICA, ABST., DEVELOPMENT AND,
B 6107 964 MOROCCO, OFFSHORE, MARGINAL BASIN, SECTION
B 6110 1729 MOROCCO, PRODUCTION, 1976
B 6210 1826 MOROCCO, PRODUCTION, 1976-1977
B 6310 1663 MOROCCO, PRODUCTION, 1978

B 6411 1771 MOROCCO, PRODUCTION, 1979
B 6407 1000 MOROCCO, SIDI AMARA FAULT
B 6407 996 MOROCCO, SIDI MANSOUR MUDSTONE
B 6407 989 MOROCCO, SOUTH ATLAS FAULT
B 6407 989 MOROCCO, SOUTHWESTERN, GEOLOGIC MAP
B 6407 992 MOROCCO, STEPHANIAN-AUTUNIAN CLASTICS
B 6407 996 MOROCCO, TADRART OUADOU SANDSTONE MEMBER
B 6407 994 MOROCCO, TANAMEURT VOLCANICLASTIC MEMBER
B 6202 228 MOROCCO, TARFAYA BASIN, SEDIMENTARY SEQUENCE
B 6407 992 MOROCCO, TIMESGADIOUINE FAULT
B 6407 990 MOROCCO, TIMESGADIOUINE FORMATION
B 6407 994 MOROCCO, TIMESGADIOUINE FORMATION
B 6407 993 MOROCCO, TIMESGADIOUINE HORST
B 6407 993 MOROCCO, TIRKOU HORST
B 6407 992 MOROCCO, TIRKOU INLIER
B 6407 993 MOROCCO, TIZI-MAACHOU GRABEN
B 6407 994 MOROCCO, TOURBIAIN CONGLOMERATE MEMBER
B 6101 83 MOROCCO, TRIASSIC-LIASSIC CORRELATIONS, TENTATIVE, NORTH
B 6101 82 MOROCCO, TRIASSIC-LIASSIC DEPOSITS
B 6407 999 MOROCCO, WEST ATLAS BASIN
B 6004 678* **MORPHOLOGIC VARIATION IN SPECIES AND PHENOTYPES OF PLANKTONIC**
B 6306 934† MORPHOLOGY AND STRUCTURE OF AMAZON UPPER CONTINENTAL
B 6206 1029† MORPHOLOGY AND STRUCTURE OF UPPER CONTINENTAL MARGIN OFF
B 6409 1553* MORPHOLOGY O TURBIDITE-CHANNEL RESERVOIRS, LOWER
M 29 226 MORPHOLOGY, SUMATRA, MARGIN
B 6405 713* MORPHOMETRY OF LATE ORDOVICIAN MICROBIAL BORINGS, ABST.
B 6207 1229 MORRIS, A. E. L., GEOLOGY AND PETROLEUM POTENTIAL OF CENTRAL
M 25 311 MORRIS, ANTHONY E. L., HYDROCARBON POTENTIAL OF OFFSHORE
B 6105 816 MORRIS, ROBERT C., HYDROCARBON POTENTIAL PREDICTED FOR
B 6108 1384 MORRIS, ROBERT C., PETROLOGY AND DIAGENESIS OF DEEP-WATER
B 6405 772 MORRIS, SIMON CONWAY, CONODONT ANIMAL—HYPOTHESES AND
B 6401 105 MORRIS-BUIE FIELD, TEXAS
B 6304 694* MORRISON FORMATION, ABST., TERTIARY OXIDATION IN; WESTWATER
S 3 94 MORRISON FORMATION, COLORADO AND NEW MEXICO
M 28 166 MORRISON FORMATION, COLORADO, JURASSIC
M 28 157 MORRISON FORMATION, COLORADO, JURASSIC
M 28 65 MORRISON FORMATION, COLORADO, JURASSIC
M 28 133 MORRISON FORMATION, COLORADO, JURASSIC
M 28 26 MORRISON FORMATION, COLORADO, JURASSIC
S 3 65 MORRISON FORMATION, COLORADO, SALT WASH SANDSTONE MEMBER
B 6407 971 MORRISON FORMATION, GREAT PLAINS
M 26 279 MORRISON FORMATION, JURASSIC
B 6211 2277 MORRISON FORMATION, MONTANA
B 6007 1010 MORRISON FORMATION, NEW MEXICO
B 6405 795* MORRISON FORMATION, SAN JUAN BASIN, AND APPLICATION OF
B 6405 795* MORRISON FORMATION, SOUTHERN SAN JUAN BASIN ABST.,
B 6304 689* MORRISON FORMATION, UPPER JURASSIC, CARRIZO MOUNTAINS,
B 6410 1718† MORRISON FORMATION, UPPER JURASSIC, GRANTS MINERAL BELT, NEW
B 6209 1673† MORRISON FORMATION, UPPER JURASSIC, GRANTS MINERAL BELT, NEW
B 6405 763* MORRISON FORMATION, UPPER JURASSIC, HENRY BASIN, UTAH, ABST.,
B 6305 837* MORRISON FORMATION, UPPER JURASSIC, OF SOUTH-CENTRA UTAH,
B 6405 796* MORRISON FORMATION, URAVAN MINERAL BELT, SOUTHWEST
M 28 56 MORRISON FORMATION, UTAH, JURASSIC
M 28 57 MORRISON FORMATION, UTAH, JURASSIC
B 6211 2277 MORRISON FORMATION, WYOMING
M 24 95 MORRISON FORMATION, WYOMING
B 6002 321 MORRISON, L. S., DEVELOPMENTS IN SOUTH-CENTRAL OKLAHOMA, ABST.
B 6405 754 MORRISON, LAWRENCE S., OIL PRODUCTION FROM FRACTURED CHERTS
B 6403 442 MORRISON, LOWELL R., ET AL, INTERMARGIN OF SOUTHERN
B 6104 595 MORROW COUNTY FIELD, OHIO
B 6311 2119* MORROW FAN-DELTA DEPOSITS OF ANADARKO BASIN, ABST., UPPER
B 6311 2119* MORROW FORMATION OF ANADARKO BASIN, ABST., DELTAIC DEPOSITS
B 6305 808 MORROW PRODUCTION, TEXAS, ATOKA-
B 6311 2119* MORROW RESERVOIRS IN REYDON-CHEYENNE AREA, WESTERN
B 6405 672* MORROW SANDSTONE DISCOVERIES IN SOUTHEASTERN COLORADO,
B 6002 326* MORROW SANDSTONE, FIELD, EDDY COUNTY, NEW MEXICO, ABST., ROCK
B 6002 318* MORROW SANDSTONE IN SOUTHERN ELLIS COUNTY, OKLAHOMA, ABST.,
M 26 453 MORROW SANDSTONE, OKLAHOMA, PENNSYLVANIAN
M 26 452 MORROW SANDSTONE, OVERPRESSURING, OKLAHOMA, ANADARKO
B 6311 2118* MORROW SANDSTONES, CIMARRON COUNTY, OKLAHOMA, ABST.,
B 6203 491* MORROW SANDSTONES IN SOUTHEASTERN COLORADO, ABST.,
M 26 453 MORROW-SPRINGER OVERPRESSURING, PREDICTABILITY
M 26 451† MORROW-SPRINGER SANDSTONES, WATONGA-CHICKASHA TREND:
B 6002 327* MORROW-SPRINGER-CHESTER RELATIONS IN ANADARKO BASIN, BY
B 6109 1513† MORROWAN UNIT IN NORTH-CENTRAL NEW MEXICO, OSHA CANYON
B 6203 566* MORROWAN-ATOKAN SERIES BOUNDARY IN NORTHWESTERN ARKANSAS
B 6203 546 MORSE, ANTHONY, SMALL ROCK OUTCROPS IN VEGETATED TERRAIN
B 6004 700 MORSE, J. W., SELECTIVE DISSOLUTION SEQUENCES FOR CALCAREOUS
B 6405 754 MORSE, ROBERT H., GETTING THE MOST OUT OF RADON
B 6207 1226 MORTON, D. J., SEISMIC AND GEOLOGIC FACIES ANALYSES IN FIELD
B 6303 496 MORTON, DENNIS J., MASSIVE MARINE-SANDSTONE FACIES, MACKEREL
B 6105 816 MORTON, R. A., AERIAL PHOTOGRAPH INTERPRETATION OF
B 6004 700 MORTON, R. A., LARGE-SCALE BED FORMS AN SEDIMENTARY
B 6309 1603 MORTON, R. A., QUEEN CITY FORMATION IN EAST TEXAS
B 6305 842* MORTON RANCH PROPERTY, SOUTHERN POWDER RIVER BASIN,
B 6303 498 MORTON, ROBERT A., CHARACTERISTICS OF SHALLOW SUBSURFACE
B 6309 1608 **MORTON, ROBERT A., DISTRIBUTION AND SIGNIFICANCE OF COARSE**
B 6109 1543 MORTON, ROBERT A., HISTORICAL SHORELINE CHANGES AND THEIR
B 6405 723 MORTON, ROBERT A., LOWER CRETACEOUS SHELF STORM DEPOSITS,
B 6303 552 MORTON, ROBERT A., MAP DISPLAY FORMATS FOR ENVIRONMENTAL
M 25 1 MORTON, ROGERS C. B., IMPORTANCE OF CIRCUM-PACIFIC ENERGY AN
S 10 196 MORUGA BOUFF MUD VOLCANO, WEST INDIES, TRINIDAD
B 6405 772* MOSBY SANDSTONE, CENTRAL MONTANA—EXAMPLE OF THIN,
M 29 465 MOSCOW BASIN, AVERAGE THICKNESS OF SEDIMENTS, USSR
B 6410 1581 MOSHEIM LIMESTONE, VIRGINIA
B 6207 1229 MOSHER, D. V., AUSTRALIA URANIUM EXPORT POTENTIAL, ABST.
B 6005 883 MOSHER, R. E., RADIOMETRIC AGE DETERMINATION OF SILURIAN
B 6004 718 MOSHER, R. E., SEDIMENTOLOGY OF SOME SILURIA ROCKS FROM
B 6405 754 MOSHRIF, MOHAMED A., CEMENTATION, DIAGENESIS AND PARAGENETIC
B 6203 546 MOSHRIF, MOHAMED A., STRATIGRAPHY, SEDIMENTATION, AND
B 6308 1301 **MOSLER, JOHN H., OIL AND GAS DEVELOPMENTS IN NORTH**
B 6409 1437 MOSLER, JOHN J., OIL AND GAS DEVELOPMENTS IN NORTH
B 6303 499 MOSLOW, THOMAS F., LATE HOLOCENE DEPOSITIONAL HISTORY OF
B 6203 547 MOSLOW, THOMAS F., RATES OF LANDWARD MIGRATION OF CORE
M 25 552 MOSQUERA C., CARLOS F., MINERAL RESOURCES OF ECUADOR, THEIR
B 6202 227 MOSSAMEDES, EVAPORITES, CUANZA, ANGOLA
B 6108 1305 MOSSLER, JOHN H., OIL AND GAS DEVELOPMENTS IN NORTH
B 6208 1477 MOSSLER, JOHN H., OIL AND GAS DEVELOPMENTS IN NORTH
B 6008 1332 MOSSLER, JOHN J., OIL AND GAS DEVELOPMENTS IN NORTH
B 6003 379 MOSSMAN, R. W., COMPLEX FAULT STRUCTURES IN VERACRUZ
B 6303 499 MOSSOP, G. D., SEDIMENTOLOGY, PETROLOGY, AND GEOTECTONIC
B 6203 547 MOSSOP, GRANT D., DEEP SINUOUS CHANNELS AND GIANT CROSS-BEDS
B 6303 499 MOSSOP, GRANT D., SEDIMENTOLOGY, PETROLOGY, AND GEOTECHNICAL
B 6209 1765* MOTAGUA FAULT ZONE IN EL PROGRESO QUADRANGLE, GUATEMALA,
B 6104 579 MOTATAN FIELD, VENEZUELA
B 6104 595 MOTATAN FIELD, VENEZUELA
M 30 235 MOTY SUITE, USSR, LENA-TUNGUSKA AREA
B 6008 1404 MOULTON, FLOYD C., LOWER MESOZOIC AND UPPER PALEOZOIC
B 6405 717 MOULTON, FLOYD C., UTAH-ARIZONA HINGE LINE—THRUST BELT—
B 6211 2357 MOULTON, FLOYD C., UTAH-ARIZONA HINGE-LINE THRUST BELT—
B 6112 2115† MOUND COMPLEX, LATE PENNSYLVANIAN PHYLLOID-ALGAL BUILDUP,
S 4 242 MOUND, CONCEPT OF, BANK, BUILDUP, OR
M 26 131 MOUND TYPES, SEISMIC, FACIES, EXTERNAL FORMS
M 26 177 MOUNDED FACIES, SEISMIC, INTERPRETATION
B 6205 767 MOUNDED ONLAPPING-FILL FACIES
M 26 180 MOUNDED ONLAPPING-FILL FACIES, SEISMIC, INTERPRETATION
B 6205 764 MOUNDED SEISMIC FACIES
B 6201 73† MOUNDS AND PRAIRIE EVAPORITE FORMATION OF SASKATCHEWAN—
B 6111 1941 MOUNDS, EVOLUTION OF, ALGAL
B 6303 555* MOUNDS FROM LOWER ORDOVICIAN WAH WAH LIMESTONE, IBEX AREA,
B 6004 733* MOUNDS FROM SHELVES AND BASINS, ABST., CARBONATE
B 6405 764* MOUNDS IN PARADOX BASIN, UTAH, ABST., SEISMIC MODELING OF,
B 6004 733* MOUNDS, LITHOHERMS, IN NORTHERN STRAITS OF FLORIDA, ABST.,
B 6405 743* MOUNDS OF FORT PAYNE FORMATION, LOWER MISSISSIPPIAN, OF
B 6303 487* MOUNDS OF FORT PAYNE FORMATION, NORTHERN TENNESSEE, ABST.,
M 25 426 MOUNT BISCHOFF, TIN DEPOSIT, TASMANIA
M 25 426 MOUNT CLEVELAND, TIN DEPOSIT, TASMANIA
B 6006 962† MOUNT DOONERAK, BROOK RANGE, NORTHERN ALASKA,
M 25 427 MOUNT ISA, LEAD-SILVER-ZINC-COPPER, QUEENSLAND
B 6410 1682 MOUNT JOHN OSMOND RANGE, LOWER CAMBRIAN FLAGS, WESTERN,
M 27 22 MOUNT LAUREL SANDSTONE, NEW JERSEY, CRETACEOUS
M 25 426 MOUNT LYELL, COPPER DEPOSIT, TASMANIA
B 6007 1113 MOUNT MITCHELL, CRUISE, 1974, RV
M 25 427 MOUNT MORGAN, GOLD-COPPER, QUEENSLAND
B 6404 568† MOUNT SEDOM SALT DIAPIR, ISRAEL, STRAIN MEASUREMENTS IN
M 25 301 MOUNT SPURR VOLCANO, ALASKA
B 6304 693* MOUNT TAYLOR URANIUM DEPOSIT, SAN MATEO, NEW MEXICO, ABST.
B 6305 839* MOUNT TAYLOR URANIUM DEPOSIT--DESCRIPTION AND COMPARISON
S 3 4 MOUNT TAYLOR VOLCANIC FIELD, NEW MEXICO, GRANTS DISTRICT
S 9 17 MOUNTAIN BUILDING, CLIMATIC CHANGES
B 6211 2270 MOUNTAIN VIEW FAULT, CALIFORNIA
B 6211 2245 MOUNTAIN VIEW PLATFORM, CALIFORNIA
B 6203 419 MOUNTENEY, NICOLA, RIFT BASINS FLANKING NORTH ATLANTIC OCEAN
B 6004 706 MOUNTENEY, S. N., NORTH SEA RIFTING AND OPENING OF NORTH
B 6105 816 MOUNTJOY, E. W., POROSITY IN REEF COMPLEXES—UPPER DEVONIAN
B 6303 492 MOUNTJOY, ERIC, BURIAL DOLOMITIZATION OF UPPER DEVONIAN
B 6303 499 MOUNTJOY, ERIC W., LATE-STAGE SUBSURFACE DOLOMITES--PROBLEMS
B 6101 85 MOUSA OU SALAH BASIN, ATLAS TREND
B 6304 699 MOUSSEAU, RICHARD J., DISSOLVED HYDROCARBONS IN COASTAL
S 10 241† MOVEMENT FROM DEPTH, NEAR-, SURFACE, EVIDENCE OF.

B 6303 496* MOVEMENT OF SUBSURFACE WATERS UNDER SABKHA, ABU DHABI,
B 6002 229 MOVEMENT UP FAULTS, FLUID
B 6303 355 MOWRY SEA, WESTERN INTERIOR
B 6405 685* MOWRY SHALE, ABST., BIOTURBATION AS FACTOR IN HYDROCARBON
B 6410 1621 MOWRY SHALE, COLORADO
B 6407 971 MOWRY SHALE, GREAT PLAINS
B 6303 354† MOWRY SHALE, LOWER CRETACEOUS, WESTERN AND CENTRAL
B 6303 356 MOWRY SHALE, THREE FACIES, WYOMING
M 28 107 MOWRY SHALE, UTAH, CRETACEOUS
M 28 72 MOWRY SHALE, UTAH, CRETACEOUS
B 6404 491 MOWRY SHALE, WYOMING
B 6410 1621 MOWRY SHALE, WYOMING
M 24 95 MOWRY SHALE, WYOMING
M 28 68 MOWRY SHALE, WYOMING, CRETACEOUS
M 24 119 MOXA ARCH, WYOMING
B 6109 1449 MOYOTES UPLIFT, MEXICO
B 6110 1749 MOZAMBIQUE, DEVELOPMENTS, 1976
B 6210 1848 MOZAMBIQUE, DEVELOPMENTS, 1977
B 6310 1693 MOZAMBIQUE, DEVELOPMENTS, 1978
B 6411 1789 MOZAMBIQUE, DEVELOPMENTS, 1979
B 6203 448 MSP MARKER, INDIA, KALOL FORMATION
M 25 141 MT. BANAHAO, LOS BANOS, GEOTHERMAL PROSPECT, PHILIPPINES, MT.
B 6408 1286* MT. CARMEL CONSOLIDATED FIELDS, GIBSON COUNTY, INDIANA, ABST.,
B 6111 1980 MT. DIABLO RANGE, CALIFORNIA
B 6402 168 MT. ENTERPRISE FAULT SYSTEM, TEXAS
M 25 360 MT. FUJI STRATOVOLCANO, JAPAN
M 25 402 MT. KEITH, NICKEL DEPOSIT, WESTERN AUSTRALIA
M 25 141 MT. MAQUILING--MT. BANAHAO, LOS BANOS, GEOTHERMAL PROSPECT,
B 6106 861 MT. MITCHELL CRUISE, MAY 17-27, 1974, RV
B 6410 1644 MT. PAINTER RIDGE, AUSTRALIA
B 6407 1007 MT. VERNON FIELD, ARKANSAS
M 27 3 MUCILAGINOUS SHEATHS
M 26 214 MUCURI-CUMURUXATIBA-JEQUITINHONHA BASIN, BRAZIL
B 6312 2215 MUD CRACKS, DESICCATION, FRACTURES
S 10 194 MUD DIAPIR WITH METHANE-HYDRATE, SEISMIC, GULF OF MEXICO
S 10 192 MUD DIAPIRS, ECHOGRAM, COLOMBIA
S 10 193 MUD DIAPIRS, SEISMIC PROFILE, WEST AFRICA
B 6303 555* MUD MOUNDS FROM LOWER ORDOVICIAN WAH WAH LIMESTONE, IBEX
B 6303 540* MUD ON SELECTED CORALS, ABST., EFFECTS OF AN OFFSHORE, DRILLING
B 6004 719* MUD TO CHALK, PLEISTOCENE, IN BAHAMIAN TROUGH. BY TITLE ONLY,
S 10 196 MUD VOLCANO, COLOMBIA, CARTAGENA, TOTUMO
S 10 195 MUD VOLCANO, COLOMBIA, CARTAGENA, GAS BUBBLE IN
S 10 195 MUD VOLCANO, COLOMBIA, VOLCANCITOS
S 10 200 MUD VOLCANO, IRAN, CASPIAN AREA, GARNIARIK TEPE
S 10 200 MUD VOLCANO, IRAN, CASPIAN AREA, NAFTLI DJEH
S 10 200 MUD VOLCANO, IRAN, MEKRAN COAST, NAPAG
S 10 199 MUD VOLCANO, PAKISTAN, MEKRAN COAST
S 10 202 MUD VOLCANO, TAIWAN
S 10 196 MUD VOLCANO, WEST INDIES, TRINIDAD, MORUGA BOUFF
S 10 203 MUD VOLCANO, DISTRIBUTION, USSR, BAKU AREA
S 10 192 MUD VOLCANOES, METHANE, PRESSURES
B 6105 832* MUD-FACIES DEFINITION FOR PALEOENVIRONMENTAL
B 6105 819* MUDBANK BUILDUPS, PENNSYLVANIAN AND PERMIAN, SACRAMENTO
B 6004 727* MUDDY AND FRONTIER FORMATIONS, WYOMING AND MONTANA, ABST.,
B 6207 1159 MUDDY CREEK FORMATION, ARIZONA AND, NEVADA, HUALAPAI
B 6405 776* MUDDY, D SANDSTONE DISCOVERY, ADAMS COUNTY, COLORADO, SECS.
M 26 432 MUDDY FORMATION, MONTANA, BELL CREEK FIELD
B 6405 676* MUDDY FORMATION, NORTHEAST POWDER RIVER BASIN, WYOMING
B 6004 699* MUDDY FORMATION OF GRIEVE FIELD, WYOMING, ABST., DEPOSITIONAL
B 6108 1384* MUDDY FORMATION OF GRIEVE FIELD, WYOMING, ABST., DEPOSITIONAL
M 26 432 MUDDY FORMATION, WYOMING, POWDER RIVER BASIN
M 26 432 MUDDY FORMATION, WYOMING, RECLUSE FIELD
B 6407 971 MUDDY SANDSTONE, GREAT PLAINS
B 6004 675* MUDDY SANDSTONE, LOWER CRETACEOUS, WIND RIVER BASIN,
B 6108 1371* MUDDY SANDSTONE, POWDER RIVER BASIN, WYOMING, ABST.,
M 24 95 MUDDY SANDSTONE, WYOMING
M 28 136 MUDDY SANDSTONE, WYOMING, CRETACEOUS
M 28 137 MUDDY SANDSTONE, WYOMING, CRETACEOUS
B 6203 457 MUDERONG GREENSAND MEMBER, AUSTRALIA
B 6203 457 MUDERONG SHALE, AUSTRALIA
B 6303 532* MUDFLOW DEPOSITS IN SUBMARINE TRENCHES, HELLENIC BASINS AND
S 10 199 MUDFLOWS, WEST INDIES, TRINIDAD, GRAND RAVINE FIELD
S 7 276 MUDFLOWS, BOTTOM, DEFORMATION
S 7 277 MUDFLOWS, DIAGRAMS, STABILITY
B 6305 836 MUDGE, MELVILLE R., STRUCTURAL PATTERN OF EASTERN PART
B 6303 550* MUDS, ABST., SEDIMENT SUSPENSION, DEWATERING, AND
B 6303 448* MUDS--VISCOSITY OF ESTUARINE SEDIMENTS, ABST., STICKY
B 6405 763* MUDSTONES AS EXPLORATION GUIDES TO TABULAR SANDSTONE-TYPE
B 6203 506* MUDSTONES FROM LIQUEFACTION, ABST., CONGLOMERATIC,
B 6203 563* MUDSTONES OF LATE JURASSIC EPICONTINENTAL SEAWAY. WESTERN

B 6008 1401 MUELLER, CLARK A., STRUCTURAL INTERPRETATION AND POROSITY
B 6105 840 MUELLER, H. W., III, POROSITY PRESERVATION IN SHALLOW-MARINE
B 6405 673 MUELLER, M., ET AL, EXPLORATION IN CLASSIC THRUST BELT AND ITS
B 6402 189 MUERTOS CONGLOMERATE MEMBER, SPAIN
M 29 368 MUERTOS TRENCH, INTERPRETATION, VENEZUELA BASIN
M 29 368 MUERTOS TRENCH, SEISMIC (FOLDOUT), VENEZUELA BASIN
B 6303 500 MUIR, MARJORIE, COORONG MODEL FOR PENECONTEMPORANEOUS
S 4 195 MUJERES, MOLLUSCA, MEXICO, ISLA
M 25 293 MULCHATNA FAULT, ALASKA
B 6102 257 MULDOON CANYON FORMATION, IDAHO
B 6004 675 MULHERN, M., HIGH-RESOLUTION STUDIES OF PALEOCLIMATIC AND
B 6403 447 MULHERN, MICHAEL E., LACUSTRINE, FLUVIAL, AND FAN
B 6207 1218 MULL, C. G., PETROLEUM EXPLORATION OF NATIONAL PETROLEUM
B 6004 727 MULL, C. G., PRUDHOE BAY SUPE GIANT FIELD AND ALASKAN ARCTIC
B 6405 754 MULL, CHARLES G., EVOLUTION OF BROOKS RANGE THRUST BELT AND
B 6103 467 MULL, CHARLES G., NORTHERN ALASKA PALEOMAGNETISM, PLATE
B 6209 1696 MULL, CHARLES G., TECTONICS OF BROOKS RANGE OPHIOLITES, ALASKA
B 6203 430 MULL VOLCANICS, BRITAIN, WEST
M 25 423 MULLAMAN BEDS, GROOTE EYLANDT
B 6004 700 MULLINS, H. T., HYDROCARBON POTENTIAL OF NORTHWESTERN
B 6410 1701 MULLINS, HENRY T., CARBONATE SEDIMENT DRIFTS IN NORTHER
B 6204 693 MULLINS, HENRY T., CHARACTERISTICS OF DEEP BAHAMA CHANNELS IN
B 6405 755 MULLINS, HENRY T., NODULAR SUBMARINE CEMENTATION ON
B 6303 500 MULLINS, HENRY T., SEISMIC FACIES AND DEPOSITIONAL PROCESSES OF
B 6007 1037 MULLINS, HENRY T., STRATIGRAPHY AND STRUCTURE OF NORTHEAST
S 4 329 MULTER, H. G., MIOCENE KINGSHILL SEAWAY--A DYNAMIC CARBONATE
B 6008 1411* MULTI-DIMENSIONAL SEISMIC DISPLAYS, ABST.
B 6105 770 * MULTIDIMENSIONAL APPROACHES TO GEOLOGIC CLASSIFICATION OF
B 6405 741* MULTI-PARAMETER REGIONAL MAPPING AND INTERPRETATIONS, ABST.
B 6211 2292 MULTICHANNEL PROFILE LINES, GULF OF MEXICO
M 29 65† MULTICHANNEL SEISMIC DEPTH SECTIONS AND INTERVAL VELOCITIES
M 29 30 MULTICHANNEL SEISMIC PROFILE, BLAKE PLATEAU
M 29 18 MULTICHANNEL SEISMIC PROFILES, GEORGIA EMBAYMENT
M 29 21 MULTICHANNEL SEISMIC PROFILES, GEORGIA EMBAYMENT
M 29 261† MULTICHANNEL SEISMIC RECORDS, STRUCTURE OF THE OUTER,
M 29 11† MULTICHANNEL SEISMIC REFLECTION PROFILES, GEOLOGY OF THE,
M 29 179 MULTICHANNEL SEISMICS, BLAKE ESCARPMENT
M 29 379 MULTICHANNEL SEISMICS, SOUTHEAST ASIA, SULU TROUGH
S 4 90 MULTICOLUMNASTREA-ACTINACIS-DICTUOPHYLLIA-DICHOCOENIA
S 4 90 MULTICOLUMNASTREA-ASTROCOENIA-ACTINACIS ASSEMBLAGE,
B 6004 651* MULTIDIMENSIONAL APPROACHES TO GEOLOGIC
B 6405 694* MULTIFOLD SEISMIC DATA ACROSS OUTER BERING SEA CONTINENTAL
B 6007 1009 MULTILATERAL SAND BODY
B 6405 799* MULTIPHASE FLUID MOVEMENTS IN GLASS MICROMODELS OF PORE
B 6405 796* MULTIPLE BARRIER-ISLAND AND DELTAIC PROGRADATIONAL
M 27 141 MULTIPLE GENERATIONS
B 6211 2353* MULTIPLE REFLECTIONS AND HEAD WAVES IN GULF OF SUEZ, ABST.
M 27 166 MULTIPLE-GENERATION CEMENTATIONS
S 5 112 MULTIPLES, SEISMIC DATA,
B 6004 705* MULTIVARIATE ANALYSES, ABST., GRANULOMETRIC RELATIONS WITH,
B 6203 562* MULTIVARIATE ANALYSIS IN CARBONATE PETROLOGY WITH EXAMPLE
B 6009 1611* MULTIVARIATE MINERAL ANALYSIS OF MIOCENE-PLIOCENE
B 6012 2188* MULTIVARIATE STATISTICAL ANALYSIS OF HEAVY-MINERAL DATA,
B 6303 481* MULTIVARIATE STUDY, ABT ST., MEGASCALE VARIABILITY OF SEDIMENT
B 6409 1559 MUMMA, MARTIN D., OSTRACODE BIOSTRATIGRAPHY OF LOWER
B 6309 1609 MUMME, S. T., GEOPRESSURE IN HOUMA AND HOLLYWOOD FIELDS,
S 10 162 MUNN, HYDRAULIC THEORY OF OIL AND GAS ACCUMULATION
B 6411 1970 MURA, DRAVA, AND SAVA SUBBASINS, YUGOSLAVIA, BANAT
B 6307 1109 MURAL LIMESTONE, FIVE DEPOSITIONAL ENVIRONMENTS, ARIZONA
B 6405 800* MURAL LIMESTONE, LOWER CRETACEOUS, ARIZONA AND SONORA, ABST.,
B 6307 1123 MURAL REEFS, MODEL, ARIZONA
B 6104 585 MURBAN, FIELD, PERSIAN GULF, BAB
B 6308 1376 MURPHY, A. G., DEVELOPMENT IN ARKANSAS, NORTH LOUISIANA, AND
B 6008 1370 MURPHY, A. G., DEVELOPMENT IN ARKANSAS, NORTH LOUISIANA, AND
B 6108 1234 MURPHY, A. G., DEVELOPMENTS IN ARKANSAS, LOUISIANA, AND EAST
B 6409 1490 MURPHY, A. G., DEVELOPMENTS IN ARKANSAS, NORTH LOUISIANA, AND
B 6208 1487 MURPHY, A. G., DEVELOPMENTS IN ARKANSAS, NORTH LOUISIANA, AND
B 6405 790 MURPHY, J., DIAGENESIS IN MONTEREY FORMATION, PISMO SYNCLINE,
M 25 250 MURPHY, R. W., STRUCTURAL EVOLUTION OF TERTIARY BASINS OF
M 25 355 MURRAY BASIN, AUSTRALIA
M 29 152 MURRAY BASIN, AUSTRALIA, CONTINENTAL MARGIN
B 6205 881 MURRAY, D. KEITH, EAGLE SPRINGS OIL FIELD, NYE COUNTY, NEVADA,
B 6008 1404 MURRAY, D. KEITH,, NEW FRONTIERS IN COLORADO COAL RESEARCH
B 6105 830 MURRAY, D. P., COAL-BEARING NARRAGANSETT BASIN, MASSACHUSETTS
B 6203 547 MURRAY, GROVER E., INDIGENOUS PRECAMBRIAN PETROLEUM
B 6109 1543 MURRAY, GROVER E., INDIGENOUS PRECAMBRIAN PETROLEUM
B 6410 1681 MURRAY, GROVER E., INDIGENOUS PRECAMBRIAN PETROLEUM
B 6203 548 MURRAY, HAYDN H., COAL CLEANING, DESULFURIZATION, BY
B 6303 500 MURRAY, JOHN C., APPLICATION OF POTENTIAL FIELD DAT TO
M 25 154 MURRAY, JOHN C., DEEP RESEARCH DRILLHOLE AT KILAUEA

M 25 355† MURRAY RIVER IN SOUTH AUSTRALIA--SUMMARY, SALINE,
B 6004 701 MURRAY, STEPHEN P., CONTROL OF SEDIMENT DISTRIBUTION BY
B 6303 501 MURRAY, STEPHEN P., DISPERSAL OF FINE-GRAINED SEDIMENT IN
B 6203 548 MURRAY, STEVEN M., EFFECT OF CLAYS ON URANIUM DISTRIBUTION IN
B 6210 1793 MURRAY, T. H., JR., FOREIGN DEVELOPMENT PAPERS, 1977
B 6310 1632 MURRAY, T. H., JR., FOREIGN DEVELOPMENT PAPERS, 1978
B 6411 1736 MURRAY, T. H., JR., FOREIGN DEVELOPMENT PAPERS, 1979
B 6405 597 MURRIS, R. J., MIDDLE EAST—STRATIGRAPHIC EVOLUTION AND OIL
B 6406 863 MURZUK BASIN, LIBYA AND, CHAD
M 27 50 MUSCAT, OMAN, JURASSIC RADIOLARITE
M 27 24 MUSCAT, OMAN, JURASSIC, RADIOLARITE
M 27 12 MUSCAT, OMAN, TERTIARY CALCARENITE
M 28 45 MUSCOVITE, DETRITAL
M 28 31 MUSCOVITE, FRAGMENTS, SCHIST, IN
M 28 8 MUSCOVITE MICROLITE INCLUSION, QUARTZ
B 6105 828 MUSIALOWSKI, F. R., DREDGE DISPOSAL IN LITTORAL ZONE, NEW RIVER
B 6105 817 MUSIALOWSKI, FRANK S., MEASUREMENT TECHNIQUE FOR
B 6303 318 MUSINIA PEAK MEMBER, UTAH, FLAGSTAFF LIMESTONE
M 24 73 MUSKEG FORMTION, ALBERTA
B 6205 782 MUSKEG SALT, ALBERTA
B 6303 433 MUSRATI, HASSAN, GEOLOGY OF A STRATIGRAPHIC GIANT--MESSLA OIL
M 30 507 MUSRATI, HASSAN, GEOLOGY OF A STRATIGRAPHIC GIANT--MESSLA OIL
B 6008 1410* MUSSELSHELL COUNTY, MONTANA, ABST., JIM COULEE-, LITTLE WALL
B 6405 797* MUSSELSHELL COUNTY, MONTANA, ABST., MASON LAKE FIELD
B 6008 1402* MUSSELSHELL COUNTY, MONTANA--FOOD FOR THOUGHT, ABST.,
B 6009 1608* MUSTANG AND PADRE ISLANDS--TIME-SERIES APPROACH, ABST., BEACH
B 6002 322* MUSTANG, CURRENT HUNTON PLAY IN MCCLAIN, GRADY, AND
M 29 151 MUTTER, JOHN, THE CRUSTAL STRUCTURE AND EVOLUTION OF THE
B 6004 701 MUTTI, EMILIANO, SEDIMENTOLOGY OF OUTER-FAN DEPOSITIONAL
B 6204 633 MV CECIL GREEN RESEARCH CRUISE
B 6007 1113 MV GULF SEAL, CRUISE, 1973
B 6203 518* MYAKKA AND PEACE RIVER ESTUARIES, CHARLOTTE HARBOR, FLORIDA,
B 6004 701 MYHR, DARYLL W., LOWER CRETACEOUS NEOCOMIAN, SANDSTONE
M 28 126 MYLONITIZATION, HEMATITE
M 22 279 MYOHOJI FIELDS, JAPAN
B 6102 193 MYS KAMCHATSKIY, CAPE KAMCHATKA
B 6102 193 MYS OLYUTORSKIY, CAPE OLYUTORSKY
B 6204 668 MZAB BASIN, ALGERIA
B 6402 158 N SANDSTONE, LOUISIANA
B 6207 1154 N-ALKANES IN OILS AND EXTRACTS
B 6211 2259 N-CHERT, CALIFORNIA
M 28 143 N.W.T., CAMBRIAN, OLD FORT ISLAND SANDSTONE, CANADA
M 28 179 N.W.T., CAMBRIAN, OLD FORT ISLAND SANDSTONE, CANADA
M 28 163 N.W.T., CRETACEOUS, SANDSTONE, CANADA
M 28 180 N.W.T., CRETACEOUS, SANDSTONE, CANADA
M 28 178 N.W.T., CRETACEOUS, SANDSTONE, CANADA
B 6409 1560* NABORTON—DOLET HILLS SEDIMENTS, TERTIARY, NORTHWESTERN
B 6409 1563* NACATOCH SAND, UPPER CRETACEOUS, EAST TEXAS BASIN AN
B 6109 1514 NACIMIENTO UPLIFT, NEW MEXICO
B 6104 603 NACO GROUP, NEW MEXICO
S 6 347 NAESER, CHARLES W., ORDOVICIAN GEOCHRONOLOGY
S 10 200 NAFTLI DJEH MUD VOLCANO, IRAN, CASPIAN AREA
B 6002 261 NAHAL HAMEAROT REEF COMPLEX, ISRAEL
B 6402 219 NAHAR-KATIYA AND MORAN FIELD, INDIA
B 6403 428 NAHARKATIYA FIELD, INDIA
B 6405 616 NAHR UMR SHALE, MIDDLE EAST
B 6303 501 NAIRN, A. E. M., STRATIGRAPHIC CORRELATION OF KEATHLEY
B 6203 508 NAIRN, K. N., COMPUTER ANALYSIS OF REGIONAL WELL-DENSITY DATA
M 25 427 NAIRNE, PYRITE, SOUTH AUSTRALIA
B 6207 1241 NAKAGAWA, SUSUMU, PRESENT STATUS OF GEOTHERMAL
M 25 148 NAKAMURA, HISAYOSHI, GEOLOGICAL AND GEOCHEMICAL
M 24 4 NAKNEK FORMATION, ALASKA
B 6408 1261 NAKNEK FORMATION, ALASKA
B 6203 377 NAKNEK FORMATION, ALASKA
B 6309 1523 NAKNEK FORMATION, ALASKA PENINSULA
B 6412 2145 NAKNEK FORMATION, BERING SEA SHELF
B 6406 947 NAKNEK FORMATION SANDSTONE, ALASKA
S 6 369 NAMA-DAMARA SYSTEM, SOUTH AFRICA
B 6412 2100 NAMAKI FORMATION, THREE MEMBER, IRAN
M 25 269 NAMHAE ARCHIPELAGO ALIGNMENT, KOREA
B 6110 1750 NAMIBIA, DEVELOPMENTS, 1976, SOUTH-WEST AFRICA
B 6210 1849 NAMIBIA, DEVELOPMENTS, 1977, SOUTH-WEST AFRICA
B 6310 1695 NAMIBIA, DEVELOPMENTS, 1978, SOUTH-WEST AFRICA
B 6411 1790 NAMIBIA, DEVELOPMENTS, 1979, SOUTHWEST AFRICA
B 6303 412* NAMORAD FIELD, MAJOR OIL DISCOVERY IN CAMPOS BASIN, BRAZIL,
B 6109 1476 NAMORADO FIELD, BRAZIL
M 30 332 NAMORADO FIELD, DISCOVERY HISTORY, BRAZIL
M 30 338 NAMORADO FIELD, GEOLOGIC CROSS SECTION, BRAZIL
M 30 329† NAMORADO OIL FIELD--A MAJOR OIL DISCOVERY IN THE CAMPOS BASIN,
M 30 331 NAMORADO SANDSTONE, BRAZIL, MACAE FORMATION
B 6004 659* NAMURIA OF CENTRAL PENNINE BASIN, ENGLAND, ABST., DELTAIC
B 6204 669 NAMURIAN, ALGERIA
B 6211 2192 NAMURIAN BASIN, IRELAND, WESTERN
B 6211 2207 NAMURIAN GROWTH FAULTS, IRELAND, WESTERN
B 6211 2194 NAMURIAN STRATIGRAPHY, IRELAND, WESTERN
B 6308 1427 NAMY, JERRY, STRATIGRAPHY AND EXPLORATION GEOLOGY OF
B 6311 2059 NANAIMO COALFIELD, BIRTISH COLUMBIA
B 6309 1584 NANCE, S. W., DEVONIAN SHALE CHARACTERIZATION--EVALUATION OF
B 6307 1094 NANGETTY FORMATION, AUSTRALIA
B 6409 1554* NANNOCONIDS OF TARAISES FORMATION, EARLY CRETACEOUS, SANTA
B 6205 892* NANNOFOSSIL ASSEMBLAGES IN ATLANTIC OCEAN—
B 6009 1562† NANNOFOSSIL BIOSTRATIGRAPH AND CLIMATIC STAGES OF
B 6009 1608* NANNOFOSSIL BIOSTRATIGRAPHY OF BOSSIER SHALE AND
B 6004 700* NANNOFOSSILS IN DEEP-SEA CARBONATE OOZE--EMPIRICAL
B 6011 2044† NANNOFOSSILS OF MIDDLE MIOCENE OF SOUTH LOUISIANA,
B 6303 443* NANNOPLANKTON BIOSTRATIGRAPHY IN EASTERN MEDITERRANEAN,
B 6012 2191* **NANNOPLANKTON BIOSTRATIGRAPHY OF UPPER SACATE AND GAVIOTA**
S 11 60† NANNOPLANKTON BIOSTRATIGRAPHY OF CENOZOIC MARINE STAGES IN
B 6303 503* **NANNOPLANKTON IN UPPER APTIAN AND ALBIAN BLACK SHALES FROM**
B 6303 423* NANNOPLANKTON PALEOBIOGEOGRAPHY AND PALEO-OCEANOGRAPHY
S 11 61 NANNOPLANKTON, REWORKING AND REDEPOSITION
S 11 61 NANNOPLANKTON ZONES, CALIFORNIA STAGES VERSUS
S 5 162 NANSEN FORMATION, SVERDRUP BASIN, HARE FIORD,
B 6111 1931 NANSEN FORMATION, SVERDRUP BASIN
B 6111 1930 NANSEN SOUND, ELLESMERE ISLAND
B 6004 672* NANTUCKET ISLAND, MASSACHUSETTS, ABST., MESOZOIC,
B 6303 492* NANUSHUK GROUP, ALBIAN-CENOMANIAN, NORTHERN ALASKA, ABST.,
B 6405 752* NANUSHUK GROUP, MIDDLE CRETACEOUS, AND ASSOCIATED STRATA,
S 10 200 NAPAG MUD VOLCANO, IRAN, MEKRAN COAST
B 6407 1091 NAPANIAN HORIZON OF KAMCHATKA, ASIA, NORTHEAST
S 10 93 NAPHTHA FRACTION, ANALYSIS, CRUDE OIL
B 6406 821 NAPIER FORMATION, AUSTRALIA
M 25 319 NAPIPI ANTICLINE, COLOMBIA
B 6201 125 NAPLES BAY GROUP, FLORIDA
S 11 91 NAPLES, CALIFORNIA, CHECK LIST OF, FORAMINIFERA, MONTEREY
S 11 94 NAPLES, CALIFORNIA, CHECK LIST OF, FORAMINIFERA, MONTEREY
B 6112 2086 NAPO FORMATION, KEROGEN, ECUADOR
B 6303 504* NAPPE STRUCTURES AROUND QUEBEC CITY, CANADA, ABST., THERMA
B 6203 503 NARASIMHAN, T. N., GEOTHERMAL RESOURCE INVESTIGATIONS AND
M 25 252 NARATHIWAT RIDGE, THAILAND
B 6212 2492 NARDIN, T. R., PLIOCENE-PLEISTOCENE DIASTROPHISM OF SANTA
B 6203 548 NARDIN, THOMAS R., HYDROCARBON SEEPAGE IN SANTA MONICA BAY—
B 6202 247 NARDIN, THOMAS R., PLIOCENE-PLEISTOCENE DIASTROPHISM OF SANTA
B 6111 1994 NARES ABYSSAL PLAIN, POSSIBLE DIAPIRS
B 6102 240 NARIZIAN ROCKS, CALIFORNIA
S 11 64 NARIZIAN STAGE
B 6405 755 NARR, WAYNE, ORIGIN OF SUBSURFACE FRACTURE SYSTEMS—EXAMPLE
B 6105 830* NARRAGANSETT BASIN, MASSACHUSETTS AND RHODE ISLAND, ABST.,
B 6203 395 NARROW CAPE FORMATION, ALASKA
B 6408 1142 NARROW CAPE FORMATION, ALASKA
B 6407 1016 NARROW CAPE FORMATION, ALASKA
B 6303 302 NARROW CAPE FORMATION, ALASKA
B 6410 1580 NARROWS BLOCK, VIRGINIA
B 6410 1576 NARROWS THRUST, VIRGINIA
B 6205 888 NASH, E. G., HYDROCARBON EXPLORATION IN GREAT SALT LAKE BASIN,
B 6405 786 NASH, J. T., GEOPHYSICAL EXPLORATION FOR PRECAMBRIAN-RELATED
B 6405 755 NASH, J. THOMAS, URANIUM OCCURRENCE IN LEADVILLE DOLOMITE A
B 6301 90 NASHWAUK BEDS, SCOTIAN BASIN
B 6301 90 NASKAPI MEMBER, SCOTIAN BASIN
M 29 239 NASU, NORIYUKI, SUBSIDENCE OF THE DAITO RIDGE AND ASSOCIATED
B 6009 1611* NATCHITOCHES PARISHES, LOUISIANA, ABST., JAMES LIMESTONE IN,
M 26 417 NATH, A. K., GEOLOGIC CONSIDERATIONS FOR STRATIGRAPHIC
B 6405 735* NATIONAL COAL RESOURCES DATA SYSTEM, NCRDS, APPLIED TO STUDY
B 6411 2036 NATIONAL DEPARTMENT OF MINERAL PRODUCTION OF MINISTRY OF
B 6207 1238* NATIONAL ENERGY POLICY, ABST., PHILIPPINE
M 25 456† NATIONAL EXPLORATION PROGRAM FOR BASE-META DEPOSITS IN
B 6207 1218* NATIONAL PETROLEUM RESERVE IN ALASKA, NPRA, ABST., PETROLEUM
B 6405 715* NATIONAL PETROLEUM RESERVE IN ALASKA, NPRA, ABST., PETROLEUM
B 6110 1802 NATIONAL RES. AUTHORITY, HASHEMITE KINGDOM OF JORDAN, ON
B 6408 1271 NATIONAL RESEARCH COUNCIL, A. W. BALLY, LEHNER, P., REVIEW OF,
B 6002 196 NATIONAL SCIENCE FOUNDATION INTERNATIONAL DECADE OF OCEAN
B 6004 702 NATLAND, M. L., CLASSIFICATION OF CLASTIC SEDIMENTS, ABST.
B 6112 2082 NATURAL CATAGENESIS OF KEROGEN TYPES
B 6109 1545* NATURAL ENVIRONMENTAL-INFORMATION AND IMPACT-ASSESSMENT
B 6312 2214† NATURAL FRACTURE SYSTEMS, DESCRIPTION AND CLASSIFICATION
S 10 95 NATURAL GAS, DISSOLVED HYDROCARBONS
S 10 102 NATURAL GAS IN SOLUTION, MIGRATION, SECONDARY
B 6303 428* NATURAL GAS, ABST., ORGANIC ACID ANIONS IN OIL-FIELD, WATERS
B 6303 515* **NATURAL GAS ACCUMULATIONS, WESTERN GULF OF MEXICO,** ABST.,

B 6212 2441† NATURAL GAS, ALIPHATIC, ACID, ANIONS IN, OIL-FIELD, WATERS—
B 6207 1207* NATURAL GAS DEPOSITS IN GULF OF THAILAND, ABST.
B 6207 1217* NATURAL GAS EXPORTER, ABST., AUSTRALIA AS
B 6412 2166† NATURAL GAS IN APPALACHIAN OVERTHRUST BELT, CROSS-STRIKE
B 6303 421* NATURAL GAS IN CALIFORNIA, ABST., INFLUENCE OF RATE ON DEMAND
B 6408 1131† NATURAL GAS IN COOK INLET BASIN, ALASKA, BIOGENIC AND,
B 6303 434* NATURA GAS IN ILLINOIS BASIN, ABST., ISOTOPE, GEOCHEMISTRY OF
B 6205 887* NATURAL GAS IN NORTHWEST COLORADO, ABST., STRATIGRAPHY OF,
B 6211 2359* NATURAL GAS IN PYRAMID LAKE AREA, NEVADA, ABST., ALTERATION
B 6303 478* NATURAL GAS IN UNITED STATES--CASE HISTORY OF POTENTIAL GAS
B 6409 1529† NATURAL GAS IN UNITED STATES—CASE HISTORY OF POTENTIAL GAS
B 6305 767 NATURAL GAS, PRIMARY MIGRATION MECHANISM
B 6309 1580* NATURAL GAS PRODUCTION IN APPLACHIAN STATES, ABST., NEW
B 6008 1413* NATURAL GAS PROSPECTS IN SOUTHERN IDAHO A DEDUCED FROM
B 6309 1577* NATURAL GAS RESOURCE--DEVONIAN SHALES, ABST., EXPLORATION
B 6409 1529 NATURAL GAS RESOURCES, CLASSIFICATION OF
B 6205 891* NATURAL GAS RESOURCES IN SHALLOW, LOW-PERMEABILITY
B 6407 969† NATURAL GAS RESOURCES, SHALLOW, LOW-PERMEABILITY, RESERVOIRS
B 6303 414* NATURAL GAS STABILITY IN DEEP SUBSURFACE, ABST.
B 6004 658* NATURAL GAS STORAGE RESERVOIRS BY ISOTOPIC ANALYSIS OF
B 6203 543* NATURAL GAS SUPPLY, ABST., FUTURE
B 6409 1566* NATURAL GASES IN OFFSHORE GULF OF MEXICO, ABST., CHEMICAL AND
B 6303 522* NATURAL GASES IN SUBALPINE TERTIARY BASINS, ABST., DUAL ORIGIN
B 6210 2119 NATURAL RESOURCES AUTHORITY, THE HASHEMITE KINGDOM OF
B 6010 1873 NATURAL RESOURCES AUTHORITY,HASHEMITE KINGDOM OF JORDAN
B 6303 482* NATURE AND FIELD APPLICATION OF PLUMOSE STRUCTURES, ABST.
B 6207 1151 NATURE AND MATURITY OF SOLID ORGANIC MATTER
B 6111 1950 NATURE AND ORIGIN OF TAR SANDS
B 6405 796* NATURE OF NON-RECOVERABLE POROSITY CHANGES IN
B 6203 521* NATURE OF SHALES AND DYNAMICS OF HYDROCARBON EXPULSION IN
B 6112 2086 NAVAJO COAL MINE, KEROGEN, NEW MEXICO
B 6203 532* NAVAJO SANDSTONE, ABST., ORIGIN OF
B 6308 1424* NAVAJO SANDSTONE ENVIRONMENTAL WAR--CAN TRACE FOSSILS HELP,
B 6303 441* NAVAJO SANDSTONE ENVIRONMENTAL WAR--CAN TRACE FOSSILS HELP.
B 6108 1386* NAVAJO SANDSTONE, JURASSIC, OF UTAH AND NORTHERN
B 6102 228 NAVAJO SANDSTONE, PERMEABILITY, UTAH
B 6102 235 NAVAJO SANDSTONE, PETROLOGY, UTAH
B 6312 2220 NAVAJO SANDSTONE, UTAH
B 6205 880* NAVAL PETROLEUM RESERVE ALASKA, ABST., NEW INFORMATION ON,
B 6109 1494 NAVAL PETROLEUM RESERVE NO. 4, ALASKA
M 25 309 NAVAL PETROLEUM RESERVES, PRESIDENTIAL ORDER IN 1912
B 6309 1468 NAVARIN BASIN, BERING SHELF
B 6002 172 NAVARIN BASIN PROVINCE, BERINGIAN SHELF
B 6002 164 NAVARIN RIDGE, BERINGIAN SHELF
B 6408 1220 NAVARRO SECTION, TEXAS
M 27 154 NAVESINK FORMATION, NEW JERSEY, CRETACEOUS
M 28 38 NAVESINK FORMATION, NEW JERSEY, CRETACEOUS
B 6406 853 NAVIDAD BASIN, SAND LAYERS
B 6206 916 NAVY SUBMARINE FAN, CALIFORNIA
B 6207 1230 NAYOAN, G. A. S., TERTIARY CARBONATE RESERVOIRS IN INDONESIA,
B 6212 2456 NAZCA AND SOUTH AMERICAN PLATES, CONVERGENCE OF
M 25 546 NAZCA AND SOUTH AMERICAN PLATES, RISE-CONVERGENT PLATE
M 25 540 NAZCA PLATE, BATHYMETRIC AND TECTONIC FEATURES, MAP
B 6403 360 NAZCA PLATE, BOUNDARY
M 25 545 NAZCA PLATE, COPPER AND NICKEL, CONCENTRATIONS OF
M 25 540 NAZCA PLATE, DYNAMIC BEHAVIOR OF
M 25 545 NAZCA PLATE, FERROMANGANESE NODULES, IRON-NICKEL RATIOS
M 25 539 NAZCA PLATE, LOCATION, PACIFIC OCEAN
M 25 544 NAZCA PLATE, MANGANESE, ALUMINUM, AND IRON, TERNARY DIAGRAM
M 25 546 NAZCA PLATE, METALLIFEROUS SEDIMENTS, GENESIS
M 25 541 NAZCA PLATE, OCEANIC CRUST, AGE OF
M 25 539† NAZCA PLATE PROJECT, METALLOGENESIS IN SOUTHEAST, PACIFIC
M 25 541 NAZCA PLATE REGION, LITHOLOGY, MINERALOGY, AND CHEMISTRY
M 25 541 NAZCA PLATE REGION, METALLIFEROUS SEDIMENTS
M 25 541 NAZCA PLATE REGION, METALS, TRANSITION
M 25 542 NAZCA PLATE REGION, MINOR AND TRACE ELEMENT GEOCHEMISTRY
M 25 543 NAZCA PLATE REGION, RADIOGENIC AND STABLE ISOTOPE
M 29 445 NAZCA PLATE, SEISMIC RECORDS IN SEDIMENTS, PERU TRENCH
M 25 543 NAZCA PLATE, SILICON, IRON, AND MANGANESE, TERNARY DIAGRAM
M 25 541 NAZCA RIDGE, CONVERGENCE WITH PERU-CHILE TRENCH, RATE OF
B 6405 735* NCRDS, APPLIED TO STUDY OF APPALACHIAN COAL BED, ABST., USCHEM,
B 6004 665* NE APPROACH TO GEOLOGIC ESTIMATES OF OIL AND GAS RESOURCES BY
B 6304 691* NE MEXICO, URANIUM DEPOSIT, ABST., GEOLOGY AND DEVELOPMENT
B 6309 1580 NEAL, DONALD W., ELEMENTAL ANALYSES OF DEVONIAN SHALES IN
B 6303 501 NEAL, DONALD W., SUBSURFACE STRATIGRAPHY OF UPPER DEVONIAN
B 6103 432 NEAL RANCH FORMATION, TEXAS
B 6103 431† NEAL RANCH-LENOX HILLS CONTACT IN GLASS MOUNTAINS, WEST
B 6410 1577 NEALMONT BEDS, VIRGINIA
B 6004 648* NEAR-SHORE REGION, SOUTHWESTERN FLORIDA, ABST., SOURCE AND
B 6203 524* NEAR-SURFACE EVIDENCE OF HYDROCARBON MOVEMENT FROM DEPTH,
S 10 241† NEAR-SURFACE EVIDENCE OF HYDROCARBON MOVEMENT FROM DEPTH

M 29 19 NEAR-TRACE SHIPBOARD MONITOR, GEORGIA EMBAYMENT
B 6409 1565* NEARSHORE BARS ALONG UNITED STATES GULF COAST, ABST.
B 6105 798* NEARSHORE CURRENT PATTERN OFF SOUTH TEXAS—INTERPRETATION FROM
B 6303 434* NEARSHORE GLACIAL-MARINE SEDIMENTATION PRELIMINARY
B 6405 740* NEARSHORE LITHOFACIES OF MANNVILLE GROUP, LLOYDMINSTER
B 6405 802* NEARSHORE MARINE AND CONTINENTAL FACIES IN EOCENE OF
B 6004 681* NEARSHORE SEDIMENTARY SEQUENCES, ABST., PHYSICAL-BIOGENIC
B 6203 552* NEARSHORE-BAR MORPHOLOGY IN EASTERN NICARAGUA, ABST.,
B 6004 681 NEASHAM, J. W., BED CONTINUITY AND PERMEABILITY VARIATIONS OF
B 6303 501 NEASHAM, J. W., CHARACTERIZATION OF ROCK MINERAL AND PORE
B 6405 756 NEAVEL, RICHARD C., COAL CHARACTERIZATION FOR OPTIMUM
B 6203 570* NEBRASKA, ABST., EARLY SUBAERIAL EXPOSURE AND, FRESHWATER
B 6002 322* NEBRASKA, ABST., GENESIS AND DISTRIBUTION OF, DESMOINESIAN,
B 6108 1392* NEBRASKA, ABST., INCIPIENT DIAGENESIS OF, ARKOSIC AND, VITRIC,
B 6303 533* NEBRASKA, ABST., ISOTOPIC COMPOSITION AN SOURCES OF, STRONTIUM
B 6105 839* NEBRASKA, ABST., LITHOFACIES OF UPPER, PENNSYLVANIAN, LANSING-,
B 6405 701* NEBRASKA, ABST., PALEOTOPOGRAPHYS INFLUENCE ON POROSITY
B 6002 320* NEBRASKA, ABST., REPORT--CENTRAL
B 6108 1393* NEBRASKA, ABST., STRATIGRAPHY AND SEDIMENTARY PETROGRAPHY
B 6107 1030 NEBRASKA, ADMIRE GROUP
B 6108 1375* NEBRASKA AND ADJACENT STATES, ABST., GEOMETRY AND
B 6107 1031 NEBRASKA, BASAL PENNSYLVANIAN SAND, PRODUCTION
B 6107 1030 NEBRASKA, CAMBRIDGE ARCH
B 6107 1030 NEBRASKA, CHASE GROUP
B 6107 1030 NEBRASKA, CHEROKEE GROUP
B 6107 1030 NEBRASKA, COUNCIL GROVE GROUP
B 6008 1402* NEBRASKA, CRETACEOUS, FRONTIER SANDSTONE, SOUTHEASTERN,
B 6410 1614 NEBRASKA, DENVER BASIN
B 6107 1032 NEBRASKA, DES MOINES, MARKER IN
B 6108 1307 NEBRASKA, DEVELOPMENTS, 1976
B 6208 1480 NEBRASKA, DEVELOPMENTS, 1977
B 6308 1305 NEBRASKA, DEVELOPMENTS, 1978
B 6409 1440 NEBRASKA, DEVELOPMENTS, 1979
B 6107 1030 NEBRASKA, DOUGLAS GROUP
B 6107 1029† NEBRASKA, GENESIS AND DISTRIBUTION OF, DESMOINESIAN,
B 6107 1030 NEBRASKA, KANSAS CITY GROUP
B 6107 1030 NEBRASKA, LANSING GROUP
B 6107 1030 NEBRASKA, MARMATON GROUP
M 28 51 NEBRASKA, MIOCENE, ARIKAREE FORMATION
B 6107 1030 NEBRASKA, NIPPEWALLA GROUP
B 6002 321* NEBRASKA OIL FIELDS AND ERTS, ABST.
B 6405 804* NEBRASKA PANHANDLE, ABST., PERMIAN-, PENNSYLVANIAN OF
B 6107 1030 NEBRASKA, PEDEE GROUP
B 6107 1038 NEBRASKA, PRECAMBRIAN CRYSTALLINE ROCKS
B 6107 1030 NEBRASKA, REAGAN SANDSTONE
B 6107 1030 NEBRASKA, RED WILLOW COUNTY, STRATIGRAPHIC COLUMN
B 6107 1030 NEBRASKA, SHAWNEE GROUP
B 6107 1030 NEBRASKA, SLEEPY HOLLOW FIELD
B 6107 1035 NEBRASKA, SLEEPY HOLLOW SANDSTONE
B 6107 1030 NEBRASKA, SUMNER GROUP
B 6107 1030 NEBRASKA, WABAUNSEE GROUP
M 29 7 NECKING , CRUSTAL THINNING
B 6010 1714 NEDERLANDSE AARDOLIE MIJ., ON NETHERLANDS, 1975
B 6105 801* NEED FOR BETTER COAL-RESOURCES ASSESSMENTS
B 6209 1754* NEED FOR REDEFINITION OF NORTH AMERICAN PLEISTOCENE STAGES,
M 28 35 NEEDLES RANGE FORMATION, NEVADA, TERTIARY
B 6209 1610 NEEDMORE SHALE, MARYLAND
B 6002 285 NEEDMORE SHALE, PENNSYLVANIA
B 6012 2186 NEEL, T. H., ALASKA EXPLORATION POTENTIAL, ABST.
B 6205 867† NEENACH VOLCANIC FORMATIONS AND THEIR BEARING ON SAN
B 6012 2128† NEENACH VOLCANIC FORMATIONS AND THEIR BEARING ON SAN
B 6303 501 NEESE, D. G., CARBONATE FACIES VARIATION ON GUADALUPIAN
B 6405 756 NEESE, DOUGLAS G., CARBON ISOTOPIC SIGNATURE AS CRITERION FOR
B 6010 1640 NEFF, C. H., PETROLEUM DEVELOPMENTS IN SOUTH AMERICA, CENTRAL
B 6304 573 NEGLIA, S., MIGRATION OF FLUIDS IN SEDIMENTARY BASINS
B 6409 1538 NEGLIA, S., MIGRATION OF FLUIDS IN SEDIMENTARY BASINS. DISCUSSION
M 25 437 NEGROS, AYA-AYA COPPER PROJECT
M 25 437 NEGROS, BASAY COPPER PROJECT
M 25 437 NEGROS, HINOBAAN COPPER PROJECT
M 25 143 NEGROS ISLAND, GEOTHERMAL PROSPECTS, PHILIPPINES
M 25 437 NEGROS, SIPALAY MINE
B 6303 502 NEGUS DE WYS, J., RELATIONS OF GAS OCCURRENCE TO GEOLOGIC
B 6309 1584 NEGUS-DE WYS, JANE, INORGANIC GEOCHEMICA STUDIES OF EASTERN
B 6309 1584 NEGUS-DE WYS, JANE, LITHOLOGY STUDIES OF UPPER DEVONIAN WELL
B 6012 2187 NEIDELL, N. S., INTERPRETIVE USES OF SEISMIC MODELING WITH
M 26 389 NEIDELL, NORMAN S., STRATIGRAPHIC MODELING AND
B 6002 326* NEILL SOUTH, ODOM LIME, FIELD, NOLAN AND FISHER COUNTIES,
S 4 53 NEKTON PROJECT, SUBMERSIBLE
B 6203 377 NELCHINA LIMESTONE, ALASKA
B 6205 890 NELLIS, JOSEPH G., FOLD SYSTEM IN SOUTHERN SIERRA NEVADA

B 6303 449 NELSON, C HANS, VIDEOTAPES OF ACTIVE PRIMARY PHYSICAL AND
B 6004 702 NELSON, C. H., SEDIMENTARY FACIES AND STRUCTURES OF BERING
B 6004 702 NELSON, C. H., THIN-BEDDED TURBIDITES IN MODERN SUBMARINE
B 6105 817 NELSON, C. HANS, COMPLEX RIDGES AND TROUGH TOPOGRAPHY ON
B 6405 706 NELSON, C. HANS, MODIFICATION OF LINEAR SAND RIDGES BY
B 6405 756 NELSON, C. HANS, SEDIMENTARY CHARACTERISTICS AND PROCESSES OF
B 6406 961 NELSON, C. HANS, VARIATION IN TURBIDITE SAND FACIES AND
B 6203 534 NELSON, HANS, HYDROCARBONS OF LOW MOLECULAR WEIGHT IN
B 6002 164 NELSON ISLAND, BERINGIAN SHELF
B 6303 502 NELSON, P. H. H., DISCOVERY AND DEVELOPMENT OF NEMBE CREEK OIL
M 30 565 NELSON, P. H. H., ROLE OF REFLECTION SEISMIC IN DEVELOPMENT OF
B 6312 2214 NELSON, R. A., NATURAL FRACTURE SYSTEMS, DESCRIPTION AND
B 6203 548 NELSON, RALPH L., LITHOFACIES AND DEPOSITIONAL ENVIRONMENTS OF
B 6102 227 NELSON, RONALD A., EXPERIMENTAL STUDY OF FRACTURE
S 4 234 NEMASTER DISCOIDEA CARPENTER, COLOMBIA
S 4 234 NEMASTER DISCOIDEA CARPENTER, GUADELOUPE
S 4 234 NEMASTER DISCOIDEA CARPENTER, JAMAICA
S 4 233 NEMASTER GRANDIS CLARK, COLOMBIA
S 4 233 NEMASTER RUBIGINOSA POURTALES, BONAIRE
S 4 233 NEMASTER RUBIGINOSA POURTALES, GUADELOUPE
S 4 236 NEMASTER RUBIGINOSA POURTALES, MARTINIQUE, PERICLIMENES ON
S 4 233 NEMASTER RUBIGINOSA POURTALES, PRUNE ISLAND
S 4 234 NEMASTER RUBIGINOSA POURTALES, SANTA LUCIA
M 30 565 NEMBE CREEK FIELD, AFRICA, NIGERIA
M 30 570 NEMBE CREEK FIELD, CROSS SECTION, NIGERIA
M 30 570 NEMBE CREEK FIELD, EXPLORATION HISTORY, NIGERIA
M 30 566 NEMBE CREEK FIELD, GEOLOGIC SETTING, NIGERIA
M 30 568 NEMBE CREEK FIELD, LOCATION, NIGERIA
M 30 565 NEMBE CREEK FIELD, NIGERIA, MIOCENE
M 30 565† NEMBE CREEK FIELD, NIGERIA, ROLE OF, REFLECTION, SEISMIC IN
B 6303 502* NEMBE CREEK OIL FIELD, NIGERIA, ABST., DISCOVERY AND
B 6204 669 NEOCOMIAN, ALGERIA
B 6201 52 NEOCOMIAN, AUSTRALIA
M 29 31 NEOCOMIAN, BLAKE PLATEAU, TOP OF
B 6012 2177* NEOCOMIAN MAP UNIT IN NORTHERN BROOKS RANGE FOLDBELT, ABST.,
B 6205 793 NEOCOMIAN, PROFILE, BLAKE PLATEAU AREA
B 6404 507 NEOCOMIAN REEFAL LIMESTONE, NEW ENGLAND MARGIN
B 6004 701* NEOCOMIAN, SANDSTONE SEQUENCE, MACKENZIE DELTA AND
B 6004 657* NEOCOMIAN STRATIGRAPHY, AND HAUTERIVIAN-EOCENE
B 6405 711* NEOGENE AND QUATERNARY SEDIMENT, ABST., HYDRAULIC PISTON,
B 6203 531* NEOGENE BACK-ARC BASIN, WESTERN HOKKAIDO ISLAND, JAPAN, ABST.,
S 6 184 NEOGENE, BADENIAN STAGE, PARATETHYS
B 6004 650* NEOGENE BASI FORMATION AND HYDROCARBON ACCUMULATION
B 6203 344† NEOGENE BASINFORMATION IN RELATION TO PLATE-TECTONIC
S 11 207 NEOGENE, BIOCHRONOLOGIC SUBDIVISIONS, HUMBOLDT BASIN
S 11 201 NEOGENE, BIOSTRATIGRAPHY, HUMBOLDT BASIN
S 11 327 NEOGENE BOLIVINIDS, KEY TO IDENTIFICATION, CALIFORNIA
B 6303 443* NEOGENE CALCAREOUS NANNOPLANKTON BIOSTRATIGRAPHY IN
B 6403 423† NEOGENE DEFORMATION OF SOUTHERN HEDIL, TUNISIAN ATLAS,
B 6405 671* NEOGENE DEPOSITIONAL AND CLIMATI CYCLES IN YAKATAGA
B 6203 368 NEOGENE DEPOSITIONAL BASINS, ONSHORE AND OFFSHORE, CALIFORNIA
B 6306 426* NEOGENE DIATOM CORRELATIONS, ABST., PAN-, PACIFIC
B 6012 2192* NEOGENE DIATOM CORRELATIONS IN CALIFORNIA--PROBLEMS AND
B 6207 1222* NEOGENE DIATOMACEOUS SEDIMENTS IN CIRCUM-PACIFIC REGION,
S 6 180 NEOGENE, EGERIAN STAGE, PARATETHYS
S 6 180 NEOGENE, EGGENBURGIAN STAGE, PARATETHYS
B 6303 473* NEOGENE EXAMPLES FROM PACIFIC, ABST., CORRELATION OF,
B 6002 271 NEOGENE FAULT SYSTEM, ISRAEL
S 11 205 NEOGENE, FORAM LINEAGES, CALIFORNIA
S 11 213 NEOGENE, FORAM NAMES, PRIOR USAGE, CALIFORNIA
B 6301 69 NEOGENE FOREDEEPS, EUROPE
B 6402 244 NEOGENE, GREECE
B 6402 213 NEOGENE, INDIA
B 6205 829 NEOGENE, ISRAEL
S 6 184 NEOGENE, KARPATHIAN STAGE, PARATETHYS
S 6 59 NEOGENE, MAGNETIC ZONES, CORRELATIONS
M 25 448 NEOGENE MINERAL DISTRIBUTION, JAPAN
B 6004 710* NEOGENE, MIOCENE, LITTORAL HARD- AND SOFT-BOTTOM
B 6012 2174† NEOGENE MOLLUSCAN STAGES OF OREGON AND WASHINGTON, ABST.
B 6407 1087† NEOGENE OF NORTHEAST ASIA, CHUKOTKA, KAMCHATKA, SAKHALIN,
S 6 179† NEOGENE OF THE PARATETHYS REGION, A, RADIOMETRIC TIME SCALE
S 6 180 NEOGENE, OTTNANGIAN STAGE, PARATETHYS
S 6 185 NEOGENE, PALEOMAGNETIC SCALES, PARATETHYS
S 6 187 NEOGENE, PANNONIAN STAGE, PARATETHYS
B 6012 2182* NEOGENE PLAEOBATHYMETRY AND PALEOENVIRONMENNS OF
S 11 111 NEOGENE PLANKTON ZONATIONS, CORRELATIONS
B 6303 514* NEOGENE RADIOLARIAN BIOSTRATIGRAPHY AND
S 6 181 NEOGENE, RADIOMETRIC DATES, TABLE, PARATETHYS
S 6 182 NEOGENE, RADIOMETRIC SAMPLES, TABLE, PARATETHYS
S 6 185 NEOGENE, RADIOMETRIC SCALES, PARATETHYS
S 4 106 NEOGENE REEFS, CARIBBEAN
S 6 189 NEOGENE, SAMPLE ANALYSIS, 12 PAGES, PARATETHYS
S 6 186 NEOGENE, SARMATIAN STAGE, PARATETHYS
B 6402 189 NEOGENE SEDIMENTARY AND OLDER BASEMENT ROCKS, SPAIN
B 6011 2019† NEOGENE SEDIMENTS OF ATLANTIC INNER CONTINENTAL SHELF OFF
S 6 59 NEOGENE, STRATIGRAPHIC ZONES, CORRELATION
B 6309 1599* NEOGENE STRATIGRAPHY OF BAY COUNTY, FLORIDA, ABST.,
B 6012 2192* NEOGENE TECTONICS OF CENTRAL VENTURA BASIN, CALIFORNIA, ABST.
B 6102 192† NEOGENE TURBIDITE SEDIMENTATION IN KOMANDORSKIY BASIN,
B 6201 87† NEOGENE VOLCANICLASTIC SEDIMENTS FROM ATKA BASIN, ALEUTIAN
S 6 172 NEOGENE-QUATERNARY BOUNDARY, A REVIEW
S 6 171 NEOGENE-QUATERNARY BOUNDARY, AGREEMENT
S 6 177 NEOGENE-QUATERNARY BOUNDARY, ARGUMENTS
S 6 173 NEOGENE-QUATERNARY BOUNDARY, ITALY
S 6 174 NEOGENE-QUATERNARY, CORRELATION, DIAGRAM
B 6012 2180* NEOGENE-QUATERNARY TECTONICS AN DEPOSITIONAL ENVIRONMENTS
B 6207 1178 NEOGONDOLELLA BITTERI ZONE, NEVADA AND, UTAH, YAKOVLEVIA
B 6207 1178 NEOSTREPTOGNATHODUS PRAYI ZONE, NEVADA AND, UTAH,
B 6207 1178 NEOSTREPTOGNATHODUS SP. C ZONE, UTAH, PENICULAURIS BASSI-
B 6207 1178 NEOSTREPTOGNATHODUS SULCOPLICATUS ZONE, NEVADA AND, UTAH,
M 30 232 NEP-BOTUOBIN ANTICLISE, USSR
B 6110 1838 NEPAL, DEVELOPMENTS, 1976
B 6210 1986 NEPAL, DEVELOPMENTS, 1977
B 6310 1824 NEPAL, DEVELOPMENTS, 1978
B 6411 1890 NEPAL, DEVELOPMENTS, 1979
B 6109 1542* NEPHELOID LAYER ON CONTINENTAL SHELF, GULF OF MEXICO, ABST.,
B 6405 690* NEREITES GROUP OF TRACE FOSSILS, ABST., SYNONYMY, ETHOLOGY,
M 26 67 NERITIC FACIES, RELATIVE RISE, SEA LEVEL FLUCTUATIONS
M 28 133 NEROLOY SANDSTONE, CALIFORNIA, MIOCENE
M 28 66 NEROLOY SANDSTONE, CALIFORNIA, MIOCENE
B 6403 444 NESBITT, ELIZABETH A., HOLOCENE MOLLUSK DISTRIBUTION PATTERNS
B 6303 450* NESQUEHONITE FROM CALCIUM-MAGNESIU BICARBONATE SOLUTIONS
B 6311 2120* NESS COUNTY, KANSAS, ABST., CHANNEL, SANDSTONE OIL RESERVOIRS
B 6407 973 NESSON ANTICLINE, MONTANA
B 6108 1382* NESSON ANTICLINE-PAST, PRESENT, AND FUTURE, ABST.
B 6301 70 NESVACILKA CANYON, CZECHOSLOVAKIA
M 30 521 NET PAY ISOPACH, MESSLA FIELD, SARIR SANDSTONE
B 6012 2150† NET SANDSTONE DETERMINATION FOR ISOPACH MAPPING OF
B 6010 1714 NETHERLANDS, 1975, NEDERLANDSE AARDOLIE MIJ., ON
B 6401 34 NETHERLANDS AND WEST GERMANY, GRONINGEN GIANT GAS FIELD
B 6401 34 NETHERLANDS AND WEST GERMANY, ROTLIEGENDES FORMATION
B 6010 1653 NETHERLANDS ANTILLES, 1975, PETROCONSULTANTS S.A., ON
B 6110 1589 NETHERLANDS ANTILLES, 1976, UMIKER, R., ON DEVELOPMENTS IN
B 6210 2056 NETHERLANDS ANTILLES, 1977, UMIKER, R., ON DEVELOPMENTS IN
B 6310 1756 NETHERLANDS ANTILLES, 1978, UMIKER, R., ON DEVELOPMENTS IN
B 6303 528* NETHERLANDS ANTILLES, ABST., DOLOMITIZATION OF, PLIOCENE-,
S 4 5 NETHERLANDS ANTILLES, ARUBA ISLAND
S 4 3† NETHERLANDS ANTILLES, CORAL, REEFS AND THEIR ZONATION IN
S 4 3 NETHERLANDS ANTILLES, LEEWARD GROUP, CORALS
S 4 155† NETHERLANDS ANTILLES, LOW-WATER EMERGENCE OF, CARIBBEAN,
S 4 3 NETHERLANDS ANTILLES, WINDWARD GROUP, CORALS
M 28 82 NETHERLANDS, CRETACEOUS, CRAIE GRISE
M 27 143 NETHERLANDS, CRETACEOUS, CRAIE GRISE
B 6110 1645 NETHERLANDS, DEVELOPMENTS, 1976
B 6210 1915 NETHERLANDS, DEVELOPMENTS, 1977
B 6310 1924 NETHERLANDS, DEVELOPMENTS, 1978
B 6411 1966 NETHERLANDS, EXCLUDING NORTH SEA, DEVELOPMENTS, 1979
B 6203 541* NETHERLANDS FROM SEISMIC DATA, ABST., PREDICTION OF, POROSITY
M 30 36 NETHERLANDS, GRONINGEN FIELD, LANDSAT IMAGE
B 6401 68 NETHERLANDS, MAESTRICHT AREA, SAMPLED SECTION
B 6107 988 NETHERLANDS, MAESTRICHTIAN CHALK
B 6411 1971 NETHERLANDS, NORTH SEA, ONSHORE, PRODUCTION, 1978-1979
B 6411 1971 NETHERLANDS, ONSHORE, PRODUCTION, 1978-1979
B 6110 1654 NETHERLANDS, PRODUCTION, 1975-1976
B 6210 1924 NETHERLANDS, PRODUCTION, 1976-1977
B 6310 1932 NETHERLANDS, PRODUCTION, 1977-1978
B 6210 1924 NETHERLANDS-NORTH SEA, PRODUCTION, 1976-1977
B 6310 1932 NETHERLANDS-NORTH SEA, PRODUCTION, 1977-1978
M 25 48 NEUMAN, LAWRENCE D., PLATE TECTONICS AND MINERAL RESOURCES
B 6004 700 NEUMANN, A. C., HYDROCARBON POTENTIAL OF NORTHWESTERN
B 6105 769 NEUMANN, A. C., LIME MUD DEPOSITION IN ENCLOSED LAGOON, BIGHT
B 6405 755 NEUMANN, A. C., NODULAR SUBMARINE CEMENTATION ON BAHAMIAN
B 6410 1701 NEUMANN, A. CONRAD, CARBONATE SEDIMENT DRIFTS IN NORTHER
B 6203 549 NEUMANN, A. CONRAD, CARBONATE SLOPES, ABST.
B 6204 693 NEUMANN, A. CONRAD, CHARACTERISTICS OF DEEP BAHAMA CHANNELS
B 6004 733 NEUMANN, A. CONRAD, PETROLOGY OF SUBSE CEMENTED CARBONATE
B 6105 841 NEUMANN, A. CONRAD, POROSITY CONTROLS IN SUBSEA CEMENTED
B 6303 500 NEUMANN, A. CONRAD, SEISMIC FACIES AND DEPOSITIONAL PROCESSES
B 6103 376 NEUMANN, A. CONRAD, SHALLOW CARBONATE-BANK-MARGIN GROWTH
B 6405 700 NEURAUTER, THOMAS W., SANDS AND SAND TRANSPORT ON

B 6010 1868 NEUTRAL ZONE, 1975, SAUDI ARABIA MINISTRY OF PETROLEUM AND
B 6110 1798 NEUTRAL ZONE, 1976, AMERICAN INDEPENDENT OIL CO., ARABIAN OIL
B 6210 2114 NEUTRAL ZONE, 1977, ARABIAN OIL CO., LTD., GETTY OIL CO., AND,
B 6310 1886 NEUTRAL ZONE, 1978, ARABIAN OIL CO., LTD., AND, GETTY OIL CO., ON
B 6210 2126 NEUTRAL ZONE, PRODUCTION, 1977, KUWAIT-, SAUDI ARABIA DIVIDED
B 6310 1895 NEUTRAL ZONE, PRODUCTION, 1978, KUWAIT-, SAUDI ARABIA DIVIDED
B 6411 1852 NEUTRAL ZONE, PRODUCTION, 1979, KUWAIT-, SAUDI ARABIA DIVIDED
B 6203 535* NEVADA, ABST., ALGAL, STROMATOLITES IN, EUREKA QUARTZITE,
B 6211 2359* NEVADA, ABST., ALTERATION OF, ALLUVIUM BY, NATURAL, GAS IN,
B 6203 571* NEVADA, ABST., DELTAIC COMPONENTS OF, VALMY AND, VININI
B 6205 881* NEVADA, ABST., EAGLE SPRINGS OIL FIELD, NYE COUNTY
B 6203 538* NEVADA, ABST., ELDER SANDSTONE—, DELTAIC SEDIMENTS OF, ROBERTS
B 6305 838* NEVADA, ABST., EVIDENCE FOR TWO MAJOR CYCLES OF PETROLEUM,
B 6205 883* NEVADA, ABST., GEOLOGIC TECHNIQUES USED IN, TRAP SPRING FIELD
B 6205 884* NEVADA, ABST., GEOMORPHIC EXPLORATION USED IN DISCOVERY OF,
B 6105 840* NEVADA, ABST., GOODWIN LIMESTONE—LOWER, ORDOVICIAN, SHELF-,
B 6403 447* NEVADA, ABST., LACUSTRINE, FLUVIAL, AND FAN
B 6403 447* NEVADA, ABST., LACUSTRINE, FLUVIAL, AND FAN SEDIMENTATION—,
B 6305 828* NEVADA, ABST., MISSISSIPPIAN, CARBONATE, SHELF, MARGIN ALONG,
B 6405 716* NEVADA, ABST., MISSISSIPPIAN SHELF MARGIN AND CARBONATE
B 6203 551* NEVADA, ABST., MODERN, LACUSTRINE, STROMATOLITES, WALKER LAKE
B 6405 767* NEVADA, ABST., OIL AND METALS IN, ORDOVICIAN AND, DEVONIAN,
B 6405 782* NEVADA, ABST., PALEOGEOGRAPHY AND MARINE COMMUNITIES OF,
B 6205 888* NEVADA, ABST., PETROLEUM GEOLOGY OF, GREAT BASIN, WESTERN,
B 6403 443* NEVADA, ABST., RESULTS OF EXPLORATORY DRILLING, NORTHERN,
B 6205 892* NEVADA, ABST., ROBERTS MOUNTAIN, THRUST—ITS SIGNIFICANCE IN OIL
B 6205 880* NEVADA, ABST., SEISMIC EXPLORATION LEADING TO DISCOVERY OF,
B 6211 2363* NEVADA, ABST., SEISMIC EXPLORATION LEADING TO DISCOVERY OF,
B 6405 737* NEVADA, ABST., TECTONICS AND SEDIMENTATION ALONG, ANTLER
B 6211 2362* NEVADA, ABST., TERTIARY GEOLOGY AND, OIL, SHALE RESOURCES OF
B 6405 703* NEVADA, ABST., TERTIARY STRUCTURAL DEVELOPMENT OF SELECTED
B 6205 886* NEVADA, ABST., TERTIARY STRUCTURAL DEVELOPMENT OF SOUTHERN,
B 6205 884* NEVADA, ABST., TERTIARY, VOLCANIC STRATIGRAPHY OF, TRAP SPRING
B 6205 883* NEVADA, ABST., TRAP SPRING OIL FIELD, NYE COUNTY
B 6405 701* NEVADA ABST., TRAP SPRING OIL FIELD, NYE COUNTY
B 6211 2354* NEVADA, ABST., TRAP SPRING OIL FIELD, NYE COUNTY
B 6203 496* NEVADA, ABST., URANIUM DEPOSITS IN, VOLCANICLASTIC ROCKS NEAR
B 6207 1172 NEVADA AND UTAH, ARCTURUS GROUP
B 6207 1174 NEVADA AND UTAH, GERSTER LIMESTONE
B 6207 1171 NEVADA AND UTAH, KAIBAB LIMESTONE
B 6207 1178 NEVADA AND UTAH, KUVELOUSIA LEPTOSA ZONE
B 6207 1172 NEVADA AND UTAH, MEADE PEAK PHOSPHATIC SHALE MEMBER,
B 6207 1178 NEVADA AND UTAH, PARK CITY GROUP, CORRELATIONS, FOSSILS
B 6207 1178 NEVADA AND UTAH, PENICULAURIS BASSI-NEOSTREPTOGNATHODUS
B 6207 1178 NEVADA AND UTAH, PENICULAURIS IVESI-NEOSTREPTOGNATHODUS
B 6207 1173 NEVADA AND UTAH, PLYMPTON FORMATION
B 6207 1178 NEVADA AND UTAH, THAMNOSIA DEPRESSA ZONE
B 6207 1172 NEVADA AND UTAH, THAYNES FORMATION
B 6207 1178 NEVADA AND UTAH, YAKOVLEVIA MULTISTRIATA-NEOGONDOLELLA
B 6008 1399* NEVADA AND WESTERN UTAH, ABST., NEW PALEOSTRUCTURAL
B 6207 1171† NEVADA AND WESTERN UTAH, STRATIGRAPHIC RELATIONS OF, PARK
B 6207 1175 NEVADA, BUTTE MOUNTAINS
B 6207 1172 NEVADA, CHERRY CREEK RANGE, SECTION
M 27 11 NEVADA, CRETACEOUS, NEWARK CANYON FORMATION
B 6207 1175 NEVADA, CURRIE RAVINE
B 6008 1239 NEVADA, DEVELOPMENTS, 1975
B 6108 1204 NEVADA, DEVELOPMENTS, 1976
B 6208 1360 NEVADA, DEVELOPMENTS, 1977
B 6308 1206 NEVADA, DEVELOPMENTS, 1978
B 6409 1340 NEVADA, DEVELOPMENTS, 1979
B 6207 1159† NEVADA DURING LATE MIOCENE TIME, GULF OF CALIFORNIA IN, LAKE,
B 6307 1139† NEVADA DURING LATE MIOCENE TIME. DISCUSSION AND REPLY, GULF
M 25 486 NEVADA, ELY COPPER DISTRICT
B 6211 2350* NEVADA, GEOTHERMAL FIELD, ABST., DISCOVERY AND GEOLOGY OF,
B 6207 1160 NEVADA, GRAND WASH CLIFFS, ARIZONA AND
B 6207 1159 NEVADA, HUALAPAI LIMESTONE MEMBER, MUDDY CREEK FORMATION,
B 6207 1175 NEVADA, LEACH MOUNTAINS
B 6207 1172 NEVADA, MEDICINE RANGE, SECTION
M 25 486 NEVADA, METALLOGENIC PROVINCES
B 6405 799* NEVADA, MONTANA, UTAH, AND WYOMING, ABST., MIDDLE-LATE,
B 6403 361 NEVADA, OROGENIC PHASES, SIERRA
B 6207 1172 NEVADA, PALOMINO RIDGE, SECTION
B 6207 1172 NEVADA, PEQUOP MOUNTAINS, SECTION
B 6207 1172 NEVADA, SPRUCE MOUNTAIN, SECTION
M 28 16 NEVADA, TERTIARY, HORSE SPRING FORMATION
M 28 34 NEVADA, TERTIARY, HORSE SPRINGS FORMATION
M 28 4 NEVADA, TERTIARY, HORSE SPRINGS FORMATION
M 28 90 NEVADA, TERTIARY, HORSE SPRINGS FORMATION
M 28 17 NEVADA, TERTIARY INTRUSIVE
M 28 35 NEVADA, TERTIARY, NEEDLES RANGE FORMATION
B 6303 442* NEVADA--EVIDENCE FOR GROUNDWATER MIXING, ABST., SHALLOW
B 6108 1378* NEVADA-IMPLICATIONS FOR PETROLEUM EXPLORATION, ABST., SHEEP
B 6405 692* NEW ALBAN SHALE GROUP, DEVONIAN-MISSISSIPPIAN, OF ILLINOIS,
B 6309 1578* NEW ALBANY SHALE GAS IN WESTERN KENTUCKY, ABST., PAST AND
B 6309 1575* **NEW ALBANY SHALE GROUP OF ILLINOIS BASIN, ABST., PETROLOGY AND,**
B 6303 344 NEW ALBANY SHALE, KENTUCKY
B 6209 1704 NEW ALBANY SHALE, UNITED STATES, EASTERN
B 6405 803* NEW AND DEVELOPING TECHNIQUES IN COAL EXPLORATION, ABST.
S 4 291 NEW BRITAIN, RABAUL, GALAXEA
B 6004 675* NEW BRUNSWICK, CANADA, ABST., MARINE-, BAR SEDIMENTATION AND
B 6108 1340 NEW BRUNSWICK PRODUCTION, 1973-1976
B 6208 1456 NEW BRUNSWICK, PRODUCTION, 1974-1977
B 6308 1349 NEW BRUNSWICK, PRODUCTION, 1975-1978
B 6409 1522 NEW BRUNSWICK, PRODUCTION, 1976-1979
M 25 460 NEW CALEDONIA, NICKEL AND COBALT, PHILIPPINES, INDONESIA, AND
S 4 291 NEW CALEDONIA, NOUMEA, GALAXEA
B 6008 1404* NEW CONCEPTS IN METHODOLOGY FOR PROCESSING U.S. GEOLOGICAL
B 6203 544* NEW DEVELOPMENTS IN METHODS OF RESOURCE ESTIMATION, GULF OF
B 6405 711* NEW DIMENSIONS IN SEISMIC EXPLORATION, ABST.
B 6309 1580* NEW ECONOMICS OF NATURAL GAS PRODUCTION IN APPLACHIAN
B 6203 559* NEW ENGLAND, ABST., SEDIMENTARY PROCESSES ON, CONTINENTAL
B 6404 508 NEW ENGLAND, BASEMENT TECTONICS
B 6404 516 NEW ENGLAND CONTINENTAL MARGIN, CRETACEOUS
B 6404 516 NEW ENGLAND CONTINENTAL MARGIN, GEOLOGY
B 6404 516 NEW ENGLAND CONTINENTAL MARGIN, JURASSIC
B 6404 516 NEW ENGLAND CONTINENTAL MARGIN, RIFTING
B 6404 516 NEW ENGLAND CONTINENTAL MARGIN, TRIASSIC
M 29 72 NEW ENGLAND, DEPTH SECTIONS, ATLANTIC COAST
B 6404 514 NEW ENGLAND, JURASSIC-CRETACEOUS BOUNDARY
B 6404 514 NEW ENGLAND, K UNCONFORMITY
B 6404 507 NEW ENGLAND MARGIN, HEEZEN CANYON
B 6404 507 NEW ENGLAND MARGIN, NEOCOMIAN REEFAL LIMESTONE
B 6404 511 NEW ENGLAND, PASSIVE MARGIN, BASEMENT TECTONIC MAP
B 6404 501† NEW ENGLAND PASSIVE MARGIN, GEOLOGY OF
B 6404 514 NEW ENGLAND, SEAMOUNT CHAIN
B 6109 1484 NEW ENGLAND SEAMOUNT CHAIN, EAST COAST
B 6404 502 NEW ENGLAND, SEISMIC REFLECTION PROFILES
B 6004 669* NEW ENVIRONMENTAL MAPS FOR GULF OF MEXICO OCS PROGRAMS,
B 6012 2185* NEW EVIDENCE FOR AGE OF CENOZOIC SAL LAKE BEDS IN
B 6203 556* NEW EVIDENCE FOR HOT, DEEP ORIGIN AND MIGRATION OF
B 6203 511* NEW EVIDENCE OF CURRENT WINNOWING ACTIVITY ON HUENEME SILL,
B 6405 672* NEW EXPLORATION AND EVALUATION OF COAL RESOURCES IN
B 6008 1404* NEW FRONTIERS IN COLORADO COAL RESEARCH ABST.
B 6206 1093* NEW GEOLOGIC FRAMEWORK FOR MINING OF OIL SANDS, ABST.
B 6207 1221* NEW GUINEA, ABST., OK TEDI, COPPER PROJECT, PAPUA
B 6207 1128 NEW GUINEAN CONTINENTAL PLATE, AUSTRALO-
M 28 32 NEW HAMPSHIRE, PALEOZOIC, ANDALUSITE SCHIST
M 28 48 NEW HAMPSHIRE, PALEOZOIC, ANDALUSITE SCHIST
M 28 49 NEW HAMPSHIRE, PALEOZOIC, ANDALUSITE SCHIST
M 28 8 NEW HAMPSHIRE, PALEOZOIC, ANDALUSITE SCHIST
M 28 55 NEW HAVEN ARKOSE, CONNECTICUT, TRIASSIC
M 28 67 NEW HAVEN ARKOSE, CONNECTICUT, TRIASSIC
M 28 164 NEW HAVEN ARKOSE, CONNECTICUT, TRIASSIC
M 28 124 NEW HAVEN ARKOSE, CONNECTICUT, TRIASSIC
M 28 135 NEW HAVEN ARKOSE, CONNECTICUT, TRIASSIC
M 28 5 NEW HAVEN ARKOSE, CONNECTICUT, TRIASSIC
M 28 136 NEW HAVEN ARKOSE, CONNECTICUT, TRIASSIC
M 28 32 NEW HAVEN ARKOSE, CONNECTICUT, TRIASSIC
M 28 6 NEW HAVEN ARKOSE, CONNECTICUT, TRIASSIC
B 6110 1882 NEW HEBRIDES AND SOLOMON ISLANDS, DEVELOPMENTS, 1976
B 6207 1225* NEW HEBRIDES ARC-TRENCH SYSTEM, ABST., SEISMICITY AND
B 6411 1779 NEW HEBRIDES, DEVELOPMENTS, 1979
B 6007 1120 NEW HOPE FIELD, TEXAS
B 6205 880* NEW INFORMATION ON LISBURNE GROUP, CARBONIFEROUS AND
B 6112 2074 NEW ISLAND, SAWU GROUP, TO AUSTRALIAN CRATON, SECTION
B 6203 495* NEW JERSEY, ABST., SLOPE, MAP DEPICTING MAJOR SUBMARINE, SLIDE
B 6004 704* NEW JERSEY, ABST., TIMING OF, TRANSGRESSIONS AND, REGRESSIONS IN,
B 6405 694* NEW JERSEY, ABST., TRACE, FOSSIL ASSEMBLAGES OF UPPER,
B 6105 836* NEW JERSEY, AND IMPLICATIONS FOR URANIUM MINERALIZATION,
B 6008 1323† NEW JERSEY AND NORTH CAROLINA IN 1975, DEVELOPMENTS ON,
B 6007 1112† NEW JERSEY ATLANTIC OUTER CONTINENTAL SHELF, EVIDENCE OF,
M 29 78 NEW JERSEY, BALTIMORE CANYON, DIAPIR, ATLANTIC COAST
B 6309 1465 NEW JERSEY, BALTIMORE CANYON, STRATIGRAPHIC COLUMN
M 27 199 NEW JERSEY, CAMBRIAN, KITTATINNY LIMESTONE
M 27 117 NEW JERSEY, CAMBRIAN, KITTATINNY LIMESTONE
B 6004 704* NEW JERSEY COAST PLAIN, ABST., CRETACEOUS AND EARLY, TERTIARY,
B 6303 505* NEW JERSEY CONTINENNAL MARGIN, ABST., OLIGOCENE TRANSGRESSIV
B 6006 932 NEW JERSEY, CONTINENTAL SHELF, SECTION
B 6004 652* NEW JERSEY COSTAL PLAIN ABST., TIMING AND, PALEOCLIMATE
M 27 22 NEW JERSEY, CRETACEOUS, MOUNT LAUREL SANDSTONE

M	28	38	NEW JERSEY, CRETACEOUS, NAVESINK FORMATION
M	27	154	NEW JERSEY, CRETACEOUS, NAVESINK FORMATION
M	27	22	NEW JERSEY, CRETACEOUS, RED BANK SANDSTONE
M	27	106	NEW JERSEY, CRETACEOUS, RED BANK SANDSTONE
M	27	90	NEW JERSEY, CRETACEOUS SEDIMENTS
M	28	38	NEW JERSEY, CRETACEOUS, TINTON SAND
B	6209	1605	NEW JERSEY, DSDP SITE 391 OFF
B	6007	1112	NEW JERSEY, FAULTS ON ATLANTIC CONTINENTAL SHELF
B	6004	705*	NEW JERSEY INTERPRETED FROM R- AND Q-MODE MULTIVARIATE
M	29	441	NEW JERSEY, MARGIN, MATURE SEDIMENTS
M	29	67	NEW JERSEY, REFRACTING HORIZONS, ATLANTIC COAST
M	29	81	NEW JERSEY, SUBSIDENCE HISTORY, ATLANTIC COAST
M	28	44	NEW JERSEY, TRIASSIC, BRUNSWICK FORMATION
M	28	152	NEW JERSEY, TRIASSIC, BRUNSWICK FORMATION
B	6001	87†	NEW JERSEY, UPPER, CRETACEOUS SUBSURFACE STRATIGRAPHY OF,
B	6305	842*	NEW LOOK AT OIL AND GAS RESOURCES OF DENVER-JULESBURG BASIN,
B	6410	1581	NEW MARKET LIMESTONE, VIRGINIA
B	6004	670*	NEW MEXIC AND WEST TEXAS, ABST., NONVADOSE ORIGIN OF, PISOLITIC,
B	6304	604	NEW MEXICO, A ZONE, NIOBRARA MEMBER, MANCOS SHALE
B	6205	781	NEW MEXICO, ABO FORMATION
B	6305	824*	NEW MEXICO, ABST., BASIN EVOLUTION OF LATE PALEOZOIC, TAOS
B	6303	553*	NEW MEXICO, ABST., DEEP-SEA DEPOSITION IN DENSITY-STRATIFIED
B	**6304**	**689***	**NEW MEXICO, ABST., DEPOSITIONAL ENVIRONMENTS, AS ORE CONTROLS**
B	6304	690*	NEW MEXICO, ABST., EFFECTS OF, URANIUM MINING ON GROUNDWATER
B	6205	881*	NEW MEXICO, ABST., ENTRADA OIL FIELDS OF SOUTHERN, SAN JUAN
B	6405	682*	NEW MEXICO, ABST., ENTRADA OIL FIELDS OF SOUTHERN, SAN JUAN
B	6108	1382*	NEW MEXICO, ABST., EVIDENCE FOR, SOURCE AREAS AND DEPOSITIONAL
B	6105	825*	NEW MEXICO, ABST., EXPLORATION AND DEVELOPMENT IN, GRANTS,
B	6304	688*	NEW MEXICO, ABST., EXPLORATION FOR, URANIUM DEPOSITS IN, GRANT
B	6304	687*	NEW MEXICO, ABST., GEOCHEMICAL STUDIES OF, GRANTS MINERAL BELT
B	6405	750*	NEW MEXICO, ABST., GEOCHEMISTRY OF REGIONALLY EXTENSIVE
B	6405	681*	NEW MEXICO, ABST., GEOCHRONOLOGIC EVIDENCE FOR, PALEOZOIC
B	6203	509*	NEW MEXICO, ABST., GEOLOGIC APPRAISAL OF UNDISCOVERED OIL AND
B	6203	558*	NEW MEXICO, ABST., GEOLOGIC APPRAISAL OF UNDISCOVERED OIL AND
B	6203	507*	NEW MEXICO, ABST., GEOLOGIC APPRAISAL OF UNDISCOVERED OIL AND
B	6305	825*	NEW MEXICO, ABST., GEOLOGIC CHARACTERISTICS OF, VALLE CALDERA,
B	6304	687*	NEW MEXICO, ABST., GEOLOGY AND ORE DEPOSITS OF, JOHNNY M MINE,
B	6303	465*	NEW MEXICO, ABST., GEOPHYSICAL INVESTIGATIONS FOR PILOT PLANT
B	6303	420*	NEW MEXICO, ABST., LITHOFACIES ANALYSIS OF, CASTILE FORMATION,
B	6304	691*	NEW MEXICO, ABST., MAGNITUDE AND VARIABILITY OF DISEQUILIBRIUM
B	**6304**	**692***	**NEW MEXICO, ABST., MINERALOGY AND GEOCHEMISTRY OF MARIANO**
B	6304	693*	NEW MEXICO, ABST., MOUNT TAYLOR, URANIUM DEPOSIT, SAN MATEO
B	6203	544*	NEW MEXICO, ABST., NEW DEVELOPMENTS IN METHODS OF RESOURCE
B	6305	822*	NEW MEXICO, ABST., OIL AND GAS POTENTIAL OF, SANTA FE
B	6105	819*	NEW MEXICO, ABST., ORIGIN OF EARLY, VUGGY, POROSITY IN,
B	6008	1405*	NEW MEXICO, ABST., PALEOCURRENT ANALYSIS OF, DAKOTA AND
B	6405	687*	NEW MEXICO, ABST., PALEOGEOGRAPHIC EVOLUTION OF LATE,
B	6203	569*	NEW MEXICO, ABST., PENNSYLVANIAN, CARBONATE AND, SILICICLASTIC
B	6305	834*	NEW MEXICO ABST., PENNSYLVANIAN COARSE-GRAINED, MEANDERING,
B	6304	687*	NEW MEXICO, ABST., PERIODS OF MINERALIZATION IN, GRANTS
B	6303	451*	NEW MEXICO, ABST., PICTURED CLIFFS SANDSTONE--UPPER,
B	6108	1373†	NEW MEXICO, ABST., RARE EARTH AND OTHER, TRACE ELEMENTS AS,
B	6405	681*	NEW MEXICO, ABST., RB-, SR DATING OF DIAGENESIS—EPIGENESIS IN,
B	6002	326*	NEW MEXICO, ABST., ROCK TANK, MORROW SANDSTONE, FIELD, EDDY
B	6304	692*	NEW MEXICO, ABST., ROLL-TYPE, URANIUM OCCURRENCE AT,
B	6303	471*	NEW MEXICO, ABST., SEAWARD PRIMARY DIP OF FALL-IN BEDS, LOWER,
B	6102	295*	**NEW MEXICO, ABST., SEDIMENTOLOGY OF, CRETACEOUS, MESA RICA**
B	6303	437*	NEW MEXICO, ABST., SEDIMENTOLOGY OF, GOAT SEEP DOLOMITE,
B	6304	693*	NEW MEXICO, ABST., SEM INVESTIGATION OF PARAGENESIS OF,
B	6405	688*	NEW MEXICO, ABST., SPECTRUM OF, DELTAIC SYSTEMS IN,
B	6004	718*	NEW MEXICO, ABST., SUBAQUEOUS, EVAPORITE DEPOSITION, PERMIAN
B	6203	541*	NEW MEXICO, ABST., SUBMARINE, CEMENTS AND FABRIC-RETENTIVE
B	6203	496*	NEW MEXICO, ABST., TURBIDITE ORIGIN FOR, ANHYDRITIC, DOLOMITES
B	6405	689*	NEW MEXICO, ABST., VARIATIONS I, CRETACEOUS COAL-BEARING
B	6104	602	NEW MEXICO, ALAMO HUECO MOUNTAINS
B	6209	1674	NEW MEXICO, AMBROSIA LAKE DISTRICT, URANIUM
B	6205	879*	NEW MEXICO AND ADJACENT SOUTHEASTERN ARIZONA, ABST.,
B	6405	761*	NEW MEXICO AND COLORADO, CLASSIC AREA OF STRATIGRAPHIC
B	6112	2141	NEW MEXICO AND COLORADO, MESAVERDE SANDSTONES
B	6112	2141	NEW MEXICO AND COLORADO, POINT LOOKOUT SANDSTONE
B	6112	2137	NEW MEXICO AND COLORADO, SAN JUAN BASIN, SUBNORMAL
B	6405	688*	NEW MEXICO AND EAGAR FORMATION AND MOGOLLON RIM GRAVELS
B	6203	526*	NEW MEXICO AND EAGAR FORMATION OF ARIZONA, ABST., GENETIC
B	6109	1456	NEW MEXICO, AND MEXICO, ALAMO HUECO BASIN, ARIZONA
B	6405	671*	NEW MEXICO AND SOUTHERN ARIZONA, ABST., MISSISSIPPIAN SYSTEM
B	6109	1449	NEW MEXICO AND TEXAS, CAPITAN REEF
B	6109	1449	NEW MEXICO AND TEXAS, ROOSEVELT UPLIFT
B	6109	1449	NEW MEXICO AND TEXAS, VICTORIO PEAK MARGIN
B	6105	843*	NEW MEXICO AND WEST TEXAS, ABST., EVOLUTION OF, CAPITAN,
B	6304	695*	NEW MEXICO AND WORLD URANIUM, ABST.
B	6104	602	NEW MEXICO, ANIMAS MOUNTAINS
B	6104	602	NEW MEXICO, APACHE HILLS
B	6109	1448	NEW MEXICO, ARIZONA, AND MEXICO, PEDREGOSA BASIN, PETROLEUM
B	6109	1515	NEW MEXICO, ARROYO PENASCO FORMATION
B	6109	1515	NEW MEXICO, ARROYO PENASCO GROUP
B	6205	855	NEW MEXICO, BASEMENT CONTOUR MAP, WEST, TEXAS AND
B	6306	887	NEW MEXICO, BASIN DAKOTA GAS FIELD
B	6004	517†	NEW MEXICO, BASINAL, SANDSTONE, FACIES, DELAWARE BASIN, WEST,
B	6007	1011	NEW MEXICO, BIG DRAW FAULT
B	6104	602	NEW MEXICO, BIG HATCHET MOUNTAINS
B	6112	2139	NEW MEXICO, BISTI FIELD
B	6112	2140	NEW MEXICO, BLANCO GAS FIELD, GREEN MARKER HORIZON
B	6104	603	NEW MEXICO, BLISS FORMATION
S	3	130	NEW MEXICO, BLUE WATER FAULT
B	6007	1011	NEW MEXICO, BLUEWATER FAULT
B	6007	1013	NEW MEXICO, BLUFF SANDSTONE
S	3	133	NEW MEXICO, BLUFF SANDSTONE
B	6007	1011	NEW MEXICO, BLUFF SANDSTONE
B	6209	1675	NEW MEXICO, BLUFF SANDSTONE
B	6205	781	NEW MEXICO, BONE SPRING FORMATION
B	6304	599	NEW MEXICO, BOULDER FIELD
B	6104	601	NEW MEXICO, BROCKMAN HILLS
B	6209	1675	NEW MEXICO, BRUSHY BASIN MEMBER
B	6007	1013	NEW MEXICO, BRUSHY BASIN MEMBER
B	6205	781	NEW MEXICO, BRUSHY CANYON FORMATION
B	6109	1449	NEW MEXICO, BURRO-FLORIDA UPLIFT, ARIZONA AND
B	6205	781	NEW MEXICO, CAPITAN FORMATION
S	5	32	NEW MEXICO, CASTILE FORMATION
B	6104	602	NEW MEXICO, CEDAR MOUNTAINS
B	6205	781	NEW MEXICO, CENTRAL BASIN PLATFORM, SEISMIC DATA
B	6109	1417	NEW MEXICO, CHA CHA FIELD
B	6205	781	NEW MEXICO, CHERRY CANYON FORMATION
B	6007	1013	NEW MEXICO, CHINLE FORMATION
B	6209	1674	NEW MEXICO, CHURCHROCK MINES
B	6209	1674	NEW MEXICO, CHUSKA DISTRICT, URANIUM
B	6104	603	NEW MEXICO, COLINA LIMESTONE
B	6206	1093*	NEW MEXICO, COLORADO, AND UTAH, ABST., PALEOCURRENTS OF,
B	6104	603	NEW MEXICO, CONCHA LIMESTONE
S	3	132	NEW MEXICO, COW SPRINGS SANDSTONE
B	6104	602	NEW MEXICO, COYOTE HILLS
B	6007	1013	NEW MEXICO, CRETACEOUS
B	6304	599	NEW MEXICO, CRETACEOUS, SECTION
B	6303	889	NEW MEXICO, DAKOTA D SANDSTONE
B	6306	887	NEW MEXICO, DAKOTA FORMATION SECTION
B	6007	1013	NEW MEXICO, DAKOTA SANDSTONE
B	6209	1675	NEW MEXICO, DAKOTA SANDSTONE
B	6205	781	NEW MEXICO, DEAN FORMATION
M	26	436	NEW MEXICO, DELAWARE BASIN, ABO FORMATION
M	26	437	NEW MEXICO, DELAWARE BASIN, SAN ANDRES FORMATION
B	6109	1449	NEW MEXICO, DELAWARE BASIN, TEXAS AND
B	6112	2116	NEW MEXICO, DRY CANYON
B	6104	603	NEW MEXICO, EARP FORMATION
B	6304	603	NEW MEXICO, EAST AND WEST PUERTO CHIQUITO FIELDS, FRACTURE
B	6008	1246	NEW MEXICO, EASTERN, PRODUCTION, 1974-1975
B	6108	1211	NEW MEXICO, EASTERN, PRODUCTION, 1975-1976
B	6208	1366	NEW MEXICO, EASTERN, PRODUCTION, 1976-1977
B	6308	1238	NEW MEXICO, EASTERN, PRODUCTION, 1976-1978
B	6409	1372	NEW MEXICO, EASTERN, PRODUCTION, 1977-1979
B	6407	1118	NEW MEXICO, EDITED BY N. M. SULLIVAN, BOUMA, ARNOLD H., REVIEW
B	6104	603	NEW MEXICO, EL PASO FORMATION
B	6401	105	NEW MEXICO, EMPIRE-ABO FIELD
B	6007	1013	NEW MEXICO, ENTRADA SANDSTONE
B	6104	603	NEW MEXICO, EPITAPH DOLOMITE
B	6104	603	NEW MEXICO, ESCABROSA GROUP
B	6109	1515	NEW MEXICO, ESPIRITU SANTO FORMATION
B	6112	2115†	NEW MEXICO, EVOLUTION OF, YUCCA, MOUND COMPLEX, LATE,
B	6112	2086	NEW MEXICO, FRUITLAND FORMATION, KEROGEN
B	6112	2139	NEW MEXICO, GALLEGOS FIELD
B	6209	1674	NEW MEXICO, GALLUP DISTRICT, URANIUM
B	6112	2139	NEW MEXICO, GALLUP SANDSTONE
B	6109	1417	NEW MEXICO, GALLUP SANDSTONES AND SHALES
B	6306	986	NEW MEXICO GEOLOGIC SOCIETY WITH ARIZONA GEOLOGICA SOCIETY
B	6111	2012†	NEW MEXICO GEOLOGICAL SOCIETY, 1947-1976, HISTORY OF
B	6111	2013	NEW MEXICO GEOLOGICAL SOCIETY, 1947-1977, OFFICERS
B	6111	2015	NEW MEXICO GEOLOGICAL SOCIETY FIELD CONFERENCE GUIDEBOOKS
B	6205	781	NEW MEXICO, GETAWAY FORMATION
B	6007	1011	NEW MEXICO, GOAT MOUNTAIN FAULT
B	6205	781	NEW MEXICO, GOAT SEEP FORMATION
S	3	4	NEW MEXICO, GRANTS DISTRICT, MOUNT TAYLOR VOLCANIC FIELD
S	3	124	NEW MEXICO, GRANTS, POISON CANYON SANDSTONE

S 3 76 NEW MEXICO, GRANTS, TREND AND STACK ORES
B 6101 38 NEW MEXICO, GRANTS URANIUM REGION
S 3 124 NEW MEXICO, GRANTS, WESTWATER SANDSTONE
B 6205 781 NEW MEXICO, GRAYBURG FORMATION
B 6306 887 NEW MEXICO, GREENHORN LIMESTONE MARKER
B 6104 602 NEW MEXICO, HATCHET GAP
B 6104 603 NEW MEXICO, HATCHITA FORMATION
B 6104 603 NEW MEXICO, HELL-TO-FINISH FORMATION
B 6104 603 NEW MEXICO, HIDALGO FORMATION
B 6304 599 NEW MEXICO, HOGBACK MONOCLINE
B 6302 158 NEW MEXICO, HOGBACK MONOCLINE, COLORADO AND
S 4 243 NEW MEXICO, HOLDER FORMATION
M 27 119 NEW MEXICO, HOLOCENE SEDIMENTS
B 6104 603 NEW MEXICO, HORQUILLA LIMESTONE
B 6304 599 NEW MEXICO, HORSESHOE FIELD, SECTION
B 6306 887 NEW MEXICO, HOSPAH-LONE PINE AREA
B 6008 1243† NEW MEXICO IN 1975, DEVELOPMENTS IN WEST, TEXAS AND EASTERN
B 6108 1208† NEW MEXICO IN 1976, DEVELOPMENTS IN WEST, TEXAS AND EASTERN
B 6208 1364† NEW MEXICO IN 1977, DEVELOPMENTS IN WEST, TEXAS AND EASTERN
B 6308 1236† NEW MEXICO IN 1978, DEVELOPMENTS IN WES, TEXAS AND EASTERN
B 6409 1369† NEW MEXICO IN 1979, DEVELOPMENTS IN WEST, TEXAS AND EASTERN
B 6209 1674 NEW MEXICO, JACKPILE MINE
B 6209 1675 NEW MEXICO, JACKPILE SANDSTONE
S 3 70 NEW MEXICO, JACKPILE SANDSTONE
B 6007 1013 NEW MEXICO, JURASSIC, COLUMN
B 6104 603 NEW MEXICO, KEATING FORMATION
B 6401 105 NEW MEXICO, KEMNITZ FIELD
B 6104 602 NEW MEXICO, KLONDIKE HILLS
S 3 125 NEW MEXICO, KM-24 MINE
B 6112 2116 NEW MEXICO, LA LUZ ANTICLINE
B 6209 1680 NEW MEXICO, LAGUNA DISTRICT, URANIUM
B 6104 602 NEW MEXICO, LARAMIDE THRUSTS, CORDILLERAN FOLDBELT
M 26 194 NEW MEXICO, LEA COUNTY, CENTRAL BASIN PLATFORM, SEISMIC
B 6205 781 NEW MEXICO, LEA COUNTY, SEISMIC DATA
B 6112 2119 NEW MEXICO, LEOPARD KNOB
B 6104 602 NEW MEXICO, LITTLE HATCHET MOUNTAINS
B 6109 1515 NEW MEXICO, LOG SPRINGS FORMATION
B 6306 899 NEW MEXICO, LONE PINE FIELD, RESERVOIR
B 6109 1515 NEW MEXICO, MADERA FORMATION
B 6302 158 NEW MEXICO, MESAVERDE FIELD
B 6302 159 NEW MEXICO, MESAVERDE SANDSTONE, COLORADO AND
M 28 23 NEW MEXICO, MIOCENE, HAYNER RANCH FORMATION
M 28 174 NEW MEXICO, MIOCENE, HAYNER RANCH FORMATION
M 28 125 NEW MEXICO, MIOCENE, HAYNER RANCH FORMATION
M 28 177 NEW MEXICO, MIOCENE, HAYNER RANCH FORMATION
B 6109 1515 NEW MEXICO, MISSISSIPPIAN, TERMINOLOGY
B 6104 603 NEW MEXICO, MOJADO FORMATION
B 6104 603 NEW MEXICO, MONTOYA DOLOMITE
B 6007 1010 NEW MEXICO, MORRISON FORMATION
B 6109 1514 NEW MEXICO, NACIMIENTO UPLIFT
B 6104 603 NEW MEXICO, NACO GROUP
B 6112 2086 NEW MEXICO, NAVAJO COAL MINE, KEROGEN
B 6004 646* NEW MEXICO, NORTHEASTERN SONORA, AND NORTHWESTERN
B 6004 723* NEW MEXICO, NORTHEASTERN SONORA, AND NORTHWESTERN
B 6306 886† NEW MEXICO, OIL AND GAS IN, DELTA-MARGIN FACIES OF DAKOTA
B 6112 2116 NEW MEXICO, OROGRANDE BASIN
B 6109 1448 NEW MEXICO, OROGRANDE BASIN
B 6109 1514 NEW MEXICO, OSHA CANYON
B 6109 1513 NEW MEXICO, OSHA CANYON FORMATION, NEW
B 6109 1513† NEW MEXICO, OSHA CANYON FORMATION, PENNSYLVANIAN, NEW,
B 6109 1517 NEW MEXICO, OSHA CANYON TYPE SECTION
B 6109 1449 NEW MEXICO, OTERO UPLIFT
B 6104 603 NEW MEXICO, PARADISE FORMATION
B 6109 1449 NEW MEXICO, PEDERNAL UPLIFT
B 6104 602 NEW MEXICO, PELONCILLO MOUNTAINS
B 6109 1514 NEW MEXICO, PENASCO CANYON
M 27 10 NEW MEXICO, PENNSYLVANIAN LIMESTONE
M 28 155 NEW MEXICO, PENNSYLVANIAN, SANGRE DE CRISTO FORMATION
M 28 19 NEW MEXICO, PENNSYLVANIAN, SANGRE DE CRISTO FORMATION
M 28 162 NEW MEXICO, PENNSYLVANIAN, SANGRE DE CRISTO FORMATION
M 28 18 NEW MEXICO, PENNSYLVANIAN, SANGRE DE CRISTO FORMATION
B 6109 1515 NEW MEXICO, PENNSYLVANIAN, TERMINOLOGY
B 6104 603 NEW MEXICO, PERCHA SHALE
B 6109 1518 NEW MEXICO, PERMIAN
M 28 43 NEW MEXICO, PERMIAN, ABO SANDSTONE
M 28 120 NEW MEXICO, PERMIAN, ABO SANDSTONE
M 28 97 NEW MEXICO, PERMIAN, ABO SANDSTONE
M 28 105 NEW MEXICO, PERMIAN, ABO SANDSTONE
M 28 119 NEW MEXICO, PERMIAN, ABO SANDSTONE
M 28 28 NEW MEXICO, PERMIAN, ABO SANDSTONE
M 28 47 NEW MEXICO, PERMIAN, ABO SANDSTONE
B 6205 781 NEW MEXICO, PERMIAN BASIN, SEISMIC DATA
B 6109 1448 NEW MEXICO, PERMIAN BASIN, TEXAS AND
M 27 5 NEW MEXICO, PERMIAN, CAPITAN FORMATION
M 27 4 NEW MEXICO, PERMIAN, CAPITAN FORMATION
M 27 6 NEW MEXICO, PERMIAN, CAPITAN LIMESTONE
M 27 67 NEW MEXICO, PERMIAN, CAPITAN LIMESTONE
M 27 77 NEW MEXICO, PERMIAN, CAPITAN LIMESTONE
M 27 174 NEW MEXICO, PERMIAN, CAPITAN LIMESTONE
M 27 47 NEW MEXICO, PERMIAN, CAPITAN LIMESTONE
M 27 9 NEW MEXICO, PERMIAN, CAPITAN LS.-TANSILL FM. TRANSITION
M 28 84 NEW MEXICO, PERMIAN, CASTILE FORMATION
M 27 133 NEW MEXICO, PERMIAN, CASTILE FORMATION
M 28 89 NEW MEXICO, PERMIAN, SALADO FORMATION
M 28 162 NEW MEXICO, PERMIAN, SANGRE DE CRISTO FORMATION
M 28 19 NEW MEXICO, PERMIAN, SANGRE DE CRISTO FORMATION
M 28 18 NEW MEXICO, PERMIAN, SANGRE DE CRISTO FORMATION
M 28 155 NEW MEXICO, PERMIAN, SANGRE DE CRISTO FORMATION
M 27 161 NEW MEXICO, PERMIAN, TANSILL FORMATION
M 27 132 NEW MEXICO, PERMIAN, TANSILL FORMATION
B 6109 1514 NEW MEXICO, PINOS CANYON
B 6304 600 NEW MEXICO, POINT LOOKOUT SANDSTONE, MESAVERDE GROUP
B 6109 1514 NEW MEXICO, PRECAMBRIAN
M 28 53 NEW MEXICO, PRECAMBRIAN, ORTEGA QUARTZITE
B 6104 602 NEW MEXICO, PYRAMID MOUNTAINS
B 6109 1518 NEW MEXICO, QUATERNARY
B 6205 781 NEW MEXICO, QUEEN FORMATION
B 6209 1675 NEW MEXICO, RECAPTURE MEMBER
B 6007 1011 NEW MEXICO, RECAPTURE MEMBER
B 6104 603 NEW MEXICO, RINGBONE FORMATION
S 3 47 NEW MEXICO, RODILTO LIMESTONE
B 6209 1673† NEW MEXICO, RUBIDIUM-, STRONTIUM MINIMUM AGES OF
S 4 243 NEW MEXICO, SACRAMENTO MOUNTAINS
B 6205 781 NEW MEXICO, SAN ANDRES FORMATION
B 6205 781 NEW MEXICO, SAN ANGELO FORMATION
M 26 102 NEW MEXICO, SAN JUAN BASIN, CRETACEOUS
M 26 106 NEW MEXICO, SAN JUAN BASIN, ELECTRIC LOG SECTIONS
B 6302 155 NEW MEXICO, SAN JUAN BASIN GAS FIELD
M 24 139 NEW MEXICO, SAN JUAN BASIN, PRODUCTIVE SANDSTONES
B 6306 887 NEW MEXICO, SAN JUAN BASIN, STRUCTURE MAP
B 6007 1011 NEW MEXICO, SAN MATEO FAULT
B 6109 1515 NEW MEXICO, SANDIA FORMATION, MEMBERS
B 6104 603 NEW MEXICO, SCHERRER FORMATION
B 6205 781 NEW MEXICO, SEVEN RIVERS FORMATION
B 6104 602 NEW MEXICO, SIERRA RICA
B 6209 1680 NEW MEXICO, SMITH LAKE DISTRICT, URANIUM
B 6104 601† NEW MEXICO, STRUCTURE O, BROCKMAN HILLS, SOUTHWESTERN
B 6007 1013 NEW MEXICO, SUMMERVILLE FORMATION
B 6109 1515 NEW MEXICO, TERERRO FORMATION
B 6109 1518 NEW MEXICO, TERTIARY
M 28 140 NEW MEXICO, TERTIARY, ASH FLOW TUFF
B 6007 1013 NEW MEXICO, TODILTO FORMATION
B 6007 1013 NEW MEXICO, TRIASSIC
B 6109 1518 NEW MEXICO, TRIASSIC
M 28 98 NEW MEXICO, TRIASSIC, DOCKUM FORMATION
B 6112 2118 NEW MEXICO, TWIN MOUNDS
B 6104 603 NEW MEXICO, U-BAR FORMATION
B 6205 854 NEW MEXICO, ULTRADEEP WELLS, LOCATION, WEST, TEXAS AND
B 6205 859 NEW MEXICO, ULTRADEEP WILDCATS, SUCCESS RATIOS, WEST, TEXAS
M 25 158 NEW MEXICO, VALLES, BACA, CALDERA FIELD, GEOTHERMAL
B 6304 599 NEW MEXICO, VERDE OIL FIELD, FRACTURE RESERVOIR
B 6104 602 NEW MEXICO, VICTORIO MOUNTAINS
B 6205 781 NEW MEXICO, VICTORIO PEAK FORMATION
B 6008 1239 NEW MEXICO, WESTERN, DEVELOPMENTS, 1975
B 6108 1204 NEW MEXICO, WESTERN, DEVELOPMENTS, 1976
B 6208 1360 NEW MEXICO, WESTERN, DEVELOPMENTS, 1977
B 6308 1206 NEW MEXICO, WESTERN, DEVELOPMENTS, 1978
B 6409 1340 NEW MEXICO, WESTERN, DEVELOPMENTS, 1979
B 6209 1675 NEW MEXICO, WESTWATER CANYON MEMBER
B 6007 1010 NEW MEXICO, WESTWATER CANYON MEMBER, FLUVIAL SHEET
S 3 75 NEW MEXICO, WESTWATER SANDSTONE
B 6205 781 NEW MEXICO, YATES FORMATION
B 6205 781 NEW MEXICO, YESO FORMATION
B 6109 1518 NEW MEXICO, YESO FORMATION
B 6112 2117 NEW MEXICO, YUCCA MOUND COMPLEX, LOWER AND UPPER MOUNDS
B 6209 1674 NEW MEXICO, ZUNI UPLIFT
B 6109 1449 NEW MEXICO, ZUNI UPLIFT, ARIZONA AND

B 6108 1126 NEW-FIELD WILDCAT
B 6308 1174 NEW-FIELD WILDCAT
B 6008 1159 NEW-FIELD WILDCAT
B 6008 1159 NEW-POOL WILDCAT
B 6308 1174 NEW-POOL WILDCAT
B 6108 1126 NEW-POOL WILDCAT
B 6208 1281 NEW-POOL WILDCAT
B 6409 1301 NEW-POOL WILDCAT
S 5 55 NEWAGO COUNTY, LOCATION, MICHIGAN,
B 6101 80 NEWARK BASIN, PENNSYLANIA-, NEW YORK
B 6105 836* NEWARK BASIN, PENNSYLVANIA-NEW JERSEY, AND IMPLICATIONS FOR
M 27 11 NEWARK CANYON FORMATION, NEVADA, CRETACEOUS
B 6101 89 NEWARK SUPERGROUP, NORTH AMERICA, EASTERN
B 6211 2327 NEWBURG SANDSTONE, WEST VIRGINIA
B 6012 2191 NEWELL, J. H., NANNOPLANKTON BIOSTRATIGRAPHY O UPPER SACATE
B 6303 516 NEWELL, NORMAN D., MARINE PERMIAN ROCKS OF TUNISIA, ABST.
B 6105 822 NEWELL, WAYNE L., OAK GROVE CORE, NORTHEASTERN VIRGINIA, ABST.,
B 6405 767* NEWFOUNDLAND, ABST., ALGAL-, METAZOAN, BIOHERMS OF LOWER,
B 6405 785* NEWFOUNDLAND, ABST., COARSENING-UPWARD SHELF SEQUENCES IN,
B 6303 474* NEWFOUNDLAND, ABST., CORRELATION AND TIMING OF
B 6105 837* NEWFOUNDLAND, ABST., GEOLOGY AND PETROLEUM POSSIBILITIES OF,
B 6109 1560* NEWFOUNDLAND, ABST., SATELLITE IMAGE ANALYSIS IN, MINERAL
B 6004 682* NEWFOUNDLAND, ABST., TIMING OF, CARBONATE AND, CLAY
B 6306 955 NEWFOUNDLAND, BRADORE FORMATION
B 6306 952 NEWFOUNDLAND, CAMBRIAN
B 6306 952 NEWFOUNDLAND, CARBONIFEROUS
B 6306 966 NEWFOUNDLAND, CONCHE TROUGH
B 6306 955 NEWFOUNDLAND, FORTEAU FORMATION
B 6306 955 NEWFOUNDLAND, GOOSE COVE FORMATION
B 6306 955 NEWFOUNDLAND, GOOSE TICKLE FORMATION
B 6306 952 NEWFOUNDLAND, GREAT NORTHERN PENINSULA, GEOLOGIC MAP
B 6306 955 NEWFOUNDLAND, HAWKE BAY FORMATION
B 6306 952 NEWFOUNDLAND, LABRADOR GROUP
B 6306 951 NEWFOUNDLAND, LANDSAT IMAGES
S 4 186 NEWFOUNDLAND, LONG POINT FORMATION, REEF
B 6306 952 NEWFOUNDLAND, LONG RANGE MOUNTAINS
B 6306 955 NEWFOUNDLAND, MAIDEN POINT FORMATION
B 6306 952 NEWFOUNDLAND, ORDOVICIAN
B 6306 955 NEWFOUNDLAND, ORDOVICIAN GRANITE
B 6306 955 NEWFOUNDLAND, PRECAMBRIAN
B 6306 951† NEWFOUNDLAND, SATELLITE IMAGE ANALYSI IN, MINERAL
B 6306 951 NEWFOUNDLAND, SPHALERITE
B 6306 955 NEWFOUNDLAND, ST. GEORGE GROUP
B 6306 955 NEWFOUNDLAND, TABLE HEAD FORMATION
B 6306 953 NEWFOUNDLAND, WESTERN, FACIES BELTS
B 6304 598† NEW MEXICO--EXPLORATION IMPLICATIONS, FRACTURES IN,
B 6102 294† NEW MEXICO. BY TITLE ONLY, OIL AND, GAS, EXPLORATION AND
B 6410 1718† NEW MEXICO. DISCUSSION AND REPLY, RUBIDIUM-, STRONTIUM
B 6203 492* NEW MEXICO—NATURE AND ORIGIN OF CEMENTS, ABST.,
B 6211 2348* NEW MIOCENE FORMATION AT COOS BAY, OREGON, ABST.
B 6206 1091* NEW MODEL FOR SOME COLORADO PLATEAU URANIUM DEPOSITS, ABST.
B 6203 501* NEW MODEL FOR SOME COLORADO PLATEAU URANIUM DEPOSITS. BY
B 6012 2179* NEW OIL AND GAS FROM MATURE PROVINCES, ABST.
B 6009 1611* NEW OIL FROM OLD FIELDS, ABST.
B 6109 1546* NEW ORLEANS AREA, LOUISIANA, ABST., DIFFERENTIAL, SUBSIDENCE OF,
B 6209 1762* NEW ORLEANS—SOUTH HANCOCK HOLOCENE BARRIER TRENDS AND
B 6008 1399* NEW PALEOSTRUCTURAL INTERPRETATION OF BASI AND RANGE
B 6009 1614* NEW PELECYPOD SPECIES FROM LOWER CRETACEOUS NEAR KERRVILLE,
B 6405 750* NEW PRECISION IN BIOSTRATIGRAPHY THROUGH GRAPHIC
B 6204 694 NEW PROVIDENCE WELL, BAHAMA PLATFORM
M 28 31 NEW RED SANDSTONE, NORTHERN IRELAND, TRIASSIC
B 6309 1579* NEW RIVER FORMATION COALS IN SOUTHERN BOONE AND NORTHERN
B 6105 828* NEW RIVER INLET, NORTH CAROLINA, ABST., ONSHORE SEDIMENT
B 6209 1610 NEW SCOTLAND FORMATION, PENNSYLVANIA
B 6209 1610 NEW SCOTLAND LIMESTONE, MARYLAND
B 6404 511 NEW SHOREHAM FAULT
B 6009 1616* NEW SMACKOVER PRODUCTION AT ANAHUAC, NORTHEASTERN MEXICO,
M 25 427 NEW SOUTH WALES, BROKEN HILL, LEAD-SILVER-ZINC DEPOSIT
M 25 426 NEW SOUTH WALES, CAPTAINS FLAT, LEAD-ZINC-COPPER DEPOSIT
M 25 426 NEW SOUTH WALES, COBAR, COPPER, ZINC, AND GOLD
B 6110 1889 NEW SOUTH WALES, DEVELOPMENTS 1976
B 6210 2151 NEW SOUTH WALES, DEVELOPMENTS, 1977
B 6310 1805 NEW SOUTH WALES, DEVELOPMENTS, 1978
B 6411 1863 NEW SOUTH WALES, DEVELOPMENTS, 1979
M 25 426 NEW SOUTH WALES, WOODLAWN, COPPER-LEAD-ZINC DEPOSIT
B 6105 814* NEW YORK, ABST., BIOGENIC STRUCTURES AND THEIR
B 6203 509* NEW YORK, ABST., FRESHWATER, ONCOLITES, ONONDAGA LAKE
B 6004 685* NEW YORK, ABST., GULL RIVER LIMESTONE--, TRANSGRESSIVE SEQUENCE
B 6203 503* NEW YORK, ABST., RECOGNITION OF, CRATONAL, BANK, SLOPE, AND,
B 6405 783* NEW YORK, ABST., SEDIMENTARY FACIES RELATIONS AND INFERRED

B 6209 1610 NEW YORK, AKRON DOLOMITE
B 6105 841* NEW YORK BIGHT INNER SHELF, ABST., GEOLOGIC EFFECTS OF, OCEAN,
B 6209 1610 NEW YORK, BRAYMAN SHALE
M 28 52 NEW YORK, CAMBRIAN, POTSDAM SANDSTONE
B 6209 1610 NEW YORK, CARLISLE CENTER SHALE
B 6209 1610 NEW YORK, COBLESKILL LIMESTONE
B 6209 1610 NEW YORK, DECEW DOLOMITE
B 6009 1570† NEW YORK, DEFINITION OF, TYPE, LITTLE FALLS, DOLOSTONE, LATE,
M 28 166 NEW YORK, DEVONIAN, BECRAFT LIMESTONE
M 27 206 NEW YORK, DEVONIAN, BECRAFT LIMESTONE
M 27 77 NEW YORK, DEVONIAN, BECRAFT LIMESTONE
M 27 42 NEW YORK, DEVONIAN, GENUNDEWA LIMESTONE
B 6412 2163 NEW YORK, DEVONIAN SHALE FACIES
B 6209 1610 NEW YORK, ESOPUS SHALE
B 6309 1506 NEW YORK, FORMATION-DENSITY LOG STUDY
B 6209 1610 NEW YORK, HELDERBERG GROUP
B 6008 1324† NEW YORK IN 1975, DEVELOPMENTS IN
B 6108 1328† NEW YORK IN 1976, DEVELOPMENTS IN
B 6208 1445† NEW YORK IN 1977, DEVELOPMENTS IN
B 6308 1318† NEW YORK IN 1978, DEVELOPMENTS IN
B 6409 1455† NEW YORK IN 1979, DEVELOPMENTS IN
B 6209 1610 NEW YORK, KALKBERG LIMESTONE
B 6209 1610 NEW YORK, LOCKPORT GROUP
B 6101 80 NEW YORK, NEWARK BASIN, PENNSYLANIA-
B 6209 1610 NEW YORK, ONONDAGA LIMESTONE
M 27 111 NEW YORK, ORDOVICIAN, DEEPKILL SHALE
B 6209 1610 NEW YORK, ORISKANY SANDSTONE
S 3 34 NEW YORK, PANTHER MOUNTAIN CIRCULAR
M 28 6 NEW YORK, PRECAMBRIAN, ADIRONDACK GNEISS
M 28 54 NEW YORK, PRECAMBRIAN, INWOOD LIMESTONE
B 6108 1333 NEW YORK, PRODUCTION, 1976
B 6208 1449 NEW YORK, PRODUCTION, 1977
B 6308 1319 NEW YORK, PRODUCTION, 1978
B 6409 1456 NEW YORK, PRODUCTION, 1979
B 6209 1610 NEW YORK, RICKARD HILL LIMESTONE
B 6209 1610 NEW YORK, RONDOUT LIMESTONE
B 6209 1610 NEW YORK, SALINA GROUP
M 27 55 NEW YORK, SILURIAN, COBLESKILL LIMESTONE
M 27 99 NEW YORK, SILURIAN, ROCHESTER SHALE
B 6209 1610 NEW YORK, SILURIAN-DEVONIAN CARBONATE SEQUENCE,
B 6209 1612 NEW YORK, SILURIAN-DEVONIAN SECTION
B 6004 660* NEW YORK STATE, ABST., INTERTIDAL AND, SUBTIDAL, STROMATOLITES
B 6309 1591* NEW YORK STATE, ABST., STRUCTURAL CONTROL OF, ONONDAGA,
B 6209 1610 NEW YORK, SYRACUSE FORMATION
B 6209 1622 NEW YORK TO OHIO, SILURIAN-DEVONIAN SECTION
B 6209 1618 NEW YORK TO PENNSYLVANIA, SILURIAN-DEVONIAN SECTION
B 6209 1610 NEW YORK, VERNON SHALE
B 6004 717* NEW YORK--TIMING AND ENVIRONMENTAL SIGNIFICANCE, ABST., THREE
B 6303 520* NEW ZEALAND, ABST., EXPLORATION RESULTS OFFSHORE SOUTH
B 6203 504* NEW ZEALAND, ABST., SEDIMENTOLOGY OF, CAPLES TERRANE, SOUTH
B 6010 1947† NEW ZEALAND AND SOUTHWEST PACIFIC ISLAND REGION,
B 6110 1880† NEW ZEALAND AND SOUTHWEST PACIFIC ISLAND REGION IN 1976,
B 6210 1900† NEW ZEALAND AND SOUTHWEST PACIFIC ISLAND REGION IN 1977,
B 6310 1680† NEW ZEALAND AND SOUTHWEST PACIFIC ISLAND REGION IN 1978,
B 6411 1776† NEW ZEALAND AND SOUTHWEST PACIFIC ISLAND REGION IN 1979,
B 6210 1901 NEW ZEALAND AND SOUTHWEST PACIFIC ISLAND REGION, PRODUCTION,
M 25 212 NEW ZEALAND, BASINS
M 25 132 NEW ZEALAND, BROADLANDS GEOTHERMAL FIELD
B 6207 1232* NEW ZEALAND, BY GEOPHYSICAL METHODS, ABST., DEFINING,
M 25 220 NEW ZEALAND, CANTERBURY-EAST OTAGO-CAMPBELL PLATEAU BASIN O
M 25 222 NEW ZEALAND, CAPE EGMONT FIELD
M 25 395 NEW ZEALAND, CHATHAM RISE, PHOSPHORITE DEPOSITS
M 25 391 NEW ZEALAND, COAL PRODUCTION
M 25 222 NEW ZEALAND, EAST COAST FOLD BELT
B 6207 1223* NEW ZEALAND GEOTHERMAL EXPLORATION AND EXPLOITATION
M 25 130 NEW ZEALAND, GEOTHERMAL FIELDS
M 25 218 NEW ZEALAND, GLOMAR CHALLENGER DSDP DRILLING SITES
B 6403 376 NEW ZEALAND, HOLOCENE FAN DELTA
B 6403 376 NEW ZEALAND, HOLOCENE FAN DELTA
M 25 391 NEW ZEALAND, IRON ORE PRODUCTION
M 25 213 NEW ZEALAND, KAPUNI FIELD
M 25 214 NEW ZEALAND, KAPUNI FORMATION
M 25 218 NEW ZEALAND, KAPUNI GAS-CONDENSATE FIELD
M 25 131 NEW ZEALAND, KAWERAU GEOTHERMAL FIELD
M 25 221 NEW ZEALAND, KOTUKU OIL SEEPAGE
M 25 225 NEW ZEALAND, LITTLE TOTARA SAND
M 25 214 NEW ZEALAND, MAHOENUI FORMATION
M 25 222 NEW ZEALAND, MANAIA FAULT
M 25 214 NEW ZEALAND, MATAMATEAONGA FORMATION
M 25 215 NEW ZEALAND, MAUI FIELD
M 25 214 NEW ZEALAND, MOHAKATINO FORMATION
M 25 214 NEW ZEALAND, MOKAU FORMATION
M 25 113 NEW ZEALAND, NORTH ISLAND, GEOTHERMAL FIELDS
S 10 201 NEW ZEALAND, NORTH ISLAND, WAIMATA MUD VOLCANO
M 25 220 NEW ZEALAND, POVERTY BASIN
B 6110 1880 NEW ZEALAND, PRODUCTION, 1976
B 6411 1781 NEW ZEALAND, PRODUCTION, 1978-1979
M 25 392 NEW ZEALAND, RAW MATERIALS, LOCATIONS
M 25 217† NEW ZEALAND, SEDIMENTARY, BASINS AND PETROLEUM PROSPECTS,
M 25 220 NEW ZEALAND, SOLANDER TROUGH
M 25 221 NEW ZEALAND, SOUTH ISLAND, STRUCTURE MAP
M 25 393 NEW ZEALAND, STEEL PRODUCTION
M 25 222 NEW ZEALAND, TARANAKI BASIN
M 25 222 NEW ZEALAND, TARANAKI FAULT
M 25 213 NEW ZEALAND, TATANAKI BASIN
M 25 214 NEW ZEALAND, URENUI FORMATION
M 25 214 NEW ZEALAND, WAIKEKIE FORMATION
M 25 131 NEW ZEALAND, WAIRAKEI GEOTHERMAL FIELD
M 25 210 NEW ZEALAND, WESTERN, MARGINAL GEOSYNCLINE
M 25 130† NEW ZEALAND--SUMMARY, GEOTHERMAL, ENERGY IN
M 25 391† NEW ZEALAND--SUMMARY, IRON ORE, COAL, AND, STEEL PRODUCTION
M 25 211† NEW ZEALAND--SUMMARY, KAPUNI AND, MAUI, GAS-, CONDENSATE
M 25 395† NEW ZEALAND--SUMMARY, SUBMARINE, PHOSPHORITE DEPOSITS OF,
B 6409 1303 NEW-FIELD DISCOVERIES, RESERVE ESTIMATES OF
B 6008 1160 NEW-FIELD DISCOVERIES, RESERVE ESTIMATES OF
B 6108 1128 NEW-FIELD DISCOVERIES, RESERVE ESTIMATES OF
B 6208 1282 NEW-FIELD DISCOVERIES, RESERVE ESTIMATES OF
B 6308 1175 NEW-FIELD DISCOVERIES, RESERVE ESTIMATES OF
B 6008 1155 NEW-FIELD DISCOVERIES--1975, 1972, 1969, SIZE ESTIMATES OF, U.S.
B 6308 1170 NEW-FIELD DISCOVERIES--1978, 1975, 1972, SIZE ESTIMATES OF, U.S.
B 6108 1123 NEW-FIELD DISCOVERIES—1976, 1973, 1970, SIZE ESTIMATES OF, U.S.
B 6208 1278 NEW-FIELD DISCOVERIES—1977, 1974, 1971, SIZE ESTIMATES OF, U.S.
B 6409 1298 NEW-FIELD DISCOVERIES—1979, 1976 1973, SIZE ESTIMATES OF, U.S.
B 6409 1301 NEW-FIELD WILDCAT
B 6208 1281 NEW-FIELD WILDCAT

B 6306 958 NEWFOUNDLAND, WHITE ARM WINDOW
B 6105 785 * NEWLY RECOGNIZED PETROLEUM SOURCE-ROCK UNITS IN EAST-CENTRAL
B 6009 1619* NEWMAN AND PENNINGTON FORMATIONS, EAST-CENTRAL KENTUCKY,
B 6205 890 NEWMAN, GARY W., EOCENE STRUCTURING IN EAST-CENTRAL
B 6103 467 NEWMAN, GARY W., NORTHERN ALASKA PALEOMAGNETISM, PLATE
B 6405 757 NEWMAN, KARL R., PALEOCENE TIME SCALE FOR ROCKY MOUNTAIN
B 6303 344 NEWMAN LIMESTONE, KENTUCKY
S 8 61 NEWPORT CANYON, CALIFORNIA, SUBMARINE CANYONS
S 8 63 NEWPORT CANYON, TOPOGRAPHY, INDEX MAP
B 6111 1982 NEWPORT-INGLEWOOD FAULT ZONE, CALIFORNIA
B 6202 255 NEWPORT-INGLEWOOD ZONE, CALIFORNIA
M 28 17 NEWTOWN GNEISS, CONNECTICUT, ORDOVICIAN
M 28 16 NEWTOWN GNEISS, CONNECTICUT, ORDOVICIAN
B 6101 83 NFIS BASIN
B 6101 85 NFIS, TISI NTEST, BASIN, ATLAS TREND
B 6002 327 NG, DAVID T. W., SUBSURFACE STUDY OF ATOKA, LOWER
B 6301 50 NG, DAVID TAI WAI, SUBSURFACE STUDY OF ATOKA, LOWER
M 25 359 NGANJUK-KERTOSONO VOLCANIC AREA, INDONESIA
S 5 1* NIAGARA AND SALINA GROUPS, NORTHERN SHELF, MICHIGAN BASIN,
S 5 2 NIAGARA GROUP, ILLINOIS, SILURIAN,
S 5 2 NIAGARA GROUP, INDIANA, SILURIAN,
S 5 2 NIAGARA GROUP, MICHIGAN, SILURIAN,
S 5 2 NIAGARA GROUP, OHIO, SILURIAN,
S 5 2 NIAGARA GROUP, ONTARIO, SILURIAN,
S 5 2 NIAGARA GROUP, WISCONSIN, SILURIAN,
S 5 4 NIAGARA GROUP, BURNT BLUFF FORMATION, MICHIGAN BASIN,
S 5 54 NIAGARA GROUP, CLINTON FORMATION, SILURIAN,
S 5 54 NIAGARA GROUP, ENGADINE FORMATION, SILURIAN,
S 5 4 NIAGARA GROUP, GUELPH FORMATION, MICHIGAN BASIN, NIAGARAN
S 5 54 NIAGARA GROUP, GUELPH FORMATION, SILURIAN,
S 5 4 NIAGARA GROUP, LOCKPORT FORMATION, MICHIGAN BASIN, NIAGARAN
S 5 54 NIAGARA GROUP, LOCKPORT FORMATION, SILURIAN,
S 5 4 NIAGARA GROUP, MANISTIQUE FORMATION, MICHIGAN BASIN,
S 5 4 NIAGARA GROUP, PINNACLE REEFS, MICHIGAN BASIN,
S 5 7 NIAGARA GROUP, PINNACLE REEFS, EVOLUTION, STAGES
S 5 54 NIAGARA GROUP, STRATIGRAPHY, MICHIGAN, SILURIAN, NIAGARAN
S 5 4 NIAGARA SERIES, NIAGARA GROUP, BURNT BLUFF FORMATION,
B 6309 1590* NIAGARAN AND LOWER CAYUGAN STRATIGRAPHY AND DEPOSITIONAL
B 6405 629† NIAGARAN BIOHERMS AND INTERBIOHERM DEPOSITS OF WESTERN WEST
B 6209 1633 NIAGARAN PALEOGEOGRAPHY, GREAT LAKES REGION
B 6304 609 NIAGARAN PALEOGEOGRAPHY, MICHIGAN BASIN
B 6405 736* NIAGARAN PINNACLE REEF IN SOUTHERN MICHIGAN BASIN, ABST.,
B 6304 608† NIAGARAN PINNACLE-REEF BELT OF NORTHERN MICHIGAN,
B 6405 739* NIAGARAN PIPE CREEK JR. REEF, GRANT COUNTY, INDIANA, ABST.,
B 6303 524* NIAGARAN REEFS BY BRINE REFLUXION AND MIXING OF FRESH WATER
B 6004 727* NIAGARAN REEFS OF WESTERN OHIO, ABST., PALEOECOLOGY AND,
B 6009 1623* NIAGARAN ROCKS, SILURIAN, IN INDIANA AND PARTS OF ILLINOIS,
S 5 100 NIAGARAN SERIES, AGE OF REEFS, MICHIGAN BASIN,
S 5 95 NIAGARAN SERIES, CORRELATION, SILURIAN,
S 5 4 NIAGARAN SERIES, NIAGARA GROUP, GUELPH FORMATION, MICHIGAN
S 5 4 NIAGARAN SERIES, NIAGARA GROUP, LOCKPORT FORMATION,
S 5 4 NIAGARAN SERIES, NIAGARA GROUP, MANISTIQUE FORMATION,
S 5 54 NIAGARAN SERIES, NIAGARA GROUP, STRATIGRAPHY, MICHIGAN,
B 6405 748* NIAGARAN THORNTON REEF, NORTHEASTERN ILLINOIS, ABST., POROSITY
M 29 239 NIAGUMO, SHOZABURO, SUBSIDENCE OF THE DAITO RIDGE AND
B 6104 601 NIALS, F. L., STRUCTURE O BROCKMAN HILLS, SOUTHWESTERN NEW
M 29 230 NIAS, SUNDA ARC, REFLECTION PROFILE
B 6010 1653 NICARAGUA, 1975, GOLDSMITH, L. H., ON DEVELOPMENTS IN
B 6110 1589 NICARAGUA, 1976, PETROCONSULTANTS S.A., ON DEVELOPMENTS IN
B 6210 2057 NICARAGUA, 1977, PETROCONSULTANTS S.A., ON DEVELOPMENTS IN
B 6310 1757 NICARAGUA, 1978, PETROCONSULTANTS S.A., ON DEVELOPMENTS IN
B 6207 1207* NICARAGUA, ABST., DEVELOPMENT AND, RESERVOIR ANALYSIS OF,
B 6203 552* NICARAGUA, ABST., EFFECTS OF VARIATIONS IN, WAVE, ENERGY ON,
B 6207 1229* NICARAGUA, ABST., GEOLOGY AND DEVELOPMENT OF, MOMOTOMBO,
M 25 164 NICARAGUA, MOMOTOMBO GEOTHERMAL PROSPECT
B 6312 2207 NICARAGUA, OFFSHORE, PROFILE
M 25 163 NICARAGUA, SAN JACINTO TISATE GEOTHERMAL PROSPECT
M 25 164 NICARAGUA, THERMAL AREAS
M 25 164 NICARAGUA, VOLCANOES
M 25 163† NICARAGUA--SUMMARY, EXPLORATION FOR, GEOTHERMAL, ENERGY IN
B 6002 299 NICHOLAS, RICHARD L., SUBSURFACE POSITIVE ELEMENTS WITHIN
B 6405 728 NICHOLS, D. J., APPLICATION OF CINGULUM INDEX TO DINOGYMNIUM
B 6303 502 NICHOLS, D. J., APPLICATION OF PALYNOLOGY TO AGE DETERMINATION,
B 6405 757 NICHOLS, K. M., DISTRIBUTION, DIAGENESIS, AND DEPOSITIONAL
B 6303 503 NICHOLS, KATHRYN M., FIELD RELATIONS OF SOME PRE-TERTIAR
B 6105 817 NICHOLS, MAYNARD M., PROBLEM OF MISPLACED SEDIMENT, ABST.
M 25 460 NICKEL AND COBALT, PHILIPPINES, INDONESIA, AND, NEW CALEDONIA
M 25 430 NICKEL, BRITISH SOLOMON ISLANDS
M 25 497 NICKEL, CANADA, ASBESTOS, CHROMITE, AND
M 25 545 NICKEL, CONCENTRATIONS OF, NAZCA PLATE, COPPER AND
M 25 402 NICKEL DEPOSIT, WESTERN AUSTRALIA, MT. KEITH
M 25 402 NICKEL DEPOSIT, WESTERN AUSTRALIA, PERSEVERANCE
M 25 402 NICKEL DEPOSIT, WESTERN AUSTRALIA, SHERLOCK BAY
M 25 403 NICKEL DEPOSITS, AUSTRALIA, LATERITE
M 25 402 NICKEL DEPOSITS, AUSTRALIA, LATERITE
M 25 402 NICKEL DEPOSITS, SILICATE AND LIMONITE ORE, LATERITE
M 25 401 NICKEL DEPOSITS, WESTERN AUSTRALIA, KALGOORLIE
M 25 401 NICKEL DEPOSITS, WESTERN AUSTRALIA, KAMBALDA
M 25 401 NICKEL DEPOSITS, WESTERN AUSTRALIA, WINDARRA
M 25 400 NICKEL MINERALIZATION, KALGOORLIE
M 25 400 NICKEL MINERALIZATION, OXIDE ZONE, AUSTRALIA
M 25 400 NICKEL MINERALIZATION, PYRITE-VIOLARITE ZONE, AUSTRALIA
M 25 403 NICKEL, RESERVES AND ECONOMICS, AUSTRALIA
M 25 399† NICKEL RESOURCES OF AUSTRALIA--SUMMARY
M 25 401 NICKEL SULFIDE DEPOSITS IN ARCHEAN, WESTERN AUSTRALIA
M 25 400 NICKEL SULFIDE DEPOSITS, WESTERN AUSTRALIA
M 25 402 NICKEL SULFIDE, WESTERN AUSTRALIA, SCOTIA-, CARR BOYD AREA
M 25 427 NICKEL SULFIDE, , WESTERN AUSTRALIA, WEST AUSTRALIAN
M 25 390† NICKEL- AND COPPER-RICH NODULES OF EQUATORIAL NORTH
M 25 402 NICKEL-ENRICHED LATERITE, WESTERN AUSTRALIA, ORA BANDA
M 25 402 NICKEL-ENRICHED LATERITE, WESTERN AUSTRALIA, WINGELLINA
M 25 402 NICKEL-ENRICHED LATERITES, AUSTRALIA, GREENVALE AND,
B 6207 1234* NICKELIFEROUS LATERITE DEPOSITS IN CIRCUM-PACIFIC REGION, ABST.
B 6310 1636 NICOD, MARC-A., PETROLEUM DEVELOPMENTS IN NORTH AFRICA
B 6010 1767 NICOD, MARC-A., PETROLEUM DEVELOPMENTS IN NORTH AFRICA IN 1975
B 6110 1698 NICOD, MARC-A., PETROLEUM DEVELOPMENTS IN NORTH AFRICA IN 1976
B 6210 1796 NICOD, MARC-A., PETROLEUM DEVELOPMENTS IN NORTH AFRICA IN 1977
B 6411 1739 NICOD, MARC-A., PETROLEUM DEVELOPMENTS IN NORTH AFRICA IN 1979
B 6311 2060 NICOLA GROUP, BRITISH COLUMBIA
M 25 509 NICOLA ROCKS, CORDILLERAN OROGEN, TAKLA-, HAZELTON-, CACHE
B 6408 1132 NICOLAI CREEK GAS FIELD, ALASKA
B 6312 2207 NICOYA PENINSULA, PROFILE, COSTA RICA
B 6405 799 NIELSON, D. L., STRATEGY OF EXPLORATION FOR HIGH TEMPERATURE
B 6305 836 NIELSON, DENNIS L., GEOLOGY AND STRUCTURAL CONTROL OF
B 6405 757 NIELSON, R. LARELL, DEPOSITIONAL ENVIRONMENTAL ANALYSIS OF
B 6201 2 NIGER DELTA, ABAKALIKI HIGH
B 6201 18 NIGER DELTA, AFIESERE-OLOMORO MACROSTRUCTURE
B 6201 2 NIGER DELTA, AFIKPO LOW
B 6202 297 NIGER DELTA, AGBADA FORMATION
B 6202 295 NIGER DELTA, AKATA FORMATION
B 6202 296 NIGER DELTA AREA, FORMATIONS, NIGERIA
B 6201 3 NIGER DELTA, BASEMENT
B 6201 2 NIGER DELTA, BENIN FLANK
B 6202 297 NIGER DELTA, BENIN FORMATION
B 6201 2 NIGER DELTA, BENUE TROUGH, SOUTHWEST EXTENSION
B 6007 1076 NIGER DELTA, BOUGUER GRAVITY MAP
B 6201 2 NIGER DELTA, CALABAR FLANK
B 6408 1252 NIGER DELTA, CAWTHORNE CHANNEL ONSHORE
B 6201 2 NIGER DELTA, CRETACEOUS
B 6007 1073 NIGER DELTA, CRETACEOUS, NORTH OF
B 6201 23 NIGER DELTA, CRUDES AND THEIR ORIGIN
B 6201 3 NIGER DELTA, EOCENE
B 6201 10 NIGER DELTA, FAULTING
B 6307 1047 NIGER DELTA, FAULTS
B 6201 30 NIGER DELTA, FIELDS, SHELL-BP
B 6201 19 NIGER DELTA, GENERATION
B 6201 19 NIGER DELTA, GEOTEMPERATURES
B 6007 1074 NIGER DELTA, GEOTHERMAL GRADIENT MAP
B 6201 18 NIGER DELTA, GREATER UGHELLI MEGASTRUCTURE
B 6211 2206 NIGER DELTA, GROWTH-FAULT FEATURES FROM, COUNTY CLARE, GULF
B 6403 429 NIGER DELTA HIGH-WAX OIL FIELD, AFRICA
B 6403 429 NIGER DELTA HIGH-WAX OIL FIELD, AFRICA
B 6201 1† NIGER DELTA, HYDROCARBON HABITAT OF, TERTIARY
B 6007 1076 NIGER DELTA, HYDROCARBON-LIQUID WINDOW
B 6007 1076 NIGER DELTA, IHENDIAGU AREA
B 6201 7 NIGER DELTA, K BLOCK, OFFSHORE
B 6201 18 NIGER DELTA, KOKORI MACROSTRUCTURE
B 6201 19 NIGER DELTA, MIGRATION
B 6201 3 NIGER DELTA, MIOCENE
B 6007 1076 NIGER DELTA, OBAMA FIELD
B 6201 2 NIGER DELTA, OBAN MASSIF
B 6007 1074 NIGER DELTA, OBETIM WELL
B 6201 3 NIGER DELTA, OLIGOCENE
B 6201 2 NIGER DELTA, ONITSHA HIGH
B 6201 3 NIGER DELTA, PALEOCENE
B 6202 298 NIGER DELTA, PETROLEUM POTENTIAL
B 6408 1251† NIGER DELTA, PETROLEUM, SOURCE-BED, EVALUATION OF, TERTIARY
B 6007 1073 NIGER DELTA PETROLIFEROUS PROVINCE, WELL-LOG RECORDS,
B 6201 4 NIGER DELTA, PLIOCENE
B 6009 1543† NIGER DELTA PROLIFIC BELT, PENDULUM EFFECT AND

B 6007 1073† NIGER DELTA SEDIMENTARY BASIN, APPROXIMATE, GEOTHERMAL,
B 6007 1073† NIGER DELTA SEDIMENTARY BASIN, APPROXIMATE, GEOTHERMAL,
B 6201 21 NIGER DELTA, SOURCE ROCKS
B 6408 1251 NIGER DELTA, SOURCE-BED EVALUATION, NIGERIA
B 6201 20 NIGER DELTA, TEMPERATURE PROFILES, SELECTED WELLS
B 6107 969 NIGER DELTA, TERTIARY
B 6202 295† NIGER DELTA, TERTIARY LITHOSTRATIGRAPHY OF
B 6201 18 NIGER DELTA, UGHELLI MACROSTRUCTURE
B 6007 1074 NIGER DELTA, UTAGBA WELL
B 6201 18 NIGER DELTA, UTOROGU MACROSTRUCTURE
B 6412 2113 NIGER DELTA WELL, HYDROCARBONS, NIGER
B 6110 1749 NIGER, DEVELOPMENTS, 1976
B 6210 1848 NIGER, DEVELOPMENTS, 1977
B 6310 1693 NIGER, DEVELOPMENTS, 1978
B 6411 1789 NIGER, DEVELOPMENTS, 1979
B 6412 2113 NIGER, NIGER DELTA WELL, HYDROCARBONS
B 6303 502* NIGERIA, ABST., DISCOVERY AND DEVELOPMENT OF, NEMBE CREEK OIL
B 6405 668* NIGERIA, ABST., PETROLEUM GEOLOGY OF, SENONIAN IN SEDIMENTS,
B 6405 672* NIGERIA, ABST., PETROLEUM PROSPECTS OF, ANAMBRA BASIN,
B 6408 1252 NIGERIA, AGBADA FORMATION
B 6202 296 NIGERIA, AJALI FORMATION
B 6202 296 NIGERIA, AMEKI FORMATION
B 6202 296 NIGERIA, ASU RIVER GROUP
B 6202 296 NIGERIA, AWGU SHALE
B 6202 296 NIGERIA, BENIN FORMATION
B 6408 1252 NIGERIA, BENIN FORMATION
B 6110 1749 NIGERIA, DEVELOPMENTS, 1976
B 6210 1848 NIGERIA, DEVELOPMENTS, 1977
B 6310 1693 NIGERIA, DEVELOPMENTS, 1978
B 6411 1789 NIGERIA, DEVELOPMENTS, 1979
B 6202 296 NIGERIA, EZE AKU SHALE
B 6202 296 NIGERIA, IMO FORMATION
B 6408 1252 NIGERIA, KATA FORMATION
B 6202 296 NIGERIA, MAMU FORMATION
M 30 565 NIGERIA, MIOCENE, NEMBE CREEK FIELD
M 30 565 NIGERIA, NEMBE CREEK FIELD, AFRICA
M 30 570 NIGERIA, NEMBE CREEK FIELD, CROSS SECTION
M 30 570 NIGERIA, NEMBE CREEK FIELD, EXPLORATION HISTORY
M 30 566 NIGERIA, NEMBE CREEK FIELD, GEOLOGIC SETTING
M 30 568 NIGERIA, NEMBE CREEK FIELD, LOCATION
B 6202 296 NIGERIA, NIGER DELTA AREA, FORMATIONS
B 6408 1251 NIGERIA, NIGER DELTA, SOURCE-BED EVALUATION
B 6202 296 NIGERIA, NKPORO SHALE
B 6202 296 NIGERIA, NSUKKA FORMATION
B 6202 296 NIGERIA, OGWASHI-ASABA FORMATION
B 6110 1761 NIGERIA, PRODUCTION, 1975-1976
B 6210 1864 NIGERIA, PRODUCTION, 1976-1977
B 6310 1705 NIGERIA, PRODUCTION, 1978
B 6411 1831 NIGERIA, PRODUCTION, 1979
M 30 565† NIGERIA, ROLE OF, REFLECTION, SEISMIC IN DEVELOPMENT OF, NEMBE
B 6004 691* NIGERIA TO CAPE OF GOOD HOPE, ABST., PETROLEUM HABITAT OF,
S 6 159 NIITSUKI FORMATION, JAPAN
B 6402 189 NIJAR FORMATION, SPAIN
B 6402 189 NIJAR FORMATION, SPAIN
B 6110 1837 NIJHUIS, H. J., ON DEVELOPMENTS IN MALAYSIA, 1976
S 6 171 NIKIFOROVA, K. V., STATUS OF THE BOUNDARY BETWEEN PLIOCENE AND
M 25 47 NIKIFOROVA, O. I., GEOLOGIC HISTORY AND PALEOZOOGEOGRAPHY OF
B 6112 2068 NIKINIKI-KOLBANO AREA, SECTION, TIMOR ISLAND
B 6106 872† NILE CONE AREA, STRUCTURE AND SEDIMENTARY HISTORY OF
B 6106 877 NILE CONE, LEVANT PLATFORM, AND ERATOSTHENES SEAMOUNT,
B 6004 716* NILE DELTA AND CONE, ABST., RECENT SEDIMENTATION ON
B 6106 877 NILE VALLEY AND DELTA, EGYPT
B 6004 650 NILSEN, T. H., NEOGENE BASI FORMATION AND HYDROCARBON
B 6203 344 NILSEN, T. H., NEOGENE BASINFORMATION IN RELATION TO
B 6405 758 NILSEN, TOR H., GEOMETRY AND DISPERSAL PATTERNS OF DEEP-SEA
B 6407 1094 NILSEN, TOR H., MODERN AND ANCIENT SUBMARINE FANS. DISCUSSION
B 6407 1094 NILSEN, TOR H., MODERN AND ANCIENT SUBMARINE FANS. DISCUSSION
B 6405 758 NILSEN, TOR H., SEDIMENTOLOGY OF FLUVIAL UPPER DEVONIAN
B 6004 701 NILSEN, TOR H., SEDIMENTOLOGY OF OUTER-FAN DEPOSITIONAL LOBES
B 6009 1607* NINE MILE POINT FIELD, ARANSAS COUNTY, TEXAS ABST., DENSITY-,
M 25 451 NINGYO-TOGE URANIUM DEPOSITS, JAPAN
M 25 453 NINGYOITE, URANIUM ORE MINERAL, JAPAN
M 30 180 NINIAN FIELD, CROSS SECTION, NORTH SEA
M 30 192 NINIAN FIELD, DEVELOPMENT, NORTH SEA
M 30 173 NINIAN FIELD, EXPLORATION HISTORY, NORTH SEA
M 30 179 NINIAN FIELD, GEOLOGY AND STRUCTURE, NORTH SEA
M 30 175 NINIAN FIELD, LICENSE BLOCKS, NORTH SEA
M 30 174 NINIAN FIELD, LOCATION, NORTH SEA
M 30 178 NINIAN FIELD, MAP, CRETACEOUS UNCONFORMITY
M 30 189 NINIAN FIELD, RESERVOIR CHARACTER, NORTH SEA
M 30 188 NINIAN FIELD, STRATIGRAPHIC COLUMN, NORTH SEA
M 30 181 NINIAN FIELD, STRATIGRAPHY, NORTH SEA
M 30 173† NINIAN FIELD, U.K. SECTOR, NORTH SEA
B 6303 503 NININGER, ROBERT D., FUTURE OF URANIUM, ABST.
M 25 488 NIOBIUM, TANTALUM AND
B 6008 1396* NIOBRARA COUNTY, WYOMING, ABST., CHARACTERISTICS OF UPPER,
M 26 279 NIOBRARA, CRETACEOUS
B 6410 1618 NIOBRARA FORMATION, DENVER BASIN
B 6407 971 NIOBRARA FORMATION, GREAT PLAINS
B 6406 943 NIOBRARA FORMATION, MINNESOTA AND, SOUTH DAKOTA
B 6403 318 NIOBRARA FORMATION, MONTANA
B 6403 318 NIOBRARA FORMATION, MONTANA
B 6407 975 NIOBRARA FORMATION, NORTH DAKOTA
B 6406 943 NIOBRARA FORMATION, SOUTH DAKOTA
B 6404 491 NIOBRARA FORMATION, WYOMING
B 6202 184 NIOBRARA FORMATION, WYOMING
B 6405 786* NIOBRARA GAS PLAY, ABST., CRETACEOUS
B 6107 997 NIOBRARA LIMESTONE, COLORADO, FORT HAYS LIMESTONE MEMBER
B 6107 983 NIOBRARA LIMESTONE, COLORADO, SMOKY HILL CHALK MEMBER
B 6107 997 NIOBRARA LIMESTONE, KANSAS, FORT HAYS LIMESTONE MEMBER
B 6107 987 NIOBRARA LIMESTONE, KANSAS, SMOKY HILL CHALK MEMBER
B 6304 604 NIOBRARA MEMBER, MANCOS SHALE, NEW MEXICO, A ZONE
B 6404 491 NIOBRARA MEMBER, WYOMING
B 6202 184 NIOBRARA SHALE, WYOMING
M 26 288 NIOBRARA STRUCTURE CONTOUR MAP, DENVER BASIN
M 26 283 NIOBRARA-PIERRE SEISMIC LINE, DENVER BASIN
B 6107 1030 NIPPEWALLA GROUP, NEBRASKA
M 25 59 NIPPON GEOSYNCLINAL REGION, USSR
M 25 463 NIPPON OROGENS, OKHOTSK-
M 25 448 NISHIWAKI, CHIKAO, RELATION BETWEEN TECTONICS AND
B 6205 837 NISSENBAUM, ARIE, DEAD SEA ASPHALTS—HISTORICAL ASPECTS
B 6206 920 NITINAT FAN
B 6012 2189* NITROGEN IN FORMATION OF PETROLEUM DEPOSITS, ABST., POSSIBLE
B 6405 679 NIXON, DONALD B., EXPLORATION IN GREAT SALT LAKE, ABST.
B 6202 296 NKPORO SHALE, NIGERIA
B 6302 218 NO PROFESSOR BESNARD CRUISES
B 6409 1552* NO. 1 PLAY IN U.S.A.—SOUTH LOUISIANA TUSCALOOSA TREND, 1975-80,
B 6405 725* NO. 11 COAL SEAM, ABST., PETROGRAPHIC VARIATION IN WESTERN,
B 6204 637 NOAA F.R.S. TOWNSEND CROMWELL CRUISE, MAY-JUNE 1975
B 6204 633 NOAA SHIP DISCOVERER RESEARCH CRUISE
B 6204 633 NOAA SHIP SURVEYOR RESEARCH CRUISE
B 6007 1078 NOAA SHIPS RESEARCHER AND DISCOVERER, CRUISES, 1971-1972
B 6303 453* NOAAS ATLANTIC OCEANOGRAPHIC AND METEOROLOGICAL
B 6105 786* NOAAS WASTE-DISPOSAL STUDIES IN NEW YORK BIGHT
B 6108 1382* NOBLE COUNTY, OKLAHOMA, ABST., ORDOVICIAN OIL, ACCUMULATION
M 26 346 NOBLE COUNTY, OKLAHOMA, SOUTH CERES FIELD
B 6203 490 NOBLE, J. E., RECENT DEVELOPMENTS AT CERRO PRIETO GEOTHERMAL
B 6405 758 NOBLE, JOHN E., STRUCTURAL AND SEDIMENTOLOGIC STUDY OF CERR
B 6405 759 NOBLE, JOHN E., STUDY OF DIAGENETIC PROCESSES IN SANDSTONE
B 6103 309 NODINE-ZELLER, D E., MISSISSIPPIAN COMBINATION TRAP, BINDLEY
S 11 234 NODOSARIA ARUNDINEA
S 11 274 NODOSARIA MONILIFORMIS
S 11 235 NODOSARIA RAPHANISTRUM
S 11 274 NODOSARIA RAPHANISTRUM
S 11 235 NODOSARIA SPINOSA
S 11 235 NODOSARIA TORNATA
S 11 274 NODOSARIA TORNATA
S 11 235 NODOSARIA TOSTA
S 11 272 NODOSARIA TOSTA
S 11 235 NODOSARIA TYMPANIPLECTRIFORMIS
S 11 272 NODOSARIA TYMPANIPLECTRIFORMIS
B 6004 688* NODULAR ANHYDRITE AND PALEOTEMPERATURE DETERMINATION,
B 6405 755* NODULAR SUBMARINE CEMENTATION ON BAHAMIAN SLOPES—POSSIBLE
M 25 382† NODULE ORIGIN--SUMMARY, MANGANESE-NODULE STRUCTURES,
M 25 397 NODULE RESERVES, CHATHAM RISE, PHOSPHATE
M 25 388 NODULES, ECONOMIC POTENTIAL OF
M 25 386† NODULES IN SOUTH PACIFIC--SUMMARY, EXPLOITATION OF,
M 25 387 NODULES, METAL CONTENTS OF
M 25 390† NODULES OF EQUATORIAL NORTH PACIFIC--SUMMARY, NICKEL- AND,
M 25 376† NODULES OF THE PACIFIC, FERROMANGANESE
M 25 535 NODULES, PACIFIC, SOUTHEASTERN, FERROMANGANESE
M 25 535 NODULES, PACIFIC, SOUTHEASTERN, PHOSPHORITE
S 11 234 NODUSARIA MONILIFORMIS
S 11 203 NOEGENE TERMINOLOGY, COMPARISON
B 6303 503 NOEL, DENISE G., CALCAREOUS NANNOPLANKTON I UPPER APTIAN AND

B 6303 503 NOEL, JAMES A., COAL RESOURCES OF PART OF SOUTHERN ILLINOIS
B 6305 846 NOEL, JAMES A., GEOLOGY OF COAL DEPOSITS IN SOUTHEASTERN SALINE
B 6004 678 NOEL, M. R., CARBONATE MASS BALANCE--CYCLING AND DEPOSITION ON
B 6203 531 NOEL P., CAVITY-DWELLING ORGANISMS, COELOBITES, IN LOWER
S 8 152 NOHILI SEAVALLEY, HAWAII, SUBMARINE CANYONS
S 8 160 NOHILI SEAVALLEY, TIME-VELOCITY CURVES, CURRENT FLOW
S 5 112 NOISE, SEISMIC DATA, COHERENT
M 26 3 NOISE, SEISMIC, REFLECTIONS, ATTENUATING
B 6002 326* NOLAN AND FISHER COUNTIES, TEXAS, ABST., NEILL SOUTH, ODOM LIME,
B 6203 539* NOLICHUCKY FORMATION, UPPER CAMBRIAN, SOUTHWEST VIRGINIA,
B 6304 652 NOLICHUCKY SHALE, VIRGINIA
B 6209 1610 NOMENCLATURE, MARYLAND, SILURIAN-, DEVONIAN CARBONATE
B 6209 1610 NOMENCLATURE, NEW YORK, SILURIAN-, DEVONIAN CARBONATE
B 6002 273 NOMENCLATURE, NOTE 44, AMERICAN, COMMISSION ON,
B 6102 248 NOMENCLATURE, NOTE 45, AMERICAN, COMMISSION ON,
B 6207 1185 NOMENCLATURE, NOTE 48, AMERICAN COMMISSION ON, STRATIGRAPHIC
B 6207 1187 NOMENCLATURE, NOTE 49, AMERICAN COMMISSION ON, STRATIGRAPHIC
B 6209 1610 NOMENCLATURE, OHIO, SILURIAN-, DEVONIAN CARBONATE SEQUENCE
B 6209 1610 NOMENCLATURE, PENNSYLVANIA, SILURIAN-, DEVONIAN CARBONATE
B 6207 1185† NOMENCLATURE TO PROVIDE GUIDELINES CONCERNING FORMAL
B 6211 2277 NOMENCLATURE, WESTERN INTERIOR, JURASSIC, CORRELATION AND
B 6105 768* NONBUOYANT DIAPIR EMPLACEMENT
B 6403 415 NONEQUILIBRIUM COMPACTION
B 6403 415 NONEQUILIBRIUM COMPACTION
B 6410 1690 NONESUCH SHALE, CRUDE OIL, MICHIGAN
S 9 19 NONGLACIAL PERIODS, PALEOCLIMATE
B 6012 2103† NONGLACIAL VARVES IN LOWER MEMBER OF ARKANSAS NOVACULITE,
S 10 97 NONHYDROCARBON ORGANIC COMPOUNDS, MIGRATION MECHANISMS
S 11 122 NONION COSTIFERUM
S 11 276 NONION COSTIFERUM
S 11 238 NONION COSTIFERUM
S 11 128 NONION GOUDKOFFI
S 11 122 NONION INCISUM
S 11 71 NONION KERNENSIS
S 11 128 NONION MONTEREYANUM CAMELOENSIS
S 11 122 NONION MONTEREYANUM CARMELOENSIS
S 11 124 NONION MONTEREYANUM CARMELOENSIS
S 11 122 NONION MONTEREYANUM MONTERAYANUM
S 11 124 NONION MONTEREYANUM MONTEREYANUM
S 11 71 NONION MULTICAMERATUM
S 11 126 NONION MULTICAMERATUM, MOHNIAN, CALIFORNIA
S 11 71 NONION PIZARRENSIS
S 11 122 NONION SCHENCKI
S 11 124 NONION SCHENCKI
S 11 71 NONION WEAVERI
S 11 239 NONIONELLA CUSHMANI
S 11 276 NONIONELLA CUSHMANI
S 11 128 NONIONELLA DAVANAENSIS
S 11 276 NONIONELLA MIOCENICA
S 11 239 NONIONELLA MIOCENICA
S 11 71 NONIONIDAE
S 11 238 NONIONIDAE
S 11 122 NONIONIDAE, CALIFORNIA, MIOCENE
S 11 124 NONIONIDAE, MOHNIAN AND DELMONTIAN STAGES, CALIFORNIA
B 6305 831* NONMARINE DEPOSITIONAL ENVIRONMENTS AND URANIUM
B 6405 764* NONMARINE ROCKS, ABST., PETROLEUM OCCURRENCES IN
B 6108 1377* NONMARINE ROCKS OF WESTERN UNITED STATES, ABST.,
B 6203 496* NONMARINE STROMATOLITES FROM CENTRAL AND SOUTHERN FRANCE
B 6408 1259† NONMARINE UPPER CRETACEOUS ROCKS, COOK INLET, ALASKA
M 26 423 NONRESERVOIR BEDS, EFFECTS OF, SEISMIC FACTORS
B 6109 1431 NONRESERVOIR ROCKS, HYDROCARBON-ABSORBING TENDENCIES OF
B 6402 145† NONSEALING FAULTS IN LOUISIANA GULF COAST SALT BASIN, SEALING
S 10 130 NONSTEADY-STATE, MIGRATION, GROUNDWATER FLOW
B 6303 431* NONSTOICHIOMETRY, ITS RELATION TO CARBONATE-ROCK FABRIC,
M 28 69 NONTRONITE, SEM VIEW, CLAY
B 6004 670* NONVADOSE ORIGIN OF PISOLITIC FACIES, CAPITAN REEF COMPLEX,
B 6103 350 NOODLE CREEK LIMESTONE, TEXAS
B 6004 659* NOPAH FORMATION, UPPER CAMBRIAN, SOUTHERN GREAT BASIN, ABST.,
B 6203 560 NORDEN, D., GREAT CANADIAN OIL SANDS—GEOLOGY AND
B 6405 759* NORDERNEYER SEEGAT, WEST GERMANY, ABST., SAND DISPERSAL AT
B 6411 1840 NORDIC PETROLEUMS, INC., ON DEVELOPMENTS IN FUJAIRAH, 1979
B 6203 435 NORDLAND RIDGE, NORTH SEA
B 6106 919 NORFOLK CANYON, EAST COAST
S 8 130 NORFOLK CANYON, TOPOGRAPHY, INDEX MAP
S 8 122 NORFOLK CANYON, U.S. EAST COAST, SUBMARINE CANYONS
B 6401 68 NORFOLK REGION, SAMPLED SECTION, ENGLAND
B 6206 1014 NORIAN, ARCTIC, CANADA, MESOZOIC, GRIESBACHIAN TO UPPER
B 6206 1014 NORIAN TO APTIAN, ARCTIC, CANADA, MESOZOIC, UPPER
M 29 7 NORMAL FAULT MECHANISMS, GRABENS
M 29 8 NORMAL FAULTING, DIAGRAM, SUBSIDENCE
B 6403 415 NORMAL FLUID PRESSURE, MODELS
B 6112 2085 NORMAL OIL-TYPE KEROGEN
S 7 230 NORMAL SEDIMENTS, SEDIMENTATION
S 7 315 NORMAL SLOPE SEQUENCE, CORE ANALYSES, DIAGRAMS
S 7 316 NORMAL SLOPE SEQUENCE, CORE ANALYSIS, EXAMPLE
S 7 315 NORMAL SLOPE SEQUENCE, NORTHWEST GULF OF MEXICO
B 6307 1043 NORMAL-FAULT ASSEMBLAGES, DETACHED
M 29 7 NORMAL-FAULT MECHANISMS, SUBSIDENCE
B 6303 536* NORMALLY AND REVERSELY GRADED BEDS WITHIN LARGE-SCALE
B 6206 967 NORMARK, W. R., CHANNEL SANDS AND PETROLEUM POTENTIAL OF
B 6203 344 NORMARK, W. R., NEOGENE BASINFORMATION IN RELATION TO
M 25 313 NORMARK, W. R., PETROLEUM POTENTIAL OF CONTINENTAL RISE OFF
B 6004 702 NORMARK, W. R., THIN-BEDDED TURBIDITES IN MODERN SUBMARINE
B 6206 912 NORMARK, WILLIAM R., FAN VALLEYS, CHANNELS, AND DEPOSITIONAL
B 6407 1094 NORMARK, WILLIAM R., MODERN AND ANCIENT SUBMARINE FANS.
B 6405 722 NORMARK, WILLIAM R., POSSIBLE GEOMETRIES OF SANDSTONE BODIES
B 6405 803 NORMARK, WILLIAM R., POTENTIAL PETROLEUM RESERVOIRS ON
B 6403 443 NORMARK, WILLIAM R., QUATERNARY STYLES OF CALIFORNIA
B 6007 1021 NORMARK, WILLIAM R., SUBLACUSTRINE FAN MORPHOLOGY IN LAKE
B 6203 549 NORMARK, WILLIAM R., TURBIDITES, TIME AND AGAIN, ABST.
M 24 184 NORPHLET FORMATION, ARKANSAS
M 26 150 NORPHLET FORMATION, TEXAS, EAST
B 6009 1610* NORPHLET IN ARK-LA-TEX REGION, ABST., CORRELATIONS AND
M 24 284 NORPHLET SANDSTONES, FLORIDA
B 6407 1005 NORPHLET SILTSTONE, ARK-LA-TEX AREA
B 6405 759 NORRIS, D. K., EASTERN CORDILLERAN FOLDBELT AND FORELAND OF
B 6105 818 NORRIS, D. K., OLD AND NEW HORIZONS IN FOR WESTERN AND ARCTIC
B 6002 184† NORT SEA SEDIMENTS, SUBSURFACE, ENVIRONMENTAL ANALYSIS OF
B 6205 780 NORTH AFRICA, CRETACEOUS
B 6205 777 NORTH AFRICA, EOCENE
B 6010 1767† NORTH AFRICA IN 1975, PETROLEUM, DEVELOPMENTS IN
B 6110 1698† NORTH AFRICA IN 1976, PETROLEUM, DEVELOPMENTS IN
B 6210 1796† NORTH AFRICA IN 1977, PETROLEUM, DEVELOPMENTS IN
B 6411 1739† NORTH AFRICA IN 1979, PETROLEUM, DEVELOPMENTS IN
B 6007 1009 NORTH AFRICA, NUBIAN
B 6205 780 NORTH AFRICA, OLIGOCENE
B 6205 777 NORTH AFRICA, PALEOCENE
B 6205 788 NORTH AFRICA, PALEOCENE
B 6310 1636† NORTH AFRICA, PETROLEUM, DEVELOPMENTS IN
B 6205 777 NORTH AFRICA, SEISMIC DATA
B 6205 788 NORTH AFRICA, SEISMIC DATA
B 6104 594 NORTH ALAZAN FIELD, TEXAS
B 6012 2179* NORTH AMERICA, ABST., APPLICATIONS OF, PLATE TECTONICS TO
B 6304 699* NORTH AMERICA, ABST., DISSOLVED, HYDROCARBONS IN COASTAL,
B 6305 827* NORTH AMERICA, ABST., ENVIRONMENTS FOR SEDIMENTARY, URANIUM
B 6309 1610* NORTH AMERICA, ABST., EVIDENCE FOR POST-JURASSIC TECTONISM IN
B 6105 824* NORTH AMERICA, ABST., MODELS OF SHALLOW-MARINE, CYCLIC
B 6405 731* NORTH AMERICA, ABST., PALEOBIOLOGIC EVIDENCE FOR, CRETACEOUS,
B 6109 1549* NORTH AMERICA, ABST., PALEOZOIC TECTONICS OF SOUTHERN MARGIN
B 6105 829* NORTH AMERICA, ABST., PETROLEUM POTENTIAL OF, BLAKE PLATEAU-,
B 6405 778* NORTH AMERICA, ABST., PHYSICAL EVIDENCE FOR, CRETACEOUS TIDES,
B 6105 832* NORTH AMERICA, ABST., SEDIMENTARY-TECTONIC EVOLUTION OF,
B 6303 467* NORTH AMERICA, ABST., SEDIMENTATION PROCESSES AND
B 6203 560* NORTH AMERICA, ABST., STRUCTURE OF, ATLANTIC SLOPE OF EASTERN
B 6101 80 NORTH AMERICA, ACADIAN OROGENE
B 6101 80 NORTH AMERICA, ALLEGHANIAN OROGENE
B 6101 83 NORTH AMERICA AND MOROCCO, TRIASSIC-LIASSIC CORRELATIONS,
B 6207 1196 NORTH AMERICA BETWEEN VANCOUVER ISLAND AND
B 6408 1233 NORTH AMERICA, CAMPANIAN SEAWAY
S 6 331 NORTH AMERICA, CARBONIFEROUS, COMPARISON
M 25 509 NORTH AMERICA, CORDILLERAN OROGEN, PORPHYRY COPPER
S 6 263 NORTH AMERICA, CRETACEOUS, ALBIAN-CENOMANIAN BOUNDARY
S 6 260 NORTH AMERICA, CRETACEOUS, BIOSTRATIGRAPHY
S 6 261 NORTH AMERICA, CRETACEOUS, CAMPANIAN-MAASTRICHTIAN
S 6 263 NORTH AMERICA, CRETACEOUS, CENOMANIAN-TURONIAN BOUNDARY
S 6 261 NORTH AMERICA, CRETACEOUS, CONIACIAN-SANTONIAN BOUNDARY
S 6 266 NORTH AMERICA, CRETACEOUS, PROPOSED TIME SCALE
S 6 261 NORTH AMERICA, CRETACEOUS, SANTONIAN-CAMPANIAN BOUNDARY
S 6 259† NORTH AMERICA, CRETACEOUS TIME SCALE FROM
S 6 262 NORTH AMERICA, CRETACEOUS, TIME-SCALES, COMPARISON
S 6 260 NORTH AMERICA, CRETACEOUS, TRICERATOPS
S 6 261 NORTH AMERICA, CRETACEOUS, TURONIAN-CONIACIAN BOUNDARY
S 6 260 NORTH AMERICA, CRETACEOUS, WESTERN INTERIOR
S 6 338 NORTH AMERICA, DEVONIAN, ABSOLUTE DATING
B 6209 1603 NORTH AMERICA, EASTERN, CONTINENTAL MARGIN
M 29 439 NORTH AMERICA, EASTERN MARGIN, MATURE SEDIMENTS
B 6101 89 NORTH AMERICA, EASTERN, NEWARK SUPERGROUP
S 9 17 NORTH AMERICA, EVIDENCES OF DRIFT, PALEOMAGNETISM

B 6302 153 NORTH AMERICA, GAS FIELDS
B 6408 1232 NORTH AMERICA, JURASSIC SAND RIDGES
B 6107 1045† NORTH AMERICA, ORIGIN OF, PHOSPHATIC, BLACK, SHALE FACIES IN,
B 6408 1233 NORTH AMERICA, OXFORDIAN SEA
B 6002 181 NORTH AMERICA PLATE
S 6 364 NORTH AMERICA, PRECAMBRIAN, RADIOMETRIC DATING
M 25 281† NORTH AMERICA, STRUCTURAL EVOLUTION OF LATE, MESOZOIC AND,
B 6405 639 NORTH AMERICA, THRUST BELT PROVINCE
M 29 79 NORTH AMERICA, TRIASSIC, RIFTING HISTORY
S 3 26 NORTH AMERICA, URANIUM AND THORIUM, CYCLES, SECTIONS
M 25 282 NORTH AMERICA, WESTERN, BASIN DISTRIBUTION
M 25 286 NORTH AMERICA, WESTERN, BASINS, PRODUCING
M 25 483 NORTH AMERICA, WESTERN CORDILLERA, METALLOGENIC PROVINCES
M 25 283 NORTH AMERICA, WESTERN, CRETACEOUS
B 6408 1232 NORTH AMERICA, WESTERN, MESOZOIC EPICONTINENTAL SEAWAYS
M 25 286 NORTH AMERICA, WESTERN, QUATERNARY
M 25 286 NORTH AMERICA, WESTERN, TERTIARY
B 6101 79† NORTH AMERICA--COMPARISON, TRIASSIC-, LIASSIC DEPOSITS OF,
B 6004 694* NORTH AMERICA--STATUS REPORT, ABST., USGS-, AAPG, DIGITIZED OIL
B 6211 2356* NORTH AMERICA. BY TITLE ONLY, MESOZOIC CONTINENTAL ACCRETION
B 6107 988 NORTH AMERICAN AND EUROPEAN CHALKS, TABLE
B 6202 265 NORTH AMERICAN AND PACIFIC PLATES, DISPLACEMENT, CALIFORNIA
B 6303 475* NORTH AMERICAN BASIN, WESTERN NORTH ATLANTIC, ABST.,
B 6309 1588* NORTH AMERICAN CONTINENT AND ITS RELATION TO FORMATION AND
B 6004 715* NORTH AMERICAN CORDILLERA, ABST., FUSULINACEAN PROVINCES OF
B 6212 2509* NORTH AMERICAN CORDILLERAS, ABST., FROM, CARIBBEAN TO, SOUTH
M 25 509 NORTH AMERICAN CRATON
B 6008 1153† NORTH AMERICAN DRILLING ACTIVITY IN 1975
B 6108 1121† NORTH AMERICAN DRILLING ACTIVITY IN 1976
B 6208 1275† NORTH AMERICAN DRILLING ACTIVITY IN 1977
B 6308 1168† NORTH AMERICAN DRILLING ACTIVITY IN 1978
B 6409 1295† NORTH AMERICAN DRILLING ACTIVITY IN 1979
M 24 1 NORTH AMERICAN OIL AND GAS FIELDS
M 24 300† NORTH AMERICAN OIL AND GAS FIELDS FROM AAPG PUBLICATIONS,
B 6101 28† NORTH AMERICAN PITCHBLENDE AND RELATED URANIUM DEPOSITS,
B 6207 1195 NORTH AMERICAN PLATE
B 6203 394 NORTH AMERICAN PLATE, ALASKA
B 6412 2152 NORTH AMERICAN PLATE, BERINGIAN MARGIN
M 29 274 NORTH AMERICAN PLATE, BOUNDARY, ALASKA
B 6211 2215 NORTH AMERICAN PLATE, CALIFORNIA
B 6103 467* NORTH AMERICAN PLATES, BERING STRAIT REGION, ALASKA, ABST.,
B 6203 346 NORTH AMERICAN PLATES, INTERACTION OF, PACIFIC, FARALLON AND
B 6209 1754* NORTH AMERICAN PLEISTOCENE STAGES, ABST., NEED FOR
B 6209 1765* NORTH AMERICAN SOUTHWEST AND GULF COAST, ABST., EARLY,
B 6004 703* NORTH AMERICAN TRANSGRESSION-REGRESSION CYCLE FROM
B 6203 351 NORTH ANTELOPE HILLS OIL FIELD, CALIFORNIA
B 6303 543* NORTH ATLANTIC, ABST., GEOLOGIC SIGNIFICANCE OF SEDIMENTAR,
B 6303 475* NORTH ATLANTIC, ABST., MESOZOIC-CENOZOIC SEDIMENTARY
B 6405 790* NORTH ATLANTIC, ABST., ORGANIC FACIES OF MID-, CRETACEOUS,
B 6004 649* NORTH ATLANTIC, ABST., PRELIMINARY RESULTS FROM, LEG, 44, DEEP
B 6303 510* NORTH ATLANTIC, ABST., TEMPORAL AND SPATIAL DISTRIBUTION OF
B 6007 1078† NORTH ATLANTIC AND BARBADOS RIDGE COMPLLEX, TECTONICS OF
B 6204 596 NORTH ATLANTIC BASIN, CRETACEOUS
B 6204 596 NORTH ATLANTIC BASIN, JURASSIC
B 6204 596 NORTH ATLANTIC BASIN, QUATERNARY
B 6204 596 NORTH ATLANTIC BASIN, SECTION
B 6204 596 NORTH ATLANTIC BASIN, TERTIARY
B 6203 419 NORTH ATLANTIC, CONTINENTAL DRIFT
M 29 437 NORTH ATLANTIC, CONTINENTAL RISE, KEROGEN MATURATION
B 6203 420 NORTH ATLANTIC, CRETACEOUS
B 6401 72 NORTH ATLANTIC, DSDP HOLES 390A, 398D, CRETACEOUS-TERTIARY
B 6401 72 NORTH ATLANTIC, DSDP HOLES 390A, 398D, CRETACEOUS-TERTIARY
B 6209 1601 NORTH ATLANTIC, DSDP SITE 391, PROFILE
B 6401 72 NORTH ATLANTIC, DSDP SITES 367 AND 391C, CARBON ISOTOPE RATIOS
B 6401 72 NORTH ATLANTIC, DSDP SITES 367 AND 391C, CARBON ISOTOPE RATIOS
B 6203 424 NORTH ATLANTIC, EOCENE
B 6105 698† NORTH ATLANTIC, GAS-HYDRATE HORIZONS DETECTED IN SEISMIC-
B 6203 420 NORTH ATLANTIC, JURASSIC
S 7 83 NORTH ATLANTIC, JURASSIC (RECONSTRUCTION) DIAGRAM
B 6312 2222† NORTH ATLANTIC KIMMERIDGIAN, COMPOSITION OF, ORGANIC MATTER
B 6312 2224 NORTH ATLANTIC KIMMERIDGIAN SHALES
B 6203 424 NORTH ATLANTIC, MIOCENE
B 6004 706* NORTH ATLANTIC OCEAN, ABST., NORTH SEA, RIFTING AND OPENING OF
B 6203 419† NORTH ATLANTIC OCEAN AND THEIR RELATION TO NORTH SEA AREA,
B 6411 2041 NORTH ATLANTIC OCEAN, EDITED BY RED H. PILGER, JR., LADD, JOHN
B 6403 364 NORTH ATLANTIC OCEAN PLATE MOVEMENTS
B 6203 424 NORTH ATLANTIC, OLIGOCENE
B 6203 420 NORTH ATLANTIC, OPENING OF
B 6203 424 NORTH ATLANTIC, PALEOCENE
B 6404 503 NORTH ATLANTIC, PLATE-TECTONIC EVENTS
B 6203 434 NORTH ATLANTIC, RIFT BASINS, HYDROCARBONS
S 7 82 NORTH ATLANTIC, SALT DIAPIRS, MAGNETIC LINEATIONS
B 6203 423 NORTH ATLANTIC, SEAFLOOR, AGE
B 6203 425 NORTH ATLANTIC, TRIASSIC
M 29 295 NORTH BALEARIC BASIN, CROSS SECTION, MEDITERRANEAN
B 6203 351 NORTH BELRIDGE OIL FIELD, CALIFORNIA
B 6202 274 NORTH BRAZILIAN RIDGE
B 6306 936 NORTH BRAZILIAN RIDGE, BATHYMETRY
B 6302 219 NORTH BRAZILIAN RIDGE, BRAZIL
B 6406 912 NORTH BURU BASIN, NEW NAME
B 6105 828* NORTH CAROLINA, ABST., ONSHORE SEDIMENT TRANSPORT—PROFILE
B 6101 80 NORTH CAROLINA AND VIRGINIA, DURHAM-DEEP RIVER BASIN
B 6309 1465 NORTH CAROLINA, BALTIMORE CANYON, STRATIGRAPHIC COLUMN
B 6008 1323† NORTH CAROLINA IN 1975, DEVELOPMENTS ON, ATLANTIC COASTAL
B 6203 547* NORTH CAROLINA—EFFECTS ON DEPOSITIONAL PATTERNS OVER PAST
B 6203 426 NORTH CELTIC SEA BASIN, EUROPE
B 6408 1132 NORTH COOK INLET GAS FIELD, ALASKA
B 6303 430* NORTH DAKOTA, ABST., DOLOMITIZATION IN UPPER, RE RIVER
B 6305 843* NORTH DAKOTA, ABST., GEOLOGY AND CURRENT DEVELOPMENT
B 6405 780* NORTH DAKOTA, ABST., HEAT-FLOW AND HEAT-PRODUCTION STUDIES
B 6405 804* NORTH DAKOTA, ABST., PROPOSED MODEL FOR DEVELOPMENT OF, RED
B 6008 1401* NORTH DAKOTA, ABST., STRATIGRAPHY AND PETROLEUM
B 6105 821* NORTH DAKOTA, ABST., STRATIGRAPHY, DEPOSITIONAL
B 6008 1398* NORTH DAKOTA, ABST., STRATIGRAPHY OF, MISSISSIPPIAN, CHARLES
S 3 119 NORTH DAKOTA AND SOUTH DAKOTA, WILLISTON BASIN
B 6407 976 NORTH DAKOTA, BELLE FOURCHE SHALE
B 6407 976 NORTH DAKOTA, CARLILE SHALE
B 6008 1207 NORTH DAKOTA, DEVELOPMENTS, 1975
B 6108 1348 NORTH DAKOTA, DEVELOPMENTS, 1976
B 6208 1501 NORTH DAKOTA, DEVELOPMENTS, 1977
B 6308 1387 NORTH DAKOTA, DEVELOPMENTS, 1978
B 6409 1501 NORTH DAKOTA, DEVELOPMENTS, 1979
B 6407 976 NORTH DAKOTA, GAMMON MEMBER
B 6407 976 NORTH DAKOTA, GREENHORN FORMATION
B 6407 975 NORTH DAKOTA, NIOBRARA FORMATION
B 6406 943 NORTH DAKOTA, PEMBINA ESCARPMENT
B 6407 975 NORTH DAKOTA, PIERRE SHALE
B 6208 1508 NORTH DAKOTA, PRODUCTION, 1976-1977
B 6308 1388 NORTH DAKOTA, PRODUCTION, 1977-1978
B 6409 1502 NORTH DAKOTA, PRODUCTION, 1978-1979
S 10 49 NORTH DAKOTA, WILLISTON BASIN, BAKKEN SHALE, MIGRATION
B 6303 485* NORTH DAKOTA--A 24-MILLION BARREL SOIL-ZONE STRATIGRAPHIC
B 6108 1372* NORTH DAKOTA-CRYPTOEXPLOSION STRUCTURE OF SUBSURFACE
B 6103 361 NORTH DUNCAN STRUCTURE, OKLAHOMA
B 6408 1134 NORTH FORK GAS FIELD, ALASKA
S 10 201 NORTH ISLAND, WAIMATA MUD VOLCANO, NEW ZEALAND
B 6404 528 NORTH, GEOLOGIC MAP, GREENLAND
S 6 209 NORTH GERMAN BASIN, CORRELATION, PALEOGENE
B 6211 2209 NORTH GOLEEN GROWTH FAULT, IRELAND, WESTERN
B 6208 1383 NORTH GUM ISLAND, JEFFERSON COUNTY, TEXAS
B 6205 885* NORTH HORN FORMATION AND LOWER PART OF GREEN RIVER
B 6303 313 NORTH HORN FORMATION, UTAH
M 25 113 NORTH ISLAND, GEOTHERMAL FIELDS, NEW ZEALAND
B 6203 443 NORTH KADI FIELD, INDIA
B 6203 453 NORTH KADI OIL FIELD, SECTIONS, INDIA
M 25 269 NORTH KIANGSU, AND SHANGHAI BASINS, KOREA, WEST, YELLOW SEA
B 6210 1986 NORTH KOREA, DEVELOPMENTS, 1977
B 6310 1824 NORTH KOREA, DEVELOPMENTS, 1978
B 6411 1890 NORTH KOREA, DEVELOPMENTS, 1979
M 24 97 NORTH LA BARGE FIELD, WYOMING
B 6104 579 NORTH LAMA FIELD, VENEZUELA
M 29 313 NORTH LEVANTINE BASIN, EVOLUTION, MEDITERRANEAN
M 29 312 NORTH LEVANTINE BASIN, SEDIMENTARY SEQUENCE, MEDITERRANEAN
B 6209 1646 NORTH LUZON TROUGH
B 6402 249 NORTH MATAPAN TROUGH, SEISMIC REFLECTION, GREECE
M 24 100 NORTH MCDONALD DRAW FIELD, WYOMING
B 6408 1132 NORTH MIDDLE GROUND SHOAL GAS FIELD, ALASKA
B 6410 1580 NORTH MOUNTAIN BLOCK, VIRGINIA
B 6410 1576 NORTH MOUNTAIN THRUST, VIRGINIA
B 6408 1279* NORTH PACIFIC AND BERING SEA, ABST., RESOURCE POTENTIAL AND,
M 25 390 NORTH PACIFIC, NICKEL- AND COPPER-RICH NODULES
B 6207 1212* NORTH PACIFIC WATER TEMPERATURES AND ANOMALOUS RUNOFF IN
M 25 390† NORTH PACIFIC--SUMMARY, NICKEL- AND, COPPER-RICH, NODULES OF
B 6108 1385* NORTH PARK, COLORADO, ABST., OIL POTENTIAL OF MESOZOIC
M 29 240 NORTH PHILIPPINE SEA, DAITO RIDGE AREA, PHYSIOGRAPHY
M 29 240 NORTH PHILIPPINE SEA, DAITO RIDGE, GEOLOGY
M 29 240 NORTH PHILIPPINE SEA, DAITO RIDGE, REFLECTION STUDIES
M 29 242 NORTH PHILIPPINE SEA, EOCENE, CARBONATE FOSSILS
M 29 242 NORTH PHILIPPINE SEA, EOCENE, SUBSIDENCE
M 29 243 NORTH PHILIPPINE SEA, KITA-DAITO BASIN, GEOLOGY

M 29 244 NORTH PHILIPPINE SEA, KITA-DAITO BASIN, STRATIGRAPHY
M 29 242 NORTH PHILIPPINE SEA, MINAMI-DAITO BASIN, GEOLOGY
M 29 243 NORTH PHILIPPINE SEA, MINAMI-DAITO BASIN, STRATIGRAPHY
M 29 240 NORTH PHILIPPINE SEA, OKI-DAITO RIDGE, GEOLOGY
M 29 241 NORTH PHILIPPINE SEA, REFLECTION SURVEY, FOLDOUT
M 29 239† NORTH PHILIPPINE SEA, SUBSIDENCE OF THE, DAITO, RIDGE AND
M 29 240 NORTH PHILIPPINE SEA, TOPOGRAPHY, INDEX MAP
B 6409 1356 NORTH PINE PRAIRIE FIELD, WALKER COUNTY, TEXAS
B 6107 1077 NORTH POINT, MICROGRAPHS, BARBADOS
M 25 236 NORTH RANKIN DISCOVERY WELL, AUSTRALIA
B 6201 43 NORTH RANKIN FIELD, AUSTRALIA
B 6207 1219* NORTH RANKIN GAS FIELD, AUSTRALIA, ABST., PRODUCTION GEOLOGY
M 29 131 NORTH SCOTIA RIDGE, ARGENTINA
M 29 135 NORTH SCOTIA RIDGE, SEISMIC REFLECTIONS, ARGENTINA
M 29 131 NORTH SCOTIA RIDGES, MAURICE EWING BANK, DESCRIBED
B 6302 183 NORTH SEA, 17-4-1 DELTA
B 6302 183 NORTH SEA, 8-3-1 DELTA
B 6303 487* NORTH SEA, ABST., BEATRIC FIELD, MORAY FIRTH
B 6405 718* NORTH SEA, ABST., BRAE FIELD AREA
B 6405 789* NORTH SEA, ABST., FAN MODELS FOR, HYDROCARBON EXPLORATION
B 6303 466* NORTH SEA, ABST., FORTIES FIELD, UNITED KINGDO, CONTINENTAL
B 6303 430* NORTH SEA, ABST., GEOTHERMAL INVESTIGATION OF NORTHERN
B 6004 721* NORTH SEA, ABST., LITHOSTRATIGRAPHY OF, NORWEGIAN
B 6203 490* NORTH SEA, ABST., REGIONAL CONTROLS ON DIAGENESIS OF,
B 6405 678* NORTH SEA AIR-GUN DATA, ABST., SIGNATURE PROCESSING OF
B 6203 419† NORTH SEA AREA, RIFT, BASINS FLANKING, NORTH, ATLANTIC OCEAN
B 6307 1053 NORTH SEA AREA, SALT FEATURES
B 6307 1031 NORTH SEA, ARGYLL FIELD, SEISMIC PROFILE
B 6405 781* NORTH SEA BASIN, ABST., CONTINENTAL STRETCHING—EXPLANATION
M 30 119 NORTH SEA, BEATRICE FIELD, EXPLORATION HISTORY
M 30 126 NORTH SEA, BEATRICE FIELD, SEISMIC SECTION REFLECTORS
M 30 121 NORTH SEA, BEATRICE FIELD, STRUCTURE MAP
M 26 350 NORTH SEA, BENTHEIM SANDSTONE, SEISMIC SECTION
B 6311 2002 NORTH SEA, BERYL EMBAYMENT
B 6312 2222 NORTH SEA, BREAM FORMATION
M 26 430 NORTH SEA, BRENT FIELD, MODEL
M 26 430 NORTH SEA, BRENT SANDSTONE
B 6311 2002 NORTH SEA, BRUCE FAN
B 6203 528* NORTH SEA BY COMPUTER, ABST., REGIONAL GEOLOGIC, MAPPING OF
B 6002 185 NORTH SEA, CARBONACEOUS DETRITUS IN DITCH CUTTINGS
B 6107 1002 NORTH SEA, CENTRAL GRABEN
B 6203 436 NORTH SEA, CENTRAL GRABEN, DANISH SECTOR, RIFT SYSTEM
M 30 196 NORTH SEA, CENTRAL GRABEN, LOCATION
B 6203 436 NORTH SEA, CENTRAL GRABEN, NETHERLANDS SECTOR, RIFT SYSTEM
B 6307 1031 NORTH SEA, CENTRAL GRABEN, ROTLIEGENDES SANDSTONE, SEISMIC
B 6302 183 NORTH SEA, CENTRAL TROUGH
B 6107 1001 NORTH SEA, CHALK, PRODUCTION FROM
M 27 149 NORTH SEA, CRETACEOUS CHALK
M 28 153 NORTH SEA, CRETACEOUS, CHALK
M 27 168 NORTH SEA, CRETACEOUS CHALK
M 27 21 NORTH SEA, CRETACEOUS CHALK
B 6107 996 NORTH SEA, CRETACEOUS CHALK, UPPER, SAMPLES
B 6306 911 NORTH SEA DEEP-WATER FACIES, LOGS
B 6010 1715 NORTH SEA, DEVELOPMENTS, 1975
B 6110 1654 NORTH SEA, DEVELOPMENTS, 1975-1976, NORWAY-
B 6110 1645 NORTH SEA, DEVELOPMENTS, 1976
B 6210 1916 NORTH SEA, DEVELOPMENTS, 1977
B 6310 1924 NORTH SEA, DEVELOPMENTS, 1978
B 6411 1966 NORTH SEA, DEVELOPMENTS, 1979
B 6002 188 NORTH SEA, DIP LOGS
B 6203 436 NORTH SEA, DOGGER HIGH
B 6302 183 NORTH SEA, DOGGER HIGH
B 6312 2222 NORTH SEA, DRAGNESET FORMATION
B 6311 2002 NORTH SEA, EAST FRIGG FAN
B 6302 183 NORTH SEA, EGERSUND SUBBASIN
B 6107 1002 NORTH SEA, EKOFISK FIELD
M 30 217 NORTH SEA, EKOFISK FIELD, BRIGHT SPOT EXAMPLE
M 30 208 NORTH SEA, EKOFISK FIELD, CROSS SECTION
M 30 221 NORTH SEA, EKOFISK FIELD, DEVELOPMENT PROGRAM
M 30 200 NORTH SEA, EKOFISK FIELD, EKOFISK FORMATION
M 30 213 NORTH SEA, EKOFISK FIELD, GEOCHEMICAL DATA
M 30 197 NORTH SEA, EKOFISK FIELD, GEOLOGIC SETTING
M 30 213 NORTH SEA, EKOFISK FIELD, GEOPRESSURES
M 30 198 NORTH SEA, EKOFISK FIELD, LITHOSTRATIGRAPHY
M 30 211 NORTH SEA, EKOFISK FIELD, SOURCE ROCK
M 30 200 NORTH SEA, EKOFISK FIELD, STRATIGRAPHY
M 30 211 NORTH SEA, EKOFISK FIELD, STRUCTURE
M 30 200 NORTH SEA, EKOFISK FIELD, TOR FORMATION
B 6401 68 NORTH SEA, EKOFISK WELL, SAMPLED SECTION, NORWEGIAN
B 6303 556* NORTH SEA EXPLORATION, ABST.
B 6302 183 NORTH SEA, FJERRITSELV FAULT
B 6311 2001 NORTH SEA, FLADEN SPUR
B 6302 183 NORTH SEA, FLEKKEFJORD HIGH
M 30 82 NORTH SEA, FORTIES FIELD, DEVELOPMENT PLAN
M 30 82 NORTH SEA, FORTIES FIELD, LITHOFACIES ANALYSIS, CORE DATA
M 30 82 NORTH SEA, FORTIES FIELD, LITHOSTRATIGRAPHY
M 30 82 NORTH SEA, FORTIES FIELD, LOCATION MAP
M 30 82 NORTH SEA, FORTIES FIELD, STRUCTURAL SETTING
B 6311 2003 NORTH SEA, FRIGG 25-1-1 DISCOVERY WELL
B 6311 2002 NORTH SEA, FRIGG FAN
M 30 59 NORTH SEA, FRIGG FIELD, HISTORY OF DISCOVERY
M 30 60 NORTH SEA, FRIGG FIELD, LOCATION MAP
M 30 63 NORTH SEA, FRIGG FIELD, STRATIGRAPHIC COLUMN
M 30 65 NORTH SEA, FRIGG FIELD, STRUCTURE
M 30 59† NORTH SEA, FRIGG FIELD--LARGE SUBMARINE-FAN TRAP IN LOWER,
B 6311 2002 NORTH SEA, FRIGG SUBBASIN
B 6203 436 NORTH SEA, FYN-RINGKOBING HIGH
B 6002 185 NORTH SEA, GAMMA-RAY LOGS
M 26 405 NORTH SEA, GEOLOGIC MODEL
M 30 81† NORTH SEA, GEOLOGY OF THE, FORTIES FIELD, U.K., CONTINENTAL
B 6303 479* NORTH SEA GIANT, ABST., STATFJORD FIELD
M 30 95† NORTH SEA GIANT, STATFJORD FIELD--A
B 6002 185 NORTH SEA, GLAUCONITE IN DITCH CUTTINGS
B 6302 184 NORTH SEA, HALDAGER FORMATION
B 6311 2001 NORTH SEA, HALIBUT HORST
B 6203 435 NORTH SEA, HELGELAND BASIN
B 6311 2001 NORTH SEA, HORDA BASIN
B 6203 436 NORTH SEA, HORDA PLATFORM
B 6302 183 NORTH SEA, HORN GRABEN
M 26 404 NORTH SEA HORST-FAULTING, MODELING
B 6203 436 NORTH SEA, INTERMEDIATE BLOCK
B 6302 183 NORTH SEA, JAEREN HIGH
B 6312 2222 NORTH SEA, JANUSFJELLET FORMATION
M 28 179 NORTH SEA, JURASSIC, SANDSTONE
B 6312 2222 NORTH SEA, KIMMERIDGE CLAY
B 6312 2223 NORTH SEA, KIMMERIDGIAN KEROGEN SAMPLE
B 6302 183 NORTH SEA, LING SUBBASIN
B 6302 183 NORTH SEA, LISTA NOSE
B 6002 184 NORTH SEA, LOG PATTERNS
B 6203 436 NORTH SEA, MAIN GRABEN
B 6302 183 NORTH SEA, MANDAL HIGH
M 30 118 NORTH SEA, MESOZOIC BASINS, LOCATION MAP
B 6302 183 NORTH SEA, MID NORTH SEA HIGH
B 6203 436 NORTH SEA, MID NORTH SEA HIGH
M 30 143 NORTH SEA, MORAY FIRTH BASIN, CALLOVIAN
M 30 134 NORTH SEA, MORAY FIRTH BASIN, MESOZOIC HISTORY
M 30 143 NORTH SEA, MORAY FIRTH BASIN, OXFORDIAN
M 30 133 NORTH SEA, MORAY FIRTH BASIN, PIPER FIELD
M 30 119 NORTH SEA, MORAY FIRTH BASIN, REGIONAL STRATIGRAPHY
M 30 123 NORTH SEA, MORAY FIRTH BASIN, REGIONAL STRUCTURE
M 30 133 NORTH SEA, MORAY FIRTH BASIN, RIFTING
M 30 139 NORTH SEA, MORAY FIRTH BASIN, TERTIARY HISTORY
M 30 120 NORTH SEA, MORAY FIRTH, REGIONAL GEOLOGY MAP
M 30 123 NORTH SEA, MORAY FIRTH, STRATIGRAPHIC COLUMN
B 6203 435 NORTH SEA, MORE-LOFOTEN AREA RIFT SYSTEM
M 30 180 NORTH SEA, NINIAN FIELD, CROSS SECTION
M 30 192 NORTH SEA, NINIAN FIELD, DEVELOPMENT
M 30 173 NORTH SEA, NINIAN FIELD, EXPLORATION HISTORY
M 30 179 NORTH SEA, NINIAN FIELD, GEOLOGY AND STRUCTURE
M 30 175 NORTH SEA, NINIAN FIELD, LICENSE BLOCKS
M 30 174 NORTH SEA, NINIAN FIELD, LOCATION
M 30 189 NORTH SEA, NINIAN FIELD, RESERVOIR CHARACTER
M 30 188 NORTH SEA, NINIAN FIELD, STRATIGRAPHIC COLUMN
M 30 181 NORTH SEA, NINIAN FIELD, STRATIGRAPHY
M 30 173† NORTH SEA, NINIAN FIELD, U.K. SECTOR
B 6203 435 NORTH SEA, NORDLAND RIDGE
B 6411 1971 NORTH SEA, ONSHORE, PRODUCTION, 1978-1979, NETHERLANDS
B 6306 905 NORTH SEA, PALEOCENE DELTA
M 28 136 NORTH SEA, PERMIAN, ROTLIEGENDES SANDSTONE
M 28 71 NORTH SEA, PERMIAN, ROTLIEGENDES SANDSTONE
M 28 132 NORTH SEA, PERMIAN, ROTLIEGENDES SANDSTONES
B 6302 182† NORTH SEA, PETROLOGY AND DIAGENESIS OF, JURASSIC SANDSTONES
M 30 134 NORTH SEA, PIPER FIELD, BASE OF CRETACEOUS FAULTING
M 30 139 NORTH SEA, PIPER FIELD, KIMMERIDGE SHALE
M 30 137 NORTH SEA, PIPER FIELD, KIMMERIDGE SHALE ISOPACH
M 30 132 NORTH SEA, PIPER FIELD, LOCATION
M 30 170 NORTH SEA, PIPER FIELD, RADIOACTIVE MARKERS
M 30 140 NORTH SEA, PIPER FIELD, STRUCTURAL CROSS SECTION

M 30 137 NORTH SEA, PIPER FIELD, TIME STRATIGRAPHIC COLUMN
M 30 136 NORTH SEA, PIPER FIELD, TOP OF PIPER SANDSTONE
M 30 163 NORTH SEA, PIPER FIELD, WELL DATA
B 6210 1924 NORTH SEA, PRODUCTION, 1976-1977, DENMARK-
B 6210 1924 NORTH SEA, PRODUCTION, 1976-1977, NETHERLANDS-
B 6210 1924 NORTH SEA, PRODUCTION, 1976-1977, NORWAY-
B 6310 1932 NORTH SEA, PRODUCTION, 1977-1978, DENMARK-
B 6310 1932 NORTH SEA, PRODUCTION, 1977-1978, NETHERLANDS-
B 6310 1932 NORTH SEA, PRODUCTION, 1977-1978, NORWAY-
B 6411 1871 NORTH SEA, PRODUCTION, 1978-1979, DENMARK
B 6411 1971 NORTH SEA, PRODUCTION, 1978-1979, NORWAY
B 6203 435 NORTH SEA, RIFT SYSTEM, SECTIONS
B 6004 706* NORTH SEA RIFTING AND OPENING OF NORTH ATLANTIC OCEAN, ABST.
B 6302 183 NORTH SEA, RINGKOBING-FYN HIGH
B 6203 436 NORTH SEA, SCHILLBANK HIGH
B 6002 184 NORTH SEA, SEDIMENTARY MODEL
B 6302 183 NORTH SEA, SELE HIGH
B 6311 2001 NORTH SEA, SHETLAND PLATFORM
B 6408 1227 NORTH SEA, SOUTHERN, SAND RIDGES
B 6002 184 NORTH SEA, SP LOGS
M 30 96 NORTH SEA, STATFJORD AREA, GEOLOGIC HISTORY
M 30 98 NORTH SEA, STATFJORD FIELD, EXPLORATION HISTORY
M 30 96 NORTH SEA, STATFJORD FIELD, LOCATION
M 30 96 NORTH SEA, STATFJORD FIELD, REGIONAL SETTING
M 30 102 NORTH SEA, STATFJORD FIELD, STRATIGRAPHIC COLUMN
M 26 430 NORTH SEA, STATFJORD SANDSTONE
B 6203 436 NORTH SEA, STORD BASIN
B 6306 905† NORTH SEA SUBMARINE-FAN SANDS, DIPMETER AND, LOG MOTIFS IN
M 30 117† NORTH SEA, THE, BEATRICE FIELD, INNER, MORAY FIRTH, U.K.
B 6203 436 NORTH SEA, UTSIRA HIGH
B 6302 183 NORTH SEA, UTSIRA HIGH
B 6311 2001 NORTH SEA, UTSIRA HIGH
B 6311 2001 NORTH SEA, VIKING BASIN
M 30 41 NORTH SEA, VIKING FIELD, GAS COMPOSITION
M 30 41 NORTH SEA, VIKING FIELD, LEMAN SANDSTONE FORMATION
M 30 40 NORTH SEA, VIKING FIELD, LOCATION MAP
M 30 41 NORTH SEA, VIKING FIELD, ROTLIEGENDES SANDSTONE
M 30 42 NORTH SEA, VIKING FIELD, STRATIGRAPHIC COLUMN
M 30 45 NORTH SEA, VIKING FIELD, WELL LOCATION MAP
M 30 41 NORTH SEA, VIKING FIELD, ZECHSTEIN SALT
B 6203 521* NORTH SEA VIKING GRABEN, ABST., FRIGG FIELD—LARGE SUBMARINE-,
B 6203 436 NORTH SEA, VIKING GRABEN, CENTRAL, RIFT SYSTEM
B 6311 1999† NORTH SEA VIKING GRABEN, FRIGG FIELD--LARGE, SUBMARINE-FAN
B 6203 435 NORTH SEA, VIKING GRABEN, NORTHERN SECTOR, RIFT SYSTEM
B 6203 436 NORTH SEA, VIKING GRABEN, SOUTHERN, RIFT SYSTEM
B 6302 183 NORTH SEA, VIKING TROUGH
B 6203 435 NORTH SEA, VORING BASIN
B 6002 189 NORTH SEA WELL LOG, ENVIRONMENTAL ANALYSES
B 6002 188 NORTH SEA, WELL LOGS
B 6002 184 NORTH SEA, WIRELINE-LOG PATTERNS
B 6311 2001 NORTH SEA, WITCH GROUND GRABEN
B 6407 1009 NORTH SIMSBORO FIELD, ARK-LA-TEX AREA
B 6211 2354* NORTH SLOPE, ABST., USE OF, VERTICAL-GRADIENT, AEROMAGNETIC
B 6405 737* NORTH SLOPE, ALASKA, ABST., BARROW GAS FIELDS
B 6405 684* NORTH SLOPE, ALASKA, ABST., UNCONFORMITIES—KEY TO MAJOR OI
M 24 23† NORTH SLOPE, ALASKA, PERMO-, TRIASSIC, RESERVOIRS OF, PRUDHO
M 30 298 NORTH SLOPE DISCOVERY WELLS, DATA TABLE, ALASKA
B 6008 1189 NORTH SLOPE, EXPLORATION, 1975, ALASKA, ARCTIC
B 6108 1157 NORTH SLOPE, EXPLORATION, 1976, ALASKA, ARCTIC
B 6208 1312 NORTH SLOPE, EXPLORATION, 1977, ALASKA, ARCTIC
B 6308 1211 NORTH SLOPE, EXPLORATION, 1978, ALASKA, ARCTIC
B 6409 1345 NORTH SLOPE, EXPLORATION, 1979, ALASKA, ARCTIC
M 30 290 NORTH SLOPE, MAJOR STRUCTURAL ELEMENTS, ALASKA
B 6405 752* NORTH SLOPE OF ALASKA, ABST., STRATIGRAPHIC RELATIONS OF,
B 6405 744* NORTH SLOPE OF ALASKA—IMPLICATIONS FOR FUTURE EXPLORATION,
M 25 303 NORTH SLOPE, SECTION, ALASKA
M 25 244 NORTH SUMATRA BASIN, INDONESIA
M 25 243 NORTH SUMATRA BASIN, INDONESIA
B 6211 2245 NORTH TEJON HIGH, CALIFORNIA
B 6108 1225 NORTH TEXAS CITY DIKE FIELD, GALVESTON COUNTY, TEXAS
M 30 48 NORTH VIKING COMPLEX POOLS, VIKING FIELD
B 6307 1093 NORTHAMPTON BLOCK, AUSTRALIA
B 6203 366 NORTHEAST BANK, CALIFORNIA
B 6103 355 NORTHEAST BLOODWORTH CANYON SANDSTONE FIELD, TEXAS
B 6109 1484 NORTHEAST CHANNEL, EAST COAST
B 6404 502 NORTHEAST CHANNEL, SEISMIC REFLECTION PROFILES
B 6101 3 NORTHEAST GREENLAND BASIN
B 6108 1382* NORTHEAST LUCIEN FIELD, NOBLE COUNTY, OKLAHOMA, ABST.,
M 25 484 NORTHEAST PACIFIC, METAL AND METALLOGENIC PROVINCES,
M 25 485 NORTHEAST PACIFIC, METAL PROVINCES
M 25 485 NORTHEAST PACIFIC, METALLIZATION, EPOCHS OF
M 25 486 NORTHEAST PACIFIC, METALLOGENIC PROVINCES
M 25 485 NORTHEAST PACIFIC, PLATE BOUNDARIES, ACTIVE
B 6204 694 NORTHEAST PROVIDENCE CHANNEL, BAHAMAS
B 6007 1037† NORTHEAST PROVIDENCE CHANNEL, BAHAMAS, STRATIGRAPHY AND
B 6007 1039 NORTHEAST PROVIDENCE CHANNEL, BATHYMETRY
B 6007 1040 NORTHEAST PROVIDENCE CHANNEL, SAMPLES, AGES
B 6007 1040 NORTHEAST PROVIDENCE CHANNEL, SEISMIC-REFLECTION PROFILES
B 6007 1050 NORTHEAST PROVIDENCE CHANNELS, BATHYMETRIC, FREE-AIR
B 6204 668 NORTHEAST SAHARA BASIN, ALGERIA
B 6109 1534* NORTHEAST THOMPSONVILLE FIELD, JIM HOGG AND WEBB COUNTIES,
B 6111 1962† NORTHEASTERN EXTENSION OF SIGSBEE SCARP, GULF OF MEXICO
B 6210 2055 NORTHEASTERN ZONE, MEXICO
B 6103 467* NORTHERN ALASKA PALEOMAGNETISM, PLATE ROTATION, AND
B 6103 467* NORTHERN ALASKA PALEOMAGNETISM, PLATE ROTATION AND
S 7 21† NORTHERN AND EASTERN GULF OF MEXICO CONTINENTAL MARGIN:
B 6406 861 NORTHERN BASIN, EGYPT
M 25 258 NORTHERN BASIN, PROSPECTS, TAIWAN
B 6006 988† NORTHERN CALIFORNIA GEOLOGICAL SOCIETY, SHORT, HISTORY OF
B 6105 831* NORTHERN COALFIELD—DIRECTION OF RESEARCH TO SUPPORT COAL
B 6211 2362* NORTHERN EAST MESA GEOTHERMAL WELLS, ABST., GEOLOGIC
B 6101 3 NORTHERN ELLESMERE AND AXEL HEIBERG ISLANDS
B 6008 1410* NORTHERN GREAT PLAINS LIGNITES, ABST., LIQUEFACTION OF
S 7 141 NORTHERN GULF OF MEXICO, CONTINENTAL SLOPE, CENTRAL SLOPE
S 7 141 NORTHERN GULF OF MEXICO, CONTINENTAL SLOPE, LOWER SLOPE
S 7 141 NORTHERN GULF OF MEXICO, CONTINENTAL SLOPE, MISSISSIPPI SLOPE
S 7 141 NORTHERN GULF OF MEXICO, CONTINENTAL SLOPE, NORTHWEST SLOPE
S 7 150 NORTHERN GULF OF MEXICO, CONTINENTAL SLOPE, SALT THICKNESS
S 7 150 NORTHERN GULF OF MEXICO, CONTINENTAL SLOPE, SEDIMENT
S 7 151 NORTHERN GULF OF MEXICO, CONTINENTAL SLOPE, STRUCTURAL
S 7 142 NORTHERN GULF OF MEXICO, CONTINENTAL SLOPE, STRUCTURAL
S 7 22 NORTHERN GULF OF MEXICO CRUSTAL FRAMEWORK
S 7 33 NORTHERN GULF OF MEXICO, DESCRIBED
S 7 22 NORTHERN GULF OF MEXICO, DESCRIBED
S 7 23 NORTHERN GULF OF MEXICO, GEOLOGIC MAP
S 7 24 NORTHERN GULF OF MEXICO MARGIN, CROSS SECTION
S 7 31 NORTHERN GULF OF MEXICO, PERIPHERAL FAULT ZONES, DESCRIBED
S 7 30 NORTHERN GULF OF MEXICO RIM, EXPOSED, ARKANSAS
S 7 30 NORTHERN GULF OF MEXICO RIM, EXPOSED, OKLAHOMA
S 7 30 NORTHERN GULF OF MEXICO RIM, EXPOSED, PALEOZOIC
S 7 27 NORTHERN GULF OF MEXICO, SALT AND DIAPIRS, JURASSIC
S 7 69† NORTHERN GULF OF MEXICO, SALT MOVEMENT ON, CONTINENTAL,
S 7 35 NORTHERN GULF OF MEXICO, SHALE STRUCTURES, DESCRIBED
S 7 28 NORTHERN GULF OF MEXICO, STRATA, CENOZOIC
S 7 25 NORTHERN GULF OF MEXICO, STRATA, JURASSIC
S 7 22 NORTHERN GULF OF MEXICO STRATIGRAPHIC FRAMEWORK
S 7 28 NORTHERN GULF OF MEXICO STRUCTURAL FRAMEWORK
S 7 141† NORTHERN GULF OF MEXICO, STRUCTURE OF, CONTINENTAL SLOPE
S 7 29 NORTHERN GULF OF MEXICO, TECTONIC MAP
M 28 78 NORTHERN IRELAND, CRETACEOUS, WHITE LIMESTONE
M 28 19 NORTHERN IRELAND, DEVONIAN, OLD RED SANDSTONE
M 28 121 NORTHERN IRELAND, TRIASSIC, BUNTER SANDSTONE
M 28 31 NORTHERN IRELAND, TRIASSIC, NEW RED SANDSTONE
B 6109 1560* NORTHERN PENINSULA, NEWFOUNDLAND, ABST., SATELLITE IMAGE
B 6101 81 NORTHERN SAHARA SALT BASIN
S 5 1* NORTHERN SHELF, MICHIGAN BASIN, DEPOSITIONAL ENVIRONMENTS OF
M 26 145 NORTHERN, SEQUENCES, GULF OF MEXICO
B 6004 733* NORTHERN STRAITS OF FLORIDA, ABST., PETROLOGY OF, SUBSE,
B 6410 1704 NORTHERN STRAITS OF FLORIDA, BATHYMETRY
B 6303 500* NORTHERN TERRITORY, AUSTRALIA, ABST., COORONG MODEL FOR
B 6203 504* NORTHERN TERRITORY, AUSTRALIA, ABST., EXPLORATION CASE
B 6110 1889 NORTHERN TERRITORY, DEVELOPMENTS, 1976
B 6411 1866 NORTHERN TERRITORY, DEVELOPMENTS, 1979
M 25 427 NORTHERN TERRITORY, EAST ALLIGATOR, URANIUM, AUSTRALIA
M 25 427 NORTHERN TERRITORY, RUM JUNGLE, URANIUM, COPPER, AND LEAD,
M 25 427 NORTHERN TERRITORY, SHERWIN IRONSTONE, IRON ORE, AUSTRALIA
B 6309 1587 NORTHROP, DAVID A., IN-SITU TESTING OF WELL-SHOOTING CONCEPTS,
B 6111 2012 NORTHROP, STUART A., HISTORY OF NEW MEXICO GEOLOGICAL
B 6203 404 NORTHSIDE RANGE, VIRGIN ISLANDS, ST. CROIX
B 6104 596 NORTHVIEW FIELD, MICHIGAN
B 6007 1050 NORTHWEST AND NORTHEAST PROVIDENCE CHANNELS, BATHYMETRIC,
M 30 429 NORTHWEST BORNEO GEOSYNCLINE, MALAYSIA
M 30 430 NORTHWEST BORNEO OROGEN, MALAYSIA
S 7 313† NORTHWEST GULF OF MEXICO, HOLOCENE, SEDIMENT, GEOCHEMISTRY
S 7 289† NORTHWEST GULF OF MEXICO, INTRASLOPE, BASIN IN
S 7 318 NORTHWEST GULF OF MEXICO, INTRASLOPE BASINS
S 7 319 NORTHWEST GULF OF MEXICO, INTRASLOPE BASINS, GYRE BASIN
S 7 320 NORTHWEST GULF OF MEXICO, INTRASLOPE BASINS, ORCA BASIN
S 7 315 NORTHWEST GULF OF MEXICO, NORMAL SLOPE SEQUENCE
S 7 303† NORTHWEST GULF OF MEXICO, ORCA, BASIN, ANOXIC DEPRESSION ON
S 7 325 NORTHWEST GULF OF MEXICO, SALT DIFFUSION
S 7 313 NORTHWEST GULF OF MEXICO, SEDIMENT GEOCHEMISTRY

M 25 244 NORTHWEST JAVA BASIN AREA, INDONESIA
B 6311 2089 NORTHWEST SHELF, AUSTRALIA
B 6201 44 NORTHWEST SHELF, AUSTRALIA
B 6207 1224* NORTHWEST SHELF GAS—ITS ROLE IN WESTERN AUSTRALIAS ENERGY
S 7 141 NORTHWEST SLOPE, NORTHERN GULF OF MEXICO, CONTINENTAL SLOPE
B 6203 523* NORTHWEST TERRITORIES, ABST., SPELEOTHEMS AND, EVAPORITE
B 6203 522* NORTHWEST TERRITORIES, ABST., TECTONIC HISTORY OF, ATHAPUSCOW,
M 24 54 NORTHWEST TERRITORIES, AKLAVIK ARCH, ALASKA
B 6203 532* NORTHWEST TERRITORIES, CANADA, ABST., CONTINENTAL SLOPE-RISE
B 6004 701* NORTHWEST TERRITORIES, CANADA, ABST., LOWER, CRETACEOUS,
B 6307 1059† NORTHWEST TERRITORIES, CANADA, DEPOSITIONAL, FACIES OF LOWER,
B 6307 1061 NORTHWEST TERRITORIES, CAPE PHILLIPS FORMATION
B 6307 1064 NORTHWEST TERRITORIES, CORNWALLIS GROUP
B 6108 1317 NORTHWEST TERRITORIES, DEVELOPMENTS, 1976
B 6308 1410 NORTHWEST TERRITORIES, DEVELOPMENTS, 1978
B 6409 1513 NORTHWEST TERRITORIES, DEVELOPMENTS, 1979
B 6307 1061 NORTHWEST TERRITORIES, IRENE BAY FORMATION
M 24 53 NORTHWEST TERRITORIES, MACKENZIE DELTA AREA
B 6108 1322 NORTHWEST TERRITORIES, PRODUCTION, 1976
B 6307 1061 NORTHWEST TERRITORIES, READ BAY FORMATION
M 24 51† NORTHWEST TERRITORIES, TAGLU GAS FIELD, BEAUFORT BASIN
M 24 65 NORTHWEST TERRITORIES, TAGLU GAS FIELD, PERMAFROST
B 6307 1061 NORTHWEST TERRITORIES, THUMB MOUNTAIN FORMATION
B 6007 1038 NORTHWESTERN BAHAMA PLATFORM, MAP
B 6101 34 NORTHWESTERN CANADIAN SHIELD, HELIKIAN URANIUM PROVINCES
B 6203 549 NORTHWOOD, E. J., HOW TO HELP SOLVE THE ENERGY PROBLEM, ABST.
B 6211 2358 NORTHWOOD, E. JOHN, EMERGENCE OF NATIONAL OIL COMPANIES,
B 6407 1078† NORTON BASIN, ALASKA, ORIGIN OF, GASOLINE-RANGE,
B 6407 1078 NORTON BASIN, GAS SEEP, ALASKA
B 6303 468* NORTON BASIN, NORTHERN BERING SEA, ABST., SONOBUOY REFRACTION
B 6203 534* NORTON SOUND, ALASKA, ABST., HYDROCARBONS OF LOW MOLECULAR
B 6309 1585 NORTON, WARREN, HOW GOOD ARE REGIONAL COAL RESERVE FIGURES.
B 6305 828* NORWAY, ABST., ALLUVIAL, FANS AND THEIR DEPOSIT IN, DEVONIAN,
B 6004 668* NORWAY, ABST., GROWTH, FAULTS AND ASSOCIATED STRUCTURES IN
M 30 196 NORWAY, EKOFISK FIELD, LOCATION
M 30 196 NORWAY, EKOFISK FIELD, LOCATION
B 6408 1245 NORWAY, JURASSIC, PIEZOMETRIC MAP, UNITED KINGDOM AND
B 6310 1927 NORWAY NORT OF 62 DEGREES, 1978, HAGEMANN, F., AND F., AL-KASIM,
B 6010 1718 NORWAY NORT OF 62 DEGREES N, 1975, HAGEMANN, F., AND F.,
B 6110 1647 NORWAY NORTH OF 62 DEGREES N, 1976, HAGEMANN, F., AND F.,
B 6210 1918 NORWAY NORTH OF 62 DEGREES N, 1977, HAGEMANN, F., AND FAROUK,
B 6411 1967 NORWAY NORTH OF 62 DEGREES N, AL-KASIM, FAROUK, AND, EGIL
B 6411 1971 NORWAY, NORTH SEA, PRODUCTION, 1978-1979
B 6408 1245† NORWAY, PRESSURE ORIGIN AND DISTRIBUTION IN, JURASSIC OF,
B 6004 714* NORWAY--CASE STUDY IN APPLICATIONS OF MARINE GEOTECHNOLOGY
B 6110 1654 NORWAY-NORTH SEA, DEVELOPMENTS, 1975-1976
B 6210 1924 NORWAY-NORTH SEA, PRODUCTION, 1976-1977
B 6310 1932 NORWAY-NORTH SEA, PRODUCTION, 1977-1978
B 6004 671* NORWAYS OIL AND GAS POTENTIAL BASED ON PRESENT EXPLORATION,
B 6303 419* NORWEGIAN CONTINENTAL SHELF, ABST., EXPLORATION RESULTS AND
B 6004 715* NORWEGIAN CONTINENTAL SHELF NORTH OF 62 DEGREES NORTH, ABST.,
B 6302 182† NORWEGIAN DANISH BASIN, NORTH SEA, PETROLOGY AND DIAGENESIS
B 6302 183 NORWEGIAN DANISH BASIN, TECTONIC MAP
B 6004 721* NORWEGIAN NORTH SEA, ABST., LITHOSTRATIGRAPHY OF
B 6401 68 NORWEGIAN NORTH SEA, EKOFISK WELL, SAMPLED SECTION
B 6203 426 NORWEGIAN PLATFORM, EUROPE
B 6203 512* NORWEGIAN SEA, ABST., MASS PHYSICAL AND ENGINEERING PROPERTIES
B 6111 1994 NORWEGIAN SEA, POSSIBLE DIAPIRS
B 6401 68 NORWEGION NORTH SEA, EKOFISK WELL, SAMPLED SECTION
B 6205 793 NOSE, BLAKE
S 6 369 NOSIB SYSTEM, SOUTH AFRICA
B 6409 1462 NOSOW, EDMUND, DEVELOPMENT IN EAST-CENTRAL STATES IN 1979
B 6008 1222 NOSOW, EDMUND, DEVELOPMENTS IN EAST-CENTRAL STATES IN 1975
B 6108 1183 NOSOW, EDMUND, DEVELOPMENTS IN EAST-CENTRAL STATES IN 1976
B 6208 1340 NOSOW, EDMUND, DEVELOPMENTS IN EAST-CENTRAL STATES IN 1977
B 6308 1324 NOSOW, EDMUND, DEVELOPMENTS IN EAST-CENTRAL STATES IN 1978
B 6002 273 NOTE 44, AMERICAN, COMMISSION ON, STRATIGRAPHIC,
B 6002 273† NOTE 44--APPLICATION FOR ADDITION TO CODE CONCERNING
B 6102 248† NOTE 45--APPLICATION FOR AMENDMENT CONCERNING TERMINOLOGY
B 6102 252† NOTE 46--APPLICATION FOR AMENDMENT OF ARTICLES 8 AND 10 OF
B 6107 1083† NOTE 47—APPLICATION FOR AMENDMENT OF ARTICLES 27 AND 34 OF
B 6207 1185† NOTE 48—APPLICATION FOR AMENDMENT OF CODE OF STRATIGRAPHIC
B 6207 1187† NOTE 49—RECORDS OF STRATIGRAPHIC COMMISSION FOR 1974-1976
B 6310 1986† NOTE 50--PROPOSAL TO CHANGE NAME OF COMMISSION
B 6401 137† NOTE 51--RECORDS OF STRATIGRAPHI NOMENCLATURE, 1976-78
B 6401 136† NOTE 51--RECORDS OF STRATIGRAPHIC COMMISSION, 1976-78, AMERICAN
B 6401 137† NOTE 51--RECORDS OF STRATIGRAPHIC NOMENCLATURE, 1976-78
S 4 291 NOUMEA, GALAXEA, NEW CALEDONIA
B 6004 697* NOVA SCOTIA, ABST., FACIES, MODE FOR, MACROTIDAL ENVIRONMENT,
B 6203 534* NOVA SCOTIA, ABST., ORDOVICIAN-, SILURIAN, GLACIO-EUSTATIC,
B 6405 730* NOVA SCOTIA, ABST., SEDIMENTOLOGIC DESCRIPTION OF PART OF,
B 6109 1483† NOVA SCOTIA AND GEORGES BANK, CONTINENTAL, SLOPE AND UPPER,
B 6405 747* NOVA SCOTIA, CANADA, ABST., TERRESTRIAL SEDIMENTATION
B 6101 80 NOVA SCOTIA, FUNDY BASIN
B 6004 675* NOVA SCOTIA SHELF, ABST., STRATIGRAPHY OF
B 6004 669* NOVA SCOTIA SHELF, CANADA--DEPOSITIONAL AND DIAGENETIC MODEL
B 6109 1484 NOVA SCOTIAN SEDIMENTARY RIDGE
B 6012 2103† NOVACULITE, DEVONIAN, ARKANSAS AND OKLAHOMA, NONGLACIAL,
B 6405 754* NOVACULITE FORMATIONS, OKLAHOMA, ABST., OIL PRODUCTION FROM,
B 6309 1578* NOVEL FORMATION-EVALUATION CONCEPT FOR SHALE RESERVOIRS
B 6102 214 NOVILLERO 9, CORE, MEXICO
B 6211 2270 NOZU SAND, CALIFORNIA
B 6405 806* NPR-ALASKA, AN UPDATE, ABST.
B 6405 715* NPRA, ABST., PETROLEUM EXPLORATION O, NATIONAL PETROLEUM
B 6207 1218* NPRA, ABST., PETROLEUM EXPLORATION OF, NATIONAL PETROLEUM
B 6211 2359* NPRA, ABST., SUMMARY OF RECENT EXPLORATION AND EVALUATION
S 10 27 NSO DISTRIBUTION, FRANCE, PARENTIS BASIN
B 6202 296 NSUKKA FORMATION, NIGERIA
B 6406 859 NUBIA FORMATION, AFRICA
B 6306 975 NUBIA FORMATION, EGYPT, EASTERN DESERT, CENTRAL
B 6306 975 NUBIA FORMATION, FIVE LITHOLOGIC UNITS, EGYPT
B 6306 975† NUBIA FORMATION OF CENTRAL EASTERN DESERT, EGYPT--MAJOR
B 6306 977 NUBIA FORMATION, STRATIGRAPHIC COLUMN, EGYPT
B 6306 974 NUBIA GROUP NOMENCLATURE AND CORRELATION, EGYPT AND,
B 6305 829* NUBIA, SOUTHWESTERN EGYPT, ABST., ALLUVIAL-PLAIN SEDIMENTS OF
B 6306 967† NUBIA STRATA, SOUTHWESTERN EGYPT, MAJOR SUBDIVISIONS AND
B 6002 258 NUBIAN MASSIF, ARABO-
B 6007 1009 NUBIAN, NORTH, AFRICA
B 6406 861 NUBIAN SHIELD, AFRICA
B 6309 1585 NUCKOLS, E. B., COMPUTER-AIDED GRAPHICS FOR ANALYSIS OF
B 6309 1588 NUCKOLS, ERNEST B., COTTAGEVILLE GAS FIELD CORRELATION
B 6008 1391* NUCLEAR ASPECTS, ABST., OKLO PHENOMENON--GEOLOGIC AND
B 6207 1234* NUCLEAR ENERGY IN FUTURE OF CIRCUM-PACIFIC, AREA, ABST., ROLE
B 6105 832* NUCLEAR FUEL, ABST., THORIUM—OTHER NATURAL SOURCE OF
M 25 522† NUCLEAR GENERATION OF ELECTRICITY IN MEXICO--SUMMARY,
M 25 522 NUCLEAR PLANTS, CALENDAR FOR ELECTRICITY GENERATION, TABLE
B 6309 1599* NUCLEAR POWER AND GEOLOGY OF URANIUM, ABST.
M 25 522 NUCLEAR POWER, DEMAND FOR
M 25 522 NUCLEAR REACTORS, HEAVY-WATER OR CANDU TYPE
B 6309 1605* NUCLEAR WASTES, ABST., SUITABILITY STUDIES OF, SALT, DOMES IN
B 6303 465* NUCLEAR-WASTE ISOLATION IN SOUTHEASTERN NEW MEXICO, ABST.,
B 6209 1756* NUECES RIVER, SOUTH TEXAS, ABST., STRATIGRAPHIC INTERPRETATION
B 6004 676* NUECES RIVER, TEXAS, ABST., BED FORMS AND STRATIFICATION TYPES
B 6409 1554* NUEVO LEON, MEXICO, ABST., CALPIONELLIDS AND, NANNOCONIDS OF,
B 6108 1376* NUGGET OIL ACCUMULATIONS AT DRY PINEY, TIP TOP, AND HOGSBACK
B 6206 1075 NUGGET SANDSTONE, IDAHO-, WYOMING, THRUST BELT
B 6307 1039 NUGGET SANDSTONE, IDAHO-, WYOMING, THRUST-FOLD BELT
M 30 281 NUGGET SANDSTONE, JURASSIC, PAINTER RESERVOIR FIELD
B 6004 708* NUGGET SANDSTONE, JURASSIC, SOUTHWESTERN WYOMING AND
B 6008 1406* NUGGET SANDSTONE, SOUTHWESTERN WYOMING AND NORTHEASTERN
B 6305 832* NUGGET SANDSTONE TREND IN ABSAROKA THRUST PLATE, UINTA
M 30 281 NUGGET SANDSTONE, TRIASSIC, PAINTER RESERVOIR FIELD
M 24 95 NUGGET SANDSTONE, WYOMING
B 6405 640 NUGGET SANDSTONE, WYOMING
B 6211 2282 NUGGET SANDSTONE, WYOMING
B 6308 1424* NUGGET-NAVAJO SANDSTONE ENVIRONMENTAL WAR--CAN TRACE
B 6303 441* NUGGET-NAVAJO SANDSTONE ENVIRONMENTAL WAR--CAN TRACE
B 6203 536* NUGSSUAQ EMBAYMENT, WEST GREENLAND, ABST., HYDROCARBON
B 6309 1585 NUHFER, EDWARD B., FABRIC-ELEMENT-BASED CLASSIFICATION FOR
B 6309 1590 NUHFER, EDWARD, PETROLOGIC EVALUATION OF SIGNIFICANCE OF
M 28 40 NUKA FORMATION, ALASKA, MISSISSIPPIAN
M 28 129 NUKA FORMATION, ALASKA, MISSISSIPPIAN
M 28 122 NUKA FORMATION, ALASKA, MISSISSIPPIAN
M 28 40 NUKA FORMATION, ALASKA, PERMIAN
M 28 129 NUKA FORMATION, ALASKA, PERMIAN
M 28 122 NUKA FORMATION, ALASKA, PERMIAN
B 6406 821 NULLARA LIMESTONE, AUSTRALIA
B 6406 821 NULLARA LIMESTONE, AUSTRALIA
B 6406 826 NULLARA REEF CYCLE, AUSTRALIA, FAMENNIAN
S 6 290 NUMERIC TIME SCALE, COMPARISON, JURASSIC
S 6 273 NUMERIC TIME SCALES, CRETACEOUS
B 6202 206 NUMERICAL AGE OF MARKERS OF TWO WELLS, GULF OF MEXICO,
B 6002 319* NUMERICAL ANALYSIS OF TRAP LOCATIONS IN MISSISSIPPIAN B,
B 6004 650* NUMERICAL ASPECTS OF CHEMICAL VARIATION AND
B 6004 663* NUMERICAL CLASSIFICATION O SANDSTONE PORE STRUCTURE, ABST.
S 6 363† NUMERICAL CORRELATION OF MIDDLE AND UPPER PRECAMBRIAN
B 6202 202 NUMERICAL STRATIGRAPHIC TECHNIQUES
B 6202 201 NUMERICAL STRATIGRAPHY
B 6202 201 NUMERICAL TIME, LINEAR
S 6 213† NUMERICAL TIME SCALE, A NEW, PALEOGENE

B 6303 429* NUMIDIAN FLYSCH COMPLEX, WESTERN MOGO MOUNTAINS, NORTHERN
B 6405 707 NUMMEDAL, DAG, EBB-TIDAL DELTA STRATIFICATION AND ITS
B 6409 1565 NUMMEDAL, DAG, ET AL, NEARSHORE BARS ALONG UNITED STATES
B 6405 759 NUMMEDAL, DAG, SAND DISPERSAL AT NORDERNEYER SEEGAT, WEST
B 6303 504 NUMMEDAL, DAG, TIDAL-INLET SEDIMENT DISPERSAL, ABST.
B 6203 550 NUMMEDAL, DAG, TRANSGRESSIVE BARRIERS OF BEAUFORT SEA COAST,
M 27 33 NUMMULITE LIMESTONE, YUGOSLAVIA, EOCENE
M 30 253 NUNAN, W. E., EUGENE ISLAND BLOCK 330 FIELD, OFFSHORE LOUISIANA
B 6002 172 NUNIVAK ARCH PROVINCE, BERINGIAN SHELF
B 6002 164 NUNIVAK ISLAND, BERINGIAN SHELF
B 6405 759 NUNN, JEFFREY A., THERMAL CONTRACTION AND PETROLEUM
B 6108 1382* NURE PROGRAM-A NATIONWIDE APPROACH TO URANIUM AND
B 6405 760 NURMI, R. D., CROSS-BEDDED SANDSTONE RESERVOIRS—GEOLOGIC
B 6004 720 NURMI, R. D., DEVELOPMENT OF POROSITY IN CARBONATE ROCKS OF
B 6303 504 NURMI, R. D., EOLIAN SEDIMENTOLOGY INTERPRETED FROM DIPMETER
S 5 23 NURMI, ROY D. SEDIMENTOLOGY AND DEPOSITIONAL ENVIRONMENTS
M 25 510 NUTZOTIN BELT, CORDILLERAN OROGEN, GRAVINA-
B 6007 1073 NWACHUKWU, SILAS O., APPROXIMATE GEOTHERMAL GRADIENTS IN
B 6305 838 NYDEGGER, GARY L., BOWDOIN DOME AREA, NORTH-CENTRAL
B 6405 760 NYDEGGER, GARY L., BOWDOIN DOME AREA, NORTH-CENTRAL
B 6008 1395 NYDEGGER, GARY L., EXPLOITATION OF GAS RESERVES AT BOWDOIN
B 6205 881* NYE COUNTY, NEVADA, ABST., EAGLE SPRINGS OIL FIELD
B 6205 883* NYE COUNTY, NEVADA, ABST., GEOLOGIC TECHNIQUES USED IN, TRAP
B 6205 884* NYE COUNTY, NEVADA, ABST., GEOMORPHIC EXPLORATION USED IN
B 6205 884* NYE COUNTY, NEVADA, ABST., TERTIARY, VOLCANIC STRATIGRAPHY OF,
B 6205 883* NYE COUNTY, NEVADA, ABST., TRAP SPRING OIL FIELD
B 6405 701* NYE COUNTY, NEVADA ABST., TRAP SPRING OIL FIELD
B 6211 2354* NYE COUNTY, NEVADA, ABST., TRAP SPRING OIL FIELD
B 6404 528 NYEBOE LAND, GREENLAND
B 6301 90 O MARKER, SCOTIAN BASIN
M 24 209 O, Q SANDSTONE, LOUISIANA, G, J, K
B 6105 822* OAK GROVE CORE, NORTHEASTERN VIRGINIA, ABST.,
B 6109 1537* OAK GROVE RIDGE CHENIER, LOUISIANA, ABST., INTERIOR STRUCTURE
B 6407 1036 OAK RIDGE FAULT, CALIFORNIA
B 6102 171 OAK RIDGE FAULT, CALIFORNIA
B 6102 139 OAK RIDGE FAULT, CALIFORNIA
B 6305 844* OAK RIDGE THRUST FAULTS, EAST-CENTRAL VENTURA BASIN,
B 6407 1005 OAKS FIELD, ARK-LA-TEX AREA
B 6309 1599* OAKS FIELD, CLAIBORNE PARISH, LOUISIANA, ABST., CLASTICITY
B 6109 1535* OAKVILLE AND GOLIAD FORMATIONS OF SOUTH TEXAS COASTAL PLAIN,
B 6303 455* OAKVILLE FORMATION OF TEXAS COASTAL PLAIN--DEPOSITIONAL
B 6206 1049 OAKVILLE FORMATION, TEXAS
B 6405 722* OAKVILLE SANDSTONE OF SOUT TEXAS—IMPLICATTONS FOR URANIUM
B 6405 711* OAKVILLE SANDSTONE OF SOUTH TEXAS COASTAL PLAIN, ABST.,
B 6303 475* OAKVILLE SANDSTONE, SOUTH TEXAS, ABST., MICROSCOPIC
S 3 105 OAKVILLE SANDSTONE, TEXAS
B 6309 1601* OAKWOOD SALT DOMES, EAST TEXAS, ABST., UPPER,
S 10 229 OB RIVER AREA, HYDROTHERMAL EFFECTS, USSR, SIBERIA
B 6007 1076 OBAMA FIELD, NIGER DELTA
B 6201 2 OBAN MASSIF, NIGER DELTA
M 27 177 OBERAUDORF SCHICHTEN, AUSTRIA, EOCENE
M 27 124 OBERAUDORF SCHICHTEN, AUSTRIA, EOCENE
B 6007 1074 OBETIM WELL, NIGER DELTA
B 6105 773* OBJECTIVES OF REGULATIONS VERSUS UNDERSTANDING OF GEOLOGIC
M 29 284 OBLIQUE CONVERGENT MARGIN, SEISMIC TRACES, GULF OF ALASKA
M 29 282 OBLIQUE CONVERGENT PLATE BOUNDARY, GULF OF ALASKA
B 6307 1021 OBLIQUE WRENCH FAULTS, IRAN
B 6205 760 OBLIQUE-PROGRADATIONAL FACIES
M 26 173 OBLIQUE-PROGRADATIONAL FACIES, SEISMIC, INTERPRETATION
B 6010 1643 OBLITAS GARCIA, J., ON DEVELOPMENTS IN BOLIVIA, 1975
B 6110 1580 OBLITAS GARCIA, J., ON DEVELOPMENTS IN BOLIVIA, 1976
B 6210 2046 OBLITAS, J., AND F. JIMENEZ, ON DEVELOPMENTS IN BOLIVIA, 1977
M 25 349 OBLITAS, J., HYDROCARBON POTENTIAL OF BOLIVIAN SEDIMENTARY
B 6212 2512 OBLITAS, JAIME, EXPLORATION OF CARBONIFEROUS SYSTEM IN
B 6209 1762 OBRIEN, BRIAN E., LOBO TREND OF SOUTH LAREDO AREA, WEBB AND
B 6009 1520† OBSCURE NATURE OF PETROLEUM MIGRATION AND ENTRAPMENT
B 6004 663* OBSERVATIONS OF CIRCULATION AN SUSPENDED-SEDIMENT TRANSPORT
B 6409 1336† OC IN 1978—ADDENDUM, DEVELOPMENTS IN, ATLANTIC
B 6104 493 OCALA ARCH, FLORIDA
M 27 164 OCALA FORMATION, FLORIDA, EOCENE
M 27 95 OCALA FORMATION, FLORIDA, EOCENE
M 27 219 OCALA FORMATION, FLORIDA, EOCENE
M 27 223 OCALA FORMATION, FLORIDA, EOCENE
M 27 217 OCALA FORMATION, FLORIDA, EOCENE
M 27 220 OCALA LIMESTONE, FLORIDA, EOCENE
M 27 31 OCALA LIMESTONE, FLORIDA, EOCENE
M 28 119 OCALA LIMESTONE, FLORIDA, EOCENE
M 27 142 OCALA LIMESTONE, FLORIDA, EOCENE
M 24 289 OCALA UPLIFT, FLORIDA
B 6010 1871 OCCIDENTAL EXPLORATION AND PRODUCTION CO., AND
B 6303 474* OCCLUSION OF POROSITY IN CARBONATE RESERVOIRS BY DICKITE, ABST.
M 25 422† OCCURRENCE AND DEVELOPMENT OF SEDIMENTARY MANGANESE ORE,
S 7 43† OCCURRENCE AND EVOLUTION OF SALT IN DEEP GULF OF MEXICO
B 6409 1567* OCCURRENCE OF HIGH GRAVITY OIL IN AN OLIGOCENE VICKSBURG
B 6203 525* OCCURRENCES OF TERTIARY POROSITY IN CARBONATE ROCKS, ABST.
S 9 4 OCEAN BASIN MODELS, CLASSIFIED EXAMPLES
B 6303 297 OCEAN BASIN REGIONS, PROSPECTS, SMALL
S 3 36 OCEAN BASINS, MANTLE-DERIVED MAGMA
B 6209 1593 OCEAN BASINS, ORGANIC CONTENT, SMALL
M 29 443 OCEAN BOTTOM SEISMOMETERS (OBS), USED IN EXPERIMENTS
B 6203 563* OCEAN CHEMISTRY IN 22D CENTURY, ABST., BURNING OF, FOSSIL FUELS
M 29 113† OCEAN CRUST BOUNDARY OFF BAHIA BLANCA, ARGENTINA, STRUCTURE
S 9 3 OCEAN CURRENTS, PRESENT DISTRIBUTION
B 6405 721* OCEAN DRILLING NEAR UNITED STATES, ABST., SCIENTIFIC
B 6105 831* OCEAN DUMPING OF DREDGED MATERIAL—MAJOR ENVIRONMENTAL
B 6105 841* OCEAN DUMPING ON NEW YORK BIGHT INNER SHELF, ABST., GEOLOGIC
B 6105 833* OCEAN DUMPING SITES SEAWARD OF COLUMBIA RIVER MOUTH, ABST.,
B 6303 419* OCEAN EVOLUTION, ABST., DEEP-SEA DRILLING AND MAJOR THEMES OF
B 6002 196 OCEAN EXPLORATION AND REMAC RECONHECIMENTO DA MARGEM
M 25 539 OCEAN EXPLORATION PROGRAM, INTERNATIONAL DECADE OF
B 6207 1220* OCEAN FLOOR AS NATURAL GUIDE TO NATIONAL-INTERNATIONAL
B 6303 546* OCEAN MARGINS, ABST., IPOD, DRILLING ON CONVERGENT
B 6408 1182 OCEAN, ORGANIC CARBON IN SURFACE LAYER OF SEDIMENTS IN WORLD
B 6408 1199 OCEAN, OXYGEN-DEPLETED LAYERS IN WORLD
B 6303 288 OCEAN REGION, PROSPECTS, CENTRAL
B 6303 298 OCEAN RESERVES AND RESOURCES
B 6408 1271 OCEAN SCIENCE BOARD OF NATIONAL RESEARCH COUNCIL, A. W. BALLY,
B 6304 700* OCEAN TEMPERATURES, ABST., TERTIARY AND, CRETACEOU
B 6212 2510* OCEAN THERMAL ENERGY, ABST., PROGRESS AND PROSPECTS IN
B 6405 704* OCEAN TO BIOSPHERE—CHEMICAL EVIDENCE FROM DEEP-WEA CORES,
M 29 76 OCEAN-CONTINENT TRANSITION ZONE, ATLANTIC COAST
B 6403 366 OCEAN-FLOOR SPREADING AND UNCONFORMITIES OF OROGENY
B 6403 359† OCEAN-FLOOR SPREADING DURING LATE JURASSIC-LATE TERTIARY
M 25 460 OCEAN-MARGIN METALLOGENIC ZONES, PACIFIDES
S 5 55 OCEANA COUNTY, LOCATION, MICHIGAN,
B 6306 870† OCEANIC ANOXIC EVENTS AS CAUSAL FACTORS IN DEVELOPMENT OF
S 6 295 OCEANIC BASEMENT AGES, JURASSIC
M 25 497 OCEANIC BASIN, MINERAL DEPOSITS, CANADIAN, CORDILLERA
M 25 497 OCEANIC BASIN, ROCKS OF, CANADIAN, CORDILLERA
B 6412 2060 OCEANIC BASINS, ANOXIC AND OXIC
B 6303 452* OCEANIC CRUST, ABST.
M 25 541 OCEANIC CRUST, AGE OF, NAZCA PLATE
M 25 459 OCEANIC METALLOGENIC ZONES, PACIFIDES
S 3 79 OCEANIC MIGRATION HISTORY OF URANIUM AND THORIUM
B 6212 2511* OCEANIC PLATES—EAST PACIFIC RISE AND BAUER BASIN, ABST.,
B 6207 1185† OCEANIC ROCKS, NOTE 48—APPLICATION FOR, AMENDMENT OF, CODE
B 6303 438* OCEANIC SEDIMENTATION RATES, ABST., PALEOENVIRONMENTAL
B 6303 484* OCEANIC SEDIMENTS, ABST., DEEP-SEA DRILLING, NEW DIMENSION IN
B 6405 688* OCEANOGRAPHIC ENVIRONMENTS, ABST., RADIOLARIAN
B 6303 483* OCEANS, ABST., RADIOLARIAN PRESERVATION IN PRESENT AND PAS
B 6303 538* OCEANS AND CLIMATE DURING CENOZOIC, ABST.
B 6405 735* OCEANS AS RELATED TO HYDROCARBON GENERATION, ABST.,
B 6111 1991† OCEANS, PIERCEMENT STRUCTURES IN, DEEP
B 6408 1171 OCHOA FACIES, TEXAS
S 10 222 OCHOA SAG, TRAP, WEST TEXAS, PERMIAN BASIN
B 6305 824 OCHS, A. M., DEFINITION OF DEPOSITIONAL FACIES OF PARACHUTE
B 6203 550 OCHS, ALLAN M., DEPOSITIONAL LITHOFACIES OF PARACHUTE CREEK
B 6305 836 OCHS, ALLAN M., PETROLOGY OF TERTIARY SANDSTONES OF SOUTHERN
B 6303 441 OCONNOR, DOLORES I., VITRINITE REFLECTANCE--WHAT, HOW, AND
B 6203 356 OCONOSTATA FAULT, CALIFORNIA
B 6405 766* OCS LEASE SALE 55, NEAR YAKUTAT, EASTER GULF OF ALASKA, ABST.,
B 6004 669* OCS PROGRAMS, ABST., NEW, ENVIRONMENTAL, MAPS FOR, GULF OF
B 6303 475* OCS TRACT EVALUATIONS, ABST., U.S. GEOLOGICAL SURVEY METHODS
B 6211 2364* OCS WELL INFORMATION, OFFSHORE CENTRAL AND NORTHERN
B 6405 690* OCYPODE CERATOPHTHALMA, PALLAS, AS POSSIBLE INDICATORS OF
S 6 247 ODIN, G. S., ISOTOPIC DATES FOR A PALEOGENE TIME SCALE
S 6 127 ODIN, G. S., RESULTS OF DATING CRETACEOUS, PALEOGENE SEDIMENTS,
B 6203 550 ODOM, I. EDGAR, LITHOSTRATIGRAPHY, PETROLOGY, AND
B 6108 1384 ODOM, I. EDGAR, PARAGENESIS OF AUTHIGENIC MINERALS IN ST. PETER
B 6303 504 ODOM, I. EDGAR,, PARAGENETIC AND STABILITY RELATIONS AMONG
B 6002 326* ODOM LIME, FIELD, NOLAN AND FISHER COUNTIES, TEXAS, ABST., NEILL
B 6207 1230 ODRISCOLL, E. S. T., FUNDAMENTAL STRUCTURAL PATTERN IN
B 6105 797 OERTEL, GEORGE F., DISTRIBUTION OF SEDIMENTARY STRUCTURES IN
B 6105 818 OERTEL, GEORGE F., SAVANNAH OCEAN-DISPOSAL SITE, GEORGIA, ABST.,
B 6207 1129 OETA ANOMALY, IRIAN JAYA
B 6012 2187 OFF, TED, ROLE OF SMALL INDEPENDENT IN TOMORROWS OIL, ABST.
B 6307 1143 OFF-SHORE PETROLEUM RESOURCES OF SOUTH-EAST ASIA, BY CORAZON
B 6108 1385 OFFIELD, TERRY W., THERMAL-INFRARED SURVEYS OF URANIUM
M 26 235 OFFLAP, DEPOSITIONAL SYSTEMS
B 6404 530 OFFLEY ISLAND FORMATION, GREENLAND

B 6203 557* OFFSHELF BUILDUPS, MIDDLE ORDOVICIAN, VIRGINIA, ABST.,
B 6402 209† OFFSHORE BASIN, INDIA, SOURCE AREAS AND, MIGRATION TRENDS OF,
B 6402 209† OFFSHORE BASIN, INDIA, SOURCE AREAS AND, MIGRATION TRENDS OF,
M 26 240 OFFSHORE BASINS, MODELS, BRAZIL
B 6207 1230* OFFSHORE CENTRAL AND NORTHERN CALIFORNIA, ABST., OIL AND GAS
B 6207 1215* OFFSHORE CENTRAL SARAWAK, ABST., GEOLOGY AND EXPLORATION
B 6308 1205 OFFSHORE CONTINENTAL SHELF, DRILLING, 1978, ATLANTIC
M 29 233 OFFSHORE, CRITICAL WELL RESULTS, SUMATRA
B 6303 297 OFFSHORE, DRILLING TECHNOLOGY, DEEP
B 6105 818* OFFSHORE EASTERN CANADA, ABST., PALEOGEOGRAPHY AND
B 6004 655* OFFSHORE GULF OF ALASKA TERTIARY PROVINCE, ABST., PRELIMINARY
B 6208 1519† OFFSHORE IN 1977, DEVELOPMENTS IN, LOUISIANA, GULF COAST
B 6009 1613* OFFSHORE LEASES, ABST., COMPETITION AND QUEST FOR
B 6105 834* OFFSHORE LOUISIANA SOUTH TIMBALIER BLOCK 54 SALT DOME, ABST.,
B 6303 286† OFFSHORE, PETROLEUM, PROSPECT OF THE, DEEP
B 6303 299 OFFSHORE, PROSPECTIVE AREA, UNITED STATES
B 6408 1225 OFFSHORE REGIMES AND SHELF ENVIRONMENTS, CLASSIFICATION
B 6408 1232 OFFSHORE SANDSTONE BODIES IN EPICONTINENTAL SEAWAYS, ANCIENT
B 6202 181† OFFSHORE SEDIMENTATION, SUSSEX SANDSTONE OF, WYOMING--
B 6206 1067† OFFSHORE, SERENDIPITY, GAS, SEEP AREA, SOUTH, TEXAS
B 6401 37† OFFSHORE SHOUTHERN KALIMANTAN, INDONESIA, STRUCTURE,
M 29 11† OFFSHORE SOUTHEAST GEORGIA EMBAYMENT, U.S. ATLANTIC
B 6207 1238* OFFSHORE SOUTHERN CALIFORNIA, ABST., OIL AND GAS EXPLORATION
B 6401 37† OFFSHORE SOUTHERN KALIMANTAN, INDONESIA, STRUCTURE,
S 7 87† OFFSHORE TEXAS-LOUISIANA, RECOGNITION OF, CONTINENTAL-SLOPE,
B 6207 1240* OFFSHORE TIN DEPOSITS OF THAILAND AND RELATED MINING
B 6007 1128 OFFSHORE WELL, LOUISIANA, SHALE, DENSITY AND, POROSITY FOR
M 26 145† OFFSHORE WEST AFRICA, SEISMIC, STRATIGRAPHY AND, GLOBAL
B 6208 1441† OFFSHORE WEST GREENLAND IN 1977, DEVELOPMENTS
B 6308 1210† OFFSHORE WEST GREENLAND IN 1978, DEVELOPMENTS
B 6108 1393* OFFSHORE-BAR SANDSTONES OF FRONTIER FORMATION, SPEARHEAD
B 6403 429 OFICINA HIGH-WAX OIL FIELD, VENEZUELA
B 6403 429 OFICINA HIGH-WAX OIL FIELD, VENEZUELA
B 6408 1165 OFICINA SANDSTONE, VENEZUELA
B 6405 743 OFTENDAHL, ORRIN G., ET AL, RESIN RODS AND WOODY ROD-LIKE
B 6303 431* OGADEN BASIN, ETHIOPIA, ABST., REGIONAL HYDROCARBON, SOURC
B 6305 841* OGALLALA FLUVIAL ENVIRONMENTS, ABST., HYDROGEOLOGIC
B 6405 677* OGALLALA VOLCANIC-ASH DEPOSITS AND THEIR POSSIBLE RELATION TO
M 30 345 OGARRIO FIELD, MEXICO, SALINE BASIN
B 6212 2510 OGAWA, KATSURO, STRUCTURAL CONTRAST BETWEEN FORE-ARC AND
M 25 269 OGCHEON, OROGENIC ZONE, KOREA
S 4 281 OGDEN, JOHN C., CARBONATE-SEDIMENT PRODUCTION BY PARROT FISH
B 6207 1230 OGLE, BURDETTE A., OIL AND GAS EXPLORATION OFFSHORE CENTRAL
M 25 475 OGNEY ORE DEPOSIT, ANTIMONY, USSR
B 6004 704 OGRADY, M. D., CRETACEOUS AND EARLY TERTIARY
B 6303 504 OGUNYOMI, O., THERMA MATURATION OF LOWER PALEOZOIC SHALES IN
B 6202 296 OGWASHI-ASABA FORMATION, NIGERIA
B 6203 532 OHARA, PATRICK, ORIGIN OF NAVAJO SANDSTONE, ABST.
B 6408 1289* OHIO, ABST., COAL RESOURCES IN
B 6408 1282* OHIO, ABST., DEPOSITIONAL TRENDS OF LOWER, SILURIAN, CLINTON
B 6309 1576* OHIO, ABST., FRACTURE PATTERN IN OUTCROP OF GAS-PRODUCING,
B 6009 1623* OHIO, ABST., LITHOFACIES RELATIONS IN, ALEXANDRIAN AND EARLY,
B 6309 1583* OHIO, ABST., OIL- AND, GAS-PRODUCING CONFIGURATIONS IN, TRENTON
B 6004 727* OHIO, ABST., PALEOECOLOGY AND, FAUNA OF SOME, NIAGARAN, REEFS
B 6305 845* OHIO, ABST., PETROLEUM GEOLOGY OF, WHIRLPOOL MEMBER, SILURIAN,
B 6009 1621* OHIO, ABST., RELATION BETWEE, SALINA, UPPER, SILURIAN, OIL
B 6408 1287* OHIO, ABST., SEISMIC REFRACTION STUDY OF, BURIED VALLEY NEAR
B 6203 528* OHIO, ABST., SUBAERIAL EXPOSURE OF, SILURIAN, SHELF-MARGIN,
B 6004 704* OHIO, ABST., TRACE, FOSSILS OF, CHAGRIN FORMATION, UPPER,
B 6209 1710 OHIO, BEDFORD SHALE
B 6304 655 OHIO, BELPRE BENTONITE, NEW NAME
B 6209 1610 OHIO, BOIS BLANC LIMESTONE
B 6209 1710 OHIO, CLEVELAND MEMBER
B 6209 1614 OHIO, COLUMBIANA COUNTY, LOG
B 6209 1610 OHIO, COLUMBUS LIMESTONE
B 6209 1610 OHIO, DELAWARE LIMESTONE
B 6104 597 OHIO, DENMARK FIELD
B 6008 1288 OHIO, DEVELOPMENTS, 1975
B 6108 1270 OHIO, DEVELOPMENTS, 1976
B 6208 1400 OHIO, DEVELOPMENTS, 1977
B 6308 1245 OHIO, DEVELOPMENTS, 1978
B 6409 1404 OHIO, DEVELOPMENTS, 1979
B 6304 655† OHIO, DEVONIAN, BENTONITES IN EASTERN
B 6304 656 OHIO, DEVONIAN BLACK SHALE
B 6412 2163 OHIO, DEVONIAN SHALE FACIES
B 6309 1506 OHIO, FORMATION-DENSITY LOG STUDY
B 6209 1710 OHIO, HURON MEMBER
B 6209 1610 OHIO, LOCKPORT AND GUELPH DOLOMITES
S 5 96 OHIO, LOCKPORT GROUP
B 6104 594 OHIO, LOGAN FIELD
B 6309 1577* OHIO, MARYLAND, AND SOUTHERN PENNSYLVANIA, ABST., REGIONAL,
B 6104 595 OHIO, MORROW COUNTY FIELD
B 6209 1710 OHIO, OHIO SHALE
B 6209 1710 OHIO, OLENTANGY SHALE
B 6308 1244† OHIO, PENNSYLVANIA, VIRGINIA, AND WEST VIRGINIA IN 1978, OIL AND,
B 6308 1244† OHIO, PENNSYLVANIA, VIRGINIA, AND WEST VIRGINIA IN 1978, OIL AND,
B 6409 1403† OHIO, PENNSYLVANIA, VIRGINIA, AND WEST VIRGINIA IN 1979, OIL AND,
B 6208 1399† OHIO, PENNSYLVANIA, VIRGINIA, AND WEST VIRGINIA, OIL AND, GAS,
B 6108 1269† OHIO, PENNSYLVANIA, VIRGINIA, AND WEST VIRGINIA, OIL AND, GAS,
B 6008 1288† OHIO, PENNSYLVANIA, VIRGINIA, AND WEST VIRGINIA, OIL AND, GAS,
B 6008 1288† OHIO, PENNSYLVANIA, VIRGINIA, AND WEST VIRGINIA, OIL AND, GAS,
M 28 9 OHIO, PENNSYLVANIAN, POTTSVILLE FORMATION
B 6008 1298 OHIO, PRODUCTION, 1974-1975
B 6108 1281 OHIO, PRODUCTION, 1975-1976
B 6208 1410 OHIO, PRODUCTION, 1976-1977
B 6308 1247 OHIO, PRODUCTION 1977-1978
B 6409 1412 OHIO, PRODUCTION, 1978-1979
B 6209 1610 OHIO, SALINA GROUP
B 6309 1576* OHIO SHALE ALONG LAKE ERIE--STRATIGRAPHY, PETROLOGY, AND
B 6209 1703† OHIO SHALE IN EASTERN KENTUCKY, DIVISION OF BLACK
B 6309 1576* OHIO, SHALE IN NORTHEASTERN OHIO, ABST., FRACTURE PATTERN IN
B 6209 1710 OHIO SHALE, OHIO
S 5 96 OHIO, SILURIAN, LOCKPORT GROUP, OUTCROP RELATIONS OF REEFS
S 5 96 OHIO, SILURIAN, LOCKPORT GROUP, STRUCTURE
S 5 96 OHIO, SILURIAN, LOCKPORT GROUP, THICKNESS
S 5 2 OHIO, SILURIAN, NIAGARA GROUP
B 6209 1610 OHIO, SILURIAN-DEVONIAN CARBONATE SEQUENCE, NOMENCLATURE
B 6209 1622 OHIO, SILURIAN-DEVONIAN SECTION, NEW YORK TO
B 6209 1614 OHIO, SILURIAN-DEVONIAN TYPE LOG
B 6209 1710 OHIO, THREE LICK BED
B 6304 656 OHIO, TIOGA BENTONITE
B 6209 1624 OHIO TO PENNSYLVANIA, SILURIAN-DEVONIAN SECTION
B 6209 1626 OHIO TO WEST VIRGINIA, SILURIAN-DEVONIAN SECTION
B 6304 656 OHIO, WASHINGTON COUNTY, STRATIGRAPHIC COLUMN
B 6003 428† OHIO--CONSTITUTION, GROWTH, AND SIGNIFICANCE, SILURIAN, REEF
B 6009 1550† OI IN MEHSANA AREA OF CAMBAY BASIN, INDIA, PATTERN OF,
M 30 267 OI SANDSTONE, RESERVOIR, EUGENE ISLAND 330 FIELD
B 6303 428* OIL ABST., GEOLOGIC DISTRIBUTION OF
B 6009 1610* OIL, ABST., ORGANIC DIAGENESIS--ITS RELATION TO, GULF COAST
B 6302 242 OIL ALREADY GENERATED, CALCULATION
B 6312 2227† OIL AND BITUMEN SATURATION ON SOURCE-ROCK PYROLYSIS, EFFECT
B 6004 711* OIL AND GA EXPLORATION ASSUMPTIONS ON PROJECTIONS OF
B 6009 1612* OIL AND GAS, ABST., INDEPENDENT AND MAJOR ROLES IN EXPLORATION
B 6008 1204† OIL AND GAS DEVELOPMENT IN NORTHERN ROCKIES IN 1975
B 6308 1301† OIL AND GAS DEVELOPMENTS I NORTH MID-CONTINENT IN 1978
B 6308 1244† OIL AND GAS DEVELOPMENTS IN MARYLAND, OHIO, PENNSYLVANIA,
B 6308 1244† OIL AND GAS DEVELOPMENTS IN MARYLAND, OHIO, PENNSYLVANIA,
B 6008 1288† OIL AND GAS DEVELOPMENTS IN MARYLAND, OHIO, PENNSYLVANIA,
B 6008 1288† OIL AND GAS DEVELOPMENTS IN MARYLAND, OHIO, PENNSYLVANIA,
B 6409 1403† OIL AND GAS DEVELOPMENTS IN MARYLAND, OHIO, PENNSYLVANIA,
B 6208 1399† OIL AND GAS DEVELOPMENTS IN MARYLAND, OHIO, PENNSYLVANIA,
B 6108 1269† OIL AND GAS DEVELOPMENTS IN MARYLAND, OHIO, PENNSYLVANIA,
B 6008 1332† OIL AND GAS DEVELOPMENTS IN NORTH MID-CONTINENT IN 1975
B 6108 1305† OIL AND GAS DEVELOPMENTS IN NORTH MID-CONTINENT IN 1976
B 6208 1477† OIL AND GAS DEVELOPMENTS IN NORTH MID-CONTINENT IN 1977
B 6308 1301† OIL AND GAS DEVELOPMENTS IN NORTH MID-CONTINENT IN 1978
B 6409 1437† OIL AND GAS DEVELOPMENTS IN NORTH MID-CONTINENT IN 1979
B 6108 1346† OIL AND GAS DEVELOPMENTS IN NORTHERN ROCKIES IN 1976
B 6208 1498† OIL AND GAS DEVELOPMENTS IN NORTHERN ROCKIES IN 1977
B 6303 513* OIL AND GAS DISCOVERY FUNCTIONS TO RESOURCE ASSESSMENT, ABST.,
B 6102 294† OIL AND GAS EXPLORATION AND DEVELOPMENT IN OIL-POTASH AREA
B 6009 1463† OIL AND GAS EXPLORATION OF CONTINENTAL MARGINS, PLATE
B 6207 1230* OIL AND GAS EXPLORATION OFFSHORE CENTRAL AND NORTHERN
B 6207 1238* OIL AND GAS EXPLORATION OFFSHORE SOUTHERN CALIFORNIA, ABST.
B 6403 450* OIL AND GAS EXPLORATION WELLS IN PEDREGOSA BASIN, ABST.
B 6004 716* OIL AND GAS FIELD DATA BANK PROJECT, UNITED STATES AND
M 24 300† OIL AND GAS FIELDS FROM AAPG PUBLICATIONS, BIBLIOGRAPHY OF,
B 6401 8† OIL AND GAS FIELDS, GEOLOGIC SIGNIFICANCE OF, LANDSAT DATA FOR
B 6401 8† OIL AND GAS FIELDS, GEOLOGIC SIGNIFICANCE OF, LANDSAT DATA FOR
B 6109 1543* OIL AND GAS FIELDS IN GULF COAST, ABST., TOPOGRAPHIC EXPRESSION
B 6408 1210† OIL AND GAS FIELDS IN HIGHLY EXPLORED, MATURE, AREAS, METHODS
M 24 1 OIL AND GAS FIELDS, NORTH AMERICAN
M 30 1 OIL AND GAS FIELDS OF THE DECADE 1968-1978,
M 30 1† OIL AND GAS FIELDS OF THE DECADE 1968-1978, GIANT
B 6405 648 OIL AND GAS FIELDS OF THE FOUR CORNERS AREA, EDITED BY JAMES E.
B 6012 2179* OIL AND GAS FROM MATURE PROVINCES, ABST., NEW
B 6207 1237* OIL AND GAS FROM SUBMARINE CHANNELS IN GREAT VALLEY,
B 6112 2082† OIL AND GAS GENERATION BY LABORATORY PYROLYSIS OF KEROGEN
B 6402 209† OIL AND GAS IN BOMBAY OFFSHORE BASIN, INDIA, SOURCE AREAS AND,

B 6305 841* OIL AND GAS IN CHINA. BY TITLE ONLY
B 6306 886† OIL AND GAS IN DELTA-MARGIN FACIES OF DAKOTA SANDSTONE, LONE
B 6303 524* OIL AND GAS IN PEOPLES REPUBLIC OF CHINA ABST.
B 6112 2134† OIL AND GAS IN RESERVOIRS WITH SUBNORMAL PRESSURES
B 6105 780* OIL AND GAS IN RESERVOIRS WITH SUBNORMAL PRESSURES
B 6201 21 OIL AND GAS KITCHENS
B 6305 761† OIL AND GAS MIGRATION--CHEMICAL AN PHYSICAL CONSTRAINTS
B 6303 552* OIL AND GAS PLAYS IN FACIES-CYCLE WEDGES, ABST., ASSESSING
B 6408 1158† OIL AND GAS PLAYS IN FACIES-CYCLE WEDGES, ASSESSING
B 6008 1401* OIL AND GAS POSSIBILITIES REDWALL LIMESTONE, BLACK MESA BASIN,
B 6004 671* OIL AND GAS POTENTIAL BASED ON PRESENT EXPLORATION, ABST.,
B 6008 1406* OIL AND GAS POTENTIAL OF NUGGET SANDSTONE, SOUTHWESTERN
B 6405 714* OIL AND GAS POTENTIAL OF SAN LUIS BASIN, SOUTH-CENTRAL
B 6305 822* OIL AND GAS POTENTIAL OF SANTA FE EMBAYMENT, SANTA FE COUNTY,
B 6002 318* OIL AND GAS POTENTIAL OF WESTER KANSAS, ABST.
B 6305 829* OIL AND GAS POTENTIAL OF WYOMING, ABST.
B 6405 767* OIL AND GAS POTENTIAL OF WYOMING-UTAH-IDAHO OVERTHRUST
B 6004 729* OIL AND GAS POTENTIALS OF MIDDLE AND UPPER ORDOVICIAN OF
B 6303 540* OIL AND GAS RESOURCE APPRAISAL--STATE OF THE ART, ABST.
B 6004 665* OIL AND GAS RESOURCES BY U.S. GEOLOGICAL SURVEY, ABST., NE
B 6203 561* OIL AND GAS RESOURCES IN GULF OF MEXICO, ABST., GEOLOGIC
B 6012 2186* OIL AND GAS RESOURCES IN THE UNITED STATES, ABST., GEOLOGIC
B 6004 698* OIL AND GAS RESOURCES IN UNITED STATES, ABST., PROBABILISTIC AND,
B 6312 2183† OIL AND GAS RESOURCES, METHODS OF, ESTIMATING
B 6305 842* OIL AND GAS RESOURCES OF DENVER-JULESBURG BASIN, ABST., NEW
B 6308 1428* OIL AND GAS TRAPS, ABST., DESIGN AND FUNCTION OF
B 6203 554* OIL AND GAS TRAPS, ABST., PHYSICAL AND CHEMICAL SIGNALS AROUND
B 6203 558* OIL AND GAS TRAPS AS CLUES TO MIGRATION, ABST., DESIGN AND
B 6405 767* OIL AND METALS IN ORDOVICIAN AND DEVONIAN KEROGENOUS
B 6004 730* OIL AND MINERAL DEPOSITS, ABST., AULACOGENS AN,
B 6203 565* OIL AND TAR SANDS OF ALBERTA. BY TITLE ONLY., HEAVY
B 6211 2358* OIL COMPANIES, ABST., EMERGENCE OF NATIONAL
B 6202 179 OIL CONSUMPTION VERSUS TIME, WORLD, POPULATION AND, CRUDE
B 6004 685* OIL DEVELOPMENT, LOS ANGELES, CALIFORNIA, ABST.,
M 30 329† OIL DISCOVERY IN THE CAMPOS BASIN, BRAZIL, THE, NAMORADO OIL
B 6305 770 OIL DROPLET EXPULSION
B 6303 412* OIL EXPLORATION IN CAMPOS BASIN, BRAZIL. MODEL FOR
B 6303 427* OIL EXPLORATION IN PRE-SALT AND POST-SALT SEQUENC OF WEST
B 6405 724* OIL EXPLORATION IN SOUTHEAST TURKEY THRUST BELT, ABST.
B 6408 1279* OIL EXPULSION—A CONSEQUENCE OF OIL GENERATION, ABST.
B 6405 739* OIL EXTRACTION METHOD TO AVAILABLE GEOLOGY, ABST., LOFRECO
B 6205 749 OIL FIELD, DEFINITION, GIANT
B 6401 12 OIL FIELD, LANDSAT DATA, IRAQ, BUZURGAN GIANT
B 6002 318* OIL FIELD WATER GEOCHEMISTRY TO EXPLORATION, ABST.,
B 6302 245† OIL FIELDS, AEROMAGNETIC DETECTION OF, DIAGENETIC, MAGNETITE
B 6306 870† OIL FIELDS, CRETACEOUS, OCEANIC, ANOXIC EVENTS AS CAUSAL
M 30 195† OIL FIELDS IN WESTERN EUROPE, EKOFISK--FIRST OF THE GIANT
B 6111 1950† OIL FIELDS, TAR, SANDS AND, SUPERGIANT
B 6309 1538† OIL FIELDS. DISCUSSION, AEROMAGNETIC DETECTION OF, DIAGENETIC,
B 6203 455† OIL FROM BARROW ISLAND, WESTERN AUSTRALIA, ISOTOPIC EVIDENCE
B 6009 1611* OIL FROM OLD FIELDS, ABST., NEW
B 6303 477* OIL GENERATION AND MIGRATION IN LUSITANIAN BASIN, PORTUGAL,
B 6201 98† OIL GENERATION IN BLACK SHALE MEMBERS OF PHOSPHORIA
B 6111 1953 OIL GENERATION, MIGRATION, ENTRAPMENT, AND DEGRADATION
B 6005 884† OIL GENESIS. DISCUSSION AND REPLY, TIME-, TEMPERATURE RELATION
B 6303 472* OIL GEOCHEMISTRY AS EXPLORATION TOOL, ABST.
S 10 94 OIL, COMPOSITION, CRUDE
B 6002 215 OIL, GHAWAR, ARABIAN, WHOLE, CRUDE
B 6009 1502† OIL HABITAT, TRINIDAD, MIOCENE, DELTAIC
B 6305 763 OIL, HYDROCARBONS DISSOLVED FROM, CRUDE
B 6002 323* OIL IN WESTERN MISSOURI--UPDATE 1975, ABST., HEAVY
B 6009 1617* OIL INDUSTRY AND COASTAL ZON MANAGEMENT IN LOUISIANA, ABST.,
B 6412 2108† OIL MIGRATION, EVIDENCES OF, PRIMARY
B 6203 545* OIL MIGRATION LIMITATIONS SUGGESTED BY GEOLOGIC AND
B 6007 1112† OIL, ORIGIN OF, ATHABASCA
B 6007 1110† OIL, ORIGIN OF, ATHABASCA
B 6209 1753* OIL POTENTIAL OF LOWER CRETACEOUS SUNNILAND FORMATION IN
B 6108 1385* OIL POTENTIAL OF MESOZOIC SEDIMENTS BENEATH INDEPENDENCE
B 6205 742 OIL PRODUCTION, 1940-2020, TOTAL, WORLD
B 6405 754* OIL PRODUCTION FROM FRACTURED CHERTS OF WOODFORD AND
B 6009 1617* OIL PROSPECTS IN GULF OF MEXICO REGION ABST.
B 6003 389† OIL PROSPECTS OF SOMALIA, EAST AFRICA, GEOLOGY AND
B 6203 463† OIL RESERVOIRS, CHADRA SANDS, GIALO FIELD, LIBYA, SECONDARY,
B 6303 459* OIL RESOURCE ASSESSMENTS AND RELATION TO UNITED STATES OIL
M 25 309† OIL RESOURCES OF THE PACIFIC FOR PEACE AND SECURITY--SUMMARY
B 6305 820 OIL SAND AND OIL SHALE CHEMISTRY, EDITED BY O. P. STRAUSZ AND E.
B 6203 547* OIL SANDS, ABST., DEEP SINUOUS, CHANNELS AND GIANT, CROSS-BEDS
B 6206 1093* OIL SANDS, ABST., NEW GEOLOGIC FRAMEWORK FOR, MINING OF
B 6303 499* OIL SANDS, ALBERTA, ABST., SEDIMENTOLOGY, PETROLOGY, AND
B 6303 499* OIL SANDS, ALBERTA. BY TITLE ONLY, SEDIMENTOLOGY, PETROLOGY,
B 6203 560* OIL SANDS—GEOLOGY AND PRODUCTION. BY TITLE ONLY., GREAT,
B 6110 1796 OIL SERVICE CO. OF IRAN, AMOCO INTERNAT. OIL CO., AMERICAN
B 6210 2112 OIL SERVICE CO. OF IRAN, AMOCO INTERNATIONAL OIL CO., SOCIETE
B 6310 1885 OIL SERVICE CO. OF IRAN, ATLANTIC RICHFIELD CO., AND AGIP
B 6010 1866 OIL SERVICE CO. OF IRAN, PHILLIPS PETROLEUM CO., DEMINEX IRA OIL
B 6405 685* OIL SHALE, ABST., CARBONATE MINERAL REACTIONS DURING,
B 6008 1412* OIL SHALE, ABST., DEPOSITIONAL ENVIRONMENNSTS OF
B 6004 666* OIL SHALE, ABST., ENVIRONMENTS OF DEPOSITION OF
B 6105 820* OIL SHALE, ABST., PREDICTIONS FOR
B 6203 505* OIL SHALE AND MARLSTONE, GREEN RIVER FORMATION, PICEANCE
B 6405 696* OIL SHALE DEPOSITS OF SOUTHERN UINTA BASIN, UTAH, ABST.,
B 6303 439* OIL SHALE ON BASIS OF CHEMICAL CHARACTERISTICS, ABST.,
B 6405 801* OIL SHALE, PICEANCE CREEK BASIN, NORTHWEST COLORADO, ABST.,
B 6211 2362* OIL SHALE RESOURCES OF SOUTH ELKO BASIN, NEVADA, ABST.,
B 6405 700* OIL SHALE TO ENERGY NEEDS OF UNITED STATES AND OTHER PARTS OF
B 6305 768 OIL SOLUBILIZED IN SURFACTANT MICELLES
B 6302 239† OIL SOURCE BED EVALUATION, SIMPLE METHOD FOR
B 6405 697* OIL SOURCE BEDS, ABST., ANOXIC ENVIRONMENTS AND
B 6205 827† OIL SOURCE ROCK, ORGANOCHEMICAL EVALUATION OF, GEVARAM
B 6004 678* OIL SPILL, ABST., GREAT, PATAGONIAN
B 6409 1554* OIL SPILLS, IXTOC I AND BURMAH AGATE, WITH CEMENTS ON AVENUES
B 6409 1554* OIL SPILLS, IXTOC I AND BURMAH AGATE, WITH COMMENTS ON AVENUE
M 28 180 OIL STAINING, EXAMPLE, FRACTURES
S 7 227† OIL TRAPPING CHARACTERISTICS OF TURBIDITES
B 6102 297* OIL TYPES IN PERMIAN BASIN BY TITLE ONLY, CHARACTERIZATION OF
S 10 91 OIL, SOLUBILITY, CRUDE
S 10 227 OIL AND COPPER IN SAME TRAP, USSR, FERGHANA
S 10 89† OIL AND GAS MIGRATION: CHEMICAL AND PHYSICAL CONSTRAINTS
S 10 217† OIL AND GAS TRAPS, DESIGN AND FUNCTICN OF
S 10 59 OIL COMPOSITION, MIGRATION IN AQUEOUS SOLUTION
S 10 11 OIL EXPULSION, MECHANISMS
S 10 114 OIL SAND, ARTIFICIAL WATER-ROCK INTERACTION
M 25 176† OIL- AND GAS-BEARING BASINS OF CIRCUM-PACIFIC BELT--SUMMARY
B 6309 1583* OIL- AND GAS-PRODUCING CONFIGURATIONS IN TRENTON LIMESTONE,
B 6009 1448† OIL-BEARING REEFS IN SALAWATI BASIN OF IRIAN JAYA, INDONESIA
B 6401 110 OIL-CARRYING CAPACITY OF GAS-SATURATED WATER
B 6401 110 OIL-CARRYING CAPACITY OF GAS-SATURATED WATER
B 6203 518* OIL-FIELD DISTRIBUTION FOR SEDIMENTARY BASINS OF UNITED STATES,
B 6212 2443 OIL-FIELD WATER SAMPLE, TITRATION CURVES FOR, SODIUM
B 6303 479* OIL-FIELD WATERS AND ORIGIN OF CARBON DIOXIDE, ABST., STABLE,
B 6303 478* OIL-FIELD WATERS FROM NORTHERN GULF OF MEXICO BASIN, ABST.,
B 6212 2441† OIL-FIELD WATERS—IMPLICATIONS FOR ORIGIN OF NATURAL GAS,
B 6108 1376* OIL-SHALE DEPOSITS, ABST., DISTRIBUTION AND, ORIGIN OF, SULFUR IN,
B 6305 823* OIL-SHALE LAND EXCHANGES, ABST., MINERAL EVALUATION OF
B 6408 1180 OIL-SOURCE BEDS, PRECURSOR SEDIMENTS TO
B 6105 836* OIL-SPILL POTENTIAL ON ATLANTIC OUTER CONTINENTAL SHELF, ABST.,
B 6405 776* OIL-TO-SOURCE CORRELATION—PINEVIEW FIELD, OVERTHRUST BELT,
B 6104 577 OILS, EXAMPLES OF, AGE CALCULATIONS OF
B 6403 427† OILS IN ASSAM BASIN, INDIA, DEPOSITIONAL ENVIRONMENT OF, SOURCE
B 6303 467* OILS IN MOLASSE BASIN, SOUTHERN GERMANY, ABST., ORIGIN AND
B 6203 572* OILS, SASKATCHEWAN AND ALBERTA. BY TITLE ONLY., HEAVY
B 6304 701* OILS--IDENTIFICATION BY CARBON ISOTOPES, ABST., BACTERIAL,
B 6104 573† OILS--PHYSICAL CHEMISTRY APPLIED T PETROLEUM GEOCHEMISTRY I,
B 6203 568* OIL—A RECONSIDERATION OF OLD PRINCIPLES, ABST., HABITAT OF
B 6102 138 OJAI VALLEY, CALIFORNIA
B 6102 170 OJAI VALLEY, CALIFORNIA
B 6207 1221* OK TEDI COPPER PROJECT, PAPUA NEW GUINEA, ABST.
B 6303 551 OKADA, H., EARLY PLEISTOCENE SUBMARINE CANYON, BOSO PENINSULA,
B 6203 531 OKADA, HAKUYU, SLOPE SEDIMENTATION PROCESSES IN NEOGENE
M 25 441 OKCHON GEOSYNCLINAL ZONE, SOUTH KOREA
M 25 440 OKCHON SYSTEM, SOUTH KOREA
M 25 443 OKCHON THRUSTS, SOUTH KOREA
M 25 468 OKHOTSK COMPLEX, USSR
B 6207 1211* OKHOTSK SEA ICE COVERS BY MICROWAVE SATELLITE IMAGERY, ABST.,
M 25 458 OKHOTSK VOLCANIC-PLUTONIC CHAIN, PACIFIDES, KHINGAN-
B 6002 163 OKHOTSK-CHUKCHI VOLCANIC AND PLUTONIC ZONE, SIBERIA
M 25 463 OKHOTSK-NIPPON OROGENS
M 30 459 OKI, T., GEOLOGY OF THE BEKAPAI FIELD
M 30 399 OKI, T., GEOLOGY OF THE HANDIL FIELD, EAST KALIMANTAN,
M 29 240 OKI-DAITO RIDGE, GEOLOGY, NORTH PHILIPPINE SEA
B 6209 1646 OKINAWA TROUGH
M 29 199† OKINAWA TROUGH, EXTENSIONAL, TECTONICS IN THE
M 29 206 OKINAWA TROUGH, HISTORY, DEPOSITION
M 29 206 OKINAWA TROUGH, HISTORY, SPREADING
M 29 200 OKINAWA TROUGH, INDEX MAP
M 29 201 OKINAWA TROUGH, SEDIMENTS, ACOUSTICS
M 29 200 OKINAWA TROUGH, SEISMIC REFLECTION DATA
M 29 203 OKINAWA TROUGH, TECTONICS, FAULTING
B 6203 491* OKLAHOMA, ABST., ATOKA FORMATION, PENNSYLVANIAN,
B 6004 727* OKLAHOMA, ABST., BASINWARD, FACIES CHANGES IN, WAPANUCKA

B 6004 719* OKLAHOMA, ABST., CHEMICAL AND, ISOTOPIC INVESTIGATION O
B 6203 543* OKLAHOMA, ABST., CONTEMPORANEOUS, FAULTS—MECHANISM FOR
B 6311 2114* OKLAHOMA, ABST., DEPOSITIONAL ENVIRONMENT OF, PENNSYLVANIA
B 6203 518* OKLAHOMA, ABST., DEPOSITIONAL ENVIRONMENTS AND, CONODONT
B 6203 526* OKLAHOMA, ABST., DEVELOPMENT OF, MILLS RANCH COMPLEX, TEXAS
B 6002 321* OKLAHOMA, ABST., DEVELOPMENTS IN SOUTH-CENTRAL
B 6002 322* OKLAHOMA, ABST., DIBBLE--, MUSTANG, CURRENT, HUNTON PLAY IN,
B 6405 677* OKLAHOMA, ABST., DISTRIBUTION AND ALTERATION OF, OGALLALA,
B 6203 566* OKLAHOMA, ABST., ECONOMIC POTENTIAL OF, CANYON-, FAN UNIT IN,
B 6203 555* OKLAHOMA, ABST., EFFECT OF, FAULT-RELATED, GRANULATION ON,
B 6311 2118* OKLAHOMA, ABST., ENVIRONMENTS AND DIAGENESIS, MORROW
B 6105 827* OKLAHOMA, ABST., FLOW PATTERNS AND, BAR MORPHOLOGY IN,
B 6311 2118* OKLAHOMA, ABST., GEOCHEMISTRY OF SMALL, LACUSTRINE, DELTA,
B 6002 320* OKLAHOMA, ABST., GEOLOGY OF, PERMIAN, BLAINE FORMATION AND
B 6002 327* OKLAHOMA, ABST., GEOTECTONIC EVOLUTION OF, WICHITA,
B 6405 754* OKLAHOMA, ABST., OIL PRODUCTION FROM, FRACTURED, CHERTS OF,
B 6108 1382* OKLAHOMA, ABST., ORDOVICIAN OIL, ACCUMULATION AT, NORTHEAST,
B 6311 2120* OKLAHOMA, ABST., PALYNOLOGIC AND PLANT COMPRESSION EVIDENCE
B 6209 1767* OKLAHOMA, ABST., PERITIDAL, CARBONATE MATERIAL AND EVIDENCE
B 6203 526* OKLAHOMA, ABST., PERMIAN, SALT DEPOSITS IN, TEXAS PANHANDLE
B 6108 1375* OKLAHOMA, ABST., PETROLEUM, SOURCE-ROCK POTENTIAL OF,
B 6108 1384* OKLAHOMA, ABST., PETROLOGY AND DIAGENESIS OF, DEEP-WATER,
B 6303 475* OKLAHOMA, ABST., POTENTIAL FOR SUBSURFACE, DISPOSAL OF
B 6002 318* OKLAHOMA, ABST., PRODUCING CHARACTERISTICS OF LOWER, MORROW
B 6203 566* OKLAHOMA, ABST., REEVALUATION OF, MORROWAN-, ATOKAN SERIES
B 6203 538* OKLAHOMA, ABST., SANDSTONE DIAGENESIS IN, BROMIDE FORMATION,
B 6311 2119* OKLAHOMA, ABST., STIMULATION DESIGN FOR UPPER, MORROW
B 6203 523* OKLAHOMA, ABST., SUBSURFACE LITHOSTRATIGRAPHY OF, HUNTON
B 6305 804 OKLAHOMA, ALEDO FIELD
B 6103 361 OKLAHOMA, ANADARKO BASIN
B 6002 233 OKLAHOMA, ANADARKO BASIN
B 6205 856 OKLAHOMA, ANADARKO BASIN, BASEMENT CONTOUR MAP, TEXAS,
M 26 447 OKLAHOMA, ANADARKO BASIN, BELLE CITY LIMESTONE
M 26 440 OKLAHOMA, ANADARKO BASIN, HASKELL LIMESTONE
M 26 445 OKLAHOMA, ANADARKO BASIN, MEDRANO SANDSTONE
M 26 452 OKLAHOMA, ANADARKO BASIN, MORROW SANDSTONE,
B 6109 1437 OKLAHOMA, ANADARKO BASIN, SEISMIC DATA
M 26 452 OKLAHOMA, ANADARKO BASIN, SPRINGER SANDSTONE
M 26 440 OKLAHOMA, ANADARKO BASIN, TONKAWA LIMESTONE
M 26 445 OKLAHOMA, ANADARKO BASIN, WADE SANDSTONE
M 26 451 OKLAHOMA, ANADARKO BASIN, WATONGA-CHICKASHA TREND
M 26 451 OKLAHOMA, ANADARKO BASIN, WICHITAN UNCONFORMITY
M 26 441 OKLAHOMA, ANADARKO, UPPER HOXBAR GROUP, MISSOURIAN
B 6203 498* OKLAHOMA, AND MODERN JAMAICAN REEFS, ABST.,
B 6108 1216† OKLAHOMA AND PANHANDLE OF TEXAS, DEVELOPMENTS IN
B 6008 1251† OKLAHOMA AND PANHANDLE OF TEXAS IN 1975, DEVELOPMENTS IN
B 6208 1372† OKLAHOMA AND PANHANDLE OF TEXAS IN 1977, DEVELOPMENTS IN
B 6308 1286† OKLAHOMA AND PANHANDLE OF TEXAS IN 1978, DEVELOPMENTS IN
B 6409 1387† OKLAHOMA AND PANHANDLE OF TEXAS IN 1979, DEVELOPMENTS IN
B 6209 1765* OKLAHOMA AND TEXAS, ABST., SEDIMENTARY FACIES, STRUCTURES,
B 6002 322* OKLAHOMA AND TEXAS PANHANDLES, ABST., REPORT--
B 6002 321* OKLAHOMA AND WESTERN ARKANSAS. BY TITLE ONLY, EASTERN
B 6103 361 OKLAHOMA, ARBUCKLE ANTICLINE
B 6307 1136 OKLAHOMA, ARBUCKLE MOUNTAINS CORRELATION
B 6103 370 OKLAHOMA, ARBUCKLE UPLIFT, ALLUVIAL FAN DEPOSITS
B 6103 361 OKLAHOMA, ARDMORE BASIN
B 6307 1025 OKLAHOMA, ARDMORE BASIN SEISMIC PROFILE
B 6103 361 OKLAHOMA, ASPHALTUM STRUCTURE
B 6405 705* OKLAHOMA AULACOGEN—IMPLICATION FOR PETROLEUM
B 6007 1009 OKLAHOMA, BARTLESVILLE, BLUEJACKET, SANDSTONE
B 6103 374 OKLAHOMA, BARTLESVILLE-BLUEJACKET SANDSTONE
B 6109 1440 OKLAHOMA, BELLE CITY LIMESTONE
B 6107 988 OKLAHOMA, BRIDGE CREEK CHALK UNIT
B 6307 1136 OKLAHOMA, BROMIDE FORMATION
B 6103 361 OKLAHOMA, CADDO ANTICLINE
B 6109 1437 OKLAHOMA, CANADIAN COUNTY, ANADARKO BASIN
M 26 439 OKLAHOMA, CANADIAN COUNTY, ANADARKO BASIN, PENNSYLVANIAN,
B 6103 361 OKLAHOMA, CARTER-KNOX STRUCTURE
B 6103 361 OKLAHOMA, CEMENT ANTICLINE
B 6302 246 OKLAHOMA, CEMENT OIL FIELD, AEROMAGNETIC SURVEY
B 6302 247 OKLAHOMA, CEMENT OIL FIELD, MAGNETITE
B 6103 360 OKLAHOMA, CEMENT URANIUM DEPOSIT
B 6103 373 OKLAHOMA, CHEROKEE DELTAIC OIL-PRODUCTIVE SANDSTONES
B 6103 361 OKLAHOMA, CHICKASHA ANTICLINE
B 6103 361 OKLAHOMA, CHITWOOD STRUCTURE
B 6104 621 OKLAHOMA CITY FIELD, OIL SAMPLES, OKLAHOMA
B 6104 622 OKLAHOMA CITY FIELD, SECTION, OKLAHOMA
B 6104 620† OKLAHOMA CITY OIL--SECOND CROP FROM PRESERVED
B 6001 115† OKLAHOMA CITY OIL--SECOND CROP FROM PRESERVED
B 6103 361 OKLAHOMA, COMANCHE STRUCTURE
B 6103 363 OKLAHOMA, COX CITY DEPOSIT
B 6103 361 OKLAHOMA, CRINER HILLS
B 6311 2115* OKLAHOMA DURING DESMOINESIAN CYCLIC SEDIMENTATION, ABST.,
B 6302 245 OKLAHOMA, EAST CEMENT DOME
B 6103 373 OKLAHOMA, ELK CITY FORMATION
B 6103 361 OKLAHOMA, EMPIRE STRUCTURE
B 6307 1135 OKLAHOMA, FITE LIMESTONE
B 6307 1135 OKLAHOMA, FITE LIMESTONE
B 6311 2117* OKLAHOMA FOR HEAVY-OIL POTENTIAL, ABST., INVESTIGATION OF,
B 6307 1135† OKLAHOMA, FORMERLY CALLED FERNVALE, WELLING FORMATION, NEW
B 6103 361 OKLAHOMA, FOX-GRAHAM STRUCTURE
M 26 460 OKLAHOMA, GEARY AREA, GEOSEISMIC MODELING
B 6103 366 OKLAHOMA, HARDEMAN-HOLLIS BASIN
B 6107 988 OKLAHOMA, HARTLAND CHALK UNIT
B 6109 1440 OKLAHOMA, HASKELL LIMESTONE
B 6103 361 OKLAHOMA, HEALDTON ANTICLINE
B 6104 583 OKLAHOMA, HEALDTON POOL
B 6104 583 OKLAHOMA, HEWITT FIELD
B 6302 246 OKLAHOMA, HOXBAR FORMATION
B 6109 1439 OKLAHOMA, HOXBAR GROUP FACIES CHANGE
B 6307 1136 OKLAHOMA, KEEL FORMATION
B 6103 361 OKLAHOMA, LOCO STRUCTURE
B 6103 361 OKLAHOMA, MARIETTA BASIN
B 6109 1440 OKLAHOMA, MEDRANO SANDSTONE
B 6305 805 OKLAHOMA, MILLS RANCH COMPLEX, TEXAS AND
B 6103 361 OKLAHOMA, MILROY STRUCTURE
M 26 453 OKLAHOMA, MISSISSIPPIAN, CHESTER SHALE
M 27 201 OKLAHOMA, MISSISSIPPIAN, HINDSVILLE LIM
M 27 76 OKLAHOMA, MISSISSIPPIAN, HINDSVILLE LIMESTONE
M 27 211 OKLAHOMA, MISSISSIPPIAN, HINDSVILLE LIMESTONE
M 27 100 OKLAHOMA, MISSISSIPPIAN, HINDSVILLE LIMESTONE
M 26 453 OKLAHOMA, MISSISSIPPIAN, PARVIN LIMESTONE
M 27 217 OKLAHOMA, MISSISSIPPIAN, PITKIN LIMESTONE
M 27 118 OKLAHOMA, MISSISSIPPIAN, PITKIN LIMESTONE
B 6012 2103† OKLAHOMA, NONGLACIAL, VARVES IN LOWER MEMBER OF, ARKANSAS,
B 6103 361 OKLAHOMA, NORTH DUNCAN STRUCTURE
S 7 30 OKLAHOMA, NORTHERN GULF OF MEXICO RIM, EXPOSED
B 6203 567* OKLAHOMA OIL COUNTRY IN CONTEXT OF PLATE TECTONICS, ABST.,
B 6104 621 OKLAHOMA, OKLAHOMA CITY FIELD, OIL SAMPLES
B 6104 622 OKLAHOMA, OKLAHOMA CITY FIELD, SECTION
M 27 119 OKLAHOMA, ORDOVICIAN, WEST SPRING CREEK FORMATION
B 6007 1125 OKLAHOMA, PALEOZOIC SHALE, ELECTRON-SCANNING
B 6103 361 OKLAHOMA, PAULS VALLEY UPLIFT
M 27 90 OKLAHOMA, PENNSYLVANIAN ASPHALTS
M 26 453 OKLAHOMA, PENNSYLVANIAN, ATOKA LIMESTONE
M 27 118 OKLAHOMA, PENNSYLVANIAN, BLOYD FORMATION
M 27 167 OKLAHOMA, PENNSYLVANIAN, BLOYD FORMATION
M 27 96 OKLAHOMA, PENNSYLVANIAN, BLOYD FORMATION
M 27 116 OKLAHOMA, PENNSYLVANIAN, BLOYD FORMATION
M 27 86 OKLAHOMA, PENNSYLVANIAN, BLOYD FORMATION
M 27 35 OKLAHOMA, PENNSYLVANIAN, BLOYD FORMATION
M 27 120 OKLAHOMA, PENNSYLVANIAN, HALE FORMATION
M 27 166 OKLAHOMA, PENNSYLVANIAN, HALE FORMATION
B 6112 2086 OKLAHOMA, PENNSYLVANIAN, KEROGEN
M 26 453 OKLAHOMA, PENNSYLVANIAN, MORROW SANDSTONE
M 26 453 OKLAHOMA, PENNSYLVANIAN, SPRINGER SANDSTONE
B 6103 366 OKLAHOMA, PERMIAN CHANNEL SANDSTONES
B 6307 1136 OKLAHOMA, PETTIT FORMATION
B 6307 1136 OKLAHOMA, POOLEVILLE MEMBER
B 6108 1218 OKLAHOMA, PRODUCTION, 1976
B 6208 1375 OKLAHOMA, PRODUCTION, 1977
B 6308 1290 OKLAHOMA, PRODUCTION, 1978
B 6409 1391 OKLAHOMA, PRODUCTION, 1979
B 6103 366 OKLAHOMA, RED RIVER UPLIFT
B 6302 247 OKLAHOMA, RUSH SPRINGS SANDSTONE
B 6109 1437† OKLAHOMA, SEISMIC STRATIGRAPHIC, MODEL OF DEPOSITIONAL,
M 26 439† OKLAHOMA, SEISMIC, STRATIGRAPHIC, MODEL OF, DEPOSITIONAL,
B 6103 361 OKLAHOMA, SHOLOM ALECHEM STRUCTURE
M 26 346 OKLAHOMA, SOUTH CERES FIELD, NOBLE COUNTY
B 6104 582 OKLAHOMA, SOUTHERN, ARBUCKLE OILS, OIL AGE
B 6104 582 OKLAHOMA, SOUTHERN, CRINER HILLS, CROSS SECTION
B 6104 583 OKLAHOMA, SOUTHWEST ENVILLE POOL
B 6307 1135 OKLAHOMA, SYLVAN SHALE
B 6307 1137 OKLAHOMA, TENKILLER-BLACKGUM FORMATIONS
B 6103 373 OKLAHOMA, TIDAL-FLAT, SABKHA, SANDSTONES-SILTSTONES
B 6109 1440 OKLAHOMA, TONKAWA LIMESTONE
B 6311 2116* OKLAHOMA TRI-STATE AREA, ABST., CORRELATION OF, CHEROKEE,
B 6110 1897 OKLAHOMA, TULSA MOUNTAINS
B 6103 361 OKLAHOMA, TUSSY-TATUMS STRUCTURE
B 6307 1136 OKLAHOMA, TYNER FORMATION

B 6205 855 OKLAHOMA, ULTRADEEP WELLS, LOCATION, TEXAS, PANHANDLE AND
B 6205 859 OKLAHOMA, ULTRADEEP WILDCATS, SUCCESS RATIOS, TEXAS,
B 6103 360† OKLAHOMA, URANIUM POTENTIAL OF, PERMIAN AND, PENNSYLVANIAN
B 6405 724* OKLAHOMA, USING SEDIMENTARY FACIES ANALYSIS, ABST.,
B 6103 361 OKLAHOMA, VELMA-CRUCE ANTICLINE
B 6307 1136 OKLAHOMA, VIOLA FORMATION
B 6109 1440 OKLAHOMA, WADE SANDSTONE
B 6103 361 OKLAHOMA, WALTERS STRUCTURE
B 6103 361 OKLAHOMA, WAURIKA-MUENSTER ARCH
B 6307 1135 OKLAHOMA, WELLING FORMATION, NEW NAME
B 6302 245 OKLAHOMA, WEST CEMENT DOME
B 6103 361 OKLAHOMA, WICHITA-CRINER HILLS UPLIFT
B 6203 508* OKLAHOMA. BY TITLE ONLY., GEOLOGY AND DEVELOPMENT OF, LEEDY
B 6405 784* OKLAHOMA—SOURCE, RESERVOIR, AND TRAP, ABST., OSWEGO
M 26 451† OKLAHOMA—T13N, R10W, A, CASE HISTORY OF, GEOSEISMIC, MODELING
B 6008 1391* OKLO PHENOMENON--GEOLOGIC AND NUCLEAR ASPECTS, ABST.
B 6408 1251 OKOYE, NDIBE V., PETROLEUM SOURCE-BED EVALUATION OF TERTIARY
M 29 239 OKUDA, YOSHIHISA, SUBSIDENCE OF THE DAITO RIDGE AND ASSOCIATED
B 6410 1644 OLARY RANGES, AUSTRALIA
B 6211 2270 OLCESE SAND, CALIFORNIA
B 6105 818* OLD AND NEW HORIZONS IN FOR WESTERN AND ARCTIC CANADA, ABST.
B 6204 694 OLD BAHAMA CHANNEL, BAHAMAS
B 6405 674 OLD BAHAMA CHANNEL, NORTH OF CUBA, ABST., EXPLORABLE
B 6009 1611* OLD FIELDS, ABST., NEW, OIL FROM
M 28 179 OLD FORT ISLAND SANDSTONE, CANADA, N.W.T., CAMBRIAN
M 28 143 OLD FORT ISLAND SANDSTONE, CANADA, N.W.T., CAMBRIAN
B 6211 2192 OLD RED SANDSTONE, IRELAND, WESTERN
M 28 19 OLD RED SANDSTONE, NORTHERN, IRELAND, DEVONIAN
B 6203 532 OLDERSHAW, A. E., CONTINENTAL SLOPE-RISE DEPOSITS FROM LOWER
B 6407 972 OLDMAN FORMATION, ALBERTA
B 6201 125 OLDSMAR FORMATION, FLORIDA
B 6104 558 OLEA, RICARDO A., REGIONALIZED VARIABLES FOR EVALUATION OF
B 6111 1951 OLENEK HEAVY OIL DEPOSIT, USSR
B 6209 1710 OLENTANGY SHALE, KENTUCKY
B 6209 1710 OLENTANGY SHALE, OHIO
M 25 463 OLGA-THERNEY VOLCANIC FIELD, USSR
B 6408 1142 OLIGOCENE, ALASKA
B 6408 1133 OLIGOCENE, ALASKA
B 6407 1016 OLIGOCENE, ALASKA
B 6309 1522 OLIGOCENE, ALASKA PENINSULA, BELKOFSKI FORMATION
S 11 144 OLIGOCENE AND MIOCENE BULIMINIDS, CALIFORNIA
B 6207 1153 OLIGOCENE, AUSTRALIA
B 6309 1458 OLIGOCENE, BALTIMORE CANYON TROUGH
S 6 254 OLIGOCENE BOUNDARY, DIAGRAM, PALEOGENE, EOCENE-
B 6309 1607* OLIGOCENE BOUNDARY IN SOUTHWEST ALABAMA, ABST., EOCENE-
S 11 146 OLIGOCENE BULIMINIDS, CALIFORNIA, MIOCENE AND
B 6105 769* OLIGOCENE CALCAREOUS NANNOFOSSIL DISTRIBUTION IN DEEP SEA
B 6205 820 OLIGOCENE, CALIFORNIA
B 6211 2223 OLIGOCENE, CALIFORNIA
S 4 105 OLIGOCENE, CARIBBEAN
M 28 105 OLIGOCENE, CATAHOULA TUFF, TEXAS
M 28 107 OLIGOCENE, CATAHOULA TUFF, TEXAS
M 28 116 OLIGOCENE, CATAHOULA TUFF, TEXAS
S 6 221 OLIGOCENE, CHATTIAN STAGE DEFINED, EUROPE
B 6104 561 OLIGOCENE, CHILE
M 25 324 OLIGOCENE, COLOMBIA
M 27 105 OLIGOCENE, CREEDE FORMATION, COLORADO
M 28 39 OLIGOCENE, CYRENENSCHICHTEN, GERMANY
M 29 341 OLIGOCENE, EVOLUTION, COLOMBIA
M 28 54 OLIGOCENE, FISH CANYON TUFF, COLORADO
M 28 71 OLIGOCENE, FRIO FORMATION, TEXAS
M 28 154 OLIGOCENE, FRIO FORMATION, TEXAS
M 28 67 OLIGOCENE, FRIO FORMATION, TEXAS
M 28 20 OLIGOCENE, FRIO FORMATION, TEXAS
M 28 173 OLIGOCENE, FRIO FORMATION, TEXAS
M 28 175 OLIGOCENE, FRIO FORMATION, TEXAS
M 28 135 OLIGOCENE, FRIO FORMATION, TEXAS
M 28 174 OLIGOCENE, FRIO FORMATION, TEXAS
M 26 216 OLIGOCENE, HACKBERRY SLOPE SYSTEM, GULF OF MEXICO
B 6108 1378* OLIGOCENE, IN EAST-CENTRAL NEVADA-IMPLICATIONS FOR PETROLEUM
B 6402 219 OLIGOCENE, INDIA
B 6402 219 OLIGOCENE, INDIA
B 6203 463 OLIGOCENE, LIBYA
B 6206 951 OLIGOCENE, LOUISIANA
B 6102 208 OLIGOCENE, MEXICO
M 27 81 OLIGOCENE, MOLASSE, GERMANY
B 6201 3 OLIGOCENE, NIGER DELTA
B 6205 780 OLIGOCENE, NORTH, AFRICA
B 6203 424 OLIGOCENE, NORTH, ATLANTIC
B 6303 370† OLIGOCENE, NORTHERN PUERTO RICO, SIDERITE, SPHERULITES OF, SAN
B 6205 793 OLIGOCENE, PROFILE, BLAKE PLATEAU AREA
S 6 227 OLIGOCENE, RADIOMETRIC AGES, CORRELATIONS
B 6209 1757* OLIGOCENE REEF COMMUNITY SUCCESSION, DAMON MOUND, TEXAS,
S 6 221 OLIGOCENE, RUPELIAN STAGE DEFINED, ERUOPE
B 6001 65† OLIGOCENE SANTA BARBARA BASIN, SANTA MONICA AND, SANTA ANA
B 6104 616† OLIGOCENE SANTA BARBARA BASIN. DISCUSSION AND REPLY, SANTA
B 6102 263† OLIGOCENE SANTA BARBARA BASIN. DISCUSSION AND REPLY, SANTA
B 6309 1473 OLIGOCENE SEQUENCE, USSR, ANADYR BASIN, SENONIAN-
B 6405 717* OLIGOCENE, SOUTH TEXAS, ABST., DELTAIC SYSTEMS AND ASSOCIATED,
B 6009 1612* OLIGOCENE, SOUTH TEXAS, ABST., LEACHED, POROSITY IN
B 6409 1553* OLIGOCENE, SOUTHEAST TEXAS, ABST., MORPHOLOGY O,
M 27 218 OLIGOCENE, SUWANNEE FORMATION, FLORIDA
M 27 165 OLIGOCENE, SUWANNEE LIMESTONE, FLORIDA
M 27 173 OLIGOCENE, SUWANNEE LIMESTONE, FLORIDA
M 27 7 OLIGOCENE, SUWANNEE LIMESTONE, FLORIDA
M 27 128 OLIGOCENE, SUWANNEE LIMESTONE, FLORIDA
M 27 223 OLIGOCENE, SUWANNEE LIMESTONE, FLORIDA
M 28 161 OLIGOCENE, TONGRIANO CONGLOMERATE, ITALY
M 28 31 OLIGOCENE, TONGRIANO FORMATION, ITALY
B 6303 505* OLIGOCENE TRANSGRESSIV SEDIMENTS OF NEW JERSEY CONTINENNAL
B 6409 1567* OLIGOCENE VICKSBURG SANDSTONE IN JIM HOGG COUNTY, TEXAS,
B 6409 1559* OLIGOCENE, VICKSBURGIAN, MISSISSIPPI AN ALABAMA, ABST.,
B 6105 839* OLIGOCENE VOLCANIC SEDIMENTARY APRON IN TRANS-PECOS
S 6 253 OLIGOCENE-EOCENE BOUNDARY, PALEOGENE
S 7 206 OLIGOCENE-LOWER MIOCENE THICKNESS, WEST FLORIDA SLOPE
S 6 221 OLIGOCENE-MIOCENE BOUNDARY DEFINED, EUROPE
S 6 255 OLIGOCENE-MIOCENE BOUNDARY, DIAGRAM, PALEOGENE
S 6 242 OLIGOCENE-MIOCENE BOUNDARY, PALEOGENE
M 29 345 OLIGOCENE-MIOCENE DIASTROPHISM, COLOMBIA, EVOLUTION
B 6303 523* OLISTOSTROMAL SEQUENCE, NORTH WALES, UNITED KINGDOM, ABST.,
B 6409 1565 OLIVER, GARY E., LATE PALEOCENE PLANKTONIC FORAMINIFERAL
B 6405 693 OLIVER, JACK E., ET AL, CRYSTALLINE OVERTHRUSTING OF PALEOZOIC
B 6409 1564* OLMOS SAN MIGUEL, AND UPSON FORMATIONS, UPPER CRETACEOUS,
B 6207 1231 OLMSTED, F. H., HYDROGEOLOGY OF LONG VALLEY GEOTHERMAL AREA,
B 6103 360 OLMSTED, RICHARD W., URANIUM POTENTIAL OF PERMIAN AND
M 25 466 OLOY-ALAZEY EUGEOSYNCLINAL SYSTEM, USSR
B 6105 772 OLSEN, CLAYTON E., RESULTS OF PILOT STUDIES OF PROJECT NURE
B 6004 723 OLSEN, C. R., RADIOCARBON DATING OF ESTUARINE CARBONATE
B 6004 703 OLSEN, C. R., SEDIMENTATION RATES IN HUDSON RIVER ESTUARY, ABST.
B 6203 551 OLSON, ALLISON C., APPLICATION OF GARNET SYSTEM FOR EVALUATION
B 6004 703 OLSON, ALLISON C., INTERACTIVE COMPUTER PROGRAM TO ANALYZE
B 6303 442 OLSON, ERIC R., SHALLOW SUBSURFACE DOLOMITIZATION OF
B 6303 505 OLSON, ERIC R., SULFUR AND OXYGEN ISOTOPE GEOCHEMISTRY OF ABU
B 6103 331 OLSON, N. K., PETROLEUM GEOLOGY AND OIL AND GAS POTENTIAL OF
B 6004 703 OLSON, WALTER S., CAUSALITY AND TIMING IN NORTH AMERICAN
B 6004 704 OLSSON, R. K., CRETACEOUS AND EARLY TERTIARY PALEOBATHYMETRIC
B 6004 704 OLSSON, R. K., TIMING OF TRANSGRESSIONS AND REGRESSIONS IN
B 6303 505 OLSSON, RICHARD K., OLIGOCENE TRANSGRESSIV SEDIMENTS OF NEW
B 6402 244 OLYMPIA BASIN, GREECE
B 6402 244 OLYMPIA BASIN, GREECE
B 6207 1233 OLYMPIA, E. V., PROGRESS OF GEOTHERMAL ENERGY EXPLORATION AND
M 25 466 OLYUTOR-KAMCHATKA GEOSYNCLINAL ZONE, USSR
M 25 448 OMACHI, HOKUICHIRO, RELATION BETWEEN TECTONICS AND
B 6310 1890 OMAN, 1978, ASIATIC PETROLEUM CORP., B.P. PETROLEUM DEVELOPMENT
B 6405 761 OMAN, CHARLES L., TRACE-ELEMENT CONTENT OF BITUMINOUS COAL
B 6405 735 OMAN, CHARLES L., USCHEM, GEOCHEMICAL DATA FILE OF NATIONAL
B 6010 1872 OMAN, INCLUDING DHOFAR, 1975, ASIATIC PETROLEUM CORP.,
B 6110 1801 OMAN, INCLUDING DHOFAR, 1976, ASIATIC PETROL. CORP., SOFIRAN, ELF
B 6210 2118 OMAN, INCLUDING DHOFAR, 1977, ASIATIC PETROLEUM CORP., SOCIETE
B 6411 1841 OMAN, INCLUDING DHOFAR, 1979, AMOCO INTERNAT. OIL CO., ET AL, ON
M 27 50 OMAN, JURASSIC RADIOLARITE, MUSCAT
M 27 24 OMAN, JURASSIC, RADIOLARITE, MUSCAT
B 6303 541* OMAN MOUNTAINS, ABST., RADIOLARIA FROM
B 6210 2133 OMAN, PRODUCTION, 1977
B 6310 1900 OMAN, PRODUCTION, 1978
B 6411 1857 OMAN, PRODUCTION, 1979
M 27 12 OMAN, TERTIARY CALCARENITE, MUSCAT
B 6112 2073 OMBAI STRAIT
M 25 500 OMINECA CRYSTALLINE BELT, CANADA
M 25 503 OMINECA CRYSTALLINE BELT, CANADIAN, CORDILLERA
B 6211 2358 OMNES, G., CONTRIBUTION OF SHEAR WAVES TO STRATIGRAPHIC
B 6305 844 OMNES, GILDAS, SHALLOW SEISMIC REFLECTION IN APPALACHIAN BASIN
M 25 471 OMOLON MASSIF, USSR
S 6 154 OMOTO FORMATION, JAPAN
S 6 205† ON DATING OF THE PALEOGENE
M 27 189 ONCOLITE, BLUE-GREEN, ALGAL
B 6203 514* ONCOLITES FROM LACUSTRINE SEDIMENTS IN CRETACEOUS OF
B 6203 509* ONCOLITES, ONONDAGA LAKE, NEW YORK, ABST., FRESHWATER
M 26 440 ONE-DIMENSIONAL, SEISMIC, MODELING
B 6211 2350 ONEIL, PHIL, DISTRIBUTION OF RECENT FORAMINIFERA IN GULF OF
B 6209 1754 ONEIL, PHILIP, DISTRIBUTION OF HOLOCENE FORAMINIFERA IN GULF OF
B 6004 704 ONEIL, THOMAS J., SOURCE IDENTIFICATION OF BEACH TARS WITHIN

Index

B 6412 2169 ONESQUETHAW GROUP, WEST VIRGINIA
B 6209 1615 ONESQUETHAW STAGE, APPALACHIAN BASIN
B 6204 657 ONGLEY, L. K., SALT DIAPIRS ON SAO PAULO PLATEAU, SOUTHEASTERN
B 6207 1129 ONIN ANTICLINORIUM FLEXURE, IRIAN JAYA, MISOOL-
B 6201 2 ONITSHA HIGH, NIGER DELTA
M 26 57 ONLAP AND DOWNLAP, DEPOSITIONAL SEQUENCES, BASELAP
M 26 235 ONLAP, DEPOSITIONAL SYSTEMS
M 26 63† ONLAP, SEISMIC, STRATIGRAPHY AND, GLOBAL CHANGES OF, SEA LEVEL,
M 26 177 ONLAPPING-FILL FACIES, SEISMIC, INTERPRETATION
M 26 180 ONLAPPING-FILL FACIES, SEISMIC, INTERPRETATION, MOUNDED
B 6205 764 ONLAPPING-FILL SEISMIC FACIES
B 6203 509* ONONDAGA LAKE, NEW YORK, ABST., FRESHWATER, ONCOLITES
B 6209 1616 ONONDAGA LIMESTONE, APPALACHIAN BASIN
B 6209 1610 ONONDAGA LIMESTONE, NEW YORK
B 6309 1591* ONONDAGA REEFS, DEVONIAN, IN SOUTH-CENTRAL NEW YORK STATE,
B 6212 2510 ONOTSUKA, TADASHI, STRUCTURAL CONTRAST BETWEEN FORE-ARC
B 6304 619 ONOWAY OIL FIELD, MICHIGAN
B 6211 2354* ONSHORE WASHINGTON AND OREGON. BY TITLE ONLY
B 6405 738* ONTARIO, ABST., APPLICATION OF, CONODONT AND, PALYNOMORPH
B 6203 510* ONTARIO, ABST., BURIED, CREVASSE FILLINGS OF, PORT HURON AGE,
B 6203 553* ONTARIO, ABST., SILURIAN, MACROBORING ORGANISMS, SOUTHERN
B 6111 1982 ONTARIO BASIN, CALIFORNIA
B 6203 539* ONTARIO, CANADA, ABST., REGRESSIVE SEDIMENTATION, MODEL FOR,
B 6108 1336 ONTARIO, DEVELOPMENTS, 1976
B 6208 1451 ONTARIO, DEVELOPMENTS, 1977
B 6308 1345 ONTARIO, DEVELOPMENTS, 1978
B 6409 1519 ONTARIO, DEVELOPMENTS, 1979
B 6101 28 ONTARIO, ELLIOTT LAKE, URANIUM DEPOSITS
S 5 99 ONTARIO, FOSSIL HILL FORMATION
S 5 34 ONTARIO, HURON COUNTY, GODERICH SALT MINE
S 5 89 ONTARIO, LAMBTON COUNTY, SARNIA, CORE SECTION
B 6108 1340 ONTARIO, PRODUCTION, 1973-1976
B 6208 1456 ONTARIO, PRODUCTION, 1974-1977
B 6308 1349 ONTARIO, PRODUCTION, 1975-1978
B 6409 1522 ONTARIO, PRODUCTION, 1976-1979
S 5 2 ONTARIO, SILURIAN, NIAGARA GROUP
B 6408 1281* ONTARIO—A PERSIAN GULF ANALOGY, ABST., MIDDLE, ORDOVICIA
M 27 207 OOID DEFORMATIONS
B 6104 511† OOID RESERVOIRS IN DOGGER, MIDDLE JURASSIC, OF FRANCE,
B 6304 676† OOID RESERVOIRS IN DOGGER, MIDDLE JURASSIC, OF FRANCE.
B 6203 520* OOID SAND-SHOAL FACIES, JOULTERS CAYS, GREAT BAHAMA BANK,
B 6303 462* OOID SHOAL, ABST., ANATOMY AND GROWTH HISTORY OF, HOLOCENE
B 6209 1758* OOID SHOAL, BAHAMAS, ABST., HOLOCENE MARINE-CEMENTED, SANDS,
B 6104 519† OOID SHOAL COMPLEX NEAR MIAMI, FLORIDA, PLEISTOCENE, BARRIER,
M 27 163 OOIDS
M 27 117 OOIDS
M 27 115 OOIDS
M 27 127 OOIDS, ALGAL BORINGS
M 27 118 OOIDS ALTERATION
M 27 211 OOIDS, GEOPETAL
B 6405 788* OOLITE DEPOSIT—DRUM LIMESTONE, PENNSYLVANIAN, MISSOURIAN,
B 6303 536* OOLITE OF SOUTHWESTERN IDAHO, ABST., NORMALLY AND REVERSELY
B 6004 679* OOLITE SAND SHOAL TO ITS PHYSICAL ENVIRONMENT AND HOLOCENE
B 6006 907† OOLITE-BAR PROGRADATION, SAN ANDRES FORMATION, MIDLAND
B 6209 1753* OOLITE-SHOAL COMPLEX—LOWER CRETACEOUS SLIGO FORMATION,
M 27 116 OOLITIC COATINGS
B 6004 667* OOLITIC ENVIRONMENT ON ELEUTHERA BANK, BAHAMAS, ABST.,
B 6204 644† OOLITIC TIDAL-BAR SEDIMENTATION IN SOUTHERN CUMBERLAND
M 27 174 OOMICRITES
M 27 179 OOSPARITE, ROUNDED
M 27 173 OOSPARITES
M 27 36 OOZE, PELAGIC
B 6004 700* OOZE--EMPIRICAL OBSERVATIONS AND EXPERIMENTAL RESULTS, ABST.,
M 27 139 OOZES
M 28 116 OPAL, IN SANDSTONE, CEMENTS
M 28 105 OPAL-CEMENTED, CHERT ARENITE
B 6405 738* OPAL-CRISTOBALITE-CEMENTED SANDS IN CATAHOULA FORMATION—
M 27 51 OPALINE SPICULES
M 28 81 OPALLINE SILICA, DIATOMITE, WITH
M 28 34 OPALLINE SILICA, FRAGMENTS
M 28 4 OPALLINE SILICA, QUARTZ
B 6306 426 OPDYKE, NEIL D., PAN-PACIFIC NEOGENE DIATOM CORRELATIONS, ABST.
B 6203 420 OPENING OF NORTH ATLANTIC
S 4 335 OPERCULINOIDES, ST. CROIX
M 25 460 OPHIOLITE DEPRESSIONS, PACIFIC
B 6209 1696† OPHIOLITES, ALASKA, TECTONICS OF, BROOKS RANGE
M 25 487 OPHIOLITIC ORIGIN, COPPER OF
B 6405 670* OPHIOMORPHA FROM UPPER BATHYAL EOCENE SUBSEA FAN FACIES,
B 6004 681* OPHIOMORPHA--GOOD NEWS AND BAD NEWS, ABST.
M 25 319 OPOGADO ANTICLINE, COLOMBIA
S 11 202 OPPELIAN ZONATION
B 6405 730* OQUIRRH BASIN, NORTHWEST UTAH, ABST., PALEOGEOGRAPHY AND
B 6303 476* OQUIRRH BASIN, UTAH, ABST., BATHYMETRIC DISTRIBUTION OF, TRACE,
M 25 402 ORA BANDA, NICKEL-ENRICHED LATERITE, WESTERN AUSTRALIA
B 6410 1577 ORANDA BEDS, VIRGINIA
B 6211 2355 ORANGE, ARNOLD, MAGNETOTELLURIC EXPLORATION—AN UPDATE,
B 6107 979 ORANGE RIVER REGION, PROFILE, AFRICA
B 6101 81 ORANIAN MESETA
B 6412 2064† ORBITAL RADAR IMAGERY FOR GEOLOGIC MAPPING—
M 27 34 ORBITOLINID FORAMS
S 4 341 ORBULINA, ST. CROIX
S 11 298 ORBULINA UNIVERSA
S 11 263 ORBULINA UNIVERSA
B 6303 428* ORCA BASIN AND DEAD SEA, ABST., COMPARISON OF, CARBONATE
S 7 303† ORCA BASIN, ANOXIC DEPRESSION ON THE CONTINENTAL SLOPE,
S 7 322 ORCA BASIN, BATHYMETRY MAP
S 7 304 ORCA BASIN, BATHYMETRY MAP, GULF OF MEXICO
S 7 310 ORCA BASIN, BRINE CHEMISTRY
S 7 310 ORCA BASIN BRINE, COMPARED TO AVERAGE SEAWATER
S 7 306 ORCA BASIN, CORE ANALYSIS
S 7 303 ORCA BASIN, GEOLOGIC SETTING, GULF OF MEXICO
S 7 305 ORCA BASIN, GEOMORPHOLOGY
B 6209 1590 ORCA BASIN, GULF OF MEXICO
S 7 320 ORCA BASIN, NORTHWEST GULF OF MEXICO, INTRASLOPE BASINS
S 7 306 ORCA BASIN, RELEASE OF GAS FROM CORES
B 6303 549* ORCA BASIN SEDIMENTS, ABST., URANIUM IN
S 7 306 ORCA BASIN, SEISMIC PROFILING
S 7 308 ORCA BASIN, SHALLOW STRUCTURAL ZONES
B 6408 1193 ORCA BASIN, SILLED DEPRESSION IN OXIC OPEN OCEAN
S 7 307 ORCA BASIN, SPARKER PROFILE, CENTRAL
S 7 307 ORCA BASIN, SPARKER PROFILE, SOUTHERN
S 7 306 ORCA BASIN, STRUCTURAL INTERPRETATION
S 7 305 ORCA BASIN, SURFICIAL SEDIMENTS
S 7 305 ORCA BASIN, THREE-DIMENSIONAL VIEW
S 7 309 ORCA BASIN, WATER-COLUMN CHEMICAL PROFILES
S 7 323 ORCA BASIN, X-RAY OF SPLIT CORE
M 29 271 ORCA FORMATION, ALASKA, KENAI PENINSULA
B 6203 395 ORCA GROUP, ALASKA
B 6405 722* ORCA GROUP, PRINCE WILLIAM SOUND, ALASKA, ABST., STRATIGRAPHY
B 6203 355 ORCUTT OIL FIELD, CALIFORNIA
B 6408 1211 ORDOS BASIN, CHINA
B 6408 1281* ORDOVICIA SHELF CARBONATE SEDIMENTATION AROUND
B 6405 767* ORDOVICIAN AGE—ST. GEORGE GROUP, WESTERN NEWFOUNDLAND,
B 6204 681 ORDOVICIAN, ALGERIA, CAMBRIAN-
M 25 341 ORDOVICIAN, AMAZON BASINS
B 6405 767* ORDOVICIAN AND DEVONIAN KEROGENOUS MARINE STRATA OF
S 6 349 ORDOVICIAN, ARGON-ARGON DATING, A REVIEW
M 27 176 ORDOVICIAN, ATHENS LIMESTONE, VIRGINIA
M 27 176 ORDOVICIAN, BAYS LIMESTONE, VIRGINIA
M 27 4 ORDOVICIAN, BAYS LIMESTONE, VIRGINIA
M 27 185 ORDOVICIAN, BAYS LIMESTONE, VIRGINIA
M 27 174 ORDOVICIAN, BEEKMANTOWN LIMESTONE, MARYLAND
M 28 7 ORDOVICIAN, BIGHORN DOLOMITE, WYOMING
M 27 72 ORDOVICIAN, BLACK RIVER LIMESTONE, PENNSYLVANIA
M 27 75 ORDOVICIAN, BLACK RIVER LIMESTONE, PENNSYLVANIA
M 27 173 ORDOVICIAN, BLACK RIVER LIMESTONE, PENNSYLVANIA
M 27 222 ORDOVICIAN, BLACK RIVER LIMESTONE, PENNSYLVANIA
M 27 200 ORDOVICIAN, BLACK RIVER LIMESTONE, PENNSYLVANIA
M 27 206 ORDOVICIAN, BLACK RIVER LIMESTONE, PENNSYLVANIA
S 6 358 ORDOVICIAN BOUNDARY, TREMADOC AGE, CAMBRIAN
M 28 113 ORDOVICIAN, CALICO ROCK SANDSTONE, ARKANSAS
S 6 347 ORDOVICIAN, CARADOC SERIES, BENTONITES
S 9 22 ORDOVICIAN, CARBONATE REEF BELT DISTRIBUTION
B 6405 777* ORDOVICIAN, CARBONATE RESERVOIR, CABIN CREEK FIELD, MONTANA,
B 6004 712* ORDOVICIAN CARBONATE ROCKS, VIRGINIA--ANCIENT ANALOGUES OF
M 27 173 ORDOVICIAN, CHAZY LIMESTONE, PENNSYLVANIA
M 27 200 ORDOVICIAN, CHAZY LIMESTONE, PENNSYLVANIA
S 6 350 ORDOVICIAN, CONFLICTING TIME SCALES
B 6304 650 ORDOVICIAN, CONODONTS, VIRGINIA
B 6209 1767* ORDOVICIAN COOL CREEK FORMATION, ARBUCKLE MOUNTAINS,
M 27 111 ORDOVICIAN, DEEPKILL SHALE, NEW YORK
M 27 48 ORDOVICIAN, EL PASO GROUP, TEXAS
B 6203 538* ORDOVICIAN ELLENBURGER DOLOMITE, PUCKETT FIELD, PECOS
M 27 132 ORDOVICIAN, ELLENBURGER LIMESTONE, TEXAS
M 28 113 ORDOVICIAN, EVERTON FORMATION, ARKANSAS
S 9 20 ORDOVICIAN, FOLDOUT, PALEOCLIMATES
B 6405 770* ORDOVICIAN FORELAND BASIN, APPALACHIAN OROGENE, VIRGINIA,
B 6101 7 ORDOVICIAN, FRANKLINIAN BASIN
B 6405 697* ORDOVICIAN GALENA GROUP, UPPER MISSISSIPPI VALLEY, ABST.,
M 28 46 ORDOVICIAN, GARNET SCHIST, CONNECTICUT

M 28 45 ORDOVICIAN, GARNET SCHIST, CONNECTICUT
S 6 347† ORDOVICIAN GEOCHRONOLOGY
S 6 347† ORDOVICIAN GEOCHRONOLOGY
S 6 347† ORDOVICIAN GEOCHRONOLOGY
S 6 347† ORDOVICIAN GEOCHRONOLOGY
S 6 352 ORDOVICIAN, GEOCHRONOLOGY, TIME SCALE CHART
B 6306 955 ORDOVICIAN GRANITE, NEWFOUNDLAND
B 6404 530 ORDOVICIAN, GREENLAND
M 28 71 ORDOVICIAN, GRIFFEL SCHIEFER, GERMANY
B 6009 1622* ORDOVICIAN HISTORY OF CINCINNATI ARCH PROVINCE, CENTRAL
B 6401 129 ORDOVICIAN, INDIANA
B 6203 514* ORDOVICIAN INTERVAL, CENTRAL PENNSYLVANIA, ABST., CARBONATE
S 6 350 ORDOVICIAN, ISOTOPIC DATING REVIEW, BRITISH ISLES
S 6 348 ORDOVICIAN, KENTUCKY, TYRONE LIMESTONE, BENTONITES
B 6407 1054 ORDOVICIAN KLIPPEN, QUEBEC
M 28 6 ORDOVICIAN, KYANITE SCHIST, CONNECTICUT
M 28 7 ORDOVICIAN, LANDER SANDSTONE, WYOMING
M 27 71 ORDOVICIAN, MARTINSBURG FORMATION, VIRGINIA
B 6405 713* ORDOVICIAN MICROBIAL BORINGS, ABST., MORPHOMETRY OF LATE
B 6410 1577 ORDOVICIAN, MIDDLE, STRATIGRAPHIC CHART, VIRGINIA
B 6306 952 ORDOVICIAN, NEWFOUNDLAND
M 28 16 ORDOVICIAN, NEWTOWN GNEISS, CONNECTICUT
M 28 17 ORDOVICIAN, NEWTOWN GNEISS, CONNECTICUT
B 6409 1566* ORDOVICIAN, NORTH ARKANSAS, ABST., PALEOENVIRONMENTAL
B 6303 430* ORDOVICIAN, NORTH DAKOTA, ABST., DOLOMITIZATION IN UPPER, RE
B 6004 685* ORDOVICIAN OF CENTRAL NEW YORK, ABST., GULL RIVER LIMESTONE--,
B 6004 717* ORDOVICIAN OF EASTERN NEW YORK--TIMING AND ENVIRONMENTAL
B 6004 729* ORDOVICIAN OF TENNESSEE, ABST., REGIONAL ENVIRONMENTA
B 6303 469* ORDOVICIAN, OF TRANS-PECOS TEXAS, ABST., CARBONATE ROCKS OF,
B 6108 1382* ORDOVICIAN OIL ACCUMULATION AT NORTHEAST LUCIEN FIELD, NOBLE
B 6203 555* ORDOVICIAN, OKLAHOMA, ABST., EFFECT OF, FAULT-RELATED,
B 6108 1375* ORDOVICIAN, OKLAHOMA, ABST., PETROLEUM, SOURCE-ROCK
B 6405 805* ORDOVICIAN PALEOGEOGRAPHY OF REGION BORDERING
M 28 166 ORDOVICIAN, PHYCODEN SCHICHTEN, GERMANY
M 28 73 ORDOVICIAN, PHYCODEN SCHICHTEN, GERMANY
B 6405 804* ORDOVICIAN, POROSITY, EASTERN MONTANA AND WESTERN NORTH
S 6 349 ORDOVICIAN, POTASSIUM-ARGON DATING, A REVIEW
B 6004 678* ORDOVICIAN, QUEBEC, ABST., SUBMARINE-, FAN COMPLEX, CAP ENRAGE
B 6407 1052† ORDOVICIAN, QUEBEC, DEPOSITIONAL FRAMEWORK OF SANDY, MID-FAN
B 6405 715* ORDOVICIAN RAMP-TO-BASIN CARBONATE ROCKS, VIRGINIA, ABST.,
B 6008 1401* ORDOVICIAN RED RIVER FORMATION, SECOND CREEK FIELD, RICHLAND
M 27 99 ORDOVICIAN, REEDSVILLE SHALE, PENNSYLVANIA
M 27 76 ORDOVICIAN, REEDSVILLE SHALE, PENNSYLVANIA
M 28 39 ORDOVICIAN, REEDSVILLE SHALE, PENNSYLVANIA
S 9 8 ORDOVICIAN, SAHARA, AFRICA, ICE FLOW DIRECTION
B 6002 322* ORDOVICIAN SANDSTONES, NORTHERN ARKANSAS, ABST., TRENDS AND
M 28 47 ORDOVICIAN SCHIST, CONNECTICUT
M 28 49 ORDOVICIAN, SCHIST, CONNECTICUT
M 28 48 ORDOVICIAN, SCHIST, MASSACHUSETTS
B 6203 517* ORDOVICIAN SEDIMENTS, ABST., TIDELESS MIDDLE
B 6105 828* ORDOVICIAN SEVIER SHALE, EASTERN TENNESSEE, ABST., DEEP-WATER
B 6105 840* ORDOVICIAN SHELF-SLOPE CARBONATE DEPOSITS IN NORTH-CENTRAL
B 6303 525* ORDOVICIAN, SOUTHERN APPALACHIANS, ABST., QUANTITATIVE, BASIN
M 28 112 ORDOVICIAN, ST. PETER SANDSTONE, ARKANSAS
M 27 131 ORDOVICIAN, STONEHENGE LIMESTONE, PENNSYLVANIA
M 27 176 ORDOVICIAN, STONEHENGE LIMESTONE, PENNSYLVANIA
M 28 50 ORDOVICIAN, STRAITS SCHIST, CONNECTICUT
B 6305 846* ORDOVICIAN STRATIGRAPHY AND SEDIMENTATION IN NORTHERN
S 6 348 ORDOVICIAN, SWEDEN, CHASMOPS LIMESTONE, BENTONITES
B 6101 9 ORDOVICIAN TO DEVONIAN, CANADIAN, ARCTIC ISLANDS
B 6101 8 ORDOVICIAN TO DEVONIAN, FRANKLINIAN BASIN
B 6203 503* ORDOVICIAN TRENTON-UTICA SEQUENCE OF NEW YORK, ABST.,
B 6307 1135† ORDOVICIAN UNIT IN EASTERN OKLAHOMA, FORMERLY CALLED
B 6107 1080 ORDOVICIAN, VERMONT, ISLE LAMOTTE
B 6304 652 ORDOVICIAN, VIRGINIA
M 27 123 ORDOVICIAN, VIRGINIA
B 6004 676* ORDOVICIAN, VIRGINIA, ABST., FENESTRAL, FABRICS IN NONCYCLIC
B 6203 557* ORDOVICIAN, VIRGINIA, ABST., GEOMETRY, FACIES, AND DEVELOPMENT
B 6410 1575† ORDOVICIAN, VIRGINIA APPALACHIANS, CARBONATE, RAMP-TO-BASIN
B 6303 555* ORDOVICIAN WAH WAH LIMESTONE, IBEX AREA, WESTERN MILLARD
M 27 119 ORDOVICIAN, WEST SPRING CREEK FORMATION, OKLAHOMA
B 6303 410* ORDOVICIAN--EARLY SILURIAN STRATA IN MID-CONTINENT AREA, ABST.,
B 6303 442* ORDOVICIAN-SILURIAN, CENTRAL NEVADA--EVIDENCE FOR
B 6203 534* ORDOVICIAN-SILURIAN GLACIO-EUSTATIC SEA-LEVEL CHANGES IN
M 25 428 ORE DEPOSITS ASSOCIATED WITH, AUSTRALIA, ARCHEAN
M 25 428 ORE DEPOSITS ASSOCIATED WITH, AUSTRALIA, MESOZOIC
M 25 428 ORE DEPOSITS ASSOCIATED WITH, AUSTRALIA, PALEOZOIC
M 25 428 ORE DEPOSITS ASSOCIATED WITH, AUSTRALIA, PROTEROZOIC
M 25 428 ORE DEPOSITS ASSOCIATED WITH, AUSTRALIA, TERTIARY
M 25 448 ORE DEPOSITS, JAPAN, KUROKO OR, BLACK
M 25 473 ORE DEPOSITS, PATTERN OF, RELATION TO CENTRAL-TYPE STRUCTURES
M 25 428 ORE DEPOSITS, RELATION TO PERIODS OF GEOLOGIC TIME
M 25 426 ORE DEPOSITS, STRATIFORM AND, STRATA-BOUND
M 25 429 ORE DEPOSITS, SURFACE EXPRESSION, PRESENCE OR ABSENCE OF
S 3 70 ORE FLUID
B 6206 1055 ORE MINERALS
M 25 435 ORE OCCURRENCES, BOUNDED IN TIME AND SPACE, PHILIPPINES
M 25 444 OREA, ORE DEPOSITS, CLASSIFICATION OF, SOUTH K
B 6107 1046 OREAD FORMATION, KANSAS
B 6203 491* OREGON, ABST., COMMUNITY SUCCESSION OF, PIONEER TERRACE,
B 6211 2358* OREGON, ABST., HOLOTHURIAN MICROFOSSILS AND, EOCENE
B 6207 1223* OREGON, ABST., LATE, MESOZOIC TECTONIC HISTORY OF NORTHERN,
B 6211 2353* OREGON, ABST., MESOZOIC PALEOGEOGRAPHY AND PALEOTECTONICS,
B 6403 439* OREGON, ABST., MIST GAS FIELD, COLUMBIA COUNTY
B 6211 2348* OREGON, ABST., NEW, MIOCENE FORMATION AT, COOS BAY
B 6108 1374* OREGON, ABST., PETROLOGY AND DIAGENETIC EFFECTS OF,
B 6303 422* OREGON, ABST., RETROGRADATIONAL SHELF SEQUENCE--UPPER,
B 6203 510* OREGON, ABST., SAND, TRANSPORT ACROSS, EOCENE SHELF AND SLOPE
B 6209 1592 OREGON AND ALASKA MARGINS, DSDP, SITES 172-182, LEG 18
B 6209 1591 OREGON AND ALASKA—CONVERGENT MARGINS
B 6303 421* OREGON AND QUEEN CHARLOTTE ISLANDS, BRITISH COLUMBIA, ABST.,
B 6403 448* OREGON AND WASHINGTON, ABST., GEOLOGICAL ASSESSMENT OF
B 6012 2174† OREGON AND WASHINGTON, ABST., NEOGENE, MOLLUSCAN STAGES OF
B 6211 2364* OREGON, AND WASHINGTON, ABST., SYNTHESIS OF, OCS WELL
B 6207 1194 OREGON AND WASHINGTON, BATEMAN FORMATION
B 6207 1194 OREGON AND WASHINGTON, CRESCENT FORMATION
B 6207 1194 OREGON AND WASHINGTON, EOCENE, RIFT HYPOTHESIS
B 6207 1194 OREGON AND WASHINGTON, FLOURNOY FORMATION
B 6207 1194 OREGON AND WASHINGTON, LOOKINGGLASS FORMATION
B 6207 1194 OREGON AND WASHINGTON, MCINTOSH FORMATION
B 6207 1194 OREGON AND WASHINGTON, RAGING RIVER FORMATION
B 6207 1194 OREGON AND WASHINGTON, ROSEBURG FORMATION
B 6207 1194 OREGON AND WASHINGTON, ROSLYN FORMATION
B 6207 1194 OREGON AND WASHINGTON, SILETZ RIVER FORMATION
B 6207 1196 OREGON AND WASHINGTON, SILETZ RIVER-CRESCENT STRATIGRAPHY
B 6207 1194 OREGON AND WASHINGTON, TEANAWAY FORMATION
B 6207 1194 OREGON AND WASHINGTON, TIGER MOUNTAIN FORMATION
B 6207 1194 OREGON AND WASHINGTON, TYEE FORMATION
B 6207 1194 OREGON AND WASHINGTON, YAMILL FORMATION
B 6408 1201 OREGON, ANOXIC OPEN OCEAN, PACIFIC OCEAN OFFSHORE,
B 6204 627 OREGON BASIN FIELD, WATER-INJECTION PROGRAM, WYOMING
B 6105 816* OREGON BASIN FIELD, WYOMING, ABST., TENSLEEP RESERVOIR STUDY
B 6204 609† OREGON BASIN FIELD, WYOMING, TENSLEEP, RESERVOIR
B 6204 622 OREGON BASIN, NORTH DOME, WYOMING
B 6204 626 OREGON BASIN, SOUTH DOME, DEVELOPMENT, WYOMING
B 6209 1592 OREGON CONTINENTAL MARGIN, ASTORIA FAN SEDIMENTS
B 6203 533* OREGON CONTINENTAL SLOPE—ACTIVE-MARGIN ENVIRONMENT, ABST.,
M 29 429 OREGON, CONVERGENT MARGIN, PACIFIC
B 6008 1196 OREGON, DEVELOPMENTS, 1975
B 6108 1166 OREGON, DEVELOPMENTS, 1976
B 6208 1322 OREGON, DEVELOPMENTS, 1977
B 6308 1292 OREGON, DEVELOPMENTS, 1978
B 6409 1393 OREGON, DEVELOPMENTS, 1979
B 6008 1394* OREGON GEOTHERMAL INVESTIGATION, ABST., ALVORD VALLEY
B 6206 949 OREGON, JURASSIC, TURBIDITES
M 29 430 OREGON, MARGIN, ORGANIC CARBON
B 6004 730* OREGON, NOT ASSOCIATED WITH CONTINENNAL MARGIN, ABST.,
M 28 139 OREGON, TERTIARY, SEDIMENTS
B 6207 1196 OREGON-CALIFORNIA BOUNDARY, PLATE BOUNDARIES IN WESTERN,
B 6207 1193† OREGON-WASHINGTON, EARLY, TERTIARY, RIFTING IN WESTERN
B 6211 2354* OREGON. BY TITLE ONLY, ONSHORE, WASHINGTON AND
B 6004 692* ORES, ABST., CHARACTERIZATION OF, INSOLUBLE, ORGANIC MATTER
B 6004 725* ORES, ABST., GLAUCONITE-, MAGHEMITE-, HEMATITE TRANSFORMATION
M 25 428 ORES, ENVIRONMENTAL PATTERN OF OCCURRENCE, STRATIFORM
M 25 488 ORES, FUTURE SEARCH FOR, METALLOGENIC
B 6303 428* ORGANIC ACID ANIONS IN OIL-FIELD WATERS AND ORIGIN OF NATURAL
B 6303 476* ORGANIC ANALYSIS AND GEOCHEMICAL MEASUREMENTS IN
B 6405 798* ORGANIC BUILDUPS—EXAMPLE FROM LOWER PALEOZOIC OF
M 29 430 ORGANIC CARBON, ALASKA, MARGIN
B 6412 2109 ORGANIC CARBON AND BITUMEN, BERING SEA
B 6401 78 ORGANIC CARBON BURIAL RATES
B 6303 293 ORGANIC CARBON CONTENT, AND OIL AND GAS SHOWS, DSDP-IPOD
B 6209 1763* ORGANIC CARBON CONTENT AND PETROLEUM SOURCE ROCK
B 6209 1589 ORGANIC CARBON IN MARINE ENVIRONMENTS
M 29 429 ORGANIC CARBON IN SEDIMENTS, ALASKA, MARGIN
B 6408 1182 ORGANIC CARBON IN SURFACE LAYER OF SEDIMENTS IN WORLD OCEAN
M 29 430 ORGANIC CARBON, OREGON, MARGIN
B 6406 921 ORGANIC CARBON RATIOS, BITUMEN AND
B 6412 2109 ORGANIC CARBON, RUSSIAN PLATFORM
B 6306 872 ORGANIC CARBON-RICH SEDIMENTS CRETACEOUS, GLOBAL

B 6002 287† ORGANIC COMPOSITION, EARLY, DIAGENETIC, GENERATION OF,
B 6404 488† ORGANIC COMPOSITION OF SOME UPPER CRETACEOUS SHALE, POWDER
B 6203 522* ORGANIC COMPOUNDS IN WATER, ABST., ORIGIN OF PETROLEUM—
B 6305 770 ORGANIC COMPOUNDS, MIGRATION OF SOLUBLE NONHYDROCARBON
S 10 170 ORGANIC COMPOUNDS, OCCURRENCE IN RIVER WATERS
S 10 169† ORGANIC COMPOUNDS IN WATER, ORIGIN OF, PETROLEUM:, IN-TRANSIT,
B 6309 1504† ORGANIC CONTENT OF APPALACHIAN DEVONIAN SHALES FROM
B 6209 1588 ORGANIC CONTENT OF CONTINENTAL-MARGIN SEDIMENTS
B 6412 2156† ORGANIC CONTENT OF DEVONIAN SHALE IN WESTERN APPALACHIAN
S 10 54 ORGANIC CONTENT OF SOURCE ROCKS, MIGRATION
B 6209 1593 ORGANIC CONTENT, SMALL, OCEAN, BASINS
M 29 427 ORGANIC CONTENT, SOURCE BEDS, CONTINENTAL MARGIN
B 6009 1610* ORGANIC DIAGENESIS--ITS RELATION TO GULF COAST OIL, ABST.
B 6405 790* ORGANIC FACIES OF MID-CRETACEOUS BLACK SHALES IN DEEP NORTH
B 6405 729* ORGANIC FACIES—STRATIGRAPHIC CONCEPT AND EXPLORATION TOOL,
B 6307 1128† ORGANIC GEOCHEMICAL SYSTEMS, ERRORS IN EXTRAPOLATION OF
B 6304 691* ORGANIC GEOCHEMISTRY AND URANIUM IN GRANTS MINERAL BELT,
B 6004 717* ORGANIC GEOCHEMISTRY AS AID TO EXPLORATION IN EASTERN JAVA
B 6201 98† ORGANIC GEOCHEMISTRY, INCIPIENT METAMORPHISM, AND OIL
B 6108 1375* ORGANIC GEOCHEMISTRY, INCIPIENT METAMORPHISM, AND OIL
B 6203 568* ORGANIC GEOCHEMISTRY OF CENOZOIC AND RECENT NONMARINE
B 6309 1610* ORGANIC GEOCHEMISTRY OF DEEP WELL IN HINDS COUNTY, MISSISSIPPI,
B 6206 1092* ORGANIC GEOCHEMISTRY OF GAS-BEARING SHALES, U.S. EASTERN
B 6412 2118† ORGANIC GEOCHEMISTRY OF PLIOCENE-PLEISTOCENE CALCAREOUS
B 6410 1690 ORGANIC GEOCHEMISTRY OF PRECAMBRIAN ROCKS
B 6303 483* ORGANIC GEOCHEMISTRY OF SEDIMENTS RECOVERE BY DSDP-IPOD,
B 6312 2135 ORGANIC GEOCHEMISTRY PARAMETERS OF BITUMEN
B 6312 2136 ORGANIC GEOCHEMISTRY PARAMETERS OF GAS
B 6312 2135 ORGANIC GEOCHEMISTRY PARAMETERS OF KEROGEN
B 6303 544* ORGANIC MATERIAL, ABST., HYDROCARBON OCCURRENCE A FUNCTION
B 6405 775* ORGANIC MATERIAL IN GREEN RIVER SHALE, ABST., SEDIMENTARY
B 6408 1180 ORGANIC MATTER ACCUMULATION IN AQUATIC ENVIRONMENTS
S 10 99 ORGANIC MATTER NETWORK, MIGRATION THROUGH
M 29 425 ORGANIC MATTER, ACCUMULATION, SOURCE BEDS
B 6203 493* ORGANIC MATTER, AND BASIN EVALUATION FOR PETROLEUM
B 6305 825* ORGANIC MATTER AND CALCIUM CARBONATE IN MODERN NORTH
B 6004 692* ORGANIC MATTER ASSOCIATED WITH URANIUM ORES, ABST.,
B 6312 2222† ORGANIC MATTER FROM NORTH ATLANTIC KIMMERIDGIAN,
B 6309 1575* ORGANIC MATTER IN NEW ALBANY SHALE GROUP OF ILLINOII BASIN,
B 6405 779* ORGANIC MATTER IN SEDIMENTS UNDERLYING ROSS ICE SHELF, ABST.
B 6203 493* ORGANIC MATTER INTERACTIONS AT SOURCE, ABST., WATER-ROCK-
B 6305 771 ORGANIC MATTER, KEROGEN, MATRIX, PETROLEUM MIGRATION FROM,
B 6207 1151 ORGANIC MATTER, NATURE AND MATURITY OF SOLID
B 6209 1586 ORGANIC MATTER, PRODUCTION AND, ACCUMULATION OF
B 6405 753* ORGANIC MATTER TRANSFORMATIONS, ABST., GENERATION OF
B 6412 2051 ORGANIC MATTER, TYPES, ATLANTIC BASINS
B 6204 594 ORGANIC MATTER, U.S. ATLANTIC SLOPE
B 6104 504† ORGANIC MATURITY, COLOR OF, KEROGEN AS INDEX OF
B 6004 729* ORGANIC METAMORPHISM IN SILICICLASTIC ROCKS, ABST., INORGANIC
B 6405 780* ORGANIC MODEL FOR ROLL-FRONT URANIUM DEPOSITS, ABST.
S 10 24 ORGANIC MOLECULES, PORE WATER
M 29 426 ORGANIC PRODUCTIVITY, CORRELATION TO DEEP-WATER UPWELLING
M 29 426 ORGANIC PRODUCTIVITY, MARINE, INDEX MAP
M 29 425 ORGANIC PRODUCTIVITY, PETROLEUM, SOURCE BEDS
B 6004 667* ORGANIC REEFS DURING JURASSIC TIME IN ATLAS MOUNTAINS OF
B 6004 608† ORGANIC RICHNESS AND EFFECT OF LITHOLOGIC VARIATION,
B 6209 1684† ORGANIC SEDIMENTATION AND GENESIS OF PETROLEUM IN MAHAKAM
M 29 431 ORGANIC SEDIMENTS, BLACK SEA, SMALL OCEAN BASIN
M 29 431 ORGANIC SEDIMENTS, MEDITERRANEAN SEA, SMALL OCEAN BASIN
B 6402 180 ORGANIC SEDIMENTS, THERMAL, METAMORPHISM OF
S 10 174 ORGANIC SULFUR COMPOUNDS, GENERATION, MODEL
S 4 372 ORGANIC TISSUE IN SUBMERGENT REEFS, MARINE, CEMENTS AND
M 29 427 ORGANIC-CARBON BY DEPTH, GULF OF MEXICO, LOUISIANA
B 6303 462* ORGANIC-RIC BLACK SHALES IN SUBSURFACE
B 6004 668* ORGANISM TRACES ON AN IN MODERN PELAGIC CARBONATE DEPOSITS,
B 6004 706* ORGANISMS, ABST., PALEOECOLOGIC, POTENTIAL OF, MICROBORING
B 6205 827† ORGANOCHEMICAL EVALUATION OF GEVARAM SHALES, LOWER
B 6405 749* ORGANOPHOSPHORITES IN CARBONIFEROUS ROCKS OF CENTRAL
S 11 257 ORIDORSALIS UMBONATUS
S 11 288 ORIDORSALIS UMBONATUS
B 6002 273 ORIEL, STEVEN S., NOTE 44--APPLICATION FOR ADDITION TO CODE
B 6107 1083 ORIEL, STEVEN S., NOTE 47—APPLICATION FOR AMENDMENT OF
B 6212 2512* ORIENTAL BASINS OF BOLIVIA, ABST., EXPLORATION OF,
M 28 107 ORIENTATION STUDIES, TEXTURE
M 25 344 ORIENTE BASIN, POTENTIAL
B 6004 699* ORIENTED HYDROCARBON MIGRATION--RESULT OF GEOLOGIC-EVENT
B 6109 1535* ORIGIN AND CHARACTER OF WASHOVER FANS ON GEORGIA COAST,
B 6303 506* ORIGIN AND CHARACTERTICS OF PENNSYLVANIAN-AGE MULTISTORY
B 6303 438* ORIGIN AND DEVELOPMENT OF BARRIER ISLANDS ON WEST-CENTRAL
B 6004 698* ORIGIN AND DISTRIBUTION OF SUSPENDED SEDIMENT IN BAY OF
B 6102 207† ORIGIN AND DISTRIBUTION OF TERTIARY CONGLOMERATES, VERACRUZ
B 6012 2117† ORIGIN AND EVOLUTION OF BIOHERMS IN WENLOCK LIMESTONES,
B 6203 556* ORIGIN AND MIGRATION OF PETROLEUM, ABST., NEW EVIDENCE FOR
B 6303 467* ORIGIN AND OCCURRENCE OF CRUDE OILS IN MOLASSE BASIN,
M 25 146† ORIGIN AND POTENTIAL OF SOM GEOTHERMAL AREAS IN
B 6002 213† ORIGIN AND PRIMARY MIGRATION, AQUEOUS, SOLUBILITY OF,
B 6202 223† ORIGIN AND SIGNIFICANCE OF EVAPORITES IN BASINS AROUND
B 6105 783* ORIGIN AND SIGNIFICANCE OF EVAPORITES IN BASINS AROUND ATLANTIC
S 10 62 ORIGIN OF OIL SANDS, ALBERTA, LOWER CRETACEOUS
S 10 61 ORIGIN OF PETROLEUM, GULF COAST TERTIARY
S 10 169† ORIGIN OF PETROLEUM: IN-TRANSIT CONVERSION OF ORGANIC
B 6207 1222* ORIGIN, DISTRIBUTION, AND DIAGENETIC MODIFICATIONS OF NEOGENE
B 6009 1607* ORIGIN FOR FRIO RESERVOIR SANDSTONES, NINE MILE POINT FIELD,
B 6108 1379* ORIGIN, MIGRATION AND CONCENTRATION OF URANIUM-CATAHOULA
B 6007 1110† ORIGIN OF ATHABASCA OIL
B 6007 1112† ORIGIN OF ATHABASCA OIL
B 6007 1110† ORIGIN OF ATHABASCA OIL
B 6204 694 ORIGIN OF BAHAMA CHANNELS
B 6004 653* ORIGIN OF BAHAMIAN OOIDS, ABST., ISOTOPIC EVIDENCE BEARING ON
B 6405 751* ORIGIN OF BIOGENIC CARBONATE SANDS OF SOUTHERN SHELF OF ST.
B 6103 466* ORIGIN OF COOK INLET OIL, ABST.
B 6203 516* ORIGIN OF CYCLICITY IN CONEMAUGH FORMATION IN APPALACHIANS
B 6004 696* ORIGIN OF DOLOMITE--MODEL BASED ON CYCLIC INTERACTION OF
B 6004 696* ORIGIN OF DOLOMITE--MODEL BASED ON CYCLIC INTERACTION OF
B 6303 532* ORIGIN OF GASES ADSORBED IN NEAR-SURFACE SEDIMENT IDENTIFIED
B 6304 701* ORIGIN OF GASES, ADSORBED IN NEAR-SURFACE SEDIMENTS IDENTIFIED
B 6407 1078† ORIGIN OF GASOLINE-RANGE HYDROCARBONS AND THEIR MIGRATION
B 6012 2162† ORIGIN OF GULF COAST SALT-DOME CAP ROCK
B 6303 471* ORIGIN OF HILLS BEACH-FLETCHER NECK TOMBOLO SYSTEM,
B 6004 733* ORIGIN OF HOLOCOCCOLITHS, ABST., MESOZOIC
B 6407 1083 ORIGIN OF HYDROCARBON GASES IN SUBMARINE SEEPS
B 6008 1407* ORIGIN OF ISOLATED REEFS AND SHELL EDGES IN MIDLAND BASIN,
B 6203 498* ORIGIN OF LATE-STAGE IRON-RICH CARBONATE CEMENTS IN
B 6011 2005† ORIGIN OF LIMESTONE BUTTES, GYPSUM PLAIN, CULBERSON COUNTY,
B 6112 2140 ORIGIN OF LOW PRESSURES
B 6303 428* ORIGIN OF NATURAL GAS, ABST., ORGANIC ACID ANIONS IN OIL-FIELD,
B 6212 2441† ORIGIN OF NATURAL GAS, ALIPHATIC, ACID, ANIONS IN, OIL-FIELD,
B 6203 532* ORIGIN OF NAVAJO SANDSTONE, ABST.
B 6203 522* ORIGIN OF PETROLEUM—IN-TRANSIT CONVERSION OF ORGANIC
B 6309 1611* ORIGIN OF POROSITY IN DEEP WOODBINE-TUSCALOOSA TREND,
B 6303 478* ORIGIN OF REGIONAL PRE-MIDDLE DEVONIAN DOLOMITIZATION IN
B 6007 1017 ORIGIN OF SHEETS OF FLUVIAL SANDSTONE
B 6305 788 ORIGIN OF SIGSBEE ESCARPMENT
B 6004 725* ORIGIN OF SOME SEDIMENTARY IRON ORES, ABST., GLAUCONITE-,
B 6405 755* ORIGIN OF SUBSURFACE FRACTURE SYSTEMS—EXAMPLE FROM
B 6108 1376* ORIGIN OF SULFUR IN COLORADO OIL-SHALE DEPOSITS, ABST.,
B 6411 2041 ORIGIN OF THE GULF OF MEXICO AND THE EARLY OPENING OF THE
B 6309 1610* ORIGINS OF PERMANEN INLETS SEPARATING BARRIER ISLANDS AND
B 6202 304 ORINOCO DELTA, REGIONAL SETTING, TRINIDAD AND
B 6202 303 ORINOCO DELTA SYSTEM, VENEZUELA
B 6111 1951 ORINOCO HEAVY OIL DEPOSIT, VENEZUELA
B 6202 303† ORINOCO , WAS IT THE
B 6209 1610 ORISKANY FORMATION, PENNSYLVANIA
B 6209 1615 ORISKANY SANDSTONE, APPALACHIAN BASIN
B 6209 1610 ORISKANY SANDSTONE, NEW YORK
B 6002 280 ORISKANY SANDSTONE, PENNSYLVANIA
M 28 165 ORISKANY SANDSTONE, VIRGINIA, DEVONIAN
M 28 97 ORISKANY SANDSTONE, VIRGINIA, DEVONIAN
M 28 65 ORISKANY SANDSTONE, VIRGINIA, DEVONIAN
B 6312 2176 OROGEN PROVENANCES, RECYCLED
S 3 85 OROGENE ROOT ZONE
B 6004 654* OROGENIC AND EPEIROGENIC EVENTS IN WESTERN ALPINE MOBILE BELT
M 25 527 OROGENIC BELT, ACTIVE, LONGEST, ANDEAN REGION
S 3 96 OROGENIC CONTINENTAL MARGIN, INCIPIENT
S 3 82 OROGENIC CONTINENTAL-MARGIN HISTORY
B 6004 693* OROGENIC EVENTS, ALTOS CUCHUMATANES MOUNTAINS, WESTERN
S 3 93 OROGENIC FORELAND MARGIN
S 3 91 OROGENIC MINERALIZATION BELTS AND METALLOTECTONIC GRADIENT
B 6403 361 OROGENIC PHASES, SIERRA, NEVADA
B 6403 361 OROGENIC PHASES, SIERRA, NEVADA
S 3 91 OROGENIC ZONES ACROSS MOBILE BELTS
S 3 136 OROGENIC-METAMORPHIC MOBILIZATION, LATEROGENESIS AT THE
S 3 115 OROGENIC-TAPHROGENIC MARGINS
B 6403 359† OROGENIES AND CHANGES IN OCEAN-FLOOR SPREADING DURING LATE
B 6403 359† OROGENIES AND CHANGES IN OCEAN-FLOOR SPREADING DURING LATE
B 6101 18 OROGENIES, CANADIAN, ARCTIC ISLANDS
B 6403 366 OROGENIES, WORLDWIDE, ALPINOTYPE
B 6403 366 OROGENIES, WORLDWIDE, ALPINOTYPE
B 6206 1021 OROGENY, CANADA, SVERDRUP BASIN, EUREKAN
B 6403 366 OROGENY DURING LATE JURASSIC TO LATE TERTIARY TIME, TIME

B 6403 365 OROGENY, KINEMATIC CONSIDERATIONS IN, PLATE MOTION AND
B 6112 2116 OROGRANDE BASIN, NEW MEXICO
B 6109 1448† OROGRANDE BASINS, PETROLEUM, POTENTIAL AND STRATIGRAPHY OF,
B 6108 1305 OROS, MARGARET O., OIL AND GAS DEVELOPMENTS IN NORTH
B 6208 1477 OROS, MARGARET O., OIL AND GAS DEVELOPMENTS IN NORTH
B 6008 1332 OROS, MARGARET O., OIL AND GAS DEVELOPMENTS IN NORTH
S 3 33 ORPHAN LODE COLLAPSED-PLUG PIPE
B 6101 80 ORPHEUS AND ABENAKI BASINS, OUTER, SCOTIAN SHELF
B 6301 89 ORPHEUS GRABEN, SCOTIAN BASIN
B 6303 505 ORR, DONALD G., AVAILABILITY OF REMOTELY SENSED DATA, ABST.
B 6303 506 ORR, DONALD G., CHARACTERISTICS OF REMOTELY SENSED DATA
B 6408 1280 ORR, R. WILLIAM, GEOLOGY OF WHEATON CONSOLIDATED OIL FIELD,
B 6211 2358 ORR, W. N., HOLOTHURIAN MICROFOSSILS AND EOCENE STRATIGRAPHY
M 28 53 ORTEGA QUARTZITE, NEW MEXICO, PRECAMBRIAN
B 6108 1385 ORTEL, WALTER H., GEOLOGY AND RESERVOIR ANALYSIS OF EAST MESA
M 28 18 ORTHOCLASE, UNTWINNED, FELDSPAR
M 28 106 ORTHOQUARTZITE, TEXTURE, SUPERMATURE SEDIMENTARY
B 6203 555* ORTHOQUARTZITES, SIMPSON GROUP, ORDOVICIAN, OKLAHOMA, ABST.,
M 25 480 ORTHOSHELVES, DEFINED
M 25 281 ORWIG, EUGENE R., JR., STRUCTURAL EVOLUTION OF LATE MESOZOIC
B 6009 1622 OSBORNE, ROBERT H., LATE MIDDLE AND EARLY LATE ORDOVICIAN
B 6203 551 OSBORNE, ROBERT H., MODERN LACUSTRINE STROMATOLITES, WALKER
B 6203 488 OSBORNE, ROBERT H.,, OXYGEN, AND CARBON ISOTOPIC VARIATION IN
B 6207 1229 OSBUN, ERIK, GEOLOGY AND DEVELOPMENT OF MOMOTOMBO
B 6112 2144 OSCEOLA LOW, FLORIDA
B 6004 704 OSGOOD, R. G., JR., TRACE FOSSILS OF CHAGRIN FORMATION, UPPER
B 6109 1513 OSHA CANYON FORMATION, NEW, NEW MEXICO
B 6109 1513† OSHA CANYON FORMATION, PENNSYLVANIAN, NEW MORROWAN UNIT
B 6109 1514 OSHA CANYON, NEW MEXICO
B 6109 1517 OSHA CANYON TYPE SECTION, NEW MEXICO
B 6004 714* OSLOFJORD, NORWAY--CASE STUDY IN APPLICATIONS OF MARINE
B 6211 2358 OSMOND, JOHN C., COAL FOR CALIFORNIA, ABST.
S 10 41 OSMOSIS, MIGRATION, AGENTS
S 10 13 OSMOTIC IMBALANCES, EXPULSION MECHANISMS
B 6105 818 OSSIAN, CLAIR RUSSELL, SCOTIAN SHELF, OFFSHORE EASTERN CANADA,
B 6303 482* OSTRACOD MICROFAUNA, ABST., EQUILIBRIUM IN MODERN, CORAL,
B 6309 1605* OSTRACOD MICROFAUNA, ABST., EQUILIBRIUM IN MODERN, CORAL,
S 4 209† OSTRACOD STANDING CROP, LAGUNA MANDINGA, VERACRUZ, MEXICO,
B 6307 1144 OSTRACODA, EDITED BY RAYMOND H. BATE AND ERIC ROBINSON,
B 6303 536* OSTRACODA FROM PENINSULAR FLORIDA, ABST., BIOSTRATIGRAPHY OF
B 6004 727* OSTRACODA FROM WELLS IN CENTRAL AND SOUTHERN FLORIDA, ABST.,
B 6209 1767* OSTRACODA OF PRAIRIE BLUFF CHALK, UPPER CRETACEOUS,
B 6405 791* OSTRACODA WITH THOSE OF WESTERN EUROPE, ABST., COMPARISON OF,
B 6409 1559* OSTRACODE BIOSTRATIGRAPHY OF LOWER OLIGOCENE, VICKSBURGIAN,
B 6409 1562* OSTRACODES AND SPECIE DIVERSITY GRADIENTS, GULF OF MEXICO,
M 27 101 OSTRACODS
M 27 100 OSTRACODS
B 6105 832* OSTRACODS, ABST., AGE OF, LAKOTA FORMATION, LOWER, CRETACEOUS,
B 6405 725* OSTRACODS, ABST., DELINEATION OF, JACKSONIAN-, VICKSBURGIAN
B 6009 1611* OSTRACODS IN H. V. HOWE COLLECTION, ABST., DIAGNOSTIC CENTRAL,
M 27 83 OSTREID MOLLUSKS
B 6303 385 OSTROFF, A. G., REVIEW OF DATA ON THE GEOCHEMISTRY OF OIL FIELD
B 6203 550 OSTROM, MEREDITH E., LITHOSTRATIGRAPHY, PETROLOGY, AND
B 6102 193 OSTROV KARAGINSKIY, KARAGINSKY ISLAND
B 6405 784* OSWEGO LIMESTONE, ALINE-LAMBERT FIELDS, OKLAHOMA—SOURCE,
B 6109 1449 OTERO UPLIFT, NEW MEXICO
B 6401 68 OTHFRESEN SAMPLED SECTION, GERMANY
B 6004 709* OTHRIS, GREECE, ABST., SEDIMENTATION AND RESEDIMENTATION OF,
B 6009 1621 OTT, R. R., RELATION BETWEE SALINA, UPPER SILURIAN, OIL
B 6009 1531† OTTAWA SAND, EFFECT OF, HIGH-, PRESSURE, DEFORMATION ON,
B 6405 761 OTTE, CAREL, OVERVIEW OF GEOTHERMAL ENERGY DEVELOPMENTS,
B 6002 164 OTTER BASIN, BERINGIAN SHELF
M 24 276 **OTTMANN, R. D., JAY FIELD, FLORIDA--A JURASSIC STRATIGRAPHIC TRAP**
S 6 180 OTTNANGIAN STAGE, PARATETHYS, NEOGENE
S 5 145* OTTO FIORD FORMATION MISSISSIPPIAN-PENNSYLVANIAN, CANADIAN
S 5 146 OTTO FIORD FORMATION, CANADA, ARCTIC ARCHIPELAGO, ELLESMERE
S 5 159 OTTO FIORD FORMATION, ALGAL MOUNDS, CANADA, SVERDRUP BASIN,
S 5 149 OTTO FIORD FORMATION, ANHYDRITE DEPOSITION, CANADA, SVERDRUP
B 6111 1930 OTTO FIORD FORMATION, ELLEF RINGNES ISLAND
B 6111 1930 OTTO FIORD FORMATION, ELLESMERE ISLAND, NORTHWESTERN
B 6105 776* OTTO FIORD FORMATION—EVIDENCE OF SUBAQUEOUS EVAPORITE
S 5 147 OTTO FIORD FORMATION, STRATIGRAPHY, CANADA, SVERDRUP BASIN,
B 6111 1930 OTTO FJORD, ELLESMERE ISLAND
B 6111 1930 OTTO FJORD FORMATION, ARCTIC ARCHIPELAGO
B 6111 1931 OTTO FJORD FORMATION, SVERDRUP BASIN
B 6209 1762 OTVOS, ERVIN G., JR., NEW ORLEANS—SOUTH HANCOCK HOLOCENE
M 25 230 OTWAY BASIN, AUSTRALIA
M 29 152 OTWAY BASIN, AUSTRALIA, CONTINENTAL MARGIN
M 29 159 OTWAY BASIN SECTION, AUSTRALIA, MARGIN
S 3 98 OUACHITA AND MARATHON MOBILE BELTS
B 6402 222 OUACHITA FACIES, TEXAS
B 6002 299† OUACHITA FOLDBELT IN TEXAS AND THEIR RELATION TO PALEOZOIC
B 6301 54 OUACHITA FOLDBELT, TEXAS
B 6105 816* OUACHITA MOUNTAINS, ABST., FAVORABLE, HYDROCARBON POTENTIAL
B 6108 1384* OUACHITA MOUNTAINS, ARKANSAS AND OKLAHOMA, ABST.,
B 6004 727* OUACHITA MOUNTAINS, OKLAHOMA, ABST., BASINWARD, FACIES
B 6405 670* OUACHITA PROBLEMS, ABST., FRESH LOOK AT SOME
B 6109 1449 OUACHITA TECTONIC BELT, TEXAS, MARATHON-
S 3 107 OUACHITA-MARATHON BELT
B 6303 409* OUARGLA REGION, ALGERIA, NORTH AFRICA, BY TITLE ONLY
M 30 540 OUARGLA REGION, BEN-KAHLA FIELD, ALGERIA
M 30 540 OUARGLA REGION, BOU-KHEZANA FIELD, ALGERIA
M 30 540 OUARGLA REGION, DRAA ET TEMRA FIELD, ALGERIA
M 30 540 OUARGLA REGION, GARET ECH CHOUF FIELD, ALGERIA
M 30 540 OUARGLA REGION, GUELLALA FIELD, ALGERIA
M 30 540 OUARGLA REGION, GUELLALA NORD-EST FIELD, ALGERIA
M 30 540 OUARGLA REGION, HANIET-EL-BEIDA FIELD, ALGERIA
M 30 540 OUARGLA REGION, HAOUD-BERKAOUI FIELD, ALGERIA
M 30 540 OUARGLA REGION, KEF EL ARGOUB FIELD, ALGERIA
M 30 540 OUARGLA REGION, N GOUSSA FIELD, ALGERIA
M 30 540 OUARGLA REGION, OUARGLA FIELD, ALGERIA
M 30 540 OUARGLA REGION, TAKHOUKT FIELD, ALGERIA
M 30 539† OUARGLA REGION TRIASSIC BASIN, ALGERIA, PETROLEUM POTENTIAL--
B 6008 1398* OUNGRE TREND IN SOUTHEASTERN SASKATCHEWAN, CANADA, ABST.,
S 5 96 OUTCROP RELATIONS OF REEFS, OHIO, SILURIAN, LOCKPORT GROUP,
B 6406 890 OUTER BANDA ARC
B 6106 904 OUTER BASIN, BAJA CALIFORNIA
B 6004 679* OUTER CONTINENTAL SHELF, ABST., BIOGENIC SEDIMENTARY
B 6004 713* OUTER CONTINENTAL SHELF, ABST., CORAL, REEFS AND, BANKS OF,
B 6004 649* OUTER CONTINENTAL SHELF, ABST., LATE, PLEISTOCENE-, HOLOCENE
B 6105 826* OUTER CONTINENTAL SHELF, ABST., PETROLEUM FRONTIER AREAS ON,
B 6405 733* OUTER CONTINENTAL SHELF, ABST., PROCESSES CONTROLLING
B 6004 722* OUTER CONTINENTAL SHELF, ABST., SEDIMENT-, DISPERSAL, PATTERNS
B 6105 836* OUTER CONTINENTAL SHELF, ABST., SEDIMENTARY MODEL PREDICTION
S 7 243† OUTER CONTINENTAL SHELF AND UPPER CONTINENTAL SLOPE, LATE,
B 6405 687* OUTER CONTINENTAL SHELF AND UPPER CONTINENTAL SLOPE,
B 6007 1114 OUTER CONTINENTAL SHELF, CDP MULTICHANNE REFLECTION PROFILE,
B 6007 1112† OUTER CONTINENTAL SHELF, EVIDENCE OF, POST-, PLEISTOCENE,
B 6409 1333 OUTER CONTINENTAL SHELF, EXPLORATORY WELLS, 1979, BALTIMORE
B 6105 830* OUTER CONTINENTAL SHELF, NORTHWEST GULF OF MEXICO, ABST.,
B 6209 1760* OUTER CONTINENTAL SHELF OF SOUTH TEXAS, ABST., LIVING, SHELLED,
B 6106 861 OUTER CONTINENTAL SHELF, SURFICIAL, SAND, SHEET ON, MIDDLE,
B 6211 2355* OUTER CONTINENTAL SHELF. BY TITLE ONLY, GEOENVIRONMENTAL
B 6007 1112† OUTER CONTINETAL SHELF, EVIDENCE OF, POST-PLEISTOCENE, FAULTS
B 6309 1606* OUTER DEEP-SEA FAN DEPOSITIONAL LOBE SEQUENCE FROM JACKFORK
M 29 45 OUTER INSULAR SHELF, ICELAND
M 29 45 OUTER INSULAR SHELF, ICELAND, PLIOCENE
B 6303 291 OUTER MARGIN APRONS, PROSPECTS
B 6410 1701 OUTER RIDGE, ABYSSAL SEDIMENT DRIFT, BLAKE-, BAHAMA
B 6301 9 OUTER RISE
B 6203 352 OUTER SANTA CRUZ BASIN, CALIFORNIA
B 6101 80 OUTER SCOTIAN SHELF, ORPHEUS AND ABENAKI BASINS
M 29 374 OUTER SULU SEA, DEEP BASIN, SOUTHEAST ASIA
M 25 454 OUTER ZONE, JAPAN
B 6004 701* OUTER-FAN DEPOSITIONAL LOBES OO UPPER MIOCENE-PLIOCENE LAGA
S 4 22 OUTER-SHELF PROVINCE, PUERTO RICO
M 25 481† OUTLINE OF KURIL-KAMCHATKA ARC--SUMMARY
B 6008 1159 OUTPOST OR EXTENSION TEST
B 6308 1174 OUTPOST OR EXTENSION TEST
B 6108 1126 OUTPOST OR EXTENSION TEST
B 6409 1301 OUTPOST OR EXTENSION TEST
B 6208 1281 OUTPOST OR EXTENSION TEST
B 6004 722 OVERBEY, W. K., JR., LINEAMENT AND STRUCTURAL ANALYSES OF
B 6309 1577 OVERBEY, WILLIAM K., EXPLORATION STRATEGY FOR
B 6102 188 OVERBURDEN AND FOLDING ON COMPACTION, EFFECT OF
M 26 37 OVERBURDEN, LABORATORY, MEASUREMENTS
B 6306 923 OVERBURDEN, SHALE MASSES BEFORE, BURIAL BY PERMEABLE
B 6207 1225* OVERCOMING EARTH RESOURCE AND HAZARD CONSTRAINTS ON
B 6008 1413* OVERPRESSURED SHALE MASSES AND GROWTH FAULTING, NEW
S 10 189 OVERPRESSURED SHALE, MIGRATION, BARRIERS
S 10 191 OVERPRESSURED SHALES, DIAPIRS, SEISMIC PROFILE
M 26 452 OVERPRESSURING, OKLAHOMA, ANADARKO BASIN, MORROW
M 26 452 OVERPRESSURING, OKLAHOMA, ANADARKO BASIN, SPRINGER
M 26 453 OVERPRESSURING, PREDICTABILITY, MORROW-, SPRINGER
B 6211 2356* OVERTHRUST BELT, ABST., EXPLORATION IN RUGGED TERRAIN IN
B 6405 806* OVERTHRUST BELT, ABST., WHITNEY CANYON—A PROVED, GIANT—ONLY
B 6412 2166† OVERTHRUST BELT, CROSS-STRIKE STRUCTURAL, DISCONTINUITIES—
B 6305 828* OVERTHRUST BELT FROM MONTANA TO NEVADA, ABST., MISSISSIPPIAN,
B 6408 1278* OVERTHRUST BELT OF ALBERTA, ABST., HYDROCARBON,

B 6405 776* OVERTHRUST BELT, UTAH, ABST., OIL-TO-SOURCE CORRELATION—
B 6205 889* OVERTHRUST BELT, UTAH, WYOMING, AND IDAHO, ABST., TWIN CREEK
B 6208 1500 OVERTHRUST BELT, WYOMING
B 6409 1503 OVERTHRUST BELT, WYOMING
B 6308 1393 OVERTHRUST BELT, WYOMING
B 6405 767* OVERTHRUST BELT—RELATION TO CANADIAN FOOTHILLS PROVINCE
M 26 391 OVERTHRUST FAULTING, MODELING
B 6405 693* OVERTHRUSTING OF PALEOZOIC SHELF IN SOUTHERN APPALACHIANS
B 6407 1005 OVERTON FIELD, ARK-LA-TEX AREA
B 6203 551 OVERTON, HAROLD L., ENVIRONMENT CREATED BY VERTICALLY
B 6205 890 OVERTON, HAROLD L., PERMEABILITY CONTROL IN CRETACEOUS
M 30 1† OVERVIEW, GIANT FIELD, DISCOVERIES 1968-1978: AN
B 6103 463* OVERVIEW OF ALASKA PLATE TECTONICS, ABST.
B 6008 1407* OVERVIEW OF AUSTRALIAN URANIUM DEPOSITS, ABST.
B 6001 3† OVERVIEW OF EXPLORATION GEOPHYSICS--RECENT BREAKTHROUGHS
B 6008 1397* OVERVIEW OF EXPLORATION GEOPHYSICS--RECENT BREAKTHROUGHS
M 25 203† OVERVIEW OF EXPLORATON GEOPHYSICS--RECENT BREAKTHROUGHS
B 6405 761* OVERVIEW OF GEOTHERMAL ENERGY DEVELOPMENTS, ABST.
M 26 51† OVERVIEW, SEISMIC, STRATIGRAPHY AND, GLOBAL, CHANGES OF, SEA
B 6203 536* OVERWASH DEPOSITS, ABST., SEDIMENT, GRADING IN
B 6008 1405 OWEN, DONALD E., PALEOCURRENT ANALYSIS OF DAKOTA AND
B 6206 1093 OWEN, DONALD E., PALEOCURRENTS OF JACKPILE SANDSTONE, BURRO
B 6405 761 OWEN, DONALD E., SAN JUAN BASIN OF NEW MEXICO AND COLORADO,
B 6007 1117 OWEN, EDGAR W., SOUTH TEXAS GEOLOGICAL SOCIETY--A
B 6203 552 OWEN, ROBERT M., FACTORS CONTROLLING GEOCHEMISTRY AND EARLY
B 6105 818 OWENS, E. H., COASTAL ENVIRONMENTS AND BEACH MORPHOLOGY OF
B 6004 704 OWENS, E. H., COMPARISON OF COASTAL DYNAMICS ON WEST- AND
B 6203 552 OWENS, EDWARD H., EFFECTS OF VARIATIONS IN WAVE ENERGY ON
B 6103 360 OWENS, RICHARD T., URANIUM POTENTIAL O PERMIAN AND
B 6305 841 OWENS, WILLARD G., HYDROGEOLOGIC SIGNIFICANCE OF OGALLALA
B 6408 1286* OWENSVILLE NORTH CONSOLIDATED AND MT. CARMEL CONSOLIDATED
B 6008 1343 OWER, JOHN R., DEVELOPMENTS IN WESTERN CANADA IN 1975
B 6108 1313 OWER, JOHN R., DEVELOPMENTS IN WESTERN CANADA IN 1976
B 6405 717 OWINGS, B. FRANK, UTAH-ARIZONA HINGE LINE—THRUST BELT—
B 6211 2357 OWINGS, B. FRANK, UTAH-ARIZONA HINGE-LINE THRUST BELT—
B 6309 1587 OWINGS, D., IMPLICATIONS OF GASEOUS HYDROCARBON GEOCHEMISTRY
B 6309 1583 OWINGS, D., PETROLEUM GEOCHEMISTRY AND GEOLOGY OF SOUTHEAST
M 30 143 OXFORDIAN, NORTH SEA, MORAY FIRTH BASIN
B 6205 793 OXFORDIAN, PROFILE, BLAKE PLATEAU AREA
B 6408 1233 OXFORDIAN SEA, NORTH AMERICA
B 6405 608 OXFORDIAN TO KIMMERIDGIAN, MIDDLE EAST
M 26 159 OXFORDIAN-KIMMERIDGIAN, AFRICA, WESTERN, SEQUENCES
B 6408 1184 OXIC ENVIRONMENT
B 6408 1188 OXIC LAKES, EXAMPLES, LARGE
B 6408 1190 OXIC NEGATIVE WATER BALANCE BASIN, MEDITERRANEAN SEA
B 6408 1190 OXIC NEGATIVE WATER BALANCE BASIN, PERSIAN GULF
B 6408 1190 OXIC NEGATIVE WATER BALANCE BASIN, RED SEA
B 6408 1192 OXIC OPEN OCEAN, CARIACO TRENCH, SILLED DEPRESSION IN
B 6408 1193 OXIC OPEN OCEAN, ORCA BASIN, SILLED DEPRESSION IN
B 6408 1192 OXIC SILLED BASINS, EXAMPLES
B 6405 671* OXIC-ANOXIC AND CARBONATE CYCLES IN CRETACEOUS ORGANI
S 3 46 OXIDATION AND MOBILIZATION, SELECTIVE
B 6108 1379* OXIDATION AT PH 5-9.5, IMPLICATIONS FOR FORMATION OF ROLL-TYPE
B 6405 722* OXIDATION-REDUCTION IN OAKVILLE SANDSTONE OF SOUT TEXAS—
M 25 400 OXIDE ZONE, AUSTRALIA, NICKEL MINERALIZATION
B 6108 1388* OXIDIZED SULFIDE MINERALS AS EXPLORATION GUIDE FOR URANIUM,
B 6407 1036 OXNARD FIELD, CALIFORNIA
B 6104 594 OXNARD FIELD, CALIFORNIA
B 6203 360 OXNARD OIL FIELD, CALIFORNIA
B 6203 360 OXNARD SHELF, CALIFORNIA
S 7 236 OXNARD SLOPE, SECTION, CALIFORNIA, VENTURA BASIN
B 6203 488* OXYGEN, AND CARBON ISOTOPIC VARIATION IN PLEISTOCENE
M 29 427 OXYGEN IN SEAWATER, INDIA, MARGIN
B 6004 696* OXYGEN ISOTOPE ANALYSES OF PLEISTOCENE FLUVIAL, ALGAL,
B 6303 505* OXYGEN ISOTOPE GEOCHEMISTRY OF ABU DHABI SABKHA, ABST.,
S 9 26 OXYGEN ISOTOPE OCEAN TEMPERATURES, CENOZOIC
B 6401 76 OXYGEN ISOTOPE SYSTEM
B 6203 489* OXYGEN ISOTOPES AS INDICATORS OF METEORIC AND MIXING-ZONE
B 6109 1554* OXYGEN ISOTOPIC COMPOSITION OF CRETACEOUS, EDWARDS GROUP,
B 6109 1555* OXYGEN ISOTOPIC EVOLUTION OF WHOLE ROCK AND CEMENTS FROM
B 6303 538* OXYGEN-DEFICIENT PALEOENVIRONMENTS IN MID-PACIFIC MOUNTAINS
B 6408 1199 OXYGEN-DEPLETED LAYERS IN WORLD OCEAN
M 25 547 OXYGEN-ISOTOPE DATA, PACIFIC OCEAN
B 6207 1231 OYAGI, NORIO, LANDSLIDE PROBLEMS IN JAPAN, ABST.
M 25 559 OYARZUN M., JORGE, METALLOGENIC BELTS AND TECTONIC EVOLUTION
B 6101 113 OYSTER RIDGE SANDSTONE MEMBER, UTAH
M 25 148 OZAWA, TAKEJIRO, GEOLOGICAL AND GEOCHEMICAL CHARACTERISTICS
B 6204 707 OZORA LIMESTONE, ILLINOIS BASIN
B 6402 149 P SANDSTONE, LOUISIANA
B 6104 597 P-8-6 WELL, LIBYA
M 26 25 P-WAVE, FLUID MOTION, ATTENUATION
B 6206 993 PABDEH MARL, IRAN
B 6405 695 PACER, JOHN C., ADVANCES IN RADON EXPLORATION TECHNIQUES FOR
M 25 518 PACHUCA Y REAL DEL MONTE MINE, SILVER AND GOLD, MEXICO
B 6004 668* PACIFIC, ABST., ABYSSAL, ORGANISM, TRACES ON AN IN MODERN,
B 6303 473* PACIFIC, ABST., CORRELATION OF, CONTINENTAL-MARGIN AND,
B 6207 1213* PACIFIC, ABST., GEOCHEMICAL EXPLORATION FOR, SUBMARINE,
B 6207 1219* PACIFIC, ABST., MINERAL RESOURCES IN SOUTHWEST
B 6405 678* PACIFIC, ABST., MIOCENE SPUMELLARIAN, RADIOLARIA FROM, SOUTH
M 29 429 PACIFIC, ALASKA, CONVERGENT MARGIN
M 25 43 PACIFIC AND ATLANTIC OCEAN MARGINS, WESTERN
M 25 44 PACIFIC AND ATLANTIC WESTERN BORDERS
B 6408 1279* PACIFIC AND BERING SEA, ABST., RESOURCE POTENTIAL AND,
S 4 76 PACIFIC AND CARIBBEAN, MAJOR REEFS AND ATOLLS, INDO-
M 25 186 PACIFIC AND INDIAN OCEANS, CONVERGENT MARGINS
B 6211 2355* PACIFIC ARCTIC OUTER CONTINENTAL SHELF. BY TITLE ONLY,
M 25 58† PACIFIC AREA, GEOLOGIC-STRUCTURAL SYNTHESIS OF
S 4 285 PACIFIC ATOLLS, GRAZING AND BROWSING FISH
S 3 79 PACIFIC BASIN
B 6207 1210* PACIFIC BASIN, ABST., CRITICAL, ENERGY PROBLEMS OF
B 6207 1213* PACIFIC BASIN, ABST., REGIONAL GEOLOGIC, STRUCTURAL, AND,
M 25 473 PACIFIC BASIN, MEGASTRUCTURES, CENTRAL-TYPE
M 25 328† PACIFIC BASINS OF ECUADOR--SUMMARY, EXPLORATION AND
M 25 176† PACIFIC BELT--SUMMARY, OIL- AND, GAS-BEARING, BASINS OF, CIRCUM-
M 25 481 PACIFIC, BENIOFF ZONE, MIGRATION OF
B 6012 2182* PACIFIC COAST, ABST., REVIEW OF ESTIMTES OF, RADIOMETRIC AGES
M 25 8† PACIFIC COAST OF ASIA, BOUNDARIES OF, SEABED, JURISDICTION OFF
M 25 318† PACIFIC COAST, POTENTIAL FOR EXPLORATION AND DEVELOPMENT OF
M 25 559† PACIFIC CONTINENTAL BORDER--SUMMARY, METALLOGENIC BELTS AND
M 25 308† PACIFIC CONTINENTAL MARGIN OF BRITISH COLUMBIA--SUMMARY,
B 6207 1216* PACIFIC CONTINENTAL MARGINS, ABST., SEAFLOOR INSTABILITY ON
M 25 461 PACIFIC COPPER AND GOLD METALLOGENIC SYSTEMS, TRANS-
M 25 460 PACIFIC COPPER-BEARING RING, GREAT
B 6405 738* PACIFIC CREEK AREA, SUBLETTE COUNTY, WYOMING, ABST.,
M 25 433† PACIFIC, DISCOVERY AND DEVELOPMENT--SUMMARY, PORPHYRY,
M 25 206 PACIFIC, EASTERN, SEISMIC STUDY
M 25 1† PACIFIC ENERGY AN MINERAL RESOURCES TO WORLD, IMPORTANCE OF,
B 6203 346 PACIFIC FARALLON AND NORTH AMERICAN PLATES, INTERACTION OF
M 25 376† PACIFIC, FERROMANGANESE, NODULES OF THE
M 25 309† PACIFIC FOR PEACE AND SECURITY--SUMMARY, OIL RESOURCES OF THE
M 25 110† PACIFIC, GEOTHERMAL PROSPECTS AROUND THE
M 25 461 PACIFIC GOLD-BEARING RING, GREAT
M 25 460 PACIFIC IRON-MANGANESE CONCRETIONS, ORIGIN OF
B 6010 1947† PACIFIC ISLAND REGION, DEVELOPMENTS IN, NEW ZEALAND AND,
B 6110 1880† PACIFIC ISLAND REGION IN 1976, DEVELOPMENTS IN, NEW ZEALAND
B 6210 1900† PACIFIC ISLAND REGION IN 1977, DEVELOPMENTS IN, NEW ZEALAND
B 6310 1680† PACIFIC ISLAND REGION IN 1978, DEVELOPMENTS IN, NEW ZEALAND
B 6411 1776† PACIFIC ISLAND REGION IN 1979, DEVELOPMENTS IN, NEW ZEALAND
B 6210 1901 PACIFIC ISLAND REGION, PRODUCTION, 1977, NEW ZEALAND AND,
B 6207 1209* PACIFIC ISLANDS—IMPORTANCE OF COOPERATION, ABST., MINE
M 25 48 PACIFIC LITHOSPHERIC PLATE BOUNDARIES
M 25 386 PACIFIC, MANGANESE NODULE DISTRIBUTION, SOUTH
B 6108 1392* PACIFIC MARGIN, ABST., WIDESPREAD OCCURRENCE OF, LAUMONTITE IN
B 6112 2077 PACIFIC MARGIN, AUSTRALIA
B 6403 444* PACIFIC MARGIN BETWEEN 32S AND 32N, ABST., BENTHIC,
B 6403 444* PACIFIC MARGIN BETWEEN 32S AND 32N, ABST., BENTHIC,
M 25 536 PACIFIC MARGIN, GEOLOGY AND METALLIC MINERAL OCCURRENCES,
B 6012 2179* PACIFIC MARGIN OF NORTH AMERICA, ABST., APPLICATIONS OF, PLATE
B 6403 448* PACIFIC MARGIN, UNITED STATES, ABST., DEPOSITIONAL ENVIRONMENTS
B 6403 448* PACIFIC MARGIN, UNITED STATES, ABST., DEPOSITIONAL ENVIRONMENTS
B 6404 467 PACIFIC, MARINE PHREATIC ARAGONITE CEMENT, INDO-
M 25 474 PACIFIC, MEGASTRUCTURES, MAP
M 25 485 PACIFIC, METAL PROVINCES, NORTHEAST
M 25 485 PACIFIC, METALLIZATION, EPOCHS OF, NORTHEAST
M 25 486 PACIFIC, METALLOGENIC PROVINCES, NORTHEAST
M 25 483† PACIFIC, METALLOGENIC PROVINCES OF NORTHEAST
M 25 461 PACIFIC, METALLOGENIC ZONATION
M 25 460 PACIFIC, MEXICO-CENTRAL AMERICA, SILVER-COPPER-POLYMETALLIC
M 25 46† PACIFIC MOBILE BELT AND PACIFIC OCEAN, GENERAL PRINCIPLES AND
M 25 458 PACIFIC MOBILE BELT OR PACIFIDES
M 25 59† PACIFIC MOBILE BELT, USSR, STRUCTURAL-FORMATION ANALYSIS OF
M 25 479† PACIFIC MOBILE BELT--SUMMARY, MEIMECHITES OF, KAMCHATKA AND,
M 25 62† PACIFIC MOBILE BELT--SUMMARY, TECTONICS OF CONTINENTAL AND
B 6306 426* PACIFIC NEOGENE DIATOM CORRELATIONS, ABST., PAN-
M 25 390 PACIFIC, NICKEL- AND COPPER-RICH NODULES, NORTH
M 25 384 PACIFIC, NORTH, MENDOCINO FRACTURE ZONE, NODULES
B 6303 538* PACIFIC OCEAN, ABST., MIDDLE, CRETACEOUS, OXYGEN-DEFICIENT
B 6004 688* PACIFIC OCEAN, ABST., RESEDIMENTATION OF, BIOGENIC, CARBONATE

M 25 46† PACIFIC OCEAN, GENERAL PRINCIPLES AND COMPILATION METHOD--,
M 25 539 PACIFIC OCEAN, NAZCA PLATE, LOCATION
M 25 539† PACIFIC OCEAN, NAZCA PLATE PROJECT, METALLOGENESIS IN
M 25 539† PACIFIC OCEAN, NAZCA PLATE PROJECT, METALLOGENESIS IN
B 6408 1201 PACIFIC OCEAN OFFSHORE WASHINGTON-OREGON, ANOXIC OPEN OCEAN
M 25 547 PACIFIC OCEAN, OXYGEN-ISOTOPE DATA
M 25 252† PACIFIC OCEAN, PETROLIFEROUS, TAIWAN BASINS IN FRAMEWOR OF
B 6111 1991 PACIFIC OCEAN, PIERCEMENT STRUCTURES, SEISMIC PROFILES
B 6111 1994 PACIFIC OCEAN, POSSIBLE DIAPIRS
M 25 546 PACIFIC OCEAN, RARE-EARTH ELEMENTS, ABUNDANCE PATTERNS
M 25 47† PACIFIC OCEAN--SUMMARY, GEOLOGIC HISTORY AND,
B 6207 1128 PACIFIC OCEANIC PLATE
B 6303 538* PACIFIC OCEEAN, ABST., MIDDLE, CRETACEOUS, OXYGEN-DEFICIENT
M 25 460 PACIFIC, OPHIOLITE DEPRESSIONS
M 29 429 PACIFIC, OREGON, CONVERGENT MARGIN
M 25 503 PACIFIC OROGEN, CANADIAN, CORDILLERA
B 6312 2206 PACIFIC, PANAMA, OFFSHORE PROFILE
B 6207 1195 PACIFIC PLATE
B 6203 394 PACIFIC PLATE, ALASKA
B 6403 360 PACIFIC PLATE, BOUNDARY
M 29 274 PACIFIC PLATE, BOUNDARY, ALASKA
B 6211 2215 PACIFIC PLATE, CALIFORNIA
B 6202 265 PACIFIC PLATES, DISPLACEMENT, CALIFORNIA, NORTH AMERICAN AND
M 27 19 PACIFIC, PLIOCENE, SEDIMENTS
M 25 379 PACIFIC, POLYGON SURVEYS
M 25 433 PACIFIC, PORPHYRY COPPER DEPOSITS, ASSOCIATED ROCKS, SOUTHWEST
M 25 51 PACIFIC REGION, BENIOFF ZONE
M 25 527† PACIFIC REGION, METALLOGENIC PROVINCES OF SOUTHEASTERN
M 25 34 PACIFIC REGION, MOOREA VOLCANIC ISLAND
M 25 49 PACIFIC REGION, PETROLEUM PRODUCTION AND POTENTIAL
M 25 32† PACIFIC REGION, TECTONIC FRAMEWORK OF
M 25 34 PACIFIC REGION, TUAMOTU ATOLLS
M 25 18† PACIFIC REGION--SUMMARY, MINERAL TRADE IN, CIRCUM-
M 25 382 PACIFIC RISE, NODULES, EAST
M 25 49 PACIFIC SEAFLOOR, AGES
M 25 459 PACIFIC SEGMENT, METALLOGENIC ZONATION, MAP
M 25 480† PACIFIC SHELVES--SUMMARY, STRUCTURAL, CLASSIFICATION OF NORTH
M 25 530 PACIFIC, SOUTHEASTERN, BATHYMETRIC MAP
M 25 535 PACIFIC, SOUTHEASTERN, FERROMANGANESE NODULES
M 25 535 PACIFIC, SOUTHEASTERN, METAL-RICH UNCONSOLIDATED PELAGIC
M 25 529 PACIFIC, SOUTHEASTERN, METALLIZATION, AGE OF DEPOSITS ASSOCIAT
M 25 530 PACIFIC, SOUTHEASTERN, MINERAL DEPOSITS
M 25 529 PACIFIC, SOUTHEASTERN, MINERAL DEPOSITS, TECTONIC AND GEOLOGIC
M 25 535 PACIFIC, SOUTHEASTERN, PHOSPHORITE NODULES
M 25 535 PACIFIC, SOUTHEASTERN, SEAFLOOR DEPOSITS BEARING MINERALS
M 29 227 PACIFIC, SUNDA TRENCH, OFF SUMATRA
B 6207 1212* PACIFIC WATER TEMPERATURES AND ANOMALOUS RUNOFF IN
M 25 386† PACIFIC--SUMMARY, EXPLOITATION OF, MANGANESE, NODULES IN,
M 25 350† PACIFIC--SUMMARY, HYDROGEOLOGY OF THE
M 25 390† PACIFIC--SUMMARY, NICKEL- AND, COPPER-RICH, NODULES OF
M 25 472† PACIFIC--SUMMARY, RING, MEGASTRUCTURES OF THE
B 6405 711* PACIFIC—PRELIMINARY RESULTS FROM DSDP SITE 503, LEG 68—INDICATE
M 25 458 PACIFIDES, ANDESITE LINE OR RING
M 25 458 PACIFIDES, ARGURI-KERULEN VOLCANIC-PLUTONIC CHAIN
M 25 460 PACIFIDES, CONTINENTAL-MARGIN METALLOGENIC ZONES
M 25 458 PACIFIDES, EASTERN ASIATIC BELT OR RHYOLITE LINE
M 25 458 PACIFIDES, GREAT KHINGAN VOLCANIC-PLUTONIC CHAIN
M 25 458 PACIFIDES, KHINGAN-OKHOTSK VOLCANIC-PLUTONIC CHAIN
M 25 460 PACIFIDES METALLOGENIC SYSTEMS, LARGE-SCALE
M 25 459 PACIFIDES, METALLOGENIC ZONES
M 25 460 PACIFIDES, METALS, TYPES OF
M 25 460 PACIFIDES, OCEAN-MARGIN METALLOGENIC ZONES
M 25 459 PACIFIDES, OCEANIC METALLOGENIC ZONES
M 25 458 PACIFIDES, PACIFIC MOBILE BELT OR
M 25 460 PACIFIDES, PERIDOTITE BELT, DUNITE-HARZBURGITE FACIES
M 25 460 PACIFIDES, PLATINUM
M 25 462 PACIFIDES, STANNIFEROUS METALLOGENIC SYSTEM
M 25 458 PACIFIDES, TECTONOSPHERE, HETEROGENEITY OF
M 25 458 PACIFIDES, VOLCANIC-PLUTONIC AREAS OR BELTS
M 25 458† PACIFIDES--SUMMARY, METALLOGENIC SYSTEMS OF
M 25 229 PACKHAM, GORDON, PHANEROZOIC HISTORY OF AUSTRALIA AND
B 6406 823 PADDYS VALLEY, AUSTRALIA
B 6105 819 PADGETT, GUY, COAL PROPERTY AND MAXIMIZING TOTAL
B 6102 167 PADRE CANYON FIELD, CALIFORNIA
B 6009 1608* PADRE ISLANDS--TIME-SERIES APPROACH, ABST., BEACH
B 6102 165 PADRE JUAN FAULT, CALIFORNIA
B 6405 762 PAGE, MARY M., ROCK-EVAL PYROLYSIS AS SOURC ROCK SCREENING
S 4 291 PAGO PAGO, GALAXEA, SAMOA
B 6109 1537 PAINE, WILLIAM R., INTERIOR STRUCTURE OF OAK GROVE RIDGE
B 6303 342 PAINT CREEK FAULT ZONE, KENTUCKY, IRVINE--
M 30 282 PAINTER RESERVOIR FIELD, LOCATION, WYOMING
M 30 281 PAINTER RESERVOIR FIELD, NUGGET SANDSTONE, JURASSIC
M 30 284 PAINTER RESERVOIR FIELD, NUGGET SANDSTONE, PHOTOMICROGRAPHY
M 30 281 PAINTER RESERVOIR FIELD, NUGGET SANDSTONE, TRIASSIC
M 30 285 PAINTER RESERVOIR FIELD, RESERVOIR CHARACTER
B 6405 639 PAINTER RESERVOIR FIELD, SECTIONS, WYOMING
M 30 284 PAINTER RESERVOIR FIELD, STRATIGRAPHY, COLUMNAR SECTION
M 30 282 PAINTER RESERVOIR FIELD, STRATIGRAPHY, WYOMING
M 30 283 PAINTER RESERVOIR FIELD, STRUCTURE SECTION
B 6205 887* PAINTER RESERVOIR FIELD, UINTA COUNTY, WYOMING, ABST.
M 30 281† PAINTER RESERVOIR FIELD--GIANT IN THE WYOMING THRUST BELT
B 6303 484* PAINTER RESERVOIR FIELD--GIANT IN WYOMING THRUST BELT, ABST.
B 6405 638† PAINTER RESERVOIR FIELD—GIANT IN WYOMING THRUST BELT
B 6305 832* PAINTER RESERVOIR--CLEAR CREEK--RYCKMA CREEK NUGGET
B 6209 1764* PAIRED FORAMINIFERAL ECOPHENOTYPES IN GULF COAST ESTUARIES—
B 6004 667 PAK, H., CONTRASTS IN CONCENTRATION OF SUSPENDED MATTER IN
B 6110 1838 PAKISTAN, 1976, HERGENRODER, J. D., ON DEVELOPMENTS IN
B 6203 542* PAKISTAN, ABST., ANCIENT, SUBMARINE CANYON AND, CHANNEL
B 6303 512* PAKISTAN, ABST., CYCLIC SEDIMENTATION OF, CRETACEOUS-,
B 6405 802* PAKISTAN, ABST., NEARSHORE MARINE AND CONTINENTAL FACIES IN,
B 6309 1490† PAKISTAN, BY COMPUTER PROCESSIN OF LANDSAT DATA,
B 6309 1494 PAKISTAN, CHAGAI HILLS, HYDROTHERMAL ALTERATION
B 6309 1494 PAKISTAN, CHAGAI HILLS, VOLCANIC FLOAT
B 6309 1494 PAKISTAN, CHAGAI INTRUSIVES
B 6209 1588 PAKISTAN CONTINENTAL MARGIN, INDIA-
B 6210 1986 PAKISTAN, DEVELOPMENTS, 1977
B 6310 1824 PAKISTAN, DEVELOPMENTS, 1978
B 6411 1890 PAKISTAN, DEVELOPMENTS, 1979
S 10 199 PAKISTAN, MEKRAN COAST, MUD VOLCANO
B 6110 1858 PAKISTAN, PRODUCTION, 1975-1976
B 6210 1990 PAKISTAN, PRODUCTION, 1975-1977
B 6310 1829 PAKISTAN, PRODUCTION, 1977-1978
B 6411 1896 PAKISTAN, PRODUCTION, 1977-1979
B 6309 1494 PAKISTAN, QUATERNARY
B 6309 1494 PAKISTAN, SINJRANI VOLCANICS
B 6407 981 PAKOWKI FORMATION, ALBERTA
B 6311 2116* PAL DURO BASIN, TEXAS, ABST., PENNSYLVANIAN, FAN-, DELTA
B 6209 1763 PALACAS, JAMES G., PRELIMINARY ASSESSMENT OF ORGANIC CARBON
B 6004 584† PALAEOAPLYSINA, HYDROZOAN, CARBONAT BUILDUPS FROM UPPER
M 28 139 PALAGONITE TUFF, PHILLIPSITE, SPHERULE, IN
S 3 99 PALANGANA DOME, TEXAS
B 6209 1654 PALAUI RIDGE
B 6406 927 PALCIAUSKAS, V. V., MICROFRACTURE DEVELOPMENT IN COMPACTING
B 6202 201 PALEO-WATER-DEPTH DETERMINATION
B 6004 646* PALEOBATHYMETR OF CALIFORNIA CONTINENTAL BORDERLAND NORTH
B 6004 704* PALEOBATHYMETRIC HISTORY OF NEW JERSEY COAST PLAIN, ABST.,
B 6202 206 PALEOBATHYMETRY CURVES AND NUMERICAL AGE OF MARKERS OF
S 7 124 PALEOBATHYMETRY, GULF OF MEXICO, MICROFAUNA
B 6205 892* PALEOBIOGEOGRAPHY AND PRESERVATIOON OF MIDDLE CRETACEOUS
B 6004 695* PALEOBIOGEOGRAPHY--IMPLICATIONS CONCERNING PLATE TECTONICS
B 6405 731* PALEOBIOLOGIC EVIDENCE FOR CRETACEOUS TIDES, WESTERN INTERIOR
B 6405 678* PALEOCALICHE IN RED BEDS O DOLORES FORMATION, SOUTHWESTERN
B 6012 2175* PALEOCEANOGRAPHIC VARIATIONS AND THEIR IMPLICATION TO
B 6408 1133 PALEOCENE, ALASKA
B 6408 1142 PALEOCENE, ALASKA
B 6207 1153 PALEOCENE, AUSTRALIA
B 6309 1456 PALEOCENE, BALTIMORE CANYON TROUGH
B 6211 2261 PALEOCENE, CALIFORNIA
B 6206 951 PALEOCENE, CALIFORNIA
B 6205 820 PALEOCENE, CALIFORNIA
B 6211 2217 PALEOCENE, CALIFORNIA, CRETACEOUS-
B 6104 561 PALEOCENE, CHILE
B 6201 125 PALEOCENE, COLUMNAR SECTION, FLORIDA
B 6303 437* PALEOCENE CONGLOMERATES IN NORTH-CENTRAL UTAH THRUST BELT,
B 6209 1767* PALEOCENE, DANIAN, ABST., OSTRACODA OF, PRAIRIE BLUFF, CHALK,
S 6 217 PALEOCENE, DANIAN STAGE DEFINED, EUROPE
B 6306 905 PALEOCENE DELTA, NORTH SEA
M 25 328 PALEOCENE, ECUADOR
M 29 341 PALEOCENE, EVOLUTION, COLOMBIA
M 27 60 PALEOCENE, FAKSE LIMESTONE, DENMARK
M 27 201 PALEOCENE, FAXEKALK, DENMARK
M 30 67 PALEOCENE, FRIGG FIELD, SEDIMENTATION HISTORY
B 6201 121† PALEOCENE, IN SOUTH FLORIDA, REBECCA, SHOAL, REEF COMPLEX,
B 6102 208 PALEOCENE, MEXICO
B 6405 777* PALEOCENE, MIDWAY, CONTINENTAL SHELF DEPOSITS, RIO GRANDE
B 6201 3 PALEOCENE, NIGER DELTA
B 6205 788 PALEOCENE, NORTH, AFRICA
B 6205 777 PALEOCENE, NORTH, AFRICA
B 6203 424 PALEOCENE, NORTH, ATLANTIC
B 6004 732* PALEOCENE, PARRAS BASIN, MEXICO, ABST., TRANSGRESSIVE-,

B 6409 1565* PALEOCENE PLANKTONIC FORAMINIFERAL BISTRATIGRAPHY OF
M 28 141 PALEOCENE, POISON CANYON FORMATION, COLORADO
S 6 227 PALEOCENE, RADIOMETRIC AGES, CORRELATIONS
B 6303 512* PALEOCENE RESERVOIR SANDSTONES IN WEST-CENTRAL PAKISTAN,
B 6004 710* PALEOCENE SANDSTONES OF CANADIAN WESTERN INTERIOR, ABST.,
B 6303 546* PALEOCENE SEDIMENTATION IN WYOMING, ABST., CONTROLS ON LATE,
M 30 543 PALEOCENE, SIRTE BASIN, INTISAR D FIELD
B 6004 716* PALEOCENE, SOUTH-CENTRAL TEXAS, ABST., SEDIMENTOLOGY AND,
M 29 173 PALEOCENE, SOUTHEAST INDIAN RIDGE, SPREADING
M 30 105 PALEOCENE, STATFJORD FIELD, ROGALAND GROUP
S 6 220 PALEOCENE, THANETIAN STAGE DEFINED, EUROPE
B 6405 757* PALEOCENE TIME SCALE FOR ROCKY MOUNTAIN REGION, ABST.
B 6405 613 PALEOCENE TO EOCENE, MIDDLE EAST
B 6305 826* PALEOCENE TONGUE RIVER MEMBER OF FORT UNION FORMATION,
B 6403 424 PALEOCENE, TUNISIA
B 6104 581 PALEOCENE, VENEZUELA
B 6303 311† PALEOCENE--LOWER EOCENE FLAGSTAFF LIMESTONE AND COEVAL
S 4 103 PALEOCENE-EARLY EOCENE, CARIBBEAN
S 6 238 PALEOCENE-EOCENE BOUNDARY, GENERAL, PALEOGENE
B 6408 1282* PALEOCHANNEL ACROSS LOUDEN ANTICLINE, FAYETTE COUNTY,
B 6004 666* PALEOCLIMATE, ABST., PALEOLATITUDE AND
S 9 22 PALEOCLIMATE, CARBONATE REEF DISTRIBUTION
S 9 14 PALEOCLIMATE, COAL, LATITUDE DISTRIBUTION
S 9 15 PALEOCLIMATE, CONTINENTAL MOBILITY
S 9 19 PALEOCLIMATE, GLACIAL PERIODS
B 6004 674* PALEOCLIMATE IN CARBONATE TIDAL FLATS, ABST., SEDIMENTARY
S 9 9 PALEOCLIMATE INDICATORS, ANCIENT WIND DIRECTION
S 9 5 PALEOCLIMATE INDICATORS, BAUXITE
S 9 6 PALEOCLIMATE INDICATORS, COAL
S 9 7 PALEOCLIMATE INDICATORS, FLORA AND FAUNA
S 9 7 PALEOCLIMATE INDICATORS, ICE FLOW DIRECTIONS
B 6004 652* PALEOCLIMATE INDICATORS IN COLOMBIA GROUP OF NEW JERSEY
S 9 9 PALEOCLIMATE INDICATORS, MEASURED PALEOTEMPERATURES
S 9 13 PALEOCLIMATE INDICATORS, PALEOMAGNETIC LATITUDES
S 9 5 PALEOCLIMATE INDICATORS, RED BEDS
S 9 7 PALEOCLIMATE INDICATORS, TILLITES
B 6004 735* PALEOCLIMATE INTERPRETATION THROUGH USE OF SIZE-DEPENDENT
S 9 19 PALEOCLIMATE, NONGLACIAL PERIODS
S 9 15 PALEOCLIMATE, NORTHERN CONTINENTS
S 9 1 **PALEOCLIMATE, PALEOMAGNETISM, AND CONTINENTAL DRIFT**
S 7 247 PALEOCLIMATE, TEXAS, SHELF-SLOPE
S 9 20 PALEOCLIMATES, CAMBRIAN, FOLDOUT
S 9 23 PALEOCLIMATES, CARBONIFEROUS, FOLDOUT
S 9 23 PALEOCLIMATES, CRETACEOUS, FOLDOUT
S 9 23 PALEOCLIMATES, DEVONIAN, FOLDOUT
S 9 23 PALEOCLIMATES, JURASSIC, FOLDOUT
S 9 20 PALEOCLIMATES, MEAN ANNUAL AIR TEMPERATURE
S 9 20 PALEOCLIMATES, MEAN ANNUAL WATER TEMPERATURE
S 9 20 PALEOCLIMATES, ORDOVICIAN, FOLDOUT
S 9 20 PALEOCLIMATES, SILURIAN, FOLDOUT
S 9 25 PALEOCLIMATES, TERTIARY, FOLDOUT
S 9 23 PALEOCLIMATES, TRIASSIC, FOLDOUT
B 6004 675* PALEOCLIMATIC AND PALEOOCEANOGRAPHIC HISTORY IN CALIFORNIA
B 6004 645* PALEOCLIMATIC IMPLICATIONS, ABST., SOME PROBLEMATIC LATE,
B 6004 664* PALEOCLIMATIC INDICATOR, SOUTH SHORE OF LOWER ST. LAWRENCE
B 6004 692* PALEOCLIMATIC INDICATORS, ABST., MARINE, COLD-, WATER,
S 4 291 PALEOCLIMATIC INFORMATION OBTAINED FROM FOSSIL GALAXEA
B 6004 678* PALEOCLIMATIC SIGNIFICANCE OF MORPHOLOGIC VARIATIO IN SPECIES
S 9 13 PALEOCLIMATIC ZONATION, GENERAL
S 7 249 PALEOCLIMATIC ZONES, PLEISTOCENE, GULF OF MEXICO
B 6004 734* PALEOCLIMATOLOGIC INTERPRETATION OF PROTEROZOIC
B 6303 519* PALEOCLIMATOLOGICAL TOOL, ABST., STABLE, CARBON, ISOTOPES A
B 6004 671* PALEOCURRENT ANALYSIS OF ADMIRE, 650 FT, SANDSTONE, PERMIAN, E
B 6008 1405* PALEOCURRENT ANALYSIS OF DAKOTA AND SUBJACENT SANDSTONES,
B 6007 1016 PALEOCURRENTS
B 6102 237† PALEOCURRENTS AND SEDIMENTATION, SA NICOLAS ISLAND,
B 6206 1093* PALEOCURRENTS OF JACKPILE SANDSTONE, BURRO CANYON
B 6405 798* PALEOECOLOGIC DETERMINATION OF BATHYMETRIC POSITION OF
B 6303 494* PALEOECOLOGIC EVALUATION OF LATE EOCENE ZONATIONS OF WEST
B 6303 548* PALEOECOLOGIC IMPLICATION OF FORAMINIFERAL DISTRIBUTIONAL
B 6303 463* PALEOECOLOGIC INTERPRETATION OF ENVIRONMENTAL STABILITY--A
B 6004 657* PALEOECOLOGIC INVESTIGATIONS O LATE HOLOCENE CARBONATE
B 6004 706* PALEOECOLOGIC POTENTIAL OF MICROBORING ORGANISMS, ABST.
B 6004 727* PALEOECOLOGY AND FAUNA OF SOME NIAGARAN REEFS OF WESTERN
B 6209 1761* PALEOECOLOGY AND POROSITY TRENDS OF ABNORMALLY HIGH
B 6405 699* PALEOECOLOGY IN BASIN ANALYSIS—HUMBOLDT BASIN OF CALIFORNIA
B 6409 1569* PALEOECOLOGY OF MIDWAY GROUP IN NORTHEAST TEXAS, ABST.
B 6004 716* PALEOECOLOGY OF TYPE LOCATION OF KINCAID FORMATION, LOWER
B 6303 542* PALEOENVIRONMENT AND PETROLEUM POTENTIAL OF
B 6412 2051† PALEOENVIRONMENT AND PETROLEUM POTENTIAL OF MIDDLE
B 6405 694* PALEOENVIRONMENT CONTROL OF BIOGENIC STRUCTURE IN UPPER
B 6004 677* PALEOENVIRONMENT OF EARLY PERMIAN, ROTLIEGENDE, VERTEBRATE
B 6303 542* PALEOENVIRONMENT OF MIDDLE DEVONIAN SANDSTONES OF UPPER
B 6105 772* **PALEOENVIRONMENT OF MOFFAT MOUND, EDWARDS FORMATION, CENTRAL**
B 6405 679* PALEOENVIRONMENTAL ANALYSIS OF DISCONFORMITY AND
B 6409 1566* PALEOENVIRONMENTAL ANALYSIS OF JOACHIM DOLOMITE, MIDDLE
B 6109 1551* PALEOENVIRONMENTAL ANALYSIS OF LOWER CRETACEOUS CUPIDO
B 6303 438* PALEOENVIRONMENTAL IMPLICATIONS OF OCEANIC SEDIMENTATION
B 6409 1559* PALEOENVIRONMENTS AND TRACE FOSSILS OF LARGE AGGRADING
B 6004 687* PALEOENVIRONMENTS AND TRACE FOSSILS OF UPPER CRETACEOUS
B 6303 354† PALEOENVIRONMENTS O MOWRY SHALE, LOWER CRETACEOUS, WESTERN
B 6004 723* PALEOENVIRONMENTS OF FRONTIER FORMATION, UPPER CRETACEOUS,
B 6004 724* PALEOENVIRONMENTS OF WESTERN INTERIOR CRETACEOUS MARINE
B 6407 1087† PALEOGENE AND NEOGENE OF NORTHEAST ASIA, CHUKOTKA,
S 6 209 PALEOGENE, AQUITAINE BASIN, CORRELATION
S 6 216 PALEOGENE, BIOSTRATIGRAPHY, A REVIEW
S 6 205 PALEOGENE BOUNDARIES, ABSOLUTE DATING, A REVIEW
B 6301 68 PALEOGENE CANYONS, CZECHOSLOVAKIA
S 6 214 PALEOGENE, DATING, A REVIEW
S 6 216 PALEOGENE, DEPOSITION, EUROPE
B 6405 698* PALEOGENE DEPOSITIONAL SYSTEMS, WESTERN TRANSVERSE RANGES,
S 6 238 PALEOGENE, EOCENE, EARLY-MIDDLE BOUNDARY
S 6 252 PALEOGENE, EOCENE, LUTETIAN STAGE, ISOTOPIC DATES
S 6 241 PALEOGENE, EOCENE, MIDDLE-LATE BOUNDARY
S 6 253 PALEOGENE, EOCENE-OLIGOCENE BOUNDARY
S 6 242 PALEOGENE, EOCENE-OLIGOCENE BOUNDARY
S 6 254 PALEOGENE, EOCENE-OLIGOCENE BOUNDARY, DIAGRAM
S 6 217 PALEOGENE, FORAMINIFERA, BERGGRENS ZONES, A REVIEW
S 6 237 PALEOGENE, GEOCHRONOLOGY, COMPARISON DIAGRAM
B 6209 1755* PALEOGENE GEOLOGY AND UNCONFORMITIES, COASTAL PLAIN OF
B 6402 213 PALEOGENE, INDIA
B 6402 213 PALEOGENE, INDIA
S 6 235 PALEOGENE, ISOTOPIC DATING, CHOICE OF A SPECIMEN
S 6 236 PALEOGENE, ISOTOPIC DATING, CONSIDERATIONS
S 6 236 PALEOGENE, ISOTOPIC DATING, DECAY OF CONSTANTS
S 6 209 PALEOGENE, NORTH GERMAN BASIN, CORRELATION
S 6 213† PALEOGENE NUMERICAL TIME SCALE, A NEW
B 6405 687* PALEOGENE, OF FOZ DO AMAZONAS BASIN, OFFSHORE NE BRAZIL, ABST.,
S 6 253 PALEOGENE, OLIGOCENE-EOCENE BOUNDARY
S 6 242 PALEOGENE, OLIGOCENE-MIOCENE BOUNDARY
S 6 255 PALEOGENE, OLIGOCENE-MIOCENE BOUNDARY, DIAGRAM
S 6 205† PALEOGENE, ON DATING OF THE
S 6 238 PALEOGENE, PALEOCENE-EOCENE BOUNDARY, GENERAL
S 6 222 PALEOGENE, PALEOMAGNETIC REVERSALS, ATLANTIC
S 6 222 PALEOGENE, PALEOMAGNETIC REVERSALS, TABLE
S 11 109 PALEOGENE PLANKTON ZONATION, CORRELATIONS
S 6 217 PALEOGENE, PLANKTONIC BIOSTRATIGRAPHY, REVIEW
S 6 139 PALEOGENE, PROPOSED RADIOMETRIC SCALE, EUROPE
S 6 256 PALEOGENE, PROPOSED TIME SCALE
B 6004 695* PALEOGENE RADIOLARIAN PALEOBIOGEOGRAPHY--IMPLICATIONS
S 6 227 PALEOGENE, RADIOMETRIC AGES, DIAGRAM
S 6 213 PALEOGENE, RADIOMETRIC DATES, RANGES
S 6 247 PALEOGENE, RADIOMETRIC DATING, A REVIEW
S 6 248 PALEOGENE, RADIOMETRIC DATING, CHOICE OF CONSTANTS
S 6 250 PALEOGENE, RADIOMETRIC DATING, COMMENTS
S 4 102 PALEOGENE REEFS, CARIBBEAN
B 6305 828* PALEOGENE ROCKS, ABST., APPLICATION OF NONMARINE, MOLLUSCA TO
M 29 132 PALEOGENE SEDIMENTS, ARGENTINA, FALKLAND PLATEAU
S 6 127† PALEOGENE SEDIMENTS, EUROPE, RESULTS OF DATING, CRETACEOUS
S 6 215 PALEOGENE, STAGE STRATOTYPES, EUROPE
S 6 240 PALEOGENE, STRATIGRAPHIC SCALE, DIAGRAM
S 6 235† PALEOGENE STRATOTYPES, CRITICAL REVIEW OF, ISOTOPIC DATES IN
S 6 214 PALEOGENE, TIME SCALE, DIAGRAM
S 6 247† PALEOGENE TIME SCALE, ISOTOPIC DATES FOR A
S 6 218 PALEOGENE, TIME SCALE, STRATOTYPES, DIAGRAM
B 6405 769* PALEOGEOGRAPH OF NORTHERN ARIZONA DURING DEPOSITION OF
B 6404 527† PALEOGEOGRAPHIC AND STRATIGRAPHIC DIFFERENTIATIO OF SILURIAN
B 6405 687* PALEOGEOGRAPHIC EVOLUTION OF LATE PALEOZOIC TAOS TROUGH,
B 6004 667* PALEOGEOGRAPHIC EXTENT O BUILDUPS AND ORGANIC REEFS DURING
B 6401 133 PALEOGEOGRAPHIC PRINCIPLES OF OI AND GAS PROSPECTING, BY N. I.
B 6211 2349* PALEOGEOGRAPHIC SIGNIFICANCE OF MIDDLE TO LATE JURASSIC
B 6405 782* PALEOGEOGRAPHY AND MARINE COMMUNITIES OF SILURIAN
B 6405 730* PALEOGEOGRAPHY AND STRUCTURAL DEVELOPMENT OF LATE
B 6405 721* PALEOGEOGRAPHY OF EUSTATIC MODEL FOR DEPOSITION OF
B 6209 1607† PALEOGEOGRAPHY OF SOME SILURIAN AND DEVONIAN REEF TRENDS,
B 6408 1218 PALEOGEOMORPHIC TRAP, DIAGRAM
B 6008 1391* PALEOGEOMORPHOLOGY OF FRED AND GEORGE CREEK AREA, TOOLE
B 6205 879* PALEOGRAVIMETER, ABST., ANALYSIS OF EFFECT ON, GRADED BEDS OF

B 6405 674* PALEOGRAVIMETRIC CAMERAS, ABST., GRADED BEDS AS
B 6305 826* PALEOKARST CONTROLS O LOCALIZATION OF URANIUM AT PITCH MINE,
B 6004 666* PALEOLATITUDE AND PALEOCLIMATE, ABST.
B 6203 537* PALEOLIMNOPHYCUS—STROMATOLITE-FORMING ALGA FROM EOCENE
S 6 146 PALEOMAGNETIC EVIDENCE, JAPAN, MESOZOIC
S 9 13 PALEOMAGNETIC LATITUDES, GENERAL
S 9 13 PALEOMAGNETIC LATITUDES, PALEOCLIMATE INDICATORS
S 6 222 PALEOMAGNETIC REVERSALS, ATLANTIC, PALEOGENE
S 6 222 PALEOMAGNETIC REVERSALS, TABLE, PALEOGENE
S 6 289 PALEOMAGNETIC SCALES, A REVIEW, JURASSIC
S 6 185 PALEOMAGNETIC SCALES, PARATETHYS, NEOGENE
S 6 15 PALEOMAGNETIC SCALES, REVIEW
S 6 311 PALEOMAGNETIC STRATIGRAPHY, A REVIEW, PERMIAN
S 9 17 PALEOMAGNETISM, AFRICA, EVIDENCES OF DRIFT
B 6103 468* PALEOMAGNETISM AND PALEOGEOGRAPHIC RECONSTRUCTION OF
S 9 17 PALEOMAGNETISM, ANTARCTICA, EVIDENCES OF DRIFT
S 9 17 PALEOMAGNETISM, EURASIA, EVIDENCES OF DRIFT
S 9 17 PALEOMAGNETISM, NORTH AMERICA, EVIDENCES OF DRIFT
S 9 16 PALEOMAGNETISM, PERIODIC RAPID POLAR SHIFTS
M 26 106 PALEONTOLOGIC CORRELATION, SEISMIC, REFLECTIONS
B 6012 2187* PALEONTOLOGIC STUDIES OF MIDDLE TERTIARY SKOONER GULCH AND
B 6004 713* PALEONTOLOGY, AND STRUCTURAL GEOLOGY IN PETROLEUM
B 6407 1117 PALEONTOLOGY, EDITED BY RHODES W. FAIRBRIDGE AND DAVID
S 11 4† PALEONTOLOGY OF WEST COAST TERTIARY, HISTORY OF
B 6203 568* PALEONTOLOGY, UNIFORMITARIANISM, AND THE
B 6004 675* PALEOOCEANOGRAPHIC HISTORY IN CALIFORNIA CONTINENTAL
B 6004 554† PALEOPORE PRESSURE, AND PALEOTEMPERATURE, SOUTHWESTERN
B 6305 812† PALEOPORE PRESSURE, AND PALEOTEMPERATURE, SOUTHWESTERN
B 6107 1087 PALEOPORE PRESSURES, EVIDENCE OF
B 6004 654* PALEOSEDIMENTOLOGIC CORRELATION OF OROGENIC AND
S 7 30 PALEOSHELF EDGES, MAP, GULF COAST BASIN
M 29 45 PALEOSHELF, ICELAND
B 6204 600 PALEOSHELF-EDGE HIGH, POTENTIAL TRAP, U.S., ATLANTIC SLOPE
B 6004 645* PALEOSOLS AND THEIR GEOTECTONIC AND PALEOCLIMATIC
B 6305 841* PALEOSOLS AS ENVIRONMENTAL INDICATORS IN NONMARINE
B 6209 1756* PALEOSTREAM CHANNEL OF ANCIENT NUECES RIVER, SOUTH TEXAS,
B 6209 1754* PALEOSTRUCTURAL MAPPING IN BATHYAL ENVIRONMENTS, ABST.
B 6104 489 PALEOSTRUCTURE TRAPS, TILTED-OUT, SALT-GENERATED, JURASSIC
B 6105 811* PALEOTECTONIC IMPLICATIONS OF SUBAQUEOUS FLOW STRUCTURES IN
B 6004 688* PALEOTEMPERATURE DETERMINATION, ABST., SABKHA FACIES,
B 6004 554† PALEOTEMPERATURE, SOUTHWESTERN PART OF WESTERN CANAD
B 6305 812† PALEOTEMPERATURE, SOUTHWESTERN PART OF WESTERN CANADA
B 6303 415* PALEOTIDAL-RANGE INDICATORS IN CARBONIFEROUS BARRIER
B 6405 701* PALEOTOPOGRAPHYS INFLUENCE ON POROSITY DISTRIBUTION IN
M 29 340 PALEOTRENCHES, BOLIVAR LINEAMENT, COLOMBIA
M 29 341 PALEOTRENCHES, COLOMBIA LINEAMENT, COLOMBIA
M 29 340 PALEOTRENCHES, ROMERAL LINEAMENT, COLOMBIA
M 29 340 PALEOTRENCHES, SINU LINEAMENT, COLOMBIA
M 25 295 PALEOZOIC, ALASKA
B 6307 1059† PALEOZOIC ALLEN BAY CARBONATE ROCKS AND CONTIGUOUS SHELF
M 25 509 PALEOZOIC ALLOCHTHON, FORMATION OF, CORDILLERAN OROGEN
M 25 94† PALEOZOIC AND MESOZOIC COAL IN REPUBLIC OF KOREA
B 6101 10 PALEOZOIC AND MESOZOIC SEDIMENTS, DEPOCENTERS, CANADIAN,
M 25 445 PALEOZOIC AND OTHER METALLOGENIC PROVINCES, SOUTH KOREA
B 6405 719* PALEOZOIC AND TRIASSIC ROCKS, CENTRAL GREAT BASIN, ABST.,
M 28 32 PALEOZOIC, ANDALUSITE SCHIST, NEW HAMPSHIRE
M 28 48 PALEOZOIC, ANDALUSITE SCHIST, NEW HAMPSHIRE
M 28 8 PALEOZOIC, ANDALUSITE SCHIST, NEW HAMPSHIRE
M 28 49 PALEOZOIC, ANDALUSITE SCHIST, NEW HAMPSHIRE
B 6107 1063 PALEOZOIC BLACK SHALE AND PHOSPHORITE, MID-CONTINENT
M 28 80 PALEOZOIC, CABALLOS NORACULITE, TEXAS
B 6004 662* PALEOZOIC CARBONATE ROCKS, ARCTIC ARCHIPELAGO, ABST., EARLY,
B 6203 558* PALEOZOIC CARBONATE SUCCESSION IN NORTHWESTERN BRITISH
B 6408 1168 PALEOZOIC, CARBONATE WEDGES, UTAH AND, COLORADO
B 6004 655* PALEOZOIC CLASTIC DEPOSITION IN MARANHAO BASIN, NORTHEAST
B 6203 503* PALEOZOIC CLASTIC ROCKS OF MIDDLE AMAZON BASIN, BRAZIL, ABST.,
M 25 496 PALEOZOIC CLASTIC WEDGE, CANADIAN, CORDILLERA, MID-
B 6405 793* PALEOZOIC CONODONT BIOSTRATIGRAPHY AND PALEOGEOGRAPHY OF
B 6405 676* PALEOZOIC CONODONT BIOSTRATIGRAPHY, BIOGEOGRAPHY, AND
B 6002 299† PALEOZOIC CRATONIC MARGIN. DISCUSSION AND REPLY, SUBSURFACE,
M 25 511 PALEOZOIC CRUST FILLS RIFT ZONE, CORDILLERAN OROGEN, NORTHERN
M 26 217 PALEOZOIC, DEPOSITIONAL SYSTEMS, FORELAND BASIN, LATE
M 26 217 PALEOZOIC, DEPOSITIONAL SYSTEMS, FORT WORTH BASIN, LATE
M 26 217 PALEOZOIC, DEPOSITIONAL SYSTEMS, MICHIGAN BASIN, LATE
B 6309 1588* PALEOZOIC DISRUPTIVE DEFORMATION IN NORTH AMERICAN
S 9 16 PALEOZOIC, EURASIA, MAGNETIC POLE POSITIONS
M 25 559 PALEOZOIC GEOSYNCLINAL CYCLE, CHILE
B 6405 807* PALEOZOIC IN YUNNAN-GUIZHOU-GUANGXI PROVINCES, CHINA, ABST.,
M 25 534 PALEOZOIC METALLIFEROUS DEPOSIT, ANDEAN REGION, PRECAMBRIAN
S 7 30 PALEOZOIC, NORTHERN GULF OF MEXICO RIM, EXPOSED
B 6405 798* PALEOZOIC OF APPALACHIANS, ABST., PALEOECOLOGIC DETERMINATION
B 6004 584† PALEOZOIC OF IDAHO, PALAEOAPLYSINA, HYDROZOAN, CARBONAT,
M 25 428 PALEOZOIC, ORE DEPOSITS ASSOCIATED WITH, AUSTRALIA
B 6405 785* PALEOZOIC PALEOGEOGRAPHY OF IDAHO, ABST., UPPER
B 6405 681* PALEOZOIC PLUTONIC ACTIVITY IN FLORIDA MOUNTAINS, NEW MEXICO,
B 6405 771* PALEOZOIC POTENTIAL OF PARADOX BASIN, ABST., LOWER AND MIDDLE
B 6205 885* PALEOZOIC ROCKS IN EASTERN GREAT BASIN, ABST., REVIEW OF,
M 28 56 PALEOZOIC, SCHIST, MASSACHUSETTS
B 6009 1620* PALEOZOIC SEDIMENTATIO IN EASTERN KENTUCKY, ABST., STRUCTURAL
B 6007 1125 PALEOZOIC SHALE, ELECTRON-SCANNING PHOTOMICROGRAPH,
B 6305 826* PALEOZOIC STRATA IN WEST-CENTRAL SASKATCHEWAN, ABST.,
B 6405 687* PALEOZOIC TAOS TROUGH, NORTHERN NEW MEXICO, ABST.,
B 6303 461* PALO DURO BASIN, TEXAS, ABST., EXPLORATION POTENTIAL OF,
B 6405 702* PALO DURO BASIN, TEXAS, ABST., SOURCE-ROCK QUALITY AND,
B 6303 442* PALO DURO BASIN, TEXAS PANHANDLE, ABST., FACIES PATTERNS AND
B 6401 88† PALO DURO BASIN, TEXAS, PENNSYLVANIAN-EARLY, PERMIAN
B 6303 511* PALO DURO BASIN, TEXAS--FACIES PATTERNS THROUGH TIME, ABST.,
B 6303 461* PALO DURO BASINS, TEXAS PANHANDLE, ABST., LOWER, PERMIAN FACIES
B 6301 54 PALO PINTO LIMESTONE, TEXAS
B 6002 327* PALO PINTO, PARKER, AND WISE COUNTIES, NORTH-CENTRAL TEXAS,
B 6102 177 PALOMA FIELD, STEVENS SAND, CALIFORNIA
B 6104 563 PALOMARES FORMATION, CHILE
B 6104 561 PALOMARES STRATIGRAPHIC UNIT, CHILE
B 6207 1172 PALOMINO RIDGE, SECTION, NEVADA
B 6202 252 PALOS VERDES FAULT, CALIFORNIA
B 6202 255 PALOS VERDES HILLS ANTICLINORIUM, CALIFORNIA
B 6202 259 PALOS VERDES HILLS AREA, COLUMN, CALIFORNIA
B 6202 248 PALOS VERDES HILLS, CALIFORNIA
B 6111 2001† PALOS VERDES HILLS, CALIFORNIA, FAULTED UPPER, PLEISTOCENE,
B 6203 361 PALOS VERDES HILLS FAULT, CALIFORNIA
B 6202 259 PALOS VERDES SAND, CALIFORNIA
B 6305 839* PALUDAL-LACUSTRINE ORIGIN IN WASATCH FORMATION OF
M 24 242 PALUXY PRODUCING SAND, MISSISSIPPI
B 6408 1220 PALUXY SECTION, TEXAS
M 28 72 PALYGORSKITE, SEM, CLAY
B 6311 2120* PALYNOLOGIC AND PLANT COMPRESSION EVIDENCE FOR
B 6405 715* PALYNOLOGY PREPARATIONS, ABST., FILAMENT-PRODUCING
B 6303 502* PALYNOLOGY TO AGE DETERMINATION, CORRELATION, AND
B 6405 738* PALYNOMORPH COLOR ALTERATION STUDIES TO THERMAL
B 6403 440* PALYNOMORPHS FROM HOLOCENE SEDIMENTS OF BASINS IN SOUTHERN
B 6004 660* PALYNOMORPHS IN PETROLEUM AND OTHER BITUMINOUS SUBSTANCES,
B 6009 1614 PAMPE, WILLIAM R., NEW PELECYPOD SPECIES FROM LOWER
B 6403 434 PAMPE, WILLIAM R., REVIEW OF ENERGY AND CLIMATE
B 6203 503 PAMPLONA, H. R. P., PETROLEUM GEOLOGY OF PALEOZOIC CLASTIC
B 6204 635 PAMPLONA RIDGE, ALASKA
B 6306 426* PAN-PACIFIC NEOGENE DIATOM CORRELATIONS, ABST.
B 6010 1654 PANAMA, 1975, PETROCONSULTANTS S.A., ON DEVELOPMENTS IN
B 6110 1590 PANAMA, 1976, CARVAJAL, J., ON DEVELOPMENTS IN
B 6210 2057 PANAMA, 1977, PETROCONSULTANTS S.A., ON DEVELOPMENTS IN
B 6310 1757 PANAMA, 1978, PETROCONSULTANTS S.A., ON DEVELOPMENTS IN
S 4 285 PANAMA, AROTHRON MELEAGRIS
B 6304 668† PANAMA CITY, FLORIDA, GAS, SEEPS AND SUBSURFAC STRUCTURE OFF
B 6303 409* PANAMA CITY, FLORIDA. ABST., SEISMIC STRATIGRAPHY AND
B 6007 1054† PANAMA, EVOLUTION OF, MODERN, CARIBBEAN, FRINGING, REEF,
B 6007 1061 PANAMA, GALETA CORES
B 6007 1055 PANAMA, GALETA POINT, DRILL HOLES
B 6007 1057 PANAMA, GALETA POINT, REEF ACROPORA PALMATA
B 6007 1056 PANAMA, GALETA POINT, REEF ASSEMBLAGES
B 6007 1054 PANAMA, GALETA REEF
B 6007 1055 PANAMA, GALETA REEF, ACANTHOPHORA ZONE
B 6007 1061 PANAMA, GALETA REEF, ACCUMULATION RATES
B 6007 1066 PANAMA, GALETA REEF, CEMENTS
B 6007 1055 PANAMA, GALETA REEF, LAURENCIA ZONE
B 6007 1069 PANAMA, GALETA REEF, OTHER COMPONENTS
B 6007 1062 PANAMA, GALETA REEF, RADIOCARBON-DATED SAMPLES
B 6007 1059 PANAMA, GALETA REEF, SECTIONS
B 6007 1055 PANAMA, GALETA REEF, THALASSIA ZONE
B 6007 1055 PANAMA, GALETA REEF, ZOANTHUS ZONE
B 6007 1058 PANAMA, GATUN FORMATION
S 4 176 PANAMA, MONTASTREA ANNULARIS AND M. CAVERNOSA
B 6312 2206 PANAMA, OFFSHORE PROFILE, PACIFIC
S 4 285 PANAMA, SCARUS CROICENSIS
B 6202 223 PANGAEA, A PROSPECT OF
B 6109 1490 PANGAEA, BREAKUP OF
B 6101 93 PANGAEA, RECONSTRUCTION OF
B 6305 835* PANHANDLE, ABST., DEPOSITIONAL FACIES AND, URANIUM
B 6303 442* PANHANDLE, ABST., FACIES PATTERNS AND DEPOSITIONAL MODELS,
B 6303 461* PANHANDLE, ABST., LOWER, PERMIAN FACIES TRACTS AND
B 6205 856 PANHANDLE AND OKLAHOMA, ANADARKO BASIN, BASEMENT CONTOUR
B 6205 855 PANHANDLE AND OKLAHOMA, ULTRADEEP WELLS, LOCATION, TEXAS

B 6205 859 PANHANDLE AND OKLAHOMA, ULTRADEEP WILDCATS, SUCCESS RATIOS,
B 6108 1216† PANHANDLE OF TEXAS, DEVELOPMENTS IN, OKLAHOMA AND
B 6008 1251† PANHANDLE OF TEXAS IN 1975, DEVELOPMENTS IN, OKLAHOMA AND
B 6208 1372† PANHANDLE OF TEXAS IN 1977, DEVELOPMENTS IN, OKLAHOMA AND
B 6308 1286† PANHANDLE OF TEXAS IN 1978, DEVELOPMENTS IN, OKLAHOMA AND
B 6409 1387† PANHANDLE OF TEXAS IN 1979, DEVELOPMENTS IN, OKLAHOMA AND
B 6308 1425* PANHANDLE TEXAS, ABST., DEPOSITIONAL MODELS AND RESOURCE
B 6002 322* PANHANDLES, ABST., REPORT--, OKLAHOMA AND, TEXAS
B 6303 370 PANNELLA, GIORGIO AND, SIDERITE SPHERULITES OF SAN SEBASTIAN
B 6405 683* PANNONIAN AND RELATED BASINS, ABST., FORELAND FOLD AND,
S 6 187 PANNONIAN STAGE, PARATETHYS, NEOGENE
B 6002 318* PANOMA GAS AREA, SOUTHWESTER KANSAS, ABST., GEOLOGY OF
S 3 34 PANTHER MOUNTAIN CIRCULAR, NEW YORK
M 28 120 PANTHER TONGUE, UTAH, CRETACEOUS
B 6004 675 PAO, G. A., HIGH-RESOLUTION STUDIES OF PALEOCLIMATIC AND
B 6310 1756 PAPALOAPAN BASIN, MEXICO
B 6102 222 PAPALOAPAN CANYON, MEXICO
B 6205 779 PAPUA, MIOCENE, GULF OF
B 6207 1221* PAPUA NEW GUINEA, ABST., OK TEDI, COPPER PROJECT
B 6110 1882 PAPUA NEW GUINEA, DEVELOPMENTS, 1976
B 6210 1901 PAPUA NEW GUINEA, DEVELOPMENTS, 1977
B 6310 1682 PAPUA NEW GUINEA, DEVELOPMENTS, 1978
B 6411 1779 PAPUA NEW GUINEA, DEVELOPMENTS, 1979
S 4 291 PAPUA, PORT MORESBY, GALAXEA
B 6205 779 PAPUA, SEISMIC DATA, GULF OF
B 6104 585 PAR-E SIAH FIELD, PERSIAN GULF
B 6202 286 PARA-MARANHAO BASIN OFF, BRAZIL, NORTHERN
B 6306 940 PARA-MARANHAO PLATFORM, AMAZON MARGIN AREA
B 6305 824* PARACHUTE CREEK MEMBER OF GREEN RIVER FORMATION,
B 6203 550* PARACHUTE CREEK MEMBER OF GREEN RIVER FORMATION, DOUGLAS
M 26 64 PARACYCLES, SEA LEVEL FLUCTUATIONS
S 4 90 PARACYCLOSERIS-TROCHOSERIS-ACTINACIS ASSEMBLAGE, JAMAICA
B 6104 603 PARADISE FORMATION, NEW MEXICO
B 6405 771* PARADOX BASIN, ABST., LOWER AND MIDDLE, PALEOZOIC POTENTIAL OF
B 6405 764* PARADOX BASIN, UTAH, ABST., SEISMIC MODELING OF, PENNSYLVANIAN,
M 27 10 PARADOX FORMATION, UTAH, PENNSYLVANIAN
M 27 34 PARADOX FORMATION, UTAH, PENNSYLVANIAN
M 27 61 PARADOX FORMATION, UTAH, PENNSYLVANIAN
M 27 133 PARADOX FORMATION, UTAH, PENNSYLVANIAN
M 27 140 PARADOX FORMATION, UTAH, PENNSYLVANIAN
B 6408 1172 PARADOX SALT, UTAH AND, COLORADO
S 10 27 PARAFFINS, DISTRIBUTION, WILLISTON BASIN
S 10 21 PARAFFINS IN WATER, MIGRATION, SOLUBILITIES OF
B 6108 1384* PARAGENESIS OF AUTHIGENIC MINERALS IN ST. PETER SANDSTONE,
B 6303 504* PARAGENETIC AND STABILITY RELATIONS AMONG AUTHIGENIC
B 6010 1654 PARAGUAY, 1975, JACOBSEN, P., JR., ON DEVELOPMENTS IN
B 6110 1590 PARAGUAY, 1976, PETROCONSULTANTS S.A., ON DEVELOPMENTS IN
B 6210 2058 PARAGUAY, 1977, PETROCONSULTANTS S.A., ON DEVELOPMENTS IN
B 6310 1758 PARAGUAY, 1978, JACOBSEN, P., JR., ON DEVELOPMENTS IN
B 6109 1471 PARAIBA BASIN, BRAZIL
B 6302 221 PARAIBA FRACTURE ZONE, BRAZIL
M 30 365 PARAJE SOLO FORMATION, CHIAPAS-TABASCO PROVINCE
M 26 123 PARALLEL, SEISMIC, REFLECTIONS, CONFIGURATION PATTERNS
B 6307 1021 PARALLEL TRANSFORM SET, GULF OF CALIFORNIA
M 26 253 PARALLEL-BEDDED REFLECTORS, STRONG-REFLECTOR FACIES
M 26 122 PARAMETERS, GEOLOGIC SIGNIFICANCE, REFLECTION
M 26 118 PARAMETERS, INTERPRETATION, SEISMIC FACIES
M 26 121 PARAMETERS, SEISMIC, FACIES
B 6104 547 PARANA BASIN, BRAZIL
B 6109 1471 PARANA BASIN, BRAZIL
B 6206 1033 PARANA BASIN, BRAZIL
S 6 105 PARANA BASIN, ESTRADA NOVA FORMATION, BRAZIL
S 6 104 PARANA BASIN, IRATI FORMATION, BRAZIL
S 6 106 PARANA BASIN, PONTA GROSSA FORMATION, BRAZIL
S 6 106 PARANA BASIN, RIO BONITO FORMATION, BRAZIL
S 6 102 PARANA BASIN, RIO DO RASTO FORMATION, BRAZIL
M 25 480 PARASHELVES, DEFINED
S 3 32 PARASHOUT WASH, SHIVWITS PLATEAU, ARIZONA, TOROWEAP
S 6 179 PARATETHYS, DESCRIPTION OF AREA, EUROPE
S 6 184 PARATETHYS, NEOGENE, BADENIAN STAGE
S 6 180 PARATETHYS, NEOGENE, EGERIAN STAGE
S 6 180 PARATETHYS, NEOGENE, EGGENBURGIAN STAGE
S 6 184 PARATETHYS, NEOGENE, KARPATHIAN STAGE
S 6 180 PARATETHYS, NEOGENE, OTTNANGIAN STAGE
S 6 185 PARATETHYS, NEOGENE, PALEOMAGNETIC SCALES
S 6 187 PARATETHYS, NEOGENE, PANNONIAN STAGE
S 6 181 PARATETHYS, NEOGENE, RADIOMETRIC DATES, TABLE
S 6 182 PARATETHYS, NEOGENE, RADIOMETRIC SAMPLES, TABLE
S 6 185 PARATETHYS, NEOGENE, RADIOMETRIC SCALES
S 6 189 PARATETHYS, NEOGENE, SAMPLE ANALYSIS, 12 PAGES
S 6 186 PARATETHYS, NEOGENE, SARMATIAN STAGE
S 6 179† PARATETHYS REGION, A, RADIOMETRIC TIME SCALE FOR THE, NEOGENE
M 25 339 PARDO, ALFREDO, HYDROCARBON POTENTIAL OF AMAZON BASINS OF
M 25 333 PARDO, ALFREDO, HYDROCARBON POTENTIAL OF COASTAL BASINS OF
B 6405 681 PARDO, GEORGES, HYDROCARBON OCCURRENCES IN NONMARINE,
M 25 135 **PARDYANTO, L., POSSIBLE ENERGY SOURCES IN INDONESIAN VOLCANIC**
S 10 154 PARENTIS ARTESIAN BASIN, PETROLEUM MIGRATION, FRANCE
S 10 27 PARENTIS BASIN, AROMATIC DISTRIBUTION, FRANCE
S 10 27 PARENTIS BASIN, NSO DISTRIBUTION, FRANCE
S 10 27 PARENTIS BASIN, SATURATE DISTRIBUTION, FRANCE
M 30 245 PARFENOVA HORIZON, USSR, MARKOVO FIELD
B 6109 1476 PARGO FIELD, BRAZIL
S 4 18 PARGUERA STUDY AREA, PUERTO RICO
S 6 132 PARIS BASIN, ALBIAN, POTASSIUM-ARGON, FRANCE
S 6 132 PARIS BASIN, CENOMANIAN, POTASSIUM-ARGON, FRANCE
B 6104 511 PARIS BASIN, CHAILLY OIL FIELD, FRANCE
B 6411 1964 PARIS BASIN, FRANCE
B 6210 1911 PARIS BASIN, FRANCE
B 6202 202 PARIS BASIN, QUALITATIVE GEOHISTORY DIAGRAM, FRANCE
B 6401 61 PARIS BASIN, TOARCIAN, KEROGEN, FRANCE
B 6110 1641 PARIS BASIN-WESTERN APPROACHES, FRANCE
B 6004 612† PARIS, FRANCE, ANTIPATHETIC RELATIONS AMON, ALGAL STRUCTURES,
B 6307 1040 PARIS THRUST, WYOMING
B 6201 100 PARK CITY FORMATION, IDAHO
B 6201 100 PARK CITY FORMATION, WYOMING
B 6207 1178 PARK CITY GROUP, CORRELATIONS, FOSSILS, NEVADA AND, UTAH
B 6207 1171† PARK CITY GROUP, PERMIAN, IN EASTERN NEVADA AND WESTERN
B 6008 1412* PARK COUNTY, MONTANA, ABST., CORWIN SPRINGS KNOWN,
B 6108 1385 PARK, GERALD M., OIL POTENTIAL OF MESOZOIC SEDIMENTS BENEATH
B 6004 705 PARK, R. K., SEDIMENTATION HISTORY OF LATE CARBONIFEROUS BASIN,
B 6302 195 PARK RANGE, COLORADO
B 6002 327* PARKER, AND WISE COUNTIES, NORTH-CENTRAL TEXAS, ABST.,
B 6301 50† PARKER, AND WISE COUNTIES, NORTH-CENTRAL TEXAS, SUBSURFACE
B 6004 705 PARKER, CALVIN A., GEOPRESSURES AND SECONDARY POROSITY IN DEEP
B 6405 682 PARKER, DOUGLAS M., UNUSUAL PONDING OF SEDIMENTS ON
B 6405 762 PARKER, JOHN M., MONDAK MISSISSIPPIAN OIL FIELD, WILLISTON BASIN,
B 6212 2511 PARKER, K. A., GEOCHEMICAL EVALUATION OF BONAPARTE GULF—
B 6311 2021 PARKER, KEVIN A., GEOCHEMICAL EVALUATION OF PETROLEUM SOURCE
B 6311 2119 PARKER, R. G., STIMULATION DESIGN FOR UPPER MORROW RESERVOIRS
B 6309 1610 PARKER, ROBERT H., ORIGINS OF PERMANEN INLETS SEPARATING
B 6211 2349 PARKINSON, GARY A., PALEOGEOGRAPHIC SIGNIFICANCE OF MIDDLE TO
B 6403 318 PARKMAN SANDSTONE MEMBER, MONTANA
B 6105 819 PARKS, DONALD A., RESERVES, ABST., FINANCING
B 6105 819 PARKS, J. M., VUGGY POROSITY IN CARBONATE-MUDBANK BUILDUPS,
B 6004 705 PARKS, JAMES M., GRANULOMETRIC RELATIONS WITH
B 6204 646 PARKWOOD FORMATION, ALABAMA
S 4 243 PARODOX BASIN, HERMOSA GROUP
B 6004 732* PARRAS BASIN, MEXICO, ABST., TRANSGRESSIVE-, REGRESSIVE DEPOSITS
B 6303 507 PARRISH, JUDITH TOTMAN,, ATMOSPHERIC CIRCULATION AND
B 6405 763 PARRISH, JUDITH TOTMAN, CLIMATE ASYMMETRY AND BIOGEOGRAPHIC
S 4 281† PARROT FISH AN SEA URCHINS ON CARIBBEAN REEFS,
B 6101 19 PARRY ISLANDS FOLDBELT, CANADIAN, ARCTIC ISLANDS
B 6406 961 PARSON, ELMER S., RED WING CREEK FIELD—COSMIC IMPACT
B 6105 840* PARSONS AND PETERSBURG STRUCTURAL LINEAMENTS, WEST VIRGINIA,
B 6412 2168 PARSONS LINEAMENT, WEST VIRGINIA
B 6203 501* PARTIAL ANALOGS, ABST., CONSTRUCTION OF PROCESS-RESPONSE,
B 6408 1223† PARTIAL ANALOGS, CONSTRUCTION OF, PROCESS-RESPONSE MODELS
B 6206 968 PARTINGTON CANYON, CALIFORNIA, LUCIA-
M 26 453 PARVIN LIMESTONE, OKLAHOMA, MISSISSIPPIAN
B 6207 1119 PASKAPOO FORMATION, ALBERTA
B 6408 1166 PASKAPOO-EDMONTON FACIES, ALBERTA
B 6303 555* PASO DEL RIO, COLIMA, MEXICO, ABST., IMMANITAS BED
B 6211 2218 PASO ROBLES HIGH, CALIFORNIA
B 6111 1959 PASSIVE ATLANTIC-TYPE MARGINS, RESERVES
M 26 246 PASSIVE BASIN-MARGINS, DEPOSITIONAL SYSTEMS
M 29 443† PASSIVE CONTINENTAL MARGINS, EXPLOSION, SEISMOLOGY STUDIES OF,
M 29 3† PASSIVE CONTINENTAL MARGINS, SUBSIDENCE MECHANISMS AT
B 6404 501† PASSIVE MARGIN, GEOLOGY OF, NEW ENGLAND
M 26 244 PASSIVE OFFLAP, MODEL, DEPOSITIONAL SYSTEMS
B 6309 1578* PAST AND CURRENT DEVELOPMENTS FOR DRILLING NEW ALBANY
B 6409 1356 PASTOR FIELD, FORT BEND COUNTY, TEXAS
B 6004 678* PATAGONIAN OIL SPILL, ABST., GREAT
S 4 186 PATCH REEF, BAHAMAS, JOULTERS CAYS
B 6004 677* PATCH REEF, BELIZE DETERMINED BY DIVER-OPERATED DRILLING
B 6405 756* PATCH REEF FACIES, ENEWETAK ATOLL, ABST., CARBON, ISOTOPIC
B 6309 1585* PATCH REEFS AND INTERREEF DEPOSITS OF SILURIAN MCKENZIE
S 4 210 PATCH REEFS, VERACRUZ
B 6109 1555* PATCH-REEF COMPLEX—BANDERA COUNTY, TEXAS, ABST., DIAGENESIS
S 4 37† PATCH-REEF ECOLOGY AND SEDIMENTOLOGY OF GLOVERS REEF ATOLL,
S 5 125 PATCHEN, D. G. RELATIONSHIPS BETWEEN DEPOSITIONAL

B	6009	1622	PATCHEN, D. G., EVOLUTION OF BASIN-SHELF RELATIONS DURING
B	6409	1403	PATCHEN, DOUGLAS G., OIL AND GAS DEVELOPMENTS IN MARYLAND,
B	6108	1269	PATCHEN, DOUGLAS G., OIL AND GAS DEVELOPMENTS IN MARYLAND,
B	6308	1244	PATCHEN, DOUGLAS G., OIL AND GAS DEVELOPMENTS IN MARYLAND,
B	6208	1399	PATCHEN, DOUGLAS G., OIL AND GAS DEVELOPMENTS IN MARYLAND,
B	6008	1288	PATCHEN, DOUGLAS G., OIL AND GAS DEVELOPMENTS IN MARYLAND,
B	6305	846	PATCHEN, DOUGLAS G., SILURIAN-DEVONIA EMBAYMENT INTO
B	6303	501	PATCHEN, DOUGLAS G., SUBSURFACE STRATIGRAPHY OF UPPER
B	6405	629	PATCHEN, DOUGLAS, NIAGARAN BIOHERMS AND INTERBIOHERM
B	6309	1585	PATCHEN, DOUGLAS, PATCH REEFS AND INTERREEF DEPOSITS OF
B	6211	2308	PATCHEN, DOUGLAS, SILURIAN EVOLUTION OF CENTRAL APPALACHIAN
M	25	161	PATHE GEOTHERMAL AREA, MEXICO
B	6201	76	PATIENCE LAKE MEMBER, SASKATCHEWAN
M	25	525	PATINO-A., M., MINERAL RESOURCES OF STATE O SINALOA,
B	6206	1043	PATOS ARCH, PROFILES, BRAZIL
B	6302	221	PATOS LINEAMENT, BRAZIL
B	6209	1768	PATRICELLI, JUDITH A., STRATIGRAPHIC RELATIONS OF ANNONA CHALK
M	25	252	PATTANI TROUGH, THAILAND
B	6001	34†	PATTERN IN PARTS OF NORTHERN EGYPT, EVOLUTION OF, EOCENE-,
B	6009	1550†	PATTERN OF ACCUMULATION AND COMPOSITION OF
B	6004	720*	PATTERNS, AND HYDROCARBON POTENTIAL OF BANGOR LIMESTONE,
B	6203	569*	PATTERNS AND SEQUENCES ACROSS SACRAMENTO SHELF, NEW MEXICO,
M	26	120	PATTERNS, BASE-DISCORDANT, SEISMIC, REFLECTIONS
M	26	127	PATTERNS, CLINOFORMS, SEISMIC, REFLECTIONS
M	26	125	PATTERNS, CLINOFORMS, SEISMIC, REFLECTIONS
B	6012	2181*	PATTERNS OF CALIFORNIA CONTINENTAL BORDERLAND
B	6004	719*	PATTERNS OF CHALKS, ABST., DIAGENETIC
B	6304	676†	**PATTERNS OF POROSITY AND CEMENT IN OOID RESERVOIRS IN DOGGER,**
B	6104	511†	**PATTERNS OF POROSITY AND CEMENT IN OOID RESERVOIRS IN DOGGER,**
B	6405	733*	PATTERNS OF SHALLOW-MARIN DEPOSITION, UPPER CRETACEOUS OF
B	6004	722*	PATTERNS OF SOUTH TEXAS OUTER CONTINENTAL SHELF, ABST.,
B	6309	1589*	PATTERNS OF SPECIES DIVERSITY IN HELDERBER GROUP OF WEST
B	6405	732*	PATTERNS OF THICK COAL DEPOSITION ACROSS POWDER RIVER BASIN
M	26	138	PATTERNS, STRATIGRAPHIC INTERPRETATION, SEISMIC, REFLECTIONS
M	26	119	PATTERNS, TOP-DISCORDANT, SEISMIC, REFLECTIONS
B	6206	1075	PATTERSON, JOHN M., PETROLEUM SOURCE-BED EVALUATION OF
S	4	381	PATTERSON, R. J., MARINE AND CONTINENTAL GROUNDWATER SOURCES
B	6211	2359	PATTERSON, S. O., SUMMARY OF RECENT EXPLORATION AND
B	6111	1980	PATTON ESCARPMENT, CALIFORNIA
B	6203	363	PATTON ESCARPMENT, CALIFORNIA
B	6203	368	PATTON RIDGE A BASIN, CALIFORNIA
B	6203	368	PATTON RIDGE B BASIN, CALIFORNIA
B	6111	1981	PATTON RIDGE, CALIFORNIA
B	6102	238	PATTON RIDGE, CALIFORNIA
B	6205	888	PATTON, T. L., HYDROCARBON EXPLORATION IN GREAT SALT LAKE
M	28	134	PATULA ARKOSE, MEXICO, CRETACEOUS
M	28	18	PATULA ARKOSE, MEXICO, CRETACEOUS
M	28	9	PATULA ARKOSE, MEXICO, CRETACEOUS
M	28	165	PATULA ARKOSE, MEXICO, CRETACEOUS
B	6012	2187	PATURET, D., PROCESSING OF NONSTRAIGHT SEISMIC LINES USING
B	6409	1437	PAUL, SHIRLEY E., OIL AND GAS DEVELOPMENTS IN NORTH
B	6208	1477	PAUL, SHIRLEY E., OIL AND GAS DEVELOPMENTS IN NORTH
B	6108	1305	PAUL, SHIRLEY E., OIL AND GAS DEVELOPMENTS IN NORTH
B	6308	1301	PAUL, SHIRLEY E., OIL AND GAS DEVELOPMENTS IN NORTH
B	6008	1332	PAUL, SHIRLEY E., OIL AND GAS DEVELOPMENTS IN NORTH
B	6307	1114	PAUL SPUR, REEFS, SECTION, ARIZONA
M	29	27	PAULL, CHARLES K., STRUCTURE AND DEVELOPMENT OF THE
B	6105	781	**PAULL, CHARLES K., STRUCTURE AND DEVELOPMENT OF SOUTHEAST**
B	6403	339	PAULL, CHARLES K., STRUCTURE, STRATIGRAPHY, AND GEOLOGIC
B	6102	256	PAULL, R. A., LITTLE COPPER FORMATION--NEW NAM FOR LOWEST
B	6103	361	PAULS VALLEY UPLIFT, OKLAHOMA
B	6109	1544	PAULSON, OSCAR L., JR., SUBSALT ORIGIN OF SALT-DOME CAP ROCK,
B	6303	495*	PAUP SPUR FIELD, MILLER COUNTY, ARKANSAS, ABST., DEPOSITIONAL
B	6203	561	PAVLAK, S. J., MAJOR URANIUM DISCOVERY IN FRONTIER AREA—
B	6304	621	PAVLAK, S. J., MAJOR URANIUM DISCOVERY IN VOLCANICLASTI
B	6405	782	PAVLAK, S. J., URANIUM DEPOSITS OF PART OF CENTRAL GREAT DIVIDE
B	6409	1566	PAVLOVIC, ROBERT, KURTEN FIELD—DISCOVERED BY STRATIGRAPHIC
B	6009	1623	PAWLOWICZ, E. F., BOWLING GREEN FAULT--CASE OF RESURGENT
B	6301	40	PAWNEE FIELD, BEE COUNTY, TEXAS
B	6107	1046	PAWNEE FORMATION, KANSAS
B	6203	523*	PAYNE, LINCOLN, AND LOGAN COUNTIES, OKLAHOMA, ABST.,
B	6004	517	PAYNE, MYRON W., BASINAL SANDSTONE FACIES, DELAWARE BASIN,
B	6203	553	PAYNE, MYRON W., PRECAMBRIAN-CAMBRIAN SEDIMENTARY ROCKS
B	6303	463	PAYNTER, DONALD D., DRAINAGE ANOMALIES IN GULF COAST
M	26	1	**PAYTON, CHARLES E., EDITED BY.SEISMIC STRATIGRAPHY--**
M	26	1	**PAYTON, CHARLES E., FOREWORD, SEISMIC STRATIGRAPHY--**
B	6105	791	**PEACE, DAVID GEORGE, HIGH-RESOLUTION REFLECTION-SEISMIC**
S	6	263	PEACE RIVER FORMATION, CANADA, CRETACEOUS
B	6111	1951	PEACE RIVER HEAVY OIL DEPOSIT, WESTERN, CANADA
B	6405	776*	PEACE RIVER OIL SANDS DEPOSIT, ALBERTA, CANADA, ABST., FACIES
M	24	73	PEACE RIVER UPLANDS, ALBERTA
S	10	60	PEAK GENERATION, MIGRATION
B	6403	359†	PEAKS IN ALPINOTYPE OROGENIES AND CHANGES IN OCEAN-FLOOR
B	6304	652	PEARIS MOUNTAIN, VIRGINIA
B	6205	882	PEARL, RICHARD HOWARD, EXPLORATION FOR NONELECTRIC
B	6108	1386	PEARL, RICHARD HOWARD, USE OF HYDROGEOLOGY, GEOCHEMISTRY,
B	6009	1615*	PEARL RIVER DELTA, LOUISIANA-MISSISSIPPI, ABST., CLAY
B	6301	46	PEARSALL FIELD, TEXAS
B	6109	1554*	PEARSALL FORMATION, LOWER CRETACEOUS, SOUTH TEXAS, ABST.,
B	6405	749	PEARSON, F. J., JR., STABLE ISOTOPE EVIDENCE FOR MODERN
B	6404	530	PEARY LAND, STRATIGRAPHIC CHART, GREENLAND
B	6206	1005	PEARYA GEANTICLINE, ARCTIC, CANADA
B	6109	1546*	PEAT AS GEOLOGIC HAZARD IN GREATER NEW ORLEANS AREA,
B	6305	801	PEBBLE CONTENT, PERMEABILITY AND
B	6305	799	PEBBLE CONTENT, POROSITY AND
B	6305	802	PEBBLE CONTENT, WATER SATURATION AND
B	6206	939	PEBBLY SANDSTONES
B	6309	1609	PECK, DOUGLAS M., STRATIGRAPHY AND PALEOECOLOGY OF TAMIAMI
B	6203	538*	PECOS COUNTY, TEXAS, ABST., DEPOSITIONAL FACIES AND OCCURRENCE
B	6303	488*	PECOS COUNTY, TEXAS, ABST., POROSITY IN, GIANT, GAS FIELD,
B	6205	860	PECOS COUNTY, ULTRADEEP WELLS, TEXAS
S	11	209	PECTEN CAURINUS
M	27	86	PECTINOID PELECYPODS
B	6107	1030	PEDEE GROUP, NEBRASKA
B	6109	1449	PEDERNAL UPLIFT, NEW MEXICO
B	6008	1405	PEDERSON, JOHN A., ECONOMIC VALUES USIN DISTRIBUTED RANGES OF
B	6105	820	PEDERSON, JOHN A., RESERVE ESTIMATES—FUEL AND ENERGY
B	6303	508	PEDRAZZINI, C. J., FACIES, ENVIRONMENTS, AND DEVELOPMENT OF
B	6109	1450	PEDREGOSA BASIN, ABRIGO FORMATION
B	6403	450*	PEDREGOSA BASIN, ABST., OIL AND GAS EXPLORATION WELLS IN
B	6109	1465	PEDREGOSA BASIN AREA, OIL AND GAS SHOWS
B	6109	1450	PEDREGOSA BASIN, BLISS SANDSTONE
B	6109	1450	PEDREGOSA BASIN, BOLSA QUARTZITE
B	6109	1461	PEDREGOSA BASIN, CRETACEOUS
B	6109	1451	PEDREGOSA BASIN, EL PASO-ELLENBURGER UNIT
B	6109	1452	PEDREGOSA BASIN, FUSSELMAN DOLOMITE
B	6109	1452	PEDREGOSA BASIN, HUNTON GROUP
B	6109	1460	PEDREGOSA BASIN, JURASSIC
B	6109	1454	PEDREGOSA BASIN, MARTIN FORMATION
B	6109	1454	PEDREGOSA BASIN, MISSISSIPPIAN
B	6109	1452	PEDREGOSA BASIN, MONTOYA GROUP
B	6109	1456	PEDREGOSA BASIN, PENNSYLVANIAN
B	6109	1454	PEDREGOSA BASIN, PERCHA UNIT
B	6109	1457	PEDREGOSA BASIN, PERMIAN
B	6109	1448	PEDREGOSA BASIN, PETROLEUM POTENTIAL, NEW MEXICO, ARIZONA,
B	6109	1451	PEDREGOSA BASIN, SIMPSON GROUP
B	6004	646*	PEDREGOSA BASIN, SOUTHEASTERN ARIZONA, SOUTHWESTERN NEW
B	6004	723*	PEDREGOSA BASIN, SOUTHEASTERN ARIZONA, SOUTHWESTERN NEW
B	6109	1454	PEDREGOSA BASIN, WOODFORD UNIT
B	6109	1448†	PEDREGOSA BASIN—COMPARISON WITH PERMIAN AND OROGRANDE
B	6203	457	PEEDAMULLA SHELF, AUSTRALIA
M	25	431	PEGMATITE TYPE, THAILAND, FLUORITE OF
B	6004	706	PEGRUM, R. M., NORTH SEA RIFTING AND OPENING OF NORTH
B	6203	419	PEGRUM, RICHARD M., RIFT BASINS FLANKING NORTH ATLANTIC OCEAN
B	6004	706	PEIRCE, F. L., COMPUTER APPROACH TO FACIES STUDIES, ABST.
B	6406	865	PELAGIAN BLOCK, AFRICA
B	6004	686*	PELAGIC AND SHALLOW-WATER EPERIC CARBONATE SEQUENCES, ABST.,
B	6004	646*	PELAGIC CALCAREOUS SEDIMENTS, ABST., PRIMARY, BEDDING,
B	6004	668*	PELAGIC CARBONATE DEPOSITS, WESTERN EQUATORIAL PACIFIC, ABST.,
B	6401	67†	PELAGIC LIMESTONES--POTENTIAL STRATIGRAPHIC AND PETROLEUM
M	27	36	PELAGIC OOZE
S	7	91	PELAGIC-SUSPENSION TRANSPORT, SEDIMENT PROCESSES
M	28	39	PELECYPOD, FRAGMENTS
B	6009	1614*	PELECYPOD SPECIES FROM LOWER CRETACEOUS NEAR KERRVILLE,
M	27	84	PELECYPODS
M	27	87	PELECYPODS
M	27	86	PELECYPODS, PECTINOID
B	6308	1225	PELICAN ISLAND FIELD, GALVESTON COUNTY, TEXAS
B	6106	896	PELISIUM LINE
S	6	97	PELITIC ROCKS, RUBIDIUM-STRONTIUM METHOD
B	6209	1763	PELLEGRIN, FREDDIE J., LATE PLEISTOCENE STRATIGRAPHY OF
B	6209	1757	PELLEGRIN, FREDDIE J., PLEISTOCENE SHORELINES AND
M	27	128	PELLETAL SEDIMENTS
M	27	4	PELLETING
M	27	112	PELLETS
M	27	175	PELMICRITES
S	4	40	PELOID, DEFINITION
B	6104	602	PELONCILLO MOUNTAINS, NEW MEXICO
B	6402	244	PELOPONNESUS AND ADJACENT ISLANDS, GEOLOGY, GREECE
B	6402	245	PELOPONNESUS CONTINENTAL MARGIN, DEEP SEA DRILLING PROJECT
B	6402	242†	PELOPONNESUS CONTINENTAL MARGIN, SEISMIC STRUCTURE OF
B	6109	1471	PELOTAS BASIN, BRAZIL
B	6206	1042	PELOTAS BASIN, PROFILES, BRAZIL

B 6002 197 PELOTAS COASTAL BASIN, BRAZIL
B 6104 549 PELOTAS SHELF AND BASIN, BRAZIL
M 27 174 PELSPARITES
M 27 111 PELSPARITES
B 6406 865 PELUSIUM LINE, AFRICA
B 6203 553 PEMBERTON, S. G., SILURIAN MACROBORING ORGANISMS, SOUTHERN
B 6303 508 PEMBERTON, S. GEORGE, ICHNOLOG OF LABRADOR GROUP, LOWER
B 6406 943 PEMBINA ESCARPMENT, NORTH DAKOTA
B 6104 594 PEMBINA FIELD, ALBERTA
B 6109 1421 PEMBINA FIELD, ALBERTA
B 6104 486 PEMBINA FIELD, TRAP, ALBERTA
B 6303 533* PENA BLANCA URANIUM DISTRICT, CHIHUAHUA, MEXICO, ABST.,
B 6405 712* PENAS ALTAS FORMATION, VENEZUELA—CASE STUDY OF CRETACEOUS
B 6109 1514 PENASCO CANYON, NEW MEXICO
B 6009 1543† PENDULUM EFFECT AND NIGER DELTA PROLIFI BELT
M 27 32 PENEROPLID FORAMS
B 6004 723 PENG, T. H., RADIOCARBON DATING OF ESTUARINE CARBONATE
B 6207 1178 PENICULAURIS BASSI-NEOSTREPTOGNATHODUS SP. C ZONE, UTAH
B 6207 1178 PENICULAURIS BASSI-NEOSTREPTOGNATHODUS SULCOPLICATUS ZONE,
B 6207 1178 PENICULAURIS IVESI-NEOSTREPTOGNATHODUS PRAYI ZONE, NEVADA
B 6109 1523 PENINSULAR ARCH, ACCUMULATION RATES, EAST COAST
B 6112 2144 PENINSULAR ARCH, FLORIDA
M 24 289 PENINSULAR ARCH, FLORIDA
B 6403 340 PENINSULAR ARCH, FLORIDA PLATFORM
B 6210 1986 PENINSULAR MALAYSIA, 1977, ARMITAGE, J., ON DEVELOPMENTS IN
B 6111 1980 PENINSULAR RANGES, CALIFORNIA
B 6203 345 PENINSULAR RANGES, CALIFORNIA
B 6309 1609 PENLAND, P. S., VARIATIONS IN LITTORAL SAND FEATURES WITH
B 6405 759 PENLAND, P. SHEA, SAND DISPERSAL AT NORDERNEYER SEEGAT, WEST
B 6303 508 PENLAND, SHEA, SEDIMENT DISPERSAL AT FORT GEORGE INLET,
B 6206 960 PENNINE BASIN, ENGLAND
B 6004 659* PENNINE BASIN, ENGLAND, ABST., DELTAIC EVOLUTION DURING, BASIN,
M 27 208 PENNINE BÜNDNERSCHIEFER, SWITZERLAND, JURASSIC
B 6206 962 PENNINE DELTA, ENGLAND
B 6009 1619* PENNINGTON FORMATIONS, EAST-CENTRAL KENTUCKY, ABST.,
B 6204 646 PENNINGTON SECTION, ALABAMA
B 6101 80 PENNSYLANIA-NEW YORK, NEWARK BASIN
B 6203 514* PENNSYLVANIA, ABST., CARBONATE ENVIRONMENTS OF MIDDLE,
B 6004 689* PENNSYLVANIA, ABST., COMPARISON OF DIAGENETIC AND,
B 6405 765* PENNSYLVANIA, ABST., DIAGENETIC MOBILIZATION OF, URANIUM AND,
B 6405 788* PENNSYLVANIA, ABST., FORMS AND ASSOCIATIONS OF, PYRITE IN UPPER,
B 6309 1592* PENNSYLVANIA, ABST., PRESSURE, CORING OF, GAS-BEARING, DEVONIAN,
B 6309 1577* PENNSYLVANIA, ABST., REGIONAL, COALIFICATION PATTERNS IN
B 6105 819* PENNSYLVANIA, AGE IN ILLINOIS, ABST., IDENTIFICATION OF MAJOR,
B 6212 2396 PENNSYLVANIA, ALLEGHENY FORMATION, WEST VIRGINIA TO
M 25 341 PENNSYLVANIA, AMAZON BASINS
B 6002 281 PENNSYLVANIA, BEAVER DAM SHALE MEMBER
B 6209 1610 PENNSYLVANIA, BLOOMSBURG RED BEDS
B 6209 1610 PENNSYLVANIA, BOSSARDVILLE LIMESTONE
B 6002 279 PENNSYLVANIA, BRALLIER FORMATION
B 6002 281 PENNSYLVANIA, BURKET MEMBER
B 6209 1610 PENNSYLVANIA, BUTTERMILK FALLS LIMESTONE
M 27 211 PENNSYLVANIA, CAMBRIAN, ALLENTOWN DOLOMITE
M 28 116 PENNSYLVANIA, CAMBRIAN, GATESBURG FORMATION
M 28 165 PENNSYLVANIA, CAMBRIAN, GATESBURG FORMATION
M 28 5 PENNSYLVANIA, CAMBRIAN, GATESBURG FORMATION
M 28 113 PENNSYLVANIA, CAMBRIAN, GATESBURG FORMATION
M 28 2 PENNSYLVANIA, CAMBRIAN, GATESBURG FORMATION
M 28 79 PENNSYLVANIA, CAMBRIAN, GATESBURG FORMATION
M 28 52 PENNSYLVANIA, CAMBRIAN, GATESBURG FORMATION
M 27 207 PENNSYLVANIA, CAMBRIAN, GATESBURG FORMATION
M 27 199 PENNSYLVANIA, CAMBRIAN, GATESBURG FORMATION
M 27 141 PENNSYLVANIA, CAMBRIAN MINES DOLOMITE
M 28 116 PENNSYLVANIA, CAMBRIAN, MINES DOLOMITE
B 6002 281 PENNSYLVANIA, CHANEYSVILLE SILTSTONE MEMBER
B 6002 281 PENNSYLVANIA, CLEARVILLE SILTSTONE
B 6209 1610 PENNSYLVANIA, COEYMANS FORMATION
B 6209 1610 PENNSYLVANIA, DECKER FORMATION
B 6008 1289 PENNSYLVANIA, DEVELOPMENTS, 1975
B 6108 1270 PENNSYLVANIA, DEVELOPMENTS, 1976
B 6208 1400 PENNSYLVANIA, DEVELOPMENTS, 1977
B 6308 1245 PENNSYLVANIA, DEVELOPMENTS, 1978
B 6409 1404 PENNSYLVANIA, DEVELOPMENTS, 1979
B 6412 2163 PENNSYLVANIA, DEVONIAN SHALE FACIES
B 6209 1610 PENNSYLVANIA, ESOPUS SHALE
B 6002 281 PENNSYLVANIA, FRAME SHALE MEMBER, RESTRICTED
B 6002 280 PENNSYLVANIA, FULTON LOBE
B 6002 281 PENNSYLVANIA, GANDER RUN SHALE MEMBER
B 6101 80 PENNSYLVANIA, GETTYSBURG BASIN
B 6003 414† PENNSYLVANIA, GROWTH FAULTS IN CAMBRIAN AND LOWER
B 6002 279 PENNSYLVANIA, HAMILTON GROUP
B 6002 285 PENNSYLVANIA, HARRELL SHALE
B 6209 1610 PENNSYLVANIA, HELDERBERG GROUP
B 6002 280 PENNSYLVANIA, HUNTERSVILLE CHERT
B 6002 279 PENNSYLVANIA, MAHANTANGO FORMATION
B 6002 284 PENNSYLVANIA, MARCELLUS SHALE
B 6002 286 PENNSYLVANIA, MILLBORO SHALE
B 6209 1610 PENNSYLVANIA, MINISINK LIMESTONE
B 6002 285 PENNSYLVANIA, NEEDMORE SHALE
B 6209 1610 PENNSYLVANIA, NEW SCOTLAND FORMATION
M 27 200 PENNSYLVANIA, ORDOVICIAN, BLACK RIVER LIMESTONE
M 27 75 PENNSYLVANIA, ORDOVICIAN, BLACK RIVER LIMESTONE
M 27 72 PENNSYLVANIA, ORDOVICIAN, BLACK RIVER LIMESTONE
M 27 206 PENNSYLVANIA, ORDOVICIAN, BLACK RIVER LIMESTONE
M 27 173 PENNSYLVANIA, ORDOVICIAN, BLACK RIVER LIMESTONE
M 27 222 PENNSYLVANIA, ORDOVICIAN, BLACK RIVER LIMESTONE
M 27 173 PENNSYLVANIA, ORDOVICIAN, CHAZY LIMESTONE
M 27 200 PENNSYLVANIA, ORDOVICIAN, CHAZY LIMESTONE
M 28 39 PENNSYLVANIA, ORDOVICIAN, REEDSVILLE SHALE
M 27 76 PENNSYLVANIA, ORDOVICIAN, REEDSVILLE SHALE
M 27 99 PENNSYLVANIA, ORDOVICIAN, REEDSVILLE SHALE
M 27 176 PENNSYLVANIA, ORDOVICIAN, STONEHENGE LIMESTONE
M 27 131 PENNSYLVANIA, ORDOVICIAN, STONEHENGE LIMESTONE
B 6209 1610 PENNSYLVANIA, ORISKANY FORMATION
B 6002 280 PENNSYLVANIA, ORISKANY SANDSTONE
B 6002 280 PENNSYLVANIA, PERRY BAY
B 6002 281 PENNSYLVANIA, POKEJOY MEMBER
B 6209 1610 PENNSYLVANIA, PORT EWEN SHALE
B 6212 2396 PENNSYLVANIA, POTTSVILLE FORMATION, WEST VIRGINIA TO
B 6209 1610 PENNSYLVANIA, POXONO ISLAND FORMATION
B 6008 1298 PENNSYLVANIA, PRODUCTION, 1974-1975
B 6108 1281 PENNSYLVANIA, PRODUCTION, 1975-1976
B 6208 1410 PENNSYLVANIA, PRODUCTION, 1976-1977
B 6308 1247 PENNSYLVANIA, PRODUCTION, 1977-1978
B 6409 1412 PENNSYLVANIA, PRODUCTION, 1978-1979
B 6002 281 PENNSYLVANIA, PURCELL MEMBER
B 6209 1610 PENNSYLVANIA, RONDOUT FORMATION
B 6209 1610 PENNSYLVANIA, SCHOHARIE FORMATION
M 27 144 PENNSYLVANIA, SILURIAN, CLINTON FORMATION
M 27 147 PENNSYLVANIA, SILURIAN, CLINTON FORMATION
M 28 45 PENNSYLVANIA, SILURIAN, CLINTON FORMATION
M 28 157 PENNSYLVANIA, SILURIAN, KEYSER LIMESTONE
M 28 157 PENNSYLVANIA, SILURIAN, TONOLOWAY LIMESTONE
M 27 90 PENNSYLVANIA, SILURIAN, TONOLOWAY-KEYSER LIMESTONE
M 27 67 PENNSYLVANIA, SILURIAN, TONOLOWAY-KEYSER LIMESTONE
M 27 94 PENNSYLVANIA, SILURIAN, TONOLOWAY-KEYSER LIMESTONE
M 27 71 PENNSYLVANIA, SILURIAN, TONOLOWAY-KEYSER LIMESTONE
M 27 4 PENNSYLVANIA, SILURIAN, TONOLOWAY-KEYSER LIMESTONE
M 27 139 PENNSYLVANIA, SILURIAN, TONOLOWAY-KEYSER LIMESTONE
M 27 164 PENNSYLVANIA, SILURIAN, TONOLOWAY-KEYSER LIMESTONE
M 27 221 PENNSYLVANIA, SILURIAN, TONOLOWAY-KEYSER LIMESTONE
M 27 75 PENNSYLVANIA, SILURIAN, TONOLOWAY-KEYSER LIMESTONE
M 27 72 PENNSYLVANIA, SILURIAN, TONOLOWAY-KEYSER LIMESTONE
B 6209 1610 PENNSYLVANIA, SILURIAN-DEVONIAN CARBONATE SEQUENCE,
B 6209 1618 PENNSYLVANIA, SILURIAN-DEVONIAN SECTION, NEW YORK TO
B 6209 1624 PENNSYLVANIA, SILURIAN-DEVONIAN SECTION, OHIO TO
B 6002 280 PENNSYLVANIA, SNYDER LOBE
B 6002 278† PENNSYLVANIA, STRATIGRAPHIC, CROS, SECTION OF, HAMILTON GROUP,
B 6002 284 PENNSYLVANIA, TIOGA BENTONITE
B 6002 285 PENNSYLVANIA, TULLY LIMESTONE
B 6311 2114* PENNSYLVANIA UPPER MARCHAND SANDSTONES, NORTHERN CADDO
B 6308 1244† PENNSYLVANIA, VIRGINIA, AND WEST VIRGINIA IN 1978, OIL AND, GAS,
B 6409 1403† PENNSYLVANIA, VIRGINIA, AND WEST VIRGINIA IN 1979, OIL AND, GAS,
B 6208 1399† PENNSYLVANIA, VIRGINIA, AND WEST VIRGINIA, OIL AND, GAS,
B 6108 1269† PENNSYLVANIA, VIRGINIA, AND WEST VIRGINIA, OIL AND, GAS,
B 6008 1288† PENNSYLVANIA, VIRGINIA, AND WEST VIRGINIA, OIL AND, GAS,
B 6303 462* PENNSYLVANIA--STRATIGRAPHY AND NATURAL GAS PRODUCTION,
B 6105 836* PENNSYLVANIA-NEW JERSEY, AND IMPLICATIONS FOR URANIUM
B 6311 2120* PENNSYLVANIAN AGE, NORTHWESTERN NESS COUNTY, KANSAS, ABST.,
B 6102 297* PENNSYLVANIAN AGE, VAL VERDE BASIN, TEXAS, BY TITLE ONLY, GAS,
B 6303 550* PENNSYLVANIAN AND CRETACEOUS RESERVOIR AND NONRESERVOIR
B 6405 677* PENNSYLVANIAN AND EARLY PERMIAN PALEOGEOGRAPHY, SOUTHERN
B 6303 476* PENNSYLVANIAN AND LOWER PERMIAN, WESTERN OQUIRRH BASIN,
B 6203 558* PENNSYLVANIAN AND MISSISSIPPIAN ROCKS OF PERMIAN BASIN, WEST
B 6002 325* PENNSYLVANIAN AND PERMIAN CARBONATE RESERVOIRS OF PERMIAN
B 6105 819* PENNSYLVANIAN AND PERMIAN, SACRAMENTO MOUNTAINS, NEW
B 6408 1283* PENNSYLVANIAN, AND THEIR RELATION LOW SULFUR COAL DEPOSITS,
M 27 90 PENNSYLVANIAN ASPHALTS, OKLAHOMA
M 26 453 PENNSYLVANIAN, ATOKA LIMESTONE, OKLAHOMA
M 28 119 PENNSYLVANIAN, AVIS SANDSTONE, TEXAS

M 28 69 PENNSYLVANIAN, BALL CLAY, MISSOURI
S 6 329 PENNSYLVANIAN, BASE OF THE MORROWAN
B 6303 533* PENNSYLVANIAN BECKLEY SEAM, ABST., ESTUARINE-COASTAL PLAIN,
B 6107 1058 PENNSYLVANIAN BLACK SHALE DEPOSITION, MODELS OF
M 27 35 PENNSYLVANIAN, BLOYD FORMATION, OKLAHOMA
M 27 86 PENNSYLVANIAN, BLOYD FORMATION, OKLAHOMA
M 27 96 PENNSYLVANIAN, BLOYD FORMATION, OKLAHOMA
M 27 167 PENNSYLVANIAN, BLOYD FORMATION, OKLAHOMA
M 27 118 PENNSYLVANIAN, BLOYD FORMATION, OKLAHOMA
M 27 116 PENNSYLVANIAN, BLOYD FORMATION, OKLAHOMA
S 6 327† PENNSYLVANIAN BOUNDARY, THE, MISSISSIPPIAN-
B 6311 2120* PENNSYLVANIAN BRAIDED STREA SEDIMENTATION, NORTHWEST
B 6111 1929† PENNSYLVANIAN, CANADIAN ARCTIC ARCHIPELAGO, CARBONATE-,
S 5 145* PENNSYLVANIAN, CANADIAN ARCTIC ARCHIPELAGO, CARBONATE-
B 6101 13 PENNSYLVANIAN, CANADIAN, ARCTIC ISLANDS, PERMO-
B 6203 569* PENNSYLVANIAN CARBONATE AND SILICICLASTIC DEPOSITIONAL
B 6405 764* PENNSYLVANIAN CARBONATE MOUNDS IN PARADOX BASIN, UTAH,
B 6303 517* PENNSYLVANIAN CARBONATE SEDIMENTATION, SOUTHEASTERN
B 6008 1407* PENNSYLVANIAN, CENNRAL WYOMING, ABST., COASTAL
B 6002 327* PENNSYLVANIAN, CLASTIC ROCKS IN PARTS OF JACK, PALO PINTO,
B 6301 50† PENNSYLVANIAN, CLASTIC ROCKS IN PARTS OF JACK, PALO PINTO,
B 6405 743* PENNSYLVANIAN COAL BEDS OF APPALACHIAN AND ILLINOIS BASINS,
B 6305 834* PENNSYLVANIAN COARSE-GRAINED MEANDERING FLUVIAL DEPOSITS,
B 6203 570* PENNSYLVANIAN CYCLIC SEDIMENTS IN NORTHERN KANSAS AND
B 6405 721* PENNSYLVANIAN CYCLOTHEMS, ABST., PALEOGEOGRAPHY OF EUSTATIC
B 6107 1045† PENNSYLVANIAN CYCLOTHEMS OF MID-CONTINENT NORTH AMERICA,
B 6107 1062 PENNSYLVANIAN DEPOSITION, MODEL FOR, MID-CONTINENT MIDDLE
B 6009 1624* PENNSYLVANIAN DEPOSITIONAL SYSTEMS IN NORTHEASTERN
B 6008 1401* PENNSYLVANIAN, DICKINSON AREA, NORTH DAKOTA, ABST.,
B 6303 485* PENNSYLVANIAN, DICKINSON AREA, NORTH DAKOTA--A 24-MILLION
B 6311 2116* PENNSYLVANIAN FAN-DELTA SANDSTONES, PAL DURO BASIN, TEXAS,
B 6303 413* PENNSYLVANIAN FRESHWATER LIMESTONES ACTUALLY INTERTIDAL
B 6203 518* PENNSYLVANIAN, FRONTAL OUACHITA MOUNTAINS, OKLAHOMA, ABST.,
B 6004 652* PENNSYLVANIAN GEOLOGIC HISTORY OF SOUTHEASTERN UNITED
M 27 48 PENNSYLVANIAN, GRAFORD FORMATION, TEXAS
M 27 62 PENNSYLVANIAN, GRAHAM FORMATION, TEXAS
M 27 70 PENNSYLVANIAN, GRAHAM FORMATION, TEXAS
M 28 119 PENNSYLVANIAN, GRAHAM FORMATION, TEXAS
M 28 128 PENNSYLVANIAN, GRAY SANDSTONE, TEXAS
M 28 157 PENNSYLVANIAN, GRAY SANDSTONE, TEXAS
M 28 112 PENNSYLVANIAN, GRAY SANDSTONE, TEXAS
B 6003 414† PENNSYLVANIAN, GROWTH, FAULTS IN, CAMBRIAN AND, LOWE
M 27 20 PENNSYLVANIAN, GUNSIGHT LIMESTONE, TEXAS
M 27 166 PENNSYLVANIAN, HALE FORMATION, OKLAHOMA
M 27 120 PENNSYLVANIAN, HALE FORMATION, OKLAHOMA
M 26 439 PENNSYLVANIAN, HOXBAR GROUP, OKLAHOMA, CANADIAN COUNTY,
B 6311 2119* PENNSYLVANIAN HYDROCARBON ACCUMULATIONS IN MID-CONTINENT,
B 6004 727* PENNSYLVANIAN, INDIAN NATIONS TURNPIKE, OUACHITA MOUNTAINS,
B 6112 2086 PENNSYLVANIAN, KEROGEN, OKLAHOMA
B 6105 839* PENNSYLVANIAN LANSING-KANSAS CITY GROUPS IN NORTHWEST
M 27 10 PENNSYLVANIAN LIMESTONE, NEW MEXICO
B 6102 295* PENNSYLVANIAN LIMESTONE RESERVOIRS, ABST., PREDICTING,
M 27 50 PENNSYLVANIAN, MARBLE FALLS LIMESTONE, TEXAS
M 27 206 PENNSYLVANIAN, MARBLE FALLS LIMESTONE, TEXAS
M 27 93 PENNSYLVANIAN, MARBLE FALLS LIMESTONE, TEXAS
M 27 105 PENNSYLVANIAN, MAZON CREEK COAL BALLS, ILLINOIS
M 28 143 PENNSYLVANIAN, MINERAL WELLS FORMATION, TEXAS
M 28 112 PENNSYLVANIAN, MINERAL WELLS FORMATION, TEXAS
M 27 10 PENNSYLVANIAN, MINTURN FORMATION, COLORADO
M 27 11 PENNSYLVANIAN, MINTURN FORMATION, COLORADO
B 6405 788* PENNSYLVANIAN, MISSOURIAN, MONTGOMERY COUNTY, KANSAS, ABST.,
M 26 453 PENNSYLVANIAN, MORROW SANDSTONE, OKLAHOMA
B 6109 1513† PENNSYLVANIAN, NEW MORROWAN UNIT IN NORTH-CENTRAL NEW
B 6102 257 PENNSYLVANIAN, NOMENCLATURE, IDAHO
B 6405 751* PENNSYLVANIAN, NORTH-CENTRAL TENNESSEE, ABST., DEPOSITIONAL
B 6109 1536* PENNSYLVANIAN, NORTH-CENTRAL TEXAS, ABST., DIAGENESIS AND,
B 6004 665* PENNSYLVANIAN OF APPALACHIAN BASIN--DETERMINATION BY FAUNAL
B 6203 498* PENNSYLVANIAN OF OKLAHOMA, AND MODERN JAMAICAN REEFS, ABST.,
B 6405 800* PENNSYLVANIAN OF WEST VIRGINIA, ABST., DEPOSITIONAL
B 6405 804* PENNSYLVANIAN OF WEST-CENTRAL NEBRASKA PANHANDLE, ABST.,
M 27 140 PENNSYLVANIAN, PARADOX FORMATION, UTAH
M 27 61 PENNSYLVANIAN, PARADOX FORMATION, UTAH
M 27 10 PENNSYLVANIAN, PARADOX FORMATION, UTAH
M 27 34 PENNSYLVANIAN, PARADOX FORMATION, UTAH
M 27 133 PENNSYLVANIAN, PARADOX FORMATION, UTAH
B 6109 1456 PENNSYLVANIAN, PEDREGOSA BASIN
B 6112 2115† PENNSYLVANIAN PHYLLOID-ALGAL BUILDUP, SACRAMENTO
B 6109 1493† PENNSYLVANIAN, POTENTIAL MAJOR HYDROCARBON OBJECTIVE OF
M 28 9 PENNSYLVANIAN, POTTSVILLE FORMATION, OHIO
M 28 143 PENNSYLVANIAN, RICKER SANDSTONE, TEXAS
M 28 112 PENNSYLVANIAN, RICKER SANDSTONE, TEXAS
B 6004 711* PENNSYLVANIAN, ROCKS IN ANADARKO BASIN, ABST., SEDIMENTARY,
B 6002 321* PENNSYLVANIAN, ROCKS IN ANADARKO BASIN, ABST., SEDIMENTARY,
B 6107 1045 PENNSYLVANIAN ROCKS OF EASTERN KANSAS, CONODONT
B 6002 322* PENNSYLVANIAN, SANDSTONE RESERVOIR AT SLEEPY HOLLOW FIELD,
B 6107 1029† PENNSYLVANIAN, SANDSTONE RESERVOIR, SLEEPY HOLLOW FIELD, RED
B 6103 360† PENNSYLVANIAN SANDSTONES IN OKLAHOMA, URANIUM POTENTIAL OF,
B 6408 1272 PENNSYLVANIAN SANDSTONES OF THE MID-CONTINENT, EDITED BY
B 6311 2115* PENNSYLVANIAN SANDSTONES, SOUTH-CENTRAL KANSAS, ABST.,
M 28 162 PENNSYLVANIAN, SANGRE DE CRISTO FORMATION, NEW MEXICO
M 28 19 PENNSYLVANIAN, SANGRE DE CRISTO FORMATION, NEW MEXICO
M 28 18 PENNSYLVANIAN, SANGRE DE CRISTO FORMATION, NEW MEXICO
M 28 155 PENNSYLVANIAN, SANGRE DE CRISTO FORMATION, NEW MEXICO
B 6311 2120* PENNSYLVANIAN SERIES BOUNDARY IN NORTHEASTERN OKLAHOMA,
M 26 453 PENNSYLVANIAN, SPRINGER SANDSTONE, OKLAHOMA
B 6107 1046 PENNSYLVANIAN, STRATIGRAPHIC COLUMN, KANSAS, EASTERN
M 28 112 PENNSYLVANIAN, STRAWN GROUP, TEXAS
M 28 143 PENNSYLVANIAN, STRAWN GROUP, TEXAS
M 28 157 PENNSYLVANIAN, STRAWN GROUP, TEXAS
M 28 128 PENNSYLVANIAN, STRAWN GROUP, TEXAS
B 6101 12 PENNSYLVANIAN, SVERDRUP BASIN
B 6412 2189 PENNSYLVANIAN SYSTEM OF THE ILLINOIS BASIN, EDITED BY JAMES E.
B 6308 1425* PENNSYLVANIAN SYSTEM, PALO DURO BASIN, PANHANDLE TEXAS, ABST.,
B 6303 442* PENNSYLVANIAN SYSTEM, PALO DURO BASIN, TEXAS PANHANDLE, ABST.,
B 6405 688* PENNSYLVANIAN TECTONICALLY ACTIVE CRATONIC BASIN,
M 28 68 PENNSYLVANIAN, TENSLEEP SANDSTONE, WYOMING
M 28 123 PENNSYLVANIAN, TENSLEEP SANDSTONE, WYOMING
M 28 129 PENNSYLVANIAN, TENSLEEP SANDSTONE, WYOMING
M 28 122 PENNSYLVANIAN, TENSLEEP SANDSTONE, WYOMING
B 6109 1515 PENNSYLVANIAN, TERMINOLOGY, NEW MEXICO
M 28 105 PENNSYLVANIAN, TEUSCHNITZER CONGLOMERATE, GERMANY
M 28 66 PENNSYLVANIAN, TEUSCHNITZER CONGLOMERATE, GERMANY
B 6406 938† PENNSYLVANIAN UPPER DELTA PLAIN COALS, SOUTHWEST WEST
M 27 68 PENNSYLVANIAN, WINCHELL LIMESTONE, TEXAS
S 9 11 PENNSYLVANIAN, WIND DIRECTION, WESTERN, UNITED STATES
B 6401 88† PENNSYLVANIAN--EARLY PERMIAN DEPOSITIONAL SYSTEMS AND
B 6112 2094 PENNSYLVANIAN-AGE COAL, KEROGEN
B 6303 506* PENNSYLVANIAN-AGE MULTISTORY FLUVIAL SANDSTONNS O ILLINOIS
B 6401 88† PENNSYLVANIAN-EARLY PERMIAN DEPOSITIONAL SYSTEMS AND
B 6405 730* PENNSYLVANIAN-EARLY PERMIAN OQUIRRH BASIN, NORTHWEST UTAH,
B 6303 461* PENNSYLVANIAN-PERMIAN CARBONATE-SHELF MARGINS AND DELTAIC
S 4 243 PENNSYLVANIAN-PERMIAN LIMESTONES, PHYLLOID ALGAE
B 6405 778* PENNSYLVANIAN—IMPLICATIONS FOR STRATIGRAPHIC-TRAP OIL
B 6408 1288* PENNSYLVANIA—A SEDIMENTOLOGIC MODEL FOR DISTRIBUTION OF
B 6408 1284* PENNSYLVANIA—GEOLOGY, CURRENT PRODUCTION, AND RESERVES,
B 6401 130 PENTAMERID ZONES, INDIANA
B 6404 530 PENTAMERUS BJERGE FORMATION, GREENLAND
B 6404 528 PENTAMERUS BJERGE, GREENLAND
S 10 257 PENTANE CONCENTRATIONS, LOUISIANA, OFFSHORE
M 25 414 PENTECOST SANDSTONE, YAMPI MEMBER, WESTERN AUSTRALIA
M 25 400 PENTLANDITE AND PYRRHOTITE, AUSTRALIA
S 11 63 PENUTIAN STAGE
B 6411 1841 PEOPLES DEMOCRATIC REPUBLIC OF YEMEN, 1979, AGIP AND, DOME
B 6303 424* PEOPLES REPUBLIC OF CHINA, ABST., EXPLORATION GEOPHYSICS IN
B 6303 524* PEOPLES REPUBLIC OF CHINA ABST., OIL AND, GAS IN
B 6207 1226* PEOPLES REPUBLIC OF CHINA, ABST., PETROLEUM GEOLOGY—
B 6405 699* PEOPLES REPUBLIC OF CHINA, ABST., SEG DELEGATION VISIT TO
B 6110 1835 PEOPLES REPUBLIC OF CHINA, DEVELOPMENTS, 1976
B 6210 1982 PEOPLES REPUBLIC OF CHINA, DEVELOPMENTS, 1977
B 6310 1825 PEOPLES REPUBLIC OF CHINA, DEVELOPMENTS, 1978
B 6411 1891 PEOPLES REPUBLIC OF CHINA, DEVELOPMENTS, 1979
B 6408 1213 PEOPLES REPUBLIC OF CHINA, GEOLOGIC BASINS
B 6210 1990 PEOPLES REPUBLIC OF CHINA, PRODUCTION, 1975-1977
B 6310 1829 PEOPLES REPUBLIC OF CHINA, PRODUCTION, 1976-1978
B 6411 1896 PEOPLES REPUBLIC OF CHINA, PRODUCTION, 1977-1979
B 6105 831 PEQUEGNAT, WILLIS E., OCEAN DUMPING OF DREDGED MATERIAL—
B 6105 805* PERCENTAGE ESTIMATE OF DISCOVERABLE RESERVES AND THICKNESS OF
B 6104 603 PERCHA SHALE, NEW MEXICO
B 6109 1454 PERCHA UNIT, PEDREGOSA BASIN
B 6209 1768 PERCIVAL, STEPHEN F., STRATIGRAPHIC RELATIONS OF ANNONA CHALK
B 6206 1068 PERDIDO SCARP, GULF OF MEXICO
B 6401 71 PEREGRINA CANYON, CARBON ISOTOPIC DATA, MEXICO
B 6303 414 PEREIRA, CHRISTOPHER P. G., DISTRIBUTION AND FACTORS
B 6303 508 PEREIRA, CHRISTOPHER P. G., GENESIS, OCCURRENCE, AND CAUSES OF
B 6008 1390* PERFORMANCE REVIEW OF TIGER RIDGE AND BULLHOOK GAS UNITS,
B 6004 678 PERHAC, RALPH M., GREAT PATAGONIAN OIL SPILL, ABST.
S 4 236 PERICLIMENES ON NEMASTER RUBIGINOSA POURTALES, MARTINIQUE
S 4 236 PERICLIMENES OR ANALCIDOMETRA ARMATA POURTALES, BARBADOS

M 25 460 PERIDOTITE BELT, DUNITE-HARZBURGITE FACIES, PACIFIDES
B 6304 687* PERIODS OF MINERALIZATION IN GRANTS MINERAL BELT, NEW MEXICO,
S 7 31 PERIPHERAL FAULT ZONES, DESCRIBED, NORTHERN, GULF OF MEXICO
S 7 269 PERIPHERAL SLUMPS, BOTTOM, DEFORMATION
B 6209 1767* PERITIDAL CARBONATE MATERIAL AND EVIDENCE FOR VANISHED
B 6405 787* PERITIDAL LIMESTONE, ABST., CARBONATE-, SULFATE MINERAL
B 6409 1559 PERKINS, BOB F., PALEOENVIRONMENTS AND TRACE FOSSILS OF LARGE
B 6305 817 PERKINS, BOB F., REVIEW OF TRACE FOSSIL CONCEPTS. EDITED BY PAUL
B 6405 743 **PERKINS, J. HUNT, COMPARISON OF SHALLOW-MARINE SHELF CARBONATE**
B 6004 669 PERKINS, R. D., EVOLUTION OF FLORIDA BAY INTERPRETED FROM
B 6004 706 PERKINS, RONALD D., PALEOECOLOGIC POTENTIAL OF MICROBORING
M 24 65 PERMAFROST, NORTHWEST TERRITORIES, TAGLU GAS FIELD
B 6305 801 PERMEABILITY AND PEBBLE CONTENT
M 26 18 PERMEABILITY ASPECTS, SEISMIC, INTERPRETATION
M 26 491 PERMEABILITY BARRIER, SEISMIC, MODELING, UPDIP
B 6205 890* PERMEABILITY CONTROL IN CRETACEOUS GAS-BEARING SANDSTONES,
S 10 44 PERMEABILITY VERSUS COMPACTION, MIGRATION
S 4 371 PERMEABILITY, CORALS, DENSITY, POROSITY, AND
B 6102 174 PERMEABILITY, EFFECT OF, CEMENTATION AND, COMPACTION ON
B 6102 227† PERMEABILITY IN POROUS ROCK, EXPERIMENTAL STUDY OF, FRACTURE
B 6405 680* PERMEABILITY OF CLAY AND SHALE, ABST.
B 6004 725* PERMEABILITY OF LOWER WILCOX, EOCENE, SANDSTONE, KARNES
B 6203 555* PERMEABILITY OF ORTHOQUARTZITES, SIMPSON GROUP, ORDOVICIAN,
B 6009 1531† PERMEABILITY OF OTTAWA SAND, EFFECT OF, HIGH-, PRESSURE,
B 6102 169† PERMEABILITY OF SANDS, STUDIES OF, VENTURA FIELD, CALIFORNIA,
B 6205 815 PERMEABILITY, POROSITY VERSUS
B 6205 891* PERMEABILITY RESERVOIR ROCKS, SOUTH-CENTRAL UINTA BASIN,
B 6405 760* PERMEABILITY RESERVOIRS, ABST., BOWDOIN DOME AREA,
B 6205 891* PERMEABILITY RESERVOIRS OF NORTHERN GREAT PLAINS, ABST.,
B 6205 848† PERMEABILITY TRAPS IN GATCHELL, EOCENE, SAND OF CALIFORNIA
B 6102 228 PERMEABILITY, UTAH, NAVAJO SANDSTONE
B 6004 681* PERMEABILITY VARIATIONS OF RECENT DELTAIC SEDIMENTS, ABST., BED
B 6305 802 PERMEABILITY VERSUS POROSITY AND SAND CONTENT IN
S 10 222 PERMIAN BASIN, OCHOA SAG, TRAP, WEST TEXAS
M 28 119 PERMIAN, ABO SANDSTONE, NEW MEXICO
M 28 120 PERMIAN, ABO SANDSTONE, NEW MEXICO
M 28 105 PERMIAN, ABO SANDSTONE, NEW MEXICO
M 28 43 PERMIAN, ABO SANDSTONE, NEW MEXICO
M 28 47 PERMIAN, ABO SANDSTONE, NEW MEXICO
M 28 28 PERMIAN, ABO SANDSTONE, NEW MEXICO
M 28 97 PERMIAN, ABO SANDSTONE, NEW MEXICO
B 6002 319* PERMIAN ADMIRE SANDSTONE, EL DORADO FIELD, BUTLER COUNTY,
M 25 344 PERMIAN, AMAZON BASINS
B 6109 1448† PERMIAN AND OROGRANDE BASINS, PETROLEUM, POTENTIAL AND
B 6103 360† PERMIAN AND PENNSYLVANIAN SANDSTONES IN OKLAHOMA, URANIUM
B 6201 50 PERMIAN, AUSTRALIA
M 25 235 PERMIAN, AUSTRALIA
B 6307 1096 PERMIAN, AUSTRALIA
B 6207 1154 PERMIAN, AUSTRALIA
B 6405 768* PERMIAN, BANK MARGIN EROSION SURFACES, GUADALUPE MOUNTAINS,
B 6002 325* PERMIAN BASIN, ABST., FACTORS CONTROLLING, POROSITY RELATIONS
B 6303 474* PERMIAN BASIN, ABST., PRECIPITATION OF, AUTHIGENIC MINERALS IN,
B 6309 1520 PERMIAN BASIN AND SURROUNDING AREAS, CORRELATION CHART
B 6102 297* PERMIAN BASIN BY TITLE ONLY, CHARACTERIZATION OF, OIL TYPES IN
M 26 200 PERMIAN BASIN, EASTERN-SHELF BUILDUPS, SEISMIC EXAMPLE, TEXAS,
B 6307 1029 PERMIAN BASIN, KEYSTONE FIELD, SECTION, TEXAS
B 6303 458* PERMIAN BASIN OF WEST TEXAS, ABST., PRELIMINARY ANALYSES OF,
B 6205 781 PERMIAN BASIN, SEISMIC DATA, NEW MEXICO
B 6205 787 PERMIAN BASIN, SEISMIC DATA, TEXAS
B 6109 1448 PERMIAN BASIN, TEXAS AND, NEW MEXICO
B 6203 507* PERMIAN BASIN, WEST TEXAS AND SOUTHEASTERN NEW MEXICO, ABST.,
B 6203 558* PERMIAN BASIN, WEST TEXAS AND SOUTHEASTERN NEW MEXICO, ABST.,
B 6203 509* PERMIAN BASIN, WEST TEXAS AND SOUTHEASTERN NEW MEXICO, ABST.,
B 6203 544* PERMIAN BASIN, WEST TEXAS AND SOUTHEASTERN NEW MEXICO, ABST.,
B 6309 1510† PERMIAN BASIN, WEST TEXAS, PROPOSED, TYPE SECTION FOR UPPER
B 6105 779* PERMIAN INFILLED SUBGLACIAL GULLY FROM BACCHUS MARSH AREA,
M 28 122 PERMIAN, BELL CANYON FORMATION, TEXAS
M 27 69 PERMIAN, BELL CANYON FORMATION, TEXAS
M 28 176 PERMIAN, BELL CANYON FORMATION, TEXAS
M 28 11 PERMIAN, BELL CANYON FORMATION, TEXAS
M 28 113 PERMIAN, BELL CANYON FORMATION, TEXAS
M 28 22 PERMIAN, BELL CANYON FORMATION, TEXAS
S 6 301 PERMIAN, BIOCHRONOLOGY, ASSELIAN STAGE
S 6 302 PERMIAN, BIOCHRONOLOGY, BAIGENDZINIAN STAGE
S 6 305 PERMIAN, BIOCHRONOLOGY, DJULFIAN STAGE
S 6 306 PERMIAN, BIOCHRONOLOGY, DORASHAMIAN STAGE
S 6 304 PERMIAN, BIOCHRONOLOGY, KAZANIAN STAGE
S 6 304 PERMIAN, BIOCHRONOLOGY, KUNGURIAN STAGE
S 6 299 PERMIAN, BIOCHRONOLOGY, MARINE FAUNAS
S 6 304 PERMIAN, BIOCHRONOLOGY, PUNJABIAN STAGE
S 6 301 PERMIAN, BIOCHRONOLOGY, SAKMARIAN STAGE
S 6 310 PERMIAN, BIOCHRONOLOGY, STANDARD NOMENCLATURE
B 6002 320* PERMIAN BLAINE FORMATION AND ASSOCIATED STRATA IN SOUTHWEST
M 27 32 PERMIAN, BONE SPRING LIMESTONE, TEXAS
M 27 78 PERMIAN, BONE SPRING LIMESTONE, TEXAS
M 28 144 PERMIAN, BRUSHY CANYON FORMATION, TEXAS
M 28 131 PERMIAN, BRUSHY CANYON FORMATION, TEXAS
M 27 5 PERMIAN, CAPITAN FORMATION, NEW MEXICO
M 27 4 PERMIAN, CAPITAN FORMATION, NEW MEXICO
M 27 174 PERMIAN, CAPITAN LIMESTONE, NEW MEXICO
M 27 6 PERMIAN, CAPITAN LIMESTONE, NEW MEXICO
M 27 47 PERMIAN, CAPITAN LIMESTONE, NEW MEXICO
M 27 67 PERMIAN, CAPITAN LIMESTONE, NEW MEXICO
M 27 77 PERMIAN, CAPITAN LIMESTONE, NEW MEXICO
M 27 9 PERMIAN, CAPITAN LS.-TANSILL FM. TRANSITION, NEW MEXICO
S 9 22 PERMIAN, CARBONATE REEF BELT DISTRIBUTION
B 6002 325* PERMIAN CARBONATE RESERVOIRS OF PERMIAN BASIN, ABST., FACTORS
B 6303 461* PERMIAN CARBONATE-SHELF MARGINS AND DELTAIC SANDSTONNS
M 27 133 PERMIAN, CASTILE FORMATION, NEW MEXICO
M 28 84 PERMIAN, CASTILE FORMATION, NEW MEXICO
B 6103 366 PERMIAN CHANNEL SANDSTONES, OKLAHOMA
S 6 299† PERMIAN, CHRONOSTRATIGRAPHY FOR THE WORLD
B 6405 799* PERMIAN CONODONT BIOSTRATIGRAPHY AND PALEOECOLOGY, ABST.
B 6303 553* PERMIAN, DELAWARE BASIN TEXAS-NEW MEXICO, ABST., DEEP-SEA
B 6203 500* PERMIAN DELAWARE MOUNTAIN GROUP, WEST TEXAS—SUBSURFACE
B 6401 88† PERMIAN DEPOSITIONAL SYSTEMS AND SHELF-MARGIN EVOLUTION,
B 6405 686 PERMIAN DEPOSITIONAL SYSTEMS UNCOMPAHGRE BASIN, EASTERN
S 6 316 PERMIAN, DEVIANT RESULTS, RADIOMETRIC DATING
S 9 12 PERMIAN, DIAGRAM, POLAR WANDERING
B 6004 671* **PERMIAN, EL DORADO MICELLAR-POLYMER PROJECT AREA, BUTLER**
B 6303 511* PERMIAN EVAPORITES AND RED BEDS OF PALO DURO BASIN,
B 6303 461* PERMIAN FACIES TRACTS AND EVOLUTIO OF CARBONATE-SHELF
S 6 308 PERMIAN, FAUNAL SUBSTAGES, CHART
B 6405 728* PERMIAN FORMATIONS IN PARTS OF COLORADO AND UTAH, ABST.,
M 25 449 PERMIAN GEOSYNCLINAL SEDIMENTATION AND MINERA
M 27 142 PERMIAN, GETAWAY LIMESTONE, TEXAS
M 27 48 PERMIAN, GETAWAY LIMESTONE, TEXAS
B 6004 670* **PERMIAN, GUADALUPE MOUNTAIN, NEW MEXICO AND WEST TEXAS,**
B 6303 471* PERMIAN, GUADALUPE MOUNTAINS, NEW MEXICO, ABST., SEAWARD
B 6203 541* PERMIAN, GUADALUPE MOUNTAINS, NEW MEXICO, ABST., SUBMARINE,
B 6303 437* PERMIAN, GUADALUPE MOUNTAINS, WEST TEXAS AND NEW MEXICO,
B 6203 573* PERMIAN, GUADALUPE MOUNTAINS, WEST TEXAS AND NEW MEXICO—
B 6303 513* PERMIAN, HOBART, TASMANIA, AUSTRALIA, ABST., GLACIOMARINE
B 6207 1171† PERMIAN, IN EASTERN NEVADA AND WESTERN UTAH, STRATIGRAPHIC
B 6205 880* PERMIAN, IN NAVAL PETROLEUM RESERVE ALASKA, ABST., NEW
S 6 309 PERMIAN, KEY FOSSIL GROUPS, AMMONOIDEA
S 6 306 PERMIAN, KEY FOSSIL GROUPS, BRACHIOPODA
S 6 310 PERMIAN, KEY FOSSIL GROUPS, CONODONTS
S 6 309 PERMIAN, KEY FOSSIL GROUPS, FUSULINACEA
S 4 243 PERMIAN LIMESTONES, PHYLLOID ALGAE, PENNSYLVANIAN-
B 6105 824* PERMIAN MARINE FAUNAS IN BIOGEOGRAPHY AND BIOSTRATIGRAPHY,
S 6 302 PERMIAN, MARINE, INTERCORRELATIONS
S 6 300 PERMIAN, MARINE, SUBDIVISIONS
B 6405 605 PERMIAN, MIDDLE EAST
S 6 313 PERMIAN, MIDDLE, RADIOMETRIC VALUES
B 6109 1518 PERMIAN, NEW MEXICO
M 28 129 PERMIAN, NUKA FORMATION, ALASKA
M 28 122 PERMIAN, NUKA FORMATION, ALASKA
M 28 40 PERMIAN, NUKA FORMATION, ALASKA
B 6105 843* PERMIAN, OF GUADALUPE MOUNTAINS, NEW MEXICO AND WEST TEXAS,
B 6405 649 PERMIAN OF PLATFORM AREAS OF POLAND, EDITED BY S. DEPOWSKI,
B 6405 730* PERMIAN OQUIRRH BASIN, NORTHWEST UTAH, ABST.,
B 6405 799* PERMIAN PALEOGEOGRAPHY OF IDAHO, NEVADA, MONTANA, UTAH,
B 6405 677* PERMIAN PALEOGEOGRAPHY, SOUTHERN COLORADO PLATEAU AND
S 6 311 PERMIAN, PALEOMAGNETIC STRATIGRAPHY, A REVIEW
B 6109 1457 PERMIAN, PEDREGOSA BASIN
M 28 83 PERMIAN, PHOSPHORIA FORMATION, IDAHO
M 28 127 PERMIAN, PHOSPHORIA FORMATION, IDAHO
M 27 120 PERMIAN, PHOSPHORIA FORMATION, IDAHO
B 6108 1375* PERMIAN PHOSPHORIA FORMATION, WESTERN INTERIOR UNITED
B 6405 746* PERMIAN PHOSPHORIA FORMATION, WESTERN UNITED STATES, ABST.,
S 6 311 PERMIAN, RADIOMETRIC, CARBONIFEROUS BOUNDARY
S 6 313 PERMIAN, RADIOMETRIC DATING, ALTERNATE VALUES
S 6 313 PERMIAN, RADIOMETRIC DATING, SCATTER
S 6 314 PERMIAN, RADIOMETRIC VALUES, CHART
S 6 315 PERMIAN, RADIOMETRIC VALUES, CORRELATION
S 6 312 PERMIAN, RADIOMETRIC VALUES, DIAGRAM
B 6303 516* PERMIAN ROCKS OF TUNISIA, ABST., MARIN
B 6004 677* PERMIAN, ROTLIEGENDE, VERTEBRATE FOOTPRINTS, ABST.,
M 28 136 PERMIAN, ROTLIEGENDES SANDSTONE, NORTH SEA

M 28 71 PERMIAN, ROTLIEGENDES SANDSTONE, NORTH SEA
M 28 132 PERMIAN, ROTLIEGENDES SANDSTONES, NORTH SEA
B 6105 819* PERMIAN, SACRAMENTO MOUNTAINS, NEW MEXICO, ABST., ORIGIN OF
M 28 89 PERMIAN, SALADO FORMATION, NEW MEXICO
B 6203 526* PERMIAN SALT DEPOSITS IN TEXAS PANHANDLE AND WESTERN
M 28 180 PERMIAN, SANDSTONE, TEXAS
B 6303 474* PERMIAN, SANDSTONES OF PERMIAN BASIN, ABST., PRECIPITATION OF,
M 28 155 PERMIAN, SANGRE DE CRISTO FORMATION, NEW MEXICO
M 28 18 PERMIAN, SANGRE DE CRISTO FORMATION, NEW MEXICO
M 28 19 PERMIAN, SANGRE DE CRISTO FORMATION, NEW MEXICO
M 28 162 PERMIAN, SANGRE DE CRISTO FORMATION, NEW MEXICO
B 6303 493* PERMIAN SEAFLOOR, ABST., IN-SITU FORMATION OF, BOTRYOIDAL,
B 6103 351 PERMIAN SHELF EDGE AND PROXIMAL SLOPE, TEXAS
B 6004 718* PERMIAN SHELF SEVEN RIVERS FORMATION, SOUTHEAST NEW MEXICO,
B 6108 1386* PERMIAN SLIP AND STRATIGRAPHIC CONSEQUENCES OF HURRICANE
S 6 316 PERMIAN, SOLAR CHRONOLOGY
S 6 318 PERMIAN, SOLAR CHRONOLOGY, CLIMATIC CHANGE
S 6 319 PERMIAN, SOLAR CHRONOLOGY, COOLINGS, TABLE
B 6205 888* PERMIAN SOURCE ROCKS, NORTHEASTERN GREAT BASIN, ABST.,
B 6308 1426* PERMIAN STRATA, PALO DURO BASIN, TEXAS, ABST., DEPOSITIONAL
B 6101 12 PERMIAN, SVERDRUP BASIN
M 27 132 PERMIAN, TANSILL FORMATION, NEW MEXICO
M 27 161 PERMIAN, TANSILL FORMATION, NEW MEXICO
B 6206 951 PERMIAN, TEXAS
B 6402 223 PERMIAN, TEXAS, CAMBRIAN THROUGH
B 6103 431 PERMIAN, TEXAS, GLASS MOUNTAINS
S 6 299 PERMIAN, TIME SCALES, A REVIEW
B 6102 257 PERMIAN TO DEVONIAN, NOMENCLATURE, IDAHO
B 6405 769* PERMIAN TOROWEAP FORMATION, ABST., PALEOGEOGRAPH OF
B 6405 669* PERMIAN TRACE FOSSILS OF WESTERN WYOMING AND ADJACENT AREAS,
S 6 306 PERMIAN, TRIASSIC BOUNDARY
S 6 307 PERMIAN, TRIASSIC BOUNDARY, ALTERNATE CLASSIFICATION
B 6109 1499 PERMIAN UNCONFORMITY, ALASKA
B 6007 1107† PERMIAN UNIT IN FINLAY MOUNTAINS, WEST TEXAS, WILKIE RANCH
S 6 313 PERMIAN, UPPER, RADIOMETRIC VALUES
B 6303 509* PERMIAN, WEST TEXAS, ABST., TRACE-ELEMENT, GEOCHEMISTRY AND
B 6303 476* PERMIAN, WESTERN OQUIRRH BASIN, UTAH, ABST., BATHYMETRIC
S 9 10 PERMIAN, WIND DIRECTION, EUROPE
S 9 11 PERMIAN, WIND DIRECTION, WESTERN, UNITED STATES
B 6002 321* PERMIAN, WOLFCAMPIAN, IN NORTHEAST KANSAS, ABST., BIOFACIES
B 6203 541* PERMIAN, ZECHSTEIN 2, CARBONATE ROCKS OF EASTERN NETHERLANDS
B 6102 297* PERMIAN-PENNSYLVANIAN AGE, VAL VERDE BASIN, TEXAS, BY TITLE
B 6405 804* PERMIAN-PENNSYLVANIAN OF WEST-CENTRAL NEBRASKA PANHANDLE,
B 6103 465* PERMIAN-TRIASSIC OF PRUDHOE BAY FIELD, ABST.
B 6101 13 PERMO-PENNSYLVANIAN, CANADIAN, ARCTIC ISLANDS
M 24 23† PERMO-TRIASSIC RESERVOIRS OF PRUDHO BAY FIELD, NORTH SLOPE,
B 6004 665* PERMOPOROSITY IN CARBONATE ROCKS, ABST., EXPERIMENTAL
B 6109 1471 PERNAMBUCO BASIN, BRAZIL
B 6302 223 PERNAMBUCO PLATEAU, BRAZIL
B 6111 1981 PERRIS BLOCK, CALIFORNIA
B 6109 1548* PERRY AND STONE COUNTIES, MISSISSIPPI, ABST., PETROLOGY OF,
B 6002 280 PERRY BAY, PENNSYLVANIA
B 6004 682 PERRY, EDWARD A., JR., TIMING OF CARBONATE AND CLAY DIAGENESIS
B 6002 321 PERRY, NORTON, EASTERN OKLAHOMA AND WESTERN ARKANSAS. BY
B 6206 1091 PERRY, R. V., NEW MODEL FOR SOME COLORADO PLATEAU URANIUM
B 6203 501 PERRY, ROBERT, NEW MODEL FOR SOME COLORADO PLATEAU URANIUM
M 24 247 PERRY SANDSTONE, MISSISSIPPI
S 5 125 PERRY, W. J., JR RELATIONSHIPS BETWEEN DEPOSITIONAL
B 6009 1623 PERRY, W. J., JR., STRATIGRAPHY AS KEY TO ARCH-RELATED ORIGIN OF
B 6304 647 PERRY, WILLIAM J., CONODONT-BASED REINTERPRETATION OF BANE
B 6004 707 PERRY, WILLIAM J., JR., POSSIBILITY OF ABNORMALLY HIGH FLUID
M 25 402 PERSEVERANCE, NICKEL DEPOSIT, WESTERN AUSTRALIA
B 6104 585 PERSIAN GULF, AB TEYMUR FIELD
B 6104 585 PERSIAN GULF, ABU HADNIYA FIELD
B 6104 585 PERSIAN GULF, ABU SAFAH FIELD
B 6104 585 PERSIAN GULF, AGHA JARI FIELD
B 6104 585 PERSIAN GULF, AHWAZ FIELD
B 6408 1281* PERSIAN GULF ANALOGY, ABST., MIDDLE, ORDOVICIA SHELF
B 6104 585 PERSIAN GULF, ASAB, ABU JIDU, FIELD
B 6104 585 PERSIAN GULF, BAB, MURBAN, FIELD
B 6104 585 PERSIAN GULF BASIN, OIL AGE
B 6104 585 PERSIAN GULF, BERRI FIELDS
B 6104 585 PERSIAN GULF, BIBI HAKIMEH FIELD
B 6104 585 PERSIAN GULF, BINAK FIELD
B 6104 585 PERSIAN GULF, BU HASA FIELD
B 6104 585 PERSIAN GULF, CHASHMEH KHUSH FIELD
S 4 381† PERSIAN GULF COASTAL SABKHA, MARINE AND, CONTINENTAL,
S 10 157 PERSIAN GULF, CRETACEOUS, POTENTIOMETRIC SURFACE
B 6104 585 PERSIAN GULF, DAMMAN FIELD
B 6312 2119 PERSIAN GULF, DARIUS FIELD, YAMAMA LIMESTONE
B 6104 585 PERSIAN GULF, DUKHAN FIELDS
S 4 285 PERSIAN GULF, ECHINOMETRA MATHEI
B 6104 585 PERSIAN GULF, GHAWAR-AIN DAR, D RESERVOIR, FIELD
B 6104 585 PERSIAN GULF, GHAWAR-HARADH, D RESERVOIR, FIELD
B 6104 585 PERSIAN GULF, GHAWAR-UTHMANIYAH, D RESERVOIR, FIELD
B 6104 585 PERSIAN GULF, HAFT KEL FIELD
M 28 121 PERSIAN GULF, HOLOCENE, SUBMARINE SANDS
B 6104 585 PERSIAN GULF, JURAYBIAT FIELDS
B 6104 585 PERSIAN GULF, KHARG FIELD
B 6104 585 PERSIAN GULF, KHURSANIYA, A B RESERVOIRS, FIELD
B 6104 594 PERSIAN GULF, KILAR KURIM FIELD
B 6104 585 PERSIAN GULF, KILUR KARIM FIELD
B 6104 585 PERSIAN GULF, MARJAN, KHAFJI RESERVOIR, FIELD
B 6104 585 PERSIAN GULF, MARUN FIELD
B 6408 1190 PERSIAN GULF, OXIC NEGATIVE WATER BALANCE BASIN
B 6104 585 PERSIAN GULF, PAR-E SIAH FIELD
S 10 156 PERSIAN GULF, PETROLEUM MIGRATION, MIDDLE EAST
B 6104 585 PERSIAN GULF, QATIF FIELDS
B 6104 585 PERSIAN GULF, RAG-E SAFID FIELD
B 6104 585 PERSIAN GULF, RAMSHIR
B 6104 585 PERSIAN GULF, RUWAIS FIELD
B 6104 585 PERSIAN GULF, SAHIL FIELD
M 28 120 PERSIAN GULF, SUBMARINE SANDS
B 6301 36 PERSON FIELD, KARNES COUNTY, TEXAS
B 6201 41 PERTH ABYSSAL PLAIN, AUSTRALIA
B 6111 1994 PERTH ABYSSAL PLAIN, POSSIBLE DIAPIRS
B 6201 41 PERTH BASIN, AUSTRALIA
B 6210 2151 PERTH BASIN, AUSTRALIA
B 6307 1092† PERTH BASIN, AUSTRALIA, GEOCHEMICAL ANALYSIS OF HYDROCARBON
B 6307 1104 PERTH BASIN, AUSTRALIA, SOURCE ROCKS, NORTH
B 6307 1093 PERTH BASIN, TECTONIC ELEMENTS, AUSTRALIA
B 6303 539* PERTH BASIN, WESTER AUSTRALIA, ABST., GENERATION AND
B 6010 1655 PERU, 1975, ZUNIGA Y RIVERO, F., ON DEVELOPMENTS IN
B 6110 1590 PERU, 1976, ZUNIGA Y RIVERO, F., ON DEVELOPMENTS IN
B 6210 2058 PERU, 1977, ZUNIGA Y RIVERO, F., AND, PETROCONSULTANTS S.A., ON
B 6310 1758 PERU, 1978, ZUNIGA Y RIVERO, J., ON DEVELOPMENTS IN
B 6405 668* PERU, ABST., EARLY, CRETACEOUS ARC SEDIMENTATION AND,
B 6004 735* PERU, ABST., PETROLEUM GEOLOGY AND, PRODUCTIVE, POTENTIAL OF
B 6112 2104 PERU, AGUA CALIENTE SECTION
M 25 339 PERU, AMAZON BASINS, POTENTIAL
M 25 533 PERU, BASE-METAL DEPOSITS, PRODUCTION
B 6112 2105 PERU, CAPAHUARI FIELD
B 6112 2105 PERU, CARMEN FIELD
B 6112 2104 PERU, CHONTA FORMATION
B 6112 2104 PERU, CHONTA LIMESTONE
M 25 531 PERU, COPPER PRODUCTION
B 6112 2104 PERU, CORRIENTES FIELD
B 6112 2105 PERU, CORRIENTES FIELD
B 6408 1196 PERU CURRENT, ANOXIC LAYERS CAUSED BY UPWELLING
B 6112 2104 PERU, CUSHABATAY SECTION
M 25 533 PERU, HUANCAVELICA DISTRICT, MERCURY
B 6112 2103† PERU, HYDROCARBON, GEOCHEMISTRY OF OILS FROM, MARANON BASIN
B 6112 2104 PERU, MARANON BASIN OILS, CLASSIFICATION
M 25 533 PERU, MOLYBDENUM
M 25 338 PERU, MOQUEGUA BASIN
B 6112 2105 PERU, PHIELIPS FIELD
M 25 338 PERU, PISCO BASIN
B 6110 1622 PERU, PRODUCTION, 1975-1976
B 6210 2090 PERU, PRODUCTION, 1976-1977
B 6310 1784 PERU, PRODUCTION, 1977-1978
M 25 335 PERU, PROGRESO BASIN
B 6112 2104 PERU, RAYA SECTION
M 25 337 PERU, SALAVERRY BASIN
M 25 335 PERU, SECHURA BASIN
M 29 443 PERU, SEISMIC EXPERIMENT, CONTINENTAL MARGINS
B 6112 2105 PERU, SHIVIYACU FIELD
B 6112 2105 PERU, SUN FIELD
M 25 336 PERU, TALARA BASIN
M 29 444 PERU TRENCH, BATHYMETRY PROFILE
M 29 444 PERU TRENCH, EARTHQUAKE HYPOCENTERS
M 29 445 PERU TRENCH, NAZCA PLATE, SEISMIC RECORDS IN SEDIMENTS
M 29 447 PERU TRENCH, SUBDUCTION ZONE, SEISMIC MODEL
M 25 533 PERU, TUNGSTEN
B 6112 2105 PERU, VALENCIA FIELD
B 6112 2104 PERU, VIVIAN SECTION
B 6112 2105 PERU, YANAYACU FIELD
B 6112 2105 PERU, YANEZ FIELD
M 25 556† PERU--SUMMARY, MINERAL RESOURCES POTENTIAL
B 6212 2470 PERU-CHILE SUBDUCTION ZONE
B 6212 2462 PERU-CHILE TRENCH

M 25 171 PERU-CHILE TRENCH
M 25 529 PERU-CHILE TRENCH
M 25 539 PERU-CHILE TRENCH
M 29 251 PERU-CHILE TRENCH, CONSTRUCTED BASIN, IQUIQUE BASIN
M 25 541 PERU-CHILE TRENCH, RATE OF, NAZCA RIDGE, CONVERGENCE WITH
B 6207 1239* PERU-CHILE TRENCH—RELATION TO MINERAL DEPOSITS, ABST.,
B 6408 1197 PERUVIAN SHELF, ANOXIC LAYERS CAUSED BY UPWELLING
B 6004 709 PERWIRA, J., INDONESIA-SOUTH SUMATRA AND JAV SEA
B 6211 2224 PESCADERO POINT, CALIFORNIA
B 6004 707 PESCATORE, T., IRPINIDS--MODEL OF TECTONICALLY CONTROLLED
S 10 232 PESCHANY FIELD, USSR, BAKU AREA
B 6303 541 PESSAGNO, EMILE A., JR., RADIOLARIA FROM OMAN MOUNTAINS, ABST.
B 6303 421 PESSAGNO, EMILE A., UPPER TRIASSIC RADIOLARIA FROM EASTERN
B 6004 727 PESSEL, G. H., PRUDHOE BAY SUPE GIANT FIELD AND ALASKAN ARCTIC
B 6012 2190 PESSEL, G. H., SUBSURFACE GEOLOGY AND SOME IMPLICATIONS
S 8 105 PETACALCO CANYON, MEXICO, SUBMARINE CANYONS
B 6007 1078 PETER, GEORGE, TECTONICS OF SOUTHWESTERN NORTH ATLANTIC AND
B 6404 530 PETERMANN HALVO FORMATION, GREENLAND
S 4 88 PETERS HILL FORMATION, UPPER SANTONIAN, JAMAICA
B 6104 504 PETERS, K. E., COLOR OF KEROGEN AS INDEX OF ORGANIC MATURITY
B 6405 763 PETERS, K. E., GEOCHEMISTRY OF ARTIFICIALLY HEATED HUMIC AND
B 6304 652 PETERS MOUNTAIN, VIRGINIA
B 6308 1301 PETERS ROGER, OIL AND GAS DEVELOPMENTS IN NORTH
B 6409 1437 PETERS, ROGER, OIL AND GAS DEVELOPMENTS IN NORTH
B 6108 1305 PETERS, ROGER, OIL AND GAS DEVELOPMENTS IN NORTH
B 6208 1477 PETERS, ROGER, OIL AND GAS DEVELOPMENTS IN NORTH
B 6412 2168 PETERSBURG LINEAMENT, WEST VIRGINIA
B 6309 1588* PETERSBURG REGION, WEST VIRGINIA, ABST., TECTONICS OF
B 6105 840* PETERSBURG STRUCTURAL LINEAMENTS. WEST VIRGINIA. ABST., FIELD
M 25 154 **PETERSON, DONALD W., DEEP RESEARCH DRILLHOLE AT KILAUEA**
M 25 371 **PETERSON, F. L., EFFECT OF SUBSURFACE WASTE-DISPOSAL PRACTICE ON**
B 6405 763 PETERSON, FRED, MUDSTONES AS EXPLORATION GUIDES TO TABULAR
B 6305 837 PETERSON, FRED, SEDIMENTARY AND TECTONIC CONTROLS OF
B 6012 2186 PETERSON, GARY L., CLAST POPULATIONS IN SESPE AND POWAY
B 6004 707 PETERSON, GARY L., SEDIMENTOLOGIC INDICATION OF EOCENE
B 6303 432 PETERSON, JAMES A., WEST SIBERIAN BASIN, ABST.
B 6203 517* PETERSON LIMESTONE, WESTERN WYOMING, ABST., LACUSTRINE,
B 6002 321 PETERSON, R. M., BIOFACIES ANALYSIS OF UPPERMOST HAMLIN SHALE
B 6303 448 **PETERSON, R. MICHAEL, GENERATION AND OCCLUSION OF POROSITY IN**
B 6405 764 PETERSON, REX M., LINEAMENT ANALYSIS FOR OIL AND GAS
B 6002 321 PETERSON, REX M., NEBRASKA OIL FIELDS AND ERTS, ABST.
B 6301 90 PETREL MEMBER, SCOTIAN BASIN
M 25 449 PETROCHEMICAL ZONING, CORRELATION OF, METAL CONCENTRATION
B 6110 1800 PETROCONSULTANTS ET AL, ON DEVELOPMENTS IN SHARJAH, 1976,
B 6210 2113 PETROCONSULTANTS S.A., AND ASIATIC PETROLEUM CO., ON
B 6210 2117 PETROCONSULTANTS S.A., AND ASIATIC PETROLEUM CORP., ON
B 6411 1838 PETROCONSULTANTS S.A. AND GETTY OIL CO., ON DEVELOPMENTS IN
B 6110 1802 PETROCONSULTANTS S.A., AND NATIONAL RES. AUTHORITY, HASHEMITE
B 6310 1886 PETROCONSULTANTS S.A., ET AL, ON DEVELOPMENTS IN SAUDI ARABIA,
B 6010 1871 PETROCONSULTANTS S.A., O DEVELOPMENTS IN RAS AL KHAIMAH, 1975,
B 6010 1770 PETROCONSULTANTS S.A., ON DEVELOPMENTS I MOROCCO, 1975, ALEM,
B 6010 1871 PETROCONSULTANTS S.A., ON DEVELOPMENTS IN AJMAN AND UMM AL
B 6110 1800 PETROCONSULTANTS S.A., ON DEVELOPMENTS IN AJMAN AND UMM AL
B 6010 1768 PETROCONSULTANTS S.A., ON DEVELOPMENTS IN ALGERIA, 1975
B 6110 1699 PETROCONSULTANTS S.A., ON DEVELOPMENTS IN ALGERIA, 1976
B 6210 1797 PETROCONSULTANTS S.A., ON DEVELOPMENTS IN ALGERIA, 1977
B 6310 1637 PETROCONSULTANTS S.A., ON DEVELOPMENTS IN ALGERIA, 1978
B 6411 1740 PETROCONSULTANTS S.A., ON DEVELOPMENTS IN ALGERIA, 1979
B 6110 1579 PETROCONSULTANTS S.A., ON DEVELOPMENTS IN ARGENTINA, 1976,
B 6110 1580 PETROCONSULTANTS S.A., ON DEVELOPMENTS IN BAHAMAS, 1976
B 6210 2046 PETROCONSULTANTS S.A., ON DEVELOPMENTS IN BAHAMAS, 1977
B 6310 1747 PETROCONSULTANTS S.A., ON DEVELOPMENTS IN BAHAMAS, 1978
B 6210 2117 PETROCONSULTANTS S.A., ON DEVELOPMENTS IN BAHRAIN, 1977,
B 6110 1580 PETROCONSULTANTS S.A., ON DEVELOPMENTS IN BARBADOS, 1976
B 6110 1580 PETROCONSULTANTS S.A., ON DEVELOPMENTS IN BELIZE, 1976
B 6210 2046 PETROCONSULTANTS S.A., ON DEVELOPMENTS IN BELIZE, 1977
B 6310 1747 PETROCONSULTANTS S.A., ON DEVELOPMENTS IN BELIZE, 1978
B 6310 1747 PETROCONSULTANTS S.A., ON DEVELOPMENTS IN BOLIVIA, 1978
B 6110 1583 PETROCONSULTANTS S.A., ON DEVELOPMENTS IN COSTA RICA, 1976
B 6210 2049 PETROCONSULTANTS S.A., ON DEVELOPMENTS IN COSTA RICA, 1977
B 6310 1750 PETROCONSULTANTS S.A., ON DEVELOPMENTS IN COSTA RICA, 1978
B 6010 1649 PETROCONSULTANTS S.A., ON DEVELOPMENTS IN CUBA, 1975
B 6210 2049 PETROCONSULTANTS S.A., ON DEVELOPMENTS IN CUBA, 1977
B 6310 1750 PETROCONSULTANTS S.A., ON DEVELOPMENTS IN CUBA, 1978
B 6010 1649 PETROCONSULTANTS S.A., ON DEVELOPMENTS IN DOMINICAN REPUBLIC,
B 6210 2050 PETROCONSULTANTS S.A., ON DEVELOPMENTS IN DOMINICAN REPUBLIC,
B 6310 1750 PETROCONSULTANTS S.A., ON DEVELOPMENTS IN DOMINICAN REPUBLIC,
B 6010 1769 PETROCONSULTANTS S.A., ON DEVELOPMENTS IN EGYPT, 1975
B 6110 1700 PETROCONSULTANTS S.A., ON DEVELOPMENTS IN EGYPT, 1976
B 6210 1797 PETROCONSULTANTS S.A., ON DEVELOPMENTS IN EGYPT, 1977
B 6310 1637 PETROCONSULTANTS S.A., ON DEVELOPMENTS IN EGYPT, 1978
B 6411 1740 PETROCONSULTANTS S.A., ON DEVELOPMENTS IN EGYPT, 1979
B 6010 1655 PETROCONSULTANTS S.A., ON DEVELOPMENTS IN EL SALVADOR, 1975
B 6210 2059 PETROCONSULTANTS S.A., ON DEVELOPMENTS IN EL SALVADOR, 1977
B 6310 1761 PETROCONSULTANTS S.A., ON DEVELOPMENTS IN EL SALVADOR, 1978
B 6110 1586 PETROCONSULTANTS S.A., ON DEVELOPMENTS IN FRENCH GUIANA, 1976
B 6310 1753 PETROCONSULTANTS S.A., ON DEVELOPMENTS IN GUATEMALA, 1978
B 6110 1586 PETROCONSULTANTS S.A., ON DEVELOPMENTS IN GUYANA, 1976
B 6210 2052 PETROCONSULTANTS S.A., ON DEVELOPMENTS IN GUYANA, 1977
B 6310 1754 PETROCONSULTANTS S.A., ON DEVELOPMENTS IN GUYANA, 1978
B 6010 1652 PETROCONSULTANTS S.A., ON DEVELOPMENTS IN HAITI, 1975
B 6210 2052 PETROCONSULTANTS S.A., ON DEVELOPMENTS IN HAITI, 1977
B 6310 1754 PETROCONSULTANTS S.A., ON DEVELOPMENTS IN HAITI, 1978
B 6110 1587 PETROCONSULTANTS S.A., ON DEVELOPMENTS IN HONDURAS, 1976
B 6411 1837 PETROCONSULTANTS S.A., ON DEVELOPMENTS IN IRAN, 1979
B 6010 1865 PETROCONSULTANTS S.A., ON DEVELOPMENTS IN IRAQ, 1975
B 6411 1836 PETROCONSULTANTS S.A., ON DEVELOPMENTS IN IRAQ, 1979
B 6110 1587 PETROCONSULTANTS S.A., ON DEVELOPMENTS IN JAMAICA, 1976
B 6310 1754 PETROCONSULTANTS S.A., ON DEVELOPMENTS IN JAMAICA, 1978
B 6010 1873 PETROCONSULTANTS S.A., ON DEVELOPMENTS IN JORDAN, 1975,
B 6210 2119 PETROCONSULTANTS S.A., ON DEVELOPMENTS IN JORDAN, 1977,
B 6411 1841 PETROCONSULTANTS S.A., ON DEVELOPMENTS IN JORDAN, 1979
B 6411 1837 PETROCONSULTANTS S.A., ON DEVELOPMENTS IN KUWAIT, 1979, KUWAIT
B 6210 2114 PETROCONSULTANTS S.A., ON DEVELOPMENTS IN KUWAIT-SAUDI
B 6010 1873 PETROCONSULTANTS S.A., ON DEVELOPMENTS IN LEBANON, 1975
B 6010 1653 PETROCONSULTANTS S.A., ON DEVELOPMENTS IN LESSER ANTILLES, 1975
B 6010 1769 PETROCONSULTANTS S.A., ON DEVELOPMENTS IN LIBYA, 1975
B 6110 1701 PETROCONSULTANTS S.A., ON DEVELOPMENTS IN LIBYA, 1976
B 6210 1798 PETROCONSULTANTS S.A., ON DEVELOPMENTS IN LIBYA, 1977
B 6310 1638 PETROCONSULTANTS S.A., ON DEVELOPMENTS IN LIBYA, 1978
B 6411 1741 PETROCONSULTANTS S.A., ON DEVELOPMENTS IN LIBYA, 1979
B 6210 2054 PETROCONSULTANTS S.A., ON DEVELOPMENTS IN MEXICO, 1977
B 6310 1755 PETROCONSULTANTS S.A., ON DEVELOPMENTS IN MEXICO, 1978
B 6110 1701 PETROCONSULTANTS S.A., ON DEVELOPMENTS IN MOROCCO, 1976, ALEM,
B 6210 1799 PETROCONSULTANTS S.A., ON DEVELOPMENTS IN MOROCCO, 1977, ALEM,
B 6310 1639 PETROCONSULTANTS S.A., ON DEVELOPMENTS IN MOROCCO, 1978, ALEM,
B 6411 1742 PETROCONSULTANTS S.A., ON DEVELOPMENTS IN MOROCCO, 1979
B 6010 1653 PETROCONSULTANTS S.A., ON DEVELOPMENTS IN NETHERLANDS
B 6110 1589 PETROCONSULTANTS S.A., ON DEVELOPMENTS IN NICARAGUA, 1976
B 6210 2057 PETROCONSULTANTS S.A., ON DEVELOPMENTS IN NICARAGUA, 1977
B 6310 1757 PETROCONSULTANTS S.A., ON DEVELOPMENTS IN NICARAGUA, 1978
B 6010 1654 PETROCONSULTANTS S.A., ON DEVELOPMENTS IN PANAMA, 1975
B 6210 2057 PETROCONSULTANTS S.A., ON DEVELOPMENTS IN PANAMA, 1977
B 6310 1757 PETROCONSULTANTS S.A., ON DEVELOPMENTS IN PANAMA, 1978
B 6110 1590 PETROCONSULTANTS S.A., ON DEVELOPMENTS IN PARAGUAY, 1976
B 6210 2058 PETROCONSULTANTS S.A., ON DEVELOPMENTS IN PARAGUAY, 1977
B 6210 2058 PETROCONSULTANTS S.A., ON DEVELOPMENTS IN PERU, 1977, ZUNIGA Y
B 6110 1590 PETROCONSULTANTS S.A., ON DEVELOPMENTS IN PUERTO RICO, 1976
B 6110 1800 PETROCONSULTANTS S.A., ON DEVELOPMENTS IN RAS AL KAHIMAH, 1976
B 6310 1888 PETROCONSULTANTS S.A. ON DEVELOPMENTS IN RAS AL KHAIMAH, 1978,
B 6411 1838 PETROCONSULTANTS S.A., ON DEVELOPMENTS IN SAUDI ARABIA, 1979,
B 6010 1871 PETROCONSULTANTS S.A., ON DEVELOPMENTS IN SHARJAH, 1975,
B 6210 2117 PETROCONSULTANTS S.A., ON DEVELOPMENTS IN SHARJAH, 1977
B 6310 1889 PETROCONSULTANTS S.A., ON DEVELOPMENTS IN SHARJAH, 1978, AMOCO
B 6210 2118 PETROCONSULTANTS S.A., ON DEVELOPMENTS IN SOUTH YEMEN, 1977,
B 6411 1742 **PETROCONSULTANTS S.A., ON DEVELOPMENTS IN SOUTH YEMEN, 1978,**
B 6010 1771 PETROCONSULTANTS S.A., ON DEVELOPMENTS IN SUDAN, 1975
B 6110 1702 PETROCONSULTANTS S.A., ON DEVELOPMENTS IN SUDAN, 1976
B 6210 1799 PETROCONSULTANTS S.A., ON DEVELOPMENTS IN SUDAN, 1977
B 6310 1639 PETROCONSULTANTS S.A., ON DEVELOPMENTS IN SUDAN, 1978
B 6110 1590 PETROCONSULTANTS S.A., ON DEVELOPMENTS IN SURINAM, 1976
B 6010 1873 PETROCONSULTANTS S.A., ON DEVELOPMENTS IN SYRIA, 1975
B 6411 1842 PETROCONSULTANTS S.A. ON DEVELOPMENTS IN SYRIA, 1979, COASTAL
B 6110 1591 PETROCONSULTANTS S.A., ON DEVELOPMENTS IN TRINIDAD, 1976
B 6010 1771 PETROCONSULTANTS S.A., ON DEVELOPMENTS IN TUNISIA, 1975
B 6110 1702 PETROCONSULTANTS S.A., ON DEVELOPMENTS IN TUNISIA, 1976
B 6210 1799 PETROCONSULTANTS S.A., ON DEVELOPMENTS IN TUNISIA, 1977
B 6310 1640 PETROCONSULTANTS S.A., ON DEVELOPMENTS IN TUNISIA, 1978
B 6411 1743 PETROCONSULTANTS S.A., ON DEVELOPMENTS IN TUNISIA, 1979
B 6010 1874 PETROCONSULTANTS S.A., ON DEVELOPMENTS IN TURKEY, 1975,
B 6110 1802 PETROCONSULTANTS S.A., ON DEVELOPMENTS IN TURKEY, 1976,
B 6310 1891 PETROCONSULTANTS S.A., ON DEVELOPMENTS IN TURKEY, 1978, ASIATIC
B 6110 1591 PETROCONSULTANTS S.A., ON DEVELOPMENTS IN URUGUAY, 1976
B 6210 2060 PETROCONSULTANTS S.A., ON DEVELOPMENTS IN URUGUAY, 1977
B 6110 1591 PETROCONSULTANTS S.A., ON DEVELOPMENTS IN VENEZUELA, 1976,
B 6310 1890 PETROCONSULTANTS S.A., ON DEVELOPMENTS IN YEMEN, 1978
B 6110 1796 PETROCONSULTANTS S.A., PHILLIPS PETROL. CO., DEMINEX IRAN OIL CO.,
B 6210 2117 PETROCONSULTANTS S.A., PRUDENTIAL INTERNATIONAL, INC., AND

Index

B 6310 1890 PETROCONSULTANTS S.A., QUINTANA OVERSEAS, INC. SUNMAR
B 6210 2117 PETROCONSULTANTS S.A., WEEKS PETROLEUM LTD., ET AL, ON
M 25 526 PETROGENIC INTERPRETATION, MEXICO, VOLCANIC PROVINCES
B 6309 1578* **PETROGRAPHIC EVALUATION OF WEST VIRGINIA COALS FOR COAL**
B 6409 1565* PETROGRAPHIC, STRATIGRAPHY, AND STRUCTURAL STUDY OF
B 6405 725* PETROGRAPHIC VARIATION IN WESTERN KENTUCKY NO. 11 COAL SEAM,
B 6209 1769* PETROGRAPHY AND DIAGENESIS OF HOSSTON SANDSTONE RESERVOIRS
B 6406 945† PETROGRAPHY AND GEOLOGIC SIGNIFICANCE OF UPPER JURASSIC
B 6309 1602* PETROGRAPHY OF SOME SUBSURFACE IGNEOUS ROCKS OF MISSISSIPPI,
B 6010 1644 PETROLEO BRASILEIRO S.A., ON DEVELOPMENTS IN BRAZIL, 1975
B 6401 107† PETROLEUM, A PROCESS FOR, PRIMARY, MIGRATION OF
B 6203 567* PETROLEUM, ABST., GRAVITY-INDUCED CROSS-FORMATIONAL, WATER,
B 6203 556* PETROLEUM, ABST., NEW EVIDENCE FOR HOT, DEEP, ORIGIN AND,
B 6001 12† PETROLEUM ACCUMULATION, LOST SOLDIER AREA, WYOMING,
B 6405 668* PETROLEUM ACCUMULATION TRENDS IN EAST TEXAS SALT DOME AREA,
B 6211 2359* PETROLEUM ACCUMULATIONS IN SANTA MARGARITA FORMATION,
B 6105 775* **PETROLEUM AND MISSISSIPPI VALLEY TYPE ORE DEPOSITS—COMMON**
B 6002 213† **PETROLEUM AS APPLIED TO ITS ORIGIN AND PRIMARY MIGRATION,**
B 6112 2146† PETROLEUM AS APPLIED TO ITS ORIGIN AND PRIMARY MIGRATION.
M 29 423 PETROLEUM, DEFINITION, SOURCE BEDS
B 6411 2040 **PETROLEUM DEVELOPMENT GEOLOGY, BY PARKE A DICKEY,**
B 6310 1689† PETROLEUM DEVELOPMENT IN CENTRAL AND SOUTHERN AFRICA IN 1978
B 6207 1211* PETROLEUM DEVELOPMENT IN COASTAL ZONE, ABST.
B 6010 1957† PETROLEUM DEVELOPMENTS IN AUSTRALIA IN 1975
B 6110 1888† PETROLEUM DEVELOPMENTS IN AUSTRALIA IN 1976
B 6210 2150† PETROLEUM DEVELOPMENTS IN AUSTRALIA IN 1977
B 6310 1804† PETROLEUM DEVELOPMENTS IN AUSTRALIA IN 1978
B 6411 1862† PETROLEUM DEVELOPMENTS IN AUSTRALIA IN 1979
B 6010 1813† PETROLEUM DEVELOPMENTS IN CENTRAL AND SOUTHERN AFRICA IN
B 6110 1746† PETROLEUM DEVELOPMENTS IN CENTRAL AND SOUTHERN AFRICA IN
B 6210 1844† PETROLEUM DEVELOPMENTS IN CENTRAL AND SOUTHERN AFRICA IN
B 6411 1785† PETROLEUM DEVELOPMENTS IN CENTRAL AND SOUTHERN AFRICA IN
B 6411 1961† PETROLEUM DEVELOPMENTS IN EUROPE IN 1979
B 6411 1880† PETROLEUM DEVELOPMENTS IN FAR EAST, 1979
B 6010 1908† PETROLEUM DEVELOPMENTS IN FAR EAST IN 1975
B 6110 1832† PETROLEUM DEVELOPMENTS IN FAR EAST IN 1976
B 6210 1978† PETROLEUM DEVELOPMENTS IN FAR EAST IN 1977
B 6310 1816† PETROLEUM DEVELOPMENTS IN FAR EAST IN 1978
B 6010 1905† PETROLEUM DEVELOPMENTS IN ISRAEL IN 1975
B 6110 1830† PETROLEUM DEVELOPMENTS IN ISRAEL IN 1976
B 6210 1898† PETROLEUM DEVELOPMENTS IN ISRAEL IN 1977
B 6310 1743† PETROLEUM DEVELOPMENTS IN ISRAEL IN 1978
B 6010 1865† **PETROLEUM DEVELOPMENTS IN MIDDLE EAST COUNTRIES IN 1975**
B 6110 1795† PETROLEUM DEVELOPMENTS IN MIDDLE EAST COUNTRIES IN 1976
B 6210 2111† PETROLEUM DEVELOPMENTS IN MIDDLE EAST COUNTRIES IN 1977
B 6310 1884† PETROLEUM DEVELOPMENTS IN MIDDLE EAST COUNTRIES IN 1978
B 6411 1836† PETROLEUM DEVELOPMENTS IN MIDDLE EAST COUNTRIES IN 1979
B 6310 1636† PETROLEUM DEVELOPMENTS IN NORTH AFRICA
B 6010 1767† PETROLEUM DEVELOPMENTS IN NORTH AFRICA IN 1975
B 6110 1698† PETROLEUM DEVELOPMENTS IN NORTH AFRICA IN 1976
B 6210 1796† PETROLEUM DEVELOPMENTS IN NORTH AFRICA IN 1977
B 6411 1739† PETROLEUM DEVELOPMENTS IN NORTH AFRICA IN 1979
B 6010 1640† PETROLEUM DEVELOPMENTS IN SOUTH AMERICA, CENTRAL AMERICA,
B 6210 2044† PETROLEUM DEVELOPMENTS IN SOUTH AMERICA, CENTRAL AMERICA,
B 6110 1578† PETROLEUM DEVELOPMENTS IN SOUTH AMERICA, CENTRAL AMERICA,
B 6310 1745† PETROLEUM DEVELOPMENTS IN SOUTH AMERICA, CENTRAL AMERICA,
B 6407 1121 PETROLEUM DICTIONARY, BY DAVI F. TVER AND RICHARD W. BERRY,
B 6303 531* PETROLEUM ENGINEERING--GEOLOGY, SYNERGISM, KEY TO
B 6010 1704† PETROLEUM EXPLORATION AND PRODUCTION IN EUROPE IN 1975
B 6110 1636† PETROLEUM EXPLORATION AND PRODUCTION IN EUROPE IN 1976
B 6210 1906† PETROLEUM EXPLORATION AND PRODUCTION IN EUROPE IN 1977
B 6310 1915† PETROLEUM EXPLORATION AND PRODUCTION IN EUROPE IN 1978
B 6405 768* PETROLEUM EXPLORATION AND WRENCHING MODEL MICHIGAN BASIN,
B 6305 836* **PETROLEUM EXPLORATION IN YOUNGER OVER OLDER THRUST PLATES IN**
B 6405 715* PETROLEUM EXPLORATION O NATIONAL PETROLEUM RESERVE IN
B 6207 1218* PETROLEUM EXPLORATION OF NATIONAL PETROLEUM RESERVE IN
M 30 225† PETROLEUM FIELDS IN PROTEROZOIC AND LOWER CAMBRIAN STRATA,
M 25 276† **PETROLEUM FIELDS OF JAPAN WITH VOLCANIC-ROCK**
B 6309 1548 PETROLEUM FORMATION AN OCCURRENCE--A NEW APPROACH TO OIL
B 6303 433* PETROLEUM GENERATION AN MIGRATION IN DENVER BASIN, ABST.
B 6410 1613† PETROLEUM GENERATION AND MIGRATION IN DENVER BASIN
B 6401 59† PETROLEUM GENERATION AND MIGRATION, ROLE OF, MINERAL,
B 6406 954 PETROLEUM GEOCHEMISTRY AND GEOLOGY, BY JOHN M. HUNT,
B 6309 1583* PETROLEUM GEOCHEMISTRY AND GEOLOGY OF SOUTHEAST GEORGIA
B 6104 573† PETROLEUM GEOCHEMISTRY I, CALCULATION OF, AGES OF,
B 6103 331† PETROLEUM GEOLOGY AND OIL AND GAS POTENTIAL OF SOUTH
B 6004 735* PETROLEUM GEOLOGY AND PRODUCTIVE POTENTIAL OF PERU, ABST.
B 6405 720* PETROLEUM GEOLOGY IN 1980S, ABST.
M 30 487† PETROLEUM GEOLOGY OF BOMBAY HIGH FIELD, INDIA
B 6304 697* PETROLEUM GEOLOGY OF CAMPECHE OFFSHOR AREA, SOUTHEASTERN
B 6207 1225* PETROLEUM GEOLOGY OF COOK INLET BASIN, ALASKA, ABST.
B 6109 1470† PETROLEUM GEOLOGY OF EASTERN BRAZILIAN CONTINENTAL MARGIN
B 6004 656* PETROLEUM GEOLOGY OF EASTERN BRAZILIAN CONTINENTAL MARGIN,
B 6008 1393* PETROLEUM GEOLOGY OF FORT UNION AND MESAVERDE
B 6205 888* PETROLEUM GEOLOGY OF GREAT BASIN, WESTERN UTAH AND NEVADA,
B 6408 1140† PETROLEUM GEOLOGY OF KODIAK SHELF, ALASKA
B 6203 503* PETROLEUM GEOLOGY OF PALEOZOIC CLASTIC ROCKS OF MIDDLE
B 6207 1219* PETROLEUM GEOLOGY OF REFORMA AREA, SOUTHEASTERN MEXICO,
B 6405 668* PETROLEUM GEOLOGY OF SENONIAN IN SEDIMENTS ANAMBRA
M 25 251† PETROLEUM GEOLOGY OF THAILAND, GULF O THAILAND, AND
B 6305 845* PETROLEUM GEOLOGY OF WHIRLPOOL MEMBER, SILURIAN, OF
B 6207 1226* PETROLEUM GEOLOGY—PEOPLES REPUBLIC OF CHINA, ABST.
B 6004 691* PETROLEUM HABITAT OF ATLANTIC MARGIN FROM NIGERIA TO CAPE
B 6212 2473 PETROLEUM IN ECUADOR SUBDUCTION SYSTEM
B 6209 1684† PETROLEUM IN MAHAKAM DELTA, BORNEO, ORGANIC SEDIMENTATION
S 10 176 PETROLEUM, ETIO PORPHYRINS RELATED TO BIOGENIC HEME
M 29 424 PETROLEUM, KEROGEN CONCENTRATION, SOURCE BEDS
M 29 424 PETROLEUM, KEROGEN MATURITY, SOURCE BEDS
M 29 424 PETROLEUM, KEROGEN TYPE, SOURCE BEDS
B 6203 520* PETROLEUM MIGRATION, ABST., METHANE GENERATION AND
B 6408 1219 PETROLEUM MIGRATION AND ACCUMULATION IN SUBTLE TRAPS
B 6009 1520† PETROLEUM MIGRATION AND ENTRAPMENT, OBSCURE NATURE OF
B 6002 232 PETROLEUM MIGRATION BY MOLECULAR SOLUTION, MODEL OF,
B 6305 771 PETROLEUM MIGRATION FROM SOURCE ROCKS THROUGH
B 6002 213 PETROLEUM MIGRATION, HYPOTHESES, PRIMARY
B 6305 775 PETROLEUM MIGRATION IN SOLUTION OR BY MICELLES, SECONDARY
B 6404 549 PETROLEUM MIGRATION STUDY, GULF COAST AREA
B 6405 764* PETROLEUM OCCURRENCES IN NONMARINE ROCKS, ABST.
B 6405 716* PETROLEUM POSSIBILITIES IN ALTAR DESERT, SONORA, MEXICO, ABST.
B 6303 446* PETROLEUM POTENTIAL OF SCOTIAN SHELF--CASE STUDY, ABST.
B 6004 708* PETROLEUM POTENTIAL AND DEPOSITIONAL ENVIRONMENTS OF
B 6203 565* PETROLEUM POTENTIAL AND STRATIGRAPHY OF SIMPSON EQUIVALENTS
M 29 400 PETROLEUM POTENTIAL, GULF OF MEXICO, MISSISSIPPI FAN
B 6207 1211* PETROLEUM POTENTIAL IN FAR EAST, ABST.
S 7 184 PETROLEUM POTENTIAL, MISSISSIPPI FAN
M 29 401 PETROLEUM POTENTIAL, MISSISSIPPI FAN, PLEISTOCENE
B 6405 707* PETROLEUM POTENTIAL OF BASIN AND RANGE PROVINCE, WESTERN
M 25 313† PETROLEUM POTENTIAL OF CONTINENTAL RISE OFF CENTRAL
B 6203 493* PETROLEUM POTENTIAL OF CONTINENTAL SLOPE DEPOSITS, ABST.
B 6304 700* PETROLEUM POTENTIAL OF EASTERN GULF OF ALASKA
B 6403 442* PETROLEUM POTENTIAL OF GREAT BASIN, ABST.
M 125 261† PETROLEUM POTENTIAL OF KOREAN OFFSHORE
B 6206 967† PETROLEUM POTENTIAL OF MONTEREY DEEP-SEA FAN, CALIFORNIA,
B 6105 811* **PETROLEUM POTENTIAL OF POSSIBLE LOWER CRETACEOUS REEF TREND**
B 6204 592† PETROLEUM POTENTIAL OF U.S. ATLANTIC SLOPE, RISE, AND ABYSSAL
B 6105 780* **PETROLEUM POTENTIAL OFF GABON**
M 30 539† PETROLEUM POTENTIAL--OUARGLA REGION TRIASSIC BASIN, ALGERIA
B 6305 844* PETROLEUM PRODUCTION FROM BASAL GREENBRIER FORMATION IN
S 10 177 PETROLEUM, ORIGIN, CONVERSION AND TRANSPORT MODEL
S 10 171 PETROLEUM, ORIGIN, GENERATION OF HYDROCARBONS
M 29 436 PETROLEUM, PRODUCTION INTERVALS, GULF COAST
M 29 126 PETROLEUM PRODUCTION, MAGELLAN BASIN, SPRINGHILL FORMATION
B 6002 233 PETROLEUM PRODUCTIVITY AND SWELLING CLAYS, CORRELATION
B 6303 286† PETROLEUM PROSPECT OF THE DEEP OFFSHORE
B 6207 1239* PETROLEUM PROSPECTIVITY OF AUSTRALIAN MARGINAL PLATEAUS,
B 6403 297† PETROLEUM PROSPECTS, CONTINENTAL, MARGINS--, CLASSIFICATION
B 6303 288 PETROLEUM PROSPECTS, ESSENTIAL REQUISITES FOR FAVORABLE
B 6405 672* PETROLEUM PROSPECTS OF ANAMBRA BASIN, SOUTHERN NIGERIA, ABST.
B 6105 795* **PETROLEUM PROSPECTS OF CONTINENTAL SLOPES AND RISES**
B 6101 43† PETROLEUM PROSPECTS OF LEBANON--REEVALUATION
B 6410 1681† PETROLEUM REVISITED, INDIGENOUS, PRECAMBRIAN
B 6303 507* PETROLEUM SOURCE BEDS, ABST., ATMOSPHERIC CIRCULATION AND
M 29 423† PETROLEUM SOURCE BEDS ON CONTINENTAL SLOPES AND RISES
B 6105 781* **PETROLEUM SOURCE BEDS ON CONTINENTAL SLOPES AND RISES**
B 6209 1584† PETROLEUM SOURCE BEDS ON CONTINENTAL SLOPES AND RISES
S 10 183 PETROLEUM, ROLE OF METHANE, MIGRATION
S 10 187 PETROLEUM, ROLE OF METHANE IN IMPEDING, MIGRATION
S 10 72 PETROLEUM IN WATER SOLUTION, MIGRATION
S 10 154 PETROLEUM MIGRATION, ALBERTA, RED EARTH REGION
S 10 156 PETROLEUM MIGRATION, CHINA, DZUNGARIAN ARTESIAN BASIN
S 10 154 PETROLEUM MIGRATION, FRANCE, PARENTIS ARTESIAN BASIN
S 10 155 PETROLEUM MIGRATION, HUNGARY, GREAT LOWLAND ARTESIAN BASIN
S 10 179† PETROLEUM MIGRATION, METHANE, GENERATION AND
S 10 156 PETROLEUM MIGRATION, MIDDLE EAST, PERSIAN GULF
S 10 121† PETROLEUM MIGRATION, THE GENERALIZED, HYDRAULIC THEORY OF
S 10 160 PETROLEUM MIGRATION, GENERALIZED THEORY
S 10 175 PETROLEUM PORPHYRINS, GENERATION, MIGRATION
S 10 154 PETROLEUM TRANSPORT AND ACCUMULATION, GROUNDWATER FLOW
M 29 425 PETROLEUM, SOURCE BEDS, ORGANIC PRODUCTIVITY
B 6206 1075† PETROLEUM SOURCE-BED EVALUATION OF JURASSIC TWIN CREEK
B 6408 1251† PETROLEUM SOURCE-BED EVALUATION OF TERTIARY NIGER DELTA
B 6405 726* PETROLEUM SOURCE-ROCK EVALUATION BY THERMAL DISTILLATION
B 6108 1375* PETROLEUM SOURCE-ROCK POTENTIAL OF ARBUCKLE GROUP,
M 29 432 PETROLEUM, THERMAL CRACKING, SOURCE BEDS
B 6004 651* PETROLEUM ZONES FOR APPRAISAL OF POTENTIAL AREAS--CLUSTER

M 29 425 PETROLEUM, ZONES OF GENERATION, SOURCE BEDS
B 6101 105 PETROLEUM-EXPLORATION CLASSROOM GAME FOR STUDENTS
S 10 169† PETROLEUM: IN-TRANSIT CONVERSION OF ORGANIC COMPOUNDS IN
M 25 252† PETROLIFEROUS TAIWAN BASINS IN FRAMEWOR OF WESTERN PACIFIC
B 6309 1590* PETROLOGIC EVALUATION OF SIGNIFICANCE OF NATURAL FRACTURES
M 25 509 PETROLOGIC MODELS, PORPHYRY, COPPER DEPOSITS
B 6309 1589* PETROLOGY AND DEPOSITIONAL ENVIRONMENTS OF BOYLE DOLOMITE,
B 6303 447* PETROLOGY AND DIAGENESIS OF ARC-DERIVED LITHIC
B 6108 1384* PETROLOGY AND DIAGENESIS OF DEEP-WATER SANDSTONES, OUACHITA
B 6302 182† PETROLOGY AND DIAGENESIS OF JURASSIC SANDSTONES FROM
B 6108 1374* PETROLOGY AND DIAGENETIC EFFECTS OF GRAYWACKE
B 6309 1575* PETROLOGY AND MATURATION OF DISPERSED ORGANIC MATTER IN
B 6203 537* PETROLOGY AND PETROGRAPHY OF GREAT BLUE FORMATION AT
B 6004 725 **PETROLOGY OF BASIN-TO-REEF SHOALING SEQUENCE IN DOGGER,**
B 6406 960* PETROLOGY OF ARBUCKLE GROUP, CENTRAL KANSAS, ABST.
B 6109 1546* PETROLOGY OF CATAHOULA SANDSTONES IN EAST TEXAS, ABST.
B 6303 493* PETROLOGY OF MIOCENE CATAHOULA FORMATION, CENTRAL TEXAS
B 6109 1548* PETROLOGY OF SMACKOVER FORMATION, JURASSIC—PERRY AND STONE
B 6004 733* **PETROLOGY OF SUBSEA CEMENTED CARBONATE MOUNDS, LITHOHERMS,**
B 6305 836* PETROLOGY OF TERTIARY SANDSTONES OF SOUTHERN PICEANCE CREEK
B 6004 671* PETROLOGY OF TRIASSIC MARINE SEDIMENTARY ROCKS, VIZCAINO
B 6305 846* PETROLOGY OF TRIASSIC-JURASSIC CULPEPER GROUP CONGLOMERATES
B 6303 456* PETROLOGY, SEDIMENTOLOGY, AND PETROLEUM POTENTIAL OF EARLY
B 6405 794* PETROPHYSICS—CASE HISTORY, ABST., APPLIED
B 6105 842* PETROS—PETROPHYSICAL LOG AND CORE-ANALYSIS SYSTEM, ABST.
B 6203 553 PETTA, T. J., EFFECTS OF SUBMARINE DIAGENESIS ON PRIMARY
B 6109 1555 PETTA, TIMOTHY J., DIAGENESIS AND GEOCHEMISTRY OF GLEN ROSE
B 6001 87 PETTERS, SUNDAY W., UPPER CRETACEOUS SUBSURFACE STRATIGRAPHY
B 6004 708 PETTIJOHN, F. J., SEDIMENTOLOGY AND EASTERN CONTINENTAL
B 6307 1136 PETTIT FORMATION, OKLAHOMA
B 6008 1196 PFEIFFER, D. H., DEVELOPMENTS IN WEST COAST AREA IN 1975
B 6108 1166 PFEIFFER, D. H., DEVELOPMENTS IN WEST COAST AREA IN 1976
B 6010 1637 PFEIFFER, D. H., FOREIGN DEVELOPMENT PAPERS, 1975
B 6110 1575 PFEIFFER, D. H., FOREIGN DEVELOPMENT PAPERS, 1976
B 6105 820 PFORZHEIMER, HARRY, OIL SHALE, ABST., PREDICTIONS FOR
B 6405 764* PHANEROZOIC CARBONATE DIAGENESIS—A NEW MODEL, ABST.
S 6 73 PHANEROZOIC, DECAY CONSTANT CALIBRATION
S 6 207 PHANEROZOIC, GLAUCONITES, ARGON DATES, DIAGRAM
M 29 470 PHANEROZOIC, GLOBAL CYCLES OF CHANGE, FIRST-ORDER CYCLES
M 29 470 PHANEROZOIC, GLOBAL CYCLES OF CHANGE, SECOND-ORDER CYCLES
M 25 229† PHANEROZOIC HISTORY OF AUSTRALIA AND SURROUNDING OCEANIC
S 6 63 PHANEROZOIC, MAGNETIC REVERSAL PATTERNS
S 6 62 PHANEROZOIC, MAGNETIC REVERSALS, EXAMPLE
S 6 372 PHANEROZOIC STRATIGRAPHY, A REVIEW, SOUTH AFRICA
B 6101 35 PHANEROZOIC TERRESTRIAL SEDIMENTS, URANIUM DEPOSITS IN
S 6 269 PHANEROZOIC TIME SCALE, A REVIEW, CRETACEOUS
S 6 73† PHANEROZOIC TIME SCALE--COMPUTER FILE OF CRITICAL DATES AND
B 6004 678* PHENOTYPES OF PLANKTONIC FORAMINIFERA, ABST., PALEOCLIMATIC
B 6112 2105 PHIELIPS FIELD, PERU
M 25 436 PHILEX MINE, LUZON
B 6403 360 PHILIPP PLATE
M 25 434 PHILIPPINE ARCHIPELAGO, MOBILE AND STABLE ZONES
B 6209 1666 PHILIPPINE BASIN, CENTRAL BASIN RIDGE
B 6209 1664 PHILIPPINE BASIN, WESTERN, STRUCTURE
S 8 158 PHILIPPINE ISLANDS, SUBMARINE CANYONS, ABRA DELTA
B 6207 1238* PHILIPPINE NATIONAL ENERGY POLICY, ABST.
M 25 436 PHILIPPINE RIFT SYSTEM
M 25 437 PHILIPPINE RIFT SYSTEM AND DIORITIC MASSES, RELATION OF
B 6203 531* PHILIPPINE SEA, ABST., LEG 58, DSDP—DEEP DRILLING IN, SHIKOKU
B 6303 413* PHILIPPINE SEA, DSDP LEGS 59 AND 60, ABST., CENOZOIC, CLAY-MINERAL
M 25 434 PHILIPPINE SEA PLATE
B 6207 1211* PHILIPPINE SEA SEDIMENTS, PARTICULARLY MANGANESE NODULES,
M 29 239† PHILIPPINE SEA, SUBSIDENCE OF THE, DAITO, RIDGE AND ASSOCIATED
M 25 438 PHILIPPINE TRENCH
B 6209 1646 PHILIPPINE TRENCH
B 6110 1839 PHILIPPINES, 1976, HICKS, G. W., ON DEVELOPMENTS IN
B 6210 1987 PHILIPPINES, 1977, MADRID, A. P., ON DEVELOPMENTS IN
B 6310 1826 PHILIPPINES, 1978, MADRID, A. P., ON DEVELOPMENTS IN
B 6212 2512* PHILIPPINES, ABST., APPRAISAL OF PRESENT PETROLEUM-EXPLORATION
B 6207 1208* PHILIPPINES, ABST., GEOLOGY AND ECONOMIC POSSIBILITIES OF SOME,
B 6303 456* PHILIPPINES, ABST., PETROLOGY, SEDIMENTOLOGY, AND PETROLEUM
B 6207 1233* PHILIPPINES, ABST., PROGRESS OF, GEOTHERMAL, ENERGY EXPLORATION
B 6207 1233* PHILIPPINES, ABST., REEF EXPLORATION IN
M 25 460 PHILIPPINES AND INDONESIA, CHROMITE
M 25 436 PHILIPPINES, COPPER, ASSOCIATED MINERALS
B 6411 1892 PHILIPPINES, DEVELOPMENTS, 1979
M 25 435 PHILIPPINES, DIORITES, INTRUSIONS OF
M 25 434 PHILIPPINES, GENERAL GEOLOGY
M 25 141 PHILIPPINES, GEOTHERMAL FIELDS
M 25 460 PHILIPPINES, INDONESIA, AND NEW CALEDONIA, NICKEL AND COBALT
M 25 144 PHILIPPINES, MAMBUCAL GEOTHERMAL PROSPECT
M 25 434 PHILIPPINES, MANILA TRENCH
M 25 434 PHILIPPINES, MINDANAO DEEP
M 25 434 PHILIPPINES, MINDANAO TRENCH
M 25 141 PHILIPPINES, MT. MAQUILING--MT. BANAHAO, LOS BANOS, GEOTHERMAL
M 25 143 PHILIPPINES, NEGROS ISLAND, GEOTHERMAL PROSPECTS
M 25 435 PHILIPPINES, ORE OCCURRENCES, BOUNDED IN TIME AND SPACE
M 25 143 PHILIPPINES, PALIMPINON-DAUIN GEOTHERMAL PROSPECT
M 25 434 PHILIPPINES, PLATE MOVEMENTS
M 25 435 PHILIPPINES, PORPHYRY COPPER DEPOSITS, GEOLOGIC ENVIRONMENT
M 25 438 PHILIPPINES, PORPHYRY COPPER, ECONOMIC POSSIBILITIES
M 25 434 PHILIPPINES, PORPHYRY COPPER, ORE RESERVE
M 25 435 PHILIPPINES, POTASSIUM-ARGON DATING OF ROCKS
B 6411 1896 PHILIPPINES, PRODUCTION, 1977-1979
S 8 161 PHILIPPINES, SUBMARINE CANYONS, PINGET SEAVALLEY
M 25 141 PHILIPPINES, TIWI GEOTHERMAL FIELD
M 25 143 PHILIPPINES, TONGONAN GEOTHERMAL PROSPECT
M 25 434† PHILIPPINES--SUMMARY, GEOLOGIC ENVIRONMENT AND ECONOMI
M 25 140† PHILIPPINES--SUMMARY, GEOTHERMAL EXPLORATION AND
B 6303 509 PHILLIPS, B. G., SMALL COMPUTER WELL-DATA SYSTEM, ABST.
B 6012 2187 PHILLIPS, F. JAY, PALEONTOLOGIC STUDIES OF MIDDLE TERTIARY
B 6005 809 PHILLIPS, J. D., CONTINENTAL MARGIN OFF WESTERN AFRICA--SENEGAL
B 6110 1796 PHILLIPS PETROL. CO., DEMINEX IRAN OIL CO., TOTAL IRAN PETROL. CO.,
B 6010 1866 PHILLIPS PETROLEUM CO., DEMINEX IRA OIL CO., HORMOZ PETROLEUM
B 6004 708 PHILLIPS, R. L., BED FORMS AND INTERNAL STRUCTURES IN DEEP
B 6004 657 PHILLIPS, R. L., MODERN AND ANCIENT ESTUARINE-FILL FACIES,
B 6303 509 PHILLIPS, R. LAWRENCE, BED FORMS AND PROCESSES ON ESTUARINE
B 6211 2359 PHILLIPS, R. LAWRENCE, PETROLEUM ACCUMULATIONS IN SANTA
B 6404 692 PHILLIPS, R. LAWRENCE, STRATIFICATION TYPES IN INTERTIDAL
S 4 87 PHILLIPSBURG LIMESTONE, JAMAICA
M 28 139 PHILLIPSITE, BISEFRINGENCE, ETC.
M 28 139 PHILLIPSITE, SPHERULE, IN NON-MARINE SEDIMENTS
M 28 139 PHILLIPSITE, SPHERULE, IN PALAGONITE TUFF
B 6405 784 PHILP, R. P., EFFECTS OF BASALT INTRUSIONS ON KEROGENS IN
B 6303 529 PHIPPS, CHARLES V. G, SLUMPS ON UPPER CONTINENTAL SLOPE,
M 28 83 PHOSPHATE, BIREFRINGENCE
M 28 84 PHOSPHATE, BONE FRAGMENTS
M 28 83 PHOSPHATE, BONE MATERIAL
M 28 40 PHOSPHATE, CONODONTS, FRAGMENTS
M 28 82 PHOSPHATE, GRAINS, WITH DETRITAL QUARTZ
B 6203 518* PHOSPHATE IN BOTTOM SEDIMENTS IN MYAKKA AND PEACE RIVER
M 28 127 PHOSPHATE, IN SANDSTONE, CEMENTS
M 28 126 PHOSPHATE, IN SANDSTONE, CEMENTS
M 25 397 PHOSPHATE NODULE RESERVES, CHATHAM RISE
M 28 127 PHOSPHATE, OOLITIC, COATINGS
M 28 83 PHOSPHATE, OOLITIC, DEPOSIT
M 28 149 PHOSPHATES, BIBLIOGRAPHY, DIAGENESIS AND POROSITY
M 25 427 PHOSPHATES, QUEENSLAND, DUCHESS AREA
B 6107 1045† PHOSPHATIC BLACK SHALE FACIES IN PENNSYLVANIAN CYCLOTHEMS OF
M 27 153 PHOSPHATIC, SHELL, FRAGMENTS
B 6207 1234* PHOSPHOGENIC PROVINCES OF CIRCUM-PACIFIC REGION, ABST.
M 28 127 PHOSPHORIA FORMATION, IDAHO, PERMIAN
M 27 120 PHOSPHORIA FORMATION, IDAHO, PERMIAN
M 28 83 PHOSPHORIA FORMATION, IDAHO, PERMIAN
B 6201 108 PHOSPHORIA FORMATION, KEROGEN
B 6207 1172 PHOSPHORIA FORMATION, NEVADA AND, UTAH, MEADE PEAK
B 6108 1375* PHOSPHORIA FORMATION, WESTERN INTERIOR UNITED STATES, ABST.,
B 6201 98† PHOSPHORIA FORMATION, WESTERN INTERIOR UNITED STATES,
B 6405 746* PHOSPHORIA FORMATION, WESTERN UNITED STATES, ABST.,
B 6204 622 PHOSPHORIA FORMATION, WYOMING
M 25 395† **PHOSPHORITE DEPOSITS OF CHATHAM RISE NEAR NEW**
B 6107 1063 PHOSPHORITE, MID-CONTINENT, PALEOZOIC, BLACK, SHALE AND
M 25 535 PHOSPHORITE NODULES, PACIFIC, SOUTHEASTERN
B 6405 746* PHOSPHORITE, ORGANIC CARBON, AND HYDROCARBON IN PERMIAN
B 6105 816* PHOTOGRAPH INTERPRETATION OF HIGH-VELOCITY WIND-DRIVEN
B 6105 835* PHOTOGRAPHIC MAPPING OF RECENT COASTAL MARINE SEDIMENTARY
B 6404 463 PHREATIC ZONE, MARINE
M 25 252 PHUKET RIDGE, THAILAND
M 28 73 PHYCODEN SCHICHTEN, GERMANY, ORDOVICIAN
M 28 166 PHYCODEN SCHICHTEN, GERMANY, ORDOVICIAN
S 11 115 PHYLETIC ORIGINS, SIPHOGENERINIDS
S 4 239† PHYLLOID ALGAE IN DEVELOPMENT OF DEPOSITIONAL
S 4 243 PHYLLOID ALGAE, PENNSYLVANIAN-, PERMIAN LIMESTONES
M 27 10 PHYLLOID ALGAL GRAINS
B 6112 2115† PHYLLOID-ALGAL BUILDUP, SACRAMENTO MOUNTAINS, NEW MEXICO,
S 11 202 PHYLOGENY, KEY TO INTERPRETATION
B 6004 650* PHYSICAL AND BIOGENI CHARACTERISTICS OF SEDIMENTS FROM UPPER
B 6203 554* PHYSICAL AND CHEMICAL SIGNALS AROUND OIL AND GAS TRAPS, ABST.
B 6203 570* PHYSICAL AND SEDIMENTOLOGIC PROCESSES IN MESOTIDAL SALT
B 6405 711* PHYSICAL CHARACTERISTICS OF SHALLOW METHANE RESERVOIRS OF
B 6105 779 * **PHYSICAL CHARACTERISTICS OF WASHOVER FANS OF GEORGIA COAST**

B 6104 573† PHYSICAL CHEMISTRY APPLIED TO PETROLEUM GEOCHEMISTRY I,
B 6207 1220* PHYSICAL CONSTRAINTS ON MINERAL DEVELOPMENT, ABST.
S 10 89† PHYSICAL CONSTRAINTS, OIL AND, GAS, MIGRATION:, CHEMICAL AND
B 6405 778* PHYSICAL EVIDENCE FOR CRETACEOUS TIDES, WESTERN INTERIOR
S 6 119† PHYSICAL TIME-SCALE STUDIES, POTASSIUM-ARGON, ISOTOPIC DATING
B 6004 681* PHYSICAL-BIOGENIC BREAK IN NEARSHORE SEDIMENTARY SEQUENCES,
M 26 15† PHYSICS FROM LABORATORY AND LOG DATA THAT ARE IMPORTANT TO
S 7 245 PHYSIOGRAPHIC FEATURES, TEXAS, SHELF-SLOPE
M 29 209 PHYSIOGRAPHY, AUSTRALIA, SHELF
S 7 157 PHYSIOGRAPHY, EASTERN GULF OF MEXICO, BATHYMETRY AND
M 29 385 PHYSIOGRAPHY, GULF OF MEXICO, MISSISSIPPI FAN
M 29 240 PHYSIOGRAPHY, NORTH PHILIPPINE SEA, DAITO RIDGE AREA
S 7 3† PHYSIOGRAPHY OF GULF OF MEXICO
S 7 155† PHYSIOGRAPHY, STRATIGRAPHY, AND SEDIMENTATIONAL PATTERNS,
M 29 354 PHYSIOGRAPHY, VIRGIN ISLANDS, ST. CROIX
B 6109 1557 PIALLI, GIAMPAOLO, LOWER CRETACEOUS SHELF MARGIN IN NORTHERN
B 6009 1617 PIALLI, GIAMPAOLO, EARLY CRETACEOUS SHELF MARGIN IN NORTHERN
B 6109 1471 PIAUI BASIN, BRAZIL
B 6305 824 PICARD, M. D., DEFINITION OF DEPOSITIONAL FACIES OF PARACHUTE
B 6108 1386 PICARD M. DANE, NAVAJO SANDSTONE, JURASSIC, OF UTAH AND
B 6008 1406 PICARD, M. DANE, OIL AND GAS POTENTIAL OF NUGGET SANDSTONE,
B 6305 841 PICARD, M. DANE, PALEOSOLS AS ENVIRONMENTAL INDICATORS IN
B 6405 764 PICARD, M. DANE, PETROLEUM OCCURRENCES IN NONMARINE ROCKS,
B 6004 708 PICARD, M. DANE, PETROLEUM POTENTIAL AND DEPOSITIONAL
B 6305 837 PICARD, M. DANE, STRATIGRAPHY OF LACUSTRINE DEPOSITS, ABST.
B 6205 890 PICARD, M. DANE, VOLCANIC SANDSTONE IN GREEN RIVER AND UINTA
B 6203 505* PICEANCE CREEK BASIN, COLORADO, ABST., IRON-, SULFIDE
B 6205 885* PICEANCE CREEK BASIN, COLORADO, ABST., MAESTRICHTIAN
B 6305 836* PICEANCE CREEK BASIN, COLORADO--IMPLICATION FOR PROVENANCE
B 6405 801* PICEANCE CREEK BASIN, NORTHWEST COLORADO, ABST., METHANE
B 6108 1387* PICEANCE CREEK BASIN, NORTHWESTERN COLORADO, ABST., GEOLOGY
B 6301 67 PICHA, FRANTISEK, ANCIENT SUBMARINE CANYONS OF TETHYAN
M 24 242 PICKENS FIELD, MISSISSIPPI
B 6405 764 PICKETT, GEORGE R., SEISMIC MODELING OF PENNSYLVANIAN
B 6405 718 PICKLES, E., ET AL, BRAE FIELD AREA, NORTH SEA, ABST.
B 6202 259 PICO FORMATION, CALIFORNIA
B 6102 177 PICO SAND, CALIFORNIA
B 6004 661 PICOU, E. B., JR., GULF COAST CENOZOIC--MODEL FOR APPLICATION OF
B 6209 1755 PICOU, E. B., JR., GULF COAST CENOZOIC—MODEL FOR APPLICATION OF
B 6303 509 PICOU, EDWARD B., JR., APPLIED BIOSTRATIGRAPHY IN GULF COAST
M 25 479 PICRITES, USSR
B 6303 451* PICTURED CLIFFS SANDSTONE--UPPER CRETACEOUS
B 6403 340 PIEDMONT COASTAL PLAIN
B 6301 69 PIENINY KLIPPEN BELT, EUROPE
B 6004 722 PIERCE, C. I., LINEAMENT AND STRUCTURAL ANALYSES OF
B 6410 1614 PIERCE FIELD, COLORADO
B 6405 750 PIERCE, R. W., NEW PRECISION IN BIOSTRATIGRAPHY THROUGH
B 6108 1386 PIERCE, WALTER H., EARLY PERMIAN SLIP AND STRATIGRAPHIC
B 6408 1280 PIERCE, WALTER H., GEOLOGY OF WHEATON CONSOLIDATED OIL FIELD,
B 6209 1564 PIERCEMENT DIAPIRISM, EMPLACEMENT HISTORIES OF
B 6209 1561† PIERCEMENT DIAPIRS, MECHANISM FOR, EMPLACEMENT OF
B 6111 2004† PIERCEMENT STRUCTURES, EXOTIC MIDDLE, MIOCENE SEDIMENT FROM,
B 6111 1991† PIERCEMENT STRUCTURES IN DEEP OCEANS
B 6111 1991 PIERCEMENT STRUCTURES, SEISMIC PROFILES, ATLANTIC OCEAN
B 6111 1991 PIERCEMENT STRUCTURES, SEISMIC PROFILES, INDIAN OCEAN
B 6111 1991 PIERCEMENT STRUCTURES, SEISMIC PROFILES, PACIFIC OCEAN
B 6410 1615 PIERRE SHALE, DENVER BASIN
B 6407 971 PIERRE SHALE, GREAT PLAINS
B 6407 975 PIERRE SHALE, NORTH DAKOTA
M 26 290 PIERRE SHALE STRUCTURE CONTOUR, DENVER BASIN
M 26 296 PIERRE SHALE, SUBSURFACE PRESSURE GRADIENT, DENVER BASIN
M 26 286 PIERRE SHALE VELOCITY-CONDUCTIVITY RELATION, DENVER BASIN
B 6108 1388* PIERRE SHALE, WEST DENVER BASIN, COLORADO, ABST., MARINE-SHELF
B 6202 184 PIERRE SHALE, WYOMING
M 26 279 PIERRE SHALE, ZONES, CRETACEOUS
B 6203 553 PIERSON, BERNARD J., CONTROL OF CATHODO-LUMINESCENCE OF
M 25 8 PIETROWSKI, ROBERT F., JR., BOUNDARIES OF SEABED JURISDICTION OFF
S 10 149 PIEZOMETRIC SURFACE, ILLINOIS, MAQUOKETA AQUITARD
B 6203 554 PIGOTT, JOHN D., INTERSTITIAL WATER CHEMISTRY OF JAMAICAN REEF
B 6405 764 PIGOTT, JOHN, PHANEROZOIC CARBONATE DIAGENESIS—A NEW MODEL,
B 6203 561 PIKE, ROBERT S., GEOLOGIC APPRAISAL OF UNDISCOVERED OIL AND GAS
B 6406 821 PIKER HILLS FORMATION, AUSTRALIA
B 6309 1468 PIKULNEY SERIES, USSR, KORYAK HIGHLANDS
M 25 520 PILARES DE NACOZARI MINE, COPPER, MEXICO, SONORA
M 25 411 PILBARA AND YILGARN BLOCKS, IRON, WESTERN AUSTRALIA
B 6201 41 PILBARA BLOCK, AUSTRALIA
B 6411 2041 PILGER, JR., LADD, JOHN W., REVIEW OF PROCEEDINGS OF THE
B 6209 1763 PILGER, REX H., JR., CLOSED GULF OF MEXICO, PRE-ATLANTIC
B 6004 692 PILKEY, O. H., MARINE COLD-WATER CARBONATE DEPOSITS AS POSSIBLE
B 6406 841 PILKEY, ORRIN H., COMPARISON OF SAND-LAYER GEOMETRY ON FLAT
B 6105 777 PILKEY, O. H., SAND PETROLOGY AND BIOCLASTIC COMPONENTS OF
B 6406 957 PILKEY, SUMMERHAYES, COLIN P., REVIEW OF, GEOLOGY OF,
B 6406 821 PILLARA LIMESTONE, AUSTRALIA
B 6406 826 PILLARA REEF CYCLE, AUSTRALIA, GIVETIAN-, FRASNIAN
B 6111 1956 PILON HEAVY OIL, VENEZUELA
B 6008 1403 PINCHOCK, JOHN, RECENT DEVELOPMENTS IN COAL EXPLORATION IN
M 26 9 PINCHOUTS, SEISMIC, REFLECTIONS, SANDSTONE
B 6402 244 PINDUS ZONE, GREECE
B 6209 1767* PINE BARREN MEMBER OF CLAYTON FORMATION, LOWER PALEOCENE,
B 6211 2277 PINE BUTTE SANDSTONE MEMBER, WYOMING
B 6205 879* PINE CANYON QUADRANGLE, CARBON COUNTY, UTAH, ABST.,
M 24 159 PINE ISLAND SHALE MEMBER, TEXAS
B 6009 1610* PINE ISLAND TO BASE OF NORPHLET IN ARK-LA-TEX REGION, ABST.,
B 6201 125 PINE KEY FORMATION, FLORIDA
B 6412 2078 PINE MOUNTAIN FAULT, APPALACHIANS
B 6203 359 PINE MOUNTAIN FAULT, CALIFORNIA
B 6103 417 PINE MOUNTAIN FAULT, VIRGINIA
B 6403 447* PINE VALLEY NEVADA, ABST., LACUSTRINE, FLUVIAL, AND FAN
B 6008 1402* PINEVIEW FIELD AREA, SUMMI COUNTY, UTAH, ABST., GEOLOGY OF
B 6206 1076 PINEVIEW FIELD, DISCOVERED JANUARY 1975, UTAH
B 6405 638 PINEVIEW FIELD, UTAH
B 6405 703 PINEZICH, A. R., TERTIARY STRUCTURAL DEVELOPMENT OF SELECTED
B 6304 698 PING, WANG, GEOLOGY OF GU-DAO OIL FIELD AN ADJACENT AREAS,
M 30 471 PING, WANG, GEOLOGY OF GUDAO OIL FIELD AND SURROUNDING
S 8 161 PINGET SEAVALLEY, PHILIPPINES, SUBMARINE CANYONS
S 8 18 PINGET SEAVALLEY, TIME-VELOCITY CURVES, CURRENT FLOW
B 6303 533 PINGITORE, NICHOLAS E., JR., CARBONATE GEOLOGY OF PENA BLANCA
B 6405 765 PINGITORE, NICHOLAS E., JR., DIFFERENTIAL DIFFUSION DURING
B 6203 554 PINGITORE, NICHOLAS E., JR., MODEL FOR TRACE-ELEMENT
B 6303 509 PINGITORE, NICHOLAS E., JR., TRACE-ELEMENT GEOCHEMISTRY AND
S 5 4 PINNACLE REEF BELT, STRATIGRAPHY, MICHIGAN BASIN,
S 5 6 PINNACLE REEF DEVELOPMENT
M 26 403 PINNACLE REEF ILLUSTRATION, MODELING
B 6405 736* PINNACLE REEF IN SOUTHERN MICHIGAN BASIN, ABST., EXPLORING FOR,
M 26 190 PINNACLE REEF, SEISMIC EXAMPLE, AFRICA, NORTH
B 6303 548* PINNACLE REEFS OF CHAPPEL LIMESTONE, FORT WORTH BASIN, TEXAS,
S 5 4 PINNACLE REEFS, MICHIGAN BASIN, NIAGARA GROUP,
S 5 7 PINNACLE REEFS, EVOLUTION, STAGES, NIAGARA GROUP,
S 5 7 PINNACLE REEFS, EVOLUTION, STAGES, SALINA GROUP,
S 5 1* PINNACLE REEFS, NIAGARA AND SALINA GROUPS, NORTHERN SHELF,
M 26 195 PINNACLE REEFS, SEISMIC EXAMPLE, CANADA, ALBERTA, RAINBOW-,
B 6304 608† PINNACLE-REEF BELT OF NORTHERN MICHIGAN, DIFFERENTIAL,
B 6205 867† PINNACLES AND NEENACH VOLCANIC FORMATIONS AND THEIR
B 6012 2128† PINNACLES AND NEENACH VOLCANIC FORMATIONS AND THEIR
B 6211 2224 PINNACLES VOLCANIC VENT AREA, CALIFORNIA
B 6109 1514 PINOS CANYON, NEW MEXICO
M 25 418 PINTER, J., MINING HEAVY-MINERAL SAND IN AUSTRALIA--SUMMARY
B 6206 968 PIONEER CANYON, CALIFORNIA
B 6405 700* PIONEER MOUNTAINS, ABST., STATUS OF, ANTLER OROGENY IN
B 6102 257 PIONEER MOUNTAINS, NOMENCLATURE, IDAHO
B 6102 256† PIONEER MOUNTAINS, SOUTH-CENTRAL IDAHO, LITTLE COPPER
B 6203 491* PIONEER TERRACE MOLLUSCAN FAUNA, CAPE BLANCO, OREGON, ABST.,
B 6303 462 PIOTROWSKI, ROBERT G., DEVONIAN ORGANIC-RIC BLACK SHALES IN
B 6308 1244 PIOTROWSKI, ROBERT G., OIL AND GAS DEVELOPMENTS IN MARYLAND,
B 6409 1403 PIOTROWSKI, ROBERT G., OIL AND GAS DEVELOPMENTS IN MARYLAND,
B 6405 739* PIPE CREE JR. REEF, GRANT COUNTY, INDIANA, ABST., POROSITY
B 6207 1239* PIPELINE, ABST., HARNESSING, ARCTIC, ENERGY—, TRANS-ALASKA
M 30 134 PIPER FIELD, BASE OF CRETACEOUS FAULTING, NORTH SEA
M 30 167 PIPER FIELD, DEVELOPMENT DRILLING, MORAY FIRTH BASIN
M 30 165 PIPER FIELD, DEVELOPMENT GEOLOGY, MORAY FIRTH BASIN
M 30 137 PIPER FIELD, KIMMERIDGE SHALE ISOPACH, NORTH SEA
M 30 139 PIPER FIELD, KIMMERIDGE SHALE, NORTH SEA
M 30 132 PIPER FIELD, LOCATION, NORTH SEA
M 30 168 PIPER FIELD, LOGGING PROCEDURES
M 30 133 PIPER FIELD, NORTH SEA, MORAY FIRTH BASIN
M 30 170 PIPER FIELD, RADIOACTIVE MARKERS, NORTH SEA
M 30 162 PIPER FIELD, RESERVOIR PERFORMANCE, MORAY FIRTH BASIN
M 30 159 PIPER FIELD, SOURCE ROCK, MORAY FIRTH BASIN
M 30 140 PIPER FIELD, STRUCTURAL CROSS SECTION, NORTH SEA
M 30 132 PIPER FIELD, TECTONIC SETTING, MORAY FIRTH BASIN
M 30 137 PIPER FIELD, TIME STRATIGRAPHIC COLUMN, NORTH SEA
M 30 136 PIPER FIELD, TOP OF PIPER SANDSTONE, NORTH SEA
M 30 163 PIPER FIELD, WELL DATA, NORTH SEA
M 30 145 PIPER FIELD, WITCH GROUND GRABEN, SEISMIC LINE
B 6211 2277 PIPER FORMATION, MONTANA
M 30 131† PIPER OIL FIELD
B 6303 489* PIPER OIL FIELD, ABST.
M 30 157 PIPER SANDSTONE, PETROGRAPHY, MORAY FIRTH BASIN
M 30 159 PIPER SANDSTONE, PETROLOGY, SEM SECTIONS
M 30 143 PIPER SANDSTONE, STRATIGRAPHY, MORAY FIRTH BASIN

B 6108 1387 PIPIRINGOS, GEORGE N., GEOLOGY AND SHALE-OIL RESOURCES OF
B 6306 942 PIRARUCU FORMATION, AMAZON BASIN
B 6405 765 PIRC, SIMON, DIAGENETIC MOBILIZATION OF URANIUM AND IRON FROM
B 6203 554 PIRSON, SYLVAIN J., PHYSICAL AND CHEMICAL SIGNALS AROUND OIL
S 4 5 PISCADERA BAY, CORALS
B 6303 510 PISCIOTTO, KENNETH A., DEDUCTION OF PAST GEOTHERMAL GRADIENTS
B 6207 1222 PISCIOTTO, KENNETH, ORIGIN, DISTRIBUTION, AND DIAGENETIC
M 25 338 PISCO BASIN, PERU
B 6203 349 PISMO BASIN, CALIFORNIA, HUASNA-
B 6403 449* PISMO SYNCLINE, CALIFORNIA, ABST., DIAGENETIC REACTIONS IN,
B 6405 790* PISMO SYNCLINE, COAST RANGES OF CALIFORNIA, ABST., DIAGENESIS IN,
B 6011 2048† PISOLITE AND CALICHE, VADOSE
M 28 74 PISOLITE, DEVELOPMENT, IN BAUXITE
M 28 74 PISOLITE, STRUCTURE, IN LATERITIC HARDPAN
M 27 119 PISOLITES
B 6004 696* PISOLITES AND PISOLITIC CALICHE IN TEXAS, ABST., CARBON AND,
M 27 120 PISOLITES, CALICHE
M 27 120 PISOLITES, PHOSPHATIC
B 6405 790* PISOLITH GROWTH IN SILURIAN REEFS OF NORTHERN INDIANA, ABST.,
B 6004 670* PISOLITIC FACIES, CAPITAN REEF COMPLEX, PERMIAN, GUADALUPE
M 27 163 PISOLITIC LIMESTONE, ITALY, JURASSIC
M 25 414 PISOLITIC LIMONITE ORE, WESTERN AUSTRALIA, IRON
B 6203 570* PITCH MINE, COLORADO. BY TITLE ONLY., URANIUM DEPOSITS OF
B 6405 755* PITCH MINE, SAGUACHE COUNTY, COLORADO, ABST., URANIUM
B 6305 826* PITCH MINE, SAWATCH RANGE, COLORADO, ABST., PALEOKARST
B 6101 28† PITCHBLENDE AND RELATED URANIUM DEPOSITS, SURFICIAL ORIGIN OF,
B 6101 29 PITCHBLENDE DEPOSITS, CANADA, BAKER LAKE
B 6101 29 PITCHBLENDE DEPOSITS, CANADA, GREAT BEAR LAKE
B 6101 29 PITCHBLENDE DEPOSITS, CANADA, RABBIT LAKE
B 6105 820 PITCHER, MAX G., EXPLORATION, ABST., TECHNOLOGICAL FOREFRONT
M 27 217 PITKIN LIMESTONE, OKLAHOMA, MISSISSIPPIAN
M 27 118 PITKIN LIMESTONE, OKLAHOMA, MISSISSIPPIAN
B 6302 259 PITMAN III, SCHLEE, JOHN S., REVIEW ON, ISLAND ARCS, DEEP SEA
B 6205 891 PITMAN, J. K., MINERALOGIC CHARACTERISTICS OF SOME LOWER
M 29 453 PITMAN, WALTER C., III, THE EFFECT OF EUSTATIC SEA LEVEL CHANGES
B 6303 539 PITTMAN, EDWARD D., APPLICATIONS OF SCANNING ELECTRON
B 6405 766 PITTMAN, EDWARD D., EFFECT OF CEMENTATION ON PHYSICAL
B 6203 555 PITTMAN, EDWARD D., EFFECT OF FAULT-RELATED GRANULATION ON
B 6405 737 PITTMAN, EDWARD D., ET AL, DEVELOPMENT OF SECONDARY POROSITY
B 6108 1387 PITTMAN, EDWARD D., POROSITY, DIAGENESIS, AND PRODUCTIVE
B 6304 692 PLACE, J. T., MINERALOGY AND GEOCHEMISTRY OF MARIANO LAKE
B 6105 815* PLACERS, SOUTH AFRICA, ABST., SEDIMENTOLOGIC SYNTHESIS OF
B 6211 2360 PLAFKER, GEORGE, FORAMINIFERAL BIOSTRATIGRAPHY IN GULF OF
B 6405 766 PLAFKER, GEORGE, GEOLOGY OF CONTINENTAL SLOPE ADJACENT TO
B 6304 700 PLAFKER, GEORGE, PETROLEUM POTENTIAL OF EASTERN GULF OF
B 6207 1231 PLAFKER, GEORGE, PLATE TECTONICS IN EVOLUTION OF SOUTHERN
B 6105 773 PLAFKER, GEORGE, STRUCTURE OF CONTINENTAL SHELF, EASTERN GULF
M 28 164 PLAGIOCLASE, COMPACTION, OF
M 28 34 PLAGIOCLASE, FRAGMENTS
M 28 97 PLAGIOCLASE, IN ARKOSE
M 28 16 PLAGIOCLASE TWINNING, FELDSPAR
B 6212 2381 PLAIN, LOWER, DELTA
B 6212 2381 PLAIN, TRANSITIONAL LOWER, DELTA
B 6209 1714† PLAIN: DISCUSSION AND REPLY, RECORRELATION OF, EOCENE-, MIOCENE
S 3 62 PLANAR SHAPES, URANIUM OREBODIES, LINEAR AND
B 6203 540* PLANETS, ABST., TERRESTRIAL EVOLUTION—INSIGHTS FROM THE, MOON
M 27 18 PLANKTON, BELIZE, HOLOCENE
S 11 109 PLANKTON ZONATION, CORRELATIONS, PALEOGENE
S 11 111 PLANKTON ZONATIONS, CORRELATIONS, NEOGENE
S 11 54† PLANKTONIC AND LARGE FORAMINIFERA IN MIDDLE TERTIARY OF
S 6 217 PLANKTONIC BIOSTRATIGRAPHY, REVIEW, PALEOGENE
B 6009 1618* PLANKTONIC CORRELATIONS--NORTHERN ITALY-JAMAICA, B.W.I., ABST.,
B 6004 678* PLANKTONIC FORAMINIFERA, ABST., PALEOCLIMATIC SIGNIFICANCE OF,
S 11 54 PLANKTONIC FORAMINIFERA, CALIFORNIA
S 11 201 PLANKTONIC FORAMINIFERA, RELATIVE ABUNDANCE
B 6012 2184* PLANKTONIC FORAMINIFERA--ADAPTIVE STRATEGY AND
M 27 35 PLANKTONIC FORAMS
B 6105 842* PLANKTONIC-BENTHONIC RATIO IN FORAMINIFERA AS
B 6306 972 PLANT BED UNIT, EGYPT
S 6 330 PLANT MEGAFOSSILS, CARBONIFEROUS, BIOSTRATIGRAPHY
B 6409 1662 PLANTAGENET FORMATION, WESTERN, ATLANTIC
B 6201 125 PLANTATION TONGUE, FLORIDA
S 11 298 PLANULINA ARIMINENSIS
S 11 265 PLANULINA ARIMINENSIS
S 11 158 PLANULINA ARIMINENSIS
S 11 300 PLANULINA MEXICANA
S 11 265 PLANULINA MEXICANA
S 11 300 PLANULINA WUELLERSTORFI
S 11 265 PLANULINA WUELLERSTORFI
B 6106 900 PLATE, AEGEAN
B 6106 900 PLATE, AFRICAN
B 6106 900 PLATE, ARABIAN
B 6207 1128 PLATE, AUSTRALO-, NEW GUINEAN, CONTINENTAL
M 25 485 PLATE BOUNDARIES, ACTIVE, NORTHEAST PACIFIC
M 29 274 PLATE BOUNDARIES, ALASKA, CENOZOIC
B 6004 697* PLATE BOUNDARIES IN USSR, ABST., EURASIAN
B 6207 1196 PLATE BOUNDARIES IN WESTERN NORTH AMERICA BETWEEN
M 29 277 PLATE BOUNDARIES, JUNCTURE, ALASKA
B 6403 360 PLATE BOUNDARIES, MAP
M 25 53 PLATE BOUNDARIES, METAL DEPOSITS AT, DIVERGENT
M 25 51 PLATE BOUNDARIES, METAL DEPOSITS AT OR NEAR, CONVERGENT
M 25 48 PLATE BOUNDARIES, PACIFIC, LITHOSPHERIC
B 6207 1193 PLATE BOUNDARIES, SEAFLOOR SPREADING ALONG, KULA-, FARALLON
M 29 376 PLATE BOUNDARIES, SOUTHEAST ASIA, SULU SEA
B 6207 1195 PLATE BOUNDARIES, TERTIARY
M 29 282 PLATE BOUNDARY, ALASKA, CROSS SOUND
B 6106 901 PLATE, CENTRAL
B 6209 1645† PLATE CONVERGENCE AND ACCRETION IN TAIWAN-LUZON REGION
M 25 546 PLATE CONVERGENCE, RELATION, TECTONIC BELTS, MOBILE, AND,
B 6106 900 PLATE, EURASIAN
B 6207 1195 PLATE, FARALLON
B 6106 900 PLATE, IONIAN
B 6207 1195 PLATE, JUAN DE FUCA
B 6207 1195 PLATE, KULA
B 6106 900 PLATE, LEVANTINE
B 6309 1602* PLATE MARGIN--NORTHERN GULF COAST, ABST., SEDIMENTATION ON
B 6403 365 PLATE MOTION AND OROGENY, KINEMATIC CONSIDERATIONS IN
B 6403 362 PLATE MOTION, EPISODIC READJUSTMENTS OF, SEAFLOOR SPREADING
B 6403 365 PLATE MOTION, KINEMATIC CONSIDERATIONS IN
M 25 448 PLATE MOVEMENT, DIRECTION OF, JAPAN
B 6403 299 PLATE MOVEMENTS ACCORDING TO MAGNETIC ANOMALIES,
B 6403 364 PLATE MOVEMENTS, NORTH, ATLANTIC OCEAN
M 25 434 PLATE MOVEMENTS, PHILIPPINES
B 6207 1195 PLATE, NORTH AMERICAN
B 6207 1195 PLATE, PACIFIC
B 6207 1128 PLATE, PACIFIC OCEANIC
B 6209 1763* PLATE RECONSTRUCTION—EVIDENCE AND IMPLICATION, ABST., CLOSED,
B 6301 8 PLATE RUPTURE, ARC-TRENCH SYSTEMS
B 6303 477* PLATE TECTONIC AND HYDROCARBON ACCUMULATION OF
B 6312 2164† PLATE TECTONIC AND SANDSTONE COMPOSITIONS
B 6403 449* PLATE TECTONIC SETTINGS, ABST., SANDSTONE COMPOSITIONS RELATED
B 6303 486* PLATE TECTONICS, ABST., LATE, MESOZOIC
B 6103 463* PLATE TECTONICS, ABST., OVERVIEW OF, ALASKA
B 6203 567* PLATE TECTONICS, ABST., SOUTHERN, OKLAHOMA OIL COUNTRY IN
B 6004 695* PLATE TECTONICS AND CLIMATIC CYCLES, ABST., PALEOGENE,
B 6105 842* PLATE TECTONICS AND EVOLUTION OF MARGINS. BY TITLE ONLY,
M 25 48† PLATE TECTONICS AND MINERAL RESOURCES OF CIRCUM-PACIFIC
B 6109 1541* PLATE TECTONICS AND SEDIMENTARY CYCLES IN CARBONATE ROCKS,
B 6406 953 PLATE TECTONICS, AND THE CHANGING ENVIRONMENT, JANE GRAY
B 6004 687* PLATE TECTONICS AS A MAJOR EXPLORATION TOOL. BY TITLE ONLY.,
B 6303 465* PLATE TECTONICS, BASIN ANALYSIS, AND EXPLORATION, ABST.
M 25 178 PLATE TECTONICS, HYPOTHESIS OF
B 6112 2063† PLATE TECTONICS IN BANDA ARCS, EASTERN INDONESIA, GEOSYNCLINES
B 6302 249† PLATE TECTONICS IN BANDA ARCS, EASTERN INDONESIA. DISCUSSION
B 6207 1231* PLATE TECTONICS IN EVOLUTION OF SOUTHERN ALASKA CONTINENTAL
B 6009 1463† PLATE TECTONICS IN OIL AND GAS EXPLORATION OF CONTINENTAL
M 25 177† PLATE TECTONICS IN PETROLEUM EXPLORATION OF CONVERGENT
B 6103 467* PLATE TECTONICS IN SOUTHERN ALASKA AND ADJACENT DEEP-SEA
B 6004 650* PLATE TECTONICS IN WESTERN CALIFORNIA, ABST., NEOGENE
M 29 274 PLATE TECTONICS, MODEL DEVELOPMENT
B 6203 493* PLATE TECTONICS, ORGANIC MATTER, AND BASIN EVALUATION FOR
B 6106 901 PLATE TECTONICS, SOUTHEASTERN, MEDITERRANEAN SEA
B 6012 2179* PLATE TECTONICS TO PETROLEUM GEOLOG ALONG PACIFIC MARGIN OF
M 25 157 PLATE TECTONICS, UNITED STATES, WESTERN
B 6009 1608* PLATE TECTONICS VIEW, ABST., GEOLOGIC EVOLUTION OF,
B 6004 694* PLATE TECTONICS. BY TITLE ONLY, STRUCTURE AN EVOLUTION OF,
B 6106 900 PLATE, TURKISH
B 6106 900 PLATE, VAN
B 6408 1279* PLATE-MARGIN GEOLOGY OF FRONTIER BASINS OF NORTH PACIFIC AND
B 6404 503 PLATE-TECTONIC EVENTS, NORTH, ATLANTIC
B 6211 2353* PLATE-TECTONIC EVOLUTION OF SACRAMENTO VALLEY, CALIFORNIA,
B 6203 344† PLATE-TECTONIC EVOLUTION OF SAN ANDREAS FAULT SYSTEM,
B 6212 2513* PLATE-TECTONIC EVOLUTION OF THE FAR EAST, ABST.
B 6307 1016† PLATE-TECTONIC HABITATS, AND HYDROCARBON TRAPS IN PETROLEUM
B 6103 464* PLATE-TECTONIC HISTORY. BY TITLE ONLY, GRAVITATIONAL AND
B 6207 1228* PLATE-TECTONIC INTERPRETATION OF MESOZOIC AND CENOZOIC ROCKS
B 6103 469* PLATE-TECTONIC MODEL BASED ON CONTINENTAL-MARGIN GEOLOGY,
M 25 547 PLATE-TECTONIC THEORY, RELATES GEOLOGIC EVENTS ON CONTINENTS
B 6007 1102 PLATE-TECTONICS HYPOTHESIS
B 6004 727* PLATE. BY TITLE ONLY, PRUDHOE BAY SUPE, GIANT FIELD AND,

B 6303 291 PLATEAU BLOCKS, PROSPECTS, MARGINAL-
B 6309 1591* PLATEAU PROVINCE, ABST., CROSS-STRIKE STRUCTURAL
B 6004 690* PLATEAU--DSDP DATA FROM SITE 356 ON SAO PAULO PLATEAU,
B 6207 1239* PLATEAUS, ABST., PETROLEUM PROSPECTIVITY OF, AUSTRALIAN,
M 25 179 PLATES AND LITHOSPHERIC SUBDUCTION, CONVERGENT
M 25 178 PLATES AND SEAFLOOR SPREADING, DIVERGENT
M 29 273† PLATES, CONTINENTAL, MARGINS OF THE EASTERN, GULF OF, ALASKA
B 6212 2456 PLATES, CONVERGENCE OF, NAZCA AND, SOUTH AMERICAN
B 6203 346 PLATES, INTERACTION OF, PACIFIC, FARALLON AND, NORTH AMERICAN
B 6209 1646 PLATES, TAIWAN-, LUZON REGION, ACTIVE AND EXTINCT
B 6212 2511* PLATES--EAST PACIFIC RISE AND BAUER BASIN, ABST., EVIDENCE FOR
B 6004 669* PLATFORM, ABST., ABENAKI FORMATION, NOVA SCOTIA SHELF,
B 6109 1552* PLATFORM, ABST., EXPLORATION HISTORY OF SOUTH, TEXAS LOWER,
B 6004 700* PLATFORM, ABST., HYDROCARBON, POTENTIAL OF NORTHWESTERN,
B 6109 1437† PLATFORM MARGIN, EASTERN ANADARKO BASIN, OKLAHOMA, SEISMIC
M 26 439† PLATFORM MARGIN, EASTERN ANADARKO BASIN, OKLAHOMA, SEISMIC,
B 6004 731* PLATFORM, MEXICO, ABST., DIAGENETIC SEQUENCE OF MIDDLE,
B 6002 258† PLATFORM OF ISRAEL, RUDISTID, FRINGING, REEFS OF, CRETACEOUS,
M 26 435 PLATFORM TRAPS, SEISMIC, MODELS
B 6203 560* PLATFORM-DERIVED SEDIMENT ON FACIES, DIAGENESIS, AND
B 6209 1758* PLATFORMS ON LOUISIANA CONTINENTAL SHELF AND SOURCE OF
M 25 533 PLATINUM, COLOMBIA
M 25 460 PLATINUM, PACIFIDES
B 6305 836 PLATT, LUCIAN B., PETROLEUM EXPLORATION I YOUNGER OVER OLDER
B 6107 1046 PLATTSBURG FORMATION, KANSAS
S 4 243 PLATTSBURG LIMESTONE, KANSAS
S 4 173 PLATYGYRA, X-RADIOGRAPH
B 6202 255 PLAYA DEL REY ANTICLINAL STRUCTURE, CALIFORNIA
S 10 51 PLAYA DEL REY FIELD, CALIFORNIA, LOS ANGELES BASIN
B 6305 835* PLAYA LAKES OF EBRO BASIN, SPAIN, ABST., RECENT, SALT
B 6406 814 PLAYFORD, PHILLIP E., DEVONIAN GREAT BARRIER REEF OF CANNING
B 6312 2188 PLAYS, ASSESSING RESOURCES, SUMMATION OF, PROSPECTS OR
B 6408 1158† PLAYS IN FACIES-CYCLE WEDGES, ASSESSING, OIL AND, GAS
B 6203 496 PLAZIAT, JEAN-CLAUDE, NONMARINE STROMATOLITES FROM CENTRAL
B 6205 850 PLEASANT VALLEY OIL FIELD, SECTION, CALIFORNIA
S 11 242 PLECTOFRONDICULARIA ADVENA
S 11 278 PLECTOFRONDICULARIA ADVENA
S 11 243 PLECTOFRONDICULARIA CALIFORNICA
S 11 278 PLECTOFRONDICULARIA CALIFORNICA
S 11 128 PLECTOFRONDICULARIA CALIFORNICA
S 11 243 PLECTOFRONDICULARIA MIOCENICA
S 11 278 PLECTOFRONDICULARIA MIOCENICA
S 11 128 PLECTOFRONDICULARIA MIOCENICA
S 11 21 PLECTOFRONDICULARIA MIOCENICA ZONE
S 11 242 PLECTOFRONDICULARIIDAE
M 27 190 PLEISTOCENE AEOLIANITE, ABU DHABI
M 27 194 PLEISTOCENE AEOLIANITE, ABU DHABI
B 6204 637 PLEISTOCENE, ALASKA
B 6201 90 PLEISTOCENE, ALASKA
B 6408 1142 PLEISTOCENE, ALASKA
S 7 134 PLEISTOCENE, ANCESTRAL MISSISSIPPI SUBMARINE CHANNEL
B 6109 1552* PLEISTOCENE AND CRETACEOUS AGE, ABST., SIGNIFICANCE OF,
B 6012 2181* PLEISTOCENE AND HOLOCENE STUDIES, ABST., PATTERNS OF,
M 27 59 PLEISTOCENE, BACK-REEF LIMESTONES, BELIZE
B 6309 1461 PLEISTOCENE, BALTIMORE CANYON TROUGH
S 4 300 PLEISTOCENE, BARBUDA
B 6104 519† PLEISTOCENE BARRIER BAR SEAWARD OF OOID SHOAL COMPLEX NEAR
S 11 201 PLEISTOCENE , BIOSTRATIGRAPHY, HUMBOLDT BASIN, PLIOCENE AND
B 6104 543 PLEISTOCENE BLUFFS, TEXAS
B 6204 665 PLEISTOCENE, BRAZIL
B 6009 1562† PLEISTOCENE BRUNHES EPOCH, NANNOFOSSIL, BIOSTRATIGRAPH AND,
B 6412 2118† PLEISTOCENE CALCAREOUS SEDIMENTS, DSDP SITE 262, TIMOR TROUGH,
B 6205 824 PLEISTOCENE, CALIFORNIA
B 6211 2267 PLEISTOCENE, CALIFORNIA
B 6102 139 PLEISTOCENE, CALIFORNIA
B 6102 171 PLEISTOCENE, CALIFORNIA
B 6202 250 PLEISTOCENE, CALIFORNIA
B 6209 1767* PLEISTOCENE CARBONATE BANK DEPOSITION, COZUMEL ISLAND,
B 6303 528* PLEISTOCENE CARBONATE SEDIMENTS, BONAIRE, NETHERLANDS
B 6410 1634† PLEISTOCENE CARBONATES, NORTHWESTERN GREAT BAHAMA BANK,
B 6303 439* PLEISTOCENE COASTAL-BARRIER COMPLEXES FROM HOLOCENE MODELS
B 6207 1132 PLEISTOCENE CONGLOMERATE FILL, BATANTA ISLAND, PLIO-
B 6407 1037 PLEISTOCENE CONTACT, CALIFORNIA, PLIOCENE-
B 6104 528 PLEISTOCENE DELTAIC MUDS AND SANDS, TEXAS
B 6109 1547* PLEISTOCENE DEPOCENTER, LOUISIANA CONTINENTAL SHELF, ABST.,
B 6009 1614* PLEISTOCENE DEPOCENTER--TEXAS-LOUISIANA CONTINENTAL SHELF,
M 26 273 PLEISTOCENE DEPOSITION, GULF OF MEXICO
M 26 249† PLEISTOCENE DEPOSITS IN NORTHWEST AND CENTRAL GULF OF MEXICO,
B 6303 434* PLEISTOCENE DEPOSITS OF PUGET LOWLANDS, ABST., NEARSHORE,
B 6212 2492† PLEISTOCENE DIASTROPHISM OF SANTA MONICA AND SAN PEDRO
B 6202 247† PLEISTOCENE DIASTROPHISM OF SANTA MONICA AND SAN PEDRO
M 27 115 PLEISTOCENE DUNES, BAHAMAS
M 27 127 PLEISTOCENE DUNES, BAHAMAS
M 27 162 PLEISTOCENE DUNES, MEXICO
M 27 163 PLEISTOCENE DUNES, MEXICO
M 24 205† PLEISTOCENE, EAST CAMERON, BLOCK 270, OFFSHORE, LOUISIANA--A
B 6303 434* PLEISTOCENE, EAST TURKANA, KENYA, ENVIRONMENTAL SIGNIFICANCE
M 29 345 PLEISTOCENE, EVOLUTION, COLOMBIA
B 6007 1112† PLEISTOCENE FAULTS ON NEW JERSEY ATLANTIC OUTER CONTINENTAL
B 6004 696* PLEISTOCENE FLUVIAL, ALGAL, PISOLITES AND PISOLITIC CALICHE IN
B 6309 1611* PLEISTOCENE FLUVIAL-DELTAIC DEPOSITION TEXAS COASTAL PLAIN
S 7 243† PLEISTOCENE GEOLOGIC HISTORY OF TEXAS OUTER CONTINENTAL
B 6109 1535* PLEISTOCENE, GULF COAST REGION, ABST., EFFECTS OF, SANDSTONE
B 6303 411* PLEISTOCENE, GULF OF ALASKA, ABST., GLACIAL-MARINE, TRACE- AND
S 7 249 PLEISTOCENE, GULF OF MEXICO, PALEOCLIMATIC ZONES
M 28 139 PLEISTOCENE, HONOLULU VOLCANICS, HAWAII
B 6004 719* PLEISTOCENE, IN BAHAMIAN TROUGH. BY TITLE ONLY, CARBONATE,
M 26 251 PLEISTOCENE ISOPACH, GULF OF MEXICO
B 6404 569 PLEISTOCENE, ISRAEL
B 6205 839 PLEISTOCENE, ISRAEL
M 27 59 PLEISTOCENE, KEY LARGO LIMESTONE, FLORIDA
M 27 218 PLEISTOCENE, KEY LARGO LIMESTONE, FLORIDA
M 27 224 PLEISTOCENE, KEY LARGO LIMESTONE, FLORIDA
M 27 7 PLEISTOCENE, KEY LARGO LIMESTONE, FLORIDA
M 27 176 PLEISTOCENE, KEY LARGO LIMESTONE, FLORIDA
M 27 71 PLEISTOCENE, KEY LARGO LIMESTONE, FLORIDA
M 27 162 PLEISTOCENE, KEY LARGO LIMESTONE, FLORIDA
B 6102 201 PLEISTOCENE, KOMANDORSKIY BASIN
B 6203 488* PLEISTOCENE LACUSTRINE STROMATOLITES FROM LAKE TURKANA,
S 4 69 PLEISTOCENE, LESSER, ANTILLES
S 4 299† PLEISTOCENE LIMESTONES ON AND AROUND BARBUDA, WEST INDIES,
B 6104 529 PLEISTOCENE LITHOLOGY, TEXAS
B 6206 951 PLEISTOCENE, LOUISIANA
M 30 253 PLEISTOCENE, LOUISIANA, EUGENE ISLAND 330 FIELD
B 6111 2001† PLEISTOCENE MARINE TERRACE, PALOS VERDES HILLS, CALIFORNIA,
B 6303 458* PLEISTOCENE METEORIC-VADOSE DIAGENESIS OF ENEWETAK ATOLL,
M 27 219 PLEISTOCENE, MIAMI OOLITES, FLORIDA
S 7 181 PLEISTOCENE, MISSISSIPPI FAN, SEISMIC PROFILES
M 28 87 PLEISTOCENE, MONTALLEGRO FORMATION, ITALY
M 27 36 PLEISTOCENE OOZE, FLORIDA
B 6303 424* PLEISTOCENE OUTCROPS ON GEORGIA COAST, ABST., ROLE OF,
B 6109 1549* PLEISTOCENE PALEOGEOGRAPHY OF FLORIDA GULF COAST
M 29 401 PLEISTOCENE, PETROLEUM POTENTIAL, MISSISSIPPI FAN
B 6011 1985† PLEISTOCENE PRODUCING SECTION OF SATICOY OIL FIELD, VENTURA
B 6303 411* PLEISTOCENE ROBINSON MOUNTAINS, GULF OF ALASKA, ABST., GLACIAL
M 28 42 PLEISTOCENE SANDSTONE, CALIFORNIA
B 6409 1555* PLEISTOCENE SANDSTONES, OFFSHORE, GULF OF MEXICO, ABST.,
M 29 396 PLEISTOCENE, SEDIMENTATION, MISSISSIPPI FAN
M 28 140 PLEISTOCENE, SEDIMENTS, CALIFORNIA
M 28 141 PLEISTOCENE, SEDIMENTS, CALIFORNIA
B 6209 1768* PLEISTOCENE SHORELINE IN SOUTH FLORIDA, ABST., HIGH-LEVEL, PLIO-
B 6209 1757* PLEISTOCENE SHORELINES AND PALEOGEOGRAPHIC FEATURES,
B 6402 189 PLEISTOCENE, SPAIN
B 6209 1754* PLEISTOCENE STAGES, ABST., NEED FOR REDEFINITION OF, NORTH
S 6 171† PLEISTOCENE, STATUS OF THE, BOUNDARY BETWEEN, PLIOCENE AND
B 6209 1763* PLEISTOCENE STRATIGRAPHY OF SOUTHERN HANCOCK COUNTY,
B 6303 551* PLEISTOCENE SUBMARINE CANYON, BOSO PENINSULA, JAPAN, ABST.,
B 6303 411* PLEISTOCENE SUBMARINE CANYONS OF NORTHWEST GULF OF
B 6209 1757* PLEISTOCENE TERRACE DEPOSITS, LOUISIANA, ABST., MINERALOGY OF
M 28 126 PLEISTOCENE, TERRACE DEPOSITS, TEXAS
S 7 257 PLEISTOCENE, TEXAS, SHELF-SLOPE, HISTORY
B 6305 792 PLEISTOCENE THICKNESS MAP, GULF COAST
S 7 79 PLEISTOCENE THICKNESS MAP, GULF OF MEXICO
S 7 214 PLEISTOCENE THICKNESS, WEST FLORIDA SLOPE
B 6405 713* PLEISTOCENE TO HOLOCENE, ABST., DEPOSITION OF, ENEWETAK, ATOLL,
B 6404 472 PLEISTOCENE, VADOSE WHISKER CALCITE, BARBADOS
M 28 2 PLEISTOCENE, YELLOWSTONE GROUP, WYOMING
M 28 90 PLEISTOCENE, YELLOWSTONE GROUP, WYOMING
M 28 3 PLEISTOCENE, YELLOWSTONE GROUP, WYOMING
B 6303 362† PLEISTOCENE, YUCATAN PENINSULA, MEXICO, STRANDLINE
B 6309 1607* PLEISTOCENE-HOLOCENE BOUNDARY, ABST., RADIOLARIAN RATIOS AN
B 6004 649* PLEISTOCENE-HOLOCENE EVOLUTION OF SOUTH TEXAS OUTER
B 6203 507* PLEISTOCENE-HOLOCENE OF BAHAMIAN TROUGH, ABST., SEDIMENTARY
S 4 94 PLEISTOCENE-HOLOCENE REEFS
S 4 107 PLEISTOCENE-HOLOCENE REEFS, CARIBBEAN
B 6403 445* PLEISTOCENE-HOLOCENE SEDIMENTARY INFILLING AND FAUNAL
S 7 210 PLEISTOCENE-RECENT, WEST FLORIDA SLOPE, NORTHERN AREA
B 6207 1132 PLIO-PLEISTOCENE CONGLOMERATE FILL, BATANTA ISLAND
B 6209 1768* PLIO-PLEISTOCENE SHORELINE IN SOUTH FLORIDA, ABST., HIGH-LEVEL
B 6004 695* PLIOCENE, ABST., SEDIMENT, LITHOFACIES IN, SOUTH, ATLANTIC OCEAN,

B 6408 1142 PLIOCENE, ALASKA
B 6408 1133 PLIOCENE, ALASKA
B 6407 1016 PLIOCENE, ALASKA
B 6201 90 PLIOCENE, ALASKA
B 6309 1524 PLIOCENE, ALASKA PENINSULA, TACHILNI FORMATION, MIOCENE AND
B 6203 533* PLIOCENE, ALTAI MOUNTAINS, ABST., LACUSTRINE STROMATOLITES
S 11 201 PLIOCENE AND PLEISTOCENE , BIOSTRATIGRAPHY, HUMBOLDT BASIN
S 6 171† PLIOCENE AND PLEISTOCENE, STATUS OF THE, BOUNDARY BETWEEN
B 6309 1461 PLIOCENE, BALTIMORE CANYON TROUGH
M 28 89 PLIOCENE, BIG SANDY FORMATION, ARIZONA
M 28 142 PLIOCENE, BIG SANDY FORMATION, ARIZONA
M 28 139 PLIOCENE, BIG SANDY FORMATION, ARIZONA
M 28 155 PLIOCENE, BIG SANDY FORMATION, ARIZONA
B 6012 2189* PLIOCENE BIOSTRATIGRAPHY AND DEPOSITIONAL ENVIRONMENT OF
S 11 1 **PLIOCENE BIOSTRATIGRAPHY OF CALIFORNIA**
S 11 183† PLIOCENE BIOSTRATIGRAPHY OF CALIFORNIA
M 28 82 PLIOCENE, BONE VALLEY FORMATION, FLORIDA
M 28 83 PLIOCENE, BONE VALLEY FORMATION, FLORIDA
M 28 126 PLIOCENE, BONE VALLEY FORMATION, FLORIDA
B 6303 427* PLIOCENE BOUSE FORMATION NEAR VIDAL, SOUTHEASTERN
B 6204 665 PLIOCENE, BRAZIL
B 6205 824 PLIOCENE, CALIFORNIA
B 6211 2247 PLIOCENE, CALIFORNIA
B 6102 139 PLIOCENE, CALIFORNIA
B 6102 171 PLIOCENE, CALIFORNIA
M 27 167 PLIOCENE, CALOOSAHATCHEE FORMATION, FLORIDA
M 27 199 PLIOCENE, CALOOSAHATCHEE FORMATION, FLORIDA
M 27 212 PLIOCENE, CALOOSAHATCHEE MARL, FLORIDA
M 27 211 PLIOCENE, CALOOSAHATCHEE MARL, FLORIDA
M 27 178 PLIOCENE, CALOOSAHATCHEE MARL, FLORIDA
M 27 175 PLIOCENE, CALOOSAHATCHEE MARL, FLORIDA
M 27 83 PLIOCENE, CALOOSAHATCHEE MARL, FLORIDA
M 27 221 PLIOCENE, CALOOSAHATCHEE MARL, FLORIDA
M 27 84 PLIOCENE, CALOOSAHATCHEE MARL, FLORIDA
S 4 106 PLIOCENE, CARIBBEAN
B 6104 561 PLIOCENE, CHILE
B 6009 1611* PLIOCENE COASTAL-PLAIN SEDIMENTS, ABST., MULTIVARIATE, MINERAL
M 25 325 PLIOCENE, COLOMBIA
M 29 345 PLIOCENE, EVOLUTION, COLOMBIA
B 6207 1169 PLIOCENE, GULF OF CALIFORNIA
B 6404 569 PLIOCENE, ISRAEL
B 6102 201 PLIOCENE, KOMANDORSKIY BASIN
B 6004 701* PLIOCENE LAGA FORMATION, EAST-CENTRAL ITALY, ABST.,
B 6112 2066 PLIOCENE, LATE, OROGENIC BELT, INDONESIA, EASTERN
S 4 69 PLIOCENE, LESSER, ANTILLES
M 27 88 PLIOCENE LIMESTONE, ITALY
M 27 89 PLIOCENE LIMESTONES, ITALY
B 6012 2175† PLIOCENE MARINE DIATO AND SILICOFLAGELLATE CORRELATIONS IN
B 6303 444* PLIOCENE MOLASSE, ZAGROS MOUNTAINS, IRAN, ABST., MIOCENE-
M 28 142 PLIOCENE, MONTEREY FORMATION, CALIFORNIA
M 28 81 PLIOCENE, MONTEREY SHALE, CALIFORNIA
B 6201 4 PLIOCENE, NIGER DELTA
M 29 45 PLIOCENE, OUTER INSULAR SHELF, ICELAND
B 6205 793 PLIOCENE, PROFILE, BLAKE PLATEAU AREA
B 6102 176 PLIOCENE SANDS, CALIFORNIA, VENTURA FIELD
B 6203 572* PLIOCENE SEDIMENTATION AND FORAMINIFERAL BIOSTRATIGRAPHY OF
B 6303 510* PLIOCENE SEDIMENTS OF NORTH ATLANTIC, ABST., TEMPORAL AND
M 27 19 PLIOCENE, SEDIMENTS, PACIFIC
B 6205 768 PLIOCENE, SEISMIC FACIES
B 6402 189 PLIOCENE, SPAIN
B 6303 444* PLIOCENE SYNDEPOSITIONAL TAR DEPOSI IN IRAN, ABST., MIOCENE-
B 6303 493* PLIOCENE TAMIAMI FORMATION, COLLIER COUNTY, FLORIDA, ABST.,
M 27 178 PLIOCENE, TAMIAMI FORMATION, FLORIDA
M 27 166 PLIOCENE, TAMIAMI FORMATION, FLORIDA
M 30 395 PLIOCENE, TEAK FIELD, GROS MORNE FORMATION
M 30 395 PLIOCENE, TEAK FIELD, MAYARO FORMATION
M 30 395 PLIOCENE, TEAK FIELD, RESERVOIR HORIZONS
M 30 387 PLIOCENE, TEAK FIELD, TRINIDAD
S 7 212 PLIOCENE THICKNESS, WEST FLORIDA SLOPE
B 6102 137† PLIOCENE TURBIDITES, STUDIES OF, VENTURA FIELD, CALIFORNIA, I--,
S 5 175 PLIOCENE, TRUBI FORMATION, ITALY,
S 7 208 PLIOCENE, WEST FLORIDA SLOPE, NORTHERN AREA
B 6303 528* PLIOCENE-PLEISTOCENE CARBONATE SEDIMENTS, BONAIRE,
B 6410 1634† PLIOCENE-PLEISTOCENE CARBONATES, NORTHWESTERN GREAT BAHAMA
B 6407 1037 PLIOCENE-PLEISTOCENE CONTACT, CALIFORNIA
B 6109 1547* PLIOCENE-PLEISTOCENE DEPOCENTER, LOUISIANA CONTINENTAL SHELF,
B 6303 434* PLIOCENE-PLEISTOCENE DEPOSITS OF PUGET LOWLANDS, ABST.,
B 6202 247† PLIOCENE-PLEISTOCENE DIASTROPHISM OF SANTA MONICA AND SAN
B 6212 2492† PLIOCENE-PLEISTOCENE DIASTROPHISM OF SANTA MONICA AND SAN
B 6303 434* PLIOCENE-PLEISTOCENE, EAST TURKANA, KENYA, ENVIRONMENTAL
B 6303 411* PLIOCENE-PLEISTOCENE, GULF OF ALASKA, ABST., GLACIAL-MARINE,
B 6109 1549* PLIOCENE-PLEISTOCENE PALEOGEOGRAPHY OF FLORIDA GULF COAST
B 6011 1985† PLIOCENE-PLEISTOCENE PRODUCING SECTION OF SATICOY OIL FIELD,
B 6303 411* PLIOCENE-PLEISTOCENE ROBINSON MOUNTAINS, GULF OF ALASKA,
B 6409 1555* PLIOCENE-PLEISTOCENE SANDSTONES, OFFSHORE, GULF OF MEXICO,
B 6211 2217 PLIOCENE-RECENT, CALIFORNIA
B 6307 1052 PLOESTI AREA, SALT FEATURES, ROMANIAN, CARPATHIAN,
B 6402 189 PLOMO FORMATION, SPAIN
B 6402 190 PLOMO REEFS, SPAIN
B 6405 747 PLUIM, SCOTT B., TERRESTRIAL SEDIMENTATION ASSOCIATED WITH
M 28 68 PLUM BENTONITE, TEXAS, EOCENE
S 3 28 PLUMES, MANTLE
B 6403 414 PLUMLEY, WILLIAM J., ABNORMALLY HIGH FLUID PRESSURE--SURVEY
B 6405 766 PLUMLEY, WILLIAM J., EFFECTIVE STRESS AND ABNORMALLY HIGH
B 6303 482* PLUMOSE STRUCTURES, ABST., NATURE AND FIELD APPLICATION OF
M 25 458 PLUTONIC AREAS OR BELTS, PACIFIDES, VOLCANIC-
B 6303 303 PLUTONIC ROCKS, ALASKA
B 6203 494* PLUTONS, ABST., TRACE ELEMENTS IN, HOLOCENE, FLUVIAL, SANDS
M 25 496 PLUTONS, CANADIAN, CORDILLERA, QUARTZ MONZONITE
B 6207 1173 PLYMPTON FORMATION, NEVADA AND, UTAH
M 30 529 PO BASIN, CROSS SECTION, ITALY
M 30 537 PO BASIN, DOLOMIA PRINCIPALE FORMATION, ITALY
M 30 525 PO BASIN, MESOZOIC, MALOSSA FIELD
M 30 527 PO BASIN, PRODUCTION HISTORY, ITALY
M 30 531 PO BASIN, STRATIGRAPHIC COLUMN, ITALY
M 30 527 PO BASIN, STRUCTURAL OUTLINE, ITALY
B 6408 1211 PO HAI BASIN, CHINA
B 6411 1964 PO VALLEY, ITALY
B 6310 1922 PO VALLEY, ITALY
B 6303 446* PO VALLEY, ITALY, ABST., MALOSSA FIELD, DEEP DISCOVERY IN
M 30 525† PO VALLEY, ITALY, MALOSSA FIELD--A DEEP DISCOVERY IN THE
M 26 343 PO VALLEY, ITALY, SERGNANO FIELD
B 6210 1913 PO VALLEY ONSHORE, ITALY
B 6110 1643 PO VALLEY ONSHORE, ITALY
B 6004 672 POAG, C. W. ET AL, MESOZOIC STRATIGRAPHIC TEST, NANTUCKET
B 6009 1614 **POAG, C. WYLIE, BIOSTRATIGRAPHY OF PLEISTOCENE**
B 6105 821 POAG, C. WYLIE, BIOSTRATIGRAPHY OF U.S. ATLANTIC CONTINENTAL
B 6303 510 POAG, C. WYLIE, IMPORTANT STRATIGRAPHIC BREAKS IN COST GE-1
B 6309 1609 POAG, C. WYLIE, LIVING FORAMINIFERS OF WEST FLOWER GARDEN
B 6209 1764 POAG, C. WYLIE, PAIRED FORAMINIFERAL ECOPHENOTYPES IN GULF
B 6405 649 POAG, C. WYLIE, REVIEW OF CURRENTS IN SUBMARINE CANYONS AND
B 6405 781 POAG, C. WYLIE, SEISMIC STRATIGRAPHY OF ATLANTIC MARGIN IN
B 6309 1452 POAG, C. WYLIE, STRATIGRAPHY AND DEPOSITIONAL ENVIRONMENTS OF
B 6212 2396 POCAHONTAS BASIN, SECTION
B 6203 548 POELCHAU, H. S., EFFECT OF CLAYS ON URANIUM DISTRIBUTION IN
M 26 389 POGGIAGLIOLMI, ELIO, STRATIGRAPHIC MODELING AND
B 6405 766 POGORSKI, LOUIS A., HELIUM SURVEYING FOR DEEPLY BURIED
M 25 269 POHANG ALIGNMENT, KOREA, KWANGJU-
M 25 269 POHANG SUBBASIN, KOREA
M 27 166 POIKILOTOPIC CEMENTS
B 6203 352 POINT ARENA BASIN, CALIFORNIA
B 6211 2215 POINT ARENA, CALIFORNIA
B 6012 2187* POINT ARENA, CALIFORNIA, ABST., PALEONTOLOGIC STUDIES OF
B 6402 265 POINT AU FER SHELL REEF, LOUISIANA
B 6409 1553* POINT BAR FACIES, AMITE RIVER, LOUISIANA, ABST., PRELIMINARY
B 6004 676* POINT BARS, NUECES RIVER, TEXAS, ABST., BED FORMS AND
B 6304 600 POINT LOOKOUT SANDSTONE, MESAVERDE GROUP, NEW MEXICO
B 6112 2141 POINT LOOKOUT SANDSTONE, NEW MEXICO AND, COLORADO
B 6204 635 POINT RIOU, ALASKA
M 28 79 POINT SAL OPHIOLITE, CALIFORNIA, JURASSIC
M 27 24 POINT SAL OPHIOLITE, CALIFORNIA, JURASSIC
M 27 23 POINT SAL OPHIOLITE, CALIFORNIA, JURASSIC
B 6109 1499 POINT STARKERSEN SECTION, ALASKA
B 6405 708* POINT-BAR SANDSTONES, ABST., APPLICATION OF, INVERSION
B 6107 1083† POINT-BOUNDARY STRATOTYPE CONCEPT, NOTE 47--APPLICATION FOR
M 30 588 POINTE CLAIRETTE SHALES, GABON BASIN, BATANGA FORMATION
B 6108 1246 POINTE COUPEE PARISH, LOUISIANA, FALSE RIVER FIELD
B 6405 795* POISON CANYO AREA, MORRISON FORMATION, SAN JUAN BASIN, AND
M 28 141 POISON CANYON FORMATION, COLORADO, PALEOCENE
B 6304 694* POISON CANYON MINE, ABST., REDISTRIBUTED OREBODIES OF
B 6304 690* POISON CANYON SANDSTONE IN GULF MARIANO LAKE MINE, ABST.,
S 3 124 POISON CANYON SANDSTONE, NEW MEXICO, GRANTS
B 6105 785 * **POISSON BED-THICKNESS DISTRIBUTION AND TURBIDITE-PELAGITE**
B 6002 281 POKEJOY MEMBER, PENNSYLVANIA
B 6010 1724 POLAND, DEVELOPMENTS, 1975
B 6110 1653 POLAND, DEVELOPMENTS, 1976
B 6405 649 POLAND, EDITED BY S. DEPOWSKI, RUHLE, EDWARD, REVIEW OF,
B 6110 1654 POLAND, PRODUCTION, 1975-1976

Index

B 6210 1924 POLAND, PRODUCTION, 1976-1977
B 6310 1932 POLAND, PRODUCTION, 1977-1978
B 6210 1923 POLAND-EAST GERMANY, DEVELOPMENTS, 1977
S 9 12 POLAR WANDERING, CRETACEOUS, DIAGRAM
S 9 12 POLAR WANDERING, PERMIAN, DIAGRAM
S 9 12 POLAR WANDERING, TERTIARY, DIAGRAM
S 9 12 POLAR WANDERING, TRIASSIC, DIAGRAM
S 6 12 POLARITY OF SEQUENCE, TIME
B 6301 8 POLARITY REVERSAL, ARC-TRENCH SYSTEMS
B 6002 275 POLARITY ROCK-STRATIGRAPHIC UNITS
M 26 303 POLARITY, SEISMIC APPLICATION
S 6 57 POLARITY, STRATIGRAPHIC CORRELATION, MAGNETIC
S 6 57† POLARITY TIME SCALE--PROSPECTS AND POSSIBILITIES IN
B 6002 276 POLARITY-CHRONOLOGIC UNITS
B 6002 276 POLARITY-CHRONOSTRATIGRAPHIC UNITS
B 6401 56 POLENG FIELD, JAVA SEA
B 6401 56 POLENG FIELD, KALIMANTAN, JAVA SEA
B 6308 1371 POLESCHOOK, DANIEL, JR., DEVELOPMENTS IN EASTERN AND
B 6205 740 POLITICAL DISTRIBUTION OF HYDROCARBONS
M 26 114 POLK COUNTY, AUSTIN CHALK, TEXAS
M 26 114 POLK COUNTY, BUDA LIMESTONE, TEXAS
M 26 114 POLK COUNTY, EAGLE FORD SHALE, TEXAS
B 6308 1225 POLK COUNTY, TEXAS, DAMASCUS FIELD
M 26 114 POLK COUNTY, WOODBINE FORMATION, TEXAS
B 6309 1601* POLK, TYLER, AND SAN JACINTO COUNTIES, TEXAS, ABST.,
S 4 239 POLLARD, WILLIAM D., IMPORTANCE OF PHYLLOID ALGAE IN
B 6405 671 POLLASTRO, RICHARD, OXIC-ANOXIC AND CARBONATE CYCLES IN
M 25 120 POLLUTION FROM HYDROGEN SULFIDE, AIR
B 6303 426* POLOCHIC FAULT BLOCKS AS DETERMINED FROM GEOMORPHIC
S 4 270 POLYCHAETE WORMS, BORERS
M 25 379 POLYGON SURVEYS, PACIFIC
B 6002 249 POLYHEDRAL PORES
B 6004 671* POLYMER PROJECT AREA, BUTLER COUNTY, KANSAS, ABST., DELTAIC
M 25 460 POLYMETALLIC ORES, CASSITERITE, EASTERN ASIATIC VOLCANOGENIC
M 25 460 POLYMETALLIC PROFILE, PACIFIC, MEXICO-, CENTRAL AMERICA, SILVER-,
S 11 274 POLYMORPHINA CHARLOTTENSIS
S 11 238 POLYMORPHINA CHARLOTTENSIS
S 11 237 POLYMORPHINIDAE
S 6 235 POMEROL, CHARLES, CRITICAL REVIEW OF ISOTOPIC DATES IN RELATION
S 6 369 PONGOLA SYSTEM, SOUTH AFRICA
B 6104 547 PONTA GROSSA ARCH, BRAZIL
S 6 106 PONTA GROSSA FORMATION, BRAZIL, PARANA BASIN
S 6 106 PONTA GROSSA FORMATION, RUBIDIUM-STRONTIUM, BRAZIL
B 6209 1762* PONTCHARTRAIN, ABST., NEW ORLEANS--, SOUTH HANCOCK, HOLOCENE,
B 6209 1758* PONTCHARTRAIN, ABST., REMOTELY SENSED SEDIMENTATION PATTERNS
B 6209 1753 PONTIGO, F. A., JR., FUTURE OIL POTENTIAL OF LOWER CRETACEOUS
B 6109 1534 PONTIGO, F. A., JR., SMACKOVER OIL PROSPECTS IN APALACHICOLA
B 6404 490 POOL CREEK MEMBER, CARLILE SHALE, WYOMING
B 6406 943 POOL CREEK MEMBER, SOUTH DAKOTA
B 6303 468* POOL DEPLETION AND GEOCHEMICAL SIGNAL DECAY. B TITLE ONLY
B 6305 838 POOLE, FORREST G., EVIDENCE FOR TWO MAJOR CYCLES OF PETROLEUM
B 6405 767 POOLE, FORREST G., OIL AND METALS IN ORDOVICIAN AND DEVONIAN
B 6307 1136 POOLEVILLE MEMBER, OKLAHOMA
B 6410 1644 POONTANA STRUCTURE, AUSTRALIA
B 6410 1644 POONTANA SUBBASIN, AUSTRALIA
B 6303 510 POORE, RICHARD Z., TEMPORAL AND SPATIAL DISTRIBUTION OF
B 6309 1611 POORT, JON M., EARLY TERTIARY LITHOSTRATIGRAPHIC
B 6106 907 POPCORN RIDGE, BAJA CALIFORNIA
B 6106 907 POPCORN RIDGE ESCARPMENT, BAJA CALIFORNIA
B 6106 904 POPCORN RIDGE FAULT, BAJA CALIFORNIA
B 6209 1752 POPE, DAVID E., RADIOMETRIC INVESTIGATION RELATIVE TO A
B 6304 693 POPP, C. J., APPLICATION OF SOLUTION-MINERAL EQUILIBRIUM
B 6202 179 POPULATION AND CRUDE OIL CONSUMPTION VERSUS TIME, WORLD
M 25 293 PORCUPINE FAULT, ALASKA
B 6203 426 PORCUPINE SEABIGHT, EUROPE
B 6411 1963 PORCUPINE TROUGH, EIRE
B 6205 884* PORE AND CEMENT VOLUMES OF CARBONATE ROCKS IN THIN SECTION,
B 6002 245† PORE CASTS AND CAPILLARY PRESSURE, PORE, GEOMETRY OF,
B 6107 1077 PORE FILLING, GRAIN BINDING AND INTERGRANULAR
M 26 422 PORE FLUID, EFFECTS OF, SEISMIC FACTORS
B 6002 297 PORE FLUID EXPELLED DURING COMPACTION
M 26 455 PORE FLUIDS, SEISMIC, MODELING, PRESSURE AND
B 6002 245† PORE GEOMETRY OF CARBONATE ROCKS AS REVEALED BY PORE CASTS
B 6107 1087 PORE PRESSURE, CHANGES IN
B 6412 2179† PORE PRESSURES IN SHALES, MECHANICAL VERSUS, THERMAL CAUSE OF,
B 6412 2179 PORE PRESSURES IN SHALES, THERMAL, HYPOTHESIS FOR, ABNORMALLY
B 6303 445* PORE SPACE IN HOLOCENE CARBONATE SEDIMENTS, ABST.
B 6004 663* PORE STRUCTURE, ABST., NUMERICAL, CLASSIFICATION
S 4 363 PORE SYSTEMS
B 6405 799* PORE SYSTEMS, ABST., MULTIPHASE FLUID MOVEMENTS IN GLASS
B 6002 249 PORE SYSTEMS IN DOLOMITES
B 6303 551* **PORE SYSTEMS IN JURASSIC CARBONATE RESERVOIRS, UNITED STATES**
B 6002 245 PORE THROATS IN DOLOMITES
B 6303 460* PORE TYPES IN SUNNILAND LIMESTONE, LOWER CRETACEOUS, ABST.
B 6401 112 PORE WATER OF SOURCE ROCKS, CONCENTRATION OF, CARBON DIOXIDE
S 10 17 PORE WATER, DIFFUSION OF WATER AND DISSOLVED SOLIDS
S 10 24 PORE WATER, HYDROGEN BONDING, STRUCTURED WATER
S 10 24 PORE WATER, ORGANIC MOLECULES
S 10 15 PORE WATER, SALINITY EQUILIBRIUM
S 10 22 PORE-CENTER NETWORK, PRIMARY MIGRATION
S 10 25 PORE-CENTER NETWORK WITH DEPTH, MIGRATION
B 6403 445* PORE-FILLING CEMENTS--PRODUCTS OF SHALE DEWATERING IN UPPER
B 6303 504* PORE-FLUID GEOCHEMISTRY, ABST., PARAGENETIC AND STABILITY
B 6002 255 PORES IN LIMESTONE
B 6002 249 PORES, INTERBOUNDARY-SHEET
B 6401 107 PORES OF SOURCE ROCKS, GENERATION OF, GAS IN
B 6002 249 PORES, POLYHEDRAL
S 4 44 PORITES ASSEMBLAGE, BELIZE, ALGAE-
S 4 33 PORITES ASSEMBLAGE, WEST FLOWER GARDEN BANK, DIPLORIA-,
S 4 362 PORITES SP., MICROGRAPH
M 27 1 **POROSITIES, A COLOR ILLUSTRATED GUIDE TO CARBONATE**
M 27 217 POROSITIES, CALCITE-FILLED INTERPARTICLE
M 27 217 POROSITIES, ENLARGED INTERPARTICLE
M 27 223 POROSITIES, ENLARGED MOLDIC
M 27 221 POROSITIES, FILLED FENESTRAL
M 27 222 POROSITIES, FILLED FRACTURE
M 27 220 POROSITIES, FILLED MOLDIC
M 27 221 POROSITIES, FILLED SHELTER
M 27 219 POROSITIES, INTERCRYSTAL
M 27 220 POROSITIES, MOLDIC
M 27 221 POROSITIES, MOLDIC
M 27 219 POROSITIES. MOLDIC
M 28 1 **POROSITIES OF SANDSTONES AND ASSOCIATED ROCKS**
M 27 222 POROSITIES, REDUCED FRACTURE
M 27 218 POROSITIES, REDUCED INTERPARTICLE
M 27 219 POROSITIES, REDUCED INTERPARTICLE
M 27 219 POROSITIES, REDUCED INTRAPARTICLE
M 27 222 POROSITIES, SECONDARY ENLARGED
M 27 217 POROSITIES, UNFILLED INTERPARTICLE
M 27 218 POROSITIES, UNFILLED INTERPARTICLE
M 27 224 POROSITIES, VUGGY
M 27 223 POROSITIES, VUGGY
B 6105 826* POROSITY ALTERATION IN CARBONATE SEDIMENTS, ABST.,
B 6104 511† **POROSITY AND CEMENT IN OOID RESERVOIRS IN DOGGER, MIDDLE**
B 6304 676† POROSITY AND CEMENT IN OOID RESERVOIRS IN DOGGER, MIDDLE
B 6305 799 POROSITY AND PEBBLE CONTENT
S 4 371 POROSITY, AND PERMEABILITY, CORALS, DENSITY
B 6004 725* POROSITY AND PERMEABILITY OF LOWER WILCOX, EOCENE, SANDSTONE,
B 6203 555* POROSITY AND PERMEABILITY OF ORTHOQUARTZITES, SIMPSON GROUP,
B 6305 802 POROSITY AND SAND CONTENT IN CONGLOMERATE AND PEBBLY
M 28 145 POROSITY, BIBLIOGRAPHY
B 6405 796* POROSITY CHANGES IN EXPERIMENTALLY DEFORMED INDIANA
M 28 170 POROSITY, CLASSIFICATION, COMPARISONS
M 28 174 POROSITY, COMPLEX, FELDSPAR GROWTH AND LEACHING
M 28 173 POROSITY, COMPLEX, IN SANDSTONE
B 6105 841* POROSITY CONTROLS IN SUBSEA CEMENTED ROCKS FROM DEEP-FLANK
B 6103 407† POROSITY DEPENDENCE ON TEMPERATURE--LIMITS ON MAXIMUM
M 28 172 POROSITY, DESTRUCTION, CEMENTATION
B 6109 1554* POROSITY DEVELOPMENT AND DISTRIBUTION INSHOAL-WATER
B 6109 1550* POROSITY DEVELOPMENT BETWEEN TWO LOWER CRETACEOUS REEFS OF
B 6109 1553* POROSITY DEVELOPMENT IN LOWER CRETACEOUS BANK COMPLEX,
B 6108 1387* POROSITY, DIAGENESIS, AND PRODUCTIVE CAPABILITY OF SANDSTONE
B 6109 1536* POROSITY DISTRIBUTION IN DELTAIC SANDSTONE, STRAWN SERIES,
B 6105 810 * POROSITY DISTRIBUTION THROUGHOUT MODERN CARIBBEAN FRINGING REEF
M 28 177 POROSITY, DUE TO GLAUCONITE SHRINKAGE
B 6203 508* POROSITY DURING BIOTURBATION, ABST., IMPROVEMENT OF,
M 26 421 POROSITY, EFFECTS OF, SEISMIC FACTORS
B 6405 739* POROSITY EVOLUTION OF NIAGARAN PIPE CREE JR. REEF, GRANT
B 6405 748* POROSITY EVOLUTION OF NIAGARAN THORNTON REEF, NORTHEASTERN
B 6402 188† POROSITY EVOLUTION OF UPPER MIOCENE REEFS, ALMERIA PROVINCE,
B 6303 457* POROSITY EXAMPLES FROM IAB FIELD, COKE COUNTY, TEXAS, ABST.
M 28 173 POROSITY, FILLED INTERPARTICLE
B 6007 1128 POROSITY FOR OFFSHORE WELL, LOUISIANA, SHALE, DENSITY AND
B 6007 1128 POROSITY FOR WELLS, VENEZUELA, SHALE
B 6109 1541* POROSITY FORMATION AND PRESERVATION, ABST., CONSOLIDATION
M 28 180 POROSITY, FRACTURE, IN INDIVIDUAL GRAINS
M 28 180 POROSITY, FRACTURE, IN SANDSTONE
M 28 169 POROSITY, GENERAL
B 6405 753* POROSITY GENERATION, AND OCCLUSION IN UPPER JURASSIC
B 6303 512* POROSITY GENERATION IN CARBONATE ROCKS MARGINAL TO

B 6404 554 POROSITY, GULF COAST SHALES, DEPTH VERSUS
M 28 180 POROSITY, HEALING OF FRACTURES
M 28 176 POROSITY, HONEYCOMBING OF GRAINS
B 6305 834* POROSITY IN SANDSTONES TO HYDROCARBON EXPLORATION, ABST.,
B 6203 553* POROSITY IN CAPRINID REEFS AND ASSOCIATED SEDIMENTS, ABST.,
B 6203 525* POROSITY IN CARBONATE ROCKS, ABST., OCCURRENCES OF, TERTIARY
B 6004 720* POROSITY IN CARBONATE ROCKS OF EVAPORITIC BASINS, ALTERNATIVE
B 6105 838* POROSITY IN CARBONATE SANDS, ABST., EFFECT OF SHAPE SORTING ON
B 6105 819* POROSITY IN CARBONATE-MUDBANK BUILDUPS, PENNSYLVANIAN AND
B 6004 705* POROSITY IN DEEP SMACKOVER OF MISSISSIPPI, ABST., GEOPRESSURES
B 6303 488* POROSITY IN GIANT GAS FIELD, ELLENBURGER FORMATION, PUCKETT
B 6105 761 * POROSITY IN LATE MIOCENE REEFS, SOUTHERN SPAIN
B 6203 538* POROSITY IN LOWER ORDOVICIAN ELLENBURGER DOLOMITE, PUCKETT
B 6205 886* POROSITY IN MISSISSIPPIAN CARBONATE RESERVOIRS OF SOUTHERN
B 6009 1612* POROSITY IN OVERPRESSURED SANDSTONES--FRIO FORMATION,
B 6203 541* POROSITY IN PERMIAN, ZECHSTEIN 2, CARBONATE ROCKS OF EASTERN
B 6105 816* POROSITY IN REEF COMPLEXES—UPPER DEVONIAN EXAMPLES FROM
B 6108 1390* POROSITY IN SANDSTONE DIAGENESIS, ABST., ROLE OF, SECONDARY
B 6109 1547* POROSITY IN SANDSTONES OF LOWER WILCOX, EOCENE, KARNES
B 6405 747* **POROSITY IN SANDSTONES TO HYDROCARBON EXPLORATION,**
M 28 178 POROSITY, LEACHING OF CEMENTS
B 6209 1761* POROSITY LOSS IN SANDSTONE BY DUCTILE GRAIN DEFORMATION
B 6002 227 POROSITY MEASUREMENTS, DEEP
M 28 179 POROSITY, MICROFRACTURE, IN SANDSTONE
M 28 172 POROSITY, PARTIAL LOSS, CEMENTATION
M 28 174 POROSITY, PARTIALLY LEACHED PLAGIOCLASE
B 6303 460* POROSITY PRESERVATION AND EARLY FRESHWATER DIAGENESIS OF
B 6105 840* POROSITY PRESERVATION IN SHALLOW-MARINE CARBONATE ROCKS,
M 28 172 POROSITY, PRESERVED PRIMARY INTERPARTICLE
M 28 175 POROSITY, PRIMARY AND SECONDARY
B 6105 809 * POROSITY REDUCTION BY CHEMICAL COMPACTION—STABLE-ISOTOPE MODEL
B 6004 730* POROSITY REDUCTION IN REEF MARGINS BY EARLY CEMENTATION,
B 6211 2329† POROSITY REDUCTION IN SHALLOW BURIED QUARTZ ARENITE,
B 6004 685* POROSITY RELATIONS, AND DIAGENESIS FROM ANALYSIS OF
B 6303 522* POROSITY RELATIONS IN CHALK RESERVOIRS, ABST.
B 6002 325* POROSITY RELATIONS IN PENNSYLVANIAN AND PERMIAN CARBONATE
B 6102 295* POROSITY RELATIONS IN PENNSYLVANIAN LIMESTONE RESERVOIRS,
M 26 15 POROSITY RELATIONS, TIME-AVERAGE RELATION, VELOCITY-
S 10 14 POROSITY, PRINCIPLES OF OSMOTIC DIFFUSION, DIAGRAM
M 28 176 POROSITY, SAMPLE ETCHED IN ACID
B 6312 2171 POROSITY, SANDSTONE
M 28 179 POROSITY, SECONDARY, DEVELOPMENT
M 28 179 POROSITY, SECONDARY, ELONGATE
M 28 176 POROSITY, SECONDARY, IN SANDSTONE
M 28 174 POROSITY, SECONDARY LEACHING OF PLAGIOCLASE
S 10 5 POROSITY, SHALE-WATER SYSTEM, GEOMETRY
S 10 2 POROSITY REDUCTION, PROCESSES
S 10 2 POROSITY REDUCTION, SHALE COMPACTION
S 10 212 POROSITY RELATED TO DEPTH, LOUISIANA, CENOZOIC
S 10 40 POROSITY VERSUS DEPTH OF CLAYS, MIGRATION
B 6107 1086† POROSITY, SIGNIFICANT GEOLOGIC PROCESSES IN DEVELOPMENT OF,
B 6004 694* POROSITY TO DEPOSITIONAL AND DIAGENETIC FACIES, SUBSURFACE
B 6209 1761* POROSITY TRENDS OF ABNORMALLY HIGH PRESSURES IN SANDSTONES
B 6404 462 POROSITY, TYPES AND TEXTURES OF, CEMENTS AND
M 28 181 POROSITY, USE OF SEM FOR STUDY
B 6205 813 POROSITY VERSUS DEPTH IN SANDSTONE RESERVOIRS
B 6205 815 POROSITY VERSUS PERMEABILITY
B 6004 709* POROSITY--NEW CONCEPT, ABST., SPATIAL, VARIATION IN
B 6004 544 POROSITY-DEPTH RELATION, SHALE
B 6102 227† POROUS ROCK, EXPERIMENTAL STUDY OF, FRACTURE, PERMEABILITY IN
B 6303 507* PORPHYRINS IN DEEP-SEA SEDIMENTS, ABST., COPPER
B 6303 507* PORPHYRINS IN SUBBITUMINOUS AND BITUMINOU COALS, ABST.
M 25 513 PORPHYRY COPPER DEPOSIT, DIAPIRIC-TYPE
M 25 434 PORPHYRY COPPER DEPOSIT, USAGE DEFINED
M 25 433 PORPHYRY COPPER DEPOSITS, ASSOCIATED ROCKS, SOUTHWEST, PACIFIC
M 25 512 PORPHYRY COPPER DEPOSITS, DIORITE-MODEL
M 25 433 PORPHYRY COPPER DEPOSITS, EXPLOITATION, SOCIAL AND POLITICAL
M 25 435 PORPHYRY COPPER DEPOSITS, GEOLOGIC ENVIRONMENT, PHILIPPINES
M 25 509 PORPHYRY COPPER DEPOSITS, GEOLOGY OF, NORTH AMERICA,
M 25 512 PORPHYRY COPPER DEPOSITS, GRANITIC-PLUTON-TYPE
M 25 519 PORPHYRY COPPER DEPOSITS, MAIN, MEXICO, SONORA
M 25 510 PORPHYRY COPPER DEPOSITS, MAP, CORDILLERAN OROGEN, NORTHERN
M 25 520 PORPHYRY COPPER DEPOSITS, MAP, MEXICO, SONORA
M 25 433† PORPHYRY COPPER DEPOSITS OF SOUTHWEST PACIFIC, DISCOVERY AND
M 25 509 PORPHYRY COPPER DEPOSITS, PETROLOGIC MODELS
M 25 512 PORPHYRY COPPER DEPOSITS, QUARTZ MONZONITE-QUARTZ DIORITE
M 25 511 PORPHYRY COPPER DEPOSITS, TABLE, CORDILLERAN OROGEN,
M 25 438 PORPHYRY COPPER, ECONOMIC POSSIBILITIES, PHILIPPINES
M 25 434 PORPHYRY COPPER, ORE RESERVE, PHILIPPINES
M 25 433 PORPHYRY COPPER PROJECTS, CAPITAL INTENSIVE
M 25 509† PORPHYRY COPPER PROVINCE OF NORTHERN CORDILLERAN
M 25 438 PORPHYRY COPPER, RELATION OF, BRECCIA PIPES AND
M 25 506 PORPHYRY DEPOSITS OF METALS, CANADIAN, CORDILLERA
M 25 506 PORPHYRY DEPOSITS, SUBCLASSES
B 6209 1610 PORT EWEN SHALE, PENNSYLVANIA
B 6303 468* PORT HUENEME AREA, CALIFORNIA, ABST., SEDIMENTARY STRUCTURES
B 6203 510* PORT HURON AGE, STOUFFVILLE, ONTARIO, ABST., BURIED, CREVASSE
S 4 291 PORT MORESBY, GALAXEA, PAPUA
B 6403 380 PORT ROYAL MOUNTAINS, JAMAICA
B 6303 550 PORTER, KAREN W., COMPARISON OF PORE-FILLING MATERIAL IN SOME
B 6303 510 PORTER, KAREN W., DIAGENESIS IN HYGIENE AND TERRY SANDSTONES,
B 6108 1388 PORTER, KAREN W., MARINE-SHELF MODEL FOR UPPER CRETACEOUS
B 6305 838 PORTER, LORNA A., LARAMIE-HANNA-SHIRLEY BASINS--FUTURE
B 6108 1208 PORTER, ROBERT B., DEVELOPMENTS IN WEST TEXAS AND EASTERN NEW
B 6408 1142 PORTLOCK ANTICLINE, ALASKA
B 6407 1024 PORTLOCK ANTICLINE, ALASKA
B 6205 882* PORTNEUF RANGE, BANNOCK AND CARIBOU COUNTIES, IDAHO, ABST.,
B 6212 2468 PORTOVIEJO FAULT, ECUADOR
B 6303 477* PORTUGAL, ABST., OIL GENERATION AND MIGRATION IN, LUSITANIAN
B 6005 809† PORTUGAL, CONTINENTAL, MARGIN OFF WESTERN, AFRICA--, SENEGAL
B 6010 1718 PORTUGAL, DEVELOPMENTS, 1975
B 6110 1648 PORTUGAL, DEVELOPMENTS, 1976
B 6210 1918 PORTUGAL, DEVELOPMENTS, 1977
B 6310 1927 PORTUGAL, DEVELOPMENTS, 1978
B 6411 1967 PORTUGAL, DEVELOPMENTS, 1979
B 6412 2057 PORTUGAL, DSDP SITE 398D, LEG 47, GEOCHEMICAL LOG
B 6110 1840 PORTUGUESE TIMOR, DEVELOPMENTS, 1976
B 6405 699 POSEY, J. S., EVALUATION OF VOLUME OF ENTRAINED METHANE IN DEEP,
B 6204 705† POSITION OF SUB-KASKASKIA UNCONFORMITY IN ILLINOIS BASIN
B 6002 299† POSITIVE ELEMENTS WITHIN OUACHITA FOLDBELT IN TEXAS AND THEIR
B 6408 1191 POSITIVE WATER BALANCE BASIN, BLACK SEA
B 6408 1191 POSITIVE WATER BALANCE BASIN, BRITISH COLUMBIA, SAANICH INLET
B 6408 1191 POSITIVE WATER BALANCE BASIN, VENEZUELA, LAKE, MARACAIBO
M 25 551† POSSIBILITIES FOR DEVELOPMENT OF MINERAL RESOURCES OF
B 6004 707* POSSIBILITY OF ABNORMALLY HIGH FLUID PRESSURES DURING
B 6303 490* **POSSIBLE BIOGENIC ORIGIN FOR SOME SEDIMENTARY DOLOMITE,**
M 25 135† POSSIBLE ENERGY SOUUCES IN INDONESIAN VOLCANIC
B 6405 722* POSSIBLE GEOMETRIES OF SANDSTONE BODIES AS REFLECTED BY
B 6105 789 * POSSIBLE PRESENCE OF ECONOMIC URANIUM DEPOSITS IN METAMORPHIC
B 6207 1212* POSSIBLE RELATIONS BETWEEN NORTH PACIFIC WATER TEMPERATURES
B 6012 2189* POSSIBLE ROLE FOR NITROGEN IN FORMATION OF PETROLEUM
B 6111 1957 POSSIBLE SETTINGS FOR SUPERGIANT RESERVES
S 4 244 POSSUM KINGDOM LAKE AREA, TEXAS, WINCHELL LIMESTONE
S 4 241 POSSUM KINGDOM MEMBER, TEXAS
B 6405 753* POST-COMPACTION, SUBSURFACE SECONDARY POROSITY GENERATION,
B 6305 827* POST-DEPOSITIONAL CONTROL OF GAS-RESERVOIR QUALITY IN EAGLE
B 6204 706 POST-EMSIAN UNCONFORMITY, ILLINOIS BASIN
B 6303 526* POST-MIDDLE CRETACEOUS SEISMIC STRATIGRAPHY AND GEOLOGIC
B 6405 796* POST-MIDDLE CRETACEOUS SEISMIC STRATIGRAPHY OF ABYSSAL
B 6105 794 * POST-PALEOZOIC CHANGES IN VOLUMES AND ORGANIC-CARBON CONTENT OF
B 6007 1112† POST-PLEISTOCENE FAULTS ON NEW JERSEY ATLANTIC OUTER
M 26 240 POST-RIFT BASIN, MODEL, BRAZIL
M 26 243 POST-RIFT CARBONATE PLATFORM, MODEL, DEPOSITIONAL SYSTEMS
B 6204 707 POST-TIPPECANOE UNCONFORMITY, ILLINOIS BASIN, SUB-KASKASKIA-
B 6102 294† POTASH AREA OF SOUTHEASTERN NEW MEXICO. BY TITLE ONLY, OIL
B 6201 81 POTASH LAYERS, SASKATCHEWAN
M 28 23 POTASSIUM, FELDSPAR, AUTHIGENIC
S 6 149 POTASSIUM-ARGON AGES, GRANITES, DIAGRAM, JAPAN
S 6 349 POTASSIUM-ARGON DATING, A REVIEW, ORDOVICIAN
S 6 147 POTASSIUM-ARGON DATING, GRANITES, JAPAN
S 6 120 POTASSIUM-ARGON DATING METHOD, A REVIEW
M 25 435 POTASSIUM-ARGON DATING OF ROCKS, PHILIPPINES
S 6 124 POTASSIUM-ARGON, DECAY CONSTANTS
S 6 132 POTASSIUM-ARGON, FRANCE, PARIS BASIN, ALBIAN
S 6 132 POTASSIUM-ARGON, FRANCE, PARIS BASIN, CENOMANIAN
S 6 119† POTASSIUM-ARGON ISOTOPIC DATING METHOD AND ITS APPLICATION
S 6 165 POTASSIUM-ARGON, QUATERNARY DATING
S 6 364 POTASSIUM-ARGON, USSR, PRECAMBRIAN
M 25 506 POTATO HILLS TUNGSTEN DEPOSIT, CANADA
B 6109 1556* POTENTIAL, ABST., DIAGENETIC PATTERNS OF, AUSTIN GROUP AND
B 6004 708* POTENTIAL AND DEPOSITIONAL ENVIRONMENTS OF NUGGET
B 6109 1448† POTENTIAL AND STRATIGRAPHY OF PEDREGOSA BASIN—COMPARISON
B 6004 671* POTENTIAL BASED ON PRESENT EXPLORATION, ABST., NORWAYS, OIL
B 6405 700* POTENTIAL CONTRIBUTION OF OIL SHALE TO ENERGY NEEDS OF UNITED
B 6303 286 POTENTIAL, DEFINITION
M 26 374 POTENTIAL DRILLING HAZARDS, RESOLUTION SYSTEMS
M 25 318† POTENTIAL FOR EXPLORATION AND DEVELOPMENT OF HYDROCARBONS
M 25 149† POTENTIAL FOR GEOTHERMAL-ENERGY DEVELOPMENT IN
B 6303 417* POTENTIAL FOR PRODUCTION OF GEOPRESSURED GEOTHERMAL

B 6303 475* POTENTIAL FOR SUBSURFACE DISPOSAL OF INDUSTRIAL WASTES IN
B 6303 459* POTENTIAL GAS COMMITTEE, ABST., CURRENT, ESTIMATES AND
B 6303 478* POTENTIAL GAS COMMITTEE, ABST., POTENTIAL RESOURCES OF,
B 6409 1529† POTENTIAL GAS COMMITTEE, POTENTIAL RESOURCES OF, NATURAL, GAS
B 6206 1059 POTENTIAL GEOCHEMICAL MECHANISMS FOR URANIUM
B 6009 1606* POTENTIAL OF ASPHALTIC HARTSELLE SANDSTONE IN NORTH ALABAMA,
B 6004 720* POTENTIAL OF BANGOR LIMESTONE, MISSISSIPPIAN, NORTHERN BLACK
B 6004 706* POTENTIAL OF MICROBORING ORGANISMS, ABST., PALEOECOLOGIC
B 6206 967† POTENTIAL OF MONTEREY DEEP-SEA FAN, CALIFORNIA, CHANNEL,
B 6203 491* POTENTIAL OF MORROW SANDSTONES IN SOUTHEASTERN COLORADO,
B 6004 700* POTENTIAL OF NORTHWESTERN BAHAMA PLATFORM, ABST.,
B 6205 885* POTENTIAL OF PALEOZOIC ROCKS IN EASTERN GREAT BASIN, ABST.,
B 6004 735* POTENTIAL OF PERU, ABST., PETROLEUM GEOLOGY AND, PRODUCTIVE
B 6204 592† POTENTIAL OF U.S. ATLANTIC SLOPE, RISE, AND ABYSSAL PLAIN,
B 6205 892* POTENTIAL OF UPPER CRETACEOUS CHALKS OF WESTERN INTERIOR,
B 6002 318* POTENTIAL OF WESTER KANSAS, ABST., OIL AND, GAS
B 6405 803* POTENTIAL PETROLEUM RESERVOIRS ON DEEP-SEA FANS OFF CENTRAL
B 6306 881 POTENTIAL RESERVOIR IN RUDIST REEF BUILDUP ON ATLANTIC-TYPE
B 6303 478* POTENTIAL RESOURCES OF NATURAL GAS IN UNITED STATES--CASE
B 6409 1529† POTENTIAL RESOURCES OF NATURAL GAS IN UNITED STATES—CASE
B 6209 1586 POTENTIAL SOURCE BED, DEFINITION
B 6409 1530 POTENTIAL SUPPLY, DEFINITION
B 6304 595 POTENTIAL USE OF ELECTRIC LOGS IN MIGRATION STUDIES
B 6004 729* POTENTIALS OF MIDDLE AND UPPER ORDOVICIAN OF TENNESSEE, ABST.,
S 10 157 POTENTIOMETRIC SURFACE, PERSIAN GULF, CRETACEOUS
B 6302 221 POTIGUAR BASIN, BRAZIL
M 26 214 POTIGUAR BASIN, BRAZIL
B 6109 1471 POTIGUAR BASIN, BRAZIL
M 28 52 POTSDAM SANDSTONE, NEW YORK, CAMBRIAN
B 6303 487 POTTER, H. C., BEATRIC FIELD, MORAY FIRTH, NORTH SEA, ABST.
M 30 117 POTTER, HENRY C., THE BEATRICE FIELD, INNER MORAY FIRTH, U.K.
B 6309 1576 POTTER, P. E., GAS-BEARIN OHIO SHALE ALONG LAKE
B 6009 1621 POTTER, PAUL EDWIN, AGE AND TECTONICS OF CINCINNATI ARCH IN
B 6209 1703 POTTER, PAUL EDWIN, DIVISION OF BLACK OHIO SHALE IN EASTERN
B 6405 769 POTTER, PAUL EDWIN, SHALE—AN OVERVIEW, ABST.
B 6408 1288* POTTSVILLE ALLUVIAL PLAIN, COALS IN NORTHERN WEST VIRGINIA,
B 6409 1555* POTTSVILLE DEPOSITIONAL SYSTEMS, OUTCROP AND SUBSURFACE, IN
B 6204 646 POTTSVILLE FORMATION, ALABAMA
M 28 9 POTTSVILLE FORMATION, OHIO, PENNSYLVANIAN
B 6212 2396 POTTSVILLE FORMATION, WEST VIRGINIA TO, PENNSYLVANIA
B 6410 1682 POUND SANDSTONE-QUARTZITE, AUSTRALIA, CAMBRIAN
B 6201 126 POURTALES ESCARPMENT FAULT, FLORIDA
B 6201 123 POURTALES ESCARPMENT, FLORIDA
S 4 235 POURTALES, GRAND CAYMAN, ANALCIDOMETRA ARMATA
B 6301 70 POUZDRANY UNIT, CZECHOSLOVAKIA
M 25 220 POVERTY BASIN, NEW ZEALAND
B 6106 911 POWAY CONGLOMERATE, BAJA CALIFORNIA
B 6012 2186* POWAY CONGLOMERATES, SOUTHERN CALIFORNIA, ABST., CLAST
B 6305 823* **POWDER RIVER BASIN OF WYOMING, ABST., COAL DEVELOPMENT IN**
B 6008 1395* POWDER RIVE BASIN, WYOMING, ABST., TEAPOT DELTA OF WESTERN
B 6305 826* **POWDER RIVER AREA, WYOMING AND MONTANA, ABST., COAL**
B 6203 528* POWDER RIVER BASIN, ABST., COMPUTER APPLICATIONS IN REGIONAL
B 6408 1233 POWDER RIVER BASIN, CRETACEOUS SAND FIELD, WYOMING
B 6405 753* POWDER RIVER BASIN HIGH AND ITS IMPLICATIONS FOR FUTURE
B 6405 732* POWDER RIVER BASIN IN NORTHEASTERN WYOMING, ABST., PATTERNS
B 6305 833* POWDER RIVER BASIN, MONTANA AND WYOMING, ABST., WRENCH,
M 26 432 POWDER RIVER BASIN, MUDDY FORMATION, WYOMING
B 6303 497* POWDER RIVER BASIN, NORTHEASTERN WYOMING AND SOUTHEASTERN
B 6305 827* POWDER RIVER BASIN OF WYOMING, ABST., COAL DEVELOPMENT IN
B 6305 830* POWDER RIVER BASIN URANIUM DEPOSITS, HISTORY AND PRODUCTION,
B 6308 1390 POWDER RIVER BASIN, WYOMING
B 6409 1503 POWDER RIVER BASIN, WYOMING
B 6208 1498 POWDER RIVER BASIN, WYOMING
B 6108 1346 POWDER RIVER BASIN, WYOMING
B 6202 183 POWDER RIVER BASIN, WYOMING
B 6008 1204 POWDER RIVER BASIN, WYOMING
B 6108 1371* POWDER RIVER BASIN, WYOMING, ABST., APPLICATIONS OF VERTICAL,
B 6405 675* POWDER RIVER BASIN, WYOMING, ABST., DEPOSITIONAL FACIES,
B 6303 491* POWDER RIVER BASIN, WYOMING, ABST., FACIES AND, RESERVOIR
B 6305 831* POWDER RIVER BASIN, WYOMING ABST., FLOOD-PLAIN SEQUENCES OF
B 6405 778* POWDER RIVER BASIN, WYOMING, ABST., MIDDLE MEMBER OF,
B 6108 1371* POWDER RIVER BASIN, WYOMING, ABST., REGIONAL DIAGENETIC
B 6004 718* POWDER RIVER BASIN, WYOMING, ABST., SYNTHETIC, SEISMIC SECTIONS
B 6008 1409* POWDER RIVER BASIN WYOMING, ABST., SYNTHETIC, SEISMIC SECTIONS
B 6008 1395* POWDER RIVER BASIN, WYOMING, ABST., THICK, COAL BEDS IN,
B 6305 842* POWDER RIVER BASIN, WYOMING, ABST., URANIUM GEOLOGY OF,
B 6405 676* POWDER RIVER BASIN, WYOMING AND MONTANA, ABST.,
B 6404 488† POWDER RIVER BASIN, WYOMING, ORGANIC COMPOSITION OF SOME
B 6405 674* POWDER RIVER BASIN, WYOMING-MONTANA, ABST., VERTICAL AN
B 6305 839* POWDER RIVER BASIN--PAST, PRESENT AND FUTURE, ABST.,
B 6203 513* POWDER RIVER-CLEAR CREEK AREA, NORTHEASTERN WYOMING, ABST.,
B 6403 429 POWDER WASH HIGH-WAX OIL FIELD, COLORADO
B 6102 234 POWELL AREA, TESTS, UTAH, LAKE
B 6311 2088 POWELL, C. MCA., SEDIMENTARY-WEDGE PROGRADATION FROM
M 25 231 POWELL, D. E., AUSTRALIAN NORTHWEST CONTINENNAL
B 6405 721* POWELL FORMATION, UPPERMOST CANADIAN, IN NORTHERN
B 6009 1607 POWELL, RAINA R., DENSITY-FLOW ORIGIN FOR FRIO RESERVOIR
B 6412 2075 POWELL RIVER DRAINAGE, TENNESSEE
M 24 250 POWELL SANDSTONE, MISSISSIPPI
B 6412 2078 POWELL VALLEY ANTICLINE, APPALACHIANS
B 6103 417 POWELL VALLEY ANTICLINE, VIRGINIA
B 6409 1553 **POWERS, BRIAN K., MORPHOLOGY OF TURBIDITE-CHANNEL RESERVOIRS,**
B 6108 1388 POWERS, RICHARD B., ASSESSMENT OF OIL AND GAS RESOURCES IN
B 6203 561 POWERS, RICHARD B., GEOLOGIC APPRAISAL OF UNDISCOVERED OIL AND
B 6105 821 POWERS, RICHARD B., IDAHO-WYOMING THRUST BELT, ABST.,
B 6004 665 POWERS, RICHARD B., NE APPROACH TO GEOLOGIC ESTIMATES OF OIL
B 6405 767 POWERS, RICHARD B., OIL AND GAS POTENTIAL OF
B 6303 515 POWERS, RICHARD B., RELATION OF SEDIMENTARY HISTORY AND
B 6303 511 POWERS, RICHARD B., RESOURCE APPRAISAL PREDICTIONS AND
B 6209 1610 POXONO ISLAND FORMATION, PENNSYLVANIA
B 6109 1552* POZA RICA TREND, MEXICO, ABST., DIAGENESIS OF A, GIANT—
M 26 477† PRACTICAL STRATIGRAPHIC MODELING AND INTERPRETATION
B 6010 1649 PRADAL, L. J., ON DEVELOPMENTS IN COSTA RICA, 1975
B 6209 1767* PRAIRIE BLUFF CHALK, UPPER CRETACEOUS, MAESTRICHTIAN, AND OF
B 6201 73† PRAIRIE EVAPORITE FORMATION OF SASKATCHEWAN—SEISMIC STUDY,
B 6201 74 PRAIRIE EVAPORITE FORMATION, SHELL LAKE MEMBER, SASKATCHEWAN
B 6201 74 PRAIRIE EVAPORITE FORMATION, UPPER SALT, SASKATCHEWAN
B 6201 74 PRAIRIE EVAPORITE FORMATION, WHITKOW SALT MEMBER,
B 6405 767 PRATT, BRIAN R., ALGAL-METAZOAN BIOHERMS OF LOWER ORDOVICIAN
B 6203 528 PRATT, F. P., COMPUTER APPLICATIONS IN REGIONAL STUDY OF
B 6405 768 PRAY, L. C., EARLY GUADALUPIAN, PERMIAN, BANK MARGIN EROSION
B 6203 555 PRAY, LLOYD C., CARBONATE SEDIMENTOLOGY AND A RETURN TO THE
B 6004 670 PRAY, LLOYD C., NONVADOSE ORIGIN OF PISOLITIC FACIES, CAPITAN
M 29 341 PRE-ANDEAN OROGENY, COLOMBIA, EVOLUTION
B 6401 44 PRE-CD LIMESTONE, KALIMANTAN
S 6 73† PRE-CENOZOIC PHANEROZOIC TIME SCALE--COMPUTER FILE OF CRITICAL
B 6203 504* PRE-CRETACEOUS GEOLOGY BENEATH GEORGIA COASTAL PLAIN, ABST.
B 6102 269† PRE-CRETACEOUS ROCKS OF NORTHWESTERN HONDURAS--BASEMENT
B 6004 566† PRE-CRETACEOUS ROCKS OF NORTHWESTERN HONDURAS--BASEMENT
B 6009 1614* PRE-FERRY LAKE EARLY CRETACEOUS DELTAS OF SOUTH MISSISSIPPI
B 6203 494* PRE-HOLOCENE CARBONATE ROCKS OF NORTHWESTERN GREAT
B 6408 1211 PRE-NAN-SHAN BASIN, CHINA
B 6209 1760* PRE-PLATFORM EXPLORATION OF HIGH ISLAND BLOCKS A-560 AND
B 6101 85 PRE-RIF, DAHAR NSOUR, BASIN, MOROCCAN MESITA
B 6303 426* PRE-STRIKE-SLIP POSITIONS OF POLOCHIC FAULT BLOCKS AS
B 6203 519* PRECAMBRIAN, ABST., UNIFORMITY AND THE
M 28 6 PRECAMBRIAN, ADIRONDACK GNEISS, NEW YORK
S 6 363 PRECAMBRIAN, AGE DETERMINATIONS, A REVIEW
M 25 534 PRECAMBRIAN AND PALEOZOIC METALLIFEROUS DEPOSIT, ANDEAN
S 6 367† PRECAMBRIAN, ASPECTS OF THE REVISED, SOUTH, AFRICAN,
S 6 109 PRECAMBRIAN, BAMBUI GROUP, BRAZIL, LATE
B 6306 976 PRECAMBRIAN BASEMENT COMPLEX, EGYPT
B 6110 1897 PRECAMBRIAN, BLACK WARRIOR BASIN
S 6 358 PRECAMBRIAN BOUNDARY, TOMMOTIAN AGE, CAMBRIAN
B 6101 6 PRECAMBRIAN, CANADIAN, ARCTIC ISLANDS
S 6 376 PRECAMBRIAN, CHRONOSTRATIGRAPHY, SOUTH AFRICA
B 6303 420* PRECAMBRIAN, CHUAR GROUP, GRAND CANYON SUPERGROUP
B 6008 1397* PRECAMBRIAN CONGLOMERATES IN GUYANA, SOUTH AMERICA, ABST.,
B 6107 1038 PRECAMBRIAN CRYSTALLINE ROCKS, NEBRASKA
B 6410 1692 PRECAMBRIAN EVOLUTION, BENCH MARKS
B 6410 1682 PRECAMBRIAN FOSSIL EVIDENCE
M 25 440 PRECAMBRIAN GEOLOGY, SOUTH KOREA
B 6009 1624* PRECAMBRIAN GEOPHYSICAL PROVINCES IN INDIANA, ABST.
B 6003 455† **PRECAMBRIAN GRANITE, LAFAYETTE COUNTY, MISSISSIPPI, SUBSURFACE** ;E
B 6203 523* PRECAMBRIAN, GREAT SLAVE LAKE, NORTHWEST TERRITORIES, ABST.,
B 6203 522* PRECAMBRIAN, GREAT SLAVE LAKE, NORTHWEST TERRITORIES, ABST.,
B 6203 443 PRECAMBRIAN, INDIA
M 28 54 PRECAMBRIAN, INWOOD LIMESTONE, NEW YORK
B 6303 529* PRECAMBRIAN IRON FORMATIONS, ABST., SEDIMENTOLOGY OF SOME
M 28 3 PRECAMBRIAN, LLANO SERIES, TEXAS
M 25 444 PRECAMBRIAN METALLOGENIC PROVINCES, SOUTH KOREA
B 6410 1682 PRECAMBRIAN MICROFOSSILS, USSR, SOUTH, KAZAKSTAN
B 6406 943 PRECAMBRIAN, MINNESOTA AND, SOUTH DAKOTA
B 6109 1514 PRECAMBRIAN, NEW MEXICO
B 6306 955 PRECAMBRIAN, NEWFOUNDLAND
B 6303 523* PRECAMBRIAN OLISTOSTROMAL SEQUENCE, NORTH WALES, UNITED
M 28 53 PRECAMBRIAN, ORTEGA QUARTZITE, NEW MEXICO
B 6109 1543* PRECAMBRIAN PETROLEUM REVISITED, ABST., INDIGENOUS
B 6203 547* PRECAMBRIAN PETROLEUM REVISITED, ABST., INDIGENOUS
B 6410 1681† PRECAMBRIAN PETROLEUM REVISITED, INDIGENOUS
S 6 364 PRECAMBRIAN, POTASSIUM-ARGON, USSR
S 6 374 PRECAMBRIAN, PROPOSED SUBDIVISIONS, SOUTH AFRICA

Index

S 6 364 PRECAMBRIAN, RADIOMETRIC DATING, AFRICA
S 6 364 PRECAMBRIAN, RADIOMETRIC DATING, NORTH AMERICA
S 6 364 PRECAMBRIAN, RADIOMETRIC DATING, RUBIDIUM CONSTANT
S 6 364 PRECAMBRIAN, RADIOMETRIC DATING, SOUTH AMERICA
B 6410 1690 PRECAMBRIAN ROCKS, ORGANIC GEOCHEMISTRY OF
S 6 364 PRECAMBRIAN, RUBIDIUM-STRONTIUM, USSR
B 6004 653* PRECAMBRIAN SEDIMENTATION IN BELT AN UINTA
S 6 363† PRECAMBRIAN SEDIMENTS, NUMERICAL, CORRELATION OF MIDDLE AND
B 6303 446* PRECAMBRIAN SEQUENCE IN SOUTH AFRICA, ABST., DOLOMITIZATION
B 6410 1682 PRECAMBRIAN, SOUTH, AUSTRALIA, EDIACARA HILLS
S 6 368 PRECAMBRIAN, STRATIGRAPHY, A REVIEW, SOUTH AFRICA
B 6203 496* PRECAMBRIAN STROMATOLITE MICROSTRUCTURES, ABST., NONMARINE,
M 26 287 PRECAMBRIAN STRUCTURE CONTOUR MAP, DENVER BASIN
B 6402 222 PRECAMBRIAN. TEXAS
S 6 377 PRECAMBRIAN, TIME-ROCK COMPARISONS, DIAGRAM
B 6201 57 PRECAMBRIAN TO CARBONIFEROUS, AUSTRALIA
B 6101 35 PRECAMBRIAN URANIUM DEPOSITS, SUMMARY OF
M 28 17 PRECAMBRIAN, WATERBURY GNEISS, CONNECTICUT
B 6203 553* PRECAMBRIAN-CAMBRIAN SEDIMENTARY ROCKS FROM EAST-CENTRAL
B 6405 786* PRECAMBRIAN-RELATED URANIUM DEPOSITS, ABST., GEOPHYSICAL
M 25 485 PRECIOUS AND BASE METALS, COLORADO
M 25 485 PRECIOUS AND BASE METALS, WASHINGTON
B 6303 474* PRECIPITATION OF AUTHIGENIC MINERALS IN ARTESIA GROUP, MIDDLE
S 5 181 PRECIPITATION OF GYPSUM, VENEZUELA, LAGO PUEBLO,
S 4 379 PRECIPITATION OF LOW-MAGNESIAN CALCITE THROUGH METEORIC
B 6408 1180 PRECURSOR SEDIMENTS TO OIL-SOURCE BEDS
S 10 20 PRECURSORS, MIGRATION, ENHANCED SOLUBILITY
B 6305 753 PREDICT ELEVATION OF WATER-FREE OIL PRODUCTION
M 26 453 PREDICTABILITY, MORROW-, SPRINGER, OVERPRESSURING
B 6303 522* PREDICTING FUTUR OIL USING THREE-DIMENSIONAL
B 6102 295* PREDICTING POROSITY RELATIONS IN PENNSYLVANIAN LIMESTONE
B 6203 541* PREDICTION OF POROSITY IN PERMIAN. ZECHSTEIN 2. CARBONATE
B 6304 699* **PREDICTIONS OF OIL OR GAS POTENTIAL BY NEAR-SURFACE**
B 6203 516* PREDICTIVE MODELS FOR CARBONATE-FACIES DISTRIBUTION, ABST.
B 6301 55 PREGNANT SHALE, TEXAS
B 6303 458* PRELIMINARY ANALYSES OF BRINES FROM PERMIAN BASIN OF WEST
B 6209 1763* PRELIMINARY ASSESSMENT OF ORGANIC CARBON CONTENT AND
B 6105 773 * **PRELIMINARY CALCAREOUS NANNOFOSSIL ZONATION OF BRIGHTSEAT AND**
B 6004 653* PRELIMINARY INVESTIGATIONS OF ELEMENTAL VARIATION AND
B 6405 748* PRELIMINARY MOLLUSCAN BIOSTRATIGRAPHY OF GULF OF ALASKA
B 6409 1553* PRELIMINARY OBSERVATIONS OF MODERN POINT BAR FACIES, AMITE
B 6409 1554* PRELIMINARY REPORT ON MICROPLANKTON AND MICROBENTHON
B 6004 649* PRELIMINARY RESULTS FROM LEG 44 DEEP SEA DRILLING PROJECT IN
B 6004 655* PRELIMINARY STUDY OF SEA-FLOOR INSTABILITY IN OFFSHORE GULF OF
B 6004 675 PRENSKY, S. E., HIGH-RESOLUTION STUDIES OF PALEOCLIMATIC AND
B 6008 1406 PRESCOTT, MAX W., SPEARHEAD RANCH FIELD, CONVERSE COUNTY
B 6207 1209* PRESENT AND FUTURE SUPPLY OF ENERGY RESOURCES OF THE WORLD,
B 6207 1241* PRESENT STATUS OF GEOTHERMAL DEVELOPMENT AND UTILIZATION IN
B 6105 812 * **PRESERVATION AND DISSOLUTION OF RADIOLARIA IN LOW-PRODUCTIVITY**
B 6004 732* PRESERVATION WINDOW AS SUBUNCONFORMIT PETROLEUM SOURCE,
B 6104 620† PRESERVED SUBUNCONFORMITY SOURCE ROCKS. DISCUSSION AND
B 6202 236 PRESIDENTE ALEMAN POOL, MEXICO
S 7 313 PRESLEY, B. J., HOLOCENE SEDIMENT GEOCHEMISTRY OF CONTINENTAL
S 7 303 PRESLEY, B. J., ORCA BASIN, ANOXIC DEPRESSION ON THE CONTINENTAL
B 6303 511 PRESLEY, M. W., PALO DURO BASIN ANALYSIS, ABST.
B 6405 717 PRESLEY, MARK W., DEPOSITIONAL AND TECTONIC EVOLUTION OF A
B 6309 1586 PRESLEY, MARK W., MAUCH CHUNK ALLUVIAL PLAIN AND MUD-FLA
B 6408 1288 PRESLEY, MARK W., POTTSVILLE ALLUVIAL PLAIN, COALS IN NORTHERN
B 6408 1288 PRESLEY, MARK W., RED-BED EVAPORITE, AND CARBONATE FACIES
B 6308 1427 PRESLEY, MARK W., SAN ANDRES FACIES PATTERNS, PALO DURO AND
B 6303 511 PRESLEY, MARK W., UPPER PERMIAN EVAPORITES AND RED BEDS OF
B 6405 766* PRESSURE, ABST., EFFECTIVE STRESS AND ABNORMALLY HIGH, FLUID
B 6305 812† PRESSURE, AND PALEOTEMPERATURE, SOUTHWESTERN PART OF
B 6004 554† PRESSURE, AND PALEOTEMPERATURE, SOUTHWESTERN PART OF
M 26 455 PRESSURE AND PORE FLUIDS, SEISMIC, MODELING
B 6107 1087 PRESSURE, CHANGES IN, PORE
B 6002 293† PRESSURE CONSOLIDATION OF KAOLINITIC CLAY, HIGH-
B 6309 1592* PRESSURE CORING OF GAS-BEARING DEVONIAN BLACK SHALES, ERIE
B 6009 1531† PRESSURE DEFORMATION ON PERMEABILITY OF OTTAWA SAND, EFFECT
B 6008 1398* PRESSURE ENVIRONMENTS AND PETROLEUM ACCUMULATION, ABST.,
B 6112 2134 PRESSURE FIELDS, SUBNORMAL
B 6102 229 PRESSURE IN FRACTURE PERMEABILITY, CONFINING
M 26 39 PRESSURE, LABORATORY, MEASUREMENTS, SATURATION EFFECTS
B 6403 415 PRESSURE, MODELS, ABNORMALLY HIGH, FLUID
B 6403 415 PRESSURE, MODELS, NORMAL, FLUID
B 6408 1245† PRESSURE ORIGIN AND DISTRIBUTION IN JURASSIC OF VIKING BASIN,
B 6002 245† PRESSURE, PORE, GEOMETRY OF, CARBONATE ROCKS AS REVEALED BY,
B 6405 749* PRESSURE PRODUCED BY HYDROCARBON GENERATION, ABST.,
B 6405 782* PRESSURE RELATIONS IN THICK SEQUENCES OF ACCUMULATING
B 6105 836* PRESSURE SHALES AND OIL-SPILL POTENTIAL ON ATLANTIC OUTER
B 6211 2329† PRESSURE SOLUTION AND POROSITY REDUCTION IN SHALLOW BURIED
B 6203 569* PRESSURE SOLUTION IN LIMESTONES, ABST., THREE TYPES OF
B 6203 527* PRESSURE, TEMPERATURE, AND WATER SALINITY, ABST., INTERACTING
B 6409 1558* PRESSURE, TEMPERATURE, SALINITY, LITHOLOGY, AND STRUCTURE IN
B 6305 738 PRESSURE TESTS. MERCURY. CAPILLARY
B 6403 414† **PRESSURE--SURVEY OF SOME BASIC PRINCIPLES, ABNORMALLY HIGH,**
B 6305 731 PRESSURE-DEPTH PLOT OF WATER PHASE IN ARTESIAN CONDITION
B 6107 1090† PRESSURE. DISCUSSION, ABNORMAL, FORMATION
B 6007 1124† PRESSURE. DISCUSSION AND REPLY, ABNORMAL, FORMATION
B 6306 918† PRESSURED SHALES, CALCULATED, COMPACTION, STATES OF THICK
B 6002 228 PRESSURES, ABNORMAL
B 6405 706* PRESSURES, ABST., GLOBAL OCCURRENCE OF, ABNORMAL FORMATION
B 6309 1527† PRESSURES AND TEMPERATURES, AQUEOUS, SOLUBILITY OF, METHANE
B 6303 480* PRESSURES ASSOCIATED WITH FRACTURE RESERVOIRS, EXAMPLE
B 6004 707* PRESSURES DURING FOLDING AND FAULTING IN EASTERN WEST
B 6007 1126 PRESSURES, EFFECT OF, COMPACTION ON, FLUID
B 6107 1087 PRESSURES, EVIDENCE OF, PALEOPORE
B 6406 929 PRESSURES IN ACCUMULATING SEDIMENTS, MODEL FOR EXCESS, FLUID
B 6303 414* PRESSURES IN GAS RESERVOIRS, ABST., ROLE OF, TEMPERATURE AND,
B 6209 1761* PRESSURES IN SANDSTONES AND SHALES, ABST., PALEOECOLOGY AND,
B 6412 2179† PRESSURES IN SHALES, MECHANICAL VERSUS, THERMAL CAUSE OF,
B 6412 2179 PRESSURES IN SHALES, THERMAL, HYPOTHESIS FOR, ABNORMALLY HIGH,
B 6112 2134† PRESSURES, OIL AND, GAS IN, RESERVOIRS WITH, SUBNORMAL
B 6112 2140 PRESSURES, ORIGIN OF LOW
B 6212 2478† PRESSURES, SOLUBILITY OF, METHANE IN, WATER AT ELEVATED,
B 6305 776 PRESSURES, SOLUBILITY OF, METHANE IN, WATER AT ELEVATED,
B 6405 753* **PRESSURES THROUGH ORGANIC MATTER TRANSFORMATIONS,**
B 6403 414† PRESSURE—SURVEY OF SOME BASIC PRINCIPLES, ABNORMALLY HIGH,
B 6403 417 PRESSURING, AQUATHERMAL
B 6403 417 PRESSURING, AQUATHERMAL
B 6209 1580 PRESSURIZED HOLES, EXPANSION OF
B 6004 663 **PRESTON, F. W., NUMERICAL CLASSIFICATION OF SANDSTONE PORE**
B 6004 709 PRESTON, F. W., SPATIAL VARIATION IN POROSITY--NEW CONCEPT, ABST.
B 6102 296 PRESTRIDGE, JEFF, SOUTH TEXAS AUSTIN CHALK, CRETACEOUS, PLAY,
B 6201 100 PREUSS RANGE, IDAHO
M 24 95 PREUSS SANDSTONE, WYOMING
B 6405 640 PREUSS-STUMP FORMATION, WYOMING
B 6108 1382* PREVENTION OF DAMAGE TO WATERFOWL ON DISPOSAL PONDS IN
B 6109 1555 PREZBINDOWSKI, D. R., CARBON AND OXYGEN ISOTOPIC EVOLUTION OF
S 6 221 PRIABONIAN STAGE DEFINED, EUROPE, EOCENE
B 6412 2142 PRIBILOF BASIN, BERING SEA SHELF
B 6002 165 PRIBILOF BASIN, BERING SHELF
B 6412 2142 PRIBILOF CANYON, BERING SEA SHELF
B 6002 162 PRIBILOF CANYON, BERINGIAN SHELF
B 6412 2142 PRIBILOF ISLAND, BERING SEA SHELF
B 6002 162 PRIBILOF ISLANDS, BERINGIAN SHELF
B 6406 945† PRIBILOF ISLANDS, SOUTHERN BERING SEA CONTINENTAL SHELF,
B 6412 2142 PRIBILOF RIDGE, BERING SEA SHELF
B 6002 164 PRIBILOF RIDGE, BERINGIAN SHELF
S 11 316 PRICE CREEK, CALIFORNIA, SAMPLE LOCALITIES
S 11 322 PRICE CREEK SECTION DESCRIPTION
S 11 194 PRICE CREEK, STRATIGRAPHIC SECTION
B 6304 652 PRICE FORMATION, VIRGINIA
B 6309 1578 PRICE, HARVEY S., CHARACTERIZATION OF PRODUCTION MECHANISM IN
B 6004 709 PRICE, I., SEDIMENTATION AND RESEDIMENTATION OF CLASTI
B 6112 2146 PRICE, L. C., AQUEOUS SOLUBILITY OF PETROLEUM AS APPLIED TO ITS
B 6309 1527 PRICE, LEIGH C., AQUEOUS SOLUBILITY OF METHANE AT ELEVATED
B 6002 213 **PRICE, LEIGH C., AQUEOUS SOLUBILITY OF PETROLEUM AS APPLIED TO ITS**
B 6309 1610 PRICE, LEIGH C., ORGANIC GEOCHEMISTRY OF DEEP WELL IN HINDS
B 6203 555 PRICE, LEIGH, CRUDE OIL AND NATURAL GAS DISSOLVED IN DEEP, HOT
B 6203 556 PRICE, LEIGH, NEW EVIDENCE FOR HOT, DEEP ORIGIN AND MIGRATION
B 6405 768 PRICE, RAYMOND A., HYDROCARBON-TRAPPING STRUCTURES IN
B 6205 885* PRICE RIVER CANYON AREA, SOUTHWESTERN UINTA BASIN, UTAH,
M 28 178 PRICE RIVER FORMATION, UTAH, CRETACEOUS
B 6309 1610 PRICE, W. ARMSTRONG, ORIGINS OF PERMANEN INLETS SEPARATING
B 6004 659* PRICE. BY TITLE ONLY, RESPONSE OF, RESERVES TO
B 6204 646 PRIDE MOUNTAIN FORMATION, ALABAMA
B 6204 654 PRIDE MOUNTAIN SANDSTONE, UNITED STATES, SOUTHEASTERN
S 5 95 PRIDOLIAN SERIES, CORRELATION, SILURIAN,
B 6004 692* PRIMARY ACCUMULATION OF HYDROCARBONS, ABST., CLAY-,
B 6409 1544 PRIMARY AND SECONDARY MIGRATION, REDEFINITION OF
B 6004 646* PRIMARY BEDDING RHYTHMS IN PELAGIC CALCAREOUS SEDIMENTS,
B 6401 76 PRIMARY CONTROLS ON CARBON ISOTOPE VARIATIONS
S 10 33† PRIMARY HYDROCARBON MIGRATION: A REVIEW, AGENTS FOR
B 6203 538* PRIMARY MIGRATION AGENTS, ABST.
B 6002 213† PRIMARY MIGRATION, AQUEOUS, SOLUBILITY OF, PETROLEUM AS
B 6304 578 PRIMARY MIGRATION, GEOCHEMICAL CONCEPT
B 6103 434† PRIMARY MIGRATION IN WESTERN CANADA BASIN, RATIO OF
B 6305 766 PRIMARY MIGRATION MECHANISM, DISSOLVED, HYDROCARBONS AS
B 6109 1428 PRIMARY MIGRATION OF HYDROCARBONS

B 6401 107† PRIMARY MIGRATION OF PETROLEUM, A PROCESS FOR
B 6304 697* PRIMARY MIGRATION OF PETROLEUM IN SEDIMENTARY BASINS, ABST.,
S 10 27 PRIMARY MIGRATION, OVERVIEW
S 10 22 PRIMARY MIGRATION, PORE-CENTER NETWORK
B 6112 2146† PRIMARY MIGRATION. DISCUSSION AND REPLY, AQUEOUS, SOLUBILITY
S 10 19† PRIMARY MIGRATION: THE IMPORTANCE OF WATER-MINERAL-ORGANIC
B 6209 1766* PRIMARY MIGRATION—NEW TACK ON AN OLD THEME, ABST.
B 6412 2108† PRIMARY OIL MIGRATION, EVIDENCES OF
B 6002 232 PRIMARY PETROLEUM MIGRATION BY MOLECULAR SOLUTION, MODEL
B 6002 213 PRIMARY PETROLEUM MIGRATION, HYPOTHESES
B 6203 553* PRIMARY POROSITY IN CAPRINID REEFS AND ASSOCIATED SEDIMENTS,
M 28 7 PRINCE CREEK FORMATION, ALASKA, CRETACEOUS
M 28 118 PRINCE CREEK FORMATION, ALASKA, CRETACEOUS
M 28 130 PRINCE CREEK FORMATION, ALASKA, CRETACEOUS
M 28 132 PRINCE CREEK FORMATION, ALASKA, CRETACEOUS
M 28 22 PRINCE CREEK FORMATION, ALASKA, CRETACEOUS
M 28 19 PRINCE CREEK FORMATION, ALASKA, CRETACEOUS
M 28 20 PRINCE CREEK FORMATION, ALASKA, CRETACEOUS
M 24 26 PRINCE CREEK-SCHRADER BLUFF FORMATION, ALASKA
B 6206 1006 PRINCE PATRICK ISLAND, ARCTIC, CANADA
B 6204 635 PRINCE WILLIAM SOUND, ALASKA
B 6405 722* PRINCE WILLIAM SOUND, ALASKA, ABST., STRATIGRAPHY AND
B 6101 3 PRINCESS MARGARET ARCH
B 6206 1013 PRINCESS MARGARET ARCH, CANADA, ARCTIC
B 6311 2059 PRINCETON COALFIELD, BRITISH COLUMBIA
B 6311 2060 PRINCETON GROUP, BRITISH COLUMBIA
B 6211 2360* PRINCETON SUBMARINE VALLEY SYSTEM BENEATH SACRAMENTO
M 25 107 PRINCETON-MERRITT COALFIELDS, CANADA
M 25 519† PRINCIPAL CHARACTERISTICS OF SOME PORPHYR COPPER DEPOSITS IN
B 6110 1749 PRINCIPE, DEVELOPMENTS, 1976, SAO TOME AND
B 6210 1848 PRINCIPE, DEVELOPMENTS, 1977, SAO TOME AND
B 6310 1694 PRINCIPE, DEVELOPMENTS, 1978, SAO TOME AND
B 6411 1790 PRINCIPE, DEVELOPMENTS, 1979, SAO TOME AND
M 26 389† PRINCIPLES AND TECHNIQUES, STRATIGRAPHIC, MODELING AND,
B 6011 2067* PRINCIPLES OF HYDROCARBON GENERATION AND APPLICATION TO OIL
B 6306 987 PRINCIPLES OF SEDIMENTOLOGY, BY GERALD M. FRIEDMAN AND JOHN
B 6203 505 PRIOR, D. B., CONTEMPORARY GRAVITY TECTONICS—AN EVERYDAY
B 6303 512 PRIOR, DAVID B., ACTIVE SUBMARINE LANDSLIDE PROCESSES, MISSISSIPPI
B 6004 698* PROBABILISTIC AND COMPUTER METHODOLOGIES USED BY U.S.
B 6312 2183 PROBABILITY CURVES, ASSESSMENT
B 6002 318* PROBABILITY METHODS OF OIL EXPLORATION IN NORTHWESTERN
B 6405 677* PROBABL RANGE CONCEPT TO BIOSTRATIGRAPHY, ABST., APPLICATIONS
B 6205 888* PROBABLE PERMIAN SOURCE ROCKS, NORTHEASTERN GREAT BASIN,
B 6303 455* PROBING BERMUDAS LAGOONS AND REEFS, ABST.
B 6105 817* PROBLEM OF MISPLACED SEDIMENT, ABST.
M 30 238 PROBLEMATICA, USSR, RIPHEAN PALEONTOLOGY
M 30 238 PROBLEMATICA, USSR, VENDIAN PALEONTOLOGY
B 6004 734* PROBLEM OF PALEOCLIMATOLOGIC INTERPRETATION OF
S 10 1 PROBLEMS OF PETROLEUM MIGRATION
B 6207 1235* PROBLEMS OF STRUCTURE AND DEVELOPMENT OF SHELF ZONES, ABST.
M 26 186 PROCEDURE, CHART, SEISMIC, REFLECTIONS, INTERPRETATION
M 26 77 PROCEDURE, SEA LEVEL FLUCTUATIONS, CORRELATION CURVE,
M 26 465 PROCEDURE, SEISMIC, MODELING
M 26 500 PROCEDURE, SEISMIC, MODELING, INTERPRETATION
M 26 187 PROCEDURE, SEISMIC, REFLECTIONS, CARBONATE, RECOGNITION
M 26 135† PROCEDURE, SEISMIC, STRATIGRAPHY AND, GLOBAL CHANGES OF SEA
M 26 479 PROCEDURE, WAVELET CORRECTION
B 6304 697* PROCESS FOR PRIMARY MIGRATION OF PETROLEUM IN SEDIMENTARY
B 6408 1223† PROCESS-RESPONSE MODELS FOR ANCIENT EPICONTINENTAL SEAWAY
M 29 139† PROCESSES ALONG THE U.S. ATLANTIC CONTINENTAL MARGIN,
B 6405 733* PROCESSES CONTROLLING CHARACTERISTICS OF SURFICIAL SAND
B 6012 2187* PROCESSING OF NONSTRAIGHT SEISMIC LINES USING SLALOM-LINE
M 26 364 PROCESSING, RESOLUTION SYSTEMS, DIGITAL
B 6203 508 PROCTER, R. M., COMPUTER ANALYSIS OF REGIONAL WELL-DENSITY
B 6004 716 PROCTER, RICHARD M., AAPG OIL AND GAS FIELD DATA BANK PROJECT,
B 6108 1384 PROCTOR, KENNETH E., PETROLOGY AND DIAGENESIS OF DEEP-WATER
B 6004 687* PRODELTA TURBIDITE-FAN APRON IN BORDE FORMTION, MISSISSIPPIAN,
B 6009 1610* PRODUCER IN SOUTH LOUISIANA--NOW WHAT, ABST., IVE GOT A NEW
B 6002 318* PRODUCING CHARACTERISTICS OF LOWER MORROW SANDSTONE IN
B 6009 1614* PRODUCING TRENDS, ABST., PRE-, FERRY LAKE EARLY, CRETACEOUS,
B 6205 742 PRODUCTION, 1940-2020, TOTAL, WORLD, OIL
B 6110 1752 PRODUCTION, 1975-1976, AFRICA, CENTRAL AND SOUTHERN
B 6409 1381 PRODUCTION, 1977-1978, MICHIGAN
B 6409 1401 PRODUCTION, 1978-1979, CALIFORNIA
B 6409 1463 PRODUCTION, 1979, ILLINOIS
B 6209 1586 PRODUCTION AND ACCUMULATION OF ORGANIC MATTER
B 6409 1529 PRODUCTION, DEFINITION, CUMULATIVE
B 6205 741 PRODUCTION FORECAST, WORLD
B 6107 1001 PRODUCTION FROM CHALKS, HYDROCARBON
B 6305 838* PRODUCTION FROM LOW-PERMEABILITY RESERVOIRS, ABST., BOWDOIN
B 6102 297* PRODUCTION FROM TIGHT SANDSTONES OF PERMIAN-PENNSYLVANIAN
B 6207 1219* PRODUCTION GEOLOGY OF NORTH RANKIN GAS FIELD, AUSTRALIA,
B 6010 1704† PRODUCTION IN EUROPE IN 1975, PETROLEUM, EXPLORATION AND
B 6110 1636† PRODUCTION IN EUROPE IN 1976, PETROLEUM, EXPLORATION AND
B 6210 1906† PRODUCTION IN EUROPE IN 1977, PETROLEUM, EXPLORATION AND
B 6310 1915† PRODUCTION IN EUROPE IN 1978, PETROLEUM, EXPLORATION AND
M 29 436 PRODUCTION INTERVALS, GULF COAST, PETROLEUM
B 6205 741 PRODUCTION RATIOS, RESERVE LIFE, OF MAJOR PRODUCING AREAS,
B 6205 862 PRODUCTION, TEXAS, BUFFALO WALLOW-, ULTRADEEP FIELD
B 6205 862 PRODUCTION, TEXAS, GOMEZ, ELLENBURGER, ULTRADEEP FIELD
B 6205 862 PRODUCTION, TEXAS, LOCKRIDGE, ULTRADEEP FIELD
B 6205 862 PRODUCTION, TEXAS, WASHITA CREEK, ULTRADEEP FIELD
B 6408 1211 PRODUCTIVE BASINS OF THE WORLD
B 6205 747 PRODUCTIVE BASINS, WORLD
B 6108 1387* PRODUCTIVE CAPABILITY OF SANDSTONE RESERVOIRS, ABST., POROSITY,
B 6004 735* PRODUCTIVE POTENTIAL OF PERU, ABST., PETROLEUM GEOLOGY AND
B 6302 218 PROFESSOR BESNARD CRUISES, NO
B 6103 468 PROFFETT, JOHN, GEOLOGY OF AMBLER MINERAL BELT, BROOKS RANG.
B 6007 1049 PROFILE, ABACO KNOLL, SEISMIC-, REFLECTION
B 6007 1114 PROFILE, ATLANTIC, OUTER, CONTINENTAL, SHELF, CDP MULTICHANNE,
B 6102 195 PROFILE, KOMANDORSKIY BASIN, GLOMAR CHALLENGER, SEISMIC
B 6211 2292 PROFILE LINES, GULF OF MEXICO, MULTICHANNEL
B 6203 490* PROFILE OF COAL-FORMING SWAMP—THE EVERGLADES, ABST.
B 6405 788* PROFILE OF UNUSUAL OOLITE DEPOSIT—DRUM LIMESTONE,
B 6108 1371* PROFILE TO STRATIGRAPHIC EXPLORATION PROBLEMS, POWDER RIVER
B 6403 300 PROFILES, CONTINENTAL, MARGINS IDENTIFIABLE FROM, CONTINUOUS,
B 6403 302 PROFILES, CONTINENTAL, RISES AND. DEEP-OCEAN. TRENCHES INFERRED
B 6403 305 PROFILES, CONTINENTAL, TERRACES CLASSIFIED USING, CONTINUOUS,
B 6306 919 PROFILES, DEFINITION, COMPACTION
M 29 48 PROFILES, DESCRIPTIONS, ICELAND, MARGIN
M 29 43† PROFILES, EROSIONAL AND, DEPOSITIONAL, STRUCTURES OF THE,
B 6305 783 PROFILES, GULF OF MEXICO, SPARKER AND SEISMIC
B 6303 326 PROFILES, ISRAEL AND, SINAI, SEISMIC
B 6007 1023 PROFILES, LAKE SUPERIOR, REFLECTION
B 6102 210 PROFILES, MEXICO, VERACRUZ BASIN, SEISMIC
B 6007 1050 PROFILES, NORTHWEST AND, NORTHEAST, PROVIDENCE CHANNELS,
B 6106 884 PROFILES, SOUTHEASTERN, MEDITERRANEAN SEA, SEISMIC
M 29 177† PROFILES, STRUCTURE AND, STRATIGRAPHY OF THE, BLAKE,
B 6106 918 PROFILES, WILMINGTON CANYON AREA, SEISMIC-REFLECTION
M 26 357 PROFILING SYSTEMS, RESOLUTION SYSTEMS, ELECTROMECHANICAL,
B 6305 847* PROFITABLE SHALLOW APPALACHIAN GA RESERVES--HARD TO FIND OR
M 26 173 PROGRADATIONAL FACIES, SEISMIC, INTERPRETATION, OBLIQUE-
M 26 172 PROGRADED-SLOPE FACIES, SEISMIC, INTERPRETATION
M 26 123 PROGRADING, SEISMIC, REFLECTIONS, CONFIGURATION PATTERNS
M 25 335 PROGRESO BASIN, PERU
B 6212 2510* PROGRESS AND PROSPECTS IN OCEAN THERMAL ENERGY, ABST.
B 6207 1233* PROGRESS OF GEOTHERMAL ENERGY EXPLORATION AND DEVELOPMENT
B 6303 445* PROGRESSIVE CEMENTATION FROM MARINE TO DEEP SUBSURFACE
S 4 353† PROGRESSIVE DIAGENETI ELIMINATION OF STRONTIUM IN
B 6403 423† PROGRESSIVE NEOGENE DEFORMATION OF SOUTHERN HEDIL, TUNISIAN
B 6309 1591* PROJECTION OF FAVORABLE GAS-PRODUCING AREAS FROM
S 4 313 PROJETO GEOMAR, GEOLOGIA MARINHA IN 1971-73
S 4 313 PROJETO REMAC, RECONHECIMENTO GLOBAL DA MARGEM CONTINENTA
B 6211 2359 PROKOPOVICH, NIKOLA P., ALTERATION OF ALLUVIUM BY NATURAL
B 6009 1543† PROLIFI BELT, PENDULUM EFFECT AND, NIGER, DELTA
B 6405 672* PROMISING MORROW SANDSTONE DISCOVERIES IN SOUTHEASTERN
B 6304 686* PROMPT FISSION NEUTRON METHO OF PULSED NEUTRON BOREHOLE
B 6405 741* PROPERTIES OF WATER IN CLAY MINERAL SYSTEMS, ABST.
B 6310 1986† PROPOSAL TO CHANGE NAME OF COMMISSION, NOTE 50--
B 6008 1411* PROPOSED GEOTHERMAL CIRCULATION PATTERN, CORWIN
B 6405 804* PROPOSED MODEL FOR DEVELOPMENT OF RED RIVER ORDOVICIAN,
S 6 291 PROPOSED TIME SCALE, JURASSIC
B 6309 1510† PROPOSED TYPE SECTION FOR UPPER SILURIAN AND LOWER DEVONIAN
B 6309 1583* PROPPANTS WITH CRACK FORMATION AND PROPAGATIO IN HYDRAULIC
M 25 469 PROPYLITIZATION
B 6309 1597* PROSPECT, ABST., HOW TO CREATE AND SUBMIT A WINNING
B 6009 1606* PROSPECT IN FRIO FORMATION OF TEXAS GULF COAST--IDEAL VERSUS
B 6303 286† PROSPECT OF THE DEEP OFFSHORE, PETROLEUM
B 6307 1040 PROSPECT THRUST, WYOMING, DARBY-
B 6401 133 PROSPECTING, BY N. I. MARKOVSKII, MCGOOKEY, D. P., REVIEW OF,
B 6304 698* PROSPECTING FOR HYDROCARBONS, ABST., MICROBIOLOGICAL
B 6108 1373† PROSPECTING GUIDES IN GRANTS MINERAL BELT, NEW MEXICO, ABST.,
B 6409 1554* PROSPECTING MODELS—WILCOX GROUP AND MERIDIAN SANDSTONE,
B 6405 693* PROSPECTIVE AND FUTURE HYDROCARBON PROVINCES OF BERING SEA.
B 6303 299 PROSPECTIVE AREA, UNITED STATES, ONSHORE
B 6303 286 PROSPECTIVE, DEFINITION
B 6303 286 PROSPECTIVENESS, DEFINITION
B 6105 777 * PROSPECTIVE HYDROCARBON PROVINCES OF THE WORLD
B 6304 700* PROSPECTS, ABST., QUANTITATIVE COMPARISONS OF GEOLOGIC, MAPS
B 6303 288 PROSPECTS, CENTRAL, OCEAN REGION

B 6303 293 PROSPECTS, CONTINENT-RELATED, DEEP-WATER, FANS AND, CONES
B 6303 289 PROSPECTS, CONTINENTAL, MARGIN REGION
B 6403 297† PROSPECTS, CONTINENTAL, MARGINS—, CLASSIFICATION AND,
B 6303 286 PROSPECTS, DEFINITION
B 6303 288 PROSPECTS, ESSENTIAL REQUISITES FOR FAVORABLE, PETROLEUM
B 6408 1283* PROSPECTS FOR COAL DEVELOPMEN IN MICHIGAN BASIN, ABST.
B 6009 1617* PROSPECTS IN GULF OF MEXICO REGION ABST., OIL
B 6303 291 PROSPECTS, MARGINAL-, PLATEAU, BLOCKS
B 6303 291 PROSPECTS, MARGINAL-, TRENCH, FILLINGS
B 6101 43† PROSPECTS OF LEBANON--REEVALUATION, PETROLEUM
B 6003 389† PROSPECTS OF SOMALIA, EAST AFRICA, GEOLOGY AND, OIL
B 6303 287 PROSPECTS, OFFSHORE, GUIDELINES TO, EVALUATION OF
B 6312 2188 PROSPECTS OR PLAYS, ASSESSING RESOURCES, SUMMATION OF
B 6303 297 PROSPECTS, SMALL, OCEAN, BASIN REGIONS
B 6004 647* PROTAGONIST--PLATE TECTONICS AS A MAJOR EXPLORATION TOOL, BY
M 30 225† PROTEROZOIC AND LOWER CAMBRIAN STRATA, LENA-TUNGUSKA
M 25 411 PROTEROZOIC BANDED IRON FORMATIONS, WESTERN AUSTRALIA
B 6203 557* PROTEROZOIC BELCHER GROUP, HUDSON BAY, CANADA, ABST.,
B 6101 6 PROTEROZOIC, FRANKLINIAN BASIN
B 6303 500* PROTEROZOIC MCARTHUR GROUP, NORTHERN TERRITORY, AUSTRALIA,
B 6004 734* PROTEROZOIC MIXTITE-BEARINN SEQUENCES, ABST., PROBLEMM OF,
M 25 428 PROTEROZOIC, ORE DEPOSITS ASSOCIATED WITH, AUSTRALIA
B 6405 691* PROTEROZOIC TIDAL-FLAT COMPLEX, ABST., EARLY DIAGENETIC,
B 6202 303 PROTO-ORINOCO DELTA SYSTEM, TRINIDAD
B 6303 450* PROTODOLOMITE, HIGH-MAGNESIAN CALCITE, HYDROMAGNESITE, AND
B 6004 709 PROUTY, C. E., IMPLICATIONS OF IMAGERY STUDIES TO TIME AND
B 6405 768 PROUTY, C. E., PETROLEUM EXPLORATION AND WRENCHING MODEL
B 6408 1283 PROUTY, C. E., PROSPECTS FOR COAL DEVELOPMEN IN MICHIGAN BASIN,
B 6409 1530 PROVED RESERVES, DEFINITION
B 6303 456* PROVENANCE STUDIES IN TUNISIA BY U-PB AGES OF DETRITAL ZIRCONS,
B 6004 710* PROVENANCE AND DISPERSAL OF UPPER CRETACEOUS AND PALEOCENE
B 6207 1208* PROVENANCE AND GEOCHEMISTRY OF CARBON DIOXIDE-RICH GROUND
B 6312 2165 PROVENANCES, TABLE, CONTINENTAL BLOCK
M 29 294 PROVENCAL BASIN, CROSS SECTION, MEDITERRANEAN
B 6411 1964 PROVENCE-MEDITERRANEAN, FRANCE
B 6007 1037† PROVIDENCE CHANNEL, BAHAMAS, STRATIGRAPHY AND STRUCTURE OF,
B 6007 1039 PROVIDENCE CHANNEL, BATHYMETRY, NORTHEAST
B 6007 1040 PROVIDENCE CHANNEL, SAMPLES, AGES, NORTHEAST
B 6007 1050 PROVIDENCE CHANNELS, BATHYMETRIC, FREE-AIR GRAVITY AND
M 26 252 PROVINCES, GULF OF MEXICO
S 7 7 PROVINCES MAP, GULF OF MEXICO
M 25 456 PROVINCIAL BUREAU OF THE MINISTRY OF INTERNATIONAL TRAD,
B 6209 1703 PROVO, LINDA J., DIVISION OF BLACK OHIO SHALE IN EASTERN
S 7 238 PROXIMAL SANDSTONE FACIES, CALIFORNIA, LOS ANGELES BASIN
S 7 236 PROXIMAL TURBIDITE FACIES, CALIFORNIA, LOS ANGELES BASIN
B 6210 2117 PRUDENTIAL INTERNATIONAL, INC., AND CANADIAN SUPERIOR OIL LTD.,
M 24 23† PRUDHO BAY FIELD, NORTH SLOPE, ALASKA, PERMO-, TRIASSIC,
B 6304 699* PRUDHOE BAY, A TEN-YEAR PERSPECTIVE, ABST.
B 6004 657* PRUDHOE BAY, ALASKA, DURING LATE CENOZOIC, ABST., CLIMATIC
B 6004 672* PRUDHOE BAY AREA, ALASKA, ABST., TRANSGRESSIVE-, REGRESSIVE,
M 30 299 PRUDHOE BAY COMPLEX, HYDROCARBON ACCUMULATIONS, ALASKA
M 30 304 PRUDHOE BAY, CRETACEOUS, PUT RIVER SANDSTONE
B 6103 465* PRUDHOE BAY FIELD, ABST., PERMIAN-, TRIASSIC OF
B 6104 594 PRUDHOE BAY FIELD, ALASKA
B 6109 1493 PRUDHOE BAY FIELD, ALASKA
M 30 290 PRUDHOE BAY FIELD, EXPLORATION HISTORY, ALASKA
M 30 297 PRUDHOE BAY FIELD, GEOLOGIC SETTING, ALASKA
M 30 22 PRUDHOE BAY FIELD, LANDSAT IMAGE, ALASKA
M 24 25 PRUDHOE BAY FIELD, LOCATION MAP, ALASKA
B 6103 439 PRUDHOE BAY FIELD, SECTION, ALASKA
B 6401 20 PRUDHOE BAY GIANT OIL AND GAS FIELD, LANDSAT DATA, ALASKA
M 30 310 PRUDHOE BAY, KUPARUK RIVER FORMATION, LOWER SAND
M 30 310 PRUDHOE BAY, KUPARUK RIVER FORMATION, MIDDLE SAND
M 30 310 PRUDHOE BAY, KUPARUK RIVER FORMATION, UPPER SAND
M 30 307 PRUDHOE BAY, LISBURNE GROUP, ALAPAH FORMATION
M 30 305 PRUDHOE BAY, LISBURNE GROUP, STRATIGRAPHY
M 30 307 PRUDHOE BAY, LISBURNE GROUP, WAHOO FORMATION
M 30 301 PRUDHOE BAY, SADLEROCHIT GROUP, IVISHAK SANDSTONE
M 30 303 PRUDHOE BAY, SAG RIVER FORMATION, RESERVOIRS
B 6109 1499 PRUDHOE BAY SECTION, ALASKA
M 26 339 PRUDHOE BAY, SEISMIC SECTION, ALASKA
M 30 303 PRUDHOE BAY, SHUBLIK FORMATION, RESERVOIRS
B 6004 727* PRUDHOE BAY SUPE GIANT FIELD AND ALASKAN ARCTIC PLATE. BY
M 30 312 PRUDHOE BAY, WEST SAK RIVER SANDSTONE, RESERVOIRS
M 30 289† PRUDHOE BAY--A 10-YEAR PERSPECTIVE
B 6008 1406 PRUIT, JOHN D., STATISTICS AS TOOL IN OIL AND GAS EXPLORATION,
S 4 233 PRUNE ISLAND, NEMASTER RUBIGINOSA POURTALES
M 25 400 PRYITE AND CHALCOPYRITE, AUSTRALIA
B 6303 512 PRYOR, WAYNE A., CYCLIC SEDIMENTATION OF
B 6309 1589 PRYOR, WAYNE A., PETROLOGY AND DEPOSITIONAL ENVIRONMENTS OF
B 6405 769 PRYOR, WAYNE A., SHALE—AN OVERVIEW, ABST.
B 6303 488 PRYOR, WAYNE A., UPPER DEVONIAN TURBIDITES OF CENTRAL AND
B 6004 659 PRZYGOCKI, R., ANALYSIS O SUBMARINE DEBRIS-FLOW DEPOSIT,
S 11 274 PSEUDOPOLYMORPHINA OVALIS
S 11 238 PSEUDOPOLYMORPHINA OVALIS
M 27 76 PSEUDOPUNCTATE BRACHIOPOD
M 27 77 PSEUDOPUNCTATE BRACHIOPODS
B 6303 489* PSEUDOSTALACTITES FROM SUBMARINE CAVE NEAR COLUMBUS CAY,
B 6203 517* PSEUDOVARVES FROM CRETACEOUS PETERSON LIMESTONE, WESTERN
B 6004 651* PSEUDOVELOCITY LOGS, ABST., COMPUTER-GENERATED, SCALE
B 6005 879† PSEUDOVELOCITY LOGS, TRANSFORMATION OF, RESISTIVITY TO
B 6203 356 PT. REYES FAULT, CALIFORNIA
M 27 105 PTERIDOSPERM SEEDS
M 27 90 PTEROPODS
B 6209 1760* PTEROPODS, AS INDICATORS OF OCEANOGRAPHIC PROCESSES IN WATER
B 6309 1597* PTEROPODS INFERRED FROM STUDIES ON LIVING FORMS IN GULF OF
B 6104 607† PUALE BAY--ALINCHAK BAY AREA, ALASKA PENINSULA, SOME LOWER
B 6303 430* PUBLIC LANDS PRIVATE, ABST., MAKING
M 24 300† PUBLICATIONS, BIBLIOGRAPHY OF, NORTH AMERICAN, OIL AND, GAS,
M 25 440 PUCHON ROCKS, SOUTH KOREA
M 25 173 PUCHULDIZA GEOTHERMAL FIELD, CHILE
B 6203 538* PUCKETT FIELD, PECOS COUNTY, TEXAS, ABST., DEPOSITIONAL FACIES
B 6303 488* PUCKETT FIELD, PECOS COUNTY, TEXAS, ABST., POROSITY IN, GIANT,
B 6104 581 PUEBLO VIEJO HIGH, VENEZUELA
B 6111 1986 PUENTE FORMATION, CALIFORNIA
B 6111 1981 PUENTE HILLS, CALIFORNIA
B 6304 603 PUERTO CHIQUITO FIELDS, FRACTURE RESERVOIRS, NEW MEXICO, EAST
B 6010 1655 PUERTO RICO, 1975, CENTONZIO, A., ON DEVELOPMENTS IN
B 6110 1590 PUERTO RICO, 1976, PETROCONSULTANTS S.A., ON DEVELOPMENTS IN
B 6210 2058 PUERTO RICO, 1977, UMIKER, R., ON DEVELOPMENTS IN
B 6310 1761 PUERTO RICO, 1978, UMIKER, R., ON DEVELOPMENTS IN
S 4 18 PUERTO RICO, AHOGADA REEF
S 4 19 PUERTO RICO, BACKREEF LAGOON
M 28 99 PUERTO RICO, CRETACEOUS, ILDEFONSO FORMATION
M 28 33 PUERTO RICO, CRETACEOUS, ILDEFONSO FORMATION
M 27 84 PUERTO RICO, CRETACEOUS, RIO YAUCO FORMATION
S 4 20 PUERTO RICO, ENRIQUE REEF ENVIRONMENT
M 27 13 PUERTO RICO, EOCENE, COAMO SPRINGS LIMESTONE
M 27 131 PUERTO RICO, EOCENE, COAMO SPRINGS LIMESTONE
M 27 33 PUERTO RICO, EOCENE, COAMO SPRINGS LIMESTONE
M 27 12 PUERTO RICO, EOCENE, COAMO SPRINGS LIMESTONE
M 27 140 PUERTO RICO, EOCENE, COAMO SPRINGS LIMESTONE
S 4 21 PUERTO RICO, FOREREEF
S 4 18 PUERTO RICO, HALIMEDA-THALASSIA MEADOWS
B 6403 376 PUERTO RICO, HOLOCENE FAN DELTA
S 4 17 PUERTO RICO, INNER-SHELF PROVINCE
B 6303 371 PUERTO RICO, KEWANEE INTERAMERICAN OIL CO. 4CPR TEST WELL
S 4 20 PUERTO RICO, MAGUEYES LIMESTONE
S 4 17 PUERTO RICO, MANGROVE-DOMINATED COAST AND SHALLOWS
S 4 22 PUERTO RICO, MEDIA LUNA REEF
S 4 22 PUERTO RICO, MIDDLE-SHELF PROVINCE
S 4 22 PUERTO RICO, OUTER-SHELF PROVINCE
S 4 18 PUERTO RICO, PARGUERA STUDY AREA
S 4 20 PUERTO RICO, REEF APRON
S 4 21 PUERTO RICO, REEF FLAT AND SHOALS
S 4 17† PUERTO RICO, SHELF, REEFS, SOUTHWESTERN
S 4 23 PUERTO RICO, SHELF-EDGE REEF
B 6303 370† PUERTO RICO, SIDERITE, SPHERULITES OF, SAN SEBASTIAN FORMATION,
S 8 124 PUERTO RICO, SUBMARINE CANYONS, RIO DE LA PLATA CANYON
B 6007 1079 PUERTO RICO TRENCH
M 25 185 PUERTO RICO TRENCH
B 6303 434* PUGET LOWLANDS, ABST., NEARSHORE, GLACIAL-MARINE
B 6303 410 PUGET SOUND, ABST., GLACIAL-MARINE SEDIMENTS OF WEST,
B 6303 520 PUJOL, P., RECENT HUGE GAS DISCOVERIES IN KANGAN AREA, IRAN,
B 6104 594 PULAI FIELD, SUMATRA
B 6103 417 PULASKI FAULT, VIRGINIA
B 6104 597 PULASKI FIELD, MICHIGAN
B 6410 1576 PULASKI THRUST, VIRGINIA
B 6410 1581 PULASKI-STAUNTON BLOCK, VIRGINIA
B 6211 2209 PULEEN BAY GROWTH FAULTS, IRELAND, WESTERN
M 26 213† PULL-APART BASINS, SEISMIC-, STRATIGRAPHIC, INTERPRETATION OF,
S 11 222 PULLEN FORMATION, ASSEMBLAGES
S 11 218 PULLEN FORMATION, FOSSIL AGGREGATE
S 11 191 PULLEN FORMATION, HUMBOLDT BASIN

S 11 294 PULLENIA BULLOIDES
S 11 261 PULLENIA BULLOIDES
S 11 120 PULLENIA, LAGENIDAE, MIOCENE, ROTALIA
S 11 120 PULLENIA MIOCENICA
S 11 294 PULLENIA MIOCENICA
S 11 261 PULLENIA MIOCENICA
S 11 261 PULLENIA SALISBURYI
S 11 294 PULLENIA SALISBURYI
S 11 261 PULLENIA SUBCARINATA
S 11 294 PULLENIA SUBCARINATA
B 6004 709 PULUNGGONO, A., INDONESIA-SOUTH SUMATRA AND
M 25 239 PULUNGGONO, A., RECENT KNOWLEDGE OF HYDROCARBON POTENTIALS
M 25 463 PUMICE LENTILS OR FIAMME
M 27 76 PUNCTATE BRACHIOPOD
S 6 304 PUNJABIAN STAGE, PERMIAN, BIOCHRONOLOGY
B 6210 1913 PUPILLI, ON DEVELOPMENTS IN ITALY, 1977, BONGIORNI, D., AND M.
B 6310 1922 PUPILLI, ON DEVELOPMENTS IN ITALY, 1978, BONGIORNI, D., M.,
M 25 357 PURBO-HADIWIDJOJO, M. M., GROUNDWATER POTENTIAL
M 25 356 PURBO-HADIWIDJOJO, M. M., GROUNDWATER POTENTIAL OF LIMESTONE
B 6301 87 PURCELL, L. P., GEOCHEMICAL CHARACTERISTICS OF SEDIMENTARY
B 6002 281 PURCELL MEMBER, PENNSYLVANIA
B 6406 868 **PURDY, G. M., ARC-CONTINENT COLLISION IN BANDA SEA REGION**
M 29 209 PURDY, G. M., SEISMIC REFRACTION AND REFLECTION STUDIES IN THE
S 3 115 PURELY TAPHROGENIC MARGINS
M 25 58 PUSHCHAROVSKY, YU. M., GEOLOGIC-STRUCTURAL SYNTHESIS OF
M 25 176 PUSHCHAROVSKY, YU. M., OIL- AND GAS-BEARING BASINS OF
B 6109 1499 PUT RIVER 9-11-13 SECTION, ALASKA
M 30 304 PUT RIVER SANDSTONE, PRUDHOE BAY, CRETACEOUS
B 6110 1582 PUTUMAYO BASIN, COLOMBIA
B 6405 700 PYLE, THOMAS E., SANDS AND SAND TRANSPORT ON PALIMPSEST
B 6402 245 PYLOS BASIN, GREECE
B 6211 2359* PYRAMID LAKE AREA, NEVADA, ABST., ALTERATION OF, ALLUVIUM BY,
B 6104 602 PYRAMID MOUNTAINS, NEW MEXICO
B 6211 2232 PYRENEES, DIAPIR
S 11 232 PYRGO OBLONGA
S 11 272 PYRGO OBLONGA
S 11 272 PYRGO ROTALARIA
S 11 232 PYRGO ROTALARIA
M 27 38 PYRGO SP., BELIZE, HOLOCENE SEDIMENTS
S 11 272 PYRGO VESPERTILIO
S 11 232 PYRGO VESPERTILIO
M 28 153 PYRITE, AUTHIGENIC, IN LIMESTONE
M 28 152 PYRITE, AUTHIGENIC, REPLACEMENT
B 6105 815* PYRITE CONCENTRATIONS IN PROTEROZOIC WITWATERSRAND PLACERS,
M 28 44 PYRITE CRYSTALS
M 27 149 PYRITE FRAMBOID
M 28 153 PYRITE, FRAMBOIDAL, REPLACEMENTS SEM
M 27 148 PYRITE GRAINS
B 6212 2401 PYRITE IN COALS, FORMS
B 6405 788* PYRITE IN UPPER FREEPORT COAL BED, HOMER CITY, PENNSYLVANIA,
M 25 448 PYRITE ORE DEPOSITS, RECENT, JAPAN, SULFUR AND
B 6108 1379* PYRITE OXIDATION AT PH 5-9.5, IMPLICATIONS FOR FORMATION OF
M 25 427 PYRITE, SOUTH AUSTRALIA, NAIRNE
M 25 426 PYRITE, TASMANIA, SAVAGE RIVER, MAGNETITE-
S 3 106 PYRITE-MARCASITE-JORDESITE-CALCITE-URANIUM ORE
M 25 400 PYRITE-VIOLARITE ZONE, AUSTRALIA, NICKEL MINERALIZATION
B 6303 432* PYRITIZATION IN SHELLS OF LIVING BIVALVES, ABST.
M 25 424 PYROLUSITE, GROOTE EYLANDT
B 6312 2227† PYROLYSIS, EFFECT OF, OIL AND, BITUMEN, SATURATION ON,
B 6309 1577* PYROLYSIS OF EASTERN GAS SHALES--EFFECTS OF TEMPERATURE AND
B 6112 2083 PYROLYSIS OF FOUR KEROGEN TYPES, LABORATORY
B 6112 2082† PYROLYSIS OF KEROGEN, OIL AND, GAS, GENERATION BY LABORATORY
B 6303 480* PYROLYSIS TECHNIQU USING WELL CUTTINGS TO MAP SOURCE ROCKS,
B 6401 59† PYROLYSIS--INFLUENCE ON PETROLEUM GENERATION AND MIGRATION,
B 6303 496* PYROLYSIS--SIMPLE TECHNIQUE FOR IDENTIFYING YIELD AND
M 25 444 PYROPHYLLITE, SOUTH KOREA
M 28 51 PYROXENE, CRYSTAL
M 25 400 PYRRHOTITE, AUSTRALIA, PENTLANDITE AND
M 24 209 Q SANDSTONE, LOUISIANA, G, J, K, O
B 6004 705* Q-MODE MULTIVARIATE ANALYSES, ABST., GRANULOMETRIC RELATIONS
B 6101 60 QARTABA STRUCTURE, LEBANON
B 6010 1872 QATAR, 1975, ASIATIC PETROLEUM CO., QATAR OIL CO., LTD., AND,
B 6110 1801 QATAR, 1976, ASIATIC PETROL. CORP. AND, WINTERSHALL A. G., ON
B 6210 2117 QATAR, 1977, WINTERSHALL A.G., PETROCONSULTANTS S.A., AND,
B 6310 1889 QATAR, 1978, WINTERSHALL A.G. AND PETROCONSULTANTS S.A., AND
B 6411 1838 QATAR, 1979, WINTERSHALL A.G. AND, DEPARTMENT OF PETROLEUM
B 6405 600 QATAR AREA, COLUMN, MIDDLE EAST
B 6310 1892 QATAR, DEVELOPMENTS, 1977-1778
B 6405 600 QATAR FORMATION, MIDDLE EAST
B 6010 1872 QATAR OIL CO., LTD., AND WINTERSHAL A.G., ON DEVELOPMENTS IN
B 6110 1811 QATAR, PRODUCTION, 1976
B 6210 2120 QATAR, PRODUCTION, 1976-1977
B 6411 1843 QATAR, PRODUCTION, 1978-1979
B 6405 602 QATAR-SOUTH FARS ARCH, MIDDLE EAST
B 6104 585 QATIF FIELDS, PERSIAN GULF
B 6106 875 QATTARA DEPRESSION, EGYPT
B 6303 512 QAZI, M. N., CYCLIC SEDIMENTATION OF CRETACEOUS-PALEOCENE
B 6412 2100 QESHM FORMATION, IRAN
B 6303 325 QISHON CANYON, ISRAEL
S 11 85 QUAILWATER CREEK, CALIFORNIA, CHECK LIST OF, FORAMINIFERA,
M 26 203 QUALIFICATIONS, SEISMIC, REFLECTIONS, RECOGNITION, LIMITATIONS
B 6009 1613* QUALITY AND QUANTITY CONTROLS ON COAL-RESOURC ASSESSMENT
M 26 493 QUANTIFICATION, SEISMIC, MODELING, STRATIGRAPHIC, TRAP
B 6303 525* QUANTITATIVE BASIN ANALYSIS AND EVOLUTION OF DEEP-MARINE
B 6304 700* QUANTITATIVE COMPARISONS OF GEOLOGIC MAPS TO PETROLEUM
S 4 195† QUANTITATIVE PALEOECOLOGICAL ASPECTS OF MODERN MARINE
B 6009 1613* QUANTITY CONTROLS ON COAL-RESOURC ASSESSMENT AND
B 6206 1058 QUARTZ
M 28 161 QUARTZ AND CALCITE FILLED, FRACTURES
M 28 98 QUARTZ ARENITE, CHERT, DETRITAL, IN
B 6211 2329† QUARTZ ARENITE, PRESSURE SOLUTION AND, POROSITY REDUCTION IN
M 27 140 QUARTZ, AUTHIGENIC
M 28 10 QUARTZ, AUTHIGENIC, IN LIMESTONE
M 28 154 QUARTZ, AUTHIGENIC, IN LIMESTONE, REPLACEMENT
M 28 154 QUARTZ, AUTHIGENIC, IN SANDSTONE SEM
M 28 10 QUARTZ, AUTHIGENIC, REPLACEMENT
M 28 153 QUARTZ, AUTHIGENIC, REPLACEMENT
M 28 12 QUARTZ, BIBLIOGRAPHY
M 28 146 QUARTZ, BIBLIOGRAPHY, DIAGENESIS AND POROSITY
M 28 2 QUARTZ BIREFRINGENCE
M 28 9 QUARTZ, BOEHM LAMELLAE
M 28 81 QUARTZ, CHALCEDONY, IN A CAVITY
M 28 7 QUARTZ, COMPLEX GRAINS
M 28 6 QUARTZ, COMPOSITE OR POLYCRYSTALLINE
M 28 80 QUARTZ, CRYSTALS, IN CAVITIES
M 28 7 QUARTZ, CRYSTALS WITH INCLUSIONS
M 28 4 QUARTZ, DETRITAL, AUTHIGENIC OVERGROWTH
M 28 7 QUARTZ, DETRITAL CHERT GRAINS
M 28 5 QUARTZ, END-PHASE EXAMPLE
M 28 8 QUARTZ, EPIDOTE MICROLITE INCLUSION
M 28 161 QUARTZ FILLED, IN LIMESTONE, FRACTURES
M 28 161 QUARTZ FILLED, IN SANDSTONE, FRACTURES
M 28 1 QUARTZ, GENERAL, DETRITAL GRAINS
M 28 164 QUARTZ GRAINS, FRACTURES OF
M 28 117 QUARTZ GROWTH, CEMENTS
M 28 11 QUARTZ, IN CHLORITE CEMENT
M 28 116 QUARTZ, IN OOLITIC LIMESTONE, CEMENTS
M 28 115 QUARTZ, MENISCUS, CEMENTS
M 28 32 QUARTZ, METAMORPHIC ORIGIN
M 25 509 QUARTZ MONZONIE COPPER DEPOSITS, TYPES OF
M 25 512 QUARTZ MONZONITE-QUARTZ DIORITE MODEL OF PORPHYRY COPPER
M 28 8 QUARTZ, MUSCOVITE MICROLITE INCLUSION
M 28 2 QUARTZ, NEGATIVE CRYSTAL
M 28 3 QUARTZ, NONDETRITAL, WITH FELDSPAR
M 28 4 QUARTZ, OPALLINE SILICA
M 28 112 QUARTZ OVERGROWTH, CEMENTS
M 28 113 QUARTZ OVERGROWTH, SEM, CEMENTS
M 28 114 QUARTZ OVERGROWTHS, CATHODOLUMINESCENCE, CEMENTS
M 28 143 QUARTZ, OVERGROWTHS, CEMENTS
M 28 114 QUARTZ OVERGROWTHS, DUST RIMS, CEMENTS
M 28 115 QUARTZ OVERGROWTHS, RE-WORKED, CEMENTS
M 28 9 QUARTZ, RUTILE INCLUSIONS
B 6004 689* QUARTZ SAND FROM MEADOWCROFT ROCK SHELTER, AVELLA,
M 28 5 QUARTZ, SEMICOMPOSITE GRAIN
M 28 8 QUARTZ, SILLIMANITE INCLUSIONS
M 28 2 QUARTZ, STRAIGHT-EXTINCTION
M 28 12 QUARTZ SURFACE TEXTURE, BIBLIOGRAPHY
M 28 2 QUARTZ, VOLCANIC, CRYSTAL
M 28 3 QUARTZ, ZONE INCLUSIONS
M 28 97 QUARTZARENITE, CLAY, AUTHIGENIC, IN
B 6408 1289* QUARTZARENITE—DOLOMITIC KEEFER SANDSTONE OF WEST VIRGINIA,
B 6004 662* QUATERNARY, ABST., CLIMATICALLY CONTROLLED, CYCLIC
B 6204 641 QUATERNARY, ALASKA
B 6203 376 QUATERNARY, ALASKA
B 6303 454* QUATERNARY ALLUVIAL-FAN DEPOSITS, BIRCH CREEK VALLEY,
M 25 236 QUATERNARY, AUSTRALIA
S 6 172 QUATERNARY BOUNDARY, A REVIEW, NEOGENE-
S 6 171 QUATERNARY BOUNDARY, AGREEMENT, NEOGENE-

S 6 177 QUATERNARY BOUNDARY, ARGUMENTS, NEOGENE-
S 6 173 QUATERNARY BOUNDARY, ITALY, NEOGENE-
B 6303 480* QUATERNARY, CALCRETES, EASTERN SPAIN, ABST., CALCIFICATION AND
M 27 189 QUATERNARY, CALICHE CRUST, TEXAS
M 28 121 QUATERNARY, CALICHE CRUST, TEXAS
M 27 120 QUATERNARY CALICHE, TEXAS
B 6202 251 QUATERNARY, CALIFORNIA
B 6403 447* QUATERNARY CLIMATE CHANGE AND TECTONISM, PINE VALLEY,
M 25 325 QUATERNARY, COLOMBIA
B 6004 722* QUATERNARY CONTINENTAL SLOPE SEDIMENTS FROM NORTHWEST
S 6 174 QUATERNARY, CORRELATION, DIAGRAM, NEOGENE-
M 27 161 QUATERNARY CRUSTS, ABU DHABI
S 6 165 QUATERNARY DATING, CARBON-14
S 6 165 QUATERNARY DATING, POTASSIUM-ARGON
S 6 166 QUATERNARY DATING, URANIUM-THORIUM, REVIEW
B 6109 1545* QUATERNARY DEPOSITS ON OUTER LOUISIANA SHELF, ABST.,
B 6306 976 QUATERNARY, EGYPT
B 6303 545* QUATERNARY FAULT ACTIVITY IN TEXAS GULF COAST, ABST.
B 6303 538* QUATERNARY FORAMINIFERAL RECORD IN EASTERN CARIBBEAN
B 6108 1379* QUATERNARY GEOLOGIC PROCESSES ASSOCIATED WITH GEOTHERMAL
S 6 165† QUATERNARY GEOLOGY, ISOTOPIC METHODS IN
B 6109 1540* QUATERNARY GEOMORPHIC EVOLUTION OF COLORADO RIVER, INNER
B 6104 563 QUATERNARY GLACIOFLUVIAL DEPOSITS, CHILE AND, ARGENTINA
M 28 11 QUATERNARY GRUS, TEXAS
S 7 31 QUATERNARY, GULF COAST BASIN, CROSS SECTIONS
M 28 42 QUATERNARY, GYPSUM DUNES, UTAH
B 6203 443 QUATERNARY, INDIA
S 6 166 QUATERNARY, ISOTOPIC DATING, A REVIEW
S 6 165 QUATERNARY, ISOTOPIC METHODS, A REVIEW
B 6101 44 QUATERNARY, LEBANON
B 6203 535* QUATERNARY MAMMOTH-BEARING SINKHOLE DEPOSITS, HOT SPRINGS,
B 6109 1518 QUATERNARY, NEW MEXICO
M 25 286 QUATERNARY, NORTH AMERICA, WESTERN
B 6204 596 QUATERNARY, NORTH, ATLANTIC BASIN
B 6309 1494 QUATERNARY, PAKISTAN
S 6 167 QUATERNARY, RADIOMETRIC DATING, PROBLEMS
B 6309 1525 QUATERNARY ROCKS, ALASKA PENINSULA
B 6405 711* QUATERNARY SEDIMENT, ABST., HYDRAULIC PISTON, CORING IN
B 6403 446* QUATERNARY SEDIMENTARY FACIE ON CONTINENTAL SHELF,
B 6203 533* QUATERNARY SEDIMENTATION AND TECTONICS ON OREGON
B 6203 542* QUATERNARY SEDIMENTATION ON UPPER CONTINENTAL SLOPE,
M 27 158 QUATERNARY SEDIMENTS, ABU DHABI
M 27 157 QUATERNARY SEDIMENTS, ABU DHABI
M 27 119 QUATERNARY SEDIMENTS, ABU DHABI
M 27 161 QUATERNARY SEDIMENTS, BAHAMAS
M 27 160 QUATERNARY SEDIMENTS, BELIZE
B 6405 734* QUATERNARY SEDIMENTS OF INDUS FAN, ABST., LITHOLOGY AND
B 6303 485* QUATERNARY SEDIMENTS OF KENYA, ABST., TRACKS AND SUBSTRATE
M 28 66 QUATERNARY SOIL, AUSTRALIA
M 28 131 QUATERNARY, SOIL, AUSTRALIA
M 28 74 QUATERNARY SOIL, AUSTRALIA
B 6403 443* QUATERNARY STYLES OF CALIFORNIA SUBMARINE FANS, ABST.
B 6101 16 QUATERNARY, SVERDRUP BASIN, TERTIARY-
B 6012 2180* QUATERNARY TECTONICS AN DEPOSITIONAL ENVIRONMENTS OF SANTA
B 6004 712* QUATERNARY TIDAL TERRACES OR SUBAERIAL KARST, ABST.,
S 4 353† QUATERNARY TO LATE TERTIARY CORAL REEFS OF RED SEA--SEQUENCE
M 25 526† QUATERNARY VOLCANISM OF BAJA CALIFORNIA--SUMMARY, MAJOR
B 6004 678* QUEBEC, ABST., SUBMARINE-, FAN COMPLEX, CAP ENRAGE FORMATION
B 6405 785* QUEBEC AND NEWFOUNDLAND, ABST., COARSENING-UPWARD SHELF
M 28 65 QUEBEC, CAMBRIAN, SILLERY GROUP, CANADA
M 28 162 QUEBEC, CAMBRIAN, SILLERY GROUP, CANADA
M 28 161 QUEBEC, CAMBRIAN, ST. ROCH FORMATION, CANADA
B 6407 1064 QUEBEC, CAP CHAT CHANNEL
B 6407 1054 QUEBEC, CAP DES ROSIERS GROUP
B 6206 938 QUEBEC, CAP ENRAGE FORMATION, MASSIVE SANDSTONE FACIES
B 6407 1062 QUEBEC, CAP STE. ANNE
B 6407 1052† QUEBEC, DEPOSITIONAL FRAMEWORK OF SANDY, MID-FAN COMPLEXES
B 6108 1335 QUEBEC, DEVELOPMENTS, 1976
B 6208 1451 QUEBEC, DEVELOPMENTS, 1977
B 6308 1344 QUEBEC, DEVELOPMENTS, 1978
B 6409 1518 QUEBEC, DEVELOPMENTS, 1979
B 6206 952 QUEBEC, GROSSES ROCHES CONGLOMERATE
B 6407 1054 QUEBEC, ORDOVICIAN KLIPPEN
B 6108 1340 QUEBEC, PRODUCTION, 1973-1976
B 6208 1456 QUEBEC, PRODUCTION, 1974-1977
B 6308 1349 QUEBEC, PRODUCTION, 1975-1978
B 6206 940 QUEBEC, ST. DAMASE FORMATION, PEBBLY SANDSTONE FACIES
B 6206 956 QUEBEC, ST. ROCH FORMATION
B 6407 1066 QUEBEC, STE. MARTHE CHANNEL
B 6407 1052 QUEBEC, SUBMARINE FANS
B 6407 1053 QUEBEC, TOURELLE FORMATION
B 6407 1055 QUEBEC, TOURELLE FORMATION, SIX FACIES
M 29 276 QUEEN CHARLOTTE ISLAND, FAULT ZONE, ALASKA
B 6303 421* QUEEN CHARLOTTE ISLANDS, BRITISH COLUMBIA, ABST., UPPER,
M 25 109 QUEEN CHARLOTTE ISLANDS, COAL, CANADA
M 29 280 QUEEN CHARLOTTE-FAIRWEATHER, CRUSTAL MODELS, ALASKA
B 6408 1169 QUEEN CITY FACIES, TEXAS
B 6309 1603* QUEEN CITY FORMATION IN EAST TEXAS EMBAYMENT--RECORD OF
B 6206 1005 QUEEN ELIZABETH ISLANDS, ARCTIC, CANADA
B 6205 781 QUEEN FORMATION, NEW MEXICO
B 6303 512 QUEEN, J. MICHAEL, CARBONATE REPLACEMENT OF SULFATE--NEW
B 6303 457 QUEEN, J. MICHAEL, MIXING ZONE DOLOMITE IN TIDAL-FLAT SEDIMENTS
B 6207 1239* QUEENSLAND COAL EXPLORATION AND DEVELOPMENT, ABST.
M 25 427 QUEENSLAND, CONSTANCE RANGE, IRON ORE
M 25 427 QUEENSLAND, DAWSONVALE, IRON ORE
B 6110 1888 QUEENSLAND, DEVELOPMENTS, 1976
B 6210 2150 QUEENSLAND, DEVELOPMENTS, 1977
B 6310 1804 QUEENSLAND, DEVELOPMENTS, 1978
B 6411 1863 QUEENSLAND, DEVELOPMENTS, 1979
M 25 427 QUEENSLAND, DUCHESS AREA, PHOSPHATES
M 25 427 QUEENSLAND, DUGALD RIVER, ZINC-LEAD
M 25 427 QUEENSLAND, HILTON DEPOSIT, LEAD-ZINC-SILVER
M 25 427 QUEENSLAND, LADY ANNIE, COPPER
M 25 427 QUEENSLAND, LADY LORETTA, LEAD-ZINC-SILVER
M 25 427 QUEENSLAND, MARY KATHLEEN, URANIUM-RARE EARTHS
M 25 427 QUEENSLAND, MOUNT ISA, LEAD-SILVER-ZINC-COPPER
M 25 427 QUEENSLAND, MOUNT MORGAN, GOLD-COPPER
B 6009 1623 QUICK, R. C., BOWLING GREEN FAULT--CASE OF RESURGENT TECTONICS,
M 29 155 QUIET ZONE, ANOMALIES, AUSTRALIA
M 29 170 QUIET ZONE AS HYBRID CRUST, AUSTRALIA
M 29 170 QUIET ZONE AS OCEANIC CRUST, AUSTRALIA
M 29 173 QUIET ZONE, COMPARED TO RIFTS, AUSTRALIA
M 29 166 QUIET ZONE, CRUSTAL ROCKS, AUSTRALIA
M 29 169 QUIET ZONE, EVOLUTION INTERPRETATION, AUSTRALIA
M 29 153 QUIET ZONE, GEOLOGY, AUSTRALIA
M 29 156 QUIET ZONE, GRAVITY, AUSTRALIA
M 29 151† QUIET ZONE ON THE MARGIN SOUTH OF AUSTRALIA, THE, CRUSTAL,
M 29 153 QUIET ZONE, PHYSIOGRAPHY, AUSTRALIA
M 29 163 QUIET ZONE, REFRACTION MEASUREMENTS, AUSTRALIA
M 29 165 QUIET ZONE, SEDIMENTARY ROCKS, AUSTRALIA
M 29 169 QUIET ZONE, SEISMIC INTERPRETATION, AUSTRALIA
M 29 157 QUIET ZONE, SEISMIC REFLECTIONS, AUSTRALIA
M 29 167 QUIET ZONE, SEISMIC SECTIONS, AUSTRALIA
B 6405 700 QUINN, JOHN, DEEP-SEA BENTHIC FORAMINIFERA AND THEIR
S 11 231 QUINQUELOCULINA AKNERIANA
S 11 270 QUINQUELOCULINA AKNERIANIA
S 11 231 QUINQUELOCULINA ELONGATA
S 11 270 QUINQUELOCULINA ELONGATA
S 11 231 QUINQUELOCULINA VULGARIS
S 11 270 QUINQUELOCULINA VULGARIS
B 6310 1890 QUINTANA OVERSEAS, INC. SUNMAR EXPLORATION CO., AND KHALIFA
B 6309 1606* QUINTANA ROO, MEXICO, ABST., CARRILLO PUERTO FORMATION OF
B 6209 1767* QUINTANA ROO, MEXICO, ABST., LATE, PLEISTOCENE, CARBONATE BANK
B 6304 688 QUINTANAR, ROBERT J., ANOMALOUS OREBODY WITHIN AMBROSIA
B 6403 447 QUINTERNO, PAULA, BENTHIC FORAMINIFERS FROM EASTERN GULF OF
B 6405 766 QUIRT, G. STEWART, HELIUM SURVEYING FOR DEEPLY BURIED
B 6107 1046 QUIVIRA, FORMATION, KANSAS, DRUM
B 6204 633 R V ACONA RESEARCH CRUISE
B 6004 705* R- AND Q-MODE MULTIVARIATE ANALYSES, ABST., GRANULOMETRIC
B 6104 597 R-2-47 WELL, LIBYA
S 4 291 RABAUL, GALAXEA, NEW BRITAIN
S 5 67 RABBIT EARS ANHYDRITE, SALINA GROUP, RUFF FORMATION,
B 6101 29 RABBIT LAKE, PITCHBLENDE DEPOSITS, CANADA
B 6405 742 RABINOWITZ, PHILLIP D., SEISMIC STRATIGRAPH AND STRUCTURE OF
B 6303 487 RACHER, D., BEATRIC FIELD, MORAY FIRTH, NORTH SEA, ABST.
M 30 117 RACHER, DAVID, THE BEATRICE FIELD, INNER MORAY FIRTH, U.K.
B 6303 432 RACHLIN, JACK, WEST SIBERIAN BASIN, ABST.
B 6411 1785 RACHWAL, C. A., PETROLEUM DEVELOPMENTS IN CENTRAL AND
B 6206 1091 RACKLEY, R. I., NEW MODEL FOR SOME COLORADO PLATEAU URANIUM
B 6203 501 RACKLEY, RUFFIN I., NEW MODEL FOR SOME COLORADO PLATEAU
B 6305 825* RADAR, ABST., LINEAMENT, MAPPING WITH ORBITING IMAGING
B 6405 619† RADAR IMAGE OF SAN ANDREAS FAULT, CALIFORNIA, SEASAT
B 6412 2064† RADAR IMAGERY FOR GEOLOGIC MAPPING—
B 6009 1623 RADER, EUGENE K., STRATIGRAPHY AS KEY TO ARCH-RELATED ORIGIN
B 6209 1570 RADIAL AND TRAP-DOOR FAULTS ASSOCIATED WITH SALT DIAPIRS
S 11 202 RADIOACTIVE DATING, BAILEY ASH BED

M 30 170 RADIOACTIVE MARKERS, NORTH SEA, PIPER FIELD
B 6004 724* RADIOACTIVE WASTE DISPOSAL, ABST., HYDROLOGIC STABILIT OF
B 6004 648* RADIOCARBON DATING AND STRATIGRAPHIC ANALYSIS OF HOLOCENE
B 6004 723* RADIOCARBON DATING OF ESTUARINE CARBONATE MATERIALS, ABST.
B 6007 1062 RADIOCARBON-DATED SAMPLES, PANAMA, GALETA REEF
S 3 12 RADIOELEMENT DISTRIBUTION, HEAT-FLOW MODEL FOR
S 3 10 RADIOELEMENT DISTRIBUTION, METEORITIC MODEL FOR
S 3 12 RADIOELEMENT DISTRIBUTION, TERRESTRIAL-LITHOLOGY MODEL FOR
S 3 41 RADIOELEMENT MOBILIZATION IN CRUSTAL ROCKS, MECHANISMS OF
S 3 22 RADIOELEMENT TRANSFER BY, MELTING, COMPLETE
S 3 23 RADIOELEMENT TRANSFER BY, MELTING, PARTIAL
S 3 14 RADIOELEMENTS, HEAT-PRODUCING CAPACITIES OF
M 27 23 RADIOLARIA
M 27 24 RADIOLARIA
M 28 79 RADIOLARIA, CROSS SECTIONS, IN OPHIOLITE
B 6303 421* RADIOLARIA FROM EASTERN OREGON AND QUEEN CHARLOTTE
B 6303 541* RADIOLARIA FROM OMAN MOUNTAINS, ABST.
B 6405 678* RADIOLARIA FROM SOUTH PACIFIC, ABST., MIOCENE SPUMELLARIAN
B 6303 531* RADIOLARIA--PRESENT AND PAST INDICATORS OF DISTANCE FROM
B 6303 514* RADIOLARIAN BIOSTRATIGRAPHY AND PALEO-OCEANOGRAPHY OF
B 6405 747* RADIOLARIAN BIOSTRATIGRAPHY AND PALEOECOLOGY OF EASTERN
B 6405 688* RADIOLARIAN CHARACTERISTICS OF MAJOR OCEANOGRAPHIC
B 6309 1598* RADIOLARIAN DISTRIBUTION, DIVERSITY, AND DENSITY IN WATER
B 6309 1597* RADIOLARIAN FAUNA IN GULF OF MEXICO--IMPLICATIONS, ABST.,
B 6004 695* RADIOLARIAN PALEOBIOGEOGRAPHY--IMPLICATIONS CONCERNING
B 6303 483* **RADIOLARIAN PRESERVATION IN PRESENT AND PAST OCEANS, ABST.,**
B 6309 1607* RADIOLARIAN RATIOS AN PLEISTOCENE-HOLOCENE BOUNDARY, ABST.
B 6209 1764* RADIOLARIANS AND PALEOOCEANOGRAPHY FROM DEEP SEA DRILLING
B 6209 1760* RADIOLARIANS, FORAMINIFERS, AND PTEROPODS, AS INDICATORS OF
B 6309 1597* RADIOLARIANS, PLANKTONIC FORAMINIFERS, AND PTEROPODS
M 27 50 RADIOLARITE, MUSCAT, OMAN, JURASSIC
M 27 24 RADIOLARITE, MUSCAT, OMAN, JURASSIC
M 28 81 RADIOLARITI, ITALY, JURASSIC
M 27 85 RADIOLITID RUDISTS
B 6005 883† RADIOMETRIC AGE DETERMINATION OF SILURIAN GLAUCONITE.
S 6 227 RADIOMETRIC AGES, CORRELATIONS, EOCENE
S 6 227 RADIOMETRIC AGES, CORRELATIONS, OLIGOCENE
S 6 227 RADIOMETRIC AGES, CORRELATIONS, PALEOCENE
S 6 227 RADIOMETRIC AGES, DIAGRAM, PALEOGENE
B 6012 2182* RADIOMETRIC AGES FOR RELIZIAN STAGE OF PACIFIC COAST, ABST.,
S 6 311 RADIOMETRIC, CARBONIFEROUS BOUNDARY, PERMIAN
S 6 337 RADIOMETRIC DATA, A REVIEW, DEVONIAN
S 6 358 RADIOMETRIC DATES, RANGES, CAMBRIAN
S 6 213 RADIOMETRIC DATES, RANGES, PALEOGENE
S 6 356 RADIOMETRIC DATES, TABLE, CAMBRIAN
S 6 181 RADIOMETRIC DATES, TABLE, PARATETHYS, NEOGENE
S 6 355 RADIOMETRIC DATING, A REVIEW, CAMBRIAN
S 6 334 RADIOMETRIC DATING, A REVIEW, CARBONIFEROUS
S 6 247 RADIOMETRIC DATING, A REVIEW, PALEOGENE
S 6 364 RADIOMETRIC DATING, AFRICA, PRECAMBRIAN
S 6 313 RADIOMETRIC DATING, ALTERNATE VALUES, PERMIAN
S 6 325 RADIOMETRIC DATING, CANTABRIAN STAGE, CARBONIFEROUS
S 6 324 RADIOMETRIC DATING, CARBONIFEROUS BOUNDARY, TABLE
S 6 248 RADIOMETRIC DATING, CHOICE OF CONSTANTS, PALEOGENE
S 6 250 RADIOMETRIC DATING, COMMENTS, PALEOGENE
S 6 359 RADIOMETRIC DATING, CONSTANTS USED, CAMBRIAN
S 6 324 RADIOMETRIC DATING, DEVONIAN BOUNDARY, TABLE
S 6 338 RADIOMETRIC DATING, NEED FOR MORE, DEVONIAN
S 6 364 RADIOMETRIC DATING, NORTH AMERICA, PRECAMBRIAN
S 6 316 RADIOMETRIC DATING, PERMIAN, DEVIANT RESULTS
S 6 167 RADIOMETRIC DATING, PROBLEMS, QUATERNARY
S 6 364 RADIOMETRIC DATING, RUBIDIUM CONSTANT, PRECAMBRIAN
S 6 313 RADIOMETRIC DATING, SCATTER, PERMIAN
S 6 364 RADIOMETRIC DATING, SOUTH AMERICA, PRECAMBRIAN
S 6 228 RADIOMETRIC, EOCENE, STAGE CORRELATIONS
B 6209 1752* RADIOMETRIC INVESTIGATION RELATIVE TO A GAMMA-RAY LOG, AND
S 6 182 RADIOMETRIC SAMPLES, TABLE, PARATETHYS, NEOGENE
S 6 339 RADIOMETRIC SCALE AGAINST ORGANIC, DEVONIAN
S 6 290 RADIOMETRIC SCALE, COMPARISON, JURASSIC
S 6 138 RADIOMETRIC SCALE, EUROPE, CRETACEOUS, PROPOSED
S 6 139 RADIOMETRIC SCALE, EUROPE, PALEOGENE, PROPOSED
S 6 275 RADIOMETRIC SCALE, GLAUCONITE, CRETACEOUS
S 6 274 RADIOMETRIC SCALE, NON-GLAUCONITE, CRETACEOUS
S 6 185 RADIOMETRIC SCALES, PARATETHYS, NEOGENE
S 6 15 RADIOMETRIC SCALES, REVIEW
S 6 179† RADIOMETRIC TIME SCALE FOR THE NEOGENE OF THE PARATETHYS
S 6 314 RADIOMETRIC VALUES, CHART, PERMIAN
S 6 315 RADIOMETRIC VALUES, CORRELATION, PERMIAN
S 6 312 RADIOMETRIC VALUES, DIAGRAM, PERMIAN
S 6 313 RADIOMETRIC VALUES, PERMIAN, MIDDLE
S 6 313 RADIOMETRIC VALUES, PERMIAN, UPPER
B 6212 2509 RADJA, VINCENT T., RECENT DEVELOPMENT OF KAMOJANG
B 6304 688* RADON EMANATION OVER OREBODY, HAS LONG-DISTANCE TRANSPORT
B 6405 695* RADON EXPLORATION TECHNIQUES FOR URANIUM, ABST., ADVANCES IN
B 6405 754* RADON GEOCHEMISTRY, ABST., GETTING THE MOST OUT OF
B 6109 1540* RADULAE—THEIR POTENTIAL IN THE FOSSIL RECORD, ABST.,
B 6004 710 RADWANSKI, A., STRUCTURE OF NEOGENE, MIOCENE, LITTORAL HARD-
B 6008 1411* RAFT RIVER, IDAHO, GEOTHERMAL WELLS--SITING, DRILLING, AND
B 6104 585 RAG-E SAFID FIELD, PERSIAN GULF
B 6207 1194 RAGING RIVER FORMATION, OREGON AND, WASHINGTON
B 6004 710 RAHMANI, R. A., PROVENANCE AND DISPERSAL OF UPPER CRETACEOUS
B 6408 1288 RAHMANIAN, VICTOR D., UPPER DEVONIAN TIDE-DOMINATED
B 6205 883* RAILROAD VALLEY, NYE COUNTY, NEVADA, ABST., GEOLOGIC
B 6004 729* RAIN SHADOWS, ABST., FLUVIAL, MODELS--, WET, SLOPES AND
B 6002 246 RAINBOW LAKE, A AND B POOLS, ALBERTA
M 26 195 RAINBOW-ZAMA AREA, PINNACLE REEFS, SEISMIC EXAMPLE, CANADA,
B 6205 782 RAINBOW-ZAMA AREA, SEISMIC DATA, ALBERTA
B 6303 513 RAINON, E., RESERVOIR STUDIES AND INVERSION OF SEISMIC TRACES,
M 29 189 RAITT, R. W., TECTONICS OF THE ANDAMAN SEA AND BURMA
B 6310 1981 RAJ, BY C. T. BARBER, LANGENHEIM, R. L., JR., REVIEW OF A GEOLOGIST
B 6108 1373* RAJAH 49 MINE, MESA COUNTY, COLORADO, ABST., GEOLOGICAL,
B 6405 769 RALEIGH, C. BARRY, DRILLING IN SAN ANDREAS FAULT, ABST.
M 25 352 RALEIGH, C. G., CONTROL OF EARTHQUAKES AT RANGELY,
B 6203 556 RAMO, ALAN O., BRIGHT SPOTS, MILLIGALS, AND GAMMAS, ABST.
B 6410 1575† RAMP-TO-BASIN TRANSITIONS AND FORELAND BASIN EVOLUTION,
B 6207 1211 RAMSEIER, R. O., STRUCTURE AND MOTION OF BERING AND OKHOTSK
B 6104 585 RAMSHIR, PERSIAN GULF
B 6405 713 RAN, W., MORPHOMETRY OF LATE ORDOVICIAN MICROBIAL BORINGS.

B 6401 74 RANCHO JACALITOS, APTIAN THROUGH CENOMANIAN CARBONATES,
B 6303 549 RANDALL, RICHARD G., DEFORMATION OF CONTINENTAL SLOPE
B 6104 492 RANDAZZO, A. F., DIAGENESIS OF MIDDLE AND UPPER EOCENE
B 6004 650 RANDS, ROBERT L., NUMERICAL ASPECTS OF CHEMICAL VARIATION AND
S 10 227 RANGELY FIELD, HYDROTHERMAL EFFECTS, COLORADO
M 25 352† RANGELY, COLORADO--SUMMARY, CONTROL OF, EARTHQUAKES AT
B 6307 1026 RANGELY OIL FIELD STRUCTURE, SEISMIC PROFILE, COLORADO
B 6405 709* RANGEL OIL FIELD, COLORADO, ABST., EOLIAN-, FLUVIATILE,
S 4 255 RANGER QUARRY, PHYLLOID ALGAE, TEXAS
B 6311 2089 RANKIN PLATFORM, AUSTRALIA
B 6201 41 RANKIN PLATFORM, AUSTRALIA
B 6105 791 * RANK OF COAL AS INDEX OF ORGANIC METAMORPHISM FOR OIL AND GAS
B 6004 710 RANNEFT, T. S. M., SEGMENTATION OF ISLAND ARCS AND IT
M 22 252 RANONG FAULT, THAILAND
B 6303 513 RAO, C. PRASADA, GLACIOMARINE CARBONATE
B 6310 1820 RAO, K. L. N., ON DEVELOPMENTS IN INDIA, 1978
M 30 487 RAO, R. P., PETROLEUM GEOLOGY OF BOMBAY HIGH FIELD, INDIA
B 6109 1522† RAPID APTIAN-ALBIAN SUBSIDENCE RATES IN EASTERN UNITED STATES
M 25 496 RAPITAN GROUP, IRON DEPOSITS, CANADA
B 6303 459 RAPOPORT, L. A., ANALYSIS OF HYDROCARBON DISTRIBUTION AND
B 6203 518 RAPOPORT, L. A., ANALYSIS OF OIL-FIELD DISTRIBUTION FOR
B 6303 513 RAPOPORT, L. A., APPLICATION OF OIL AND GAS DISCOVERY FUNCTIONS
B 6004 711 RAPOPORT, L. A., IMPACT OF OIL AND GAS EXPLORATION ASSUMPTIONS
B 6203 531 RAPP, GEORGE, JR., HOLOCENE EPOCH ALLUVIAL-DELTAIC-COASTAL
B 6203 534 RAPP, JOHN B., HYDROCARBONS OF LOW MOLECULAR WEIGHT IN
B 6303 433 RAPP, JOHN B., SPILLE OIL AND INFAUNAL ACTIVITY--MODIFICATION OF
B 6403 448 RAPPEPORT, M. L., DEPOSITIONAL ENVIRONMENTS WITHIN
B 6012 2188 RAPPEPORT, MELVYN L., HOLOCENE SHELF SEDIMENTATION PATTERNS
B 6108 1373† RARE EARTH AND OTHER TRACE ELEMENTS AS PROSPECTING GUIDES IN
M 25 427 RARE EARTHS, QUEENSLAND, MARY KATHLEEN, URANIUM-
M 25 546 RARE-EARTH ELEMENTS, ABUNDANCE PATTERNS, PACIFIC OCEAN
B 6309 1453 RARITAN EMBAYMENT, MID-ATLANTIC
B 6110 1800 RAS AL KAHIMAH, 1976, PETROCONSULTANTS S.A., ON DEVELOPMENTS IN
B 6010 1871 RAS AL KHAIMAH, 1975, VITOL EXPLORATION B.V. AND,
B 6210 2117 RAS AL KHAIMAH, 1977, PETROCONSULTANTS S.A., WEEKS PETROLEUM
B 6310 1888 RAS AL KHAIMAH, 1978, GULF EXPLORATION AND PRODUCTION CO.
B 6411 1840 RAS AL KHAIMAH, 1979, GULF OIL EXPLORATION AND PRODUCTION CO.
B 6309 1602 RASCHILLA, STEPHEN N., SEDIMENTATION ON TRAILING PLATE
B 6311 2119 RASCOE, BAILEY, JR., CONTROLS ON PENNSYLVANIAN HYDROCARBON
B 6002 321 RASCOE, BAILEY, JR., SEDIMENTARY CYCLES AND DEPOSITIONAL
B 6004 711 RASCOE, BAILEY, JR., SEDIMENTARY CYCLES IN VIRGILIAN, UPPER
B 6301 87 RASHID, M. A. AND, GEOCHEMICAL CHARACTERISTICS OF SEDIMENTARY
B 6303 448 RATCLIFE, BRETT C., RECENT ARTHROPO LEBENSSPUREN AS INDICATORS
B 6004 703* RATES IN HUDSON RIVER ESTUARY, ABST., SEDIMENTATION
B 6303 513* RATES OF CREEP OF ACTIVE FAULTS IN NORTHWESTERN HOUSTON,
B 6105 764 * RATES OF ICE GOUGING AND SEDIMENT REWORKING, BEAUFORT SEA,
B 6203 547* RATES OF LANDWARD MIGRATION OF CORE BANKS, NORTH CAROLINA—
B 6103 434† RATIO OF PETROLEUM TO WATER DURING PRIMARY MIGRATION IN
B 6405 792* RATIONALIZATION OF KOYUKUK CRUNCH, NORTHERN AND CENTRAL
B 6402 210 RATNAGIRI SHELF, INDIA, BOMBAY-
B 6104 596 RATTLESNAKE FIELD, FOUR CORNERS
B 6211 2360 RAU, WELDON W., FORAMINIFERAL BIOSTRATIGRAPHY IN GULF OF
B 6309 1575 RAUCH, HENRY W., LINEAMENTS AND WATER WELLS AS EXPLORATION
B 6004 711 RAUTMAN, C. A., LATE JURASSIC BARRIER ISLAND COMPLEX IN LOWER
B 6211 2275 RAUTMAN, C. A., SEDIMENTOLOGY OF LATE JURASSIC BARRIER-ISLAND
M 25 392 RAW MATERIALS, LOCATIONS, NEW ZEALAND
B 6008 1401 RAWSON, RICHARD,, OIL AND GAS POSSIBILITIES REDWALL LIMESTONE,
B 6405 769 RAWSON, RICHARD R., PALEOGEOGRAPH OF NORTHERN ARIZONA
B 6008 1406 RAWSON, RICHARD R., SABKHA ENVIRONMENT--NEW FRONTIER FOR
B 6304 692 RAWSON, RICHARD R., URANIUM IN TODILTO LIMESTONE--SABKHA-LIKE
B 6005 794 RAY, K. K., CONCEALED MESOZOIC-CENOZOIC ALPINE HIMALAYAN
B 6203 556 RAY, PULAK K., BACK-BARRIER SEDIMENTATION ON TIDE-DOMINATED
B 6405 734 RAY, PULAK K., LITHOLOGY AND STRUCTURES OF QUATERNARY
B 6112 2104 RAYA SECTION, PERU

M 25 485 RAYLIKE DISTRIBUTION, METAL OCCURRENCES
B 6203 361 RAYMOND HILL FAULT, CALIFORNIA, SANTA MONICA-
B 6004 712 RAYMOND, R., JR., EARLY CAMBRIAN SHALLOW SUBTIDAL
B 6108 1246 RAYNE FIELD, ACADIA PARISH, LOUISIANA
B 6412 2099 RAZAK FORMATION, IRAN
B 6405 681* RB-SR DATING OF DIAGENESIS—EPIGENESIS IN MESOZOIC ROCKS, SAN
B 6211 2352* RB-SR EVALUATION, ABST., AGE OF SEDIMENTATION-DIAGENESIS FOR,
B 6209 1676 RB-SR GEOCHRONOLOGY
B 6303 430* RE RIVER FORMATION, UPPER ORDOVICIAN, NORTH DAKOTA, ABST.,
B 6303 538 REA, D. K., MIDDLE CRETACEOUS OXYGEN-DEFICIENT
B 6108 1394* REACTION MODELS OF DIAGENETIC PROCESSES IN VOLCANOGENIC
B 6307 1061 READ BAY FORMATION, NORTHWEST TERRITORIES
B 6405 770 READ, J. F., CARBONATE DEPOCENTERS AND FACIES DISTRIBUTION ON
B 6410 1575 READ, J. F., CARBONATE RAMP-TO-BASIN TRANSITIONS AND FORELAND
B 6203 539 READ, J. F., DIGITATE STROMATOLITIC BIOHERMS AND ASSOCIATED
B 6203 557 READ, J. F., GEOMETRY, FACIES, AND DEVELOPMENT OF LARGE
B 6004 712 READ, J. F., SCALLOPED SOLUTION SURFACES, MIDDLE ORDOVICIAN
B 6407 1117 READING, HARMS, J. C., REVIEW OF, SEDIMENTARY ENVIRONMENTS AND
B 6107 1030 REAGAN SANDSTONE, NEBRASKA
B 6402 156 REALTY A ZONE, LOUISIANA
B 6203 548 REASER, DONALD F., LITHOFACIES AND DEPOSITIONAL ENVIRONMENTS
B 6203 571 REASER, DONALD F., STRUCTURAL GEOLOGY OF SOUTHWEST EDWARDS
B 6405 770 REAUGH, A. B., DEPOSITIONAL ENVIRONMENTS AND KEROGEN TYPES. BY
B 6405 770 REAUGH, A. B., FLUORESCENCE OF ACRITARCHS IN STUDY OF MARINE
B 6201 124 REBECCA SHOAL DOLOMITE, FLORIDA
B 6201 124 REBECCA SHOAL DOLOMITE TONGUES, FLORIDA
B 6201 121† REBECCA SHOAL REEF COMPLEX, UPPER CRETACEOUS AND PALEOCENE,
B 6012 2188 REBER, S. J., USE OF IR COLOR PHOTOGRAPHY AND CANYON PHOTOS IN
B 6004 669 REBMAN, J., NEW ENVIRONMENTAL MAPS FOR GULF OF MEXICO OCS
B 6209 1675 RECAPTURE MEMBER, NEW MEXICO
B 6007 1011 RECAPTURE MEMBER, NEW MEXICO
B 6405 745* RECENT ACTIVITIES IN UNITED STATES TAR SAND, ABST.
B 6303 527* **RECENT ADVANCES IN PASSIVE-MARGIN RESEARCH,**
B 6405 771* RECENT ADVANCES IN HELIUM ANALYSIS AS EXPLORATION TOOL FOR
B 6303 527* RECENT ADVANCES IN PASSIVE-MARGIN RESEARCH, ABST.
B 6303 516* RECENT AND POTENTIAL ADVANCES IN DSDPBIOSTRATIGRAPHY, ABST.
B 6004 677* RECENT AND RELICT TOPOGRAPHY OF BOO BEE PATCH REEF, BELIZE
B 6403 445* RECENT ARCTIC FORAMINIFERA--AN OVERVIEW, ABST.
B 6303 448* RECENT ARTHROPO LEBENSSPUREN AS INDICATORS OF ANCIENT
B 6211 2357* RECENT BENTHONIC FORAMINIFERAL BIOFACIES IN ARCTIC OCEAN,
B 6211 2217 RECENT, CALIFORNIA, PLIOCENE-
B 6004 681* RECENT DELTAIC SEDIMENTS, ABST., BED CONTINUITY AND,
B 6203 499* RECENT DEPOSITIONAL HISTORY OF UPPER CONTINENTAL SLOPE,
B 6212 2509* RECENT DEVELOPMENT OF KAMOJANG GEOTHERMAL PROJECT, WEST
B 6203 490* RECENT DEVELOPMENTS AT CERRO PRIETO GEOTHERMAL FIELD, BAJA
B 6008 1403* RECENT DEVELOPMENTS IN COAL EXPLORATION IN SOUTHEASTERN
B 6009 1609* RECENT DEVELOPMENTS IN SUNNILAND EXPLORATION O SOUTH
B 6303 520* RECENT HUGE GAS DISCOVERIES IN KANGAN AREA, IRAN, ABST.
M 25 239† RECENT KNOWLEDGE OF HYDROCARBON POTENTIALS IN SEDIMENTARY
B 6405 610 RECENT, MIDDLE EAST, SENONIAN TO
B 6012 2182* RECENT RESULTS IN EARTHQUAKE PREDICTION, ABST.
B 6305 835* RECENT SALT SEDIMENTATION IN PLAYA LAKES OF EBRO BASIN, SPAIN,
B 6102 176 RECENT SAND, MISSISSIPPI RIVER
B 6102 176 RECENT SANDS, FLORIDA BEACHES
B 6102 176 RECENT SANDS, TEXAS BEACHES
B 6004 716* RECENT SEDIMENTATION ON NILE DELTA AND CONE, ABST.
B 6002 290 RECENT SEDIMENTS, GAS COMPOSITION OF
B 6203 552* RECENT SEDIMENTS IN LAKE HURON, ABST., FACTORS CONTROLLING,
B 6002 197 RECIFE PLATEAU, BRAZIL MARGIN
B 6302 221 RECIFE-JOAO PESSOA BASIN, BRAZIL
B 6109 1474 RECIFE-JOAO PESSOA BASIN, BRAZIL
B 6002 197 RECIFE-JOAO PESSOA COASTAL BASIN, BRAZIL
M 26 432 RECLUSE FIELD, MUDDY FORMATION, WYOMING
M 26 185 RECOGNITION, BANKS, SEISMIC, REFLECTIONS
M 26 185 RECOGNITION, BUILDUPS, SEISMIC, REFLECTIONS
M 26 203 RECOGNITION, CRITERIA, CHART, SEISMIC, REFLECTIONS, CARBONATE
M 26 203 RECOGNITION, EXAMPLES, SEISMIC, REFLECTIONS, CARBONATE
M 26 136 RECOGNITION, FACIES, ANALYSIS
M 26 187 RECOGNITION, GEOPHYSICAL CRITERIA, SEISMIC, REFLECTIONS,
M 26 203 RECOGNITION, INTERVAL VELOCITY, CARBONATES, SEISMIC,
M 26 185† RECOGNITION OF CARBONATE BUILDUPS, SEISMIC, STRATIGRAPHY AND,
S 7 87† RECOGNITION OF CONTINENTAL-SLOPE SEISMIC FACIES, OFFSHORE
B 6203 503* RECOGNITION OF CRATONAL BANK, SLOPE, AND BASINAL FACIES,
B 6108 1388* RECOGNITION OF OXIDIZED SULFIDE MINERALS AS EXPLORATION
M 26 187 RECOGNITION, PROCEDURE, SEISMIC, REFLECTIONS, CARBONATE
M 26 185 RECOGNITION, REEFS, SEISMIC, REFLECTIONS
B 6109 1471 RECONCAVO BASIN, BRAZIL
B 6307 1033 RECONCAVO BASIN, SANTO AMARO GROUP, BRAZIL
B 6307 1034 RECONCAVO BASIN, SECTION, BRAZIL
B 6002 197 RECONCAVO COASTAL BASIN, BRAZIL
B 6008 1399* RECONNAISSANCE FOR URANIUM IN APPALACHIAN BASIN--SOME INITIAL
B 6309 1490† RECONNAISSANCE GEOLOGIC MAPPING IN CHAGAI HILLS, BALUCHISTAN,
B 6101 93 RECONSTRUCTION OF PANGAEA
B 6401 137† RECORDS OF STRATIGRAPHI NOMENCLATURE, 1976-78, NOTE 51--
B 6401 136 RECORDS OF STRATIGRAPHIC COMMISSION, 1976-78, AMERICAN
B 6401 136† RECORDS OF STRATIGRAPHIC COMMISSION, 1976-78, AMERICAN
B 6401 137† RECORDS OF STRATIGRAPHIC NOMENCLATURE, 1976-78, NOTE 51--
B 6209 1714† RECORRELATION OF EOCENE-MIOCENE BOUNDARY OF DELAWARE
B 6105 723† RECORRELATION OF EOCENE-MIOCENE BOUNDARY OF DELAWARE COASTAL
B 6105 831* RECOVERY, ABST., APPLICATION OF, RESERVOIR GEOLOGIC METHODS TO
B 6105 835* RECOVERY FROM KNOWN FIELDS BY ENHANCED RECOVERY METHODS,
B 6105 835* RECOVERY METHODS, ABST., INCREMENTAL, ULTIMATE, RECOVERY
B 6309 1590* RECOVERY OF ENERGY FROM MICHIGAN ANTRI SHALE BY IN-SITU
B 6206 1092* RECOVERY PROJECTS, ABST., INSTRUMENTATION FOR, EVALUATION OF,
B 6203 529* RECOVERY PROJECTS. BY TITLE ONLY., INSTRUMENTATION FOR,
M 27 201 RECRYSTALLIZATION, STRAIN
M 27 208 RECRYSTALLIZATION, STRAIN
M 27 200 RECRYSTALLIZATION, STRAIN
B 6001 12† RECURRENT LARAMIDE STRUCTURAL GROWTH ON SEDIMENTATION ON
B 6312 2176 RECYCLED OROGEN PROVENANCES
M 27 176 RED ALGAL BIOLITHITE
M 27 176 RED ALGAL ENCRUSTATION
M 27 13 RED ALGAL ENCRUSTING
M 27 13 RED ALGAL FRAGMENT, DOLOMITE REPLACEMENT
M 27 12 RED ALGAL GRAIN, GONIOLITHON
M 27 131 RED ALGAL GRAINS
M 27 22 RED BANK SANDSTONE, NEW JERSEY, CRETACEOUS
M 27 106 RED BANK SANDSTONE, NEW JERSEY, CRETACEOUS
S 9 6 RED BED, DEPOSITIONAL MODEL, FRAMEWORK
B 6405 678* **RED BEDS OF DOLORES FORMATION, SOUTHWESTERN COLORADO,** ABST.,
B 6405 765* RED BEDS OF CATSKILL FORMATION IN EASTERN PENNSYLVANIA, ABST.,
B 6303 511* RED BEDS OF PALO DURO BASIN, TEXAS--FACIES PATTERNS THROUGH
S 9 5 RED BEDS, PALEOCLIMATE INDICATORS
B 6202 187 RED BIRD SILTY MEMBER, WYOMING
B 6207 1118 RED DEER AREA STUDY, ALBERTA
B 6203 561* RED DESERT, WYOMING, ABST., VERTICAL DISTRIBUTION OF, HE4, RN222,
S 10 141 RED EARTH REGION, GROUNDWATER FLOW, ALBERTA
S 10 125 RED EARTH REGION, HYDRAULIC DISTRIBUTION, ALBERTA
S 10 141 RED EARTH REGION, HYDROGEOLOGIC CROSS SECTION, ALBERTA
S 10 154 RED EARTH REGION, PETROLEUM MIGRATION, ALBERTA
S 10 140 RED EARTH REGION, STRATIGRAPHY, CANADA, ALBERTA
B 6203 399 RED GLACIER FORMATION, ALASKA
B 6211 2220 RED HILLS FAULT, CALIFORNIA
B 6102 138 RED MOUNTAIN THRUST, CALIFORNIA
B 6102 170 RED MOUNTAIN THRUST, CALIFORNIA
B 6405 674* RED RIVER FORMATION, POWDER RIVER BASIN, WYOMING-MONTANA,
B 6008 1401* RED RIVER FORMATION, SECOND CREEK FIELD, RICHLAND COUNTY,
B 6209 1765* RED RIVER IN OKLAHOMA AND TEXAS, ABST., SEDIMENTARY FACIES,
B 6405 804* RED RIVER ORDOVICIAN, POROSITY, EASTERN MONTANA AND WESTERN
B 6105 827* RED RIVER, TEXAS AND OKLAHOMA, ABST., FLOW PATTERNS AND, BAR
B 6103 366 RED RIVER UPLIFT, OKLAHOMA
B 6405 777* RED RIVER, UPPER ORDOVICIAN, CARBONATE RESERVOIR, CABIN CREEK
B 6303 453* RED SEA, ABST., DOLOMITE IS EVAPORITE MINERAL--EVIDENCE FROM
B 6203 536* RED SEA AND GULF OF SUEZ AREA—TALE OF FOUR TROUGHS, ABST.,
S 7 83 RED SEA, CROSS SECTIONS (GENERALIZED)
B 6403 376 RED SEA, HOLOCENE FAN DELTA
B 6004 669* RED SEA MARGIN, ABST, GEOLOGY OF
B 6305 795 RED SEA MODEL FOR GULF OF MEXICO
S 7 82 RED SEA MODEL, GULF OF MEXICO
B 6408 1190 RED SEA, OXIC NEGATIVE WATER BALANCE BASIN
S 4 237 RED SEA, REEFS
S 4 353 RED SEA, REEFS
S 3 117 RED SEA REGION, MINERALIZATION TEMPERATURES
B 6412 2125 RED SEA, SITE 229, GEOCHEMISTRY
S 4 353† RED SEA--SEQUENCE AND TIME SCALE, PROGRESSIVE, DIAGENETI
S 9 5 RED SOILS, PRESENT DISTRIBUTION
M 26 348 RED WASH FIELD, UINTAH COUNTY, UTAH
B 6403 429 RED WASH HIGH-WAX OIL FIELD, UTAH
B 6002 322* RED WILLOW COUNTY, NEBRASKA, ABST., GENESIS AND DISTRIBUTION
B 6107 1029† RED WILLOW COUNTY, NEBRASKA, GENESIS AND DISTRIBUTION OF,
B 6107 1030 RED WILLOW COUNTY, STRATIGRAPHIC COLUMN, NEBRASKA
B 6108 1372* RED WING CREEK FIELD, NORTH DAKOTA-CRYPTOEXPLOSION
B 6406 961* RED WING CREEK FIELD—COSMIC IMPACT STRUCTURE, ABST.
B 6408 1288* RED-BED EVAPORITE, AND CARBONATE FACIES ASSOCIATIONS IN
B 6209 1759* REDDELL SALT DOME OF LOUISIANA—GEOTEMPERATURES AND
B 6105 806 REDDEN, GEORGE D., HYDROCARBON GASES IN SEDIMENTS OF EASTERN
B 6311 2070 REDDEN, GEORGE D., HYDROCARBON POTENTIAL OF ALEUTIAN BASIN,
B 6409 1544 REDEFINITION OF PRIMARY AND SECONDARY MIGRATION
B 6304 694* REDISTRIBUTED OREBODIES OF POISON CANYON MINE, ABST.
B 6202 248 REDONDO CANYON, CALIFORNIA

S 8 64 REDONDO CANYON, CALIFORNIA, SUBMARINE CANYONS
B 6202 250 REDONDO CANYON FAULT SCARP, CALIFORNIA
S 8 66 REDONDO CANYON, TIME-VELOCITY CURVES, CURRENT FLOW
S 8 65 REDONDO CANYON, TOPOGRAPHY, INDEX MAP
B 6202 253 REDONDO PLATFORM ANTICLINORIUM, CALIFORNIA
B 6202 248 REDONDO PLATFORM, CALIFORNIA
B 6206 927 REDONDO SUBMARINE FAN
B 6202 248 REDONDO SUBMARINE FAN, CALIFORNIA
B 6203 374 REDOUBT BAY, ALASKA
B 6408 1132 REDOUBT SHOAL OIL FIELD, ALASKA
B 6203 374 REDOUBT VOLCANO, ALASKA
B 6203 554* REDOX MODEL FOR SUBMARINE CEMENTATION, ABST., INTERSTITIAL,
B 6203 513* REDOX POTENTIAL IN URANIUM ORE DEPOSITS, ABST., BACTERIAL
B 6004 730* REDUCTION IN REEF MARGINS BY EARLY CEMENTATION, GOLDEN SPIKE
B 6408 1172 REDWALL FACIES, UTAH AND, COLORADO
M 28 82 REDWALL LIMESTONE, ARIZONA, MISSISSIPPIAN,
B 6008 1401* REDWALL LIMESTONE, BLACK MESA BASIN, ARIZONA, ABST., OIL AND
B 6104 597 REDWATER FIELD, ALBERTA
B 6211 2277 REDWATER SHALE MEMBER, WYOMING
B 6211 2360 REDWINE, LOWELL E., TERTIARY PRINCETON SUBMARINE VALLEY
B 6405 770 REDWINE, LOWELL, HYPOTHESIS INVOLVING DILATION AND NATURA
B 6303 435 REED, J. T., SURFACE SEISMIC METHODS APPLIED TO COAL-MINING
B 6004 608 REED, P. ROBERT, THERMAL-ANALYSIS TECHNIQU FOR SOURCE-ROCK
M 27 76 REEDSVILLE SHALE, PENNSYLVANIA, ORDOVICIAN
M 28 39 REEDSVILLE SHALE, PENNSYLVANIA, ORDOVICIAN
M 27 99 REEDSVILLE SHALE, PENNSYLVANIA, ORDOVICIAN
B 6002 240 REEDY CREEK FIELD, MISSISSIPPI
B 6002 220 REEDY CREEK TOPPED OIL, MISSISSIPPI
S 5 6 REEF DEVELOPMENT, PINNACLE
B 6004 683* REEF, ABST., AUTOPSY OF A DEAD, CORAL
B 6405 682* REEF, ABST., UNUSUAL PONDING OF SEDIMENTS ON DEEP-WATER
B 6007 1057 REEF ACROPORA PALMATA, PANAMA, GALETA POINT
M 24 85 REEF, ALBERTA, SWAN HILLS
B 6205 789 REEF AND BANK COMPLEX, SEISMIC DATA, ALBERTA, JUDY CREEK
M 25 204 REEF ANOMALY, MICHIGAN BASIN, SILURIAN
S 4 20 REEF APRON, PUERTO RICO
B 6401 116 REEF ARCHIPELAGO, GREAT LAKES AREA
B 6007 1056 REEF ASSEMBLAGES, PANAMA, GALETA POINT
B 6401 115† REEF AT DELPHI, INDIANA, STRATIGRAPHY, STRUCTURE, AND,
S 4 37† REEF ATOLL, BELIZE, PATCH-REEF ECOLOGY AND SEDIMENTOLOGY OF,
M 25 27 REEF, AUSTRALIA, GREAT BARRIER
S 4 187 REEF, BAHAMAS, ANDROS ISLAND
S 4 186 REEF, BAHAMAS, JOULTERS CAYS, PATCH
B 6004 677* REEF, BELIZE DETERMINED BY DIVER-OPERATED DRILLING DEVICE,
S 9 22 REEF BELT DISTRIBUTION, CAMBRIAN, CARBONATE
S 9 22 REEF BELT DISTRIBUTION, CARBONIFEROUS, CARBONATE
S 9 22 REEF BELT DISTRIBUTION, CRETACEOUS, CARBONATE
S 9 22 REEF BELT DISTRIBUTION, DEVONIAN, CARBONATE
S 9 22 REEF BELT DISTRIBUTION, HOLOCENE, CARBONATE
S 9 22 REEF BELT DISTRIBUTION, JURASSIC, CARBONATE
S 9 22 REEF BELT DISTRIBUTION, ORDOVICIAN, CARBONATE
S 9 22 REEF BELT DISTRIBUTION, PERMIAN, CARBONATE
S 9 22 REEF BELT DISTRIBUTION, SILURIAN, CARBONATE
S 9 22 REEF BELT DISTRIBUTION, TERTIARY, CARBONATE
S 9 22 REEF BELT DISTRIBUTION, TRIASSIC, CARBONATE
B 6304 608† REEF BELT OF NORTHERN MICHIGAN, DIFFERENTIAL, ENTRAPMENT OF
S 4 187 REEF, BERMUDA, TOBACCO BAY
S 4 263 REEF, BORERS, JAMAICA, DISCOVERY BAY
S 4 127† REEF BUILDERS AND SHAPERS, SPONGES AS
B 6306 881 REEF BUILDUP ON ATLANTIC-TYPE PASSIVE MARGIN, POTENTIAL,
M 26 197 REEF BUILDUP, SEISMIC EXAMPLE, CANADA, ALBERTA, INNISFAIL
B 6303 519* **REEF CARBONATE ROCKS--EXAMPLE FROM CAPITAN COMPLEX,**
B 6405 782* REEF COMMUNITIES IN GULF COAST, ABST., EARLY, CRETACEOUS
B 6209 1757* REEF COMMUNITY SUCCESSION, DAMON MOUND, TEXAS, ABST.,
B 6105 839* REEF COMPLEX, ABST., BIOFACIES RELATIONS IN LOWER, DEVONIAN,
M 24 83 REEF COMPLEX, ALBERTA, KAYBOB
M 24 80 REEF COMPLEX, ALBERTA, KAYBOB
B 6107 1080 REEF COMPLEX, ALBERTA, MIETTE
B 6004 670* REEF COMPLEX, PERMIAN, GUADALUPE MOUNTAIN, NEW MEXIC AND
B 6003 428† REEF COMPLEX, ROCKFORD, OHIO--CONSTITUTION, GROWTH, AND
B 6201 121† REEF COMPLEX, UPPER CRETACEOUS AND PALEOCENE, IN SOUTH
B 6406 827 REEF COMPLEXES, AUSTRALIA, FRASNIAN
B 6406 820 REEF COMPLEXES, CANNING BASIN, AUSTRALIA
B 6105 816* REEF COMPLEXES—UPPER DEVONIAN EXAMPLES FROM ALBERTA, ABST.,
B 6109 1555* REEF COMPLEX—BANDERA COUNTY, TEXAS, ABST., DIAGENESIS AND,
B 6303 489* REEF COMPLLX--EVIDENCE OF EXTENSIVE SUBMARINE LITHIFICATION,
S 4 178 REEF CORAL, SAIPAN, MIOCENE
S 4 171† REEF CORALS MONTASTREA ANNULARIS AND MONTASTREA
S 4 158 REEF CORALS OBSERVED DURING EMERGENCE, LIVING, CARIBBEAN
B 6205 772 REEF, DEFINITION
M 25 236 REEF DISCOVERY WELL, AUSTRALIA, SCOTT
S 9 22 REEF DISTRIBUTION, PALEOCLIMATE, CARBONATE
S 4 261† REEF ECOLOGY AND PRESERVATION, CARBONATE, BORERS--THEIR ROLE
S 4 96 REEF ECOSYSTEMS, TROPHIC ANALYSIS OF, ANCIENT
B 6405 800* REEF, ENEWETAK ATOLL, MARSHALL ISLANDS, ABST.,
S 4 20 REEF ENVIRONMENT, PUERTO RICO, ENRIQUE
B 6004 653* REEF ENVIRONMENTS, ABST., ANALYSIS OF, BORERS AND THEIR,
B 6203 525* REEF EXPLORATION IN MICHIGAN BASIN, ABST., USE OF, SEISMIC,
B 6207 1233* REEF EXPLORATION IN PHILIPPINES, ABST.
S 4 21 REEF FLAT AND SHOALS, PUERTO RICO
S 4 187 REEF, FLORIDA, KEY LARGO LIMESTONE
B 6007 1054† REEF, GALETA POINT, PANAMA, EVOLUTION OF, MODERN, CARIBBEAN,
B 6405 739* REEF, GRANT COUNTY, INDIANA, ABST., POROSITY EVOLUTION OF,
M 30 554 REEF HISTORY, DEPOSITIONAL HISTORY, SIRTE BASIN
B 6309 1609* REEF IN GULF OF MEXICO, ABST., LIVING, FORAMINIFERS OF, WEST
B 6405 736* REEF IN SOUTHERN MICHIGAN BASIN, ABST., EXPLORING FOR,
B 6401 127 REEF, INDIANA, DELPHI
S 4 129 REEF, JAMAICA, DISCOVERY BAY
S 4 129 REEF, JAMAICA, EAST PALISADES
S 4 128 REEF, JAMAICA, RUNAWAY BAY
B 6402 265 REEF, LOUISIANA, POINT AU FER SHELL
B 6004 730* REEF MARGINS BY EARLY CEMENTATION, GOLDEN SPIKE OIL FIELD,
B 6405 713* REEF, MIDDLE PLEISTOCENE TO HOLOCENE, ABST., DEPOSITION OF,
B 6004 684* REEF MODEL--GREA LAKES AREA, ABST., SEARCH FOR, SILURANIAN
B 6303 427* REEF MORPHOGENESIS, ST. CROIX, UNITED STATES VIRGIN ISLANDS,
B 6004 682* REEF MORPHOLOGY AND SEDIMENT TRANSPORT, LUCAYA, GRAND
S 4 186 REEF, NEWFOUNDLAND, LONG POINT FORMATION
B 6405 748* REEF, NORTHEASTERN ILLINOIS, ABST., POROSITY EVOLUTION OF,
B 6406 814† REEF OF CANNING BASIN, WESTERN AUSTRALIA, DEVONIAN, GREAT,
B 6403 400† REEF OIL FIELD—WILFRED POOL, SULLIVAN COUNTY, INDIANA,
B 6405 709* REEF PALEOECOLOGY, ABST., UNIFORMITARIANISM AND, TERTIARY
B 6007 1054 REEF, PANAMA, GALETA
B 6203 541* REEF, PERMIAN, GUADALUPE MOUNTAINS, NEW MEXICO, ABST.,
B 6004 732* REEF PRESERVATION--ROL OF CORAL BORERS, ABST., EVENT SEQUENCES
S 4 18 REEF, PUERTO RICO, AHOGADA
S 4 22 REEF, PUERTO RICO, MEDIA LUNA
S 4 23 REEF, PUERTO RICO, SHELF-EDGE
B 6004 674* REEF RESERVOIRS, ABST., SOURCE-, RESERVOIR-, SEAL RELATIONS IN
B 6211 2256 REEF RIDGE SAND, CALIFORNIA
B 6211 2247 REEF RIDGE SHALE, CALIFORNIA
M 27 49 REEF SEDIMENT, JAMAICA, HOLOCENE
B 6203 554* REEF SEDIMENTS—REDOX MODEL FOR SUBMARINE CEMENTATION,
B 6205 783 REEF, SEISMIC DATA, ALBERTA, GOLDEN SPIKE
M 26 190 REEF, SEISMIC EXAMPLE, AFRICA, NORTH, PINNACLE
M 26 192 REEF, SEISMIC EXAMPLE, GULF OF PAPUA, SHELF-MARGIN
B 6004 725* REEF SHOALING SEQUENCE IN DOGGER, MIDDLE JURASSIC, OF CENTRAL
S 4 127 REEF SPONGES, KINDS AND DISTRIBUTION
S 4 145 REEF, ST. CROIX, ISAAC BAY
S 4 93† REEF SYSTEMS OF CARIBBEAN--PROSPECTS FOR PALEOECOLOGIC
S 4 186 REEF, TENNESSEE, HOLSTON LIMESTONE
M 24 80 REEF, TERMINOLOGY
B 6205 783 REEF TREND, ALBERTA, LEDUC-RIMBEY, BARRIER
M 26 199 REEF TREND, SHELF-MARGIN BUILDUP, SEISMIC EXAMPLE, TEXAS,
B 6209 1607† REEF TRENDS, CENTRAL APPALACHIAN BASIN, PALEOGEOGRAPHY OF
S 4 301 REEF, WEST INDIES, BARBUDA
B 6102 294* REEF--EVOLUTION OF A CONCEPT, ABST., CAPITAN
B 6004 661* REEF--EVOLUTION OF CONCEPT, ABST., CAPITAN
S 4 223 REEF-BUILDING CORALS
B 6303 517* REEF-CREST WAVE AND CURRENT INTERACTIONS AND SEDIMENT
S 4 219† REEF-DWELLING FORAMINIFERA, CALCIUM CARBONATE FIXATION BY
B 6309 1582* REEF-FLANK DEPOSITS, MIDDLE SILURIAN, OF INDIANA, ABST.,
B 6306 870† REEF-RESERVOIRED GIANT OIL FIELDS, CRETACEOUS, OCEANIC, ANOXIC
S 4 183 REEFAL BRYOZOAN GROUPS, CLASSIFICATION
B 6306 877 REEFAL BUILDUPS, DISTRIBUTION, CRETACEOUS, RUDIST-, CORAL
S 5 189* REEFS AND EVAPORITES: A SUMMARY
S 5 1 **REEFS AND EVAPORITES--CONCEPTS AND DEPOSITIONAL MODELS**
S 4 1 **REEFS AND RELATED CARBONATES**
S 5 4 REEFS, MICHIGAN BASIN, NIAGARA GROUP, PINNACLE
B 6303 455* REEFS, ABST., PROBING, BERMUDAS, LAGOONS AND
B 6203 498* REEFS, ABST., STROMATOPOROIDS—EPIC STRUGGLE FOR SURVIVAL,
B 6303 528* REEFS, ABST., THREE-DIMENSIONAL ASPECTS OF, BELIZE PATCH
B 6402 188† REEFS, ALMERIA PROVINCE, SOUTHERN SPAIN, POROSITY EVOLUTION OF
B 6203 553* REEFS AND ASSOCIATED SEDIMENTS, ABST., EFFECTS OF, SUBMARINE
S 4 76 REEFS AND ATOLLS, INDO-, PACIFIC AND, CARIBBEAN, MAJOR
B 6004 713* REEFS AND BANKS OF TEXAS OUTER CONTINENTAL SHELF, ABST., CORAL
B 6205 786 REEFS AND BANKS, TEXAS, WASHITA-, FREDERICKSBURG RUDISTID
S REEFS, MODEL, ARIZONA, MURAL
S 5 1* REEFS, NIAGARA AND SALINA GROUPS, NORTHERN SHELF, MICHIGAN
S 5 111* REEFS, NORTHERN MICHIGAN, SEISMIC DATA-PROCESSING TECHNIQUES
S 5 93* REEF-BEARING ROCKS ON WABASH PLATFORM WITH CYCLC EVAPORITES
B 6203 528* REEFS, NORTHWESTERN OHIO, ABST., SUBAERIAL EXPOSURE OF,
S 4 102 REEFS OF CARIBBEAN AREA, CENOZOIC FOSSIL
B 6303 548* REEFS OF CHAPPEL LIMESTONE, FORT WORTH BASIN, TEXAS, ABST.,
B 6002 258† REEFS OF CRETACEOUS SHALLOW CARBONATE PLATFORM OF ISRAEL,
B 6303 508* REEFS OF MOMBASA, KENYA, ABST., GENESIS, OCCURRENCE, AND CAUSES
B 6405 790* REEFS OF NORTHERN INDIANA, ABST., SUBMARINE CARBONATE
S 4 353† REEFS OF RED SEA--SEQUENCE AND TIME SCALE, PROGRESSIVE,
B 6109 1550* REEFS OF TEXAS, ABST., CONTRASTS IN, CEMENTATION, DISSOLUTION,
B 6004 727* REEFS OF WESTERN OHIO, ABST., PALEOECOLOGY AND, FAUNA OF SOME,
S 4 94 REEFS, PLEISTOCENE-, HOLOCENE
S 4 353 REEFS, RED SEA
S 4 237 REEFS, RED SEA
S 4 3 REEFS, SABA BANK, CORAL
M 26 195 REEFS, SEISMIC EXAMPLE, CANADA, ALBERTA, RAINBOW-, ZAMA AREA,
M 26 185 REEFS, SEISMIC, REFLECTIONS, RECOGNITION
S 4 357† REEFS, SEQUENCE OF PROGRESSIVE, DIAGENESIS IN, CORAL
S 4 17† REEFS, SOUTHWESTERN PUERTO RICO, SHELF
B 6402 190 REEFS, SPAIN, PLOMO
S 4 94 REEFS, UNIFORMITARIANISM AND
S 4 113 REEFS, VENEZUELA, CHICHIRIVICHE AREA, COASTAL

Index

S 4 210 REEFS, VERACRUZ, PATCH
B 6007 1064 REEFS, VERTICAL DEVELOPMENT OF, YUCATAN SHELF
B 6303 482* REEFS, WESTERN GULF OF MEXICO--ROLE OF ECOLOGY AND OSTRACOD
B 6309 1605* REEFS, WESTERN GULF OF MEXICO--ROLE OF ECOLOGY AND OSTRACOD
S 4 111† REEFS--A EXAMPLE FROM VENEZUELA, MANS IMPACT ON COASTAL
B 6008 1239 REESE, D. L., DEVELOPMENTS IN FOUR CORNERS--INTERMOUNTAIN ARE
B 6108 1389 REESE, D. L., THRUST BELT OF WYOMING, IDAHO, AND NORTHERN
B 6009 1614 REESE, DALE, PRE-FERRY LAKE EARLY CRETACEOUS DELTAS OF SOUTH
B 6209 1769* REEVALUATING FLORIDA BASEMENT, ABST.
B 6405 796* REEVALUATION OF DEPOSITIONAL ENVIRONMENTS OF SALT WASH
B 6203 566* REEVALUATION OF MORROWAN-ATOKAN SERIES BOUNDARY IN
B 6104 612 REEVES, C. C., JR., THE FILL IN MARFA BASIN, TEXAS
B 6405 698 REEVES, H. NEAL, HOW GEOLOGIC OBJECTIVES SHOULD DETERMINE
M 29 143 REFLECTING HORIZONS, ATLANTIC MARGIN
M 29 367 REFLECTING HORIZONS, VENEZUELA BASIN, CURACAO RIDGE
M 26 455 REFLECTION COEFFICIENTS, SEISMIC, MODELING
M 26 166 REFLECTION CONFIGURATION, CHART, SEISMIC, INTERPRETATION,
B 6007 1088 REFLECTION DATA, ATLANTIC BASIN, SEISMIC-
B 6007 1096 REFLECTION DATA, BARBADOS RIDGE COMPLEX, SEISMIC-
B 6002 163 REFLECTION DATA, BERINGIAN SHELF, SEISMIC-
M 26 122 REFLECTION PARAMETERS, GEOLOGIC SIGNIFICANCE
M 26 117† REFLECTION PATTERNS IN DEPOSITIONAL SEQUENCES, SEISMIC,
M 26 330 REFLECTION PATTERNS IN EXPLORATION
S 7 91 REFLECTION PATTERNS, SEDIMENT PROCESSES
B 6007 1049 REFLECTION PROFILE, ABACO KNOLL, SEISMIC-
B 6007 1114 REFLECTION PROFILE, ATLANTIC, OUTER, CONTINENTAL, SHELF, CDP
M 29 216 REFLECTION PROFILE, AUSTRALIA, ARU BASIN
M 29 215 REFLECTION PROFILE, AUSTRALIA, TANIMBAR TROUGH
M 29 215 REFLECTION PROFILE, AUSTRALIA, TIMOR TROUGH
M 29 418 REFLECTION PROFILE, GULF OF MEXICO, CAMPECHE BANK
M 29 418 REFLECTION PROFILE, GULF OF MEXICO, FLORIDA STRAITS
M 29 418 REFLECTION PROFILE, GULF OF MEXICO, WEST FLORIDA BANKS
M 29 230 REFLECTION PROFILE, NIAS, SUNDA ARC
M 29 228 REFLECTION PROFILE, SIBERUT, SUNDA ARC
M 29 232 REFLECTION PROFILE, SUMATRA, SUNDA ARC
M 29 11† REFLECTION PROFILES, GEOLOGY OF THE, OFFSHORE, SOUTHEAST,
B 6007 1023 REFLECTION PROFILES, LAKE SUPERIOR
B 6007 1040 REFLECTION PROFILES, NORTHEAST PROVIDENCE CHANNEL, SEISMIC-
M 29 177† REFLECTION PROFILES, STRUCTURE AND, STRATIGRAPHY OF THE,
M 29 264 REFLECTION RECORDS (24-CHANNEL), ALASKA, KODIAK
M 29 417 REFLECTION SECTION, GULF OF MEXICO, FLORIDA ESCARPMENT
M 29 409 REFLECTION SECTION, GULF OF MEXICO, MISSISSIPPI CONE
M 29 411 REFLECTION SECTION, MEXICO, MEXICAN RIDGES
M 30 565† REFLECTION SEISMIC IN DEVELOPMENT OF NEMBE CREEK FIELD,
M 26 221 REFLECTION SIGNATURES, CORRELATION, SEISMIC
M 26 221 REFLECTION SIGNATURES, INTERPRETATION, SEISMIC
M 26 302 REFLECTION STRENGTH, SEISMIC APPLICATION
M 29 214 REFLECTION STUDIES, AUSTRALIA, SHELF
M 29 209† REFLECTION STUDIES IN THE TIMOR-ARU TROUGH SYSTEM AND
M 29 240 REFLECTION STUDIES, NORTH PHILIPPINE SEA, DAITO RIDGE
M 29 241 REFLECTION SURVEY, FOLDOUT, DAITO RIDGE
M 29 241 REFLECTION SURVEY, FOLDOUT, NORTH PHILIPPINE SEA
M 26 128 REFLECTION-FREE AREAS, SEISMIC, REFLECTIONS, CONFIGURATION
S 5 112 REFLECTIONS, SEISMIC DATA,
M 26 6 REFLECTIONS, ACOUSTIC IMPEDANCE, SEISMIC
M 26 3† REFLECTIONS AND GEOLOGIC DETAIL DERIVABLE FROM THEM,
M 26 3 REFLECTIONS, ATTENUATING NOISE, SEISMIC
M 26 203 REFLECTIONS, CARBONATE RECOGNITION, CRITERIA, CHART, SEISMIC
M 26 203 REFLECTIONS, CARBONATE RECOGNITION, EXAMPLES, SEISMIC
M 26 187 REFLECTIONS, CARBONATE RECOGNITION, GEOPHYSICAL CRITERIA,
M 26 187 REFLECTIONS, CARBONATE RECOGNITION, PROCEDURE, SEISMIC
M 26 188 REFLECTIONS, CARBONATES, DIRECT RECOGNITION, SEISMIC
M 26 188 REFLECTIONS, CARBONATES, INDIRECT RECOGNITION, SEISMIC
M 26 115 REFLECTIONS, CHRONOSTRATIGRAPHIC CORRELATION, ACCURACY,
M 26 116 REFLECTIONS, CHRONOSTRATIGRAPHIC CORRELATIONS, PROBLEMS,
M 26 109 REFLECTIONS, CHRONOSTRATIGRAPHY, STRATIGRAPHIC TRAP
M 26 128 REFLECTIONS, CONFIGURATION PATTERNS, CHAOTIC, SEISMIC
M 26 123 REFLECTIONS, CONFIGURATION PATTERNS, DIVERGENT, SEISMIC
M 26 123 REFLECTIONS, CONFIGURATION PATTERNS, PARALLEL, SEISMIC
M 26 123 REFLECTIONS, CONFIGURATION PATTERNS, PROGRADING, SEISMIC
M 26 128 REFLECTIONS, CONFIGURATION PATTERNS, REFLECTION-FREE AREAS,
M 26 123 REFLECTIONS, CONFIGURATION PATTERNS, SUB-PARALLEL, SEISMIC
M 26 123 REFLECTIONS, CONFIGURATION PATTERNS, TYPES, SEISMIC
M 26 122 REFLECTIONS, CONFIGURATION, SEISMIC
M 26 132 REFLECTIONS, CONFIGURATIONS, MIGRATING-WAVE, SEISMIC
M 26 100 REFLECTIONS, CONTINUOUS, SEISMIC
M 26 100 REFLECTIONS, DISCONTINUOUS, SEISMIC
M 26 121 REFLECTIONS, FACIES DEFINITION, SEISMIC
M 26 130 REFLECTIONS, FACIES UNITS, EXTERNAL FORMS, TYPES, SEISMIC
M 26 122 REFLECTIONS, FACIES UNITS, SEISMIC
M 26 115 REFLECTIONS, FLUID CONTACTS, HYDRATE ZONES, SEISMIC
M 26 6 REFLECTIONS, INTERFERENCE, SEISMIC
M 26 186 REFLECTIONS, INTERPRETATION PROCEDURE, CHART, SEISMIC
M 26 100 REFLECTIONS, LITHOSTRATIGRAPHY, SEISMIC
M 26 106 REFLECTIONS, PALEONTOLOGIC CORRELATION, SEISMIC
M 26 120 REFLECTIONS, PATTERNS, BASE-DISCORDANT, SEISMIC
M 26 127 REFLECTIONS, PATTERNS, CLINOFORMS, SEISMIC
M 26 125 REFLECTIONS, PATTERNS, CLINOFORMS, SEISMIC
M 26 138 REFLECTIONS, PATTERNS, STRATIGRAPHIC INTERPRETATION, SEISMIC
M 26 119 REFLECTIONS, PATTERNS, TOP-DISCORDANT, SEISMIC
M 26 99 REFLECTIONS, PHYSICAL SURFACES, SEISMIC
M 26 185 REFLECTIONS, RECOGNITION, BANKS, SEISMIC
M 26 185 REFLECTIONS, RECOGNITION, BUILDUPS, SEISMIC
M 26 203 REFLECTIONS, RECOGNITION, INTERVAL VELOCITY, CARBONATES,
M 26 203 REFLECTIONS, RECOGNITION, LIMITATIONS AND QUALIFICATIONS,
M 26 185 REFLECTIONS, RECOGNITION, REEFS, SEISMIC
M 26 4 REFLECTIONS, RESOLUTION, VERTICAL, SEISMIC
M 26 9 REFLECTIONS, SANDSTONE PINCHOUTS, SEISMIC
M 26 141 REFLECTIONS, SECTION DESCRIPTION, SEISMIC
M 26 101 REFLECTIONS, SEISMIC MODELS, SEISMIC
M 26 99† REFLECTIONS, SEISMIC, STRATIGRAPHY AND, GLOBAL CHANGES OF SEA
M 26 117 REFLECTIONS, SEQUENCES, SEISMIC
M 26 6 REFLECTIONS, THIN BEDS, SEISMIC
M 26 100 REFLECTIONS, UNCONFORMITIES, CHRONOSTRATIGRAPHY, SEISMIC
M 26 101 REFLECTIONS, VELOCITY RELATIONS, SEISMIC
M 26 4 REFLECTIONS, WAVELENGTH, SEISMIC
M 29 30 REFLECTORS, BLAKE PLATEAU, AGE OF
M 29 35 REFLECTORS, BLAKE PLATEAU, STRUCTURE CONTOURS
M 29 30 REFLECTORS, GEORGIA EMBAYMENT, AGE OF
B 6303 543* REFLECTORS IN DEEP WESTERN NORTH ATLANTIC, ABST., GEOLOGIC
M 26 255 REFLECTORS, STRONG, CLASSIFICATION CHART
M 26 255 REFLECTORS, WEAK, CLASSIFICATION CHART
B 6207 1219* REFORMA AREA, SOUTHEASTERN MEXICO, AND EXPLORATORY EFFORT
M 29 67 REFRACTING HORIZONS, ATLANTIC COAST, NEW JERSEY
M 29 209† REFRACTION AND REFLECTION STUDIES IN THE TIMOR-ARU TROUGH
M 29 285 REFRACTION DATA, ALASKA, ICY BAY
M 29 285 REFRACTION DATA, ALASKA, YAKUTAT BAY
B 6203 566* REFRACTION SEISMIC DATA IN WOLLASTON FOLDBELT REGION OF
M 29 284 REFRACTION STATION DATA, GULF OF ALASKA
M 29 210 REFRACTION STUDIES, AUSTRALIA, SHELF
M 29 268 REFRACTION VELOCITY DATA, ALASKA, KODIAK
S 11 16 REFUGIAN STAGE
S 11 65 REFUGIAN STAGE
B 6012 2190* REFUGIAN STAGE, UPPERMOST EOCENE OF CALIFORNIA, ABST.,
B 6303 526* REGIONAL ASPECT OF SEDIMENT DISTRIBUTION ACROSS MIDDLE EAST
B 6105 813* REGIONAL CEMENTATION PATTERNS IN MISSISSIPPIAN LIMESTONES OF
B 6309 1577* REGIONAL COALIFICATION PATTERNS IN EASTERN KENTUCKY,
B 6203 490* REGIONAL CONTROLS ON DIAGENESIS OF ROTLIEGENDES FORMATION IN
B 6108 1371* REGIONAL DIAGENETIC TRENDS IN LOWER CRETACEOUS MUDDY
B 6004 729* REGIONAL ENVIRONMENTAL STRATIGRAPHY AND OIL AND GAS
B 6006 926† REGIONAL GEOLOGIC FRAMEWORK OFF NORTHEASTERN UNITED
B 6105 742† REGIONAL GEOLOGIC FRAMEWORK OFF NORTHEASTERN UNITED STATES:
B 6105 741† REGIONAL GEOLOGIC FRAMEWORK OFF NORTHEASTERN UNITED STATES:
B 6203 528* REGIONAL GEOLOGIC MAPPING OF NORTH SEA BY COMPUTER, ABST.
B 6207 1213* REGIONAL GEOLOGIC, STRUCTURAL, AND MINERAL TRENDS IN
B 6303 431* REGIONAL HYDROCARBON SOURCE ROCK AND THERMAL MATURITY
B 6303 425* REGIONAL HYDROCARBON SOURCE-ROCK EVALUATION OF ATLANTIC
B 6108 1381* REGIONAL IMPLICATIONS OF OIL ENTRAPMENT AT COTTONWOOD
B 6105 794* REGIONAL LOWER MANNVILLE LITHOFACIES COMPUTER MAPS
B 6409 1552* REGIONAL PATTERNS OF WOODBINE-TUSCALOOSA, NORTHERN COASTAL
B 6008 1398* REGIONAL PRESSURE ENVIRONMENTS AND PETROLEUM
B 6004 699* REGIONAL SEISMIC INTERPRETATION USING SEQUENCES AND EUSTATIC
B 6303 464* REGIONAL STRATIGRAPHY OF LIMESTONE MARKER BED IN BRIDGE
B 6409 1563* REGIONAL STRATIGRAPHY OF UPPER JURASSIC SMACKOVER
B 6103 465* REGIONAL TECTONIC IMPLICATIONS OF ALASKA BLUESCHIST BELTS, BY
B 6409 1569* REGIONAL TECTONIC FEATURES OF INNER GULF COAST BASIN AND
B 6303 477* REGIONAL TECTONICS AND PETROLEUM GEOLOGY OF
B 6203 563* REGIONAL VARIATION OF DEPOSITIONAL ENVIRONMENTS IN THEBES
B 6012 2183* REGIONAL VELOCITY REDUCTION, ABST.
B 6104 558† REGIONALIZED VARIABLES FOR EVALUATION OF PETROLEUM
B 6405 681 REGISTER, J. K., RB-SR DATING OF DIAGENESIS—EPIGENESIS IN MESOZOIC
B 6004 703* REGRESSION CYCLE FROM CAMBRIAN TO PRESENT, ABST., CAUSALITY
M 26 66 REGRESSION, SEA LEVEL FLUCTUATIONS
B 6004 704* REGRESSIONS IN CRETACEOUS AND TERTIARY OF NEW JERSEY, ABST.,
B 6004 728* REGRESSIONS OF DELAWARE COASTAL PLAIN, ABST., MIOCENE,
B 6004 732* REGRESSIVE DEPOSITS OF DIFUNTA GROUP, UPPER
B 6406 857† REGRESSIVE FACIES, NORTHEAST AFRICA CRATON, LATEST,
B 6004 665* REGRESSIVE RATES IN PENNSYLVANIAN O APPALACHIAN
B 6203 539* REGRESSIVE SEDIMENTATION MODEL FOR JAMES BAY LOWLANDS,
B 6004 672* REGRESSIVE SHELF DEPOSITION, SHUBLIK FORMATION, PRUDHOE BAY
B 6203 515* REGRESSIVE-DELTA MODEL WITH EXAMPLE FROM UPPER DEVONIAN—
B 6101 80 REGUIBAT MASSIF
M 25 420 REHABILITATION OF MINED LANDS, AUSTRALIA, HEAVY-, MINERAL
B 6105 821 REHBEIN, EDWARD A., LIGNITE RESOURCES OF FORT UNION
B 6405 771 REID, FRED S., LOWER AND MIDDLE PALEOZOIC POTENTIAL OF PARADOX
B 6308 1386 REID, S. G., DEVELOPMENTS IN NORTHERN ROCKIES IN 1978
B 6208 1498 REID, S. G., OIL AND GAS DEVELOPMENTS IN NORTHERN ROCKIES IN 1977
B 6211 2360 REID, STEPHEN A., EARLY MIOCENE SUBMARINE DUNE FIELDS—
B 6303 513 REID, WILLIAM M., RATES OF CREEP OF ACTIVE FAULTS IN
B 6105 821 REIJERS, T. J. A., CRETACEOUS COGOLLO GROUP, MARACAIBO
B 6309 1578 REIKE, H. H., NOVEL FORMATION-EVALUATION CONCEPT FOR SHALE
B 6405 771 REIMER, G. M., RECENT ADVANCES IN HELIUM ANALYSIS AS
B 6008 1407 REIMER, LOUIS R., OVERVIEW OF AUSTRALIAN URANIUM DEPOSITS,
B 6012 2176 REIMNITZ, ERK, DISPERSAL, DISTRIBUTION, AND THICKNESS OF
B 6105 764 REIMNITZ, ERK, RATES OF ICE GOUGING AND SEDIMENT REWORKING,
B 6004 712 REIMNITZ, ERK, SLOPE VALLEYS AND SEDIMENTARY STRUCTURES OFF
M 25 269 REINAN, MASSIF, KOREA, SOBAEKSAN, SOBAEGON OR, YONGNAM
M 25 441 REINAN MASSIF, SOUTH KOREA, FUKIEN-
B 6405 692 REINBOLD, M. L., RELATION OF DEPOSITIONAL FACIES AND HISTORY TO
B 6004 681 REINECK, H.-E., PHYSICAL-BIOGENIC BREAK I NEARSHORE SEDIMENTARY
B 6303 468 REINECK, HANS-ERICH, SEDIMENTARY STRUCTURES OF HIGH-ENERGY

B 6105 822 REINHARDT, JUERGEN, OAK GROVE CORE, NORTHEASTERN VIRGINIA,
B 6303 518 REINHART, PETER J., HOLOCENE STRATIGRAPHY OF RAPIDLY
B 6303 514 REINSON, GERALD E., DEPOSITIONAL SEQUENCE IN SUBARCTIC SANDY
B 6405 771 REINSON, GERALD E., SEABED CHARACTERISTICS AND SAND DISPERSAL
B 6004 656 REIS FONSECA, J., PETROLEUM GEOLOGY OF EASTERN BRAZILIAN
B 6108 1373 REISE, WALTER C., RARE EARTH AND OTHER TRACE ELEMENTS AS
B 6006 962 REISER, H. N., CARBONIFEROUS SECTION AND UNCONFORMITY AT
B 6006 952 REISER, H. N., GEOLOGIC SIGNIFICANCE OF DOONERAK STRUCTURAL
B 6405 771 REITER, MARSHALL, DEEP TERRESTRIAL HEAT-FLOW STUDIES IN
B 6009 1621* **RELATION BETWEEN SALINA, UPPER SILURIAN, OIL PRODUCTION AND**
M 25 448† RELATION BETWEEN TECTONICS AND METALLOGENESIS IN
B 6405 692* RELATION OF DEPOSITIONAL FACIES AND HISTORY TO HYDROCARBON
B 6004 694* RELATION OF EARLY CARBONATE POROSITY TO DEPOSITIONAL AND
B 6303 515* RELATION OF SEDIMENTARY HISTORY AND TECTONICS TO NATURAL
B 6004 653* **RELATION OF SEISMICITY AND TECTONIC PROVINCES, CENTRAL STABLE**
B 6108 1379* RELATIONS AMONG DEPOSITIONAL-SYSTEM, GROUNDWATER FLOW
S 5 125* RELATIONS BETWEEN DEPOSITIONAL ENVIRONMENTS, TONOLOWAY
B 6305 831* RELATIONS BETWEEN SEDIMENTARY FACIES AND DIAGENESIS IN
B 6405 803* RELATIONS BETWEEN STROMATOLITES AND BURROWING ORGANISMS,
B 6105 761* **RELATIONS BETWEEN TRIASSIC CARBONATE SABKHA AND KENNECOTT-TYPE**
B 6303 502* RELATIONS OF GAS OCCURRENCE TO GEOLOGIC PARAMETERS IN
M 26 15 RELATIONS, TIME-AVERAGE RELATION, VELOCITY-, POROSITY
M 26 63† RELATIVE CHANGES OF SEA LEVEL FROM COASTAL ONLAP, SEISMIC,
M 29 469† RELATIVE CHANGES OF SEA LEVEL FROM SEISMIC STRATIGRAPHY,
M 26 71 RELATIVE FALLS, SEA LEVEL FLUCTUATIONS
B 6004 665* RELATIVE MARINE TRANSGRESSIVE AND REGRESSIVE RATES IN
S 6 338 RELATIVE PALEONTOLOGIC UNITS, INDEX, DEVONIAN
B 6404 551 RELATIVE POSITIONS OF ISOTHERMS DURING COMPACTION
M 26 67 RELATIVE RISE, SEA LEVEL FLUCTUATIONS, NERITIC FACIES
M 26 65 RELATIVE RISES, SEA LEVEL FLUCTUATIONS
B 6105 764* **RELATIVE STABILITIES OF TIDAL-INLET COMPONENT FEATURES—COMPUTER**
M 26 68 RELATIVE STILLSTAND, SEA LEVEL FLUCTUATIONS
B 6401 74 RELIABILITY OF ISOTOPE DATA--EFFECTS OF DIAGENESIS
B 6401 74 RELIABILITY OF ISOTOPE DATA--EFFECTS OF DIAGENESIS, CARBO
B 6309 1597* RELICT AND EXPATRIATE RADIOLARIAN FAUNA IN GULF OF
B 6309 1597* RELICT AND EXPATRIATE RADIOLARIAN FAUNA IN GULF OF
B 6004 691* RELICT FORESTS OFF NORTHWEST FLORIDA
B 6004 734* RELICT SHORELINES, ABST., DEPOSITION, DEFORMATION, AND, HEAVY,
B 6109 1549* RELICT SHORELINES, ABST., PLIOCENE-, PLEISTOCENE PALEOGEOGRAPHY
B 6211 2219 RELIZ CANYON FORMATION, CALIFORNIA
B 6211 2219 RELIZ CANYON SECTION, CALIFORNIA
B 6211 2218 RELIZ FAULT SYSTEM, CALIFORNIA, RINCONADA-
B 6211 2221 RELIZIAN BOUNDARY PALEOGEOGRAPHY, CALIFORNIA, SAUCESIAN-
S 11 66 RELIZIAN STAGE
B 6211 2246 RELIZIAN STAGE, CALIFORNIA
B 6202 259 RELIZIAN STAGE, CALIFORNIA
B 6012 2182* RELIZIAN STAGE OF PACIFIC COAST, ABST., REVIEW OF ESTIMTES OF,
S 11 23 RELIZIAN STAGE, SUPERPOSITION AND ZONATION
B 6211 2221 RELIZIAN-LUISIAN BOUNDARY PALEOGEOGRAPHY, CALIFORNIA
B 6206 1029 REMAC, COOPERATIVE INVESTIGATION OF BRAZILIAN CONTINENTAL
B 6206 1030 REMAC CRUISES, 1972-73
B 6002 196 REMAC RECONHECIMENTO DA MARGEM BRASILEIRO, NATIONAL
S 4 313 REMAC, RECONHECIMENTO GLOBAL DA MARGEM CONTINENTA
S 6 3 REMANENT MAGNETISM, REVIEW OF PRINCIPLE
M 25 321 REMOLINO GRANDE ANTICLINE, COLOMBIA
B 6207 1236* REMOTE SENSING, ABST., INTRODUCTION TO, GEOLOGIC
B 6211 2349* REMOTE SENSING FOR ENERGY RESOURCES, ABST.
B 6207 1220* REMOTE SENSING FROM SPACE—POTENTIAL AND PROBLEMS, ABST.,
B 6412 2092 REMOTE SENSING, GLOSSARY
B 6004 677* REMOTE SENSING LANDSAT DATA TO PETROLEUM EXPLORATION, ABST.,
B 6207 1222* REMOTE SENSING TECHNIQUES TO MINERAL RESOURCES EXPLORATION
B 6109 1559* REMOTE-SENSING SYSTEMS OF FUTURE, ABST., GEOSAT—
B 6303 505* REMOTELY SENSED DATA, ABST., AVAILABILITY OF
B 6303 506* REMOTELY SENSED DATA ACQUIRED FROM AIRCRAFT, ABST.,
B 6209 1758* REMOTELY SENSED SEDIMENTATION PATTERNS IN LAKE
B 6004 554† REMOVED SEDIMENTARY ROCKS, PALEOPORE PRESSURE, AND
B 6305 812† REMOVED SEDIMENTARY ROCKS. PALEOPORE PRESSURE, AND
B 6303 380 REMSON, HAYWARD, O. T., REVIEW OF, GEOLOGY IN ENVIRONMENTAL
M 27 5 RENALCIS, EXAMPLE
B 6304 689 RENAULT, J. R., THERMOLUMINESCENCE OF URANIUM HOST ROCKS IN
B 6207 1218* RENEWABLE ENERGY SYMPOSIUM, INTRODUCTION, ABST.
B 6008 1407 RENFRO, ARTHUR R., ORIGIN OF ISOLATED REEFS AND SHELL EDGES IN
B 6101 3 RENS FIORD UPLIFT
B 6405 800 RENTON, J. J., DEPOSITIONAL ENVIRONMENTS OF LACUSTRINE
B 6105 775 **RENTON, JOHN J., GAS-PRODUCING SUBMARINE-FAN DEPOSITS OF**
B 6309 1581 RENTON, JOHN J., GEOLOGIC AGE VERSUS MAJOR MINOR, AND
B 6309 1584 RENTON, JOHN J., INORGANIC GEOCHEMICA STUDIES OF EASTERN
S 11 227 REOPHACIDAE
S 11 227 REOPHAX
B 6405 772 REPETSKI, JOHN E., CONODONT ANIMAL—HYPOTHESES AND
B 6405 671 REPETSKI, JOHN E., MISSISSIPPIAN SYSTEM OF NEW MEXICO AND
S 11 215 REPETTIAN GUIDE FOSSILS
S 11 51 REPETTIAN STAGE
S 11 68 REPETTIAN STAGE
B 6202 254 REPETTIAN STRATA, CALIFORNIA
B 6206 958 REPETTO FIELD, CALIFORNIA
B 6202 259 REPETTO FORMATION, CALIFORNIA
B 6102 138 REPETTO FORMATION, CALIFORNIA
B 6102 177 REPETTO SAND, CALIFORNIA
B 6212 2493 REPETTO STRUCTURE, UPPER, CALIFORNIA, EAST WILMINGTON FIELD
M 28 156 REPLACEMENT, DOLOMITE, IN CHERT CLAST
M 28 157 REPLACEMENT OF A FOSSIL, CHALCEDONY
M 28 155 REPLACEMENT OF ANALCIME, FELDSPAR, AUTHIGENIC
M 28 157 REPLACEMENT OF CLAY, KAOLINITE
M 28 156 REPLACEMENT OF DOLOMITE, CHERT
M 28 155 REPLACEMENT OF LIMESTONE, FELDSPAR, AUTHIGENIC
M 28 157 REPLACEMENT OF QUARTZ, COFFINITE
M 28 157 REPLACEMENT OF QUARTZ, URANINITE
M 28 151 REPLACEMENT OR DISPLACEMENT, GENERAL, FABRICS
S 3 90 REPLACEMENT POWER VERSUS TEMPERATURE
M 28 152 REPLACEMENT, PYRITE, AUTHIGENIC
M 28 153 REPLACEMENT, QUARTZ, AUTHIGENIC
M 28 154 REPLACEMENT, QUARTZ, AUTHIGENIC, IN LIMESTONE
M 25 432 REPLACEMENT TYPE, THAILAND, FLUORITE OF
M 28 153 REPLACEMENTS SEM, PYRITE, FRAMBOIDAL
B 6207 1115 REPLOTTED LOGS
S 6 323† REPORT ON ISOTOPIC DATING OF ROCKS IN THE CARBONIFEROUS
B 6002 322* REPORT--ANADARKO BASIN. BY TITLE ONLY
B 6002 320* REPORT--CENTRALNEBRASKA, ABST.
B 6002 318 REPORT--EASTERN KANSAS. BY TITLE ONLY
B 6002 322* REPORT--OKLAHOMA AND TEXAS PANHANDLES, ABST.
B 6002 322* REPORT--WESTERN KANSAS. BY TITLE ONLY
B 6210 1988 REPUBLIC OF CHINA, 1977, YANG, Y. F., ON DEVELOPMENTS IN, TAIWAN
M 25 94† REPUBLIC OF KOREA, PALEOZOIC AND, MESOZOIC, COAL IN
B 6311 1996† RESEARCH CONFERENCE GUIDELINES, AAPG
B 6207 1236* RESEARCH PROGRAM, ABST., CANADIAN, GEOTHERMAL
B 6405 694* RESEARCH SYMPOSIUM—SOURCE BEDS—DEPOSITIONAL ENVIRONMENTS
B 6007 1078 RESEARCHER AND DISCOVERER, CRUISES, 1971-1972, NOAA SHIPS
B 6004 688* RESEDIMENTATION OF BIOGENIC CARBONATE SEDIMENTS IN DSDP
B 6206 933 RESEDIMENTED COARSE-CLASTIC FAMILY
B 6303 465* RESEDIMENTED CONGLOMERATES IN SUBMARINE CHANNEL, MIOCENE
B 6407 1034† RESEDIMENTED FACIES IN VENTURA BASIN, CALIFORNIA, AND MODEL
B 6310 1889 **RESERVE OIL, INC., AND PETROCONSULTANTS S.A., ON DEVELOPMENTS IN**
B 6105 819* RESERVE, ABST., EVALUATING, COAL PROPERTY AND MAXIMIZING
B 6108 1128 RESERVE ESTIMATES OF NEW-FIELD DISCOVERIES
B 6308 1175 RESERVE ESTIMATES OF NEW-FIELD DISCOVERIES
B 6208 1282 RESERVE ESTIMATES OF NEW-FIELD DISCOVERIES
B 6409 1303 RESERVE ESTIMATES OF NEW-FIELD DISCOVERIES
B 6008 1160 RESERVE ESTIMATES OF NEW-FIELD DISCOVERIES
B 6105 820* RESERVE ESTIMATES—FUEL AND ENERGY MINERALS—INTRODUCTION,
B 6007 1024 RESERVE FAN AND DELTA, LAKE SUPERIOR
B 6206 918 RESERVE FAN, LAKE, SUPERIOR
B 6007 1023 RESERVE FAN, NEW TERM, LAKE SUPERIOR
B 6105 787* **RESERVE INFORMATION—INVALUABLE TOOL OF PORTFOLIO MANAGER**
B 6007 1021 RESERVE MINING CO., TACONITE ORE TAILINGS, DEPOSITION,
B 6010 1871 RESERVE OIL AND GAS CO., AND PETROCONSULTANTS S.A., ON
B 6010 1871 RESERVE OIL AND GAS CO., ON DEVELOPMENTS IN FUJAIRAH, 1975
B 6110 1800 RESERVE OIL AND GAS CO., ON DEVELOPMENTS IN FUJAIRAH, 1976
B 6210 2117 RESERVE OIL AND GAS CO., ON DEVELOPMENTS IN FUJAIRAH, 1977
B 6110 1800 RESERVE OIL AND GAS CO., PETROCONSULTANTS ET AL, ON
B 6310 1888 RESERVE OIL, INC., ON DEVELOPMENTS IN FUJAIRAH, 1978
B 6408 1284* RESERVES, ABST., COAL IN, PENNSYLVANIA—GEOLOGY, CURRENT
B 6203 508* RESERVES, ABST., COMPUTER ANALYSIS OF REGIONAL, WELL-DENSITY
B 6105 819* RESERVES, ABST., FINANCING
B 6408 1134 RESERVES, ALASKA, COOK INLET BAY
B 6407 983 RESERVES, ALBERTA, SUFFIELD BLOCK GAS
M 25 403 RESERVES AND ECONOMICS, AUSTRALIA, NICKEL
B 6008 1405* RESERVES AND IN-SITU MARKET VALUES, ABST., ECONOMIC VALUES
B 6304 623 RESERVES, AND POTENTIAL, REGIONAL, URANIUM PRODUCTION
B 6303 298 RESERVES AND RESOURCES, OCEAN
B 6008 1395* RESERVES AT BOWDOIN AREA, MONTANA--LOW-PERMEABILITY,
B 6004 698* RESERVES BY COMPUTER METHODS, ABST., CALCULATION OF, COAL
M 25 105 RESERVES, CANADA, COAL
B 6101 25 RESERVES, CANADIAN, ARCTIC ISLANDS
B 6409 1530 RESERVES, DEFINITION, PROVED
B 6105 823* RESERVES ESTIMATES, ABST., DATA NEEDS FOR, HYDROCARBON
B 6303 497* RESERVES FORECASTING METHOD AND ASSESSMENT OF REGIONAL
B 6111 1957 RESERVES, FORELAND, BASINS
B 6002 319* RESERVES, GAS RESOURCE BASE, AND PROSPECT FOR FUTURE GAS
S 3 3 RESERVES, HISTORICAL ESTIMATES, URANIUM ORE

M 6004 726* RESERVES IN GREGOIRE LAKE AREA, ATHABASCA TAR DEPOSITS,
B 6303 531* RESERVES IN MATURE BASINS. BY TITLE ONLY, PETROLEUM
M 25 86 RESERVES, INDONESIA, COAL
M 25 99 RESERVES, JAPAN, COAL
M 25 96 RESERVES, KOREA, SOUTHERN, COAL
B 6111 1957 RESERVES, POSSIBLE SETTINGS FOR, SUPERGIANT
B 6203 534* RESERVES PREDICTION, ABST., STATISTICAL APPROACH FOR FUTURE
B 6109 1481 RESERVES, RECOVERABLE, BRAZIL
B 6111 1957 RESERVES, RIFT, BASINS
M 25 505 RESERVES, TABLE, CANADIAN, CORDILLERA, METAL PRODUCTION AND
M 25 522 RESERVES, TABLE, MEXICO, URANIUM ORE
M 25 92 RESERVES, TAIWAN, COAL
B 6004 659* RESERVES TO PRICE. BY TITLE ONLY, RESPONSE OF
B 6205 741 RESERVES TO PRODUCTION RATIOS, RESERVE LIFE, OF MAJOR
B 6305 847* RESERVES--HARD TO FIND OR EASY TO MISS, ABST., PROFITABLE
M 30 414 RESERVOIR 19-7, INDONESIA, HANDIL FIELD
B 6405 780* RESERVOIR, ABST., GEOLOGIC-SEISMIC EXPLORATION MODEL FOR
B 6108 1385* RESERVOIR ANALYSIS OF EAST MESA GEOTHERMAL FIELD, IMPERIAL
B 6207 1207* RESERVOIR ANALYSIS OF MOMOTOMBO GEOTHERMAL PROJECT,
B 6211 2362* RESERVOIR AND FLUID-FLOW RATE VARIATIONS IN NORTHERN EAST
B 6303 550* RESERVOIR AND NONRESERVOIR SANDSTONES, ABST., COMPARISON OF
B 6105 844* RESERVOIR AND SOURCE-BED HISTORY IN GREAT VALLEY, CALIFORNIA,
B 6012 2192* RESERVOIR AND SOURCE-BED HISTORY IN GREAT VALLEY, CALIFORNIA,
B 6205 813† RESERVOIR AND SOURCE-BED HISTORY OF GREAT VALLEY, CALIFORNIA
B 6002 322* RESERVOIR AT SLEEPY HOLLOW FIELD, RED WILLOW COUNTY,
B 6206 974 RESERVOIR BEDS AND TRAPS, CALIFORNIA, MONTEREY FAN
B 6002 179 RESERVOIR BEDS, BERING SEA SHELF
M 30 189 RESERVOIR CHARACTER, NORTH SEA, NINIAN FIELD
B 6303 491* RESERVOIR CHARACTERISTICS OF SHELF SANDSTONE, HARTZOG DRAW
B 6408 1146 RESERVOIR DATA, ALASKA, KODIAK SHELF, SOURCE AND
B 6305 730 RESERVOIR DENSITY OF GAS CONDENSATE
B 6203 517* RESERVOIR DEPOSITS, ABST., DEVONIAN SLOPE SEDIMENTS IN CENTRAL,
B 6303 501* RESERVOIR DESCRIPTION AND FORMATION EVALUATION, GULF COAST.
B 6105 800* RESERVOIR DESCRIPTION AND POOL PERFORMANCE OF JUDY CREEK AND
B 6303 484* RESERVOIR FIELD--GIANT IN WYOMING THRUST BELT, ABST., PAINTER
B 6203 494* RESERVOIR FLUID CONTENT BY SEISMIC MODELING, ABST.,
B 6303 439* RESERVOIR FROM SEISMIC AND BOREHOLE DATA, ABST., KRIGING OF
B 6105 831* RESERVOIR GEOLOGIC METHODS TO IMPROVE ULTIMATE RECOVERY,
B 6007 1009† RESERVOIR GEOMETRY OF A FLUVIAL SHEET SANDSTONE
M 26 453 RESERVOIR GEOMETRY, SEISMIC, MODELING
M 30 395 RESERVOIR HORIZONS, PLIOCENE, TEAK FIELD
B 6306 881 RESERVOIR IN RUDIST REEF BUILDUP ON ATLANTIC-TYPE PASSIVE
B 6303 528* RESERVOIR IN WOODBINE--EAGLE FORD INTERVAL, UPPER CRETACEOUS,
B 6408 1276* RESERVOIR MANAGEMENT, ABST., IMPACT OF, DIAGENESIS ON
B 6309 1588* RESERVOIR MODELING, ABST., COTTAGEVILLE GAS FIELD CORRELATION
B 6303 418* RESERVOIR MORPHOLOGIES FOR GULF COAST TURBIDITE SANDSTONES,
B 6306 899 RESERVOIR, NEW MEXICO, LONE PINE FIELD
B 6304 599 RESERVOIR, NEW MEXICO, VERDE OIL FIELD, FRACTURE
B 6204 609† RESERVOIR, OREGON BASIN FIELD, WYOMING, TENSLEEP
B 6309 1578* RESERVOIR PARAMETERS, ABST., CHARACTERIZATION OF PRODUCTION
B 6309 1581* RESERVOIR PERFORMANCE, ABST., MODELING OF, DEVONIAN, SHALE,
M 30 162 RESERVOIR PERFORMANCE, MORAY FIRTH BASIN, PIPER FIELD
B 6406 854 RESERVOIR POTENTIAL, FLAT, BASIN FLOORS
S 7 184 RESERVOIR POTENTIAL, MISSISSIPPI FAN
B 6312 2193† RESERVOIR POTENTIAL OF BARENTS SHELF, SANDSTONE IN LOWER,
B 6209 1755* RESERVOIR POTENTIAL OF RECENT CLASTIC DEPOSITIONAL SYSTEMS OF
B 6203 508* RESERVOIR POTENTIAL OF VOLCANOGENIC SANDSTONES, ABST.
B 6203 417 RESERVOIR POTENTIAL, VIRGIN ISLANDS, KINGSHILL LIMESTONE
B 6305 799† RESERVOIR PROPERTIES OF CONGLOMERATES AND CONGLOMERATIC
B 6108 1378* RESERVOIR QUALITY IN ARC-DERIVED SANDSTONES-IMPLICATIONS FOR
B 6305 827* RESERVOIR QUALITY IN EAGLE SANDSTONE OF BEARPAW MOUNTAINS,
B 6303 488* RESERVOIR QUALITY IN TERTIARY SANDSTONES ALONG TEXAS GULF
B 6409 1555* RESERVOIR QUALITY, PLIOCENE-PLEISTOCENE SANDSTONES, OFFSHORE,
B 6109 1535* RESERVOIR QUALITY, TERTIARY-PLEISTOCENE, GULF COAST REGION,
B 6405 803* RESERVOIR ROCK, SOURCE ROCK, AND TRAPPING MECHANISM OF
B 6205 891* RESERVOIR ROCKS, SOUTH-CENTRAL UINTA BASIN, UTAH, ABST.,
B 6002 322* RESERVOIR SANDS, KINCAID OIL FIELD, SOUTHEAST KANSAS, ABST.,
B 6303 293 RESERVOIR SANDS, ORGANIC CARBON CONTENT, AND OIL AND GAS
B 6009 1621* RESERVOIR SANDSTONE, ABST., WILDCAT VALLEY SANDSTONE IN
B 6103 344† RESERVOIR SANDSTONE IN WEST-CENTRAL TEXAS, DEPOSITIONAL
B 6405 737* RESERVOIR SANDSTONES B DISSOLUTION OF SILICATE MINERAL
B 6303 512* RESERVOIR SANDSTONES IN WEST-CENTRAL PAKISTAN, ABST., CYCLIC
B 6009 1607* RESERVOIR SANDSTONES, NINE MILE POINT FIELD, ARANSAS COUNTY,
B 6107 1029† RESERVOIR, SLEEPY HOLLOW FIELD, RED WILLOW COUNTY, NEBRASKA,
M 30 127 RESERVOIR STRATIGRAPHY, BEATRICE FIELD
B 6303 513* RESERVOIR STUDIES AND INVERSION OF SEISMIC TRACES, ABST.
B 6407 1004† RESERVOIR TRENDS, EAST TEXAS, NORTH LOUISIANA, AND SOUTH
B 6309 1602* RESERVOIR--INTERPRETATION CASE STUDY OF WATER SATURATIONS
B 6309 1582* RESERVOIR-FORMING PROCESS, ABST., METEORITIC IMPACT--
M 26 427 RESERVOIR-SEAL CONTACT, ABRUPT VERTICAL CONTACTS
M 26 427 RESERVOIR-SEAL CONTACT, ACOUSTIC IMPEDANCE CONTRAST
M 26 427 RESERVOIR-SEAL CONTACT, LATERAL CONTACTS
B 6004 674* RESERVOIR-SEAL RELATIONS IN PRODUCTIVE EVAPORITE-ENCASED REEF
B 6306 870† RESERVOIRED GIANT OIL FIELDS, CRETACEOUS, OCEANIC, ANOXIC
B 6303 460* RESERVOIRS, ABST., AUTHIGENIC, ILLITES IN, SANDSTONE
B 6305 838* RESERVOIRS, ABST., BOWDOIN DOME AREA, NORTH-CENTRAL,
B 6105 815* RESERVOIRS, ABST., DIAGENETIC CONSTRAINTS ON USE OF PREDICTIVE
B 6303 409* RESERVOIRS, ABST., DRILLING AND PRODUCTION PROBLEMS IN
B 6408 1276* RESERVOIRS, ABST., ENERGY RESOURCES OF, WATER-BEARING
B 6203 560* RESERVOIRS, ABST., EXPLORATION MODELS AND CRITERIA FOR
B 6309 1585* RESERVOIRS, ABST., FABRIC-ELEMENT-BASED CLASSIFICATION FOR
B 6303 448* RESERVOIRS, ABST., GENERATIO AND OCCLUSION OF POROSITY IN,
B 6108 1387* RESERVOIRS, ABST., POROSITY, DIAGENESIS, AND, PRODUCTIVE
B 6303 522* RESERVOIRS, ABST., POROSITY RELATIONS IN, CHALK
B 6102 295* RESERVOIRS, ABST., PREDICTING, POROSITY RELATIONS IN,
B 6303 414* RESERVOIRS, ABST., ROLE OF, TEMPERATURE AND, BURIAL DEPTH I
B 6004 674* RESERVOIRS, ABST., SOURCE-, RESERVOIR-, SEAL RELATIONS IN
B 6405 699* RESERVOIRS ALONG ONSHORE TEXAS GULF COAST, ABST., EVALUATION
B 6309 1606* RESERVOIRS ALONG TEXAS GULF COAST, ABST., IMPORTANCE OF
B 6309 1578* RESERVOIRS AND BLACK SHALE FORMATIONS, ABST., NOVEL
B 6209 1769* RESERVOIRS AT BASSFIELD, JEFFERSON DAVIS COUNTY, MISSISSIPPI,
B 6201 70 RESERVOIRS, AUSTRALIA, EXMOUTH PLATEAU
B 6303 474* RESERVOIRS BY DICKITE, ABST., OCCLUSION OF POROSITY IN,
B 6004 658* RESERVOIRS BY ISOTOPIC ANALYSIS OF METHANE, ABST.,
B 6104 484 RESERVOIRS, CARBONATE
B 6203 463† RESERVOIRS, CHADRA SANDS, GIALO FIELD, LIBYA, SECONDARY,
B 6104 484 RESERVOIRS, CLASTIC
B 6303 480* RESERVOIRS, EXAMPLE FROM ANADARKO BASIN, ABST., SIMPLE,
B 6305 822* RESERVOIRS IN BIGHORN BASIN, WYOMING, ABST., BOREHOLE GRAVITY
B 6308 1425* RESERVOIRS IN CANYO SANDSTONE, ROUNDTOP AREA, FISHER COUNTY,
B 6002 324* RESERVOIRS IN CANYON SANDSTONE, VAL VERDE BASIN, TEXAS, ABST.,
B 6104 511† RESERVOIRS IN DOGGER, MIDDLE JURASSIC, OF FRANCE, PATTERNS OF,
B 6304 676† RESERVOIRS IN DOGGER, MIDDLE JURASSIC, OF FRANCE.
B 6207 1230* RESERVOIRS IN INDONESIA, ABST., TERTIARY, CARBONATE
B 6405 770* RESERVOIRS IN MONTEREY SHALE, SANTA MARIA AREA, CALIFORNIA,
B 6309 1598* RESERVOIRS IN VALENTINE AND MENKING FIELDS, LAVACA COUNTY,
B 6205 845 RESERVOIRS, ISOPACH, MAPPING OF HYDROCARBON
B 6309 1596* RESERVOIRS, MCALLEN RANCH FIELD, HIDALGO COUNTY, TEXAS, ABST.,
B 6304 603 RESERVOIRS, NEW MEXICO, EAST AND, WEST, PUERTO CHIQUITO FIELDS,
B 6109 1534* RESERVOIRS, NORTHEAST THOMPSONVILLE FIELD, JIM HOGG AND WEBB
B 6405 712* RESERVOIRS OF EAGLE SANDSTONE, MONTANA, ABST., DIAGENESIS IN
B 6205 891* RESERVOIRS OF NORTHERN GREAT PLAINS, ABST., NATURAL, GAS
B 6405 711* RESERVOIRS OF NORTHERN GREAT PLAINS, ABST., PHYSICAL
B 6407 969† RESERVOIRS OF NORTHERN GREAT PLAINS—ASSESSMENT OF THEIR
B 6002 325* RESERVOIRS OF PERMIAN BASIN, ABST., FACTORS CONTROLLING,
M 24 23† RESERVOIRS OF PRUDHOE BAY FIELD, NORTH SLOPE, ALASKA, PERMO-,
B 6205 886* RESERVOIRS OF SOUTHERN SASKATCHEWAN, ABST., INFLUENCE OF
B 6104 586 RESERVOIRS, OIL AGE, LOUISIANA SHELF, TERTIARY
B 6205 813 RESERVOIRS, POROSITY VERSUS, DEPTH IN, SANDSTONE
B 6305 840* RESERVOIRS, ROCKY MOUNTAIN BASINS, ABST., SEISMIC MODEL OF
B 6405 794* RESERVOIRS, ROCKY MOUNTAINS, ABST., EVOLUTION AND STRESS
B 6303 551* RESERVOIRS, UNITED STATES GULF COAST, ABST., PORE SYSTEMS IN,
B 6309 1595* RESERVOIRS, VALENTINE FIELD, LAVACA COUNTY, TEXAS, ABST.,
B 6012 2150† RESERVOIRS, VERTICAL, NET, SANDSTONE DETERMINATION FOR,
M 24 177† RESERVOIRS, WALKER CREEK FIELD AREA, LAFAYETTE AND COLUMBIA
B 6112 2134† RESERVOIRS WITH SUBNORMAL PRESSURES, OIL AND, GAS IN
M 25 276† RESERVOIRS--SUMMARY, PETROLEUM FIELDS OF, JAPAN
* RESERVOIRS, ALBIAN-CENOMANIAN, CAMPOS BASIN, OFFSHORE RIO DE
B 6203 495* RESERVOIRS—CANYON GROUP OF VAL VERDE BASIN, WILCOX GROUP OF
B 6405 773* RESERVOIRS—EXAMPLES FROM SOUTHEASTERN ALBERTA AND
B 6105 825* RESERVOIRS—EXPLORATION MODELS AND RECOGNITION CRITERIA,
B 6405 760* RESERVOIRS—GEOLOGIC MODELING OF GEOMETRY OF RESERVOIR
B 6109 1541* RESERVOIRS—POROSITY FORMATION AND PRESERVATION, ABST.,
B 6408 1280* RESERVOIRS—SUBMARINE FANS AND FANTASIES, ABST., DEEP-WATER
M 29 253 RESIDUAL BASIN, EVOLUTION, STRUCTURAL HIGH
B 6102 297* RESIDUAL BOUGUER GRAVITY ANOMALY ANALYSIS OF CHIHUAHUA
S 10 101 RESIDUAL OIL SATURATION, MIGRATION, SECONDARY
M 29 259 RESIDUAL, EVOLUTION DIAGRAM, FOREARC BASIN
B 6405 743* RESIN RODS AND WOODY ROD-LIKE STRUCTURES IN PENNSYLVANIAN
B 6004 651* RESISTIVITY TO PSEUDOVELOCITY LOGS, ABST., COMPUTER-GENERATED,
B 6005 879† RESISTIVITY TO PSEUDOVELOCITY LOGS, TRANSFORMATION OF
B 6303 498* RESOLUTION, ABST., IN SEARCH OF
M 26 11 RESOLUTION, FRESNEL ZONE, SEISMIC
M 26 419 RESOLUTION, LATERAL, SEISMIC
M 26 3† RESOLUTION OF SEISMIC REFLECTIONS AND GEOLOGIC DETAIL
M 26 352† RESOLUTION SEISMIC DATA, ANALYSIS OF, HIGH
M 26 354 RESOLUTION SYSTEMS, ACOUSTIC SYSTEMS, CLASSIFICATION CHART
M 26 374 RESOLUTION SYSTEMS, BATHYMETRIC MAPPING
M 26 364 RESOLUTION SYSTEMS, DIGITAL PROCESSING
M 26 357 RESOLUTION SYSTEMS ELECTROMECHANICAL SUBBOTTOM PROFILING

Index

M 26 374 RESOLUTION SYSTEMS, ENGINEERING ANALYSES
M 26 369 RESOLUTION SYSTEMS, EXPLORATION ANALYSES
M 26 379 RESOLUTION SYSTEMS, GLOSSARY
M 26 371 RESOLUTION SYSTEMS ISOPACH MAPPING, SHALLOW
M 26 371 RESOLUTION SYSTEMS, LOCATING SEEPS AND SHALLOW
M 26 374 RESOLUTION SYSTEMS, POTENTIAL DRILLING HAZARDS
M 26 369 RESOLUTION SYSTEMS, SEAFLOOR MAPPING
M 26 356 RESOLUTION SYSTEMS SIDE-SCAN SONAR SYSTEMS
M 26 358 RESOLUTION SYSTEMS, SPARKER SYSTEMS
M 26 371 RESOLUTION SYSTEMS, STRUCTURAL MAPPING, SHALLOW
M 26 371 RESOLUTION SYSTEMS, STRUCTURE FORM CONTOUR MAPPING
M 26 371 RESOLUTION SYSTEMS, SUBCROP MAPPING
M 26 357 RESOLUTION SYSTEMS TUNED TRANSDUCER SYSTEMS
M 26 355 RESOLUTION SYSTEMS WATER-DEPTH SYSTEMS
M 26 418 RESOLUTION THICKNESS, SEISMIC RESOLUTION, CRITICAL
M 26 418 RESOLUTION, THIN OR THICK, SEISMIC
M 26 4 RESOLUTION, VERTICAL, SEISMIC, REFLECTIONS
B 6303 497* RESOURCE APPRAISAL METHODS IN U.S. GEOLOGICAL SURVEY AND
B 6303 523* RESOURCE APPRAISAL PREDICTIONS AND EXPLORATION PERFORMANCE
B 6303 511* RESOURCE APPRAISAL PREDICTIONS AND EXPLORATION
B 6303 540* RESOURCE APPRAISAL--STATE OF THE ART, ABST., OIL AND, GAS
B 6303 513* RESOURCE ASSESSMENT, ABST., APPLICATION OF, OIL AND, GAS
B 6303 517* RESOURCE ASSESSMENT TO IMPROVE EXPLORATION PERFORMANCE,
B 6303 459* RESOURCE ASSESSMENTS AND RELATION TO UNITED STATES OIL
B 6409 1531 RESOURCE ESTIMATES, EVALUATION OF
B 6405 774* **RESOURCE EVALUATION OF GAS-BEARING COAL BEDS,**
B 6408 1279* RESOURCE POTENTIAL AND PLATE-MARGIN GEOLOGY OF FRONTIER
B 6303 418* RESOURCE POTENTIAL OF ANTARCTICA, ABST., SPECULATIONS ON
B 6004 691* RESOURCE-APPRAISAL ESTIMATES RESULTING FROM AREAL
B 6305 818 RESOURCES, BY BERNARDO F. GROSSLING, MEYERHOFF, A. A., REVIEW
B 6004 698* RESOURCES IN UNITED STATES, ABST., PROBABILISTIC AND, COMPUTER
B 6312 2183† RESOURCES, METHODS OF, ESTIMATING, OIL AND, GAS
B 6303 298 RESOURCES, OCEAN, RESERVES AND
B 6305 842* RESOURCES OF DENVER-JULESBURG BASIN, ABST., NEW LOOK AT, OIL
B 6004 679* RESPONSE OF ACTIVE OOLITE SAND SHOAL TO ITS PHYSICAL
B 6004 659* RESPONSE OF RESERVES TO PRICE. BY TITLE ONLY
M 26 490 RESPONSE, SHALE DISTRIBUTION, SEISMIC
B 6303 466* RESPONSIBILITIES OF TEXAS RAILROAD COMMISSION FOR URANIUM
S 10 76 RESTORATION OF SEDIMENTARY COLUMN, UNCOMPACTION
S 6 127† RESULTS OF DATING CRETACEOUS, PALEOGENE SEDIMENTS, EUROPE
B 6403 443* RESULTS OF EXPLORATORY DRILLING, NORTHERN FALLON BASIN,
B 6105 772* RESULTS OF PILOT STUDIES OF PROJECT NURE FROM NEW MEXICO:
B 6201 108 RETORT MEMBER, GAS CHROMATOGRAPHIC ANALYSES
B 6201 99 RETORT PHOSPHATIC SHALE MEMBER, IDAHO
B 6201 99 RETORT PHOSPHATIC SHALE MEMBER, MONTANA
B 6201 107 RETORT PHOSPHATIC SHALE MEMBER, WESTERN INTERIOR, UNITED
B 6201 99 RETORT PHOSPHATIC SHALE MEMBER, WYOMING
B 6004 653* RETRIEVAL FROM MEGA-BYTE DATA BASES, ABST., SET-THEORETIC,
B 6405 674* RETROGRADATIONAL SHELF SEDIMENTATION, VIKING SANDSTONE,
B 6303 422* RETROGRADATIONAL SHELF SEQUENCE--UPPER CRETACEOUS,
B 6212 2489† REVERSAL OF BASEMENT-BLOCK MOTIONS IN CAMBAY BASIN, INDIA,
B 6203 518* REVERSALS, ABST., GEOPHYSICAL AND GEOLOGICAL SIGNIFICANCE OF,
S 6 222 REVERSALS, ATLANTIC, PALEOGENE, PALEOMAGNETIC
S 6 222 REVERSALS, TABLE, PALEOGENE, PALEOMAGNETIC
B 6105 813* REVIEW AND APPLICATION OF PETROLEUM FINDING-RATE
B 6108 1383* REVIEW AND APPLICATION OF PETROLEUM-FINDING RATE
B 6105 813* REVIEW AND APPLICATION OF PETROLEUM FINDING-RATE METHODOLOGIES
S 6 1† REVIEW, DATING AND CORRELATION, A
S 6 9 REVIEW, GEOCHRONOLOGIC SCALES
B 6407 1122 REVIEW OF 1. SEISMIC, SAFETY AND LAND-USE PLANNING. 2. FLATLAND
B 6211 2351* REVIEW OF 30 SEISMIC CASE HISTORIES. BY TITLE ONLY
B 6310 1981 REVIEW OF A COLOR ILLUSTRATED GUIDE TO CONSTITUENTS,
B 6410 1723 REVIEW OF A FIRST COURSE IN GEOPHYSICAL EXPLORATION AND
B 6310 1981 REVIEW OF A GEOLOGIST IN THE SERVICE OF THE RAJ, BY C. T. BARBER,
B 6309 1545 REVIEW OF ANCIENT SEDIMENTARY ENVIRONMENTS, B R. C. SELLEY,
B 6401 133 REVIEW OF ARCYANA FAMOUS--PHOTOGRAPHIC ATLAS OF THE
B 6308 1421 REVIEW OF ATLAS OF ECONOMIC MINERAL DEPOSITS, BY COLIN J.
B 6412 2190 REVIEW OF BEITRAGE ZUR BIOSTRATIGRAPHIE DER TETHYS-TRIAS,
B 6309 1547 REVIEW OF BIBLIOGRAPHY OF THE GEOLOGY OF LIBYA BY J. AZZOUZ
B 6310 1983 REVIEW OF COAL RESOURCES OF THE AMERICAS--SELECTED PAPERS,
B 6306 989 REVIEW OF COASTAL SOUTHERN CALIFORNIA, BY ROBERT SHARP, BEAR,
B 6304 682 REVIEW OF COMPACTION AND FLUID MIGRATION, PRACTICAL
B 6408 1271 REVIEW OF CONTINENTAL MARGINS—GEOLOGICAL AND GEOPHYSICAL
B 6411 2038 REVIEW OF CONTINENTAL TECTONICS, STUDIE IN GEOPHYSICS
B 6407 1119 REVIEW OF CRITERIA IN CORRELATION—RELEVANT PRINCIPLES OF
B 6405 649 REVIEW OF CURRENTS IN SUBMARINE CANYONS AND OTHER
B 6303 385 REVIEW OF DATA ON THE GEOCHEMISTRY OF OIL FIELD WATERS, BY
B 6412 2189 REVIEW OF DEPOSITIONAL AND STRUCTURAL HISTORY OF THE
B 6410 1726 REVIEW OF DIAGENESIS IN SEDIMENTS AND SEDIMENTARY ROCKS,
B 6409 1550 REVIEW OF DIAGRAPHIES DIFFEREES—BASES DE L'INTERPRETATION,
B 6305 820 REVIEW OF ENCYCLOPEDIA OF SEDIMENTOLOGY, EDITED BY RHODES W.
B 6403 434 REVIEW OF ENERGY AND CLIMATE, PAMPE, WILLIAM R.
B 6012 2182* REVIEW OF ESTIMTES OF RADIOMETRIC AGES FOR RELIZIAN STAGE OF
B 6408 1274 REVIEW OF FACIES ANATOMY AND DIAGENESIS O A BAHAMIAN OOID
B 6404 585 REVIEW OF FACIES MODELS, EDITED B ROGER G. WALKER, VAN HOUTEN,
B 6307 1144 REVIEW OF FLUVIAL SEDIMENTOLOGY, EDITED BY ANDREW D. MIALL, LE
B 6307 1143 REVIEW OF GEOCHEMISTRY, BY A. H. BROWNLOW, HUNT, JOHN M.
B 6411 2037 REVIEW OF GEOCHEMISTRY IN MINERAL EXPLORATION BY ARTHUR W.
B 6411 2039 REVIEW OF GEOCHEMISTRY OF ORGANIC MOLECULES, EDITED BY KEITH
B 6412 2189 REVIEW OF GEOLOGICAL FIELD TRIPS COLOMBIA 1959-1978, BY
B 6403 434 REVIEW OF GEOLOGY AND HYDROGEOLOGY OF NORTHEASTERN
B 6403 434 REVIEW OF GEOLOGY AND HYDROGEOLOGY OF NORTHEASTERN
B 6303 380 REVIEW OF GEOLOGY IN ENVIRONMENTAL PLANNING. BY ARTHUR D.
B 6406 957 REVIEW OF GEOLOGY OF CONTINENTAL SLOPES, EDITED BY LARRY J.
B 6303 384 REVIEW OF GEOLOGY OF MANITOULIN AREA, INCLUDING THE ROAD
B 6310 1984 REVIEW OF GEOREF THESAURUS AND GUIDE T INDEXING, BY CAROL
B 6407 1118 REVIEW OF GUADALUPIAN DELAWARE MOUNTAI GROUP OF WEST
B 6405 650 REVIEW OF GUIDEBOOK OF SANTA FE COUNTRY EDITED BY RAYMOND
B 6406 953 REVIEW OF HISTORICAL BIOGEOGRAPHY, PLATE TECTONICS, AND THE
B 6205 885* REVIEW OF HYDROCARBON POTENTIAL OF PALEOZOIC ROCKS IN
B 6309 1546 REVIEW OF ICE AGES, SOLVING THE MYSTERY BY JOHN IMBRIE AND
B 6303 380 REVIEW OF INDIAN OCEAN GEOLOGY AND BIOSTRATIGRAPHY. STUDIES
B 6410 1725 REVIEW OF INTRODUCTION TO EXPLORATION GEOCHEMISTRY, BY A. A.
B 6306 986 REVIEW OF LAND OF COCHISE, SOUTHEASTERN ARIZONA, 1978, NEW
B 6406 956 REVIEW OF LECTURES IN ISOTOPE GEOLOGY EDITED BY E. JAGER AND J.
B 6405 649 REVIEW OF LITHOFACIES-PALEOGEOGRAPHICAL ATLAS OF THE PERMIAN
B 6406 954 REVIEW OF LITHOSTRATIGRAPHIC ANALYSIS OF SEDIMENTARY BASINS,
B 6412 2191 REVIEW OF MANUAL FOR TEPHROCHRONOLOGY, BY VIRGINIA
B 6306 987 REVIEW OF MARINE EVAPORITES, EDITED BY WALTER E. DEAN AND B.
B 6407 1120 REVIEW OF MARINE GRAVITY, BY PETER DEHLINGER, BOWIN, CARL
B 6309 1545 REVIEW OF MODERN PETROLEUM, A BASIC PRIME OF THE INDUSTRY, BY
B 6307 1143 REVIEW OF OFF-SHORE PETROLEUM RESOURCES OF SOUTH-EAST ASIA,
B 6405 648 REVIEW OF OIL AND GAS FIELDS OF THE FOUR CORNERS AREA, EDITED
B 6305 820 REVIEW OF OIL SAND AND OIL SHALE CHEMISTRY, EDITED BY O. P.
B 6412 2191 REVIEW OF OLD ELECTRICAL LOG INTERPRETATION, BY DOUGLAS W.
B 6401 133 **REVIEW OF PALEOGEOGRAPHIC PRINCIPLES OF OIL AND GAS**
B 6408 1272 REVIEW OF PENNSYLVANIAN SANDSTONES OF THE MID-CONTINENT,
B 6411 2040 **REVIEW OF PETROLEUM DEVELOPMENT GEOLOGY, BY PARKE A. DICKEY,**
B 6309 1548 REVIEW OF PETROLEUM FORMATION AN OCCURRENCE--A NEW
B 6406 954 REVIEW OF PETROLEUM GEOCHEMISTRY AND GEOLOGY, BY JOHN M.
B 6309 1467† REVIEW OF PETROLEUM GEOLOGY OF ANADYR AND KHATYRKA BASINS,
B 6306 987 REVIEW OF PRINCIPLES OF SEDIMENTOLOGY, BY GERALD M. FRIEDMAN
B 6411 2041 REVIEW OF PROCEEDINGS OF THE SYMPOSIUM ON THE ORIGIN OF THE
B 6105 801* REVIEW OF RECENT CARBONATE LAGOONAL SEDIMENTATION
B 6409 1548 REVIEW OF ROCKS AND ROCK MINERALS, BY RICHARD V. DIETRICH AND
B 6408 1272 REVIEW OF SALT DOMES OF GULF REGION, UNITED STATES AND
B 6407 1117 REVIEW OF SEDIMENTARY ENVIRONMENTS AND FACIES, EDITED BY H. G.
B 6410 1723 REVIEW OF STATISTICAL ANALYSIS IN GEOLOGY, BY JOHN M. CUBITT
B 6307 1144 REVIEW OF STRATIGRAPHIC INDEX OF BRITISH OSTRACODA, EDITED BY
B 6411 2036 REVIEW OF TECTONIC MAP OF SOUTH AMERICA BY COMMISSION FOR
B 6304 682 REVIEW OF TECTONICS OF ASIA, BY EMILE ARGAND, TRANSLATED AND
B 6409 1548 REVIEW OF TECTONOMECHANICS—AN INTRODUCTION TO STRUCTURAL
B 6409 1549 REVIEW OF THE ANDES, A GEOLOGICAL REVIEW, BY WERNER ZEIL, VAN
B 6407 1117 REVIEW OF THE ENCYCLOPEDIA O PALEONTOLOGY, EDITED BY RHODES
B 6310 1982 REVIEW OF THE IMPACT OF THE GEOSCIENCES ON CRITICAL ENERGY
B 6407 1121 REVIEW OF THE PETROLEUM DICTIONARY, BY DAVI F. TVER AND
B 6410 1722 REVIEW OF THE SOVIET ENERGY SYSTEM, BY LESLIE DIENES AND
B 6302 258 REVIEW OF THE STRUCTURE OF WESTERN EUROPE, BY J. G. C.
B 6305 817 REVIEW OF TRACE FOSSIL CONCEPTS, EDITED BY PAUL B. BASAN,
B 6305 816 REVIEW OF TRACE FOSSILS 2, EDITED BY T. P CRIMES AND J. C. HARPER,
B 6408 1273 REVIEW OF TURKIYE DORDUNCU PETROL KONGRESI, FOURTH TURKISH
B 6306 988 REVIEW OF URANIUM DEPOSITS, THEIR MINERALOGY AND ORIGIN,
B 6404 585 REVIEW OF VOLCANOLOGY, BY HOWEL WILLIAMS AND ALEXANDER R.
B 6305 818 REVIEW OF WINDOW ON OIL, A SURVEY OF WORLD PETROLEUM
B 6302 259 REVIEW ON ISLAND ARCS, DEEP SEA TRENCHES, AND BACK-ARC BASINS,
B 6302 257 REVIEW ON SEDIMENTATION IN SUBMARINE CANYONS FANS, AND
B 6302 259 REVIEW ON SEISMIC EXPLORATION FUNDAMENTALS, BY J A. COFFEEN,
B 6309 1605* REVISION OF TAMPA FORMATION, WEST-CENTRAL FLORIDA, ABST.
B 6309 1591* REVISION OF U.S GEOLOGICAL SURVEY AND U.S. BUREAU OF MINES
B 6201 100 REX CHERT MEMBER, IDAHO AND, WYOMING
B 6107 1014 REX MEMBER, ALBERTA AND, SASKATCHEWAN
B 6009 1623 REXROAD, CARL B., LITHOFACIES RELATIONS IN ALEXANDRIAN AND
B 6311 2119* REYDON-CHEYENNE AREA, WESTERN OKLAHOMA, ABST., STIMULATION
M 29 45 REYKJANES RIDGE, ICELAND
B 6303 514 REYNAUD, R., DELTAIC FACIES RECOGNITIO ON SEISMIC DATA, ABST.
B 6008 1407 REYNOLDS, MITCHELL W., CHANGING SITES AND STYLES OF MESOZOIC
B 6008 1407 REYNOLDS, MITCHELL W., COASTAL SEDIMENTATION MODEL, TENSLEEP
B 6001 12 REYNOLDS, MITCHELL W., INFLUENCE OF RECURRENT LARAMIDE
B 6004 713 REYNOLDS, MITCHELL W., INTEGRATION OF SEDIMENTOLOGY,
B 6407 1005 REYNOLDS OOLITE, ARK-LA-TEX AREA

B 6405 772 REYNOLDS, R. L., HISTORY OF SULFIDIZATION OF SOUTH TEXAS
B 6209 1764 REYNOLDS, RICHARD A., COSMOPOLITAN BIOZONATION FOR LATE
B 6303 514 REYNOLDS, RICHARD A., NEOGENE RADIOLARIAN BIOSTRATIGRAPHY
B 6303 483 REYNOLDS, RICHARD A., RADIOLARIAN PRESERVATION IN PRESENT AND
B 6108 1391 REYNOLDS, RICHARD, ASSESSMENT OF MAGNETIC MEASUREMENTS IN
B 6309 1597 REYNOLDS, RICHARD, ECOLOGIC NICHES OF RADIOLARIANS,
B 6108 1379 REYNOLDS, RICHARD L., EXPERIMENTAL STUDY OF PYRITE OXIDATION
B 6303 514 REYNOLDS, RICHARD L., MARCOSITE IN ROLL-TYPE URANIUM DEPOSITS,
B 6108 1388 REYNOLDS, RICHARD L., RECOGNITION OF OXIDIZED SULFIDE MINERALS
B 6309 1598 REYNOLDS, RICHARD, RADIOLARIAN DISTRIBUTION, DIVERSITY, AND
S 4 27 REZAK, R., WEST FLOWER GARDEN BANK, GULF OF MEXICO
B 6004 713 REZAK, RICHARD, CORAL REEFS AND BANKS OF TEXAS OUTER
B 6112 2115 REZAK, RICHARD, EVOLUTION OF YUCCA MOUND COMPLEX, LATE
B 6109 1542 REZAK, RICHARD, INTERNAL WAVES AND NEPHELOID LAYER ON
M 25 189 RF GULFREX SEISMIC SURVEY
S 11 268 RHABDAMMINA ABYSSORUM
S 11 227 RHABDAMMINA ABYSSORUM
M 27 19 RHABDOLITH, COCCOLITH
B 6405 607 RHAETIAN, MIDDLE EAST
M 30 127 RHAETIAN-LIAS, STRATIGRAPHY, BEATRICE FIELD
B 6101 83 RHARB BASINS, MOROCCO, DOUKKALA-, BERRICHID-
B 6105 828* RHIZOCORALLIUM IN CRETACEOUS DAKOTA SANDSTONE, COLORADO,
B 6105 830* RHODE ISLAND, ABST., RENEWED EXPLORATION OF, COAL-BEARING,
B 6303 440* RHODE RIVER, MARYLAND, ABST., SEDIMENTATION RATES IN
B 6004 714 RHODEHAMEL, EDWARD C., SCOTIAN SHELF GEOTHERMAL GRADIENTS,
B 6008 1408 RHODES, ROBERT B., EAST KEITH GAS AREA, LIBERTY COUNTY,
B 6410 1690 RHODESIA, BULAWAYAN STROMATOLITIC LIMESTONE
B 6210 1849 RHODESIA, RWANDA, SWAZILAND, UPPER VOLTA, ZAMBIA,
B 6203 463† RHOMBS IN OIL RESERVOIRS, CHADRA SANDS, GIALO FIELD, LIBYA,
B 6206 920 RHONE FAN
M 25 458 RHYOLITE LINE, PACIFIDES, EASTERN ASIATIC BELT OR
B 6303 525* RHYTHMS IN DEEP-MARINE TURBIDITIC SHALE AND FINE-GRAINED
B 6004 646* RHYTHMS IN PELAGIC CALCAREOUS SEDIMENTS, ABST., PRIMARY,
B 6109 1476 RIACHUELO FIELD, BRAZIL
B 6303 514 RICART Y MENENDEZ, FERNANDO O., SUBAERIAL DIAGENESIS OF AXIAL
B 6004 701 RICCI LUCCHI FRANCO, SEDIMENTOLOGY OF OUTER-FAN DEPOSITIONAL
B 6111 2001 RICCIO, JOSEPH F., FAULTED UPPER PLEISTOCENE MARINE TERRACE,
B 6305 838 RICE, DUDLEY D., BOWDOIN DOME AREA, NORTH-CENTRAL
B 6405 760 RICE, DUDLEY D., BOWDOIN DOME AREA, NORTH-CENTRAL MONTANA—
B 6409 1566 RICE, DUDLEY D., CHEMICAL AND ISOTOPIC EVIDENCE OF ORIGINS OF
B 6403 316 RICE, DUDLEY D., COASTAL AND DELTAIC SEDIMENTATION OF UPPER
B 6008 1408 RICE, DUDLEY D., DEPOSITIONAL ENVIRONMENTS OF EAGLE SANDSTONE,
B 6405 773 RICE, DUDLEY D., DEVELOPMENT OF BIOGENIC GAS FROM SHALLOW,
B 6205 891 RICE, DUDLEY D., NATURAL GAS RESOURCES IN SHALLOW,
B 6303 515 RICE, DUDLEY D., RELATION OF SEDIMENTARY HISTORY AND
B 6004 714 RICE, DUDLEY D., SHALLOW GAS ACCUMULATIONS--THEIR ORIGIN AND
B 6407 969 RICE, DUDLEY D., SHALLOW, LOW-PERMEABILITY RESERVOIRS OF
B 6405 772 RICE, DUDLEY D., UPPER CRETACEOUS MOSBY SANDSTONE, CENTRAL
B 6303 515 **RICE, G. W., GEOSEISMIC MODELING--AN INTERACTIVE COMPUTER**
B 6304 700 RICE, G. W., WAVE-FORM FACTOR ANALYSIS--QUANTITATIVE APPROACH
B 6004 714 RICE, L. F., COMPUTER ANALYSIS OF LINEAMENTS TO INFER EXISTENCE
B 6310 1984 RICE, PEGGY J., REVIEW OF GEOREF THESAURUS AND GUIDE
B 6004 714 RICE, ROBERT B., GEOPHYSICAL-GEOLOGIC SYNERGISM. BY TITLE ONLY
B 6004 665 RICH, DAVID W., EXPERIMENTAL DEVELOPMENT OF PERMOPOROSITY IN
B 6105 781 **RICH, DAVID W., FABRIC-SELECTIVE PERMOPOROSITY EXPERIMENTALLY**
B 6012 2188 RICH, ERNEST I., SATELLITE LOOK AT REGIONAL GEOLOGY OF
M 25 63 RICH, ERNEST I., SURVEYING EARTH
B 6203 557 RICHARDS, ADRIAN F., GEOTECHNICAL PROPERTIES OF UPPER
B 6004 714 RICHARDS, ADRIAN F., OSLOFJORD, NORWAY--CASE STUDY IN
B 6105 822 RICHARDS, ADRIAN F., SHIPBOARD GEOTECHNICAL RESULTS FROM U.S.
M 24 52 RICHARDSON MOUNTAINS, ALASKA
B 6004 701* RICHARDSON MOUNTAINS, NORTHWEST TERRITORIES, CANADA, ABST.,
B 6008 1401* RICHLAND COUNTY, MONTANA, ABST., STRUCTURAL INTERPRETATION
B 6409 1566 RICHMANN, D. L., ET AL, MINERALOGY, DIAGENESIS, AND POROSITY IN
B 6303 449 RICHMOND, WILLIAM C., VIDEOTAPES OF ACTIVE PRIMARY PHYSICAL
B 6004 608† RICHNESS AND EFFECT OF LITHOLOGIC VARIATION, THERMAL-ANALYSIS
M 30 30 RICINUS FIELD, LANDSAT IMAGE, ALBERTA
B 6303 469* RICINUS GAS FIELD, ALBERTA, ABST., STRACHAN-
B 6401 28 RICINUS GIANT GAS FIELD, LANDSAT DATA, ALBERTA
M 30 315 RICINUS WEST, ALBERTA, DEVONIAN
M 30 327 RICINUS WEST FIELD, RESERVES AND PRODUCTION, ALBERTA
M 30 315† RICINUS WEST GAS FIELDS, ALBERTA, CANADA, STRACHAN AND
B 6209 1610 RICKARD HILL LIMESTONE, NEW YORK
M 28 112 RICKER SANDSTONE, TEXAS, PENNSYLVANIAN
M 28 143 RICKER SANDSTONE, TEXAS, PENNSYLVANIAN
B 6203 557 RICKETTS, BRIAN D., EXAMPLE OF CARBONATE BANKS IN LOWER
M 25 519 RICKETTS FAULT, MEXICO
B 6309 1605 RICKLES, SUE E., EQUILIBRIUM IN MODERN CORAL REEFS, WESTERN
B 6303 482 RICKLES, SUSAN E., EQUILIBRIUM IN MODERN CORAL REEFS, WESTERN
B 6109 1544 RICOY, JOSE ULISES, DEPOSITIONAL SYSTEMS IN SPARTA FORMATION,
B 6211 2191 RIDER, M. H., GROWTH FAULTS IN CARBONIFEROUS OF WESTERN
B 6105 830 RIDER, T. H., COAL-BEARING NARRAGANSETT BASIN, MASSACHUSETTS
M 29 239† RIDGE AND ASSOCIATED BASINS, NORTH PHILIPPINE SEA, SUBSIDENCE
B 6203 369 RIDGE BASIN, CALIFORNIA
B 6007 1078† RIDGE COMPLEX, TECTONICS OF SOUTHWESTERN, NORTH, ATLANTIC
M 29 455 RIDGE SPREADING, DIAGRAM, ATLANTIC MARGINS
B 6004 705* RIDGE-AND-SWALE TOPOGRAPHY ON INNER CONTINENNAL SHELF OFF
B 6209 1610 RIDGELEY SANDSTONE, MARYLAND
B 6303 293 RIDGES, PROSPECTS, TRANSMARGINAL
B 6106 929† RIDGES—EVIDENCE FROM SEDIMENTS ON RIO GRAND RISE, SOUTHWEST
B 6304 692 RIDGLEY, JENNIE L., ROLL-TYPE URANIUM OCCURRENCE AT
B 6405 773 RIEDEL W., LOW-COST MICROPROCESSOR SYSTEM FOR PALEONTOLOGIC
B 6303 516 RIEDEL, WILLIAM R., RECENT AND POTENTIAL ADVANCES IN
B 6405 774 RIEKE, H. H., RESOURCE EVALUATIO OF GAS-BEARING COAL BEDS, ABST.
B 6211 2277 RIERDON FORMATION, MONTANA
B 6304 682 RIES, EDWARD R., REVIEW OF TECTONICS OF ASIA, BY EMILE ARGAND,
B 6304 693 RIESE, A. C., APPLICATION OF SOLUTION-MINERAL EQUILIBRIUM
B 6304 693 RIESE, WALTER C., MOUNT TAYLOR URANIUM DEPOSIT, SAN MATEO,
B 6305 839 RIESE, WALTER C., MOUNT TAYLOR URANIUM DEPOSIT--DESCRIPTION
B 6304 693 RIESE, WALTER C., SEM INVESTIGATION OF PARAGENESIS OF URANIUM
M 28 28 RIESELBERGER SANDSTONE, GERMANY, CRETACEOUS
B 6405 773 RIESTER, DEBRA, LOCATION OF LITTORAL ENERGY FENCE AND
M 26 213† RIFT AND PULL-APART BASINS, SEISMIC-, STRATIGRAPHIC,
B 6203 419† RIFT BASINS FLANKING NORTH ATLANTIC OCEAN AND THEIR RELATION
B 6203 434 RIFT BASINS, HYDROCARBONS, NORTH, ATLANTIC
B 6111 1957 RIFT BASINS, RESERVES
B 6207 1194 RIFT HYPOTHESIS, OREGON AND, WASHINGTON, EOCENE
B 6207 1144 RIFT HYPOTHESIS, TESTING, IRIAN JAYA
B 6101 79 RIFT, MID-ATLANTIC
B 6203 429 RIFT SYSTEM, SECTIONS, BRITAIN, WEST
B 6203 435 RIFT SYSTEM, SECTIONS, NORTH SEA
B 6101 83 RIFT VALLEY, CENTRAL, ATLANTIC
B 6404 569 RIFT VALLEY, ISRAEL, DEAD SEA
B 6202 226 RIFT VALLEYS, AFRICA
M 25 511 RIFT ZONE, CORDILLERAN OROGEN, NORTHERN, PALEOZOIC, CRUST
M 26 240 RIFT-BASIN, MODEL, BRAZIL
B 6405 732* RIFT-RELATED BASINS, ABST., INTEGRATED GEOLOGIC AND
B 6004 706* RIFTING AND OPENING OF NORTH ATLANTIC OCEAN, ABST., NORTH SEA
B 6405 714* RIFTING GEOMETRY OF ARCTIC BASIN, ABST., IMPLICATIONS OF
M 29 79 RIFTING HISTORY, AFRICA, TRIASSIC
M 29 79 RIFTING HISTORY, NORTH AMERICA, TRIASSIC
B 6207 1193† RIFTING IN WESTERN OREGON-WASHINGTON, EARLY, TERTIARY
B 6402 175 RIFTING MODELS, SIMPLE
B 6404 516 RIFTING, NEW ENGLAND CONTINENTAL MARGIN
B 6405 680* RIFTING PHASES OF SOUTH ATLANTIC, ABST., LOWER, CRETACEOUS
M 25 466 RIFTOGENIC ZONES, USSR, NORTHEAST, MESOZOIC, TAPHROGENIC OR
B 6105 822 RIGASSI, DANILO A., USSR, ABST., FUTURE OIL AND GAS RESOURCES OF
B 6303 516 RIGBY, J. KEITH, MARIN PERMIAN ROCKS OF TUNISIA, ABST.
B 6402 225 RIGGS GRABEN, TEXAS
B 6303 516 RIGGS, KARL A., LIGNITES OF TOMBIGBEE AND HOLLY SPRINGS
B 6003 455 RIGGS, KARL A., SUBSURFACE PRECAMBRIAN GRANITE, LAFAYETT
B 6405 774 RIGHTMIRE, C. T., RESOURCE EVALUATIO OF GAS-BEARING COAL BEDS,
B 6402 222 RILEY FORMATION, TEXAS
M 28 164 RILEY FORMATION, TEXAS, CAMBRIAN
M 28 126 RILEY FORMATION, TEXAS, CAMBRIAN
M 28 106 RILEY FORMATION, TEXAS, CAMBRIAN
B 6102 166 RINCON ANTICLINE, CALIFORNIA
B 6104 594 RINCON FIELD, CALIFORNIA
B 6407 1039 RINCON FIELD, CALIFORNIA
B 6203 353 RINCONADA FAULT, CALIFORNIA
B 6211 2218 RINCONADA-RELIZ FAULT SYSTEM, CALIFORNIA
B 6004 731 **RINE, J. M., TIMING, CHARACTER, AND PRESERVABILITY OF**
M 25 474 RING MEGASTRUCTURES AND METALLOGENIC PROSPECTING, RELATION
M 25 472† RING MEGASTRUCTURES OF THE PACIFIC--SUMMARY
M 25 472 RING STRUCTURES
M 25 473 RING STRUCTURES, DISTINCT ON EARTH, MOON, MARS, AND VENUS
B 6104 603 RINGBONE FORMATION, NEW MEXICO
B 6302 183 RINGKOBING-FYN HIGH, NORTH SEA
S 8 108 RIO BALSAS CANYON, DESCRIBED, TURBIDITY CURRENTS
S 8 31 RIO BALSAS CANYON, TIME-VELOCITY CURVES, CURRENT FLOW
S 8 110 RIO BALSAS CANYON, TIME-VELOCITY CURVES, CURRENT FLOW
S 8 34 RIO BALSAS CANYON, TIME-VELOCITY RECORD, CURRENT FLOW
S 8 109 RIO BALSAS CANYON, TOPOGRAPHY, INDEX MAP
S 8 35 RIO BALSAS CANYON, VELOCITY AVERAGES, CURRENT FLOW
S 8 105 RIO BALSAS DELTA CANYON, MEXICO, SUBMARINE CANYONS
S 8 16 RIO BALSAS DELTA, TIME-VELOCITY CURVES, CURRENT FLOW
B 6004 712* RIO BALSAS OF WESTERN MEXICO, ABST., SLOPE, VALLEYS AND
B 6008 1393* RIO BLANCO COUNTY, COLORADO, ABST., PETROLEUM GEOLOGY OF,

B 6008 1393* RIO BLANCO UNIT, RIO BLANCO COUNTY, COLORADO, ABST.,
S 6 106 RIO BONITO FORMATION, BRAZIL, PARANA BASIN
S 6 106 RIO BONITO FORMATION, RUBIDIUM-STRONTIUM, BRAZIL
B 6102 177 RIO BRAVO SAND, CALIFORNIA
S 8 129 RIO DE LA PLATA CANYON, DESCRIBED, TURBIDITY CURRENTS
S 8 124 RIO DE LA PLATA CANYON, PUERTO RICO, SUBMARINE CANYONS
S 8 38 RIO DE LA PLATA CANYON, TIME-VELOCITY CURVES, CURRENT FLOW
S 8 131 RIO DE LA PLATA CANYON, TOPOGRAPHY, INDEX MAP
S 11 196 RIO DELL FORMATION
S 11 223 RIO DELL FORMATION, ASSEMBLAGES
S 11 219 RIO DELL FORMATION, FOSSIL AGGREGATE
S 6 102 RIO DO RASTO FORMATION, BRAZIL, PARANA BASIN
S 6 102 RIO DO RASTO FORMATION, RUBIDIUM-STRONTIUM, BRAZIL
B 6409 1564* RIO ESCONDIDO BASIN, COAHUILA, MEXICO, ABST., DIAGENESIS OF,
B 6106 929† RIO GRAND RISE, SOUTHWEST ATLANTIC OCEAN, SUBSIDENCE OF,
B 6302 228 RIO GRANDE DO NORTE PLATEAU, BRAZIL
B 6009 1616* RIO GRANDE EMBAYMENT, ABST., NEW, SMACKOVER PRODUCTION AT,
S 10 208 RIO GRANDE EMBAYMENT, GEOLOGIC SECTION, TEXAS
B 6206 1051 RIO GRANDE EMBAYMENT, TEXAS
B 6405 777* RIO GRANDE EMBAYMENT, TEXAS, ABST., PALEOCENE, MIDWAY,
B 6104 547 RIO GRANDE FAULT, BRAZIL
B 6002 197 RIO GRANDE RISE, BRAZIL MARGIN
B 6106 931 RIO GRANDE RISE SAMPLES, LOCATION OF
B 6106 930 RIO GRANDE RISE, WALVIS RIDGE, DEEP SEA DRILLING SITES ON
S 7 11 RIO GRANDE SLOPE, DESCRIBED
B 6206 1044 RIO GRANDE-SANTOS FAULT, BRAZIL
B 6206 1034 RIO GRANDE-SANTOS FAULT SYSTEM, BRAZIL
B 6110 1748 RIO MUNI-FERNANDO POO, DEVELOPMENTS, 1976, EQUATORIAL GUINEA
B 6310 1691 RIO MUNI-FERNANDO POO, DEVELOPMENTS, 1978, EQUATORIAL GUINEA
B 6411 1787 RIO MUNI-FERNANDO POO, DEVELOPMENTS, 1979, EQUATORIAL GUINEA
S 4 87 RIO NUEVO FORMATION, ALBIAN, JAMAICA
B 6109 1479 RIO PARAIBA PLATFORM, BRAZIL
B 6206 1068 RIO PERDIDO CANYON, GULF OF MEXICO
M 25 554 RIO PLAYAS, COPPER AND MOLYBDENUM DEPOSITS, ECUADOR
M 25 554 RIO PLAYAS, COPPER-MOLYBDENUM, ECUADOR
M 27 84 RIO YAUCO FORMATION, PUERTO RICO, CRETACEOUS
B 6210 1846 RIO-MUNI-FERNANDO POO, DEVELOPMENTS, 1977, EQUATORIAL GUINEA
M 29 113 RIOS, F. F., STRUCTURE OF COLORADO BASIN AND CONTINENT-OCEAN
M 30 238 RIPHEAN PALEONTOLOGY, PROBLEMATICA, USSR
M 30 238 RIPHEAN PALEONTOLOGY, SPOROPHYTES, USSR
B 6004 679* RISE, ABST., RESPONSE OF ACTIVE, OOLITE, SAND, SHOAL TO ITS
B 6204 592† RISE, AND ABYSSAL PLAIN, PETROLEUM, POTENTIAL OF, U.S., ATLANTIC,
B 6405 691* RISE AND FALL OF TRIASSIC CONODONTS, ABST.
B 6106 930 RISE, CEARA
B 6303 438* RISE OF EASTERN SOUTH AMERICA, ABST., MASS WASTING ON,
M 25 313† RISE OFF CENTRAL CALIFORNIA--SUMMARY, PETROLEUM POTENTIAL OF,
B 6303 467* RISE OFF NORTH AMERICA, ABST., SEDIMENTATION PROCESSES AND
B 6109 1483† RISE OFF WESTERN NOVA SCOTIA AND GEORGES BANK, CONTINENTAL,
M 26 67 RISE, SEA LEVEL FLUCTUATIONS, NERITIC FACIES, RELATIVE
B 6106 929† RISE, SOUTHWEST ATLANTIC OCEAN, SUBSIDENCE OF, ASEISMIC,
B 6105 837* RISES, ABST., SEISMIC RECOGNITION OF DEPOSITIONAL FACIES ON,
B 6403 302 RISES AND DEEP-OCEAN TRENCHES INFERRED FROM BATHYMETRIC AND
M 29 423† RISES, PETROLEUM, SOURCE BEDS ON, CONTINENTAL, SLOPES AND
B 6209 1584† RISES, PETROLEUM, SOURCE BEDS ON, CONTINENTAL, SLOPES AND
B 6312 2204† RISES, SEISMIC EVIDENCE FOR WIDESPREAD POSSIBLE, GAS, HYDRATE,
B 6312 2211 RISES, THERMAL GRADIENTS, CONTINENTAL, SLOPES AND
B 6207 1232 RISK, G. F., DEFINING BOUNDARIES AND STRUCTURE OF BROADLANDS
B 6203 553 RISK, M. J., SILURIAN MACROBORING ORGANISMS, SOUTHERN ONTARIO,
B 6107 1069 RISK, MICHAEL J., MICRITIZATION AND CARBONATE-GRAIN BINDING BY
B 6304 693 RISTORCELLI, STEVEN J., GEOLOGY OF EASTERN SMITH LAKE ORE
B 6012 2188 RITCH, H. J., EVIDENCE FOR LOW FREE-GAS SATURATION IN
B 6303 516 RITCHIE, SUSAN H., FACIES RECOGNITION FROM WELL LOGS TO PREDICT
B 6309 1609 RITCHIE, W., VARIATIONS IN LITTORAL SAND FEATURES WITH SPECIAL
B 6309 1538 RITZMA, HOWARD R., AEROMAGNETIC DETECTION OF DIAGENETIC
B 6405 774 RITZMA, HOWARD R., MIGRATION AND ENTRAPMENT O PETROLEUM—
B 6008 1392 RITZMA, HOWARD R., STRUCTURAL GEOLOGY AND PETROLEUM
B 6105 822 RITZMA, HOWARD R., UNITED STATES AND UTAH OIL-IMPREGNATED
M 25 558 RIVAS, S., DEPOSITS OF BOLIVIAN TIN BELT--SUMMARY
B 6004 655* RIVER, ABST., EFFECTS OF, ICE ON SEDIMENTS ON, SANDY, BRAIDED
B 6107 1075 RIVER BAY, MICROGRAPHS, BARBADOS
B 6002 295 RIVER FAL KAOLINITIC DEPOSITS, ENGLAND
B 6109 1561 RIVEREAU, JEAN-CLAUDE, CONTRIBUTION OF LANDSAT IMAGERY TO
B 6303 451* RIVER NEAR CHARLESTON SOUTH CAROLINA, ABST.
B 6409 1566 RIVES, JOHN S., II, PALEOENVIRONMENTAL ANALYSIS OF JOACHIM
B 6105 827* RN222 AND HE4-AR36 RATIOS IN SOIL GAS OVERLYING TWO URANIUM
B 6105 823 ROADIFER, ROY E., HYDROCARBON RESOURCES AND RESERVES
B 6301 106 ROAN, VIRGIL H., HISTORY OF ARDMORE GEOLOGICAL SOCIETY
B 6006 926 ROBB, JAMES M., REGIONAL GEOLOGIC FRAMEWORK OFF
B 6105 742 ROBB, JAMES M., REGIONAL GEOLOGIC FRAMEWORK OFF NORTHEASTERN
B 6303 480 ROBB, JAMES M., SLUMPING IN INTERCANYON AREAS, MIDDLE ATLANTIC
B 6310 1746 ROBBIANO, J. A., ON DEVELOPMENTS IN ARGENTINA, 1978, YRIGOYEN, M.
B 6405 783 ROBBIN, DANIEL M., EXPERIMENTAL COMPACTION OF LIME SEDIMENT,
B 6303 528 ROBBIN, DANIEL M., THREE-DIMENSIONAL ASPECTS OF BELIZE PATC
B 6004 714 ROBBINS, E. I., SCOTIAN SHELF GEOTHERMAL GRADIENTS, ABST.
B 6105 775 ROBBINS, ELEANORA I., GEOLOGIC FACTORS CONTROLLING
B 6405 687 ROBBINS, RAYMOND K., SACHA FIELD OF ECUADORIAN ORIENTE, ABST.
M 25 414 ROBE PISOLITE, WESTERN AUSTRALIA
B 6002 196 ROBERT D. CONRAD CRUISES THROUGH MAY 1974, RV
B 6405 771 ROBERTS, A. A., RECENT ADVANCES IN HELIUM ANALYSIS AS
B 6302 245 ROBERTS, ALAN A., AEROMAGNETIC DETECTION OF DIAGENETIC
B 6008 1408 ROBERTS, BILLY W., MESAVERDE GAS POTENTIAL IN DRY CREEK AREA,
B 6303 527 ROBERTS, D. G., RECENT ADVANCES IN PASSIVE-MARGIN RESEARCH,
B 6402 264 ROBERTS, H. H., EVOLUTION OF SAND-DOMINANT SUBAERIAL PHASE,
B 6409 1568 ROBERTS, HARRY H., ATCHAFALAYA DELTA—LOUISIANAS NEW
B 6303 501 ROBERTS, HARRY H., DISPERSAL OF FINE-GRAINED SEDIMENT IN
B 6303 517 ROBERTS, HARRY H., REEF-CREST WAVE AND CURRENT INTERACTIONS
B 6409 1567 ROBERTS, HARRY H., SEDIMENTARY STRUCTURES ASSOCIATED WITH
S 4 239 ROBERTS, J. WAYLAND, IMPORTANCE OF PHYLLOID ALGAE IN
B 6311 2117 ROBERTS, JOHN F., INVESTIGATION OF DESMOINESIAN ROCKS IN
B 6303 475 ROBERTS, JOHN F., POTENTIAL FOR SUBSURFACE DISPOSAL OF
M 24 292 ROBERTS LIMESTONE PRODUCING ZONE, FLORIDA
B 6203 538* ROBERTS MOUNTAIN ALLOCHTHON, NORTH-CENTRAL NEVADA, ABST.,
B 6205 892* ROBERTS MOUNTAIN THRUST—ITS SIGNIFICANCE IN OIL EXPLORATION
S 10 217 ROBERTS, W. H., DESIGN AND FUNCTION OF OIL AND GAS TRAPS
B 6405 774 ROBERTS, W. H., III, COMMON FACTORS AMONG ATYPICAL FIELDS, ABST.
B 6308 1428 ROBERTS, W. H., III, DESIGN AND FUNCTION OF OIL AND GAS TRAPS,
B 6203 558 ROBERTS, W. H., III, DESIGN AND FUNCTION OF OIL AND GAS TRAPS AS
S 10 1 ROBERTS, W. H., III, EDITED BY, PROBLEMS OF PETROLEUM MIGRATION
S 10 1 ROBERTS, W.H., III, INTRODUCTION, PROBLEMS OF PETROLEUM
B 6308 1344 ROBERTSON, D. C., DEVELOPMENTS IN EASTERN CANADA IN 1978
B 6409 1518 ROBERTSON, D. C., DEVELOPMENTS IN EASTERN CANADA IN 1979
B 6105 804 ROBERTSON, D. J., SILICIFICATION OF CALCIUM PHOSPHATE IN
M 25 231 ROBERTSON, G. A., AUSTRALIAN NORTHWEST CONTINENNAL
B 6303 483 ROBESON, HERMAN E., SILICA DISSOLUTION FROM MONTMORILLONITE,
M 25 526 ROBIN, CLAUDE, MAJOR CHEMICAL CHARACTERISTICS OF TERTIARY
B 6209 1764 ROBINSON, GRAY, COMPUTER-ASSISTED PALEOECOLOGIC ANALYSES AND
B 6405 774 ROBINSON, GRAY S., COMPUTER ASSISTED PALEOECOLOGIC ANALYSES
B 6004 715 ROBINSON, J. C., COMBINATION ENTRAPMENT IN SMACKOVER
B 6009 1614 ROBINSON, J. D., COMBINATION ENTRAPMENT IN SMACKOVER
B 6105 823 ROBINSON, J. E., GRAY-LEVEL MAPPING OF GEOLOGIC DATA, ABST.
B 6304 700 ROBINSON, J. E., QUANTITATIVE COMPARISONS OF GEOLOGIC MAPS TO
B 6303 517 ROBINSON, J. E., STATISTICAL COMPARISON OF MAPPED DATA, ABST.
B 6309 1591 ROBINSON, J. E., STRUCTURAL CONTROL OF ONONDAGA REEFS,
B 6203 558 ROBINSON, KEITH, GEOLOGIC APPRAISAL OF UNDISCOVERED OIL AND
B 6307 1144 ROBINSON, LUNDIN, ROBERT F., REVIEW OF, STRATIGRAPHIC INDEX OF,
B 6303 411* ROBINSON MOUNTAINS, GULF OF ALASKA, ABST., GLACIAL ORIGIN OF
B 6405 775 ROBINSON, W. E., SEDIMENTARY ASPECTS ORGANIC MATERIAL IN
S 11 120 ROBULUS MIOCENSICUS
S 11 120 ROBULUS MOHNENSIS
S 11 120 ROBULUS SMILEYI
B 6209 1610 ROCHESTER SHALE, MARYLAND
M 27 99 ROCHESTER SHALE, NEW YORK, SILURIAN
B 6211 2315 ROCHESTER SHALES, WEST VIRGINIA
M 28 25 ROCK FRAGMENTS, GENERAL
B 6203 522* ROCK INTERACTION,ABST., SOME ECONOMIC ASPECTS OF, WATER-
S 10 47 ROCK PROPERTIES, MIGRATION
B 6205 864 ROCK, OR FOSSIL COLLECTIONS AVAILABLE FROM FORTY-EIGHT STATES,
M 26 15† ROCK PHYSICS FROM LABORATORY AND LOG DATA THAT ARE
M 26 39 ROCK PROPERTIES, LABORATORY, MEASUREMENTS, SATURATION
M 26 32 ROCK PROPERTIES, TEMPERATURE EFFECTS, LABORATORY,
B 6405 732* ROCK PROPERTIES—INFLUENCE ON HYDROCARBON ACCUMULATION,
B 6004 689* ROCK SHELTER, AVELLA, PENNSYLVANIA, ABST., COMPARISON OF
B 6002 326* ROCK TANK, MORROW SANDSTONE, FIELD, EDDY COUNTY, NEW
B 6405 762* **ROCK-EVAL PYROLYSIS AS SOURCE ROCK SCREENING TECHNIQUE,** ABST.
B 6102 252 ROCK-STRATIGRAPHIC TERM, VOLCANIC ROCKS, FORMAL
B 6102 252† ROCK-STRATIGRAPHIC UNIT, NOTE 46--, APPLICATION FOR AMENDMENT
M 25 402 ROCKHAMTON, NICKEL-ENRICHED LATERITES, AUSTRALIA, GREENVALE
B 6008 1204† ROCKIES IN 1975, OIL AND, GAS, DEVELOPMENT IN NORTHERN
B 6108 1346† ROCKIES IN 1976, OIL AND, GAS, DEVELOPMENTS IN NORTHERN
B 6208 1498† ROCKIES IN 1977, OIL AND, GAS, DEVELOPMENTS IN NORTHERN
B 6308 1386† ROCKIES IN 1978, DEVELOPMENTS IN NORTHERN
B 6409 1501† ROCKIES IN 1979, DEVELOPMENTS IN NORTHERN
B 6405 768* ROCKIES SEGMENT OF CORDILLERAN FORELAND THRUST BELT, ABST.,
B 6409 1548 ROCKS AND ROCK MINERALS, BY RICHARD V. DIETRICH AND BRIAN J.
B 6412 2078 ROCKY FACE FAULT, APPALACHIANS
B 6304 652 ROCKY GAP SANDSTONE, VIRGINIA
B 6108 1372* ROCKY MOUNTAIN BASINS, ABST., MEGAEXPLORATION FOR NEW
B 6405 778* ROCKY MOUNTAIN BASINS, ABST., SEISMIC MODEL OF 15

B 6305 840* ROCKY MOUNTAIN BASINS, ABST., SEISMIC MODEL OF FIFTEEN
B 6405 761* ROCKY MOUNTAIN COAL PROVINCE—DATA AS OF 1979, ABST.,
B 6008 1408* ROCKY MOUNTAIN CRETACEOUS SOURCE-ROCK TYPES, MATURITY, AND
B 6008 1395* ROCKY MOUNTAIN FORELANDS, ABST., ANALYTIC SOLUTIONS APPLIED
B 6008 1411* ROCKY MOUNTAIN FORELANDS, ABST., FIELD EVIDENCE FOR
B 6008 1409* ROCKY MOUNTAIN FORELANDS, ABST., SEISMIC EVIDENCE FOR
B 6405 751* ROCKY MOUNTAIN FRONT RANGES, CANMORE, ALBERTA, ABST.,
B 6405 806* ROCKY MOUNTAIN OVERTHRUST BELT, ABST., WHITNEY CANYON—A
B 6004 651* ROCKY MOUNTAIN REGION, ABST., ALTERED, VOLCANIC, ASH PARTINGS
B 6004 713* ROCKY MOUNTAIN REGION, ABST., INTEGRATION OF SEDIMENTOLOGY,
B 6405 757* ROCKY MOUNTAIN REGION, ABST., PALEOCENE TIME SCALE FOR
B 6303 502* ROCKY MOUNTAINS, ABST., APPLICATION OF, PALYNOLOGY TO AGE
B 6405 794* ROCKY MOUNTAINS, ABST., EVOLUTION AND STRESS HISTORY OF
B 6304 701* ROCKY MOUNTAINS, ABST., STRUCTURAL ANALYSIS OF NORTHERN
B 6007 1009 ROCKY MOUNTAINS, CHINLE FORMATION
B 6112 2076 ROCKY MOUNTAINS, FRONT RANGES
M 26 296 ROCKY MOUNTAINS, NEW, TECTONIC STYLE
M 24 159 RODESSA FORMATION, TEXAS
M 24 242 RODESSA PRODUCING SAND, MISSISSIPPI
B 6411 2038 RODGERS, DONALD A., REVIEW OF CONTINENTAL TECTONICS, STUDIE IN
B 6102 296 RODGERS, ELTON, BROOKS FIELD, SOUTHEAST MIDLAND BASIN, IRION
S 3 47 RODILTO LIMESTONE, NEW MEXICO
B 6405 775 RODRIGUEZ-PARADINAS, ANGEL, CASTILLO GAS FIELD, SPAIN, ABST.
B 6405 775 ROEDER, DIETRICH, APPALACHIAN THRUST BELT BETWEEN TRENTON,
B 6405 775 ROEDER, DIETRICH, CENTRAL MEDITERRANEAN THRUST BELTS, ABST.
B 6209 1696 ROEDER, DIETRICH, TECTONICS OF BROOKS RANGE OPHIOLITES, ALASKA
B 6309 1586 ROEHL, P. O., DILATION BRECCIATION--PROPOSED MECHANISM OF
B 6305 839 ROEHLER, HENRY W., VERMILLION CREEK COAL BED, HIGH-SULFUR,
B 6305 839 ROEHRS, ROBERT C., MINNELUSA--POWDER RIVER BASIN--PAST, PRESENT
B 6109 1545 ROEMER, LAMAR B., STRUCTURE AND STRATIGRAPHY OF LATE
B 6303 551 ROEN, JOHN B., CORRELATION OF SUBSURFACE MIDDLE AND UPPER
M 30 105 ROGALAND GROUP, PALEOCENE, STATFJORD FIELD
M 30 105 ROGALAND GROUP, TERTIARY, STATFJORD FIELD
B 6107 1029 ROGERS, JAMES P., DESMOINESIAN, PENNSYLVANIAN, SANDSTONE
B 6002 322 ROGERS, JAMES P., GENESIS AND DISTRIBUTION OF DESMOINESIAN,
B 6303 411 ROGERS, JOHN, GLACIAL ORIGIN OF MEGACHANNELS OF UPPER
B 6004 646 ROGGENTHEN, W. M., PRIMARY BEDDING RHYTHMS IN PELAGIC
B 6405 763 ROHRBACK, B. G., GEOCHEMISTRY OF ARTIFICIALLY HEATED HUMIC
B 6204 687 ROHRBACK, B. G., HYDROCARBON GENERATION BY THERMAL
B 6409 1550 ROKSANDIC, MIODRAG M., REVIEW OF DIAGRAPHIES DIFFEREES—BASES
B 6303 429* ROL OF BRINES AND FERROUS IRON IN FORMATION OF DOLOSTONE,
B 6303 424* ROLE OF BIOEROSION IN MASS-WASTING OF PLEISTOCENE OUTCROPS ON
B 6203 567* ROLE OF ELECTROCHEMICAL CELLS IN FORMATION OF
S 10 207† ROLE OF GEOPRESSURE IN THE HYDROCARBON AND WATER SYSTEM
B 6401 59† ROLE OF MINERAL MATRIX IN KEROGEN PYROLYSIS--INFLUENCE ON
B 6207 1234* ROLE OF NUCLEAR ENERGY IN FUTURE OF CIRCUM-PACIFIC, AREA,
B 6303 547* ROLE OF PHYSICAL SEDIMENTATION I CARBONATE-BANK GROWTH,
M 30 565† ROLE OF REFLECTION SEISMIC IN DEVELOPMENT OF NEMBE CREEK
B 6211 2214† ROLE OF SALINIAN BLOCK IN EVOLUTION OF SAN ANDREAS FAULT
B 6108 1390* ROLE OF SECONDARY POROSITY IN SANDSTONE DIAGENESIS, ABST.
B 6012 2187* ROLE OF SMALL INDEPENDENT IN TOMORROWS OIL, ABST.
B 6203 541* ROLE OF SOLUBILITY IN MIGRATION OF PETROLEUM FROM SOURCE,
B 6303 414* ROLE OF TEMPERATURE AND BURIAL DEPTH I DEVELOPMENT OF
B 6008 1355 ROLIFF, W. A., DEVELOPMENTS IN EASTERN CANADA IN 1975
B 6108 1335 ROLIFF, W. A., DEVELOPMENTS IN EASTERN CANADA IN 1976
B 6208 1451 ROLIFF, W. A., DEVELOPMENTS IN EASTERN CANADA IN 1977
B 6308 1344 ROLIFF, W. A., DEVELOPMENTS IN EASTERN CANADA IN 1978
B 6409 1518 ROLIFF, W. A., DEVELOPMENTS IN EASTERN CANADA IN 1979
B 6305 839* ROLL FRONTS, ABST., MOUNT TAYLOR, URANIUM
S 3 60 ROLL FRONTS, URANIUM
B 6108 1373* ROLL WITHIN A TABULAR URANIUM BODY-RAJAH 49 MINE, MESA
B 6405 780* ROLL-FRONT URANIUM DEPOSITS, ABST., ORGANIC MODEL FOR
B 6405 772* ROLL-TYPE URANIUM DEPOSIT, ABST., HISTORY OF, SULFIDIZATION OF
B 6108 1379* ROLL-TYPE URANIUM DEPOSITS, ABST., EXPERIMENTAL STUDY OF,
B 6303 514* ROLL-TYPE URANIUM DEPOSITS, ABST., MARCOSITE IN
B 6304 694* ROLL-TYPE URANIUM DEPOSITS--MARIANO LAKE TYPE, ABST.,
B 6304 692* ROLL-TYPE URANIUM OCCURRENCE AT DENNISON-BUNN CLAIM AND
B 6004 665 ROLLINS, HAROLD B., RELATIVE MARINE TRANSGRESSIVE AND
B 6008 1408 ROLLINS, RICHARD A., ROCKY MOUNTAIN CRETACEOUS SOURCE-ROCK
B 6304 701 ROLLINSON, DON, LOCATIN STRATIGRAPHIC FEATURES AND
B 6202 274 ROMANCHE FRACTURE ZONE
B 6302 219 ROMANCHE FRACTURE ZONE, BRAZIL
B 6107 969 ROMANCHE-TAKORADI FRACTURES, AFRICA
B 6110 1652 ROMANIA, DEVELOPMENTS, 1976
B 6210 1923 ROMANIA, DEVELOPMENTS, 1977
B 6110 1654 ROMANIA, PRODUCTION, 1975-1976
B 6210 1924 ROMANIA, PRODUCTION, 1976-1977
B 6310 1932 ROMANIA, PRODUCTION, 1977-1978
B 6307 1052 ROMANIAN CARPATHIAN THRUST-FOLD BELT, PLOESTI AREA, SALT
B 6303 344 ROME FORMATION, KENTUCKY
B 6110 1899 ROME FORMATION, MISSISSIPPI
B 6304 652 ROME FORMATION, VIRGINIA
B 6009 1621* ROME TROUGH AND THEIR RELATION TO CINCINNATI ARCH, ABST.,
B 6303 345 ROME TROUGH, COMPUTER MODELING OF GRAVITY DATA, KENTUCKY
B 6303 343 ROME TROUGH, GRAVITY MAP, KENTUCKY
B 6303 341† ROME TROUGH IN EASTERN KENTUCKY BY GRAVITY AND DEEP
B 6303 344 ROME TROUGH, STRATIGRAPHY, KENTUCKY
M 29 340 ROMERAL LINEAMENT, COLOMBIA, PALEOTRENCHES
B 6108 1389 ROMERO, JOHN C., GEOTHERMAL EXPLORATION IN CENTRAL
B 6303 498 ROMINE, KAREN, IN SEARCH OF RESOLUTION, ABST.
M 25 48 RONA, PETER A., PLATE TECTONICS AND MINERAL RESOURCES OF
M 27 194 RONDA UNIT, SPAIN, JURASSIC
M 27 221 RONDA UNIT, SPAIN, JURASSIC
M 27 88 RONDA UNIT, SPAIN, JURASSIC
M 27 111 RONDA UNIT, SPAIN, JURASSIC
B 6209 1615 RONDOUT FORMATION, APPALACHIAN BASIN
B 6209 1610 RONDOUT FORMATION, PENNSYLVANIA
B 6209 1610 RONDOUT LIMESTONE, NEW YORK
B 6004 715 RONNEVIK, H. C., GEOLOGY OF NORWEGIAN CONTINENTAL SHELF
B 6212 2403 ROOF CONDITIONS, MINES
B 6203 523* ROOF PROBLEMS IN COAL MINES, ABST., USE OF DEPOSITIONAL, MODELS
B 6209 1753 ROOKE, J. H., FUTURE OIL POTENTIAL OF LOWER CRETACEOUS
B 6109 1534 ROOKE, J. H., SMACKOVER OIL PROSPECTS IN APALACHICOLA
B 6305 836* **ROOSEVELT HOT SPRINGS KGRA, BEAVER COUNTY, UTAH,**
B 6008 1390* ROOSEVELT KGRA, UTAH, ABST., GEOTHERMAL EXPLORATION OF
B 6109 1449 ROOSEVELT UPLIFT, NEW MEXICO AND, TEXAS
B 6303 434* ROOT CAST FROM KOOBI FORA FORMATION, PLIOCENE-PLEISTOCENE,
B 6105 782 ROOT, D. J., STATISTICAL HISTORY OF PETROLEUM EXPLORATION IN
B 6409 1484 ROOT, MICHAEL R., DEVELOPMENTS IN EASTERN AND NORTHWESTERN
B 6405 776 ROOT, MICHAEL R., ZENITH FIELD—SIGNIFICANT DAKOTA, MUDDY, D
S 3 85 ROOT ZONE, OROGENE
M 27 163 ROOT-HAIR CEMENTS
B 6309 1610 ROPER, PAUL J., EVIDENCE FOR POST-JURASSIC TECTONISM IN EASTERN
B 6209 1765 ROPER, PAUL S., STRUCTURAL FABRIC OF SERPENTINITES AND
B 6203 370 ROSA BANK, BAJA CALIFORNIA
S 11 286 ROSALINA COLUMBIENSIS
S 11 255 ROSALINA COLUMBIENSIS
B 6104 579 ROSARIO FIELD, VENEZUELA
B 6405 765 ROSE, ARTHUR W., DIAGENETIC MOBILIZATION OF URANIUM AND IRON
B 6402 225 ROSE GRABEN, TEXAS
B 6402 225 ROSE GRABEN, TEXAS
B 6411 2037 ROSE, HERBERT E. HAWKES, AND JOHN S. WEBB, BRIGHT, M. J., REVIEW
M 28 29 ROSE HILL FORMATION, VIRGINIA, SILURIAN
B 6211 2309 ROSE HILL FORMATION, WEST VIRGINIA
B 6304 652 ROSE HILL SHALE, VIRGINIA
B 6405 635 ROSE HILL SHALE, WEST VIRGINIA
B 6105 823 ROSE, PETER R., MISSISSIPPIAN CARBONATE-SHELF MARGINS, WESTERN
B 6109 1552 ROSE, PETER R., SIGNIFICANCE OF FRESHWATER LIMESTONES IN MARINE
B 6303 517 ROSE, PETER R., USING PETROLEUM RESOURCE ASSESSMENT TO IMPROVE
B 6303 344 ROSE RUN SANDSTONE, KENTUCKY
B 6412 2191 ROSE, W. I., JR., REVIEW OF MANUAL FOR TEPHROCHRONOLOGY, BY
B 6309 1586 ROSE, WALTER, CONSTRAINTS AND PROBLEMS IN PRODUCING GAS FROM
B 6309 1581 ROSE, WALTER, MODELING OF DEVONIAN SHALE GAS-RESERVOIR
M 25 426 ROSEBERY, LEAD-ZINC-COPPER DEPOSIT, TASMANIA
B 6207 1194 ROSEBURG FORMATION, OREGON AND, WASHINGTON
B 6206 951 ROSEDALE CHANNEL, CALIFORNIA
B 6211 2243 ROSEDALE CYCLE, INFORMAL NAME, CALIFORNIA
B 6211 2245 ROSEDALE FAN, CALIFORNIA
B 6211 2247 ROSEDALE FAN, CALIFORNIA
B 6211 2256 ROSEDALE SAND, CALIFORNIA
B 6211 2247 ROSEDALE-BELLEVUE TREND, CALIFORNIA
B 6105 788 ROSEN, P. S., WAVE-CLIMATE STUDIES IN BALTIMORE CANYON TROUGH
B 6303 514 ROSEN, PETER S., DEPOSITIONAL SEQUENCE IN SUBARCTIC SANDY BEACH
B 6405 776 ROSENFELD, JEFFREY K., OIL-TO-SOURCE CORRELATION—PINEVIEW
B 6303 411 ROSENMEIER, FRED, GLACIAL ORIGIN OF MEGACHANNELS OF UPPER
B 6106 880 ROSETTA FAN, EGYPT
B 6407 1005 ROSEWOOD FIELD, ARK-LA-TEX AREA
B 6008 1332 ROSHARDT, MARY A., OIL AND GAS DEVELOPMENTS IN NORTH
B 6207 1194 ROSLYN FORMATION, OREGON AND, WASHINGTON
B 6004 715 ROSS, C. A., FUSULINACEAN PROVINCES OF NORTH AMERICAN
B 6105 824 ROSS, C. A., PERMIAN MARINE FAUNAS IN BIOGEOGRAPHY AND
S 4 219 ROSS, CHARLES A., CALCIUM CARBONATE FIXATION BY LARGE
B 6303 517 ROSS, CHARLES A., CONTRASTING PENNSYLVANIAN CARBONATE
B 6203 558 ROSS, CHARLES A., DEPOSITIONAL HISTORY OF UPPER PALEOZOIC
B 6311 2058 ROSS, CHARLES A., DEPOSITIONAL SETTING AND COAL PETROLOGY OF
B 6106 872 ROSS, DAVID A., MEDITERRANEAN SEA-NILE CONE AREA, STRUCTURE
B 6004 716 ROSS, DAVID A., RECENT SEDIMENTATION ON NILE DELTA AND CONE,
B 6203 559 ROSS, DAVID A., SEDIMENTARY PROCESSES ON CONTINENTAL SLOPE OFF
B 6405 799 ROSS, H. P., STRATEGY OF EXPLORATION FOR HIGH TEMPERATURE
B 6406 823 ROSS HILL, AUSTRALIA

B 6405 779* ROSS ICE SHELF, ABST., ORGANIC MATTER IN SEDIMENTS UNDERLYING
B 6105 824 ROSS, J. R. P., PERMIAN MARINE FAUNAS IN BIOGEOGRAPHY AND
M 24 1 ROSS, R. F., MIDDLE GROUND SHOAL OI FIELD, ALASKA
S 6 347 ROSS, REUBEN J., JR., ORDOVICIAN GEOCHRONOLOGY
B 6203 541 ROSS, REUBEN J., JR., STEEPLY INCLINED ALGAL COLUMNS IN LATE
S 11 257 ROTALIA BECCARII
S 11 288 ROTALIA BECCARII
S 11 120 ROTALIA GARVEYENSIS
S 11 120 ROTALIA, PULLENIA, LAGENIDAE, MIOCENE
B 6207 1114 ROTATION ASSOCIATED WITH BREAKOUTS, ASPECTS OF, TOOL
B 6303 423 ROTH, P. H., MID-CRETACEOUS CALCAREOUS NANNOPLANKTON
B 6205 892 ROTH, PETER H., PALEOBIOGEOGRAPHY AND PRESERVATIOON OF
B 6004 677* ROTLIEGENDE, VERTEBRATE FOOTPRINTS, ABST., PALEOENVIRONMENT
B 6203 490* ROTLIEGENDES FORMATION IN DUTCH SECTOR OF NORTH SEA, ABST.,
B 6401 34 ROTLIEGENDES FORMATION, NETHERLANDS AND, WEST, GERMANY
M 30 50 ROTLIEGENDES SANDSTONE, EOLIAN SANDS, VIKING FIELD
M 30 44 ROTLIEGENDES SANDSTONE FACIES DISTRIBUTION, VIKING FIELD
M 30 51 ROTLIEGENDES SANDSTONE, FACIES ISOPACH, VIKING FIELD
M 30 43 ROTLIEGENDES SANDSTONE ISOPACH, VIKING FIELD
M 28 71 ROTLIEGENDES SANDSTONE, NORTH SEA, PERMIAN
M 28 136 ROTLIEGENDES SANDSTONE, NORTH SEA, PERMIAN
M 30 41 ROTLIEGENDES SANDSTONE, NORTH SEA, VIKING FIELD
B 6307 1031 ROTLIEGENDES SANDSTONE, SEISMIC PROFILE, NORTH SEA, CENTRAL
M 28 132 ROTLIEGENDES SANDSTONES, NORTH SEA, PERMIAN
M 25 479 ROTMAN, V. K., MEIMECHITES OF KAMCHATKA AND KIMBERLITE
M 25 481 ROTMAN, V. K., OUTLINE OF KURIL-KAMCHATKA ARC--SUMMARY
B 6405 776 ROTTENFUSSER, BRIAN A., FACIES CONTROL ON BITUMEN SATURATION,
M 28 102 ROUNDNESS, ANGULARITY, CLASSIFICATION
B 6308 1425* ROUNDTOP AREA, FISHER COUNTY, TEXAS, ABST., TURBIDITE CHANNEL,
B 6108 1389 ROUSE, JIM V., APPLICABLE REGULATIONS OF IN-SITU MINING OF
B 6004 716 ROUSE, JOHN T., AAPG OIL AND GAS FIELD DATA BANK PROJECT,
B 6004 654 ROWE, G. T., SEDIMENT PROCESSES CONTROLLED BY BOTTOM CURRENTS
B 6203 559 ROWE, GILBERT T., BIOTA AND BIOLOGIC PROCESSES OF CONTINENTAL
B 6109 1545 ROWE, PETER G., NATURAL ENVIRONMENTAL-INFORMATION AND
B 6201 1 ROWLANDS, P. H., HYDROCARBON HABITAT OF TERTIARY NIGER DELTA
B 6303 518 ROWLEY, PETER D., METALLIC MINERAL RESOURCES OF
B 6010 1658 ROWLINSON, N. R., ON DEVELOPMENTS IN COLOMBIA, 1975
B 6110 1582 ROWLINSON, N. R., ON DEVELOPMENTS IN COLOMBIA, 1976
B 6004 716 ROY, E. C., JR., SEDIMENTOLOGY AND PALEOECOLOGY OF TYPE
B 6405 777 ROY, EDWARD C., JR., PALEOCENE, MIDWAY, CONTINENTAL SHELF
B 6303 413 ROY, KENNETH J., SVERDRUP BASIN RESPONSE TO CRATONIC EVENTS,
B 6002 197 ROYAL CHARLOTTE BANK, BRAZIL MARGIN
B 6007 1080 ROYAL DEEP, BATHYMETRIC PROFILE
B 6405 683 ROYDEN, L., FORELAND FOLD AND THRUST BELT OF CARPATHIANS AND
B 6402 173 ROYDEN, LEIGH, CONTINENTAL MARGIN SUBSIDENCE AND HEAT
B 6108 1389 ROYSE, FRANK, JR., THRUST BELT OF WYOMING, IDAHO, AND
B 6002 299 ROZENDAL, ROGER A., SUBSURFACE POSITIVE ELEMENTS WITHIN
B 6104 597 RUAGA FIELD, LIBYA
M 24 97 RUBEN FIELD, WYOMING
S 6 364 RUBIDIUM CONSTANT, PRECAMBRIAN, RADIOMETRIC DATING
S 6 109 RUBIDIUM-STRONTIUM, BRAZIL, BAMBUI GROUP
S 6 108 RUBIDIUM-STRONTIUM, BRAZIL, ESTANCIA FORMATION
S 6 105 RUBIDIUM-STRONTIUM, BRAZIL, ESTRADA NOVA FORMATION
S 6 104 RUBIDIUM-STRONTIUM, BRAZIL, IRATI FORMATION
S 6 106 RUBIDIUM-STRONTIUM, BRAZIL, PONTA GROSSA FORMATION
S 6 106 RUBIDIUM-STRONTIUM, BRAZIL, RIO BONITO FORMATION
S 6 102 RUBIDIUM-STRONTIUM, BRAZIL, RIO DO RASTO FORMATION
S 6 107 RUBIDIUM-STRONTIUM, BRAZIL, SEPOTUBA FORMATION
S 6 101 RUBIDIUM-STRONTIUM, BRAZIL, TROMBETAS FORMATION
S 6 152 RUBIDIUM-STRONTIUM DATA, JAPAN, MIYAKO GRANITE
S 6 152 RUBIDIUM-STRONTIUM DATA, JAPAN, TARO GRANITE
S 6 131 RUBIDIUM-STRONTIUM, ISOCHRON DIAGRAM, CENOMANIAN
S 6 156 RUBIDIUM-STRONTIUM ISOCHRON DIAGRAM, JAPAN, MIYAKO GRANITE
S 6 157 RUBIDIUM-STRONTIUM ISOCHRON PLOT, JAPAN, TARO GRANITE
S 6 93 RUBIDIUM-STRONTIUM METHOD, A REVIEW
S 6 97 RUBIDIUM-STRONTIUM METHOD, PELITIC ROCKS
S 6 93† RUBIDIUM-STRONTIUM METHOD TO SHALES AND RELATED ROCKS,
B 6410 1718† RUBIDIUM-STRONTIUM MINIMUM AGES OF SEDIMENTATION, URANIUM
B 6209 1673† RUBIDIUM-STRONTIUM MINIMUM AGES OF SEDIMENTATION, URANIUM
S 6 100 RUBIDIUM-STRONTIUM, SHALES, CASE HISTORIES
S 6 364 RUBIDIUM-STRONTIUM, USSR, PRECAMBRIAN
B 6008 1392 RUBIN, BRUCE, URANIUM IN TERTIARY CHANNEL SOUTHEAST OF LAKE
B 6004 717 RUBIN, D. M., THREE DIAGENETIC FEATURES FROM WHITEHALL
B 6405 777 RUBIN, DAVID M., TWO CONTROLS OF SAND-WAVE SIZE—DYNAMIC
B 6203 490 RUBIN, MEYER, PROFILE OF COAL-FORMING SWAMP—THE EVERGLADES,
S 6 205 RUBINSTEIN, M., ON DATING OF THE PALEOGENE
B 6004 716 RUBY, C. H., LITTORAL PROCESSES AND GEOMORPHIC VARIABILITY ON
B 6203 512 RUBY, CHRISTOPHER H., DEPOSITIONAL MODEL OF MESOTIDAL
B 6303 518 RUBY, CHRISTOPHER H., HOLOCENE STRATIGRAPHY OF RAPIDLY
B 6306 877 RUDIST BUILDUPS AND SOURCE ROCK
B 6306 876 RUDIST BUILDUPS AS STRATIGRAPHIC TRAPS, CRETACEOUS
S 4 83† RUDIST FRAMEWORKS AND THEI GEOLOGIC SETTING, JAMAICAN,
B 6306 881 RUDIST REEF BUILDUP ON ATLANTIC-TYPE PASSIVE MARGIN,
B 6307 1108† RUDIST REEFS, ARIZONA, DEPOSITIONAL, MODEL OF EARLY,
B 6306 877 RUDIST-CORAL REEFAL BUILDUPS, DISTRIBUTION, CRETACEOUS
B 6002 258† RUDISTID FRINGING REEFS OF CRETACEOUS SHALLOW CARBONATE
M 27 84 RUDISTID MOLLUSKS
M 27 14 RUDISTID REEF
M 27 85 RUDISTS, CAPRINID
B 6109 1551* RUDISTS, OF GULF COAST, ABST., EARLY AND MIDDLE, CRETACEOUS,
M 27 85 RUDISTS, RADIOLITID
M 27 85 RUDISTS, TOUCASIAID
B 6004 651 RUDMAN, A. J., COMPUTER-GENERATED SCALE FUNCTIONS FOR
B 6303 518 RUDMAN, ALBERT J., COMPUTER PREDICTION OF CHANGES IN
B 6009 1624 RUDMAN, ALBERT J., PRECAMBRIAN GEOPHYSICAL PROVINCES IN
B 6005 879 RUDMAN, ALBERT J., TRANSFORMATION OF RESISTIVITY TO
B 6303 519 RUDOLPH, KURT W., DIAGENESIS OF BACK-REEF CARBONAT
B 6403 428 RUDRASAGAR FIELD, INDIA
B 6312 2222 RUESLATTEN, HAKON G., COMPOSITION OF ORGANIC MATTER FROM
S 5 53* RUFF FORMATION (UPPER SILURIAN) IN SOUTHEAST MICHIGAN,
S 5 16 RUFF FORMATION, MICHIGAN BASIN, SALINA GROUP,
S 5 66 RUFF FORMATION, DEPOSITIONAL HISTORY, SILURIAN, SALINA GROUP,
S 5 58 RUFF FORMATION, LITHOFACIES DESCRIPTIONS, SILURIAN, SALINA
S 5 67 RUFF FORMATION, RABBIT EARS ANHYDRITE, SALINA GROUP,
M 27 62 RUGOSE CORALS, COLONIAL
M 27 62 RUGOSE, CORALS, SOLITARY
M 27 63 RUGOSE CORALS, SOLITARY
B 6405 649 RUHLE, EDWARD, REVIEW OF LITHOFACIES-PALEOGEOGRAPHICAL ATLAS
M 25 427 RUM JUNGLE, URANIUM, COPPER, AND LEAD, AUSTRALIA, NORTHERN
B 6309 1610 RUMEN, LINDA L., ORGANIC GEOCHEMISTRY OF DEEP WELL IN HINDS
S 4 128 RUNAWAY BAY, REEF, JAMAICA
B 6105 809* RUNDLE OIL SHALE DEPOSIT, QUEENSLAND, AUSTRALIA
B 6105 824 RUNNELS, T. D., COMPUTER GEOLOGY IN SMALL COMPANY, ABST.,
B 6207 1212* RUNOFF IN CIRCUM-PACIFIC REGION, ABST., POSSIBLE RELATIONS
S 6 221 RUPELIAN STAGE DEFINED, ERUOPE, OLIGOCENE
B 6405 798 RUPPEL, S. C., PALEOECOLOGIC DETERMINATION OF BATHYMETRIC
B 6312 2194 RURIKFJELLET MEMBER, SVALBARD
B 6002 326* RUSH CREEK MEMBER OF WOODBINE FORMATION, NORTH-CENTRAL
B 6302 247 RUSH SPRINGS SANDSTONE, OKLAHOMA
B 6004 566 RUSHKAR, PAUL, PRE-CRETACEOUS ROCKS OF NORTHWESTERN
M 25 59 RUSS, V. V., STRUCTURAL-FORMATION ANALYSIS OF NORTHWESTERN
B 6203 569* RUSSELL COUNTY, KANSAS, ABST., GORHAM OIL FIELD
B 6103 417 RUSSELL FORK FAULT, VIRGINIA
B 6303 428 RUSSELL, K. L., GEOLOGIC DISTRIBUTION OF OIL ABST.
B 6004 717 RUSSELL, K. L., ORGANIC GEOCHEMISTRY AS AID TO EXPLORATION IN
B 6203 353 RUSSELL RANCH OIL FIELD, CALIFORNIA
B 6202 178 RUSSIA, CHINA, ET AL, ENERGY RESOURCES
B 6412 2109 RUSSIAN PLATFORM, ORGANIC CARBON
B 6009 1609 RUSSO, MARTIN, SOUTH SAREPTA FIELD, GRAY SANDSTONE,
B 6405 719 RUST, CLAUDE C., ET AL, CONODONT-BASED ASSESSMENT OF THERMAL
B 6303 536 RUST, CLAUDE C., SOUTH HALLETTSVILLE, GAS FIELD IN LOWER WILCOX,
M 25 418 RUTILE, COMPOSITION AND USES
M 28 59 RUTILE, CRYSTALS, IN GRAIN MOUNT
M 28 9 RUTILE INCLUSIONS, QUARTZ
B 6303 519 RUTLEDGE, C., IN-SITU URANIUM SOLUTION MINING--ENVIRONMENTAL
B 6104 585 RUWAIS FIELD, PERSIAN GULF
B 6405 777 RUZYLA, KENNETH, MECHANISMS CONTROLLING POROSITY IN RED
B 6111 1962 RV ALAMINOS 1973 CRUISE AL-11
B 6206 1067 RV ALAMINOS CRUISE 11-73
B 6406 871 RV ATLANTIS II, 1976 GEOPHYSICAL INVESTIGATIONS
B 6404 501 RV ATLANTIS II CRUISE 91
B 6109 1483 RV ATLANTIS II CRUISE 91, JULY-AUGUST 1975
B 6106 861 RV ATLANTIS II CRUISE, JUNE 2-10, 1975
B 6306 934 RV BESNARD CRUISE, 1973
B 6106 872 RV CHAIN CRUISE 119, FEBRUARY-APRIL 1975
B 6412 2126 RV CHAIN CRUISE 161
B 6302 218 RV CHAIN CRUISE, 1974
B 6209 1647 RV CHIU LIEN, TAIJUNT CRUISE, 1975
B 6007 1050 RV CONRAD, CRUISE 10--LEG 12
B 6204 660 RV CONRAD CRUISE 15
B 6104 546 RV CONRAD CRUISES 15 AND 16
B 6410 1658 RV CONRAD SEISMIC REFLECTION PROFILES
B 6204 698 RV EASTWARD CRUISE E-16A-74
B 6204 701 RV EASTWARD CRUISE E-1B-74
B 6204 702 RV EASTWARD CRUISE E-1C-77
B 6410 1703 RV EASTWARD, SEISMIC REFLECTION PROFILES

Index

S 4 54 RV EASTWARD SOUNDINGS
B 6007 1039 RV EASTWOOD CRUISE, JANUARY, 1974, DUKE UNIVERSITY
B 6403 339 RV FAY CRUISES 017, 018, AND 025
B 6410 1660 RV FAY CRUISES 19, 20, 21, AND 25, USGS
B 6204 700 RV GILLIS CRUISE
B 6007 1042 RV GLOMAR CHALLENGER, DSDP SITE 98
B 6205 794 RV IDA GREEN CRUISE
B 6206 1067 RV IDA GREEN CRUISES 1G-7 AND 1G-8
B 6109 1483 RV LULU CRUISE 33, SEPTEMBER 1970
B 6007 1113 RV MOUNT MITCHELL, CRUISE, 1974
B 6106 861 RV MT. MITCHELL CRUISE, MAY 17-27, 1974
B 6002 196 RV ROBERT D. CONRAD CRUISES THROUGH MAY 1974
B 6407 1014 RV S.P. LEE, CRUISES 1976 AND 1977
B 6406 945 RV SEA SOUNDER CRUISE S6-77
B 6204 633 RV SEA SOUNDER RESEARCH CRUISE
B 6204 636 RV THOMAS G. THOMPSON CRUISE, SEPTEMBER-OCTOBER 1974
B 6406 871 RV THOMAS WASHINGTON, 1976 GEOPHYSICAL INVESTIGATIONS
B 6111 2004 RV VEMA 1975 CRUISE 3205
B 6406 906 RV VEMA CRUISE 28
B 6204 658 RV VEMA CRUISE 31, STATION 114
B 6002 196 RV VEMA CRUISES THROUGH MAY 1974
B 6210 1849 RWANDA, SWAZILAND, UPPER VOLTA, ZAMBIA, DEVELOPMENTS, 1977,
B 6004 717 RYAN, J. D., EOCENE PROVENANCE OF HANNA AND CARBON BASINS
B 6203 559 RYAN, J. DONALD, VOLCANOGENIC SEDIMENTATION ON ANDEAN
B 6305 832* RYCKMA CREEK NUGGET SANDSTONE TREND IN ABSAROKA THRUST
B 6206 1076 RYCKMAN CREEK FIELD, WYOMING
B 6405 638 RYCKMAN CREEK FIELD, WYOMING
B 6108 1371 RYDER, R. T., APPLICATIONS OF VERTICAL SEISMIC PROFILE TO
B 6004 718 RYDER, R. T., SYNTHETIC SEISMIC SECTIONS OF SELECTED
B 6108 1389 RYDER, ROBERT T., HYDROCARBON POTENTIAL OF ARCHULETA
B 6405 778 RYDER, ROBERT T., MIDDLE MEMBER OF MINNELUSA FORMATION,
B 6405 778 RYDER, ROBERT T., SEISMIC MODEL OF 15 STRATIGRAPHICALLY
B 6305 840 RYDER, ROBERT T., SEISMIC MODEL OF FIFTEEN STRATIGRAPHICALLY
B 6008 1409 RYDER, ROBERT T., SYNTHETIC SEISMIC SECTIONS OF SELECTED
B 6101 112 RYER, THOMAS A., AGE OF FRONTIER FORMATION IN NORTH-CENTRAL
B 6105 824 RYER, THOMAS A., CYCLIC SEDIMENTATION, WESTERN INTERIOR
B 6303 519 RYER, THOMAS A., DELTAIC COALS OF FERRON SANDSTONE MEMBER OF
B 6305 840 RYER, THOMAS A., DEPOSITIONAL SETTING OF COALS IN UP
B 6405 731 RYER, THOMAS A., PALEOBIOLOGIC EVIDENCE FOR CRETACEOUS TIDES,
B 6405 778 RYER, THOMAS A., PHYSICAL EVIDENCE FOR CRETACEOUS TIDES,
M 25 441 RYONGNAM MASSIF, SOUTH KOREA
M 25 440 RYONGNAM SYSTEM, SOUTH KOREA
B 6209 1670 RYUKYU ISLAND ARC AND TRENCH, STRUCTURE
M 25 47 RZHONSNITSKAYA, M. A., GEOLOGIC HISTORY AND
M 26 22 S-WAVE MEASUREMENT LABORATORY SYSTEMS, BROADBAND VELOCITY
M 30 530 S. GIOVANNI BIANCO FORMATION, TRIASSIC, MALOSSA FIELD
B 6109 1476 S. M. CAMPOS FIELD, BRAZIL
B 6407 1014 S.P. LEE, CRUISES 1976 AND 1977, RV
B 6004 668* SA ANDREAS FAULT IN SOUTHERN CALIFORNIA AND ITS PALINSPASTIC
B 6102 237† SA NICOLAS ISLAND, CALIFORNIA, EOCENE, PALEOCURRENTS AND
B 6408 1191 SAANICH INLET, POSITIVE WATER BALANCE BASIN, BRITISH COLUMBIA
S 4 3 SABA BANK, CORAL REEFS
B 6411 1896 SABAH, PRODUCTION, 1977-1979, MALAYSIA EAST
M 30 434 SABAH, SAMARANG FIELD, STRATIGRAPHY AND PRODUCTION
M 27 42 SABELARIID WORM TUBES
M 30 550 SABIL FORMATION, LIBYA, SIRTE BASIN
B 6206 1051 SABINE ARCH, TEXAS
B 6206 1006 SABINE PENINSULA, ARCTIC, CANADA
B 6408 1220 SABINE UPLIFT, SECTION, TEXAS
B 6206 1051 SABINE UPLIFT, TEXAS
B 6407 1006 SABINE UPLIFT, TEXAS AND, LOUISIANA
B 6207 1232 SABINS, FLOYD F., APPLICATIONS OF LANDSAT IMAGERY IN
B 6405 619 SABINS, FLOYD F., JR., SEASAT RADAR IMAGE OF SAN ANDREAS FAULT,
B 6303 505* SABKHA, ABST., SULFUR AND, OXYGEN, ISOTOPE, GEOCHEMISTRY OF,
B 6303 496* SABKHA, ABU DHABI, UNITED ARAB EMIRATES, AND ITS RELATION TO
S 4 384 SABKHA BELT, SECTION, ABU DHABI COASTAL-CONTINENTAL
B 6404 479 SABKHA BRINES
B 6111 1934 SABKHA CYCLE, IDEALIZED
B 6008 1406* SABKHA ENVIRONMENT--NEW FRONTIER FOR URANIUM EXPLORATION,
B 6004 688* SABKHA FACIES NODULAR ANHYDRITE AND PALEOTEMPERATURE
S 4 382 SABKHA HYDROLOGY
S 4 381† SABKHA, MARINE AND, CONTINENTAL, GROUNDWATER SOURCES IN A,
S 4 381 SABKHA MORPHOLOGY AND STRATIGRAPHY
M 28 129 SABKHA SANDS, ABU DHABI, HOLOCENE
B 6103 373 SABKHA, SANDSTONES-SILTSTONES, OKLAHOMA, TIDAL-FLAT
B 6304 692* SABKHA-LIKE DEPOSIT, ABST., URANIUM IN, TODILTO LIMESTONE--
B 6004 665 SABLE, EDWARD G., NE APPROACH TO GEOLOGIC ESTIMATES OF OIL AND
B 6301 90 SABLE MEMBER, SCOTIAN BASIN
B 6301 89 SABLE SUBBASIN, SCOTIAN BASIN
B 6012 2191* SACATE AND GAVIOTA FORMATIONS, ARROYO EL BULITO, SANTA
S 11 227 SACCAMMINA SPHAERICA
S 11 268 SACCAMMINA SPHAERICA
S 11 227 SACCAMMINIDAE
B 6405 687* SACHA FIELD OF ECUADORIAN ORIENTE, ABST.
B 6304 694 SACHDEV, SURESH C., MINERALOGICAL AND GEOCHEMICAL ZONATION
B 6405 779 SACKETT, WILLIAM M., ORGANIC MATTER IN SEDIMENTS UNDERLYING
B 6303 519 SACKETT, WILLIAM M., STABLE CARBON ISOTOPES A
B 6303 549 SACKETT, WILLIAM M., URANIUM IN ORCA BASIN SEDIMENTS, ABST.
B 6208 1355 SACKREITER, D. K., DEVELOPMENTS IN EASTERN AND NORTHWESTERN
S 4 243 SACRAMENTO MOUNTAINS, NEW MEXICO
B 6105 819* SACRAMENTO MOUNTAINS, NEW MEXICO, ABST., ORIGIN OF EARLY,
B 6112 2115† SACRAMENTO MOUNTAINS, NEW MEXICO, EVOLUTION OF, YUCCA,
B 6203 569* SACRAMENTO SHELF, NEW MEXICO, ABST., PENNSYLVANIAN,
B 6206 951 SACRAMENTO VALLEY, CALIFORNIA
B 6211 2244 SACRAMENTO VALLEY, CALIFORNIA
B 6301 23 SACRAMENTO VALLEY, CALIFORNIA
B 6203 515* SACRAMENTO VALLEY, CALIFORNIA, ABST., DEPOSITIONAL SYSTEMS AND
B 6211 2353* SACRAMENTO VALLEY, CALIFORNIA, ABST., PLATE-TECTONIC
B 6211 2360* SACRAMENTO VALLEY, CALIFORNIA, ABST., TERTIARY, PRINCETON
B 6004 645* SACRAMENTO VALLEY, CALIFORNIA, ABST., TIMING OF, TERTIARY,
B 6012 2179* **SACRAMENTO VALLEY, CALIFORNIA, ABST., UPPER CRETACEOUS**
B 6301 27 SACRAMENTO VALLEY PETROLEUM OCCURRENCE SUMMARY DIAGRAM,
B 6205 818 SACRAMENTO VALLEY, SOURCE ROCKS, CALIFORNIA
B 6008 1409 SACRISON, WILLIAM R., SEISMIC EVIDENCE FOR STRUCTURAL STYLE OF
B 6305 832 **SACRISON, WILLIAM R., WHITNEY CANYON FIELD--POTENTIAL GAS GIANT**
B 6405 730 SACRISON, WILLIAM R., WHITNEY CANYON FIELD—POTENTIAL GAS
B 6408 1261 SADDLE MOUNTAIN AREA, ALASKA
M 24 97 SADDLE RIDGE FIELD, WYOMING
B 6406 821 SADLER FORMATION, AUSTRALIA
B 6401 20 SADLEROCHIT FORMATION, ALASKA
M 24 26 SADLEROCHIT GROUP, ALASKA
B 6109 1496 SADLEROCHIT GROUP, ALASKA
M 30 301 SADLEROCHIT GROUP, IVISHAK SANDSTONE, PRUDHOE BAY
B 6103 438 SADLEROCHIT SANDSTONE, ALASKA
B 6109 1499 SAG DELTA SECTION, ALASKA
M 24 40 SAG RIVER FORMATION, ALASKA
M 30 303 SAG RIVER FORMATION, RESERVOIRS, PRUDHOE BAY
M 24 26 SAGAVANIRTOK FORMATION, ALASKA
B 6404 490 SAGE BREAKS MEMBER, CODY SHALE, WYOMING
B 6406 943 SAGE BREAKS MEMBER, SOUTH DAKOTA
B 6108 1373† SAGEBRUSH AND GEOCHEMICAL EXPLORATION IN EAST TINTIC
S 5 80 SAGINAW COUNTY LOCATION, MICHIGAN,
B 6004 666 SAGRI, M., SPECTRAL AND MARKOVIAN ANALYSIS OF BED THICKNESS
B 6307 1035 SAGS, BASEMENT, WARPS--, ARCHES, DOMES, AND
B 6405 755* SAGUACHE COUNTY, COLORADO, ABST., URANIUM OCCURRENCE IN,
B 6110 1901 SAHAI, SURINDER K., CLAY, MINERALS AND PETROLEUM-FORMING
S 9 8 SAHARA, AFRICA, ICE FLOW DIRECTION, ORDOVICIAN
B 6204 668 SAHARA BASIN, ALGERIA, NORTHEAST
M 29 463 SAHARA BASIN, VOLUME OF SEDIMENT PRESERVED
B 6204 669 SAHARA, EASTERN, HERCYNIAN UNCONFORMITY, ALGERIAN
B 6101 81 SAHARA SALT BASIN, NORTHERN
B 6101 81 SAHARAN ATLAS SALT BASIN
B 6204 667† SAHARA—INFLUENCE ON HYDROCARBON OCCURRENCE,
B 6104 585 SAHIL FIELD, PERSIAN GULF
B 6311 2024 SAHUL PLATFORM, AUSTRALIA
M 25 241 SAHUL SHELF, INDONESIA, ARAFURA PLATFORM OR
B 6311 2024 SAHUL SYNCLINE, AUSTRALIA
B 6403 427 SAIKIA, M. M., DEPOSITIONAL ENVIRONMENT OF SOURCE BEDS OF
B 6207 1131 SAILEEN BAY, IRIAN JAYA
B 6103 467 SAINSBURY, C. L., COMPRESSIVE INTERACTION OF SIBERIAN AND NORTH
B 6104 513 SAINT BRICE WELL, FRANCE
B 6304 652 SAINT CLAIR FAULT SYSTEM, VIRGINIA
S 4 291 SAIPAN, GALAXEA, MARIANA ISLANDS
S 4 178 SAIPAN, MIOCENE, REEF CORAL
B 6407 1087† SAKHALIN, STRATIGRAPHY OF MARINE, PALEOGENE AND, NEOGENE OF
S 6 301 SAKMARIAN STAGE, PERMIAN, BIOCHRONOLOGY
B 6012 2185* SAL LAKE BEDS IN NORTHEASTERN GREAT BASIN, ABST., NEW EVIDENCE
M 28 89 SALADO FORMATION, NEW MEXICO, PERMIAN
S 5 96 SALAMONIE DOLOMITE, INDIANA,
B 6401 129 SALAMONIE DOLOMITE, INDIANA
M 25 515 SALAS, GUILLERMO P., METALLOGENETIC CHART OF MEXICO--SUMMARY
M 25 337 SALAVERRY BASIN, PERU
B 6207 1127† SALAWATI AREA, IRIAN JAYA, INDONESIA, TERTIARY TECTONIC HISTORY
M 25 248 SALAWATI BASIN, IRIAN JAYA
B 6009 1448† SALAWATI BASIN OF IRIAN JAYA, INDONESIA, OIL-BEARING, REEFS IN
B 6207 1138 SALAWATI ISLAND, IRIAN JAYA
B 6207 1233 SALDIVAR-SALI, ARTHUR, PROGRESS OF GEOTHERMAL ENERGY
B 6207 1233 SALDIVAR-SALI, ARTHUR, REEF EXPLORATION IN PHILIPPINES, ABST.
B 6211 2361 SALEEBY, J., EARLY MESOZOIC PALEOTECTONIC-PALEOGEOGRAPHIC
B 6303 520 SALEHI, IRAJ, RECENT HUGE GAS DISCOVERIES IN KANGAN AREA, IRAN,

M 27 90 SALEM LIMESTONE, INDIANA, MISSISSIPPIAN
B 6001 34 SALEM, RAFIK, EVOLUTION OF EOCENE-MIOCENE SEDIMENTATION
B 6408 1286* SALEM STRATIGRAPHY AND OIL PRODUCTION, OWENSVILLE NORTH
S 5 169* SALEMI BASIN (SICILY) AND A MODERN ANALOG, AN EVAPORITIC
S 5 175 SALEMI BASIN, ITALY, SICILY,
B 6209 1637 SALINA EVAPORITE UNITS, APPALACHIAN BASIN
B 6211 2309 SALINA EVAPORITES, WEST VIRGINIA
B 6405 636 SALINA EVAPORITES, WEST VIRGINIA
B 6401 129 SALINA FORMATION, KENNETH LIMESTONE MEMBER, INDIANA
S 5 94 SALINA FORMATION, KENNETH LIMESTONE
B 6401 128 SALINA FORMATION, KOKOMA LIMESTONE MEMBER, INDIANA
S 5 94 SALINA FORMATION, KOKOMO LIMESTONE
M 28 88 SALINA FORMATION, MICHIGAN, SILURIAN
S 5 130 SALINA FORMATION, TONOLOWAY LIMESTONE, MARYLAND, CAYUGAN
S 5 127 SALINA FORMATION, TONOLOWAY LIMESTONE, WEST VIRGINIA,
S 5 23* SALINA GROUP (UPPER SILURIAN), MICHIGAN BASIN, SEDIMENTOLOGY
B 6209 1611 SALINA GROUP, APPALACHIAN BASIN
B 6209 1610 SALINA GROUP, NEW YORK
B 6209 1610 SALINA GROUP, OHIO
B 6209 1611 SALINA GROUP, UNITS, MICHIGAN BASIN
S 5 1* SALINA GROUPS, NORTHERN SHELF, MICHIGAN BASIN, DEPOSITIONAL
S 5 4 SALINA GROUP, MICHIGAN BASIN, CAYUGAN SERIES,
S 5 3 SALINA GROUP, MICHIGAN, SILURIAN,
S 5 3 SALINA GROUP, A-0 CARBONATE
S 5 3 SALINA GROUP, A-1 CARBONATE
S 5 3 SALINA GROUP, A-1 EVAPORITE
S 5 30 SALINA GROUP, A-1 EVAPORITE, MICHIGAN BASIN,
S 5 4 SALINA GROUP, A-2 CARBONATE
S 5 4 SALINA GROUP, A-2 EVAPORITE
S 5 4 SALINA GROUP, B-SALT
S 5 4 SALINA GROUP, C-SHALE
S 5 4 SALINA GROUP, D-SALT
S 5 4 SALINA GROUP, E-UNIT
S 5 4 SALINA GROUP, F-SALT
S 5 4 SALINA GROUP, G-UNIT
S 5 24 SALINA GROUP, LITHOFACIES, MICHIGAN BASIN, LOWER
S 5 7 SALINA GROUP, PINNACLE REEFS, EVOLUTION, STAGES
S 5 16 SALINA GROUP, RUFF FORMATION, MICHIGAN BASIN,
S 5 66 SALINA GROUP, RUFF FORMATION, DEPOSITIONAL HISTORY, SILURIAN,
S 5 58 SALINA GROUP, RUFF FORMATION, LITHOFACIES DESCRIPTIONS,
S 5 67 SALINA GROUP, RUFF FORMATION, RABBIT EARS ANHYDRITE
S 5 54 SALINA GROUP, STRATIGRAPHY, MICHIGAN, SILURIAN, CAYUGAN
S 5 24 SALINA GROUP, TERMINOLOGY, SILURIAN,
B 6009 1621* SALINA, UPPER SILURIAN, OIL PRODUCTION AND SHALLOW STRUCTURE
B 6203 349 SALINAS BASIN, CALIFORNIA
S 5 177 SALINAS, ITALY, COMMERCIAL
S 5 178 SALINA, CALIFORNIA, COMMERCIAL
S 5 171 SALINA, ITALY, ADRIATIC COAST, SANTA MARGHERITA DI SAVOIA,
B 6212 2512 SALINAS, CARLOS,, EXPLORATION OF CARBONIFEROUS SYSTEM IN
B 6207 1207 SALINAS, O., DEVELOPMENT AND RESERVOIR ANALYSIS OF MOMOTOMBO
B 6211 2217 SALINAS VALLEY, CALIFORNIA
S 11 8 SALINAS VALLEY SEQUENCE, CALIFORNIA, MIOCENE
B 6012 2174† SALINAS VALLEY TODAY AND TOMORROW, ABST.
M 25 349 SALINAS-E., C., HYDROCARBON POTENTIAL OF BOLIVIAN SEDIMENTARY
M 30 343 SALINE BASIN, CINCO PRESIDENTES FIELD, MEXICO
M 30 342 SALINE BASIN, EL PLAN FIELD, MEXICO
M 30 345 SALINE BASIN, MAGALLANES FIELD, MEXICO
M 30 345 SALINE BASIN, OGARRIO FIELD, MEXICO
M 30 342 SALINE BASIN, TONALA-EL BURRO FIELD, MEXICO
B 6305 846* SALINE COUNTY, ILLINOIS, ABST., GEOLOGY OF, COAL DEPOSITS IN
S 5 125* SALINE FORMATION, WEST VIRGINIA, RELATIONS BETWEEN
M 25 355† SALINE GROUNDWATER INFLOWS TO MURRAY RIVER IN SOUTH
B 6211 2363* SALINE LAGOON COMPLEX, UPPER SESPE CREEK, VENTURA COUNTY,
B 6405 801* SALINE ZONE OIL SHALE, PICEANCE CREEK BASIN, NORTHWEST
B 6211 2217 SALINIAN BASEMENT, CALIFORNIA, SIERRAN-
B 6203 349 SALINIAN BLOCK, CALIFORNIA
B 6211 2214† SALINIAN BLOCK IN EVOLUTION OF SAN ANDREAS FAULT SYSTEM,
B 6211 2355* SALINIAN BLOCK OF CALIFORNIA, AN ENIGMA, ABST.
B 6002 230 SALINITIES, BASIN
S 10 223 SALINITY CONCENTRATION, TRAP FUNCTION
S 10 15 SALINITY EQUILIBRIUM, PORE WATER
S 10 212 SALINITY WITH DEPTH, LOUISIANA, CALCASIEU PARISH, MANCHESTER
B 6203 527* SALINITY, ABST., INTERACTING DYNAMICS OF, PRESSURE,
B 6203 571* SALINITY, CRACKING, AND MIGRATION, ABST., SHALE, DEHYDRATION
B 6207 1235* SALINITY IN DARLING RANGE, WESTERN AUSTRALIA, ABST., BAUXITE,
B 6304 590 SALINITY MAPS, GEOCHEMICAL TOOL
B 6004 676* SALINITY TIDAL-FLAT CARBONATE ROCKS, MIDDLE ORDOVICIAN,
B 6309 1453 SALISBURY EMBAYMENT, MID-ATLANTIC
B 6105 824 SALLENGER, A. H., JR., GRADING IN GRAIN FLOWS, ABST., DEVELOPMENT
B 6311 2118* SALLYARDS FIELD, GREENWOOD COUNTY, KANSAS, ABST.,
M 25 507 SALMO TYPE OF LEAD-ZINC DEPOSIT, CANADA
S 11 317 SALMON CREEK, CALIFORNIA, SAMPLE LOCALITIES
B 6402 222 SALMON PEAK FORMATION, TEXAS
B 6303 520 SALSMAN, CHARLES W., CONQUISTA PROJECT, URANIUM MINING AN
S 7 27 SALT AND DIAPIRS, JURASSIC, NORTHERN, GULF OF MEXICO
B 6303 427* SALT AND POST-SALT SEQUENC OF WEST AFRICA--RESULTS IN CABINDA,
B 6407 1007 SALT ANTICLINES, TEXAS, LOUISIANA, AND, ARKANSAS
B 6405 797* SALT BASIN, ABST., GEOLOGY OF, CALVIN FIELD—DEEP BASINAL,
B 6303 529* SALT BASIN, ABST., SEALING AND NONSEALING, FAULTS IN, GULF COAST
B 6407 1006 SALT BASIN, LOUISIANA
B 6102 207 SALT BASIN, MEXICO, ISTHMUS
B 6101 81 SALT BASIN, NORTHERN, SAHARA
B 6101 81 SALT BASIN, SAHARAN, ATLAS
B 6402 145† SALT BASIN, SEALING AND, NONSEALING, FAULTS IN, LOUISIANA, GULF
B 6407 1006 SALT BASIN, TEXAS, EAST TEXAS
B 6002 199 SALT BOUNDARY, BRAZIL COASTAL MARGIN
S 5 84 SALT FIELD, MICHIGAN, MIDLAND COUNTY, MIDLAND
S 7 49 SALT, CAMPECHE SCARP, GULF OF MEXICO
S 7 49 SALT, CHALLENGER KNOLL, GULF OF MEXICO
S 7 70 SALT, CONTINENTAL SLOPE, SEDIMENT DEPOSITION
B 6405 793* SALT CREEK AREA, WYOMING, ABST., SHANNON SANDSTONE, UPPER,
B 6202 188 SALT CREEK DOME, WYOMING
B 6211 2262 SALT CREEK SECTION, CALIFORNIA
S 7 50 SALT DEPOSITION, DISCUSSED, GULF OF MEXICO
B 6203 526* SALT DEPOSITS IN TEXAS PANHANDLE AND WESTERN OKLAHOMA, ABST.,
B 6206 997 SALT DIAPIR, IRAN, HORMUZ
B 6404 568† SALT DIAPIR, ISRAEL, STRAIN MEASUREMENTS IN EASTERN MARGINAL
B 6003 458† SALT DIAPIRISM IN SOUTHERN IRAN. DISCUSSION
B 6206 999 SALT DIAPIRS AND BUCKLE FOLDING, IRAN
S 7 82 SALT DIAPIRS, MAGNETIC LINEATIONS, NORTH ATLANTIC
B 6204 657† SALT DIAPIRS ON SAO PAULO PLATEAU, SOUTHEASTERN BRAZILIAN
B 6209 1570 SALT DIAPIRS, RADIAL AND, TRAP-DOOR, FAULTS ASSOCIATED WITH
B 6311 2097† SALT DIAPIRS, SEISMIC RECOGNITION OF
S 7 325 SALT DIFFUSION, NORTHWEST GULF OF MEXICO
B 6002 323* SALT DISSOLUTION IN OIL AND GAS TEST HOLE IN CENTRAL KANSAS,
B 6004 660* SALT DOM AREA, LOUISIANA, ABST., FOSSIL SALT STRUCTURES IN,
B 6105 834* SALT DOME, ABST., PALEONTOLOGIC TECHNIQUES FOR DETERMINING
B 6209 1768* SALT DOME AND SALT DOME GROWTH MODELS, ABST., DEPOSITIONAL
B 6405 668* SALT DOME AREA, ABST., PETROLEUM ACCUMULATION TRENDS IN, EAST,
S 7 72 SALT DOME GROWTH, CONTINENTAL SLOPE
S 7 76 SALT DOME GROWTH, DIAGRAM, CONTINENTAL SLOPE
B 6305 785 SALT DOME GROWTH ON CONTINENTAL SLOPE
B 6101 67 SALT DOME, LOUISIANA, COTE BLANCHE ISLAND
B 6209 1759* SALT DOME OF LOUISIANA—GEOTEMPERATURES AND HALOKINETICS,
B 6404 573 SALT DOME, SECTIONS, GERMANY, BENTHE
B 6409 1556* SALT DOME, SOUTH TEXAS, ABST., CAP-ROCK FORMATION AND
B 6109 1540* SALT DOME, WOOD COUNTY, TEXAS, ABST., GROWTH HISTORY OF,
B 6309 1601* SALT DOMES, EAST TEXAS, ABST., UPPER, CRETACEOUS--LOWE, EOCENE
B 6309 1605* SALT DOMES IN EAST TEXAS BASIN FOR GEOLOGIC ISOLATION OF
B 6211 2240 SALT DOMES IN PRESENT DEEP-SEA ENVIRONMENTS, COMPARISON OF,
B 6408 1272 SALT DOMES OF GULF REGION, UNITED STATES AND MEXICO, BY
B 6209 1574 SALT DOMES, ONSHORE, GULF COAST
B 6004 724* SALT DOMES--POTENTIAL SITES FOR RADIOACTIVE WASTE DISPOSAL,
B 6002 198 SALT EDGE, SOUTH ATLANTIC
B 6002 211 SALT EDGES, ALBIAN, MODEL, MATCHING, WEST, AFRICAN AND,
B 6307 1053 SALT FEATURES, NORTH SEA AREA
B 6307 1052 SALT FEATURES, ROMANIAN, CARPATHIAN, THRUST-FOLD BELT, PLOESTI
S 7 46 SALT, GULF OF MEXICO, CHALLENGER KNOLL, REFLECTION PROFILE
B 6305 790 SALT, GULF OF MEXICO, DIAPIRIC
S 7 58 SALT, GULF OF MEXICO, MISSISSIPPI CONE
S 7 47 SALT IDENTIFICATION, CRITERIA, GULF OF MEXICO
S 7 43† SALT IN DEEP GULF OF MEXICO, OCCURRENCE AND, EVOLUTION OF
B 6205 888* SALT LAKE BASIN, ABST., HYDROCARBON EXPLORATION IN, GREAT
B 6203 570* SALT MARSH, ABST., PHYSICAL AND SEDIMENTOLOGIC PROCESSES IN,
B 6303 556* SALT MARSH AND IT EFFECT ON ESTUARINE SEDIMENT TRANSPORT AND
B 6107 1091 SALT MINE, LOUISIANA, BELLE ISLE
B 6009 1612* SALT MOVEMENT, HALOKINETICS, IN NORTH LOUISIANA, ABST.
B 6305 782† SALT MOVEMENT ON CONTINENTAL SLOPE NORTHERN GULF OF MEXICO
S 7 69† SALT MOVEMENT ON CONTINENTAL SLOPE, NORTHERN GULF OF
B 6004 683* SALT MOVEMENT ON CONTINENTAL SLOPE NORTHERN GULF OF
S 3 136 SALT OR SHALE DIAPIRISM, MOBILIZATION OR LOCALIZATION BY
B 6311 2118* SALT PLAINS, ALFALFA COUNTY, OKLAHOMA, ABST., GEOCHEMISTRY OF
S 7 59 SALT RIDGES, DIAGRAM, MISSISSIPPI CONE, BURIED
B 6305 833* SALT RIVER RANGE, IDAHO-WYOMING THRUST BELT--PRELIMINARY
S 8 19 SALT RIVER CANYON, TIME-VELOCITY CURVES, CURRENT FLOW
S 8 136 SALT RIVER CANYON, TIME-VELOCITY CURVES, CURRENT FLOW
S 8 135 SALT RIVER CANYON, TOPOGRAPHY, INDEX MAP
S 8 129 SALT RIVER CANYON, WEST INDIES, ST. CROIX

S	8	129	SALT RIVER CANYON, WEST INDIES, SUBMARINE CANYONS
B	6004	674*	SALT RIVER ESTUARY, ST. CROIX, U.S.V.I.--MIXED CARBONATE AND
B	6305	835*	SALT SEDIMENTATION IN PLAYA LAKES OF EBRO BASIN, SPAIN, ABST.,
B	6008	1390*	SALT SOLUTION-COLLAPSE, WILLISTON BASIN--WHAT, WHERE, WHY,
B	6009	1434†	SALT STOCK HELP TO DELINEATE SPINES OF MOVEMENT, SHEAR ZONES
B	6107	1090†	SALT STOCKS HELP TO DELINEATE SPINES OF MOVEMENT. DISCUSSION
B	6307	1048	SALT STRUCTURES
S	7	50	SALT STRUCTURES, DISCUSSED, GULF OF MEXICO
B	6303	491*	SALT STRUCTURES IN GULF O MEXICO, ABST., DISTRIBUTION OF
B	6004	660*	SALT STRUCTURES IN VACHERIE SALT DOM AREA, LOUISIANA, ABST.,
M	30	261	SALT TECTONICS, GULF OF MEXICO, DISTRIBUTION OF
B	6004	679*	SALT TECTONICS OF WINN PARISH, LOUISIANA, ABST.
B	6307	1050	SALT, TEXAS, LOUANN
S	7	150	SALT THICKNESS, NORTHERN GULF OF MEXICO, CONTINENTAL SLOPE
B	6209	1577	SALT TO SHALE DIAPIRISM, COMPARISON OF
B	6405	763*	SALT WASH MEMBER, MORRISON FORMATION, UPPER JURASSIC, HENRY
B	6304	689*	SALT WASH MEMBER OF MORRISON FORMATION, UPPER JURASSIC,
B	6405	796*	SALT WASH MEMBER OF MORRISON FORMATION, URAVAN MINERAL
S	3	124	SALT WASH SANDSTONE, COLORADO, URAVAN MINERAL BELT
S	3	65	SALT WASH SANDSTONE MEMBER, MORRISON FORMATION, COLORADO
B	6109	1544*	SALT-DOME CAP ROCK, ABST., SUBSALT ORIGIN OF
B	6012	2162†	SALT-DOME CAP ROCK, ORIGIN OF, GULF COAST
B	6104	489	SALT-GENERATED JURASSIC PALEOSTRUCTURE TRAPS, TILTED-OUT
B	6203	494*	SALT-WITHDRAWAL BASIN, FRIO FORMATION, BRAZORIA COUNTY,
B	6004	728*	SALTON SEA GEOTHERMAL FIELD, ABST., GEOLOGY OF
B	6203	345	SALTON TROUGH, CALIFORNIA
B	6111	1980	SALTON TROUGH, CALIFORNIA
B	6012	2081†	SALTON TROUGH, CALIFORNIA, TECTONIC TRANSPRESSION AND,
B	6002	231	SALTS IN OIL-FIELD WATERS VS. DEPTH, ARKANSAS
B	6002	231	SALTS IN OIL-FIELD WATERS VS. DEPTH, LOUISIANA
B	6410	1580	SALTVILLE BLOCK, VIRGINIA
B	6412	2078	SALTVILLE FAULT, APPALACHIANS
B	6103	417	SALTVILLE FAULT, VIRGINIA
B	6410	1576	SALTVILLE THRUST, VIRGINIA
B	6303	520	SALVADOR, AMOS,, LATE TRIASSIC-JURASSIC PALEOGEOGRAPHY AND
B	6207	1233	SALVESON, J. O., FUTURE POTENTIAL FOR DEVELOPMENT OF
B	6405	779	SALVESON, JAMES O., GENERATION, MIGRATION, AND ENTRAPMENT OF
M	25	506	SAM GOOSLY DEPOSIT, CANADA
M	30	434	SAMARANG FIELD, STRATIGRAPHY AND PRODUCTION, SABAH
M	25	449	SAMBAGAWA SYSTEM, STRATA-BOUND SULFIDE DEPOSITS, JAPAN
S	4	291	SAMOA, PAGO PAGO, GALAXEA
M	30	24	SAMOTLOR FIELD, LANDSAT IMAGE, SIBERIA
M	30	24	SAMOTLOR FIELD, LANDSAT IMAGE, USSR
B	6401	22	SAMOTLOR, SAMOTLORSKOYE, GIANT OIL FIELD, LANDSAT DATA,
B	6401	22	SAMOTLORSKOYE, GIANT OIL FIELD, LANDSAT DATA, U.S.S.R.,
S	11	309	SAMPLE LOCALITIES, BEAR RIVER VALLEY, CALIFORNIA
S	11	310	SAMPLE LOCALITIES, CENTERVILLE, CALIFORNIA
S	11	314	SAMPLE LOCALITIES, FIELDS LANDING, CALIFORNIA
S	11	314	SAMPLE LOCALITIES, FRESHWATER CREEK, CALIFORNIA
S	11	305	SAMPLE LOCALITIES, HUMBOLDT COUNTY, CALIFORNIA
S	11	315	SAMPLE LOCALITIES, LARABEE CREEK, CALIFORNIA
S	11	316	SAMPLE LOCALITIES, PRICE CREEK, CALIFORNIA
S	11	317	SAMPLE LOCALITIES, SALMON CREEK, CALIFORNIA
S	11	318	SAMPLE LOCALITIES, SCOTIA-, EEL RIVER, CALIFORNIA
S	11	319	SAMPLE LOCALITIES, TABLE BLUFF COAST, CALIFORNIA
S	11	312	SAMPLE LOCALTIES, ELK RIVER VALLEY, CALIFORNIA
B	6209	1766*	SAMPLE ORDERING—NEW STATISTICAL TECHNIQUE FOR
B	6004	685*	SAMPLES, ABST., INTERPRETATION OF DEPOSITIONAL ENVIRONMENTS,
B	6007	1040	SAMPLES, AGES, NORTHEAST, PROVIDENCE CHANNEL
B	6007	1062	SAMPLES, PANAMA, GALETA REEF, RADIOCARBON-DATED
S	6	120	SAMPLING FOR TIME STUDIES, A REVIEW
B	6303	488	SAMUELS, NEIL D., UPPER DEVONIAN TURBIDITES OF CENTRAL AND
B	6012	2178*	SAN ANDREAS FAULT, ABST., DEPOSITIONAL AND TECTONIC HISTORY OF,
B	6405	769*	SAN ANDREAS FAULT, ABST., DRILLING IN
B	6405	626	SAN ANDREAS FAULT, AIRCRAFT RADAR IMAGE, CALIFORNIA
M	25	486	SAN ANDREAS FAULT, APPARENT RELATION TO METAL OCCURRENCE
M	25	41	SAN ANDREAS FAULT, CALIFORNIA
B	6205	819	SAN ANDREAS FAULT, CALIFORNIA
B	6211	2244	SAN ANDREAS FAULT, CALIFORNIA
B	6111	1980	SAN ANDREAS FAULT, CALIFORNIA
B	6405	619†	SAN ANDREAS FAULT, CALIFORNIA, SEASAT, RADAR IMAGE OF
B	6405	624	SAN ANDREAS FAULT, LANDSAT IMAGE, CALIFORNIA
B	6012	2128†	SAN ANDREAS FAULT PROBLEM, CORRELATION OF, PINNACLES AND,
B	6205	867†	SAN ANDREAS FAULT PROBLEM. DISCUSSION AND REPLY, CORRELATION
B	6405	626	SAN ANDREAS FAULT, SKYLAB PHOTOGRAPH, CALIFORNIA
B	6202	265	SAN ANDREAS FAULT, SLIP, CALIFORNIA
B	6004	650*	SAN ANDREAS FAULT SYSTEM AN PLATE TECTONICS IN WESTERN
B	6203	344†	SAN ANDREAS FAULT SYSTEM, CALIFORNIA, NEOGENE BASINFORMATION
B	6211	2214†	SAN ANDREAS FAULT SYSTEM, CALIFORNIA, ROLE OF, SALINIAN BLOCK
B	6401	25	SAN ANDREAS FAULT ZONE, CALIFORNIA
B	6012	2081†	SAN ANDREAS FAULT ZONE, SALTON TROUGH, CALIFORNIA, TECTONIC
M	25	485	SAN ANDREAS FAULT ZONE, TRANSCURRENT MOVEMENTS
B	6003	356†	SAN ANDREAS FAULTING, SAN JOAQUIN VALLEY, CALIFORNIA,
M	25	526	SAN ANDREAS-GULF OF CALIFORNIA FAULT SYSTEM, MOVEMENTS
B	6308	1427*	SAN ANDRES FACIES PATTERNS, PALO DURO AND DALHART BASINS,
B	6408	1171	SAN ANDRES FACIES, TEXAS
B	6202	244	SAN ANDRES FORMATION, MEXICO
B	6006	907†	SAN ANDRES FORMATION, MIDLAND BASIN, TEXAS, OOLITE-, BAR
B	6205	781	SAN ANDRES FORMATION, NEW MEXICO
M	26	437	SAN ANDRES FORMATION, NEW MEXICO, DELAWARE BASIN
B	6205	781	SAN ANGELO FORMATION, NEW MEXICO
B	6405	749*	SAN ANTONIO AREA, TEXAS, ABST., STABLE, ISOTOPE EVIDENCE FOR
B	6211	2220	SAN ANTONIO FAULT, CALIFORNIA
B	6007	1117	SAN ANTONIO GEOLOGICAL SOCIETY, HISTORY
B	6304	691*	SAN ANTONIO VALLEY URANIUM DEPOSIT, VALENCIA COUNTY NEW
B	6211	2219	SAN ARDO OIL FIELD, CALIFORNIA
M	25	553	SAN BARTOLOME DEPOSIT, SILVER-ZINC-LEAD, ECUADOR
M	25	553	SAN BARTOLOME, SILVER, ZINC, AND LEAD DEPOSITS, ECUADOR
B	6203	349	SAN BENITO BASIN, CALIFORNIA
M	27	153	SAN CARLOS FORMATION, TEXAS, CRETACEOUS
M	27	82	SAN CARLOS FORMATION, TEXAS, CRETACEOUS
B	6102	138	SAN CAYENTO FAULT, CALIFORNIA
B	6102	170	SAN CAYENTO FAULT, CALIFORNIA
B	6305	844*	SAN CAYETANO AND OAK RIDGE THRUST FAULTS, EAST-CENTRAL
B	6111	1981	SAN CLEMENTE BASIN, CALIFORNIA
B	6203	363	SAN CLEMENTE ESCARPMENT, CALIFORNIA
B	6111	1981	SAN CLEMENTE ISLAND, CALIFORNIA
B	6102	238	SAN CLEMENTE ISLAND, CALIFORNIA
B	6203	367	SAN CLEMENTE ISLAND PLATFORM, CALIFORNIA
B	6111	1984	SAN CLEMENTE RIDGE, CALIFORNIA
S	8	91	SAN CLEMENTE RIFT, CALIFORNIA, SUBMARINE CANYONS
S	8	96	SAN CLEMENTE RIFT, TIME-VELOCITY CURVES, CURRENT FLOW
S	8	95	SAN CLEMENTE RIFT, TOPOGRAPHY, INDEX MAP
B	6111	1981	SAN DIEGO TROUGH, CALIFORNIA
B	6203	364	SAN DIEGO TROUGH, CALIFORNIA
B	6102	152	SAN DIEGO TROUGH, CALIFORNIA
B	6102	238	SAN DIEGO TROUGH, CALIFORNIA
M	25	518	SAN DIMAS MINE, GOLD, MEXICO, TAYOLTITA-
M	25	329	SAN EDUARDO FORMATION, ECUADOR
B	6211	2270	SAN EMIDIO MOUNTAINS, CALIFORNIA
B	6212	2447	SAN EMIDIO NOSE, CALIFORNIA
B	6107	988	SAN FELIPE CHALK UNIT, MEXICO, AUSTIN-
M	30	363	SAN FELIPE FORMATION, CHIAPAS-TABASCO PROVINCE
B	6403	448*	SAN FRANCISCO BAY, ABST., FORAMINIFERA OF, SANGAMON ESTUARY,
B	6012	2178*	SAN FRANCISCO BAY REGION EAST OF SAN ANDREAS FAULT, ABST.,
B	6004	689*	SAN FRANCISCO BAY SYSTEM, ABST., CLAY-MINERAL VARIABILITY IN,
B	6111	1982	SAN GABRIEL BASIN, CALIFORNIA
B	6202	248	SAN GABRIEL CANYON, CALIFORNIA
B	6203	359	SAN GABRIEL FAULT, CALIFORNIA
B	6111	1982	SAN GABRIEL MOUNTAINS, CALIFORNIA
B	6203	357	SAN GREGORIO FAULT, CALIFORNIA
B	6211	2218	SAN GREGORIO-HOSGRI FAULT, CALIFORNIA
B	6106	904	SAN ISIDRO BASIN, BAJA CALIFORNIA
M	29	337	SAN JACINTO BELT, STRUCTURE, COLOMBIA
B	6309	1601*	SAN JACINTO COUNTIES, TEXAS, ABST., DEPOSITIONAL ENVIRONMENT OF
B	6208	1383	SAN JACINTO COUNTY, TEXAS, SHEPARD FIELD
B	6111	1981	SAN JACINTO FAULT, CALIFORNIA
M	29	335	SAN JACINTO FRAGMENTED BELT, COLOMBIA, NORTHWESTERN
M	25	163	SAN JACINTO TISATE GEOTHERMAL PROSPECT, NICARAGUA
B	6203	350	SAN JOAQUIN BASIN, CALIFORNIA
M	26	74	SAN JOAQUIN BASIN, FACIES SEQUENCES, STRATIGRAPHY, CALIFORNIA
M	26	72	SAN JOAQUIN BASIN, FRUITVALE FORMATION, CALIFORNIA
B	6211	2267	SAN JOAQUIN FORMATION, CALIFORNIA
B	6111	1982	SAN JOAQUIN HILLS, CALIFORNIA
B	6307	1036	SAN JOAQUIN VALLEY BASIN, EAST, CONTOUR MAP, CALIFORNIA
B	6211	2224	SAN JOAQUIN VALLEY, CALIFORNIA
B	6212	2442	SAN JOAQUIN VALLEY, CALIFORNIA
B	6211	2351*	SAN JOAQUIN VALLEY, CALIFORNIA, ABST., LATE, MIOCENE, TURBIDITE
B	6211	2243†	SAN JOAQUIN VALLEY, CALIFORNIA, SEDIMENTATION AND, TRAPPING
B	6003	356†	SAN JOAQUIN VALLEY, CALIFORNIA, TECTONIC SIGNIFICANCE AND,
B	6403	446*	SAN JOAQUIN VALLEY, CALIFORNIA--CONFIGURATION AND
B	6403	440*	SAN JOAQUIN VALLEY, CALIFORNIA. BY TITLE ONLY, BASIN-EDGE
B	6403	446*	SAN JOAQUIN VALLEY, CALIFORNIA—CONFIGURATION AND
M	26	173	SAN JOAQUIN VALLEY, MIOCENE, CALIFORNIA
B	6205	820	SAN JOAQUIN VALLEY, SOURCE ROCKS, CALIFORNIA
B	6206	922	SAN JOSE CANYON
S	8	101	SAN JOSE CANYON, MEXICO, SUBMARINE CANYONS
B	6111	1982	SAN JOSE FAULT, CALIFORNIA

B 6104 579 SAN JOSE FIELD, VENEZUELA
M 25 328 SAN JOSE FORMATION, ECUADOR
B 6111 1982 SAN JOSE HILLS, CALIFORNIA
M 25 319 SAN JUAN AREA, POTENTIAL, COLOMBIA
B 6405 795* SAN JUAN BASIN ABST., SEDIMENTOLOGY OF, WESTWATER CANYON
B 6405 795* SAN JUAN BASIN, AND APPLICATION OF LACUSTRINE-HUMATE MODEL,
M 26 102 SAN JUAN BASIN, CRETACEOUS, NEW MEXICO
M 26 169 SAN JUAN BASIN, CRETACEOUS, SHELF FACIES
M 26 106 SAN JUAN BASIN, ELECTRIC LOG SECTIONS, NEW MEXICO
B 6302 155 SAN JUAN BASIN GAS FIELD, NEW MEXICO
B 6205 757 SAN JUAN BASIN, LEWIS SHALE
B 6205 757 SAN JUAN BASIN, MANCOS SHALE
B 6205 881* SAN JUAN BASIN, NEW MEXICO, ABST., ENTRADA OIL FIELDS OF
B 6405 682* SAN JUAN BASIN, NEW MEXICO, ABST., ENTRADA OIL FIELDS OF
B 6008 1405* SAN JUAN BASIN, NEW MEXICO, ABST., PALEOCURRENT ANALYSIS OF,
B 6303 451* SAN JUAN BASIN, NEW MEXICO, ABST., PICTURED CLIFFS
B 6405 681* SAN JUAN BASIN, NEW MEXICO, ABST., RB-, SR DATING OF DIAGENESIS—
B 6304 598† SAN JUAN BASIN, NEW MEXICO--EXPLORATION IMPLICATIONS,
B 6405 761* SAN JUAN BASIN OF NEW MEXICO AND COLORADO, CLASSIC AREA OF
M 24 139 SAN JUAN BASIN, PRODUCTIVE SANDSTONES, NEW MEXICO
B 6206 1093* SAN JUAN BASIN REGION, NEW MEXICO, COLORADO, AND UTAH, ABST.,
B 6008 1392* SAN JUAN BASIN, RELATED TO SOURCE-ROCK TYPE, THERMAL
B 6205 758 SAN JUAN BASIN, SEISMIC SECTION
B 6306 887 SAN JUAN BASIN, STRUCTURE MAP, NEW MEXICO
B 6112 2137 SAN JUAN BASIN, SUBNORMAL PRESSURES, NEW MEXICO AND,
B 6304 692* SAN JUAN BASIN,NEW MEXICO, ABST., ROLL-TYPE, URANIUM
B 6004 653* SAN JUAN COUNTY, UTAH, ABST., PRELIMINARY INVESTIGATIONS OF,
S 3 30 SAN JUAN MOUNTAINS, COLORADO
S 8 106 SAN LUCAS CANYON, TIME-VELOCITY CURVES, CURRENT FLOW
S 8 104 SAN LUCAS CANYON, TOPOGRAPHY, INDEX MAP
B 6206 922 SAN LUCAS FAN, BAJA CALIFORNIA
B 6407 1073 SAN LUCAS FAN, BAJA CALIFORNIA
B 6405 714* SAN LUIS BASIN, SOUTH-CENTRAL COLORADO, ABST., OIL AND GAS
B 6206 1051 SAN MARCOS ARCH, TEXAS
B 6301 46 SAN MARCOS ARCH, TEXAS
B 6301 34 SAN MARCOS PLATFORM, NOMENCLATURE AND CORRELATION, TEXAS
B 6007 1011 SAN MATEO FAULT, NEW MEXICO
M 25 329 SAN MATEO FORMATION, ECUADOR
B 6304 693* SAN MATEO, NEW MEXICO, ABST., MOUNT TAYLOR, URANIUM DEPOSIT
B 6409 1564* SAN MIGUEL, AND UPSON FORMATIONS, UPPER CRETACEOUS,
B 6309 1611* SAN MIGUEL FORMATION, MAVERICK BASIN, SOUTH TEXAS, ABST.,
B 6102 238 SAN MIGUEL ISLAND, EOCENE, CALIFORNIA
B 6405 787* SAN MIGUEL LIGNITE DEPOSIT, JACKSON GROUP, SOUTH TEXAS, ABST.,
B 6308 1426* SAN MIGUEL SANDS OF SOUTHWEST TEXAS, ABST., STRATIGRAPHY AND
B 6109 1540* SAN MIGUEL SANDSTONES OF SOUTHWEST TEXAS, ABST., STRATIGRAPHY
M 24 147 SAN MIGUEL SANDSTONES, TEXAS
B 6102 167 SAN MIGUELITO FIELD, CALIFORNIA
B 6104 594 SAN MIGUELITTO FIELD, CALIFORNIA
B 6102 238 SAN NICOLAS BASIN, CALIFORNIA
B 6203 363 SAN NICOLAS BASIN, CALIFORNIA
B 6111 1981 SAN NICOLAS BASIN, CALIFORNIA
B 6203 544* SAN NICOLAS BASIN, CALIFORNIA, ABST., BOTTOM SEDIMENTS OF
B 6111 1981 SAN NICOLAS ISLAND, CALIFORNIA
B 6102 237 SAN NICOLAS ISLAND, EOCENE, CALIFORNIA
B 6102 238 SAN PEDRO BASIN, CALIFORNIA
B 6202 248 SAN PEDRO BASIN, CALIFORNIA
B 6111 1984 SAN PEDRO BASIN, CALIFORNIA
B 6203 363 SAN PEDRO BASIN, CALIFORNIA
B 6202 255 SAN PEDRO BAY ANTICLINORIUM, CALIFORNIA
B 6202 248 SAN PEDRO ESCARPMENT, CALIFORNIA
B 6202 255 SAN PEDRO ESCARPMENT FAULT ZONE, CALIFORNIA
B 6202 259 SAN PEDRO FORMATION, CALIFORNIA
B 6202 248 SAN PEDRO SEA VALLEY, CALIFORNIA
B 6202 249 SAN PEDRO SHELF, PROFILES, CALIFORNIA
B 6202 249 SAN PEDRO SHELF, SEISMIC DATA, CALIFORNIA
B 6202 247† SAN PEDRO SHELVES, CALIFORNIA CONTINENTAL BORDERLAND,
B 6212 2492† SAN PEDRO SHELVES, CALIFORNIA CONTINENTAL BORDERLAND.
B 6106 904 SAN QUENTIN BASIN, BAJA CALIFORNIA
B 6203 355 SAN RAFAEL MOUNTAINS, CALIFORNIA
B 6203 360 SAN RAFAEL UPLIFT, CALIFORNIA, UPPER PIRU-
B 6204 694 SAN SALVADOR WELL, BAHAMA PLATFORM
B 6303 370† SAN SEBASTIAN FORMATION, OLIGOCENE, NORTHERN PUERTO RICO,
B 6303 305 SANAK BASIN, ALASKA
B 6303 538 SANCETTA, C. A., MIDDLE CRETACEOUS OXYGEN-DEFICIENT
B 6101 65† SAND AREA, SOUTH LOUISIANA--IMPLICATIONS, GEOTHERMAL AND,
B 6209 1755* SAND BODY GEOMETRY AND RESERVOIR POTENTIAL OF RECENT
B 6007 1009 SAND BODY, MULTILATERAL
B 6405 759* SAND DISPERSAL AT NORDERNEYER SEEGAT, WEST GERMANY, ABST.
B 6004 661* SAND FLAT-SHALLOW SUBTIDAL ENVIRONMENT, ABST., TRACE-, FOSSIL
S 7 117 SAND FRACTION, GULF OF MEXICO, CONTINENTAL SLOPE
B 6004 689* SAND FROM MEADOWCROFT ROCK SHELTER, AVELLA, PENNSYLVANIA,
S 10 246 SAND PERCENTAGE CONCENTRATIONS, ALABAMA, FLOMATON FIELD
S 10 246 SAND PERCENTAGE CONCENTRATIONS, FLORIDA, JAY FIELD
S 10 246 SAND PERCENTAGE CONCENTRATIONS, FLORIDA, BLACKJACK CREEK
B 6105 777* **SAND PETROLOGY AND BIOCLASTIC COMPONENTS OF BLACK SHELL**
B 6408 1229 SAND RIDGES, LONGITUDINAL
B 6405 792* SAND RIDGES ON STORM-DOMINATED SHELVES—STATUS REPORT, ABST.,
B 6106 861† SAND SHEET, BALTIMORE CANYON TROUGH AREA, THICKNESS AND AGE
B 6106 861 SAND SHEET ON MIDDLE ATLANTIC OUTER CONTINENTAL SHELF,
B 6004 679* SAND SHOAL TO ITS PHYSICAL ENVIRONMENT AND HOLOCENE
B 6203 510* SAND TRANSPORT ACROSS EOCENE SHELF AND SLOPE IN
B 6302 198 SAND WASH BASIN, COLORADO AND, WYOMING
B 6405 696* SAND WAVES, ABST., INTERNAL STRUCTURES OF SHALLOW-MARINE,
B 6408 1229 SAND WAVES AND RIPPLES, TRANSVERSE
B 6405 681* SAND WAVES IN VINEYARD SOUND, MASSACHUSETTS, ABST.,
B 6402 264† SAND-DOMINANT SUBAERIAL PHASE, ATCHAFALAYA DELTA, EVOLUTION
B 6402 264† SAND-DOMINANT SUBAERIAL PHASE, ATCHAFALAYA DELTA, LOUISIANA,
B 6004 683* SAND-FLOW CROSS-STRATA, ABST., COMPARISON OF, EOLIAN AND,
B 6406 841† SAND-LAYER GEOMETRY ON FLAT FLOORS OF 10 MODERN
B 6203 520* SAND-SHOAL FACIES, JOULTERS CAYS, GREAT BAHAMA BANK, ABST.,
B 6405 777* SAND-WAVE SIZE—DYNAMIC EQUILIBRIUM PROCESSES AND KINEMATIC
B 6305 828 SANDBERG, CHARLES A., MISSISSIPPIAN CARBONATE SHELF MARGIN
B 6405 716 SANDBERG, CHARLES A., MISSISSIPPIAN SHELF MARGIN AND CARBONATE
B 6405 779 SANDBERG, CHARLES A., SEDIMENTATION, BIOSTRATIGRAPHY, AND
B 6405 780 SANDBERG, CHARLES A., USE OF DEVONIAN CONODONTS IN PETROLEUM
B 6004 652 SANDBERG, P. A., ISOTOPIC EVIDENCE BEARING ON ORIGIN OF
B 6306 987 SANDERS, BUSCH, DANIEL A., REVIEW OF, PRINCIPLES OF,
B 6203 560 SANDFORD, G. E. R., GREAT CANADIAN OIL SANDS—GEOLOGY AND
B 6211 2219 SANDHOLDT MEMBER, CALIFORNIA
B 6109 1515 SANDIA FORMATION, MEMBERS, NEW MEXICO
B 6305 828 SANDO, WILLIAM J., MISSISSIPPIAN CARBONATE SHELF MARGIN ALONG
B 6405 716 SANDO, WILLIAM J., MISSISSIPPIAN SHELF MARGIN AND CARBONATE
B 6303 541* SANDS, ABST., DISTRIBUTION OF, CLAY IN RECENT
B 6105 838* SANDS, ABST., EFFECT OF SHAPE SORTING ON PREDICTION OF, POROSITY
B 6004 647* SANDS, ABST., ENERGY FROM, TAR
B 6206 1093* SANDS, ABST., NEW GEOLOGIC FRAMEWORK FOR, MINING OF, OIL
B 6303 499* SANDS, ALBERTA, ABST., SEDIMENTOLOGY, PETROLOGY, AND
B 6004 735* SANDS AN SANDSTONES, ABST., PALEOCLIMATE INTERPRETATION
B 6206 967† SANDS AND PETROLEUM POTENTIAL OF MONTEREY DEEP-SEA FAN,
B 6405 700* SANDS AND SAND TRANSPORT ON PALIMPSEST CARBONATE SHELF, ABST.
B 6111 1950† SANDS AND SUPERGIANT OIL FIELDS, TAR
M 25 420 SANDS, AUSTRALIA, BLACK
B 6203 494* SANDS DERIVED FROM GRANITIC PLUTONS, ABST., TRACE ELEMENTS IN,
B 6306 908 SANDS, DIPMETER MOTIFS OF, DEEP-SEA
B 6306 906 SANDS, GAMMA, LOG MOTIFS OF, DEEP-SEA
B 6209 1758* SANDS, JOULTERS OOID SHOAL, BAHAMAS, ABST., HOLOCENE
B 6007 1110 SANDS, MISNAMED, TAR
B 6111 1950 SANDS, NATURE AND ORIGIN OF, TAR
B 6203 565* SANDS OF ALBERTA. BY TITLE ONLY., HEAVY, OIL AND, TAR
B 6102 169† SANDS, STUDIES OF, VENTURA FIELD, CALIFORNIA, II--, LITHOLOGY,
B 6403 449* SANDSTON COMPOSITIONS RELATED TO PLATE TECTONIC SETTINGS,
B 6303 541* SANDSTONE, ADMIRE 650-FOOT SANDSTONE, EL DORADO FIELD, KANSAS,
M 28 123 SANDSTONE, ATLANTIC, CRETACEOUS
B 6405 722* SANDSTONE BODIES AS REFLECTED BY GEOMORPHIC FEATURES ON
B 6408 1232 SANDSTONE BODIES IN EPICONTINENTAL SEAWAYS, ANCIENT, OFFSHORE
B 6209 1761* SANDSTONE BY DUCTILE GRAIN DEFORMATION DURING COMPACTION,
M 28 172 SANDSTONE, CANADA, CRETACEOUS
M 28 180 SANDSTONE, CANADA, N.W.T., CRETACEOUS
M 28 178 SANDSTONE, CANADA, N.W.T., CRETACEOUS
M 28 163 SANDSTONE, CANADA, N.W.T., CRETACEOUS
B 6203 497* SANDSTONE CEMENTATION, ABST., CLAY DIAGENESIS IN, WILCOX
B 6303 533* SANDSTONE CEMENTS IN HIGH PLAINS SEQUENCE OF WYOMING AND
B 6108 1392* SANDSTONE, CENOZOIC HIGH PLAINS SEQUENCE, WYOMING AND
M 28 93 SANDSTONE CLASSIFICATION, GENERAL
M 28 94 SANDSTONE CLASSIFICATIONS, DIAGRAMS
B 6109 1535* SANDSTONE COMPOSITION AND DIAGENESIS ON RESERVOIR QUALITY,
B 6312 2164† SANDSTONE COMPOSITIONS, PLATE, TECTONIC AND
B 6403 449* SANDSTONE COMPOSITIONS RELATED TO PLATE TECTONIC SETTINGS,
B 6105 843* SANDSTONE DEPOSITIONAL SYSTEMS IN URANIFEROUS SANDSTONES OF
B 6012 2150† SANDSTONE DETERMINATION FOR ISOPACH MAPPING OF
B 6108 1390* SANDSTONE DIAGENESIS, ABST., ROLE OF, SECONDARY, POROSITY IN
B 6203 538* SANDSTONE DIAGENESIS IN BROMIDE FORMATION, SOUTH-CENTRAL
B 6405 741* SANDSTONE DIAGENESIS IN GEOPRESSURED TERTIARY GULF COAST
B 6108 1380* SANDSTONE DIAGENESIS-THE HOLE TRUTH, ABST.
B 6209 1760* SANDSTONE DISTRIBUTION AND POTENTIAL FOR GEOPRESSURED
B 6305 810† SANDSTONE FACIES AND ANCIENT SUBMARINE FANS. MODELS FOR
B 6206 932† SANDSTONE FACIES AND ANCIENT SUBMARINE FANS: MODELS FOR
B 6004 517† SANDSTONE FACIES, DELAWARE BASIN, WEST TEXAS AND SOUTHEAST
B 6312 2193† SANDSTONE IN LOWER CRETACEOUS HELVETIAFJELLET FORMATION,
B 6304 688* SANDSTONE MINE, ABST., ANOMALOUS OREBODY WITHIN, AMBROSIA

B 6004 680* SANDSTONE MODELS FROM APPALACHIAN CARBONIFEROUS DELTAS,
B 6107 1030 SANDSTONE, NEBRASKA, REAGAN
M 28 179 SANDSTONE, NORTH SEA, JURASSIC
B 6311 2120* SANDSTONE OIL RESERVOIRS OF PENNSYLVANIAN AGE, NORTHWESTERN
M 26 445 SANDSTONE, OKLAHOMA, ANADARKO BASIN, MEDRANO
M 26 445 SANDSTONE, OKLAHOMA, ANADARKO BASIN, WADE
B 6007 1017 SANDSTONE, ORIGIN OF, SHEETS OF, FLUVIAL
M 26 9 SANDSTONE PINCHOUTS, SEISMIC, REFLECTIONS
B 6004 663* SANDSTONE PORE STRUCTURE, ABST., NUMERICAL, CLASSIFICATION O
B 6312 2171 SANDSTONE POROSITY
B 6203 508* SANDSTONE POROSITY DURING BIOTURBATION, ABST., IMPROVEMENT
B 6007 1009† SANDSTONE, RESERVOIR, GEOMETRY OF A, FLUVIAL, SHEET
B 6107 1029† SANDSTONE RESERVOIR, SLEEPY HOLLOW FIELD, RED WILLOW COUNTY,
B 6303 460* SANDSTONE RESERVOIRS, ABST., AUTHIGENI, ILLITES IN
B 6203 560* SANDSTONE RESERVOIRS, ABST., EXPLORATION MODELS AND CRITERIA
B 6108 1387* SANDSTONE RESERVOIRS, ABST., POROSITY, DIAGENESIS, AND,
B 6309 1606* SANDSTONE RESERVOIRS ALONG TEXAS GULF COAST, ABST.,
B 6205 813 SANDSTONE RESERVOIRS, POROSITY VERSUS, DEPTH IN
B 6203 495* SANDSTONE RESERVOIRS—CANYON GROUP OF VAL VERDE BASIN,
B 6105 825* SANDSTONE RESERVOIRS—EXPLORATION MODELS AND RECOGNITION
B 6109 1541* SANDSTONE RESERVOIRS—POROSITY FORMATION AND PRESERVATION,
M 26 480 SANDSTONE, SEISMIC, MODELING, BARRIER-BAR
M 28 180 SANDSTONE, TEXAS, PERMIAN
B 6007 1009 SANDSTONE, USAGE OF TERM, SHEET
B 6408 1164 SANDSTONE WEDGES, ALBERTA, MESOZOIC
B 6408 1163 SANDSTONE WEDGES, EXAMPLES
B 6408 1167 SANDSTONE WEDGES, TEXAS, GULF COAST, TERTIARY
B 6408 1163 SANDSTONE WEDGES, VENEZUELA, EASTERN, TERTIARY
B 6109 1537* SANDSTONES, ABST., CRITICAL, WATER ESTIMATES FOR, GULF COAST
B 6108 1393* SANDSTONES, ABST., DIAGENESIS OF, VOLCANOGENIC
B 6108 1372* SANDSTONES, ABST., DIAGENETIC PROCESSES IN
B 6304 686* SANDSTONES, ABST., DISSOLUTION AND AUTHIGENESIS IN HOST
B 6203 520* SANDSTONES, ABST., FACIES ASSOCIATIONS AND, FABRIC ANALYSES OF,
B 6303 409* SANDSTONES, ABST., MECHANISMS OF FORMATION DAMAGE IN
B 6004 735* SANDSTONES, ABST., PALEOCLIMATE INTERPRETATION THROUGH USE
B 6205 890* SANDSTONES, ABST., PERMEABILITY CONTROL IN, CRETACEOUS,
B 6203 508* SANDSTONES, ABST., RESERVOIR POTENTIAL OF, VOLCANOGENIC
B 6108 1394* SANDSTONES, ABST., TRANSFER-, REACTION MODELS OF DIAGENETIC
M 28 1 **SANDSTONES AND ASSOCIATED ROCKS, A COLOR ILLUSTRATED**
B 6310 1981 SANDSTONES AND ASSOCIATED ROCKS, BY PETER A. SCHOLLE, BOLES,
B 6209 1761* SANDSTONES AND SHALES, ABST., PALEOECOLOGY AND, POROSITY
B 6203 511* SANDSTONES IN SOUTHEASTERN KANSAS, ABST., SUBSURFACE,
B 6004 730* SANDSTONES, JURASSIC OF SOUTHWESTERN OREGON, NOT ASSOCIATED
B 6206 936 SANDSTONES, MASSIVE
B 6009 1607* SANDSTONES, NINE MILE POINT FIELD, ARANSAS COUNTY, TEXAS ABST.,
B 6004 710* SANDSTONES OF CANADIAN WESTERN INTERIOR, ABST., PROVENANCE
B 6105 835* SANDSTONES OF CRETACEOUS GREAT VALLEY SEQUENCE, CALIFORNIA,
B 6108 1393* SANDSTONES OF FRONTIER FORMATION, SPEARHEAD RANCH FIELD,
B 6109 1547* SANDSTONES OF LOWER WILCOX, EOCENE, KARNES COUNTY, TEXAS,
B 6203 521* SANDSTONES OF NORTH SEA VIKING GRABEN, ABST., FRIGG FIELD—
B 6102 297* SANDSTONES OF PERMIAN-PENNSYLVANIAN AGE, VAL VERDE BASIN,
B 6305 836* SANDSTONES OF SOUTHERN PICEANCE CREEK BASIN,
B 6303 542* SANDSTONES OF UPPER MISSISSIPPI VALLEY, ABST., PALEOENVIRONMENT
B 6108 1384* SANDSTONES, OUACHITA MOUNTAINS, ARKANSAS AND OKLAHOMA,
B 6206 939 SANDSTONES, PEBBLY
B 6305 802 SANDSTONES, PERMEABILITY VERSUS, POROSITY AND SAND CONTENT
B 6305 799† SANDSTONES, RESERVOIR PROPERTIES OF, CONGLOMERATES AND
M 26 479 SANDSTONES, SEISMIC, MODELING, CHANNEL
M 26 480 SANDSTONES, SEISMIC, MODELING, SHORELINE
B 6305 834* SANDSTONES TO HYDROCARBON EXPLORATION, ABST., IMPORTANCE OF,
B 6009 1612* SANDSTONES--FRIO FORMATION, OLIGOCENE, SOUTH TEXAS, ABST.,
B 6303 447* SANDSTONES--WAGWATER TROUGH, EAST-CENTRAL JAMAICA, ABST.,
B 6108 1378* SANDSTONES-IMPLICATIONS FOR PETROLEUM EXPLORATION, ABST.,
B 6203 498* SANDSTONES—ACTIVE ANKERITE CEMENTATION IN WILCOX OF
B 6305 840* SANDSTONE RESERVOIRS, ROCKY MOUNTAIN BASINS, ABST., SEISMIC
B 6303 506* SANDSTONES ILLINOIS BASIN, ABST., ORIGIN AND CHARACTERTICS
B 6303 461* **SANDSTONES PALO DURO BASIN, TEXAS, ABST., EXPLORATION**
B 6203 560* SANDS—GEOLOGY AND PRODUCTION, BY TITLE ONLY., GREAT,
B 6008 1265 **SANDT, JAMES E., DEVELOPMENTS IN LOUISIANA GULF COAST IN 1975**
B 6108 1244 SANDT, JAMES E., DEVELOPMENTS IN LOUISIANA GULF COAST IN 1976
B 6004 655* SANDY BRAIDED RIVER, ABST., EFFECTS OF, ICE ON SEDIMENTS ON
B 6005 883 SANFORD, J. T., RADIOMETRIC AGE DETERMINATION OF SILURIAN
B 6004 718 SANFORD, J. T., SEDIMENTOLOGY OF SOME SILURIA ROCKS FROM
B 6303 520 SANFORD, R. M., EXPLORATION RESULTS OFFSHORE SOUTH NEW
B 6403 448* SANGAMON ESTUARY, CENTRAL SAN FRANCISCO BAY, ABST.,
M 28 18 SANGRE DE CRISTO FORMATION, NEW MEXICO, PENNSYLVANIAN
M 28 155 SANGRE DE CRISTO FORMATION, NEW MEXICO, PENNSYLVANIAN
M 28 19 SANGRE DE CRISTO FORMATION, NEW MEXICO, PENNSYLVANIAN
M 28 162 SANGRE DE CRISTO FORMATION, NEW MEXICO, PENNSYLVANIAN
M 28 162 SANGRE DE CRISTO FORMATION, NEW MEXICO, PERMIAN
M 28 18 SANGRE DE CRISTO FORMATION, NEW MEXICO, PERMIAN
M 28 19 SANGRE DE CRISTO FORMATION, NEW MEXICO, PERMIAN
M 28 155 SANGRE DE CRISTO FORMATION, NEW MEXICO, PERMIAN
S 7 87 SANGREE, J. B., RECOGNITION OF CONTINENTAL-SLOPE SEISMIC FACIES,
B 6004 699 SANGREE, J. B., REGIONAL SEISMIC INTERPRETATION USING SEQUENCES
M 26 49 SANGREE, J. B., SEISMIC STRATIGRAPHY AND GLOBAL CHANGES OF SEA
M 26 99 SANGREE, J. B., SEISMIC STRATIGRAPHY AND GLOBAL CHANGES OF SEA
M 26 117 SANGREE, J. B., SEISMIC STRATIGRAPHY AND GLOBAL CHANGES OF SEA
B 6205 752 SANGREE, J. B., SEISMIC STRATIGRAPHY AND GLOBAL CHANGES OF SEA
M 26 165 SANGREE, J. B., SEISMIC STRATIGRAPHY AND GLOBAL CHANGES OF SEA
M 28 17 SANIDINE, FELDSPAR
S 5 55 SANILAC COUNTY, LOCATION, MICHIGAN,
B 6308 1428 SANNES, TORSTEIN, WAVELAND FIELD, UNIQU STRUCTURAL AND
B 6309 1610 SANNESS, TORSTEIN, WAVELAND FIELD, UNIQU STRUCTURAL AND
B 6004 670* SANT ROSA-CORTES RIDGE, SOUTHERN CALIFORNIA CONTINENTAL
B 6111 1982 SANTA ANA MOUNTAINS, CALIFORNIA
B 6202 248 SANTA ANA MOUNTAINS, CALIFORNIA
B 6011 2038† SANTA ANA MOUNTAINS, CALIFORNIA, ALLOCHTHONOUS, CARBONATE
B 6211 2363* SANTA ANA MOUNTAINS, SOUTHERN CALIFORNIA, ABST., LATE,
B 6111 1983 SANTA ANA MOUNTAINS, STRATIGRAPHIC COLUMN, CALIFORNIA
B 6001 65† SANTA ANA MOUNTAINS--RELATION TO OLIGOCENE SANTA BARBARA
B 6102 263† SANTA ANA MOUNTAINS--RELATION TO OLIGOCENE SANTA BARBARA
B 6104 616† SANTA ANA MOUNTAINS--RELATION TO OLIGOCENE SANTA BARBARA
B 6102 207 SANTA ANA UPLIFT, MEXICO
B 6303 465* SANTA BARBARA AND VENTURA COUNTIES, CALIFORNIA, ABST.,
B 6303 436* SANTA BARBARA BASIN, ABST., CRUDE OIL-LIKE HYDROCARBONS IN
B 6102 238 SANTA BARBARA BASIN, CALIFORNIA
B 6203 364 SANTA BARBARA BASIN, CALIFORNIA
B 6111 1981 SANTA BARBARA BASIN, CALIFORNIA
B 6206 973 SANTA BARBARA BASIN, CALIFORNIA
B 6012 2180* SANTA BARBARA BASIN, CALIFORNIA, ABST., LATE, NEOGENE-,
B 6405 792* SANTA BARBARA BASIN, CALIFORNIA BORDERLAND, ABST.,
B 6204 688 SANTA BARBARA BASIN, KEROGEN, CALIFORNIA
B 6001 65† SANTA BARBARA BASIN, SANTA MONICA AND, SANTA ANA
B 6104 616† SANTA BARBARA BASIN. DISCUSSION AND REPLY, SANTA MONICA AND,
B 6102 263† SANTA BARBARA BASIN. DISCUSSION AND REPLY, SANTA MONICA AND,
B 6303 473* SANTA BARBARA COAST, CALIFORNIA, ABST., LATERAL DIAGENESIS IN,
B 6403 444* SANTA BARBARA COAST, CALIFORNIA, ABST., MONTEREY ROCKS ALONG
B 6012 2191* SANTA BARBARA COUNTY, CALIFORNIA, ABST., NANNOPLANKTON
B 6102 238 SANTA BARBARA ISLAND, CALIFORNIA
B 6111 1981 SANTA BARBARA ISLAND, CALIFORNIA
B 6102 238 SANTA CATALINA BASIN, CALIFORNIA
B 6206 973 SANTA CATALINA BASIN, CALIFORNIA
B 6102 238 SANTA CATALINA ISLAND, CALIFORNIA
B 6202 248 SANTA CATALINA ISLAND, CALIFORNIA
B 6111 1981 SANTA CATALINA ISLAND, CALIFORNIA
B 6203 359 SANTA CLARA FAULT, CALIFORNIA
B 6203 359 SANTA CLARA TROUGH, CALIFORNIA
B 6303 471* SANTA CRUZ BASIN, BOLIVIA--CASE STUDY OF
B 6111 1981 SANTA CRUZ BASIN, CALIFORNIA
B 6203 346 SANTA CRUZ BASIN, CALIFORNIA
B 6102 238 SANTA CRUZ BASIN, CALIFORNIA
B 6401 108 SANTA CRUZ BASIN RECENT SEDIMENTS, GENERATION OF
S 8 73 SANTA CRUZ CANYON, CALIFORNIA, SUBMARINE CANYONS
S 8 24 SANTA CRUZ CANYON, TIDAL EFFECTS, CURRENT FLOW
S 8 79 SANTA CRUZ CANYON, TIME-VELOCITY CURVES, CURRENT FLOW
S 8 78 SANTA CRUZ CANYON, TOPOGRAPHY, INDEX MAP
B 6203 356 SANTA CRUZ HIGH, CALIFORNIA
B 6102 238 SANTA CRUZ ISLAND, EOCENE, CALIFORNIA
B 6012 2177* SANTA CRUZ ISLAND FAULT, ABST., MIOCENE VOLCANIC AND
B 6012 2177* SANTA CRUZ ISLAND--EVIDENCE FOR OFFSET ALONG SANTA CRUZ
B 6211 2224 SANTA CRUZ MOUNTAINS, CALIFORNIA
B 6211 2359* SANTA CRUZ MOUNTAINS, CALIFORNIA, ABST., PETROLEUM
B 6212 2461 SANTA ELENA OIL FIELD, ECUADOR
B 6405 650 SANTA FE COUNTRY EDITED BY RAYMOND V. INGERSOLL, LEE A.
B 6305 822* SANTA FE COUNTY, NEW MEXICO, ABST., OIL AND GAS POTENTIAL OF,
B 6305 822* SANTA FE EMBAYMENT, SANTA FE COUNTY, NEW MEXICO, ABST., OIL
B 6010 1872 SANTA FE MINERALS, INC., ON DEVELOPMENTS IN YEMEN, 1975, ASIATIC
B 6203 357 SANTA LUCIA BANK FAULT, CALIFORNIA
S 4 234 SANTA LUCIA, NEMASTER RUBIGINOSA POURTALES
B 6211 2217 SANTA LUCIA RANGE, CALIFORNIA
S 11 9 SANTA MARGARITA FORMATION
B 6211 2245 SANTA MARGARITA FORMATION, CALIFORNIA
B 6211 2359* SANTA MARGARITA FORMATION, WESTERN SANTA CRUZ MOUNTAINS,
S 5 171 SANTA MARGHERITA DI SAVOIA, COMMERCIAL SALINA, ITALY, ADRIATIC
B 6403 450* **SANTA MARIA BASIN, ABST., STRATIGRAPHIC AND STRUCTURAL**
B 6403 450* SANTA MARI BASIN, ABST., STRATIGRAPHIC AND STRUCTURAL
B 6405 770* SANTA MARIA AREA, CALIFORNIA, ABST., HYPOTHESIS INVOLVING
B 6203 350 SANTA MARIA BASIN, CALIFORNIA
B 6203 355 SANTA MARIA OIL DISTRICT AND BASIN, CALIFORNIA

B 6203 355 SANTA MARIA VALLEY OIL FIELD, CALIFORNIA
B 6212 2492† SANTA MONICA AND SAN PEDRO SHELVES, CALIFORNIA CONTINENTAL
B 6202 247† SANTA MONICA AND SAN PEDRO SHELVES, CALIFORNIA CONTINENTAL
B 6001 65† SANTA MONICA AND SANTA ANA MOUNTAINS--RELATION TO
B 6102 263† SANTA MONICA AND SANTA ANA MOUNTAINS--RELATION TO
B 6104 616† SANTA MONICA AND SANTA ANA MOUNTAINS--RELATION TO
B 6102 238 SANTA MONICA BASIN, CALIFORNIA
B 6111 1981 SANTA MONICA BASIN, CALIFORNIA
B 6202 248 SANTA MONICA BASIN, CALIFORNIA
B 6203 364 SANTA MONICA BASIN, CALIFORNIA
B 6406 853 SANTA MONICA BASIN, SAND LAYERS, CALIFORNIA
B 6203 548* SANTA MONICA BAY—CALIFORNIA CONTINENTAL BORDERLAND, ABST.,
B 6202 248 SANTA MONICA CANYON, CALIFORNIA
S 8 65 SANTA MONICA CANYON, CALIFORNIA, SUBMARINE CANYONS
S 8 70 SANTA MONICA CANYON, TIME-VELOCITY CURVES, CURRENT FLOW
B 6202 255 SANTA MONICA FAULT ZONE, CALIFORNIA
B 6202 248 SANTA MONICA MOUNTAINS, CALIFORNIA
B 6211 2351* SANTA MONICA MOUNTAINS, CALIFORNIA, ABST., LATE, CRETACEOUS
B 6102 266 SANTA MONICA MOUNTAINS, NORTHWESTERN, GEOLOGIC MAP,
B 6202 248 SANTA MONICA SHELF, CALIFORNIA
B 6202 249 SANTA MONICA SHELF, PROFILES, CALIFORNIA
B 6202 249 SANTA MONICA SHELF, SEISMIC DATA, CALIFORNIA
B 6211 2352* SANTA MONICA SLATE IN SOUTHERN CALIFORNIA; RB-SR EVALUATION,
B 6203 361 SANTA MONICA-RAYMOND HILL FAULT, CALIFORNIA
B 6409 1554* SANTA ROSA CANYON, SIERRA DE SANTA ROSA, NUEVO LEON, MEXICO,
B 6102 238 SANTA ROSA ISLAND, EOCENE, CALIFORNIA
B 6108 1382* SANTA ROSA SANDSTONE, LATE TRIASSIC, OF EAST-CENTRAL NEW
B 6001 66 SANTA ROSA VALLEY, CALIFORNIA
B 6203 363 SANTA ROSA-CORTES RIDGE, CALIFORNIA
B 6111 1981 SANTA ROSA-CORTES RIDGE, CALIFORNIA
B 6203 359 SANTA SUSANA FAULT, CALIFORNIA
B 6203 359 SANTA YNEZ FAULT, CALIFORNIA
B 6303 465* SANTA YNEZ MOUNTAINS, SANTA BARBARA AND VENTURA COUNTIES,
B 6203 360 SANTA YNEZ UPLIFT, CALIFORNIA
M 30 339 SANTIAGO ACEVEDO, JOSE, GIANT FIELDS OF THE SOUTHERN
M 25 330 SANTIAGO AND ZAPALLO FORMATIONS, ECUADOR
B 6111 1986 SANTIAGO PEAK INTRUSIVE, CALIFORNIA
B 6109 1478 SANTIAGO SANDSTONE, BRAZIL
B 6307 1033 SANTO AMARO GROUP, BRAZIL, RECONCAVO BASIN
B 6106 906 SANTO TOMAS FAULT, OFFSHORE, CALIFORNIA
B 6401 75 SANTONIAN, ENGLAND, ISLE OF WIGHT, ALBIAN THROUGH
S 4 88 SANTONIAN, JAMAICA, ARTHURS SEAT FORMATION
S 4 88 SANTONIAN, JAMAICA, PETERS HILL FORMATION, UPPER
B 6109 1480 SANTONIAN PETROLEUM RESERVOIR SYSTEM, BRAZIL, ALBIAN-
B 6202 184 SANTONIAN, WYOMING
S 6 261 SANTONIAN-CAMPANIAN BOUNDARY, NORTH AMERICA, CRETACEOUS
B 6109 1471 SANTOS BASIN, BRAZIL
M 26 214 SANTOS BASIN, BRAZIL
B 6206 1035 SANTOS BASIN, PROFILES, BRAZIL
B 6002 197 SANTOS COASTAL BASIN, BRAZIL
B 6206 1034 SANTOS FAULT SYSTEM, BRAZIL, RIO GRANDE-
B 6206 1033 SAO GABRIEL ARCH, BRAZIL
B 6104 547 SAO GABRIEL ARCH, BRAZIL
B 6306 940 SAO LUIS BASIN, AMAZON MARGIN AREA
B 6109 1476 SAO MATEUS FIELD, BRAZIL
B 6106 930 SAO PAULO PLATEAU
B 6204 658 SAO PAULO PLATEAU, BATHYMETRIC MAP, BRAZIL
B 6002 197 SAO PAULO PLATEAU, BRAZIL MARGIN
B 6004 690* SAO PAULO PLATEAU, BRAZILIAN CONTINENTAL MARGIN, ABST.,
B 6204 657 SAO PAULO PLATEAU, GEOTHERMAL MEASUREMENTS, BRAZIL
B 6204 661 SAO PAULO PLATEAU, SEISMIC REFLECTION PROFILES, BRAZIL
B 6204 657† SAO PAULO PLATEAU, SOUTHEASTERN BRAZILIAN CONTINENTAL
B 6110 1749 SAO TOME AND PRINCIPE, DEVELOPMENTS, 1976
B 6210 1848 SAO TOME AND PRINCIPE, DEVELOPMENTS, 1977
B 6310 1694 SAO TOME AND PRINCIPE, DEVELOPMENTS, 1978
B 6411 1790 SAO TOME AND PRINCIPE, DEVELOPMENTS, 1979
B 6412 2064 SAR IMAGERY, APPALACHIANS, SOUTHERN, VALLEY AND RIDGE AREA
B 6412 2064 SAR, IMAGERY, SEASAT, SYNTHETIC-APERTURE RADAR
M 25 554 SARAGURO TIN AND TUNGSTEN DEPOSITS, ECUADOR
B 6303 422* SARATOGA FORMATION, ABST., ICHNOLOGY AND DEPOSITIONAL
B 6207 1215* SARAWAK, ABST., GEOLOGY AND EXPLORATION HISTORY OF, OFFSHORE
M 30 431 SARAWAK, BARAM DELTA, BARONIA FIELD
M 30 436 SARAWAK, CENTRAL LUCONIA FIELD, MALAYSIA
B 6411 1896 SARAWAK, PRODUCTION, 1977-1979, MALAYSIA EAST
M 30 433 SARAWAK SHELF, RECENT SEDIMENTATION, MALAYSIA
B 6210 1915 SARDINIA, DEVELOPMENTS, 1977, SICILY AND
B 6310 1924 SARDINIA, DEVELOPMENTS, 1978, SICILY AND
B 6004 718 SARG, J. F., SUBAQUEOUS EVAPORITE DEPOSITION, PERMIAN SHELF
B 6004 719 SARGENT, K. A., CHEMICAL AND ISOTOPIC INVESTIGATION O
M 30 517 SARIR GROUP, STRATIGRAPHY, MESSLA FIELD
M 30 507 SARIR SANDSTONE, CRETACEOUS, MESSLA FIELD
B 6401 16 SARIR SANDSTONE, LIBYA
M 30 521 SARIR SANDSTONE, NET PAY ISOPACH, MESSLA FIELD
M 30 514 SARIR SANDSTONE, SEISMIC LINE, MESSLA FIELD
B 6104 594 SARITA FIELD, TEXAS
B 6105 778 SARJEANT, WILLIAM A. S., DINOSAUR TRACKS FROM CRETACEOUS
S 6 186 SARMATIAN STAGE, PARATETHYS, NEOGENE
B 6212 2455 SARMIENTO RIDGE
S 5 89 SARNIA, CORE SECTION, ONTARIO, LAMBTON COUNTY,
B 6203 538 SARNIAK, T., ELDER SANDSTONE—DELTAIC SEDIMENTS OF ROBERTS
B 6203 571 SARNIAK, T. M., DELTAIC COMPONENTS OF VALMY AND VININI
B 6004 719 SARNTHEIN, M., EOLIAN-SAND TURBIDITES, ABST.
B 6104 492 SAROOP, H. C., DIAGENESIS OF MIDDLE AND UPPER EOCENE CARBONATE
M 24 242 SARTARTIA FIELD, MISSISSIPPI
B 6203 530* SASKATCHEWAN, ABST., CARBONATE-, EVAPORITE RELATIONS IN
B 6305 841* SASKATCHEWAN, ABST., GEOLOGY AND HYDROCARBON ACCUMULATION
B 6305 826* SASKATCHEWAN, ABST., GEOLOGY OF, PALEOZOIC STRATA IN
B 6205 886* SASKATCHEWAN, ABST., INFLUENCE OF DIAGENETIC AND
B 6405 740* SASKATCHEWAN, ABST., NEARSHORE LITHOFACIES OF, MANNVILLE
B 6305 831* SASKATCHEWAN, ABST., RELATIONS BETWEEN SEDIMENTARY FACIES
B 6101 30 SASKATCHEWAN, ACE MINE
B 6008 1413* SASKATCHEWAN AND ALBERTA, ABST., TRAPPING OF OIL AT MIDDLE,
B 6203 572* SASKATCHEWAN AND ALBERTA. BY TITLE ONLY., HEAVY, OILS
B 6201 81 SASKATCHEWAN, ASHERN CORRELATION
B 6101 29 SASKATCHEWAN, BEAVERLODGE AREA, PITCHBLENDE
B 6101 31 SASKATCHEWAN, BEAVERLODGE AREA, URANIUM
B 6201 76 SASKATCHEWAN, BELLE PLAINE MEMBER
B 6201 74 SASKATCHEWAN, BIRDBEAR LIMESTONE-DOLOMITE
B 6201 74 SASKATCHEWAN, BLAIRMORE SAND-SHALE-SILT
B 6008 1398* SASKATCHEWAN, CANADA, ABST., GEOLOGY AND HYDROCARBON
B 6107 1014 SASKATCHEWAN, COLONY-MCLAREN-WASECA MEMBERS, ALBERTA AND
B 6201 81 SASKATCHEWAN, COLORADO, LOWER, CORRELATION
B 6201 74 SASKATCHEWAN, CRETACEOUS
B 6107 1014 SASKATCHEWAN, CUMMINGS MEMBER, ALBERTA AND
B 6201 74 SASKATCHEWAN, DAVIDSON EVAPORITES
B 6201 74 SASKATCHEWAN, DAWSON BAY LIMESTONE
B 6108 1316 SASKATCHEWAN, DEVELOPMENTS, 1976
B 6308 1409 SASKATCHEWAN, DEVELOPMENTS, 1978
B 6409 1512 SASKATCHEWAN, DEVELOPMENTS, 1979
B 6107 1014 SASKATCHEWAN, DINA MEMBER, ALBERTA AND
B 6201 74 SASKATCHEWAN, DUPEROW LIMESTONE-DOLOMITE
B 6107 1016 SASKATCHEWAN, EPPING FIELD
B 6107 1025 SASKATCHEWAN, EVESHAM AREA
B 6107 1024 SASKATCHEWAN, EYEHILL FIELD
B 6101 30 SASKATCHEWAN, FAY MINE
B 6201 83 SASKATCHEWAN, FIRST RED BED
B 6201 74 SASKATCHEWAN, FIRST RED CLAYSTONE
B 6107 1014 SASKATCHEWAN, GENERAL PETROLEUM MEMBER, ALBERTA AND
B 6101 33 SASKATCHEWAN, HELIKIAN URANIUM DEPOSITS
B 6201 81 SASKATCHEWAN, INTERLAKE CORRELATION
B 6107 1010 SASKATCHEWAN, LLOYDMINSTER AREA, HISTORY
B 6107 1014 SASKATCHEWAN, LLOYDMINSTER MEMBER, ALBERTA AND
B 6107 1010 SASKATCHEWAN, MANNVILLE GROUP
B 6101 30 SASKATCHEWAN, MARTIN FORMATION
B 6407 981 SASKATCHEWAN, MEDICINE HAT FIELD, ALBERTA AND
B 6201 76 SASKATCHEWAN, PATIENCE LAKE MEMBER
B 6201 81 SASKATCHEWAN, POTASH LAYERS
B 6201 74 SASKATCHEWAN, PRAIRIE EVAPORITE FORMATION, SHELL LAKE MEMBER
B 6201 74 SASKATCHEWAN, PRAIRIE EVAPORITE FORMATION, UPPER SALT
B 6201 74 SASKATCHEWAN, PRAIRIE EVAPORITE FORMATION, WHITKOW SALT
B 6108 1322 SASKATCHEWAN, PRODUCTION, 1976
B 6107 1014 SASKATCHEWAN, REX MEMBER, ALBERTA AND
B 6201 76 SASKATCHEWAN, SECOND RED BED
B 6201 74 SASKATCHEWAN, SECOND RED CLAYSTONE
B 6201 74 SASKATCHEWAN, SECOND SPECKS CALCAREOUS SHALE
B 6201 74 SASKATCHEWAN, SOURIS RIVER, LOWER, LIMESTONE
B 6201 74 SASKATCHEWAN, SOURIS RIVER, UPPER,
B 6107 1014 SASKATCHEWAN, SPARKY MEMBER, ALBERTA AND
B 6305 767 SASKATCHEWAN, STOUGHTON CRUDE OIL
B 6101 30 SASKATCHEWAN, TAZIN GNEISS
B 6107 1010† SASKATCHEWAN, TRAPPING OF OIL AT INTRA-, MANNVILLE, LOWER,
B 6101 30 SASKATCHEWAN, VERNA MINE
B 6201 74 SASKATCHEWAN, WATROUS MUDSTONE-SILTSTONE
B 6201 75 SASKATCHEWAN, WINNIPEGOSIS FORMATION
B 6205 885* SASKATCHEWAN—REINTERPRETATION, ABST., CARBONATE, BUILDUP-,
B 6201 73† SASKATCHEWAN—SEISMIC STUDY, WINNIPEGOSIS, MOUNDS AND,
M 30 531 SASS DE LA LUNA FORMATION, CRETACEOUS, MALOSSA FIELD
M 25 319 SATAGUI ANTICLINE, COLOMBIA

B 6306 951† SATELLITE IMAGE ANALYSI IN MINERAL EXPLORATION IN COMPLICATED
B 6109 1560* SATELLITE IMAGE ANALYSIS IN MINERAL EXPLORATION IN
B 6004 663* SATELLITE IMAGERY, ABST., OBSERVATIONS OF, CIRCULATION AN,
B 6207 1211* SATELLITE IMAGERY, ABST., STRUCTURE AND MOTION OF, BERING AND,
B 6012 2188* SATELLITE LOOK AT REGIONAL GEOLOGY OF NORTHERN CALIFORNIA,
B 6207 1237* SATELLITE-ASSISTED MINERAL EXPLORATION IN BAJA CALIFORNIA—
B 6102 139 SATICOY FIELD, CALIFORNIA
B 6102 171 SATICOY FIELD, CALIFORNIA
B 6407 1036 SATICOY FIELD, CALIFORNIA
B 6102 177 SATICOY FIELD, PICO SAND, CALIFORNIA
B 6203 360 SATICOY OIL FIELD, CALIFORNIA
B 6011 1985† SATICOY OIL FIELD, VENTURA BASIN, CALIFORNIA, MAGNETIC-REVERSAL
B 6212 2509 SATO, SHUNJI, CONTRAST OF HYDROCARBON POTENTIAL BETWEEN
B 6207 1234 SATSUMA, A., NICKELIFEROUS LATERITE DEPOSITS IN CIRCUM-PACIFIC
B 6012 2183 SATTLEGGER, JOHANN, REGIONAL VELOCITY REDUCTION, ABST.
M 25 252 SATUN RIDGE, THAILAND
S 10 27 SATURATE DISTRIBUTION, FRANCE, PARENTIS BASIN
M 26 39 SATURATION EFFECTS, BULK DENSITY, LABORATORY, MEASUREMENTS
M 26 39 SATURATION EFFECTS, PRESSURE, LABORATORY, MEASUREMENTS
M 26 39 SATURATION EFFECTS, ROCK PROPERTIES, LABORATORY,
M 26 39 SATURATION EFFECTS, SHEAR VELOCITY, LABORATORY, MEASUREMENTS
B 6312 2227† SATURATION ON SOURCE-ROCK PYROLYSIS, EFFECT OF, OIL AND,
B 6402 165 SATURDAY ISLAND FIELD, LOUISIANA
B 6211 2221 SAUCESIAN PALEOGEOGRAPHY, CALIFORNIA
S 11 65 SAUCESIAN STAGE
S 11 20 SAUCESIAN STAGE, SUPERPOSITION
S 11 21 SAUCESIAN STAGE, ZONATION
B 6211 2221 SAUCESIAN-RELIZIAN BOUNDARY PALEOGEOGRAPHY, CALIFORNIA
B 6405 780 SAUCIER, A. E., ORGANIC MODEL FOR ROLL-FRONT URANIUM DEPOSITS,
B 6304 694 SAUCIER, A. E., TERTIARY OXIDATION IN WESTWATER CANYON MEMBER
B 6105 825 SAUCIER, ALVA E., GRANTS URANIUM REGION, NEW MEXICO, ABST.,
B 6010 1868 SAUDI ARABIA, 1975, SAUDI ARABIA MINISTRY OF PPTROLEUM AND
B 6110 1798 SAUDI ARABIA, 1976, MINISTRY OF PETROL. AND MINERAL RESOURCES,
B 6210 2114 SAUDI ARABIA, 1977, ARABIAN AMERICAN OIL CO., ON DEVELOPMENTS IN
B 6310 1886 SAUDI ARABIA, 1978, PETROCONSULTANTS S.A., ET AL, ON
B 6411 1838 SAUDI ARABIA, 1979, ARABIAN AMERICAN OIL CO. AND,
B 6405 754* SAUDI ARABIA, ABST., CEMENTATION, DIAGENESIS AND PARAGENETIC
B 6203 546* SAUDI ARABIA, ABST., STRATIGRAPHY, SEDIMENTATION, AND
B 6401 9 SAUDI ARABIA, ARAB FORMATION
B 6310 1892 SAUDI ARABIA, DEVELOPMENTS, 1977-1978
B 6010 1868 SAUDI ARABIA DIVIDED NEUTRAL ZONE, 1975, SAUDI ARABIA MINISTRY
B 6110 1798 SAUDI ARABIA DIVIDED NEUTRAL ZONE, 1976, AMERICAN INDEPENDENT
B 6210 2114 SAUDI ARABIA DIVIDED NEUTRAL ZONE, 1977, ARABIAN OIL CO., LTD.,
B 6310 1886 SAUDI ARABIA DIVIDED NEUTRAL ZONE, 1978, ARABIAN OIL CO., LTD.,
B 6411 1838 SAUDI ARABIA DIVIDED NEUTRAL ZONE, 1979, PETROCONSULTANTS S.A.
B 6210 2126 SAUDI ARABIA DIVIDED NEUTRAL ZONE, PRODUCTION, 1977, KUWAIT-
B 6310 1895 SAUDI ARABIA DIVIDED NEUTRAL ZONE, PRODUCTION, 1978, KUWAIT-
B 6411 1852 SAUDI ARABIA DIVIDED NEUTRAL ZONE, PRODUCTION, 1979, KUWAIT-
B 6401 9 SAUDI ARABIA, EN NALA ANTICLINAL AXIS
M 30 9 SAUDI ARABIA, GHAWAR FIELD, LANDSAT IMAGE
B 6401 9 SAUDI ARABIA, GHAWAR GIANT OIL FIELD, LANDSAT DATA
B 6401 9 SAUDI ARABIA, JUBAILA FORMATION
B 6010 1868 SAUDI ARABIA MINISTRY OF PETROLEUM AND MINERAL RESOURCES
B 6210 2120 SAUDI ARABIA, PRODUCTION, 1976-1977
B 6411 1843 SAUDI ARABIA, PRODUCTION, 1978-1979
S 4 1 **SAUNDERS, JOHN B., EDITED BY, REEFS AND RELATED CARBONATES--**
S 4 1 **SAUNDERS, JOHN B., INTRODUCTION, REEFS AND RELATED CARBONATES--**
S 4 1 **SAUNDERS, JOHN B., PREFACE, REEFS AND RELATED CARBONATES--**
B 6303 380 SAUNDERS, AND J. G. SCHLATER, BECKMANN, J. P., REVIEW OF, INDIAN
B 6411 1970 SAVA SUBBASINS, YUGOSLAVIA, BANAT, MURA, DRAVA, AND
M 25 426 SAVAGE RIVER, MAGNETITE-PYRITE, TASMANIA
B 6105 818* SAVANNAH OCEAN-DISPOSAL SITE, GEORGIA, ABST., DEPOSITIONAL
B 6303 446 SAVELLI, S., MALOSSA FIELD, DEEP DISCOVERY IN PO VALLEY, ITALY,
M 30 525 SAVELLI, S., MALOSSA FIELD--A DEEP DISCOVERY IN THE PO VALLEY,
B 6304 700 SAVIN, SAMUEL M., TERTIARY AND CRETACEOU OCEAN TEMPERATURES,
B 6310 1920 SAVOIE, FRANCE, JURA-, BRESSE-
B 6112 2070 SAVU BASIN, SECTION
B 6305 826* SAWATCH RANGE, COLORADO, ABST., PALEOKARST CONTROLS O
B 6303 445 SAWATSKY, L. H., PORE SPACE IN HOLOCENE CARBONATE SEDIMENTS,
B 6008 1409 SAWATZKY, HENRY B., WHY PETROLEUM EXPLORATIONIST SHOULD BE
B 6112 2074 SAWU GROUP, TO AUSTRALIAN CRATON, SECTION, NEW ISLAND
B 6112 2068 SAWU ISLAND SECTION
B 6207 1151 SAXBY, J. D., HYDROCARBON GENERATION IN GIPPSLAND BASIN,
B 6405 780 SAXENA, R. S., GEOLOGIC-SEISMIC EXPLORATION MODEL FOR REWORKED
B 6105 825 SAXENA, RAM S., DELTAIC SANDSTONE RESERVOIRS—EXPLORATION
B 6203 560 SAXENA, RAM S., EXPLORATION MODELS AND CRITERIA FOR
B 6303 538 SAYRE, W. O., MIDDLE CRETACEOUS OXYGEN-DEFICIENT
B 6004 651† SCALE FUNCTIONS FOR CONVERSION OF RESISTIVITY TO
B 6203 497* SCALE MODELING OF BED CONFIGURATIONS, ABST., DYNAMIC
B 6411 1841 SCALLOP CORP., ON DEVELOPMENTS IN YEMEN, 1979
B 6004 712* SCALLOPED SOLUTION SURFACES, MIDDLE ORDOVICIAN CARBONATE
M 28 188 SCANNING ELECTRON MICROSCOPY, TECHNIQUES
B 6303 539* SCANNING ELECTRON MICROSCOPY TO HYDROCARBON EXPLOITATION,
B 6405 794* SCANNING ELECTRON MICROSCOPY, X-RAY DIFFRACTION TECHNIQUES,
B 6007 1125 SCANNING PHOTOMICROGRAPH, OKLAHOMA, PALEOZOIC SHALE,
B 6002 297 SCANNING-ELECTRON MICROGRAPH, CLAY
S 7 318 SCANNING-ELECTRON MICROGRAPHS, GYRE BASIN
S 7 317 SCANNING-ELECTRON MICROGRAPHS, GYRE BASIN
M 27 89 SCAPHOPODS
M 27 88 SCAPHOPODS
S 4 285 SCARUS CROICENSIS, PANAMA
B 6301 90 SCATARIE MEMBER, SCOTIAN BASIN
S 6 313 SCATTER, PERMIAN, RADIOMETRIC DATING
S 4 155 SCATTERDAY, JAMES W., LOW-WATER EMERGENCE OF CARIBBEAN REEFS
B 6405 780 SCATTOLINI, R., HEAT-FLOW AND HEAT-PRODUCTION STUDIES IN
S 4 245 SCENIC POINT MUD BANK, TEXAS
B 6303 538 SCHAAF, A., MIDDLE CRETACEOUS OXYGEN-DEFICIENT
B 6405 800 SCHAFERSMAN, S D., CONTEMPORANEOUS RECONSTRUCTION AND
B 6405 781 SCHAFERSMAN, S. D., BIOSTRATIGRAPHIC PROBLEMS GENERATED BY DEE
B 6209 1757 SCHAFERSMAN, STEVEN D., OLIGOCENE REEF COMMUNITY SUCCESSION,
S 4 37 SCHAFERSMAN, STEVEN D., PATCH-REEF ECOLOGY AND
B 6303 483 SCHAFERSMAN, STEVEN D., RADIOLARIAN PRESERVATION IN PRESENT
B 6303 521 SCHAFERSMAN, STEVEN D., TROPICAL CENOZOIC PALEO-OCEANOGRAPHY
B 6303 477 SCHAMEL, S., TUNISIA--PLATE TECTONIC AND HYDROCARBON
B 6303 531 SCHAMEL, STEVEN, KARASEK, RICHARD M., KEF ANTICLINE--BOX
B 6403 423 SCHAMEL, STEVEN, PROGRESSIVE NEOGENE DEFORMATION OF
B 6303 521 SCHAMEL, STEVEN, STRUCTURE OF TUNISIAN ATLAS, ABST.
B 6209 1753 SCHATZINGER, R. A., DISTRIBUTION AND GEOMETRY OF AN
M 25 444 SCHEELITE DEPOSITS, SOUTH KOREA
M 25 426 SCHEELITE, TUNGSTEN, TASMANIA, KING ISLAND
B 6010 1908 SCHEIDECKER, WILLIAM R., PETROLEUM DEVELOPMENTS IN FAR EAST IN
B 6110 1832 SCHEIDECKER, WILLIAM R., PETROLEUM DEVELOPMENTS IN FAR EAST IN
B 6203 533 SCHEIDEGGER, K. F., QUATERNARY SEDIMENTATION AND TECTONICS ON
B 6004 657 SCHEIHING, J. E., MODERN AND ANCIENT ESTUARINE-FILL FACIES,
B 6109 1546 SCHELDT, JOHN, PETROLOGY OF CATAHOULA SANDSTONES IN EAST
B 6203 562* SCHEME FOR MULTIVARIATE ANALYSIS IN CARBONATE PETROLOGY
B 6105 771 SCHENCK, DAN R., COMPUTER APPROACH TO VERTICAL-VARIABILITY
B 6203 534 SCHENK, PAUL E., ORDOVICIAN-SILURIAN GLACIO-EUSTATIC SEA-LEVEL
B 6108 1390 SCHERBA, MARSHAL G., HIGH-RESOLUTION SEISMIC EXPLORATION FOR
B 6409 1567 SCHERER, DONALD R., HOUSTON TREND, MISSISSIPPI, ABST.
M 30 423 SCHERER, F. C., EXPLORATION IN EAST MALAYSIA OVER THE PAST
B 6012 2153 SCHERER, MICHAEL, INFLUENCES OF DIAGENETIC ENVIRONMENT ON
B 6104 603 SCHERRER FORMATION, NEW MEXICO
S 8 6 SCHICK-SAVONIUS, MODEL, CURRENT METER
B 6209 1765 SCHIEBOUT, JUDITH A., EARLY TERTIARY TERRESTRIAL
B 6203 536 SCHIENER, E. J., HYDROCARBON GENERATION POTENTIAL OF
B 6203 436 SCHILLBANK HIGH, NORTH SEA
B 6405 650 SCHILLING, JOHN, REVIEW OF GUIDEBOOK OF SANTA FE COUNTRY
B 6304 690 SCHIPPER, M. R., EFFECTS OF URANIUM MINING ON GROUNDWATER IN
M 28 47 SCHIST, CONNECTICUT, ORDOVICIAN
M 28 49 SCHIST, CONNECTICUT, ORDOVICIAN
M 28 48 SCHIST, MASSACHUSETTS, ORDOVICIAN
M 28 56 SCHIST, MASSACHUSETTS, PALEOZOIC
M 28 48 SCHIST, MASSACHUSETTS, SILURIAN
B 6202 255 SCHIST RIDGE, CALIFORNIA
M 28 156 SCHISTI GALESTRINI, ITALY, CRETACEOUS
B 6303 521 SCHLAEFER, JILL T., SOIL REGIMES OF TAZIRBU REGION OF CENTRA
B 6303 522 SCHLAEFER, JILL T., SUBSURFACE GEOLOGY OF HONOR RANCHO AREA,
B 6105 825 SCHLAEPFER, C. J., JURASSIC SEDIMENTATION, EASTERN SIERRA MADRE
B 6004 719 SCHLAGER, W., CARBONATE MUD TO CHALK, PLEISTOCENE, IN
B 6203 560 SCHLAGER, WOLFGANG, INFLUENCE OF PLATFORM-DERIVED SEDIMENT
B 6412 2190 SCHLAGER, WOLFGANG, REVIEW OF BEITRAGE ZUR BIOSTRATIGRAPHIE
B 6306 870 SCHLANGER, SEYMOUR O., CRETACEOUS OCEANIC ANOXIC EVENTS AS
B 6105 762 SCHLANGER, SEYMOUR O., MIDDLE CRETACEOUS OCEANIC ANOXIC EVENTS
B 6303 380 SCHLATER, BECKMANN, J. P., REVIEW OF, INDIAN OCEAN GEOLOGY AND
B 6105 826 SCHLEE, JOHN, ATLANTIC OUTER CONTINENTAL SHELF, ABST.,
B 6105 790 SCHLEE, JOHN, DEPTH CONVERSION OF MULTICHANNEL SEISMIC-
B 6105 742 SCHLEE, JOHN, REGIONAL GEOLOGIC FRAMEWORK OFF NORTHEASTERN
B 6006 926 SCHLEE, JOHN, REGIONAL GEOLOGIC FRAMEWORK OFF NORTHEASTERN
B 6405 715 SCHLEE, JOHN S., ET AL, DEEP STRATIGRAPHY AND EVOLUTION OF
M 29 65 SCHLEE, JOHN S., MULTICHANNEL SEISMIC DEPTH SECTIONS AND
B 6302 259 SCHLEE, JOHN S., REVIEW ON ISLAND ARCS, DEEP SEA TRENCHES, AND
B 6405 781 SCHLEE, JOHN S., SEISMIC STRATIGRAPHY OF ATLANTIC MARGIN IN
B 6203 560 SCHLEE, JOHN S., STRUCTURE OF ATLANTIC SLOPE OF EASTERN NORTH
B 6303 521* SCHLERACTINIAN CORALS, ABST., TROPICAL, CENOZOIC
B 6207 1112 SCHLUMBERGER 4-ARM HIGH-RESOLUTION DIPMETER LOG
B 6011 2069 SCHMALZ, ROBERT F., EVAPORITES, SULFIDES, AND PETROLEUM, ABST.

B 6303 538 SCHMIDT, R. R., MIDDLE CRETACEOUS OXYGEN-DEFICIENT
B 6105 826 SCHMIDT, RICHARD A., FUEL-RESERVE INFORMATION, ABST., ELECTRIC
B 6309 1587 SCHMIDT, RICHARD A., IN-SITU TESTING OF WELL-SHOOTING CONCEPTS,
B 6105 826 SCHMIDT, VOLKMAR, POROSITY ALTERATION IN CARBONATE
B 6108 1390 SCHMIDT, VOLKMAR, ROLE OF SECONDARY POROSITY IN SANDSTONE
B 6309 1504 SCHMOKER, JAMES W., DETERMINATION OF ORGANIC CONTENT OF
B 6412 2156 SCHMOKER, JAMES W., ORGANIC CONTENT OF DEVONIAN SHALE IN
B 6205 848 SCHNEEFLOCK, ROBERT, PERMEABILITY TRAPS IN GATCHELL, EOCENE,
B 6303 496 SCHNEIDER, JEAN, MOVEMENT OF SUBSURFACE WATERS UNDER
B 6405 800 SCHNEIDERMANN, N., CONTEMPORANEOUS RECONSTRUCTION AND
S 4 17 SCHNEIDERMANN, NAHUM, SHELF REEFS, SOUTHWESTERN PUERTO RICO
B 6303 522 SCHOELL, MARTIN, DUAL ORIGIN OF NATURAL GASES IN SUBALPINE
B 6103 435 SCHOENNAGEL, F. A., MISSISSIPPIAN UNCONFORMITIES IN NORTHER
B 6105 826 SCHOETTLE, M., GEOCHEMISTRY OF RECENT CONTINENTAL-MARGIN
B 6209 1610 SCHOHARIE FORMATION, PENNSYLVANIA
M 25 280 SCHOLL, D. W., CENOZOIC AN MESOZOIC PETROLEUM PROSPECTS,
B 6405 745 SCHOLL, D. W., HYDROCARBON PROSPECTS OF BERING SEA SHELF SOUTH
B 6303 490 SCHOLL, D. W., SHALLOW-WATER UPPER JURASSIC ROCKS DREDGED
B 6004 694 SCHOLL, D. W., STRUCTURE AN EVOLUTION OF BERING SEA MARGIN
B 6303 436 SCHOLL, D. W., THICK SEDIMENT ACCUMULATIONS BENEATH
B 6403 439 SCHOLL, DAVID W., HYDROCARBON POTENTIAL OF ALEUTIAN BASIN,
B 6311 2070 SCHOLL, DAVID W., HYDROCARBON POTENTIAL OF ALEUTIAN BASIN,
B 6103 467 SCHOLL, DAVID W., INDICATOR OF PLATE TECTONICS IN SOUTHERN
B 6405 694 SCHOLL, DAVID W., MULTIFOLD SEISMIC DATA ACROSS OUTER BERING
B 6405 693 SCHOLL, DAVID W., PROSPECTIVE AND FUTURE HYDROCARBON
B 6408 1279 SCHOLL, DAVID W., RESOURCE POTENTIAL AND PLATE-MARGIN
B 6002 161 SCHOLL, DAVID W., STRUCTURE AND EVOLUTION OF BERING SEA SHELF
B 6212 2481 SCHOLL, DAVID W., VAMPS—POSSIBLE HYDROCARBON-BEARING
B 6310 1981 SCHOLLE, BOLES, JAMES R., REVIEW OF A, COLOR ILLUSTRATED GUIDE
B 6004 719 SCHOLLE, P. A., DIAGENETIC PATTERNS OF CHALKS, ABST.
B 6204 592 SCHOLLE, P. A., PETROLEUM POTENTIAL OF U.S. ATLANTIC SLOPE, RISE,
B 6401 67 SCHOLLE, PETER A., CARBON ISOTOPE FLUCTUATIONS IN CRETACEOUS
B 6107 982 SCHOLLE, PETER A., CHALK DIAGENESIS AND ITS RELATION TO
M 27 1 SCHOLLE, PETER A., A COLOR ILLUSTRATED GUIDE TO CARBONATE ROCK
M 28 1 SCHOLLE, PETER A., A COLOR ILLUSTRATED GUIDE TO CONSTITUENTS,
B 6105 827 SCHOLLE, PETER A., DEEP-WATER CARBONATE SEDIMENTOLOGY,
B 6109 1556 SCHOLLE, PETER A., DIAGENETIC PATTERNS OF AUSTIN GROUP AND
B 6105 811 SCHOLLE, PETER A., PETROLEUM POTENTIAL OF POSSIBLE LOWER
B 6303 522 SCHOLLE, PETER A., POROSITY RELATIONS IN CHALK RESERVOIRS, ABST.
B 6305 820 SCHOLLE, PETER A., REVIEW OF ENCYCLOPEDIA OF SEDIMENTOLOGY,
B 6205 892 SCHOLLE, PETER, DIAGENESIS AND HYDROCARBON POTENTIAL OF
B 6405 764 SCHOONMAKER, JANE, PHANEROZOIC CARBONATE DIAGENESIS—A NEW
B 6209 1645 SCHOUTEN, HANS, PLATE CONVERGENCE AND ACCRETION IN
B 6305 840 SCHOWALTER, TIM T., HYDROCARBON SHOWS--PROXIMITY INDICATORS
B 6305 723 SCHOWALTER, TIM T., MECHANICS OF SECONDARY HYDROCARBON
B 6108 1391 SCHOWALTER, TIM T., MECHANICS OF SECONDARY MIGRATION AND
M 26 477 SCHRAMM, M. W., JR., PRACTICAL STRATIGRAPHIC MODELING AND
B 6303 487 SCHRAMM, MARTIN W., JR., SEISMIC STRATIGRAPHIC, AND ECONOMIC
B 6309 1602 SCHRANK, J., SMACKOVER RESERVOIR--INTERPRETATION CASE STUDY OF
B 6004 720 SCHREIBER, B. C., DEVELOPMENT OF POROSITY IN CARBONATE ROCKS OF
S 5 169 SCHREIBER, B. CHARLOTTE AN EVAPORITIC LITHOFACIES CONTINUUM:
S 5 169 SCHREIBER, EDWARD AN EVAPORITIC LITHOFACIES CONTINUUM:
B 6306 987 SCHREIBER, EVANS, ROBERT, REVIEW OF, MARINE, EVAPORITES, EDITED
B 6411 1969 SCHRODER, L., AND DEVELOPMENTS IN WEST GERMANY, 1979
B 6310 1920 SCHRODER, L., ON DEVELOPMENTS IN FEDERAL REPUBLIC OF GERMANY,
B 6110 1642 SCHRODER, L., ON DEVELOPMENTS IN GERMAN FEDERAL REPUBLIC, 1976
B 6210 1911 SCHRODER, L., ON DEVELOPMENTS IN GERMAN FEDERAL REPUBLIC, 1977
B 6303 522 SCHUENEMEYER, JOHN H., PREDICTING FUTUR OIL USING
B 6105 782 SCHUENEMEYER, J. H., STATISTICAL HISTORY OF PETROLEUM
B 6007 1119 SCHULER FORMATION, TEXAS
B 6407 1005 SCHULER RED BEDS, ARK-LA-TEX AREA
M 24 79 SCHULTHEIS, N. H., KAYBOB OIL FIELD, ALBERTA, CANADA
B 6309 1595 SCHULTZ, ALVIN L., IMPACT OF DIAGENESIS ON LOG INTERPRETATION,
B 6105 814 SCHULTZ, D. M., GEOCHEMISTRY OF LIGHT HYDROCARBONS IN
B 6105 814 SCHULTZ, D. M., HYDROCARBONS IN SURFACE SEDIMENTS OF
B 6309 1587 SCHULTZ, D. M., IMPLICATION OF GASEOUS HYDROCARBON
B 6309 1583 SCHULTZ, D. M., PETROLEUM GEOCHEMISTRY AND GEOLOGY OF
B 6303 527 SCHULTZ, PHILIP, DEPTH MIGRATION OF SEISMIC DATA, ABST.
B 6211 2362 SCHULTZ, PHILIP, EQUALIZING STACKING VELOCITIES OF DIPPING
B 6012 2177 SCHULTZ, PHILIP S., SLANT-WAVE STACK, ABST.
B 6305 840 SCHUMM, STANLEY A., EVOLUTION AND RESPONSE OF FLUVIAL
B 6303 523 SCHUSTER, DAVID, GWNA MELANGE, UPPER PRECAMBRIAN
B 6008 1410 SCHUSTER, J. ERIC,, HEAT-FLOW STUDIES IN STEAMBOAT
B 6105 827 SCHUTZ, DONALD F., RN222 AND HE4-AR36 RATIOS IN SOIL GAS
B 6203 561 SCHUTZ, DONALD F., VERTICAL DISTRIBUTION OF HE4, RN222, AND PB210
B 6203 490 SCHWAB, ROLAND, REGIONAL CONTROLS ON DIAGENESIS OF
B 6309 1578 SCHWALB, HOWARD, PAST AND CURRENT DEVELOPMENTS FOR
B 6403 359 SCHWAN, WERNER, GEODYNAMIC PEAKS IN ALPINOTYPE OROGENIES
B 6303 505 SCHWARCZ, HENRY P., SULFUR AND OXYGEN ISOTOPE GEOCHEMISTRY
M 25 268 SCHWARTZ, D., MARIN PETROLEUM EXPLORATION OF HUKSAN
B 6105 827 SCHWARTZ, DANIEL E., FLOW PATTERNS AND BAR MORPHOLOGY IN
B 6209 1765 SCHWARTZ, DANIEL E., SEDIMENTARY FACIES, STRUCTURES, AND
B 6303 523 SCHWARTZ, M. H., IMPACT OF COAL PROPERTIES ON COMBUSTION
B 6105 828 SCHWARTZ, R. K., DREDGE DISPOSAL IN LITTORAL ZONE, NEW RIVER
B 6405 791 SCHWARTZ, ROBERT K., LATE MESOZOIC TO EARLY CENOZOIC FORELAND
B 6105 817 SCHWARTZ, ROBERT K., MEASUREMENT TECHNIQUE FOR
B 6208 1399 SCHWARZ, KENNETH A., OIL AND GAS DEVELOPMENTS IN MARYLAND,
B 6409 1403 SCHWARZ, KENNETH A., OIL AND GAS DEVELOPMENTS IN MARYLAND,
B 6308 1244 SCHWARZ, KENNETH A., OIL AND GAS DEVELOPMENTS IN MARYLAND,
B 6309 1597 SCHWARZER, RUDY, RELICT AND EXPATRIATE RADIOLARIAN FAUNA IN
B 6004 720 SCHWARZER, THERESA FLYNN,, WOMEN GEOSCIENTISTS IN PETROLEUM
B 6104 573 SCHWEISBERGER, R T., CALCULATION OF AGES OF HYDROCARBONS IN
B 6405 721* SCIENTIFIC OCEAN DRILLING NEAR UNITED STATES, ABST.
B 6402 173 SCLATER, J. G., CONTINENTAL MARGIN SUBSIDENCE AND HEAT
B 6405 683 SCLATER, J. G., FORELAND FOLD AND THRUST BELT OF CARPATHIANS
B 6405 781 SCLATER, JOHN G., CONTINENTAL STRETCHING—EXPLANATION OF
S 4 289 SCLERACTINIAN CORAL, GALAXEA CLAVUS DANA, HERMATYPIC
M 27 61 SCLERACTINIAN CORALS
M 27 59 SCLERACTINIAN CORALS
S 4 361 SCLERACTINIAN CORALS, ANATOMY OF
S 4 95 SCLERACTINIAN CORALS, DIVERSITY AND ABUNDANCE OF
S 4 375 SCLERODERMITES UNDER SUBAERIAL CONDITIONS, LEACHING OF
M 27 49 SCLEROSPONGES
S 4 95 SCLEROSPONGES AS FRAMEWORK BUILDERS
B 6004 673 SCOFFIN, TERENCE P., BERMUDAS NORTHWEST REEFS COLONIZE LAST
B 6310 1689 SCORCELLETTI, P. GIORGIO, PETROLEUM DEVELOPMENT IN CENTRAL
B 6210 1844 SCORCELLETTI, P. GIORGIO, PETROLEUM DEVELOPMENTS IN CENTRAL
B 6102 257 SCORPION MOUNTAIN FORMATION, IDAHO
M 25 535 SCOTIA ARC, MINERAL OCCURRENCES
S 11 224 SCOTIA BLUFFS AND CARLOTTA FORMATION, ASSEMBLAGES
S 11 196 SCOTIA BLUFFS SANDSTONE
M 29 133 SCOTIA SEA BASIN, SEAFLOOR SPREADING, MAGNETICS
B 6202 228 SCOTIA SHELF, SEDIMENTARY SEQUENCE
M 25 402 SCOTIA-CARR BOYD AREA, NICKEL SULFIDE, WESTERN AUSTRALIA
S 11 318 SCOTIA-EEL RIVER, CALIFORNIA, SAMPLE LOCALITIES
S 11 323 SCOTIA-EEL RIVER SECTION DESCRIPTION
B 6301 90 SCOTIAN BASIN, ABENAKI FORMATION
B 6301 89 SCOTIAN BASIN, ABENAKI SUBBASIN
B 6203 526* SCOTIAN BASIN, ABST., COST B-2 WELL, BALTIMORE CANYON, AND ITS
B 6301 90 SCOTIAN BASIN, ARGO FORMATION
B 6301 90 SCOTIAN BASIN, BACCARO MEMBER
B 6301 90 SCOTIAN BASIN, BANQUEREAU FORMATION
B 6301 90 SCOTIAN BASIN, DAWSON CANYON FORMATION
B 6109 1484 SCOTIAN BASIN, EAST COAST
B 6301 90 SCOTIAN BASIN, ESPERANTO BEDS
B 6301 90 SCOTIAN BASIN, EURYDICE FORMATION
B 6301 87† SCOTIAN BASIN, GEOCHEMICAL CHARACTERISTICS OF SEDIMENTARY
B 6301 89 SCOTIAN BASIN, HYDROCARBON SHOWS
B 6301 90 SCOTIAN BASIN, IROQUOIS FORMATION
B 6301 89 SCOTIAN BASIN, LAHAVE PLATFORM
B 6301 90 SCOTIAN BASIN, LAURENTIAN FORMATION
B 6301 90 SCOTIAN BASIN, LOGAN CANYON FORMATION
B 6301 90 SCOTIAN BASIN, MANHASSET BEDS
B 6301 90 SCOTIAN BASIN, MASKONOMET BEDS
B 6301 90 SCOTIAN BASIN, MIC MAC FORMATION
B 6301 90 SCOTIAN BASIN, MISAINE MEMBER
B 6301 90 SCOTIAN BASIN, MISSISAUGA FORMATION
B 6301 90 SCOTIAN BASIN, MOHICAN FORMATION
B 6301 90 SCOTIAN BASIN, NASHWAUK BEDS
B 6301 90 SCOTIAN BASIN, NASKAPI MEMBER
B 6301 90 SCOTIAN BASIN, O MARKER
B 6301 89 SCOTIAN BASIN, ORPHEUS GRABEN
B 6301 90 SCOTIAN BASIN, PETREL MEMBER
B 6301 90 SCOTIAN BASIN, SABLE MEMBER
B 6301 89 SCOTIAN BASIN, SABLE SUBBASIN
B 6301 90 SCOTIAN BASIN, SCATARIE MEMBER
B 6301 89 SCOTIAN BASIN, SEDIMENTARY RIDGE PROVINCES
B 6404 502 SCOTIAN BASIN, SEISMIC REFLECTION PROFILES
B 6301 90 SCOTIAN BASIN, SHORTLAND SHALE
B 6301 90 SCOTIAN BASIN, STRATIGRAPHIC COLUMN
B 6301 90 SCOTIAN BASIN, VERRILL CANYON FORMATION
B 6301 90 SCOTIAN BASIN, WYANDOT FORMATION
B 6202 230 SCOTIAN SHELF, EVAPORITES
B 6004 714* SCOTIAN SHELF GEOTHERMAL GRADIENTS, ABST.
B 6105 818* SCOTIAN SHELF, OFFSHORE EASTERN CANADA, ABST., PALEOGEOGRAPHY
B 6101 80 SCOTIAN SHELF, ORPHEUS AND ABENAKI BASINS, OUTER
B 6107 988 SCOTIAN SHELF, WYANDOT CHALK UNIT, CANADA
B 6303 446* SCOTIAN SHELF--CASE STUDY, ABST., PETROLEUM
B 6306 905 SCOTLAND, HELMSDALE FAULT

Index

B 6203 430 SCOTLAND, RIFT SYSTEM WEST OF
B 6405 688 SCOTT, ALAN J., SPECTRUM OF DELTAIC SYSTEMS IN PENNSYLVANIAN
B 6203 561 SCOTT, EDWARD W., GEOLOGIC APPRAISAL OF UNDISCOVERED OIL AND
B 6303 515 SCOTT, EDWARD W., RELATION OF SEDIMENTARY HISTORY AND
B 6303 523 SCOTT, EDWARD W., RESOURCE APPRAISAL PREDICTIONS AND
B 6004 720 SCOTT, G. L., DEPOSITION, FACIES PATTERNS, AND HYDROCARBON
B 6008 1410 SCOTT, G. L., STRETCHING THE SEISMIC DOLLAR, ABST.
B 6108 1391 SCOTT, JAMES,, ASSESSMENT OF MAGNETIC MEASUREMENTS IN
B 6201 41 SCOTT PLATEAU, AUSTRALIA
B 6004 650 SCOTT, R. M., PHYSICAL AND BIOGENI CHARACTERISTICS OF SEDIMENTS
B 6405 782 SCOTT, R. W., EARLY CRETACEOUS REEF COMMUNITIES IN GULF COAST,
B 6109 1557 SCOTT, R. W., LOWER CRETACEOUS DEPOSITIONAL SYSTEMS, WEST
B 6105 828 SCOTT, R. W., RHIZOCORALLIUM IN CRETACEOUS DAKOTA SANDSTONE,
M 25 236 SCOTT REEF DISCOVERY WELL, AUSTRALIA
B 6004 720 SCOTT, RICHARD M., SEDIMENTARY ENVIRONMENTS OF TIDAL FLAT
B 6109 1546 SCOTT, ROBERT J., AUSTIN CHALK—BUDA TREND OF SOUTH TEXAS,
B 6102 296 SCOTT, ROBERT J., SOUTH TEXAS AUSTIN CHALK, CRETACEOUS, PLAY,
B 6307 1108 SCOTT, ROBERT W., DEPOSITIONAL MODEL OF EARLY CRETACEOUS
B 6305 841 SCOTT, ROBERT W., OIL AND GAS IN CHINA. BY TITLE ONLY
B 6303 524 SCOTT, ROBERT W., OIL AND GAS IN PEOPLES REPUBLIC OF CHINA ABST.
M 25 411 SCREE ORE, CONGLOMERATE ORE, DETRITAL ORE, OR CANGA
S 8 58 SCRIPPS CANYON, CALIFORNIA, SUBMARINE CANYONS
S 8 61 SCRIPPS CANYON, DESCRIBED, TURBIDITY CURRENTS
S 8 59 SCRIPPS CANYON, TOPOGRAPHY, INDEX MAP
B 6206 972 SCRIPPS CRUISE 7 TOW, LEG 10, CALIFORNIA
B 6206 972 SCRIPPS CRUISE ANTIPODE, LEG 1, CALIFORNIA
B 6206 972 SCRIPPS CRUISE MONTCANYON II, CALIFORNIA
B 6004 695* SCRIPPS SUBMARINE CANYON TRIBUTARY, ABST., LARGE SEDIMENT,
B 6004 721 SCULL, B. J., LITHOSTRATIGRAPHY OF NORWEGIAN NORTH SEA, ABST.
B 6303 524 SCULLY, R. TUCKER, LEGAL AND POLITICAL STATUS OF ANTARCTICA,
B 6303 462* SEA GULLIES AND DEVELOPMENT OF LINEAR CONGLOMERATIC UNITS,
M 26 138 SEA LEVEL ANALYSIS, TOOLS
M 29 461† SEA LEVEL CHANGE: A VIEW FROM THE CRATON, GLOBAL
M 29 453† SEA LEVEL CHANGES ON STRATIGRAPHIC SEQUENCES AT ATLANTIC
M 26 139 SEA LEVEL, CHART, CORRELATION, CHRONOSTRATIGRAPHIC WITH
M 26 81 SEA LEVEL FLUCTUATIONS, CHRONOCORRELATIONS, AFRICA, WEST
M 26 66 SEA LEVEL FLUCTUATIONS, COASTAL ONLAP
M 26 77 SEA LEVEL FLUCTUATIONS, CORRELATION CURVE CONSTRUCTION
M 26 64 SEA LEVEL FLUCTUATIONS, CYCLES
M 26 89 SEA LEVEL FLUCTUATIONS, EUSTATIC ESTIMATION
M 26 83 SEA LEVEL FLUCTUATIONS, GLOBAL CYCLES
M 26 92 SEA LEVEL FLUCTUATIONS, GLOBAL CYCLES, CAUSES
M 26 88 SEA LEVEL FLUCTUATIONS, GLOBAL CYCLES, CHART CONSTRUCTION
M 26 92 SEA LEVEL FLUCTUATIONS, GLOBAL HIGHSTANDS
M 26 92 SEA LEVEL FLUCTUATIONS, GLOBAL LOWSTANDS
M 26 65 SEA LEVEL FLUCTUATIONS, HINTERLAND SEQUENCES
M 26 92 SEA LEVEL FLUCTUATIONS, INTERREGIONAL UNCONFORMITIES
M 26 65 SEA LEVEL FLUCTUATIONS, MARITIME SEQUENCES
M 26 67 SEA LEVEL FLUCTUATIONS, NERITIC FACIES, RELATIVE RISE
M 26 64 SEA LEVEL FLUCTUATIONS, PARACYCLES
M 26 66 SEA LEVEL FLUCTUATIONS, REGRESSION
M 26 71 SEA LEVEL FLUCTUATIONS, RELATIVE FALLS
M 26 65 SEA LEVEL FLUCTUATIONS, RELATIVE RISES
M 26 68 SEA LEVEL FLUCTUATIONS, RELATIVE STILLSTAND
M 26 66 SEA LEVEL FLUCTUATIONS, TRANSGRESSION
M 29 469† SEA LEVEL FROM SEISMIC STRATIGRAPHY, GLOBAL, CYCLES OF,
M 26 71 SEA LEVEL INTERPRETATION, TEXAS, EAST TEXAS FIELD, WOODBINE
B 6203 509* SEA LEVEL ON SOUTHEASTERN UNITED STATES COAST, ABST.,
B 6205 772† SEA LEVEL, PART 10: SEISMIC RECOGNITION OF CARBONATE BUILDUPS,
M 26 185† SEA LEVEL, PART 10: SEISMIC RECOGNITION OF CARBONATE BUILDUPS,
M 26 205† SEA LEVEL, PART 11: GLOSSARY OF TERMS USED IN SEISMIC
M 26 51† SEA LEVEL, PART 1: OVERVIEW, SEISMIC, STRATIGRAPHY AND, GLOBAL,
M 26 53† SEA LEVEL, PART 2: THE DEPOSITIONAL SEQUENCE AS A BASIC UNIT FOR
M 26 63† SEA LEVEL, PART 3: RELATIVE CHANGES OF SEA LEVEL FROM COASTAL
M 26 165† SEA LEVEL, PART 9: SEISMIC INTERPRETATION OF CLASTIC
B 6205 752† SEA LEVEL, PART 9: SEISMIC INTERPRETATION OF CLASTIC
M 26 49† SEA LEVEL, SEISMIC, STRATIGRAPHY AND, GLOBAL, CHANGES OF
B 6109 1538* SEA LEVEL—EVIDENCE FROM CHOCTAWHATCHEE FORMATION, FLORIDA
M 25 263 SEA OF JAPAN, KOREAN OFFSHORE, EAST SEA
B 6406 945 SEA SOUNDER CRUISE S6-77, RV
B 6204 633 SEA SOUNDER RESEARCH CRUISE, RV
S 4 281† SEA URCHINS ON CARIBBEAN REEFS, CARBONATE-SEDIMENT
B 6004 696* SEA WATER, ABST., ORIGIN OF, DOLOMITE--, MODEL BASED ON, CYCLIC
B 6004 655* SEA-FLOOR INSTABILITY IN OFFSHORE GULF OF ALASKA TERTIARY
S 9 19 SEA-LEVEL CHANGES, CLIMATIC CHANGES
B 6109 1553* SEA-LEVEL CHANGES, GULF COAST, ABST., DEPOSITION AND DIAGENESIS
B 6109 1555* SEA-LEVEL CHANGES, GULF COAST, ABST., LOWER, CRETACEOUS
B 6203 534* SEA-LEVEL CHANGES IN EASTERNMOST APPALACHIANS, SOUTHERN
B 6303 529* SEA-LEVEL MOVEMENTS AROUND SOUTHERN AFRICA, ABST., TERTIARY
B 6004 679* SEA-LEVEL RISE, ABST., RESPONSE OF ACTIVE, OOLITE, SAND, SHOAL TO
B 6309 1584 SEABAUGH, P. W., DEVONIAN SHALE CHARACTERIZATION--EVALUATION
B 6309 1587 **SEABAUGH, P. W., INTERPRETATION AND STATISTICAL ANALYSIS OF**
B 6405 771* SEABED CHARACTERISTICS AND SAND DISPERSAL ON
M 25 16 SEABED JURISDICTION, GULF OF, THAILAND
M 25 8† SEABED JURISDICTION OFF PACIFIC COAST OF ASIA, BOUNDARIES OF
M 25 15 SEABED JURISDICTION, SOUTH, CHINA SEA
M 24 26 SEABEE FORMATION, ALASKA
S 4 87 SEAFIELD LIMESTONE, JAMAICA
M 25 535 SEAFLOOR DEPOSITS BEARING MINERALS, PACIFIC, SOUTHEASTERN
B 6207 1216* SEAFLOOR INSTABILITY ON PACIFIC CONTINENTAL MARGINS, ABST.
M 26 369 SEAFLOOR MAPPING, RESOLUTION SYSTEMS
M 29 355 SEAFLOOR MORPHOLOGY, VIRGIN ISLANDS, ST. CROIX
B 6207 1193 SEAFLOOR SPREADING ALONG, KULA-, FARALLON, PLATE BOUNDARIES
B 6403 362 SEAFLOOR SPREADING AND PLATE MOTION, EPISODIC READJUSTMENTS
M 25 178 SEAFLOOR SPREADING, DIVERGENT, PLATES AND
M 25 485 SEAFLOOR SPREADING, EAST PACIFIC RISE
M 25 485 SEAFLOOR SPREADING, EXPLORER RISE
M 25 485 SEAFLOOR SPREADING, GORDA RISE
M 25 485 SEAFLOOR SPREADING, JUAN DE FUCA RISE
M 29 133 SEAFLOOR SPREADING, MAGNETICS, SCOTIA SEA BASIN
M 25 547 SEAFLOOR SPREADING, PLATE-TECTONIC THEORY, RELATES GEOLOGIC
M 25 487 SEAFLOOR SPREADING, RELATION OF, MINERAL DEPOSITS AND
B 6305 747 SEAL CAPACITY
B 6004 674* SEAL RELATIONS IN PRODUCTIVE EVAPORITE-ENCASED REEF
B 6204 635 SEAL ROCKS, ALASKA
B 6303 529* SEALING AND NONSEALING FAULTS IN GULF COAST SALT BASIN, ABST.
B 6402 145† SEALING AND NONSEALING FAULTS IN LOUISIANA GULF COAST SALT
B 6303 533* SEAM, ABST., ESTUARINE-COASTAL PLAIN, COAL DEPOSITION IN
B 6405 795* SEAM IN ILLINOIS, ABST., SEELYVILLE, COAL—MAJOR UNEXPLOITED
B 6404 514 SEAMOUNT CHAIN, NEW ENGLAND
B 6106 877 SEAMOUNT, CONTINENTAL RISE, NILE CONE, LEVANT PLATFORM, AND,
B 6002 197 SEAMOUNTS, BRAZIL MARGIN
B 6108 1390* SEAMS, ABST., HIGH-RESOLUTION, SEISMIC EXPLORATION FOR
B 6004 684* **SEARCH FOR SILURANIAN REEF MODEL--GREAT LAKES AREA,** , ABST.
B 6405 741 SEARCY, WILLIAM P., III, MULTI-PARAMETER REGIONAL MAPPING AND
M 28 141 SEARLESITE, BIREFRINGENCE, ETC.
M 28 141 SEARLESITE, FILLING VOIDS
B 6303 524 SEARS, S. O., DOLOMITIZATION OF NORTHERN MICHIGAN NIAGARAN
B 6412 2064† SEASAT ORBITAL RADAR IMAGERY FOR GEOLOGIC MAPPING—
B 6405 619† SEASAT RADAR IMAGE OF SAN ANDREAS FAULT, CALIFORNIA
B 6412 2064 SEASAT, SYNTHETIC-APERTURE RADAR, SAR, IMAGERY
S 8 40 SEAVALLEYS, CONGO CANYON, AFRICA
S 8 42 SEAVALLEYS, CROSS-VALLEY CURRENTS
S 8 22 SEAVALLEYS, CURRENT FLOW, SEDIMENT TRANSPORT
S 8 40 SEAVALLEYS, FRASER DELTA, CANADA
S 8 4 SEAVALLEYS, INDEX MAP
S 8 42 SEAVALLEYS, INTERNAL WAVES ALONG AXES
S 8 41 SEAVALLEYS, ORIGIN AND CURRENTS
B 6303 471* SEAWARD PRIMARY DIP OF FALL-IN BEDS, LOWER SEVEN RIVERS
S 4 289† SEAWATER, USE OF, CORALS IN DETERMINING, GLACIAL-, INTERGLACIAL
B 6203 501* SEAWAY DEPOSITIONAL SYSTEMS USING PARTIAL ANALOGS, ABST.,
B 6408 1223† SEAWAY DEPOSITIONAL SYSTEMS USING PARTIAL ANALOGS,
B 6203 563* SEAWAY, WESTERN UNITED STATES, ABST., MOLLUSCAN, BIOCLASTIC,
B 6408 1232 SEAWAYS, ANCIENT, OFFSHORE, SANDSTONE BODIES IN,
B 6203 362 SEBASTIAN VIZCAINO BASIN, BAJA CALIFORNIA
B 6401 38 SEBUKU PLATFORM, INDONESIA
M 25 329 SECE FORMATION, ECUADOR
M 25 335 SECHURA BASIN, PERU
B 6008 1401* SECOND CREEK FIELD, RICHLAND COUNTY, MONTANA, ABST.,
B 6201 76 SECOND RED BED, SASKATCHEWAN
B 6201 74 SECOND RED CLAYSTONE, SASKATCHEWAN
B 6211 2239 SECOND SANDSTONE UNIT, BAY OF BISCAY AREA
B 6201 74 SECOND SPECKS CALCAREOUS SHALE, SASKATCHEWAN
B 6407 981 SECOND WHITE SPECKS, ALBERTA
M 26 84 SECOND-ORDER, CORRELATION WITH FIRST-ORDER, GLOBAL CYCLES
M 29 470 SECOND-ORDER CYCLES, PHANEROZOIC, GLOBAL CYCLES OF CHANGE
M 26 84 SECOND-ORDER, GLOBAL CYCLES, FIRST-ORDER, CORRELATION WITH
B 6303 416* SECONDARY CARBONATE POROSITY, ZELTEN FIELD LIBYA, ABST.
B 6203 463† SECONDARY FERROAN DOLOMITE RHOMBS IN OIL RESERVOIRS, CHADRA
B 6305 723† SECONDARY HYDROCARBON MIGRATION AND ENTRAPMENT,
B 6108 1391* SECONDARY MIGRATION AND ENTRAPMENT OF HYDROCARBONS, ABST.,
B 6304 581 SECONDARY MIGRATION, GEOCHEMICAL CONCEPT
B 6409 1544 SECONDARY MIGRATION, REDEFINITION OF, PRIMARY AND
B 6305 775 SECONDARY PETROLEUM MIGRATION IN SOLUTION OR BY MICELLES
B 6405 753* SECONDARY POROSITY GENERATION, AND OCCLUSION IN UPPER
B 6305 834* **SECONDARY POROSITY IN SANDSTONES TO HYDROCARBON**
B 6004 705* SECONDARY POROSITY IN DEEP SMACKOVER OF MISSISSIPPI, ABST.,
B 6405 737* SECONDARY POROSITY IN RESERVOIR SANDSTONES B DISSOLUTION OF
B 6108 1390* SECONDARY POROSITY IN SANDSTONE DIAGENESIS, ABST., ROLE OF
B 6109 1547* SECONDARY POROSITY IN SANDSTONES OF LOWER WILCOX, EOCENE,

Index

B 6405 747* SECONDARY POROSITY IN SANDSTONES TO HYDROCARBON
S 10 104 SECONDARY, HYDROCARBONS IN SOLUTION, MIGRATION
S 10 105 SECONDARY, MICELLAR MIGRATION, MIGRATION
S 10 102 SECONDARY, NATURAL GAS IN SOLUTION, MIGRATION
S 10 59 SECONDARY MIGRATION, MIGRATION IN AQUEOUS SOLUTION
S 3 136 SECRETION, TRANSPORT, AND FIXATION OF URANIUM BY NATURAL
M 26 141 SECTION DESCRIPTION, SEISMIC, REFLECTIONS
B 6002 278† SECTION OF HAMILTON GROUP, DEVONIAN, AND ADJACENT STRATA
M 29 65† SECTIONS AND INTERVAL VELOCITIES OVER OUTER CONTINENTAL
B 6310 1888 SEDCO ENERGY CO., ON DEVELOPMENTS IN DUBAI, 1978, DUBAI
B 6411 1840 SEDCO ENERGY CORP., ON DEVELOPMENTS IN DUBAI, 1979, DUBAI
S 7 299 SEDIMENT ACCUMULATION, HOLOCENE, GYRE BASIN
B 6202 212 SEDIMENT ACCUMULATION RATES
S 7 297 SEDIMENT ANOMALY MAP, GYRE BASIN
M 28 104 SEDIMENT COMPARISON, SORTING
S 7 293 SEDIMENT CORE, GYRE BASIN
B 6007 1025 SEDIMENT CORES, LAKE SUPERIOR
B 6403 307 SEDIMENT DAMS BENEATH CONTINENTAL MARGINS
S 7 70 SEDIMENT DEPOSITION, SALT, CONTINENTAL SLOPE
S 7 89 SEDIMENT DEPOSITION, TEXAS-LOUISIANA SLOPE
B 6303 508* SEDIMENT DISPERSAL AT FORT GEORGE INLET, FLORIDA, ABST.
M 29 395 SEDIMENT DISTRIBUTION, GULF OF MEXICO, MISSISSIPPI FAN
S 7 294 SEDIMENT DISTRIBUTION, GYRE BASIN
B 6202 273† SEDIMENT DISTRIBUTION IN WESTERN ATLANTIC OFF NORTHERN
B 6105 806* SEDIMENT DYNAMICS IN MACROTIDAL AVON RIVER ESTUARY, NOVA SCOTIA
B 6004 674* SEDIMENT FACIES OF SALT RIVER ESTUARY, ST. CROIX, U.S.V.I.--MIXED
S 7 313 SEDIMENT GEOCHEMISTRY, NORTHWEST GULF OF MEXICO
S 7 313† SEDIMENT GEOCHEMISTRY OF CONTINENTAL SLOPE AND INTRASLOPE
B 6203 536* SEDIMENT GRADING IN OVERWASH DEPOSITS, ABST.
M 29 454 SEDIMENT INFLUX AND REMOVAL, MECHANISMS
B 6004 695* SEDIMENT LITHOFACIES IN SOUTH ATLANTIC OCEAN, EARLY
S 7 150 SEDIMENT LOADING, NORTHERN GULF OF MEXICO, CONTINENTAL
M 28 58 SEDIMENT, MINNESOTA, HOLOCENE
S 7 91 SEDIMENT PROCESSES, BOTTOM-FLOWING CURRENT TRANSPORT
B 6004 654* SEDIMENT PROCESSES CONTROLLED BY BOTTOM CURRENTS AND
S 7 89 SEDIMENT PROCESSES, FLUVIAL TRANSPORT
S 7 91 SEDIMENT PROCESSES, HIGH-ENERGY TURBIDITY-CURRENT TRANSPORT
S 7 90 SEDIMENT PROCESSES, LOW-ENERGY TURBIDITY-CURRENT TRANSPORT
S 7 91 SEDIMENT PROCESSES, MASS TRANSPORT
S 7 91 SEDIMENT PROCESSES, PELAGIC-SUSPENSION TRANSPORT
S 7 91 SEDIMENT PROCESSES, REFLECTION PATTERNS
S 7 112 SEDIMENT PROCESSES, SEISMIC IDENTIFICATION, CHAOTIC PATTERNS
S 7 108 SEDIMENT PROCESSES, SEISMIC IDENTIFICATION, COMPLEX PATTERNS
S 7 92 SEDIMENT PROCESSES, SEISMIC IDENTIFICATION, SIMPLE PATTERNS
S 7 90 SEDIMENT PROCESSES, WAVE TRANSPORT
M 28 103 SEDIMENT SORTING, CLASSIFICATION
B 6303 550* SEDIMENT SUSPENSION, DEWATERING, AND MASS-MOVEMENT
S 7 250 SEDIMENT TEXTURE, TEXAS, SHELF-SLOPE
M 29 400 SEDIMENT THICKNESS, MISSISSIPPI FAN, HOLOCENE
S 8 22 SEDIMENT TRANSPORT, SEAVALLEYS, CURRENT FLOW
S 7 89 SEDIMENT TRANSPORT, TEXAS-LOUISIANA SLOPE
B 6004 722* SEDIMENT-DISPERSAL PATTERNS OF SOUTH TEXAS OUTER
B 6109 1537* SEDIMENT-DISPERSAL TRENDS OF CAMINADA-MOREAU CHENIER-BEACH
B 6305 837* SEDIMENTARY AND TECTONIC CONTROLS OF URANIUM
B 6105 839* SEDIMENTARY APRON IN TRANS-PECOS VOLCANIC FIELD OF TEXAS,
B 6405 775* SEDIMENTARY ASPECTS OF ORGANIC MATERIAL IN GREEN RIVER SHALE,
S 10 172 SEDIMENTARY BASIN, GENERAL, IDEALIZED VERTICAL SECTION
M 25 217† SEDIMENTARY BASINS AND PETROLEUM PROSPECTS, ONSHORE AND
M 29 28 SEDIMENTARY BASINS, ATLANTIC MARGIN
S 10 152 SEDIMENTARY BASINS, HYDRODYNAMIC TYPES, EXAMPLES
B 6304 573† SEDIMENTARY BASINS, MIGRATION OF, FLUIDS IN
M 25 441 SEDIMENTARY BASINS, SOUTH KOREA, KYONGSANG
B 6409 1538† SEDIMENTARY BASINS. DISCUSSION AND REPLY, MIGRATION OF, FLUIDS
B 6405 756* SEDIMENTARY CHARACTERISTICS AND PROCESSES OF
B 6002 321* SEDIMENTARY CYCLES AND DEPOSITIONAL ENVIRONMENTS IN
B 6004 711* SEDIMENTARY CYCLES IN VIRGILIAN, UPPER PENNSYLVANIAN, ROCKS IN
B 6004 644* SEDIMENTARY ENVIRONMENTS AND DIAGENESIS OF E ABRA LIMESTONE
B 6407 1117 SEDIMENTARY ENVIRONMENTS AND FACIES, EDITED BY H. G. READING,
B 6004 720* SEDIMENTARY ENVIRONMENTS OF TIDAL FLAT CREATED BY SHORELINE
B 6004 702* SEDIMENTARY FACIES AND STRUCTURES OF BERING EPICONTINENTAL
B 6405 783* SEDIMENTARY FACIES RELATIONS AND INFERRED DYNAMICS OF A
B 6209 1765* SEDIMENTARY FACIES, STRUCTURES, AND GRAIN-SIZE DISTRIBUTION—
M 29 455 SEDIMENTARY HISTORY, ATLANTIC-TYPE MARGINS
B 6002 184 SEDIMENTARY MODEL, NORTH SEA
M 25 414 SEDIMENTARY ORE, WESTERN AUSTRALIA, IRON
B 6403 442* SEDIMENTARY PROCESSES ACTIVE ON SLOPES OF CALIFORNIA
B 6004 686* SEDIMENTARY PROCESSES IN SUBMARINE CANYONS OF NORTHWESTERN
B 6405 731* SEDIMENTARY PROCESSES ON CONTINENTAL SHELF OFF SOUTHERN
B 6203 559* SEDIMENTARY PROCESSES ON CONTINENTAL SLOPE OFF NEW ENGLAND,
B 6203 507* SEDIMENTARY RECORD OF LONGITUDINAL TRANSPORT IN LATE
B 6004 674* SEDIMENTARY RECORD OF PALEOCLIMATE IN CARBONATE TIDAL FLATS,
B 6301 89 SEDIMENTARY RIDGE PROVINCES, SCOTIAN BASIN
M 28 26 SEDIMENTARY ROCK FRAGMENTS (SRF)
B 6101 100† SEDIMENTARY ROCKS, DISTRIBUTION OF, CARBON AS, HYDROCARBONS
M 29 297 SEDIMENTARY SEQUENCE, MEDITERRANEAN, ADRIATIC BASINS
M 29 310 SEDIMENTARY SEQUENCE, MEDITERRANEAN, AEGEAN BASINS
M 29 312 SEDIMENTARY SEQUENCE, MEDITERRANEAN, ANTALYA BASIN
M 29 312 SEDIMENTARY SEQUENCE, MEDITERRANEAN, CILICIA-ADANA BASIN
M 29 299 SEDIMENTARY SEQUENCE, MEDITERRANEAN, HERODOTUS BASIN
M 29 299 SEDIMENTARY SEQUENCE, MEDITERRANEAN, IONIAN BASIN
M 29 299 SEDIMENTARY SEQUENCE, MEDITERRANEAN, LEVANTINE BASIN
M 29 312 SEDIMENTARY SEQUENCE, MEDITERRANEAN, NORTH LEVANTINE BASIN
M 29 297 SEDIMENTARY SEQUENCE, MEDITERRANEAN, TYRRHENIAN BASIN
M 29 313 SEDIMENTARY SEQUENCE, USSR, BLACK SEA BASIN
M 29 293 SEDIMENTARY SEQUENCE, WESTERN, MEDITERRANEAN BASIN
B 6012 2177* SEDIMENTARY SERPENTINITE OF THE MIOCENE BIG BLUE FORMATION
B 6007 1015 SEDIMENTARY STRUCTURES AND TEXTURES, SEQUENCE OF
B 6409 1567* SEDIMENTARY STRUCTURES ASSOCIATED WITH MISSISSIPPI RIVER
B 6303 468* SEDIMENTARY STRUCTURES OF HIGH-ENERGY BEACH-TO-OFFSHORE
S 7 220 SEDIMENTARY STRUCTURES, WEST FLORIDA SLOPE
S 7 221 SEDIMENTARY STRUCTURES, WEST FLORIDA SLOPE
B 6303 530* SEDIMENTARY SUBENVIRONMENTS OF WILKINS PEA MEMBER OF GREEN
M 25 509 SEDIMENTARY TROUGH, ALASKA, VOLCANIC-
B 6311 2088† SEDIMENTARY-WEDGE PROGRADATION FROM TRANSFORM-FAULTED
M 25 503 SEDIMENTATION AND MINERALIZATION, CYCLES OF, CANADIAN,
B 6008 1412* SEDIMENTATION AND MULTIPLE COAL-SEA CORRELATIONS IN UPPER
B 6004 709* SEDIMENTATION AND RESEDIMENTATION OF CLASTI CARBONATE
M 25 308† SEDIMENTATION AND TECTONICS OF PACIFIC CONTINENTAL MARGIN OF
B 6211 2243† SEDIMENTATION AND TRAPPING MECHANISM IN UPPER MIOCENE
B 6405 779* SEDIMENTATION, BIOSTRATIGRAPHY, AND SOURCE-ROCK POTENTIAL OF
S 7 230 SEDIMENTATION, ENVIRONMENTAL CYCLE CONCEPT
S 7 230 SEDIMENTATION, GRAVITY SEDIMENTS
S 7 117† SEDIMENTATION, GULF OF MEXICO, CONTINENTAL-SLOPE, SEDIMENTS
S 7 188 SEDIMENTATION HISTORY, MISSISSIPPI FAN
B 6004 705* SEDIMENTATION HISTORY OF LATE CARBONIFEROUS BASIN,
B 6105 779* SEDIMENTATION IN LAKE REEVE, VICTORIA, AUSTRALIA—MODERN BARRIER
B 6303 449* SEDIMENTATION IN MISSISSIPPI TROUGH, GULF OF MEXICO, ABST.
M 29 398 SEDIMENTATION, MISSISSIPPI FAN, HOLOCENE
M 29 396 SEDIMENTATION, MISSISSIPPI FAN, PLEISTOCENE
S 7 230 SEDIMENTATION, NORMAL SEDIMENTS
B 6203 490* SEDIMENTATION ON ANTARCTIC CONTINENTAL SLOPE, ABST.
B 6203 510* SEDIMENTATION ON EASTERN UNITED STATES CONTINENTAL SLOPE,
B 6309 1602* SEDIMENTATION ON TRAILING PLATE MARGIN--NORTHERN GULF
B 6303 467* SEDIMENTATION PROCESSES AND HYDROCARBON POTENTIAL OF
B 6405 725* SEDIMENTATION RATES AND ILLITE-SMECTITE DIAGENESIS, ABST.
B 6203 515* SEDIMENTATION RATES FROM CONTINUUM OF
B 6303 440* SEDIMENTATION RATES IN COASTAL-PLAIN ESTUARY, RHODE RIVER,
B 6004 703* SEDIMENTATION RATES IN HUDSON RIVER ESTUARY, ABST.
S 7 178 SEDIMENTATION RATES, TABLE, MISSISSIPPI FAN
S 7 228 SEDIMENTATION, TURBIDITES
S 7 155† SEDIMENTATIONAL PATTERNS, MISSISSIPPI FAN, GULF OF MEXICO—,
B 6405 730* SEDIMENTOLOGIC DESCRIPTION OF PART OF COAL-BEARING
B 6303 483* SEDIMENTOLOGIC FACIES I MODERN GLACIAL-MARINE ENVIRONMENT,
B 6403 442* SEDIMENTOLOGIC HISTORY AND CHARACTERISTICS OF CONTINENTAL
B 6004 707* SEDIMENTOLOGIC INDICATION OF EOCENE CLIMATIC CHANGE,
B 6404 582† SEDIMENTOLOGIC PROBLEMS OF URANIUM DEPOSITS OF TEXAS GULF
B 6303 553* SEDIMENTOLOGIC STUDIES IN JURASSIC OF TUNISIA, ABST.
S 5 23† SEDIMENTOLOGY AND DEPOSITIONAL ENVIRONMENTS OF BASIN-CENTER
B 6004 708* SEDIMENTOLOGY AND EASTERN CONTINENTAL MARGIN, 1926-1976, ABST.
B 6004 708* SEDIMENTOLOGY AND EASTERN CONTINENTAL MARGIN, ABST.
B 6004 716* SEDIMENTOLOGY AND PALEOECOLOGY OF TYPE LOCATION OF KINCAID
B 6405 750* SEDIMENTOLOGY AND PALEOGEOGRAPHY OF CAMBRIAN
B 6308 1426* SEDIMENTOLOGY AND SOURCE-BED GEOCHEMISTRY OF SPRABERRY
B 6303 541* SEDIMENTOLOGY AND SYNERGY OF DELTAIC SANDSTONE, ADMIRE
B 6306 987 SEDIMENTOLOGY, BY GERALD M. FRIEDMAN AND JOHN E. SANDERS,
B 6307 1144 SEDIMENTOLOGY, EDITED BY ANDREW D. MIALL, LE BLANC, RUFUS J.,
B 6305 820 SEDIMENTOLOGY, EDITED BY RHODES W. FAIRBRIDGE AND JEANNE
B 6203 504* SEDIMENTOLOGY OF CAPLES TERRANE, SOUTH ISLAND, NEW ZEALAND.
B 6102 295* SEDIMENTOLOGY OF CRETACEOUS MESA RICA SANDSTONE, TUCUMCARI
B 6405 758* SEDIMENTOLOGY OF FLUVIAL UPPER DEVONIAN KANAYUT
B 6303 437* SEDIMENTOLOGY OF GOAT SEEP DOLOMITE GUADALUPIAN, PERMIAN,
B 6211 2275† SEDIMENTOLOGY OF LATE JURASSIC BARRIER-ISLAND COMPLEX—
B 6203 535* SEDIMENTOLOGY OF LATE QUATERNARY MAMMOTH-BEARING
B 6405 728* SEDIMENTOLOGY OF LOWER CRETACEOUS SUBTIDAL SAND COMPLEX,
B 6004 644* SEDIMENTOLOGY OF MELANGE, FRANCISCAN OF TRINIDAD,
B 6004 701* SEDIMENTOLOGY OF OUTER-FAN DEPOSITIONAL LOBES OO UPPER
B 6303 529* SEDIMENTOLOGY OF SOME PRECAMBRIAN IRON FORMATIONS, ABST.
B 6004 718* SEDIMENTOLOGY OF SOME SILURIAN ROCKS FROM GOTLAND, SWEDEN,

M 26 249† SEDIMENTOLOGY OF TERRIGENOUS PLEISTOCENE DEPOSITS IN
B 6305 843* SEDIMENTOLOGY OF VOLCANICLASTIC DEPOSITS FROM 1971-1974
B 6405 795* SEDIMENTOLOGY OF WESTWATER CANYON MEMBER, MORRISON
B 6303 499* SEDIMENTOLOGY, PETROLOGY, AND GEOTECTONIC PROPERTIES OF
B 6004 672* SEDIMENTOLOGY--RETROSPECT AN PROSPECT, ABST., 50 YEARS OF
S 7 117† SEDIMENTS AND SEDIMENTATION, GULF OF MEXICO,
S 10 180 SEDIMENTS BENEATH THE OCEANS, METHANE, OCCURRENCES
M 28 141 SEDIMENTS, CALIFORNIA, PLEISTOCENE
M 28 140 SEDIMENTS, CALIFORNIA, PLEISTOCENE
M 28 57 SEDIMENTS, MINNESOTA, HOLOCENE
S 6 363† SEDIMENTS, NUMERICAL, CORRELATION OF MIDDLE AND UPPER,
M 28 139 SEDIMENTS, OREGON, TERTIARY
M 27 19 SEDIMENTS, PACIFIC, PLIOCENE
S 6 94 SEDIMENTS, STRONTIUM COMPOSITION
B 6404 571 SEDOM COVE AND QUARRY, ISRAEL
B 6404 569 SEDOM FORMATION, ISRAEL
M 27 106 SEEDS, MOSS SPORES
M 27 106 SEEDS, POLLEN GRAINS
M 27 108 SEEDS, TRICOLPATE ANGIOSPERM POLLEN GRAIN
M 27 107 SEEDS, TRICOLPATE ANGIOSPERM POLLEN GRAINS
M 27 107 SEEDS, TRILETE FERN SPORES
M 27 106 SEEDS, TRILETE SPORES
B 6105 781 SEELY, DONALD R., STRATIGRAPHY AND STRUCTURE OF CONPRESSIONAL
B 6301 2 SEELY, D. R., STRUCTURE AND STRATIGRAPHY OF FOREARC REGIONS
M 29 245 SEELY, D. R., THE EVOLUTION OF STRUCTURAL HIGHS BORDERING
B 6004 721 SEELY, D. R., VARIATIONS IN TRENCH-SLOPE MODEL, ABST.
B 6405 795* SEELYVILLE COAL—MAJOR UNEXPLOITED SEAM IN ILLINOIS, ABST.
B 6206 1067† SEEP AREA, SOUTH TEXAS OFFSHORE, SERENDIPITY, GAS
B 6203 548* SEEPAGE IN SANTA MONICA BAY—CALIFORNIA CONTINENTAL
B 6304 668† SEEPS AND SUBSURFACE STRUCTURE OFF PANAMA CITY, FLORIDA, GAS
B 6407 1083 SEEPS, ORIGIN OF, HYDROCARBON, GASES IN, SUBMARINE
B 6405 699* SEG DELEGATION VISIT TO PEOPLES REPUBLIC OF CHINA, ABST.
B 6004 710* SEGMENTATION OF ISLAND ARCS AND ITS APPLICATION TO PETROLEUM
B 6303 527 SEIBOLD, E., RECENT ADVANCES IN PASSIVE-MARGIN RESEARCH,
B 6303 538 SEIFERT, K. E., MIDDLE CRETACEOUS OXYGEN-DEFICIENT
B 6303 525 SEIGLIE, G. A., SIGNIFICANCE OF MIDDLE TERTIARY LARGE
B 6303 370 SEIGLIE, GEORGE A.., SIDERITE SPHERULITES OF SAN SEBASTIAN
B 6012 2175* SEISCROP MAP TECHNIQUES, ABST., SEISMIC DATA ANALYSIS WITH, TIME
B 6008 1391* SEISCROP MAP TECHNIQUES, ABST., SEISMIC DATA ANALYSIS WITH, TIME-
B 6203 564* SEISLOG APPLICATIONS, ABST., EXPANDING FIELD OF
B 6012 2184* SEISLOG METHOD, ABST., STRATIGRAPHIC EXPLORATION WITH
M 29 368 SEISMIC (FOLDOUT), VENEZUELA BASIN, MUERTOS TRENCH
M 25 120 SEISMIC ACTIVITY FROM FLUID REINJECTION, INDUCTION OF
M 26 53 SEISMIC ANALYSIS, DEPOSITIONAL SEQUENCE
M 26 55 SEISMIC ANALYSIS, DEPOSITIONAL SEQUENCE, CHRONOSTRATIGRAPHIC
B 6105 843* SEISMIC ANALYSIS OF GLIDE-PLANE FAULT STRUCTURES, STYLES,
B 6207 1226* SEISMIC AND GEOLOGIC FACIES ANALYSES IN FIELD DEVELOPMENT,
M 26 303 SEISMIC APPLICATION, FREQUENCY
M 26 302 SEISMIC APPLICATION, INSTANTANEOUS PHASE
M 26 303 SEISMIC APPLICATION, POLARITY
M 26 302 SEISMIC APPLICATION, REFLECTION STRENGTH
M 26 303 SEISMIC APPLICATION, USE OF COLOR DISPLAY
B 6203 537* SEISMIC CALIBRATION TECHNIQUES FOR IMPROVED SUBSURFACE
B 6211 2351* SEISMIC CASE HISTORIES, BY TITLE ONLY, REVIEW OF 30
S 7 263 SEISMIC CHARACTER, TEXAS, UPPER CONTINENTAL SLOPE
B 6205 753 SEISMIC CHARACTERISTICS AND ENVIRONMENTAL FACIES
M 29 324 SEISMIC CHARACTERISTICS, GULF OF MEXICO, SOUTHWESTERN
B 6303 514* SEISMIC DATA, ABST., DELTAIC FACIES RECOGNITIO ON
B 6303 527* SEISMIC DATA, ABST., DEPTH MIGRATION OF
B 6012 2180* SEISMIC DATA, ABST., GEOLOGY OF OFFSHORE LOWER, COOK INLET,
B 6304 701* SEISMIC DATA, ABST., LOCATION STRATIGRAPHIC FEATURES AND,
B 6203 541* SEISMIC DATA, ABST., PREDICTION OF, POROSITY IN, PERMIAN,
B 6405 694* SEISMIC DATA ACROSS OUTER BERING SEA CONTINENTAL MARGIN,
B 6203 380 SEISMIC DATA, ALASKA, COOK INLET
B 6212 2415 SEISMIC DATA, ALASKA, ICY BAY-, MALASPINA GLACIER SLUMP
B 6205 783 SEISMIC DATA, ALBERTA, GOLDEN SPIKE, REEF
B 6205 784 SEISMIC DATA, ALBERTA, INNISFAIL FIELD
B 6205 789 SEISMIC DATA, ALBERTA, JUDY CREEK, REEF AND BANK COMPLEX
B 6205 782 SEISMIC DATA, ALBERTA, RAINBOW-, ZAMA AREA
B 6205 785 SEISMIC DATA, ALBERTA, YEKAU LAKE REEF
M 26 352† SEISMIC DATA, ANALYSIS OF, HIGH, RESOLUTION
B 6012 2175* SEISMIC DATA ANALYSIS WITH TIME SEISCROP MAP TECHNIQUES, ABST.
B 6008 1391* SEISMIC DATA ANALYSIS WITH TIME-SEISCROP MAP TECHNIQUES, ABST.
B 6410 1661 SEISMIC DATA, ATLANTIC MAGNETIC QUIET ZONE
B 6201 48 SEISMIC DATA, AUSTRALIA, EXMOUTH PLATEAU
B 6202 249 SEISMIC DATA, CALIFORNIA, SAN PEDRO SHELF
B 6202 249 SEISMIC DATA, CALIFORNIA, SANTA MONICA SHELF
M 26 8 SEISMIC DATA, CONVOLUTION
B 6205 779 SEISMIC DATA, GULF OF, PAPUA
B 6203 566* SEISMIC DATA IN WOLLASTON FOLDBELT REGION OF ATHABASCA BASIN,
B 6004 528† SEISMIC DATA, INFERRING, STRATIGRAPHY FROM
M 26 249 SEISMIC DATA, INTERPRETATION
M 26 3 SEISMIC DATA, MODELING
B 6205 781 SEISMIC DATA, NEW MEXICO, CENTRAL BASIN PLATFORM
B 6205 781 SEISMIC DATA, NEW MEXICO, LEA COUNTY
B 6205 781 SEISMIC DATA, NEW MEXICO, PERMIAN BASIN
B 6205 777 SEISMIC DATA, NORTH, AFRICA
B 6205 788 SEISMIC DATA, NORTH, AFRICA
M 26 10 SEISMIC DATA, SEISMIC LOG PROCESS
M 26 9 SEISMIC DATA, SEISMIC TRACES
M 26 11 SEISMIC DATA, SPATIAL RESOLUTION
M 26 9 SEISMIC DATA, SYNTHETIC SEISMOGRAMS,
M 26 3 SEISMIC DATA, SYNTHETIC SEISMOGRAMS
M 26 9 SEISMIC DATA, SYNTHETIC SEISMOGRAMS
B 6205 787 SEISMIC DATA, TEXAS, CROSBY COUNTY
B 6205 787 SEISMIC DATA, TEXAS, EASTERN SHELF AREA
B 6205 787 SEISMIC DATA, TEXAS, PERMIAN BASIN
B 6205 786 SEISMIC DATA, TEXAS, WASHITA REEF TREND
M 26 5 SEISMIC DATA, WAVE EQUATION MIGRATION
B 6205 778 SEISMIC DATA, WEST, AFRICA, OFFSHORE
S 5 111* SEISMIC DATA-PROCESSING TECHNIQUES IN EXPLORATION FOR REEFS,
S 5 119 SEISMIC DATA, AUTOMATIC STATICS CORRECTION
S 5 112 SEISMIC DATA, COHERENT NOISE
S 5 118 SEISMIC DATA, DECONVOLUTION
S 5 118 SEISMIC DATA, ELEVATION CORRECTIONS
S 5 118 SEISMIC DATA, FILTERING
S 5 121 SEISMIC DATA, HAND-STATICS CORRECTION
S 5 112 SEISMIC DATA, MULTIPLES
S 5 112 SEISMIC DATA, REFLECTIONS
S 5 117 SEISMIC DATA, STACKING
B 6405 703* SEISMIC DATA—BASIN AND RANGE PROVINCE NORTHERN NEVADA,
S 5 113 SEISMIC EXAMPLE, MICHIGAN, MANISTEE COUNTY, BEAR LAKE
M 26 224 SEISMIC, DEPOSITIONAL SYSTEMS, DELTAS
M 26 231 SEISMIC, DEPOSITIONAL SYSTEMS, SHELFS
M 26 232 SEISMIC, DEPOSITIONAL SYSTEMS, SLOPES
M 29 65† SEISMIC DEPTH SECTIONS AND INTERVAL VELOCITIES OVER OUTER
M 26 301 SEISMIC DETECTION, WAVES AS SIGNALS
B 6008 1411* SEISMIC DISPLAYS, ABST., MULTI-DIMENSIONAL
B 6004 670* SEISMIC DISPLAYS AS INTERPRETATION AID, ABST., THREE-DIMENSIONAL
B 6008 1410* SEISMIC DOLLAR, ABST., STRETCHING THE
M 26 277† SEISMIC EVIDENCE FOR LATE CRETACEOUS GROWTH FAULTING, DENVER
B 6008 1409* SEISMIC EVIDENCE FOR STRUCTURAL STYLE OF LARAMIDE TRAPS IN
B 6312 2204† SEISMIC EVIDENCE FOR WIDESPREAD POSSIBLE GAS HYDRATE HORIZONS
M 26 190 SEISMIC EXAMPLE, AFRICA, NORTH, PINNACLE, REEF
M 26 191 SEISMIC EXAMPLE, AFRICA, WEST, SHELF-MARGIN, CARBONATE
M 26 196 SEISMIC EXAMPLE, CANADA, ALBERTA, GOLDEN SPIKE, CARBONATE
M 26 197 SEISMIC EXAMPLE, CANADA, ALBERTA, INNISFAIL, REEF BUILDUP
M 26 202 SEISMIC EXAMPLE, CANADA, ALBERTA, JUDY CREEK, SHALLOW-WATER,
M 26 195 SEISMIC EXAMPLE, CANADA, ALBERTA, RAINBOW-, ZAMA AREA,
M 26 198 SEISMIC EXAMPLE, CANADA, ALBERTA, YEKAN LAKE, CARBONATE
M 26 192 SEISMIC EXAMPLE, GULF OF PAPUA, SHELF-MARGIN, REEF
M 26 194 SEISMIC EXAMPLE, NEW MEXICO, LEA COUNTY, CENTRAL, BASIN
M 26 200 SEISMIC EXAMPLE, TEXAS, CROSBY COUNTY, PERMIAN BASIN,
S 7 114 SEISMIC EXAMPLE, TEXAS, MASS-TRANSPORT SCAR
M 26 199 SEISMIC EXAMPLE, TEXAS, WASHITA, REEF TREND, SHELF-MARGIN
M 29 448 SEISMIC EXPERIMENT, CONTINENTAL MARGINS, COLOMBIA
M 29 446 SEISMIC EXPERIMENT, CONTINENTAL MARGINS, MEXICO
M 29 443 SEISMIC EXPERIMENT, CONTINENTAL MARGINS, PERU
M 29 448 SEISMIC EXPERIMENT, CONTINENTAL MARGINS, U.S. EAST COAST
B 6008 1390* SEISMIC EXPERIMENTS, CLINKERS AREA, WYOMING, ABST.
B 6405 711* SEISMIC EXPLORATION, ABST., NEW DIMENSIONS IN
B 6207 1224* SEISMIC EXPLORATION FOR COAL IN AUSTRALIA, ABST.,
B 6108 1390* SEISMIC EXPLORATION FOR GEOLOGIC FAULTS IN COAL SEAMS, ABST.,
M 26 329† SEISMIC EXPLORATION FOR STRATIGRAPHIC TRAPS
B 6302 259 SEISMIC EXPLORATION FUNDAMENTALS, BY J A. COFFEEN, HUH, JOHN
B 6102 294* SEISMIC EXPLORATION IN CENTRAL TEXAS. BY TITLE ONLY, EVOLUTION
B 6211 2363* SEISMIC EXPLORATION LEADING TO DISCOVERY OF TRAP SPRING FIELD,
B 6205 880* SEISMIC EXPLORATION LEADING TO DISCOVERY OF TRAP SPRING OIL
B 6307 1054 SEISMIC EXPRESSION OF SHALE DIAPIRS WITHIN DELTAIC SEQUENCE
M 26 165 SEISMIC FACIES, ANALYSIS
B 6205 752 SEISMIC FACIES ANALYSIS, DEFINITION
S 7 87 SEISMIC FACIES ANALYSIS, STRATIGRAPHY
B 6303 500* SEISMIC FACIES AND DEPOSITIONAL PROCESSES OF MOOERN
M 26 249† SEISMIC FACIES AND SEDIMENTOLOGY OF TERRIGENOUS PLEISTOCENE
B 6205 762 SEISMIC FACIES, BASIN-SLOPE AND BASIN FLOOR
S 7 100 SEISMIC FACIES DIAGRAM, LOUISIANA, MIDSLOPE
S 7 99 SEISMIC FACIES EXAMPLE, ALABAMA, MOBILE BAY
S 7 103 SEISMIC FACIES EXAMPLE, FLORIDA, SHELF-UPPER SLOPE
S 7 96 SEISMIC FACIES EXAMPLE, LOUISIANA, MARSH ISLAND
S 7 101 SEISMIC FACIES EXAMPLE, LOUISIANA, MIDSLOPE

S 7 107 SEISMIC FACIES EXAMPLE, LOUISIANA, MIDSLOPE
S 7 102 SEISMIC FACIES EXAMPLE, LOUISIANA, OUTER SHELF EDGE
S 7 96 SEISMIC FACIES EXAMPLE, MISSISSIPPI DELTA
S 7 99 SEISMIC FACIES EXAMPLE, MISSISSIPPI RIVER DELTA
S 7 97 SEISMIC FACIES, EXAMPLE, TEXAS, MIDSLOPE
S 7 107 SEISMIC FACIES EXAMPLE, TEXAS, MIDSLOPE
S 7 104 SEISMIC FACIES EXAMPLE, TEXAS, SHELF-SLOPE MARGIN
S 7 110 SEISMIC FACIES EXAMPLE, TEXAS, UPPER SLOPE
M 26 131 SEISMIC FACIES, EXTERNAL FORMS, FILL TYPES
M 26 131 SEISMIC FACIES, EXTERNAL FORMS, MOUND TYPES
B 6205 770 SEISMIC FACIES INTERPRETATION, FUTURE OF
M 26 142 SEISMIC FACIES, MAPPING
B 6205 764 SEISMIC FACIES, MOUNDED
S 7 87† SEISMIC FACIES, OFFSHORE TEXAS-LOUISIANA, RECOGNITION OF,
B 6205 764 SEISMIC FACIES, ONLAPPING-FILL
M 26 121 SEISMIC FACIES, PARAMETERS
M 26 118 SEISMIC FACIES, PARAMETERS, INTERPRETATION
B 6205 756 SEISMIC FACIES, SHELF
B 6205 759 SEISMIC FACIES, SHELF-MARGIN AND PROGRADED-, SLOPE
S 7 113 SEISMIC FACIES, THICKNESS, EXAMPLE, GULF OF MEXICO, CHAOTIC
S 7 94 SEISMIC FACIES TYPE, DIVERGENT, DIAGRAM
S 7 93 SEISMIC FACIES TYPE, EXAMPLES
S 7 106 SEISMIC FACIES TYPE, MOUNDED-CHAOTIC, DIAGRAM
S 7 104 SEISMIC FACIES TYPE, OBLIQUE-PROGRADATIONAL, DIAGRAM
S 7 100 SEISMIC FACIES TYPE, ONLAPPING FILL, DIAGRAM
S 7 94 SEISMIC FACIES TYPE, PARALLEL, DIAGRAM
S 7 97 SEISMIC FACIES TYPE, SHEET DRAPE, DIAGRAM
S 7 103 SEISMIC FACIES TYPE, SIGMOID-PROGRADATIONAL, DIAGRAM
B 6205 753 SEISMIC FACIES UNITS, SEISMIC CHARACTERISTICS AND,
M 26 423 SEISMIC FACTORS, NONRESERVOIR BEDS, EFFECTS OF
M 26 422 SEISMIC FACTORS, PORE FLUID, EFFECTS OF
M 26 421 SEISMIC FACTORS, POROSITY, EFFECTS OF
M 26 423 SEISMIC FACTORS, STRESS, EFFECTS OF
B 6405 698* SEISMIC FIELD DESIGN, ABST., HOW GEOLOGIC OBJECTIVES SHOULD
B 6303 305 SEISMIC HORIZONS, ALASKA, TUGIDAK BASIN
S 7 112 SEISMIC IDENTIFICATION, CHAOTIC PATTERNS, SEDIMENT PROCESSES
S 7 108 SEISMIC IDENTIFICATION, COMPLEX PATTERNS, SEDIMENT PROCESSES
S 7 92 SEISMIC IDENTIFICATION, SIMPLE PATTERNS, SEDIMENT PROCESSES
M 30 565† SEISMIC IN DEVELOPMENT OF NEMBE CREEK FIELD, NIGERIA, ROLE OF,
B 6004 644* SEISMIC INTERPRETATION, ABST., USE OF, COLOR, DISPLAYS IN
M 26 18 SEISMIC INTERPRETATION, AGE-VELOCITY-DEPTH ASPECTS
M 26 15† SEISMIC INTERPRETATION, ASPECTS OF, ROCK, PHYSICS FROM,
M 26 175 SEISMIC INTERPRETATION, BASIN-FLOOR FACIES
M 26 175 SEISMIC INTERPRETATION, BASIN-SLOPE FACIES
M 26 183 SEISMIC INTERPRETATION, CHAOTIC-FILL FACIES
M 26 304 SEISMIC INTERPRETATION, COLOR DISPLAYS, EXAMPLES
M 26 166 SEISMIC INTERPRETATION, DEPOSITIONAL FRAMEWORK, REFLECTION
M 26 165 SEISMIC INTERPRETATION, DEPOSITIONAL SETTINGS
M 26 169 SEISMIC INTERPRETATION, HIGH-AMPLITUDE FACIES
M 26 169 SEISMIC INTERPRETATION, HIGH-CONTINUITY FACIES
M 26 170 SEISMIC INTERPRETATION, LOW-AMPLITUDE FACIES
M 26 170 SEISMIC INTERPRETATION, LOW-CONTINUITY FACIES
B 6303 424* SEISMIC INTERPRETATION METHODS, ABST., THREE-DIMENSIONAL
S 7 170 SEISMIC INTERPRETATION, MISSISSIPPI FAN, STRATIGRAPHY
M 26 177 SEISMIC INTERPRETATION, MOUNDED FACIES
M 26 180 SEISMIC INTERPRETATION, MOUNDED ONLAPPING-FILL FACIES
M 26 173 SEISMIC INTERPRETATION, OBLIQUE-PROGRADATIONAL FACIES
M 26 165† SEISMIC INTERPRETATION OF CLASTIC DEPOSITIONAL FACIES, SEISMIC,
M 26 177 SEISMIC INTERPRETATION, ONLAPPING-FILL FACIES
M 26 18 SEISMIC INTERPRETATION, PERMEABILITY ASPECTS
M 26 172 SEISMIC INTERPRETATION, PROGRADED-SLOPE FACIES
M 26 176 SEISMIC INTERPRETATION, SHEET-DRAPE FACIES
M 26 169 SEISMIC INTERPRETATION, SHELF FACIES
M 26 172 SEISMIC INTERPRETATION, SHELF-MARGIN FACIES
M 26 175 SEISMIC INTERPRETATION, SIGMOID-PROGRADATIONAL FACIES
M 26 177 SEISMIC INTERPRETATION, SLOPE-FRONT FILL FACIES
B 6004 699* SEISMIC INTERPRETATION USING SEQUENCES AND EUSTATIC CYCLES,
M 26 170 SEISMIC INTERPRETATION, VARIABLE-AMPLITUDE FACIES
M 26 18 SEISMIC INTERPRETATION, VELOCITY-DENSITY-LITHOLOGY ASPECTS
B 6205 798 SEISMIC INTERVALS, BLAKE PLATEAU
B 6205 804 SEISMIC INTERVALS, BLAKE PLATEAU AREA, DEEP OCEAN
M 29 17 SEISMIC INTERVALS, GEORGIA EMBAYMENT, MAJOR
B 6104 546† SEISMIC INVESTIGATIONS, SOUTHERN BRAZIL MARGIN, MARINE
M 26 333 SEISMIC LINE, BALTIMORE CANYON
M 26 283 SEISMIC LINE, DENVER BASIN, NIOBRARA-PIERRE
M 26 261 SEISMIC LINE, GULF OF MEXICO, MISSISSIPPI TROUGH
B 6303 544* SEISMIC LINES, ABST., GEOLOGIC HISTORY OF DEEP, GULF ON MEXICO
B 6012 2187* SEISMIC LINES USING SLALOM-LINE METHOD, ABST., PROCESSING OF
M 26 421 SEISMIC, LITHOLOGY, EFFECTS OF
M 26 10 SEISMIC LOG PROCESS, SEISMIC, DATA
B 6303 438* SEISMIC METHODS IN COAL EXPLORATION, ABST., HIGH-RESOLUTION
B 6203 533* SEISMIC METHODS—CASE HISTORY, ABST., FIELD DEVELOPMENT WITH,
B 6405 778* SEISMIC MODEL OF 15 STRATIGRAPHICALLY CONTROLLED OIL AND GAS
B 6305 840* SEISMIC MODEL OF FIFTEEN STRATIGRAPHICALLY CONTROLLED OIL
M 29 447 SEISMIC MODEL, PERU TRENCH, SUBDUCTION ZONE
B 6203 494* SEISMIC MODELING, ABST., DETERMINATION OF, RESERVOIR, FLUID
M 26 455 SEISMIC MODELING, ACOUSTIC STRATIGRAPHY
M 26 482 SEISMIC MODELING, AMPLITUDE ANOMALY,
M 26 484 SEISMIC MODELING, APPLICATIONS
M 26 480 SEISMIC MODELING, BARRIER-BAR SANDSTONE
M 26 479 SEISMIC MODELING, CHANNEL SANDSTONES
B 6203 525* SEISMIC MODELING FOR REEF EXPLORATION IN MICHIGAN BASIN, ABST.,
M 26 453 SEISMIC MODELING, INPUT PARAMETERS
M 26 500 SEISMIC MODELING, INTERPRETATION PROCEDURE
M 26 453 SEISMIC MODELING, LITHOLOGIC CHARACTER
B 6405 764* SEISMIC MODELING OF PENNSYLVANIAN CARBONATE MOUNDS IN
M 26 440 SEISMIC MODELING, ONE-DIMENSIONAL
M 26 455 SEISMIC MODELING PRESSURE AND PORE FLUIDS
M 26 465 SEISMIC MODELING, PROCEDURE
M 26 455 SEISMIC MODELING, REFLECTION COEFFICIENTS
M 26 453 SEISMIC MODELING, RESERVOIR GEOMETRY
M 26 480 SEISMIC MODELING, SHORELINE SANDSTONES
M 26 493 SEISMIC MODELING, STRATIGRAPHIC TRAP, QUANTIFICATION
M 26 444 SEISMIC MODELING, TWO-DIMENSIONAL
M 26 491 SEISMIC MODELING, UPDIP PERMEABILITY BARRIER
M 26 478 SEISMIC MODELING, VERTICAL LITHOLOGY, SEISMIC RESPONSE
M 26 482 SEISMIC MODELING, WATER SANDSTONE, GAS CAP
B 6012 2187* SEISMIC MODELING WITH EMPHASIS ON STRATIGRAPHY AND SEISMIC
M 26 435 SEISMIC MODELS, CARBONATE SHELF MARGIN TRAPS
M 26 431 SEISMIC MODELS, DELTAIC TRAPS
M 26 435 SEISMIC MODELS, PLATFORM TRAPS
M 26 101 SEISMIC MODELS, SEISMIC, REFLECTIONS
M 26 428 SEISMIC MODELS, SUBSURFACE TRAPS
M 26 432 SEISMIC MODELS, THIN RESERVOIR SYSTEMS
M 26 430 SEISMIC MODELS, UNCONFORMITY TRAP
M 29 393 SEISMIC PROFILE, (FOLDOUT), MISSISSIPPI FAN
M 29 30 SEISMIC PROFILE, BLAKE PLATEAU, MULTICHANNEL
B 6111 1993 SEISMIC PROFILE, CANARY BASIN, VEMA-30
B 6111 2006 SEISMIC PROFILE, CAPE VERDE RISE
B 6111 1993 SEISMIC PROFILE, CAPE VERDE RISE, VEMA-30
B 6111 1992 SEISMIC PROFILE, DEEP SEA DRILLING, SITE 141
S 7 215 SEISMIC PROFILE, DESOTO CANYON
S 7 199 SEISMIC PROFILE, DESOTO CANYON
M 29 121 SEISMIC PROFILE, GOLFO SAN JORGE
B 6102 195 SEISMIC PROFILE, KOMANDORSKIY BASIN, GLOMAR CHALLENGER
B 6111 1994 SEISMIC PROFILE LIBRARY, DEEP-SEA SEDIMENT, DIAPIRS FROM,
M 29 391 SEISMIC PROFILE, MISSISSIPPI FAN, DESOTO DEEP SEA CHANNEL
S 7 257 SEISMIC PROFILE, TEXAS, SHELF-SLOPE
S 7 256 SEISMIC PROFILE, TEXAS, SHELF-SLOPE
S 7 255 SEISMIC PROFILE, TEXAS, SHELF-SLOPE
B 6108 1371* SEISMIC PROFILE TO STRATIGRAPHIC EXPLORATION PROBLEMS, POWDER
S 7 211 SEISMIC PROFILE, WEST FLORIDA SLOPE
S 7 208 SEISMIC PROFILE, WEST FLORIDA SLOPE
S 7 201 SEISMIC PROFILE, WEST FLORIDA SLOPE
S 7 218 SEISMIC PROFILE, WEST FLORIDA SLOPE
B 6111 1991 SEISMIC PROFILES, ATLANTIC OCEAN, PIERCEMENT STRUCTURES
B 6212 2481 SEISMIC PROFILES, BERING SEA BASIN
B 6403 302 SEISMIC PROFILES, CONTINENTAL, RISES AND, DEEP-OCEAN, TRENCHES
M 29 18 SEISMIC PROFILES, GEORGIA EMBAYMENT, MULTICHANNEL
B 6204 634 SEISMIC PROFILES, GULF OF, ALASKA, NORTHERN
B 6111 1991 SEISMIC PROFILES, INDIAN OCEAN, PIERCEMENT STRUCTURES
B 6303 326 SEISMIC PROFILES, ISRAEL AND, SINAI
B 6103 382 SEISMIC PROFILES, LITTLE BAHAMA BANK
B 6405 674* SEISMIC PROFILES MAP AQUIFERS IN MADISON GROUP AND RED RIVER
B 6102 210 SEISMIC PROFILES, MEXICO, VERACRUZ BASIN
B 6107 961 SEISMIC PROFILES OFF, AFRICA
B 6202 282 SEISMIC PROFILES OFF, BRAZIL, NORTHERN
B 6111 1991 SEISMIC PROFILES, PACIFIC OCEAN, PIERCEMENT STRUCTURES
S 7 181 SEISMIC PROFILES, PLEISTOCENE, MISSISSIPPI FAN
B 6106 884 SEISMIC PROFILES, SOUTHEASTERN, MEDITERRANEAN SEA
B 6406 871 SEISMIC PROFILING, BANDA SEA REGION
S 7 306 SEISMIC PROFILING, ORCA BASIN
M 26 185† SEISMIC RECOGNITION OF CARBONATE BUILDUPS, SEISMIC,
B 6205 772† SEISMIC RECOGNITION OF CARBONATE BUILDUPS, SEISMIC
B 6105 837* SEISMIC RECOGNITION OF DEPOSITIONAL FACIES ON SLOPES AND RISES,
B 6311 2097† SEISMIC RECOGNITION OF SALT DIAPIRS
M 29 445 SEISMIC RECORDS IN SEDIMENTS, PERU TRENCH, NAZCA PLATE
M 29 261† SEISMIC RECORDS, STRUCTURE OF THE OUTER, CONVERGENT, MARGIN
B 6303 426* SEISMIC REFLECTIO PROFILES, ABST., MIOCENE SUBMARINE, FAN IN DEEP
B 6206 970 SEISMIC REFLECTION DATA, CALIFORNIA, MONTEREY FAN

Index

M 29 200 SEISMIC REFLECTION DATA, OKINAWA TROUGH
B 6412 2141 SEISMIC REFLECTION DATA, PROFILES, BERING SEA SHELF
M 26 117† SEISMIC REFLECTION PATTERNS IN DEPOSITIONAL SEQUENCES, SEISMIC,
B 6204 702 SEISMIC REFLECTION PROFILE, ABACO KNOLL
B 6204 694 SEISMIC REFLECTION PROFILES, BAHAMA PLATFORM
B 6204 661 SEISMIC REFLECTION PROFILES, BRAZIL, SAO PAULO PLATEAU
B 6403 300 SEISMIC REFLECTION PROFILES, CONTINENTAL, MARGINS IDENTIFIABLE
B 6403 305 SEISMIC REFLECTION PROFILES, CONTINENTAL, TERRACES CLASSIFIED
M 29 11† SEISMIC REFLECTION PROFILES, GEOLOGY OF THE, OFFSHORE,
B 6404 502 SEISMIC REFLECTION PROFILES, GREAT SOUTH CHANNEL
B 6404 502 SEISMIC REFLECTION PROFILES, NEW ENGLAND
B 6404 502 SEISMIC REFLECTION PROFILES, NORTHEAST CHANNEL
B 6404 502 SEISMIC REFLECTION PROFILES, SCOTIAN BASIN
M 29 177† SEISMIC REFLECTION PROFILES, STRUCTURE AND, STRATIGRAPHY OF
B 6405 715* SEISMIC REFLECTION, REFRACTION, GRAVITY, AND MAGNETIC DATA,
M 26 221 SEISMIC, REFLECTION SIGNATURES, CORRELATION
M 26 221 SEISMIC, REFLECTION SIGNATURES, INTERPRETATION
B 6202 249 SEISMIC REFLECTION, SOUTHERN, CALIFORNIA
M 26 6 SEISMIC REFLECTIONS, ACOUSTIC IMPEDANCE
M 26 3† SEISMIC REFLECTIONS AND GEOLOGIC DETAIL DERIVABLE FROM THEM,
M 29 135 SEISMIC REFLECTIONS, ARGENTINA, FALKLAND PLATEAU
M 29 135 SEISMIC REFLECTIONS, ARGENTINA, FALKLAND TROUGH
M 29 135 SEISMIC REFLECTIONS, ARGENTINA, NORTH SCOTIA RIDGE
M 26 3 SEISMIC REFLECTIONS, ATTENUATING NOISE
M 26 203 SEISMIC REFLECTIONS, CARBONATE RECOGNITION, CRITERIA, CHART
M 26 203 SEISMIC REFLECTIONS, CARBONATE RECOGNITION, EXAMPLES
M 26 187 SEISMIC REFLECTIONS, CARBONATE RECOGNITION, GEOPHYSICAL
M 26 187 SEISMIC REFLECTIONS, CARBONATE RECOGNITION, PROCEDURE
M 26 188 SEISMIC REFLECTIONS, CARBONATES, DIRECT RECOGNITION
M 26 188 SEISMIC REFLECTIONS, CARBONATES, INDIRECT RECOGNITION
M 26 115 SEISMIC REFLECTIONS, CHRONOSTRATIGRAPHIC CORRELATION,
M 26 116 SEISMIC REFLECTIONS, CHRONOSTRATIGRAPHIC CORRELATIONS,
M 26 109 SEISMIC REFLECTIONS, CHRONOSTRATIGRAPHY, STRATIGRAPHIC TRAP
M 26 122 SEISMIC REFLECTIONS, CONFIGURATION
M 26 128 SEISMIC REFLECTIONS, CONFIGURATION PATTERNS, CHAOTIC
M 26 123 SEISMIC REFLECTIONS, CONFIGURATION PATTERNS, DIVERGENT
M 26 123 SEISMIC REFLECTIONS, CONFIGURATION PATTERNS, PARALLEL
M 26 123 SEISMIC REFLECTIONS, CONFIGURATION PATTERNS, PROGRADING
M 26 128 SEISMIC REFLECTIONS, CONFIGURATION PATTERNS, REFLECTION-FREE
M 26 123 SEISMIC REFLECTIONS, CONFIGURATION PATTERNS, SUB-PARALLEL
M 26 123 SEISMIC REFLECTIONS, CONFIGURATION PATTERNS, TYPES
M 26 132 SEISMIC REFLECTIONS, CONFIGURATIONS, MIGRATING-WAVE
M 26 100 SEISMIC REFLECTIONS, CONTINUOUS
M 26 100 SEISMIC REFLECTIONS, DISCONTINUOUS
M 26 121 SEISMIC REFLECTIONS, FACIES DEFINITION
M 26 122 SEISMIC REFLECTIONS, FACIES UNITS
M 26 130 SEISMIC REFLECTIONS, FACIES UNITS, EXTERNAL FORMS, TYPES
M 26 115 SEISMIC REFLECTIONS, FLUID CONTACTS, HYDRATE ZONES
M 26 6 SEISMIC REFLECTIONS, INTERFERENCE
M 26 186 SEISMIC REFLECTIONS, INTERPRETATION PROCEDURE, CHART
M 26 100 SEISMIC REFLECTIONS, LITHOSTRATIGRAPHY
M 26 106 SEISMIC REFLECTIONS, PALEONTOLOGIC CORRELATION
M 26 120 SEISMIC REFLECTIONS, PATTERNS, BASE-DISCORDANT
M 26 125 SEISMIC REFLECTIONS, PATTERNS, CLINOFORMS
M 26 127 SEISMIC REFLECTIONS, PATTERNS, CLINOFORMS
M 26 138 SEISMIC REFLECTIONS, PATTERNS, STRATIGRAPHIC INTERPRETATION
M 26 119 SEISMIC REFLECTIONS, PATTERNS, TOP-DISCORDANT
M 26 99 SEISMIC REFLECTIONS, PHYSICAL SURFACES
M 26 185 SEISMIC REFLECTIONS, RECOGNITION, BANKS
M 26 185 SEISMIC REFLECTIONS, RECOGNITION, BUILDUPS
M 26 203 SEISMIC REFLECTIONS, RECOGNITION, INTERVAL VELOCITY,
M 26 203 SEISMIC REFLECTIONS, RECOGNITION, LIMITATIONS AND
M 26 185 SEISMIC REFLECTIONS, RECOGNITION, REEFS
M 26 4 SEISMIC REFLECTIONS, RESOLUTION, VERTICAL
M 26 9 SEISMIC REFLECTIONS, SANDSTONE PINCHOUTS
M 26 141 SEISMIC REFLECTIONS, SECTION DESCRIPTION
M 26 101 SEISMIC REFLECTIONS, SEISMIC MODELS
M 26 99† SEISMIC REFLECTIONS, SEISMIC, STRATIGRAPHY AND, GLOBAL CHANGES
M 26 117 SEISMIC REFLECTIONS, SEQUENCES
M 26 6 SEISMIC REFLECTIONS, THIN BEDS
M 26 100 SEISMIC REFLECTIONS, UNCONFORMITIES, CHRONOSTRATIGRAPHY
M 26 101 SEISMIC REFLECTIONS, VELOCITY RELATIONS
M 26 4 SEISMIC REFLECTIONS, WAVELENGTH
M 29 181 SEISMIC REFLECTORS, ATLANTIC, BLAKE-BAHAMA BASIN
M 29 182 SEISMIC REFLECTORS, BLAKE ESCARPMENT, MIOCENE
M 29 405 SEISMIC REFLECTORS, GULF OF MEXICO, SOUTHWESTERN
M 29 209† SEISMIC REFRACTION AND REFLECTION STUDIES IN THE TIMOR-ARU
B 6408 1287* SEISMIC REFRACTION STUDY OF BURIED VALLEY NEAR PENINSULA,
M 29 118 SEISMIC REFRACTIONS, ARGENTINA, MARGIN
M 26 418 SEISMIC RESOLUTION, CRITICAL RESOLUTION THICKNESS
M 26 11 SEISMIC RESOLUTION, FRESNEL ZONE
M 26 419 SEISMIC RESOLUTION, LATERAL
M 26 418 SEISMIC RESOLUTION, THIN OR THICK
M 26 478 SEISMIC RESPONSE, SEISMIC, MODELING, VERTICAL LITHOLOGY
M 26 490 SEISMIC RESPONSE, SHALE DISTRIBUTION
M 29 373 SEISMIC RESULTS, SOUTHEAST ASIA
B 6407 1122 SEISMIC, SAFETY AND LAND-USE PLANNING. 2. FLATLAND DEPOSITS—
M 29 323 SEISMIC SECTION (24-FOLD), GULF OF MEXICO, MEXICAN RIDGES
M 29 324 SEISMIC SECTION (FOLDOUT), GULF OF MEXICO, MEXICAN RIDGES
M 26 339 SEISMIC SECTION, ALASKA, PRUDHOE BAY
B 6206 1072 SEISMIC SECTION, DEFINITION, SMEARING OF
B 6206 1072 SEISMIC SECTION, DEFINITION, WIPEOUT OF
M 26 338 SEISMIC SECTION, GERMANY, HOHNE FIELD
M 29 258 SEISMIC SECTION, GUATEMALA, SHELF
M 29 407 SEISMIC SECTION, GULF OF MEXICO, CAMPECHE SCARP
M 29 407 SEISMIC SECTION, GULF OF MEXICO, CHALLENGER KNOLL
M 29 407 SEISMIC SECTION, GULF OF MEXICO, MEXICAN RIDGES
M 29 406 SEISMIC SECTION, GULF OF MEXICO, SIGSBEE KNOLLS
M 30 552 SEISMIC SECTION, INTISAR D FIELD, INTISAR D REEF
B 6311 2091 SEISMIC SECTION, ISRAEL, MEDITERRANEAN MARGIN
M 26 256 SEISMIC SECTION, LOUISIANA, TIMBALIER CHANNEL
M 26 350 SEISMIC SECTION, NORTH SEA, BENTHEIM SANDSTONE
M 29 365 SEISMIC SECTION, VENEZUELA BASIN, CURACAO RIDGE
M 29 366 SEISMIC SECTION, VENEZUELA BASIN, LOS ROQUES TRENCH
M 26 336 SEISMIC SECTION, WEST, TEXAS, HORSESHOE ATOLL
M 29 276 SEISMIC SECTIONS, ALASKA, CROSS SOUND
M 29 167 SEISMIC SECTIONS, AUSTRALIA, QUIET ZONE
B 6004 718* SEISMIC SECTIONS OF SELECTED STRATIGRAPHIC TRAPS AN AQUIFERS IN
B 6008 1409* SEISMIC SECTIONS OF SELECTED STRATIGRAPHIC TRAPS AND AQUIFERS
B 6104 547 SEISMIC SECTIONS, SOUTHERN MARGIN, BRAZIL
M 29 358 SEISMIC SECTIONS, VIRGIN ISLANDS, FREDERICKSTED PLATEAU
B 6303 423* **SEISMIC SEQUENCE ANALYSIS, NEW APPROACH FOR EXPLORATION IN**
M 26 145† SEISMIC SEQUENCES IN GULF OF MEXICO AND OFFSHORE WEST AFRICA,
M 26 153 SEISMIC SEQUENCES, TEXAS, SOUTH
B 6410 1658† **SEISMIC STRATIGRAPHY AND BASEMENT STRUCTURE OF WESTERN**
B 6405 742* **SEISMIC STRATIGRAPHY AND STRUCTURE OF FALKLAND PLATEAU,**
M 26 220 SEISMIC STRATIGRAPHIC ANALYSIS, DEPOSITIONAL SYSTEMS
B 6303 487* SEISMIC STRATIGRAPHIC, AND ECONOMIC ANALYSIS OF POTENTIAL GAS
B 6405 726* SEISMIC, STRATIGRAPHIC, AND STRUCTURAL ANALYSIS OF NORTHEAST
B 6405 682* SEISMIC STRATIGRAPHIC INTERPRETATION AHEAD OF DRILL, ABST.,
M 26 135† SEISMIC STRATIGRAPHIC INTERPRETATION PROCEDURE, SEISMIC,
S 7 193† SEISMIC STRATIGRAPHIC INVESTIGATION OF WEST FLORIDA SLOPE,
B 6409 1568 SEISMIC-STRATIGRAPHIC MAPPING OF GULF COAST STRATIGRAPHIC
M 26 439† SEISMIC STRATIGRAPHIC MODEL OF DEPOSITIONAL PLATFORM MARGIN,
B 6109 1437† SEISMIC STRATIGRAPHIC MODEL OF DEPOSITIONAL PLATFORM MARGIN,
B 6405 793* SEISMIC STRATIGRAPHY, ABST., JURASSIC SEA-LEVEL CHANGES FROM
B 6304 700* SEISMIC STRATIGRAPHY, ABST., WAVE-FORM FACTOR
B 6205 792† SEISMIC STRATIGRAPHY AND GEOLOGIC HISTORY OF BLAKE PLATEAU
B 6303 487* SEISMIC STRATIGRAPHY AND GEOLOGIC HISTORY OF CAMPECHE
B 6004 690* SEISMIC STRATIGRAPHY AND GEOLOGIC HISTORY OF DEEP WESTERN
M 26 185† SEISMIC STRATIGRAPHY AND GLOBAL CHANGES OF SEA LEVEL, PART 10:
B 6205 772† SEISMIC STRATIGRAPHY AND GLOBAL CHANGES OF SEA LEVEL, PART 10:
M 26 205† SEISMIC STRATIGRAPHY AND GLOBAL CHANGES OF SEA LEVEL, PART 11:
M 26 51† SEISMIC STRATIGRAPHY AND GLOBAL CHANGES OF SEA LEVEL, PART 1:
M 26 53† SEISMIC STRATIGRAPHY AND GLOBAL CHANGES OF SEA LEVEL, PART 2:
M 26 63† SEISMIC STRATIGRAPHY AND GLOBAL CHANGES OF SEA LEVEL, PART 3:
M 26 83† SEISMIC STRATIGRAPHY AND GLOBAL CHANGES OF SEA LEVEL, PART 4:
M 26 99† SEISMIC STRATIGRAPHY AND GLOBAL CHANGES OF SEA LEVEL, PART 5:
M 26 117† SEISMIC STRATIGRAPHY AND GLOBAL CHANGES OF SEA LEVEL, PART 6:
M 26 135† SEISMIC STRATIGRAPHY AND GLOBAL CHANGES OF SEA LEVEL, PART 7:
M 26 145† SEISMIC STRATIGRAPHY AND GLOBAL CHANGES OF SEA LEVEL, PART 8:
M 26 165† SEISMIC STRATIGRAPHY AND GLOBAL CHANGES OF SEA LEVEL, PART 9:
B 6205 752† SEISMIC STRATIGRAPHY AND GLOBAL CHANGES OF SEA LEVEL, PART 9:
M 26 49† SEISMIC STRATIGRAPHY AND GLOBAL CHANGES OF SEA LEVEL
B 6303 409* SEISMIC STRATIGRAPHY AND STRUCTURE OFF PANAMA CITY, FLORIDA.
B 6205 799 SEISMIC STRATIGRAPHY, BLAKE PLATEAU
M 29 16 SEISMIC STRATIGRAPHY, GEORGIA EMBAYMENT
M 29 469† SEISMIC STRATIGRAPHY, GLOBAL, CYCLES OF, RELATIVE, CHANGES OF,
M 29 321 SEISMIC STRATIGRAPHY, GULF OF MEXICO, CENTRAL GULF
M 29 403 SEISMIC STRATIGRAPHY, GULF OF MEXICO, MARGINS
M 29 393 SEISMIC STRATIGRAPHY, GULF OF MEXICO, MISSISSIPPI FAN
B 6405 798* SEISMIC STRATIGRAPHY IN LOW-ENERGY CARBONATE DEPOSITIONAL
B 6405 796* SEISMIC STRATIGRAPHY OF ABYSSAL SOUTHWESTERN GULF OF MEXICO
B 6405 781* SEISMIC STRATIGRAPHY OF ATLANTIC MARGIN IN VICINITY OF
B 6105 773* **SEISMIC STRATIGRAPHY, SEDIMENTARY HISTORY, AND PETROLEUM**
M 26 205† SEISMIC STRATIGRAPHY, SEISMIC, STRATIGRAPHY AND, GLOBAL
M 26 205 SEISMIC STRATIGRAPHY, TERMINOLOGY, GLOSSARY
M 26 1 **SEISMIC STRATIGRAPHY--APPLICATIONS TO HYDROCARBON EXPLORATION**
B 6012 2179* SEISMIC STRATIGRAPHY--NEW KIND OF GEOPHYSICAL INTERPRETATION,
B 6402 242† SEISMIC STRUCTURE OF SOUTHWESTERN PELOPONNESUS CONTINENTAL

B 6008 1389* SEISMIC STUDIES OF BOISE FRONT, IDAHO, FOR GEOTHERMAL RESOURCE
B 6403 347 SEISMIC STUDY, FLORIDA-, HATTERAS SHELF AND INNER, BLAKE
B 6201 73† SEISMIC STUDY, WINNIPEGOSIS, MOUNDS AND, PRAIRIE EVAPORITE
B 6405 729* SEISMIC SURVEY, ABST., UNDERSTANDING FIELD DEVELOPMEN HISTORY
S 7 259 SEISMIC, TEXAS, SHELF-SLOPE, GAS MIGRATION
B 6105 829* SEISMIC TRACE, ABST., GEOLOGIC INTERPRETATION OF, COLOR DISPLAYS
M 26 302 SEISMIC TRACE, COMPLEX, DESCRIPTION
B 6303 513* SEISMIC TRACES, ABST., RESERVOIR STUDIES AND INVERSION OF
M 29 284 SEISMIC TRACES, GULF OF ALASKA, OBLIQUE CONVERGENT MARGIN
M 26 9 SEISMIC TRACES, SEISMIC, DATA
B 6211 2291 SEISMIC UNITS, GULF OF MEXICO, SOUTHWESTERN
M 29 324 SEISMIC UNITS, GULF OF MEXICO, SOUTHWESTERN
S 7 48 SEISMIC UNITS, TABLE, WESTERN, GULF OF MEXICO
M 29 13 SEISMIC VELOCITIES, GEORGIA EMBAYMENT
B 6203 562* SEISMIC-ATTRIBUTE MEASUREMENTS, ABST., GEOLOGIC
B 6105 698† SEISMIC-PROFILER DATA FROM THE WESTERN NORTH ATLANTIC, GAS-
B 6007 1088 SEISMIC-REFLECTION DATA, ATLANTIC BASIN
B 6007 1096 SEISMIC-REFLECTION DATA, BARBADOS RIDGE COMPLEX
B 6002 163 SEISMIC-REFLECTION DATA, BERINGIAN SHELF
B 6007 1049 SEISMIC-REFLECTION PROFILE, ABACO KNOLL
B 6007 1040 SEISMIC-REFLECTION PROFILES, NORTHEAST PROVIDENCE CHANNEL
B 6106 918 SEISMIC-REFLECTION PROFILES, WILMINGTON CANYON AREA
M 26 213† SEISMIC-STRATIGRAPHIC INTERPRETATION OF DEPOSITIONAL SYSTEMS:
S 7 170 SEISMIC-STRATIGRAPHIC UNITS, MISSISSIPPI FAN
B 6203 506* SEISMICALLY DERIVED IMPEDANCE LOGS, ABST., UNDERSTANDING
B 6004 653* SEISMICITY AN TECTONIC PROVINCES, CENTRAL STABLE REGION, ABST.,
B 6207 1225* SEISMICITY AND STRUCTURE OF NEW HEBRIDES ARC-TRENCH SYSTEM,
B 6212 2458 SEISMICITY, ECUADOR
M 29 446 SEISMICITY, MEXICO, MIDDLE AMERICAS TRENCH
M 29 95 SEISMICS, ATLANTIC MARGIN, MAGNETICS
M 26 390 SEISMOGRAM SYNTHESIS, MODELING
M 26 3 SEISMOGRAMS, SEISMIC, DATA, SYNTHETIC
M 26 9 SEISMOGRAMS, SEISMIC, DATA, SYNTHETIC
M 29 443† SEISMOLOGY STUDIES OF ACTIVE AND PASSIVE CONTINENTAL MARGINS,
M 25 456 SEKINE, YOSHIHIRO, NATIONAL EXPLORATION PROGRAM FOR
B 6203 532* SEKWI FORMATION, MACKENZIE MOUNTAINS, NORTHWEST
M 30 531 SELCIFERO FORMATION, JURASSIC, MALOSSA FIELD
B 6203 384 SELDOVIA ARCH, ALASKA, AUGUSTINE-
B 6203 376 SELDOVIA FAULT, ALASKA
B 6207 1132 SELE HIGH, IRIAN JAYA
B 6302 183 SELE HIGH, NORTH SEA
B 6207 1129 SELE STRAIT—MISOOL FAULT, IRIAN JAYA
B 6010 1900† SELECTED REFERENCES ON MIDDLE EAST, 1970 TO 1975
S 3 47 SELECTIVE DISSOLUTION
B 6004 700* SELECTIVE DISSOLUTION SEQUENCES FOR CALCAREOUS NANNOFOSSILS
S 3 46 SELECTIVE OXIDATION AND MOBILIZATION
M 26 352 SELF, GEORGE W., ANALYSIS OF HIGH RESOLUTION SEISMIC DATA
B 6012 2189 SELLERS, GEORGE A., POSSIBLE ROLE FOR NITROGEN IN FORMATION OF
B 6309 1545 SELLEY, KUMAR, NARESH, REVIEW OF, ANCIENT SEDIMENTARY
B 6306 905 SELLEY, RICHARD C., DIPMETER AND LOG MOTIFS IN NORTH SEA
B 6002 184 SELLEY, RICHARD C., SUBSURFACE ENVIRONMENTAL ANALYSIS OF NORT
M 24 242 SELMA CHALK, MISSISSIPPI
B 6107 988 SELMA CHALK UNIT, ALABAMA
M 25 493 SELWYN BASIN, CANADA
M 25 497 SELWYN BASIN, ZINC-LEAD, CANADA
M 28 188 SEM, EXPLANATION, TECHNIQUES
B 6304 693* SEM INVESTIGATION OF PARAGENESIS OF URANIUM DEPOSITS, GRANTS
B 6303 538 SEN GUPTA, BARUN K., LATE QUATERNARY FORAMINIFERAL RECORD IN
M 25 539 SENECHAL, RONALD G., METALLOGENESIS IN SOUTHEAST PACIFIC
B 6101 80 SENEGAL BASIN
B 6107 965 SENEGAL BASIN, AFRICA
B 6110 1749 SENEGAL, DEVELOPMENTS, 1976
B 6210 1848 SENEGAL, DEVELOPMENTS, 1977
B 6310 1694 SENEGAL, DEVELOPMENTS, 1978
B 6411 1790 SENEGAL, DEVELOPMENTS, 1979
B 6104 595 SENEGAL, DKM-2 WELL
B 6104 597 SENEGAL, DKM-2 WELL
B 6202 230 SENEGAL, EVAPORITES
B 6110 1752 SENEGAL, PRODUCTION, 1976
B 6210 1852 SENEGAL, PRODUCTION, 1976-1977
B 6005 809† SENEGAL TO PORTUGAL, CONTINENTAL, MARGIN OFF WESTERN,
B 6109 1542 SENI, S. J., DEPOSITIONAL FRAMEWORK OF LOWER DOCKUM GROUP,
S 6 160 SENNAN ACID PYROCLASTICS, JAPAN
B 6204 669 SENONIAN, ALGERIA
S 6 277 SENONIAN, FORAMINIFERAL SCHEME, CRETACEOUS
B 6405 668* SENONIAN IN SEDIMENTS ANAMBRA SYNCLINE, SOUTHEASTERN
B 6205 829 SENONIAN, ISRAEL
B 6101 44 SENONIAN, LEBANON
B 6405 610 SENONIAN TO RECENT, MIDDLE EAST
B 6201 50 SENONIAN-MAESTRICHTIAN, AUSTRALIA
B 6309 1473 SENONIAN-OLIGOCENE SEQUENCE, USSR, ANADYR BASIN
B 6309 1583 SENSENY, PAUL E., INTERACTION OF PROPPANTS WITH CRACK
B 6007 1015 SEPARATE CHANNEL SYSTEM, SEQUENCE
S 10 21 SEPARATE OIL PHASE, MIGRATION, MECHANISMS
B 6405 694* SEPM RESEARCH SYMPOSIUM—SOURCE BEDS—DEPOSITIONAL
S 6 107 SEPOTUBA FORMATION, BRAZIL, ALTO PARAGUAY GROUP
S 6 107 SEPOTUBA FORMATION, RUBIDIUM-STRONTIUM, BRAZIL
B 6311 2059 SEQUASH COALFIELD, BRITISH COLUMBIA
M 26 135 SEQUENCE ANALYSIS, INTERPRETATION
M 26 53† SEQUENCE AS A BASIC UNIT FOR STRATIGRAPHIC ANALYSIS, SEISMIC,
S 4 357† SEQUENCE OF PROGRESSIVE DIAGENESIS IN CORAL REEFS
B 6007 1015 SEQUENCE OF SEDIMENTARY STRUCTURES AND TEXTURES
M 29 453† SEQUENCES AT ATLANTIC MARGINS, THE EFFECT OF, EUSTATIC, SEA
M 26 145 SEQUENCES, GULF OF MEXICO, NORTHERN
M 26 155 SEQUENCES, IDENTIFICATION, AFRICA, WESTERN
M 26 145† SEQUENCES IN GULF OF MEXICO AND OFFSHORE WEST AFRICA, SEISMIC,
M 26 158 SEQUENCES, JURASSIC, BAJOCIAN, AFRICA
M 26 158 SEQUENCES, JURASSIC, BATHONIAN, AFRICA
M 26 158 SEQUENCES, JURASSIC, CALLOVIAN, AFRICA
M 26 159 SEQUENCES, OXFORDIAN-KIMMERIDGIAN, AFRICA, WESTERN
M 26 117 SEQUENCES, SEISMIC, REFLECTIONS
M 26 117† SEQUENCES, SEISMIC, STRATIGRAPHY AND, GLOBAL CHANGES OF SEA
M 26 153 SEQUENCES, TEXAS, GULF COAST
M 26 160 SEQUENCES, TITHONIAN-BERRIASIAN, AFRICA, WESTERN
S 7 84 SEQUENTIAL DEVELOPMENT, GULF OF MEXICO
B 6003 356† SEQUENTIAL FOLDING SYNCHRONOUS WITH SAN ANDREAS FAULTING,
B 6406 884 SERAM TROUGHS, TIMOR, ARU, AND
B 6206 1067† SERENDIPITY GAS SEEP AREA, SOUTH TEXAS OFFSHORE
B 6206 1068 SERENDIPITY GAS SEEPS, LOCATION, TEXAS
B 6109 1471 SERGIPE-ALAGOAS BASIN, BRAZIL
B 6405 705* SERGIPE-ALAGOAS BASIN, BRAZIL—SOURCE ROCK CHARACTERIZATION
B 6202 229 SERGIPE-ALAGOAS BASIN, SEDIMENTARY SEQUENCE, BRAZIL
B 6002 197 SERGIPE-ALAGOAS COASTAL BASIN, BRAZIL
B 6202 230 SERGIPE-ALAGOAS, EVAPORITES, BRAZIL
M 26 343 SERGNANO FIELD, PO VALLEY, ITALY
B 6402 219 SERIA FIELD, MALAYSIA
B 6004 728* SERIAL SECTIONS, ABST., COMPUTERIZED, MODELING IN
M 28 70 SERICITE, COARSE, IN CLAYSTONE
M 28 67 SERICITE, IN ARKOSE, CLAY
M 28 18 SERICITIZATION, FELDSPAR
B 6104 562 SERIE TOBIFERA STRATIGRAPHIC UNIT, CHILE
S 11 7 SERIES SUBDIVISIONS, FORMATIONAL APPROACH, CALIFORNIA
M 26 214 SERIGIPE-ALAGOAS BASIN, BRAZIL
B 6209 1765* SERPENTINITES AND AMPHIBOLITES ALONG MOTAGUA FAULT ZONE IN
M 28 87 SERPULID WORM TUBE, ANHYDRITE, CRYSTALS, IN A
M 27 41 SERPULID WORM TUBES
B 6206 1033 SERRA GERAL BASALTS, BRAZIL
B 6002 206 SERRA GERAL LAVA, BRAZIL COASTAL MARGIN
B 6104 551 SERRA GERAL LAVAS, BRAZIL
B 6409 1550 SERRA, ROKSANDIC, MIODRAG M., REVIEW OF, DIAGRAPHIES
B 6109 1477 SERRARIA FORMATION, BRAZIL
B 6010 1710 SERVICE CONS. GISEMENTS, ON DEVELOPMENTS IN FRANCE, 1975
B 6110 1640 SERVICE CONS. GISEMENTS, ON DEVELOPMENTS IN FRANCE, 1976
B 6210 1910 SERVICE CONS. GISEMENTS, ON DEVELOPMENTS IN FRANCE, 1977
B 6310 1919 SERVICE CONS. GISEMENTS, ON DEVELOPMENTS IN FRANCE, 1978
B 6308 1174 SERVICE WELL
B 6208 1281 SERVICE WELL
B 6108 1127 SERVICE WELL
B 6409 1302 SERVICE WELL
B 6008 1159 SERVICE WELL
B 6012 2186* SESPE AND POWAY CONGLOMERATES, SOUTHERN CALIFORNIA, ABST.,
B 6211 2360* SESPE CREEK, VENTURA COUNTY, CALIFORNIA, ABST., EARLY, MIOCENE
B 6211 2363* SESPE CREEK, VENTURA COUNTY, CALIFORNIA, ABST., UPPER, MIOCENE,
B 6111 1986 SESPE FORMATION, CALIFORNIA
B 6004 653* SET-THEORETIC INFORMATION RETRIEVAL FROM MEGA-BYTE DATA
B 6205 781 SEVEN RIVERS FORMATION, NEW MEXICO
B 6303 471* SEVEN RIVERS FORMATION, PERMIAN, GUADALUPE MOUNTAINS, NEW
B 6004 718* SEVEN RIVERS FORMATION, SOUTHEAST NEW MEXICO, ABST.,
B 6008 1410 SEVERSON, DONALD E., LIQUEFACTION OF NORTHERN GREAT PLAINS
B 6105 828* SEVIER SHALE, EASTERN TENNESSEE, ABST., DEEP-WATER
S 10 180 SEWAGE SLUDGE, METHANE, OCCURRENCES
B 6407 1037 SEXTON CANYON, CALIFORNIA
B 6102 170 SEXTON CONGLOMERATE OUTCROPS, CALIFORNIA
B 6102 138 SEXTON CONGLOMERATE OUTCROPS, CALIFORNIA
B 6204 706 SEXTON CREEK LIMESTONE, ILLINOIS BASIN
B 6401 129 SEXTON CREEK LIMESTONE, INDIANA
B 6407 1042 SEXTON LAKE CONGLOMERATE, CALIFORNIA
B 6110 1749 SEYCHELLES, DEVELOPMENTS, 1976
B 6210 1848 SEYCHELLES, DEVELOPMENTS, 1977
B 6310 1694 SEYCHELLES, DEVELOPMENTS, 1978

B 6411 1790 SEYCHELLES, DEVELOPMENTS, 1979
B 6410 1722 SHABAD, MEYERHOFF, A. A., REVIEW OF THE, SOVIET ENERGY SYSTEM,
B 6303 344 SHADY TOMSTOWN FORMATION, KENTUCKY
B 6009 1608 SHAFFER, B. L., NANNOFOSSIL BIOSTRATIGRAPHY OF BOSSIER SHALE
B 6309 1579 SHAFFER, R. J., THEORETICAL AND EXPERIMENTAL ANALYSES OF
M 30 472 SHAHEJIE FORMATION, CHINA, ZHANHUA BASIN
S 10 2 SHALE COMPACTION, POROSITY REDUCTION
S 10 10 SHALE COMPACTION SEQUENCE, GULF COAST
S 10 192 SHALE DIAPIRS, METHANE, PRESSURES
S 10 6 SHALE MINERALOGY EXAMPLE, MIOCENE, LOUISIANA
S 10 10 SHALE POROSITY, CHANGES WITH DEPTH
S 10 2 SHALE POROSITY, CHANGES WITH DEPTH
S 10 3 SHALE POROSITY, COMPACTION STAGES
S 10 4 SHALE POROSITY, COMPACTION STAGES, GULF COAST
S 10 4 SHALE POROSITY, COMPACTION STAGES, MIOCENE, GULF COAST
S 10 8 SHALE POROSITY, COMPACTION VERSUS CLAY FRACTION
S 10 7 SHALE POROSITY, EFFECTS ON ADSORPTION
S 10 7 SHALE POROSITY, GRAIN SIZE
S 10 7 SHALE POROSITY, PORE SIZE
S 10 5 SHALE POROSITY, PROCESSES, GENERAL
S 10 12 SHALE POROSITY, RELATION BETWEEN TEMPERATURE AND AGE
S 10 9 SHALE POROSITY, RELATION OF SURFACE AREAS TO SAND
S 10 13 SHALE POROSITY, RELATION TO BULK DENSITY
S 10 13 SHALE POROSITY, RELATION TO KEROGEN COMPOSITION
S 10 9 SHALE POROSITY, THERMAL STABILITY OF WATERS
S 10 43 SHALE POROSITY-CHLORINITY RELATION, MIGRATION
B 6008 1412* SHALE, ABST., DEPOSITIONAL ENVIRONMENNSTS OF, OIL
B 6004 666* SHALE, ABST., ENVIRONMENTS OF DEPOSITION OF, OIL
B 6405 680* SHALE, ABST., PERMEABILITY OF, CLAY AND
B 6105 820* SHALE, ABST., PREDICTIONS FOR, OIL
B 6309 1580* SHALE, ABST., UNCONVENTIONAL EVALUATION OF, COTTAGEVILL FIELD,
M 28 160 SHALE, ALASKA, CRETACEOUS
M 28 160 SHALE, ALASKA, JURASSIC
B 6303 525* SHALE AND FINE-GRAINED DEBRIS FLOW, ABST., RHYTHMS IN
B 6309 1578* SHALE AND ITS SENSITIVITY TO CHANGE IN VARIOUS RESERVOIR
B 6203 505* SHALE AND MARLSTONE, GREEN RIVER FORMATION, PICEANCE CREEK
B 6107 1063 SHALE AND PHOSPHORITE, MID-CONTINENT, PALEOZOIC, BLACK
M 28 97 SHALE, ARENITE
B 6303 525* SHALE BASIN, MIDDLE ORDOVICIAN, SOUTHERN APPALACHIANS, ABST.,
M 28 75 SHALE, BIBLIOGRAPHY
B 6309 1584* SHALE CHARACTERIZATION--EVALUATION OF SOME SIGNIFICANT
B 6305 820 SHALE CHEMISTRY, EDITED BY O. P. STRAUSZ AND E. M. LOWN, COPELIN,
B 6311 2115* SHALE CLAST, ABST., DISTRIBUTION OF, BITUMENS AND, KEROGEN IN
M 28 65 SHALE CLASTS, CLAY
B 6007 1124 SHALE COMPACTION
B 6304 573 SHALE COMPACTION, GEOCHEMICAL CONCEPT
B 6306 922 SHALE COMPACTION, MECHANISM OF
B 6203 571* SHALE DEHYDRATION, SALINITY, CRACKING, AND MIGRATION, ABST.
B 6007 1128 SHALE DENSITY AND POROSITY FOR OFFSHORE WELL, LOUISIANA
B 6107 1058 SHALE DEPOSITION, MODELS OF, PENNSYLVANIAN, BLACK
B 6405 696* SHALE DEPOSITS OF SOUTHERN UINTA BASIN, UTAH, ABST., SHALLOW,
B 6209 1577 SHALE DIAPIRISM, COMPARISON OF, SALT TO
S 3 136 SHALE DIAPIRISM, MOBILIZATION OR LOCALIZATION BY, SALT OR
B 6307 1054 SHALE DIAPIRS WITHIN DELTAIC SEQUENCE, SEISMIC EXPRESSION OF
M 26 490 SHALE DISTRIBUTION, SEISMIC, RESPONSE
B 6107 1045† SHALE FACIES IN PENNSYLVANIAN CYCLOTHEMS OF MID-CONTINENT
B 6309 1578* SHALE GAS IN WESTERN KENTUCKY, ABST., PAST AND CURRENT
B 6305 847* SHALE GAS PRODUCTION, PHOTOLINEAMENTS, AND GEOLOGY IN
B 6309 1585* SHALE GAS RESERVOIRS, ABST., FABRIC-ELEMENT-BASED CLASSIFICATION
B 6309 1590* SHALE GAS RESERVOIRS--RESULTS OF INVESTIGATION IN UPPER
B 6309 1581* SHALE GAS-RESERVOIR PERFORMANCE, ABST., MODELING OF, DEVONIAN
B 6206 959 SHALE GRIT: A COMPLETE ANCIENT SUBMARINE FAN, ENGLAND
B 6412 2156† SHALE IN WESTERN APPALACHIAN BASIN, ORGANIC CONTENT OF,
B 6008 1413* SHALE MASSES AND GROWTH FAULTING, NEW TECTONIC STYLE IN
B 6306 923 SHALE MASSES BEFORE BURIAL BY PERMEABLE OVERBURDEN
B 6306 925 SHALE MASSES DURING BURIAL
B 6405 709* SHALE MINERALOGY OF GENERAL CRUDE OIL AND DEPARTMENT OF
B 6408 1285* SHALE OF EASTERN UNITED STATES, ABST., DEVONIA OIL
B 6304 656 SHALE, OHIO, DEVONIAN, BLACK
B 6405 739* SHALE OIL EXTRACTION METHOD TO AVAILABLE GEOLOGY, ABST.,
B 6303 439* SHALE ON BASIS OF CHEMICAL CHARACTERISTICS, ABST., STATISTICAL
S 11 99 SHALE POINT, CALIFORNIA, CHECK LIST OF, MIOCENE, FORAMINIFERA
B 6007 1128 SHALE POROSITY FOR WELLS, VENEZUELA
B 6004 544 SHALE POROSITY-DEPTH RELATION
B 6309 1578* SHALE RESERVOIRS AND BLACK SHALE FORMATIONS, ABST., NOVEL
B 6211 2362* SHALE RESOURCES OF SOUTH ELKO BASIN, NEVADA, ABST., TERTIARY
M 28 73 SHALE, SOLUTION SEAMS
B 6307 1052 SHALE STRUCTURES
S 7 35 SHALE STRUCTURES, DESCRIBED, NORTHERN, GULF OF MEXICO
B 6405 700* SHALE TO ENERGY NEEDS OF UNITED STATES AND OTHER PARTS OF THE
B 6011 2044† SHALE TYPES BY ALLOCHTHONOU AND AUTOCHTHONOUS
B 6108 1387* SHALE-OIL RESOURCES OF WHITE RIVER-GRAY HILLS AREA, PICEANCE
B 6103 332 SHALERS LINE, SOUTH CAROLINA
S 10 1† SHALES AND THE DYNAMICS OF HYDROCARBON EXPULSION IN THE
S 10 5 SHALE-WATER SYSTEM, GEOMETRY, POROSITY
B 6203 489* SHALES, ABST., ANALOGIES BETWEEN, HOLOCENE, COASTAL-, MARSH
B 6309 1581* SHALES, ABST., COMPOSITIONAL VARIATION IN FABRIC-ELEMENT-BASED
B 6309 1586* SHALES, ABST., CONSTRAINTS AND PROBLEMS IN PRODUCING, GAS FROM
B 6309 1577* SHALES, ABST., EXPLORATION STRATEGY FOR UNCONVENTIONAL,
B 6209 1761* SHALES, ABST., PALEOECOLOGY AND, POROSITY TRENDS OF
B 6309 1576* SHALES, ABST., SUGGESTED ORIGIN FOR, GAS IN EASTERN, DEVONIAN
B 6309 1579* SHALES, ABST., THEORETICAL AND EXPERIMENTAL ANALYSES OF,
B 6203 521* SHALES AND DYNAMICS OF HYDROCARBON EXPULSION IN GULF COAST
B 6105 836* SHALES AND OIL-SPILL POTENTIAL ON ATLANTIC OUTER CONTINENTAL
S 6 93† SHALES AND RELATED ROCKS, APPLICABILITY OF THE,
B 6009 1425† SHALES AT HAMILTON DOME, BIGHORN BASIN, WYOMING,
B 6412 2051 SHALES, ATLANTIC OCEAN, DEEP SEA DRILLING PROJECT, BLACK
B 6306 918† SHALES, CALCULATED, COMPACTION, STATES OF THICK ABNORMALLY,
S 6 100 SHALES, CASE HISTORIES, RUBIDIUM-STRONTIUM
B 6007 1125 SHALES, DENSITY VERSUS, DEPTH RELATION FOR
B 6309 1592* SHALES, ERIE COUNTY PENNSYLVANIA, ABST., PRESSURE, CORING OF,
B 6309 1504† SHALES FROM FORMATION-DENSITY LOGS, DETERMINATION OF,
B 6303 503* SHALES FROM SOUTH ATLANTIC, DSDP LEGS 36, 40--SEDIMENTOLOGIC
M 28 63 SHALES, GENERAL, CLAYS AND
B 6303 542* SHALES IN ATLANTIC BASINS, ABST., PALEOENVIRONMENT AND
B 6412 2051† SHALES IN ATLANTIC BASINS, PALEOENVIRONMENT AND PETROLEUM
B 6303 504* SHALES IN NORTHERN APPALACHIAN NAPPE STRUCTURES AROUND
B 6309 1580* SHALES IN SOUTHERN WEST VIRGINIA, ABST., ELEMENTAL ANALYSES OF,
B 6303 462* SHALES IN SUBSURFACE PENNSYLVANIA--STRATIGRAPHY AND NATURAL
B 6412 2179† SHALES, MECHANICAL VERSUS, THERMAL CAUSE OF, ABNORMALLY
B 6412 2179 SHALES, THERMAL, HYPOTHESIS FOR, ABNORMALLY HIGH, PORE,
B 6309 1577* SHALES--EFFECTS OF TEMPERATURE AND ATMOSPHERE, ABST.,
B 6004 651 SHALEY, J. F., COMPUTER-GENERATED SCALE FUNCTIONS FOR
B 6405 769* SHALE—AN OVERVIEW, ABST.
B 6103 376† **SHALLOW CARBONATE-BANK-MARGIN GROWTH AND STRUCTURE, LITTLE**
B 6004 687* **SHALLOW CONTINENTAL SHELF OFF NORTHWESTERN FLORIDA AND**
B 6002 258† SHALLOW CARBONATE PLATFORM OF ISRAEL, RUDISTID, FRINGING,
B 6105 82* **SHALLOW FRAMEWORK OF INNER CONTINENTAL SHELF OF CAPE FEAR**
B 6403 316† SHALLOW GAS ACCUMULATIONS, NORTH-CENTRAL MONTANA, COASTAL
B 6004 714* SHALLOW GAS ACCUMULATIONS--THEIR ORIGIN AND SIGNIFICANCE FOR
S 7 271 SHALLOW GRABEN FAULTS, BOTTOM, DEFORMATION
B 6105 814* SHALLOW HOLES, ATLANTIC MARGIN CORING PROJECT—PRELIMINARY
B 6407 969† SHALLOW, LOW-PERMEABILITY RESERVOIRS OF NORTHERN GREAT
B 6405 696* SHALLOW OIL SHALE DEPOSITS OF SOUTHERN UINTA BASIN, UTAH, ABST.
B 6004 647* SHALLOW PETROLEUM GENERATION. ATTEMPT TO RESOLVE PARADOX,
M 24 96 SHALLOW RIDGE FIELD, WYOMING
B 6105 808* **SHALLOW SEDIMENTARY FRAMEWORK OF GEORGES BANK**
B 6305 844* SHALLOW SEISMIC REFLECTION IN APPALACHIAN BASIN WITH
B 6004 670* SHALLOW STRUCTURE OF NORTHERN SANT ROSA-CORTES RIDGE,
B 6303 442* SHALLOW SUBSURFACE DOLOMITIZATION OF SUBTIDALLY DEPOSITED
B 6004 661* SHALLOW SUBTIDAL ENVIRONMENT, ABST., TRACE-, FOSSIL ASSEMBLAGE
B 6004 712* SHALLOW SUBTIDAL ENVIRONMENT--TIME AND PLACE FOR TAXONOMIC
S 3 40 SHALLOW URANIUM MOBILIZATION PROCESSES
S 3 56 SHALLOW URANIUM-BEARING FLUID
B 6105 840* SHALLOW-MARINE CARBONATE ROCKS, ABST., POROSITY PRESERVATION
B 6303 460* SHALLOW-MARINE ENVIRONMENTS, ABST., HUMMOCKY
B 6405 701* SHALLOW-MARINE TRACE ASSEMBLAGES IN A CAMBRIAN TIDAL SAND
B 6409 1301 SHALLOW-POOL TEST
B 6303 542* SHALLOW-SUBMARINE SEISMIC STRATIGRAPHY, MISSISSIPPI RIVER
B 6303 462* SHALLOW-WATER CARBONATE AND EVAPORITE SEDIMENTATION
M 26 202 SHALLOW-WATER CARBONATE BUILDUP, SEISMIC EXAMPLE, CANADA,
B 6004 686* SHALLOW-WATER EPERIC CARBONATE SEQUENCES, ABST., SIMILARITIES
B 6303 490* SHALLOW-WATER UPPER JURASSIC ROCKS DREDGED FROM BERING SE
B 6308 1174 SHALLOWER POOL TEST
B 6108 1126 SHALLOWER POOL TEST
B 6008 1159 SHALLOWER POOL TEST
B 6208 1281 SHALLOWER POOL TEST
B 6102 295* SHALLOWS, SABKHAS, SHELVES, AND SUMPS, ABST., ENVIRONMENTS OF,
B 6304 649 SHALY DOLOMITE, VIRGINIA
M 25 269 SHANGHAI BASINS, KOREA, WEST, YELLOW SEA, NORTH KIANGSU, AND
B 6411 2040 SHANMUGAM, G., REVIEW OF PETROLEUM DEVELOPMENT GEOLOGY, B
B 6105 828 SHANMUGAM, GANAPATHY, DEEP-WATER SEDIMENTATION AND
B 6303 525 SHANMUGAM, GANAPATHY, QUANTITATIVE BASIN ANALYSIS AND
B 6303 525 SHANMUGAM, GANAPATHY, RHYTHMS IN DEEP-MARINE TURBIDITIC
B 6211 2194 SHANNON GROUP, IRELAND, WESTERN
M 24 141 SHANNON HYGIENE DISCOVERY WELLS, COLORADO, SUSSEX-
B 6202 184 SHANNON INTERVAL, WYOMING
B 6408 1235 SHANNON SAND SEQUENCE, WYOMING
B 6405 793* SHANNON SANDSTONE, UPPER CRETACEOUS, OFFSHORE BAR FACIES
B 6202 185 SHANNON SANDSTONE, WYOMING

B 6405 783* SHANNON-EQUIVALENT SANDSTONE IN NORTHERN BLACK HILLS,
B 6303 526 SHARIEF, FAROOQ A., REGIONAL ASPECT OF SEDIMENT DISTRIBUTION
B 6010 1871 SHARJAH, 1975, CRYSTAL OIL CO., RESERVE OIL AND GAS CO., AND,
B 6110 1800 SHARJAH, 1976, RESERVE OIL AND GAS CO., PETROCONSULTANTS ET AL,
B 6210 2117 SHARJAH, 1977, PETROCONSULTANTS S.A., ON DEVELOPMENTS IN
B 6310 1889 SHARJAH, 1978, AMOCO INTERNATIONAL OIL CO., FORMAN
B 6411 1840 SHARJAH, 1979, AMOCO INTERNATIONAL OIL CO. AND, FORMAN
B 6310 1892 SHARJAH, DEVELOPMENTS, 1977-1978
B 6210 2120 SHARJAH, PRODUCTION, 1976-1977
B 6411 1843 SHARJAH, PRODUCTION, 1978-1979
B 6202 187 SHARON SPRINGS MEMBER, WYOMING
B 6306 989 SHARP, BEAR, TED L., REVIEW OF, COASTAL SOUTHERN, CALIFORNIA, BY
B 6105 829 SHARP, J. M., JR., ENERGY-TRANSPORT PROCESSES IN SEDIMENTARY
B 6405 782 SHARP, JOHN M., JR., TEMPERATURE AND PRESSURE RELATIONS IN
B 6211 2361 SHARP, W., EARLY MESOZOIC PALEOTECTONIC-PALEOGEOGRAPHIC
S 6 264 SHASTA BALLY BATHOLITH, CALIFORNIA, CRETACEOUS
M 29 319 SHAUB, F. JEANNE, ANATOMY OF THE MEXICAN RIDGES, SOUTHWESTERN
B 6303 544 SHAUB, F. JEANNE, GEOLOGIC HISTORY OF DEEP GULF ON MEXICO
B 6405 682 SHAUB, F. JEANNE, GEOLOGIC HISTORY OF DEEP SOUTHEASTERN GULF
B 6303 426 SHAUB, F. JEANNE, GRAVITY--SLIDE ORIGIN FOR MEXICAN RIDGES
S 7 43 SHAUB, F. JEANNE, OCCURRENCE AND EVOLUTION OF SALT IN DEEP
B 6303 526 SHAUB, F. JEANNE, POST-MIDDLE CRETACEOUS SEISMIC STRATIGRAPHY
B 6312 2204 SHAUB F. JEANNE, SEISMIC EVIDENCE FOR WIDESPREAD POSSIBLE GAS
B 6203 501 SHAUB, F. JEANNE, STRUCTURE AND EARLY GEOLOGIC HISTORY OF DEEP
B 6404 501 SHAUGHNESSY, D. R., III, GEOLOGY OF NEW ENGLAND PASSIVE MARGIN
B 6004 684 SHAVER, ROBERT H, SEARCH FOR SILURANIAN REEF MODEL--GREA
S 5 93 SHAVER, ROBERT H. SYNCHRONIZATION OF DEPOSITION: SILURIAN
B 6408 1283 SHAVER, ROBERT H., STRUCTURAL INTERPRETATION OF BURIED
B 6207 1234 SHAW, H. E., ROLE OF NUCLEAR ENERGY IN FUTURE OF CIRCUM-PACIFIC,
B 6107 1030 SHAWNEE GROUP, NEBRASKA
M 26 39 SHEAR VELOCITY, LABORATORY, MEASUREMENTS, SATURATION EFFECTS
B 6211 2358* SHEAR WAVES TO STRATIGRAPHIC INTERPRETATION, ABST.,
B 6009 1434† SHEAR ZONES INSIDE GULF COAST SALT STOCK HELP TO DELINEATE
B 6107 1090† SHEAR ZONES INSIDE GULF COAST SALT STOCKS HELP TO DELINEATE
B 6405 771 SHEARER, CHARLES, DEEP TERRESTRIAL HEAT-FLOW STUDIES IN
B 6405 782 SHEEHAN, PETER M., PALEOGEOGRAPHY AND MARINE COMMUNITIES OF
M 25 108 SHEEP CREEK, AND DAWSON COALFIELDS, CANADA, LIARD RIVER
B 6108 1378* SHEEP PASS, CRETACEOUS TO EOCENE, AND ASSOCIATED CLOSED-BASIN
B 6106 861† SHEET, BALTIMORE CANYON TROUGH AREA, THICKNESS AND AGE OF,
B 6106 861 SHEET ON MIDDLE ATLANTIC OUTER CONTINENTAL SHELF, SURFICIAL,
B 6007 1010 SHEET SANDSTONE, NEW MEXICO, WESTWATER CANYON MEMBER,
B 6007 1009† SHEET SANDSTONE, RESERVOIR, GEOMETRY OF A, FLUVIAL
B 6007 1009† SHEET SANDSTONE, RESERVOIR, GEOMETRY OF A, FLUVIAL
B 6007 1009 SHEET SANDSTONE, USAGE OF TERM
B 6205 763 SHEET-DRAPE FACIES
M 26 178 SHEET-DRAPE FACIES, EXAMPLE, GULF OF MEXICO
M 26 176 SHEET-DRAPE FACIES, SEISMIC, INTERPRETATION
B 6205 892* SHEET-FLOOD DEPOSITION AND DEPOSITS, ABST.
B 6007 1017 SHEETS OF FLUVIAL SANDSTONE, ORIGIN OF
B 6303 526 SHEFFIELD, F. C., FUTURE EXPLORATION OPPORTUNITIES IN GULF OF
B 6002 322 SHELBY, JERRY M., REPORT--OKLAHOMA AND TEXAS PANHANDLES,
B 6311 2119 SHELBY, JERRY M., UPPER MORROW FAN-DELTA DEPOSITS OF
B 6207 1234 SHELDON, RICHARD P., PHOSPHOGENIC PROVINCES OF CIRCUM-PACIFIC
B 6004 679* SHELF, ABST., BIOGENIC SEDIMENTARY STRUCTURES OF SOUTH, TEXA,
B 6009 1614* SHELF, ABST., BIOSTRATIGRAPHY O, PLEISTOCENE DEPOCENTER--TEXAS-,
B 6004 713* SHELF, ABST., CORAL, REEFS AND, BANKS OF, TEXAS, OUTER,
B 6209 1759* SHELF, ABST., DISTRIBUTION OF LIVING, BENTHONIC, FORAMINIFERA AS
B 6303 419* SHELF, ABST., EXPLORATION RESULTS AND POTENTIAL OF, NORWEGIAN,
B 6405 675* SHELF, ABST., FRACTIONAL PRESERVATION OF TRANSGRESSIVE COASTAL,
B 6105 841* SHELF, ABST., GEOLOGIC EFFECTS OF, OCEAN, DUMPING ON, NEW YORK
B 6004 649* SHELF, ABST., LATE, PLEISTOCENE-, HOLOCENE EVOLUTION OF SOUTH,
B 6105 826* SHELF, ABST., PETROLEUM FRONTIER AREAS ON, ATLANTIC, OUTER,
B 6405 733* SHELF, ABST., PROCESSES CONTROLLING CHARACTERISTICS OF
B 6004 722* SHELF, ABST., SEDIMENT-, DISPERSAL, PATTERNS OF SOUTH, TEXAS,
B 6004 702* SHELF, ABST., SEDIMENTARY, FACIES AND STRUCTURES OF, BERING,
B 6105 836* SHELF, ABST., SEDIMENTARY MODEL PREDICTION OF HIGH-, PRESSURE,
B 6004 675* SHELF, ABST., STRATIGRAPHY OF, NOVA SCOTIA
B 6109 1545* SHELF, ABST., STRUCTURE AND STRATIGRAPHY OF LATE, QUATERNARY
B 6109 1547* SHELF, ABST., STRUCTURE AND STRATIGRAPHY OF SMALL, PLIOCENE-,
B 6307 1059† SHELF AND BASIN STRATA, CORNWALLIS AND GRIFFITH ISLANDS,
B 6203 557* SHELF AND OFFSHELF BUILDUPS, MIDDLE ORDOVICIAN, VIRGINIA, ABST.,
S 7 267† SHELF AND SLOPE OFF LOUISIANA AND TEXAS: GEOTECHNICAL ASPECTS,
B 6303 481* SHELF AND SLOPE--MULTIVARIATE STUDY, ABT ST., MEGASCALE
B 6209 1758* SHELF AND SOURCE OF TURBID BOTTOM WATER LAYER, ABST., WATER
M 29 65† SHELF AND UPPER CONTINENTAL SLOPE BETWEEN CAPE HATTERAS AND
S 7 243† SHELF AND UPPER CONTINENTAL SLOPE, LATE, PLEISTOCENE, GEOLOGIC
B 6405 687* SHELF AND UPPER CONTINENTAL SLOPE, NORTHWEST GULF OF MEXICO,
M 29 213 SHELF, ARAFURA SEA SECTION, AUSTRALIA
M 29 212 SHELF, BANDA SEA, AUSTRALIA
S 4 321 SHELF, BENTHIC FORAMINIFERA, ASSEMBLAGES, BRAZIL, CONTINENTAL
S 7 254 SHELF BREAK PROFILE, TEXAS, SHELF-SLOPE
S 4 317 SHELF, CARBONATE SEDIMENTATION, BRAZIL, CONTINENTAL
B 6007 1114 **SHELF, CDP MULTICHANNEL REFLECTION PROFILE, ATLANTIC, OUTER,**
B 6004 672* SHELF DEPOSITION, SHUBLIK FORMATION, PRUDHOE BAY AREA,
B 6308 1205 SHELF, DRILLING, 1978, ATLANTIC, OFFSHORE, CONTINENTAL
B 6004 697* SHELF EDGE AND ITS CONTRIBUTION TO AUTOKINETI BASINAL
M 29 79 SHELF EDGE, ATLANTIC COAST, BALTIMORE CANYON
M 29 72 SHELF EDGE, ATLANTIC COAST, CRETACEOUS
B 6408 1224 SHELF ENVIRONMENTS AND PROCESSES, REVIEW, CONTINENTAL-
B 6408 1225 SHELF ENVIRONMENTS, CLASSIFICATION, OFFSHORE REGIMES AND
B 6007 1112† SHELF, EVIDENCE OF, POST-PLEISTOCENE, FAULTS ON, NEW JERSEY,
M 26 169 SHELF FACIES, SAN JUAN BASIN, CRETACEOUS
M 26 169 SHELF FACIES, SEISMIC, INTERPRETATION
B 6004 714* SHELF GEOTHERMAL GRADIENTS, ABST., SCOTIAN
B 6109 1542* SHELF, GULF OF MEXICO, ABST., INTERNAL, WAVES AND, NEPHELOID
B 6303 527* SHELF, GULF OF MEXICO-TIME-SEQUENCE STUDY, ABST., SURFACE,
M 29 66 SHELF INDEX MAP, ATLANTIC COAST
B 6305 828* SHELF MARGIN ALONG OVERTHRUST BELT FROM MONTANA TO
B 6109 1557* SHELF MARGIN IN NORTHERN MEXICO, ABST., LOWER, CRETACEOUS
B 6009 1617* SHELF MARGIN IN NORTHERN MEXICO AND SOUTH TEXAS, ABST.,
M 26 435 SHELF MARGIN TRAPS, SEISMIC, MODELS, CARBONATE
B 6105 823* SHELF MARGINS, WESTERN UNITED STATES, ABST., MISSISSIPPIAN,
M 29 213 SHELF, MELVILLE ISLAND SECTION, AUSTRALIA
S 4 329† SHELF MODEL, ST. CROIX, U.S VIRGIN ISLANDS, MIOCENE, KINGSHILL
M 26 231 SHELF MODELS, DEPOSITIONAL SYSTEMS
B 6004 715* SHELF NORTH OF 62 DEGREES NORTH, ABST., GEOLOGY OF, NORWEGIAN,
B 6303 449* SHELF, NORTHERN BERING SEA, ABST., VIDEOTAPES OF ACTIVE PRIMARY
B 6012 2176* SHELF, NORTHERN GULF OF ALASKA, ABST., DISPERSAL, DISTRIBUTION,
B 6105 830* SHELF, NORTHWEST GULF OF MEXICO, ABST., SUSPENDED-SEDIMENT
S 4 315 SHELF, OCEANOGRAPHY, BRAZIL, CONTINENTAL
B 6209 1760* SHELF OF SOUTH TEXAS, ABST., LIVING, SHELLED, MICROZOOPLANKTON,
B 6212 2412† SHELF OFF MALASPINA GLACIER, GULF OF ALASKA, HOLOCENE, SLUMP
B 6004 705* SHELF OFF NEW JERSEY INTERPRETED FROM R- AND Q-MODE
B 6011 2019† SHELF OFF NORTHERN FLORIDA, NEOGENE SEDIMENTS OF, ATLANTIC,
B 6004 687* SHELF OFF NORTHWESTERN FLORIDA AND ALABAMA, ABST.,
S 4 313† SHELF OFF SOUTHERN BRAZIL, WITH SPECIAL REFERENCE TO BENTHIC
B 6406 945† SHELF, PETROGRAPHY AND GEOLOGIC SIGNIFICANCE OF UPPER,
M 29 209 SHELF, PHYSIOGRAPHY, AUSTRALIA
S 4 17† SHELF REEFS, SOUTHWESTERN PUERTO RICO
M 29 214 SHELF, REFLECTION STUDIES, AUSTRALIA
M 29 210 SHELF, REFRACTION STUDIES, AUSTRALIA
B 6408 1224 SHELF REGIMES
B 6105 814* SHELF REGION—INITIAL SURVEY, ABST., HYDROCARBONS IN SURFACE
B 6006 932 SHELF, SECTION, NEW JERSEY, CONTINENTAL
B 6203 542* SHELF SEDIMENT AND LOCAL FLOW PHENOMENA—IN SITU
B 6105 789* **SHELF-SEDIMENT DYNAMICS AND SOLID-WASTE DISPOSAL**
B 6205 756 SHELF SEISMIC FACIES
M 29 209† SHELF, SEISMIC, REFRACTION AND, REFLECTION STUDIES IN THE,
M 29 258 SHELF, SEISMIC SECTION, GUATEMALA
B 6007 1113 SHELF, SHALLOW FAULTS, ATLANTIC, OUTER, CONTINENTAL
B 6004 724* SHELF, SILURIAN MCKENZIE FORMATION, WEST VIRGINIA, ABST.,
B 6002 161† SHELF SOUTH OF ST LAWRENCE ISLAND, STRUCTURE AND EVOLUTION
B 6303 301† SHELF SOUTHWEST OF KODIAK ISLAND, ALASKA, STRUCTURE AND
S 7 269 SHELF STABILITY, DEFORMATIONAL FEATURES
S 7 8 SHELF SUBPROVINCES MAP, GULF OF MEXICO
B 6204 633† SHELF, SURFACE SEDIMENTARY UNITS OF NORTHERN, GULF OF, ALASKA,
B 6106 861 SHELF, SURFICIAL, SAND, SHEET ON, MIDDLE, ATLANTIC, OUTER,
B 6205 880* SHELF TAR SAND DEPOSITS, CENTRAL AND SOUTHERN UTAH, ABST.,
M 29 211 SHELF, VELOCITY SOLUTIONS, AUSTRALIA
B 6207 1235* SHELF ZONES, ABST., PROBLEMS OF STRUCTURE AND DEVELOPMENT OF
M 25 231† SHELF--RESULTS OF TEN YEARS OF EXPLORATION, AUSTRALIAN
S 7 262 SHELF-EDGE OUTBUILDING, TEXAS
S 4 23 SHELF-EDGE REEF, PUERTO RICO
B 6205 759 SHELF-MARGIN AND PROGRADED-SLOPE SEISMIC FACIES
M 26 199 SHELF-MARGIN BUILDUP, SEISMIC EXAMPLE, TEXAS, WASHITA, REEF
M 26 191 SHELF-MARGIN CARBONATE, SEISMIC EXAMPLE, AFRICA, WEST
B 6401 88† SHELF-MARGIN EVOLUTION, PALO DURO BASIN, TEXAS,
M 26 172 SHELF-MARGIN FACIES, SEISMIC, INTERPRETATION
M 26 192 SHELF-MARGIN REEF, SEISMIC EXAMPLE, GULF OF PAPUA
B 6203 528* SHELF-MARGIN REEFS, NORTHWESTERN OHIO, ABST., SUBAERIAL
M 26 269 SHELF-PLATFORM FLUVIAL-DELTAIC SYSTEMS
B 6105 840* SHELF-SLOPE CARBONATE DEPOSITS IN NORTH-CENTRAL NEVADA, ABST.,
S 7 282 SHELF-SLOPE, STABILITY EVALUATION, ENGINEERING ANALYSIS
S 7 279 SHELF-SLOPE, STABILITY EVALUATION, PROCEDURES
B 6211 2355* SHELF. BY TITLE ONLY, GEOENVIRONMENTAL STUDIES OF, PACIFIC,
M 26 231 SHELFS, SEISMIC, DEPOSITIONAL SYSTEMS
B 6303 306 SHELIKOF BASIN, ALASKA, COOK-
B 6104 608 SHELIKOF FORMATION, ALASKA
B 6203 374 SHELIKOF STRAIT, ALASKA
B 6303 493* SHELL BORINGS IN UPPER MIOCENE-PLIOCENE TAMIAMI FORMATION,

Index

B 6104 579 SHELL LAMAR FIELD, VENEZUELA
B 6102 156 SHELL TAYLOR LEASE, ISOPACH MAP, CALIFORNIA
B 6201 30 SHELL-BP, NIGER DELTA, FIELDS
B 6209 1760* SHELLED MICROZOOPLANKTON, RADIOLARIANS, FORAMINIFERS, AND
B 6004 689* SHELTER, AVELLA, PENNSYLVANIA, ABST., COMPARISON OF DIAGENETIC
M 24 239 SHELTON, J. F., JR., TINSLEY OIL FIELD, YAZOO COUNTY, MISSISSIPPI
B 6408 1272 SHELTON, JOHN W., REVIEW OF SALT DOMES OF GULF REGION, UNITED
B 6203 463 SHELTON, JOHN W., SECONDARY FERROAN DOLOMITE RHOMBS IN OIL
B 6103 360 SHELTON, JOHN W., URANIUM POTENTIAL OF PERMIAN AND
B 6004 733* SHELVES AND BASINS, ABST., CARBONATE, MOUNDS FROM
B 6004 678* SHELVES AND IN DEEP SEA, ABST., CARBONATE MASS BALANCE--,
B 6102 295* SHELVES, AND SUMPS, ABST., ENVIRONMENTS OF, EVAPORITI
M 25 480 SHELVES, MINERALIZATION TYPICAL OF
M 25 480 SHELVES, STRUCTURAL CLASSIFICATION
M 29 355 SHEPARD CANYON, VIRGIN ISLANDS, ST. CROIX
B 6004 721 SHEPARD, F. P., CURRENTS IN SUBMARINE CANYON AND THEIR
B 6004 712 SHEPARD, F. P., SLOPE VALLEYS AND SEDIMENTARY STRUCTURES OFF
B 6208 1383 SHEPARD FIELD, SAN JACINTO COUNTY, TEXAS
S 8 1 **SHEPARD, FRANCIS P., CURRENTS IN SUBMARINE CANYONS AND**
B 6203 561 SHEPARD, FRANCIS P., CURRENTS IN SUBMARINE CANYONS AND OTHER
B 6405 649 SHEPARD, NEIL F. MARSHALL, PATRICK A. MCLOUGHLIN, AND GARY G.
B 6305 841 SHEPHERD, RUSSELL G., HYDROGEOLOGIC SIGNIFICANCE OF OGALLALA
B 6403 317 SHERARD FIELD, MONTANA
B 6203 561 SHERBORNE, J. E., JR., MAJOR URANIUM DISCOVERY IN FRONTIER
B 6304 621 SHERBORNE, J. E., JR., MAJOR URANIUM DISCOVERY IN VOLCANICLASTIC
B 6405 782 SHERBORNE, J. E., JR., URANIUM DEPOSITS OF PART OF CENTRAL GREAT
B 6303 526 SHERER, F. C., EXPLORATION IN EAST MALAYSIA OVER PAST DECADE,
B 6007 1112 SHERIDAN, R. E., EVIDENCE OF POST-PLEISTOCENE FAULTS ON NEW
B 6004 649 SHERIDAN, R. E., PRELIMINARY RESULTS FROM LEG 44 DEEP SEA
B 6303 527 SHERIDAN, R. E., RECENT ADVANCES IN PASSIVE-MARGIN RESEARCH,
M 29 177 SHERIDAN, R. E., STRUCTURE AND STRATIGRAPHY OF THE BLAKE
B 6105 829 SHERIDAN, ROBERT E., BLAKE PLATEAU-BAHAMA REGION OF ATLANTIC
B 6303 445 SHERIDAN, ROBERT E., MIOCENE CARBONATE GRAVITY FLOW IN
B 6410 1723 SHERIFF, BACKUS, MILO M., REVIEW OF A FIRST COURSE IN,
M 26 301 SHERIFF, R. E., APPLICATION OF AMPLITUDE, FREQUENCY, AND OTHER
B 6105 829 SHERIFF, R. E., COLOR DISPLAYS OF COMPLEX SEISMIC TRACE, ABST.,
B 6203 562 SHERIFF, R. E., GEOLOGIC INTERPRETATION OF SEISMIC-ATTRIBUTE
B 6304 699 SHERIFF, R. E., HIGH-RESOLUTION LANDSAT FOR GEOLOGIC STUDIES,
B 6405 698 SHERIFF, R. E., HOW GEOLOGIC OBJECTIVES SHOULD DETERMINE SEISMIC
B 6004 528 SHERIFF, R. E., INFERRING STRATIGRAPHY FROM SEISMIC DATA
B 6304 701 **SHERIFF, R. E., LOCATING STRATIGRAPHIC FEATURES AND**
M 26 3 SHERIFF, ROBERT E., LIMITATIONS ON RESOLUTION OF SEISMIC
M 25 402 SHERLOCK BAY, NICKEL DEPOSIT, WESTERN AUSTRALIA
B 6405 764 SHERRILL, FRANCES G., SEISMIC MODELING OF PENNSYLVANIAN
M 25 427 SHERWIN IRONSTONE, IRON ORE, AUSTRALIA, NORTHERN TERRITORY
B 6303 527 SHERWOOD, JOHN, DEPTH MIGRATION OF SEISMIC DATA, ABST.
B 6012 2183 SHERWOOD, JOHN, REGIONAL VELOCITY REDUCTION, ABST.
B 6211 2362 SHERWOOD, JOHN W., EQUALIZING STACKING VELOCITIES OF DIPPING
M 30 549 SHETERAT SHALE, LIBYA, SIRTE BASIN
M 30 105 SHETLAND GROUP, CRETACEOUS, STATFJORD FIELD
B 6203 426 SHETLAND PLATFORM, EUROPE
B 6311 2001 SHETLAND PLATFORM, NORTH SEA
B 6203 429 SHETLAND-FAEROE TROUGH, BRITAIN, WEST
B 6408 1247 SHETLANDS PLATFORM, UNITED KINGDOM
B 6405 695 SHETTEL, DON L., JR., APPLICATION OF REGIONAL GEOCHEMICAL DATA
B 6207 1151 SHIBAOKA, M., HYDROCARBON GENERATION IN GIPPSLAND BASIN,
S 6 143 SHIBATA, KEN, ISOTOPIC AGES AND STRATIGRAPHIC CONTROL OF
B 6105 708 **SHIDELER, GERALD L., LATE HOLOCENE SEDIMENTARY PROVINCES, SOUTH**
B 6004 722 SHIDELER, GERALD L., SEDIMENT-DISPERSAL PATTERNS OF SOUTH
B 6303 527 SHIDELER, GERALD L., SURFACE TURBIDITY AND HYDROGRAPHI
B 6105 830 SHIDELER, GERALD L., TEXAS OUTER CONTINENTAL SHELF, NORTHWEST
B 6209 1766 SHIER, DANIEL E., SAMPLE ORDERING—NEW STATISTICAL TECHNIQUE
B 6111 1962 SHIH, TAI-CHANG, NORTHEASTERN EXTENSION OF SIGSBEE SCARP, GULF
B 6203 531* SHIKOKU BASIN AND NORTHEASTERN PHILIPPINE SEA, ABST., LEG 58,
M 25 466 SHILO, N. A., GEOLOGIC FRAMEWORK OF METALLIC MINERAL DEPOSITS
B 6207 1235 SHILO, N. A., PROBLEMS OF STRUCTURE AND DEVELOPMENT OF SHELF
M 25 451 **SHIMAZAKI, YOSHIHIKO, GEOLOGIC ASPECTS OF URANIUM RESOURCES OF**
B 6110 1901 SHIMOYAMA, A., CLAY, MINERALS AND PETROLEUM-FORMING
M 25 98 SHIMOYAMA, TOSHIO, INFLUENCE OF TEMPERATURE ON COALIFICATION
S 3 124 SHINARUMP CONGLOMERATE, UTAH, MONUMENT VALLEY
B 6212 2512 SHINDO, SHIZUO, GROUNDWATER IN VOLCANIC ISLANDS, ABST.
S 3 90 SHINKOLOBWE VEIN DISTRICT, CONGO
B 6205 892 SHINN, E. A., DEPOSITION AND DIAGENESIS OF HOLOCENE REEFS, BELIZE,
B 6004 683 SHINN, EUGENE A., AUTOPSY OF A DEAD CORAL REEF, ABST.
B 6405 783 SHINN, EUGENE A., EXPERIMENTAL COMPACTION OF LIME SEDIMENT,
B 6104 519 SHINN, EUGENE A., PLEISTOCENE BARRIER BAR SEAWARD OF OOID
B 6004 677 SHINN, EUGENE A., RECENT AND RELICT TOPOGRAPHY OF BOO BEE
B 6303 528 SHINN, EUGENE A., THREE-DIMENSIONAL ASPECTS OF BELIZE PATCH
B 6208 1383 SHIP CHANNEL FIELD, CHAMBERS COUNTY, TEXAS
B 6104 587 SHIP SHOAL BLOCK FIELD, LOUISIANA
B 6105 822* SHIPBOARD GEOTECHNICAL RESULTS FROM U.S. GEOLOGICAL SURVEYS
B 6312 2204 SHIPLEY, THOMAS H., SEISMIC EVIDENCE FOR WIDESPREAD POSSIBLE GAS
B 6205 792 SHIPLEY, THOMAS H., SEISMIC STRATIGRAPHY AND GEOLOGIC HISTORY
B 6105 773 **SHIPLEY, THOMAS H., SEISMIC STRATIGRAPHY, SEDIMENTARY HISTORY,**
B 6405 783 SHIPP, R. CRAIG, SEDIMENTARY FACIES RELATIONS AND INFERRED
B 6208 1382 SHIPWRECK FIELD, GALVESTON AREA, TEXAS
B 6305 838* SHIRLEY BASINS--FUTURE PETROLEUM POTENTIAL, ABST., LARAMIE-,
B 6102 193 SHIRSHOV RIDGE
B 6311 2071 SHIRSHOV RIDGE, BERING SEA
B 6112 2105 SHIVIYACU FIELD, PERU
S 3 32 SHIVWITS PLATEAU, ARIZONA, TOROWEAP SANDSTONE, PARASHOUT
B 6303 462* SHOAL, ABST., ANATOMY AND GROWTH HISTORY OF, HOLOCENE, OOID
B 6104 519† SHOAL COMPLEX NEAR MIAMI, FLORIDA, PLEISTOCENE, BARRIER, BAR
B 6004 685* SHOAL DEPOSITS IN MEDIAL ORDOVICIAN OF CENTRAL NEW YORK,
B 6201 121† SHOAL REEF COMPLEX, UPPER CRETACEOUS AND PALEOCENE, IN SOUTH
B 6004 679* SHOAL TO ITS PHYSICAL ENVIRONMENT AND HOLOCENE SEA-LEVEL RISE,
B 6109 1554* SHOAL-WATER CARBONATE COMPLEXES--SUBSURFACE
B 6004 725* SHOALING SEQUENCE IN DOGGER, MIDDLE JURASSIC, OF CENTRAL HIGH
B 6405 802 SHOEMAKER, EUGENE M., URANIUM MINERALIZATION IN HOPI BUTTES,
B 6105 765 **SHOEMAKER, J. W., GEOMETRY OF COAL BEDS IN SAN JUAN BASIN, NEW**
B 6309 1596 SHOEMAKER, PHILIP W., CHARACTERISTICS OF LOWER VICKSBURG
B 6211 2362 SHOHAM, YORAM, MAGNETOTELLURIC GEOPHYSICAL EXPLORATION
B 6103 361 SHOLOM ALECHEM STRUCTURE, OKLAHOMA
B 6303 536* SHOOFLY OOLITE OF SOUTHWESTERN IDAHO, ABST., NORMALLY AND
M 29 209 SHOR, G. G., JR., SEISMIC REFRACTION AND REFLECTION STUDIES IN THE
B 6406 868 **SHOR, GEORGE, ARC-CONTINENT COLLISION IN BANDA SEA REGION**
M 29 273 SHOR, GEORGE G., JR., CONTINENTAL MARGINS OF THE EASTERN GULF
B 6109 1543* SHORELINE CHANGES AND THEIR CAUSES, TEXAS GULF COAST, ABST.,
B 6109 1536* SHORELINE CHANGES BY ARCHEOLOGICAL DATING, GEORGIA COAST,
B 6004 663* SHORELINE CHANGES ON GEORGIA COAST, ABST., ARCHAELOGIC,
B 6004 720* SHORELINE EROSION IN TIDAL ESTUARY, ABST., SEDIMENTARY
B 6209 1768* SHORELINE IN SOUTH FLORIDA, ABST., HIGH-LEVEL, PLIO-, PLEISTOCENE
B 6207 1228* SHORELINE PROCESSES ON RESOURCE DEVELOPMENT, ABST., IMPACT OF
M 26 480 SHORELINE SANDSTONES, SEISMIC, MODELING
B 6104 492† SHORELINE SEQUENCES, CENTRAL FLORIDA, DIAGENESIS OF MIDDLE
B 6004 734* SHORELINES, ABST., DEPOSITION, DEFORMATION, AND, HEAVY, MINERAL
B 6109 1549* SHORELINES, ABST., PLIOCENE-, PLEISTOCENE PALEOGEOGRAPHY OF,
B 6209 1757* SHORELINES AND PALEOGEOGRAPHIC FEATURES, MISSISSIPPI GULF
B 6006 988† SHORT HISTORY OF NORTHERN CALIFORNIA GEOLOGICAL SOCIETY
B 6009 1624 SHORT, MICHAEL R., EARLY PENNSYLVANIAN DEPOSITIONAL SYSTEMS IN
B 6301 90 SHORTLAND SHALE, SCOTIAN BASIN
B 6409 1562* SHOULD WE PERMIT MISSISSIPPI-ATCHAFALAYA DIVERSION, ABST.
B 6209 1610 SHRIVER CHERT, MARYLAND
B 6012 2117† SHROPSHIRE, ENGLAND, ORIGIN AND EVOLUTION OF, BIOHERMS IN,
B 6303 512 SHUAIB, S M., CYCLIC SEDIMENTATION OF CRETACEOUS-PALEOCENE
B 6405 616 SHUAIBA FORMATION, MIDDLE EAST
B 6405 616 SHUAIBA LIMESTONE, MIDDLE EAST
M 24 26 SHUBLIK FORMATION, ALASKA
M 24 38 SHUBLIK FORMATION, ALASKA
B 6004 672* SHUBLIK FORMATION, PRUDHOE BAY AREA, ALASKA, ABST.,
M 30 303 SHUBLIK FORMATION, RESERVOIRS, PRUDHOE BAY
B 6303 305 SHUMAGIN BASIN, ALASKA
M 28 34 SHUMAGIN FORMATION, ALASKA, CRETACEOUS
B 6309 1523 SHUMAGIN FORMATION, ALASKA PENINSULA
B 6207 1228* SHUMAGIN KODIAK SHELF, ABST., PLATE-TECTONIC INTERPRETATION
B 6004 722 SHUMAKER, R. C., LINEAMENT AND STRUCTURAL ANALYSES OF
B 6303 502 SHUMAKER, R. C., RELATIONS OF GAS OCCURRENCE TO GEOLOGIC
B 6309 1588 SHUMAKER, ROBERT C., PALEOZOIC DISRUPTIVE DEFORMATION IN
B 6207 1235 SHUPE, JOHN W., GEOTHERMAL ENERGY IN HAWAII—PRESENT AND
B 6104 612 SHURBET, D. H., THE FILL IN MARFA BASIN, TEXAS
B 6406 942 SHURR, GEORGE W., EXPOSURES OF GREENHORN FORMATION AND
B 6405 783 SHURR, GEORGE W., GEOMETRY OF SHELF SANDSTONE BODIES IN
B 6205 891 SHURR, GEORGE W., NATURAL GAS RESOURCES IN SHALLOW,
B 6407 969 SHURR, GEORGE W., SHALLOW, LOW-PERMEABILITY RESERVOIRS OF
M 25 507 SHUSWAP TYPE OF LEAD-ZINC DEPOSIT, CANADA
B 6407 1016 SHUYAK FORMATION, ALASKA
B 6303 446 SHVEIMA, J. S., PETROLEUM POTENTIA OF SCOTIAN SHELF--CASE STUDY,
B 6305 836 SIBBETT, BRUCE S., GEOLOGY AND STRUCTURAL CONTROL OF
B 6309 1467 SIBERIA, ANADYR BASIN
B 6002 163 SIBERIA, KAMCHATKA-KORYAK EUGEOSYNCLINAL COMPLEX
B 6309 1467 SIBERIA, KHATYRKA BASIN
S 10 230 SIBERIA, KOLPASHEVO, GEOTHERMAL PROFILE, USSR
S 10 229 SIBERIA, OB RIVER AREA, HYDROTHERMAL EFFECTS, USSR
B 6002 163 SIBERIA, OKHOTSK-CHUKCHI VOLCANIC AND PLUTONIC ZONE
M 30 24 SIBERIA, SAMOTLOR FIELD, LANDSAT IMAGE
M 25 42 SIBERIA, SIKHOTE-ALIN COAST, GEOLOGIC MAP
M 25 42 SIBERIA, SIKHOTE-ALIN FAULT
M 30 225† SIBERIA, USSR, GEOLOGY AND, PETROLEUM FIELDS IN, PROTEROZOIC
B 6002 164 SIBERIA, WESTERN, GEOLOGIC MAP
B 6103 467* SIBERIAN AND NORTH AMERICAN PLATES, BERING STRAIT REGION,

B 6303 432* SIBERIAN BASIN, ABST., WEST
B 6401 22 SIBERIAN BASIN, U.S.S.R., WEST
B 6410 1691 SIBERIAN PLATFORM, OIL AND GAS FIELDS, USSR
M 29 228 SIBERUT, SUNDA ARC, REFLECTION PROFILE
B 6104 597 SIBIRCARA FIELD, VENEZUELA
B 6405 720 SIBLEY, DUNCAN F., AUTHIGENIC QUARTZ IN DEVONIAN BLACK SHALE,
B 6303 528 SIBLEY, DUNCAN F., DOLOMITIZATION OF PLIOCENE-PLEISTOCENE
B 6211 2329 SIBLEY, DUNCAN F., PRESSURE SOLUTION AND POROSITY REDUCTION IN
S 6 165 SIBRAVA, VLADIMIR, ISOTOPIC METHODS IN QUATERNARY GEOLOGY
B 6104 579 SIBUCARA FIELD, VENEZUELA
B 6210 1915 SICILY AND SARDINIA, DEVELOPMENTS, 1977
B 6310 1924 SICILY AND SARDINIA, DEVELOPMENTS, 1978
S 5 175 SICILY, CATTOLICA BASIN, ITALY,
S 5 175 SICILY, CIMMINA BASIN, ITALY,
B 6411 1965 SICILY, DEVELOPMENTS, 1979
B 6304 589 SICILY, GELA OIL FIELD
S 5 174 SICILY, MESSINIAN STRATIGRAPHY, ITALY,
S 5 175 SICILY, SALEMI BASIN, ITALY,
B 6307 1143 SIDDAYAO, DURKEE, E. F., REVIEW OF, OFF-SHORE PETROLEUM
B 6405 619 SIDE-LOOKING AIRBORNE RADAR, SLAR
M 26 356 SIDE-SCAN SONAR RECORD, ALASKA, CAPE, YAKATAGA
M 26 356 SIDE-SCAN SONAR SYSTEMS, RESOLUTION SYSTEMS
B 6012 2183* SIDE-SCAN SONAR--APPLICATIONS IN OIL EXPLORATION AND
M 28 124 SIDERITE, IN SANDSTONE, CEMENTS
M 28 123 SIDERITE, IN SANDSTONE, CEMENTS
M 28 163 SIDERITE NODULES AND PELLETS, COMPACTION, OF
B 6303 370† SIDERITE SPHERULITES OF SAN SEBASTIAN FORMATION, OLIGOCENE,
B 6405 714* SIDERITE TEXTURES IN CARDIUM FORMATION, FERRIER FIELD,
B 6407 1000 SIDI AMARA FAULT, MOROCCO
B 6407 996 SIDI MANSOUR MUDSTONE, MOROCCO
B 6004 722 SIDNER, B. R., GEOCHEMICAL CONTROLS ON VERTICAL DISTRIBUTION OF
S 7 289 SIDNER, BRUCE R., INTRASLOPE BASIN IN NORTHWEST GULF OF MEXICO
S 7 243 SIDNER, BRUCE R., LATE PLEISTOCENE GEOLOGIC HISTORY OF TEXAS
B 6310 1890 SIEBENS OIL AND GAS LTD., AGIP S.P.A., AND PETROCONSULTANT S.A., ON
B 6010 1873 SIEBENS OIL AND GAS LTD., ON DEVELOPMENTS IN SOUTH YEMEN, 1975
B 6110 1802 SIEBENS OIL AND GAS, LTD., ON DEVELOPMENTS IN SOUTH YEMEN, 1976
B 6210 2118 SIEBENS OIL GAS LTD., AGIP S.P.A., AND PETROCONSULTANTS S.A., ON
M 26 352 SIECK, HERMAN C., ANALYSIS OF HIGH RESOLUTION SEISMIC DATA
B 6004 723 SIEMERS, CHARLES T., PALEOENVIRONMENTS OF FRONTIER FORMATION,
B 6303 528 SIEMERS, CHARLES T., DESCRIPTION AND SEDIMENTOLOGY O
B 6209 1766 SIEMERS, CHARLES T., SUBMARINE-FAN DEPOSITION OF WOODBINE
B 6109 1548* SIERRA BALUARTES DIAPIR AND SIERRA DE PAJAROS AZULES,
B 6102 207 SIERRA DE CHIAPAS, MEXICO
B 6409 1550 SIERRA DE JUAREZ, CHIHUAHUA, MEXICO. STRUCTURE AND
B 6004 566† SIERRA DE OMOA, PRE-CRETACEOUS ROCKS OF NORTHWESTERN,
B 6102 269† SIERRA DE OMOA. DISCUSSION AND REPLY, PRE-CRETACEOUS ROCKS OF
B 6109 1548* SIERRA DE PAJAROS AZULES, COAHUILA, MEXICO, ABST., GEOLOGY OF,
B 6211 2218 SIERRA DE SALINAS SCHIST-TYPE BASEMENT, CALIFORNIA
B 6409 1554* SIERRA DE SANTA ROSA, NUEVO LEON, MEXICO, ABST., CALPIONELLIDS
B 6210 1849 SIERRA LEONE, DEVELOPMENTS, 1977
B 6310 1694 SIERRA LEONE, DEVELOPMENTS, 1978
B 6411 1790 SIERRA LEONE, DEVELOPMENTS, 1979
M 25 518 SIERRA MADRE DEL SUR METALLOGENETIC PROVINCE, MEXICO
B 6203 361 SIERRA MADRE FAULT, CALIFORNIA, CUCAMONGA-
M 25 515 SIERRA MADRE OCCIDENTAL METALLOGENETIC PROVINCE, MEXICO
M 25 518 SIERRA MADRE ORIENTAL METALLOGENETIC PROVINCE, MEXICO
B 6102 207 SIERRA MADRE ORIENTAL, MEXICO
B 6105 825* SIERRA MADRE ORIENTAL, MEXICO, ABST., LATE, JURASSIC
B 6303 430* SIERRA MADRE ORIENTAL, SOUTHWEST OF VERACRUZ PORT, ABST.,
B 6205 890* SIERRA NEVADA BATHOLITH, CALIFORNIA, ABST., FOLD SYSTEM IN
B 6211 2244 SIERRA NEVADA, CALIFORNIA
B 6203 345 SIERRA NEVADA, CALIFORNIA
B 6303 493* SIERRA NEVADA FOOTHILLS, CALIFORNIA, ABST., DIAGENETIC
B 6211 2349* SIERRA NEVADA FOOTHILLS, CALIFORNIA, ABST., PALEOGEOGRAPHIC
B 6403 361 SIERRA NEVADA, OROGENIC PHASES
B 6211 2361* SIERRA NEVADA REGION, ABST., EARLY, MESOZOIC
B 6111 1982 SIERRA PEAK, CALIFORNIA
B 6104 602 SIERRA RICA, NEW MEXICO
B 6104 613 SIERRA VIEJA MOUNTAINS, TEXAS
B 6211 2217 SIERRAN-SALINIAN BASEMENT, CALIFORNIA
B 6110 1749 SIERRE LEONE, DEVELOPMENTS, 1976
B 6303 529 SIESSER, WILLIAM G., TERTIARY SEA-LEVEL MOVEMENTS AROUND
B 6004 687* SIEVERING FORMATION, VIENNA, AUSTRIA, ABST., TRAILS FROM, THE
M 25 500 SIFTON BASIN, CANADA
M 25 501 SIFTON BASINS, CROSS SECTION, CANADA, SLOKO, SUSTUT, AND
B 6205 762 SIGMOID-PROGRADATIONAL FACIES
M 26 175 SIGMOID-PROGRADATIONAL FACIES, SEISMIC, INTERPRETATION
S 11 232 SIGMOILINA CELATA
S 11 270 SIGMOILINA CELATA
S 11 270 SIGMOILINA TENUIS
S 11 232 SIGMOILINA TENUIS
B 6104 579 SIGNAL-CENTRO FIELD, VENEZUELA
B 6405 678* SIGNATURE PROCESSING OF NORTH SEA AIR-GUN DATA, ABST.
M 26 221 SIGNATURES, CORRELATION, SEISMIC, REFLECTION
M 26 221 SIGNATURES, INTERPRETATION, SEISMIC, REFLECTION
B 6004 686* **SIGNIFICANCE OF CLAST SIZE, SHAPE, AND COMPOSITION IN ANCIENT AND**
B 6109 1552* SIGNIFICANCE OF FRESHWATER LIMESTONES IN MARINE CARBONATE
B 6303 525* **SIGNIFICANCE OF MIDDLE TERTIARY LARGE FORAMINIFERA COMMON TO**
M 26 122 SIGNIFICANCE, REFLECTION, PARAMETERS, GEOLOGIC
M 26 55 SIGNIFICANCE, SEISMIC ANALYSIS, DEPOSITIONAL SEQUENCE,
S 7 57 SIGSBEE ESCARPMENT, DISCUSSED, GULF OF MEXICO, CENTRAL
S 7 57 SIGSBEE ESCARPMENT, DISCUSSED, GULF OF MEXICO, WESTERN
B 6209 1590 SIGSBEE ESCARPMENT, GULF OF MEXICO
S 7 75 SIGSBEE ESCARPMENT, ORIGIN, GULF OF MEXICO
B 6305 788 SIGSBEE ESCARPMENT, ORIGIN OF
S 7 60 SIGSBEE ESCARPMENT, REFLECTION PROFILE, GULF OF MEXICO
S 7 78 SIGSBEE ESCARPMENT, REFLECTION PROFILE, GULF OF MEXICO
S 7 71 SIGSBEE ESCARPMENT, SPARKER PROFILE, GULF OF MEXICO
S 7 73 SIGSBEE ESCARPMENT, SPARKER PROFILE, GULF OF MEXICO
S 7 74 SIGSBEE ESCARPMENT, SPARKER PROFILE, GULF OF MEXICO
B 6211 2232† SIGSBEE KNOLL IN EARLY TERTIARY BAY OF BISCAY AND ASSOCIATED
S 7 53 SIGSBEE KNOLLS, CHALLENGER UNIT, GULF OF MEXICO
B 6305 783 SIGSBEE KNOLLS, GULF OF MEXICO
M 26 258 SIGSBEE KNOLLS, GULF OF MEXICO
S 7 52 SIGSBEE KNOLLS, REFLECTION PROFILE, GULF OF MEXICO
B 6211 2294 SIGSBEE KNOLLS REGION, GULF OF MEXICO
M 29 406 SIGSBEE KNOLLS, SEISMIC SECTION, GULF OF MEXICO
S 7 17 SIGSBEE PLAIN, DESCRIBED
M 29 408 SIGSBEE SALT DOME PROVINCE, SECTION, GULF OF MEXICO
B 6111 1973 SIGSBEE SCARP, GAS-ASSOCIATED FEATURES
B 6206 1068 SIGSBEE SCARP, GULF OF MEXICO
B 6111 1962† SIGSBEE SCARP, GULF OF MEXICO, NORTHEASTERN EXTENSION OF
B 6111 1965 SIGSBEE SCARP, SEGMENTS
B 6111 1966 SIGSBEE SCARP, SEISMIC SECTIONS
B 6211 2291 SIGSBEE UNIT, GULF OF MEXICO
M 26 272 SIGSBEE WEDGE SYSTEM
B 6009 1615 SIGSBY, R. J., PALEOENVIRONMENTAL ANALYSIS OF BIG ESCAMBIA
M 25 321 SIGUIRISUA ANTICLINE, COLOMBIA
M 25 440 SIHUNG ROCKS, SOUTH KOREA
M 25 465 SIKHOTE ALIN, GEOTHERMAL GRADIENT
M 25 463 SIKHOTE ALIN, VOLCANIC BELTS, CHARACTERISTICS OF, USSR
M 25 42 SIKHOTE-ALIN COAST, GEOLOGIC MAP, SIBERIA
M 25 42 SIKHOTE-ALIN FAULT, SIBERIA
B 6405 784 SILBERLING, N. J., MESOZOIC ACCRETIONARY TECTONICS OF ALASKA,
B 6211 2356 SILBERLING, N. J., MESOZOIC CONTINENTAL ACCRETION OF NORTH
B 6405 702* SILESIAN, ABST., BIVALVE, TRACE, FOSSILS IN, BRITISH
B 6207 1194 SILETZ RIVER FORMATION, OREGON AND, WASHINGTON
B 6207 1196 SILETZ RIVER-CRESCENT STRATIGRAPHY, OREGON AND, WASHINGTON
M 28 146 SILICA, BIBLIOGRAPHY, DIAGENESIS AND POROSITY
B 6303 483* SILICA DISSOLUTION FROM MONTMORILLONITE, EFFECT OF SOLUTION
M 27 141 SILICA, REPLACEMENT
M 27 140 SILICA, REPLACEMENT
M 28 80 SILICA REPLACEMENT, OF LIMESTONE
M 28 79 SILICA REPLACEMENT, OF OOLITIC LIMESTONE
M 25 402 SILICATE AND LIMONITE ORE, LATERITE, NICKEL DEPOSITS
B 6405 737* SILICATE MINERAL CONSTITUENTS, ABST., DEVELOPMENT OF,
M 27 26 SILICEOUS ALGAE, DIATOMS
M 27 25 SILICEOUS ALGAE, DIATOMS
B 6303 510* SILICEOUS ROCKS IN CIRCUM-PACIFIC REGION, ABST., DEDUCTION OF
S 4 128 SILICEOUS SPICULES IN SEDIMENTS
B 6203 569* SILICICLASTIC DEPOSITIONAL PATTERNS AND SEQUENCES ACROSS
B 6004 729* SILICICLASTIC ROCKS, ABST., INORGANIC AND, ORGANIC,
B 6205 885* SILICICLASTIC ROCKS FROM MESAVERDE GROUP, NORTH HORN
B 6105 804* SILICIFICATION OF CALCIUM PHOSPHATE IN CARBONATE ENVIRONMENT,
B 6012 2175† SILICOFLAGELLATE CORRELATIONS IN CALIFORNIA, ABST., MIDDLE,
B 6403 438* SILICOFLAGELLATES FROM HOLOCENE SEDIMENTS, SOUTHERN
M 25 543 SILICON, IRON, AND MANGANESE, TERNARY DIAGRAM, NAZCA PLATE
B 6111 1951 SILIGIR HEAVY OIL DEPOSIT, USSR
B 6311 2103† SILL, CALHOUN COUNTY, MISSISSIPPI, VARDAMAN, TRIASSIC-, JURASSIC
B 6203 511* SILL, CALIFORNIA CONTINENTAL BORDERLAND, ABST., NEW EVIDENCE
M 25 447 SILLA SERIES, UPPER, MINERALS, SOUTH KOREA
B 6408 1189 SILLED BASINS, EXAMPLES, ANOXIC
B 6408 1192 SILLED DEPRESSION IN OXIC OPEN OCEAN, CARIACO TRENCH
B 6408 1193 SILLED DEPRESSION IN OXIC OPEN OCEAN, ORCA BASIN
M 28 162 SILLERY GROUP, CANADA, QUEBEC, CAMBRIAN
M 28 65 SILLERY GROUP, CANADA, QUEBEC, CAMBRIAN
M 28 48 SILLIMANITE, CRYSTALLINE
M 28 8 SILLIMANITE INCLUSIONS, QUARTZ

B 6102 196 SILT LAYERS, KOMANDORSKIY BASIN, TURBIDITE SAND AND
M 28 73 SILTSTONE, BURROWED
B 6203 492* SILTSTONE IN DEVONIAN BRALLIER FORMATION OF WEST VIRGINIA AND
B 6004 684* SILURANIAN REEF MODEL--GREA LAKES AREA, ABST., SEARCH FOR
B 6004 718* SILURIA ROCKS FROM GOTLAND, SWEDEN, ABST., SEDIMENTOLOGY OF
S 5 93* SILURIAN REEF-BEARING ROCKS ON WABASH PLATFORM WITH CYCLC
S 5 53* SILURIAN) IN SOUTHEAST MICHIGAN, DEPOSITIONAL ENVIRONMENT OF
S 5 23* SILURIAN), MICHIGAN BASIN, SEDIMENTOLOGY AND DEPOSITIONAL
M 25 341 SILURIAN, AMAZON BASINS
B 6209 1607† SILURIAN AND DEVONIAN REEF TRENDS, CENTRAL APPALACHIAN BASIN,
B 6009 1618* SILURIAN AND EARLY DEVONIAN EVENTS IN SOUTHWESTERN INDIANA,
B 6309 1510† SILURIAN AND LOWER DEVONIAN SUBSURFACE UNITS IN PERMIAN
S 6 341 SILURIAN, BIOSTRATIGRAPHIC DATING, A REVIEW
S 6 343 SILURIAN, BIOSTRATIGRAPHY, CORRELATION WITH ISOTOPIC
M 27 50 SILURIAN, BROWNSPORT GROUP, TENNESSEE
B 6404 527† SILURIAN CARBONATE BUILDUPS AND BIOSTROMES OF NORTH
S 9 22 SILURIAN, CARBONATE REEF BELT DISTRIBUTION
B 6405 782* SILURIAN CARBONATE SHELF IN UTAH AND NEVADA, ABST.,
S 5 54 SILURIAN, CAYUGAN SERIES, SALINA GROUP, STRATIGRAPHY,
S 5 95 SILURIAN, CAYUGN SERIES, CORRELATION
B 6303 442* SILURIAN, CENTRAL NEVADA--EVIDENCE FOR GROUNDWATER MIXING,
M 27 147 SILURIAN, CLINTON FORMATION, PENNSYLVANIA
M 27 144 SILURIAN, CLINTON FORMATION, PENNSYLVANIA
M 28 45 SILURIAN, CLINTON FORMATION, PENNSYLVANIA
M 28 5 SILURIAN, CLINTON FORMATION, VIRGINIA
M 28 106 SILURIAN, CLINTON FORMATION, VIRGINIA
B 6408 1282* SILURIAN CLINTON SANDSTONE, NORTHEASTERN OHIO, ABST.,
M 27 55 SILURIAN, COBLESKILL LIMESTONE, NEW YORK
B 6401 119 SILURIAN CORRELATION CHART, GREAT LAKES AREA
S 6 344 SILURIAN, DEVONIAN BOUNDARY DISCUSSED
B 6211 2308† SILURIAN EVOLUTION OF CENTRAL APPALACHIAN BASIN
S 9 20 SILURIAN, FOLDOUT, PALEOCLIMATES
B 6304 610 SILURIAN FORMATIONS, MICHIGAN BASIN
B 6101 4 SILURIAN, FRANKLINIAN BASIN
B 6203 534* SILURIAN GLACIO-EUSTATIC SEA-LEVEL CHANGES IN EASTERNMOST
B 6005 883† SILURIAN GLAUCONITE. DISCUSSION AND REPLY, RADIOMETRIC, AGE
S 5 130 **SILURIAN, HELDERBERG GROUP, KEYSER LIMESTONE, WEST VIRGINIA,**
B 6009 1623* SILURIAN, IN INDIANA AND PARTS OF ILLINOIS, KENTUCKY AND OHIO,
B 6401 128 SILURIAN, INDIANA
S 6 342 SILURIAN, ISOTOPIC DATING, A REVIEW
S 6 343 SILURIAN, ISOTOPIC DATING, CORRELATION WITH BIOSTRATIGRAPHY
S 6 344 SILURIAN, ISOTOPIC DATING, USE OF BENTONITES DISCUSSED
M 28 157 SILURIAN, KEYSER LIMESTONE, PENNSYLVANIA
B 6203 553* SILURIAN MACROBORING ORGANISMS, SOUTHERN ONTARIO, ABST.
B 6004 724* SILURIAN MCKENZIE FORMATION, WEST VIRGINIA, ABST., EVOLUTION
B 6309 1585* SILURIAN MCKENZIE FORMATION, WEST VIRGINIA, ABST., PATCH, REEFS
S 5 96 SILURIAN, LOCKPORT GROUP, OUTCROP RELATIONS OF REEFS, OHIO,
S 5 96 SILURIAN, LOCKPORT GROUP, STRUCTURE, OHIO,
S 5 96 SILURIAN, LOCKPORT GROUP, THICKNESS, OHIO,
S 5 95 SILURIAN, LUDLOVIAN SERIES, CORRELATION
S 5 2 SILURIAN, NIAGARA GROUP, ILLINOIS,
S 5 2 SILURIAN, NIAGARA GROUP, INDIANA,
S 5 2 SILURIAN, NIAGARA GROUP, MICHIGAN,
S 5 2 SILURIAN, NIAGARA GROUP, OHIO,
S 5 2 SILURIAN, NIAGARA GROUP, ONTARIO,
S 5 2 SILURIAN, NIAGARA GROUP, WISCONSIN,
S 5 54 SILURIAN, NIAGARA GROUP, CLINTON FORMATION
S 5 54 SILURIAN, NIAGARA GROUP, ENGADINE FORMATION
S 5 54 SILURIAN, NIAGARA GROUP, GUELPH FORMATION
S 5 54 SILURIAN, NIAGARA GROUP, LOCKPORT FORMATION
S 5 95 SILURIAN, NIAGARAN SERIES, CORRELATION
S 5 54 SILURIAN, NIAGARAN SERIES, NIAGARA GROUP, STRATIGRAPHY,
B 6309 1582* SILURIAN, OF INDIANA, ABST., SUBMARINE, SLUMPING IN, REEF-FLANK
B 6305 845* SILURIAN, OF MCCONNELSVILLE FIELD, MORGAN COUNTY, OHIO, ABST.,
B 6004 660* SILURIAN, OF NEW YORK STATE, ABST., INTERTIDAL AND, SUBTIDAL,
B 6012 2117† SILURIAN, OF SHROPSHIRE, ENGLAND, ORIGIN AND EVOLUTION OF,
B 6009 1621* SILURIAN, OIL PRODUCTION AND SHALLOW STRUCTURE IN
S 6 344 SILURIAN, ORDOVICIAN BOUNDARY DISCUSSED
M 25 204 SILURIAN REEF ANOMALY, MICHIGAN BASIN
B 6401 115† SILURIAN REEF AT DELPHI, INDIANA, STRATIGRAPHY, STRUCTURE, AND,
B 6401 121 SILURIAN REEF COMPLEX, FOSSILS, INDIANA
B 6401 121 SILURIAN REEF COMPLEX, FOSSILS, INDIANA
B 6003 428† SILURIAN REEF COMPLEX, ROCKFORD, OHIO--CONSTITUTION, GROWTH,
B 6408 1283* SILURIAN REEFS IN SOUTHWESTERN INDIANA, ABST., STRUCTURAL
B 6405 790* SILURIAN REEFS OF NORTHERN INDIANA, ABST., SUBMARINE
B 6204 672 SILURIAN RESERVOIRS, F6 UNITS, ALGERIA, DEVONIAN-
S 5 95 SILURIAN, PRIDOLIAN SERIES, CORRELATION
M 27 99 SILURIAN, ROCHESTER SHALE, NEW YORK
M 28 29 SILURIAN, ROSE HILL FORMATION, VIRGINIA
M 28 88 SILURIAN, SALINA FORMATION, MICHIGAN
S 5 3 SILURIAN, SALINA GROUP, MICHIGAN,
S 5 66 SILURIAN, SALINA GROUP, RUFF FORMATION, DEPOSITIONAL HISTORY
S 5 58 SILURIAN, SALINA GROUP, RUFF FORMATION, LITHOFACIES
S 5 24 SILURIAN, SALINA GROUP, TERMINOLOGY
M 28 48 SILURIAN, SCHIST, MASSACHUSETTS
B 6203 528* SILURIAN SHELF-MARGIN REEFS, NORTHWESTERN OHIO, ABST.,
B 6303 410* SILURIAN STRATA IN MID-CONTINENT AREA, ABST., LATE,
S 6 341† SILURIAN SYSTEM, THE
B 6009 1622* SILURIAN TIME IN WEST VIRGINIA, ABST., EVOLUTION OF, BASIN-SHELF
B 6203 562* SILURIAN TONOLOWAY LIMESTONE, ABST., SCHEME FOR, MULTIVARIATE
M 28 157 SILURIAN, TONOLOWAY LIMESTONE, PENNSYLVANIA
M 27 164 SILURIAN, TONOLOWAY-KEYSER LIMESTONE, PENNSYLVANIA
M 27 67 SILURIAN, TONOLOWAY-KEYSER LIMESTONE, PENNSYLVANIA
M 27 71 SILURIAN, TONOLOWAY-KEYSER LIMESTONE, PENNSYLVANIA
M 27 72 SILURIAN, TONOLOWAY-KEYSER LIMESTONE, PENNSYLVANIA
M 27 90 SILURIAN, TONOLOWAY-KEYSER LIMESTONE, PENNSYLVANIA
M 27 94 SILURIAN, TONOLOWAY-KEYSER LIMESTONE, PENNSYLVANIA
M 27 139 SILURIAN, TONOLOWAY-KEYSER LIMESTONE, PENNSYLVANIA
M 27 221 SILURIAN, TONOLOWAY-KEYSER LIMESTONE, PENNSYLVANIA
M 27 75 SILURIAN, TONOLOWAY-KEYSER LIMESTONE, PENNSYLVANIA
M 27 4 SILURIAN, TONOLOWAY-KEYSER LIMESTONE, PENNSYLVANIA
B 6304 652 SILURIAN, VIRGINIA
S 5 95 SILURIAN, WENLOCKIAN SERIES, CORRELATION
B 6305 846* SILURIAN-DEVONIA EMBAYMENT INTO APPALACHIA, ABST.
B 6209 1609 SILURIAN-DEVONIAN CARBONATE SEQUENCE, APPALACHIAN BASIN
B 6209 1609 SILURIAN-DEVONIAN CARBONATE SEQUENCE, KENTUCKY
B 6209 1609 SILURIAN-DEVONIAN CARBONATE SEQUENCE, MARYLAND
B 6209 1610 SILURIAN-DEVONIAN CARBONATE SEQUENCE, NOMENCLATURE,
B 6209 1610 SILURIAN-DEVONIAN CARBONATE SEQUENCE, NOMENCLATURE, OHIO
B 6209 1609 SILURIAN-DEVONIAN CARBONATE SEQUENCE, VIRGINIA
B 6209 1612 SILURIAN-DEVONIAN SECTION, NEW YORK
B 6209 1622 SILURIAN-DEVONIAN SECTION, NEW YORK TO, OHIO
B 6209 1618 SILURIAN-DEVONIAN SECTION, NEW YORK TO, PENNSYLVANIA
B 6209 1624 SILURIAN-DEVONIAN SECTION, OHIO TO, PENNSYLVANIA
B 6209 1626 SILURIAN-DEVONIAN SECTION, OHIO TO, WEST VIRGINIA
B 6209 1614 SILURIAN-DEVONIAN TYPE LOG, OHIO
B 6406 851 SILVER ABYSSAL PLAIN, SAND LAYERS
M 25 552 SILVER, AND COPPER, ECUADOR, TOACHI DEPOSITS, LA PLATA MINE,
M 25 553 SILVER, AND GOLD DEPOSITS, ECUADOR, ANGAS, COPPER, ZINC, LEAD
M 25 518 SILVER AND GOLD, MEXICO, PACHUCA Y REAL DEL MONTE MINE
B 6007 1023 SILVER BAY, TAILINGS, DEPOSITION, MINNESOTA
B 6405 756 SILVER, BURR A., CARBON ISOTOPIC SIGNATURE AS CRITERION FOR
B 6004 646 SILVER, BURR A., DEPOSITIONAL ENVIRONMENTS OF TAMAROA
B 6004 723 SILVER, BURR A., DIAGENETIC HISTORY OF MISSISSIPPIAN CARBONATE
B 6303 419 SILVER, BURR A., ISOTOPIC COMPOSITIONS OF BIOGENIC CARBONATE
B 6405 784 SILVER, BURR A., OSWEGO LIMESTONE, ALINE-LAMBERT FIELDS,
M 25 483 SILVER CHANNEL, GREAT
B 6203 344 SILVER, E. A., NEOGENE BASINFORMATION IN RELATION TO
M 25 460 SILVER, EASTERN ASIATIC VOLCANOGENIC BELT, GOLD AND
B 6203 562 SILVER, ELI A., UNIFORMITARIANISM AND GLOBAL TECTONICS, ABST.
M 25 515 SILVER, LEAD, ZINC, AND ARSENIC, MEXICO
B 6405 786 SILVER, LEON T., BASINS, BASEMENT, AND THE DRILL, ABST.
M 25 505 SILVER, PRODUCTION AND RESERVES, CANADIAN, CORDILLERA
M 25 427 SILVER, QUEENSLAND, HILTON DEPOSIT, LEAD-, ZINC-
M 25 427 SILVER, QUEENSLAND, LADY LORETTA, LEAD-, ZINC-
M 25 444 SILVER, SOUTH KOREA, GOLD-
M 25 469 SILVER, USSR, NORTHEAST
M 25 553 SILVER, ZINC, AND LEAD DEPOSITS, ECUADOR, SAN BARTOLOME
M 25 439 SILVER-COPPER DEPOSITS, TAIWAN, GOLD-
M 25 460 SILVER-COPPER-POLYMETALLIC PROFILE, PACIFIC, MEXICO-, CENTRAL
M 25 496 SILVER-LEAD-ZINC, CANADIAN, CORDILLERA
M 25 500 SILVER-RICH VEINS, CANADA
M 25 427 SILVER-ZINC DEPOSIT, NEW SOUTH WALES, BROKEN HILL, LEAD-
M 25 427 SILVER-ZINC-COPPER, QUEENSLAND, MOUNT ISA, LEAD-
M 25 553 SILVER-ZINC-LEAD, ECUADOR, SAN BARTOLOME DEPOSIT
M 25 84 SIMANJUNTAK, MANGARA, COAL RESOURCES AND POTENTIAL IN
B 6004 686* SIMILARITIES IN EARLY DIAGENETIC HISTORIES OF CONDENSED
B 6209 1766 SIMMONS, JACK, PRIMARY MIGRATION—NEW TACK ON AN OLD THEME,
B 6109 1546 SIMMONS, WM. B., DIFFERENTIAL SUBSIDENCE OF MARSHLAND PEAT AS
B 6203 501 SIMMS, R. J., COMPUTER SYSTEM FOR LITHOSTRATIGRAPHIC ANALYSIS,
B 6311 2119 SIMON, D. E., STIMULATION DESIGN FOR UPPER MORROW RESERVOIRS IN
B 6405 784 SIMONEIT, BERND R. T., EFFECTS OF BASALT INTRUSIONS ON KEROGENS
B 6405 785 SIMONSON, BRUCE M., COARSENING-UPWARD SHELF SEQUENCES IN
B 6303 529 SIMONSON, BRUCE M., SEDIMENTOLOGY OF SOME PRECAMBRIAN IRON
B 6302 239† SIMPLE METHOD FOR OIL SOURCE BED EVALUATION
S 7 92 SIMPLE PATTERNS, DIVERGENT FACIES
S 7 102 SIMPLE PATTERNS, ONLAPPING-FILL SEISMIC FACIES
S 7 92 SIMPLE PATTERNS, PARALLEL-LAYERED FACIES
S 7 92 SIMPLE PATTERNS, SEDIMENT PROCESSES, SEISMIC IDENTIFICATION
S 7 98 SIMPLE PATTERNS, SHEET-DRAPE SEISMIC FACIES

B 6303 480* SIMPLE PYROLYSIS TECHNIQUE USING WELL CUTTINGS TO MAP SOURCE
B 6402 175 SIMPLE RIFTING MODELS
B 6203 565* SIMPSON EQUIVALENTS IN ARKANSAS AND ADJACENT AREAS, ABST.,
B 6203 555* SIMPSON GROUP, ORDOVICIAN, OKLAHOMA, ABST., EFFECT OF,
B 6109 1451 SIMPSON GROUP, PEDREGOSA BASIN
B 6402 222 SIMPSON GROUP, TEXAS
B 6004 723 SIMPSON, H. J., RADIOCARBON DATING OF ESTUARINE CARBONATE
B 6004 703 SIMPSON, H. J., SEDIMENTATION RATES IN HUDSON RIVER ESTUARY,
B 6303 427 SIMS, J. D., DEFORMATIONAL STRUCTURES IN PLIOCENE BOUSE
B 6004 673* SIMULATED GEOTHERMAL MATURATION OF ATHABASCA BITUMEN,
M 25 366† SIMULATION MODEL OF HONOLULU BASAL AQUIFER--SUMMARY
B 6203 573* SIMULATION OF GEOLOGIC, HYDRODYNAMIC, AND THERMODYNAMIC
B 6004 654* SIMULATION TECHNIQUES, ABST., EXAMINATION OF RELATIONS
B 6303 325 SINAI, GAZA DISTURBANCE, ISRAEL AND
B 6303 326 SINAI, SEISMIC PROFILES, ISRAEL AND
B 6303 324† SINAI, SUBMARINE, SLUMPING IN, CONTINENTAL, MARGIN OF, ISRAEL
M 25 525 SINALOA, METALLIC CONCENTRATION, TIMES OF, MEXICO
M 25 525† SINALOA, MEXICO--SUMMARY, MINERAL RESOURCES OF STATE O
M 25 525 SINALOA, MINERAL DEPOSITS, GROUPS OF, MEXICO
B 6203 442 SINGH, LAKSHMAN, EARLY EOCENE SUBAERIAL EROSIONAL VALLEYS IN
B 6103 468 SINGH, SUKI, WATONA DAM PROJECT. BY TITLE ONLY
B 6305 841 SINGLER, CHARLES R., PALEOSOLS AS ENVIRONMENTAL INDICATORS IN
B 6410 1615 SINGLETREE FIELD, DENVER BASIN
B 6206 1093 SINHA, R. P., NEW GEOLOGIC FRAMEWORK FOR MINING OF OIL SANDS,
B 6209 1697 SINIKTAMEYAK MOUNTAIN, ALASKA
B 6405 602 SINJAR FOLDBELT, MIDDLE EAST, PALMYRA-
B 6309 1494 SINJRANI VOLCANICS, PAKISTAN
B 6203 535* SINKHOLE DEPOSITS, HOT SPRINGS, SOUTH DAKOTA, ABST.,
M 29 339 SINU BELT, COLOMBIA, NORTHWESTERN
B 6212 2510* SINU BELT, NORTHWESTERN COLOMBIA, ABST., GEOLOGY AND,
M 29 340 SINU LINEAMENT, COLOMBIA, PALEOTRENCHES
M 25 437 SIPALAY MINE, NEGROS
S 11 251 SIPHOGENERINA BRANNERI
S 11 150 SIPHOGENERINA BRANNERI
S 11 148 SIPHOGENERINA BRANNERI
S 11 152 SIPHOGENERINA BRANNERI
S 11 152 SIPHOGENERINA, CALIFORNIA, MIOCENE
S 11 150 SIPHOGENERINA, CALIFORNIA, MIOCENE
S 11 148 SIPHOGENERINA, CALIFORNIA, MIOCENE
S 11 152 SIPHOGENERINA COLLOMI
S 11 150 SIPHOGENERINA COLLOMI
S 11 148 SIPHOGENERINA CYMRICENSIS
S 11 146 SIPHOGENERINA CYMRICENSIS
S 11 146 SIPHOGENERINA FORTICOSTATA
S 11 251 SIPHOGENERINA HUGHESI
S 11 148 SIPHOGENERINA HUGHESI
S 11 152 SIPHOGENERINA KLEINPELLI
S 11 146 SIPHOGENERINA KLEINPELLI
S 11 150 SIPHOGENERINA KLEINPELLI
S 11 148 SIPHOGENERINA KLEINPELLI
S 11 146 SIPHOGENERINA NODIFERA
S 11 144 SIPHOGENERINA NODIFERA
S 11 152 SIPHOGENERINA NUCIFORMIS
S 11 150 SIPHOGENERINA NUCIFORMIS
S 11 150 SIPHOGENERINA OBISPOENSIS
S 11 77 SIPHOGENERINA OBISPOENSIS
S 11 152 SIPHOGENERINA OBISPOENSIS
S 11 150 SIPHOGENERINA REEDI
S 11 152 SIPHOGENERINA REEDI
S 11 144 SIPHOGENERINA TENUA
S 11 146 SIPHOGENERINA TENUA
S 11 146 SIPHOGENERINA TRANSVERSA
S 11 148 SIPHOGENERINA TRANSVERSA
S 11 144 SIPHOGENERINA TRANSVERSA
S 11 21 SIPHOGENERINA TRANSVERSA ZONE
S 11 77 SIPHOGENERINID FORAMINIFERA, NEW GENERA, WEST COAST
S 11 115 SIPHOGENERINIDS, PHYLETIC ORIGINS
B 6307 1046 SIPHONINA DAVISI ZONE, LOUISIANA
S 11 251 SIPHONODOSARIA INSECTA
S 11 282 SIPHONODOSARIA INSECTA
S 4 270 SIPUNCULID WORMS, BORERS
B 6101 61 SIR ED DANIEH DOME, LEBANON
B 6207 1230 SIREGAR, E. M., TERTIARY CARBONATE RESERVOIRS IN INDONESIA, ABST.
B 6109 1476 SIRIRIZINHO FIELD, BRAZIL
M 30 511 SIRTE BASIN, GEOLOGIC HISTORY, LIBYA
M 30 548 SIRTE BASIN, GIALO FORMATION, EL GIZA MEMBER
M 30 543 SIRTE BASIN, INTISAR D FIELD, PALEOCENE
M 30 523 SIRTE BASIN, INTRA-CRETACEOUS UNCONFORMITY, LIBYA
M 30 550 SIRTE BASIN, KHEIR FORMATION, LIBYA
B 6203 464 SIRTE BASIN, LIBYA
B 6401 16 SIRTE BASIN, LIBYA
B 6406 860 SIRTE BASIN, LIBYA
B 6401 16 SIRTE BASIN, LIBYA
M 30 508 SIRTE BASIN, LOCATION MAP, LIBYA
M 30 546 SIRTE BASIN, MARSA BREGA TROUGH, LIBYA
M 30 507 SIRTE BASIN, MESSLA FIELD, LIBYA
M 30 515 SIRTE BASIN, MESSLA FIELD, STRUCTURE
M 30 554 SIRTE BASIN, REEF HISTORY, DEPOSITIONAL HISTORY
M 30 550 SIRTE BASIN, SABIL FORMATION, LIBYA
M 30 549 SIRTE BASIN, SHETERAT SHALE, LIBYA
B 6102 170 SISAR FAULT, CALIFORNIA
B 6102 138 SISAR FAULT, CALIFORNIA
B 6204 596 SITE 100, JOIDES, DSDP
B 6204 596 SITE 101, JOIDES, DSDP
B 6204 596 SITE 105, JOIDES, DSDP
B 6204 596 SITE 106, JOIDES, DSDP
B 6111 1992 SITE 141, SEISMIC PROFILE, DEEP SEA DRILLING
B 6412 2125 SITE 147, GEOCHEMISTRY, CARIACO TRENCH
B 6201 88 SITE 186, ALASKA, DEEP SEA DRILLING PROJECT
B 6212 2486 SITE 190, DSDP
B 6102 193 SITE 191, BERING SEA, KOMANDORSKIY BASIN, DEEP SEA DRILLING
B 6412 2125 SITE 229, GEOCHEMISTRY, RED SEA
B 6201 62 SITE 260, AUSTRALIA, DEEP SEA DRILLING PROJECT
B 6112 2068 SITE 262, DEEP SEA DRILLING PROJECT
B 6004 690* SITE 356 ON SAO PAULO PLATEAU, BRAZILIAN CONTINENTAL MARGIN,
B 6309 1608* SITE 386, ABST., MERIDIONALLY ORNAMENTED HEDBERGELLID,
B 6204 597 SITE 391, LEG 44, BLAKE BASIN, JOIDES, DSDP
B 6209 1605 SITE 391 OFF, NEW JERSEY, DSDP
B 6209 1601 SITE 391, PROFILE, NORTH ATLANTIC, DSDP
B 6405 711* SITE 503, LEG 68—INDICATE CONTINUOUS SECTION OF UNDISTURBED
B 6211 2306 SITE 97, DEEP SEA DRILLING PROJECT
B 6007 1041 SITE 98, DEEP SEA, DRILLING PROJECT
B 6204 596 SITE 99, CAT GAP, JOIDES, DSDP
B 6402 245 SITES 127 AND 128, PELOPONNESUS CONTINENTAL MARGIN, DEEP SEA
B 6209 1592 SITES 172-182, LEG 18, OREGON AND, ALASKA MARGINS, DSDP
B 6105 840 SITES, ROY S., PARSONS AND PETERSBURG STRUCTURAL LINEAMENTS,
B 6105 830 SITES, ROY S., SMOKE HOLES AREA, VALLEY AND RIDGE, WEST
B 6309 1588 SITES, ROY S., TECTONICS OF PETERSBURG REGION, WEST VIRGINIA,
M 30 365 SITIO GRANDE FIELD, CHIAPAS-TABASCO PROVINCE
B 6408 1142 SITKALIDAK FORMATION, ALASKA
B 6407 1016 SITKALIDAK FORMATION, ALASKA
B 6303 302 SITKALIDAK FORMATION, ALASKA
B 6405 671* SITKALIDAK FORMATION, EOCENE SUBMARINE FAN COMPLEX, KODIAK
B 6408 1142 SITKINAK FORMATION, ALASKA
B 6407 1016 SITKINAK FORMATION, ALASKA
B 6203 395 SITKINAK FORMATION, ALASKA
B 6303 302 SITKINAK FORMATION, ALASKA
B 6408 1142 SITKINAK ISLAND, ALASKA
B 6009 1618* SIX PROPOSED FORMATIONS IN UNDESCRIBED PART OF LOWER
B 6008 1155 SIZE ESTIMATES OF U.S. NEW-FIELD DISCOVERIES--1975, 1972, 1969
B 6308 1170 SIZE ESTIMATES OF U.S. NEW-FIELD DISCOVERIES--1978, 1975, 1972
B 6208 1278 SIZE ESTIMATES OF U.S. NEW-FIELD DISCOVERIES—1977, 1974, 1971
B 6409 1298 SIZE ESTIMATES OF U.S. NEW-FIELD DISCOVERIES—1979, 1976 1973
B 6004 735* SIZE-DEPENDENT COMPOSITIONAL TRENDS IN SANDS
B 6203 519* SIZE-FREQUENCY DISTRIBUTIONS AT LAMINAR AND INTRALAMINAR
M 30 471 SIZHONG, CHEN, GEOLOGY OF GUDAO OIL FIELD AND SURROUNDING
B 6008 1410* SKAMANIA COUNTY, WASHINGTON, ABST., HEAT-FLOW STUDIES IN,
M 25 505 SKARN DEPOSITS OF METALS, CANADIAN, CORDILLERA
M 25 504 SKEENA ARCH, CANADA
M 25 500 SKEENA ARCH, CANADA
B 6105 830 SKEHAN, J. W., COAL-BEARING NARRAGANSETT BASIN, MASSACHUSETTS
S 4 172 SKELETAL CARBONATE, DENSITY VARIATIONS IN
S 4 171† SKELETAL X-RADIOGRAPHY, CARIBBEAN, REEF, CORALS, MONTASTREA
B 6408 1284 SKEMA, VIKTORAS W., COAL IN PENNSYLVANIA—GEOLOGY, CURRENT
B 6410 1688 SKILLOGALEE DOLOMITE, CELLULAR FILAMENTS AND SHEATHS
B 6405 785 SKINNER, BRIAN J., DRILL AS TOOL FOR STUDYING MINERAL DEPOSITS
B 6409 1548 SKINNER, SUCZEK, CHIRSTOPHER A., REVIEW OF, ROCKS AND ROCK
B 6405 785 SKIPP, BETTY, UPPER PALEOZOIC PALEOGEOGRAPHY OF IDAHO, ABST.
B 6012 2187* SKOONER GULCH AND GALLAWAY FORMATIONS AT POINT ARENA,
M 25 376 SKORNYAKOVA, N. S., FERROMANGANESE NODULES OF THE PACIFIC
B 6410 1623 SKULL CREEK EQUIVALENT, DENVER BASIN
B 6407 971 SKULL CREEK SHALE, GREAT PLAINS
B 6405 626 SKYLAB PHOTOGRAPH, CALIFORNIA, SAN ANDREAS FAULT
B 6012 2187* SLALOM-LINE METHOD, ABST., PROCESSING OF NONSTRAIGHT, SEISMIC
B 6012 2177* SLANT-WAVE STACK, ABST.
B 6303 455* SLAR, ABST., A NEW LOOK AT GEOLOGY OF, TOGO BY
B 6002 325* SLAR IMAGERY IS NOT JUST A PICTURE, ABST.
B 6405 619 SLAR, SIDE-LOOKING AIRBORNE RADAR

M 28 30 SLATE, FRAGMENTS
B 6309 1609 SLATER, DAVID H., STRATIGRAPHY AND PALEOECOLOGY OF TAMIAMI
B 6303 529 SLATER, RICHARD A., SLUMPS ON UPPER CONTINENTAL SLOPE,
B 6402 236 SLAUGHTER BEND FAULT ZONE, TEXAS
M 24 73 SLAVE POINT FORMATION, ALBERTA
B 6205 782 SLAVE POINT FORMATION, ALBERTA
B 6405 759 SLEEP, NORMAN H., THERMAL CONTRACTION AND PETROLEUM
B 6002 322* SLEEPY HOLLOW FIELD, RED WILLOW COUNTY, NEBRASKA, ABST.,
B 6107 1029† SLEEPY HOLLOW FIELD, RED WILLOW COUNTY, NEBRASKA, GENESIS AND
B 6107 1035 SLEEPY HOLLOW SANDSTONE, NEBRASKA
B 6207 1235 SLESSAR, GRAHAM, BAUXITE MINING AND STREAM SALINITY IN
B 6109 1539* SLICE TECHNIQUE APPLIED TO URANIUM FAVORABILITY IN UPPERMOST
M 26 263 SLIDE MODIFICATION
M 25 509 SLIDE MOUNTAIN- NICOLA ROCKS, CORDILLERAN OROGEN, TAKLA-,
B 6203 495* SLIDE ON ATLANTIC CONTINENTAL SLOPE EAST OF CAPE MAY, NEW
B 6303 435* SLIDES, ABST., GENERATION OF DEBRIS, FLOWS AND,
B 6303 498* SLIDES, ABST., SUBMARINE
B 6303 444* SLIDES ON EASTERN CONTINENTA MARGIN--CAPE COD TO FLORIDA,
B 6109 1551* SLIGO AND HOSSTON DEPOSITIONAL PATTERNS, SUBSURFACE OF SOUTH
B 6209 1753* SLIGO FORMATION, SOUTH TEXAS, ABST., DISTRIBUTION AND
B 6205 786 SLIGO INTERVAL, TEXAS, WASHITA-
B 6301 48 SLIGO LIMESTONE, TEXAS
M 24 242 SLIGO PRODUCING SANDS, MISSISSIPPI
B 6108 1386* SLIP AND STRATIGRAPHIC CONSEQUENCES OF HURRICANE AND GRAND
S 4 88 SLIPPERY ROCK FORMATION, MAESTRICHTIAN, JAMAICA
B 6403 448 SLOAN, DORIS, FORAMINIFERA OF SANGAMON ESTUARY, CENTRAL SAN
M 25 500 SLOKO BASIN, CANADA
M 25 501 SLOKO, SUSTUT, AND SIFTON BASINS, CROSS SECTION, CANADA
B 6304 700* SLOPE, ABST., PETROLEUM POTENTIAL OF EASTERN, GULF OF, ALASKA
B 6303 480* SLOPE, ABST., SLUMPING IN INTERCANYON AREAS, MIDDLE, ATLANTIC,
B 6203 503* SLOPE, AND BASINAL FACIES, ORDOVICIAN TRENTON-UTICA SEQUENCE
S 7 313† SLOPE AND INTRASLOPE BASIN AREAS, NORTHWEST GULF OF MEXICO,
B 6109 1483† SLOPE AND UPPER RISE OFF WESTERN NOVA SCOTIA AND GEORGES
M 29 65† SLOPE BETWEEN CAPE HATTERAS AND CAPE COD, MULTICHANNEL,
B 6105 840* SLOPE CARBONATE DEPOSITS IN NORTH-CENTRAL NEVADA, ABST.,
M 29 268 SLOPE DATA, INTERPRETATION, ALASKA, KODIAK
B 6203 493* SLOPE DEPOSITS, ABST., PETROLEUM POTENTIAL OF, CONTINENTAL
B 6206 1092* SLOPE DEPOSITS, NORTHEASTERN MEXICO, ABST., MIDDLE, CRETACEOUS,
B 6203 506* SLOPE DEVELOPMENT, ABST., ANCIENT, CONTINENTAL SLOPE DEPOSITS
B 6002 165 SLOPE, DREDGE SAMPLES, BERING SEA, CONTINENTAL
S 7 276 SLOPE FAILURES, DIAGRAM, STABILITY
M 29 321 SLOPE, GOLDEN LANE CARBONATE BANK, MEXICO
S 7 243† SLOPE, LATE, PLEISTOCENE, GEOLOGIC HISTORY OF, TEXAS, OUTER
B 6212 2427 SLOPE MAP CONCEPT
B 6203 495* SLOPE MAP DEPICTING MAJOR SUBMARINE SLIDE ON ATLANTIC
B 6212 2427† SLOPE MAPPING
B 6004 665* SLOPE MAPPING, ABST., TECHNIQUES AND USES OF
B 6212 2428 SLOPE MAPS, COMPUTER-MADE
B 6212 2429 SLOPE MAPS, DATACOLOR
B 6212 2427 SLOPE MAPS, MANUAL
M 26 232 SLOPE MODELS, DEPOSITIONAL SYSTEMS
B 6303 529* SLOPE, NORTHEASTERN UNITED STATES--OBSERVATIONS FROM
B 6004 683* SLOPE NORTHERN GULF OF MEXICO, ABST., SALT MOVEMENT ON,
B 6305 782† SLOPE NORTHERN GULF OF MEXICO, SALT MOVEMENT ON,
S 7 69† SLOPE, NORTHERN GULF OF MEXICO, SALT MOVEMENT ON,
B 6405 687* SLOPE, NORTHWEST GULF OF MEXICO, ABST., GEOLOGIC AND
S 7 303† SLOPE, NORTHWEST GULF OF MEXICO, ORCA, BASIN, ANOXIC
S 7 267† SLOPE OFF LOUISIANA AND TEXAS: GEOTECHNICAL ASPECTS, STABILITY
B 6204 592† SLOPE, RISE, AND ABYSSAL PLAIN, PETROLEUM, POTENTIAL OF, U.S.,
B 6305 785 SLOPE, SALT, DOME GROWTH ON, CONTINENTAL
B 6106 918† SLOPE SEDIMENT INSTABILITY NORTHEAST OF WILMINGTON CANYON,
B 6203 531* SLOPE SEDIMENTATION PROCESSES IN NEOGENE BACK-ARC BASIN,
B 6004 722* SLOPE SEDIMENTS FROM NORTHWEST GULF OF MEXICO, ABST.,
B 6303 549* SLOPE SEDIMENTS RESULTING FROM LARGE-SCALE DOWNSLOPE CREEP,
B 6205 759 SLOPE SEISMIC FACIES, SHELF-MARGIN AND PROGRADED-
S 7 269 SLOPE STABILITY, DEFORMATIONAL FEATURES
S 7 10 SLOPE SUBPROVINCES MAP, GULF OF MEXICO
M 26 216 SLOPE SYSTEM, GULF OF MEXICO, OLIGOCENE, HACKBERRY
B 6203 561* SLOPE VALLEYS, ABST., CURRENTS IN, SUBMARINE CANYONS AND OTHER
B 6004 712* SLOPE VALLEYS AND SEDIMENTARY STRUCTURES OFF RIO BALSAS OF
B 6303 481* SLOPE--MULTIVARIATE STUDY, ABT ST., MEGASCALE VARIABILITY OF
B 6203 510* SLOPE-BASIN BENTHIC FORAMINIFERA OF CALIFORNIA BORDERLAND,
B 6405 792* SLOPE-CENTERED PROCESSES IN SANTA BARBARA BASIN, CALIFORNIA
B 6205 764 SLOPE-FRONT FILL FACIES
M 26 177 SLOPE-FRONT FILL FACIES, SEISMIC, INTERPRETATION
B 6207 1238* SLOPE-STABILITY PROBLEMS OF CIRCUM-PACIFIC REGION AS RELATED
B 6203 549* SLOPES, ABST., CARBONATE
B 6004 729* SLOPES AND RAIN SHADOWS, ABST., FLUVIAL, MODELS--, WET
B 6105 837* SLOPES AND RISES, ABST., SEISMIC RECOGNITION OF DEPOSITIONAL
B 6209 1584† SLOPES AND RISES, PETROLEUM, SOURCE BEDS ON, CONTINENTAL
M 29 423† SLOPES AND RISES, PETROLEUM, SOURCE BEDS ON, CONTINENTAL
B 6312 2204† SLOPES AND RISES, SEISMIC EVIDENCE FOR WIDESPREAD POSSIBLE, GAS,
B 6312 2211 SLOPES AND RISES, THERMAL GRADIENTS, CONTINENTAL
M 26 232 SLOPES, SEISMIC, DEPOSITIONAL SYSTEMS
S 5 189 SLOSS, L. L. REEFS AND EVAPORITES: A SUMMARY
B 6405 786 SLOSS, L. L., BASINS, BASEMENT, AND THE DRILL, ABST.
B 6105 832 SLOSS, L. L., BRAZILIAN CRATON—COMPARISON WITH NORTH AMERICA,
M 29 461 SLOSS, L. L., GLOBAL SEA LEVEL CHANGE: A VIEW FROM THE CRATON
B 6212 2413 SLUMP AREA, GULF OF ALASKA
B 6211 2198 SLUMP FEATURES, IRELAND, WESTERN
M 26 259 SLUMP MODIFICATION
B 6212 2412† SLUMP ON CONTINENTAL SHELF OFF MALASPINA GLACIER, GULF OF
B 6004 695* SLUMP REOPENS UNKNOWN, SCRIPPS SUBMARINE CANYON TRIBUTARY,
B 6303 324† SLUMPING IN CONTINENTAL MARGIN OF ISRAEL AND NORTHERN SINAI,
B 6303 480* SLUMPING IN INTERCANYON AREAS, MIDDLE ATLANTIC CONTINENTAL
B 6309 1582* SLUMPING IN REEF-FLANK DEPOSITS, MIDDLE SILURIAN, OF INDIANA,
B 6303 529* SLUMPS ON UPPER CONTINENTAL SLOPE, NORTHEASTERN UNITED
B 6009 1609* SMACKOVER, BOSSIER AND WEBSTER PARISHES, LOUISIANA, ABST.,
B 6409 1563* SMACKOVER CARBONATE ROCKS OF SOUTHWEST ALABAMA, ABST.,
B 6303 544* SMACKOVER CASE HISTORY, ABST., WALKER CREEK FIELD, ARKANSAS--
B 6009 1607* SMACKOVER DIAGENETIC TRAP, ABST., WALKER CREEK FIELD--
B 6409 1568 SMACKOVER, FIELD, MISSISSIPPI, ABST., PACHUTA CREEK
B 6303 445* SMACKOVER FORMATION, ABST., PROGRESSIVE, CEMENTATION FROM,
M 27 112 SMACKOVER FORMATION, ALABAMA, JURASSIC
B 6009 1614* SMACKOVER FORMATION AT CHALYBEAT SPRINGS FIELD, JURASSIC OF
B 6004 715* SMACKOVER FORMATION AT CHALYBEAT SPRINGS FIELD, JURASSIC OF
B 6404 466 SMACKOVER FORMATION, CEMENTS, LOUISIANA
B 6401 30 SMACKOVER FORMATION, FLORIDA
B 6303 495* SMACKOVER FORMATION, JURASSIC, PAUP SPUR FIELD, MILLER COUNTY,
B 6109 1548* SMACKOVER FORMATION, JURASSIC—PERRY AND STONE COUNTIES,
M 24 252 SMACKOVER FORMATION, MISSISSIPPI
B 6303 534* SMACKOVER FORMATION OF GULF COAST REGION, ABST.
B 6007 1119 SMACKOVER FORMATION, TEXAS
M 26 150 SMACKOVER FORMATION, TEXAS, EAST
M 26 147 SMACKOVER FORMATION, TEXAS, EAST TEXAS BASIN
B 6409 1565* SMACKOVER GRAY SAND, JURASSIC, NORTH LOUISIANA, ABST.,
B 6405 753* SMACKOVER IN SOUTHERN ARKANSAS, ABST., POST-COMPACTION,
B 6004 705* SMACKOVER OF MISSISSIPPI, ABST., GEOPRESSURES AND, SECONDARY,
B 6109 1534* SMACKOVER OIL PROSPECTS IN APALACHICOLA EMBAYMENT, ABST.
B 6009 1616* SMACKOVER PRODUCTION AT ANAHUAC, NORTHEASTERN MEXICO,
B 6407 1004† SMACKOVER RESERVOIR TRENDS, EAST TEXAS, NORTH LOUISIANA, AND
B 6309 1602* SMACKOVER RESERVOIR--INTERPRETATION CASE STUDY OF WATER
B 6309 1599* SMACKOVER RESERVOIRS, OAKS FIELD, CLAIBORNE PARISH, LOUISIANA,
M 24 177† SMACKOVER RESERVOIRS, WALKER CREEK FIELD AREA, LAFAYETTE AND
M 24 184 SMACKOVER, SUBMEMBER, BASAL UPPER, ARKANSAS
B 6405 786 SMAGALA, TOM, CRETACEOUS NIOBRARA GAS PLAY, ABST.
B 6303 509* SMALL COMPUTER WELL-DATA SYSTEM, ABST.
B 6303 297 SMALL OCEAN BASIN REGIONS, PROSPECTS
B 6209 1593 SMALL OCEAN BASINS, ORGANIC CONTENT
B 6203 546* SMALL ROCK OUTCROPS IN VEGETATED TERRAIN IDENTIFIED BY
B 6006 973 SMALLEY, WILLARD C., FLOW OF WATER AN SOLUTES THROUGH
B 6206 1072 SMEARING OF SEISMIC SECTION, DEFINITION
M 28 134 SMECTITE (AUTHIGENIC), IN SANDSTONE, CEMENTS
B 6405 735* SMECTITE AND CARBONATE CEMENT, ABST., EARLY POROSITY AND
M 28 66 SMECTITE, AUTHIGENIC, CLAY, MONTMORILLONITE-
M 28 68 SMECTITE, BIREFRINGENCE, CLAY
M 28 130 SMECTITE, COATING QUARTZ
B 6405 725* SMECTITE DIAGENESIS, ABST., SEDIMENTATION RATES AND, ILLITE-
M 28 68 SMECTITE, IN A BENTONITE, CLAY
B 6405 794* SMECTITE MIXED-LAYERED CLAY USING COMBINED SCANNING
M 28 133 SMECTITE, ON DETRITAL SAND, COATINGS
B 6203 497* SMECTITE-ILLITE REACTION FOR SANDSTONE CEMENTATION, ABST.,
B 6405 708* SMECTITE-ILLITE TRANSFORMATION—ROLE IN GENERATING AND
B 6004 673 SMILEY G. T., SIMULATED GEOTHERMAL MATURATION OF ATHABASCA
B 6303 370 SMITH, ALA L., SIDERITE SPHERULITES OF SAN SEBASTIAN FORMATION,
B 6405 786 SMITH, B. D., GEOPHYSICAL EXPLORATION FOR PRECAMBRIAN-RELATED
B 6108 1391 SMITH, BRUCE, ASSESSMENT OF MAGNETIC MEASUREMENTS IN
B 6004 724 SMITH, C. G., HYDROLOGIC STABILIT OF NORTH LOUISIANA SALT
B 6309 1577 SMITH, CARL J., REGIONAL COALIFICATION PATTERNS IN EASTERN
B 6105 831 SMITH, CARL J., WEST VIRGINIAS NORTHERN COALFIELD—DIRECTION OF
B 6105 821 SMITH, CHARLES C., BIOSTRATIGRAPHY OF U.S. ATLANTIC CONTINENTAL
B 6303 494 SMITH, D. F., COMPUTER DRAFTING--ITS APPLICATIO IN PETROLEUM
B 6105 831 SMITH, DAVID D., OCEAN DUMPING OF DREDGED MATERIAL—MAJOR
B 6403 439 SMITH, DENNIS R., COMPUTER SAND VOLUME DETERMINATIONS IN
B 6303 529 SMITH, DERRELL A., SEALING AND NONSEALING FAULTS IN GULF COAST
B 6402 145 SMITH, DERRELL A., SEALING AND NONSEALING FAULTS IN LOUISIANA
B 6308 1286 SMITH, DOROTHY J., DEVELOPMENTS IN OKLAHOMA AND PANHANDLE
B 6409 1387 SMITH, DOROTHY J., DEVELOPMENTS IN OKLAHOMA AND PANHANDLE
B 6112 2143 SMITH, DOUGLAS L., FLORIDA BASEMENT—AN ISOMETRIC VIEW
B 6209 1769 SMITH, DOUGLAS L., REEVALUATING FLORIDA BASEMENT, ABST.

B 6303 530 SMITH, G. W., ELLENBURGER GROUP, DELAWARE BASIN, WEST TEXAS
B 6405 722 SMITH, GARY E., ET AL, OXIDATION-REDUCTION IN OAKVILLE
B 6405 711 SMITH, GARY E., HYDROSTRATIGRAPHIC FRAMEWORK AND FLOW
B 6405 778 SMITH, GERALD N., SEISMIC MODEL OF 15 STRATIGRAPHICALLY
B 6305 840 SMITH, GERALD N., SEISMIC MODEL OF FIFTEEN STRATIGRAPHICALLY
B 6303 536 SMITH, GERALD R., NORMALLY AND REVERSELY GRADED BEDS WITHIN
B 6203 527 SMITH, GERALD R., TERTIARY ALGAL CARBONATE REEFS FROM SNAKE
B 6012 2188 SMITH, J. T., EVIDENCE FOR LOW FREE-GAS SATURATION IN
B 6203 455 SMITH, J. W., ISOTOPIC EVIDENCE FOR MICROBIOLOGIC ROLE IN GENESIS
B 6405 696 SMITH, J. WARD, SHALLOW OIL SHALE DEPOSITS OF SOUTHERN UINTA
B 6211 2362 SMITH, J. WARD, TERTIARY GEOLOGY AND OIL SHALE RESOURCES OF
B 6209 1767 SMITH, JAMES K., OSTRACODA OF PRAIRIE BLUFF CHALK, UPPER
B 6008 1243 SMITH, JEFFRY A., DEVELOPMENTS IN WEST TEXAS AND EASTERN NEW
B 6211 2362 SMITH, JOHN L., GEOLOGIC INTERPRETATIONS FOR RESERVOIR AND
B 6409 1567 SMITH, L. W., OCCURRENCE OF HIGH GRAVITY OIL IN AN OLIGOCENE
B 6304 692* SMITH LAKE DISTRICT, NEW MEXICO, ABST., MINERALOGY AND
B 6209 1680 SMITH LAKE DISTRICT, URANIUM, NEW MEXICO
B 6304 693* SMITH LAKE ORE TREND, ABST., GEOLOGY OF EASTERN
S 7 289 SMITH, LESTER B., INTRASLOPE BASIN IN NORTHWEST GULF OF MEXICO
B 6004 724 SMITH, M. B., MINICOMPUTERS FOR MAXIANALYSIS, ABST.
B 6105 760 SMITH, MICHAEL A., GEOLOGIC AND OPERATIONAL SUMMARY OF COST B-2
B 6207 1227 SMITH, ROBERT L., GEOTHERMAL EXPLORATION ON ADAK ISLAND,
B 6012 2081 SMITH, ROBERT R., TECTONIC TRANSPRESSION AND
B 6305 841 SMITH, SCOTT R., GEOLOGY AND HYDROCARBON ACCUMULATION IN
B 6103 468 SMITH, THOMAS E., GEOLOGY OF AMBLER MINERAL BELT, BROOKS
B 6104 594 SMITHDALE FIELD, MISSISSIPPI
B 6105 830* SMOKE HOLES AREA, VALLEY AND RIDGE, WEST VIRGINIA—EXPOSED
B 6107 983 SMOKY HILL CHALK MEMBER, NIOBRARA LIMESTONE, COLORADO
B 6107 987 SMOKY HILL CHALK MEMBER, NIOBRARA LIMESTONE, KANSAS
B 6303 530 SMOOT, JOSEPH P., SEDIMENTARY SUBENVIRONMENTS OF WILKINS PEA
S 5 125 SMOSNA, R. A. RELATIONSHIPS BETWEEN DEPOSITIONAL ENVIRONMENTS,
B 6105 839 SMOSNA, R. A., BIOFACIES RELATIONS IN LOWER DEVONIAN
B 6009 1622 SMOSNA, R. A., EVOLUTION OF BASIN-SHELF RELATIONS DURING
B 6004 724 SMOSNA, R. A., EVOLUTION OF CARBONATE SHELF, SILURIAN MCKENZIE
B 6405 800 SMOSNA R., DEPOSITIONAL ENVIRONMENTS OF LACUSTRINE LIMESTONES
B 6305 846 SMOSNA, RICHARD A., SILURIAN-DEVONIA EMBAYMENT INTO
B 6408 1289 SMOSNA, RICHARD, CARBONATE DIAGENESIS IN A QUARTZARENITE—
B 6405 787 SMOSNA, RICHARD, CARBONATE-SULFATE MINERAL REPLACEMENTS IN
B 6303 530 SMOSNA, RICHARD, COMPACTION EFFECTS IN FUSULINID LIMESTONE
B 6405 629 SMOSNA, RICHARD, NIAGARAN BIOHERMS AND INTERBIOHERM
B 6309 1585 SMOSNA, RICHARD, PATCH REEFS AND INTERREEF DEPOSITS OF
B 6203 562 SMOSNA, RICHARD, SCHEME FOR MULTIVARIATE ANALYSIS IN
B 6211 2308 SMOSNA, RICHARD, SILURIAN EVOLUTION OF CENTRAL APPALACHIAN
M 29 46 SNAEFELLSNES ARCH, ICELAND
M 25 157 SNAKE RIVER DOWNWARP, GEOTHERMAL POTENTIAL, UNITED STATES
M 25 508 SNAKE RIVER IRON DEPOSIT, CANADA
B 6008 1392* SNAKE RIVER PLAIN, IDAHO, ABST., HEAT-FLOW STUDY OF
B 6203 527* SNAKE RIVER PLAIN, IDAHO, ABST., TERTIARY, ALGAL, CARBONATE,
B 6104 595 SNAPPER FIELD, AUSTRALIA
B 6203 563 SNAVELY, P. D., III, REGIONAL VARIATION OF DEPOSITIONAL
B 6402 188 SNAVELY, PARKE D., JR., POROSITY EVOLUTION OF UPPER MIOCENE
B 6105 761 SNAVELY, PARKE D., POROSITY IN LATE MIOCENE REEFS, SOUTHERN
B 6402 188 SNAVELY, PARKE D., JR., POROSITY EVOLUTION OF UPPER MIOCENE
B 6405 787 SNEDDEN, JOHN W., COASTAL SWAMP ORIGIN OF SAN MIGUEL LIGNITE
B 6303 531 SNEIDER, R. M., PETROLEUM ENGINEERING--GEOLOGY, SYNERGISM, KEY
B 6105 831 SNEIDER, ROBERT M., RESERVOIR GEOLOGIC METHODS TO IMPROVE
B 6303 531 SNOKE, ARTHUR W., KEF ANTICLINE--BOX FOLDING IN TUNISIAN ATLAS,
B 6009 1615 SNOWDEN, JESSE O., CLAY SEDIMENTATION IN PEARL RIVER DELTA,
B 6109 1546 SNOWDEN, JESSE O., DIFFERENTIAL SUBSIDENCE OF MARSHLAND PEAT
B 6307 1128 SNOWDON, LLOYD R., ERRORS IN EXTRAPOLATION OF EXPERIMENTA
B 6002 280 SNYDER LOBE, PENNSYLVANIA
B 6109 1547 SNYDER, RICHARD H., EVALUATION OF AUSTIN AND BUDA FORMATIONS
B 6004 647* SO4 DIAGENESIS AND TIMING OF HYDROCARBON MIGRATION, ABST.,
B 6105 832 SOARES, PAULO C., BRAZILIAN CRATON—COMPARISON WITH NORTH
M 25 269 SOBAEGON OR YONGNAM, REINAN, MASSIF, KOREA, SOBAEKSAN
M 25 443 SOBAEKSAN ANTICLINE, SOUTH KOREA
M 25 441 SOBAEKSAN MASSIF, SOUTH KOREA
M 25 269 SOBAEKSAN, SOBAEGON OR YONGNAM, REINAN, MASSIF, KOREA
B 6210 2112 SOCIETE NATIONALE ELF AQUITAINE, AGIP MINERARIA S.P.A., DEMINEX
B 6210 2118 SOCIETE NATIONALE ELF AQUITAINE, ET AL, ON DEVELOPMENTS IN
B 6102 253† SOCIETY, HISTORY OF, EL PASO, GEOLOGICAL
B 6007 1117 SOCIETY, HISTORY, SAN ANTONIO, GEOLOGICAL
B 6007 1119 SOCIETY, HONORARY MEMBERS, 1935-1975, SOUTH, TEXAS, GEOLOGICAL
B 6102 255 SOCIETY, OFFICERS, 1967-1976, EL PASO, GEOLOGICAL
B 6006 988† SOCIETY, SHORT, HISTORY OF, NORTHERN, CALIFORNIA, GEOLOGICAL
B 6007 1117† SOCIETY--A FRAGMENTARY INFORMAL HISTORY, SOUTH, TEXAS,
M 25 329 SOCORRO FORMATION, ECUADOR
M 25 269 SOCOTRA SUBBASIN, KOREA
B 6004 696* SODA-RICH WATERS WITH CONCENTRATED SEA WATER, ABST., ORIGIN
B 6009 1621 SODERBERG, R. K., GEOPHYSICAL AND TECTONIC STUDY OF MOORMAN
B 6307 1059 SODERO, DARIO E., DEPOSITIONAL FACIES OF LOWER PALEOZOIC ALLEN
B 6212 2443 SODIUM BICARBONATE, SODIUM ACETATE, AND OIL-FIELD WATER
M 25 357 SOEKARDI, R., GROUNDWATER POTENTIAL OF AREAS UNDERLAIN BY
M 25 356 SOEKARDI, R., GROUNDWATER POTENTIAL OF LIMESTONE TERRANES,
B 6009 1448 SOEPARJADI, R. A., OIL-BEARING REEFS IN SALAWATI BASIN OF IRIAN
B 6207 1236 SOERJADJI, R. SOERJO, GEOTHERMAL WELL COMPLETION AND CASING
B 6110 1801 SOFIRAN, ELF AQUITAINE, ET AL, ON DEVELOPMENTS IN OMAN,
B 6010 1866 SOFIRAN, IRAN NIPPON PETROLEUM CO., AGI S.P.A., AMOCO, AND
B 6010 1712 SOGARO, ON DEVELOPMENTS IN ITALY, 1975, BONGIORNI, D., AND L.
B 6207 1187 SOHL, N. F., NOTE 49—RECORDS OF STRATIGRAPHIC COMMISSION FOR
B 6102 248 SOHL, NORMAN F., NOTE 45--APPLICATION FOR AMENDMENT
B 6102 252 SOHL, NORMAN F., NOTE 46--APPLICATION FOR AMENDMENT OF
B 6207 1185 SOHL, NORMAN F., NOTE 48—APPLICATION FOR AMENDMENT OF CODE
B 6406 851 SOHM AND HATTERAS ABYSSAL PLAINS, SAND LAYERS
B 6101 53 SOHMOR 1, LOG, LEBANON
B 6101 53 SOHMOR ANTICLINE, LEBANON
B 6105 832 SOHN, I. G., LAKOTA FORMATION, LOWER CRETACEOUS, IN SOUTH
B 6109 1560* SOIL ANOMALIES FROM LANDSAT, ABST., DETECTION OF
M 28 74 SOIL, AUSTRALIA, QUATERNARY
M 28 131 SOIL, AUSTRALIA, QUATERNARY
B 6105 827* SOIL GAS OVERLYING TWO URANIUM ORE DEPOSITS, ABST.,
B 6303 521* SOIL REGIMES OF TAZIRBU REGION OF CENTRA LIBYA DETERMINED
B 6303 485* SOIL-ZONE STRATIGRAPHIC TRAP, ABST., TYLER SANDSTONES,
B 6309 1604* SOILS, ABST., CHEMICAL DIFFERENTIATION OF TEMPERATURE AND
M 25 58 SOKOLOV, B. A., GEOLOGIC-STRUCTURAL SYNTHESIS OF PACIFIC AREA
M 25 176 SOKOLOV, B. A., OIL- AND GAS-BEARING BASINS OF CIRCUM-PACIFIC
M 25 220 SOLANDER TROUGH, NEW ZEALAND
S 6 318 SOLAR CHRONOLOGY, CLIMATIC CHANGE, PERMIAN
S 6 319 SOLAR CHRONOLOGY, COOLINGS, TABLE, PERMIAN
S 6 317 SOLAR CHRONOLOGY, GLACIATION DATING
S 6 316 SOLAR CHRONOLOGY, PERMIAN
M 25 73† SOLAR DESALINATION EXPERIENCE OF UNIVERSITY OF
B 6309 1599* SOLAR ENERGY. WHAT IS PRACTICAL IN THE NEAR TERM, ABST.,
B 6207 1215* SOLAR ENERGY—CENTRAL RECEIVER PROGRAM, BARSTOW POWER
B 6106 904 SOLEDAD BASIN, BAJA CALIFORNIA
B 6203 369 SOLEDAD BASIN, CALIFORNIA
B 6106 904 SOLEDAD GAP, BAJA CALIFORNIA
B 6106 904 SOLEDAD RIDGE, BAJA CALIFORNIA
B 6206 1053 SOLEDAD VOLCANIC CONGLOMERATE MEMBER, TEXAS
M 26 22 SOLENHOFEN LIMESTONE, S-WAVE MEASUREMENT LABORATORY
B 6309 1479 SOLENHOFEN LIMESTONE, TEXTURE AND COMPOSITION DATA
M 27 14 SOLENOPOROID GRAIN
S 10 180 SOLID GAS HYDRATES, METHANE, OCCURRENCES
M 27 175 SOLNHOFEN LIMESTONE, GERMANY, JURASSIC
M 27 222 SOLNHOFEN LIMESTONE, GERMANY, JURASSIC
M 27 128 SOLNHOFEN LIMESTONE, GERMANY, JURASSIC
B 6211 2362 SOLOMON, BARRY J., TERTIARY GEOLOGY AND OIL SHALE RESOURCES
M 25 430† SOLOMON ISLANDS AND FIJI--SUMMARY, BACKGROUND TO PRESENT
M 25 430 SOLOMON ISLANDS, BAUXITE, BRITISH
B 6110 1882 SOLOMON ISLANDS, DEVELOPMENTS, 1976, NEW HEBRIDES AND
B 6310 1682 SOLOMON ISLANDS, DEVELOPMENTS, 1978
B 6411 1779 SOLOMON ISLANDS, DEVELOPMENTS, 1979
M 25 472 SOLOVYEV, V. V., RING MEGASTRUCTURES OF THE PACIFIC--SUMMARY
M 25 480 SOLOVYEV, V. V., STRUCTURAL CLASSIFICATION OF NORTH PACIFIC
S 10 90 SOLUBILITIES IN WATER, MIGRATION, CONSTRAINTS
S 10 21 SOLUBILITIES OF ACIDS IN WATER, MIGRATION
S 10 21 SOLUBILITIES OF ALCOHOLS IN WATER, MIGRATION
B 6305 762 SOLUBILITIES OF HYDROCARBONS IN WATER
S 10 21 SOLUBILITIES OF PARAFFINS IN WATER, MIGRATION
S 10 37 SOLUBILITY AS TEMPERATURE FUNCTION, MIGRATION
B 6203 541* SOLUBILITY IN MIGRATION OF PETROLEUM FROM SOURCE, ABST., ROLE
S 10 103 SOLUBILITY IN WATER AT TEMPERATURE, METHANE
S 10 38 SOLUBILITY OF GAS IN FORMATION WATER, MIGRATION
S 10 20 SOLUBILITY OF HYDROCARBON IN WATER, MIGRATION
B 6309 1527† SOLUBILITY OF METHANE AT ELEVATED PRESSURES AND
B 6212 2478† SOLUBILITY OF METHANE IN WATER AT ELEVATED TEMPERATURES AND
B 6305 776 SOLUBILITY OF METHANE IN WATER AT ELEVATED TEMPERATURES AND
B 6002 213† SOLUBILITY OF PETROLEUM AS APPLIED TO ITS ORIGIN AND PRIMARY
B 6112 2146† SOLUBILITY OF PETROLEUM AS APPLIED TO ITS ORIGIN AND PRIMARY
B 6006 973† SOLUTES THROUGH COMPACTED CLAYS, FLOW OF, WATER AN
B 6002 214 SOLUTION HYPOTHESIS, GASEOUS
B 6002 214 SOLUTION HYPOTHESIS, MICELLAR
B 6002 215 SOLUTION HYPOTHESIS, MOLECULAR
S 10 20 SOLUTION, MIGRATION, MECHANISMS
M 28 166 SOLUTION, IN A SANDSTONE, STYLOLITIC
M 28 166 SOLUTION, IN SILTSTONE, STYLOLITIC
B 6305 843* SOLUTION MINING OF URANIUM AND ITS EFFECT ON EXPLORATION,
B 6304 693* SOLUTION MINING OF URANIUM ORES, ABST., APPLICATION OF
B 6002 232 SOLUTION, MODEL OF, PRIMARY, PETROLEUM, MIGRATION BY,

M 28 166 SOLUTION, OF FOSSILS, INTERGRANULAR
M 28 73 SOLUTION SEAMS, SHALE
B 6004 712* SOLUTION SURFACES, MIDDLE ORDOVICIAN CARBONATE ROCKS,
M 27 201 SOLUTION-CAVITY FILLS
B 6111 1994 SOMALI BASIN, POSSIBLE DIAPIRS
B 6110 1749 SOMALI REPUBLIC, DEVELOPMENTS, 1976
B 6210 1849 SOMALI REPUBLIC, DEVELOPMENTS, 1977
B 6310 1694 SOMALI REPUBLIC, DEVELOPMENTS, 1978
B 6411 1790 SOMALI REPUBLIC, DEVELOPMENTS, 1979
B 6003 401 SOMALIA, COTTON FORMATION
B 6003 389† SOMALIA, EAST AFRICA, GEOLOGY AND, OIL, PROSPECTS OF
B 6004 677* SOME APPROACHES TO CLUSTER-ANALYSIS INTERPRETATION OF
M 25 463† SOME CHARACTERISTICS OF ACID VOLCANISM OF SOUTHERN FAR
B 6203 522* SOME ECONOMIC ASPECTS OF WATER-ROCK INTERACTION,ABST.
S 10 109† SOME ECONOMIC ASPECTS OF WATER-ROCK INTERACTION
B 6104 607† SOME LOWER AND MIDDLE JURASSI BEDS IN PUALE BAY--ALINCHAK BAY
B 6109 1550* SOME LOWER CRETACEOUS STRATIGRAPHIC MODELS FROM TEXAS AND
B 6203 528* SOME MASS BALANCE AND GEOLOGIC CONSTRAINTS ON MIGRATION
S 10 47† SOME MASS BALANCE AND GEOLOGICAL CONSTRAINTS ON MIGRATION
B 6205 885* SOME PETROGRAPHIC CHARACTERISTICS OF SEQUENCE OF SILICICLASTIC
B 6004 645* SOME PROBLEMATIC LATE CENOZOIC CARIBBEAN PALEOSOLS AND THEIR
B 6405 732* SOME RELATIONS BETWEEN DIAGENESIS AND POROSITY REAL AND
M 25 189† SOME SHALLOW TECTONIC CONSEQUENCES O SUBDUCTION AND THEIR
M 26 356 SONAR RECORD, ALASKA, CAPE, YAKATAGA, SIDE-SCAN
M 26 356 SONAR SYSTEMS, RESOLUTION SYSTEMS, SIDE-SCAN
M 25 252 SONGKHLA BASIN, THAILAND
B 6401 9 SONGLIAO BASIN, CHINA
M 25 442 SONGRIM DISTURBANCE, KOREA
M 25 442 SONGRIM DISTURBANCE, SOUTH KOREA, TRIASSIC DEFORMTION
B 6309 1579* SONIC LOGGING IN GAS SHALES, ABST., THEORETICAL AND
B 6405 806* SONIC LOGS FROM EXPLORATIONISTS VIEWPOINT, ABST.
M 29 70 SONIC VELOCITIES, COST B-2, WELL DATA
B 6212 2512 SONIDO, ERNESTO P., APPRAISAL OF PRESENT PETROLEUM-EXPLORATION
B 6009 1615 SONNENBERG, STEPHEN A., INTERPRETATION OF COTTON VALLEY
B 6405 801 SONNENBERG, STEPHEN A., WATTENBERG AND SPINDLE FIELDS—
S 5 181 SONNENFELD, P. BASE- METAL CONCENTRATION IN A DENSITY-
B 6303 468* SONOBUOY REFRACTION MEASUREMENT FROM NORTON BASIN,
M 29 164 SONOBUOY VELOCITIES, AUSTRALIA, MARGIN
B 6405 800* SONORA, ABST., DOLOMITE AND DEDOLOMITE IN, MURAL LIMESTONE,
B 6004 646* SONORA, AND NORTHWESTERN CHIHUAHUA, ABST., DEPOSITIONAL
B 6004 723* SONORA, AND NORTHWESTERN CHIHUAHUA, ABST., DIAGENETIC
M 25 519 SONORA, CANANEA MINE, COPPER, MEXICO
M 25 520 SONORA, COPPER MINERALIZATION, MEXICO
M 25 521 SONORA, LA CARIDAD MINE, COPPER, MEXICO
B 6303 547* SONORA, MEXICO, ABST., LOWER, CRETACEOUS, CARBONATE SHELF IN
B 6405 716* SONORA, MEXICO, ABST., PETROLEUM POSSIBILITIES IN, ALTAR DESERT
M 25 519† SONORA, MEXICO--SUMMARY, PRINCIPAL CHARACTERISTICS OF SOME
B 6203 545* SONORA PASS, CALIFORNIA, ABST., URANIUM IN, TUFFACEOUS, FLUVIAL
M 25 520 SONORA, PILARES DE NACOZARI MINE, COPPER, MEXICO
M 25 519 SONORA, PORPHYRY COPPER DEPOSITS, MAIN, MEXICO
M 25 520 SONORA, PORPHYRY COPPER DEPOSITS, MAP, MEXICO
B 6111 1986 SOQUEL SANDSTONE, CALIFORNIA
B 6402 189 SORBAS BASIN, SPAIN
S 4 335 SORITES MARGINALIS, ST. CROIX
M 25 242 SORONG FAULT ZONE, INDONESIA
B 6207 1127 SORONG FAULT ZONE, IRIAN JAYA
M 28 103 SORTING, CLASSIFICATION, SEDIMENT
M 28 104 SORTING, SEDIMENT COMPARISON
M 24 259 SOSO FIELD, MISSISSIPPI
B 6004 724 SOUDER, W. W., MINICOMPUTERS FOR MAXIANALYSIS, ABST.
B 6303 500 SOULE, RALPH, APPLICATION OF POTENTIAL FIELD DAT TO STRUCTURAL
B 6311 2077 SOUNDER RIDGE, BERING SEA
B 6303 431* **SOURCE ROCK AND THERMAL MATURITY EVALUATION OF OGADEN**
B 6405 762* **SOURCE ROCK SCREENING TECHNIQUE, ABST., ROCK-EVAL PYROLYSIS AS**
S 10 19† SOURCE ROCK, PRIMARY, MIGRATION: THE IMPORTANCE OF,
S 10 57 SOURCE ROCK DRAINAGE EFFICIENCY, MIGRATION
B 6004 732* SOURCE, ABST., PRESERVATION, WINDOW AS, SUBUNCONFORMIT
B 6004 648* SOURCE AND DYNAMICS OF SUSPENDED SEDIMENTS IN NEAR-SHORE
B 6405 800* SOURCE AND TIME OF GENERATION OF HYDROCARBONS IN FOSSIL
B 6108 1382* SOURCE AREAS AND DEPOSITIONAL ENVIRONMENTS OF SANTA ROSA
B 6402 209† SOURCE AREAS AND MIGRATION TRENDS OF OIL AND GAS IN BOMBAY
B 6209 1586 SOURCE BED, DEFINITION
B 6209 1586 SOURCE BED, DEFINITION, POTENTIAL
B 6302 241 SOURCE BED EVALUATION, EXPULSION EFFICIENCY IN
B 6302 239† SOURCE BED EVALUATION, SIMPLE METHOD FOR, OIL
B 6408 1179† SOURCE BED GENESIS, ANOXIC ENVIRONMENTS AND, OI
B 6405 697* SOURCE BEDS, ABST., ANOXIC ENVIRONMENTS AND, OIL
B 6303 507* SOURCE BEDS, ABST., ATMOSPHERIC CIRCULATION AND UPWELLING IN
B 6405 668* SOURCE BEDS, ABST., EVALUATION OF TEMPERATE ZONE COASTAL-,
M 29 427 SOURCE BEDS, CONTINENTAL MARGIN, ORGANIC CONTENT
M 29 432 SOURCE BEDS, CONTINENTAL MARGIN, THERMAL MATURITY
B 6209 1584 SOURCE BEDS, DEFINITION OF
B 6209 1584 SOURCE BEDS, KEROGEN IN
B 6403 427† SOURCE BEDS OF HIGH-WAX OILS IN ASSAM BASIN, INDIA, DEPOSITIONAL
B 6008 1412* SOURCE BEDS OF PETROLEUM IN DENVER BASIN, ABST.
B 6209 1584† SOURCE BEDS ON CONTINENTAL SLOPES AND RISES, PETROLEUM
M 29 423† SOURCE BEDS ON CONTINENTAL SLOPES AND RISES, PETROLEUM
M 29 425 SOURCE BEDS, ORGANIC MATTER, ACCUMULATION
M 29 425 SOURCE BEDS, ORGANIC PRODUCTIVITY, PETROLEUM
M 29 423 SOURCE BEDS, PETROLEUM, DEFINITION
M 29 424 SOURCE BEDS, PETROLEUM, KEROGEN CONCENTRATION
M 29 424 SOURCE BEDS, PETROLEUM, KEROGEN MATURITY
M 29 424 SOURCE BEDS, PETROLEUM, KEROGEN TYPE
M 29 432 SOURCE BEDS, PETROLEUM, THERMAL CRACKING
M 29 425 SOURCE BEDS, PETROLEUM, ZONES OF GENERATION
B 6408 1180 SOURCE BEDS, PRECURSOR SEDIMENTS TO, OIL-
B 6004 704* SOURCE IDENTIFICATION OF BEACH TARS WITHIN SOUTHERN
B 6303 437* SOURCE OF OILS IN GULF COAST TERTIARY. WHY LOO A GIFT HORSE IN
B 6401 60 SOURCE ROCK AND ISOLATED KEROGEN, EXPERIMENTAL EVOLUTION OF
B 6405 705* SOURCE ROCK CHARACTERIZATION AND EVALUATION, ABST.,
B 6311 2033 SOURCE ROCK, GEOTHERMAL, MATURATION OF POTENTIAL
B 6311 2021† SOURCE ROCK IN BONAPARTE GULF--TIMOR SEA REGION,
M 30 159 SOURCE ROCK, MORAY FIRTH BASIN, PIPER FIELD
B 6205 827† SOURCE ROCK, ORGANOCHEMICAL EVALUATION OF, GEVARAM SHALES,
B 6209 1763* SOURCE ROCK POTENTIAL OF CRETACEOUS AND LOWER TERTIARY
B 6306 877 SOURCE ROCK, RUDIST BUILDUPS AND
B 6303 489* SOURCE ROCKS, ABST., ENVIRONMENTAL AND DIAGENETIC CONTROLS
B 6405 744* SOURCE ROCKS, ABST., ENVIRONMENTAL AND DIAGENETIC CONTROLS
B 6303 496* SOURCE ROCKS, ABST., TEST-TUBE, PYROLYSIS--SIMPLE TECHNIQUE FOR
B 6205 818 SOURCE ROCKS AND HYDROCARBON GENERATION
B 6401 112 SOURCE ROCKS, CONCENTRATION OF, CARBON DIOXIDE IN, PORE,
B 6303 480* SOURCE ROCKS, GAS-CONDENSATE MATURITY, AND ABNORMAL FLUID
B 6401 107 SOURCE ROCKS, GENERATION OF, GAS IN, PORES OF
B 6108 1377* SOURCE ROCKS IN EAST-CENTRAL UTAH-IMPLICATIONS FOR
B 6303 431* SOURCE ROCKS IN GULF COAST AREA--THEIR IDENTIFICATION AND
B 6205 819 SOURCE ROCKS, LOPATIN CALCULATIONS ON
S 7 187 SOURCE ROCKS, MISSISSIPPI FAN
B 6307 1104 SOURCE ROCKS, NORTH PERTH BASIN, AUSTRALIA
B 6205 888* SOURCE ROCKS, NORTHEASTERN GREAT BASIN, ABST., PROBABLE,
B 6001 115† SOURCE ROCKS, OKLAHOMA CITY OIL--SECOND CROP FROM PRESERVED,
M 30 109 SOURCE ROCKS, STATFJORD FIELD
B 6305 771 SOURCE ROCKS THROUGH THREE-DIMENSIONAL ORGANIC MATTER,
B 6104 620† SOURCE ROCKS. DISCUSSION AND REPLY, OKLAHOMA CITY OIL--SECOND
B 6206 1075† SOURCE-BED EVALUATION OF JURASSIC TWIN CREEK LIMESTONE,
B 6408 1251† SOURCE-BED EVALUATION OF TERTIARY NIGER DELTA, PETROLEUM
B 6205 813† SOURCE-BED HISTORY OF GREAT VALLEY, CALIFORNIA, RESERVOIR AND
B 6004 674* SOURCE-RESERVOIR-SEAL RELATIONS IN PRODUCTIVE
B 6405 798* SOURCE-ROCK ANALYSIS, ABST., EVALUATION OF, URANIUM POTENTIAL
B 6405 726* SOURCE-ROCK EVALUATION BY THERMAL DISTILLATION AND
B 6410 1619 SOURCE-ROCK EVALUATION, DENVER BASIN
B 6303 425* SOURCE-ROCK EVALUATION OF ATLANTIC COASTAL PLAIN ADJACENT
B 6004 608† SOURCE-ROCK EVALUATION--QUANTITATIVE ESTIMATE OF ORGANIC
B 6311 2025 SOURCE-ROCK POTENTIAL, CLASSIFICATION
B 6108 1375* SOURCE-ROCK POTENTIAL OF ARBUCKLE GROUP,
B 6405 733* SOURCE-ROCK POTENTIAL OF EVAPORITIC ENVIRONMENT, ABST.
B 6312 2227† SOURCE-ROCK PYROLYSIS, EFFECT OF, OIL AND, BITUMEN, SATURATION
B 6405 702* SOURCE-ROCK QUALITY AND THERMAL MATURITY PALO DURO BASIN,
M 25 488 SPOURCES, CORDILLERAN
B 6303 434* SOURCES OF LIABILITY FOR GEOLOGIC HAZARDS, ABST.
B 6201 74 SOURIS RIVER, LOWER, LIMESTONE, SASKATCHEWAN
B 6201 74 SOURIS RIVER, UPPER, LIMESTONE-DOLOMITE-ANHYDRITE,
B 6303 446* SOUTH AFRICA, ABST., DOLOMITIZATION MODELS FROM EARLY,
B 6105 815* SOUTH AFRICA, ABST., SEDIMENTOLOGIC SYNTHESIS OF CARBONATE
S 6 369 SOUTH AFRICA, CAPE SYSTEM
S 6 369 SOUTH AFRICA, CHRONOLITHOSTRATIGRAPHIC COLUMN
B 6110 1749 SOUTH AFRICA, DEVELOPMENTS, 1976
B 6210 1849 SOUTH AFRICA, DEVELOPMENTS, 1977
B 6310 1694 SOUTH AFRICA, DEVELOPMENTS, 1978
B 6411 1790 SOUTH AFRICA, DEVELOPMENTS, 1979
S 6 369 SOUTH AFRICA, DOMINION REEF SYSTEM
S 9 8 SOUTH AFRICA, ICE FLOW DIRECTION, CARBONIFEROUS
S 6 369 SOUTH AFRICA, KAROO SYSTEM
S 6 369 SOUTH AFRICA, KHEIS SYSTEM
S 6 369 SOUTH AFRICA, MOODIES SYSTEM
S 6 369 SOUTH AFRICA, NAMA-DAMARA SYSTEM
S 6 369 SOUTH AFRICA, NOSIB SYSTEM
S 6 372 SOUTH AFRICA, PHANEROZOIC STRATIGRAPHY, A REVIEW

Index

S 6 369 SOUTH AFRICA, PONGOLA SYSTEM
S 6 376 SOUTH AFRICA, PRECAMBRIAN, CHRONOSTRATIGRAPHY
S 6 374 SOUTH AFRICA, PRECAMBRIAN, PROPOSED SUBDIVISIONS
S 6 368 SOUTH AFRICA, PRECAMBRIAN, STRATIGRAPHY, A REVIEW
S 6 370 SOUTH AFRICA, STRATIGRAPHIC COLUMN
S 6 367 SOUTH AFRICA, STRATIGRAPHY, A REVIEW
S 6 369 SOUTH AFRICA, SWAZILAND SYSTEM
S 6 369 SOUTH AFRICA, TRANSVAAL SYSTEM
S 6 369 SOUTH AFRICA, WATERBERG SYSTEM
S 6 369 SOUTH AFRICA, WITWATERSRAND SYSTEM
S 6 367† SOUTH AFRICAN STRATIGRAPHIC CLASSIFICATION AND A PROPOSAL
B 6303 438* SOUTH AMERICA, ABST., MASS WASTING ON, CONTINENTAL, RISE OF
B 6004 734* SOUTH AMERICA, ABST., SUBMARINE-, FAN SEDIMENTATION IN,
B 6008 1397* SOUTH AMERICA, ABST., URANIUM EXPLORATION IN, PRECAMBRIAN
B 6206 917 SOUTH AMERICA, AMAZON SUBMARINE FAN
B 6207 1213* SOUTH AMERICA AS RELATED TO PACIFIC BASIN, ABST., REGIONAL
B 6411 2036 SOUTH AMERICA BY COMMISSION FOR GEOLOGICAL MAP OF THE
B 6010 1640† SOUTH AMERICA, CENTRAL AMERICA, AND CARIBBEAN AREA IN 1975,
B 6210 2044† SOUTH AMERICA, CENTRAL AMERICA, MEXICO, AND CARIBBEAN AREA IN
B 6310 1745† SOUTH AMERICA, CENTRAL AMERICA, MEXICO, AND CARIBBEAN AREA IN
B 6110 1578† SOUTH AMERICA, CENTRAL AMERICA, MEXICO, AND CARIBBEAN AREA IN
S 6 364 SOUTH AMERICA, PRECAMBRIAN, RADIOMETRIC DATING
B 6104 558† SOUTH AMERICA, REGIONALIZED, VARIABLES FOR EVALUATION OF
S 3 79 SOUTH AMERICAN ANDEAN MARGIN
M 25 540 SOUTH AMERICAN PLATE
B 6212 2456 SOUTH AMERICAN PLATES, CONVERGENCE OF, NAZCA AND
M 25 548 SOUTH AMERICAN PLATES, PLATE-TECTONIC SUMMARY, MAZCA AND
M 25 546 SOUTH AMERICAN PLATES, RISE-CONVERGENT PLATE SYSTEM, NAZCA
B 6212 2509* SOUTH AND NORTH AMERICAN CORDILLERAS, ABST., FROM, CARIBBEAN
B 6405 680* SOUTH ATLANTIC, ABST., LOWER, CRETACEOUS LACUSTRINE SOURCE
B 6002 202 SOUTH ATLANTIC, COASTAL WELLS
B 6002 196† SOUTH ATLANTIC DIAPIRIC STRUCTURES
B 6303 503* SOUTH ATLANTIC, DSDP LEGS 36, 40--SEDIMENTOLOGIC IMPLICATIONS,
B 6004 695* SOUTH ATLANTIC OCEAN, EARLY CRETACEOUS THROUGH PLIOCENE,
B 6002 198 SOUTH ATLANTIC, SALT EDGE
B 6407 989 SOUTH ATLAS FAULT, MOROCCO
B 6204 668 SOUTH ATLAS FLEXURE, ALGERIA
B 6101 80 SOUTH ATLAS FRACTURE ZONE
B 6101 88 SOUTH ATLAS FRACTURE ZONE, CORNWALL-
B 6004 654* SOUTH AUGUSTINE ISLAND, LOWER COOK INLET, ALASKA, ABST.,
B 6008 1392* SOUTH AUSTRALIA, ABST., URANIUM IN, TERTIARY CHANNEL
M 25 427 SOUTH AUSTRALIA, BEVERLEY, URANIUM
B 6110 1889 SOUTH AUSTRALIA, DEVELOPMENTS, 1976
B 6210 2151 SOUTH AUSTRALIA, DEVELOPMENTS, 1977
B 6411 1864 SOUTH AUSTRALIA, DEVELOPMENTS, 1979
B 6410 1643† SOUTH AUSTRALIA, DISTRIBUTION AND GENESIS OF SEDIMENTARY,
B 6410 1682 SOUTH AUSTRALIA, EDIACARA HILLS, PRECAMBRIAN
M 25 399 SOUTH AUSTRALIA, GARNIERITE
M 25 427 SOUTH AUSTRALIA, MIDDLEBACK RANGES, IRON ORE
M 25 427 SOUTH AUSTRALIA, NAIRNE, PYRITE
M 25 355† SOUTH AUSTRALIA--SUMMARY, SALINE, GROUNDWATER INFLOWS TO,
B 6111 1994 SOUTH AUSTRALIAN ABYSSAL PLAIN, POSSIBLE DIAPIRS
M 29 295 SOUTH BALEARIC BASIN, CROSS SECTION, MEDITERRANEAN
M 29 307 SOUTH BALEARIC-ALGERIAN BASIN, EVOLUTION, MEDITERRANEAN
B 6402 218 SOUTH BASSEIN FIELD, INDIA
B 6203 351 SOUTH BELRIDGE OIL FIELD, CALIFORNIA
B 6107 1046 SOUTH BEND FORMATION, KANSAS
B 6203 556* SOUTH CAROLINA, ABST., BACK-, BARRIER SEDIMENTATION ON,
B 6303 518* SOUTH CAROLINA, ABST., HOLOCENE STRATIGRAPHY OF RAPIDLY,
B 6303 499* SOUTH CAROLINA, ABST., LATE HOLOCENE DEPOSITIONAL HISTORY OF
B 6303 451* SOUTH CAROLINA, ABST., RIVER NEAR CHARLESTON
B 6103 337 SOUTH CAROLINA AND GULF COAST EQUIVALENTS, STRATIGRAPHIC
B 6103 332 SOUTH CAROLINA, BEAUFORT HIGH
B 6403 340 SOUTH CAROLINA, CAPE ROMAIN
B 6303 451* SOUTH CAROLINA COAST, ABST., ACCRETIONAL SHORELINE PROCESSES
B 6103 333 SOUTH CAROLINA COASTAL PLAIN, OIL AND GAS TESTS
M 28 70 SOUTH CAROLINA, CRETACEOUS, MIDDENDORF FORMATION
B 6103 332 SOUTH CAROLINA, DUNBARTON BASIN
B 6207 1164 SOUTH CAROLINA, EOCENE, OPALINE CLAYSTONES
B 6103 332 SOUTH CAROLINA, FLORENCE BASIN
B 6403 339 SOUTH CAROLINA, FLORIDA, HATTERAS SHELF, FLORIDA, GEORGIA, AND
B 6103 331† SOUTH CAROLINA, PETROLEUM GEOLOGY AND OIL AND GAS POTENTIAL
B 6103 332 SOUTH CAROLINA, SHALERS LINE
B 6103 332 SOUTH CAROLINA, STONO ARCH
B 6103 332 SOUTH CAROLINA, YAMACRAW RIDGE
B 6203 426 SOUTH CELTIC SEA BASIN, EUROPE
M 26 346 SOUTH CERES FIELD, NOBLE COUNTY, OKLAHOMA
B 6209 1654 SOUTH CHINA BASIN, MAGNETIC ANOMALIES
M 25 15 SOUTH CHINA SEA SEABED JURISDICTION
M 25 474 SOUTH CHINESE MEGASTRUCTURE
B 6203 353 SOUTH CUYAMA OIL FIELD, CALIFORNIA
B 6405 783* SOUTH DAKOTA, ABST., GEOMETRY OF SHELF SANDSTONE BODIES IN,
B 6405 679* SOUTH DAKOTA, ABST., INDIAN CREEK FIELD, FALL RIVER COUNTY
B 6203 535* SOUTH DAKOTA, ABST., SEDIMENTOLOGY OF LATE, QUATERNARY,
B 6105 832* SOUTH DAKOTA AND WYOMING AS DETERMINED BY NONMARINE
B 6406 943 SOUTH DAKOTA, BELLE FOURCHE SHALE
B 6211 2275 SOUTH DAKOTA, BLACK HILLS
S 3 93 SOUTH DAKOTA, BLACK HILLS
B 6406 943 SOUTH DAKOTA, BLACK HILLS, NORTHERN, UPPER CRETACEOUS
B 6406 943 SOUTH DAKOTA, BLUE HILL MEMBER, MINNESOTA AND
B 6406 943 SOUTH DAKOTA, CODELL SANDSTONE MEMBER, MINNESOTA AND
M 27 89 SOUTH DAKOTA, CRETACEOUS, FOX HILLS SANDSTONE
B 6008 1208 SOUTH DAKOTA, DEVELOPMENTS, 1975
B 6108 1349 SOUTH DAKOTA, DEVELOPMENTS, 1976
B 6208 1502 SOUTH DAKOTA, DEVELOPMENTS, 1977
B 6308 1388 SOUTH DAKOTA, DEVELOPMENTS, 1978
B 6409 1502 SOUTH DAKOTA, DEVELOPMENTS, 1979
B 6406 943 SOUTH DAKOTA, FAIRPORT MEMBER, MINNESOTA AND
B 6406 943 SOUTH DAKOTA, GRANEROS SHALE, MINNESOTA AND
B 6406 943 SOUTH DAKOTA, NIOBRARA FORMATION
B 6406 943 SOUTH DAKOTA, NIOBRARA FORMATION, MINNESOTA AND
B 6406 943 SOUTH DAKOTA, POOL CREEK MEMBER
B 6406 943 SOUTH DAKOTA, PRECAMBRIAN, MINNESOTA AND
B 6208 1508 SOUTH DAKOTA, PRODUCTION, 1976-1977
B 6308 1388 SOUTH DAKOTA, PRODUCTION, 1977-1978
B 6409 1502 SOUTH DAKOTA, PRODUCTION, 1978-1979
B 6406 943 SOUTH DAKOTA, SAGE BREAKS MEMBER
B 6406 943 SOUTH DAKOTA, TURNER SANDY MEMBER
S 3 119 SOUTH DAKOTA, WILLISTON BASIN, NORTH DAKOTA AND
B 6108 1226 SOUTH DUBINA FIELD, COLORADO COUNTY, TEXAS
B 6008 1256 SOUTH EXTENSION FEDERAL BLOCK 341, TEXAS, HIGH ISLAND EAST
B 6201 122 SOUTH FLORIDA BASIN
B 6209 1763* SOUTH FLORIDA BASIN, ABST., PRELIMINARY ASSESSMENT OF, ORGANIC,
M 24 289 SOUTH FLORIDA EMBAYMENT, FLORIDA
M 24 289 SOUTH FLORIDA SHELF, FLORIDA
S 7 15 SOUTH FLORIDA SLOPE, DESCRIBED
B 6301 21 SOUTH FORK MOUNTAIN THRUST, CALIFORNIA
B 6211 2268 SOUTH FORK VALLEY, CALIFORNIA
S 10 28 SOUTH GLENROCK FIELD, SHALE CORE, WYOMING
B 6409 1557* SOUTH HALLETSVILLE FIELD, LAVACA COUNTY, TEXAS, ABST., CLAY
B 6303 536* SOUTH HALLETTSVILLE, GAS FIELD IN LOWER WILCOX, LAVACA
B 6209 1762* SOUTH HANCOCK HOLOCENE BARRIER TRENDS AND ORIGINS OF LAKE
B 6203 504* SOUTH ISLAND, NEW ZEALAND, ABST., SEDIMENTOLOGY OF, CAPLES
M 25 221 SOUTH ISLAND, STRUCTURE MAP, NEW ZEALAND
B 6410 1682 SOUTH KAZAKSTAN, PRECAMBRIAN MICROFOSSILS, USSR
M 25 443 SOUTH KOREA, ANDONG ANTICLINE
M 25 443 SOUTH KOREA, BONGHWAJAE THRUST
M 25 440 SOUTH KOREA, BULKUKSA GRANITE
M 25 444 SOUTH KOREA, BULKUKSA GRANITES
M 25 443 SOUTH KOREA, CHUGARYONG RIFT VALLEY
M 25 440 SOUTH KOREA, CHUNSONG GROUP
M 25 444 SOUTH KOREA, COPPER
M 25 443 SOUTH KOREA, CRETACEOUS, LATE, TO TERTIARY DEFORMATION
M 25 446 SOUTH KOREA, CRETACEOUS, LATE, TO TERTIARY, EARLY,
M 25 440 SOUTH KOREA, DAEBO GRANITE
M 25 444 SOUTH KOREA, DAEBO GRANITES
M 25 443 SOUTH KOREA, DAEDONG SEDIMENTARY ROCKS
B 6110 1837 SOUTH KOREA, DEVELOPMENTS, 1976
B 6210 1987 SOUTH KOREA, DEVELOPMENTS, 1977
B 6310 1826 SOUTH KOREA, DEVELOPMENTS, 1978
B 6411 1893 SOUTH KOREA, DEVELOPMENTS, 1979
M 25 445 SOUTH KOREA, DONGJOM QUARTZITE
M 25 444 SOUTH KOREA, FLUORITE
M 25 441 SOUTH KOREA, FUKIEN-REINAN MASSIF
M 25 261 SOUTH KOREA, FUKIEN-REINAN MASSIF
M 25 442 SOUTH KOREA, GEOLOGIC AND TECTONIC MAP
M 25 441 SOUTH KOREA, GEOLOGIC PROVINCES
M 25 441 SOUTH KOREA, GEOLOGIC SEQUENCE, GENERALIZED, TABLE
M 25 440 SOUTH KOREA, GEOLOGIC SETTING
M 25 444 SOUTH KOREA, GOLD-SILVER
M 25 440 SOUTH KOREA, GRANITE GNEISS SYSTEM
M 25 443 SOUTH KOREA, GRANITES
M 25 444 SOUTH KOREA, GRAPHITE
M 25 441 SOUTH KOREA, HWANGGANGRI FORMATION, TILLITIC ORIGIN
M 25 443 SOUTH KOREA, ILWOLSAN UPTHRUST
M 25 444 SOUTH KOREA, IRON
M 25 440 SOUTH KOREA, JANGRAK GROUP
M 25 443 SOUTH KOREA, JOMCHON THRUST
M 25 443 SOUTH KOREA, JURASSIC DEFORMATION, DAEBO OROGENY
M 25 446 SOUTH KOREA, JURASSIC TO CRETACEOUS, EARLY, METALLOGENI

M 25 444 SOUTH KOREA, KAOLIN
M 25 441 SOUTH KOREA, KYONGGI MASSIF
M 25 447 SOUTH KOREA, KYONGSANG BASIN, MINERALS
M 25 441 SOUTH KOREA, KYONGSANG SEDIMENTARY BASINS
M 25 444 SOUTH KOREA, LEAD
M 25 440 SOUTH KOREA, MESOZOIC GRANITES
M 25 445 SOUTH KOREA, METALLOGENIC EPOCHS
M 25 446 SOUTH KOREA, METALLOGENIC PROVINCES
M 25 447 SOUTH KOREA, MINERAL DEPOSITS, ZONAL DISTRIBUTION
M 25 444 SOUTH KOREA, MINERALS
M 25 444 SOUTH KOREA, MOLYBDENUM
M 25 441 SOUTH KOREA, OKCHON GEOSYNCLINAL ZONE
M 25 440 SOUTH KOREA, OKCHON SYSTEM
M 25 443 SOUTH KOREA, OKCHON THRUSTS
M 25 444 SOUTH KOREA, ORE DEPOSITS, CLASSIFICATION OF
M 25 445 SOUTH KOREA, PALEOZOIC AND OTHER METALLOGENIC PROVINCES
M 25 442 SOUTH KOREA, PRE-TRIASSIC DEFORMATION
M 25 440 SOUTH KOREA, PRECAMBRIAN GEOLOGY
M 25 444 SOUTH KOREA, PRECAMBRIAN METALLOGENIC PROVINCES
M 25 440 SOUTH KOREA, PUCHON ROCKS
M 25 444 SOUTH KOREA, PYROPHYLLITE
M 25 441 SOUTH KOREA, RYONGNAM MASSIF
M 25 440 SOUTH KOREA, RYONGNAM SYSTEM
M 25 444 SOUTH KOREA, SCHEELITE DEPOSITS
M 25 440 SOUTH KOREA, SIHUNG ROCKS
M 25 447 SOUTH KOREA, SILLA SERIES, UPPER, MINERALS
M 25 443 SOUTH KOREA, SOBAEKSAN ANTICLINE
M 25 441 SOUTH KOREA, SOBAEKSAN MASSIF
M 25 440 SOUTH KOREA, STRATIGRAPHY
M 25 444 SOUTH KOREA, TALC
M 25 442 SOUTH KOREA, TECTONISM AND RELATED GRANITES
M 25 441 SOUTH KOREA, TERTIARY BASINS
M 25 443 SOUTH KOREA, TONGRAE FAULT, HEAT FLOW, HIGH
M 25 442 SOUTH KOREA, TRIASSIC DEFORMTION, SONGRIM DISTURBANCE
M 25 444 SOUTH KOREA, TUNGSTEN
M 25 444 SOUTH KOREA, URANIUM
M 25 440 SOUTH KOREA, YANGPYONG ROCKS
M 25 443 SOUTH KOREA, YANGSAN FAULT, HEAT FLOW, HIGH
M 25 440 SOUTH KOREA, YONCHON SYSTEM
M 25 443 SOUTH KOREA, YULRI SYNCLINE
M 25 440 SOUTH KOREA, YULRI SYSTEM
M 25 444 SOUTH KOREA, ZINC
M 24 99 SOUTH LA BARGE FIELD, WYOMING
B 6209 1762* SOUTH LAREDO AREA, WEBB AND ZAPATA COUNTIES, TEXAS, ABST.,
B 6104 587 SOUTH MARSH ISLAND FIELDS, LOUISIANA
B 6402 256 SOUTH MATAPAN TROUGH, SEISMIC REFLECTION, GREECE
B 6407 1036 SOUTH MOUNTAIN FIELD, CALIFORNIA
B 6203 360 SOUTH MOUNTAIN OIL FIELD, CALIFORNIA
B 6405 678* SOUTH PACIFIC, ABST., MIOCENE SPUMELLARIAN, RADIOLARIA FROM
M 25 386 SOUTH PACIFIC, MANGANESE NODULE DISTRIBUTION
M 25 386† SOUTH PACIFIC--SUMMARY, EXPLOITATION OF, MANGANESE, NODULES
S 3 124 SOUTH PARK, EOCENE SANDSTONE, COLORADO
B 6402 155 SOUTH PASS BLOCK 24 FIELD, LOUISIANA
B 6402 165 SOUTH PASS BLOCK 27 FIELD, LOUISIANA
B 6108 1226 SOUTH PEACH POINT, BRAZORIA COUNTY, TEXAS
B 6112 2086 SOUTH PECAN LAKE FIELD, KEROGEN, LOUISIANA
B 6303 413* SOUTH PHILIPPINE SEA, DSDP LEGS 59 AND 60, ABST., CENOZOIC,
B 6104 587 SOUTH PLASS BLOCK FIELD, LOUISIANA
B 6104 595 SOUTH PYOTE FIELD, TEXAS
B 6009 1609* SOUTH SAREPTA FIELD, GRAY SANDSTONE, SMACKOVER, BOSSIER AND
M 25 263 SOUTH SEA, EAST CHINA SEA, KOREA OFFSHORE
B 6004 709* SOUTH SUMATRA AND JAV SEA BASINS--MECHANISM OF BASIN
M 25 244 SOUTH SUMATRA BASINS, INDONESIA, CENTRAL AND
B 6402 218 SOUTH TAPTI FIELD, INDIA
B 6102 296* SOUTH TEXAS AUSTIN CHALK, CRETACEOUS, PLAY, ABST.
B 6109 1535* SOUTH TEXAS COASTAL PLAIN, ABST., LITHOFACIES ANALYSIS AND,
B 6007 1119 SOUTH TEXAS GEOLOGICAL SOCIETY, HONORARY MEMBERS, 1935-1975
B 6007 1117† SOUTH TEXAS GEOLOGICAL SOCIETY--A FRAGMENTARY INFORMAL
B 6008 1365† **SOUTH TEXAS IN 1975, DEVELOPMENTS IN**
B 6208 1511† SOUTH TEXAS IN 1977, DEVELOPMENTS IN
B 6308 1338† SOUTH TEXAS IN 1978, DEVELOPMENTS IN
B 6409 1476† SOUTH TEXAS IN 1979, DEVELOPMENTS IN
B 6105 795* SOUTH TEXAS MINERAL BELT—REVIEW
B 6105 834* SOUTH TIMBALIER BLOCK 54 SALT DOME, ABST., PALEONTOLOGIC
B 6209 1768* SOUTH TIMBALIER BLOCK 54 SALT DOME AND SALT DOME GROWTH
M 30 49 SOUTH VIKING COMPLEX POOLS, VIKING FIELD
B 6010 1873 SOUTH YEMEN, 1975, SIEBENS OIL AND GAS LTD., ON DEVELOPMENTS IN
B 6110 1802 SOUTH YEMEN, 1976, SIEBENS OIL AND GAS, LTD., ON DEVELOPMENTS IN
B 6210 2118 SOUTH YEMEN, 1977, SIEBENS OIL, GAS LTD., AGIP S.P.A., AND,
B 6301 1890 SOUTH YEMEN, 1978, SIEBENS OIL AND GAS LTD., AGIP S.P.A., AND
B 6411 1841 SOUTH YEMEN, PEOPLES DEMOCRATIC REPUBLIC OF YEMEN, 1979, AGIP
B 6110 1750 SOUTH-WEST AFRICA NAMIBIA, DEVELOPMENTS, 1976
B 6210 1849 SOUTH-WEST AFRICA, NAMIBIA, DEVELOPMENTS, 1977
B 6310 1695 SOUTH-WEST AFRICA, NAMIBIA, DEVELOPMENTS, 1978
B 6408 1195 SOUTH-WEST AFRICAN SHELF, BENGUELA CURRENT, ANOXIC LAYER
B 6004 678 SOUTHAM, J. R., CARBONATE MASS BALANCE--CYCLING AND DEPOSITION
B 6203 520 SOUTHAM, JOHN R., BINARY SIMILARITY COEFFICIENTS AND
B 6203 563 SOUTHAM, JOHN R., BURNING OF FOSSIL FUELS AND OCEAN CHEMISTRY
B 6105 794 SOUTHAM, JOHN R., POST-PALEOZOIC CHANGES IN VOLUMES AND ORGANIC
B 6405 681 SOUTHARD, J. B., TIDAL-CURRENT SAND WAVES IN VINEYARD SOUND,
B 6203 497 SOUTHARD, JOHN B., DYNAMIC SCALE MODELING OF BED
M 29 376 SOUTHEAST ASIA, INNER SULU SEA, OCEANIC BASIN
M 29 374 SOUTHEAST ASIA, OUTER SULU SEA, DEEP BASIN
M 29 373 SOUTHEAST ASIA, SEISMIC RESULTS
M 29 373 SOUTHEAST ASIA, STRUCTURAL FRAMEWORK
M 29 374 SOUTHEAST ASIA, STRUCTURAL FRAMEWORK, INDEX
M 29 380 SOUTHEAST ASIA, STRUCTURE, SUBDUCTION ZONES
M 29 375 SOUTHEAST ASIA, SULU SEA, BATHYMETRY
M 29 376 SOUTHEAST ASIA, SULU SEA, PLATE BOUNDARIES
M 29 376 SOUTHEAST ASIA, SULU SEA, SCHEMATIC SECTION
M 29 379 SOUTHEAST ASIA, SULU TROUGH, MULTICHANNEL SEISMICS
M 29 373† SOUTHEAST ASIA, THE, SULU SEA: A, MARGINAL, BASIN IN
B 6403 340 SOUTHEAST GEORGIA EMBAYMENT
B 6205 793 SOUTHEAST GEORGIA EMBAYMENT
B 6103 336 SOUTHEAST GEORGIA EMBAYMENT
B 6303 452* SOUTHEAST GEORGIA EMBAYMENT, ABST., GEOLOGY OF
B 6303 510* SOUTHEAST GEORGIA EMBAYMENT, ABST., IMPORTANT STRATIGRAPHIC
B 6109 1523 SOUTHEAST GEORGIA EMBAYMENT, ACCUMULATION RATES, EAST
B 6309 1583* SOUTHEAST GEORGIA EMBAYMENT AND FLORIDA-HATTERAS SLOPE,
M 29 27† SOUTHEAST GEORGIA EMBAYMENT AND NORTHERN BLAKE PLATEAU:
M 29 29 SOUTHEAST GEORGIA EMBAYMENT, INDEX
M 29 33 SOUTHEAST GEORGIA EMBAYMENT, REFLECTORS, INTERPRETATION
M 29 11 SOUTHEAST GEORGIA EMBAYMENT, SEE GEORGIA EMBAYMENT
M 29 11† SOUTHEAST GEORGIA EMBAYMENT, U.S. ATLANTIC CONTINENTAL
M 29 173 SOUTHEAST INDIAN RIDGE, SPREADING, PALEOCENE
B 6409 1558* SOUTHEAST LITTLE PECAN LAKE FIELDS, CAMERON PARISH, LOUISIANA,
B 6104 587 SOUTHEAST PASS FIELD, LOUISIANA
S 7 15 SOUTHEASTERN GULF FLOOR, DESCRIBED
B 6106 901 SOUTHEASTERN MEDITERRANEAN SEA, PLATE TECTONICS
B 6106 884 SOUTHEASTERN MEDITERRANEAN SEA, SEISMIC PROFILES
B 6411 2026† SOUTHEASTERN STATES, DEVELOPMENTS IN
B 6008 1277† SOUTHEASTERN STATES IN 1975, DEVELOPMENTS IN
B 6108 1259† SOUTHEASTERN STATES IN 1976, DEVELOPMENTS IN
B 6208 1390† SOUTHEASTERN STATES IN 1977, DEVELOPMENTS IN
B 6308 1399† SOUTHEASTERN STATES IN 1978, DEVELOPMENTS IN
B 6210 2055 SOUTHEASTERN ZONE, CAMPECHE BAY, MEXICO
B 6310 1756 SOUTHEASTERN ZONE, MEXICO
B 6207 1236 SOUTHER, J. G., CANADIAN GEOTHERMAL RESEARCH PROGRAM, ABST.
B 6406 862 SOUTHERN BASIN, EGYPT
M 25 257 SOUTHERN BASIN, PROSPECTS, TAIWAN
B 6203 346 SOUTHERN CALIFORNIA BASIN, CALIFORNIA
B 6212 2513* SOUTHERN CALIFORNIA—TECTONICS AND OIL ACCUMULATION, ABST.
B 6002 266 SOUTHERN CARMEL REEF, ISRAEL
B 6209 1753* SOUTHERN FLORIDA SUBSURFACE FEATURES RELATED TO OIL
B 6203 567* SOUTHERN OKLAHOMA OIL COUNTRY IN CONTEXT OF PLATE
S 3 80 SOUTHERN ROCKY MOUNTAIN OROGENY
B 6402 221† SOUTHERN VAL VERDE BASIN, SOUTHWEST TEXAS, STRUCTURAL
B 6305 832* SOUTHMAN CANYON GAS FIELD, UINTA BASIN, UTAH, ABST., INFLUENCE
B 6411 1790 SOUTHWEST AFRICA, NAMIBIA, DEVELOPMENTS, 1979
B 6104 583 SOUTHWEST ENVILLE POOL, OKLAHOMA
B 6403 340 SOUTHWEST GEORGIA EMBAYMENT
M 29 43† SOUTHWEST ICELAND INSULAR MARGIN: THIRTEEN GEOPHYSICAL
B 6207 1213* SOUTHWEST PACIFIC, ABST., GEOCHEMICAL EXPLORATION FOR,
B 6010 1947† SOUTHWEST PACIFIC ISLAND REGION, DEVELOPMENTS IN, NEW ZEALAND
B 6110 1880† SOUTHWEST PACIFIC ISLAND REGION IN 1976, DEVELOPMENTS IN, NEW
B 6210 1900† SOUTHWEST PACIFIC ISLAND REGION IN 1977, DEVELOPMENTS IN, NEW
B 6310 1680† SOUTHWEST PACIFIC ISLAND REGION IN 1978, DEVELOPMENTS IN, NEW
B 6411 1776† SOUTHWEST PACIFIC ISLAND REGION IN 1979, DEVELOPMENTS IN, NEW
B 6210 1901 SOUTHWEST PACIFIC ISLAND REGION, PRODUCTION, 1977, NEW ZEALAND
M 25 433 SOUTHWEST PACIFIC, PORPHYRY COPPER DEPOSITS, ASSOCIATED ROCKS
B 6008 1402* SOUTHWEST RAGGED POINT FIELD, MUSSELSHELL COUNTY,
B 6410 1722 SOVIET ENERGY SYSTEM, BY LESLIE DIENES AND THEODORE SHABAD,
B 6110 1650 SOVIET UNION, DEVELOPMENTS, 1976
B 6210 1921 SOVIET UNION, DEVELOPMENTS, 1977
B 6310 1929 SOVIET UNION, DEVELOPMENTS, 1978
B 6002 184 SP LOGS, NORTH SEA
B 6212 2510* SPACE, ABST., WHICH WAY IS UP IN
B 6207 1220* SPACE—POTENTIAL AND PROBLEMS, ABST., FUTURE GEOLOGIC, REMOTE

B 6010 1718 SPAIN, 1975, MINISTERIO DE INDUSTRIA, ON DEVELOPMENTS IN
B 6110 1648 SPAIN, 1976, MINISTERIO DE INDUSTRIA, ON DEVELOPMENTS IN
B 6210 1918 SPAIN, 1977, MINISTERIO DE INDUSTRIA, ON DEVELOPMENTS IN
B 6310 1927 SPAIN, 1978, DIRECCION GENERAL DE LA ENERGIA, ON DEVELOPMENTS
B 6303 480* SPAIN, ABST., CALCIFICATION AND SIGNIFICANCE OF SOI FILAMENTOUS
B 6405 775* SPAIN, ABST., CASTILLO GAS FIELD
B 6405 675* SPAIN, ABST., DEVONIAN, REEFS EXPOSED ALONG CENTRAL, CANTABRIC
B 6203 514* SPAIN, ABST., ONCOLITES FROM, LACUSTRINE SEDIMENTS IN,
B 6305 835* SPAIN, ABST., RECENT, SALT SEDIMENTATION IN, PLAYA LAKES OF, EBRO
B 6004 705* SPAIN, ABST., SEDIMENTATION HISTORY OF LATE, CARBONIFEROUS,
B 6411 1968 SPAIN, ALBORAN SEA
B 6411 1968 SPAIN, BALEARIC OFFSHORE
B 6411 1968 SPAIN, BAY OF BISCAY
B 6211 2232 SPAIN, BAY OF BISCAY AREA
B 6402 189 SPAIN, CABO DE GATA VOLCANIC COMPLEX
B 6402 189 SPAIN, CAMPO DE NIJAR SECTION
B 6402 189 SPAIN, CUEVAS VIEJAS FORMATION
B 6402 189 SPAIN, CUEVAS VIEJOS FORMATION
B 6411 1967 SPAIN, DEVELOPMENTS, 1979
B 6411 1968 SPAIN, GULF OF CADIZ
B 6411 1968 SPAIN, GULF OF VALENCIA
M 27 221 SPAIN, JURASSIC, RONDA UNIT
M 27 88 SPAIN, JURASSIC, RONDA UNIT
M 27 111 SPAIN, JURASSIC, RONDA UNIT
M 27 194 SPAIN, JURASSIC, RONDA UNIT
B 6402 189 SPAIN, MORALES FORMATION
B 6402 189 SPAIN, MORALES FORMATION
B 6402 189 SPAIN, MUERTOS CONGLOMERATE MEMBER
B 6402 189 SPAIN, MUERTOS CONGLOMERATE MEMBER
B 6402 189 SPAIN, NEOGENE SEDIMENTARY AND OLDER BASEMENT ROCKS
B 6402 189 SPAIN, NEOGENE SEDIMENTARY AND OLDER BASEMENT ROCKS
B 6402 189 SPAIN, NIJAR FORMATION
B 6402 189 SPAIN, NIJAR FORMATION
B 6402 189 SPAIN, PLEISTOCENE
B 6402 189 SPAIN, PLEISTOCENE
B 6402 189 SPAIN, PLIOCENE
B 6402 189 SPAIN, PLIOCENE
B 6402 189 SPAIN, PLOMO FORMATION
B 6402 190 SPAIN, PLOMO REEFS
B 6402 188† SPAIN, POROSITY EVOLUTION OF UPPER, MIOCENE, REEFS, ALMERIA
B 6110 1654 SPAIN, PRODUCTION, 1975-1976
B 6210 1924 SPAIN, PRODUCTION, 1976-1977
B 6310 1932 SPAIN, PRODUCTION, 1977-1978
B 6411 1971 SPAIN, PRODUCTION, 1978-1979
B 6402 189 SPAIN, SORBAS BASIN
B 6402 189 SPAIN, VICAR FORMATION
B 6405 751 SPANG, J. H., GEOMETRY AND MECHANISMS OF TRANSVERSE FAULTING,
B 6405 796 SPANG, J. H., NATURE OF NON-RECOVERABLE POROSITY CHANGES IN
B 6304 701 SPANG, JOHN H., STRUCTURAL ANALYSIS OF NORTHERN TERMINATION
B 6211 2199 SPANISH POINT GROWTH FAULT, IRELAND, WESTERN
B 6107 1070 SPAR SUBSTRATES, MICRITIZATION OF, ICELAND
S 7 307 SPARKER PROFILE, CENTRAL, ORCA BASIN
S 7 307 SPARKER PROFILE, SOUTHERN, ORCA BASIN
M 26 358 SPARKER SYSTEMS, RESOLUTION SYSTEMS
B 6403 448 SPARKS, DENNIS M., GEOLOGICAL ASSESSMENT OF PETROLEUM
B 6107 1014 SPARKY MEMBER, ALBERTA AND, SASKATCHEWAN
M 30 71 SPARNACIAN, FRIGG FIELD, PALEOGEOGRAPHY
B 6408 1169 SPARTA FACIES, TEXAS
B 6109 1544* SPARTA FORMATION, EOCENE GULF COAST BASIN OF TEXAS, ABST.,
B 6402 169 SPARTA SANDSTONE, TEXAS
M 27 207 SPASTOLITHS
M 26 11 SPATIAL RESOLUTION, SEISMIC, DATA
B 6004 709* SPATIAL VARIATION IN POROSITY--NEW CONCEPT, ABST.
B 6303 531 SPAULDING A. O., GEOLOGISTS AND POLITICS--A MATTER OF SURVIVAL,
B 6012 2189 SPAULDING, ARTHUR O., LEGISLATION AND TOMORROWS OIL FROM
B 6309 1597 SPAW, JOAN MUSSLER, ECOLOGIC NICHES OF RADIOLARIANS,
B 6303 531 SPAW, JOAN MUSSLER, RADIOLARIA--PRESENT AND PAST INDICATORS OF
B 6405 688 SPAW, JOAN MUSSLER, RADIOLARIAN CHARACTERISTICS OF MAJOR
B 6309 1598 SPAW, JOAN MUSSLER, RADIOLARIAN DISTRIBUTION, DIVERSITY, AND
B 6303 483 SPAW, JOAN MUSSLER,, RADIOLARIAN PRESERVATION IN PRESENT AND
B 6309 1597 SPAW, JOAN MUSSLER, RELICT AND EXPATRIATE RADIOLARIAN FAUNA
B 6209 1767 SPAW, RICHARD H., LATE PLEISTOCENE CARBONATE BANK DEPOSITION,
B 6203 505* SPEAKEASY—COMPUTER LANGUAGE FOR GEOLOGISTS, ABST.
B 6211 2280 SPEARFISH FORMATION, BLACK HILLS
B 6008 1406* SPEARHEAD RANCH FIELD, CONVERSE COUNTY WYOMING, ABST.
B 6108 1393* SPEARHEAD RANCH FIELD, WYOMING, ABST., DIAGENESIS OF,
B 6409 1552* SPEARY GAS FIELD, KARNES COUNTY, TEXAS, ABST., ENVIRONMENTS OF
B 6203 563 SPECHT, RALPH W., MOLLUSCAN BIOCLASTIC ACCUMULATIONS IN
B 6004 678* SPECIES AND PHENOTYPES OF PLANKTONIC FORAMINIFERA, ABST.,
S 10 41 SPECIFIC VOLUME AT DEPTH, MIGRATION
B 6004 666* SPECTRAL AND MARKOVIAN ANALYSIS OF BED THICKNESS AND
B 6305 845* SPECTROMETER DATA FOR URANIUM EXPLORATION IN EASTER
B 6007 1112 SPECTROMETRIC ANALYSIS, GRAND RAPIDS OIL, MASS
B 6007 1112 SPECTROMETRIC ANALYSIS, LEDUC OIL, MASS
B 6007 1112 SPECTROMETRIC ANALYSIS, LLOYDMINSTER OIL, MASS
B 6007 1112 SPECTROMETRIC ANALYSIS, MCMURRAY OIL, MASS
B 6405 688* SPECTRUM OF DELTAIC SYSTEMS IN PENNSYLVANIAN TECTONICALLY
B 6303 464* SPECTRUM OF DEPOSITIONAL MODELS FOR CLASTIC STRANDLINE
B 6303 498* SPECULATIONS ON BASEMENT MOBILITY IN GULF OF MEXICO, ABST.
B 6403 440* SPECULATIONS ON HYDROCARBON POTENTIAL OF DEEP-WATER BASINS
B 6303 418* SPECULATIONS ON PETROLEUM RESOURCE POTENTIAL OF ANTARCTICA,
M 24 23 SPEERS, R. G., PERMO-TRIASSIC RESERVOIRS OF PRUDHO BAY FIELD,
B 6203 523* SPELEOTHEMS AND EVAPORITE SOLUTION COLLAPSE IN ATHAPUSCOW
B 6403 446* SPENCER, ABST., TWENTIETH-CENTURY HISTORY OF, GULF OF, ALASKA
B 6405 738 SPENCER, C. W., EVALUATION OF ORGANIC MATTER AND SUBSURFACE
B 6004 724 SPENCER, CHARLES W., MODELS USING SUBSURFACE DATA FO
M 27 56 SPHAERACTINIDS
S 11 261 SPHAEROIDINA BULLOIDES
B 6306 951 SPHALERITE, NEWFOUNDLAND
S 11 294 SPHAREROIDINA BULLOIDES
M 28 60 SPHENE, CRYSTALS, IN GRAIN MOUNT
M 28 54 SPHERE, CRYSTALS, IN CALCITIC MARBLE
B 6303 370† SPHERULITES OF SAN SEBASTIAN FORMATION, OLIGOCENE, NORTHERN
S 4 128 SPICULES IN SEDIMENTS, SILICEOUS
B 6106 861 SPIKER, ELLIOTT, SURFICIAL SAND SHEET, BALTIMORE CANYON TROUGH
B 6004 678* SPILL, ABST., GREAT, PATAGONIAN, OIL
B 6105 836* SPILL POTENTIAL ON ATLANTIC OUTER CONTINENTAL SHELF, ABST.,
B 6303 433* SPILLE OIL AND INFAUNAL ACTIVITY--MODIFICATION OF BURROWING
B 6409 1554* SPILLS, IXTOC I AND BURMAH AGATE, WITH CEMENTS ON AVENUES OF
B 6409 1554* SPILLS, IXTOC I AND BURMAH AGATE, WITH COMMENTS ON AVENUE OF
B 6303 510* SPINDLE FIELD, COLORADO, ABST., DIAGENESIS IN, HYGIENE AND, TERRY
B 6405 801* SPINDLE FIELDS—PALEOSTRUCTURAL AND STRATIGRAPHIC TRAPS,
M 26 278 SPINDLE OIL FIELD, COLORADO
M 28 17 SPINDLE TWINNING, FELDSPAR, MICROCLINE
B 6109 1547 SPINDLER, WILLIAM MATHAIS, III, STRUCTURE AND STRATIGRAPHY OF
B 6009 1434† SPINES OF MOVEMENT, SHEAR ZONES INSIDE, GULF COAST, SALT, STOCK
B 6107 1090† SPINES OF MOVEMENT. DISCUSSION AND REPLY, SHEAR ZONES INSIDE,
B 6405 787 SPIRAKIS, CHARLES S., ANOMALOUS THERMOLUMINESCENCE AROUND
B 6204 705 SPIRAKIS, CHARLES S., POSITION OF SUB-KASKASKIA UNCONFORMITY IN
M 26 60 SPIRIT RIVER FORMATION, CANADA
S 11 270 SPIROLOCULINA DEPRESSA
S 11 232 SPIROLUCULINA DEPRESSA
S 6 341 SPJELDNAES, NILS, THE SILURIAN SYSTEM
B 6303 532 SPLETTSTOESSER, JOHN F., COAL IN ANTARCTICA, ABST.
B 6009 1616* SPLUNGE FIELDS, BLACK WARRIOR BASIN, MONROE COUNTY,
B 6004 725 SPOLJARIC, NENAD,, GLAUCONITE-MAGHEMITE-HEMATITE
M 27 184 SPONGE BIOEROSION
M 27 184 SPONGE BORINGS
M 27 51 SPONGE SPICULES
M 27 50 SPONGE SPICULES
M 28 78 SPONGE SPICULES, IN CHERT MATRIX
S 4 132 SPONGES AND REEFS, BURROWING
S 4 127† SPONGES AS REEF BUILDERS AND SHAPERS
S 4 270 SPONGES, BORERS
M 27 49 SPONGES, CALCAREOUS
S 4 127 SPONGES, WESTERN, ATLANTIC REGION, SPECIES OF
B 6009 1616 SPOONER, HARRY, JR., FOURMILE CREEK AND SPLUNGE FIELDS, BLACK
M 30 238 SPOROPHYTES, USSR, RIPHEAN PALEONTOLOGY
M 30 238 SPOROPHYTES, USSR, VENDIAN PALEONTOLOGY
S 7 117 SPOTTS, J. H., GULF OF MEXICO CONTINENTAL-SLOPE SEDIMENTS AND
B 6012 2192 SPOTTS, J. H., RESERVOIR AND SOURCE-BED HISTORY IN GREAT VALLEY,
B 6105 844 SPOTTS, J. H., RESERVOIR AND SOURCE-BED HISTORY IN GREAT VALLEY,
B 6205 813 SPOTTS, J. H., RESERVOIR AND SOURCE-BED HISTORY OF GREAT VALLEY,
B 6308 1426* SPRABERRY SANDSTONE, MIDLAND BASIN, WEST TEXAS, ABST.,
M 25 354 SPRAGG, W. T., ARTIFICIAL RECHARGE OF GR
M 29 355 SPRAT HALL CANYON, VIRGIN ISLANDS, ST. CROIX
B 6403 366 SPREADING AND UNCONFORMITIES OF OROGENY DURING LATE
B 6403 359† SPREADING DURING LATE JURASSIC-LATE TERTIARY TIME,
M 29 194 SPREADING EXTENSION, BURMA, ANDAMAN SEA
M 29 173 SPREADING, PALEOCENE, SOUTHEAST INDIAN RIDGE
B 6101 113 SPRING CANYON MEMBER, UTAH
M 26 453 SPRINGER OVERPRESSURING, PREDICTABILITY, MORROW-
M 26 453 SPRINGER SANDSTONE, OKLAHOMA, PENNSYLVANIAN
M 26 452 SPRINGER SANDSTONE OVERPRESSURING, OKLAHOMA, ANADARKO
M 26 451† SPRINGER SANDSTONES, WATONGA-CHICKASHA TREND: GEARY,
B 6309 1585 SPRINGER, THOMAS E., COMPUTER-AIDED GRAPHICS FOR ANALYSIS OF
B 6309 1588 SPRINGER, THOMAS E., COTTAGEVILLE GAS FIELD CORRELATION
B 6002 327* SPRINGER-CHESTER RELATIONS IN ANADARKO BASIN. BY TITLE ONLY,
B 6104 562 SPRINGHILL FORMATION, CHILE
M 29 126 SPRINGHILL FORMATION, PETROLEUM PRODUCTION, MAGELLAN BASIN

B 6111 1951 SPRINGS HEAVY OIL DEPOSIT, UNITED STATES
B 6403 449 SPROUSE, DONALD W., YOWLUMNE OIL FIELD. BY TITLE ONLY
B 6207 1172 SPRUCE MOUNTAIN, SECTION, NEVADA
S 7 278 SQUEEZING AND DIAPIRISM, BOTTOM, DEFORMATION
S 7 278 SQUEEZING AND DIAPIRISM, DIAGRAM, STABILITY
B 6303 416 SQUYRES, C. H., MODERN EXPLORATION STRATEGY IN FRONTIER AREAS,
B 6405 681* SR DATING OF DIAGENESIS—EPIGENESIS IN MESOZOIC ROCKS, SAN JUAN
B 6209 1676 SR GEOCHRONOLOGY, RB-
M 30 244 SREDNE-BOTUOBIN FIELD, EXPLORATION HISTORY, USSR
B 6410 1691 SREDNE-BUTUOBIN DISCOVERY, USSR
B 6110 1840 SRI LANKA, DEVELOPMENTS, 1976
B 6210 1988 SRI LANKA, DEVELOPMENTS, 1977
B 6310 1827 SRI LANKA, DEVELOPMENTS, 1978
B 6411 1893 SRI LANKA, DEVELOPMENTS, 1979
B 6002 161† ST LAWRENCE ISLAND, STRUCTURE AND EVOLUTION OF, BERING SEA,
S 4 85 ST. ANNS INLIER, JAMAICA
S 4 78 ST. BARTS, CARBONATE SHELF
B 6409 1569* ST. BERNARD PARISH, LOUISIANA, ABST., BLOCK 25 FIELD, CHANDELEUR
B 6104 511 ST. BRICE-VILLENEUVE WELLS, FRANCE
B 6410 1580 ST. CLAIR BLOCK, VIRGINIA
B 6204 706 ST. CLAIR LIMESTONE, ILLINOIS BASIN
B 6410 1576 ST. CLAIR THRUST, VIRGINIA
S 4 335 ST. CROIX, AMPHISTEGINA GRAINSTONES
S 4 345 ST. CROIX, ANEGADA PASSAGE
B 6406 853 ST. CROIX BASIN, SAND LAYERS
M 29 356 ST. CROIX, BATHYMETRY, VIRGIN ISLANDS
S 4 283 ST. CROIX, BOILER BAY, ECHINOMETRA LUCUNTER
S 4 283 ST. CROIX, BOILER BAY, PORITES ASTREOIDES
B 6203 414 ST. CROIX, CALEDONIA FORMATION, VIRGIN ISLANDS
B 6203 404 ST. CROIX, CENTRAL LIMESTONE PLAIN, VIRGIN ISLANDS
S 8 139 ST. CROIX, CHRISTIANSTED CANYON, WEST INDIES
S 4 341 ST. CROIX, CORONOCYCLUS
B 6203 416 ST. CROIX, CRETACEOUS, VIRGIN ISLANDS
S 4 339 ST. CROIX, CYCLICARGOLITHUS FLORIDANUS
B 6203 414 ST. CROIX, EAST END RANGE, VIRGIN ISLANDS
B 6203 404 ST. CROIX, EAST RANGE, VIRGIN ISLANDS
S 4 336 ST. CROIX, FACIES, VIRGIN ISLANDS
B 6203 404 ST. CROIX, FOUNTAIN VALLEY, VIRGIN ISLANDS
M 29 355 ST. CROIX, FREDERICKSTED CANYON, VIRGIN ISLANDS
S 4 330 ST. CROIX, GEOLOGIC MAP, VIRGIN ISLANDS
S 4 341 ST. CROIX, GLOBOROTALIA MAYERI
B 6203 414 ST. CROIX, HAMS BLUFF, VIRGIN ISLANDS
S 4 145 ST. CROIX, ISAAC BAY, REEF
S 4 331 ST. CROIX, JEALOUSY FORMATION
B 6203 414 ST. CROIX, JUDITH FANCY FORMATION, VIRGIN ISLANDS
S 4 332 ST. CROIX, KINGSHILL MARL
B 6203 403 ST. CROIX, KINGSHILL MARL, VIRGIN ISLANDS
B 6203 404 ST. CROIX, NORTHSIDE RANGE, VIRGIN ISLANDS
S 4 335 ST. CROIX, OPERCULINOIDES
S 4 341 ST. CROIX, ORBULINA
M 29 354 ST. CROIX, PHYSIOGRAPHY, VIRGIN ISLANDS
S 8 129 ST. CROIX, SALT RIVER CANYON, WEST INDIES
M 29 355 ST. CROIX, SEAFLOOR MORPHOLOGY, VIRGIN ISLANDS
M 29 355 ST. CROIX, SHEPARD CANYON, VIRGIN ISLANDS
S 4 335 ST. CROIX, SORITES MARGINALIS
M 29 355 ST. CROIX, SPRAT HALL CANYON, VIRGIN ISLANDS
M 29 359 ST. CROIX, STRATIGRAPHIC SECTIONS, VIRGIN ISLANDS
M 29 359 ST. CROIX, SUBSURFACE GEOLOGY, VIRGIN ISLANDS
B 6203 406 ST. CROIX, TECTONICS, VIRGIN ISLANDS
S 4 329† ST. CROIX, U.S VIRGIN ISLANDS, MIOCENE, KINGSHILL SEAWAY--A
M 29 353† ST. CROIX, U.S. VIRGIN ISLANDS, GEOMORPHOLOGY AND, SUBSURFACE,
B 6203 403† ST. CROIX, U.S. VIRGIN ISLANDS, STRATIGRAPHY AND DEPOSITIONAL
B 6203 509* ST. CROIX, U.S. VIRGIN ISLANDS. BY TITLE ONLY, GEOLOGY SEEN WHILE
B 6004 674* ST. CROIX, U.S.V.I.--MIXED CARBONATE AND CLASTIC DEPOSITS, ABST.,
B 6303 427* ST. CROIX, UNITED STATES VIRGIN ISLANDS, ABST., BANK-BARRIER, REEF
B 6203 502* ST. CROIXAN, ABST., ENIGMAS IN, WISCONSIN, CAMBRIAN AND NEW
B 6206 940 ST. DAMASE FORMATION, PEBBLY SANDSTONE FACIES, QUEBEC
M 25 305 ST. ELIAS BASIN, SECTION
B 6412 2142 ST. GEORGE BASIN, BERING SEA SHELF
B 6002 170 ST. GEORGE BASIN PROVINCE, BERINGIAN SHELF
B 6306 955 ST. GEORGE GROUP, NEWFOUNDLAND
B 6405 767* ST. GEORGE GROUP, WESTERN NEWFOUNDLAND, ABST., ALGAL-,
B 6203 430 ST. GEORGES CHANNEL BASIN, BRITAIN, WEST
B 6203 430 ST. GEORGES CHANNEL, RIFT SYSTEM
B 6209 1767 ST. JOHN, JACK W., JR., PERITIDAL CARBONATE MATERIAL AND
S 4 79 ST. KITTS, CARBONATE SHELF
B 6004 664* ST. LAWRENCE ESTUARY, ABST., ICE-DRIFTED, BOULDERS AS,
B 6405 693* ST. LAWRENCE ISLAND, ABST., PROSPECTIVE AND FUTURE
B 6408 1286* ST. LOUIS AND SALEM STRATIGRAPHY AND OIL PRODUCTION,
S 4 236 ST. LUCIA, TROPIOMETRA CARINATA LAMARCK
S 4 78 ST. MARTIN, CARBONATE SHELF
B 6407 972 ST. MARY RIVER FORMATION, MONTANA
B 6002 165 ST. MATTHEW BASIN, BERING SHELF
B 6002 162 ST. MATTHEW ISLAND, BERINGIAN SHELF
B 6107 963 ST. PAUL FRACTURE ZONE, AFRICA
B 6302 219 ST. PAUL FRACTURE ZONE, BRAZIL
B 6107 963 ST. PAUL-LAS PALMAS FRACTURES, AFRICA
B 6202 274 ST. PAULS FRACTURE ZONE
B 6306 939 ST. PAULS FRACTURE ZONE, AMAZON CONTINENTAL MARGIN
B 6405 805* ST. PETER SANDSTONE, ABST., MODEL OF AUTOCEMENTATION OF
B 6108 1384* ST. PETER SANDSTONE, ABST., PARAGENESIS OF, AUTHIGENIC, MINERALS
M 28 112 ST. PETER SANDSTONE, ARKANSAS, ORDOVICIAN
M 28 161 ST. ROCH FORMATION, CANADA, QUEBEC, CAMBRIAN
B 6206 956 ST. ROCH FORMATION, QUEBEC
B 6405 751* ST. THOMAS, U.S. VIRGIN ISLANDS, ABST., ORIGIN OF BIOGENIC
S 4 79 ST. VINCENT, CARBONATE SHELF
B 6007 1094 STA. LUCIA-BARBADOS TRANSVERSE VALLEY
B 6105 832 STAATZ, MORTIMER H., THORIUM—OTHER NATURAL SOURCE OF
S 7 284 STABILITY, BOTTOM-PRESSURE ANOMALY, STRESS
S 7 276 STABILITY, CREEP BEHAVIOR IN CLAY SOILS
S 7 269 STABILITY, DEFORMATIONAL FEATURES, SHELF
S 7 269 STABILITY, DEFORMATIONAL FEATURES, SLOPE
S 7 279 STABILITY EVALUATION, CONTINENTAL SHELF-SLOPE
S 7 282 STABILITY EVALUATION, ENGINEERING ANALYSIS, SHELF-SLOPE
S 7 279 STABILITY EVALUATION, PROCEDURES, SHELF-SLOPE
S 7 277 STABILITY, MUDFLOWS, DIAGRAMS
S 7 267† STABILITY OF CONTINENTAL SHELF AND SLOPE OFF LOUISIANA AND
S 7 276 STABILITY, SLOPE FAILURES, DIAGRAM
S 7 278 STABILITY, SQUEEZING AND DIAPIRISM, DIAGRAM
S 7 283 STABILITY, STRUCTURAL LOADING AND CREEP, DIAGRAM
S 7 262 STABILITY, TEXAS, UPPER CONTINENTAL SLOPE
S 7 274 STABILITY, WAVE-INDUCED BOTTOM-PRESSURE ANOMALY
B 6303 519* STABLE CARBON ISOTOPES A PALEOCLIMATOLOGICAL TOOL, ABST.
B 6303 479* STABLE CARBON ISOTOPES IN OIL-FIELD WATERS AND ORIGIN OF
B 6405 749* STABLE ISOTOPE EVIDENCE FOR MODERN FRESHWATER DIAGENESIS OF
B 6004 653* STABLE REGION, ABST., RELATION OF, SEISMICITY AN TECTONIC
B 6009 1616 STABLER, COLIN L., NEW SMACKOVER PRODUCTION AT ANAHUAC,
B 6405 678 STACEY, A. P., SIGNATURE PROCESSING OF NORTH SEA AIR-GUN DATA,
S 3 62 STACK COMPOUND, URANIUM OREBODY
B 6211 2362* STACKING VELOCITIES OF DIPPING EVENTS, ABST., EQUALIZING
S 5 117 STACKING, SEISMIC DATA,
B 6002 319* STAFFORD COUNTY, KANSAS, ABST., NUMERICAL ANALYSIS OF, TRAP
S 11 15 STAGE BOUNDARIES, CALIFORNIA, TERTIARY
S 5 7 STAGES, NIAGARA GROUP, PINNACLE REEFS, EVOLUTION,
S 5 7 STAGES, SALINA GROUP, PINNACLE REEFS, EVOLUTION,
B 6304 698 STAHL, W. J., CARBON ISOTOPE MEASUREMENTS OF HYDROCARBONS
B 6303 532 STAHL, W. J., ORIGIN OF GASES ADSORBED IN NEAR-SURFACE SEDIMENT
B 6304 701 STAHL, WOLFGANG J., BACTERIAL DEGRADATION OF CRUDE
B 6303 532 STAHL, WOLFGANG J., CARBON ISOTOPES--NEW TOOL IN HYDROCARBON
B 6304 701 STAHL, WOLFGANG J., ORIGIN OF GASES. ADSORBED IN NEAR-SURFACE
B 6009 1564 STAINFORTH, R. M., UNRELIABILITY OF GLOBIGERINOIDES DATUM
B 6004 690 STAINFORTH, R. M., UNRELIABILITY OF GLOBIGERINOIDES DATUM, ABST.
B 6202 303 STAINFORTH, R. M., WAS IT THE ORINOCO
M 28 185 STAINING, EXPLANATION, TECHNIQUES
B 6306 951 STANCIOFF, ANDREW, SATELLITE IMAGE ANALYSIS IN MINERAL
B 6109 1560 STANCIOFF, ANDREW, SATELLITE IMAGE ANALYSIS IN MINERAL
B 6302 257 STANLEY AND GILBERT KELLING, BALLY, A. W., REVIEW ON
B 6405 787 STANLEY, DANIEL J., DEPOSITIONAL MODELS FOR FINE-GRAINED
B 6203 564 STANLEY, DANIEL JEAN, EROSION VERSUS DEPOSITION ON UPPER
B 6303 532 STANLEY, DANIEL JEAN, GIANT MUDFLOW DEPOSITS IN SUBMARINE
B 6004 725 STANLEY, DANIEL JEAN, LATERAL INFILL AS MAJOR FACTOR IN
B 6105 832 STANLEY, DANIEL JEAN, MUD-FACIES DEFINITION FOR
B 6002 326 STANLEY, JACK R., ROCK TANK, MORROW SANDSTONE, FIELD, EDDY
B 6303 311 STANLEY, K. O. AND, DEPOSITIONAL HISTORY OF PALEOCENE--LOWER
B 6303 535 STANLEY, K. O., EVOLUTION OF HYDROGRAPHIC BASINS AND
B 6108 1392 STANLEY, KENNETH O., INCIPIENT DIAGENESIS OF ARKOSIC AND VITRIC
B 6303 533 STANLEY, KENNETH O., ISOTOPIC COMPOSITION AN SOURCES OF
B 6004 725 STANLEY, R. G., CARBONATE PETROLOGY OF BASIN-TO-REEF SHOALING
B 6203 564 STANLEY, STEVE M., EVOLUTION AND UNIFORMITARIANISM, ABST.
M 25 462 STANNIFEROUS METALLOGENIC SYSTEM, PACIFIDES
M 25 460 STANNIFEROUS SYSTEM, BOLIVIAN
M 25 460 STANNIFEROUS ZONE, AUSTRALIAN-, ASIATIC
B 6107 1046 STANTON FORMATION, KANSAS
B 6004 725 STANTON, GEORGE D., FACTORS INFLUENCING POROSITY AND
B 6109 1547 STANTON, GEORGE, SECONDARY POROSITY IN SANDSTONES OF LOWER
B 6012 2189 STANTON, R. J., JR., PLIOCENE BIOSTRATIGRAPHY AND DEPOSITIONAL
B 6405 788 STANTON, R. W., FORMS AND ASSOCIATIONS OF PYRITE IN UPPER
B 6405 689 STANTON, R. W., GEOLOGIC CONTROLS ON MINERAL-MATTER CONTENT

Index

B	6405	689	STANTON, R. W., GEOLOGIC CONTROLS ON SULFUR CONTENT IN COAL,
B	6405	702	STANTON, R. W., X-RAY MINERALOGY OF UPPER FREEPORT COAL, ABST.
B	6405	699	STANTON, ROBERT J., JR., PALEOECOLOGY IN BASIN ANALYSIS—
B	6105	775	**STANTON, RONALD W., GEOLOGIC FACTORS CONTROLLING COALIFICATION**
B	6109	1548	STAPP, WILFORD LEE, GEOLOGY OF FRACTURED AUSTIN CHALK AND
M	24	97	STAR CORRAL FIELD, WYOMING
M	28	180	STAR POINT SANDSTONE, UTAH, CRETACEOUS
M	28	120	STAR POINT SANDSTONE, UTAH, CRETACEOUS
M	28	156	STAR POINT SANDSTONE, UTAH, CRETACEOUS
B	6104	594	STARICHKOFF FIELD, ALASKA
B	6008	1410	STARK, DAN, LEASE OF FEDERAL GEOTHERMAL ENERGY--ROCKY ROAD
B	6105	833	STARK, P. H., DRILLING STATISTICS IN BASIN EVALUATION, ABST.,
B	6108	1392	STARK, PETE, APPLICATION OF DRILLING STATISTICS IN BASIN
B	6405	788	STARK, PHILIP H., WHATS NEW IN U.S. INDUSTRY ACTIVITY, ABST.
M	29	383	STARKE, G. W., INVESTIGATION OF MISSISSIPPI FAN, GULF OF MEXICO
S	7	155	STARKE, G. W., MISSISSIPPI FAN, GULF OF MEXICO—PHYSIOGRAPHY,
S	3	107	STARKS DOME, LOUISIANA
S	3	99	STARKS DOME, LOUISIANA
B	6405	705	STARKS, T. H., GEOSTATISTICAL APPROAC IN COAL RESOURCE
B	6008	1410	STARKWEATHER, JACK A., JIM COULEE-LITTLE WALL FIELD,
B	6405	779*	STARVED BASIN, MISSISSIPPIAN, WESTERN UNITED STATES, ABST.,
B	6311	2034	STATE OF PRESERVATION INDEX
B	6405	791*	STATE OF THE ART IN CONODONT TAXONOMY, ABST.
B	6205	863†	STATE SURVEYS, AVAILABILITY OF, GEOLOGIC MATERIALS FROM
B	6104	597	STATE UNIVERSITY FIELD, TEXAS
B	6205	864	STATES, MINERAL, ROCK, OR, FOSSIL, COLLECTIONS AVAILABLE FROM
B	6306	918†	STATES OF THICK ABNORMALLY PRESSURED SHALES, CALCULATED,
M	30	96	STATFJORD AREA, GEOLOGIC HISTORY, NORTH SEA
M	30	102	STATFJORD FIELD, BRENT FORMATION, JURASSIC
M	30	111	STATFJORD FIELD, BRENT FORMATION, STRUCTURE CONTOUR
M	30	105	STATFJORD FIELD, CROMER KNOLL GROUP, CRETACEOUS
M	30	102	STATFJORD FIELD, DUNLIN FORMATION, JURASSIC
M	30	99	STATFJORD FIELD, EAST SHETLAND BASIN, CROSS SECTION
M	30	98	STATFJORD FIELD, EXPLORATION HISTORY, NORTH SEA
M	30	104	STATFJORD FIELD, HEATHER FORMATION, JURASSIC
M	30	104	STATFJORD FIELD, HUMBER GROUP, JURASSIC
M	30	109	STATFJORD FIELD, HYDROCARBON RESERVOIRS
M	30	105	STATFJORD FIELD, KIMMERIDGE CLAY FORMATION, JURASSIC
M	30	106	STATFJORD FIELD, LITHOFACIES COMPARISONS
M	30	96	STATFJORD FIELD, LOCATION, NORTH SEA
B	6303	479*	STATFJORD FIELD, NORTH SEA GIANT, ABST.
M	30	96	STATFJORD FIELD, REGIONAL SETTING, NORTH SEA
M	30	115	STATFJORD FIELD, RESERVOIR INFORMATION
M	30	105	STATFJORD FIELD, ROGALAND GROUP, PALEOCENE
M	30	105	STATFJORD FIELD, ROGALAND GROUP, TERTIARY
M	30	105	STATFJORD FIELD, SHETLAND GROUP, CRETACEOUS
M	30	109	STATFJORD FIELD, SOURCE ROCKS
M	30	101	STATFJORD FIELD, STATFJORD FORMATION, JURASSIC
M	30	112	STATFJORD FIELD, STATFJORD FORMATION, STRUCTURE CONTOUR
M	30	102	STATFJORD FIELD, STRATIGRAPHIC COLUMN, NORTH SEA
M	30	101	STATFJORD FIELD, STRATIGRAPHY, TRIASSIC
M	30	109	STATFJORD FIELD, STRUCTURE, BRENT RESERVOIR
M	30	109	STATFJORD FIELD, STRUCTURE, STATFJORD FORMATION
M	30	95†	STATFJORD FIELD--A NORTH SEA GIANT
M	30	101	STATFJORD FORMATION, JURASSIC, STATFJORD FIELD
M	30	109	STATFJORD FORMATION, STATFJORD FIELD, STRUCTURE
M	30	112	STATFJORD FORMATION, STRUCTURE CONTOUR, STATFJORD FIELD
M	26	430	STATFJORD SANDSTONE, NORTH SEA
M	26	30	STATIC MODULI, LABORATORY, MEASUREMENT
B	6109	1550*	STATICS AND DYNAMICS OF NATURAL SYSTEMS—GENETIC BASIS FOR
S	5	119	STATICS CORRECTION, SEISMIC DATA, AUTOMATIC
S	5	121	STATICS CORRECTION, SEISMIC DATA, HAND-
B	6410	1723	STATISTICAL ANALYSIS IN GEOLOGY, BY JOHN M. CUBITT AND STEPHEN
B	6203	534*	STATISTICAL APPROACH FOR FUTURE RESERVES PREDICTION, ABST.
B	6303	517*	STATISTICAL COMPARISON OF MAPPED DATA, ABST.
B	6203	495*	STATISTICAL CURVATURE ANALYSIS TECHNIQUES FOR STRUCTURAL
B	6211	2350*	STATISTICAL CURVATURE ANALYSIS TECHNIQUES FOR STRUCTURAL
B	6105	782*	**STATISTICAL HISTORY OF PETROLEUM EXPLORATION IN DENVER BASIN**
B	6004	691*	STATISTICAL MAPS AS AID TO GEOLOGIC INTERPRETATION, ABST.
B	6205	853†	STATISTICAL SUMMARY OF WELLS DRILLED BELOW 18,000 FEET, 5,500
B	6102	297*	STATISTICAL SUMMARY OF WELLS DRILLED BELOW 18,000 FT, 5,486 M, IN
B	6303	439*	STATISTICAL ZONATION OF OIL SHALE ON BASIS OF CHEMICAL
B	6008	1406*	STATISTICS AS TOOL IN OIL AND GAS EXPLORATION, ABST.
B	6108	1392*	STATISTICS IN BASIN EVALUATION, ABST., APPLICATION OF, DRILLING
B	6208	1340	STATLER, ANTHONY, DEVELOPMENTS IN EAST-CENTRAL STATES IN 1977
B	6008	1222	STATLER, ANTHONY T., DEVELOPMENTS IN EAST-CENTRAL STATES IN
B	6108	1183	STATLER, ANTHONY T., DEVELOPMENTS IN EAST-CENTRAL STATES IN
B	6405	700*	STATUS OF ANTLER OROGENY IN CENTRAL IDAHO—CLARIFICATIONS
M	25	80†	STATUS OF COAL EXPLORATION AND MINE DEVELOPMENT IN
S	6	171†	STATUS OF THE BOUNDARY BETWEEN PLIOCENE AND PLEISTOCENE
B	6303	533	STAUB, JAMES R., ESTUARINE-COASTAL PLAIN COAL DEPOSITION IN
B	6303	533	STAUFFER, K. W., GEOCHEMICAL EVALUATION OF ENSENADA DE LA
B	6410	1580	STAUNTON BLOCK, VIRGINIA
B	6410	1581	STAUNTON BLOCK, VIRGINIA, PULASKI-
B	6410	1576	STAUNTON THRUST, VIRGINIA
M	28	49	STAUROLITE, CRYSTAL
B	6303	431*	STE. GENEVIEVE FORMATION, ILLINOIS BASIN, ABST., MISSISSIPPIAN
M	27	163	STE. GENEVIEVE LIMESTONE, INDIANA, MISSISSIPPIAN
M	27	207	STE. GENEVIEVE LIMESTONE, TENNESSEE, MISSISSIPPIAN
B	6407	1066	STE. MARTHE CHANNEL, QUEBEC
S	10	123	STEADY-STATE, MIGRATION, GROUNDWATER FLOW
B	6008	1410*	STEAMBOAT MOUNTAIN-LEMEI ROCK AREA, SKAMANIA COUNTY,
B	6008	1411	STEARNS, D. W., FIELD EVIDENCE FOR STRUCTURAL STYLE OF LARAMIDE
B	6008	1395	STEARNS, DAVID W., ANALYTIC SOLUTIONS APPLIED TO FAULTING IN
B	6304	702	STECKLER, M. S., MECHANISMS OF BASIN SUBSIDENCE, ABST.
M	25	391†	STEEL PRODUCTION IN NEW ZEALAND--SUMMARY, IRON ORE, COAL,
M	25	393	STEEL PRODUCTION, NEW ZEALAND
B	6305	828	STEEL, RON, ALLUVIAL FANS AND THEIR DEPOSIT IN DEVONIAN
B	6202	184	STEELE SHALE, WYOMING
B	6412	2191	STEEN-MCINTYRE, ROSE, W. I., JR., REVIEW OF, MANUAL FOR,
B	6105	741	**STEENLAND, NELSON C., REGIONAL GEOLOGIC FRAMEWORK OFF**
B	6002	299	STEENLAND, NELSON C., SUBSURFACE POSITIVE ELEMENTS WITHIN
B	6203	541*	STEEPLY INCLINED ALGAL COLUMNS IN LATE TERTIARY FURNACE
B	6303	533	STEGE, BRUCE R., CARBONATE GEOLOGY OF PENA BLANCA URANIUM
B	6203	564	STEIDL, PETER F., FOAM STIMULATION FOR ENHANCED RECOVERY FROM
B	6108	1383	STEIDTMANN, JAMES R., DEPOSITION AND DIAGENESIS OF TENSLEEP
B	6405	750	STEIDTMANN, JAMES R., SEDIMENTOLOGY AND PALEOGEOGRAPHY OF
S	6	114	STEIGE SERIES, ISOCHRON DIAGRAM, FRANCE
S	6	67	STEIGER, R. H., SUBCOMMISSION ON GEOCHRONOLOGY--CONVENTION
B	6105	833	STEINEN, R. P., HOLOCENE DOLOMITE LOCALITY IN FLORIDA BAY, ABST.
B	6405	788	STEINEN, RANDOLPH P., CEMENTATION OF LIME-MUD AND PELLET MU
B	6405	783	STEINEN, RANDOLPH P., EXPERIMENTAL COMPACTION OF LIME
B	6303	431	STEINEN, RANDOLPH P., MISSISSIPPIAN NON-SUPRATIDAL DOLOMITE STE.
B	6303	457	STEINEN, RANDOLPH P., MIXING ZONE DOLOMITE IN TIDAL-FLAT
M	24	229	STEINER, RICHARD J., GRAND ISLE BLOCK 16 FIELD, OFFSHORE
B	6303	548	STEINKER, DON C., PALEOECOLOGIC IMPLICATION OF FORAMINIFERAL
B	6405	682	STEINMETZ, JOHN C., ET AL, UNUSUAL PONDING OF SEDIMENTS ON
M	6004	726	STEINMETZ, RICHARD, APPLICATION OF MONTE CARLO METHOD FOR
B	6102	297	STEINMETZ, RICHARD, STATISTICAL SUMMARY OF WELLS DRILLED
B	6205	853	STEINMETZ, RICHARD, STATISTICAL SUMMARY OF WELLS DRILLED
B	6211	2271	STENDERUP SAND, CALIFORNIA
B	6111	1930	STEPANOW CREEK, ELLESMERE ISLAND
B	6105	795*	**STEPHANIAN A AND B ALLUVIAL-FAN AND LACUSTRINE SEDIMENTARY**
B	6407	992	STEPHANIAN-AUTUNIAN CLASTICS, MOROCCO
B	6004	726	STEPHEN, M. F., LITTORAL PROCESSES AND GEOMORPHIC VARIABILITY
B	6303	534	STEPHENS, DANIEL G., GEOLOGY OF GABES BASIN, TUNISIA, ABST.
B	6103	468	STEPHENS, JACKIE, LARGE NEW PORPHYRY MOLYBDENUM DISCOVERY
B	6009	1613	STEPHENS, R. W., COMPETITION AND QUEST FOR OFFSHORE LEASES,
B	6109	1546	STEPHENS, RAYMOND W., DIFFERENTIAL SUBSIDENCE OF
B	6309	1589	STEPHENSON, J. TODD, PETROLOGY AND DEPOSITIONAL ENVIRONMENTS
B	6103	407	STEPHENSON, LEE P., POROSITY DEPENDENCE ON TEMPERATURE--LIMITS
B	6304	686	STEPHENSON, WILLIAM A., DIRECT MEASUREMENT OF URANIUM BY
B	6309	1523	STEPOVAK FORMATION, ALASKA PENINSULA
B	6203	377	STERLING FORMATION, ALASKA
M	24	11	STERLING FORMATION, ALASKA
B	6408	1133	STERLING FORMATION, ALASKA
B	6408	1132	STERLING GAS FIELD, ALASKA
B	6105	833	STERNBERG, R. W., DREDGED MATERIAL DEPOSITED AT OCEAN DUMPING
B	6405	710*	STEROIDS IN MARINE ENVIRONMENT, ABST., EARLY TRANSFORMATION
B	6211	2243†	STEVENS AND OLDER TURBIDITE FANS OF SOUTHEASTERN SAN JOAQUIN
B	6303	474	STEVENS, ROBERT K., CORRELATION AND TIMING OF
B	6102	177	STEVENS SAND, CALIFORNIA
B	6211	2246	STEVENS SANDSTONE, CALIFORNIA
B	6403	445*	STEVENS SANDSTONE, FLK HILLS FIELD, KERN COUNTY, CALIFORNIA,
M	24	250	STEVENS SANDSTONE, MISSISSIPPI
B	6211	2264	STEVENS TURBIDITE FAN, CALIFORNIA
B	6211	2257	STEVENS TURBIDITES, CALIFORNIA
B	6311	2070	STEVENSON, ANDREW J., HYDROCARBON POTENTIAL OF ALEUTIAN
B	6408	1142	STEVENSON BASIN, ALASKA
B	6407	1023	STEVENSON BASIN, ALASKA
B	6109	1538	STEVENSON, JACK E., JFS FIELD, DIMMIT COUNTY, TEXAS—SOME UNIQUE
B	6409	1568	STEVENSON, JOHN, LOG EVALUATION OF WELLS IN TUSCALOOSA TREND
B	6105	767	**STEWARD, ROBERT G., HURRICANE IMPACT ON CONTINENTAL-SHELF**
B	6408	1272	STEWART, G. F., REVIEW OF PENNSYLVANIAN SANDSTONES OF THE
B	6305	825	STEWART, HARRY E., LINEAMENT MAPPING WITH ORBITING IMAGING
B	6008	1411	STEWART, KENNETH A., MULTI-DIMENSIONAL SEISMIC DISPLAYS, ABST.
B	6405	736*	STEWART PEAK CULMINATION, IDAHO-WYOMIN THRUST BELT, AS
B	6310	1819	STEWART, R. D., ON DEVELOPMENTS IN BANGDALESH, 1978,
B	6110	1833	STEWART, R. D., ON DEVELOPMENTS IN BANGLADESH, 1976

B 6210 1980 STEWART, R. D., ON DEVELOPMENTS IN BANGLADESH, 1977
B 6102 192 STEWART, RICHARD J., NEOGENE TURBIDITE SEDIMENTATION IN
B 6201 87 STEWART, RICHARD J., NEOGENE VOLCANICLASTIC SEDIMENTS FROM
B 6403 446 STEWART, RICHARD J., SUBSURFACE INTERFACE BETWEEN ZEOLITIZED
B 6108 1392 STEWART, RICHARD J., WIDESPREAD OCCURRENCE OF LAUMONTITE IN
B 6308 1236 STICHTENOTH, CRAIG W., DEVELOPMENTS IN WEST TEXAS AND EASTERN
B 6303 448* STICKY MUDS--VISCOSITY OF ESTUARINE SEDIMENTS, ABST.
M 25 500 STIKINE ARCH, CANADA
M 25 504 STIKINE ARCH, CANADA
M 26 68 STILLSTAND, SEA LEVEL FLUCTUATIONS, RELATIVE
S 11 282 STILOSTOMELLA ADVENA
S 11 252 STILOSTOMELLA ADVENA
S 11 252 STILOSTOMELLA LEPIDULA
S 11 282 STILOSTOMELLA LEPIDULA
B 6002 320* STIMULATING RECOVERY FROM HEAVY-OIL RESOURCES,
B 6311 2119* STIMULATION DESIGN FOR UPPER MORROW RESERVOIRS IN
B 6203 564* STIMULATION FOR ENHANCED RECOVERY FROM COAL-GAS WELLS,
M 25 436 STO. NINO MINE, LUZON
B 6009 1434† STOCK HELP TO DELINEATE SPINES OF MOVEMENT, SHEAR ZONES INSIDE,
B 6211 2279 STOCKADE BEAVER SHALE MEMBER, BLACK HILLS
B 6211 2277 STOCKADE BEAVER SHALE MEMBER, WYOMING
B 6304 701 STOCKMAL, GLEN S., STRUCTURAL ANALYSIS OF NORTHERN
B 6002 197 STOCKS, BRAZIL MARGIN
B 6107 1090† STOCKS HELP TO DELINEATE SPINES OF MOVEMENT. DISCUSSION AND
B 6211 2244 STOCKTON ARCH, CALIFORNIA
B 6406 939 STOCKTON COAL SEAM, WEST VIRGINIA, LOWER
M 29 177 STOFFA, P. L., STRUCTURE AND STRATIGRAPHY OF THE BLAKE
B 6008 1411 STOKER, ROGER C., RAFT RIVER, IDAHO, GEOTHERMAL WELLS--SITING,
B 6205 892 STOKES, WILLIAM LEE, SHEET-FLOOD DEPOSITION AND DEPOSITS, ABST.
B 6207 1212 STOLLER, HAROLD M., UTILIZATION OF MAGMA ENERGY, ABST.
B 6206 1013 STOLZ FAULT, CANADA, ARCTIC
B 6109 1548* STONE COUNTIES, MISSISSIPPI, ABST., PETROLOGY OF, SMACKOVER
B 6103 468 STONE, D. B., PALEOMAGNETISM AND PALEOGEOGRAPHIC
B 6104 620 STONE, DONALD S., OKLAHOMA CITY OIL--SECOND CROP FROM
B 6104 492 STONE, G. C., DIAGENESIS OF MIDDLE AND UPPER EOCENE CARBONATE
B 6206 1049 STONE SWITCH FORMATION, TEXAS
B 6405 788 STONE, WILLIAM P., JR., PROFILE OF UNUSUAL OOLITE DEPOSIT—DRUM
M 27 131 STONEHENGE LIMESTONE, PENNSYLVANIA, ORDOVICIAN
M 27 176 STONEHENGE LIMESTONE, PENNSYLVANIA, ORDOVICIAN
B 6103 332 STONO ARCH, SOUTH CAROLINA
B 6004 658* STORAGE RESERVOIRS BY ISOTOPIC ANALYSIS OF METHANE, ABST.,
B 6203 436 STORD BASIN, NORTH SEA
B 6303 534 STOREY, SIDNEY R., CLAY-CARBONATE DIAGENESIS OF DELTAIC
B 6405 723* STORM DEPOSITS, NORTH TEXAS, ABST., LOWER, CRETACEOUS SHELF
B 6207 1229 STORM, PAUL V., GEOLOGY AND DEVELOPMENT OF MOMOTOMBO
B 6004 716* STORM-DOMINATED, GLACIAL SHORELINE, MALASPINA FORELAND,
B 6405 792* STORM-DOMINATED SHELVES—STATUS REPORT, ABST., GENESIS OF,
B 6105 764* STORM-GENERATED SHELL RIDGE ON OPEN LAGOONAL COAST OF BULLS BAY,
B 6405 745 STOSUR, J. J., RECENT ACTIVITIES IN UNITED STATES TAR SAND, ABST.
B 6303 534 STOUDT, EMILY L., SMACKOVER FORMATION OF GULF COAST REGION,
B 6305 767 STOUGHTON CRUDE OIL, SASKATCHEWAN
B 6303 535 STOUT, JOHN G., HUDSON BAY BASIN, ABST.
B 6305 842 STOUT, JOHN L., NEW LOOK AT OIL AND GAS RESOURCES OF
B 6405 789 STOUT, JOHN, UNITED STATES PROVINCE OVERVIEWS, ABST.
B 6405 789 STOW, DORRIK A. V., FAN MODELS FOR HYDROCARBON EXPLORATION
B 6303 535 STOW, DORRIK A. V., LAURENTIAN FAN--DEEP-SEA FAN MODELS,
M 30 315† STRACHAN AND RICINUS WEST GAS FIELDS, ALBERTA, CANADA
M 30 315 STRACHAN FIELD, ALBERTA, DEVONIAN
M 30 30 STRACHAN FIELD, LANDSAT IMAGE, ALBERTA
M 30 324 STRACHAN FIELD, LEDUC REEF, ALBERTA
M 30 322 STRACHAN FIELD, RESERVES AND PRODUCTION, ALBERTA
M 30 322 STRACHAN FIELD, SEISMIC SECTION, ALBERTA
B 6401 28 STRACHAN GIANT GAS FIELD, LANDSAT DATA, ALBERTA
B 6303 469* STRACHAN-RICINUS GAS FIELD, ALBERTA, ABST.
M 30 315 STRACHAN-RICINUS PRODUCTION, ALBERTA, LEDUC FORMATION
M 30 317 STRACHAN-RICINUS WEST FIELDS, HISTORY OF DISCOVERY
M 30 316 STRACHAN-RICINUS WEST FIELDS, LOCATION MAP
M 30 317 STRACHAN-RICINUS WEST FIELDS, REEF HORIZONS, DEVONIAN
M 30 321 STRACHAN-RICINUS WEST FIELDS, RESERVOIR CHARACTERISTICS
M 30 316 STRACHAN-RICINUS WEST FIELDS, STRATIGRAPHY
B 6008 1412* STRAIGHT CLIFFS FORMATION, SOUTHERN KAIPAROWITS REGION,
B 6404 568† STRAIN MEASUREMENTS IN EASTERN MARGINAL SHEAR ZONE OF
M 27 201 STRAIN RECRYSTALLIZATION
M 27 208 STRAIN RECRYSTALLIZATION
M 27 200 STRAIN RECRYSTALLIZATION
B 6410 1704 STRAITS OF FLORIDA, BATHYMETRY, NORTHERN
B 6201 123 STRAITS OF FLORIDA, BATHYMETRY OF SOUTHERN
B 6410 1701† STRAITS OF FLORIDA, CARBONATE SEDIMENT, DRIFTS IN NORTHER
M 28 50 STRAITS SCHIST, CONNECTICUT, ORDOVICIAN
B 6102 176 STRAND FIELD, MIOCENE SANDS, CALIFORNIA
B 6102 177 STRAND FIELD, RIO BRAVO SAND, CALIFORNIA
B 6211 2260 STRAND OIL FIELD, SECTIONS, CALIFORNIA
B 6211 2247 STRAND TREND, CALIFORNIA, TEN SECTION-
B 6203 500* STRAND-PLAIN LIMESTONES OF EASTERN YUCATAN PENINSULA,
B 6303 362† STRANDLINE SEDIMENTATION OF CARBONATE GRAINSTONES, UPPER
B 6303 464* STRANDLINE SYSTEMS, ABST., SPECTRUM OF DEPOSITIONAL MODELS
M 26 227 STRANDPLAIN DEPOSITS, DEPOSITIONAL SYSTEMS, FACIES ASSOCIATIONS
M 25 486 STRATA-BOUND COPPER DEPOSITS, GENETIC DISTINCTIONS
M 25 487 STRATA-BOUND DEPOSITS, METALS
M 25 426 STRATA-BOUND ORE DEPOSITS, STRATIFORM AND
M 25 449 STRATA-BOUND SULFIDE DEPOSITS, JAPAN, SAMBAGAWA SYSTEM
B 6405 799* STRATEGY OF EXPLORATION FOR HIGH TEMPERATURE
B 6404 692* STRATIFICATION TYPES IN INTERTIDAL SEDIMENT, WILLAPA BAY,
S 5 181* STRATIFIED EVAPORITE PAN, BASE- METAL CONCENTRATION IN A
M 25 426† STRATIFORM AND STRATA-BOUND METAL CONCENTRATIONS IN
M 25 426 STRATIFORM AND STRATA-BOUND ORE DEPOSITS
M 25 426 STRATIFORM BASE-METAL DEPOSITS, PRODUCTION AND RES. AUSTRALIA
M 25 475 STRATIFORM DEPOSITS OF CINNABAR-FLUORITE-ANTIMONITE, USSR,
M 25 507 STRATIFORM DEPOSITS OF METALS, CANADIAN, CORDILLERA
M 25 428 STRATIFORM ORES, ENVIRONMENTAL PATTERN OF OCCURRENCE
B 6401 137† STRATIGRAPHI NOMENCLATURE, 1976-78, NOTE 51--, RECORDS OF
M 26 220 STRATIGRAPHIC ANALYSIS, DEPOSITIONAL SYSTEMS, SEISMIC
M 26 53† STRATIGRAPHIC ANALYSIS, SEISMIC, STRATIGRAPHY AND, GLOBAL
M 26 301† STRATIGRAPHIC AND HYDROCARBON DETERMINATION, APPLICATION
B 6305 823* STRATIGRAPHIC AND SEDIMENTOLOGIC RELATION AMONG THREE
M 26 277† STRATIGRAPHIC AND SEISMIC EVIDENCE FOR LATE CRETACEOUS
S 7 21† STRATIGRAPHIC AND STRUCTURAL FRAMEWORK, NORTHERN AND,
B 6403 450* STRATIGRAPHIC AND STRUCTURAL RELATIONS TO PRE-TERTIARY
B 6403 450* STRATIGRAPHIC AND STRUCTURAL RELATIONS TO PRE-TERTIARY
B 6205 845† STRATIGRAPHIC AND VERTICAL THICKNESSES IN DEVIATED WELLS,
S 6 367† STRATIGRAPHIC CLASSIFICATION AND A PROPOSAL FOR THE
B 6107 1083† STRATIGRAPHIC CODE TO INTRODUCE POINT-BOUNDARY STRATOTYPE
B 6304 656 STRATIGRAPHIC COLUMN, OHIO, WASHINGTON COUNTY
B 6301 90 STRATIGRAPHIC COLUMN, SCOTIAN BASIN
S 6 370 STRATIGRAPHIC COLUMN, SOUTH AFRICA
B 6401 136† STRATIGRAPHIC COMMISSION, 1976-78, AMERICAN COMMISSION ON,
B 6401 136 STRATIGRAPHIC COMMISSION, 1976-78, AMERICAN COMMISSION ON,
B 6207 1187† STRATIGRAPHIC COMMISSION FOR 1974-1976, NOTE 49—RECORDS OF
B 6203 502* STRATIGRAPHIC CONCEPTS AND TERMINOLOGY, ABST.
B 6304 690* STRATIGRAPHIC CONTROL OF URANIUM DEPOSITS ABST.
S 6 57 STRATIGRAPHIC CORRELATION, MAGNETIC, POLARITY
B 6303 501* STRATIGRAPHIC CORRELATION OF KEATHLEY ANOMALY,
M 26 408 STRATIGRAPHIC CORRELATIONS, MODELING
B 6002 278† STRATIGRAPHIC CROS SECTION OF HAMILTON GROUP, DEVONIAN, AND
B 6012 2184* STRATIGRAPHIC EXPLORATION WITH SEISLOG METHOD, ABST.
S 7 22 STRATIGRAPHIC FRAMEWORK, NORTHERN, GULF OF MEXICO
M 30 507† STRATIGRAPHIC GIANT--MESSLA OIL FIELD, LIBYA, GEOLOGY OF A
B 6307 1144 STRATIGRAPHIC INDEX OF BRITISH OSTRACODA, EDITED BY RAYMOND
B 6209 1756* STRATIGRAPHIC INTERPRETATION OF A POSSIBLE PALEOSTREAM
M 26 213† STRATIGRAPHIC INTERPRETATION OF DEPOSITIONAL SYSTEMS:
M 26 117† STRATIGRAPHIC INTERPRETATION OF SEISMIC REFLECTION PATTERNS
M 26 135† STRATIGRAPHIC INTERPRETATION PROCEDURE, SEISMIC,
M 26 138 STRATIGRAPHIC INTERPRETATION, SEISMIC, REFLECTIONS, PATTERNS
S 7 193† STRATIGRAPHIC INVESTIGATION OF WEST FLORIDA SLOPE, GULF OF
B 6004 651* STRATIGRAPHIC MARKER BEDS IN COALS OF ROCKY MOUNTAIN
M 26 439† STRATIGRAPHIC MODEL OF DEPOSITIONAL PLATFORM MARGIN,
B 6105 842* STRATIGRAPHIC MODELING, ABST., INTERPRETIVE METHODS OF
M 26 417† STRATIGRAPHIC MODELING AND INTERPRETATION, GEOLOGIC
M 26 477† STRATIGRAPHIC MODELING AND INTERPRETATION, PRACTICAL
M 26 389† STRATIGRAPHIC MODELING AND INTERPRETATION—GEOPHYSICAL
B 6303 554* STRATIGRAPHIC MODELS FOR MODERN BACK-BARRIER ENVIRONMENTS,
B 6401 137† STRATIGRAPHIC NOMENCLATURE, 1976-78, NOTE 51--, RECORDS OF
B 6002 273 STRATIGRAPHIC NOMENCLATURE, NOTE 44, AMERICAN, COMMISSION ON
B 6102 248 STRATIGRAPHIC NOMENCLATURE, NOTE 45, AMERICAN, COMMISSION ON
B 6207 1185 STRATIGRAPHIC NOMENCLATURE, NOTE 48, AMERICAN COMMISSION ON
B 6207 1187 STRATIGRAPHIC NOMENCLATURE, NOTE 49, AMERICAN COMMISSION ON
B 6310 1986 STRATIGRAPHIC NOMENCLATURE, NOTE 50, AMERICAN COMMISSION OF
B 6401 136 STRATIGRAPHIC NOMENCLATURE NOTE 51--RECORDS OF
B 6401 136† STRATIGRAPHIC NOMENCLATURE NOTE 51--RECORDS OF
B 6207 1185† STRATIGRAPHIC NOMENCLATURE TO PROVIDE GUIDELINES
B 6009 1619* STRATIGRAPHIC, PALEOENVIRONMENTAL, AND STRUCTURALLY
B 6209 1768* STRATIGRAPHIC RELATIONS OF ANNONA CHALK AND GOBER CHALK,
B 6405 752* STRATIGRAPHIC RELATIONS OF NANUSHUK GROUP, MIDDLE
B 6207 1171† STRATIGRAPHIC RELATIONS OF PARK CITY GROUP, PERMIAN, IN
B 6405 728* STRATIGRAPHIC RELATIONS OF PERMIAN FORMATIONS IN PARTS OF
S 6 240 STRATIGRAPHIC SCALE, DIAGRAM, PALEOGENE
S 11 194 STRATIGRAPHIC SECTION, BEAR RIVER
S 11 194 STRATIGRAPHIC SECTION, CENTERVILLE COASTAL

Index

S 11 194 STRATIGRAPHIC SECTION, PRICE CREEK
M 29 359 STRATIGRAPHIC SECTIONS, VIRGIN ISLANDS, ST. CROIX
M 29 453† STRATIGRAPHIC SEQUENCES AT ATLANTIC MARGINS, THE EFFECT OF,
B 6105 799 * STRATIGRAPHIC SYNTHESIS AND HYDROCARBON POTENTIAL OF CENTRAL
B 6008 1159 STRATIGRAPHIC TEST
B 6208 1281 STRATIGRAPHIC TEST
B 6409 1302 STRATIGRAPHIC TEST
B 6308 1174 STRATIGRAPHIC TEST
B 6108 1127 STRATIGRAPHIC TEST
B 6004 672* STRATIGRAPHIC TEST, NANTUCKET ISLAND, MASSACHUSETTS, ABST.,
B 6303 485* STRATIGRAPHIC TRAP, ABST., TYLER SANDSTONES, PENNSYLVANIAN,
B 6309 1610* STRATIGRAPHIC TRAP, ABST., WAVELAND FIELD, UNIQU STRUCTURAL
B 6308 1428* STRATIGRAPHIC TRAP, ABST., WAVELAND FIELD, UNIQU STRUCTURAL
B 6004 645* STRATIGRAPHIC TRAP ANALYSIS BY COMPUTER, ABST., CLASTIC AND
B 6408 1218 STRATIGRAPHIC TRAP, DIAGRAM
M 26 109 STRATIGRAPHIC TRAP EXPLORATION, SEISMIC, REFLECTIONS,
B 6405 709* STRATIGRAPHIC TRAP FOR PETROLEUM IN WEBER FORMATION, RANGEL
M 24 276† STRATIGRAPHIC TRAP, JAY FIELD, FLORIDA--A, JURASSIC
M 24 72† STRATIGRAPHIC TRAP, MITSUE OIL FIELD, ALBERTA--A RICH
B 6203 516* STRATIGRAPHIC TRAP POSSIBILITIES IN ARBUCKLE GROUP, ABST.
B 6002 325* STRATIGRAPHIC TRAP POSSIBILITIES IN ARBUCKLE GROUP. BY TITLE
B 6203 516* STRATIGRAPHIC TRAP POSSIBILITIES IN ARBUCKLE GROUP, ABST.
M 26 493 STRATIGRAPHIC TRAP, QUANTIFICATION, SEISMIC, MODELING
B 6104 489 STRATIGRAPHIC TRAP, TILTED-OUT
M 24 121† STRATIGRAPHIC TRAP, UINTA BASIN, UTAH, ALTAMONT-, BLUEBELL--A
B 6305 748 STRATIGRAPHIC TRAPS
B 6409 1568 STRATIGRAPHIC TRAPS, ABST., SEISMIC-STRATIGRAPHIC, MAPPING OF,
B 6311 2117* STRATIGRAPHIC TRAPS, ABST., GEOCHEMICAL PROSPECTING FOR
B 6303 448* STRATIGRAPHIC TRAPS, ABST., HIGH-RESOLUTION SEISMIC WORK AS
B 6305 840* STRATIGRAPHIC TRAPS, ABST., HYDROCARBON SHOWS--PROXIMITY
B 6004 718* STRATIGRAPHIC TRAPS AN AQUIFERS IN SOUTHEAST POWDER RIVER
M 25 314 STRATIGRAPHIC TRAPS, CALIFORNIA, CENTRAL
B 6206 978 STRATIGRAPHIC TRAPS, CHANNELS AND, SUPRAFANS AS
B 6306 876 STRATIGRAPHIC TRAPS, CRETACEOUS, RUDIST BUILDUPS AS
B 6206 932† STRATIGRAPHIC TRAPS, DEEP-, WATER, SANDSTONE FACIES AND,
B 6405 801* STRATIGRAPHIC TRAPS, DENVER BASIN COLORADO, ABST., WATTENBERG
B 6104 485 STRATIGRAPHIC TRAPS, MISINTERPRETATION OF, DIAGENETIC, TRAPS AS
M 26 329† STRATIGRAPHIC TRAPS, SEISMIC, EXPLORATION FOR
B 6305 810† STRATIGRAPHIC TRAPS. DISCUSSION AND REPLY, DEEP-, WATER,
S 6 59 STRATIGRAPHIC ZONES, CORRELATION, NEOGENE
B 6405 778* STRATIGRAPHIC-TRAP OIL ACCUMULATIONS AS IN POWDER RIVER
S 5 147 STRATIGRAPHY, CANADA, SVERDRUP BASIN, OTTO FIORD FORMATION,
S 5 79 STRATIGRAPHY, DEVONIAN, DETROIT RIVER GROUP, LUCAS
S 5 4 STRATIGRAPHY, MICHIGAN BASIN, PINNACLE REEF BELT,
S 5 54 STRATIGRAPHY, MICHIGAN, SILURIAN, CAYUGAN SERIES, SALINA
S 5 54 STRATIGRAPHY, MICHIGAN, SILURIAN, NIAGARAN SERIES, NIAGARA
S 6 367 STRATIGRAPHY, A REVIEW, SOUTH AFRICA
S 6 372 STRATIGRAPHY, A REVIEW, SOUTH AFRICA, PHANEROZOIC
S 6 368 STRATIGRAPHY, A REVIEW, SOUTH AFRICA, PRECAMBRIAN
B 6108 1372* STRATIGRAPHY AND CLAY DIAGENESIS OF TRINIDAD SANDSTONE AND
B 6305 824* STRATIGRAPHY AND COAL DEPOSITS OF UPPER CRETACEOUS
B 6309 1452† STRATIGRAPHY AND DEPOSITIONAL ENVIRONMENTS OF BALTIMORE
B 6004 659* STRATIGRAPHY AND DEPOSITIONAL ENVIRONMENTS OF LOWER PART
B 6203 513* STRATIGRAPHY AND DEPOSITIONAL ENVIRONMENTS OF TONGUE RIVER
B 6405 721* STRATIGRAPHY AND DEPOSITIONAL HISTORY OF POWELL FORMATION,
B 6203 403† STRATIGRAPHY AND DEPOSITIONAL SETTING, KINGSHILL LIMESTONE,
B 6308 1426* STRATIGRAPHY AND ENTRAPMENT OF HYDROCARBONS IN SAN MIGUEL
B 6109 1540* STRATIGRAPHY AND ENTRAPMENT OF HYDROCARBONS IN SAN MIGUEL
B 6412 2095† STRATIGRAPHY AND ENVIRONMENTAL SEDIMENTOLOG OF LOWER FARS
B 6308 1427* STRATIGRAPHY AND EXPLORATION GEOLOGY OF MARBLE FALLS
B 6203 500* STRATIGRAPHY AND GEOLOGIC HISTORY OF CENTRAL FLORIDA GULF
M 26 185† STRATIGRAPHY AND GLOBAL CHANGES OF SEA LEVEL, PART 10: SEISMIC
M 26 205† STRATIGRAPHY AND GLOBAL CHANGES OF SEA LEVEL, PART 11:
M 26 51† STRATIGRAPHY AND GLOBAL CHANGES OF SEA LEVEL, PART 1:
M 26 53† STRATIGRAPHY AND GLOBAL CHANGES OF SEA LEVEL, PART 2: THE
M 26 63† STRATIGRAPHY AND GLOBAL CHANGES OF SEA LEVEL, PART 3: RELATIVE
M 26 83† STRATIGRAPHY AND GLOBAL CHANGES OF SEA LEVEL, PART 4: GLOBAL
M 26 99† STRATIGRAPHY AND GLOBAL CHANGES OF SEA LEVEL, PART 5:
M 26 117† STRATIGRAPHY AND GLOBAL CHANGES OF SEA LEVEL, PART 6:
M 26 135† STRATIGRAPHY AND GLOBAL CHANGES OF SEA LEVEL, PART 7: SEISMIC
M 26 145† STRATIGRAPHY AND GLOBAL CHANGES OF SEA LEVEL, PART 8:
M 26 165† STRATIGRAPHY AND GLOBAL CHANGES OF SEA LEVEL, PART 9: SEISMIC
M 26 49† STRATIGRAPHY AND GLOBAL CHANGES OF SEA LEVEL, SEISMIC
B 6309 1609* STRATIGRAPHY AND PALEOECOLOGY OF TAMIAMI FORMATION IN LEE
B 6008 1401* STRATIGRAPHY AND PETROLEUM ACCUMULATION, TYLER SANDSTONES,
B 6108 1393* STRATIGRAPHY AND SEDIMENTARY PETROGRAPHY OF SUBSURFACE
B 6004 684* STRATIGRAPHY AND SEDIMENTATION OF FRESHWATER,
S 7 155† STRATIGRAPHY, AND SEDIMENTATIONAL PATTERNS, MISSISSIPPI FAN,
B 6211 2357* STRATIGRAPHY AND SEDIMENTOLOGY OF TERTIARY STRATA IN COLD
B 6202 235† STRATIGRAPHY AND STRUCTURE OF CHICONTEPEC TURBIDITES,
B 6105 781 * STRATIGRAPHY AND STRUCTURE OF COMPRESSIONAL CONTINENTAL MARGINS
B 6405 722* STRATIGRAPHY AND STRUCTURE OF EARLY TERTIARY ORCA GROUP,
B 6007 1037† STRATIGRAPHY AND STRUCTURE OF NORTHEAST PROVIDENCE
B 6105 782 * STRATIGRAPHY AND STRUCTURE OF PULL-APART MARGINS
B 6009 1623* STRATIGRAPHY AS KEY TO ARCH-RELATED ORIGIN OF LITTLE NORTH
M 26 74 STRATIGRAPHY, CALIFORNIA, SAN JOAQUIN BASIN, FACIES SEQUENCES
B 6004 528† STRATIGRAPHY FROM SEISMIC DATA, INFERRING
M 29 469† STRATIGRAPHY, GLOBAL, CYCLES OF, RELATIVE, CHANGES OF, SEA
B 6205 886* STRATIGRAPHY OF CEDAR MOUNTAIN AND DAKOTA FORMATIONS,
B 6311 2042† STRATIGRAPHY OF INTERMONTANE, LACUSTRINE DELTA, CATATUMBO
B 6305 837* STRATIGRAPHY OF LACUSTRINE DEPOSITS, ABST.
B 6407 1087† STRATIGRAPHY OF MARINE PALEOGENE AND NEOGENE OF NORTHEAST
B 6205 887* STRATIGRAPHY OF MESAVERDE GROUP AND OCCURRENCE OF NATURAL
B 6008 1398* STRATIGRAPHY OF MISSISSIPPIAN CHARLES FORMATION, WILLISTON
B 6004 675* STRATIGRAPHY OF NOVA SCOTIA SHELF, ABST.
M 29 177† STRATIGRAPHY OF THE BLAKE ESCARPMENT BASED ON SEISMIC
B 6203 546* STRATIGRAPHY, SEDIMENTATION, AND PALEOENVIRONMENTS OF
S 7 87 STRATIGRAPHY, SEISMIC FACIES ANALYSIS
S 7 170 STRATIGRAPHY, SEISMIC INTERPRETATION, MISSISSIPPI FAN
M 26 205† STRATIGRAPHY, SEISMIC, STRATIGRAPHY AND, GLOBAL CHANGES OF,
B 6401 115† STRATIGRAPHY, STRUCTURE, AND ZONATION OF LARGE SILURIAN REEF
M 26 205 STRATIGRAPHY, TERMINOLOGY, GLOSSARY, SEISMIC
M 26 150 STRATIGRAPHY, TEXAS, EAST
B 6107 1083† STRATOTYPE CONCEPT, NOTE 47—APPLICATION FOR AMENDMENT OF,
S 6 35 STRATOTYPE, DEFINITION
S 6 33† STRATOTYPES AND AN INTERNATIONAL GEOCHRONOLOGIC SCALE
S 6 235† STRATOTYPES, CRITICAL REVIEW OF, ISOTOPIC DATES IN RELATION TO,
S 6 218 STRATOTYPES, DIAGRAM, PALEOGENE, TIME SCALE
S 6 215 STRATOTYPES, EUROPE, PALEOGENE, STAGE
B 6105 806 *ˉ STRAT-SEIS—SIMPLE EXPLANATION WITH FIELD EXAMPLES
B 6305 820 STRAUSZ AND E. M. LOWN, COPELIN, EDWARD C., REVIEW OF, OIL SAND
B 6301 54 STRAWN CLASTIC SEQUENCE, TEXAS
B 6002 324* STRAWN DEPOSITIONAL SYSTEMS, LATE DESMOINESIAN, OF
B 6309 1601* STRAWN, DESMOINESIAN, CRATONIC DELTA SYSTEMS, CONCHO
B 6408 1171 STRAWN FACIES, TEXAS
M 28 157 STRAWN GROUP, TEXAS, PENNSYLVANIAN
M 28 112 STRAWN GROUP, TEXAS, PENNSYLVANIAN
M 28 128 STRAWN GROUP, TEXAS, PENNSYLVANIAN
M 28 143 STRAWN GROUP, TEXAS, PENNSYLVANIAN
B 6109 1536* STRAWN SERIES, PENNSYLVANIAN, NORTH-CENTRAL TEXAS, ABST.,
B 6207 1235* STREAM SALINITY IN DARLING RANGE, WESTERN AUSTRALIA, ABST.,
B 6203 542* STREAMS IN CARBONIFEROUS OF MARITIME PROVINCES, CANADA, ABST.,
B 6203 564 STREET, A. V., EXPANDING FIELD OF SEISLOG APPLICATIONS, ABST.
B 6403 414 STRESS, DEFINITION, EFFECTIVE
M 26 423 STRESS, EFFECTS OF, SEISMIC FACTORS
B 6007 1126 STRESS IN SEDIMENTS, VERTICAL
S 7 284 STRESS, STABILITY, BOTTOM-PRESSURE ANOMALY
B 6008 1410* STRETCHING THE SEISMIC DOLLAR, ABST.
B 6004 727 STRICKLAND, K. E., PALEOECOLOGY AND FAUNA OF SOME NIAGARAN
B 6002 175 STRIKE-SLIP FAULTS, BERINGIAN SHELF
B 6203 565 STROM, N. A., HEAVY OIL AND TAR SANDS OF ALBERTA. BY TITLE ONLY.
M 27 6 STROMATOLITE, BLUE-GREEN, ALGAL
M 27 4 STROMATOLITE, BLUE-GREEN, ALGAL
B 6203 537* STROMATOLITE-FORMING ALGA FROM EOCENE BRIDGER AND GREEN
B 6203 568* STROMATOLITES, ABST., ORGANIC, GEOCHEMISTRY OF, CENOZOIC AND
B 6405 803* STROMATOLITES AND BURROWING ORGANISMS, ABST., RELATIONS
B 6203 496* STROMATOLITES FROM CENTRAL AND SOUTHERN FRANCE
B 6203 488* STROMATOLITES FROM LAKE TURKANA, KENYA, ABST., OXYGEN, AND,
B 6203 565* STROMATOLITES IN EOCENE GREEN RIVER FORMATION, ABST.
B 6203 535* STROMATOLITES IN EUREKA QUARTZITE, SOUTHEASTERN NEVADA,
B 6004 660* STROMATOLITES IN LOCKPORT FORMATION, MIDDLE SILURIAN, OF NEW
B 6203 551* STROMATOLITES, WALKER LAKE, NEVADA, ABST., MODERN, LACUSTRINE
B 6203 539* STROMATOLITIC BIOHERMS AND ASSOCIATED FACIES, NOLICHUCKY
B 6105 839* STROMATOPOROID REEF COMPLEX, ABST., BIOFACIES RELATIONS IN
B 6401 129 STROMATOPOROID-CORAL ZONES, INDIANA
M 27 55 STROMATOPOROIDS
B 6203 498* STROMATOPOROIDS—EPIC STRUGGLE FOR SURVIVAL, EXAMPLES FROM
M 26 253 STRONG-REFLECTOR FACIES, PARALLEL-BEDDED REFLECTORS
S 6 94 STRONTIUM COMPOSITION, SEDIMENTS
S 4 353† STRONTIUM IN QUATERNARY TO LATE TERTIARY CORAL REEFS OF RED
B 6303 533* STRONTIUM IN SANDSTONE CEMENTS IN HIGH PLAINS SEQUENCE OF
B 6209 1673† STRONTIUM MINIMUM AGES OF SEDIMENTATION, URANIUM
B 6410 1718† STRONTIUM MINIMUM AGES OF SEDIMENTATION, URANIUM
M 25 544 STRONTIUM-ISOTOPE ABUNDANCE DATA OF METALLIFEROUS
B 6402 249 STROPHADES RISE, SEISMIC REFLECTION, GREECE
B 6402 221† STRUCTURAL ANALYSIS OF DEVILS RIVER UPLIFT--SOUTHERN VAL
B 6304 701* STRUCTURAL ANALYSIS OF NORTHERN TERMINATION OF LEWIS
B 6405 758* STRUCTURAL AND SEDIMENTOLOGIC STUDY OF CERR PRIETO
M 25 480† STRUCTURAL CLASSIFICATION OF NORTH PACIFIC SHELVES--SUMMARY
B 6212 2510* STRUCTURAL CONTRAST BETWEEN FORE-ARC AND BACK-ARC BASINS OF
B 6309 1591* STRUCTURAL CONTROL OF ONONDAGA REEFS, DEVONIAN, IN
B 6009 1620* STRUCTURAL CONTROL ON PALEOZOIC SEDIMENTATIO IN EASTERN
M 29 46 STRUCTURAL DIAGRAM, ICELAND, INSULAR MARGIN

M 29 329† STRUCTURAL ELEMENTS AND EVOLUTION OF NORTHWESTERN
B 6012 2176* STRUCTURAL ELEMENTS OF OFFSHORE GULF OF ALASKA TERTIARY
M 26 263 STRUCTURAL EROSION, MODIFICATIONS
M 25 281† STRUCTURAL EVOLUTION OF LATE MESOZOIC AND CENOZOIC BASINS IN
B 6303 450* STRUCTURAL EVOLUTION OF MICHIGAN BASIN AND ITS PETROLEUM
M 25 250† STRUCTURAL EVOLUTION OF TERTIARY BASINS OF SOUTHEAST
B 6209 1765* STRUCTURAL FABRIC OF SERPENTINITES AND AMPHIBOLITES ALONG
S 7 246 STRUCTURAL FEATURES, TEXAS, SHELF-SLOPE
S 7 21† STRUCTURAL FRAMEWORK, NORTHERN AND, EASTERN, GULF OF
S 7 28 STRUCTURAL FRAMEWORK, NORTHERN, GULF OF MEXICO
S 7 151 STRUCTURAL GENESIS, NORTHERN GULF OF MEXICO, CONTINENTAL
B 6008 1392* STRUCTURAL GEOLOGY AND PETROLEUM POTENTIAL OF SOUTH FLANK
B 6305 833* STRUCTURAL GEOLOGY OF NORTHERN SALT RIVE RANGE,
B 6203 571* STRUCTURAL GEOLOGY OF SOUTHWEST EDWARDS PLATEAU REGION,
M 29 252 STRUCTURAL HIGH, ACCRETIONARY BASIN, EVOLUTION
M 29 252 STRUCTURAL HIGH, CONSTRUCTED BASIN, EVOLUTION
M 29 253 STRUCTURAL HIGH, RESIDUAL BASIN, EVOLUTION
M 29 245† STRUCTURAL HIGHS BORDERING MAJOR FOREARC BASINS, THE,
B 6008 1401* STRUCTURAL INTERPRETATION AND POROSITY VARIATIONS OF
B 6408 1283* STRUCTURAL INTERPRETATION OF BURIED SILURIAN REEFS IN
S 7 306 STRUCTURAL INTERPRETATION, ORCA BASIN
S 7 283 STRUCTURAL LOADING AND CREEP, DIAGRAM, STABILITY
M 29 292 STRUCTURAL MAP, MEDITERRANEAN, MARGINS
S 4 135† STRUCTURAL MODIFICATION OF CORALS IN RELATION TO WAVES AND
B 6305 836* **STRUCTURAL PATTERN OF EASTERN PART OF DISTURBED BELT OF**
M 25 230† STRUCTURAL STYLE AND HYDROCARBON OF BASS, GIPPSLAND, AND
B 6307 1018 STRUCTURAL STYLES AND HABITATS, CLASSIFICATION
B 6307 1016† STRUCTURAL STYLES, THEIR PLATE-TECTONIC HABITATS, AND
B 6305 748 STRUCTURAL TRAP
M 25 59† STRUCTURAL-FORMATION ANALYSIS OF NORTHWESTERN PART OF
B 6004 694* STRUCTURE AN EVOLUTION OF BERING SEA MARGIN AND RELATION TO
M 29 223† STRUCTURE AND CENOZOIC EVOLUTION OF THE SUNDA ARC IN THE
M 29 27† STRUCTURE AND DEVELOPMENT OF THE SOUTHEAST GEORGIA
B 6105 781 STRUCTURE AND DEVELOPMENT OF SOUTHEAST GEORGIA EMBAYMENT–BLAKE
B 6203 501* STRUCTURE AND EARLY GEOLOGIC HISTORY OF DEEP CENTRAL GULF
B 6002 161† STRUCTURE AND EVOLUTION OF BERING SEA SHELF SOUTH OF ST
M 29 151† STRUCTURE AND EVOLUTION OF THE AREA UNDERLYING THE
B 6405 706* STRUCTURE AND HYDROCARBON POTENTIAL OF KODIAK SHELF,
B 6207 1211* STRUCTURE AND MOTION OF BERING AND OKHOTSK SEA ICE COVERS BY
B 6302 218† STRUCTURE AND ORIGIN OF THREE CONTINENTAL-MARGIN PLATEAUS.
B 6301 2† **STRUCTURE AND STRATIGRAPHY OF FOREARC REGIONS**
B 6109 1545* STRUCTURE AND STRATIGRAPHY OF LATE QUATERNARY DEPOSITS ON
B 6109 1547* STRUCTURE AND STRATIGRAPHY OF SMALL PLIOCENE-PLEISTOCENE
M 29 177† STRUCTURE AND STRATIGRAPHY OF THE BLAKE ESCARPMENT BASED
B 6303 301† STRUCTURE AND TECTONIC SETTING OF CONTINENTAL SHELF
B 6203 513* STRUCTURE AND TECTONIC SETTING OF CONTINENTAL SHELF
B 6004 693* STRUCTURE AND TIMING OF OROGENIC EVENTS, ALTOS
S 5 96 STRUCTURE, OHIO, SILURIAN, LOCKPORT GROUP,
M 29 36 STRUCTURE, ANGULAR UNCONFORMITY, BLAKE PLATEAU
M 29 34 STRUCTURE, BLAKE PLATEAU, REGIONAL
M 29 35 STRUCTURE CONTOURS, REFLECTORS, BLAKE PLATEAU
S 7 270 STRUCTURE, FEATURES IN GEOPHYSICAL RECORDS
M 26 371 STRUCTURE FORM CONTOUR MAPPING, RESOLUTION SYSTEMS
B 6104 601† **STRUCTURE OF BROCKMAN HILLS, SOUTHWESTERN NEW MEXICO**
B 6203 535* STRUCTURE OF ALISO CANYON FIELD, EASTERN VENTURA BASIN,
B 6203 560* STRUCTURE OF ATLANTIC SLOPE OF EASTERN NORTH AMERICA, ABST.
B 6105 766* STRUCTURE OF BALTIMORE CANYON TROUGH, U.S. ATLANTIC CONTINENTAL
B 6105 805* STRUCTURE OF COASTAL WASHOVER FANS AND THEIR ROLE IN INTERNAL
M 29 113† STRUCTURE OF COLORADO BASIN AND CONTINENT-OCEAN CRUST
B 6105 773* STRUCTURE OF CONTINENTAL SHELF, EASTERN GULF OF ALASKA
S 7 141† STRUCTURE OF CONTINENTAL SLOPE, NORTHERN GULF OF MEXICO
M 29 125† STRUCTURE OF FALKLAND PLATEAU AND OFFSHORE TIERRA DEL
B 6004 710* STRUCTURE OF NEOGENE, MIOCENE, LITTORAL HARD- AND
B 6305 844* STRUCTURE OF SAN CAYETANO AND OAK RIDGE THRUST FAULTS,
M 29 261† STRUCTURE OF THE OUTER CONVERGENT MARGIN OFF KODIAK ISLAND,
M 29 85† STRUCTURE OF THE U.S. ATLANTIC MARGIN, BASIN
B 6303 521* STRUCTURE OF TUNISIAN ATLAS, ABST.
B 6407 1014† **STRUCTURE OF UPPER CENOZOIC STRATA BENEATH KODIAK SHELF,**
M 29 193 STRUCTURE SECTIONS, BURMA, ANDAMAN SEA
B 6403 339† STRUCTURE, STRATIGRAPHY, AND GEOLOGIC HISTORY OF
B 6401 37† STRUCTURE, STRATIGRAPHY, AND HYDROCARBONS OFFSHORE
M 29 226 STRUCTURE, SUMATRA, MARGIN
M 29 34 STRUCTURE, TOP OF CRETACEOUS, BLAKE PLATEAU
B 6303 555* STRUCTURE-CONTOUR MAPS COMPUTER CONSTRUCTED FROM
S 10 24 STRUCTURED WATER, PORE WATER, HYDROGEN BONDING
M 25 472 STRUCTURES, CENTRAL-TYPE, DEVELOPMENT OF
M 25 472 STRUCTURES, CENTRAL-TYPE, GROUPS OF
M 25 473 STRUCTURES, DISTINCT ON EARTH, MOON, MARS, AND VENUS, RING
M 25 472 STRUCTURES, DOME
B 6307 1037 STRUCTURES, MAP, WILLISTON BASIN AREA
M 29 43† STRUCTURES OF THE SOUTHWEST ICELAND INSULAR MARGIN:
M 25 472 STRUCTURES, RING
B 6008 1411 STRUHSACKER, ERIC M., PROPOSED GEOTHERMAL CIRCULATION
B 6105 834 STRUHSACKER, ERIC MITCHELL, GEOTHERMAL SYSTEMS OF CORWIN
B 6207 1153 STRZELECKI GROUP, AUSTRALIA
B 6009 1616 STUART, CHARLES J., FORM AND COMPOSITION OF MISSISSIPPI FAN,
M 26 249 STUART, CHARLES J., SEISMIC FACIES AND SEDIMENTOLOGY OF
B 6301 38 STUART CITY FIELD, LASALLE COUNTY, TEXAS
B 6409 1552* STUART CITY FORMATION, CRETACEOUS, SPEARY GAS FIELD, KARNES
B 6301 38 STUART CITY REEF TREND, TEXAS
B 6109 1555* STUART CITY TREND, LOWER CRETACEOUS, SOUTH-CENTRAL TEXAS,
B 6101 1 STUART SMITH, J. H., GEOLOGY AND HYDROCARBON DISCOVERIES OF
B 6405 792 STUBBLEFIELD, WILLIAM, GENESIS OF SAND RIDGES ON
B 6209 1767 STUCKEY, CHAS. W., JR., MILESTONES IN GULF COAST ECONOMIC
B 6405 789 STUCKLESS, JOHN S., INTERPRETATION OF THORIUM TO URANIUM
B 6209 1768 STUDE, GERALD R., DEPOSITIONAL ENVIRONMENTS OF GULF OF MEXICO
B 6105 834 STUDE, GERALD R., OFFSHORE LOUISIANA SOUTH TIMBALIER BLOCK 54
B 6101 105 STUDENTS, PETROLEUM-, EXPLORATION CLASSROOM GAME FOR
B 6411 2038 STUDIE IN GEOPHYSICS RESEARCH BOARD, RODGERS, DONALD A.,
B 6102 137† STUDIES OF VENTURA FIELD, CALIFORNIA, I--FACIES GEOMETRY AND
B 6102 169† STUDIES OF VENTURA FIELD, CALIFORNIA, II--LITHOLOGY, COMPACTION
B 6207 1117 STUDY, ALBERTA, CLARESHOLM AREA
B 6002 328* STUDY OF CARBONATE ENVIRONMENTS USING WELL CUTTINGS, ABST.
B 6405 759* STUDY OF DIAGENETIC PROCESSES IN SANDSTONE FROM CERRO PRIETO
B 6303 458* **STUDY OF FOSSIL MICROBIAL BORINGS--NEW APPROACH,**
B 6002 288 STUDY OF WHOLE FREE-GAS PHASE OF HYDROCARBONS
B 6405 640 STUMP FORMATION, WYOMING, PREUSS-
B 6002 322 STUMP, FRED, REPORT--WESTERN KANSAS. BY TITLE ONLY
M 24 95 STUMP SANDSTONE, WYOMING
B 6311 2024 STURT BLOCK, AUSTRALIA
M 28 166 STYLOLITIC SOLUTION, IN A SANDSTONE
M 28 166 STYLOLITIC SOLUTION, IN SILTSTONE
B 6411 1963 STYNE TROUGH, DONEGAL AND FASTNET BASINS, EIRE
M 27 206 STYOLITE SOLUTION
B 6411 1963 STYRIAN BASIN, AUSTRIA
B 6210 1909 STYRIAN BASIN, AUSTRIA
B 6310 1918 STYRIAN BASIN, AUSTRIA
B 6110 1639 STYRIAN BASIN, AUSTRIA
B 6112 2068 SUAI AREA, SECTION, TIMOR ISLAND
M 30 399 SUARDY, ATIK, GEOLOGY OF THE HANDIL FIELD, EAST KALIMANTAN,
B 6210 2045 SUAREZ LYNCH, ON DEVELOPMENTS IN ARGENTINA, 1977, YRIGOYEN, M.
B 6204 705† SUB-KASKASKIA UNCONFORMITY IN ILLINOIS BASIN, POSITION OF
B 6204 707 SUB-KASKASKIA-POST-TIPPECANOE UNCONFORMITY, ILLINOIS BASIN
M 26 123 SUB-PARALLEL, SEISMIC, REFLECTIONS, CONFIGURATION PATTERNS
B 6303 514* SUBAERIAL DIAGENESIS OF AXIAL CORALLITE OF ACROPORA
B 6203 528* SUBAERIAL EXPOSURE OF SILURIAN SHELF-MARGIN REEFS,
B 6004 712* SUBAERIAL KARST, ABST., SCALLOPED, SOLUTION SURFACES, MIDDLE,
B 6402 264† SUBAERIAL PHASE, ATCHAFALAYA DELTA, EVOLUTION OF,
B 6402 264† SUBAERIAL PHASE, ATCHAFALAYA DELTA, LOUISIANA, EVOLUTION OF,
B 6303 522* SUBALPINE TERTIARY BASINS, ABST., DUAL ORIGIN OF, NATURAL, GASES
B 6004 718* SUBAQUEOUS EVAPORITE DEPOSITION, PERMIAN SHELF SEVEN RIVERS
B 6303 453* SUBAQUEOUS GRAVITY-DISPLACEMENT PRODUCTS, ABST.
B 6303 426* SUBAQUEOUS LANDSLIDES AND DIAPIRS, GULF OF MEXICO. BY TITLE
B 6004 683* SUBAQUEOUS SAND-FLOW CROSS-STRATA, ABST., COMPARISON OF,
B 6303 507* SUBBITUMINOUS AND BITUMINOU COALS, ABST., PORPHYRINS IN
M 26 357 SUBBOTTOM PROFILING SYSTEMS, RESOLUTION SYSTEMS,
S 6 67† SUBCOMMISSION ON GEOCHRONOLOGY--CONVENTION ON THE USE OF
M 26 371 SUBCROP MAPPING, RESOLUTION SYSTEMS
M 25 485 SUBDUCTION, ALEUTIAN TRENCH
M 25 189† SUBDUCTION AND THEIR MEANING TO THE HYDROCARBON
B 6301 18 SUBDUCTION COMPLEX
B 6211 2349* SUBDUCTION COMPLEX, MENDOCINO COAST, NORTHERN CALIFORNIA,
B 6312 2176 SUBDUCTION COMPLEX PROVENANCE
M 29 224 SUBDUCTION HISTORY, SUMATRA, CENOZOIC
M 25 526 SUBDUCTION, MEXICO, MIOCENE
M 25 485 SUBDUCTION, MIDDLE AMERICA TRENCH
B 6212 2454† SUBDUCTION SYSTEM, ECUADORIAN
B 6212 2473 SUBDUCTION SYSTEM, PETROLEUM IN, ECUADOR
B 6402 280† SUBDUCTION SYSTEM. DISCUSSION AND REPLY, ECUADORIAN
S 3 82 SUBDUCTION ZONE
B 6212 2470 SUBDUCTION ZONE, PERU-, CHILE
M 25 448 SUBDUCTION ZONE, RELATION STUDIED, JAPAN, MINERAL
M 29 447 SUBDUCTION ZONE, SEISMIC MODEL, PERU TRENCH
M 25 330 SUBIBAJA FORMATION, ECUADOR
B 6007 1021† SUBLACUSTRINE FAN MORPHOLOGY IN LAKE SUPERIOR
B 6201 100 SUBLETT RANGE, IDAHO
M 24 91† SUBLETTE AND LINCOLN COUNTIES, WYOMING, BIG PINEY--, LA BARGE
B 6405 738* SUBLETTE COUNTY, WYOMING, ABST., EVALUATION OF ORGANIC
B 6108 1376* SUBLETTE COUNTY, WYOMING, ABST., NUGGET OIL, ACCUMULATIONS

Index

B 6004 654* SUBMARINE CANYON, ABST., SEDIMENT PROCESSES CONTROLLED BY,
B 6203 542* SUBMARINE CANYON AND CHANNEL MORPHOLOGY, INDUS CONE,
B 6004 721* SUBMARINE CANYON AND THEIR RELATION TO DEEP-SEA FANS, ABST.,
B 6303 551* SUBMARINE CANYON, BOSO PENINSULA, JAPAN, ABST., EARLY,
B 6004 650* SUBMARINE CANYON, CALIFORNIA COAST, ABST., PHYSICAL AND,
B 6004 695* SUBMARINE CANYON TRIBUTARY, ABST., LARGE SEDIMENT, SLUMP
B 6004 686* SUBMARINE CANYON-FAN SYSTEMS, ABST., SIGNIFICANCE OF, CLAST
S 8 158 SUBMARINE CANYONS, ABRA DELTA, PHILIPPINE ISLANDS
B 6004 667* SUBMARINE CANYONS, ABST., CONTRASTS IN CONCENTRATION OF,
B 6004 702* SUBMARINE CANYONS AND FANS, ABST., THIN-BEDDED, TURBIDITES IN,
B 6004 645* SUBMARINE CANYONS AND MARINE CYCLES OF DEPOSITION IN
B 6405 649 SUBMARINE CANYONS AND OTHER SEAVALLEYS, BY FRANCIS P.
B 6203 561* SUBMARINE CANYONS AND OTHER TYPES OF SLOPE VALLEYS, ABST.,
S 8 101 SUBMARINE CANYONS, CAPE SAN LUCAS CANYON, MEXICO
S 8 78 SUBMARINE CANYONS, CARMEL CANYON, CALIFORNIA
S 8 139 SUBMARINE CANYONS, CHRISTIANSTED CANYON, WEST INDIES
S 8 161 SUBMARINE CANYONS, CONGO CANYON, AFRICA
B 6302 257 SUBMARINE CANYONS FANS, AND TRENCHES, EDITED BY DANIEL JEAN
S 8 97 SUBMARINE CANYONS, FRASER SEAVALLEY, CANADA
S 8 152 SUBMARINE CANYONS, HANAKAPIAI CANYON, HAWAII
S 8 121 SUBMARINE CANYONS, HUDSON CANYON, U.S. EAST COAST
S 8 67 SUBMARINE CANYONS, HUENEME CANYON, CALIFORNIA
S 8 111 SUBMARINE CANYONS, HYDROGRAPHER CANYON, U.S. EAST COAST
S 8 4 SUBMARINE CANYONS, INDEX MAP
S 8 152 SUBMARINE CANYONS, KALALAU CANYON, HAWAII
S 8 147 SUBMARINE CANYONS, KAUAI CANYONS, HAWAII
S 8 58 SUBMARINE CANYONS, LA JOLLA CANYON, CALIFORNIA
S 8 87 SUBMARINE CANYONS, MONTEREY CANYON, CALIFORNIA
S 8 61 SUBMARINE CANYONS, NEWPORT CANYON, CALIFORNIA
S 8 152 SUBMARINE CANYONS, NOHILI SEAVALLEY, HAWAII
S 8 122 SUBMARINE CANYONS, NORFOLK CANYON, U.S. EAST COAST
B 6303 411* SUBMARINE CANYONS OF NORTHWES GULF OF MEXICO--STUDY IN
B 6004 686* SUBMARINE CANYONS OF NORTHWESTERN ATLANTIC MARGIN, ABST.,
B 6301 67† SUBMARINE CANYONS OF TETHYAN CONTINENTAL MARGINS,
S 8 105 SUBMARINE CANYONS, PETACALCO CANYON, MEXICO
S 8 161 SUBMARINE CANYONS, PINGET SEAVALLEY, PHILIPPINES
S 8 64 SUBMARINE CANYONS, REDONDO CANYON, CALIFORNIA
S 8 105 SUBMARINE CANYONS, RIO BALSAS DELTA CANYON, MEXICO
S 8 124 SUBMARINE CANYONS, RIO DE LA PLATA CANYON, PUERTO RICO
S 8 129 SUBMARINE CANYONS, SALT RIVER CANYON, WEST INDIES
S 8 91 SUBMARINE CANYONS, SAN CLEMENTE RIFT, CALIFORNIA
S 8 101 SUBMARINE CANYONS, SAN JOSE CANYON, MEXICO
S 8 73 SUBMARINE CANYONS, SANTA CRUZ CANYON, CALIFORNIA
S 8 65 SUBMARINE CANYONS, SANTA MONICA CANYON, CALIFORNIA
S 8 58 SUBMARINE CANYONS, SCRIPPS CANYON, CALIFORNIA
S 8 122 SUBMARINE CANYONS, WASHINGTON CANYON, U.S. EAST COAST
S 8 122 SUBMARINE CANYONS, WILMINGTON CANYON, U.S. EAST COAST
B 6405 790* SUBMARINE CARBONATE CEMENTATION AND PISOLITH GROWTH IN
B 6203 554* SUBMARINE CEMENTATION, ABST., INTERSTITIAL, WATER CHEMISTRY
B 6203 573* SUBMARINE CEMENTATION OF CAPITAN LIMESTONE, PERMIAN,
B 6203 541* SUBMARINE CEMENTS AND FABRIC-RETENTIVE DIAGENESIS, CAPITAN
B 6207 1237* SUBMARINE CHANNELS IN GREAT VALLEY, CALIFORNIA, ABST., OIL AND
B 6004 659* SUBMARINE DEBRIS-FLOW DEPOSIT, NORTHERN TUNISIA, ABST.,
B 6405 697* SUBMARINE DIAGENESIS, ARAGONITE DISSOLUTION, CEMENTATION BY
B 6105 808 * SUBMARINE DIAGENESIS IN RELICT HOLOCENE CORAL REEF, SOUTHEAST
B 6203 553* SUBMARINE DIAGENESIS ON PRIMARY POROSITY IN CAPRINID REEFS
B 6303 495* SUBMARINE FAN, ABST., DOWNCURRENT CHANGE IN, MIOCENE
B 6405 671* SUBMARINE FAN COMPLEX, KODIAK ISLAND ARCHIPELAGO, ALASKA,
M 30 86 SUBMARINE FAN DEPOSITIONAL MODEL, FORTIES FIELD
B 6206 959 SUBMARINE FAN, ENGLAND, SHALE, GRIT: A COMPLETE, ANCIENT
B 6407 1096 SUBMARINE FAN MODELS AND SUPRAFAN CONCEPT
B 6405 722* SUBMARINE FANS, ABST., POSSIBLE GEOMETRIES OF, SANDSTONE BODIES
B 6403 443* SUBMARINE FANS, ABST., QUATERNARY STYLES OF, CALIFORNIA
B 6206 950 SUBMARINE FANS, EXAMPLES OF, ANCIENT
B 6306 907 SUBMARINE FANS, NOMENCLATURE
B 6407 1052 SUBMARINE FANS, QUEBEC
B 6407 1094† SUBMARINE FANS. DISCUSSION AND REPLIES, MODERN AND, ANCIENT
B 6305 810† SUBMARINE FANS. MODELS FOR EXPLORATION FOR STRATIGRAPHIC
B 6206 912† SUBMARINE FANS: CHARACTERS FOR RECOGNITION OF SANDY
B 6206 932† SUBMARINE FANS: MODELS FOR EXPLORATION FOR STRATIGRAPHIC
B 6203 521* SUBMARINE FANS--IMPORTANT REGIONS FOR OIL AND GAS GENESIS
B 6207 1213* SUBMARINE METALLIFEROUS DEPOSITS IN SOUTHWEST PACIFIC, ABST.,
M 25 395† SUBMARINE PHOSPHORITE DEPOSITS OF CHATHAM RIS NEAR NEW
M 28 120 SUBMARINE SANDS, PERSIAN GULF
M 28 121 SUBMARINE SANDS, PERSIAN GULF, HOLOCENE
B 6407 1083 SUBMARINE SEEPS, ORIGIN OF, HYDROCARBON, GASES IN
B 6303 498* SUBMARINE SLIDES, ABST.
B 6303 324† SUBMARINE SLUMPING IN CONTINENTAL MARGIN OF ISRAEL AND
B 6309 1582* SUBMARINE SLUMPING IN REEF-FLANK DEPOSITS, MIDDLE SILURIAN, OF
B 6105 783* SUBMARINE SLUMPS AND SEDIMENT FLOWS OFF SAN NICOLAS ISLAND,
S 7 5 SUBMARINE TOPOGRAPHY, OUTLINE CHART, GULF OF MEXICO
B 6004 725* SUBMARINE-CANYON AND FAN-VALLEY SEDIMENTATION, ABST.,
B 6004 678* SUBMARINE-FAN COMPLEX, CAP ENRAGE FORMATION MIDDLE
B 6004 730* SUBMARINE-FAN CONGLOMERATES AND SANDSTONES, JURASSIC OF
B 6209 1766* SUBMARINE-FAN DEPOSITION OF WOODBINE FORMATION, UPPER
B 6303 528* SUBMARINE-FAN GAS RESERVOIR IN WOODBINE--EAGLE FORD
B 6306 905† SUBMARINE-FAN SANDS, DIPMETER AND, LOG MOTIFS IN, NORTH SEA
B 6004 734* SUBMARINE-FAN SEDIMENTATION IN MESOZOIC ARC-REAR MARGINAL
B 6311 1999† SUBMARINE-FAN TRAP IN LOWER EOCENE ROCKS OF NORTH SEA VIKING
B 6011 2068* SUBMARINE-SLOPE FAILURES AND DOWNSLOPE MASS MOVEMENT OF
B 6105 774 * SUBMARINE-SLUMP FEATURES SEAWARD OF ICY BAY AND MALASPINA
M 28 105 SUBMATURE SEDIMENTARY, (ARKOSE), TEXTURE
B 6004 691* SUBMERGED RELICT FORESTS OFF NORTHWEST FLORIDA
B 6004 732* SUBMERSIBLE, ABST., GEOLOGIC EXPLORATION OF SOUTHERN,
B 6303 529* SUBMERSIBLE, ABST., SLUMPS ON UPPER, CONTINENTAL, SLOPE,
S 4 53 SUBMERSIBLE NEKTON PROJECT
B 6112 2134 SUBNORMAL PRESSURE FIELDS
B 6112 2134† SUBNORMAL PRESSURES, OIL AND, GAS IN, RESERVOIRS WITH
B 6109 1544* SUBSALT ORIGIN OF SALT-DOME CAP ROCK, ABST.
B 6004 733* SUBSE CEMENTED CARBONATE MOUNDS, LITHOHERMS, IN NORTHERN
B 6304 702* SUBSIDENCE, ABST., MECHANISMS OF, BASIN
B 6402 173† SUBSIDENCE AND HEAT FLOW--IMPORTANT PARAMETERS IN
B 6405 705* SUBSIDENCE AND THERMAL HISTORY OF SOUTHERN OKLAHOMA
M 29 6 SUBSIDENCE, CRUSTAL FLOW, DIAGRAM
M 29 81 SUBSIDENCE HISTORY, ATLANTIC COAST, COST B-2 WELL
M 29 81 SUBSIDENCE HISTORY, ATLANTIC COAST, NEW JERSEY
M 25 121 SUBSIDENCE IN GEOTHERMAL FIELDS
B 6303 454* SUBSIDENCE IN HOUSTON-GALVESTON REGION, TEXAS, ABST.,
M 25 362† SUBSIDENCE IN JAPAN RESULTING FROM FLUID
M 29 3† SUBSIDENCE MECHANISMS AT PASSIVE CONTINENTAL MARGINS
M 29 8 SUBSIDENCE, NORMAL FAULTING, DIAGRAM
M 29 7 SUBSIDENCE, NORMAL-FAULT MECHANISMS
M 29 242 SUBSIDENCE, NORTH PHILIPPINE SEA, EOCENE
B 6106 929† SUBSIDENCE OF ASEISMIC RIDGES--EVIDENCE FROM SEDIMENTS ON RIO
B 6109 1546* SUBSIDENCE OF MARSHLAND PEAT AS GEOLOGIC HAZARD IN GREATER
M 29 239† SUBSIDENCE OF THE DAITO RIDGE AND ASSOCIATED BASINS, NORTH
B 6202 215 SUBSIDENCE RATES
B 6109 1522 SUBSIDENCE RATES, EAST COAST
B 6109 1522† SUBSIDENCE RATES IN EASTERN UNITED STATES, RAPID, APTIAN-,
B 6402 174 SUBSIDENCE, UNITED STATES, EAST COAST
S 7 272 SUBSURFACE CONVEX FAULT, BOTTOM, DEFORMATION
B 6004 724* SUBSURFACE DATA OF INTERPRETATION OF PALEOENVIRONMENTS OF
B 6004 677* SUBSURFACE DATA. BY TITLE ONLY., SOME APPROACHES TO,
B 6002 184† SUBSURFACE ENVIRONMENTAL ANALYSIS OF NORT SEA SEDIMENTS
B 6006 981† SUBSURFACE ENVIRONMENTS, HYDRATE FORMATION IN
B 6207 1111† SUBSURFACE FRACTURES FROM DIPMETER LOGS, MEASUREMENT OF
B 6004 714* SUBSURFACE GEOLOGIC STRUCTURES, ABST., COMPUTER ANALYSIS OF,
B 6012 2190* SUBSURFACE GEOLOGY AND SOME IMPLICATIONS CONCERNING FUTURE
B 6303 522* SUBSURFACE GEOLOGY OF HONOR RANCHO AREA, VENTURA COUNTY,
M 29 359 SUBSURFACE GEOLOGY, VIRGIN ISLANDS, ST. CROIX
M 29 353† SUBSURFACE GEOLOGY WEST OF ST. CROIX, U.S. VIRGIN ISLANDS,
B 6207 1216* SUBSURFACE GEOTECHNICAL HAZARDS IN PETROLEUM DEVELOPMENT,
B 6203 511* SUBSURFACE HEAVY-OIL-BEARING SANDSTONES IN SOUTHEASTERN
B 6403 446* SUBSURFACE INTERFACE BETWEEN ZEOLITIZED AND OVERLYING
B 6004 657* SUBSURFACE JURASSIC-CRETACEOUS BOUNDARY, NEOCOMIAN
B 6203 523* SUBSURFACE LITHOSTRATIGRAPHY OF HUNTON GROUP IN PARTS OF
B 6309 1599* SUBSURFACE NEOGENE STRATIGRAPHY OF BAY COUNTY, FLORIDA,
B 6002 299† SUBSURFACE POSITIVE ELEMENTS WITHIN OUACHITA FOLDBELT IN
B 6003 455† SUBSURFACE PRECAMBRIAN GRANITE,
M 26 296 SUBSURFACE PRESSURE GRADIENT, DENVER BASIN, PIERRE SHALE
B 6203 488* SUBSURFACE STRATIGRAPHIC ANALYSIS, ABST., INTERACTIVE, GRAPHICS
B 6309 1596* SUBSURFACE STRATIGRAPHY OF MIDWAY-WILCOX, ZAPATA COUNTY,
B 6303 501* SUBSURFACE STRATIGRAPHY OF UPPER DEVONIAN CLASTICS IN
B 6301 50† SUBSURFACE STUDY OF ATOKA, LOWER PENNSYLVANIAN, CLASTIC
B 6002 327* SUBSURFACE STUDY OF ATOKA, LOWER PENNSYLVANIAN, CLASTIC
B 6205 893* SUBSURFACE THRUSTS ALONG WASATCH HINGE LINE IN CENTRAL
S 10 180 SUBSURFACE WATERS, METHANE, OCCURRENCES
B 6409 1555* SUBSURFACE WILCOX LIGNITE IN WEST-CENTRAL LOUISIANA, ABST.
B 6004 685* SUBTIDAL AND SHOAL DEPOSITS IN MEDIAL ORDOVICIAN OF CENTRAL
B 6004 661* SUBTIDAL ENVIRONMENT, ABST., TRACE-, FOSSIL ASSEMBLAGE OF
B 6004 712* SUBTIDAL ENVIRONMENT--TIME AND PLACE FOR TAXONOMIC
B 6004 660* SUBTIDAL STROMATOLITES IN LOCKPORT FORMATION, MIDDLE
B 6004 727* SUBTIDAL TRACE FOSSILS, MUDDY AND FRONTIER FORMATIONS,
B 6408 1218 SUBTLE PETROLEUM TRAPS, TYPES
B 6004 732* SUBUNCONFORMITY PETROLEUM SOURCE, ABST., PRESERVATION,
B 6408 1174 SUBUNCONFORMITY PLAYS
B 6001 115† SUBUNCONFORMITY SOURCE ROCKS, OKLAHOMA CITY OIL--SECOND
B 6104 620† SUBUNCONFORMITY SOURCE ROCKS. DISCUSSION AND REPLY,

M 25 499 SUCCESSOR BASIN, CROSS SECTION, CANADIAN, CORDILLERA, JURASSIC
M 25 500 SUCCESSOR BASINS, MINERAL DEPOSITS, CANADA
M 25 500 SUCCESSOR BASINS, ROCKS OF, CANADIAN, CORDILLERA
B 6405 790 SUCHECKI, ROBERT K., DIAGENESIS IN VOLCANOGENIC ROCKS OF GREAT
B 6004 682 SUCHECKI, ROBERT K., TIMING OF CARBONATE AND CLAY DIAGENESIS
B 6305 842 SUCHOMEL, THOMAS J., URANIUM GEOLOGY OF MORTON RANCH
B 6403 446* SUCKLING TO CAPE SPENCER, ABST., TWENTIETH-CENTURY HISTORY OF,
B 6409 1548 SUCZEK, CHIRSTOPHER A., REVIEW OF ROCKS AND ROCK MINERALS, BY
B 6312 2164 SUCZEK, CHRISTOPHER A., PLATE TECTONIC AND SANDSTONE
B 6403 449 SUCZEK, CHRISTOPHER A., SANDSTON COMPOSITIONS RELATED TO
B 6010 1771 SUDAN, 1975, PETROCONSULTANTS S.A., ON DEVELOPMENTS IN
B 6110 1702 SUDAN, 1976, PETROCONSULTANTS S.A., ON DEVELOPMENTS IN
B 6210 1799 SUDAN, 1977, PETROCONSULTANTS S.A., ON DEVELOPMENTS IN
B 6310 1639 SUDAN, 1978, PETROCONSULTANTS S.A., ON DEVELOPMENTS IN
B 6411 1742 SUDAN, 1979, PETROCONSULTANTS S.A., ON DEVELOPMENTS IN
B 6406 863 SUDAN, KUFRA BASIN, LIBYA AND
B 6306 974 SUDAN, NUBIA GROUP NOMENCLATURE AND CORRELATION, EGYPT
B 6211 2353* SUEZ, ABST., MULTIPLE REFLECTIONS AND HEAD WAVES IN, GULF OF
B 6203 536* SUEZ AREA—TALE OF FOUR TROUGHS, ABST., CRETACEOUS-, HOLOCENE
B 6407 981 SUFFIELD BLOCK, ALBERTA
B 6407 983 SUFFIELD BLOCK GAS RESERVES, ALBERTA
B 6303 528* SUGAR CREEK FIELD, TYLER COUNTY, TEXAS, ABST., DESCRIPTION AND
B 6309 1576* SUGGESTED ORIGIN FOR GAS IN EASTERN DEVONIAN SHALES, ABST.
B 6207 1224 SUHEIMI, S., CORRELATION OF MAJOR UNCONFORMITIES IN WESTERN
B 6203 565 SUHM, RAYMOND W., PETROLEUM POTENTIAL AND STRATIGRAPHY OF
B 6002 322 SUHM, RAYMOND W., TRENDS AND ORIGINS OF ORDOVICIAN
B 6309 1605* SUITABILITY STUDIES OF SALT DOMES IN EAST TEXAS BASIN FOR
M 25 138 SULAWESI, THERMAL AREA, PROSPECTIVE
B 6203 530* SULFATE DIAGENESIS, MISSISSIPPIAN OF EASTERN WILLISTON BASIN,
B 6405 787* SULFATE MINERAL REPLACEMENTS IN DIAGENESIS OF PERITIDAL
B 6303 420* SULFATE REDUCTION AND METHANE PRODUCTION, ABST., MODEL FOR
B 6303 512* SULFATE--NEW MECHANISM FOR POROSITY GENERATION IN CARBONATE
M 25 487 SULFIDE ASSOCIATIONS, METALS, VOLCANOGENIC
M 25 449 SULFIDE DEPOSITS, JAPAN, SAMBAGAWA SYSTEM, STRATA-BOUND
M 25 506 SULFIDE DEPOSITS OF METALS, CANADIAN, CORDILLERA, MASSIVE
B 6203 505* SULFIDE MINERALOGY AND MORPHOLOGY IN OIL SHALE AND
B 6108 1388* SULFIDE MINERALS AS EXPLORATION GUIDE FOR URANIUM, ABST.,
B 6011 2069* SULFIDES, AND PETROLEUM, ABST., EVAPORITES
M 25 554 SULFIDES, SURVEYS, ECUADOR, GOLD AND, BASE-, METAL
B 6405 772* SULFIDIZATION OF SOUTH TEXAS ROLL-TYPE URANIUM DEPOSIT, ABST.,
B 6303 505* SULFUR AND OXYGEN ISOTOPE GEOCHEMISTRY OF ABU DHABI SABKHA,
M 25 448 SULFUR AND PYRITE ORE DEPOSITS, RECENT, JAPAN
B 6408 1283* SULFUR COAL DEPOSITS, SOUTHERN ILLINOIS, ABST., FACIES,
S 10 174 SULFUR COMPOUNDS, GENERATION, MIGRATION
B 6004 692* SULFUR CONTENT, BIG BROWN LIGNITE DEPOSIT, FREESTONE COUNTY,
B 6405 689* SULFUR CONTENT IN COAL, ABST., GEOLOGIC CONTROLS ON
M 25 448 SULFUR DEPOSITS, RELATION TO VOLCANIC FRONT, JAPAN
M 25 439 SULFUR DEPOSITS, TAIWAN
B 6108 1376* SULFUR IN COLORADO OIL-SHALE DEPOSITS, ABST., DISTRIBUTION AND,
B 6305 824* SULFUR ISOTOPES AND WHOLE-ROCK MINERALOGIC DATA, ABST.,
B 6212 2400 SULFUR PROBLEMS, COAL QUALITY
B 6305 839* SULFUR, RADIOACTIVE COAL OF PALUDAL-LACUSTRINE ORIGIN IN
B 6407 1118 SULLIVAN, BOUMA, ARNOLD H., REVIEW OF, GUADALUPIAN, DELAWARE
B 6403 400† SULLIVAN COUNTY, INDIANA, ANALYSIS OF, GRAVITY, ANOMALY OVER,
M 25 507 SULLIVAN DEPOSIT, CANADA
B 6405 750 SULLIVAN, F. R., NEW PRECISION IN BIOSTRATIGRAPHY THROUGH
S 8 1 SULLIVAN, GARY G., CURRENTS IN SUBMARINE CANYONS AND
M 25 496 SULLIVAN MINE, CANADA
B 6405 649 SULLIVAN, POAG, C. WYLIE, REVIEW OF, CURRENTS IN, SUBMARINE
B 6407 1037 SULPHUR CANYON, CALIFORNIA
B 6205 782 SULPHUR POINT FORMATION, ALBERTA
M 29 375 SULU SEA, BATHYMETRY, SOUTHEAST ASIA
M 29 376 SULU SEA, PLATE BOUNDARIES, SOUTHEAST ASIA
M 29 376 SULU SEA, SCHEMATIC SECTION, SOUTHEAST ASIA
M 29 373† SULU SEA: A MARGINAL BASIN IN SOUTHEAST ASIA, THE
M 29 379 SULU TROUGH, MULTICHANNEL SEISMICS, SOUTHEAST ASIA
M 28 30 SULZBACHER, GERMANY, CARBONIFEROUS
B 6004 709* SUMATRA AND JAV SEA BASINS--MECHANISM OF BASIN FORMATION. BY
M 25 244 SUMATRA BASIN, INDONESIA, NORTH
M 25 243 SUMATRA BASIN, INDONESIA, NORTH
M 25 244 SUMATRA BASINS, INDONESIA, CENTRAL AND, SOUTH
M 29 235 SUMATRA, CENOZOIC, HISTORY
M 29 224 SUMATRA, CENOZOIC, SUBDUCTION HISTORY
M 29 225 SUMATRA, CENOZOIC, SUNDA CONVERGENCE ZONE
M 29 226 SUMATRA, INDIAN OCEAN BASIN, RELATIONS
B 6104 594 SUMATRA, LIRIK FIELD
M 29 224 SUMATRA, MARGIN, KINEMATIC FRAMEWORK
M 29 226 SUMATRA, MARGIN, MORPHOLOGY
M 29 226 SUMATRA, MARGIN, STRUCTURE
B 6402 219 SUMATRA, MINAS FIELD
M 29 233 SUMATRA, OFFSHORE, CRITICAL WELL RESULTS
B 6104 594 SUMATRA, PULAI FIELD
M 29 223† SUMATRA REGION, STRUCTURE AND, CENOZOIC, EVOLUTION OF THE,
M 29 231 SUMATRA, SUNDA ARC, BASEMENT MELANGE
M 29 229 SUMATRA, SUNDA ARC, EVOLUTION OF STRUCTURE
M 29 224 SUMATRA, SUNDA ARC, INDEX MAP
M 29 232 SUMATRA, SUNDA ARC, REFLECTION PROFILE
M 29 226 SUMATRA, SUNDA ARC, SHALLOW STRUCTURES
M 29 234 SUMATRA, TAPANULI DISTRICT, GEOLOGIC MAP
M 25 138 SUMATRA, THERMAL AREAS, PROSPECTIVE
M 25 462 SUMATRA-CLIPPERTON-TEHUANTEPEC-BARTLETT FAULT, HUTANOPAN
B 6112 2068 SUMBA ISLAND, SECTION
M 25 148 SUMI, KIYOSHI, GEOLOGICAL AND GEOCHEMICAL CHARACTERISTICS OF
B 6207 1224 SUMINTA, CORRELATION OF MAJOR UNCONFORMITIES IN WESTERN
B 6207 1236 SUMITRAMI-HARDJA, A., GEOTHERMAL WELL COMPLETION AND CASING
B 6302 173 SUMMARY OF LOG ANALYSIS BY FORMATION
B 6101 35 SUMMARY OF PRECAMBRIAN URANIUM DEPOSITS
B 6211 2359* SUMMARY OF RECENT EXPLORATION AND EVALUATION WORK ON
S 5 189* SUMMARY, REEFS AND EVAPORITES: A
B 6312 2188 SUMMATION OF PROSPECTS OR PLAYS, ASSESSING RESOURCES
S 4 88 SUMMERFIELD FORMATION, MAESTRICHTIAN, JAMAICA
B 6405 790 SUMMERHAYES, C. P., ORGANIC FACIES OF MID-CRETACEOUS BLACK
B 6406 957 SUMMERHAYES, COLIN P., REVIEW OF GEOLOGY OF CONTINENTAL
B 6004 716 SUMMERHAYES, COLIN, RECENT SEDIMENTATION ON NILE DELTA AND
B 6007 1013 SUMMERVILLE FORMATION, NEW MEXICO
B 6008 1402* SUMMIT COUNTY, UTAH, ABST., GEOLOGY OF, PINEVIEW FIELD AREA
B 6002 320* SUMNER COUNTY, KANSAS, ABST., YARNELL TREND
B 6107 1030 SUMNER GROUP, NEBRASKA
B 6102 295* SUMPS, ABST., ENVIRONMENTS OF, EVAPORITI SEDIMENTATION--,
B 6112 2105 SUN FIELD, PERU
B 6104 579 SUN LAMA FIELD, VENEZUELA
B 6104 579 SUN LAMAR FIELD, VENEZUELA
B 6010 1872 SUN OIL CO., AN ELF R.E., ON DEVELOPMENTS IN OMAN, INCLUDING
B 6407 971 SUNBURST SANDSTONE MEMBER, GREAT PLAINS
B 6209 1705 SUNBURY SHALE, KENTUCKY
M 29 231 SUNDA ARC, BASEMENT MELANGE, SUMATRA
M 29 229 SUNDA ARC, EVOLUTION OF STRUCTURE, SUMATRA
M 29 223† SUNDA ARC IN THE CENTRAL SUMATRA REGION, STRUCTURE AND,
M 29 224 SUNDA ARC, INDEX MAP, SUMATRA
M 29 230 SUNDA ARC, REFLECTION PROFILE, NIAS
M 29 228 SUNDA ARC, REFLECTION PROFILE, SIBERUT
M 29 232 SUNDA ARC, REFLECTION PROFILE, SUMATRA
M 29 226 SUNDA ARC, SHALLOW STRUCTURES, SUMATRA
M 29 225 SUNDA CONVERGENCE ZONE, SUMATRA, CENOZOIC
B 6401 37 SUNDA SHELF, STRATIGRAPHY
M 29 227 SUNDA TRENCH, OFF SUMATRA, PACIFIC
B 6211 2275 SUNDANCE FORMATION, CALLOVIAN, BLACK HILLS
B 6004 711* SUNDANCE FORMATION OF BLACK HILLS REGION, ABST., LATE,
B 6211 2275† SUNDANCE FORMATION OF BLACK HILLS, SEDIMENTOLOGY OF LATE,
B 6211 2277 SUNDANCE FORMATION, WYOMING
B 6211 2363 SUNDBERG, FREDERICK A., LATE CRETACEOUS DEPOSITIONAL
S 4 85 SUNDERLAND INLIER, JAMAICA
B 6405 790 SUNDERMAN, JACK A., SUBMARINE CARBONATE CEMENTATION AND
B 6408 1211 SUNG-LIAO BASIN, CHINA
B 6310 1890 SUNMAR EXPLORATION CO., AND KHALIFA M. AL-HINAI, ON
B 6009 1609* SUNNILAND EXPLORATION O SOUTH FLORIDA, ABST., RECENT
B 6209 1753* SUNNILAND FORMATION IN SOUTH FLORIDA, ABST., FUTURE, OIL
B 6303 460* SUNNILAND LIMESTONE, LOWER CRETACEOUS, ABST., PORE TYPES IN
B 6010 1870 SUNNINGDALE OILS, U.K., LTD., AND TOTAL ABU AL BU KHOOSH, ON
B 6111 1951 SUNNYSIDE HEAVY OIL DEPOSIT, UNITED STATES
B 6009 1613* SUNOCO-FELDA AND WEST SUNOCO-FELDA FIELDS HENDRY AND
M 24 287† SUNOCO-FELDA FIELD, HENDRY AND COLLIER COUNTIES, FLORIDA
M 29 223 SUPARKA S., STRUCTURE AND CENOZOIC EVOLUTION OF THE SUNDA
B 6309 1534† SUPER OVERPRESSURED GAS
B 6004 700* SUPERCRITICAL FLOW DURING HURRICANE FLOODING, ABST.,
B 6111 1950† SUPERGIANT OIL FIELDS, TAR, SANDS AND
B 6111 1957 SUPERGIANT RESERVES, POSSIBLE SETTINGS FOR
B 6210 2117 SUPERIOR OIL CO., AND PETROCONSULTANTS S.A., ON DEVELOPMENTS IN
B 6310 1889 SUPERIOR OIL CO., ET AL, ON DEVELOPMENTS IN BAHRAIN, 1978,
B 6010 1872 SUPERIOR OIL CO., ON DEVELOPMENTS IN BAHRAIN, 1975, CHEVRON
B 6110 1801 SUPERIOR OIL CO., ON DEVELOPMENTS IN BAHRAIN, 1976, CHEVRON
B 6411 1838 SUPERIOR OIL CO., ON DEVELOPMENTS IN BAHRAIN, 1979, CHEVRON
B 6206 918 SUPERIOR, RESERVE FAN, LAKE
B 6007 1021† SUPERIOR, SUBLACUSTRINE, FAN MORPHOLOGY IN, LAKE
M 28 106 SUPERMATURE SEDIMENTARY, ORTHOQUARTZITE, TEXTURE
B 6408 1282 SUPHASIN, C., DEPOSITIONAL TRENDS OF LOWER SILURIAN CLINTON
B 6004 690 SUPKO, P. R., EVOLUTION OF MARGINAL PLATEAU--DSDP DATA FROM
B 6004 695 SUPKO, P. R., SEDIMENT LITHOFACIES IN SOUTH ATLANTIC OCEAN,

B 6211 2353* SUPLEE-IZEE AND ALDRICH MOUNTAINS INLIER, CENTRAL OREGON,
B 6004 711* SUPPLIES AND ECONOMICS, ABST., IMPACT OF, OIL AND, GA,
B 6407 1096 SUPRAFAN CONCEPT, SUBMARINE, FAN MODELS AND
B 6206 914 SUPRAFAN, DEFINITION
B 6206 952 SUPRAFAN LOBE, ANCIENT, MID-, FAN:, BRAIDED, CHANNELS ON
B 6206 947 SUPRAFAN LOBES, SUBMARINE FAN
B 6206 978 SUPRAFANS AS STRATIGRAPHIC TRAPS, CHANNELS AND
B 6004 685* SUPRATIDAL TO SUBTIDAL AND SHOAL DEPOSITS IN MEDIAL
B 6203 357 SUR BASIN, CALIFORNIA
B 6206 968 SUR CANYON, CALIFORNIA
B 6211 2222 SUR-NACIMIENTO FAULT, CALIFORNIA
B 6203 345 SUR-NACIMIENTO FAULT, CALIFORNIA
B 6203 451 SURAJ EROSIONAL VALLEYS, INDIA
S 10 232 SURAKHANY FIELD, USSR, BAKU AREA
B 6405 790 SURDAM, R. C., DIAGENESIS IN MONTEREY FORMATION, PISMO
B 6205 881 SURDAM, RONALD C., CARBONATE CYCLES IN GREEN RIVER FORMATION,
B 6108 1393 SURDAM, RONALD C., DIAGENESIS OF VOLCANOGENIC SANDSTONES,
B 6403 449 SURDAM, RONALD C., DIAGENETIC REACTIONS IN MONTEREY
B 6303 535 SURDAM, RONALD C., EVOLUTION OF HYDROGRAPHIC BASINS AND
B 6203 565 SURDAM, RONALD C., STROMATOLITES IN EOCENE GREEN RIVER
B 6108 1394 SURDAM, RONALD C., TRANSFER-REACTION MODELS OF DIAGENETIC
B 6008 1412 SURDAM, RONALL C., DEPOSITIONAL ENVIRONMENNSTS OF OIL SHALE,
S 10 245 SURFACE EVIDENCE, ETHANE INTERPRETATION, MIGRATION
S 10 241† SURFACE EVIDENCE OF HYDROCARBON MOVEMENT FROM DEPTH, NEAR-
B 6004 689* SURFACE FAULTING, TEXAS COASTAL ZONE, ABST., ACTIVE
S 9 21 SURFACE MEAN TEMPERATURES, PRESENT
B 6303 466* SURFACE MINING AND RECLAMATION ACT OF 1975, ABST.,
S 7 271 SURFACE MUDFLOW, BOTTOM, DEFORMATION
B 6204 633† SURFACE SEDIMENTARY UNITS OF NORTHERN GULF OF ALASKA
B 6303 435* SURFACE SEISMIC METHODS APPLIED TO COAL-MINING PROBLEMS, ABST.
B 6303 527* SURFACE TURBIDITY AND HYDROGRAPHI VARIABILITY ON SOUTH
S 10 95 SURFACTANT MICELLES, MIGRATION MECHANISMS
B 6305 768 SURFACTANT MICELLES, OIL SOLUBILIZED IN
S 10 96 SURFACTANT SOLUTIONS, MICELLE CONCENTRATIONS
B 6101 28† SURFICIAL ORIGIN OF NORTH AMERICAN PITCHBLENDE AND RELATED
B 6106 861† SURFICIAL SAND SHEET, BALTIMORE CANYON TROUGH AREA,
B 6106 861 SURFICIAL SAND SHEET ON MIDDLE ATLANTIC OUTER CONTINENTAL
S 7 305 SURFICIAL SEDIMENTS, ORCA BASIN
S 10 180 SURFICIAL WATERS, METHANE, OCCURRENCES
B 6110 1590 SURINAM, 1976, PETROCONSULTANTS S.A., ON DEVELOPMENTS IN
B 6210 2059 SURINAM, 1977, JACOBSEN, P., JR., ON DEVELOPMENTS IN
B 6310 1761 SURINAM, 1978, JACOBSEN, P., JR., ON DEVELOPMENTS IN
B 6010 1656 SURINAM, DEVELOPMENTS, 1975
M 25 174 SURIRE THERMAL AREA, CHILE
M 25 63† SURVEYING EARTH
B 6209 1593 SURVEYOR CHANNEL, ALASKA CONTINENTAL MARGIN
B 6204 633 SURVEYOR RESEARCH CRUISE, NOAA SHIP
B 6205 863† SURVEYS, AVAILABILITY OF, GEOLOGIC MATERIALS FROM, STATE
B 6203 566 SURYAM, J. V., COMPUTER APPLICATIONS IN GEOLOGIC
B 6004 667* SUSPENDED MATTER IN HYDROGRAPHER, HUDSON, AND WILMINGTON
B 6004 698* SUSPENDED SEDIMENT IN BAY OF FUNDY, CANADA, ABST., ORIGIN AND
B 6004 693* SUSPENDED SEDIMENT IN STREAMS DRAINING CURRENTLY GLACIATED
B 6004 648* SUSPENDED SEDIMENTS IN NEAR-SHORE REGION, SOUTHWESTERN
B 6004 689* SUSPENDED SEDIMENTS OF SAN FRANCISCO BAY SYSTEM, ABST.,
B 6004 663* SUSPENDED-SEDIMENT TRANSPORT OVER PART OF SOUTHERN
B 6202 184 SUSSEX INTERVAL, WYOMING
B 6405 675* SUSSEX SANDSTONE AT HOUSE CREEK FIELD, POWDER RIVER BASIN,
B 6202 197 SUSSEX SANDSTONE, HYDROCARBON ACCUMULATIONS, WYOMING
B 6202 181† SUSSEX SANDSTONE OF WYOMING—EXAMPLE OF CRETACEOUS
B 6408 1237 SUSSEX SANDSTONE, WYOMING
M 24 141 SUSSEX-SHANNON HYGIENE DISCOVERY WELLS, COLORADO
M 25 501 SUSTUT, AND SIFTON BASINS, CROSS SECTION, CANADA, SLOKO
M 25 500 SUSTUT BASIN, CANADA
M 25 507 SUSTUT COPPER DEPOSIT, CANADA
B 6305 831 SUTHERLAND, G. N., RELATIONS BETWEEN SEDIMENTARY FACIES AND
B 6004 727 SUTHERLAND, PATRICK K., BASINWARD FACIES CHANGES IN
B 6203 566 SUTHERLAND, PATRICK K., REEVALUATION OF MORROWAN-ATOKAN
M 24 205 SUTLEY, CLARKE E., EAST CAMERON BLOCK 270, OFFSHORE LOUISIANA--A
B 6004 735 SUTTNER, L. J., PALEOCLIMATE INTERPRETATION THROUGH USE OF
B 6405 791 SUTTNER, LEE J., LATE MESOZOIC TO EARLY CENOZOIC FORELAND
B 6004 717 SUTTON, C., ORGANIC GEOCHEMISTRY AS AID TO EXPLORATION IN
M 27 218 SUWANNEE FORMATION, FLORIDA, OLIGOCENE
M 27 173 SUWANNEE LIMESTONE, FLORIDA, OLIGOCENE
M 27 128 SUWANNEE LIMESTONE, FLORIDA, OLIGOCENE
M 27 165 SUWANNEE LIMESTONE, FLORIDA, OLIGOCENE
M 27 223 SUWANNEE LIMESTONE, FLORIDA, OLIGOCENE
M 27 7 SUWANNEE LIMESTONE, FLORIDA, OLIGOCENE
B 6403 340 SUWANNEE SADDLE
B 6112 2144 SUWANNEE STRAITS, FLORIDA
B 6403 449 SUZCEK, CHRISTOPHER A., SANDSTONE COMPOSITIONS RELATED TO
B 6312 2194 SVALBARD, ADVENTDALEN GROUP
B 6312 2193† SVALBARD, BEARING ON RESERVOIR POTENTIAL OF BARENTS SHELF,
B 6312 2194 SVALBARD, CAROLINEFJELLET FORMATION
B 6312 2194 SVALBARD, DALKJEGLA MEMBER
B 6010 1719 SVALBARD, DEVELOPMENTS, 1975
B 6110 1649 SVALBARD, DEVELOPMENTS, 1976
B 6210 1919 SVALBARD, DEVELOPMENTS, 1977
B 6310 1927 SVALBARD, DEVELOPMENTS, 1978
B 6411 1967 SVALBARD, DEVELOPMENTS, 1979
B 6312 2194 SVALBARD, FESTNINGEN MEMBER
B 6312 2194 SVALBARD, GLITREFJELLET MEMBER
B 6003 341† SVALBARD, GROWTH, FAULTS IN UPPER, TRIASSIC, DELTAIC SEDIMENTS
B 6312 2194 SVALBARD, HELVETIAFJELLET FORMATION
B 6312 2194 SVALBARD, INNKJEGLA MEMBER
B 6312 2194 SVALBARD, JANUSFJELLET FORMATION
B 6004 668* SVALBARD, NORWAY, ABST., GROWTH, FAULTS AND ASSOCIATED
B 6312 2194 SVALBARD, RURIKFJELLET MEMBER
B 6111 1930 SVARTFJELD PENINSULA, ELLESMERE ISLAND
B 6111 1931 SVERDRUP BASIN, ANTOINETTE FORMATION
B 6206 1004† SVERDRUP BASIN, ARCTIC CANADA, EVOLUTION OF
B 6111 1931 SVERDRUP BASIN, BELCHER CH FORMATION
B 6111 1931 SVERDRUP BASIN, BORUP FJORD FORMATION
B 6111 1930 SVERDRUP BASIN, CANADIAN ARCTIC
B 6101 11 SVERDRUP BASIN, CANADIAN, ARCTIC ISLANDS
B 6111 1931 SVERDRUP BASIN, CANYON FJORD FORMATION
B 6101 16 SVERDRUP BASIN, CRETACEOUS
B 6101 5 SVERDRUP BASIN, ELLSMERIAN OROGENY
B 6101 5 SVERDRUP BASIN, EUREKAN OROGENY
B 6206 1021 SVERDRUP BASIN, EUREKAN OROGENY, CANADA
B 6206 1016 SVERDRUP BASIN, HALOKINETIC STRUCTURES, CANADA
S 5 162 SVERDRUP BASIN, HARE FIORD, NANSEN FORMATION
B 6111 1931 SVERDRUP BASIN, HARE FJORD FORMATION
B 6206 1025 SVERDRUP BASIN, HYDROCARBONS, CANADA
B 6101 16 SVERDRUP BASIN, IGNEOUS ACTIVITY
B 6101 15 SVERDRUP BASIN, JURASSIC
B 6101 15 SVERDRUP BASIN, JURASSIC-CRETACEOUS
B 6206 1019 SVERDRUP BASIN, MAFIC INTRUSIVE AND VOLCANIC ROCKS, CANADA
B 6101 5 SVERDRUP BASIN, MELVILLIAN DISTURBANCE
B 6101 12 SVERDRUP BASIN, MESOZOIC
B 6101 11 SVERDRUP BASIN, MISSISSIPPIAN
B 6111 1931 SVERDRUP BASIN, NANSEN FORMATION
B 6206 1006 SVERDRUP BASIN, OIL AND GAS, ARCTIC, CANADA
S 5 159 SVERDRUP BASIN, OTTO FIORD FORMATION, ALGAL MOUNDS, CANADA,
S 5 149 SVERDRUP BASIN, OTTO FIORD FORMATION, ANHYDRITE DEPOSITION,
S 5 147 SVERDRUP BASIN, OTTO FIORD FORMATION, STRATIGRAPHY, CANADA,
B 6111 1931 SVERDRUP BASIN, OTTO FJORD FORMATION
B 6101 12 SVERDRUP BASIN, PENNSYLVANIAN
B 6101 12 SVERDRUP BASIN, PERMIAN
B 6303 413* SVERDRUP BASIN RESPONSE TO CRATONIC EVENTS, ABST.
B 6111 1931 SVERDRUP BASIN, SECTION, ELLESMERE ISLAND
B 6101 5 SVERDRUP BASIN, STRATIGRAPHIC COLUMN
B 6206 1007 SVERDRUP BASIN, STRATIGRAPHY, CANADA
B 6111 1931 SVERDRUP BASIN, TANQUARY FJORD FORMATION
B 6101 16 SVERDRUP BASIN, TERTIARY
B 6101 16 SVERDRUP BASIN, TERTIARY-QUATERNARY
B 6101 12 SVERDRUP BASIN, TRIASSIC
B 6101 13 SVERDRUP BASIN, TRIASSIC-JURASSIC
B 6101 3 SVERDRUP RIDGE
B 6101 20 SVERDRUP RIDGE, CANADIAN, ARCTIC ISLANDS
B 6303 484 SVETLICHNY, MICHAEL, DETAILED TEMPERATURE LOGGING AS USEFUL
B 6405 791 SWAIN, F. M., COMPARISON OF GULF COAST LATE JURASSIC OSTRACODA
B 6004 727 SWAIN, F. M., MARINE AND BRACKISH-WATE CRETACEOUS OSTRACODA
B 6303 536 SWAIN, FREDERICK M., BIOSTRATIGRAPHY OF EARLY AND EARLIEST
B 6004 705* SWALE TOPOGRAPHY ON INNER CONTINENNAL SHELF OFF NEW JERSEY
B 6008 1412* SWAMP COMPLEX, JOHN HENRY MEMBER OF STRAIGHT CLIFFS
M 24 81 SWAN HILLS FORMATION, ALBERTA
M 24 80 SWAN HILLS PLATFORM, ALBERTA
M 24 85 SWAN HILLS, REEF, ALBERTA
B 6309 1611 SWANN, CHARLES, EARLY TERTIARY LITHOSTRATIGRAPHIC
B 6311 2119 SWANSON, DONALD C., DELTAIC DEPOSITS IN UPPER MORROW
M 24 6 SWANSON RIVER FIELD, ALASKA
B 6401 26 SWANSON RIVER GIANT OIL FIELD, LANDSAT DATA, ALASKA
M 30 28 SWANSON RIVER, LANDSAT IMAGE, ALASKA
B 6408 1132 SWANSON RIVER OIL FIELD, ALASKA
B 6004 699* SWATCH OF NO GROUND CANYON, NORTHEAST INDIAN OCEAN, ABST.,
S 6 369 SWAZILAND SYSTEM, SOUTH AFRICA
B 6210 1849 SWAZILAND, UPPER VOLTA, ZAMBIA, DEVELOPMENTS, 1977, TERRITORY
S 6 348 SWEDEN, CHASMOPS LIMESTONE, BENTONITES, ORDOVICIAN
B 6010 1719 SWEDEN, DEVELOPMENTS, 1975
B 6110 1649 SWEDEN, DEVELOPMENTS, 1976

B 6210 1919 SWEDEN, DEVELOPMENTS, 1977
B 6310 1928 SWEDEN, DEVELOPMENTS, 1978
B 6411 1968 SWEDEN, DEVELOPMENTS, 1979
B 6110 1654 SWEDEN, PRODUCTION, 1975-1976
B 6210 1924 SWEDEN, PRODUCTION, 1976-1977
B 6310 1932 SWEDEN, PRODUCTION, 1977-1978
B 6411 1971 SWEDEN, PRODUCTION, 1978-1979
B 6210 1984 SWEENEY, P. J., ON DEVELOPMENTS IN INDONESIA, 1977
B 6110 1836 SWEENEY, PAT, ON DEVELOPMENTS IN INDONESIA, 1976
B 6405 791 SWEET, WALTER C., STATE OF THE ART IN CONODONT TAXONOMY, ABST.
B 6407 973 SWEETGRASS HILLS, MONTANA
B 6203 537* SWEETWATER COUNTIES, WYOMING, ABST., PALEOLIMNOPHYCUS--,
S 3 94 SWEETWATER GRANITE, WYOMING
M 26 218 SWEETWATER SLOPE SYSTEM, MIDLAND BASIN
B 6002 233 SWELLING CLAYS, CORRELATION BETWEEN, PETROLEUM PRODUCTIVITY
B 6303 433 SWETLAND, P. J., PETROLEUM GENERATION AN MIGRATION IN DENVER
B 6410 1613 SWETLAND, P. J., PETROLEUM GENERATION AND MIGRATION IN DENVER
B 6206 1075 SWETLAND, PAUL J., PETROLEUM SOURCE-BED EVALUATION OF
B 6008 1412 SWETLAND, PAUL J., SOURCE BEDS OF PETROLEUM IN DENVER BASIN,
B 6303 453 SWIFT, D. J. P., ENVIRONMENTAL STUDIES AT NOAAS ATLANTIC
B 6405 792 SWIFT, DONALD J. P., GENESIS OF SAND RIDGES ON STORM-DOMINATED
B 6405 706 SWIFT, DONALD J. P., MODIFICATION OF LINEAR SAND RIDGES BY
B 6211 2277 SWIFT FORMATION, MONTANA
B 6105 897* SWIFT, J. P., NOAAS WASTE-DISPOSAL STUDIES IN NEW
B 6108 1393 SWINEHART, JAMES B, STRATIGRAPHY AND SEDIMENTARY
B 6303 536 SWIRYDCZUK, KRYSTYNA,, NORMALLY AND REVERSELY GRADED BEDS
B 6303 536 SWITEK, MICHAEL J., JR., SOUTH HALLETTSVILLE, GAS FIELD IN LOWER
B 6010 1719 SWITZERLAND, 1975, BUCHI, U. P., AND G., WIENER, ON DEVELOPMENTS
B 6110 1649 SWITZERLAND, 1976, BUCHI, U. P., AND G., WIENER, ON DEVELOPMENTS
B 6210 1919 SWITZERLAND, 1977, BUCHI, U. P., AND G., WIENER, ON DEVELOPMENTS
B 6310 1928 SWITZERLAND, 1978, BUCHI, U. P., AND G., WIENER, ON DEVELOPMENTS
B 6411 1968 SWITZERLAND, 1979, BUCHI, U. P., AND GABRIEL, WIENER, ON
S 8 40 SWITZERLAND, DOWNSLOPE FLOW, CURRENT FLOW, LAKE WALENSEE
B 6411 1968 SWITZERLAND, JURA MOUNTAINS
M 27 208 SWITZERLAND, JURASSIC, PENNINE BÜNDNERSCHIEFER
B 6411 1968 SWITZERLAND, MOLASSE BASIN, NORTHERN
B 6107 1046 SWOPE FORMATION, KANSAS
B 6111 1986 SYCAMORE CANYON SANDSTONE AND CONGLOMERATE, CALIFORNIA
B 6402 237 SYCAMORE CREEK MONOCLINE, TEXAS
B 6307 1135 SYLVAN SHALE, OKLAHOMA
B 6012 2081 SYLVESTER, ARTHUR G., TECTONIC TRANSPRESSION AND
M 25 497 SYLVESTER TROUGH, CANADA
B 6105 808 SYLWESTER, R. E., SHALLOW SEDIMENTARY FRAMEWORK OF GEORGES BANK
M 28 89 SYLVITE, CRYSTALLOGRAPHIC
B 6112 2144 SYMAP AND SYMVU, USED FOR FLORIDA BASEMENT,
B 6112 2144 SYMVU, USED FOR FLORIDA BASEMENT, COMPUTER-PROGRAM
S 5 93* SYNCHRONIZATION OF DEPOSITION: SILURIAN REEF-BEARING ROCKS ON
B 6405 742* SYNCRUDE CANADA LIMITED MINE SITE, ATHABASCA OIL SAND AREA,
B 6203 573* SYNDEPOSITIONAL SUBMARINE CEMENTATION OF CAPITAN LIMESTONE,
B 6203 492* SYNDEPOSITIONAL SUBMARINE CEMENTATION OF CAPITAN LIMESTONE,
B 6312 2216 SYNERESIS FRACTURES
B 6004 714* SYNERGISM. BY TITLE ONLY, GEOPHYSICAL-, GEOLOGIC
B 6405 690* SYNONYMY, ETHOLOGY, AND ENVIRONMENTAL IMPLICATIONS OF
B 6405 740* SYNSEDIMENTARY DEFORMATION IN FOSSIL ACCRETIONARY PRISM,
B 6004 655* SYNSEDIMENTARY TECTONIC CONTROL OF PALEOZOIC CLASTIC
M 26 390 SYNTHESIS, MODELING, SEISMOGRAM
B 6211 2364* SYNTHESIS OF OCS WELL INFORMATION, OFFSHORE CENTRAL AND
B 6303 450* SYNTHESIS OF PROTODOLOMITE, HIGH-MAGNESIAN CALCITE,
B 6303 409* SYNTHETIC FUEL FORECASTS. BY TITLE ONLY
M 29 195 SYNTHETIC MAGNETICS, BURMA, ANDAMAN SEA
B 6004 718* SYNTHETIC SEISMIC SECTIONS OF SELECTED STRATIGRAPHIC TRAPS AN
B 6008 1409* SYNTHETIC SEISMIC SECTIONS OF SELECTED STRATIGRAPHIC TRAPS AND
M 26 3 SYNTHETIC SEISMOGRAMS, SEISMIC, DATA
M 26 9 SYNTHETIC SEISMOGRAMS, , SEISMIC, DATA
B 6209 1610 SYRACUSE FORMATION, NEW YORK
B 6010 1873 SYRIA, 1975, PETROCONSULTANTS S.A., ON DEVELOPMENTS IN
B 6110 1802 SYRIA, 1976, JAVOQUES, J.-P., ON DEVELOPMENTS IN
B 6210 2119 SYRIA, 1977, JAVOGUES, J.-P., ON DEVELOPMENTS IN
B 6310 1891 SYRIA, 1978, JAVOGUES, J. P., ET AL, ON DEVELOPMENTS IN
B 6411 1842 SYRIA, 1979, COASTAL CORP., MARATHON OIL CORP.,AND,
B 6310 1892 SYRIA, DEVELOPMENTS, 1977-1978
B 6210 2120 SYRIA, PRODUCTION, 1976-1977
B 6411 1843 SYRIA, PRODUCTION, 1978-1979
M 29 304 SYRTE GULF, MAGNETIC ANOMALIES, MEDITERRANEAN
M 26 273 SYSTEM, ABYSSAL PLAIN
M 26 272 SYSTEM, CAMPECHE FAN
B 6105 776* SYSTEM FOR REGIONAL LITHOFACIES MAPPING BY COMPUTER
M 26 271 SYSTEM, MISSISSIPPI FAN
M 26 270 SYSTEM, MISSISSIPPI TROUGH
M 26 272 SYSTEM, SIGSBEE WEDGE
M 26 272 SYSTEM, VERACRUZ FAN
M 26 22 SYSTEMS, BROADBAND VELOCITY MEASUREMENT, GERMANY,
M 26 354 SYSTEMS, CLASSIFICATION CHART, RESOLUTION SYSTEMS, ACOUSTIC
M 26 270 SYSTEMS, CONTINENTAL SLOPE
M 26 432 SYSTEMS, SEISMIC, MODELS, THIN RESERVOIR
M 26 269 SYSTEMS, SHELF-PLATFORM, FLUVIAL-DELTAIC
S 6 24 SYSTEMS, TIME SCALE MODEL
M 26 19 SYSTEMS, VELOCITY MEASUREMENT, LABORATORY
M 26 213† SYSTEMS: EXAMPLES FROM BRAZILIAN RIFT AND PULL-APART BASINS,
B 6408 1211 SZECHWAN BASIN, CHINA
B 6105 834 SZUMILAS, D., COMPUTER FOR TIME-TO-DEATH CONVERSION AND
B 6402 152 T ZONE, LOUISIANA
B 6204 682 TABANKORT DEVONIAN OIL FIELD, ALGERIA, TIN FOUYE-
S 11 319 TABLE BLUFF COAST, CALIFORNIA, SAMPLE LOCALITIES
M 28 64 TABLE, FINE-GRAINED CLASTIC NOMENCLATURE
B 6306 955 TABLE HEAD FORMATION, NEWFOUNDLAND
B 6108 1373* TABULAR URANIUM BODY-RAJAH 49 MINE, MESA COUNTY, COLORADO,
B 6004 653* TABULAR URANIUM DEPOSIT LA SAL MINE, SAN JUAN COUNTY, UTAH,
B 6405 795* TABULAR URANIUM ORE IN POISON CANYO AREA, MORRISON
M 27 61 TABULATE CORALS, SILICIFIED
B 6109 1476 TABULEIRO DOS MARTINS FIELD, BRAZIL
B 6309 1524 TACHILNI FORMATION, MIOCENE AND PLIOCENE, ALASKA PENINSULA
B 6401 9 TACHING, DAQING, GIANT OIL FIELD, LANDSAT DATA, CHINA
M 30 10 TACHING FIELD, LANDSAT IMAGE, CHINA
B 6403 429 TACHING HIGH-WAX OIL FIELD, CHINA
B 6405 718 TACKENBERG, T., BRAE FIELD AREA, NORTH SEA, ABST.
M 25 411 TACONITE, BENEFICIABLE, IRON FORMATION OR
B 6007 1021 TACONITE ORE TAILINGS, DEPOSITION, MINNESOTA, RESERVE MINING
B 6407 996 TADRART OUADOU SANDSTONE MEMBER, MOROCCO
M 25 440 TAEBO, SEE DAEBO
B 6203 351 TAFT OIL DISTRICT, CALIFORNIA
M 25 319 TAGACHI ANTICLINE, COLOMBIA
M 24 51† TAGLU GAS FIELD, BEAUFORT BASIN, NORTHWEST TERRITORIES
M 24 65 TAGLU GAS FIELD, PERMAFROST, NORTHWEST TERRITORIES
M 24 61 TAGLU SHALE, ALASKA
B 6209 1647 TAIJUNT CRUISE, 1975, RV, CHIU LIEN
B 6007 1022 TAILINGS DEPOSITION, LAKE SUPERIOR
B 6007 1021 TAILINGS, DEPOSITION, MINNESOTA, RESERVE MINING CO., TACONITE
B 6007 1023 TAILINGS, DEPOSITION, MINNESOTA, SILVER BAY
B 6004 727 TAILLEUR, I. L., PRUDHOE BAY SUPE GIANT FIELD AND ALASKAN ARCTIC
B 6405 792 TAILLEUR, I. L., RATIONALIZATION OF KOYUKUK CRUNCH, NORTHERN
B 6012 2190 TAILLEUR, I. L., SUBSURFACE GEOLOGY AND SOME IMPLICATIONS
B 6207 1218 TAILLEUR, IRVIN L., PETROLEUM EXPLORATION OF NATIONAL
B 6109 1476 TAINHA FIELD, BRAZIL
B 6310 1827 TAIWAN, 1978, YANG, S., AND G. L., FLETCHER, ON DEVELOPMENTS IN
B 6207 1221* TAIWAN, ABST., EXPLORATION AND DEVELOPMENT OF, GEOTHERMAL
B 6209 1667 TAIWAN, BACKBONE RANGE AND YU SHAN
M 25 259 TAIWAN BASIN, PROSPECTS
M 25 252† TAIWAN BASINS IN FRAMEWOR OF WESTERN PACIFIC OCEAN,
M 25 439 TAIWAN, CARBONATE DEPOSITS
B 6209 1667 TAIWAN, CENTRAL RANGE PROVINCE
M 25 92 TAIWAN, COAL RESERVES
M 25 89 TAIWAN, COALFIELDS
B 6209 1667 TAIWAN, COASTAL RANGE PROVINCE
M 25 439 TAIWAN, CUPRIFEROUS IRON SULFIDE DEPOSITS, BEDDED
B 6411 1893 TAIWAN, DEVELOPMENTS, 1979
B 6209 1667 TAIWAN, EASTERN SUBPROVINCE
M 25 90 TAIWAN, GEOLOGIC MAP
M 25 146 TAIWAN, GEOTHERMAL AREAS
M 25 439 TAIWAN, GEOTHERMS
M 25 439 TAIWAN, GOLD-SILVER-COPPER DEPOSITS
B 6209 1667 TAIWAN, HSUEH SHAN RANGE
B 6209 1670 TAIWAN, ILAN PLAIN
S 10 202 TAIWAN, MUD VOLCANO
M 25 258 TAIWAN, NORTHERN BASIN, PROSPECTS
M 25 259 TAIWAN, PETROLEUM PRODUCTION
B 6110 1858 TAIWAN, PRODUCTION, 1975-1976
B 6210 1990 TAIWAN, PRODUCTION, 1975-1977
B 6310 1829 TAIWAN, PRODUCTION, 1977-1978
B 6411 1896 TAIWAN, PRODUCTION, 1977-1979
B 6110 1840 TAIWAN, REPUBLIC OF CHINA, 1976, YANG, Y. F., ON DEVELOPMENTS IN
B 6210 1988 TAIWAN, REPUBLIC OF CHINA, 1977, YANG, Y. F., ON DEVELOPMENTS IN
M 25 257 TAIWAN, SOUTHERN BASIN, PROSPECTS
B 6209 1666 TAIWAN, STRUCTURE
M 25 439 TAIWAN, SULFUR DEPOSITS
B 6209 1667 TAIWAN, WESTERN FOOTHILLS PROVINCE
B 6209 1667 TAIWAN, WESTERN SUBPROVINCE
M 25 89† TAIWAN--SUMMARY, COAL RESOURCES OF
M 25 439† TAIWAN--SUMMARY, GEOLOGIC CONTROLS OF MINERAL DEPOSITS IN
M 25 146† TAIWAN--SUMMARY, ORIGIN AND POTENTIAL OF SOME GEOTHERMAL

Index

B 6209 1646 TAIWAN-LUZON REGION, ACTIVE AND EXTINCT PLATES
B 6209 1645† TAIWAN-LUZON REGION, PLATE, CONVERGENCE AND ACCRETION IN
B 6209 1646 TAIWAN-LUZON REGION, VOLCANOES
B 6303 551 TAKANO, H., EARLY PLEISTOCENE SUBMARINE CANYON, BOSO
M 25 371 **TAKASAKI, K. J., EFFECT OF SUBSURFACE WASTE-DISPOSAL PRACTICE ON**
M 30 540 TAKHOUKT FIELD, ALGERIA, OUARGLA REGION
B 6004 727 TAKIN, M., GEOLOGIC HISTORY AND TECTONICS OF ZAGROS MOUNTAINS
B 6409 1387 TAKKEN, SUZANNE, DEVELOPMENTS IN OKLAHOMA AND PANHANDLE OF
B 6308 1286 TAKKEN, SUZANNE, DEVELOPMENTS IN OKLAHOMA AND PANHANDLE OF
B 6208 1372 TAKKEN, SUZANNE, DEVELOPMENTS IN OKLAHOMA AND PANHANDLE OF
B 6108 1216 TAKKEN, SUZANNE, DEVELOPMENTS IN OKLAHOMA AND PANHANDLE OF
B 6002 327 TAKKEN, SUZANNE, MORROW-SPRINGER-CHESTER RELATIONS IN
B 6002 322 TAKKEN, SUZANNE, REPORT--ANADARKO BASIN. BY TITLE ONLY
M 25 509 TAKLA-HAZELTON-CACHE CREEK-SLIDE MOUNTAIN- NICOLA ROCKS,
B 6107 969 TAKORADI FRACTURES, AFRICA, ROMANCHE-
M 25 336 TALARA BASIN, PERU
M 25 444 TALC, SOUTH KOREA
M 24 4 TALKEETNA FORMATION, ALASKA
B 6203 376 TALKEETNA FORMATION, ALASKA
S 3 129 TALLAHASSEE CREEK, COLORADO
B 6303 537 TALLEY, BOBBY J., LITHOLOGY DATA SYSTEMS--ROCKS TO
B 6002 259 TALME YAFE FORMATION, ISRAEL
M 30 487 TALUKDAR, S. N., PETROLEUM GEOLOGY OF BOMBAY HIGH FIELD, INDIA
S 4 56 TALUS CLIFFS, DEFINITION
B 6302 259 TALWANI AND WALTER C. PITMAN III, SCHLEE, JOHN S., REVIEW ON,
B 6303 537 TALWANI, MANIK, DEEP-SEA DRILLING AND GLOBAL TECTONICS, ABST.
M 29 151 TALWANI, MANIK, THE CRUSTAL STRUCTURE AND EVOLUTION OF THE
B 6202 244 TAMABRA FORMATION, MEXICO
M 27 190 TAMABRA LIMESTONE, MEXICO, CRETACEOUS
M 27 14 TAMABRA LIMESTONE, MEXICO, CRETACEOUS
M 27 13 TAMABRA LIMESTONE, MEXICO, CRETACEOUS
M 27 195 TAMABRA LIMESTONE, MEXICO, CRETACEOUS
M 27 85 TAMABRA LIMESTONE, MEXICO, CRETACEOUS
M 27 212 TAMABRA LIMESTONE, MEXICO, CRETACEOUS
M 27 9 TAMABRA LIMESTONE, MEXICO, CRETACEOUS
B 6004 646* TAMAROA SEQUENCE, UPPER DEVONIAN AND MISSISSIPPIAN,
B 6202 244 TAMAULIPAS FORMATION, MEXICO
B 6101 85 TAMDAFELT BASIN, ATLAS TREND
B 6402 209 TAMHANE, D. M., SOURCE AREAS AND MIGRATION TRENDS OF OIL AND
M 27 166 TAMIAMI FORMATION, FLORIDA, PLIOCENE
M 27 178 TAMIAMI FORMATION, FLORIDA, PLIOCENE
B 6309 1609* TAMIAMI FORMATION IN LEE AND HENDRY COUNTIES, FLORIDA, ABST.,
B 6105 798 TAMM, ALAN H., GEOTHERMAL POTENTIAL OF THREE AREAS IN
B 6309 1605* TAMPA FORMATION, WEST-CENTRAL FLORIDA, ABST., REVISION OF
B 6102 207 TAMPICO MISANTLA BASIN, MEXICO
B 6202 235† TAMPICO-MISANTLA BASIN, MEXICO, STRATIGRAPHY AND STRUCTURE
B 6407 994 TANAMEURT VOLCANICLASTIC MEMBER, MOROCCO
B 6401 56 TANDJUNG FIELD, KALIMANTAN
M 26 301 TANER, M. T., APPLICATION OF AMPLITUDE, FREQUENCY, AND OTHER
B 6105 829 TANER, M. T., COLOR DISPLAYS OF COMPLEX SEISMIC TRACE, ABST.,
B 6203 562 TANER, M. T., GEOLOGIC INTERPRETATION OF SEISMIC-ATTRIBUTE
B 6304 701 **TANER, M. T., LOCATION STRATIGRAPHIC FEATURES AND**
B 6012 2181 TANER, M. TURHAN, APPLICATIONS OF SURFACE-CONSISTENT
B 6408 1187 TANGANYIKA, LARGE ANOXIC LAKE, LAKE
M 29 215 TANIMBAR TROUGH, REFLECTION PROFILE, AUSTRALIA
B 6102 238 TANNER BANK, CALIFORNIA
B 6203 366 TANNER BANK, CALIFORNIA
B 6203 368 TANNER BASIN, CALIFORNIA
B 6111 1981 TANNER BASIN, CALIFORNIA
B 6102 238 TANNER BASIN, CALIFORNIA
B 6104 504 TANNER BASIN, KEROGEN, CALIFORNIA
B 6204 688 TANNER BASIN, KEROGEN, CALIFORNIA
B 6009 1616 TANNER, WILLIAM F., GEOTHERMAL EXPLORATION FROM DEEP-WELL
B 6209 1768 TANNER, WILLIAM F., HIGH-LEVEL PLIO-PLEISTOCENE SHORELINE IN
B 6009 1617 TANNER, WILLIAM F., OIL PROSPECTS IN GULF OF MEXICO REGION ABST.
B 6111 1931 TANQUARY FJORD FORMATION, SVERDRUP BASIN
M 27 132 TANSILL FORMATION, NEW MEXICO, PERMIAN
M 27 161 TANSILL FORMATION, NEW MEXICO, PERMIAN
B 6303 501* TANSILL FORMATIONS, GUADALUPE MOUNTAINS, NEW MEXICO, ABST.,
M 25 488 TANTALUM AND NIOBIUM
M 25 500 TANTALUS BASIN, CANADA
B 6110 1750 TANZANIA, DEVELOPMENTS, 1976
B 6210 1849 TANZANIA, DEVELOPMENTS, 1977
B 6310 1695 TANZANIA, DEVELOPMENTS, 1978
B 6411 1791 TANZANIA, DEVELOPMENTS, 1979
B 6305 834* TAOS TROUGH, NORTH-CENTRAL NEW MEXICO ABST., PENNSYLVANIAN
B 6305 824* TAOS TROUGH, NORTHERN NEW MEXICO, ABST., BASIN EVOLUTION OF
B 6405 687* TAOS TROUGH, NORTHERN NEW MEXICO, ABST., PALEOGEOGRAPHIC
M 29 234 TAPANULI DISTRICT, GEOLOGIC MAP, SUMATRA
S 3 135 TAPHROGENIC ALKALIC VOLCANIC CENTERS, HYDROTHERMAL-VEIN
S 3 115 TAPHROGENIC CONTINENTAL-MARGIN HISTORY
S 3 135 TAPHROGENIC HYDROTHERMAL FEEDING, AT TECTONIC INTERSECTIONS
S 3 115 TAPHROGENIC MARGINS, OROGENIC-
S 3 115 TAPHROGENIC MARGINS, PURELY
M 25 466 TAPHROGENIC OR RIFTOGENIC ZONES, USSR, NORTHEAST, MESOZOIC
B 6109 1476 TAQUIPE FIELD, BRAZIL
B 6307 1034 TAQUIPE FIELD, SECTION, BRAZIL
B 6303 444* TAR DEPOSI IN IRAN, ABST., MIOCENE-, PLIOCENE SYNDEPOSITIONAL
M 6004 726* TAR DEPOSITS, ALBERTA, ABST., APPLICATION OF, MONTE CARLO
B 6405 745* TAR SAND, ABST., RECENT ACTIVITIES IN, UNITED STATES
B 6205 880* TAR SAND DEPOSITS, CENTRAL AND SOUTHERN UTAH, ABST., LOWER,
B 6105 822* TAR SAND, RESOURCES, ABST., UNITED STATES AND, UTAH
B 6004 647* TAR SANDS, ABST., ENERGY FROM
B 6111 1950† TAR SANDS AND SUPERGIANT OIL FIELDS
B 6012 2178* TAR SANDS GIVE CLUES TO SEARCH FOR SUPERGIANT OIL FIELDS, ABST.
B 6007 1110 TAR SANDS, MISNAMED
B 6111 1950 TAR SANDS, NATURE AND ORIGIN OF
B 6203 565* TAR SANDS OF ALBERTA. BY TITLE ONLY., HEAVY, OIL AND
M 28 142 TAR SANDS, WITH HYDROCARBONS
B 6111 1951 TAR TRIANGLE HEAVY OIL DEPOSIT, UNITED STATES
B 6409 1554* TARAISES FORMATION, EARLY CRETACEOUS, SANTA ROSA CANYON,
M 25 222 TARANAKI BASIN, NEW ZEALAND
M 25 222 TARANAKI FAULT, NEW ZEALAND
B 6303 537 TARANIK, JAMES V., CHARACTERISTICS OF LANDSAT SYSTEM
B 6303 537 TARANIK, JAMES V., COMPUTER PROCESSING OF LANDSAT DATA
B 6207 1236 TARANIK, JAMES V., INTRODUCTION TO GEOLOGIC REMOTE SENSING,
M 25 170 TARAPACA HYDROTHERMAL AREA, CHILE
B 6402 218 TARAPUR FIELD, INDIA
B 6203 444 TARAPUR SHALE, INDIA
B 6306 973 TAREF SANDSTONE UNIT, EGYPT
B 6207 1127 TARERA-AIDUNA FAULT ZONE, IRIAN JAYA
B 6101 85 TARFAYA, BASIN, AFRICAN PLATFORM, AAIUN
B 6202 228 TARFAYA BASIN, SEDIMENTARY SEQUENCE, MOROCCO
B 6408 1211 TARIM BASIN, CHINA
S 6 152 TARO GRANITE, RUBIDIUM-STRONTIUM DATA, JAPAN
S 6 157 TARO GRANITE, RUBIDIUM-STRONTIUM ISOCHRON PLOT, JAPAN
B 6204 635 TARR BANK, ALASKA
B 6004 704* TARS WITHIN SOUTHERN CALIFORNIA BORDERLAND, ABST., SOURCE
B 6207 1207 TASAN, RUMBOKO, EXPLORATION AND DEVELOPMENT OF GEOTHERMAL
B 6111 1994 TASMAN BASIN, POSSIBLE DIAPIRS
B 6303 513* TASMANIA, AUSTRALIA, ABST., GLACIOMARINE CARBONATE
B 6110 1889 TASMANIA, DEVELOPMENTS, 1976
B 6210 2151 TASMANIA, DEVELOPMENTS, 1977, VICTORIA AND
B 6310 1805 TASMANIA, DEVELOPMENTS, 1978, VICTORIA AND
B 6411 1864 TASMANIA, DEVELOPMENTS, 1979
M 25 426 TASMANIA, KING ISLAND SCHEELITE, TUNGSTEN
M 25 426 TASMANIA, MOUNT BISCHOFF, TIN DEPOSIT
M 25 426 TASMANIA, MOUNT CLEVELAND, TIN DEPOSIT
M 25 426 TASMANIA, MOUNT LYELL, COPPER DEPOSIT
M 25 426 TASMANIA, RENISON BELL, TIN DEPOSIT
M 25 426 TASMANIA, ROSEBERY, LEAD-ZINC-COPPER DEPOSIT
M 25 426 TASMANIA, SAVAGE RIVER, MAGNETITE-PYRITE
B 6203 566 TASSONE, JEFF A., ECONOMIC POTENTIAL OF CANYON-FAN UNIT IN
M 25 213 TATANAKI BASIN, NEW ZEALAND
B 6409 1355 TAVENER FIELD, FT. BEND COUNTY, TEXAS
B 6201 125 TAVERNIER TONGUE, FLORIDA
B 6407 1007 TAWAKONI FIELD, TEXAS
B 6004 712* TAXONOMIC EXPANSION, ABST., EARLY, CAMBRIAN, SHALLOW,
S 11 70† TAXONOMY
B 6203 353 TAYLOR CANYON TREND, CALIFORNIA, MORALES AND
B 6103 346 TAYLOR COUNTY, LOWER PERMIAN, TEXAS
B 6207 1237 TAYLOR, DON S., OIL AND GAS FROM SUBMARINE CHANNELS IN GREAT
B 6207 1151 TAYLOR, G. H., HYDROCARBON GENERATION IN GIPPSLAND BASIN,
B 6405 679* TAYLOR GROUPS, UPPER CRETACEOUS, EAST-CENTRAL AND
B 6008 1412 TAYLOR, H. C. JIM, CORWIN SPRINGS KNOWN GEOTHERMAL RESOURCE
B 6004 727 TAYLOR, JAMES C., CALIFORNIAS DEVELOPED AND UNDEVELOPED
B 6002 322 TAYLOR, JOHN ALLEN, DIBBLE--MUSTANG, CURRENT HUNTON PLAY IN
B 6102 156 TAYLOR LEASE, ISOPACH MAP, CALIFORNIA, SHELL
S 11 1 TAYLOR, MARY E., MIOCENE STRATIGRAPHY OF CALIFORNIA
B 6105 742 TAYLOR, P. T., REGIONAL GEOLOGIC FRAMEWORK OFF NORTHEASTERN
B 6006 926 TAYLOR, P. T., REGIONAL GEOLOGIC FRAMEWORK OFF NORTHEASTERN
B 6408 1220 TAYLOR SECTION, TEXAS
B 6102 154 TAYLOR THRUST FAULT, CALIFORNIA
B 6102 189 TAYLOR THRUST FAULT, CALIFORNIA
M 25 518 TAYOLTITA-SAN DIMAS MINE, GOLD, MEXICO
B 6410 1580 TAZEWELL ARCH, VIRGINIA
B 6101 30 TAZIN GNEISS, SASKATCHEWAN
B 6303 521* TAZIRBU REGION OF CENTRA LIBYA DETERMINED FROM LANDSAT

B 6407 1005 TEAGUE FIELD, ARK-LA-TEX AREA
M 30 389 TEAK FIELD, DEVELOPMENT HISTORY, TRINIDAD
M 30 395 TEAK FIELD, GROS MORNE FORMATION, PLIOCENE
M 30 395 TEAK FIELD, MAYARO FORMATION, PLIOCENE
M 30 389 TEAK FIELD, REGIONAL GEOLOGY, TRINIDAD
M 30 398 TEAK FIELD, RESERVOIR CHARACTERISTICS, TRINIDAD
M 30 395 TEAK FIELD, RESERVOIR HORIZONS, PLIOCENE
M 30 391 TEAK FIELD, SEISMIC SECTION, TRINIDAD
M 30 393 TEAK FIELD, STRATIGRAPHY, TRINIDAD
M 30 396 TEAK FIELD, STRUCTURE, TRINIDAD
M 30 387 TEAK FIELD, TRINIDAD, PLIOCENE
B 6303 414* TEAK OIL FIELD, TRINIDAD, WEST INDIES, ABST., GEOLOGY AND
M 30 387† TEAK OIL FIELD, TRINIDAD, WEST INDIES, GEOLOGY AND DEVELOPMENT
B 6207 1194 TEANAWAY FORMATION, OREGON AND, WASHINGTON
B 6008 1395* TEAPOT DELTA OF WESTERN POWDER RIVE BASIN. WYOMING, ABST.
B 6105 671† TEAPOT DOME—PAST, PRESENT, AND FUTURE
M 28 175 TEAPOT SANDSTONE, WYOMING, CRETACEOUS
M 28 178 TEAPOT SANDSTONE, WYOMING, CRETACEOUS
M 28 179 TEAPOT SANDSTONE, WYOMING, CRETACEOUS
M 28 176 TEAPOT SANDSTONE, WYOMING, CRETACEOUS
B 6407 1007 TEAS FIELD, TEXAS
B 6004 665* TECHNIQUES AND USES OF SLOPE MAPPING, ABST.
S 5 111* TECHNIQUES IN EXPLORATION FOR REEFS, NORTHERN MICHIGAN,
M 28 191 TECHNIQUES, BIBLIOGRAPHY
M 28 186 TECHNIQUES, CATHODOLUMINESCENCE, EXPLANATION
M 28 190 TECHNIQUES, ELECTRON MICROPROBE ANALYSIS
M 28 189 TECHNIQUES, ENERGY DISPERSIVE ANALYSIS
M 28 187 TECHNIQUES, FLUID INCLUSION STUDIES, EXPLANATION
M 28 183 TECHNIQUES, GENERAL
M 28 188 TECHNIQUES, SCANNING ELECTRON MICROSCOPY
M 28 188 TECHNIQUES, SEM, EXPLANATION
M 28 185 TECHNIQUES, STAINING, EXPLANATION
M 26 389† TECHNIQUES, STRATIGRAPHIC, MODELING AND, INTERPRETATION—,
B 6303 508* TECOLUTLA CARBONATE PLATFORM, GULF COAST, MEXICO, ABST.,
S 6 16 TECTOGENETIC SCALES, REVIEW
M 26 278 TECTONIC AND GEOLOGIC MAP, DENVER BASIN
B 6312 2164† TECTONIC AND SANDSTONE COMPOSITIONS, PLATE
M 25 503 TECTONIC BELTS, CANADIAN, CORDILLERA
M 25 546 TECTONIC BELTS, MOBILE, AND LITHOSPHERIC PLATE CONVERGENCE,
B 6203 537* TECTONIC CONTROL OF SEDIMENTATION IN TRIASSIC-JURASSIC
M 25 483 TECTONIC CROSSROADS CONCEPT
M 29 74 TECTONIC DEFORMATION, ATLANTIC COAST, JURASSIC
M 29 363† TECTONIC DEVELOPMENT OF TRENCH-ARC COMPLEXES ON THE
M 29 349 TECTONIC EVOLUTION, COLOMBIA, NORTHWESTERN
B 6405 807* TECTONIC FEATURES OF OIL AND GAS BASINS OF EASTERN CHINA, ABST.
B 6212 2508* TECTONIC FRAMEWORK AND METALLOGENESIS IN CALIFORNIA, ABST.
B 6207 1210* TECTONIC FRAMEWORK OF MIDDLE AMERICA TRENCH, ABST.
M 25 32† TECTONIC FRAMEWORK OF PACIFIC REGION
M 25 291† TECTONIC FRAMEWORK OF PETROLIFEROUS ROCKS IN ALASKA
B 6203 522* TECTONIC HISTORY OF ATHAPUSCOW AULACOGEN, MIDDLE
B 6105 763* TECTONIC IMPLICATIONS OF CROWN POINT SECTION, NEW YORK
M 29 191 TECTONIC MAP, BURMA, ANDAMAN SEA
M 29 253 TECTONIC MAP, CENTRAL AMERICA, GUATEMALA
B 6411 2036 TECTONIC MAP OF SOUTH AMERICA BY COMMISSION FOR GEOLOGICAL
M 29 273† TECTONIC PLATES, CONTINENTAL, MARGINS OF THE EASTERN, GULF OF,
M 25 486 TECTONIC PROVINCES, MOLYBDENUM DEPOSITS, IGNEOUS-
M 30 132 TECTONIC SETTING, MORAY FIRTH BASIN, PIPER FIELD
B 6207 1239* TECTONIC SETTING OF PERU-CHILE TRENCH—RELATION TO MINERAL
B 6003 356† TECTONIC SIGNIFICANCE AND HYDROCARBON TRAPPING
B 6405 669* TECTONIC SIGNIFICANCE OF MICROSTRUCTURES IN IDAHO-WYOMING
M 26 296 TECTONIC STYLE, ROCKY MOUNTAINS, NEW
B 6012 2081† TECTONIC TRANSPRESSION AND BASEMENT-CONTROLLED
B 6004 695* TECTONICS AND CLIMATIC CYCLES, ABST., PALEOGENE, RADIOLARIAN,
B 6105 842* TECTONICS AND EVOLUTION OF MARGINS. BY TITLE ONLY,
M 25 448† TECTONICS AND METALLOGENESIS IN PERIPHERAL-SEA - ISLAND-AR
B 6405 737* TECTONICS AND SEDIMENTATION ALONG ANTLER OROGENIC BELT OF
B 6004 647* TECTONICS AS A MAJOR EXPLORATION TOOL, BY TITLE ONLY,
B 6004 687* TECTONICS AS A MAJOR EXPLORATION TOOL. BY TITLE ONLY.,
M 29 197 TECTONICS EVALUATED, BURMA, ANDAMAN SEA
M 29 203 TECTONICS, FAULTING, OKINAWA TROUGH
M 29 199† TECTONICS IN THE OKINAWA TROUGH, EXTENSIONAL
B 6004 650* TECTONICS IN WESTERN CALIFORNIA, ABST., NEOGENE BASI FORMATION
M 29 330 TECTONICS, MAJOR ELEMENTS, COLOMBIA
B 6103 465* TECTONICS, MIDDLE PALEOZOIC VOLCANISM, AND SULFIDE
B 6304 682 TECTONICS OF ASIA, BY EMILE ARGAND, TRANSLATED AND EDITED BY
B 6209 1696† TECTONICS OF BROOKS RANGE OPHIOLITES, ALASKA
M 25 62† TECTONICS OF CONTINENTAL AND OCEANIC STRUCTURES OF NORTH
B 6309 1588* TECTONICS OF PETERSBURG REGION, WEST VIRGINIA, ABST.
B 6007 1078† TECTONICS OF SOUTHWESTERN NORTH ATLANTIC AND BARBADOS
M 29 189† TECTONICS OF THE ANDAMAN SEA AND BURMA
B 6004 679* TECTONICS OF WINN PARISH, LOUISIANA, ABST., SALT
B 6004 694* TECTONICS. BY TITLE ONLY, STRUCTURE AN EVOLUTION OF, BERING
B 6408 1211 TECTONISM IN PRINCIPAL ONSHORE REGIONS, CHINA
M 25 473 TECTONOMAGMATIC PROCESSES WITHIN MEGASTRUCTURES
B 6409 1548 TECTONOMECHANICS—AN INTRODUCTION TO STRUCTURAL ANALYSIS
B 6109 1534 TEDFORD, FREDERICK J., CHARACTERISTICS OF WILCOX GAS
B 6012 2175 TEGLAND, EDWARD R., SEISMIC DATA ANALYSIS WITH TIME SEISCROP
B 6008 1391 TEGLAND, EDWARD R., SEISMIC DATA ANALYSIS WITH TIME-SEISCROP
B 6211 2245 TEHACHAPI MOUNTAINS, CALIFORNIA
M 25 462 TEHUANTEPEC-BARTLETT FAULT, HUTANOPAN OR, SUMATRA-,
B 6406 823 TEICHERT HILLS, AUSTRALIA
B 6205 819 TEJON DEPOCENTER, CALIFORNIA
B 6211 2246 TEJON PLATFORM, CALIFORNIA
B 6407 971 TELEGRAPH CREEK FORMATION, GREAT PLAINS
B 6403 319 TELEGRAPH CREEK FORMATION, MONTANA
B 6105 817 TELEKI, PAUL G., MEASUREMENT TECHNIQUE FOR PROCESS-RESPONSE
B 6311 2059 TELKWA COALFIELD, BRITISH COLUMBIA
M 25 108 TELKWA COALFIELD, CANADA
B 6101 51 TELL ZNOUB 1B, LOG, LEBANON
B 6412 2074 TELLICO SANDSTONE, TENNESSEE
B 6111 1956 TEMBLADOR OIL FIELD, VENEZUELA
S 11 88 TEMBLOR-MONTEREY SEQUENCE, EDWARDS CANYON, CALIFORNIA,
S 11 82 TEMBLOR-MONTEREY SEQUENCE, INDIAN CREEK, CALIFORNIA, CHECK
S 11 85 TEMBLOR-MONTEREY SEQUENCE, QUAILWATER CREEK, CALIFORNIA,
S 10 73 TEMPERATURE EFFECTS ON PORE FLUIDS, MIGRATION
S 10 42 TEMPERATURE VERSUS DEPTH, MIGRATION
B 6307 1129 TEMPERATURE, AND ACTIVATION ENERGIES FOR VARIOUS
B 6303 414* TEMPERATURE AND BURIAL DEPTH I DEVELOPMENT OF SUBNORMAL
B 6402 184 TEMPERATURE, AND DEPTH PLOT, CAPE, HATTERAS, TIME
B 6402 182 TEMPERATURE, AND DEPTH PLOT, FALKLAND PLATEAU, TIME
B 6402 184 TEMPERATURE, AND DEPTH PLOT, HATTERAS ABYSSAL PLAIN, TIME
S 4 289† TEMPERATURE AND ISOTOPIC COMPOSITION OF SEAWATER, USE OF,
B 6405 782* TEMPERATURE AND PRESSURE RELATIONS IN THICK SEQUENCES OF
B 6203 527* TEMPERATURE, AND WATER SALINITY, ABST., INTERACTING DYNAMICS
S 9 24 TEMPERATURE CHANGES, CENOZOIC
M 26 32 TEMPERATURE EFFECTS CLASSIFICATION CHART, LABORATORY,
M 26 32 TEMPERATURE EFFECTS, LABORATORY, MEASUREMENTS, ROCK
B 6303 447* TEMPERATURE FROM MAGNETIC DATA, BLIGH WATER, FIJI, ABST.,
B 6102 230 TEMPERATURE IN FRACTURE PERMEABILITY
B 6406 916† TEMPERATURE IN PETROLEUM FORMATION—APPLICATION OF LOPATINS
B 6406 921 TEMPERATURE INDEX GEOCHEMICAL DATA, CORRELATION OF, TIME-
B 6406 919 TEMPERATURE INDEX OF MATURITY VERSUS VITRINITE REFLECTANCE,
B 6303 421* TEMPERATURE, LAUMONTITE OCCURRENCE, AND BURIAL HISTORY IN
B 6303 484* TEMPERATURE LOGGING AS USEFUL TOOL FOR LITHOLOGIC
B 6304 594 TEMPERATURE MAP, EGYPT, ABU MADI FIELD
B 6304 591 TEMPERATURE MAPS, GEOCHEMICAL TOOL
M 25 98† TEMPERATURE ON COALIFICATION OF TERTIARY COAL IN
S 9 20 TEMPERATURE, PALEOCLIMATES, MEAN ANNUAL AIR
S 9 20 TEMPERATURE, PALEOCLIMATES, MEAN ANNUAL WATER
B 6201 20 TEMPERATURE PROFILES, SELECTED WELLS, NIGER DELTA
B 6307 1131 TEMPERATURE RANGE, EXTRAPOLATION OF, ARRHENIUS PLOTS OVER
B 6101 100 TEMPERATURE RANGES, HYDROCARBONS AND ASPHALTIC COMPOUNDS
B 6005 884† TEMPERATURE RELATION IN OIL GENESIS. DISCUSSION AND REPLY,
S 3 90 TEMPERATURE, REPLACEMENT POWER VERSUS
B 6112 2085 TEMPERATURE VERSUS CARBONIZATION
B 6103 407† TEMPERATURE--LIMITS ON MAXIMUM POSSIBLE EFFECT, POROSITY
S 7 320 TEMPERATURE-SALINITY DIAGRAM, GYRE BASIN
B 6304 700* TEMPERATURES, ABST., TERTIARY AND, CRETACEOU, OCEAN
B 6207 1212* TEMPERATURES AND ANOMALOUS RUNOFF IN CIRCUM-PACIFIC REGION,
B 6304 578 TEMPERATURES AND OCCURRENCE OF OIL AND GAS, EARTH
B 6212 2478† TEMPERATURES AND PRESSURES, SOLUBILITY OF, METHANE IN, WATER
B 6305 776 TEMPERATURES AND PRESSURES, SOLUBILITY OF, METHANE IN, WATER
B 6309 1527† TEMPERATURES, AQUEOUS, SOLUBILITY OF, METHANE AT ELEVATED,
S 9 26 TEMPERATURES, CENOZOIC, OXYGEN ISOTOPE OCEAN
B 6002 231 TEMPERATURES, DEEP-, BASIN
B 6007 1073 TEMPERATURES, NIGER DELTA PETROLIFEROUS PROVINCE, WELL-, LOG
S 9 21 TEMPERATURES, PRESENT, SURFACE MEAN
B 6303 538 TEMPLES, TOMMY J., LATE QUATERNARY FORAMINIFERAL RECORD IN
B 6303 510* TEMPORAL AND SPATIAL DISTRIBUTION OF ICE-RAFTED MINERAL
B 6405 727* TEMPORAL CHANGES IN DEPOSITIONAL FACIES IN GREAT VALLEY
B 6204 668 TEN FOUYE OIL FIELD, ALGERIA
B 6211 2247 TEN SECTION-STRAND TREND, CALIFORNIA
B 6307 1137 TENKILLER-BLACKGUM FORMATIONS, OKLAHOMA
B 6110 1798 TENNECO, AND ARABIAN AMERICAN OIL CO., DEVELOPMENTS IN SAUDI
B 6010 1868 TENNECO OIL, ON DEVELOPMENNS IN SAUDI ARABIA, 1975, SAUDI
B 6303 487* TENNESSEE, ABST., CONTROLS OF SEDIMENTATIO OF LOWER,
B 6105 828* TENNESSEE, ABST., DEEP-WATER SEDIMENTATION AND TECTONICS IN
B 6405 751* TENNESSEE, ABST., DEPOSITIONAL ENVIRONMENTS AND BIOGENIC
B 6408 1283* TENNESSEE, ABST., DEVONIAN SHALE EXPLORATION RATIONALE TESTED
B 6305 845* TENNESSEE, ABST., GEOLOGY AND PRODUCTION POTENTIAL OF,

B 6009 1622* TENNESSEE, ABST., LATE, MIDDLE AND EARLY LATE, ORDOVICIAN
B 6203 544* TENNESSEE, ABST., NON-SABKHA, EVAPORITES, SILICIFIED, FROM,
B 6004 729* TENNESSEE, ABST., REGIONAL ENVIRONMENTA STRATIGRAPHY AND, OIL
B 6212 2383 TENNESSEE, BARRIER MODEL
B 6110 1898 TENNESSEE, CAMBRIAN TYPE LOCALITIES, ALABAMA, GEORGIA, AND
S 6 348 TENNESSEE, CARTERS LIMESTONE, CORRELATION
B 6408 1286* TENNESSEE COAL, ABST., GEOLOGY AND, MINING O
B 6008 1222 TENNESSEE, DEVELOPMENTS, 1975
B 6108 1184 TENNESSEE, DEVELOPMENTS, 1976
B 6208 1341 TENNESSEE, DEVELOPMENTS, 1977
B 6308 1324 TENNESSEE, DEVELOPMENTS, 1978
B 6409 1462 TENNESSEE, DEVELOPMENTS, 1979
S 3 61 TENNESSEE, FORT PAYNE CHERT
S 4 186 TENNESSEE, HOLSTON LIMESTONE, REEF
M 27 207 TENNESSEE, MISSISSIPPIAN, STE. GENEVIEVE LIMESTONE
B 6204 644 TENNESSEE, MONTEAGLE LIMESTONE
B 6412 2075 TENNESSEE, POWELL RIVER DRAINAGE
B 6008 1223 TENNESSEE, PRODUCTION, 1975
B 6108 1184 TENNESSEE, PRODUCTION, 1976
B 6208 1341 TENNESSEE, PRODUCTION, 1977
B 6308 1325 TENNESSEE, PRODUCTION, 1978
B 6409 1463 TENNESSEE, PRODUCTION, 1979
M 27 50 TENNESSEE, SILURIAN, BROWNSPORT GROUP
B 6412 2074 TENNESSEE, TELLICO SANDSTONE
S 3 61 TENNESSEE, WELLS CREEK CRYPTOEXPLOSIVE STRUCTURE
B 6405 743* TENNESSEE WITH WAULSORTIAN MOUNDS OF WESTERN EUROPE, ABST.,
B 6412 2064† TENNESSEE-KENTUCKY-VIRGINIA, SEASAT, ORBITAL, RADAR, IMAGERY
S 7 278 TENSIONAL STRESS, BOTTOM, DEFORMATION
B 6305 822* TENSLEEP, AND MADISON RESERVOIRS IN BIGHORN BASIN, WYOMING,
B 6204 630 TENSLEEP G ZONE, WYOMING
B 6204 609† TENSLEEP RESERVOIR, OREGON BASIN FIELD, WYOMING
B 6105 816* TENSLEEP RESERVOIR STUDY, OREGON BASIN FIELD, WYOMING, ABST.
B 6108 1383* TENSLEEP SANDSTONE, EASTERN BIG HORN BASIN, WYOMING, ABST.,
B 6008 1407* TENSLEEP SANDSTONE, PENNSYLVANIAN, CENNRAL WYOMING, ABST.,
B 6201 100 TENSLEEP SANDSTONE, WYOMING
B 6204 610 TENSLEEP SANDSTONE, WYOMING
M 28 122 TENSLEEP SANDSTONE, WYOMING, PENNSYLVANIAN
M 28 68 TENSLEEP SANDSTONE, WYOMING, PENNSYLVANIAN
M 28 129 TENSLEEP SANDSTONE, WYOMING, PENNSYLVANIAN
M 28 123 TENSLEEP SANDSTONE, WYOMING, PENNSYLVANIAN
M 27 42 TENTACULITEN KNOTTENKALK, GERMANY, DEVONIAN
B 6004 545 TEPETATE-BATON ROUGE FAULT SYSTEM, LOUISIANA
B 6412 2191 TEPHROCHRONOLOGY, BY VIRGINIA STEEN-MCINTYRE, ROSE, W. I., JR.,
B 6101 51 TERBOL, 1A, LOG, LEBANON
B 6101 61 TERBOL NO. 1, SECTION, LEBANON
B 6109 1515 TERERRO FORMATION, NEW MEXICO
B 6212 2513 TERMAN, MAURICE J., PLATE-TECTONIC EVOLUTION OF THE FAR EAST,
S 5 24 TERMINOLOGY, SILURIAN, SALINA GROUP,
B 6203 502* TERMINOLOGY, ABST., STRATIGRAPHIC CONCEPTS AND
B 6102 248† TERMINOLOGY FOR IGNEOUS AND HIGH-GRADE METAMORPHIC ROCKS,
M 26 205 TERMINOLOGY, GLOSSARY, SEISMIC, STRATIGRAPHY
B 6102 248 TERMINOLOGY, NOTE 45, METAMORPHIC ROCKS
B 6209 1757* TERRACE DEPOSITS, LOUISIANA, ABST., MINERALOGY OF, PLEISTOCENE
M 28 126 TERRACE DEPOSITS, TEXAS, PLEISTOCENE
B 6111 2001† TERRACE, PALOS VERDES HILLS, CALIFORNIA, FAULTED UPPER,
B 6403 305 TERRACES CLASSIFIED USING CONTINUOUS SEISMIC REFLECTION
B 6004 712* TERRACES OR SUBAERIAL KARST, ABST., SCALLOPED, SOLUTION
B 6203 546* TERRAIN IDENTIFIED BY DIGITAL PROCESSING OF LANDSAT DATA,
B 6109 1560* TERRANES, GREAT NORTHERN PENINSULA, NEWFOUNDLAND, ABST.,
B 6409 1557* TERREBONNE PARISH, LOUSIANA, ABST., DIAGENESIS OF MIDDLE,
M 24 231 TERREBONNE TROUGH, LOUISIANA
B 6203 540* TERRESTRIAL EVOLUTION—INSIGHTS FROM THE MOON AND PLANETS,
B 6405 747* TERRESTRIAL SEDIMENTATION ASSOCIATED WITH STRIKE-SLIP FAULT
B 6303 485* TERRESTRIAL VERTEBRATES IN QUATERNARY SEDIMENTS OF KENYA,
S 3 12 TERRESTRIAL-LITHOLOGY MODEL FOR RADIOELEMENT DISTRIBUTION
M 24 157 TERRIERE, ROBERT T., GEOLOGY OF FAIRWAY FIELD, EAST TEXAS
M 26 249† TERRIGENOUS PLEISTOCENE DEPOSITS IN NORTHWEST AND CENTRAL
B 6210 1849 TERRITORY OF AFARS AND, ISSAS, BOTSWANA, BURUNDI, MALAWI,
B 6303 538 TERRY, GLENN A., ENVIRONMENTAL ASSESSMENT OF IN-SITU LEACHING
M 26 279 TERRY SANDSTONE, CRETACEOUS
B 6410 1615 TERRY SANDSTONE, DENVER BASIN
B 6303 510* TERRY SANDSTONES, UPPER CRETACEOUS, SPINDLE FIELD, COLORADO,
B 6407 1005 TERRYVILLE SANDS, ARK-LA-TEX AREA
B 6303 525* TERTIAR LARGE FORAMINIFERA COMMON TO WEST AFRICA AND
S 10 1† TERTIARY SECTION, THE NATURE OF, SHALES AND THE DYNAMICS OF,
B 6303 509* TERTIARY, ABST., APPLIED BIOSTRATIGRAPHY IN, GULF COAST
B 6303 303 TERTIARY, ALASKA
B 6407 1016 TERTIARY, ALASKA
B 6408 1142 TERTIARY, ALASKA
B 6408 1133 TERTIARY, ALASKA
B 6407 1016 TERTIARY, ALASKA
M 24 4 TERTIARY, ALASKA
M 25 297 TERTIARY, ALASKA
B 6204 637 TERTIARY, ALASKA
B 6203 380 TERTIARY, ALASKA
B 6203 527* TERTIARY ALGAL CARBONATE REEFS FROM SNAKE RIVER PLAIN, IDAHO,
M 25 344 TERTIARY, AMAZON BASINS
B 6304 700* TERTIARY AND CRETACEOU OCEAN TEMPERATURES, ABST.
M 25 526† TERTIARY AND QUATERNARY VOLCANISM OF BAJA
B 6303 465* TERTIARY ARKOSES, SANTA YNEZ MOUNTAINS, SANTA BARBARA AND
M 28 140 TERTIARY, ASH FLOW TUFF, NEW MEXICO
M 25 236 TERTIARY, AUSTRALIA
M 28 69 TERTIARY, BASALT WEATHERING, WASHINGTON
M 30 388 TERTIARY BASIN, REGIONAL TECTONIC MAP, VENEZUELA
B 6403 448* TERTIARY BASIN, WESTERN OREGON AND WASHINGTON, ABST.,
B 6303 522* TERTIARY BASINS, ABST., DUAL ORIGIN OF, NATURAL, GASES IN,
B 6108 1392* TERTIARY BASINS OF PACIFIC MARGIN, ABST., WIDESPREAD
M 25 250† TERTIARY BASINS OF SOUTHEAST ASIA--SUMMARY, STRUCTURAL
M 25 441 TERTIARY BASINS, SOUTH KOREA
M 28 74 TERTIARY, BAUXITE, ARKANSAS
B 6211 2232† TERTIARY BAY OF BISCAY AND ASSOCIATED TURBIDITY CURRENTS, A,
B 6211 2233 TERTIARY, BAY OF BISCAY AREA
B 6109 1427 TERTIARY BEDS, LOUISIANA, OFFSHORE
B 6205 805 TERTIARY, BLAKE PLATEAU
S 6 260 TERTIARY BOUNDARY, A REVIEW, CRETACEOUS
B 6401 72 TERTIARY BOUNDARY, ISOTOPE RATIOS, NORTH ATLANTIC, DSDP HOLES
M 27 12 TERTIARY CALCARENITE, MUSCAT, OMAN
B 6211 2223 TERTIARY, CALIFORNIA
S 9 22 TERTIARY, CARBONATE REEF BELT DISTRIBUTION
B 6207 1230* TERTIARY CARBONATE RESERVOIRS IN INDONESIA, ABST.
B 6209 1763* TERTIARY CARBONATE ROCKS, SOUTH FLORIDA BASIN, ABST.,
B 6008 1392* TERTIARY CHANNEL SOUTHEAST OF LAKE FROME, SOUTH AUSTRALIA,
M 25 98† TERTIARY COAL IN JAPAN--SUMMARY, INFLUENCE OF, TEMPERATURE
M 25 323 TERTIARY, COLOMBIA
B 6102 207† TERTIARY CONGLOMERATES, VERACRUZ BASIN, MEXICO, ORIGIN AND
S 4 353† TERTIARY CORAL REEFS OF RED SEA--SEQUENCE AND TIME SCALE,
B 6305 825* TERTIARY CORRELATIONS IN BIGHORN AN UINTA BASINS, WYOMING
M 28 41 TERTIARY, CREEDE FORMATION, COLORADO
M 25 443 TERTIARY DEFORMATION, SOUTH KOREA, CRETACEOUS, LATE, TO
B 6004 697* TERTIARY DELTAIC CYCLES ON UNSTABLE CRATO MARGIN, BANKS
S 9 12 TERTIARY, DIAGRAM, POLAR WANDERING
M 25 446 TERTIARY, EARLY, METALLOGENIC PROVINCES, SOUTH KOREA,
B 6306 976 TERTIARY, EGYPT
M 30 64 TERTIARY, FAUNAL ASSEMBLAGES, FRIGG FIELD
S 9 25 TERTIARY, FOLDOUT, PALEOCLIMATES
M 30 67 TERTIARY, FRIGG FIELD, SEDIMENTATION HISTORY
B 6203 541* TERTIARY FURNACE CREEK FORMATION, DEATH VALLEY, CALIFORNIA,
B 6211 2362* TERTIARY GEOLOGY AND OIL SHALE RESOURCES OF SOUTH ELKO BASIN,
M 26 85 TERTIARY, GLOBAL CYCLES
B 6405 741* TERTIARY GULF COAST BASIN, ABST., SANDSTONE DIAGENESIS IN
S 7 31 TERTIARY, GULF COAST BASIN, CROSS SECTIONS
S 7 84 TERTIARY, GULF OF MEXICO, END OF MIOCENE
M 30 139 TERTIARY HISTORY, NORTH SEA, MORAY FIRTH BASIN
S 11 4† TERTIARY, HISTORY OF STRATIGRAPHIC, PALEONTOLOGY OF, WEST
M 28 16 TERTIARY, HORSE SPRING FORMATION, NEVADA
M 28 4 TERTIARY, HORSE SPRINGS FORMATION, NEVADA
M 28 90 TERTIARY, HORSE SPRINGS FORMATION, NEVADA
M 28 34 TERTIARY, HORSE SPRINGS FORMATION, NEVADA
B 6402 219 TERTIARY, INDIA
B 6203 443 TERTIARY, INDIA
M 28 17 TERTIARY INTRUSIVE, NEVADA
B 6207 1136 TERTIARY, IRIAN JAYA
M 28 172 TERTIARY, KOGUSUKRUK FORMATION, ALASKA
M 25 406 TERTIARY LATERITES, AUSTRALIAN, BAUXITE DEPOSITS, CONFINED TO
B 6101 44 TERTIARY, LEBANON
S 4 69 TERTIARY, LESSER, ANTILLES
B 6309 1611* TERTIARY LITHOSTRATIGRAPHIC INTERPRETATION OF SOUTHWEST
B 6202 295† TERTIARY LITHOSTRATIGRAPHY OF NIGER DELTA
B 6205 891* TERTIARY LOW-PERMEABILITY RESERVOIR ROCKS, SOUTH-CENTRAL
M 29 133 TERTIARY, LOW-VELOCITY SEDIMENTS, MALVINAS BASIN
B 6111 1989 TERTIARY METAVOLCANIC CLASTS, CALIFORNIA
B 6102 208 TERTIARY, MEXICO
M 28 35 TERTIARY, NEEDLES RANGE FORMATION, NEVADA
B 6109 1518 TERTIARY, NEW MEXICO
B 6107 969 TERTIARY, NIGER DELTA
B 6201 1† TERTIARY NIGER DELTA, HYDROCARBON HABITAT OF
B 6408 1251† TERTIARY NIGER DELTA, PETROLEUM, SOURCE-BED, EVALUATION OF
M 25 286 TERTIARY, NORTH AMERICA, WESTERN
B 6204 596 TERTIARY, NORTH, ATLANTIC BASIN
B 6302 194† TERTIARY, NORTHWEST COLORADO, AND THEIR RELATION TO COAL,

Index

B 6409 1560* TERTIARY, NORTHWESTERN LOUISIANA, AND THEIR RELATION TO
S 11 54† TERTIARY OF CALIFORNIA, ZEMORRIAN THROUGH DELMONTIAN,
B 6109 1539* TERTIARY OF EAST TEXAS AND WEST LOUISIANA, ABST., SLICE
B 6004 704* TERTIARY OF NEW JERSEY, ABST., TIMING OF, TRANSGRESSIONS AND,
B 6405 722* TERTIARY ORCA GROUP, PRINCE WILLIAM SOUND, ALASKA, ABST.,
M 25 428 TERTIARY, ORE DEPOSITS ASSOCIATED WITH, AUSTRALIA
B 6304 694* TERTIARY OXIDATION IN WESTWATER CANYON MEMBER OF MORRISON
B 6004 704* TERTIARY PALEOBATHYMETRIC HISTORY OF NEW JERSEY COAST PLAIN,
B 6303 554* TERTIARY PALEOCLIMATIC HISTORY, ABST., DEEP-SEA, DRILLING IN,
B 6109 1480 TERTIARY PETROLEUM RESERVOIR SYSTEM, BRAZIL, CRETACEOUS, LATE-
B 6207 1195 TERTIARY, PLATE BOUNDARIES
B 6203 525* TERTIARY POROSITY IN CARBONATE ROCKS, ABST., OCCURRENCES OF
B 6211 2360* TERTIARY PRINCETON SUBMARINE VALLEY SYSTEM BENEATH
B 6211 2360* TERTIARY PROVINCE, ABST., FORAMINIFERAL BIOSTRATIGRAPHY IN,
B 6405 748* TERTIARY PROVINCE, ABST., PRELIMINARY, MOLLUSCAN
B 6004 655* TERTIARY PROVINCE, ABST., PRELIMINARY STUDY OF, SEA-FLOOR
B 6012 2176* TERTIARY PROVINCE, ABST., STRUCTURAL ELEMENTS OF OFFSHORE,
B 6405 709* TERTIARY REEF PALEOECOLOGY, ABST., UNIFORMITARIANISM AND
B 6104 586 TERTIARY RESERVOIRS, OIL AGE, LOUISIANA SHELF
B 6207 1193† TERTIARY RIFTING IN WESTERN OREGON-WASHINGTON, EARLY
B 6108 1393* TERTIARY ROCKS IN WESTERN NEBRASKA, ABST., STRATIGRAPHY AND
B 6405 699* TERTIARY SANDSTONE RESERVOIRS ALONG ONSHORE TEXAS GULF
B 6309 1606* TERTIARY SANDSTONE RESERVOIRS ALONG TEXAS GULF COAST, ABST.,
B 6109 1541* TERTIARY SANDSTONE RESERVOIRS—POROSITY FORMATION AND
B 6408 1167 TERTIARY, SANDSTONE WEDGES, TEXAS, GULF COAST
B 6408 1163 TERTIARY, SANDSTONE WEDGES, VENEZUELA, EASTERN
B 6303 463* TERTIARY SANDSTONES, ABST., DRAINAGE ANOMALIES IN, GULF COAST
B 6303 440* TERTIARY SANDSTONES ALONG TEXAS GULF COAST, ABST.,
B 6303 488* TERTIARY SANDSTONES ALONG TEXAS GULF COAST, ABST., RESERVOIR
B 6405 738* TERTIARY SANDSTONES IN PACIFIC CREEK AREA, SUBLETTE COUNTY,
B 6305 836* TERTIARY SANDSTONES OF SOUTHERN PICEANCE CREEK BASIN,
B 6303 529* TERTIARY SEA-LEVEL MOVEMENTS AROUND SOUTHERN AFRICA, ABST.
B 6203 521* TERTIARY SECTION, ABST., NATURE OF, SHALES AND DYNAMICS OF
B 6405 698* TERTIARY SEDIMENTS IN ALASKA, ABST., URANIUM IN
M 28 139 TERTIARY, SEDIMENTS, OREGON
M 27 153 TERTIARY SEDIMENTS, TEXAS
B 6205 767 TERTIARY, SEISMIC FACIES
B 6012 2187* TERTIARY SKOONER GULCH AND GALLAWAY FORMATIONS AT POINT
S 11 15 TERTIARY, STAGE BOUNDARIES, CALIFORNIA
S 11 7 TERTIARY STAGES AND ZONES, EUROPEAN
M 30 105 TERTIARY, STATFJORD FIELD, ROGALAND GROUP
B 6203 536* TERTIARY STRATA FROM NUGSSUAQ EMBAYMENT, WEST GREENLAND,
B 6211 2357* TERTIARY STRATA IN COLD BAY AREA OF ALASKA PENINSULA, ABST.,
B 6012 2181* TERTIARY STRATIGRAPHIC AND DEPOSITIONAL ENVIRONMENTS NEAR
B 6309 1522† TERTIARY STRATIGRAPHY AND PETROLEUM POTENTIAL O COLD
B 6405 703* TERTIARY STRUCTURAL DEVELOPMENT OF SELECTED VALLEYS BASED
B 6205 886* TERTIARY STRUCTURAL DEVELOPMENT OF SOUTHERN EGAN RANGE,
B 6211 2349* TERTIARY SUBDUCTION COMPLEX, MENDOCINO COAST, NORTHERN
B 6004 645* TERTIARY SUBMARINE CANYONS AND MARINE CYCLES OF DEPOSITION
B 6101 16 TERTIARY, SVERDRUP BASIN
B 6207 1127† TERTIARY TECTONIC HISTORY OF SALAWATI AREA, IRIAN JAYA,
B 6209 1765* TERTIARY TERRESTRIAL PALEOGEOGRAPHY OF NORTH AMERICAN
B 6403 359† TERTIARY TIME, GEODYNAMIC, PEAKS IN, ALPINOTYPE, OROGENIES AND
B 6403 366 TERTIARY TIME, TIME RELATIONS BETWEEN, DISCONTINUITIES OF,
B 6111 1987 TERTIARY UNCONFORMITY, CALIFORNIA, CRETACEOUS-
S 11 164 TERTIARY VALVULINERIDS, CALIFORNIA
B 6405 724* TERTIARY VERSUS CARBONIFEROUS DEPOSITIONAL SETTINGS—FACTORS
M 28 4 TERTIARY, VIEJA GROUP, TEXAS
M 28 83 TERTIARY, VIEJA GROUP, TEXAS
M 28 3 TERTIARY, VIEJA GROUP, TEXAS
M 28 80 TERTIARY, VIEJA GROUP, TEXAS
M 28 115 TERTIARY, VIEJA GROUP, TEXAS
M 28 138 TERTIARY, VIEJA GROUP, TEXAS
B 6205 884* TERTIARY VOLCANIC STRATIGRAPHY OF TRAP SPRING FIELD, NYE
B 6108 1375* TERTIARY VOLCANICLASTICS, GUATEMALA, ABST., DIAGENESIS OF
B 6109 1535* TERTIARY-PLEISTOCENE, GULF COAST REGION, ABST., EFFECTS OF,
S 4 299† TERTIARY-PLEISTOCENE LIMESTONES ON AND AROUND BARBUDA, WEST
B 6101 16 TERTIARY-QUATERNARY, SVERDRUP BASIN
B 6303 437* **TERTIARY. WHY LOOK A GIFT HORSE IN THE MOUTH, ABST., SOURCE OF**
B 6308 1386 TESELLE, R. D., DEVELOPMENTS IN NORTHERN ROCKIES IN 1978
B 6409 1501 TESELLE, R. D., DEVELOPMENTS IN NORTHERN ROCKIES IN 1979
B 6304 694 TESSENDORF, TIMOTHY NEAL, REDISTRIBUTED OREBODIES OF POISON
B 6308 1174 TEST, DEEPER POOL
B 6108 1126 TEST, DEEPER POOL
B 6208 1281 TEST, DEEPER POOL
B 6008 1159 TEST, DEEPER POOL
B 6409 1301 TEST, DEEPER-POOL
B 6002 323* TEST HOLE IN CENTRAL KANSAS, ABST., SALT, DISSOLUTION IN OIL AND
B 6004 672* TEST, NANTUCKET ISLAND, MASSACHUSETTS, ABST., MESOZOIC,
B 6409 1301 TEST, SHALLOW-POOL
B 6008 1159 TEST, SHALLOWER POOL
B 6208 1281 TEST, SHALLOWER POOL
B 6108 1126 TEST, SHALLOWER POOL
B 6308 1174 TEST, SHALLOWER POOL
B 6105 765* TEST SITE FOR GEOPRESSURED GEOTHERMAL WELL—FRIO SANDSTONE,
B 6308 1174 TEST, STRATIGRAPHIC
B 6108 1127 TEST, STRATIGRAPHIC
B 6208 1281 TEST, STRATIGRAPHIC
B 6409 1302 TEST, STRATIGRAPHIC
B 6203 515* TEST USE OF ERTS-LANDSAT IMAGERY IN URANIUM EXPLORATION,
B 6303 496* TEST-TUBE PYROLYSIS--SIMPLE TECHNIQUE FOR IDENTIFYING YIELD
B 6309 1595* TESTIN GEOPRESSUREDD GEOTHERMAL RESOURCE, FRIO FORMATION,
B 6101 81 TETHYS
B 6109 1490 TETHYS SEA
B 6112 2079 TETHYS SEA
B 6002 258 TETHYS SEA
B 6412 2190 TETHYS-TRIAS, EDITED BY H. ZAPFE, SCHLAGER, WOLFGANG, REVIEW OF,
M 28 66 TEUSCHNITZER CONGLOMERATE, GERMANY, PENNSYLVANIAN
M 28 105 TEUSCHNITZER CONGLOMERATE, GERMANY, PENNSYLVANIAN
B 6405 745 TEWALT, S. J., DEVELOPMENT OF CONCEPTUAL MODEL TO
B 6309 1600* TEXAS, ABST, LOG EVALUATION OF TIGHT ROCKS OF SOUTH
B 6308 1425* TEXAS, ABST., ATOKA GAS IN SOUTHERN, COTTLE AND NORTHER, KING
B 6102 294* TEXAS, ABST., ATOKA, GAS IN SOUTHERN, COTTLE AND NORTHERN,
B 6109 1546* TEXAS, ABST., AUSTIN, CHALK–, BUDA TREND OF SOUTH
B 6004 676* TEXAS, ABST., BED FORMS AND STRATIFICATION TYPES OF MODERN
B 6004 672* TEXAS, ABST., BIOHERMAL DEVELOPMENT IN, WOLFCAMPIAN,
B 6002 327* TEXAS, ABST., BLOCK 16 FIELD, WARD COUNTY
B 6102 296* TEXAS, ABST., BROOKS FIELD, SOUTHEAST, MIDLAND BASIN, IRION
B 6405 678* TEXAS, ABST., CALCIUM BUDGET IN, FRIO SANDSTONES, SOUTHWEST
B 6409 1556* TEXAS, ABST., CAP-ROCK FORMATION AND DIAGENESIS, GYP HILL, SALT,
B 6004 696* TEXAS, ABST., CARBON AND, OXYGEN, ISOTOPE ANALYSES OF,
B 6109 1555* TEXAS, ABST., CARBON AND, OXYGEN, ISOTOPIC EVOLUTION OF WHOLE
B 6004 694* TEXAS, ABST., CARBONATE DIAGENESIS, AND FACIES
B 6303 469* TEXAS, ABST., CARBONATE ROCKS OF, MONTOYA GROUP, MIDDLE AN
B 6309 1596* TEXAS, ABST., CHARACTERISTICS OF LOWER, VICKSBURG, RESERVOIRS,
B 6309 1595* TEXAS, ABST., CHARACTERISTICS OF LOWER, WILCOX, RESERVOIRS,
B 6109 1534* TEXAS, ABST., CHARACTERISTICS OF, WILCOX, GAS, RESERVOIRS,
B 6409 1557* TEXAS, ABST., CLAY MINERALS AS INDICATORS FOR DEPOSITIONAL
B 6405 787* TEXAS, ABST., COASTAL SWAMP ORIGIN OF, SAN MIGUEL LIGNITE
B 6303 520* TEXAS, ABST., CONQUISTA PROJECT, URANIUM, MINING AN MILLING IN
B 6109 1550* TEXAS, ABST., CONTRASTS IN, CEMENTATION, DISSOLUTION, AND,
B 6203 494* TEXAS, ABST., DELTAIC DEPOSITIONAL STYLE IN, SALT-WITHDRAWAL,
B 6405 717* TEXAS, ABST., DELTAIC SYSTEMS AND ASSOCIATED, GROWTH, FAULTING
B 6009 1607* TEXAS ABST., DENSITY-, FLOW, ORIGIN FOR, FRIO, RESERVOIR,
B 6109 1553* TEXAS, ABST., DEPOSITION AND DIAGENESIS OF, FORT TERRETT
B 6303 548* TEXAS, ABST., DEPOSITION AND DIAGENESIS OF, MISSISSIPPIAN,
B 6303 479* TEXAS, ABST., DEPOSITION OF, MARBLE FALLS FORMATION, CENTRAL
B 6405 717* TEXAS, ABST., DEPOSITIONAL AND TECTONIC EVOLUTION OF A
B 6109 1541* TEXAS, ABST., DEPOSITIONAL ENVIRONMENT OF, GRAYSON FORMATION,
B 6309 1601* TEXAS, ABST., DEPOSITIONAL ENVIRONMENT OF WOODBINE
B 6203 566* TEXAS, ABST., DEPOSITIONAL ENVIRONMENTS OF, MAXON FORMATION,
B 6203 538* TEXAS, ABST., DEPOSITIONAL FACIES AND OCCURRENCE OF, POROSITY IN
B 6311 2116* TEXAS, ABST., DEPOSITIONAL HISTORY OF HIGH-CONSTRUCTIVE, DELTA
B 6308 1425* TEXAS, ABST., DEPOSITIONAL MODELS AND RESOURCE POTENTIAL OF,
B 6308 1426* TEXAS, ABST., DEPOSITIONAL SYSTEMS AND PETROLEU POTENTIAL OF
B 6109 1544* TEXAS, ABST., DEPOSITIONAL SYSTEMS IN, SPARTA FORMATION, EOCENE,
B 6303 528* TEXAS, ABST., DESCRIPTION AND SEDIMENTOLOGY O, SUBMARINE-, FAN,
B 6303 440* TEXAS, ABST., DETAILED STRATIGRAPHY AND SEDIMENTOLOGY OF,
B 6004 688* TEXAS, ABST., DEVELOPMENT OF EARLY, CRETACEOUS, BANK COMPLEX,
B 6109 1555* TEXAS, ABST., DIAGENESIS AND, GEOCHEMISTRY OF, GLEN ROSE, PATCH-,
B 6109 1536* TEXAS, ABST., DIAGENESIS AND, POROSITY DISTRIBUTION IN, DELTAIC
B 6209 1753* TEXAS, ABST., DISTRIBUTION AND GEOMETRY OF AN, OOLITE-SHOAL
B 6303 472* TEXAS, ABST., DOLOMITIZATION OF OFFSHORE, CARBONATE DEPOSITS
B 6009 1617* TEXAS, ABST., EARLY, CRETACEOUS, SHELF, MARGIN IN NORTHERN,
B 6405 768* TEXAS, ABST., EARLY, GUADALUPIAN, PERMIAN, BANK MARGIN EROSION
B 6203 548* TEXAS, ABST., EFFECT OF, CLAYS ON, URANIUM DISTRIBUTION IN SOUTH
B 6303 530* TEXAS ABST., ELLENBURGER GROUP, DELAWARE BASIN, WEST
B 6409 1556* TEXAS, ABST., ENVIRONMENT OF DEPOSITION OF UPPER, WILCOX
B 6409 1552* TEXAS, ABST., ENVIRONMENTS OF DEPOSITION AND DIAGENESIS OF,
B 6004 692* TEXAS, ABST., EOCENE DEPOSITIONAL ENVIRONMENTS AND RELATED,
B 6308 1428* TEXAS, ABST., EVIDENCE FOR, WRENCH, FAULTING, SOUTHERN, VAL
B 6105 843* TEXAS, ABST., EVOLUTION OF, CAPITAN, MASSIVE LIMESTONE, PERMIAN,
B 6309 1598* TEXAS, ABST., EXPLORATION METHODS OF DISCOVERY AND
B 6303 461* TEXAS, ABST., EXPLORATION POTENTIAL OF, PENNSYLVANIAN-, PERMIAN,
B 6109 1553* TEXAS, ABST., FACIES, DIAGENESIS, AND, POROSITY DEVELOPMENT IN
B 6004 725* TEXAS, ABST., FACTORS INFLUENCING POROSITY AND, PERMEABILITY OF
B 6303 432* TEXAS, ABST., FAULTS OFFSETTING LAND SURFACE IN SOUTHEASTERN,
B 6309 1607* TEXAS ABST., FIELD RELATIONS AND PETROLOGY OF, CATAHOULA
B 6209 1761* TEXAS, ABST., FORAMINIFERAL PALEOECOLOGY OF, GRAYSON

B 6109 1548* TEXAS, ABST., GEOLOGY OF, FRACTURED, AUSTIN, CHALK AND, BUDA
B 6409 1563* TEXAS, ABST., GEOLOGY OF, GUM ISLAND NORTH AND, FRENCH ISLAND
B 6109 1539* TEXAS, ABST., GEOMETRY OF, GULF COAST, AQUIFERS, HOUSTON-,
B 6405 703* TEXAS, ABST., GROWTH, FAULTING AND SEDIMENTATION IN UPPER,
B 6109 1540* TEXAS, ABST., GROWTH HISTORY OF, HAINESVILLE, SALT, DOME, WOOD
B 6309 1602* TEXAS, ABST., HYDROLOGY AND, WATER QUALITY OF, EOCENE, WILCOX
B 6303 545* TEXAS, ABST., LAND DEVELOPMENT AND, FAULTING NEA HOUSTON
B 6109 1539* TEXAS, ABST., LAND RESOURCES OF
B 6303 454* TEXAS, ABST., LAND-SURFACE, SUBSIDENCE IN, HOUSTON-, GALVESTON
B 6009 1612* TEXAS, ABST., LEACHED, POROSITY IN OVERPRESSURED, SANDSTONES--,
B 6203 526* TEXAS, ABST., LEONARDIAN STRATA IN NORTHERN, MIDLAND BASIN OF
B 6203 548* TEXAS, ABST., LITHOFACIES AND DEPOSITIONAL ENVIRONMENTS OF,
B 6002 326* TEXAS, ABST., LITHOFACIES AND DEPOSITIONAL ENVIRONMENTS OF,
B 6409 1556* TEXAS, ABST., LIVE OAK, DELTA COMPLEX—UNSTABLE, SHELF-EDGE
B 6209 1760* TEXAS, ABST., LIVING, SHELLED, MICROZOOPLANKTON, RADIOLARIANS,
B 6209 1762* TEXAS, ABST., LOBO TREND OF, SOUTH, LAREDO AREA, WEBB AND,
B 6109 1557* TEXAS, ABST., LOWER, CRETACEOUS, CARBONATE ROCKS OF CENTRAL
B 6109 1557* TEXAS, ABST., LOWER, CRETACEOUS DEPOSITIONAL SYSTEMS, WEST
B 6405 723* TEXAS, ABST., LOWER, CRETACEOUS SHELF, STORM DEPOSITS, NORTH
B 6303 475* TEXAS, ABST., MICROSCOPIC DISTRIBUTION OF, URANIU IN, OAKVILLE
B 6109 1557* TEXAS, ABST., MIDDLE, CRETACEOUS ROCKS OF, MEXICO AND
B 6109 1551* TEXAS, ABST., MIDDLE, GLEN ROSE, CRETACEOUS, FACIES MOSAIC,
B 6309 1601* TEXAS, ABST., MIDDLE, STRAWN, DESMOINESIAN, CRATONIC, DELTA
B 6409 1566* TEXAS, ABST., MINERALOGY, DIAGENESIS, AND POROSITY IN, VICKSBURG
B 6209 1761* TEXAS, ABST., MIOCENE GEOLOGY OF, MATAGORDA ISLAND, CALHOUN
B 6409 1553* **TEXAS, ABST., MORPHOLOGY OF TURBIDITE-CHANNEL RESERVOIRS,**
B 6002 326* TEXAS, ABST., NEILL SOUTH, ODOM LIME, FIELD, NOLAN AND, FISHER
B 6009 1614* TEXAS, ABST., NEW, PELECYPOD SPECIES FROM LOWER, CRETACEOUS
B 6305 831* TEXAS, ABST., NONMARINE DEPOSITIONAL ENVIRONMENTS AND,
B 6004 670* TEXAS, ABST., NONVADOSE ORIGIN OF, PISOLITIC, FACIES, CAPITAN, REEF
B 6409 1567* TEXAS, ABST., OCCURRENCE OF HIGH GRAVITY OIL IN AN, OLIGOCENE,
B 6209 1757* TEXAS, ABST., OLIGOCENE, REEF COMMUNITY SUCCESSION, DAMON
B 6105 839* TEXAS, ABST., OLIGOCENE, VOLCANIC, SEDIMENTARY, APRON IN,
B 6203 498* TEXAS, ABST., ORIGIN OF LATE-STAGE, IRON-RICH, CARBONATE,
B 6405 777* TEXAS, ABST., PALEOCENE, MIDWAY, CONTINENTAL SHELF DEPOSITS, RIO
B 6409 1569* TEXAS, ABST., PALEOECOLOGY OF, MIDWAY GROUP IN NORTHEAST
B 6405 679* TEXAS, ABST., PALEOENVIRONMENTAL ANALYSIS OF DISCONFORMITY
B 6409 1559* TEXAS, ABST., PALEOENVIRONMENTS AND, TRACE FOSSILS OF LARGE
B 6311 2116* TEXAS, ABST., PENNSYLVANIAN, FAN-, DELTA SANDSTONES, PAL DURO
B 6109 1546* TEXAS, ABST., PETROLOGY OF, CATAHOULA SANDSTONES IN EAST
B 6109 1554* TEXAS, ABST., POROSITY DEVELOPMENT AND DISTRIBUTION IN, SHOAL-,
B 6303 457* TEXAS, ABST., POROSITY EXAMPLES FROM, IAB FIELD, COKE COUNTY
B 6303 488* TEXAS, ABST., POROSITY IN, GIANT, GAS FIELD, ELLENBURGER
B 6303 458* TEXAS, ABST., PRELIMINARY ANALYSES OF, BRINES FROM, PERMIAN
B 6303 513* TEXAS, ABST., RATES OF, CREEP OF ACTIVE, FAULTS IN NORTHWESTERN,
B 6004 694* TEXAS, ABST., RELATION OF EARLY, CARBONATE, POROSITY TO
B 6303 418* TEXAS, ABST., RESERVOIR MORPHOLOGIES FOR, GULF COAST, TURBIDITE
B 6308 1427* TEXAS, ABST., SAN ANDRES FACIES PATTERNS, PALO DURO AND,
B 6109 1547* TEXAS, ABST., SECONDARY, POROSITY IN, SANDSTONES OF LOWER,
B 6209 1765* TEXAS, ABST., SEDIMENTARY FACIES, STRUCTURES, AND GRAIN-SIZE
B 6004 716* TEXAS, ABST., SEDIMENTOLOGY AND, PALEOECOLOGY OF TYPE
B 6308 1426* TEXAS, ABST., SEDIMENTOLOGY AND SOURCE-BED, GEOCHEMISTRY OF,
B 6405 709* TEXAS, ABST., SHALE MINERALOGY OF, GENERAL CRUDE OIL AND,
B 6109 1551* TEXAS, ABST., SLIGO AND, HOSSTON DEPOSITIONAL PATTERNS,
B 6405 702* TEXAS, ABST., SOURCE-ROCK QUALITY AND, THERMAL MATURITY, PALO
B 6303 536* TEXAS, ABST., SOUTH HALLETTSVILLE, GAS FIELD IN LOWER, WILCOX,
B 6405 749* TEXAS, ABST., STABLE, ISOTOPE EVIDENCE FOR MODERN FRESHWATER
B 6209 1756* TEXAS, ABST., STRATIGRAPHIC INTERPRETATION OF A POSSIBLE,
B 6109 1540* TEXAS, ABST., STRATIGRAPHY AND, ENTRAPMENT OF, HYDROCARBONS
B 6308 1426* TEXAS, ABST., STRATIGRAPHY AND ENTRAPMENT OF, HYDROCARBONS
B 6203 571* TEXAS, ABST., STRUCTURAL GEOLOGY OF SOUTHWEST, EDWARDS
B 6309 1596* TEXAS, ABST., SUBSURFACE STRATIGRAPHY OF, MIDWAY-WILCOX,
B 6002 327* TEXAS, ABST., SUBSURFACE STUDY OF, ATOKA, LOWER, PENNSYLVANIAN,
B 6108 1385* TEXAS, ABST., THERMAL-, INFRARED SURVEYS OF, URANIUM
B 6303 509* TEXAS, ABST., TRACE-ELEMENT, GEOCHEMISTRY AND DIAGENESIS IN,
B 6203 525* TEXAS, ABST., TRANSITION BETWEEN A SANDY AND MUDDY
B 6308 1425* TEXAS, ABST., TURBIDITE CHANNEL, RESERVOIRS IN,
B 6002 324* TEXAS, ABST., TURBIDITE, RESERVOIRS IN, CANYON SANDSTONE, VAL
B 6309 1601* TEXAS, ABST., UPPER, CRETACEOUS--LOWE, EOCENE STRATA,
B 6002 324* TEXAS, ABST., UPPER, STRAWN DEPOSITIONAL SYSTEMS, LATE,
B 6309 1611* TEXAS, ABST., WAVE-DOMINATED DELTAIC SYSTEMS OF UPPER,
B 6203 570* TEXAS, ABST., WHEELER-PAN HUNTON FIELD, WHEELER COUNTY
B 6303 417* TEXAS, ABST., WORLDS FIRST GEOPRESSURED, GEOTHERMAL TEST WELL,
B 6402 225 TEXAS, ALMOND GRABEN
B 6108 1225 TEXAS, ALVIN FIELD, BRAZORIA COUNTY
B 6305 804 TEXAS, ANADARKO BASIN, MILLS RANCH COMPLEX
B 6408 1169 TEXAS, ANAHUAC SHALE
B 6102 297* TEXAS AND ANADARKO BASIN, ABST., STATISTICAL SUMMARY OF, WELLS
B 6205 853† TEXAS AND ANADARKO BASIN, STATISTICAL SUMMARY OF, WELLS
B 6203 495* TEXAS, AND BELL CANYON FORMATION OF DELAWARE BASIN, ABST.,
B 6008 1243† TEXAS AND EASTERN NEW MEXICO IN 1975, DEVELOPMENTS IN WEST
B 6108 1208† TEXAS AND EASTERN NEW MEXICO IN 1976, DEVELOPMENTS IN WEST
B 6208 1364† TEXAS AND EASTERN NEW MEXICO IN 1977, DEVELOPMENTS IN WEST
B 6308 1236† TEXAS AND EASTERN NEW MEXICO IN 1978, DEVELOPMENTS IN WES
B 6409 1369† TEXAS AND EASTERN NEW MEXICO IN 1979, DEVELOPMENTS IN WEST
B 6407 1006 TEXAS AND LOUISIANA, SABINE UPLIFT
B 6109 1550* TEXAS AND MEXICO, ABST., SOME LOWER, CRETACEOUS STRATIGRAPHIC,
B 6303 420* TEXAS AND NEW MEXICO, ABST., LITHOFACIES ANALYSIS OF, CASTILE
B 6303 437* TEXAS AND NEW MEXICO, ABST., SEDIMENTOLOGY OF, GOAT SEEP
B 6205 855 TEXAS AND NEW MEXICO, BASEMENT CONTOUR MAP, WEST
B 6109 1449 TEXAS AND NEW MEXICO, DELAWARE BASIN
B 6109 1448 TEXAS AND NEW MEXICO, PERMIAN BASIN
B 6205 854 TEXAS AND NEW MEXICO, ULTRADEEP WELLS, LOCATION, WEST
B 6205 859 TEXAS AND NEW MEXICO, ULTRADEEP WILDCATS, SUCCESS RATIOS,
B 6203 573* TEXAS AND NEW MEXICO—CEMENT EVOLUTION, ABST.,
B 6203 492* TEXAS AND NEW MEXICO—NATURE AND ORIGIN OF CEMENTS, ABST.,
B 6203 526* TEXAS AND OKLAHOMA, ABST., DEVELOPMENT OF, MILLS RANCH
B 6105 827* TEXAS AND OKLAHOMA, ABST., FLOW PATTERNS AND, BAR
B 6305 805 TEXAS AND OKLAHOMA, MILLS RANCH COMPLEX
B 6004 517† TEXAS AND SOUTHEAST NEW MEXICO, BASINAL, SANDSTONE, FACIES,
B 6407 1118 TEXAS AND SOUTHEAST NEW MEXICO, EDITED BY N. M. SULLIVAN,
B 6203 558* TEXAS AND SOUTHEASTERN NEW MEXICO, ABST., GEOLOGIC APPRAISAL
B 6203 507* TEXAS AND SOUTHEASTERN NEW MEXICO, ABST., GEOLOGIC APPRAISAL
B 6203 509* TEXAS AND SOUTHEASTERN NEW MEXICO, ABST., GEOLOGIC APPRAISAL
B 6203 544* TEXAS AND SOUTHEASTERN NEW MEXICO, ABST., NEW DEVELOPMENTS
B 6209 1768* TEXAS AND SOUTHWEST ARKANSAS, ABST., STRATIGRAPHIC RELATIONS
B 6002 299† TEXAS AND THEIR RELATION TO PALEOZOIC CRATONIC MARGIN.
B 6109 1539* TEXAS AND WEST LOUISIANA, ABST., SLICE TECHNIQUE APPLIED TO,
B 6305 806 TEXAS, ARBUCKLE DEVELOPMENT, MILLS RANCH COMPLEX
B 6305 804 TEXAS, ARBUCKLE PRODUCTION
S 4 251 TEXAS, ARCHEOLITHOPHYLLUM SEGMENT
B 6303 447* TEXAS, AREA A OBSERVED ON LANDSAT IMAGERY, ABST., ACTIVE,
B 6301 35 TEXAS, ATASCOSA TROUGH
B 6301 54 TEXAS, ATOKA CLASTICS
B 6305 808 TEXAS, ATOKA-MORROW PRODUCTION
B 6102 296* TEXAS AUSTIN CHALK, CRETACEOUS, PLAY, ABST., SOUTH
B 6107 988 TEXAS, AUSTIN CHALK UNIT
B 6409 1356 TEXAS, AUSTIN COUNTY
B 6402 231 TEXAS, AUSTIN FORMATION
B 6408 1220 TEXAS, AUSTIN SECTION
B 6208 1382 TEXAS, BAFFLE POINT FIELD, GALVESTON COUNTY
B 6301 35 TEXAS, BANDERA COUNTY, CRETACEOUS
B 6408 1220 TEXAS BASIN, CROSS SECTION, EAST
B 6309 1605* TEXAS BASIN FOR GEOLOGIC ISOLATION OF NUCLEAR WASTES, ABST.,
B 6102 176 TEXAS BEACHES, RECENT SANDS
B 6104 486 TEXAS, BELL CANYON SANDSTONES TRAPS
B 6301 54 TEXAS, BEND ARCH, ORIGIN
B 6301 54 TEXAS, BEND CONGLOMERATES
M 24 159 TEXAS, BEXAR SHALE MEMBER
B 6104 594 TEXAS, BIG CAESAR FIELD
B 6301 46 TEXAS, BIG WELLS FIELD
M 24 145† TEXAS, BIG WELLS FIELD, DIMMIT AND, ZAVALA COUNTIES
M 24 149 TEXAS, BIG WELLS, LO EAST, FIELD
M 24 149 TEXAS, BIG WELLS, SAN MIGUEL FIELD
B 6103 346 TEXAS, BLACKWELL FIELD
B 6401 105 TEXAS, BLACO FIELD
B 6409 1355 TEXAS, BLOCK 154, HIGH ISLAND AREA
B 6208 1382 TEXAS, BLOCK 27-L FIELD, HIGH ISLAND AREA
B 6104 613 TEXAS, BOFECILLOS MOUNTAINS
B 6104 594 TEXAS, BORREGAS FIELD
B 6007 1119 TEXAS, BOSSIER, SHALE, FORMATION
B 6407 1007 TEXAS, BRANTLEY JACKSON FIELD
B 6007 1107 TEXAS, BRIGGS FORMATION
B 6402 237 TEXAS, BROTHERTON ARCH
B 6402 237 TEXAS, BROTHERTON ARCH
B 6007 1119 TEXAS, BUCHNER FORMATION
B 6007 1119 TEXAS, BUCKNER FORMATION
B 6402 230 TEXAS, BUDA FORMATION
B 6301 35 TEXAS, BUDA LIMESTONE
B 6305 804 TEXAS, BUFFALO WALLOW FIELD
B 6205 862 TEXAS, BUFFALO WALLOW-ULTRADEEP FIELD, PRODUCTION
B 6102 297* **TEXAS. BY TITLE ONLY, COAL DEPOSITS IN, THURBER AREA**
B 6301 55 TEXAS, CADDO LIMESTONE
B 6408 1171 TEXAS, CADDO LIMESTONE
B 6206 1053 TEXAS, CALLIHAN SANDSTONE MEMBER
B 6402 222 TEXAS, CAMBRIAN
M 28 164 TEXAS, CAMBRIAN, HICKORY SANDSTONE
M 28 106 TEXAS, CAMBRIAN, HICKORY SANDSTONE

M 28 126 TEXAS, CAMBRIAN, HICKORY SANDSTONE
M 28 164 TEXAS, CAMBRIAN, RILEY FORMATION
M 28 126 TEXAS, CAMBRIAN, RILEY FORMATION
M 28 106 TEXAS, CAMBRIAN, RILEY FORMATION
B 6402 223 TEXAS, CAMBRIAN THROUGH PERMIAN
M 27 212 TEXAS, CAMBRIAN, WILBERNS FORMATION
B 6203 369 TEXAS CANYON BASIN, CALIFORNIA
S 10 209 TEXAS, CAMERON COUNTY, TEMPERATURES SHOWING GEOPRESSURE
B 6408 1171 TEXAS, CANYON REEF
B 6109 1449 TEXAS, CAPITAN REEF, NEW MEXICO AND
B 6402 168 TEXAS, CARRIZO SANDSTONE
B 6402 234 TEXAS, CARTA VALLEY FAULT ZONE
B 6402 222 TEXAS, CARTA VALLEY QUADRANGLE
B 6407 1011 TEXAS, CARTHAGE FIELD
S 5 32 TEXAS, CASTILE FORMATION
B 6206 1050 TEXAS, CATAHOULA TUFFACEOUS ROCKS, URANIUM SOURCE
B 6402 224 TEXAS, CENOZOIC
B 6205 787 TEXAS, CENTRAL BASIN PLATFORM
B 6007 1120 TEXAS, CHEROKEE TRACE FIELD
B 6112 2086 TEXAS, CHERRY CANYON FORMATION, KEROGEN
S 4 243 TEXAS, CHICO RIDGE AREA
B 6109 1449 TEXAS, CHIHUAHUA TECTONIC BELT
B 6104 613 TEXAS, CHINITI MOUNTAINS
B 6212 2444 TEXAS, CHOCOLATE BAYOU
B 6206 1053 TEXAS, CHUSA TUFF MEMBER
B 6408 1171 TEXAS, CISCO FACIES
B 6205 787 TEXAS, CISCO-CANYON SEISMIC DATA
B 6008 1256 TEXAS, CITY DIKE FIELD, GALVESTON COUNTY
B 6408 1220 TEXAS, CLAIBORNE SECTION
B 6112 2091 TEXAS, CLEARFORK OIL, KEROGEN
B 6209 1755* TEXAS COAST, ABST., SAND BODY GEOMETRY AND, RESERVOIR
B 6405 711* TEXAS COASTAL PLAIN, ABST., HYDROSTRATIGRAPHIC FRAMEWORK
B 6109 1540* TEXAS COASTAL PLAIN, ABST., LATE, QUATERNARY GEOMORPHIC
B 6109 1535* TEXAS COASTAL PLAIN, ABST., LITHOFACIES ANALYSIS AND, URANIUM
B 6303 493* TEXAS COASTAL PLAIN, ABST., PETROLOGY OF, MIOCENE, CATAHOULA
B 6108 1379* TEXAS COASTAL PLAIN, ABST., RELATIONS AMONG
B 6309 1611* TEXAS COASTAL PLAIN AND SHELF, ABST., LATE, PLEISTOCENE
B 6303 455* TEXAS COASTAL PLAIN--DEPOSITIONAL SYSTEMS, COMPOSITION,
B 6004 689* TEXAS COASTAL ZONE, ABST., ACTIVE, SURFACE, FAULTING
B 6009 1609* TEXAS COASTAL ZONE, ABST., INTERPRETATION OF UNENHANCED,
B 6402 277 TEXAS, COLORADO DELTA
B 6301 34 TEXAS, COMANCHE SERIES, NOMENCLATURE
B 6301 54 TEXAS, COMYN LIMESTONE
B 6301 52 TEXAS, CONCHO ARCH
B 6209 1759* TEXAS CONTINENTAL SHELF, ABST., DISTRIBUTION OF LIVING,
B 6303 527* TEXAS CONTINENTAL SHELF, GULF OF MEXICO-TIME-SEQUENCE STUDY,
S 7 291 TEXAS, CONTINENTAL SLOPE, GYRE BASIN
B 6206 951 TEXAS, COOK CHANNEL
B 6103 352 TEXAS, COOK DELTAS
B 6103 345 TEXAS, COOK PALEODRAINAGE SYSTEMS
B 6103 348 TEXAS, COOK POINT-BAR STRATIGRAPHIC TRAP
B 6103 346 TEXAS, COOK SANDSTONE
B 6007 1119 TEXAS, COTTON VALLEY LIMESTONE
B 6407 1008 TEXAS, COTTON VALLEY LIMESTONE
M 27 153 TEXAS, CRETACEOU, SAN CARLOS FORMATION
B 6402 225 TEXAS, CRETACEOUS
M 28 153 TEXAS, CRETACEOUS, ATCO FORMATION
M 28 84 TEXAS, CRETACEOUS, ATCO FORMATION
M 27 17 TEXAS, CRETACEOUS, ATCO FORMATION
M 27 18 TEXAS, CRETACEOUS, ATCO FORMATION
M 27 193 TEXAS, CRETACEOUS, AUSTIN GROUP
M 27 223 TEXAS, CRETACEOUS, AUSTIN GROUP
M 27 220 TEXAS, CRETACEOUS, AUSTIN GROUP
M 27 21 TEXAS, CRETACEOUS, BUDA LIMESTONE
M 27 33 TEXAS, CRETACEOUS, DEL RIO CLAY
M 27 205 TEXAS, CRETACEOUS, EDWARDS LIMESTONE
M 28 87 TEXAS, CRETACEOUS, FERRY LAKE ANHYDRITE
M 27 134 TEXAS, CRETACEOUS, FERRY LAKE ANHYDRITE
M 27 41 TEXAS, CRETACEOUS, GLEN ROSE LIMESTONE
M 27 144 TEXAS, CRETACEOUS, GLEN ROSE LIMESTONE
M 27 135 TEXAS, CRETACEOUS, GLEN ROSE LIMESTONE
M 27 147 TEXAS, CRETACEOUS, GLEN ROSE LIMESTONE
M 27 87 TEXAS, CRETACEOUS, GLEN ROSE LIMESTONE
M 27 134 TEXAS, CRETACEOUS, GLEN ROSE LIMESTONE
M 27 124 TEXAS, CRETACEOUS, GLEN ROSE LIMESTONE
S 4 262 TEXAS, CRETACEOUS LIMESTONE, BORERS
M 27 82 TEXAS, CRETACEOUS, SAN CARLOS FORMATION
M 28 27 TEXAS, CRETACEOUS, TRAVIS PEAK CONGLOMERATE
M 28 115 TEXAS, CRETACEOUS, TRAVIS PEAK FORMATION
M 28 117 TEXAS, CRETACEOUS, TRAVIS PEAK FORMATION
M 28 9 TEXAS, CRETACEOUS, TRAVIS PEAK FORMATION
M 28 78 TEXAS, CRETACEOUS, TRAVIS PEAK FORMATION
B 6404 471 TEXAS, CRETACEOUS, VADOSE PENDANT CEMENTS
M 28 124 TEXAS, CRETACEOUS, WOODBINE FORMATION
M 26 200 TEXAS, CROSBY COUNTY, PERMIAN BASIN, EASTERN-SHELF BUILDUPS,
B 6205 787 TEXAS, CROSBY COUNTY, SEISMIC DATA
B 6103 348 TEXAS, CRYSTAL FALLS LIMESTONE
B 6407 1007 TEXAS, CURRIE FIELD
B 6308 1225 TEXAS, DAMASCUS FIELD, POLK COUNTY
B 6208 1382 TEXAS, DANBURY DOME, BRAZORIA COUNTY
B 6301 38 TEXAS, DARST CREEK FIELD
B 6104 613 TEXAS, DAVIS MOUNTAINS
B 6007 1119† TEXAS, DEFINITION OF, GILMER LIMESTONE, UPPER, JURASSIC
B 6402 230 TEXAS, DEL RIO FORMATION
B 6104 527 TEXAS, DELTAIC SEDIMENTATION, LAVACA BAY
B 6103 344† TEXAS, DEPOSITIONAL ENVIRONMENTS OF A, RESERVOIR SANDSTONE IN
B 6108 1216† TEXAS, DEVELOPMENTS IN, OKLAHOMA AND, PANHANDLE OF
B 6008 1257 TEXAS, DEVILLIER FIELD, CHAMBERS COUNTY
B 6402 222 TEXAS, DEVILS LAKE QUADRANGLE
B 6402 230 TEXAS, DEVILS RIVER FORMATION
B 6206 1053 TEXAS, DEWEESVILLE SANDSTONE MEMBER
B 6301 35 TEXAS, DEWITT COUNTY, CRETACEOUS
B 6109 1449 TEXAS, DIABLO UPLIFT
B 6206 1053 TEXAS, DILWORTH URANIUM HOST ROCK
S 4 262 TEXAS, DIPLOTHYRA SMYTHI
B 6108 1225 TEXAS, DOLLAR POINT FIELD, GALVESTON COUNTY
B 6402 222 TEXAS, DRY DEVIL QUADRANGLE
B 6402 230 TEXAS, EAGLE FORD GROUP
S 3 100 TEXAS, EAGLE HILL SILICIFIED MUD VOLCANO
B 6408 1220 TEXAS, EAGLEFORD SECTION
B 6301 34 TEXAS, EAST AND NORTH-CENTRAL, NOMENCLATURE AND
S 10 262 TEXAS, EAST BERNARD FIELD, LEVERIDGE ESTATE WELL, LOG
S 10 223 TEXAS, EAST TEXAS FIELD, TRAP
M 26 150 TEXAS, EAST, BUCKNER FORMATION
M 26 150 TEXAS, EAST, EAGLE MILLS FORMATION
M 26 310 TEXAS, EAST, EDWARDS REEF
M 26 150 TEXAS, EAST, GILMER FORMATION
M 26 150 TEXAS, EAST, LOUANN-WERNER FORMATION
M 26 150 TEXAS, EAST, NORPHLET FORMATION
B 6208 1382 TEXAS, EAST OLD OCEAN FIELD, BRAZORIA COUNTY
M 26 150 TEXAS, EAST, SMACKOVER FORMATION
B 6409 1356 TEXAS, EAST SOUR LAKE FIELD, HARDIN COUNTY
M 26 150 TEXAS, EAST, STRATIGRAPHY
M 26 146 TEXAS, EAST TEXAS BASIN, BUCKNER FORMATION
M 26 151 TEXAS, EAST TEXAS BASIN, CORRELATION OF UNITS TO A CYCLE CHART
M 26 147 TEXAS, EAST TEXAS BASIN, COTTON VALLEY FORMATION
M 26 147 TEXAS, EAST TEXAS BASIN, EAGLE MILLS FORMATION
M 26 147 TEXAS, EAST TEXAS BASIN, GILMER FORMATION
M 26 147 TEXAS, EAST TEXAS BASIN, HOSSTON FORMATION
M 26 147 TEXAS, EAST TEXAS BASIN, LOUANN FORMATION
M 26 147 TEXAS, EAST TEXAS BASIN, SMACKOVER FORMATION
M 26 147 TEXAS, EAST TEXAS BASIN, WERNER FORMATION
B 6408 1220 TEXAS, EAST TEXAS FIELD, SECTION
M 26 71 TEXAS, EAST TEXAS FIELD, WOODBINE SEQUENCE, SEA LEVEL
B 6407 1006 TEXAS, EAST TEXAS SALT BASIN
M 26 70 TEXAS, EAST, WILCOX SANDSTONE
M 26 70 TEXAS, EAST, WOODBINE SANDSTONE
B 6103 346 TEXAS, EASTERN SHELF
B 6205 787 TEXAS, EASTERN SHELF AREA, SEISMIC DATA
B 6402 221 TEXAS, EDWARDS COUNTY, MAPPING
B 6301 37 TEXAS, EDWARDS LIMESTONE
B 6402 238 TEXAS, EDWARDS PLATEAU, ERTS IMAGE
B 6402 233 TEXAS, ELLENBURGER DOLOMITE
B 6308 1225 TEXAS, ELWOOD FIELD, CHAMBERS COUNTY
B 6309 1603* TEXAS EMBAYMENT--RECORD OF RIVERINE, TIDAL, AND
B 6206 951 TEXAS, EOCENE
M 28 68 TEXAS, EOCENE, PLUM BENTONITE
B 6104 527† TEXAS ESTUARY, LAVACA BAY--, TRANSGRESSIVE, DELTAIC
M 24 157 TEXAS, FAIRWAY FIELD
B 6206 1053 TEXAS, FANT TUFF MEMBER
B 6301 47 TEXAS, FASHING CARBONATE PRODUCTION
M 24 159 TEXAS, FERRY LAKE ANHYDRITE
B 6402 222 TEXAS, FLAT ROCK QUADRANGLE
B 6408 1169 TEXAS, FLEMING TONGUE
B 6103 348 TEXAS, FLIPPEN LIMESTONE
B 6212 2444 TEXAS, FORMATION WATERS
B 6309 1600* TEXAS FORMATION-EVALUATION PROBLEMS, ABST., INTERPRETIVE,
B 6402 222 TEXAS, FORT LANCASTER FORMATION
B 6402 222 TEXAS, FORT TERRETT FORMATION
B 6301 50 TEXAS, FORT WORTH BASIN

B 6308 1225 TEXAS, FOUR CORNERS FIELD, BRAZORIA COUNTY
B 6309 1511 TEXAS, FRAME MEMBER, NEW NAME
B 6402 222 TEXAS, FREDERICKSBURG GROUP
B 6408 1220 TEXAS, FREDERICKSBURG SECTION
B 6305 804 TEXAS, FREEPORT OIL CO. 1 SIDNEY FABIAN, MILLS RANCH FIELD
B 6206 1053 TEXAS, FRIO CLAY
B 6408 1169 TEXAS, FRIO SAND
S 4 249 TEXAS, GAINES BEND MOUND
B 6007 1119 TEXAS GEOLOGICAL SOCIETY, HONORARY MEMBERS, 1935-1975, SOUTH
B 6007 1117† TEXAS GEOLOGICAL SOCIETY--A FRAGMENTARY INFORMAL HISTORY,
M 24 157† TEXAS, GEOLOGY OF, FAIRWAY FIELD, EAST
B 6308 1227 TEXAS, GEOTHERMAL ACTIVITY, 1978
B 6409 1356 TEXAS, GEOTHERMAL EXPLORATION
B 6407 1004 TEXAS, GILMER FIELD
B 6007 1119 TEXAS, GILMER, INDIAN ROCK, FIELD
B 6309 1535 TEXAS, GILMER LIMESTONE GAS-CONDENSATE FIELD
B 6007 1120 TEXAS, GILMER LIMESTONE, TYPE SECTION
B 6007 1119 TEXAS, GILMER SOUTH FIELD
B 6407 1005 TEXAS, GINGER FIELD
B 6103 431 TEXAS, GLASS MOUNTAINS, PERMIAN
B 6402 232 TEXAS, GLEN ROSE FORMATION
B 6301 33 TEXAS, GLEN ROSE LIMESTONE
B 6408 1220 TEXAS, GLEN ROSE SECTION
B 6011 2058† TEXAS, GLOBOTRUNCANA FORNICATA ZONE OF UPPER, AUSTIN GROUP,
B 6402 225 TEXAS, GOBBLE GRABEN
B 6206 1053 TEXAS, GOLIAD SAND
B 6205 862 TEXAS, GOMEZ, ELLENBURGER, ULTRADEEP FIELD, PRODUCTION
B 6112 2086 TEXAS, GOMEZ FIELD, KEROGEN
B 6301 39 TEXAS, GONZALES TROUGH
M 24 149 TEXAS, GOOD LUCK FIELD
B 6305 808 TEXAS, GRANITE WASH PRODUCTION
B 6103 346 TEXAS, GROUP 4000 FIELD
S 3 103 TEXAS, GROWTH-FAULT SYSTEMS
B 6402 277 TEXAS, GUADALUPE DELTA
B 6408 1171 TEXAS, GUADALUPE FACIES
B 6007 1109 TEXAS, GULF 1 BURNER-STATE B WELL
B 6405 699* TEXAS GULF COAST, ABST., EVALUATION OF VOLUME OF ENTRAINED,
B 6109 1543* TEXAS GULF COAST, ABST., HISTORICAL, SHORELINE CHANGES AND
B 6309 1606* TEXAS GULF COAST, ABST., IMPORTANCE OF SECONDARY, LEACHED
B 6303 440* TEXAS GULF COAST, ABST., MINERALOGIC COMPOSITION AND
B 6309 1610* **TEXAS GULF COAST, ABST., ORIGINS OF PERMANENT INLETS SEPARATING,**
B 6303 417* TEXAS GULF COAST, ABST., POTENTIAL FOR PRODUCTION OF
B 6303 545* TEXAS GULF COAST, ABST., QUATERNARY, FAULT ACTIVITY IN
B 6303 488* TEXAS GULF COAST, ABST., RESERVOIR QUALITY IN, TERTIARY
B 6209 1760* TEXAS GULF COAST, ABST., SANDSTONE DISTRIBUTION AND POTENTIAL
B 6309 1595* **TEXAS GULF COAST, ABST., TESTING GEOPRESSURED GEOTHERMAL**
B 6307 1043 TEXAS, GULF COAST BASIN, SEISMIC PROFILE
B 6307 1045 TEXAS, GULF COAST BASIN, SEISMIC PROFILE, LOUISIANA AND
M 26 153 TEXAS, GULF COAST, CHRONOSTRATIGRAPHIC CORRELATION TO CYCLES
M 26 153 TEXAS, GULF COAST SEQUENCES
B 6408 1167 TEXAS, GULF COAST, TERTIARY, SANDSTONE WEDGES
M 26 162 TEXAS GULF COAST, WEST AFRICA, COMPARISON, CYCLE CHARTS
B 6009 1606* TEXAS GULF COAST--IDEAL VERSUS ACTUAL MODELS, ABST.,
B 6303 470* TEXAS GULF COASTA PLAIN--TREND, EXPLORATION, AND PRODUCTION,
B 6203 524* TEXAS GULF COASTAL PLAIN, ABST., ANALYSIS OF, GEOCHEMICAL AND
B 6206 1049† TEXAS GULF COASTAL PLAIN, GEOCHEMICAL AND SEDIMENTOLOGIC
B 6404 582† TEXAS GULF COASTAL PLAIN. DISCUSSION, GEOCHEMICAL AND,
B 6109 1536* TEXAS GULF COASTAL PLAIN--INFLUENCE OF GENETIC FACIES ON
M 26 260 TEXAS, GULF OF MEXICO, MEXICAN RIDGE, SOUTH
B 6307 1050 TEXAS, HAINESVILLE DOME, SEISMIC SECTION
S 4 249 TEXAS, HAIRPIN CURVE EXPOSURE
B 6108 1225 TEXAS, HALF MOON SHOAL FIELD, GALVESTON COUNTY
B 6208 1383 TEXAS, HANNAH ISLAND FIELD, GALVESTON COUNTY
B 6007 1119 TEXAS, HAYNESVILLE LIMESTONE
B 6402 225 TEXAS, HICKMAN GRABEN
B 6008 1256 TEXAS, HIGH ISLAND AREA EAST ADDITION SOUTH EXTENSION,
B 6008 1256 TEXAS, HIGH ISLAND AREA SOUTH, FEDERAL BLOCK A-532
B 6008 1256 TEXAS, HIGH ISLAND EAST ADDITION, SOUTH EXTENSION FEDERAL
B 6108 1225 TEXAS, HIGH ISLAND SOUTH, FEDERAL BLOCK A-492
B 6108 1225 TEXAS, HIGH ISLAND SOUTH, FEDERAL BLOCK A-560
B 6008 1256 TEXAS, HIGH ISLAND SOUTH, FEDERAL BLOCK A-561
B 6403 376 TEXAS, HOLOCENE FAN DELTA
M 28 80 TEXAS, HOLOCENE, SEDIMENTS
M 27 116 TEXAS, HOLOCENE SEDIMENTS
B 6205 787 TEXAS, HORSESHOE ATOLL
M 26 336 TEXAS, HORSESHOE ATOLL, SEISMIC SECTION, WEST
M 24 159 TEXAS, HOSSTON FORMATION
B 6206 1051 TEXAS, HOUSTON EMBAYMENT
B 6007 1107 TEXAS, HUECO LIMESTONE
B 6109 1449 TEXAS, HUECO UPLIFT
B 6305 805 TEXAS, HUNTON GEOLOGY, MILLS RANCH COMPLEX
B 6402 225 TEXAS, HUTTO GRABEN
B 6301 47 TEXAS, IMOGENE CARBONATE PRODUCTION
B 6008 1370† TEXAS IN 1975, DEVELOPMENT IN, ARKANSAS, NORTH, LOUISIANA, AND
B 6008 1340† TEXAS IN 1975, DEVELOPMENTS IN NORTH-CENTRAL
B 6008 1251† TEXAS IN 1975, DEVELOPMENTS IN, OKLAHOMA AND, PANHANDLE OF
B 6008 1365† **TEXAS IN 1975, DEVELOPMENTS IN, SOUTH**
B 6008 1254† TEXAS IN 1975, DEVELOPMENTS IN UPPER, GULF COAST OF
B 6108 1234† TEXAS IN 1976, DEVELOPMENTS IN, ARKANSAS, LOUISIANA, AND EAST
B 6108 1220† TEXAS IN 1976, DEVELOPMENTS IN NORTH-CENTRAL
B 6108 1356† TEXAS IN 1976, DEVELOPMENTS IN SOUTH
B 6108 1223† TEXAS IN 1976, DEVELOPMENTS IN UPPER, GULF COAST OF
B 6208 1487† TEXAS IN 1977, DEVELOPMENTS IN, ARKANSAS, NORTH, LOUISIANA, AND
B 6208 1377† TEXAS IN 1977, DEVELOPMENTS IN NORTH-CENTRAL
B 6208 1372† TEXAS IN 1977, DEVELOPMENTS IN, OKLAHOMA AND, PANHANDLE OF
B 6208 1511† TEXAS IN 1977, DEVELOPMENTS IN, SOUTH
B 6208 1380† TEXAS IN 1977, DEVELOPMENTS IN UPPER, GULF COAST OF
B 6308 1376† TEXAS IN 1978, DEVELOPMENT IN, ARKANSAS, NORTH, LOUISIANA, AND
B 6308 1233† TEXAS IN 1978, DEVELOPMENTS IN NORTH-CENTRAL
B 6308 1286† TEXAS IN 1978, DEVELOPMENTS IN, OKLAHOMA AND, PANHANDLE OF
B 6308 1338† TEXAS IN 1978, DEVELOPMENTS IN, SOUTH
B 6308 1223† TEXAS IN 1978, DEVELOPMENTS IN UPPER, GULF COAST OF
B 6409 1490† TEXAS IN 1979, DEVELOPMENTS IN, ARKANSAS, NORTH, LOUISIANA, AND
B 6409 1366† TEXAS IN 1979, DEVELOPMENTS IN NORTH-CENTRAL
B 6409 1387† TEXAS IN 1979, DEVELOPMENTS IN, OKLAHOMA AND, PANHANDLE OF
B 6409 1476† TEXAS IN 1979, DEVELOPMENTS IN, SOUTH
B 6409 1354† TEXAS IN 1979, DEVELOPMENTS IN UPPER, GULF COAST OF
B 6303 420* TEXAS IN ATLAS FORMAT, ABST., ENVIRONMENTAL GEOLOGY OF,
B 6303 498* TEXAS INNER SHELF, ABST., CHARACTERISTICS OF SHALLOW
B 6309 1608* TEXAS INNER SHELF, ABST., DISTRIBUTION AN SIGNIFICANCE OF COARSE
M 24 159 TEXAS, JAMES LIMESTONE MEMBER
M 24 158 TEXAS, JAMES LIMESTONE MEMBER
B 6103 355 TEXAS, JAMESON STRAWN SANDSTONE FIELD
B 6301 36 TEXAS, JOURDANTON FIELD, ATASCOSA COUNTY
B 6007 1120 TEXAS, JURASSIC, UPPER, CORRELATION CHART
B 6301 42 TEXAS, KAINER FORMATION
B 6301 35 TEXAS, KARNES TROUGH
B 6307 1029 TEXAS, KEYSTONE FIELD, SECTION
B 6402 221 TEXAS, KINNEY COUNTY, MAPPING
B 6108 1226 TEXAS, KURTEN FIELD, BRAZOS COUNTY
B 6104 528 TEXAS, LAVACA VALLEY
B 6103 432 TEXAS, LENOX HILLS FORMATION
B 6408 1171 TEXAS, LEONARD DOLOMITE
B 6408 1171 TEXAS, LEONARD REEF
B 6007 1107 TEXAS, LEONARD SERIES, UNDIVIDED
B 6105 801* **TEXAS LIGNITE**
B 6205 862 TEXAS, LOCKRIDGE ULTRADEEP FIELD, PRODUCTION
B 6307 1050 TEXAS, LOUANN SALT
B 6007 1119 TEXAS, LOUARK GROUP
B 6407 1007 TEXAS, LOUISIANA, AND ARKANSAS, BASEMENT STRUCTURES
B 6407 1011 TEXAS, LOUISIANA, AND ARKANSAS, COTTON VALLEY SANDSTONES
B 6407 1008 TEXAS, LOUISIANA, AND ARKANSAS, GRABEN-FAULT STRUCTURES
B 6407 1007 TEXAS, LOUISIANA, AND ARKANSAS, MEXIA-TALCO FAULT SYSTEM
B 6407 1007 TEXAS, LOUISIANA, AND ARKANSAS, SALT ANTICLINES
B 6109 1552* TEXAS LOWER CRETACEOUS CARBONATE PLATFORM, ABST.,
B 6301 32† TEXAS LOWER CRETACEOUS CARBONATE PLATFORM, EXPLORATION
B 6301 34 TEXAS, LULING FAULT ZONES
B 6109 1449 TEXAS, MARATHON-OUACHITA TECTONIC BELT
B 6301 55 TEXAS, MARBLE FALLS LIMESTONE
B 6408 1171 TEXAS, MARBLE FALLS LIMESTONE
B 6104 612 TEXAS, MARFA BASIN, FILL
S 7 114 TEXAS, MASS-TRANSPORT SCAR, SEISMIC EXAMPLE
B 6104 544 TEXAS, MATAGORDA BAY, SEDIMENTATION
B 6301 41 TEXAS, MAVERICK BASIN
B 6402 225 TEXAS, MCBEE GRABEN
B 6402 222 TEXAS, MCKNIGHT FORMATION
B 6208 1382 TEXAS, MIDDLE BANK REEF, BRAZOS AREA
S 7 107 TEXAS, MIDSLOPE, SEISMIC FACIES EXAMPLE
S 7 97 TEXAS, MIDSLOPE, SEISMIC FACIES, EXAMPLE
B 6408 1220 TEXAS, MIDWAY SECTION
B 6402 225 TEXAS, MIERS GRABEN
B 6402 225 TEXAS, MIERS GRABEN
B 6305 804 TEXAS, MILLS RANCH FIELD, DISCOVERED 1972
B 6401 105 TEXAS, MOBEETIE FIELD
B 6401 105 TEXAS, MORRIS-BUIE FIELD
B 6402 168 TEXAS, MT. ENTERPRISE FAULT SYSTEM
B 6408 1220 TEXAS, NAVARRO SECTION
B 6103 432 TEXAS, NEAL RANCH FORMATION
B 6103 431† TEXAS, NEAL RANCH-, LENOX HILLS CONTACT IN, GLASS MOUNTAINS,
B 6007 1120 TEXAS, NEW HOPE FIELD

Index

B 6103 350 TEXAS, NOODLE CREEK LIMESTONE
B 6104 594 TEXAS, NORTH ALAZAN FIELD
B 6208 1383 TEXAS, NORTH GUM ISLAND, JEFFERSON COUNTY
B 6407 1004† TEXAS, NORTH LOUISIANA, AND SOUTH ARKANSAS, JURASSIC, COTTON
B 6409 1356 TEXAS, NORTH PINE PRAIRIE FIELD, WALKER COUNTY
B 6108 1225 TEXAS, NORTH TEXAS CITY DIKE FIELD, GALVESTON COUNTY
B 6108 1221 TEXAS, NORTH-CENTRAL, PRODUCTION, 1975-1976
B 6208 1378 TEXAS, NORTH-CENTRAL, PRODUCTION, 1976-1977
B 6308 1234 TEXAS, NORTH-CENTRAL, PRODUCTION, 1977-1978
B 6409 1368 TEXAS, NORTH-CENTRAL, PRODUCTION, 1978-1979
B 6301 58 TEXAS, NORTH-CENTRAL, SECTIONS
B 6103 355 TEXAS, NORTHEAST BLOODWORTH CANYON SANDSTONE FIELD
B 6206 1049 TEXAS, OAKVILLE FORMATION
S 3 105 TEXAS, OAKVILLE SANDSTONE
B 6408 1171 TEXAS, OCHOA FACIES
B 6206 1067† TEXAS OFFSHORE, SERENDIPITY, GAS, SEEP AREA, SOUTH
B 6111 1960 TEXAS, OIL ACCUMULATION, EAST
M 28 105 TEXAS, OLIGOCENE, CATAHOULA TUFF
M 28 116 TEXAS, OLIGOCENE, CATAHOULA TUFF
M 28 107 TEXAS, OLIGOCENE, CATAHOULA TUFF
M 28 67 TEXAS, OLIGOCENE, FRIO FORMATION
M 28 154 TEXAS, OLIGOCENE, FRIO FORMATION
M 28 71 TEXAS, OLIGOCENE, FRIO FORMATION
M 28 20 TEXAS, OLIGOCENE, FRIO FORMATION
M 28 135 TEXAS, OLIGOCENE, FRIO FORMATION
M 28 173 TEXAS, OLIGOCENE, FRIO FORMATION
M 28 174 TEXAS, OLIGOCENE, FRIO FORMATION
M 28 175 TEXAS, OLIGOCENE, FRIO FORMATION
B 6006 907† TEXAS, OOLITE-, BAR PROGRADATION, SAN ANDRES FORMATION,
M 27 48 TEXAS, ORDOVICIAN, EL PASO GROUP
M 27 132 TEXAS, ORDOVICIAN, ELLENBURGER LIMESTONE
B 6011 2005† TEXAS, ORIGIN OF, LIMESTONE, BUTTES, GYPSUM PLAIN, CULBERSON
B 6402 222 TEXAS, OUACHITA FACIES
B 6301 54 TEXAS, OUACHITA FOLDBELT
B 6004 713* TEXAS OUTER CONTINENTAL SHELF, ABST., CORAL, REEFS AND, BANKS
B 6004 649* TEXAS OUTER CONTINENTAL SHELF, ABST., LATE, PLEISTOCENE-,
B 6004 722* TEXAS OUTER CONTINENTAL SHELF, ABST., SEDIMENT-, DISPERSAL,
S 7 243† TEXAS OUTER CONTINENTAL SHELF AND UPPER CONTINENTAL SLOPE,
B 6105 830* TEXAS OUTER CONTINENTAL SHELF, NORTHWEST GULF OF MEXICO,
B 6105 708† TEXAS OUTER CONTINENTAL SHELF, LATE HOLOCENE SEDIMENTARY
B 6210 2116 TEXAS PACIFIC OIL CO., INC., ON DEVELOPMENTS IN DUBAI, 1977, DUBAI
B 6010 1870 TEXAS PACIFIC OIL CO., ON DEVELOPMENT IN DUBAI, 1975,
B 6110 1799 TEXAS PACIFIC OIL CO., ON DEVELOPMENTS IN DUBAI, 1976, DUBAI
S 3 99 TEXAS, PALANGANA DOME
M 28 80 TEXAS, PALEOZOIC, CABALLOS NORACULITE
B 6401 89 TEXAS, PALO DURO BASIN, STRATIGRAPHIC CHART
B 6301 54 TEXAS, PALO PINTO LIMESTONE
B 6408 1220 TEXAS, PALUXY SECTION
B 6305 835* TEXAS PANHANDLE, ABST., DEPOSITIONAL FACIES AND, URANIUM
B 6109 1542* TEXAS PANHANDLE, ABST., DEPOSITIONAL FRAMEWORK OF LOWER,
B 6303 442* TEXAS PANHANDLE, ABST., FACIES PATTERNS AND DEPOSITIONAL
B 6203 499* TEXAS PANHANDLE, ABST., LACUSTRINE DEPOSITIONAL SYSTEMS OF,
B 6303 461* TEXAS PANHANDLE, ABST., LOWER, PERMIAN FACIES TRACTS AND
B 6205 856 TEXAS PANHANDLE AND OKLAHOMA, ANADARKO BASIN, BASEMENT
B 6205 855 TEXAS PANHANDLE AND OKLAHOMA, ULTRADEEP WELLS, LOCATION
B 6205 859 TEXAS PANHANDLE AND OKLAHOMA, ULTRADEEP WILDCATS, SUCCESS
B 6203 526* TEXAS PANHANDLE AND WESTERN OKLAHOMA, ABST., PERMIAN, SALT
B 6108 1218 TEXAS PANHANDLE, PRODUCTION, 1976
B 6208 1375 TEXAS PANHANDLE, PRODUCTION, 1977
B 6308 1290 TEXAS PANHANDLE, PRODUCTION, 1978
B 6409 1391 TEXAS PANHANDLE, PRODUCTION, 1979
B 6002 322* TEXAS PANHANDLES, ABST., REPORT--, OKLAHOMA AND
B 6409 1356 TEXAS, PASTOR FIELD, FORT BEND COUNTY
B 6301 40 TEXAS, PAWNEE FIELD, BEE COUNTY
B 6301 46 TEXAS, PEARSALL FIELD
B 6205 860 TEXAS, PECOS COUNTY, ULTRADEEP WELLS
B 6308 1225 TEXAS, PELICAN ISLAND FIELD, GALVESTON COUNTY
M 28 119 TEXAS, PENNSYLVANIAN, AVIS SANDSTONE
M 27 48 TEXAS, PENNSYLVANIAN, GRAFORD FORMATION
M 27 70 TEXAS, PENNSYLVANIAN, GRAHAM FORMATION
M 27 62 TEXAS, PENNSYLVANIAN, GRAHAM FORMATION
M 28 119 TEXAS, PENNSYLVANIAN, GRAHAM FORMATION
M 28 112 TEXAS, PENNSYLVANIAN, GRAY SANDSTONE
M 28 157 TEXAS, PENNSYLVANIAN, GRAY SANDSTONE
M 28 128 TEXAS, PENNSYLVANIAN, GRAY SANDSTONE
M 27 20 TEXAS, PENNSYLVANIAN, GUNSIGHT LIMESTONE
M 27 206 TEXAS, PENNSYLVANIAN, MARBLE FALLS LIMESTONE
M 27 50 TEXAS, PENNSYLVANIAN, MARBLE FALLS LIMESTONE
M 27 93 TEXAS, PENNSYLVANIAN, MARBLE FALLS LIMESTONE
M 28 143 TEXAS, PENNSYLVANIAN, MINERAL WELLS FORMATION
M 28 112 TEXAS, PENNSYLVANIAN, MINERAL WELLS FORMATION
M 28 143 TEXAS, PENNSYLVANIAN, RICKER SANDSTONE
M 28 112 TEXAS, PENNSYLVANIAN, RICKER SANDSTONE
M 28 143 TEXAS, PENNSYLVANIAN, STRAWN GROUP
M 28 128 TEXAS, PENNSYLVANIAN, STRAWN GROUP
M 28 157 TEXAS, PENNSYLVANIAN, STRAWN GROUP
M 28 112 TEXAS, PENNSYLVANIAN, STRAWN GROUP
M 27 68 TEXAS, PENNSYLVANIAN, WINCHELL LIMESTONE
B 6401 88† TEXAS, PENNSYLVANIAN-EARLY, PERMIAN DEPOSITIONAL SYSTEMS AND,
B 6206 951 TEXAS, PERMIAN
S 3 119 TEXAS, PERMIAN BASIN
B 6307 1029 TEXAS, PERMIAN BASIN, KEYSTONE FIELD, SECTION
B 6205 787 TEXAS, PERMIAN BASIN, SEISMIC DATA
M 28 11 TEXAS, PERMIAN, BELL CANYON FORMATION
M 28 22 TEXAS, PERMIAN, BELL CANYON FORMATION
M 28 113 TEXAS, PERMIAN, BELL CANYON FORMATION
M 28 176 TEXAS, PERMIAN, BELL CANYON FORMATION
M 27 69 TEXAS, PERMIAN, BELL CANYON FORMATION
M 28 122 TEXAS, PERMIAN, BELL CANYON FORMATION
M 27 32 TEXAS, PERMIAN, BONE SPRING LIMESTONE
M 27 78 TEXAS, PERMIAN, BONE SPRING LIMESTONE
M 28 4 TEXAS, PERMIAN, BRUSHY CANYON FORMATION
M 28 144 TEXAS, PERMIAN, BRUSHY CANYON FORMATION
M 28 131 TEXAS, PERMIAN, BRUSHY CANYON FORMATION
M 27 48 TEXAS, PERMIAN, GETAWAY LIMESTONE
M 27 142 TEXAS, PERMIAN, GETAWAY LIMESTONE
M 28 180 TEXAS, PERMIAN, SANDSTONE
B 6103 351 TEXAS, PERMIAN SHELF EDGE AND PROXIMAL SLOPE
B 6301 36 TEXAS, PERSON FIELD, KARNES COUNTY
M 24 159 TEXAS, PINE ISLAND SHALE MEMBER
B 6104 543 TEXAS, PLEISTOCENE BLUFFS
B 6104 528 TEXAS, PLEISTOCENE DELTAIC MUDS AND SANDS
B 6104 529 TEXAS, PLEISTOCENE LITHOLOGY
M 28 126 TEXAS, PLEISTOCENE, TERRACE DEPOSITS
M 26 114 TEXAS, POLK COUNTY, AUSTIN CHALK
M 26 114 TEXAS, POLK COUNTY, BUDA LIMESTONE
M 26 114 TEXAS, POLK COUNTY, EAGLE FORD SHALE
M 26 114 TEXAS, POLK COUNTY, WOODBINE FORMATION
S 4 241 TEXAS, POSSUM KINGDOM MEMBER
B 6402 222 TEXAS, PRECAMBRIAN
M 28 3 TEXAS, PRECAMBRIAN, LLANO SERIES
B 6301 55 TEXAS, PREGNANT SHALE
B 6309 1510† TEXAS, PROPOSED, TYPE SECTION FOR UPPER, SILURIAN AND LOWER,
M 27 120 TEXAS, QUATERNARY CALICHE
M 27 189 TEXAS, QUATERNARY, CALICHE CRUST
M 28 121 TEXAS, QUATERNARY, CALICHE CRUST
M 28 11 TEXAS, QUATERNARY GRUS
B 6408 1169 TEXAS, QUEEN CITY FACIES
B 6303 466* TEXAS RAILROAD COMMISSION FOR URANIUM MINING PURSUANT TO
S 4 255 TEXAS. RANGER QUARRY. PHYLLOID ALGAE
B 6402 225 TEXAS, RIGGS GRABEN
B 6402 222 TEXAS, RILEY FORMATION
S 3 103 TEXAS, RIO GRANDE EMBAYMENT
B 6206 1051 TEXAS, RIO GRANDE EMBAYMENT
S 10 208 TEXAS, RIO GRANDE EMBAYMENT, GEOLOGIC SECTION
M 24 159 TEXAS, RODESSA FORMATION
B 6405 772* TEXAS ROLL-TYPE URANIUM DEPOSIT, ABST., HISTORY OF,
B 6109 1449 TEXAS, ROOSEVELT UPLIFT, NEW MEXICO AND
B 6402 225 TEXAS, ROSE GRABEN
B 6308 1225 TEXAS, S. W. RUGELY FIELD, MATAGORDA COUNTY
B 6206 1051 TEXAS, SABINE ARCH
B 6408 1220 TEXAS, SABINE UPLIFT, SECTION
B 6402 222 TEXAS, SALMON PEAK FORMATION
B 6405 668* TEXAS SALT DOME AREA, ABST., PETROLEUM ACCUMULATION TRENDS
B 6408 1171 TEXAS, SAN ANDRES FACIES
B 6206 1051 TEXAS, SAN MARCOS ARCH
B 6301 46 TEXAS, SAN MARCOS ARCH
B 6301 34 TEXAS, SAN MARCOS PLATFORM, NOMENCLATURE AND CORRELATION
M 24 147 TEXAS, SAN MIGUEL SANDSTONES
B 6104 594 TEXAS, SARITA FIELD
S 4 245 TEXAS, SCENIC POINT MUD BANK
B 6007 1119 TEXAS, SCHULER FORMATION
B 6206 1068 TEXAS, SERENDIPITY GAS SEEPS, LOCATION
S 7 262 TEXAS, SHELF-EDGE OUTBUILDING
S 7 244 TEXAS, SHELF-SLOPE, BATHYMETRY
S 7 250 TEXAS, SHELF-SLOPE, BIOGENIC CALCIUM CARBONATE
S 7 260 TEXAS, SHELF-SLOPE, CHAOTIC FACIES FROM UPLIFT
S 7 258 TEXAS, SHELF-SLOPE, CHAOTIC SEISMIC FACIES
S 7 246 TEXAS, SHELF-SLOPE, COLORADO DELTA BULGE
S 7 259 TEXAS, SHELF-SLOPE, GAS MIGRATION, SEISMIC

S 7 247 TEXAS, SHELF-SLOPE, GROWTH FAULTS, DIAGRAM
S 7 257 TEXAS, SHELF-SLOPE, HISTORY, PLEISTOCENE
S 7 104 TEXAS, SHELF-SLOPE MARGIN, SEISMIC FACIES EXAMPLE
S 7 247 TEXAS, SHELF-SLOPE, PALEOCLIMATE
S 7 245 TEXAS, SHELF-SLOPE, PHYSIOGRAPHIC FEATURES
S 7 250 TEXAS, SHELF-SLOPE, SEDIMENT TEXTURE
S 7 253 TEXAS, SHELF-SLOPE, SEISMIC FACIES ANALYSES
S 7 257 TEXAS, SHELF-SLOPE, SEISMIC PROFILE
S 7 256 TEXAS, SHELF-SLOPE, SEISMIC PROFILE
S 7 255 TEXAS, SHELF-SLOPE, SEISMIC PROFILE
S 7 254 TEXAS, SHELF-SLOPE, SHELF BREAK PROFILE
S 7 246 TEXAS, SHELF-SLOPE, STRUCTURAL FEATURES
B 6208 1383 TEXAS, SHEPARD FIELD, SAN JACINTO COUNTY
B 6208 1383 TEXAS, SHIP CHANNEL FIELD, CHAMBERS COUNTY
B 6208 1382 TEXAS, SHIPWRECK FIELD, GALVESTON AREA
B 6104 613 TEXAS, SIERRA VIEJA MOUNTAINS
B 6402 222 TEXAS, SIMPSON GROUP
B 6402 236 TEXAS, SLAUGHTER BEND FAULT ZONE
B 6301 48 TEXAS, SLIGO LIMESTONE
B 6007 1119 TEXAS, SMACKOVER FORMATION
B 6206 1053 TEXAS, SOLEDAD VOLCANIC CONGLOMERATE MEMBER
B 6108 1226 TEXAS, SOUTH DUBINA FIELD, COLORADO COUNTY
B 6108 1226 TEXAS, SOUTH PEACH POINT, BRAZORIA COUNTY
B 6104 595 TEXAS, SOUTH PYOTE FIELD
M 26 153 TEXAS, SOUTH SEISMIC SEQUENCES
B 6408 1169 TEXAS, SPARTA FACIES
B 6402 169 TEXAS, SPARTA SANDSTONE
B 6104 597 TEXAS, STATE UNIVERSITY FIELD
B 6206 1049 TEXAS, STONE SWITCH FORMATION
B 6301 54 TEXAS, STRAWN CLASTIC SEQUENCE
B 6408 1171 TEXAS, STRAWN FACIES
B 6402 221† TEXAS, STRUCTURAL ANALYSIS OF, DEVILS RIVER UPLIFT--, SOUTHERN,
B 6301 38 TEXAS, STUART CITY FIELD, LASALLE COUNTY
B 6301 38 TEXAS, STUART CITY REEF TREND
B 6301 50† TEXAS, SUBSURFACE STUDY OF, ATOKA, LOWER, PENNSYLVANIAN,
B 6402 237 TEXAS, SYCAMORE CREEK MONOCLINE
B 6409 1355 TEXAS, TAVENER FIELD, FT. BEND COUNTY
B 6407 1007 TEXAS, TAWAKONI FIELD
B 6103 346 TEXAS, TAYLOR COUNTY, LOWER PERMIAN
B 6408 1220 TEXAS, TAYLOR SECTION
B 6407 1007 TEXAS, TEAS FIELD
M 27 153 TEXAS, TERTIARY SEDIMENTS
M 28 115 TEXAS, TERTIARY, VIEJA GROUP
M 28 80 TEXAS, TERTIARY, VIEJA GROUP
M 28 138 TEXAS, TERTIARY, VIEJA GROUP
M 28 3 TEXAS, TERTIARY, VIEJA GROUP
M 28 83 TEXAS, TERTIARY, VIEJA GROUP
M 28 4 TEXAS, TERTIARY, VIEJA GROUP
B 6104 612† TEXAS, THE, FILL IN, MARFA BASIN
B 6309 1513 TEXAS, THIRTYONE FORMATION, NEW NAME
B 6002 233 TEXAS, THOMPSONVILLE, NORTHEAST, FIELD
B 6206 1049 TEXAS, TORDILLA MEMBER, WHITSETT FORMATION
B 6408 1220 TEXAS, TRAVIS PEAK SECTION
B 6104 595 TEXAS, TRES ENCINOS FIELD
M 28 137 TEXAS, TRIASSIC, DOCKUM GROUP
M 28 28 TEXAS, TRIASSIC, DOCKUM GROUP
M 28 173 TEXAS, TRIASSIC, DOCKUM GROUP
B 6402 277 TEXAS, TRINITY DELTA
B 6402 222 TEXAS, TRINITY GROUP
B 6401 99 TEXAS, TUBB FORMATION
B 6402 168 TEXAS, TYLER BASIN
S 7 263 TEXAS, UPPER CONTINENTAL SLOPE, SEISMIC CHARACTER
S 7 262 TEXAS, UPPER CONTINENTAL SLOPE STABILITY
S 7 95 TEXAS, UPPER SLOPE, SEISMIC DIAGRAM
S 7 110 TEXAS, UPPER SLOPE, SEISMIC FACIES EXAMPLE
B 6309 1535 TEXAS, UPSHUR COUNTY, GULF ENERGY AND MINERALS CO., U.S.A., D. A.
B 6102 296* TEXAS URANIUM, ABST.
B 6303 472* TEXAS URANIUM DEPOSITS, ABST., GEOCHEMISTRY OF SOUTH
B 6206 1050 TEXAS, URANIUM SOURCE ROCKS
B 6402 221 TEXAS, VAL VERDE COUNTY, MAPPING
B 6104 613 TEXAS, VAN HORN MOUNTAINS
S 3 102 TEXAS, VICKSBURG FLEXURE
B 6109 1449 TEXAS, VICTORIO PEAK MARGIN, NEW MEXICO AND
B 6103 350 TEXAS, WALDRIP LIMESTONE
B 6402 236 TEXAS, WARDLAW GRABEN AREA
B 6305 804 TEXAS, WASHITA CREEK FIELD
B 6205 862 TEXAS, WASHITA CREEK ULTRADEEP FIELD, PRODUCTION
B 6402 222 TEXAS, WASHITA GROUP
B 6205 786 TEXAS, WASHITA REEF TREND, SEISMIC DATA
M 26 199 TEXAS, WASHITA REEF TREND, SHELF-MARGIN BUILDUP, SEISMIC
B 6408 1220 TEXAS, WASHITA SECTION
B 6205 786 TEXAS, WASHITA-FREDERICKSBURG RUDISTID REEFS AND BANKS
B 6205 786 TEXAS, WASHITA-SLIGO INTERVAL
B 6402 168 TEXAS, WECHES SHALE
B 6402 222 TEXAS, WEST NUECES FORMATION
B 6408 1168 TEXAS, WEST, PALEOZOIC, CARBONATE WEDGES
B 6008 1246 TEXAS, WEST, PRODUCTION, 1974-1975
B 6108 1211 TEXAS, WEST, PRODUCTION, 1975-1976
B 6208 1366 TEXAS, WEST, PRODUCTION, 1976-1977
B 6308 1238 TEXAS, WEST, PRODUCTION, 1976-1978
B 6409 1372 TEXAS, WEST, PRODUCTION, 1977-1979
B 6008 1257 TEXAS, WEST RAYWOOD FIELD, LIBERTY COUNTY
S 10 261 TEXAS, WHARTON COUNTY, LEVERIDGE ESTATE WELL, HYDROCARBON
B 6305 804 TEXAS, WHEELER COUNTY, MILLS RANCH COMPLEX
B 6402 225 TEXAS, WHITEHEAD GRABEN
B 6205 787 TEXAS, WICHITA SEISMIC DATA
M 24 151 TEXAS, WICKHAM SANDSTONES
B 6402 222 TEXAS, WILBERNS FORMATION
B 6408 1169 TEXAS, WILCOX FACIES
B 6007 1009 TEXAS, WILCOX GROUP
B 6408 1220 TEXAS, WILCOX SECTION
B 6007 1107† TEXAS, WILKIE RANCH FORMATION, NEW, PERMIAN UNIT IN, FINLAY
S 4 244 TEXAS, WINCHELL LIMESTONE, POSSUM KINGDOM LAKE AREA
B 6309 1511 TEXAS, WINK MEMBER, NEW NAME
B 6402 236 TEXAS, WITTENBURG DRAW FAULT ZONE
B 6402 236 TEXAS, WITTENBURG DRAW FAULT ZONE AREA
S 4 246 TEXAS, WOLF MOUNTAIN SHALE FORMATION
B 6408 1171 TEXAS, WOLFCAMP LIMESTONE
B 6205 787 TEXAS, WOLFCAMP SEISMIC DATA
B 6401 88 TEXAS, WOLFCAMPAIN SERIES, PALO DURO BASIN
B 6409 1558* TEXAS WOODBINE, ABST., LOUISIANA, TUSCALOOSA VERSUS SOUTHEAST
B 6408 1220 TEXAS, WOODBINE OUTCROP
B 6408 1169 TEXAS, WOODBINE SANDSTONE
B 6408 1220 TEXAS, WOODBINE SECTION
B 6402 225 TEXAS, WORD GRABEN
B 6309 1510 TEXAS, WRISTEN FORMATION, NEW NAME
B 6206 951 TEXAS, YOAKUM CHANNEL
B 6303 511* TEXAS--FACIES PATTERNS THROUGH TIME, ABST., UPPER, PERMIAN,
B 6309 1607* TEXAS--IMPLICATIONS LOW-TEMPERATURE GEOTHERMAL WATERS,
B 6303 409* TEXAS--STRATIGRAPHY. DEPOSITIONAL GEOMETRY, AND DIAGENESIS,
S 7 87† TEXAS-LOUISIANA, RECOGNITION OF, CONTINENTAL-SLOPE, SEISMIC
S 7 4 TEXAS-LOUISIANA SHELF, DESCRIBED
S 7 12 TEXAS-LOUISIANA SLOPE, DESCRIBED
S 7 56 TEXAS-LOUISIANA SLOPE, DISCUSSED, GULF OF MEXICO
S 7 314 TEXAS-LOUISIANA SLOPE, INTRASLOPE BASINS, MAP
S 7 183 TEXAS-LOUISIANA SLOPE, RELATION TO MISSISSIPPI FAN
S 7 89 TEXAS-LOUISIANA SLOPE, SEDIMENT DEPOSITION
S 7 89 TEXAS-LOUISIANA SLOPE, SEDIMENT TRANSPORT
B 6303 553* TEXAS-NEW MEXICO, ABST., DEEP-SEA DEPOSITION IN
B 6102 295* TEXAS. BY TITLE ONLY, COAL DEPOSITS IN, THURBER AREA
B 6102 294* TEXAS. BY TITLE ONLY, EVOLUTION OF PRESENT-DAY, SEISMIC
S 7 267† TEXAS: GEOTECHNICAL ASPECTS, STABILITY OF, CONTINENTAL, SHELF
B 6109 1554* TEXAS,ABST., HYDROGEN AND, OXYGEN, ISOTOPIC COMPOSITION OF,
B 6203 497* TEXAS—IMPLICATIONS OF SMECTITE-ILLITE REACTION FOR SANDSTONE
B 6405 722* TEXAS—IMPLICATTONS FOR URANIUM MINERALIZATION, ABST.,
B 6405 804* TEXAS—RELATION TO GROWTH FAULTING AND GEOTHERMAL
B 6109 1538* TEXAS—SOME UNIQUE ASPECTS OF EDWARDS-MCKNIGHT DIAGENESIS,
B 6209 1766* TEXAS—SUBSURFACE CORE STUDY, ABST., SUBMARINE-FAN DEPOSITION
B 6203 500* TEXAS—SUBSURFACE EVIDENCE, ABST., FLUID DENSITY, CURRENT
S 11 229 TEXTULARIA
S 11 268 TEXTULARIA FLINTII
M 24 235 TEXTULARIA L ZONE, LOUISIANA
S 11 229 TEXTULARIIDAE
B 6309 1478† TEXTURE AND COMPOSITION ON MECHANICAL BEHAVIOR OF
M 28 108 TEXTURE, BIBLIOGRAPHY
M 28 105 TEXTURE, IMMATURE SEDIMENTARY
M 28 106 TEXTURE, INVERSION, BIMODAL GRAIN SIZE DISTRIBUTION
M 28 106 TEXTURE, INVERSION, WELL-ROUNDED BUT POORLY SORTED
M 28 105 TEXTURE, MATURE SEDIMENTARY
M 28 107 TEXTURE, ORIENTATION STUDIES
M 28 105 TEXTURE, SUBMATURE SEDIMENTARY, (ARKOSE)
M 28 106 TEXTURE, SUPERMATURE SEDIMENTARY, ORTHOQUARTZITE
B 6004 650* TEXTURE--CONTRIBUTION FROM MAYA ARCHAEOLOGY, ABST.,
M 28 101 TEXTURES, GENERAL
B 6007 1015 TEXTURES, SEQUENCE OF, SEDIMENTARY STRUCTURES AND
M 25 431 THAI-MALAYA GEOSYNCLINE
B 6110 1841 THAILAND, 1976, MILNE, A. D., ON DEVELOPMENTS IN
B 6210 1988 THAILAND, 1977, MILNE, A. D., ON DEVELOPMENTS IN
B 6207 1223* THAILAND, ABST., EVAPORITE DEPOSITS OF NORTHEAST
B 6207 1207* THAILAND, ABST., NATURAL, GAS DEPOSITS IN, GULF OF

B 6207 1240* THAILAND AND RELATED MINING PROBLEMS, ABST., OFFSHORE, TIN
M 25 252 THAILAND, ANDAMAN BASIN
M 25 252 THAILAND, CHAO PHRAYA DEPRESSION
M 25 252 THAILAND, CHUMPHON BASIN, GULF OF
B 6310 1827 THAILAND, DEVELOPMENTS, 1978
B 6411 1894 THAILAND, DEVELOPMENTS, 1979
M 25 431 THAILAND, FLUORITE DEPOSITS, OCCURRENCE
M 25 432 THAILAND, FLUORITE, GEOPHYSICAL EXPLORATION
M 25 432 THAILAND, FLUORITE, MINING AND PROCESSING
M 25 432 THAILAND, FLUORITE OF CAVITY-FILLING TYPE
M 25 431 THAILAND, FLUORITE OF FISSURE-VEIN TYPE
M 25 431 THAILAND, FLUORITE OF PEGMATITE TYPE
M 25 432 THAILAND, FLUORITE OF REPLACEMENT TYPE
M 25 432 THAILAND, FLUORITE, RESERVES
M 25 431 THAILAND, FLUORITE, USE FOR GEMSTONES
M 25 67 THAILAND, GEOLOGIC MAPPING
M 25 431 THAILAND, GEOLOGY
M 25 251† **THAILAND, GULF OF THAILAND, AND ANDAMAN SEA--SUMMARY,**
M 25 252 THAILAND, KHORAT PLATEAU
M 25 252 THAILAND, KLONG MARUI FAULT
M 25 252 THAILAND, KO KRA RIDGE
M 25 252 THAILAND, KWAE NOI FAULT
M 25 252 THAILAND, MALAY BASIN
M 25 252 THAILAND, NARATHIWAT RIDGE
M 25 252 THAILAND, PATTANI TROUGH
M 25 252 THAILAND, PHUKET RIDGE
M 22 252 THAILAND, RANONG FAULT
M 25 252 THAILAND, SATUN RIDGE
M 25 16 THAILAND, SEABED JURISDICTION, GULF OF
M 25 252 THAILAND, SONGKHLA BASIN
M 25 252 THAILAND, STRATIGRAPHIC SECTION, GULF OF
M 25 431† THAILAND--SUMMARY, FLUORITE RESOURCES OF
S 4 18 THALASSIA MEADOWS, PUERTO RICO, HALIMEDA-
B 6007 1055 THALASSIA ZONE, PANAMA, GALETA REEF
B 6409 1501 THAMES, D. B., JR.,, DEVELOPMENTS IN NORTHERN ROCKIES IN 1979
B 6207 1178 THAMNOSIA DEPRESSA ZONE, NEVADA AND, UTAH
M 30 67 THANETIAN, FRIGG FIELD, PALEOGEOGRAPHY, MONTIAN
S 6 220 THANETIAN STAGE DEFINED, EUROPE, PALEOCENE
B 6211 2353 THAYER, THOMAS P., MESOZOIC PALEOGEOGRAPHY AND
B 6207 1172 THAYNES FORMATION, NEVADA AND UTAH
B 6406 960* THE ALSANDS PROJECT—A CHALLENGE IN PETROLEUM AND MINING
M 30 117† THE BEATRICE FIELD, INNER MORAY FIRTH, U.K. NORTH SEA
S 6 355† THE CAMBRIAN SYSTEM
M 29 151† THE CRUSTAL STRUCTURE AND EVOLUTION OF THE AREA UNDERLYING
M 29 453† THE EFFECT OF EUSTATIC SEA LEVEL CHANGES ON STRATIGRAPHIC
M 29 245† THE EVOLUTION OF STRUCTURAL HIGHS BORDERING MAJOR FOREARC
B 6104 612† THE FILL IN MARFA BASIN, TEXAS
S 10 121† THE GENERALIZED HYDRAULIC THEORY OF PETROLEUM MIGRATION
M 25 158 THE GEYSERS FIELD, GEOTHERMAL, CALIFORNIA
B 6211 2356* THE GUS-BUS, ABST.
M 25 4† THE MACROPROBLEM OF MODERNIZATION
S 6 57† THE MAGNETIC POLARITY TIME SCALE--PROSPECTS AND POSSIBILITIES
M 29 403† THE MARGINS OF THE GULF OF MEXICO
S 6 327† THE MISSISSIPPIAN-PENNSYLVANIAN BOUNDARY
M 30 329† THE NAMORADO OIL FIELD--A MAJOR OIL DISCOVERY IN THE CAMPOS
S 10 1† THE NATURE OF SHALES AND THE DYNAMICS OF HYDROCARBON
B 6009 1617* THE OIL INDUSTRY AND COASTAL ZON MANAGEMENT IN LOUISIANA,
B 6211 2219 THE ROCKS MEMBER, CALIFORNIA
S 6 341† THE SILURIAN SYSTEM
M 29 373† THE SULU SEA: A MARGINAL BASIN IN SOUTHEAST ASIA
B 6004 687* THE VIENNA WOODS--PALEOENVIRONMENTS AND TRACE FOSSILS OF
B 6203 563* THEBES FORMATION, EOCENE, OF EASTERN EGYPT—RESPONSE TO
B 6309 1579* THEORETICAL AND EXPERIMENTAL ANALYSES OF HYDRAULIC
B 6209 1579 THEORY OF DIAPIRISM VERSUS BUOYANCY THEORY
B 6406 918 THEORY OF LOPATINS METHOD
B 6405 707 THERIOT, J. C., DIPMETER VALIDITY IN DEVIATED BORE HOLES, ABST.
B 6303 504* THERMA MATURATION OF LOWER PALEOZOIC SHALES IN NORTHERN
B 6204 687† THERMAL ALTERATION OF KEROGEN FROM DIFFERENT SEDIMENTS,
B 6303 544* THERMAL ALTERATION OF ORGANIC MATERIAL, ABST., HYDROCARBON
M 25 164 THERMAL AREAS, NICARAGUA
M 25 138 THERMAL AREAS, PROSPECTIVE, JAVA
B 6412 2179† THERMAL CAUSE OF ABNORMALLY HIGH PORE PRESSURES IN SHALES,
B 6304 592 THERMAL CONDUCTIVITY OF ROCKS AND FLUIDS
B 6405 759* THERMAL CONTRACTION AND PETROLEUM MATURATION IN MICHIGAN
B 6312 2216 THERMAL CONTRACTIONAL FRACTURES
M 29 432 THERMAL CRACKING, SOURCE BEDS, PETROLEUM
B 6002 236 THERMAL DESTRUCTION OF HYDROCARBONS
B 6112 2148 THERMAL DESTRUCTION OF HYDROCARBONS
B 6405 726* THERMAL DISTILLATION AND PYROLYSIS, ABST., PETROLEUM,
B 6212 2510* THERMAL ENERGY, ABST., PROGRESS AND PROSPECTS IN, OCEAN
B 6405 670* **THERMAL EVOLUTION OF SEDIMENTARY BASINS ALONG ATLANTIC**
S 10 12 THERMAL EXPANSION, EXPULSION MECHANISMS
B 6312 2211 THERMAL GRADIENTS, CONTINENTAL, SLOPES AND, RISES
B 6303 540* THERMAL HISTORY, ABST., LIGHT HYDROCARBONS OF PETROLEUM,
B 6405 705* THERMAL HISTORY OF SOUTHERN OKLAHOMA AULACOGEN—
B 6412 2179 THERMAL HYPOTHESIS FOR ABNORMALLY HIGH PORE PRESSURES IN
B 6405 738* THERMAL MATURATION HISTORY, SOUTHERN ONTARIO, ABST.,
B 6312 2128† THERMAL MATURATION INDICATORS, COMPILATION AND CORRELATION
B 6312 2139 THERMAL MATURATION INDICATORS, GLOSSARY
B 6303 547* THERMAL MATURATION ON CHEMICAL COMPOSITION OF GULF COAS
B 6008 1392* THERMAL MATURITY, AND MIGRATION, ABST., VARIATIONS IN
M 29 437 THERMAL MATURITY, DIVERGENT CONTINENTAL MARGINS
B 6303 431* THERMAL MATURITY EVALUATION OF OGADEN BASIN, ETHIOPIA, ABST.,
M 29 433 THERMAL MATURITY, GULF OF MEXICO, LOUISIANA
B 6209 1595 THERMAL MATURITY OF CONTINENTAL MARGIN SEDIMENTS
B 6302 240 THERMAL MATURITY OF KEROGENS
M 29 435 THERMAL MATURITY OF SEDIMENTS, CONVERGENT MARGINS
B 6405 702* THERMAL MATURITY PALO DURO BASIN, TEXAS, ABST., SOURCE-ROCK
M 29 432 THERMAL MATURITY, SOURCE BEDS, CONTINENTAL MARGIN
B 6406 916 THERMAL MATURITY, TIME-TEMPERATURE INDEX OF
B 6402 180 THERMAL METAMORPHISM OF ORGANIC SEDIMENTS
B 6402 186 THERMAL MODELS
M 25 152 THERMAL SPRINGS, ALASKA
S 10 9 THERMAL STABILITY OF WATERS, SHALE POROSITY
B 6004 608† **THERMAL-ANALYSIS TECHNIQUE FOR SOURCE-ROCK**
B 6108 1385* THERMAL-INFRARED SURVEYS OF URANIUM EXPLORATION IN SOUTH
B 6003 452† THERMALLY INDUCED DIAGENESIS OF ALIPHATIC FATTY ACIDS IN
B 6405 806* THERMODYNAMIC DEVELOPMENT OF A SEDIMENTARY BASIN—NEW
B 6203 573* THERMODYNAMIC DEVELOPMENT OF A SEDIMENTARY BASIN—
B 6405 736* THERMOGENIC HYDROCARBON GASES IN UNCONSOLIDATED SEAFLOOR
B 6408 1131† THERMOGENIC ORIGINS OF NATURAL GAS IN COOK INLET BASIN,
B 6405 787* THERMOLUMINESCENCE AROUND URANIUM DEPOSITS, ABST.,
B 6304 689* THERMOLUMINESCENCE OF URANIUM HOST ROCKS IN AMBROSIA LAKE
M 24 95 THERMOPOLIS SHALE, WYOMING
M 25 463 THERNEY VOLCANIC FIELD, USSR, OLGA-
B 6008 1395* THICK COAL BEDS IN WASATCH FORMATION, WESTERN POWDER RIVER
B 6303 436* THICK SEDIMENT ACCUMULATIONS BENEATH CONTINENTAL MARGIN OF
M 26 418 THICK, SEISMIC, RESOLUTION, THIN OR
B 6004 666* THICKNESS AND LITHOLOGIC PERIODICITIES IN TURBIDITE SUCCESSIONS
B 6004 554† THICKNESS OF REMOVED SEDIMENTARY ROCKS, PALEOPORE PRESSURE,
B 6305 812† THICKNESS OF REMOVED SEDIMENTARY ROCKS. PALEOPORE PRESSURE,
S 5 96 THICKNESS, OHIO, SILURIAN, LOCKPORT GROUP,
M 26 418 THICKNESS, SEISMIC RESOLUTION, CRITICAL, RESOLUTION
B 6205 845† THICKNESSES IN DEVIATED WELLS, GRAPHIC DETERMINATION OF,
B 6303 538 THIEDE, JORN, MIDDLE CRETACEOUS OXYGEN-DEFICIENT
B 6303 538 THIEDE, JORN, OCEANS AND CLIMATE DURING CENOZOIC, ABST.
B 6106 929 THIEDE, JORN, SUBSIDENCE OF ASEISMIC RIDGES—EVIDENCE FROM
M 25 163 THIGPEN, JOHN B., EXPLORATION FOR GEOTHERMAL ENERGY IN
M 26 6 THIN BEDS, SEISMIC, REFLECTIONS
M 26 418 THIN OR THICK, SEISMIC, RESOLUTION
M 26 432 THIN RESERVOIR SYSTEMS, SEISMIC, MODELS
B 6205 884* THIN SECTION, ABST., ESTIMATING, PORE AND, CEMENT VOLUMES OF,
B 6105 759* **THIN SECTION ANALYSIS OF COASTAL-MARSH SEDIMENTS AND ITS USE IN**
B 6004 702* THIN-BEDDED TURBIDITES IN MODERN SUBMARINE CANYONS AND
B 6203 363 THIRTY MILE BANK, CALIFORNIA
B 6102 152 THIRTY MILE BANK, CALIFORNIA
B 6102 238 THIRTY MILE BANK, CALIFORNIA
B 6309 1513 THIRTYONE FORMATION, NEW NAME, TEXAS
M 25 448 THOLEIITIC SUITE, JAPAN
B 6405 648 THOMAIDIS, AND MARVIN L. MATHENY, WENGERD, SHERMAN A.,
B 6307 1092 THOMAS, B. M., GEOCHEMICAL ANALYSIS OF HYDROCARBON
B 6303 539 THOMAS, B.M., GENERATION AND OCCURRENCE OF HYDROCARBONS
B 6105 835 THOMAS, G. M., COMPUTER-GENERATED BASE MAPS IN EXPLORATION
B 6204 636 THOMAS G. THOMPSON CRUISE, SEPTEMBER-OCTOBER 1974, RV
B 6303 539 THOMAS, JOHN B., APPLICATIONS OF SCANNING ELECTRON MICROSCOPY
B 6405 737 THOMAS, JOHN B., DEVELOPMENT OF SECONDARY POROSITY IN
M 30 195 THOMAS, OWEN D., EKOFISK--FIRST OF THE GIANT OIL FIELDS IN
B 6304 701 THOMAS, OWEN D., EKOFISK--FIRST OF WESTERN EUROPEAN GIANT OIL
B 6406 871 THOMAS WASHINGTON, 1976 GEOPHYSICAL INVESTIGATIONS, RV
B 6105 835 THOMASSON, M. R., U.S. AND WORLD ENERGY OUTLOOK FOR 1980S, ABST.
B 6303 539 THOMASSON, M. RAY, FORECASTING--FACT OR FOLLY, ABST.
B 6305 842 THOMASSON, M. RAY, UNITED STATES AND WORLD ENERGY OUTLOOK
B 6105 762 **THOMPSON, ALLAN M., DISCRIMINATION OF SURFICIAL MATERIALS IN**
B 6207 1237 THOMPSON, ALLAN M., SATELLITE-ASSISTED MINERAL EXPLORATION IN
B 6203 506 THOMPSON, D. D., UNDERSTANDING SEISMICALLY DERIVED IMPEDANCE
B 6304 695 THOMPSON D. T., GEOPHYSICAL EXPERIMENTS AT THE MARIANO LAK
B 6203 566 THOMPSON, GARY, DEPOSITIONAL ENVIRONMENTS OF MAXON
B 6405 742 THOMPSON, GEORGE A., DEPTH MIGRATION AND INTERPRETATION OF
B 6303 540 THOMPSON, JACK H., JR., EFFECTS OF AN OFFSHORE DRILLING MUD ON
B 6303 540 THOMPSON, K. F. M., LIGHT HYDROCARBONS OF PETROLEUM, INTERNAL

Index

B 6209 1768 THOMPSON, LAIRD B., STRATIGRAPHIC RELATIONS OF ANNONA CHALK
B 6109 1448 THOMPSON, S., III, POTENTIAL AND STRATIGRAPHY OF PEDREGOSA
M 26 49 THOMPSON, S., III, SEISMIC STRATIGRAPHY AND GLOBAL CHANGES OF
M 26 63 THOMPSON, S., III, SEISMIC STRATIGRAPHY AND GLOBAL CHANGES OF
M 26 53 THOMPSON, S., III, SEISMIC STRATIGRAPHY AND GLOBAL CHANGES OF
M 26 83 THOMPSON, S., III, SEISMIC STRATIGRAPHY AND GLOBAL CHANGES OF
B 6403 450 THOMPSON, SAM, III, OIL AND GAS EXPLORATION WELLS IN PEDREGOSA
B 6403 450 THOMPSON, SAM, III, OIL AND GAS EXPLORATION WELLS IN PEDREGOSA
B 6405 705 THOMPSON, THOMAS L., ET AL, SUBSIDENCE AND THERMAL HISTORY OF
B 6303 540 THOMPSON, THOMAS L., EXPLORATION RESEARCH ALONG
B 6009 1463 THOMPSON, THOMAS L., PLATE TECTONICS IN OIL AND GAS
M 25 177 THOMPSON, THOMAS L., PLATE TECTONICS IN PETROLEUM
B 6203 567 THOMPSON, THOMAS L., SOUTHERN OKLAHOMA OIL COUNTRY IN
B 6109 1534* THOMPSONVILLE FIELD, JIM HOGG AND WEBB COUNTIES, TEXAS, ABST.,
B 6002 233 THOMPSONVILLE, NORTHEAST, FIELD, TEXAS
B 6004 724 THOMS, R. L., HYDROLOGIC STABILIT OF NORTH LOUISIANA SALT
B 6405 792 THOMS, RICHARD E., INTERPRETATION OF FOSSI FLUVIAL BIVALVE
B 6012 2186 THOMSEN, HARRY L., GEOLOGIC ESTIMATES OF UNDISCOVERED OIL AND
B 6303 540 THOMSEN, HARRY L., OIL AND GAS RESOURCE APPRAISAL--STATE OF
B 6004 698 THOMSEN, HARRY L., PROBABILISTIC AND COMPUTER METHODOLOGIES
B 6309 1611 THOMSON, ALAN, ORIGIN OF POROSITY IN DEEP
B 6209 1769 THOMSON, ALAN, PETROGRAPHY AND DIAGENESIS OF HOSSTON
B 6211 2363 THOR, DEVIN R., UPPER MIOCENE BARRIER BEACH-SALINE LAGOON
B 6101 24 THOR FIELD, CANADIAN, ARCTIC ISLANDS
S 3 26 THORIUM, CYCLES, SECTIONS, NORTH AMERICA, URANIUM AND
S 3 10 THORIUM, FUNDAMENTAL SOURCES OF, URANIUM AND
S 3 19 THORIUM, GEOCHEMISTRY OF, URANIUM AND
S 3 12 THORIUM, LITHOPHILE ELEMENTS, URANIUM AND
S 3 79 THORIUM, OCEANIC, MIGRATION HISTORY OF, URANIUM AND
S 3 83 THORIUM, SECTION, CYCLES OF, URANIUM AND
S 3 77 THORIUM, SECTION, CYCLES OF, URANIUM AND
B 6405 789* THORIUM TO URANIUM RATIO IN GRANITIC ROCKS AND IMPLICATIONS
S 3 22 THORIUM TRANSFER TO THE CRUST, MECHANISMS OF, URANIUM AND
S 3 36 THORIUM WITHIN CRUST, MIGRATION OF, MANTLE-DERIVED, URANIUM
S 3 135 THORIUM-URANIUM MINERALS, CRYSTALLIZATION DISSEMINATION OF
S 3 135 THORIUM-URANIUM MINERALS, DEUTERIC CRYSTALLIZATION OF
B 6105 832* THORIUM—OTHER NATURAL SOURCE OF NUCLEAR FUEL, ABST.
B 6304 702 THORNE, J., MECHANISMS OF BASIN SUBSIDENCE, ABST.
B 6303 541 THORNHILL, RON R. D., DISTRIBUTION OF CLAY IN RECENT SANDS, ABST.
B 6405 748* THORNTON REEF, NORTHEASTERN ILLINOIS, ABST., POROSITY
B 6405 792 THORNTON, SCOTT E., SLOPE-CENTERED PROCESSES IN SANTA BARBARA
B 6105 835 THRASHER, L. W., ULTIMATE RECOVERY FROM KNOWN FIELDS BY
B 6004 717* THREE DIAGENETIC FEATURES FROM WHITEHALL FORMATION,
B 6102 257 THREE FORKS FORMATION, IDAHO
B 6305 823* THREE FRONTIER-TURNER DEPOCENTERS IN WYOMING PORTION OF
B 6209 1711 THREE LICK BED CORRELATION IN, APPALACHIAN BASIN
B 6209 1711 THREE LICK BED CORRELATION IN, ILLINOIS BASIN
B 6209 1705 THREE LICK BED, NEW MARKER, KENTUCKY
B 6209 1710 THREE LICK BED, OHIO
B 6203 569* THREE TYPES OF PRESSURE SOLUTION IN LIMESTONES, ABST.
B 6303 528* THREE-DIMENSIONAL ASPECTS OF BELIZE PATC REEFS, ABST.
B 6303 522* THREE-DIMENSIONAL DISCOVERY-PROCESS MODEL, ABST., PREDICTING
B 6004 670* THREE-DIMENSIONAL SEISMIC DISPLAYS AS INTERPRETATION AID, ABST.
B 6303 424* THREE-DIMENSIONAL SEISMIC INTERPRETATION METHODS, ABST.
B 6405 806* THREE-DIMENSIONAL SIMULATION OF GEOLOGIC, HYDRODYNAMIC,
S 7 305 THREE-DIMENSIONAL VIEW, ORCA BASIN
M 25 230 THRELFALL, W. R., STRUCTURAL STYLE AND HYDROCARBON OF BASS,
B 6408 1131 THRELKELD, CHARLES N., BIOGENIC AND THERMOGENIC ORIGINS OF
B 6002 245 THROATS IN DOLOMITES, PORE
B 6312 2222 THRONDSEN, TORBJORN, COMPOSITION OF ORGANIC MATTER FROM
B 6303 437* THRUST BELT, ABST., ALLUVIAL-, FAN DEPOSITION AND TECTONIC
B 6303 500* THRUST BELT, ABST., APPLICATION OF POTENTIAL FIELD DAT TO
B 6105 821* THRUST BELT, ABST., ASSESSMENT OF OIL AND GAS RESOURCES IN,
B 6108 1388* THRUST BELT, ABST., ASSESSMENT OF OIL AND GAS RESOURCES IN,
B 6405 768* THRUST BELT, ABST., HYDROCARBON-, TRAPPING STRUCTURES IN
B 6109 1559* THRUST BELT, ABST., LANDSAT—USEFUL PETROLEUM EXPLORATION
B 6405 724* THRUST BELT, ABST., OIL EXPLORATION IN SOUTHEAST, TURKEY
B 6303 484* THRUST BELT, ABST., PAINTER, RESERVOIR FIELD--GIANT IN, WYOMING
B 6405 800* THRUST BELT, ABST., SOURCE AND TIME OF GENERATION OF
B 6305 832* **THRUST BELT, ABST., WHITNEY CANYON FIELD--POTENTIAL, GAS, GIANT**
B 6405 730* THRUST BELT, ABST., WHITNEY CANYON FIELD—POTENTIAL, GAS, GIANT
B 6405 754* THRUST BELT AND ARCTIC SLOPE, ALASKA, ABST., EVOLUTION OF,
B 6405 669* THRUST BELT AND HINTERLAND, ABST., TECTONIC SIGNIFICANCE OF
B 6405 673* THRUST BELT AND ITS FORELAND--BAVARIAN ALPS, GERMANY, ABST.,
B 6405 736* THRUST BELT, AS COMPARED WITH OTHER FOLD-AND-THRUST BELT
B 6405 775* THRUST BELT BETWEEN TRENTON, GEORGIA, AND TUSCALOOSA,
B 6206 1075 THRUST BELT, NUGGET SANDSTONE, IDAHO-, WYOMING
B 6405 683* THRUST BELT OF CARPATHIANS AND ITS RELATION TO PANNONIAN AND
B 6108 1389* THRUST BELT OF WYOMING, IDAHO, AND NORTHERN
M 30 281† THRUST BELT, PAINTER RESERVOIR FIELD--GIANT IN THE, WYOMING
B 6405 638† THRUST BELT, PAINTER RESERVOIR FIELD—, GIANT IN, WYOMING
B 6206 1075† THRUST BELT, PETROLEUM, SOURCE-BED, EVALUATION OF, JURASSIC,
B 6405 639 THRUST BELT PROVINCE, NORTH AMERICA
B 6305 833* THRUST BELT--PRELIMINARY REPORT, ABST., STRUCTURAL GEOLOGY OF
B 6405 775* THRUST BELTS, ABST., CENTRAL, MEDITERRANEAN
B 6405 704* THRUST BELTS FORM, ABST., HOW DO
B 6405 684* THRUST BELTS—REVIEW, ABST., FORELAND FOLD AND
B 6405 717* THRUST BELT—POTENTIAL NEW HYDROCARBON PROVINCE, ABST.,
B 6211 2357* THRUST BELT—POTENTIAL NEW HYDROCARBON PROVINCE, ABST.,
B 6108 1385* THRUST FAULT, NORTH PARK, COLORADO, ABST., OIL POTENTIAL OF
B 6410 1576 THRUST FAULTS, VIRGINIA
B 6405 805* THRUST GEOMETRY OF NORTHERN ABSAROKA SHEET, IDAHO AND
B 6305 832* THRUST PLATE, UINTA COUNTY, WYOMING, ABST., PAINTER
B 6305 836* THRUST PLATES IN SOUTHEASTERN IDAHO, ABST., PETROLEUM
B 6307 1039 THRUST, SEISMIC PROFILE, WYOMING, HOGSBACK
B 6307 1040 THRUST, WYOMING, ABSAROKA
B 6307 1040 THRUST, WYOMING, DARBY-, PROSPECT
B 6307 1040 THRUST, WYOMING, MEADE
B 6307 1040 THRUST, WYOMING, PARIS
M 25 182 THRUST-FAULT HYPOTHESIS FOR GENERATION OF OIL AND GAS
B 6307 1038 THRUST-FOLD ASSEMBLAGES, DECOLLEMENT
B 6307 1039 THRUST-FOLD BELT, FRONTIER FORMATION, IDAHO-, WYOMING
B 6307 1039 THRUST-FOLD BELT, HILLIARD SHALE, IDAHO-, WYOMING
B 6307 1038 THRUST-FOLD BELT, MAP, IDAHO-, WYOMING
B 6307 1039 THRUST-FOLD BELT, NUGGET SANDSTONE, IDAHO-, WYOMING
B 6307 1052 THRUST-FOLD BELT, PLOESTI AREA, SALT FEATURES, ROMANIAN,
B 6307 1040 THRUST-FOLD BELT SEQUENTIAL EVOLUTION, IDAHO-, WYOMING
B 6205 893* THRUSTS ALONG WASATCH HINGE LINE IN CENTRAL UTAH, ABST.,
B 6307 1026 THRUSTS, COMPRESSIVE, FAULT BLOCKS AND, BASEMENT
B 6205 892* THRUST—ITS SIGNIFICANCE IN OIL EXPLORATION IN CENTRAL NEVADA,
B 6308 1386 THUESEN, R. A., DEVELOPMENTS IN NORTHERN ROCKIES IN 1978
B 6409 1501 THUESEN, R. A., DEVELOPMENTS IN NORTHERN ROCKIES IN 1979
B 6101 6 THULE BASIN
B 6307 1061 THUMB MOUNTAIN FORMATION, NORTHWEST TERRITORIES
B 6102 295* **THURBER AREA, TEXAS. BY TITLE ONLY, COAL DEPOSITS IN**
S 4 87 TIBER FORMATION, ALBIAN, JAMAICA
B 6406 858 TIBESTI MASSIF, AFRICA
B 6004 720* TIDAL ESTUARY, ABST., SEDIMENTARY ENVIRONMENTS OF, TIDAL, FLAT
B 6004 720* TIDAL FLAT CREATED BY SHORELINE EROSION IN TIDAL ESTUARY,
B 6004 674* TIDAL FLATS, ABST., SEDIMENTARY RECORD OF, PALEOCLIMATE IN,
B 6303 422* TIDAL FLATS OF WESTERN ANDROS, BAHAMAS, ABST., HURRICANES AND
B 6405 707* TIDAL INLET PROCESSES, ABST., EBB-TIDAL, DELTA STRATIFICATION AND
B 6405 701* TIDAL SAND BODY, ABST., SHALLOW-MARINE, TRACE ASSEMBLAGES IN A,
B 6405 696* TIDAL SAND WAVES, ABST., INTERNAL STRUCTURES OF
B 6004 712* TIDAL TERRACES OR SUBAERIAL KARST, ABST., SCALLOPED, SOLUTION
B 6204 644† TIDAL-BAR SEDIMENTATION IN SOUTHERN CUMBERLAND PLATEAU,
B 6303 509* TIDAL-CURRENT RIDGE, WILLAPA BAY, WASHINGTON, ABST., BED FORMS
B 6405 681* TIDAL-CURRENT SAND WAVES IN VINEYARD SOUND, MASSACHUSETTS,
B 6004 676* TIDAL-FLAT CARBONATE ROCKS, MIDDLE ORDOVICIAN, VIRGINIA, ABST.,
B 6405 691* TIDAL-FLAT COMPLEX, ABST., EARLY DIAGENETIC, GLAUCONITIZATION
M 26 227 TIDAL-FLAT DEPOSITS, DEPOSITIONAL SYSTEMS, FACIES ASSOCIATIONS
B 6405 801* TIDAL-FLAT MUDS, ABST., DYNAMICS OF UNVEGETATED
B 6103 373 TIDAL-FLAT, SABKHA, SANDSTONES-SILTSTONES, OKLAHOMA
B 6105 783* **TIDAL-FLAT SEDIMENTATION ON TIDE-DOMINATED SHORELINE, BAHIA DE**
B 6303 504* TIDAL-INLET SEDIMENT DISPERSAL, ABST.
B 6203 556* TIDE-DOMINATED COASTLINE OF SOUTH CAROLINA, ABST., BACK-,
B 6203 517* TIDELESS MIDDLE ORDOVICIAN SEDIMENTS, ABST.
B 6405 731* TIDES, WESTERN INTERIOR BASIN, NORTH AMERICA, ABST.,
M 24 300 TIDWELL, E. M., BIBLIOGRAPHY OF NORTH AMERICAN OIL AND GAS
B 6405 738 TIEH, THOMAS T., OPAL-CRISTOBALITE-CEMENTED SANDS IN
M 29 125† TIERRA DEL FUEGO, ARGENTINA, STRUCTURE OF, FALKLAND PLATEAU
M 29 126 TIERRA DEL FUEGO, OFFSHORE, GEOLOGY, ARGENTINA
M 25 308 TIFFIN, D. L., SEDIMENTATION AND TECTONICS OF PACIFIC
B 6207 1194 TIGER MOUNTAIN FORMATION, OREGON AND, WASHINGTON
B 6008 1390* TIGER RIDGE AND BULLHOOK GAS UNITS, ABST., PERFORMANCE REVIEW
B 6403 317 TIGER RIDGE FIELD, MONTANA
B 6407 981 TIGER RIDGE FIELD, MONTANA
B 6106 882 TIGER SCARP, BALTIMORE CANYON TROUGH AREA
B 6102 297* TIGHT SANDSTONES OF PERMIAN-PENNSYLVANIAN AGE, VAL VERDE
B 6204 669 TIGUENTOURINE, ALGERIA
M 25 154 **TILLING, ROBERT I., DEEP RESEARCH DRILLHOLE AT KILAUEA**
S 9 7 TILLITES, PALEOCLIMATE INDICATORS
M 25 441 TILLITIC ORIGIN, SOUTH KOREA, HWANGGANGRI FORMATION
B 6309 1589 TILLMAN, J. EDWARD, HYDROTHERMAL AND UPLIF HISTORIES OF
B 6004 671 TILLMAN, R. W., DELTAIC SUBENVIRONMENTS AND PALEOCURRENT
B 6105 759 TILLMAN, R. W., DIAGENESIS IN FRONTIER FORMATION, SPEARHEAD
B 6105 835 TILLMAN, R. W., FLUVIAL TO BATHYAL SANDSTONES OF CRETACEOUS
B 6203 540 TILLMAN, R. W., HARTZOG DRAW, NEW GIANT OIL FIELD, ABST.
B 6205 888 TILLMAN, R. W., HARTZOG DRAW, NEW GIANT OIL FIELD, ABST.
B 6004 727 TILLMAN, R. W., SUBTIDAL TRACE FOSSILS, MUDDY AND FRONTIER

B 6303 491 TILLMAN, ROD W., FACIES AND RESERVOIR CHARACTERISTICS OF SHELF
B 6303 541 TILLMAN, ROD W., SEDIMENTOLOGY AND SYNERGY OF DELTAIC
B 6108 1393 TILLMAN, RODERICK W., DIAGENESIS OF OFFSHORE-BAR SANDSTONES OF
B 6405 793 TILLMAN, RODERICK W., SHANNON SANDSTONE, UPPER CRETACEOUS,
B 6207 1235 TILMAN, S. M., PROBLEMS OF STRUCTURE AND DEVELOPMENT OF SHELF
B 6203 567 TILSLEY, JAMES A., ROLE OF ELECTROCHEMICAL CELLS IN FORMATION
B 6104 486 TILTED OIL-WATER CONTACTS, HYDRODYNAMICS CONNRIBUTING TO
B 6104 489 TILTED-OUT SALT-GENERATED JURASSIC PALEOSTRUCTURE TRAPS
B 6104 489 TILTED-OUT STRATIGRAPHIC TRAP
B 6104 489 TILTED-OUT TRUNCATION TRAP
M 26 256 TIMBALIER CHANNEL, SEISMIC SECTION, LOUISIANA
M 29 71 TIME AND DEPTH SECTIONS, ATLANTIC COAST, CRETACEOUS
B 6004 709* TIME AND ORIGIN OF MICHIGAN BASIN LINEAR STRUCTURES, ABST.,
B 6406 916† TIME AND TEMPERATURE IN PETROLEUM FORMATION—APPLICATION OF
S 10 57 TIME OF PRIMARY MIGRATION, MIGRATION
S 6 12 TIME, CONCEPT OF SEQUENCE
S 6 11 TIME, GEOLOGICAL
S 6 11 TIME, GEOSCIENTIST
B 6202 201 TIME, LINEAR, NUMERICAL
S 6 12 TIME, POLARITY OF SEQUENCE
B 6403 366 TIME RELATIONS BETWEEN DISCONTINUITIES OF OCEAN-FLOOR
S 6 11 TIME SCALE (1893), WILLIAMS, H. S.
B 6004 498† TIME SCALE, A, CRETACEOUS
B 6004 489† TIME SCALE, A, JURASSIC
S 6 2 TIME SCALE, LORD KELVIN
S 6 24 TIME SCALE MODEL, SYSTEMS
S 4 353† TIME SCALE, PROGRESSIVE, DIAGENETI ELIMINATION OF, STRONTIUM IN,
S 6 218 TIME SCALE, STRATOTYPES, DIAGRAM, PALEOGENE
S 6 1 TIME SCALE, THE GEOLOGIC
S 6 73† TIME SCALE--COMPUTER FILE OF CRITICAL DATES AND CONSEQUENCES
B 6012 2175* TIME SEISCROP MAP TECHNIQUES, ABST., SEISMIC DATA ANALYSIS WITH
B 6307 1129 TIME, TEMPERATURE, AND ACTIVATION ENERGIES FOR VARIOUS
M 26 15 TIME-AVERAGE RELATION, VELOCITY-, POROSITY, RELATIONS
S 6 377 TIME-ROCK COMPARISONS, DIAGRAM, PRECAMBRIAN
B 6008 1391* TIME-SEISCROP MAP TECHNIQUES, ABST., SEISMIC DATA ANALYSIS WITH
B 6009 1608* TIME-SERIES APPROACH, ABST., BEACH SEDIMENTOLOGY OF, MUSTANG
B 6107 1084 TIME-STRATIGRAPHIC UNITS, PROCEDURE IN ESTABLISHING
B 6406 921 TIME-TEMPERATURE INDEX GEOCHEMICAL DATA, CORRELATION OF
B 6406 919 TIME-TEMPERATURE INDEX OF MATURITY VERSUS VITRINITE
B 6406 916 TIME-TEMPERATURE INDEX OF THERMAL MATURITY
B 6005 884† TIME-TEMPERATURE RELATION IN OIL GENESIS. DISCUSSION AND REPLY
B 6105 834* TIME-TO-DEATH CONVERSION AND STRUCTURE MAPPING IN
B 6407 992 TIMESGADIOUINE FAULT, MOROCCO
B 6407 990 TIMESGADIOUINE FORMATION, MOROCCO
B 6407 994 TIMESGADIOUINE FORMATION, MOROCCO
B 6407 993 TIMESGADIOUINE HORST, MOROCCO
S 10 78 TIMING OF FLUID SOURCES, COMPUTER MODELING
B 6004 699* TIMING, ABST., ORIENTED, HYDROCARBON, MIGRATION--RESULT OF
B 6004 652* TIMING AND PALEOCLIMATE INDICATORS IN COLOMBIA GROUP OF NEW
B 6004 731* TIMING, CHARACTER, AND PRESERVABILIT OF SEDIMENTATION EVENTS
B 6004 688* TIMING FACTOR IN ENERGY EXPLORATION. BY TITLE ONLY
B 6004 703* TIMING IN NORTH AMERICAN TRANSGRESSION-REGRESSION CYCLE
B 6004 682* TIMING OF CARBONATE AND CLAY DIAGENESIS IN
B 6004 668* TIMING OF DISPLACEMENT ON SAN ANDREAS FAULT IN SOUTHERN
B 6004 647* TIMING OF HYDROCARBON MIGRATION, ABST., MIDDLE, GUADALUPIAN
B 6004 693* TIMING OF OROGENIC EVENTS, ALTOS CUCHUMATANES MOUNTAINS,
B 6004 645* TIMING OF TERTIARY SUBMARINE CANYONS AND MARINE CYCLES OF
B 6004 704* TIMING OF TRANSGRESSIONS AND REGRESSIONS IN CRETACEOUS AND
B 6004 685* TIMING--GEOLOGIC AN POLITICAL--IN URBAN OIL DEVELOPMENT, LOS
B 6112 2079 TIMOR, AND NORTHWESTERN SHELF OF AUSTRALIA, SECTION, FLORES
B 6112 2073 TIMOR AREA, SECTIONS
B 6112 2078 TIMOR AREA, TECTONIC EVOLUTION, TENTATIVE
B 6112 2067 TIMOR AREA, WELLS, SEISMIC CONTROL, AND CROSS-SECTION LINES
B 6406 884 TIMOR, ARU, AND SERAM TROUGHS
B 6112 2068 TIMOR ISLAND, NIKINIKI-KOLBANO AREA, SECTION
B 6112 2068 TIMOR ISLAND, SUAI AREA, SECTION
B 6112 2070 TIMOR, KOLBANO AREA, SECTION
B 6112 2076 TIMOR RIDGE
B 6212 2511* TIMOR SEA REGION, NORTHWESTERN AUSTRALIA, ABST., GEOCHEMICAL
B 6311 2021† TIMOR SEA REGION, NORTHWESTERN AUSTRALIA, GEOCHEMICAL
B 6302 251 TIMOR, STRATIGRAPHY AND STRUCTURE
B 6112 2070 TIMOR TROUGH
B 6311 2024 TIMOR TROUGH, AUSTRALIA
B 6412 2134 TIMOR TROUGH, HYDROCARBON POTENTIAL
B 6412 2121 TIMOR TROUGH, KEROGEN
B 6412 2118† TIMOR TROUGH, ORGANIC, GEOCHEMISTRY OF PLIOCENE-, PLEISTOCENE
M 29 215 TIMOR TROUGH, REFLECTION PROFILE, AUSTRALIA
M 29 209† TIMOR-ARU TROUGH SYSTEM AND AUSTRALIAN CONTINENTAL SHELF,
M 29 214 TIMOR-TANIMBAR-ARU TROUGH, CRUSTAL SECTION, AUSTRALIA
B 6105 835 TIMSON, BARRY S., PHOTOGRAPHIC MAPPING OF RECENT COASTAL
M 25 533 TIN AND ASSOCIATED DEPOSITS, ANDEAN REGION
M 25 554 TIN AND TUNGSTEN DEPOSITS, ECUADOR, SARAGURO
M 25 554 TIN AND TUNGSTEN, ECUADOR, SARAGURD DEPOSIT
M 25 558† TIN BELT--SUMMARY, DEPOSITS OF, BOLIVIAN
M 25 533 TIN, BOLIVIA
M 25 426 TIN DEPOSIT, TASMANIA, MOUNT BISCHOFF
M 25 426 TIN DEPOSIT, TASMANIA, MOUNT CLEVELAND
M 25 426 TIN DEPOSIT, TASMANIA, RENISON BELL
B 6207 1240* TIN DEPOSITS OF THAILAND AND RELATED MINING PROBLEMS, ABST.,
B 6204 682 TIN FOUYE-TABANKORT DEVONIAN OIL FIELD, ALGERIA
M 25 462 TIN, METALLOGENIC ZONATION, SINGLE-ELEMENT
M 25 518 TIN, MEXICO
M 25 487 TIN, OCCURRENCE OF
M 24 239† TINSLEY OIL FIELD, YAZOO COUNTY, MISSISSIPPI
M 28 38 TINTON SAND, NEW JERSEY, CRETACEOUS
B 6002 284 TIOGA BENTONITE, PENNSYLVANIA
B 6403 424 TIOUR EL LIL-ANNTRA, TUNISIA, DJEBEL
B 6108 1376* TIP TOP, AND HOGSBACK FIELDS, SUBLETTE COUNTY, WYOMING, ABST.,
B 6405 793 TIPNIS, R. S., EARLY PALEOZOIC CONODONT BIOSTRATIGRAPHY AND
B 6004 728 TIPPER, J. C., COMPUTERIZED MODELING IN RECONSTRUCTION O
B 6303 541 TIPPIT, PHYLLIS,, RADIOLARIA FROM OMAN MOUNTAINS, ABST.
B 6012 2190 TIPTON, ANN, FORAMINIFERAL ZONATION OF REFUGIAN STAGE,
S 11 70 TIPTON, ANN, TAXONOMY
B 6407 993 TIRKOU HORST, MOROCCO
B 6407 992 TIRKOU INLIER, MOROCCO
M 25 163 TISATE GEOTHERMAL PROSPECT, NICARAGUA, SAN JACINTO
B 6101 85 TISI NTEST, BASIN, ATLAS TREND, NFIS
B 6309 1548 TISSOT AND DIETRIC H. WELTE, MOMPER, JAMES A., REVIEW OF,
B 6303 542 TISSOT, B., PALEOENVIRONMENT AND PETROLEUM POTENTIAL OF
B 6412 2051 TISSOT, B., PALEOENVIRONMENT AND PETROLEUM POTENTIAL OF
B 6401 59 TISSOT, B., ROLE OF MINERAL MATRIX IN KEROGEN
B 6303 542 TISSUE, J. STEPHEN,, PALEOENVIRONMENT OF MIDDLE DEVONIAN
M 25 418 TITANIUM BOOM
M 29 31 TITHONIAN, BLAKE PLATEAU, TOP OF
B 6405 609 TITHONIAN, MIDDLE EAST
M 26 160 TITHONIAN-BERRIASIAN, AFRICA, WESTERN, SEQUENCES
B 6212 2443 TITRATION CURVES FOR SODIUM BICARBONATE, SODIUM ACETATE, AND
M 25 141 TIWI GEOTHERMAL FIELD, PHILIPPINES
B 6407 993 TIZI-MAACHOU GRABEN, MOROCCO
M 25 552 TOACHI DEPOSITS, LA PLATA MINE, GOLD, SILVER, AND COPPER,
M 30 127 TOARCIAN-BAJOCIAN, STRATIGRAPHY, BEATRICE FIELD
S 4 187 TOBACCO BAY, REEF, BERMUDA
B 6010 1656 TOBAGO, 1975, MINISTRY OF PETROLEUM AND MINES, ON
B 6210 2093 TOBAGO, PRODUCTION, 1976-1977, TRINIDAD AND
B 6310 1784 TOBAGO, PRODUCTION, 1977-1978, TRINIDAD AND
B 6007 1094 TOBAGO TROUGH
B 6110 1643 TOCCO, ON DEVELOPMENTS IN ITALY, 1976, BONGIORNI, D., AND R.
B 6104 594 TOCITO DOME FIELD, FOUR CORNERS
B 6104 597 TOCITO DOME FIELD, FOUR CORNERS
B 6405 793 TODD, R. G., JURASSIC SEA-LEVEL CHANGES FROM SEISMIC
B 6004 699 TODD, R. G., REGIONAL SEISMIC INTERPRETATION USING SEQUENCES
M 26 49 TODD, R. G., SEISMIC STRATIGRAPHY AND GLOBAL CHANGES OF SEA
M 26 99 TODD, R. G., SEISMIC STRATIGRAPHY AND GLOBAL CHANGES OF SEA
M 26 145 TODD, R. G., SEISMIC STRATIGRAPHY AND GLOBAL CHANGES OF SEA
B 6006 907 TODD, ROBERT G., OOLITE-BAR PROGRADATION, SAN ANDRES
B 6007 1013 TODILTO FORMATION, NEW MEXICO
B 6304 692* TODILTO LIMESTONE--SABKHA-LIKE DEPOSIT, ABST., URANIUM IN
B 6303 455* TOGO BY SLAR, ABST., A NEW LOOK AT GEOLOGY OF
B 6110 1750 TOGO, DEVELOPMENTS, 1976
B 6210 1849 TOGO, DEVELOPMENTS, 1977
B 6310 1695 TOGO, DEVELOPMENTS, 1978
B 6411 1791 TOGO, DEVELOPMENTS, 1979
B 6212 2509* TOHOKU ARC, JAPAN, ABST., CONTRAST OF HYDROCARBON POTENTIAL
B 6105 764 TOIMIL, LAWRENCE, RATES OF ICE GOUGING AND SEDIMENT REWORKING,
B 6004 712 TOIMIL, L. J., SLOPE VALLEYS AND SEDIMENTARY STRUCTURES OFF RIO
M 25 452 TOKI LIGNITE-BEARING FORMATION, URANIUM HOST ROCKS, JAPAN
M 25 436 TOLEDO COPPER OREBODIES, CEBU
B 6303 302 TOLSTOI FORMATION, ALASKA
B 6309 1523 TOLSTOI FORMTION, ALASKA PENINSULA
B 6303 516* TOMBIGBEE AND HOLLY SPRINGS NATIONAL FORESTS, MISSISSIPPI,
S 6 358 TOMMOTIAN AGE, CAMBRIAN, PRECAMBRIAN BOUNDARY
B 6105 792* TOMORROWS ENERGY—WHERE WILL IT COME FROM
B 6012 2180* TOMORROWS OIL FROM SOUTHERN CALIFORNIA BASINS, ABST.
B 6405 794 TOMPKINS, ROBERT E., CHARACTERIZATION OF REGULARLY
M 30 342 TONALA-EL BURRO FIELD, MEXICO, SALINE BASIN
B 6110 1881 TONGA, DEVELOPMENTS, 1976
B 6210 1901 TONGA, DEVELOPMENTS, 1977
B 6310 1681 TONGA, DEVELOPMENTS, 1978
B 6411 1778 TONGA, DEVELOPMENTS, 1979
S 4 291 TONGA, GALAXEA

Index

M 25 143 TONGONAN GEOTHERMAL PROSPECT, PHILIPPINES
M 28 161 TONGRIANO CONGLOMERATE, ITALY, OLIGOCENE
M 28 31 TONGRIANO FORMATION, ITALY, OLIGOCENE
B 6204 694 TONGUE OF THE OCEAN, BAHAMAS
B 6203 560* TONGUE OF THE OCEAN, BAHAMAS, ABST., INFLUENCE OF,
B 6406 853 TONGUE OF THE OCEAN, SAND LAYERS
B 6203 513* TONGUE RIVER MEMBER OF FORT UNION FORMATION, EOCENE, POWDER
B 6305 826* TONGUE RIVER MEMBER OF FORT UNION FORMATION, POWDER RIVER
B 6109 1440 TONKAWA LIMESTONE, OKLAHOMA
M 26 440 TONKAWA LIMESTONE, OKLAHOMA, ANADARKO BASIN
B 6203 386 TONNIE SYNCLINE, ALASKA
B 6208 1498 TONNSEN, J. J., OIL AND GAS DEVELOPMENTS IN NORTHERN ROCKIES IN
M 25 454 TONO AREA, URANIUM CONTENT OF GROUNDWATERS, JAPAN
M 25 452 TONO URANIUM DEPOSITS, JAPAN
S 5 130 TONOLOWAY LIMESTONE, MARYLAND, CAYUGAN GROUP, SALINA
S 5 127 TONOLOWAY LIMESTONE, WEST VIRGINIA, CAYUGAN GROUP, SALINA
B 6203 562* TONOLOWAY LIMESTONE, ABST., SCHEME FOR, MULTIVARIATE ANALYSIS
S 5 125* TONOLOWAY LIMESTONE, AND DISTRIBUTION OF EVAPORITES IN THE
S 5 126 TONOLOWAY LIMESTONE, DEPOSITIONAL ENVIRONMENT, WEST
B 6209 1610 TONOLOWAY LIMESTONE, MARYLAND
M 28 157 TONOLOWAY LIMESTONE, PENNSYLVANIA, SILURIAN
B 6211 2309 TONOLOWAY LIMESTONE, WEST VIRGINIA
M 27 72 TONOLOWAY-KEYSER LIMESTONE, PENNSYLVANIA, SILURIAN
M 27 71 TONOLOWAY-KEYSER LIMESTONE, PENNSYLVANIA, SILURIAN
M 27 139 TONOLOWAY-KEYSER LIMESTONE, PENNSYLVANIA, SILURIAN
M 27 4 TONOLOWAY-KEYSER LIMESTONE, PENNSYLVANIA, SILURIAN
M 27 164 TONOLOWAY-KEYSER LIMESTONE, PENNSYLVANIA, SILURIAN
M 27 90 TONOLOWAY-KEYSER LIMESTONE, PENNSYLVANIA, SILURIAN
M 27 94 TONOLOWAY-KEYSER LIMESTONE, PENNSYLVANIA, SILURIAN
M 27 221 TONOLOWAY-KEYSER LIMESTONE, PENNSYLVANIA, SILURIAN
M 27 67 TONOLOWAY-KEYSER LIMESTONE, PENNSYLVANIA, SILURIAN
M 27 75 TONOLOWAY-KEYSER LIMESTONE, PENNSYLVANIA, SILURIAN
B 6207 1114 TOOL ROTATION ASSOCIATED WITH BREAKOUTS, ASPECTS OF
B 6008 1391* TOOLE COUNTY, MONTANA, ABST., PALEOGEOMORPHOLOGY OF, FRED
B 6203 457 TOOLONGA CALCILUTITE, AUSTRALIA
M 26 138 TOOLS, SEA LEVEL, ANALYSIS
B 6112 2115 TOOMEY, DONALD FRANCIS, EVOLUTION OF YUCCA MOUND COMPLEX,
M 26 119 TOP-DISCORDANT, SEISMIC, REFLECTIONS, PATTERNS
B 6111 1986 TOPANGA FORMATION, CALIFORNIA
B 6203 360 TOPATOPA UPLIFT, CALIFORNIA
B 6107 1046 TOPEKA FORMATION, KANSAS
M 26 58 TOPLAP, DEPOSITIONAL SEQUENCES
B 6109 1543* TOPOGRAPHIC EXPRESSION OF OIL AND GAS FIELDS IN GULF COAST,
S 7 5 TOPOGRAPHY, OUTLINE CHART, GULF OF MEXICO, SUBMARINE
M 30 200 TOR FORMATION, NORTH SEA, EKOFISK FIELD
B 6206 1049 TORDILLA MEMBER, WHITSETT FORMATION, TEXAS
M 28 33 TOROK FORMATION, ALASKA, CRETACEOUS
B 6405 769* TOROWEAP FORMATION, ABST., PALEOGEOGRAPH OF NORTHERN,
B 6405 757* TOROWEAP FORMATIONS IN SOUTHWESTERN UTAH, ABST.,
S 3 32 TOROWEAP SANDSTONE, PARASHOUT WASH, SHIVWITS PLATEAU,
B 6203 362 TORRANCE OIL FIELD, CALIFORNIA
B 6202 255 TORRANCE-WILMINGTON ANTICLINAL STRUCTURE, CALIFORNIA
B 6206 1033 TORRES ARCH, BRAZIL
B 6104 552 TORRES ARCH, BRAZIL
S 4 291 TORRES STRAIT, GALAXEA
B 6104 547 TORRES SYNCLINE, BRAZIL
B 6206 1033 TORRES SYNCLINE, BRAZIL
B 6203 346 TORTUGAS BASIN, CALIFORNIA
B 6201 100 TOSI CHERT MEMBER, WYOMING
B 6010 1870 TOTAL ABU AL BU KHOOSH, ON DEVELOPMENTS IN ABU DHABI, 1975,
B 6110 1796 TOTAL IRAN PETROL. CO., MOBIL IRAN ONSHORE, INC., AND ATLANTIC
B 6210 2112 TOTAL IRANIAN PETROLEUM CO., AMERICAN ULTRAMAR LTD., AND
B 6302 242 TOTAL OIL SOURCE POTENTIAL, CALCULATION
S 10 59 TOTAL ORGANIC CONTENT, MIGRATION IN AQUEOUS SOLUTION
B 6205 742 TOTAL WORLD OIL PRODUCTION, 1940-2020
S 10 121 TOTH, JOZSEF, CROSS-FORMATIONAL GRAVITY-FLOW OF
B 6203 567 TOTH, JOZSEF, GRAVITY-INDUCED CROSS-FORMATIONAL WATER FLOW—
S 10 121 TOTH, JOZSEF, THE GENERALIZED HYDRAULIC THEORY OF PETROLEUM
B 6012 2191 TOTTEN, DAVID K., JR., COCCOLITHUS PELAGICUS--CLARIFICATION,
S 10 196 TOTUMO MUD VOLCANO, COLOMBIA, CARTAGENA
M 27 85 TOUCASIAID RUDISTS
B 6407 994 TOURBIAIN CONGLOMERATE MEMBER, MOROCCO
B 6407 1052† TOURELLE FORMATION, ORDOVICIAN, QUEBEC, DEPOSITIONAL
B 6407 1053 TOURELLE FORMATION, QUEBEC
B 6407 1055 TOURELLE FORMATION, SIX FACIES, QUEBEC
M 28 60 TOURMALINE, CRYSTALS, IN GRAIN MOUNT
M 28 52 TOURMALINE, CRYSTALS IN LIMESTONE
B 6303 456* TOURMALINE CRYSTALS IN PORE SPACE OF UPPER CRETACEOUS
M 28 52 TOURMALINE, DETRITAL
M 28 53 TOURMALINE, DETRITAL, FIRST-CYCLE ORIGIN
B 6403 450 TOVAR, JORGE C., OIL AND GAS EXPLORATION WELLS IN PEDREGOSA
B 6204 637 TOWNSEND CROMWELL CRUISE, MAY-JUNE 1975, NOAA F.R.S.
B 6108 1385 TOWNSEND, TIMOTHY E., THERMAL-INFRARED SURVEYS OF URANIUM
B 6405 794 TOWSE, DONALD F., EVOLUTION AND STRESS HISTORY OF
B 6004 728 TOWSE, DONALD, GEOLOGY OF SALTON SEA GEOTHERMAL FIELD, ABST.
B 6105 796 TOWSE, DONALD, HISTORY OF EXPLOITATION AT SALTON SEA GEOTHERMAL
S 7 303 TRABANT, P. K., ORCA BASIN, ANOXIC DEPRESSION ON THE
B 6303 542 TRABANT, P., SHALLOW-SUBMARINE SEISMIC STRATIGRAPHY, MISSISSIPPI
B 6405 701* TRACE ASSEMBLAGES IN A CAMBRIAN TIDAL SAND BODY, ABST.,
B 6108 1373† TRACE ELEMENTS AS PROSPECTING GUIDES IN GRANTS MINERAL BELT,
B 6203 494* TRACE ELEMENTS IN HOLOCENE FLUVIAL SANDS DERIVED FROM
B 6305 823* TRACE ELEMENTS IN SANDSTONE-TYPE URANIUM DEPOSITS--THEORY
B 6405 694* TRACE FOSSIL ASSEMBLAGES OF UPPER CRETACEOUS SAND UNITS,
B 6305 817 TRACE FOSSIL CONCEPTS, EDITED BY PAUL B. BASAN, PERKINS, BOB F.,
B 6405 733* TRACE FOSSIL MORPHOLOGY, ABST., ENVIRONMENTAL CONTROL OF
B 6305 816 TRACE FOSSILS 2, EDITED BY T. P CRIMES AND J. C. HARPER, WARME,
B 6303 423* TRACE FOSSILS, ABST., CHANGING PERSPECTIVES IN PALEOECOLOGIC USE
B 6405 690* TRACE FOSSILS, ABST., SYNONYMY, ETHOLOGY, AND ENVIRONMENTAL
B 6405 704* TRACE FOSSILS AND STAGNATION OF DEEP-SEA BASINS, ABST.
B 6405 729* TRACE FOSSILS AND STRATIGRAPHY OF DEVONIAN BLACK SHALE IN
B 6405 697* TRACE FOSSILS FOR INTERPRETATION OF TRIASSIC DEPOSITIONAL
B 6405 691* TRACE FOSSILS FROM PROTEROZOIC SEDIMENTS OF GODAVARI VALLEY,
B 6308 1424* TRACE FOSSILS HELP, ABST., NUGGET-, NAVAJO SANDSTONE
B 6303 441* TRACE FOSSILS HELP. BY TITLE ONLY, NUGGET-, NAVAJO SANDSTONE
B 6405 702* TRACE FOSSILS IN BRITISH SILESIAN, ABST., BIVALVE
B 6105 811* TRACE FOSSILS IN RELATION TO WATER DEPTH IN PERMIAN DELTA,
B 6303 476* TRACE FOSSILS IN UPPER PENNSYLVANIAN AND LOWER PERMIAN,
B 6004 727* TRACE FOSSILS, MUDDY AND FRONTIER FORMATIONS, WYOMING AND
B 6004 704* TRACE FOSSILS OF CHAGRIN FORMATION, UPPER DEVONIAN,
B 6409 1559* TRACE FOSSILS OF LARGE AGGRADING DELTA MARGIN EMBAYMENT—
B 6004 687* TRACE FOSSILS OF UPPER CRETACEOUS SIEVERING FORMATION, VIENNA,
B 6405 669* TRACE FOSSILS OF WESTERN WYOMING AND ADJACENT AREAS, ABST.,
B 6303 411* TRACE- AND BODY-FOSSIL ASSOCIATIONS, UPPER YAKATAGA
B 6405 761* TRACE-ELEMENT CONTENT OF BITUMINOUS COAL FROM APPALACHIAN
B 6303 509* TRACE-ELEMENT GEOCHEMISTRY AND DIAGENESIS IN CAPITAN REEF,
B 6303 485* TRACE-ELEMENT GEOCHEMISTRY OF DOLOMITE--STATE OF THE ART,
B 6203 554* TRACE-ELEMENT PARTITIONING IN LIMESTONES, ABST., MODEL FOR
B 6004 661* TRACE-FOSSIL ASSEMBLAGE OF EARLY CRETACEOUS SAND
B 6303 443* TRACE-FOSSIL EVIDENCE FOR DEEP-WATER SEDIMENTATIO IN
B 6405 729* TRACE-MAKERS AS HISTORIANS IN LARGE-SCALE CYCLES OF WESTERN
B 6004 668* TRACES ON AN IN MODERN PELAGIC CARBONATE DEPOSITS, WESTERN
B 6007 1079 TRACKLINE COVERAGE EAST OF LESSER ANTILLES ISLAND ARC
B 6303 485* TRACKS AND SUBSTRATE REWORKING BY TERRESTRIAL VERTEBRATES
B 6303 542 TRACY, PATRICIA A., COMPUTER APPLICATIONS OF PETROLEUM DATA
M 25 6 TRADING BAY FIELD, ALASKA
B 6408 1132 TRADING BAY OIL FIELD, ALASKA
B 6004 687* TRAILS FROM THE VIENNA WOODS--PALEOENVIRONMENTS AND TRACE
B 6311 2060 TRANQUILLE FORMATION, BRITISH COLUMBIA
S 10 231 TRANS CAUCASUS, REGIONAL GEOTHERMAL MAP, USSR
S 10 229 TRANS CAUCASUS AND BAKU, HYDROTHERMAL EFFECTS, USSR
B 6207 1239* TRANS-ALASKA PIPELINE, ABST., HARNESSING, ARCTIC, ENERGY—
B 6101 79 TRANS-ATLANTIC SIMILARITIES, CONTINENTAL DRIFT
M 25 461 TRANS-PACIFIC COPPER AND GOLD METALLOGENIC SYSTEMS
B 6102 297* TRANS-PECOS, ABST., RESIDUAL, BOUGUER, GRAVITY, ANOMALY
B 6303 469* TRANS-PECOS TEXAS, ABST., CARBONATE ROCKS OF, MONTOYA GROUP,
B 6105 839* TRANS-PECOS VOLCANIC FIELD OF TEXAS, ABST., OLIGOCENE, VOLCANIC,
B 6405 805* TRANSCONTINENTAL ARCH, ABST., MIDDLE AND UPPER, ORDOVICIAN
B 6303 550* TRANSCONTINENTAL ARCH O CRETACEOUS SEDIMENTATION AND
M 25 512 TRANSCURRENT FAULTS, LARGE, ORIGIN OF, ALASKA AND, BRITISH
M 26 357 TRANSDUCER SYSTEMS, RESOLUTION SYSTEMS, TUNED
B 6108 1394* TRANSFER-REACTION MODELS OF DIAGENETIC PROCESSES IN
B 6307 1021 TRANSFORM SET, GULF OF CALIFORNIA, PARALLEL
B 6307 1021 TRANSFORM SYSTEM, DEAD SEA, ASYMMETRIC MASTER
B 6311 2088† TRANSFORM-FAULTED CONTINENTAL RIM, SOUTHERN EXMOUTH
B 6005 879† TRANSFORMATION OF RESISTIVITY TO PSEUDOVELOCITY LOGS
B 6004 648* TRANSGRESSION, NORTHWEST FLANK OF BALTIMORE CANYON
M 26 66 TRANSGRESSION, SEA LEVEL FLUCTUATIONS
B 6004 703* TRANSGRESSION-REGRESSION CYCLE FROM CAMBRIAN TO PRESENT,
B 6004 704* TRANSGRESSIONS AND REGRESSIONS IN CRETACEOUS AND TERTIARY OF
B 6004 728* TRANSGRESSIONS AND REGRESSIONS OF DELAWARE COASTAL PLAIN,
B 6004 665* TRANSGRESSIVE AND REGRESSIVE RATES IN PENNSYLVANIAN O
B 6303 518* TRANSGRESSIVE BARRIER ISLAND, CAPE ROMAIN, SOUTH CAROLINA,
B 6312 2145† TRANSGRESSIVE BARRIER, LATERAL AND, VERTICAL, FACIES RELATIONS
B 6203 550* TRANSGRESSIVE BARRIERS OF BEAUFORT SEA COAST, ABST.
B 6104 527† TRANSGRESSIVE DELTAIC SEDIMENTATION IN CENTRAL TEXAS
B 6004 648† TRANSGRESSIVE SANDSTONE--CARBONATE SEQUENCE--HOSSTON--SLIGO
B 6004 685* TRANSGRESSIVE SEQUENCE OF SUPRATIDAL TO SUBTIDAL AND SHOAL
B 6312 2160 TRANSGRESSIVE WASHOVER BARRIERS, ORIGIN
B 6004 732* TRANSGRESSIVE-REGRESSIVE DEPOSITS OF DIFUNTA GROUP, UPPER
B 6004 672* TRANSGRESSIVE-REGRESSIVE SHELF DEPOSITION, SHUBLIK FORMATION,

B 6203 525* TRANSITION BETWEEN A SANDY AND MUDDY SHORELINE—HOLOCENE
M 29 77 TRANSITION ZONE, ATLANTIC COAST, CRUST
B 6212 2381 TRANSITIONAL LOWER DELTA PLAIN
B 6303 293 TRANSMARGINAL RIDGES, PROSPECTS
B 6203 510* TRANSPORT ACROSS EOCENE SHELF AND SLOPE IN SOUTHWESTERN
S 10 86 TRANSPORT BY BYPASSING GEOPRESSURED SHALE, MIGRATION
S 3 136 TRANSPORT, AND FIXATION OF URANIUM BY NATURAL FLUI OR
B 6004 682* TRANSPORT, LUCAYA, GRAND BAHAMAS ISLAND, ABST., REEF
B 6004 663* TRANSPORT OVER PART OF SOUTHERN CALIFORNIA BORDERLAND
S 3 54 TRANSPORTATION IN METEORIC GROUNDWATER SYSTEM
B 6004 681* TRANSPORTATIONAL AND DEPOSITIONAL HISTORY OF LAMOTTE
B 6410 1690 TRANSVAAL STROMATOLITIC LIMESTONE, AFRICA
S 6 369 TRANSVAAL SYSTEM, SOUTH AFRICA
B 6111 1980 TRANSVERSE RANGES, CALIFORNIA
B 6203 345 TRANSVERSE RANGES, CALIFORNIA
B 6405 698* TRANSVERSE RANGES, SOUTHERN CALIFORNIA, ABST., PALEOGENE
B 6408 1229 TRANSVERSE SAND WAVES AND RIPPLES
B 6303 485* TRAP, ABST., TYLER SANDSTONES, PENNSYLVANIAN, DICKINSON AREA,
B 6009 1607* TRAP, ABST., WALKER CREEK FIELD--, SMACKOVER, DIAGENETIC
B 6309 1610* TRAP, ABST., WAVELAND FIELD, UNIQUE STRUCTURAL AND,
B 6308 1428* TRAP, ABST., WAVELAND FIELD, UNIQUE STRUCTURAL AND,
B 6104 486 TRAP, ALBERTA, PEMBINA FIELD
B 6004 645* TRAP ANALYSIS BY COMPUTER, ABST., CLASTIC AND, STRATIGRAPHIC
B 6002 319* TRAP, BINDLEY FIELD, HODGEMAN COUNTY, KANSAS, ABST., GEOLOGY
B 6103 309† TRAP, BINDLEY FIELD, HODGEMAN COUNTY, KANSAS, MISSISSIPPIAN
S 10 220 TRAP, DEFINED, ANTICLINAL SUPREMACY
S 10 220 TRAP, DEFINED, DEEP-WATER DISCHARGE
S 10 218 TRAP, DEFINED, GUIDE PLANES
S 10 219 TRAP, DEFINED, LOCAL WEAKNESS
S 10 219 TRAP, DEFINED, MEMBRANES
S 10 220 TRAP, DEFINED, PRESSURE CHANGE
S 10 221 TRAP, DEFINED, SALINITY
S 10 221 TRAP, DEFINED, TEMPERATURE CHANGE
B 6104 486 TRAP, DEFINITION, DIAGENETIC
B 6408 1218 TRAP, DIAGRAM, PALEOGEOMORPHIC
B 6408 1218 TRAP, DIAGRAM, STRATIGRAPHIC
B 6408 1218 TRAP, DIAGRAM, UNCONFORMITY
B 6209 1576 TRAP EXISTENCE IN DIAPIRS
M 26 109 TRAP EXPLORATION, SEISMIC, REFLECTIONS, CHRONOSTRATIGRAPHY,
B 6405 709* TRAP FOR PETROLEUM IN WEBER FORMATION, RANGEL OIL FIELD,
B 6203 521* TRAP IN LOWER EOCENE SANDSTONES OF NORTH SEA VIKING GRABEN,
S 10 218 TRAP, FUNCTION DEFINED
S 10 223 TRAP FUNCTION, SALINITY CONCENTRATION
S 10 234 TRAP FUNCTIONS, ACCUMULATION
S 10 235 TRAP FUNCTIONS, DIFFERENTIAL ENTRAPMENT
S 10 234 TRAP FUNCTIONS, MATURATION AND RESIDENCE TIME
S 10 234 TRAP FUNCTIONS, TRAP CONTENT
M 24 276† TRAP, JAY FIELD, FLORIDA--A, JURASSIC, STRATIGRAPHIC
B 6002 319* TRAP LOCATIONS IN MISSISSIPPIAN B, STAFFORD COUNTY, KANSAS,
M 24 72† TRAP, MITSUE OIL FIELD, ALBERTA--A RICH, STRATIGRAPHIC
B 6405 778* TRAP OIL ACCUMULATIONS AS IN POWDER RIVER BASIN, WYOMING,
B 6203 516* TRAP POSSIBILITIES IN ARBUCKLE GROUP, ABST., STRATIGRAPHIC
B 6002 325* TRAP POSSIBILITIES IN ARBUCKLE GROUP. BY TITLE ONLY,
M 26 493 TRAP, QUANTIFICATION, SEISMIC, MODELING, STRATIGRAPHIC
M 26 430 TRAP, SEISMIC, MODELS, UNCONFORMITY
B 6205 883* TRAP SPRING FIELD DISCOVERY, RAILROAD VALLEY, NYE COUNTY,
B 6211 2363* TRAP SPRING FIELD, NEVADA, ABST., SEISMIC EXPLORATION LEADING TO
B 6205 884* TRAP SPRING FIELD, NYE COUNTY, NEVADA, ABST., TERTIARY,
B 6205 880* TRAP SPRING OIL FIELD, NEVADA, ABST., SEISMIC EXPLORATION
B 6205 883* TRAP SPRING OIL FIELD, NYE COUNTY, NEVADA, ABST.
B 6211 2354* TRAP SPRING OIL FIELD, NYE COUNTY, NEVADA, ABST.
B 6405 701* TRAP SPRING OIL FIELD, NYE COUNTY, NEVADA ABST.
B 6205 884* TRAP SPRING OIL FIELD, NYE COUNTY, NEVADA, ABST., GEOMORPHIC
B 6305 748 TRAP, STRUCTURAL
B 6104 489 TRAP, TILTED-OUT, STRATIGRAPHIC
B 6104 489 TRAP, TILTED-OUT, TRUNCATION
M 24 121† TRAP, UINTA BASIN, UTAH, ALTAMONT-, BLUEBELL--A MAJOR,
B 6302 152† TRAP, WESTERN CANADA, DEEP BASIN, GAS
B 6209 1570 TRAP-DOOR FAULTS ASSOCIATED WITH SALT DIAPIRS, RADIAL AND
B 6405 686* TRAPPING AND ACCRETION OF AEOLIAN SEDIMENT BY CYANOPHYTES,
S 7 227† TRAPPING CHARACTERISTICS OF TURBIDITES, OIL
B 6003 356† TRAPPING CONSEQUENCES OF SEQUENTIAL FOLDING SYNCHRONOUS
B 6211 2243† TRAPPING MECHANISM IN UPPER MIOCENE STEVENS AND OLDER
B 6107 1010† TRAPPING OF OIL AT INTRA-MANNVILLE, LOWER CRETACEOUS,
B 6008 1413* TRAPPING OF OIL AT MIDDLE MANNVILLE UNCONFORMITY IN
B 6405 768* TRAPPING STRUCTURES IN SOUTHERN CANADIAN ROCKIES SEGMENT OF
S 10 217† TRAPS, DESIGN AND FUNCTION OF, OIL AND, GAS
B 6308 1428* TRAPS, ABST., DESIGN AND FUNCTION OF, OIL AND, GAS
B 6311 2117* TRAPS, ABST., GEOCHEMICAL PROSPECTING FOR, STRATIGRAPHIC
B 6303 448* TRAPS, ABST., HIGH-RESOLUTION SEISMIC WORK AS TOOL TO LOCATE,
B 6305 840* TRAPS, ABST., HYDROCARBON SHOWS--PROXIMITY INDICATORS FOR,
B 6203 554* TRAPS, ABST., PHYSICAL AND CHEMICAL SIGNALS AROUND, OIL AND,
B 6004 718* TRAPS AN AQUIFERS IN SOUTHEAST POWDER RIVER BASIN, WYOMING,
B 6008 1409* TRAPS AND AQUIFERS IN SOUTHEAST POWDER RIVER BASIN WYOMING,
B 6203 558* TRAPS AS CLUES TO MIGRATION, ABST., DESIGN AND FUNCTION OF, OIL
B 6104 485 TRAPS AS STRATIGRAPHIC TRAPS, MISINTERPRETATION OF, DIAGENETIC
B 6002 179 TRAPS, BERING SEA SHELF
M 25 314 TRAPS, CALIFORNIA, CENTRAL, STRATIGRAPHIC
B 6206 974 TRAPS, CALIFORNIA, MONTEREY FAN, RESERVOIR BEDS AND
M 26 334 TRAPS, CARBONATE
B 6206 978 TRAPS, CHANNELS AND, SUPRAFANS AS, STRATIGRAPHIC
B 6306 876 TRAPS, CRETACEOUS, RUDIST BUILDUPS AS, STRATIGRAPHIC
B 6206 932† TRAPS, DEEP-, WATER, SANDSTONE FACIES AND, ANCIENT, SUBMARINE,
M 26 334 TRAPS, DELINEATION, HYDROCARBON
B 6405 801* TRAPS, DENVER BASIN COLORADO, ABST., WATTENBERG AND, SPINDLE
B 6205 848† TRAPS IN GATCHELL, EOCENE, SAND OF CALIFORNIA, PERMEABILITY
B 6307 1016† TRAPS IN PETROLEUM PROVINCES, STRUCTURAL STYLES, THEIR,
B 6008 1409* TRAPS IN ROCKY MOUNTAIN FORELANDS, ABST., SEISMIC EVIDENCE FOR
B 6104 485 TRAPS, MISINTERPRETATION OF, DIAGENETIC, TRAPS AS,
S 7 187 TRAPS, MISSISSIPPI FAN
M 26 329† TRAPS, SEISMIC, EXPLORATION FOR, STRATIGRAPHIC
M 26 435 TRAPS, SEISMIC, MODELS, CARBONATE, SHELF MARGIN
M 26 431 TRAPS, SEISMIC, MODELS, DELTAIC
M 26 435 TRAPS, SEISMIC, MODELS, PLATFORM
M 26 428 TRAPS, SEISMIC, MODELS, SUBSURFACE
B 6305 748 TRAPS, STRATIGRAPHIC
B 6104 486 TRAPS, TEXAS, BELL CANYON SANDSTONES
M 26 335 TRAPS, TRUNCATIONAL
B 6408 1218 TRAPS, TYPES, SUBTLE PETROLEUM
B 6204 600 TRAPS, U.S., ATLANTIC SLOPE, POTENTIAL
B 6104 483† TRAPS--EXPLORATION TARGETS, FROZEN-IN, HYDROCARBON,
B 6305 810† TRAPS. DISCUSSION AND REPLY, DEEP-, WATER, SANDSTONE FACIES AND,
B 6408 1289 TRASK, BRIAN C., ILLINOIS COAL—A MAJOR BITUMINOUS COAL
B 6303 506 TRASK, C. BRIAN, ORIGIN AND CHARACTERTICS OF
B 6109 1546 TRAUGHBER, E. BERKLEY, DIFFERENTIAL SUBSIDENCE OF
B 6409 1354 TRAUTMAN, CYNTHIA HAYNES, DEVELOPMENTS IN UPPER GULF COAST
B 6308 1223 TRAUTMAN, TIMOTHY A., DEVELOPMENTS IN UPPER GULF COAST OF
B 6208 1380 TRAUTMAN, TIMOTHY A., DEVELOPMENTS IN UPPER GULF COAST OF
B 6303 543 TRAUTWEIN, CHARLES M., ANALYTIC AND INTERPRETIVE PROCEDURES
B 6207 1236 TRAUTWEIN, CHARLES M., INTRODUCTION TO GEOLOGIC REMOTE
B 6105 836 TRAVERS, WILLIAM B., PRESSURE SHALES AND OIL-SPILL POTENTIAL ON
B 6401 128 TRAVERSE FORMATION, INDIANA
B 6406 942† TRAVERSE, WESTERN MINNESOTA, EXPOSURES OF, GREENHORN
M 25 553 TRAVERTINE MARBLE, SURVEYS, ECUADOR, CLAYS AND
B 6309 1589 TRAVIS, JACK W., PATTERNS OF SPECIES DIVERSITY IN HELDERBERG
M 28 27 TRAVIS PEAK CONGLOMERATE, TEXAS, CRETACEOUS
M 28 117 TRAVIS PEAK FORMATION, TEXAS, CRETACEOUS
M 28 78 TRAVIS PEAK FORMATION, TEXAS, CRETACEOUS
M 28 9 TRAVIS PEAK FORMATION, TEXAS, CRETACEOUS
M 28 115 TRAVIS PEAK FORMATION, TEXAS, CRETACEOUS
B 6408 1220 TRAVIS PEAK SECTION, TEXAS
B 6205 845 TRAVIS, RUSSELL B., GRAPHIC DETERMINATION OF STRATIGRAPHIC AND
M 25 331 TRAVIS, RUSSELL B., HYDROCARBON POTENTIAL OF COASTAL BASINS OF
S 6 358 TREMADOC AGE, CAMBRIAN, ORDOVICIAN BOUNDARY
B 6405 794 TREMPER, LAUREN R., APPLIED PETROPHYSICS—CASE HISTORY, ABST.
B 6207 1210* TRENCH, ABST., TECTONIC FRAMEWORK OF, MIDDLE, AMERICA
B 6403 438* TRENCH AN TRENCH-SLOPE SEDIMENTATION ALONG ACTIVE
B 6403 438* TRENCH AND TRENCH-SLOPE SEDIMENTATION ALONG ACTIVE
B 6301 13 TRENCH AXIS OR FLOOR
B 6303 291 TRENCH FILLINGS, PROSPECTS, MARGINAL-
B 6301 14 TRENCH INNER SLOPE
B 6301 11 TRENCH OUTER SLOPE
M 25 171 TRENCH, PERU-, CHILE
B 6207 1225* TRENCH SYSTEM, ABST., SEISMICITY AND STRUCTURE OF, NEW,
M 29 369 TRENCH-ARC COMPLEXES, CONVERGENCE, VENEZUELA BASIN
M 29 363† TRENCH-ARC COMPLEXES ON THE NORTHERN AND SOUTHERN MARGINS
B 6004 721* TRENCH-SLOPE MODEL, ABST., VARIATIONS IN
B 6403 302 TRENCHES INFERRED FROM BATHYMETRIC AND SEISMIC PROFILES,
M 25 410 TRENDALL, A. F., IRON-ORE DEPOSITS OF WESTERN
B 6002 322* TRENDS AND ORIGINS OF ORDOVICIAN SANDSTONES, NORTHERN
B 6401 118 TRENTON LIMESTONE, INDIANA
B 6309 1583* TRENTON LIMESTONE, NORTHWESTERN OHIO, ABST., OIL- AND,
B 6203 503* TRENTON-UTICA SEQUENCE OF NEW YORK, ABST., RECOGNITION OF,
B 6104 595 TRES ENCINOS FIELD, TEXAS
B 6309 1609 TRESSLAR, R. CHRISTOPHER, LIVING FORAMINIFERS OF WEST FLOWER
B 6011 2070 TREVES, SAMUEL B., DRILLING IN ANTARCTICA, ABST.
B 6203 552 TREWORGY, COLIN G., COAL RESOURCES OF ILLINOIS, ABST.
B 6405 795 TREWORGY, COLIN G., SEELYVILLE COAL—MAJOR UNEXPLOITED SEAM
B 6311 2116* TRI-STATE AREA, ABST., CORRELATION OF, CHEROKEE, DESMOINESIAN,
B 6407 1016 TRIASSIC, ALASKA

M	25	295	TRIASSIC, ALASKA
B	6104	608	TRIASSIC, ALASKA
B	6203	378	TRIASSIC, ALASKA
B	6204	667	TRIASSIC, ALGERIA
M	29	34	TRIASSIC, ANGULAR UNCONFORMITY, BLAKE PLATEAU
M	28	53	TRIASSIC, ARKOSE, MASSACHUSETTS
B	6201	58	TRIASSIC, AUSTRALIA
B	6307	1096	TRIASSIC, AUSTRALIA
M	25	235	TRIASSIC, AUSTRALIA
B	6204	696	TRIASSIC, BAHAMA PLATFORM
M	30	539†	TRIASSIC BASIN, ALGERIA, PETROLEUM POTENTIAL--, OUARGLA REGION
S	6	307	TRIASSIC BOUNDARY, ALTERNATE CLASSIFICATION, PERMIAN
S	6	306	TRIASSIC BOUNDARY, PERMIAN
M	28	152	TRIASSIC, BRUNSWICK FORMATION, NEW JERSEY
M	28	44	TRIASSIC, BRUNSWICK FORMATION, NEW JERSEY
M	28	121	TRIASSIC, BUNTER SANDSTONE, NORTHERN IRELAND
B	6101	14	TRIASSIC, CANADIAN, ARCTIC ISLANDS
S	9	22	TRIASSIC, CARBONATE REEF BELT DISTRIBUTION
M	28	26	TRIASSIC, CHINLE FORMATION, ARIZONA
M	28	97	TRIASSIC, CHINLE FORMATION, ARIZONA
M	28	20	TRIASSIC, CHINLE FORMATION, UTAH
B	6405	691*	TRIASSIC CONODONTS, ABST., RISE AND FALL OF
M	27	3	TRIASSIC DACHSTEIN LIMESTONE, AUSTRIA
M	27	164	TRIASSIC, DACHSTEIN LIMESTONE, AUSTRIA
M	25	442	TRIASSIC DEFORMATION, SOUTH KOREA, PRE-
M	25	442	TRIASSIC DEFORMTION, SONGRIM DISTURBANCE, SOUTH KOREA
B	6003	341†	TRIASSIC DELTAIC SEDIMENTS, SVALBARD, GROWTH, FAULTS IN UPPER
B	6004	668*	TRIASSIC DELTAIC SEDIMENTS, SVALBARD, NORWAY, ABST., GROWTH,
B	6405	697*	TRIASSIC DEPOSITIONAL ENVIRONMENTS, NORTHEAST BORDER OF
S	9	12	TRIASSIC, DIAGRAM, POLAR WANDERING
M	28	98	TRIASSIC, DOCKUM FORMATION, NEW MEXICO
M	28	173	TRIASSIC, DOCKUM GROUP, TEXAS
M	28	137	TRIASSIC, DOCKUM GROUP, TEXAS
M	28	28	TRIASSIC, DOCKUM GROUP, TEXAS
B	6305	835*	TRIASSIC, DOCKUM GROUP, TEXAS PANHANDLE, ABST., DEPOSITIONAL
S	9	23	TRIASSIC, FOLDOUT, PALEOCLIMATES
M	27	86	TRIASSIC, HALLSTATTER LIMESTONE, GERMANY
B	6205	880*	TRIASSIC HINGE LINE AND SHELF TAR SAND DEPOSITS, CENTRAL AND
B	6207	1128	TRIASSIC, IRIAN JAYA
M	26	145†	TRIASSIC, JURASSIC, AND LOWER CRETACEOUS SEISMIC SEQUENCES IN
B	6112	2086	TRIASSIC, KEROGEN, ALBERTA
M	30	529	TRIASSIC, MALOSSA FIELD, ESINO FORMATION
M	30	537	TRIASSIC, MALOSSA FIELD, HAUPT DOLOMITE FORMATION
M	30	530	TRIASSIC, MALOSSA FIELD, S. GIOVANNI BIANCO FORMATION
B	6004	671*	TRIASSIC MARINE SEDIMENTARY ROCKS, VIZCAINO PENINSULVA, BAJA
B	6405	606	TRIASSIC, MIDDLE EAST
M	28	125	TRIASSIC, MOENKOPI FORMATION, ARIZONA
B	6404	516	TRIASSIC, NEW ENGLAND CONTINENTAL MARGIN
M	28	55	TRIASSIC, NEW HAVEN ARKOSE, CONNECTICUT
M	28	164	TRIASSIC, NEW HAVEN ARKOSE, CONNECTICUT
M	28	6	TRIASSIC, NEW HAVEN ARKOSE, CONNECTICUT
M	28	67	TRIASSIC, NEW HAVEN ARKOSE, CONNECTICUT
M	28	32	TRIASSIC, NEW HAVEN ARKOSE, CONNECTICUT
M	28	5	TRIASSIC, NEW HAVEN ARKOSE, CONNECTICUT
M	28	135	TRIASSIC, NEW HAVEN ARKOSE, CONNECTICUT
M	28	124	TRIASSIC, NEW HAVEN ARKOSE, CONNECTICUT
M	28	136	TRIASSIC, NEW HAVEN ARKOSE, CONNECTICUT
B	6109	1518	TRIASSIC, NEW MEXICO
B	6007	1013	TRIASSIC, NEW MEXICO
M	28	31	TRIASSIC, NEW RED SANDSTONE, NORTHERN IRELAND
B	6203	425	TRIASSIC, NORTH, ATLANTIC
B	6108	1382*	TRIASSIC, OF EAST-CENTRAL NEW MEXICO, ABST., EVIDENCE FOR,
B	6103	465*	TRIASSIC OF PRUDHOE BAY FIELD, ABST., PERMIAN-
M	30	281	TRIASSIC, PAINTER RESERVOIR FIELD, NUGGET SANDSTONE
B	6405	678*	**TRIASSIC PALEOCALICHE IN REF BEDS OF DOLORES FORMATION,**
B	6303	421*	TRIASSIC RADIOLARIA FROM EASTERN OREGON AND QUEEN
M	24	23†	TRIASSIC RESERVOIRS OF PRUDHO BAY FIELD, NORTH SLOPE, ALASKA,
M	29	79	TRIASSIC, RIFTING HISTORY, AFRICA
M	29	79	TRIASSIC, RIFTING HISTORY, NORTH AMERICA
B	6405	719*	TRIASSIC ROCKS, CENTRAL GREAT BASIN, ABST., CONODONT-BASED
B	6407	988†	TRIASSIC ROCKS OF ARGANA VALLEY, SOUTHERN MOROCCO, AND THEIR
B	6101	58	TRIASSIC, SEDIMENTARY ROCKS, LEVANT
M	30	101	TRIASSIC, STATFJORD FIELD, STRATIGRAPHY
B	6101	12	TRIASSIC, SVERDRUP BASIN
B	6109	1542*	TRIASSIC, TEXAS PANHANDLE, ABST., DEPOSITIONAL FRAMEWORK OF
B	6303	526*	TRIASSIC TIME, ABST., REGIONAL ASPECT OF SEDIMENT DISTRIBUTION
B	6403	424	TRIASSIC, TUNISIA
M	30	55	TRIASSIC, VIKING FIELD, TIME-VELOCITY
M	24	117	TRIASSIC, WYOMING
M	25	344	TRIASSIC-JURASSIC, AMAZON BASINS
B	6305	827*	TRIASSIC-JURASSIC BASINS, EASTERN NORTH AMERICA, ABST.,
B	6101	15	TRIASSIC-JURASSIC, CANADIAN, ARCTIC ISLANDS
B	6203	537*	TRIASSIC-JURASSIC CULPEPER BASIN, VIRGINIA, ABST., TECTONIC
B	6305	846*	TRIASSIC-JURASSIC CULPEPER GROUP CONGLOMERATES OF VIRGINIA,
B	6303	520*	TRIASSIC-JURASSIC PALEOGEOGRAPHY AND ORIGIN OF GULF OF
B	6311	2103†	TRIASSIC-JURASSIC SILL, CALHOUN COUNTY, MISSISSIPPI, VARDAMAN
B	6101	13	TRIASSIC-JURASSIC, SVERDRUP BASIN
B	6101	83	TRIASSIC-LIASSIC CORRELATIONS, TENTATIVE, NORTH AMERICA AND,
B	6101	82	TRIASSIC-LIASSIC DEPOSITS, MOROCCO
B	6101	79†	TRIASSIC-LIASSIC DEPOSITS OF MOROCCO AND EASTERN NORTH
B	6004	695*	TRIBUTARY, ABST., LARGE SEDIMENT, SLUMP REOPENS UNKNOWN,
S	6	260	TRICERATOPS, NORTH AMERICA, CRETACEOUS
B	6203	568	TRICHET, J. G., ORGANIC GEOCHEMISTRY OF CENOZOIC AND RECENT
M	27	99	TRILOBITE FRAGMENTS
M	28	39	TRILOBITE, FRAGMENTS
B	6110	1591	TRINIDAD, 1976, PETROCONSULTANTS S.A., ON DEVELOPMENTS IN
B	6210	2059	TRINIDAD, 1977, TRINIDAD MINISTRY OF PETROLEUM AND MINES, ON
B	6310	1761	TRINIDAD, 1978, TRINIDAD MINISTRY OF PETROLEUM AND MINES, ON
B	6202	304	TRINIDAD AND ORINOCO DELTA, REGIONAL SETTING
B	6010	1656	TRINIDAD AND TOBAGO, 1975, MINISTRY OF PETROLEUM AND MINES, ON
B	6210	2093	TRINIDAD AND TOBAGO, PRODUCTION, 1976-1977
B	6310	1784	TRINIDAD AND TOBAGO, PRODUCTION, 1977-1978
B	6004	644*	TRINIDAD, CALIFORNIA, ABST., SEDIMENTOLOGY OF, MELANGE,
S	10	198	TRINIDAD, CHATHAM MUD ISLAND, WEST INDIES
M	25	185	TRINIDAD, CONVERGENT MARGIN
B	6202	303	TRINIDAD, CRUSE-FOREST OIL SANDSTONES
M	30	396	TRINIDAD, EASTERN OFFSHORE BASIN, STRATIGRAPHIC COLUMN
M	30	393	TRINIDAD, EASTERN OFFSHORE BASIN, STRUCTURAL BOUNDARIES
S	10	199	TRINIDAD, GRAND RAVINE FIELD, MUDFLOWS, WEST INDIES
S	10	198	TRINIDAD, LABREA FORMATION, GAS-CUT MUDFLOW, WEST INDIES
B	6205	840	TRINIDAD, MANJAK, SOLID HYDROCARBON
B	6310	1761	TRINIDAD MINISTRY OF PETROLEUM AND MINES, ON DEVELOPMENTS IN
B	6210	2059	TRINIDAD MINISTRY OF PETROLEUM AND MINES, ON DEVELOPMENTS IN
B	6009	1502†	TRINIDAD, MIOCENE, DELTAIC, OIL HABITAT
B	6202	303	TRINIDAD, MIOCENE OIL-BEARING SANDSTONES
S	10	196	TRINIDAD, MORUGA BOUFF MUD VOLCANO, WEST INDIES
M	30	387	TRINIDAD, PLIOCENE, TEAK FIELD
B	6110	1626	TRINIDAD, PRODUCTION, 1976
B	6202	303	TRINIDAD, PROTO-ORINOCO DELTA SYSTEM
B	6108	1372*	TRINIDAD SANDSTONE AND VERMEJO FORMATION, WALSENBURG AREA,
B	6007	1094	TRINIDAD SHELF, VENEZUELA-
M	30	389	TRINIDAD, TEAK FIELD, DEVELOPMENT HISTORY
M	30	388	TRINIDAD, TEAK FIELD, LOCATION MAP
M	30	389	TRINIDAD, TEAK FIELD, REGIONAL GEOLOGY
M	30	398	TRINIDAD, TEAK FIELD, RESERVOIR CHARACTERISTICS
M	30	391	TRINIDAD, TEAK FIELD, SEISMIC SECTION
M	30	393	TRINIDAD, TEAK FIELD, STRATIGRAPHY
M	30	396	TRINIDAD, TEAK FIELD, STRUCTURE
B	6303	414*	TRINIDAD, WEST INDIES, ABST., GEOLOGY AND DEVELOPMENT OF, TEAK
M	30	387†	TRINIDAD, WEST INDIES, GEOLOGY AND DEVELOPMENT OF THE, TEAK
B	6402	277	TRINITY DELTA, TEXAS
M	24	298	TRINITY GROUP, FLORIDA
B	6402	222	TRINITY GROUP, TEXAS
B	6012	2191	TRIPLEHORN, DON M., VOLCANIC-ASH PARTINGS I
S	11	233	TROCHAMMINA
S	11	233	TROCHAMMINIDAE
S	4	90	TROCHOSERIS-ACTINACIS ASSEMBLAGE, JAMAICA, PARACYCLOSERIS-
B	6004	715	TROELL, A. R., COMBINATION ENTRAPMENT I SMACKOVER FORMATION
B	6009	1614	TROELL, A. R., COMBINATION ENTRAPMENT IN SMACKOVER FORMATION
B	6109	1559	TROLLINGER, W. V., LANDSAT—USEFUL PETROLEUM EXPLORATION
S	6	101	TROMBETAS FORMATION, BRAZIL, AMAZON BASIN
S	6	101	TROMBETAS FORMATION, RUBIDIUM-STRONTIUM, BRAZIL
S	4	96	TROPHIC ANALYSIS OF ANCIENT REEF ECOSYSTEMS
B	6303	521*	TROPICAL CENOZOIC PALEO-OCEANOGRAPHY AND CORRELATED EVENTS
B	6203	488*	TROPICAL DELTA, ABST., HEAVY-MINERAL RESPONSE TO
S	4	236	TROPIOMETRA CARINATA LAMARCK, BARBADOS
S	4	233	TROPIOMETRA CARINATA LAMARCK, COLOMBIA
S	4	236	TROPIOMETRA CARINATA LAMARCK, ST. LUCIA
B	6105	817*	TROUGH TOPOGRAPHY ON SHALLOW, CURRENT-DOMINATED SHELF,
M	25	466	TROUGHS, USSR, NORTHEAST, MESOZOIC
S	5	175	TRUBI FORMATION, ITALY, PLIOCENE,
M	29	199	TRUCHAN, MAREK, EXTENSIONAL TECTONICS IN THE OKINAWA
B	6405	696	TRUDELL, LAURENCE G., SHALLOW OIL SHALE DEPOSITS OF SOUTHERN
B	6212	2492	TRUEX, JOHN N., PLIOCENE-PLEISTOCENE DIASTROPHISM OF SANTA
B	6001	65	TRUEX, JOHN N., SANTA MONICA AND SANTA ANA
B	6102	263	TRUEX, JOHN N., SANTA MONICA AND SANTA ANA
B	6104	616	TRUEX, JOHN N., SANTA MONICA AND SANTA ANA
M	26	59	TRUNCATION, DEPOSITIONAL SYSTEMS
B	6104	489	TRUNCATION TRAP, TILTED-OUT
M	26	335	TRUNCATIONAL TRAPS

M 25 309 TRUNZ, J. P., OIL RESOURCES OF THE PACIFIC FOR PEACE AND
B 6201 52 TRYAL ROCKS AREA, AUSTRALIA
B 6408 1211 TSAIDAM BASIN, CHINA
B 6405 763 TSCHUDY, R. H., MUDSTONES AS EXPLORATION GUIDES TO TABULAR
B 6105 785 TSCHUDY, ROBERT H., NEWLY RECOGNIZED PETROLEUM SOURCE-ROCK
B 6212 2423 TSUNAMIS, GULF OF ALASKA
M 25 269 TSUSHIMA BASIN AND CONTINENTAL BORDERLAND, KOREA
M 25 34 TUAMOTU ATOLLS, PACIFIC REGION
B 6401 99 TUBB FORMATION, TEXAS
B 6004 592 TUBULAR-FORAM FACIES, IDAHO
M 27 6 TUBULES EXAMPLE, ALGAL NODULE
S 3 117 TUCANO BASIN
B 6109 1471 TUCANO BASIN, BRAZIL
B 6105 698 TUCHOLKE, BRIAN E., GAS-HYDRATE HORIZONS DETECTED
B 6303 543 TUCHOLKE, BRIAN E., GEOLOGIC SIGNIFICANCE OF SEDIMENTAR
B 6203 568 TUCKER, LEWIS R., HABITAT OF OIL—A RECONSIDERATION OF OLD
B 6102 295* TUCUMCARI BASIN, NEW MEXICO, ABST., SEDIMENTOLOGY OF,
B 6111 1956 TUCUPITA OIL FIELD, VENEZUELA
B 6306 942 TUCURARE FORMATION, AMAZON BASIN
B 6203 514* TUFA FROM SHORES OF ANCESTRAL DEAD SEA, ABST., ALGAL
M 28 140 TUFF, RHYOLITIC VITRIC, ALTERED
B 6203 545* TUFFACEOUS FLUVIAL SANDSTONE, SONORA PASS, CALIFORNIA, ABST.,
B 6303 307 TUGIDAK ANTICLINE, ALASKA
B 6408 1142 TUGIDAK BASIN, ALASKA
B 6303 301 TUGIDAK BASIN, FORE-ARC BASIN, ALASKA
B 6303 305 TUGIDAK BASIN, SEISMIC HORIZONS, ALASKA
B 6303 302 TUGIDAK FORMATION, ALASKA
B 6408 1142 TUGIDAK FORMATION, ALASKA
B 6407 1016 TUGIDAK FORMATION, ALASKA
B 6408 1142 TUGIDAK ISLAND, ALASKA
B 6408 1142 TUGIDAK UPLIFT, ALASKA
M 24 55 TUK FLEXURE ZONE, ALASKA
M 28 30 TUKTU FORMATION, ALASKA, CRETACEOUS
M 28 99 TUKTU FORMATION, ALASKA, CRETACEOUS
B 6311 2058† TULAMEEN COALFIELD, SOUTH-CENTRAL BRITISH COLUMBIA,
B 6211 2220 TULARCITOS FAULT, CALIFORNIA
B 6211 2267 TULARE FORMATION, CALIFORNIA
B 6203 499* TULE CANYON, TEXAS PANHANDLE, ABST., LACUSTRINE DEPOSITIONAL
B 6105 814* TULLY CLASTIC CORRELATIVES, DEVONIAN, EAST-CENTRAL NEW YORK,
B 6002 285 TULLY LIMESTONE, PENNSYLVANIA
B 6110 1897 TULSA MOUNTAINS, OKLAHOMA
M 25 321 TUMACO AREA, POTENTIAL, COLOMBIA
B 6307 1094 TUMBLAGOODA SANDSTONE, AUSTRALIA
B 6211 2262 TUMEY GULCH SECTION, CALIFORNIA
B 6207 1153 TUNA-FLOUNDER CHANNEL, AUSTRALIA
M 26 357 TUNED TRANSDUCER SYSTEMS, RESOLUTION SYSTEMS
M 25 449 TUNGSTEN AND MOLYBDENUM DEPOSITS, ZONAL DISTRIBUTION, JAPAN
M 25 535 TUNGSTEN, ANDEAN REGION
M 25 496 TUNGSTEN, CANADIAN, CORDILLERA
M 25 506 TUNGSTEN DEPOSIT, CANADA, POTATO HILLS
M 25 554 TUNGSTEN DEPOSITS, ECUADOR, SARAGURO, TIN AND
M 25 554 TUNGSTEN, ECUADOR, SARAGURD DEPOSIT, TIN AND
M 25 533 TUNGSTEN, PERU
M 25 444 TUNGSTEN, SOUTH KOREA
M 25 426 TUNGSTEN, TASMANIA, KING ISLAND, SCHEELITE
B 6408 1211 TUNGTING-HU BASIN, CHINA
M 30 239 TUNGUSKA SYNCLISE, CROSS SECTION, USSR
M 30 239 TUNGUSKA SYNCLISE, PETROLEUM POTENTIAL, USSR
M 27 21 TUNICATE SPICULE
B 6010 1771 TUNISIA, 1975, PETROCONSULTANTS S.A., ON DEVELOPMENTS IN
B 6110 1702 TUNISIA, 1976, PETROCONSULTANTS S.A., ON DEVELOPMENTS IN
B 6210 1799 TUNISIA, 1977, PETROCONSULTANTS S.A., ON DEVELOPMENTS IN
B 6310 1640 TUNISIA, 1978, PETROCONSULTANTS S.A., ON DEVELOPMENTS IN
B 6411 1743 TUNISIA, 1979, PETROCONSULTANTS S.A., ON DEVELOPMENTS IN
B 6303 429* TUNISIA, ABST., ALLOCHTHONOUS NATURE OF, NUMIDIAN FLYSCH
B 6004 659* TUNISIA, ABST., ANALYSIS OF SUBMARINE DEBRIS-FLOW DEPOSIT,
B 6303 534* TUNISIA, ABST., GEOLOGY OF, GABES BASIN
B 6303 516* TUNISIA, ABST., MARINE PERMIAN ROCKS OF
B 6203 572* TUNISIA, ABST., MIDDLE, MIOCENE TO EARLY, PLIOCENE
B 6303 553* TUNISIA, ABST., SEDIMENTOLOGIC STUDIES IN, JURASSIC OF
B 6303 462* TUNISIA, ABST., SHALLOW-WATER, CARBONATE AND, EVAPORITE
B 6303 501* TUNISIA, ABST., STRATIGRAPHIC CORRELATION OF, KEATHLEY
B 6406 860 TUNISIA, AND ALGERIA, GHADAMES BASIN, LIBYA
B 6303 456* TUNISIA BY U-PB AGES OF DETRITAL ZIRCONS, ABST., PROVENANC
B 6202 204 TUNISIA, CENTRAL, QUALITATIVE GEOHISTORY DIAGRAMS
B 6403 424 TUNISIA, CRETACEOUS
B 6403 424 TUNISIA, DJEBEL EL HARA
B 6403 424 TUNISIA, DJEBEL HASSANE
B 6403 424 TUNISIA, DJEBEL TIOUR EL LIL-ANNTRA
M 30 20 TUNISIA, EL BORMA FIELD, LANDSAT IMAGE
B 6401 18 TUNISIA, EL BORMA GIANT OIL FIELD, LANDSAT DATA
B 6403 424 TUNISIA, EOCENE
B 6401 18 TUNISIA, GHADAMES BASIN
B 6403 424 TUNISIA, HENNCHIR EN NEMRA
B 6403 424 TUNISIA, KEF EL KATTOUS
B 6401 18 TUNISIA, KIRCHAOU SANDSTONE FORMATION
B 6403 424 TUNISIA, MIOCENE
B 6403 424 TUNISIA, PALEOCENE
B 6110 1735 TUNISIA, PRODUCTION, 1976
B 6210 1832 TUNISIA, PRODUCTION, 1976-1977
B 6310 1668 TUNISIA, PRODUCTION, 1978
B 6411 1775 TUNISIA, PRODUCTION, 1979
B 6303 544* TUNISIA, RECORD OF LATE ALPINE ACTIVITY ON CRATONIC MARGIN OF
B 6403 424 TUNISIA, TRIASSIC
B 6303 477* TUNISIA--INTRODUCTION AND OVERVIEW, ABST., REGIONAL TECTONICS
B 6303 477* TUNISIA--PLATE TECTONIC AND HYDROCARBON ACCUMULATION OF
B 6303 521* TUNISIAN ATLAS, ABST., STRUCTURE OF
B 6303 531* TUNISIAN ATLAS, CLUE TO REGIONAL TECTONIC STYLE, ABST., KEF
B 6403 423† TUNISIAN ATLAS, PROGRESSIVE, NEOGENE DEFORMATION OF
B 6203 506* TUNISIA—LINK IN GENESIS OF PEBBLY MUDSTONES FROM
B 6209 1758* TURBID BOTTOM WATER LAYER, ABST., WATER CLARITY NEAR OIL
B 6303 548* TURBIDITE AND CANYON-RELATED POOLS, ABST., ENTRAPMENT
B 6308 1425* TURBIDITE CHANNEL RESERVOIRS IN CANYO SANDSTONE, ROUNDTOP
B 6405 694* TURBIDITE DEPOSIT, ABST., PALEOENVIRONMENT CONTROL OF
B 6211 2351* TURBIDITE DEPOSITION, CROCKER CANYON AREA, SOUTHWEST SAN
B 6206 912† TURBIDITE ENVIRONMENTS, FAN, VALLEYS, CHANNELS, AND
B 6206 934 TURBIDITE FACIES, CALIFORNIA, CRETACEOUS
B 6211 2243† TURBIDITE FANS OF SOUTHEASTERN SAN JOAQUIN VALLEY,
B 6203 496* TURBIDITE ORIGIN FOR ANHYDRITIC DOLOMITES IN BELL CANYON
B 6002 324* TURBIDITE RESERVOIRS IN CANYON SANDSTONE, VAL VERDE BASIN,
B 6102 196 TURBIDITE SAND AND SILT LAYERS, KOMANDORSKIY BASIN
B 6406 961* TURBIDITE SAND FACIES AND APPLICATION TO PETROLEUM GEOLOGY,
B 6203 495* TURBIDITE SANDSTONE RESERVOIRS—CANYON GROUP OF VAL VERDE
B 6303 418* TURBIDITE SANDSTONES, TEXAS, ABST., RESERVOIR MORPHOLOGIES FOR,
B 6105 767* TURBIDITE SEDIMENTATION IN CAMBRIAN OF NORTH WALES
B 6102 192† TURBIDITE SEDIMENTATION IN KOMANDORSKIY BASIN, WESTERN
B 6203 492* TURBIDITE SILTSTONE IN DEVONIAN BRALLIER FORMATION OF WEST
B 6004 666* TURBIDITE SUCCESSIONS OF MONTE ANTOLA FORMATION, NORTHERN
B 6409 1553* TURBIDITE-CHANNEL RESERVOIRS, LOWER HACKBERRY, OLIGOCENE,
B 6004 687* TURBIDITE-FAN APRON IN BORDE FORMTION, MISSISSIPPIAN, KENTUCKY
B 6004 719* TURBIDITES, ABST., EOLIAN-SAND
S 7 233 TURBIDITES, BEHAVIOR PATTERNS, ANCIENT
B 6206 935 TURBIDITES, BOUMA, MODEL FOR, CLASSIC
B 6206 934 TURBIDITES, CALIFORNIA, MATILIJA FORMATION
B 6206 933 TURBIDITES, CLASSIC
B 6202 237 TURBIDITES, CROSS SECTIONS, MEXICO, CHICONTEPEC
S 7 228 TURBIDITES, DISTAL FACIES DEFINED
B 6004 702* TURBIDITES IN MODERN SUBMARINE CANYONS AND FANS, ABST.,
B 6303 488* TURBIDITES OF CENTRAL AND SOUTHERN APPALACHIAN BASIN--
S 7 227† TURBIDITES, OIL, TRAPPING CHARACTERISTICS OF
B 6206 949 TURBIDITES, OREGON, JURASSIC
S 7 228 TURBIDITES, PROXIMAL FACIES DEFINED
S 7 228 TURBIDITES, SEDIMENTATION
B 6202 235† TURBIDITES, SOUTHEASTERN TAMPICO-MISANTLA BASIN, MEXICO,
B 6102 137† TURBIDITES, STUDIES OF, VENTURA FIELD, CALIFORNIA, I--, FACIES
B 6203 549* TURBIDITES, TIME AND AGAIN, ABST.
B 6303 525* TURBIDITIC SHALE AND FINE-GRAINED DEBRIS FLOW, ABST., RHYTHMS
B 6303 527* TURBIDITY AND HYDROGRAPHI VARIABILITY ON SOUTH TEXAS
B 6211 2232† TURBIDITY CURRENTS, A, SIGSBEE, KNOLL IN EARLY, TERTIARY, BAY OF,
S 8 161 TURBIDITY CURRENTS, ABRA SEAVALLEY, DESCRIBED
S 8 32 TURBIDITY CURRENTS, LA JOLLA CANYON, CURRENT FLOW
S 8 61 TURBIDITY CURRENTS, LA JOLLA CANYON, DESCRIBED
B 6407 1034† TURBIDITY CURRENTS, RESEDIMENTED FACIES IN, VENTURA BASIN,
S 8 29 TURBIDITY CURRENTS, REVIEW, CURRENT FLOW
S 8 108 TURBIDITY CURRENTS, RIO BALSAS CANYON, DESCRIBED
S 8 129 TURBIDITY CURRENTS, RIO DE LA PLATA CANYON, DESCRIBED
S 8 61 TURBIDITY CURRENTS, SCRIPPS CANYON, DESCRIBED
S 8 21 TURBIDITY CURRENTS, VAR CANYON, FRANCE
S 7 278 TURBIDITY FLOWS, BOTTOM, DEFORMATION
B 6303 435* TURBIDITY-CURRENT FLOWS FROM SUBMARINE SLIDES, ABST.,
S 7 91 TURBIDITY-CURRENT TRANSPORT, SEDIMENT PROCESSES, HIGH-ENERGY
S 7 90 TURBIDITY-CURRENT TRANSPORT, SEDIMENT PROCESSES, LOW-ENERGY
B 6105 698 TURCHOLKE, BRIAN E., GAS-HYDRATE HORIZONS DETECTED IN SEISMIC-
B 6405 670 TURCOTTE, D. I., THERMAL EVOLUTION OF SEDIMENTARY BASINS ALONG
B 6303 543 TURCOTTE, DONALD L., MODELS FOR EVOLUTION OF INTERIOR BASINS
S 5 181 TUREK, A. BASE- METAL CONCENTRATION IN A DENSITY- STRATIFIED
B 6408 1211 TURFAN BASIN, CHINA
B 6203 488* TURKANA, KENYA, ABST., OXYGEN, AND, CARBON, ISOTOPIC VARIATION
B 6010 1874 TURKEY, 1975, TURKISH PETROLEUM CORP., TPAO, ASIATIC PETROLEUM
B 6110 1802 TURKEY, 1976, TURKISH PETROL. CORP., ASIATIC PETROL. CORP., AND,
B 6210 2119 TURKEY, 1977, ASIATIC PETROLEUM CORP. AND J.P., JAVOGUES, ON

B 6310 1891 TURKEY, 1978, ASIATIC PETROLEUM CO., MOBIL EXPLORATION
B 6411 1842 TURKEY, 1979, ASIATIC PETROLEUM CO., SHELL, AND, MOBIL
B 6110 1815 TURKEY, PRODUCTION, 1976
B 6210 2135 TURKEY, PRODUCTION, 1976-1977
B 6310 1901 TURKEY, PRODUCTION, 1978
B 6411 1860 TURKEY, PRODUCTION, 1979
B 6405 724* TURKEY THRUST BELT, ABST., OIL EXPLORATION IN SOUTHEAST
B 6110 1802 TURKISH PETROL. CORP., ASIATIC PETROL. CORP., AND
B 6408 1273 TURKISH PETROLEUM CONGRESS PROCEEDINGS, EDITED BY GOKSENIN
B 6010 1874 TURKISH PETROLEUM CORP., TPAO, ASIATIC PETROLEUM CORP., AND
B 6106 900 TURKISH PLATE
B 6408 1273 TURKIYE DORDUNCU PETROL KONGRESI, FOURTH TURKISH
B 6310 1891 TURKIYE PETROLLERI A.O., AND PETROCONSULTANTS S.A., ON
B 6206 1093 TURMELLE, JOHN M., PALEOCURRENTS OF JACKPILE SANDSTONE, BURRO
B 6105 836 TURNER, A. KEITH, COMPUTER-ASSISTED COMPOSITE MAPPING SYSTEM
B 6103 469 TURNER, D. F., GEOCHRONOLOGY OF SOUTHWESTERN BROOKS RANGE.
B 6305 823* TURNER DEPOCENTERS IN WYOMING PORTION OF POWDER RIVE BASIN,
B 6303 543 TURNER, MORT D., ANTARCTIC LOGISTICS FOR EARTH SCIENCES, ABST.
B 6404 490 TURNER SANDY MEMBER, CARLILE SHALE, WYOMING
B 6406 943 TURNER SANDY MEMBER, SOUTH DAKOTA
M 30 173 TURNER, W. L., NINIAN FIELD, U.K. SECTOR, NORTH SEA
B 6405 795 TURNER-PETERSON, C. E., SEDIMENTOLOGY OF WESTWATER CANYON
B 6405 795 TURNER-PETERSON, C. E., TABULAR URANIUM ORE IN POISON CANYO
B 6105 836 TURNER-PETERSON, CHRISTINE E., LACUSTRINE SEDIMENTATION IN
B 6405 769 TURNER-PETERSON, CHRISTINE E., PALEOGEOGRAPH OF NORTHERN
B 6305 843 TURNER-PETERSON, CHRISTINE, LACUSTRINE HUMATE
B 6204 669 TURONIAN, ALGERIA, CENOMANIAN-
B 6211 2233 TURONIAN, BAY OF BISCAY AREA
B 6405 802* TURONIAN, DEPOSITIONAL ENVIRONMENTS ALONG A PART OF EASTERN
B 6101 44 TURONIAN, LEBANON, CENOMANIAN-
S 6 278 TURONIAN, MACROFOSSIL ZONES, CRETACEOUS
B 6405 606 TURONIAN, MIDDLE EAST, JURASSIC TO
B 6412 2059 TURONIAN, ORGANIC MATTER, ATLANTIC BASINS
B 6309 1473 TURONIAN SEQUENCE, USSR, ANADYR BASIN, APTIAN-
B 6002 258 TURONIAN STRATA, ISRAEL, ALBIAN-
B 6406 860 TURONIAN TRANSGRESSIVE DEPOSITS, AFRICA, CENOMANIAN-
S 6 261 TURONIAN-CONIACIAN BOUNDARY, NORTH AMERICA, CRETACEOUS
B 6207 1153 TURRUM FORMATION, AUSTRALIA
B 6307 1093 TURTLE DOVE RIDGE, AUSTRALIA
B 6407 1092 TUSATUVAYAMIAN HORIZON, ASIA, NORTHEAST
B 6409 1565* TUSCAHOMA MARLS, SOUTHWEST ALABAMA, ABST., LATE, PALEOCENE
B 6309 1598* TUSCALOOSA, ABST., YOU AINT SEEN NOTHING TILL YOUVE SEEN THE
B 6109 1421 TUSCALOOSA FORMATION, HYDROCARBONS, MISSISSIPPI
M 28 144 TUSCALOOSA FORMATION, LOUISIANA, CRETACEOUS
M 28 176 TUSCALOOSA FORMATION, LOUISIANA, CRETACEOUS
M 28 138 TUSCALOOSA FORMATION, LOUISIANA, CRETACEOUS
M 28 135 TUSCALOOSA FORMATION, LOUISIANA, CRETACEOUS
M 28 70 TUSCALOOSA FORMATION, LOUISIANA, CRETACEOUS
B 6405 710* TUSCALOOSA GAS TREND OF SOUTH LOUISIANA, ABST., DEEP
B 6409 1552* TUSCALOOSA, NORTHERN COASTAL REGION, ABST., REGIONAL
M 24 242 TUSCALOOSA PRODUCING SAND, MISSISSIPPI
M 24 265 TUSCALOOSA SANDSTONE, ALABAMA
B 6409 1552* TUSCALOOSA TREND, 1975-80, ABST., NO. 1 PLAY IN U.S.A.—SOUTH,
B 6309 1611* TUSCALOOSA TREND, LOUISIANA, ABST., ORIGIN OF POROSITY IN DEEP,
B 6409 1568* TUSCALOOSA TREND OF SOUTH LOUISIANA, ABST., LOG EVALUATION OF
B 6409 1558* TUSCALOOSA VERSUS SOUTHEAST TEXAS WOODBINE, ABST., LOUISIANA
B 6209 1635 TUSCARORA SANDSTONE, APPALACHIAN BASIN
B 6304 652 TUSCARORA SANDSTONE, VIRGINIA
B 6211 2309 TUSCARORA SANDSTONE, WEST VIRGINIA
B 6204 646 TUSCUMBIA SECTION, ALABAMA
B 6103 361 TUSSY-TATUMS STRUCTURE, OKLAHOMA
B 6405 600 TUWAIQ FORMATION, MIDDLE EAST
B 6203 374 TUXEDNI BAY, ALASKA
B 6203 377 TUXEDNI GROUP, ALASKA
M 24 4 TUXEDNI GROUP, ALASKA
B 6303 508* TUXPAN-TECOLUTLA CARBONATE PLATFORM, GULF COAST, MEXICO,
B 6102 207 TUXTLAS UPLIFT, MEXICO
B 6407 1121 TVER AND RICHARD W. BERRY, HOWER, S. GRACE, REVIEW OF THE,
B 6403 446* TWENTIETH-CENTURY HISTORY OF GULF OF ALASKA COASTLINE, CAPE
B 6303 470* TWENTY YEARS OF UNITED STATES PETROLEUM ESTIMATES, ABST.
B 6303 480 TWICHELL, DAVID C., SLUMPING IN INTERCANYON AREAS, MIDDLE
B 6303 529 TWICHELL, DAVID C., SLUMPS ON UPPER CONTINENTAL SLOPE,
B 6205 889* TWIN CREEK FORMATION AND EQUIVALENT UNITS, HINGE LINE—
B 6405 640 TWIN CREEK FORMATION, WYOMING
B 6206 1075† TWIN CREEK LIMESTONE, IDAHO-WYOMING THRUST BELT, PETROLEUM,
B 6206 1077 TWIN CREEK LIMESTONE, SAMPLES AND GEOCHEMICAL DATA, UTAH
B 6206 1077 TWIN CREEK LIMESTONE, SAMPLES AND GEOCHEMICAL DATA, WYOMING
M 24 95 TWIN CREEK SANDSTONE, WYOMING
B 6112 2118 TWIN MOUNDS, NEW MEXICO
M 28 16 TWINNING, FELDSPAR, ALBITE
M 28 17 TWINNING, FELDSPAR, ALBITE
M 28 17 TWINNING, FELDSPAR, MICROCLINE
M 28 16 TWINNING, FELDSPAR, PLAGIOCLASE
B 6405 777* TWO CONTROLS OF SAND-WAVE SIZE—DYNAMIC EQUILIBRIUM
B 6405 673* TWO HILLS COLONY GAS FIELD OF ALBERTA, ABST., GEOPHYSICAL CASE
B 6407 972 TWO MEDICINE FORMATION, MONTANA
B 6405 744* TWO OIL TYPES ON NORTH SLOPE OF ALASKA—IMPLICATIONS FOR
M 26 444 TWO-DIMENSIONAL, SEISMIC, MODELING
B 6405 796 TWOMBLY, G., NATURE OF NON-RECOVERABLE POROSITY CHANGES IN
M 25 500 TYAUGHTON TROUGH, CANADA
B 6303 554 TYE, ROBERT S., STRATIGRAPHIC MODELS FOR MODERN BACK-BARRIER
B 6207 1194 TYEE FORMATION, OREGON AND, WASHINGTON
M 24 287 TYLER, A. N., SUNOCO-FELDA FIELD, HENDRY AND COLLIER COUNTIES,
B 6309 1601* TYLER, AND SAN JACINTO COUNTIES, TEXAS, ABST., DEPOSITIONAL
B 6402 168 TYLER BASIN, TEXAS
B 6105 777 TYLER, THEODORE F., PROSPECTIVE HYDROCARBON PROVINCES OF THE
B 6303 528* TYLER COUNTY, TEXAS, ABST., DESCRIPTION AND SEDIMENTOLOGY
B 6209 1766* TYLER COUNTY, TEXAS—SUBSURFACE CORE STUDY, ABST.,
B 6405 796 TYLER, NOEL, REEVALUATION OF DEPOSITIONAL ENVIRONMENTS OF
B 6008 1401* TYLER SANDSTONES, PENNSYLVANIAN, DICKINSON AREA, NORTH
B 6303 485* TYLER SANDSTONES, PENNSYLVANIAN, DICKINSON AREA, NORTH
B 6307 1136 TYNER FORMATION, OKLAHOMA
B 6104 594 TYONEK FIELD, ALASKA
M 24 10 TYONEK FORMATION, ALASKA
B 6408 1133 TYONEK FORMATION, ALASKA
B 6203 377 TYONEK FORMATION, ALASKA
B 6109 1549* TYPE AND COTYPE WELLS FOR SOME CRETACEOUS-EOCENE FORMATIONS
B 6009 1570† TYPE LITTLE FALLS DOLOSTONE, LATE CAMBRIAN, EAST-CENTRAL NEW
B 6309 1510† TYPE SECTION FOR UPPER SILURIAN AND LOWER DEVONIAN
B 6405 772* TYPE URANIUM DEPOSIT, ABST., HISTORY OF, SULFIDIZATION OF SOUTH,
S 6 328 TYPE-MISSISSIPPIAN REGION, CARBONIFEROUS
S 6 329 TYPE-MORROWAN SERIES, CARBONIFEROUS
S 6 328 TYPE-PENNSYLVANIAN REGION, CARBONIFEROUS
B 6404 462 TYPES AND TEXTURES OF CEMENTS AND POROSITY
M 26 123 TYPES, SEISMIC, REFLECTIONS, CONFIGURATION PATTERNS
S 6 348 TYRONE LIMESTONE, BENTONITES, ORDOVICIAN, KENTUCKY
M 29 297 TYRRHENIAN BASIN, EVOLUTION, MEDITERRANEAN
M 29 297 TYRRHENIAN BASIN, SEDIMENTARY SEQUENCE, MEDITERRANEAN
B 6104 603 U-BAR FORMATION, NEW MEXICO
B 6303 456* U-PB AGES OF DETRITAL ZIRCONS, ABST., PROVENANC STUDIES IN,
B 6309 1591* U.S GEOLOGICAL SURVEY AND U.S. BUREAU OF MINES COAL RESOURCES
S 4 329† U.S VIRGIN ISLANDS, MIOCENE, KINGSHILL SEAWAY--A DYNAMIC,
B 6105 835* U.S. AND WORLD ENERGY OUTLOOK FOR 1980S, ABST.
B 6105 821* U.S. ATLANTIC CONTINENTAL MARGIN, ABST., PRELIMINARY,
M 29 11† U.S. ATLANTIC CONTINENTAL MARGIN, BASED ON MULTICHANNEL
M 29 139† U.S. ATLANTIC CONTINENTAL MARGIN, VARIATIONS IN, BOTTOM,
M 29 85† U.S. ATLANTIC MARGIN, BASIN, STRUCTURE OF THE
M 29 85 U.S. ATLANTIC MARGIN, SEE ATLANTIC MARGIN
B 6405 733* U.S. ATLANTIC OUTER CONTINENTAL SHELF, ABST., PROCESSES
B 6204 604 U.S. ATLANTIC SLOPE, CARBONATE FACIES
B 6204 603 U.S. ATLANTIC SLOPE, JURASSIC
B 6204 594 U.S. ATLANTIC SLOPE, ORGANIC MATTER
B 6204 600 U.S. ATLANTIC SLOPE, PALEOSHELF-EDGE HIGH, POTENTIAL TRAP
B 6204 600 U.S. ATLANTIC SLOPE, POTENTIAL TRAPS
B 6204 592† U.S. ATLANTIC SLOPE, RISE, AND ABYSSAL PLAIN, PETROLEUM,
B 6204 596 U.S. ATLANTIC SLOPE, SEDIMENT DISTRIBUTION
B 6008 1154 U.S. DRILLING SUMMARY--1975
B 6308 1169 U.S. DRILLING SUMMARY--1978
B 6108 1122 U.S. DRILLING SUMMARY—1976
B 6208 1277 U.S. DRILLING SUMMARY—1977
B 6409 1297 U.S. DRILLING SUMMARY—1979
M 29 448 U.S. EAST COAST, SEISMIC EXPERIMENT, CONTINENTAL MARGINS
S 8 121 U.S. EAST COAST, SUBMARINE CANYONS, HUDSON CANYON
S 8 111 U.S. EAST COAST, SUBMARINE CANYONS, HYDROGRAPHER CANYON
S 8 122 U.S. EAST COAST, SUBMARINE CANYONS, NORFOLK CANYON
S 8 122 U.S. EAST COAST, SUBMARINE CANYONS, WASHINGTON CANYON
S 8 122 U.S. EAST COAST, SUBMARINE CANYONS, WILMINGTON CANYON
S 8 112 U.S. EAST COAST, TOPOGRAPHY, CANYON INDEX
B 6206 1092* U.S. EASTERN INTERIOR BASIN, ABST., ORGANIC, GEOCHEMISTRY OF,
B 6004 665* U.S. GEOLOGICAL SURVEY, ABST., NE APPROACH TO GEOLOGIC,
B 6303 497* U.S. GEOLOGICAL SURVEY AND ROLE OF APPRAISAL GROUP IN
B 6407 1122 U.S. GEOLOGICAL SURVEY, ANONYMOUS, REVIEW OF 1., SEISMIC, SAFETY
B 6004 698* U.S. GEOLOGICAL SURVEY FOR GEOLOGIC ESTIMATES OF UNDISCOVERED
B 6303 475* U.S. GEOLOGICAL SURVEY METHODS FOR OCS TRACT EVALUATIONS,
B 6008 1404* U.S. GEOLOGICAL SURVEY PROBABILISTIC ASSESSMENTS OF
B 6105 822* U.S. GEOLOGICAL SURVEYS ATLANTIC MARGIN CORING PROJECT, ABST.,
B 6105 842* U.S. GULF COAST, ABST., CONTINENTAL, MARGIN—
B 6405 788* U.S. INDUSTRY ACTIVITY, ABST., WHATS NEW IN

B 6410 1660 U.S. NAVAL OCEANOGRAPHIC OFFICE CRUISES LYNCH 73-1 AND 73-2
B 6008 1155 U.S. NEW-FIELD DISCOVERIES--1975, 1972, 1969, SIZE ESTIMATES OF
B 6308 1170 U.S. NEW-FIELD DISCOVERIES--1978, 1975, 1972, SIZE ESTIMATES OF
B 6108 1123 U.S. NEW-FIELD DISCOVERIES—1976, 1973, 1970, SIZE ESTIMATES OF
B 6208 1278 U.S. NEW-FIELD DISCOVERIES—1977, 1974, 1971, SIZE ESTIMATES OF
B 6409 1298 U.S. NEW-FIELD DISCOVERIES—1979, 1976 1973, SIZE ESTIMATES OF
B 6405 751* U.S. VIRGIN ISLANDS, ABST., ORIGIN OF BIOGENIC CARBONATE SANDS OF
M 29 353† U.S. VIRGIN ISLANDS, GEOMORPHOLOGY AND, SUBSURFACE, GEOLOGY
B 6203 403† U.S. VIRGIN ISLANDS, STRATIGRAPHY AND DEPOSITIONAL SETTING,
B 6203 509* U.S. VIRGIN ISLANDS. BY TITLE ONLY, GEOLOGY SEEN WHILE LIVING IN
B 6110 1654 U.S.S.R., PRODUCTION, 1975-1976
B 6210 1924 U.S.S.R., PRODUCTION, 1976-1977
B 6310 1932 U.S.S.R., PRODUCTION, 1977-1978
B 6401 22 U.S.S.R., SAMOTLOR, SAMOTLORSKOYE, GIANT OIL FIELD, LANDSAT
B 6401 22 U.S.S.R., WEST SIBERIAN BASIN
B 6109 1476 UBARANA FIELD, BRAZIL
B 6109 1480 UBATUBA MEMBER, BRAZIL
M 25 344 UCAYALI BASIN, POTENTIAL
B 6005 809 UCHUPI, ELAZAR, CONTINENTAL MARGIN OFF WESTERN
B 6109 1483 UCHUPI, ELAZAR, CONTINENTAL SLOPE AND UPPER RISE OFF WESTERN
B 6404 501 UCHUPI, ELAZAR, GEOLOGY OF NEW ENGLAND PASSIVE MARGIN
B 6106 872 UCHUPI, ELAZAR, MEDITERRANEAN SEA-NILE CONE AREA, STRUCTURE
B 6004 716 UCHUPI, ELAZAR, RECENT SEDIMENTATION ON NILE DELTA AND CONE,
S 4 31 UDOTEA SP., CORAL-SAND SUBFACIES
B 6110 1950 UGANDA, DEVELOPMENTS, 1976
B 6210 1849 UGANDA, DEVELOPMENTS, 1977
B 6310 1695 UGANDA, DEVELOPMENTS, 1978
B 6411 1791 UGANDA, DEVELOPMENTS, 1979
B 6201 18 UGHELLI MACROSTRUCTURE, NIGER DELTA
B 6004 653* UINTA AULACOGENS--INITIAL DEPOSITS OF CORDILLERAN
B 6401 61 UINTA BASIN, EOCENE, KEROGEN, UTAH
S 10 53 UINTA BASIN, GREEN RIVER FORMATION, MIGRATION, UTAH
B 6305 832* UINTA BASIN, UTAH, ABST., INFLUENCE OF DIAGENETIC REACTIONS ON
B 6205 891* UINTA BASIN, UTAH, ABST., MINERALOGIC CHARACTERISTICS OF SOME
B 6405 755* UINTA BASIN, UTAH, ABST., ORIGIN OF SUBSURFACE, FRACTURE
B 6405 696* UINTA BASIN, UTAH, ABST., SHALLOW, OIL, SHALE DEPOSITS OF
B 6205 885* UINTA BASIN, UTAH, ABST., SOME PETROGRAPHIC CHARACTERISTICS OF
B 6405 732* UINTA BASIN, UTAH, ABST., SOME RELATIONS BETWEEN DIAGENESIS AND
B 6205 890* UINTA BASIN, UTAH, ABST., VOLCANIC SANDSTONE IN, GREEN RIVER
M 24 121† UINTA BASIN, UTAH, ALTAMONT-, BLUEBELL--A MAJOR, NATURALLY,
B 6305 825* UINTA BASINS, WYOMING AND UTAH, ABST., TERTIARY CORRELATIONS
B 6205 887* UINTA COUNTY, WYOMING, ABST., PAINTER RESERVOIR FIELD
B 6305 832* UINTA COUNTY, WYOMING, ABST., PAINTER RESERVOIR--, CLEAR
B 6305 835* UINTA COUNTY, WYOMING, ABST., YELLOW CREEK FIELD
B 6205 890* UINTA FORMATIONS, UINTA BASIN, UTAH, ABST., VOLCANIC SANDSTONE
B 6008 1392* UINTA MOUNTAIN UPLIFT, ABST., STRUCTURAL GEOLOGY AND
B 6302 195 UINTA MOUNTAINS, COLORADO
M 26 348 UINTAH COUNTY, UTAH, RED WASH FIELD
M 25 276 UKAI, M., PETROLEUM FIELDS OF JAPAN WIT VOLCANIC-ROCK
B 6102 240 ULATISIAN ROCKS, CALIFORNIA
S 11 64 ULATISIAN STAGE
B 6408 1287* ULLIN LIMESTONE AND FORT PAYNE FORMATION, MISSISSIPPIAN,
B 6004 704 ULRICH, B. C., TIMING OF TRANSGRESSIONS AND REGRESSIONS IN
B 6105 835* ULTIMATE RECOVERY FROM KNOWN FIELDS BY ENHANCED RECOVERY
B 6409 1529 ULTIMATELY DISCOVERABLE VOLUME, DEFINITION
M 25 460 ULTRABASITES, METAL CONCENTRATIONS IN, EAST PACIFIC RISE
B 6205 862 ULTRADEEP FIELD, PRODUCTION, TEXAS, BUFFALO WALLOW-
B 6205 862 ULTRADEEP FIELD, PRODUCTION, TEXAS, GOMEZ, ELLENBURGER
B 6205 862 ULTRADEEP FIELD, PRODUCTION, TEXAS, LOCKRIDGE
B 6205 862 ULTRADEEP FIELD, PRODUCTION, TEXAS, WASHITA CREEK
B 6205 854 ULTRADEEP WELLS AS OF MARCH 1976, UNITED STATES
B 6205 862 ULTRADEEP WELLS, AVERAGE COSTS
B 6205 855 ULTRADEEP WELLS, LOCATION, TEXAS, PANHANDLE AND, OKLAHOMA
B 6205 854 ULTRADEEP WELLS, LOCATION, WEST, TEXAS AND, NEW MEXICO
B 6205 860 ULTRADEEP WELLS, TEXAS, PECOS COUNTY
B 6205 859 ULTRADEEP WILDCATS, SUCCESS RATIOS, TEXAS, PANHANDLE AND,
B 6205 859 ULTRADEEP WILDCATS, SUCCESS RATIOS, WEST, TEXAS AND, NEW
M 25 479 ULTRAMAFIC VOLCANIC ROCKS, KALIMANTAN
M 25 479 ULTRAMAFIC VOLCANIC ROCKS, MALAITA
M 25 479 ULTRAMAFIC VOLCANIC ROCKS, USSR, KAMCHATKA
M 27 212 UMBRELLA-VOID FILLS
B 6210 2046 UMIKER, R., ON DEVELOPMENTS IN BARBADOS, 1977
B 6310 1747 UMIKER, R., ON DEVELOPMENTS IN BARBADOS, 1978
B 6210 2054 UMIKER, R., ON DEVELOPMENTS IN LESSER ANTILLES, 1977
B 6310 1755 UMIKER, R., ON DEVELOPMENTS IN LESSER ANTILLES, 1978
B 6110 1589 UMIKER, R., ON DEVELOPMENTS IN NETHERLANDS ANTILLES, 1976
B 6210 2056 UMIKER, R., ON DEVELOPMENTS IN NETHERLANDS ANTILLES, 1977
B 6310 1756 UMIKER, R., ON DEVELOPMENTS IN NETHERLANDS ANTILLES, 1978
B 6210 2058 UMIKER, R., ON DEVELOPMENTS IN PUERTO RICO, 1977
B 6310 1761 UMIKER, R., ON DEVELOPMENTS IN PUERTO RICO, 1978
B 6010 1871 UMM AL QAIWAIN, 1975, UNITED REFINING CO., OCCIDENTAL
B 6110 1800 UMM AL QAIWAIN, 1976, HOUSTON OIL AND MINERALS CORP. AND,
B 6210 2117 UMM AL QAIWAIN, 1977, PETROCONSULTANTS S.A., PRUDENTIAL
B 6411 1841 UMM AL QAIWAIN, 1979, HOUSTON OIL AND MINERALS OF DUBAI, INC.,
B 6310 1888 UMM AL-QAIWAIN, 1978, HOUSTON OIL AND MINERALS CORP AND,
B 6002 162 UMNAK PLATEAU, BERINGIAN SHELF
B 6108 1374* UMPQUA FORMATION, SOUTHWEST OREGON, ABST., PETROLOGY AND
S 10 76 UNCOMPACTION, RESTORATION OF SEDIMENTARY COLUMN
B 6405 686 UNCOMPAHGRE BASIN, EASTERN UTAH AND SOUTHWESTERN
M 26 100 UNCONFORMITIES, CHRONOSTRATIGRAPHY, SEISMIC, REFLECTIONS
B 6209 1755* UNCONFORMITIES, COASTAL PLAIN OF GEORGIA, ABST., UPPER,
M 26 56 UNCONFORMITIES, DEPOSITIONAL SEQUENCES
M 26 93 UNCONFORMITIES, GLOBAL CYCLES, INTERREGIONAL
B 6103 435† UNCONFORMITIES IN NORTHERN ALASKA RELATED TO ANTLER TECTONIC
B 6207 1224* UNCONFORMITIES IN WESTERN INDONESIA, ABST., CORRELATION OF
B 6403 366 UNCONFORMITIES OF OROGENY DURING LATE JURASSIC TO LATE
B 6405 684* UNCONFORMITIES—KEY TO MAJOR OI ACCUMULATIONS, NORTH SLOPE,
B 6109 1534* UNCONFORMITIES—NEGLECTED EXPLORATION TOOL, ABST.
B 6006 962† UNCONFORMITY AT MOUNT DOONERAK, BROOK RANGE, NORTHERN
M 29 75 UNCONFORMITY, ATLANTIC COAST, CONTINENTAL SLOPE
B 6004 657* UNCONFORMITY IN BEJUCO FIELD, EASTERN MEXICO, ABST.,
B 6408 1218 UNCONFORMITY TRAP, DIAGRAM
M 26 430 UNCONFORMITY TRAP, SEISMIC, MODELS
B 6209 1569 UNCONSTRAINED AND CONSTRAINED DIAPIRISM
B 6309 1580* UNCONVENTIONAL EVALUATION OF COTTAGEVILLE FIELD, DEVONIAN
S 3 135 UNDEREMPHASIZED PROCESSES OF URANIUM MINERALIZATION
B 6004 658* UNDERGROUND NATURAL GAS STORAGE RESERVOIRS BY ISOTOPIC
B 6405 729* UNDERSTANDING FIELD DEVELOPMEN HISTORY USING 3-D SEISMIC
B 6203 506* UNDERSTANDING SEISMICALLY DERIVED IMPEDANCE LOGS, ABST.
B 6406 945 UNDERWOOD, M. B., PETROGRAPHY AND GEOLOGIC SIGNIFICANCE OF
M 28 162 UNICOI FORMATION, VIRGINIA, CAMBRIAN
S 8 18 UNIDIRECTIONAL FLOW, CURRENT FLOW
S 6 2 UNIDIRECTIONAL HISTORY
B 6203 564* UNIFORMITARIANISM, ABST., EVOLUTION AND
B 6203 514* UNIFORMITARIANISM, ABST., ICHNOLOGY AND
B 6203 562* UNIFORMITARIANISM AND GLOBAL TECTONICS, ABST.
S 4 94 UNIFORMITARIANISM AND REEFS
B 6405 709* UNIFORMITARIANISM AND TERTIARY REEF PALEOECOLOGY, ABST.
B 6203 568* UNIFORMITARIANISM, AND THE HYPOTHETICO-DEDUCTIVE METHOD,
B 6203 492* UNIFORMITARIANISM—DOCTRINE WHICH NEEDS RETHINKING, ABST.
B 6203 519* UNIFORMITY AND THE PRECAMBRIAN, ABST.
B 6412 2142 UNIMAK ISLAND, BERING SEA SHELF
B 6211 2355* UNION ISLAND GAS FIELD, ABST.
B 6002 220 UNION MOONIE TOPPED OIL, AUSTRALIA
M 27 34 UNISERIAL FORAMS
B 6303 496* UNITED ARAB EMIRATES, AND ITS RELATION TO DOLOMITE GENESIS,
B 6210 2115 UNITED ARAB EMIRATES, DEVELOPMENTS, 1977
B 6310 1887 UNITED ARAB EMIRATES, DEVELOPMENTS, 1978
B 6010 1720 UNITED KINGDOM, 1975, B.P. PETROL. DEVEL. LTD., BRITISH GAS CORP.,
B 6110 1649 UNITED KINGDOM, 1976, BRITISH PETROL. DEV. LTD., BRITISH GAS CORP.,
B 6303 549* UNITED KINGDOM ABST., COMPUTATION OF INITIAL WELL
B 6303 523* UNITED KINGDOM, ABST., GWNA MELANGE, UPPER, PRECAMBRIAN,
B 6303 552* UNITED KINGDOM AND EIRE, ABST., HYDROCARBON PROSPECTS IN,
B 6408 1245 UNITED KINGDOM AND NORWAY, JURASSIC, PIEZOMETRIC MAP
B 6408 1247 UNITED KINGDOM, BERYL EMBAYMENT
B 6210 1920 UNITED KINGDOM, DEVELOPMENTS, 1977
B 6310 1928 UNITED KINGDOM, DEVELOPMENTS, 1978
B 6411 1969 UNITED KINGDOM, EXCLUDING NORTH SEA, DEVELOPMENTS, 1979
B 6408 1247 UNITED KINGDOM, FRIGG FIELD
B 6408 1247 UNITED KINGDOM, FRIGG SUBBASIN
B 6110 1654 UNITED KINGDOM, PRODUCTION, 1975-1976
B 6210 1924 UNITED KINGDOM, PRODUCTION, 1976-1977
B 6610 1932 UNITED KINGDOM, PRODUCTION, 1977-1978
B 6411 1971 UNITED KINGDOM, PRODUCTION, 1978-1979
B 6408 1247 UNITED KINGDOM, SHETLANDS PLATFORM
B 6408 1245† UNITED KINGDOM-NORWAY, PRESSURE ORIGIN AND DISTRIBUTION IN,
B 6010 1871 UNITED REFINING CO., OCCIDENTAL EXPLORATION AND PRODUCTION
B 6203 518* UNITED STATES, ABST., ANALYSIS OF, OIL-FIELD DISTRIBUTION FOR
B 6207 1222* UNITED STATES, ABST., AVAILABILITY AND QUALITY OF, WATER
B 6405 703* UNITED STATES, ABST., DEEP DRILLING AND CURRENT MODELS OF,
B 6405 771* UNITED STATES, ABST., DEEP TERRESTRIAL, HEAT-FLOW STUDIES IN
B 6403 448* UNITED STATES, ABST., DEPOSITIONAL ENVIRONMENTS WITHIN
B 6408 1285* UNITED STATES, ABST., DEVONIA OIL, SHALE OF EASTERN
B 6004 652* UNITED STATES, ABST., EPISODIC UPLIFT AND SEDIMENTATION IN POST-,
B 6207 1233* UNITED STATES, ABST., FUTURE POTENTIAL FOR DEVELOPMENT OF,
B 6405 762* UNITED STATES, ABST., FUTURE TRENDS IN, COAL RESOURCE AND
B 6012 2186* UNITED STATES, ABST., GEOLOGIC ESTIMATES OF UNDISCOVERED, OIL
B 6203 557* UNITED STATES, ABST., GEOTECHNICAL PROPERTIES OF UPPER,
B 6108 1377* UNITED STATES, ABST., INTERDISCIPLINARY ANALYSIS OF SOME
B 6105 823* UNITED STATES, ABST., MISSISSIPPIAN, CARBONATE-, SHELF, MARGINS,

B 6203 563* UNITED STATES, ABST., MOLLUSCAN, BIOCLASTIC, ACCUMULATIONS IN
B 6108 1375* UNITED STATES, ABST., ORGANIC, GEOCHEMISTRY, INCIPIENT,
B 6405 749* UNITED STATES, ABST., ORGANOPHOSPHORITES IN, CARBONIFEROUS
B 6405 707* UNITED STATES, ABST., PETROLEUM POTENTIAL OF, BASIN AND RANGE
B 6405 746* UNITED STATES, ABST., PHOSPHORITE, ORGANIC CARBON, AND
B 6004 698* UNITED STATES, ABST., PROBABILISTIC AND, COMPUTER
B 6303 464* UNITED STATES, ABST., REGIONAL STRATIGRAPHY OF, LIMESTONE
B 6303 511* UNITED STATES, ABST., RESOURCE APPRAISAL PREDICTIONS AND
B 6303 523* UNITED STATES, ABST., RESOURCE APPRAISAL PREDICTIONS AND
B 6105 813* UNITED STATES, ABST., REVIEW AND APPLICATION OF PETROLEUM
B 6108 1383* UNITED STATES, ABST., REVIEW AND APPLICATION OF PETROLEUM-,
B 6405 721* UNITED STATES, ABST., SCIENTIFIC, OCEAN DRILLING NEAR
B 6405 779* UNITED STATES, ABST., SEDIMENTATION, BIOSTRATIGRAPHY, AND
B 6405 670* **UNITED STATES, ABST., THERMAL EVOLUTION OF SEDIMENTARY, BASINS**
B 6405 780* UNITED STATES, ABST., USE OF, DEVONIAN, CONODONTS IN PETROLEUM
B 6004 716* UNITED STATES AND CANADA, ABST., AAPG, OIL AND, GAS, FIELD, DATA
B 6408 1272 UNITED STATES AND MEXICO, BY MICHEL T. HALBOUTY, SHELTON, JOHN
B 6405 700* UNITED STATES AND OTHER PARTS OF THE WORLD, ABST., POTENTIAL
B 6105 822* UNITED STATES AND UTAH OIL-IMPREGNATED ROCK, TAR SAND,
B 6305 842* UNITED STATES AND WORLD ENERGY OUTLOOK FOR THE 1980S, ABST.
B 6212 2509* UNITED STATES ANTITRUST LAWS, ABST., WHAT FOREIGN COMPANIES
B 6111 1951 UNITED STATES, ASPHALT RIDGE HEAVY OIL DEPOSIT
B 6303 481* UNITED STATES ATLANTIC CONTINENTAL SHELF AND
B 6405 706* UNITED STATES ATLANTIC SHELF, ABST., MODIFICATION OF LINEAR
B 6111 1951 UNITED STATES, CIRCLE CLIFFS HEAVY OIL DEPOSIT
B 6203 509* UNITED STATES COAST, ABST., ARCHAEOLOGICAL EVIDENCE FOR
B 6405 746* UNITED STATES CONTINENTAL MARGIN NORTH OF CAPE HATTERAS,
B 6203 510* UNITED STATES CONTINENTAL SLOPE, ABST., SEDIMENTATION ON
B 6202 177 UNITED STATES, CRUDE OIL RESOURCE, TWO ESTIMATES
B 6402 174 UNITED STATES, EAST COAST, SUBSIDENCE
B 6209 1704 UNITED STATES, EASTERN, ANTRIM SHALE
B 6209 1704 UNITED STATES, EASTERN, ARKANSAS NOVACULITE
B 6209 1709 UNITED STATES, EASTERN, BEREA SANDSTONE
B 6209 1704 UNITED STATES, EASTERN, BURKET BLACK SHALE
B 6209 1704 UNITED STATES, EASTERN, CHATTANOOGA SHALE
B 6209 1704 UNITED STATES, EASTERN, GENESEO SHALE
B 6209 1704 UNITED STATES, EASTERN, MILLBORO SHALE
B 6209 1704 UNITED STATES, EASTERN, NEW ALBANY SHALE
B 6202 178 UNITED STATES, ENERGY RESOURCES
M 25 65 UNITED STATES, ERTS IMAGES
M 25 485 UNITED STATES, GREAT BASIN, METAL PROVINCE
B 6409 1565* UNITED STATES GULF COAST, ABST., NEARSHORE, BARS ALONG
B 6303 551* UNITED STATES GULF COAST, ABST., PORE SYSTEMS IN, JURASSI,
B 6409 1557* UNITED STATES GULF COAST, ABST., WELL LOG ANALYSIS CONCEPTS IN,
B 6111 1951 UNITED STATES, HILL CREEK HEAVY OIL DEPOSIT
B 6201 105 UNITED STATES, MEADE PEAK PHOSPHATIC SHALE MEMBER, WESTERN
B 6303 299 UNITED STATES, OFFSHORE, PROSPECTIVE AREA
B 6303 459* UNITED STATES OIL HISTORY, ABST., WORLD, OIL, RESOURCE
B 6303 299 UNITED STATES, ONSHORE, PROSPECTIVE AREA
B 6201 98† UNITED STATES, ORGANIC, GEOCHEMISTRY, INCIPIENT, METAMORPHISM,
S 9 11 UNITED STATES, PENNSYLVANIAN, WIND DIRECTION, WESTERN
S 9 11 UNITED STATES, PERMIAN, WIND DIRECTION, WESTERN
B 6303 470* UNITED STATES PETROLEUM ESTIMATES, ABST., TWENTY YEARS OF
B 6303 473* UNITED STATES PETROLEUM IN WORLD ENERGY PERSPECTIVE, ABST.
B 6008 1404* UNITED STATES PETROLEUM PROVINCES, ABST., NEW CONCEPTS IN
B 6405 789* UNITED STATES PROVINCE OVERVIEWS, ABST.
B 6109 1522† UNITED STATES, RAPID, APTIAN-, ALBIAN, SUBSIDENCE RATES IN
B 6006 926† UNITED STATES, REGIONAL GEOLOGIC, FRAMEWORK OFF
B 6201 107 UNITED STATES, RETORT PHOSPHATIC SHALE MEMBER, WESTERN
M 25 157 UNITED STATES, SNAKE RIVER DOWNWARP, GEOTHERMAL POTENTIAL
B 6204 654 UNITED STATES, SOUTHEASTERN, MISSISSIPPIAN, LATE
B 6204 654 UNITED STATES, SOUTHEASTERN, PRIDE MOUNTAIN SANDSTONE
B 6111 1951 UNITED STATES, SPRINGS HEAVY OIL DEPOSIT
B 6111 1951 UNITED STATES, SUNNYSIDE HEAVY OIL DEPOSIT
B 6405 745* UNITED STATES TAR SAND, ABST., RECENT ACTIVITIES IN
B 6111 1951 UNITED STATES, TAR TRIANGLE HEAVY OIL DEPOSIT
B 6205 854 UNITED STATES, ULTRADEEP WELLS AS OF MARCH 1976
B 6205 862 UNITED STATES, WELL COSTS, PER FOOT DRILLED
M 25 156 UNITED STATES, WESTERN, GEOTHERMAL PROVINCES
M 25 486 UNITED STATES, WESTERN, MERCURY DEPOSITS
M 25 157 UNITED STATES, WESTERN, PLATE TECTONICS
M 25 157 UNITED STATES, WESTERN, VOLCANIC ROCKS
B 6303 478* UNITED STATES--CASE HISTORY OF POTENTIAL GAS COMMITTEE, ABST.,
B 6303 529* UNITED STATES--OBSERVATIONS FROM SUBMERSIBLE, ABST., SLUMPS ON
M 25 156† UNITED STATES--SUMMARY, GEOTHERMAL POTENTIAL OF WESTERN
B 6211 2355* UNITED STATES, BY TITLE ONLY, MESOZOIC TECTONICS OF WESTERN
B 6409 1529† UNITED STATES—CASE HISTORY OF POTENTIAL GAS COMMITTEE,
B 6405 773* UNITED STATES—FOURIER GRAIN SHAPE ANALYSIS, ABST., LOCATION OF
M 25 73† UNIVERSITY OF ARIZONA--SUMMARY, SOLAR, DESALINATION
B 6009 1564† UNRELIABILITY OF GLOBIGERINOIDES DATUM
B 6004 690* **UNRELIABILITY OF GLOBIGERINOIDES DATUM, ABST**
B 6405 682* UNUSUAL PONDING OF SEDIMENTS ON DEEP-WATER REEF, ABST.
B 6203 524* UP THE DECAY CHAIN—PERSPECTIVE ON MULTIELEMENT URANIUM
B 6002 323* UPDATE 1975, ABST., HEAVY, OIL IN WESTERN, MISSOURI--
B 6012 2174† UPDATE OF GEOTHERMAL RESOURCES AN DEVELOPMENTS IN
B 6305 830* UPDATE ON EXPLORATION FOR DIAMONDS IN COLORADO-WYOMING
M 26 491 UPDIP PERMEABILITY BARRIER, SEISMIC, MODELING
M 26 264 UPLIFT, LOUISIANA, WESTERN, DIAPIR
B 6012 2179* UPPE CRETACEOUS LITHOFACIES MODEL, SACRAMENTO VALLEY,
M 25 474 UPPER AMAZON MEGASTRUCTURE
M 28 78 UPPER CHALK, ENGLAND, CRETACEOUS
M 27 95 UPPER CHALK, ENGLAND, CRETACEOUS
M 27 82 UPPER CHALK, ENGLAND, CRETACEOUS
M 27 35 UPPER CHALK, ENGLAND, CRETACEOUS
S 7 243† UPPER CONTINENTAL SLOPE, LATE, PLEISTOCENE, GEOLOGIC HISTORY
S 7 262 UPPER CONTINENTAL SLOPE STABILITY, TEXAS
M 29 19 UPPER CRETACEOUS INTERVAL, GEORGIA EMBAYMENT
B 6405 772* UPPER CRETACEOUS MOSBY SANDSTONE, CENTRAL MONTANA—
B 6001 87† UPPER CRETACEOUS SUBSURFACE STRATIGRAPHY OF ATLANTIC
B 6309 1601* **UPPER CRETACEOUS--LOWER EOCENE STRATA, HAINESVILLE, KEECHI, AND**
B 6408 1288* UPPER DEVONIAN TIDE-DOMINATED DELTAIC-INTRADELTAIC
B 6303 488* UPPER DEVONIAN TURBIDITES OF CENTRAL AND SOUTHERN
B 6405 702* UPPER FREEPORT COAL, ABST., X-RAY MINERALOGY OF
B 6209 1755* UPPER MESOZOIC AND PALEOGENE GEOLOGY AND UNCONFORMITIES,
S 7 209 UPPER MIDDLE MIOCENE THICKNESS, WEST FLORIDA SLOPE
B 6211 2363* UPPER MIOCENE BARRIER BEACH-SALINE LAGOON COMPLEX, UPPER
S 7 210 UPPER MIOCENE THICKNESS, WEST FLORIDA SLOPE
B 6309 1596* UPPER MISSISSIPPIAN DELTAS IN BLACK WARRIOR BASIN OF MISSISSIPPI
B 6311 2119* UPPER MORROW FAN-DELTA DEPOSITS OF ANADARKO BASIN, ABST.
B 6309 1590* UPPER NIAGARAN AND LOWER CAYUGAN STRATIGRAPHY AND
B 6405 785* UPPER PALEOZOIC PALEOGEOGRAPHY OF IDAHO, ABST.
B 6303 511* UPPER PERMIAN EVAPORITES AND RED BEDS OF PALO DURO BASIN,
B 6203 360 UPPER PIRU-SAN RAFAEL UPLIFT, CALIFORNIA
M 30 310 UPPER SAND, PRUDHOE BAY, KUPARUK RIVER FORMATION
M 24 177† UPPER SMACKOVER RESERVOIRS, WALKER CREEK FIELD AREA,
B 6002 324* UPPER STRAWN DEPOSITIONAL SYSTEMS, LATE DESMOINESIAN, OF
B 6211 2277 UPPER SUNDANCE FORMATION, WYOMING
B 6303 421* UPPER TRIASSIC RADIOLARIA FROM EASTERN OREGON AND QUEEN
B 6210 1849 UPPER VOLTA, ZAMBIA, DEVELOPMENTS, 1977, TERRITORY OF, AFARS
B 6309 1535 UPSHUR COUNTY, GULF ENERGY AND MINERALS CO., U.S.A., D. A. COLLIE
B 6409 1564* UPSON FORMATIONS, UPPER CRETACEOUS, NORTHERN RIO ESCONDIDO
B 6101 113 UPTON SANDSTONE MEMBER, UTAH
B 6111 1955 URAL HEAVY OIL, USSR, VOLGA-
B 6405 711* URANIFEROUS AQUIFER-OAKVILLE SANDSTONE OF SOUTH TEXAS
M 25 454 URANIFEROUS GLAUCONITE AND APATITE, FUTURE POTENTIAL, JAPAN
S 3 135 URANIFEROUS MANTLE VOLATILES, DIATREME FEEDING OF
B 6105 843* URANIFEROUS SANDSTONES OF COLORADO PLATEAU, ABST., SANDSTONE
B 6206 1059 URANINITE OR COFFINITE
M 28 157 URANINITE, REPLACEMENT OF QUARTZ
M 25 453 URANINITE, URANIUM ORE MINERAL, JAPAN
B 6405 695* URANIUM, ABST., ADVANCES IN, RADON EXPLORATION TECHNIQUES
B 6303 538* URANIUM, ABST., ENVIRONMENTAL ASSESSMENT OF, IN-SITU, LEACHING
B 6303 503* URANIUM, ABST., FUTURE OF
B 6304 695* URANIUM, ABST., NEW MEXICO AND WORLD
B 6309 1599* URANIUM, ABST., NUCLEAR POWER AND GEOLOGY OF
B 6108 1388* URANIUM, ABST., RECOGNITION OF, OXIDIZED, SULFIDE MINERALS AS
B 6102 296* URANIUM, ABST., TEXAS
B 6206 1053 URANIUM ACCUMULATION
B 6206 1059 URANIUM ACCUMULATION, POTENTIAL, GEOCHEMICAL MECHANISMS
M 25 453 URANIUM ADSORBED ON ZEOLITES, JAPAN
M 25 451 URANIUM AND GRANITIC MASSES, RELATION OF, JAPAN
M 25 454 URANIUM AND GRANITIC ROCKS, JAPAN
B 6405 765* URANIUM AND IRON FROM RED BEDS OF CATSKILL FORMATION IN
B 6305 843* URANIUM AND ITS EFFECT ON EXPLORATION, ABST., SOLUTION, MINING
M 25 522† URANIUM AND ITS PROJECTED USE IN NUCLEAR GENERATION OF
B 6108 1382* URANIUM AND MINERAL EXPLORATION, ABST., NURE PROGRAM-A
B 6105 815* URANIUM, AND PYRITE CONCENTRATIONS IN PROTEROZOIC
S 3 26 URANIUM AND THORIUM, CYCLES, SECTIONS, NORTH AMERICA
S 3 10 URANIUM AND THORIUM, FUNDAMENTAL SOURCES OF
S 3 19 URANIUM AND THORIUM, GEOCHEMISTRY OF
S 3 12 URANIUM AND THORIUM, LITHOPHILE ELEMENTS
S 3 79 URANIUM AND THORIUM, OCEANIC, MIGRATION HISTORY OF
S 3 77 URANIUM AND THORIUM, SECTION, CYCLES OF
S 3 83 URANIUM AND THORIUM, SECTION, CYCLES OF
S 3 22 URANIUM AND THORIUM TRANSFER TO THE CRUST, MECHANISMS OF
S 3 36 URANIUM AND THORIUM WITHIN CRUST, MIGRATION OF,
B 6303 440* URANIUM AREA OF WESTERN KARNES COUNTY, TEXAS, ABST., DETAILED
B 6305 826* URANIUM AT PITCH MINE, SAWATCH RANGE, COLORADO, ABST.,

M 25 427 URANIUM, AUSTRALIA, NORTHERN TERRITORY, EAST ALLIGATOR
B 6108 1373* URANIUM BODY-RAJAH 49 MINE, MESA COUNTY, COLORADO, ABST.,
B 6101 39 URANIUM, BOUNDARY-ROLL DEPOSITS
S 3 136 URANIUM BY NATURAL FLUI OR GASEOUS HYDROCARBONS, SECRETION,
B 6304 686* URANIUM BY PROMPT FISSION NEUTRON METHO OF PULSED NEUTRON
M 25 452 URANIUM CONCENTRATION AND GRANITIC BODIES, DISTRIBUTION,
M 25 453 URANIUM CONCENTRATION, GEOCHEMISTRY OF, JAPAN
M 25 454 URANIUM CONTENT OF GROUNDWATERS, JAPAN, TONO AREA
M 25 427 URANIUM, COPPER, AND LEAD, AUSTRALIA, NORTHERN TERRITORY,
M 25 523 URANIUM DEMAND AND SERVICES, MEXICO
B 6304 691* URANIUM DEPOSIT, ABST., GEOLOGY AND DEVELOPMENT OF, MARQUEZ,
B 6405 772* URANIUM DEPOSIT, ABST., HISTORY OF, SULFIDIZATION OF SOUTH,
B 6004 653* URANIUM DEPOSIT LA SAL MINE, SAN JUAN COUNTY, UTAH, ABST.,
B 6203 504* URANIUM DEPOSIT, NORTHERN TERRITORY, AUSTRALIA, ABST.,
B 6103 360 URANIUM DEPOSIT, OKLAHOMA, CEMENT
B 6203 561* URANIUM DEPOSIT, RED DESERT, WYOMING, ABST., VERTICAL
B 6203 573* URANIUM DEPOSIT REDISCOVERED. BY TITLE ONLY., COPPER
B 6304 693* URANIUM DEPOSIT, SAN MATEO, NEW MEXICO, ABST., MOUNT TAYLOR
B 6304 692* URANIUM DEPOSIT, SMITH LAKE DISTRICT, NEW MMXICO, ABST.,
B 6304 691* URANIUM DEPOSIT, VALENCIA COUNTY NEW MEXICO, ABST.,
B 6305 839* URANIUM DEPOSIT--DESCRIPTION AND COMPARISON WITH WYOMING
B 6304 686* URANIUM DEPOSITION IN GRANTS MINERAL BELT, ABST., MECHANISMS
B 6405 787* URANIUM DEPOSITS, ABST., ANOMALOUS, THERMOLUMINESCENCE
B 6004 664* URANIUM DEPOSITS, ABST., ENVIRONMENTS OF FORMATION OF
B 6108 1379* URANIUM DEPOSITS, ABST., EXPERIMENTAL STUDY OF, PYRITE,
B 6303 435* URANIUM DEPOSITS, ABST., EXPLOITATION OF VERY LOW-GRADE
B 6303 472* URANIUM DEPOSITS, ABST., GEOCHEMISTRY OF SOUTH, TEXAS
B 6405 786* URANIUM DEPOSITS, ABST., GEOPHYSICAL EXPLORATION FOR,
B 6405 766* URANIUM DEPOSITS, ABST., HELIUM SURVEYING FOR DEEPLY BURIED
B 6305 843* **URANIUM DEPOSITS, ABST., LACUSTRINE, HUMATE**
B 6303 514* URANIUM DEPOSITS, ABST., MARCOSITE IN, ROLL-TYPE
B 6206 1091* URANIUM DEPOSITS, ABST., NEW, MODEL FOR SOME, COLORADO
B 6405 780* URANIUM DEPOSITS, ABST., ORGANIC MODEL FOR, ROLL-FRONT
B 6008 1407* URANIUM DEPOSITS, ABST., OVERVIEW OF, AUSTRALIAN
B 6203 567* URANIUM DEPOSITS, ABST., ROLE OF, ELECTROCHEMICAL, CELLS IN
B 6304 690* URANIUM DEPOSITS ABST., STRATIGRAPHIC CONTROL OF
B 6101 35 URANIUM DEPOSITS, COLORADO PLATEAU
B 6304 693* URANIUM DEPOSITS, GRANTS MINERAL BELT, NEW MEXICO, ABST., SEM
B 6305 830* URANIUM DEPOSITS, HISTORY AND PRODUCTION, ABST., POWDER RIVER
B 6207 1217* URANIUM DEPOSITS IN CALCRETES OF WESTERN AUSTRALIA, ABST.,
B 6304 688* URANIUM DEPOSITS IN GRANT MINERAL BELT, NEW MEXICO, ABST.,
B 6101 35 URANIUM DEPOSITS IN PHANEROZOIC TERRESTRIAL SEDIMENTS
B 6405 763* URANIUM DEPOSITS IN SALT WASH MEMBER, MORRISON FORMATION,
B 6203 524* URANIUM DEPOSITS IN SOUTH TEXAS GULF COASTAL PLAIN, ABST.,
B 6203 496* URANIUM DEPOSITS IN VOLCANICLASTIC ROCKS NEAR MOUNTAIN CITY,
M 25 451 URANIUM DEPOSITS, JAPAN, NINGYO-TOGE
M 25 452 URANIUM DEPOSITS, JAPAN, TONO
B 6405 782* URANIUM DEPOSITS OF PART OF CENTRAL GREAT DIVIDE BASIN,
B 6203 570* URANIUM DEPOSITS OF PITCH MINE, COLORADO. BY TITLE ONLY.
B 6303 470* URANIUM DEPOSITS OF TEXAS GULF COASTA PLAIN--TREND,
B 6206 1049† URANIUM DEPOSITS OF TEXAS GULF COASTAL PLAIN, GEOCHEMICAL
B 6404 582† URANIUM DEPOSITS OF TEXAS GULF COASTAL PLAIN. DISCUSSION,
B 6105 765* **URANIUM DEPOSITS OF NORTHERN SASKATCHEWAN**
B 6101 28 URANIUM DEPOSITS, ONTARIO, ELLIOTT LAKE
B 6101 33 URANIUM DEPOSITS, SASKATCHEWAN, HELIKIAN
B 6101 35 URANIUM DEPOSITS, SUMMARY OF, PRECAMBRIAN
B 6101 28† URANIUM DEPOSITS, SURFICIAL ORIGIN OF, NORTH AMERICAN,
B 6306 988 URANIUM DEPOSITS, THEIR MINERALOGY AND ORIGIN, EDITED BY M. M.
M 25 427 URANIUM DEPOSITS, WESTERN AUSTRALIA, YEELIRRIE
B 6304 694* URANIUM DEPOSITS--MARIANO LAKE TYPE, ABST., MINERALOGICAL
B 6305 823* URANIUM DEPOSITS--THEORY AND APPLICATION, ABST., LESS COMMON,
B 6203 501* URANIUM DEPOSITS. BY TITLE ONLY., NEW, MODEL FOR SOME,
S 3 30 URANIUM, DIATREME FEEDING OF
B 6203 561* URANIUM DISCOVERY IN FRONTIER AREA—ANDERSON MINE, YAVAPAI
B 6304 621† URANIUM DISCOVERY IN VOLCANICLASTIC SEDIMENTS, BASIN AND
S 3 5 URANIUM DISCOVERY-EFFICIENCY INDEX, 1948-72
M 25 451 URANIUM, DISTRIBUTION AND OCCURRENCE, JAPAN
B 6203 548* URANIUM DISTRIBUTION IN SOUTH TEXAS, ABST., EFFECT OF, CLAYS ON
B 6303 533* URANIUM DISTRICT, CHIHUAHUA, MEXICO, ABST., CARBONATE
B 6405 695* URANIUM EXPLORATION, ABST., APPLICATION OF REGIONAL,
B 6108 1391* URANIUM EXPLORATION, ABST., ASSESSMENT OF, MAGNETIC
B 6108 1376* URANIUM EXPLORATION, ABST., HELIUM DETECTION IN
B 6008 1406* URANIUM EXPLORATION, ABST., SABKHA ENVIRONMENT--NEW
B 6203 515* URANIUM EXPLORATION, ABST., TEST USE OF, ERTS-, LANDSAT IMAGERY
S 3 2 URANIUM EXPLORATION ACTIVITY, 1948-72
B 6011 1993† URANIUM EXPLORATION, EXPECTATIONS FROM
S 3 1 URANIUM EXPLORATION, HISTORICAL ANALYSIS
B 6305 845* **URANIUM EXPLORATION IN EASTERN APPALACHIANS,** ABST.,
B 6305 831* URANIUM EXPLORATION IN LOWER CRETACEOUS ANTLER FORMATION,
B 6008 1397* URANIUM EXPLORATION IN PRECAMBRIAN CONGLOMERATES IN
B 6108 1385* URANIUM EXPLORATION IN SOUTH TEXAS, ABST., THERMAL-, INFRARED
B 6203 524* URANIUM EXPLORATION RATIONALE, ABST., UP THE, DECAY CHAIN—
B 6405 679* URANIUM EXPLORATION SYSTEMS CASE STUDY—COPPER MOUNTAIN,
B 6207 1229* URANIUM EXPORT POTENTIAL, ABST., AUSTRALIA
B 6109 1539* URANIUM FAVORABILITY IN UPPERMOST TERTIARY OF EAST TEXAS AND
B 6109 1535* URANIUM FAVORABILITY OF OAKVILLE AND GOLIAD FORMATIONS OF
M 25 454 URANIUM, FIXATION FROM GROUNDWATER
S 3 75 URANIUM, FIXATION OF
B 6304 640 URANIUM, GEOCHEMISTRY, ARIZONA
B 6305 842* URANIUM GEOLOGY OF MORTON RANCH PROPERTY, SOUTHERN
B 6109 1536* URANIUM HOST DEPOSITIONAL SYSTEMS OF TEXAS GULF COASTAL
B 6206 1053 URANIUM HOST ROCKS
B 6304 689* URANIUM HOST ROCKS IN AMBROSIA LAKE AREA, ABST.,
M 25 452 URANIUM HOST ROCKS, JAPAN, TOKI, LIGNITE-BEARING FORMATION
S 3 124 URANIUM HOST SANDSTONES, GRAIN-SIZE ANALYSIS
B 6008 1399* URANIUM IN APPALACHIAN BASIN--SOME INITIAL RESULTS, ABST.,
B 6203 512* URANIUM IN GRANITES. BY TITLE ONLY.
B 6304 691* URANIUM IN GRANTS MINERAL BELT, ABST., ORGANIC, GEOCHEMISTRY
B 6303 549* URANIUM IN ORCA BASIN SEDIMENTS, ABST.
B 6008 1389* URANIUM IN PLUTONIC ROCKS, ABST.
B 6008 1392* URANIUM IN TERTIARY CHANNEL SOUTHEAST OF LAKE FROME, SOUTH
B 6405 698* URANIUM IN TERTIARY SEDIMENTS IN ALASKA, ABST.
B 6304 692* URANIUM IN TODILTO LIMESTONE--SABKHA-LIKE DEPOSIT, ABST.
B 6305 827* URANIUM IN TRIASSIC-JURASSIC BASINS, EASTERN NORTH AMERICA,
B 6203 545* URANIUM IN TUFFACEOUS FLUVIAL SANDSTONE, SONORA PASS,
B 6303 486* URANIUM IN-SITU LEACH MINING--A THIRD ALTERNATIVE, ABST.
B 6206 1051 URANIUM LEACHING
B 6304 625 URANIUM MINE, PRODUCTION, ARIZONA, ANDERSON
B 6105 836* URANIUM MINERALIZATION, ABST., LACUSTRINE SEDIMENTATION IN,
B 6303 455* URANIUM MINERALIZATION, ABST., OAKVILLE FORMATION OF, TEXAS
B 6405 722* URANIUM MINERALIZATION, ABST., OXIDATION-REDUCTION IN,
B 6209 1673† URANIUM MINERALIZATION, AND PROVENANCE, MORRISON
B 6410 1718† URANIUM MINERALIZATION, AND PROVENANCE, MORRISON
B 6410 1650 URANIUM MINERALIZATION, AUSTRALIA
S 3 123 URANIUM MINERALIZATION, ELEMENTS, VANADIUM-
B 6405 802* URANIUM MINERALIZATION IN HOPI BUTTES, ARIZONA, ABST.
B 6305 837* URANIUM MINERALIZATION IN MORRISON FORMATION, UPPER
B 6405 677* URANIUM MINERALIZATION IN WESTERN OKLAHOMA, ABST.,
S 3 121 URANIUM MINERALIZATION PROCESSES, GEOCHEMICAL DISTINCTION OF
S 3 135 URANIUM MINERALIZATION, UNDEREMPHASIZED PROCESSES OF
B 6206 1055 URANIUM MINERALS
S 3 135 URANIUM MINERALS, DEUTERIC CRYSTALLIZATION OF, THORIUM-
B 6303 520* **URANIUM MINING AND MILLING IN SOUTH TEXAS, ABST., CONQUISTA**
B 6303 445* URANIUM MINING, FROM EXPLORATION TO RESTORATION, ABST.,
B 6304 690* URANIUM MINING ON GROUNDWATER IN AMBROSIA LAKE AREA, NEW
B 6303 466* URANIUM MINING PURSUANT TO SURFACE MINING AND RECLAMATION
S 3 40 URANIUM MOBILIZATION PROCESSES, SHALLOW
M 25 524 URANIUM, NEEDS OF, TABLE, MEXICO
B 6209 1674 URANIUM, NEW MEXICO, AMBROSIA LAKE DISTRICT
B 6209 1674 URANIUM, NEW MEXICO, CHUSKA DISTRICT
B 6209 1674 URANIUM, NEW MEXICO, GALLUP DISTRICT
B 6209 1680 URANIUM, NEW MEXICO, LAGUNA DISTRICT
B 6209 1680 URANIUM, NEW MEXICO, SMITH LAKE DISTRICT
B 6304 689* URANIUM OCCURRENCE, ABST., DISCONFORMITIES IN, GRANTS MINERAL
B 6304 692* URANIUM OCCURRENCE AT DENNISON-BUNN CLAIM AND POSSIBILITY
B 6405 755* URANIUM OCCURRENCE IN LEADVILLE DOLOMITE A PITCH MINE,
B 6305 835* URANIUM OCCURRENCE TRIASSIC, DOCKUM GROUP, TEXAS
B 6203 513* URANIUM ORE DEPOSITS, ABST., BACTERIAL MEDIATION OF, REDOX
B 6105 827* URANIUM ORE DEPOSITS, ABST., MEASUREMENT OF, RN222 AND, HE4-,
B 6405 795* **URANIUM ORE IN POISON CANYON AREA, MORRISON FORMATION, SAN**
M 25 453 URANIUM ORE MINERAL, JAPAN, AUTUNITE
M 25 453 URANIUM ORE MINERAL, JAPAN, CARNOTITE
M 25 453 URANIUM ORE MINERAL, JAPAN, COFFINITE
M 25 453 URANIUM ORE MINERAL, JAPAN, NINGYOITE
M 25 453 URANIUM ORE MINERAL, JAPAN, URANINITE
S 3 106 URANIUM ORE, PYRITE-, MARCASITE-, JORDESITE-, CALCITE-
S 3 3 URANIUM ORE RESERVES, HISTORICAL ESTIMATES
M 25 453 URANIUM ORE RESERVES, JAPAN
M 25 522 URANIUM ORE RESERVES, TABLE, MEXICO
S 3 62 URANIUM OREBODIES, LINEAR AND PLANAR SHAPES
S 3 62 URANIUM OREBODIES, TYPICAL SHAPES
B 6304 695* URANIUM OREBODY, ABST., GEOPHYSICAL EXPERIMENTS AT THE,
S 3 62 URANIUM OREBODY, C ROLL
S 3 62 URANIUM OREBODY, STACK COMPOUND
B 6304 693* URANIUM ORES, ABST., APPLICATION OF SOLUTION-MINERAL
B 6004 692* URANIUM ORES, ABST., CHARACTERIZATION OF, INSOLUBLE, ORGANIC
B 6101 33 URANIUM ORES, CHINLE FORMATION
B 6103 360† **URANIUM POTENTIAL OF PERMIAN AND PENNSYLVANIAN SANDSTONES**

B 6405 798* URANIUM POTENTIAL OF FRONTIER AREAS BY FUNCTIONAL
B 6103 360† URANIUM POTENTIAL OF PERMIAN AND PENNSYLVANIAN SANDSTONES
B 6304 623 URANIUM PRODUCTION, RESERVES, AND POTENTIAL, REGIONAL
B 6101 34 URANIUM PROVINCES, NORTHWESTERN, CANADIAN SHIELD, HELIKIAN
B 6405 789* URANIUM RATIO IN GRANITIC ROCKS AND IMPLICATIONS FOR
B 6105 825* URANIUM REGION, NEW MEXICO, ABST., EXPLORATION AND
B 6101 38 URANIUM REGION, NEW MEXICO, GRANTS
B 6304 687* URANIUM REGION SINCE 1963, ABST., EXPLORATION IN, GRANTS
M 25 451 URANIUM, RESERVES, JAPAN
B 6303 441* **URANIUM RESOURCE EVALUATION IN ANTARCTICA,** ABST.
B 6305 822* URANIUM RESOURCE OF GREAT DIVIDE BASIN, WYOMING, ABST.,
B 6207 1210* URANIUM RESOURCES IN AMERICAN WEST—WORLDWIDE IMPLICATIONS,
B 6008 1403* URANIUM RESOURCES IN NONCOMMUNIST WORLD, ABST.,
M 25 451† URANIUM RESOURCES OF JAPAN--SUMMARY, GEOLOGI ASPECTS OF
S 3 60 URANIUM ROLL FRONTS
B 6101 31 URANIUM, SASKATCHEWAN, BEAVERLODGE AREA
B 6109 1560* URANIUM SEARCH IN LAKE ATHABASKA REGION BY IMAGE PROCESSING
B 6303 519* URANIUM SOLUTION MINING--ENVIRONMENTAL AND LICENSING
B 6405 693* URANIUM SOLUTION MINING—INTEGRATION OF EXPLORATION AND
B 6206 1050 URANIUM SOURCE ROCKS, TEXAS
B 6206 1050 URANIUM SOURCE, TEXAS, CATAHOULA TUFFACEOUS ROCKS
B 6008 1396* URANIUM SOURCES, ABST., HELIUM DETECTION EXPLORATION GUIDE
M 25 427 URANIUM, SOUTH AUSTRALIA, BEVERLEY
M 25 444 URANIUM, SOUTH KOREA
M 25 523 URANIUM SUPPLY AND SERVICES, MEXICO
M 25 523 URANIUM, UNIT DATA, TABLE
S 3 56 URANIUM-BEARING FLUID, SHALLOW
B 6304 635 URANIUM-BEARING INTERVAL, ARIZONA, ANDERSON MINE
B 6108 1379* URANIUM-CATAHOULA FORMATION, TEXAS COASTAL PLAIN, ABST.,
M 25 427 URANIUM-RARE EARTHS, QUEENSLAND, MARY KATHLEEN
S 6 166 URANIUM-THORIUM, REVIEW, QUATERNARY DATING
B 6108 1389* URANIUM. BY TITLE ONLY., APPLICABLE REGULATIONS OF, IN-SITU,
B 6105 840* URANIUM—DEVELOPING TECHNOLOGY, ABST., IN-SITU, LEACHING OF
M 25 453 URANOPHANE, URANIUM ORE MINERAL, JAPAN, B-
S 3 124 URAVAN MINERAL BELT, SALT WASH SANDSTONE, COLORADO
B 6405 796* URAVAN MINERAL BELT, SOUTHWEST COLORADO, ABST.,
B 6303 544 URBAN, JAMES B., HYDROCARBON OCCURRENCE A FUNCTION OF
B 6211 2355* URBAN OIL AND GAS DEVELOPMENT, ABST., WORKSHOP ON
B 6004 685* URBAN OIL DEVELOPMENT, LOS ANGELES, CALIFORNIA, ABST.,
B 6104 579 URDANETA FIELD, VENEZUELA
B 6104 581 URDANETA HIGH, VENEZUELA
B 6307 1093 URELLA FAULT, AUSTRALIA
M 25 214 URENUI FORMATION, NEW ZEALAND
M 25 554 URITOHUASER DEPOSIT, LEAD-ZINC, ECUADOR
M 25 554 URITOHUASER, LEAD AND ZINC DEPOSITS, ECUADOR
M 25 319 URODO ANTICLINE, COLOMBIA
B 6010 1657 URUGUAY, 1975, MACKINNON, J. L., ON DEVELOPMENTS IN
B 6110 1591 URUGUAY, 1976, PETROCONSULTANTS S.A., ON DEVELOPMENTS IN
B 6210 2060 URUGUAY, 1977, PETROCONSULTANTS S.A., ON DEVELOPMENTS IN
M 29 120 URUGUAY, MARGIN, STRUCTURE SECTION
B 6007 1009 USAGE OF TERM SHEET SANDSTONE
B 6405 735* USCHEM, GEOCHEMICAL DATA FILE OF NATIONAL COAL RESOURCES
B 6004 644* USE OF COLOR DISPLAYS IN SEISMIC INTERPRETATION, ABST.
S 4 289† USE OF CORALS IN DETERMINING GLACIAL-INTERGLACIAL CHANGES IN
B 6203 523* USE OF DEPOSITIONAL MODELS IN PREDICTING ROOF PROBLEMS IN
B 6405 780* USE OF DEVONIAN CONODONTS IN PETROLEUM EXPLORATION,
B 6108 1386* USE OF HYDROGEOLOGY, GEOCHEMISTRY, AND GEOTHERMOMETER
B 6008 1389* USE OF INTERACTIVE COMPUTER GRAPHICS FOR EXPLORATION, ABST.
B 6012 2188* USE OF IR COLOR PHOTOGRAPHY AND CANYON PHOTOS IN
B 6105 770* USE OF RESERVE-RESOURCE ESTIMATES BY NATIONAL ERDA LABORATORY
B 6203 525* USE OF SEISMIC MODELING FOR REEF EXPLORATION IN MICHIGAN
B 6405 697* USE OF TRACE FOSSILS FOR INTERPRETATION OF TRIASSIC
B 6211 2354* USE OF VERTICAL-GRADIENT AEROMAGNETIC DATA FOR GEOLOGIC
B 6309 1453 USGS ATLANTIC MARGIN CORING PROJECT, AMCOR, BALTIMORE
B 6410 1660 USGS RV FAY CRUISES 19, 20, 21, AND 25
B 6004 694* USGS-AAPG DIGITIZED OIL AND GAS FIELD MA OF NORTH
M 30 235 USHAKOVA SUITE, USSR, LENA-TUNGUSKA AREA
B 6303 517* USING PETROLEUM RESOURCE ASSESSMENT TO IMPROVE EXPLORATION
B 6405 796 USMANI, TARIQ U., POST-MIDDLE CRETACEOUS SEISMIC STRATIGRAPHY
B 6212 2481 USNS BARTLETT AND BENT CRUISES, 1972
B 6305 794 USNS ELISHA KANE SPARKER PROFILES
B 6111 1962 USNS KEATHLEY T-AGS 35 CRUISE
M 30 236 USOLYE SUITE, USSR, LENA-TUNGUSKA AREA
B 6203 449 USP MARKER, INDIA, KALOL FORMATION
B 6004 697* USSR, ABST., EURASIAN, PLATE BOUNDARIES IN
B 6105 822* USSR, ABST., FUTURE OIL AND GAS RESOURCES OF
M 25 59 USSR, AMUR FOLDED REGION
B 6309 1473 USSR, ANADYR BASIN, APTIAN-TURONIAN SEQUENCE
B 6309 1474 USSR, ANADYR BASIN, MIOCENE-HOLOCENE SEQUENCE
B 6309 1473 USSR, ANADYR BASIN, SENONIAN-OLIGOCENE SEQUENCE
M 25 466 USSR, ANADYR-KORYAK GEOSYNCLINAL ZONE
M 30 232 USSR, ANGARA-LENA STRUCTURAL TERRACE
S 10 232 USSR, BAKU AREA, GOUSANY FIELD
S 10 232 USSR, BAKU AREA, KARACHUKHUR FIELD
S 10 203 USSR, BAKU AREA, MUD VOLCANO, DISTRIBUTION
S 10 232 USSR, BAKU AREA, PESCHANY FIELD
S 10 232 USSR, BAKU AREA, SURAKHANY FIELD
S 10 230 USSR, BAKU AREA, ZYKH FIELD, HYDROTHERMAL EFFECTS
M 29 305 USSR, BLACK SEA BASIN, CROSS SECTION
M 29 313 USSR, BLACK SEA BASIN, EVOLUTION
M 29 313 USSR, BLACK SEA BASIN, SEDIMENTARY SEQUENCE
M 25 463 USSR, BUREYA STABLE MASSIF
S 6 333 USSR, CARBONIFEROUS, COMPARISON
S 10 233 USSR, DON MEDVEDICA AREA (VOLGA), HYDROTHERMAL EFFECTS
M 25 59 USSR, EAST ASIAN VOLCANOGENIC BELT
M 25 475 USSR, EASTERN, ANTIMONY-MERCURY MINERALIZATION, TIME OF
M 25 476 USSR, EASTERN, GOLD-ANTIMONITE MINERALIZATION, TIME OF
M 25 477 USSR, EASTERN, LODE DEPOSITS OF ANTIMONITE AN
M 25 476 USSR, EASTERN LODE DEPOSITS OF GOLD-ANTIMONITE
M 25 475 USSR, EASTERN, STRATIFORM DEPOSITS O
M 25 463 USSR, FAR EAST, ACID VOLCANISM
S 10 227 USSR, FERGHANA, OIL AND COPPER IN SAME TRAP
S 10 227 USSR, FERGHANA FIELD, TRAP
M 25 479 USSR, GARNETS
B 6002 292 USSR, GAS FIELDS OF THE
M 25 466† USSR, GEOLOGIC FRAMEWORK OF METALLIC, MINERAL DEPOSITS OF
M 30 225† USSR, GEOLOGY AND, PETROLEUM FIELDS IN, PROTEROZOIC AND,
M 30 241 USSR, IRKUTSK AMPHITHEATER, PETROLEUM POTENTIAL
B 6410 1691 USSR, IRKUTSK, BASIN, OIL AND GAS FIELDS
M 25 479 USSR, KAMCHATKA, ULTRAMAFIC VOLCANIC ROCKS
B 6309 1468 USSR, KAMCHATKA-KORYAK FOLDBELT
M 25 468 USSR, KAVRALYAN COMPLEX
M 25 475 USSR, KELYAN DEPOSITS, ANTIMONY
M 25 466 USSR, KOLYNA MASSIF, MIDDLE, RECLASSIFIED
B 6309 1468 USSR, KORYAK HIGHLANDS, PIKULNEY SERIES
M 25 475 USSR, LENA STAGE, DOLOMITES
M 30 237 USSR, LENA-TUNGUSKA AREA, ANGARA SUITE
M 30 236 USSR, LENA-TUNGUSKA AREA, BELYA SUITE
M 30 237 USSR, LENA-TUNGUSKA AREA, BULAY SUITE
M 30 237 USSR, LENA-TUNGUSKA AREA, LITVINTSEV SUITE
M 30 226 USSR, LENA-TUNGUSKA AREA, LOCATION MAP
M 30 235 USSR, LENA-TUNGUSKA AREA, MOTY SUITE
M 30 233 USSR, LENA-TUNGUSKA AREA, STRATIGRAPHY
M 30 235 USSR, LENA-TUNGUSKA AREA, USHAKOVA SUITE
M 30 236 USSR, LENA-TUNGUSKA AREA, USOLYE SUITE
M 30 237 USSR, LENA-TUNGUSKA AREA, VERKHOLENSK SUITE
M 30 231 USSR, LENA-TUNGUSKA PROVINCE, EXPLORATION HISTORY
M 30 228 USSR, LENA-TUNGUSKA PROVINCE, STRUCTURE CONTOURS
M 30 243 USSR, MARKOVA FIELD, EXPLORATION HISTORY
B 6410 1691 USSR, MARKOVO DISCOVERY
M 30 234 USSR, MARKOVO FIELD, CROSS SECTION
M 30 232 USSR, MARKOVO FIELD, LITHOLOGIC COLUMN
M 30 245 USSR, MARKOVO FIELD, PARFENOVA HORIZON
B 6111 1951 USSR, MELEKESS HEAVY OIL DEPOSIT
M 25 466 USSR, MESOZOIC GEOSYNCLINAL DEVELOPMENT
M 29 465 USSR, MOSCOW BASIN, AVERAGE THICKNESS OF SEDIMENTS
M 30 232 USSR, NEP-BOTUOBIN ANTICLISE
M 25 59 USSR, NIPPON GEOSYNCLINAL REGION
M 25 466 USSR, NORTHEAST, BASINS, SUPERIMPOSED OR SUCCESSOR
M 25 470 USSR, NORTHEAST, CASSITERITE
M 25 467 USSR, NORTHEAST, COAL DEPOSITS
M 25 469 USSR, NORTHEAST, COPPER, LEAD, AND ZINC
M 25 467 USSR, NORTHEAST, GOLD AND CASSITERITE
M 25 468 USSR, NORTHEAST, GOLD DEPOSITS
M 25 467 USSR, NORTHEAST, HEAVY METALS
M 25 467 USSR, NORTHEAST, MAGMATISM
M 25 470 USSR, NORTHEAST, MERCURY
M 25 466 USSR, NORTHEAST, MESOZOIC TAPHROGENIC OR RIFTOGENIC ZONES
M 25 466 USSR, NORTHEAST, MESOZOIC TROUGHS
M 25 471 USSR, NORTHEAST, METALLOGENESIS, MAJOR, TIME OF
M 25 469 USSR, NORTHEAST, METASOMATIC REWORKING
M 25 468 USSR, NORTHEAST, MINERAL DEPOSITS
M 25 469 USSR, NORTHEAST, SILVER
M 25 475 USSR, OGNEY ORE DEPOSIT, ANTIMONY
M 25 468 USSR, OKHOTSK COMPLEX
M 25 466 USSR, OKHOTSK-CHUKOTSK VOLCANIC BELT
B 6111 1951 USSR, OLENEK HEAVY OIL DEPOSIT
M 25 463 USSR, OLGA-THERNEY VOLCANIC FIELD
M 25 466 USSR, OLOY-ALAZEY EUGEOSYNCLINAL SYSTEM
M 25 466 USSR, OLYUTOR-KAMCHATKA GEOSYNCLINAL ZONE

M 25 471 USSR, OMOLON MASSIF
M 25 479 USSR, PICRITES
S 6 364 USSR, PRECAMBRIAN, POTASSIUM-ARGON
S 6 364 USSR, PRECAMBRIAN, RUBIDIUM-STRONTIUM
B 6309 1467† USSR, REVIEW OF PETROLEUM GEOLOGY OF, ANADYR AND, KHATYRKA
M 30 238 USSR, RIPHEAN PALEONTOLOGY, PROBLEMATICA
M 30 238 USSR, RIPHEAN PALEONTOLOGY, SPOROPHYTES
M 30 24 USSR, SAMOTLOR FIELD, LANDSAT IMAGE
S 10 230 USSR, SIBERIA, KOLPASHEVO, GEOTHERMAL PROFILE
S 10 229 USSR, SIBERIA, OB RIVER AREA, HYDROTHERMAL EFFECTS
B 6410 1691 USSR, SIBERIAN PLATFORM, OIL AND GAS FIELDS
M 25 463 USSR, SIKHOTE ALIN, VOLCANIC BELTS, CHARACTERISTICS OF
B 6111 1951 USSR, SILIGIR HEAVY OIL DEPOSIT
B 6410 1682 USSR, SOUTH KAZAKSTAN, PRECAMBRIAN MICROFOSSILS
M 30 244 USSR, SREDNE-BOTUOBIN FIELD, EXPLORATION HISTORY
B 6410 1691 USSR, SREDNE-BUTUOBIN DISCOVERY
M 25 59† USSR, STRUCTURAL-FORMATION ANALYSIS OF NORTHWESTERN PART
S 10 231 USSR, TRANS CAUCASUS, REGIONAL GEOTHERMAL MAP
S 10 229 USSR, TRANS CAUCASUS AND BAKU, HYDROTHERMAL EFFECTS
M 30 239 USSR, TUNGUSKA SYNCLISE, CROSS SECTION
M 30 239 USSR, TUNGUSKA SYNCLISE, PETROLEUM POTENTIAL
M 30 238 USSR, VENDIAN PALEONTOLOGY, PROBLEMATICA
M 30 238 USSR, VENDIAN PALEONTOLOGY, SPOROPHYTES
M 30 250 USSR, VERKHNEVILYUY FIELD, STRUCTURE CONTOURS MAP
M 25 466 USSR, VERKHOYANSK-CHUKOTSK FOLDED AREA
M 25 59 USSR, VERKHOYANSK-CHUKOTSK FOLDED REGION
M 29 463 USSR, VOLGA-URAL BASIN, VOLUME OF SEDIMENT PRESERVED
B 6111 1955 USSR, VOLGA-URAL HEAVY OIL
M 29 463 USSR, WEST SIBERIAN BASIN, VOLUME OF SEDIMENT PRESERVED
M 25 466 USSR, YANA-KOLYMA MIOGEOSYNCLINAL SYSTEM
M 30 244 USSR, YARAKTIN FIELD, EXPLORATION HISTORY
M 30 251 USSR, YARAKTIN FIELD, STRUCTURE CONTOUR MAP
S 10 233 USSR, ZYKH FIELD, MINERAL CONTENT OF WATER
M 25 475† USSR--SUMMARY, EVOLUTIONARY SERIES OF, ANTIMONY DEPOSITS OF
B 6407 1092 UST-LIMIMTEVAYAMIAN HORIZON, ASIA, NORTHEAST
B 6007 1074 UTAGBA WELL, NIGER DELTA
B 6303 476* UTAH, ABST., BATHYMETRIC DISTRIBUTION OF, TRACE, FOSSILS IN
B 6303 555* UTAH, ABST., CARBONATE, MUD, MOUNDS FROM LOWER, ORDOVICIAN,
B 6405 757* UTAH, ABST., DEPOSITIONAL ENVIRONMENTAL ANALYSIS OF, KAIBAB
B 6305 840* UTAH, ABST., DEPOSITIONAL SETTING OF, COALS IN UPPE, CRETACEOUS,
B 6205 879* UTAH, ABST., ENVIRONMENTAL RECONSTRUCTION OF MIDDLE,
B 6405 712* UTAH, ABST., FACIES AND DEPOSITIONAL TECTONICS OF MIDDLE,
B 6305 836* UTAH, ABST., GEOLOGY AND STRUCTURAL CONTROL OF, GEOTHERMAL
B 6008 1402* UTAH, ABST., GEOLOGY OF, PINEVIEW FIELD AREA, SUMMI COUNTY
B 6008 1390* UTAH, ABST., GEOTHERMAL EXPLORATION OF, ROOSEVELT, KGRA
B 6205 883* UTAH, ABST., HOLOCENE, ICHNOFACIES FROM NORTHERN, GULF OF,
B 6305 832* UTAH, ABST., INFLUENCE OF DIAGENETIC REACTIONS ON NONMARINE
B 6008 1404* UTAH, ABST., LOWER MESOZOIC AND UPPER PALEOZOIC PETROLEUM
B 6205 880* UTAH, ABST., LOWER, TRIASSIC, HINGE LINE AND, SHELF, TAR SAND
B 6205 891* UTAH, ABST., MINERALOGIC CHARACTERISTICS OF SOME LOWER,
B 6405 763* UTAH, ABST., MUDSTONES AS EXPLORATION GUIDES TO TABULAR
B 6405 796* UTAH, ABST., MULTIPLE BARRIER-ISLAND AND DELTAIC
B 6008 1399* UTAH, ABST., NEW PALEOSTRUCTURAL INTERPRETATION OF, BASI AND
B 6008 1406* UTAH, ABST., OIL AND GAS POTENTIAL OF, NUGGET SANDSTONE,
B 6405 776* UTAH, ABST., OIL-TO-SOURCE CORRELATION—PINEVIEW FIELD,
B 6405 755* UTAH, ABST., ORIGIN OF SUBSURFACE, FRACTURE SYSTEMS—EXAMPLE
B 6206 1093* UTAH, ABST., PALEOCURRENTS OF, JACKPILE SANDSTONE, BURRO
B 6405 730* UTAH, ABST., PALEOGEOGRAPHY AND STRUCTURAL DEVELOPMENT OF
B 6004 708* UTAH, ABST., PETROLEUM, POTENTIAL AND DEPOSITIONAL
B 6203 537* UTAH, ABST., PETROLOGY AND PETROGRAPHY OF, GREAT BLUE
B 6004 653* UTAH, ABST., PRELIMINARY INVESTIGATIONS OF, ELEMENTAL,
B 6108 1373† UTAH, ABST., SAGEBRUSH AND, GEOCHEMICAL EXPLORATION IN, EAST
B 6305 837* UTAH, ABST., SEDIMENTARY AND TECTONIC CONTROLS OF, URANIUM
B 6008 1412* UTAH, ABST., SEDIMENTATION AND MULTIPLE, COAL-SEA
B 6405 764* UTAH, ABST., SEISMIC MODELING OF, PENNSYLVANIAN, CARBONATE,
B 6405 696* UTAH, ABST., SHALLOW, OIL, SHALE DEPOSITS OF SOUTHERN, UINTA
B 6205 885* UTAH, ABST., SOME PETROGRAPHIC CHARACTERISTICS OF SEQUENCE OF,
B 6405 732* UTAH, ABST., SOME RELATIONS BETWEEN DIAGENESIS AND POROSITY
B 6405 728* UTAH, ABST., STRATIGRAPHIC RELATIONS OF, PERMIAN FORMATIONS IN
B 6305 824* UTAH, ABST., STRATIGRAPHY AND, COAL DEPOSITS OF UPPER,
B 6205 893* UTAH, ABST., SUBSURFACE, THRUSTS ALONG, WASATCH, HINGE LINE IN
B 6305 825* UTAH, ABST., TERTIARY CORRELATIONS IN, BIGHORN AN, UINTA BASINS,
B 6205 890* UTAH, ABST., VOLCANIC SANDSTONE IN, GREEN RIVER AND, UINTA
B 6101 112† UTAH, AGE OF, FRONTIER FORMATION IN NORTH-CENTRAL
B 6101 113 UTAH, ALLEN HOLLOW SHALE MEMBER
M 24 121 UTAH, ALTAMONT-BLUEBELL FIELD
M 24 121† UTAH, ALTAMONT-, BLUEBELL--A MAJOR, NATURALLY, FRACTURED,
B 6408 1172 UTAH AND COLORADO, CUTLER ARKOSE
B 6408 1172 UTAH AND COLORADO, HERMOSA CARBONATE ROCK
B 6408 1172 UTAH AND COLORADO, MOLAS CARBONATE ROCK
B 6408 1168 UTAH AND COLORADO, PALEOZOIC, CARBONATE WEDGES
B 6408 1172 UTAH AND COLORADO, PARADOX SALT
B 6408 1172 UTAH AND COLORADO, REDWALL FACIES
B 6408 1172 UTAH AND COLORADO, WHITE RIM TAR
B 6405 782* UTAH AND NEVADA, ABST., PALEOGEOGRAPHY AND MARINE
B 6205 888* UTAH AND NEVADA, ABST., PETROLEUM GEOLOGY OF, GREAT BASIN,
B 6108 1386* UTAH AND NORTHERN ARIZONA-EOLIAN OR MARINE IN ORIGIN, ABST.,
B 6405 686 UTAH AND SOUTHWESTERN COLORADO, ABST., LOWER, PERMIAN
B 6405 799* UTAH, AND WYOMING, ABST., MIDDLE-LATE, PERMIAN
B 6207 1172 UTAH, ARCTURUS GROUP, NEVADA AND
B 6101 113 UTAH, ASPEN SHALE
B 6303 314 UTAH, CANYON RANGE THRUST
B 6101 113 UTAH, CHALK CREEK MEMBER
B 6201 102 UTAH, CHARLESTON OVERTHRUST FAULT
B 6101 113 UTAH, COALVILLE MEMBER
B 6101 114 UTAH, COALVILLE, TO CUMBERLAND GAP, WYOMING, CORRELATION
B 6303 313 UTAH, COLTON FORMATION
B 6207 1172 UTAH, CONFUSION RANGE, SECTION
M 28 107 UTAH, CRETACEOUS, MOWRY SHALE
M 28 72 UTAH, CRETACEOUS, MOWRY SHALE
M 28 120 UTAH, CRETACEOUS, PANTHER TONGUE
M 28 178 UTAH, CRETACEOUS, PRICE RIVER FORMATION
B 6408 1236 UTAH, CRETACEOUS SAND BARS
M 28 180 UTAH, CRETACEOUS, STAR POINT SANDSTONE
M 28 156 UTAH, CRETACEOUS, STAR POINT SANDSTONE
M 28 120 UTAH, CRETACEOUS, STAR POINT SANDSTONE
S 3 58 UTAH, CUTLER FORMTION
B 6303 311† UTAH, DEPOSITIONAL HISTORY OF, PALEOCENE--LOWER, EOCENE,
B 6008 1239 UTAH, DEVELOPMENTS, 1975
B 6108 1204 UTAH, DEVELOPMENTS, 1976
B 6208 1360 UTAH, DEVELOPMENTS, 1977
B 6308 1206 UTAH, DEVELOPMENTS, 1978
B 6409 1340 UTAH, DEVELOPMENTS, 1979
B 6101 113 UTAH, DRY HOLLOW MEMBER
B 6408 1238 UTAH, FERRON SANDSTONES
B 6303 316 UTAH, FLAGSTAFF LIMESTONE, COVE MOUNTAIN MEMBER
B 6303 316 UTAH, FLAGSTAFF LIMESTONE FERRON MOUNTAIN MEMBER
B 6303 318 UTAH, FLAGSTAFF LIMESTONE, MUSINIA PEAK MEMBER
B 6101 113 UTAH, FRONTIER FORMATION
B 6207 1174 UTAH, GERSTER LIMESTONE, NEVADA AND
B 6205 840 UTAH, GILSONITE, SOLID HYDROCARBON
B 6207 1172 UTAH, GOLD HILL, SECTION
B 6101 113 UTAH, GRASS CREEK MEMBER
B 6303 314 UTAH, GREEN RIVER FORMATION
M 24 123 UTAH, GREEN RIVER FORMATION
B 6111 1955 UTAH, HEAVY OIL
B 6101 113 UTAH, HENEFER FORMATION
M 28 85 UTAH, HOLOCENE, DUNES
M 27 217 UTAH, HOLOCENE OOLITES
M 27 115 UTAH, HOLOCENE SEDIMENTS
B 6101 113 UTAH, JUDD SHALE MEMBER
M 28 39 UTAH, JURASSIC, CURTIS FORMATION
M 28 115 UTAH, JURASSIC, CURTIS FORMATION
M 28 27 UTAH, JURASSIC, CURTIS FORMATION
M 28 114 UTAH, JURASSIC, CURTIS FORMATION
M 28 7 UTAH, JURASSIC, CURTIS FORMATION
M 28 107 UTAH, JURASSIC, ENTRADA SANDSTONE
M 28 106 UTAH, JURASSIC, ENTRADA SANDSTONE
M 28 56 UTAH, JURASSIC, MORRISON FORMATION
M 28 57 UTAH, JURASSIC, MORRISON FORMATION
B 6207 1171 UTAH, KAIBAB LIMESTONE, NEVADA AND
B 6101 113 UTAH, KELVIN FORMATION
B 6207 1178 UTAH, KUVELOUSIA LEPTOSA ZONE, NEVADA AND
B 6303 311 UTAH, LAKE FLAGSTAFF LACUSTRINE-ALLUVIAL
B 6102 234 UTAH, LAKE POWELL AREA, TESTS
S 3 129 UTAH, LISBON VALLEY
S 3 58 UTAH, LISBON VALLEY URANIUM DISTRICT
B 6101 113 UTAH, LONGWALL SANDSTONE MEMBER
B 6101 115 UTAH, LOWER CRETACEOUS-UPPER CRETACEOUS BOUNDARY
B 6201 99 UTAH, MEADE PEAK PHOSPHATIC SHALE MEMBER
B 6207 1172 UTAH, MEADE PEAK PHOSPHATIC SHALE MEMBER, PHOSPHORIA
S 3 129 UTAH, MONUMENT VALLEY
S 3 124 UTAH, MONUMENT VALLEY, SHINARUMP CONGLOMERATE
B 6312 2217 UTAH, NAVAJO SANDSTONE
B 6312 2220 UTAH, NAVAJO SANDSTONE
B 6102 228 UTAH, NAVAJO SANDSTONE, PERMEABILITY
B 6102 235 UTAH, NAVAJO SANDSTONE, PETROLOGY
B 6303 313 UTAH, NORTH HORN FORMATION
B 6105 822* UTAH OIL-IMPREGNATED ROCK, TAR SAND, RESOURCES, ABST., UNITED
B 6405 774* UTAH OIL-IMPREGNATED SANDSTONE DEPOSITS, ABST., MIGRATION AND

B 6101 113 UTAH, OYSTER RIDGE SANDSTONE MEMBER
B 6207 1178 UTAH, PARK CITY GROUP, CORRELATIONS, FOSSILS, NEVADA AND
B 6207 1178 UTAH, PENICULAURIS BASSI-NEOSTREPTOGNATHODUS SP. C ZONE
B 6207 1178 UTAH, PENICULAURIS BASSI-NEOSTREPTOGNATHODUS SULCOPLICATUS
B 6207 1178 UTAH, PENICULAURIS IVESI-NEOSTREPTOGNATHODUS PRAYI ZONE,
M 27 34 UTAH, PENNSYLVANIAN, PARADOX FORMATION
M 27 61 UTAH, PENNSYLVANIAN, PARADOX FORMATION
M 27 10 UTAH, PENNSYLVANIAN, PARADOX FORMATION
M 27 133 UTAH, PENNSYLVANIAN, PARADOX FORMATION
M 27 140 UTAH, PENNSYLVANIAN, PARADOX FORMATION
B 6405 638 UTAH, PINEVIEW FIELD
B 6206 1076 UTAH, PINEVIEW FIELD, DISCOVERED JANUARY 1975
B 6207 1173 UTAH, PLYMPTON FORMATION, NEVADA AND
M 28 42 UTAH, QUATERNARY, GYPSUM DUNES
M 26 348 UTAH, RED WASH FIELD, UINTAH COUNTY
B 6403 429 UTAH, RED WASH HIGH-WAX OIL FIELD
B 6303 312 UTAH, SEVIER THRUST BELT
B 6101 113 UTAH, SPRING CANYON MEMBER
B 6207 1171† UTAH, STRATIGRAPHIC RELATIONS OF, PARK CITY GROUP, PERMIAN, IN
B 6207 1178 UTAH, THAMNOSIA DEPRESSA ZONE, NEVADA AND
B 6303 437* UTAH THRUST BELT, ABST., ALLUVIAL-, FAN DEPOSITION AND TECTONIC
M 28 20 UTAH, TRIASSIC, CHINLE FORMATION
B 6206 1077 UTAH, TWIN CREEK LIMESTONE, SAMPLES AND GEOCHEMICAL DATA
B 6401 61 UTAH, UINTA BASIN, EOCENE, KEROGEN
S 10 53 UTAH, UINTA BASIN, GREEN RIVER FORMATION, MIGRATION
B 6101 113 UTAH, UPTON SANDSTONE MEMBER
B 6303 317 UTAH, WASATCH PLATEAU, COLUMNS
B 6205 889* UTAH, WYOMING, AND IDAHO, ABST., TWIN CREEK FORMATION AND
B 6207 1178 UTAH, YAKOVLEVIA MULTISTRIATA-NEOGONDOLELLA BITTERI ZONE,
B 6405 717* UTAH-ARIZONA HINGE LINE—THRUST BELT—POTENTIAL NEW
B 6211 2357* UTAH-ARIZONA HINGE-LINE THRUST BELT—POTENTIAL NEW
B 6405 767* UTAH-IDAHO OVERTHRUST BELT—RELATION TO CANADIAN FOOTHILLS
B 6108 1377* UTAH-IMPLICATIONS FOR HYDROCARBON EXPLORATION IN
B 6108 1389* UTAH-STRUCTURAL GEOMETRY AND RELATED STRATIGRAPHIC
B 6205 882* UTAH—A FIRST APPRAISAL, ABST., CARBON DIOXIDE RESOURCES OF
B 6408 1283 UTGAARD, JOHN, FACIES, PALEOECOLOGY, AND DEPOSITIONAL
B 6203 503* UTICA SEQUENCE OF NEW YORK, ABST., RECOGNITION OF, CRATONAL,
B 6303 554* UTILITY OF DIATOMS IN PETROLEUM EXPLORATION, ABST.
B 6108 1380* UTILIZATION OF GEOPHYSICS IN COAL EXPLORATION AND
B 6207 1212* UTILIZATION OF MAGMA ENERGY, ABST.
B 6407 1091 UTKHOLOKIAN-VIVENTEKIAN HORIZON, ASIA, NORTHEAST
B 6201 18 UTOROGU MACROSTRUCTURE, NIGER DELTA
B 6203 436 UTSIRA HIGH, NORTH SEA
B 6311 2001 UTSIRA HIGH, NORTH SEA
B 6302 183 UTSIRA HIGH, NORTH SEA
S 11 146 UVIGERINA ATWILLI
S 11 142 UVIGERINA, CALIFORNIA, MIOCENE
S 11 142 UVIGERINA CARMELOENSIS
S 11 146 UVIGERINA COCOAENSIS
S 11 146 UVIGERINA GALLOWAYI
S 11 20 UVIGERINA GALLOWAYI ZONE
S 11 146 UVIGERINA GESTERI
S 11 144 UVIGERINA GESTERI
S 11 142 UVIGERINA HANNAI
S 11 144 UVIGERINA HANNAI
S 11 282 UVIGERINA HISPIDA
S 11 249 UVIGERINA HISPIDA
S 11 142 UVIGERINA HOOTSI
S 11 249 UVIGERINA HOOTSI
S 11 280 UVIGERINA HOOTSI
S 11 144 UVIGERINA HOOTSI
S 11 142 UVIGERINA JOAQUINENSIS
S 11 142 UVIGERINA MODELOENSIS
S 11 76 UVIGERINA OBESA IMPOLITA
S 11 142 UVIGERINA OBESA IMPOLITA
S 11 146 UVIGERINA OBESA IMPOLITA
S 11 76 UVIGERINA OBESA OBESA
S 11 142 UVIGERINA OBESA OBESA
S 11 21 UVIGERINA OBESA ZONE
S 11 250 UVIGERINA PEREGRINA
S 11 282 UVIGERINA PEREGRINA
S 11 280 UVIGERINA PEREGRINA
S 11 142 UVIGERINA SEGUNDOENSIS
S 11 146 UVIGERINA SPARSICOSTATA
S 11 76 UVIGERINA SPARSICOSTATA
S 11 20 UVIGERINA SPARSICOSTATA ZONE
S 11 144 UVIGERINA SUBPEREGRINA
S 11 251 UVIGERINA SUBPEREGRINA
S 11 142 UVIGERINA SUBPEREGRINA
S 11 777 UVIGERINA SUBPEREGRINA
S 11 140 UVIGERINE BULMINIDAE, CALIFORNIA, MIOCENE
S 11 252 UVIGERINELLA CALIFORNICA
S 11 140 UVIGERINELLA CALIFORNICA CALIFORNICA
S 11 140 UVIGERINELLA CALIFORNICA GRACILIS
S 11 140 UVIGERINELLA CALIFORNICA ORNATA
S 11 140 UVIGERINELLA NUDOCOSTATA
S 11 75 UVIGERININAE
S 11 282 UVIGERNA SUBPEREGRINA
B 6405 600 UWAINAT MEMBER, MIDDLE EAST
B 6407 1016 UYAK FORMATION, ALASKA
B 6303 302 UYAK FORMATION, ALASKA
B 6004 660* VACHERIE SALT DOM AREA, LOUISIANA, ABST., FOSSIL, SALT
M 28 18 VACUOLIZATION, FELDSPAR
B 6203 500* VADOSE CEMENT PATTERNS IN STRAND-PLAIN LIMESTONES OF EASTERN
S 5 11 VADOSE DEPOSITS, MICHIGAN, BELLE RIVER MILLS FORMATION,
M 27 195 VADOSE DIAGENESIS
B 6011 2048† VADOSE PISOLITE AND CALICHE
B 640 468 VADOSE ZONE
B 6303 446 VAGHI, G. C., MALOSSA FIELD, DEEP DISCOVERY IN PO VALLEY, ITALY,
M 30 525 VAGHI, G. C., MALOSSA FIELD--A DEEP DISCOVERY IN THE PO VALLEY,
M 29 469 VAIL, P. R., GLOBAL CYCLES OF RELATIVE CHANGES OF SEA LEVEL FROM
B 6405 793 VAIL, P. R., JURASSIC SEA-LEVEL CHANGES FROM SEISMIC
B 6303 527 VAIL, P. R., RECENT ADVANCES I PASSIVE-MARGIN RESEARCH, ABST.
B 6004 699 VAIL, P. R., REGIONAL SEISMIC INTERPRETATION USING SEQUENCES AND
B 6105 837 VAIL, P. R., SEISMIC RECOGNITION OF DEPOSITIONAL FACIES ON SLOPES
M 26 49 VAIL, P. R., SEISMIC STRATIGRAPHY AND GLOBAL CHANGES OF SEA
M 26 51 VAIL, P. R., SEISMIC STRATIGRAPHY AND GLOBAL CHANGES OF SEA
M 26 63 VAIL, P. R., SEISMIC STRATIGRAPHY AND GLOBAL CHANGES OF SEA
M 26 53 VAIL, P. R., SEISMIC STRATIGRAPHY AND GLOBAL CHANGES OF SEA
M 26 117 VAIL, P. R., SEISMIC STRATIGRAPHY AND GLOBAL CHANGES OF SEA
M 26 99 VAIL, P. R., SEISMIC STRATIGRAPHY AND GLOBAL CHANGES OF SEA
M 26 83 VAIL, P. R., SEISMIC STRATIGRAPHY AND GLOBAL CHANGES OF SEA
M 26 135 VAIL, P. R., SEISMIC STRATIGRAPHY AND GLOBAL CHANGES OF SEA
B 6308 1428* VAL VERDE BASIN, SOUTHWEST TEXAS, ABST., EVIDENCE FOR, WRENCH,
B 6402 221† VAL VERDE BASIN, SOUTHWEST TEXAS, STRUCTURAL ANALYSIS OF,
B 6002 324* VAL VERDE BASIN, TEXAS, ABST., TURBIDITE, RESERVOIRS IN, CANYON
B 6102 297* VAL VERDE BASIN, TEXAS, BY TITLE ONLY, GAS, PRODUCTION FROM,
B 6203 495* VAL VERDE BASIN, WILCOX GROUP OF SOUTH TEXAS, AND BELL
B 6402 221 VAL VERDE COUNTY, MAPPING, TEXAS
B 6405 609 VALANGINIAN, MIDDLE EAST
B 6303 472 VALASTRO, S., JR., GEOCHEMISTRY OF SOUTH TEXAS URANIUM
B 6102 297 VALDER, CLAYTON, GAS PRODUCTION FROM TIGHT SANDSTONES OF
B 6210 2049 VALDERRAMA, B., AND J. VELANDIA, ON DEVELOPMENTS IN COLOMBIA,
B 6310 1749 VALDERRAMA, R., AND VELANDIA, J., ON DEVELOPMENTS IN COLOMBIA,
B 6203 395 VALDEZ GROUP, ALASKA
M 25 339 VALDIVIA, HUGO, HYDROCARBON POTENTIAL OF AMAZON BASINS OF
B 6304 691* VALENCIA COUNTY NEW MEXICO, ABST., MAGNITUDE AND VARIABILITY
B 6112 2105 VALENCIA FIELD, PERU
B 6309 1598* VALENTINE AND MENKING FIELDS, LAVACA COUNTY, TEXAS, ABST.,
B 6409 1557* VALENTINE AREA OF SOUTHEASTERN LOUISIANA, ABST., EXPLORATION
B 6309 1595* VALENTINE FIELD, LAVACA COUNTY, TEXAS, ABST., CHARACTERISTICS
B 6203 568 VALENTINE, JAMES W., PALEONTOLOGY, UNIFORMITARIANISM, AND THE
B 6407 1034 VALENTINE, JAMES W., RESEDIMENTED FACIES IN VENTURA BASIN,
B 6009 1614 **VALENTINE, PAGE C., BIOSTRATIGRAPHY OF PLEISTOCENE**
B 6105 821 VALENTINE, PAGE C., BIOSTRATIGRAPHY OF U.S. ATLANTIC
B 6004 728 VALIA, HARDARSHAN S., MIOCENE TRANSGRESSIONS AND REGRESSIONS
B 6209 1714 VALIA, HARDARSHAN S., RECORRELATION OF EOCENE-MIOCENE
B 6105 723 VALIA, HARDARSHAN S., RECORRELATION OF EOCENE-MIOCENE BOUNDARY
B 6305 825* VALLE CALDERA GEOTHERMAL SYSTEM IN NEW MEXICO, ABST.,
M 25 158 VALLES, BACA, CALDERA FIELD, GEOTHERMAL, NEW MEXICO
B 6412 2064 VALLEY AND RIDGE AREA, SAR IMAGERY, APPALACHIANS, SOUTHERN
B 6309 1579* VALLEY AND RIDGE BENEATH METAMORPHIC SEQUENCES OF
B 6410 1576 VALLEY AND RIDGE PROVINCE, GEOLOGIC SKETCH, APPALACHIANS
B 6410 1576 VALLEY AND RIDGE PROVINCE, GEOLOGIC SKETCH, VIRGINIA
B 6412 2076 VALLEY AND RIDGE PROVINCE, STRUCTURE, APPALACHIANS
B 6103 417 VALLEY AND RIDGE PROVINCE, VIRGINIA
B 6105 830* VALLEY AND RIDGE, WEST VIRGINIA—EXPOSED MODEL FOR COMPLEX
M 24 242 VALLEY PARK FIELD, MISSISSIPPI
B 6004 725* VALLEY SEDIMENTATION, ABST., LATERAL, INFILL AS MAJOR FACTOR IN,
B 6203 561* VALLEYS, ABST., CURRENTS IN, SUBMARINE CANYONS AND OTHER TYPES
B 6004 712* VALLEYS AND SEDIMENTARY STRUCTURES OFF RIO BALSAS OF WESTERN
B 6206 912† VALLEYS, CHANNELS, AND DEPOSITIONAL LOBES ON MODERN
B 6303 538 VALLIER, T. L., MIDDLE CRETACEOUS OXYGEN-DEFICIENT
B 6406 945 VALLIER, T. L., PETROGRAPHY AND GEOLOGIC SIGNIFICANCE OF UPPER
B 6408 1281* VALMEYERAN, EWING AREA, JEFFERSON ANDFRANKLIN COUNTIES,
B 6202 259 VALMONTE DIATOMITE MEMBER, CALIFORNIA
B 6203 571* VALMY AND VININI FORMATIONS, NORTH-CENTRAL NEVADA, ABST.,
B 6211 2245 VALPREDO FAN, CALIFORNIA
S 11 255 VALVULINERIA ARAUCANA

S	11	284	VALVULINERIA ARAUCANA
S	11	128	VALVULINERIA ARAUCANA
S	11	170	VALVULINERIA, CALIFORNIA, MIOCENE
S	11	166	VALVULINERIA, CALIFORNIA, MIOCENE
S	11	168	VALVULINERIA, CALIFORNIA, MIOCENE
S	11	164	VALVULINERIA CALIFORNICA APPRESSA
S	11	170	VALVULINERIA CALIFORNICA APRESSA
S	11	164	VALVULINERIA CALIFORNICA CALIFORNICA
S	11	166	VALVULINERIA CALIFORNICA CALIFORNICA
S	11	168	VALVULINERIA CALIFORNICA CALIFORNICA
S	11	170	VALVULINERIA CALIFORNICA CALIFORNICA
S	11	166	VALVULINERIA CALIFORNICA OBESA
S	11	168	VALVULINERIA CALIFORNICA OBESA
S	11	164	VALVULINERIA CALIFORNICA OBESA
S	11	164	VALVULINERIA CASITASENSIS CASITASENSIS
S	11	164	VALVULINERIA DEPRESSA
S	11	166	VALVULINERIA DEPRESSA
S	11	170	VALVULINERIA MIOCENICA MIOCENICA
S	11	166	VALVULINERIA MIOCENICA MIOCENICA
S	11	166	VALVULINERIA ORNATA
S	11	164	VALVULINERIA WILLIAMI
S	11	164	VALVULINERIDS, CALIFORNIA, TERTIARY
S	11	117	VALVULINERIDS, PHYLETIC ORIGINS
S	11	230	VALVULINIDAE
B	6212	2483	VAMP, VELOCITY-AMPLITUDE FEATURE, OR
B	6212	2483	VAMPS, BERING ABYSSAL PLAIN
B	6212	2486	VAMPS, DEEP-WATER BRIGHT SPOTS
B	6212	2481†	VAMPS—POSSIBLE HYDROCARBON-BEARING STRUCTURES IN BERING SEA
S	6	39	VAN COUVERING, J. A., BIOCHRONOLOGY
B	6004	729	VAN DE KAMP, PETER C., INORGANIC AND ORGANIC METAMORPHISM IN
M	30	195	VAN DEN BARK, EDWIN, EKOFISK--FIRST OF THE GIANT OIL FIELDS IN
B	6304	701	VAN DEN BARK, EDWIN, EKOFISK--FIRST OF WESTERN EUROPEAN GIANT
B	6008	1222	**VAN DEN BERG, JACOB, DEVELOPMENTS IN EAST-CENTRAL STATES IN 1975**
B	6409	1462	VAN DEN BERG, JACOB, DEVELOPMENT IN EAST-CENTRAL STATES IN 1979
B	6108	1183	VAN DEN BERG, JACOB, DEVELOPMENTS IN EAST-CENTRAL STATES IN
B	6208	1340	VAN DEN BERG, JACOB, DEVELOPMENTS IN EAST-CENTRAL STATES IN
B	6308	1324	VAN DEN BERG, JACOB, DEVELOPMENTS IN EAST-CENTRAL STATES IN
B	6405	759	VAN DER HAR, STEPHEN, STUDY OF DIAGENETIC PROCESSES IN
B	6004	698	VAN DYKE, L. H., CALCULATION OF COAL RESERVES BY COMPUTER
B	6002	322	VAN DYKE, RICHARD, DEPOSITIONAL ENVIRONMENTS OF CHEROKE
B	6110	1585	VAN HART, D., ON DEVELOPMENTS IN GUATEMALA, 1976
B	6111	1930	VAN HAUEN PASS, CANADIAN ARCTIC
B	6409	1568	VAN HEERDEN, IVOR LL., ATCHAFALAYA DELTA—LOUISIANAS NEW
B	6004	498	VAN HINTE, J. E., A CRETACEOUS TIME SCALE
S	6	269	VAN HINTE, J. E., A CRETACEOUS TIME SCALE
S	6	289	VAN HINTE, J. E., A JURASSIC TIME SCALE
B	6004	489	VAN HINTE, J. E., A JURASSIC TIME SCALE
B	6202	201	VAN HINTE, J. E., GEOHISTORY ANALYSIS—APPLICATION OF
B	6104	613	VAN HORN MOUNTAINS, TEXAS
B	6408	1289	VAN HORN, ROBERT G., COAL RESOURCES IN OHIO, ABST.
B	6303	425	VAN HOUTEN, F. B., EARLY MESOZOIC TECTONIC FRAMEWORK AND
B	6406	857	VAN HOUTEN, FRANKLYN B., LATEST JURASSIC-EARLY CRETACEOUS
B	6303	544	VAN HOUTEN, FRANKLYN B., MID-CENOZOIC FORTUNA FORMATION,
B	6404	585	VAN HOUTEN, FRANKLYN B., REVIEW OF FACIES MODELS, EDITED B
B	6409	1549	VAN HOUTEN, FRANKLYN B., REVIEW OF THE ANDES, A GEOLOGICAL
B	6101	79	VAN HOUTEN, FRANKLYN B., TRIASSIC-LIASSIC DEPOSITS OF MOROCCO
B	6405	763	VAN LOENEN, S. D., MUDSTONES AS EXPLORATION GUIDES TO TABULAR
B	6009	1617	**VAN LOPIK, JACK R., THE OIL INDUSTRY AND COASTAL ZONE**
B	6303	544	VAN OVERMEEREN, F. A., WALKER CREEK FIELD,
B	6106	900	VAN PLATE
B	6303	545	VAN SICLEN, DEWITT, C., LAND DEVELOPMENT AND FAULTING NEA
B	6204	657	VAN STEVENICK, W., SALT DIAPIRS ON SAO PAULO PLATEAU,
B	6008	1324	VAN TYNE, ARTHUR M., DEVELOPMENTS IN NEW YORK IN 1975
B	6108	1328	VAN TYNE, ARTHUR M., DEVELOPMENTS IN NEW YORK IN 1976
B	6208	1445	VAN TYNE, ARTHUR M., DEVELOPMENTS IN NEW YORK IN 1977
B	6308	1318	VAN TYNE, ARTHUR M., DEVELOPMENTS IN NEW YORK IN 1978
B	6409	1455	VAN TYNE, ARTHUR M., DEVELOPMENTS IN NEW YORK IN 1979
B	6203	569	VAN WAGONER, JOHN C., PENNSYLVANIAN CARBONATE AND
B	6203	541	VAN WIJHE, D. H., PREDICTION OF POROSITY IN PERMIAN, ZECHSTEIN 2,
S	3	123	VANADIUM-URANIUM MINERALIZATION, ELEMENTS
B	6207	1196	VANCOUVER ISLAND AND OREGON-CALIFORNIA BOUNDARY, PLATE
M	25	109	VANCOUVER ISLAND, COAL, CANADA
M	25	497	VANCOUVER ISLAND, WESTERN MINES COPPER-ZINC-LEAD DEPOSIT
B	6405	796	VANINETTI, GERALD E., MULTIPLE BARRIER-ISLAND AND DELTAIC
B	6008	1412	VANINETTI, JERRY, SEDIMENTATION AND MULTIPLE COAL-SEA
B	6010	1658	VANMIDDLESWORTH, N. R., ON DEVELOPMENTS IN VENEZUELA, 1975
B	6210	2060	VAN MIDDLESWORTH, N. R., ON DEVELOPMENTS IN VENEZUELA, 1977
B	6310	1763	VANMIDDLESWORTH, N. R., ON DEVELOPMENTS IN VENEZUELA, 1978
B	6110	1591	VANMIDDLESWORTH, N. R., PETROCONSULTANTS S.A., ON
B	6211	2220	VAQUEROS FAULT, CALIFORNIA
B	6211	2360*	VAQUEROS FORMATION ALONG SESPE CREEK, VENTURA COUNTY,
B	6211	2219	VAQUEROS FORMATION, CALIFORNIA
S	11	8	VAQUEROS SANDSTONE
S	8	21	VAR CANYON, FRANCE, TURBIDITY CURRENTS
B	6311	2103†	VARDAMAN TRIASSIC-JURASSIC SILL, CALHOUN COUNTY, MISSISSIPPI
B	6212	2512	VARGAS, CARLOS, EXPLORATION OF CARBONIFEROUS SYSTEM IN
B	6205	757	VARIABLE-AMPLITUDE FACIES, LOW-CONTINUITY AND
B	6403	438*	VARIABLES AFFECTING TRENCH AND TRENCH-SLOPE SEDIMENTATION
B	6104	558†	VARIABLES FOR EVALUATION OF PETROLEUM ACCUMULATION IN
B	6109	1410	VARIABLES, TECHNIQUES FOR MEASURING
B	6004	653*	VARIATION AND DIAGENETIC ASPECTS OF TABULAR URANIUM DEPOSIT
B	6004	709*	VARIATION IN POROSITY--NEW CONCEPT, ABST., SPATIAL
B	6406	961*	VARIATION IN TURBIDITE SAND FACIES AND APPLICATION TO
B	6405	689*	**VARIATIONS IN CRETACEOUS COAL-BEARING STRATA, GALLUP COAL**
M	29	139†	VARIATIONS IN BOTTOM PROCESSES ALONG THE U.S. ATLANTIC
B	6309	1609*	VARIATIONS IN LITTORAL SAND FEATURES WITH SPECIAL REFERENCE
B	6008	1392*	VARIATIONS IN NATURAL GAS COMPOSITION I BLANCO FIELD, SAN JUAN
B	6004	721*	VARIATIONS IN TRENCH-SLOPE MODEL, ABST.
B	6303	450*	VARIATIONS OF MORPHOLOGY AND SEDIMEN TRANSPORT WITHIN
B	6306	973	VARIEGATED SHALE UNIT, EGYPT
M	29	293†	VARIETY OF MARGINS AND DEEP BASINS IN THE MEDITERRANEAN
B	6301	69	VARISCAN PLATFORM, CZECHOSLOVAKIA
B	6104	601	VARNELL, R. J., STRUCTURE O BROCKMAN HILLS, SOUTHWESTERN NEW
B	6207	1238	VARNES, D. J., SLOPE-STABILITY PROBLEMS OF CIRCUM-PACIFIC REGION
B	6203	507	VARNES, KATHARINE L., GEOLOGIC APPRAISAL OF UNDISCOVERED OIL
B	6012	2103†	VARVES IN LOWER MEMBER OF ARKANSAS NOVACULITE, DEVONIAN,
S	6	179	VASS, DIONYZ, A RADIOMETRIC TIME SCALE FOR THE NEOGENE OF THE
B	6407	972	VAUGHN MEMBER, MONTANA
B	6205	892	VEAL, HARRY K., ROBERTS MOUNTAIN THRUST—ITS SIGNIFICANCE IN
B	6004	646	VEDDER, J. G., CHANGES IN MIOCENE PALEOBATHYMETR OF CALIFORNIA
B	6203	344	VEDDER, J. G., NEOGENE BASINFORMATION IN RELATION TO
B	6211	2355	VEDDER, J. G., SALINIAN BLOCK OF CALIFORNIA, AN ENIGMA, ABST.
B	6403	450	VEDDER, J. G., STRATIGRAPHIC AND STRUCTURAL RELATIONS TO
B	6012	2174	VEDDER, JOHN G., LATE MIOCENE PALEOBATHYMETR OF CALIFORNIA
M	25	539	VEEH, HERBERT H., METALLOGENESIS IN SOUTHEAST PACIFIC OCEAN,
M	25	431	VEERABURUS, MANAS, FLUORITE RESOURCES OF THAILAND--SUMMARY
B	6311	2088	VEEVERS, J. J., SEDIMENTARY-WEDGE PROGRADATION FROM
B	6203	546*	VEGETATED TERRAIN IDENTIFIED BY DIGITAL PROCESSING OF LANDSAT
M	25	507	VEIN DEPOSITS OF METALS, CANADIAN, CORDILLERA
B	6310	1749	VELANDIA, J., ON DEVELOPMENTS IN COLOMBIA, 1978, VALDERRAMA, R.,
B	6210	2049	VELANDIA, ON DEVELOPMENTS IN COLOMBIA, 1977, VALDERRAMA, B.,
M	25	339	VELARDE, PEDRO, HYDROCARBON POTENTIAL OF AMAZON BASINS OF
B	6102	208	VELASCO FORMATION, MEXICO
B	6202	244	VELASCO FORMATION, MEXICO
B	6207	1238	VELASCO, GERONIMO ZAMAN, PHILIPPINE NATIONAL ENERGY POLICY,
M	25	522	VELEZ, C., URANIUM AND ITS PROJECTED USE IN NUCLEAR GENERATION
B	6103	361	VELMA-CRUCE ANTICLINE, OKLAHOMA
M	29	65†	VELOCITIES OVER OUTER CONTINENTAL SHELF AND UPPER
M	29	182	VELOCITY ANALYSES, BLAKE ESCARPMENT
M	29	267	VELOCITY DATA, ALASKA, KODIAK
M	26	39	VELOCITY, LABORATORY, MEASUREMENTS, SATURATION EFFECTS,
M	26	22	VELOCITY MEASUREMENT, GERMANY, SOLENHOFEN LIMESTONE,
M	26	19	VELOCITY MEASUREMENT, LABORATORY, SYSTEMS
B	6012	2183*	VELOCITY REDUCTION, ABST., REGIONAL
M	26	101	VELOCITY RELATIONS, SEISMIC, REFLECTIONS
B	6203	518*	VELOCITY REVERSALS, ABST., GEOPHYSICAL AND GEOLOGICAL
M	29	69	VELOCITY SECTIONS, EXAMPLES, ATLANTIC COAST
M	29	211	VELOCITY SOLUTIONS, AUSTRALIA, SHELF
B	6212	2483	VELOCITY-AMPLITUDE FEATURE, OR VAMP
M	26	286	VELOCITY-CONDUCTIVITY RELATION, DENVER BASIN, PIERRE SHALE
M	26	18	VELOCITY-DENSITY-LITHOLOGY ASPECTS, SEISMIC, INTERPRETATION
M	26	18	VELOCITY-DEPTH ASPECTS, SEISMIC, INTERPRETATION, AGE-
M	26	15	VELOCITY-POROSITY RELATIONS, TIME-AVERAGE RELATION
B	6410	1660	VEMA 18, 19 30.01, AND 31.08, LAMONT-DOHERTY GEOLOGICAL
B	6111	2004	VEMA 1975 CRUISE 3205, RV
B	6106	930	VEMA CHANNEL
B	6002	197	VEMA CHANNEL, BRAZIL MARGIN
B	6406	906	VEMA CRUISE 28, RV
B	6104	546	VEMA CRUISE 31
B	6204	658	VEMA CRUISE 31, STATION 114, RV
B	6002	196	VEMA CRUISES THROUGH MAY 1974, RV
B	6202	274	VEMA FRACTURE ZONE
B	6007	1089	VEMA FRACTURE ZONE
B	6111	1993	VEMA-30, SEISMIC PROFILE, CANARY BASIN
B	6111	1993	VEMA-30, SEISMIC PROFILE, CAPE VERDE RISE
M	30	238	VENDIAN PALEONTOLOGY, PROBLEMATICA, USSR
M	30	238	VENDIAN PALEONTOLOGY, SPOROPHYTES, USSR
B	6108	1223	VENDITTI, ANTHONY R., DEVELOPMENTS IN UPPER GULF COAST OF
B	6010	1658	VENEZUELA, 1975, VANMIDDLESWORTH, N. R., ON DEVELOPMENTS IN
B	6110	1591	VENEZUELA, 1976, VANMIDDLESWORTH, N. R., PETROCONSULTANTS S.A.,

B 6210 2060 VENEZUELA, 1977, VAN MIDDLESWORTH, N. R., ON DEVELOPMENTS IN
B 6310 1763 VENEZUELA, 1978, VANMIDDLESWORTH, N. R., ON DEVELOPMENTS IN
B 6105 821* VENEZUELA, ABST., DEPOSITIONAL PATTERNS AND DIAGENETIC
B 6303 516* VENEZUELA, ABST., FACIES RECOGNITION FROM, WELL, LOGS TO
B 6111 1956 VENEZUELA, ALTAMIRA-SUATA HEAVY OIL
B 6104 595 VENEZUELA, ALTURITAS FIELD
B 6104 579 VENEZUELA, BACHAQUERO FIELD
B 6104 579 VENEZUELA, BACHAQUERO, SHELL, FIELD
M 29 364 VENEZUELA BASIN, CURACAO RIDGE, INTERPRETATION
M 29 367 VENEZUELA BASIN, CURACAO RIDGE, REFLECTING HORIZONS
M 29 365 VENEZUELA BASIN, CURACAO RIDGE, SEISMIC SECTION
M 29 366 VENEZUELA BASIN, LOS ROQUES TRENCH, SEISMIC SECTION
M 29 368 VENEZUELA BASIN, MUERTOS TRENCH, INTERPRETATION
M 29 368 VENEZUELA BASIN, MUERTOS TRENCH, SEISMIC (FOLDOUT)
M 29 363† VENEZUELA BASIN, TECTONIC DEVELOPMENT OF, TRENCH-ARC
M 29 369 VENEZUELA BASIN, TRENCH-ARC COMPLEXES, CONVERGENCE
M 29 368 VENEZUELA BASIN, VENEZUELA TRENCH, CARIB BEDS
M 29 364 VENEZUELA BASIN, VENEZUELA TRENCH, INTERPRETATION
M 26 340 VENEZUELA, BOSCAN FIELD
M 26 342 VENEZUELA, BOSCAN SANDSTONE
B 6104 579 VENEZUELA, BOSEAN FIELD
B 6311 2054 VENEZUELA, CATATUMBO RIVER DELTA, CHARACTERISTICS, TABLE
S 4 121 VENEZUELA, CAYO DE LOS MUERTOS
S 4 123 VENEZUELA, CAYO PERAZA
S 4 122 VENEZUELA, CAYO SAL
B 6104 581 VENEZUELA, CENTRAL LAKE HIGH
B 6104 579 VENEZUELA, CENTRO FIELD
B 6111 1956 VENEZUELA, CERRO NEGRO-MORICHAL-JOBO HEAVY OIL
S 4 113 VENEZUELA, CHICHIRIVICHE AREA, COASTAL REEFS
B 6104 579 VENEZUELA, CREOLE CENTRAL LAKE FIELD
B 6104 579 VENEZUELA, CREOLE LA ROSA FIELD
B 6104 579 VENEZUELA, CREOLE LAGUNILLAS FIELD
B 6104 579 VENEZUELA, CREOLE PUNTA BENITEZ FIELD
B 6104 579 VENEZUELA, CREOLE TIA JUANA FIELD
B 6104 581 VENEZUELA, CRETACEOUS
B 6111 1953 VENEZUELA, EASTERN, HEAVY OIL
B 6408 1163 VENEZUELA, EASTERN, TERTIARY, SANDSTONE WEDGES
B 6104 581 VENEZUELA, EOCENE
B 6408 1165 VENEZUELA, FREITES SHALE
B 6111 1956 VENEZUELA, GORRIN-MACHETE HEAVY OIL
S 5 182 VENEZUELA, GRAN ROQUE, LAGO PUEBLO, LOCATION
B 6111 1956 VENEZUELA, GREATER ANACO OIL FIELD
B 6111 1956 VENEZUELA, GREATER MERCEDES OIL FIELD
B 6111 1956 VENEZUELA, GREATER OFICINA OIL FIELD
B 6111 1956 VENEZUELA, GUARICO OIL FIELD
B 6111 1956 VENEZUELA, HAMACA-SANTA CLARA HEAVY OIL
B 6104 581 VENEZUELA, ICOTEA HIGH
B 6104 579 VENEZUELA, LA PAZ FIELD
B 6408 1165 VENEZUELA, LA PICA SANDSTONE
S 5 184 VENEZUELA, LAGO PUEBLO, CONCENTRATIONS IN BRINE
S 5 181 VENEZUELA, LAGO PUEBLO, PRECIPITATION OF GYPSUM
B 6109 1424 VENEZUELA, LAKE MARACAIBO AREA, RESERVOIRS
B 6408 1191 VENEZUELA, LAKE MARACAIBO, POSITIVE WATER BALANCE BASIN
B 6104 579 VENEZUELA, LAMA FIELD
B 6104 581 VENEZUELA, LAMA TREND
B 6104 579 VENEZUELA, LAMAR FIELD
B 6104 581 VENEZUELA, LAMAR HIGH
B 6408 1165 VENEZUELA, LAS PIEDRAS SANDSTONE
B 6104 579 VENEZUELA, LOWER LAGUNILLAS FIELD
S 4 111† VENEZUELA, MANS IMPACT ON COASTAL, REEFS--A EXAMPLE FROM
B 6104 579 VENEZUELA, MARA FIELD
B 6104 579 VENEZUELA, MARACAIBO BASIN, OIL AGE
B 6104 579 VENEZUELA, MENE GRANDE FIELD
B 6104 595 VENEZUELA, MENEG-CENTA FIELD
B 6104 579 VENEZUELA, MENEG. CUETA FIELD
B 6104 579 VENEZUELA, MENEG. LAMAR FIELD
B 6104 581 VENEZUELA, MIOCENE
B 6007 1009 VENEZUELA, MISOA-MIRADOR FORMATION
B 6104 595 VENEZUELA, MOTATAN FIELD
B 6104 579 VENEZUELA, MOTATAN FIELD
B 6104 579 VENEZUELA, NORTH LAMA FIELD
B 6403 429 VENEZUELA, OFICINA HIGH-WAX OIL FIELD
B 6408 1165 VENEZUELA, OFICINA SANDSTONE
B 6202 303 VENEZUELA, ORINOCO DELTA SYSTEM
B 6111 1951 VENEZUELA, ORINOCO HEAVY OIL DEPOSIT
B 6104 581 VENEZUELA, PALEOCENE
B 6111 1956 VENEZUELA, PILON HEAVY OIL
B 6110 1627 VENEZUELA, PRODUCTION, 1976
B 6210 2096 VENEZUELA, PRODUCTION, 1976-1977
B 6310 1786 VENEZUELA, PRODUCTION, 1977-1978
B 6104 581 VENEZUELA, PUEBLO VIEJO HIGH
B 6104 579 VENEZUELA, ROSARIO FIELD
B 6104 579 VENEZUELA, SAN JOSE FIELD
B 6007 1128 VENEZUELA, SHALE POROSITY FOR WELLS
B 6104 579 VENEZUELA, SHELL LAMAR FIELD
B 6104 597 VENEZUELA, SIBIRCARA FIELD
B 6104 579 VENEZUELA, SIBUCARA FIELD
B 6104 579 VENEZUELA, SIGNAL-CENTRO FIELD
B 6311 2042† VENEZUELA, STRATIGRAPHY OF, INTERMONTANE, LACUSTRINE, DELTA,
B 6104 579 VENEZUELA, SUN LAMA FIELD
B 6104 579 VENEZUELA, SUN LAMAR FIELD
B 6111 1956 VENEZUELA, TEMBLADOR OIL FIELD
M 30 388 VENEZUELA, TERTIARY BASIN, REGIONAL TECTONIC MAP
M 29 368 VENEZUELA TRENCH, CARIB BEDS, VENEZUELA BASIN
M 29 364 VENEZUELA TRENCH, INTERPRETATION, VENEZUELA BASIN
B 6111 1956 VENEZUELA, TUCUPITA OIL FIELD
B 6104 579 VENEZUELA, URDANETA FIELD
B 6104 581 VENEZUELA, URDANETA HIGH
B 6007 1094 VENEZUELA-TRINIDAD SHELF
B 6405 712* VENEZUELA—CASE STUDY OF CRETACEOUS SHALLOW-MARINE SHELF,
B 6102 182 VENTURA ANTICLINE, CALIFORNIA
B 6102 165 VENTURA ANTICLINE, CALIFORNIA
B 6104 594 VENTURA AVENUE FIELD, CALIFORNIA
B 6407 1036 VENTURA AVENUE FIELD, CALIFORNIA
B 6203 360 VENTURA AVENUE OIL FIELD, CALIFORNIA
B 6203 356 VENTURA BASIN, CALIFORNIA
B 6012 2192* VENTURA BASIN, CALIFORNIA, ABST., NEOGENE TECTONICS OF CENTRAL
B 6203 535* VENTURA BASIN, CALIFORNIA, ABST., STRUCTURE OF, ALISO CANYON
B 6305 844* VENTURA BASIN, CALIFORNIA, ABST., STRUCTURE OF, SAN CAYETANO
B 6407 1034† VENTURA BASIN, CALIFORNIA, AND MODEL OF LONGITUDINAL
B 6011 1985† VENTURA BASIN, CALIFORNIA, MAGNETIC-REVERSAL STRATIGRAPHY OF,
S 7 232 VENTURA BASIN, CROSS SECTION, CALIFORNIA
S 7 235 VENTURA BASIN, DISTAL TURBIDITE FACIES, CALIFORNIA
B 6102 170 VENTURA BASIN, GEOLOGIC MAP, CALIFORNIA, WESTERN
B 6102 138 VENTURA BASIN, GEOLOGIC MAP, CALIFORNIA, WESTERN
B 6407 1036 VENTURA BASIN, OIL FIELDS, CALIFORNIA
S 7 236 VENTURA BASIN, OXNARD SLOPE, SECTION, CALIFORNIA
B 6303 421* VENTURA BASINS, CALIFORNIA, ABST., VITRINITE REFLECTANCE,
B 6303 465* VENTURA COUNTIES, CALIFORNIA, ABST., DIAGENESIS OF, TERTIARY,
B 6211 2360* VENTURA COUNTY, CALIFORNIA, ABST., EARLY, MIOCENE SUBMARINE
B 6303 522* VENTURA COUNTY, CALIFORNIA, ABST., SUBSURFACE GEOLOGY OF,
B 6211 2363* VENTURA COUNTY, CALIFORNIA, ABST., UPPER, MIOCENE, BARRIER
B 6102 154 VENTURA FIELD, A, B, C, D PRODUCING BLOCKS, CALIFORNIA
B 6206 958 VENTURA FIELD, CALIFORNIA
B 6102 137† VENTURA FIELD, CALIFORNIA, I--FACIES GEOMETRY AND GENESIS OF
B 6102 169† VENTURA FIELD, CALIFORNIA, II--LITHOLOGY, COMPACTION AND
B 6102 176 VENTURA FIELD, PLIOCENE SANDS, CALIFORNIA
B 6102 177 VENTURA FIELD, REPETTO AND PICO SANDS, CALIFORNIA
B 6102 172 VENTURA FIELD, TYPE ELECTRIC LOG, CALIFORNIA
B 6102 140 VENTURA FIELD, TYPE ELECTRIC LOG, CALIFORNIA
B 6102 172 VENTURA FIELD, ZONES 1 TO 8, CALIFORNIA
B 6102 140 VENTURA FIELD, ZONES 1 TO 8, CALIFORNIA
B 6102 149 VENTURA FIELD, ZONES 5, 6, 7, 8, CALIFORNIA
B 6303 468* VENTURA--PORT HUENEME AREA, CALIFORNIA, ABST., SEDIMENTARY
S 11 68 VENTURIAN STAGE
S 11 216 VENTURIAN STAGE
B 6202 259 VENTURIAN STAGE, CALIFORNIA
M 25 473 VENUS, RING, STRUCTURES, DISTINCT ON, EARTH, MOON, MARS, AND
B 6102 207† VERACRUZ BASIN, MEXICO, ORIGIN AND DISTRIBUTION OF, TERTIARY,
B 6102 210 VERACRUZ BASIN, SEISMIC PROFILES, MEXICO
M 26 272 VERACRUZ FAN SYSTEM
S 4 209† VERACRUZ, MEXICO, CHANGE IN, OSTRACOD STANDING CROP, LAGUNA,
S 4 210 VERACRUZ, PATCH REEFS
B 6003 379† VERACRUZ PROVINCE OF MEXICO, COMPLEX, FAULT STRUCTURES IN
B 6303 432 VERBEEK, EARL R., FAULTS OFFSETTING LAND SURFACE IN
B 6303 545 VERBEEK, EARL R., QUATERNARY FAULT ACTIVITY IN TEXAS GULF
B 6304 599 VERDE OIL FIELD, FRACTURE RESERVOIR, NEW MEXICO
M 30 399 VERDIER, A. C., GEOLOGY OF THE HANDIL FIELD, EAST KALIMANTAN,
B 6303 545 VERDIER, A., ET AL, HANDIL FIELD, EAST KALIMANTAN, INDONESIA,
B 6102 207 VERDUGO V., RODOLFO, ORIGIN AND DISTRIBUTION OF TERTIARY
B 6211 2217 VERGELLES FAULT, CALIFORNIA
M 30 250 VERKHNEVILYUY FIELD, STRUCTURE CONTOURS MAP, USSR
M 30 237 VERKHOLENSK SUITE, USSR, LENA-TUNGUSKA AREA
M 25 466 VERKHOYANSK-CHUKOTSK FOLDED AREA, USSR
M 25 59 VERKHOYANSK-CHUKOTSK FOLDED REGION, USSR
B 6108 1372* VERMEJO FORMATION, WALSENBURG AREA, COLORADO, ABST.,
M 28 137 VERMICULAR, PORE FILLING, KAOLINITE
B 6104 587 VERMILION BLOCK FIELD, LOUISIANA
B 6305 839* VERMILLION CREEK COAL BED, HIGH-SULFUR, RADIOACTIVE COAL OF
B 6107 1080 VERMONT, ISLE LAMOTTE, ORDOVICIAN

B 6101 30 VERNA MINE, SASKATCHEWAN
S 11 268 VERNEUILINA SCABRA
S 11 229 VERNEUILINA SCABRA
S 11 229 VERNEUILINIDAE
B 6209 1610 VERNON SHALE, NEW YORK
B 6405 797 VERRALL, P., GEOLOGIC AND STRUCTURAL HISTORY OF ZAGROS
B 6301 90 VERRILL CANYON FORMATION, SCOTIAN BASIN
B 6303 485* VERTEBRATES IN QUATERNARY SEDIMENTS OF KENYA. ABST., TRACKS
B 6405 674* **VERTICAL AND SURFACE SEISMIC PROFILES MAP AQUIFERS IN MADISON**
B 6007 1064 VERTICAL DEVELOPMENT OF YUCATAN SHELF REEFS
B 6203 561* VERTICAL DISTRIBUTION OF HE4, RN222, AND PB210 OVER
B 6312 2145† VERTICAL FACIES RELATIONS OF TRANSGRESSIVE BARRIER, LATERAL
M 26 478 VERTICAL LITHOLOGY, SEISMIC RESPONSE, SEISMIC, MODELING
B 6012 2150† VERTICAL NET SANDSTONE DETERMINATION FOR ISOPACH MAPPING OF
B 6002 294 VERTICAL SECTION THROUGH COMPACTION CELL
M 26 4 VERTICAL, SEISMIC, REFLECTIONS, RESOLUTION
B 6007 1126 VERTICAL STRESS IN SEDIMENTS
B 6205 845† VERTICAL THICKNESSES IN DEVIATED WELLS, GRAPHIC DETERMINATION
B 6211 2354* VERTICAL-GRADIENT AEROMAGNETIC DATA FOR GEOLOGIC MAPPING
B 6012 2159 VERTREES, CHARLES D., WEST TEXAS GEOLOGICAL SOCIETY, 1926 TO 1976
B 6112 2091 VERY HIGH HYDROGEN KEROGEN
B 6305 843 VESSELL, RICHARD K., SEDIMENTOLOGY OF VOLCANICLASTIC DEPOSITS
B 6402 189 VICAR FORMATION, SPAIN
S 3 102 VICKSBURG FLEXURE, TEXAS
B 6209 1760* VICKSBURG FORMATION ALONG TEXAS GULF COAST, ABST., SANDSTONE
B 6405 717* VICKSBURG FORMATION, OLIGOCENE, SOUTH TEXAS, ABST., DELTAIC
B 6309 1596* VICKSBURG RESERVOIRS, MCALLEN RANCH FIELD, HIDALGO COUNTY,
B 6409 1567* VICKSBURG SANDSTONE IN JIM HOGG COUNTY, TEXAS, ABST.,
B 6409 1566* VICKSBURG SANDSTONES, MCALLEN RANCH FIELD, HIDALGO COUNTY,
B 6109 1538* VICKSBURGIAN BOUNDARY IN EAST-CENTRAL GULF COAST EVALUATED
B 6405 725* VICKSBURGIAN BOUNDARY IN EAST-CENTRAL GULF COAST USING
B 6409 1559* VICKSBURGIAN, MISSISSIPPI AN ALABAMA, ABST., OSTRACODE
B 6009 1611* VICKSBURGIAN OSTRACODS IN H. V. HOWE COLLECTION, ABST.,
B 6210 2151 VICTORIA AND TASMANIA, DEVELOPMENTS, 1977
B 6310 1805 VICTORIA AND TASMANIA, DEVELOPMENTS, 1978
B 6110 1889 VICTORIA, DEVELOPMENTS, 1976
B 6411 1863 VICTORIA, DEVELOPMENTS, 1979
B 6404 528 VICTORIA FJORD, GREENLAND
B 6206 1005 VICTORIA ISLAND, ARCTIC, CANADA
B 6104 602 VICTORIO MOUNTAINS, NEW MEXICO
B 6205 781 VICTORIO PEAK FORMATION, NEW MEXICO
B 6109 1449 VICTORIO PEAK MARGIN, NEW MEXICO AND, TEXAS
B 6303 546 VIDAL, J. P., GRONDIN OIL FIELD, GABON, ABST.
M 30 577 VIDAL, JACQUES, GEOLOGY OF GRONDIN FIELD
B 6105 833 VIDELOCK, SHARI L., HOLOCENE DOLOMITE LOCALITY IN FLORIDA BAY,
B 6303 449* VIDEOTAPES OF ACTIVE PRIMARY PHYSICAL AND BIOGENIC
B 6303 489 VIDETICH, PATRICIA E., PSEUDOSTALACTITES FROM SUBMARINE CAVE
M 28 4 VIEJA GROUP, TEXAS, TERTIARY
M 28 80 VIEJA GROUP, TEXAS, TERTIARY
M 28 83 VIEJA GROUP, TEXAS, TERTIARY
M 28 3 VIEJA GROUP, TEXAS, TERTIARY
M 28 138 VIEJA GROUP, TEXAS, TERTIARY
M 28 115 VIEJA GROUP, TEXAS, TERTIARY
S 7 49 VIEJO UNIT, GULF OF MEXICO
B 6004 687* VIENNA, AUSTRIA, ABST., TRAILS FROM, THE VIENNA WOODS--,
B 6411 1894 VIETNAM, 1979, FLOOD, LLOYD, ON DEVELOPMENTS IN
B 6110 1841 VIETNAM, DEVELOPMENTS, 1976
B 6210 1989 VIETNAM, DEVELOPMENTS, 1977
B 6310 1828 VIETNAM, DEVELOPMENTS, 1978
B 6203 561 VIETTI, BARBARA T., GEOLOGIC APPRAISAL OF UNDISCOVERED OIL AND
B 6105 837 VIGRASS, L. W., GRAND BANKS, NEWFOUNDLAND, ABST., GEOLOGY AND
B 6107 1010 VIGRASS, LAURENCE W., TRAPPING OF OIL AT INTRA-MANNVILLE,
B 6008 1413 VIGRASS, LAURENCE W., TRAPPING OF OIL AT MIDDLE MANNVILLE
M 30 53 VIKING B HORST, VIKING FIELD, SEISMIC LINE
M 30 61 VIKING BASIN, EFFECTS OF CIMMERIAN OROGENY
B 6311 2001 VIKING BASIN, NORTH SEA
M 30 61 VIKING BASIN, STRUCTURE MAP, TOP OF CRETACEOUS
B 6408 1245† VIKING BASIN, UNITED KINGDOM-NORWAY, PRESSURE ORIGIN AND
M 30 41 VIKING FIELD, GAS COMPOSITION, NORTH SEA
M 30 41 VIKING FIELD, LEMAN SANDSTONE FORMATION, NORTH SEA
M 30 40 VIKING FIELD, LOCATION MAP, NORTH SEA
M 30 48 VIKING FIELD, NORTH VIKING COMPLEX POOLS
M 30 50 VIKING FIELD, ROTLIEGENDES SANDSTONE, EOLIAN SANDS
M 30 44 VIKING FIELD, ROTLIEGENDES SANDSTONE FACIES DISTRIBUTION
M 30 51 VIKING FIELD, ROTLIEGENDES SANDSTONE, FACIES ISOPACH
M 30 43 VIKING FIELD, ROTLIEGENDES SANDSTONE ISOPACH
M 30 41 VIKING FIELD, ROTLIEGENDES SANDSTONE, NORTH SEA
M 30 56 VIKING FIELD, ROTLIEGENDES, STRUCTURE MAP
M 30 53 VIKING FIELD, SEISMIC LINE, VIKING B HORST
M 30 49 VIKING FIELD, SOUTH VIKING COMPLEX POOLS
M 30 42 VIKING FIELD, STRATIGRAPHIC COLUMN, NORTH SEA
M 30 55 VIKING FIELD, TIME-VELOCITY, TRIASSIC
M 30 45 VIKING FIELD, WELL LOCATION MAP, NORTH SEA
M 30 41 VIKING FIELD, ZECHSTEIN SALT, NORTH SEA
B 6008 1391* VIKING FORMATION, ALBERTA, ABST.
B 6109 1420 VIKING FORMATION, HYDROCARBONS, ALBERTA
M 30 39† VIKING GAS FIELD, A REVIEW OF THE
B 6203 521* VIKING GRABEN, ABST., FRIGG FIELD—LARGE SUBMARINE-, FAN, TRAP
B 6203 436 VIKING GRABEN, CENTRAL, RIFT SYSTEM, NORTH SEA
B 6203 426 VIKING GRABEN, EUROPE
B 6311 1999† VIKING GRABEN, FRIGG FIELD--LARGE, SUBMARINE-FAN TRAP IN LOWER
M 30 59† VIKING GRABEN, NORTH SEA, FRIGG FIELD--LARGE SUBMARINE-FAN
B 6203 435 VIKING GRABEN, NORTHERN SECTOR, RIFT SYSTEM, NORTH SEA
B 6203 436 VIKING GRABEN, SOUTHERN, RIFT SYSTEM, NORTH SEA
B 6408 1166 VIKING SANDSTONE, ALBERTA
B 6405 674* VIKING SANDSTONE, LOWER CRETACEOUS, CENTRAL ALBERTA, ABST.,
B 6112 2134 VIKING SANDSTONES, SUBNORMAL PRESSURES, ALBERTA
B 6302 183 VIKING TROUGH, NORTH SEA
B 6405 771 VILKS, GUSTAV, SEABED CHARACTERISTICS AND SAND DISPERSAL ON
S 6 173 VILLAFRANCHIAN CORRELATION, ITALY, CALABRIAN-
S 6 114 VILLE SERIES, ISOCHRON DIAGRAM, FRANCE
B 6011 2070 VINCELETTE, RICHARD R., CONCEPTS AND TECHNIQUES UTILIZED IN
B 6009 1448 VINCELETTE, RICHARD R., OIL-BEARING REEFS IN SALAWATI BASIN OF
B 6303 538 VINCENT, E, MIDDLE CRETACEOUS OXYGEN-DEFICIENT
B 6105 837 VINE, JAMES D., ENERGY TECHNOLOGY TO CREATE SURGE IN DEMAND
B 6105 838 VINE, JAMES D., LITHIUM IN BRINES AND SEDIMENTARY ROCKS, ABST.,
B 6109 1548 VINET, MARSHALL J., GEOLOGY OF SIERRA BALUARTES DIAPIR AND
B 6008 1370 VINEYARD, JOHN W., DEVELOPMENT IN ARKANSAS, NORTH LOUISIANA,
B 6108 1234 VINEYARD, JOHN W., DEVELOPMENTS IN ARKANSAS, LOUISIANA, AND
B 6405 681* VINEYARD SOUND, MASSACHUSETTS, ABST., TIDAL-CURRENT, SAND,
B 6003 379 VINIEGRA, FRANCISCO, COMPLEX FAULT STRUCTURES IN VERACRUZ
B 6303 546 VINIEGRA-O., FRANCISCO,, GREAT CARBONATE BANK OF YUCATAN AND
B 6203 571* VININI FORMATIONS, NORTH-CENTRAL NEVADA, ABST., DELTAIC
B 6203 518 VINKOVETSKY, Y., ANALYSIS OF OIL-FIELD DISTRIBUTION FOR
B 6203 569 VINKOVETSKY, YAKOV, WHY ARE LARGE DEPOSITS DISCOVERED FIRST
B 6105 838 VINOPAL, R. J., POROSITY IN CARBONATE SANDS, ABST., EFFECT OF
B 6309 1585 VINOPAL, ROBERT J., FABRIC-ELEMENT-BASED CLASSIFICATION FOR
B 6309 1590 VINOPAL, ROBERT, PETROLOGIC EVALUATION OF SIGNIFICANCE OF
B 6307 1136 VIOLA FORMATION, OKLAHOMA
M 25 400 VIOLARITE ZONE, AUSTRALIA, NICKEL MINERALIZATION, PYRITE-
B 6407 972 VIRGELLE SANDSTONE, GREAT PLAINS
B 6403 319 VIRGELLE SANDSTONE MEMBER, FIVE LITHOFACIES, MONTANA
B 6403 324 VIRGELLE SANDSTONE MEMBER, UPPER MEMBER, MONTANA
B 6403 320 VIRGELLE SANDSTONE, MIDDLE MEMBER, MONTANA
B 6002 321* VIRGILIAN, PENNSYLVANIAN, ROCKS IN ANADARKO BASIN, ABST.,
B 6004 711* VIRGILIAN, UPPER PENNSYLVANIAN, ROCKS IN ANADARKO BASIN, ABST.,
B 6406 821 VIRGIN HILLS FORMATION, AUSTRALIA
B 6203 404 VIRGIN ISLAND BASIN
B 6303 427* VIRGIN ISLANDS, ABST., BANK-BARRIER, REEF MORPHOGENESIS, ST.
B 6405 751* VIRGIN ISLANDS, ABST., ORIGIN OF BIOGENIC CARBONATE SANDS OF
S 4 283 VIRGIN ISLANDS, DIADEMA ANTILLARUM
S 4 283 VIRGIN ISLANDS, DIPLORIA STRIGOSA
M 29 360 VIRGIN ISLANDS, FREDERICKSTED PLATEAU, BATHYMETRY
M 29 362 VIRGIN ISLANDS, FREDERICKSTED PLATEAU, GEOMORPHOLOGIC
M 29 358 VIRGIN ISLANDS, FREDERICKSTED PLATEAU, SEISMIC SECTIONS
M 29 353† VIRGIN ISLANDS, GEOMORPHOLOGY AND, SUBSURFACE, GEOLOGY WEST
M 27 159 VIRGIN ISLANDS, HOLOCENE SEDIMENTS
B 6203 406 VIRGIN ISLANDS, JEALOUSY FORMATION
B 6203 407 VIRGIN ISLANDS, KINGSHILL LIMESTONE, FIVE MAJOR FACIES
B 6203 417 VIRGIN ISLANDS, KINGSHILL LIMESTONE, RESERVOIR POTENTIAL
B 6203 405 VIRGIN ISLANDS, KINGSHILL LIMESTONE, STRATIGRAPHIC COLUMN
B 6203 407 VIRGIN ISLANDS, KINGSHILL LIMESTONE, TYPE SECTION
S 4 329† VIRGIN ISLANDS, MIOCENE, KINGSHILL SEAWAY--A DYNAMIC,
M 29 356 VIRGIN ISLANDS, ST. CROIX, BATHYMETRY
B 6203 414 VIRGIN ISLANDS, ST. CROIX, CALEDONIA FORMATION
B 6203 404 VIRGIN ISLANDS, ST. CROIX, CENTRAL LIMESTONE PLAIN
B 6203 416 VIRGIN ISLANDS, ST. CROIX, CRETACEOUS
B 6203 414 VIRGIN ISLANDS, ST. CROIX, EAST END RANGE
B 6203 404 VIRGIN ISLANDS, ST. CROIX, EAST RANGE
S 4 336 VIRGIN ISLANDS, ST. CROIX, FACIES
B 6203 404 VIRGIN ISLANDS, ST. CROIX, FOUNTAIN VALLEY
M 29 355 VIRGIN ISLANDS, ST. CROIX, FREDERICKSTED CANYON
S 4 330 VIRGIN ISLANDS, ST. CROIX, GEOLOGIC MAP
B 6203 414 VIRGIN ISLANDS, ST. CROIX, HAMS BLUFF
B 6203 414 VIRGIN ISLANDS, ST. CROIX, JUDITH FANCY FORMATION
B 6203 403 VIRGIN ISLANDS, ST. CROIX, KINGSHILL MARL
B 6203 404 VIRGIN ISLANDS, ST. CROIX, NORTHSIDE RANGE
M 29 354 VIRGIN ISLANDS, ST. CROIX, PHYSIOGRAPHY
M 29 355 VIRGIN ISLANDS, ST. CROIX, SEAFLOOR MORPHOLOGY
M 29 355 VIRGIN ISLANDS, ST. CROIX, SHEPARD CANYON

M 29 355 VIRGIN ISLANDS, ST. CROIX, SPRAT HALL CANYON
M 29 359 VIRGIN ISLANDS, ST. CROIX, STRATIGRAPHIC SECTIONS
M 29 359 VIRGIN ISLANDS, ST. CROIX, SUBSURFACE GEOLOGY
B 6203 406 VIRGIN ISLANDS, ST. CROIX, TECTONICS
B 6203 403† VIRGIN ISLANDS, STRATIGRAPHY AND DEPOSITIONAL SETTING,
B 6203 509* VIRGIN ISLANDS. BY TITLE ONLY, GEOLOGY SEEN WHILE LIVING IN A
B 6009 1623* VIRGINI AND WEST VIRGINIA, ABST., STRATIGRAPHY AS KEY TO,
B 6405 770* VIRGINIA, ABST., CARBONATE DEPOCENTERS AND FACIES DISTRIBUTION
B 6405 715* VIRGINIA, ABST., CEMENT TYPES AND CEMENTATION PATTERNS
B 6408 1284* VIRGINIA ABST., COAL RESOURCE STUDIES IN
B 6203 539* VIRGINIA, ABST., DIGITATE, STROMATOLITIC, BIOHERMS AND
B 6203 492* VIRGINIA, ABST., DISTINCTIVE, TURBIDITE, SILTSTONE IN, DEVONIAN,
B 6004 676* VIRGINIA, ABST., FENESTRAL, FABRICS IN NONCYCLIC LOW, SALINITY,
B 6203 557* VIRGINIA, ABST., GEOMETRY, FACIES, AND DEVELOPMENT OF LARGE
B 6105 822* VIRGINIA, ABST., LITHOSTRATIGRAPHY OF, OAK GROVE CORE,
B 6309 1589* VIRGINIA, ABST., PATTERNS OF SPECIES DIVERSITY IN, HELDERBER
B 6305 846* VIRGINIA, ABST., PETROLOGY OF, TRIASSIC-, JURASSIC, CULPEPER
B 6203 537* VIRGINIA, ABST., TECTONIC CONTROL OF SEDIMENTATION IN, TRIASSIC-,
B 6309 1590* **VIRGINIA AND WEST VIRGINIA, ABST., PETROLOGIC EVALUATION OF**
B 6308 1244† VIRGINIA, AND WEST VIRGINIA IN 1978, OIL AND, GAS, DEVELOPMENTS
B 6409 1403† VIRGINIA, AND WEST VIRGINIA IN 1979, OIL AND, GAS, DEVELOPMENTS
B 6208 1399† VIRGINIA, AND WEST VIRGINIA, OIL AND, GAS, DEVELOPMENTS IN,
B 6108 1269† VIRGINIA. AND WEST VIRGINIA. OIL AND. GAS. DEVELOPMENTS IN,
B 6008 1288† VIRGINIA, AND WEST VIRGINIA, OIL AND, GAS, DEVELOPMENTS IN,
B 6410 1575† VIRGINIA APPALACHIANS, CARBONATE, RAMP-TO-BASIN TRANSITIONS
B 6410 1582 VIRGINIA, APPALACHIANS, SECTIONS
B 6309 1465 VIRGINIA, BALTIMORE CANYON, STRATIGRAPHIC COLUMN
B 6410 1605 VIRGINIA, BAYS FORMATION
B 6410 1577 VIRGINIA, BEEKMANTOWN FORMATION
B 6410 1577 VIRGINIA, BENBOLT FORMATION
B 6410 1590 VIRGINIA, BLACKFORD FORMATION
B 6410 1576 VIRGINIA, BLUE RIDGE THRUST
B 6304 652 VIRGINIA, BLUEFIELD FORMATION
B 6410 1577 VIRGINIA, BOTETOURT FORMATION
B 6410 1605 VIRGINIA, BOWEN FORMATION
B 6304 652 VIRGINIA, BRAILLER AND MILLBORO SHALES
B 6304 649 VIRGINIA, CALIFORNIA CO. F. P. STRADER NO. 1, SECTION
B 6304 652 VIRGINIA, CAMBRIAN
M 28 160 VIRGINIA, CAMBRIAN, CONOCOCHEAGUE FORMATION
M 28 154 VIRGINIA, CAMBRIAN, CONOCOCHEAGUE LIMESTONE
M 27 141 VIRGINIA, CAMBRIAN, CONOCOCHEAGUE LIMESTONE
M 27 139 VIRGINIA, CAMBRIAN, CONOCOCHEAGUE LIMESTONE
M 27 140 VIRGINIA, CAMBRIAN, COPPER RIDGE DOLOMITE
M 27 141 VIRGINIA, CAMBRIAN, COPPER RIDGE DOLOMITE
M 27 139 VIRGINIA, CAMBRIAN, COPPER RIDGE DOLOMITE
M 28 162 VIRGINIA, CAMBRIAN, UNICOI FORMATION
B 6410 1577 VIRGINIA, CHATHAM HILL FORMATION
B 6304 649 VIRGINIA, CHEPULTEPEC DOLOMITE
B 6103 417 VIRGINIA, CLINCHPORT FAULT
B 6410 1577 VIRGINIA, COLLIERSTOWN BEDS
B 6410 1580 VIRGINIA, COPPER CREEK BLOCK
B 6103 417 VIRGINIA, COPPER CREEK FAULT
B 6410 1576 VIRGINIA, COPPER CREEK THRUST
B 6101 80 VIRGINIA, CULPEPER BASIN
B 6103 417 VIRGINIA, CUMBERLAND OVERTHRUST
B 6103 417 VIRGINIA, CUMBERLAND PLATEAU
B 6008 1291 VIRGINIA, DEVELOPMENTS, 1975
B 6108 1272 VIRGINIA, DEVELOPMENTS, 1976
B 6208 1402 VIRGINIA, DEVELOPMENTS, 1977
B 6308 1248 VIRGINIA, DEVELOPMENTS, 1978
B 6409 1406 VIRGINIA, DEVELOPMENTS, 1979
B 6304 652 VIRGINIA, DEVONIAN
M 28 165 VIRGINIA, DEVONIAN, ORISKANY SANDSTONE
M 28 65 VIRGINIA, DEVONIAN, ORISKANY SANDSTONE
M 28 97 VIRGINIA, DEVONIAN, ORISKANY SANDSTONE
B 6101 80 VIRGINIA, DURHAM-DEEP RIVER BASIN, NORTH CAROLINA AND
B 6410 1586 VIRGINIA, EDINBURG BEDS
B 6410 1577 VIRGINIA, EFFNA BEDS
B 6410 1602 VIRGINIA, EGGLESTON FORMATION
B 6410 1578 VIRGINIA, ELBROOK CARBONATE ROCKS
B 6410 1590 VIRGINIA, ELWAY FORMATION
B 6410 1590 VIRGINIA, FIVE OAKS FORMATION
B 6309 1506 VIRGINIA, FORMATION-DENSITY LOG STUDY
B 6304 652 VIRGINIA, GILES COUNTY, CROSS SECTIONAL MODEL
B 6304 652 VIRGINIA, GREENBRIER LIMESTONE
B 6103 417 VIRGINIA, GREENDALE SYNCLINE
B 6304 652 VIRGINIA, HINTON FORMATION
B 6304 652 VIRGINIA, HONAKER DOLOMITE
B 6103 417 VIRGINIA, HUNTER VALLEY FAULT
B 6304 652 VIRGINIA, HUNTERSVILLE CHERT
B 6304 652 VIRGINIA, JUNIATA FORMATION
B 6304 652 VIRGINIA, KEEFER SANDSTONE
B 6304 649 VIRGINIA, KINGSPORT DOLOMITE
B 6304 652 VIRGINIA, KNOX DOLOMITE OR GROUP
B 6410 1577 VIRGINIA, KNOX GROUP
B 6410 1577 VIRGINIA, LANTZ MILLS BEDS
B 6410 1577 VIRGINIA, LIBERTY HALL BEDS
B 6410 1577 VIRGINIA, LINCOLNSHIRE LIMESTONE
B 6304 652 VIRGINIA, MACCRADY FORMATION
B 6410 1578 VIRGINIA, MARTINSBURG FORMATION
B 6304 652 VIRGINIA, MARTINSBURG SHALE
B 6304 649 VIRGINIA, MASCOT DOLOMITE
B 6103 417 VIRGINIA, MIDDLESBORO SYNCLINE
B 6304 652 VIRGINIA, MISSISSIPPIAN
B 6410 1605 VIRGINIA, MOCCASIN FORMATION
B 6410 1581 VIRGINIA, MOSHEIM LIMESTONE
B 6410 1580 VIRGINIA, NARROWS BLOCK
B 6410 1576 VIRGINIA, NARROWS THRUST
B 6410 1577 VIRGINIA, NEALMONT BEDS
B 6410 1581 VIRGINIA, NEW MARKET LIMESTONE
B 6304 652 VIRGINIA, NOLICHUCKY SHALE
B 6410 1580 VIRGINIA, NORTH MOUNTAIN BLOCK
B 6410 1576 VIRGINIA, NORTH MOUNTAIN THRUST
B 6410 1577 VIRGINIA, ORANDA BEDS
B 6304 652 VIRGINIA, ORDOVICIAN
M 27 123 VIRGINIA, ORDOVICIAN
M 27 176 VIRGINIA, ORDOVICIAN, ATHENS LIMESTONE
M 27 4 VIRGINIA, ORDOVICIAN, BAYS LIMESTONE
M 27 176 VIRGINIA, ORDOVICIAN, BAYS LIMESTONE
M 27 185 VIRGINIA, ORDOVICIAN, BAYS LIMESTONE
B 6304 650 VIRGINIA, ORDOVICIAN, CONODONTS
M 27 71 VIRGINIA, ORDOVICIAN, MARTINSBURG FORMATION
B 6410 1577 VIRGINIA, ORDOVICIAN, MIDDLE, STRATIGRAPHIC CHART
B 6304 652 VIRGINIA, PEARIS MOUNTAIN
B 6304 652 VIRGINIA, PETERS MOUNTAIN
B 6103 417 VIRGINIA, PINE MOUNTAIN FAULT
B 6103 417 VIRGINIA, POWELL VALLEY ANTICLINE
B 6304 652 VIRGINIA, PRICE FORMATION
B 6008 1298 VIRGINIA, PRODUCTION, 1974-1975
B 6108 1281 VIRGINIA, PRODUCTION, 1975-1976
B 6208 1410 VIRGINIA, PRODUCTION, 1976-1977
B 6308 1247 VIRGINIA, PRODUCTION, 1977-1978
B 6409 1412 VIRGINIA, PRODUCTION, 1978-1979
B 6103 417 VIRGINIA, PULASKI FAULT
B 6410 1576 VIRGINIA, PULASKI THRUST
B 6410 1581 VIRGINIA, PULASKI-STAUNTON BLOCK
B 6304 652 VIRGINIA, ROCKY GAP SANDSTONE
B 6304 652 VIRGINIA, ROME FORMATION
B 6304 652 VIRGINIA, ROSE HILL SHALE
B 6103 417 VIRGINIA, RUSSELL FORK FAULT
B 6304 652 VIRGINIA, SAINT CLAIR FAULT SYSTEM
B 6410 1580 VIRGINIA, SALTVILLE BLOCK
B 6103 417 VIRGINIA, SALTVILLE FAULT
B 6410 1576 VIRGINIA, SALTVILLE THRUST
B 6412 2064† VIRGINIA, SEASAT, ORBITAL, RADAR, IMAGERY FOR GEOLOGIC,
B 6304 649 VIRGINIA, SHALY DOLOMITE
B 6304 652 VIRGINIA, SILURIAN
M 28 5 VIRGINIA, SILURIAN, CLINTON FORMATION
M 28 106 VIRGINIA, SILURIAN, CLINTON FORMATION
M 28 29 VIRGINIA, SILURIAN, ROSE HILL FORMATION
B 6209 1609 VIRGINIA, SILURIAN-DEVONIAN CARBONATE SEQUENCE
B 6410 1580 VIRGINIA, ST. CLAIR BLOCK
B 6410 1576 VIRGINIA, ST. CLAIR THRUST
B 6410 1580 VIRGINIA, STAUNTON BLOCK
B 6410 1576 VIRGINIA, STAUNTON THRUST
B 6410 1580 VIRGINIA, TAZEWELL ARCH
B 6410 1576 VIRGINIA, THRUST FAULTS
B 6410 1590 VIRGINIA, TUMBEZ FORMATION
B 6304 652 VIRGINIA, TUSCARORA SANDSTONE
B 6103 417 VIRGINIA, VALLEY AND RIDGE PROVINCE
B 6410 1576 VIRGINIA, VALLEY AND RIDGE PROVINCE, GEOLOGIC SKETCH
B 6103 417 VIRGINIA, WALLEN VALLEY FAULT
B 6410 1602 VIRGINIA, WARDELL FORMATION
B 6410 1602 VIRGINIA, WASSUM FORMATION
B 6309 1577* VIRGINIA, WEST VIRGINIA OHIO, MARYLAND, AND SOUTHERN
B 6103 422 VIRGINIA, WILDCAT VALLEY SANDSTONE, MAP
B 6410 1602 VIRGINIA, WITTEN FORMATION
B 6004 712* VIRGINIA--ANCIENT ANALOGUES OF QUATERNARY TIDAL TERRACES OR
B 6009 1621* VIRGINIA--POSSIBLE RESERVOIR SANDSTONE, ABST., WILDCAT VALLEY
B 6103 416† VIRGINIA--POSSIBLE RESERVOIR SANDSTONE, WILDCAT VALLEY

S 11 138 VIRGULINA CALIFORNIENSIS GRANDIS
S 11 138 VIRGULINA CALIFORNIENSIS TICEENSIS
S 11 138 VIRGULINELLA MIOCENICA
S 11 138 VIRULINA DELMONTEENSIS
B 6405 747 VISCONTI, ROBERT, RADIOLARIAN BIOSTRATIGRAPHY AND
B 6303 448* VISCOSITY OF ESTUARINE SEDIMENTS, ABST., STICKY, MUDS--
S 10 39 VISCOSITY VERSUS QUANTITY OF WATER IN CLAY, MIGRATION
B 6204 669 VISEAN, ALGERIA
B 6206 1007 VISEAN TO DZHULFIAN, ARCTIC, CANADA, PALEOZOIC
B 6004 729 VISHER, G. S., FLUVIAL MODELS--WET SLOPES AND RAIN SHADOWS, ABST.
B 6303 546 VISHER, GLENN S., CONTROLS ON LATE CRETACEOUS-PALEOCENE
B 6203 566 VISHER, GLENN S., ECONOMIC POTENTIAL OF CANYON-FAN UNIT IN
S 4 291 VITI LEVU, GALAXEA, FIJI
B 6010 1871 VITOL EXPLORATION B.V. AND PETROCONSULTANTS S.A.,
M 30 330 VITORIA ARCH, BRAZIL, CAMPOS BASIN
B 6002 197 VITORIA ARCH, BRAZIL MARGIN
B 6002 197 VITORIA-TRINDADE CHAIN, BRAZIL MARGIN
B 6108 1392* VITRIC SANDSTONE, CENOZOIC HIGH PLAINS SEQUENCE, WYOMING AND
B 6302 242 VITRINITE REFLECTANCE
B 6312 2223 VITRINITE REFLECTANCE, MEASUREMENTS OF
B 6303 421* VITRINITE REFLECTANCE, PRESENT ROCK TEMPERATURE, LAUMONTITE
B 6207 1157 VITRINITE REFLECTANCE PROFILES, AUSTRALIA
B 6406 919 VITRINITE REFLECTANCE, TIME-, TEMPERATURE INDEX OF MATURITY
B 6303 441* VITRINITE REFLECTANCE--WHAT, HOW, AND WHY, ABST.
B 6402 242 VITTORI, JACQUES, SEISMIC STRUCTURE OF SOUTHWESTERN
B 6407 1091 VIVENTEKIAN HORIZON, ASIA, NORTHEAST, UTKHOLOKIAN-
B 6112 2104 VIVIAN SECTION, PERU
B 6004 671* VIZCAINO PENINSULVA, BAJA CALIFORNIA SUR, MEXICO, ABST.,
B 6307 1093 VLAMING SUBBASIN, AUSTRALIA
M 27 166 VOID-FILL CEMENTS
M 27 162 VOIDS, LEACHED
S 3 45 VOLATILIZATION, FRACTIONAL
S 10 195 VOLCANCITOS, MUD VOLCANO, COLOMBIA
B 6309 1603* VOLCANIC ACTIVITY IN TIME AND SPACE IN GULF COASTAL PROVINCE,
S 10 180 VOLCANIC ACTIVITY, METHANE, OCCURRENCES
M 25 269 VOLCANIC ACTIVITY, KOREA, CRETACEOUS
B 6301 3 VOLCANIC ARC, DEFINITION
B 6406 903 VOLCANIC ARC, INDONESIAN ISLAND INNER
M 25 360† VOLCANIC AREAS IN JAPAN--SUMMARY, GROUNDWATER IN
M 25 135† VOLCANIC AREAS--SUMMARY, POSSIBLE, ENERGY SOUUCES IN,
B 6004 651* VOLCANIC ASH PARTINGS AS STRATIGRAPHIC MARKER BEDS IN COALS
B 6205 793 VOLCANIC BASEMENT, PROFILE, BLAKE PLATEAU AREA
M 25 269 VOLCANIC BELT, KOREA, CHEJU
M 25 463 VOLCANIC BELTS, CHARACTERISTICS OF, JAPAN, SOUTHWEST
M 25 463 VOLCANIC BELTS, CHARACTERISTICS OF, USSR, SIKHOTE ALIN
S 3 135 VOLCANIC CENTERS, HYDROTHERMAL-VEIN FORMATION AROUND,
M 25 526 VOLCANIC ERUPTIONS, RECENT, BAJA CALIFORNIA
B 6105 839* VOLCANIC FIELD OF TEXAS, ABST., OLIGOCENE, VOLCANIC,
B 6309 1494 VOLCANIC FLOAT, PAKISTAN, CHAGAI HILLS
M 25 531 VOLCANIC FLOW, CHILE, EL LACO, MAGNETITE INTERPRETED AS A
B 6012 2128† VOLCANIC FORMATIONS AND THEIR BEARING ON SAN ANDREAS FAULT
B 6205 867† VOLCANIC FORMATIONS AND THEIR BEARING ON SAN ANDREAS FAULT
M 25 448 VOLCANIC FRONT, JAPAN, SULFUR DEPOSITS, RELATION TO
M 25 137 VOLCANIC GEOTHERMAL POTENTIAL, INDONESIA
B 6212 2512* VOLCANIC ISLANDS, ABST., GROUNDWATER IN
M 25 448 VOLCANIC OR GREEN TUFF AREA, JAPAN, MIOCENE SUBMARINE
M 25 526 VOLCANIC PROVINCES, PETROGENIC INTERPRETATION, MEXICO
B 6206 1019 VOLCANIC ROCKS, CANADA, SVERDRUP BASIN, MAFIC INTRUSIVE AND
B 6303 430* VOLCANIC ROCKS CLOSE TO GULF COAST, SIERRA MADRE ORIENTAL,
B 6102 252 VOLCANIC ROCKS, FORMAL ROCK-STRATIGRAPHIC TERM
B 6209 1680 VOLCANIC ROCKS, ISOTOPIC AGES; ARIZONA
M 25 479 VOLCANIC ROCKS, KALIMANTAN, ULTRAMAFIC
M 25 479 VOLCANIC ROCKS, MALAITA, ULTRAMAFIC
M 25 149 VOLCANIC ROCKS, RECENT, ALASKA
M 25 157 VOLCANIC ROCKS, UNITED STATES, WESTERN
M 25 479 VOLCANIC ROCKS, USSR, KAMCHATKA, ULTRAMAFIC
B 6205 890* VOLCANIC SANDSTONE IN GREEN RIVER AND UINTA FORMATIONS,
B 6105 839* VOLCANIC SEDIMENTARY APRON IN TRANS-PECOS VOLCANIC FIELD OF
B 6004 662* VOLCANIC SEDIMENTS AND THEIR SEDIMENTATION, ABST.,
M 25 526 VOLCANIC SERIES, CYCLES OF, MEXICO, BAJA CALIFORNIA
B 6205 884* VOLCANIC STRATIGRAPHY OF TRAP SPRING FIELD, NYE COUNTY,
B 6405 677* VOLCANIC-ASH DEPOSITS AND THEIR POSSIBLE RELATION TO URANIUM
B 6012 2191* **VOLCANIC-ASH PARTINGS IN COALS--CHARACTERISTICS AND**
M 25 458 VOLCANIC-PLUTONIC CHAIN, PACIFIDES, ARGURI-, KERULEN
M 25 276† VOLCANIC-ROCK RESERVOIRS--SUMMARY, PETROLEUM FIELDS OF,
M 25 509 VOLCANIC-SEDIMENTARY TROUGH, ALASKA
B 6304 621† VOLCANICLASTIC SEDIMENTS, BASIN AND RANGE PROVINCE, YAVAPAL
B 6305 843* VOLCANICLASTIC DEPOSITS FROM 1971-1974 ERUPTION CLUSTER OF
B 6211 2356* VOLCANICLASTIC ROCKS ADJACENT TO LOWER COOK INLET,
M 25 357† VOLCANICLASTIC ROCKS, EXAMPLES FROM INDONNSIA--SUMMARY,
B 6203 496* VOLCANICLASTIC ROCKS NEAR MOUNTAIN CITY, NEVADA, ABST.,
B 6304 621† **VOLCANICLASTIC SEDIMENTS, BASIN AND RANGE PROVINCE, YAVAPAL**
B 6201 87† VOLCANICLASTIC SEDIMENTS FROM ATKA BASIN, ALEUTIAN RIDGE,
B 6108 1375* VOLCANICLASTICS, GUATEMALA, ABST., DIAGENESIS OF, TERTIARY
B 6107 972 VOLCANICS, AFRICA, CAMEROON
M 25 464 VOLCANISM, CHARACTERISTICS OF, FAR EAST, SOUTHERN, ACID
M 25 169 VOLCANISM, CHILE
B 6405 668* VOLCANISM IN COASTAL RANGES, CENTRAL PERU, ABST., EARLY,
M 25 465 VOLCANISM, JAPAN, NORTHEAST, IGNIMBRITE
M 25 463 VOLCANISM, LESSER, KHINGAN RANGE, IGNIMBRITE
B 6211 2351* VOLCANISM, LOWER COOK INLET, ALASKA. BY TITLE ONLY, MOBILE BED
M 25 526† VOLCANISM OF BAJA CALIFORNIA--SUMMARY, MAJOR CHEMICAL
M 25 463† VOLCANISM OF SOUTHERN FAR EAST--SUMMARY, SOME
M 25 463 VOLCANISM, RELATED TO ARC-FAULT-BLOCK MOVEMENT
M 25 463 VOLCANISM, USSR, FAR EAST, ACID
B 6004 694* VOLCANO, DIMMI COUNTY, TEXAS, ABST., CARBONATE DIAGENESIS, AND
B 6305 843* VOLCANO FUEGO, GUATEMALA, ABST., SEDIMENTOLOGY OF,
B 6009 1607* VOLCANO IN SOUTHEAST LOUISIANA, ABST., DOOR POINT--, BURIED
M 25 136 VOLCANOES, ACTIVE, INDONESIA
M 25 357 VOLCANOES, INDONESIA
M 25 164 VOLCANOES, NICARAGUA
B 6209 1646 VOLCANOES, TAIWAN-, LUZON REGION
B 6405 790* VOLCANOGENIC ROCKS OF GREAT VALLEY SEQUENCE, NORTHERN
B 6108 1393* VOLCANOGENIC SANDSTONES, ABST., DIAGENESIS OF
B 6203 508* VOLCANOGENIC SANDSTONES, ABST., RESERVOIR POTENTIAL OF
B 6108 1394* VOLCANOGENIC SANDSTONES, ABST., TRANSFER-, REACTION MODELS OF
B 6203 559* VOLCANOGENIC SEDIMENTATION ON ANDEAN PLATEAU, NORTHERN
M 25 487 VOLCANOGENIC SULFIDE ASSOCIATIONS, METALS
B 6404 585 VOLCANOLOGY, BY HOWEL WILLIAMS AND ALEXANDER R. MCBIRNEY,
M 29 463 VOLGA-URAL BASIN, VOLUME OF SEDIMENT PRESERVED, USSR
B 6111 1955 VOLGA-URAL HEAVY OIL, USSR
M 25 354 VOLKER, R. E., ARTIFICIAL RECHARGE OF GR
B 6302 182 VOLLSET, JAN, PETROLOGY AND DIAGENESIS OF JURASSIC SANDSTONES
B 6210 1849 VOLTA, ZAMBIA, DEVELOPMENTS, 1977, TERRITORY OF, AFARS AND,
M 29 454 VOLUME CHANGES IN MID-OCEAN RIDGE, MECHANISMS
B 6312 2186 VOLUMETRIC YIELD, ASSESSING RESOURCES
B 6303 500 VON DER BORCH, CHRIS C., COORONG MODEL FOR
B 6402 173 VON HERZEN, R. P., CONTINENTAL MARGIN SUBSIDENCE AND HEAT
B 6105 838 VON HUENE, ROLAND, ALEUTIAN TRENCH AND ADJACENT
M 29 273 VON HUENE, ROLAND, CONTINENTAL MARGINS OF THE EASTERN GULF
B 6303 546 VON HUENE, ROLAND, IPOD DRILLING ON CONVERGENT OCEAN
B 6103 469 VON HUENE, ROLAND, LATE TERTIARY GULF OF ALASKA
M 29 261 VON HUENE, ROLAND, STRUCTURE OF THE OUTER CONVERGENT
B 6407 1014 VON HUENE, ROLAND, STRUCTURE OF UPPER CENOZOIC STRATA
B 6405 712 VOORHEES, BRENT J., FACIES AND DEPOSITIONAL TECTONICS OF MIDDLE
B 6203 435 VORING BASIN, NORTH SEA
B 6203 512* VORING PLATEAU, NORWEGIAN SEA, ABST., MASS PHYSICAL AND
B 6405 797 VOSS, JAMES D., GEOLOGY OF CALVIN FIELD—DEEP BASINAL JURASSIC
B 6204 669 VRACONIAN, ALGERIA
B 6301 70 VRANOVICE CANYON, CZECHOSLOVAKIA
B 6303 423* VRAZILIAN COASTAL BASINS, ABST., SEISMIC SEQUENC ANALYSIS, NEW
B 6205 880 VREELAND, JOHN H., SEISMIC EXPLORATION LEADING TO DISCOVERY OF
B 6211 2363 VREELAND, JOHN H., SEISMIC EXPLORATION LEADING TO DISCOVERY OF
B 6105 819* VUGGY POROSITY IN CARBONATE-MUDBANK BUILDUPS,
B 6311 2024 VULCAN SUB-BASIN, AUSTRALIA
B 6111 1951 WABASCA HEAVY OIL DEPOSIT, WESTERN, CANADA
B 6406 960* WABASCA OIL SANDS DEPOSIT, ALBERTA, ABST., GEOLOGY OF, IN SITU
B 6401 128 WABASH FORMATION, INDIANA
S 5 99 WABASH FORMATION, MISSISSINEWA SHALE, INDIANA,
S 5 93* WABASH PLATFORM WITH CYCLC EVAPORITES OF MICHIGAN BASIN,
B 6107 1030 WABAUNSEE GROUP, NEBRASKA
B 6103 437 WACHSMUTH LIMESTONE, ALASKA
B 6303 421 WADDELL, C. T., VITRINITE REFLECTANCE, PRESENT ROCK
M 25 74 WADE, F. ALTON, ANTARCTICA, AN UNPROSPECTED, UNEXPLOITED
B 6203 526 WADE, J. A., COST B-2 WELL, BALTIMORE CANYON, AND ITS RELATION TO
B 6406 824 WADE KNOLL PINNACLE REEF, AUSTRALIA
B 6109 1440 WADE SANDSTONE, OKLAHOMA
M 26 445 WADE SANDSTONE, OKLAHOMA, ANADARKO BASIN
B 6408 1286 **WADE, WILLIAM J., GEOLOGY AND MINING OF TENNESSEE COAL,** ABST.
M 29 273 WAGEMAN, JOHN, CONTINENTAL MARGINS OF THE EASTERN GULF OF
B 6307 1094 WAGINA FORMATION, AUSTRALIA
B 6004 705 WAGNER, R. H., SEDIMENTATION HISTORY OF LATE CARBONIFEROUS
B 6105 795 **WAGNER, ROBERT H., STEPHANIAN A AND B ALLUVIAL-FAN AND**
B 6003 414 WAGNER, WALTER R.,, GROWTH FAULTS IN CAMBRIAN AND LOWER
B 6303 447* WAGWATER TROUGH, EAST-CENTRAL JAMAICA, ABST., PETROLOGY AND
B 6403 380 WAGWATER TROUGH, JAMAICA
B 6303 555* WAH WAH LIMESTONE, IBEX AREA, WESTERN MILLARD COUNTY, UTAH,
M 30 307 WAHOO FORMATION, PRUDHOE BAY, LISBURNE GROUP
M 25 214 WAIKEKIE FORMATION, NEW ZEALAND
S 10 201 WAIMATA MUD VOLCANO, NEW ZEALAND, NORTH ISLAND

B 6111 1955 WAINWRIGHT HYDROCARBONS, WESTERN, CANADA
M 25 114 WAIRAKEI FIELD, GEOTHERMAL, CHILE
M 25 131 WAIRAKEI GEOTHERMAL FIELD, NEW ZEALAND
B 6405 797 WAKEFIELD, W. W., MASON LAKE FIELD, MUSSELSHELL COUNTY,
B 6109 1548 WAKELYN, BRIAN D., PETROLOGY OF SMACKOVER FORMATION,
B 6103 350 WALDRIP LIMESTONE, TEXAS
B 6401 128 WALDRON FORMATION, INDIANA
B 6303 523* WALES, UNITED KINGDOM, ABST., GWNA MELANGE, UPPER,
B 6207 1132 WALIO OIL FIELD, IRIAN JAYA
B 6012 2162 WALKER, CHARLES W., ORIGIN OF GULF COAST SALT-DOME CAP ROCK
B 6409 1356 WALKER COUNTY, TEXAS, NORTH PINE PRAIRIE FIELD
M 24 177† WALKER CREEK FIELD AREA, LAFAYETTE AND COLUMBIA COUNTIES,
M 24 177 WALKER CREEK FIELD, ARKANSAS
B 6407 1008 WALKER CREEK FIELD, ARKANSAS
B 6303 544* WALKER CREEK FIELD, ARKANSAS--SMACKOVER CASE HISTORY, ABST.
B 6104 594 WALKER CREEK FIELD, FOUR CORNERS
B 6009 1607* WALKER CREEK FIELD--SMACKOVER DIAGENETIC TRAP, ABST.
B 6407 1004 WALKER CREEK STRATIGRAPHIC FIELD, ARKANSAS
B 6407 1004 WALKER FIELD, ARKANSAS
B 6004 729 WALKER, K. R., REGIONAL ENVIRONMENTAL STRATIGRAPHY AND OIL
B 6105 828 WALKER, KENNETH R., DEEP-WATER SEDIMENTATION AND TECTONICS
B 6405 798 WALKER, KENNETH R., PALEOECOLOGIC DETERMINATION OF
B 6303 525 WALKER, KENNETH R., QUANTITATIVE BASIN ANALYSIS AND EVOLUTION
B 6203 551* WALKER LAKE, NEVADA, ABST., MODERN, LACUSTRINE, STROMATOLITES
B 6303 460 WALKER, R. G., HUMMOCKY CROSS-STRATIFICATION--INDICATOR OF
B 6108 1346 WALKER, ROBERT E., OIL AND GAS DEVELOPMENTS IN NORTHERN
B 6408 1280 WALKER, ROGER G., DEEP-WATER RESERVOIRS—SUBMARINE FANS AND
B 6206 932 WALKER, ROGER G., DEEP-WATER SANDSTONE FACIES AND ANCIENT
B 6305 810 WALKER, ROGER G., DEEP-WATER SANDSTONE FACIES AND ANCIENT
B 6407 1094 WALKER, ROGER G., MODERN AND ANCIENT SUBMARINE FANS.
B 6004 730 WALKER, ROGER G., SUBMARINE-FAN CONGLOMERATES AND
B 6211 2268 WALKER VALLEY, CALIFORNIA
B 6404 585 WALKER, VAN HOUTEN, FRANKLYN B., REVIEW OF, FACIES MODELS,
B 6404 490 WALL CREEK MEMBER, FRONTIER FORMATION, WYOMING
B 6201 41 WALLABY PLATEAU, AUSTRALIA
S 4 37 WALLACE, ROBERT JAMES, PATCH-REEF ECOLOGY AND SEDIMENTOLOGY
M 27 20 WALLED CALCISPHERES, GREEN ALGAE
B 6103 417 WALLEN VALLEY FAULT, VIRGINIA
B 6409 1556* WALLER COUNTY, TEXAS, ABST., ENVIRONMENT OF DEPOSITION OF
B 6101 24 WALLIS FIELD, CANADIAN, ARCTIC ISLANDS
B 6207 1238 WALLIS, WILLIAM S., OIL AND GAS EXPLORATION OFFSHORE SOUTHERN
B 6105 816 WALLS, R. A., POROSITY IN REEF COMPLEXES—UPPER DEVONIAN
B 6004 730 WALLS, R. A., POROSITY REDUCTION IN REEF MARGINS BY EARLY
B 6004 730 WALPER, JACK L., AULACOGENS AN MEGASHEARS--NATURAL HABITAT
B 6002 327 WALPER, JACK L., GEOTECTONIC EVOLUTION OF WICHITA AULACOGEN,
B 6109 1549 WALPER, JACK L., PALEOZOIC TECTONICS OF SOUTHERN MARGIN OF
B 6309 1602 WALPER, JACK L., SEDIMENTATION ON TRAILING PLATE
B 6002 164 WALRUS BASIN, BERINGIAN SHELF
B 6108 1372* WALSENBURG AREA, COLORADO, ABST., STRATIGRAPHY AND, CLAY
B 6408 1282 WALTERS, G. M., DEPOSITIONAL TRENDS OF LOWER SILURIAN CLINTON
B 6205 886 WALTERS, KENNETH L., INFLUENCE OF DIAGENETIC AND
B 6105 803 WALTERS, K. L., EFFECTS OF BURIAL STAGES ON POROSITY IN
B 6303 547 WALTERS, LESTER J., JR., EFFECTS OF SOURCE MATERIAL AND THERMAL
B 6008 1405 WALTERS, LESTER J., JR., PALEOCURRENT ANALYSIS OF DAKOTA AND
B 6206 1093 WALTERS, LESTER J., JR., PALEOCURRENTS OF JACKPILE SANDSTONE,
B 6311 2120 WALTERS, ROBERT F., CHANNEL SANDSTONE OIL RESERVOIRS OF
B 6203 569 WALTERS, ROBERT F., GORHAM OIL FIELD, RUSSELL COUNTY, KANSAS,
B 6002 323 WALTERS, ROBERT F., SALT DISSOLUTION IN OIL AND GAS TEST HOLE IN
B 6103 361 WALTERS STRUCTURE, OKLAHOMA
B 6405 798 WALTON, ANTHONY W., EVALUATION OF URANIUM POTENTIAL OF
B 6105 839 WALTON, ANTHONY W., OLIGOCENE VOLCANIC SEDIMENTARY APRON IN
B 6106 930 WALVIS RIDGE DEEP SEA DRILLING SITES ON RIO GRANDE RISE
B 6002 210 WALVIS RIDGE TO BENUE TROUGH, WEST, AFRICA
B 6107 975 WALVIS RIDGE—CAPE GOOD HOPE SEGMENT
B 6307 1093 WALYERING FIELD, AUSTRALIA
B 6405 718* WAMSUTTER ARCH TIGHT GAS PLAY, SOUTHERN WYOMING—NEW LOOK
B 6101 11 WANDEL SEA BASIN, CANADIAN, ARCTIC ISLANDS
B 6004 731 WANLESS, H. R., TIMING, CHARACTER, AND PRESERVABILITY OF
B 6303 547 WANLESS, HAROLD R., ROLE OF PHYSICAL SEDIMENTATION
B 6203 569 WANLESS, HAROLD R., THREE TYPES OF PRESSURE SOLUTION IN
B 6203 518* WAPANUCKA FORMATION, PENNSYLVANIAN, FRONTAL OUACHITA
B 6004 727* WAPANUCKA LIMESTONE, LOWER PENNSYLVANIAN, INDIAN NATIONS
B 6411 2039 WAPLES, DOUGLAS, REVIEW OF GEOCHEMISTRY OF ORGANIC
B 6005 884 WAPLES, DOUGLAS, TIME-TEMPERATURE RELATION IN OIL GENESIS.
B 6302 239 WAPLES, DOUGLAS W., SIMPLE METHOD FOR OIL SOURCE BED
B 6406 916 WAPLES, DOUGLAS W., TIME AND TEMPERATURE IN PETROLEUM
B 6403 434 WARD AND A. E. 'WEIDIE, GUZMAN, EDWARD J., REVIEW OF, GEOLOGY
B 6002 327* WARD COUNTY, TEXAS, ABST., BLOCK 16 FIELD
B 6004 731 WARD, J. A., DIAGENETIC SEQUENCE OF MIDDLE CRETACEOUS
B 6405 798 WARD, J. A., SEISMIC STRATIGRAPHY IN LOW-ENERGY CARBONATE
B 6409 1568 WARD, JOHN A., SEISMIC-STRATIGRAPHIC MAPPING OF GULF COAST
B 6405 708 WARD, JOHN A., APPLICATION OF INVERSION PROCESSING TO
B 6203 512 WARD, LARRY G., DEPOSITIONAL MODEL OF MESOTIDAL BACK-BARRIER
B 6203 570 WARD, LARRY G., PHYSICAL AND SEDIMENTOLOGIC PROCESSES IN
B 6004 682 WARD, LARRY G., REEF MORPHOLOGY AND SEDIMENT TRANSPORT,
B 6203 570 WARD, MERSCH, URANIUM DEPOSITS OF PITCH MINE, COLORADO. BY
B 6207 1238 WARD, ROSCOE F., BIOMASS AS ENERGY SOURCE, ABST.
B 6405 799 WARD, S. H., STRATEGY OF EXPLORATION FOR HIGH TEMPERATURE
B 6409 1552 WARD, SUZANNE, ENVIRONMENTS OF DEPOSITION AND DIAGENESIS OF
B 6303 362 WARD, W. C. AND, STRANDLINE SEDIMENTATION OF CARBONATE
B 6309 1606 WARD, W. C., CARRILLO PUERTO FORMATION OF NORTHEASTERN
B 6306 975 WARD, W. C., NUBIA FORMATION OF CENTRAL EASTERN DESERT,
B 6004 731 WARD, WILLIAM C., INDICATORS OF CLIMATE IN CARBONATE DUNE
B 6109 1546 WARD, WILLIAM C., PETROLOGY OF CATAHOULA SANDSTONES IN EAST
B 6203 500 WARD, WILLIAM C., VADOSE CEMENT PATTERNS IN STRAND-PLAIN
B 6410 1602 WARDELL FORMATION, VIRGINIA
B 6405 719 WARDLAW, BRUCE R., CONODONT-BASED ASSESSMENT OF THERMAL
B 6405 718 WARDLAW, BRUCE R., CONODONTS—MODELS OF PRAGMATIC
B 6405 799 WARDLAW, BRUCE R., MIDDLE-LATE PERMIAN PALEOGEOGRAPHY OF
B 6405 799 WARDLAW, BRUCE R., PERMIAN CONODONT BIOSTRATIGRAPHY AND
B 6207 1171 WARDLAW, BRUCE R., STRATIGRAPHIC RELATIONS OF PARK CITY
B 6402 236 WARDLAW GRABEN AREA, TEXAS
B 6105 776 WARDLAW, N. C., OTTO FIORD FORMATION—EVIDENCE OF SUBAQUEOUS
B 6002 245 WARDLAW, N. C., PORE GEOMETRY OF CARBONATE ROCKS AS REVEALED
B 6405 799 WARDLAW, NORMAN C., MULTIPHASE FLUID MOVEMENTS IN GLASS
B 6004 725 WARME, J. E., CARBONATE PETROLOGY OF BASIN-TO-REEF SHOALING
B 6405 800 WARME, J. E., CONTEMPORANEOUS RECONSTRUCTION AND
B 6004 732 WARME, J. E., EVENT SEQUENCES IN GEOLOGIC REEF PRESERVATION--ROL
B 6004 732 WARME, J. E., GEOLOGIC EXPLORATION OF SOUTHERN CALIFORNIA
S 4 261 WARME, JOHN E., CARBONATE BORERS--THEIR ROLE IN REEF ECOLOGY
B 6305 816 WARME, JOHN E., REVIEW OF TRACE FOSSILS 2, EDITED BY T. P CRIMES
B 6307 1094 WARNBRO GROUP, AUSTRALIA
B 6309 1590 WARNER, ALBERT J., UPPER NIAGARAN AND LOWER CAYUGAN
B 6405 800 WARNER, M. A., SOURCE AND TIME OF GENERATION OF HYDROCARBONS
B 6108 1389 WARNER, M. A., THRUST BELT OF WYOMING, IDAHO, AND NORTHERN
B 6008 1413 WARNER, MONT M., NATURAL GAS PROSPECTS IN SOUTHERN IDAHO A
B 6004 732 WARNING, KARL R., TRANSGRESSIVE-REGRESSIVE DEPOSITS OF DIFUNTA
B 6309 1587 WARPINSKI, NORMAN R., IN-SITU TESTING OF WELL-SHOOTING
B 6307 1035 WARPS--ARCHES, DOMES, AND SAGS, BASEMENT
S 11 60 WARREN, A. D., CALCAREOUS NANNOPLANKTON BIOSTRATIGRAPHY OF
B 6012 2191 WARREN, A. D., NANNOPLANKTON BIOSTRATIGRAPHY
B 6211 2364 WARREN, JOSEPH E., EXPLORATION RISK ANALYSIS—DEVELOPMENT
B 6412 2078 WARREN VALLEY FAULT, APPALACHIANS
B 6103 312 WARSAW FORMATION, KANSAS
S 5 125 WARSHAUER, S. M. RELATIONSHIPS BETWEEN DEPOSITIONAL
B 6105 839 WARSHAUER, S. M., BIOFACIES RELATIONS IN LOWER DEVONIAN
B 6405 800 WARSHAUER, S. M., DEPOSITIONAL ENVIRONMENTS OF LACUSTRINE
B 6004 724 WARSHAUER, S. M., EVOLUTION OF CARBONATE SHELF, SILURIAN
B 6203 562 WARSHAUER, STEVEN M., SCHEME FOR MULTIVARIATE ANALYSIS IN
B 6401 56 WARUKIN FIELD, KALIMANTAN
B 6203 570 WARWICK, RICHARD L., WHEELER-PAN HUNTON FIELD, WHEELER
B 6405 800 WARZESKI, E. ROBERT, DOLOMITE AND DEDOLOMITE IN MURAL
B 6303 547 WARZESKI, E. ROBERT, LOWER CRETACEOUS CARBONATE SHELF IN
B 6202 303† WAS IT THE ORINOCO
B 6305 831* WASATCH AND UPPER FORT UNION FORMATIONS, CENTRAL POWDER
B 6302 197 WASATCH FORMATION, COLORADO
B 6305 839* WASATCH FORMATION OF VERMILLION CREEK BASIN, WYOMING AND
B 6008 1395* WASATCH FORMATION, WESTERN POWDER RIVER BASIN, WYOMING,
M 24 92 WASATCH FORMATION, WYOMING
B 6205 893* WASATCH HINGE LINE IN CENTRAL UTAH, ABST., SUBSURFACE, THRUSTS
B 6303 317 WASATCH PLATEAU, COLUMNS, UTAH
B 6305 824* WASATCH PLATEAU, UTAH, ABST., STRATIGRAPHY AND, COAL DEPOSITS
B 6405 640 WASATCH-GREEN RIVER FORMATIONS, WYOMING
B 6107 1014 WASECA MEMBERS, ALBERTA AND, SASKATCHEWAN, COLONY-,
B 6303 548 WASHBURN, JUDITH R., DEPOSITION AND DIAGENESIS OF MISSISSIPPIAN
B 6004 708* WASHINGTON, ABST., BED FORMS AND, INTERNAL STRUCTURES IN, DEEP,
B 6405 718* WASHINGTON, ABST., DEPOSITIONAL SETTING OF MIDDLE DOLOMITE
B 6403 448* WASHINGTON, ABST., GEOLOGICAL ASSESSMENT OF PETROLEUM
B 6008 1410* WASHINGTON, ABST., HEAT-FLOW STUDIES IN, STEAMBOAT MOUNTAIN-,
B 6004 657* WASHINGTON, ABST., MODERN AND, ANCIENT, ESTUARINE-, FILL, FACIES,
B 6012 2174† WASHINGTON, ABST., NEOGENE, MOLLUSCAN STAGES OF, OREGON AND
B 6405 670* WASHINGTON, ABST., OPHIOMORPHA FROM UPPER BATHYAL, EOCENE
B 6404 692* WASHINGTON, ABST., STRATIFICATION TYPES IN INTERTIDAL SEDIMENT,
B 6211 2364* WASHINGTON, ABST., SYNTHESIS OF, OCS WELL INFORMATION,
B 6211 2354* WASHINGTON AND OREGON. BY TITLE ONLY, ONSHORE
B 6207 1194 WASHINGTON, BATEMAN FORMATION, OREGON AND
B 6106 919 WASHINGTON CANYON, EAST COAST
S 8 130 WASHINGTON CANYON, TOPOGRAPHY, INDEX MAP

S 8 122 WASHINGTON CANYON, U.S. EAST COAST, SUBMARINE CANYONS
B 6304 656 WASHINGTON COUNTY, STRATIGRAPHIC COLUMN, OHIO
B 6207 1194 WASHINGTON, CRESCENT FORMATION, OREGON AND
B 6008 1196 WASHINGTON, DEVELOPMENTS, 1975
B 6108 1166 WASHINGTON, DEVELOPMENTS, 1976
B 6208 1322 WASHINGTON, DEVELOPMENTS, 1977
B 6308 1292 WASHINGTON, DEVELOPMENTS, 1978
B 6409 1393 WASHINGTON, DEVELOPMENTS, 1979
B 6207 1193† WASHINGTON, EARLY, TERTIARY, RIFTING IN WESTERN, OREGON-
M 28 160 WASHINGTON, EOCENE, CRESCENT FORMATION
B 6207 1194 WASHINGTON, EOCENE, RIFT HYPOTHESIS, OREGON AND
B 6207 1194 WASHINGTON, FLOURNOY FORMATION, OREGON AND
B 6309 1590 WASHINGTON, L. J., RECOVERY OF ENERGY FROM MICHIGAN ANTRI
B 6404 528 WASHINGTON LAND, GREENLAND
B 6404 530 WASHINGTON LAND, STRATIGRAPHIC CHART, GREENLAND
B 6207 1194 WASHINGTON, LOOKINGGLASS FORMATION, OREGON AND
B 6207 1194 WASHINGTON, MCINTOSH FORMATION, OREGON AND
M 25 485 WASHINGTON, PRECIOUS AND BASE METALS
B 6207 1194 WASHINGTON, RAGING RIVER FORMATION, OREGON AND
B 6207 1194 WASHINGTON, ROSEBURG FORMATION, OREGON AND
B 6207 1194 WASHINGTON, ROSLYN FORMATION, OREGON AND
B 6207 1194 WASHINGTON, SILETZ RIVER FORMATION, OREGON AND
B 6207 1196 WASHINGTON, SILETZ RIVER-CRESCENT STRATIGRAPHY, OREGON AND
B 6207 1194 WASHINGTON, TEANAWAY FORMATION, OREGON AND
M 28 69 WASHINGTON, TERTIARY, BASALT WEATHERING
B 6207 1194 WASHINGTON, TIGER MOUNTAIN FORMATION, OREGON AND
B 6207 1194 WASHINGTON, TYEE FORMATION, OREGON AND
B 6207 1194 WASHINGTON, YAMILL FORMATION, OREGON AND
B 6408 1201 WASHINGTON-OREGON, ANOXIC OPEN OCEAN, PACIFIC OCEAN
B 6305 804 WASHITA CREEK FIELD, TEXAS
B 6205 862 WASHITA CREEK ULTRADEEP FIELD, PRODUCTION, TEXAS
B 6303 490* WASHITA GROUP, ABST., MODELS FOR INTERPRETATION OF,
B 6402 222 WASHITA GROUP, TEXAS
M 24 242 WASHITA PRODUCING SAND, MISSISSIPPI
B 6205 786 WASHITA REEF TREND, SEISMIC DATA, TEXAS
M 26 199 WASHITA REEF TREND, SHELF-MARGIN BUILDUP, SEISMIC EXAMPLE,
B 6205 786 WASHITA-FREDERICKSBURG RUDISTID REEFS AND BANKS, TEXAS
B 6205 786 WASHITA-SLIGO INTERVAL, TEXAS
B 6312 2160 WASHOVER BARRIERS, ORIGIN, TRANSGRESSIVE
B 6109 1535* WASHOVER FANS ON GEORGIA COAST, ABST., ORIGIN AND CHARACTER
B 6407 1005 WASKOM FIELD, ARK-LA-TEX AREA
B 6410 1602 WASSUM FORMATION, VIRGINIA
B 6004 724* **WASTE DISPOSAL, ABST., HYDROLOGIC STABILITY OF NORTH, LOUISIANA,**
M 25 371† WASTE-DISPOSAL PRACTICE ON GROUNDWATER RESOURCES IN
B 6303 475* WASTES IN OKLAHOMA, ABST., POTENTIAL FOR SUBSURFACE, DISPOSAL
M 25 365† WASTEWATER DISPOSAL IN HONOLULU--SUMMARY, IMPACT OF,
B 6303 548 WASZCZAK, RONALD F., PALEOECOLOGIC IMPLICATION OF
B 6004 727* WATE CRETACEOUS OSTRACODA FROM WELLS IN CENTRAL AND
S 10 169† WATER, ORIGIN OF, PETROLEUM:, IN-TRANSIT, CONVERSION OF,
B 6004 696* WATER, ABST., ORIGIN OF, DOLOMITE--, MODEL BASED ON, CYCLIC
B 6203 522* WATER, ABST., ORIGIN OF PETROLEUM—IN-TRANSIT CONVERSION OF,
B 6006 973† WATER AN SOLUTES THROUGH COMPACTED CLAYS, FLOW OF
B 6212 2478† WATER AT ELEVATED TEMPERATURES AND PRESSURES, SOLUBILITY OF,
B 6305 776 WATER AT ELEVATED TEMPERATURES AND PRESSURES, SOLUBILITY OF,
B 6302 168 WATER BLOCK, DOWNDIP WATER FLOW, OR
B 6109 1554* WATER CARBONATE COMPLEXES--SUBSURFACE PEARSALLFORMATION,
B 6004 692* WATER CARBONATE DEPOSITS AS POSSIBLE PALEOCLIMATIC
B 6203 554* WATER CHEMISTRY OF JAMAICAN REEF SEDIMENTS—REDOX MODEL FOR
B 6209 1758* WATER CLARITY NEAR OIL PRODUCTION PLATFORMS ON LOUISIANA
S 7 319 WATER COLUMN ANALYSIS, GYRE BASIN
B 6103 434† WATER DURING PRIMARY MIGRATION IN WESTERN CANADA BASIN,
B 6109 1537* WATER ESTIMATES FOR GULF COAST SANDSTONES, ABST., CRITICAL
B 6004 543† WATER EXPULSION FROM CLASTIC SEDIMENTS DURING COMPACTION
B 6203 567* WATER FLOW—POSSIBLE MECHANISM FOR TRANSPORT AND
B 6002 318* WATER GEOCHEMISTRY TO EXPLORATION, ABST., APPLICATIONS OF, OIL,
B 6209 1760* WATER OF OUTER CONTINENTAL SHELF OF SOUTH TEXAS, ABST., LIVING,
B 6401 112 WATER OF SOURCE ROCKS, CONCENTRATION OF, CARBON DIOXIDE IN,
S 10 11 WATER, MODEL, DYNAMIC STRUCTURING
S 10 58 WATER MOVEMENT, MIGRATION IN AQUEOUS SOLUTION
S 10 207† WATER SYSTEM, ROLE OF, GEOPRESSURE IN THE, HYDROCARBON AND
B 6401 110 WATER, OIL-CARRYING CAPACITY OF, GAS-SATURATED
B 6004 686* WATER PELAGIC AND SHALLOW-WATER EPERIC CARBONATE SEQUENCES,
B 6305 731 WATER PHASE IN ARTESIAN CONDITION, PRESSURE-DEPTH PLOT OF
B 6309 1602* WATER QUALITY OF EOCENE WILCOX GROUP IN RELATION TO LIGNITE
B 6207 1222* WATER RELATED TO ENERGY IN WESTERN UNITED STATES, ABST.,
B 6203 527* WATER SALINITY, ABST., INTERACTING DYNAMICS OF, PRESSURE,
B 6212 2443 WATER SAMPLE, TITRATION CURVES FOR, SODIUM BICARBONATE,
B 6305 810† WATER SANDSTONE FACIES AND ANCIENT SUBMARINE FANS. MODELS
B 6206 932† WATER SANDSTONE FACIES AND ANCIENT SUBMARINE FANS: MODELS
M 26 482 WATER SANDSTONE, GAS CAP, SEISMIC, MODELING
B 6305 802 WATER SATURATION AND PEBBLE CONTENT
B 6309 1602* **WATER SATURATIONS VERSUS PRODUCTION, ABST., SMACKOVER,**
B 6305 762 WATER, SOLUBILITIES OF, HYDROCARBONS IN
B 6207 1213* WATER SUPPLIES IN LIMESTONE TERRANES OF MICRONESIA, ABST.
M 25 351† WATER SUPPLIES ON BASALTIC ISLANDS--SUMMARY, DEVELOPING,
B 6207 1212* WATER TEMPERATURES AND ANOMALOUS RUNOFF IN CIRCUM-PACIFIC
B 6303 490* **WATER UPPER JURASSIC ROCKS DREDGED FROM BERING SEA**
B 6408 1276* WATER-BEARING GEOPRESSURED RESERVOIRS, ABST., ENERGY
B 6202 201 WATER-DEPTH DETERMINATION, PALEO-
M 26 355 WATER-DEPTH SYSTEMS, RESOLUTION SYSTEMS
B 6305 753 WATER-FREE OIL PRODUCTION, PREDICT ELEVATION OF
B 6204 627 WATER-INJECTION PROGRAM, WYOMING, OREGON BASIN FIELD
S 10 19† WATER-MINERAL-ORGANIC MATTER INTERACTIONS IN THE SOURCE
B 6405 726* WATER-ROCK INTERACTION DURING CLASTIC DIAGENESIS IN BOTH
S 10 109† WATER-ROCK INTERACTION, SOME, ECONOMIC ASPECTS OF
B 6203 522* WATER-ROCK INTERACTION,ABST., SOME ECONOMIC ASPECTS OF
B 6203 493* WATER-ROCK-ORGANIC MATTER INTERACTIONS AT SOURCE, ABST.
S 6 369 WATERBERG SYSTEM, SOUTH AFRICA
M 28 55 WATERBURY GNEISS, CONNECTICUT, CAMBRIAN
M 28 46 WATERBURY GNEISS, CONNECTICUT, CAMBRIAN
M 28 17 WATERBURY GNEISS, CONNECTICUT, PRECAMBRIAN
B 6108 1382* WATERFOWL ON DISPOSAL PONDS IN NORTHEAST COLORADO, ABST.,
S 6 299 WATERHOUSE, J. B., CHRONOSTRATIGRAPHY FOR THE WORLD PERMIAN
B 6309 1607* WATERS, ABST., FACIES DISTRIBUTION OF, HOSSTON SAND, CENTRAL,
B 6303 479* WATERS AND ORIGIN OF CARBON DIOXIDE, ABST., STABLE, CARBON,
B 6303 428* WATERS AND ORIGIN OF NATURAL GAS, ABST., ORGANIC ACID ANIONS
B 6303 385 WATERS, BY KARL KREJCI-GRAF, OSTROFF, A. G., REVIEW OF, DATA ON
B 6212 2444 WATERS, CALIFORNIA, FORMATION
B 6303 478* WATERS FROM NORTHERN GULF OF MEXICO BASIN, ABST.,
B 6203 519* **WATERS FROM SOUTH LOUISIANA WELL, ABST., CHEMICAL ANALYSES OF,**
B 6203 529* WATERS IN SOUTH LOUISIANA WELL, ABST., CHEMICAL ANALYSIS OF,
B 6304 699* WATERS OF NORTH AMERICA, ABST., DISSOLVED, HYDROCARBONS IN
B 6203 555* WATERS OF PETROLEUM BASINS—POSSIBLE SIGNIFICANT NEW ENERGY
S 4 379 WATERS, PRECIPITATION OF LOW-, MAGNESIAN, CALCITE THROUGH,
B 6212 2444 WATERS, TEXAS, FORMATION
B 6303 496* WATERS UNDER SABKHA, ABU DHABI, UNITED ARAB EMIRATES, AND ITS
B 6004 696* WATERS WITH CONCENTRATED SEA WATER, ABST., ORIGIN OF,
B 6212 2441† WATERS—IMPLICATIONS FOR ORIGIN OF NATURAL GAS, ALIPHATIC,
M 24 82 WATERWAYS FORMATION, ALBERTA
M 30 59 WATHNE, E., FRIGG FIELD--LARGE SUBMARINE-FAN TRAP IN LOWER
B 6311 1999 WATHNE, E., FRIGG FIELD--LARGE SUBMARINE-FAN TRAP IN LOWER
B 6203 521 WATHNE, E., FRIGG FIELD—LARGE SUBMARINE-FAN TRAP IN LOWER
S 7 267 WATKINS, DAVID J., STABILITY OF CONTINENTAL SHELF AND SLOPE OFF
M 29 383 WATKINS, J. S., INVESTIGATION OF MISSISSIPPI FAN, GULF OF MEXICO
B 6004 690 WATKINS, J. S., SEISMIC STRATIGRAPHY AND GEOLOGIC HISTORY OF
M 29 319 WATKINS, JOEL S., ANATOMY OF THE MEXICAN RIDGES, SOUTHWESTERN
M 29 1 **WATKINS, JOEL S., EDITED BY, GEOLOGICAL AND GEOPHYSICAL**
B 6303 544 WATKINS, JOEL S., GEOLOGIC HISTORY OF DEEP GULF ON MEXICO BASED
M 29 11 WATKINS, JOEL S., GEOLOGY OF THE OFFSHORE SOUTHEAST GEORGIA
B 6303 426 WATKINS JOEL S., GRAVITY--SLIDE ORIGIN FOR MEXICAN RIDGES
B 6111 1962 WATKINS, JOEL S., NORTHEASTERN EXTENSION OF SIGSBEE SCARP, GULF
S 7 43 WATKINS, JOEL S., OCCURRENCE AND EVOLUTION OF SALT IN DEEP
B 6105 773 **WATKINS, JOEL S., SEISMIC STRATIGRAPHY, SEDIMENTARY HISTORY,**
B 6205 792 WATKINS, JOEL S., SEISMIC STRATIGRAPHY AND GEOLOGIC HISTORY OF
B 6206 1067 WATKINS, JOEL S., SERENDIPITY GAS SEEP AREA, SOUTH TEXAS
B 6203 501 WATKINS, JOEL S., STRUCTURE AND EARLY GEOLOGIC HISTORY OF DEEP
M 29 363 WATKINS, JOEL S., TECTONIC DEVELOPMENT OF TRENCH-ARC
M 29 1 **WATKINS, JOEL, FOREWORD, GEOLOGICAL AND GEOPHYSICAL**
B 6308 1233 WATKINS, JOHNNY W., DEVELOPMENTS IN NORTH-CENTRAL TEXAS IN
B 6105 797 **WATKINS, N. D., DEPOSITION AND RESEDIMENTATION OF ABYSSAL TEPHRA**
B 6203 570 WATNEY, W. LYNN, EARLY SUBAERIAL EXPOSURE AND FRESHWATER
B 6105 839 WATNEY, W. LYNN, PENNSYLVANIAN LANSING-KANSAS CITY GROUPS IN
B 6103 468* WATONA DAM PROJECT. BY TITLE ONLY
M 26 451 WATONGA-CHICKASHA TREND, OKLAHOMA, ANADARKO BASIN
M 26 451† WATONGA-CHICKASHA TREND: GEARY, OKLAHOMA—T13N, R10W, A,
B 6201 74 WATROUS MUDSTONE-SILTSTONE, SASKATCHEWAN
B 6004 709 WATSON, G., SPATIAL VARIATION IN POROSITY--NEW CONCEPT, ABST.
B 6008 1329 WATSON H. J., DEVELOPMENTS IN WEST GREENLAND IN 1975
B 6108 1326 WATSON, HILREY J., DEVELOPMENTS IN GREENLAND IN 1976
M 25 354 WATSON, K. K, ARTIFICIAL RECHARGE OF GR
M 24 73 WATT MOUNTAIN FORMATION, ALBERTA
B 6405 801* WATTENBERG AND SPINDLE FIELDS—PALEOSTRUCTURAL AND
B 6410 1631 WATTENBERG FIELD, COLORADO
M 24 136 WATTENBERG FIELD, COLORADO
B 6302 155 WATTENBERG FIELD, COLORADO
M 24 136† WATTENBERG FIELD, DENVER BASIN, COLORADO
M 26 278 WATTENBERG OIL FIELD, COLORADO
B 6004 672 WATTERS, G. E., BIOHERMAL DEVELOPMENT IN WOLFCAMPIAN
B 6304 702 WATTS, A. B., MECHANISMS OF BASIN SUBSIDENCE, ABST.
M 27 179 WAUL SORTIAN LIMESTONE, IRELAND, MISSISSIPPIAN

Index

M 27 69 WAULSORTIAN LIMESTONE, IRELAND, MISSISSIPPIAN
B 6405 743* WAULSORTIAN MOUNDS OF WESTERN EUROPE, ABST., COMPARISON
B 6303 487* WAULSORTIAN-TYPE MOUNDS OF FORT PAYNE FORMATION, NORTHERN
B 6103 361 WAURIKA-MUENSTER ARCH, OKLAHOMA
B 6203 552* WAVE ENERGY ON BEACH AND NEARSHORE-BAR MORPHOLOGY IN
M 26 5 WAVE EQUATION MIGRATION, SEISMIC DATA
B 6405 777* WAVE SIZE—DYNAMIC EQUILIBRIUM PROCESSES AND KINEMATIC
M 26 398 WAVE THEORY, BOX CAR REEF, MODELING
S 7 90 WAVE TRANSPORT, SEDIMENT PROCESSES
B 6105 788* **WAVE-CLIMATE STUDIES IN BALTIMORE CANYON TROUGH OCS—**
B 6309 1611* WAVE-DOMINATED DELTAIC SYSTEMS OF UPPER CRETACEOUS SAN
B 6405 739* WAVE-DOMINATED DELTAS—IMPORTANT ECONOMIC DEPOSITIONAL
B 6105 807* **WAVE-EQUATION MIGRATION—TWO APPROACHES**
B 6012 2184* WAVE-EQUATION MIGRATION--TWO APPROACHES, ABST.
B 6304 700* WAVE-FORM FACTOR ANALYSIS--QUANTITATIVE APPROACH TO SEISMIC
S 7 274 WAVE-INDUCED BOTTOM-PRESSURE ANOMALY, STABILITY
M 26 392 WAVEFORMS, MODELING, USING, AMPLITUDES
B 6308 1428* **WAVELAND FIELD, UNIQUE STRUCTURAL AND STRATIGRAPHIC TRAP,**
B 6309 1610* **WAVELAND FIELD, UNIQUE STRUCTURAL AND STRATIGRAPHIC TRAP,**
M 26 4 WAVELENGTH, SEISMIC, REFLECTIONS
M 26 479 WAVELET CORRECTION, PROCEDURE
B 6405 696* WAVES, ABST., INTERNAL STRUCTURES OF SHALLOW-MARINE, TIDAL,
S 4 135† WAVES AND CURRENTS, STRUCTURAL MODIFICATION OF, CORALS IN
B 6109 1542* WAVES AND NEPHELOID LAYER ON CONTINENTAL SHELF, GULF OF
M 26 301 WAVES AS SIGNALS, SEISMIC, DETECTION
B 6405 681* WAVES IN VINEYARD SOUND, MASSACHUSETTS, ABST., TIDAL-CURRENT,
B 6211 2358* WAVES TO STRATIGRAPHIC INTERPRETATION, ABST., CONTRIBUTION OF,
B 6403 429 WAX OIL FIELDS, TABLE 1, HIGH-
B 6403 427† WAX OILS IN ASSAM BASIN, INDIA, DEPOSITIONAL ENVIRONMENT OF,
S 7 87 WAYLETT, D. C., RECOGNITION OF CONTINENTAL-SLOPE SEISMIC FACIES,
M 26 259 WEAK-REFLECTOR FACIES
B 6004 689* WEATHERING MICROTEXTURES ON QUARTZ SAND FROM
B 6203 571 WEAVER, CHARLES E., SHALE DEHYDRATION, SALINITY, CRACKING, AND
B 6203 490 WEAVER, FRED M., SEDIMENTATION ON ANTARCTIC CONTINENTAL
B 6209 1762* WEBB AND ZAPATA COUNTIES, TEXAS, ABST., LOBO TREND OF, SOUTH,
B 6411 2037 WEBB, BRIGHT, M. J., REVIEW OF, GEOCHEMISTRY IN MINERAL
B 6109 1534* WEBB COUNTIES, TEXAS, ABST., CHARACTERISTICS OF, WILCOX, GAS,
B 6303 548 WEBB, GREGORY W., ENTRAPMENT FACTORS IN CALIFORNIA TURBIDITE
B 6104 628 WEBB, GREGORY W., OKLAHOMA CITY OIL--SECOND CROP FROM
B 6001 115 WEBB, GREGORY W., OKLAHOMA CITY OIL--SECOND CROP FROM
B 6004 732 **WEBB, GREGORY W., PRESERVATION WINDOW AS SUBUNCONFORMITY**
B 6203 542 WEBB, J. E., ANCIENT SUBMARINE CANYON AND CHANNEL
B 6303 495 WEBB, JAMES E., DOWNCURRENT CHANGE IN MIOCENE
B 6409 1568 WEBER, ANTHONY, PACHUTA CREEK, SMACKOVER, FIELD, MISSISSIPPI,
B 6405 709* WEBER FORMATION, RANGEL OIL FIELD, COLORADO, ABST., EOLIAN-,
B 6303 549 WEBER, FREDERICK F., URANIUM IN ORCA BASIN SEDIMENTS, ABST.
S 4 171 WEBER, JON N., CARIBBEAN REEF CORALS MONTASTREA ANNULARIS
S 4 289 WEBER, JON N., USE OF CORALS IN DETERMINING
B 6303 549 WEBER, K. J., COMPUTATION OF INITIAL WELL PRODUCTIVITIES IN
B 6406 893 WEBER TROUGH
B 6406 894 WEBER TROUGH AND DEEP
B 6002 287 WEBER, V. V., EARLY DIAGENETIC GENERATION OF HYDROCARBON
B 6009 1609* WEBSTER PARISHES, LOUISIANA, ABST., SOUTH SAREPTA FIELD, GRAY
B 6308 1428 WEBSTER, ROBERT E., EVIDENCE FOR WRENCH FAULTING, SOUTHERN
B 6402 221 WEBSTER, ROBERT E., STRUCTURAL ANALYSIS OF DEVILS RIVER
B 6203 571 WEBSTER, ROBERT E., STRUCTURAL GEOLOGY OF SOUTHWEST EDWARDS
B 6402 168 WECHES SHALE, TEXAS
B 6303 467* WEDDELL SEA, ANTARCTICA--700,000 YEARS TO PRESENT, ABST., FLOW OF
B 6408 1159 WEDGE ANALYSIS
B 6109 1489 WEDGE, EAST COAST, CONTINENTAL, MARGIN, EUGEOCLINAL
B 6109 1486 WEDGE, EAST COAST, CONTINENTAL, MARGIN, MIOGEOCLINAL
B 6311 2088† WEDGE PROGRADATION FROM TRANSFORM-FAULTED CONTINENTAL
B 6408 1170 WEDGE-BASE PLAYS
B 6408 1174 WEDGE-EDGE PLAYS
B 6408 1172 WEDGE-MIDDLE PLAYS
B 6408 1173 WEDGE-TOP PLAYS
B 6303 552* WEDGES, ABST., ASSESSING, OIL AND, GAS PLAYS IN FACIES-CYCLE
B 6408 1164 WEDGES, ALBERTA, MESOZOIC, SANDSTONE
B 6408 1158† WEDGES, ASSESSING, OIL AND, GAS, PLAYS IN, FACIES-CYCLE
B 6408 1167 WEDGES, EXAMPLES, CARBONATE
B 6408 1163 WEDGES, EXAMPLES, SANDSTONE
B 6408 1167 WEDGES, TEXAS, GULF COAST, TERTIARY, SANDSTONE
B 6408 1168 WEDGES, UTAH AND, COLORADO, PALEOZOIC, CARBONATE
B 6211 2258 WEDGING, BASIN-MARGIN
B 6007 1110 WEEKS, LEWIS G., ORIGIN OF ATHABASCA OIL
B 6210 2117 WEEKS PETROLEUM LTD., ET AL, ON DEVELOPMENTS IN RAS AL
B 6203 550 WEGRZYN, RICHARD S., LITHOSTRATIGRAPHY, PETROLOGY, AND
B 6109 1545 WEICHERT, PHIL, NATURAL ENVIRONMENTAL-INFORMATION AND
B 6405 801 WEICHMAN, BERNARD E., METHANE MEASUREMENT FROM SALINE ZONE
B 6309 1606 WEIDIE, A. E., CARRILLO PUERTO FORMATION OF NORTHEASTERN
B 6109 1548 WEIDIE, ALFRED E., GEOLOGY OF SIERRA BALUARTES DIAPIR AND
B 6403 434 WEIDIE, GUZMAN, EDUARDO J., REVIEW OF, GEOLOGY AND
B 6303 549 WEIL, CHARLES B., DEFORMATION OF CONTINENTAL SLOPE SEDIMENTS
B 6303 550 WEIMER, ROBERT J., COMPARISON OF PORE-FILLING MATERIAL IN SOME
B 6303 510 WEIMER, ROBERT J., DIAGENESIS IN HYGIENE AND TERRY SANDSTONES,
B 6303 550 WEIMER, ROBERT J., INFLUENCE OF BASEMENT TECTONICS ON
B 6303 550 WEIMER, ROBERT J., INFLUENCE OF TRANSCONTINENTAL ARCH
B 6008 1413 WEIMER, ROBERT J., OVERPRESSURED SHALE MASSES AND GROWTH
M 26 277 WEIMER, ROBERT J., STRATIGRAPHIC AND SEISMIC EVIDENCE FOR LATE
B 6405 801 WEIMER, ROBERT J., WATTENBERG AND SPINDLE FIELDS—
B 6303 515 **WEINBERG, D. M., GEOSEISMIC MODELING--AN INTERACTIVE COMPUTER**
M 25 405 WEIPA, BAUXITE RESERVES, AUSTRALIA
B 6309 1611 WEISE, BONNIE R., WAVE-DOMINATED DELTAIC SYSTEMS OF UPPER
B 6110 1899 WEISNER SANDSTONE, MISSISSIPPI
S 4 111 WEISS, M. P., MANS IMPACT ON COASTAL REEFS--A EXAMPLE FROM
B 6401 136 WEISS, MALCOLM P., AMERICAN COMMISSION ON STRATIGRAPHIC
S 4 1 **WEISS, MALCOLM P., EDITED BY, REEFS AND RELATED CARBONATES--**
S 4 1 **WEISS, MALCOLM P., INTRODUCTION, REEFS AND RELATED CARBONATES--**
B 6310 1986 WEISS, MALCOLM P., NOTE 50--PROPOSAL TO CHANGE NAME OF
B 6401 137 WEISS, MALCOLM P., NOTE 51--RECORDS OF STRATIGRAPHI
S 4 1 **WEISS, MALCOLM P., PREFACE, REEFS AND RELATED CARBONATES--**
B 6105 843 WEISSER, G. H., SEISMIC ANALYSIS OF GLIDE-PLANE FAULT STRUCTURES,
M 26 277 WELD, AND ADAMS COUNTIES, COLORADO, BOULDER
B 6102 153 WELDON CANYON, CALIFORNIA
B 6207 1236* WELL COMPLETION AND CASING PROGRAMS IN INDONESIA, ABST.,
B 6205 862 WELL COSTS, PER FOOT DRILLED, UNITED STATES
B 6002 328* WELL CUTTINGS, ABST., STUDY OF, CARBONATE ENVIRONMENTS USING
B 6009 1616* WELL DATA, ABST., GEOTHERMAL EXPLORATION FROM, DEEP-
M 29 70 WELL DATA, SONIC VELOCITIES, COST B-2
B 6208 1281 WELL, DEVELOPMENT
B 6308 1174 WELL, DEVELOPMENT
B 6108 1127 WELL, DEVELOPMENT
B 6409 1557* WELL LOG ANALYSIS CONCEPTS IN CLASTIC AND ARGILLACEOUS
B 6002 189 WELL LOG, ENVIRONMENTAL ANALYSES, NORTH SEA
B 6405 760* WELL LOGS, ABST., CROSS-BEDDED SANDSTONE, RESERVOIRS—GEOLOGIC
B 6405 705* WELL LOGS, ABST., GEOTHERMAL RESOURCES EVALUATED BY
B 6002 188 WELL LOGS, NORTH SEA
B 6303 516* WELL LOGS TO PREDICT PERMEABILITY OF EOCENE DELTAIC
B 6409 1302 WELL, SERVICE
B 6108 1127 WELL, SERVICE
B 6308 1174 WELL, SERVICE
B 6208 1281 WELL, SERVICE
B 6209 1756* WELL STIMULATION, ABST., CLAY TECHNOLOGY AND
B 6308 1203 WELL TICKET, API--AAPG, CSD, INDIVIDUAL
B 6008 1188 WELL TICKET, API--, AAPG, CSD, INDIVIDUAL
B 6208 1310 WELL TICKET, API-, AAPG, CSD, INDIVIDUAL
B 6108 1156 WELL TICKET, API-, AAPG, CSD, INDIVIDUAL
B 6409 1320 WELL TICKET, API—AAPG, CSD, INDIVIDUAL
B 6409 1301 WELL-CLASSIFICATION GUIDELINES
B 6108 1126 WELL-CLASSIFICATION GUIDELINES
B 6208 1281 WELL-CLASSIFICATION GUIDELINES
B 6008 1158 WELL-CLASSIFICATION GUIDELINES
B 6308 1174 WELL-CLASSIFICATION GUIDELINES
B 6004 685* WELL-CUTTING SAMPLES, ABST., INTERPRETATION OF DEPOSITIONAL
B 6004 651* WELL-DATA DISPLAY SYSTEM--INTERACTIVE EXPLORATION TOOL, ABST.
B 6303 509* WELL-DATA SYSTEM, ABST., SMALL, COMPUTER
S 3 79 WELL-DEFINED BENIOFF SUBDUCTION ZONES
B 6203 508* WELL-DENSITY DATA FOR PREDICTING UNDISCOVERED HYDROCARBON
B 6008 1393* WELL-LOG ANALYSIS IN CRETACEOUS GAS SAND IN NORTHERN
B 6007 1073 WELL-LOG RECORDS, TEMPERATURES, NIGER DELTA PETROLIFEROUS
B 6309 1600* WELL-LOGGING CONCEPTS SOLVE SOUTH TEXAS
B 6309 1587* WELL-SHOOTING CONCEPTS, ABST., IN-SITU TESTING OF
B 6203 571 WELLAND, M. J. P., DELTAIC COMPONENTS OF VALMY AND VININI
B 6203 538 WELLAND, M. J. P., ELDER SANDSTONE—DELTAIC SEDIMENTS OF
B 6105 840 WELLAND, MICHAEL J. P., GOODWIN LIMESTONE—LOWER ORDOVICIAN
B 6307 1135† WELLING FORMATION, NEW NAME
B 6308 1202 WELLS, AAPG AND, API, CLASSIFICATION OF
B 6108 1155 WELLS, AAPG AND, API, CLASSIFICATION OF
B 6208 1309 WELLS, AAPG AND, API, CLASSIFICATION OF
B 6409 1319 WELLS, AAPG AND, API, CLASSIFICATION OF
B 6008 1187 WELLS, AAPG AND, API, CLASSIFICATION OF
B 6203 564* WELLS, ABST., FOAM, STIMULATION FOR, ENHANCED RECOVERY FROM,
B 6208 1309 WELLS, AS APPLIED BY CSD, LAHEE, CLASSIFICATION OF
B 6008 1187 WELLS, AS APPLIED BY CSD, LAHEE, CLASSIFICATION OF
B 6409 1319 WELLS, AS APPLIED BY CSD, LAHEE, CLASSIFICATION OF
B 6108 1155 WELLS, AS APPLIED BY CSD, LAHEE, CLASSIFICATION OF
B 6308 1202 WELLS, AS APPLIED BY CSD, LAHEE, CLASSIFICATION OF
B 6205 854 WELLS AS OF MARCH 1976, UNITED STATES, ULTRADEEP
B 6205 862 WELLS, AVERAGE COSTS, ULTRADEEP
B 6002 233 WELLS, BOTTOM-HOLE TEMPERATURES, HUNGARY

B 6002 202 WELLS, BRAZIL, COASTAL
S 3 61 WELLS CREEK CRYPTOEXPLOSIVE STRUCTURE, TENNESSEE
B 6108 1356 WELLS, DAVID R., DEVELOPMENTS IN SOUTH TEXAS IN 1976
B 6208 1511 WELLS, DAVID R., DEVELOPMENTS IN SOUTH TEXAS IN 1977
B 6205 845 WELLS, DEVIATED, THICKNESSES IN
B 6205 743 WELLS DRILLED ANNUALLY IN NON-COMMUNIST WORLD, 1970-1976,
B 6205 853† WELLS DRILLED BELOW 18,000 FEET, 5,500 METERS, IN WEST TEXAS AND
B 6102 297* WELLS DRILLED BELOW 18,000 FT, 5,486 M, IN WEST TEXAS AND
B 6109 1549* WELLS FOR SOME CRETACEOUS-EOCENE FORMATIONS IN FLORIDA,
B 6201 100 WELLS FORMATION, IDAHO
B 6205 845† WELLS, GRAPHIC DETERMINATION OF, STRATIGRAPHIC AND, VERTICAL,
B 6004 727* WELLS IN CENTRAL AND SOUTHERN FLORIDA, ABST., MARINE AND,
B 6002 323 WELLS, JACK, HEAVY OIL IN WESTERN MISSOURI--UPDATE 1975, ABST.
B 6409 1437 WELLS, JACK S., OIL AND GAS DEVELOPMENTS IN NORTH
B 6208 1477 WELLS, JACK S., OIL AND GAS DEVELOPMENTS IN NORTH
B 6108 1305 WELLS, JACK S., OIL AND GAS DEVELOPMENTS IN NORTH
B 6008 1332 WELLS, JACK S., OIL AND GAS DEVELOPMENTS IN NORTH
B 6308 1301 WELLS, JACK S., OIL AND GAS DEVELOPMENTS IN NORTH
B 6405 801 WELLS, JOHN T., DYNAMICS OF UNVEGETATED TIDAL-FLAT MUDS, ABST.
B 6303 550 WELLS, JOHN T., SEDIMENT SUSPENSION, DEWATERING, AND
B 6205 855 WELLS, LOCATION, TEXAS, PANHANDLE AND, OKLAHOMA, ULTRADEEP
B 6205 854 WELLS, LOCATION, WEST, TEXAS AND, NEW MEXICO, ULTRADEEP
B 6405 802 WELLS, NEIL, NEARSHORE MARINE AND CONTINENTAL FACIES IN
B 6002 202 WELLS, SOUTH ATLANTIC, COASTAL
B 6205 860 WELLS, TEXAS, PECOS COUNTY, ULTRADEEP
B 6007 1128 WELLS, VENEZUELA, SHALE, POROSITY FOR
B 6203 537* WELLSVILLE MOUNTAIN, UTAH, ABST., PETROLOGY AND PETROGRAPHY
B 6205 893 WELSH, JOHN E., SUBSURFACE THRUSTS ALONG WASATCH HINGE LINE IN
B 6203 426 WELSH PLATFORM, EUROPE
B 6405 806 WELTE, D. H., THREE-DIMENSIONAL SIMULATION OF GEOLOGIC,
B 6309 1548 WELTE, MOMPER, JAMES A., REVIEW OF, PETROLEUM FORMATION AN
B 6012 2187 WELTON, BRUCE, PALEONTOLOGIC STUDIES OF MIDDLE TERTIARY
B 6405 740 WELTON, JOANN E., HYDROCARBON POTENTIAL OF MATILIJA
B 6012 2187 WELTON, JOANN, PALEONTOLOGIC STUDIES OF MIDDLE TERTIARY
B 6303 551 **WENDTE, J. C., PORE SYSTEMS IN JURASSIC CARBONATE RESERVOIRS,**
B 6105 840 WENDTE, J. C., POROSITY PRESERVATION IN SHALLOW-MARINE
B 6405 648 WENGERD, SHERMAN A., REVIEW OF OIL AND GAS FIELDS OF THE FOUR
B 6012 2117† WENLOCK LIMESTONES, SILURIAN, OF SHROPSHIRE, ENGLAND, ORIGIN
S 5 95 WENLOCKIAN SERIES, CORRELATION, SILURIAN,
B 6203 572 WENNEKERS, HENRY, HEAVY OILS, SASKATCHEWAN AND ALBERTA. BY
B 6101 1 WENNEKERS, J. H. N., GEOLOGY AND HYDROCARBON DISCOVERIES OF
B 6405 802 WENRICH-VERBEEK, KAREN J., URANIUM MINERALIZATION IN HOPI
B 6004 723 WERMIEL, DAN E., DIAGENETIC HISTORY OF MISSISSIPPIAN CARBONATE
B 6305 847 WERNER, EBERHARD, DEVONIAN SHALE GAS PRODUCTION,
M 26 150 WERNER FORMATION, TEXAS, EAST, LOUANN-
M 26 147 WERNER FORMATION, TEXAS, EAST TEXAS BASIN
B 6203 572 WESCOTT, WILLIAM A., DEPOSITIONAL ENVIRONMENTS OF YALLAHS FAN
B 6403 374 WESCOTT, WILLIAM A., FAN-DELTA SEDIMENTOLOGY AND TECTONIC
B 6303 447 WESCOTT, WILLIAM A., PETROLOGY AND DIAGENESIS OF ARC-DERIVED
S 7 227 WESER, OSCAR E., OIL TRAPPING CHARACTERISTICS OF TURBIDITES
B 6204 635 WESSELS REEF, ALASKA
B 6303 476 WESSMAN, F. H., INTERACTIVE MODELING AND INTERPRETATION OF
S 10 193 WEST AFRICA, MUD DIAPIRS, SEISMIC PROFILE
B 6205 778 WEST AFRICA, OFFSHORE, JURASSIC
B 6205 778 WEST AFRICA, OFFSHORE, SEISMIC DATA
M 26 145† WEST AFRICA, SEISMIC, STRATIGRAPHY AND, GLOBAL CHANGES OF SEA
B 6002 210 WEST AFRICA, WALVIS RIDGE TO BENUE TROUGH
B 6303 427* WEST AFRICA--RESULTS IN CABINDA, ABST., OIL EXPLORATION IN PRE-,
B 6002 211 WEST AFRICAN AND BRAZILIAN SALT EDGES, ALBIAN, MODEL,
B 6407 999 WEST ATLAS BASIN, MOROCCO
B 6104 587 WEST CAMERON BLOCK FIELD, LOUISIANA
B 6302 245 WEST CEMENT DOME, OKLAHOMA
B 6303 494* WEST COAST, ABST., PALEOECOLOGIC EVALUATION OF LATE, EOCENE
B 6008 1196† WEST COAST AREA IN 1975, DEVELOPMENTS IN
B 6108 1166† WEST COAST AREA IN 1976, DEVELOPMENTS IN
B 6208 1322† WEST COAST AREA IN 1977, DEVELOPMENTS IN
B 6308 1292† WEST COAST AREA IN 1978, DEVELOPMENTS IN
B 6409 1393† WEST COAST AREA IN 1979, DEVELOPMENTS IN
S 11 77 WEST COAST SIPHOGENERINID FORAMINIFERA, NEW GENERA
S 11 4† WEST COAST TERTIARY, HISTORY OF STRATIGRAPHIC, PALEONTOLOGY
B 6111 1984 WEST CORTES BASIN, CALIFORNIA
B 6109 1429 WEST DELTA BLOCK 73 FIELD, LOUISIANA
B 6104 587 WEST DELTA BLOCK FIELDS, LOUISIANA
B 6105 807 WEST, F. G., HOT, DRY ROCK, HDR, GEOTHERMAL ENERGY—EXPLOITATION
B 6211 2306 WEST FLORIDA BANKS, GULF OF MEXICO
M 29 418 WEST FLORIDA BANKS, REFLECTION PROFILE, GULF OF MEXICO
S 7 63 WEST FLORIDA PLATFORM, GULF OF MEXICO
B 6211 2303 WEST FLORIDA SCARP, GULF OF MEXICO
S 7 9 WEST FLORIDA SHELF, DESCRIBED
S 7 220 WEST FLORIDA SLOPE, BANDING
S 7 197 WEST FLORIDA SLOPE, BATHYMETRY AND CORE HOLES
S 7 220 WEST FLORIDA SLOPE, BURROWS
S 7 217 WEST FLORIDA SLOPE, CARBONATE TEXTURES
S 7 219 WEST FLORIDA SLOPE, CARBONATE TEXTURES
S 7 217 WEST FLORIDA SLOPE, CHARACTERISTICS OF SEDIMENTS
S 7 198 WEST FLORIDA SLOPE, CHRONOSTRATIGRAPHY, CHART
S 7 203 WEST FLORIDA SLOPE, CRETACEOUS THICKNESS
S 7 200 WEST FLORIDA SLOPE, DEPOSITION
S 7 13 WEST FLORIDA SLOPE, DESCRIBED
S 7 213 WEST FLORIDA SLOPE, DESOTO CANYON AREA
S 7 218 WEST FLORIDA SLOPE, DOLOMITE
S 7 220 WEST FLORIDA SLOPE, GRADED TURBIDITES( )
S 7 193† WEST FLORIDA SLOPE, GULF OF MEXICO, SEISMIC, STRATIGRAPHIC
S 7 217 WEST FLORIDA SLOPE, MINERALOGY
S 7 207 WEST FLORIDA SLOPE, MIOCENE THICKNESS
S 7 202 WEST FLORIDA SLOPE, NORTHERN AREA, EOCENE-LOWER MIOCENE
S 7 200 WEST FLORIDA SLOPE, NORTHERN AREA, LOWER CRETACEOUS
S 7 205 WEST FLORIDA SLOPE, NORTHERN AREA, MIDDLE AND UPPER MIOCENE
S 7 210 WEST FLORIDA SLOPE, NORTHERN AREA, PLEISTOCENE-RECENT
S 7 208 WEST FLORIDA SLOPE, NORTHERN AREA, PLIOCENE
S 7 202 WEST FLORIDA SLOPE, NORTHERN AREA, UPPER CRETACEOUS
S 7 200 WEST FLORIDA SLOPE, NORTHERN SLOPE AREA
S 7 206 WEST FLORIDA SLOPE, OLIGOCENE-LOWER MIOCENE THICKNESS
S 7 214 WEST FLORIDA SLOPE, PLEISTOCENE THICKNESS
S 7 212 WEST FLORIDA SLOPE, PLIOCENE THICKNESS
S 7 220 WEST FLORIDA SLOPE, SEDIMENTARY STRUCTURES
S 7 221 WEST FLORIDA SLOPE, SEDIMENTARY STRUCTURES
S 7 201 WEST FLORIDA SLOPE, SEISMIC PROFILE
S 7 211 WEST FLORIDA SLOPE, SEISMIC PROFILE
S 7 208 WEST FLORIDA SLOPE, SEISMIC PROFILE
S 7 218 WEST FLORIDA SLOPE, SEISMIC PROFILE
S 7 217 WEST FLORIDA SLOPE, SOUTHERN SLOPE AREA
S 7 200 WEST FLORIDA SLOPE, STRATIGRAPHY
S 7 209 WEST FLORIDA SLOPE, UPPER MIDDLE MIOCENE THICKNESS
S 7 210 WEST FLORIDA SLOPE, UPPER MIOCENE THICKNESS
S 4 33 WEST FLOWER GARDEN BANK, DIPLORIA-MONTASTREA-PORITES
S 4 27† WEST FLOWER GARDEN BANK, GULF OF MEXICO
B 6309 1609* WEST FLOWER GARDEN BANK, NORTHERNMOST CORAL REEF IN GULF
S 4 33 WEST FLOWER GARDEN BANK, REEFS, GULF OF MEXICO
B 6203 377 WEST FORELAND FORMATION, ALASKA
M 24 10 WEST FORELAND FORMATION, ALASKA
B 6408 1261 WEST FORELAND FORMATION, ALASKA
B 6408 1133 WEST FORELAND FORMATION, ALASKA
B 6408 1132 WEST FORELAND GAS FIELD, ALASKA
B 6408 1132 WEST FORK GAS FIELD, ALASKA
B 6411 1969 WEST GERMANY, 1979, SCHRODER, L., AND DEVELOPMENTS IN
B 6405 759* WEST GERMANY, ABST., SAND DISPERSAL AT, NORDERNEYER SEEGAT
M 26 337 WEST GERMANY, DOGGER SANDSTONE, HOHNE FIELD
M 30 36 WEST GERMANY, GRONINGEN FIELD, LANDSAT IMAGE
B 6401 34 WEST GERMANY, GRONINGEN GIANT GAS FIELD LANDSAT DATA,
B 6411 1971 WEST GERMANY, PRODUCTION, 1978-1979
B 6401 34 WEST GERMANY, ROTLIEGENDES FORMATION, NETHERLANDS AND
B 6303 342 WEST HICKMAN CREEK-BRYAN STATION FAULT, KENTUCKY
S 4 231† WEST INDIAN CORAL REEFS, CRINOIDS OF
S 4 231 WEST INDIAN CRINOIDS
B 6303 414* WEST INDIES, ABST., GEOLOGY AND DEVELOPMENT OF, TEAK OIL FIELD,
S 4 301 WEST INDIES, BARBUDA, REEF
S 4 299† WEST INDIES, FACIES ANALYSIS OF, HOLOCENE, CARBONATE SEDIMENTS
M 30 387† WEST INDIES, GEOLOGY AND DEVELOPMENT OF THE, TEAK OIL FIELD,
S 8 139 WEST INDIES, ST. CROIX, CHRISTIANSTED CANYON
S 8 129 WEST INDIES, ST. CROIX, SALT RIVER CANYON
S 8 139 WEST INDIES, SUBMARINE CANYONS, CHRISTIANSTED CANYON
S 8 129 WEST INDIES, SUBMARINE CANYONS, SALT RIVER CANYON
S 10 198 WEST INDIES, TRINIDAD, CHATHAM MUD ISLAND
S 10 199 WEST INDIES, TRINIDAD, GRAND RAVINE FIELD, MUDFLOWS
S 10 198 WEST INDIES, TRINIDAD, LABREA FORMATION, GAS-CUT MUDFLOW
S 10 196 WEST INDIES, TRINIDAD, MORUGA BOUFF MUD VOLCANO
B 6212 2509* WEST JAVA, INDONESIA, ABST., RECENT DEVELOPMENT OF, KAMOJANG,
B 6402 159 WEST LAKE VERRET FIELD, LOUISIANA
B 6209 1646 WEST LUZON TROUGH
B 6303 551 WEST, MARETA,, CORRELATION OF SUBSURFACE MIDDLE AND UPPER
B 6004 701* WEST MISSISSIPPI DELTA AREA, ABST., CONTROL OF SEDIMENT
B 6104 594 WEST MONTALVO FIELD, CALIFORNIA
B 6407 1036 WEST MONTALVO FIELD, CALIFORNIA
B 6407 1036 WEST MOUNTAIN FIELD, CALIFORNIA
B 6203 426 WEST NETHERLANDS BASIN, EUROPE
B 6309 1604* WEST NUECES CARBONATE ROCKS OF MAVERICK BASIN, ABST., FACIES,
B 6402 222 WEST NUECES FORMATION, TEXAS
B 6304 603 WEST PUERTO CHIQUITO FIELDS, FRACTURE RESERVOIRS, NEW MEXICO,
B 6008 1257 WEST RAYWOOD FIELD, LIBERTY COUNTY, TEXAS

M 29 463 WEST SIBERIAN BASIN, VOLUME OF SEDIMENT PRESERVED, USSR
B 6203 426 WEST SOLE TROUGH, EUROPE
M 27 119 WEST SPRING CREEK FORMATION, OKLAHOMA, ORDOVICIAN
B 6405 679 WEST, STEVE C., EXPLORATION IN GREAT SALT LAKE, ABST.
B 6009 1613* WEST SUNOCO-FELDA FIELDS HENDRY AND COLLIER COUNTIES,
B 6004 670* WEST TEXAS, ABST., NONVADOSE ORIGIN OF, PISOLITIC, FACIES,
B 6205 853† WEST TEXAS AND ANADARKO BASIN, STATISTICAL SUMMARY OF, WELLS
B 6205 855 WEST TEXAS AND NEW MEXICO, BASEMENT CONTOUR MAP
B 6205 854 WEST TEXAS AND NEW MEXICO, ULTRADEEP WELLS, LOCATION
B 6205 859 WEST TEXAS AND NEW MEXICO, ULTRADEEP WILDCATS, SUCCESS RATIOS
B 6004 517† WEST TEXAS AND SOUTHEAST NEW MEXICO, BASINAL, SANDSTONE,
M 26 219 WEST TEXAS BASIN, DEPOSITIONAL SYSTEMS
B 6012 2159† WEST TEXAS GEOLOGICAL SOCIETY, 1926 TO 1976
M 26 336 WEST TEXAS, HORSESHOE ATOLL, SEISMIC SECTION
B 6007 1107† WEST TEXAS, WILKIE RANCH FORMATION, NEW, PERMIAN UNIT IN,
B 6201 43 WEST TRYAL ROCKS FIELD, AUSTRALIA
B 6408 1289* WEST VIRGINIA, ABST., CARBONATE DIAGENESIS IN A, QUARTZARENITE–
B 6309 1585* WEST VIRGINIA, ABST., COMPUTER-AIDED GRAPHICS FOR ANALYSIS OF,
B 6405 800* WEST VIRGINIA, ABST., DEPOSITIONAL ENVIRONMENTS OF LACUSTRINE
B 6305 847* WEST VIRGINIA, ABST., DEVONIAN, SHALE, GAS PRODUCTION,
B 6309 1580* WEST VIRGINIA, ABST., ELEMENTAL ANALYSES OF, DEVONIAN, SHALES IN
B 6009 1622* WEST VIRGINIA, ABST., EVOLUTION OF, BASIN-SHELF RELATIONS
B 6004 724* WEST VIRGINIA, ABST., EVOLUTION OF, CARBONATE, SHELF, SILURIAN,
B 6105 840* WEST VIRGINIA, ABST., FIELD STUDIES OF, PARSONS AND, PETERSBURG
B 6309 1575* WEST VIRGINIA, ABST., LINEAMENTS AND WATER WELLS AS
B 6309 1586* WEST VIRGINIA, ABST., MAUCH CHUNK ALLUVIAL PLAIN AND MUD-FLA
B 6309 1577* WEST VIRGINIA, ABST., MIDDLE AND UPPER, DEVONIAN STRATIGRAPHY
B 6203 516* WEST VIRGINIA, ABST., ORIGIN OF, CYCLICITY IN, CONEMAUGH
B 6309 1585* WEST VIRGINIA, ABST., PATCH, REEFS AND INTERREEF DEPOSITS OF,
B 6309 1590* WEST VIRGINIA, ABST., PETROLOGIC EVALUATION OF SIGNIFICANCE OF
B 6004 707* WEST VIRGINIA, ABST., POSSIBILITY OF ABNORMALLY HIGH, FLUID,
B 6408 1288* WEST VIRGINIA, ABST., POTTSVILLE ALLUVIAL PLAIN, COALS IN
B 6009 1623* WEST VIRGINIA, ABST., STRATIGRAPHY AS KEY TO, ARCH-RELATED
B 6303 501* WEST VIRGINIA, ABST., SUBSURFACE STRATIGRAPHY OF UPPER,
B 6309 1588* WEST VIRGINIA, ABST., TECTONICS OF, PETERSBURG REGION
B 6203 492* WEST VIRGINIA AND VIRGINIA, ABST., DISTINCTIVE, TURBIDITE,
B 6309 1589* WEST VIRGINIA AND VIRGINIA, ABST., PATTERNS OF SPECIES DIVERSITY
B 6211 2309 WEST VIRGINIA, BASS ISLANDS FORMATION
B 6212 2399 WEST VIRGINIA, BECKLEY COAL AND SURROUNDING LITHOLOGIES
B 6405 631 WEST VIRGINIA, BIG SIX SAND
B 6211 2309 WEST VIRGINIA, BLOOMSBURG FORMATION
B 6211 2312 WEST VIRGINIA, CACAPON SANDSTONE MEMBER
B 6212 2385 WEST VIRGINIA, CARBONIFEROUS BACK-BARRIER DEPOSITS, KENTUCKY
S 5 127 WEST VIRGINIA, CAYUGAN GROUP, SALINA FORMATION, TONOLOWAY
B 6212 2408 WEST VIRGINIA, CEDAR GROVE COAL
B 6412 2169 WEST VIRGINIA, CHEMUNG GROUP
B 6309 1581* WEST VIRGINIA COALS, ABST., GEOLOGIC AGE VERSUS MAJOR MINOR,
B 6309 1578* **WEST VIRGINIA COALS FOR COAL UTILIZATION SCHEMES,** ABST..
B 6412 2168 WEST VIRGINIA, COTTAGEVILLE FIELD
B 6412 2169 WEST VIRGINIA, DEER PARK ANTICLINE
B 6008 1293 WEST VIRGINIA, DEVELOPMENTS, 1975
B 6108 1274 WEST VIRGINIA, DEVELOPMENTS, 1976
B 6208 1405 WEST VIRGINIA, DEVELOPMENTS, 1977
B 6308 1253 WEST VIRGINIA, DEVELOPMENTS, 1978
B 6409 1407 WEST VIRGINIA, DEVELOPMENTS, 1979
B 6412 2169 WEST VIRGINIA, ELKINS VALLEY ANTICLINE
B 6412 2169 WEST VIRGINIA, ETAM ANTICLINE
B 6406 938† WEST VIRGINIA, EXPLORATION MODEL FOR, PENNSYLVANIAN UPPER,
B 6309 1506 WEST VIRGINIA, FORMATION-DENSITY LOG STUDY
B 6412 2169 WEST VIRGINIA, HAMPSHIRE FORMATION
B 6412 2169 WEST VIRGINIA, HANNASVILLE SYNCLINE
B 6308 1244† WEST VIRGINIA IN 1978, OIL AND, GAS, DEVELOPMENTS IN, MARYLAND,
B 6409 1403† WEST VIRGINIA IN 1979, OIL AND, GAS, DEVELOPMENTS IN, MARYLAND,
B 6211 2309 WEST VIRGINIA, JUNIATA FORMATION
B 6406 938 WEST VIRGINIA, KANAWHA FORMATION
B 6405 629 WEST VIRGINIA, KEEFER SANDSTONE
B 6211 2309 WEST VIRGINIA, KEEFER SANDSTONE
B 6211 2309 WEST VIRGINIA, KEYSER LIMESTONE
S 5 135 WEST VIRGINIA, KEYSER LIMESTONE, BIG MOUNTAIN SHALE
S 5 135 WEST VIRGINIA, KEYSER LIMESTONE, MANDATA SHALE
B 6405 631 WEST VIRGINIA, LOCKPORT BIOHERMS
B 6405 635 WEST VIRGINIA, LOCKPORT DOLOMITE, GAS POTENTIAL
B 6406 939 WEST VIRGINIA, LOWER STOCKTON COAL SEAM
B 6405 634 WEST VIRGINIA, MCKENZIE FORMATION
B 6211 2309 WEST VIRGINIA, MCKENZIE FORMATION
B 6211 2327 WEST VIRGINIA, NEWBURG SANDSTONE
B 6405 629† WEST VIRGINIA, NIAGARAN, BIOHERMS AND, INTERBIOHERM DEPOSITS
B 6309 1577* WEST VIRGINIA OHIO, MARYLAND, AND SOUTHERN PENNSYLVANIA,
B 6208 1399† WEST VIRGINIA, OIL AND, GAS, DEVELOPMENTS IN, MARYLAND, OHIO,
B 6108 1269† WEST VIRGINIA, OIL AND, GAS, DEVELOPMENTS IN, MARYLAND, OHIO,
B 6008 1288† WEST VIRGINIA, OIL AND, GAS, DEVELOPMENTS IN, MARYLAND, OHIO,
B 6412 2169 WEST VIRGINIA, ONESQUETHAW GROUP
B 6309 1587* WEST VIRGINIA OPEN-FLOW DATA, INCORPORATING GEOCHEMICAL
B 6412 2168 WEST VIRGINIA, PARSONS LINEAMENT
B 6303 533* WEST VIRGINIA, PENNSYLVANIAN BECKLEY SEAM, ABST.,
B 6412 2168 WEST VIRGINIA, PETERSBURG LINEAMENT
B 6008 1298 WEST VIRGINIA, PRODUCTION, 1974-1975
B 6108 1281 WEST VIRGINIA, PRODUCTION, 1975-1976
B 6208 1410 WEST VIRGINIA, PRODUCTION, 1976-1977
B 6308 1247 WEST VIRGINIA, PRODUCTION, 1977-1978
B 6409 1412 WEST VIRGINIA, PRODUCTION, 1978-1979
S 5 125* WEST VIRGINIA, RELATIONS BETWEEN DEPOSITIONAL ENVIRONMENTS,
B 6211 2315 WEST VIRGINIA, ROCHESTER SHALES
B 6211 2309 WEST VIRGINIA, ROSE HILL FORMATION
B 6405 635 WEST VIRGINIA, ROSE HILL SHALE
B 6405 636 WEST VIRGINIA, SALINA EVAPORITES
B 6211 2309 WEST VIRGINIA, SALINA EVAPORITES
B 6209 1626 WEST VIRGINIA, SILURIAN-DEVONIAN SECTION, OHIO TO
B 6212 2396 WEST VIRGINIA TO PENNSYLVANIA, ALLEGHENY FORMATION
B 6212 2396 WEST VIRGINIA TO PENNSYLVANIA, POTTSVILLE FORMATION
B 6211 2309 WEST VIRGINIA, TONOLOWAY LIMESTONE
S 5 126 WEST VIRGINIA, TONOLOWAY LIMESTONE, DEPOSITIONAL
B 6211 2309 WEST VIRGINIA, TUSCARORA SANDSTONE
S 5 130 WEST VIRGINIA, UPPER SILURIAN, HELDERBURG GROUP, KEYSER
B 6211 2309 WEST VIRGINIA, WILLIAMSPORT SANDSTONE
B 6211 2309 WEST VIRGINIA, WILLS CREEK FORMATION
B 6309 1579* WEST VIRGINIA--POSSIBLE NEW COAL AND-OR METHANE RESOURCE,
B 6105 831* WEST VIRGINIAS NORTHERN COALFIELD–DIRECTION OF RESEARCH TO
B 6105 830* WEST VIRGINIA–EXPOSED MODEL FOR COMPLEX SUBSURFACE
B 6408 1287* WEST VIRGINIA–GEOLOGY AND CURRENT MINING TRENDS, ABST., COAL
B 6408 1285* WEST VIRGINIA–STATISTICAL COMPARISON AND DEPOSITIONAL
M 25 269 WEST YELLOW SEA, NORTH KIANGSU, AND SHANGHAI BASINS, KOREA
B 6007 1078 WESTBROOK, GRAHAM K., TECTONICS OF SOUTHWESTERN NORTH
B 6007 1078 WESTBROOK, GRAHAM K., TECTONICS OF SOUTHWESTERN NORTH
B 6004 654* WESTERN ALPINE MOBILE BELT AND FORELANDS, ABST.,
B 6203 428 WESTERN APPROACHES BASIN, BRITAIN, SOUTHWEST
B 6203 426 WESTERN APPROACHES, EUROPE
B 6411 1964 WESTERN APPROACHES, FRANCE, ENGLISH CHANNEL-
B 6110 1641 WESTERN APPROACHES, FRANCE, PARIS BASIN-
B 6405 740* WESTERN APPROACHES, OFFSHORE ENGLAND, ABST., HYDROCARBON
B 6409 1662 WESTERN ATLANTIC, BLAKE-BAHAMA FORMATION
B 6410 1659 WESTERN ATLANTIC, CONTINENTAL OFFSHORE STRATIGRAPHIC TEST
B 6410 1659 WESTERN ATLANTIC, DEEP SEA DRILLING PROJECT, SITES
B 6409 1662 WESTERN ATLANTIC JURASSIC INNER MAGNETIC QUIET ZONE, PROFILE
B 6409 1662 WESTERN ATLANTIC JURASSIC OUTER MAGNETIC QUIET ZONE, PROFILE
B 6410 1676 WESTERN ATLANTIC, KNAUSS KNOLL
B 6410 1677 WESTERN ATLANTIC MAGNETIC QUIET ZONE, JURASSIC SEDIMENTATION
B 6409 1662 WESTERN ATLANTIC, PLANTAGENET FORMATION
S 4 127 WESTERN ATLANTIC REGION, SPECIES OF SPONGES
B 6207 1235* WESTERN AUSTRALIA, ABST., BAUXITE, MINING AND, STREAM, SALINITY
M 25 411 WESTERN AUSTRALIA, ARCHEAN BANDED IRON FORMATIONS OR BIFS
M 25 427 WESTERN AUSTRALIA, ARCHEAN IRON ORE
M 25 415 WESTERN AUSTRALIA, BOOLGEEDA IRON FORMATION
M 25 411 WESTERN AUSTRALIA, BROCKMAN IRON FORMATION
M 25 411 WESTERN AUSTRALIA, DALES GORGE MEMBER
B 6110 1889 WESTERN AUSTRALIA, DEVELOPMENTS, 1976
B 6210 2151 WESTERN AUSTRALIA, DEVELOPMENTS, 1977
B 6310 1806 WESTERN AUSTRALIA, DEVELOPMENTS, 1978
B 6411 1864 WESTERN AUSTRALIA, DEVELOPMENTS, 1979
B 6406 814† WESTERN AUSTRALIA, DEVONIAN, GREAT, BARRIER, REEF OF, CANNING
M 25 411 WESTERN AUSTRALIA, FORTESCUE GROUP
B 6201 40† WESTERN AUSTRALIA, GEOLOGY AND PETROLEUM POTENTIAL OF,
M 25 411 WESTERN AUSTRALIA, GREENSTONE BELTS
M 25 411 WESTERN AUSTRALIA, HAMERSLEY BASIN, IRON
M 25 411 WESTERN AUSTRALIA, HAMERSLEY GROUP
M 25 427 WESTERN AUSTRALIA, HAMERSLEY IRON PROVINCE, BANDED IRO
M 25 419 WESTERN AUSTRALIA, HEAVY-MINERAL SAND
M 25 411 WESTERN AUSTRALIA, IRON, HEMATITE ENRICHMENT ORE
M 25 415 WESTERN AUSTRALIA, IRON ORE, ECONOMICS
M 25 411 WESTERN AUSTRALIA, IRON ORE, GEOLOGY AND DISTRIBUTION
M 25 410 WESTERN AUSTRALIA, IRON ORE, HISTORICAL BACKGROUND
M 25 415 WESTERN AUSTRALIA, IRON ORE, PRODUCTION
M 25 411 WESTERN AUSTRALIA, IRON ORE, TYPES OF
M 25 414 WESTERN AUSTRALIA, IRON, PISOLITIC LIMONITE ORE
M 25 414 WESTERN AUSTRALIA, IRON, SEDIMENTARY ORE
M 25 415 WESTERN AUSTRALIA, IRON-ORE DEPOSITS, EXPLORATION
M 25 412 WESTERN AUSTRALIA, IRON-ORE DEPOSITS, MAP

M 25 416 WESTERN AUSTRALIA, IRON-ORE PRODUCTION BY COMPANIES
M 25 416 WESTERN AUSTRALIA, IRON-ORE PRODUCTION, FUTURE OUTLOOK
B 6203 455† WESTERN AUSTRALIA, ISOTOPIC EVIDENCE FOR, MICROBIOLOGIC ROLE
M 25 411 WESTERN AUSTRALIA, JOFFRE MEMBER
M 25 401 WESTERN AUSTRALIA, KALGOORLIE, NICKEL DEPOSITS
M 25 401 WESTERN AUSTRALIA, KAMBALDA, NICKEL DEPOSITS
M 25 399 WESTERN AUSTRALIA, KAMBALDA, NICKEL ORE, SULFIDE
M 25 414 WESTERN AUSTRALIA, KIMBERLEY BASIN
M 25 414 WESTERN AUSTRALIA, KIMBERLEY GROUP
M 25 411 WESTERN AUSTRALIA, MARRA MAMBA IRON FORMATION
B 6410 1682 WESTERN AUSTRALIA, MOUNT JOHN OSMOND RANGE, LOWER
M 25 402 WESTERN AUSTRALIA, MT. KEITH, NICKEL DEPOSIT
M 25 400 WESTERN AUSTRALIA, NICKEL SULFIDE DEPOSITS
M 25 401 WESTERN AUSTRALIA, NICKEL SULFIDE DEPOSITS IN ARCHEAN
M 25 402 WESTERN AUSTRALIA, ORA BANDA, NICKEL-ENRICHED LATERITE
M 25 414 WESTERN AUSTRALIA, PENTECOST SANDSTONE, YAMPI MEMBER
M 25 402 WESTERN AUSTRALIA, PERSEVERANCE, NICKEL DEPOSIT
M 25 411 WESTERN AUSTRALIA, PILBARA AND YILGARN BLOCKS, IRON
M 25 411 WESTERN AUSTRALIA, PROTEROZOIC BANDED IRON FORMATIONS
M 25 414 WESTERN AUSTRALIA, ROBE PISOLITE
M 25 402 WESTERN AUSTRALIA, SCOTIA-CARR BOYD AREA, NICKEL SULFIDE
B 6311 2088† WESTERN AUSTRALIA, SEDIMENTARY-, WEDGE PROGRADATION FROM,
M 25 402 WESTERN AUSTRALIA, SHERLOCK BAY, NICKEL DEPOSIT
M 25 427 WESTERN AUSTRALIA, WEST AUSTRALIAN NICKEL SULFIDE
M 25 401 WESTERN AUSTRALIA, WINDARRA, NICKEL DEPOSITS
M 25 402 WESTERN AUSTRALIA, WINGELLINA, NICKEL-ENRICHED LATERITE
M 25 411 WESTERN AUSTRALIA, WYLOO GROUP
M 25 427 WESTERN AUSTRALIA, YAMPI SOUND, IRON ORE
M 25 414 WESTERN AUSTRALIA, YARRALOOLA CONGLOMERATE
M 25 427 WESTERN AUSTRALIA, YEELIRRIE URANIUM DEPOSITS
M 25 410† WESTERN AUSTRALIA--GEOLOGY AND DEVELOPMENT, IRON-ORE
B 6207 1224* WESTERN AUSTRALIAS ENERGY ECONOMY, ABST., NORTHWEST SHELF,
B 6410 1702 WESTERN BOUNDARY CURRENT
B 6410 1701 WESTERN BOUNDARY UNDERCURRENT
B 6004 554† WESTERN CANAD BASIN, THICKNESS OF, REMOVED SEDIMENTARY
B 6111 1951 WESTERN CANADA, ATHABASCA HEAVY OIL DEPOSIT
B 6103 434† WESTERN CANADA BASIN, RATIO OF PETROLEUM TO, WATER DURING,
B 6305 812† WESTERN CANADA BASIN. DISCUSSION AND REPLY, THICKNESS OF,
B 6111 1951 WESTERN CANADA, COLD LAKE HEAVY OIL DEPOSIT
B 6111 1953 WESTERN CANADA, HEAVY OIL
B 6008 1343† WESTERN CANADA IN 1975, DEVELOPMENTS IN
B 6108 1313† WESTERN CANADA IN 1976, DEVELOPMENTS IN
B 6308 1407† WESTERN CANADA IN 1978, DEVELOPMENTS IN
B 6409 1510† WESTERN CANADA IN 1979, DEVELOPMENTS IN
B 6111 1954 WESTERN CANADA, LLOYDMINSTER HEAVY OIL
B 6111 1951 WESTERN CANADA, PEACE RIVER HEAVY OIL DEPOSIT
B 6111 1951 WESTERN CANADA, WABASCA HEAVY OIL DEPOSIT
B 6111 1955 WESTERN CANADA, WAINWRIGHT HYDROCARBONS
B 6105 775* WESTERN CANADIAN OIL AND GAS DATA BASE AS WORKING MODEL FOR
M 25 486 WESTERN CORDILLERA, COPPER PROVINCES
B 6202 274 WESTERN EQUATORIAL ATLANTIC, PHYSIOGRAPHIC PROVINCE MAP
M 30 195† WESTERN EUROPE, EKOFISK--FIRST OF THE GIANT, OIL FIELDS IN
B 6306 953 WESTERN, FACIES BELTS, NEWFOUNDLAND
B 6209 1667 WESTERN FOOTHILLS PROVINCE, TAIWAN
S 7 48 WESTERN GULF OF MEXICO, SEISMIC UNITS, TABLE
S 7 15 WESTERN GULF RISE, DESCRIBED
B 6207 1224* WESTERN INDONESIA, ABST., CORRELATION OF MAJOR,
B 6307 1021 WESTERN INDONESIA, LONGITUDINAL WRENCH FAULTS
B 6303 519* WESTERN INTERIOR, ABST., DELTAIC, COALS OF, FERRON SANDSTONE
B 6205 892* WESTERN INTERIOR, ABST., DIAGENESIS AND, HYDROCARBON,
B 6303 550* WESTERN INTERIOR, ABST., INFLUENCE OF, TRANSCONTINENTAL ARCH
B 6004 710* WESTERN INTERIOR, ABST., PROVENANCE AND, DISPERSAL OF UPPER,
B 6405 739* WESTERN INTERIOR, ABST., WAVE-DOMINATED, DELTAS—IMPORTANT
B 6405 731* WESTERN INTERIOR BASIN, NORTH AMERICA, ABST., PALEOBIOLOGIC
B 6405 778* WESTERN INTERIOR BASIN, NORTH AMERICA, ABST., PHYSICAL EVIDENCE
M 29 463 WESTERN INTERIOR BASIN, VOLUME OF SEDIMENT PRESERVED
B 6004 724* WESTERN INTERIOR CRETACEOUS MARINE AND COASTAL DEPOSITS,
B 6105 824* WESTERN INTERIOR CRETACEOUS, NORTH AMERICA, ABST., MODELS OF
B 6405 729* WESTERN INTERIOR CRETACEOUS STRATA, ABST., TRACE-MAKERS AS
B 6211 2277 WESTERN INTERIOR, JURASSIC CORRELATION AND NOMENCLATURE
B 6303 355 WESTERN INTERIOR, MOWRY SEA
S 6 260 WESTERN INTERIOR, NORTH AMERICA, CRETACEOUS
B 6108 1375* WESTERN INTERIOR UNITED STATES, ABST., ORGANIC, GEOCHEMISTRY,
B 6303 464* WESTERN INTERIOR UNITED STATES, ABST., REGIONAL STRATIGRAPHY
B 6201 105 WESTERN INTERIOR UNITED STATES, MEADE PEAK PHOSPHATIC SHALE
B 6201 98† WESTERN INTERIOR UNITED STATES, ORGANIC, GEOCHEMISTRY,
B 6201 107 WESTERN INTERIOR UNITED STATES, RETORT PHOSPHATIC SHALE
B 6007 1022 WESTERN LAKE SUPERIOR, BATHYMETRIC MAP
M 25 210 WESTERN, MARGINAL GEOSYNCLINE, NEW ZEALAND
M 29 293 WESTERN MEDITERRANEAN BASIN, SEDIMENTARY SEQUENCE
M 25 486 WESTERN, MERCURY DEPOSITS, UNITED STATES
M 25 497 WESTERN MINES COPPER-ZINC-LEAD DEPOSIT, VANCOUVER ISLAND
M 25 43 WESTERN PACIFIC AND ATLANTIC OCEAN MARGINS
B 6102 297 WESTERN, STEPHEN K., RESIDUAL BOUGUER GRAVITY ANOMALY
M 25 269 WESTERN SUBBASIN, KOREA
B 6209 1667 WESTERN SUBPROVINCE, TAIWAN
S 9 11 WESTERN UNITED STATES, PENNSYLVANIAN, WIND DIRECTION
S 9 11 WESTERN UNITED STATES, PERMIAN, WIND DIRECTION
M 25 157 WESTERN, VOLCANIC ROCKS, UNITED STATES
M 25 29 WESTERNPORT BAY, AUSTRALIA
B 6104 597 WESTEROSE FIELD, ALBERTA
B 6112 2135 WESTLOCK FIELD, ALBERTA
B 6105 840 WESTPHAL, WARREN H., IN-SITU LEACHING OF URANIUM—DEVELOPING
B 6305 843 WESTPHAL, WARREN H., SOLUTION MINING OF URANIUM AND ITS
B 6204 669 WESTPHALIAN, ALGERIA
S 6 325 WESTPHALIAN-STEPHANIAN BOUNDARY, CARBONIFEROUS
B 6007 1010 WESTWATER CANYON MEMBER, FLUVIAL SHEET SANDSTONE, NEW
B 6405 795* WESTWATER CANYON MEMBER, MORRISON FORMATION, SOUTHERN SAN
B 6209 1675 WESTWATER CANYON MEMBER, NEW MEXICO
B 6304 694* WESTWATER CANYON MEMBER OF MORRISON FORMATION, ABST.,
B 6304 689* WESTWATER CANYON WET ALLUVIAL-FAN SYSTEM, ABST., DEPOSITION
S 3 75 WESTWATER SANDSTONE, NEW MEXICO
S 3 124 WESTWATER SANDSTONE, NEW MEXICO, GRANTS
B 6004 729* WET SLOPES AND RAIN SHADOWS, ABST., FLUVIAL, MODELS--
B 6112 2092 WET-GAS-GENERATING KEROGEN
B 6009 1609* WETLAND AND LAND-USE DELINEATION IN TEXAS COASTAL ZONE,
S 3 102 WHALE BACKS, DEFINITION
B 6203 428 WHALE BASIN, GRAND BASIN
B 6004 651 WHALEY, J. F., COMPUTER-GENERATED SCALE FUNCTIONS FOR
B 6009 1624 WHALEY JOSEPH F., PRECAMBRIAN GEOPHYSICAL PROVINCES IN
B 6005 879 WHALEY, JOSEPH F., TRANSFORMATION OF RESISTIVITY TO
B 6201 44 WHARTON BASIN, AUSTRALIA
S 10 261 WHARTON COUNTY, LEVERIDGE ESTATE WELL, HYDROCARBON LOG,
B 6212 2509* WHAT FOREIGN COMPANIES SHOULD KNOW ABOUT UNITED STATES
B 6207 1216* WHAT LAW SHALL GOVERN DEEP-SEA MINING , ABST.
B 6405 788* WHATS NEW IN U.S. INDUSTRY ACTIVITY, ABST.
B 6408 1280* WHEATON CONSOLIDATED OIL FIELD, GIBSON COUNTY, INDIANA, ABST.,
B 6305 804 WHEELER COUNTY, MILLS RANCH COMPLEX, TEXAS
B 6203 570* WHEELER COUNTY, TEXAS, ABST., WHEELER-PAN HUNTON FIELD
B 6212 2447 WHEELER RIDGE, CALIFORNIA
B 6309 1591 WHEELER, RUSSELL, L., CROSS-STRIKE STRUCTURAL
B 6412 2166 WHEELER, RUSSELL L., CROSS-STRIKE STRUCTURAL DISCONTINUITIES—
B 6105 840 WHEELER, RUSSELL L., PARSONS AND PETERSBURG STRUCTURAL
B 6105 830 WHEELER, RUSSELL L., SMOKE HOLES AREA, VALLEY AND RIDGE, WEST
B 6203 570* WHEELER-PAN HUNTON FIELD, WHEELER COUNTY, TEXAS, ABST.
S 11 68 WHEELERIAN STAGE
S 11 216 WHEELERIAN STAGE
B 6202 259 WHEELERIAN STAGE, CALIFORNIA
B 6303 551 WHELAN, THOMAS, EFFECT OF BIOGENIC METHANE ON SEDIMENT
B 6212 2510* WHICH WAY IS UP IN SPACE, ABST.
M 26 439 WHIPPLE, ARTHUR P., SEISMIC STRATIGRAPHIC MODEL OF
B 6109 1437 WHIPPLE, ARTHUR P., SEISMIC STRATIGRAPHIC MODEL OF
B 6305 845* WHIRLPOOL MEMBER, SILURIAN, OF MCCONNELSVILLE FIELD, MORGAN
B 6404 472 WHISKER CALCITE
M 27 162 WHISKER-CRYSTAL CEMENTS
B 6303 551 WHITAKER, J. H. M., EARLY PLEISTOCENE SUBMARINE CANYON, BOSO
B 6303 552 WHITBREAD, D. R., HYDROCARBON PROSPECTS IN BASINS WEST OF
B 6306 958 WHITE ARM WINDOW, NEWFOUNDLAND
B 6408 1158 WHITE, DAVID A., ASSESSING OIL AND GAS PLAYS IN FACIES-CYCLE
B 6303 552 WHITE, DAVID A., ASSESSING OIL AND GAS PLAYS IN FACIES-CYCLE
B 6312 2183 WHITE, DAVID A., METHODS OF ESTIMATING OIL AND GAS RESOURCES
B 6303 552 WHITE, DAVID L., BRAIDED-STREAM PROCESSES AND FACIES IN MODERN
S 4 171 WHITE, EUGENE W., CARIBBEAN REEF CORALS MONTASTREA ANNULARIS
B 6102 257 WHITE KNOB LIMESTONE GROUP, IDAHO
B 6107 988 WHITE LIMESTONE, CHALK UNIT, IRELAND
M 27 184 WHITE LIMESTONE, IRELAND, CRETACEOUS
M 27 160 WHITE LIMESTONE, IRELAND, CRETACEOUS
M 27 51 WHITE LIMESTONE, IRELAND, CRETACEOUS
M 27 205 WHITE LIMESTONE, IRELAND, CRETACEOUS
M 28 78 WHITE LIMESTONE, NORTHERN IRELAND, CRETACEOUS
B 6412 2078 WHITE MOUNTAIN FAULT ZONE, APPALACHIANS
B 6408 1172 WHITE RIM TAR, UTAH AND, COLORADO
B 6302 195 WHITE RIVER PLATEAU, COLORADO
B 6108 1387* WHITE RIVER-GRAY HILLS AREA, PICEANCE CREEK BASIN,
M 26 281 WHITE ROCKS AREA, CROSS SECTION, COLORADO
B 6303 552 WHITE, WILLIAM A., MAP DISPLAY FORMATS FOR ENVIRONMENTAL
B 6211 2245 WHITE WOLF FAULT, CALIFORNIA
B 6004 717* WHITEHALL FORMATION, CAMBRIAN-ORDOVICIAN OF EASTERN NEW
B 6402 225 WHITEHEAD GRABEN, TEXAS
M 25 108 WHITEHORSE COAL AREAS, CANADA

M 25 499 WHITEHORSE TROUGH, CANADA
S 7 313 WHITEHOUSE, U. G., II, HOLOCENE SEDIMENT GEOCHEMISTRY OF
B 6407 972 WHITEMUD FORMATION, ALBERTA
B 6309 1599 WHITESIDES, VIRGIL S., JR., CLASTICITY INDEX--KEY TO CORRELATING
B 6405 802 WHITLEY, DONALD L., MIDDLE CRETACEOUS, ALBIAN-TURONIAN,
B 6305 832* WHITNEY CANYON FIELD--POTENTIAL GAS GIANT IN WYOMING THRUST
B 6405 730* WHITNEY CANYON FIELD—POTENTIAL GAS GIANT IN WYOMING THRUST
B 6405 806* WHITNEY CANYON—A PROVED GIANT—ONLY SMALL PART OF ROCKY
B 6103 465 WHITNEY, O. T., CENOZOIC SEDIMENTATION AND TECTONICS OF HOPE
B 6303 440* WHITSETT FORMATION IN URANIUM AREA OF WESTERN KARNES
B 6206 1049 WHITSETT FORMATION, TEXAS, TORDILLA MEMBER
B 6109 1522 WHITTEN, E. H. TIMOTHY, RAPID APTIAN-ALBIAN SUBSIDENCE RATES IN
B 6202 255 WHITTIER FAULT, CALIFORNIA
B 6203 361 WHITTIER FAULT, CALIFORNIA
B 6111 1982 WHITTIER FAULT ZONE, CALIFORNIA
B 6111 1988 WHITTIER-ELSINORE FAULT ZONE, CALIFORNIA
B 6002 215 WHOLE CRUDE OIL, GHAWAR, ARABIAN
B 6002 288 WHOLE FREE-GAS PHASE OF HYDROCARBONS, STUDY OF
B 6203 569* WHY ARE LARGE DEPOSITS DISCOVERED FIRST GEOLOGIC KNOWLEDGE
B 6008 1409* WHY PETROLEUM EXPLORATIONIST SHOULD BE ON LOOKOUT FOR
S 6 5 WHY TIME SCALES, REVIEW OF PRINCIPLE
B 6002 327* WICHITA AULACOGEN, OKLAHOMA, ABST., GEOTECTONIC EVOLUTION OF
B 6205 787 WICHITA SEISMIC DATA, TEXAS
B 6103 367 WICHITA UPLIFT, ALLUVIAL-FAN DEPOSITS, ALLUVIAL-FAN DEPOSIT
B 6103 361 WICHITA-CRINER HILLS UPLIFT, OKLAHOMA
M 26 451 WICHITAN UNCONFORMITY, OKLAHOMA, ANADARKO BASIN
B 6209 1769 WICKER, RUSSEL A., REEVALUATING FLORIDA BASEMENT, ABST.
B 6112 2143 WICKER, RUSSELL A., FLORIDA BASEMENT—AN ISOMETRIC VIEW
M 24 151 WICKHAM SANDSTONES, TEXAS
B 6004 667* WIDESPREAD GENERATION OF HARDGROUNDS IN HIGH-ENERGY
B 6108 1392* WIDESPREAD OCCURRENCE OF LAUMONTITE IN LATE MESOZOIC AND
M 26 49 WIDMIER, J. M., SEISMIC STRATIGRAPHY AND GLOBAL CHANGES OF SEA
M 26 165 WIDMIER, J. M., SEISMIC STRATIGRAPHY AND GLOBAL CHANGES OF SEA
B 6205 752 WIDMIER, J. M., SEISMIC STRATIGRAPHY AND GLOBAL CHANGES OF SEA
B 6405 803 WIECZOREK, JOZEF, RELATIONS BETWEEN STROMATOLITES AND
B 6010 1719 WIENER, ON DEVELOPMENTS IN SWITZERLAND, 1975, BUCHI, U. P., AND G.
B 6110 1649 WIENER, ON DEVELOPMENTS IN SWITZERLAND, 1976, BUCHI, U. P., AND G.
B 6210 1919 WIENER, ON DEVELOPMENTS IN SWITZERLAND, 1977, BUCHI, U. P., AND G.
B 6310 1928 WIENER, ON DEVELOPMENTS IN SWITZERLAND, 1978, BUCHI, U. P., AND G.
B 6411 1968 WIENER, ON DEVELOPMENTS IN SWITZERLAND, 1979, BUCHI, U. P., AND
B 6405 803 WIER, CHARLES E., NEW AND DEVELOPING TECHNIQUES IN COAL
B 6303 428 WIESENBURG, DENIS A., COMPARISON OF CARBONATE CHEMISTRY OF
B 6012 2177 WIGGINS, V. D., IPEWICK FORMATION, SIGNIFICANT
S 4 299 WIGLEY, PETER, FACIES ANALYSIS OF HOLOCENE CARBONATE
M 25 84 WIJAYA, SOELISTIYO U., COAL RESOURCES AND POTENTIAL IN
B 6002 293 WIJEYESEKERA, DEVAPRIYA C., HIGH-PRESSURE CONSOLIDATION OF
B 6405 755 WILBER, R. J., ET AL, NODULAR SUBMARINE CEMENTATION ON
B 6410 1701 WILBER, R. JUDE, CARBONATE SEDIMENT DRIFTS IN NORTHERN STRAITS
B 6105 841 WILBER, R. JUDE, POROSITY CONTROLS IN SUBSEA CEMENTED ROCKS
B 6004 733 WILBER, ROBERT J., PETROLOGY OF SUBSE CEMENTED CARBONATE
B 6402 222 WILBERNS FORMATION, TEXAS
M 27 212 WILBERNS FORMATION, TEXAS, CAMBRIAN
B 6405 703* WILCOX DELTA SYSTEM, SOUTH TEXAS, ABST., GROWTH, FAULTING AND
B 6109 1547* WILCOX, EOCENE, KARNES COUNTY, TEXAS, ABST., SECONDARY,
B 6004 725* WILCOX, EOCENE, SANDSTONE, KARNES COUNTY TEXAS, ABST., FACTORS
B 6408 1169 WILCOX FACIES, TEXAS
B 6109 1534* WILCOX GAS RESERVOIRS, NORTHEAST THOMPSONVILLE FIELD, JIM
B 6309 1602* WILCOX GROUP IN RELATION TO LIGNITE DEVELOPMENT IN EAST
B 6203 495* WILCOX GROUP OF SOUTH TEXAS, AND BELL CANYON FORMATION OF
B 6007 1009 WILCOX GROUP, TEXAS
B 6303 536* WILCOX, LAVACA COUNTY, TEXAS, ABST., SOUTH HALLETTSVILLE, GAS
B 6409 1555* WILCOX LIGNITE IN WEST-CENTRAL LOUISIANA, ABST., SUBSURFACE
B 6203 498* WILCOX OF SOUTHWEST TEXAS, ABST., ORIGIN OF LATE-STAGE,
B 6309 1598* WILCOX RESERVOIRS IN VALENTINE AND MENKING FIELDS, LAVACA
B 6309 1595* WILCOX RESERVOIRS, VALENTINE FIELD, LAVACA COUNTY, TEXAS,
M 26 70 WILCOX SANDSTONE, TEXAS, EAST
B 6409 1556* WILCOX SANDSTONES, KATY GAS FIELD, WALLER COUNTY, TEXAS, ABST.,
B 6203 497* WILCOX SANDSTONES OF SOUTHWEST TEXAS—IMPLICATIONS OF
B 6408 1220 WILCOX SECTION, TEXAS
B 6405 804* WILCOX SHELF EDGE IN TEXAS—RELATION TO GROWTH FAULTING AND
B 6409 1556* WILCOX TREND OF SOUTH TEXAS, ABST., LIVE OAK, DELTA COMPLEX—
B 6405 725 WILD, GARRY D., PETROGRAPHIC VARIATION IN WESTERN KENTUCKY
S 11 224 WILDCAT GROUP, FAUNULES
S 11 220 WILDCAT GROUP UNDIFFERENTIATED, FAUNAS
B 6108 1126 WILDCAT, NEW-FIELD
B 6308 1174 WILDCAT, NEW-FIELD
B 6008 1159 WILDCAT, NEW-FIELD
B 6208 1281 WILDCAT, NEW-FIELD
B 6208 1281 WILDCAT, NEW-POOL
B 6008 1159 WILDCAT, NEW-POOL
B 6308 1174 WILDCAT, NEW-POOL
B 6108 1126 WILDCAT, NEW-POOL
B 6409 1301 WILDCAT, NEW-POOL
S 11 185 WILDCAT SERIES, CALIFORNIA
B 6009 1621* WILDCAT VALLEY SANDSTONE IN SOUTHWEST VIRGINIA--POSSIBLE
B 6103 416† WILDCAT VALLEY SANDSTONE IN SOUTHWEST VIRGINIA--POSSIBLE
B 6103 422 WILDCAT VALLEY SANDSTONE, MAP, VIRGINIA
B 6311 2117* WILDCATS MUST WE DRILL, ABST., HOW MANY
B 6205 859 WILDCATS, SUCCESS RATIOS, TEXAS, PANHANDLE AND, OKLAHOMA,
B 6205 859 WILDCATS, SUCCESS RATIOS, WEST, TEXAS AND, NEW MEXICO,
B 6206 967 WILDE, P., CHANNEL SANDS AND PETROLEUM POTENTIAL OF MONTEREY
M 25 313 WILDE, P., PETROLEUM POTENTIAL OF CONTINENTAL RISE OFF CENTRAL
B 6405 803 WILDE, P., POTENTIAL PETROLEUM RESERVOIRS ON DEEP-SEA FANS OFF
B 6208 1441 WILDFANG, BO, DEVELOPMENTS OFFSHORE WEST GREENLAND IN 1977
B 6405 751 WILEY, GEORGE N., ORIGIN OF BIOGENIC CARBONATE SANDS OF
B 6403 400† WILFRED POOL, SULLIVAN COUNTY, INDIANA, ANALYSIS OF, GRAVITY,
B 6002 323 WILHITE, G. PAUL, ENGINEERING AND GEOLOGIC CHALLENGES IN
B 6007 1107† WILKIE RANCH FORMATION, NEW PERMIAN UNIT IN FINLAY
B 6303 530* WILKINS PEA MEMBER OF GREEN RIVER FORMATION, EOCENE,
B 6104 527 WILKINSON, B. H., LAVACA BAY--TRANSGRESSIVE DELTAIC
B 6303 425 WILKINSON, BRUCE H., DRANEY LIMESTONE--EARLY CRETACEOUS
B 6203 534 WILKINSON, BRUCE H., EGGSHELL DIAGENESIS AND PRECOMPACTIONAL
B 6203 517 WILKINSON, BRUCE H., LACUSTRINE PSEUDOVARVES FROM CRETACEOUS
B 6303 536 WILKINSON, BRUCE H., NORMALLY AND REVERSELY GRADED BEDS
B 6203 527 WILKINSON, BRUCE H., TERTIARY ALGAL CARBONATE REEFS FROM
B 6108 1384 WILLAND, T. M., PARAGENESIS OF AUTHIGENIC MINERALS IN ST. PETER
B 6004 708* WILLAPA BAY, WASHINGTON, ABST., BED FORMS AND, INTERNAL
B 6303 509* WILLAPA BAY, WASHINGTON, ABST., BED FORMS AND PROCESSES ON,
B 6004 657* WILLAPA BAY, WASHINGTON, ABST., MODERN AND, ANCIENT,
B 6404 692* WILLAPA BAY, WASHINGTON, ABST., STRATIFICATION TYPES IN
B 6201 40 WILLCOX, J. B., GEOLOGY AND PETROLEUM POTENTIAL OF EXMOUTH
B 6207 1239 WILLCOX, JOHN BARRY, PETROLEUM PROSPECTIVITY OF AUSTRALIAN
B 6304 700 WILLETTE, P. D., QUANTITATIVE COMPARISONS OF GEOLOGIC MAPS TO
B 6303 517 WILLETTE, P. D., STATISTICAL COMPARISON OF MAPPED DATA, ABST.
B 6309 1591 WILLETTE, P. D., STRUCTURAL CONTROL OF ONONDAGA REEFS,
B 6404 585 WILLIAMS AND ALEXANDER R. MCBIRNEY, MORAN, W. R., REVIEW OF,
B 6309 1597 WILLIAMS, DAMON, ECOLOGIC NICHES OF RADIOLARIANS, PLANKTONIC
B 6408 1290 WILLIAMS, DAVID A., COAL AND COAL DEVELOPMENTS IN WESTERN
B 6309 1577 WILLIAMS, DAVID A., REGIONAL COALIFICATION PATTERNS IN EASTERN
B 6203 526 WILLIAMS, G. L., COST B-2 WELL, BALTIMORE CANYON, AND ITS
S 6 11 WILLIAMS, H. S. TIME SCALE (1893)
B 6102 297 WILLIAMS, J. A., CHARACTERIZATION OF OIL TYPES IN PERMIAN BASIN
B 6303 497 WILLIAMS, JACK A., GEOCHEMICAL EXPLORATION IN POWDER RIVER
B 6304 699 WILLIAMS, JOHN C., DISSOLVED HYDROCARBONS IN COASTAL WATERS
B 6411 1887 WILLIAMS, ON DEVELOPMENTS IN INDONESIA, 1979, JAMES, R., AND W.
B 6303 518 WILLIAMS, PAUL L., METALLIC MINERAL RESOURCES OF
B 6004 723 WILLIAMS, S. C., RADIOCARBON DATING OF ESTUARINE CARBONATE
B 6004 703 WILLIAMS, S. C., SEDIMENTATION RATES IN HUDSON RIVER ESTUARY,
B 6004 687 WILLIAMS, S. J., DISTRIBUTION AND RELATION OF PHYSICAL AND
B 6105 841 WILLIAMS, S. JEFFRESS, OCEAN DUMPING ON NEW YORK BIGHT INNER
B 6311 2058 WILLIAMS, V. EILEEN, DEPOSITIONAL SETTING AND COAL PETROLOGY OF
B 6204 694 WILLIAMS WELL, BAHAMA PLATFORM
B 6408 1290 WILLIAMSON, ALLEN D., COAL AND COAL DEVELOPMENTS IN WESTERN
B 6303 553 WILLIAMSON, CHARLES R., DEEP-SEA DEPOSITION IN
B 6405 803 WILLIAMSON, CHARLES R., RESERVOIR ROCK, SOURCE ROCK, AND
M 30 173 WILLIAMSON, K. R., NINIAN FIELD, U.K. SECTOR, NORTH SEA
B 6211 2309 WILLIAMSPORT SANDSTONE, WEST VIRGINIA
B 6008 1400* WILLISTON BASIN, ABST., DOLOMITIZATION PATTERNS IN PALEOZOIC
B 6405 762* WILLISTON BASIN, ABST., MONDAK, MISSISSIPPIAN OIL FIELD
B 6303 478* WILLISTON BASIN, ABST., ORIGIN OF REGIONAL PRE-MIDDLE, DEVONIAN,
B 6008 1403* WILLISTON BASIN, AND ITS RELATION TO PETROLEUM GENERATION,
B 6307 1037 WILLISTON BASIN AREA, STRUCTURES, MAP
M 29 465 WILLISTON BASIN, AVERAGE THICKNESS OF SEDIMENTS
S 10 49 WILLISTON BASIN, BAKKEN SHALE, MIGRATION, NORTH DAKOTA
B 6407 972 WILLISTON BASIN, CRETACEOUS, MONTANA
B 6211 2277 WILLISTON BASIN, JURASSIC, MONTANA
S 3 119 WILLISTON BASIN, NORTH DAKOTA AND, SOUTH DAKOTA
B 6008 1398* WILLISTON BASIN, NORTH-CENTRAL NORTH DAKOTA, ABST.,
S 10 27 WILLISTON BASIN, PARAFFINS, DISTRIBUTION
B 6008 1390* WILLISTON BASIN--WHAT, WHERE, WHY, HOW, AND WHEN, ABST., SALT
B 6405 723* WILLISTON BASIN—REAWAKENING OF GIANT PETROLEUM PROVINCE,
B 6407 972 WILLOW CREEK FORMATION, MONTANA
B 6211 2309 WILLS CREEK FORMATION, WEST VIRGINIA
B 6209 1610 WILLS CREEK SHALE, MARYLAND
B 6202 255 WILMINGTON ANTICLINAL STRUCTURE, CALIFORNIA, TORRANCE-
B 6106 918 WILMINGTON CANYON AREA, SEISMIC-REFLECTION PROFILES
B 6106 918 WILMINGTON CANYON AREA, SUBMARINE SLIDE, DELAWARE
B 6106 918† WILMINGTON CANYON, CONTINENTAL, SLOPE SEDIMENT INSTABILITY

S 8 128 WILMINGTON CANYON, TIME-VELOCITY CURVES, CURRENT FLOW
S 8 127 WILMINGTON CANYON, TOPOGRAPHY, INDEX MAP
S 8 122 WILMINGTON CANYON, U.S. EAST COAST, SUBMARINE CANYONS
B 6403 439* WILMINGTON OIL FIELD, ABST., COMPUTER SAND VOLUME
B 6004 667* WILMINGTON SUBMARINE CANYONS, ABST., CONTRASTS IN
M 24 1 WILSON, B. W., MIDDLE GROUND SHOAL OI FIELD, ALASKA
B 6102 269 WILSON, H. H., PRE-CRETACEOUS ROCKS OF NORTHWESTERN
B 6104 483 WILSON, H. HUGH, FROZEN-IN HYDROCARBON ACCUMULATIONS
B 6107 1090 WILSON, H. HUGH, SHEAR ZONES INSIDE GULF COAST SALT STOCKS HELP
B 6004 733 WILSON, J. L., CARBONATE MOUNDS FROM SHELVES AND BASINS, ABST.
B 6303 526 WILSON, J. L., REGIONAL ASPECT OF SEDIMENT DISTRIBUTION ACROSS
B 6002 159 WILSON, JAME E., IN THE PUBLIC INTEREST
B 6409 1550 WILSON, JAMES E., REVIEW OF SIERRA DE JUAREZ, CHIHUAHUA, MEXICO.
B 6405 682 WILSON, JAMES L., DIAGENESIS OF LOWER CORALLINE LIMESTONE,
B 6009 1617 WILSON, JAMES LEE, EARLY CRETACEOUS SHELF MARGIN IN NORTHERN
B 6112 2115 WILSON, JAMES LEE, EVOLUTION OF YUCCA MOUND COMPLEX, LATE
B 6109 1557 WILSON, JAMES LEE, LOWER CRETACEOUS SHELF MARGIN IN NORTHERN
B 6303 553 WILSON, JAMES LEE, SEDIMENTOLOGIC STUDIES IN JURASSIC OF TUNISIA,
B 6105 728 WILSON, JOSEPH M., HISTORY OF DALLAS GEOLOGICAL SOCIETY
B 6002 273 WILSON, JOHN A., NOTE 44--APPLICATION FOR ADDITION TO CODE
B 6405 804 WILSON, JOHN M., PERMIAN-PENNSYLVANIAN OF WEST-CENTRAL
B 6311 2120 WILSON, LEONARD R., PALYNOLOGIC AND PLANT COMPRESSION
B 6004 716 WILSON, S. J., LITTORAL PROCESSES AND GEOMORPHIC VARIABILITY ON
B 6211 2329 WILSON, THOMAS V., PRESSURE SOLUTION AND POROSITY REDUCTION
B 6105 653 WILTSCHKO, D. V., FLOW OF WEAK ROCKS IN APPALACHIAN PLATEAU
B 6203 572 WIMAN, STEPHEN K., MIDDLE MIOCENE TO EARLY PLIOCENE
S 4 244 WINCHELL LIMESTONE, POSSUM KINGDOM LAKE AREA, TEXAS
M 27 68 WINCHELL LIMESTONE, TEXAS, PENNSYLVANIAN
B 6105 841 WINCHESTER, PAUL D., COMPACTION IN LIME GRAINSTONES, ABST.,
S 9 11 WIND DIRECTION, BRAZIL, JURASSIC
S 9 10 WIND DIRECTION, EUROPE, PERMIAN
S 9 11 WIND DIRECTION, WESTERN, UNITED STATES, PENNSYLVANIAN
S 9 11 WIND DIRECTION, WESTERN, UNITED STATES, PERMIAN
B 6004 733 WIND, F. H., MESOZOIC ORIGIN OF HOLOCOCCOLITHS, ABST.
B 6004 700 WIND, F. H., SELECTIVE DISSOLUTION SEQUENCES FOR CALCAREOUS
B 6409 1503 WIND RIVER BASIN, WYOMING
B 6108 1347 WIND RIVER BASIN, WYOMING
B 6311 2052 WIND RIVER BASIN, WYOMING
B 6208 1499 WIND RIVER BASIN, WYOMING
B 6308 1391 WIND RIVER BASIN, WYOMING
B 6008 1205 WIND RIVER BASIN, WYOMING
B 6004 675* WIND RIVER BASIN, WYOMING, ABST.,
S 3 71 WIND RIVER FORMATION, WYOMING
B 6201 100 WIND RIVER RANGE, WYOMING
M 28 85 WIND-DEPOSITED, GYPSUM
B 6203 457 WINDALIA OIL ZONE, AUSTRALIA
B 6203 457 WINDALIA RADIOLARITE, AUSTRALIA
B 6203 457 WINDALIA SAND MEMBER, AUSTRALIA
M 25 401 WINDARRA, NICKEL DEPOSITS, WESTERN AUSTRALIA
M 29 177 WINDISCH, C. C., STRUCTURE AND STRATIGRAPHY OF THE BLAKE
M 29 113 WINDISCH, C. C., STRUCTURE OF COLORADO BASIN AND
M 29 125 WINDISCH, C. C., STRUCTURE OF FALKLAND PLATEAU AND OFFSHORE
B 6406 821 WINDJANA LIMESTONE, AUSTRALIA
B 6404 483 WINDLEY KEY QUARRY, PHREATIC ENVIRONMENT, FLORIDA
B 6303 538 WINDOM, K. E., MIDDLE CRETACEOUS OXYGEN-DEFICIENT
B 6004 732* WINDOW AS SUBUNCONFORMIT PETROLEUM SOURCE, ABST.,
B 6007 1076 WINDOW, NIGER DELTA, HYDROCARBON-LIQUID
B 6305 818 WINDOW ON OIL, A SURVEY OF WORLD PETROLEUM RESOURCES, BY
S 4 3 WINDWARD GROUP, CORALS, NETHERLANDS, ANTILLES
B 6103 384 WINDWARD-BANK MARGINS, LITTLE BAHAMA BANK
B 6211 2277 WINDY HILL SANDSTONE MEMBER, WYOMING
B 6207 1215* WIND—OLD IDEA WITH NEW POTENTIAL, ABST., ENERGY FROM
B 6305 831 WING, S. J. C., RELATIONS BETWEEN SEDIMENTARY FACIES AND
B 6403 317 WINIFRED FIELD, MONTANA
B 6309 1585* WINIFREDE COAL, BOONE COUNTY, WEST VIRGINIA, ABST., HOW GOOD
B 6309 1511 WINK MEMBER, NEW NAME, TEXAS
B 6004 734 WINKER, C. D., DEPOSITION, DEFORMATION, AND HEAVY MINERAL
B 6309 1608 WINKER, CHARLES D., DISTRIBUTION AN SIGNIFICANCE OF COARSE
B 6405 804 WINKER, CHARLES D., LOWER WILCOX SHELF EDGE IN TEXAS—RELATION
B 6109 1549 WINKER, CHARLES D., PLIOCENE-PLEISTOCENE PALEOGEOGRAPHY OF
B 6309 1611 WINKLER, C. D., LATE PLEISTOCENE FLUVIAL-DELTAIC DEPOSITION
B 6211 2360 WINKLER, G. R., FORAMINIFERAL BIOSTRATIGRAPHY IN GULF OF
B 6304 700 WINKLER, G. R., PETROLEUM POTENTIAL OF EASTERN GULF OF ALASKA
B 6405 766 WINKLER, GARY, GEOLOGY OF CONTINENTAL SLOPE ADJACENT TO OCS
B 6207 1231 WINKLER, GARY, PLATE TECTONICS IN EVOLUTION OF SOUTHERN
B 6009 1611* WINN AND NATCHITOCHES PARISHES, LOUISIANA, ABST., JAMES
B 6004 679* WINN PARISH, LOUISIANA, ABST., SALT, TECTONICS OF
B 6004 734 WINN, R. D., JR., SUBMARINE-FAN SEDIMENTATION IN MESOZOIC
B 6201 50 WINNING GROUP, AUSTRALIA
B 6203 457 WINNING GROUP, AUSTRALIA
B 6201 75 WINNIPEGOSIS FORMATION, SASKATCHEWAN
B 6201 73† WINNIPEGOSIS MOUNDS AND PRAIRIE EVAPORITE FORMATION OF
B 6004 654 WINNOCK, ETIENNE, PALEOSEDIMENTOLOGIC CORRELATION OF
B 6203 511* WINNOWING ACTIVITY ON HUENEME SILL, CALIFORNIA CONTINENTAL
B 6006 992 WINSTON, GEORGE O., FLORIDAS OCALA UPLIFT IS NOT AN UPLIFT
B 6201 121 WINSTON, GEORGE O., REBECCA SHOAL REEF COMPLEX, UPPER
B 6009 1618 WINSTON, GEORGE O., SIX PROPOSED FORMATIONS IN UNDESCRIBED
B 6109 1549 WINSTON, GEORGE O., TYPE AND COTYPE WELLS FOR SOME
B 6303 553 WINTERER, EDWARD L., BIOGENOUS SEDIMENTS RECOVERED BY
B 6403 441* WINTERS FORMATION, ABST., MEMBER SANDS OF
B 6010 1872 WINTERSHAL A.G., ON DEVELOPMENTS IN QATAR, 1975, ASIATIC
B 6310 1889 WINTERSHALL A. G. AND PETROCONSULTANTS S.A., AND DEVELOPMENTS
B 6110 1801 WINTERSHALL A. G., ON DEVELOPMENTS IN QATAR, 1976, ASIATIC
B 6411 1838 WINTERSHALL A.G. AND DEPARTMENT OF PETROLEUM AFFAIRS, ON
B 6210 2117 WINTERSHALL A.G., PETROCONSULTANTS S.A., AND ASIATIC PETROLEUM
B 6010 1872 WINTERSHALL A.G., SUN OIL CO., AN ELF R.E., ON DEVELOPMENTS IN
B 6412 2149 WIPE-OUT ZONES, BERING SEA SHELF, ACOUSTIC ANOMALIES, BRIGHT
B 6206 1072 WIPEOUT OF SEISMIC SECTION, DEFINITION
B 6002 184 WIRELINE-LOG PATTERNS, NORTH SEA
B 6405 685* WISCONSIN, ABST., DEPOSITIONAL ENVIRONMENTS IN UPPER, CAMBRIAN,
B 6211 2330 WISCONSIN, BARABOO QUARTZITE
B 6211 2330 WISCONSIN, BARABOO RANGE
B 6203 502* WISCONSIN CAMBRIAN AND NEW DEPOSITIONAL MODEL FOR TYPE ST.
B 6211 2329 WISCONSIN, GALESVILLE QUARTZ ARENITE
B 6211 2331 WISCONSIN, GALESVILLE SANDSTONE
B 6104 529 WISCONSIN LOWSTAND OF GULF OF MEXICO
S 5 2 WISCONSIN, SILURIAN, NIAGARA GROUP
B 6002 327* WISE COUNTIES, NORTH-CENTRAL TEXAS, ABST., SUBSURFACE STUDY OF,
B 6301 50† WISE COUNTIES, NORTH-CENTRAL TEXAS, SUBSURFACE STUDY OF,
B 6004 733 WISE, S. W., JR., MESOZOIC ORIGIN OF HOLOCOCCOLITHS, ABST.
B 6009 1618 WISE, S. W., JR., MIDDLE EOCENE PLANKTONIC
B 6004 700 WISE, S. W., SELECTIVE DISSOLUTION SEQUENCES FOR CALCAREOUS
B 6303 554 WISE, SHERWOOD W., DEEP-SEA DRILLING IN ANTARCTIC--LATE
B 6309 1609 WISE, SHERWOOD W., JR., STRATIGRAPHY AND PALEOECOLOGY OF
B 6004 701 WISEMAN, WM. J., JR., CONTROL OF SEDIMENT DISTRIBUTION BY
B 6405 785* WISHART QUARTZITE, KNOB LAKE AREA, QUEBEC AND
B 6203 377 WISHBONE FORMATION, ALASKA
B 6311 2001 WITCH GROUND GRABEN, NORTH SEA
M 30 145 WITCH GROUND GRABEN, SEISMIC LINE, PIPER FIELD
B 6405 710 WITHJACK, MARTHA O., BASEMENT TECTONICS OF CHINA—
B 6405 727 WITHJACK, MARTHA O., BASINS OF EAST INDIA—
B 6410 1602 WITTEN FORMATION, VIRGINIA
B 6207 1239 WITTEN, WESLEY, HARNESSING ARCTIC ENERGY—TRANS-ALASKA
B 6402 236 WITTENBURG DRAW FAULT ZONE AREA, TEXAS
B 6105 842 WITTICK, THOMAS R., STRATIGRAPHIC MODELING, ABST., INTERPRETIVE
B 6305 843 WITTSTROM, MARTIN D., GEOLOGY AND CURRENT DEVELOPMENT
B 6405 804 WITTSTROM, MARTIN D., PROPOSED MODEL FOR DEVELOPMENT OF RED
B 6105 815* WITWATERSRAND PLACERS, SOUTH AFRICA, ABST., SEDIMENTOLOGIC
S 6 369 WITWATERSRAND SYSTEM, SOUTH AFRICA
B 6405 805 WITZKE, BRIAN J., MIDDLE AND UPPER ORDOVICIAN PALEOGEOGRAPHY
B 6405 728* WOBURN SANDS, SOUTHERN ENGLAND, ABST., SEDIMENTOLOGY OF
B 6303 554 WOJTAL, AILEEN M., STRATIGRAPHIC MODELS FOR MODERN
S 4 246 WOLF MOUNTAIN SHALE FORMATION, TEXAS
B 6408 1171 WOLFCAMP LIMESTONE, TEXAS
B 6205 787 WOLFCAMP SEISMIC DATA, TEXAS
B 6401 88 WOLFCAMPAIN SERIES, PALO DURO BASIN, TEXAS
B 6004 672* WOLFCAMPIAN CARBONATES, LEONARD MOUNTAI AREA, GLASS
B 6002 321* WOLFCAMPIAN, IN NORTHEAST KANSAS, ABST., BIOFACIES ANALYSIS OF
B 6401 88 WOLFCAMPIAN SERIES, PALO DURO BASIN, TEXAS
B 6311 2116* WOLFCAMPIAN, SOUTHEASTERN PALO DURO BASIN, TEXAS, ABST.,
B 6012 2185 WOLFE, JACK A., ET AL, GEOLOGIC FRAMEWORK OF LOWER COOK INLET,
B 6203 566* WOLLASTON FOLDBELT REGION OF ATHABASCA BASIN, ABST.,
B 6203 490 WOLLENBERG, H. A., RECENT DEVELOPMENTS AT CERRO PRIETO
B 6401 68 WOLTWIESCHE SAMPLED SECTION, GERMANY
B 6409 1569 WOLTZ, DAVID, BLOCK 25 FIELD, CHANDELEUR SOUND, ST. BERNARD
B 6108 1244 WOLTZ, DAVID, DEVELOPMENTS IN LOUISIANA GULF COAST IN 1976
B 6308 1357 WOLTZ, DAVID, DEVELOPMENTS IN LOUISIANA GULF COAST ONSHORE IN
B 6411 2012 WOLTZ, DAVID, DEVELOPMENTS IN LOUISIANA GULF COAST ONSHORE IN
B 6208 1462 WOLTZ, DAVID, DEVELOPMENTS IN LOUISIANA GULF COAST ONSHORE IN
B 6201 44 WOMBAT PLATEAU, AUSTRALIA
B 6004 720* WOMEN GEOSCIENTISTS IN PETROLEUM INDUSTRY, ABST.
B 6203 510 WOO, C. C., SEDIMENTATION ON EASTERN UNITED STATES CONTINENTAL
B 6109 1540* WOOD COUNTY, TEXAS, ABST., GROWTH HISTORY OF, HAINESVILLE,
B 6405 668 WOOD, D. H., PETROLEUM ACCUMULATION TRENDS IN EAST TEXAS SALT
B 6009 1613 WOOD, G. H., JR., EVALUATION OF QUALITY AND QUANTITY CONTROLS
B 6303 466 WOOD, G. V., FORTIES FIELD, UNITED KINGDOM CONTINENTAL SHELF,

B 6105 801 **WOOD, GORDON H., NEED FOR BETTER COAL-RESOURCES ASSESSMENTS**
M 30 81 WOOD, G. V., GEOLOGY OF THE FORTIES FIELD, U.K. CONTINENTAL
B 6309 1591 WOOD, GORDON H., JR., REVISION OF U.S GEOLOGICAL SURVEY AND U.S.
B 6405 805 WOOD, J. R., MODEL OF AUTOCEMENTATION OF QUARTZ SANDS AS
B 6108 1394 WOOD, JAMES R., TRANSFER-REACTION MODELS OF DIAGENETIC
B 6102 257 WOOD RIVER FORMATION, IDAHO
B 6307 1094 WOODADA FORMATION, AUSTRALIA
M 25 399 WOODALL, R., NICKEL RESOURCES OF AUSTRALIA--SUMMARY
M 24 83 WOODBEND GROUP, ALBERTA
M 30 318 WOODBEND GROUP, FACIES DISTRIBUTION MAP, ALBERTA
B 6409 1558* WOODBINE, ABST., LOUISIANA, TUSCALOOSA VERSUS SOUTHEAST, TEXAS
B 6002 326* WOODBINE FORMATION, NORTH-CENTRAL TEXAS, ABST., LITHOFACIES
B 6409 1559* WOODBINE FORMATION OF NORTHEAST TEXAS, ABST.,
M 28 124 WOODBINE FORMATION, TEXAS, CRETACEOUS
M 26 114 WOODBINE FORMATION, TEXAS, POLK COUNTY
B 6209 1766* WOODBINE FORMATION, UPPER CRETACEOUS, CENTRAL TYLER COUNTY,
B 6408 1220 WOODBINE OUTCROP, TEXAS
B 6408 1169 WOODBINE SANDSTONE, TEXAS
M 26 70 WOODBINE SANDSTONE, TEXAS, EAST
B 6408 1220 WOODBINE SECTION, TEXAS
M 26 71 WOODBINE SEQUENCE, SEA LEVEL INTERPRETATION, TEXAS, EAST TEXAS
B 6303 528* WOODBINE--EAGLE FORD INTERVAL, UPPER CRETACEOUS, SUGAR CREEK
B 6409 1552* WOODBINE-TUSCALOOSA, NORTHERN COASTAL REGION, ABST.,
B 6309 1611* WOODBINE-TUSCALOOSA TREND, LOUISIANA, ABST., ORIGIN OF
S 7 117 WOODBURY, H. O., GULF OF MEXICO CONTINENTAL-SLOPE SEDIMENTS
M 29 383 WOODBURY, H. O., INVESTIGATION OF MISSISSIPPI FAN, GULF OF MEXICO
B 6203 546 WOODBURY, H. O., MISSISSIPPI FAN, GULF OF MEXICO, ABST.
S 7 155 WOODBURY, H. O., MISSISSIPPI FAN, GULF OF MEXICO—PHYSIOGRAPHY,
B 6111 1979 WOODFORD, A. O., LOS ANGELES EROSION SURFACE OF MIDDLE
B 6405 754* WOODFORD AND ARKANSAS NOVACULITE FORMATIONS, OKLAHOMA,
B 6109 1454 WOODFORD UNIT, PEDREGOSA BASIN
M 30 253 WOODHAMS, R. L., EUGENE ISLAND BLOCK 330 FIELD, OFFSHORE
M 25 426 WOODLAWN, COPPER-LEAD-ZINC DEPOSIT, NEW SOUTH WALES
B 6409 1569 WOODRUFF, C. M., JR., REGIONAL TECTONIC FEATURES OF INNER GULF
B 6109 1550 WOODRUFF, C. M., JR., STATICS AND DYNAMICS OF NATURAL SYSTEMS—
B 6309 1607 WOODRUFF, CHARLES M., JR., FACIES DISTRIBUTION OF HOSSTON SAND,
B 6405 700 WOODRUFF, F., DEEP-SEA BENTHIC FORAMINIFERA AND THEIR
M 24 243 WOODRUFF SANDSTONE, MISSISSIPPI
B 6410 1660 WOODS HOLE OCEANOGRAPHIC INSTITUTION CRUISES CHAIN 57 AND 73
B 6207 1239 WOODS, JACK T., QUEENSLAND COAL EXPLORATION AND DEVELOPMENT,
B 6405 650 WOODWARD, AND H. L. JAMES, SCHILLING, JOHN, REVIEW OF,
B 6304 598 WOODWARD, LEE A., FRACTURES IN CRETACEOUS ROCKS FROM
B 6109 1513 WOODWARD, LEE A., OSHA CANYON FORMATION, PENNSYLVANIAN, NEW
B 6405 805 WOODWARD, NICHOLAS B., THRUST GEOMETRY OF NORTHERN
B 6305 830 WOODZICK, T. L., UPDATE ON EXPLORATION FOR DIAMONDS IN
B 6207 1239 WOOLLARD, G. P., TECTONIC SETTING OF PERU-CHILE TRENCH—
B 6109 1557 WOOTEN, J. W., LOWER CRETACEOUS CARBONATE ROCKS OF CENTRAL
B 6402 225 WORD GRABEN, TEXAS
B 6405 806 WORK, DAVID F., WHITNEY CANYON—A PROVED GIANT—ONLY SMAL
B 6405 806 WORK, PHILLIP L., NPR-ALASKA, AN UPDATE, ABST.
B 6207 1223 WORKMAN, DAVID R., EVAPORITE DEPOSITS OF NORTHEAST THAILAND,
B 6211 2355* WORKSHOP ON URBAN OIL AND GAS DEVELOPMENT, ABST.
B 6403 303 WORLD DISTRIBUTION OF STAGES OF CONTINENTAL MARGINS
B 6305 842* WORLD ENERGY OUTLOOK FOR THE 1980S, ABST., UNITED STATES AND
B 6303 473* WORLD ENERGY PERSPECTIVE, ABST., UNITED STATES PETROLEUM IN
M 25 515 WORLD, METALLOGENETIC CHART OF THE
M 25 18 WORLD MINERAL TRADE
B 6205 742 WORLD OIL PRODUCTION, 1940-2020, TOTAL
B 6303 459* WORLD OIL RESOURCE ASSESSMENTS AND RELATION TO UNITED STATES
B 6305 818 WORLD PETROLEUM RESOURCES, BY BERNARDO F. GROSSLING,
B 6202 179 WORLD POPULATION AND CRUDE OIL CONSUMPTION VERSUS TIME
B 6303 417* WORLDS FIRST GEOPRESSURED GEOTHERMAL TEST WELL, BRAZORIA
M 27 42 WORM TUBES, SABELARIID
M 27 41 WORM TUBES, SERPULID
S 4 270 WORMS, BORERS, POLYCHAETE
S 4 270 WORMS, BORERS, SIPUNCULID
B 6012 2192 WORNARDT, W. W., NEOGENE DIATOM CORRELATIONS IN
B 6303 554 WORNARDT, WALTER W., JR., UTILITY OF DIATOMS IN PETROLEUM
B 6105 769 **WORSLEY, THOMAS R., OLIGOCENE CALCAREOUS NANNOFOSSIL**
M 29 383 WORZEL, J. L., INVESTIGATION OF MISSISSIPPI FAN, GULF OF MEXICO
B 6004 690 WORZEL, J. L., SEISMIC STRATIGRAPHY AND GEOLOGIC HISTORY OF DEEP
M 29 319 WORZEL, J. LAMAR, ANATOMY OF THE MEXICAN RIDGES,
B 6105 842 WORZEL, J. LAMAR, CONTINENTAL MARGIN—U.S. GULF COAST, ABST.
B 6304 668 **WORZEL, J. LAMAR, GAS SEEPS AND SUBSURFACE STRUCTURE OFF**
B 6303 544 WORZEL, J. LAMAR, GEOLOGIC HISTORY OF DEEP GULF ON MEXICO
B 6303 426 WORZEL, J. LAMAR, GRAVITY--SLIDE ORIGIN FOR MEXICAN RIDGES
B 6211 2290 WORZEL, J. LAMAR, MARGINS OF GULF OF MEXICO
B 6111 1962 WORZEL, J. LAMAR, NORTHEASTERN EXTENSION OF SIGSBEE SCARP,
S 7 43 WORZEL, J. LAMAR, OCCURRENCE AND EVOLUTION OF SALT IN DEEP
B 6303 526 **WORZEL, J. LAMAR, POST-MIDDLE CRETACEOUS SEISMIC STRATIGRAPHY**
B 6312 2204 WORZEL, J. LAMAR, SEISMIC EVIDENCE FOR WIDESPREAD POSSIBLE GAS
B 6303 409 WORZEL, J. LAMAR, SEISMIC STRATIGRAPHY AND STRUCTURE OFF
B 6206 1067 WORZEL, J. LAMAR, SERENDIPITY GAS SEEP AREA, SOUTH TEXAS
B 6203 501 WORZEL, J. LAMAR, STRUCTURE AND EARLY GEOLOGIC HISTORY OF
M 29 403 WORZEL, J. LAMAR, THE MARGINS OF THE GULF OF MEXICO
M 25 149 WRANGELL MOUNTAINS VOLCANIC PILE, ALASKA
B 6203 565 WRAY, J. L., STROMATOLITES IN EOCENE GREEN RIVER FORMATION,
B 6305 833* WRENCH FAULTING AND HYDROCARBON OCCURRENCE IN
B 6308 1428* WRENCH FAULTING, SOUTHERN VAL VERDE BASIN, SOUTHWEST TEXAS,
B 6307 1021 WRENCH FAULTS, IRAN, OBLIQUE
B 6307 1021 WRENCH FAULTS, WESTERN, INDONESIA, LONGITUDINAL
B 6307 1018 WRENCH-FAULT ASSEMBLAGES, BASEMENT-INVOLVED STRUCTURES
B 6405 768* WRENCHING MODEL MICHIGAN BASIN, ABST., PETROLEUM
B 6406 961 WRIGHT, AUDREY A., VARIATION IN TURBIDITE SAND FACIES AND
B 6002 328 WRIGHT, BILLY J., STUDY OF CARBONATE ENVIRONMENTS USING WELL
B 6203 548 WRIGHT, BRUCE, COAL CLEANING, DESULFURIZATION, BY
B 6210 2053 WRIGHT, R. M., ON DEVELOPMENTS IN JAMAICA, 1977
B 6109 1538 WRIGHT, RAMIL C., LATE MIOCENE GLACIO-EUSTATIC LOWERING OF SEA
B 6309 1605 WRIGHT, RAMIL C., REVISION OF TAMPA FORMATION, WEST-CENTRAL
B 6309 1599 WRIGHT, RAMIL C., SUBSURFACE NEOGENE STRATIGRAPHY OF BAY
B 6105 842 WRIGHT, RAMIL G., PLANKTONIC-BENTHONIC RATIO IN FORAMINIFERA
B 6304 695 WRIGHT, ROBERT J., NEW MEXICO AND WORLD URANIUM, ABST.
B 6303 410 WRIGHT, ROBYN, GLACIAL-MARINE SEDIMENTS OF WEST ANTARCTICA
B 6207 1240 WRIGHT, THOMAS L., INTEGRATING ENVIRONMENTAL ASSESSMENT INTO
B 6303 555 WRIGHTSON, C. B., STRUCTURE-CONTOUR MAPS COMPUTER
B 6008 1251 WROBLEWSKI, E. F., DEVELOPMENTS IN OKLAHOMA AND PANHANDLE OF
M 25 89 WU, PATZEN, COAL RESOURCES OF TAIWAN--SUMMARY
M 25 146 **WU, YUNG-JUH, ORIGIN AND POTENTIAL OF SOME GEOTHERMAL AREAS**
B 6405 806 WUENSCHEL, P. C., SONIC LOGS FROM EXPLORATIONISTS VIEWPOINT,
B 6101 3 WULFF LAND ARCH
B 6404 530 WULFF LAND, STRATIGRAPHIC CHART, GREENLAND
B 6203 509 WUNDER, JOHN M., GEOLOGIC APPRAISAL OF UNDISCOVERED OIL AND
B 6107 988 WYANDOT CHALK UNIT, CANADA, SCOTIAN SHELF
B 6301 90 WYANDOT FORMATION, SCOTIAN BASIN
S 4 243 WYANDOTTE FORMATION, KANSAS
B 6107 1046 WYANDOTTE FORMATION, KANSAS
B 6303 555 WYATT, DANNY J., CARBONATE MUD MOUNDS FROM LOWER
B 6105 842 WYLIE, BOB, PETROS—PETROPHYSICAL LOG AND CORE-ANALYSIS
M 25 411 WYLOO GROUP, WESTERN AUSTRALIA
B 6108 1373 WYNN, JEFFREY C., GEOLOGICAL, GEOCHEMICAL, AND PETROPHYSICAL
B 6305 826* WYOMIN AND MONTANA, ABST., COAL VARIATIONS IN FLUVIAL
B 6405 736* WYOMIN THRUST BELT, AS COMPARED WITH OTHER FOLD-AND-THRUST
B 6405 639 WYOMING, ABSAROKA THRUST
B 6307 1040 WYOMING, ABSAROKA THRUST
B 6004 723* WYOMING, ABST., PALEOENVIRONMENTS OF, FRONTIER FORMATION,
B 6405 728* WYOMING, ABST., APPLICATION OF CINGULUM INDEX TO,
B 6108 1371* WYOMING, ABST., APPLICATIONS OF VERTICAL, SEISMIC, PROFILE TO
B 6305 822* WYOMING, ABST., APPLIED EXPLORATION GEOLOGY AND, URANIUM
B 6305 822* WYOMING, ABST., BOREHOLE GRAVITY STUDY OF DENSITY AND
B 6205 881* WYOMING, ABST., CARBONATE, CYCLES IN, GREEN RIVER FORMATION
B 6008 1407* WYOMING, ABST., CHANGING SITES AND STYLES OF MESOZOIC AND
B 6008 1396* WYOMING, ABST., CHARACTERISTICS OF UPPER, CRETACEOUS, LANCE
B 6305 827* WYOMING, ABST., COAL DEVELOPMENT IN, POWDER RIVER BASIN
B 6008 1407* WYOMING, ABST., COASTAL SEDIMENTATION MODEL, TENSLEEP
B 6303 546* WYOMING, ABST., CONTROLS ON LATE, CRETACEOUS-, PALEOCENE
B 6108 1383* WYOMING, ABST., DEPOSITION AND DIAGENESIS OF, TENSLEEP
B 6004 699* WYOMING, ABST., DEPOSITIONAL ENVIRONMENT OF LOWER,
B 6108 1384* WYOMING, ABST., DEPOSITIONAL ENVIRONMENT OF LOWER,
B 6205 887* WYOMING, ABST., DEPOSITIONAL ENVIRONMENTS AND DIAGENESIS OF,
B 6405 675* WYOMING, ABST., DEPOSITIONAL FACIES, GEOMETRY, AND GENESIS OF
B 6108 1393* WYOMING, ABST., DIAGENESIS OF, OFFSHORE-, BAR, SANDSTONES OF,
B 6004 717* WYOMING, ABST., EOCENE PROVENANCE OF, HANNA AND, CARBON,
B 6405 738* WYOMING, ABST., EVALUATION OF ORGANIC MATTER AND SUBSURFACE
B 6303 491* WYOMING, ABST., FACIES AND, RESERVOIR CHARACTERISTICS OF SHELF
B 6305 831* WYOMING ABST., FLOOD-PLAIN SEQUENCES OF FINE-GRAINED.
B 6004 675* **WYOMING, ABST., INFLUENCE OF SOURCE AND PROCESSES IN**
B 6203 517* WYOMING, ABST., LACUSTRINE, PSEUDOVARVES FROM, CRETACEOUS,
B 6405 764* WYOMING, ABST., LINEAMENT ANALYSIS FOR OIL AND GAS EXPLORATIO
B 6405 778* WYOMING, ABST., MIDDLE MEMBER OF, MINNELUSA FORMATION,
B 6405 799* WYOMING, ABST., MIDDLE-LATE, PERMIAN PALEOGEOGRAPHY OF,
B 6108 1376* WYOMING, ABST., NUGGET OIL, ACCUMULATIONS AT, DRY PINEY, TIP
B 6305 829* WYOMING, ABST., OIL AND, GAS POTENTIAL OF
B 6205 887* WYOMING, ABST., PAINTER RESERVOIR FIELD, UINTA COUNTY
B 6305 832* WYOMING, ABST., PAINTER RESERVOIR--, CLEAR CREEK--, RYCKMA
B 6203 537* WYOMING, ABST., PALEOLIMNOPHYCUS—, STROMATOLITE-FORMING,
B 6405 732* WYOMING, ABST., PATTERNS OF THICK, COAL DEPOSITION ACROSS,
B 6008 1403* WYOMING, ABST., RECENT DEVELOPMENTS IN, COAL EXPLORATION IN
B 6108 1371* WYOMING, ABST., REGIONAL DIAGENETIC TRENDS IN LOWER,
B 6108 1381* WYOMING, ABST., REGIONAL IMPLICATIONS OF OIL, ENTRAPMENT AT,

B 6303 530* WYOMING, ABST., SEDIMENTARY SUBENVIRONMENTS OF, WILKINS PEA
B 6405 750* WYOMING, ABST., SEDIMENTOLOGY AND PALEOGEOGRAPHY OF,
B 6008 1390* WYOMING, ABST., SEISMIC EXPERIMENTS, CLINKERS AREA
B 6405 793* WYOMING, ABST., SHANNON SANDSTONE, UPPER, CRETACEOUS,
B 6008 1406* WYOMING, ABST., SPEARHEAD RANCH FIELD, CONVERSE COUNTY
B 6203 513* WYOMING, ABST., STRATIGRAPHY AND DEPOSITIONAL ENVIRONMENTS
B 6004 718* WYOMING, ABST., SYNTHETIC, SEISMIC SECTIONS OF SELECTED,
B 6008 1409* WYOMING, ABST., SYNTHETIC, SEISMIC SECTIONS OF SELECTED
B 6105 816* WYOMING, ABST., TENSLEEP RESERVOIR STUDY, OREGON BASIN FIELD
B 6008 1395* WYOMING, ABST., THICK, COAL BEDS IN, WASATCH FORMATION,
B 6405 805* WYOMING, ABST., THRUST GEOMETRY OF NORTHERN, ABSAROKA SHEET,
B 6405 782* WYOMING, ABST., URANIUM DEPOSITS OF PART OF CENTRAL, GREAT
B 6405 679* WYOMING ABST., URANIUM EXPLORATION SYSTEMS CASE STUDY–
B 6305 842* WYOMING, ABST., URANIUM GEOLOGY OF, MORTON RANCH PROPERTY,
B 6203 561* WYOMING, ABST., VERTICAL DISTRIBUTION OF, HE4, RN222, AND PB210
B 6305 833* WYOMING, ABST., WRENCH, FAULTING AND HYDROCARBON
B 6305 835* WYOMING, ABST., YELLOW CREEK FIELD, UINTA COUNTY
B 6101 114 WYOMING, ADAVILLE FORMATION
B 6204 622 WYOMING, AMSDEN FORMATION
B 6405 669* WYOMING AND ADJACENT AREAS, ABST., PERMIAN, TRACE, FOSSILS OF
B 6305 839* WYOMING AND COLORADO, ABST., VERMILLION CREEK, COAL BED,
B 6205 889* WYOMING, AND IDAHO, ABST., TWIN CREEK FORMATION AND
B 6405 676* WYOMING AND MONTANA, ABST., HYDRODYNAMIC FLOW IN LOWER,
B 6004 727* WYOMING AND MONTANA, ABST., SUBTIDAL, TRACE, FOSSILS, MUDDY
B 6108 1392* WYOMING AND NEBRASKA, ABST., INCIPIENT DIAGENESIS OF, ARKOSIC
B 6303 533* WYOMING AND NEBRASKA, ABST., ISOTOPIC COMPOSITION AN SOURCES
B 6004 708* WYOMING AND NORTHEASTERN UTAH, ABST., PETROLEUM, POTENTIAL
B 6008 1406* WYOMING AND NORTHEASTERN UTAH, ABST., OIL AND GAS POTENTIAL
B 6108 1383* WYOMING AND NORTHWESTERN COLORADO, ABST., BIOSTRATIGRAPHIC,
B 6303 425* WYOMING AND SOUTHEASTERN IDAHO, ABST., DRANEY
B 6303 497* WYOMING AND SOUTHEASTERN MONTANA, ABST., GEOCHEMICAL
B 6305 825* WYOMING AND UTAH, ABST., TERTIARY CORRELATIONS IN, BIGHORN
B 6008 1402* WYOMING AND WESTERN NEBRASKA, CRETACEOUS, FRONTIER
B 6202 185 WYOMING, ARDMORE BENTONITE BED
B 6105 832* WYOMING AS DETERMINED BY NONMARINE OSTRACODS, ABST., AGE OF,
M 24 115 WYOMING, ASPEN SHALE
B 6101 114 WYOMING, ASPEN SHALE
M 24 92 WYOMING, BAXTER SHALE
B 6202 190 WYOMING, BEAR CREEK AREA, CROSS SECTION
B 6101 114 WYOMING, BEAR RIVER FORMATON
B 6404 490 WYOMING, BELLE FOURCHE MEMBER, FRONTIER FORMATION
B 6404 490 WYOMING, BELLE FOURCHE SHALE
B 6108 1347 WYOMING, BIG HORN BASIN
B 6409 1503 WYOMING, BIG HORN BASIN
B 6208 1499 WYOMING, BIG HORN BASIN
B 6008 1205 WYOMING, BIG HORN BASIN
M 24 119 WYOMING, BIG PINEY ANTICLINE
M 24 96 WYOMING, BIG PINEY FIELD
M 24 91† WYOMING, BIG PINEY--, LA BARGE PRODUCING COMPLEX, SUBLETTE
B 6202 185 WYOMING, BIGHORN BASIN
B 6308 1390 WYOMING, BIGHORN BASIN
B 6307 1028 WYOMING, BIGHORN BASIN
B 6202 185 WYOMING, BIGHORN MOUNTAINS
M 24 99 WYOMING, BIRCH CREEK FIELD
M 24 99 WYOMING, BIRCH CREEK SANDSTONE
B 6202 184 WYOMING, BLACK HILLS
B 6211 2275 WYOMING, BLACK HILLS
B 6410 1617 WYOMING, BORIE FIELD
B 6405 639 WYOMING, BRIDGER HILL THRUST
M 24 119 WYOMING, C CANYON ANTICLINE
S 3 60 WYOMING, C ORE ROLL
M 27 118 WYOMING, CAMBRIAN, GALLATIN FORMATION
B 6202 184 WYOMING, CAMPANIAN
B 6202 182 WYOMING, CAMPANIAN INTERIOR SEAWAY
B 6211 2277 WYOMING, CANYON SPRINGS SANDSTONE MEMBER
B 6202 184 WYOMING, CASPER ARCH
M 24 103 WYOMING, CHAPPO MEMBER
B 6202 185 WYOMING, CODY SHALE
B 6202 184 WYOMING, CONIACIAN
B 6101 114 WYOMING, CORRELATION, UTAH, COALVILLE, TO, CUMBERLAND GAP
M 24 109 WYOMING, CRETACEOUS
M 28 144 WYOMING, CRETACEOUS, FRONTIER FORMATION
M 28 174 WYOMING, CRETACEOUS, FRONTIER FORMATION
M 28 134 WYOMING, CRETACEOUS, FRONTIER FORMATION
M 28 130 WYOMING, CRETACEOUS, FRONTIER FORMATION
M 28 21 WYOMING, CRETACEOUS, FRONTIER FORMATION
B 6202 182 WYOMING, CRETACEOUS, LATE, PALEOGEOGRAPHY
M 28 178 WYOMING, CRETACEOUS, MESAVERDE FORMATION
M 28 179 WYOMING, CRETACEOUS, MESAVERDE FORMATION
M 28 176 WYOMING, CRETACEOUS, MESAVERDE FORMATION
M 28 175 WYOMING, CRETACEOUS, MESAVERDE FORMATION
M 28 68 WYOMING, CRETACEOUS, MOWRY SHALE
M 28 137 WYOMING, CRETACEOUS, MUDDY SANDSTONE
M 28 136 WYOMING, CRETACEOUS, MUDDY SANDSTONE
M 28 175 WYOMING, CRETACEOUS, TEAPOT SANDSTONE
M 28 176 WYOMING, CRETACEOUS, TEAPOT SANDSTONE
M 28 179 WYOMING, CRETACEOUS, TEAPOT SANDSTONE
M 28 178 WYOMING, CRETACEOUS, TEAPOT SANDSTONE
M 24 95 WYOMING, DAKOTA SANDSTONE
B 6307 1040 WYOMING, DARBY-PROSPECT THRUST
B 6009 1425† WYOMING, DEFORMATION OF, MESOZOIC, SHALES AT, HAMILTON, DOME,
B 6308 1393 WYOMING, DENVER BASIN
B 6410 1614 WYOMING, DENVER BASIN
B 6409 1503 WYOMING, DENVER BASIN
B 6008 1206 WYOMING, DENVER BASIN
B 6208 1500 WYOMING, DENVER BASIN
B 6008 1204 WYOMING, DEVELOPMENTS, 1975
B 6108 1346 WYOMING, DEVELOPMENTS, 1976
B 6208 1498 WYOMING, DEVELOPMENTS, 1977
B 6308 1388 WYOMING, DEVELOPMENTS, 1978
B 6409 1502 WYOMING, DEVELOPMENTS, 1979
B 6312 2218 WYOMING, DEVILS TOWER, COLUMNAR JOINTING
B 6201 100 WYOMING, DINWOODY FORMATION, IDAHO AND
M 24 93 WYOMING, DISTURBED BELT
M 24 114 WYOMING, DRY PINEY FIELD
M 27 6 WYOMING, EOCENE, GREEN RIVER FORMATION
M 27 142 WYOMING, EOCENE, GREEN RIVER FORMATION
M 28 118 WYOMING, EOCENE, GREEN RIVER FORMATION
M 27 100 WYOMING, EOCENE, GREEN RIVER FORMATION
M 27 189 WYOMING, EOCENE, INDIAN MEADOWS FORMATION
B 6201 100 WYOMING, ERVAY CARBONATE ROCK MEMBER
B 6405 640 WYOMING, EVANSTON FORMATION
B 6002 220 WYOMING FARMERS WHOLE OIL
B 6202 185 WYOMING, FISHTOOTH SANDSTONE
B 6311 2052 WYOMING, FORT UNION FORMATION
M 24 106 WYOMING, FORT UNION MARKER
B 6201 100 WYOMING, FRANSON MEMBER, IDAHO AND
M 24 111 WYOMING, FRONTIER FORMATION
B 6101 114 WYOMING, FRONTIER FORMATION
B 6202 185 WYOMING, GAMMON FERRUGINOUS SHALE
M 24 115 WYOMING, GANNETT GROUP
B 6405 640 WYOMING, GANNETT GROUP
B 6410 1650 WYOMING, GEOCHEMICAL CELL CONCEPT
S 3 76 WYOMING, GHOST ROLL FRONTS
B 6201 100 WYOMING, GOOSE EGG FORMATION
B 6410 1621 WYOMING, GRANEROS SHALE
B 6201 100 WYOMING, GRANITE MOUNTAINS
B 6008 1205 WYOMING, GREATER GREEN RIVER BASIN
B 6208 1499 WYOMING, GREATER GREEN RIVER BASIN, GREATER
M 24 91 WYOMING, GREEN RIVER BASIN
B 6409 1503 WYOMING, GREEN RIVER BASIN, GREATER
B 6108 1347 WYOMING, GREEN RIVER BASIN, GREATER
B 6308 1391 WYOMING, GREEN RIVER BASIN, GREATER
M 24 92 WYOMING, GREEN RIVER FORMATION
B 6404 490 WYOMING, GREENHORN FORMATION
B 6211 2277 WYOMING, GYPSUM SPRING FORMATION
B 6409 1503 WYOMING, HANNA-LARAMIE BASIN
B 6308 1391 WYOMING, HANNA-LARAMIE BASIN
B 6208 1500 WYOMING, HANNA-LARAMIE BASIN
B 6108 1347 WYOMING, HANNA-LARAMIE-DENVER BASIN
B 6101 114 WYOMING, HILLIARD FORMATION
M 24 105 WYOMING, HILLIARD SHALE
M 24 103 WYOMING, HOBACK FORMATION
B 6405 639 WYOMING, HOGSBACK THRUST
B 6307 1039 WYOMING, HOGSBACK THRUST, SEISMIC PROFILE
B 6410 1617 WYOMING, HORSE CREEK FIELD
B 6202 184 WYOMING, HOUSE CREEK FIELD
B 6211 2277 WYOMING, HULETT SANDSTONE MEMBER
B 6108 1389* WYOMING, IDAHO, AND NORTHERN UTAH-STRUCTURAL GEOMETRY
B 6001 12† WYOMING, INFLUENCE OF, RECURRENT, LARAMIDE STRUCTURAL
B 6211 2282 WYOMING, JELM FORMATION
B 6305 830* WYOMING KIMBERLITE PROVINCE, ABST., UPDATE ON EXPLORATION
M 24 106 WYOMING, LA BARGE ANTICLINE
M 24 92 WYOMING, LA BARGE AREA, GEOLOGIC MAP
M 24 105 WYOMING, LA BARGE FIELD
M 24 103 WYOMING, LA BARGE MEMBER
B 6211 2277 WYOMING, LAK MEMBER
M 24 95 WYOMING, LAKOTA SANDSTONE
B 6104 595 WYOMING, LANCE CREEK FIELD
B 6008 1206 WYOMING, LARAMIE-HANNA BASINS

B 6211 2277 WYOMING, LOWER SUNDANCE FORMATION
M 24 97 WYOMING, MCDONALD DRAW FIELD
B 6201 99 WYOMING, MEADE PEAK PHOSPHATIC SHALE MEMBER
B 6307 1040 WYOMING, MEADE THRUST
B 6405 639 WYOMING, MEDICINE BUTTE THRUST
B 6307 1039 WYOMING, MERIDIAN ANTICLINE, SEISMIC PROFILE
B 6202 185 WYOMING, MESAVERDE FORMATION
M 24 109 WYOMING, MESAVERDE FORMATION
M 28 50 WYOMING, MIOCENE, ARIKAREE FORMATION
B 6202 187 WYOMING, MITTEN BLACK SHALE MEMBER
M 24 119 WYOMING, MONUMENT-BLUE RIM ARCH
M 24 95 WYOMING, MORRISON FORMATION
B 6211 2277 WYOMING, MORRISON FORMATION
B 6410 1621 WYOMING, MOWRY SHALE
M 24 95 WYOMING, MOWRY SHALE
B 6404 491 WYOMING, MOWRY SHALE
B 6303 356 WYOMING, MOWRY SHALE, THREE FACIES
M 24 119 WYOMING, MOXA ARCH
M 24 95 WYOMING, MUDDY SANDSTONE
B 6404 491 WYOMING, NIOBRARA FORMATION
B 6202 184 WYOMING, NIOBRARA FORMATION
B 6404 491 WYOMING, NIOBRARA MEMBER
B 6202 184 WYOMING, NIOBRARA SHALE
M 24 97 WYOMING, NORTH LA BARGE FIELD
M 24 100 WYOMING, NORTH MCDONALD DRAW FIELD
M 24 95 WYOMING, NUGGET SANDSTONE
B 6405 640 WYOMING, NUGGET SANDSTONE
B 6211 2282 WYOMING, NUGGET SANDSTONE
M 28 7 WYOMING, ORDOVICIAN, BIGHORN DOLOMITE
M 28 7 WYOMING, ORDOVICIAN, LANDER SANDSTONE
B 6204 627 WYOMING, OREGON BASIN FIELD, WATER-INJECTION PROGRAM
B 6204 622 WYOMING, OREGON BASIN, NORTH DOME
B 6204 626 WYOMING, OREGON BASIN, SOUTH DOME, DEVELOPMENT
B 6208 1500 WYOMING, OVERTHRUST BELT
B 6409 1503 WYOMING, OVERTHRUST BELT
B 6308 1393 WYOMING, OVERTHRUST BELT
M 30 282 WYOMING, PAINTER RESERVOIR FIELD, LOCATION
B 6405 639 WYOMING, PAINTER RESERVOIR FIELD, SECTIONS
M 30 282 WYOMING, PAINTER RESERVOIR FIELD, STRATIGRAPHY
B 6303 354† **WYOMING, PALEOENVIRONMENTS OF MOWRY SHALE, LOWER,**
B 6307 1040 WYOMING, PARIS THRUST
B 6201 100 WYOMING, PARK CITY FORMATION
M 28 68 WYOMING, PENNSYLVANIAN, TENSLEEP SANDSTONE
M 28 122 WYOMING, PENNSYLVANIAN, TENSLEEP SANDSTONE
M 28 123 WYOMING, PENNSYLVANIAN, TENSLEEP SANDSTONE
M 28 129 WYOMING, PENNSYLVANIAN, TENSLEEP SANDSTONE
B 6204 622 WYOMING, PHOSPHORIA FORMATION
B 6202 184 WYOMING, PIERRE SHALE
B 6211 2277 WYOMING, PINE BUTTE SANDSTONE MEMBER
M 28 90 WYOMING, PLEISTOCENE, YELLOWSTONE GROUP
M 28 3 WYOMING, PLEISTOCENE, YELLOWSTONE GROUP
M 28 2 WYOMING, PLEISTOCENE, YELLOWSTONE GROUP
B 6404 490 WYOMING, POOL CREEK MEMBER, CARLILE SHALE
B 6305 823* WYOMING PORTION OF POWDER RIVE BASIN, ABST., STRATIGRAPHIC
B 6308 1390 WYOMING, POWDER RIVER BASIN
B 6108 1346 WYOMING, POWDER RIVER BASIN
B 6202 183 WYOMING, POWDER RIVER BASIN
B 6208 1498 WYOMING, POWDER RIVER BASIN
B 6409 1503 WYOMING, POWDER RIVER BASIN
B 6008 1204 WYOMING, POWDER RIVER BASIN
B 6408 1233 WYOMING, POWDER RIVER BASIN, CRETACEOUS SAND FIELD
M 26 432 WYOMING, POWDER RIVER BASIN, MUDDY FORMATION
M 24 95 WYOMING, PREUSS SANDSTONE
B 6405 640 WYOMING, PREUSS-STUMP FORMATION
B 6008 1213 WYOMING, PRODUCTION, 1968-1975
B 6108 1352 WYOMING, PRODUCTION, 1969-1976
B 6208 1503 WYOMING, PRODUCTION, 1970-1977
B 6308 1394 WYOMING, PRODUCTION, 1971-1978
B 6409 1508 WYOMING, PRODUCTION, 1972-1979
M 26 432 WYOMING, RECLUSE FIELD, MUDDY FORMATION
B 6202 187 WYOMING, RED BIRD SILTY MEMBER
B 6211 2277 WYOMING, REDWATER SHALE MEMBER
B 6201 99 WYOMING, RETORT PHOSPHATIC SHALE MEMBER
B 6201 100 WYOMING, REX CHERT MEMBER, IDAHO AND
B 6305 839* WYOMING ROLL FRONTS, ABST., MOUNT TAYLOR, URANIUM
M 24 97 WYOMING, RUBEN FIELD
B 6206 1076 WYOMING, RYCKMAN CREEK FIELD
B 6405 638 WYOMING, RYCKMAN CREEK FIELD
M 24 97 WYOMING, SADDLE RIDGE FIELD
B 6404 490 WYOMING, SAGE BREAKS MEMBER, CODY SHALE
B 6202 188 WYOMING, SALT CREEK DOME
B 6302 198 WYOMING, SAND WASH BASIN, COLORADO AND
B 6202 184 WYOMING, SANTONIAN
B 6405 742* WYOMING, SEISMIC REFLECTION DATA, ABST., DEPTH MIGRATION AND
M 24 96 WYOMING, SHALLOW RIDGE FIELD
B 6202 184 WYOMING, SHANNON INTERVAL
B 6408 1235 WYOMING, SHANNON SAND SEQUENCE
B 6202 185 WYOMING, SHANNON SANDSTONE
B 6202 187 WYOMING, SHARON SPRINGS MEMBER
S 10 28 WYOMING, SOUTH GLENROCK FIELD, SHALE CORE
M 24 99 WYOMING, SOUTH LA BARGE FIELD
M 24 97 WYOMING, STAR CORRAL FIELD
B 6202 184 WYOMING, STEELE SHALE
B 6211 2277 WYOMING, STOCKADE BEAVER SHALE MEMBER
M 24 95 WYOMING, STUMP SANDSTONE
B 6211 2277 WYOMING, SUNDANCE FORMATION
B 6202 184 WYOMING, SUSSEX INTERVAL
B 6408 1237 WYOMING, SUSSEX SANDSTONE
B 6202 197 WYOMING, SUSSEX SANDSTONE, HYDROCARBON ACCUMULATIONS
S 3 94 WYOMING, SWEETWATER GRANITE
B 6204 630 WYOMING, TENSLEEP G ZONE
B 6204 609† WYOMING, TENSLEEP, RESERVOIR, OREGON BASIN FIELD
B 6204 610 WYOMING, TENSLEEP SANDSTONE
B 6201 100 WYOMING, TENSLEEP SANDSTONE
M 24 95 WYOMING, THERMOPOLIS SHALE
B 6303 500* WYOMING THRUST BELT, ABST., APPLICATION OF POTENTIAL FIELD DAT
B 6108 1388* WYOMING THRUST BELT, ABST., ASSESSMENT OF OIL AND GAS
B 6105 821* WYOMING THRUST BELT, ABST., ASSESSMENT OF OIL AND GAS
B 6109 1559* WYOMING THRUST BELT, ABST., LANDSAT–USEFUL PETROLEUM
B 6303 484* WYOMING THRUST BELT, ABST., PAINTER, RESERVOIR FIELD--GIANT IN
B 6405 800* WYOMING THRUST BELT, ABST., SOURCE AND TIME OF GENERATION OF
B 6305 832* **WYOMING THRUST BELT, ABST., WHITNEY CANYON FIELD--POTENTIAL,**
B 6405 730* WYOMING THRUST BELT, ABST., WHITNEY CANYON FIELD–POTENTIAL,
B 6405 669* WYOMING THRUST BELT AND HINTERLAND, ABST., TECTONIC
B 6206 1075 WYOMING THRUST BELT, NUGGET SANDSTONE, IDAHO-
M 30 281† WYOMING THRUST BELT, PAINTER RESERVOIR FIELD--GIANT IN THE
B 6405 638† WYOMING THRUST BELT, PAINTER RESERVOIR FIELD–, GIANT IN
B 6206 1075† WYOMING THRUST BELT, PETROLEUM, SOURCE-BED, EVALUATION OF,
B 6305 833* WYOMING THRUST BELT--PRELIMINARY REPORT, ABST., STRUCTURAL
B 6307 1039 WYOMING THRUST-FOLD BELT, FRONTIER FORMATION, IDAHO-
B 6307 1039 WYOMING THRUST-FOLD BELT, HILLIARD SHALE, IDAHO-
B 6307 1038 WYOMING, THRUST-FOLD BELT, MAP, IDAHO-
B 6307 1039 WYOMING THRUST-FOLD BELT, NUGGET SANDSTONE, IDAHO-
B 6307 1040 WYOMING THRUST-FOLD BELT SEQUENTIAL EVOLUTION, IDAHO-
B 6201 100 WYOMING, TOSI CHERT MEMBER
M 24 117 WYOMING, TRIASSIC
B 6404 490 WYOMING, TURNER SANDY MEMBER, CARLILE SHALE
B 6405 640 WYOMING, TWIN CREEK FORMATION
B 6206 1077 WYOMING, TWIN CREEK LIMESTONE, SAMPLES AND GEOCHEMICAL DATA
M 24 95 WYOMING, TWIN CREEK SANDSTONE
B 6211 2277 WYOMING, UPPER SUNDANCE FORMATION
B 6203 573* WYOMING, URANIUM DEPOSIT REDISCOVERED. BY TITLE ONLY., COPPER
B 6404 490 WYOMING, WALL CREEK MEMBER, FRONTIER FORMATION
M 24 92 WYOMING, WASATCH FORMATION
B 6405 640 WYOMING, WASATCH-GREEN RIVER FORMATIONS
B 6108 1347 WYOMING, WESTERN, OVERTHRUST BELT
B 6108 1347 WYOMING, WIND RIVER BASIN
B 6409 1503 WYOMING, WIND RIVER BASIN
B 6208 1499 WYOMING, WIND RIVER BASIN
B 6308 1391 WYOMING, WIND RIVER BASIN
B 6311 2052 WYOMING, WIND RIVER BASIN
B 6008 1205 WYOMING, WIND RIVER BASIN
S 3 71 WYOMING, WIND RIVER FORMATION
B 6201 100 WYOMING, WIND RIVER RANGE
B 6211 2277 WYOMING, WINDY HILL SANDSTONE MEMBER
B 6312 2218 WYOMING, YELLOWSTONE NATIONAL PARK, COLUMNAR JOINTING
B 6405 674* WYOMING-MONTANA, ABST., VERTICAL AN SURFACE, SEISMIC PROFILES
B 6405 767* WYOMING-UTAH-IDAHO OVERTHRUST BELT–RELATION TO CANADIAN
B 6202 181† WYOMING–EXAMPLE OF CRETACEOUS OFFSHORE SEDIMENTATION,
B 6405 718* WYOMING–NEW LOOK AT OLD AREA, ABST., WAMSUTTER ARCH TIGHT
S 4 173 X-RADIOGRAPH, PLATYGYRA
S 4 171† X-RADIOGRAPHY, CARIBBEAN, REEF, CORALS, MONTASTREA ANNULARIS
B 6311 2036 X-RAY ANALYSIS OF CLAY MINERALS
B 6303 488* **X-RAY DIFFRACTION ESTIMATES OF DOLOMITE IN CARBONATE**
B 6405 794* X-RAY DIFFRACTION TECHNIQUES, ABST., CHARACTERIZATION OF
B 6405 702* X-RAY MINERALOGY OF UPPER FREEPORT COAL, ABST.
S 7 323 X-RAY OF SPLIT CORE, ORCA BASIN
B 6403 438* YAKATAGA DISTRICT CONTINENTAL SHELF, GULF OF ALASKA. BY TITLE

B 6405 671* YAKATAGA FORMATION, GULF OF ALASKA, ABST., LATE, NEOGENE
B 6303 411* YAKATAGA FORMATION, PLIOCENE-PLEISTOCENE, GULF OF ALASKA,
B 6303 411* YAKATAGA FORMATION, PLIOCENE-PLEISTOCENE ROBINSON
M 26 356 YAKATAGA, SIDE-SCAN SONAR RECORD, ALASKA, CAPE
B 6403 440* YAKATAGA—YAKUTAT AREA NORTHERN GULF OF ALASKA, ABST.,
B 6207 1178 YAKOVLEVIA MULTISTRIATA-NEOGONDOLELLA BITTERI ZONE, NEVADA
B 6403 440* YAKUTAT AREA, NORTHERN GULF OF ALASKA, ABST., HOLOCENE,
B 6204 635 YAKUTAT BAY, ALASKA
M 26 355 YAKUTAT BAY, DEPTH SOUNDER RECORD, ALASKA
M 29 285 YAKUTAT BAY, REFRACTION DATA, ALASKA
B 6212 2417 YAKUTAT SEA VALLEY, ALASKA
B 6204 635 YAKUTAT SEA VALLEY, ALASKA
B 6403 374† YALLAHS FAN DELTA, SOUTHEAST JAMAICA, FAN-, DELTA
B 6203 572* YALLAHS FAN DELTA, SOUTHEASTERN JAMAICA, ABST., DEPOSITIONAL
B 6103 332 YAMACRAW RIDGE, SOUTH CAROLINA
S 6 159 YAMADORI FORMATION, JAPAN, HAUTERIVIAN
B 6312 2119 YAMAMA LIMESTONE, PERSIAN GULF, DARIUS FIELD
M 25 360 YAMAMOTO, SOKI, GROUNDWATER IN VOLCANIC AREAS IN
B 6212 2512 YAMAMOTO, SOKI, GROUNDWATER IN VOLCANIC ISLANDS, ABST.
B 6207 1194 YAMILL FORMATION, OREGON AND, WASHINGTON
B 6101 44 YAMMOUNEH FAULT, LEBANON
M 25 414 YAMPI MEMBER, WESTERN AUSTRALIA, PENTECOST SANDSTONE
M 25 427 YAMPI SOUND, IRON ORE, WESTERN AUSTRALIA
M 25 474 YANA-KOLYMA MEGASTRUCTURE
M 25 466 YANA-KOLYMA MIOGEOSYNCLINAL SYSTEM, USSR
S 6 143 YANAGI, TAKERU, ISOTOPIC AGES AND STRATIGRAPHIC CONTROL OF
B 6112 2105 YANAYACU FIELD, PERU
M 26 439 YANCEY, MARSHALL S., SEISMIC STRATIGRAPHIC MODEL OF
B 6109 1437 YANCEY, MARSHALL S., SEISMIC STRATIGRAPHIC MODEL OF
B 6112 2105 YANEZ FIELD, PERU
B 6310 1827 YANG, S., AND G. L. FLETCHER, ON DEVELOPMENTS IN TAIWAN, 1978
B 6110 1840 YANG, Y. F., ON DEVELOPMENTS IN TAIWAN, REPUBLIC OF CHINA, 1976
B 6210 1988 YANG, Y. F., ON DEVELOPMENTS IN TAIWAN, REPUBLIC OF CHINA, 1977
M 25 440 YANGPYONG ROCKS, SOUTH KOREA
M 25 443 YANGSAN FAULT, HEAT FLOW, HIGH, SOUTH KOREA
M 30 244 YARAKTIN FIELD, EXPLORATION HISTORY, USSR
M 30 251 YARAKTIN FIELD, STRUCTURE CONTOUR MAP, USSR
B 6105 842 YARBOROUGH, H., CONTINENTAL-MARGIN TYPES RELATED TO PLATE
B 6307 1093 YARDARINO FIELD, AUSTRALIA
B 6404 511 YARMOUTH ARCH
B 6109 1484 YARMOUTH ARCH, EAST COAST
B 6002 320* YARNELL TREND, SUMNER COUNTY, KANSAS, ABST.
B 6307 1097 YARRAGADEE FORMATION, AUSTRALIA
M 25 414 YARRALOOLA CONGLOMERATE, WESTERN AUSTRALIA
B 6410 1644 YARRAMBA CHANNEL, AUSTRALIA
B 6203 544 YASUDA, Y., BOTTOM SEDIMENTS OF SAN NICOLAS BASIN, CALIFORNIA,
B 6303 501* YATES AND LOWER TANSILL FORMATIONS, GUADALUPE MOUNTAINS,
S 10 221 YATES FIELD, TRAP, WEST TEXAS
B 6205 781 YATES FORMATION, NEW MEXICO
B 6203 561* YAVAPAI COUNTY, ARIZONA, ABST., MAJOR, URANIUM DISCOVERY IN,
B 6304 621† YAVAPAI COUNTY, ARIZONA, MAJOR, URANIUM DISCOVERY IN,
M 24 239† YAZOO COUNTY, MISSISSIPPI, TINSLEY OIL FIELD
B 6011 1985 YEATS, ROBERT S., MAGNETIC-REVERSAL STRATIGRAPHY OF
B 6212 2513 YEATS, ROBERT S., SOUTHERN CALIFORNIA—TECTONICS AND OIL
B 6012 2192 YEATS, ROBERTS S., NEOGENE TECTONICS OF CENTRAL VENTURA BASIN,
B 6305 847 YEDLOSKY, R. J., PROFITABLE SHALLOW APPALACHIAN GA
M 25 427 YEELIRRIE URANIUM DEPOSITS, WESTERN AUSTRALIA
M 26 198 YEKAN LAKE, CARBONATE BUILDUP, SEISMIC EXAMPLE, CANADA,
B 6205 785 YEKAU LAKE REEF, SEISMIC DATA, ALBERTA
B 6203 573 YELLICH, JOHN, COPPER MOUNTAIN, WYOMING, URANIUM DEPOSIT
M 25 523 YELLOW CAKE, DEMAND FOR, MEXICO
B 6305 835* YELLOW CREEK FIELD, UINTA COUNTY, WYOMING, ABST.
M 25 262 YELLOW SEA, KOREAN OFFSHORE, WEST SEA
M 25 269 YELLOW SEA, NORTH KIANGSU, AND SHANGHAI BASINS, KOREA, WEST
M 25 13 YELLOW SEA, SEABED JURISDICTION
M 28 3 YELLOWSTONE GROUP, WYOMING, PLEISTOCENE
M 28 2 YELLOWSTONE GROUP, WYOMING, PLEISTOCENE
M 28 90 YELLOWSTONE GROUP, WYOMING, PLEISTOCENE
B 6312 2218 YELLOWSTONE NATIONAL PARK, COLUMNAR JOINTING, WYOMING
B 6010 1872 YEMEN, 1975, ASIATIC PETROLEUM CORP. AND, SANTA FE MINERALS,
B 6110 1802 YEMEN, 1976, ASIATIC PETROL. CORP., ON DEVELOPMENTS IN
B 6210 2118 YEMEN, 1977, ASIATIC PETROLEUM CORP., ON DEVELOPMENTS IN
B 6310 1890 YEMEN, 1978, PETROCONSULTANTS S.A., ON DEVELOPMENTS IN
B 6411 1841 YEMEN, 1979, SCALLOP CORP., ON DEVELOPMENTS IN
B 6411 1841 YEMEN, PEOPLES DEMOCRATIC REPUBLIC OF YEMEN, 1979, AGIP AND,
M 25 439 YEN, T. P., GEOLOGIC CONTROLS OF MINERAL DEPOSITS IN
B 6402 219 YENANGYAUNG FIELD, BURMA
B 6203 553 YEO, R., SILURIAN MACROBORING ORGANISMS, SOUTHERN ONTARIO,
B 6102 263 YERKES, R. F., SANTA MONICA AND SANTA ANA MOUNTAINS--RELATION
B 6205 781 YESO FORMATION, NEW MEXICO
B 6109 1518 YESO FORMATION, NEW MEXICO
B 6201 41 YILGARN BLOCK, AUSTRALIA
M 25 411 YILGARN BLOCKS, IRON, WESTERN AUSTRALIA, PILBARA AND
S 11 62 YNEZIAN STAGE
B 6206 951 YOAKUM CHANNEL, TEXAS
M 25 357 YOGYAKARTA VOLCANIC AREA, INDONESIA
B 6101 53 YOHMOR 1, LOG, LEBANON
M 25 440 YONCHON SYSTEM, SOUTH KOREA
S 4 291 YONGE REEF, GALAXEA
M 25 269 YONGNAM, REINAN, MASSIF, KOREA, SOBAEKSAN, SOBAEGON OR
B 6207 1240 YONGSAKUL, KANIT, OFFSHORE TIN DEPOSITS OF THAILAND AND
M 25 442 YONIL DISTURBANCE, KOREA
B 6207 1241 YORA, MITSUO, PRESENT STATUS OF GEOTHERMAL DEVELOPMENT AND
B 6111 1986 YORBA SILTSTONE, CALIFORNIA
B 6309 1600 YORK, MARK E., BAYOU MIDDLE FORK FIELD, CLAIBORNE PARISH,
B 6405 707 YORK, P. L., DIPMETER VALIDITY IN DEVIATED BORE HOLES, ABST.
B 6105 843 YORSTON, H. J., SEISMIC ANALYSIS OF GLIDE-PLANE FAULT STRUCTURES,
M 25 279 YOSHII GAS FIELD, JAPAN
B 6309 1598* YOU AINT SEEN NOTHING TILL YOUVE SEEN THE TUSCALOOSA, ABST.
B 6104 573 YOUNG, ALLEN, CALCULATION OF AGES OF HYDROCARBONS IN
B 6109 1407 YOUNG, ALLEN, HYDROCARBONS BETWEEN OILS AND ASSOCIATED
B 6004 701 YOUNG, FREDERICK G., LOWER CRETACEOUS NEOCOMIAN, SANDSTONE
B 6004 734 YOUNG, G. M., PROBLEMM OF PALEOCLIMATOLOGIC INTERPRETATION
B 6303 555 YOUNG, KEITH, IMMANITAS BED, PASO DEL RIO, COLIMA, MEXICO, ABST.
B 6109 1557 YOUNG, KEITH, MIDDLE CRETACEOUS ROCKS OF MEXICO AND TEXAS,
B 6105 786 **YOUNG, R. A., NOAAS WASTE-DISPOSAL STUDIES IN NEW YORK BIGHT**
B 6105 843 YOUNG, ROBERT G., SANDSTONE DEPOSITIONAL SYSTEMS IN
B 6004 735 YOUNG, S. W., PALEOCLIMATE INTERPRETATION THROUGH USE OF
B 6305 831 YOUNGBERG, ALV D., FLOOD-PLAIN SEQUENCES OF FINE-GRAINED
B 6004 699 YOUNT, J. C., DYNAMIC PROCESSES OF UPPER BENGAL FAN AND SWATCH
B 6409 1569 YOUSSEFNIA, IRADJ, PALEOECOLOGY OF MIDWAY GROUP IN NORTHEAST
B 6403 449* YOWLUMNE OIL FIELD. BY TITLE ONLY
M 30 71 YPRESIAN, FRIGG FIELD, PALEOGEOGRAPHY
S 6 220 YPRESIAN STAGE DEFINED, EUROPE, EOCENE
B 6210 2045 YRIGOYEN, M. R., AND J. A. SUAREZ LYNCH, ON DEVELOPMENTS IN
B 6110 1579 YRIGOYEN, M. R., AND PETROCONSULTANTS S.A., ON DEVELOPMENTS IN
B 6310 1746 YRIGOYEN, M. R., AND ROBBIANO, J. A., ON DEVELOPMENTS IN
B 6010 1642 YRIGOYEN, ON DEVELOPMENTS IN ARGENTINA, 1975, LESTA, P. H., AND
B 6303 546* YUCATAN AND ITS PETROLEUM POTENTIAL, ABST., GREAT, CARBONATE
S 3 107 YUCATAN AXIS, ANTILLES-
S 4 205 YUCATAN, CHIONE CANCELLATA
S 4 195† YUCATAN COAST, MEXICO, QUANTITATIVE PALEOECOLOGICAL ASPECTS
B 6403 434 YUCATAN, EDITED BY W. C. WARD AND A. E. WEIDIE, GUZMAN,
S 4 195 YUCATAN, MOLLUSCA
B 6211 2293 YUCATAN PENINSULA AREA, MULTICHANNEL PROFILE LINES
B 6203 500* YUCATAN PENINSULA, MEXICO, ABST., VADOSE, CEMENT PATTERNS IN,
B 6303 362† YUCATAN PENINSULA, MEXICO, STRANDLINE SEDIMENTATION
S 7 9 YUCATAN SHELF, DESCRIBED
B 6007 1064 YUCATAN SHELF REEFS, VERTICAL DEVELOPMENT OF
B 6211 2293 YUCATAN STRAIT, MULTICHANNEL PROFILE LINES
B 6112 2115† YUCCA MOUND COMPLEX, LATE PENNSYLVANIAN PHYLLOID-ALGAL
B 6112 2117 YUCCA MOUND COMPLEX, LOWER AND UPPER MOUNDS, NEW MEXICO
M 25 365 YUEN, GEORGE, IMPACT OF ECONOMIC DEVELOPMENT ON DEMANDS
B 6207 1235 YUEN, PAUL C., GEOTHERMAL ENERGY IN HAWAII—PRESENT AND
B 6411 1970 YUGOSLAVIA, ADRIATIC OFFSHORE AND DINARIC ONSHORE ZONE
B 6411 1970 YUGOSLAVIA, BANAT, MURA, DRAVA, AND SAVA SUBBASINS
B 6110 1650 YUGOSLAVIA, DEVELOPMENTS, 1976
B 6210 1921 YUGOSLAVIA, DEVELOPMENTS, 1977
B 6310 1929 YUGOSLAVIA, DEVELOPMENTS, 1978
B 6411 1970 YUGOSLAVIA, DEVELOPMENTS, 1979
M 27 33 YUGOSLAVIA, EOCENE, NUMMULITE LIMESTONE
B 6110 1654 YUGOSLAVIA, PRODUCTION, 1975-1976
B 6210 1924 YUGOSLAVIA, PRODUCTION, 1976-1977
B 6310 1932 YUGOSLAVIA, PRODUCTION, 1977-1978
B 6411 1971 YUGOSLAVIA, PRODUCTION, 1978-1979
B 6405 806 YUKLER, A. M., THREE-DIMENSIONAL SIMULATION OF GEOLOGIC,
B 6203 573 YUKLER, M. ARIF, SIMULATION OF GEOLOGIC, HYDRODYNAMIC, AND
M 25 462 YUKON FAULT ZONE, ANYUY-
B 6002 172 YUKON-KOYUKUK GEOSYNCLINE, ALASKA
B 6405 796* YULE MARBLE, ABST., NATURE OF NON-RECOVERABLE, POROSITY
B 6309 1479 YULE MARBLE, TEXTURE AND COMPOSITION DATA
M 25 443 YULRI SYNCLINE, SOUTH KOREA
M 25 440 YULRI SYSTEM, SOUTH KOREA
B 6405 807* YUNNAN-GUIZHOU-GUANGXI PROVINCES, CHINA, ABST., LITHOFACIES
B 6203 573 YUREWICZ, D. A., SYNDEPOSITIONAL SUBMARINE CEMENTATION OF
B 6203 492 YUREWICZ, D. A., SYNDEPOSITIONAL SUBMARINE CEMENTATION OF
B 6105 843 YUREWICZ. DONALD A., CAPITAN MASSIVE LIMESTONE, PERMIAN, OF
M 25 154 **ZABLOCKI, CHARLES J., DEEP RESEARCH DRILLHOLE AT KILAUEA**

B 6203 355 ZACA OIL FIELD, CALIFORNIA
B 6311 2120 ZACHRY, DOY L., EARLY PENNSYLVANIAN BRAIDED STREA
B 6405 797* ZAGROS FOLDBELT, IRAN, ABST., GEOLOGIC AND STRUCTURAL HISTORY
B 6405 602 ZAGROS FOLDBELT, MIDDLE EAST
B 6401 10 ZAGROS FOOTHILLS THRUST BELT, IRAN
B 6401 12 ZAGROS FOOTHILLS THRUST BELT, IRAQ
B 6412 2096 ZAGROS GEOSYNCLINE, MAP, IRAN
B 6412 2098 ZAGROS GEOSYNCLINE, STRATIGRAPHIC CHART, IRAN
B 6206 985 ZAGROS MOUNTAINS, BASEMENT, IRAN
B 6206 987 ZAGROS MOUNTAINS, COMPETENT GROUP, IRAN
B 6004 727† ZAGROS MOUNTAINS IN SOUTHWEST IRAN. BY TITLE ONLY, GEOLOGIC
B 6206 987 ZAGROS MOUNTAINS, INCOMPETENT GROUP, IRAN
B 6303 444* ZAGROS MOUNTAINS, IRAN, ABST., MIOCENE-, PLIOCENE MOLASSE
B 6206 987 ZAGROS MOUNTAINS, MOBILE GROUP, IRAN
B 6307 1022 ZAGROS OROGENIC BELT, IRAN
B 6206 984† ZAGROS SIMPLY FOLDED BELT, SOUTHWEST IRAN, FOLD DEVELOPMENT
B 6206 986 ZAGROS SIMPLY FOLDED BELT, STRATIGRAPHIC COLUMN, IRAN
B 6110 1750 ZAIRE, DEVELOPMENTS, 1976
B 6310 1695 ZAIRE, DEVELOPMENTS, 1978
B 6411 1791 ZAIRE, DEVELOPMENTS, 1979
B 6110 1764 ZAIRE, PRODUCTION, 1975-1976
B 6210 1866 ZAIRE, PRODUCTION, 1976-1977
B 6310 1709 ZAIRE, PRODUCTION, 1978
B 6411 1794 ZAIRE, PRODUCTION, 1978-1979
B 6404 568 ZAK, ISRAEL, STRAIN MEASUREMENTS IN EASTERN MARGINAL SHEAR
M 26 195 ZAMA AREA, PINNACLE REEFS, SEISMIC EXAMPLE, CANADA, ALBERTA,
B 6205 782 ZAMA AREA, SEISMIC DATA, ALBERTA, RAINBOW-
B 6210 1849 ZAMBIA, DEVELOPMENTS, 1977, TERRITORY OF, AFARS AND, ISSAS,
B 6401 32 ZANDOBBIO DOLOMITE, ITALY
M 30 530 ZANDOBBIO FORMATION, JURASSIC, MALOSSA FIELD
M 25 330 ZAPALLO FORMATIONS, ECUADOR, SANTIAGO AND
B 6209 1762* ZAPATA COUNTIES, TEXAS, ABST., LOBO TREND OF, SOUTH, LAREDO
B 6309 1596* ZAPATA COUNTY, TEXAS, ABST., SUBSURFACE STRATIGRAPHY OF,
B 6310 1888 ZAPEX, U.K., LTD., ON DEVELOPMENTS IN UMM AL-QAIWAIN, 1978,
B 6411 1841 ZAPEX, U.K., ON DEVELOPMENTS IN UMM AL QAIWAIN, 1979, HOUSTON
B 6412 2190 ZAPFE, SCHLAGER, WOLFGANG, REVIEW OF, BEITRAGE ZUR
B 6010 1643 ZAPPATERRA, E., ON DEVELOPMENTS IN BELIZE, 1975
B 6211 2233 ZARAUZ DIAPIR, BAY OF BISCAY AREA
B 6211 2240 ZARAUZ DIAPIR WITH SALT DOMES IN PRESENT DEEP-SEA
B 6303 556 ZARILLO, GARY A., ENERGY DISSIPATION OVER SALT MARSH AND IT
M 24 145† ZAVALA COUNTIES, TEXAS, BIG WELLS FIELD, DIMMIT AND
B 6301 75 ZDANICE NAPPE, CZECHOSLOVAKIA
B 6203 541* ZECHSTEIN 2, CARBONATE ROCKS OF EASTERN NETHERLANDS FROM
M 30 41 ZECHSTEIN SALT, NORTH SEA, VIKING FIELD
B 6409 1549 ZEIL, VAN HOUTEN, FRANKLYN B., REVIEW OF THE, ANDES, A
B 6206 1092 ZEILINSKI, R. E., ORGANIC GEOCHEMISTRY OF GAS-BEARING SHALES, U.S.
B 6310 1821 ZELIFF, C. W., ON DEVELOPMENTS IN INDONESIA, 1978, JAUNCEY, W., AND
B 6002 319 ZELLER, DORIS E. N., GEOLOGY OF MISSISSIPPIAN COMBINATION TRAP,
B 6303 441 ZELLER, EDWARD J., URANIUM RESOURCE EVALUATIO IN ANTARCTICA,
B 6104 597 ZELTEN FIELD, LIBYA
B 6303 416* ZELTEN FIELD LIBYA, ABST., SECONDARY, CARBONATE POROSITY
B 6002 196 ZEMBRUSCKI, S., SOUTH ATLANTIC DIAPIRIC STRUCTURES
S 11 65 ZEMORRIAN STAGE
S 11 17 ZEMORRIAN STAGE, SUPERPOSITION
S 11 19 ZEMORRIAN STAGE, ZONATION AND CONGREGATIONS
S 11 54† ZEMORRIAN THROUGH DELMONTIAN, PLANKTONIC AND LARGE,
B 6203 490 ZEN, C. S., PROFILE OF COAL-FORMING SWAMP—THE EVERGLADES, ABST.
B 6207 1241 ZEN, M. T., ENERGY, RESOURCES, AND ENVIRONMENT MANAGEMENT—
B 6009 1570 ZENGER, DONALD H., DEFINITION OF TYPE LITTLE FALLS DOLOSTONE,
B 6405 776* ZENITH FIELD—SIGNIFICANT DAKOTA, MUDDY, D SANDSTONE
B 6405 708* ZEOLITE CEMENTATION IN UPPER JURASSIC SANDSTONES, LOWER COOK
M 28 138 ZEOLITE, IN SANDSTONE, CEMENTS
B 6206 1057 ZEOLITES
M 28 149 ZEOLITES, BIBLIOGRAPHY, DIAGENESIS AND POROSITY
M 25 101 ZEOLITES IN TERTIARY COALFIELDS, JAPAN, AUTHIGENIC
M 25 453 ZEOLITES, JAPAN, URANIUM ADSORBED ON
B 6405 807 ZHAN, XIE, LITHOFACIES AND BIOHERMS OF UPPER PALEOZOIC IN
M 30 476 ZHANHUA BASIN, GUANTAO FORMATION, CHINA
M 30 472 ZHANHUA BASIN, SHAHEJIE FORMATION, CHINA
M 30 473 ZHANHUA BASIN, STRATIGRAPHIC COLUMN, CHINA
B 6002 164 ZHEMCHUG BASIN, BERINGIAN SHELF
B 6002 162 ZHEMCHUG CANYON, BERINGIAN SHELF
B 6405 807 ZHI, TANG, TECTONIC FEATURES OF OIL AND GAS BASINS OF EASTERN
B 6012 2192 ZIEGLAR, D. L., RESERVOIR AND SOURCE-BED HISTORY IN GREAT
B 6105 844 ZIEGLAR, D. L., RESERVOIR AND SOURCE-BED HISTORY IN GREAT
B 6205 813 ZIEGLAR, D. L., RESERVOIR AND SOURCE-BED HISTORY OF GREAT
B 6211 2364 ZIEGLAR, D. L., SYNTHESIS OF OCS WELL INFORMATION, OFFSHORE
B 6405 763 ZIEGLER, A. M., CLIMATE ASYMMETRY AND BIOGEOGRAPHIC
B 6303 507 ZIEGLER, ALFRED M., ATMOSPHERIC CIRCULATION AND UPWELLING IN
M 24 276 ZIEGLER, M. A., JAY FIELD, FLORIDA--A JURASSIC STRATIGRAPHIC TRAP
B 6303 556 ZIEGLER, PETER A., NORTH SEA EXPLORATION, ABST.
S 6 337 ZIEGLER, WILLI, DEVONIAN
B 6309 1584 ZIELINSKI, R. E., DEVONIAN SHALE CHARACTERIZATION--EVALUATION
B 6309 1592 ZIELINSKI, R. E., PRESSURE CORING OF GAS-BEARING DEVONIAN BLACK
B 6309 1591 ZIELINSKI, R. E., PROJECTION OF FAVORABLE GAS-PRODUCING AREAS
B 6309 1587 ZIELINSKI, R. R., INTERPRETATION AND STATISTICAL ANALYSIS OF
B 6002 263 ZIKHRON REEF, ISRAEL
B 6203 566 ZIMBRICK, GRANT D., REEVALUATION OF MORROWAN-ATOKAN SERIES
B 6004 695 ZIMMERMAN, H. B., SEDIMENT LITHOFACIES IN SOUTH ATLANTIC OCEAN,
M 25 515 ZINC, AND ARSENIC, MEXICO, SILVER, LEAD
M 25 426 ZINC, AND GOLD, NEW SOUTH WALES, COBAR, COPPER
M 25 553 ZINC, AND LEAD DEPOSITS, ECUADOR, SAN BARTOLOME, SILVER
M 25 427 ZINC, AUSTRALIA, NORTHERN TERRITORY, H.Y.C. DEPOSIT, LEAD-
M 25 496 ZINC, CANADIAN, CORDILLERA, SILVER-, LEAD-
M 25 535 ZINC DEPOSIT, ARGENTINA, AGUILAR, LEAD-
M 25 507 ZINC DEPOSIT, CANADA, SALMO TYPE OF, LEAD-
M 25 507 ZINC DEPOSIT, CANADA, SHUSWAP TYPE OF, LEAD-
M 25 427 ZINC DEPOSIT, NEW SOUTH WALES, BROKEN HILL, LEAD-, SILVER-
M 25 426 ZINC DEPOSIT, NEW SOUTH WALES, WOODLAWN, COPPER-, LEAD-
M 25 554 ZINC DEPOSITS, ECUADOR, URITOHUASER, LEAD AND
M 25 554 ZINC, ECUADOR, URITOHUASER DEPOSIT, LEAD-
M 25 554 ZINC, LEAD, AND MOLYBDENUM DEPOSITS, ECUADOR, FIERRO URCO,
M 25 553 ZINC, LEAD, SILVER, AND GOLD DEPOSITS, ECUADOR, ANGAS, COPPER
M 25 457 ZINC ORES, JAPAN, LEAD-
M 25 505 ZINC, PRODUCTION AND RESERVES, CANADIAN, CORDILLERA
M 25 444 ZINC, SOUTH KOREA
M 25 426 ZINC-COPPER DEPOSIT, NEW SOUTH WALES, CAPTAINS FLAT, LEAD-
M 25 426 ZINC-COPPER DEPOSIT, TASMANIA, ROSEBERY, LEAD-
M 25 427 ZINC-COPPER, QUEENSLAND, MOUNT ISA, LEAD-, SILVER-
M 25 497 ZINC-LEAD, CANADA, SELWYN BASIN
M 25 553 ZINC-LEAD, ECUADOR, SAN BARTOLOME DEPOSIT, SILVER-
M 25 427 ZINC-LEAD, QUEENSLAND, DUGALD RIVER
M 25 427 ZINC-SILVER, QUEENSLAND, HILTON DEPOSIT, LEAD-
M 25 427 ZINC-SILVER, QUEENSLAND, LADY LORETTA, LEAD-
B 6405 695 ZINKL, RICHARD J., APPLICATION OF REGIONAL GEOCHEMICAL DATA TO
M 28 57 ZIRCON, BIREFRINGENCE
M 25 418 ZIRCON, COMPOSITION AND USES
M 28 56 ZIRCON, CRYSTALS, DETRITAL
M 28 60 ZIRCON, CRYSTALS, IN GRAIN MOUNT
B 6303 456* ZIRCONS, ABST., PROVENANC STUDIES IN, TUNISIA BY, U-PB AGES OF
B 6007 1055 ZOANTHUS ZONE, PANAMA, GALETA REEF
B 6009 1531 ZOBACK, MARK D., EFFECT OF HIGH-PRESSURE DEFORMATION ON
B 6008 1414 ZOERB, RICHARD M., LANDSAT IMAGERY--EVOLVING EXPLORATION
B 6012 2192 ZOERB, RICHARD M., SATELLITE EXPLORATION BY GEOPHYSICISTS, ABST.
B 6104 562 ZONA GLAUCONITICA STRATIGRAPHIC UNIT, CHILE AND, ARGENTINA,
B 6401 115† ZONATION OF LARGE SILURIAN REEF AT DELPHI, INDIANA,
M 30 443 ZONE A, RESERVOIR, BADAK FIELD
M 30 456 ZONE B, STRATIGRAPHY, BADAK FIELD
M 30 457 ZONE C, RESERVOIRS, BADAK FIELD
M 30 457 ZONE D, RESERVOIRS, BADAK FIELD
M 30 457 ZONE E, RESERVOIRS, BADAK FIELD
M 30 457 ZONE F, RESERVOIRS, BADAK FIELD
M 28 3 ZONE INCLUSIONS, QUARTZ
B 6102 140 ZONES 1 TO 8, CALIFORNIA, VENTURA FIELD
B 6102 172 ZONES 1 TO 8, CALIFORNIA, VENTURA FIELD
B 6102 149 ZONES 5, 6, 7, 8, CALIFORNIA, VENTURA FIELD
B 6410 1614 ZORICHAK FIELD, COLORADO
M 25 24 ZRALY, ARTHUR J., ECONOMICS OF ENERGY AND MINERAL
B 6401 12 ZUBAIR SANDSTONE FORMATION, IRAQ
B 6401 12 ZUBAIR SANDSTONE FORMATION, IRAQ
B 6405 761 ZUBOVIC, PETER, TRACE-ELEMENT CONTENT OF BITUMINOUS COAL
M 27 141 ZULOAGO FORMATION, MEXICO, JURASSIC
B 6109 1449 ZUNI UPLIFT, ARIZONA AND, NEW MEXICO
B 6209 1674 ZUNI UPLIFT, NEW MEXICO
B 6004 735 ZUNIGA, FERNANDO, PETROLEUM GEOLOGY AND PRODUCTIVE
B 6210 2058 ZUNIGA Y RIVERO, F., AND PETROCONSULTANTS S.A., ON
M 25 339 ZUNIGA Y RIVERO, F., HYDROCARBON POTENTIAL OF AMAZON BASINS
B 6010 1655 ZUNIGA Y RIVERO, F., ON DEVELOPMENTS IN PERU, 1975
B 6110 1590 ZUNIGA Y RIVERO, F., ON DEVELOPMENTS IN PERU, 1976
B 6310 1758 ZUNIGA Y RIVERO, F., ON DEVELOPMENTS IN PERU, 1978
S 10 233 ZYKH FIELD, MINERAL CONTENT OF WATER, USSR